21st Innovative Manufacturing Engineering & Energy International Conference (IManE&E 2017)

MATEC Web of Conferences Volume 112 (2017)

Iasi, Romania
24 - 27 May 2017

Part 1 of 2

Editors:

L. Slatineanu
G. Nagit
O. Dodun
V. Merticaru
M. Coteata
M.I. Ripanu
A.M. Mihalache
M. Boca
R. Ibanescu
C.E. Panait
G. Oancea
P. Kyratsis

ISBN: 978-1-5108-4430-8

Printed from e-media with permission by:

Curran Associates, Inc.
57 Morehouse Lane
Red Hook, NY 12571

Some format issues inherent in the e-media version may also appear in this print version.

Printed by Curran Associates, Inc. (2017)

For additional information, please contact EDP Sciences – Web of Conferences at the address below.

EDP Sciences – Web of Conferences
17, Avenue du Hoggar
Parc d'Activité de Courtabœuf
BP 112
F-91944 Les Ulis Cedex A
France

Phone: +33 (0) 1 69 18 75 75
Fax: +33 (0) 1 69 28 84 91

contact-edps@webofconferences.org

Additional copies of this publication are available from:

Curran Associates, Inc.
57 Morehouse Lane
Red Hook, NY 12571 USA
Phone: 845-758-0400
Fax: 845-758-2633
Email: curran@proceedings.com
Web: www.proceedings.com

TABLE OF CONTENTS

PART 1

PART 2

ISSN (electronic edition): 2261-236X

Preface

The Machine Manufacturing Technology Department from the Gheorghe Asachi Technical University of Iasi organizes since 1996 an International Conference. The first edition has taken place in 1996 under the title of *Modern Technologies in Machine Manufacturing Technology*. Over the past 20 years the conference had various titles due to ever changing educational and industrial environments and to collaborations between similar departments from other countries: *Modern Technologies, Quality, Reorganization – TMCR* (1999-2008), *Modern Technologies, Quality and Innovation - New face of TMCR – Modtech* (2009-2012) and *Innovative Manufacturing Engineering International Conference-IManE* (2012- 2015). Starting from 2016 the conference's title is *Innovative Manufacturing Engineering & Energy - IManE&E*. Starting from 2009, excepting the year 2015, all editions have been ISI indexed.

2017 marks the 21th anniversary edition since our first conference in manufacturing engineering was organized. Over the years, these scientific events were seen as excellent opportunities to present researches results and exchange professional opinions among participants. These scientific manifestations encourage young researchers to find new solutions and receive valuable suggestions to their problems.

The conference editions were yearly organized by the Department of Machine Manufacturing Technology from the Gheorghe Asachi Technical University of Iasi, Romania in partnership with both Romanian and foreign manufacturing departments of other universities: the Manufacturing Engineering Department from Technical University of Cluj - Napoca, Romania, the Manufacturing Engineering Department from Transilvania University of Brasov, Romania, the Manufacturing Engineering Department from Politechnica University of Bucharest, Romania, the Manufacturing Engineering Department from University of Pitesti, Romania, the Engineering and Technology Management Department from North University Center, Baia Mare, Romania and the Department of Machine Manufacturing Technology from Technical University of Moldova, Chisinau, Republic of Moldova and the Department of Mechanical Engineering and Industrial Design from Technological Education from University of Applied Sciences, Institute of West Macedonia, Kozani, Greece.

These events turned into great occasions for researchers to get to better know each other and join their efforts for mutual progress in science and manufacturing technologies! It is our belief that this year's edition has continued this wonderful tradition and it has represented a step further in the development of our conference with benefits for everyone involved!

Opening address

The 21st edition of the *Innovative Manufacturing Engineering & Energy international Conference – IMANE&E 2017* took place in Iasi, Romania from 24th till 27th of May 2017.

The conference's proceedings have been divided into ten sections as follows:

- Advanced Machining and Surface Engineering;
- Assembling Technologies, Forming Technologies;
- Non-Conventional Technologies in Manufacture and Industry, Welding Technologies;
- Advanced Materials;
- Flexible Manufacturing, Automation and Robotics in Technological Processes;
- CAD/CAM/CAE/CAx Technologies, Manufacturing Optimization;
- Mechanical and Manufacturing Equipment, Devices and Instrumentation, Design and Analysis;
- Innovation, Creativity, Learning and Education in Engineering;
- Industrial and Product Management, Quality and Evaluation;
- Energy (Smart Grids, Energy Efficiency in Buildings and Energy systems and energy management, Energy policies, Environmental technologies and studies, Renewable energy technologies and energy saving materials).

Each paper was presented and sent for publishing to MATEC Web of Conferences for an Open Access Mode and will be submitted for indexation and ratings.

Here are some raw numbers which complete the image of the amplitude of this year's event: people from over 24 countries were involved as authors or co-authors; more than 200 authors and co-authors from 14 countries have attended the conference's proceedings; over 260 abstracts submitted on the official conference's website www.imane.ro, 218 works were received and 196 paperwork were accepted to be presented in the conference's proceedings; there were 8 papers in plenary session and over 188 in sections of the conference. Each paper has been reviewed by two, three or even four reviewers. The conference has benefited of support of more than 94 reviewers in the evaluation of it scientific level.

We hope that in this year also the plenary debates and those from each section have constituted a solid ground for excellent exchange of ideas which will pave the way for new collaborations between specialists. The organizing committee would like to take the opportunity to thank all our collaborators, reviewers, authors and especially participants for their extraordinary contribution brought to the increase of the scientific level of the 21st edition of the conference. The organizing committee wishes all the participants to be successful and expresses conviction that we will have the opportunity to meet again at future IManE&E conference editions.

We consider the present edition to be a real success looking at the number of papers and participants who have attended this wonderful event as well as the variety of topics that were approached in each section. Authors have done their best to deliver important research related results and highly valuable information from a scientific point of view thus upgrading the level of significance of IManE&E conference. Topics were widely spread in an array of ideas, concepts and result based conclusions or confirmations which make up an undeniable source of valuable scientific content for all researchers, students, teachers or any type of entrepreneurs at the present moment.

The level of the scientific content presented at this conference oblige us to step up and find new ways to improve our organizational skills at all levels thus providing a solid ground for significant content to be drawn towards us in order for it to be displayed, well known and spread by researchers from all over the world.

The conference's official website is www.imane.ro and the participants as well as the organizers have enjoyed the benefits of the online platform provided by EasyChair (www.easychair.org) for their paper submissions, reviews, comments, acceptance or rejections and other publishing related issues.

Professor Laurentiu Slatineanu,
Honorary President of IManE&E 2017 International Conference
Machine Manufacturing Technology Department
Gheorghe Asachi Technical University of Iasi, Romania

INTERNATIONAL SCIENTIFIC COMMITTEE

REVIEWERS FOR THE IMANE&E2017 CONFERENCE

edp sciences

MATEC Web of Conferences

Statement of Peer review

In submitting conference proceedings to MATEC Web of Conferences, I certify to the Publisher that I adhere to the **Policy on Publishing Integrity** of the journal in order to safeguard good scientific practice in publishing.

1. All articles have been subjected to peer review administered by the proceedings editors.
2. Reviews have been conducted by expert referees, who have been requested to provide unbiased and constructive comments aimed, whenever possible, at improving the work.
3. Proceedings editors have taken all reasonable steps to ensure the quality of the materials they publish and their decision to accept or reject a paper for publication has been based only on the merits of the work and the relevance to the journal.

Title, date and place of the conference

21st Innovative Manufacturing Engineering & Energy International Conference, IManE&E 2017
24th – 27th May 2017, Iasi, Romania

Proceedings editor(s):

Laurentiu SLATINEANU, "Gheorghe Asachi" Technical University of Iaşi, România
Gheorghe NAGIT, "Gheorghe Asachi" Technical University of Iaşi, România
Oana DODUN, "Gheorghe Asachi" Technical University of Iaşi, România
Vasile MERTICARU, "Gheorghe Asachi" Technical University of Iaşi, România
Margareta COTEATA, "Gheorghe Asachi" Technical University of Iaşi, România
Marius Ionut RIPANU, "Gheorghe Asachi" Technical University of Iaşi, România
Andrei Marius MIHALACHE, "Gheorghe Asachi" Technical University of Iaşi, România
Mihai BOCA, "Gheorghe Asachi" Technical University of Iaşi, România
Radu IBANESCU, "Gheorghe Asachi" Technical University of Iaşi, România
Carmen Ema PANAIT "Gheorghe Asachi" Technical University of Iaşi, România
Gheorghe OANCEA, "Transilvania" University of Brasov, Romania
Panagiotis KYRATSIS, Technological Education Institute of Western Macedonia, Greece

Date and editor's signature

04.06.2017

17, avenue du Hoggar - PA de Courtabœuf – BP 112 - 91944 Les Ulis Cedex A (France)
Tél. : 33 (0)1 69 18 75 75 - Fax : 33(0)1 69 07 45 17 – www.edpsciences.org

Portable selfcentred device with calibrated rollers

Adriana Munteanu*, *Florentin* Cioata, and *Emilian* Paduraru

Technical University Gheorghe Asachi of Iaşi, Department of Machining Tool, Bdul Mangeron, 59A, 700050, Iaşi, România

Abstract. The critical analysis of the self-centering devices for inner cylindrical surfaces highlights a number of advantages and disadvantages of these solutions. The key benefit is the precision but the most important disadvantage it is that doesn't covers a sufficiently range of values for the inner diameters and are very complex. In the present paper a solution of centering mechanism with calibrated rollers that eliminates the disadvantages mentioned above, with a medium complexity but ensuring a sufficient range of values for measured surfaces is proposed. The method used to obtain the specified range of values is graphic method.

1 Introduction

The inspection of geometrical deviations (especially the coaxial deviation and concentricity deviation) is an issue developed in specified standards [1, 2]. There are several methods for inspecting but these methods are more or less precise [3].

In industrial practice the measurement of geometrical deviations is an economic aspect that imposes specified inspection methods if the production is mass one. So because of this are used inspection methods more precise that are normally more time consuming and more costly. When the manufacturing production is small the less precise inspection methods that are normally less time consuming and less costly are used [4].

In the coaxiality and concentricity deviation measurement the literature propose in various cases the deviations measuring with the coordinate measuring machines [5] that are time consuming so the development was focused on optical technology, high-speed electronics and computers. But those are expensive. In exclusive production type is necessary a device control that satisfies the following conditions:

- High manoeuvrability, so as to ensure quick measurement of the considered deviation;
- Zeroing quick and easy, without requiring costly additional accessories;
- High degree of universality that would ensure a measuring range as large as possible;
- The indicator instrument: the control device to be fitted with indicator instrument;
- The degree of mobility: control device to be portable;
- Fixing and positioning elements: the control device must have in its structure, fixing and positioning elements for instrument indicator, respectively its orientation relative to the controlled workpiece [6, 7].

*Corresponding author: adycypmunteanu@yahoo.com

2 The schematic diagram of the portable device

In any design process when it is establish a schematic diagram for a control device based on adopted measurement scheme following aspects must be resolved [7, 8, 9]:
• The method of difference application, as a method of measuring of coaxial deviation of inner cylindrical surfaces;
• The compliance of adopted measuring scheme;
• The orientation and the positioning of control device on the inner cylindrical surface specified as datum, in order to materialized its axis;
• To establishing a simple solution for displacement in the radial direction, of the contact elements to the inspected surface;
• The existence, in the device structure, of an element for the realization of inner cylindrical surface tolerated axis;
•The device must be fitted with some indicator whose feeler can perform a circular motion, or if this feeler has a linear movement to establish a variant of displacement transmission of the feeler at the measuring tip;
• To ensure for a sufficiently range values of inner diameters of the two cylindrical surfaces (adjusting solution);
• The possibility to zeroing the control device with calibrated elements accurate mounting on the device;

The proposed solution of selfcentering mechanism (presented in a previous paper [3]) provide greater accuracy of centering and covers a sufficiently typo dimensional values for the cylindrical inner surfaces diameter and is characterized by a high degree of universality (Figure 1).

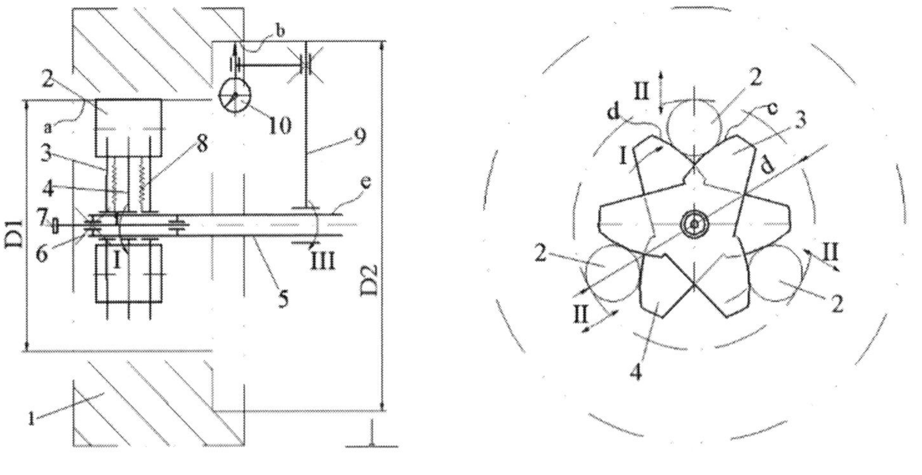

Fig. 1. Portable device

For generation of schematic diagrams of this portable control device, adopted by the measuring scheme, it has been identified a number of requirements to be met (Figure1):
• It is necessary to materialize the axis of inner tolerated cylindrical surface (*a* and *b* surfaces); this requirement is imposed by the adopted scheme measurement;
• It is necessary to materialize the axis of inner cylindrical specified as datum (by the calibrated rollers 2, and the mobile and fix disks 3 and 4);

• The measuring of deviation from coaxiality of inner cylindrical surfaces is achieved with a portable control device (with the cylindrical body 5) that must provide both orientation and positioning relative to the workpiece (1) and the measurement;
• Performing measurement of deviation from coaxiality (distance between the materialized rotation axes) using an indicating instrument (the indicating instrument 10 and its support 9);
• Providing guidance of the control device on inner cylindrical surfaces of different diameters (the active surface *e*);
•Ensure the measuring deviation from coaxiality of inner cylindrical surfaces of parts with different distances between the cylindrical surfaces considered (by movement of calibrated rollers by hand wheels 7).

3 The graphic solution for control device dimensioning

Dimensional element of the workpiece that is decisive in determining the characteristic dimensions of the control device is the diameter *D* of the inner cylindrical surface of the workpiece because it determines the overall dimensions of the control device. To establish the characteristic dimensions of the device control it was considered the nominal values for diameter *D* of the inner cylindrical surface of the workpiece. This range of values is set between 90 and 130 mm.

The decisive size in determining the overall dimensions of the self centering mechanism with calibrated roller from the control device structure is the minimum diameter of the cylindrical inner surface, in which the centering mechanism should be introduced. Thus, the diameter of this mechanism must have the diameter less than the diameter of the cylindrical inner surface.

The characteristics dimensions of the control device are dimensions that determine the overall of all assemble sizes which determine the structure of the device subassemblies, the lengths of running components who execute displacements, etc. After the obtaining mode, the characteristic dimensions of the device control fall into two distinct categories:
• The characteristic dimensions adopted;
• The characteristics dimensions calculated.

In the considered situation an important place in the control device designed the following adopted characteristic dimensions are relevant:
• The diameter of centering mechanism with calibrated roller (the three fixed and mobile discs, will have a smaller diameter than the minimum diameter of the surface to be inspected;
• The length of the cylindrical body of the self centring device with calibrated rollers (the length of the cylindrical body will be greater than the maximum value of the length of the control);
• The diameter of calibrated rollers. Calibrated rollers, when their generators are brought into contact with the inner cylindrical surface of control, are designed to materialize the cylinder adjacent thereto; by moving their radials to come into contact with the inner cylindrical surface.

But for this selfcentering device the calculated characteristics dimensions depends on the size characteristic of the device control, for ensuring that measurement functions, the orientation of control device for limiting the course of moving parts that perform rotational motion or translation.

To designed control device must considered the following characteristic dimensions:
• Travel length of calibrated rollers - To materialize the adjacent cylinder of the inner cylindrical surface with different values of diameter, the calibrated rollers must move on

radial direction with the distance l_c, called the length of stroke; it is calculated by the relationship 1:

$$l_c = \frac{D_{max} - D_{min}}{2} \quad (1)$$

where D_{max} and D_{min} are the extreme diameters of the controlled workpiece.

• The angle of rotation of the mobile disc – The mobile disc determine the displacement on radial direction of the calibrated rollers so as to cover the specified value range for the diameter of the inner cylindrical surface with which they come into contact.

In Table 1 are presented the successive position of the calibrated rollers, corresponding to the rotation of the mobile disc with values from 0 ° to 40 °.β is the angle between the fix disc 2 and mobile disc 3 in his initial position , when α (the rotation angle of mobile disc) have the value zero (Fig. 2,a).

The solution for determining the rotation angle can be obtained by calculation or graphic representation and in the present paper was chosen the graphics solution. Graphical method for obtaining the angle shown in the Table 1 can be seen in Figure 2.

Table 1. Device dimensioning

Graphical representation	**Calculation**
Fig. 2 a)	α- rotation angle of mobile disc 3; α=0°; β –angle between the fix disc 2 and mobile disc 3; β= 46°; d- diameter of materialized adjacent cylinder ; d=130 mm.
Fig. 2 b)	α- rotation angle of mobile disc 3; α=5°; β –angle between the fix disc 2 and mobile disc 3; β= 46°; d- diameter of materialized adjacent cylinder ; d=125 mm.
Fig. 2 c)	α- rotation angle of mobile disc 3; α=10°; β –angle between the fix disc 2 and mobile disc 3; β= 46°; d- diameter of materialized adjacent cylinder ; d=118 mm
Fig. 2 d)	α- rotation angle of mobile disc 3; α=15°; β –angle between the fix disc 2 and mobile disc 3; β= 46°; d- diameter of materialized adjacent cylinder ; d=111 mm
Fig. 2 e)	α- rotation angle of mobile disc 3; α=20°; β –angle between the fix disc 2 and mobile disc 3; β= 46°; d- diameter of materialized adjacent cylinder ; d=106 mm
Fig. 2 f)	α- rotation angle of mobile disc 3; α=25°; β –angle between the fix disc 2 and mobile disc 3; β= 46°; d- diameter of materialized adjacent cylinder ; d=99 mm
Fig. 2 g)	α- rotation angle of mobile disc 3; α=30°; β –angle between the fix disc 2 and mobile disc 3; β= 46°; d- diameter of materialized adjacent cylinder ; d=96 mm
Fig. 2 h)	α- rotation angle of mobile disc 3; α=35°; β –angle between the fix disc 2 and mobile disc 3; β= 46°; d- diameter of materialized adjacent cylinder ; d=93 mm
Fig. 2 i)	α- rotation angle of mobile disc 3; α=40°; β –angle between the fix disc 2 and mobile disc 3; β= 46°; d- diameter of materialized adjacent cylinder ; d=90 mm

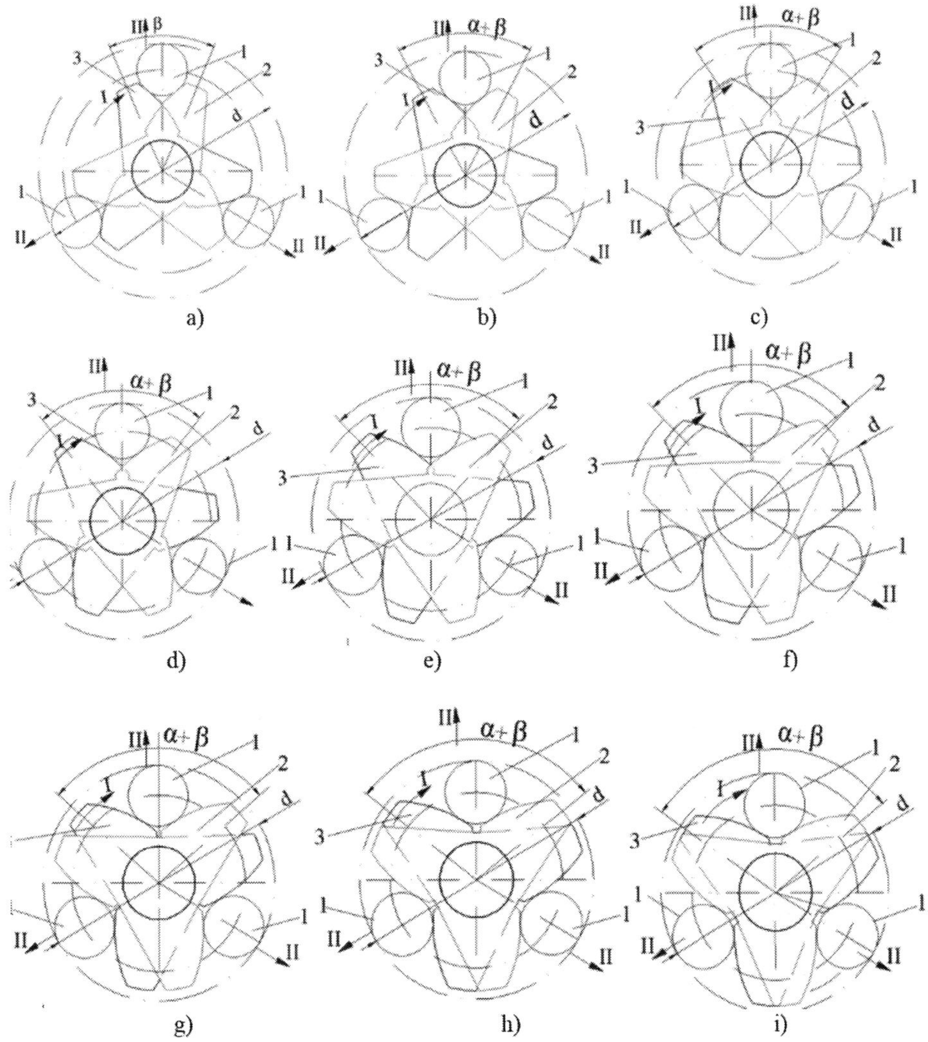

Fig. 2. Graphical dimensioning.

These dimensional characteristics are determined by workpiece dimensions. So must consider:

• Inner diameter cylindrical surface whose axis is specified as a datum axis and determining dimensions orientation- positioning mechanism (subassembly adjacent cylinder to materialize), respectively calibrated roller centering mechanism;

• Inner diameter of the cylindrical surface that measured deviation from concentricity / coaxial and determines the size and positioning mechanism orientation- fixing indicator instrument;

• The values for the total length of the inner cylindrical surfaces of the socket type parts (datum line and tolerated) which determine the length of the cylindrical body of the controller;

• The values for carcass length (distance between walls are charged cylindrical inner surfaces of skeletal parts (baseline and tolerated), which determines the length of the cylindrical body of the controller.

Considering the fact that the control device is portable, it isn't necessary fixing the workpiece with a mechanism of the control device because it installs (it is oriented and positioned) and fixed on the controlled workpiece, within the area / cylindrical surface / thereof. Therefore, the overall dimensions of the piece to control, respectively, the maximum outer diameter and length of the piece (for parts bushing type) are not decisive in determining the characteristic dimensions of the system.

Conclusion

In this paper a new approach has been proposed to verify and determine the geometrical specifications. Conceived and designed controller has the following operational and constructive characteristics: the controlled parameter is the form deviation, the measured technique is the direct one, the zeroing is realised on the workpiece, the number of the controlled parameters is one, the device is portable, have the indicating instrument and the mass about half a kilogram.

The focus was on the dimensioning the control device so that to obtain specified values for measured surfaces and the method is the graphic method. In a future work the authors will study and will determine the measurement uncertainties of the control portable device. Crucial issue given that any control equipment is the measurement error which can certify the validity of measurements performed or not, it is essential to determine the error limits with which it will measure. The sources of error will be identified and will be established the specified relationships and the physical model it will be designed. The experimental results regarding the applicability of the developed device will be the purpose of a future work too.

References

1. ISO 1101:2004 *Geometrical product specifications and verification. Geometrical tolerancing. Tolerancing of form, orientation, location and run-out.*
2. *** *The ISO System of Limits and Fits* — Tolerances and Deviations
3. *Geometrical Dimensioning and Tolerancing for Design, Manufacturing and Inspection (Second Edition)*. A Handbook for Geometrical Product Specification using ISO and ASME Standards. Edited by: Georg Henzold, ISBN: 978-0-7506-6738-8
4. Yuhe Li, Yongrong Qiu, Yanxiang Chen, Kaisen Guan. Opt. and Lasers in Eng. **62,** 31–37 (2014)
5. X.B. Zhang, X.Q. Jiang, Z. Li.A. Key Eng. Mat., **295-296**, 283-288 (2005)
6. Munteanu A., Cioata F. Bul. of Pol. Ins. Iaşi, **62** (2016)
7. T. Damian, *Constructive elements for fine mechanics (*in Romanian*)* (Didactic and Pedagogical Publishing House, Bucharest, 1980)
8. N. Seghedin, *Technological work devices fine mechanics*, (Tehnopress Publishing House, Iasi, 2006)
9. W. Gao, S.W. Kim, H. Bosse, H. Haitjema, Y.L. Chen, X.D. Lu, W. Knapp, A. Weckenmann, W.T. Estler, H. Kunzmann, CIRP Annals – Man.Tech., **64**, **2**, 773-796 (2015)
10. Y.-P. Cang, G.-J. Wu, A.-M. Xie, W.-J. Wang, Ac. Phot. Sinica **38**(11), 2932-2936 (2009)

Application of optical scanning system to determine the machining allowances

Jolanta B. Krolczyk[1,*], *Gabriel* Pihan[2], and *Stanislaw* Legutko[2]

[1]Opole University of Technology, Faculty of Mechanical Engineering, 76 Proszkowska Street, 45-758 Opole, Poland
[2]Poznan University of Technology, Faculty of Mechanical Engineering and Management, 3 Piotrowo Street, 60-965 Poznan, Poland

Abstract. The paper contains a description of replacing a process of marking-out iron casts with 3D scanning operation and the profits it implies. It presents some of the available scanning methods which is a part of surface metrology. The paper includes a description of scanning operation using structured light method and all information necessary to perform such operation, from the equipment construction up to creation of a report for the CNC machinist.

1 Introduction

Monitoring of machine part condition is an important research field in reliability analysis of mechanical components [1-3]. Many metrological techniques have been developed to analyse the surface quality, defect analysis and measurement of process temperature [4-6]. These techniques can be successful in various practical applications [7]. For example, surface profilometry as a method of topography inspection enables the identification of singular components for fault diagnosis of machine parts [8–11]. The defects of parts are very often caused by the inappropriately selected technological input parameters, condition and properties of a machine, as well as the process dynamics [12-14].

Nowadays, many technologies are available making it possible to collect data concerning the shape of an object or to measure whole surfaces necessary to monitor the individual processes of production. The existing measurement systems allow for the determination of the location of the object points in space with the accuracy of micrometers, x-raying and projection of an element inside without the necessity do destroy it, as well as for analysis of stresses and temperature distribution in objects.

Tracing is one of the basic operations in the technology of machine construction, long used as the method of positioning the machining tool in relation to the object being machined. Even today, tracing is used in many enterprises whose production profile requires such operations. The major problem with this technology is its total dependence on the operator performing the task and its susceptibility to human errors (fatigue, oppression, routine) [15].

* Corresponding author: j.krolczyk@po.opole.pl

 DOI: 10.1051/matecconf/201711201002

Direct setting of a casting on the machine tool is often impossible for various reasons, such as too complex shape of the part under machining, dimensional repeatability, even within one series (low quality of casting plant products) or lack of points on the model which could be a base ensuring correct machining of all the required surfaces. In such a case, the first operation to be done in the technological process used so far is tracing in which an adequately trained employee draws lines which will be a basis for the machine tool operator to adopt machining bases on the machine. The operation of tracing is more and more often avoided to reduce the cost of labour. The solution which is expected to replace tracing is collection of data on the geometrical properties of the actual casting and comparison of them to the 3D CAD model [16, 17]. Then, basing on the comparison, one can determine the values of machining allowances and optimum distribution of them on the casting.

The purpose of the present work was to demonstrate the application of the optical scanning system for the determination of machining allowances in the technology of machining machine tool bodies as an example.

2 Scanning technology

At the moment, there are many technologies of collecting data concerning the shape of an object available in the market. Unfortunately, most of them will not be applied in practice in the case of the castings under discussion due, for example, to their too big overall sizes (x-ray tomography) or to too little data collected in a unit of time (CMM). Fig. 1 shows the division of shape measurement methods, in Fig. 2, one can see the division of optical methods.

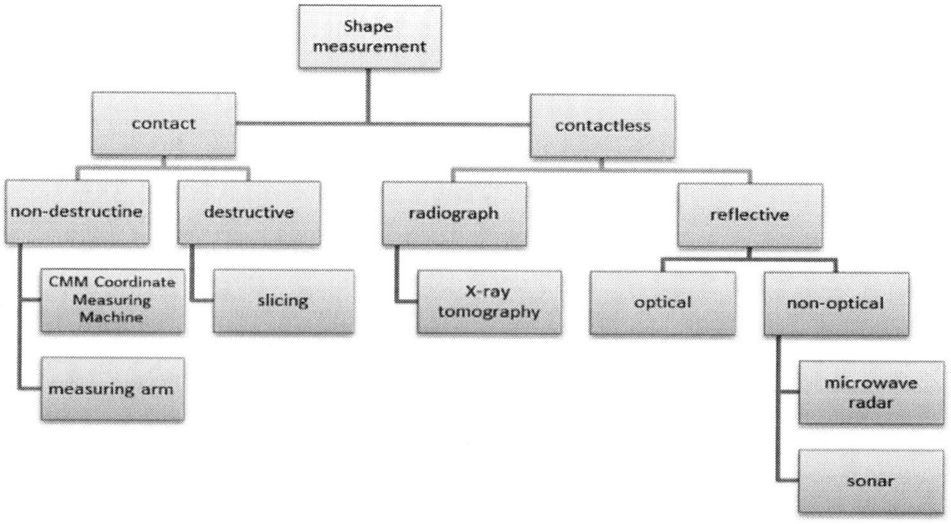

Fig. 1. The division of methods of shape measurement [based on 18, 19].

The optical method, lighting with structural light, has been selected for the scanning system to determine the values of machining allowances on the example of machining the bodies of machine tools. The method has a number of advantages, i.e.:

- non-contact measurement – the casting can be partially dirty or oiled without the hazard of damaging the measurement device,
- possibility of collecting even as many as 4 million points as result of one measurement [20],
- high quantity of data collected at a time reduces the time of operation,

- possibility of automating the measurement - no need to have a trained employee to perform the measurement,
- precision of measurement by an order better than in the case of tracing [20-22] – measurement tolerance lower than actually necessary.

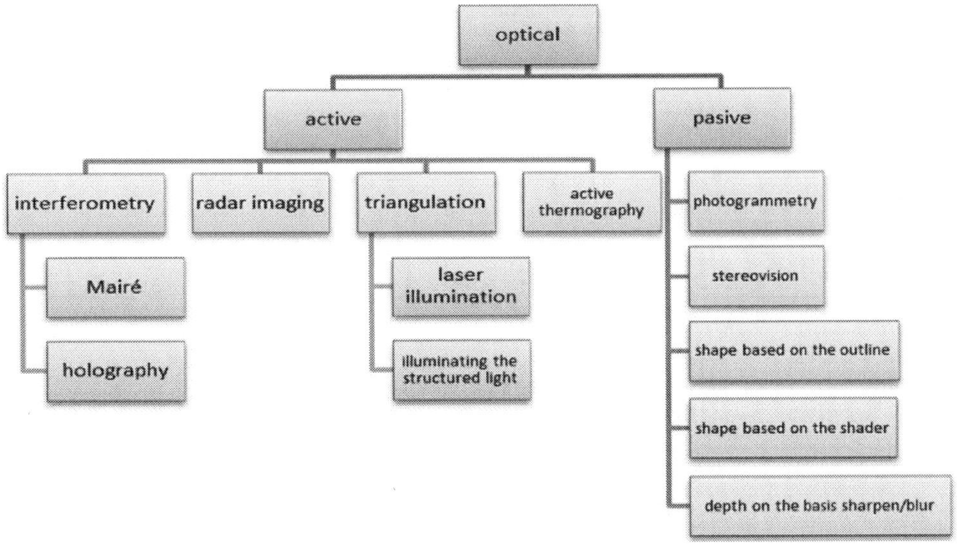

Fig. 2. The division of optical methods [based on 18, 19].

3 Materials and methods

The scanning stand consists of a number of elements:
- overhead crane with the capacity of 5 tons,
- separated space for storing scanned castings and ones to be scanned,
- measuring cell, ATOS ScanBox 6130 (Fig. 3) which comprises industrial robot, Fanuc M-20iA [23], rotating table with provided with a special plate with holes for mounting the device and a measurement probe, ATOS II Triple Scan [24],
- light curtain and sensors of the object presence in the prohibited zone,
- PC computer with ATOS Professional 2016 software.

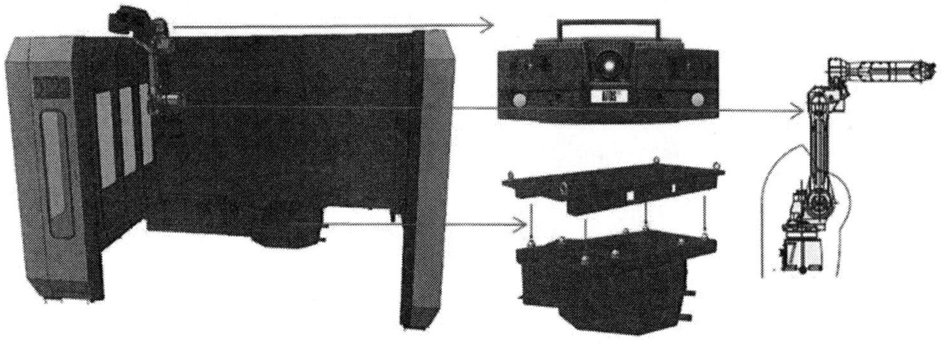

Fig. 3. Measuring cell, ATOS ScanBox 6130 (left): industrial robot, rotating table and measurement probe.

The rotating table (Fig. 3), a part of the stand, is also the seventh controllable axis, without which execution of correct and complete scanning process would be, in most cases impossible. The maximum load of the table determined by the producer is 2000 kg; its range of motion is ±180°.The stand is secured against a man's entering into the scanning zone. The parameters of the ATOS II Triple Scan probe are: cameras – 2x5 Megapixels GigE CCD with the resolving power of 2448 x 2050 pixels, impact resistant casing made of carbon fibre, dimensions about 570 mm x 340 mm x 240 mm, weight about 12 kg, LED lighting – blue light technology. The measuring gate of the measuring cell has dimensions which allow for passing objects with the dimensions of up to 3 metres, however, for practical reasons, the biggest scanned casting had just a little more than 2300 mm in the longest axis, which caused problems with moving it and setting on the rotating table.

4 Technological operations of scanning

4.1. Basing the casting on the device on the scanner stand

Adequate setting of the casting is crucial for the reasons of safety and the process correctness. Incorrect setting can result in a collision because there is a hazard of the scanner probe striking the casting or the device.

4.2. Construction of the scanning device

The next stage is construction of the device for scanning. The device must ensure the object stability and adequate precision of basing. It is also necessary to ensure the possibility of scanning the bottom of the casting and structure covering possibly smallest part of the scanned object.

4.3. Distribution of supports, determinatio0n of the casting position and locating the casting on the supports

The most often used method of distributing supports is setting them experimentally while creating the program. Of course, computer models of the device and the object to be scanned can be used, but one must consider the resulting discrepancies between the theoretical and the actual state. Castings always have allowances, which makes them differ from the CAD models which show a ready-made part. Determination of the centre of gravity is very important due to the centrifugal forces acting in the casting during the table rotation, particularly in case of very bug castings. In such a case, one can assume that the centre of gravity of the casting and the 3D model coincide. Location of the element on the supports, however, should be determined in the experimental way. Frequently, castings have significant allowances resulting from the reduction of the form complexity by the casting plant for technological reasons.

The required accuracy of positioning the model on the scanning device supports is adopted with the tolerance of ±5 mm. This is due to practical reasons – most of the castings machined in the plant have masses of the order of several hundred kilograms and more and are carried by means of the overhead crane located under the ceiling. Very precise setting of an object with such a mass and in such conditions actually infeasible for the crane operator or senselessly time consuming.

4.4. Distribution of reference points

The reference points (RP) allow for combining the individual photos (scans) made by the scanner into an image of the whole. Although there is a technology nowadays which allows us to eliminate the use of reference points, the utilised stand does not possess adequate hardware for it. When the object is so small that can be contained in the measuring space of the scanner, the use of reference points is often not necessary. In other circumstances, particularly in the case of big objects, it is, unfortunately, inevitable. The software can combine the individual scans if there are at least three common reference points on each two scans. The software has the possibility of matching the scans basing on shapes, but the method is less precise and more unreliable.

DOI: 10.1051/matecconf/201711201002

One of the two major factors to be considered when distributing the p[points on the scanned element is the above mentioned required visibility of the common points. Too long distance between them can make it difficult to ensure the presence of 3 reference points in the space of an individual scan. The other factor is masking of the casting surface by the reference points. The points should be located in places which are not subjected to machining because those surfaces are most important in the process.

When distributing the reference points, we can use two methods: gluing (Fig. 4) or magnetic beams (Fig. 5). The simplest method of distributing the points is gluing them on the element (Fig. 4). Fig. 6 shows an example of distribution of magnetic beams on the castings.

Fig. 4. Distribution of glued reference points on casting.

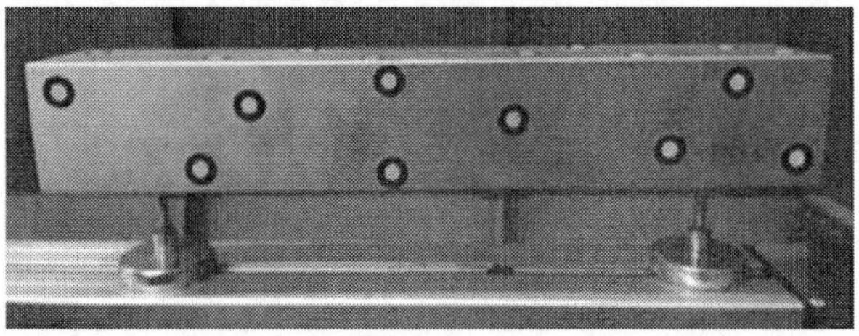

Fig. 5. Magnetic beam.

Fig. 6. Positioning the magnetic beams on casts.

Another solution of distributing RP is the use of previously marked and additionally made magnetic beams (Fig. 5) set on the casting and kept on it thanks to the magnets built into them. The solution shows many advantages and, in a longer period, a more cost-effective option:

- once marked beam can be put on the element and taken off it as many times as necessary,
- much easier ensuring of RP distribution repeatability,
- much shorter time of RP distributing.

Magnetic beams have one important drawback as compared to the RP glued directly on the casting; they mask much larger surface of it thus limiting the number of useful data collected by the scanner. However, the problem can be eliminated by adequate positioning of the scanner probe while creating the program.

4.5 Creation of the scanning program

The possibility of automation of the scanning process is one of its most important advantages as it reduces the task of the stand operator to merely adequate setting of the device, putting the casting on the device and taking it off the device, putting the beams on the casting and taking them off it and actuation of the prepared program.

The program controlling the robot's motions can be created in automatic and manual way. Both in the case of automatic and manual process, writing the program begins with entering the ATOS SCABOX cell into the program or opening of a pattern already possessing the cell. Next models of the scanned object and the scanning device are imported. The device is imported as a VMR element (Visual Measuring Room) while the object as a network of medium solution. The program enables a test scan to be performed in order to match the 3D model with the actual object. In Fig 7, one can see the arrangement of the probe measuring space on the scanned model. It is also visible, basing on the test scan performed, that shifting in relation to the actual object is insignificant.

Fig. 7. Model with visible imposed scanned surfaces.

Although the option of automatic creation of measuring points is available, most measurement series created by the program require manual corrections. When the program is created in the offline mode, the subsequent positions of the scanner probe are created basing on a CAD model located in the measuring space. The program makes it possible to determine which surfaces will theoretically be subjected to scanning (Fig. 8 left) in the given position of the probe. The problem is the discrepancies between the actual state (Fig. 8 right) and the theoretical one (Fig. 8 left).

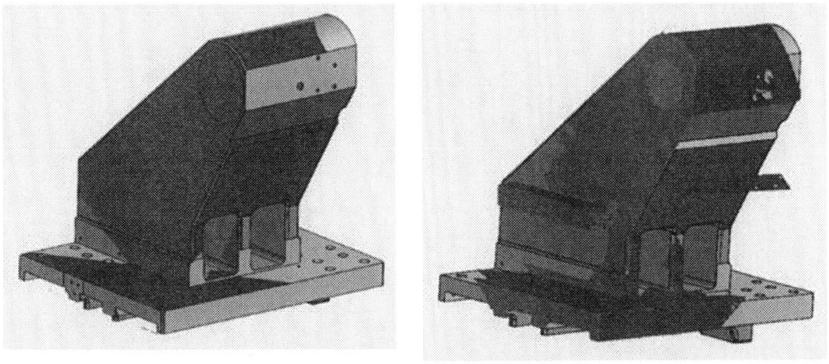

Fig. 8 Surfaces which should theoretically be scanned (marked red) – left; surfaces actually scanned (marked red) – right.

The reason might have been various kinds of physical disturbances, such as, for example light reflexes or ordinary differences between the casting and the theoretical model. The right approach, under the circumstances, is verification of the program written offline on a number of actual castings prior to proceeding to further activities.

When the program is already prepared (regardless of the method applied), the scanning parameters are set and the proper scan is performed in order to collect the data for further processing.

4.6. Traversing

The subsequent stages of the process involve data processing. The data obtained as result of scanning must be adequately processed in order to determine the allowance values and to prepare a report for the machine tool operator.

Scanning renders data in the form of a cloud of points. Often, the device elements or even the rotating table are scanned. Correct performance of further operations requires removal of the unnecessary data. At the same time, it also reduces the file size and the quantity of information for the processor to cope with, which positively influences the time of calculations. Once the unnecessary data are removed, the next step is traversing of the possessed cloud of points.

4.7. Scan basing on a 3D model

The network obtain after traversing can be "described" on the 3D model. In order to perform basing on the basis of definite geometries, it is necessary to mark them on the model and to create separate surfaces. Next, a part of the network corresponding to the defined geometry is marked (Fig. 9). Selection of adequate basing parameters is a very important issue. Inadequate zone of tolerance or the upper or lower limit can result in incorrect basing of the network on the model and, consequently, result in negative allowances on the surfaces not machined. This is inadmissible, particularly in the case when the casting can be set in a way enabling all the surfaces to be correctly machined. Unfortunately those parameters are selected by the method of trial and error checking the allowance values obtained with the subsequent combinations of the parameters. Setting optimum values is often very difficult, particularly when scanning castings from different casting plants.

 DOI: 10.1051/matecconf/201711201002

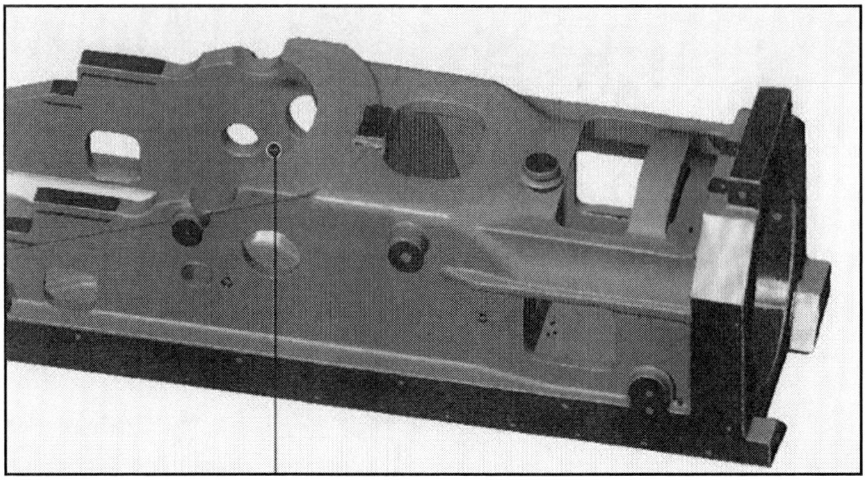

Fig. 9. Surfaces admitted to basing - marked in red.

4.8. Preparation of report

The objective of scanning process is the determination and presentation of the machining allowance values distribution onto the mold. On the basis of the obtained data one can ascertain if the machining of the mold would cause the generation of shortage. Therefore, it is necessary to prepare a report including all required information. The coloristic dimension deviations maps – visual representation of machining allowances on the model are being created (Fig. 10).

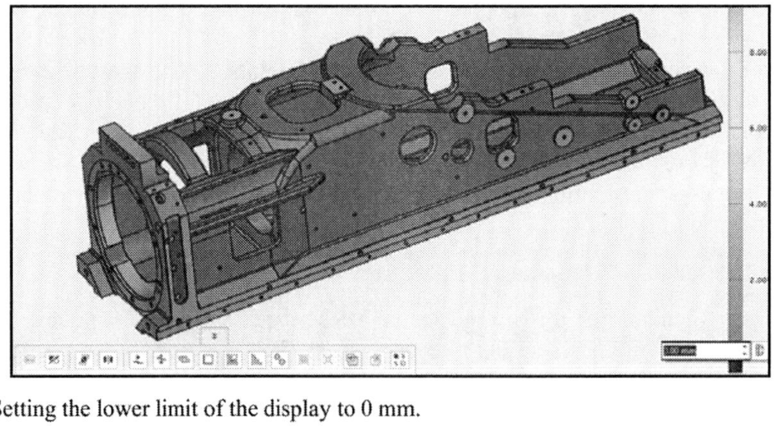

Fig. 10. Setting the lower limit of the display to 0 mm.

Since the zero (or negative) machining allowance values are unacceptable, the lower tolerance boundary is set to 0 mm. It results in a contrast occurring between the positive and non-positive values. It is applied in order to quick identification of a shortage. Despite the common visibility, as well as the possibility of approximate valuation of machining allowance on the basis of the legend located on the right side, the precise determination of allowances in a few points is advisable. It is made by the creation of the allowance labels (Fig. 11).

 DOI: 10.1051/matecconf/201711201002

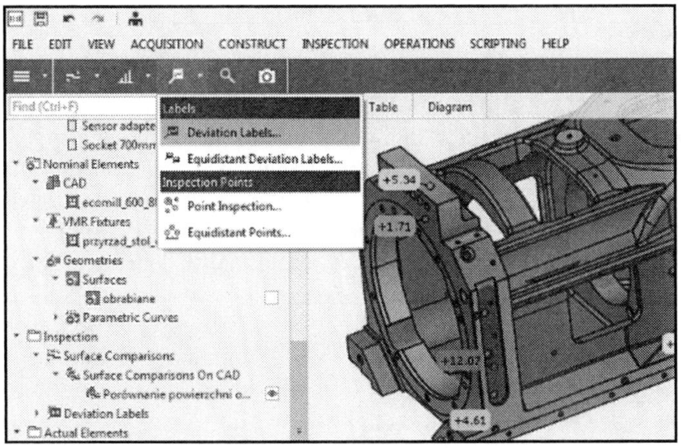

Fig. 11 Creating deviation labels.

The next stage is the determination of allowance values in the cantilever points. The most important areas are the cantilever points located on the machine tool's device. They are often located on the machined surfaces (they are being machined in the subsequent fix). The determination of these values is essential for the appropriate positioning of the mold in the device.

The last parts of creating the report involve the determination of control and fixing points. The control points are being created in the same way as cantilever points. They are prepared in order to define the machining base and validation of fixing carried out on the device. Knowing the machining program, the definition of values which have to be entered during fixing of machine base is possible. The exemplary fixing and control points together with the allowance values are presented in Fig. 12 left and Fig. 12 right respectively. The visible negative allowance is not a problem, because the PBX1 and PBY1 fixing points are not located on the machined surfaces.

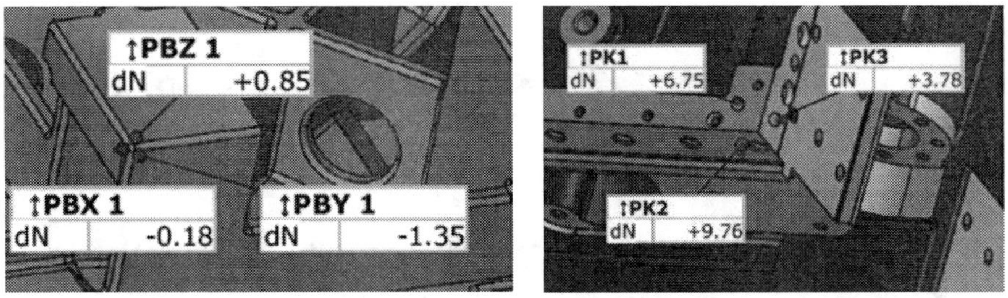

Fig. 12 Basing points (left) and control points (right).

The prepared report should contain any information necessary for the appropriate determination of mold machining validity, as well as its positioning on the machining device.

4.9 Fixing on the machining device

Hitherto the positioning of the mold on the device has been carried out on the basis of line-marking. The adjusting of the level has been checked with the application of altimeter, whereas the rotation of the machine table has been set on the basis of line-marking in the two points. In the process which applies the report generated by scanner, the device is set to the appropriate dimensions without the fixed mold.

5 Conclusions

The introduction of scanner is accompanied by the two primary advantages for company:
- shortening of technological times in relation to line-marking. The time savings obtained on the different mold models differ significantly because of drawing time. In case of some parts, the line-marking in the three different positions is necessary. Furthermore, the change of molds positioning on the line-marking plate is time consuming, especially in case of objects characterized by the mass in range of hundreds of kilograms. The scanning is always carried out in one position, which additionally decreases the preparation and completion times.
- the higher scanning accuracy and possibility of shortage detection. The elimination of such mold during the scanning stage allows the complaint in the casting division and generates substantial savings resulting i.e. from the lack of tool wear during machining and limitation of non-productive machine's application.

References

1. S. Jun, O. Kochan, Meas Sci Rev, **14** (1), 29-34 (2014).
2. A. Kujawińska, M. Rogalewicz, M. Piłacińska, A. Kochański, A. Hamrol, M. Diering, Metalurgija, **55** (4), 821-824 (2016).
3. A. Glowacz, Meas. Sci. Rev., **15**, 167–75 (2015).
4. G.M. Krolczyk, J.B. Krolczyk, R.W. Maruda, S. Legutko, M. Tomaszewski, Measurement, **88**, 176–85 (2016).
5. M. Niemczewska-Wojcik. Measurement, **96**, 8–17 (2017).
6. G.M. Krolczyk, R.W. Maruda, P. Nieslony, M. Wieczorowski, Measurement, **94**, 464–70 (2016).
7. J. B. Królczyk. Int. Agrophys., **30** (2), 193-202 (2016).
8. M. Merola, A. Ruggiero, J.S. De Mattia, S. Affatato, Measurement, **90**, 365–71 (2016).
9. P. Nieslony, G.M. Krolczyk, K. Zak K, R.W. Maruda, S. Legutko, Precis. Eng., 47, 104–10 (2017).
10. C. Zhang, Z. Li, C. Hu, S. Chen, J. Wang, X. Zhang, Meas. Sci. Technol., **28**, 35102 (2017).
11. D. Lehocka, J. Klich, J. Foldyna, S. Hloch, J.B. Krolczyk, J. Carach, G.M. Krolczyk, Measurement, **82**, 375–383 (2016).
12. S. Wojciechowski, P. Twardowski, M. Pelic, Procedia CIRP, **14**, 113-118 (2014).
13. S. Wojciechowski, R.W. Maruda, P. Nieslony, G.M. Krolczyk, Int. J. Mech. Sci., **119**, 360-369 (2016).
14. S. Wojciechowski, P. Twardowski, M. Pelic, P.W. Maruda, S. Barrans, S., G.M. Krolczyk, Precis. Eng., **46**, 158-165 (2016).
15. S. Legutko. **42** (2), 8-16 (2009).
16. P. Hakala, *Direct Modeling in Global CAD Environment,* (Master of Science Thesis, Tampere University of Technology, 2015).
17. J. Liu, X. Liu, Z. Ni, H. Zhou, J. Intell. Manuf., **1-18** (2016).
18. 1B. Curless, *Overview of active vision techniques* (SIGGRAPH 99 Course on 3D Photography, 1999).
19. J. Geng, *Structured-light 3D surface imaging: a tutorial* (Optical Society of America, 2011).

20. A. Gessner, *Photogrammetry and scanning in machine tool body technology* (Publishing House of Poznan University of Technology, 2015).
21. K. Derejczyk, P. Siemiński, Mechanik, **4**, 312-313 (2016).
22. J. Erbel, *Encyclopaedia of manufacturing techniques used in the machinery industry* (Publishing House of the Warsaw University of Technology, 2001).
23. http://www.fanuc.eu/pl/pl/roboty/robot-strona-filtrowania/m-20-series/m-20ia
24. http://www.gom.com/pl/systemy-pomiarowe/atos/atos-triple-scan.html

MATEC Web of Conferences 112, 01003 (2017) DOI: 10.1051/matecconf/201711201003
IManE&E 2017

Productivity improvement of the water meter's body manufacturing process

Jolanta B. Krolczyk[1*], *Stanislaw* Legutko[2], and *Agnieszka* Soberska[2]

[1]Opole University of Technology, Faculty of Mechanical Engineering, 76 Proszkowska Street, 45-758 Opole, Poland
[2]Poznan University of Technology, Faculty of Mechanical Engineering and Management, 3 Piotrowo Street, 60-965 Poznan, Poland

Abstract. In the paper, the water meter's body manufacturing process, in the aspect of its productivity improvement was analyzed. The analysis was focused on the identification of the fundamental problems of manufacturing company, related to the productivity, and subsequently on the ideas of their elimination (alternatively limitation). The paper was focused on the analysis of problems influencing productivity and ideas of their elimination. The experimental research results related to the productivity, generation of production shortages and also the dimensional deviations of the water meter's body were also presented.

1 Introduction

Most contemporary manufacturers are making efforts, the essence of which is the determination of the ways of acquisition or maintenance of their position in the competitive environment [1]. More efficiently process that the greater will be the companies' productivity [2]. The main fault of companies is a too high net cost of product due defects [3]. The defects of manufactured products can be caused by the inappropriately selected technological input parameters, condition and properties of a machine, as well as the process dynamics [4-6]. Several techniques have been used successfully aiming at reducing the manufacturing defects [6, 7]. With increased dimension accuracy of the products, the methods of defects analysis of machine components are ever demanding [8, 9].

Quality is an important thing in analysing the aspect of competitiveness, productivity and costs [1]. Increase of productivity can be achieved by improvement of the operational activity and increase of the employees commitment [10]. Operational activity can be improved by increasing the expenses on investigations and development works. More expenses on R&D will contribute to identification of new products and new applications for the existing ones and new methods of manufacturing them [11].

The major purpose of the present work is an analysis of the process of production of water meter body in a factory of fittings in order to increase productivity. The scope of the work includes identification of the major problems of the enterprise influencing the

* Corresponding author: j.krolczyk@po.opole.pl

productivity of the water meter body manufacturing. Concept of eliminating or reducing the problems is presented also in this paper.

2 Materials and methods

The production process analysis has been performed in an enterprise located in Poland producing fittings and devices for liquid propane-butane gas, as well as for technical gases. The factory cooperates with companies of all over the world, exporting over 50% of its production to the countries of Europe and USA to the automotive branch (40%), building market (30%), installation (15%) and gas markets (15%). The main assortment offered by the companies includes: rings of gear box synchronizer; machined water meter bodies (Figure 1); gas bottle valves; etc. Since 2010, the strategic product manufactured by the enterprise is the water meter body (Figure 1) the sales of which is growing each year.

Fig. 1. Water meter body.

Fig. 2. A forging with flash.

The constructional elements of the water meter are made of brass, MO 58, stainless steel and plastics: PP, PPO, EPDM, PPA. The technological process includes modification of the form, chemical and physical properties of the material being processed. The blank in the process is an extruded brass rod with the diameter of 40 mm provided by an external supplier. The rod is cut with a transverse saw, ADIGE CM 601, into sections of 46 mm length and mass of 0.485 kg with the tolerance up to 0.005 kg. Disk formed blanks are cut out at intervals of four seconds. The next stage is transport of the blanks from the store to the forging plant. The process of forging is performed on the crank press, ROVETTA PRESSE FO 300. In one 8 hour shift, an employee should make 3478 parts of forgings. The specific time of execution of one part is 6.1 seconds while heating takes 6 seconds. Crank presses are intended for hot forging. In the heater, the blank is heated up to the temperature of about 720°C – 740°C. Improperly heated material is shifted by the feeder to a selected chest. Hot semi product is taken by the manipulator from the discharge channel and put into the die. Next, the process of forging follows resulting in forgings (Figure 2). The next operation is trimming which is performed on the eccentric press, KD 2128 with the force of 63 tons. The next process is cleaning, also referred to as shot peening. The operation is performed on a strip cleaner, OWT-400. It consists in covering the forgings with shot and takes about an hour. The last stage is "final finishing" on a transfer machine tool, GNUTTI. During a shift, the operator should make 2000 parts of the water meter bodies. The company has two such machines working in parallel. After machining, the bodies are dried to get rid of the coolant remaining after machining and to remove chips. Next, assembling is executed. The last auxiliary procedures of the production process include actions related to: preparation of production, transport, quality control and storing.

3 Problems limiting productivity

Productivity is a serious problem of the company because there is high demand of the water meter bodies, the percentage of defective parts is relatively high and production capability is limited. The survey performed in the company shows that over 9500 defective parts were made in August 2014). The percentage of body water meter manufacturing defects comes from metal forming (55%) and machining (45%). This means that 114 000 defective parts can be made in a year; with the annual production determined at the level of 1 200 000, this will generate as much as 9.5% defective parts in the whole production.

In connection with the above, the basic factor increasing productivity would be significant reduction of the number of defects in the products. It has been found that product defects appearing during other stages of processing result from incorrect forging. When the material does not flow correctly during the process of forging, fractures may occur on the collar, as well as too thin or not full fillings on the collar (Figure 3) or on the thread, material delaminations, air cavities on the thread (Figure 4) or in the inlet.

Fig. 3. Defects of machined surfaces due to improper material flow: on the collar.

Fig. 4. Defects as a result of improper flow of material: air cavity

It has been determined during the investigation that machining causes more oscillations in the bottom thickness values for various tested bodies as compared to the proves of forging.

Defective products can also be made on the assembling stand due to employee's inattention, e.g. improper position of the plate, too weak clamping of the plate by the press, fixing of cracked strainers or deflection of the collar. Productivity problems include not only generation of many defective products, but also organization of the working stands. Much time intended for machining is lost for unplanned idling, for example pouring oil, water, cleaning jammed water pipes enabling chips removal, unplanned tool exchange, cleaning, drum failures.

There are also problems with material transport. There are two fork lift operators in a shift. One of them works in the store, the other in the production hall and the forging plant. They must ensure raw material circulation not only for the water meter body, but also for many other products. Additionally, they have to carry out waste and ready made products. 15 minutes of waiting for material on one GNUTTI machine tool results in the drop of productivity by about 64 parts in a shift. Figure 5 shows a diagram presenting the problems influencing the productivity of the water meter body manufacturing. The most important problems are marked with red.

4 Proposals to increase productivity

Recently, the company has bought a modern, twice as productive hydraulic press, MECOL PRESS ME 250, in order to increase the rate and, consequently, productivity of forging while maintaining high quality level. The new press also makes it possible to eliminate trimming of the flash, which will save time and accelerate the whole process. The forgings of the new press are more accurate and more repeatable. The high quality of the forgings influences improvement of the further processes.

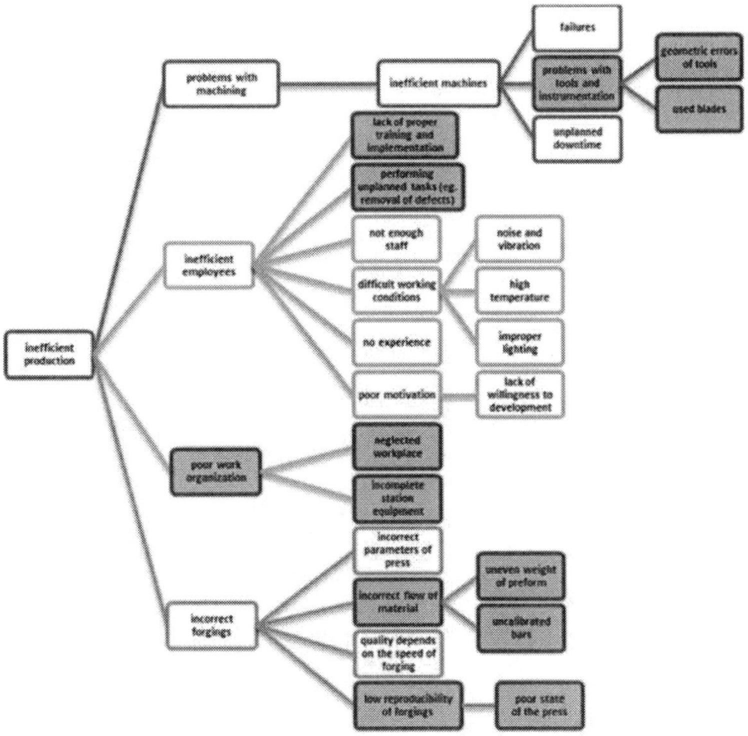

Fig. 5. The problems affecting the production capacity of the body water meter.

The water meter body is a strategic product and that is why actions involved in the production of it have priority. Earlier observation suggests that the bottleneck is caused by the machining process. In order to eliminate unplanned idling on the machines (e.g. waiting for the material, seeking the fork lift operator) one can provide employees in short wave radio or a pager to accelerate the transfer of information from the machine concerning the necessity of transporting the chests with the material.

Another solution to rationalize the machine tool operator's work is to provide him with exchangeable tools. There are several tools working on the machine simultaneously; they get worn and, consequently blunt. In order to analyse the influence of the machine idling time, investigation concerning the number of machined bodies in a period of time has been performed (Figure 6).

The data show that the least number of parts has been machined in the time period of 8.00 a.m to 10.00 a.m. This is due to the machine idling from 8.38 till 9.26 caused by the necessity of regeneration of the cutting tool edges. For this purpose, the operator disassembles the blunt tool from the machine, carries it to the tool-room, regenerates it on a grinding machine and finally, with the help of the foreman, fixes the sharpened tool in the machine. The whole operation takes a lot of time while the machine is not functioning. In order to minimize idling times, the machine could be provided with a set of spare tools, so that the tool can just be replaced during idling, which will take about 21 minutes instead of

DOI: 10.1051/matecconf/201711201003

48 minutes. A responsible and competent person should also be selected, e.g. a foreman whose task will be to make sure that the spare tools are always sharpened and adequately marked to avoid problems with identification of the tool intended for the given operation.

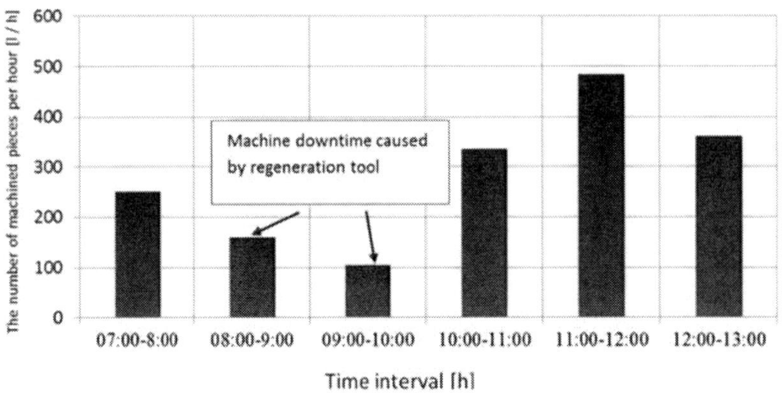

Fig. 6. Number of machined bodies for an hour in the time period from 7:00 to 13:00.

Insufficient training of the employees and their poor experience are also serious trouble. Employees often unwillingly generate defective parts. They cannot check quality in the correct way, due to which defective parts often pass through the whole production process. Additional training of the production employees should be performed.

One of the major causes of defects in the machined bodies (e.g. burrs on the edges, incorrectly machined thread) is excessive wear (blunt edges) of the cutting tools. In some cases, the phenomenon of wear is not identified by the operator at the right time and, consequently, the machines (often many parts) with a blunt edge. In order to reduce the influence of the cutting edge wear on generation of defective parts, one of the methods of the edge condition monitoring could be adopted. Cutting wedge condition can be realized by the following methods: direct ones or indirect ones. From among the indirect methods, the one practically applied is the one based on the measurement of the components of the total force. In order to reduce the number of defective part resulting from blunting of the cutting tool edge, purchase of a stand for supervising the edge condition has been suggested. This involves significant costs of force gauges, amplifiers, software for the analysis and supervision of the edges, as well as modification of the machine tool. Considering that almost half of the defective parts result from the problems of machining, the investment will be paid back in a short time.

Another problem influencing the occurrence of defective parts are dimension errors (mostly on the internal surfaces) which appear after machining. One of the reasons why they arise is run-out of the tool fixed in the spindle of the machine. The value of the run-out depends on geometrical errors of the tool itself, the tool holder, the spindle, as well as the run-out resulting from inaccurate joining of the above elements. This kind of error can be minimized by the application of a DIHART holder with the possibility of elimination of the tool run-out after fixing. The holder is intended for correcting the tool axis in such a way as to align it with spindle axis.

The application of the above mentioned solutions aiming to reduce the dimensional errors of the machined surfaces involves the necessity of purchasing new tools and holder. The involved costs, however, can be compensated by improvement of the dimensional accuracy of the machined surfaces and, consequently, reduction of the number of defective parts.

5 Conclusions

1. The major factors influencing productivity of manufacturing the water meter body are: high percentage of defective parts (estimated at almost 10% of the whole year's production), organization of production and the degree of the production personnel training.
2. It has been shown that defective parts result mainly from the process of forging and machining. The problem of defective parts resulting from forging can be partially solved by implementing a new press type MECOL PRESS ME 250. Defective parts resulting from machining can be significantly minimized by introduction of: a stand for supervising the cutting edge condition during machining, application of special tool holders enabling radial tool run-out to be eliminated or purchase of a modern tool set.
3. In order to increase the productivity the water meter body manufacture, a number of production organization modifications can be performed. Those modifications consists in: the inclusion of a wave transmitter or pager provided for the fork lift operators to accelerate information flow, the provision of spare regenerated tools to the operator and training of production personnel with respects to the identification of the most frequent product defects and correct utilization of the measurement devices.
4. Some of the suggested modifications, e. g. introduction of the stand for supervision of the cutting edge condition during machining, the use of special tool holders compensating the radial run-out of the tool or purchase of modern tool sets will also make possible the improvement of the quality related to the manufacture productivity the quality an manufacture productivity.
5. In the operation of a transfer machine tool, idling as result of failures due to the machine overload take place. The reason of those failures is probably machining overload. In connection with that, investigation to modify the cutting parameters for reduces the machining power should also be performed.
6. The productivity of the water meter body manufacture could also be increased by introducing adequate actions to increase the staff motivation and remuneration of the personnel.

References

1. J.B. Krolczyk, G.M. Krolczyk, S. Legutko, J. Napiorkowski, S. Hloch, J. Foltys, E. Tama, Teh. Vjesn., **22**, (6), 1447-1456 (2015).
2. S. Jun, O. Kochan, Meas. Sci. Rev., **14** (1), 29-34 (2014).
3. J.B. Krolczyk, B. Gapinski, G.M. Krolczyk, I. Samardzic, R.W. Maruda, K. Soucek, S. Legutko, P. Nieslony, Y. Javadi, L. Stas, Teh. Vjesn., **23** (1), 301-306 (2016).
4. S. Wojciechowski, P. Twardowski, M. Pelic, Procedia CIRP, **14**, 113-118 (2014).
5. S. Wojciechowski, R.W. Maruda, P. Nieslony, G.M. Krolczyk, Int. J. Mech. Sci., **119**, 360-369 (2016).
6. S. Wojciechowski, P. Twardowski, M. Pelic, P.W. Maruda, S. Barrans, S., G.M. Krolczyk, Precis. Eng., **46**, 158-165 (2016).
7. A. Glowacz, Meas. Sci. Rev., **15**, 167-75 (2015).
8. P. Hreha, A. Radvanska, L. Knapcikova, G. M. Królczyk, S. Legutko, J.B. Królczyk, S. Hloch, P. Monka, Metrol. Meas. Syst., **22** (2), 315-326 (2015).
9. D. Lehocka, J. Klich, J. Foldyna, S. Hloch, J.B. Krolczyk, J. Carach, G.M. Krolczyk, Measurement, **82**, 375-383 (2016).
10. R.W. Griffin, *Fundamentals of management* (Cengage Learning, 2013).

The application of response surface method to optimization of precision ball end milling

Szymon Wojciechowski[1,*], *Radosław* W. Maruda[2], *Grzegorz* M. Królczyk[3]

[1] Poznan University of Technology, Faculty of Mechanical Engineering and Management, Institute of Mechanical Technology, Poznan, Poland
[2] University of Zielona Gora, 4 Prof. Z. Szafrana street, 65-516 Zielona Gora, Poland
[3] Opole University of Technology, 76 Proszkowska St., Opole 45-758, Poland

Abstract. This paper is focused on the multi criteria optimization of precision ball end milling process of hardened 55NiCrMoV6 steel. The proposed method enables the selection of optimal input parameters which affect the minimization of cutting forces and vibrations signals, as well as the maximization of process efficiency. The experiment includes the measurement of forces and vibrations during the milling tests with variable input parameters. Ultimately, the optimization of the ball end milling process with the application of response surface method is carried out.

1 Introduction

Ball end milling of molds and dies made of hardened alloy steels is very often conducted as a finish process, which consequently imposes restrictive requirements on machined surface's quality and tool's condition. The surface roughness of the machined surfaces can be affected by the kinematic-geometric parameters, frictional effects in the tool-work material inter-face [1], process stability [2], plastic-elastic deformations of work material induced by the ploughing mechanism [3, 4], as well as the thermal phenomena in the cutting zone [5-7]. However, one of the most significant factors affecting milled surface quality is tool-work piece relative displacement (vibrations) [8].

The vibrations are mainly induced by the cutting forces and milling process kinematics, and they are sensitive to the selected machining parameters and tool's slenderness [9]. In addition, during ball end milling of curvilinear surfaces, the process dynamics is significantly affected by the surface inclination angle [10, 11]. Thus, the selection of cutting parameters and milling strategy which enable the reduction of forces and vibrations with the simultaneous maximization of process productivity is a significant task. In order to achieve it, the machining process optimization procedure can be applied.

The state of the art reveals that machining process' optimization objectives include mainly surface roughness [12], cutting forces [13], tool life [14] and tool wear [15]. The most popular optimization approaches are based on the determination of Taguchi's signal-to-noise ratio, grey relational analysis and response surface method. Durakbasa et al. [16] applied the Taguchi method for the optimization of process parameters and different

* Corresponding author: sjwojciechowski@o2.pl

coating materials, combined with different tool radiuses to obtain the minimum surface roughness values during end milling process of AISI H13 hot work steel in dry cutting conditions. Zhou et al. [17] carried out the optimization of multi-axis ball-end milling of Ni-based superalloy Inconel 718 with the application of grey relational analysis. The purpose of the study was to simultaneously obtain minimum surface roughness and maximum compressive residual stress by optimizing the inclination angle, cutting speed, and feed. Masmiati et al. [18] applied the response surface method (RSM) to the optimization of surface integrity after ball end milling of S50C steel. Through the analysis, it was found that minimum residual stress and cutting force can be achieved during up milling, by adopting the MQL-SiO2 nanolubrication system.

Despite the comprehensive research regarding the optimization of milling, the minimization of forces and vibrations with the simultaneous maximization of process efficiency has not been reported in the literature. Therefore, this study is focused on the optimal selection of surface inclination angle α and feed per tooth f_z, in order to minimize the forces and vibrations generated during ball end milling and maximize the process efficiency. The proposed optimization procedure is carried out with the application of response surface method.

2 Experimental details

2.1 Work and tool materials

The experiments have been conducted on hardened alloy steel 55NiCrMoV6 sample with average hardness of 58 HRC. The selected tool was monolithic ball end mill made of fine-grained sintered tungsten carbide (WC) with diameter D=10 mm and number of teeth z=2. The tool's geometry was as follows: γ_o = -15°, λ_s = 30°, α_o = 6°. Experiments were carried out on 5-axes CNC milling workstation (DMU 60monoBLOCK).

2.2 Research range and method

Experimental milling tests involved the measurement of cutting force components and acceleration of vibrations carried out in the range of variable surface inclination angles α and feed per tooth f_z. In order to obtain surface inclination during the milling process, tool's rotational axis was inclined against the work piece in the range of various α angles (Fig. 1).

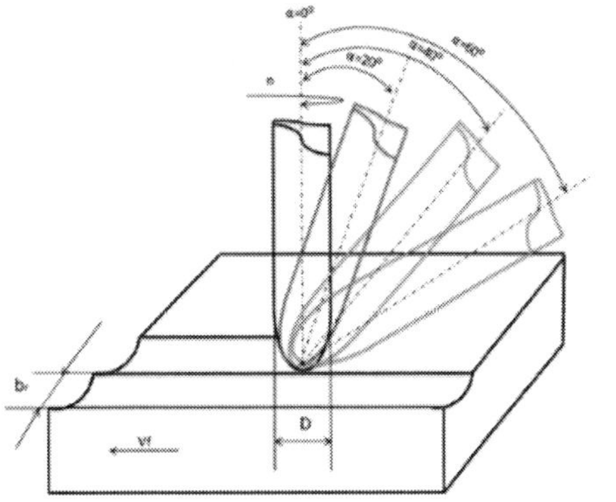

Fig. 1. Ball end milling kinematics.

In all investigated cases tool's effective diameter was lower than the value of pick feed – D_{ef}<b_r. The milling parameters applied in the research were depicted in table 1.

Table 1. Design of experiment.

Cutting speed v_c [m/min]	Depth of cut a_p [mm]	Pick feed b_r [mm]	Inclination angle α [°]	Feed per tooth f_z [mm/tooth]	Overhang l [mm]
200	0.3	3.5–3.9	0 – 60 20 interval	0.05 – 0.2 0.05 interval	63

The forces and accelerations of vibrations were measured in machine tool's coordinates, in the three directions (A_x [m/s^2], F_x [N] –feed normal direction, A_y [m/s^2], F_y [N] - feed direction, A_z [m/s^2], F_z [N] - thrust direction) with the application of the 3-directional piezoelectric Kistler 9257B force dynamometer and Brüel&Kjaer 4321 accelerometer. The piezoelectric sensors were connected through amplifiers and band-pass filters to a data acquisition computer. The dedicated software was applied for the calculation of signal's peak values and filtration carried out with a low-pass filter.

2.3 Optimization based on response surface method

The multi-criteria optimization procedure was applied in the research. This optimization is based on the transformation of the approximated output factors' values into the singular value of the resultant total utility (desirability) U_{tot}. This function is based on the idea that the "quality" of a process that has multiple quality characteristics (multi-criteria), with one of them outside of some "desired" limits, is completely unacceptable. The method finds operating conditions that provide the "most desirable" response values. Thus, the problem of simultaneous optimization of many output factors is simplified to the determination of one output parameter, which maximizes total response utility.

In the carried out research, the following optimization criteria were adopted:

- cutting force and acceleration of vibration components (A_x, A_y, A_z, F_x, F_y, F_z),
- material removal rate Q_V.

In the first part, the milling tests conducted on the basis of the central composite experimental design (with 16 repetitions and 1 block) were carried out. The material removal rate was calculated on the basis of equation:

$$Q_v = a_p \cdot b_r \cdot v_f \tag{1}$$

The obtained milling tests' results were the starting point to the definition of response utility profile, formulated for the investigated output factors (F_i, A_i, Q_V). During this optimization stage, all output factors' values were assigned to the partial utilities, which values were contained in the range: <0, 1>. The equal weights of all responses have been selected. The response utility profiles for the responses were formulated in accordance with the recommendations presented in [19], on the basis of the following expressions:

$$Q_v \leq 3000 \text{ mm}^3/\text{min} \rightarrow U_{(Qv)} = 0; \tag{2}$$

$$3000 \text{ mm}^3/\text{min} < Q_v \leq 4000 \text{ mm}^3/\text{min} \rightarrow 0 < U_{(Qv)} \leq 0.5; \tag{3}$$

$$4000 \text{ mm}^3/\text{min} < Q_v < 5000 \text{ mm}^3/\text{min} \rightarrow 0.5 < U_{(Qv)} < 1; \tag{4}$$

$$Q_v \geq 5000 \text{ mm}^3/\text{min} \rightarrow U_{(Qv)} = 1. \tag{5}$$

$$A_{x,y,z} \leq 50 \text{ m/s}^2 \rightarrow U_{(A)} = 1; \tag{6}$$

$$50 < A_{x,y,z} \leq 100 \text{ m/s}^2 \rightarrow 0.5 \leq U_{(A)} < 1; \tag{7}$$

$$100 < A_{x,y,z} < 150 \text{ m/s}^2 \rightarrow 0 < U_{(A)} < 0.5; \tag{8}$$

$$A_{x,y,z} \geq 150\ \mathrm{m/s^2} \quad \rightarrow \quad U_{(A)} = 0; \tag{9}$$

$$F_{x,y,z} \leq 100\ \mathrm{N} \quad \rightarrow \quad U_{(F)} = 1; \tag{10}$$

$$100\ \mathrm{N} < F_{x,y,z} \leq 150\ \mathrm{N} \quad \rightarrow \quad 0.5 \leq U_{(F)} < 1; \tag{11}$$

$$150 < F_{x,y,z} < 200\ \mathrm{N} \rightarrow \quad 0 < U_{(F)} < 0.5; \tag{12}$$

$$F_{x,y,z} \geq 200\ \mathrm{N} \quad \rightarrow \quad U_{(F)} = 0. \tag{13}$$

After the definition of response utility profiles, the optimal solution's searching procedure was applied. This procedure included the mesh definition of the input variables (f_z, α). Subsequently, the output factors' predictive values were calculated for the all mesh points and as a result, the partial utility values were received. In order to determine the total utility value, the mean geometric values of partial utilities were calculated in the each mesh point. The next step involved the formulation of the total utility model, with the application of spline function. After the generation of this model, the optimal solution searching algorithm was adopted. The optimal solution determines the values of input variables, corresponding to maximal value of the total utility. In order to achieve it, the simplex method was applied. Ultimately, the optimal input parameters' (f_{z_opt}, $\alpha_{\_opt}$) values were received.

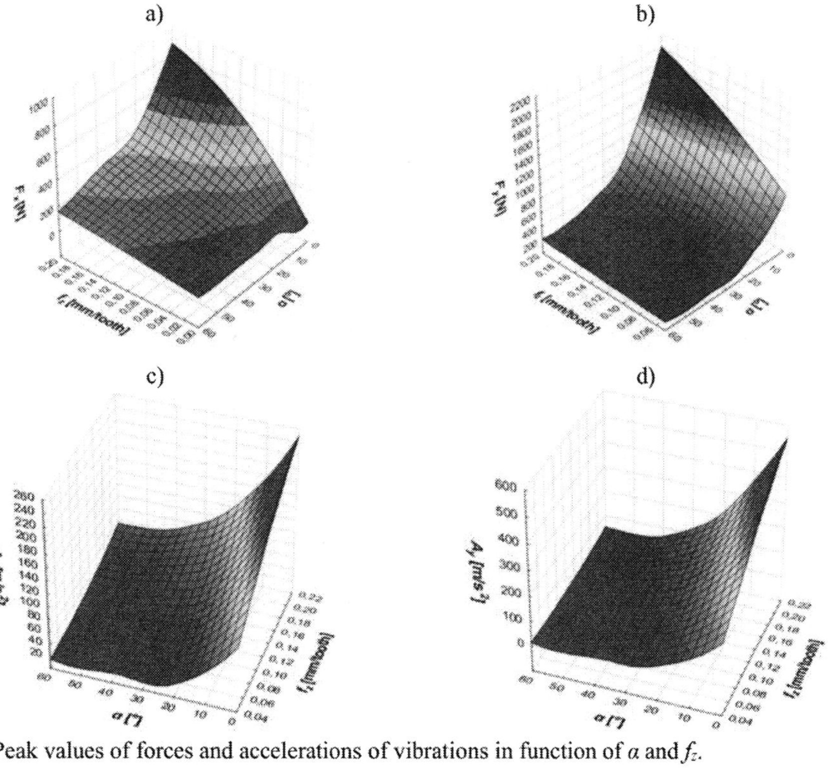

Fig. 2. Peak values of forces and accelerations of vibrations in function of α and f_z.

3 Results and discussion

The Figure 2 and Table 2 depict the measured peak values of forces (F_x, F_y) and accelerations of vibrations (A_x, A_y) in function of surface inclination angle α and feed per tooth f_z. Figures 2a,b reveal that increasing feed per tooth, f_z, induces an increase in cutting forces, which is a typical relation found during metal cutting. Furthermore, the most intense effect of f_z factor on forces is found for the lower investigated surface inclinations ($\alpha \leq 20°$).

It can be also seen that the growth of the α angle induces the decline in force values. According to [13], during slot milling ($\alpha = 0$), cutting speed near ball end mill's rotational axis is very low. Consequently, the large volume of the material which flows toward the cutter is not being transformed into the chip, but as a result of the large elastic-plastic deformations is pressed under the tool's flank face. This phenomenon is followed by the generation of the relatively high ploughing forces, which consequently contribute to the growth of cutting force components. Figures 2c and 2d reveal also that the qualitative influence of the α and f_z factors on the accelerations of vibrations is similar to that observed for the cutting forces. This observation is attributed to the proportionality of tool's displacements to the value of force acting on the tool's working part, according to the milling process' dynamic vibration model [19].

The resultant total utility function for the surface inclination angle α and feed per tooth f_z is presented in Figure 3. It can be concluded (Figure 3) that the optimum cutting parameters are for the higher ranges of inclination angle and the feed per tooth. The maximal value of utility function is achieved with feed per tooth f_{z_opt} = 0.163 mm/tooth and surface inclination angle $\alpha_{\_opt} = 45°$. The application of these optimal cutting parameters gave cutting forces lower than 170 N, acceleration of vibrations lower than 55 m/s^2 and material removal rate higher than 3820 mm^3/min.

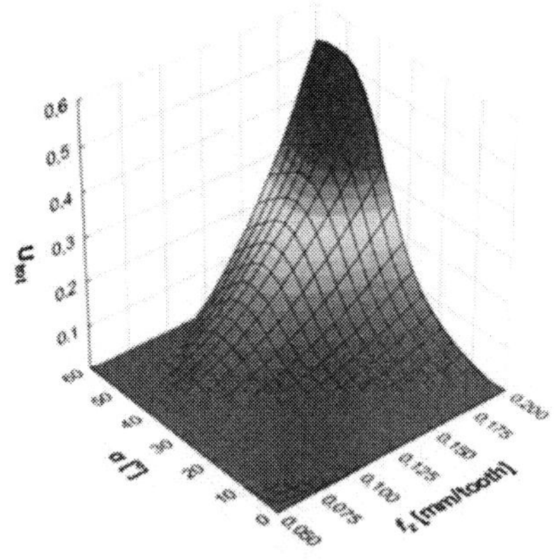

Fig. 3. Total utility function for the α and f_z.

Table 2. Experimental results.

α, °	f_z, mm	F_x, N	F_y, N	F_z, N	A_x, m/s^2	A_y, m/s^2	A_z, m/s^2
0	0.05	292	812	541	111	207	188
0	0.1	560	1276	970	162	318	252
0	0.15	771	1728	1337	200	424	335
0	0.2	881	2000	1656	237	517	383
20	0.05	168	220	81	24	50	59
20	0.1	230	274	149	46	91	99
20	0.15	307	403	266	76	142	142
20	0.2	363	663	451	107	186	182
40	0.05	91	120	62	20	20	28
40	0.1	153	186	65	30	30	37
40	0.15	211	244	87	45	39	65
40	0.2	277	292	120	57	53	88
60	0.05	65	119	93	15	14	23
60	0.1	94	143	89	19	15	27
60	0.15	130	159	92	33	26	47
60	0.2	156	213	98	47	44	48

4 Conclusions

On the basis of the carried out investigations, the following conclusions are formulated:

- The α and f_z factors have qualitatively similar effect on forces and vibrations' values generated during precision ball end milling of hardened steel. Furthermore, the growth of the α angle induced the decline in force and vibration values, whereas the increase of f_z caused the growth of the measured quantities.
- On the basis of the carried out optimization procedure, the following optimal milling parameters were obtained: f_{z_opt} = 0.163 mm/tooth, $\alpha_{\_opt} = 45°$.
- The selection of the optimized α and f_z can result in the growth of productivity and reduction of dynamical load during milling, and thus improvement of machined surface quality and tool's life.

References

1. R. Kumar, S. Chattopadhyaya, S. Hloch, G.M. Krolczyk, S. Legutko, Maint. Reliab. **18**, 128–135 (2016)
2. R. Rusinek, M. Wiercigroch, P. Wahi, Procedia CIRP **31**, 429–434 (2015)
3. S. Wojciechowski, R.W. Maruda, P. Nieslony, G.M. Krolczyk, International Journal of Mechanical Sciences **119**, 360–369 (2016)
4. S. Wojciechowski, P. Twardowski, Procedia CIRP **1**, 289 – 294 (2012)
5. D. Przestacki, P. Szymanski, S. Wojciechowski, Composites Part A: Applied Science and Manufacturing, **91**, (1), 370–379 (2016)
6. D. Przestacki, M. Jankowiak, Journal of Physics: Conference Series **483** (2014)
7. D. Przestacki, Procedia CIRP **14**, 229–233 (2014)
8. S. Wojciechowski, P. Twardowski, M. Pelic, R.W. Maruda, S. Barrans, G. Krolczyk, Precision Eng **46**, 158-165 (2016)
9. L.N. López de Lacalle, A. Lamikiz, J.A. Sanchez, M.A. Salgado, Int J Adv Manuf Technol **24** 621–631 (2004)
10. S. Wojciechowski, P. Twardowski, M. Pelic, Procedia CIRP **14**, 113 – 118 (2014)
11. S. Wojciechowski, Int J Mach Tools Manuf **89**, 110-123 (2015)
12. T. Kivak, Measurement **50**, 19–28 (2014)
13. S. Karabulut, Measurement **66**, 139–149 (2015)
14. G.M. Krolczyk, P. Niesłony, S. Legutko, Arch Civil Mech Eng **15** (2) 347-354 (2015)
15. X. Wang, I. S. Jawahir, International Journal of Production Research **43**, 3543–3559 (2005)
16. M. N. Durakbasa, A. Akdogan, A. S. Vanli, A. G. Bulutsuz, Measurement **68**, 92–99 (2015).
17. J. Zhou, J. Ren, C. Yao, Measurement, doi: 10.1016/j.measurement.2017.01.057 (2017)
18. N. Masmiati, A.A.D. Sarhan, M.A.N. Hassan, M. Hamdi, Measurement **86**, 253–265 (2016)
19. S. Wojciechowski, T. Chwalczuk, P. Twardowski, G.M. Krolczyk, Arch Civil Mech Eng, **15**, Issue 4, 798-805 (2015)

Requirements in designing a device for experimental investigation of threading accuracy

Laurențiu Slătineanu[1], *Miroslav* Radovanovic[2], *Margareta* Coteață[1,*], *Irina* Beşliu[3], *Oana* Dodun[1], *Ionel* Coman[1], and *Sergiu-Constantin* Olaru[1]
[1]"Gheorghe Asachi" Technical University of Iaşi, Department of Machine Manufacturing Technology, Blvd. D. Mangeron, 59A, Iaşi, Romania
[2]University of Niš, Department of Production Engineering, Aleksandra Medvedeva 14, 18000 Niš, Serbia
[3] "Ştefan cel Mare" University of Suceava, Department of Mechanics and Technology, Universității Street, 13, Suceava, Romania

Abstract. When the threads are obtained by using turning tools, various factors are able to affect the machining accuracy. In order to reveal the way in which these factors exert influence on the threading accuracy, an analysis of some proper situations of obtaining threads on universal lathes and using turning tools was developed. The analysis showed that the thread profile error could be essentially affected by the position of the cutting tool active edges relative to the workpiece. Taking into consideration the results of analysis, some aspects concerning the requirements to be met during the design of a device that could be used in experimental investigation of influence exerted by some factors on the threading accuracy are presented.

1 Introduction

Threads are surfaces obtained by moving a certain profile along a helix curve. They are used in order to achieve a motion (threads of moving), a demountable assembly (clamping threads) or to meet special objectives (special threads). There are various criteria applicable when classifying the threads (measuring system, surface on which they are placed, the position of this surface (internal or external), direction of winding, type of profile moved along the helix curve, number of thread spirals etc.

If the problem of obtaining an external thread is stated, one could take into consideration the plastic deformation processes (rolling), the cutting processes (turning, milling, grinding), the use of distinct nonconventional machining processes etc.

The main geometrical parameters of metric external cylindrical thread are considered the average diameter, the pitch and the half of the profile angle. Some other secondary parameters could be also additionally considered: major diameter, minor diameter, the profile theoretical height, the profile real height, the angle of helicoid curve, the radius of

* Corresponding author: mcoteata@tcm.tuiasi.ro

rounding profile contour etc. [1, 2, 3, 4]. For each of the above mentioned parameters, adequate values for the tolerances could be established.

There are many factors able to exert influence on the thread machining accuracy: the wear of technological system components, the geometrical accuracy of the technological system components, the elastic deformation of the technological system components under the action of mechanical forces generated by the machining process, the positioning of the tool point relative to the workpiece surface etc.

Two of the thread profile errors that are analyzed when the problem of thread accuracy is formulated are the flank profile error and the error of the so-called pitch diameter. There are some groups of factors able the affect the sizes of the above mentioned two errors:

- the position of the tool cutting edges relative to the axial plane of workpiece;
- the angular position of the turning tool axis;
- the errors of the cutting tool profile etc.

Distinct aspects concerning the obtaining of threads on lathe and the ways of analyzing their accuracy constituted preoccupation subjects for the researchers interested by the problem of the manufacturing threaded parts.

Thus, Khoshdarregi and Altintas investigated the multi-point thread turning operation [3]. They considered that the threads accuracy and surface quality are crucial aspects that must be considered in order to avoid the fatigue failure and shearing of threads flanks. A generalized and systemic approach was proposed to predict the cutting forces in turning threads with multi-point inserts.

Stanciu developed an ample research aiming to highlight the influence exerted by some technological factors (tool profile errors, machine tool geometrical errors, elastic deformation of the technological system components etc.) on the dimensional and profile accuracy of threads obtained on lathes [5]. Some problems concerning the machining accuracy in the case of threaded surfaces of screws gears obtained by thread whirling were investigated by G. Crețu [2], by taking into consideration the influence exerted by the cutting conditions and geometrical parameters of cutting tool on the accuracy of machined surface. Machining errors appeared during the thread whirling process were also investigated by V.V. Merticaru, who addressed the problem of modeling the influence exerted by the cutting tools accuracy and setting the position of cutting tools on the machining accuracy [4].

In this paper, the authors took into consideration the analysis of the possibility to design a device for experimental study of influence exerted by the position of the turning tool active edges on the thread geometrical and profile errors.

2 Hypotheses concerning the influence of various factors on the thread profile accuracy obtained by using a turning tool

In Figure 1, a schematic presentation of the external metric thread profile and some of main its characteristics are presented. The main geometrical characteristics are the nominal (basic) diameter, major and minor diameter, the pitch, the angle of thread flank inclination. If the thread profile is examined, one could notice that this profile has essentially the shape of an equilateral triangle; a rounded zone is placed at the bottom of the thread open space. In such a case, in the axial section of the threaded cylindrical part, the thread flank is represented by a line segment.

In the case of the threaded surface, one of the machining accuracy indicators could be the error relative to the given profile and the error relative to the given shape of the surface, respectively. The error to the given shape of the profile is defined as the maximum distance between the real profile and the adjacent profile of given shape, this distance being measured between the limits of the reference length.

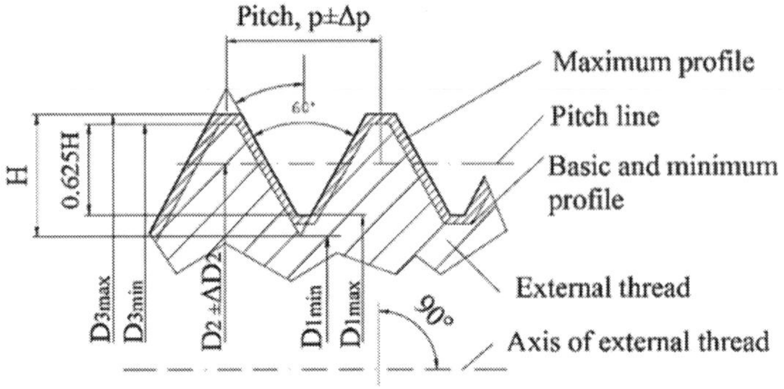

Fig. 1. Characteristics of metric external thread.

On the other hand, the error relative to the given shape of the surface could be defined as the maximum distance between the real surface and the adjacent surface of given shape, between the limits of the reference length.

As above mentioned, a common way of obtaining external threads is based on a machining process developed on a lathe. In such a case, the threaded surface is gradually generated using an apron longitudinal movement strictly correlated with the main shaft rotation. It is known that in the case of universal lathes, the position of the cutting tool in the tool holder is usually established by the lathe operator; in the case of a computer numerical controlled lathe, the position of the threading tool is established out of the machine tool, using a device for pre-setting the position of threaded tool.

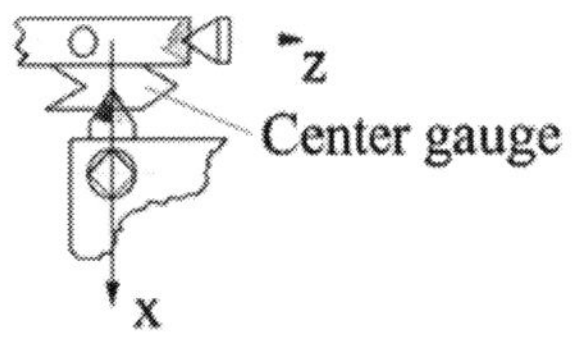

Fig. 2. Positioning the threading tool on universal lathe by means of a center gauge.

When a metric threaded surface is obtained on a universal lathe, in order to correctly position the tool body in tool holder relative to the workpiece axis before starting the threading process, a center gauge could be used (Figure 2). One could mention that even in such a case, however there are various factors deriving from the position of the tool edges able to affect the profile error of the machined thread. Some such situations could be examined.

In a coordinate system *xOyz* (Figure 3), the correctness of thread profile could be determined essentially by the position of the threading tool along the *Oy* axis, and by the possible rotations of the tool around the axes *Ox*, *Oy* and *Oz*. One could mention that a tool positioning error along the *Oz* axis and respectively along the *Oy* axis practically does not affect the thread profile error, but they could exert influence on the sizes of other characteristics corresponding to threaded surface.

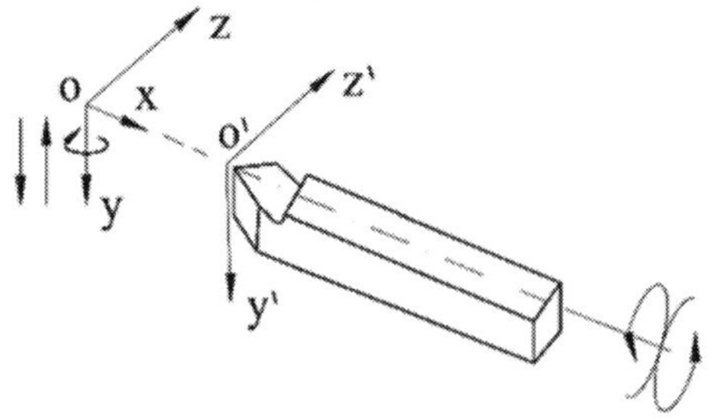

Fig. 3. Main positioning errors of the threading tool.

In the case of a tool lathe characterized by a value equal to zero of the side rake angle, the two tool active edges are placed in the workpiece axial plane. Due to the fact that the rectilinear line segment corresponding to one tool active edge intersects the workpiece axis, if there is not the threading tool longitudinal

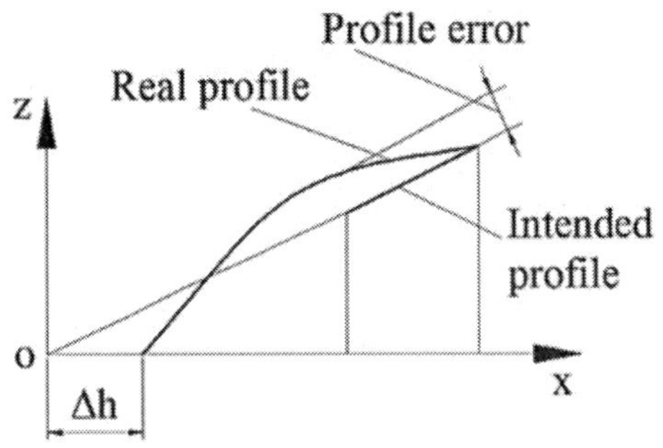

Fig. 4. Error generated by placing the threading tool over or under the horizontal plane of the main shaft rotation axis.

movement, a conical surface could be generated during the transversal feed motion.

If there is only a wrong threading tool position along the *Oy* axis (this meaning that the tool corner is placed over or under the horizontal plane of the workpiece axis), the threading tool rectilinear edge will not intersect the workpiece axis and a hyperbolic surface will be generated (Figure 4). The minimum distance Δh between the line corresponding to the threading tool cutting edge and the lathe main shaft rotation axis will correspond to the distance between the tool corner and the horizontal plane of the main shaft rotation axis (Figure 4).

If in the *xOyz* coordinate system origin *O* is placed on the workpiece axis (Figure 4) and the threading tool has a position rotated relative to the *Oz* axis, practically profile errors will not appear; but one could mention that in such a situation, an error of the thread basic and minor diameters could be generated. From theoretical point of view, the major diameter of the external threaded surface could be determined by the diameter of the initial cylindrical surface corresponding to the workpiece. Practically, due to a process of plastic deformation of workpiece material, a certain increase of the threaded surface major diameter could be observed.

If the origin *O'* of the coordinate system is placed in a position corresponding to the bottom of the future threaded surface and the threading tool is rotated around the new *Oz'* axis (Figure 3), the threading tool active edge does not intersect the workpiece rotation axis and a hyperbolic surface is also generated.

If the threading tool has a position rotated relative to the *Ox* axis, one could notice that the rectilinear line segment corresponding to one of the threading tool active edges also does not intersect the workpiece axis and a hyperbolic surface could be generated as a consequence of a transversal movement achieved by the threading tool and when there is not an apron longitudinal movement.

It is known that during the process of setting the position of threading tool, some errors could affect each of the above mentioned aspects of positioning the threading tool.

In all this cases, an experimental research could show the influence exerted by the positioning errors corresponding to the threading tool and their weight in the total positioning error, but a theoretical appreciation of these errors could be also useful.

In the case of the threading tools whose cutting edges are not placed in the plane which contains the rotation axis of lathe main shaft, the two cutting edges do not intersect the rotation axis *Oz* and, when there is not the apron longitudinal movement, a hyperbolic surface could be

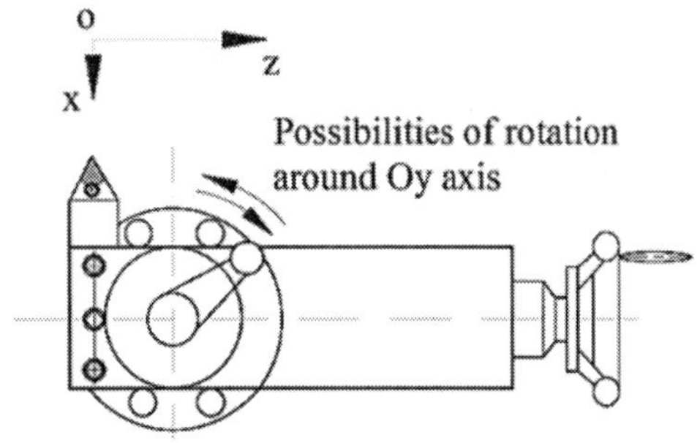

Fig. 5. Rotation of threading tool around the Oy axis ensured by the rotation of universal lathe usual tool holder.

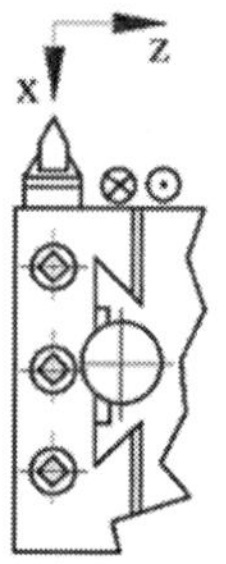

Fig. 6. Tool holder for positioning the threading tool along the Oy axis.

generated instead of a conical one. Taking into consideration this situation, an estimated theoretical evaluation of the machined thread shape error could be developed, by determining the distance from the hyperbolic surface to the conical surface, in the space interval corresponding to the threaded surface of interest (fig. 4).

One of the frequent errors able to affect the thread parameters is generated by the position of the turning tool.

3 Problems and solutions in the case of a device for experimental investigation of the profile error when using a turning tool

In order to conceive and materialize a device for experimental research of the profile errors corresponding to an external metric thread in accordance with the above mentioned aspects, some general requests could be addressed.

As essential requests concerning this device, one could take into consideration the possibilities of changing the position of the threading tool along the *Oy* axis and of rotating the threading tool around the axes *Ox*, *Oy*, *Oz*, respectively.

If the conditions offered by a turning process developed on a universal lathe are analyzed, one could notice that the rotation around *Oy* axis could be relatively easily ensured by rotating the slide of the tool holder around the vertical axis (Figure 5). Usually, this operation needs the unscrewing some nuts and the rotation of the guide slide following the angular indication existing on the apron. In this way, one could notice that the searched device could ensure the positioning of the threading tool along the *Oy* axis and the two rotations around the axis *Oz* and *Ox*, respectively.

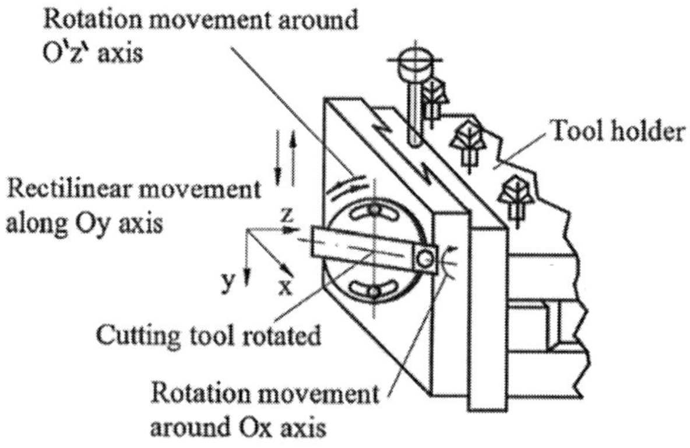

Fig. 7. Schematic representation of the device for the experimental study of influence exerted by some positioning errors of the threading tool on the thread profile.

It is known that usually, to position the turning tool so that the tool corner is placed in the horizontal plane of the main shaft rotation axis, some calibrated blades could be placed under the turning tool up to obtaining the desired position of the turning tool. The operation is sometimes achieved by using the position of the live center corner placed in the conical hole from the tailstock. This solution could be also applied in the case of the device for investigation the influence exerted by the position of the threading tool on the machined thread accuracy.

An easier possibility to position the threading tool along the *Oy* axis could include a slide able to be moved and fixed in the desired position, eventually using a measuring scale drawn on the slide or on its guide (Figure 6 and Figure 7).

The rotation of the threading tool around an axis parallel to the *Oz* axis could not be achieved by the common solutions existing on the universal lathe and an adequate constructive solution could be included in the device able to facilitate the investigation of the thread profile errors (Figure 7).

The third rotation of the treading tool that around the *Ox* axis could be inspired by the solution of tools holder necessary in the case of tools used on drilling and milling machines and such a solution could be applied in the case of the device for investigating the machining errors generated by the position of the threading tool when using a universal lathe.

Thus, the main requirements for a device utilizable to experimentally evaluate the influence exerted by the threading tool position on the profile errors of the threaded surfaces, the following ways of changing the position of the threading tool have to be ensured:

- positioning along the *Oy* axis;
- rotations around the axes *Ox*, *Oy*, *Oz*.

4 Conclusions

In principle, cylindrical threaded surfaces are achieved by moving a certain profile along a cylindrical helix. If this profile is an equilateral triangle, a so-called metric thread could be obtained. On the accuracy of the threaded surface, many factors could exert influence. If only the position of the threading tool is considered, one could notice that the profile error of threaded surface could be affected by the position of the threading tool cutting edges relative to the workpiece. In order to highlight the influence exerted by the threading tool position on the threaded surfaces accuracy, an analysis of some specific positions of the threading tool were considered. This analysis showed that the position of the threading tool along *Oy* axis and respectively the rotation of the threading tool around the axes *Ox*, *Oy* and *Oz* or around lines parallel relative to these axes could affect the profile accuracy of the threaded surface. The rotation of the threading tool around *Oy* axis could be easily achieved by rotating the lathe common tool holder. In order to materialize the threading tool positioning along the *Oy* axis and the rotation of the threading tool around *Oz* and *Ox* axes, a device that could be clamped in the lathe common tool holder was conceived. In the future, there is the intention to achieve a device that could be used to experimentally evaluate the influence of the threading tool position on the profile errors of the threaded surfaces.

References

1. Advanced Mechanical Engineering Solutions. Unified screw threads and tolerances calculator (inch), available: http://www.amesweb.info/Screws/AsmeUnifiedInchScrewThread.aspx, accessed: 25.02.2017 (2013)
2. G. Crețu, *Theoretical and practical contributions in the field of gear screws using the principle of thread whirling. Doctoral thesis* (in Romanian) (Iași, Universitatea Tehnică "Gheorghe Asachi", 1997)
3. M.R. Khoshdarregi, Y. Altintas, Generalized modeling of chip geometry and cutting forces in multi-point thread turning, Int. J. Mach. Tools Manuf. **98** (2015)
4. V.V. Merticaru, *Theoretical and experimental researches concerning the accuracy at thread whirling. Doctoral thesis* (in Romanian) (Iași, Universitatea Tehnică "Gh. Asachi", 2003)

5. A.R. Stanciu, *Contributions to the study of accuracy corresponding to threads obtained by turning tools and multipoint turning tools. Doctoral thesis* ("Gheorghe Asachi" Technical University of Iasi, Romania, 2008)

Researches regarding the reducing of burr size by optimising the cutting parameters on a CNC milling machine

Cristina Biriş*, and *Gabriel* Racz

Lucian Blaga University of Sibiu, 10. B-dul Victoriei, Sibiu, România

Abstract. This paper presents some experimental researches regarding burrs dimensions reduction that appear after the milling process together with an approach to reduce or eliminate the burrs resulted after this process. In order to reduce burrs dimensions, the milling process was executed with different cutting parameters and strategies then the results were evaluated.

1 Introduction

Milling is a machining operation performed with a rotating, multi-edge cutting tool. The tool performs programmed feed movements in almost any direction for removing material from a workpiece, achieving a prismatic, polyhedrical or free-form shape [1]. Milling is much more versatile than drilling, turning, slotting or any other material removal operation. Each of the milling tool cutting edges removes a small amount of material, with a limited engagement into the raw piece, generally producing small chips which it is easy to remove from the cutting area. As in all machining processes, the harder the material, the more difficult it is to remove by cutting [2].

A literature survey has shown that many researches were focused upon improving the accuracy of machined parts, by altering the control parameters of the CNC milling machines using comprehensive modeling and simulation approaches [3, 4], or even virtual machine-tools [5] or by using new solutions for actuating the main drives or the feed drives [6], but only few papers had tackled the problem of reducing the burr size [7].

Depending on the destination of parts, burrs can significantly affect the overall functioning of an assembly. There is always a danger that a burr to be detachable and come off affecting the operation and cleanliness, where a cleaner working environment frees of impurities is necessary and sometimes imperative.

2 Experimental researches

The experimental researches were carried out on a CNC cutting machine Chiron FZ12 S equipped with Sinumerik 840D CNC control.

Some of the CNC machine characteristics are shown in Table 1.

The parts used for the experiment were cylinder shaped of 20 mm in diameter of alloy steel 16MnCr5 (1.7131). The dimensions which must result after milling are shown in Figure 1 represented by the thick black contour.

* Corresponding author: cristina.biris@ulbsibiu.ro

Table 1. CNC machine characteristics.

Parameter	Unit	Value
Maximum path [X–Y–Z]	[mm]	550–400–420
Maximum electrical power	[kW]	40
Maximum spindle speed	[rpm]	40000
Chip-to-chip time starting from	[s]	2.2
Maximum axis acceleration [X–Y–Z]	[g]	1-1.5-2
Rapid feed as fast as	[m/min]	75
Max. number of tools	-	206
Tool taper	-	HSK 40 / HSK 50 / HSK 63
Maximum tool weight	[kg]	[5.0]
Maximum tool diameter	[mm]	125
Maximum tool length	[mm]	250

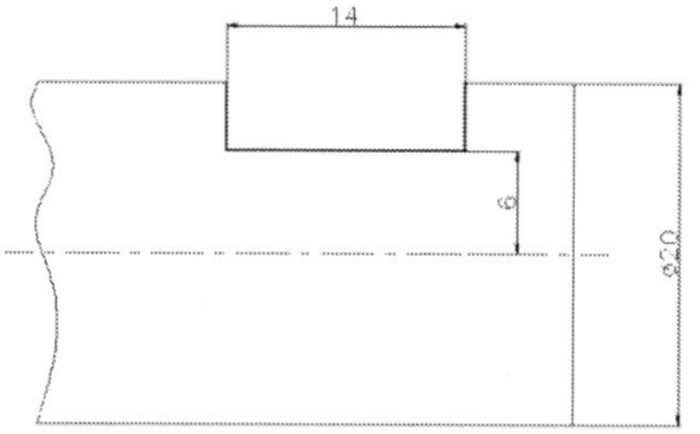

Fig.1 Dimensions after milling.

For the experiment, two types of tools were used shown in Figures 2 and 3 respectively. First was a 4 flute Guhring milling cutter and the second a 6 flute Sandvik milling cutter.

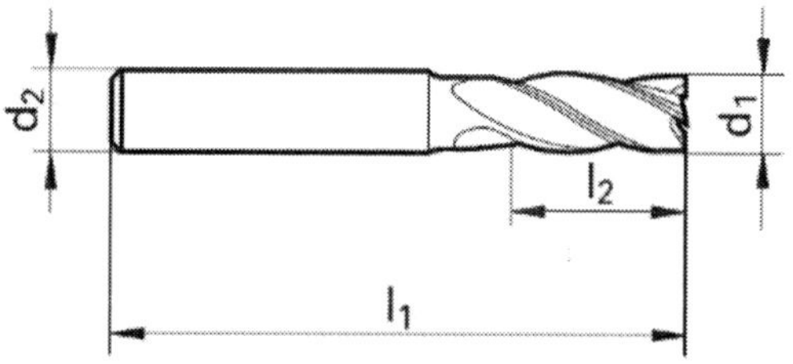

Fig. 2. Guhring cutter 4 flute.

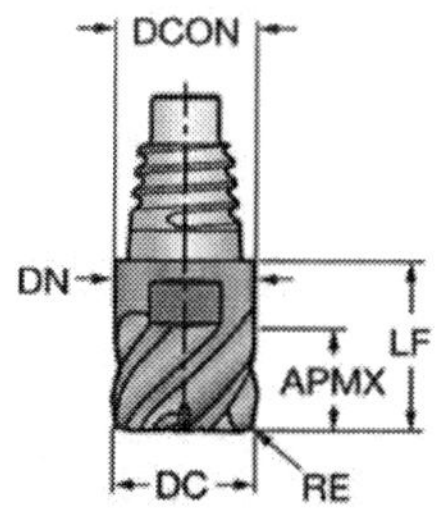

Fig. 3. Sandvik cutter 6 flute.

Both tools characteristics are shown in Table 2 and 3 respectively.

Table 2. Guhring cutter characteristics.

Parameter	Unit	Value
Total length l1	[mm]	73
Depth of cut maximum l2	[mm]	16
Cutting diameter d1	[mm]	12
Shank diameter d2	[mm]	12
Corner chamfer	[mm]	0.15
Flute helix angle	[deg]	30

Table 3. Sandvik cutter characteristics.

Parameter	**Unit**	**Value**
Connection diameter DCON	[mm]	12.3
Neck Diameter DN	[mm]	12.3
Cutting diameter DC	[mm]	12.7
Depth of cut maximum APMX	[mm]	7
Flute helix angle	[deg]	50
Functional length LF	[mm]	14.6
Corner chamfer RE	[mm]	0.38

For the experiment, climb milling was used because according to the specialty literature the surface finish is better by working with this strategy.

The first tool used for the experiment was Guhring cutter 4 teeth.

The milling strategy was established as follows: one pass roughing and two passes finishing. Material left on bottom (a_p) and sides (a_e) was 1 mm for finishing.

The cutting parameters are shown in Table 4.

Figures 4 and 5 respectively shows the burrs resulted after the test with Guhring cutter 4 flute 30° helix angle. As seen, it is almost impossible to have a precise measure of the burrs.

Table 4. Cutting parameters Guhring cutter, one pass roughing, two passes finishing.

Parameter	Roughing	Finishing	Unit
Spindle speed S	2000	3000	[rpm]
Feed per turn f	0.2	0.25	[mm/rot]
Feed per tooth fz	0.05	0.0625	[mm]
Axial depth of cut a_p	4	1	[mm]
Radial depth of cut a_e	12	1	[mm]
Cutting speed vf	400	750	[mm/min]

As shown in Table 4, the cutting parameters for finishing were increased because after roughing there is much less material to remove and the cutting forces are reduced.

Fig. 4. Burrs resulted after one pass roughing and two passes finishing (a_p=1; a_e=1) 4 flute 30 ° helix angle.

Fig. 5. Burrs resulted after one pass roughing and two passes finishing (a_p=1; a_e=1) 4 flute 30° helix angle.

Considering specialty literature, the ae and ap are influencing the surface quality and burrs dimensions. Therefore in order to reduce the burrs dimensions, the milling strategy was changed to two passes roughing and two passes finishing. Considering the influencing factors, the material left for finishing was a_p = a_e=0.2. The speed and feed of the tool were not changed. Figure 6 shows the results after changing the axial (a_p) and radial (a_e) depth of cut.

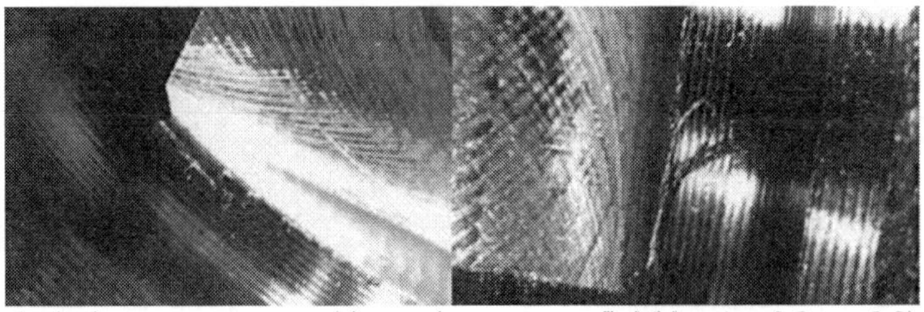

Fig. 6. Burrs resulted after two passes roughing and two passes finishing (a_p=0.2; a_e=0.2) 4 flute 30° helix angle.

As seen, by reducing the a_e and a_p for finishing, the burrs dimensions are reduced. In Figure 7 are shown the burrs dimensions resulted after working with a_e =0.2 and a_p =0.2, where it can be seen that the root of the burr x and the height z can be measured now.

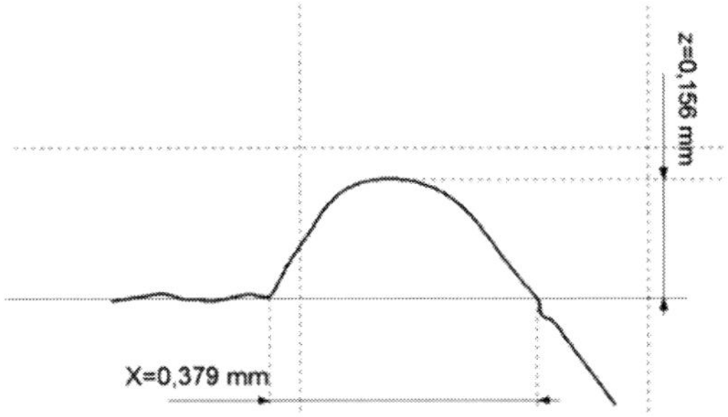

Fig. 7. Burrs dimensions after two passes roughing and two passes finishing (a_p=0.2; a_e=0.2) 4 flute 30° helix angle.

In the next stage of the experiment, Sandvik cutter 6 flute 50° helix angle was tested.

First, the milling strategy was established as follows: one pass roughing and two passes finishing. Because of the 6 tooth milling cutter, all the cutting parameters were increased for this tool. Increasing the cutting parameters also allows a reduced cycle time, which is very important in a serial production resulting more parts in a shorter time. The cutting parameters for this strategy are shown in Table 5.

Table 5. Cutting parameters Sandvik cutter, one pass roughing, two passes finishing.

Parameter	Roughing	Finishing	Unit
Spindle speed S	3000	3500	[rpm]
Feed per turn f	0.25	0.3	[mm/rot]
Feed per tooth fz	0.041	0.05	[mm]
Axial depth of cut a_p	4	0.65	[mm]
Radial depth of cut a_e	12.7	0.65	[mm]
Cutting speed vf	750	1050	[mm/min]

Figure 8 and 9 respectively shows the burrs resulted after the test with Sandvik cutter 6 flute 50° helix angles. Figure 9 shows the measurements of the burr's root x and height z.

As can be seen, burrs resulted are smaller compared to the burrs resulted by using the first cutter.

In order to further reduce the burrs dimensions, the cutting strategy was changed again by working with two passes roughing and two passes finishing, reducing the material left for finishing to a_e=a_p=0.2 mm.

Fig. 8. Burrs resulted after one pass roughing and two passes finishing (a_p=0.65; a_e=0.65)
6 flute 50° helix angle

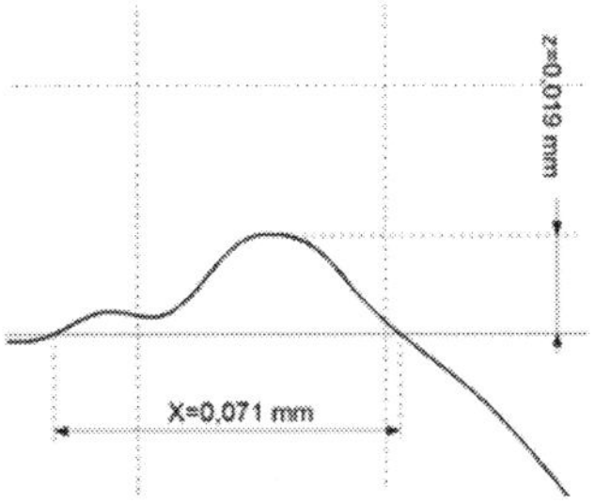

Fig. 9. Burrs dimensions after one pass roughing and two passes finishing (a_p=0.65; a_e=0.65)
6 flute 50° helix angle

Cutting parameters for this strategy are shown in Table 6.

Table 6. Cutting parameters Sandvik cutter, two passes roughing, two passes finishing.

Parameter	Roughing	Finishing	Unit
Spindle speed S	3000	3500	[rpm]
Feed per turn f	0.25	0.3	[mm/rot]
Feed per tooth fz	0.041	0.05	[mm]
Axial depth of cut a_p	4	0.2	[mm]
Radial depth of cut a_e	12.7	0.2	[mm]
Cutting speed vf	750	1050	[mm/min]

Burrs dimensions resulted after working with this strategy is shown in Figure 10.

Fig. 10. Burrs dimensions after two passes roughing and two passes finishing (a_p=0.2; a_e=0.2)
6 flute 50° helix angle.

As can be seen in Figure 10, the burrs are much smaller and can be removed if necessary by a brushing operation.

To further reduce or eliminate the burrs, the cutting parameters were changed as shown in Table 7.

Table 7. Cutting parameters Sandvik cutter, two passes roughing, two passes finishing.

Parameter	Roughing	Finishing	Unit
Spindle speed S	3000	3800	[rpm]
Feed per turn f	0.25	0.25	[mm/rot]
Feed per tooth fz	0.041	0.041	[mm]
Axial depth of cut a_p	4	0.2	[mm]
Radial depth of cut a_e	12.7	0.2	[mm]
Cutting speed vf	750	950	[mm/min]

The results after milling with these parameters for finishing pass are shown in Figures 11 and 12. In Figure 11 it can be seen that the root x and height z of the burrs are greatly reduced after optimizing the cutting parameters. Figure 12 shows that no burr was detected during measuring, after implementing the parameters shown in Table 7.

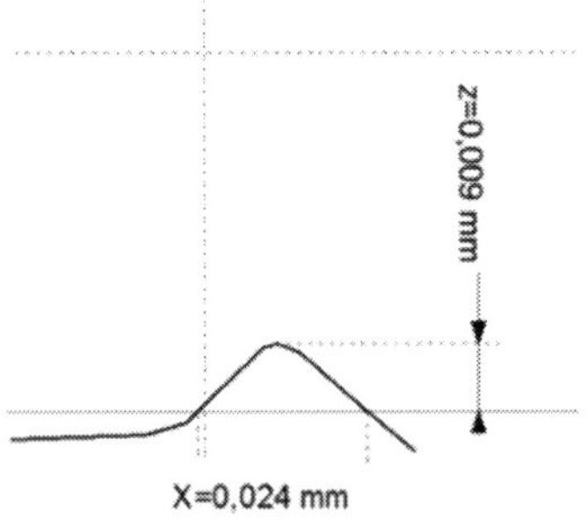

Fig. 11. Without burrs resulted after two passes roughing and two passes finishing (a_p= a_e=0.2 mm) and after changing the parameters for finishing
6 flute 50° helix angle.

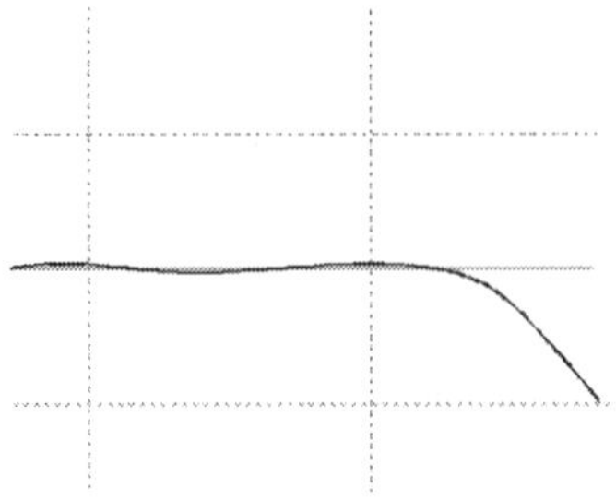

Fig. 12. Without burrs resulted after two passes roughing and two passes finishing (a_p= a_e=0.2 mm) and after changing the parameters for finishing
6 flute 50° helix angle.

As seen above, the burrs disappeared by cutting with these parameters using 6 flute cutter.

3 Conclusions

The experiment shows that the cutting parameters such as speed and feed have a major influence on the burrs dimensions, which is why these parameters must be chosen depending on the application and maintaining a balance between cycle time and tool capabilities. The material left for finishing pass after roughing also influences the size of burrs. Also, regarding the cutting tool, the experiment shows that choosing a tool with more teeth and a greater helix angle help to significantly reduce the burrs dimensions. Considering that in order to reduce the burrs dimensions were needed to increase the speed and to reduce the feed, further researches can take place on the cutting parameters influence on the tool durability.

References

1. L.N.L. de Lacalle, F.J. Campa, A. Lamikiz, *Milling*, (University of the Basque Country, Spain, (2001)
2. J.P. Davim, *Modern Machining Technology. A practical guide*, (Woodhead Publishing Limited, Cambridge, (2011)
3. C. Biriş, B. Breaz, C., Gîrjob, A., Chicea, Applied Mechanics and Materials **555**, 580-585, (2014)
4. K, Erkorkmaz, Y. Altintas, Int. J. of Mach. Tools and Manuf., **41**, 1487–1509, (2001)
5. Y. Altintas, C. Brecher, M. Weck, S. Witt, Annals of CIRP, **54**, 651-673 (2005)
6. R.E. Breaz, O.C. Bologa, C. Biriş, G. Racz, C. Gîrjob, V. Oleksik, IEEE Catalog Number: CFP09INI-CDR, 813 – 818, (2009)
7. K. Sanjib, D. Santanu, P.S. Partha, GCCM, **97**, 230-240, (2014)

TPM implantation in manufacturing systems: A survey of the state-of-the-art and extensions

Kenza Chaabane*, *Jérémie* Schutz, and *Sofiène* Dellagi

Université de Lorraine, LGIPM, EA 3096, Metz, F-57045, France

Abstract. Maintenance activities represent a significant pillar in industry. Its real goal consists at improving the production systems productivity. Specially, TPM represents one of the main evolutions of the maintenance methods in order to improve the machines efficiency by a proactive approach. This paper presents an overview of several TPM implantations and especially efficiency and effectiveness measurement of the TPM in several industries. In a second hand, a proposition of possible extension about analytical assessment of the TPM performance, based on both technical and economic criteria, is presented.

1 Introduction

Companies are distinguished in terms of their industrial performance by the availability, productivity and flexibility of their resources as well as by ensuring lower expenses costs [1]. In today's dynamic environment, improving industrial performances is crucial for competitiveness, thus, companies should remove the unexpected (breakdowns, defects, waste of . . .). In this context, maintenance has a very strategically role to play.

We note that several researchers have explored ways of integrating maintenance and production after decades during which those two activities had been studied separately. Works on maintenance policies started with Barlow et al. [2] and continued with a vast number of contributions as it can be seen in a survey on maintenance models by Wang [3]. During the last two periods, numerous companies have applied that strategies dissociating maintenance and production were unsuccessful. The real need for developing integrated maintenance-production strategies became compulsory. Brandolese et al. [4] proposed a strategy for maintaining a multiple machines production system. His planning consists in programming the execution date of each task and the machine that should perform it. They integrated preventive maintenance activities in the planning placing them as close as possible to the optimal maintenance periods. In the same framework, some researchers have taken into ac- count several external constraints. In fact, integrated maintenance-production strategies which take into consideration subcontracting have been studied by Dellagi et al. [5]

* Corresponding author: kenza.chaabane9@etu.univ- lorraine.fr, jeremie.schutz@univ-lorraine.fr, sofiene.dellagi@univ-lorraine.fr

The TPM approach proposed in this survey deals with the frame of integrated production maintenance. In fact, works on TPM started in 1970's with Nakajima [6], he developed the concept of TPM when he improved the American method PM (Productive Maintenance) in Japanese company (Nippon Denso); since that, TPM Has refined over time, evolved according to the different experiments in different companies around the world. The purpose of this article is to present "the state of the art" in definition and implementation of TPM in manufacturing plant, and suggest a possible extension for measuring the assessment of the method which is rarely treated in literature. The reminder of this paper is organized as follows, the first section will review briefly some history, the target and expected results of the method. A second section will present some recent cases of study of the implementation of TPM in different companies. section 3 is dedicated to the comparison of the different ways to assess TPM improvement and propose a possible extensions based on both technical and economic criteria. Finally, conclusions and potential future work are provided in Section 4.

2 Presentation of the TPM method

The original TPM was mediated by Seichi Nakajima, Director of JIPM (Japan Institute of Plan Maintenance). His concept was "Creating small groups to take ownership of equipment and improve its performance" [6]. This concept stems from big changes in the company, namely that we are no longer in the basic scheme "I produce, you repair ", but in a scheme where maintenance is everyone's busi- ness. This is reflected in the fact that the operator performs some of the so-called "repair level" such as lubrication, monitoring, cleaning and control; in the 70'th, this vision was innovative and encountered a lot of opposition but since then, success has swept away all fears.

Nakajima [6] define TPM as "productive maintenance involving total participation in addition to maximizing equipment effectiveness and establishing a through system of planned maintenance. Also, the JIPM defined TPM as a system of Maintenance covering the entire life of the equipment including manufacturing, planning and maintenance. The world "total" in TPM has three principal meanings which are the feather of the TPM: the first means total effectiveness indicates TPM's pursuit of economic efficiency, the second is for total maintenance system including maintenance prevention, maintainability improvement and preventive maintenance. The last meaning and the most important is for total participation of all employees, from the top management to workers on the shop floor.

Thus, TPM is closely tied to "just in time", 5S and "total quality management"; also, it is an extension of "PM", where the machine works at high productivity and efficiency, and where the maintenance is all employee's responsibility, focusing to prevent the problems before it may occur. Referring to bufferne [1], the principal goals of the TPM are: Improve equipment effectiveness by looking into the six big losses, to make the equipment work at it highest performance; reduce production's direct and indirect costs; enhance the life cycle of equipment and reduce it cost by improving maintenance efficiency and effectiveness and create an operational value for the company by educating and training personnel witch is an important value added of the TPM approach.

Notice that despite the efforts made, some companies fail to deploy TPM within their production systems. The reasons are numerous and constitute barriers to the implementation of this proactive strategy. In his study, Kumar et al. [7] identified various causes and proposed a model to analyze and eliminate them later. One of the most important causes is the lack of involvement of the manager and human resources. We must also cite the resistance to change of employees who sometimes have trouble getting out of the traditional pattern (I produce, maintenance repairs). The lack of communication between the departments also constitutes a barrier to the implementation of the method, hence the need to set up pyramidal structures.

3 Effectiveness and efficiency measurement: cases of study

3.1 Effectiveness measurement

Industrials affirm that the performance measurement system is a key element in order to motivate the whole staff (We progress only if we measure a performance). In the TPM approach, the OEE (Overall Equipment Effectiveness) is the basic indicator to which other indicators can be added. Since the aim of the TPM is to achieve ideal performance and zero losses, quantifying the accumulation of time waste is the best way to measure the equipment effectiveness, then by definition:

$$\text{OEE} = \text{Valuable operating time} / \text{Loading time [1]} \tag{1}$$

Where loading time is the total time available to produce in a given period and the valuable operating time is only the time use to fabric good products.
It is also expressed by the relation [1]:

$$\text{OEE} = \text{Availability} \times \text{Performance rate} \times \text{Quality rate [1]} \tag{2}$$

The availability is calculated as:

$$\text{Availability} = (\text{Required availability - Downtime}) / \text{Required availability [1]} \tag{3}$$

The formula to compute the performance rate can be expressed as:

$$\text{Performance rate (P)} = (\text{Design cycle time x Output}) / \text{Operating time [1]} \tag{4}$$

The design cycle time or the production output will be in the unit of production, like parts per hour and the output will be the total output in the given time period interval.
The quality rate can be expressed as:

$$\text{Quality rate (Q)} = (\text{Production input} - \text{Quality defects}) / \text{Production input [1]} \tag{5}$$

The production input means that the unit of product being feeds into the production process. The quality defects mean the number of products which are below the quality standards i.e. the rejected items after the production process.

In 2016, Kumar et al. [7] studies the relation between OEE, A, P and Q. By looking at the values of correlation, he came to conclusion that OEE is highly correlated with Availability then Performance rate and then Quality rate. The paper also compares the effectiveness of the TPM implementation between two plastic industries in India using the OEE; it shows that both industries are lagging in effective implementation of important pillars of the TPM when comparing there OEE to the world standard OEE; Still Company A has got a little bit better implementation of TPM than Company B, same time OEE value of Company A is higher than Company B by 12%.

Singh et al. [8] study the implementation of the TPM in a Machine Shop in Indian automotive company. The paper shows detailed steps of the implementation of the method, it also compares OEE before and after TPM implementation, it has improved from 63% to 79% indicating the improvement in productivity and improvement in quality of products.

In 2016, Vittaleshwar et al. [9] establish an empirical study of effect of Total Productive Maintenance on OEE in a Water Bottling Industry. Data was collected on daily activities and was used for calculation of OEE for Water Treatment Plant, Bottle Blowing Unit, and Bottle Filling Machine 30

Bottles per Minute, Pouch Packing Machine and Inkjet Jet Coders before and after TPM implementation. The average value of OEE were found fall between the ranges of 0.14 and 0.7. The world-class benchmarks for OEE is 0.85. TPM and OEE helped the Water Bottling Company to identify previously hidden manufacturing losses and inefficiency.

We can deduct from various cases of study seen in this part, that the OEE is not a simple indicator. it is a tool of progress which allows the identification of possible hidden losses, to better eliminate them.

3.2 Efficiency measurement: DEA method

When it comes to performance evaluation in TPM, OEE has commonly been used as a performance measure since TPM aims to maximize equipment effectiveness. Although OEE has been considered as a standard measure for equipment performance what it captures is only effectiveness of TPM, not efficiency. However, the performance evaluation of TPM should include an objective and a method based on multiple different factors (inputs and outputs) [10]. The DEA method is a linear program- ming model for evaluating the relative efficiency of comparable decision-making units (DMUs) in transforming inputs into outputs in a direct fashion without recourse to any underlying model [11]. It is a non-parametric approach that does not require any assumptions about the functional form of a production function or a priori information on importance of inputs and outputs [12]. There are two models of the DEA, the first (CCR model for Charnes, Cooper, and Rhodes, 1978) assumes that production exhibits constant returns to scale; that is, if an activity (x, y) is feasible under a given production frontier, then, for every positive scalar t, the activity (tx, ty) is also feasible.

The second model (BCC for Banker, Charnes, and Cooper (1984)) for encompassing variable returns to scale along the efficiency frontier. DEA models are distinguished by the objective of a model: maximize outputs (output-oriented) or minimize inputs (input-oriented). Since it is implicitly assumed that the objective of TPM lies in increasing outputs rather than decreasing inputs, also it cannot be assumed that TPM activities exhibit constant returns to scale [13]. More precisely, it is the output-oriented BCC model which is used to measure TPM efficiency, it is formulated as the following [14]:

$$\min \theta_p \tag{6}$$

$$\text{st} \sum_{j=1}^{n} \lambda_j \cdot x_{ij} \leq \theta_p \cdot x_{ip} \quad \forall i, \tag{7}$$

$$\sum_{j=1}^{n} \lambda_j \cdot y_{rj} \geq y_{rp} \quad \forall r, \tag{8}$$

$$\sum_{j=1}^{n} \lambda_j = 1, \quad \lambda_j \geq 0, \forall j. \tag{9}$$

Jeon et al.[13] study measured the efficiency of TPM using DEA where SDWT was the DMUs using DEA. It considered that TPM implementation process is captured in a three-stage model: stage 1 from TPM inputs (employee, expense) to TPM intermediate outputs (autonomous maintenance, planned maintenance, education and training), stage 2 from TPM intermediate outputs to TPM final outputs (availability, performance efficiency), and stage 3 from TPM inputs to TPM final outputs. Every SDWT in every team was evaluated by DEA for each stage. Based on the efficiency scores of DEA, the relationships between stage-wise efficiency scores of SDWTs is analyzed by correlation analysis. It finds that SDWTs efficient transforming intermediate outputs into final outputs are also efficient at transforming inputs into final outputs.

Bekar et al. [10] go further to deal with situations where some of the input and output data for TPM are imprecise or not available, they used a newly method where fuzzy logic is introduced to DEA for measuring efficiency of DMUs under uncertainty. It's called Fuzzy data envelopment analysis (FDEA).

In the study, the performance efficiency of TPM for four production lines (DMUs) was evaluated (Rail Machining, Rail Assembly and HPV, NHB, Beginning of Line and NHB End of Line) with respect to seven outputs and ten inputs, the fuzzy weights of inputs and outputs are obtained by fuzzy COPRAS. After estimation, the study shows that the rail machine was on efficient frontiers, it is 100% efficient, unlike rail assembly and HPV which was the lowest efficient. This study is the first introducing FUZZY COPRA-FUZZY DEA method to evaluate performance efficiency of TPM.

4 Possible extensions

Looking at the literature related to TPM, detailed in last section, we have seen different methods to evaluate TPM implementation in companies, depending on what we want to improve (efficiency or effectiveness). In the other hand, those methods still too relatives results.

For example, the JIPM precise that it's not the number itself that count for OEE, but how much we improve it, it depends on the ideal objective that we fixed and how much did we approach it, that's the meaning of the effectiveness measuring. For example, Buffern [1] says that in a Japanese company he visited, the OEE presented was 110%, it's explained by the fact that progress realized by the company during the 4 years of TPM implementation exceeds the initial too pessimist objectives. It proves that OEE still a relative factor.

The DAE method, as for her, aims to compare different units (DMUs) in same company relative to their best peers (efficient frontier). DMUs which belong to this frontier are 100% efficient, even if the TPM wasn't well implemented in the company.

Also, the OEE measurement, which is a function of availability, performance and quality rate need large amount of data collection. As Hammar et al. [15] says that "The data collection should be at such a detailed level that it fulfils its objectives without being unnecessarily demanding resources. Too detailed data collection may result in unmotivated personnel and reaction against the measurement". Moreover, it's mentioned that "To succeed with data collection, it is necessary to find a less time consuming method that is also precise". These statements show that a simple measure of OEE and co-operation of production personnel are the crucial factors for successful data collection.

A new measuring method or economical or technical estimation of the efficiency of the TPM is compulsory at this stage using less data measurement, and giving non relative assessment. Looking to the evolution in studies related to preventive maintenance and new integrated maintenance production strategies. These studies started by simple description or proposition of maintenance strategies [3] in order to reach recently a hay level research evolving a real estimation of economical or technical strategies. More then, in several studies the authors developed an analytical model representing an objective function related to economical or technical profit. The authors optimize analytically or by simulation methods theses developed functions in order to obtain an optimal variable decisions helpful for industrial (Schutz et al. [16], Hajej et al. [17]).

Dealing with this context, our idea consists at introducing a new variable which related to the time spent by operator on doing first level maintenance. In fact, one of the most important added value of the TPM is first level maintenance. We find the optimal time spending by that operator on checking the machine in its work-time, noted TPM period, ensuring compromise between system reliability and economical profit related to selling costs, production costs and others relative dispenses.

The benefice related to maintenance after TPM implementation will be the new measurement of its efficiency and depends on this new variable, it could be estimated analytically or by simulation, according to the complexity of the developed model, in the aim to maximize it as an objective function.

5 Conclusion

The purpose of this report was to present a state-of-the-art of the TPM approach, in order to better identify the method. We first considerate the origin of the approach and its objectives. Then, we focused at measuring the effectiveness and efficiency of TPM; the various indicators that we used, the OEE for efficiency and the DEA method for performance measurement, were presented based on the latest studies on TPM over the last decade.

After reflection, it was possible to say that the indicators that were used were limited; also, in the literature, rare are the documents which are interested in measuring the effectiveness of the TPM using economic criteria.

Following in this study, it is proposed to develop an economical way to evaluate the performance of the TPM, which will have as objective function the gain maximization.

References

1. J. Bufferne, *Le guide de la TPM: Total Productive Maintenance* (Editions Eyrolles, 2006), ISBN 978-2-7081-3723-3, google-Books-ID: F5A8y88nn7cC
2. R.E. Barlow, F. Proschan, Studies in Applied Probability and Management Science **7**, 63 (1962)
3. H. Wang, European Journal of Operational Research **139**, 469 (2002)
4. M. Brandolese, M. Franci, A. Pozzetti, International Journal of Production Research **34**, 2059 (1996)
5. S. Dellagi, N. Rezg, X. Xie, International Journal of Production Research **45**, 1233 (2007)
6. S. Nakajima, *Introduction to TPM: Total Productive Maintenance* (Productivity Press, 1988), ISBN 978-0-915299-23-2, google-Books-ID: XKc28H3JeUUC
7. A. Kumar, R.K. Singh, T.A. Khan, International Journal of Advance Research and Innovation **3**, 82 (2016)
8. R. Singh, A.M. Gohil, D.B. Shah, S. Desai, Procedia Engineering **51**, 592 (2013)
9. A. Vittaleshwar, K.S. Dasharathraj, P. Prajual, International Journal of Applied Engineering Research **11**, 5573 (2016)
10. E.T. Bekar, C. Kahraman, Measuring Efficiency of Total Productive Maintenance (TPM) with Newly Developed Performance Measures using Fuzzy Data Envelopment Analysis, in Proceed- ings of the World Congress on Engineering (2016), Vol. 1
11. W.W. Cooper, L.M. Seiford, J. Zhu, Socio-Economic Planning Sciences **34**, 1 (2000)
12. W.D. Cook, L.M. Seiford, European journal of operational research **192**, 1 (2009)

13. J. Jeon, C. Kim, H. Lee, Total Quality Management & Business Excellence **22**, 911 (2011)
14. A. Charnes, W.W. Cooper, Accounting, Organizations and Society **5**, 87 (1980)
15. P. Hammar, B. Bennett, M. Yang, M. Johnson, Physical Review Letters **83**, 203 (1999)
16. J. Schutz, *Joint optimization of production-maintenance plans based on optimal production rates*, in 2014 IEEE International Conference on Industrial Engineering and Engineering Management (2014), pp. 933–937
17. H. Zied, D. Sofiene, R. Nidhal, International Journal of Production Research **49**, 5695 (2011)

MATEC Web of Conferences **112**, 01008 (2017) DOI: 10.1051/matecconf/201711201008
IManE&E 2017

The energetic load smoothing along the rack tool cutting edge in toothing process

Mihai Bordeanu, *Gabriel* Frumuşanu *, and *Nicolae* Oancea

Dunărea de Jos University of Galaţi, Department of Manufacturing Engineering, Domnească Street 111, 800201 - Galaţi, Romania

Abstract. The method of generating the gear tooth with a rack tool is applied on MAAG machine tools. Their specific cutting scheme leads to an irregular energetic loading of the tool cutting edge, during the rolling process. In this paper, we present a new method for analyzing the energetic load along the cutting edges of the rack shaped cutter, based on graphical modelling of the cutting process performed in CATIA. The energetic load smoothing is proposed to be obtained by finding an appropriate variation law for the circular feed communicated to the workpiece by toothing machine kinematical chain, during the rolling motion that permanently respects the rolling condition established between the centrodes attached to both tool and workpiece. A numerical solution in a concrete case of analysis is also included.

1 Introduction

Generating the tooth of cylindrical gear having involute or non-involute profile is a machining process that can be performed on different machine tools by using a multitude of tool types: slotting with rack shaped tool or with pinion cutter, hobbing with worm tool, milling with profiled end mill or profiled disc tool.

The process of generating the tooth space of machined gear takes place with high non-uniformity due to the rolling process between the centrodes to which they are associated the tool cutting edges (usually straight lines in the case of rack with trapezoidal teeth), on one side, and the enwrapped flanks of workpiece teeth (with involute profile, in the mentioned case), on the other side. At the same time, one should not neglect that rack shaped cutter teeth must cut from full material during the process of tool engagement in workpiece [1, 2].

Analysis of tooth space generating process have been already realized and presented in dedicated literature [3-5]. Most of them address to the case of a mono-tooth rack shaped cutter and they all prove that, during the toothing process, the area of the chip detached by tool cutting edge shows high variation when determined for successive rolling positions considered for the couple of centrodes attached to both tool and workpiece. This lack of uniformity leads to occurrence of highly variable cutting specific stress, depending on the tool position relative to the workpiece, with direct impact on the magnitude of main cutting force [6].

* Corresponding author: gabriel.frumusanu@ugal.ro

There are also presented effective solutions suggesting constructive modifications of the generating rack gear in order to diminish the mentioned lack of uniformity [7].

In this paper, we present an analysis of the tooth space generating process, based on graphical modelling in CATIA and approaching the case of cylindrical gear with involute profile machined by mono-tooth tool. This is specific mainly to the gears with high module manufactured on MAAG-type machine tools. The tooth space generating process is tried to be uniformed by adopting a non-uniform process of rolling between the conjugated centrodes attached to both tool and workpiece, achieved by discrete variation of workpiece circular feed and aiming to uniform the area of the chip detached at each double stroke of the rack shaped cutter.

Next section is about modelling the variation of detached chip area. Third section presents the procedure to find the variation law of workpiece circular feed, if targeting to uniform the chip area during gear machining, sampled in a concrete case. The last section is for conclusion.

2 Detached chip area

A solution for measuring the detached chip area when machining a gear with mono-tooth rack tool, based on graphical modelling in CATIA, is further proposed.

2.1 Kinematics of tooth generating by rolling

In Figure 1, there is depicted the kinematical scheme of generating the gear tooth with a rack shaped cutter [8, 9].

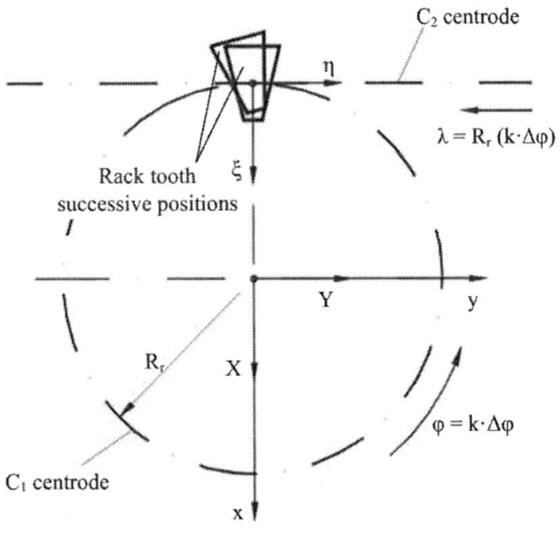

Fig. 1. Reference systems and rolling centrodes

By C_1 is denoted the centrode associated to workpiece, which is a circle having rotation motion of φ angular parameter, while C_2 is the tool centrode – a straight line in translation motion of λ linear parameter.

The following reference systems have to be considered in order to find the equation of the relative motion between workpiece and tool:

- xy, which is the global system, having the origin in workpiece centre
- XY – relative system, attached to the workpiece and initially overlapped to xy
- $\xi\eta$ – relative system, attached to the tool.

The equation of workpiece absolute motion is:

$$x = \omega_3^{T}(k \cdot \Delta\varphi) \cdot X\,, \tag{1}$$

while the equation of tool absolute motion is:

$$x = \xi + a\,,\ \text{with}\ a = \begin{pmatrix} -R_r \\ -R_r \cdot (k \cdot \Delta\varphi) \end{pmatrix}. \tag{2}$$

The equation of relative motion between workpiece and tool results from (1) and (2), after eliminating x:

$$X = \omega_3(k \cdot \Delta\varphi)[\xi + a]. \tag{3}$$

In relations from above, $\Delta\varphi$ means the increment in workpiece's rotation motion and k is the order number of tool double stroke. In up to now approach, the increment remains constant all along the toothing process. In our approach, smoothing of detached chip area should be obtained by transforming $\Delta\varphi$ in a variable parameter, its variation law following to be determined. We also must observe that to $\Delta\varphi$ increment in workpiece's rotation motion corresponds $\Delta\lambda$ increment in tool rectilinear motion,

$$\Delta\lambda = R_r \cdot \Delta\varphi\,. \tag{4}$$

Note Obviously, kinematics of the rolling process taking part on MAAG machine tool needs to be modified by introducing a stepper motor and an adequate numerical control system in order to actually obtain the aimed rolling motion with variable speed.

2.2 CATIA modelling of tooth generating process

The area of the chip detached at each active stroke of the rack shaped cutter can be found by performing a graphical modelling in CATIA of the toothing process. In this purpose, two stages need to be covered.

In first stage, the solid models of both workpiece and tooth representing the active part of the generating tool are built by using *Sketch* module of CATIA. First, the workpiece is sketched and geometrical / dimensional constrains are applied to it. Workpiece model is a cylindrical disc having the exterior diameter equal to head diameter of the gear to be generated.

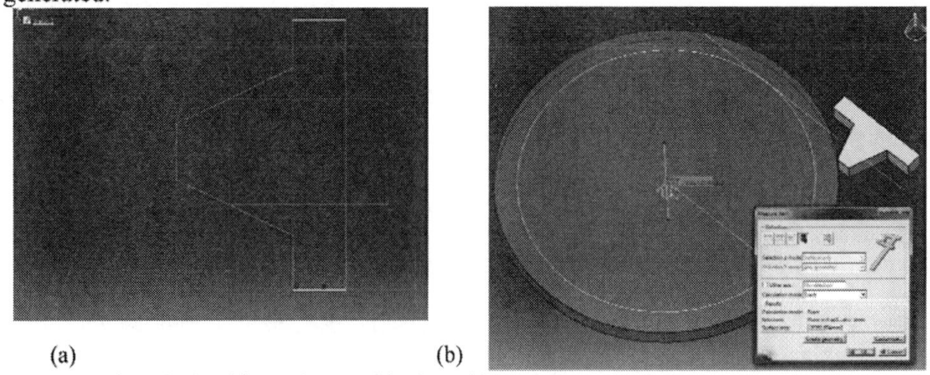

(a) (b)

Fig. 2. Tool tooth sketching – a) & positioning – b)

Then, the tool tooth is sketched in a separate file (Figure 2-a) and submitted to geometrical / dimensional constrains, being positioned in contact with the projection of gear head circle. The condition of tangency between rack shaped cutter rolling line and workpiece rolling circle is also imposed (Figure 2-b). After sketching, the profiles are extruded at the required dimensions with the help of *Pad* modelling tool (available in *Sketch Based Features* toolbar), hereby two solids being generated.

In second stage, the motions following to be imposed to each of solids from above are established such as reproducing the process kinematics (Figure 3). The motions are supposed to be discrete, having $\Delta\varphi$ and $\Delta\lambda$ increments, which respects the rolling condition (4).

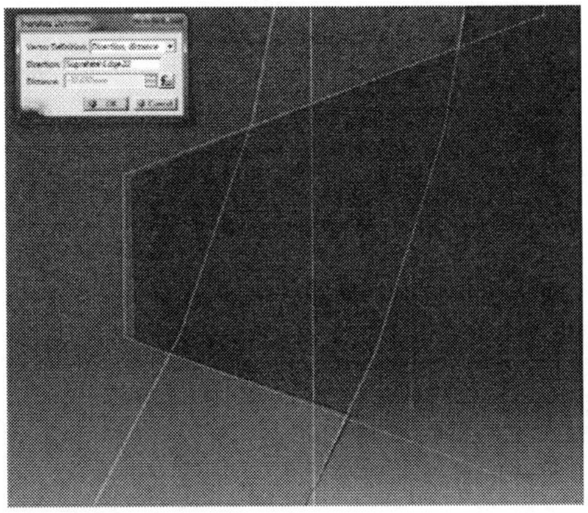

Fig. 3. Imposition of the relative kinematics

2.3 Measurement of detached chip area

For finding the areas of successively detached chips, the tooth model is imported in the file where workpiece disk was sketched. After import, the workpiece is rotated with $\Delta\varphi$ angle, while the tooth is translated along the rolling line with $\Delta\lambda$ distance. By using *Remove* command of *Operations* modelling tool, the common region of the two elements is extracted. The workpiece area is measured before and after extraction (with *Measure Item* tool from *Measure* toolbar) and the removed chip area results by subtraction (Figure 4). This procedure requires importing tooth model before each new intersection.

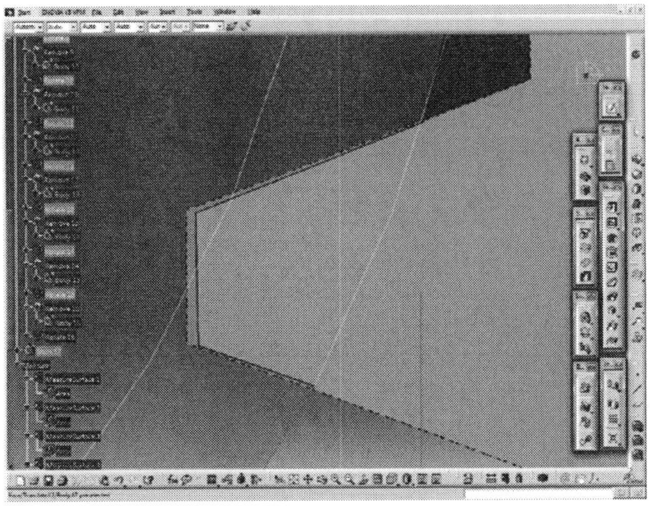

Fig. 4. Measurement of removed chip area

3 Algorithm to find the variation law of workpiece circular feed aiming detached chip area equalisation

The purpose of research here presented was to find the variation law of workpiece circular feed (obviously, in strict correlation with rack shaped cutter linear feed) such as the energetic load along tool cutting edge is smoothened. The smoothening can be achieved by equalising the areas of successively detached chips. A dedicated two-steps algorithm has been developed in order to reach the mentioned target. For better understanding the way it works, we further present algorithm implementation in a concrete case.

3.1 Variation law of detached chip area

Before finding the feed variation law, the variation law of detached chip area need to be determined. The method for measurement of detached chip area A_c presented in previous section is applied when machining a gear with following actual parameters: module $m = 10$ mm and number of teeth $z = 60$, see also Figure 5.

Fig. 5. Machined gear geometry

Table 1. Removed chips area

Crt. no.	Angle, φ [°]	Displacement of tool, λ [mm]	Workpiece area, [mm^2]	Removed chip area, A_c [mm^2]
1	0	0	301907.054	0.000
2	0.394	2.063	301906.131	0.923
3	0.694	3.634	301904.211	1.920
4	0.994	5.205	301901.264	2.947
5	1.294	6.775	301897.316	3.948
…………………………………				
25	5.494	28.767	301813.531	6.394
26	5.794	30.337	301807.128	6.403
27	6.094	31.908	301800.716	6.412
28	6.394	33.479	301794.306	6.410
29	6.694	35.050	301787.906	6.400
…………………………………				
92	27.394	143.435	301585.398	0.021
93	27.694	145.006	301585.385	0.013
94	27.994	146.577	301585.378	0.007
95	28.294	148.147	301585.375	0.003
96	28.594	149.718	301585.375	0.000

The circular feed value was considered as constant, $\Delta\varphi_0 = 0.3$ deg/dbl stroke (starting from the second effective double stroke), corresponding to a rack tool displacement $\Delta\lambda_0 = 1.571$ mm. Excerpts from the table containing the results obtained for all 96 dbl strokes needed for generating the entire tooth space are presented in Table 1 for sampling. By graphical representing the variation of removed chip area during tooth space machining, the curve from Figure 6 resulted.

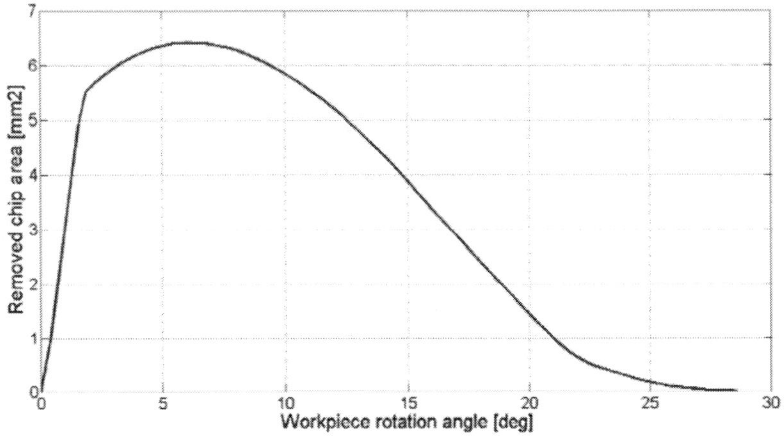

Fig. 6. Variation of detached chip area

As it can be easily noticed, detached chip area presents a high variability during toothing process, this leading to an irregular energetic loading of the tool cutting edge. In order to find the analytical expression of the chip variation law, an approximation curve fitted to the points (φ, A_c) from Table 1 can be found with the help of *Curve Fitting Tool* toolbox from MatLab soft. If choosing to approximate by a 9th degree polynomial function:

$$A(\varphi) = p_1 \cdot \varphi^9 + p_2 \cdot \varphi^8 + \ldots + p_9 \cdot \varphi + p_{10}, \quad (5)$$

then fitting session results are: p_1 = 2.997e-10, p_2 = -4.473e-8, p_3 = 2.87e-6, p_4 = -0.0001038, p_5 = 0.002322, p_6 = -0.03313, p_7 = 0.2987, p_8 = -1.659, p_9 = 5.18, and p_{10} = -0.4928, with a satisfactory goodness of fit (the root mean square error being 0.1238).

3.2 Variation law of workpiece circular feed

By accepting a local proportionality (when workpiece position is defined by angle φ) between A_c and $\Delta\varphi$, one can write:

$$A_c = \frac{\Delta\varphi}{\Delta\varphi_0} \cdot A(\varphi). \quad (6)$$

The knowledge of chip area A variation law in analytical form (5) enabled the development of a MatLab application aiming to find in any moment the actual value of $\Delta\varphi$ (this time considered as variable), corresponding to which the area A_c of next chip detached will have an imposed value, A_{c0}. Hereby, the application determines the variation law of workpiece circular feed leading to a constant value of chip area. The input data are the limits of angle φ variation domain (φ_{min} and φ_{max}), the imposed value of chip area (A_{c0}), the expression of area approximation function (A – imported from *Curve Fitting Tool* toolbox), the value of angular increment ($\Delta\varphi_0$) based on which the chip area variation law was determined. There are also two auxiliary variables involved in numerically solving the equation (6), namely an incremental parameter $d\varphi$ ($\Delta\varphi = i \cdot d\varphi$, $i = 1, 2, \ldots$) and the maximum admissible value of the error, ε.

In the implementation example here addressed, $\varphi_{min} = 0°$, $\varphi_{max} = 28.594°$, $A_{c0} = 2$ mm^2, the values of A function coefficients are the ones from above, $\Delta\varphi_0 = 0.3°$, $d\varphi = 0.0001°$, $\varepsilon = 0.01$. By plotting the calculated values of $\Delta\varphi$ relative to double stroke current number and by joining the points, the variation law of workpiece circular feed results in graphical form (Figure 7).

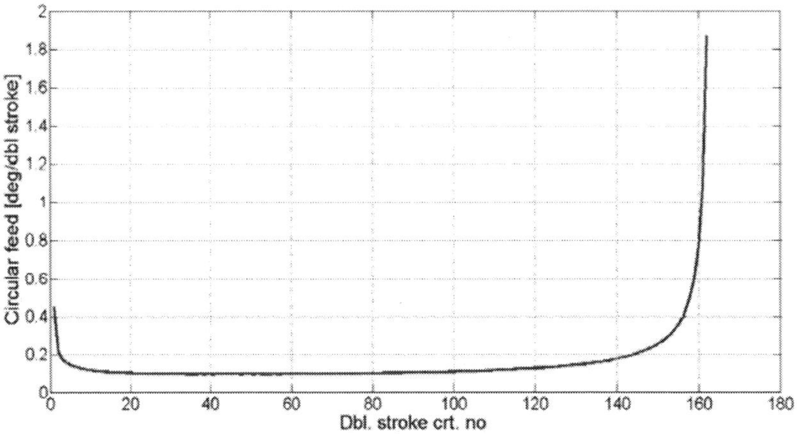

Fig. 7. Circular feed variation law

After examining the diagram from Figure 7, two remarks should be made:

- The number of tool double strokes required for machining the tooth space by removing chips with constant area is different from the one resulted if machining with constant feed (in the addressed case, higher)
- At both beginning and end of machining process, the circular feed values resulted very high, unacceptable in practice because by applying them, the generated surface geometry could be affected. This can be solved by limiting the maximum value of the circular feed, at a preset value (e.g. 0.4°/dbl stroke), see Figure 8. The only disadvantage is, in this case, a higher number of required double strokes.

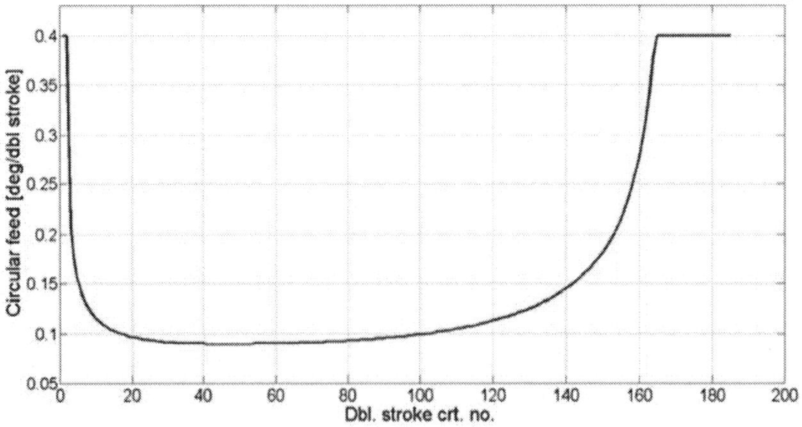

Fig. 8. Circular feed variation law (modified)

The variation of removed chip area resulted if applying the determined variation law of the circular feed is presented in Figure 9.

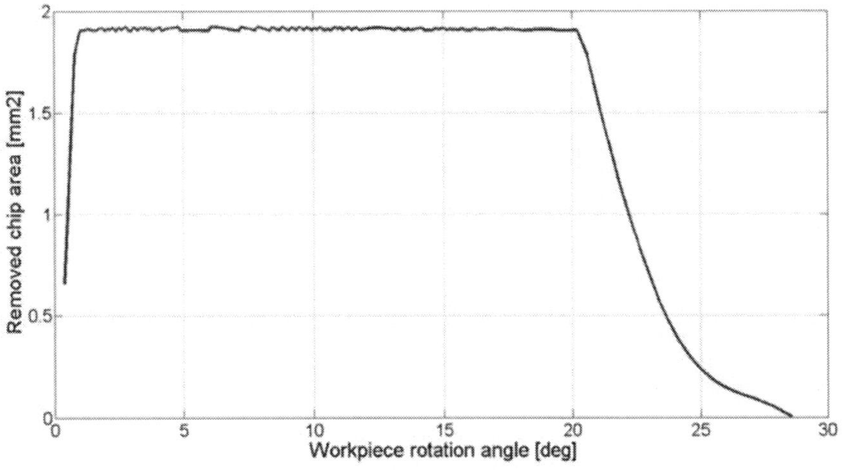

Fig. 9. Smoothened variation of removed chip area

4 Conclusion

This paper analyses the possibility of smoothing the energetic load along the rack tool cutting edge, when machining gear teeth. In this purpose, a graphical modelling of the toothing process performed with a mono-tooth rack shaped cutter has been realized in CATIA. The variation law of removed chip area has been deducted on this base, as the analytical expression of a polynomial function fitted to the points representing the areas of successively detached chips, with the help of MatLab soft facilities. A MatLab application was also written in order to determine the variation law of the workpiece circular feed, if aiming to maintain a constant area of the removed chips, hence to smooth the energetic load along the rack tool cutting edge. The results obtained by solving the approached problem in a particular case prove that, from theoretical point of view, the proposed methodology works fast and precisely and removed chip area may be considered constant for a significant number of tool double strokes. The remaining challenges are to extend the application domain to other types of tools & generated surfaces and to develop a control system enabling to actually implement in practice the imagined solution.

References

1. I. Lăzărescu, *Calculation and construction of cutting tools*, (Tehnică Publishing House, Bucureşti, 1961).
2. I.I. Semencenko, V.M. Matiushin and G.N. Saharov, *Cutting tools design*, (Science and technology for machine building, Moscow, 1962).
3. M. Dima, N. Oancea and V. Teodor, *Modelling the cutting schemes of the toothing processes*, (Cermi, Iaşi, 2007).
4. A. Antoniadis, N. Vidakis and N. Bilalis, J Manuf Sci Eng, **124**, pp.784-791, (2002).
5. V. Georgescu, N. Oancea and E. Constantin, Buletinul Universităţii din Galaţi, Fasc. V, pp.33-42, (1982).
6. J. Ma, X. Ge, S.I. Chang and S. Lei, Int J Adv Manuf Technol, **74**, pp.1701–1708, (2014).
7. M Dima and N. Oancea, Romanian Journal of Technical Sciences, Applied Sciences, **49**, pp.237-241, (2004).

8. S.P. Radzevich, *Kinematics Geometry of Surface Machining* (CRC Press, London, 2008).
9. N. Oancea, *Surface Generation by Enwrapping, Vol. I, Fundamental theorems* („Dunărea de Jos" Publishing House, Galaţi 2004).

Using regression analysis method to model and optimize the quality of chip-removing processed metal surfaces

Mihail Aurel Țîțu[1,*], and *Alina Bianca* Pop[2]

[1]Lucian Blaga University of Sibiu, Victoriei Street no. 10, code 550024, Sibiu, România
[2]SC TechnoCAD SA, Vasile Alecsandri no. 72, code 430351, Baia Mare, România

Abstract. The paper aim is to identify based on regression analysis, the quantitative relationship between the cutting process parameters (cutting speed, cutting depth and feed per tooth) and the arithmetic mean deviation of the surface profile, measured longitudinally and transversely on the cutting feed direction, which describe the system at any point in the chosen experimentally studied range. The equations coefficients means the influence of the variables on the pursued answer.

1 Introduction

Among the previous research are the investigation of the effect of the feed rate on the surface roughness and the reduction of processing time [1], and also the mathematical modeling of a regression equation of the surface roughness obtained by end milling [2]. This paper is a continuation of the researches conducted in [2].

In the scientific literature, there are studies based on the mathematical modeling such: [3] and [4], in which the implementation of the Taguchi method was analyzed in order to improve the product quality; [5] which used the Response Surface Methodology (RSM) as optimization technique, [6] used the regression analysis and [7] which used ANOVA and the regression analysis. The regression analysis allows to use the experimental data to calculate the equation coefficients that are describing the studied system at any point in the experimentally range.

2 Research method

Any experimenter begins an experimental study based on a specific working plan, which will use an experimental plan or some experimental project. The experimenter can reach the desired solution, based on his knowledge in the field, or based on his intuition - but this is rarely achievable. Doing so, it may lose time and money, up to reach an appropriate result, or it may not ever reach the desired objective. Therefore it is necessary to use mathematical statistics methods and optimization methods for a rational choice of the experimental measurements to be carried out and to get reliable and consistent information [4]. In the

* Corresponding author: mihail.titu@ulbsibiu.ro

first phase to conduct the research it is necessary to establish the process input factors: cutting speed, cutting depth and feed per tooth. Given that the full factorial experiment was used, the number of factors corresponding levels are shown in table 1. Based on these values it is necessary to conduct a number of 6x5x5=150 experiments. Thus, the 150 cutting regimes obtained under the combinations, were tested by processing of 21 Al7136 blocks with the following dimensions: 500 x 101 x 24.5 mm. All of the three parameters takes variables values according to the tool manufacturer recommendation for aluminum machining [8]. The experiment will be performed using a standard tools set of aluminum machining - 16 mm End milling cutter - SECO R217.69-1616.0-09-2AN, holding two indexable cutting inserts XOEX090308FR-E05, H15. The machine used for the milling tests is a HAAS VF2 CNC. Abundant amounts of Blasocut BC 35 Kombi SW mineral coolant were provided at the cutting zone throughout the experiment. To analyze the surface roughness for different machining conditions it was used the portable surface roughness tester - Mitutoyo SURFTEST SJ-210. The machined surface roughness R_a will be measured longitudinally (R_a long) and transversely (R_a transv) on the cutting feed direction. Then, the regression equations will be obtained using the least squares method, because the experiment aim is to determine the most accurate constants values for any equation type, assigning equal or unequal weights (as needed), to various experimental data. After the regression equations obtaining, it is necessary to assess the associated experimental data, based on the variance analysis ANOVA. To use the ANOVA test to evaluate the quality match, it is necessary to repeat at least once the experiment in the center of the experimental range. In this research, all the experiments were repeated for seven times. It is very important to know the precision which the coefficient values were predicted, not only their absolute value. Finally, based on the proposed regression model, the arithmetic mean deviation of the surface profile will be determined, and it will be compared with the experimental results.

Table 1. Cutting regime

Cutting regime	**Cutting speed [m/min] – 6 levels**
	495 530 570 610 660 710
	Cutting depth [mm] – 5 levels
	2 2.5 3 3.5 4
	Feed per tooth [mm/tooth] – 5 levels
	0.04 0.06 0.08 0.11 0.14

3 Regression analysis

3.1 R_a longitudinal regression analysis

The entire analysis starts with the studied research factors and the interactions among them, and also the R_a long and R_a transv measurements (each with 1050 measurements). From all experimental data tested, the values affected by aberrant error were removed and the rest of data were used further to determinate the regression equations. To facilitate the necessary calculations to achieve the objective, the statistical application MINITAB 17 was used. In this application the experimental data were introduced and by its predefined calculation method, it was determined the squares sum and the percentage distribution for each factor, respectively factors interaction. The freedom degrees were calculated for each factor/interaction, using Taguchi's method. For Ra long situation it was found that the cutting depth B influence is 0.02% - a negligible value of. It is the same case for interaction between B and C factors, whose influence is felt in 0.14% -a small percentage. Therefore, to obtain the regression equation the influence percentage of the other factors were recalculated (Table 2). It was found that the model errors, aggregates accounting for 13.93% and therefore, to determine the regression equation it will be consider the parameters and the interactions among them.

Table 2. The percentage presentation of the cutting parameters influence exerted on the R_a long [2]

Source	Freedom degrees		Square sum		Percentage contribution	
	R_a long	R_a long recalculated	R_a long	R_a long recalculated	R_a long	R_a long recalculated
Regression model	7	5	199.482	199.38	86.12%	86.07%
Cutting speed A	1	1	194.322	194.322	83.89%	83.89%
Cutting depth B	1	-	0.040	-	0.02%	-
Feed per tooth C	1	1	1.157	1.194	0.50%	0.52%
A x B	1	1	3.093	0.569	1.34%	0.25%
A x C	1	1	0.316	2.795	0.16%	1.21%
B x C	1	-	0.366	-	0.14%	-
A x B x C	1	1	0.188	0.501	0.08%	0.22%
Error	827	829	32.156	32.258	13.88%	13.93%
Fitting error	143	145	18.735	18.837	8.09%	8.13%
Pure error	684	684	13.421	13.421	5.79%	5.79%
Total	834	834	231.638	231.638	100%	100%

Thus, by MINITAB 17 using, and its predefined calculation procedure, the regression coefficients were determined and the regression equation related to Ra long, has the following form:

$$R_a long = 0.000287 \cdot A - 5.808 \cdot C + 0.000196 \cdot A \cdot B + 0.01345 \cdot A \cdot C - 0.001582 \cdot A \cdot B \cdot C \quad (1)$$

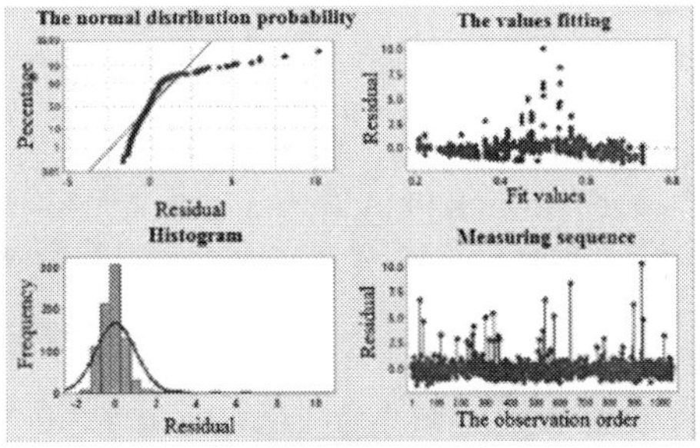

Fig. 1. The R_a long equation residuals diagram.

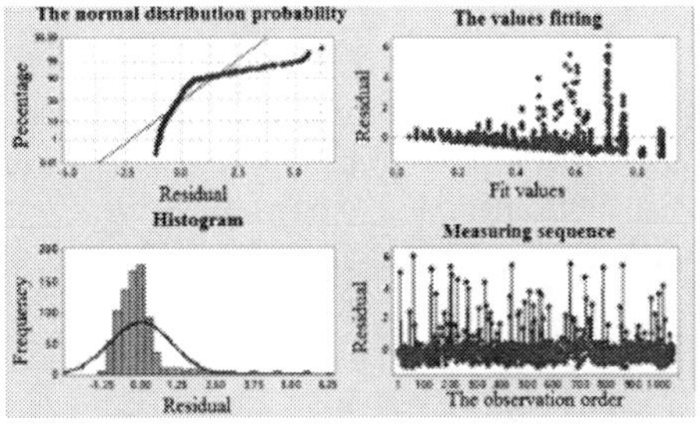

Fig. 2. The R_a transv equation residuals diagram.

In Figure 1 the residual diagrams corresponding to the obtained regression equation are plotted. Analyzing the normal distribution it was found that the residuals does not approximate a straight line, and the distribution histogram of the residuals indicates some values that exceed a Gaussian distribution curve. This is due to the roughness values jumps recorded when the cutting speed is in the middle of the studied interval - 570-610 [m/min]. Regarding the values matching diagrams and the diagram with the measurements order, it was found that the residuals have a uniform distribution to the zero line but, in some cases, there are some big jumps from one value to another due to the recorded values at 570-610 [m/min] cutting speeds.

3.2 R_a transversal regression analysis

Similar to the previous paragraph, Table 3 shows the R_a transv situation. Also, the cutting speed is the parameter with the greatest influence on the surface roughness with 39.11%. At the opposite side is the feed per tooth, whose influence on R_a transv, is only 0.62%.

Table 3. The percentage presentation of the cutting parameters influence exerted on the R_a transv

Source	Freedom degrees	Square sum	Percentage contribution
Regression model	7	229.728	46.24%
Cutting speed A	1	194.328	39.11%
Cutting depth B	1	27.813	5.60%
Feed per tooth C	1	3.078	0.62%
A x B	1	3.598	0.72%
A x C	1	0.007	0,001%
B x C	1	0.063	0,01%
A x B x C	1	0.842	0.17%
Error	846	267.089	53.76%
Fitting error	143	163.840	32.98%
Pure error	703	103.250	20.78%
Total	853	496.818	100%

The regression equation related Ra transv, has the following form (MINITAB 17):

$$R_a transv = -0.000416 \cdot A - 0.027 \cdot B - 12.23 \cdot C + 0.000502 \cdot A \cdot B + 0.0182 \cdot A \cdot C + 4.67 \cdot B \cdot C - 0.00767 \cdot A \cdot B \cdot C \quad (2)$$

In Figure 2 the residual diagrams corresponding to the obtained regression equation are plotted and it is found that the residuals does not approximate a straight line, and the distribution histogram of the residuals indicates the some values that exceed a Gaussian distribution curve. The residuals have not a uniform distribution to the zero line because of the big jumps of the measured values at 570-610 [m/min] cutting speeds. Next for the each determinate regression equation, the variance analysis ANOVA will be used.

4 The variance analysis ANOVA of the regression equation

The ANOVA analysis terms were determined based on the MINITAB 17 calculation procedure, and they were centralized in Tables 4.

For R_a long, given that the F calculated value is higher than F tubular's value (F_{calc}> F_{tab}) and always greater than 1, it can be considered that the relationship between the adjusting variables and the responses is statistically significant. Therefore, the mathematical model is acceptable and it permitted to describe the studied phenomenon evolution in the experimentally field, with a strong correlation between the studied factors, responses and the regression equation between the proposed and the experimental data obtained. But in the R_a transv case, F_{calc}> F_{tab} and it is not greater than 1, because the Ra transv data have not a normal distribution. Conducting the statistical analysis of the regression equations coefficients it was obtained the confidence significance of 85.92% for R_a long and 45.53% for R_a transv. This percentage has not a higher value due to the recorded data values distribution in the fields of 570-610 [m/min] of cutting speed.

Table 4. ANOVA from the R_a regression equations

Source	F_{calc}		F_{tab}	
	R_a long	R_a transv	R_a long	R_a transv
Regression model	1025.09	103.954	$(F_{0.05,5,834})$ 2.225	$(F_{0.05,7,853})$ 2.020
A	5.054	1.293	$(F_{0.05,1,834})$ 3.853	$(F_{0.05,1,853})$ 3.853
B	-	0.039	-	$(F_{0.05,1,853})$ 3.853
C	94.717	2.783	$(F_{0.05,1,834})$ 3.853	$(F_{0.05,1,853})$ 3.853
A x B	22.429	3.903	$(F_{0.05,1,834})$ 3.853	$(F_{0.05,1,853})$ 3.853
A x C	64.997	2.063	$(F_{0.05,1,834})$ 3.853	$(F_{0.05,1,853})$ 3.853
B x C	-	2.864	-	$(F_{0.05,1,853})$ 3.853
A x B x C	12.871	2.666	$(F_{0.05,1,834})$ 3.853	$(F_{0.05,1,853})$ 3.853
Fitting error	6.628	7.799	$(F_{0.05,145,834})$ 1.222	$(F_{0.05,143,853})$ 1.223

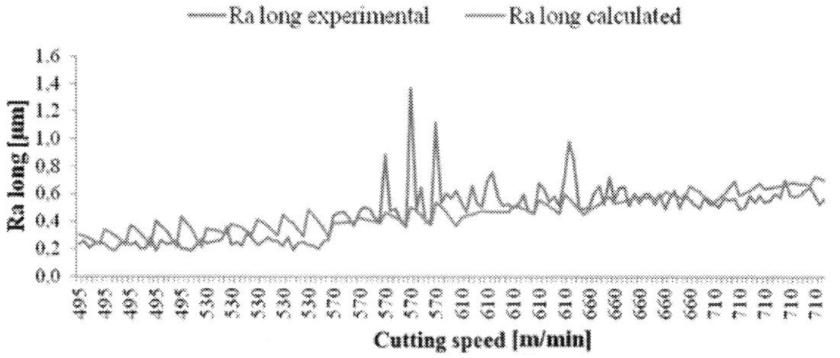

Fig. 3. The Ra long experimental measurements versus Ra long calculated measurements diagram

The last step that must be met to complete this research lies in results predicting at any point in the obtained experimentally range based on the regression model. Specifically, the arithmetic mean deviation of surface profile will be determined using the obtained regression equations. The graphical representations to compare the experimental measurements and the calculated values by using the equation, were made (fig. 3 and fig. 4). In both graphs, were introduced the R_a long and R_a transv evolution values, depending on the cutting speed, because this parameter has the greatest influence on the response. The problem of the higher errors values occurred in both cases, is caused by the cutting speeds of 570 m/min and 610 m/min, where the arithmetic mean deviation of surface profile measured, recorded significant jumps distributions.

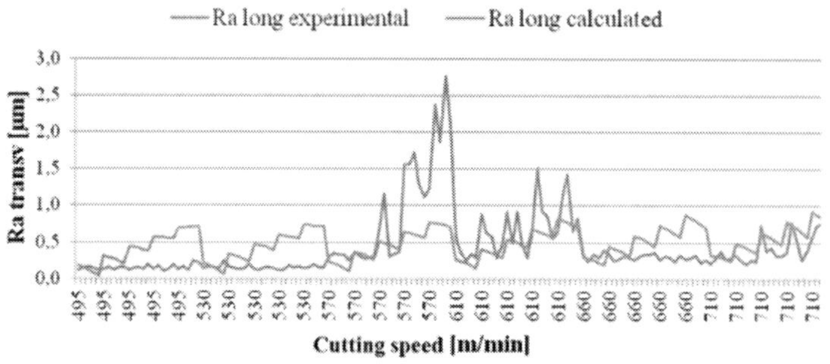

Fig. 4. The Ra transv experimental measurements versus Ra transv calculated measurements diagram.

Conclusions

After the performed regression analysis were identified the quantitative relationship between the cutting process parameters and the arithmetic mean deviation of surface profile. In both cases for R_a long and for R_a transv, cutting speed has the greatest influence on the arithmetic mean deviation of the machined surface profile. The obtained regression equations, approximates the studied system at any point in the chosen experimental range, with a grade of 85.91% in matching longitudinal Ra and 45.52% for transverse Ra.

The increases of the Ra values on 570 m/min and 610 m/min cutting speed, are caused by the vibrations occurred during the end milling process. The vibrations emergence, leads to a rough surface. These vibrations are the effect of the resonance phenomenon given by the contact between the tool and the workpiece and the materials used. The causes, the producing manner of and the vibration effect could be detailed studied in further research.

References

1. A.B. Bonţiu Pop, *Proceedings of the International Conference of the Carpathian Euro-Region specialists in industrial systems-Ceursis 2014*, (Technical University of Cluj Napoca, România, North University Center in Baia Mare, 2014)
2. A.B. Pop, M.A. Ţîţu, IOP Conf. Series: Materials Science and Engineering **161** (2016)
3. A.B. Bonţiu Pop, Applied Mechanics and Materials, Vols. **809-810**, pp 123-128 (2015)
4. M. Ţîţu, C. Oprean, Al. Boroiu, *Applied experimental research for increasing the quality of products and services (in Romanian)* (AGIR Publishing House, Bucharest, 2011)
5. A. Kannan, K. Esakkiraja, M. Nataraj, Journal of Mechanical and Civil Engineering **9**, 4, 59-64 (2013)
6. Z.A. Shekh, T.A. Khan, International Journal of Scientific Engineering and Technology **2**, 7, 705-709 (2013)
7. D. Vakondios, Kyratsis, P., Yaldiz, A. Measurement, **45**, 6, 1480-1488 (2012)
8. A.B. Bonţiu Pop, *Aluminum alloys surface quality by cylindrical-frontal milling (in Romanian)*, (PhD Dissertation, Technical University of Cluj Napoca, România, North University Center in Baia Mare, 2015)

The experimental research strategy of the end-milled aluminum alloys

Alina Bianca Pop[1], and *Mihail Aurel* Ţîţu[2,*]

[1]SC TechnoCAD SA, Vasile Alecsandri no. 72, code 430351, Baia Mare, România
[2]Lucian Blaga University of Sibiu, Victoriei Street no. 10, code 550024, Sibiu, România

Abstract. In this paper, the most important aspects of the research methodology of the resulted surface quality by aluminum alloys end-milling, are presented. The first part of the paper includes a short presentation of the cutting processes aluminum alloys evolution. In the second part, the research problem was defined by setting a fundamental objective and identifying the specific objectives which contribute to achieve the fundamental objective. In this context an experimental study, was performed, starting with the organization and conducting the end-milling process, in which the cutting regime were established, then the characteristic measured parameters were determined, so that the research factors and parameters will result.

1 Introduction

Each science has its own history that marks its development stages. Paul H. Black's book "Theory of Metal Cutting" has a starting point the paragraph "Metal Cutting - Art to Science" in which assumes that cutting history began with 600000-1000000 years ago, somewhere in Tanganyika, Africa. The cutting processes development started since the 19th century [1], when the first experimental tests were conducted aimed to explain the chip formation [2]. Later, in the 1930s, the sintered metal or mineral-ceramic materials were discovered, than in 1950 the indexable cutting inserts were carried which were widespread in the production, leading to the new tools types development. The machining process has a considerable age with a continuous evolution derived by the progresses occurred in the new materials developing, also by the cutting tools and machine tools improvement. The machining process is closely related to the costs, it is desired to raise the machine tools and machining centers performance, and also the new products continuous development, which also contributes to increase the machined parts precision. Choi [3] presents a fairly detailed theory accompanied by a number of applications related to surface modeling techniques used by different software for computer aided design/computer-aided machining. Following a literature review, conducted on the materials used in the cutting processes [4], it was found that over the last decade, the researches have focused particularly on steels. This statement is supported by the Table 1 [5], in which the percentage comparison of the workpiece materials use frequency is presents.

* Corresponding author: mihail.titu@ulbsibiu.ro

Table. 1. A percentage comparison on frequency of raw material use [5].

Steel	Cast iron	Aluminium	Plastic	Other material
64%	1%	23%	1%	11%

In this paper the attention will be focused on aluminum, which has been studied only in 23%. Today, aluminum and its alloys are considered to be the most practical metals due their low cost, low weight and modern appearance. Aluminum has a god electrical conductivity, is a good heat conductor, it is magnetic, it is reflexive and resistant to chemicals. It is commonly used in the construction, marine, and aircraft industry because of the ease in processing, non-toxicity and corrosion resistance. The anodizing increases his strength and also allows iridescent finishes in different colors. Due to the aluminum alloy properties of that make it one worthwhile material, the variety of its applications is continuously growing. The aluminum alloys typically contain alloying elements such as copper, magnesium, manganese, silicon and zinc, the aluminum (Al) is predominantly metal. In [4], the research workability of aluminum alloys depending on the year and the number of papers identified through a literature review (period 2000 to 2014), is highlighted (Figure 1). As it can be seen in Figure 1, the higher percentage research, were performed on Al6061 alloy which occupies a proportion of 32%. The aluminum alloys machinability has been studied and tested most often through milling – 46%, drilling – 7% and turning – 4% [4]. Another important aspect is the main research directions pursued in aluminum cutting processes. Following an analysis conducted by [4], it was found that the surface roughness was the most frequently studied (Table. 2).

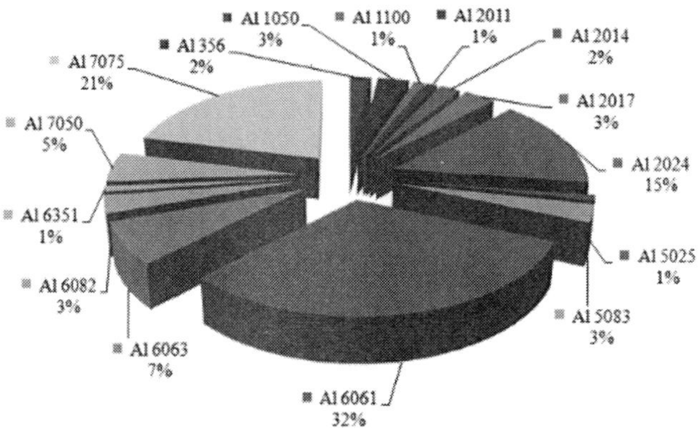

Fig. 1. A percentage comparison on frequency of research on aluminum alloy machinability [5].

Table. 2. Percentage comparison on the research directions.

Friction	7%	Deformations	9%	Temperature	10%	Forces	19%
Tool wear	7%	Stress	13%	Chip formation	14%	Roughness	21%

On the other hand, the most of the identified research paper in order to study these alloys machinability, use for optimization, mathematical methods. The percentage of this method use is shown in table 3 [4]. Starting with this information, the main problem of this scientific work is to present and develop the research methodology of the resulted surface quality by aluminum alloys end-milling.

Table. 3. Percentage comparison on mathematical methods used to optimize the cutting process.

RBFN	Taguchi	Regression	ANOVA	RSM
2%	24%	14%	38%	22%

2 Research Methodology

2.1 Experimental Methods Used in Research

The scientific research involves the theoretical studies development combined with the experimental research. The theoretical studies enable to underline the laws of the studied phenomena by using the mathematical equipment and the achievements in the respective field or in other fields [6], [7]. The methods used in the experimental research are [6]: - *The hypothesis testing* using: χ^2 test, Fisher - Snedecor test, Cochran test or Bartlett test.- *The anomalous result elimination* involves the measurement results reading and it allows highlighting the data that are inconsistent with reality - Chauvenet test, Romanovski test, Grubbs test, Student test etc. - *Check the randomness data string* is often carried out using test Young. - *To check the normal distribution of experimental data,* the Gauss's law in the experimental results processing obtained by measurements, is more accepted, χ^2 test, Massey test, Shapiro-Wilk test, Kolmogorov test. - *When determining the range and level of trust* - the arithmetic mean and the dispersion are the main parameters describing a statistical set. - *To choose the confidence level value* the measurements made in: industry, laboratories, some operations control, reception, testing, expertise, etc., take the 95% confidence level; the measurement of high quality and responsibility in science and technology, one of the values: 99%, 99.5%, 99.73% and 99.99%. In this context, the research presented in this paper were conducted using the above mentioned methods of covering the requirements of processing / modeling of experimental data held.

2.2 Research Problems Defining

For the beginning, the *fundamental research objective* was defined: the study of the arithmetic mean deviation of the surface profile Ra, resulted from the 7136 aluminum alloys end milling process, depending on the cutting process parameters.

Specific objectives helps to fulfill the fundamental objective. These must identified and consists in answers to the following research questions [4]:

Question 1: What is the influences percentage of each cutting parameter on the arithmetic mean deviation of the machined surface profile Ra?

Question 2: How is the arithmetic mean deviation of the surface profile Ra, influenced by the each cutting process parameter?

Question 3: It is possible to get a valid mathematical model of Ra, depending on the cutting process parameters?

Question 4: It is possible to make some determined settings on the cutting process parameters in order to obtain a predetermined Ra?

The answers of these questions represent the following specific objectives:

Specific objective 1: Calculation of the percentages influences of the cutting process parameters to attain the performance on obtaining an arithmetic mean deviation of the machined surface profile Ra, as good as possible and also, to have a minimum sensitivity of the uncontrollable factors.

Specific objective 2: The Ra experimental determination of the Al7136 end milled surface profile, and the data analysis and the interpretation of the measured results.

Specific objective 3: The establishing of the mathematical model to determine the arithmetic mean deviation Ra according to the studied cutting parameters in one established experimentally field.

Specific objective 4: The mathematical model validation and the experimental data optimization.

2.3 Research Design

The research design stages summarized and adapted by [8] are: first to conducting a rigorous bibliographic study, then the research problem definition, setting the input data entry: raw material, the cutting operation, the research factors and their variation levels, cutting tools, CNC Machine, instruments and measurement devices; setting the output data: the choice of the responses tracked in research and the methods of their determination; choice of the experimental method so the information progressive acquiring is allowed, and finally the system knowledge (phenomenon).

The experimental method must to be efficient, and allows to obtaining the maximum information with a minimum experiments number. Also, the required number of measured values, samples and tools used will be established; the experiment; analyzing, interpreting and validating the results analysis, is facilitated by the initial choice of the experimental method, which also must achieve the objectives; the information gradual acquisition is necessary because if a phenomenon is studied, the experimenter did not know the results, so he must forward progressively and reorient the research based of the obtained results. First experiments lead to primary conclusions and based on these conclusions can be make another series of experiments. The experiments series allow obtaining more comprehensive information about the phenomenon. If it is necessary, depending on the obtained results in the first experiments series, it can be carried out another experiments series. In this way the experimenter seeks the needed answers and he stops when he obtained them; knowledge of the studied system.

2.4 Research Planning

The research planning, pursue the expected results in the maximum efficiency terms. The measurements volume will be based: on the accuracy, on the timing of the instruments for measuring purchase, on the results processing timing and on the research cost. A low measurement volume has a negative effect on their accuracy, and a too big volume charges the research program useless. The excessive repetition of the measurement data may lead to measure some parameters variation tendencies as a result of the unavoidable data dispersion, resulting incomplete or erroneous conclusions.

Table 4. Al7136 Chemical Composition Standard AMS4415A.

Element	Min	Max	Element	Min	Max	Element	Min	Max
Silicon	-	0,12	Magnesium	1,8	2,5	Zirconium	0,10	0,20
Iron	-	0,15	Chromium	-	0,05	Other elements, Each	-	0,05
Copper	1,9	2,5	Zinc	8,4	9,4	Other elements, Total	-	0,15
Manganese	-	0,05	Titanium	-	0,10	Aluminum	difference	

The effective research planning is based on the material used in the study - 7136 aluminum alloy with the chemical composition indicated in Table 4. The effective experiments will go towards to detailed study of the behavior of Al 7136 according with the proposed objectives taking into account the variation of the cutting regime parameters.

The experiment will be performed using a standard tools set on aluminum machining - SECO R217.69-1616.0-09-2AN - 16 mm End milling cutter with 100% tool engagement - holding two indexable cutting inserts XOEX090308FR-E05, H15.The machine used for the milling tests is a HAAS VF2 CNC. The optical microscope Micro-Vu VERTEX 310 was used to make an optical determination and evaluation of the machined surface. To analyze the surface roughness for different machining conditions it was used the portable surface roughness tester - Mitutoyo SURFTEST SJ-210, with a resolution of 0.002 μm.

2.5 Research Strategy

The research strategy (Figure 2) has as a starting point the fundamental objective initially set. To conduct the research strategy, it is required an experimental study on the defined research problem, based on an experimental plan. To make a rational choice of the experiment to carry out and also, to obtain the most clear and precise results it is necessary to use modeling and optimization techniques of the experimental research.

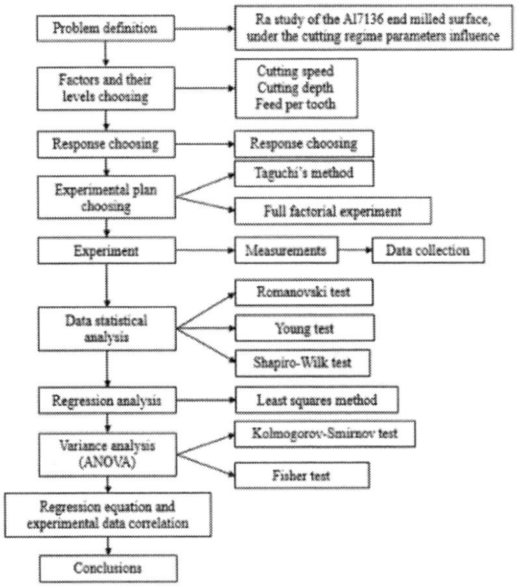

Fig. 2. Research strategy [4].

The Figure 2 requirements will be accomplished according to the producer's recommendations and technological possibilities of the tool, CNC Machine and workpiece materials. In order to identify the influence of cutting speed; cutting depth and feed per tooth on the arithmetic mean deviation of the surface profile Ra, in the chosen experimental field, it is necessary to conduct preliminary screening programs [8]. Then the optimization is used to predict the response value in the experimental field of any factors combination to identifying the optimum to increase the cutting process performance and obtaining a minimum sensitivity to the uncontrollable factors and to determine the influence percentage of the studied factors on the followed response - finally to be able to get a regression model. In this way the methods used are Taguchi, because it can be use a small number of experiments, and a full factorial experiment based on a 7 times experiments replication, to obtain more precise results, according to Montgomery [9] which says that this repetitions can variate between 3 and 7 times for each set of input parameter values to determine the consistency of measurements. A statistical analysis will be performed to certify that the obtained values are real for the studied process. Finally, the regression analysis using the least squares method [6] will be performed to identify a quantitative relationship between the cutting process parameters and the response, which describe the system at any point in the studied experimentally range. The experimental data evaluation related to the regression equation must be done with the Variance Analysis ANOVA, which is focused on the estimation of a various variability types of responses made by Fisher test. By Fisher test the coefficients significance of the regression equation is tested by comparing the ratio of two variances and show the probability that between them exist or not a statistical difference [8]. But the ANOVA analysis of the regression equation can be carried out only under a normal data distribution. To confirm that, the Kolmogorov-Smirnov test. Finally the proposed regression model will be compared with the experimental results.

Conclusion

Starting from the fundamental objective and specific objectives of this research paper defined in the current chapter, the whole activity is based on the experimental plan settled.

The design, planning, programming and control of processing activities conducted are taking into account the whole range of technologies and methods used to define how products must be manufactured.

The machining activities such: design, planning, programming and control that transform the inputs (workpiece) into finished products was done taking into account the whole range of the technologies and methods used to define how products must be manufactured.

With the experimental stand, it have been fulfilled the main necessary requirements to achieve the sets objectives and also to obtain the experimental data on which can be analyzed the cutting parameters influence (cutting speed, cutting depth and feed per tooth) exerted on the roughness of the end milled surface.

The research strategy was established, beginning with a study that was conducted on the defined research problem, with an experimental design plan using Taguchi's method. For this, it was used modelling and optimizing methods and techniques for the experimental research.

References

1. I. Time, *Resistance of metals and Wood to Cutting (in Russian)*, (St. Petersburg, Russia, 1870).
2. H. Treska, Bulletin de la Sciete d'Ecouragement pour l'Industrie Nationale **15**, 585-685 (1873).
3. Y. K. Choi, *Tool path generation and 3D tolerance analysis for free form surfaces.* Thesis for Doctor of Philosophy, (Texas A&M University, 2004).
4. A.B. Bonţiu Pop, *Aluminum alloys surface quality by cylindrical-frontal milling (in Romanian)*, (PhD Dissertation, Technical University of Cluj Napoca, România, North University Center in Baia Mare, 2015)
5. A.B. Bonţiu Pop, M Lobonţiu, *Proceedings of the International Conference of the Carpathian Euro-Region specialists in industrial systems-Ceursis 2014*, p. 35-39 (Technical University of Cluj Napoca, România, North University Center in Baia Mare, 2014)
6. V. Năsui, A. Coteţiu, R. Coteţiu, M. Lobonţiu, N. Ungureanu, *Basis of experimental research of electromechanical actuators (in Romanian)* (Universităţii de Nord Publishing House, Baia Mare, 2007).
7. G. Creţu, *The basis of experimental research (in Romanian)*, (Gheorghe Asachi Tehnical University of Iaşi, 1998).
8. M. Ţîţu, C. Oprean, A. Boroiu, *Applied experimental research for increasing the quality of products and services (in Romanian)* (AGIR Publishing House, Bucharest, 2011).
9. D. Montgomery, *Design and Analysis of Experiments.* (Eighth Edition, Hoboken: John Wiley & Sons, Inc, 2013).

Defining the machining process for complex parts made from aluminium alloy extruded profiles by correcting raw material deviations

Ştefan Adrian Moldovan*, *Mihai* Banica, and *Vasile* Năsui

Technical University of Cluj-Napoca - North University Center of Baia Mare, Victor Babes, 62A, Baia Mare, 430083, România

Abstract. In this paper we present technical solutions to define the machining process to create good products specific for aerospace industry made of aluminium alloy extruded profile. Those parts have very tight tolerances and it is necessary to create a robust system to ensure repeatability of machining processes.
The major factor which affects our machining process is internal stress effect of the aluminium alloy extrusions. To determinate this influence were made several tests, using parts for aerospace industry, those parts were measured through coordinate measuring machine and all data was recorded for statistical analysis. Based on input data was determinate a technical solution to eliminate deviations caused by raw material internal stress for obtaining parts in tolerance.
Through several tests and recording all dimensions changes during the milling process, was obtained good products and was ensured repeatability of the machining process.

1 Introduction

The major concern for our production is to maintain the production repeatability. In aerospace industry it's not space to make mistakes for this reason we need to create a robust system to can ensure our products quality. Starting a new project on S.C. Universal Alloy Corporation Europe S.R.L. factory, were determined a machining strategy to can machine parts made from aluminium alloy extrusions. Through several tests which was measured on the coordinate measuring machine and all data was recorded for statistical analysis. Creating the Ishikava diagram was determined all factors which affect the machining process in this factory (Figure 1). The aim for this research is to track and control the extrusion internal stress using input data's to determinate the proper machining strategy [1].

* Corresponding author: moldovan.stefanadrian@gmail.com

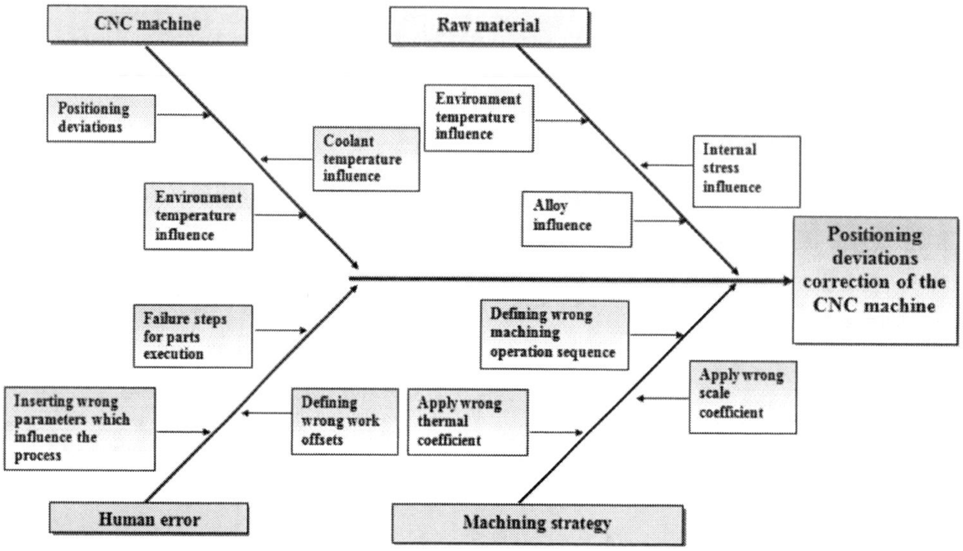

Fig. 1. Ishikawa diagram.

2 Manufacturing methodology

The machining process is performed in a normal processing unit, with climate control and the CNC (computer numerical control) machine used is Handtmann PBZ HD 600 (Figure 2). The CNC machine provides temperature compensation based on environment temperature fluctuation and also it is equipped with chiller system to maintain constant the coolant temperature [2].

Fig. 2. Handtmann PBZ HD 600.

Developing optimum machining results with best accuracy, high material removal rates with 63 kW spindle power a maximum 30000 rpm. As well as short machining times and reduced cost per part are the outcomes. The HD machine provides full 5-axis simultaneous machining [3]. All machined parts specific for aerospace industry are validated by CMM (coordinate measuring machine), placed in a room without temperature control environment.

CMM is a device for measuring the physical geometrical characteristics of an object. The typical 3 "bridge" CMM is composed of three axes, X, Y and Z [4]. These axes are orthogonal to each other in a typical three-dimensional coordinate system. Each axis has a scale system that indicates the location of that axis. The machine will read the input from the touch probe, as directed by the operator or programmer. The machine then uses the X, Y, Z coordinates of each of these points to determine size and position with micrometre precision typically.

The research was conducted on extruded aluminium alloys work pieces with 5600 mm length, used in aerospace industry, with a wide range of alloys and treatments. In Figure 3 is showing the test part placed inside the subassembly. With thin walls and the section represents a high resistance to shock and breakage, develop from advanced alloys tailored for specific applications to help find the right balance of strength, damage tolerance, and corrosion resistance.

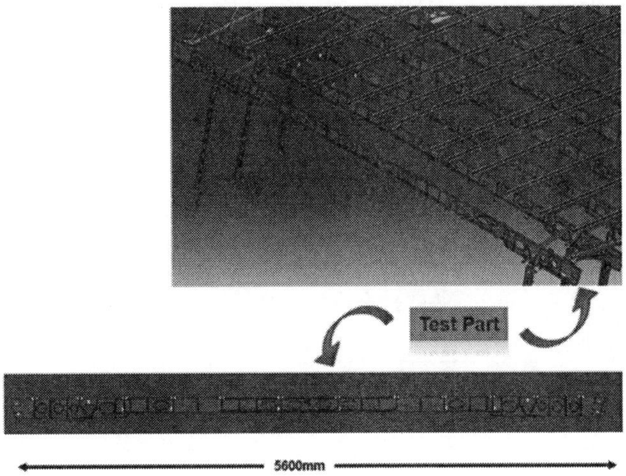

Fig. 3. Aerospace part assembled view.

The problem appears after machining process for those parts on the holes position, which have deviation out of tolerance on X-axis based on the datum reference (Figure 4). The tolerance for the all holes is ± 0.2 mm.

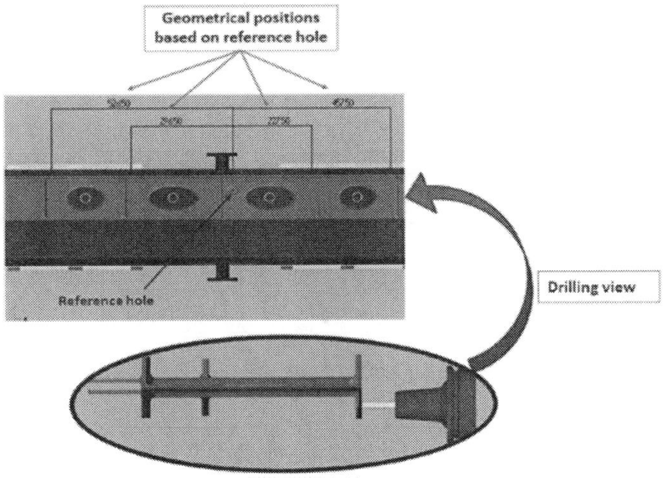

Fig. 4. Machined part view.

3 Machining results

The parts geometry is constrained by one middle hole. After machining four identical pieces, were measured in CMM. In Table 1 are represented the measurement values of the holes geometrical position based on reference hole.

Table 1. First four identical parts machined results.

Hole position on X-axis [mm]	P1	P2	P3	P4
-2316.7	-0.157	-0.275	-0.302	-0.277
-2162.5	-0.190	-0.255	-0.292	-0.265
-1991.5	-0.139	-0.244	-0.280	-0.255
-1727.5	-0.125	-0.196	-0.224	-0.233
-1585.0	-0.114	-0.173	-0.202	-0.208
-1503.7	-0.096	-0.163	-0.201	-0.200
-1261.5	-0.077	-0.137	-0.185	-0.169
-787.5	-0.058	-0.081	-0.111	-0.092
-457.5	-0.024	-0.040	-0.066	-0.048
-227.5	0.007	-0.009	-0.018	-0.023
296.5	-0.093	-0.066	-0.056	-0.067
526.5	-0.121	-0.097	-0.084	-0.104
856.5	-0.143	-0.131	-0.116	-0.136
1330.5	-0.239	-0.193	-0.172	-0.199
1572.7	-0.213	-0.235	-0.205	-0.211
1654.0	-0.220	-0.237	-0.219	-0.226
1796.5	-0.287	-0.258	-0.246	-0.248
2060.5	-0.306	-0.283	-0.282	-0.286
2231.5	-0.327	-0.303	-0.296	-0.309
2385.7	-0.354	-0.330	-0.314	-0.302

By analysing the results it is visible a material contraction to the middle of the part. The maximum deviation value which is on the ends of the part is -0.354 mm, on chart analysing is visible a polynomial growth (Figure 5).

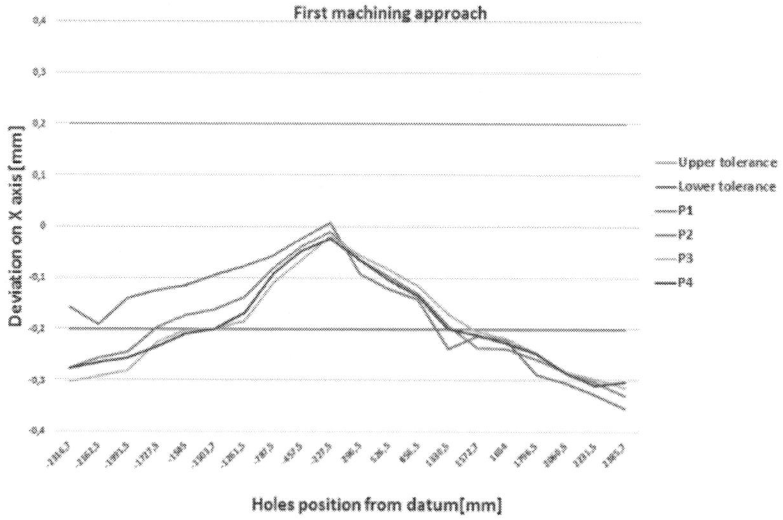

Fig. 5. First four identical parts machined results.

It is visible the final part contraction at both ends. Through elimination of other factors from Ishikava diagram, which can influence the holes position, was determinate the root cause to by internal stress (Figure 6).

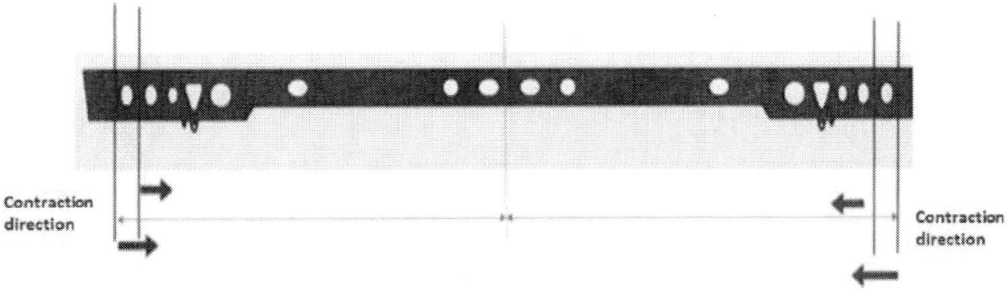

Fig. 6. Part contraction.

The solution for this issue is to eliminate the material stress before drilling operation, that means is to release the part to be on the free state from the clamping device and clamped again after that. Using this approach was made other four identical pieces. In Table 2 are represented the measurement values of the holes position based on the reference hole.

Table 2. Second four identical parts machined results.

Hole position on X-axis [mm]	P5	P6	P7	P8
-2316.7	0.006	0.008	0.022	0.007
-2162.5	0.013	0.008	0.002	-0.009
-1991.5	0.011	-0.001	-0.018	-0.017
-1727.5	0.093	0.004	-0.021	-0.021
-1585.0	0.004	-0.001	-0.023	-0.018
-1503.7	0.001	-0.003	-0.021	-0.026
-1261.5	0.032	-0.014	-0.021	-0.027
-787.5	0.032	0.006	-0.015	-0.020
-457.5	0.032	0.035	0.004	0.009
-227.5	0.033	0.045	0.010	0.017
296.5	-0.060	-0.054	-0.033	-0.032
526.5	-0.078	-0.057	-0.040	-0.037
856.5	-0.082	-0.054	-0.029	-0.030
1330.5	-0.074	-0.035	-0.028	-0.026
1572.7	-0.088	-0.046	-0.034	-0.045
1654.0	-0.084	-0.047	-0.037	-0.044
1796.5	-0.09	-0.054	-0.039	-0.042
2060.5	-0.087	-0.066	-0.030	-0.034
2231.5	-0.084	-0.057	-0.015	-0.022
2385.7	-0.085	-0.069	-0.007	-0.020

It is visible an improvement on the holes positions. If the internal stress from extrusion is eliminated by releasing the part, the holes are in tolerance (Figure 7).

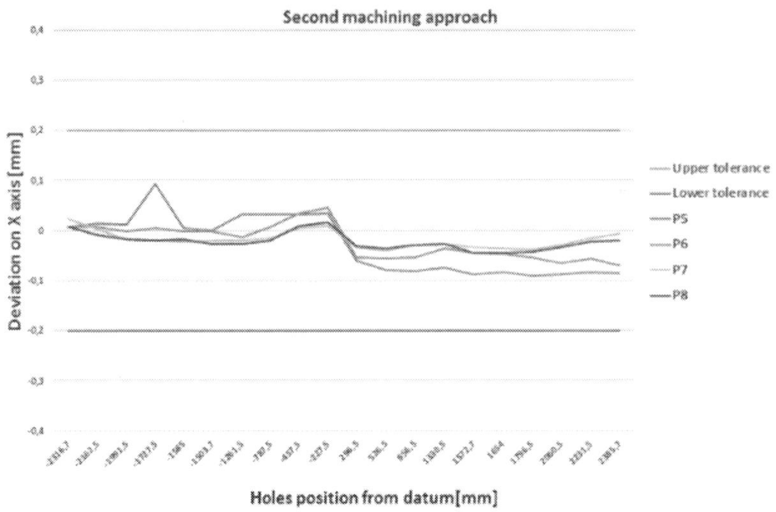

Fig. 7. Second four identical parts machined results.

4 Conclusions

This experimental research bring a solution using a simple procedure by releasing the part to be on free state before drilling operation, through this was eliminated the deviations created by internal stress of the extruded profiles. This methodology improves the machining process control and creates a manufacturing repeatability.

Future improvement which creates better precision on the final part will be to apply a temperature coefficient to eliminate the environment temperature fluctuation [5].

The original contribution for this research is: creating a solution to solve the deviations problem.

The experimental research conducted with the support of the machining department of S.C. Universal Alloy Corporation Europe S.R.L., Dumbravita.

References

1. Ş.A. Moldovan, Research on Increasing Efficiency in machining on Numerically Controlled Machine at S.C. Universal Alloy Corporation Europe S.R.L., No.3657/09.02.2016 (in Romanian)
2. ***Training Guide PBZ HD Ver. (2015)
3. ***https://www.handtmann.de/en/machining-centres/products/pbz-profile-machining-centres/pbz-hd/ (2017)
4. ***Metris Bridge CMM User Guide (2016)
5. Ş.A. Moldovan, V. Năsui, Applied Mechanics and Materials **809-810**, 33-38 (2015)

Study of a additive device which measures the angled walls of injection mold components on a three axis CNC

Bogdan Pralea* and *Gheorghe* Nagîţ

Gheorghe Asachi Technical University of Iaşi, Department of Machine Manufacturing Technology, Faculty of Machine Manufacturing and Industrial Management, Bdul Mangeron, 59A, 700050, Iaşi, România

Abstract. The paper attempts to demonstrate an efficient on machine measurement system. During this study the authors will develop a measurement device, in order to accurately analyze the dimensional precision of the angled vertical walls used to center active cavities inside an injection mold, on a three axis CNC. The designed device will work in strong connection with a touch trigger probe. The final results will show an improved accuracy of the measurement system.

1. Introduction

Recent studies show that the measurement probes part of CMM (Coordinate-measuring machines) represents one of the most important system of dimensional control and they are responsible for the precision of the measurements. The usage of electronic devices and modern algorithms are also improving the quality of measurements which are taken by CMMs. Some studies show that, despite the fact that they have been continuously improving and companies developed different types of probes, the most common and used one remains the triggering probes, which prevail on the industrial market, due to their lower costs and a good precision [1]. This types of probes generate a binary signal when the tip of the probe touches the surface that has to be measured. During this time the tip of the probe sends the values which can be seen and further collected. However, studies are underlining the fact that these types of probes are generating some type of errors, which can lead to damage parts in manufacturing. Some of the errors that appear could be solved by calibrating the probes, before the actual measurement takes part. In present in all types of probes a compensation of future displacement takes part [2]. During calibration stage which is done by using a master sphere with a known diameter that is placed on the tip of the probe, the effective radius is being calculated, considering various points of measuring, generating this way an average of the values for future movements [3].

Some researchers are underlining the fact that the measuring probe is the source of more than 55% of measurement errors done by a coordinate measuring machine. So in order to

* Corresponding author: bpralea@gmail.com

improve the precision, devices or mathematical models can be developed. The fast and correct answers of the probe can be controlled by a computer and this opens new trends and possibilities for future generations of measurements. A large variety of types of probes is available and it is compatible with CNC machines and CMM machines. The dimensional inspections on the components that are manufactured and have free form surfaces, need accurate measurements on a large variety of points. That is why some papers show that planning the measurement strategy can be done by starting from the base CAD model and selecting the points which have to be measurement [3, 4].

Through analyses of different papers we establish that there are two types of measurement probes. The first one is represented by hard probes. This probes are available in a large variety and they continue to have a large horizon inside the field of coordinate measurement machines and devices. This types of probes are used manual, as they are best suited for medium or low requirements in precision of manufactured parts. They are easy to use, but their repeatability in precision of dimensional measurement is strongly connected to the operator's experience [4]. Due to the fact that each individual operator has different touch the measurement values will also differ and this will result in lots of errors. The second type of probes is represented by the triggering probes, which engage to measure when they touch the surface that has to be evaluated by means of dimensional control. Like stated above, this category of devices are very sensible and they conduct the actual control of dimensions trough the tip of the probe, generating an electronic impulse [5]. This moment is signaled by a LED. The capacity of this probe and the technology behind it makes it versatile and they are best suited for high precision jobs (die, molds and large series of manufacturing).

A conclusion for what was stated before and through analyses of paper regarding this particular field the CMMs are opened to a series of errors. Based on their components it can be shown a classification of errors source that affect the accuracy of the probes. First of all the mechanical components of the CMM's can produces errors due to the manufacturing of their entire assembly, such as table, carriages, guide ways, bearings and the carriages. This can involve differences in thermal expansion, stiffness and adjustment. All these marker are resulting in static, quasi-static and dynamic error. [6, 7].

Secondly the drive components, which are reserved for CNC operated CMM's. The axis, in this particular case are equipped with electronic units and servo control units. The errors can appear due to an uneven measuring speed, the mechanical load can also be uneven and vibrations of the entire structure could also generate them [8].

Thirdly the software that collects the data. This involves strictly the algorithms which collect the data and also calculated it.

In this study the authors will try to develop an experimental device for reducing the errors which appear in measuring the angular surfaces of, in this case, the cavity walls which centers the active parts of an injection mold. This particular surface requires a high level of accuracy. So the object of study will involve a close look on CNC operated CMM's and designing an experimental, which hopefully will eliminate or reduce the error level in measuring [9].

2. Methodology and experimental procedure

As it was stated before the object of the study will be, for the authors, for this stage the developing of an experimental device that will reduce the number of errors in measuring the angular surfaces, on a three axis CNC, with the help of a triggering probe. The device will be manual and will function as an additive form to the triggering probe.

Till now there were created numerous solutions for mechanical measurement of the vertical and plane surfaces on three axis CNC machines, but almost none of them took into consideration the possibility of milling free form and angular surfaces and obtain measurement data from on machine measurement devices. The problem of creating new solutions that can work and be easy to use, appears when environment request it. Here we can mention the machines and the parts that are being manufactured together with the possibility of lower the costs. In order to put it in an adequate form, the companies that have as main domain the manufacturing of complex parts, in our case injection molds, need to developed devices for what it is in essence the job completion. These types of devices are usually designed and manufactured, in a short number, sometimes they are unique, and in order to answer indirectly to client needs. So to remain at the subject at hand, the problem of designing solutions for measuring devices which are versatile, is to answer to some major criteria. First of all the complexity level of the device, secondly the time and costs involving the financial resources that are allocated for the outlook and testing and thirdly the number of devices that are manufactured.

The final stage of designing the correct model of the experimental device, was to choose the final ideas that are based on value analysis, of some multi-criteria methods, using basic points, which start from advantages and disadvantages of proposed solutions. Another step was the identification of a simple way to use for the operator, this means in particular that it needs to be manipulated on and off the measured surface with ease, never the less the active surfaces that enter in contact with the surfaces should be smooth and easy to clean form any residual material, obtained from the actual manufacturing stage. After that the stage of material selection. In this case the authors considered the factors that have an impact for this particular experimental device. So in this way the authors have chosen a material which has good machinability properties and can have good precision in time. Another stage is represented by the dimension of the experimental device. The authors first made an evaluation of the dimensions and on which we can work on. The final dimensions needed to answer the request of most surfaces for which was designed. From previous historic data we manage to extract an angle which appears always in terms of centering the active parts of the mold. It needs to answer the give geometric tolerances due to the fact that it is an additive system which is helping to reduce errors regarding measurement. Because the dimensions of the device determine the rigidity of the part and the rigidity has a strong impact on maintaining initial position and also accuracy.

To keep the design on track, it were considered the fits of involved parts. If you take into consideration that the precision of the device has a direct impact on the action of measurement that we can assume that the fits which take place must be small or even better tight (H/h). Also the roughness for the entire device needs to be very small. Another stage will be represented by choosing different technical activities, other than previous ones, conditions that are extracted from the manufacturing stage of the experimental device, the assembly and also the way we operate with. Between this conditions, is to be mentioned the fact that the device surfaces, which make contact with the part that needs to be measured, must be grinded.

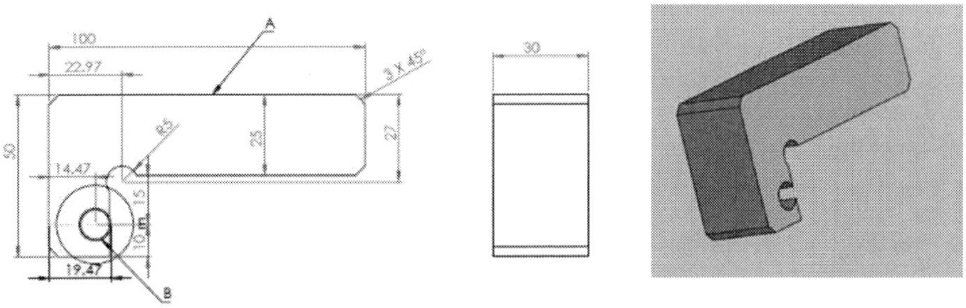

Fig. 1. Experimental device.

Taken into consideration the facts which were discussed above and the research, the authors developed the device which is represented in Figure 1. It is characterized through a simple design and represents an additive mechanism to the triggering probe installed on the three axis CNC.

The device is an assembly, and it is composed from two parts:

- Part A represents the main body, made from 1.2312 (40CrMnNiMo8-6-4), because it is delivered in a quenched and tempered conditions at around 950-1100N/mm^2, which makes it rigid and stable, also the composition makes it suitable for better machinability capacity. It was milled from two fixtures on a DOOSAN 750 L, which is a vertical machining center which is designed for high speed and precision of small to large work pieces such as dies molds and other complex parts [10]. It operates on three axis, X and Y movement is generated by the table, Z axis controls the head movement. In the first fixture had several operations like roughing, pre-finishing and finishing of the outside contour. Both hole (Φ10mm) and open cavity (R5mm) which appear in the drawing, were subject to drilling and afterwards reaming. It was very important keeping the geometric tolerances tight during milling and drilling process. At the end the part resulted according to initial specifications.
- Part B represents a shaft which was cold mounted into the main body A. The material used for this part is from a broken widia end mill, which had a diameter of 10 mm. We cut the shank of the mill and afterwards made two chamfers, one at each end. Afterwards we inserted tightly into part A. We choose the shank of a high accuracy end mill, due to its high precision and small roughness, being the most important part of the assembly.

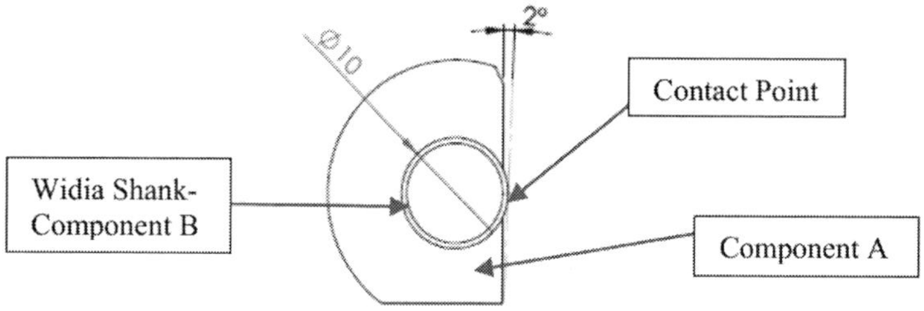

Fig. 2. Detail E from the Experimental device, represented in Figure 1.

As suggested above, in the Figure 2 is represented more clearly component B of the assembly, tightly fitted into component A. The construction line on the left which has an angular value of two degrees signify the theoretical surface that needs to be measured. The surface is the one part of the cavity (Figure 3.a.) which centers the active part of the injection mold (Figure 3.b.). Both components where milled on the same three axis CNC machine as the experimental device and both are parts subject to high precision. Assuming the accurate circularity of the shaft, it is known that a circle will touch/tangent a straight line in only one point. Starting from this we can assume that putting the shank tangent with the surface will eventually generate a line which will be characterize in terms of coordinates. The coordinates will be define according to the part main origin from which the triggering probe will also start the measuring cycle.

Basically the experimental device will translate the measuring line from the angular surface with a fixed given value into a simple vertical surface which can be measured by the triggering probe, in our case the Renishaw RMP60, which is placed on the CNC machine together with a software package Inspection Plus which is generating the cycle measuring programs [11]. It has several operating cycles such as: X or Y or Z single surface measure, web or pocket, bore/boss, external corner, single point (radial) measurement, two-point (diameter) measurement. As it can be observed none of which include angular measurement of surfaces. This is where the experimental device takes places, being an additive mechanism to the entire process and assuring a lower cost with good accuracy. As it was discussed during the first part of the paper, many error that can appear and affect the accuracy of collected values is given by drive system for the CMM's that are installed on CNC's, which is the case and also the algorithm inside the software, which again is missing, part because the installed software is suited for a three axis CNC and part because that particular macro would have higher costs and the precision of the measurement will be affected by the random movement of the probe tip (sphere).

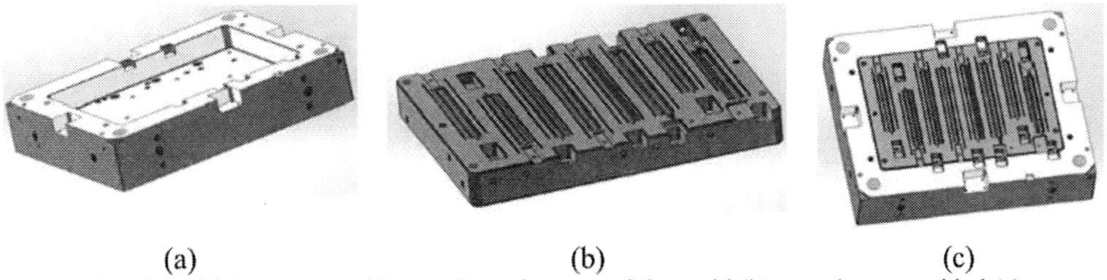

(a) (b) (c)

Fig. 3. Plate (a) which centers and keeps the active part of the mold (b); together assembled (c).

The process is illustrated in Figure 4 and it can be observed that the probe touches the experimental device, which has translated the angular surface into a vertical one. The value that resulted and appears on the screen of machine interface needs now to be summed or subtracted according to the direction of coordinate measurement.

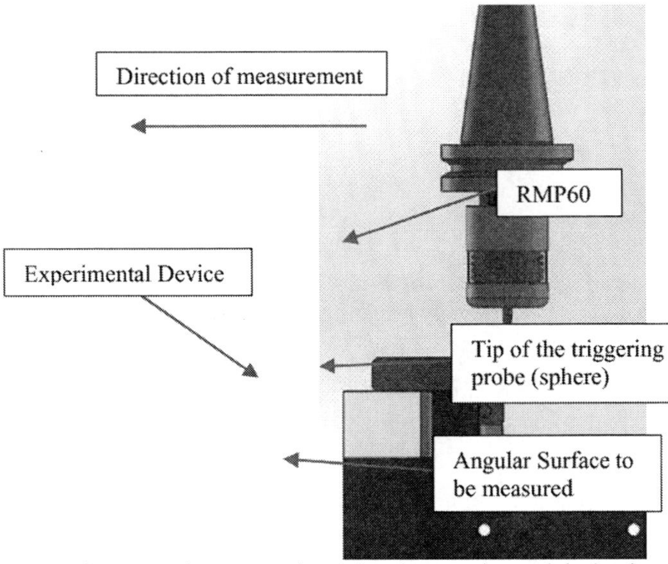

Fig. 4. The Measuring process involving the experimental device for angular surface measuring.

In order to find the value that is going to be added or subtracted we can apply the tangent in the virtual right triangle formed between the angular surface and the surface of the experimental device (Figure 5). The tangent of 2 degrees will result in a value of 0.035 and this multiply by 15 mm will result into a constant value x, given by equations number 2, of 0.530mm. So if we sum that value to the known one from the vertical wall 19.47mm, it will give us exactly 20mm. That value is the one to be added or subtracted from the measured coordinate. The basic mathematical equations for this procedure are:

$$|A| + x = |B| \quad (1)$$

$$x = \frac{tan\,2°}{15}[mm] \quad (2)$$

where: $|A|$-is the measured value given by the triggering probe. We will use the absolute value of it. $|B|$-represents the value of the angular surface.

Fig. 5. The actual experimental device and the virtual right triangle that is created between it and the angular surface.

3. Conclusions

The present study shows that the quality and the dimensional control of the angled walls, which are machined on a three axis CNC, are better measured using the proposed experimental device. It should also reduce the errors in measurement between the triggering probe and angular surfaces part of high precision components from injection molds.

The authors would like to thank TREVIS GRUP ROMANIA and UNITEAM ITALY, for their assistance and for having them enabled the factory equipment.

References

1. C.K. Gohel, A.H. Makwana, , Int. J. Adv. Manuf. Technol. **4**, pp 21-26 (2014)
2. Abhijit Sahasrabudhe, Xiambing Liu, *Theoretical and Experimental Study on Surface Finish for Multi-Axis CNC Milling,* University of California, Davis CA 95916, (2003)
3. Salah H. R. Ali, Measurement Science Review **10** (2010)
4. W.G. Weekers, *Compensation for dynamic errors of coordinate measuring machines,* (Eindhoven University of Technology, 1996)
5. Gyula Hermann, Acta Polytechnica Hugarica **4** (2007)
6. Zhenjiu Zhang, Hong Hu, Int. J. Adv. Manuf. Technol. **69**, 1483-1497 (2013)
7. H. B. Liu, Y. Q. Wang, Int. J. Adv. Manuf. Technol. **80**, 1721-1731 (2015)
8. Gaiyun He, Xin Huang, Wenkui Ma, Guanmin Yu, Int. J. Adv. Manuf. Technol. doi:10.1007/s00170-016-9932-2 December (2016)
9. http://www.doerrenberg.de
10. http://www.doosanmachinetools.com/en/main.do
11. http://www.renishaw.com/en/machine-tool-probes-and-software--6073

Investigation on hole quality of cutting conditions in drilling of CFRP composite

Nafiz Yaşar[1,*], *Mehmet Erdi* Korkmaz[2], and *Mustafa* Günay[2]

[1] Karabük University, TOBB Vocational School, Bilimler Meslek Yüksekokulu, Demir Çelik Kampüsü, 78050, Karabük, Turkey
[2] Karabük University, Department of Mechanical Engineering, Balıklarkayası, Mevkii, 78050, Karabük, Turkey

Abstract. Surface damages and delamination occurred in drilling of the carbon fiber reinforced polymer composite materials, which change depending on their physical and mechanical properties and cutting conditions, and so affect the hole quality. Therefore, it is requires to be selective with regards to drill geometry and drilling parameters during drilling of CFRP. This study investigated the effects of cutting parameters and drill geometry on the hole quality in terms of hole diameter and surface roughness during the drilling of CFRP composite laminate. Drilling experiments were performed with three different levels of drilling parameters that are point angle, cutting speed and feed rate. At the end of the drilling tests, it has determined that if the feed rate increases, the surface roughness value increases while the hole diameter decreases. The feed rate was determined as the most important parameter with a PCR of 78.02% according to analysis of variance. The better hole quality was obtained with 130° drill among the three drill geometry.

1 Introduction

Beyond aerospace, automotive, marine and sport industry [1], the demand for composite materials in implant and its prosthesis parts, manufacturing of chemical, biological and electronic equipment is gradually increasing [2]. CFRP and CFRP composite materials having much more advantages than other materials are preferred and qualified with their excellent properties such as long life, lightness, high strength and corrosion resistance [3]. (Fiber reinforced polymer) FRP materials may display different mechanical properties depending on location of fiber in epoxy matrix. Fiber reinforcements and polymer matrix are existed within FRP materials [4]. Their mechanical properties may be different according to location of fiber in epoxy matrix [5]. Fiber reinforced polymer (FRP) composite materials are usually manufactured in geometry of final form in order to provide excellent assembly performance. Therefore, metal removal processes like milling, drilling etc. are needed to meet the requirements of quality and dimension in assembly [6]. For example, structural weight of above 50% of Boeing 787 airplane is of composite materials and about 55000 drilling processes are performed in manufacturing of structural

* Corresponding author: nafizyasar@karabuk.edu.tr

components in airplane [7,8]. Thus, drilling is a very common process applied to carbon fiber reinforced polymers (CFRP) [9].

Machinability of FRP composites are harder than conventional metal cutting processes [4] due to poor machinability of fiber reinforced polymer (FRP) laminates usually presenting heterogenic characteristic, anisotropic structure and brittle behavior [10]. Therefore, it requires more attention for selection of cutting tool (tool material, coating type and tool geometry) and cutting parameters (feed rate and cutting speed) when machinability of CFRP composite materials become current issue [9,11]. Moreover, delamination that is defined as layer segregation in entry and exit of hole is the most common problem in drilling of composites [12].

Merino-Perez et. al presented the influence of the material properties and the effect of the cutting speed variation on the heat dissipation in the drilling of the studied CFRP composites with SEM analysis. They determined that high cutting speeds caused more severe matrix cracking and hole entry/exit splintering than low cutting speeds. They also suggested using of temperatures not exceeding the decomposition temperature of the matrix within the machining time [13]. Mohan et. al stated that the most important parameters in drilling of GFRP composites are cutting speed and material thickness for entry delamination, feed rate and material thickness for exit delamination according to their experimental and numerical studies [14]. On the other hand, it was emphasized that delamination factor (Fd) is proportional with cutting speed and feed rate, similarly remarked that delamination increases with carbide drills having high point of angles [15]. SenthilKumar et. al emphasized that less progressive tool wear and better chip evacuation was achievable with larger point angle drills in their study of optimizing cutting parameters with genetic algorithm [16]. El-Sonbaty et al. determined that cutting speed has negligible effect on thrust force (*Fz*), but torque values increase with increasing cutting speed [17]. İşbilir ve Ghassemieh indicated that average surface roughness value increase with increasing feed rate while it decreases with increasing cutting speed [18]. It was reported that anisotropic CFRP composites exhibit different machinability characteristics according to different fiber directions during drilling. It is emphasized that various surface defects like resin loss, remove from fiber and matrix degradation may arise on hole surface, causing irregular surface forms [19].

Literature mentioned that CFRP materials are generally machined in dry cutting conditions due to the fact that their dehumidifying properties effect shear fatigue toughness negatively. In this study, the effects of cutting parameters (drill point angle, cutting speed and feed rate) on machinability criteria (Ra and Diameter) under dry cutting condition have been investigated experimentally and statistically in drilling of CFRP composites.

2 Material and method

As the workpiece material, for drilling, carbon fiber reinforced polymer laminate (CFRP) are used which are produced by vacuum bagging method, laying 245 g/m^2 2x2-3K structured carbon fiber and prepregs containing Huntsman XU3508 resin to a glass surface symmetrically with suitable fiber angles. In this material, twill textured carbon fibers form with directions of 0° and 90° angles. Therefore, only laying angles of 0° and 45° are used in production. Reinforcement materials are used 40 times for 10 mm laminate and the sides of the laminates are vacuumed by bagging with vacuum equipment. Reinforcements compressed by vacuum were placed into furnace and a temperature of 120 °C was applied to the system by four hours.

Drilling experiments were performed in CNC vertical machining center (Johnford VMC–550) with three various levels of point angle, cutting speed and feed rate in dry machining conditions. Karcan branded tungsten carbide (WC) drills were used as cutting tools. Drills have a diameter of 5 mm, helical angle of 30° and geometry of two cutting edges. Feed rate (f), cutting speed (V) and drill point angle (φ) were chosen as factors to apply full factorial experimental design in drilling of CFRP. The levels of these factors were determined by considering tool manufacturer's recommendations and studies on drilling of CFRP composites in literature. Drilling parameters with levels are given in Table 1.

Table 1. Drilling factors and levels.

Factors	Levels
Point Angle P (°)	120-130-140
Cutting Speed V (m/min)	60-100-140
Feed Rate f (mm/rev)	0.05-0.1-0.15

Drilling processes were repeated as three times and evaluations were done by taking arithmetic averages of machinability criteria (Ra and Diameter). Measurements of diameters (D) were performed in 5 points (Figure 1a) from entry, middle and exit of the hole. Hexagon Global Performance device was used in measuring of hole diameters. Surface roughness on hole surfaces were measured with portative Mahr Perthometer M300 device. Three surface roughness measurements were applied for each hole surfaces with a cut-off length of 0.25 mm and sampling length of 1.75 mm according to ISO 4288 standard. The arithmetic average of surface roughness values were calculated and evaluations were performed with regards to average surface roughness (*Ra*) value.

In the light of experimental results, the effects of factors on Ra and D were determined by analysis of variance (ANOVA) with 95% confidence level. Since the spindle speed of the CNC milling machine used in the experiments is insufficient, suitable spindle speeds matching the appropriate cutting speeds were obtained using spindle speeder. The experimental setup for drilling experiments (b) and drill geometry (c) were given in Figure 1.

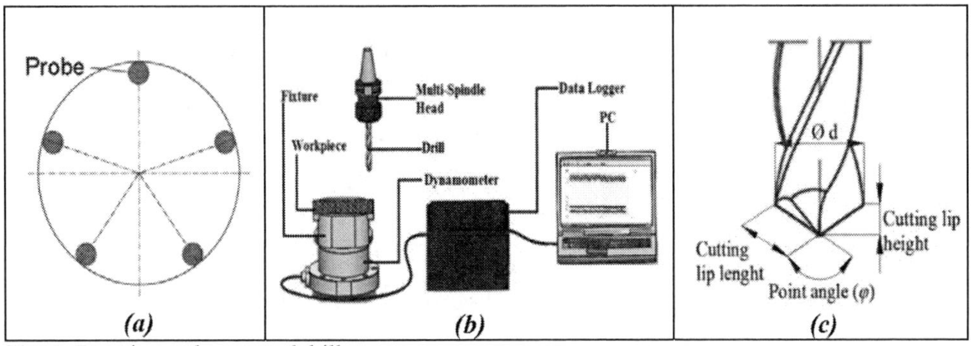

Fig. 1. Experimental setup and drill geometry.

3 Results and discussion

The variations of surface roughness (Ra) and diameter (D) in drilling of CFRP with different cutting parameters were given in Figure 2 and Figure 3.

3.1 Analysis of surface roughness (Ra)

Surface roughness is one of the surface integrity criteria for all parts manufactured by machining and also is a parameter to be considered in the analysis of the hole quality in

CFRP drilling. In this study, variation of average surface roughness (Ra) occurred in drilling of CFRP according to drill point angle and drilling parameters is given in Figure 2.

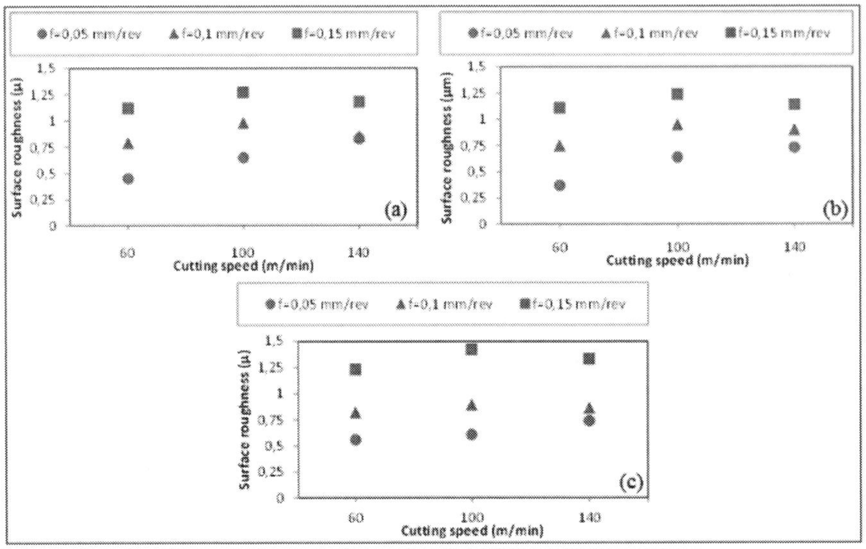

Fig. 2. Variation of surface roughness; point angle of a) 120°, b) 130°, c) 140.

Examining Figure 3, similar tendencies for average surface roughness (Ra) are observed for each drill point angle. Surface roughness values obtained in drilling process increase with increasing feed rate and cutting speed. It is generally known that surface roughness increases with increasing feed rate. This result is similar with literature [18]. On the other hand, increasing surface roughness values can be explained with increasing thrust force by increasing feed rate [20]. This increase can be referred to increasing chip cross-section area by increasing feed rate. Moreover, it is possible to say that drill cannot perform better due to adhesion on it in higher feed rates [15] and this situation effects increasing surface roughness.

It is understood from the Figure 3 that variation of surface roughness according to drilling parameters with increasing cutting speed. Surface roughness increases with increasing cutting speed from 60 m/min to 100 m/min, subsequently, roughness tends to

Table 2. ANOVA results for surface roughness

Factor	**DF**	**SS**	**MS**	**F**	**P**	**%PCR**
Φ (°)	2	2.382	1.191	2.74	0.124	1.1
V (m/min)	2	21.807	10.904	25.1	0.000	10.05
f (mm/rev)	2	169.404	84.702	194.99	0.000	78.02
φ*V	4	2.564	0.6409	1.48	0.296	1.18
φ*f	4	2.052	0.5129	1.18	0.388	0.95
V* f	4	15.426	3.8565	8.88	0.005	7.10
Error	8	3.475	0.4344			1.6
Total	26	217.109				100
R^2 =98.4%						

decrease. As mentioned in recent studies [13], this is mostly originated from the incomplete cutting of fibers and their removal from the matrix with the increasing cutting speed and/or the surface defects on drilled hole that occur as a result of matrix-fiber debonding. Surface roughness decreased 6-8 % in cutting speed of 140 m/min. This result can be referred to the probable remaining of the hole surface temperature below the glass transition temperature of matrix material as a result of decreasing cutting time with high cutting speed.

The lowest Ra was measured as 0.37 μm with drill point angle of 130°, feed rate of 0.05 mm/rev and cutting speed of 60 m/min. The highest Ra value was obtained as 1.42 μm in drill point angle of 140°, feed rate of 0.15 mm/rev and cutting speed of 100 m/min. Analysis of variance (ANOVA) was performed to determine the effects of drilling parameters and drill point angle on Ra that is a significant factor for performance life of structural part manufactured from CFRP (Table 2). As seen in the ANOVA table, the most important drilling parameter that effects the surface roughness is feed rate with a PCR of 78.02%. The other statistically significant parameters over Ra are cutting speed (V) with PCR of 10.05%, 7.10% and 1.18%, interaction between point angle (φ) and point angle-feed rate (φ*f) respectively.

3.2 Hole size

Another significant factor taken into account in determining the hole quality is the hole size. Especially, the need to drilling in desired tolerance with low roughness and suitable roundness due to necessity of high strength in assembly parts is an important safety problem in aerospace industry [10]. In this study, variation of hole diameter values occurred in drilling of CFRP according to point angle and drilling parameters was given in Figure 3.

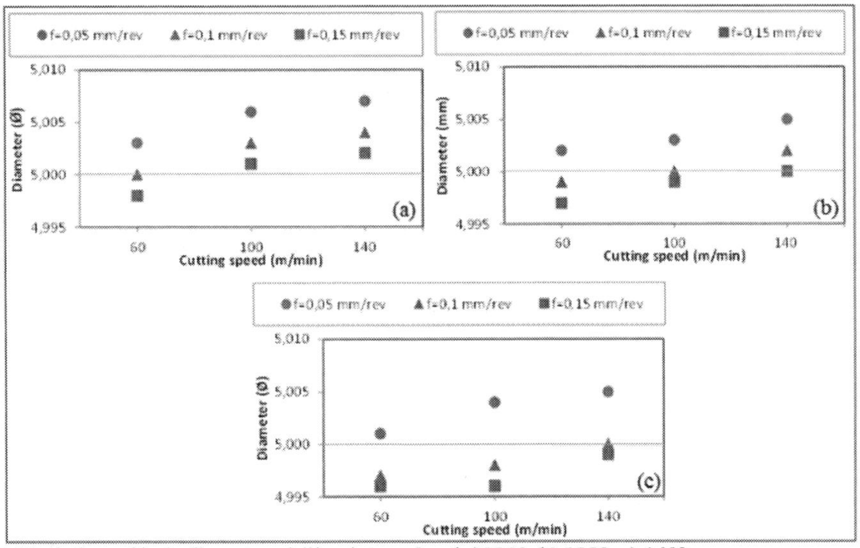

Fig. 3. Variation of hole diameter; drill point angle of a)120°, b) 130°, c) 140°.

As seen from Figure 4, the hole diameters decreased with increasing of feed rate for all drill point angle. The main reason of higher hole size than nominal value can be explained with cutting temperature varying according to feed rate and cutting speed. It is generally known that the cutting temperature increases with lower feed conditions due to the low thermal conductivity of CFRP [21]. It is generally known that increasing cutting speed leads to increase in vibration [22]. This situation explains the increase in hole diameters with increasing cutting speed. On the other hand, the highest hole size was measured as 5.07 mm in drill point angle of 120, cutting speed of 140 m/min and feed rate of 0.05 mm/rev. The lowest hole size was measured as 4.996 mm in drill point angle of 140, cutting speed of 60 m/min and feed rate of 0.15 mm/rev. This result can be explained with decreasing in tool-chip contact area due to increasing drill point angle. Decreasing in cutting lip length and height with increasing drill point angle (see Figure 1.c) will lead to reduce of friction through tool-chip interface, which decreases generated heat, thereby result in CFRP hole sizes closer to the drill diameter.

4 Conclusion

In this study, the effects of drill point angle, cutting speed and feed rate on machinability factors (Ra, D) when drilling of CFRP composites with uncoated WC drill bits were experimentally and statistically analyzed.

The lowest Ra was measured as 0.37 μm with drill point angle of 130°, feed rate of 0.05 mm/rev and cutting speed of 60 m/min. The highest Ra value was obtained as 1.42 μm in drill point angle of 140°, feed rate of 0.15 mm/rev and cutting speed of 100 m/min. Lower cutting speed and feed rate are suggested for better surface roughness. The most important drilling parameter that effects the surface roughness is feed rate with a PCR of 78.02%.

The highest hole size was measured as 5.07 mm in drill point angle of 120°, cutting speed of 140 m/min and feed rate of 0.05 mm/rev. The lowest hole size was measured as 4.996 mm in drill point angle of 140°, cutting speed of 60 m/min and feed rate of 0.15 mm/rev. This result can be explained with decreasing in tool-chip contact area due to increasing drill point angle.

This study is supported by Scientific Research Project Unit of Karabük University (KBÜ-BAP-15/2-DR-022) and the authors express their appreciation for this support.

References

1. O. İsbilir, E. Ghassemieh, Compos Struct, **105**, 126-133, (2013)
2. E.D. Eneyew, M. Ramulu, J of Materials Research and Techn, **3**, 4 354–362 (2014)
3. O. Pecat, E. Brinksmeier, Procedia CIRP, **13**, 1-7 (2014)
4. J. Xu, A. Mkaddem, M.E. Mansori, Compos Struct, **135**, 316–338. (2016)
5. D. Liu, Y. Tang, W.L. Cong, Compos Struct, **94**(4) 1265–1279 (2014)
6. I. Singh, N. Bhatnagar, P. Viswanath, Mater Design, **29(2)**, 546-553 (2008)
7. V. Krishnaraj, A. Prabukarthi, A. Ramanathan, N. Elanghovan, M.S. Kumar, R. Zitoune, J.P. Davim, Compos: Part B, **43**, 1791–1799 (2012)
8. Y. Fan, L. Zhang, Advanced of Aircraft Manufacture Techn, **30(3)**, 534-543 (2009)
9. Jamal Y. Sheikh-Ahmad, *Machining of Polymer Composites*, (Springer, 2009)
10. A.M. Abrão, P.E. Faria, J.C.C. Rubio, P. Reis, J.P. Davim, J Mater Process Tech, **186(1–3)**, 1–7 (2007)
11. C.R. Dandekar, Y.C. Shin, J Manuf Sci E-T ASME, **130(5)**, 051016 (2008)
12. G.V.G. Rao, P. Mahajana, N. Bhatnagar, Compos Sci Techn, **67(3–4)**, 579-593 (2007)
13. J.L. Merino-Pérez, R. Royer, S. Ayvar-Soberanis, E. Merson, A. Hodzic, Composite Structures, **123**, 161-168 (2015)
14. N.S. Mohan, S.M. Kulkarni, J Mater Process Tech, **186**, 265-271 (2007)
15. S.R. Karnik, et al., Mater Design, **29**, 1768-1776 (2008)
16. M.SenthilKumar, A. Prabukarthi, V. Krishnaraj, International Conference On Design And Manufacturing, IConDM, Procedia Engineering, **64**, 582-592 (2013)

17. I. El-Sonbaty, U.A. Khashaba, T. Machaly, Compos Struct, **63(3–4)**, 329–338 (2004)
18. O. İsbilir, E. Ghassemieh, Mach Sci Technol: An Int. Journal, **17(3)** 380-409 (2013)
19. C.Y. Wang, Y.H. Chen, Q.L. An, X.J. Cai, W.W. Ming, M. Chen, International Journal of Precision Engineerıng And Manufacturing, **16(8)** 1689-1697 (2015)
20. B. Latha, V.S. Senthilkumar, J Reinf Plast Comp, **28** 951–964 (2009)
21. S.A. Ashrafi, P.W. Miller, K.M. Wandro, D. Kim, Materials **9,** 828 (2016)
22. R. Rusinek, Int J Nonlinear Mech, **45**, 458-462 (2010)

The investigation of the chip behaviour during the end milling cutting process

Ovidiu Virgil Vereş[1,*], *Marcel Sabin* Popa[1], *Stefan* Sattel[2], *Claudiu Ioan* Jugrestan[1], and *Elena Adina* Cotargă[1]

[1]Technical University of Cluj-Napoca, Faculty of Mechanical Engineering, Department of Manufacturing Engineering, Bulevardul Muncii 103-105, 400461, Cluj-Napoca, jud. Cluj, Romania
[2]Gühring KG, Herderstraße 50-54, 72458, Albstadt, Germany

Abstract. Generally the study, evaluation and general understanding of end milling cutters are complicated by the complexity of the cutting process and of the cutting tool. The main performance factor of the end milling cutters is given by the durability of the tool. In practice the sound analysis of the tool engaged in the cutting process is used as a simple predictability mean to forecast the future durability of the tool. An alternative or complementary tool to predict future durability is the investigation of the chip behaviour. Chip behaviour investigations can also give clues concerning the possible improvement of the cutting tools. The current article presents a detailed chip behaviour investigation for 5-fluted end milling cutters with different performances and various geometries when approaching different types of metal cutting operations in 42CrMo4 alloy steel. Current study can be applied in future researches regarding the evaluation and the development of end milling cutters. Furthermore the current approach can be used on other types of cutting tools.

1 Introduction

The metal cutting technology continues to be the main and most precise manufacturing technique in the world today. Current trends and research in ultra-precision metal cutting anticipate an even greater role for this technology in the future.

The first scientific approaches to analyse the dynamics of the cutting process and behaviour of the metal chip was made by Merchant [1], which introduced the shear angle model for orthogonal metal cutting, and by Lee and Shaffer [2], which introduced the slip line field theory in the chip formation process, as an attempt to apply the plasticity theory to the problem of orthogonal metal cutting. From these early models, more precise and more detailed models have been developed by taking into consideration the influences of friction, temperature, work-hardening and strain-rate [3-5]. Although this early chip formation models are the base for the metal cutting theory, they are not in full correspondence with the applied experiments.

In the field of milling cutting process, first attempts to create a model for a milling process, were made by Young [6], whose research was based on Oxley's predictive

* Corresponding author: veres_ovidiu_virgil@yahoo.com

machining theory [7]. Young used Oxley's predictive theory for forces, temperatures and stresses in relation with the material flow, which he applied on the chip segments of a divided milling chip with varying thickness. In his study he considered the chip segments as constant geometrical elements. Young's model has been validated trough similar results of amplitude and geometry of the force profile in applied experiments.

Oxley's generalized predictive machine theory has been widely used to define specific cutting situations and specific materials. Ekanayake et al. [8] applied the theory to predict cutting forces in high speed milling of AISI 1045 and AISI 4140, while Li et al. [9] predicted forces in helical end milling. Masmali et al. applied Oxley's theory to predict the forces of a non-uniform helix angle milling cutter [10] and conducted a geometric chip analysis for helical end milling tools [11]. Masmali's model results show significant similarities with the models of the experimental results. However there are still inconsistencies that can be improved, like the force variation between the analytical model and the experimental results on the tool-axis direction.

Other studies in the field of milling cutting process have focused on the effort to investigate chip formation, chip morphology and tool wear in the case of high speed milling of hardened steels [12-14]. Meanwhile Ning et al. [15] and Wu et al. [16] have classified the chip obtained through milling in different groups. By interpolation of the previous two chip classifications, four major chip types are obtained: stable (complete), unstable, critical and severe chips. More recent Patwari et al. [17] developed a mathematical model to predict the instability process of the chip formation by using the concept of chip serration frequencies.

Although much research has been done in the direction of studying, understanding and theorizing the end milling cutting process, the scientific effort is far from a universal solution for the complex process. Firstly the complexity of the process, which distinguishes itself from simple metal cutting procedures like turning and drilling, is given the interconnected cut along two cutting edges: primary or major cutting edge, located on the tools diameter, and secondary or minor cutting edge, located perpendicular to the tools axis. This aspect complicates the effort to apply orthogonal cutting theory in milling cutting. Furthermore, in previous studies there have been found no considerations regarding the dynamic non-linear chip behaviour in the cutting process, the influence of the tools geometry on overall stability and the influence of the secondary surfaces (as seen in Figure 1), surfaces that are not in direct cutting contact with the chip, on the chips removal success and tool stability. Therefore, the current article will present a different approach to the study, evaluation and understanding of the end milling cutting process, which is based on simple practical observations, resulting from the investigation of the chip behaviour of 5-fluted milling cutters with different geometries and performances when approaching different types of milling cutting operations in 42CrMo4 (EN 10269:2008) alloy steel.

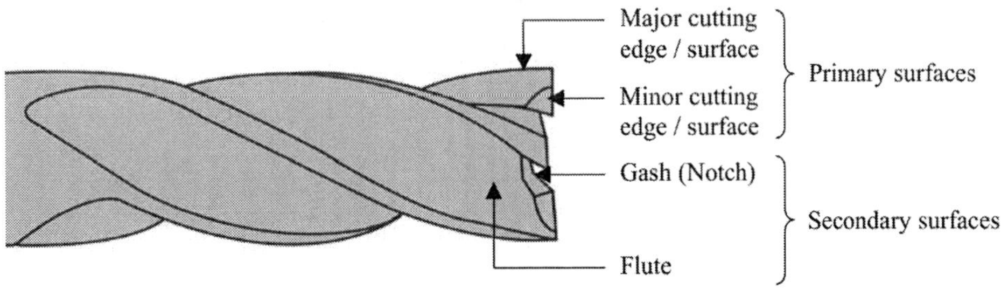

Fig. 1. End milling cutters surfaces and cutting edges.

2 Experimental setup

The experimental setup, as shown schematically in Figure 2.a., consists of the following major components: the machine tool, the tool clamping device (tool holder), the end milling cutter, the workpiece, the workpiece clamping unit and the high-speed camera.

The machine tool for the experiments is the universal milling machine for 5-sided / 5-axis machining DMU 85 monoBLOCK® (Figure 2.b.) manufactured by DMG MORI.

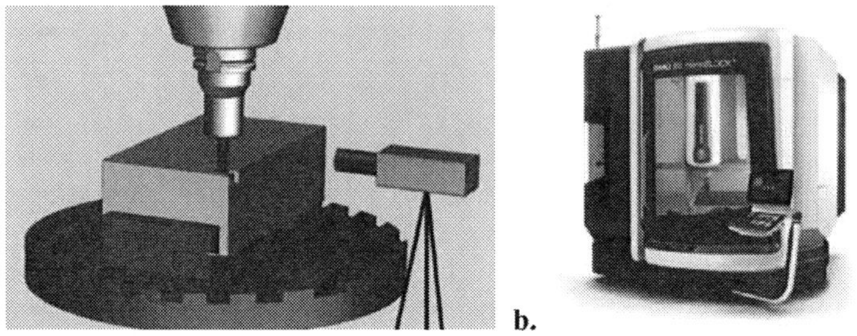

a. **b.**

Fig. 2. Experimental setup: a. schematic representation and b. universal milling mach. DMU 85 monoBLOCK® [Courtesy of DMG MORI]

The tools tested are 5-fluted end milling cutters with Ø12mm cutting diameter (*abbr.* D) and with different geometrical characteristics and performances. The geometrical differences consist in the following parameters: spiral angle (38° / 45°), minor cutting edge position (under and above center), rake angle of major cutting edge (7° / 9°), rake angle of minor cutting edge (5° / 10°), core diameter (between 6 mm - 7,8 mm), corner geometry (chamfer / radius) and geometry of center cutting (with and without center-cut). Another geometrical aspect that should be taken into consideration is the use of different gash and flute profiles. The tools used were clamped with the help of a Gühring HPC precision clamping chuck.

The workpiece used for the testing is a 42CrMo4 (EN 10269:2008) alloy steel block.

The cutting operations and parameters are shown in table 1. Additionally, it should be mentions that the experiments were made using the down milling method and were conducted in a dry environment, the cooling been done only with air.

The major component which makes the investigation of the chip behaviour possible is the high-speed camera. Although the experiments were conducted using an additional normal video camera, no conclusive observations have been made, using this method.

Table 1. The machining parameters for the experiments (*abbr.* exp.): cutting speed (*abbr.* v_c), spindle speed (*abbr.* n), feed per tooth (*abbr.* f_z), axial depth of cut (*abbr.* a_p), radial depth of cut (*abbr.* a_e)

Experiment 1: HPC Roughing	
Exp. 1.1	v_c = 200 m/min (n = 5305 RPM); f_z = 0,12 mm/rev; a_p = 1xD; a_e = 0,1(6)xD
Exp. 1.2	v_c = 200 m/min (n = 5305 RPM); f_z = 0,12 mm/rev; a_p = 1xD; a_e = 0,25xD
Experiment 2: Slotting	
Exp. 2.1	v_c = 135 m/min (n = 3581 RPM); f_z = 0,08 mm/rev; a_p = 0,25xD; a_e = 1xD
Exp. 2.2	v_c = 135 m/min (n = 3581 RPM); f_z = 0,08 mm/rev; a_p = 1xD; a_e = 1xD
Exp. 2.3	v_c = 160 m/min (n = 4244 RPM); f_z = 0,08 mm/rev; a_p = 1xD; a_e = 1xD
Experiment 3: 5°/10°/15° Ramping	
Exp. 3.1	5° ramping; v_c= 160 m/min (n= 4244 RPM); f_z= 0,08 mm/rev; a_p= 0-0,5xD; a_e=1xD
Exp. 3.2	10° ramping; v_c= 160 m/min (n= 4244 RPM); f_z= 0,08 mm/rev; a_p= 0-0,5xD; a_e=1xD
Exp. 3.3	15° ramping; v_c= 160 m/min (n= 4244 RPM); f_z= 0,08 mm/rev; a_p= 0-0,5xD; a_e=1xD

3 Observations, results and discussions

The goal of the experiments was to observe and determine the chip behaviour during the end milling cutting process, when using 5-fluted end milling cutters, and to look for any significant correlation between the milling process stability and the chip behaviour.

3.1 The investigation of the chip behaviour for HPC roughing

The experiment 1.1 shows several differences between the stable and the unstable milling process. The chip behaviour can be observed through the release pattern of the chip from the gash and flute and the movement of the chip after the contact with the tool has been interrupted.

In the stable situation (Figure 3.a) a orderly movement of the chip without showing restraint behaviour to the tools flute can be observed, whereas in a unstable situation (Figure 3.b) the movement is slightly erratic with restrained behaviour to the flute. The bottom part of the chip, the part which is cut by the minor cutting edge, twists after bottom cut and leaves a more pronounced scratch on the machined surface in the unstable operation. The release of the chip is done synchronized in stable and disorganized in unstable situations. In the last case, after the tool contact release, a greater twist of the chips can be noticed.

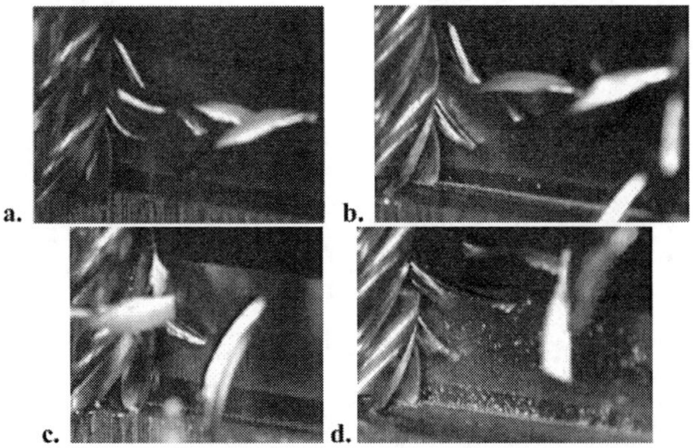

Fig. 3. Chip behaviour in HPC roughing operations: a. stable, b. unstable process (exp. 1.1), c. stable, d. unstable process (exp.1.2)

Furthermore, the chip from the stable situation is intact and no supplementary elemental chips (punctiform shape) are observed, while in unstable conditions there are serrated chips and a few elemental chips detected.

The experiment 1.2 is similar to the first experiment, but the chip behaviour differences are more pronounced. The stable situation (Figure 3.c) shows a further orderly movement of the chips with no restrains to the tools flute, while the unstable situation (Figure 3.d) shows a extreme erratic movement with a more visible restrained behaviour to the tools flute. The scratched surface of the workpiece becomes more visible in unstable conditions. The release of the chip is done more disorganized in unstable situations. The chips serrations are bigger in the unstable situations and more elemental chips are observed, which adhere to the machined surface.

Additional there has been observed scratches in tools flute, which are more visible for tools with big core diameter. The release trajectory of the chips is bigger for tools with bigger spiral angle and the scratches on the surface of the workpiece are less wider for tools with small spiral angle.

3.2 The investigation of the chip behaviour for slotting

The experiment 2.1 shows a bigger chip agglomeration in the machined slot between the stable (Figure 4.a) and the unstable (Figure 4.b) processes. In various unstable conditions, there have been observed chips that are stopped in the flute, being transported back in the cutting process and composing clusters of stacked chips.

In the experiment 2.2 and 2.3 similar observations have been made. The chip agglomeration has in this operation a more visible influence. In unstable conditions (Figure 4.d) this chip behaviour is more pronounced in comparison to stable situations (Figure 4.c). The extreme cases of this behaviour, as seen in Figure 4.e, can lead to tool

DOI: 10.1051/matecconf/201711201014

failure. Although agglomerations occur, in stable conditions the chips are evacuated much faster.

In unstable conditions the chips have a tendency to be stopped in the flute, which together with the chips from the near and dense agglomeration are being cut several times. The chips from this event are chips with multiple longitudinal cuts or long and thin chips. Additional there have been observed short, needle and elemental chips.

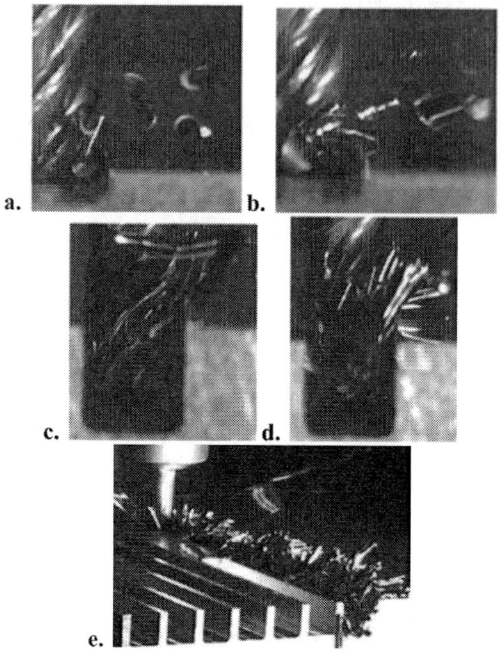

Fig. 4. Chip behaviour in slotting operations: a. stable, b. unstable process (exp.2.1), c. stable, d. and e. unstable process (exp.2.3)

3.3 The investigation of the chip behaviour for 5°/10°/15° ramping

The performance of the milling cutters and chip behaviour in ramping is very similar to the slotting operation.

The experiment 3.1 shows significant differences between the chip behaviour in stable (Figure 5.a) and unstable (Figure 5.b) situations. Right at the beginning of the cut ($a_p \approx 0$mm) the chips have the tendency not to detach themselves from the machined material, becoming so very long spiral chips. Under stable situation as the tool progresses in depth, these long chips are removed and the process behaves similar to slotting. The initial formation of these long chips gives clues for understanding the processes failure. In some cases this long chip consists of a cluster of attached chips, which slide on the flute as more chips are added, and in other cases the chips are welded on the flutes surface, building a long chip formed by the pressing of the welded chips. This second type, if not removed during the cut, can destabilize the process. As observed, the simultaneous filling of more clearance surfaces (gas and flute) can have a catastrophic result by leading to tools failure.

The experiments 3.2 (Figure 5.c - stable; Figure 5.d - unstable situation) and 3.3 show similar observations as previous. The chance of multiple simultaneous filling of the clearance surfaces increases and so the failure rate becomes greater.

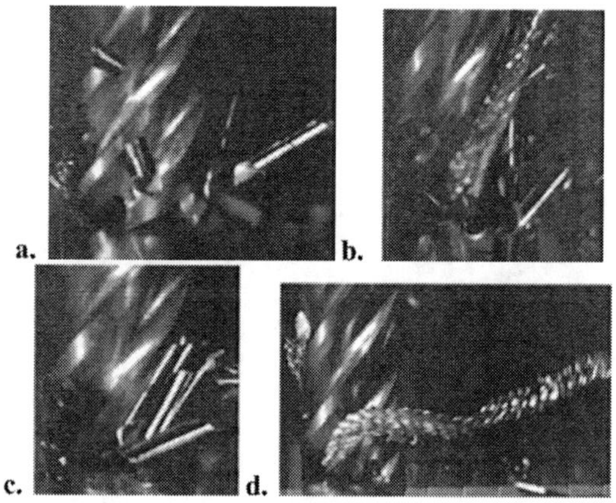

Fig. 5. Chip behaviour in 5°/10° ramping operations: a. stable, b. unstable process (exp. 3.1), c. stable, d. unstable process (exp. 3.2)

4 Conclusions

In the current article was accomplished the investigation of the chip behaviour during the cutting process with 5-fluted end milling cutters. It has been shown that the stability of the cutting process is in direct correlation with the chip behaviour. The chip behaviour and thus the performance of the process are conditioned by the geometry of the tool.

Furthermore, it has been noticed that the secondary surfaces have a great role in the chip behaviour and tool performance. Influences of these surfaces are subject for further studies.

The investigation of the chip behaviour is a quick and more direct method, which can be used to inspect the formation and the removal succes of the chip. In comparison to traditional techniques is provides more information about the cutting process and about the performance of a tool's geometry. This method can be used for further academic researches and can help the private sectors need of improving cutting tool faster.

References

1. M. E. Merchant, J. Appl. Phys. **16**, 267-275, (1945)
2. E. H. Lee, B. W. Schaffer, ASME J. Appl. Mech. **73**, 405-413, (1951)
3. G. Boothroyd, J. A. Bailey, J. Mech. Eng. Sci. **8** (3), 264-275, (1966)
4. V. Gorianis, S. Kobayashi, Ann. CIRP **15** (4), 425-431, (1967)
5. R. G. Fenton, P. L.B. Oxley, Proc. Inst. Mech. Engrs. **183**, 417-428, (1969)
6. H. T. Young, , *Application of predictive machining theory to more complicated processes including single-point oblique cutting tools with other than single straight cutting edges and milling, and experimental verification* (Ph.D. dissertation, The University of New South Wales, 1986)
7. P. L. B. Oxley, *Mechanics of machining: An analytical approach to assessing machinability*, (Ellis Horwood Ltd, 1989)
8. R. A. Ekanayake, P. Mathew, N. Amer. Manuf. R. Conf. **36**, 429-436, (2008)
9. H. Z. Li, B. Zhang, X. P. Li, Int. J. of Mech. Sci. **43** (8), 1711-1730, (2001)
10. M. Masmali, P. Mathew, *Variable flow stress machining theory applied to predict cutting forces in end milling with a variable helix angle end mills*, (Int. Conf. on Ind., Syst. and Manuf. Eng. 1, 2014)
11. M. Masmali, P. Mathew, *Geometric chip analysis of helical end milling based on a variable flow stress machining theory*, 1515-1526, (Int. Conf. on Ind. Eng. And Op. Mgt. 6, 2016)
12. M. A. Elbestawi, L. Chen, C. E. Becze, T. I. El-Wardany, CIRP Ann. **46** (1), 57-62, (1997)
13. E. G. Ng, D. W. Lee, A. R. C. Sharman, R. C. Dewes, D. K. Aspinwall, CIRP Ann. **49** (1), 41-46, (2000)
14. C. E. Becze, P. Clayton, L. Chen, T. I. El-Wardany, M .A. Elbestawi, Int. J. Mach. T. Manuf. **40** (6), 869-885, (2000)
15. Y. Ning, M. Rahman, Y. S. Wong, J. Mater. Process. Tech. **113**, 360-367, (2001)
16. X. Wu, Y. Li, Asia Int. Symp. On Mechatronics **1**, 225-230, (2004)
17. A. U. Patwari, N. Amin, W. Faris, Amer. J. of Eng. and App. Sci. **3** (1), 193-200, (2010)

Tool material, tool wear and machined hole quality evaluation at CFRP drilling

Claudiu-Ioan Jugrestan[1,*], *Marcel-Sabin* Popa[1], *Stefan* Sattel[2], *Adina-Elena* Cotarga[1], *Ovidiu-Virgil* Veres[1] and *Camil-Octav* Chetreanu Don[3]

[1]Technical University of Cluj-Napoca, Faculty of Mechanical Engineering, Department of Manufacturing Engineering, BulevardulMuncii 103-105, Cluj-Napoca, Romania
[2]Gühring KG, Herderstraße 50-54, Albstadt, Germany
[3]Gühring S.R.L., Transilvaniei 12, Baciu, Romania

Abstract. Because of increasingly stricter safety rules, requirements regarding the quality of the components used in the airline industry have increased. Besides the materials used in aircraft components and areas where they are machined, these components must comply certain standards for quality and tolerances of the machined surfaces.Due to high hardness of carbon fibers, hole machining in carbon fiber reinforced plastics has always been a challenge. Wear that occurs in the cutting tool leads to the appearance of workpiece defects such as delamination, peeling, fiber pulling out, etc. Besides cutting tool geometry, a very important role on tool wear, it is owned by the material from which it is made. This paper will make a quantitative assessment of defects that occur in the process of CFRP (carbon fiber reinforced plastics) drilling and a cutting tool wear evaluation. In order to obtain conclusive results for this study will be used three drill bits with the same geometry but different carbides.The aim of this paper is to find which one of this three carbide used is more feasible for holes machining in composite materials reinforced with carbon fibers.

1 Introduction

In the entire world, the growing demand for new materials with new or better chemical and physical properties has increased a lot in the last 20 years. Due to their weight-specific properties, low thermal expansion, high chemical resistance, etc. the CFRP are used in a wide range.

The main fields where these materials are usedare those that are based on weight reduction. Some of those are aerospace and automotive industry, but also the wind energy industry. In the Figure 1, in a report made by Kraus and Kuehnel, are shown the quantities, in tones, used in different areas in 2014 [1]. In the same report is made a forecast regarding the produced quantities of CFRP, annually, by 2021. Can be observed that there are a lot of fields, where, the composite materials are replacing conventional materials. A good example is the manufacturing of a plane fuselage, where until last years were used aluminium alloys that now are replaced by composite materials, especially composite materials reinforced with carbon fibers or fiberglass.

*Corresponding author: jugrestan_claudiu@yahoo.com

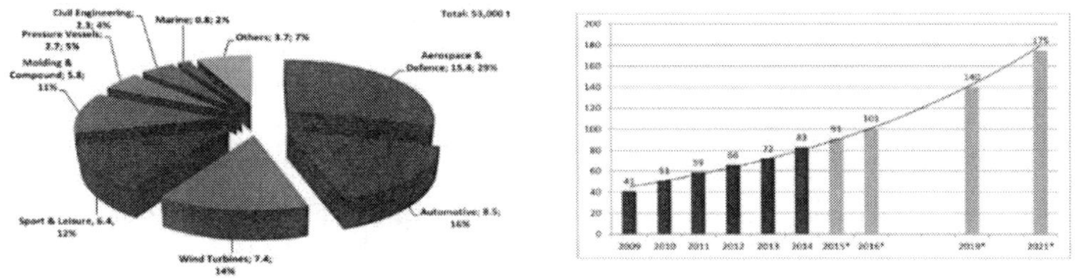

Fig.1. CFRP application fields and CFRP demand forecast [1].

Accordingly with Islam [2], drilling is one of the most used operation in carbon fiber reinforced plastics machining. In his doctoral thesis, he showed that are acoustic panels that can incorporate up to 100.000 holes with diameters between 3-6 mm.

Because of the use on a raising scale of these types of materials, also the cutting tools industry is forced to come up with new solutions for their machining. Compared to conventional materials, at CFRP machining must be taken into account other defects that can occur, as delamination, pull outs, fibers projection, pyrolysis, etc., partially represented in Figure 2 [3].

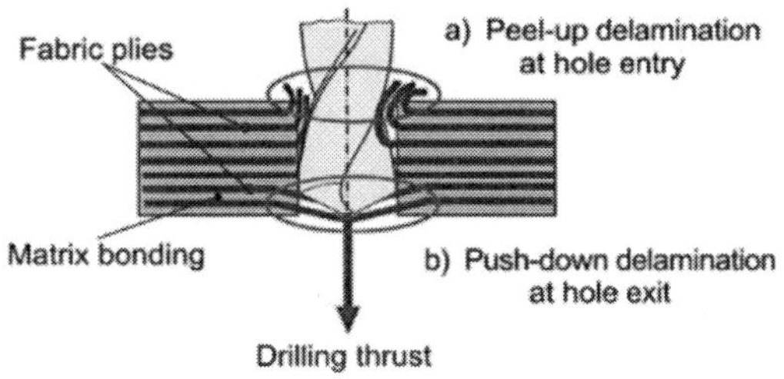

Fig. 2. Hole quality defects [3].

These problems occur, primarily, due to inhomogeneity that composite materials reinforced with carbon fibers, due tothe cutting tools geometry, their wear and due tocutting parameters that are used.

2 Experimental work

2.1 Equipment

For the experiments was used a Matsuura VX-1000 CNC machining center with the characteristics shown in Table 1.

Table 1. CNC characteristics

Spindle motor power	kW	15-22	X-Axis Travel	mm	1020
Spindle speedrange	min $^{-1}$	50-15000	Y-Axis Travel	mm	610
Feedrate	mm/min	1-20000	X-Axis Travel	mm	610

For delamination measurement was used a Keyence VHX-5000 digital microscope, equipped with a 4800x3600 pixels super high resolution camera. Tool wear was measured with an Aliconanon-contact, optical, three-dimensional scanning, based on focus-variation equipment.

2.2 Workpiece material

For test, the M21/34%/UD194/IMA-12K material, produced by Firma Hexcel was selected. In present, this material is used for the Airbus A380 civilian airplane fuselage construction. The main physical and mechanical properties are shown in Table 2.

Table 2. Material properties [4]

Properties	Units	M21/34%/UD194/IMA-12K
Fiber Weave/DU Fiber Mass	g/m^2	194
TheoreticalCalculatedCuredPlyThickness	mm	0,184
TheoreticalCalculated Fiber Volume	%	59.2
ResinDensity	g/cm^3	1.28
Fiber Desnsity	g/cm^3	1.78
TheoreticalCalculated Laminate Density	g/cm^3	1.58
Tension Strength	MPa	3050
Tension Modulus	GPa	178
CompressionStrength	MPa	1500
CompressionModulus	GPa	146

2.3 Cutting tools

In order to have conclusive results, for this study will be used three drills with the same geometry but made from three different carbides. Alsothe cutting parameters will be the same for all tests: vc=100m/min; f=0.06mm/U; vf=300mm/min; n=5013 rev/min.

2.3.1 Drill geometry

In his dissertation thesis, Barbagallo has shown that the step drills have better result in CFRP drilling. He has explained that the drill chisel edge and the axial force, which acts on the last material layers, have an important role on delamination produced by the first drill diameter [5].

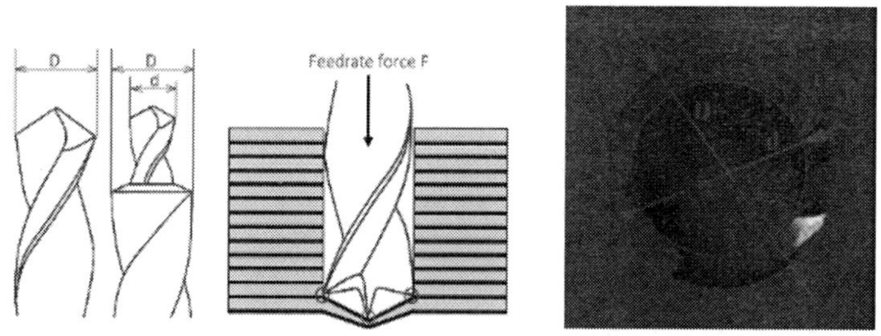

Fig. 3. Drill geometry (left) [5] and machining dynamics (center and right).

The tested drills will have two steps, with d=4.75 and D=6.35. The results will be compared with previously tests made with an RT100U, article 7924 drill, provided by the Gühring Company, used for metallic materials machining.

2.3.2 Cutting tools materials

The tested drills are produced from three different WC-Co carbide grades, Type 1 (0.7μmgrainsize), Type 2(0.6 μm), and Type 3(0.5 μm), , produced by Firma Gühring[6]. In table 3 are described the characteristics of these three carbide grades.

Table 3. Carbide grades [6]

Carbide grade	Chemical composition (%)		WC grainsize (μm)	Density (g/cm³)	Hardness (HV 30)
	WC	Co			
Type 1	92	8	0.7	14.55	1710
Type 2	92	8	0.6	14.55	1790
Type 3	95	5	0.5	15	2010

3 Tests and results

In this paper was aimed to obtain holes where the delaminated area (A_2), calculated with relation 2, must be maximum 2% from the machined hole area (A_1), calculated with relation 3. For that, relation 1 will be taken into account:

$$2\% \geq (A_2/A_1)*100 \tag{1}$$

$$A_2=a_1+a_2+\dots a_n \tag{2}$$

$$A_1=\pi*r^2=31.6\ mm^2 \tag{3}$$

where r is the machined hole radius (3.175 mm) and $a_1,a_2,\dots a_n$ are the errors surfaces.

Also the number of machined holes with the RT100U drill will be taken as a percentage of 100% and depending on it the others three drills will be evaluated. In the Figure 4 are pictures made with the first machined hole and with the hole where the test was stopped, for each drill.

In Figure 4 are represented the errors that has occurred during a test made by Gauggel in his dissertation thesis[7]. The tests were made in the same workpiece material and with the same Gühring RT100U drill[8]. These results will be taken as reference for future tests.

In Figure 5 are shown, in the left side, the machined holes, together with the surfaces where the errors has occurred. Also, in the same figure, in the right side, by 3D scanning, was made a comparison between the initial drill cutting edge and the cutting edge after the tests were stopped.

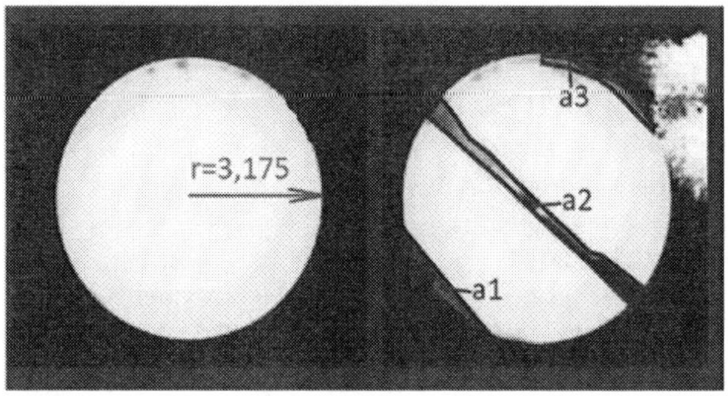

Fig. 4. First hole / last hole machined with RT100U [7].

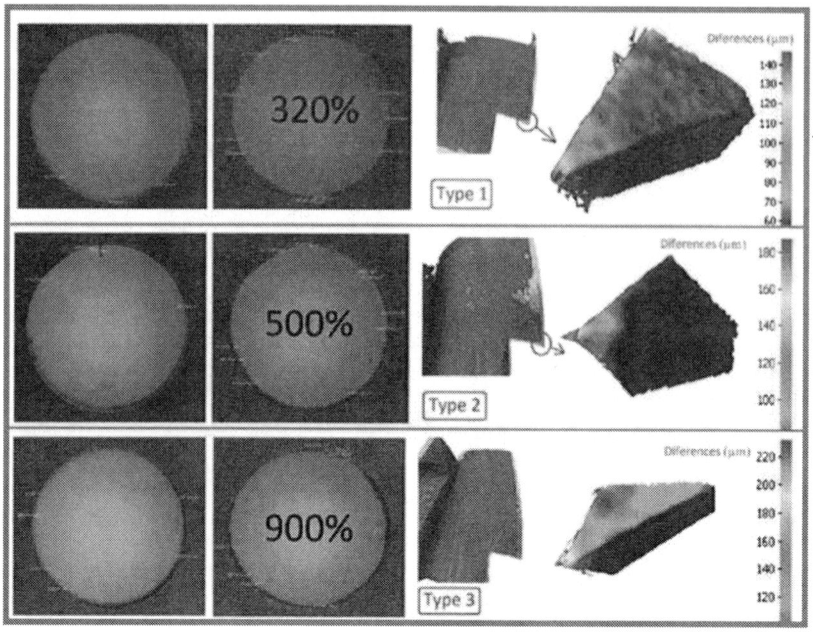

Fig. 5. Machined holes errors (left) and tool wear (right).

The 3D scan was made with the Alicona equipment. Before starting the tests, the drills were 3D scanned, obtaining the green contour. The obtained image was overlapped on the scan made at the end of the tests (blue contour) thus obtaining the wear values "W", represented in Figure 6.

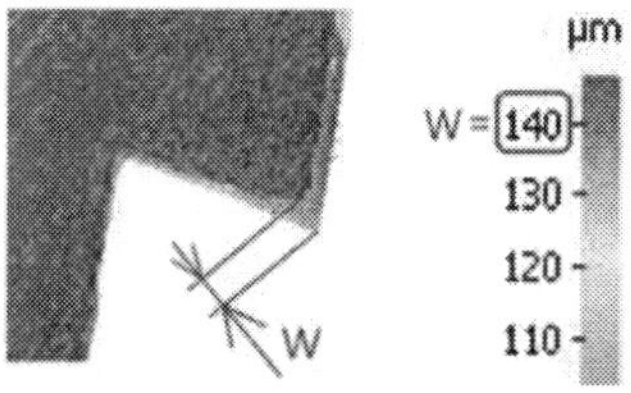

Fig. 6. Alicona 3D

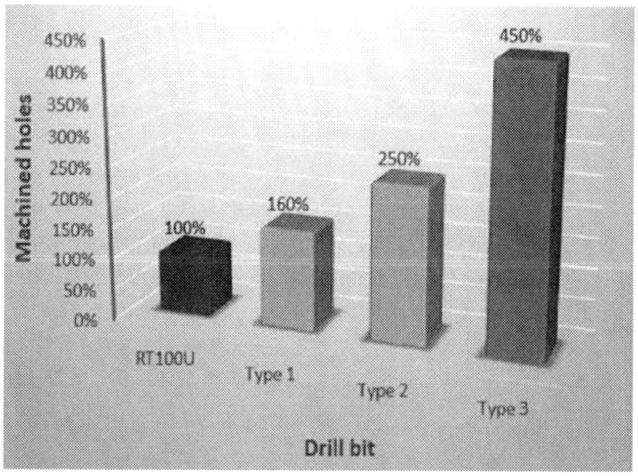

Fig. 7. Comparison between conventional and step drills.

After the chart from Figure 7 analyzation, can be seen that with step drills can be obtain results with up to 450% better than using conventional drills used for the metallic materials machining. Also can be observed that the carbide grade has an important role on the number of machined holes. Although all three carbide grades have the same components, the percentage of those components has a significant influence on final results.The Type 3 drill has the WC content bigger than the other two carbides, and that makes the cutting edge tougher. Because of this, the carbon fibers abrasiveness do not wear so fast the drill cutting edges. In the same Figure 6 is visible that hole number made with the Type 3 drill is almost double, compared with the Type 2. The Type 1 and Type 2 are two carbides with the same properties and the same percentage of components, the only difference between them being only the WC grain size. Accordingly with Faraz, as main cutting edge radius is smaller, the errors surface will be reduced [3]. Also it makes sense that as the smaller the WC grains are, a better cutting edge sharpening can be obtain. From the chart above, between the results obtained with these two drills is a big difference. The Type 2 grain size is with 0,1μm smaller that the Type 1, but this small difference had substantially influenced the final results as is visible in the chart from Figure 6.

4 Conclusions

In the last 20 years, the carbon fiber reinforced plastics had started to replace the conventional materials in many industrial applications. Because of that in this paper was made, firstly, an assessment of the problems that appear during the holes machining regarding the workpiece quality and then regarding the cutting tool wear. In order to reduce the problems regarding the workpiece and the tool, special drill geometries were tested and compared with classical geometries. Was observed that the classical drills are not feasible for the CFRP machining because they produce large delaminated areas compared with the step drills. For the step drills was noticed that the delaminated area produced by the first diameter (d) is cut off by the tool nominal diameter (D).Also this paper reveals that the tool material is critical for the tool life. The Type 3 drill, that have a bigger concentration of WC, have performed better than the others, the conclusion being that the tool life is strictly related to the tool material hardness. A correlation was also found between the WC grain size and the machined hole quality. The Type 1 and Type 2 drills have the same chemical composition, density and hardness, the only difference between them being only the WC grain size. After the tests, is visible the fact that the Type 2 drill , with 0.6 μm grain size, gave with more than 150% better results than the Type 1 drill, with 0.7 μm grain size. The explanation is that the smaller the carbide grains are, the better the cutting edges can be sharpened, in the drilling process the carbon fibers being cut easier.

References

1. T. Kraus, M. Kühnel, E. Witten, *Composites market report 2015*, (Federation of Reinforced Plastic, 2015)
2. S.E.M.S. Islam, *Drilling of carbon fibre reinforced plastic composites*, (Doctoral Thesis, University of Birmingham, 2010)
3. A. Faraz, D. Biermann, K. Weinert, International Journal of Machine Tools & Manufacture, **49** (2009)
4. Information brochure, Hexcel, www.hexcel.com, (2017)
5. S.F. Barbagallo, *Optimization of the cutting geometries of a solid carbide drilling tool for the processing of CFRP composite materials (in German)*, (Technology and Management Faculty, Ravensburg-Weingarten, 2014)
6. Gühring Catalog, *HARTMETALLE for precision tools. Price list no.42. Hard metal grades for basic tools (in German)*, (2011)
7. C.W. Gaugel, *Development of high-performance cutting tool for the processing of CFRP composite materials (in German)*, (Albstadt-Sigmaringen University, 2013)
8. Gühring Catalog, *Gühring drilling*, (2015).

Influence of the cutting edge microgeometry on the tool life in austenitic stainless steel machining with carbide end mill

Marius-Bogdan Bozga[1*], *Marcel Sabin* Popa[1], *Stefan* Sattel[2], and *Vlad-Bogdan* Tomoiagă[1]

[1] Technical University of Cluj-Napoca Romania, Faculty of Mechanical Engineering, Department of Manufacturing Engineering, Blvd. Muncii, No. 103-105, Cluj-Napoca, Romania
[2] Gühring KG, Head of Research and Development Department, Winterlingen Street, No. 12, 72488, Sigmaringen-Laiz, Germany

Abstract. The goal of this paper is to determine the optimal cutting edge microgeometry for a carbide end mill used in stainless steel machining. The optimal cutting edge microgeometry will be determined through conducting more tool life tests. Tools with three different cutting edge radiuses and one without edge preparation will be tested in order to determine the influence of the cutting edge microgeometry on the wear and tool life. The edge microgeometry is obtained through wet blasting preparation.

1 Introduction

This study aims to determine the influence of the tools cutting edge microgeometry obtained in the end milling process of a 1.4301 grade austenitic stainless steel through conducting more tool life tests. In order to determine the optimal cutting edge microgeometry, tools with three different cutting edge rounding were prepared through wet blasting with different parameters. The tools were then tested in order to determine de influence of the cutting edge microgeometry obtained through wet blasting on the tool life.

The 1.4301 grade is an austenitic stainless steel, which is widely used in many industries such as automotive industry, food industry, chemical industry, petrochemical industry, decorative items and kitchen utensils for its good corrosion and heat resistance, ease of formability, good cryogenic properties and its aesthetic appearance. This grade of austenitic stainless steel is also known for its poor machinability.

The end milling process is often used in many industries for machining stainless steel parts. The milling process is a hard cutting process due to the interrupted contact between the cutting tool and the work piece and due to the variable chip thickness, therefore an extensive optimisation of the process is necessary [1].

An optimisation of the austenitic stainless steel cutting process is possible through choosing the right tool geometry and cutting parameters [2]. Another efficient way of increasing the tool life is the usage of an efficient cutting fluid, which decreases the cutting

* Corresponding author: marius.bozga@yahoo.com

edge temperature, reduces the cutting forces, helps to reduce the friction coefficient between the chip/workpiece and cutting tool, inhibits adhesion and flushes the chips away from the tool and workpiece.

Ciftci and Endrino [2,3] showed that choosing the right coating for the tool also helps to increase the tool life. High temperature hardness, oxidation resistance, low friction coefficient and low thermal conductivity are the main properties of a good coating for a cutting tool used in machining austenitic stainless steel.

The researches regarding cutting edge microgeometry from the last two decades show that the tool life can be increased through preparation of the tools cutting edge with a specific rounding. For each machining process and machined material there is an optimal cutting edge rounding [4].

The present tests were carried out with a Gühring RF100VA solid carbide end mill with four cutting edges and a specific geometry for machining stainless steel. The main purpose of the tool life tests was to determine the influence of the cutting edge microgeometry prepared through wet blasting on the tool life and performance.

2 Cutting edge microgeometry and preparation

Currently there are no international standards defining the microgeometry of the cutting edge. The characterisation of a rounded edge defined by a radius is given in DIN 6582, but there are no specific details about the criteria used to measure this radius. Proposals for a classification were made but none of these are universally used [4].There are currently different methods being used for the measuring of the microgeometry and this often leads to significantly different results.

2.1 Microgeometry classification by form

The classification of the cutting edge by form is done by analysing the edge profile in a perpendicular section on the cutting edge.

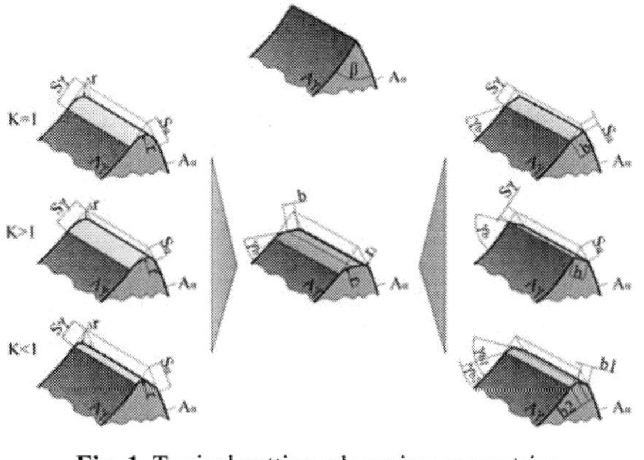

Fig. 1. Typical cutting edge microgeometries.

2.1.1 Round

The profile of the perpendicular section to the edge has a round shape defined by a radius r and two lengths $S\gamma$ and $S\alpha$. $S\gamma$ and $S\alpha$ are the length from the tangent point of the radius with rake face ($S\gamma$) and flank face ($S\alpha$) up to the imaginary intersection of the rake face with the flank face. The length Δr helps defining the profile flattening [4].

DOI: 10.1051/matecconf/201711201016

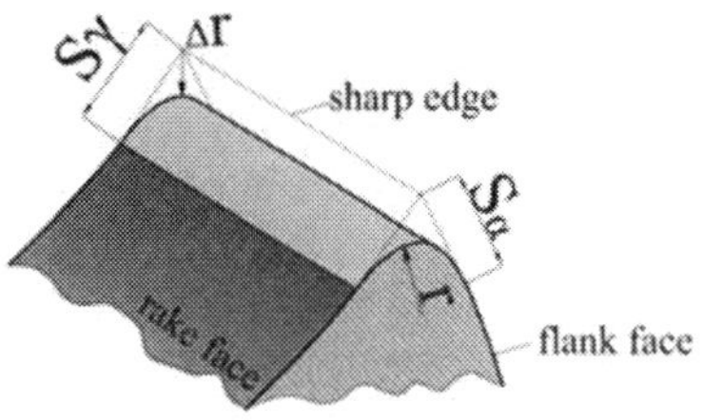

Fig. 2. Characterisation of the round cutting edge.

The *K*-factor is the ratio between $S\gamma$ and $S\alpha$, which shows the orientation of the rounded edge towards the flank face or the rake face [4].

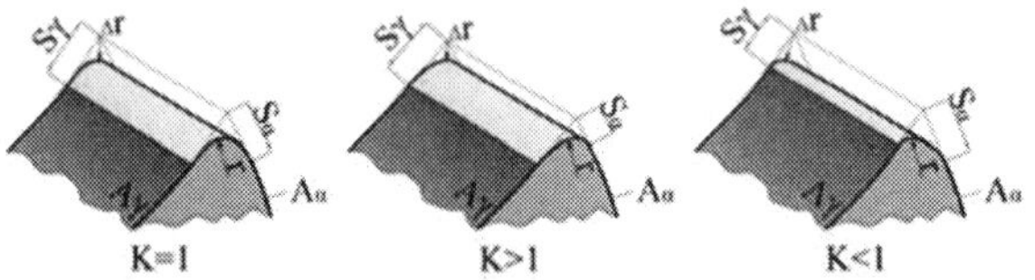

Fig. 3. Different types of cutting edge rounding.

2.1.2 Chamfered

It is defined as an intermediary flat surface in between flank and rake face and it is characterised by an angle γ_b and a width b. It can have a single or more flat surfaces and can have a greater inclination towards the flank face or towards the rake face. This inclination is characterised by the *K*-factor.

2.1.3 Chamfered with rounded edges

This form is a combination between the rounded and chamfered cutting edge.

2.2 Microgeometry used in the tests

In the present tests, tools with cutting edges prepared through wet blasting were used. Wet blasting produced a cutting edge with a round form with a *K*-Factor lower than 1 on the peripheral cutting edge and bigger than 1 on the end cutting edge. The parameters for the wet blasting of the solid carbide end mill were established after intense research and with the help of the specialised personnel from Gühring Company. The interval of the cutting edge rounding values, that were obtained after wet blasting, were chosen after conducting more FEM simulations in order to restrict the interval.

3 The preparation of the cutting edge microgeometry through wet blasting

Wet blasting (or Vapormatting) is a process similar to conventional blasting in the way of using abrasive media and compressed air. The differences, however, is that the wet blasting uses a third element, a liquid to enhance the surface finish [5]. The abrasive is suspended in a liquid projected at high velocity by a high pressured flow of compressed air [6]. The benefit of including water in the process is that it offers a much smoother and more consistent finish.

In this experiment, the blasting installation uses as abrasive medium a corundum granulate and water as mixing medium. The concentration of abrasive in water is 15%. The pressure of the compressed air is adjustable between 1 and 7 bar. Time is an important factor that determines the size of the cutting edge rounding. The distance between blasting nozzle and tool is 120mm. In order to obtain different sizes of the cutting edge rounding, the following blasting parameters were used:

- For the tool with the small cutting edge radius, an air pressure of 2 bar and 31 seconds as the total time of blasting was used.
- To obtain tools with a medium rounding, they were blasted for 130 second at 2 bar.
- The large cutting edge rounding was obtained by blasting the tools at 3.5bar for 60 seconds.

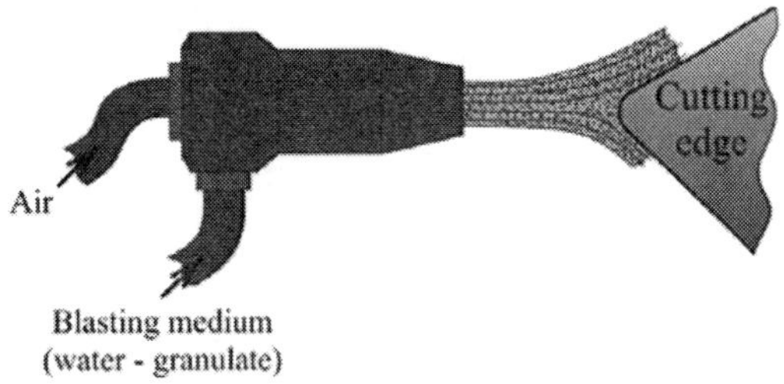

Fig. 4. Schematic representation of the cutting edge wet basting.

The tools cutting edge blasting process aims to increase the edge stability and rigidity, to eliminate the cutting edge defects and to smooth the edge surfaces.

In the following images, the uncoated tools with the cutting edge preparation are presented. From left to right the tool with a small cutting edge radius, the tool with a medium cutting edge radius and the tool with the large cutting edge radius.

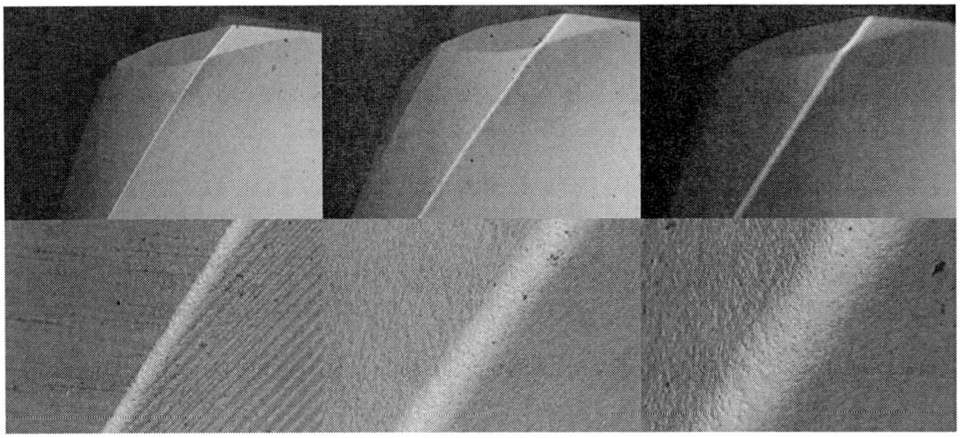

Fig. 5. Tools cutting edge after wet blasting.

4 Coating

After the wet blasting process the tools were coated with a multilayer coating of TiN and SiN which enhances the tool cutting performance. Guhring developed the new Signum coating that has a hardness of 5500 HV one of the hardest nitride coatings on the market. Guhring could achieve this extraordinary coating hardness thanks to the special nano composite structure with a layer build-up of TiAlN and SiN. Thus, the Signum coating offers particularly high heat resistance and at the same time high diffusion resistance [7].

5 Tool life test

The durability tests were done on a 3 axis vertical CNC milling centre. The tools were analysed at predefined intervals to follow the edge wear and deterioration. At this intervals the wear was measured and the cutting edge was analysed at an electron microscope to see the edge wear, material adhesion and coating deterioration.

The intervals of documentation were done at first after 1m and 3m of cutting length. After that the documentation was done every 5m until the maximum wear width of 360μm was achieved.

The cutting parameters were chosen after a functional cutting test through which the stability of the process was analysed. The following parameters were chosen: cutting speed V_c=130m/min and the feed per teeth f_z=0.03mm. The cutting depth a_p is 12mm and the cutting width a_e is 2mm. These parameters are in the range of semi finishing cutting.

6 Results and discussion

The results of the durability tests showed that the tools with the small cutting edge radius show the greatest cutting performance with a 54.8% increase in tool life in comparison with the reference tool with the sharp edge. The tools with the medium and large cutting show a decrease in the tool life with 38.3% in the case of the tool with medium cutting edge rounding and with 72.7% in the case of the tool with small cutting edge rounding.

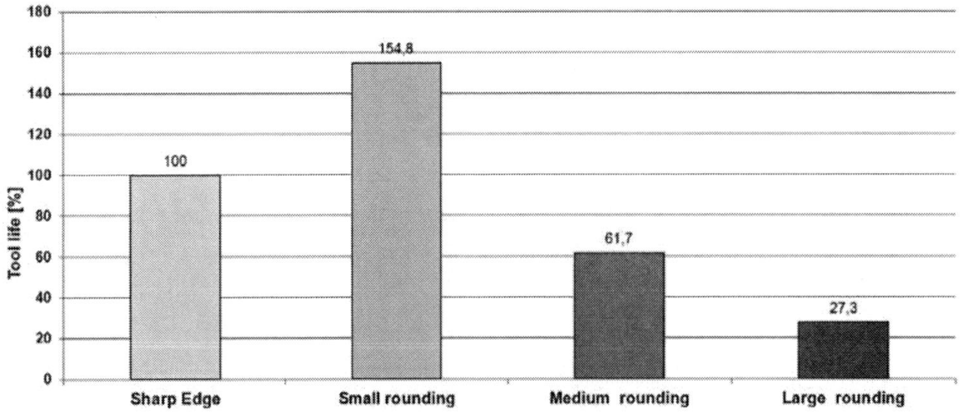

Fig. 6. Results of the tool life cutting tests

The tool with the sharp edge showed an excessive wear of the mill corner chamfer (Figure 7-a) which lead to an increase in tool vibration and a decrease of the surface quality of the machined workpiece and therefore the end of the cutting test.

In the case of the tool with the small cutting edge rounding (Figure 7-b) the wear is localised evenly alongside the cutting edge with a small increase on the mill corner chamfer. The surface quality of the machined workpiece was constant through all the tool life range and with no increase in the tool vibration.

The wear of the tool with the medium and the big cutting edge radius (Figure 7-c and d) is distributed mostly on the cutting edge. This happened due to increased material deformation, increased temperature and material adhesion.

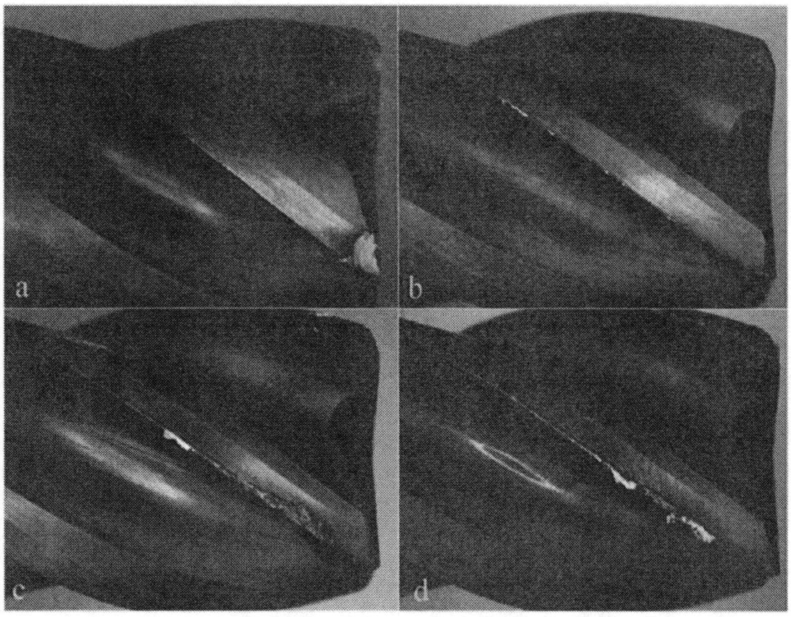

Fig. 7. The cutting edge wear at the end of the test.

7 Conclusion

It can be concluded that the Guhring RF100VA carbide end mill with a small cutting edge rounding that have been obtained by wet blasting the tool at 3.5 bar for 60 seconds using water and corundum as abrasive medium, have the best cutting performances and tool life in the process of machining a 1.4301 grade austenitic stainless steel.

The authors would like to thank the Gühring Company for the technical and material support and the Technical University of Cluj-Napoca for knowledge and information support.

References

1. H. Shao, L. Liu, H.L. Qu, Wear, **263**, 736-744, (2007)
2. J.L. Endrino, G.S. Fox-Rabinovich, C. Gey, Surface & Coatings Technology, **200**, 6840-6845, (2006)
3. I. Ciftci, Tribology International, **39**, 565-569, (2006)
4. B. Denkena, D. Biermann, CIRP Annals – Manufacturing Technology, **63**, 631-653, (2014)
5. https://www.vapormatt.com/news/what-wet-blasting
6. http://www.azom.com/article.aspx?ArticleID=7894
7. http://pvdcoating.com.au/signum-coating/

Behavior under contact loading of surface asperities obtained by turning

Cristina- Florentina Savu[1], *Vasile* Ermolai[1,*], *Claudiu- Ioan* Guriţă[1], *Vlad-Constantin* Cazacu[1], and *Gabriele* Taormina[2]

[1]"Gheorghe Asachi" Technical Univeristy of Iaşi, Department of Machine Manufacturing Technology, Blvd. D. Mangeron, 59A, Iaşi, Romania
[2]University of Parma, Department of Industrial Engineering, University street, 12 – I, 43121 Parma, Italy

Abstract. The quality of metallic surfaces is described by the physic-mechanical and chemical state (microhardness, microstructure) and geometrical state. Contact interaction between two solid bodies has a very important role in many physical phenomena and engineering applications. The main idea of the research was to experimentally study the behavior at pressure of a surface obtained by turning. For this experiment, cylindrical parts made of aluminum were used. The test samples were obtained by turning using tools having distinct corner radius and distinct feeding rates. Experimental method was applied in order to evaluate and model the influence exerted by some input factors (force, turning tool corner radius, machining feed) on the *Ra* surface roughness parameter. By using the data obtained after the experimental tests, an empirical mathematical model able to describe the evolution of the roughness under pressure contact was determined.

1 Introduction

The surface roughness consists in an assembly of micro-irregularities with a reduced pitch in comparison with its depth, which forms a certain relief of surface. The surface roughness is significant for knowing how an object will interact with its environment. Unlike the smooth surfaces, the higher friction coefficients of the rough surfaces will cause more wear on this type of surface. Roughness frequently is a right benchmark to determinate the performance of the mechanical element, because the irregularities of the surface may "thrive" in time and form cracks or corrosion. [1]

Various parameters could be used in order to characterize the surface roughness. One of them is arithmetic mean deviation of profile, which is universally recognized and most used parameter of surface roughness. It is defined as an arithmetic mean of the absolute deviations of the roughness profile from the mean line *Ra*.

The mechanical contact is an important concept in the field of engineering. It is agreed that two or more distinct entities (solid bodies) are in contact as long as there is a common

* Corresponding author: ermolai.vasile@gmail.com

DOI: 10.1051/matecconf/201711201017

surface that is separating them without any material transfer from one solid to another. Two rough surfaces are in contact on a small area that is composed by all the contacts between the asperities (which will deform).

According to the rigidity of the surfaces, the contact between them can be rigid-flexible. A rigid-flexible contact is defined by the fact that one of the surfaces that are in contact is more rigid than the other. One of the main applications of the contact analysis refers to the subject of quality [2].

Many researchers have addressed issue of contact with or without friction between two elastic solids. Hertz was the first that solved this issue for a frictionless contact between smooth solids, claiming that the point of contact between two spheres of radius r becomes an area of contact. The explanation of Bowden and Tabor (in 1939) was simple, claiming that the asperities of two rough surfaces reach the field of plastic deformation as soon as the asperities get in contact, noting the existence of a direct proportionality between the load and the contact area of the bodies get in contact [3].

The main objective of the present paper was to highlight the influence exerted by several factors (feed, tool corner radius and force) on the surface roughness, respectively on the behavior under contact loading of surface asperities.

2 Hypotheses concerning the influence of various factors on the surface asperities accuracy obtained by turning

When two surfaces are in contact, the real area of contact is not the nominal one, but the resulting area of asperities found in contact [2].

It can be observed that when they are pressed against a rough surface with another cylindrical body that has a higher hardness, the asperities of the initial surface are reduced/ flattened. The main factors considered in this study that have influence on the roughness in the turning machining are the following: the processed material, the pressure between bodies found in contact, the feed and the turning tool corner radius. These factors can be taken into consideration as input data. In order to observe the surface roughness modification under contact loading, the test pieces surface were pressed with another rigid body. The force used for pressing against the surface is also an input factor that will influence the flattening of the initial surface asperities. The required result is to obtain a surface with "lower" asperities, compared with the initial ones that is expressed as an output data and that consists of the modified value of Ra surface roughness parameter. For illustrating the correlation between input and output parameters, it can be used a systemic analysis (Figure 1).

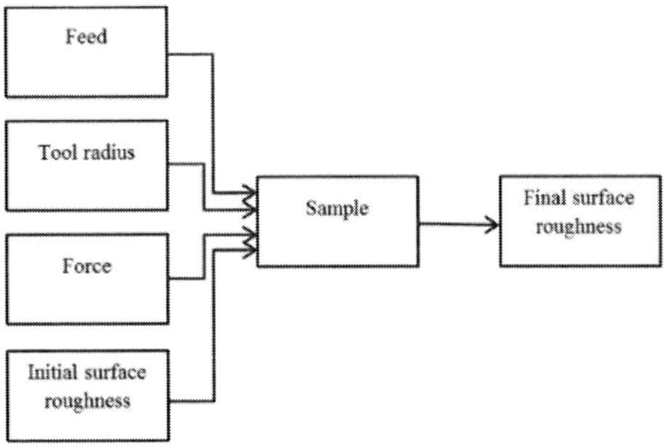

Fig. 1. Systemic analysis of input and output.

As known, a higher tool corner radius and a smaller turning feed will determine a smaller roughness at facing the test pieces.

By summing the above mentioned statements, three cases of material behavior can be described by considering the compression force and the value of roughness parameter:

- The first case, where the compression force has a reduced value, and the asperities are stressed in the elastic domain;

- The second case, where the compression force is highly enough to deform the asperities in the elastoplastic domain, producing a flattening of the superficial layer. Much higher strains are produced if the force goes in the plastic domain;

- The third case and the most "hazardous" where the compression force is high enough to crush the asperities of the sample and mark the form and roughness of the punch (pressing tool).

The way in which a rigid body presses on rough surfaces is represented in Figure 2.

One can assume that once with the increasing of the contact loading, respectively of the pressure between the rigid body and the asperities of the test piece surface, a smoother roughness will be obtained.

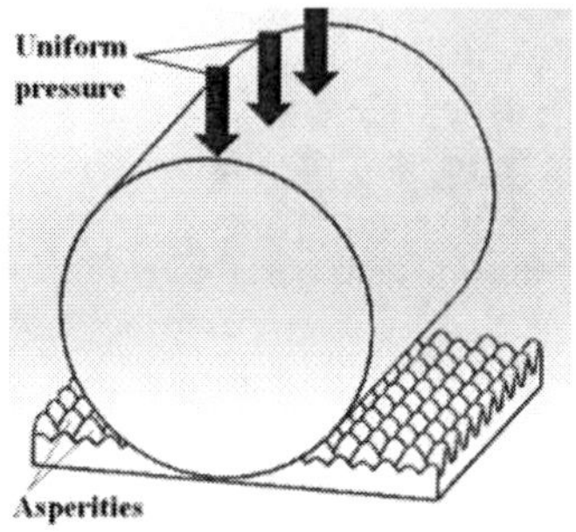

Fig. 2. Pressing with a uniform force on a flat surface.

3 Experimental research

The experimental research aimed to evaluate the contact behavior at pressure of a surface asperities obtained by turning.

For the experimental study, it was used an upright drilling machine as an equipment able to offer the necessary force/pressure (Figure 3). One considered initially that during experimental research, a force up to 500 [daN] could be necessary in order to generate plastic deformation of the surface asperities. Four test samples of aluminum and a cylindrical pressing tool made of high speed steel were also used. The aluminum was selected as test samples material due to its high plasticity, adopting the hypothesis that under the pressure action, the surface asperities will be affected by a process of plastic deformation. The cylindrical test samples had thickness s=10 [mm] and an external diameter 25[mm]. An hole with a diameter 5 [mm] was achieved in each test sample, in order to simplify the process of frontal turning.

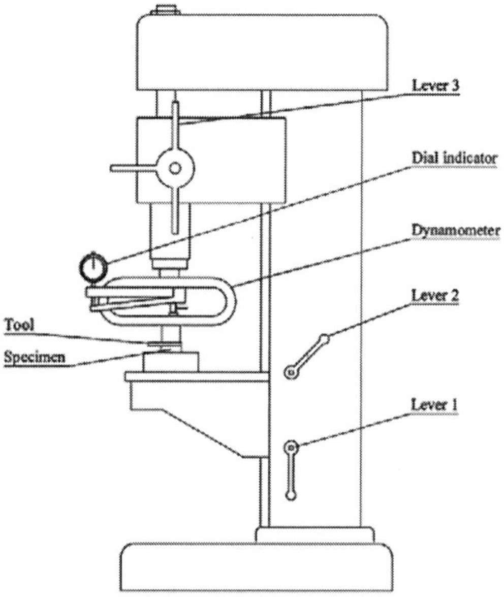

Fig. 3. Drilling machine used as a pressing machine.

The samples were prepared on the frontal surface using two feed values (0.24 mm / rev and 0.48 mm / rev) during turning process and two cutting tools with distinct tool corner radii (0.4 mm and 0.8 mm). Firstly, it was measured the roughness of the initial surface of test samples by using an electronic roughness tester (Mitutoyo SJ 201) to determinate the value of this parameter. The second stage was the pressing stage, where in first instance pressing forces with values of 250 [daN] and then another one with 462 [daN] were used. In order to generate the plastic deformation of the test samples surface asperities, a cylindrical pressing tool having a diameter of 2.95 mm was used. It was expected that using the hardened high speed steel as material for pressing tool and not very high value of the pressing force, the pressing tool material will not be affected by plastic deformations.

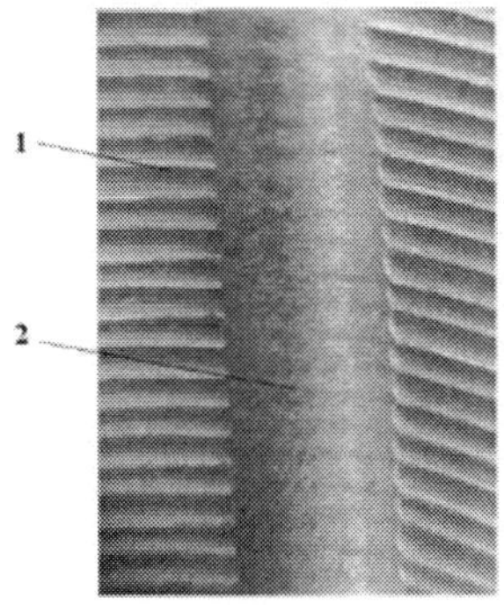

Fig. 4. Image concerning surface roughness of the sample: 1- asperities corresponding to the feed movement of turning process; 2- surface generated by pressing a cylindrical body with a known force.

The result pressure exerted by the cylindrical body determined a visible change in the frontal aspect of each test sample (Figure 4). All forces were generated by using a dynamometer. Finally, using the same electronic surface roughness tester, the roughness corresponding to the surface generated by plastic deformation of the test sample material was measured.

The values of feed *f*, radius tool r_ε and force *F* were inscribed in Table 1, for each of the eight experiments. The last columns from this table included the sizes of the surface roughness parameter determined by measurements and corresponding to the initial and final surface states, respectively. Every roughness parameter was measured three times and the final value was settled as arithmetic mean of those three values.

To determine the empirical mathematical model able to show the influence exerted by the three independent variables on the *Ra* surface roughness parameter, specialized software was used [4]. This program uses the least square roots method to determine the most suitable equation for certain experimental research. The output of using the program consists in several types of equations that are possible mathematical expressions corresponding to the given input.

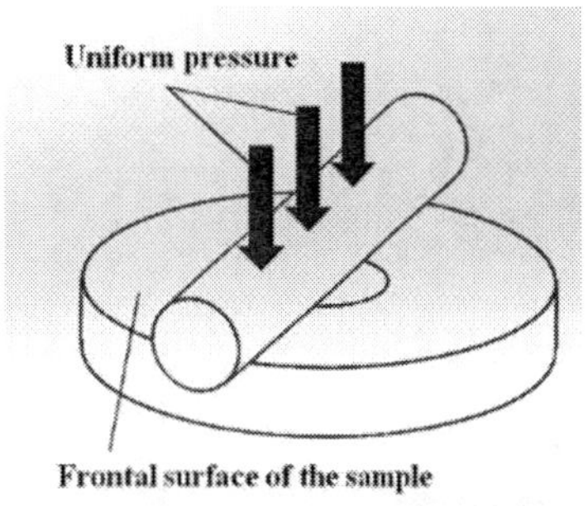

Fig. 5. Pressing with a cylindrical tool on the frontal surface of the sample.

Table 1 Experimental conditions and results

Exp. no.	Input parameters				Output parameters
	Feed *f* [mm/rev]	Tool radius r_ε [*mm*]	Force *F* [daN]	Initial surface roughness parameter R_a [μm]	Final surface roughness parameter R_a [μm]
1	0.24	0.4	250	5.12	2.32
2	0.24	0.4	462		0.84
3	0.24	0.8	250	11.28	1.42
4	0.24	0.8	462		1.20
5	0.48	0.4	250	17.87	2.79
6	0.48	0.4	462		2.11
7	0.48	0.8	250	22.35	3.89
8	0.48	0.8	462		2.93

Considering that in machine manufacturing there is a real preference to use power functions when elaborating empirical mathematical models, we selected such a function to model the influence exerted by the machining feed *f*, tool corner radius r_ε and force *F* on the value of *Ra* surface roughness parameter:

$$R_a = 30.65\, f^{1.41} r_\varepsilon^{\,0.518} F^{-0.178} \qquad (1)$$

This mathematical empirical model was used to elaborate the graphical representations from Figure 6,7 and 8.

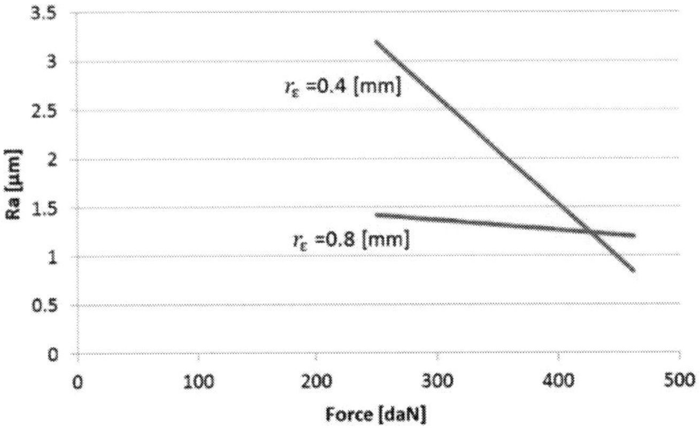

Fig. 6. Influence exerted by force on *Ra* surface roughness parameter (f=0.24 [daN]).

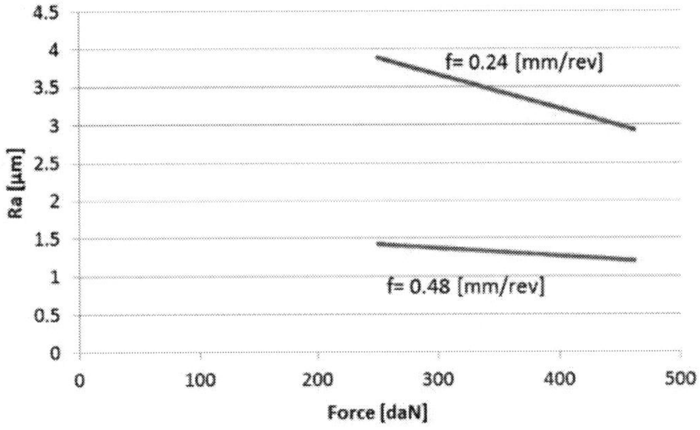

Fig. 7. Influence exerted by force on *Ra* surface roughness parameter ($r_\varepsilon = 0.8\ [mm]$).

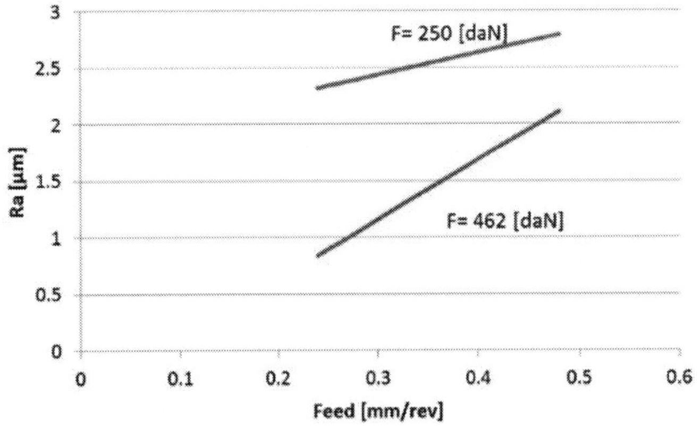

Fig. 8. Influence exerted by feed on *Ra* surface roughness parameter ($r_\varepsilon = 0.4\ [mm]$).

The analysis of these graphical representations and of the empirical model constituted by the relation (1) shows that the most important factor able to influence the value of the *Ra* surface roughness parameter is the machining feed *f*, since the value of the exponent attached to this factor has the maximum value compared to the values of the exponents attached to the other factors. One could notice that as expected, when the machining feed *f* increase, the value of the *Ra* roughness parameter corresponding to the plastically deformed surface increase also, since the value of the exponent attached to the size *f* is positive. The increase of the value of tool corner radius determines a growth of the rough parameter value in that the narrowing between the peaks of asperities is reduced.

4 Conclusions

We could say that character of a produced surface is highlighted by the precision of the process with fulfill of the dimensions defined in the technical documentation. Each machining process leaves particular pattern of the machined outer. This pattern is known as surface roughness. When two manufactured surfaces are in contact, surface finish of the mating elements is a significant factor in the performance and behavior of the matching elements. The roughness of a machined part occurs as a result of the following factors: cutting conditions parameters (feed, cutting speed, cutting depth), tool geometry, vibrations generated in the technological system which consists of machine tool, cutting tool and workpiece. Thus, it can be assumed that surface roughness determines the quality of contact interaction between two solid bodies which plays an important role in many physical phenomena and engineering applications.

Within this paper, we presented some research results concerning the contact behavior between a cylindrical pressing tool made of high speed steel and test samples made of aluminum on 8 experiments, where we settled as a combination of three main process input factors (machining feed, tool corner radius, force), each of them taking, one by one, two values. We measured the initial roughness of the front surface of test samples, then we pressed the pressing tool on this surface with two values of force and after that we measured again the roughness of the new surfaces created by plastic deformation of the asperities (Figure 5). We found out the function that expresses the influence exerted by the input factors on the *Ra* roughness parameter valid to the plastically deformed zone of the test sample. In this way, the power function determined as empirical mathematical model revealed us that all input factors are able to significantly affect the values of the Ra roughness parameter of the deformed zone.

In order to illustrate the above mentioned influence, we realized three graphical representations, where f, r_ε, and F were used, one by one, as independent variables in the power type function valid for the *Ra* surface roughness parameter. We have observed that the main influence on roughness is exerted by the machining feed. If the feed increases, the roughness increases too (Figure 8). This variation of the *Ra* surface roughness parameter is also valid for the tool corner radius. Another thing that we have observed refers to value of the force. Thus, as much as the pressing force grows, the values corresponding to the *Ra* surface roughness parameter reduces, also (Figures 6,7).

In the future, we could study the influence of the same or other input factors on other surface roughness parameters, like *Rz*, *Rq*, *Ry* are or even on factors able to better characterize the contact surfaces.

References

1. G. Carbone, F. Bottiglione, Journal of the mechanics and physics of solids **56** (8), 2555-2572 (2008)
2. D. Prodan, *Research concerning the circular concentrated contact of roughness surfaces in elastoplastic field* (2008)
3. S. Brown, C. H. Scholz, Journal of geophysical research atmospheres (1985)
4. G. Crețu, G. Varvara and V. Merticaru, *Experimental research methods-applications (in Romanian)*, (Chișinău, 2000)

MATEC Web of Conferences **112**, 01018 (2017) DOI: 10.1051/matecconf/201711201018
IManE&E 2017

The effect of drilling parameters for surface roughness in drilling of AA7075 alloy

Nafiz Yaşar[1,*], *Mehmet* Boy[1], and *Mustafa* Günay[2]

[1] Karabük University, TOBB Vocational School, Bilimler Meslek Yüksekokulu, Demir Çelik Kampüsü, 78050, Karabük, Turkey
[2] Karabük University, Department of Mechanical Engineering, Balıklarkayası, Mevkii, 78050, Karabük, Turkey

Abstract. AA7075 aluminum alloy has been very popular significantly interest in the production of structural components in automotive and aviation applications due to its high strength, low density, good plasticity and better machinability comparable to many metals. Particularly, final products must have uniformly high quality to ensure essential safety standards in the aircraft industry. The optimization of hole quality which can variable according to tool geometry and drilling parameters is important in spite of high machinability rate of AA7075 alloy. In this study, the effects of drilling parameters on average surface roughness (Ra) has been investigated in drilling of AA7075 with tungsten carbide drills. Machining experiments were performed with three different drill point angles and three different levels of cutting parameters (feed rate, cutting speed). The effects of drilling parameters on thrust force has been determined with ANOVA in %95 confidence level. Feed rate was determined as the most important factor on Ra according to ANOVA results. Moreover, it was shown that increasing feed rate leads to increase of Ra while increasing drill point angle leads to decrease of Ra. The optimum surface roughness was obtained with point angle of 130°, cutting speed of 40 m/min and feed rate of 0.1 mm/rev, thereby the validity of optimization was confirmed with Taguchi method.

1 Introduction

Aluminum and its many alloys are commonly used in aerospace, computer and medical industries due to their lower weight, height strength and electrical, thermal conductivity [1]. Machining of aluminum became an important research manner because of its behavior during its machining. Tool wear, burr formation and hole surface quality are the most common problems in drilling of aluminum and its alloys. These materials adhere to cutting tool due to their smearing properties. This situation cause to change in cutting tool geometry. Thus, thrust force and surface roughness increase, too. The surface of machined part is one of the most important criteria in terms of quality. The cutting parameters used

* Corresponding author: nafizyasar@karabuk.edu.tr

during machining of materials are important factors affecting surface accuracy [2]. Surface quality can be obtained by right choice of machining parameters.

Cutting force is known as related with tool-chip contact length in machining process. According to that adhering chip on tool change tool nose geometry. Thus, that cause to a change of force on cutting tool. Although aluminum is a ductile material, machining of these alloys will affect tool life more than expected. In recent years, many research are done in order to measure the force on tool in adequate accuracy. The forces on tool is an important part of machining. Measuring of cutting forces is useful in optimizing of tool design and necessary for scientific analysis of cutting [3].

Drilling of hole is one of the most common machining processes applied to aluminum alloys [4]. Drilling is the most problematic machining process in removal of chips from cutting region. The leading reason is that chip forming is in closed area and not seen [5]. Rivero et al. described an experimental research study on machinability in the dry drilling of aluminum alloys (AA7075-T6) and on the potential of the new design of tools and coatings. Dry drilling tests were performed using uncoated drills and two different coatings produced by means of an arc evaporation PVD process. Experiments consisted of machining with a 10 mm diameter three-edged drill to produce 25 mm deep holes. Tool wear evolution and burr size were analyzed, as well as the impact of the process parameters on torque, power, feed force and tool temperature. The authors stated that drilling of aluminum alloys effectively in dry condition is possible [6]. Kılıçkap [7] investigated the influence of cutting parameters, such as cutting speed and feed rate, and point angle on burr height and surface roughness produced when drilling Al-7075. All tests were run without coolant at three different levels of cutting speed, feed rate and point angle. The orthogonal array, signal-to-noise ratio, and analysis of variance (ANOVA) were employed to investigate the optimal drilling parameters of AA7075. The optimization results showed that the combination of low cutting speed, low feed rate, and high point angle is necessary to minimize burr height. The best results of the surface roughness were obtained at lower cutting speed and feed rates while at higher point angle. The predicted values and measured values are quite close to each other. The author indicated that the developed models can be effectively used to predict the burr height and the surface roughness on drilling of AA7075. Kannan et al. [8] observed that AA7075 alloy has less burring when comparing with pure aluminum in their study. Hanyu et al. [9] studied that shaft tools with new coatings were applied to dry and semi-dry machining of highly adherent aluminum alloys whose efficient means of dry machining have not yet been established for. In comparison to the case of conventional diamond coatings with rough surfaces, more than four times longer durability of drills was realized in perfectly dry condition. In the case of minimum quantity of lubricant condition, the lifetime was found again to be more than four times longer and successfully achieved the largest improvement of diamond-coated drills ever reported in semi-dry condition. Kurt et al. investigated the effects of cutting parameters on drilling temperature, cutting force and surface roughness in drilling of AA2024 with diamond like carbon coated drills. In their study, it was determined that the most significant parameters on surface quality is feed rate and drill diameter. Also, author observed that the change in feed rate and drill diameter affect average surface roughness considerably in high cutting speed [10].

It is noticed that there is not sufficient study about 7000 series of aluminum alloys whose usage is increasingly spreading. In this study, the effects of drilling parameters (point angle, cutting speed and feed rate) on thrust force were examined in dry drilling of AA7075 aluminum alloy.

DOI: 10.1051/matecconf/201711201018

2 Material and method

In this study, drilling processes of AA7075 alloy in three different levels of point angle (120°, 130° and 140°), cutting speed (40, 80 and 120 m/min) and feed rate (0.05, 0.1 and 0.15 mm/rev) were performed in CNC vertical drilling machine (Johnford VMC–550). Karcan branded tungsten carbide drills were used as cutting tool in experiments. Drills had diameters of 5 mm, helical angle of 30° and geometry of 2 cutting edge. Each drilling process was repeated three times and evaluations were done by taking arithmetic average of drilling outputs. The drilling parameters (factors) were given in Table 1.

Table 1. Drilling parameters and levels

Factors	Levels
Point angle, P (°)	120-130-140
Cutting speed, V (m/min)	60-100-140
Feed rate, f (mm/rev)	0.05-0.1-0.15

Proper number of revolutions according to cutting speed by using speed head were obtained due to the fact that CNC milling machine in experiments has 6000 rev/min as a maximum number of revolution. Surface roughness (Ra) values in drilling experiments were measured by Mahr Perthometer M300 device. The experimental setup for drilling tests was given in Figure 1.

Fig. 1. The experimental setup for drilling tests.

Three different levels of feed rate, cutting speed and point angle were chosen as drilling parameters in experimental design according to full factorial. The levels of these parameters were determined by considering cutting tool company and other studies about AA7075 in literature. In the light of experimental results, the effects of parameters on Ra were determined by analysis of variance (ANOVA) performed with 95% confidence level. Finally, drilling parameters for average surface roughness were optimized based on Taguchi method according to L_{27} orthogonal array. Therefore, the-smaller-the-better approach were applied due to desire of average surface roughness (Ra). The drilling parameters giving optimum surface roughness were determined in optimization study according to S/N ratio. In the Eq. (1), n and y shows the number of experiments and performance characteristic, respectively [11].

$$S/N = -10\log\left(\frac{1}{n}\sum_{i=1}^{n} y_i^2\right) \quad (1)$$

3 Results and Discussion

3.1 Analysis of surface roughness (Ra)

Captions should be typed in 9-point Times. They should be centred above the tables and The variations of Ra values according to drilling parameters in experimental results were given in Figure 2. Ra values increased with increasing feed rate when examining of graph and this result show similar tendency with literature. The reason of that can be explained by increasing chip volume with increasing feed rate. Because, increasing feed rate causes high thrust force, low shear angle and thick chip formation [12,13].

It is known that low temperatures cause BUE formation in machining of aluminum and its alloys due to having multiple phases. Figure 2 show that increasing of Ra values with increasing point angle can be referred to that tool-chip contact area decrease with increasing point angle. Decreasing tool-chip contact area leads to decrease in temperature because it causes less friction. Increase of BUE can be explained with decreasing temperature [13]. Increasing Ra value can be referred that BUE formation is unstable form.

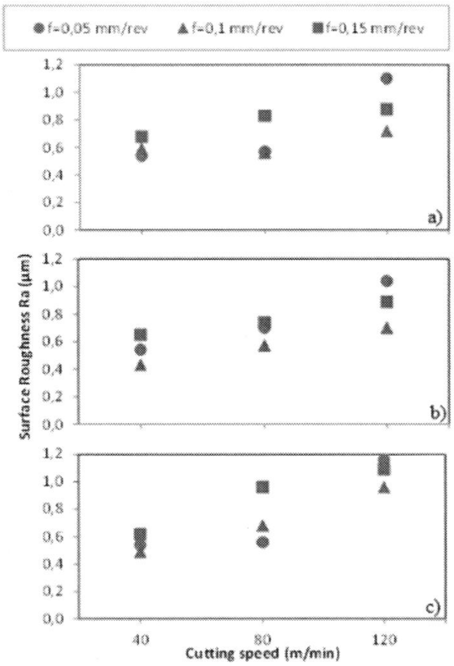

Fig. 2. The variation of Ra according to drilling parameters a)120°, b) 130°, c) 140° .

As literature examining, it is known that Ra values decrease with increasing cutting speed. However, previous studies were performed in lower cutting speeds. It is seen from Figure 2 that Ra values increase with increasing cutting speed. Increase in cutting speed is known as important parameter on increasing chatter effecting Ra values negatively in drilling operations [14]. Amplitude increase on torque is seen with increasing cutting speed by examining torque graphs obtained during drilling operations (Figure 3). It can be said that these amplitudes generated in this torque values cause increase in Ra values.

The smallest Ra value was obtained as 0.43 µm (feed rate of 0.1 mm/rev, cutting speed of 40 m/min and point angle of 130°) while highest value was obtained as 1.16 µm (feed rate of 0.05 mm/rev, cutting speed of 120 m/min and point angle of 140°).

Additionally, analysis of variance (ANOVA) results performed on the 95% confidence level in order to determine the effects of factors on the Ra are given in Table 2. Here, the probability (P) values indicating the importance level of each factor, degree of freedom (DF), the sum of squares (SS), F-ratio and the percent contribution ratio (PCR) is shown. P value should be lower than 0.05 in 95% confidence level in order to determine that any factor acting on the force is effective on it. Table 2 indicates that the most effective factor on the surface roughness is cutting speed with 62.03 % PCR. The feed rate and point angle have effect on thrust force with 13.68 % and 3.40 % PCR, respectively. According to that, statistical analysis results confirm the experimental results.

 DOI: 10.1051/matecconf/201711201018

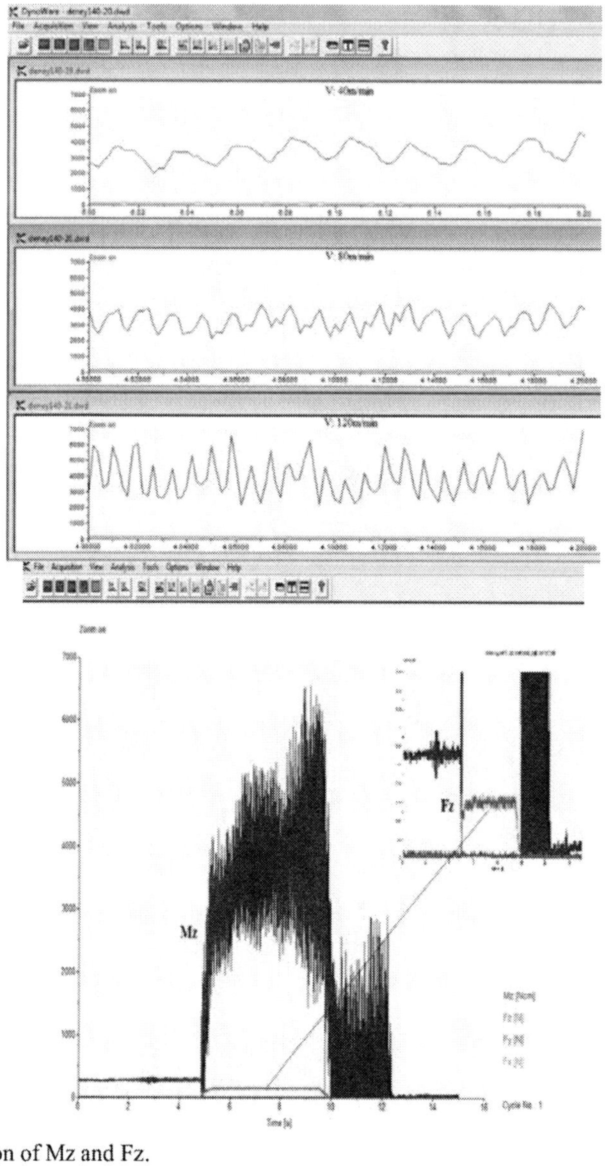

Fig. 3. The variation of Mz and Fz.

Table 2. ANOVA result for Ra

Factor	DF	SS	MS	F-ratio	P	%PCR
Φ (°)	2	0.03823	0.019115	4.75	0.044	3.40
V (m/min)	2	0.69543	0.347715	86.37	0.000	62.03
f (mm/rev)	2	0.15334	0.076670	19.04	0.001	13.68
φ*V	4	0.04624	0.011559	2.87	0.095	4.12
φ*f	4	0.02086	0.005215	1.30	0.349	1.87
V* f	4	0.13486	0.033715	8.37	0.006	12.03
Error	8	0.03221	0.004026			2.87
Total	26	1.12116				100
R^2 =98.4%						

The surface roughness values were obtained from drilling experiments and their S/N ratios calculated according to Eq. (1) are given in Table 3.

Table 3. Surface roughness and their S/N ratios.

Exp. no	Point angle	Cutting speed	Feed rate	Ra	S/N ratio	Exp. no	Point angle	Cutting speed	Feed rate	Ra	S/N ratio
1	1	1	1	0.540	5.3521	15	2	2	3	0.740	2.6153
2	1	1	2	0.590	4.5829	16	2	3	1	1.040	-0.3406
3	1	1	3	0.680	3.3498	17	2	3	2	0.700	3.0980
4	1	2	1	0.570	4.8825	18	2	3	3	0.890	1.0122
5	1	2	2	0.560	5.0362	19	3	1	1	0.540	5.3521
6	1	2	3	0.830	1.6184	20	3	1	2	0.490	6.1960
7	1	3	1	1.100	-0.8278	21	3	1	3	0.620	4.1521
8	1	3	2	0.720	2.8533	22	3	2	1	0.560	5.0362
9	1	3	3	0.880	1.1103	23	3	2	2	0.680	3.3498
10	2	1	1	0.540	5.3521	24	3	2	3	0.960	0.3545
11	2	1	2	0.430	7.3306	25	3	3	1	1.160	-1.2891
12	2	1	3	0.650	3.7417	26	3	3	2	0.960	0.3545
13	2	2	1	0.700	3.0980	27	3	3	3	1.090	-0.7485
14	2	2	2	0.570	4.8825						

For optimizing of process factor, S/N ratios are considered in Taguchi optimization method. The highest S/N ratio shows the optimum level of parameters according to the-smaller-the-better approach. Moreover, distribution of calculated average S/N ratios according to parameters for Ra is given in Table 4. It is seen that the most important parameter on Ra is cutting speed (V) when examining maximum and minimum points of S/N ratios in Table 4. Consequently, the optimum levels of parameters were determined as P2, V1 and f2 according to calculated S/N ratios as can be seen from Table 4.

Table 4. Response table of S/N ratios for surface roughness

Factors	S/N ratios			
	Level 1	Level 2	Level 3	Max.-Min.
Point angle	3.1064	3.4211*	2.5287	0.8925
Cutting speed	5.0455*	3.4304	0.5803	4.4653
Feed rate	2.9573	4.1871*	2.2753	2.2753
*Optimum levels of parameters				

Average of drilling experiment result performed at optimal levels of factors are evaluated by Eq. (2) to forecast the mean for the improvement conditions. Eq. (1) which is the expression of calculated surface roughness (Ra) is derived from Eq. (3).

$$\eta_G = \bar{\eta}_G + (\bar{P}_O - \bar{\eta}_G) + (\bar{V}_O - \bar{\eta}_G) + (\bar{f}_O - \bar{\eta}_G) \tag{2}$$

$$Ra_{cal} = 10^{-\eta G/20} \tag{3}$$

Where, η_G is the S/N ratio calculated at optimal level of factors (dB), $\overline{\eta}_G$ is the mean S/N ratio of all parameters (dB), $\overline{P}_O$, $\overline{V}_O$ and $\overline{f}_O$ are the mean S/N ratio when point angle, cutting speed and feed rate are at optimum levels, respectively and Ra_{cal} is the calculated value. Consequently, η_G and Ra_{cal} for optimum drilling parameters were determined as 6.9800 dB and 0.44, respectively. Lastly, confirmation experiments is done by using the optimum drilling parameters after the determination of these factors for Ra and thus reliability of the optimization has been confirmed. However, the Ra (0.43 μm) and S/N ratio were obtained for surface roughness is very close to the predicted or calculated values. Thus, there is no need to carry out confirmation experiments if the drilling parameters found with the optimization procedure are included in the drilling conditions within the Taguchi experimental design.

4 Conclusion

In this study, the effects of drilling parameters (point angle, cutting speed and feed rate) on Ra were examined in dry drilling of AA7075 aluminum alloy based on Taguchi method. It was determined that Ra values increased with increasing feed rate in drilling of AA7075 aluminum alloys. The most effective factor on the Ra is cutting speed with 62.03 % PCR according to ANOVA results. The smallest Ra value was obtained as 0.43 μm (feed rate of 0.1 mm/rev, cutting speed of 40 m/min and point angle of 130°) while highest value was obtained as 1.16 μm (feed rate of 0.05 mm/rev, cutting speed of 120 m/min and point angle of 140°). The results obtained have shown one more time that Taguchi method is a design and analysis of experiment technique that can be applied in machining investigates successfully.

References

1. ASM Handbook, (1999), Machining, Printed in USA, Volume: 16, 761-804.
2. B.J. Griffiths, *Surface Integrity and Functional Performance*, (Manufacturing Surface Technology, Penton Press, London, 2001)
3. A. Çakır, *Investigation of Cutting Parameters During Drilling Operations of Al 7075 and Al 6013 Aluminum Materials (in Turkish),* (Gazi University Graduate School of Natural and Applied Sciences, Ankara, Turkey, 2009)
4. J. Kim, D.A. Dornfeldd, Trans ASME, **124**, 192–198, (2002)
5. M. Akkurt, *Cutting Methods and Machine Tools (in Turkish),* (Birsen Publishing House, Istambul, 1998)
6. A. Rivero, G. Aramendi, S. Herranz, L.N. Lopez de Lacella, Int J Adv Manuf Technol **28** (1), 1-11, (2006)
7. E. Kılıçkap, Int J Adv Manuf Technol **49**, 911–923, (2010)
8. I. Suresh Kannan, A. Ghosh, Procedia Materials Science **5**, 2615-2621, (2014)
9. H. Hanyu, S. Kamiya, Y. Murakami, M. Saka, Surface and Coatings Technology, **173-174**, 992-995, (2003)
10. M. Kurt, Y. Kaynak, B. Bakır, U. Köklü, G. Atakök, L. Kutlu, *Proceedings of the International Advanced Technologies Symposium - IATS'09*, (Karabuk University, 2009)
11. M. Günay, J Fac Eng Archit Gaz, **28** (3), 437-444 (2013)

12. C. Çakır, *Principles of Modern Machining (in Turkish)*, (Nobel Publishing, Ankara, 2006)
13. E.M. Trent, *Metal Cutting*, (Butterworths Pres, London, 1989)
14. M. Kurt, Y. Kaynak, E. Bagci, Int J Adv Manuf Technol **37**, 1051-1060 (2008)
15. M. Boy, I. Ciftci, M. Gunay, F. Ozhan, Materials and Technology **49**(5), 765–772, (2015)

Grinding of the gears with high depth processing

Sergiu Mazuru, *Serghei* Scaticailov, and *Ion* Stingaci*

Technical University of Moldova, 9, Studenţilor str. Chişinău, Republic of Moldova

Abstract. The goal research consists in the analysis and synthesis of technological schemes, processes, tools and equipment for processing of conical gears and precession gears with medium module. A new processing method which allows to intensify the process of processing as a result is proposed. It ensures a required accuracy of processing. The theoretical researches are based on the theory gear drives, theory of involute profiles generation and precession gears, generating machine, tool design theory machine building technology. Methods of mathematical analysis, analytic geometry, mathematical model and 3D computer model are used. The experimental researches were conducted in laboratories of departments of machine manufacturing technology and fundamentals of machines design in the Technical University of Moldova, by using machines and technological equipment for machining and testing. Experimental data processing were accomplished using approximation methods and mathematical statistics.

1 Background of profile deep grinding of gear elements

By its kinematic, the deep grinding differs substantially from traditional grinding methods and in particular from pendular processing scheme. Larger material layer thickness - up to 10 mm or more, removable in a single pass, in conjunction with low workpiece feed - up to 20 - 40 mm/min, leads to the fact that the contact length of the abrasive tool with a workpiece and the lengths of the zones characterized by thermodynamic effects are one - two orders higher than those corresponding to traditional forms of pendular grinding.

The characteristics of deep grinding gears made of hardened steel deep grinding, due to the exact shape and large size of the contact area of the grinding wheel with the surface to be processed and its elevated sensitivity to thermal injury involves the unconditional fulfilment of the following requirements:

a) during grinding process, the cavity between the adjacent tines machined in a single pass must provide a high efficiency of material removal, this being of 3 – 4 times higher than, for example, creep feed grinding of the blade roots made of nickel alloys;

b) the maximum allowable temperature in the cutting zone with respect to time of one deep grinding cavity should be about 1.5 – 2 times less than that corresponding to the creep feed grinding, applied to heat-resistant nickel and titanium alloys;

* Corresponding author: ion.stingaci@gmail.com

c) the need to maintain the original profile of the grinding wheel while processing a maximum number of gears and multiple cavities in the period between revisions, as used in gear grinding machines, with just cyclical corrections.

Thus, effective implementation of profiled deep grinding process is possible, it ensuring a high productivity with minimum concomitant heating temperature (700°C). Conditions of forming the cavity found between two teeth in one or more passes with the grinding wheel are determined mainly by its cross-sectional area and the size of the contact surfaces between tool and workpiece [1-6].

The problem of finishing conical gears was a preoccupation for the researchers in the field of manufacturing technologies. Thus, Lopatin and Plotnikova investigated the possibilities of positioning the tools and the workpiece in the case of using shaving process and grinding process to finish helical bevel gears teeth flanks [7]. They noticed that the common devices used for obtaining teeth or devices similar to those could be also applied in the finishing processes.

Komatsubara et al. studied the concave conical gear used in marine transmissions. They took into consideration a theoretical method of generating the concave conical gears, aiming to establish the essential dimensions of gear and tool [8]. The experiments proved a good correspondence of the principal normal radii of test gears with the values theoretically determined. Another remark was that the Hertzian contact ellipse generated in the case of concave conical gears is larger than that corresponding to the conventional conical gear.

The problem of tooth contact analysis in the case of conical involute gears was investigated by Jingliang et al. [9]. They developed a mathematical model corresponding to the conical involute gears, considering the theory of gearing and the generating mechanism.

Various aspects concerning the conical gears and methods for their machining were investigated by Radzevich [10]. The use of form grinding wheel is discussed, but especially in the case of grinding cylindrical gears.

The objective of this paper was to highlight some aspects corresponding to the deep grinding process applied in the case of conical gears, by considering the conditions specific to these gears.

When optimizing the parameters of the of creep feed grinding mode range Vk, the speed of longitudinal movement of the workpiece or grinding wheel and the cutting depth are considered; the main difficulty is to solve the problems of single -pass profiling, and imposing cuts at multipass profiling gear wheels. Figure 1 shows the schema of gears deep profiled grinding.

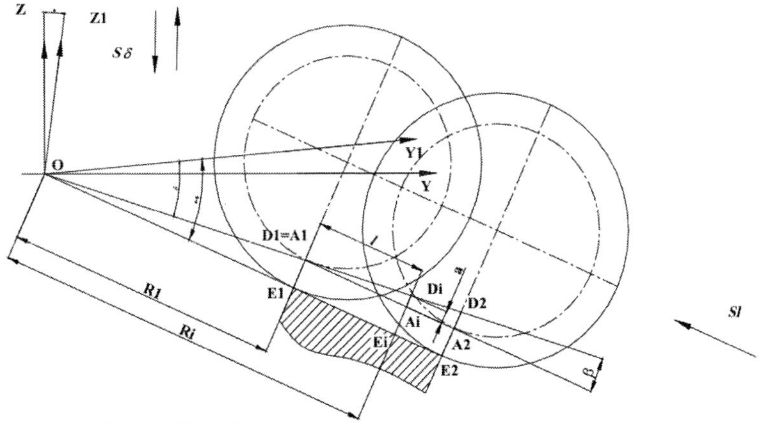

Fig. 1. The deep profile grinding of gear.

The most important kinematic parameter of the process is the area of cavity section F_c, which is affected by the tool movement in a single pass. F_c depends on the size of the workpiece material cutting resistance, which is proportional to the cross sectional area of the shear layer, and accordingly, to the intensity of the thermodynamic tension in the grinding zone. In addition, the workpiece material determines the performance of the profiling process speed during the material removal.

$$W = F_c \cdot S_d \quad (1)$$

Analysis of calculation of cavity section area formed in a single pass shown that F_c section increases with the number of the teeth z. The extent of this effect is negligible: with unlimited including of the teeth, the decrease of the F_c is only of 5.7%. This small difference makes possible the analysis of the process of wheels formation, in a first approximation, without taking into account the number of teeth.

In forming cavity between the adjacent teeth in a single pass, the sectional area of F_c increases from 14.95 to 229.4 mm^2, this meaning of about 15.3 times.

Accordingly, the cutting resistance increases, this being necessary to be considered when evaluating the possibility of profiling gears with large modules in a single pass.

The creep feed grinding for the first pass of the grinding wheel is advisable to appoint a cutting depth, which will provide the final shapes of the teeth, or taking into account the retained allowance for finish grinding after thermal or chemical-thermal treatment. However, this option is largely due to the technical capabilities of the machine and tools, as well as to the restrictions derived from the thermodynamic conditions of the current process without forming defects.

As an example, one can see the way of changing the sectional area F_c depending on the teeth height h and cutting depth of the first pass t (Figure 2).

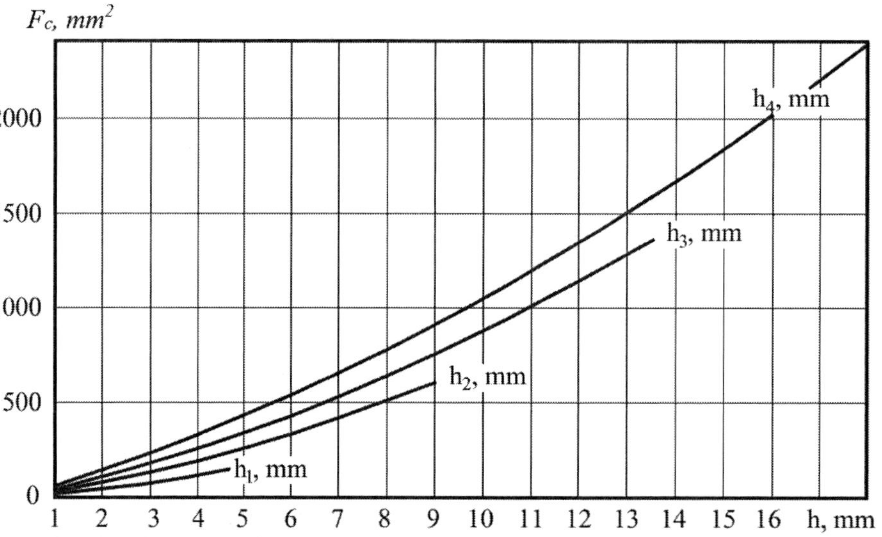

Fig. 2. Changing the cross sectional area slice F_c, depending on the height of the teeth h and the depth grinding.

In its turn, the required power shaping as thermodynamic processing quality conditions depends not only on the component due to the removed slice section F_c, but on the friction work, able to contribute to the overall grinding energy consumption and which could be quite noticeable.

Frictional work is proportional to the contact area of the grinding wheel with F_k treated surface, which for the deep grinding of gear is determined by the scheme shown in Figure 1.

The total area of the contact zone of the friction develops along the bottom cavity width and with a length of the contact *s*, equal to the contact arc AB, and two inclined side surfaces, the dimensions of which depend on the ratio of the width *B* and the length AC. Figure 3 shows the change in cross-sectional area F_c cavity and F_k contact area of the grinding wheel diameter $D = 85$ mm. Let us note that the estimated value of F_k significantly

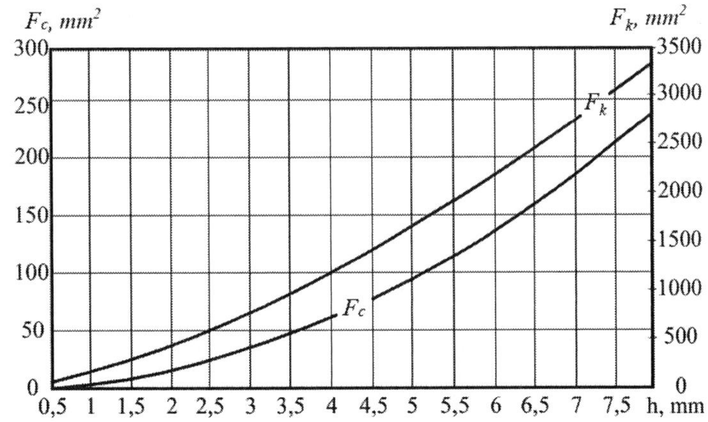

Fig. 3. Comparison of the calculated values of the area F_c the slice section and F_k contact with the grinding gear wheels

exceeds the pre-sectional area F_c: *h1* at F_k value is greater than F_c of about 55.4 times, while the excess is stored h4 great, but is reduced up to 13.8 times.

The most closely linked feed rate and cross-sectional area (the simple correlation coefficient is of 0.838), as well as the grinding speed, weakly correlated with each other kinematic and other operating parameters of the process, indicating their statistical independence.

Thus, we can talk about the decisive influence exerted by work profile creep feed grinding gear wheels on the parameter assignment of movement speeds. It is appropriate to stress that, in turn, is a function of the magnitude of the cutting depth, of the module and number of the teeth of the gear being processed.

The stability of the thermodynamic effects on the machined gears, due to dispersion conditions with vigorous rubbing, has, as shown by the results of the study, a decisive influence on the quality and accuracy of the profiled of the teeth.

From the analysis of the experimental data, one notices that it is important in relation to pragmatic conclusion: to stabilize the thermodynamic effect of processing of the gear, it is necessary to stabilize any controlled input parameter shaping process.

Such a parameter can be the value of the removed section of cavity, which is regulated by the depth of the passage. In conclusion, it should be noted that on the basis of theoretical analysis and preliminary experimental studies, there was established a fundamental possibility of implementing the process of forming gear wheels profile by creep feed grinding method.

The methodology for optimizing profiling in one or more passes of the grinding wheel depends on the characteristics of the toothed wheel and its handling process conditions.

It is also important to emphasize that the fundamental difference between the new process of forming gear wheels known as an analogy of profile creep feed grinding is that it provides a high performance material removal at a lower heating of surface layer to be treated to prevent the appearance of sanding defects.

Kinematic forming of cavity implies an increased thermodynamic process intensity with an intense heating. The need to obtain an accurate profile and the minimum accumulated pitch error of machined wheels with circular grinding wheel dressing imposes stringent requirements for its durability and evenness of wear resistance for the service period.

Under these requirements, it is advisable to use high hardness tool that becomes an additional factor able to intensify the heating process. In this connection, the abrasive tool for profile creep feed grinding gear wheels have special requirements that are to be the optimum combination of necessary values of its strength and durability, together with very high cutting properties, able to minimize the stresses generated by the thermodynamic process.

When profile creep feed grinding, to avoid overheating of the workpiece material, usually the soft abrasive tools with abundant cooling of the cutting area are used.

To ensure high accuracy to the formed profile, tool is subject to continuous compulsory revision with special diamond profiled rollers. In this case, when using the cyclic changes, it is possible to use the tools with higher hardness and which should be reformed quickly and easily to the desired cavity profile of machined gears.

The problem in grinding is mainly the work of friction of the abrasive grains on the surface to be processed, the weight of which is up to 70 - 80% of the total energy costs for material removal. This is due to the fact that most of the abrasive grains, which are at the working surface of the grinding wheel do not participate in the cutting process, and only deform (crushes) the processed material, being a significant source of heat during grinding. The abrasive tool is the determining element in the process of grinding system.

The correct assignment of the grinding wheel characteristics largely depend on character of the grinding process, the level of performance, the intensity of heating, tool life, machining accuracy and quality.

2 Assigning characteristics highly porous wheel for gear grinding

In the grinding interaction of each abrasive grain and a grinding wheel as a whole with the cutting material takes place under extreme conditions. Under the high pressures and temperatures in the cutting zone the adhesive grasp the abrasive with the work material, the mutual diffusion of chemical elements, increase work bundle friction and abrasive grains of the grinding wheel with the workpiece surface and, as a consequence - the softening and breakdown of the abrasive grains and the formation of the surface layer parts with unfavorable conditions for the operation of the properties. Figure 1 shows a block diagram of the relationship of the constituent elements of the grinding wheel - grinding grain and bundle it with properties that need to be considered when appointing the tool characteristics. Under these conditions, complex thermodynamic efficiency of the grinding wheel is determined by the ability of abrasive grains on the working surface of the tool to resist said contact physicochemical phenomena and, above all, their softening and thermal degradation, as well as the ability to bundle hold the abrasive grains to the loss of cutting properties. In accordance with the block diagram, you can define the characteristics of grinding wheels for gear grinding optimal conditions. These include: a) optimal assignment modification of alumina as an abrasive material under high temperatures and high resistance to brittle fracture; b) grit selection range to provide the desired grinding performance with a minimum roughness of the processed surface; c) to find the optimal ratio of the structure (porosity) of the grinding wheel and its hardness to ensure a minimum intensity of heat and maximum durability of the tool; d) selection of a ceramic bond, which firmly holds the abrasive grains under conditions of intense heat generation at a minimum of friction with the workpiece surface.

3 Conclusion

The theoretic al analysis showed that the main technical difficulty of implementing profiling gearwheels grinding process is extremely unfavorable combination of large sections of the removed material Fc with more than an order of magnitude of larger contact area Fk wheel with machined surface profile. Dimensions Fc and Fk determine respectively the two main characteristic components of the cutting operation: removal of the workpiece material and the work of the friction. Both kinematic components responsible for the thermodynamic strength of the cutting process and the stability of its course are very dependent on the gear unit element m and the number of passes.

The optimal choice of the characteristics of the grinding wheel provides the related appointment of grain, hardness and structure of the instrument based on unconditional restrictions to ensure the set parameters on material removal rate, the roughness of the treated surface of the part and the thermal stress in the cutting zone. Optimization of other characteristics of a highly porous material on the tool abrasive grain, hardness and operational properties ligament takes into account, in addition to these restrictions, the need for increased durability and efficiency of the grinding process.

If you have been on the working surface of the grinding wheel, which can collect the chips, increased cutting power tool, as it increases the time may come when his "brining", that is, the loss of cutting properties by filling the pores chips. In addition, the selection of the optimal ratio of "grain-hardness-structure" by optimizing the composition of the abrasive mass and ceramic binder can provide grinding conditions for grinding wheel wear process takes place in a self-sharpening mode.

References

1. I. Bostan, S. Mazuru, S. Scaticailov, TEHNOMUS Journal **20** (2013)
2. S. Mazuru, A. Vlase and S. Scaticailov, *Processing technologies on machine tools for machining gears (in Romanian)* (Tehnica-U.T.M Publishing House, Chişinău, 2014))
3. I. Bostan, S. Mazuru, M. Vaculenco, S. Scaticailov, Proceedings of the IX International Congress "Machines,Technologies, Materials 2012", (Varna, Bulgaria, 2012)
4. V. Botnari and S. Mazuru, Applied Mechanics and Materials **657** (2014)
5. S. Mazuru S. and M. Casian, Advanced Materials Research **1036** (2014).
6. M. Casian and S. Mazuru, Advanced Materials Research **1036** (2014)
7. B.A. Lopatin, S.V. Plotnikova, Procedia Engineering, **150** (2016)
8. H. Komatsubara, K. Mitome, T. Ohmachi, JSME International Journal **45** 2 (2002)
9. HE Jingliang, WU Xutang, CUI Yahui, NIE Gang, Chinese Journal Of Mechanical Engineering **19** 1 (2006)
10. S.P. Radzevich, *Gear cutting tools. Fundamentals of design and computation* (Taylor and Francis Group, Broken Sound Parkway, U.S.A., 2010)

Aspects concerning the circularity and cylindricity deviation of a work piece processed by turning

Mihai Boca[1,*], *Gheorghe* Nagit[1], and *Florentin* Cioată[2]

[1] Machine Manufacturing Department, Machine Manufacturing and Industrial Management Faculty, Gheorghe Asachi Technical University, D. Mangeron Blvd, No. 59A, Iaşi, România
[2] Machine Tools and Production Systems Department, Machine Manufacturing and Industrial Management Faculty, Gheorghe Asachi Technical University, D. Mangeron Blvd, No. 59A, Iaşi, România

Abstract. All the cutting processes impose certain specifications for the work pieces obtained. In such conditions, the parameters as tolerance, roughness and geometrical deviations must be well known in order to determine the obtained accuracy of the processed work pieces. During the cutting process, due to some influences from the MTDW (Machine Tool, Device Work piece), a work piece can result with processing errors. In order to compensate this errors or, if it is possible to eliminate them, some measurement of the work piece dimensions are taken. In this way, the present study propose an evaluation of such aspects in order to establish, based on the work piece shape on radial and longitudinal direction, the imposed measures to compensate the processing errors. For high accuracy of the obtained data, an automated device will be used. The results will highlight the influenced areas of the processed work pieces and the possible factors which conduct to generating errors.

1 Introduction

In most cutting processes the obtained accuracy of the manufactured parts depends on influences exerted, during the processing, by several factors as cutting parameters, cutting force, work piece stiffness, machine-tool behaviour and others. Not in all cases the accuracy expected is similar with the predicted one. From the cutting processes, the turning process is one of most used processes in industry and implies a high level of work piece accuracy. From this reason, it is mandatory to find the entire error sources that can modify the expectations concerning the obtained accuracy of the manufactured work pieces.

During the cutting processing, due to some unexpected influences produced in the inside of the MTDW system (Machine, Tool, Device, Work piece), the accuracy might suffer some changes. Concerning the work piece shape obtained into a turning process, excluding the roughness parameter, the errors which occur can be associated to the radial deviations measured in transversal sections of the work piece and on the work piece profile

* Corresponding author: mihaitzaboca@yahoo.com

generator along the work piece length. Depending on the differences between the expected profile and real one, the differences should be included into the tolerances prescripts. In such conditions, a well knowing of the real work piece profile and of the mentioned influences, both in radial and in longitudinal direction, it can improve the obtained accuracy of the manufactured products. From this reason the present paper wants to present several aspects concerning the measurement made on some work pieces processed into a turning process, in order to highlight the aspects mentioned earlier and to furnish some measures to avoid work piece errors.

2 General considerations

Usually, the accuracy of a work piece is expressed by his dimensional values, which contain the specification of tolerances. Beside the large dimensions of the work piece, the cylindricity deviation, both with the profile cilindricity represents the main aspects which furnish an idea about the wok piece accuracy. The cylindricity deviation appears in case of a cylindrical work piece processed by turning and it is caused by the cutting force variation during processing. This type of error consists by a non-uniform shape profile founded on entire work piece length. For measuring such kind of error it is necessary to measure the work piece shape in several transversal sections along the work piece length [1]. This method is known as the measuring method on radial direction, made in several sections perpendiculars to the work piece axis. Depending on the technique of the applied method, two ways of measurement can be distinguished: the measurement which is made with contact in points of the measured section and measurement which is made with continuous contact in measured section. In this way it is obtained the cylindricity deviation that includes the circularity error in each considered section perpendicular to the work piece axis and considers the values measured in all the sections disposed on the entire length of the work piece. Depending on the used device, the measurements can be taken by contact with work piece as in our case, or without contact when is used a laser device.

For the present study it was chosen the method of measuring with contact in points. Also, the work piece profiles can highlight the effective cylindrical shape of these, supposing that a number of measured sections can give an image of the profile shape of the work piece in axial sections. This fact is assumed by the hypothesis of concentricity of the real circular sections from each measured section with the longitudinal axis of the processed surface.

The studies made until now on this topic presents as errors sources the following factors: the work piece deflection, initial geometry of the raw part [1], vibration, thermal deformations, machine tool and tool wear [3] and others. In the last year a sustained activity was developed on errors compensation [2]. The first step to make an error compensation is to know the values of the errors appearance and the factors which conduct to their presence [1]. Beside these, the technical literature presents a series of studies related to MTDW stiffness, which can justify the cilindricity and cilindricity deviations. Between them, the study made by Olvera [2] presents some aspects concerning the evaluation of the MTDW components stiffness along his kinematic chain. He noted that a percentage of 21% of the errors are generated by the uncompensated thermal error, tool flexibility, material instability and axis acceleration, all these in case of classical cutting machines and CNC machines. Another study [3], using a FEM analysis, presents the deflection of the tool, due to the cutting force influences, as main parameter which can cause cylindricity deviations. Also, the Renjith author [3], purposes a Taguchi optimization plan for minimize the cutting tool deflection. Another approach concerning the cylindricity and cilindricity deviations presents an effective measurement of a rod, using a two points measurement, in order to reduce the errors caused by the spindle rotary movement [4].

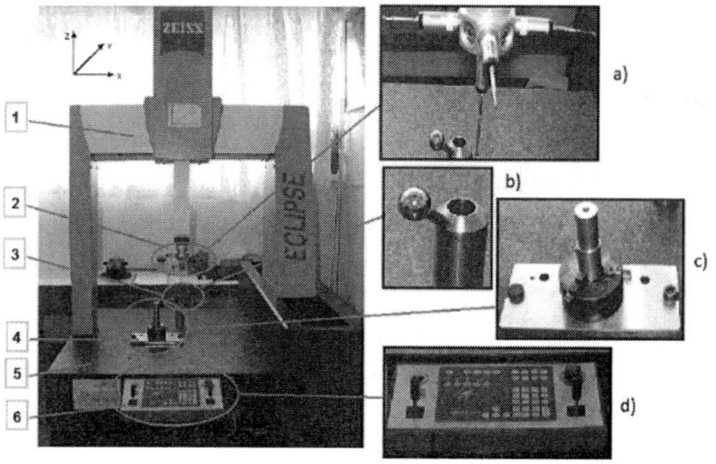

Fig. 1. Zeiss Eclipse Measurement Coordinate Machine: 1 – pillar machine, 2 – measurement head, 3 – calibration sphere, 4 – work piece clamping system; 5 – granite table; 6 – command panel.

3 Experimental setup

For the experimental study it was used a Zeiss Eclipse equipment presented in Figure 1,which allows to make measurements on three axes. The measurement head (Figure 2.a.) has possibility to use multiple spheres with diameters between 3 and 5 mm. The head used in this study has a diameter of 3.97 mm, dimension specified by the technical book of machine in case of the measured work pieces and which depends by the surface roughness obtained after processing. The clamping system chosen (Figure 2.) was represented by a classical tailstock in order to maintain the initial processing conditions. The measurement accuracy of the machine is 0.001µm, reported to the displacement error of the machine axis table.

In order to determine the cilindricity error, both with cylindricity deviation, an experimental methodology was used. The experimental measurement procedure uses the coordinate measurement system of the measurement equipment and consists into a couple of measurement made on several radial sections of the work piece, together with measurement of the work piece profile along his length. The experimental method applied for measuring the diameter of the cylindrical surface it is an absolute method which supposed to evaluate directly the sectors chosen for reconstructed the real shape of the processed work piece. Due to the necessity of knowing the dimensional variations of the processed surface, reported to the measured work piece length, the work piece diameter it

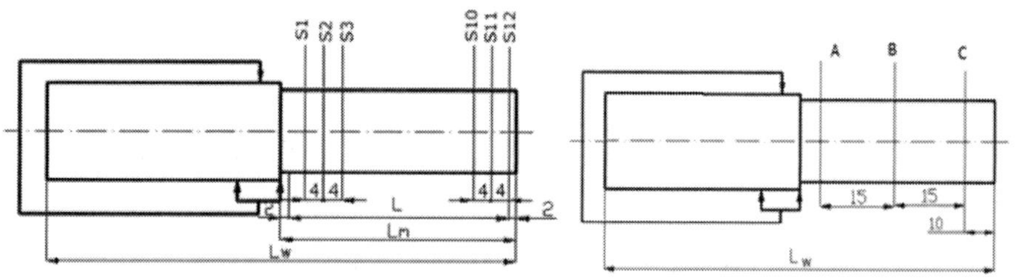

a) b)

Fig. 2. The measuring strategy for determine the work piece diameter, along the work piece axis and transverse to the work piece.

Table 1. The measuring strategy for determining the workpiece diameter

Measured section	Measured nominal diameter values d [mm]				Cilindricity error [mm]			
	Work piece number				Work piece code			
	1	2	3	4	1	2	3	4
0 (SI)	15.743	25.0298	14.6019	25.4704	0.0803	0.0989	0.0728	0.0633
4 (SII)	15.2739	25.0424	14.9434	25.4684	0.0835	0.0939	0.063	0.0731
8 (SIII)	15.3422	25.0512	14.9702	25.4732	0.0824	0.0715	0.0948	0.0743
12 (SIV)	15.2379	25.0487	14.9472	25.4711	0.069	0.0673	0.0731	0.0746
16 (SV)	15.2249	25.0561	14.9763	25.4769	0.0676	0.0622	0.065	0.0819
20 (SVI)	15.2191	25.0569	14.9735	25.4765	0.0712	0.0554	0.0625	0.078
24 (SVII)	15.2934	25.059	14.9092	25.4785	0.0762	0.0483	0.0702	0.0767
28 (SVIII)	15.2961	25.0353	14.8923	25.4736	0.0744	0.0479	0.0817	0.0796
32 (SIX)	15.2974	25.0278	14.93	25.4776	0.0704	0.0409	0.0499	0.0717
36 (SX)	15.3912	25.0398	14.9461	25.4756	0.0779	0.0367	0.0611	0.0763
40 (SXI)	15.3898	25.0324	14.9462	25.4765	0.077	0.0459	0.0736	0.0729
44 (SXII)	15.4054	25.0612	14.8314	25.4785	0.0672	0.0371	0.0777	0.0653
48 (SXIII)			14.7999	25.4788			0.0659	0.0762
52 (SXIV)			14.813	25.4774			0.089	0.0702
56 (SXV)			14.9022	25.4798			0.0893	0.00710

will be measure in several sections perpendicular on the work piece axis. More precisely, it was measured eight work piece, but only four of them were chosen as presenting the significant results to be discussed, other work pieces do not show relevant information. In this way each section was placed on a distance established at 4 mm one from another, in order to offer complete measurement informations concerning the work piece shape. The first and the last measurement sections were placed at 2 mm distance between the measuring limits of the end / beginning of the work piece (Figure 2b). In the same figure it was presented the strategy to determine the cilindricity error, measuring the work piece in four positions. In this way it was measured 12 sectors on the short wok pieces and 15 sectors on long wok pieces. This fact is considered as reason for eliminating the possible errors which could be inserted into the measurement results, possible dimensional or shape errors, which regularly are founded to the reference end lengths. With the symbol L_w was noted the work piece length, with L_m the theoretical, respectively with L the effective measurement length. The orientation, positioning and clamping of the measured work piece it was succeeded using the clamping system of the measurement device used. The measured diameters which characterize the cylindrical surface of the work piece processed allow determining directly the variation of the dimensional characteristic of the processed surface.

A steel alloy 41Cr4 it was used for the processing work pieces. The sample dimensions have two values: 18x100mm and 25x125 mm. It is considered that the used material is homogeneous and isotropic, to assure a uniform deformation of the work piece material during processing. As cutting parameters was chosen depth of cut a_p=1.5 mm, speed s=35.83 m/s and feed f= 0.023 mm/rot. The presented values was obtained considering a research experiment planning furnish by a Anova simulation, made in Datafit software, taking into account two values for each parameter of interest (a_p= 1.5 and 2.5 mm, work piece diameter D= 18 and 30 mm and work piece length L_W=100 and 125 mm, with L= 48 mm measured length, respectively L= 60 mm, and L_m=48 mm and 61 mm).

In the Table 1 are presented the values corresponding to the nominal measured diameter of the considered work pieces and the cilindricity errors for each measured sector (given by mean of measured values). The graphical results exposed in the Figure 3 presents four of the eight work pieces considered. It can be observed the cylindrical shape corresponding both to the short work pieces and for the biggest ones. In the Figure 4 are presented graphical representations of the cilindricity error corresponding to the work piece no. 3,

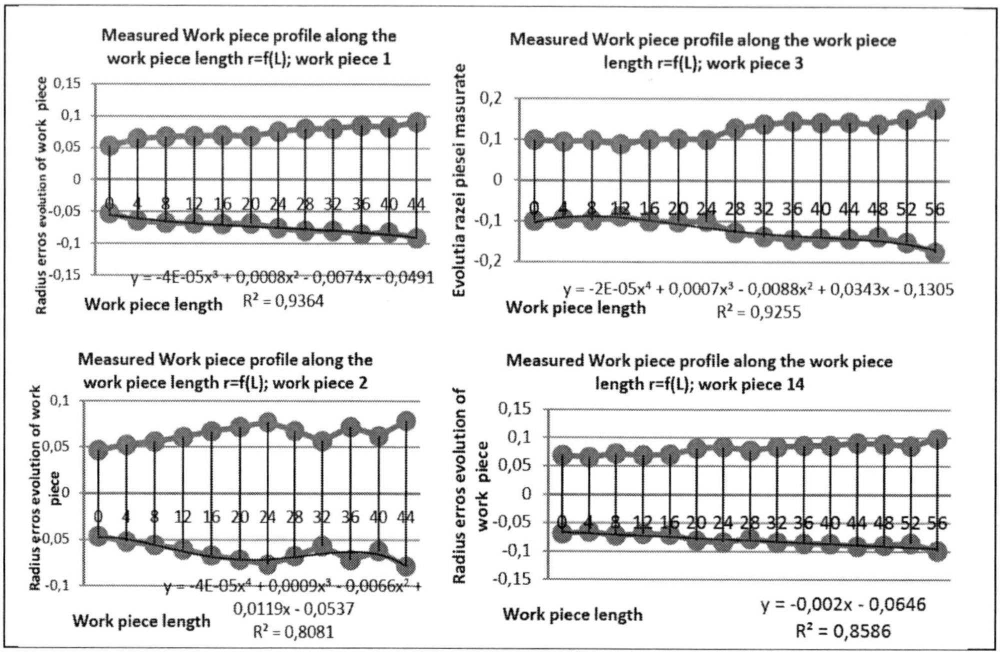

Fig. 3. Graphical representation of the cylindrical error of the most significant processed work pieces

assumed as the work piece with greater deviations justified by a weak stiffness of the work piece and greater cutting force in comparison with these. In this case, depending by the tolerances specified, the manufactured work pieces can be accepted or not for the following stages of research.

4 Conclusions

After processing all the work pieces concerning the cylindricity deviation, some observations can be highlighted:

- all parts dimensions of the work pieces processed by turning with feed, on sections perpendicular to the axis diameter, varies incrementally over tolerances prescript, starting at its clamped end into the tailstock to the free end of the processed part; this evolution of the diameter varies along the length of the processed surface, evolution materialized by a double concavity (customized by the funnel-shaped) of the work piece, normal evolution characteristic to the work pieces clamped at one end;
- it is observed a pronounced variation of the processed section diameter, this situation is highlighted in case of increased cutting length, both with maintaining a work piece constant diameter; it is noticed that of 25% cutting length increasing on the same depth of cut (from 100 mm to 125 mm), the diameter of the processed work piece deviation increases almost with 10% (from 0.038 mm to 0.045 mm);
- the processing of the work pieces with a diameter of 18 mm, on the same cutting length reveal that the diameter is changing between higher limits than in case of the work pieces with diameters of 28 mm; also, at an increasing depth of cut (from 1.5 mm to 3 mm) the difference between diameters becomes double (from 0.059 mm to 0.120 mm);
- once with 25% of length variation (from 100 mm to 125 mm), for a constant depth of cut, it is noted that the variation of the processed diameter has a much quicker evolution than for the work piece with the diameter of 28 mm; in this case, the limit becomes larger than

66% (from 0.059 mm to 0.090 mm) to a 1.5 mm depth of cut and the same, with 60% (from 0.120 mm to 0.200 mm) to 3 mm depth of cut; due to the high stiffness of the cross section

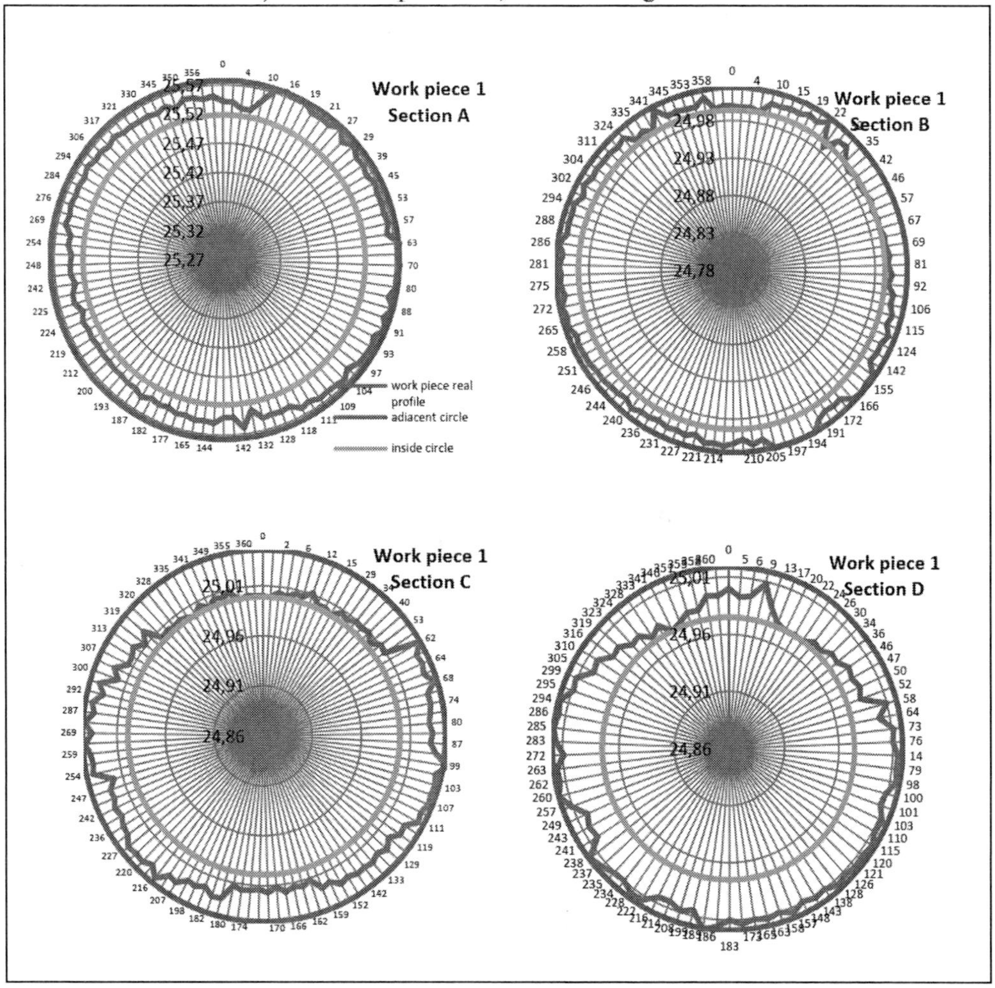

Fig. 4. Graphical representations of the diameter sections of the work piece no. 3.

placed near of the work piece chuck, the error value is lower compared with the section located at the free end of the work piece which present a significant changing of the profile.

References

1. *** *Kalsi Seals Handbook, D4 - Shaft deflection, runout, vibration, and axial motion*, (Kalsi Engineering Inc. 2014)
2. D. Olvera, L. N. López de Lacalle, F. I. Compeán, Fz-Valdivielso, A. Lamikiz, F. J. Campa, Int J Adv Manuf Technol **60** (2012)
3. V. B. Renjith, M. Baby, K. R. Jayadevan, Int. J. of Scientific and Research Publications **3**, 7, (2013)
4. Q. Chen, X. Tao, J. Lu, X. Wang, Hindawi Publishing Corporation Adv. in Materials Science and Engineering, **1** (2016)

DOI: 10.1051/matecconf/201711201021

Experimental research on the durability of the cutting tools for cutting-off steel profiles

Alexandru Cristea, *Cătălin* Fetecău* and *Nicoleta Violeta* Stanciu

Center of Excellence Polymer Processing, *Dunărea de Jos* University of Galaţi, 111 Domneasca Street, Galaţi, România

Abstract. The production lines used for manufacturing U-shaped profiles are very complex and they must have high productivity. One of the most important stages of the fabrication process is the cutting-off. This paper presents the experimental research and analysis of the durability of the cutting tools used for cutting-off U-shaped metal steel profiles. The results of this work can be used to predict the durability of the cutting tools.

1 Introduction

Cold-formed steel profiles are found in a wide range of products, with a wide variety of shapes and sizes, in almost all aspects of contemporary life [1,2,3,4]. The industrial use of these profiles started only in the second half of past century. In recent years the use of these profiles also expanded to achieve proper strength structures of buildings.

Light profile metal systems with complex sections are also commonly used for the production of curtain walls. Advanced technologies for corrosion protection have led to increased competitiveness of cold-formed steel profiles in domains where, until recently, their use was restricted due to high corrosion risk [5,6].

The development of manufacturing technologies by cold plastic deformation allowed to increase the thickness from 3 mm to 25 mm [1]. Increasingly used to manufacture profiles, the steels with flow limits of which these profiles are obtained have flow limits within the (250 ÷ 550) MPa range [1, 2].

The shapes of cross sections for cold-formed profiles are usually more complex than those of hot rolled and welded sections, such as sections I or U. Usually, cold-formed sections have symmetric, or asymmetric shapes, usually with additional stiffeners at the end areas, or even intermediate stiffeners [6]. As shown in Figure 1, cold forming can be used to manufacture different simple or complex sections.

One of the very important operations in the production flow of cold-formed profiles is their cutting-off, which is affected by the tool wear. Generally, the wear should progress gradually and should be easily monitored. However, during the cutting-off process, the tool wear is hard to predict. Thus, in this paper, since productivity is important, the durability of a tool for cutting-off U-shaped metal profiles is investigates.

* Corresponding author: catalin.fetecau@ugal.ro

Fig. 1. Cold-formed sections [2].

2 Experimental methodology

The tool for cutting-off U-shaped metal profiles from zinc plating metal sheets investigated in this paper (Figure 2) is part of the length cutting subassembly of a production line for U-shaped metal profiles. This line manufactures UD-30 metal profiles used for dry-wall, at 80 mm width and 0.5 or 0.6 mm thickness, at standard lengths of 3 or 4 meters.

The tool is made of hardened HS 6-5-2 and tempered to 63~65 HRC, and the metal profile is made of S250 zinc plating metal sheet.

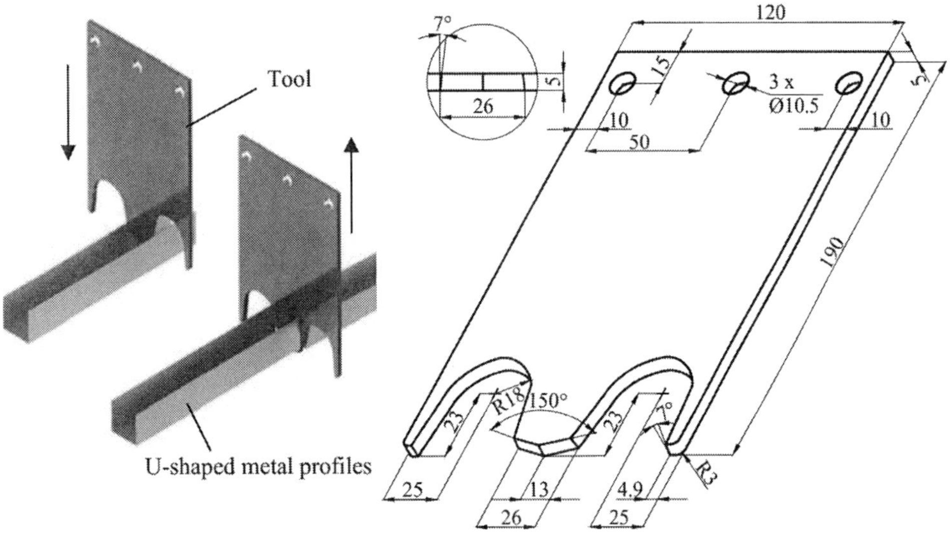

Fig. 2. Tool for cutting-off U-shaped metal profiles.

Figure 3 shows photos of profiles cut with the tool. The profiles in Figure 3.a have been cut-off when the tool was new, and those shown in Figure 3.b have visible defects after cutting-off with a used tool, as a result of pronounced wear of the tool.

The production line (Figure 4) has a maximum work speed of 60 m/min, the rolls train being driven by a SIEMENS DC engine with an output of 12.5 kW.

The actual speed of the profile is translated using a digital encoder that sends impulses to a specialized computer which synchronizes the tool with the speed of the line.

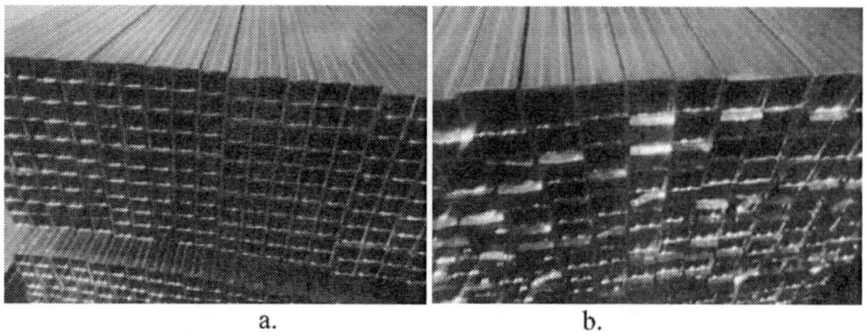

Fig. 3. U-shaped profiles cut: a. at the beginning of the cutting process; b. with tool with a high degree of wear.

Fig. 4. The production line.

The tool holder assembly is supported by two linear guides. The translational motion is performed by a ball bolt-and-nut assembly, which it's driven by a 3.2 kW servomotor.

The tool is placed in this assembly, fixed by a ram port connected to a hydraulic cylinder making a back-and-forth race in about 0.7 seconds, which results in the actual cutting-off of the U-shaped profile.

Figure 5 shows schematically the stages of the experimental research.

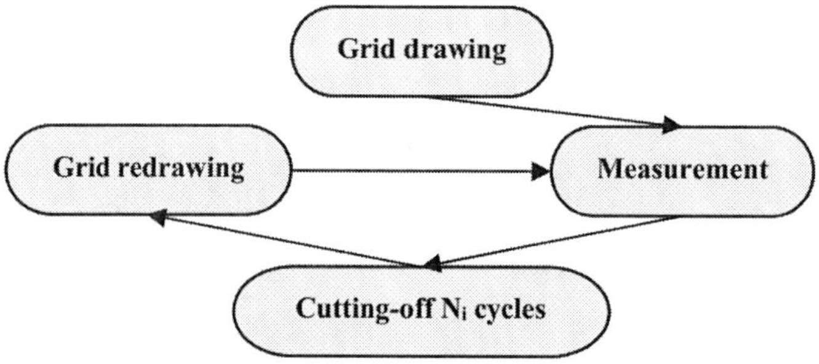

Fig. 5. Stages of experimental research.

To measure the tool wear, initially, a grid size of 3×3 mm^2 was designed on the rake surface and several points were marked, where maximal wear was expected (Figure 6). The measurements were performed 3 times for each marked zones and were used 4 tools.

The cutting-off process is done at high speed which leads to forming crater wear along the cutting edge of the tool, h_α, or flank wear on the rake surface.

To determine the wear progress, after each N_i (i=1 ÷ 4) cutting cycle, the production line was stopped and the crater depth along the cutting edge of the tool was measured using a SINOWON WMM 3D video microscope (Sinowon, China) as shown in Figure 7.

DOI: 10.1051/matecconf/201711201021

The values of the wear in the four most important zones of the tool, resulted from the measurements, are presented in Table 1. It should be noted that the values presented in Table 1 represents the average for three measurements and the corresponding standard deviation.

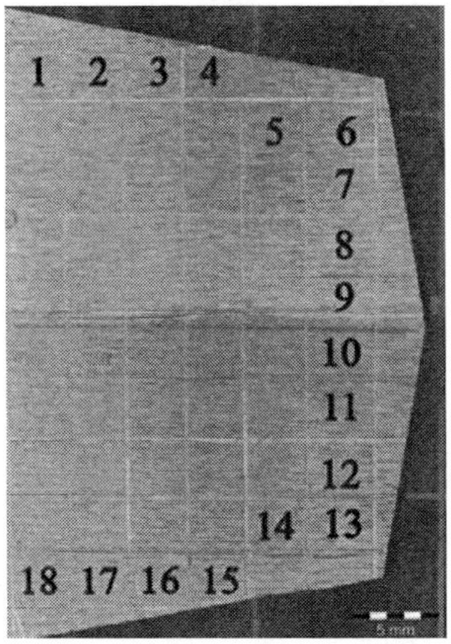

Fig. 6. Initial picture of the grid on the tool.

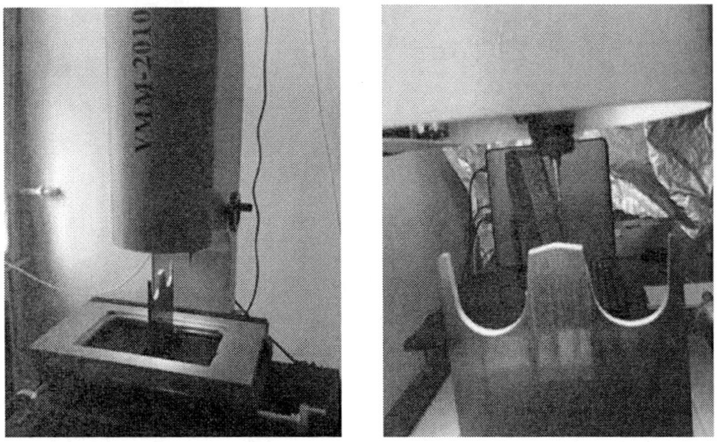

Fig. 7. Wear measurement with a SENOWON 3D microscope.

Table 1. Wear in four zones of the tool

Number of cycles		**Wear [mm]**			
		Zone 1 (grid 6)	**Zone 2 (grid 8)**	**Zone 3 (grid 11)**	**Zone 4 (grid 13)**
N1	188000	0.104±0.001	0.012±0.001	0.052±0.001	0.194±0.002
N2	298000	0.165±0.001	0.012±0.001	0.07±0.001	0.275±0.002
N3	415000	0.196±0.001	0.02±0.002	0.093±0.002	0.294±0.001
N4	726000	0.212±0.002	0.036±0.001	0.106±0.001	0.301±0.001

 DOI: 10.1051/matecconf/201711201021

Figure 8 shows the main effects plot for the tool wear and the corresponding measurement zone. As shown in Figure 8 the wear increases with increasing number of cycles; on the other hand, the wear decreases in the zones approaching the tip of the tool, which is used to cut the bottom of the metal profile. As shown in Figure 8 and Table 2, the greatest impact on the tool wear is the zone 4 wear, used for cutting-off the side of the U-shaped profiles.

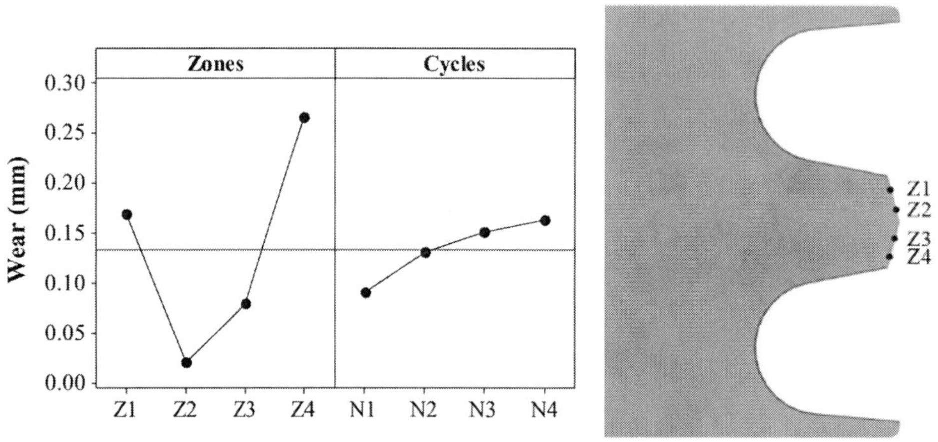

Fig. 8. Main effect plot for the wear of the tool.

Table 2. Wear of the tool in the 4 zones (average values for four tools)

Level	Zone	Cycles
1	0.16925	0.0905
2	0.02025	0.13075
3	0.08025	0.15075
4	0.266	0.16375
Delta	0.24575	0.07325
Rank	1	2

Table 3. Analysis of Variance for Means

Source	DF	Seq SS	Adj SS	MS	F	P
Zone	3	**0.137979**	**0.137979**	**0.045993**	**105.69**	**0**
Cycles	3	**0.012274**	**0.012274**	**0.004091**	**9.4**	**0.004**
Residual	9	**0.003917**	**0.003917**	**0.000435**		
Total	15	**0.154169**				

3 Conclusions

This paper shows the wear that occurs using a tool for cutting-off U-shaped metal profiles at different cycles.

On the active surfaces of the tool it develops abrasion wear because of the existent oxides on the surfaces of metal profiles and chipping wear as a result of the "hits" made at the cutting-off.

The wear is more pronounced at the sides of the tool tip, where the cutting-off of the edges of the profile is done, and less obvious in the zones where it cuts the bottom of the profile.

The analysis of the wear process showed that, after about $(700{\sim}800)\cdot 10^3$ work, fatal wear occurs, which requires, after stopping the production line, that the tool is sharpened or replaced. The cutting knives can be sharpened on the rake surface until its active height is enough for the cutting-off process.

The research results provides the necessary information to reduce the number of waste and to ensure a good productivity of the manufacturing process by the knowledge of the tool durability for cutting-off U-shaped metal profiles.

Many thanks to Associate prof. eng. Teodor Virgil for his contribution in this research by measuring the wear of the tools analysed.

References

1. P.J. Trebilcock, *Building design using cold formed steel sections: an architect's guide* (Berkshire, Steel Construction Institute Publication, 1994);
2. V. P. Astakhov, *Metal cutting mechanics* (CRC Press, 1998);
3. N. E. Dowling, *Mechanical behavior of materials: Engineering Methods for Deformation, Fracture, and Fatigue* (Prentice Hall, Inc, USA, 1999).
4. E. Alfredo, *The complete part design handbook* (Hanser publishers, Munich, 2006);
5. E. Rabinowich, *Friction and wear materials*; (Wiley, 2013);
6. R. Kobayashi, S. Xu, K. Shimada, M. Mizutani, T. Kuriyagawa, *Defining the effects of cutting parameters on burr formation and minimization in ultra-precision grooving of amorphous alloy* (Elsevier, Precision Engineering, 2017).

Computational and experimental analysis of machine tool vibrations in micro-milling

Dimitrios Sagris[1], *Constantine* David[1], *Evlampia* Stergianni[2], *Christos* Tsiafis[2], and *Ioannis* Tsiafis[2,*]

[1]Laboratory of Manufacturing Technology and Production Systems, Mechanical Engineering Department, Institute of Technology & Education of Central Macedonia, Serres, Greece.
[2]Laboratory of Machine Tools and Manufacturing Engineering, Mechanical Engineering Department, Aristotle University of Thessaloniki, Greece.

Abstract. In machining processes, it is necessary to study the vibration phenomena which take place in cutting zone between the cutting tool and the workpiece in order to ascertain the factors that affect them. The subject of this paper is to identify the influence of the cutting parameters and the dynamic characteristics of the machine tool spindle and the cutting tool on the vibrations by means of FEM analysis. A FEM modal analysis of the machine tool spindle in combination with cutting force measurements are used to build up a transient FEM model in time domain in order to study the vibrations induced by the cutting process. The results of the transient FEM model will serve for further analysis to find out the optimum cutting parameters in micro-milling.

1 Introduction

During the cutting process, it is crucial to know the vibration behaviour of the "Machine Tool- Cutting Tool- Workpiece" system in order to ascertain the factors that affect it [1]. The subject of this paper is to develop a numerical model using the Finite Element Method (FEM) to simulate the vibration behaviour of the system. To succeed that, it is necessary to model the system in such a way taking into account the particular stiffness of the system components which take an effect in the machine tool dynamics. The current work aims at the development of a FEM simulation model to

- assess the natural frequencies and eigenmodes of the "Machine Tool Spindle - Cutting Tool" system
- calculate the cutting tool deformation, results from the cutting forces during the micro-milling process

A FEM modal analysis in combination with experimentally measured cutting forces results are taken in order to develop a transient FEM model in time domain to describe the vibration phenomena in milling. The results of the transient FEM model will serve for further analysis to find out numerically the optimum cutting parameters in micro-milling.

* Corresponding author: tsiafis@eng.auth.gr

2 Experimental determination of cutting forces

The experiments to measure the cutting forces during the milling process were conducted in the CNC machining center DECKEL MAHO MH600C. The measuring equipment includes a dynamometer (KISTLER 9257B), a charge amplifier (KISTLER 5233A) and an Analog/Digital converter (NATIONAL INTRUMENTS NI PCI-MIO-16E) connected to a computer. The dynamometer is capable to measure the cutting forces in three axes simultaneously. In Figure 1 are presented the cutting force's components. The analog force signals are amplified and consequently digitized and recorded.

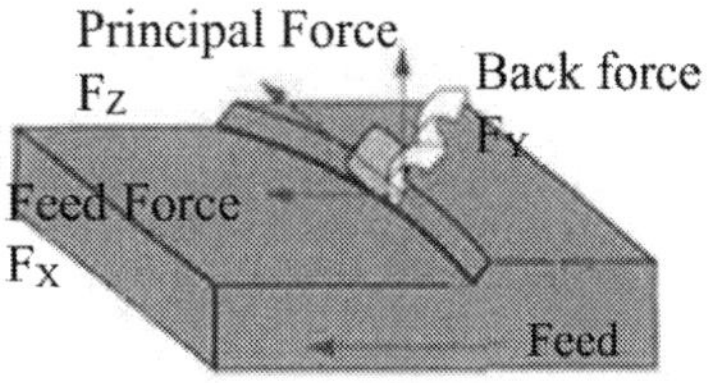

Fig. 1. Cutting force's components during milling process.

The cutting parameters of the conducted experiments are shown in Table 1. The material of the specimens used to machining was aluminium alloy 7075 T651.

Table 1. Cutting parameters

Symbol	Parameter (Units)		Value
S	Spindle Speed	(rpm)	5500
Vc	Cutting Speed	(m/min)	34.5
f	Feed Speed	(mm/min)	220
f_x	Feed per cut	(mm/cut)	0.01
a_e	Radial depth of cut	(mm)	0.5
a_{xz}	Axial depth of cut	(mm)	0.2/0.4/0.6
d	Cutter diameter	(mm)	2

The data of the force components are recorded within a time frame of 0.04 sec with a sampling resolution of 4e-05 sec. In Figure 2 a typical force component measurement is shown. It can be assumed that during the cutting process the main cutting force component is developed in ZZ direction and the feed force component in XX direction. On the other hand, the force component in YY direction is negligible.

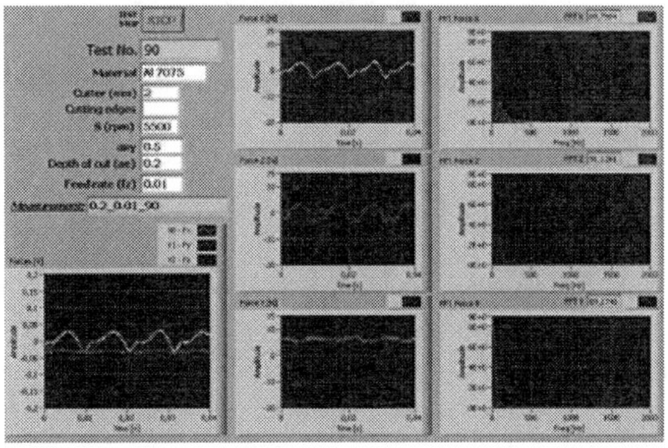

Fig. 2. Cutting force measurement during milling process.

3 Experimental determination of the machine spindle dynamic characteristics

In order to determine the natural frequencies and the damping factor of the "Machine Tool-Cutting Tool" system, there were conducted specific measurements of the vibration response of the cutting tool mounted on the machine spindle with the aid of the cutting tool holder [2]. The vibration behavior of the assembly in cutting position is mainly affected by the machine tool spindle and the cutting tool. Based on excitation-response measurements of the machine spindle - cutting tool assembly at various critical positions it was verified that a subsystem is substantially isolated when mounted in rolling elements (bearings). For this reason, it is necessary to determine just only the dynamic characteristics of the Spindle-Tool holder- Cutting tool compound.

The experimental modal analysis was conducted in the following manner [3]. An impact excitation produced by a hammer hit is applied on the cutting tool edge and the response (acceleration) is measured. The analog signals (impact force and response) are undergoing further processing and the vibration characteristics (natural frequencies and damping factor) are determined. The analog signals taken with the aid of a piezoelectric accelerometer, after being amplified, were digitized and recorded for further processing and analysis. The results of the experiments are shown in Table 2 and 3 for the XX and ZZ direction accordingly. After the experimental measurement of the cutting forces and the determination of the dynamic characteristics of the vibratory system, it is possible to setup the 3D simulation model of the machine spindle in order to develop the transient model in time domain, because of the time-dependent self-excited vibrations induced by the cutting process.

Table 2. Dynamic characteristics on XX direction

Frequency w_q (Hz)	Damping ratio j_q	Stiffness k_q (N/m)	Mass m_q(kg)
4817	**0.098**	**172866**	**0.0076**

Table 3. Dynamic characteristics on ZZ direction

Frequency w_q (Hz)	Damping ratio j_q	Stiffness k_q (N/m)	Mass m_q(kg)
4650	0.064	227941	0.01

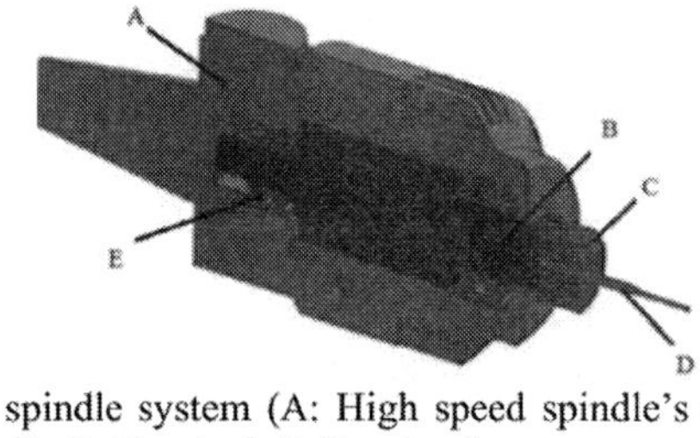

Fig. 3. 3D modelling of the spindle system (A: High speed spindle's shell, B: Machine tool high speed spindle, C: Tool holder, D: Cutting tool, E: Bearings).

4 High speed spindle modelling

In order to develop the 3D simulation model of the "Machine - Cutting Tool" assembly (Figure 3), it is necessary to design, as accurately as possible, the geometry of each part of the system. To facilitate the modeling work, specific experiments were conducted. It should be mentioned that while there were not available the mechanical drawings and therefore many of the geometry details of the vibratory system were missing, special attention was paid in proper distribution of the mass of the various components, in order to develop a consistent equivalent model. However, the drawing of the high-speed spindle assembly allowed us to extract some useful information about the main dimensions of the parts geometry.

The cutting tool in the 3D graphical representation in Fig. 3 has the same geometry as the cutting tool used in the experiments, in order to be possible to compare the simulation results with the experimental ones. In particular, an end-mill tool was used with a diameter of 3 mm and a shaft diameter of 4 mm. In order to simplify the FEM model, special consideration was paid on the fact that the bearings due to their stiffness dynamically separate the shaft vibrations from the rest of the machine tool. Therefore, it was not necessary to include the whole machine tool spindle as well as the bearings into the FEM model. As a result of that in the 3D-model, which was built for the FEM analysis, the shaft surfaces where the bearings are located were also defined as pivot joints and are indicated with light grey color in Fig. 4.

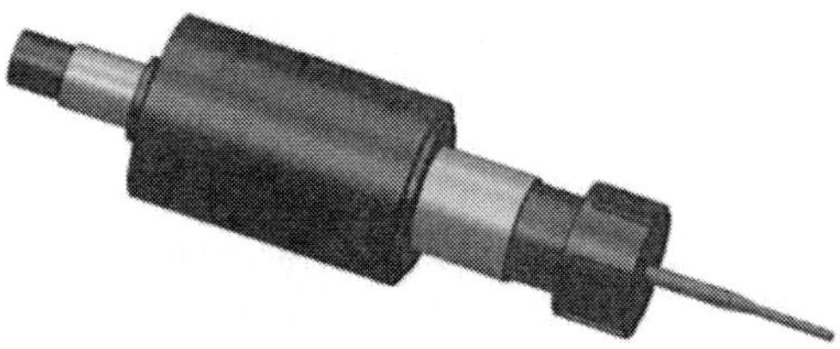

Fig. 4. Equivalent 3D-model used for the FEM Analysis.

5 FEM modal analysis

The Finite Element Method (FEM) simulation of the machine tool spindle was performed with the aid of the ANSYS WorkBench software. It concerns the modal analysis to find out the natural frequencies and eigenmodes of the spindle. This numerical simulation was based on the machine tool spindle equivalent model presented in Figure 4, as well as on the appropriate model characteristics, such as material properties, boundary conditions and contact conditions between the parts of the spindle assembly. In order to simulate the dynamic behavior of the spindle model, it is necessary to define the boundary conditions of the problem. In particular, the bearing positions described in section 4, were considered as cylindrical joints in order to allow the free rotation of the shaft and its axial movement, preventing in this way the development of virtual forces and stresses. In Fig. 5 it is shown the meshing into finite elements of the 3D model on the left and the boundary conditions in FEM modal analysis in order to proceed with the FEM Analysis on the right.

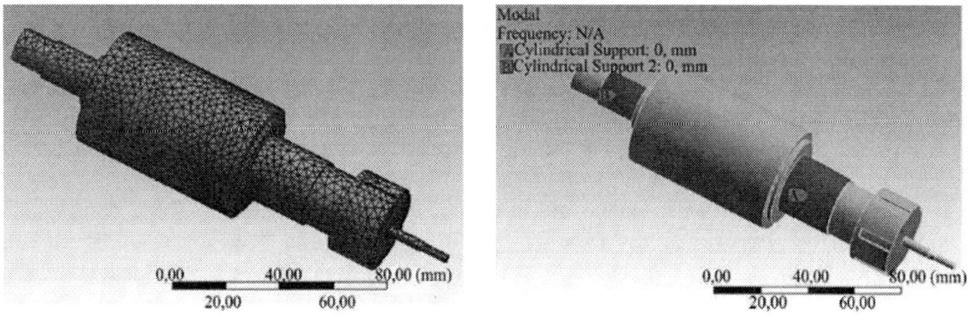

Fig. 5. Meshing of the spindle model and boundary conditions in FEM modal analysis.

With the aid of the FEM modal analysis it was possible to calculate the first two natural frequencies of the spindle and the relevant eigenmodes. The first natural frequency was calculated at 4794,5 Hz and the first eigenmode in XX direction is shown in Fig. 6 (left). The second eigenmode in ZZ direction corresponding to the second natural frequency calculated at 4837,8 Hz is shown in Fig. 6 (right).

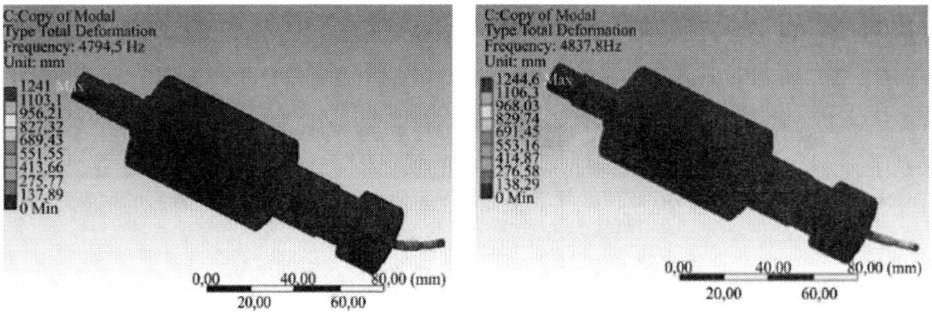

Fig. 6. Graphical representation of the eigenmode corresponding to the first natural frequency of the machine tool spindle (left) and the eigenmode for the second natural frequency of the machine tool spindle (right)

The values of the natural frequencies (4794.5 Hz and 4837.8 Hz) is substantially the first main natural frequency of the assembly. The deviation between the two eigenmodes is due to the small geometrical deviation of the device in two directions XX and ZZ, mainly because of the cutting tool since the elements of the tool holder and the spindle are rotationally symmetrical. It can be assumed that the spindle natural frequencies calculated with the aid of the FEM simulation model are very closed to the natural frequencies, which were determined experimentally, so that the accuracy of the FEM simulation model can be validated. After the approval of the FEM simulation model with the experimentally performed modal analysis of the machine tool spindle it was possible to go forward and to develop the FEM transient model in time domain. This is useful to numerically evaluate the time-dependent self-excited vibrations induced by the cutting process.

6 FEM transient analysis

Transient analysis is used to predict the response of structures where time depend excitation is implemented [4,5]. In order to describe the excitation, it is necessary to know the mathematical description of it. In situations, where no mathematical function exists for the implemented excitation, an empirical model from experimental measurements is required. The subsequent analysis of the data in time domain is used to obtain a frequency response model of the vibratory system.

Thus, in this paper, in order to develop a consistent transient FEM model a simulation model validated with FEM modal analysis was firstly developed. The self-excited vibration phenomena which take place during the milling process were introduced in the model as an implemented force at the cutting tool edge. In this case, the implemented force represents the phenomena occurring in an up-milling process. The values of the force components were determined experimentally for specific cutting conditions (feed per cut 0.01 mm/cut and cutting depth 0.2 mm) and introduced in the transient analysis as numerical data. The FEM analysis results are derived for a time step equal to 4e-05 sec, where 1000 steps of force history for each experiment were applied. In Figure 7 the deformation course of the cutting tool edge estimated by means of the FEM transient analysis is presented. By comparing this result with the experimentally determined deformation a good correlation

 DOI: 10.1051/matecconf/201711201022

can be established. More specifically, according to the conducted damped forced vibration experiments the defined cutting tool stiffness amounted to 172866 N/m for 4 N cutting force (see Figure 2 and Table 2) and the tool deformation was calculated at about 2,3e-05 m, which is in a good agreement with the FEM calculated deformation of about 2,0e-05 m, at the specific time point 0.011s of the deformation course.

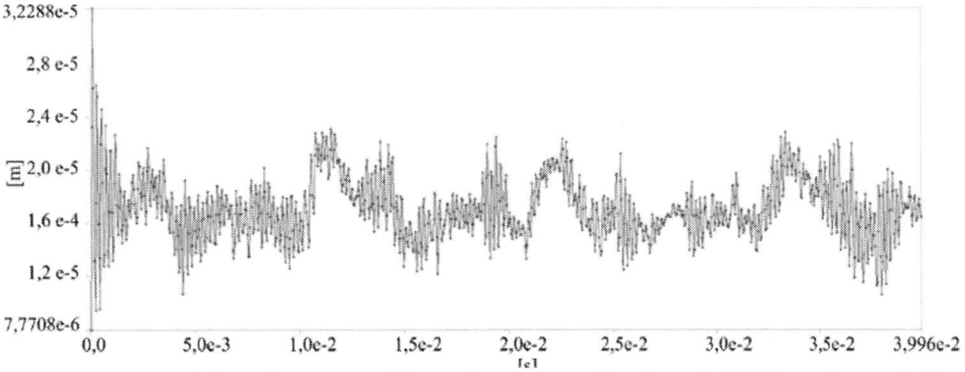

Fig. 7. Total deformation course of the cutting edge resulting from the FEM transient analysis.

7 Conclusions

In milling processes it is necessary to study the vibration phenomena which take place in cutting area between the cutting tool and the workpiece in order to ascertain the factors that affect them. In this paper, the influence of the cutting process parameters on the machine tool spindle vibration by means of FEM analysis is investigated.

Firstly, by means of a wide spectrum of end-milling experiments the cutting forces that occur during the cutting process under various cutting conditions were determined. Moreover, there were also specific experiments conducted in order to determine the machine tool dynamic characteristics of the machine tool spindle. A simulation model of the machine tool spindle was developed and it was validated by FEM modal analysis.

Furthermore, the experimentally determined cutting forces were used in combination with the cutting tool and the machine tool spindle stiffness, in order to develop the transient FEM model in time domain. The results of the transient FEM model are validated with experimental results. This FEM model will serve in future for further analysis to find out the optimum cutting parameters in micro-milling.

References

1. Y. Altintas, *Manufacturing Automation – Metal cutting mechanics, machine tool vibrations and CNC design*, second edition, (University of British Columbia, Cambridge University Press, 2012).
2. T.L. Schmitz, K.S. Smith, *Machining Dynamics Frequency Response to Improved Productivity* (Springer Science + Business Media, LLC, 2009).
3. Y. Altintaş, E. Budak, CIRP Annals-Manufacturing Technology **44**, 1, 357-362 (1995).
4. J.K. Rai, *FEM-MILL: A finite element based 3d transient milling simulation Environment for process plan verification and optimization,* (Ecole Polytechnique Federale de Lausanne, These No 4190, 2008).
5. A. Burgess, *Transient response of mechanical structures using modal analysis techniques* (University of London, 1988).

MATEC Web of Conferences **112**, 01023 (2017) DOI: 10.1051/matecconf/201711201023
IManE&E 2017

Innovative approach of drills acquisition using electrical current. A case study

Raluca Daicu*, *Valentin* Dițu, and *Gheorghe* Oancea

Transilvania University of Brașov, Department of Manufacturing Engineering, Mihai Viteazu No. 5, Brașov, Romania

Abstract. During the process of drilling metals there are used cutting tools that can be acquisitioned directly from the producers, from distributors or from supermarkets. Because the offer is varied and there are different producers, the acquisition must be made after an economical analysis that includes experimental tests. The paper presents an innovative methodology of drills acquisition based on electrical current at cutting process. The researches were developed using HS 6-5-2 steel drills of 8 mm diameter, in two batches, acquisitioned from two different suppliers, at different prices. The drills were tested on 1C45 material using the same values for cutting parameters and because the electrical current at cutting is sensible to any imperfection of the tools, the batches were analysed from qualitative and economical point of view.

1 Introduction

Due to high frequency of reaming processing, the correct acquisition of cutting tools can lead to significant economies in industry and it also guides the small business to avoid bad quality tools that can be found, for example in supermarkets at lower prices. The supermarkets do not give any information about the tool life of drills in order to make a quality/price ratio. For this reason it is opportune to analyse different brands of drills acquisitioned from different sources.

An approach of the outlined issue is presented in other researches [1], where it was used the electrical current at cutting to analyse the quality of drills. Previous researches [2-11] conducted to the fact that the state of a cutting edge can be appreciated by the value of cutting electrical current voltage, because the higher this value is, the higher the initial temperature is.

The previous researches developed by the authors demonstrate a connection between the electrical current at cutting and the quality of the cutting edges, by microscopically examination of the drills edges and identifying causes of the high electrical current voltage [1]. The conclusion was that the electrical current is sensible to any imperfection of the drill edge. In this research is used the electrical current at cutting to develop an innovative acquisition methodology for drills.

*Corresponding author: raluca.daicu@unitbv.ro

2 Qualitative analysis of two drill batches, from two suppliers, using electrical current at cutting

There were acquisitioned two HS 6-5-2 steel drill batches, from two different suppliers, at different prices, each batch having 10 drills.

For the qualitative analysis of the two batches it will be used the voltage of electrical current at cutting. The experimental steps are:

• voltage measurement of the electrical current at cutting, for each drill, with the installation presented in Figure 1;

• microscopic visualisation of drill edges to obtain additional information about the connection between the electrical current voltage at drilling and the state of the cutting edge;

• for the drill with the lowest voltage and for the drill with the highest voltage there will be done tool life tests;

• tool life estimation for all the drills from the batch.

The experiments were done on 1C45 material which has the following chemical composition: C = 0.45%, Si = 0.22%, Mn = 0.56%, P = 0.025%, S = 0.034. The tensile strength of 1C45 steel is 660 MPa (66 daN/mm^2) and the hardness is 224 HB.

Fig. 1. Experimental stand used in measuring process of the electrical current at normal drilling.

The experimental stand is composed of mandrel for fixing the drill (1), collector brushes (2), measuring instrument for electrical current voltage at cutting (3), and clamping device for multiple drillings (4).

The tests made for batch I from the first supplier and for batch II from the second supplier are centralized in Table 1.

Based on the obtained results shown in Table 1, it was generated the Table 2 that shows, comparatively, for the two batches, the number of drills that had the same value of the electrical current voltage, and also the lowest and highest values for each batch.

Table 1. Obtained values for cutting electrical current voltage at drilling 1C45 steel with drills from batch I and batch II

No. of drill	Drill diameter [mm]	Spindle speed [rev/min]	Feed [mm/rev]	Voltage "U_I" for drills from batch I [mV]	Voltage "U_{II}" for drills from batch II [mV]
1	8	560	0.25	0.4	0.4
2	8	560	0.25	0.7	0.3
3	8	560	0.25	0.6	0.4
4	8	560	0.25	0.6	0.6
5	8	560	0.25	0.3	0.3
6	8	560	0.25	0.3	0.4
7	8	560	0.25	0.5	0.5
8	8	560	0.25	0.4	0.3
9	8	560	0.25	0.2	0.2
10	8	560	0.25	0.4	0.3

Table 2. Number of drills from batch I and II that had the same value of the electrical current voltage

Crt. No.	Electrical current voltage U [mV]	No. of drills from batch I	No. of drills from batch II
1	0.2	1	1
2	0.3	1	4
3	0.4	3	3
4	0.5	2	1
5	0.6	2	1
6	0.7	1	0
Total		10	10

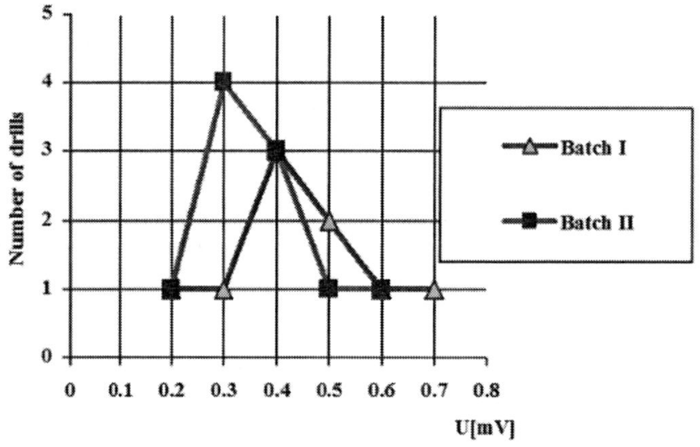

Fig. 2. Number of drills from each batch that had the same value of the electrical current voltage.

From Figure 2 where the data from Table 2 are represented, it can be observed that the batch II has the highest number of drills with voltage of 0.3 mV (4 drills) and batch I has the highest number of drills with voltage of 0.4 mV. Also, it can be observed that the difference between the highest and the lowest voltage is smaller for batch II, and this mean that the process of drills manufacturing is better handled.

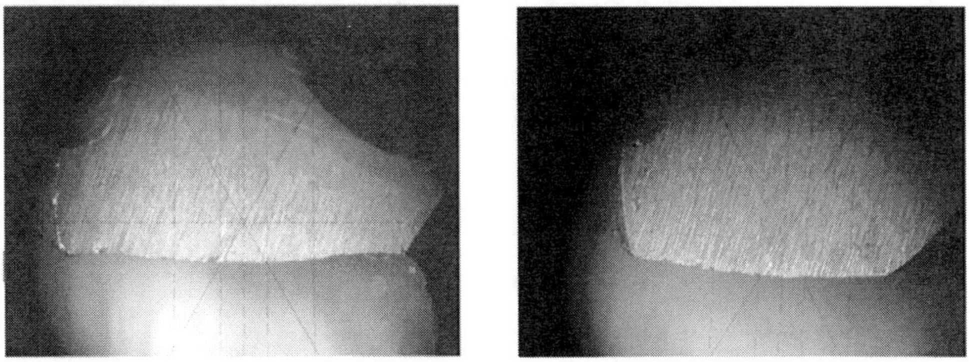

Fig. 3. Microscopic image of cutting edges at drill no. 9 from batch I.

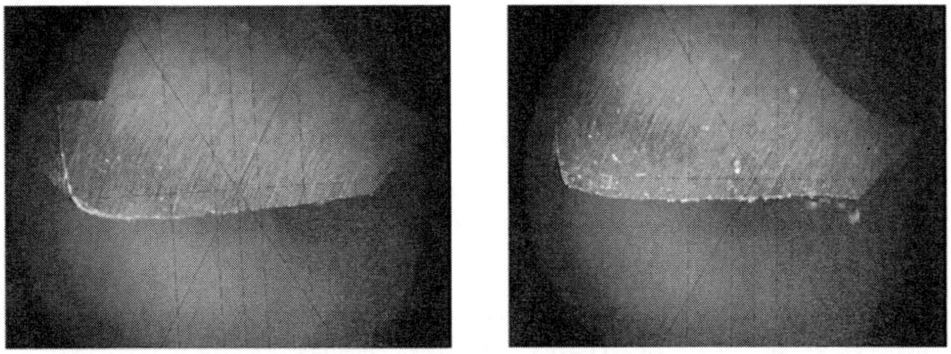

Fig. 4. Microscopic image of cutting edges at drill no. 9 from batch II.

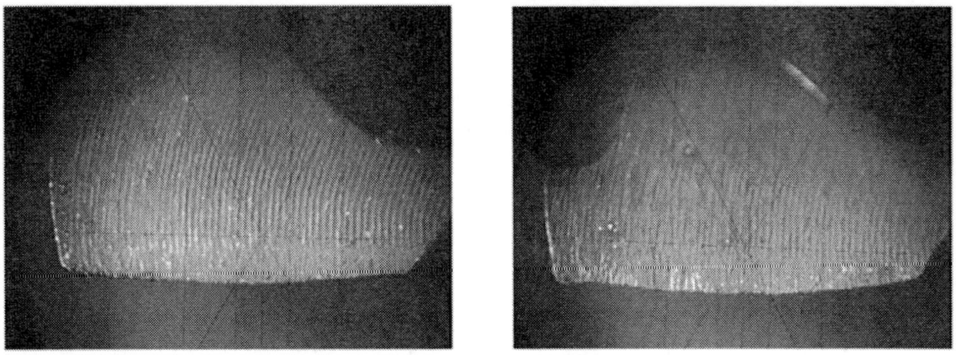

Fig. 5. Microscopic image of cutting edges at drill no. 2 from batch I.

The lowest value of voltage, 0.2 mV, was obtained with drills no. 9 from the two batches. In Figure 3 is presented the microscopic image of drill no. 9 from batch I, and in Figure 4 is presented the image of drill no. 9 from batch II. It can be observed that both of them have a very good state of the cutting edges.

The highest value of the voltage, 0.7 mV, was given by drill no. 2 from batch I, and the voltage of 0.6 mV was obtained for drill no. 4 from batch II. The Figures 5 and 6 show the state of the cutting edges for these two drills.

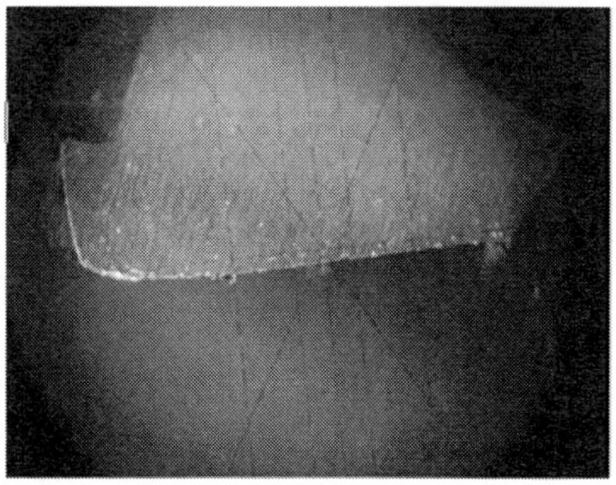

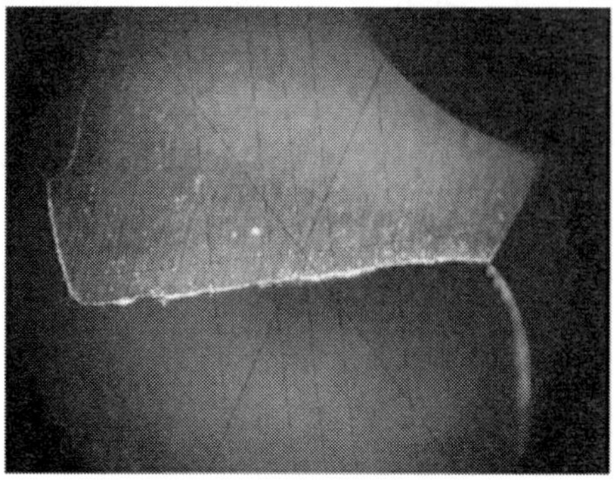

Fig. 6. Microscopic image of cutting edges at drill no. 4 from batch II.

From these microscopic images it can be observed that the high voltage of 0.7 mV (Figure 5) was obtained due to the high roughness of the land surfaces fact that lead to imperfections on the cutting lips. The voltage of 0.6 mV (Figure 6) was due to the notches from one cutting lip, and a higher roughness on the land surfaces than drill no. 9 from same batch (Figure 4).

3 Economical analysis of two drill batches

The economical analysis of the two batches starts from the qualitative analysis presented in section 2. The steps for the economical analysis are:

- tests for tool life will be made on the drills from the two batches, on the 1C45 material, until it appears a 0.4 mm wear at corners;
- tool life for each drill will be estimated based on the voltage of the electrical current;
- total machined length with all 10 drills from a batch has to be calculated;
- acquisition cost of the drill batch is divided by the total machined length with all drills from the batch and it will result the average cost to process 1 mm length;
- obtained value for this indicator for the two analysed batches will be compared and a decision will be taken.

Table 3. Tool life tests for the drills from each batch

Drill No.	Drill diameter [mm]	Spindle speed [rev/min]	Feed [mm/rev]	Voltage "U_I" for batch I [mV]	Machined length batch I [mm]	Voltage "U_{II}" for batch II [mV]	Machined length batch II [mm]
1				0.4	168	0.4	171
2				0.7	17	0.3	231
3				0.6	54	0.4	228
4				0.6	56	0.6	63
5	8	560	0.25	0.3	224	0.3	225
6				0.3	226	0.4	166
7				0.5	112	0.5	115
8				0.4	163	0.3	224
9				0.2	282	0.2	285
10				0.4	166	0.3	230
Total					1468		1938

In Table 3 are presented the results obtained for tool life test for all drills from each batch. The admissible wear at the corner of the cutting lip was 0.4 mm (not 0.6 mm as recommended) in order to shorten the time for experiments and reduce the material consumption.

The first batch was acquisitioned with 11.1 EUR and the second batch with 19.15 EUR. The price per mm length for the first batch will be 11.1 EUR divided to 1468 mm, thus the result is 0.0075 EUR/mm. For the second batch, the price per mm length is 0.0098 EUR/mm, obtained from the drills price (19.15 EUR) and the associated machined length (1938 mm).

4 Conclusions

Due to the experiments, the next conclusions are drawn:

• from the qualitative analysis resulted that the drills from the second batch acquisitioned from the second supplier, had a better quality than the drills from the first batch;

• from the economical analysis resulted that for personal and occasional use, it is better to acquire drills from the first supplier, because they are cheaper;

• for industrial purposes, it resulted that is better to acquire drills from the second supplier, because even if the cost per mm length is higher, the quality of the processed reaming is better, due to a good state of the cutting edges, fact observed also from microscopic analysis;

• electrical current at cutting is sensible to any imperfection of the cutting edge and it represents a simple, fast and low cost method to appreciate the quality of a drill batch.

References

1. V. Dițu, G. Oancea, R. Daicu, MATEC Web Conf. **94**, 02005 (2017)
2. V. Dițu, AJME **1**, 63 (2008)
3. R. Daicu, G. Oancea, AMM **808**, 40 (2015)
4. H. Bagchi, S.K. Basu, Wear **26**, 39 (1973)
5. V. Dițu, WSEAS **2**, 322 (2009)
6. A.L. Plotnikov, E.G. Krylov, E.M. Frolov, STIN **1**, 2 (2009)
7. M. Murata, S. Kurokawa, O. Ohnishi, CSME **79**, 2546 (2013)
8. M. Murata, S. Kurokawa, O. Ohnishi, M. Uneda, T. Doi, CSME **6**, 958 (2012)
9. H. Hirota, M. Murata, A. Kojima, K. Yanagihara, JSPE **64**, 1191 (1998)
10. V. Dițu, AJME **8**, 18 (2010)
11. R. Tanaka, A. Hosokawa, K. Yamada, T. Ueda, *Proceedings of the Seventh International Conference on Progress of Machining Technology,* 786 (2004)

Experimental studies on drilling in cryogenic conditions the industrial plastic parts

Sandor Ravai-Nagy[1,*], *Mircea* Lobonțiu[1], *Nicolae* Medan[1] and *Ferenc* Szigeti[2]

[1] Technical University of Cluj-Napoca, North University Center Baia Mare, Dr. V. Babeş str., 62A, 430083, Baia Mare, România
[2] University of Nyíregyháza, Institute of Engineering and Agricultural Sciences, Sóstói út 31/B, 4400, Hungary

Abstract. Starting from an identified industrial need, therefore to process plastics in certain accuracy classes with machining time as low as possible, we studies performed in the field of plastic processing in cryogenic conditions. To parameterize the cutting potential of the industrial plastics with low rigidity in cryogenic conditions, the tests where we made was an exploratory character. During the experiments, we followed the variation of feed forces and cutting torque as well as the machining precision. We noticed an improvement in chip remove capacity of the tool and increased machining accuracy, due to low temperatures used during the processing.

1 Introduction

Diversification of the polymers (industrial plastics) applied to a wide range of industry fields, creates a constant challenge in determining the optimal conditions for machining by chip removing.

The dimensional accuracy and surface quality, prescribed by designers, must be achieved after processing at a low cost.

Studying scientific literature relating to processing materials in cryogenic conditions, we identified three areas of research, characterized by:

- cooling of the cutting tool [1], [2];
- cooling of the work piece [3], [4], [5];
- cooling of the cutting tool and the work piece [6].

In the present paper, we present the results of exploratory experiments achieved in the field of processing by drilling certain industrial plastics, used in the car industry components in the framework of the car body manufacturing lines.

Industrial plastics being sensitive to temperature, will limit the intensity of cutting parameters due to thermal effect caused by the chip removing and friction between piece and tool respectively between tool and chip. Due to the interaction between the lubricating and cooling liquid and the polymers, it is recommended the dry cutting manufacturing method.

* Corresponding author: Ravai.Nagy.Sandor@cunbm.utcluj.ro

The industrial plastic change the mechanical properties depending on temperature. This was the reason we carried out the research in terms of cooling the work piece.

Thus, it was solved the cooling in cutting process and also the changing the mechanism of chip removal.

In the study, we follow the works piece cooling effects on the feed force, on the cutting torque and on the machining precision in the drilling process.

2 Research layout

The technological assembly used to perform the experiments shown in Figure 1, and consists of:

- Universal drilling machine P=3.2kW, n_{max}=1400 rmp, s_{max}=1.72 mm/rev;
- Clamping device for works piece, equipped with force and torque measuring sensor Kistler 9345B;
- Thermometer with K Type probe, measuring range -200 °C to 1200 °C, 0,1 °C accuracy.

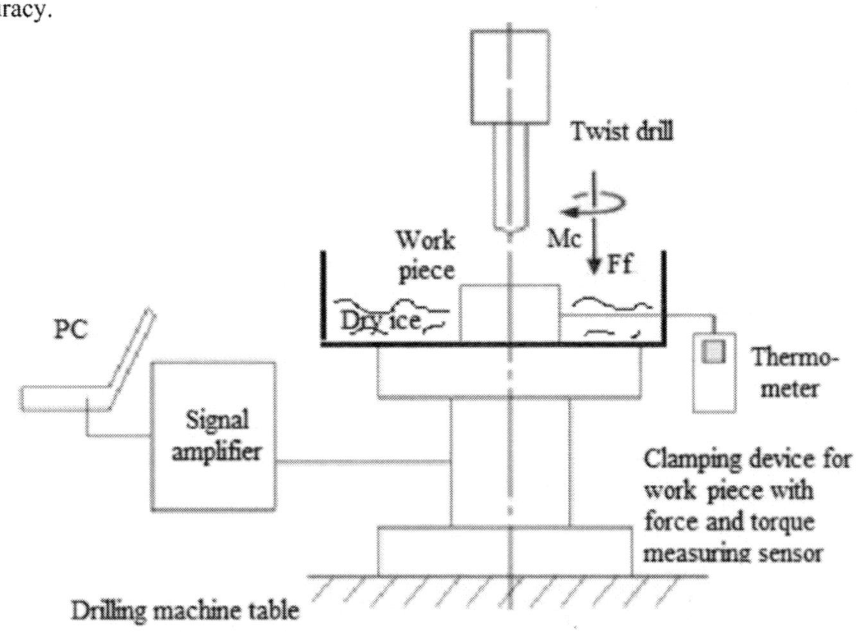

Fig. 1. Technological assembly used to perform the experiments

3 Approach of the experiment

In the experiment were performed holes using HSS twist drills, Ø 12 mm diameter, in concordance with DIN 338.

Materials used in the experiment: SIKA Block M960, PA6oil, PA6SA

Feed: 0.32 mm/rev; 0.62 mm/rev

Spindle speed: 180 rpm (Cutting speed: 6,5 m/min);

Work piece temperature: 20 °C, -30 °C

The coolant: dry ice, solid carbon dioxide.

Data that will result from the experiment:

- Feed force F_f (N);
- Cutting torque M_c (Nm);

- Out of Roundness for the machined holes (mm).

Based on experimental data we track:

- The influence of temperature on the feed force and cutting torque;
- The influence of temperature on removed chip form;
- The influence of temperature on machined surface precision.

4 Results

4.1 The influence of temperature on the feed force and cutting torque

The variation of registered feed force "F_f" and cutting torque "M_c" for each material are presented in Figures 2, 3, 4 and 5.

In the Table 1 we present, as a synthesis, the average of the feed force and cutting torque depending on temperature. The feed and cutting speed (spindle speed) was constant under the experiment.

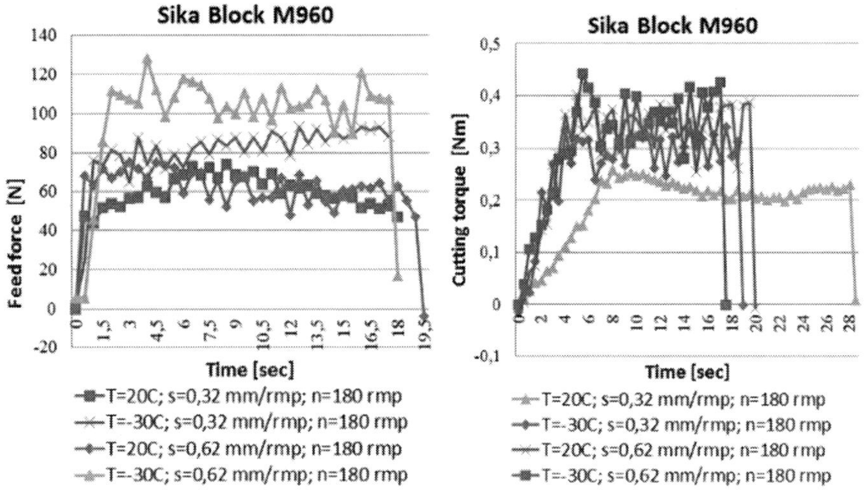

Fig. 2. The variation of feedd force and cutting torque for M960 material.

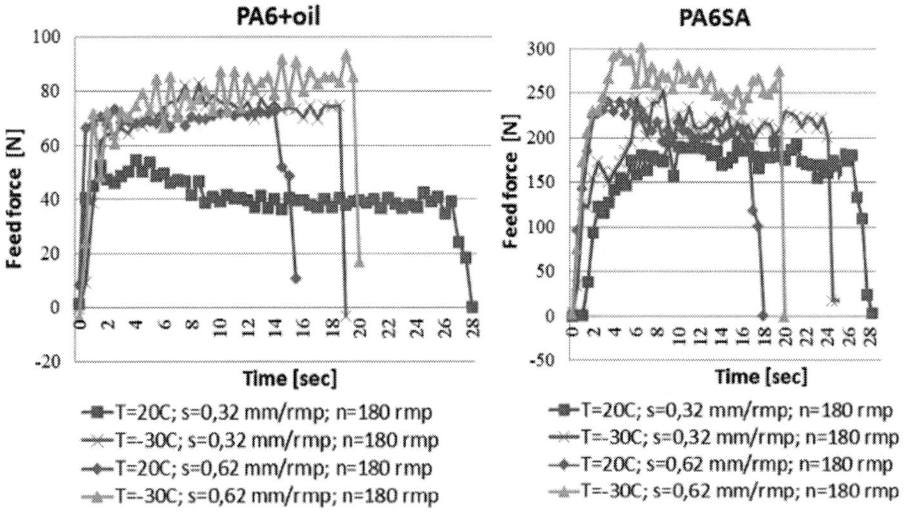

Fig. 3. The variation of feed force for PA6oil and PA6SA material.

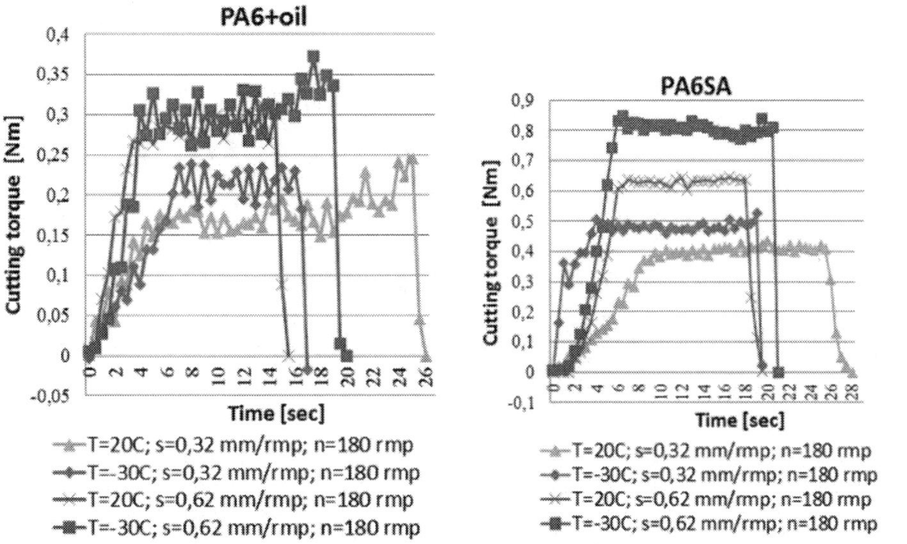

Fig. 4. The variation of cutting torque for PA6oil and PA6SA material.

Table 1. Experimental results. Average values of feed force F_f and cutting torque M_c

		Sika Block M960				PA6 oil		PA6SA	
		n=180 [rmp]		n=500 [rmp]		n=180 [rmp]		n=180 [rmp]	
		s=0.32 [mm/ rev]	s=0.62 [mm/ rev]	s=0.32 [mm/ rev]	s=0.62 [mm/ rev]	s=0.32 [mm/ rev]	s=0.62 [mm/ rev]	s=0.32 [mm/ rev]	s=0.62 [mm/ rev]
Feed force, F_f [N]	20 °C	59.49	63.33	65.92	75.99	41.30	70.99	177.1	214.8
	-30 °C	84.36	107.1	80.40	113.2	74.94	82.47	216.4	263.1
Cutting torque, M_c [Nm]	20 °C	0.222	0.346	0.237	0.344	0.191	0.286	0.411	0.629
	-30 °C	0.299	0.365	0.283	0.367	0.216	0.307	0.484	0.801

4.2 The influence of temperature on the shape of detached chips

In the case of M960 materials the chips become smaller with decreasing the temperature and drill ejects easily from hole, Figure 5.

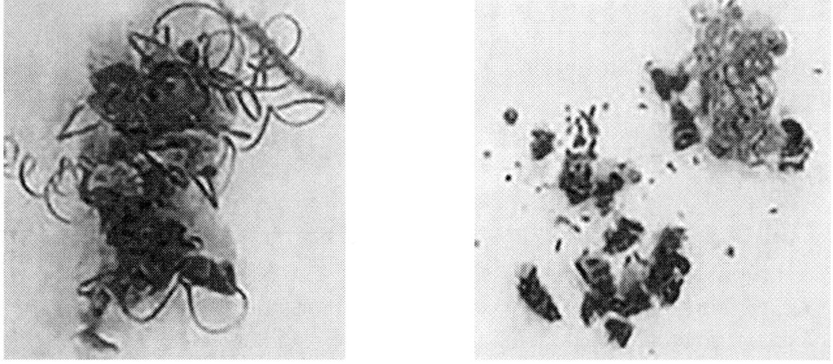

Fig. 5. The shape of the chip at 20 °C respectively at -30 °C for M960 material (feed 0.62 mm/rev).

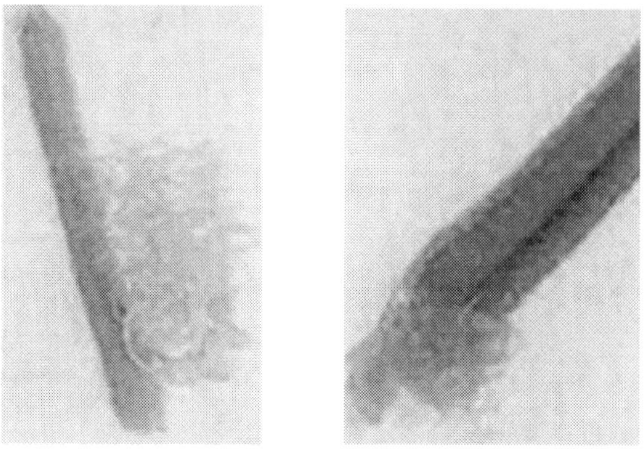

Fig. 6. The shape of the chip at 20 °C respectively at -30 °C for PA6oil material (feed 0,62 mm/rev)

With decreasing the temperature, the removed chips of PA6SA and PA6oil materials became more rigids and with larger volume. Contrary to the formed chip at 20 °C with a tendency of clogging of the drill at -30 °C it is easily discharged from the machined hole, Figure 6.

4.3 The influence of temperature on machined surface precision

A stage of the study was to verify the accuracy of drilled holes. Diameter holes were measured and in Table 2 were introduced the values of deviations.

Table 2. Deviations from the hole diameter

	M960				PA6AS			
Temperature	20 °C		-30 °C		20 °C		-30 °C	
Feed [mm/rev]	0.32	0.62	0.32	0.62	0.32	0.62	0.32	0.62
Deviation	0.03	0.03	0.02	0.01	0.02	0.02	0.01	0.01

It can be observed that deviations from circularity of the hole are lower in case of machining at low temperatures.

5 Conclusions

Analysing the results of the experiments have found the following:

- At temperatures of -30 °C increases the cutting forces (feed force) values;
- At temperatures of -30 °C increases the cutting torque values;
- The chips are more rigid at temperatures -30 °C and it evacuated more easily from the drilled hole and not remain locked in the drill channel;

- The holes machined at -30 °C presents a 50% reduced deviation from circularity, which for certain conditions in industrial manufacturing requires no boring;
- In summary, the behaviour of industrial plastics, when processing in cryogenic conditions by chip remove methods is different from the normal temperature.

Considering the results of exploratory research of the industrial plastics machining in low temperature conditions, we consider it appropriate development of this research for various cutting parameters, various plastics and various cooling temperatures.

References

1. A.A. Khan, M. I. Ahmed, Journal of Materials Processing Technology **196**, 149–154 (2008)
2. F. Pušavec, J, Kopač, Journal of Mechanical Engineering **57/9**, 637-647 (2011)
3. B. Aldwella, J. O'Mahonyb, G., E. O'Donnella, Procedia CIRP **33**, 305–310 (2014)
4. F. Ambrosy, F. Zanger, V., Schulze, I.S., Jawahir, Procedia CIRP **13**, 169–174 (2014)
5. B. E. Mocerneac, M. Lobontiu, Academic Journal of Manufacturing Engineering **7(1)** 51-56 (2009)
6. Y. Kakinuma, S. Kidani, T. Aoyama, CIRP Annals - Manufacturing Technology, **61(1)**, 79-82 (2012)

DOI: 10.1051/matecconf/201711201025

Investigation of the sound intensity level in the case of a universal lathe

Sergiu-Constantin Olaru, *Laurențiu* Slătineanu, *Manuela* Silitră, *Andrei-Marius* Mihalache, and *Margareta* Coteață

"Gheorghe Asachi" Technical University of Iași, Department of Machine Manufacturing Technology, Blvd. D. Mangeron, 59A, Iași, România

Abstract. The phonic pollution is a problem of interest in the mechanical workshops. Various solutions could be applied in order to diminish the intensity level of the sound generated by various noise sources. The research presented in this paper aimed to highlight some aspects concerning the measuring of the sound intensity level in a mechanical workshop and some possibilities of decreasing the sound intensity level by using fences made of distinct materials. The sound intensity level corresponding to the use of a universal lathe was experimentally measured by means of an adequate apparatus. The results of experimental research were mathematically processed and some empirical mathematical models were determined. The analysis of the experimental results and empirical mathematical models highlighted the influence exerted by some factors on the sound intensity level and some possibilities of diminishing the phonic pollution by using various phonic fences.

1 Introduction

According to the literature, the phonic pollution represents the process of surrounding medium contamination with materials that interfere with human health, the quality of life or the natural function of ecosystems (live organisms and their habitats). Even though sometimes the pollution of the surrounding medium results from natural causes like volcanic eruptions, the most of polluting substances comes from human related activities.

Acoustic pollution, also known as *phonic pollution* or *sonic pollution,* represents a significant component of the pollution process of our surroundings by its harmful nature. By its presence in all modern life compartments, the sonic pollution represents a major issue for all highly economically developed countries and for the emergent ones [1-5].

Phonic pollution materializes by a continuous aggression determined by distinct sound that are produced by cars, equipment, industrial or domestic machines, inside or outside constructions, buildings, chambers, etc. The intensity of noise may be influenced by the way by which noise sources are put into place and how they should be contained.

Studies about acoustic pollution regarding the determination of any type of sound, especially for the industrial environment protection etc. related domains use sonic meters

* Corresponding author: olarusergiu@gmail.com

(equipment that measures the acoustic pressure level).

By the help of such an equipment and by repeated measurements, we may obtain a so called noise map, which may be available for a certain area or surface.

Fighting against noise represents a problem for each work place that may be considered as a real system. In this case, by *system* we refer to the assembly that was formed by the noise source, the environment (the path on which the acoustic energy is propelled) and the receiver. The source is that part of the system where acoustic energy is generated. In general, the source should be considered as a group of noise generators that may have distinct physical characteristics and with a certain distribution in time and space.

Fighting noise does not mean only reducing its intensity level, as temperature adjustment does not always mean that it will decrease. But we all can agree upon the fact that a lot of noise fighting related issues are solved by reducing the acoustic power or acoustic pressure with a certain degree of confidence. There are situations in which the recommended solution focuses, for example, on modifying only the noise frequency spectrum.

In paper [1], the authors show that diagrams of stability must be developed in combination with the development of a FEA analysis, in order to be able to determine the dynamic performance of machines. The authors present two different ways to obtain such diagrams. They have conducted experimental test on a horizontal lathe. The results they have recorded are similar to those obtained by numerical simulation in the time frame of the proposed analysis.

In another paper [2], the authors base themselves on conducting experimental studies also on results obtained from the simulation and by doing this, they hope to achieve a better stiffness for the tools they use in order withstand the intense vibrations generated by the machining processes. They have succeeded to obtain an improvement in terms of machined surface quality. According to other researches, the authors have highlighted vibration phenomena that appear between the tool and the tool holder; such vibrations influence in a negative way the roughness state of the machined surfaces [3].

The present study aims to highlight the way in which several noise producing factors interact and behave during cutting processes performed on a universal machine tool as part as machine manufacturing processes. The paper also highlights the need for new and improved techniques and technologies to reduce phonic pollution. Such solutions may require additional investments, new materials from civil and industrial engineering as part of the machine manufacturing industry, but also they may require rethinking new procedures as well as machining processes.

2 Theoretical considerations about sonic vibrations generation in case of a lathe machining

The main sources of vibration in the case of developing a turning process on a lathe are the electric engine, the gearing between the engine and the gearbox, the existence of some worn parts inside the gearbox and even the cutting process itself.

Figure 1 shows a upper view of a lathe, highlighting the most significant noise sources and how the noise intensity level measurement process takes place, but also the how fences may be put into place in order to diminish the intensity of the noise propagation process in the neighborhood of the machine tool equipment.

Vibration appearance is inevitable during lathe functioning. Its presence is more than often harmful to the lathe functionality and machining process. It is considered that inside the lathe's internal structure may appear free vibrations, forced vibrations and auto vibrations (self-maintained), thus the vibrating phenomenon being frequently accompanied by noises with distinct patterns.

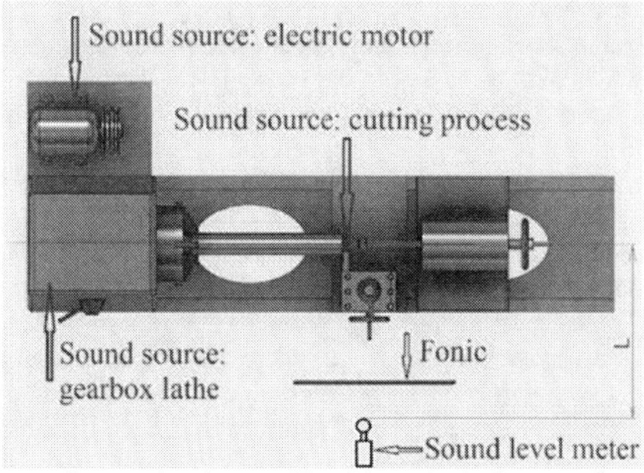

Fig. 1. Sound sources and measuring the sound level intensity in the case of a universal lathe.

Free vibrations have a relative short time when they do manifest, being amortized by different joints between lathe sub-assemblies. The kinematic and dynamic factors that correspond to a proper functioning of the lathe generate forced vibrations that display a continuous presence during the machine tool functioning.

Forced vibrations that appear during functioning may be classified taking into consideration all excitement sources: a) forced vibrations which do not depend on the cutting process; b) forced vibrations which depend on the cutting process.

Vibrations that do not depend only on the cutting process may be highlighted experimentally during machine tool idle state. They can be generated also by other equipment found in the neighborhood or other functioning machine tools and may appear due to inertia forces that intervene as result of the revolution movements of machine parts or unbalanced subassemblies. These vibrations may amplify sometimes due to an inappropriate fixation of equipment onto foundation or because of technological imperfections in terms of machining and assembling the parts of machine tool. The main sources of vibrations and noises in the case of machine tools are their mechanical transmissions, which are composed out of gears, bearing housing, electrical motors, but also the cutting process itself. The noise produced by bearings may appear due to direct and indirect propagation of vibration that appear as a result of the relative movement of components (rolling holders, rings, cage) during functioning conditions. Based on the analyses of bearing vibration causes, few general recommendations were formulated regarding the decrease of noise, such as: using silent bearings with a higher accuracy level, diminishment of impurities from contact surfaces, elimination or reduction of clearances when the axial bearing is being used (by axial preloading).

Globally studies show that the most important noise and vibration sources inside machines and mechanical equipment are the gears. One of the main reason is considered to be the ever increasing need for high power and speed of modern equipment, as they become lighter and smaller. These imposed conditions to worsen the acoustic vibrating behavior of gears and gearing, because even from the drawing board, but also during execution and fitting, no compliances were made towards optimization criteria of equipment's acoustic behavior. Reducing noise and vibrations produced by gearing becomes one of the main elements in fighting sonic pollution.

As a result of analyzing the statistical information by regulations, one notice that there have been imposed some admissible threshold noise levels, established for each domain: medical, technical, economical and professional. For the technical domain which includes machine tools, the admissible values for noise levels have been presented in the older standard STAS 8857-87 and they can reach up to 80-81 dB, depending on the machine tool that is used.

Vibrations are transmitted through environment as the machine tool is surrounded by a complex acoustic field. This acoustic filed is characterized by the energy of vibrating sources and by the acoustic properties of the chamber in which the machine tool performs.

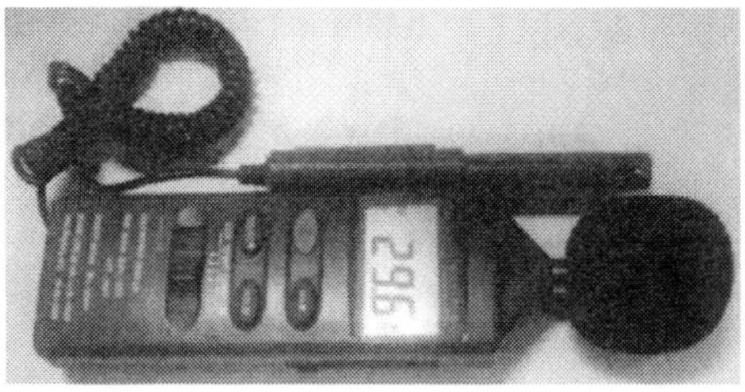

Fig. 2. Soundmeter type Voltcraft model DT 8820.

The main factors that may influence the intensity level of vibrations during cutting processes are the cutting parameters, the length of the workpiece zone which is not additionally supported or fixed, the tool's length found cantilever, the tool's active wear state, the type of metallic carbide tool tip and the way in which it gets assembled in the tool holder.

3 Experimental research

The objective of the experimental research was to measure the intensity of the vibration phenomenon corresponding to the lathe functioning, both during the machining process and idle operation. In accordance with the accepted definition, the L_i sound intensity level or acoustic intensity level (a logarithmic quantity, expressed in dB) is the level of the intensity of a sound relative to a reference value I_0

$$L_i = 10 \cdot \log_{10}\left(\frac{I}{I_0}\right) \quad (1)$$

As a reference sound intensity level in air, the value I_0=1 pW/m^2 is used. The measurements were made on a universal lathe type SN 400x1000, for which the workpiece maximum diameter turned over the lathe bad is of 400 mm and the distance between centers is of 1000 mm. The main shaft rotation is obtained by means of an electric motor able to offer a power of 15 kW at a speed rotation of 1480 rev/min.

In order to measure the intensity of vibrations, a multifunctional apparatus type Voltcraft model DT 8820 was used (Figure 2). The apparatus allows determining of some microclimate parameters, having also functions of luxmeter, hydrometer and thermometer.

To materialize a turning process, a turning carbide-tipped tool was used. The hard material of the tool active part was the metallic carbide P30. As a test piece, a cylindrical laminated bar made of a steel alloyed with molybdenum and chromium was used. All the measurements were made in front of the lathe, at a height of 1 m relative to the flooring. In some cases, fences made of distinct materials were placed in front of the sound level meter, in order to see the effect of sound damping. Some of the experimental results obtained by means of the sound level meter were included in Tables 1, 2 and 3.

The experimental results were

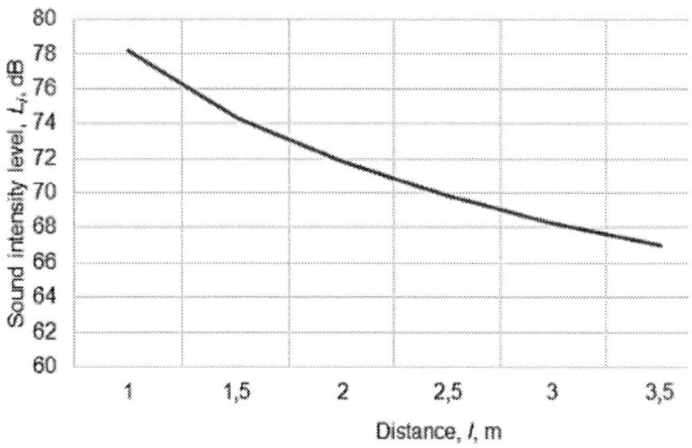

Fig. 3. Influence exerted by the distance l on the sound intensity level L_i during the idle operation (in accordance with the relation (2)).

mathematically processed by means of a specialized software based on the least square method [4]; power functions were preferred in order to model the studied phenomena, due to the possibility to easier highlighting the effects exerted by some independent variables on the intensity of sounds generated by the lathe found in distinct situations. The adequacy of the power type empirical models to the experimental results was evaluated by means of the Gauss's criterion.

Thus, in the Table 1, the results obtained during the electric motor rotation were inscribed; one must mention that the lathe main shaft was not trained in rotation movement.

Table 1. Intensity of the sound generated by the electric motor during the idle operation (the lathe main shaft was not trained in rotation movement)

Sound intensity L_i (idle operation, only the rotation of the electric motor)	**Measuring distance, *l*, m**		
	1	**2**	**3**
Measured values, dB	78	72.2	68.5

The empirical mathematical model corresponding to the measured values included in table 1 is the following:

$$L_i = 78.174 \cdot l^{-0.123} \quad (2)$$

for which the Gauss criterion has the value S_G=9.616448*10^{-2}. For the experimental results include in Table 2, the following empirical model was determined:

$$L_i = 62.180 \cdot n^{0.040} \cdot l^{-0.0249}, \quad (3)$$

the Gauss criterion having the value S_G=2.670875.

Table 2. Sound level intensity when the lathe main shaft is trained in rotation movement with distinct rotation speeds

Rotation speed, *n*, rev/min	**Intensity of sound level intensity, L_i, in dB, relative to distance to the universal lathe, *l*, m**		
	1	2	3
380	80.1	79.2	77.7
765	79.8	77.8	76.8
1500	84.1	83.8	82.8

In the case of this dependence L_i=*f(n, l)*, the software showed that a more adequate empirical model is that based on a polynomial of second degree:

$$L_i = 84.422 - 0.0121 \cdot n + 0.0000086 \cdot n^2 - 0.915 \cdot l - 0.05 \cdot l^2, \quad (4)$$

for which the Gauss criterion has the value S_G=0.1308351.

Considering that there is a possible correlation between the sound intensity level L_i and the roughness of the machined surface, the values of the *Ra* surface roughness parameter were measured; one noticed that relatively high values were obtained on the machined surfaces (*Ra*=7.02-8.87 μm).

Some experiments aimed to highlight the influence exerted by the distance *l* between the soundmeter and lathe and the depth of cut a_p on the sound level intensity L_i, this meaning just during the process of machining of the test piece made of alloyed steel. The experimental results obtained in this case are presented in table 3.

The power type empirical mathematical model corresponding to the information included in table 3 is the following:

$$L_i = 81.62 l^{-0.0472} a_p^{0.00855}, \quad (5)$$

the Gauss's criterion having the value S_G=0.1206107.

Table 3. Sound intensity level in case of turning and when using various materials as fences, respectively

Dis-tance l, m	Depth of cut a_p, mm				Card-board	Paper A4	Cellophane	Plywood	Synthetic material
	1	2	5	10	300x190	simple	1000x500		
1	81.7	82.1	82.7	83.5	81	80.2	71.2	74.2	78.8
2	78.9	79.1	79.3	81.1	79.1	78.4	68.1	69.1	77.5
3	77.9	78.1	78.6	78.9	77.9	76.9	67.7	70	76.3

On the base of the empirical mathematical models, the graphical representations from Figures 3 and 4 were elaborated. The analysis of the empirical mathematical models and of the graphical representations allowed establishing of some general remarks. Thus, one noticed that, as expected, the increase of the distance l between the sound source and the soundmeter determines a diminishing of the L_i sound intensity level. At the same time, the increase of the rotation speed n has as a result an increase of the L_i intensity sound level.

4 Conclusions

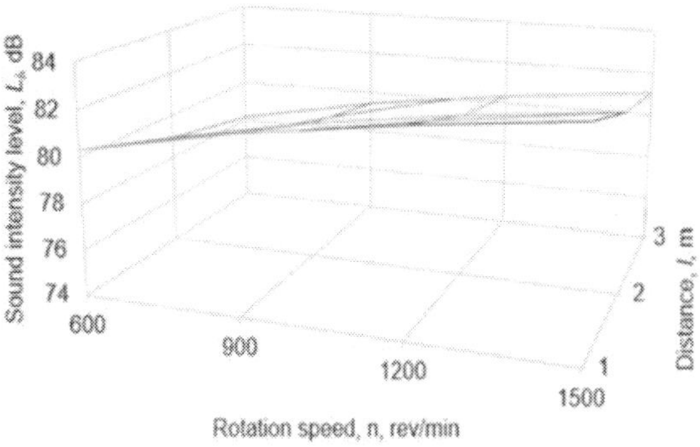

Fig. 4. Influence exerted by the speed rotation n and distance l on the L_i sound intensity level when the lathe main shaft is trained in rotation movement

The vibration generated during the functioning of the lathe, both in idle operation and during the machining process, could be considered as a phonic pollution source. The analysis of the turning process developed on a lathe highlighted the main sources of vibrations. An experimental research was planned and materialized, in order to evaluate the influence exerted by some turning process input factors and also by some materials of fences placed in order to diminish the sound intensity level. By processing the experimental results, empirical mathematical models were determined. On noticed the diminishing of the sound intensity level when the rotation speed of the lathe main shaft reduces and when the distance from the machine tool is higher. Some fences materials proved a good capacity of dumping the sound intensity.

References

1. G. Urbikain, F.-J. Campa, Journal of Sound and Vibration **340**, 317-330, (2015)
2. V. Khoroshailo, V. Kovolav, Procedia Technology **22**, 153-160, (2016)
3. M. Thomas, Y. Beauchamp, Computers & Industrial Engineering **31**, 3-4, 637-644, (1996)
4. G. Crețu, *Fundamentals of experimental research. Laboratory handbook* (Technical University of Iași, Romania, 1992)

MATEC Web of Conferences **112**, 01026 (2017) DOI: 10.1051/matecconf/201711201026
IManE&E 2017

The processing accuracy of the gear

Sergiu Mazuru*, *Maxim* Casian, and *Serghei* Scaticailov

Technical University of Moldova, 9, Studenţilor str. Chişinău, Republic of Moldova

Abstract. The paper presents analysis of processing quality of gears. A special attention was drawn to detecting and removing burns, the existence of which greatly reduced operating properties of gears. The study emphases the influence of schemes, processing modes and abrasive tools features on the physico-mechanical layer of teeth processed and phase changes, micro-cracks appearances, depth and degree of hardening, meaning and character of residual stresses distribution.

1 Introduction

The formation of the surface layer during deep grinding of gears from alloyed steels occur under the influence of an uneven thermodynamic influence on the work-piece and, as a consequence, the appearance of inhomogeneous elasto-plastic deformations along the section of cavity profile between the adjacent teeth and the structure-phase transformations from heating and cooling of metal. This leads to distortion of the teeth profile, to the appearance of a step error and a decrease in the kinematic accuracy of the gear and to the appearance of grinding defects on the treated surfaces. Experience shows that the accuracy of machining of gears depends not only on the rigidity and kinematic accuracy of the machine, but also on the accepted grinding modes, tool characteristics and the conditions of its correctness.

The following describes an example of the profile deep grinding of identical gears on the same P600G machine with a grinding wheel with one characteristic. With the option of processing in a softer profiling mode, when the total allowance was divided into 5 passes, the accuracy of the gear wheel was achieved no less than with the optimal version in 2 passes.

As an example illustrating the accuracy of this processing process, measurements are given of spur gears manufactured by ARP GmbH Alpirsbah (Germania) with the following specifications: module - 5 mm, angle of engagement - 20 °, number of teeth - 23, width of sprocket -80 mm. Accuracy of gears according to DIN 3062 - 6th qualitet.

The gears were machined by the method of profile deep grinding on a 53A30P machine model with a highly porous disk with the characteristic 25A16PM212K. The preliminary profiling of the teeth was carried out by the method of deep grinding in 2 passes, and after carburizing with quenching, fine grinding was also performed in 2 passes. The meshing parameters were measured on a special-purpose gear-measuring center model P 100 CNC from company Klingelnberg. Measurement protocols containing graphical and digital information on the accuracy of profiles and tooth lines, as well the step errors and radial runout of the sprocket are shown in Figure 1 in accordance with GOST 1643-81.

*Corresponding author: s_mazuru@mail.utm.md

The received high accuracy of gear wheel processing is provided due to the kinematics of profile grinding process (absence of the long and difficult kinematic chains, control from highly developed system of CNC) and high dimensional stability of the abrasive tool. The requirements of the drawing and the results of measuring the processing accuracy parameters after grinding with a high porosity circle are given in Table 1.

Table 1. Precision parameters of gears

Precision parameters	Notation	Tolerance by the drawing, μm	The real precision, μm
Total profile error	F_a	12	6.4 ÷ 7.9
Profile shape deviation	F_{fa}	10	6.4÷6.8
The deviation of gear angle	F_{ka}	±7	-1.5÷2.6
Form error of the tooth straightness	F_P	9	0.4÷0.6
Angle deviation of the tooth straightness	F_{fp}	10	1.2÷5.8

Thermal action during grinding is often accompanied by structural changes in the surface layer of the treated material in the form of burns. Depending on the capacity of the heat source, the time of its action, and also the heat resistance of the processed steel, structural changes can develop into different depths and have a different nature. After the most intense thermal action, the surface layer represents secondary hardening austenite-martensite, under which is located a zone of the secondarily tempered metal with the structure of troostomartensite and troostite. With a decrease of heating of the surface layer, a secondary release structure is formed in it, proceeding through all the stages of tempering into the basic structure of martensite.

The likelihood of the burns appearance during tooth grinding depends not only on the processing conditions of processing, but also on the grade of the steel being processed. For heat-resistant steels of the type AISI 415, 16Cr3NiWMo this probability is lower than for non-heat-resistant steels such as 14NiCr14 and others [1].

During etching, the areas with a modified microstructure after tooth-grinding are usually extended along the tooth or revealed as a dashed grid, depending on the processing method. The greatest number of burns strokes and their maximum intensity, as a rule, is observed on the heads of ground teeth. Closer to the teeth legs, their number and intensity decrease or they are generally absent. Structural changes cause concomitant changes in the hardness and stress of the surface layer of the treated teeth. The micro-hardness in the burn zones is reduced, and residual tensile stresses are formed in the surface layer. Cogwheels with burns have bending endurance 1.4 to 1.6 times less than ground wheels without burnings. Accordingly, the durability of contact fatigue in gears with burns is 3.5 times lower [1, 2, 3, 4, 5, 6 and 7].

The appearance and development of grinding nature defects, before all, of burns and micro-cracks, caused by structural-phase changes in the surface layer, depends on the temperature and the time of its action. In the gear grinding schemes used, the heating intensity and the degree of its localization greatly differ and are related to the kinematic features of contacting grinding wheel with the profiled surface of the teeth. Under optimal conditions of gear grinding by various rolling patterns or profile processing by highly porous abrasive wheels, as experimental studies have shown, production tests and experience of industrial application, burns and micro-cracks on the machined surfaces of the teeth are not observed.

A general recommendation to prevent grinding defects in all gear grinding schemes is the decrease of the heat source intensity and the heating time of the treated zone. The use of new highly porous circles already by itself reduces the heating temperature by 300 to 400 ° C compared to grinding wheels with a normal structure. In combination with the optimization of the processing mode due to the speed increase of the relative movement relative to tool and to work-piece, as well as the depth of cutting, their application makes it possible to increase the removal rate of the material and to reduce the time of action of the heat source on the machined surface, that is, to create the prerequisites for high-performance and defect-free grinding. The level of intensification of the processing parameters depends on the selected characteristic of the highly porous grinding wheel.

Thus, taking into account a significant reduction in the operational properties of gears by the formation of defective layers with burns, the introduction of new technologies of non-burn gear grinding is an important condition for ensuring high reliability and durability of tooth wheels in exploitation.

2 Results of metallographic analysis

Analyzing the obtained data on the influence of various conditions for the profile deep grinding of the gear wheel from hardened steel 14NiCr14 (HRC 40) on the formation of surface layer properties. The cavity between the neighboring teeth 2 or 4 mm in width was profiled in 4-6 passes at a constant processing speed of 20 m / s with varying parameters of the grinding mode t and B_d with grinding wheels with characteristics 25A12PBM214K5, 25A12PM2.16K5 and 25A16PM2.12K5 for various kinematic schemes of interaction between the tool and the work-piece. The total nominal depth of the cavity after multi-pass treatment varied from 5 to 6.1 mm. The quality of the surface layer by the change in microstructure and micro-hardness was estimated after the final passage. In the process of shaping the profile, after each pass on the machined surfaces of the teeth, defects of the grinding character were visually recorded: burns, traces of crushing and also.

Practically in all cases of profile deep grinding of a conical gear wheel made from 16Cr3NiWMo steel its initial structure of tempering martensite is transformed into a secondary quenching structure in a thin near-surface layer with an increase in micro hardness followed by a transition to a secondary tempering zone. Under unfavorable conditions of deep grinding in the surface layer, a structure of secondary hardening with increased micro hardness along the valleys and to tooth profile in the form of an ellipse with a width at the center of 0.2 to 0.86 mm and a length of 1.28 to 5 mm is observed.

The results of the metallographic study of the grooves microstructure after profile grinding testify to the fact that the groove geometry (cross-sectional area), the scheme and processing regimes, as well as the characteristics of the grinding wheel, have a great influence on the formation nature of the surface layer - the depth of the altered layer and the degree of hardening.

With multi-pass groove profiling of the gear wheel, the technological inheritance from previous passages and their number have a big influence on the quality of the treatment. In this regard, not always the final finishing passes of the grinding wheel can ensure the formation of the surface layer with minimal distortion of its structure. As an example, we can compare the results of processing two different options. In the first variant, in 4 passes, despite the high energy density of 16.2 J / mm, the quality of the treatment is better than in the other version in 6 passes at a lower energy level of 7.68 J / mm, this conclusion is confirmed by the example described in [5].

As an additional factor in analyzing the treatment options considered, it is necessary to take into account the scheme of fixing the work-piece, and the direction of movement and rotation of the grinding wheel relative to it. Experiments were carried out on a conical spur gear wheel, which was cantilevered in the mandrel, enabling the implementation of horizontal and vertical profiling schemes. Undoubtedly, for these conditions, a horizontal layout is more preferable for deep grinding, in which higher rigidity of the technological cutting system is achieved.

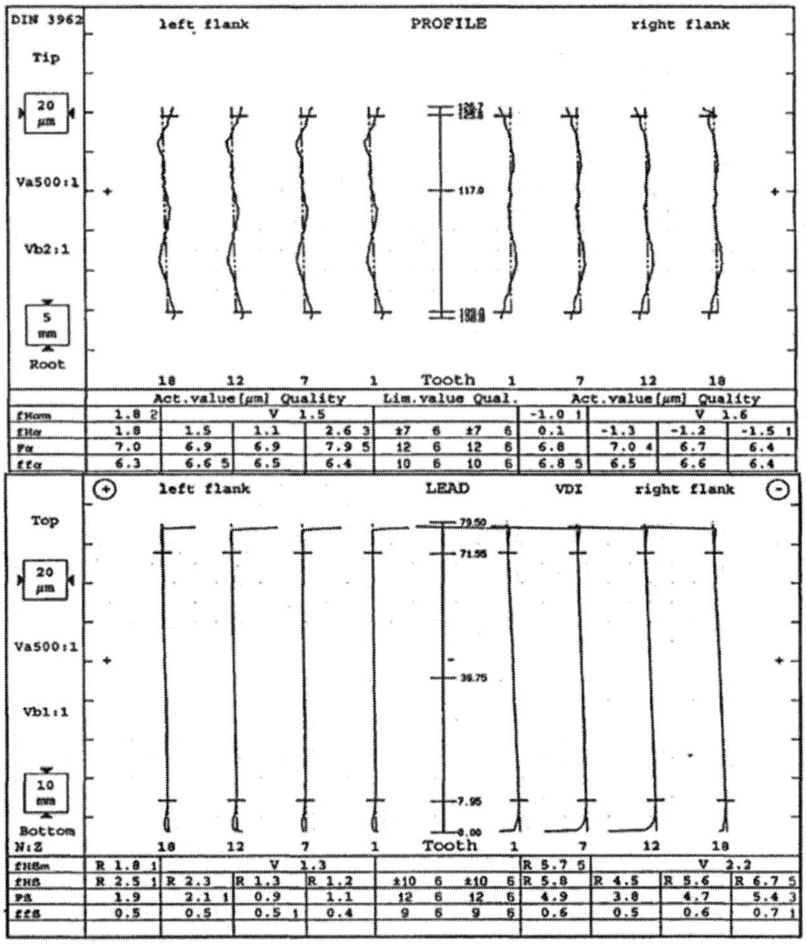

Fig. 1. Results of measuring the accuracy of the profile and the direction of the tooth line.

3 Elimination of burns

When assigning the characteristics of a highly porous grinding wheel, it is more expedient to use a tool with a larger grain size and a smaller structure than with a smaller grain and a larger number of the structure. Analyzing the obtained results, it can be concluded that the optimal conditions for the gears profile deep grinding, from the point of view of the surface layer formation with minimal structural changes is the horizontal counter processing scheme with a minimum number of passes with a grinding wheel with the characteristic 25A16PM2.12K5.

In addition to multi-pass groove profiling of the gear wheel, experiments were conducted to evaluate the direct influence of the deep grinding regime on the formation of the surface layer properties. Profiling of the groove with a width of 2 mm was carried out under optimal grinding conditions with a circle with the characteristic 25A16PM2.12K5 with a horizontal counter processing scheme. The depth of the passage varied from 0.5 to 3 mm, the speed of the longitudinal movement of the tool was chosen in such a way as to ensure the constancy of the item $t \times S_{fl}$ equal to 60 mm / min.

The results of the metallographic study of the surface layer after processing show that the item $t * SA$ is kept constant, the depth of the altered layer h and the degree of adhesion

DOI: 10.1051/matecconf/201711201026

of the treated surface N increases with increasing depth of the passage t. This is explained by the fact that the depth of cutting during the teeth profiling implicitly affects the actual removal of material inconsistently, the thermodynamic intensity of the process of deep grinding. If the depth t is increased from 0.5 to 3 mm or 6 times, the calculated sectional area of the profile groove of the gear rises from 1.09 mm to 9.28 mm or 8.5 times, and the actual metal removal speed is 130, 9 mm / min (at $t = 0.5$ mm, $S_{fl} = 120$ mm / min) to 185.5 mm / min (at $t = 3$ mm, $B_d = 20$ mm / min) or 42%.

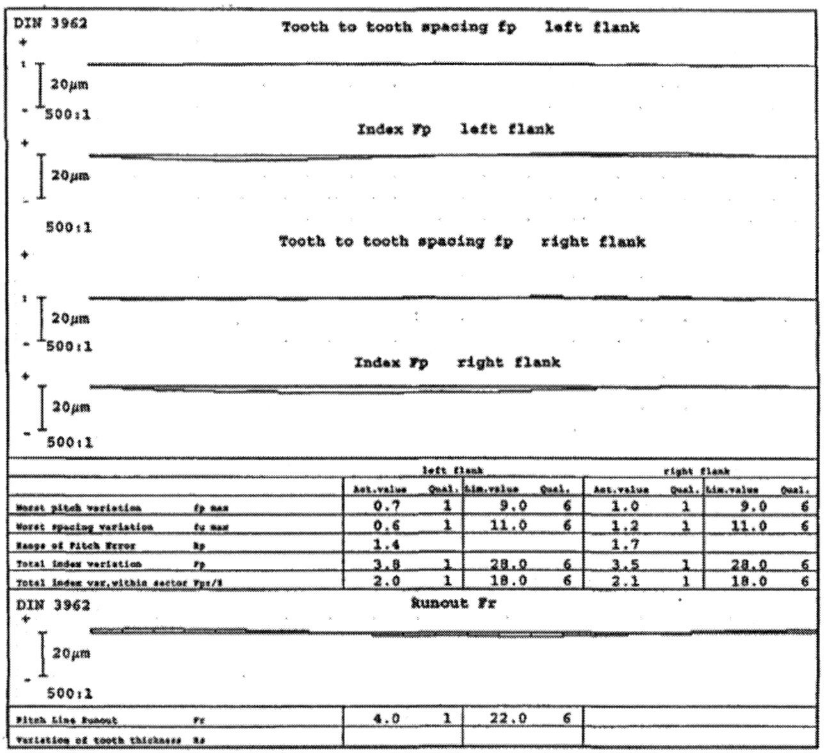

		left flank				right flank			
		Act.value	Qual.	Lim.value	Qual.	Act.value	Qual.	Lim.value	Qual.
Worst pitch variation	fp max	0.7	1	9.0	6	1.0	1	9.0	6
Worst spacing variation	fu max	0.6	1	11.0	6	1.2	1	11.0	6
Range of Pitch Error	Rp	1.4				1.7			
Total index variation	Fp	3.8	1	28.0	6	3.5	1	28.0	6
Total index var.within sector	Fpz/8	2.0	1	18.0	6	2.1	1	18.0	6
Pitch Line Runout	Fr	4.0	1	22.0	6				
Variation of tooth thickness	Rs								

Fig. 2. Results of measuring the stepwise error of the gear.

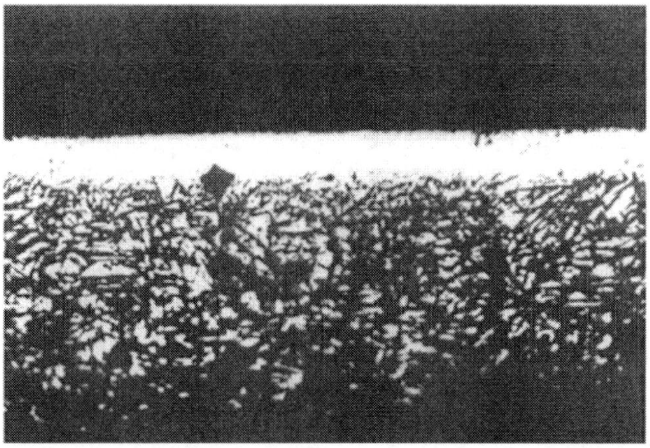

Fig. 3. Gear tooth microstructure, made from steel 14NiCr14 after grinding with traces of burns (x500).

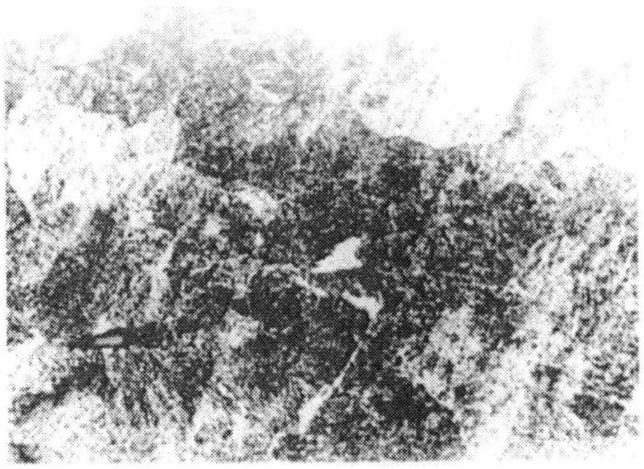

Fig. 4. Initial structure of steel 37Cr4 (x500).

Fig. 5. Outer layer of gear with structural changes.

Figures 4 and 5 show examples of structural changes in the surface layer of the 14NiCr14 gearwheel grooves before and after the profile grinding. Under optimal conditions for profile deep grinding of the gear wheel, structural changes in the surface layer are not observed (Figure 5).

4 Conclusions

Precision of processing gears was rated in accordance with GOST 1643-81 and European standard DIN 3062 after the achieving precision of the profile and tooth line, step error and gear runout. To indicate the optimal constant processing conditions are ensured a level of precision 4-5 depending on the type and condition of the machine used.

Lifetime of the gear wheels, processed by new technology increases about 3 times. This effect is due to the fact that in the surface layer of the work-piece are ensured minimal deviations micro and macro structural at a depth of up to 10 micrometers or excluding them with the forming of compression residual stress of 240 MPa up value.

References

1. S. Mazuru, A. Vlase, S. Scaticailov, *Processing technologies for gear cutting machines* (Tehnica-U.T.M. Publishing House, Chişinău, Republic of Moldova, 2014)
2. I. Bostan, V. Dulgheru, C. Gluşco, S. Mazuru, M. Vaculenco, *Anthalogy of Inventions*. (Chişinău: Bons Offices, 2011)
3. I. Bostan, S. Mazuru, S. Scaticailov, M. Casian, *Precession transmission*. Patent nr. 1116, BOPI nr. 1/2017
4. I. Bostan, S. Mazuru, S. Scaticailov, *Proceedings of TEHNOMUS Conference* (Ştefan cel Mare University of Suceava, România, 2013)
5. V. Botnari and S. Mazuru, Applied Mechanics and Materials **657** 147-153 (2014)
6. S. Mazuru and M. Casian, Advanced Materials Research **1036** 286-291 (2014)
7. M. Casian and S. Mazuru, Advanced Materials Research **1036** 292-297 (2014)

DOI: 10.1051/matecconf/201711202001

Experimental research on the effect of the ball burnishing process, using new kinematical scheme on hardness and phase composition of surface layer of AISI 304L stainless steel

Diyan M. Dimitrov[1,*], *Stoyan* D. Slavov[2], and *Zhivko* Dimitrov[3]

[1]TU-Varna, Department of Mechanics and Machine Elements, 9010 Varna, Bulgaria
[2]TU-Varna, Department of Technology of Machine Tools and Manufacturing, 9010 Varna, Bulgaria
[3] HTP TU-Varna, 9010 Varna, Studentska 1 str., Bulgaria

Abstract. In this paper, experimental results about hardness and phase composition of surface layer of AISI 304L SS, machined by using ball burnishing process, conducted under a new kinematic scheme, are presented. The effect of different combinations of the process regime parameters on the amount of strain induced martensitic phase, is discussed. The amounts of austenitic and strain induced martensitic phases are identified by x-ray diffractometer. Micro hardness along the depth of the hardened layer is measured. Conclusions about the influence of the ball burnishing process on strain induced martensitic phase are given.

1 Introduction

Besides producing a specific surface finish (roughness with regular microshape), the ball burnishing process has additional advantages in comparison to other finishing processes, such as providing increased hardness, corrosion resistance and fatigue life, caused by compressive residual stress in the surface layer. Previously published works indicated that ball-burnished surfaces have many advantages in comparison with machined, milled or ground surfaces [1-5, 12]. In general, the regime parameters, affecting the plastic deformation process, are variable and they can include the burnishing speed, the burnishing feed, the burnishing compressive force as well as the number of passes of the burnishing tool over the same region of the processed surface. Thus, by appropriate adjustment of the regime parameters of ball burnishing process, it is possible to obtain compressive residual stresses in the surface region that could lead to considerable improvement in the component life.

Considering some known drawbacks of the classical kinematical schemes for implementation of vibrational ball-burnishing process, some new schemes are developed, based on the advantages of contemporary CNC machines [3]. In the process of ball burnishing, implemented under a new kinematical scheme, the ball tool also performs motion, following a complex trajectory. In this case, however, the tool path is obtained by means of a linear (or circular) interpolation, using the pre-calculated in a suitable CAD (or

* Corresponding author: dimitrov.diyan@gmail.com

CAM) software curve (polyline) [3], and not by compelled oscillations as in the classical scheme of the process. Furthermore, using CNC machines, some other parameters of burnishing process, such as feedrate can be changed steplessly and more widely, allowing far more precise adjustments of the processing.

In most of the austenitic stainless steel grades strain hardening effect is combined with a phase transformation. The formation and the amount of strain-induced martensite depends on the austenite stability (the chemical composition and the initial austenite grain size) and the strain hardening conditions (the deformation temperature, strain and strain rate) [6, 7].

The main objective of the current research is to investigate the correlation between phase composition of the hardened layer, obtained by a ball burnishing process under a new kinematic scheme and fatigue life of stainless steel grade 304L.

2 Experimental

2.1 Material and specimen preparation

Beam like specimens with a hemispherical stress concentrator, Fig.1.a, are laser cut of stainless steel grade ASTM 304L (Cr18Ni8) sheet with a thickness of 4 mm. The chemical composition of the steel is given in Table 1.

Table 1. Chemical composition of the steel grade ASTM 304L (Cr18Ni8).

C, %	Si, %	Mn, %	P, %	S, %	Cr, %	Ni, %	N, %
0.021	0.38	1.55	0.031	< 0.001	18.2	8.1	0,059

Before cutting, on both sides of the plate in rectangular bands with a width of 10 and a length of 40 mm regular relief (RR) by a ball burnishing under a new scheme, are formed.The regimes of the ball burnishing process are chosen, following four factors, full factorial experiment design with two levels per factor [1]. A low and a high level of each factor are given in tabl.2, where: **F [N]**-the external force on the ball tool; **i**– number of the sinusoids within the length of the toolpath, determining the parameter $(\pi \cdot D)/i$,[mm], related to the height of the obtained RR cells; **e [mm]**–half of the amplitude of the sine wave, related to the width of the obtained RR cells; **s [mm/min]** – feeding speed of the ball tool, following the sinusoidal trajectory.

Table 2. Regime parameters of the ball burnishing process.

Level	Regime parameters			
	F, N	I	e, mm	s, mm/min
Low	1865	600.15	1	150
High	3040	1200.15	2.5	300

2.2 Vibration fatigue resonance test

For the fatigue life test, the specimens are positioned on a shaker as a cantilever beam with a mounted accelerometer at the tip end, Fig.1.b. The accelerometer acts as a measuring device and a tip mass.

Tests are carried out at a frequency, equal to the first resonance frequency of a cantilever, which is ~90Hz.

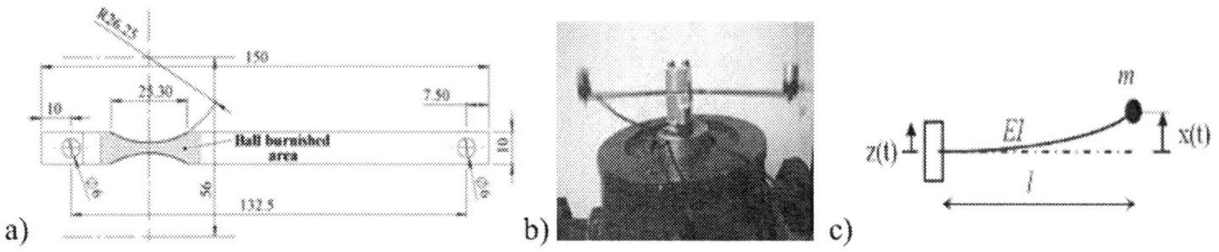

Fig. 1. a) Configuration and dimensions of the experimental specimens; b) Positioning of the specimens on the shaker; c) Equivalent dynamics model.

The loading amplitude is kept equal for all tests. Two specimens are tested simultaneously to shorten the experimental time. Detailed explanation of vibration test is given in [1].

2.3 Phase composition and metallography analysis

The phase composition of the surface layer is studied with an X-rays diffractometer (XRD) with Brag-Brentano geometry in Cr-Kα radiation. According to the rolling direction longitudinal and transversal sections are cut, polished and etched from some of the specimens for metallography analysis. A light optical microscope Neophot 32 is used. A microhardness Vickers test is carried out with a 50g loading to measure the depth of the hardened layer.

3 Results and discussion

The X-ray diffractogram of the row material shows only the austenite γ peaks. In the ball burnished specimens, peaks of strain induced α' martensite also appears. Since in some diffractograms only α'(110) peak is clear, as an indicator for strain hardening effect after a ball burnishing, ratio "r" between intensities of α'(110) and γ(111) diffraction peaks, is used. According to this ratio the structures are divided into 5 groups. The typical diffraction patterns for each group are given in Fig. 2.

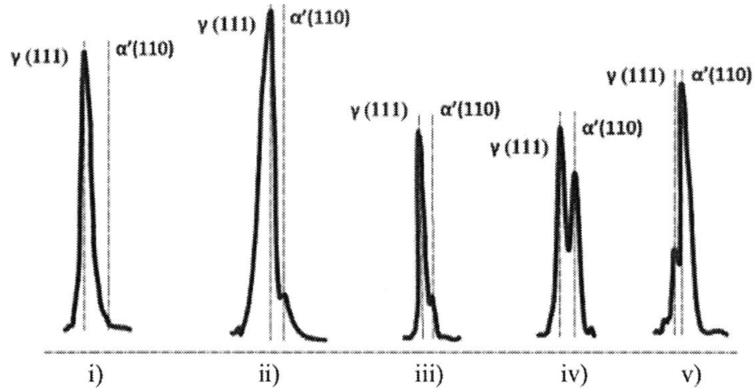

Fig. 2. Typical X-ray diffraction for γ(111) - 2θ=67.5° α'(110) - 2θ=69.5° each structural group i) r=0 ii) r=0.03-0.10 iii) r=0.12-0.25 iv) r=0.50-1.00 v) r>2.

The first group is characterised with no α'(110) peak r=0. The austenite γ(111) peak is broadened with a high intensity. The second group is characterised with a small α'(110) peak, partly hindered from broad γ(111) peak r=0.03-0.10. The third group is characterised with a more clear weak intensity α'(110) peak r=0.12-0.25. The next group is characterised by similar intensities of the two peaks r=0.5-1.00. And the last group is characterised with a strong α'(110) peak r>2.

Such scattering of the phase composition can be explained with a particular chemical composition of the raw material. On the base of the chemical composition, given in tabl.1 temperature M_{d30} =13,75°C (temperature at which 50% of A is transformed to M under 30% true strain) is calculated. As stated in [8] at such high value of M_{d30} austenite phase is very unstable. Authors in [9] shows X-ray difractogram after a deep rolling of 304L steel, which according to α'(110)/ γ(111) ratio can be classified in group (iv), and estimated the martensitic volume fraction on 27,5%. Similar results are obtained from Altenberger and all [10] after deep rolling.

The registered cycles to failure from fatigue test of specimens from each structure group are shown on Fig. 4 and a summary statistic of each group (mean and coefficient of variation CV) is presented in Table.3.

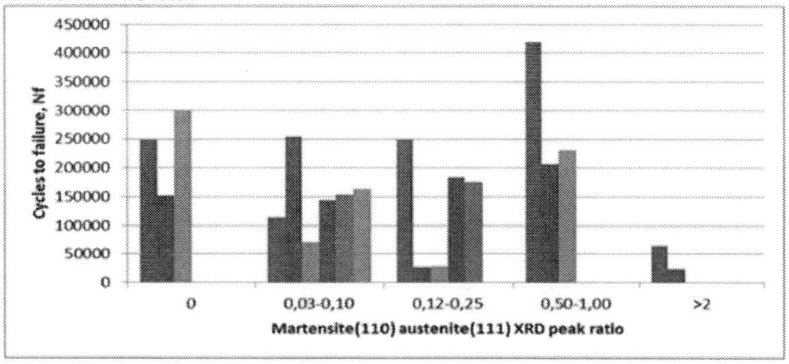

Fig. 3. Cycles to failure of specimens fall in each structural group.

Table 3. Summary statistic of specimens in each group from Fig.3.

M/A peak ratio "r"	0	0,03-0,10	0,12-0,25	0,50-1,00	>2
mean	233802	149794	132647	285598	44156
CV, %	32	41	76	41	65

It can be seen that the fourth (r=0.5-1.0) structure group give the highest mean fatigue life and the fifth group with r>2 gives a very short fatigue life, similar to non-plastically deformed specimens, which is about 20000 cycles [1]. As a general rule, the fatigue life increases with the increase of the martensite content in the volume but, as it is mentioned in [11], at 54 vol-% martensite more brittle behaviour and higher notch sensitivity of the martensite phase become predominant and lead to crack initiation at inclusions in the very high cycle fatigue (VHCF) regime. The fatigue tests of predeformed specimens with martensite content up to 36%, made in [8] show optimum for martensite volume fraction below 26%.

In the third group (r=0.12-0.25) there are two specimens with a short fatigue life, similar to those in r>2 group. We suppose that fatigue life of these specimens is affected by other factors (random micro stress concentrators or inclusions), which have to be additionally investigated on broken surfaces.

3.2 Hardness and microstructure

The surface hardness of all specimens increases with about 50% from 180HV10 of raw sheet to 260-280HV10 on plastically deformed surfaces. For a microstructure analysis two specimens, coded as "5" and "15" are chosen. These specimens fall in the third structural group. The regime parameters of the ball burnishing process differ only in parameter "i". It has a low value for specimen 5 (N_f=*182826*) and a high value for specimen 15(N_f=*250202*), tabl.2. A higher value of "i" means that the burnishing tool passes over the same region more frequently. Other parameters of the regime are F=3040N, e=1 mm; s=150mm/min. The microstructures in the near surface area, presented on Fig. 4 and 5, are similar.

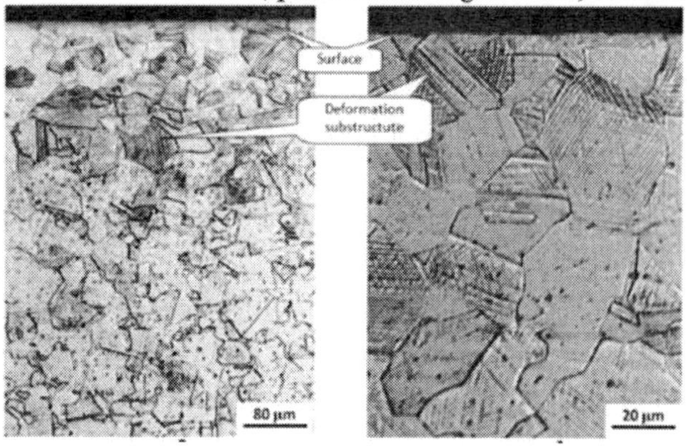

Fig. 4. Microstructures of specimen “5” near the surface.

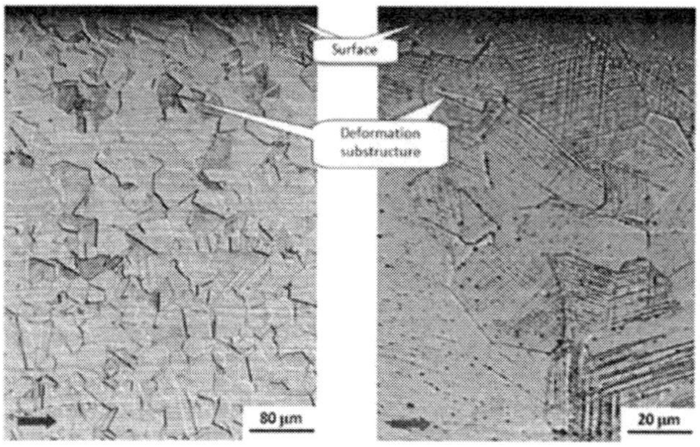

Fig. 5. Microstructures of specimen “15” near the surface (red arrow –rolling direction).

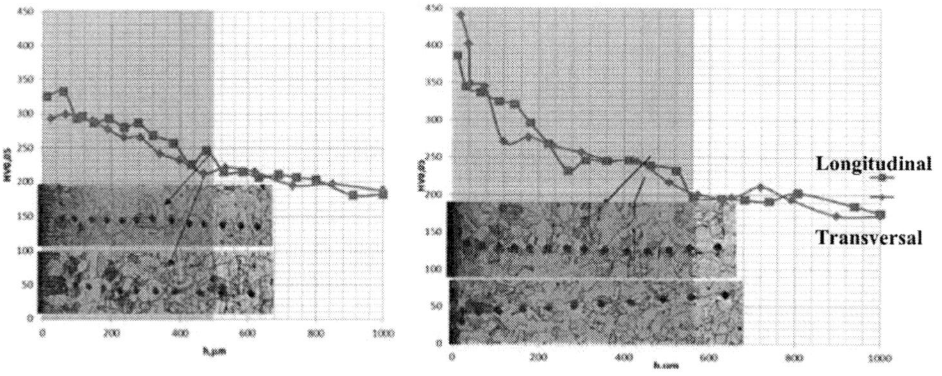

Fig. 6. Microhardness profile in depth of the deformed layer (left -5; right -15).

The deformation substructure with martensite needles, crossing each other in some of the austenitic grains, is clearly visible. The microhardness measurements, Fig. 6, show that the depth of the hardened layer in both specimens is about 500µm, but in specimen “15” higher hardness values are measured close to the surface, which results in slightly higher fatigue life.

4 Conclusion

On specimens from sheet stainless steel 304L grade, a hardened layer with a depth of about 500μm was obtained by ball burnishing process under a new kinematic scheme. Obviously, the fatigue life is extended but careful control on regime parameters of burnishing is needed, since the austenite phase in 304L steel is unstable and sensitive to the particular chemical composition of the alloy, different amounts of strain induced martensite are registered. As a result from the experiment, it is evident that there is a threshold value of the amount of martensite content, beyond which the fatigue life drastically shortens. It can be recommended to use a low level of the loading force, combined with high values of "i", which guaranties that the burnishing tool passes over the same region more frequently and obtaining near surface sublayer with a high hardness.

Our future work will be focused on investigation on the influence of the regime parameters of the ball-burnishing process on operating characteristics and phase composition for different grades of austenitic stainless steels, like 316L for example.

References

1. S.D. Slavov, D.M. Dimitrov. ЕСУ. **4-2 (25),** 11-22 (2016)
2. L. G. Odintsov, Hardening and Finishing of Parts by Surface Plastic Deformation. Mashinostroenie, Moscow (1987).
3. S. D. Slavov. International virtual journal "Machines, Technologies, Materials", **8,** 29-33 (2011)
4. Schneider Yu. G. Operational properties of parts with regular microrelief, publishing IVA, St. Petersburg, (2001)
5. I. Altenberger, Proceedings of 9th international conference on shot peening. IITT-International, 6-9, (2005)
6. J.Y. Choi, W. Jin,Scr. Mater., **36 (1)**, p. 99-104 (1997)
7. H Mirzadeh, A Najafizadeh Materials characterization **59 (11)**, 1650-1654 (2008)
8. Müller-Bollenhagen, C., Zimmermann, M., & Christ, H. J. Int. J. Fatigue, **32(6)**, 936-942. (2010)
9. J. Muñoz-Cubillos, J.J. Coronado, S.A. Rodríguez, , Int. J. Fatigue, **95** (2017), pp. 120–131
10. I. Altenberger, B. Scholtes, U. Martin, and H. Oettel, Mater. Sci. Eng. A, 1999, 264, p. 1–16
11. Müller-Bollenhagen, C., Zimmermann, M., & Christ, H. J. Procedia Engineering, **2(1)**, 1663-1672. (2010)
12. Amdouni, H., Bouzaiene, H., Montagne, A. et al. Int J Adv Manuf Technol (2016). doi: 10.1007/s00170-016-9529-9

Removal of the nonconformities in the drug boxes packaging industry

Emilia Bălan*, and *Florin Marian* Dobrea

Politehnica University of Bucharest, Machines and Production Systems Department, Splaiul Independenţei 313, District 6, Bucharest, Romania

Abstract. The paper presents the specific quality aspects of cardboard drug boxes (folding boxes) used as packaging in pharmaceutical industry. The types of defects and nonconformities that occur during offset printing and finishing of the packaging products are being identified and analyzed, such as: differences in color printing, scratches on the printed sheets, cracks during creasing, unparalleled gluing in respect to the closing flaps, ungluing, successive drug boxes stick to each other. The paper also focuses on aspects regarding the nonconformities removal of the drug boxes by establishing a control plan and preventive and corrective methods applicable in different technological stages of the production flow. Monitoring and analyzing activities for quality improvement of the drug boxes, in accordance with the quality specifications required by customers were performed for 20 months on a production line with Heidelberg machines. Nonconformities considered in this paper are also encountered in advertising printing.

1 Introduction

The materials used in packaging are paper, cardboard, wood, plastics or textiles, metal, glass etc. They are used for packaging products obtained in different industrial fields: pharmaceutical and cosmetic industry, food and beverage industry, consumer goods, industrial machinery and equipment. The package protects products and isolates them from the environment, preparing them for transport, warehousing, logistics, sale, in the end reaching the consumer.

Printing the packages and labels (included or made separate from the packaging product) is a field of activity continuously growing. The European Union has allocated a budget of 454 billion euros in 2014-2020 for Southern East Europe, which is situated on the fourth place in the packaging industry. The incomes obtained from printing packages and labels were of approximately 76 billion euros in 2015 [1].

Currently, paper and cardboard packaging are used widely in various industries because they can be reused and recycled. In Europe, their recycling rate reached 72% from a consumption of 80 million tons in 2014 [1]. They can also be incinerated for energy recovery and most of them are biodegradable.

* Corresponding author: emilia.balan59@yahoo.com

Paper and cardboard are made as sheets. They can be differentiated by thickness, the type of fiber pulp (bleached/unbleached, separate mechanical/chemical, fiber first use/recycled), the amount of the fiber, after treatment and additives used in the preparation of the pulp, but also by manufacturing processes (arrangement into layers with other ingredients, the number of layers or laminating operations, etc.). The characteristics of paper and cardboard must provide both the necessary technical performance of the packaging, as well as its visual aspect. This can be achieved by using printing and finishing processes of the packaging [2-3].

In packaging, print runs are large and very large. The strategy of product quality assurance of specialized organizations should be oriented towards identifying and eliminating defects and nonconformities that occur during offset printing and finishing packaging products because they directly affect the performance of the organization [4].

Frequently, the following major nonconformities are found in packaging [2, 5-8]: differences in color after printing, scratches on the printed sheets, cracks during creasing, unparalleled gluing in respect to the closing flaps, ungluing, successive packages stick to each other.

2 Case study

The experimental work presented in this paper was performed in a company with a significant portfolio of clients. The company is certified for integrated management systems according to SR EN ISO: 9001: 2015, 14001: 2005, 18001: 2008.

According to their quality documented information, for treating a nonconforming product, a number of activities should be undertaken, these being included in a strategic plan for ensuring the products quality. Their aim is to control the nonconformities effect, to limit their occurrence and to remove their recurrence [9].

2.1 Heidelberg production line

The experimental part was performed on an integrated Heidelberg production line, consisting of equipment which works in a semiautomatic-automatic system: flat offset printing, hot thermal sheet laminating, flat-bed die-cutter and off-line gluing system. The print runs can vary between 500-5,000,000 products per model, with an efficiency of up to 60,000 products per hour.

The production line consists of: • Computer-to-Plate - *Suprasetter* A74; • offset printing machine - *Heidelberg Speed Master* SM74-5P2 LH; • cutting machines - *Heidelberg Polar* 92NT and 78XT; • hot thermal film laminating machine - *Laminator*; • flat-bed die-cutter machine - *Heidelberg Varimatrix* CS 82; • cylindrical die-cutter machine - *Heidelberg Cylinder*; • gluing machine - *Heidelberg Easygluer* 100.

2.2 Specific nonconformities of printed drug boxes

The aim of this study was to identify defects and nonconformities of the printed drug boxes, to determine the causes of their appearance and to propose preventive and corrective methods applicable in different technological stages of the production flow. Monitoring and analyzing activities for quality improvement of the drug boxes in accordance with the quality specifications required by customers were performed for 20 months (December 2014 - February 2016). During this period, 14,237,650 pieces of drug boxes were produced. Print runs with nonconformities totaled 2.652 million pieces, but only 47,058 pieces have nonconformities. They were found in 49 orders.

Nonconformities encountered due to the use of *GC1* cardboards (coated paper with mixed pulp and white back; chemical and mechanical pulp bleaching; codification according to DIN 19303:2011 standard) are shown in Table 1.

Table 1. Nonconformities generated by improper characteristics of cardboard.

Characteristic of cardboard	Nonconformity generated	Place in technological flow
Cardboard excessive curving	Blocking the equipment and destruction of the cardboard	On printing machine, hot thermal film laminating machine, die-cutter machine, gluing machine
Cardboard with variable porosity or thickness		
Cardboard electrostatically charged		
Cardboard with variable roughness	Poor quality of the printing process	On printing machine

GC1 cardboards are supplied by four manufacturers: Avanta Prima and Charter Integra (Europe), Ningbo Fold and Zenith (Asia). Even if when contracting the cardboard suppliers they provide quality certificates, in practice many deviations from the technical characteristics occur. This generates nonconformities on the technological production flow.

Preventive measures implemented to avoid such nonconformities were:

- renegotiation with customers in order for them to accept a superior quality of the raw materials;
- identification of other potential suppliers of cardboards;
- requesting of quality specifications of cardboards for each run supplied.

Nonconformities frequently occurring due to the offset printing were as follows: variation of the color density; picking of the cardboard fibers or picking its coating layer due to improper transfer of inks (Fig. 1); scratches on the printed sheet (Fig. 2); apparition of spots due to damage on the printing plate surface (Fig. 3).

The corrective measures implemented in order to remove these nonconformities were:

- guiding sticks have been ordered and were mounted in the removal part of the printing machine - in this way the cardboard sheets is no longer touching the crossbar;
- replacement of printed plates, which were scratched.

The major nonconformity that may occur after varnishing is picking the varnish of the printed sheet (Fig. 4). Protective coating of the varnish is recommended to be done in order to avoid unnecessary inconvenience that may occur in further processing the printed material (due to insufficient time allowed for the ink to dry) and for giving an attractive aspect of the final product.

The implemented corrective measures to overcome this nonconformity were:

- performing the settings for applying the varnish layer on the printed sheet more carefully;
- adjusting the temperature of the drying oven or the thickness of the varnish layer.

Nonconformities occurring in the die-cutting operation have two main causes [6]:

- improper behavior of cardboard during compression-folding operations (the compression of cardboard depends on its elastic-plastic characteristics);
- mismatching mold-puncheon used to crease (improper pressure on cardboard by misuse of creasing groove width).

The appearance of cracks on the creases of the boxes is the nonconformity generated by these two reasons (Fig. 5 and Fig. 6). Figure 7 illustrates the microscope image of nonconformities arising when using nonconforming creases. Nonconformities due to the improper use of the fiber of the cardboard are shown in figure 8.

 DOI: 10.1051/matecconf/201711202002

Fig. 1. Nonconformity "lint" due to the improper transfer of inks.

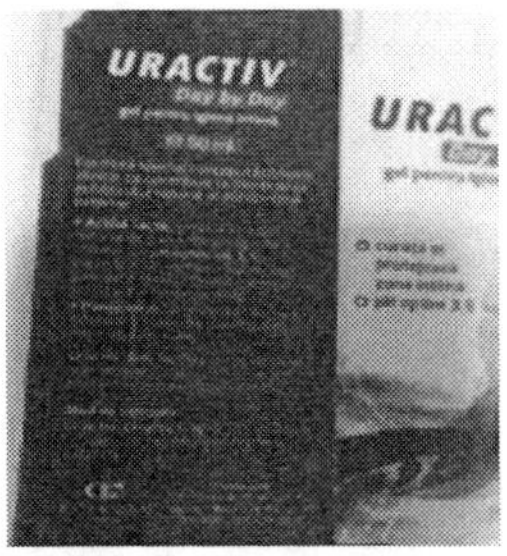

Fig. 2. Scratches on the printed sheet.

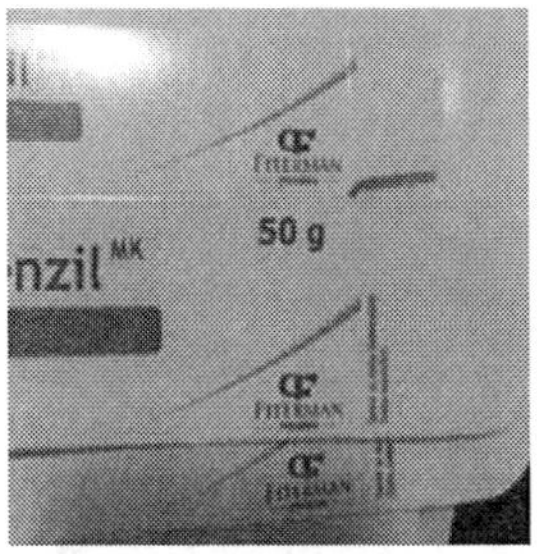

Fig. 3. Stripes due to the damage on the printing plate surface.

Fig. 4. Picking of the varnish from the printed sheet.

Fig. 5. Cracks on the surface of printed boxes, cracks on creasing side edges due to the fiber of the cardboard.

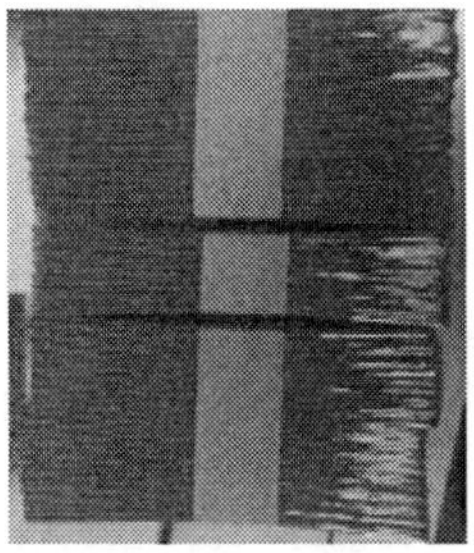

Fig. 6. Cracks on the surface of printed boxes, cracks on creasing side edges due to mold-puncheon assembly.

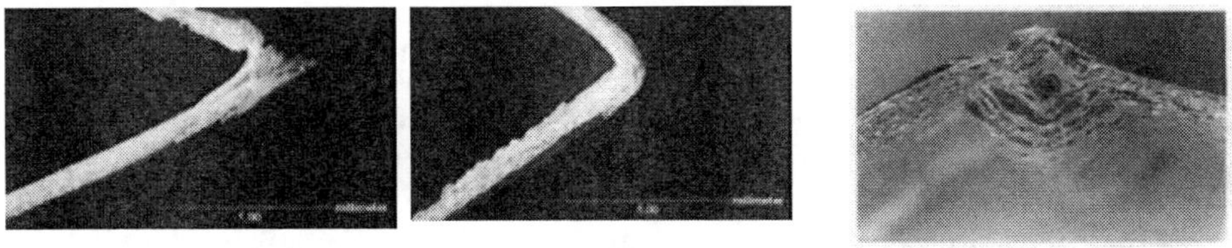

Fig. 7. Nonconformities arising when using nonconforming creases

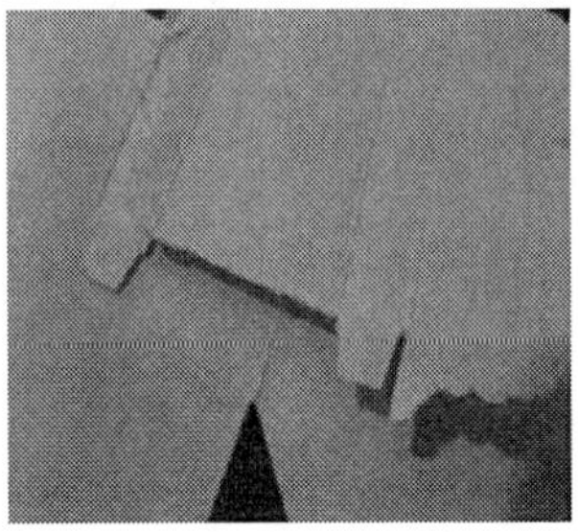

Fig. 8. Nonconformities due to the improper use of the fiber of the cardboard.

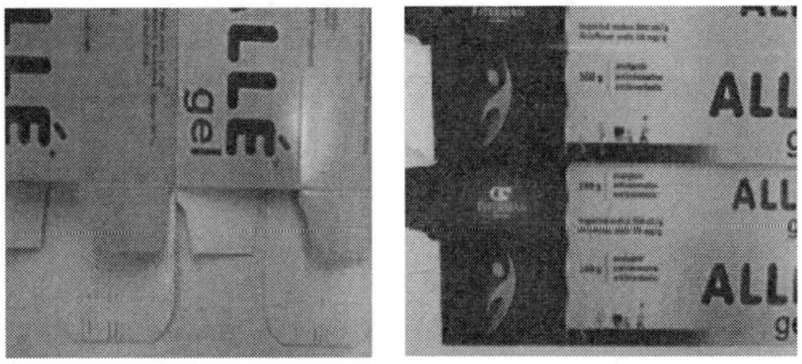

Fig. 9. Traces of ink on the unprinted surface (inside) of the drug boxes.

Taking into account the potential causes for these nonconformities, upon identifying them, a number of tests were performed:

a. ***Die-cutting tests on the same type of GC1 cardboard provided by the four producers.*** After completing the die-cutting of the four cardboard specimens under identical conditions, the following results were obtained:
- on Avanta Prima and Charter Integra cardboard - there were not detected essentially cracks after creasing;
- on NingBo Fold and Zenith cardboard - cracks persisted, blocking the major variations of the main characteristics (thickness of cardboard [μm] and specific weight [g/m^2]) of the same run; major variations were more than +/- 5%.

b. ***Die-cutting tests for the same GC1 cardboard from NingBo Fold in two types of die-cutter machines.*** There was used a flat-bed die-cutter machine, respectively a cylinder die-cutter machine. The cardboard was fed on machines in two different ways: • oriented on fiber direction; • oriented versus fiber direction.
In this study, the occurrence of cracks was observed on the creases perpendicular to the feed direction of the cardboard only for the cylinder die-cutter machine; the cardboard was fed against fiber direction.

c. ***Die-cutting tests for the same GC1 cardboard from NingBo Fold on a flat-bed die-cutter machine.*** In this study, there were used various thicknesses of grooves creases (direct and versus creases). The following aspects have been observed:
- the cracks did not appear when the thickness of grooves crease was decreased, but the quality decreased (the depth of creasing);
- according to the thickness of the cardboard used the recalibration of the die-cutting form with versus creases type "long life" was required;
- a new die-cutting technology involving milled versus creases (Pertinax puncheon) was used; it was obtained a more accurate die-cutting and creasing throughout the entire process.

Corrective and preventive measures applied in order to avoid generating these nonconformities were:
- the die-cutter workers were retrained on how to use the versus crease correlated with the thickness of cardboard, but also correlated with the fiber direction;
- the use of Pertinax puncheon in the die-cutting process became the new norm.

Following these actions, the quality problems such as cracks of drug boxes (cracks due to die-cutting and creasing process) have been decreased with more than 80%.

In the gluing stage of the boxes, the following nonconformities occurred:
1. poor gluing of the drug boxes;
2. successive boxes stick to each other;
3. occurrence of traces of ink on the surface of the boxes (Fig. 9).

The corrective and preventive measures applied were the following:
1. the glue lubrication with a rotating disk was removed, using only the glue lubrication with a digital controlled nozzle;
2. the boxes that remained on the line between the folding-gluing zone and the pressing zone were eliminated if the machine was stopped by accident or deliberately.

Regarding the third nonconformity identified, the gluing equipment producer was consulted and it has been found that the gates, which limited the supply of the machine, were positioned in the same area with the braille codes. Therefore, the ink on the braille points (embossed) was transmitted to the strings of the next package. Consequently, it was proposed to change the position of the gate which limited the supply on a side where there is no braille code (embossed). Thus, the ink transfer was removed.

In addition, the following corrective measures were applied: • reducing the power of vacuum; • increasing the separation distance between the box and the conveyor belt.

A preventive measure proposed was to disassemble and clean the conveyor belts weekly.

In figure 10 there is presented the report of changes, in percentage, of nonconformities identified on the production line for the 49 orders during the monitoring period.

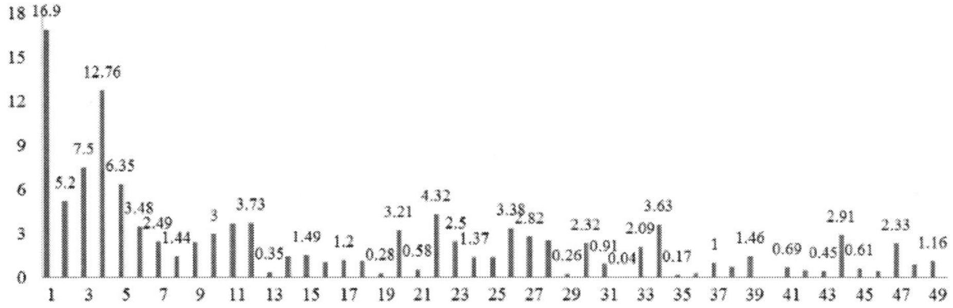

Fig. 10. Evolution of nonconformities of drug boxes.

3 Conclusions

This paper presented specific quality aspects of printed cardboard drug boxes. These may not always be aligned with the quality specifications required by customers.

All nonconformities that occurred during the launching in manufacturing of 49 orders have been studied and analyzed. It was considered that all these nonconformities are significantly affecting the quality of the final product. The originality of the presented research consists in the solutions that were proposed by the authors in order to prevent the amount of nonconformities.

The made research has led to the following conclusions: • a great number of nonconformities was generated at the beginning of the monitoring period, but the situation gradually decreased due to the implementation of corrective and preventive measures • the design of the removal part of the printing machine needs to be improved; • the use of Pertinax puncheon in the die-cutting process became the new norm; • the glue lubrication must be done with a digital controlled nozzle.

Using a high grade cardboard on the production line proved to be the best solution for removing the nonconformities that occur in the manufacturing process of drug boxes. The company has changed its procedure for selecting the suppliers of raw materials in consequence.

References

1. D. Calin, Print Magazin, **65**, (2015) - http://www.print-magazin.ro
2. D. Twede, S.E. Selke, D.P. Kamdem, D. Shires, *Cartons, crates and corrugated board: handbook of paper and wood packaging technology* (DEStech Publications, Inc., 2014)
3. www.sappi.com
4. A. Nagyova, M. Palko, H. Pacaiova, IQC, pp. 33-42, (2015)
5. G. Müller, Paper Technology, **52** (3), pp. 24-25, (2011)
6. E.D. Demaine, *Folding and Unfolding*, PhD thesis, University of Waterloo, Ontario, Canada, (2001)
7. B. Frank, Packaging Technology and Science, **27** (2), pp. 105-128, (2014)
8. B.Y. Sun, Y.L. Li, Packaging Engineering, **4**, 019, (2008)
9. V. Bendic, D. Tilina, *WSEAS*, **7**, pp. 133-136, (2010)

MATEC Web of Conferences **112**, 02003 (2017) DOI: 10.1051/matecconf/201711202003
IManE&E 2017

Contact statuses distributions of flat forming tool profile effect in incremental forming process by finite element analysis

Khalil Ibrahim Abass*

AL-Mustansiriya University, Engineering Collage, Mechanical Engineering Department, Baghdad, Iraq.

Abstract: Single Point Incremental Forming process (SPIF) is a modern forming technique of sheet material. Principles of the process are based on manufacturing by layers-two dimensions layers with step size depth of the tool path. The movement of a hemispherical forming tool is using Computer Numerical Control (CNC) machine to produce the profile of the final product. The description of the process is more complicated by highly nonlinear boundary conditions. The paper presents a study of the effect of the flat forming tool profile through FEA on SPIF that permits the modelling of complex product geometries, material behaviour and boundary conditions. The results showed that the model of simulation can predict the behaviour of contact tool-blank, and the accuracy of product.

1 Introduction

Single Point Incremental Forming (SPIF) is a flexible sheet metal forming process with potential for small production and for rapid forming tools. Numerical Control (NC) milling machine is used for execution of the product and the forming tool path is programmed by Computer Aided Manufacturing (CAM) software. The geometry of the product is included into two-dimensional (2D) layers and the designed product profile is deformed in multi-layer movement of a hemispherical forming tool [1] (Fig. 1, [2]).

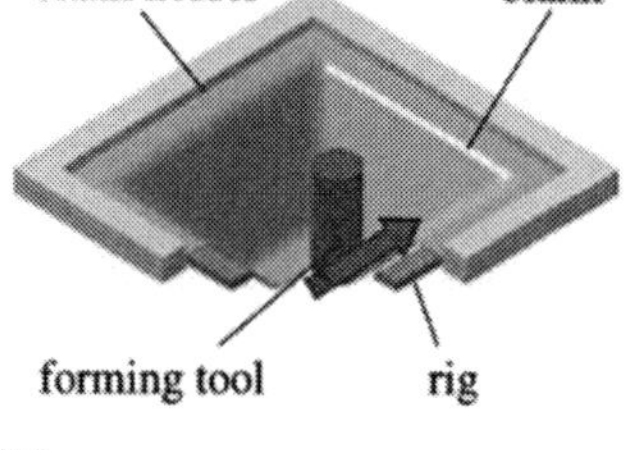

Fig. 1. SPIF technology principle [2].

The deformation of the sheet metal occurs only in the processing zone and is a combination of shearing and stretching processes. Finally, the thinning occurs during SPIF process and the wall thickness along the product section is lower than the blank thickness [4].

* Corresponding author: Khalil1969a@yahoo.com

Main parameters of the process that influence the quality and the accuracy of the formed blank are: spindle speed (ω), feed rate (v), tool vertical step (Δz) and tool diameter (D_p) [3].

Advantages of SPIF process: highly flexible process, low hardware costs, enabling production of complex shapes, shorter time-to-market. SPIF process would be useful for prototyping and small-batches production for automobile, biomedical and aerospace industries [5]. Disadvantages of SPIF process: long forming time when compared to deep drawing process, small-lot production, low accuracy of the geometry, high springback, in bending edges and convex radius areas [6].

The forming tool radius effect on accuracy and ability of SPIF deforming process is studied with wide ranges (between 2-25 mm) of radiuses of the round head. The interaction between the tool step size and sheet metal thickness was researched [7]. They concluded that the maximum forming wall angle is influenced by this interaction.

The effect of forming tool radius, step size, and speed of forming on the formability were studied [8]. The formability was observed through a surface response method called central composite rotational design. The authors concluded that the forming tool radius and the step size have an important effect on the formability, and it can be decreased or increased depending on the step size.

Five hemispherical forming tools with 4-25 mm radius to deform aluminum AA1050-H111 were used in the experiment [9]. As a result, the stability effect is not significant to ensure localization, and the failure mechanism will change to promote fracture with suppression of necking when using the small forming tool.

A range of the forming tool diameter between 5-10 mm was selected, to predict the effect with low rotating speed by FEM [10]. While, selected a range between 4-6 mm forming tool diameter to study the forming tool radius effect on polymer products [11].

Two types of the forming tools, flat and hemispherical heads were studied [12]. The paper showed that the flat head forming tool required lower forming force, and can give a better accuracy of the product profile and formability than the others. A lot of researchers didn't take into account the effect of the forming tool radius and selected one value of the radius, depending upon previous papers [13, 14, 15], who selected a 5 mm forming tool radius, or [16, 17] who selected a 15 mm forming tool radius.

This research focuses on using a flat forming tool profiles and on evaluating the product profile, thickness and strain distribution, compared to other profiles. A new type of evaluation was used - contact status distributions by Auto CAD application and FEA method to observe the behaviour of the entire blank surface and the forming tool surface.

2 FE simulation model development

The FE simulation model was undertaken to represent the metal forming process. The input stage involves physical SPIF model geometric, material properties and boundary conditions. The processing stage takes into account the analysis of FEM and evaluated shape, for the failure product or success. The output stage includes the final product for calculation and conclusions. FEM analysis application program, ANSYS 11 of linear and nonlinear general purpose, is used. Script files of the commands were used to improve specific tedious data response tasks during pre-processing and post-processing, by using the FEM in a system for an integrated engineering methodology to sheet metal forming.

The FE simulation model of SPIF process was developed. The model contains the effect of direct and indirect contact between the tool set - blank, in addition to the elastic-plastic behavior of the blank material. To develop the FE model, the effect of flat forming tools was studied with constant values (dimension and materials) of fixture (die and blank holder) and blank, as shown in Tables 1 and 2.

Symmetric components to model blanks are used, that are rotated about an axis symmetrically. The analysis of the two dimensional sector of the deformed blank is carried out in order to evaluate the complete shape of the product. A pure Aluminum AA1050 as a workpiece was used, and the mechanical properties of the material are shown in Table 2.

Table 1. Dimensions and variables of SPIF tools.

Variable	Value
Dia. of active forming tool, D_p	Ø8 – 16 (mm)
Radius of Die, R_d	5 (mm)
Radius of forming tool edge, R_P	4 (mm)
Dia. of blank, D_b	Ø220 (mm)
Blank thickness, t	0.9 (mm)
Dia. of die hole, D_d	Ø180 (mm)
Step size, x and y directions	0.1 (mm)
Friction coefficient, μ	0.05
Dim. of start tool position	Ø76 – 80 (mm)

Table 2. Material properties of the blank.

Material, Al (AA1050)	
Yield stress, σ_y, MPa	80
Young's modulus, **E**, GPa	75
Poisson's ratio, **ν**	0.3
Density, ρ, g/m^3	2700
Tangent modules, E_T, Pa	0.5

The elements PLANE42-2D and V15C0106-2D were chosen to represent the forming tool set (fixture, die and forming tool) and blank respectively. Both elements have the nodal directions x and y translation. For the blank meshing is used a quadrilateral mapped mesh.

The boundary conditions include loading, displacement, friction coefficient and (direct and indirect) contact and are represented as following (Fig. 4):

1. The constraint of the die was held fixed in two directions x and y, N1 and N2, by nodal;
2. The constraint of the blank holder was as to allow in one direction y movement, N3 and N4;
3. The constraint of the blank was along the centerline, N5 with zero displacement in y direction;
4. Movement of forming tool was identified in a profile with constant feed rate (Fig. 4);
5. The specific incremental loads and load steps in two directions are applied.

The pair Point-to-Surface contact elements CONTAC169 and CONTAC171-2D were used to represent the contact conditions, like a Rigid-to-flexible contact between the tool set and the blank respectively. The contact surfaces, a real constant set were used for each interface. The convergence criteria and non-linear analysis are applied. The convergence tolerance was based on minimization of the residual force.

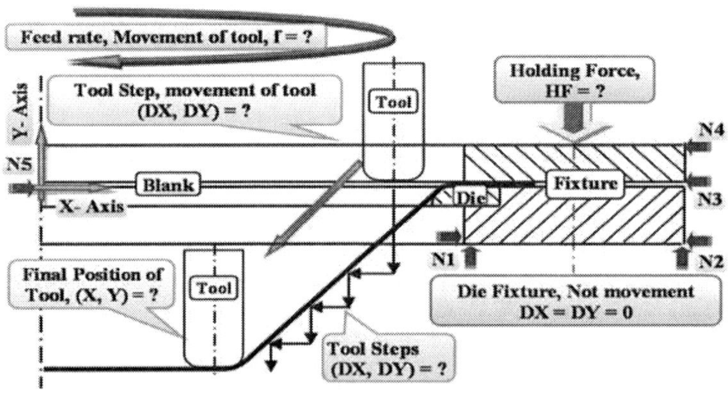

Fig. 4. SPIF process boundary conditions.

3 Results and discussion

The results evaluation included the following: a) forming tool movement examination, resulting from Auto CAD application; b) analysis of the simulated product profile based on the forming tool movement; c) thickness and strain distributions comparison; d) contact statuses distributions assessment at the interface between tool and workpiece.

The contact statuses distributions assessment at the interface between tool and workpiece evaluation is representing in Fig. 5. In this case, the forming tool moves at a 45° at three stages (2x2, 20x20, 40x40) mm in x and y directions.

The contact in case of the spherical profile case begins with a point, until the shape of the blank surface becomes identical to the forming tool edge profile shape, which occurs gradually, with increasing the depth, more precisely at final stroke. The highest value for the contact distance was 2.64 mm, around the edge.

In case of the flat forming tool, the observation was that at the beginning of the stroke, the contact takes the form of a line, and as the depth increases, it takes the form of the edge of the forming tool. At the end of the stroke, the maximum contact distance was 10.64 mm. Another observation was that in all the cases, the contact area increases as depth of deformation increases as follows: 0 mm to 2.64 mm, 2 mm to 4.64 mm, 4 mm to 6.64 mm, 6 mm to 8.64 mm, 8 mm to 10.64 mm. Also, the shape of the product end edge is going to emulate the shape of the edge of the forming tool profile.

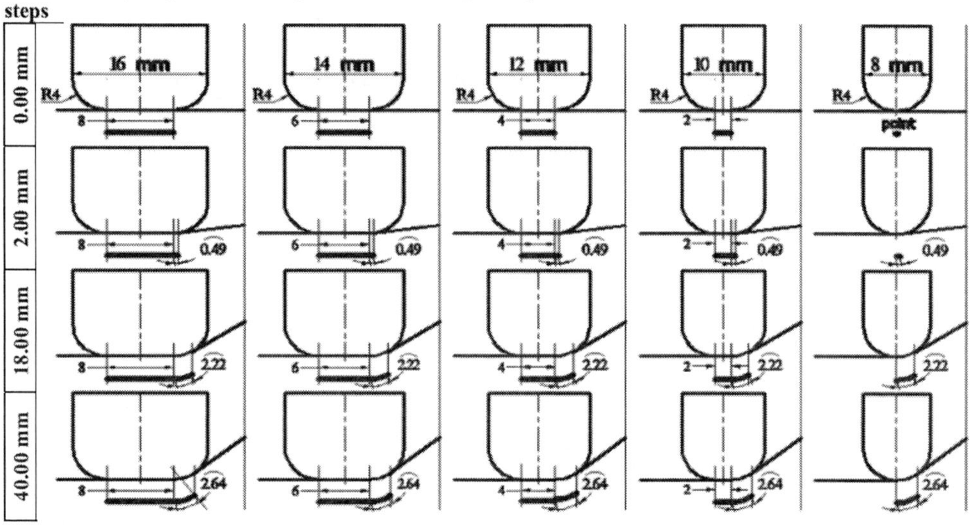

Fig. 5. Forming tool profile movement and a comparing of contact areas of spherical and flat.

The contact behaviour of each forming tool profile (simulation and Auto CAD models) shows differences, as illustrated Figs. 5 and 6. The regions are defined depending on the contact state (forming tool surface – deformed blank surface), the simulation prediction shows four regions (far open, near contact, sliding and sticking), with different shapes and areas according to the deformation depth (Fig. 6).

Far open contact occurs between the flat region of the flat forming tool, which indicates small contact area. It also indicates the fact that the real contact only occurs at the forming tool edge contour, while two factors have impact on it: revolving speed and forming tool active diameter.

In case of the spherical forming tool, springback has a great impact, resulting in a larger contact area, higher friction and plastic region, where sticking also occurs.

Based on the contact statuses for successive depths (0.2, 10, 20, 30 and 40 mm) shown in Fig. 6, the following remarks can be made:

a. The contact has higher effects when the area is increased, and when depth increases
b. Sticking region increases and is transferred around the forming tool when depth of deformation increases, raising the probability of fracture, notably in the case of spherical forming tool, due to recurring deformation in wider areas.

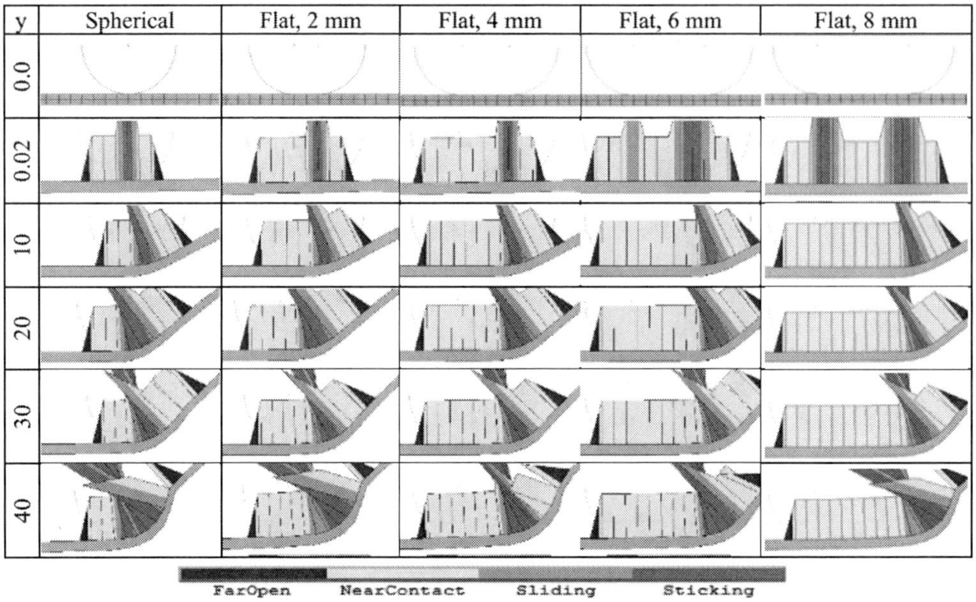

Fig. 6. Forming tool profile structure and contact status of flat forming tool profiles.

The stages of the deformation process and the forming tool strokes from the FEM simulation are presented in Fig. 7. The evaluation shows the results of the forming stages:

a. The end shape of the product has the same shape as the forming tool.
b. The final predicted shape of the spherical profile tool was quite different (Fig. 5).
c. Springabck is concentrated around the edges of the forming tool edge, at end stroke.

The thickness distribution along the section of the deformed shape section was evaluated at 2, 20 and 40 mm depths, as shown in Fig. 7. Thickness is constant at 2 mm and 20 mm depths. The changes in thickness occur at 40 to 50 mm distance from the product center, at 40 mm depth, around the edge of the forming tool, at final stroke. This phenomenon has no significant effect in case of the flat forming tool shape, even if the area is increased.

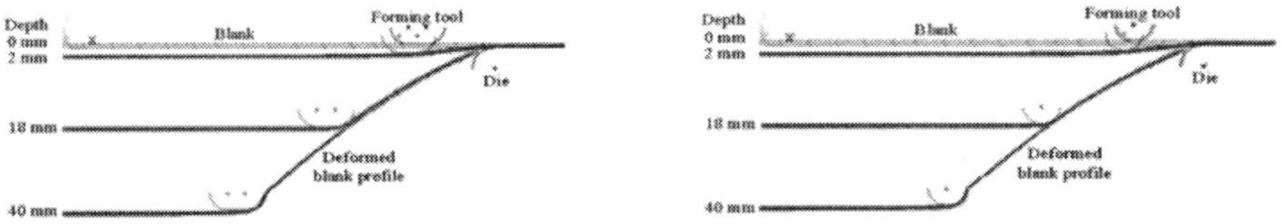

Fig. 7. Forming tool movement and product profiles and thickness distributions steps of Spherical and Flat.

The thickness distribution is influenced by the forming tool profile with radius from 2 to 6 mm is presented in Fig 8. The analysis is performed for a 40 mm forming tool stroke.

The highest thinning occurred at 40 mm depth, at 45 mm to 50 mm from the product center, in case of all products. Severe metal bending is occurring in case of the spherical forming tool lead to maximum values for thinning, which enables the formation of neck. At this point, product failure is clearly indicated. Thickness distribution varies according to forming tool changes, only in the case of the forming tool edge profile.

The edge of the forming tool has an obvious effect detected along the section of the product, even in the areas which have not been in direct contact (from 0 mm to 85 mm from the product center), as is presented in Fig. 8-a. Another observation is that between 45 mm to 85 mm from the product center, the thickness distribution is improved when the tool radius is increased. Thickness is reduced in the area from 0 to 45 mm from the product center (end of stroke).

Thickness distribution is not influenced by the flat forming tool area, from 0 mm to 110 mm from the product center. As shown in Fig. 8-b, the only exception is the last position of the forming tool, where an improvement can be noticed, but it is not incidental.

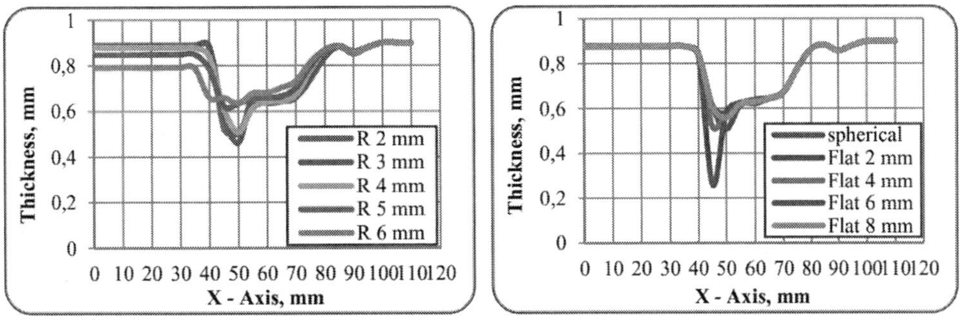

Fig. 8. Thickness distribution of spherical and flat forming tool.

4 Conclusions

The present work establishes the fact that any adjustment of the forming tool profile influences the SPIF formability based on surface shape of the interface. Product accuracy is not necessarily improved if the contact area is increased.

Defects appear as a consequence of the interaction between forming tool profile and the sheet thickness. As a result, the tool edge radius has impact on the thickness distribution.

There is no deformation effect of the flat forming tool area during the deformation steps except increasing the stiffness of the forming tool used. And the contact tool - blank only occurs at the edge of the tool. The wall and corner fold appear around the forming tool as a result of modifications in the flat forming tool edge.

Using 2D simulation model with high number of boundary conditions is easier, and the evaluation of contact statues is easier and quicker indication.

References

1. R. P. Singh, G. Goyal, Int. J. Eng. Research and Applications (IJERA), pp. 33-37, (2014).
2. M. Bambach , M. Cannamela , M. Azaouzi , G. Hirt and J. L. Batoz, Adv. Meth. Mate. Form, pp. 233-250, (2007).
3. R. Crina, Vasile Alecsandri Uni. of Bacau, Romania, J. Eng. Stud. Res., **16**, 4, pp. 33-39, (2010).
4. G. Hussain, N. Hayat, G. Lin, J. Mech. Sci. and Tech., **26**, 8, pp. 2337-2345, (2012).
5. M. A. Dittrich, T. G. Gutowski, J. Cao, J. T. Roth, J. Prod. Eng., **6**, 2, pp. 169-177, (2012).
6. R. Benmessaoud, Int. J. Adv. Res. Comp. Sci. and Soft. Eng., **4,** 5, pp. 1035-1044, (2014).
7. M. Ham, J. Jeswiet, CIRP Annals - Manufacturing Technology, **55**, 1, pp. 241-244, (2006).
8. G. Hussain, L. Gao, N. Hayat, Int. J. Adv. Manuf. Tech., **46**, 5-8, pp. 543-549, (2010).
9. M. B. Silva, P. S. Nielsen, N. Bay, Int. J. Adv. Manuf. Technol., **56**, 9-12, pp. 893-903, (2011).
10. P. Eyckens, B. Belkassem, C. Henrard, Int. J. Mater. Form., **4**, 1, pp. 55-71, (2011).
11. T. A. Marques, M. B. Silva, P. A. F. Martins, Int. J. Adv. Manuf. Tech., **60**, 1-4, 75-86, (2012).
12. X. Ziran, L. Gao, G. Hussain, Int. J. Adv. Manuf. Tech., **46**, 9-12, pp. 1113-1118, (2010).
13. M. Fiorotto, M. Sorgente, G. Lucchetta, Int. J. Mater. Form., **3**, 1, pp. 951-954, (2010).
14. A. K. Behera, J. Verbert, J. R. Duflou, Computer-Aided Design, **45**, 3, pp. 575-590, (2013).
15. C. F. Guzmán, J. Gu, J. Duflou, Int. J. Solids and Struct., **49**, 25, pp. 3594-3604, (2012).
16. B. T. Araghi, G. L. Manco, M. Bambach, CIRP Annals-Manuf. Tech., **58**, 1, pp. 225-228, (2009).
17. C. Robert, A. Delameziere, J. Mate. Pro. Tech., **212**, 5, pp. 1123-1131, (2012).

Using two tools movement in incremental forming process by finite element analysis

Khalil Ibrahim Abass*

Al Mustansiriyah University, Engineering Collage, Mechanical Engineering Department, Baghdad, Iraq

Abstract. Single Point Incremental Forming, SPIF and Two Point Incremental Forming, TPIF are two types of Incremental Sheet Forming ISF processes with high economic potential payoff for rapid prototyping applications. The process requires a CNC machining centre, a spherical tip tool and a simple support to fix the workpiece. The process is more complicated by highly nonlinear boundary conditions, namely contact and frictional effects. However, this process, numerical approach, dominated by the Finite Element Method, FEM is being used. The paper presents the data and main results of a study on the effect of using two forming tool movements through FE analysis. The certain process conditions referring to the test blank, tool, etc., applying ANSYS application program. The results show that the upper TPIF process is more stable during the deformation process steps by chattering statuses evaluation and improve the strain distributions of the deformed blank.

1 Introduction

The advanced Incremental Sheet Forming or ISF process includes the local (point) deformation of the blank sheet by using a simple geometry (spherical or hemispherical) tooling system [1]. The tool is mounted on a CNC machine. A forming tool path is moved accordingly over a clamped blank with step size (vertical feed) [1, 2, 3]. The complicated geometry information is resolved into a series of 2D-dimensional layers, and then the plastic deformation is carried out layer-by-layer through the CNC controlled movements of the forming tool to obtain the desired part [4]. A sketch of ISF types is presented in Fig. 1 [3].

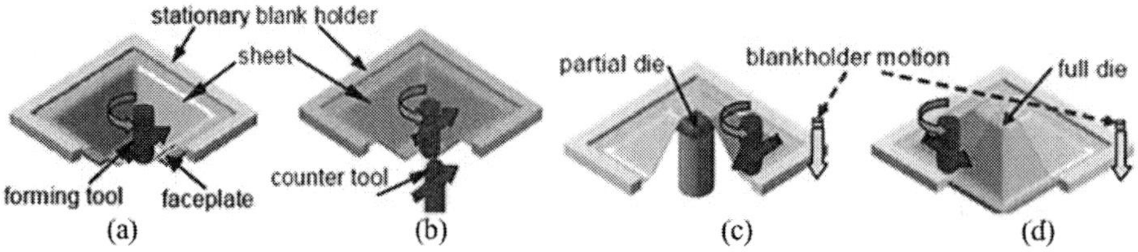

Fig. 1. Configurations of asymmetrical ISF [3].

The applications of the Single Point Incremental Forming process (SPIF) include production of small quantities or fast prototyping, as it is a highly flexible forming process. Sheet metal parts can be obtained without using special dies [5].

* Corresponding author: Khalil1969a@yahoo.com

The technology of SPIF has several benefits, which include high flexibility, low costs for hardware and high formability. Complicated geometries can also be obtained through SPIF without using expensive dies. The time to market is also reduced due to the fact that there are no dies needed for the deformation. SPIF could be successfully employed in automobile, aerospace and biomedical industries for prototyping and small-lot productions [6].

The main drawbacks of SPIF are low predictability due to springback, wall angle restraints, thinning of the material and the product surface quality. There have been employed some approaches to reduce the negative effects of the process: warm forming [7], laser assisted heating [8] and electric heating through the tool [9]. The ISF process is influenced by several factors, such as type of tool, use of lubrication, plane anisotropy, feed rate [10].

Two Point Incremental Forming (TPIF) with partial die was introduced [11]. According to the analytical model obtained, the mechanical deformation of SPIF and TPIF are handled. Also, the partial die prevents the formation of neck in TPIF. The parts obtained with partial die by TPIF have higher geometric accuracy than those obtained through SPIF. This can be explained by smaller elastic recovery at loading.

Asymmetric Incremental Sheet Forming (AISF) was mixed with stretch forming to obtain a new hybrid process [12]. Compared to AISF, the new process requires less time, especially when identifying in advance the processed areas. The thickness distribution was more homogenous and the maximum thinning was improved.

The use of different deformation steps on both surfaces in order to test the process self-capabilities to correct inaccuracies was analyzed [13]. At the second repeated step, the accuracy increases. At new steps, the accuracy is not improved.

This research focuses on using a new type of forming tools movement (upper TPIF) and evaluating the product profile, thickness and strain distribution after usage. Using the simple way to simulation upper TPIF model (2D model), a new type of evaluation was employed - chattering status - to observe the behavior of the entire blank surface as a reflection of the forming tool movement. Some of the applications of Aluminum material used are: buildings; roofing, cladding, signage, road signs and furniture.

2 FE model development

The FE model was developed to simulate the forming process using ANSYS 11, purpose of linear and nonlinear FEM analysis program. The FE simulation was incorporated into a combined system of sheet metal forming. The model developed to simulate the SPIF process incorporates the effects of contact between the tool set and workpiece. In the FE model, constant forming tool, fixture and blank characteristics were used in order to analyze the effects of pre-forming tool movement, as shown in Fig. 2.

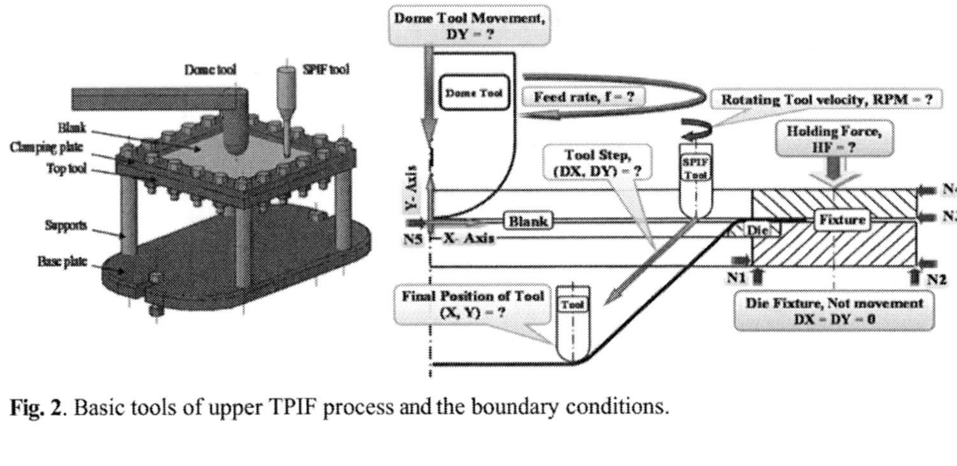

Fig. 2. Basic tools of upper TPIF process and the boundary conditions.

The blank is represented through a quadrilateral mapped mesh. Fig. 2 presents the boundary conditions - displacement, loading, contact and the forming tool speed was constant.

For the model, the blank type is of 220 mm outside diameter (Db) and 0.9 mm thickness (t). The forming tool profile diameter (Dp) is 10 mm, and all six cases are with the same die profile (Dd) of 5 mm. The interface tool-blank is represented with the friction coefficient (μ), 0.05. For the material, Aluminum EN AW 1050 (Al99,5) was selected, the specific mechanical properties of a tensile test resulted from load-extension curve are shown in Table 1.

Table 1. Material properties of the blank.

Variable	Density, ρ, g/m^3	Young's modulus, **E**, GPa	Poisson's ratio, **v**	Yield stress, σ_y, MPa	Tangent modules, **E$_T$**, GPa
Value	2700	75	0.3	80	0.5

For the simulation, the element PLANE42-2D was selected to represent the tool set and the element V15C0106-2D to represent the blank, and both have x and y translation directions.

The element representing the contact tool set, which includes the forming tool, die and blank holder, is CONTAC169 and for the blank is CONTAC171-2D Point-to-Surface Contact, like a Rigid-to-Flexible contact. The real constant set was used for each contact surface. A non-linear analysis, convergence criteria, incremental load and specified load steps are applied. The convergence tolerance was based on minimization of the force residual.

For the tool path, the CAD model built includes the properties of the product profile, all dimensions related with the fixture parts and the forming tool coordinates.

3 Results and Discussion

Five predicted AutoCAD models to produce the product profile with pre-movement in depths, h = 0, 0.2, 1, 5 and 10 mm, and the details of the model are presented in the Fig. 3.

These procedures are used as a primary movement to start producing a new profile of SPIF process with forming tool path movement 50 mm depth and with 45° degree.

In Fig. 3 the FEM models of the products are presented. The dome forming tool starts to move vertically on the y-axis to distance h mm (Fig. 3 a-1), followed by the SPIF forming tool according to the tool path profile (45° degree) (Fig. 3 a-2) to be at that time move together until product profile complete (Fig. 3 a-3).

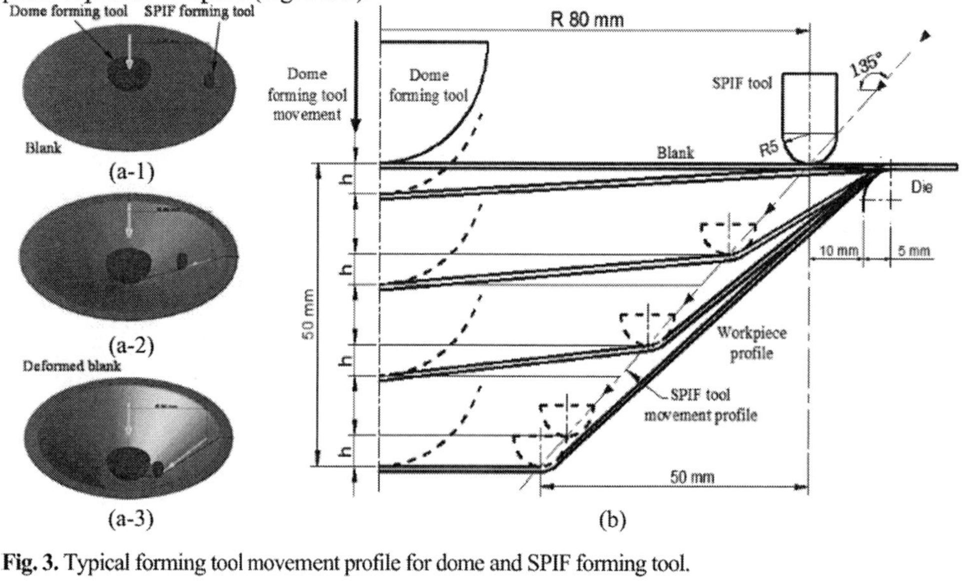

Fig. 3. Typical forming tool movement profile for dome and SPIF forming tool.

The thickness distribution evaluation of the upper TPIF deformed product is shown in the Fig. 4. The section of the product is divided into four regions, as follows:

- 100 to 110 mm from blank center (fixed region), there is not effect of the tool movement;
- 90 to 100 mm from blank center (backing plate profile effect region), maximum reduction is 0.08 mm from 0.9 mm blank thickness, when using 10 mm h value;
- 40 to 90 mm from blank center, (SPIF forming tool movement effect region), clear regular reducing in thickness distribution of all h values;
- 0 to 40 mm from blank center, (dome forming tool movement effect region), where the thinning decreases with increasing the h value, as illustrated in the Fig. 6.

The results for strain distribution include the strain values for each node of the numerical model and the values of all nodes at each forming tool step size. As a consequence, a very high number of data is included in the final results.

To analyse the effects of pre-forming tool movement, Fig. 5 includes strain distribution for the upper surfaces, for 50 mm vertical stroke and 45° angle of forming tool.

The strain peak values are concentrated along the blank section between 0-45 mm distances from the blank center and increase towards the blank center. The strain increases uniformly and the strain values are acceptable at distances 45-100 mm from the blank center. There is no deformation occurring under the blank holder (100-110 mm from the blank center).

The maximum strain values are concentrated on the places between the forming tool movements, as well as on the places under the center of the forming tool dome (0-10 mm from the blank center). The strain values increase with increasing the h value.

The evaluation of the strains along the deformed blank section is illustrated in the Fig. 6. The section of the product can be divided into four regions - blank holder, die edge, tool path and between tools. In the first and second regions there is no effect of the forming tool movement, while in the third and the fourth regions the highest strain distributions appear in the areas that are in direct contact or near the forming tool position than others areas along product section.

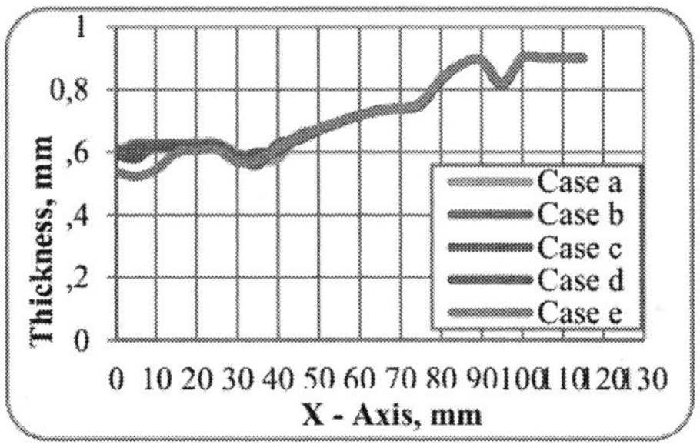

Fig 4. Thickness distribution of different ways.

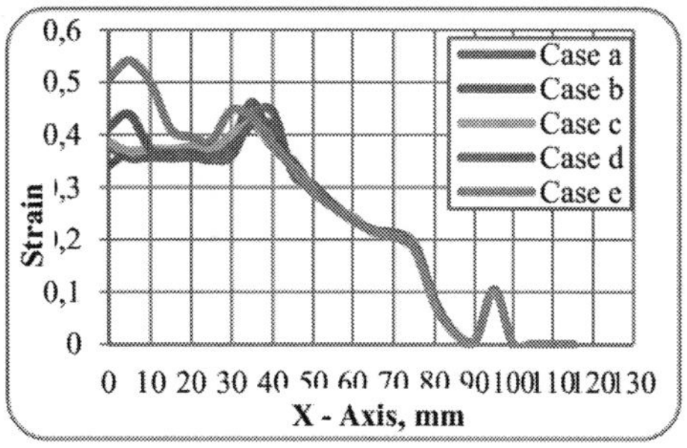

Fig. 5. Strain distribution of different ways.

The strain distributions are concentrating on the zone between 30-50 mm from the blank center. This is obviously improved on two levels. The first level is to reduce the surface strain on the upper blank surface and on blank section, while the second level is to reduce thinning with increasing the h value (cases a, b, c, d and e respectively). At same time, the maximum strain value increased from 0.495 to 0.542 and moved from the zone near the forming tool edge (in case a), to be under the dome tool (in case e) (Fig. 6). Other important notes: the profile of the product is the same in all the cases, which is clearly a good indication that any changes in h value will not affect the product profile resulting from this processing model.

The rotating tool velocity, the constant feed rate along the contour of the product and the support-free wide blank surface produce an operation similar to continuous forming with pulses. The operation is producing a continuous vibration of the blank. The level of vibration is affected by depth of deformation and product shape. Large support-free area and increasing vibration lead to chattering effect as is shown in Fig. 7.

Successive stages that are used in pre-forming tool movement to produce the SPIF product profile are shown in the Fig. 7. It is clear that the forming tool stroke influences greatly the chattering values, in which these values increase with depth of stroke. The peak height appears to be a good indicator to assess whether a forming operation will be stable or not. The severe of the chattering statuses distribution is located at the positions where the forming tools are.

Six steps, 0.2, 10, 20, 30, 40 and 50 mm depth, are used to evaluate the chattering statuses during simulation process (Fig. 7). At two positions the chattering appears under SPIF and dome forming tool, where the direct contact occurs.

At the first step, there is no chattering status because of the elastic properties of the blank material. The same situation appears at the end stroke, as well, but because of high tension and reducing the unsupported areas.

Using the dome forming tool caused clear reducing in the chattering statuses at all the steps under the SPIF forming tool, and transfer it to be under the dome forming tool with low levels. That means the dome forming tool absorbed the vibrations, first by reducing the support free-area and giving the blank stability during blank metal deformation by the SPIF forming tool with continuous direct contact between tool and blank.

Evaluating the chattering statuses under the SPIF forming tool in all cases with six steps is represented and detailed in Fig. 8. At the first steps (0.2 mm depth) low values were registered and without chattering in cases d and e, where there is no direct contact because the h value which is higher than the step size.

At step 10 mm depth, there are no chattering statuses in all cases, and that is a good indication of improving the conditions of the process. At step 20 mm depth, the chattering values increased at 5 and 10 mm h values, slightly decreased at the step 30 mm depth and at 10 mm h value very low chattering values were recorded.

Chattering values levels continued to decrease during the step 40 mm depth with all h values, as well as until the latest step 50 mm depth of all h values. Depending on the chattering statuses, increasing the h value with more than 1 mm depth makes the process more stable, especially at the SPIF forming tool position.

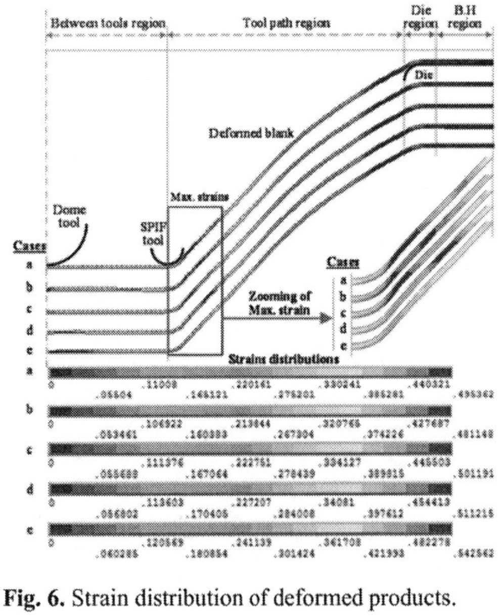

Fig. 6. Strain distribution of deformed products.

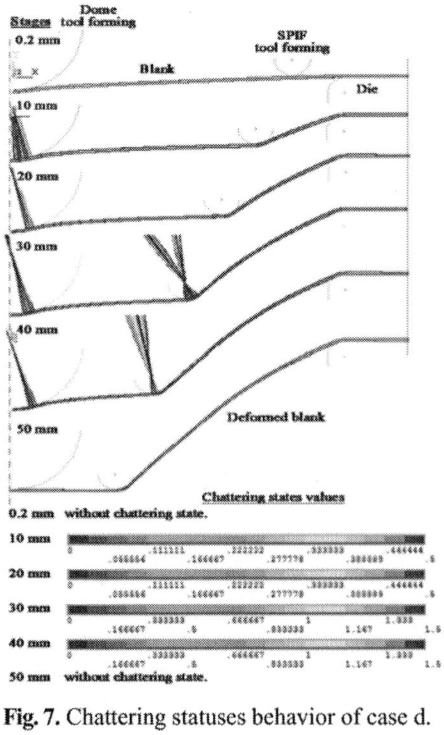

Fig. 7. Chattering statuses behavior of case d.

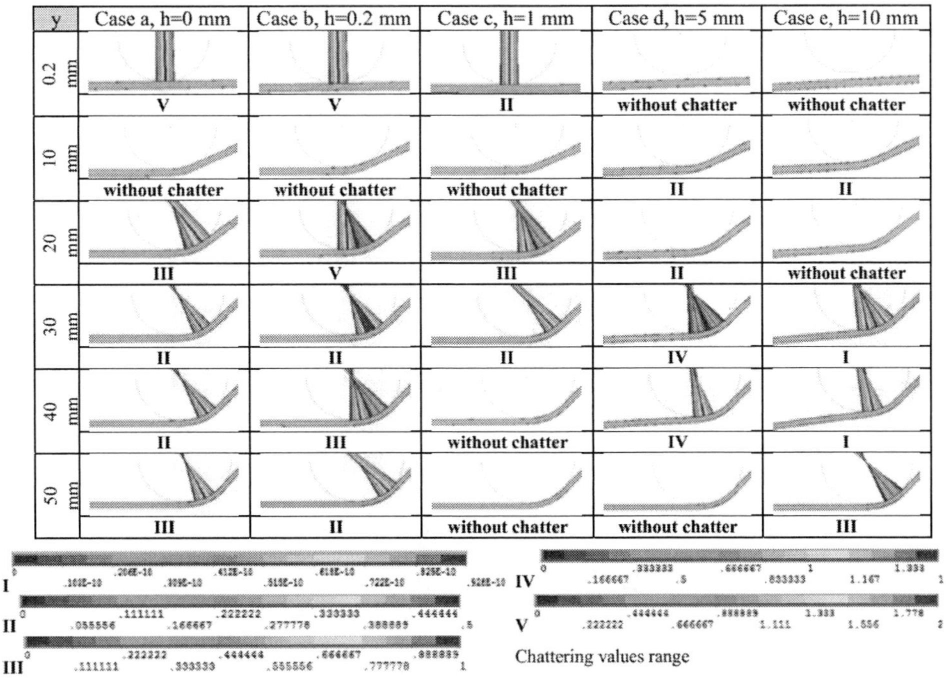

Fig. 8. Chattering statuses behavior at SPIF forming tool.

4 Conclusions

The 2D simulation model of ISF by FEM with pre-forming tool movement (upper TPIF) can predict an ideal profile of processing trajectory. The high values of strain distributions along the section of the product profile appear to be good indicators to assess whether a product will be successful or not. The results also showed improvement in the strain distributions along the deformed blank section by using upper TPIF process.

Based on the chattering statuses evaluations the upper TPIF process appears to be more stable during the deformation process steps than traditional SPIF.

In addition, decreasing the chattering statuses means reducing the pulses, keeping continuously deformation contact tool-blank and reducing the surface roughness.

References

1. J. Kopac, Z. Kampus, J. of mate. Pro. techno., 162-163 622-628 (2005)
2. A. Hadoush, A. H. van den Boogaard, Int. J. Mater. Form. **2** pp 181-189 (2009)
3. U. Khare , M. Pandagale, ICAET-2014, IOSR J. Mech. Civil Eng. PP 42-46 (2014)
4. R. P. Singh, G. Goyal, Int. J. Eng. Research and Applications (IJERA) pp 33-37 (2014)
5. R. Benmessaoud, Int. J. Adv. Research in Comp. Sci. and Soft. Eng. **4** pp 1035-1044 (2014)
6. M. A. Dittrich, T. G. Gutowski, J. Cao, Prod. Eng. Res. Devel. **6** pp 169-177 (2012)
7. G. Ambrogio, L. Filice, G. L. Manco, CIRP Annal – Manuf. Tehcn. **57** 1 pp 257-260 (2008)
8. J. R. Duflou et al, Int. J. Mach. Tools and Manufacture **48** 5 pp 543-549 (2008)
9. F. Guoqiang, S. Fentao, M. Xiangguo, Int. J. Adv. Manuf.Techno. **49** 9 pp 941-947 (2010)
10. J. Sheoran , I. Monga, Int. J. for Tech. Res.In Eng. **2**, 8, pp 1365-1368 (2015)
11. M. B. Silva, P. A. F. Martins, JMEPEG **22** pp 1018-1027 (2013)
12. B. T. Araghi, G. L. Manco, M. Bambach, CIRP Annals-Manuf. Techno. **58** pp 225-228 (2009)
13. G. Ambrogio, L. Filice, Int. J. Mate. Form. **5** pp 269-274 (2012)

Study regarding the springback of metal sheet parts after bending

Mihail Purcarea*, and *Andrei* Dumitrescu

Politehnica University of Bucharest, Department of Manufacturing Engineering, Bucharest, Romania

Abstract. Plastic deformation of materials is accompanied by elastic deformation. After the bending process, the elastic deformation disappears and the part is changing its dimensions comparing with those of the tool. The change of bent part's dimensions due to the elastic properties of material is called *elastic springback*. It influences the precision of bent parts, especially the bending angle, and this is why its value should be known in order to take the necessary measures in order to reduce or even annihilate its effects. Its value can be determined with analytical relations, from various nomograms or with experimental relations. Because the analytical relations offer results far from reality, this paper proposes a multivariable relation of springback (dependent variable) as a function of several independent variables obtained on the basis of experimental results, in real working conditions and using the method of regression functions. This method is based on simultaneous variation of considered variables, using a limited number of values in the experimentation domain. A split experimental program was used. The obtained relation is true for any combination of values for considered variables, but, from their variation domains, it can be obtained separately the effects of independent variables on dependent variables.

1 Introduction

When bending the parts, the plastic deformation is accompanied by an elastic deformation that disappear, after removing the forces which has produced the bending. The part has the dimensions α_r, r_r, φ_r, different from those of the tool (α_0, r_0, φ_0), as displayed in Figure 1, where α, r and φ are the bending angle, bending radius and supplementary bending angle in the two situations. The modification of the bent part dimensions, because of the material's elastic properties, is called *elastic springback* or *elastic recovery*. Springback has influence on the bending angle but also on the bending radius, the biggest influence being on the bending angle.

The value of elastic springback can be determined by analytical relations or is obtained usually with empirical relations or with graphics and nomograms [1, 2, 3]. In all cases, the obtained results are more or less satisfactory, meaning that the results obtained with these relations, graphics or nomograms are different from those obtained by measuring the real processed parts.

* Corresponding author: mhlpurcarea@yahoo.com

DOI: 10.1051/matecconf/201711202005

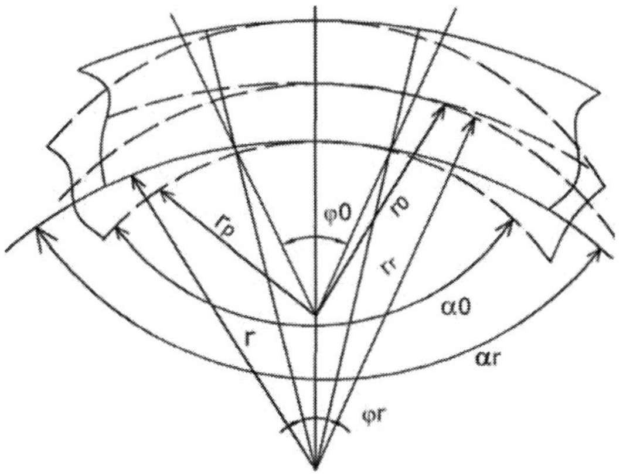

Fig.1. Parameters of bent part before and after removing from the work zone.

These differences arise mainly because of the simplifying hypotheses. For example, in the analytical determination of elastic springback, it is considered that the bending moment characterizes a state of total plastic deformation, when in fact, the state of deformation is an elastic-plastic one with hardening. The importance of analytical-established relations is a theoretical one by establishing parameters that influence the elastic springback, and by establishing the direction effect of their influence on elastic springback.

The elastic springback is defined as the difference between the remnant angle of the part α_r, after it comes out of the work area, and the angle α_0 of the tool's active elements.

2 Relations of elastic springback determined on the basis of experimental results

In order to eliminate the previously indicated disadvantage, respectively the low precision of analytical relations, the present paper determines some relations obtained from measured values in established working conditions. Experiments carried out in real working conditions should generally be the basis of the validation of process simulation results [4], and therefore of the bending process - the springback phenomenon in this case.

The relations that will be determined will be of multivariable type, respectively dependencies like:

$$\Delta\alpha = f(x_1, x_2, \ldots, x_n) \tag{1}$$

Where:

$\Delta\alpha$ is the dependent variable – elastic springback;

x_i – independent variables, with a direct influence upon the elastic springback;

f – the dependency between dependent variable and the independent variables.

Knowledge of quantitative relations between process variables (dependent variables) and their significant variables (independent variables) is very important for studying, researching and optimal conducting of manufacturing processes.

3 Employed experimental research method

The experimental research method that was used is the response surface method which considers the relationship between process parameters and characteristic responses of them as surfaces in the dimensional space of variables [5]. In experiments conducted using this method, the independent variables are varied simultaneously taking a limited number of values in the considered domain, called levels. This enables the highlighting of existing interactions between independent variables contributing to a more accurate determination of the overall optimum. With this method, the main effects and of higher order and the interactions of independent variables can be determined separately though more independent variables are varied simultaneously. It can be established the order of variables' influence and be excluded those without significant influence. It can be verified the accuracy of the determined mathematical model and the model can be perfected in order to increase the precision of reproduction of experimental data.

Using response surface method in establishing mathematical models of a process generally requires the following steps: choosing the form and the complexity of the proposed mathematical model; programming the experiment; establishing the experimental conditions; conducting the experiment; determining the model's coefficients.

The determined relation (if adequate) displays the following characteristics:

- Can be used (it is consistent) for any combination of values of the independent variables considered in their domain of variation;
- Although the independent variables are varied simultaneously, their effects on the dependent variable can be determined separately when other variables are constant;
- The order of independent variables' influence can be determined and also the variables with no significant influence can be established;
- The precision of the established mathematical model can be determined and the model can be refined in order to increase the precision of reproduction of experimental data;
- The established model can be visualized as a hyper surface in the multidimensional space of independent variables, this being the response surface whose points have as coordinates the corresponding values of process variables.

4 Variables of the studied process

The purpose is to establish a relation in the following type:

$$\Delta\alpha = f(\alpha, g, b, \beta) \tag{2}$$

Where: $\Delta\alpha$ is the elastic springback and represents the dependent variable [°];

α – bending angle [°];

g – thickness of the processed material [mm];

b – width of the bent part [mm];

β – angle between the bending line and the direction of part's rolling lines [°].

Variables α, g, b, β are the dependent variables of the studied process.

Tables 1 shows the variation levels of independent variables.

Table 1. Independent variables and their levels of variation.

Independent Variables		α [°]	g [mm]	b [mm]	β [°]
Level of variation	min (-1)	30	0.5	10	0
	mean (0)	90	1	20	45
	max (+1)	150	1.5	30	90

The form of dependency type (2) can be, depending on the chosen forms for independent and dependent variables, as follows:

$$\Delta\alpha = e^{a0} \cdot \alpha^{a1} \cdot g^{a2} \cdot b^{a3} \cdot e^{a4 \cdot \beta} \tag{3}$$

Where the variables are polytrophic function type and

$$\Delta\alpha = a_0 + a_1\alpha + a_2 g + a_3 b + a_4\beta \tag{4}$$

Where the variables are of linear multivariable function type, and coefficients a_i, $i = 0, 1, 2, 3, 4$ are the models' coefficients that will be determined using regression program.

5 Experimental programs and obtained results

Fractional experimental programs have been used for four variables [5, 6]. The tests were performed in the Cold Pressing Lab of Manufacturing Engineering Department, on 6tf Press, using sets of active plates and punches for "V" bending on TDA sheets.

Table 2. Experimental program measured and calculated results, regression analysis.

No. exp.	(α) x1	(g) x2	(b) x3	(β) x4	Measured values	Calculated values – rel. (3)	Calculated values – rel. (4)
1	-1	-1	-1	-1	2.63	2.97	3.19
2	1	-1	-1	1	1.47	1.64	1.58
3	-1	1	-1	1	4.45	4.83	4.55
4	1	1	-1	-1	1.55	1.72	1.71
5	-1	-1	1	1	4.37	4.86	4.53
6	1	-1	1	-1	1.6	1.74	1.7
7	-1	1	1	-1	4.55	5.09	4.66
8	1	1	1	1	2.5	2.82	3.06
9	0	0	0	0	3.47	2.89	3.12
10	0	0	0	0	3.25	2.89	3.12
11	0	0	0	0	3.42	2.89	3.12
12	0	0	0	0	4.22	2.89	3.12
Regression coefficients					a_0	2.395	3.006
					a_1	-0.503	-0.019
					a_2	0.242	0.745
					a_3	0.249	0.036
					a_4	0.002	0.007
Indicators of regression analysis					R^*	0.557	0.236
					R_0	55.3	68.22
					R_1	5.35	5.74
					R_2	0.58	0.65
					R_3	0.61	0.62
					R_4	0.39	0.44

The elastic springback was calculated as the difference between the actual angle of active plate and the part's angle after processing. The part's angle was measured with a dial goniometer possessing a precision of 5'. The final value was the mean of three readings.

The experimental program structure, the elastic springback value for each experience, and the regression analysis coefficients and indicators are presented in Table 2.

6 Interpretation of results

It can be observed that:

- both models are adequate, R^* coefficients being smaller than 1 ($R^* < 1$);
- relation (4), by its calculated values, is more close to the measured values ($R^*_4 < R^*_3$);
- α variable (bending angle) is highly significant in both cases ($R_1 > 1$), the angle increase leading to the decrease of elastic springback ($a_1 < 0$);
- variables g, b, β (with the meanings presented above) are not statistically significant (R_2, R_3, $R_4 < 1$), but their increase leads to the increase of elastic springback (a_2, a_3, $a_4 > 0$).

As was mentioned before, the determined relations (type 3 or 4) are adequate for any combination of values of the independent variables in their experimentation domain. Also, it can be determined the individual influences of each independent variable or group of independent variables on the dependent variable, keeping other variables constant, characterized by analytical or graphical dependencies, as in the following examples:

Considering the relation type (4):

$$\Delta\alpha = a_0 + a_1\alpha + a_2 g + a_3 b + a_4\beta \qquad (4)$$

For values at mean level (see Table 1), it is obtained the following invariable mathematical dependency for independent variables g, b, β (variable α possessing the highest significance):

$$\Delta\alpha = f(\alpha) = 4{,}786 - 0{,}019 \cdot \alpha \qquad (5)$$

The corresponding graphical dependency (increasing the bending angle leads to the decrease of elastic springback) is displayed in Figure 2.

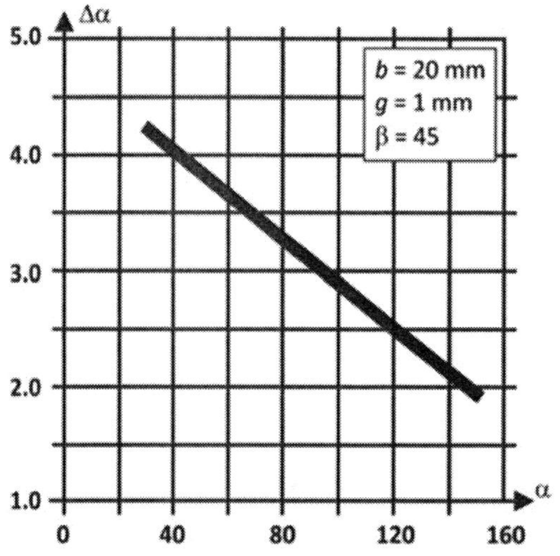

Fig. 2. Variation of springback $\Delta\alpha$ versus bending angle α.

For the same relation (4), it can be determined the mathematical and graphical dependencies of two variables, the others having constant values. For example, on the basis of relation (6), in which the independent variables b, β have mean values in their variation domain, it was obtained the graphical dependency displayed in Figure 3.

$$\Delta\alpha = f(\alpha, g) = 4{,}041 - 0{,}019 \cdot \alpha + 0{,}745 \cdot g \qquad (6)$$

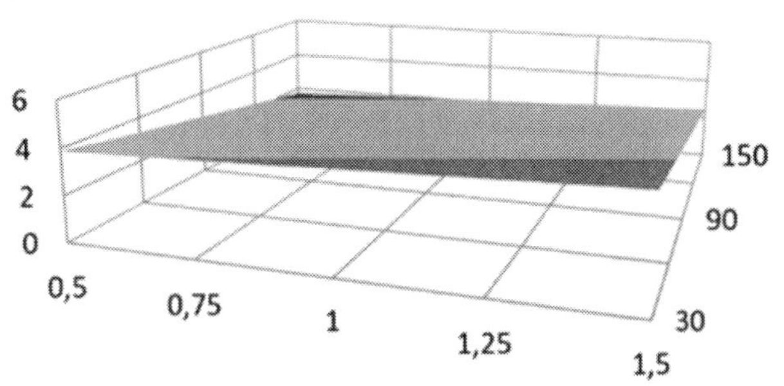

Fig. 3. Variation of springback $\Delta\alpha$ versus bending angle α and thickness of processed material g.

Such dependencies (mathematical and graphical) can be obtained for any constant values for the independent variables, provided that these values remain in the domain of variation.

7 Objectives and Conclusions

It is envisaged the development of a normative for direct determination of elastic springback in given conditions for the independent variables (in the considered variation domains).

Also, there will be developed annexes with graphical dependencies, with one or two parameters, for elastic springback.

The variation domain for statistically insignificant variables will be enlarged in order to establish more precisely their significance degree. Other influences (material's type, bending radius) on the elastic springback will be studied.

Knowing the precise value of elastic springback is important for part's manufacture with the prescribed bending angle. If elastic springback makes the bending angle out-of-tolerance, then constructive and technological measures are taken in order to reduce, even eliminate, the effects of elastic springback.

References

1. C. Ciocardia, et.al., *Cold pressing technology (in Romanian),* (EDP, Bucharest, 1991)
2. C. Iliescu, (ed.), *Cold Pressing Technology*, Elsevier, 1990
3. Gh. Sindila, et.al., *Technological systems for cold pressing technology (vol.1) (in Romanian)*, (BREN Publishing House, Bucharest, 2014)
4. Demeri, M., Lou, M., and Saran, M., SAE Technical Paper 2000-01-2657 (2000)
5. Gh. Sindila, et.al., *Manufacturing by cold pressing technology (in Romanian),* (BREN Publishing House, Bucharest, 2009)
6. R. Mihail, *Introduction in strategy of experimentation with applications in chemical technology (in Romanian),* (Stiintifica si Enciclopedica Publishing House, 1976)

A study on the variability of the pressing force during processing of regular shaped roughness by using new kinematical scheme for ball-burnishing process

Stoyan Slavov*, and *Iliyan* Iliev

Technical University of Varna, Department of Mechanical Engineering and Machine Tools, Varna, Bulgaria

Abstract. In many studies of the ball-burnishing processing often assumes that the pressing force applied to a ball tool remains a constant during the whole processing. The present study focused on measuring the compression force continuously during conducting ball-burnishing process, using new kinematical scheme. The combinations of values of the regime parameters are according to four factors full factorial experiment design with two levels per factor of type 2^4 and 4 replications per run. The experimental results are processed statistically and using techniques such as Pareto and ANOVA, the regime parameters of the process are sorted by degree of significance.

1 Introduction

In the cold forming processes like burnishing, vibratory ball burnishing, thread or spline rolling, there takes place the deformation of the surface layer, which modifies its physical and mechanical characteristics. The strain-hardening phenomenon introduces some compressive stresses into the surface layer, which results with the increase of surface hardness and endurance strength [2,3,5]. The process of burnishing also uses as finishing machining operation after which, the workpiece has improvements such as better surface roughness, decrease of the formation of fractures, fatigue strength, and increase of the wear resistance [3,5]. These positive improvements are dependent on the regime parameters of the (vibratory) burnishing process such as feed rate, burnishing (compressive) force, ball diameter, etc. The influence of these parameters, used while conducting the ball burnishing process for improving surface roughness, micro hardness, fatigue strength, etc., have been examined in numerous experimental studies over the years. As a result, there are obtained many functional dependencies between the regime parameters and the roughness, strength and physical and mechanical characteristics of workpieces (made of different types of steels, non-ferrous metals, etc.), processed by different types of cold forming processes. Significant part of conducted experiments are focused on determination of the impact of the pressing force applied to the ball tool, which is one of the most significant regime parameters of these processes [5]. When describing the experimental studies, however, many authors consider

* Corresponding author: sdslavov@tu-varna.bg

that once the required nominal value of the pressing force has been set, it remains constant during the whole processing. That assumption may be not always correct even when processing flat surfaces from the prismatic workpieces for example. This is because the burnishing process actually performs on real workpiece, by using real instruments and real machines, which usually have certain surface irregularities, deviations in its dimensional tolerances, and deformations caused by external loads. Therefore, it is interesting to study what is actually the variability of the pressing force during ball burnishing process, as this may have an impact on the degree of uniformity of resulting plastic deformation in surface layer, and can reduce the effect of the improvements in the surface layer of the burnished workpieces. In this regard, the main objective of the present work is to measure experimentally what is the variability of the pressing force when processing flat surfaces from a steel plate by ball burnishing process (using a new kinematical scheme), and how the main regime parameters influenced on that variability.

2 Methodology of experimental investigation

2.1 Setup for experimental measuring the pressing force during ball burnishing process

Fig. 1 shows a measurement setup, used for determining the nominal value and deviations of the pressing force, during ball vibratory burnishing processing of flat surfaces. It is consist of specially designed deforming ball tool [4] whit integrated miniature force sensor, type FCM (F_{nom}=5 kN), SIKA, Dr. Siebert&Kühn GmbH & Co (pos. 3). A standard manufactured helical compression spring with ground ends provides the needed pressing force of the tool. The spring, which is used, gives maximum force up to 3200 N at maximum deflection length of 48 mm. The helical spring (pos. 2) presses the force sensor (pos. 3).

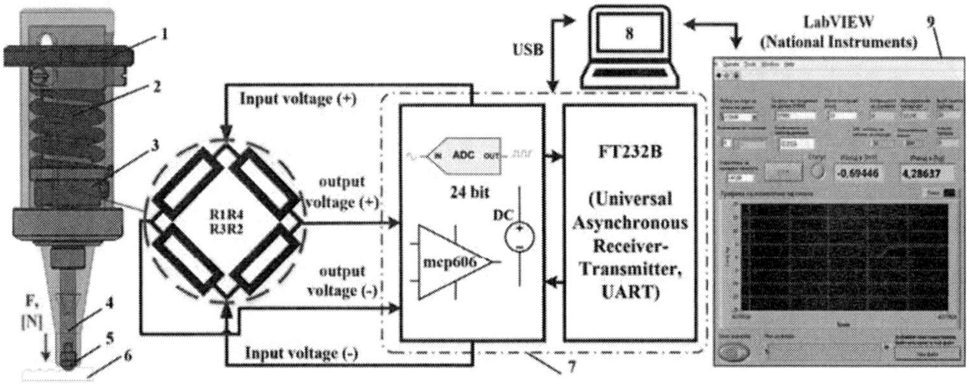

Fig. 1. Setup for measurement the pressing force during ball burnishing process.

The front part of the force sensor (pos. 3) presses the head of the plunger (pos. 4). The plunger transfers the force from the spring to the ball burnishing element (pos. 5), which rolls along a complex trajectory on processed surface of the workpiece (pos. 6). The complex trajectory provides by using appropriate CNC milling machine [3,5]. This is necessary for obtaining the desired surface roughness with a regular micro shape. Position 7 of Fig. 1, shows a principal block diagram of a device for amplification and conversion the signal from the force sensor to a PC. The device provides a stabilized DC input of 10 V, 50 mA, and thereby the measuring range of the output signal of the sensor is varied from 0 to 20 mV when changing the pressing force in the range between 0 and 5 kN. The output signal from the sensor is amplified by precise low-noise amplifier type MCP606 (Microchip) with adjustable amplification factor to an appropriate level for converting from analogue to digital form, using the built-in ADC. The converted signal is transmitted to the universal asynchronous receiver-transmitter (type FT232B on FTDI) for conversion into serial data for transmission to a PC (pos. 8), through the USB 2.0 interface. For adjusting the operating mode of the microcontroller, as well as to visualize and record the measurement data of the force, a specialized software product (pos. 9) based on the integrated program environment LabVIEW (NI) is developed. Thus, it is possible to balance the force sensor, to adjust the factor of amplification, to change the rate of sampling, etc. Moreover, it is possible to show the measured instantaneous values of the pressing force digitally and in graphic form. The software allows recording the measured data from the force sensor in CSV (comma-separated values) file format, supported by MS Excel for further statistical analysis of the data.

2.2 Design of the experimental research

Four factors full factorial experiment design with two levels per factor of type 2^4 and four replications per run [1] is carried out to determine the degree of influence of the main regime factors of the process. The design of the experiment is shown in "Experimental design" section of the Table 1.

Table 1. Full factorial design plan and results for coefficient of variation of the pressing force F.

Run	Experimental design				Experimental results (coefficient of variation)			
	F, N	i	e, mm	s, mm/min	CV 1	CV 2	CV 3	CV 4
1	1059.1	600.15	1	300	0.737	0.514	0.778	0.556
2	1059.1	600.15	2.5	150	0.606	0.654	0.599	0.510
3	1059.1	1200.15	1	150	0.522	0.608	0.587	0.407
4	1059.1	1200.15	2.5	300	0.531	0.721	0.569	0.421
5	1735.8	600.15	1	150	1.531	1.702	1.443	1.298
6	1735.8	600.15	2.5	300	1.605	1.564	1.285	1.213
7	1735.8	1200.15	1	300	1.406	1.543	1.286	1.185
8	1735.8	1200.15	2.5	150	1.716	1.619	1.430	1.410
9	1059.1	600.15	1	150	1.461	1.433	0.539	0.412
10	1059.1	600.15	2.5	300	1.401	1.530	0.510	0.453
11	1059.1	1200.15	1	300	1.201	1.394	0.552	0.434
12	1059.1	1200.15	2.5	150	1.641	1.716	0.571	0.477
13	1735.8	600.15	1	300	1.306	1.466	1.234	1.155
14	1735.8	600.15	2.5	150	1.516	1.508	1.216	1.188
15	1735.8	1200.15	1	150	1.258	1.463	1.254	1.129
16	1735.8	1200.15	2.5	300	1.424	1.551	1.334	1.325

The diameter of the used ball tool (a steel ball from rolling bearing) remains the same (8 mm) in all of the16 runs and four replications of the experimental study. Due to conducted preliminary studies of the ball burnishing process, carried out using CNC milling machines [3], the significant (adjustable) regime parameters, which mainly affect the characteristics of the processed surface, have been established. They are as follows:

– F, N– applied external force (by helical spring) on the ball tool;

– i – the set number of the sinusoids along the path travelled by instrument, which determines the height of the obtained cells from roughness with regular microshape;

– e, mm – the set half of the amplitude of the sine wave, related to the width of the obtained cells from roughness with regular microshape;

– s, mm/min – the set feed rate of the ball tool following the sinusoidal trajectory, provided by the CNC milling machine;

The experimental specimen is made of sheet stainless steel, grade 316L ASTM (Cr17 Ni11Mo2) with a thickness δ= 4 mm, which has physical and mechanical properties, as follows: ρ=8000 kg/m^3; E=193 GPa, $Rp_{0.2}$=353 MPa, $Rp_{1.0}$=394 MPa, Rm=628 MPa, A5=49%, A50= 47.2% and HRB=82. The plate has a square shape and dimensions 400x400 mm. The sections, corresponding to the rows of the experimental design from Table 1 which processed by ball burnishing, have rectangular shape and dimensions 40X10 mm.

2.3 Statistical processing of measured data

Measured data for the pressing force, obtained during processing by vibratory ball-burnishing for all runs and replications according to the experimental design plan, shown in Table 1, are imported into MS Excel, where displayed through XY line (or scatter) charts. Fig. 2.a shows graphics of two measurements, captured under the nominal values of the pressing force F, respectively 1059.1, and 1735.8 N. As seen in Fig. 2.a, the duration of measurements is different for different rows of the experimental design ($\Delta t_1 \neq \Delta t_2$). This is because the processing time depends on the regime parameters *e* and *i* and feed rate of the ball tool *s*. According to the experimental plan from Table 1, the shortest duration will be obtained at row №13: Δt_{min}=6:52 min. Thus, at rate 16 SPS will be accumulated 6,500 measured values. Respectively, the longest time will be obtained at row №12: Δt_{max}=26:28 min and 25,414 values will be accumulated. The Excel XY charts are used to identify and clipping those measured data, when the ball tool is not processing the workpiece, i.e. when it reposition from an already processed area of the specimen to the next.

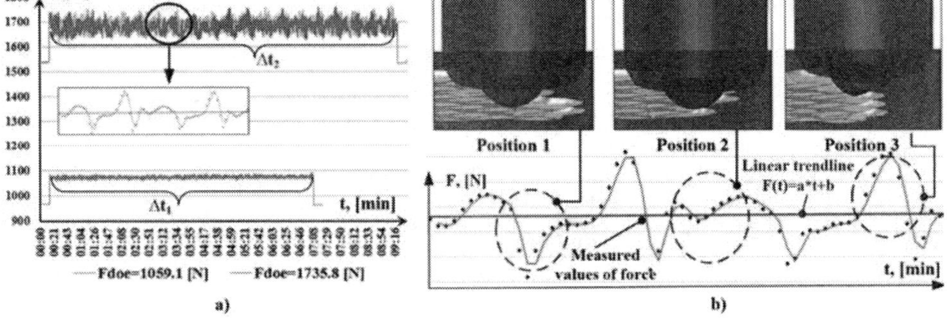

Fig. 2. a) Examples of the XY line graphs of the measured data in Ms Excel; b) Manner of changing the pressing force depending on the current position of the ball tool.

For more accurately assess of the force variability, a least squares linear regression is used for calculate the slope and the intercept to determine whether the measured data for the pressing force have significant slope trend. Where it is found significant slope, the magnitudes of the deviations of pressing force is assessed against the least squares fitted regression line $F(t) = a \cdot t + b$ (see Fig. 2.b). Thus, the effect of the irregularities in the geometric shape and dimensions of the specimen and the working bodies of the used milling machine is reduced. After that, the Descriptive Statistics analysis tool add-in of MS Excel is used for every run and replication of the experimental design. Because for different rows of the experimental plan the samples have different sizes and different mean values, the variability of the force is assessed by using the coefficient of variation (CV) [1]. The CV is the ratio of the standard deviation (σ) to the mean ($\overline{X}$), calculated using the following formula:

$$CV = (\sigma/\overline{X}) \cdot 100 \qquad (1)$$

Calculated CVs using (1) are given in the section "Experimental results" of the Table 1. To build the effects plot (see Fig. 3.a), the average effects Avg_{low} and Avg_{high} are determined for the four regime parameters, using the following formulas:

$$Avg_{low} = \frac{1}{8} \cdot \left(\sum_{i=1}^{8} CV_i^{low} \right) \tag{2}$$

$$Avg_{high} = \frac{1}{8} \cdot \left(\sum_{i=9}^{16} CV_i^{high} \right), \tag{3}$$

Where: CV_i^{low} – are those eight results from Table 1, where the regime parameters have low level values; CV_i^{high} – the other eight results, where the parameters have high level values.

Based on (2) and (3) the effect of each of the four regime parameters on the variability of the pressing force can be calculated using the formula:

$$Eff_{(F),(i),(e),(s)} = Avg_{high} - Avg_{low} \tag{4}$$

To visualize the effects of a design experiment, Pareto analysis technique is used for sorting the regime parameters according to their levels of significance (see Fig. 3.b) based on calculated by using (4) values of effects. Additionally an analysis of variance [1] (see Fig. 3.c) is carried out to calculate the Fisher-values of the factors. For confidence level 95% and total degrees of freedom $DF = (Runs \cdot Repplications) - 1 = (16 \cdot 4) - 1 = 63$, the critical Fisher value is 3.993.

3 Results and discussion

Comparing the ranges of variation, when the pressing force has low nominal value F=1059.1 N, the variability is amended from ΔF_{min}=25.3 N to ΔF_{max}=105.8 N. At the high level of the force F=1735.8 N, this amendment is from ΔF_{min}=84.2 N to ΔF=171.5 N. Thus, the range of variability is amended from 5 to 10% of the nominally assigned compression forces. The main cause of variability of the pressing force is represented in Fig. 2.b. It is seen how the pressing force varies within one full reciprocation of the ball-tool following the toolpath during the process of ball burnishing of flat surface of the specimen. In Position 1, the ball-tool is located over already burnished region of the specimen. In this position, the pressing force has a minimum value because the ball falls within the valleys of regular roughness, and the spring expands slightly in comparison to its preliminary tensioning. In Position 2, the pressing force has an average value and slightly fluctuates because the ball passes over the heights, which surrounding the cells of the regular roughness. In Position 3, the spring is shrinking additionally in comparison to its preliminary tensioning.

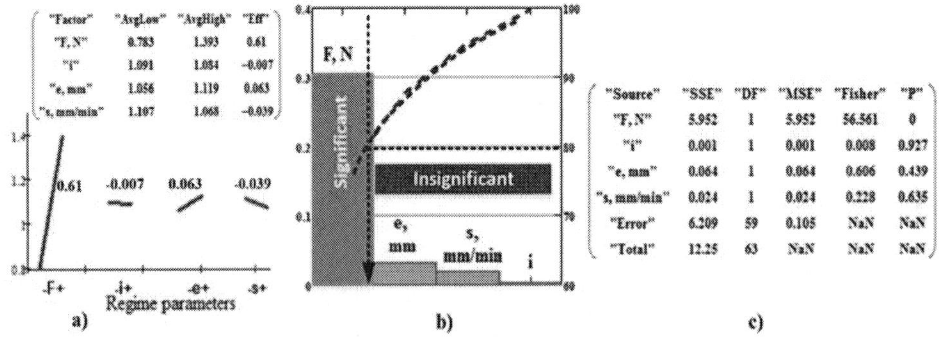

Fig. 3. a) Effects plot of the regime parameters; b) Pareto histogram and cumulative graphs, based on the effects; c) ANOVA table with calculated statistics and Fisher-values of the process factors.

This is because the ball goes up over not yet deformed region of processed surface. In addition, this is the main reason for observed short-term pulsations in the values of the pressing force.

Obtained data after evaluation of the effects and carried out Pareto analysis (see Fig. 3.a and 3.b) shows that the variation mainly depends on the magnitude of the nominal value of the pressing force F. With increasing the magnitude of the pressing force, proportionally increases observed variability and vice versa. The other three regime parameters (i, e, and s) have a much smaller impact on the variability of the pressing force. After conducted ANOVA analysis (see Fig. 3.c), the factor F was assesses as a significant at the 95% level of confidence since his Fisher-value =56.561 is greater than critical Fisher value=3.993. The F-values of the other regime parameters are less than the critical Fisher value. Thus, conducted analysis of variance reinforces the subjective conclusion derived from the Pareto histogram plot.

4 Conclusions

Based on conducted experimental study and the undertaken analysis of the obtained results could be made following conclusions:

- The established experimental setup (see Fig. 1) including miniature force sensor, device for amplification and signal conversion, and specially created software, allows precise measuring the pressing force (in real time) as well as visualization and recording the measurement data in appropriate file format. This significantly facilitates the subsequent processing of the measurement data in suitable software such as Excel, Mathcad etc.
- The observed variation of the pressing force F, during ball-burnishing process carried out by new kinematical scheme (using CNC milling machine) on flat surfaces mainly depends on the pre-set magnitude of this regime parameter. The remaining three regimes parameters (i, e, and s) have a negligible impact on the variability of the pressing force. This gives grounds they do not take into consideration for future research.
- The measured variability reaches a maximum of 10% (± 5%), compared to the nominal pre-set level of the pressing force. Largely, this irregularity is due to cyclical crossing of the ball-tool between the already deformed and have not yet processed surface of the workpiece (see Fig. 2.b), resulting in a periodic shortening and extending of the helical spring, which provides the pressing force.

Further research of the collective will be focused on examining the degree of variability of the pressing force when processing workpieces with a more complex shape of its functional surfaces, using ball-burnishing process.

References

1. Michael H. Kutner, John Neter, Christopher J. Nachtsheim, William Li, Applied Linear Statistical Models, McGraw-Hill Education, 827-829, (2004)
2. L. G. Odintsov, Hardening and Finishing of Parts by Surface Plastic Deformation. Mashinostroenie, Moscow (1987).
3. Slavov Stoyan Dimitrov, Dimitrov Diyan Minkov, ECУ **4-2 (25),** 11-22 (2016)
4. Slavov S. D., Iliev I.V., Fiability& Durability, "Academica Brancusi", 3–9 (2016)
5. Schneider Yu. G. Operational properties of parts with regular microrelief, publishing IVA, St. Petersburg, (2001)

Study of the influence of punched contour geometry over the punch force using finite element analysis

Stelian Uțuleanu[1, 2,*], *Aurelian* Vlase[1], *Gheorghe* Sindilă[1], and *Nicolae* Căpățână[1]

[1]University POLITEHNICA of Bucharest, Faculty of Engineering and Management of Technological Systems, Splaiul Independenței 313, 060042 Bucharest, Romania
[2]ICPE SA – Special Electric Machines, Splaiul Unirii 313, 030138 Bucharest, Romania

Abstract. When is calculated the punching force is taken into account only the perimeter and not its form. In this case, the pressing force for the contours with the same length, but with different geometries (circle, square, hexagon, triangle and shearing) may have different values. To demonstrate this hypothesis were made experimental tests and they confirm this supposition. Now this hypothesis is verified using the finite element analysis in order to show the possible differences and to view the material behavior during punching process. The analyzed materials are AISI 304, DC01 (AISI 1008), Al99.8 and Brass CDA 110. The software used to make the finite element analysis is Deform 3D version 6.1 (service pack 2).

1 Introduction

Punching is a group of cold plastic processing methods based on wholly or partially material separation, after a closed or open contour [1, 2].

It is used both for obtaining finished parts and semi-finished products for further processing operations by plastic deformation such as bending, drawing, cold forming.

Punching contains several methods like: cutting, punching, shearing, notching, trimming [2, 3].

The software used for finite element analysis (FEA) is Deform 3D Service Pack 2. This system consists of three major components [4]:

1. A *pre-processor* for creating, assembling, or modifying the data required to analyze the simulation, and for generating the required database file.

2. A *simulation engine* for performing the numerical calculations required to analyze the process, and writing the results to the database file. The simulation engine reads the database file, performs the actual solution calculation, and appends the appropriate solution data to the database file. The simulation engine also works seamlessly with the Automatic Mesh Generation (AMG) system to generate a new FEM mesh on the workpiece whenever necessary. While the simulation engine is running, it writes status information, including any error messages, to the message (.MSG) and log (.LOG) files.

* Corresponding author: stelian_u_86@yahoo.com

3. A *post-processor* for reading the database file from the simulation engine and displaying the results graphically and for extracting numerical data.

The analyzed contour length is 60 mm, and the geometries are presented in the next figures (Fig. 1 to Fig. 5).

Fig. 1. Circle contour.

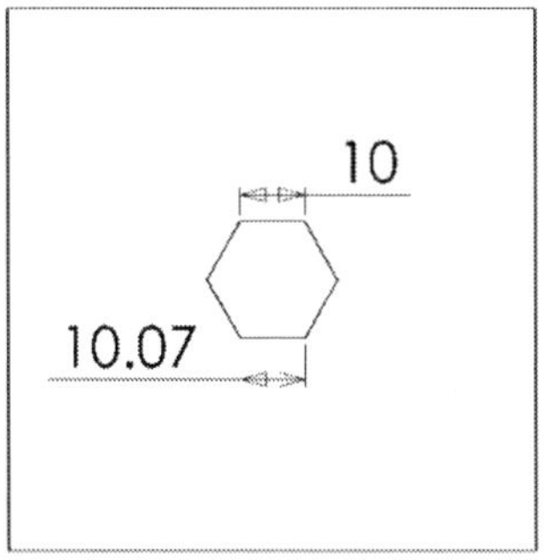

Fig. 2. Hexagon contour.

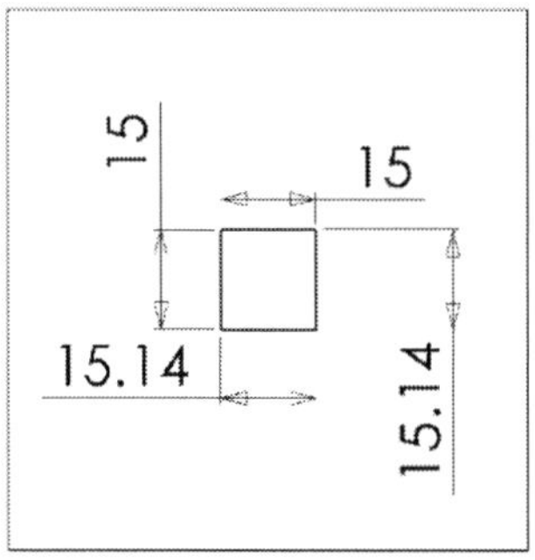

Fig. 3. Square contour.

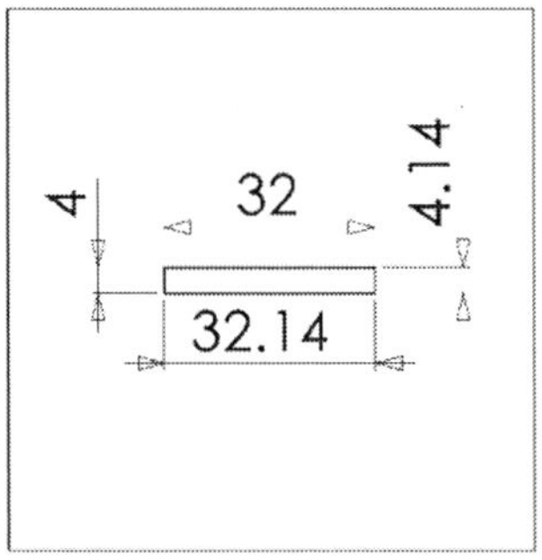

Fig. 4. Rectangular contour.

Fig. 5. Triangle contour.

2 Setup

To set up a new simulation it must be selected the "New Problem" under the "File" menu. In the wizard window that appears is selected "Forming" as guided template and after that should be chosen the problem location and its name. This window will be automatically closed and will be opened the pre-processor component.

After is selected the unit system [System International (SI)], should be entered a name for operation, followed by "Cold forming" process type, "Shape Complexity" as "Complex" and "Accuracy" as "Accurate", "Workpiece Shape" as "Whole part" and "Number of Objects" as "1 workpiece + 2 dies".

Now can be started the defining of objects: workpiece, top die and bottom die.

These elements were created in SolidWorks 2010 x64 Edition and exported to STL format in order to be imported in Deform 3D.

The first object defined is the workpiece and only it will have a mesh network, because the dies are considered as rigid elements. Can be let the default name for workpiece or can be choose a new one. To add the geometry in the pre-processor can be selected one option from: "Import geometry", "Extract from mesh" and "Define primitive geometry".

After is added the geometry of workpiece should be set the number of mesh elements. In our case was selected the maximum number of 100.000 elements, and the "Size ratio" as 3, followed by the mesh generation, the selection of material and the boundary conditions as velocity 0 mm / sec. The materials used were imported from the library (AISI 304, AISI 1008, Al99.8 and Brass CDA 110).

In the next step will be imported the top die and will be set the movement for "Press / Hammer" on Z axis for a mechanical machine with 8 mm of "Total stroke" and with a speed of 2 Stroke/sec.

Also, will be inserted the bottom die and all of these objects will be positioning manually. Friction coefficient will be let as default (0.08), the "Total primary die travel" will be set as 8 mm and will be not checked any "Stopping Controls".

After that will be define the number of steps (400) and step increment (10), followed by the checking of data, generation of database, switching to simulation engine and with starting of this process.

3 Results

After the FEA studies had been finished were created comparative graphs for each type of material, with all five types of punched geometry.

Considering that the top die is 4 mm above the bottom die, the registration of force will start at 3.19 mm stroke for brass, at 3.49 mm stroke for AISI 304 and AISI 1008, respectively at 3.68 mm stroke for Al99.8. The first registered value is when the top die hits the material, and the maximum force is registered when the material is completely ruptured. To highlight the interval where the force presents variation, there was fixed the interval of stroke between 3 and 6 mm. The following figures show these graphs.

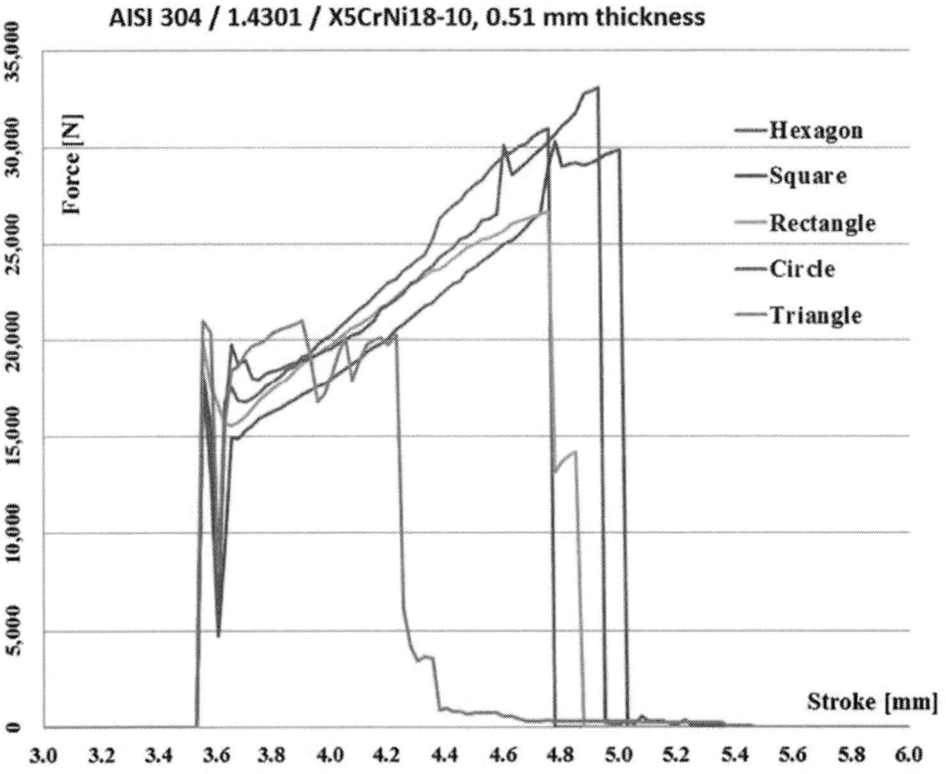

Fig. 6. FEA results for AISI 304 material.

The graphs for DC01 and Al99.8 materials are shown in the figures 7 and 8.

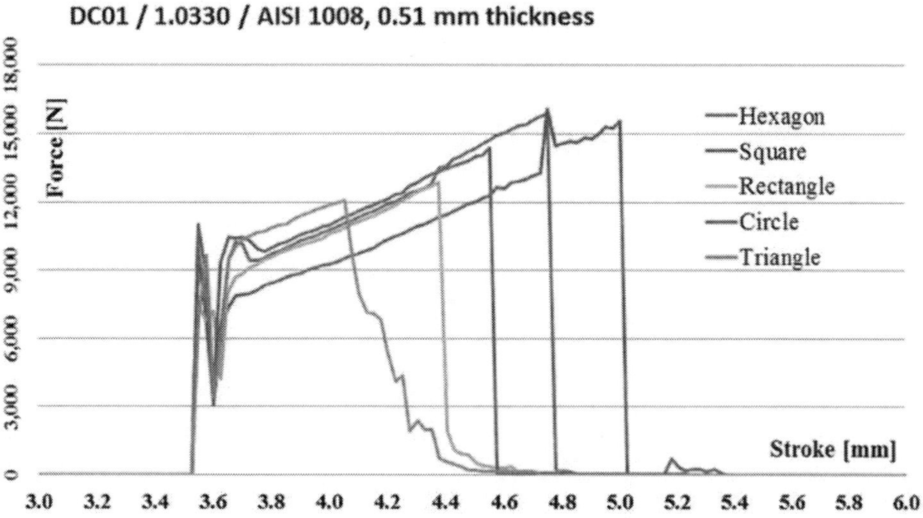

Fig. 7. FEA results for AISI 1008 material.

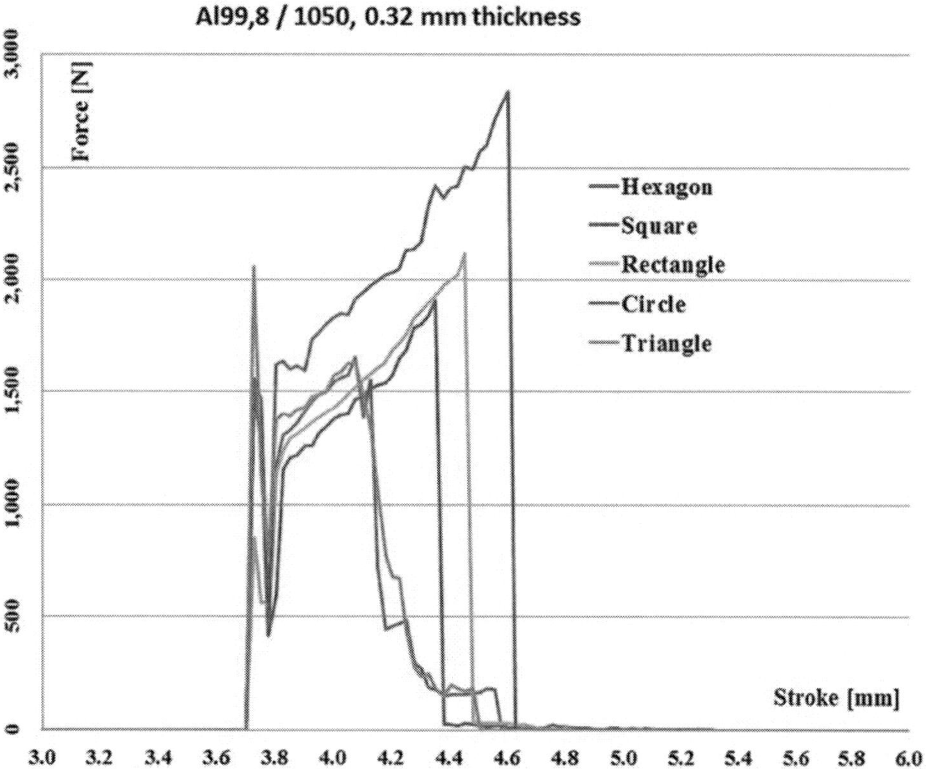

Fig. 8. FEA results for Al99.8 material.

As we can see in these graphs, the geometry of punched contour influence the punching force because the material behavior is different for each type of geometry and also the flow of it is modified with impact over the punching force and over the interval of stroke needed to achieve the full rupture of material (when the force registration is stopped).

The graphs for Brass CDA 110 are shown in the Figure 9.

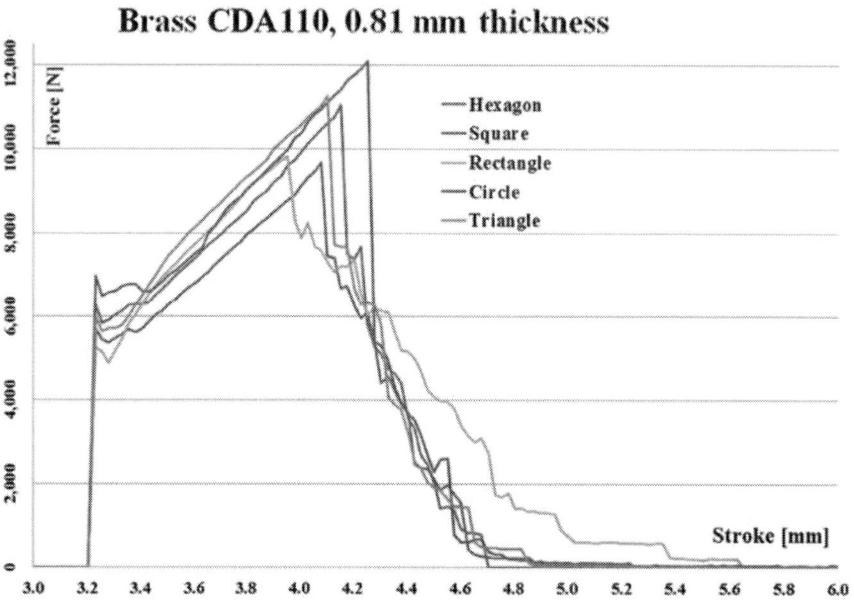

Fig. 9. FEA results for Brass CDA 110 material.

To easily observe the differences between the forces obtained for each type of contour shapes and materials was made a comparative table and was determined the percentage difference between the maximum force and the minimum force.

Table 1. Comparison of influence of punched contour geometry over the punch force.

		AISI 304	AISI 1008	Al99.8	Brass
Hexagon	**F con**	18,366.31	9,408.24	1,540.58	6,269.08
	F max	30,944.10	15,870.65	1,654.75	11,050.73
Square	**F con**	17,945.79	9,938.70	1,506.11	5,712.40
	F max	30,295.54	16,108.86	1,910.74	9,693.75
Shearing / cutback	**F con**	19,980.57	7,416.46	853.24	5,253.49
	F max	26,666.41	12,880.79	2,115.74	9,826.86
Circle	**F con**	17,917.21	11,029.96	1,561.13	6,999.85
	F max	33,089.05	14,386.40	2,838.76	12,095.67
Triangle	**F con**	21,030.08	6,854.26	2,058.01	6,002.73
	F max	21,030.08	12,089.80	2,058.01	11,261.86
Max – min [%]	**F con**	14.80	37.86	58.54	24.95
	F max	36.44	24.95	41.71	19.86

The "F con" represent the force recorded when the top die hit the workpiece and the "F max" represent the maximum force recorded during simulation. Because it is hard to observe in images the differences of material deformation for different geometry contours, in figures 10 and 11 are presented as examples two steps of deformation for AISI 304 material, for circle geometry.

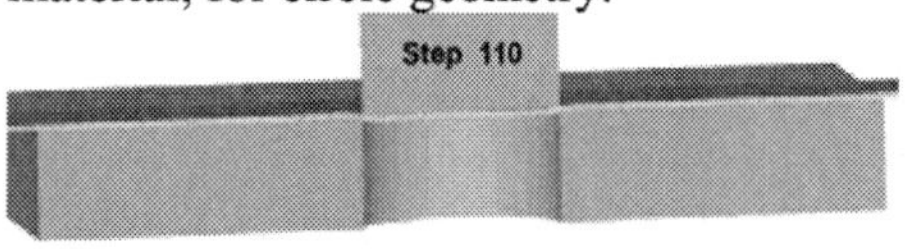

Fig. 10. AISI 304 material deformation at step 110 for circle geometry, 0.51 mm thickness.

Fig. 11. AISI 304 material deformation at step 150 for circle geometry, 0.51 mm thickness.

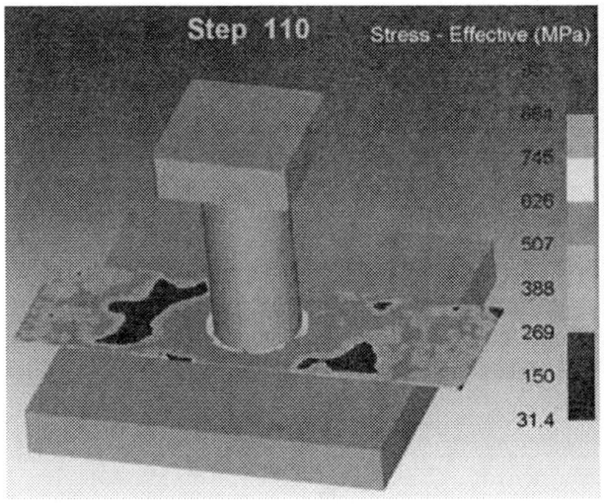

Fig. 12. Stress representation at step 110 for round geometry on AISI 304 material.

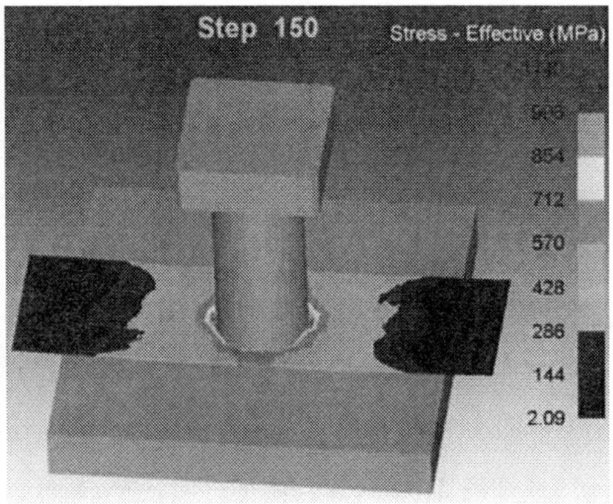

Fig. 13. Stress representation at step 150 for round geometry on AISI 304 material.

Stress representation is more interesting, but due to the limitations to 6 pages of paper, in the figures 12 and 13 were presented the stress analyses at step 110 and 150 for AISI 304.

4 Conclusion

Observing the variations of punched force in the graphs presented in figures 6 – 9 and after the comparison of forces centralized in table 1, we can conclude that our hypothesis about the influence of punched contour geometry over the punch force is confirmed and gives us the opportunity to continue the research.

In further researches will be analyzed the influence of different thickness of material over the punched force, over the material deformation and its flow. Also, there will be used other setups (such as modification of: shape complexity and accuracy, number of elements, minimum element size and size ratio of elements).

Also, will be made a comparison between calculated, experimental and FEA punch forces for the same length of punched contours but with different geometry.

To achieve FEA results very close to the real deformation of material, must be correlated the experimental tests with simulation results and to determine the optimal setup for this kind of process.

References

1. Gh. Sindilă, Cold plastic deformation manufacturing technologies, Ed. Bren, Bucharest (2007)
2. S. Uțuleanu, A. Vlase, Gh. Sindilă, N. Căpățână, ISCAME 1, **212**, Debrecen (2013)
3. C. Ciocârdia, Cold pressing technology, Ed. Did. and Ped., R.A., Bucharest (1991)
4. *DEFORM 3D Version 6.1(sp2) User's Manual,* SFTC , Columbus, Ohio (2008)

A numerical solution for a closed die forging process

Dorin Luca*

"Gheorghe Asachi" Technical University of Iaşi, Department of Technologies and Equipments for Materials Processing, 51, Prof.dr.doc. Dimitrie Mangeron Street, 700050 Iaşi, Romania

Abstract. One of the manufacturing processes that can be permanent improved is plastic deformation of metallic materials, as incorporating reserves on the manufacture of products with reduced material and energy consumptions. This paper presents finite element analysis for a closed die forging process, showing the stresses, strains and temperature into deformed part and stresses in the working tools. The analysis of obtained results for different flash dimensions of the working tools has enabled optimization of the forging process studied. To be able to validate the numerical results obtained, experimental tests were conducted. The compared data series show a good agreement between the numerical and experimental data.

1 Introduction

Closed die forging procedure continues to show particular interest because it can produce parts of various shapes and sizes, with high complexity, from different metals and alloys with moderate costs. For these reasons, research of the forging process in the closed die is current, aiming embodiments, investigation aspects of the interdependence of the various working parameters, the optimum temperature for forging and optimization of the die cavity for a good flow of the metal, increasing the life of the dies.

Companies that manufacture parts in closed die forging are interested in improving the quality and manufacture of net-shape parts [1]. A particular importance for achieving these goals has the designing process of the forging dies. Modern designing is a complex process which requires, among other things, the application of numerical methods for modelling and simulation based on the use of performant software. Finite element modelling offers solutions for metal forming processes, to identify fields of variation of displacements and stresses, mainly, and viewing the flow behaviour of the material.

The main elements that need to be considered when designing the dies are: flash dimensions, internal and external draft angles, fillet and corner radii, shrinkage and parting line of dies [2]. Closed forging dies can be made with one or more cavities depending on the shape and dimensions of the parts to be forged. Cavities fulfil the different roles in the forging process, and therefore can be classified into several types: forging finishing; pre-forging; cutting. Workpiece shapes obtained in each cavity and processes necessary to

* Corresponding author: dluca@tuiasi.ro

obtain these shapes represent forging steps.

Radev and Tomov [3] describe in their paper the role of shape complexity factor of part in the steps established of forging in closed die: multiple pre-forging or only a single forging. Closed die forging for parts with simple shape, generally symmetrical, can be made using billet with constant section. In the case of forgings with complex shape and asymmetrical, the use of workpieces with variable section is recommended to obtain the correct part with all details intact (e.g., ribs). Wei et al. [4] were concerned with unequal thickness billet design for isothermal forging of rib-web components from titanium alloys. This research aims to provide guidelines for designing the billet with unequal thickness, used to manufacture a wide range of precision components (without any under-filling defects) of titanium alloys.

Friction and wear forging dies are elements that must be considered for increasing the life of dies and avoidance of their failure. Chavoshi [5] performed a study on the influence of friction on the closed die forging of aluminium alloy AA7075. The process was simulated by finite volume method to compare the effect of two laws of friction (Coulomb's Amonton and Prandtl's) on the forging load, stresses and strains. Abachi et al. [6] have analyzed the hot forging die wear. By comparing the simulation results by the finite volume method with measurements made on the worn die, it was determined a value of the wear coefficient. Die wear is the main cause which leads to failure in hot forging. The failure mechanism of hot forging dies has been studied by Gronostajski et al. [7]. They found that thermomechanical fatigue is the factor affecting most die because it produces fine cracks that extend over the entire contact surface. This phenomenon, together with oxide abrasive particles from hot material, contributes to a sharp rise in abrasive wear.

Knowing and respecting the forging optimum temperature for each material is another decisive factor for the success forging process. Ghanbari et al. [8] were occupied of manufacturing a nanostructured Ti-6Al-4V alloy by closed die forging along several axes, in isothermal conditions. The same material is studied in their paper by Gupta et al. [9]. Two manufacturing variants were investigated, closed die forging of pressure vessels from titanium alloy Ti6Al4V: one with several stages of forging and mill annealing and one by one / two step forging and mill annealing.

Sedighi and Pourbashiri [10, 11] were concerned to reduce material waste in closed die forging. They proposed a new method, called variable gutter technique, which resulted in a decrease of about 50% of waste material from one forged part.

This paper presents a case study for closed die forging of a gear. The study shows finite element analysis, performed in order to determine the influence of channel flash dimensions on forging pressure and force to optimize the forging process.

2 Process description

A numerical simulation of a closed die forging process in one step, to a gear (Fig. 1.a) is presented. Flash dimensions have great influence on: forging load, forging energy and durability dies. Channel flash plays an important role in achieving a proper flow of the material in the finisher cavity. Depending on the complexity of the forged part, there are different constructive channels forms that can be used, most used flash channel being the one with the shape shown in Fig. 1.b. Flash channel consists of two main parts: the flash land (with the dimensions w_1 and h_1) and flash gutter (with the dimensions w_2 and h_2).

To establish the dimensions of the flash channel, flash land height h_1 is first determined, as follows:

- For the forging hammers,

$$h_1 = 0.015\,(A_p)^{1/2} \tag{1}$$

DOI: 10.1051/matecconf/201711202008

- For the forging presses,

$$h_1 = 0.020\,(A_p)^{1/2} \quad (2)$$

A_p being part's area in the parting line.

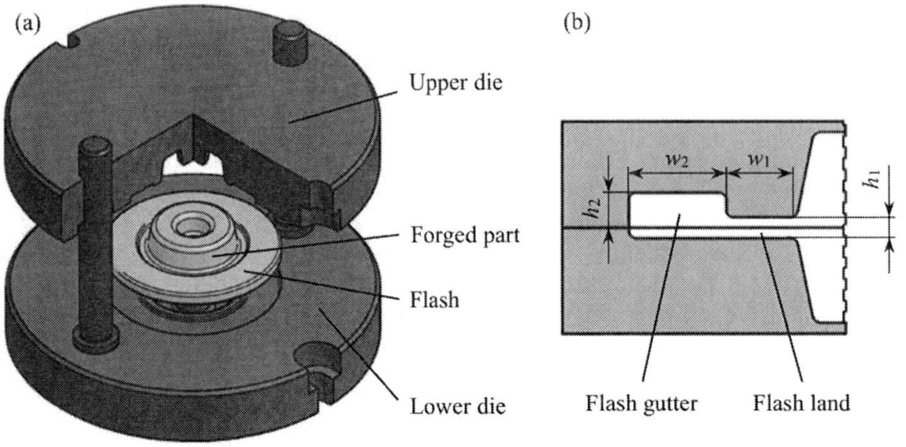

Fig. 1. 3D model of the studied forged part (a) and the dimensions of a flash channel (b).

The relationships (1) and (2), empirically determined and used in practice, can be improved by numerical simulation. The values calculated by these relationships can be corrected (by increasing or decreasing the value of h_1) after examining how the flow of the material for any type of forged part, in order to optimize the forging process. After calculating the flash land height h_1, other dimensions of the flash channel w_1, w_2 and h_2 are extracted from tables after previously was provided the flow mode of material into cavity.

The type of finite element analysis chosen was coupled elastic-plastic analysis, which enables establishing the dependency between stress and the temperature from process. Forging process occurs with the transfer of heat to the environment, which is calculated based on the updated shape (workpiece) after each increment and contact conditions. A fraction of the energy for plastic deformation is converted into heat and the friction between the dies and the workpiece, also produces heat. Coupled analysis considers the effect of temperature changes on the forged material properties. Because of the axial-symmetric geometry of the forged part, the simulation can be more efficient in terms of time analysis by adopting a 2D model analysis. The material model used is the elastic-plastic with hardening, the material is assumed to be isotropic and von Mises plasticity criterion was used in the analysis. The workpiece was defined as a deformable body, was meshed by a number of 300 quadrilateral finite elements and an adaptive remeshing method was selected. The two dies, the lower fixed and upper mobile, were defined as rigid bodies and meshed with a number of 2146 triangular finite elements. Material data for the workpiece used in experiment, and for dies are shown in Table 1.

Table 1. Material data for workpiece and dies.

Mechanical characteristics	Work-piece	Dies	Thermal characteristics	Work-piece	Dies
Young's modulus [MPa]	14000	217000	Thermal expansion [1/K]	2.93×10^{-5}	1.05×10^{-5}
Poisson's ratio	0.3	0.3	Conductivity [W/m·K]	24	45.1×10^{3}
Mass density [kg/m^3]	11360	7850	Specific heat [J/kg·K]	129	371×10^{3}

The initial temperature of the workpiece at the start of the forging process is 20°C. Contact conditions were established by specifying the coefficient of heat transfer with the environment for the workpiece, the coefficient of contact heat transfer and the initial temperature (20°C) for dies and the coefficient of friction between the dies and the workpiece ($\mu = 0.3$). The friction coefficient between the dies and the workpiece is assumed to be constant along the contact surfaces.

3 Results and discussions

The first results and analyzes undertaken followed the flow behaviour of the material and variations stresses, strains and temperature in the workpiece volume during the forging.

Some results regarding the stress, strain and temperature fields from the forged part are graphically illustrated in Fig. 2. In it is observed that all dies corners are well filled with material before it flows in the flash through the land. This shows that the flash channel dimensions were well designed so that it is ensured that firstly filling of the cavity corresponding to the forged part is complete and then excess material passes to the flash.

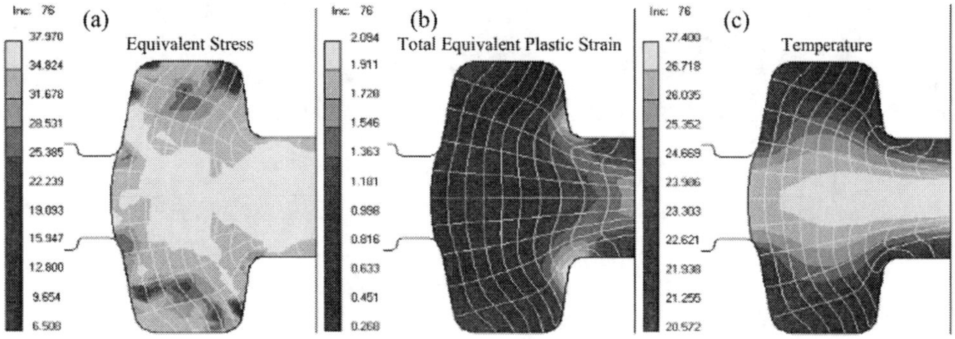

Fig. 2. Variation fields: (a) equivalent stress [MPa]; (b) equivalent plastic strain; (c) temperature [°C].

Images concerning the fields of stress and strain were also obtained for both the equivalent and their components, for any increment of time from the forging process.

Next, equivalent stresses from closed forging dies were analyzed, for various dimensions of the flash channel in order to optimize the process. Three different cases of flash channel dimensions were chose to be analyzed, shown in Table 2.

Table 2. The cases studied for flash dimensions.

Dimensions [mm]	Case I	Case II	Case III
Height, h_1	1	0.5	2
Height, h_2	4	3	6
Width, w_1	6	6	6
Width, w_2	20	20	20

The maximum values of stresses from dies are recorded at the end of the forging process. In Fig. 3 it can be observed that stresses have maximum values at the outer corners of the dies, but do not exceed the allowable strength at break of the steel from which dies are made (42CrMo4). This shows that the corner radius may be increased to ensure easy flow of the material in that area. The maximum value of forging force is achieved with the dies' closure.

DOI: 10.1051/matecconf/201711202008

An analysis of data obtained for the three simulated cases can be done by examining the graph from Fig. 4. The forging forces values show that by halving the flash land height (h_1) the force increases by about 71% (case II) and through doubling the height, forging force decreases by about 13% (case III). Maximum equivalent stresses of dies, compared to the case I which is considered a reference, increase by about 66% in case II and have an insignificantly change in case III (about 3%).

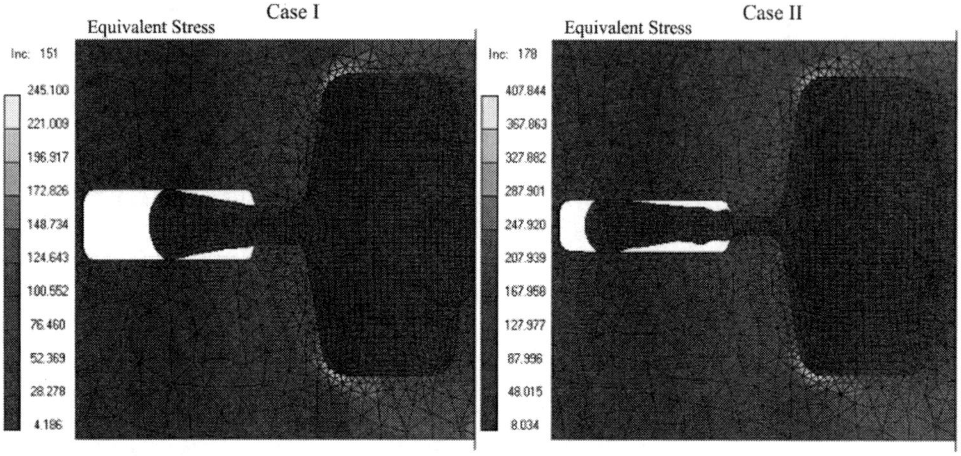

Fig. 3. Variation field of equivalent stress [MPa] from dies in two studied cases.

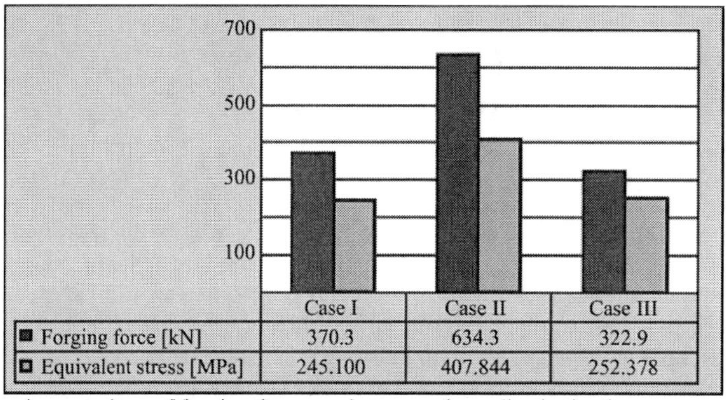

	Case I	Case II	Case III
Forging force [kN]	370.3	634.3	322.9
Equivalent stress [MPa]	245.100	407.844	252.378

Fig. 4. The maximum values of forging forces and stresses from dies in the three cases.

After obtaining finite element model it is important to check the agreement between this and the experimental model that it represents. Verification is required for validation of the model obtained and this is made by comparing the some parameters obtained by measuring from experiments, with values of the same parameters determined by simulation. Experimental tests were carried out on a hydraulic press of 75 tonnes and a lead (Pb) workpiece with the dimensions Ø40×75 mm was used. Experimental data acquired were: forging force [kN], forging pressure [MPa] and press ram stroke [mm]. The experimental obtained part can be seen in Fig. 5.a.

Comparison of numerical data obtained by simulation with experimental data was done for forging force. The two variations of forging force with press ram stroke show a good agreement between the numerical data and those from the experiments (Fig. 5.b).

In finite element modelling, the compatibility between material model, physical model and type of analysis is very important. If simulations are conducted with proper models, it can generate results sufficiently accurate for the modelling to be taken into account.

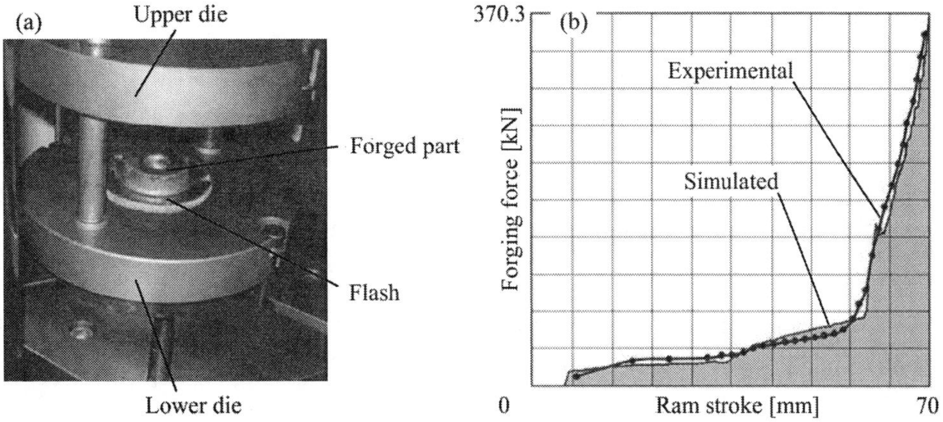

Fig. 5. Forged part obtained experimentally (a) and forging force vs. ram stroke for case I (b).

The results presented in Fig. 5.b show that the finite element model developed provide accurate data for numerical analysis performed in this paper.

4 Conclusions

In the case of closed die forging process numerically analyzed, it was noticed a good agreement between the data obtained by means of the finite element model developed and data acquired from experiments. Thus, forging force values extracted from finite element modelling results show a deviation of about 12% from the forging force recorded in the laboratory.

The main conclusions are: (*i*) Equivalent stresses from forging tools vary inversely with the flash land height; (*ii*) Forging force decreases with increasing of the flash land height however, increasing the flash land height can be made only within the limits in which ensures a complete filling of final cavity corners; (*iii*) Equivalent elastic strains in the dies, for all three cases considered, are low, below 1 µm (the maximum is 0.4 than 0.5 µm); it shows that the dies were well designed, are sufficiently rigid and provide the necessary precision for the forged part.

References

1. C. Brecher, W. Klein, M. Tannert, Prod. Eng. Res. Devel. **4**, 279 (2010)
2. T. Altan, M. Shirgaokar, *Process Design in Impression Die Forging*, Chapter 14, https://ercnsm.osu.edu/sites/ercnsm.osu.edu/files/uploads/F-hotforging/467.pdf, accessed 05.02.2017
3. B. Tomov, R. Radev, Int. J. Mater. Form., **3** (1), 319 (2010)
4. K. Wei, H. Yang, X. Fan, P. Gao, Int. J. Adv. Manuf. Technol., **81**, 729 (2015)
5. S.Z. Chavoshi, Int. J. Interact. Des. Manuf., Published online: 20 November 2014, http://link.springer.com/article/10.1007/s12008-014-0256-x, accessed 21.02.2017
6. S. Abachi, M. Akkok, M.I. Gokler, Tribol. Int., **43**, 467 (2010)
7. Z. Gronostajski, M. Kaszuba, S. Polak et al., Mat. Sci. Eng. A-Struct, **657**, 147 (2016)
8. B.F. Ghanbari, H. Arabi, S.M. Abbasi, S.M. Ali Boutorabi, Int. J. Adv. Manuf. Technol., **87**, 755 (2016)
9. R.K. Gupta, V.A. Kumar, P.R. Kumar, J. Mater. Eng. Perform., **25**, 2551 (2016)
10. M. Sedighi, M. Pourbashiri, J. Mech. Sci. Technol., **28** (12), 5129 (2014)
11. M. Pourbashiri, M. Sedighi, J. Mech. Sci. Technol., **30** (4), 1851 (2016)

Considerations about the influence of lubricant in different machining mechanical processes

Gheorghe Nagit*, *Andrei Marius* Mihalache, *Marius Ionut* Ripanu, and *Madalin* Pista
Gheorghe Asachi Technical University of Iasi, Department of Machine Manufacturing and Industrial Management, Blvd. Dimitrie Mangeron no. 59A, Iasi, Romania

Abstract. There are numerous researches that highlight the fact that wear related processes of machine components are one of the main causes which leads to their destruction. The main reasons for wear are related to the initial state of contact surfaces described by mechanical and physical properties of their superficial layer. Thus friction surface's state depends on previous mechanical machining done on this surface especially the last one. This study presents a method which proposes and creates premises for a micro-geometry favorable to maintaining the lubricant on part surfaces for as long as possible dependent on different machining processes. Comparisons are made between values specific to the wear state of the considered part obtained by various machining by measuring material mass removed during functioning in similar working conditions.

1 Introduction

As part of this research test probes were manufactured having their outer surface resulted through cutting and superficial plastic deformation. Final machining operations on part surfaces were by turning, rectification, rolling, diamond polishing and vibration rolling. The last process was then divided into ball vibration rolling and ball vibration rolling with channels. For the machining of parts obtained through cold superficial plastic deformation a device was manufactured which had interchangeable deforming elements. Wear resistance estimation was carried out by means of a device which allowed friction moment measurement and the friction coefficient between the test probe and the machine shoe.

2 Investigation technique and instruments

The Machine Manufacturing Technology laboratory from the Gheorghe Asachi Technical University of Iasi holds a device, designed and developed inside the lab, for cold plastic deformation which allows rolling, vibrating rolling and polishing by means of diamond spherical tips [1]. Figure 1 shows the device working principle as figure 2 depicts a photography taken during experimental attempts. We are attaching the cold superficial plastic deformation device to the tool holder sledge of a SN400x1000 Romanian lathe. Pushing the deforming element onto the probe has been done by the arc based system (6).

* Corresponding author: nagit@tcm.tuiasi.ro

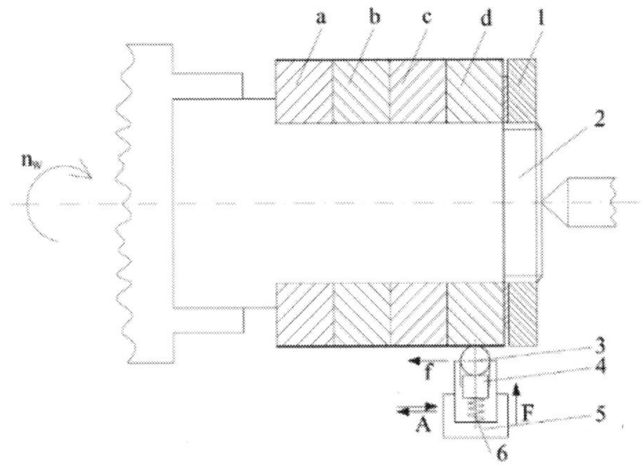

Fig. 1. The working principle of the used superficial plastic deformation device.

The device allows to inflict on the deforming element (3) oscillations (A) with amplitudes between 0 and 7 mm by means of a revolution variation device and a rod-crank type of tappet mechanism which allows oscillation frequency adjustment. The support (5) and the backing mechanism of the deforming element may be interchanged and replaced with other types; for example diamond tips. The revolution movement (n_w) of the mandrel support (2) makes the testing probes (a, b, c and d) revolve which overlaps with the feed of the lathe. The device allows the adjustment of the pressing force with values between 20 and $1000N$ by adequately replacing the arc mechanism (6).

To appreciate the wear resistance of parts whose surfaces were machined by different processes the following criteria was taken into consideration [2]:

- Part dimensions decrease during functioning;
- Mass reduction of the two elements which make up the friction couple (the part and the machine shoe);
- Time or revolutions executed by the part until friction moment stabilization;
- Values and variations of the friction moment and the friction coefficient inside the friction couple during functioning.

To adequately study those values and parameters a device (figure 3) was designed and developed which adapts on to a fine mechanics designed lathe.

The studied friction couples have been formed from a cylindrical part which had its outer surface machined by means of cold superficial plastic deformation processes or by cutting and a machine shoe with its inner cylindrical surface machined by rectification [3].

Fig. 2. Image taken during experimental attempts.

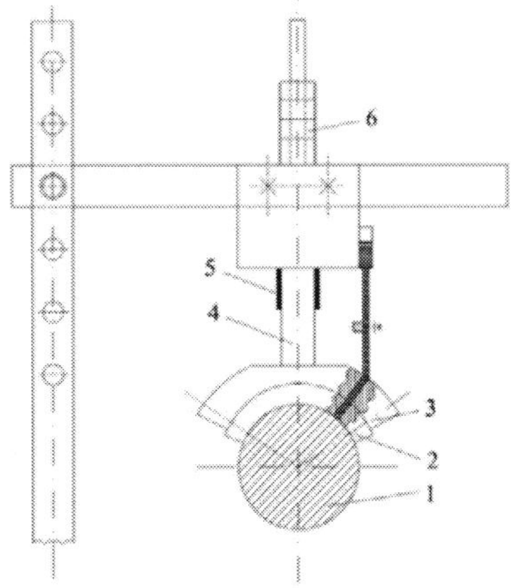

Fig. 3. The working principle of the device that estimates wear resistance of outer cylindrical parts.

Between friction couple's elements there is a relative motion obtained when the part revolves as the machine shoe stays fixed. Figure 3 presents a sketch of the designed device which has the interchangeable machine shoe (2) fixed on the shoe holder part (3). Inside the shoe holder part we have drilled a hole for the lubricant to be injected in the friction area between the part and the machine's shoe.

The shoes were made out of wear resistant materials such as the X210Cr12 steel which complies with the EN ISO 4957 standard. The shoe holder part is being fixed with the rod (4) on top of which dependent to its axis are mounted in a semi-deck format two 2359/TH 120-20 type of tensometric transmitters (5) having a resistance of $R = 128\Omega \pm 0.2\%$ and a constant of $2.39 \pm 3\%$.

By means of rod (4) the system has been loaded with the weights (6). The friction moment produced between the part (1) and the machine shoe (2) is being transformed into a bending moment at rod's (4) level and is being sensed by the transmitters.

The obtained information is then overtaken and recorded by a tensometric bridge. The lubrication system's adjustment is made through H20 oil flow control which complies with the STAS 9691-74 standard. An image taken during functioning is presented in figure 4.
The designed device allows the measurement of the following parameters:

- The friction moment between the part and the machine shoe;
- The friction coefficient and the friction force between the part and the machine shoe.

Fig. 4. Image taken during experiments.

The friction moment is given by [4]:

$$Mf = r \cdot F_f \quad [Nm] \tag{1}$$

Where r represents the radius of outer cylindrical test probe used in the experiment and Ff is the friction force between the two elements.
This force is expressed as [5]:

$$Ff = \mu \cdot N_a \quad [N] \tag{2}$$

Where μ represents the friction coefficient between the part and the machine shoe and Na is the force of normal pressing on the part's surface imposed by the friction shoe.

This force is determined as a product between the mass of the weights mounted on the vertical rod (4) and the gravitational acceleration g.
It results that by measuring the moment we can compute the friction coefficient as [4]:

$$\mu = \frac{M_f}{(m \cdot g)} \cdot \frac{1}{r} \tag{3}$$

Where g represents the gravitational acceleration and m the mass of the parts used as weights.

After experiments we can establish:

- Friction moment values and the values of the friction coefficient;
- The time needed for the grinding type of wear by drawing out the correlation graphical representations between the friction moment and the functioning time;
- The optimum type of micro-geometry in friction conditions;
- Comparison of different technological processes from the point of view of wear resistance of the machined parts;
- The optimum type of micro-geometry that is more suitable for energy dissipation which translates into lower values recorded for the friction moments.

3 Experimental results and interpretation

The experimental research has been focused on the comparison of friction moments between a part obtained by means of various machining and the machine's shoe made out of X210Cr12 steel which complies with the SR EN ISO 4957 standard depending on the pressing pressure between the part and the machine shoe and the oiling regime.

Test probes from ST60 steel have been manufactured having a 60 mm diameter and a widening of 22 mm machined through various processes up to 1 μm for the Ra roughness parameter for surface quality. A revolution speed of 800 rot/min has been used and lubricant up to $12 - 14$ $drops\ per\ minute$ for one set of experiments and no oiling at all for another set of experiments.

The pressing force was of 550 N for a pressure of $p = \frac{550}{\pi \cdot 2.2 \cdot 6} = 13.26 \quad N / cm^2$. The friction moment was periodically measured. By analyzing the experimental data we can observe the friction moment is greater for the scenario which does not uses lubricant between the part and the machine shoe. The friction forces are greater for the cutting machined parts.

Table 1. Friction moments between the part and the machine shoe with lubricant.

Part characteristics	$N \cdot mm$											
The number of part revolutions expressed in thousands	0	12	24	36	48	60	72	84	96	108	120	132
$Ra = 1.69\mu m$ (turning)	275	289	300	325	340	355	360	375	390	400	405	410
$Ra = 0.82\mu m$ (rectification)	250	268	283	310	329	346	355	370	385	400	410	410
$F = 350N, f = 0.059\ mm/rot, Ra = 0.65\mu m$ (ball rolling)	140	180	215	245	270	295	315	330	340	347	350	350
$f = 0.059, Ra = 0.1\mu m$ (ball vibration rolling)	120	155	185	210	230	245	260	270	276	280	281	281
(ball vibration rolling with channels)	90	120	145	170	190	205	215	217	220	220	220	220

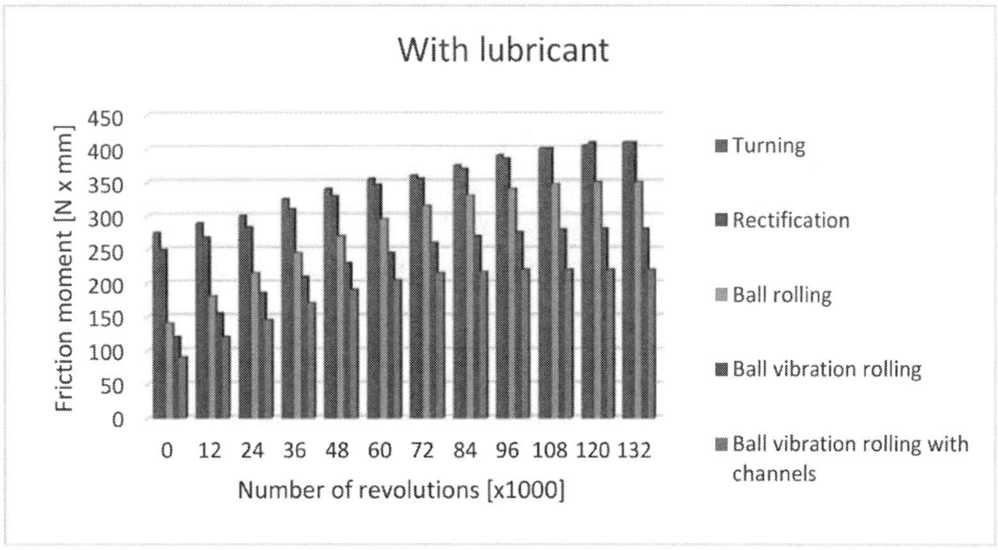

Fig. 5. Friction moments variations dependent on different machining types using lubricant.

The explanation points out the modifications of superficial plastic deformation surface characteristics. We can see that the parts machined by vibration rolling have a much smaller friction moment than the ones machined by other cold superficial plastic deformation processes. That is because we are achieving a much more pronounced hardening at vibration rolling obtaining a complete new topology as opposed to rolling or diamond based polishing. Rounding radiuses for the micro-irregularities in case of vibration rolling present greater values and have a much more obvious regular disposition which result in the so called "elementary cells of regular micro-relief". In the case of the micro-relief with a pressed channels system a significant role is attributed to the channels and their behavior as true "*oiling pockets*" in which the oil is maintained and/or transported in the friction zone.

Table 2. Friction moments between the part and the machine shoe without lubricant.

Part characteristics	$N \cdot mm$											
The number of part revolutions expressed in thousands	0	12	24	36	48	60	72	84	96	108	120	132
$Ra = 1.69\mu m$ (turning)	330	370	400	425	450	470	490	505	515	525	530	535
$Ra = 0.82\mu m$ (rectification)	300	340	370	400	425	445	460	475	485	495	500	504
$F = 350N, f = 0.059\ mm/rot, Ra = 0.65\mu m$ (ball rolling)	180	175	210	240	265	285	300	315	330	340	355	355
$f = 0.059, Ra = 0.1\mu m$ (ball vibration rolling)	150	180	210	230	246	261	273	280	287	292	295	295
(ball vibration rolling with channels)	125	150	172	195	206	220	226	229	235	235	235	235

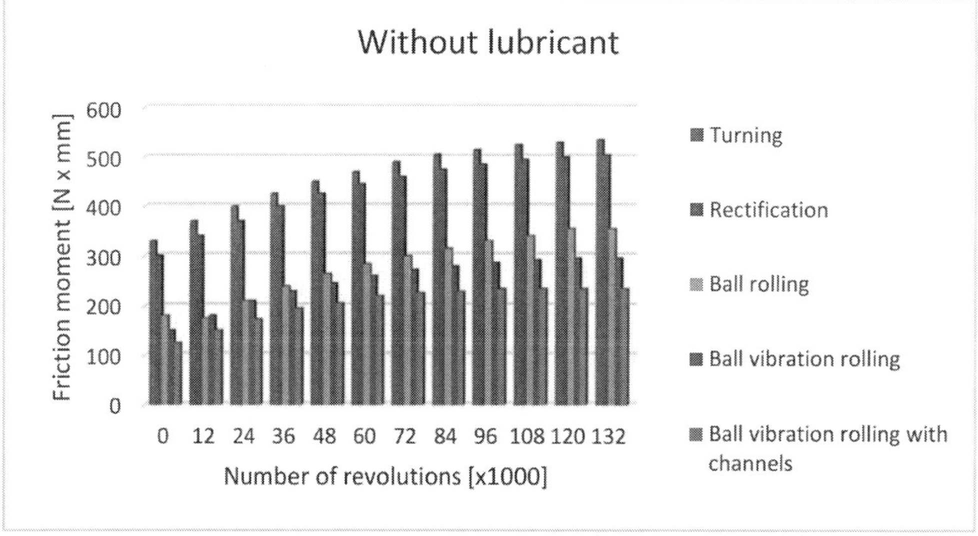

Fig. 6. Friction moments variations dependent on different machining types without lubricant.

A comparison between the two cases was carried out in order to highlight the main differences that occur in terms of friction moments related values. For all cases, as expected the use of lubricant decreases the friction process that occurs between the part and the machine shoe. Higher values are recorded for turning and rectification as the rolling processes do manage to record about half of those values. As well as the number of revolutions are getting bigger the friction moments are significant larger as result.

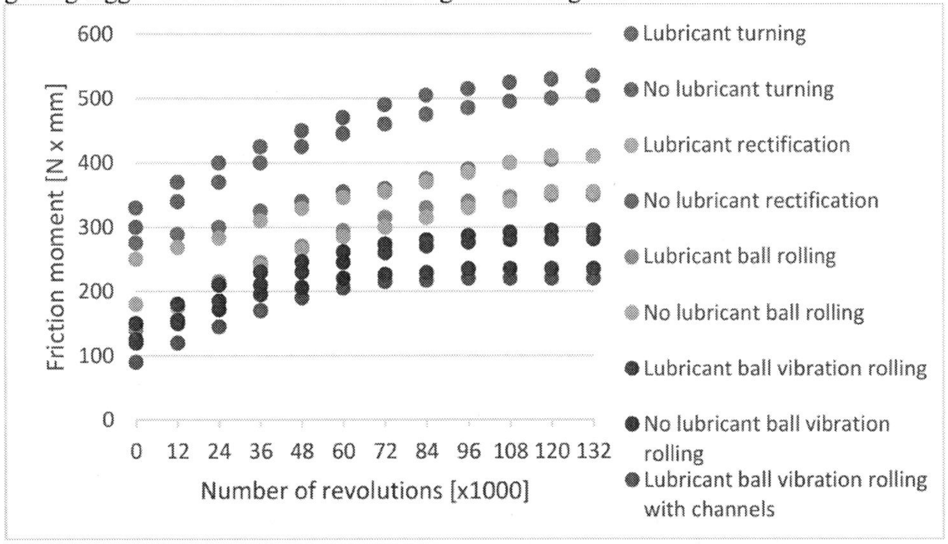

Fig. 7. Comparison between friction moments variations dependent on different machining types with versus without lubricant.

4 Conclusions

In all cases we can observe that that using lubricant produces better results in terms of friction moments between the part and the machine shoe. The best results are obtained for the ball vibration rolling with channels. The channels dissipate the energy produced as moments of friction in a very timely and accurate manner. The results obtained for this particular process do overlap all other ones even if no lubricant are used highlighting the importance of energy dissipation by means of channels.

In a brief presentation based on the sustained experiments and results in case of vibration rolling we are noticing:

- A double increase of wear resistance for the cold superficial plastic deformation machined parts and especially for the vibration rolled ones; the case of the frication couples;
- A decrease of the friction moment and that of the friction coefficient for the vibration rolling obtained parts;
- A decrease of the grinding time up to two times for the superficial deformed parts and especially at those subjected to vibration rolling.

References

1. A.V. Antsupov, V.P. Antsupov, M.G. Slobodianskii, Procedia Engineering, 2nd International Conference on Industrial Engineering (ICIE-2016), **150**, 411-415 (2016)
2. I.P. Troyanovskaya. B.M. Pozin, Procedia Engineering, International Conference on Industrial Engineering (ICIE-2015), **129**, 156-160 (2015)
3. L. Duana, Z. An, R. Yanga, Z. Fub, Tribology International, **103**, 218-227 (2016)
4. N. Banerjee, A. Sharma, Journal of Materials Processing Technology, **238**, 181-194 (2016)
5. M. Korzynski, J. Lubas, *Nonconventional finishing technologies* (Polish Scientific Publishers PWN, Warsaw, 2013)

Electrochemical and corrosion behavior of two chromium dental alloys in artificial bioenvironments

Alexandra Banu[1,*], *Cristina* Juganaru[1] , and *Maria* Marcu[2]

[1]UPB, Manufacturing Engineering Department, 313 Splaiul Independentei, Romania
[2]"Ilie Murgulescu" Institute of Physical Chemistry of the Romanian Academy, 202 Splaiul Independentei, Romania

Abstract. The purpose of this study is to compare the corrosion and tarnish behavior of NiCrMo and CoCrMo cast dental alloys in artificial bio environments. The cobalt chromium alloys are known and used in dentistry for many years, but its difficult machinability because of the strength and hardness, is an argument for scientists to study alternative materials with comparable biocompatibility. On the other hand, for dentistry devices beside corrosion behavior is important the aesthetic so, the used alloys have to preserve their shining and do not stain. The corrosion resistance has been evaluated using the Atomic mass spectroscopy method for ion release determination, the anodic polarization curves and the open circuit potential – time monitoring for corrosion behavior evaluation and optical microscopy for the structure analysis. The tarnish tendency of alloys was estimated using the method of cyclic immersion with frequency of 10 seconds for each minute during 72 hours in Na_2S containing solution. The most important conclusion is that the alloys are comparable from corrosion and tarnish point of view, but we recommend to use the nickel base alloy only for orthodontic devices implanted for short periods of time, because of higher quantity of released ions.

1 Introduction

Mechanical properties of alloys used in dental crown and bridges applications differ from those of porcelain devices [1-3]. The first materials have to be tough and rigid and the second one, on the contrary, because these properties make difficult the finishing steps as grinding and polishing. Dental alloys must have excellent corrosion resistance to the oral environment, which is highly important for maintaining biocompatibility throughout the time of use of the device. From this point of view, is not so important and dangerous the ions generation because of corrosion, how is their toxicity, their poisoning power for human body [4, 5]. Negative effect of dental alloys in the oral cavity is their capability to produce local or systematic changes that may be short term or long term or systematic. Interactions between the oral environment and dental alloys can also generate some discomforts like metallic taste,

* Corresponding author: alexandrabanu14@yahoo.com

discoloration of teeth, galvanic currents, oral lesions, dermatitis hypersensitivity, dental implant rejection, and generating tumors and decay [5-6].

Because the noble metals (gold, palladium, silver) became less and less accessible for many people, in the recent decades, new alternative and cheaper materials came to dental market. The materials used in oral environment must be completely tarnish-resistant, they must be chemically and electrochemically inert in contact with a very big variety of foods, that means a large pH domain [7, 8]. While corrosion behavior of a material is generally understood, the tarnishing is a less used notion. Tarnish can be defined as a surface discoloration, in fact color changes, or decrease in luster of a dental alloy because of the formation of a surface product.

The main task of this work is to evaluate how two nickel and cobalt chromium alloys behave in oral cavity from metallic ions release and tarnish point of view.

2 Experimental details

The experimental samples have been cut from two chromium molybdenum cast dental alloys with the chemical composition shown in Table 1.

Table 1. Chemical composition (% weight) of studied alloys.

Sample label	Alloy	Cr %	Mo %	Fe %	Si %	Mn %	Al %	Ti %	Co %	Ni %
1.	NiCr	19.2	5.2	2	2.09	1.5	0.6	0.34	0.4	base
2.	CoCr	28.1	4.2	-	1.01	0.6	-	0.45	base	-

The CoCrMo alloy (sample 2) is close to a known dental material, Stellite 21, but with less Molybdenum content (4% of Molybdenum instead of 6%).

Before each test, the samples were mechanically prepared according with the practical instruction from dentistry laboratories.

2.1 Techniques, solutions, samples

For structural characterization of materials was used optical microscopy after electrochemical etching of surfaces using 2% chromic acid solution. Have been used the following tests for characterization of corrosion behavior of studied alloys:

1. The Stationary immersion test during 168 h at (37±1)°C was used in order to evaluate the corrosion rates and pitting corrosion resistance [2]. For corrosion rate have been used two ways: mass loss and the electrolyte's analysis after immersion time. As corroding solution was used, according to ISO 10271:2011, lactic acid $C_3H_6O_3$ (0.1 M); Natrium chloride, NaCl (0.1 M); H_2O grade 2, according to ISO 3696. All reagents were pro analysis grade of quality. The samples were cut at 32 x 10 x 1.5 mm of dimension. Corrosion rates were done by weighing using the analytical balance (10^{-4} % accuracy) before and after immersion and by solution analysis using the Atomic mass spectroscopy method for ion release determination.

2. Time monitoring of open circuit potential (OCP) during 168 h and the cyclic potentiodynamic polarization method were done in artificial saliva with chemical composition according [2] (NaCl, KCl, $CaCl_2x2H_2O$, NaH_2PO_4 H_2O, Na_2Sx9H_2O, Urea; H_2O grade 2); all reagents were pro analysis grade of quality; as reference electrode was used the saturated calomel electrode (sce). Were used cylindrical samples with a total surface area of 1 cm^2; the electrical contact was made by screw (a threaded hole was made at one end of the

sample). Potentiodynamic polarization test was performed using a Radiometer potentiostat/ galvanostat PGP 201 and acquired data were processed with Voltalab specialized software. The potential was scanned between –250mV and +1000mV forward and back with a scan rate of 0.6V/h.

3. Sulfide tarnish test (cyclic immersion) [2], was performed using cylindrical samples (L=0.5 mm; Φ=10 mm) with immersion frequency of 10 seconds for each minute, during 72 hours at room temperature. As a staining solution was used Na_2S x$7H_2O$ and distilled water grade 2. The device for cyclic immersion was built in our laboratory. A vertical wheel is driven by a step-by-step electric motor so that the specimen is immersed in the test solution 10 seconds for each minute (a cycle consists in 10 seconds of immersion and 50 seconds of air exposure).

3 Results and Discussions

3.1 Structure

As is known, the chemical composition of an alloy is not determining factor for the physical properties of the material [9] and from this point of view very important is the material's structure. In such alloys, the cobalt element affects the modulus, strength and hardness of the alloy, the carbon increases the hardness because a lot of elements such as chromium, molybdenum and cobalt react with carbon to form carbides, affecting the properties of alloys. In addition, the silicon and manganese increase the liquidity and stability and the molybdenum and chromium are benefic for corrosion resistance in the chloride ions presence [10]. As shown in the Fig. 1, both alloys present dendritic structure with an important carbides quantity. The nickel base alloy (Fig. 1a) seems to have a bigger quantity of carbides compared with the cobalt alloy and the structure shows fine elements, oriented in the solidification direction. The cobalt alloy (Fig. 1b) has an austenitic matrix composed of the solid solution of cobalt and chromium, and the dendritic structure formed by different chemical segregations. Despite the fact that it is the most studied non precious dental alloy, yet many works are devoted to the study of its structure [11-13]. Most of segregated phases are carbides with different formal oxidation number, compounds from MC to $M_{23}C_6$ with specific morphologies such as spherical with discontinuous arrangement or agglomerated within grain boundaries.

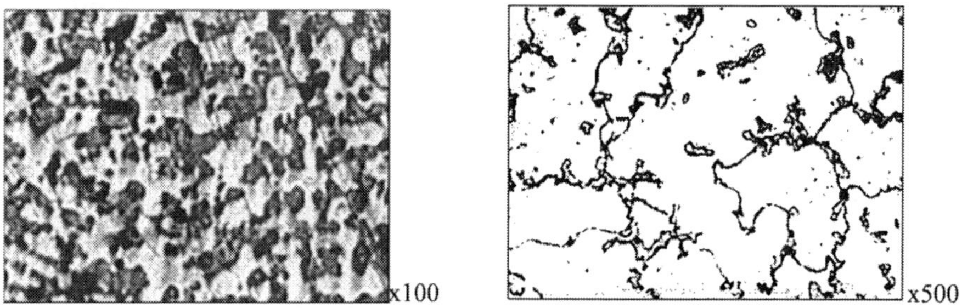

Fig. 1. Microstructure of samples: (a) NiCrMo alloy and (b) CoCrMo alloy, after electrochemical etching using 2% of chromium acid solution.

3.2 Stationary immersion test

The Stationary immersion test consists, according to ISO 10271, in determination of mass loss of the samples by weighing before and after immersion, and in determination by analyze of electrolyte of metallic ions released. After 168 h of immersion in synthetic bio fluid at

(37±1) °C, the quantities of metallic ions released, expressed as μg cm^{-2} of immersed surface are presented in Table 2. As is shown in the table, the nickel alloy released an important quantity of ions in 168 h of immersion (about 93 μg cm^{-2}). Despite of the fact that the nickel has an important toxicity potential, some studies [4, 14, 15] reveal that nickel is a moderate allergen and no evidence of toxicity associated with the intraoral use of nickel-containing dental alloys is found. It is concluded that, subject to use it according to established techniques, the dental alloys with nickel do not pose a risk to patients or members of the dental team. On the other hand, the quantity of nickel ions ingested from food or water is higher than that from oral devices or from dental reconstruction [4].

Table 2. The ions release, expressed in μg cm^{-2}, after 168 h of immersion in lactic acid solution at (37±1) °C - Atomic mass spectroscopy.

Alloy	Release, μg cm^{-2}			
	Ni	Co	Cr	Mo
NiCrMo	77.85	-	4.234	4.4
CoCrMo	-	0.314	0	0

Compared with nickel alloy, the cobalt alloy loses a very small quantity of metallic ions (less than 1 μgcm^{-2} in 168 h) and this shows a better behavior as biocompatible material for cobalt chromium alloy. Table 3 presents the corrosion rates ($gm^{-2}h^{-1}$) determined by weight loss of the samples. This values, are in good agreement with those obtained by Okazaki and Gotoh [16] in similar conditions.

Table 3. Corrosion rates after 168 h of immersion in lactic acid solution.

Alloy	Immersed surface cm^2	Weight loss $mgcm^{-2}$	Corrosion rate $gm^{-2}h^{-1}$	Penetration rate mmpy
NiCrMo	7.8	0.395	0.0235	0.02445
CoCrMo	7.08	0.0876	0.00522	0.00574

From practical point of view, an important design parameter is the corrosion rate expressed as the rate of decreasing of the oral implant thickness, in millimeters per year (penetration rate). The average values of penetration rates, calculated as arithmetic average of three values, highlighted superior behavior of cobalt alloy. The value of 0.005 millimeters per year (mmpy) falls this alloy in class II, "Very resistant", according to STAS 9684-82, while the nickel alloy presents a corrosion rate of five times higher and it falls in class III, "Resistant". The pitting corrosion evaluation was made by optical inspection at a magnification of 10 times and we found that the cobalt alloy was perfect resistant (no pits), while the nickel alloy is sensible, an incipient corrosion attack was observed.

3.3 Electrochemical tests

Have been run two types of electrochemical tests: time evolution of open circuit potential (OCP) for 90 hours and cyclic polarization, toward anodic direction and back, in order to calculate corrosion rates and estimate the sensitivity to pitting corrosion related to passivation - repassivation tendency. Electrochemical tests were performed in artificial saliva environment. Figure 2 shows the time evolution of OCP for both alloys and highlights the following aspects:

• The instantaneous values of mixed potentials are positives at about +0,24V/sce and +0,48V/sce, respectively, this important difference is given by the higher content of chromium in CoCrMo alloy (28% Cr) compared with 18%Cr in NiCrMo alloy.

• In the first 20 hours, both alloys presented an activation tendency, OCP moved toward electronegative values which mean that the native passive superficial film is quasi-stable in the presence of chlorine ions. As was shown [16] in aqueous environments, the chromium is subject of so called hydro oxolation reactions which in the absence of chloride ions reinforce the passive film, but when the chloride ions are present, they damage this film. This behavior of damage and reinforce of passive film is highlighted by the oscillation of OCP toward electronegative and electropositive values during immersion period.

• After about 70 hours of immersion the OCP values are approximately stabilized around 0 V/sce for NiCrMo alloy and around +0,150mV/sce for CoCrMo alloy, respectively. Correlating this behavior with cyclic anodic polarization test, both alloys tend to passivity.

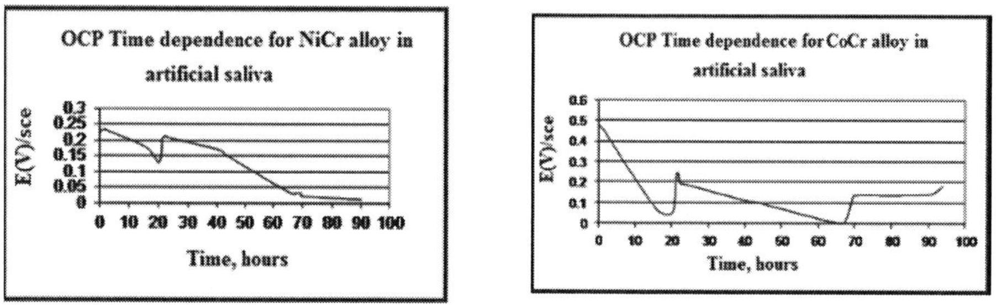

(a) (b)

Fig. 2. Time evolution of the open circuit potential in artificial saliva environment for (a) NiCrMo alloy and (b) CoCrMo alloy .

The Figure 3 shows potentiodynamic polarization curves performed in artificial saliva.

During anodic polarization sequence of the curve, the surface of sample oxidizes and, as can be seen in the figure 3, the corrosion density current does not rises exponentially according Buttler Volmer equation, in fact it is limited diffusion density current with values less than $1mAcm^{-2}$. During the reverse scanning of potentials, toward the electronegative values, despite of the reduction processes on the surface, the current density remains at low values, meaning that the surface of sample is still passive. It is the effect of oxygen affinity of chromium and hydro oxolation processes, as was described above and in reference [17]. Despite of chloride ions presence, both alloys are passive in a wide range of potentials, the most important differences between its are the values of difusion limited current density (passive current density).

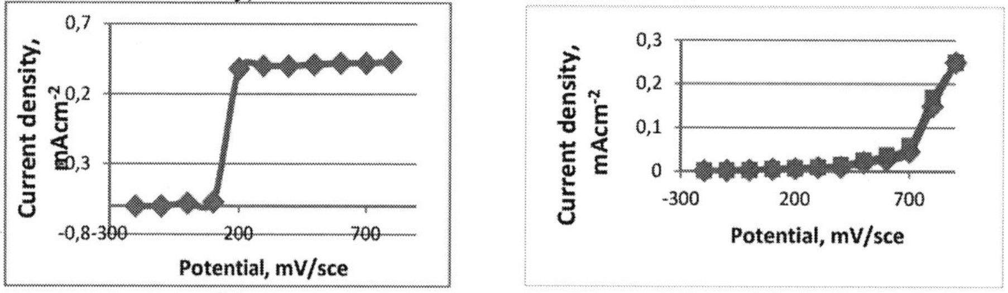

(a) (b)

Fig. 3. Potentiodynamic polarization curves performed in artificial saliva for (a) NiCrMo alloy and (b) CoCrMo alloy .

A higher chromium concentration in the CoCrMo alloy is benefic for an easier passivation, the limited current density is less than $50\mu Acm^{-2}$ while for NiCrMo alloy this parameter is about 8 time higher, around of 400 μAcm^{-2}. Concerning to the corrosion rates obtained by Tafel slopes, these values are in agreement with that obtained by weighing of samples, at about 5-6 µm per year for CoCrMo alloy and at 15-20µm per year for NiCrMo alloy.

3.4 Tarnish test

Tarnish can be defined as a surface discoloration, or decrease in lustre of a dental alloy because of the corrosion product formation on the surface. Several synthetic saliva solutions, as well as aggressive solutions have been used in attempts to identify in accelerated tests the conditions under which an alloy tarnishes [8]. Since chlorides are present in saliva and certain foods such as fish and eggs have high sulphur content, both chloride and sulphide solutions are commonly chosen as immersion test media. Ranking of alloys in terms of susceptibility to such tarnishing has been recorded subjectively, by comparing colour changes visually. After 72 hours of cyclic immersion in chloride and sulphide containing solution, the surfaces of tested samples are not modified, no color changes or loss of shine. The tested alloys are similar from tarnish resistance point of view.

4 Conclusions

Chromium dental alloys used in the oral environment may be affected by electrochemical corrosion processes, so their corrosion behavior can be predicted from results of electrochemical testing. The cobalt chromium alloy revealed an excellent behavior. The nickel chromium alloy tends in time, in artificial saliva, to the electrochemical activity, by movement of OCP toward electronegative potential range, but the corrosion rate is relatively low related to the quantity of released ions. Because of that and correlated with their good tarnish behavior, the nickel chromium alloys can be used for production of dental devices which remain for short time in contact with the oral environment.

References

1. ASTM F75, *Book of Standards Volume*: 13.01 F75-12
2. ***ISO 10271:2011, Dentistry - Corrosion Test Methods For Metallic Materials
3. C. Manaranche, H. Hornberger, Dental Materials **23**, 1428-1437 (2007)
4. J. C. Setcos, A.B. Mahani, L. Silvio, et all, Dental Materials **22,** 1163–1168 (2006)
5. E. Vasilescu, P. Drob, I. Cinca, et all, Rev. Chim. **61**, 168-171 (2010)
6. M. Mikulewicz, K. Chojnacka, Biol Trace Elem Res **142**, 865–889 (2011)
7. N. Rin, I. Bau, S. Miko, M. Papi, et all, Coll. Antropol. **27** (2), 99–106 (2003)
8. W. L. Leon, G.W. Stanford, GoldBull. **14** (1), 13-18 (1981)
9. K. Asgar, F. A. Peyton, J. Dental Research **40**, 63-72 (1961)
10. J.C. Wataha In Craig RG, JM Powers editor. *Restorative dental materials. The 11th edition.* (St Louis: Mosby, 163-184, 2002)
11. J. Marciniak, Marciniak J., Biomaterials in bone surgery. (in Polish), (Silesian University of Technology, Gliwice, 1992)
12. A. Woźna, Current Problems of Biomechanics (in Polish), **5**, 183 (2011)
13. A. Łukaszczyk, J. Augustyn-Pieniazek, Archives of metal and materials, **60** (2015)
14. ***Toxicological profile for nickel U.S. Dep. of health and human services Public Health Service Agency for Toxic Substances and Disease Registry, (2005)
15. R.D. Barrett, S.E. Bishara, Am. J. Orthod. Dentofacial Orthop. **103**, 8–14 (1993)
16. Y Ocazaki, E Gotoh, Biomaterials **26**, 11-21 (2005)
17. A. Banu, O. Radovici, M. Marcu, Roum. Biotechnol. Lett. **9**, 1603-1608 (2004)

MATEC Web of Conferences **112**, 03002 (2017) DOI: 10.1051/matecconf/201711203002
IManE&E 2017

Kerf variation analysing for abrasive water jet cutting of a steel square part

Adrian-Paul Basarman* and *Mircea* Lobonțiu

Technical University Cluj Napoca, North University Center Baia Mare, Dr. Victor Babeș Street 62A, 430083 Baia Mare, Romania

Abstract. The abrasive water jet cutting method is a modern method for cutting materials. It is used for cutting different type of materials, from glass, rocks, and even metals like titanium. This method has a reached a high level of usability in the nowadays modern production. In order to obtain the class of precision needed for different requirements in production, the surface quality, the kerf aspect, the shape and respectively the form of the obtained part have to be researched and analyzed. This paper presents the results obtained after cutting one square shaped part, made of S355 material. This paper presents the study regarding both the inside and the outside of the cut, the kerf width, the aspect of the taper and the profile deviation.

1 Introduction

The water jet is a modern method for cutting materials and there is no relevant literature for calculation of the stock left for machining for metallic parts, so there is a need of some research in order to have a more accurate view regarding this matter. This paper presents a continuation of a study started with two steel pieces made from 1C45 cutted using the same principles [1,2]. The present study was made on a S355 material. Researches in this field, regarding the water jet, have been made, by international authors like Fowler [3], studying the of milling the surface using the abrasive water jet technology, Hloch et. al. [4] and Valíček et.al. [5] which researched the way of turning materials using the same technology and recent studies regarding the surface roughness for steel parts cutted with AWJ was made by Ravai-Nagy et.al.[6] and recent studies by Hergheleghiu et. al. [7] on aluminium alloy, Filip et. al [8] on machinability and Medan et.al. [9] for parameters influence.

2 Experimental Setup

2.1 Preparation for experiments

The experiments were conducted using an abrasive water jet (AWJ) cutting machine - Bystronic ByJet Pro L - as presented in Fig 1.

*Corresponding author: adrian.basarman@gmail.com

The method of cutting materials using water, which has in it's consistency abrasive particles is a new and very different technology compared to the more commonly used conventional technology in which, in order to pierce or cut the material, a tool is needed. In this case, the tool takes contact with the surface, but using the AWJ technology, that doesn't happen anymore, so no more tools that wear out, but the parts that restrain and guide the abrasive water jet are the ones that take the extreme pressure and in the end break down.

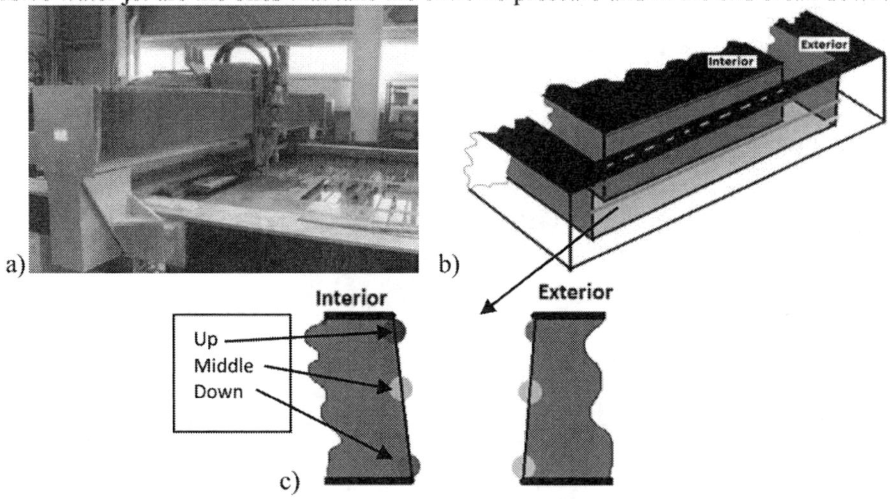

Fig. 1. a) Bystronic ByJet Pro L, b) the general aspect of the AWJ cut and the way of measuring, c)detail regarding the way of measuring the surface after cutting.

In order to take notice of how the surface has modified after cutting, we had to analyse both, the interior and the exterior of the kerf, as presented in Fig. 1b) and c). By measuring the kerf width, the values obtained can be used in order to determine the stock left for machining, needed in order to have a more precise cut and a more precise resulted surface.

These experiments were conducted on a S355 material plate, of 30 mm thickness. The experiment was made by cutting a 50x50 mm square part in a 70x70 square part, in order to study both the inside and outside of the cutting area.

From the list of cutting regimes offered by the machine, regarding the speed of cut, the parameters selected are presented in Table 1.

Table 1. The parameters used for the cutting regime

Parameters	Value selected
Breakthrough time	19 s
Breakthrough pressure	3600 bar
Abrasive material used	GMA Garnet 80 Mesh (300-150 micron)
Quantity of abrasive material	342 g/min
Cutting pressure	3600 bar
Cutting speed	57 mm/min
Interior sapphire nozzle	0,28 mm
Exterior nozzle	0,8 mm

2.2 Conducting the experiments

The first step for cutting the part was using the machine's clamping system to hold the base plate in place. After this step, the program of cut was loaded in the software of the machine and after that was selecting the regime of cutting, presented in Table 1.

After this, the part is being cut and the result is presented in Fig. 2.

2.3 Measuring the profile deviations

After cutting the steel part, for both parts resulted, as one can observe in Fig 2, every of the surfaces were measured separately one after another, but analysed one with each other, both for the interior and for the exterior of the cut.

Because it's a square part, it will have four exterior surfaces (E1-4) and four interior surfaces (I1-4) which were analysed together (example- exterior E1 with interior I1).

To determine the exact aspect of the surfaces, the surface was divided in three rows, as presented in Fig. 1b) and c) and using a 3D measuring arm, they were taken a number of 10 measurements on each row, 30 points per surface. Those values were set to be measured from the ideal path of cut which is exactly the theoretically path for the AWJ cutting.

Table 2. Values measured on every surface on the steel square part cutted using AWJ.

Area/ Part	*Row*	*M 1*	*M 2*	*M 3*	*M 4*	*M 5*	*M 6*	*M 7*	*M 8*	*M 9*	*M 10*
I/Int	Up	0.907	0.857	0.874	0.868	0.897	0.899	0.874	0.876	0.869	0.884
	Mid.	0.729	0.686	0.673	0.676	0.741	0.731	0.681	0.673	0.685	0.718
	Down	0.518	0.528	0.494	0.513	0.581	0.547	0.517	0.487	0.484	0.554
I/Ext	Up	-0.021	-0.008	-0.008	-0.016	-0.041	-0.014	-0.003	0.001	-0.004	-0.032
	Mid.	-0.132	-0.054	-0.058	-0.075	-0.153	-0.130	-0.052	-0.061	-0.060	-0.149
	Down	-0.161	-0.104	-0.120	-0.129	-0.274	-0.192	-0.128	-0.140	-0.147	-0.235
II/Int	Up	0.898	0.877	0.878	0.882	0.885	0.886	0.886	0.897	0.895	0.915
	Mid.	1.043	0.964	0.966	0.968	0.974	0.979	0.969	0.970	0.980	1.061
	Down	1.143	1.028	0.997	1.006	1.038	1.025	0.951	1.009	1.016	1.101
II/Ext	Up	-0.059	-0.026	-0.012	0.149	0.100	0.028	-0.040	-0.025	-0.058	-0.078
	Mid.	0.047	0.131	0.141	0.139	0.160	0.151	0.153	0.132	0.126	0.113
	Down	0.236	0.311	0.392	0.301	0.321	0.318	0.322	0.360	0.340	0.335
III/Int	Up	0.935	0.915	0.902	0.907	0.899	0.893	0.887	0.880	0.872	0.882
	Mid.	1.042	0.949	0.940	0.946	0.925	0.933	0.922	0.929	0.916	0.990
	Down	1.008	0.887	1.014	0.939	0.964	0.957	1.048	0.944	0.968	1.081
III/Ext	Up	-0.014	0.010	0.043	-0.009	-0.006	0.014	0.006	-0.029	-0.013	-0.046
	Mid.	0.126	0.148	0.141	0.161	0.183	0.165	0.139	0.114	0.121	0.062
	Down	0.347	0.297	0.217	0.236	0.318	0.298	0.149	0.320	0.271	0.223
IV/Int	Up	0.858	0.821	0.823	0.834	0.835	0.843	0.847	0.862	0.894	0.921
	Mid.	0.714	0.657	0.647	0.641	0.650	0.656	0.673	0.678	0.702	0.790
	Down	0.537	0.418	0.398	0.443	0.368	0.461	0.384	0.472	0.512	0.455
IV/Ext	Up	-0.114	-0.107	-0.100	-0.026	-0.071	-0.067	-0.056	-0.048	-0.027	-0.024
	Mid.	-0.206	-0.168	-0.154	-0.135	-0.125	-0.138	-0.133	-0.106	-0.078	-0.122
	Down	-0.345	-0.313	-0.271	-0.261	-0.246	-0.218	-0.221	-0.220	-0.126	-0.240

a)

Fig. 2. The part cutted using AWJ.

3 Analysing the results

After measuring every surface, the values were calculated making reference to an ideal path of cut. The results in table 2 position the measurements at some distance from the ideal path. Both, the interior or the exterior measurements, can be on same side of the path in some cases, or some of the values resulted, were with minus, positioning the result on the other side of the ideal path.

Regarding the kerf width, the values from each side of the ideal path were summed together, to form the resulted measurement of the kerf width and the values are presented in the Table 3.

This values are correspondent directly proportional with the kerf width. After analysing the values, one can observe that the smallest kerf width from all the surfaces analysed is 0,590 mm and the biggest kerf width was 0,996 mm.

4 Analysing the aspect of the surface

The principles of cutting were the same all around the square part, and in that case, the values obtained should be almost the same, so in this care we will only present two cases:
-first one - when the cutting head moved longitudinally
-second one- when the cutting head moved transversally.

Table 3. The kerf width on every surface.

Area/ Row	M 1	M 2	M 3	M 4	M 5	M 6	M 7	M 8	M 9	M 10
I/Up	0.928	0.865	0.882	0.884	0.938	0.913	0.877	0.875	0.873	0.916
I/Mid.	0.861	0.740	0.731	0.751	0.894	0.861	0.733	0.734	0.745	0.867
I/Down	0.679	0.632	0.614	0.642	0.855	0.739	0.645	0.627	0.631	0.789
II/Up	0.957	0.903	0.890	0.733	0.785	0.858	0.926	0.922	0.953	0.993
II/Mid.	0.996	0.833	0.825	0.829	0.814	0.828	0.816	0.838	0.854	0.948
II/Down	0.907	0.717	0.605	0.705	0.717	0.707	0.629	0.649	0.676	0.766
III/Up	0.949	0.905	0.859	0.916	0.905	0.879	0.881	0.909	0.885	0.928
III/Mid.	0.916	0.801	0.799	0.785	0.742	0.768	0.783	0.815	0.795	0.928
III/Down	0.661	0.590	0.797	0.703	0.646	0.659	0.899	0.624	0.697	0.858
IV/Up	0.972	0.928	0.923	0.860	0.906	0.910	0.903	0.910	0.921	0.945
IV/Mid.	0.920	0.825	0.801	0.776	0.775	0.794	0.806	0.784	0.780	0.912
IV/Down	0.882	0.731	0.669	0.704	0.614	0.679	0.605	0.692	0.638	0.695

In Fig 3a) it can be seen the aspect of the cut, and also of the surfaces, both for the interior of the cut and also for the exterior. One can see that the surface has an irregular shape, and it presents, by observing the values in the table 2, a narrowing of the surface in the down area, and the whole surface that was cut, it's not straight, from top to bottom, it has an inclination of cut towards the exterior.

Putting the values together, it was generated the aspect of the width of cut, shown in the Fig 3b).

As it can be observed in Fig 3a), the bulging aspect of the surface is accentuated in the middle and even more up in the corners and the more different is measurement 5 which is opposite to measurements 3 and 9 (counting starts from the first left vertical line). The aspect from the corners may result because of the machine error while switching the movement direction from one axis to another. On this surface, the smallest kerf width resulted here is 0,614mm and the biggest kerf width is 0,938 mm, as presented in Table 3 and Fig 3b).

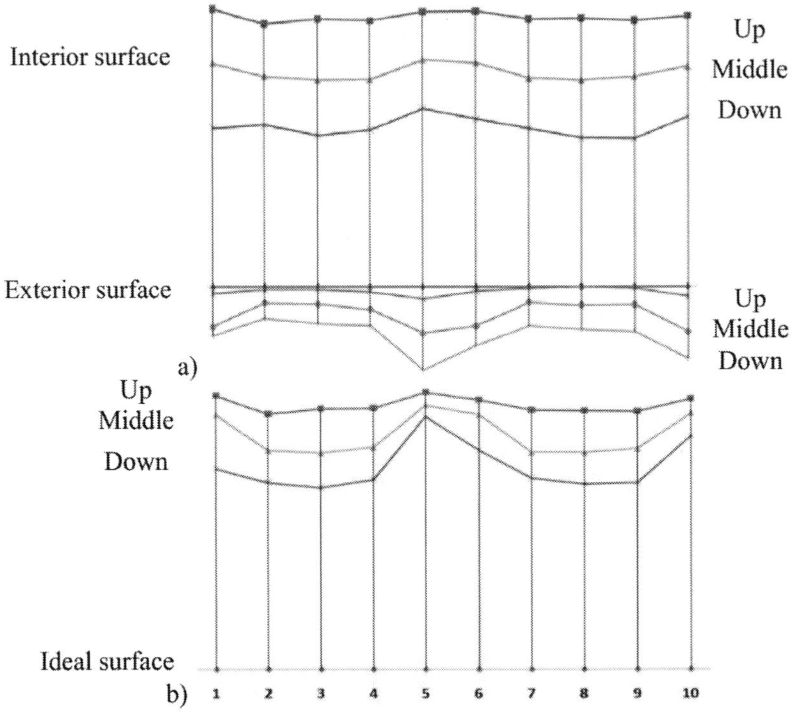

Fig. 3. a)Top view for the aspect of the surface I; b)The width of cut for surface I.

Analysing the next surface, presented in Fig 4, similar aspects can be seen, like in the ones in Fig 3, but here, some of the exterior, and also some of the interior lines, intersect.

As one can see, in Fig 4, as presented, it can be observed a little difference from the surface I. The surfaces are a bit more straight but presents also a V shaped taper, and in this case, the sliding is towards the interior, opposite to the first surface. In Fig 5 one can observe the kerf width for the second surface.

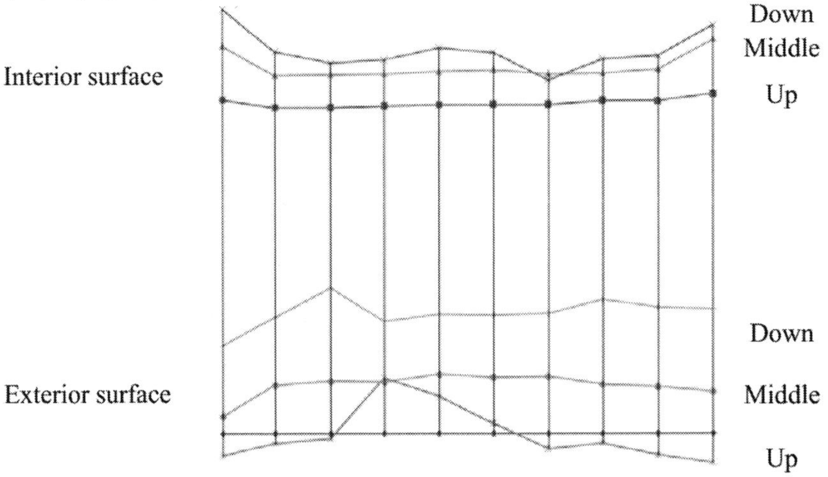

Fig. 4. Top view for the aspect of surface II.

DOI: 10.1051/matecconf/201711203002

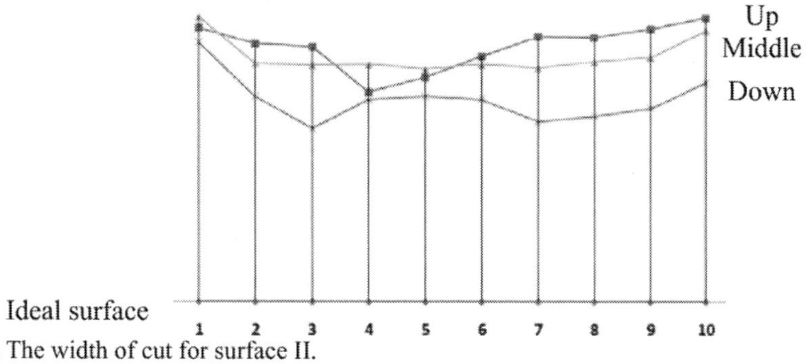

Fig. 5. The width of cut for surface II.

As can be seen in Fig 5, the surface a bulging aspect in the middle and more up in the corners because the water jet kicked back, the surface has irregular shape, and the lines intersect. The smallest kerf width resulted is 0,605 mm and the biggest one is 0,996 mm. Rezults that the error of the machine is the problem, while switching direction. Analyzing the results, both, the smallest kerf width and the biggest kerf width are very similar.

5 Conclusions

1)The surface aspect has an inclination because of the error of the machine.

2)The maximum stock needed to be left for machining with abrasive water jet (AWJ) is 0,996 mm for a 30 mm thickness steel S355 part, cutted at a speed of 57 mm/min.

3)The surface cut using AWJ technology presents a V-shape tapper.

References

A.P. Basarman, M. Lobonţiu, Nonconventional Technologies Review **20**, 2 (2016)

A.P. Basarman, D. Damjanovic, M. Lobonţiu, N. Medan, *TEAM 2016 Proceedings* (2016)

G. Fowler, *Abrasive water jet: controlled depth milling of titanium alloys*, PhD. Thesis, University of Nottingham (2003)

S. Hloch, J. Hlaváček, K. Vasilko, J. Cárach, et al, METABK **53** (4), 537-540 (2014)

J. Valíček, M. Harničárová, I. Hlavatý, R. Grznárik, M. Kušnerová, Z. Hutyrová, A. Panda, Materialwissen-schaft und Werkstoff-technik, **46**, 4-5 (2015)

S. Ravai-Nagy, N. Medan, Hidraulica **4**, 2-17 (2016)

E.Herghelegiu, M.C. Radu, C. Schnakovszky, C.N. Tampu., *CoSME'16 MATEC Web of Conferences* 94, 03007 (2017)

A.C. Filip, M.A. Vasiloni, L. A. Mihail, *CoSME'16 MATEC Web of Conferences 94*, 03003, (2017)

N. Medan, M. Banica, *IManEE2016, IOP Conf. Series: Materials Science and Engineering 161, 012016 (2016)*

An overview on the conventional and non-conventional methods for manufacturing the metallic glasses

Eugen Axinte [1,*], *Andrei* Bofu[1], *Yan* Wang[2], *Ahmad Majdi* Abdul-Rani[3], and *Abdul Azeez* Abdu Aliyu[3]

[1] Gheorghe Asachi Technical University of Iasi, Romania
[2] School of Materials Science and Engineering, University of Jinan, Jinan. PR China
[3] Universiti Teknologi Petronas, Seri Iskandar, Perak Darul Ridzuan, Malaysia

Abstract. Metallic glasses (MGs), first discovered in 1959 at Caltech are currently among the most studied metallic materials. MGs called also glassy metals, amorphous metals, liquid metals, are considered to be among the materials of the future. The "classic" methods for industrialization of MGs are : end-casting in copper molds and protected environment, die forging , atomization for obtaining MG powder ,selective laser melting , imprinting in molds, thermoplastic shaping in the super-cooled temperature region. These methods are suitable for producing high value-added precision components but the problems still exists: expensive tools, limited lifetime of tools and the occurring of crystallization. Actually methods (thermoplastic shaping, casting and die forging) are limited by the low flexibility of production and by higher costs of tools and accessories. More suitable methods are greatly desired to machine MGs for their wider applications.

1 Introduction

Metallic glasses (MGs) are a class of metallic alloys with an isotropic amorphous structure that is rapidly quenched from liquid melts [1].The first reported metallic glass was the alloy Au75Si25 produced at Caltech by Klement, Willens and Duwez in 1959 [2]. In "Metallic glasses-historical background",Duwez describes the first experiment,based on ''gun technique'' as ''a success combined with a failure''. The success was given by sufficient metallic glass obtained, the failure was the destruction of experimental apparatus. Duwez says: ''the shock pressure was too high and about half of apparatus disintegrated, sending hot broken pieces into the laboratory'' [3]. In 1980s, Inoue and Johnson (Caltech) have discovered strongly glass forming multi-component La-, Mg-, Zr-, Pd-, Fe-, Cu-, and Ti-based alloys with large under-cooling and low critical cooling rates of 106 oC/s, comparable to oxide glasses [4]. In 1992, the first commercial amorphous alloy, Vitreloy 1 – Vit1 (41.2% Zr, 13.8% Ti, 12.5% Cu, 10% Ni, and 22.5% Be), was discovered by Johnson and Peker in findings of some new aerospace materials. In 2001, Kündig at ETH Zürich has been

* Corresponding author: axintee@tuiasi.ro

investigating Zr–Ti–Cu–Ni–Al alloys focusing on those similar to Vit105 (Zr52.5Ti5Cu17.9Ni14.6Al10) – one of the best glass-forming alloys [5].

In 2010 Demetriou and colleagues developed a palladium based metallic glass (with formula $Pd_{79}Ag_{3.5}P_6Si_{9.5}Ge_2$) that is not only strong, but also tough as steel. The damage tolerance of this metallic glass, its combination of strength and toughness, is higher than any known and studied material [6].

2 Mechanical properties of metallic glasses

The absence of a crystalline micro-structure endows them with a portfolio of properties such as high strength, high elasticity, and excellent corrosion resistance. All studies summarizes that MGs have much higher tensile strengths and much lower Young's moduli. The difference in these values between the MGs and crystalline alloys is as large as 60%. The significant difference in the mechanical properties is thought to be a reflection of the difference in the deformation and fracture mechanisms between MGs and crystalline alloys. Plastic deformation in metallic glasses is generally associated with inhomogeneous flow in highly localized shear bands [5-10]. Metallic glasses cannot have the crystallographic defined slip-systems of polycrystalline metals. In the absence of dislocation-mediated slip, they show high yield stresses, much closer to the theoretical limit than their crystalline counterparts – see Fig. 1a (reproduced with permission from [4] 2010 Elsevier). The Ashby plot for materials selection exhibit a large range of common engineering materials along with selected metallic glass ribbons and MGs, as it can see in Fig. 1b [8] A typical fracture of a MG specimen (commercial Vitreloy1)is shown in Fig. 1c [9]. Hardness is the most influential mechanical property that dictate the wear resistance capabilities of a material. Since hardness is understood to be a measure of flow stress, it correlates linearly with the material yield strength. The Vickers hardness plotted from different materials (1.d) shows that the amorphous metals demonstrate an advantage over crystalline metals in terms of hardness [10].

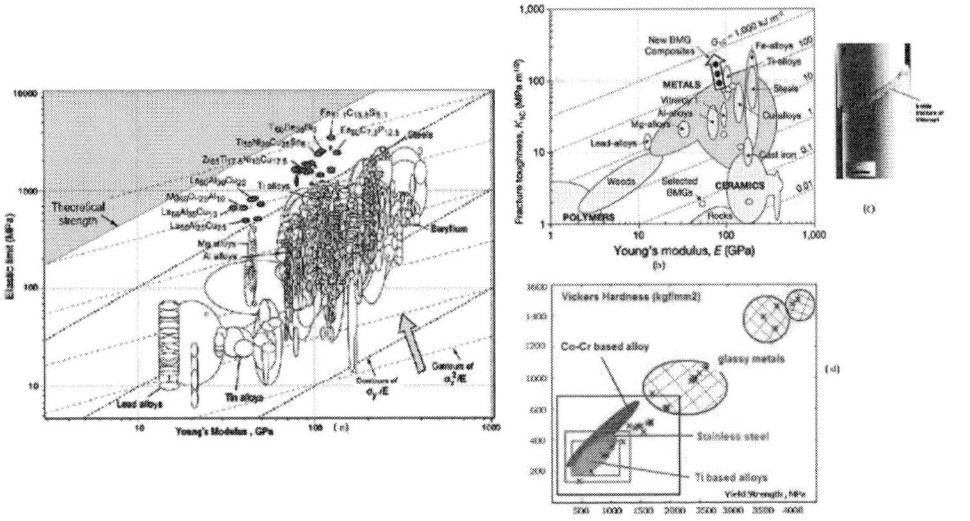

Fig.1. The elasticity and fracture toughness of MGs. a. Yield stress σ_y plotted against the Young modulus E for 1507 materials (reproduced with permission from [4]© 2010 Elsevier) b. Ashby plot for materials selection (adapted with permission from [8] © 2008NPG) c. brittle fracture of MGs [9] © 2013 Elsevier) d. Vickers hardness plotted from different materials. (adapted with permission from [10] © 2011 Elsevier)

3 Conventional (classic) methods for fabrication of MGs parts

The main "classic" methods for industrialization of MGs are: end-casting in copper molds and protected environment, die forging, atomization for obtaining MG bubbles, selective laser melting, imprinting in molds. Thermoplastic shaping in the supercooled temperature region. These methods are suitable for producing high value-added precision components but the problems still exists: high cost and limited lifetime of tools, the crystallization of MGs Actually methods (thermoplastic shaping, casting and die forging) are limited by low flexibility and higher costs of tools and accessories.

A synthesis of classical methods for manufacturing the MGs parts is illustrated in Fig.2 [5, 11,12]

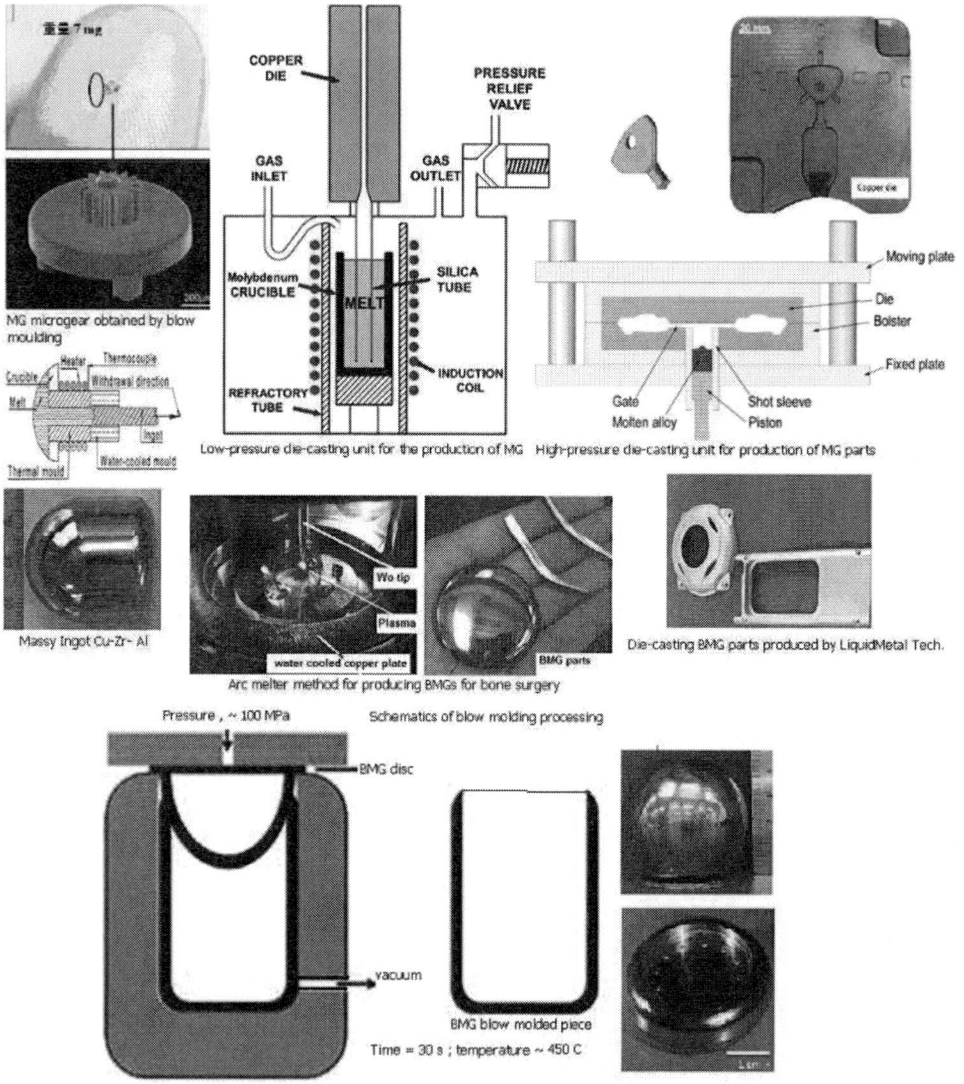

Fig.2. Conventional methods for manufacturing MGs parts. [5, 11, 12]

Another usual method for manufacturing MGs micro-parts is the hot embossing method, schematized in Fig.3 [1]

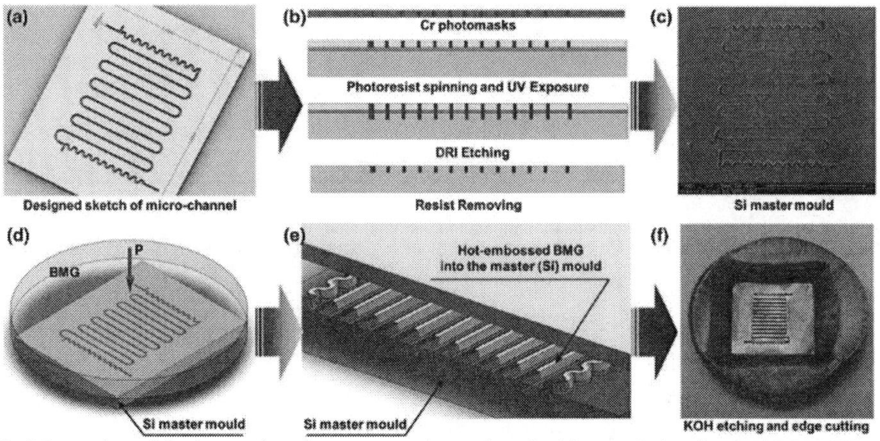

Fig.3. Schematics of the hot-embossing process. (reproduced with permission from[1]© 2016Elsevier)

Recent research shows that by application of an electric current pulse in the presence of a normally directed magnetic field can heat (by electrical resistance) a metallic glass to a softened state, while simultaneously inducing a large enough magnetic body force to plastically shape it. The heating and shaping is performed on millisecond timescales, effectively bypassing crystallization producing fully amorphous- shaped parts. This electromagnetic forming process may become a promising manufacturing platform for strong metals [13-15].

4 Non-conventional methods for fabrication of MGs parts

4.1 The cutting of MGs

It exists some sporadically studies of the machining characteristics of MGs during cutting and drilling and found that although mechanical machining of MGs can obtain high dimensional accuracy and surface quality, some problems, such as oxidation and crystallization during high speed machining, still exists [16-20].

In [16] the ultra-precision turning of a Zr-based MG was investigated under various machining parameters (depth of cutting, feed rate, and spindle rate) .The influences of machining parameters on the turned surface of a Zr-based MG after observing the 3D morphologies of this surface were characterized. The results showed that the influence of the spindle rate on the surface morphologies is more substantial as compared to the depth of cutting and the feed rate. Scratch tests were conducted to further characterize the separation mechanism of the chips.

The chips are torn off the surface of a MG because of inhomogeneous localized maximum shear stress. According to the sketch in Fig. 4 [16], three factors (depth of cut, feed rate, and spindle rate) can essentially affect the turning of a metallic glass . Unfortunately, the excellent comprehensively mechanical properties of the Zr -based metallic glass lead to a severe abrasion of the diamond tool.

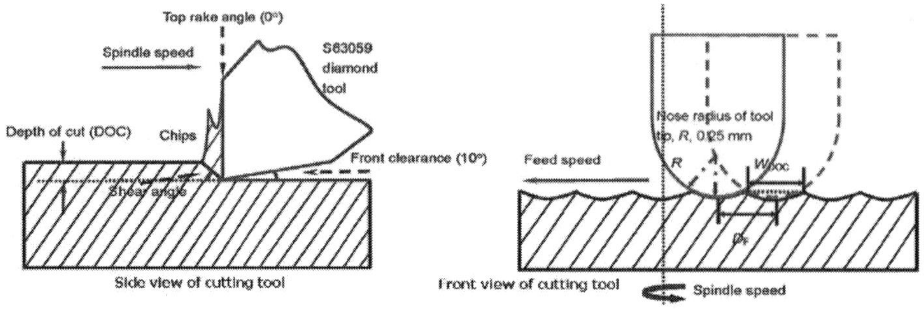

Fig.4. The schematic of the ultra-precision turning of a Zr-based MG. (adapted with permission from [16]© 2015 Elsevier)

The cut-ability of Zr&Pd-based MGs by lathe was examined using different tool tip materials (diamond, CBN, ceramics and cermet), nose radii of the tip 0.4, 0.8 and 1.2 mm , feed speed (f =0.05mm/rev) and cutting speeds (V=5-150 m/min).

From the chip observation results, the mechanisms that the roughness on the turned surface of the MGs exhibited the precise finishing level. The chips slipped off at planar and evenly fine spaced slip planes controlled only by a maximum shearing stress. The cutting force became constant and chattering was not generated. As the MGs are generally difficult to produce the plastic deformation in addition to the slipping off behavior of the MGs, the MGs were easily cut off along the shape of the tool tip. A built-up edge did not occur in the MG workpieces - Fig.5 [17].

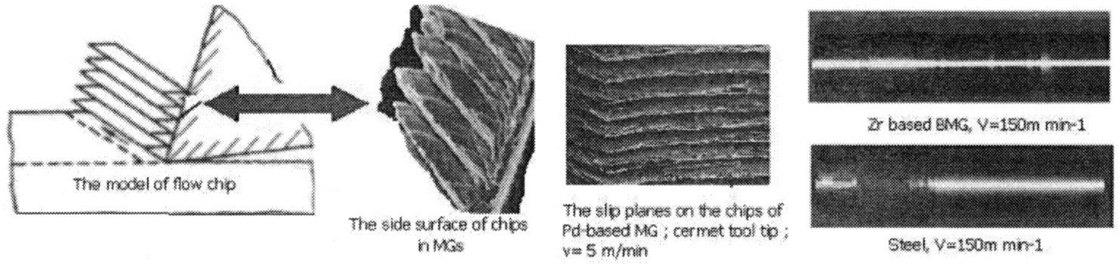

Fig.5. Chip formation mechanism of MG at turning operation. (adapted with permission [17] © 2005 Elsevier)

The severe tool wear because of the high hardness of MGs and the formation of built-up edge due to the adhesion between the chip and tool in the contact region affected the machining quality. Furthermore, for hard-brittle MGs with hardness over 10 GPa but fracture toughness less that 10 MPa , mechanical machining is more challenging. More suitable methods are greatly desired to machine MGs for their wider applications.

4.2 Fabrication of metallic glass ordered at large scale

A recent and totally unconventional method for obtain a metallic glass that is ordered at a large scale was reported in a study conducted by Geophysical Laboratory (Carnegie Institution for Science) [21]. It was created a single crystal by applying 25 GPa of pressure (equivalent of 1800 tons per square inch) to the cerium–aluminum MG and the new order formed is preserved even when the glass is restored to ambient pressure. In Fig. 6 is schematized the experiment in where was fabricated a long range ordered (LRO) metallic glass.

 DOI: 10.1051/matecconf/201711203003

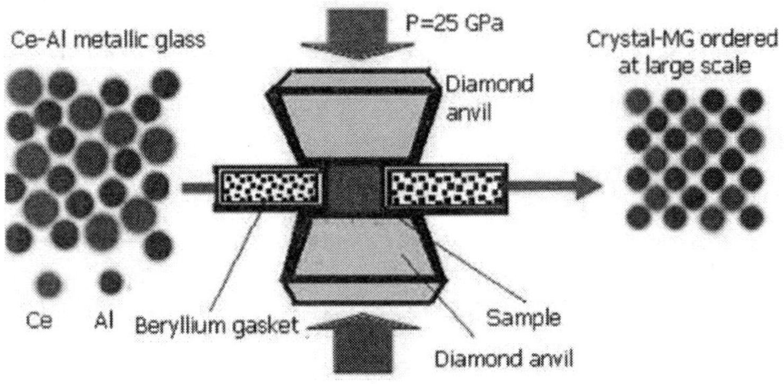

Fig.6. The schematic of fabrication of a large range ordered (LRO) metallic glass. [21]

4.3 The welding of metallic glass parts

Because amorphous metal formation requires specific critical cooling rates, the part size and thickness are limited; however, these limitations can be overcome with welding. Welding possibilities and the weldability of MGs was recently studied in [22-23]. In [22], a Zr-based metal glassy alloy was successfully welded by small-scale resistance spot welding (RSW).

In [23], the similar and dissimilar friction welding of tubular Zr-based MGs to amorphous and crystalline metals have been performed and compared with the cases of MG rods. A successful joining of the bulk metallic glass (BMG) to crystalline metals could be obtained for certain pairs of the material combination through the precise control of friction conditions. Electron beam welding method is industrially used by Liquidmetal Technologies -Fig. 7[24].

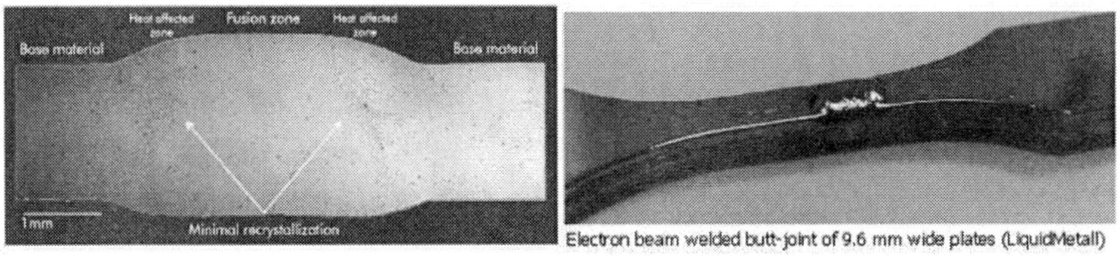

Fig.7. Electron beam welded butt-joint used by Liquidmetal Technologies Inc. [24]

4.4 The painting and coating of metallic glass parts

MG parts can be painted with a range of traditional techniques and can also be processed through Physical Vapor Deposition (PVD) processes.

Painting requires the same cleaning and surface preparation similarly to the other metallic surfaces. The paint baking cycle temperatures above 250°C could affect the amorphous structure of the material. PVD processes are used to apply high-purity coatings such as titanium, chromium, and aluminum to MG parts. In the PVD process, these coating materials are either evaporated by heat or bombarded with ions (sputtering).

While the PVD coating metal is being vaporized, a reactive gas is introduced which forms a compound with the metal vapor and is deposited on the metal parts as a thin and highly adherent coating, Fig.8 [24].

DOI: 10.1051/matecconf/201711203003

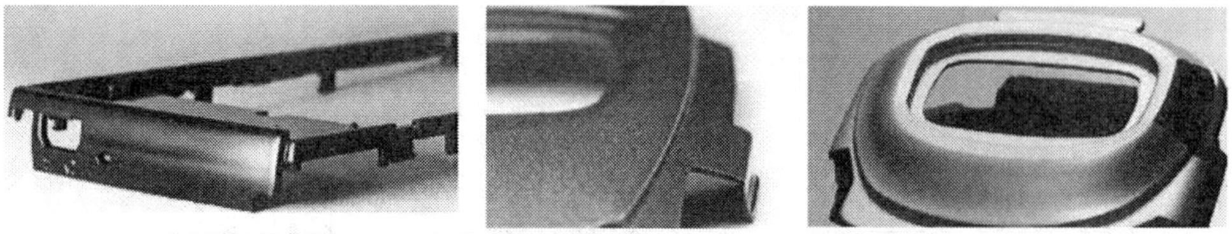

Painted metallic glass components (by Liquidmetal) PVD coated component (by Liquidmetal)

Fig.8. Metallic glass components industrially painted and coated. [24]

4.5 The electrical discharge machining (EDM) of metallic glass parts

Considering the limited size of MGs and their main applications in precision and micro-components, the micro-electrical discharge machining (M-EDM) may be a good choice for machining MGs. However, M-EDM is an electro-thermal process and MGs are very sensitive to temperature. A small number of recent researches [25-29] indicated that severe crystallization happened during EDM of MG because of high electrical discharge energy. It was revealed that crystallization of MGs occurs on the top surface in EDM. The penetration depth of the thermal damage into the bulk material is actually less studied and less known. The microstructural changes of MGs in heat affected zone during M-EDM have not been clarified. Also, as a disadvantage, the EDM machining is a low productivity process.

5 Conclusions

The machinability of MGs has been given less attention, although it is seems to become very important from the perspective of industrialization. It were sporadically studied the machining characteristics of MGs during cutting and drilling and found that although mechanical machining of MGs can obtain high dimensional accuracy and surface quality, some problems, such as oxidation and crystallization during high speed machining, still exists. Some recent researches shows that micro-electrical discharge machining (M-EDM) is suitable for machining hard-brittle materials and can be used for production of micro-parts and three dimensional structures such as micro-holes, micro-gears, micro-array, and micro-tools.

References

1. N. Li, W. Chen, L. Liu, JOM, **68** (4), 1246-1261 (2016)
2. W. Klement, R.H. Willens, P. Duwez, Nature **187** (4740), 869–70 (1960)
3. P. Duwez, Top. Appl. Phys. **46** 19–23 (1981)
4. M. Telford, Materials today, **7** (3), 36-43 (2004)
5. E. Axinte, MaterDesign, **35**, 518-556 (2012).
6. M.D. Demetriou, M.E. Launey, G. Garrett, J.P. Schramm, D.C. Hofmann, W.L. Johnson, R.O. Ritchie, Nature Materials **10** (2), 123-128 (2011)
7. M.M. Trexler, N.N. Thadhani, Progr Mater Sci. **55**, 759–839 (2010)
8. D.C. Hofmann, J.-Y. Suh, A. Wiest, G. Duan, M.-L. Lind, M.D. Demetriou, W.L. Johnson, Nature, **451**, 1085–9 (2008)
9. K.L. Edwards, E. Axinte, L.L. Tabacaru, MaterDesign, **50**, 713-723 (2013)
10. E. Axinte, Y. Wang, L.Tabacaru, N. Radwan, Recent Patents on Mat. Sci. J., **9** (1), 2-19 (2016)
11. K.J. Laws, B. Gun, M. Ferry, Mat. Sci. & Eng A, **475** (1), 348-354 (2008)
12. P. Ramasamy, A. Szabo, S. Borzel, J. Eckert et al., Scientific Rep. **6** (2016)

13. G. Kaltenboeck, M.D. Demetriou, S. Roberts, W.L. Johnson, NatComm **7** (2016)
14. G. Kaltenboeck et al, Sci. Rep. **4** 6441 (2014)
15. W.L. Johnson et al, Science **332**, 828-833 (2011)
16. D.X. Han, G. Wang, J. Li, K.C. Chan et al, J. Mater. Sci. Technol. **31**, 153-158 (2015)
17. K. Fujita, Y. Morishita, N. Nishiyama, H. Kimura, A. Inoue, Mater. Trans. **46**, 2856-2863 (2005)
18. M. Bakkal et al, Intermetallics **12**, 195-204 (2004)
19. H. Huang, H.W. Zhao, C.L. Shi, B.D. Wu et al, AIP Adv. **2** /042193 (2012)
20. T. H. Li, P. H. Tsai, K. T. Hsu, Y. C. Liu, J. S. C. Jang, J. C. Huang, Intermetallics, **78,** 17-20 (2016)
21. Q. Zeng, Y. Ding, W. L. Mao, W. Yang, S. V. Sinogeikin, J. Shu, H. Mao, J. Z. Jiang, Phys Rev Lett. **104** (10), 105702 (2010)
22. K. Fujiwara, S. Fukumoto, Y .Yokoyama, M. Nishijima, A.Yamamoto, Mater Sci Eng A, **498**, 302–307 (2008)
23. H.S. Shin, J.S. Park, Y. Yokoyama, J Alloys Compd **504S**, 275–278 (2010)
24. *Liquidmetal® Design Guide 3.0* ,(Liquidmetal Technologies, 2015)
25. D. Wang, W.S. Zhao, L. Gu, X.M. Kang, J. Mater. Process. Tech. **2113-11** (2011)
26. H. Huang, J.W. Yan, Appl. Surf. Sci. **355**, 1306-1315 (2015)
27. S.F. Hsieh, S.L. Chen, M.H. Lin, S.F. Ou et al, J. Mater.Res. **28,** 3177-3184 (2013)
28. D. Wang, W.S. Zhao, L. Gu, X.M. Kang, J. Mater. Process. Tech. **2113-11** (2011)
29. J.W. Yan, K. Watanabe, T. Aoyama, CIRP Ann.-Manuf. Techn. **63209-212,** (2014)

Experiments for WAAM Technologies

Laurentiu Dan Ghenghea*

Technical University "Gheorghe Asachi" TCM Dep., D.Mangeron Blvd. 59, B201, Iasi, Romania

Abstract. The most recent group of technologies added to Additive Manufacturing (AM) field, Wire and Arc Additive Manufacturing (WAAM) use some well known welding processes, generally gas metal arc welding (GMAW codification at American Welding Society and 135 at International Institute of Welding/Institute International de Soudure) or Tungsten Inert Gas to manufacture products used in top ranged production field like aeronautics. Remarkable results in this new method for industrial production have been achieved by research groups from Cranfield, Delft and many other universities around the world and were published in prestigious research journals. Together with our students from Welding Engineering Program and our limited resources for powerful researches an approach to these new working technologies have been done and our main objective was to show that strong researches could be done with more resources during masters or doctoral studies. In this proposed paper some experiments with shielded metal arc welding process will be presented which have been done to investigate what are the internal structure of welded products made when bead over bead have been deposed, also will be presented some hardness tests of manufactured structure and machining tests for milling process.

1 Introduction

The welding processes together with metallic coatings (metallisation), thermal cutting both groups could be done with mixed oxygen and gases as flames or thermal plasma, bonding processes with metallic alloys or with polymers and special welding processes like underwater or space welding are in the front view of researchers and producers. The utilisation of welding procedures could be identified from the beginning of humanity speaking about cold pressing welding or forget welding for some metals that are native on the Earth like: gold, tin, silver. The modern welding processes have been developed as we know them today at the end of the nineteenth century by scientists like: Slavianov, Bernandos, Kjilberg and many other that used the developments from physics, chemistry and other scientific fields to materialise welding and connected to welding processes, the examples could be utilisation of: lasers, thermal plasma, CNC, robotic, induction, high frequency currents, electronic devices for electric current frequency change (invertors).

In the recent years, no more than fifty, a new way to manufacture industrial products has been given to producers by scientific research and it is additive manufacturing (AM) that is a way to construct products layer upon layer and the first encouraging result have been made

* Corresponding author: lghenghe@tcm.tuiasi.ro

with polymers which have been solidified in thin layers with heating source like ultraviolet light or lasers. This production method is totally different from known methods that subtract or redistribute the excess from raw materials and it could not be utilised. A great advantage of additive manufacturing is that it could use electronic files of models which are very quick transmitted to the execution parts, this is the reason them have been called "rapid prototyping manufacturing processes". In Figure 1 some simplified drawings for manufacturing processes are presented the first line (a and b) shows subtractive (the milling process) and redistributive process (cold deep drawing process) the final line (c and d) shows joining process (welding) and additive manufacturing using layer upon layer solidification.

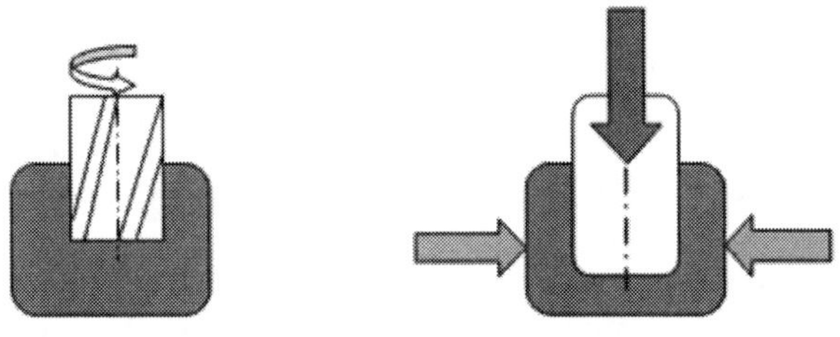

a Subtracting technology (milling process) *b* Redistributing technology (cold deep drawing)

c Joining technology (corner welding) *d* Additive technology, layer upon layer deposition

Fig. 1. The manufacturing methods: ***a*** - cutting excess of material; ***b*** - redistribution of raw material; ***c*** - joining base materials; ***d*** - additive manufacturing.

2 Technological processes related to additive manufacturing

A very good survey of additive manufacturing processes it is offered by Wohler Report published each year, but in 2014 it described The History of Additive Manufacturing [1] focussed on producers of industrial installations for lot of specific technologies used in this scientific and industrial area.

In a short summary of the report it could be outlined that very different materials have been used starting with known liquid polymers (polyamides, polycarbonate, ABS) also special compositions made by producers like DuPont, powders of different types bonded with liquid binders, papers to be cut in different shapes and bond together to manufacture pieces from cross sections, metallic particle or wire and some processes use sand. Having this in view, the additive manufacturing processes could be classified in following classes counting on materials used:

1 Polymer Based Additive Manufacturing Processes;

2 Metals Based Additive Manufacturing Processes;

3 Exotic Materials Based Additive Manufacturing Processes.

The additive manufacturing equipments prices start from few hundred dollars and for complex manufacturing equipments for metal casting arrive at six hundred thousand dollars, important world manufacturers like 3 D Systems, Z Corp, Mitsubishi, EOS have been involved in equipments development and introducing its on world market.

Important scientific initiative to develop knowledge of additive manufacturing have been done in United States of America where American Society for test and Materials founded the committee F-42 to standardize additive manufacturing terminology and to develop industry standards.

In 2009 Edison Welding Institute organised Additive Manufacturing Consortium (AMC) consisting in twenty two industrial members and partner organizations representing both large and small companies, government agencies and important universities with researches in additive manufacturing field.

The European Union have identified the additive manufacturing field as a key emergent technology fields and cooperative projects have been founded, just two examples, Custom Fit [2] and Repolac (Rapid Production of Large Aerospace Components).

The director of AMC [3] considers that for metal based additive manufacturing two main components are important: the raw material used and the energy source used to forms it and from this point of view only few technologies could be used and those are: powder bed systems where electron beam or laser are used to give the needed shape to layers from powders in enclosed environments, laser powder injection where a nozzle directed powder to laser waves and free form fabrication with electron beam melting metallic wires or arc deposition of wire and powders sometimes together with ultrasonic consolidation of metallic layers.

3 The wire and arc additive manufacturing (WAAM)

In the paper review about metal additive manufacturing [4,5] the author William E. Frazeir presents the group of metals used for researches and production using additive manufacturing processes and those are: Titanium (Ti-6Al-4V, ELI Ti, CP Ti, γ-TiAl), Aluminium (Al-Si-Mg, 6061), Tool steels (H13, Cermets), Super alloys (IN625, IN718, Stellite), Stainless steels (316 & 316L, 420, 347, PH 17-4), Refractory metals (MoRe, Ta-W, CoCr, Alumina).

In the scientific meeting Titanium Europe 2015 organised by International Titanium Association at Birmingham, UK [7] researchers from Cranfield (Adrian Addison, Jialuo Ding, Filomeno Martina, Helen Lockett, Stewart Williams) and Coventry University (Xiang Zhang) presented solid results about complex Titanium parts manufacturing using wire and arc additive manufacturing considering it capability to uses available welding equipments and manipulators to produce parts of medium or high complexity at different scales.

The authors consider origin of this technology in Baker\s patent (1929) titled "The use of an electric arc as a heat source to generate 3D objects by depositing molten metal in superimposed layers", in their researches wires of Ti-6Al-4V have been deposited layer by layer onto a base support, in single bead without weaving producing two straight walls 6.8 mm thickness, the first 925 mm long and 195 mm high and the second 975 mm long and 155 mm high.

For the manufactured walls using wire and arc additive manufacturing in the communication experimental results are presented related to: mechanical proprieties (the average yield strength (YS), ultimate tensile strength (UTS), and strain to failure from the baseline tests were 950±150 MPa, 1033±100 MPa, and 11.7%), fatigue properties (twenty-one samples had a fatigue life well above three million cycles), fracture toughness testing, crack propagation at WAAM to wrought interface, the effect of cold rolling for each layer deposited using wire and arc additive manufacturing test pieces made in Ti-6Al-4V alloy and in the final part of scientific communication the researchers present complex parts manufactured for aeronautical industry like: landing gear rib, a frame demonstrator for BAE Systems and flap rib demonstrator for Fokker Aeronautics.

4 Objectives, methodology and procedure of this research

At Iasi TCM department some experimental researches have been done for additive manufacturing using a self made device for fused object manufacturing of pieces for the car presented in formula Students, also same colleague uses a Z Corp 3D printing device to manufacture a small model of the car in powder and liquid binder. The pieces made using fused object manufacturing are presented in Figure 2, with the permission of our colleague.

In this research the main objective was to investigate if it is possible to weld layer upon layer of welded bead to form a vertical wall with good mechanical characteristics for further mechanical operation like milling.

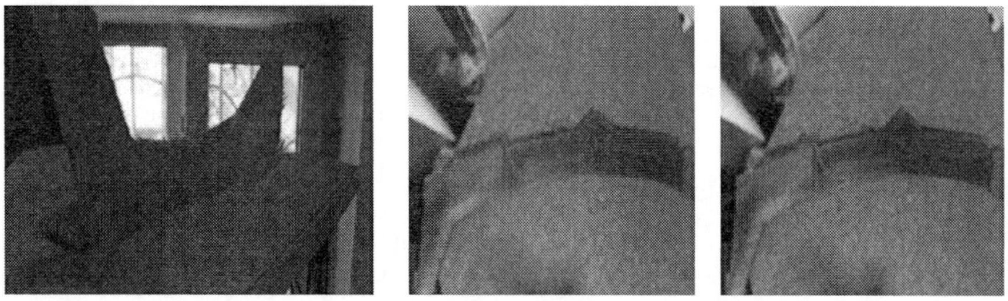

Fig. 2. Pieces manufacturing in plastic materials (PVC) made with the self made devices for Fused Object Manufacturing (FOM) with utilization in Formula Student car manufactured by the team lead by PhD Ciofu Ciprian (posted with his kind permission).

To achieve the objective the shielded metal arc welding process have been used with Technology Inverter 200 welding device, a 2.5 mm diameter shielded electrode type E 43 2 RR 22 (ISO 2560) or E 6013 (AWS A 5.1) with chemical composition of deposed material in %: C 0.06-1; Mn 0.4-0.7; Si 0.2-0.6; S max. 0.04; P max. 0.04 and mechanical characteristic of deposed metal: $Rp_{0.2}$ 430-490 N/mm^2; Rm 490-550 N/mm^2; A5% min.24; KV-J 78 at 20 ^{0}C, producer recommended welding current 60-80 A, electrode manufactured with authorisation from Lloyd's Register of Shipping, American Bureau of Shipping, Bureau VERITAS.

The welded bead have been deposed on a plate with dimensions 100x20x8 mm, polished to metallic shine with the speed of electrode around 20 mm/min, shot waving of electrode half diameter using 70 A, each bead has been carefully cleaned for welding slag and quickly cooled at around 150 ^{0}C, ten layer upon layer of welded bead have been deposed. The phases of experimental examination are presented in Figure 3 where could be seen: the support and fixing devices, the deposition of the first layer of welded bead and the colour of a layer of welded bead given by electric arc utilisation.

During experimental research preparation the definition of additive manufactured processes proposed by ASTM [6] stated that: "additive manufacturing (AM) a process of joining materials to make objects from 3D model data, usually layer upon layer, as opposed to subtractive manufacturing methodologies. Synonyms: additive fabrication, additive processes, additive techniques, additive layer manufacturing, layer manufacturing, and freeform fabrication'' to make object from 3D model is very important part of this definition and this mean that the 3D model data controlled by a computer must be used as a design precursor and the further development of this experiment will be in this direction, but the objective of this research was to observe if important changes in internal structure of deposed welding bead will occur and this will be investigate by hardness and milling tests to understand if constructed material could be used for pieces manufacturing using machining technologies.

 DOI: 10.1051/matecconf/201711203004

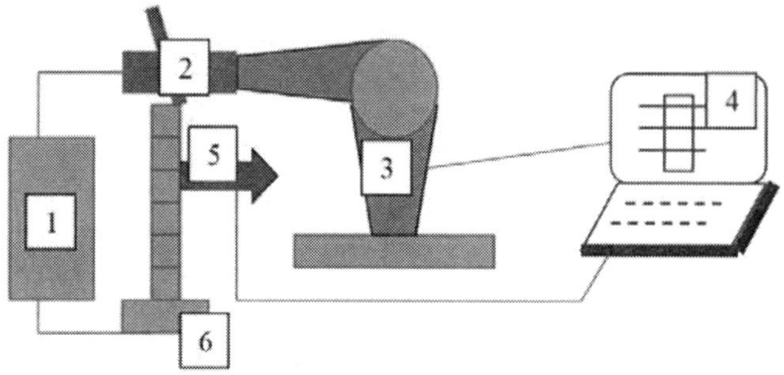

Fig. 3. Experimental stand for waam: 1- Welding machine; 2 - robotic catching device; 3 - robot IRR 120; 4 - process computer; 5 - temperature measurement device; 6 - test piece manufactured using waam technique; lines represent connectors.

5 Results and Conclusions

The phases of welding deposition layers are presented in Figure 4, could be seen the initial catching of metallic support in specific welding devices, the first layer deposition and attentive cleaning of surface and the colour of the other welding deposition layer because of the heat developed by electric arc used like heating source to melt the electrode

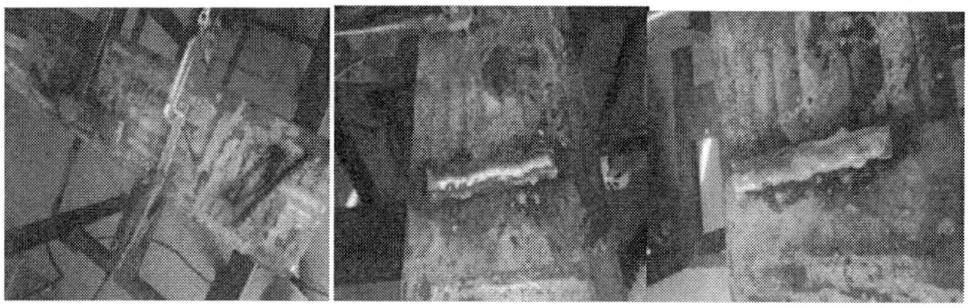

Fig. 4. Manufacturing phases for preliminary experiments using Shaped Deposition Metal (SDM) method our approach to wire and arc additive manufacturing.

After test piece manufacturing using wire and arc manufacturing (wam) in this case, the cutting of cross sections have been done for grinding surfaces to do micro-hardness tests, this operation give the first information that resulted pieces have not fragile internal structures. The resulted piece, the cross sections and micro-hardness measurement are shown in Fig. 5.

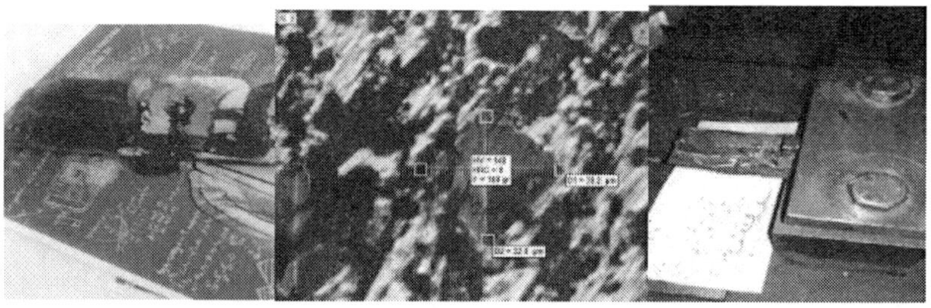

Fig. 5. The results of experimental wire and arc object manufacturing (WAOM).

In this experimental research the prove that manufacturing objects using shielded wires with 2.5 mm diameter by adding one upon other layers consisting of welded bead it is shown and the method could be named wire and arc object manufacturing the processed piece was 80x40x8 mm.

The measurements of micro-hardness in cross sections of deposited material from wire core of the shielded electrode did not change it values showing that internal structure is not composed by fragile part, in the experiments a high cooling speed of welded material has been used wishing to avoid repeated thermal cycles, using milling technology on test piece the process was smooth, with small cutting forces and the resulted cheeps has metallic colour with no burned zone showing that it is possible to use welded test pieces to manufacture precise mechanical elements.

References

1. Wohlers, T., *Wohlers Report 2010*, Wohlers Associates, Ft. Collins (2010)
2. Custom-fit.org, http://ec.europa.eu/research/headlines/news/article_09_02_16_en.html (2016)
3. Harris, I., *Development and Implementation of Metals Additive Manufacturing Welding* Institute presentation to the U.S. Department of Transportation (2011)
4. W.E. Frazier, J. Materials Engineering and Performance, **23** (6), 1917- 1928 (2014)
5. W.E. Frazier, *Solid Free Form Fabrication Proceedings,* Austin TX, 717–732 (2010)
6. E. Herderick, *Proceedings of MS&T, Additive Manufacturing of Metals*, Columbus, OH, (2011)
7. Adrian Addison, *TITANIUM EUROPE 2015* (Birmingham, 11-13, 2015)

Some aspects of finite element modelling of ultrasonically aided micro-EDM of CoCr alloys

*Daniel Ghiculescu**, *Niculae* Marinescu and *Claudiu* Pirnau
"Politehnica" University of Bucharest, Manufacturing Engineering Department, 313 Splaiul Independentei, Bucharest, Romania

Abstract. The paper deals with finite element modelling of micro-machining CoCr alloys by ultrasonically aided electrical discharge machining. This hybrid machining process has two components: a thermal one due to EDM, and a mechanical one to ultrasonic assistance. Both components were modelled using Thermal and Structural Mechanics time dependent modules of Comsol Multiphysics. The results were compared with the experimental data obtained in our laboratories, proving a good agreement and offering some solutions for machining optimization.

1 Introduction

The mechanical properties of CoCr alloys, extremely wear resistant and also at corrosion, and biocompatibility as well, recommend their use in medical applications, injection molding and engine components [1, 2]. The stellite superalloys have similar using areas but also at cutting tools, components for guns etc. The C content and existence of carbides play an important part in their functional behavior [3, 4]. Conventional casting of CoCr alloys determines existence of metallic carbides, affecting later machining, in comparison with other nonconventional techniques like selective laser melting (SLM) or Electron Beam Melting (EBM) that assures a homogenous structure [1, 2].

All increased properties of CoCr alloys determine high difficulties at conventional machining of micro-surfaces [5]. Ultrasonically aided micro-electrical discharge machining (μEDM+US) is an appropriate response in terms of machining rate, precision and surface quality, supported by relative low costs [6]. This hybrid process was applied for other advanced materials difficult to machine by cutting technologies [7, 8].

At micro-EDM+US application, CoCr alloys thermo-physical characteristics as well as mechanical ones become critical, corresponding to their two components of this hybrid machining: thermal one due to EDM, and ultrasonically induced cavitation owed to ultrasonic assistance. In this respect, values of properties needed for modelling of EDM part are: density 8290 kg/m^3 for usual alloys, or much higher (9120–9240 kg/m^3) at additional alloying elements, and 8690 kg/m^3 for stellite, thermal conductivity 33W/m °K at 1000 °C, which is lower than at usual steels, heat capacity 420-460 J/kg °K, which is slight lower than at usual steels, melting range 1380-1430°C [5, 10, 11], and boiling temperature around 3000 °C, taking into account the boiling points of alloy main components (Co, Cr, Mo).

* Corresponding author: daniel.ghiculescu@upb.ro

About mechanical characteristics of CoCr alloys, necessary for ultrasonic material removal modelling, modulus of elasticity varies between 200 – 220 GPa, much higher than other alloys used in medical applications, based on Ti and Au, and ultimate tensile strength is 1300 MPa [1,2,9]. Stellite alloys have higher elasticity modulus at 230 GPa, and ultimate tensile strength at 1195 MPa [3,4], in comparison with CoCr Arcam alloys with 960 MPa [2]. The fatigue limit is 610 MPa at rotating beam fatigue test [2]. Other values of fatigue limits of the alloys are: at bending, this is estimated in the region of 480 MPa [9]; for stents CoCr alloys, it is predicted much lower, at 207 MPa [10].

2 Experimental data

An experimental setup for preliminary studying the micro-EDM+US process on CoCr alloy was created, presented in fig. 1. This comprises basically an ultrasonic chain, integrating the tool its end, connected to an ultrasonic generator of 20 kHz, producing tool amplitude of around 2 µm during EDM process. CoCr alloy rectangular samples of 30 x 50 x 1 mm were machined comparatively on Romanian ELER 01 machine with US vibrations of tool (EDM+US) and without US (classic EDM).

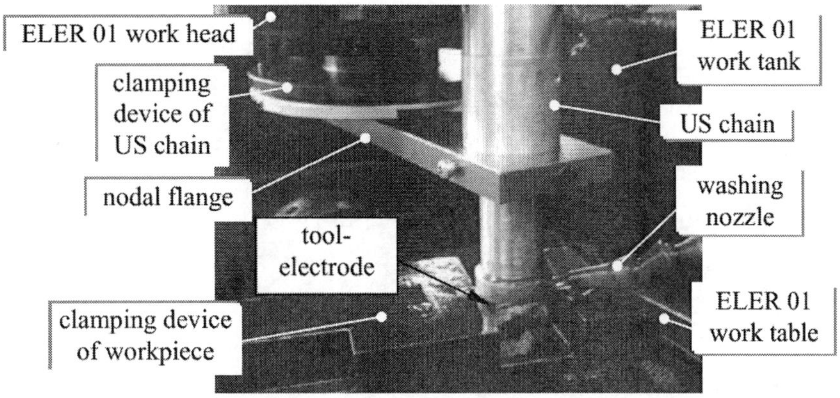

Fig. 1. Technological system elements for EDM+US micromachining of CoCr alloy samples.

Both commanded and relaxation pulses were used provided by a special generator for finishing and micro-machining. The working parameters at commanded pulses were: current step, I=3A, positive (tool) polarity; pulse time, t_i=48 µs, pause time, t_0=24 µs; t_i=8µs t_0=8µs. At relaxation pulses, the specific working parameters were: negative polarity, capacitor step, C=10 nF, resistance step, R=0.74 kΩ (for adjusting supply current). For flushing, the lateral one was chosen with pressure, p_{hl}=0.04 MPa. The values of consumed power on ultrasonic chain (P_{cUS}) were: 90 and 120 W.

The variation of total hydraulic pressure (p_{ht}) is calculated with the following relation, and presented in fig. 2:

$$p_{ht} = 2\pi\, c\, \rho\, f_{US}\, z + p_{hl} \quad \text{[Pa]}, \tag{1}$$

where: f_{US} is ultrasonic frequency [Hz]; z - elongation, $z=A\sin\,\omega t$; $\omega = 2\pi f_{US}$ [s^{-1}] A - amplitude [m]; c - sound velocity in dielectric liquid [m/s]; ρ - dielectric liquid density, ρ=840 kg/m^3 [kg/m^3]; $c=(K/\rho)^{1/2}$=1267.7 m/s; K=1.35x10^9 Pa (K-bulk modulus); p_{hl} - local hydraulic pressure [Pa], p_{hl}=0.04MPa. At each final of ultrasonic period T_{US}, when pressure increases, *cumulative microjets stage* (*CMS*) occurs, due to the collective implosion of gas bubbles, generating shock waves. This produces pressure of 100 MPa order of magnitude, oriented along the gap, frontal in this case - fig. 2.

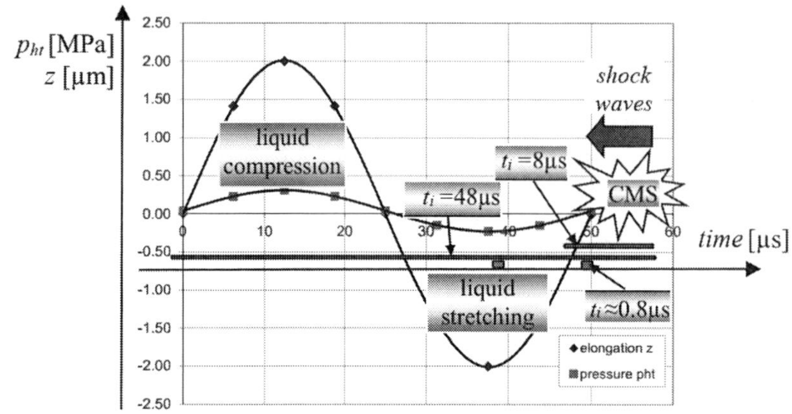

Fig. 2. Variation of tool elongation and total pressure in the gap within T_{US} during EDM+US tests.

During T_{US}, commanded and relaxation pulses are delivered. At overlapping the pulses on *CMS*, it was noticed experimentally that the discharge is stopped due to high pressure generated by collective gas bubble implosion. Then the dielectric liquid is allowed to access the melted material by discharge, and therefore increased volume of material is removed.

The microtopographies provided by scanning electron microscope (SEM) QUANTA INSPECT F50, with field emission electron gun and resolution of 1 nm, and corresponding profile of craters average dimensions, obtained by mediation of craters diameters with their depth (Rz) given by surface roughness tester SJ411, are presented in fig. 3 – 8.

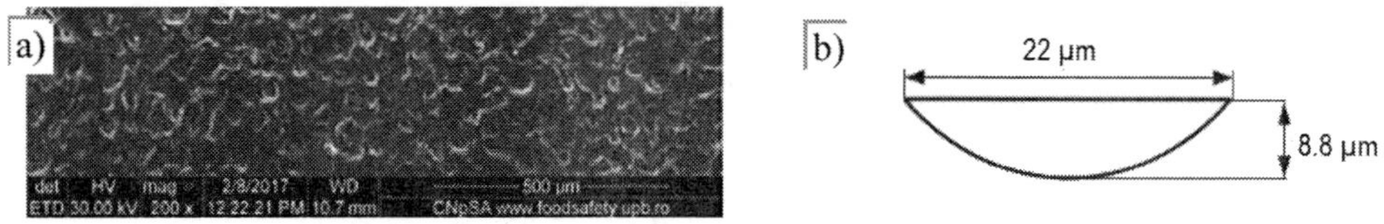

Fig. 3. Surface microgeometry at EDM with commanded pulses, I=3A, t_i=48 µs, t_0=24 µs: (a) SEM view; b) craters averaged dimensions.

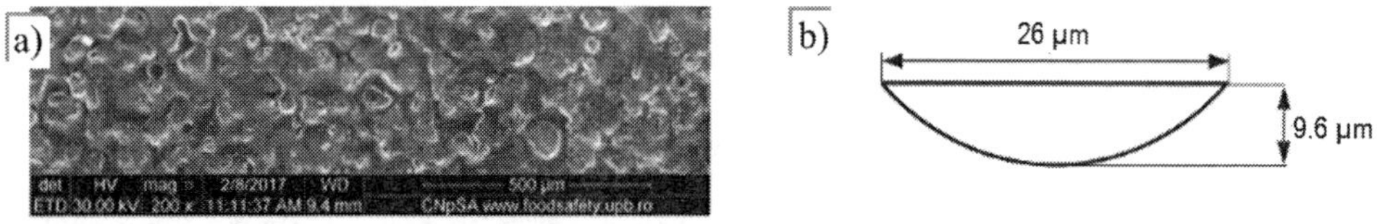

Fig. 4. Surface microgeometry at µEDM+US with commanded pulses, P_{cUS} = 120 W, I=3A, t_i=48 µs, t_0=24 µs: (a) SEM view; b) craters averaged dimensions.

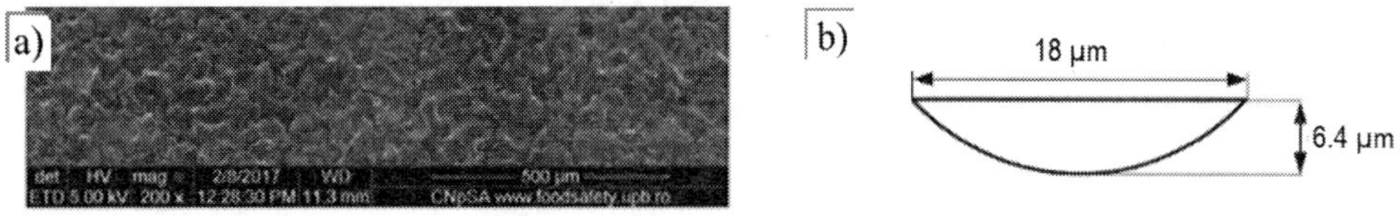

Fig. 5. Surface microgeometry at µEDM with commanded pulses, I=3A, t_i=8 µs, t_0=8 µs: (a) SEM view; b) craters averaged dimensions

Fig. 6. Surface microgeometry at µEDM+US with commanded pulses, P_{cUS} = 90 W, I=3A, t_i=8 µs, t_0=8 µs: (a) SEM view; b) craters averaged dimensions.

DOI: 10.1051/matecconf/201711203005

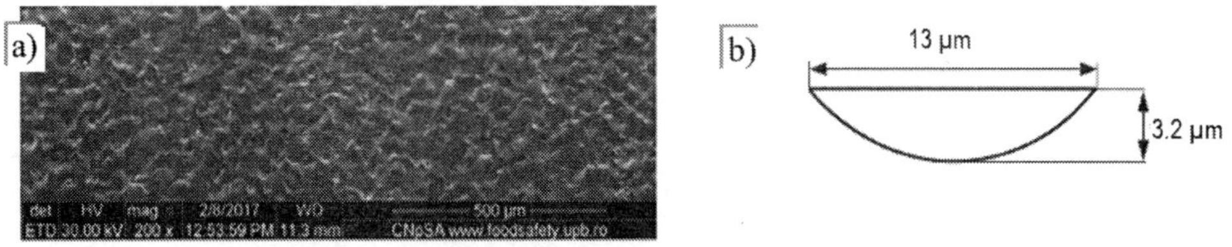

Fig. 7. Surface microgeometry at µEDM with relaxation pulses, C=10 nF, R=0.74 kΩ: (a) SEM view; b) craters averaged dimensions.

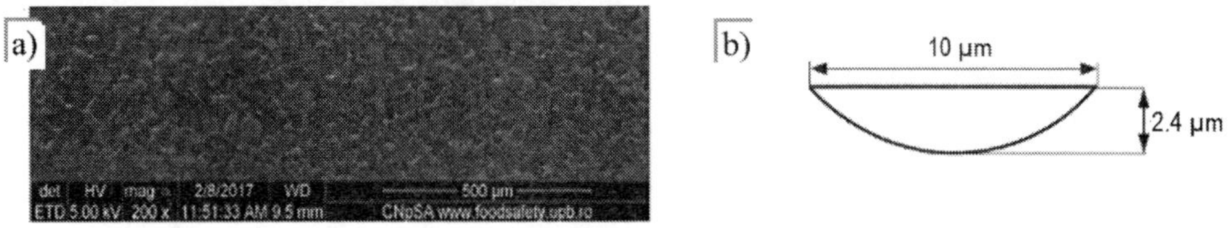

Fig. 8. Surface microgeometry at µEDM+US with relaxation pulses, P_{cUS} =90W, C=10 nF, R=0.74 kΩ: (a) SEM view; b) craters averaged dimensions.

3 Finite element modelling of micro-EDM+US process

The material removal process of µEDM+US on CoCr alloys was modelled by finite element method using Comsol Multiphysics, time dependent modules of Heat Transfer in Solids and Solid Mechanics, corresponding to EDM, and ultrasonic components. Some parametric models were created for each pulse type, using the assignation shown in fig. 9.

▾ Parameters

Name	Expression	Value	Description
lp	10[mm]	0.01 m	workpiece length
hp	1[mm]	0.001 m	workpiece height
acr	11e-6	1.1E-5	radius of initial craters
bcr	4*2.2e-6	8.8E-6	depth of initial craters
rms	0.1e-6	1.0E-7	radius of resolified material
rgb	0.1[mm]	1.0E-4 m	radius of gas bubble
ti	10e-6	1.0E-5	discharge duration
modulE	220[GPa]	2.2E11 Pa	Young's modulus
PR	0.29	0.29	Poisson ratio
pus	150[MPa]	1.5E8 Pa	ultrasonic shock pressure
tb	0.8e-6	8.0E-7	bubbles implosion time

Fig. 9. Example of assigned parameters for material removal process of µEDM+US on CoCr alloy.

The Van Dijck and Snoeys's overheating model was considered for EDM component modelling. In this case, boiling is the main mean to remove overheated material after pulse end when pressure drops suddenly; the gas bubble formed around plasma channel lasts very long after the end of pulse, so the material melted by discharge is already solidified [6].

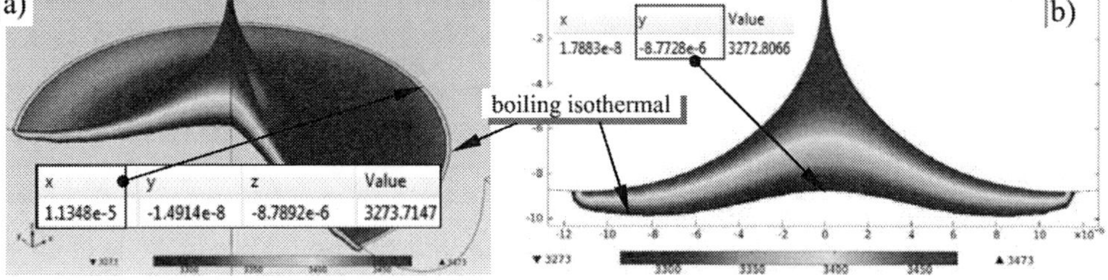

Fig. 10. Removed material, boiling isothermal [K] disposal at µEDM after t_i=48 µs: (a) 3D; b) section view.

Constant 3473 K temperature on EDM spot, thermal isolation on area covered by gas bubble, and convection cooling on workpiece periphery in contact with dielectric liquid with 313 K were setup as boundary conditions. The temperature distribution obtained around EDM spot with t_i=48 μs and t_0=24 μs is presented in fig. 10, indicating close value to experimental ones (see fig. 3.b), validating the model.

At ultrasonic assistance, due to relative long pulse time t_i=48 μs, comparable with T_{US}, it is very high probability to its overlapping on *CMS* [14]. So, melting becomes the main mechanism of material removal since the gas bubble around plasma channel collapses at *CMS*, and dielectric liquid accesses the melted material. The temperature distribution obtained in this case is presented in fig. 11.a:

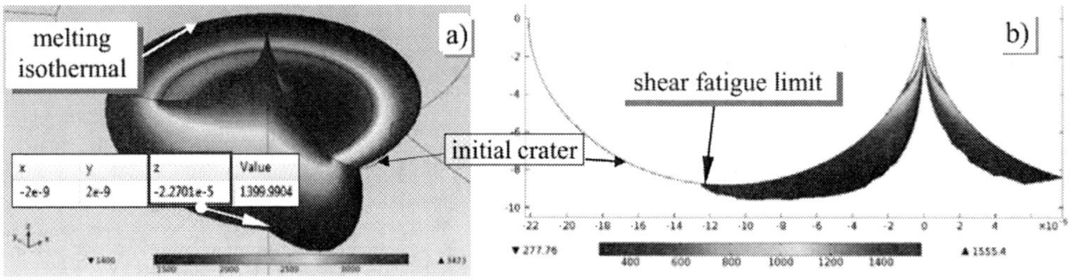

Fig. 11. Removed material at μEDM+US: a) melting isothermal [K] disposal after t_i=10 μs from pulse start, at *CMS*; b) Von Mises stress [MPa] at 150 MPa ultrasonic pressure shock wave.

During EDM removal, another mechanism is produced due to ultrasonic cavitation. Crater margins are removed by shear of shock waves oriented parallel to machined surface. In order to underline the volume of material removed by hydraulic mechanical effect of ultrasonic cavitation, the fatigue limit at shear pulsing cycle (τ_0) was calculated [11]:

$$\tau_0 = 1.12(40 + 0.16\,\sigma_r), \qquad [\text{MPa}] \tag{2}$$

where: σ_r is the usual ultimate tensile strength; in this case, σ_r=1300 MPa, and the parameter of interest is τ_o=277.76 MPa. In fig. 11.b, it can be observed a massive removal also from the bottom of crater, which determines the roughness increase, i.e. the depth of crater. This is confirmed by experimental data (see fig. 4.b), explained by relative high power, P_{cUS} = 120 W used on ultrasonic chain. Both removal mechanism determined increase of craters dimensions in comparison with pure micro-EDM, experimentally emphasized by fig. 4.

At μEDM+US with t_i=8 μs, t_0=8 μs, the overlapping probability of pulse time on CMS, decreases very much, under 20% [6]. So, classic thermal removal like in classic μEDM occurs dominantly, and the temperature distribution after pulse end is presented in fig. 12:

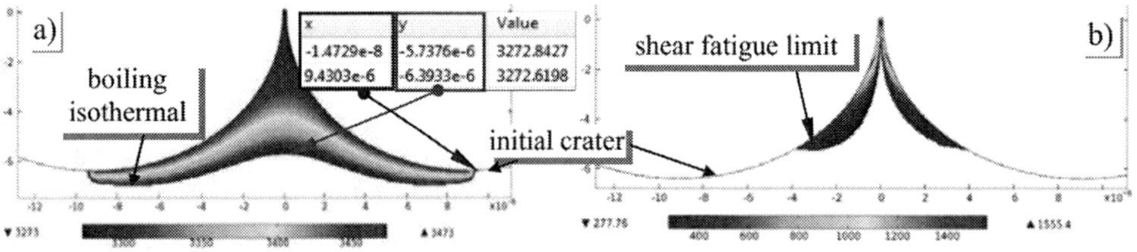

Fig. 12. Material removed at μEDM+US, t_i=8 μs commanded pulse: a) boiling isothermal [K] disposal; b) Von Mises stress [MPa] at pulse cycling at 110 MPa ultrasonic pressure shock wave.

The boiling isothermal position is in agreement with experimental data from fig. 5, validating the model. Over this thermal removal mechanism is superimposed, the ultrasonic removal due to shock wave produced by *CMS* parallel to machined surface [6]. In fig. 12.b, distribution of Von Mises stress is presented, higher than shear fatigue limit, which shows the material that is removed. The decrease of craters depth is produced, and implicitly the surface roughness, confirming the experimental data from fig. 6.b, at lower P_{cUS}=90W.

At μEDM+US with relaxation pulses, the nature of material removal mechanism is very similar with the previous working mode. The ultra-short discharge duration of relaxation pulses makes a very low probability to overlap them on *CMS*. So, classic thermal removal is dominant, as presented in fig. 13.a, associated with ultrasonic shear of craters margins, producing lowering of craters dimensions, confirmed experimentally by data from fig. 8.

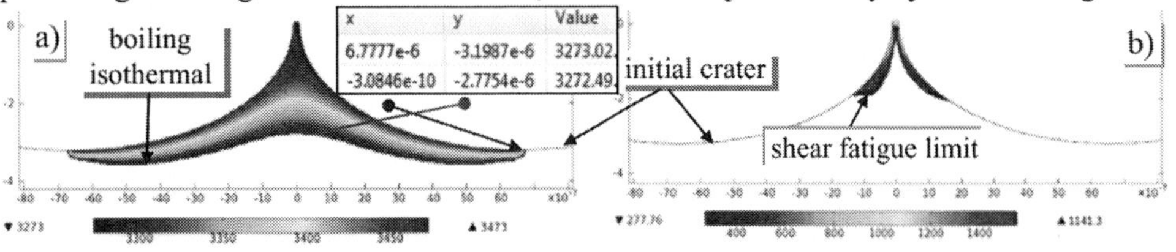

Fig. 13. Material removed at μEDM+US, relaxation pulse: a) boiling isothermal [K] disposal after pulse end; b) Von Mises stress [MPa] at pulse cycling at 110 MPa ultrasonic pressure shock wave.

4 Conclusions

The FEM study of μEDM+US on CoCr alloy, not systematically treated in the state of the art yet, emphasized two types of material removal mechanism, confirmed by experimental data. At longer pulse time, comparable with ultrasonic period, the thermal mechanism of removal is dominant, and position of melting isothermal shows a spectacular increase of machining rate. At shorter pulse time, especially at relaxation pulses using, the ultrasonic hydraulic removal becomes dominant by removing the craters margins – the debris are not spherical like at melting and resolidification. Finding an optimum value of consumed power on ultrasonic chain, correlated with CoCr alloy fatigue limit, is of utmost importance in order to decrease the surface roughness in comparison with pure μEDM.

The paper was achieved in Joint Applied Research Project supported by MEN-UEFISCDI, project no. PN-II-PT-PCCA-2013-4-0236, Contract no. 222/2014. The SEM analyzes/images obtained on the samples were possible due to EU-funding project POSCCE-A2-O2.2.1-2013-1, SMIS-code 48652.

References

1. Y. S. Al Jabbari, J. Adv. Prosthodont. **6**, 138-145 (2014)
2. * * * Arcam AB, ASTM F75 CoCr Alloy, 3 (2010)
3. S. Kapoor, R. Liu, X. J. Wu, M. X. Yao, Int. J. Ch. M. Nucl. Mat. Met. Eng. **6**, 592-601 (2012)
4. * * * Hardfacing Alloys 22, kennametal.com/stellite (2016)
5. S. Baron, E. Ahearne, P. Connolly, S. Keaveney, G. Byrne, Proc. Mach. Tool Tech. Res. Found., 1-7 (2015)
6. D. Ghiculescu, N.I. Marinescu, O. Alupei, *Applied Mechanics and Materials*, **834**, 126-131 (2016)
7. M. Iwai, S. Ninomiya, K.Suzuki, *Procedia CIRP*, **6**, 146–150 (2013)
8. Z. Qinjian, Z. Luming, L. Jianyong, et al., *Procedia CIRP*, **6**, 589–593 (2013)
9. * * * Cobalt Chrome Alloy Co28Cr6Mo, Material specifications, 3T RPD Ltd. UK (2012)
10. R. V. Marreya, R. Burgermeistera, R. B. Grishabera, R.O. Ritchie, Biomaterials **27**, 1988–2000 (2006)
11. V. Drobota, *Resistance of Materials*, (Didactic and Pedagogical, Bucharest, 1982)

MATEC Web of Conferences **112**, 03006 (2017) DOI: 10.1051/matecconf/201711203006
IManE&E 2017

Experimental issues at ultrasonically aided micro-EDM of CoCr alloys

Niculae Marinescu, *Daniel Ghiculescu**, and *Claudiu* Pirnau
"Politehnica" University of Bucharest, Manufacturing Engineering Department, 313 Splaiul Independentei, Bucharest, Romania

Abstract. The paper deals with researches of microtopography obtained at different working modes at ultrasonically aided micro-electrical discharge machining (μEDM+US) and pure micro-EDM, using both commanded and relaxation pulses. Images of machined surface taken with scanning electron microscope were analyzed in correlation with chemical composition of samples detected by X rays spectrometer. This led to some explanations of material removal mechanism, aiming at improvement of process modelling and performances increase of μEDM+US.

1 Introduction

The exceptional properties of CoCr alloys made them matter of utilization in many areas. High stiffness, high surface quality, extremely wear-resistant and corrosion resistant in lactic acid and sodium chloride and also biocompatibility [1, 2] recommend their use in medical applications as material for knee implants, metal-to-metal hip joints, fixation devices for fractured bones, and dental prosthetics. Under these conditions, fractures should be attributed generally to incorrectly designed clasps, or to stress concentration areas, and not to material mechanical characteristics [1].

The existence of Ni in CoCr alloys composition allows the utilization as fuel nozzles and vanes for industrial gas turbines. High hardness and excellent surface quality permit utilization at injection molding of plastic parts [2].

There are also several types of stellite superalloys, belonging to the CoCr ones, having variable content of: Ti, Si, S, P, Mo, W, Mg, Cr, Co, B, Al, Fe, and Ni. Due to their high mechanical properties, these alloys are used in different industrial applications like: valves, valve seats for internal combustion engines, machine gun and machine gun barrels, artificial hip joints, other bone replacements, dental prosthesis, and cutting tools. The C content and as a result, carbides volume within stellite alloys decisively influences the material performances. Therefore, the following classification of these alloys is possible: high C content - for high wear applications; low C content - for high temperature use; low C / higher Cr content for corrosion resistance [3, 4].

But CoCr alloys mechanical properties determine consistent difficulties at machining, especially when carbides are present in their structure, and complex surfaces are generated by micromachining. Ultrasonically aided electrical discharge machining (EDM+US) could

* Corresponding author: daniel.ghiculescu@upb.ro

be a competitive response of all these, in terms of machining rate, precision and surface quality in relation with manufacturing costs [5]. Other advanced materials, difficult to process with other means, were machined by EDM+US with spectacular increase of main technological performances [6, 7, 8], but no reported references for CoCr alloys yet [9].

2 Experimental setup

In order to study the micro-EDM+US process on CoCr alloy, an experimental setup was created, presented in Fig. 1. Sheet samples of 30 x 50 x 1 mm were machined comparatively on Romanian ELER 01 machine with 20 kHz longitudinal ultrasonic oscillations of tool-electrode (EDM+US), and without US (classic EDM). The device for clamping the ultrasonic chain, which included the tool at its end, is a granted patent [10].

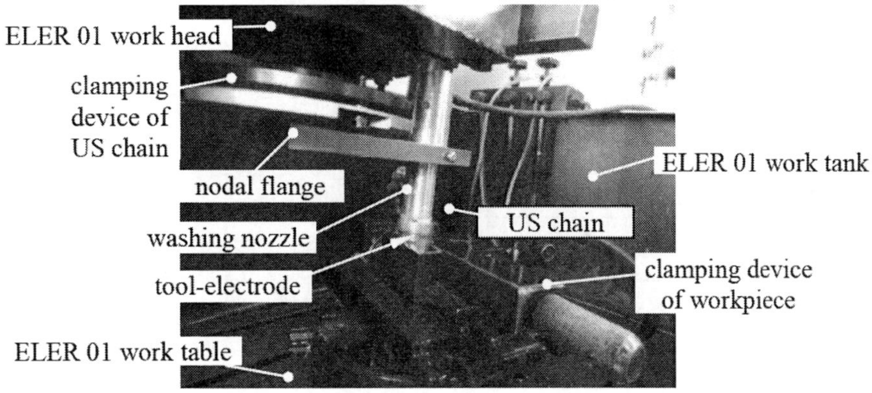

Fig. 1. Technological system elements for EDM+US micromachining of CoCr alloy samples.

Some preliminary tests on this CoCr alloy were done using both commanded and relaxation pulses provided by a special generator dedicated to finishing and micro-machining, connected to ELER 01 machine. The working parameters at commanded pulses were: current step, I=3 A, positive (tool) polarity, lateral flushing with pressure, p_{hl}=0.04 MPa, pulse time t_i=48 µs, pause time, t_0=24 µs - working mode no.1; t_i=8 µs, t_0=8 µs -working mode no. 2. At relaxation pulses, the specific working parameters were used: negative polarity, capacitor step, C=10 nF, resistance step, R=0.74 kΩ (for adjusting supply current) - working mode no. 3. At µEDM+US, the consumed power on ultrasonic chain (P_{cUS}) was 90, 120 W.

The variation of total hydraulic pressure (p_{ht}) in the gap was determined by elongation (z), according to relation:

$$p_{ht} = 2\pi \, c \, \rho \, f_{US} \, z + p_{hl} \quad \text{[Pa]}, \tag{1}$$

where: c is sound velocity in dielectric liquid [m/s]; ρ - dielectric liquid density [kg/m^3]; f_{US} - ultrasonic frequency [Hz]; z - elongation, $z=A\sin \omega t$; A - amplitude [m]; $\omega = 2\pi f_{US}$ [s^{-1}]; p_{hl} - local hydraulic pressure [Pa]. In this case, the values of parameters were: p_{hl}=0.04MPa, ρ=840 kg/m^3 (P3 dielectric liquid), K=1.35x10^9 Pa (K-bulk modulus), $c=(K/\rho)^{1/2}$=1267.7 m/s, A=2µm, f_{US}=20kHz.

A collective implosion of gas bubbles, generating shock waves occurs at each final of ultrasonic period T_{US}. This is called the *cumulative microjets stage* (CMS), generating pressure of 100 MPa order of magnitude, oriented horizontally along the frontal gap. This is able to reduce the surface roughness by removing by shearing the peaks of microgemetry. The work parameter related to ultrasonic pressure is the consumed power on ultrasonic chain, including the tool (P_{cUS}).

The commanded or relaxation pulses are delivered during ultrasonic oscillations of tool-electrode. Experimentally, it was noticed that at overlapping the pulses on CMS, the discharge is stopped, hence dielectric liquid accesses the melting material by discharge since the gas bubble formed around plasma channel collapses.

3 Analysis of chemical composition and microtopography

The chemical composition of machined samples (cast and cold-worked) was determined by X rays spectrometer energy dispersive (EDS) with resolution at MnK of 133 eV - table 1. It can be noticed close values of weight % for Co, Cr and Mo in case of all used samples, the C content being under 0.1%. Therefore, three different working modes, EDM and EDM+US, were applied on very close chemical composition of the samples no. 1, 2 and 3 because their surfaces were so small to have three different machined marks on each of them.

Table 1. Chemical composition of CoCr alloy samples.

Element	Sample no. 1, weight %	Sample no. 2, weight %	Sample no. 3, weight %
Si	0.7	0.68	0.68
Cr	26.72	27.48	27.61
Co	58.09	59.6	57.28
Mo	14.49	12.42	14.12

Investigation of machined samples microtopography was made by scanning electron microscope QUANTA INSPECT F50, provided with field emission electron gun with resolution of 1 nm.

The images of machined microtopography obtained at pure μEDM with working mode no.1, and sample no.1 are presented in Fig. 2:

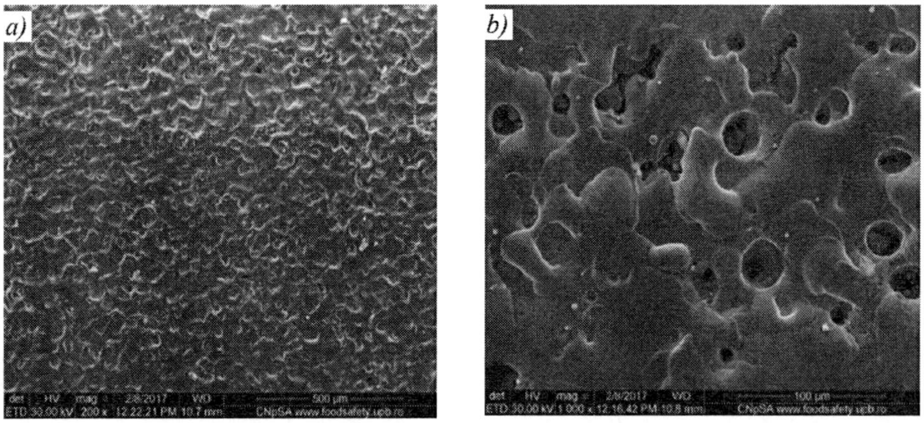

Fig. 2. Microtopography at μEDM, I=3A, t_i=48 μs, t_0=24 μs, polarity +; (a) general; (b) detail view.

The roughness of machined surface was Ra=2.2 μm, determined by SJ411 surface instrument. The average of craters transversal dimensions was D_{cr}=22 μm.

At μEDM+US with working mode no.1, on sample no.1, with consumed power on US chain, P_{cUS} = 120 W, the images of machined microtopography are presented in Fig. 3.

DOI: 10.1051/matecconf/201711203006

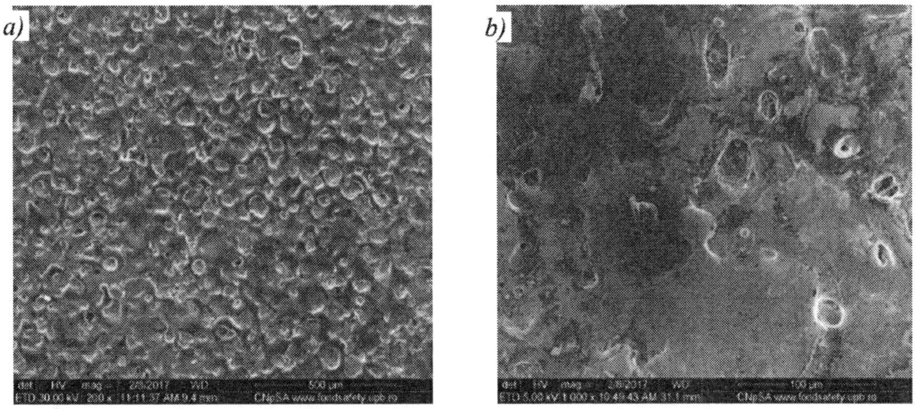

Fig. 3. Microtopography at μEDM+US, I=3A, t_i=48 μs, t_0=24 μs, polarity +; P_{cUS} = 120 W; (a) general view; (b) detail view.

In this case, the sample roughness was Ra=2.4 μm, and the average of craters transversal dimensions was D_{cr}=26 μm. As it can be noticed, at relative high value P_{cUS} = 120 W, an increase of crater volume is produced due to ultrasonic aiding not only in terms of crater diameter but also in depth.

A specific phenomena also occurred: a smaller crater is often developed in a greater crater as it is emphasized in Fig. 4, probably as a result of cavitational phenomenon ultrasonically induced – a gas bubble collapsing inside of a main crater produced by electrical discharge.

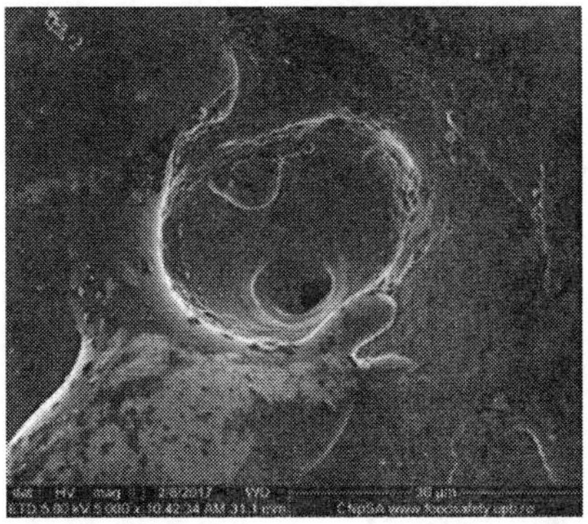

Fig. 4. Microtopography of smaller crater within a main crater at μEDM+US, I=3A, t_i=48 μs, t_0=24 μs, polarity +; P_{cUS} = 120 W.

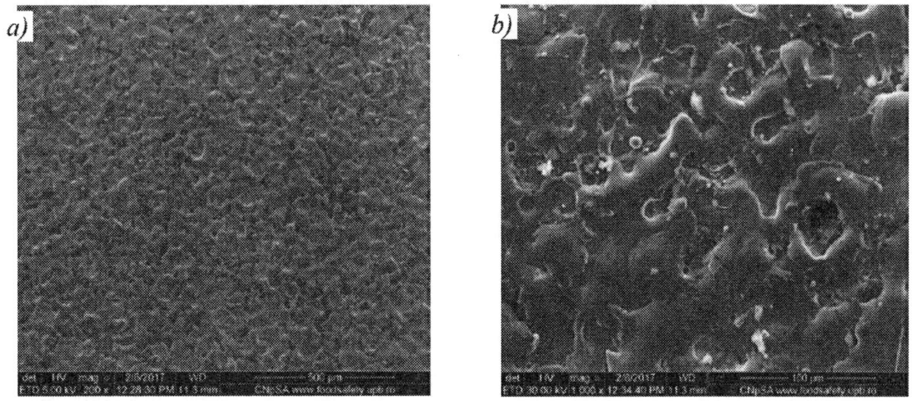

Fig. 5. Microtopography at μEDM, I=3A, t_i=8 μs, t_0=8 μs, polarity +; (a) general; (b) detail view.

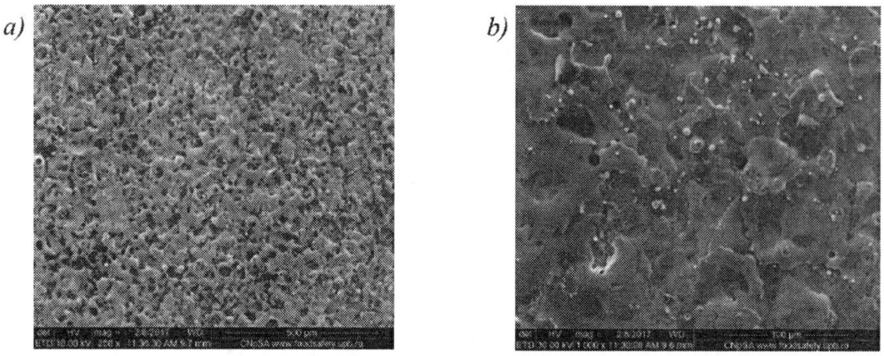

Fig. 6. Microtopography at μEDM+US, I=3A, t_i=8 μs, t_0=8 μs, polarity +, P_{cUS}=90 W: (a) general; (b) detail view.

The images of machined microtopography obtained at pure μEDM with working mode no. 2, on sample no. 2, are presented in Fig. 5. The sample roughness was Ra=1.6 μm, and the average of craters transversal dimensions was D_{cr}=18 μm.

At μEDM+US and the same working mode, no. 2, sample no. 2, with lower power consumed on the US chain, P_{cUS} = 90 W, than in previous case, the microtopography is shown in fig. 7. The sample roughness was Ra=1.2μm, and the average of craters transversal dimensions was D_{cr}=16 μm. As it can be noticed comparing figures 5.b and 6.b a certain levelness of machined surface was obtained through US assistance at lowering of P_{cUS}.

At μEDM with relaxation pulses, working mode no. 3, on sample no. 3, the images of microtopography are presented in Fig. 7. The sample roughness was Ra=0.8μm, and the average of craters transversal dimensions was D_{cr}=13 μm.

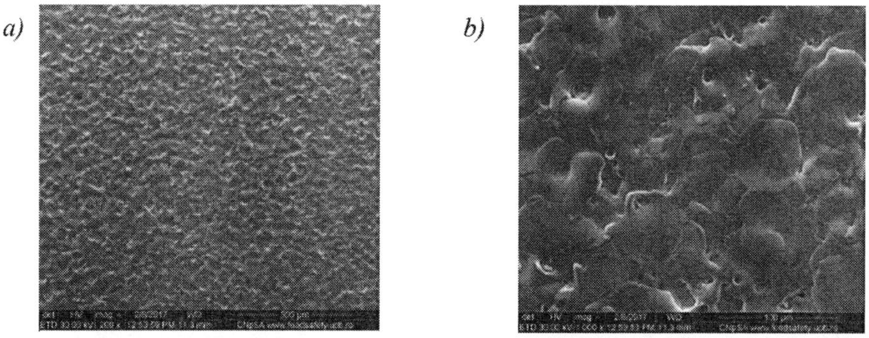

Fig. 7. Microtopography at μEDM, C=10 nF, R=0.74 kΩ, polarity - ; (a) general; (b) detail view.

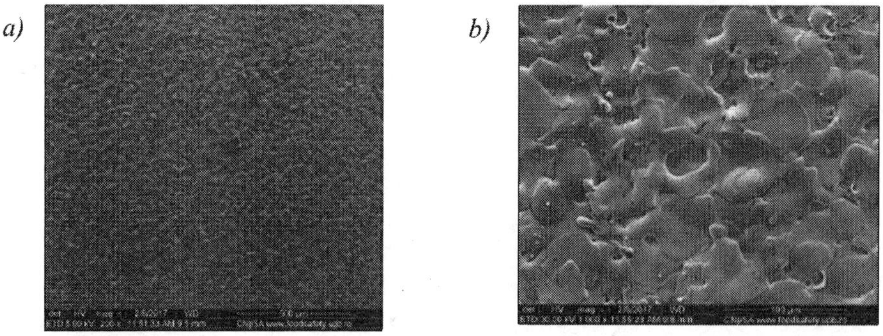

Fig. 8. Microtopography at μEDM+US, C=10 nF, R=0.74 kΩ, polarity -, P_{cUS} = 90 W: (a) general view; (b) detail view.

At μEDM+US with relaxation pulses, working mode no. 3, on sample no. 3, with P_{cUS} = 90 W, the images of microtopography are presented in Fig. 8. In this last case, the sample roughness was *Ra*=0.6 μm, and the average of craters transversal dimensions was D_{cr}=10 μm. As it can be noticed, keeping relative low value of P_{cUS}, the surface roughness was also decreased as well as craters transversal dimensions.

4 Conclusions

Analyzing the microtopography of CoCr alloy samples, one can observe that crater average dimensions are increased as the discharge energy increases (in this order, working modes no. 1, 2, 3), the variation of samples chemical composition, which is very small, having insignificant influence.

At lower values of the power consumed on ultrasonic chain (P_{cUS}) in case of ultrasonically aided micro-electrical discharge machining (μEDM+US), the craters transversal dimensions are decreased. On the contrary, at higher values, these are increased. This effect is the result of crater margins removal, since they are more sensitive of ultrasonic shock waves produced by cavitation, oriented parallel to the working gap direction.

The ultrasonic aiding is able to reduce the roughness of machined surface of CoCr alloy samples, resulted from micro-EDM if the power P_{cUS} has optimum value. At higher P_{cUS} values, the surface roughness is increased, together with craters transversal dimensions. In this case, it was also observed a specific microgeometry, e. g. smaller craters in main EDM craters, probably produced by ultrasonically induced cavitation within the working gap, which could also increase the machining rate spectacularly.

After these preliminary researches of μEDM+US applied on CoCr alloy, further researches will be developed in order to draw the variation of craters dimensions with respect to consumed power on ultrasonic chain, and on this base, to find the optimum value of this parameter, leading to minimum surface roughness.

The paper was achieved in Joint Applied Research Project supported by MEN-UEFISCDI, project no. PN-II-PT-PCCA-2013-4-0236, Contract no. 222/2014. The SEM analyzes/images obtained on the samples were possible due to EU-funding project POSCCE-A2-O2.2.1-2013-1/Prioritary Axe 2, Project No. 638/12.03.2014, ID 1970, SMIS-CSNR code 48652.

References

1. Y. S. Al Jabbari, J. Adv. Prosthodont. **6**, 138-145 (2014)
2. * * * Arcam AB, ASTM F75 CoCr Alloy, 3 (2010)
3. S. Kapoor, R. Liu, X. J. Wu, M. X. Yao, *Int. J. Ch. M. Nucl. Mat. Met. Eng.* **6**, 592-601 (2012)
4. * * * Hardfacing Alloys 22, kennametal.com/stellite (2016)
5. M. P. Jahan, et al., *Electrical Discharge Machining (EDM). Types, Technologies and Applications* (Nova, New York, 2015)
6. C. Praneetpongrung, Y. Fukuzawa, S. Nagasawa, et al., *Mat.Trans.*, **51**, 2113–2120 (2010)
7. M. Iwai, S. Ninomiya, K.Suzuki, *Procedia CIRP*, **6**, 146–150 (2013)
8. Z. Qinjian, Z. Luming, L. Jianyong, et al., *Procedia CIRP*, **6**, 589–593 (2013)
9. B. C. Khatri, P. Rathod, J. B. Valaki, *J. Eng.Manuf.*, **1**, 1-12 (2015)
10. D. Ghiculescu, N.I. Marinescu, S. Nanu, Romanian granted patent, RO-126191 (2012)

Taguchi versus full factorial design to determine the equation of impact forces produced by water jets used in sewer cleaning

Nicolae Medan[1,*], *Mircea* Lobontiu[1], *Sandor* Ravai Nagy[1] and *Gergely* Dezső[2]

[1]Technical University of Cluj-Napoca, Department of Engineering and Management of Technology, Victor Babes 62A, Baia Mare, Romania
[2]University of Nyiregyhaza, Department of Physics and Production Engineering, Sóstói 31/B, Nyiregyhaza, Hungary

Abstract. The regular cleaning up of the materials deposed in sewer networks is realised with equipment that uses pressure water jets. The functioning of this equipment is dependent on certain process parameters that can vary, causing variations of the impact forces. The research method of this study is the experiment. The purpose of this study is to determine the equation of impact forces produced by water jets in two ways: once using the Taguchi design of experiments and secondly using the full factorial design of experiments. For the experimental determination of the impact forces a stand for generating water jets and a device for measuring the impact forces are used. Finally, an analysis of the advantages and disadvantages of the two methods used is presented.

1 Introduction

Phenomena that occur in the cleaning water jets are complex. Adler [1] describes mechanisms occurring at the impact of a jet with a surface. Leach et al [2], Leu et al [3] and Guha et al [4] analysed pressure distribution along the centreline of the water jet.

The aim of this paper is to determine the equation of impact forces produced by water jets from the sewerage system cleaning head, impact forces that depend on the process parameters. The research method of this study is the experiment.

To determine the equation of impact forces produced by water jets are used two type of design of experiment: once using the Taguchi design of experiments and secondly using the full factorial design of experiments.

To measure the impact forces between the water jet and a flat and rigid surface, a stand for generating pressure water jets, as well as a device to measure the impact forces have been designed and built. In the measurements, water jets using pressure between 100 and 200 bars have been used. The values of these pressures correspond to low pressure water jets [5].

* Corresponding author: nicolae.medan@cunbm.utcluj.ro

2 Equipment used and methodology of the measurements

To measure the impact forces between the water jet and a flat and rigid surface, a stand for generating pressure water jets, as well as a device to measure the impact forces were designed and built [6].

2.1 The stand to generate pressure water jet

Schematic diagram of the stand to generate a pressure jet is shown in Figure 1.

The component parts of the stand are: 1) electric motor 2) flexible coupling; 3) high pressure pump, 4) pressure regulator, 5) pressure gauge, 6) nozzle, 7) tap water, 8) water tank, 9) chassis. Water coming out of the high-pressure pump (3) goes into the pressure regulator (4). The pressure regulator adjusts the pressure and flow of the water in the path of the high-pressure water. This pressure corresponds to the one at the outlet of the nozzle.

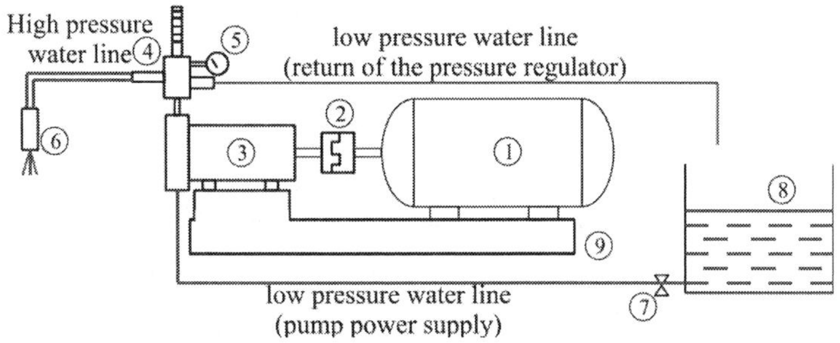

Fig. 1. Diagram of the stand to generate pressure jet.

2.2 The device to measure the impact forces

Figure 2 represents the principle diagram of the device for the measurement of the impact force between the water jet and a flat and rigid surface.

The main component parts of the device are:1) high-pressure water hose, 2) support nozzle, 3) nozzle block, 4) nozzle, 5) water jet, 6) flat and rigid target plate, 7) collection path water, 8) scaled container, 9) piezoelectric sensor mounting, 10) piezoelectric sensor, 11) data acquisition Personal Daq/3000, 12) computer for the processing of data; 13) support plate, 14) acrylic tube, 15) rods for adjusting distance x.

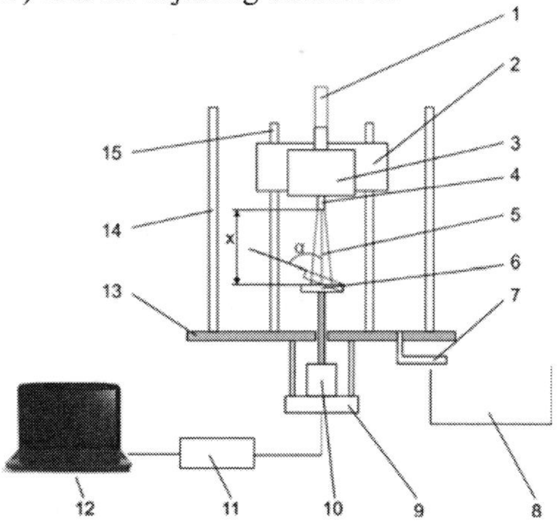

Fig. 2. Diagram of the device for the measurement of the impact force of the water jet.

2.3 The process parameters

In the water jet cleaning process a series of parameters are involved [5]. These parameters can be divided into two major groups, namely: 1) target parameters which shall be defined according to the contact area between the water jet and the surface to be cleaned and 2) process parameters. In the measurement of the impact forces of a stationary water jet and flat and rigid surface the process parameters are involved (figure 3).

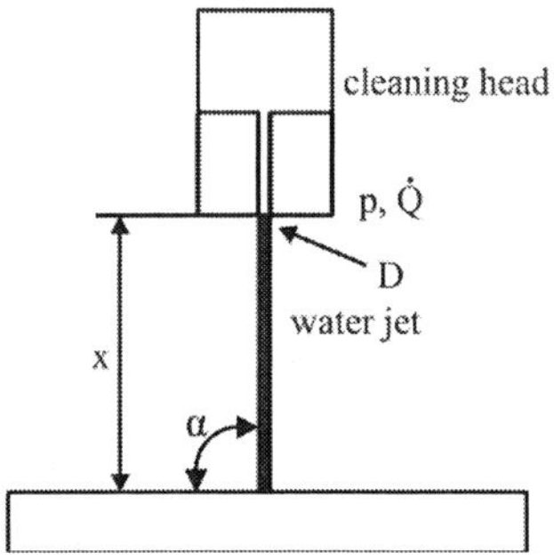

Fig. 3. Process parameters.

They can be divided into two groups, namely hydraulic parameters and the performance parameters. Hydraulic parameters characterized by the system's high pressure pump-nozzle are represented by work pressure (p), volume flow ($\dot{Q}$) and nozzle diameter (D). The performance parameters are more related to the process performance and include the following: cleaning distance (x), impact angle (α) and traverse rate (v_T). For a stationary water jet v_T= 0 m/s.

2.3.1 The values of the used parameters

In table 1 are presented the values of parameters for Taguchi and full factorial methods.

For Taguchi method, each parameter has 2 values, the minim and the maxim, in concordance with the experimental domain. For full factorial method, each parameter is assigned all the values set in experimental field [7].

Table 1. The values of the parameters for Taguchi and full factorial methods.

Parameter	Taguchi		Full factorial
	Value 1	Value 2	Values
Nozzle diameter D [mm]	1	2	1, 1.5, 2
Pressure p [bar]	100	200	100, 120,140,160,180,200
Impact angle α [^{0}]	60	90	60, 75, 90
Distance x [mm]	25	200	25, 50,75,100,125,150,175,200

Based on the parameters established in accordance with Table 1 the impact forces were determined. For each experiment was performed three measures of impact force and still it has worked with F_{med}, representing the arithmetic average of the three forces measured.

For Taguchi method results 16 experiments and a total of 48 measurements. For full factorial method results. For full factorial method results 144 experiments and a total of 432 measurements.

For brief of this paper, we dare not to present the table with the values of F_{med} obtained, for the two used methods (Taguchi and full factorial).

3 Results

For determine the equation of impact forces, in the first step it is necessary to determine the contribution of each parameter and their interactions.

3.1 Contribution of parameters and their interactions

Using the Taguchi method and full factorial method has conducted an analysis of the variance to determine the influence of each parameter and their interactions on the impact force. In Table 2 is presented the analysis of the variance, using Minitab 17.

Table 2. Analysis of Variance (Taguchi method and full factorial method).

Source	DF (degree of freedom)		Contribution	
	Taguchi	**Full factorial**	**Taguchi**	**Full factorial**
Regression	10	10	99.76%	99.14%
Diameter (A)	1	1	64.38%	71.32%
Pressure (B)	1	1	22.71%	18.05%
Angle (C)	1	1	2.84%	3.63%
Distance (D)	1	1	0.54%	0.24%
Diameter*Pressure	1	1	7.63%	4.63%
Diameter* Angle	1	1	1.04%	0.93%
Diameter* Distance	1	1	0.08%	0.09%
Pressure*Angle	1	1	0.42%	0.24%
Pressure*Distance	1	1	0.07%	0.01%
Angle*Distance	1	1	0.04%	0.00%
Error	421	5	0.24%	0.86%
Total	431	15	100%	100%

From table 2 results that 3 parameters (diameter, pressure, angle) and 2 interactions (diameter*pressure and diameter*angle) have a significant contribution to the impact forces in both cases Taguchi method and full factorial method.

3.2 Equation of impact forces

Based on the results obtains in table 2, was determined the equation of impact forces in both cases Taguchi method and full factorial method.

Using Minitab 17, the equation (1) of impact forces for Taguchi method was determined

$$F_{Taguchi} = 23.0 - 33.6 \cdot A - 0.2101 \cdot B - 0.254 \cdot C + 0.2992 \cdot A \cdot \mathrm{B} + 0.375 \cdot A \cdot C \quad (1)$$

Using Minitab 17, the equation (2) of impact forces for full factorial method was determined:

$$F_{full\ fact.} = 22.91 - 32.97 \cdot A\ 0.2088 \cdot B - 0{,}26 \cdot C + 0.30072 \cdot A \cdot \mathrm{B} + 0.3755 \cdot A \cdot C \quad (2)$$

3.3 Comparison between the two equations determined

To compare the accuracy of two equations determined using Taguchi and full factorial designs of experiment, in table 3 are presented the impact forces measured. Was used the values of the parameters for full factorial methods, less the parameter distance, who was fixed at 50 mm (the influence of this parameter it is insignificant).

It was calculated relative error of the force measured and calculated force $F_{Taguchi}$. Also, was calculated relative error of the force measured and calculated force $F_{full\ factorial}$.

Table 3. Measured and calculated impact forces. Relative errors.

Exp. No.	Diameter [mm]	Pressure [bar]	Angle [α]	F_{det}	$F_{taguchi}$	F_{full}	Relative error ε_x[%]	
							$F_{Taguchi}$	$F_{full\ factorial}$
1	1	100	60	6.59	6.51	6.06	1.24	7.98
2	1	100	75	8.49	8.34	7.80	1.83	8.17
3	1	100	90	9.95	10.16	9.53	2.13	4.23
4	1	120	60	8.48	8.29	7.90	2.17	6.75
5	1	120	75	10.54	10.12	9.64	4.04	8.61
6	1	120	90	11.83	11.94	11.37	0.96	3.91
7	1	140	60	9.04	10.07	9.74	11.46	7.80
8	1	140	75	11.24	11.90	11.47	5.84	2.06
9	1	140	90	12.60	13.73	13.21	8.91	4.79
10	1	160	60	11.41	11.86	11.58	3.94	1.53
11	1	160	75	14.18	13.68	13.31	3.52	6.12
12	1	160	90	15.25	15.51	15.05	1.73	1.31
13	1	180	60	13.54	13.64	13.42	0.73	0.87
14	1	180	75	15.09	15.46	15.15	2.50	0.44
15	1	180	90	16.55	17.29	16.89	4.45	2.00
16	1	200	60	15.51	15.42	15.26	0.57	1.61
17	1	200	75	17.03	17.25	16.99	1.29	0.21
18	1	200	90	18.25	19.07	18.72	4.51	2.60
19	1.5	100	60	15.95	15.92	15.88	0.20	0.45
20	1.5	100	75	20.18	20.56	20.43	1.91	1.25
21	1.5	100	90	24.23	25.20	24.98	4.02	3.09
22	1.5	120	60	21.23	20.69	20.73	2.52	2.37
23	1.5	120	75	25.09	25.34	25.28	0.98	0.74
24	1.5	120	90	28.3	29.98	29.82	5.93	5.39
25	1.5	140	60	23.83	25.47	25.57	6.87	7.31
26	1.5	140	75	29.64	30.11	30.12	1.59	1.62
27	1.5	140	90	32.77	34.75	34.67	6.05	5.80
28	1.5	160	60	29.01	30.24	30.42	4.25	4.86
29	1.5	160	75	35.22	34.88	34.97	0.95	0.72
30	1.5	160	90	37.74	39.53	39.52	4.73	4.71
31	1.5	180	60	33.29	35.02	35.27	5.19	5.93
32	1.5	180	75	40.88	39.66	39.81	2.99	2.61
33	1.5	180	90	43.81	44.30	44.36	1.12	1.26
34	1.5	200	60	37.34	39.79	40.11	6.56	7.42
35	1.5	200	75	45.36	44.43	44.66	2.04	1.54
36	1.5	200	90	48.62	49.07	49.21	0.93	1.21
37	2	100	60	26.01	25.33	25.70	2.61	1.21
38	2	100	75	32.36	32.79	33.06	1.33	2.17
39	2	100	90	40.26	40.25	40.43	0.03	0.41
40	2	120	60	34.62	33.10	33.55	4.40	3.09
41	2	120	75	39.78	40.56	40.91	1.96	2.86
42	2	120	90	49.25	48.01	48.28	2.51	1.97
43	2	140	60	40.94	40.86	41.40	0.19	1.13
44	2	140	75	47.27	48.32	48.77	2.23	3.18
45	2	140	90	57.91	55.78	56.13	3.68	3.07
46	2	160	60	48.15	48.63	49.26	1.00	2.31
47	2	160	75	57.43	56.09	56.62	2.33	1.40
48	2	160	90	64.04	63.54	63.99	0.78	0.09
49	2	180	60	57.24	56.39	57.11	1.48	0.23
50	2	180	75	64.29	63.85	64.48	0.68	0.29
51	2	180	90	70.79	71.31	71.84	0.73	1.48

Table 3. Measured and calculated impact forces. Relative errors (continued).

Exp. No.	Diameter [mm]	Pressure [bar]	Angle [α]	F_{det}	$F_{taguchi}$	F_{full}	Relative error ε_x[%]	
							$F_{Taguchi}$	$F_{full\ factorial}$
52	2	200	60	64.99	64.16	64.96	1.28	0.04
53	2	200	75	71.88	71.62	72.33	0.37	0.62
54	2	200	90	79.06	79.08	79.69	0.02	0.80

4 Conclusions

The purpose of this study is to determine the equation of impact forces produced by water jets in two ways: once using the Taguchi design of experiments and secondly using the full factorial design of experiments.

Using the Taguchi method and full factorial method has conducted an analysis of the variance to determine the influence of parameters and their interactions on the impact force.

In both cases the conclusion was the same: only 3 parameters (diameter, pressure and angle) and two interactions (diameter-pressure, diameter-angle) have an important contribution.

The parameter distance has an insignificant contribution (less to 1%) and can be neglected. Further value of distance parameter was set at 50 mm.

Comparing the relative errors between the two equations relative to measured forces can be seen:

- For $F_{Taguchi}$ the maximum value of relative error is 11.46% and for $F_{full\ factorial}$ the maximum value of relative error is 7.98%.
- In generally the values obtained from $F_{full\ factorial}$ are closer to the measured forces.

The advantages of Taguchi method are the reduced number of experiments, reduced costs and reduced time. The disadvantage is that the method is not as accurate like full factorial method.

The advantage of full factorial method is the accurate compare with Taguchi method. The disadvantages of this method are the large number of experiments, the costs of the experiments and more time to perform the experiments.

It is hard to say which method it is better. It depends on the experience of the researcher, on the time and resources, how accurate must be the results.

References

1. W. F. Adler, *The Mechanics of Liquid Impact*. Treatise on Materials Science and Technology, (16) editor C.M.Preece, (Academic Press, 127-183, 1979)
2. S. J. Leach, G. L. Walker, Phil. Trans. Roy. Soc. **260**, (A), 295-308 (1966)
3. M. Leu, C.P. Meng, E.S. Geskin, L. Tismeneskiy, J. Manuf. Sci. Eng **120** (3), 571-579 (1998)
4. A. Guha, R.M. Barron, R. Balachandar, J. Mater. Process. Technol. **211** (4) 610-618 (2011)
5. A. W. Momber, *Hydroblasting and Coating of Stell structures* (Oxford: Elsevier Ltd, 2003)
6. N. Medan, *Research and Contributions to the Functioning of Cleaning Head-Sewer System*, *PhD Thesis* (Baia Mare, 2014)
7. D.C. Montgomery, *Design and analysis of experiments*, 8th edition (Wiley, 2013)

Advanced product design principles applied for developing a reconfigurable multi-station welding workbench

Vasile Merticaru*, *Alexandru-Catalin* Paraschiv and *Marius-Ionut* Ripanu

"Gheorghe Asachi" Technical University of Iasi, Romania

Abstract. The research results presented in the paper are part of a larger approach concerning the improvement of product design process sustainability by enhancing the effectiveness and efficiency of Product Design activities. As case study, an applied educational project is exemplified as result sample, particularly referring to applying such advanced design principles in developing a reconfigurable multi-station welding workbench, intended to solve the problem of the lack of space which usually most of the welding workshops have to confront with.

1 Introduction. Problem statement

There is a good while since the concept and principles of Agile Manufacturing have become a valuable strategic philosophy for industrial production survival and development within a high frequency changing market, [1], bringing fundamental and irreversible structural changes in mastering the uncertainty and redefining the mass production bases, where Flexible Manufacturing only is not any more able to solve the complex competition problems.

In order to deal with the continuously increasing product variety and individualization, Mass Customization has come to replace Mass Production as industrial strategy, [2], and Reconfigurable Manufacturing Systems are nowadays bringing more effective responsiveness to renewing circumstances, being designed for rapidly adjustable capabilities, in terms of functionality and production capacity, [3]. The essence in this direction is represented by modularity and by changing or rearranging the system components.

On the other hand, Sustainability of Product and Process Development is another major theme for advanced industrial and manufacturing engineering research, a lot of new concepts, principles and strategies evolving also in this direction.

But some of the principles of reconfiguration, as they are nominated for integrated manufacturing systems in large production enterprises, [4], are sometimes useful and effective also for product or equipment design for smaller industrials or even for manufacturing workshops, in order to enhance their flexibility and sustainability. We hereby talk about principles such as modularity, scalability, convertibility and not only.

* Corresponding author: merticaru@tcm.tuiasi.ro

The particular research results presented further on in the paper are referring to such a problem of integrating and applying some advanced design principles within product and process development activities. The research approach is part of a larger approach concerning the improvement of product design process sustainability by enhancing the effectiveness and efficiency of Product Design activities, by improving the product flexibility, functionality and reliability, by reducing the resources consumption and, not at last, by reducing the product lifecycle environmental impact, all of these being based on integrating some advanced Product Design principles, theories and tools, [5-8].

2 Research approach substantiation. Conceptual models

Within the particular research approach being discussed in the paper, a conceptual model has been first of all structured, based on a conceptual analysis, and it is hereby proposed in Fig. 1, referring to the importance of product specifications correctly defining, as core condition for providing product development sustainability.

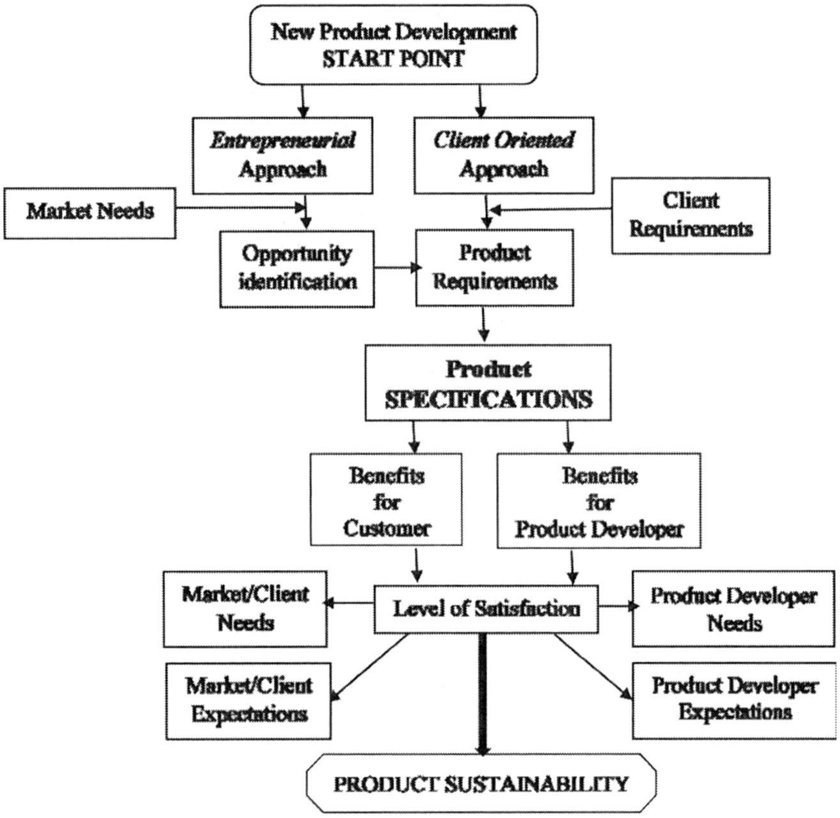

Fig. 1. Conceptual model of product specifications defining for sustainable product development.

Another conceptual model has been further on developed and is proposed in Fig.2, where some ideas related to the principles of Reconfigurable Product Design and Modular Product Design are mainly considered, starting from the need to provide easiness in change for the product or equipment physical structure and functionality, [4], and from the fact that modularity in product structure helps a lot in this direction by the possibility of components rearrangement or change, [3].

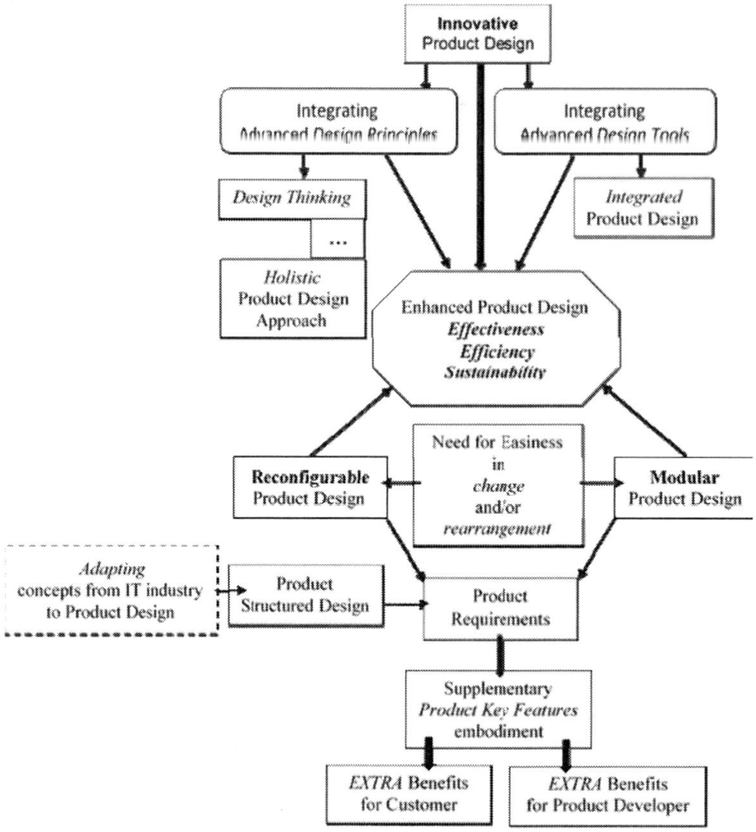

Fig. 2. Conceptual model for convergent integration of advanced design principles within Reconfigurable Product Design.

More than that, the idea of adapting some concepts from IT industry to Product Design activities, have determined to consider also some discussions upon Product Structured Design, [9], within the above mentioned model.

As start point in innovative Product Development, the idea that product conceptualisation must respond to a particular production need identified as an opportunity, [10], has been considered.

Further on, a set of general product requirements and also a specific one have been considered to be identified, in relation to all the advanced design principles included in the above model, Design Thinking, [11], and Holistic Design, [12], being among them.

Some of the identified and further on considered general product requirements are as following: reconfigurable structure; modular structure; product decomposition principles embedment; enhanced features and functionality; layout space optimization (compact design); durable construction; modules reusability for further applications development; ergonomic principles embedment; easy to assemble, install and relocate; easiness in access and use; safety in use; reduced material resource consumption (dematerialization principle); reduced carbon footprint (product lifecycle environmental impact); cleaner work environment.

All the above presented requirements will provide key features for product sustainability, which will become competitive advantages for it.

3 Case study. Product design approach and results

As case study, an applied educational project is exemplified as result sample, particularly referring to applying such advanced design principles in developing a reconfigurable multi-station welding workbench, intended to solve the problem of the lack of space which usually most of the welding workshops have to confront with, [13].

Advanced capabilities of Solid Edge as CAD solution, [14], referring mainly to the performances of providing dynamic generation of product model structure and to the advanced parameterizing functions have been used within the case study.

By combining innovation tools and simplicity considerations, the structure of the reconfigurable multi-station welding workbench has been conceived based on two types of main functional modules which can be combined, rearranged and eventually changed for obtaining configurations able to respond to different particular needs of a welding workshop, dimensioned for four up to ten individual working stations. The designed solution also provides optimized spaces for equipment and personal stuff storage.

The main functional module no.1, shown in Fig.3, provides two individual working stations endowed with equipment for welding, cutting and drilling operations, welding gas recipients, fixing devices and additional storage spaces. The module is designed to be combined at least two of them in tandem, for integrating four welding equipment of invertor type, in order to support WIG, MIG/MAG welding processes and also electric arc welding processes using coated electrodes. The modular structure of this module, consisting of a frame and of some auxiliary components, is shown in Fig. 4.

The solution of steel profiles framework, built to withstand the hard conditions of production environment, provides enhanced durability for the mechanical construction.

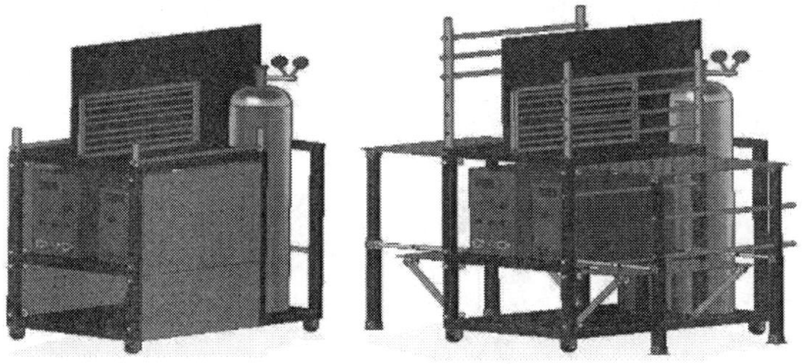

Fig. 3. Folded and unfolded view of main functional module no.1.

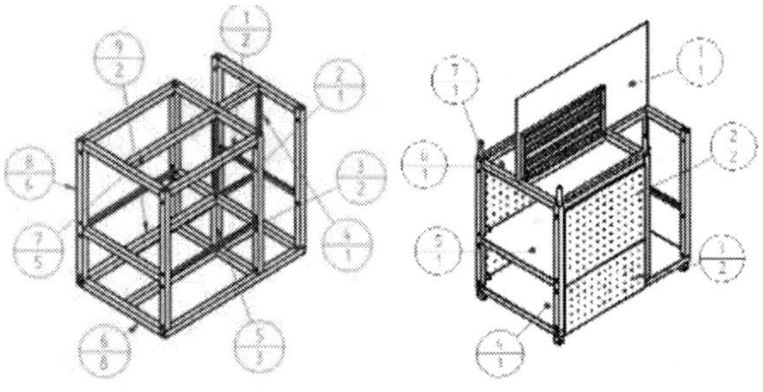

Fig. 4. Modular structure (frame and auxiliary components) of main module no.1

The main functional module no.2, shown in Fig.5, provides a welding fume extraction hood, two foldable tables for creating two additional working stations, two lockers and an auxiliary storage space. The modular structure of this module, consisting also of a frame and of some auxiliary components, is shown in Fig. 6.

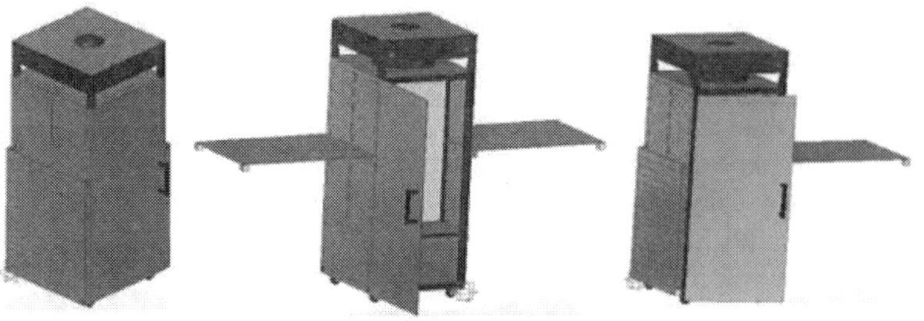

Fig. 5. Folded, unfolded and half-folded views of main module no. 2.

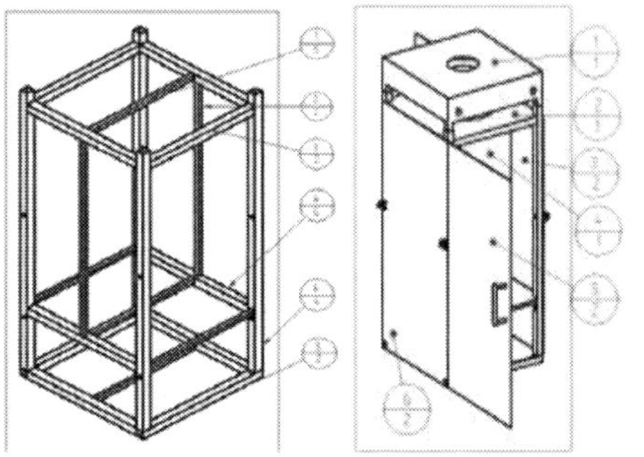

Fig. 6. Modular structure (frame and auxiliary components) of main module no.2.

In Fig.7 and Fig.8, two particularly designed configurations, respectively those for four and for ten individual working stations are shown.

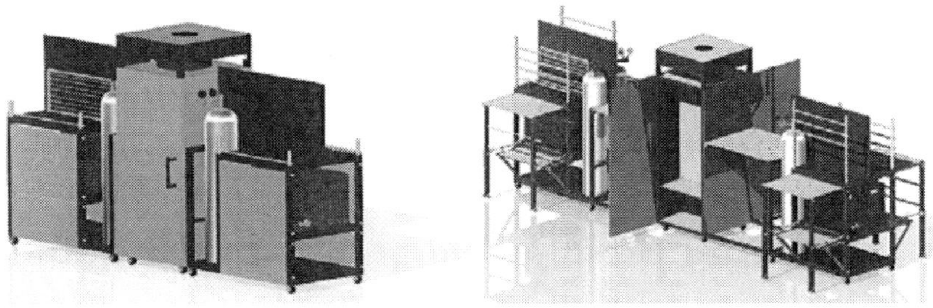

Fig. 7. Folded and unfolded view of configuration with four working stations.

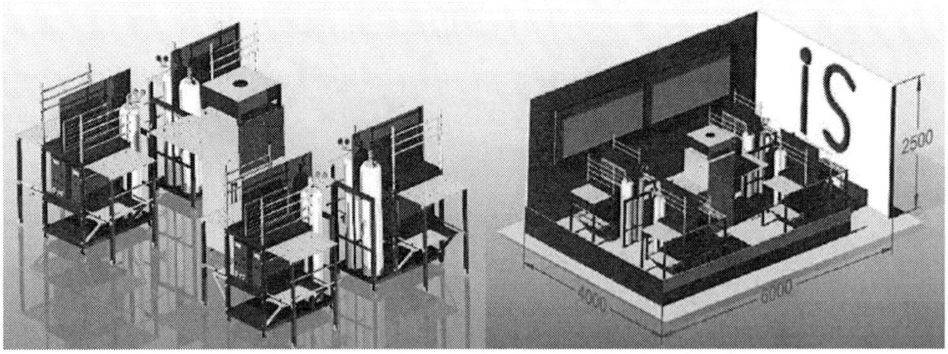

Fig. 8. Unfolded view and workshop layout view of configuration with ten working stations.

4 Conclusions

As a conclusion, we can firstly say that all the above presented conceptual considerations and the results of the case study come to sustain the great importance of appropriately identifying, selecting, combining and aligning the best options to the needs, within Product Design and Product Development activities, starting with an analytic and detailed establishment of product requirements and with their translation into product specifications.

Some general product requirements intended to become product key features able to provide supplementary benefits for the product developer and for the final user have been identified and some of them have been applied and analysed within the case study, for providing effectiveness and efficiency through the reconfigurable product development activities.

An innovative and flexible solution of a reconfigurable multi-station welding workbench, intended to solve the problem of layout optimisation, which usually most of the welding workshops have to deal with, has been designed and analysed within this work. Several benefits of that solution can be nominated: custom engineered to meet welding workshops facilities and application requirements; space optimization; low cost production and installation; modular design for ease of assembly, installation or rearranging; cleaner work environment.

Solid Edge parameterized modelling capabilities have enhanced the flexibility of the reconfigurable product model and have created premises for increasing the agility in technological changes implementation.

As another conclusion of the research approach, a few directions of further research development are identified for enhancing the Efficiency and Effectiveness of Product Development Process and the New Product Development Sustainability: integrating also an Economic Efficiency Analysis and a Life Cycle Impact Assessment approach as Concurrent Engineering activities will be considered.

References

1. A. Gunasekaran, Int. J. Prod. Res. **36**, 5, 1223-1247 (1998)
2. B. Xing, G. Bright, N.S. Tlale, J. Potgieter, *CAD/CAM, Robotics and Factories of the Future*, 473 – 482 (2006)
3. A.S.Yigit, A.G. Ulsoy, A. Allahverdi, Journal of Intelligent Manufacturing **13**, 309-316 (2002)
4. Y. Koren, M. Shpitalni, JMSY **29**, 130-141 (2010)
5. V. Merticaru, M.I. Ripanu, M.A. Mihalache, M.M. Cucos, *Applied Mechanics and Materials* **809-810**, 1492-1497 (2015)
6. V. Merticaru, G. Nagit, B. Pralea, R. Oana, AJME **12**, 43-48 (2014)
7. V. Merticaru, M.I. Ripanu, T.M. Runcanu, M.M. Nechita, S.C. Minciu, *Applied Mechanics and Materials* **657**, 996-1000 (2014)
8. V. Merticaru, M.I. Ripanu, *Applied Mechanics and Materials* **371**, 499-503 (2013)
9. E. Pelchat, *A Brief Introduction to Structured Design* (Coll. of Charleston, USA, 2004)
10. J. Gubau, Temes de Disseny **32**, 57-71 (2016)
11. A.M. Del Corral, J. Fernández, M. González, X. Riudor, Temes de disseny **32**, 81-97 (2016)
12. K. Scharwath, *Yves Béhar's 7 Principles of Holistic Product Design* (Triple Pundit, 2011)
13. A.C. Paraschiv, *Bachelor Thesis* (TUIASI, 2015)
14. SIEMENS, *SolidEdge* (2017)

The influence of build orientation on the mechanical properties of medical implants made from PA 2200 by Selective Laser Sintering

Răzvan Păcurar*, *Ancuța* Păcurar, and *Anna* Petrilak
Technical University of Cluj-Napoca, Faculty of Machine Building, Department of Manufacturing Engineering, B-dul Muncii no. 103-105, 400641, Cluj-Napoca, Romania

Abstract. This paper presents a series of research that have been made by the authors in the field of medical implants realized from PA 2200 powder material using the Selective Laser Sintering (SLS) method. Several sets of samples differently oriented into the working area of the machine were manufactured from PA 2200 by SLS, in order to determine which is the recommended orientation to be used during the manufacturing process, in such a way that at the end the mechanical properties (e.g. Compressive strength, Young's modulus, etc.) will be as closed as possible to the ones of the human bones. The obtained results were further on used in the case of a medical implant (Acetabular liner) that was manufactured at the Technical University of Cluj-Napoca (TUCN) from PA 2200 material using the DTM Sinterstation 2000 SLS equipment.

1 Introduction

Additive manufacturing (AM) technologies are widely used in medical field to produce human bone replacements and different types of prostheses, dental or hip joint replacement, etc [1,2]. Selective Laser Sintering (SLS) technology is one of the most developed AM technologies which nowadays are widely used for producing medical implants directly from biomaterials (PA 2200, PEEK, etc) [3,4]. A significant condition to produce medical implants by Selective Laser Sintering technology is to use a biocompatible material with the closest mechanical and material properties to those of the bone structure which must be replaced by surgical intervention [5,6]. Biocompatibility is related to the behavior of the biomaterials . The biocompatibility of scaffolds has to be similar with the one of the human bone, from the mechanical, chemical and biological point of view [7]. Also the biocompatible material must have the possibility to perform as a substrate that will support the appropriate cellular activity and also to optimize tissue generation [8, 9]. Other important characteristics are the biomechanical properties of the total hip replacement produced by SLS technology which should be the closest to the human bone properties. The mechanical properties of the hip joint prostheses produced by SLS depend on the process parameters and the part orientation used during the manufacturing process. In the last years,

* Corresponding author: razvan.pacurar@tcm.utcluj.ro

a significant research effort has been focused on the optimization of the SLS process parameters (laser power, scanning speed, layer thickness, etc.) for each type of material tested for producing different types of customized medical implants by using this technology [10,11]. In order to determine the influence of the process parameters on the mechanical properties of the SLS parts, researchers have tried to use different approaches to obtain better results regarding the process [12]. Many researchers have varied the laser power or scanning speed and maintained constant the other parameters to determine their influence on the mechanical properties of the manufactured parts [13,14]. Another approach was to determine the influence of different position and part building orientations on the mechanical characteristics of different types of customized medical implants made from biocompatible materials by SLS [15,16]. This paper presents results that were obtained at the Technical University of Cluj-Napoca (TUC-N) in the case of testing the influence of build orientations of samples made from PA2200 powder by SLS, on the mechanical properties of these samples (compressive strength, Young's modulus, etc.)

2 SLS manufacturing setup

In order to determine the mechanical properties of the PA 2200 powder under compressive loads, 50 compressive test samples were designed using Solidworks CAD program. According to the Standard Test Method for Compressive Properties of Rigid Plastics - ISO 604, the 50 samples have been realized with the same rectangular shape and dimensions: 12.7 x 12.7 x 25.4 mm. In order to easily identify the parts after the manufacturing process, the samples have been numbered from 1 to 50. In the current research four different types of orientations of the samples have been used: first batch of 10 samples placed in the building chamber were oriented at 0° along the Y-axis, the second batch of 10 samples were oriented at 90° (perpendicular to the Y- axis). In order to avoid collisions between the first two batches, the second one was translated along the z-axis with 19 mm from the building platform. The third batch orientation was similar to the first one (oriented at 0° along the Y-axis) , but having a 36 mm translation along the Z-axis in this case, while the fourth batch orientation was similar to the second one (at 90° - perpendicular to the Y- axis) but having a 51 mm translation along the Z-axis in this case. The last batch of 10 samples was split in two categories: the first five samples being oriented at -45° with respects to the Z-axis, the other five ones being oriented at +45° with respects to the Z-axis. The last batch was translated with a value of 70 mm along the Z-axis.

The mechanical properties of the parts produced by Selective Laser Sintering depend not only on the part orientation, but also on the process parameters (laser power [W]; scanning speed [mm/s] and layer thickness [mm]). In the case of the made research, only the laser power [W] was modified during the manufacturing process of the samples. The first, second and fifth batch of 10 samples were manufactured by using P=4,5 [W] laser power, the third and the fourth set of samples being realized with a laser power of P=4 [W]. The layer thickness of 0.1 [mm] and the scanning speed S=1257.3 [mm/s] were maintained constant in all cases. All samples were scaled before being manufactured by SLS with the following factors: 1.03510 (along X-axis); 1.03310 (along Y-axis) and 1.01800 (along Z-axis).

3 Samples manufactured by SLS and compressive tests

The selective laser sintering (SLS) process requires three stages to be strictly followed during the manufacturing process, while temperature values are varied in the building chamber: warm up stage, when the temperature gradually increases from 60°C to the

DOI: 10.1051/matecconf/201711203009

melting point of the material, which in this case was setup to 170°C; the build stage, when the temperature is maintained fixed at 170 °C during the entire manufacturing process and the cool down stage, when the temperature gradually decreases from 170°C to 95°C in order to avoid deformations and shrinkage of the manufactured parts.

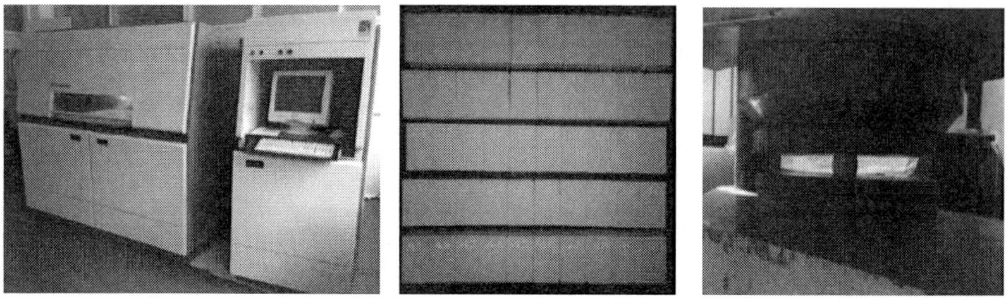

Fig.1. DTM Sinterstation 2000 equipment, samples and compressive tests performed using the ZD 40 Tension-Compressive Testing Machine from TUC-N.

In order to determine which is the correlation between the laser power and part orientation with the mechanical properties of the samples manufactured from PA2200 powder using the DTM Sinterstation 2000 equipment from TUC-N presented in Fig. 1, and under compressive loads of the biocompatible PA 2200 samples, compressive mechanical test were performed using the ZD 40 Tension-Compressive Testing Machine from TUC-N which is also presented in the last image of Fig.1.

The compressive strain was calculated for each sample tested with the following equation:

$$\varepsilon = dl/l_0 \tag{1}$$

where: dl-change of length [mm], l_0- initial dimension (25.4 mm) and ε -engineering strain.

The curves of the PA2200 samples determined by compressive tests are shown for each batch in Fig. 2. The compressive stress was calculated with the following equation:

$$\sigma = F_n/A \tag{2}$$

where: σ- compressive stress [N/mm^2], F_n-normal compressive force [N] and A is the cross-sectional area of the samples (12.7x12.7=161.29 mm^2)

As shown in Figure 2, the curves show a state of deformation in the range 0 - 0.1 mm in which the stress is zero corresponds to the self-positioning of the samples in the grips of the testing machine. As one may notice in Fig. 2, the results obtained in the case of samples 6-10, 16-20, 26-30 and 36-40 were not presented, because the obtained values were very similar to the ones shown in the diagrams and the intention was to reduce the volume of the output data. This is the main reason why, in the case of each batch, only the most 5 representative values were displayed in the diagrams.

As regarding the obtained data, according to Fig. 2, the highest values of the mechanical strength (before failure) were obtained in the case of the first batch of samples, which was manufactured with a P=4.5W laser power and with an orientation of 0° along the Y-axis (Fig. 2, a). The highest value of the compressive stress has been determined in the case of Sample no 2 (Rm= 128.87 N/mm^2). The lowest value of the mechanical strength was obtained in the case of the fourth batch of samples which were manufactured with a P=4 W laser power and with an orientation of 90° along the y-axis. The value obtained in the case of sample no. 34 was approximately six times lower as compared to the most resistant sample (Rm= 21.78 N/mm^2). Values in the range 125-160 [N/mm^2] according to the age of the patient are reference values for the compressive strength of the human bones [17]. It can be concluded that for manufacturing parts from PA2200 powder by selective laser sintering technology it is more suitable to use a laser power P=4.5 [W] combined with a build orientation of 0° along Y-axis in the case of parts made by SLS, so as a higher mechanical strength will be obtained at the end.

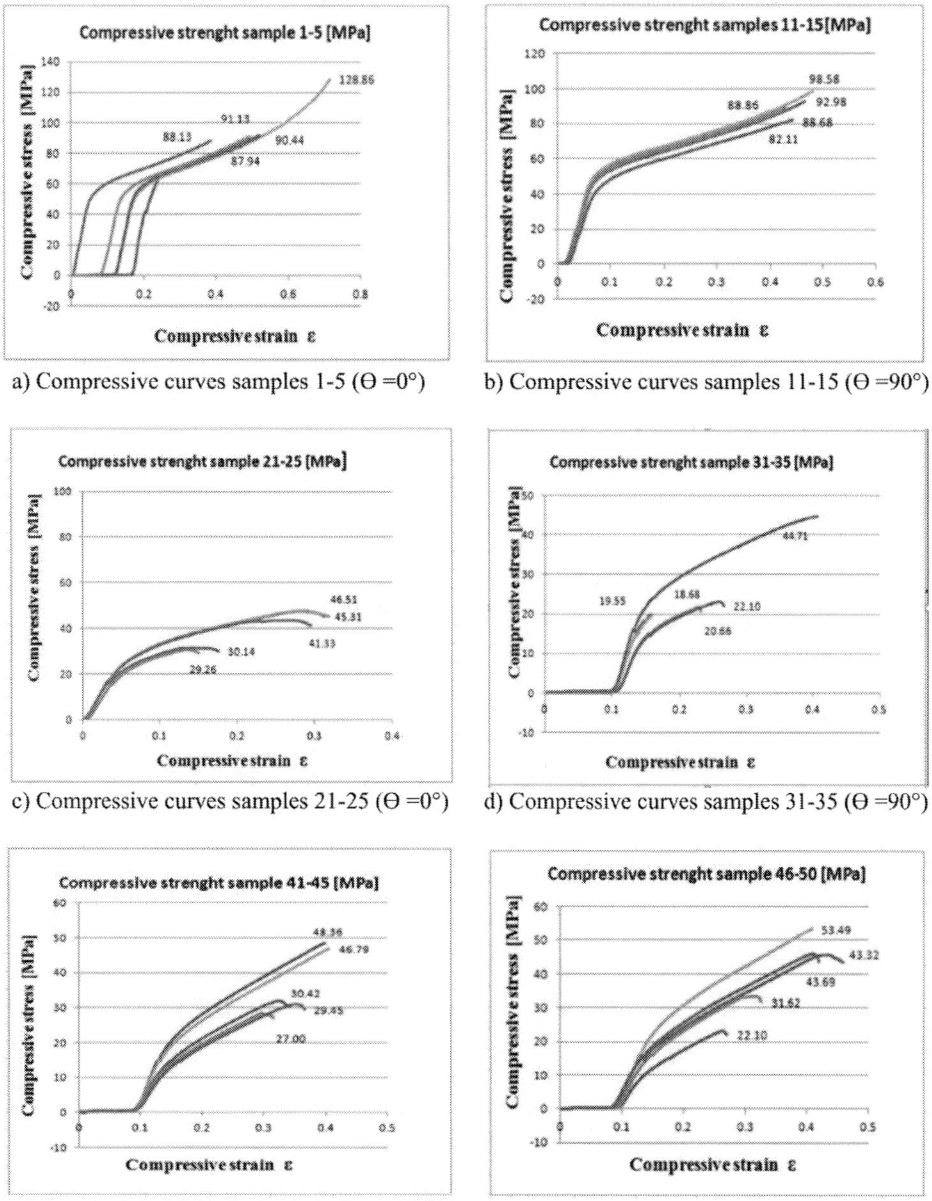

a) Compressive curves samples 1-5 (Θ =0°) b) Compressive curves samples 11-15 (Θ =90°)

c) Compressive curves samples 21-25 (Θ =0°) d) Compressive curves samples 31-35 (Θ =90°)

e) Compressive curves samples 41-45 (Θ =-45°) f) Compressive curves samples 46-50 (Θ =+45°)

Fig.2. Stress-strain compressive curves of PA 2200 samples made by SLS.

The Young's modulus determined by the compressive tests for each batch of PA2200 samples manufactured by SLS, are shown in Fig. 3. These values were calculated from the tangent at the beginning of the curvilinear region of the stress-strain curve using the Test

DOI: 10.1051/matecconf/201711203009

and Motion software of the ZD 40 Tension-Compressive Testing Machine. As noticeable in Fig. 3, the lowest value of Young's modulus was obtained in the case of the last batch of five samples manufactured from PA2200 by Selective Laser Sintering (SLS) technology (manufactured with an orientation +45° with respects to the Z-axis), while the highest values were obtained in the case of the first batch manufactured by SLS with an orientation of 0° along the Y-axis. It can be concluded that the first batch presents a more pronounced elastic behavior than plastic behavior during the axial loading. The lowest value (32.42 GPa – see Fig.3, e) obtained in the case when the sample was manufactured at +45° with respects to the Z-axis was the closest one to the value of Young's modulus reported in the case of the human bone (18-21 GPa) [18].

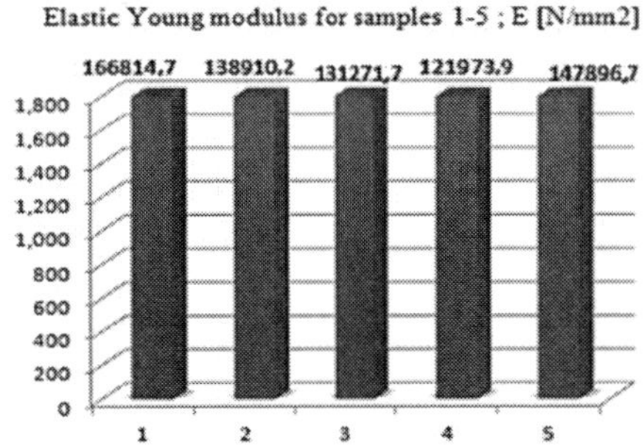

a) Young's Modulus for samples 1-5

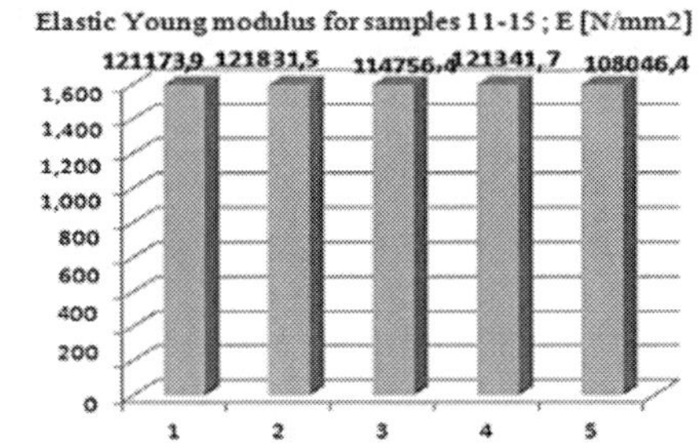

b) Young's Modulus for samples 11-15

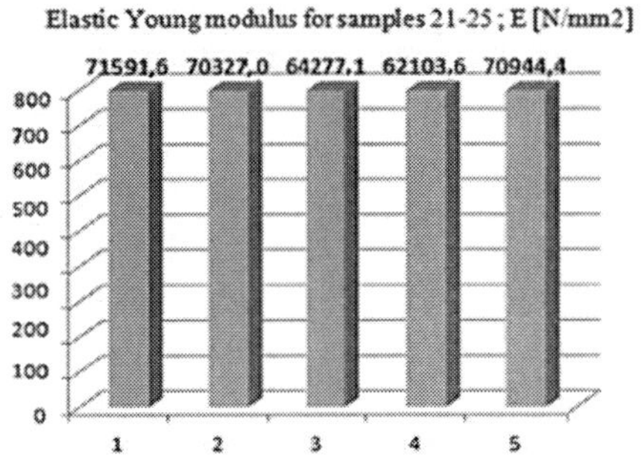

c) Young's Modulus for samples 21-25

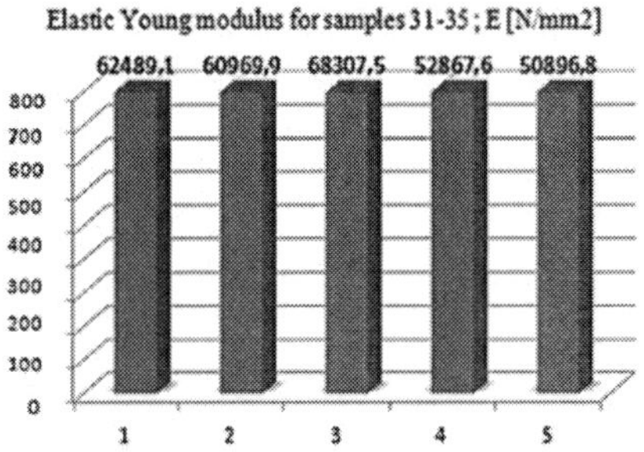

d) Young's Modulus for samples 31-35

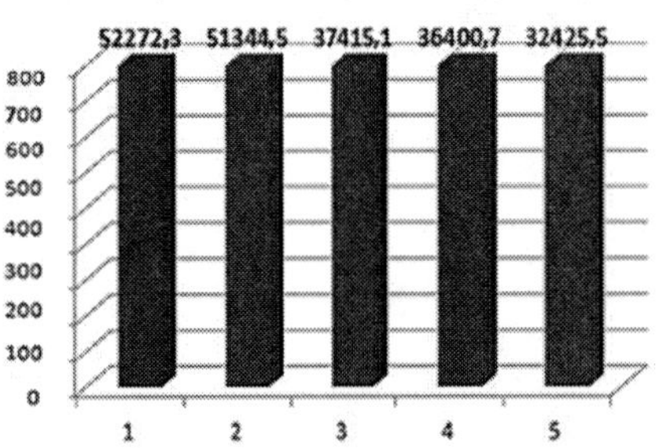

e) Young's Modulus for samples 41-45

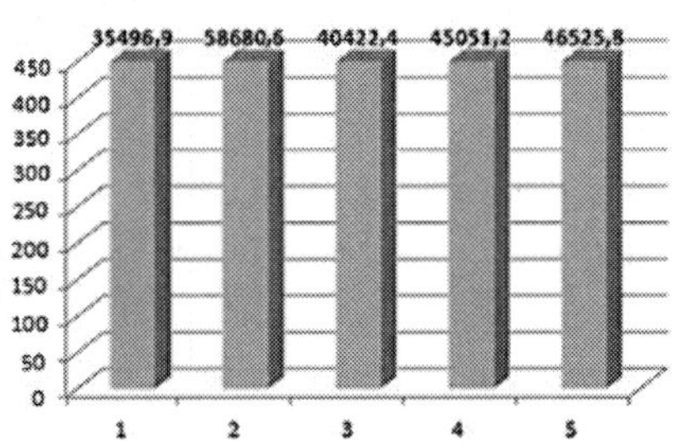

f) Young's Modulus for samples 46-50

Fig.3. Young's modulus of the PA 2200 samples determined by compressive tests.

4 Conclusions

A total batch of 50 compressive test samples was manufactured from PA2200 biocompatible material by selective laser sintering (SLS) technology using a DTM Sinterstation 2000 SLS equipment from the Technical University of Cluj-Napoca (TUC-N), with the aim of performing compressive tests according to ISO 604 standard using a Universal Testing Machine which is available also at TUC-N. The manufacturing process of the samples was realized with different laser power values (4-4.5 W) and different build orientations in the working area of the machine. The compressive tests that were realized further on emphasized that a laser power having the 4.5 W value and an orientation of +45° along the Y-axis is recommended to be used in the case of parts (customized medical implants) made from PA 2200 powder by SLS with a Young's modulus value as closed as possible to the one of the human bone. If a higher value of the compressive strength is required to be achieved in the case of a customized implant to be manufactured from PA2200 powder by SLS, a laser power of 4.5W and an orientation of 0° along the Y-axis is recommended to be used due to the fact that in this case the compressive stress determined (Rm= 128.87 N/mm^2) will be closest to the one of the human bones (125-160 N/mm^2). The obtained results were further on used in the case of a medical implant (Acetabular liner) that was manufactured at the Technical University of Cluj-Napoca from PA 2200 material using the DTM Sinterstation 2000 SLS equipment. This customized medical implant was manufactured with a laser power of 4.5W and an orientation of 0° along the Y-axis.

References

1. A.L. Jardini, M.A. Larosa, A. Kaasi, P. Kharmandayan, Reference Module in Materials Science and Materials Engineering (2017)
2. I. Gibson,A. Srinath, Procedia Technology, **20** (2015)
3. J.P. Singh, Pulak M. Pandey, A.K. Verma, Rapid Prototyping Journal, **22**, 4 (2016)
4. S. Berretta, O. Ghita, K.E. Evans, European Polymer Journal, **59** (2014)
5. A. Cerardi, M. Caneri, R. Meneghello, G. Concheri, M. Ricotta, Materials & Design, **46** (2013)
6. K.H. Hussein, K.M. Park, K.S. Kang, H.M. Woo, Materials Science and Engineering: C, **67** (2016)
7. M. Salmi, K.S. Paloheimo, J. Tuomi, J. Wolff, A. Mäkitie, Journal of Cranio-Maxillofacial Surgery, **41**, 7 (2013)
8. S. Lohfeld, V. Barron, P.E. McHugh, Ann. Biomed. Eng. **33** (10) (2005)
9. S. Cahill, S. Lohfeld, PE. McHugh, Journal of Materials Science Materials in Medicine **20** (6) (2009)
10. D. Drummer, K. Wudy, F. Kühnlein, M. Drexler, Physics Procedia, **39** (2012)
11. P. Delfs, M. Töws, H.J. Schmid, Additive Manufacturing, **12**, B (2016)
12. C.S. Miron Borzan, M.C. Dudescu, V. Ceclan, A. Trif, M. Ridzon, P. Berce, Materiale Plastice, **53**, 1 (2016)
13. K. Senthilkumaran, Pulak M. Pandey, P.V.M. Rao, Materials & Design, **30**, 8 (2009)
14. A. Gupta, P.S. Kumar, Int. Journal of Tech. Research and Applications, **2**, 2 (2014)
15. C.S. Borzan, P. Berce, H. Chezan, E. Sabău, S.A. Radu, M. Ridzon, Academic Journal of Manufacturing Engineering, **11**, 4 (2013)
16. J.P. Singh, P.M. Pandey, Procedia Engineering, **59** (2013)
17. R. Havaldar, S. C. Pilli, B.B. Putti, Advanced Biomedical Research (2014)
18. J. Rho, R. Ashman, C. Turner, Journal of Biomechanics, **26**, 2 (1993)

MATEC Web of Conferences **112**, 03010 (2017) DOI: 10.1051/matecconf/201711203010
IManE&E 2017

The Process of weld brazing CMT (Cold Metal Transfer pulse) for assembly of the different thin sheets

Gheorghe Sima*, *Elena* Muncut, *Adriana* Motica and *Aurelia* Tanasoiu

Aurel Vlaicu University of Arad, Department of Automation, Industrial Engineering, Textiles and Transport, Street Elena Drăgoi no. 2, RO 310068

Abstract. This paper describes the experimental program realized for the research of the behavior at the assembly through weld-brazing CMTP (Cold Metal Transfer pulse) of thin sheets of copper with aluminum. Four variants were chosen and studied, settling the best conditions for the assembly. They were made welded samples which were analyzed in volume terms of metallurgical and mechanical structure. It was also analyzed the influence of cooling velocity on the structure.

1 Introduction

To solve corrosion problems and simultaneously achieve light weight for metallic constructions, a combination of dissimilar alloyed galvanized sheet steel and aluminium is commonly used. The protective layer must satisfy the following conditions: be continuously adhering to the metal support, have uniform thickness and sufficient mechanical strength so as not to be destroyed during operation [1].

In the case of weld-brazing aluminium steel sheet, only aluminium melts in the joint. Thus, the filler material is chosen to be similar to the base material which is melted, considering the following characteristics: melting point, hardness, durability, electrical conductivity, and chemical composition, the flow of filler material and granulation of the filler material compared to the base material which does not melt. Such joints occur in electronics industry, in vehicles construction, and in supplies field [2].

Getting lighter structures using materials that until now could not be welded and therefore achieving lower consumption in operation (steel galvanized + steel galvanized, steel galvanized + Cu).

2 Experimental program weld-brazing CMTP and CMT of thin sheets of galvanized steel with aluminium

2.1 Base materials

While conducting the experimental program following materials were used [3]:

* Corresponding author: gheorghe.sima@uav.ro

- Galvanized steel sheet DX51D + Z150-NAC (SR EN 10327: 2004) 1mm thick.
- Aluminium alloy sheet EN AW 1200 (SR EN 1706: 2000) thick 1 mm.

The main alloying elements that are found in aluminium alloys: copper, silicon, manganese, magnesium, lithium and zinc [4]. Elements such as nickel, chromium, titanium, scandium and zirconium can be added in very small amounts to achieve certain properties. Other elements can be found as impurities and these elements still have a significant contribution in changing or thermo-mechanical properties.

Filler alloy electrode was used, with wire diameter 1.2mm AlSi5 using cc+ polarity, the wire electrode EL-AlSi5 DIN 1732.

Pure argon was used as a protective gas (Ar 100%), since aluminium alloys are suitable for the inert gas welding [5].

Using these base materials and filler materials with aluminium chemically predominant, Al and Zn will have the greatest influence in joints.

2.2 Determining the type of dissimilar joints and weld-brazing variants

The experimental program consisted of dissimilar joints making corner "overlap" because of the many applications that can be found in the field of machine manufacturing (auto, rail, etc.) [6].

The experimental program aimed to obtain a structural and mechanical characterization of intermetallic layer formed from dissimilar joints made with different variants of weld-brazing conducted on two levels and the choice of specific parameters of weld-brazing technologies for joining galvanized aluminium alloy.

Table 1 presents the parameter values used to achieve heterogeneous galvanized steel joints - aluminium alloy.

Table 1 Values of parameters used to achieve joints.

Weld-brazing variant number	I_s[A]	U_a [V]	v_{as}[m/min]	CMT * CMTP**	v_s[mm/min]	I_{na}
0	70.0	12.9	3.80	CMT	1000	5
1	67.2	11.7	3.70	CMT	800	5
2	88.6	15.2	4.37	CMTP	800	5
3	88.6	15.2	4.35	CMTP	1000	5
4a	62.6	11.3	3.64	CMT	1000	5
4b	62.6	11.3	3.64	CMT	1000	5
5	61.0	11.3	3.47	CMT	800	-5
6	69.3	14.3	4.35	CMTP	800	-5
7	70.1	14.5	4.39	CMTP	1000	-5
8	63.6	11.6	3.69	CMT	1000	-5

* CMT Cold Metal Transfer ** CMTP Cold Pulse Transfer

In the experimental program were used 150x250x1mm table sizes.

To optimize the process of weld-brazing with CMT process of galvanized aluminium sheets were selected the following influencing factors:

- $x_1 \rightarrow$ CMT transfer mode * Cold Metal Transfer **CMTP Cold Steel Pulse Transfer
- $x_2 \rightarrow$Welding speed v_s[mm/min]
- $x_3 \rightarrow$Dynamic correction factor of spring I_{na}.

It has been selected as objective function correlated analysis of the intermetallic layer by a factor of measurable tensile strength of the joint, while maintaining the function of the layer of zinc anticorrosion.

The experimental program includes sampling mode and specimens needed to assess the structural and mechanical properties of heterogeneous connections (galvanized steel, aluminium alloy).

Figure 1 represents the procedure of the experiment on two levels with changing parameters determined in preliminary experiments.

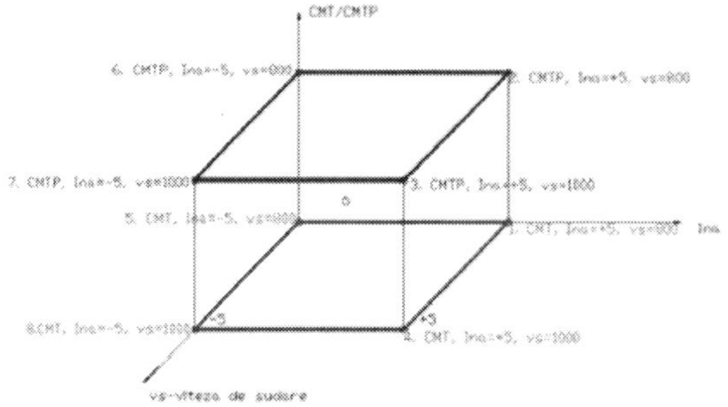

Fig. 1. Implementation plan on two levels factorial experiment. [2]

Positioning on samples plates weld-brazing micro-macro structural examination and tensile specimens (TRAC1, TRAC 2), bending (IND1 and IND2) for mechanical testing is performed according to standardized methodology.

2.3 Macro and microscopic examination of the heterogeneous joint aluminium-zinc coated steel

Macroscopic examination the cross sections of heterogeneous connections analysis showed no welding defects such as cracks, but 0 and 1 variants showed fine pores with a maximum diameter of 0.3 mm in the welded region (Fig. 2 and Table 2).

Fig. 2. Variant 0 of weld-brazing, made in the central point [Attack Nital 10%]

Table 2 Macrostructures samples of the joints.

Sample Test	Sample macro CMT	Sample Test	Sample macro CMTP
Version 1 weld-brazing [Attack Nital 10%]	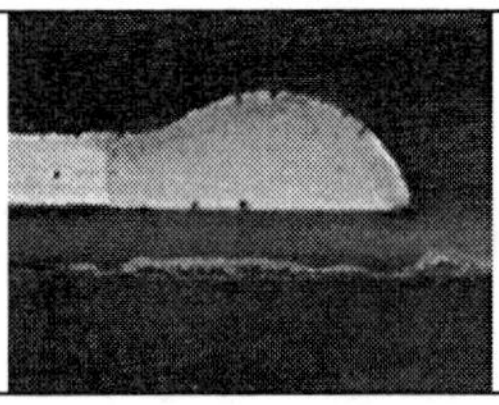	Version 2 weld-brazing [Attack Nital 10%]	

Version 6 weld-brazing [Attack Nital 10%]		Version 3 weld-brazing [Attack Nital 10%]	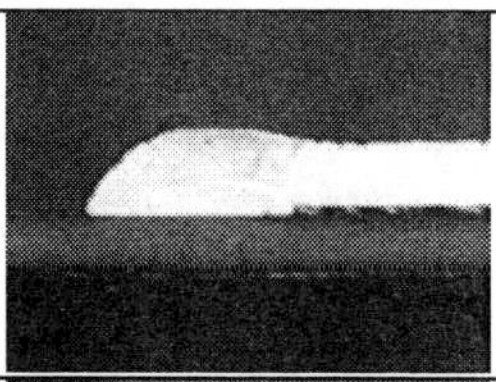
Version 8 weld-brazing [Attack Nital 10%]		Version 4 weld-brazing [Attack Nital 10%]	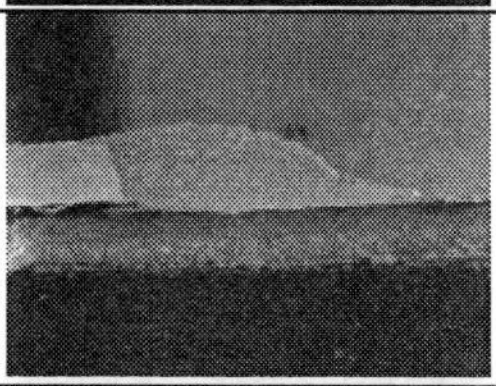
Version 8b weld-brazing [Attack Nital 10%]		Version 4b weld-brazing [Attack Nital 10%]	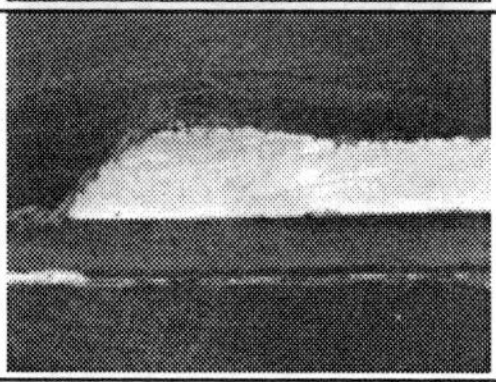

Over small beginning and final lengths of the weld-brazing process arc instability occurs resulting in uneven weld geometry, consequently it was considered necessary to remove the zone length of 50 mm using mechanical processes without excessive heating.

Figure 2 and table 2 present images of samples made in the 8 variants weld-brazing heterogeneous joints aluminium-zinc coated steel.

Analysing the visual connections made, it appears that the weld-brazing version 5 and weld-brazing version 8 heterogeneous presents defects in solder joints (sharp discontinuities) which are considered inappropriate, other variants showed no major defects, so are considered appropriate.

Microscopic examinations were performed in specific weld-brazing areas of heterogeneous connections (SUD, MB, ZIT) according to EN 1321: 2006 distinguishing the following structures (Fig. 3 and Table 3).

Fig. 3. Weld-brazing variant 0 [Attack Nital 2% + 5% NaOH], 100x.

Table 3 Microstructure samples of the joints.

Sample test	Macro sample CMT	Sample test	Macro sample CMTP
Variant 1 weld-brazing [Attack Nital 2% + 5% NaOH], 100x	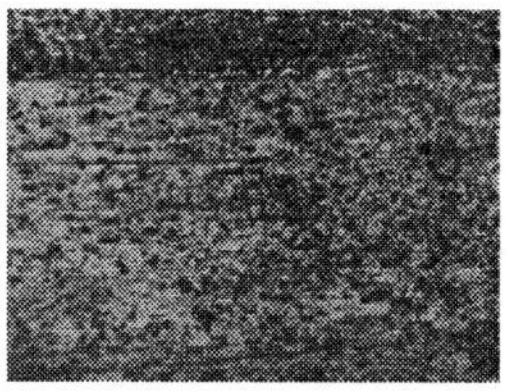	Variant 2 weld-brazing [Attack Nital 2% + 5% NaOH], 100x	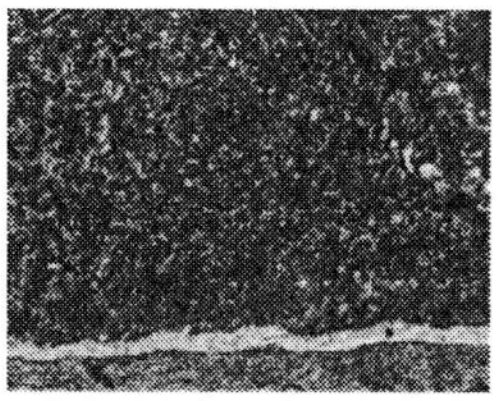

Variant 6 weld-brazing [Attack Nital 2% + 5% NaOH], 100x	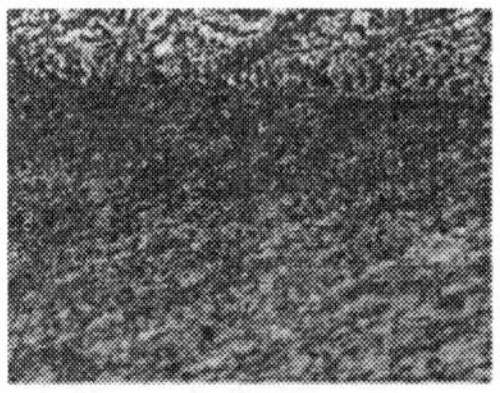	Variant 3 weld-brazing [Attack Nital 2% + 5% NaOH], 100x	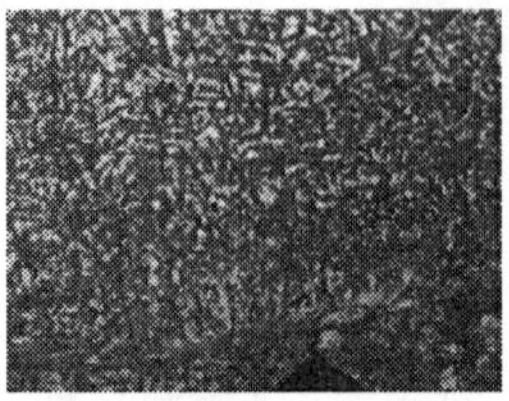
Variant 8a weld-brazing [Attack Nital 2% + 5% NaOH], 100x	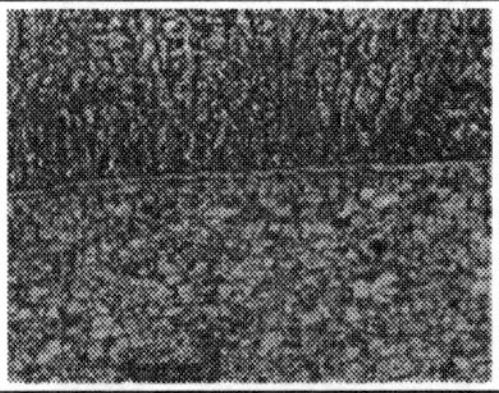	Variant 4a weld-brazing [Attack Nital 2% + 5% NaOH], 100x	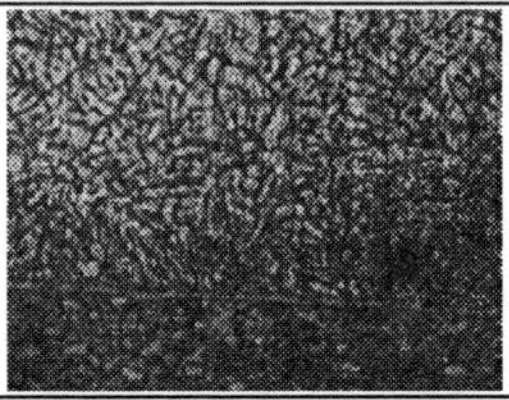
Variant 8b weld-brazing [Attack Nital 2% + 5% NaOH], 100x	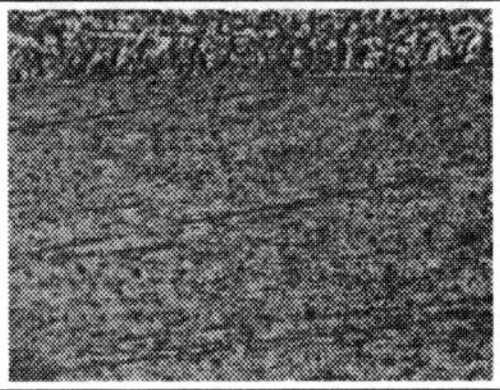	Variant 4b weld-brazing [Attack Nital 2% + 5% NaOH], 100x	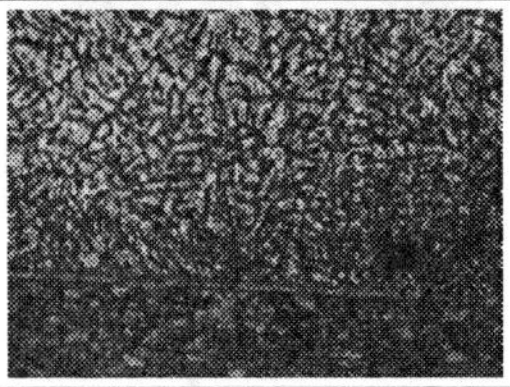

Conclusions result in the comparative study of the two levels: Weld-brazing the joint is achieved with the flattest CMTP compared with CMT. The same CMT process by changing only the dynamic correction factor I_{na} at +5 samples 4a and 4b the cord is flatter, to -5 8a and 8b the cord more convex, in circumstances where the cord thickness does not change significantly.

As a result of this stage is noted that the CMT process gives flatter cords than CMTP and correction factor and it makes it less tall.

ZIT (heat affected zone) Areas examined microscopically showed no welding defects such as micro cracks and heat affected zones were clearly defined in the base material.

Primary dendrite is perpendicular oriented on interface at the CMT process, this show that the process structure cools faster than the CMTP.

At the process CMTP there is a greater flow of zinc in the aluminium, which explains the start fracture in this area.

Joint Some pores appear, but not outstanding defects. The structure is detrital, homogeneous, and harder than base material - aluminium. The cord ruptures are caused by initial cracks, due to a defect in the zinc layer. This is a common phenomenon at welded structure and galvanized later.

MB generally the basic material does not suffer structural diffusion; phenomenon encountered is zinc-silicon diffusion especially in aluminium.

3 Conclusions

Analysing the thickness of intermetallic is observed that they vary between 10-26 μm, it appears that the CMT process intermetallic layer thickness is thinner than at the process CMTP (Sample 1 CMT, sample 2 CMTP).

As important conclusion of this work, thickness intermetallic layer must be maintained around, because the spacing of the aluminium atom allows migration of atoms of zinc. As a future conclusion it will be possible weld-brazing of those dissimilar materials on which atoms of one material can migrate on the network of the other material.

It is observed from the thickness study and from analysis of the breaking strength at the transfer CMT the intermetallic layer is thinner and breaking strength [4] is increased. As a result when using materials with brittle intermetallic layers give recommended procedure weld-brazing CMT. If we weld-brazing similar materials that do not give CMTP preferred intermetallic layer which introduces a superior linear energy and make mechanical connections formed to be stronger and joints have required mechanical characteristics.

In the base metal (carbon steel) fine ferrite-pearlite structure with oxide inclusions, perlite is placed it in general within grains. In the base metal (aluminium alloy) structure composed of α-rich solid solution in aluminium, results Al-Si intermetallic compounds and oxides of aluminium.

In the interface between the galvanized steel and weld are noted intermetallic layers of different thicknesses that separate the two areas.

The welds (SUD) made by the process of CMT or CMTP dendritic structure grows with specific structures zinc-aluminium alloys, α-solid solution and intermetallic compounds of silicon.

Acknowledgements: Research Contract ISIM Timisoara, PN 09-160103, 01.03.2009 *Electric arc weld brazing in shielded gases*, coordinator Dragut, L

References

1. S. Kroger, R. Killing, *Software for creating and managing parameters for MIG/MAG welding.* German Welding Assoc. Yearbook, DVS-Verlag, 150-161 (2004)
2. R. A. Rosu, D. R. Pascu, E. Muncut, C. Toma, *ModTech 2011 Intern. Conf. Proceedings*, Republic of Moldova, 929 (2011)
3. G. Trommer, Sudura Review, **20,** 1 (2010)
4. E. Muncut, R. Lile, Gh. Sima and D. Glavan, Pensee Journal, **76**, 5 (2014)
5. G. Mathers, *The welding of aluminium and its alloys,* Ch.3 (Woodhead Publishing Limited, Abington Hall, 2002)
6. ***. *CMT: Cold Metal Transfer MIG/MAG dip-transfer process for automated applications.* (Fronius International, 2005)
7. S. Kodoma, Y. Ishida, K. Asai, M. Mizumoto, T. Namekata, H. Nagasaki, Welding in the World, **54**, 328-336 (2010)

Polyjet technology applications for rapid tooling

Razvan Udroiu*, and *Ion Cristian* Braga

Transilvania University of Brasov, Department of Manufacturing Engineering, Mihai Viteazu No.5, Brasov, Romania

Abstract. Polymer Jetting (PolyJet) has proved to be one of the most accurate additive manufacturing technologies, in order to manufacture rapid tools. Rapid Tooling (RT) is different from conventional tooling as follow: manufacturing time is shorter, the cost is much less, but the tool life is shorter and tolerances are wider. The purpose of this paper is to make a comparative study between the soft tools (silicon moulds) and hard tools (acrylic thermoplastic moulds) based on the Polymer Jetting technology. Thus, two types of moulds have been made in order to manufacture a test part. Reaction injection moulding (RIM) and casting techniques were used to fill these moulds with resins that simulate the plastic injection materials. Rapid tooling applications, such as indirect tooling and direct tooling, based on PolyJet technology were experimentally investigated.

1 Introduction

Rapid X [1] is a generic term based on additive manufacturing, which includes: Rapid Prototyping (RP), Rapid Tooling (RT), Rapid Manufacturing (RM) and so on. Rapid prototyping allows making, in an additive way, of physical objects, directly from computer. Rapid Tooling, based on RP techniques, allows to build tools and moulds in a rapid way. RT technologies are classified in soft tooling (silicon rubber, epoxy resins, low-melting-point alloys) and hard tooling (hard plastic, metal alloys). In indirect tooling a mould, die or tool is manufactured using a master model made by RP. In direct tooling, the tool, mould or die is created directly on the additive manufacturing system.

Different approaches for RT applications [1-5] are found in the literature. Chua presents [3] a comparison of rapid tooling technologies based on tool life, tool development time and cost. In [4] applications of vacuum forming tooling using Fused Deposition Modelling (FDM) technology were presented. Singh [5] investigated and compared PolyJet printing and silicon moulding using Alchemix VC3340 material. The companies that provide additive manufacturing systems also develop RT application based on their AM techniques. Thus, Stratasys Ltd proposes some solutions for manufacturing tools [6] based on fused deposition modelling (FDM) and polymer jetting (PolyJet) technology.

Indirect rapid tooling applications based on PolyJet technology are: silicon moulding, investment casting, sand casting and vacuum forming. All of these required a master model made by 3D printing. Direct rapid tooling with PolyJet technology requires 3D printed master mould or tool, and comprise the following applications: blow moulding, jigs and fixtures,

* Corresponding author: udroiu.r@unitbv.ro

reaction injection moulding (RIM), thermoplastic injection moulding, thermoforming, composite tooling, liquid silicon casting and investment casting mould. All of these applications focused on manufacturing tooling, are under development and needs improvements based on experimental researches.

The purpose of this paper is to explain how are made the tools or moulds based on the Polymer Jetting (PolyJet) technology. Quite a few researches have studied the problem of manufacturing of tools and mould by direct and indirect methods using PolyJet technology. For a small series production, 3D printing offers an effective cost compared with milled mould by computer numerical control (CNC).

In the first part of the paper are investigated the rapid tooling techniques based on a master model made by PolyJet 3D printing. The second part presents direct rapid tooling techniques focused on PolyJet 3D printing of moulds. These rapid tooling techniques have applications for resins casting, RIM, wax pressing in a mould and vacuum casting.

2 PolyJet 3D printing technology and materials

The first Polyjet 3D printing system was launched by Objet Geometries Ltd. in 2000. Stratasys has acquired Objet Geometries Ltd. in 2012. The Polyjet 3D printing technique [6] consists in building of parts, layer by layer, combining inkjet technology with photo-polymerization process. The main steps are the following: pre-processing, processing and post-processing. In the pre-processing step, the part orientation on the build tray is optimized using Objet Studio software. In the next step, droplets of resin are deposited, in 16 micron layers, onto a built tray via an inkjet print head (Fig. 1), each layer being cured by UV light. The building tray moves down and the process continues, layer by layer, until the part is complete. Two different photopolymer materials are used, model and support. In the last step, using a water jet recycling station the support material is removed. The accuracy of parts obtained by PolyJet technique is 0,1mm [6].

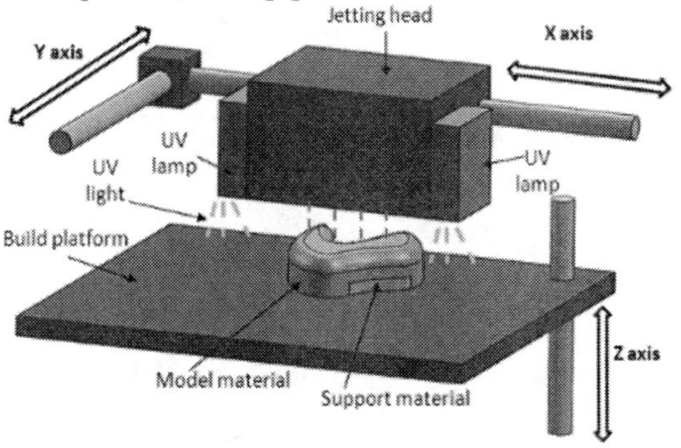

Fig. 1. Schematic representation of PolyJet printing process.

The test parts, including the master model and the master mould, were made by Polyjet 3D printing technology on the Objet Eden 350 machine, from FullCure™720 material by Objet Geometries (Statasys). FullCure materials family is an acrylic-based photopolymer.

3 Resins that simulate the injection plastic materials

The resins from Sika, having similar properties with the injection plastic materials are presented in Fig.2 and Fig. 3. Some of these resins are experimentally investigated in this paper.

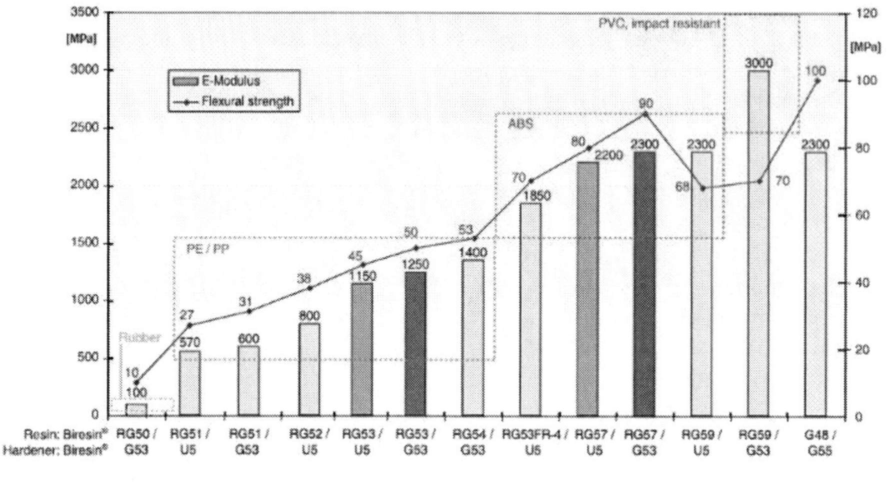

Fig. 2. RIM resins properties from Sika. [7]

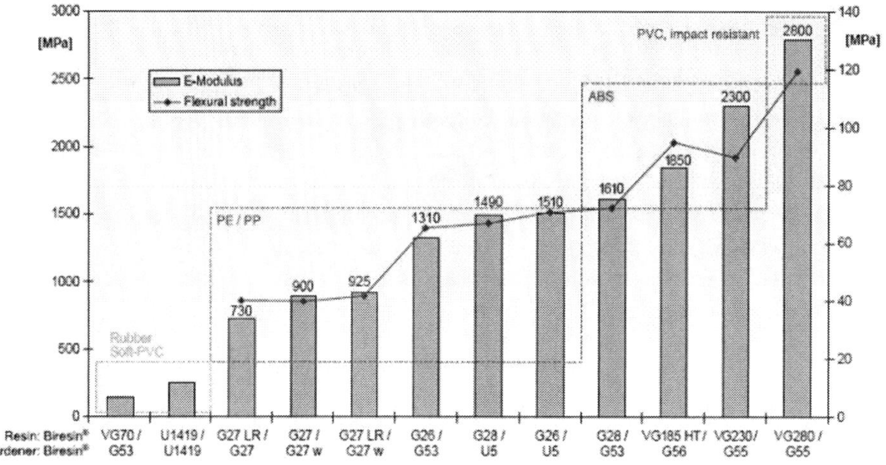

Fig. 3. Cast resin properties from Sika. [7]

The mechanical characteristics of the RIM resins (Fig. 2) and casting resins (Fig. 3), indicated on the horizontal axis, are presented on the left vertical axis (E-modulus), the right axis (flexural strength) and the doted rectangles show the equivalent injection plastic materials used in serial production.

4 Indirect rapid tooling based on PolyJet technology

In the first case study, the development of silicon mould is investigated. The test parts (Fig. 4b) are manufactured by casting or reaction injection moulding (RIM) from a thermoset resin, into the silicon mould (Fig. 4a).

4.1 Silicon moulding for casting

Silicon rubber moulding is an indirect rapid soft tooling technique that consists in the following main steps: master model manufacturing, mould making, part casting and part finishing. A master model built by rapid prototyping reduces significantly the manufacturing time and allows making complex shape. The master pattern needs to be prepared for the next step, by marking the parting lines and shut-offs. The silicon mould is built following the steps:

- Make a rigid box from plastic (or wood);
- Suspend the master pattern into the mould box and add gates and vents;
- Prepare the model master by marking the parting lines and shut-offs and then apply a mould release on its surface;
- Prepare the silicon by mixing the two components of the silicon rubber kit and vacuum degassing the liquid silicon;
- Pour liquid silicone rubber over the master model until it fills the mould box and repeat the vacuum degassing;
- Cure the silicon mould;
- Remove the mould box, cut with a knife the silicon mould in order to define the parting surface and finally open the mould and extract the master model.

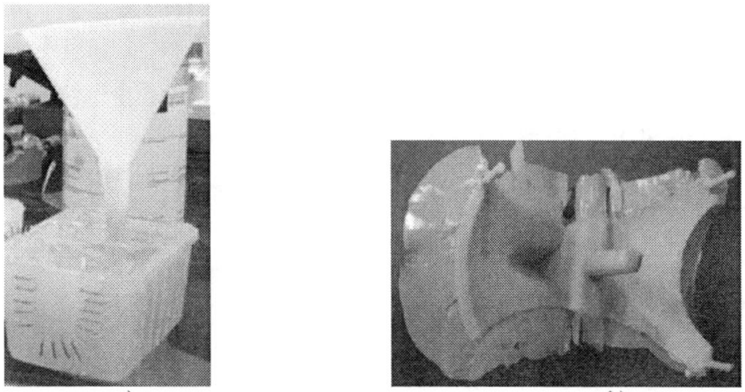

a) b)

Fig. 4. a) Silicon mould, b) Polyurethane cast part.

Two components (A + B) addition silicon rubber from Alpina Technische Produkte GmbH, having a very low linear shrinkage (around 0.1%) was used, obtaining a transparent mould. The mould is prepared for fill, by applying a release agent on its internal surface, and then closes the mould and binds it to prevent separation. Two fill variants of the silicon mould were experimented: cast by gravity (Fig. 4a) and injection using a syringe. The resin is inserted into the mould until it begins to exit from the vents. A polyurethane resin G26 / G53, similar with high density polyethylene, was casted into the mould. The resin potlife is around 7-10 minutes and demoulding time 20-30 minutes. The final step consists in extracting the part from the mould and make post processing operations (remove the gates, vents and parting surface).

4.2 Low pressure Reaction injection moulding (RIM) with silicon moulds

Reaction injection molding (RIM) is a technique that basically consists in mixing two thermosetting polymers (resin A-polyol and hardener B-isocyanate) with a curing reaction followed by the rapid injection of the mixture at low pressure and temperature into the mould. RIM is used in the development and preproduction stages of plastic components, allowing to simulate the thermoplastic materials for serial production. RIM technology applications are focused on the field of technical parts for aerospace, automotive etc.

The advantages of RIM technique compared with other plastic technologies are the following:

 DOI: 10.1051/matecconf/201711203011

- a quick cycle times compared to typical cast materials;
- a high-density skin is formed with a low-density core;
- the bi-component mixture has a much lower viscosity than molten thermoplastic polymers;
- less clamping forces are required, which leads to smaller equipment.

A UNIDOS 100 RIM machine was used to make this experiment. The RIM machine has two separate containers, two pumps for each one and a mixing head. The mixing ratio between resin A and hardener B is adjusted by the pumps volume adjustment. After the resin and hardener were mixed, there is a very short potlife (timeframe) for injection the material and fill-up the mould. In this case study, a polyurethane resin RG53/ G53 from Sika, similar with thermoplastic injection mouldable material like HDPE (high density polyethylene), was low pressure injected into the mould. The potlife is 55 seconds and demoulding time around 15-20 minutes.

5 Direct rapid tooling made by PolyJet technology

Direct rapid tooling is a manufacturing process of tools using additive manufacturing. The additive manufactured tools can be made from different materials, such as plastics, metal alloys or ceramics. In this case study the tools are manufactured from polymeric material by Polyjet 3D printing technology. 3D printing of mould is an important method used in plastic parts manufacturing allowing shortening of the manufacturing time of the mould and decreasing the manufacturing cost, comparing with CNC milling of the mould. The mould can be designed using software tools such as CATIA, SolidWorks, Creo etc. In this case study the mould was designed using SolidWorks by Dassault Systemes.

5.1 Low pressure Reaction injection moulding (RIM) with 3D printed moulds

PolyJet technology having main advantages high quality printing, smooth detailed parts and highly accuracy [5, 6], is suitable technique to manufacture 3D printed moulds and tools.

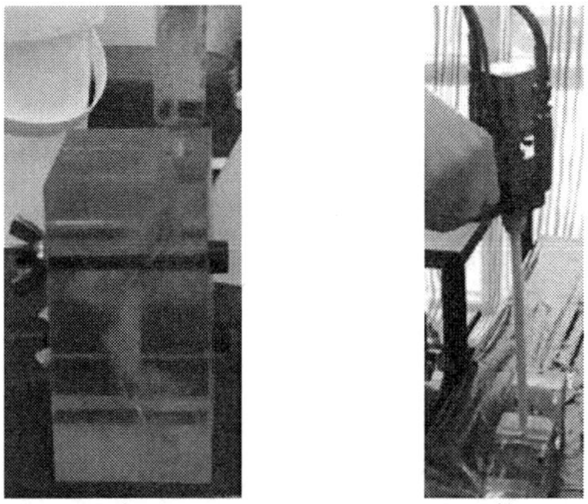

a) b)

Fig. 5. a) Fill the mould with a syringe, b) RIM with PolyJet 3D printed mould.

Both components of the mould (Fig. 5) were 3D printed by PolyJet, having the cavities oriented upwards and using glossy surface finish. Smooth surfaces with the roughness around of 3.8 μm [10] are obtained. In this way the cavities of the mould does not need any post processing operations. The components of the mould (Fig. 5a) are assembled using two screws. An important preparing step, before assembling the components of the mould, is applying of a release agent on all surfaces of the mould cavity. Using the UNIDOS 100 RIM machine, the polyurethane resin RG53/ G53 from Sika, was low pressure injected into the 3D printed mould (Fig. 5b). The potlife and demoulding time were the same like in the silicon mould reaction injection moulding.

5.2 Plastic injection moulding with 3D printed moulds

In this case, PolyJet 3D printed moulds needs to replace machined injection moulds. The moulding temperature of the thermoplastic materials (PE, PP, ABS, etc) is up to 300°C. Thus, PolyJet injection mould tooling (PIMT) needs to be cool between shots, at a target temperature around 50°C. Digital ABS material [6] is recommended to be used for PolyJet printing of the injection plastic mould. Its mechanical and thermal properties will offer longer mould life and reduced cycle times [9]. Depending on material and part complexity, PIMT are suitable for the injection moulding of up to 100 parts.

6 Conclusions

Rapid tooling based on additive manufacturing is an important method utilized in small series production where the conventional tooling like as CNC milled and spark erosion moulds are expensive. RT integrated into CAD/ CAE/ CAM systems allows rapid product development, low cost optimizing the product in a short period of time, reduce time-consuming mould corrections and increase product innovation.

The paper presents solution for prototyping, functional testing and short-run production applications. PolyJet 3D printed moulds are not production tools. Two different techniques to make moulds with PolyJet technology were experimented: silicon moulding (indirect rapid tooling) and direct 3D printing of mould (direct rapid tooling). Reaction injection moulding and casting techniques were used to fill with polyurethane base resins the moulds. The RIM materials have good mechanical properties comparable with thermoplastic mouldable materials but the potlife is very short. Cast resins have longer potlife than RIM materials. Direct rapid mould is more durable than silicon mould. Thus, using RIM technique can be manufacture up to 50-80 parts in the first case, and 300-500 parts in the second one. The Sika resins can simulate injection mouldable materials. The productivity of RIM is smaller than Injection Mould process but the tooling cost is lower.

References

1. R. Udroiu, *3rd WSEAS International Conference on Manufacturing Engineering, Quality And Production Systems*, Romania, 177 (2011)
2. P. Bere, C.Neamtu, Centr. Europ. J. of Eng., **4**, 148 (2014)
3. C.K. Chua, K.H. Hong, S.L. Ho, Int. J. of Adv. Manuf. Techn., **15**, 604 (1999)
4. F. Rayegani, G. C. Onwubolu, A.Nagy and H. Singh, *ASME International Mechanical Engineering Congress and Exposition*, 810 (2014)
5. R. Singh, Int. J. of Autom. and Mech. Eng., **6**, 777 (2012)
6. ***, *Stratasys Ltd*, http://www.stratasys.com (2017)
7. ***, *Biresin® Low pressure RIM-system and casting*, Sika Group (2010)
8. R. Udroiu, A. Nedelcu, *Optimization of Additive Manufacturing Processes Focused on 3D Printing in Rapid Prototyping Technology - Principles* (Intech, 2011)
9. B.O. Thorsten, The AM Magazine, **6**, 8 (2015)
10. R. Udroiu, L.A. Mihail, *Proceedings of the 8th WSEAS international conference on circuits, systems, electronics, control & signal processing*, Spain, 283 (2009)

A welding technology using RobotStudio

Viorel Cohal*

Technical University of Iaşi, Department of TCM, Iaşi, Romania

Abstract. The paper presents the results obtained in designing a welding workstation by using offline programming in RobotStudio. The developed model was used to simulate welding elements of a welded assembly. Welding routes are obtained offline. By using ArcWelding module and VirtualArc there were obtained parameters of welding technology, which are automatically in WPS.

1. Introduction

This paper presents an embodiment of an assembly welding technology welding using a welding robot a robot. In this case, the model used is the IRB 140 robot programmed using software RobotStudio. To illustrate the simplicity of use of robot software and Virtual Studio Arc, programming and simulation of welding to achieve a welded assembly (Fig.1).

In this design phase welding technology was used a restraint system consists of several components that help the assembly welded assembly, welding operations in the order execution, using the robot. The system is constructed so as not to impede penetration of the robot arm with welding torch places for the welding process.

Finally both technological descriptions of the welding process and economic returns will be listed using Virtual Arc program, resulting in production price of welded subassembly according to production capacity and welding sequence of operations.

The final results on the computer and robot motion simulation and in the real environment during the welding process. This simulation is needed to determine the best angles for both the welding process and an economic use of the robot.

Programming the robot.

To program the robot is used IRB140 Robot Studio program, produced by ABB company, this is intended for off-line programming of robots from ABB range and simulation them within the workplace. RobotStudio is useful for programming, training in the use of robots and optimization program offering a number of advantages:

- eliminating the risk that exists when using real robots manual control.
- commissioning controller in a shorter time.
- opportunity to work with different types of effectors.
- ability to test a wider range of ABB robots.
- possibility to create and import your own objects, which can be saved in our database.
- interfacing with Visual Basic and Visual C ++

* Corresponding author: cohal@tcm.tuiasi.ro

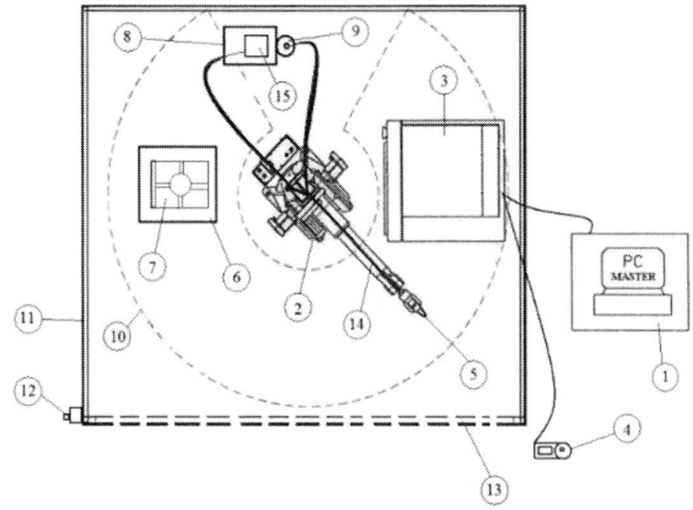

Fig. 1. Workstation 1. P.C. Master - computer connected to the controller; 2. Robot IRB 140; 3. Controller; 4. Console; 5. Gun; 6. Table; 7. Clamping system fixing piece; 8. Source; 9. Inert Gas Cylinder; 10. The robot operating area; 11. Fencing; 12. Control Panel-emergency; 13. Safety zone; 14. Cable.

Disadvantages simulation program:

- any simulator, it will not completely create real environment
- some positions must be redefined real robot,
- workstations component may have only one active robot with a controller

2. Programming stages in Robotstudio

For using RobotStudio in order to design and simulate an welding workstation, some specific requirements had to be fulfilled with the following stages:

1. Creating the workstation
   - ✓ Robot Choice

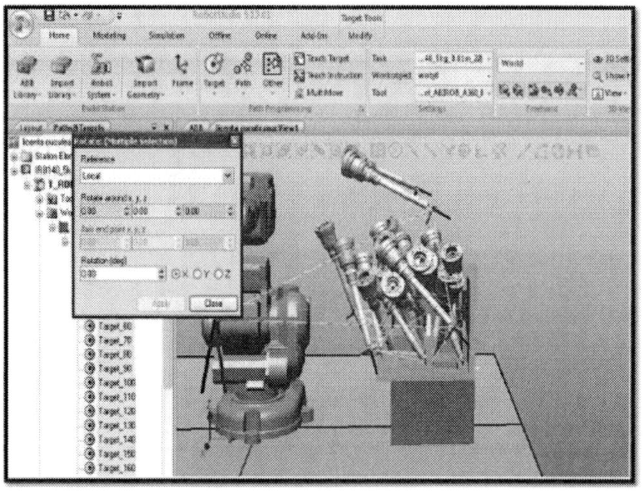

Fig. 2. Target.

DOI: 10.1051/matecconf/201711203012

- ✓ milestones welding program has been projected (ex. Solid Edge, Solid Works, CATIA)
- ✓ Import-device mounting fixture and equipment.

2. Programming the robot
   - ☑ creation targets (Fig. 2)
   - ☑ create paths (Fig.3)
   - ☑ Check the collision, accessibility, self-configuration.
   - ☑ offline simulation (Fig. 4)
   - ☑ Saving console program

3. Creating work program online:

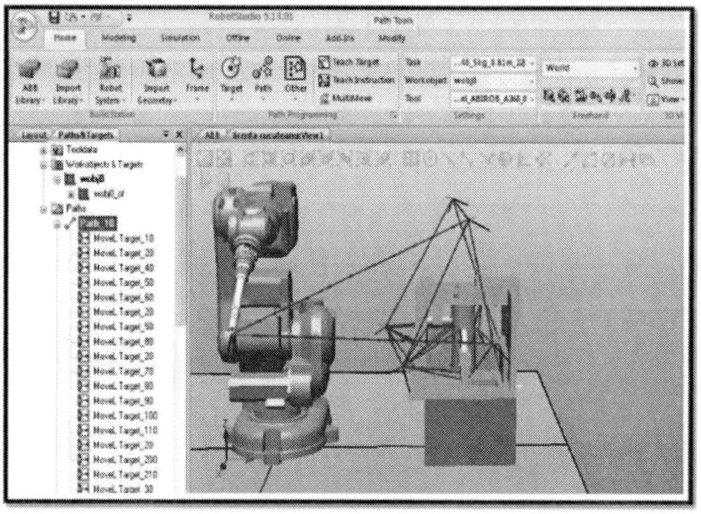

Fig.3. Path.

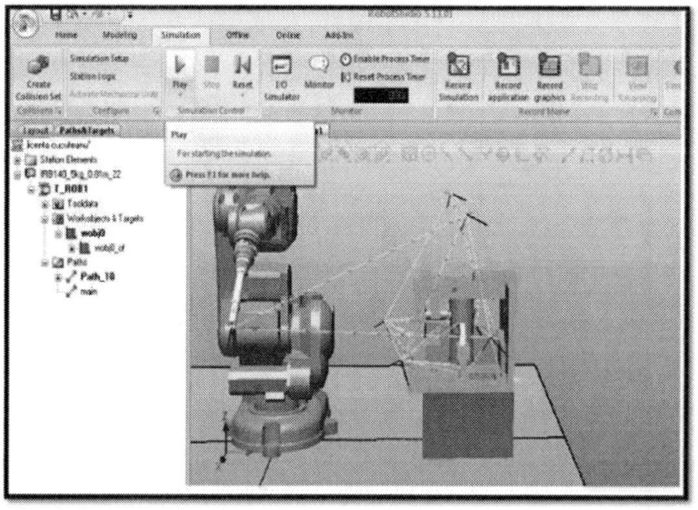

Fig.4. Offline simulation.

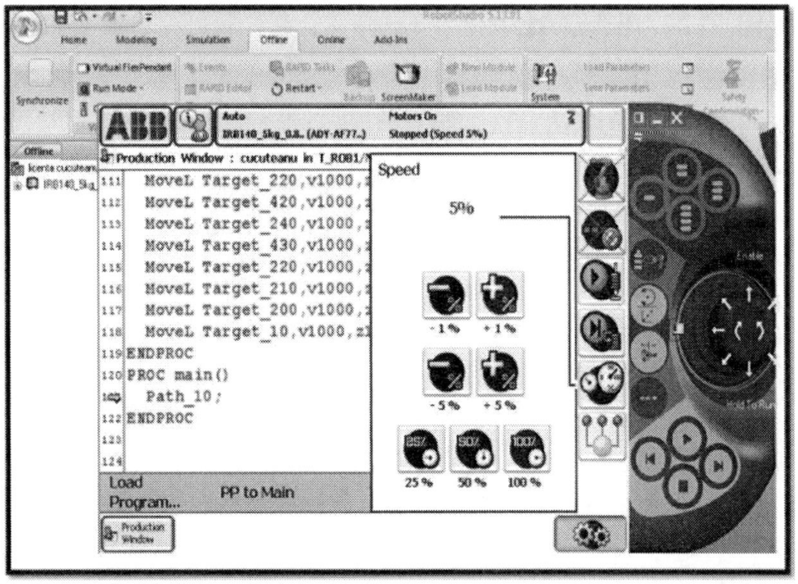

Fig.5. Speed control with flexpendant.

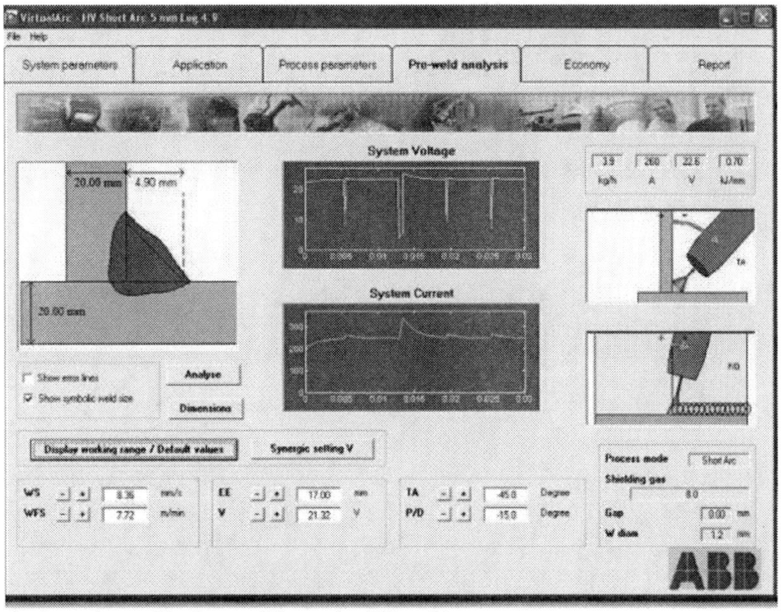

Fig.6. Analysis performed before welding using VirtualArc.

Creating the program is done using the console to manipulate the robot which is present in the virtual option program.

Creating software to be installed on the robot is entering the console menu and then select the "Program Editor". After loading will save the program so created that can be identified and stored in an easily accessible place to be copied to stick or to be charged directly from your PC to console. The next step is to load the program back into the console to be checked if it will work in real console. Robot speed is set depending on whether you are going to execute robot (Fig.5).

The last step is moving the robot program thus created by pressing the "play" of the console.

3. USING THE VIRTUAL ARC

After the end stages of programming, engineering calculations on the robot and management can be done by software, a software module of Robot Studio, which is Virtual Arc.

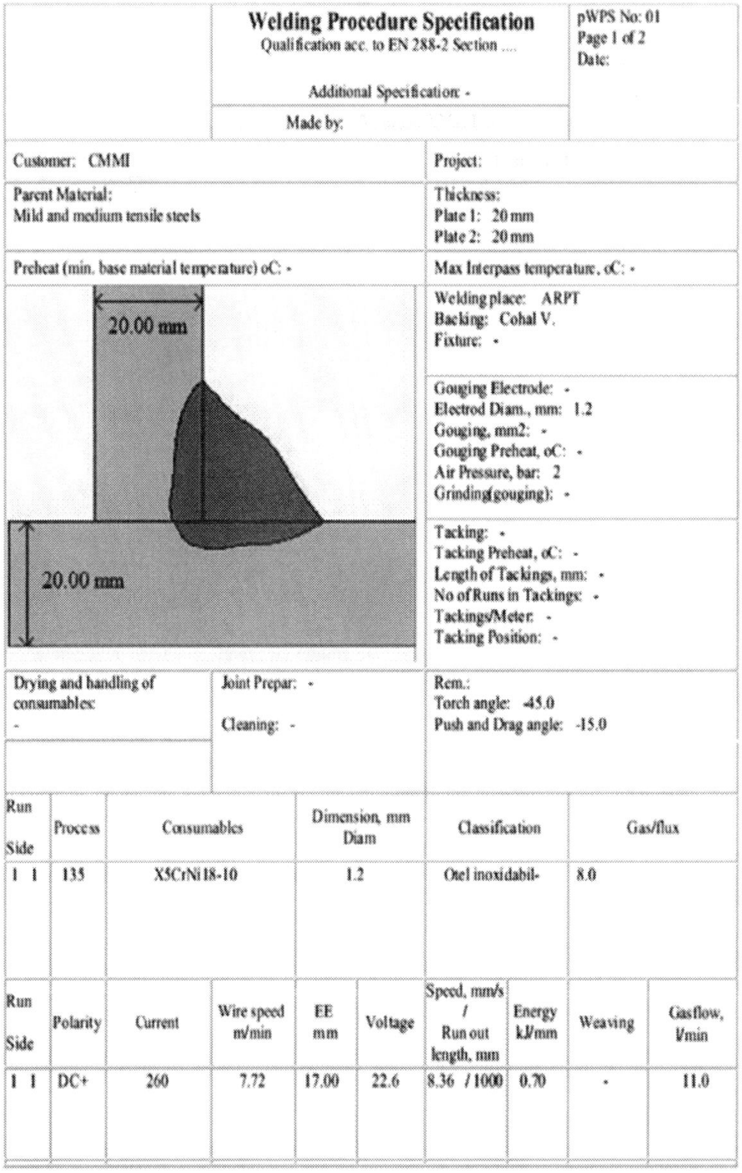

Welding Procedure Specification
Qualification acc. to EN 288-2 Section ....
Additional Specification: -
Made by:

pWPS No: 01
Page 1 of 2
Date:

Customer: CMMI	Project:
Parent Material: Mild and medium tensile steels	Thickness: Plate 1: 20 mm; Plate 2: 20 mm
Preheat (min. base material temperature) oC: -	Max Interpass temperature, oC: -

Welding place: ARPT
Backing: Cohal V.
Fixture: -

Gouging Electrode: -
Electrod Diam., mm: 1.2
Gouging, mm2: -
Gouging Preheat, oC: -
Air Pressure, bar: 2
Grinding(gouging): -

Tacking: -
Tacking Preheat, oC: -
Length of Tackings, mm: -
No of Runs in Tackings: -
Tackings/Meter: -
Tacking Position: -

Drying and handling of consumables: -
Joint Prepar: -
Cleaning: -
Rem.:
Torch angle: -45.0
Push and Drag angle: -15.0

Run Side	Process	Consumables	Dimension, mm Diam	Classification	Gas/flux
1 1	135	X5CrNi18-10	1.2	Otel inoxidabil-	8.0

Run Side	Polarity	Current	Wire speed m/min	EE mm	Voltage	Speed, mm/s / Run out length, mm	Energy kJ/mm	Weaving	Gasflow, l/min
1 1	DC+	260	7.72	17.00	22.6	8.36 / 1000	0.70	-	11.0

Fig.7. WPS using VirtualArc.

Opening Arc virtual program is in the "Add-Ins" module "ArcWelding" (fig.6).
Further complement the required values selected from the database or calculated:

- ☑ System parameters
- ☑ Regimes welding process parameters
- ☑ Pre-weld analysis
- ☑ Elements on economic calculations

☑ Report

Economic application will be completed as desired, depending on the number of pieces of operator costs, etc. At the end on the right will show the calculation results of the robot expenses and costs and the company that manufactures welded assembly.

Finally sheet resulting technological WPS (fig. 7). A Welding Procedure Specification (WPS) is the formal written document describing welding procedures, which provides direction to the welder or welding operators for making quality production welds as per the requirements. The purpose of the document is to guide welders to the accepted procedures so that repeatable and trusted welding techniques are used. A WPS is developed for each material and for each welding type used.

4. CONCLUSIONS

Using software RobotStudio is done offline work program the robot in a short time, being accomplished online testing reducing costs and development time of technological design. Arc program virtual appearance in Robot Studio enables a detailed view of the data sheet, which is displayed, and the formation of weld material inside out and eliminating errors.

References

1. V. Cohal, *Workstation Design for Robot Welding using RobotStudio* (in Romanian), (Ed. Politehnium, Iaşi, Romania, 2008)
2. V. Cohal, *Programming robots for welding and simulation in RobotStudio*, (in Romanian), (Ed. Politehnium, Iaşi, 2008)
3. *VirtualArc® 1.1, ABB AB Robotics Products, S-721-68 Västerås*, Sweden (2016)
4. *Operating Manual ArcWelding PowerPac RobotStudio 6.02*, Document ID: 3HAC028931-001, Revision: F, ABB AB Robotics Products, SE-721 68, Västerås, Sweden (2016)
5. *Operating manual RobotStudio 6.02*, Document ID: 3HAC032104-001, Revision: C, ABB AB Robotics Products, SE-721 68, Västerås, Sweden (2016)

Effect of friction speed on wear intensity in abrasive soil mass

Magdalena Lemecha[1,*], *Jerzy* Napiórkowski[1] and *Mariusz* Rosiak[2]

[1]University of Warmia nad Mazury in Olsztyn, Faculty of Technical Sciences Chair of Building, Operation of Vechicles and Machines, M. Oczapowskiego 11, 10-079 Olsztyn, Poland
[2]Opole University of Technology, Faculty of Mechanical Engineering, Department of Manufacturing Engineering and Production Automation, S.Mikołajczyka 5, 45-271 Opole, Poland

Abstract. One of the major problems when machining soil mass is wear of working tools. The process consists of many factors related to, amongst other things, the work item, the properties of the soil and the working process parameters. During surgeries working big impact on the consumption process parameters are working, such as the type of cutting, machining depth and speed of its implementation. So far, there is no work that would have recognized the issue of the impact of cutting speed on the course wear. Processes which occur in the surface layer material are associated with changes in its mechanical properties, physical and chemical properties. The paper analyses the process of wear steel 38GSA and the same steel weld electrode EstelMn60. Studies were conducted in laboratory conditions by the "spinning bowl" method. During the tests carried out etering of samples depending on the type of road friction and the treated pulp. The results found that with increasing speed decreases wear intensity weight steel 38GSA. And for the deposit EstelMn60 there was no clear effect of speed on wear intensity. It is also noted that a significant impact on the amount of abrasion is the percentage of alloying elements..

1 Introduction

Wear is defined as a process in which the top layer is changed by gradually removing material from the surface of a solid which is characterised by a permanent deformation or mass loss [1]. Abrasive wear is the most common type of machine working part wearing [2,3]. It usually consists in moving a hardened or loose abrasive material against the surface of a working part, which is in direct contact with the material being processed. It is difficult to control since it is a natural process.

Considering the high intensity reported by many authors [4-6], it is a serious scientific and technical issue. In most papers, the wear of machine working parts in soil mass has been regarded as wear caused by single-cycle separation of material from the surface. Properties of whole soil as abrasive mass are disregarded. Such system-creating factors as the disorderly arrangement of a large number of abrasive grains on the active cutting surface, non-ductility of the abrasion surface, free movement of abrasive grains depending on the condition of soil

* Corresponding author: magdalena.lemecha@uwm.edu.pl

mass and the action on the working part of organic mass whose properties depend on the natural surroundings are not taken into account. It has been found that wear of construction materials in soil mass is affected by physicochemical properties of soil and of the construction material, as well as by the parameters of the work process [7-9]. The most important of them include mechanics of cooperation of the working part and soil mass, cutting depth and working speed. Especially the latter of these parameters has not been addressed to a sufficient extent in the literature.

The effect of the cutting speed on wear is usually considered equivalent to friction work [10]. Friction work alone can bring about various effects of wear. The heat which is associated with the friction speed in the surface layer of material brings about changes in the chemical, mechanical and physical properties. Hardness decreases with increasing temperature, which brings about structural changes in the surface layer and affects strains. Friction work can generate various effects, which are caused by such components as: friction speed and friction coefficient, as well as normal pressure during the cutting process. It has been found for steel-steel friction that the wear intensity increases up to the speed of 0.5 m^{-s}, and then decreases rapidly to 2 m^{-s} [11].

Therefore, the relationship which defines the friction force as a function of the speed of the bodies in contact against each other is important for bodies in movement with friction. It is usually assumed in engineering calculations that the relationship is constant, although in fact it is not. It was shown in a study conducted by Stribeck that friction force decreases with increasing relative speed of bodies (within a range of small sliding speed). It is of particular importance in initiating *stick-slip* movements [12].

Considering the fact that no studies have been found regarding the relationship between the cutting speed and friction force, or wear intensity in soil mass, the objective of this study was to evaluate the effect of cutting speed on wear of various construction materials in diverse abrasive soil masses

2 Study methodology

The study was conducted in laboratory conditions on a "centrifugal bowl" machine. 30 x 20 x 10 mm samples were made of 38GSA steel and of the same steel pad-welded with an EstelMn60 electrode. The structure of the surface layers was assessed metalographically under an optical microscope – Neophot 52 – coupled with a Visitron Systems digital camera and by electron scanning microscopy. A microanalysis of the chemical composition was conducted with a JEOL JSM-5800 LV scanning microscope coupled with an Oxford LINK ISIS-300 X-Ray microanalyser.

38GSA is a steel of martensitic structure, containing perlite and bainite (Fig. 1). EstelMn60 welded pad is a material of a plate structure, containing residual austenite and primary carbides [α +$(Fe,Cr)_7C_3$)] with numerous carbides $(Fe,Cr_3)C$ (Fig. 2). The chemical composition of the steel and the welded pad is shown in Table 1. Hardness was measured with a Vickers hardness tester, type HM010u, in accordance with PN-EN ISO 6507-1:1999. The indenting tool was loaded with a weight of 98 N, loading time - 10 s. The steel hardness was 436 HV10, and that of the welded pad – 697 HV10. Sand and clay were used as the abrasive mass. Granulation was classified in accordance with PN- EN ISO 14668-2(2004).

The abrasive soil mass taken for the study and its parameters are shown in Table 2.

Weight wear was determined from the formula:

$$Z_w = m_w - m_i \tag{1}$$

where: m_w- starting weight of a sample [g], m_i - weight of a sample after it completed the friction distance [g].

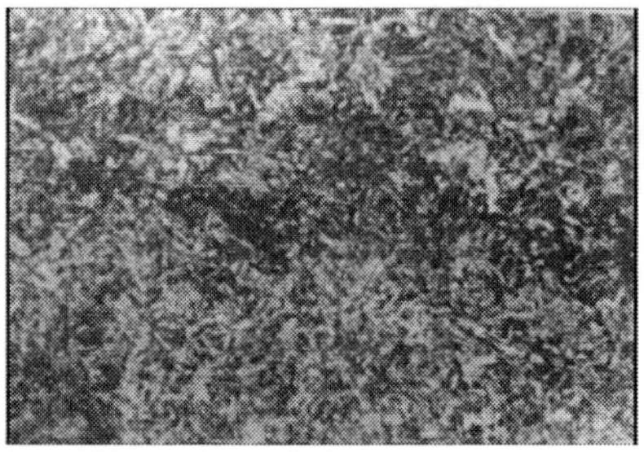

Fig. 1. Steel 38GSA. Martensite and bainite with perlite fine grained. Magnification 500x.

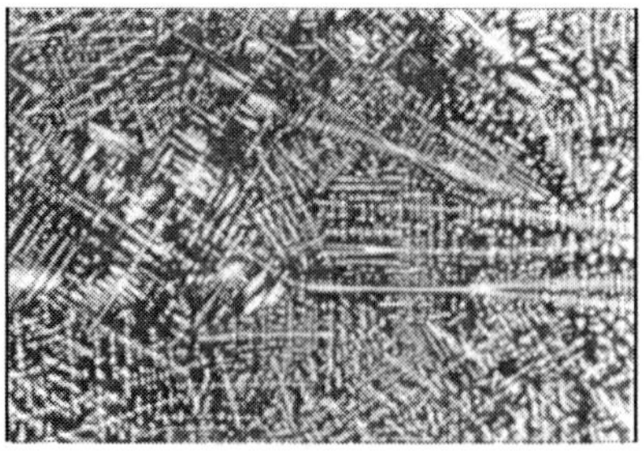

Fig. 2. EStelMn60 surfacing weld: The distribution of primary carbides in a matrix of eutectic structure of the leaf + retained austenite. Magnification 200x.

Table 1. Chemical composition of the materials tested.

Element (symbol)	38GSA etceteras [%]	EStelMn 60 etceteras [%]
C	0.35	2.99
Mn	1.07	6.28
Si	1.17	1.68
P	0.03	0.05
S	0.02	0.14
Cr	0.18	31.6
Ni	0.09	0.30
Cu	0.16	0.09
Al	0.022	0.04
Mo	-	0.04
V	-	0.07
Ti	-	0.10
B	-	0.002

Table 2. Characteristics of soil particle size.

Soil	% the fraction with a diameter in mm			
	gravel > 1.0	sand 0.1-1.0	dust 0.02-0.1	loam <0.02
Sand	0.8	71.6	21.0	7.4
Clay	0.4	28.4	25.6	46.0

Three friction speeds were taken for the experiment, which usually occur in cutting soil mass, i.e. 0.56, 1.39 and 2.00 m^{-s}. Unit pressure 57 kPa. Total friction distance was 2,000 m. The samples were weighed every 2,000 m and the soil mass was replaced. The soil moisture content was determined by measuring the weight of the solid phase dried at 105°C, which ranged from 10% to 12%. Before the test, the samples were washed in an ultrasonic washer with a potassium hydroxide-based washing agent.

DOI: 10.1051/matecconf/201711203013

3 Test results and analysis

Figures 3-4 show the wear of the construction material at various speeds.

a)

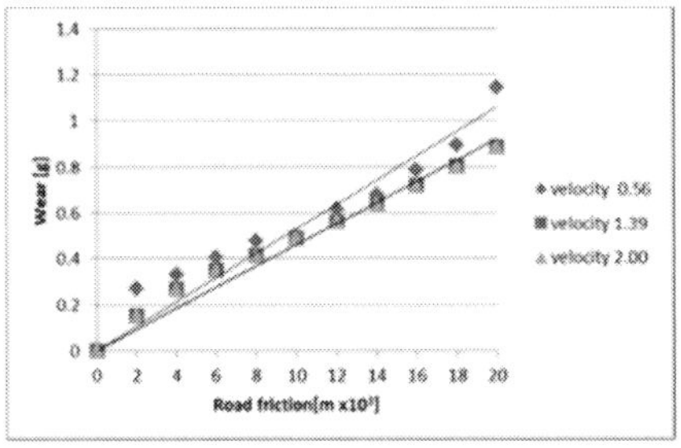

b)

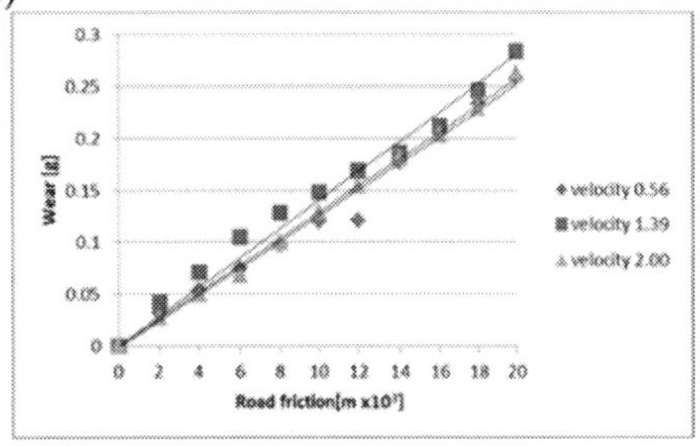

Fig. 3. The course of wear in the sand: a) 38 GSA steel, b) EStelMn60 welded.

a)

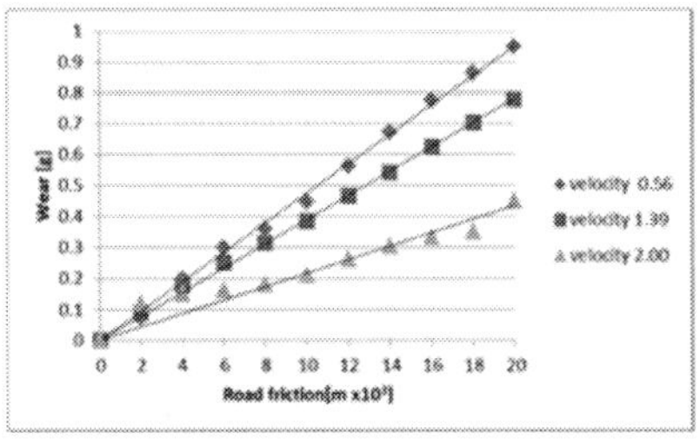

b)

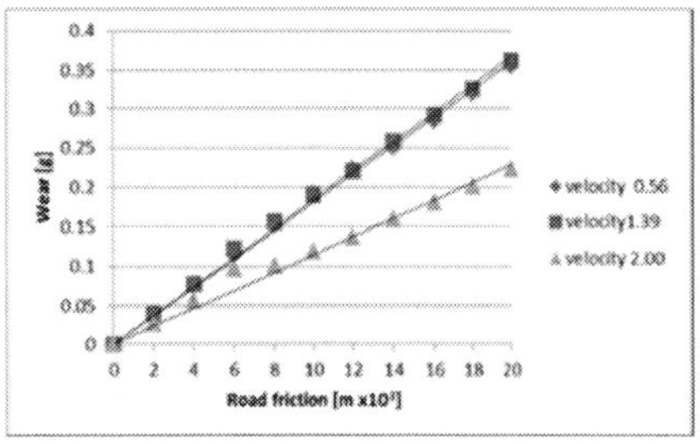

Fig. 4. The course of wear in clay: a) 38 GSA steel, b) EStelMn60 welded.

The results have shown no significant effect of the cutting speed on the wearing intensity of 38GSA steel in clayey soil mass. The most intensive wear was recorded at the speed of $V=0.56\ ms^{-1}$, and the least intensive was at the speed of $V=2.00ms^{-1}$. The wearing analysis of the EStelMn60 welded pad did not show any significant differences in the wear intensity (see Fig. 5).

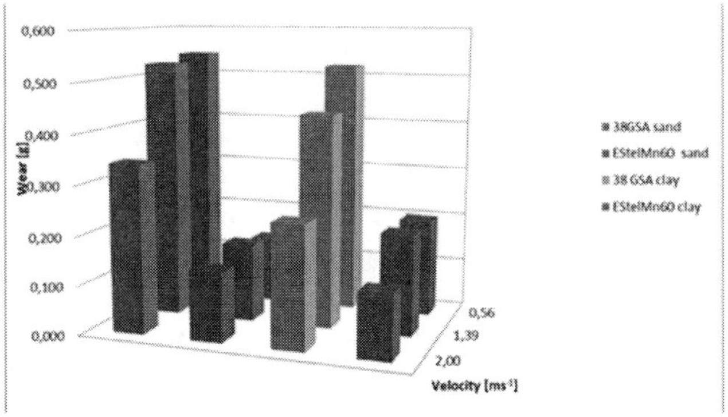

Fig. 5. Compilation of total consumption of materials in studied soils.

The results of the statistical analysis conducted to identify significant wearing intensity differences between the materials under study in sand and clay showed that cutting speed has a significant effect on the wearing intensity, both for 38GSA steel and for the EStelMn60 welded pad (Table 3, 4).

Table 3. Duncan's statistics differ in the intensity of wear depending on the cutting speed.

Soil	Duncan test			
	Materials	**Average consumption**	**1**	**2**
Sand	EStelMn60	0.1455	****	
Sand	38GSA	0.4888		****

**** - it illustrates the statistically significant difference $p < 0.05$ between different grades of steel in a particular soil species.

Tabele 4. Duncan's statistics differ in the intensity of wear depending on the cutting speed.

Soil	Duncan test			
	Materials	**Average consumption**	**1**	**2**
Clay	EStelMn60	0.1782	****	
Clay	38GSA	0.4002		****

**** - it illustrates the statistically significant difference $p < 0.05$ between different grades of steel in a particular soil species.

This is caused by the smallest percentage of silt and clay fractions in this type of abrasive mass. These finest fractions of abrasive mass, in combination with moisture, are responsible for removing the matrix from the surface of the sintered carbide samples. A higher percentage of fine fractions in abrasive mass increases the wearing intensity of sintered carbides.

4 Summary

This study showed that the cutting speed affects the wear of working parts, depending on the material properties and type of soil. The process is affected by the soil contact with the surface layer of a construction material. The contact time of soil grains in sandy soil is much shorter than in clayey soil. Abrasive grains in sandy soil have a considerable number of degrees of freedom. The contact with a construction material is discrete regardless of the friction speed. For clayey soils, a high percentage of silty and clayey fractions (71.6%) makes a soil under load acquire properties of a continuous medium. Soil is under plastic deformations at low speeds and adhesion on the friction surface takes place. With increasing speed, the elastic strain of soil mass begins to dominate, which changes the wearing intensity.

The same relationships were found for both construction materials, despite different surface layer characteristics. The wear of the EStelMn60 welded pad was equal to half of that of 38 GSA steel, regardless of the speed or the type of soil mass.

Apart from the cutting speed, the wear process is affected by the susceptibility of the surface layer of steel and the welded pad. It is noteworthy that no significant effect of temperature was observed as the processed soil consistently took away the energy generated in the process.

References

1. A. Zmitrowicz, JTAM, **44**, 219-253 (2006)
2. R.W. Maruda, G.M. Krolczyk, E. Feldshtein, P. Nieslony, B. Tyliszczak, F. Pusavec, Wear, **372-373**, 54-67 (2017)

3. R.W. Maruda, G.M. Krolczyk, P. Nieslony, S. Wojciechowski, M. Michalski, S. Legutko, J Manuf Process, **24**, 107-115 (2016)
4. A. Gierek, *Tribological consumption.* Gliwice: The Silesian Technical University (2005)
5. R. Maruda, E. Feldshtein, S. Legutko, G.M. Krolczyk, J Frict Wear+, **36**, 548 – 553 (2015)
6. J. Napiórkowski, (ed.), P. Szczyglak, P. Drożyner, K. Ligier, P. Mikolajczak, A. Rychlik, K. Kolakowski. *Research and modeling of abrasive and fatigue wear processes.* (in Polish) (Olsztyn: University of Warmia and Mazury, 2014)
7. M. Nalbant, A.T. Palali, TURJAF, **35**, 215-223 (2011)
8. J. Napórkowski, *Spatial impact of soil on working elements of agricultural tools* (habilitation thesis). Agricultural Engineering, **12,** 72 (2005)
9. P. Kostencki, B. Letkowska, R. Nowowiejski, (in Polish), Tribologia **44** (3), 49-79 (2013)
10. Y. Bayhan, Tribology International **39** (6), 570-574 (2006)
11. U. König, H. Grewe *Investigations on the production and properties of hard material layers produced by reactive vapor deposition.* 197 –250 Tribology **1,** (Berlin: Springer 1981)
12. M. Wiercigroch, JSV, **175(5)**, 700-704 (1994)

Case study of residual stresses distribution in steel welded parts using ultrasound

Radu Baciu Lupascu, *Petru* Dusa* and *Eugen* Purice

Technical University "Gheorghe Asachi" Iasi, Faculty of Machine Manufacturing and Industrial Management, Bd. D. Mangeron, 59A, Iasi, Romania

Abstract. Residual stresses occur in every welded manufactured structure. Different studies aimed to classify the methods of investigating the residual stresses and to highlight the advantages and disadvantages/limits of each of them. The conclusion is that ultrasonic investigation permits the evaluation of bulk stress state of welded components with acceptable accuracy. Our past research has addressed various aspects of ultrasonic investigation of residual stresses in welded steel parts: different cases of the analysis of residual stresses, edge effect on analyses of residual stresses. The purpose of this paper is to continue the previous research by introducing a new case: comparative study of equal size samples obtained by splitting an initial sample. The probe was conceived as a frame with a detachable side. After welding, the assembly is processed on a grinding machine to ensure parallel surfaces. The sample, as obtained, is ultrasonic investigated as part. After that, the detachable side is removed and the resulted part is split in six equal samples. Every so obtained sample was investigated and results are presented. New direction for future research is proposed.

1 Introduction

Residual stresses occur in every welded manufactured structure. Many investigations have been carried out to study this phenomenon. Different studies aimed to classify the methods of investigating the residual stresses and to highlight the advantages and disadvantages/limits of each of them [1-10].

Our past research has addressed various aspects of ultrasonic investigation of residual stresses in welded steel parts: different cases of the analysis of residual stresses, edge effect on analyses of residual stresses [11, 12]. Observations made during experiments opened up new investigative directions. By studying, for example, edge effect we used samples of different thicknesses and we noticed an interesting distribution of stresses in the samples obtained by splitting an initial sample.

This paper aims to investigate the mode of distribution of residual stresses in identical samples obtained by splitting, perpendicular to the welding line, of an initially welded sample.

* Corresponding author: pdusa@tcm.tuiasi.ro

2 Theoretical backgrounds

There are several methods for the analysis of the phenomena and relationships arising from residual stresses and propagation of ultrasounds in metals. Our previous research [11, 12], presented the theoretical principles used for experimentation. A synthesis is presented below. The approach is based on identifying a relationship between Young's modulus and velocity of a longitudinal wave.

Longitudinal wave velocity is given by Eq. 1 [1]:

$$\rho V_L^2 = \lambda + \mu \tag{1}$$

where:

ρ - Density of the material, [kg/m^3];

V_L - Velocity of a longitudinal wave in an isotropic solid, [m/s];

λ, μ - Lamé moduli of an isotropic solid.

According with [2] between Lamé moduli and Young's modulus there are the relationship presented by equation 2 and equation 3:

$$\lambda = \frac{\upsilon E}{(1+\upsilon)(1-2\upsilon)} \tag{2}$$

where:
E - Young modulus;
υ - Poisson ratio.

$$\mu = \frac{E}{2(1+\upsilon)} \tag{3}$$

Using into equation 1 the values of Lamé moduli from equations 2 and 3 results:

$$\rho V_L^2 = \frac{\upsilon E}{(1+\upsilon)(1-2\upsilon)} + \frac{E}{2(1+\upsilon)} = \frac{E(1-\upsilon)}{(1+\upsilon)(1-2\upsilon)} \tag{4}$$

In Eq. 5 is explicit relation between Young's modulus, sound velocity, density of the material and Poisson ratio.

$$E = \frac{\rho V_L^2 (1+\upsilon)(1-2\upsilon)}{(1-\upsilon)} \tag{5}$$

Relation between stress, strain and Young's modulus is assumed to be governed by Hooke's law.

3 Method

Experiments were conducted using a previews equipment and methodology [11, 12] and a short description is presented in the following. The probe was conceived as a frame with a detachable side (see Figure 1). After welding, the assembly is processed on a grinding machine to ensure parallel surfaces. The sample, as obtained, is ultrasonic investigated as Part P (see Figure 7). After that, the detachable side is removed (see Figure 3) and the resulted part is splitting in six equal samples denominate as P1, P2, P3, P4, P5, P6 (see Figures 2 and 4). Every sample P1-P6 was processed on a grinding machine to ensure parallel surfaces (in same manner as the part P). Each of the P1-6 samples was marked with divisions to locate the measurement points (see # 5).

The sample material is USt.37-2 [13] C% 0.25; P% 0.065; S% 0.065; N% 0.009; tensile strength Rm N/mm^2:340 up to 470; upper yield point ReH N/mm^2 225.

Testing equipment consists of ultrasonic measuring device USLT 2000 (Krautkramer), transducer K5K, frequency 5 MHz, 5 mm in diameter.

For each sample to be brought to the initial request position a screw extension was used (see Figures 6 and 8).

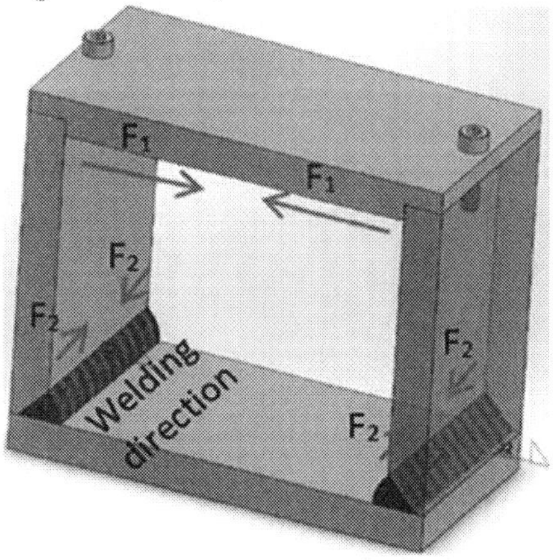

Fig. 1. The sample P conceived as a frame with a detachable side.

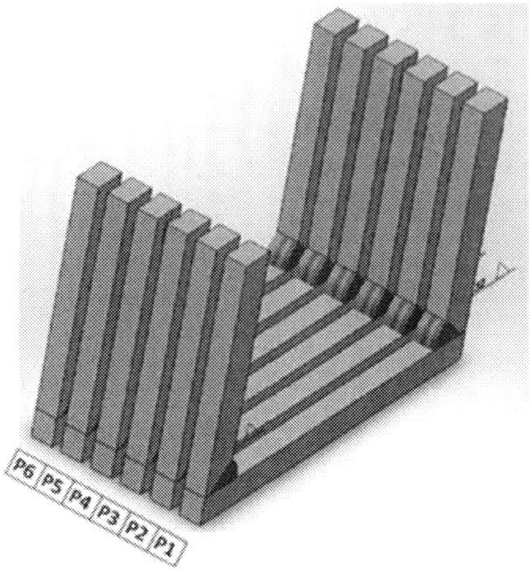

Fig. 2. The samples resulted by splitting of initial P sample (into samples P1-P6).

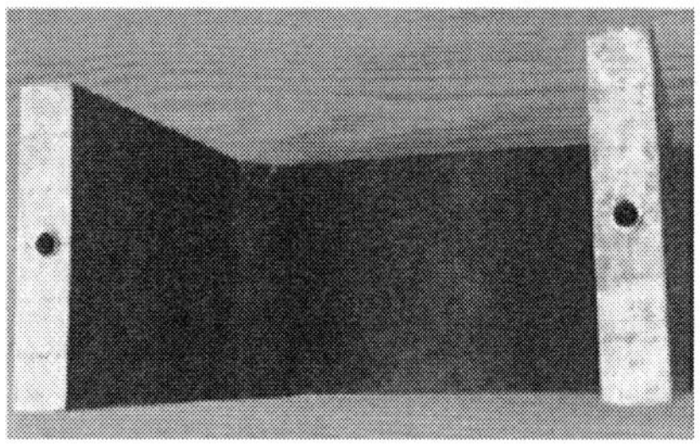

Fig. 3. View of the sample P.

Fig. 4. View of the samples P1-P6.

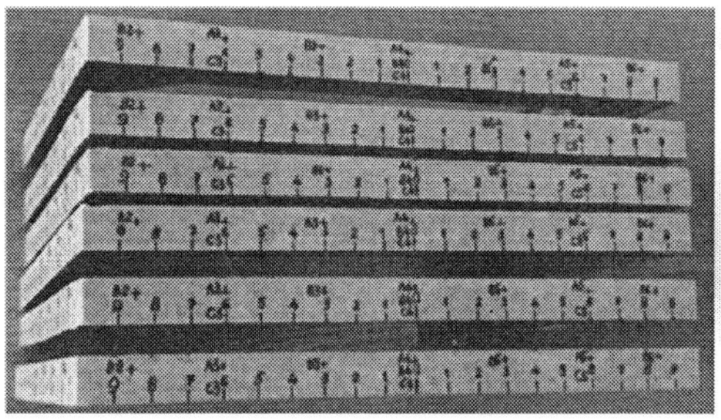

Fig. 5. Marked samples with divisions to locate the measurement points.

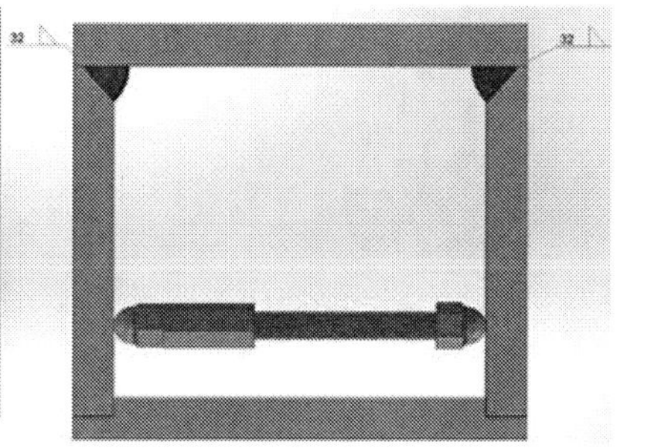

Fig. 6. The screw extension to be brought to the initial request position.

The points of interest are marked as series of points A1, A2, A3, A4, A5, A6, A7; B1, B2, B3, B4, B5, B6, B7; C1, C2, C3, C4, C5, C6, C7 (see figure 9) and was established in our previous research [11,12] . These points were investigated for every sample (P, P1, P2, P3, P4, P5 and P6) ultrasonic in the state charged of the sample (with constriction), condition obtained by using the extension screw (as shown in the figure 6) and preserved by fastening on the base piece (as shown in the figure 8).

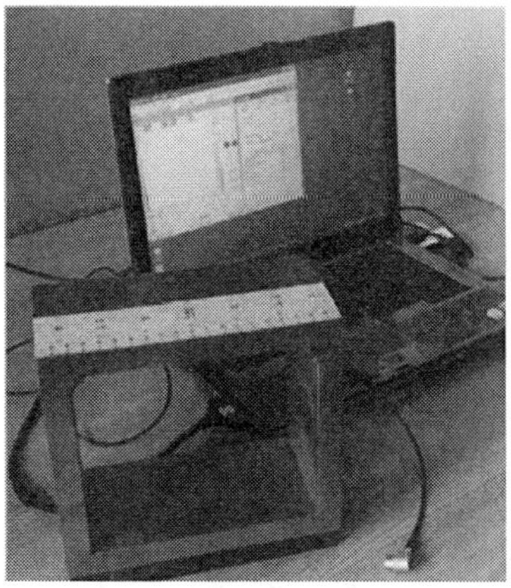

Fig. 7. Overview of ultrasonic investigation of sample P.

Fig. 8. Overview of ultrasonic investigation of samples P1-P6.

4 Results and discussion

As mentioned above the points of interest marked as series of points A1, A2, A3, A4, A5, A6, A7; B1, B2, B3, B4, B5, B6, B7; C1, C2, C3, C4, C5, C6, C7 were investigated for every samples P1-P6 in state charged (with constriction). For every point several measurements (minimum five) were done. To assessing whether one piece of experimental data from a set of observations, is likely to be spurious Chauvenet's criterion was applied.

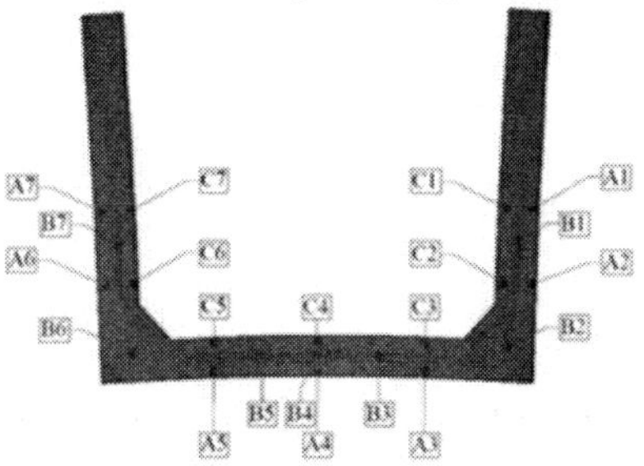

Fig. 9. Measuring point location.

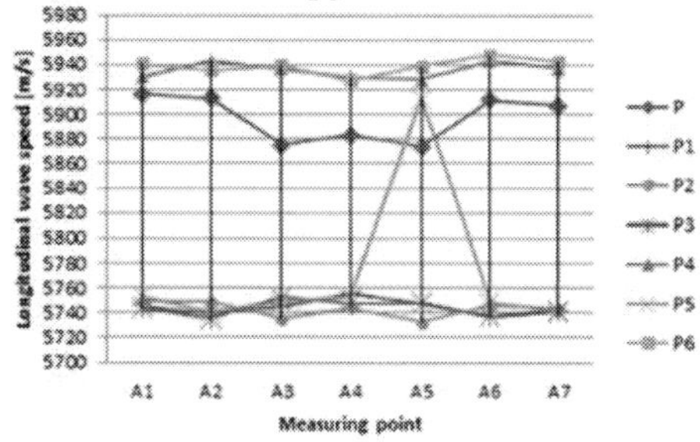

Fig. 10. Wave longitudinal velocity measured in *A* points (samples P, P1, P2, P3, P4, P5, P6).

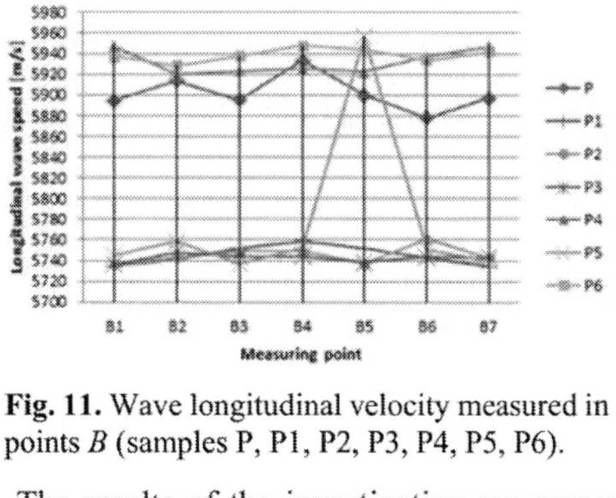

Fig. 11. Wave longitudinal velocity measured in points *B* (samples P, P1, P2, P3, P4, P5, P6).

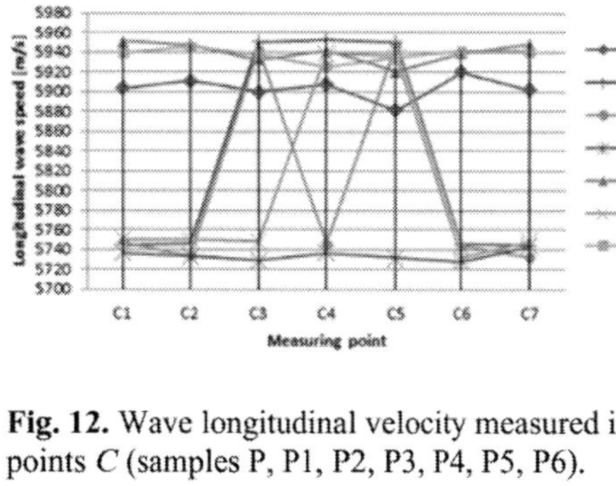

Fig. 12. Wave longitudinal velocity measured in points *C* (samples P, P1, P2, P3, P4, P5, P6).

The results of the investigation are presented in tables 1-7, figures 10-12 and refer to points of interest A1, A2, A3, A4, A5, A6, A7 B1, B2, B3, B4, B5, B6, B7, C1, C2, C3, C4, C5, C6, C7 for each of the P, P1, P2, P3, P4, P5, P6 samples. On the tables below $\Delta\sigma_A$, $\Delta\sigma_B$, $\Delta\sigma_C$ represent stresses variation related to standard sample for the measuring points.

Table 1. Wave longitudinal velocity and stress state analysis on part P.

Measuring Point	*Longitudinal Wave speed [m/s]*	*Measuring Point*	*Longitudinal Wave speed [m/s]*	*Measuring Point*	*Longitudinal Wave speed [m/s]*	$\Delta\sigma_A$ *[N/mm2]*	$\Delta\sigma_B$ *[N/mm2]*	$\Delta\sigma_C$ *[N/mm2]*
A1	*5916,40*	**B1**	*5895,00*	**C1**	*5903,40*	-3	-18	-12
A2	*5913,20*	**B2**	*5913,80*	**C2**	*5910,60*	-5	-4	-7
A3	*5875,20*	**B3**	*5895,20*	**C3**	*5899,60*	-32	-18	-15
A4	*5883,00*	**B4**	*5933,00*	**C4**	*5907,40*	-26	9	-9
A5	*5874,20*	**B5**	*5900,00*	**C5**	*5880,80*	-33	-14	-28
A6	*5911,40*	**B6**	*5878,00*	**C6**	*5919,60*	-6	-30	0
A7	*5907,00*	**B7**	*5897,20*	**C7**	*5901,40*	-9	-16	-13

Table 2. Wave longitudinal velocity and stress state analysis on part P1.

Measuring Point	*Longitudinal Wave speed [m/s]*	*Measuring Point*	*Longitudinal Wave speed [m/s]*	*Measuring Point*	*Longitudinal Wave speed [m/s]*	$\Delta\sigma_A$ *[N/mm2]*	$\Delta\sigma_B$ *[N/mm2]*	$\Delta\sigma_C$ *[N/mm2]*
A1	*5745,00*	**B1**	*5734,40*	**C1**	*5744,20*	-123	-130	-123
A2	*5740,00*	**B2**	*5741,60*	**C2**	*5745,60*	-126	-125	-122
A3	*5744,80*	**B3**	*5751,40*	**C3**	*5950,20*	-123	-118	22
A4	*5755,80*	**B4**	*5759,00*	**C4**	*5952,60*	-115	-113	23
A5	*5747,60*	**B5**	*5751,80*	**C5**	*5950,20*	-121	-118	22
A6	*5737,20*	**B6**	*5743,00*	**C6**	*5745,40*	-128	-124	-123
A7	*5742,00*	**B7**	*5734,40*	**C7**	*5744,00*	-125	-130	-124

Table 3. Wave longitudinal velocity and stress state analysis on part P2.

Measuring Point	*Longitudinal Wave speed [m/s]*	*Measuring Point*	*Longitudinal Wave speed [m/s]*	*Measuring Point*	*Longitudinal Wave speed [m/s]*	$\Delta\sigma_A$ *[N/mm2]*	$\Delta\sigma_B$ *[N/mm2]*	$\Delta\sigma_C$ *[N/mm2]*
A1	*5748,40*	**B1**	*5736,60*	**C1**	*5747,00*	-121	-129	-122
A2	*5748,40*	**B2**	*5743,40*	**C2**	*5733,80*	-121	-124	-131
A3	*5735,60*	**B3**	*5749,40*	**C3**	*5942,20*	-129	-120	16
A4	*5744,20*	**B4**	*5749,40*	**C4**	*5746,60*	-123	-120	-122
A5	*5733,00*	**B5**	*5737,00*	**C5**	*5948,00*	-131	-128	20
A6	*5746,60*	**B6**	*5761,20*	**C6**	*5743,60*	-122	-112	-124
A7	*5743,00*	**B7**	*5743,00*	**C7**	*5732,00*	-124	-124	-132

Table 4. Wave longitudinal velocity and stress state analysis on part P3.

Measuring Point	*Longitudinal Wave speed [m/s]*	*Measuring Point*	*Longitudinal Wave speed [m/s]*	*Measuring Point*	*Longitudinal Wave speed [m/s]*	$\Delta\sigma_A$ *[N/mm2]*	$\Delta\sigma_B$ *[N/mm2]*	$\Delta\sigma_C$ *[N/mm2]*
A1	*5743,60*	**B1**	*5736,80*	**C1**	*5736,40*	-124	-129	-129
A2	*5735,60*	**B2**	*5748,20*	**C2**	*5733,00*	-129	-121	-131
A3	*5752,60*	**B3**	*5744,20*	**C3**	*5728,60*	-118	-123	-134
A4	*5748,00*	**B4**	*5744,60*	**C4**	*5736,00*	-121	-123	-129
A5	*5747,80*	**B5**	*5738,40*	**C5**	*5731,20*	-121	-127	-132
A6	*5737,20*	**B6**	*5743,80*	**C6**	*5727,60*	-128	-124	-135
A7	*5741,60*	**B7**	*5743,40*	**C7**	*5742,80*	-125	-124	-124

Table 5. Wave longitudinal velocity and stress state analysis on part P4.

Measuring Point	*Longitudinal Wave speed [m/s]*	*Measuring Point*	*Longitudinal Wave speed [m/s]*	*Measuring Point*	*Longitudinal Wave speed [m/s]*	$\Delta\sigma_A$ *[N/mm2]*	$\Delta\sigma_B$ *[N/mm2]*	$\Delta\sigma_C$ *[N/mm2]*
A1	*5930,40*	**B1**	*5947,20*	**C1**	*5951,00*	7	19	22
A2	*5942,40*	**B2**	*5920,00*	**C2**	*5947,40*	16	0	20
A3	*5936,80*	**B3**	*5923,00*	**C3**	*5932,20*	12	2	9
A4	*5930,00*	**B4**	*5925,80*	**C4**	*5941,20*	7	4	15
A5	*5928,00*	**B5**	*5922,40*	**C5**	*5920,40*	6	2	0
A6	*5943,60*	**B6**	*5937,00*	**C6**	*5939,20*	17	12	14
A7	*5937,20*	**B7**	*5947,00*	**C7**	*5947,80*	12	19	20

Table 6. Wave longitudinal velocity and stress state analysis on part P5.

Measuring Point	*Longitudinal Wave speed [m/s]*	*Measuring Point*	*Longitudinal Wave speed [m/s]*	*Measuring Point*	*Longitudinal Wave speed [m/s]*	$\Delta\sigma_A$ *[N/mm2]*	$\Delta\sigma_B$ *[N/mm2]*	$\Delta\sigma_C$ *[N/mm2]*
A1	*5753,00*	**B1**	*5745,20*	**C1**	*5750,20*	-117	-123	-119
A2	*5740,80*	**B2**	*5758,40*	**C2**	*5749,60*	-126	-114	-120
A3	*5748,40*	**B3**	*5737,20*	**C3**	*5747,60*	-121	-128	-121
A4	*5752,80*	**B4**	*5751,80*	**C4**	*5938,00*	-118	-118	13
A5	*5912,00*	**B5**	*5956,00*	**C5**	*5935,60*	-6	26	11
A6	*5748,40*	**B6**	*5750,60*	**C6**	*5731,80*	-121	-119	-132
A7	*5740,60*	**B7**	*5739,80*	**C7**	*5748,40*	-126	-127	-121

Table 7. Wave longitudinal velocity and stress state analysis on part P6.

Measuring Point	*Longitudinal Wave speed [m/s]*	*Measuring Point*	*Longitudinal Wave speed [m/s]*	*Measuring Point*	*Longitudinal Wave speed [m/s]*	$\Delta\sigma_A$ *[N/mm2]*	$\Delta\sigma_B$ *[N/mm2]*	$\Delta\sigma_C$ *[N/mm2]*
A1	*5940,40*	**B1**	*5937,20*	**C1**	*5938,80*	15	12	13
A2	*5935,20*	**B2**	*5928,20*	**C2**	*5945,40*	11	6	18
A3	*5939,20*	**B3**	*5937,60*	**C3**	*5937,60*	14	13	13
A4	*5927,80*	**B4**	*5947,40*	**C4**	*5925,20*	6	20	4
A5	*5939,00*	**B5**	*5943,60*	**C5**	*5934,40*	14	17	10
A6	*5948,20*	**B6**	*5933,60*	**C6**	*5941,00*	20	10	15
A7	*5943,00*	**B7**	*5941,80*	**C7**	*5940,80*	16	16	15

The state of stresses in the P-piece is caused by the simultaneous action of forces F1 and F2 (see Figure 1). For the whole piece P the effects of the forces F1 and F2 determine the stress state shown in Table 1 and represented by line P on the graphs in Figures 10-12. It can be seen that by splitting the P-piece into six equal parts the stress distribution in the six P1-P6 pieces is different from the P-piece. For parts P1, P2, P3, P5 there is a decrease in the longitudinal wave speed, so a relaxation of the material by stretching.

It is also noted for the P4, P6 parts an increase in longitudinal wave speed, as compared to the P piece, which can be interpreted as a compression.

There are also local points P1 (C3, C4, C5), P2 (C3, C5), P5 (A5, B5, C4, C5) for which the speed is very different from the other points of the sample. These particular observed values may be linked with weld or fusion zone and heat-affected zone.

The distribution of most of the points for samples P1, P2, P3 and P4, P6 on either side of the sample P can be interpreted in relation to the direction in which the welding lines were made (see Figure 1).

It can be hypothesized that an influence on the distribution of the stresses in the measured points on each sample P1, P2, P3, P4, P5, P6 has the split order and the slot size.

5 Conclusions

By splitting the P-piece into six equal parts the stress distribution in the six P1-P6 pieces is different from the P-piece. For samples P1, P2, P3, P5 there is a decrease in the longitudinal wave speed, so a relaxation of the material by stretching. It is noted for the P4, P6 parts an increase in longitudinal wave speed, as compared to the P piece, which can be interpreted as a compression.

Weld or fusion zone and heat-affected zone have a particular and differentiated influence on the state of the stresses at the neighboring points. An example of this is the distribution of the stresses in the points P1 (C3, C4, C5), P2 (C3, C5), P5 (A5, B5, C4, C5).

The direction in which the welding layers are performed has an important influence. For the experiments the welding layers were performed in the direction of the arrow (see figure 1). It can be said that the sense of welding was from P6 to P1. It can be noticed that after splitting, the P1, P2, P3 pieces (located at the end of the welding cord) relaxed while the P4, P6 pieces (located towards the start of the welding cord) contracted.

The stress variation in the measured points, in absolute values, is between 0 and 132 N / mm2. Similar variations are identified in similar works in the literature [5]. Case studies are, however, particular, and the establishment of similarities is not always easy.

The behavior of the P5 piece rather similar to P1, P2, P3 and not similar to P4, P6 as expected would lead to the hypothesis that the order of splitting the original piece in the P1-P6 parts is important.

It can also be suspected as part of stress distribution in parts after splitting has the size of the cutting slit.

Taking into account the aforementioned, new research may be continued by taking into account the order of the splitting into parts, the parameters of the welding regime, and the characteristics of the heat-affected zone.

References

1. V. Hauk, *Structural and Residual Stress Analysis by Nondestructive Methods. Evaluation - Application – Assessment* (Elsevier, 1997)
2. A. F. Bower, *Applied Mechanics of Solids*, (CRC Press, Taylor&Francis Group, 2010)
3. G. Totten, M. Howes, T. Inoue, *Handbook of Residual Stress and Deformation of Steel* (ASM International, 2002)
4. N.S. Rossini, M. Dassisti, K.Y. Benyounis, A.G. Olabi, Materials and Design **35**, 572-588 (2012)
5. Y. Javadi, M. A. Najafabadi, Materials and Design **47**, 473-482 (2013)
6. NDT Resource Center, Transducer Beam Spread, accessed at https://www.nde-ed.org/EducationResources/CommunityCollege/Ultrasonics/EquipmentTrans/beamspread.htm
7. Signal Processing, The Ultrasonic Field, accessed at http://www.signal-processing.com/us_field.html
8. J. Krautkramer, H. Krautkramer, *Ultrasonic Testing of Materials, 4th Edition* (Springer-Verlang, Berlin, 1990)
9. E 1065 – 99 *Standard Guide for Evaluating Characteristics of Ultrasonic Search Units ASTM Standard* (2003)
10. P. J. Shull *Non-destructive Evaluation: Theory, Techniques, and Applications*, (CRC Press, 2002)
11. P. Dusa, R. B. Lupascu, *Applied Mechanics and Materials,* **657**, 326-331 (2014)
12. P. Dusa, R. B. Lupascu, E. Purice, *Applied Mechanics and Materials,* **809-810**, 479-485 (2015)
13. DIN 17100 Grade classification and chemical composition of the steels, accessed at http://www.mesteel.com/info/german.htm

Integrity evaluation concerning vibrations of a welded structure

Emilian Stanciu[1,2], *Codrin* Nitescu[1], *Zeno-Iosif* Praisach[1,3], and *Gilbert-Rainer* Gillich[1,*]

[1]Universitatea "Eftimie Murgu" din Resita, P-ta Taian Vuia 1-4, 320085 Resita, Romania
[2]Windkraft Simonsfeld RO S.R.L., Str. Sportului B 26, 320110 Resita, Romania
[3]S.C. UCM Reşiţa S.A., Str. Golului Nr. 1, 320053 Resita, Romania, Romania

Abstract. In this paper, a predictive model that describes the structural behavior in the absence and presence of damage, and anticipates the modal parameter changes, is introduced for weld structures. Degradation of engineering structures in the welds or HAZ due to damage affects their safety by reducing the stiffeners of structural components. In this paper was considered the kinetic energy distribution which reflects the mass participation, and the modal strain energy distribution which take into consideration the stiffeners reduction. The relation is successfully proved against FEM simulation results.

1 Introduction

Structural integrity assessment of engineering structures by means of vibration-based techniques had become a growing attention in the last decades [1,2]. The relation between stiffness and mass changes and natural frequency changes, coupled with ease of measurement of the natural frequencies by using only a single sensor, was the impetus for using modal methods to identify damage [3]. Generally, in such procedures, the stiffness decrease due to damage is correlated with the natural frequency drop, in this way being found the damage location and severity [4-6]. The greatest success in the use of natural frequency shifts for damage identification, as evidenced by the number of published examples, is in small simple laboratory structures with only single damage locations [7]. For weld structures, this approach is not possible, since here a relevant mass increase is expected. In addition, physical-mechanical parameters in the weld area, especially in the fusion zone, can significantly differ from that of the base material. This makes involving of more complex models necessary.

This paper, continuing our research in damage assessment of simple and complex structures [8-10], introduces a predictive model that describes the weld structure dynamical behavior in the absence and presence of damage and anticipates the modal parameter changes. The specific but weld joint zones have associated finite elements with tailored parameters. Based on this theoretical model a mathematical relation linking the changes introduced by the weld joint and the modal parameter shifts is contrived.

* Corresponding author: gr.gillich@uem.ro

2 Weld beam behavioral model

Let us consider a clamped-hinged prismatic steel beam, with the density ρ=7850 kg/m^3, the Young modulus $E=2*10^{11}$ N/m^2 and Poisson coefficient $\nu=0.3$. The beam has the width $b=0.020$ mm and the height $h=0.005$ mm, and the length $L=1$ m. The weld is located at distance L_1 taken from the left beam end, as shown in Fig. 1. The length of the weld in longitudinal beam direction is $L_w=20$ mm. From literature is known that the heat affected zone, denoted HAZ, can be divided into three zones, depicted in Fig. 2. Starting from the base plate, these are the intercritical zone, the fine-grained zone, and the coarse-grained zone. Each of these zones has different characteristics, as follows: the intercritical zone has lower values of hardness, strength, and resilience and higher values of plasticity properties, the fine-grained zone has higher values of hardness and strength and the coarse-grained zone has increased mass density and good corrosion resistance. Due to its shape, see Fig. 3, the weld height is bigger than the beam thickness.

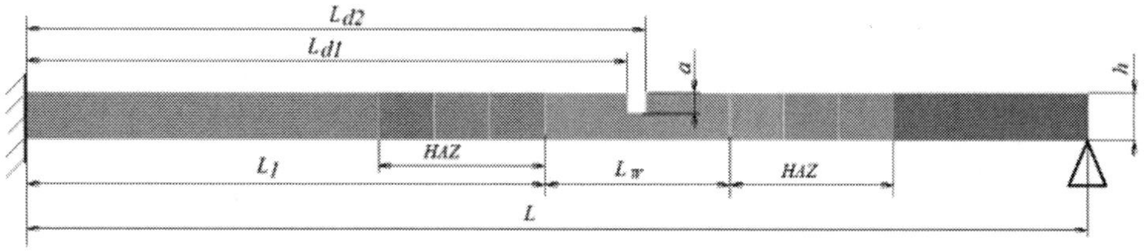

Fig. 1. The clamped-hinged beam in welded construction.

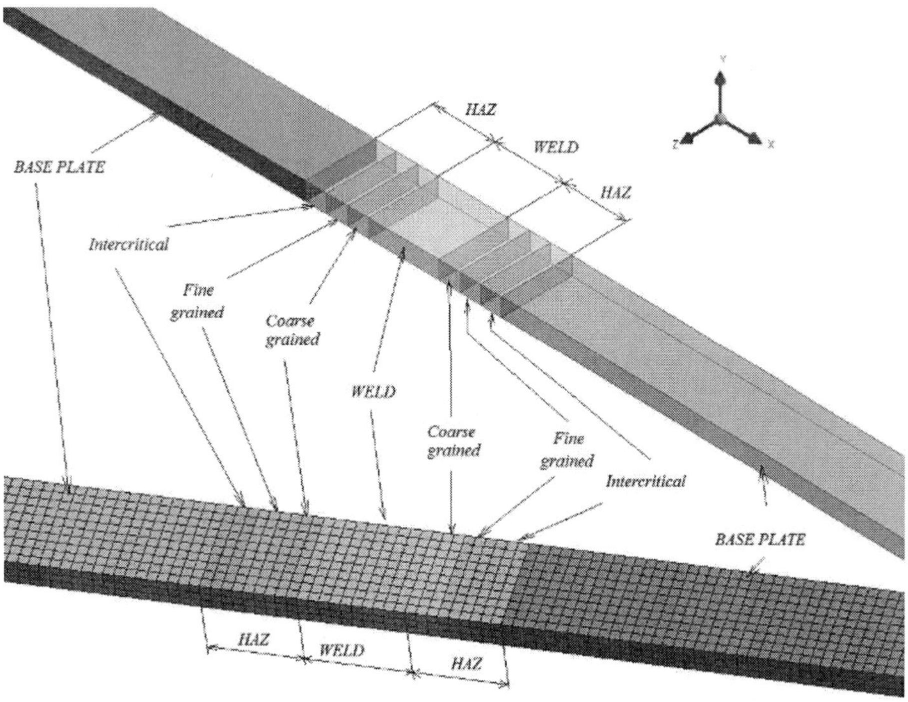

Fig. 2. 3D model and meshing of the model of a welded structure.

 DOI: 10.1051/matecconf/201711203015

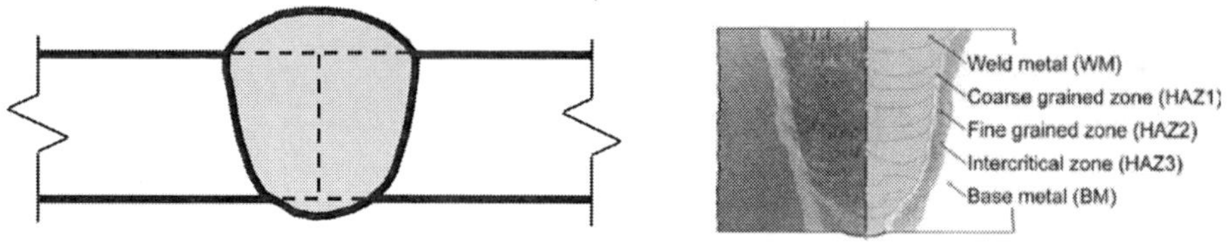

Fig. 3. Shape of weld (left image) and specific zones of the weld structure (right image).

For a clamped-hinged beam, the frequency equation is according to eq. (1) and the bending vibration mode shape is defined in eq. (2).

$$\tanh \alpha_n - \tan \alpha_n = 0 \tag{1}$$

$$W(x) = D\left\{\frac{\cos \alpha_n + \cosh \alpha_n}{\sin \alpha_n + \sinh \alpha_n}\left[\sin(\alpha_n x) - \sinh(\alpha_n x)\right] - \cos(\alpha_n x) + \cosh(\alpha_n x)\right\} \tag{2}$$

where α is the dimensionless wave number and $D=0.5$ is an integration constant, that get the value 0.5 to normalize the mode shape.

To simplify the analytic model, the weld in Fig. 3 will be considered with the same height as the beam, see Fig. 1 and 2. Thus, the weld density has been artificially increased in order not to affect the real weld beam mass. Finally, the resulting model is a welded beam with a constant cross-sectional area and different mass densities, one for the base plate and one for the weld.

Starting from kinetic energy conservation [11] we can write the equivalent mass contribution for each segment of the weld beam:

$$\begin{cases} m_{L_1}{}^{eq} = \bar{m}_{b1} L \int\limits_{0}^{L_1} \left[W_n(x)\right]^2 dx \\ m_{L_w}{}^{eq} = \bar{m}_w L \int\limits_{L_1}^{L_1+L_w} \left[W_n(x)\right]^2 dx \\ m_{L_2}{}^{eq} = \bar{m}_{b2} L \int\limits_{L_1+L_w}^{L-L_1-L_w} \left[W_n(x)\right]^2 dx \\ m_L{}^{eq} = \bar{m} L \int\limits_{0}^{L} \left[W_n(x)\right]^2 dx = 0.25 \bar{m} L \end{cases} \tag{3}$$

where $m_{L_1}{}^{eq}$, $m_{L_w}{}^{eq}$, $m_{L_2}{}^{eq}$, $m_L{}^{eq}$ is the equivalent specific mass corresponding to the nth bending vibration mode for the left beam segment with length L_1, for the right beam segment with length L_2 , for the weld L_w, respectively for or whole beam length L, and $\bar{m}_{b1}$, $\bar{m}_{b1}$, $\bar{m}_{b1}$, $\bar{m}_{b1}$ are the specific masses for respective lengths.

For α_n derived from eq. (1) and taking in consideration eq. (3), the natural frequencies of the weld beam can be calculated from:

$$f_i = \frac{\alpha_i^2}{2\pi}\sqrt{\frac{EI_z}{\bar{m}_b L^4}} = \frac{\alpha_i^2}{2\pi}\sqrt{\frac{EI_z}{4 m_n{}^{ech} L^3}} = \frac{\alpha_i^2}{2\pi}\sqrt{\frac{EI_z}{4\left(m_{L_1}{}^{ech} + m_{L_{cs}}{}^{ech} + m_{L_2}{}^{ech}\right)L^3}} \tag{4}$$

If a damage located in the middle of the weld ($L_w/2$) and having the depth $a=1$ mm respectively width L_{d2}-$L_{d1}=1$ mm is considered, see Fig. 1, the stiffness in this section is modified. The moment of inertia will decrease from $I=bh^3/12$ to $I_D=b(h-a)^3/12$ and taking into consideration the ratio between loss energy distribution and total energy distribution along the beam, we have:

$$\frac{\bar{U}_{iD}(x)}{\bar{U}_i}=\frac{\frac{1}{2EI}\int_0^L\left[\frac{d^2W(x)}{dx^2}\right]^2dx-\frac{I-I_D}{I}\frac{1}{2EI}\int_{L_{d1}}^{L_{d2}}\left[\frac{d^2W(x)}{dx^2}\right]^2dx}{\frac{1}{2EI}\int_0^L\left[\frac{d^2W(x)}{dx^2}\right]^2dx} \tag{5}$$

thus

$$\frac{\bar{U}_{iD}(x)}{\bar{U}_i}=1-4\frac{I-I_D}{I}\int_{L_{d1}}^{L_{d2}}\left[\frac{d^2W(x)}{dx^2}\right]^2dx \tag{6}$$

It is known that the ratio of the natural frequencies between damaged and healthy beam lead us to expression:

$$\frac{f_{iD}}{f_i}=\sqrt{\frac{\bar{U}_{iD}(x)}{\bar{U}_i}}=\sqrt{1-4\frac{I-I_D}{I}\int_{L_{d1}}^{L_{d2}}\left(\frac{d^2W(x)}{dx^2}\right)^2dx} \tag{7}$$

and finally, the relationship which permit us to calculate the natural frequency for a welded beam with damage:

$$f_{di}=f_i\sqrt{1-4\frac{I-I_d}{I}\int_{L_{d1}}^{L_{d2}}\left(\frac{d^2W(x)}{dx^2}\right)^2dx} \tag{8}$$

For the first six vibration modes, the results are presented in tables 1, 2 and 3.

3 Numerical investigations

The investigations by finite element method were performed via ANSYS software, by using the modal analysis module. The medium size of the mesh element is *2* mm, as shown in Fig. 2. For the HAZ were used following parameters:

- intercritical zone length of *6* mm with mass density of $\rho_{ic}=7700$ kg/m³, and Young modulus $E_{ic}=2.0*10^{11}$ N/m²;
- fine grained zone of *6* mm with mass density of $\rho_{fg}=7850$ kg/m³, and Young modulus $E_{fg}=2.1*10^{11}$ N/m²;
- coarse grained zone of *6* mm with mass density of $\rho_{cg}=8000$ kg/m³, and Young modulus $E_{cg}=2.05*10^{11}$ N/m².

For the weld was considered an artificially increased mass density of $\rho_w=10990$ kg/m³ corresponding for a weld with *2* mm over beam height and Young modulus $E_w=2.0*10^{11}$ N/m²; and for the base plate the mass density of $\rho_{bp}=7850$ kg/m³ and Young modulus $E_{bp}=2.0*10^{11}$ N/m².

The weld is stepwise replaced from $L_1+L_w/2=0.5L$ to $L_1+L_w/2=0.1L$. For the first six vibration modes, the results attained from numerical investigations are presented in tables 1 to 3. In these tables, the natural frequency values obtained from the analytic evaluation by employing eq. (4) and (8) are also presented. In Figure 4 the results are presented in graphical form, making them easily comparable; the fit is obvious.

Table 1. Natural frequencies for welded beam with damage at $L_1+L_w/2=0.5L$ and $L_1+L_w/2=0.4L$

i	α	$L_1+L_w/2=0.5L$			$L_1+L_w/2=0.4L$		
		f_{di} [Hz]	f_{FEM} [Hz]	Deviation [%]	f_{di} [Hz]	f_{FEM} [Hz]	Deviation [%]
1	3.9266	17.8658	17.7330	*0.74*	17.8721	17.7898	*0.46*
2	7.0686	57.9280	57.9031	*0.04*	57.8937	57.5068	*0.67*
3	10.2102	120.7760	119.9339	*0.70*	120.8743	120.9362	*0.05*
4	13.3518	206.6783	206.4900	*0.09*	206.5049	204.7770	*0.84*
5	16.4934	315.1636	312.7717	*0.76*	315.3611	314.7990	*0.18*
6	19.6350	446.9689	446.2430	*0.16*	446.8095	444.6690	*0.48*

Table 2. Natural frequencies for welded beam with damage at $L_1+L_w/2=0.3L$ and $L_1+L_w/2=0.2L$

i	α	$L_1+L_w/2=0.3L$			$L_1+L_w/2=0.2L$		
		f_{di} [Hz]	f_{FEM} [Hz]	Deviation [%]	f_{di} [Hz]	f_{FEM} [Hz]	Deviation [%]
1	3.9266	17.8773	17.8489	*0.16*	17.8759	17.8743	*0.01*
2	7.0686	57.8898	57.3852	*0.87*	57.9254	57.7056	*0.38*
3	10.2102	120.8137	120.2281	*0.48*	120.7884	119.7365	*0.87*
4	13.3518	206.7063	206.7814	*0.04*	206.5397	204.8341	*0.83*
5	16.4934	315.1943	313.1465	*0.65*	315.3269	314.3157	*0.32*
6	19.6350	446.6495	443.1368	*0.79*	447.0328	446.8960	*0.03*

Table 3. Natural frequencies for welded beam with damage at $L_1+L_w/2=0.1L$

i	α	$L_1+L_w/2=0.1L$		
		f_{di} [Hz]	f_{FEM} [Hz]	Deviation [%]
1	3.9266	17.8646	17.8494	*0.09*
2	7.0686	57.9249	57.9152	*0.02*
3	10.2102	120.8776	120.7347	*0.12*
4	13.3518	206.6846	205.9404	*0.36*
5	16.4934	315.3033	313.1942	*0.67*
6	19.6350	446.7288	442.5647	*0.93*

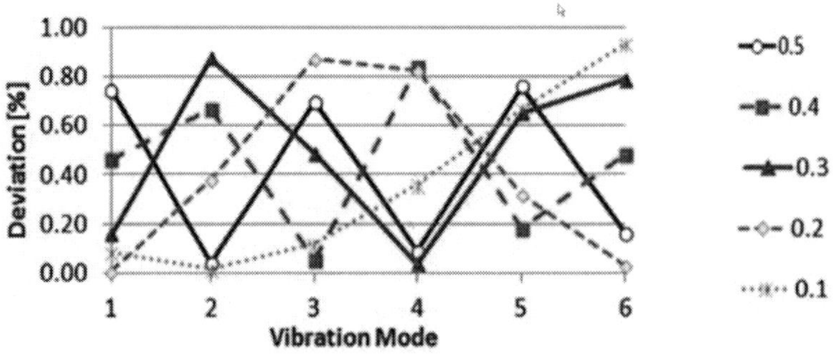

Fig. 4. Deviation of the natural frequency values attained analytically and from numerical simulations for a welded beam with damage.

5 Conclusions

The paper introduces a mathematical model which allows predicting the vibrational behavior of weld structures with damage. It takes into account the mass increase, due to weld and the changes of mass density and elasticity occurred in the HAZ Results obtained by using this relation were tested against FEM simulation results. Analyzing the results one can observe that natural frequency deviation between analytical approach and numerical investigation, for the first six vibration mode, is less than 1%. This qualifies the proposed relation for predicting frequency changes due to damage and finds its application in damage detection procedures.

Next approaches will take into consideration other types of defects occurred in welding processes, and will be extended to real structures.

References

1. G.-R. Gillich, Z. Praisach, D.M. Onchis, *Proceedings of the International Conference on Engineering Mechanics, Structures, Engineering Geology, International Conference on Geography and Geology*, 204-209 (2010)
2. G.-R. Gillich, Z.-I. Praisach, M. Abdel Wahab, N. Gillich, I.C. Mituletu, C. Nitescu, Free Vibration of a Perfectly Clamped-Free Beam with Stepwise Eccentric Distributed Masses, *Shock and Vibration*, article 2086274 (2016)
3. J.S. Sakellariou, S.D. Fassois, Journal of Sound and Vibration, **297**(3–5), 1048-1067 (2006)
4. Y. Hui, S.S. Law, C.J. Ku, Journal of Sound and Vibration, **389**, 168-182 (2017)
5. G.R. Gillich, M. Abdel Wahab, Z.I. Praisach, J.L. Ntakpe, *Proceedings of ISMA 2014*, 485-498 (2014)
6. M. Chandrashekhar, R. Ganguli, *Mechanical Systems and Signal Processing*, 75, 75-93 (2016)
7. G.-R. Gillich, E.D. Birdeanu, N. Gillich, D. Amariei, V. Iancu, C.S. Jurcau, *Annals of DAAAM and Proceedings of the International DAAAM Symposium*, 623-624 (2011)
8. G.-R. Gillich, N.M.M. Maia, I.-C. Mituletu, Z.-I. Praisach, M. Tufoi, I. Negru, *Latin American Journal of Solids and Structures*, 12(12), 2311-2329 (2015)
9. Z. Praisach, G.-R. Gillich, D.E. Birdeanu, *Proceedings of the International Conference on Engineering Mechanics, Structures, Engineering Geology, International Conference on Geography and Geology*,, 2014-2019 (2010)
10. G.-R. Gillich, Z.-I. Praisach, I. Negru, *Materiale Plastice*, **49** (3), 186-191 (2012)
11. G.-R. Gillich, Z.I. Praisach, V. Iancu, H. Furdui, I. Negru, *Strojniski Vestnik/Journal of Mechanical Engineering*, **61** (12), 721-730 (2015)

Measurement of defects in carbon fiber reinforced polymer drilled

Víctor Pascual*, *Manuel* San Juan, *Óscar* Martín, and *Francisco* Santos

Manufacturing-UVa, Escuela de Ingenierías Industriales, Universidad de Valladolid, España

Abstract. Increasingly, fiber-reinforced materials are more widely used because of their good mechanical properties. It is usual to join pieces of these materials through screws and rivets, for which it is necessary to make a hole in the piece, usually by drilling. One of the problems of use CFRP resides in the appearance of defects due to the machining. The main defect to be taken into account is the delamination. Delamination implies poor tolerance when assembling parts, reducing the structural integrity of the part, and areas with high wear, as a series of stresses arise when mounting the screws. Much has been published about delamination and the factors that influence its appearance, so we are not going to focus on it. The present study aims to quantify and measure the defects associated with the drilling of compounds reinforced with carbon fibers, in relation to the cutting parameters used in each case. For this purpose, an optical measurement system and a posterior digital image processing will be used through Deltec Vision software.

1 Introduction

The union of pieces made with the materials of the type carbon fiber reinforced polymer (CFRP) is made mainly by screws and rivets, for which it is necessary to drill a hole in these pieces. It is of vital importance to study how these drills are made and what their quality and final geometries are, because the tolerances of the assembly process and the structural integrity of the assembly will depend on the appearance of possible stresses in the joining areas, the screws or the rivets [1].

2 Experimental Procedure

For the development of the study, a carbon fiber reinforced polymer plate was used, similar to the material used in the aeronautical industry. The plate consists of three layers of fabric, of the taffeta type, with fibers at 45° between the different layers. The last layer has a special treatment to reduce the delamination at the exit (push-out), reason why in this study we will focus on the delamination to the entrance (peel-up). Holes were practiced in that plate, using three drills: two of them were equal, and the other one was different. All three were of the Gühring brand, 6 mm in diameter and were HSS cobalt-coated. The difference between them resided in the geometry: the angles of the points were 118° and 130° respectively [2]. The

* Corresponding author: victor.pascual@uva.es

commercial references of these drills are HSCO5524 and HSCO1261. These tools were not specific to drilling carbon fiber, but were used to observe the results obtained, and to be able to study if their use is profitable, because the cost is five times inferior to a specific one for carbon fiber.

The machine used to perform the machining was a numerical control mill, Nicolás Correa brand, model A-16, equipped with a control Heidenhain brand, model TC 355 [3].

To simulate the working conditions, the drills were made at a minimum distance from each other of twice the diameter, as is the case in the aeronautical industry. With this, we avoid the influence of some holes on others. The sequence of drills carried out was as follows: keeping constant the speed of rotation of the spindle, five holes were realized with an intermediate feed speed. Subsequently, five more holes were drilled with a feed speed 50% higher, and finally, five more holes were drilled with a feed speed 50% lower than the first one. The sequence was repeated twice with each drill. That is, with each drill used, thirty drills were made, in series of five and with three different feed speeds. The cutting conditions were taken based on previous experiences, because the tools were not specific to mechanize this type of materials [4].

Table 1. Cutting conditions.

Parameter	HSCO5524	HSCO5524	HSCO1261
Diameter (mm)	6	6	6
Tip angle (°)	118	118	130
Cutting speed (m/min)	61,07	16,30	16,30
Spindle speed (rpm)	3240	865	865
Feed rate (mm/min)	300	80	80
Feed rate (mm/rev)	0,092	0,092	0,092

As for the measurement of the diameters and the affected areas, they were realized by means of an optical microscope, coupled to a computer to process the image through the software Deltec Vision and its utilities to determine longitudinal magnitudes. Each diameter was determined by taking 10 points, and the software itself implemented its algorithm to determine the diameter by least squares, as well as the roundness error. The delamination factor (F_d) was defined as the quotient of the maximum diameter of the zone affected by the drill (D_{max}) between the nominal diameter of the hole drilled (D_o), that is, the diameter of the bit, as indicated by equation 1 [5].

$$F_d = D_{max} / D_o \quad (1)$$

We only focused on the peel-up delamination, that is, the one produced when the tool was inserted in the material, because as we mentioned, the plate was treated to avoid push-out delamination (at the exit) [6].

3 Results

In figures 1, 2 and 3 we represent the values obtained for the delamination factor in the initial and final bores of each of the two series performed for the three feed ratios used, that is, we have four data in each line represented in each graph.

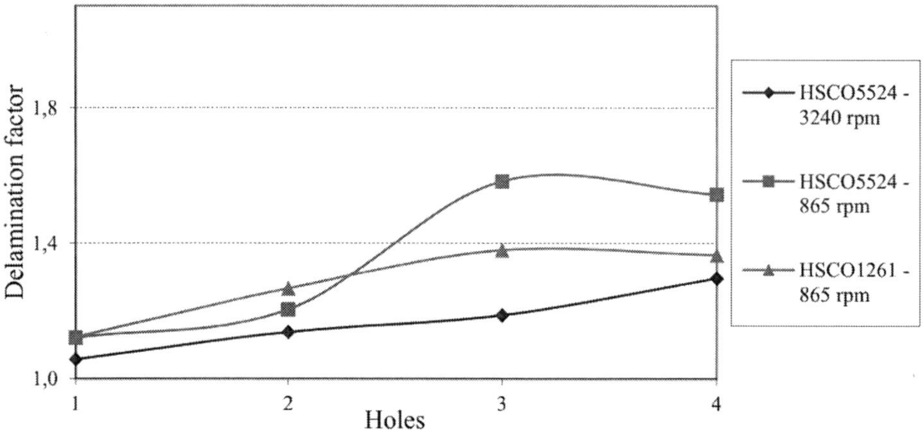

Fig. 1. Drills performed at 0,10 mm/rev.

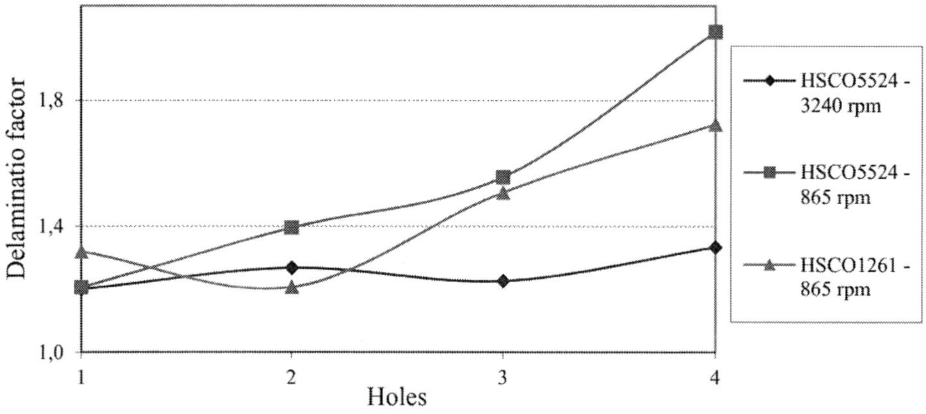

Fig. 2. Drills performed at 0,15 mm/rev.

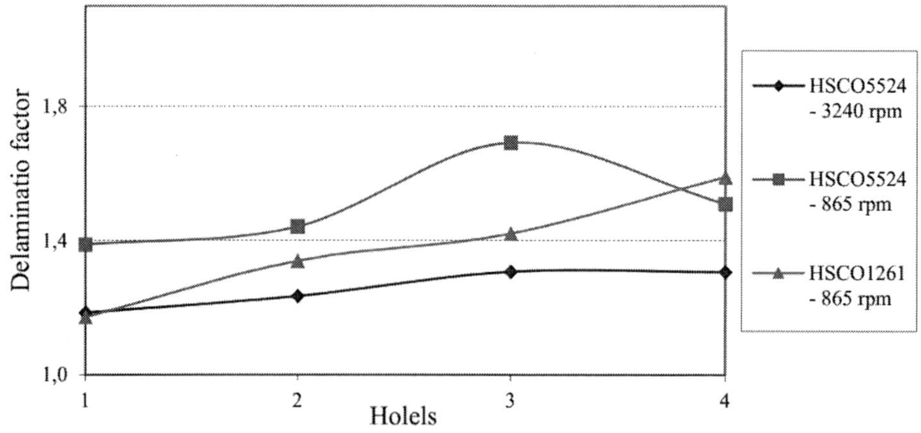

Fig. 3. Drills performed at 0,05 mm/rev.

In figure 4 we can see how the delamination factor evolves as the holes were produced with each drill. The wear of the tool was appearing, in some cases more progressively than in others.

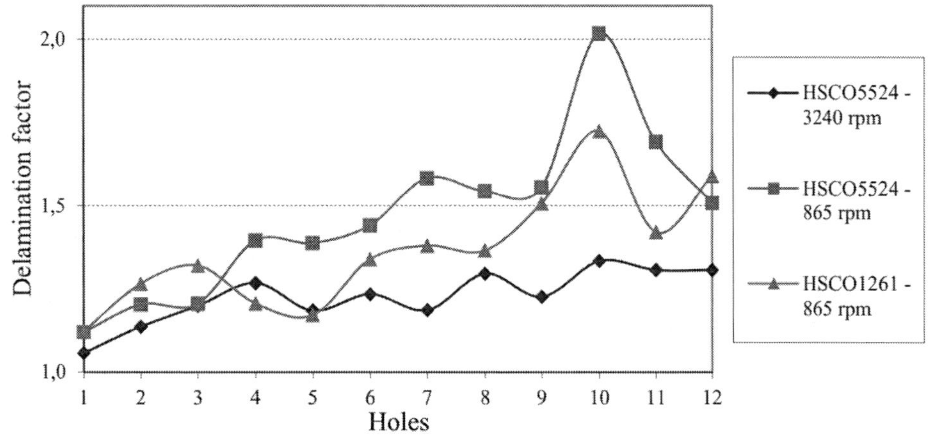

Fig. 4. Delamination in successive holes.

The different qualities obtained for the drills practiced are visible in the following images (see figures 5 and 6), corresponding to the first bore of a series and to the last, respectively. It can be seen how the wear of the tool has repercussions in obtaining a drill of poor quality only having made 30 holes with that tool.

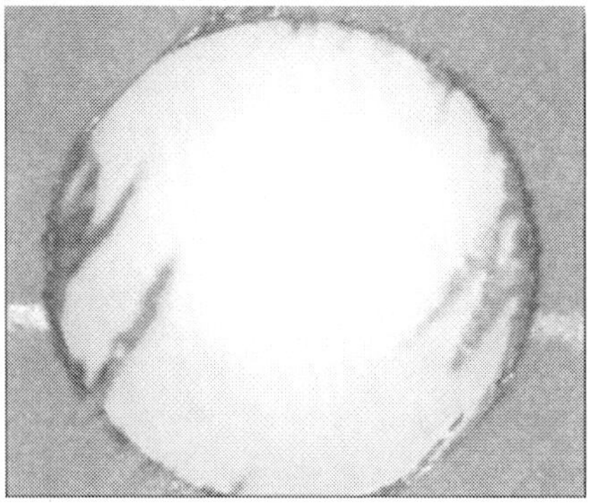

Fig. 5. First hole in a series.

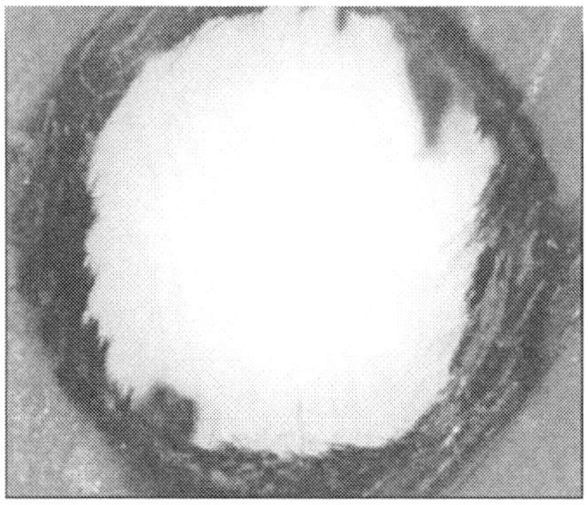

Fig. 6. Last hole in a series.

4 Conclusions

If we focused on the HSCO5524 drill bit, we could clearly see that the tool wears obtained for 3240 rpm were much lower than those obtained for 865 rpm, as well as the delamination data were also lower for the higher cutting speed.

If we focus on the cutting speed (16 m/min), we obtained better results with the HSCO1261 drill, probably due to the more progressive wear suffered by this tool. Both drills followed a similar behavior, but the delamination values obtained for HSCO1261, for the same cutting conditions, were generally lower than those obtained for HSCO5524.

We can say that the tool tip angle has great influence when machining carbon fiber, because under equal conditions, we obtain smaller delamination defects with the 130º bit.

The lower delamination obtained at high cutting speeds may be due to the increase in the temperature in the plate, which would reduce its resistance to machining. On the contrary, internal damages could have been generated that we cannot observe through these techniques.

In the HSCO5524 drill bit at 3240 rpm, it is observed that, in the last drills performed, the delamination factor remains almost constant, due to the lesser wear suffered by the tool.

When we worked with low feeds ratios and the tools already worn, we did not obtain higher values of delamination, reason why, at lower feed speeds, less delamination, although the tool wear was greater.

It is worth noting the excessive wear suffered by the tools in all cases, so it will not be profitable to mechanize these materials with tools of this type if we want to obtain a minimum of quality.

He would fail to do the experiment with a HSCO1261 drill at 3240 rpm, where it would be expected to obtain lower delamination factors.

References

1. M. San-Juan, Ó. Martín, and F. Santos, Int. J. Mach. Tools Manuf., **50**, no. 7, 591–600, (2010).
2. M. San Juan, Ó. Martín, F. J. Santos, P. De Tiedra, and A. Sánchez, Advanced Materials Research, **498**, 43-48, (2012).
3. N. Feito, J .López-Puente, C. Santiuste and M.H. Miguélez, Comp. Struct., **108**, 677–683, (2014).
4. A. Faraz, D. Biermann, and K. Weinert, Int. J. Mach. Tools Manuf., **49**, 1185-1196, (2009).
5. I.S. Shyha, D.K. Aspinwall, S.L. Soo, S. Bradley, Int. J. Mach. Tools Manuf., **49**, 1008–1014, (2009).
6. M. San Juan, O. Martín, F.J. Santos, J.A. Cabezudo, A. Sánchez, AIP Conference Proceedings, **1431**, Issue 1, 495-500, (2012).

Evaluation of sintered gears in operational conditions

Mariusz Rosiak[1,*], *Jerzy* Napiórkowski[2], *Magdalena* Lemecha[2]
[1]Opole University of Technology, Poland
[2]University of Warmia and Mazury, Poland

Abstract. This article presents the description of tested gears that were made from sintered material, as well as of gears made of grade 16MnCr5 steel (number 1.7131), which are installed in transmissions of power tools. The procedure for the gear tests was described, and a test bench and the methodology for carrying out the tests were presented. The aim of the article was to examine whether sintered gears could replace gears made of steel 16MnCr5 while maintaining the required stability in operating conditions. The results of the evaluation of geometry changes and the results of metallographic examinations were presented. The results achieved were analyzed and the sintered gear technology was evaluated. Recommendations were formulated concerning the enhancement of the technology with which sintered gears are produced in order to obtain products with even higher operational properties.

1 Introduction

Sintered products made from metal powders are winning even greater recognition from machine manufacturers. The automotive industry, among others, is showing its exceptionally high interest in the technology. A possibility to manufacture cheap parts that can work under high and complex loads, manufacturing without waste, reduction or elimination of machining in the technological process, no fibrousness of a material, lower mass, adjustable density size and distribution constitute only a part of the characteristics which render manufacturers highly interested in sintered products made from metal powders. It is very popular to manufacture small sintered parts in mass production, such as pistons, rings, plain bearing sleeves, gears in transmissions [1], while the automotive industry is the industry that is introducing sintered elements in their products at the highest pace [2]. The global market for powder metallurgy (PM) parts is expected to reach US$ 15.7 billion by 2020, driven by emerging applications in transport, construction and engineering. It is also expected to benefit from the recovery of the global automotive industry and an optimistic outlook for the commercial aerospace market growth. It is important to try to test the stability of sintered components in operating conditions. In order to prevent and limit possible damage to them in operating conditions, such parts are subjected to various tests in order to evaluate their stability [3]. Machines as entire technical systems are also subjected to the stability evaluation

* Corresponding author: m.rosiak@po.opole.pl

tests. Nowadays, modern methods for testing the mechanical condition of machines, such as thermal analysis, vibration analysis or acoustic effect analysis, are available. Some of these methods are characterized by high effectiveness; however, the results of such stability tests have a significant risk of uncertainty. Such a state of affairs gives an incentive to devise more effective methods for evaluating damage [4] and to approach individually the evaluation of operational stability of parts with physical modelling methods, which reproduce the actual operating conditions as closely as possible.

2 Test methods

The aim of the test was to examine the operating stability of sintered gears in conditions similar to operating and to evaluate the feasibility of their application as an alternative to steel 16MnCr5 (1.7131) gears that have been used in transmissions of power tools so far. Sintered gears were made from DISTALOY AE metal powder, which was selected based on the following criteria: predisposition of metal powder to produce parts with high strength properties, good compatibility, and chemical composition that provides high strength properties of a part after heat treatment, availability on the market and economic relationships. In order to enhance strength properties of a heat-treated part, DISTALOY AE metal powder was enriched with 0.3% carbon by adding graphite during a charge preparation phase. The sintered gears were made in typical conditions of powder metallurgy for this kind of material, which involved the following operations: preparation of a powder charge, compaction, sintering, and calibration. The chemical composition of DISTALOY AE metal powder is shown in Table 1[5]. The geometry of the tested gears is shown in figure 1.

Table 1. Chemical composition of the powder used to make sinters.

DISTALOY AE			
Element	Element content %		
	Min.	Max.	Actual
H2 loss	-	0.15	0.09
C	-	0.01	0.01
Ni	3.60	4.40	3.98
Cu	1.35	1.65	1.50
Mo	0.45	0.55	0.50
Fe			Rest*

Rest* - ASC 100.29 powder, which is the basis for the production of DISTALOY-type metal powders.

Performance tests of the test gears were carried out on a test bench, which was built for this purpose and aimed at obtaining the actual operational conditions of drills [6]. The tested gears were installed in transmissions of electric drills. The selected gear made of steel 16MnCr5 (1.7131) and the sintered gear for tests are shown in figure 2. The diagram of the test bench is shown in figure 3.

Total operating time of each gear was 800 hours. The gears installed in transmissions of drills operated in a cyclical manner at a rated voltage of 230 V (U=230 V) [6].

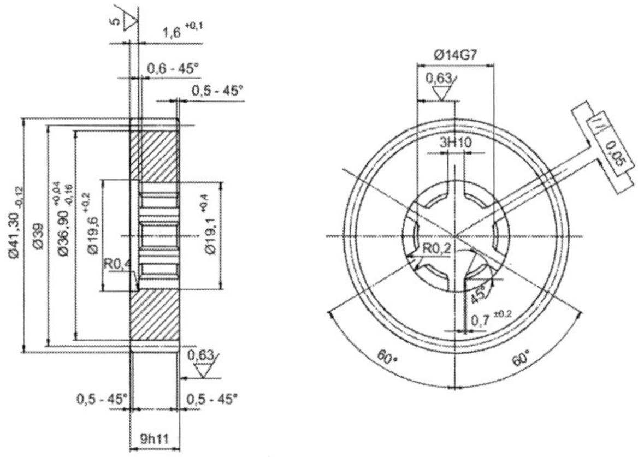

Type of teeth	Straight		Mutual cooperation error	Fi	0.070
Module	m	1	Single mutual cooperation error	fi	0.025
Pressure angle	α	200	Side tooth profile error	Δe	0.018
Addendum	ha	1*m	Radius of base circle	rz	18.324
Dedendum	hf	1.2*m	Base circle error	Δrz	0.016
Number of teeth	z	39	Pressure pitch	tp	2.952
Profile shift	X*m	+0.15	Pressure pitch error	Δtp	0.018
Tooth thickness	s	1.68	Circular pitch evenness error	Δo	0.018
Base tangent length	W5	$13{,}93_{-0{,}04}$	Deviation of profile per tooth width		0.018

Fig. 1. Drawing of gears used in tests.

Fig. 2. The selected gears for operational tests on the test bench: steel 16MnCr5 (1.7131) gear on the left, sintered gear made from DISTALOY AE + 0.3%C metal powder on the right.

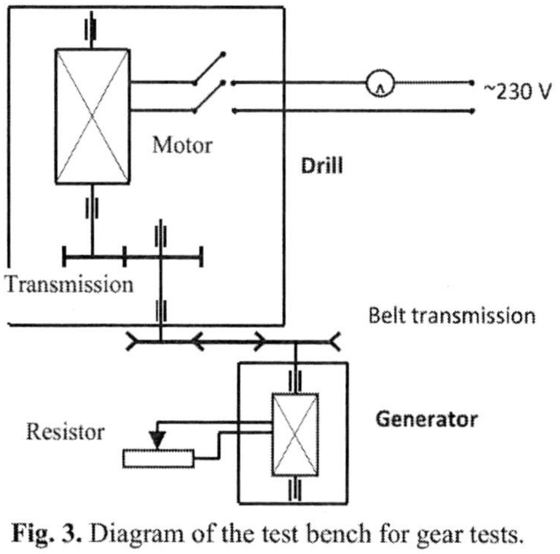

Fig. 3. Diagram of the test bench for gear tests.

Drill tests on the test bench were performed according to the following plan: 2-hour working cycles (involving the following phases: drill work time under load for 100 s, then a 20-s break), and 1-hour breaks for the drill to cool off. The repeated 2-hour working cycles followed by 1-hour breaks were carried out over the next 50 hours. Then, the drill in the grip was rotated 900 [6], and the tests according to the above plan were resumed. The work time included 20-s breaks but it did not include 1-hour breaks for the drill to cool off. Before the tests began, the gears were subjected to geometry measurements and a general assessment. The geometry measurements included dedendum diameter, addendum diameter, whole diameter, groove width, base tangent length W5, single cooperation error fi, total cooperation error Fi (see figure 1). After each 200 hours of operation (i.e. after 200 h, 400 h, 600 h, 800 h), the drills were disassembled, the gears inside removed and washed in paraffin. Then, the gears were subjected to measurements, which at the same time involved checking whether they were damaged with the naked eye.

3 Test results

The values of the selected geometry parameters of the gears after the testing periods on the test bench are shown in Figure 4-5. Importantly, in the process of gear wear, there are changes of the condition of the surface and dimensions where the cooperating parts, which bear the load, have a direct contact with each other. The toothed-wheel rim and spline grooves were especially exposed to the change of the surface and geometry. The evaluation of geometry changes of the toothed-wheel rim was made on the basis of the results of the base tangent length W5 measurements through five teeth.

The sintered gears underwent metallographic examinations. Structure photographs were taken in the cross section of the tooth in zooms from 50 to 500 times. After heat treatment (austenitization combined with quenching in oil) of the gear, metallographic examinations of the structure of the gear material, which was made from the sinter based on Distaloy AE powder, revealed that it had a very complex structure. Inhomogeneity of the chemical composition and the presence of pores (voids) contributed to a diverse process of heat transfer phenomenon in local areas. Across the entire section, evenly distributed areas of coarse acicular martensite, surrounded by fine acicular martensite, which transform into bainine, were revealed. In the structure, there was a very small number of white ferrite grains surrounded by narrow areas of fine pearlite and bainine. Areas with coarse acicular martensite reach a maximum hardness of approx. 1000 ÷ 1050 HV0.1.

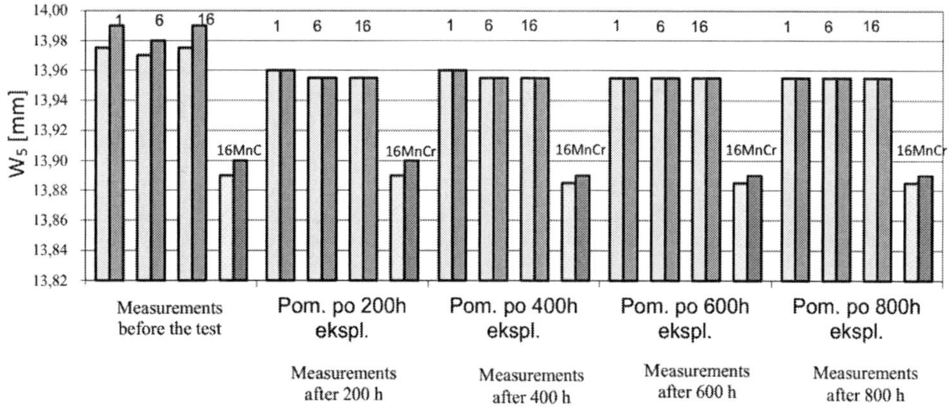

Fig. 4. Values of a base tangent length measurement W5 [mm] (Figure 2) "base tangent length through 5 teeth", measured in sintered gears and steel 16MnCr5 (16HG) gears after operational periods (1,6,16 - marks of the tested gears made from Distaloy AE + 0.3%C powder, □□minimum value, ■■ maximum value, allowed difference between maximum and minimum length W5 is 0.04 mm).

The evaluation of the geometry change of spline grooves was made on the basis of the measurements results of the spline groove width (Fig. 5).

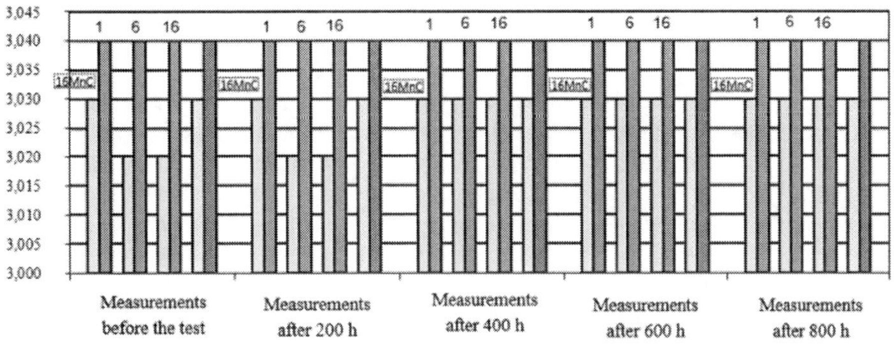

Fig. 5. Values of a "spline groove width" measurement L [mm] measured in sintered gears and steel 16MnCr5 (16HG) gears after the operational periods (1,6,16 - marks of the tested sintered gears, □□ minimum value, ■■ maximum values, 3H10 size and tolerance zone of the spline groove width).

The presence of alloying components and 0.3%C, which involved adding graphite (during the charge powder preparation phase) to Distaloy AE powder, enabled the gear material to obtain so high hardness after austenitization and quenching in oil. Metallographic images of the structure from the area of the sintered gear tooth are shown in figure 6.

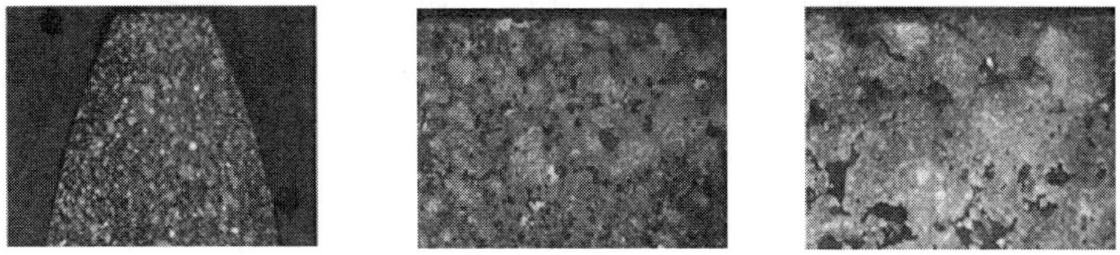

a. x50 zoom **b.** x200 zoom c. x500 zoom

Fig. 6. Images of the structure taken on the cross section of the sintered gear tooth (material -- Distaloy AE+0.3%C, heat treatment - austenitization and quenching in oil).

3 Discussion of the results

The measured mean value of length W5 (base tangent length through 5 teeth) on the gears made from powdered material before the tests using the test bench was 0.05 mm higher than the basic length. After 200 h of operation, there was a significant change in the value of length W5. Further hours of operation of the gear did not have a significant impact on the value of the base tangent length. The maximum change of length W5, which was measured after 800 hours of operation, was 0.035 mm, so it was lower than the minimum tolerance of 0.04 mm. The mean value of base tangent length W5, which was measured on the steel 16MnCr5 gears before the tests, was 13.89 mm. After 400 hours of operation, the value changed noticeably to 13.887 mm. Further operation of the steel 16MnCr5 gears on the test bench until the end of test, i.e. until 800 hours passed, did not have an effect on the change of the base tangent length value. The gears based on metal powders showed no difference between the minimum and maximum values of the base tangent length after first 200 hours of operation and further operation hours. The steel 16MnCr5 gears showed small differences between the minimum and maximum values of the base tangent length after each operational period.

The spline is the other essential element in the gear structure, which is directly involved in bearing loads. The width of spline grooves in the gears based on metal powders before the operational tests showed higher differences than the steel 16MnCr5 gears (see figure 4). After 400 hours of operation on the test bench, both gears based on metal powders and steel 16MnCr5 gears were characterized by a small difference between the measured values of the spline groove width, which did not exceed the tolerance zone. The analysis of the results achieved indicated high dimensional stability of the tested gears. Examinations of the surface condition of the tested gears with the naked eye after the operational periods did not reveal any changes. No cracks, cavities or damage were noticed. The measurements of the tested gears in respective operational periods, as well as after 800 hours of operation, showed that the geometry dimensions of the gears changed to a smaller extent than the allowed dimensional tolerance (the stability of solid steel gears is assured for 400 hours of operation).

Due to the dimensional stability, the tested gears could have been further used. The performed tests enabled the formulation of the following statements:

- the results of the geometry measurements of the tested gears after consecutive operational phases showed no changes of the geometry that exceeded the dimensional tolerances;
- this new technology can be used to manufacture gears for transmissions of power tools due to high dimensional stability;
- no damage to the surface of the tested gears was observed, even after their normative operation time was increased by another 400 hours;
- the applied technology of manufacturing gears, based on iron powder as charge material, enabled the products that meet the stability requirements to be obtained;
- the presence of porosity in the sintered material of the gears did not initiate the development of micro cracks in the structure, which could have led to premature damage and the separation of material particles from the surface.

References:

1. http://www.inop.poznan.pl/
2. http://www.strategyr.com/MarketResearch/Powder_Metallurgy_Market_Trends.asp
3. J. Krolczyk, G. Krolczyk, S. Legutko, J. Napiorkowski, S. Hloch, J. Foltys, E.Tama, Tehnicki Vjesnik – Technical Gazette, **22** (6), 1447 – 1456 (2015)
4. R. Kumar, S. Chattopadhyaya, S. Hloch, G. Krolczyk, S. Legutko, *Maintenance and Reliability*, **18** (1), 128 – 135 (2016)
5. www.hoganas.com
6. PN-85/E-08400/02, PN-92/E08409,

Digital Direct Metallization (DDM) by a plasma coating process on polymeric components and its applications

Alexandru Sover*, *Pilipp* Popp, and *Tamer* Aydin

Ansbach University of Applied Sciences, Faculty of Engineering Sciences, Residenzstr. 8, Ansbach, Germany

Abstract. Digital Direct Metallization (DDM) is an innovative process for the production of metallic layers on plastic substrates, e.g. to generate electrical circuit path. This paper describes the manufacturing process of Digital Direct Metallization (DDM) in the case of plasma based copper coating on polymeric components. The process-temperature has a significant impact on the metal deposition and can be controlled by the distance between the plasma head and plastic part or/and its cross-runs. The deposition of the copper powder on the polymeric surface was substantial improved by a temperature close to the melting point of the polymer.

1 Introduction

Digital Direct Metallization (DDM) is an innovative plasma process for the metal deposition on different substrates materials, planar or three-dimensional. The flexibility of this technology offers a great solution for the currently existing technologies of metal deposition on plastic materials and open new ways for different applications, i.e. design, electronic, sensors, telecommunication etc. [1-3].

The combination of plastic and metal on the surface of the parts can also improve the quality and functionality of the components concerning design, thermal and electrical properties, which are presented in different papers [1-4]. One of the new applications is the production of circuit layers, which can replace the conventional chemical metallization processes of the printed circuit board (PCB). The former is presented in papers [2-4] and applied by companies like Plasma Innovation.

Plasmadust® is a thermal spray technology using cold active plasma to melt a metal powder such as copper, which can be deposited directly on a surface [3]. This can be described as additive metallization technology.

Joerg Franke *et. all* present in different papers [1, 3] the importance of the particle size and the surface treatment with e.g. solder pastes and the interconnections between the substrate and the components in electronics [3]. Also the particle sizes of the metal powder of the process can highly influence the diffusion distances of the plasma jet [3].

* Corresponding author: a.sover@hs-ansbach.de

2 DDM Technology and used materials

The main advantages of the Plasmadust® technologies are the variable process temperatures ranging between 90°C and 180°C, the good adhesion of the metal layer, the high process speed of about 100 mm/s and the fast growing metallic thickness with about 15 µm thickness per cycle [4, 5]. The settings of parameters like temperature, process speed or powder feed rate in the process have an important influence on the thickness of the metallic layer that can be achieved on thermoplastic or thermosets substrates [4-5]. Plastic specimens with or without surface treatment can be metallized with no or minimal thermal damage of the plastic surface, depending on these plasma process parameters [4].

The Plasmadust® technology can combine the surface treatment and metal deposition or can use plastic parts with an already mechanical prepared surface, e.g. one that is structured by laser (Laser-Direct-Structuring LDS). This provides the necessary surface structure for a better adhesion of the melted or partial melted metal particles [3]. The following picture presents the Plasmadust® technology by Plasma Innovations GmbH.

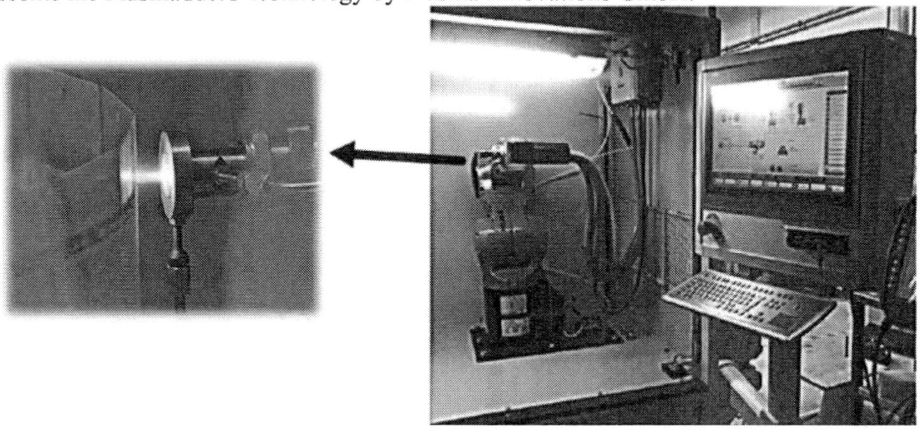

Fig. 1. Plasmadust® Equipment for Digital Direct Metallization (DDM).

The devices for active atmospheric plasma-metallization consists of a plasma head, installed on a robot arm, which can move in a small chamber in different directions, following the surface geometry of a product. For this machine argon gas is used as it can provide the desired plasma qualities but also due to its low costs and wide availability. The robot arm can be programmed to perform several move cycles to increase the thickness of the deposition and also for keeping a definite distance between the plasma head and surface, and the speed of movements can be adjusted. The following figure shows the plasma head with the powder feeders of the DDM equipment.

Fig. 2. Plasma head (left) for the DDM-Technique and the copper feeder (right).

To avoid the thermal damage of the surface a certain distance between the plasma jet and surface is necessary, otherwise the polymeric surface can be deteriorated and the adhesion of the "printed" structures is negatively affected. The metallic powder used for the metallization is a special high tech copper powder produced by ECKART GmbH (member of ALTANA Group) with the trade name Plasmavario 1001 Copper [6]. This powder is a high purity copper (99.9%) with very fine spherical particles (average particle size 10µm) and narrow particle size distribution at a high electrical conductivity [6]. The copper powder is added to the plasma head where under high temperature it melts and from where it is projected with the argon gas onto the surface which should be metallized.

As polymer materials with different melting points were used in the experiments, the distance to the surface needed to be adjusted for each polymer to avoid the melting of the material but also depending of the surface treatment and copper dosage. At a distance between 10 and 30 mm the temperature on the specimen surface can vary between 80 to 120°C by per cycle (Fig. 2).

In the first test Polypropylen (PPH 9069) and Polyamide 6 (TECHNYL® C 206F) were used. The dimension of the prepared specimens was 100x150x4 mm. The melting point of the both materials are 165°C for PP respectively 222°C for PA6.

The used parameters for the plasma deposition on the both polymer materials were:

- Distance plasma head to surface (30, 20 and 10 mm)
- Moving speed of the plasma head (0,3 mm/s)
- Gas supply (25 l/min)
- carrier gas (5 l/min) for the feed rate copper
- feed rate 10, 20 and 40%

The plastic specimens were fixed on a device and half of the surface was pre-treated with plasma, immediately before the metal deposition, to improve the adhesion of the powder particles.

The third polymeric material tested was a Polyphenylen Sulfid (PPS) (Therma-Tech™ TT9200-5010 EC), a high temperature polymer with the melting point at 310°C and very good thermal management. The Therma-Tech™ compounds by the PolyOne Corporation improved thermal conductivity up to 100-times that of conventional plastics [7].

The PPS parts were coated with a thermoset varnish (PLASMACOAT PU41) which acted as an electric insulation layer between the metal layer and the circuit carrier. The electric insulation and non-stick varnish provided also the possibility to protect some areas from the metal deposition and allows the generation of the conductor lines for logic circuits. The coated layer can be structured or removed by a laser, to provide certain surface characteristics (roughness) for the metal deposition, Figure 7. The roughened substrate offers the necessary environment for a mechanical bond between the first melted or softened metal particles and the polymer. The coating thickness increased with each deposed metallic layer.

3 Results

The PP and PA plastic specimens were metallized at different parameters, as described above. The procedure is presented in the following figure in which two different prepared PA surfaces where metallized: a- no treatment, b- with plasma treatment.

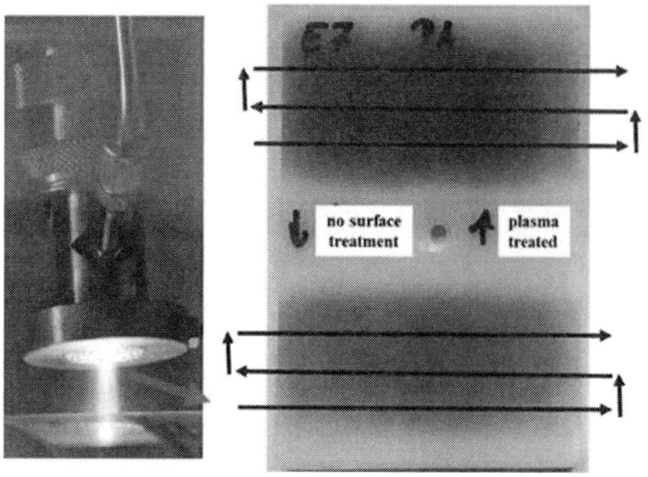

Fig. 3. PA specimen with (top) and without plasma surface treatment (down) at plasma metallization with the path of the plasma head.

The arrows show the path of the plasma head for the metal deposition. The temperature on the specimen surface showed values between 90°C (first run) and 124°C (third run) at a distance of 30 mm [6]. The deposition of the copper on the PP and PA specimens was hardly influenced by the distance between the plasma head and the surface.

At a distance of 30 mm only a thin cooper layer with the dimension of ca. 10 µm was observed, which is also the average copper particle size. Also, the particles kept their spherical form, which indicates that they were not completely melted before the deposition. The adhesion of the metallic layer was very poor; it could be partially removed with a finger. When the distance between the plasma head and specimen surface from 30 to 10 mm was decreased an improvement of the deposition was observed for both samples (PP and PA).The PP sample showed even better deposition compared to the PA sample (Fig. 5).

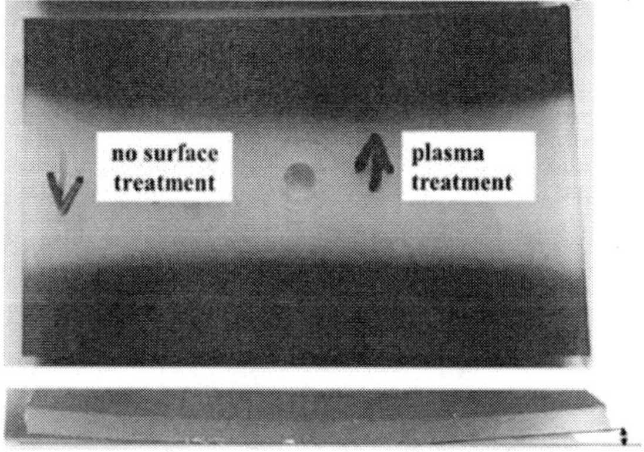

Fig. 4. Local copper metallized PP samples with and without surface treatment at 10 mm distance between plasma head and surface [5].

The adhesion of the copper particles on the PP surface was much better compared to the adhesion to the PA samples. A reason for this effect can be the lower melting point of the PP material (ca. 165°C), which facilitates a local melting of the PP polymer and the copper particles can be better enclosed in the surface. An evidence of the high temperature exposure of the surface is the warpage of the plate specimens, with was more than 1 mm, Figure 5 above. Also, the microscopically comparison of the copper layer on the PP surface at 30 and

10 mm plasma head distances shows different types of adhesion of the copper particles on the material surface.

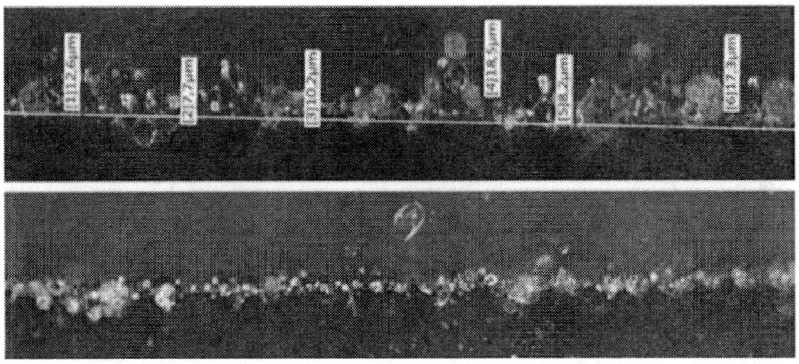

Fig. 5. Copper Layer on PP specimen by 30 and 10 mm distance of the plasma head to the surface [5].

In the upper image of Figure 6 one can see that at higher temperatures on the polymer surfaces the copper particles are enclosed in the plastic material. This increases the adhesion of the first metallic layer, but it can also change the dimensions of the plastic parts Figure 5. When the powder feed rate was increased up to 40% the metallic layer became thicker - between 20 and 40 μm.

The next figure shows the circuit carrier after the copper depositing processes. The PPS material was coated with an anti-adherent varnish and after drying the circuit path was structured / removed by IR-laser.

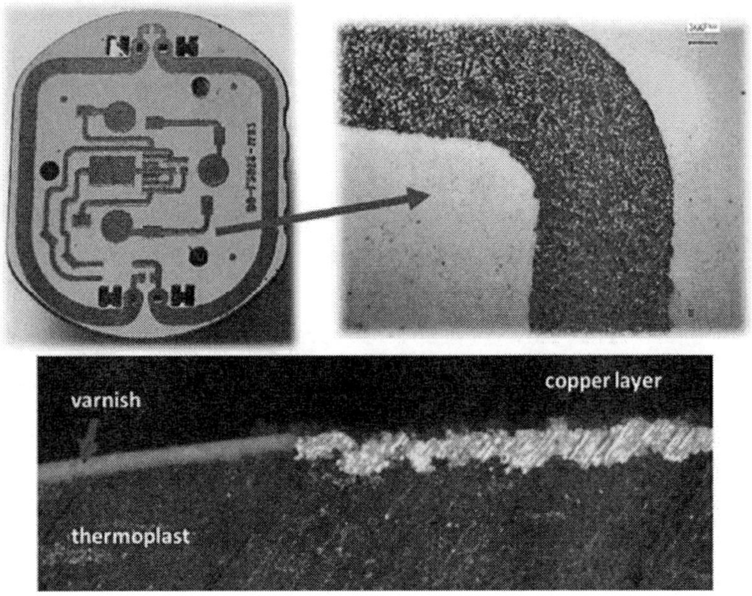

Fig. 6. LED plastic carrier produced with MID technique by DDM metallization [8].

Figure 7 shows a cross section through the part, where the varnish and copper layer on the PPS polymer can be seen. The copper particles are adhering on the surface structure without thermal damage of the substrate and form a continuous coating. The particles are melted by plasma and accelerated to the surface by the gas jet where they solidify and build a mechanically bond, firstly with the roughened surface and then with each other as the coating thickness is increased. At a higher optical magnification, the spherical geometry of the particles can be perceived, which shows that the particles are not completely melted. The layer is electrically conductive and the high thermal resistance of the plastic materials allows a higher process temperature which results in a better adhesion. With these types of polymers, the metallization of the surface can be done more easily without other warpage effects on the components.

4 Conclusion

The deposition of copper on three different polymeric substrates (PP, PA, and PPS) was tested under different conditions. Without any surface structuring the adhesion of the copper substrate depends mostly on the thermal proprieties of the polymeric material. A higher temperature on the polymer surface, close to his melting point (e.g. 165°C for PP) leads to a better deposition of the copper powder on the surface. At higher temperatures the copper particles can easily be integrated in the melted polymer film, but the geometry of the plastic specimen can suffer a highly (more millimeters) warpage. On the other hand, a structured surface can provide a mechanical bond for the melted or partially melted copper particles. The plasma pre-treatment of the surfaces did not significantly improve the adhesion of the copper layer on the examined PP and PA polymers.

A special gratitude I give to the students of the HS Ansbach who contributed in different steps with their work to this research project. Special thanks for her support go to Ms Antje Sover.

References

1. J. Franke, Materials, Manufacturing, Assembly and Applications for Injection Molded Circuit Carriers, Carl Hanser Verlag (2014)
2. M. Mueller, J. Franke, *Spring Seminar on Electronics Technology* (2016)
3. J. Franke, A. Syed-Khajaa, R. Schramm, R. Ochs, *Packaging through Additive Plasma Technology*, Procedia CIRP, **37** (2015)
4. R. Schramm, J. Franke, *Electronics Packaging Technology Conference* (2013)
5. P. Popp, T. Aydin, Plasma Oberflächenbehandlung, Projektarbeit an der HS Ansbach, (2017)
6. U.-A. Hirth, High Tech Copper Powder, for Digital Direct Metallization, Technical Seminar at kunststoffcampus Bayern (2016)
7. J. Kuntermann, Thermally conductive plastics: Challenges, experience and the road ahead, Technical Seminar at kunststoffcampus Bayern (2016)
8. M. Horn, H. Hanson, Voruntersuchung und Testverfahren zur Beurteilung der Eigenschaften von plasmabeschichteten MID-Bauteilen, Projektarbeit HS-Ansbach (2016)

Simulation of extrusion of high density polyethylene tubes

Antonios Koutelieris[1], *Kyriaki* Kioupi[1], *Onoufrios* Haralampous[1], *Konstantinos* Kitsakis[1,2,*], *Nikolaos* Vaxevanidis[3], and *John* Kechagias[1]

[1]Technological Educational Institute of Thessaly, Mechanical Engineering Department, 41110 Larissa, Greece
[2]University of Western Macedonia, Mechanical Engineering Department, 50100 Kozani, Greece
[3]School of Pedagogical and Technological Education, Mechanical Engineering Educators Department, 14121 N. Heraklion Attikis, Greece

Abstract. The production of extruded polyethylene films rods, tubes and pipes is a typical industrial process that has been extensively investigated over many years. In this study the extrusion of High Density Polyethylene (HDPE) tubes was simulated by using the 3D finite element simulation software StarCCM+®. The simulation was applied in order to examine the influence of design and operational parameters on the characteristics and the soundness of the tube and for optimizing the overall quality of the product. Two alternative die configurations were studied; one with a four inputs head and the other with an eight inputs head. From the results obtained it is concluded that extrusion with the eight inputs head die design results in better overall pipe quality; this design was implemented successfully in an industrial production line.

1 Introduction

High-density polyethylene (HDPE) tubes are typically used in many applications such as to convey potable water, waste water, chemicals, etc. HDPE has high strength, toughness, durability and resistance to chemicals. In order to manufacture solid wall pipes the following steps should be followed: heat, melt, mix, and convey the raw material into a mold to achieve the final form.

Finally, the shape should be kept uniform during the cooling process. All these phases were implemented within a special-purpose designed pipe manufacturing line (Fig. 1).

Figure 2 shows the extruder set-up.

* Corresponding author: kitsakis@teithessaly.gr

Fig. 1. A typical pipe manufacturing line [1].

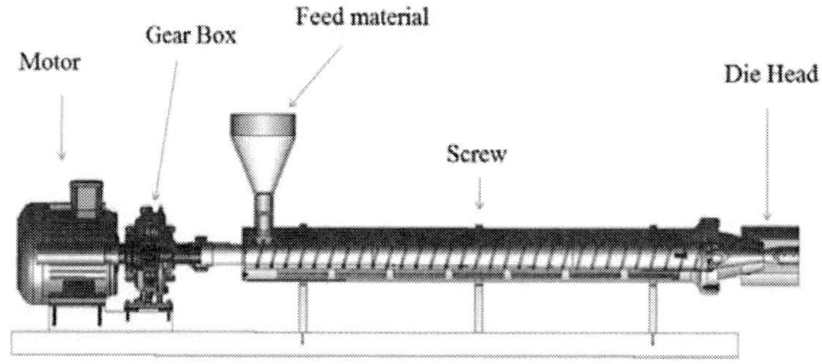

Fig. 2. Extruder set-up.

The design of the die head is one of the most influencing factor that affects the pipe shape quality. The main characteristics of die heads are the number of screw inputs and the geometry of the screw [2-6].

This work studies the extrusion of HDPE tubes (equivalently called pipes) by applying 3D simulation of the extrusion process. Two alternative die configurations were studied; one with a four inputs head and the other with an eight inputs head. From the results obtained it is concluded that extrusion with the eight inputs head die design results in better overall pipe quality; this design was implemented successfully in an industrial production line.

2 Simulation set-up

In order to simulate the process at three dimensions (3D), as the first step the two configurations of head die in a 3D CAD program had to be designed; Solidworks® was used for this job. Figure 3 shows the head components, all of them being designed at 1/1 scale (final dimensions).

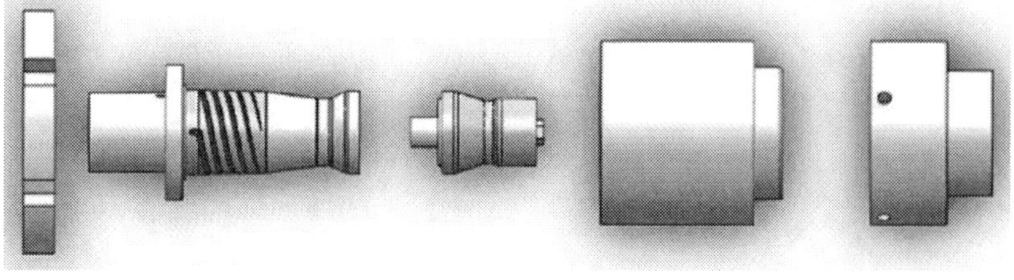

Fig. 3. Components of the four input head die (by Solidworks®).

Figure 4 shows the die with raw material inside (red color) as it placed in extruder line.

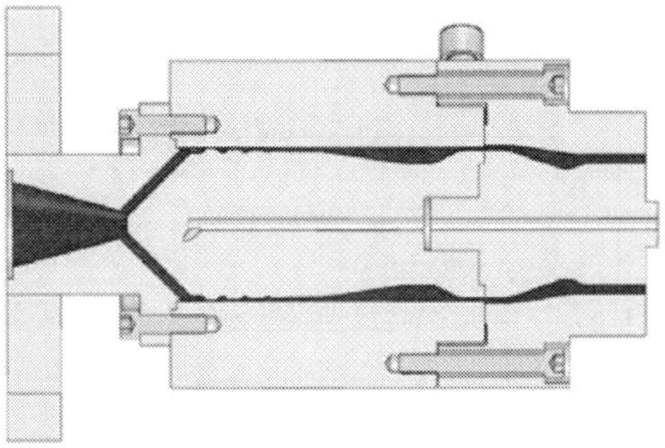

Fig. 4. Head cross section view.

In Figure 5 and 6 the volume of raw material is represented as 3D volume models for the two configurations (four inputs and eight inputs heads) examined.

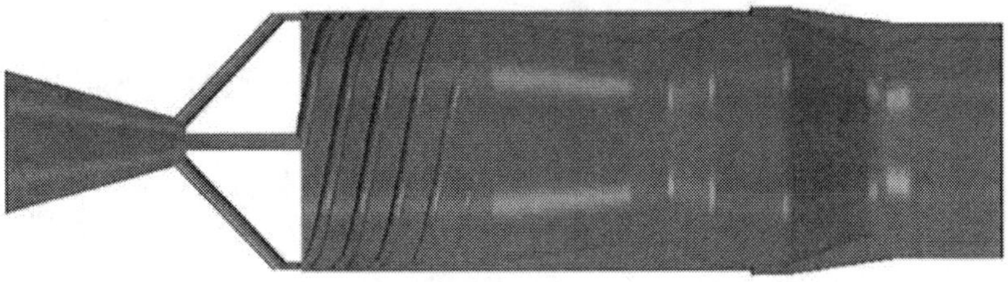

Fig. 5. Volume of raw material inside the head with four inputs.

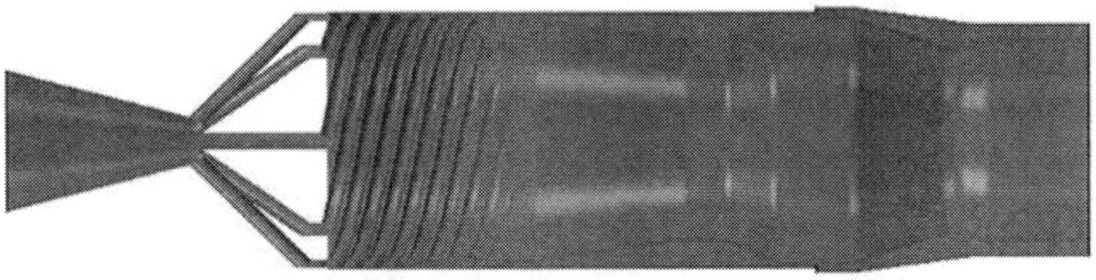

Fig. 6. Volume of raw material inside the head with eight inputs.

Then, the 3D CAD representations were transfer in the StarCCM+ software in order to configure the tessellation parameters and to produce the tessellations models (Fig. 7). The models that used for the tessellations were: polyhedral mesher, prism layer mesher and surface mesher. The "base size" for the tessellation process was selected to 5μm.

For solving the problem, a number of assumptions were made:

- first that the flow is in a steady state and second that
- the heat flow is adiabatic (adiabatic head wall; (see figure 8).

The initial conditions for the simulation were:

- Mass Flow Rate at the entrance: m = 0.05556 kg/s
- Fluid Temperature: T = 200°C
- Physical Model: isothermal process.
- At the exit the pressure is set at one atmosphere.

The material properties were set as follows:

- Density: 950 kg/m^3
- Viscosity follows the Carreau-Yasuda model:

$$\mu = \mu_\infty + (\mu_0 - \mu_\infty)\,(1 + \lambda\gamma)^\wedge[(v - 1)/2] \tag{1}$$

Where μ_∞=0.
Therefore, the viscosity was calculated by the formula:

$$\mu = \mu_0(1 + \lambda\gamma)^\wedge[(v - 1)/2] \tag{2}$$

It has the following parameter values:

- μ_0 = 2100 Pa-s; Zero Shear Viscosity
- μ_∞ = 0.0 Pa-s; Infinite Shear Viscosity
- v = 0.28612; Power Constant:
- λ = 0.058 s; Relaxation Time
- γ: Shear Rate: This value was obtained by the program in order to calculate the viscosity

Fig 7. Tessellation model of the two models (four and eight inputs respectively).

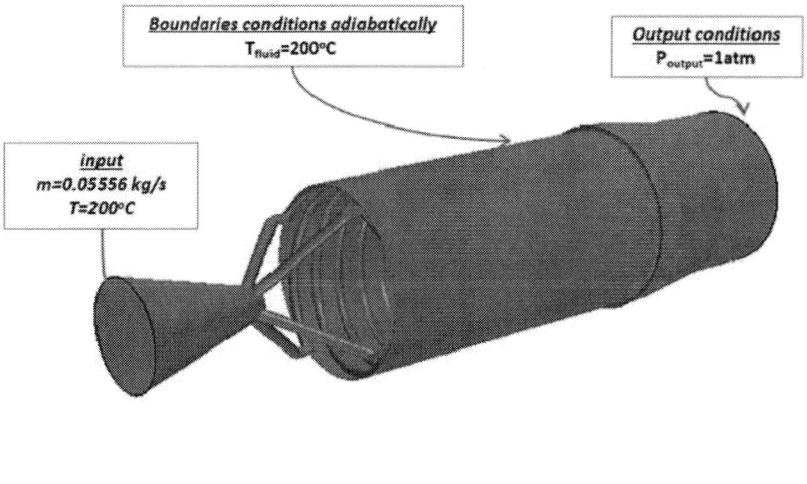

Fig. 8. Boundary conditions of the FE model.

3 Simulation results and discussion

After the FEA simulations of the two different setup the following are observed:

- The use of the 4 inputs head resulted in higher values of the pressures and the velocities in comparison to the ones estimated for the 8 inputs head (see Figs 9, 10 &11).

- The topology of these values forms a polygon (see Figs 10 & 11). This, is an indication that the shape of the pipe at the exit should be more precise at 8 input head than that of 4 input head.
- At 8 input head the pressure increased in a general way with uniformity along the head; moreover, we observe lower velocities of the fluid material in this case that the ones with the 4 inputs head.

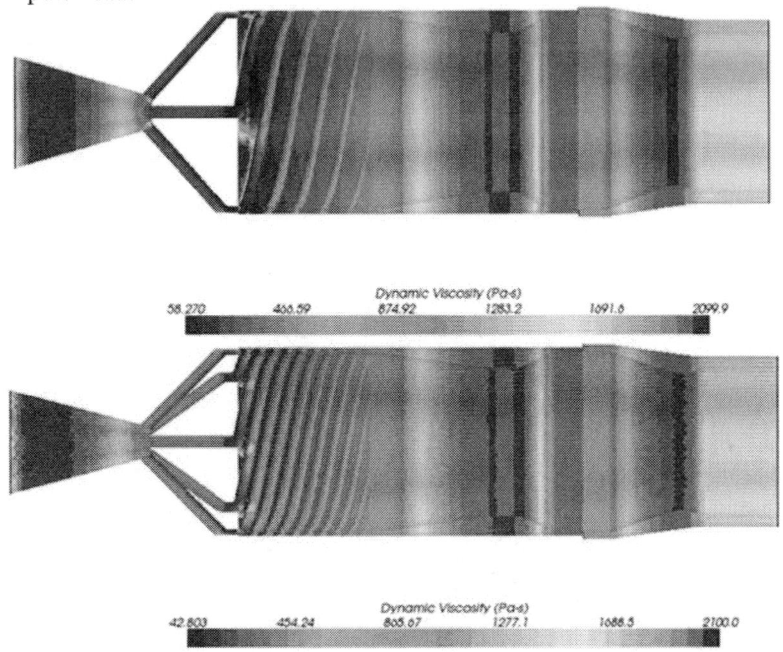

Fig. 9. Dynamic viscosity of fluid material along the die length (4 and 8 inputs head respectively).

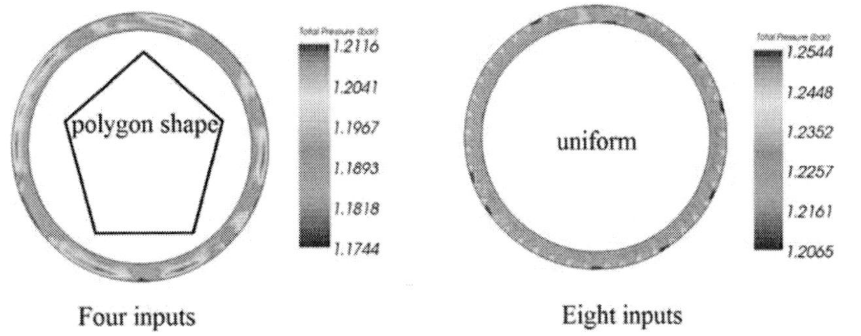

Fig. 10. Output pressure.

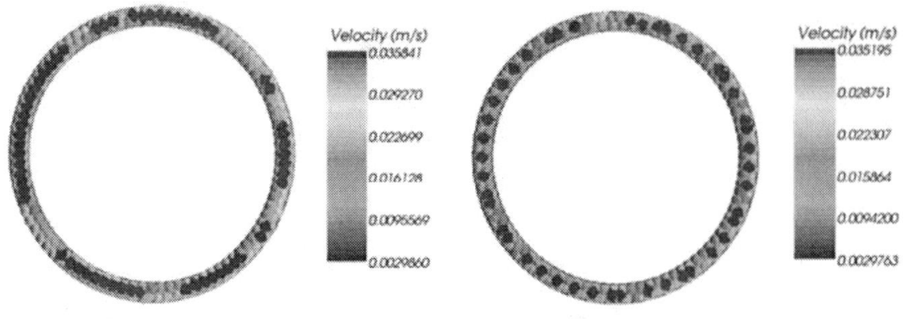

Four inputs Eight inputs

Fig. 11. Velocity distribution.

The 8 inputs head is manufactured and gave better quality pipes in the production line. The quality indicators measured were the wall thickness of the pipes and the precision of the profile.

4 Conclusions

This paper studies the thermo-mechanical behavior of HDPE plastic pipes extrusion. The conclusions drawn are:

The rheological results were better when the 8 inputs head was used for plastic HDPE pipe extrusion since, in general, lower pressures, velocities, and viscosity were obtained. The polymeric fluid material flux possessed better characteristics than that of 4 inputs head.

The quality and the soundness of the pipe product was improved when the 4 input head was replaced by the 8 inputs head in the production line.

The authors are grateful to Zisis Buzoukis Plastic Manufacturer Co for kindly providing the equipment and the material for dies manufacturing.

References

1. *Handbook Of Polyethylene Pipe*, 2nd Ed., http://plasticpipe.org/publications/pe-handbook.html
2. G. Raju, M. L. Sharma, M. L. Meena, Int. J. Adv. Mech. Eng., *Recent Methods for Optimization of Plastic Extrusion Process: A Literature Review*, **4**, 583-588 (2014)
3. A. G. Mamalis, A. K. Vortselas, G. Kouzilos, J. App. Polym. Sc., *Tube extrusion of polymeric materials: Optimization of the processing parameters*, **126**, 186-193 (2012)
4. E. Taliadorou, G. C. Georgiou, E. Mitsoulis, Rheologica Acta, *Numerical simulation of the extrusion of strongly compressible Newtonian liquids*, **47**, 49-62 (2008)
5. Y. Huang, C. R. Gentle, M. Lacey, P. Prentice, Mat. & Des., *Analysis and improvement of die design for the processing of extruded plastic pipes*, **21**, 465-475 (2000)
6. B. D. Freeman, M.M. Denn, R. Keunings, G.E. Molau, J. Ramos, J., Polym. Eng., *Profile Development in Drawn Hollow Tubes,* **6**, 171-186 (1986)

DOI: 10.1051/matecconf/201711204005

Preliminary study on the microstructure and mechanical properties of dissimilar joints of aluminum alloy and pure copper by FSW

Daniela-Monica Iordache*, *Cătălin-Marian* Ducu, *Eduard-Laurentiu* Nițu, *Doina* Iacomi, and *Adriana-Gabriela* Plăiașu

University of Pitești, Manufacturing and Industrial Management Department, Târgu din Vale Street No.1, Pitești, Romania

Abstract. Aluminum and copper are two common metals and Al-Cu joints are commonly used in the power industry. Overlay joints of 3105 aluminum alloy and commercially pure copper were produced by FSW, microstructure and mechanical properties was investigated. A combination of optical and scanning electron microscopy is used to characterize the microstructures in the copper and heat affected zone in the aluminum near the fusion line. The hardness distribution was analyzed and residual stresses were measured at the stir zone, copper side and aluminum alloy side.

1 Introduction

Joining of metals and materials is an important stage in manufacturing and assembling of industrial and engineering parts. Friction stir welding is a variant of friction welding and represents the newest welding process introduced in practice [1-2]. This process was invented in Cambridge, at the Welding Institute in 1991. The procedure allows joining without melting the parts, without their relative pressing and without filling. Welds are the result of a complex heating action by friction and plastic deformation by means of a tool moving in rotation and translation. Friction stir welding is mainly used to weld aluminum and its alloys, resulting strong joints with good plasticity, although aluminum is difficult to weld by conventional methods [3]. Current research in the field are oriented towards welding dissimilar materials (welded plates have different materials) used especially in the construction of parts specific to the aerospace industry, transport, automotive industry, electric power industry [4-5]. Aluminum and copper are two common metals in the electric power industry, but the dissimilar combination of aluminum and copper is generally difficult for fusion welding because of the wide difference in their physical, chemical and mechanical properties [6-8].

Considering current research, joining Al-Cu is of particular importance in future electric cars as electrodes for batteries or electrical conductor [7-9].

In the present work, the microstructure and mechanical properties of dissimilar friction stir welded overlay joints of aluminum alloy (EN AW 3105) and pure copper (Cu) were evaluated.

*Corresponding author: monica.iordache@upit.ro

Thus, the macrostructure and the microstructure of the joint were analyzed using electron microscopy and the HV micro-hardness was measured.

2 Experimental procedure

In the experimental part of this investigation, the welded samples are overlay joints between two plates, where the dimensions are 160 x 80 [mm^2] and the thickness is 2 [mm]. The friction stir welds have been carried out on samples of Cu 99 and EN-AW-3005 with the composition listed in Tables 1 and 2.

Table 1. Chemical composition [%] of material EN-AW-3005 (in weight percent).

Alloy	Si	Fe	Cu	Mn	Mg	Cr	Zn	Ti	other	Al
3005	0.6	0.7	0.3	1-1.5	0.2-0.6	0.1	0.25	0.1	0.05	Rest

Table 2. Chemical composition [%] of material Cu 99 (in weight percent).

Alloy	Ag	As	Bi	Cu+Ag	Fe	Pb
Cu99	0.15	0.2	0.005	0.99	0.1	0.03

Here, we used a cylindrical tool where the diameter of the shoulder is 20 mm and the pin is conical, Figure 1. The length of the pin was 2.7 [mm]. The values of the process parameters were the same in both cases: (i) the rotation speed of the pin 1200 [rpm] (rotation per minute), (ii) feed rate 70 [mm / min].

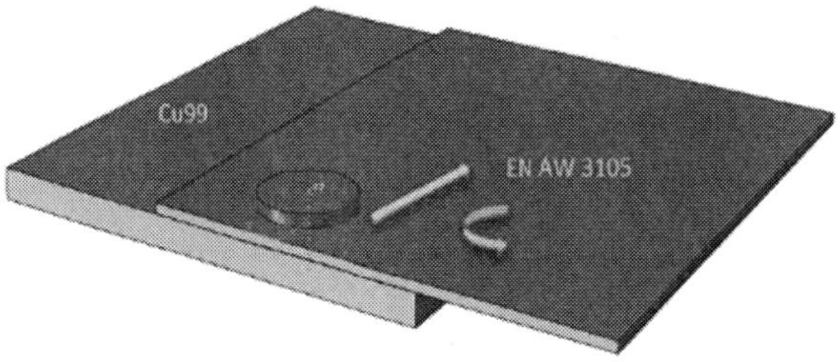

Fig. 1. Schematic illustration of dissimilar FSW.

Welded test specimens were cut perpendicular to the welding direction and prepared by standard methods for the analysis of microstructure. Microstructure analysis was performed by Transmission Electron Microscopy (TEM) using the 5000-SU Hitachi microscope coupled with an EDS spectrometer for the elemental analysis of the joined area.

The voltage was 25 kV and the working distance varied between 9-15 mm in High Vacuum MODE. The hardness was measured using the Vickers method by the equipment Falcon 500 Innovatest. It has achieved a line of hardness with the step between two fingerprints of 1 mm.

To determine the residual stress there was used the method with the Rigaku AutoMate II diffractometer, the tensions being determined on the crystallographic direction by normal perpendicular on the section of the test samples.

3 Results and discussions

3.1 Macroscopic and microscopic analysis

The macroscopic appearance of the welded joint was analyzed at different times / sequences of the welding process and it was noticed that the core has a time-varying evolution in terms of configuration of sizes and distribution of compounds characteristic of the two materials. For example:

- at 60 mm from the start of the welding process, the portion of the mixing zone (core) expands mainly from the aluminum alloy to copper (Figure 2),

- after 150 mm of welding, along with the increase of temperature developed during the process, it is observed that the mixing zone expands to a greater extent in the copper alloy (Figure 3).

The core portion of aluminum alloy consists of the following regions:

- an area of the core, lighter in color and located at the top in Al,

- an area under the previous one which contains many particles rich in Cu and expanding in the basic material Cu (see figure 3).

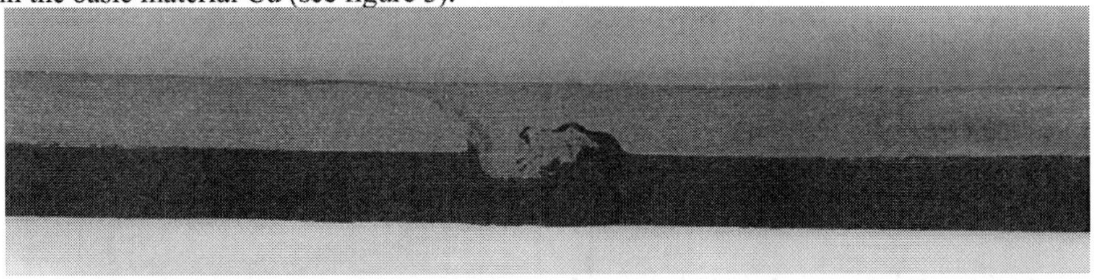

Fig. 2. Macroscopic appearance in samples taken at 60 mm from the start of the weld.

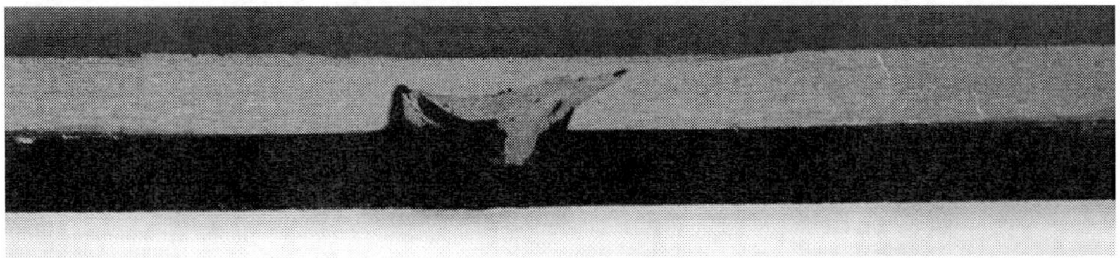

Fig. 3. Macroscopic appearance in samples taken at 150 mm from the start of the weld.

There were investigated three areas of joint, Figure 4, images of the weld nugget, acquired by EDS technique, are shown in Figure 5.

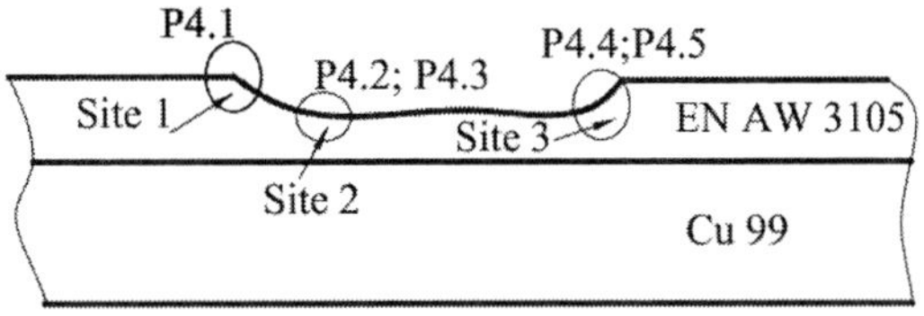

Fig. 4. Position of analyzed areas.

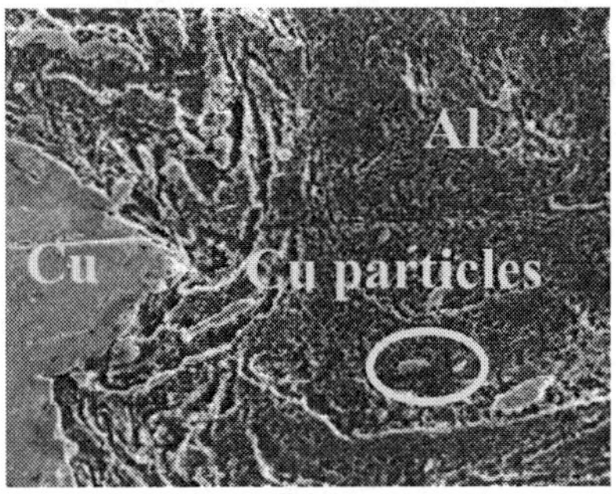

P4.1

DOI: 10.1051/matecconf/201711204005

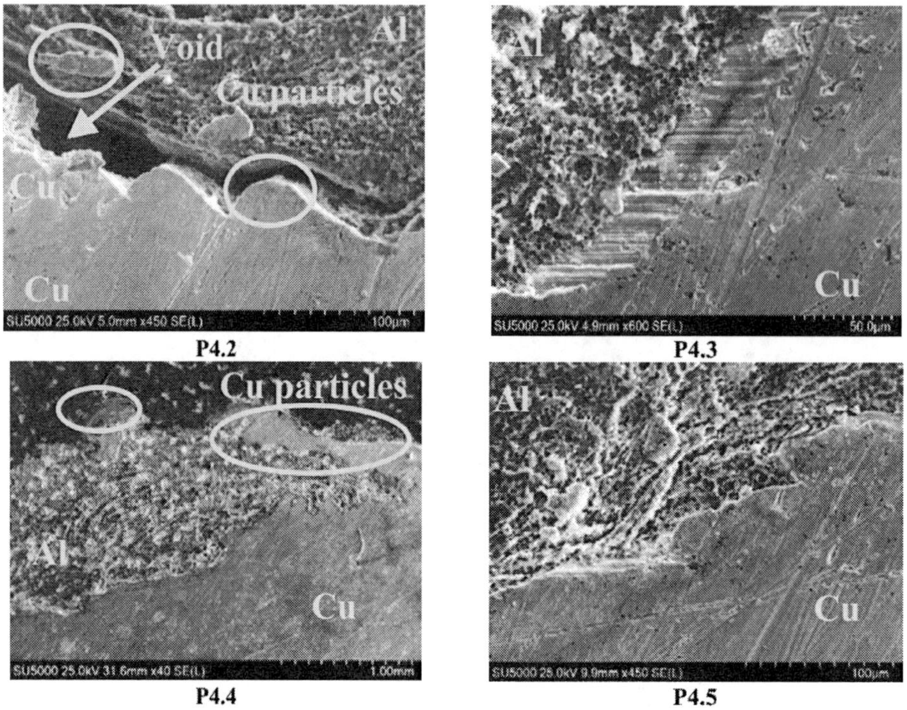

Fig. 5. Morphology of joining area.

From the analysis of the joining area it can be seen that at the interface between the two materials there appear slip bands, micro-pores and voids. Some defects like voids can be associated with the insufficient material flow because at very low welding temperature. In fact, different melting temperatures and contraction coefficients of dissimilar sheets materials lead to this lap joint cavity defect.

Cu fragment were brought in the aluminum side. The distribution of Cu particles with irregular shapes and various sizes was inhomogeneous in the welding zone and a particles rich zone was formed.

Joint microstructure and chemical composition in areas (sites), marked in Figure 4 are shown in Figures 6-8.

In the analysis of the images it is observed that pieces of Cu are brought in the material made of Al, obtaining a mixture of the two materials. In all joining areas analyzed, in the chemical composition besides aluminum there is also copper.

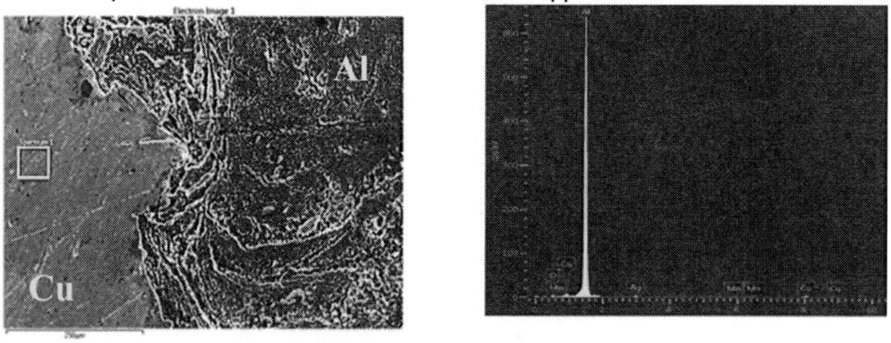

Fig. 6. Microstructure and chemical composition – Site 1, P4.1.

DOI: 10.1051/matecconf/201711204005

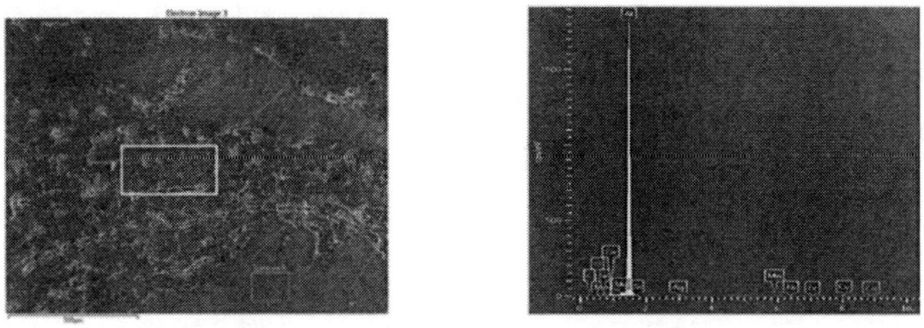

Fig. 7. Microstructure and chemical composition – Site 2, P4.2.

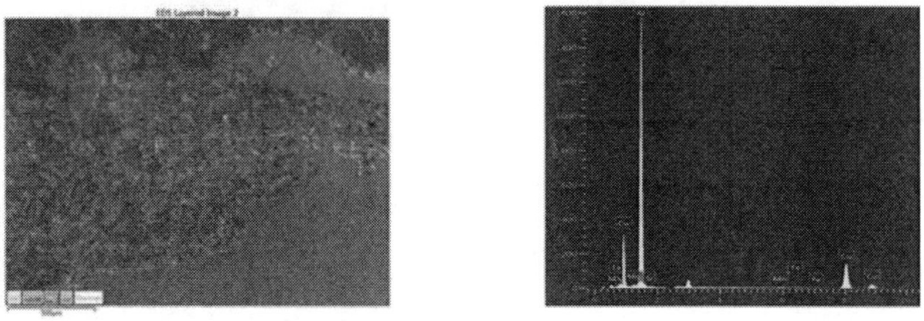

Fig. 8. Microstructure and chemical composition – Site 3, P4.4.

3.2 Microhardness of the joint

There was performed a line of microhardness, in the cross section of the joined parts, in the direction parallel to the outer surfaces of the joined parts at their mid-thickness. The load used was 300 grams, the step was of 1 mm and the determinations were made in 10 points. The values of microhardness thus obtained are shown in Figure 9.

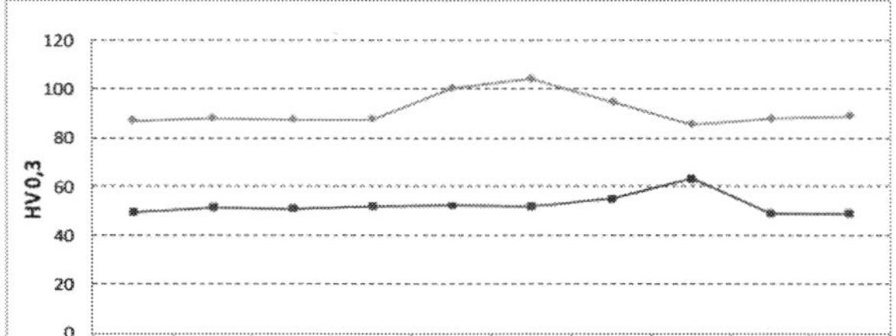

	1	2	3	4	5	6	7	8	9	10
Al	49,5	51,5	51,2	52,1	52,4	52,1	55,3	63,5	49,2	49,3
Cu	87,5	88,3	87,9	88	100,7	104,4	95	85,9	88,1	89,3

Fig. 9. Values and evolution of microhardness.

There are slight increases of microhardness in the joining area (measurement points 5, 6 and 7) for the Cu material. The values of microhardness in this area, in the Cu material are approximately 100 HV0.3, being somewhat larger than the base material (about 87 HV0.3). Hardness value increases significantly in welding zone relative to base metals because is fine grain microstructure caused by the dynamic recrystallization resulting from the combined role of heat and stirring force. Such increases were not identified for the Al material.

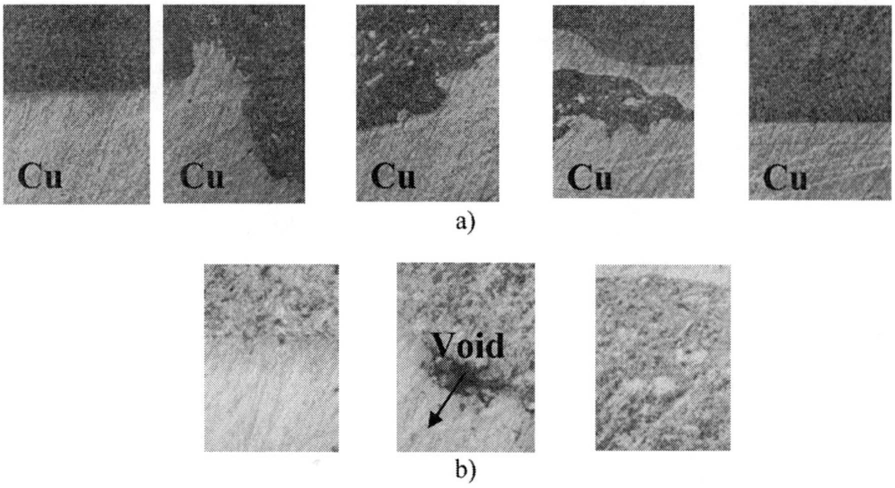

Fig. 10. Micrographic aspects of the welded joint: a) increase x10, b) increase x50.

4 Conclusions

A preliminary investigation has shown that FSW is an effective method of joining dissimilar materials. The FSW joint appeared to contain some evidence of micro-pores and voids at the interface of the aluminum and copper. Some defects like voids that show up at a region near the sheets interface especially in the Cu sheet because very low welding temperature the particle rich zone formed in the welding zone had a structure with variously sized particles dispersed in the Al. As the pin is introduced into the area of Al in the core predominates Al, and Cu is distributed in the form of pieces of irregular shapes. The portion of the core of the aluminum alloy is made of an area of the core, light-colored and located at the top of Al and an area under the previous one that contains many particles rich in Cu and which expands in the base material Cu. The value of Cu micro-hardness in the core increases to 100 HV and that of aluminum up to 63 HV.

This work was accomplished within the "Partnerships in priority areas – PN II" program, implemented with the support of Executive Agency for Higher Education, Research, Development and Innovation Funding (UEFISCDI) and Romanian Ministry of Education and Scientific Research, project no. PN II–PT–PCCA–2013–4–185. All authors are grateful to UEFISCDI for research funding.

References

1. Z. Shen, Y. Chena, M. Haghshenas, T. Nguyenb, J. Gallowayc, A.P. Gerlicha, Mater.Charact., **104** (2015)
2. X. Li, D. T.Zhang, C.Qiu, W.Zhang,Trans. Nonferrous Met. Soc. China, **22** (2012)
3. V.C. Sinha, S. Kundu, S. Chatterjee Perspect. Sci, **8** (2016)
4. P. Xue, D.R. Ni, D. Wang, B.L. Xiao, Z.Y. Ma. Mater Sci. Eng A **528** (2011)
5 Y.Demmouche, These pour obtenir le grade de docteur délivré par l'École Nationale Supérieure d'Arts et. Métiers Spécialité Mécanique - Matériaux (le 17 Décembre 2012)
6. J.E. Lee, D.H. Bae, W.S. Chung, K.H. Kim, J.H. Lee, Y.R. Cho, J. Mater. Process. Technol. **187–188** (2007)
7. Y.J. Guo, G.W. Liu, H.Y. Jin, Z.Q. Shi, G.J. Qiao, J. Mater. Sci. **46** (8) (2010).
8. L. Yang, B.X. Mi, L. Lin, H.J. Huang, X.P. Lin, X.G. Yuan,. Mater. Sci. For. **749** (2013)
9. J.Q. Zhang, Y.F. Shen, X. Yao, H.S. Xu, B. Li, Mat.& Design. **64** (2014)

Influence of the stacking sequence on the mechanical proprieties of glass fiber reinforced polymer

Paul Bere, *Adrian* Popescu, *Cristian* Dudescu*, and *Liana* Hancu

Technical University of Cluj-Napoca, Department of Manufacturing Engineering B-dul Muncii 103-105, Cluj-Napoca, Romania

Abstract. The reinforced composite materials are in a very impressive development in the last decades. In this paper the influence of the ply-stacking sequence of glass fiber reinforced polymer on mechanical properties is investigated. Composite material structures consist of layers from glass mat and fabrics with different disposal sequence. The mechanical properties of the manufactured composite materials have been determined by tensile and bending tests. The obtained results are used to design the optimal materials architecture.

1 Introduction

Composite materials, especially those with a polymeric matrix, are of considerable interest to researchers and engineers due to the use of them in a broad spectrum of applications, particularly for high-performance aerospace and automotive components and a wide range of possible combinations to get specific material properties. In almost all engineering applications requiring strength, high stiffness and fatigue resistance, composites are reinforced with continuous fibers rather than small particles or whiskers. Fiber reinforced plastic (FRP) composites can be simply described as multi-constituent materials that consist of reinforcing fibers embedded in a rigid polymer matrix. Most composites used in engineering applications contain fibers made of glass, carbon or aramid. A wide range of polymers can be used as the matrix to FRP composites, and these are generally classified as thermoset (eg. epoxy, polyester) or thermoplastic (eg. polyether-ether-ketone, polyamide) resins.

Glass fiber reinforced polymeric composite (GFRP) is a composite material made of a plastic (resin) matrix reinforced by fine fibers made of glass. GFRP is a lightweight, strong material with very many uses, including boats, automobiles, water tanks, roofing and pipes. The mechanical behavior of a GFRP composite basically depends on the fiber strength and modulus, their amount, orientation and type, the chemical stability, matrix strength and the interface bonding between the fiber/matrix to enable stress transfer [1, 2]. Various GF reinforcements like long longitudinal, woven mat, chopped fiber (distinct) and chopped mat

* Corresponding author: mircea.dudescu@rezi.utcluj.ro

in the composites have been produced to enhance the mechanical and tribological properties of the composites.

Influence of the stacking sequence on the mechanical properties of composites is investigated the literature. Amico S C et al. [3] are focused on the mechanical properties of pure sisal, pure glass, and hybrid sisal/glass compression-molded composites, in which various stacking sequences of fiber mat layers were used, the importance of controlling the stacking sequence to enhance properties was evident. Similar study of the tensile and flexural properties of glass fiber epoxy composites hybridized with basalt, flax or jute fibers is presented in [4] and for cotton/glass fiber reinforced polyester composites in [5]. Mechanical properties of hybrid composites made from glass fiber (plain-woven and stitched bi-axial ±45°) and plain-woven carbon fiber was analyzed by Irina M [6]. Effect of fiber orientation and stacking sequence on mechanical and thermal characteristics is investigated also for hybrid natural composites based on banana-kenaf fibers [7]. Davallo M [8] studied the effects of laminate thickness and ply-stacking sequence on the mechanical properties and failure mechanism of unidirectional glass-polyester composites.

The primary goal of this study is to evaluate the effect of the stacking sequence on the mechanical properties of GFRP laminates, and to assess the best combination that can be applied to design the optimal materials architecture. The first part of the paper presents manufacturing process of the material. The mechanical proprieties of the material, presented in the second part of the paper, were determined by standard tensile tests and three point bending test according to ASTM standards. The main results and conclusion are included at the end of the paper.

2 Materials and manufacturing method

2.1 Materials

To investigate the stacking sequence on mechanical properties of composite materials two types of plates were manufactured. The same reinforcement material and the same matrix were used, the difference between them being the architecture of the reinforcement material. In the both cases of glass/epoxy composite plate the reinforced ratio obtained was 40% and was calculated as ratio between weight of reinforcement material and weight of composite material.

Reinforcement material was a roving fabrics type Combimat RT 300/S300 (Keltecs, Croatia) with a density of 600 g/m^2. The base materials are E glass and ECR glass. This type of material is made by a layer of 300 g/m^2 glass mat and balanced biaxial glass woven fabrics with 300 g/m^2. These two types of materials are assembled together like a single layer. The glass woven fabrics has the role to increase the mechanical properties of composite materials. In the same time, it also reduces the workmanship for the laminating of another layer of reinforcing material. The matrix impregnation in case of Combimat materials is simultaneous for both materials that composed the roving fabrics.

The epoxy matrix was Epiphen RE 4020/De 4020, with the ratio between resin and hardener of 100:30 in weight fraction. The matrix properties are: transparent aspect; pot life at 20 °C - 45 min; Brookfield viscosity at 25 °C – 300 mPa.s; gelling time (thin layer in hours at 20°C) - 8 to 9 hours; density at 25 °C - 1150 kg/m^3; curing at 20 °C - 24 hours.

2.2 Manufacturing method

The Hand Lay-up and pressured processes were used to obtain the composite materials plates [9, 10]. Two 10 mm thick steel plates were used as a mold. The surface of the mold was milled and polished to prevent the bonding of composite material on the mold surface. Additionally, five layers of wax were applied.

Both plates (denoted A and B) were manufacturing by two layers of Combimat materials and epoxy matrix. The difference between them is the stacking sequence, respectively the position of reinforcement materials. For the material "A" the balanced biaxial glass woven fabrics was applied on outer position. The glass mat layers are positioning in the interior of the plates like presented in Figure 1.

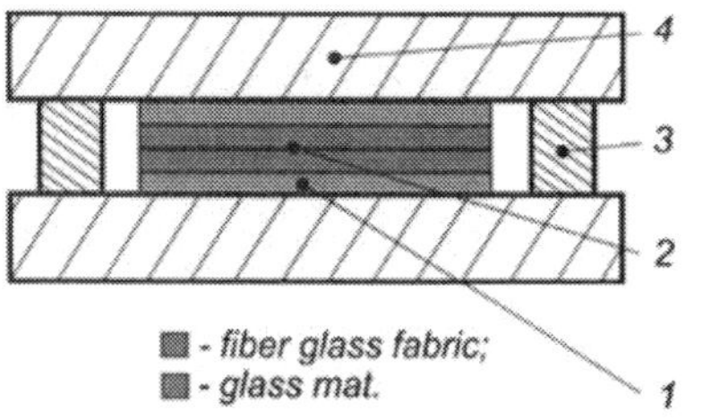

Fig. 1. Manufacturing set-up (1- glass woven fabrics, 2- mat layers, 3- path spacers, 4- mold).

Fig. 2. Composite plate laminating process.

The fiber glass mat layers are outer positioned and the biaxial glass woven fabrics layers in inner position for the composite material "B". After reinforcement materials impregnation, the composite materials were covered by metallic steel mold and pressed at 9 kPa using a screw press. This operation was performed to eliminate the matrix excess from composite material. On the border of the mold path spacers by 1.8 mm were placed to obtain the composite materials limitation of this dimensions. Polymerizations of composite plates were made 24 hours at 24 °C. Post-polymerization heat treatment temperature at 80° C for 24 hours it was carried out in an oven. After this time the composite plates were kept to full cooling in the oven for more 6 hours.

3 Mechanical characterization

To determinate the mechanical properties of the manufactured materials standard samples were cut out from the plates by a water jet machine. To prevent the samples delamination during the water jet penetration time of the composite plates, the penetration water jet area was positioned at 30 mm distance from the samples border. For tensile test five specimens from each material were manufactured according to ASTM standard. At the ends of the specimens the TABs were glued by a structural two-compound epoxy adhesive 3M Scotch-Welt 9323 B/A.

Mechanical proprieties of the above described materials were determined by standard tensile tests and three point bending test according to ASTM standards D638-10 and D790-02 respectively. All measurements were carried out at room temperature (23°C) and humidity in the range of 45-50%. The specimens for tensile test 1.8 mm thick and 25 mm wide (Figure 3a), were loaded to failure with constant crosshead speed (5 mm/min) using an Instron 8801 (100 KN) servo hydraulic testing system. The gauge length was 180 mm and the engineering strain was calculated based on the crosshead displacement. For the bending test the specimens have 1.8 mm thickness and 15 mm width (Figure 3b), the calculated test speed being 0.948 mm/min.

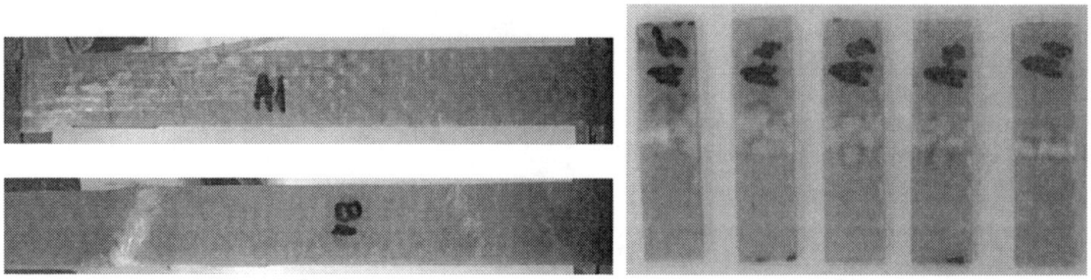

a) b)

Fig. 3. Composite samples subjected to tensile test (a) and bending test (b).

Comparatively in Figure 4 is presented the strain-stress diagrams obtained from the tensile test for the two types of composites. Similarly, Figure 5 shows the flexural behavior of the GFR composite material.

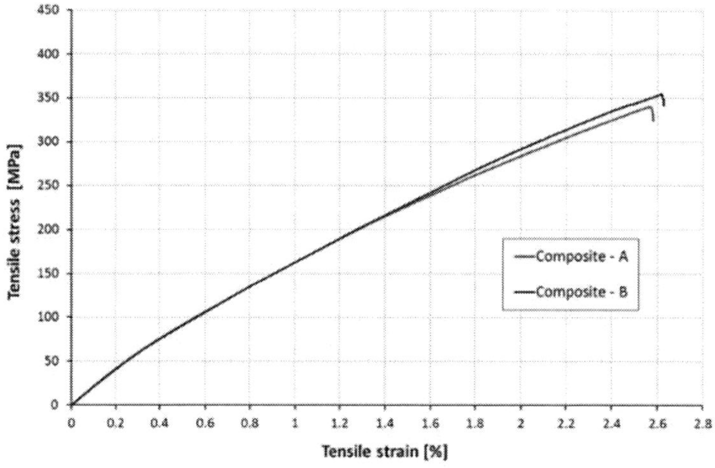

Fig. 4. Tensile strain-stress diagrams of GFR composites.

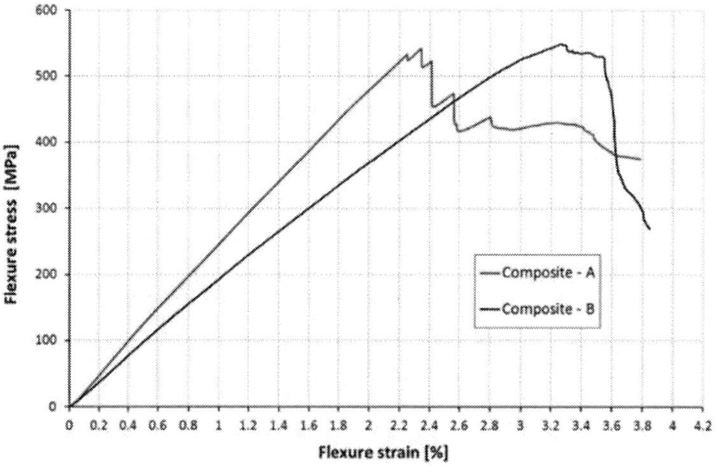

Fig. 5. Flexure strain – stress diagrams of GFR composites.

The measured mechanical data for the tensile test are presented in Table 1 and for the three-point bending test in the Table 2. For each stacking sequence five specimens were tested, the reported data are the measured mean values, the standard deviation appearing in the brackets.

Table 1. The mean values of maximum tensile stress and strain of the GFR composite.

Composite type	Load at Brake [N]	Tensile strain at tensile strength (mm/mm)	Tensile strength (MPa)	Young's modulus (MPa)
A	15004 (575)	0.0253 (0.0016)	334 (13)	18655 (598)
B	16135 (1923)	0.0260 (0.0030)	359 (43)	18294 (216)

Table 2. The mean values of maximum flexure stress and strain of the GFR composite.

Composite type	Flexure Load [N]	Flexure strain at flexure strength (mm/mm)	Flexure strength (MPa)	Flexural modulus (MPa)
A	549 (28)	0.026 (0.0024)	542 (28)	26124 (321)
B	673 (45)	0.032 (0.0011)	539 (36)	19683 (1075)

Flexure stress is more than 500 MPa for both composite materials and tensile stress is around 350 MPa, the obtained values being comparable with the metals but in condition of 6 times reduced density.

To highlight the compatibility between fiber and matrix the microstructure of damaged samples area was investigated. The morphological analyses were carried out by electronic microscope type Quanta 200 3D DUAL BEAM. Microstructures of damaged samples area presented in Figure 6 indicate a good impregnation of glass monofilaments and good adhesion between matrix and glass monofilaments surfaces.

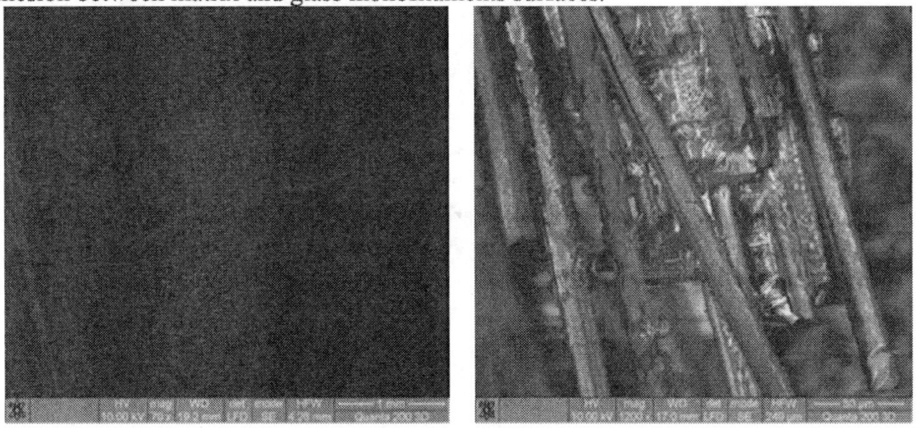

Fig. 6. Microstructure of damaged samples (material type A).

4 Conclusions

The hand lay-up and pressure is a simple and very used technology of composite materials that allows to obtain short manufacturing time and a minimal investment. In our case the manufactured GFR composite plates had a constant thickness and good surface quality. For "A" type composite plate, where the glass woven fabrics are applied on outer surfaces, a resin insufficiency was observed on the surface of the plates, the warp and weft glass wires being visible. For "B" type material where we have the glass mat on exterior the surface is smoother.

Analyzing the obtained mechanical characteristics, it can be noticed that the tensile proprieties of the composite material is not influenced by the stacking sequence. Positioning of the reinforcements closer to the neutral axis or the outer fibers will strongly affect the flexural proprieties, composite A being much stiffer than B. The importance of controlling the stacking sequence to enhance properties was evident. For instance, to optimize flexural behavior, there must be glass fibers mainly on the top and bottom surfaces.

This work was supported by a grant of the Romanian National Authority for Scientific Research and Innovation, CNCS/CCCDI–UEFISCDI, project number PN-III-P2-2.1-BG-2016-0210, within PNCDI III.

References

1. T. Sathishkumar, S. Satheeshkumar, J. Naveen, J. Reinf. Plast. Compos., **33** 1258-1275 (2014).
2. J. M. Stickel, M. Nagarajan, Int. J. Appl. Glass Sci., **3** 122-136 (2012).
3. S. C. Amico, C. C. Angrizani, M. L. Drummond, J. Reinf. Plast. Compos., **29** 179-189 (2010).
4. M. S. M. Jusoh, C. Santulli, M. Y. M. Yahya, N. S. Hussein, H. A. I. Ahmad, Mater. Sci. Eng. Adv. Res, **1** 19-25 (2016).
5. E. H. Portella, D. Romanzini, C. C. Angrizani, S. C. Amico, A. J. Zattera, Mater. Res., **19** 542-547 (2016).
6. M. M. W. Irina, A. I. Azmi, C. L. Tan, C. C. Lee, A. N. M. Khalil, Procedia Manuf., **2** 236-240 (2015).
7. P. Sathish, R. Kesavan, B. V. Ramnath, C. Vishal, Silicon, 1-9 (2015).
8. M. Davallo, H. Pasdar, M. Mohseni, Islamic Azad University, North Tehran Branch, **2** 2118-2124 (2010).
9. P. Bere, *Theoretical and experimental research on the manufacturing and mechanical behavior of the tubes made from polymeric composite materials*, PhD, Technical University of Cluj-Napoca, Romania (2009).
10. G. L. Enachescu, M. F. Stefanescu, Materiale plastice **53** 198-201 (2016).

Electromagnetic configurable architectures for assessment of Carbon Fiber Reinforced Plastics

Rozina Steigmann[1,2], *Adriana* Savin[1,*], *Nicoleta* Iftimie[1], *Gabriel Silviu* Dobrescu[1], *Zoran* Bergant[3], and *Janez* Grum[3]

[1]National Institute of R&D for Technical Physics, NDT Department, Iasi, Romania
[2]Alexandru Ioan Cuza University, Faculty of Physics, Iasi, Romania
[3]University of Ljubljana, Faculty of Mechanical Engineering, Ljubljana, Slovenia

Abstract. Carbon Fiber Reinforced Plastics are used in most wide domains due their low density, lack of mechanical fatigue phenomena and high strength –to weight ratio. From electromagnetic point of view, Carbon Fiber Reinforced Plastics structure represents an inhomogeneous structure of electric conductive fibers embedded into a dielectric material, thus an electromagnetic configurable architecture can be used to evaluate above mentioned defects. The paper proposes a special sensor, send receiver type and the obtaining of electromagnetic image by post-processing each coil signals in each point of scanning, using a sub-encoding image reconstruction algorithm and super-resolution procedures. The layout of fibers can be detected interrogating only diagonal reception coils.

1 Introduction

Carbon Fiber Reinforced Plastics (CFRP) started to be used as usually materials in aerospace industry since the middle of the last century [1]. Nowadays, AIRBUS A350 is 52% CFRP, including wing spars and fuselage components [2]. The high costs of CFRP have been surpassed by their strength-to-weight ratio and feasibility. The CFRP structures in aviation/aerospace industry make compromise between advantages – weight reduction, the mechanical properties can be tailored by layout design, high resistance to corrosion at interfaces with metallic joints and disadvantages as high recurring costs, barely visible defects appearance and their propagation, difficulties in repairing, raising the requirement of testing during the fabrication of structures as well as in service monitoring. CFRP structures can be evaluated by many non-destructive techniques (NDT) using conventional strain gages [3], ultrasound using different types of sensors and methods [4,5], Bragg fibers [6], PZT sensors [7], thermography and acoustic emission [8,9] and also sensors for electrical properties, wireless sensing with low sampling rate, wireless strain and temperature sensors, especially knowing that carbon fibers have electrical properties [10,11] and these properties are modified by the effect of impact with low energies [12].

Due to the numerous types of layups and layouts in configuring the composites in function of layers number, orientation of fibers and woven, type of matrix, etc., the full effect of design

* Corresponding author: asavin@phys-iasi.ro

over the behavior of the composites at impact isn't yet elucidated, different designs leading to the development of different control methods [13, 14]. Taking into studies samples of CFRP, usually used in aerospace engineering, carbon fiber embedded in Polyphenylene sulphide (PPS) matrix, this paper present a novel type of electromagnetic sensors defined as CA due to the possibility to interrogate only a part of the sensors from the architecture or the entire assembly, to detect damages induced by low energy impact, increasing the spatial resolution compared to the classical sensors. The electromagnetic image is obtained by post-processing each coil signals in each point of scanning, using a sub-encoding image reconstruction algorithm and super-resolution procedures. The novelty consist in the possibility to interrogate diagonal reception coils in order to detect the fiber breaks and delamination that are tilted from the direction of scanning. The interrogation is sequentially made by programming the acquisition data board. The method will be further developed for semi-automatic control of complex structures made from composites.

2 Materials and methods

2.1 Studied samples

The strength in using CFRP is given by the volume ratio of fibers and matrix. A good ratio means a qualitative composite. Plates of CFRP made by Tencate, the Netherlands having the dimensions 150mmx100mmx4.2mm [15] have been taken into study. The reinforcement consists in 12 biaxial woven [0°/90°] harness satin 5H-T300JB type. The successive laminas are disposed by rotation at 45°, obtaining a quasi-isotropic material, the matrix is PPS. This offers exceptional chemical resistance at high temperatures, comparable to PEEK, along with excellent thermal properties for long-term (over 200°C) and short-term use (up to 260°C) [16]. The fiber volume ratio is 50±3% and the composite density is ρ=1520 kg/m^3. This composite has been used in the construction of Airbus and Boeing aircrafts [17]. The properties of weave weft for input in simulation are yarn width 1.31 mm, yarn thickness 0.162 mm, yarn spacing 1.50 mm, yarn filament count 3000, filament diameter 0.007 mm, carbon fiber density 1.75g/cm^3 [18]. The microstructure of the cross-section was determined using microscope Olympus BX51 coupled with digital camera using QuickPhoto Micro3.1 in order to see that the material is very compact, having no voids or mislaying of tows. These kind of flaws can influence the mechanical behavior of the composite (Figure 1).

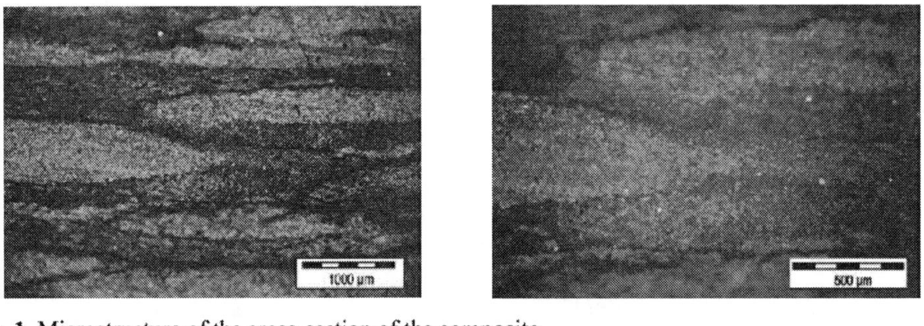

Fig. 1. Microstructure of the cross-section of the composite.

Using equipment FRACTOVIS PLUS 9350-CEAST-Instron USA [12], [15], the samples were impacted with different energies to obtain delamination. Figure 2a present the 5 Harness Satin woven layout and Figure 2b shows the front side of impacted sample, at 10J energy, which creates a concavity at surface and delamination inside.

DOI: 10.1051/matecconf/201711204007

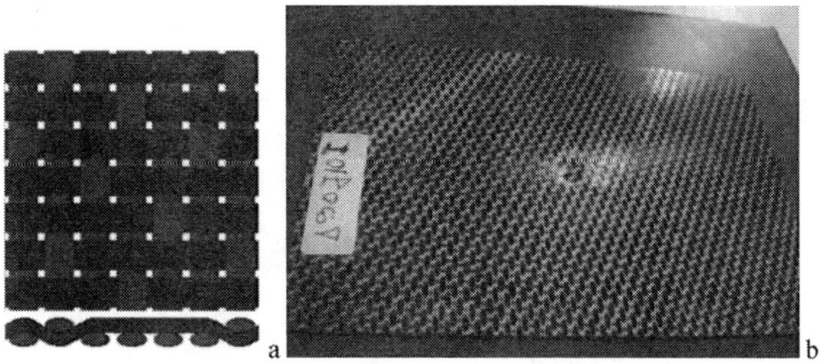

Fig. 2. Studied samples: a) 5H satin layout; b) front side of impacted sample.

2.2 Sensor concept

The sensor consists in a 2D architecture of 9 identical reception coils having inner radii 1mm and outer radii 1.2 mm, having 18 turns with 0.03 mm from CuEm placed into a polyimide support with 7x7x2 mm dimensions. The emission coil has 1 turn from CuEm wire with 1mm diameter, has average side of 7.1 mm and is placed around the half height of the support. For CFRP plate with above characteristics; the moduli of the magnetic field components generated by a current through the emission coil at frequency of 300 kHz and amplitude of 0.1A is presented in Figure 3, in the middle, the field assures a uniform value. The design of the CA with 2D sensor array of receiver coils is presented in Figure 4.

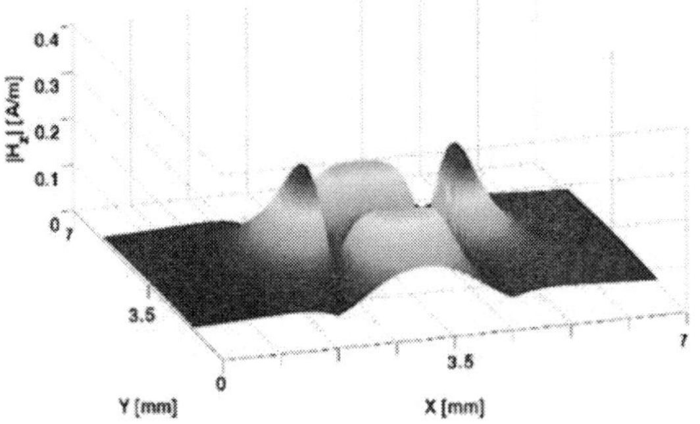

Fig. 3. Magnetic field provided by the emission coil.

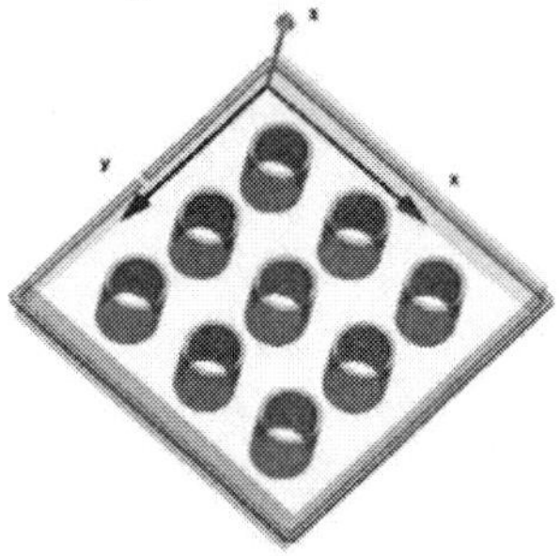

Fig. 4. Configurable architecture.

The geometry was tested based on simulation using Dyadic Green function and volume integral method described in [19, 20]. The functioning of 2D sensor array [21] assures good lateral resolution and emphasizes hidden defects by simultaneous interrogation of the reception coils by means a multiplexor, improving the scan time and the response resolution. An internal multiplexing system reduces the mutual inductance and programs the actual time when each coil is interrogated. The architecture is configurable, being able to interrogate only few of reception coils by programming the data acquisition board.

The reception coils detect the reflections of electromagnetic field (EM) created by the emission part in composite, the system measuring the received signal both in amplitude and phase. The behavior of the CA has been simulated using XFDTD software, Remcom, placing a simulated delamination under the reception coils denoted (3.2, 3.3), Figure 5. Finite Difference Time Domain (FDTD) finds approximate solutions to Maxwell's equation in differential form, discretizing using central-difference by 30 approximations to the space and time partial derivatives.

The equations are implemented in the software and solved in a cyclic manner: the electric field vector components in a volume of space are solved in time, the magnetic field vector components in the same spatial volume are solved at the next instant in time. The process is repeated until the desired transient or steady-state EM field behavior is fully achieved. The FDTD method is able to cover a range of frequency in a single simulation using Fourier transform, treating the simulated structures with a wide variety of EM material properties. Both time domain and frequency domain information are offered for complete characterization of a lot of problems and applications in electromagnetics.

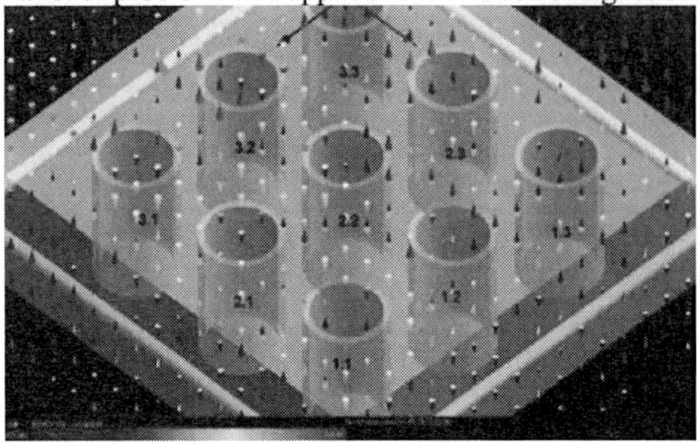

Fig. 5. FDTD simulation.

The 2D array allows parallel acquisition of the signals, reducing the data storage and processing requirements. The signals are reconstructed into separate electromagnetic images and can be improved using super-resolution (SR) procedures [21, 22]. SR is a technique in terms of its ability to resolve closely spaced scatters in a solid, measuring the near field of scattered waves. Using notations from [21] SR uses the maximum likelihood estimation procedure [23] for coherent sources. The source locations are located where the function $(\bar{S} * \bar{R}_{xx})$ is maximal, where $\overline{R_{xx}}$ is the autocorrelation matrix of sensor signal response. The SR method can lead to improvement of the signal to noise region (SNR). The SNR in dB was measured in the time domain $SNR = 20\log_{10} A_{signalpeak} / A_{noiseRMS}$ where $A_{signalpeak}$ represent the peak amplitude of signal reflected of the discontinuities and $A_{noiseRMS}$ represent the RMS of noise measured away of the signal. The array gain G is defined as $G = SNR_{array} / SNR_{sensor}$ and here G=0.7*N*M, where M and N represent the number of rows and respectively columns of reception coils from the array. Applying SR method on the results obtained by numerical simulation, Figure 6, the sensor array response is obtained (Figure 7).

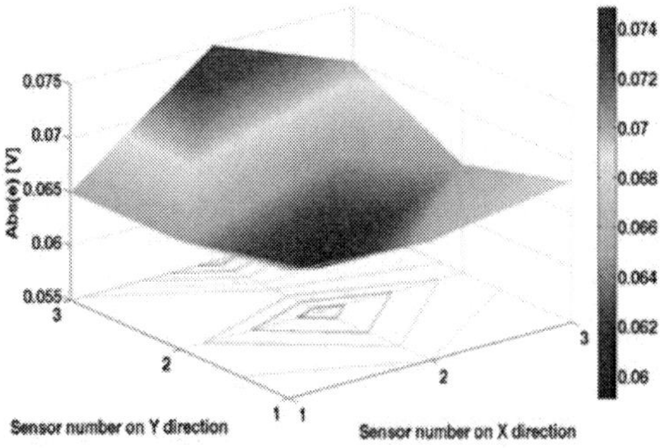

Fig. 6. Amplitude of the signal on each of the 9 receiver coils placed on a simulated delamination.

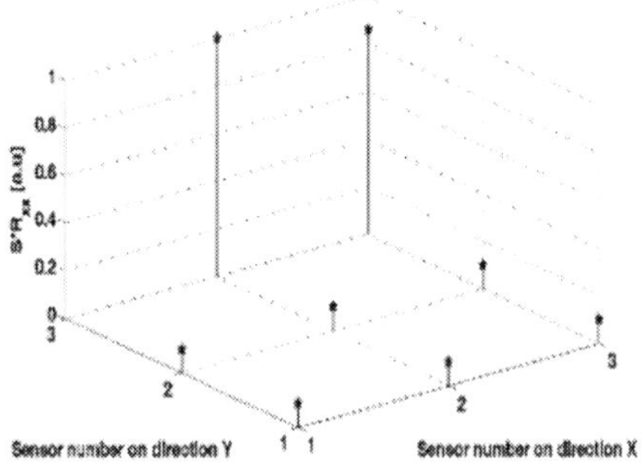

Fig.7. Simulation of signals delivered by the sensors from array, using the SR procedure.

The maximum value corresponds to position where the delamination is placed under the coils indexed (3.2, 3.3).

2.3 Experimental set-up

The experimental set-up is presented in Figure 8. The emission coil is fed through a function generator WW1074 Tabor Ltd, the signal is amplified by a power amplifier RF A1012 AG-T&C Power Conversion Inc. The multiplexer send the received signal to the data acquisition towards PC.

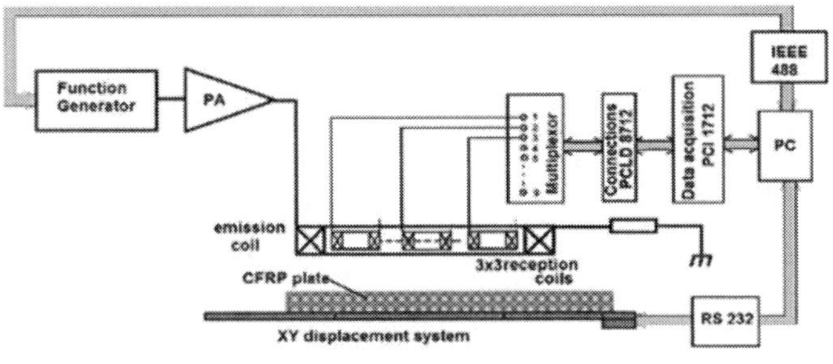

Fig. 8. Experimental set-up.

The samples to be examined are displaced under the sensor, being raster scanned with a XY automated displacement system, Newmark USA, the acquisition of signal being synchronized with the buffering of it. The experimental devices are controlled by PC though algorithms developed in Matlab 2014a.

3 Results and discussions

The answer of sensor array for experimental measurements of CFRP of the plate with impacted zone is presented in Figure 9a after the signal processing. The sensor array is placed in the same region for which the simulation has been carried on.

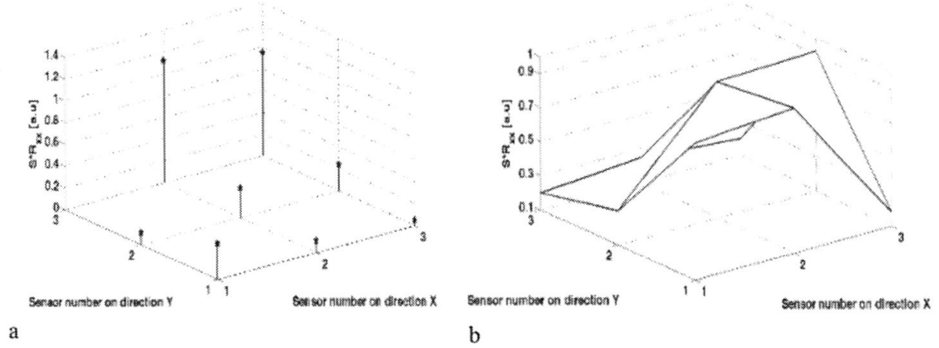

Fig. 9. a) Signals delivered by the sensors from array, using the super-resolution procedure; b) signals received when the 45^0 fibers are displaced by delamination.

Comparing the results from Figure 7 with Figure 9a, it can be observed that the position of maxima are clearly improved for experimental data. Figure 9b presents the response of the sensor array using the SR method for signal processing, in the region where the position of carbon fibers at 45° have been disturbed from the layout. A surface of 60x60mm of the composite is scanned using above mentioned system. The EM image [24] of the surface is obtained by post-processing the acquired signals in each coil in each point of scan using a sub encoding image reconstruction algorithm. A sub encoding image reconstruction starts at the beginning scanning point with the set of component signals received using a reduced number of phase-encoding gradient steps. Fourier transformation of these signal sets results in aliased component images. Further, the sub encoding reconstruction operates entirely in the image domain based on the fact that each pixel in an aliased image is a superposition of multiple pixels [25]. The results obtained at the reading of the middle columns sensors, as numbered in Figure 5, are presented in Figure 10. The method allows the visualization of delamination induced in the composite plate due to impact with low energy.

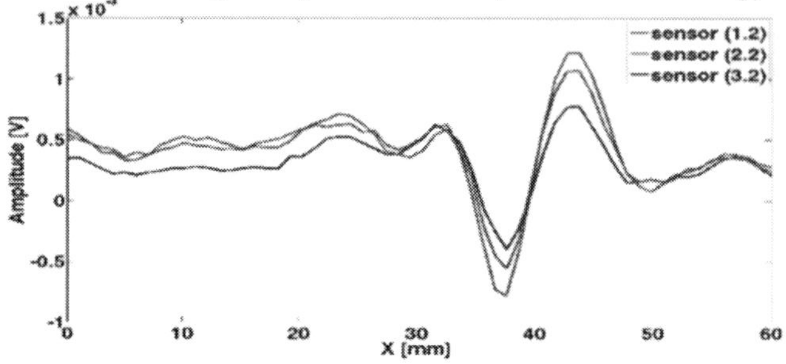

Fig. 10. Signals recorded by the middle column sensors.

In Figure 11a is presented the electromagnetic image of the plate impacted with 10J. It can be observed the area of delamination and the breaks of few fibers. Interrogating only the diagonal reception coils, the layout of 45^0 fiber orientation can be also emphasized. The results are presented in Figure 11b.

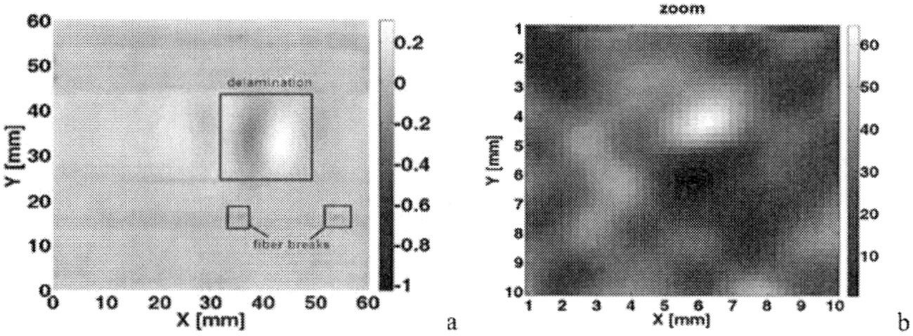

Fig. 11. Electromagnetic image: a) entire scanned region; b) zoom on 10x10mm around the central region of the delamination.

4 Conclusions

Using a suitable CA of sensors and the SR procedures as the signal post processing method, will lead to improvement of the SNR reported to the situation in which only one sensor is used. The simulations and their validations using real measurements prove the ability of the CA of sensors to improve the resolution of control. Also, the evaluation become more efficient, the scanning time being reduced and parallel acquisition of the signals digitized and recorded, reduces the data storage and processing requirements. Delamination's due to impact with low energies can be emphasized and localized, as well as displacement of the fibers from the layout are determined. The electromagnetic image is obtained by post-processing each coil signals in each point of scanning, using a sub-encoding image reconstruction algorithm and super-resolution procedures. The layout of fibers can be detected interrogating only diagonal reception coils. The method can be applied for rapid detection of flaws or misalignment of fibers/layers after a certain functioning time in any domain where CFRP composites are employed, especially for semi-automatic control of complex components made from CFRP.

Further tests for curve surfaces can be made embedding the sensors on a flexible support, allowing the enveloping of a surface as turbine blades or corners and joints of composite products.

This paper is partially supported by Romanian Ministry of Research and Innovation under project PN-II-PCCA-2013-4-0656 and Nucleus Program PN 16-37 01 01.

References

1. K. Potter, *An Introduction to Composite Products: Design, Development and Manufacture* (Springer, 5th Ed. Chapman & Hall, London 1996)
2. G. Marsh, Reinforced plastics **54**, 6, 20-24 (2010)
3. R. Steigmann, A. Savin, V. Goanta, P. D. Barsanescu, B. Leitoiu, N. Iftimie, M D. Stanciu, I. Curtu, *IOP Conf. Ser. Mat. Sci.* **147,** 012140 (2016)
4. R.Grimberg, A. Savin, R. Steigmann, A. Bruma, P.D. Barsanescu, P. D. Salavastru, *Proceedings of NDT in Progress 2009*, 65-74 (2009)
5. A.A. Karabutov, N.B. Podymova, I.O. Belyaev, Acoustical Physics **59**, 6, 667–673 (2013)
6. G. Luyckx, E. Voet, W. De Waele, J. Degrieck, Smar. Mat. St. **19**, 10, 105017 (2010)
7. G. Dib, E. Koricho, O. Karpenko, M. Haq, L. Udpa, *AIP Conference* **1650**, 1, 1072-1080 (2015)
8. J. Kober, Z.Prevorovsky, M. Chlada, Sensors and Actuators A: Physical, **240**, 50-56 (2016)
9. E.Z. Kordatos, D.G. Aggelis, T.E Matikas, Compos. Part B: Eng. **43**, 6, 2676-2686 (2012)
10. R. Grimberg, D. Premel, A. Savin, Y. Le Bihan, D. Placko, Insight **43**, 4, 260-264 (2001)
11. Y. He, G. Tian, M. Pan, D. Chen, Compos. Struct.**109**, 1-7 (2014)
12. R. Steigmann, A. Savin, *Advanced Sensor for Enhancement of Electromagnetic Imaging of Impacted, Carbon Fibers-Pps Composites*, Chapt. 52 in *DAAAM International Scientific Book,* 633-644 (Katalinic, B., Ed.; DAAAM International: Vienna, Austria, 2014)
13. W. Yin, P.J. Withers, U. Sharma, A.J .Peyton, IEEE T.Instrum. Meas. **58**, 3, 738-743 (2009)
14. G. Mook, J. Pohl, F. Michel, Smart Mat. Struct, **12**, 6, 997-1004 (2003)
15. A. S. T. M. D7136, *Standard test method for measuring the damage resistance of a fiber-reinforced polymer matrix composite to a drop-weight impact event* (2005)
16. S.Zhou, Q. Zhang, C. Wu, J. Huang, Mater. Design, **44**, 493-499 (2013)
17. J. Chacksfield, *Aircraft Design, Systems and Operations Meeting* (1987)
18. S. Daggumati, E. Voet, W. Van Paepegem, J. Degrieck, J. Xub, S.V. Lomov, I. Verpoest, Compos. Sci. Technol. **71**, 1171-1179 (2011)
19. O. Mihalache, R. Grimberg, E. Radu, A. Savin, Sensor.Actuat. A: Phys., **59**, 1-3 (1997).
20. E. Radu, R. Grimberg, A. Savin, O. Mihalache, Sensor.Actuat. A: Phys. **59**, 1, 213-218 (1997)
21. R. Grimberg, L. Udpa, A. Savin, R. Steigmann, NDT & E Int, **39** (4), 264-271 (2006)
22. S.V. Shell, W.A. Gardner, *High resolution direction finding* (Bose NK, Rao CR, editors. Handbook of statistics, vol.10. Amsterdam: Elsevier; 1993)
23. L.Zhao, P. Krishnaiah, Z. Boi, J. Mutivariate Anal, **20**, 1-25 (1986)
24. http://www.solvay.com/en/binaries/Composites-Specialty-Polymers_EN-220508.pdf
25. D. K.Sodickson, *Spatial encoding using multiple RF coils: SMASH imaging and parallel MRI*, eMagRes (2000).

The structure and formation of functional hard coatings: a short review

Vlad Diciuc[1,*], and *Alicja* Kazek-Kęsik [2,*]

[1] Technical University of Cluj-Napoca, North University Center of Baia Mare, Romania
[2] Faculty of Chemistry, Silesian University of Technology, Poland

Abstract. Turning tools come in different shapes and sizes, geometry, base material and coating, according to their destination. They are widely used both for obtaining parts and for machinability tests. In this paper a short review about high-speed steel (HSS) turning tools and their coatings is presented. Hard coatings formed on the tool material should be functional depending on the tool final application. Requirements for hard coatings and technological problems for layer formation on the real cutting tool are discussed.

1 Techniques and production of hard coatings

In a statistical study made for the year 2013 [1] the 16.33 billion US-Dollars estimated industry of cutting tools includes 39% milling tools followed by 30% of turning tools, with an annual growth assumed to be 4-5%. The same study shows that in 53% of the cases, the cutting tool substrate is represented by cemented carbide and in 20% of the cases by high-speed steel (HSS).

Coatings on cutting tools are widely spread due to the improvement they provide during the cutting process. In the early 2000's made statistic 50% of HSS tools, 85% of carbide tools and 40% of super-hard tools used in industry were coated [2]. Due to the more complex phenomenology which accompanies the intermittent cutting process, in this study only coatings of HSS turning tools will be treated.

The use of coatings is expected to yield a better protection of the tool against abrasion and adhesion, diffusion, oxidation and to produce a heat barrier for the intensive heat flowing from the cutting area into the tool material [3], thus increasing productivity and tool's life. However, regarding the heat insulation the opinions are divided, as in the paper [4] it was proven that the coatings of continuous cutting tools are not able to insulate the substrate.

Mainly, depending on the method and the coating structure, the same coatings can be applied to both HSS and carbide tools, each substrate having its advantages and disadvantages according to the cutting application. The properties of the tools surface should be improved according their novelty, application and to achieve higher productivity for the manufacturing process. Depending on the turning tools destination various hard coatings can be formed on its surface. The shape of the turning tool is usually not complicated, thus techniques like physical or chemical vapor deposition for layer formation are applied. Hard

* Corresponding author: vlad.diciuc@gmail.com, alicja.kazek-kesik@polsl.pl

coatings are not well defined, but it might be assumed that their hardness is between 2300-5500 (HV, 0.05). The final mechanical properties of the coatings significantly depend on their structure, phase composition and homogeneity. Besides the mechanical properties, the coatings formed on the material should exhibit oxidation and thermal resistance.

Chemical vapor deposition (CVD) or physical vapor deposition (PVD), arc ion plating, ion implantation, plasma-enhanced chemical vapor deposition (PECVD), magnetron sputtering are widely used for hard coatings formation [5]. By using PVD technique the user is allowed to play with a combination of different materials (Ti, Al, Cr, Ni, N, O, Si and C) and is able to obtain layers of 1-5μm whereas CVD normally uses a narrower range of materials (Ti(C,N), TiN and Al_2O_3) for a 4-20 μm thick coating but it lays the emphasis on control parameters such as stress state, phase composition and orientation of crystals [6].

Limitation of CVD and PVD techniques is high temperature of layer deposition (usually around 950°C), poor adhesion of the layer to the substrate and sometimes poor adhesion between interlayers in the coating. The tools market is dynamically increasing and new shapes of the turning tool are desirable. For tools with irregular shape the formed layer cannot be homogenized thought the layer or thickness. Surface roughness might be also different in some places of the material.

The coatings cannot be also brittle, due to the decrease of their working life time [7]. Thus, PVD and CVD techniques are very often combined. Nano - sized and more plastic layer may be formed on the substrate by PVD technique. Coated materials should be suitable for minimum quantity lubrication (MQL) cutting, high-speed turning and should be chemically inert, phase and thermal stable. The surface of the treated materials should be smooth, homogenous, compact and wear and abrasive resistant.

Even though carbide tools are more used, HSS tools are still the choice of professionals in many cutting applications due to their low cost, high performance in machining, especially for mild steels and due to the possibility of easy re-sharpening / re-shaping operations. Any re-sharpening or re-shaping operation must be followed by re-coating so as to achieve the best results.

2 Formation of hard coatings on HSS

2.1 Single ceramic coating

The coatings adhesion to the metal substrate can be improved by pre- and/or post-treatment techniques. Typical protocols of surface pretreatment include micro-blasting, grinding, sand-blasting, shot penning, etching or also for selected materials electropolishing. Procedures of the surface pretreatment play a key role for further single or multiple layer formation [8]. Post treatment includes usually thermal coating treatment. The most popular single, hard layers formed on high speed steel are composed of TiC, TiN, Al_2O_3, TiAlN [1,9]. Titanium-based materials are characterized by high wear resistance to abrasion whereas Al_2O_3 ceramic material exhibits good chemical and thermal stability. Pure aluminum dioxide (99.99%) is often used as a reference in thermal analysis techniques. These layers are formed during one step coating formation. Duplex layers composed of TiN and TiC phase are also desirable, due to high hardness of titanium carbide and toughness and plastic properties of the titanium nitride. The coatings composed of TiC/Ti(CN)/TiN enhance also corrosion resistance of steel substrate in chloride environments [10]. Multilayer coatings might exhibit much better mechanical properties, especially hardness. Thus, in recent years the new generation of tools coated by functional multilayer ceramic materials has been widely studied.

2.2 Multilayer ceramic coating

2.2.1. Titanium-based layers

Microstructure and mechanical properties of multilayer Ti/TiAlN/TiAlCN material was investigated by M.A. AL-Bukhaiti et al. [11]. The coated materials showed lower-friction coefficients than ceramic material (Al_2O_3) and higher hardness compared to the same uncoated materials. The coatings showed good cohesion within the multiple layers. The layers weren`t brittle, only spallation-chipping failure was registered according to the typical cracking or delamination during scratch-test measurement.

Tribological and mechanical properties of the material during the working of the tools are important. During the scratch measurements it was noticed that heat influences the plastic and chip deformation of the layer. Thermal stability of multilayer $Ti_{1-x}Al_xN$/TiN formed on two PM-HSS (S390PM, S290PM) substrates were investigated by T. Weirather et al. [12]. Temperature of the substrate during multilayer coating formation was between 373°C-575°C. Chemical composition of the substrate slightly influenced the thermal stability of the whole material. The difference in the hardness value was 1 HRC (66 HRC for S390PM and 67 HRC for S290PM) when coatings were formed at temperature up to 530°C. A higher temperature of the substrate during the coating process has caused the hardness to increase for both materials. Heating of the substrate also influenced on the total thickness of the layer and its tribological properties. All layers showed excellent adhesion to the substrates with Rockwell nanoindentation test. Thickness of the coatings formed on metal substrates can be between 10 nm to around 20 μm.

Depending on the applied techniques the chemical composition of the substrate could strongly influence the final effects of substrate treatment. One of the investigations about the influence of the steel's hardness over the surface treatment and the tribological behavior was presented in paper [13].

Comparison of the mechanical properties of single TiC_xN_{1-x} (x=0-1.0) layer with TiC/$TiC_{0.5}N_{0.5}$/TiN layers formed on high-speed steel was presented by M. Yasuoka et al. [14]. The layers were formed using hollow cathode discharge (HCD) ion plating method. Multilayer coatings showed higher wear and dynamic resistance compared with a single layered material. However, the success of the multilayer coating was connected with the first single layer - TiN, which exhibits excellent adhesion to the HSS substrate. TiC layer led to a much better wear resistance of the material. It was found that coatings composed of 50-70% carbon showed the highest hardness, thus the tools life should be longer. Similar investigations with TiN layer as an intermediate layer between substrate and upper ceramic layer was reported by K. Kawata et. al. [15].

The thickness of the lower layer was higher (2.5 μm) compared to the thickness of the upper (Ti,Al)N layer (1.0 μm). The layers were homogenize and well adhered. The upper layer was composed of 21.10 at.% of aluminum and 28.65 at.% of titanium (determined by EDX analysis). Addition of the aluminum caused that coating hardness increased, and coating exhibited low friction coefficient (0.47-0.55). The layers were also resistant to oxidation at air at 1073 K.

To enhance adhesion of two layers TiN and TiAlN, the surface of the HSS was additionally pretreated using plasma nitriding [16]. Then the ceramic layers were deposited by physical vapor deposition technique. The formed layer was smooth, compact with average surface roughness below 5 nm, and total layer thickness 3.1 μm. Plasma-nitrided surface was more favorable for hard coatings formation, which weren`t delaminated during surface mechanical testing.

2.2.2. Chromium-based layers

Another group of the multilayer ceramic coatings for preventing premature steel material degradation contains layers with chromium nitride. Coatings composed of Cr-X-N, where X is Ti, Si, Al or C have been investigated as a functional layer on various metal substrates. Basic chromium nitride (CrN) layer is considered as a material which prevents premature consumption of steel substrates due to their chemical inert properties. Besides increasing mechanical properties of the substrate (mainly hardness), chromium nitride also prevents corrosion and wear.

K. Valleti et al. [17] presented difference in microstructure and mechanical properties of single CrN and multilayer CrN coatings formed on HSS using cathodic arc physical vapor deposition. Hardness of the coatings was similar, i.e. around 22-14 GPa, but the differences were observed in the coating microstructure and adhesion to the substrate. Monolayer CrN was more compact and regular compared with multilayer CrN_x. Adhesion of the coatings wasn`t improved by formation of additional CrN_x layers. It was also reported that increasing thickness of the layer resulted in the increase of the internal stresses and could led to reduced adhesion strength of the coating to the HSS substrate. Similar results were presented by C.-M. Shi et al. [18]. The HSS was coated by various thickness ratios of CrN and Cr_2O_3 layer using an arc ion plating method. The main goal of the CrN interlayer was improvement adhesion of the outer ceramic layer to the metal substrate. It was found that CrN layer indeed improved additional ceramic layer adhesion, but also influenced on their final average surface roughness. The multilayer coatings composed of the thickness ratio 2:1 of CrN and Cr_2O_3 showed the best wear resistance. The average surface roughness of these coatings was 0.13 μm, when for the coatings composed of the layers in thickness ratio 1:4 was much higher 0.26 μm. Thus the significant difference in friction coefficient was observed. It was proved that the CrN underlayer reduced plastic deformation of the multilayer ceramic coatings.

Thermal stability of the chromium-based coating was improved by yttrium. S. Liu et al. [19] characterized the CrAlYN layers formed on the HSS by a hybrid high-power impulse magnetron sputtering – magnetron sputtering disposition method. CrN/ AlN layers were enriched by Y up to 9 at.%. Yttrium was preferentially adsorbed into the AlN layer. The layers composed of Y were more compact and better integrated after heating at 1000°C compared with layers without Y. The thermal stability of the layers was enhanced by formation of yttrium nitride (YN) which was considered as a diffusion barrier layer.

Addition of carbon in chromium nitride composite decreases layer adhesion to the substrate but increases their hardness. However the amount of carbon in the layer is limited. It was reported that more than 19 at.% of carbon content decreases wear rate at higher temperature. It was found that Cr-X-N coating with carbon content between 5 and 15 at.% reduced the friction coefficient more than 0.15 [20]. Thus, the investigation of the multilayer CrCN/CrN coatings composed of carbon (10 at.%) was investigated by A. Gilewicz et al. [21]. It was found that multilayer coatings obtained by cathodic arc evaporation technique showed much better mechanical properties and better adherence to the substrate compared to the single layer. The critical load depends on the structure of the coating, its thickness, microstructure and adhesion between each interlayer. Another multilayer coatings formed on HSS composed of Cr/CrN/CrTiN was proposed by M.S. Kabir et al. [22]. The layers were obtained using close field unbalanced magnetron sputtering (CF-UBMS). Microstructure of the coatings strongly depends on the content of titanium columnar microstructure decreased when coatings were composed of a higher content of titanium. The decrease of the columnar grain size influenced the top layer hardness. The top layer of the ceramic composite was very smooth and the average surface roughness was 21.0 nm ± nm (determined by atomic force microscopy). The influence of TiN layers on micromechanical properties of the TiN/CrN bilayer coating formed on high-speed steel by physical vapor deposition was presented by J.J. Roa et al. [23].

Chromium-based coatings may also be composed of additional elements like aluminum and/or silica. Procedures for multilayer formation and nanocomposite AlCrN and AlCrSiN obtained on the high-speed steel by arc ion plating were proposed by W. Wu et al. [24]. Addition of Si into layers caused the formation of nanocomposite solid-solution (Cr,Al)N crystallites phases and amorphous Si_3N_4. This kind of ceramic layers resulted in higher hardness 48 GPa, lower surface roughness and lower friction coefficients compared to layers without silica compounds. Physical vapor deposition technique is excellent to obtain duplex layers on nitrided substrate (CrAlN layer) during one step process [25]. The hardness, high temperature stability and extreme wear resistance significantly increased after the steel's surface treatment. The layers were composed also by intermediate Cr layer, CrN layer with the thickness of 0.3 µm. It was reported that TiAlN/CrN layer composed of chromium up to 30% reduced low oxidation rate at temperatures up to 900°C [26].

3 Summary and conclusions

Various techniques are used to obtain hard coatings on high-speed steel surface. Chemical composition of the layers is varied, but mainly composed of elements such as Ti, Cr, Al, N, C, Si. The formed ceramic layers are usually crystals, and phase composition includes compounds like TiC, TiN, TiAlN, CrN, CrCN. These layers are usually chemical and thermal stable in the temperature range of the cutting tools. Single layer formed on the substrate exhibits usually good adhesion. It is desirable to obtain tools with much longer tool life. Thus, investigations on the multilayer hard coatings have been extensively carried out. Indeed some single interlayers such as TiN, CrN enhance adhesion between harder outer layer with the metal substrate. Multilayer coatings are characterized by higher thickness and this may influence their fragility. As it was mentioned, chemical and phase stability is one of the main requirements of the hard coatings formed on the cutting tools. The ceramic layer might be formed as a separate layer (one by one) or as a layer with diffusion phase between these layers.

The chemical composition of course influences on the final mechanical properties of the material, however the microstructure and surface roughness play the main role in the cutting tool's life. In order to better synthesize the amount of information above, Figure 1 is proposed to illustrate some key aspects of the 2 type of coatings, i.e. single layer and multilayer.

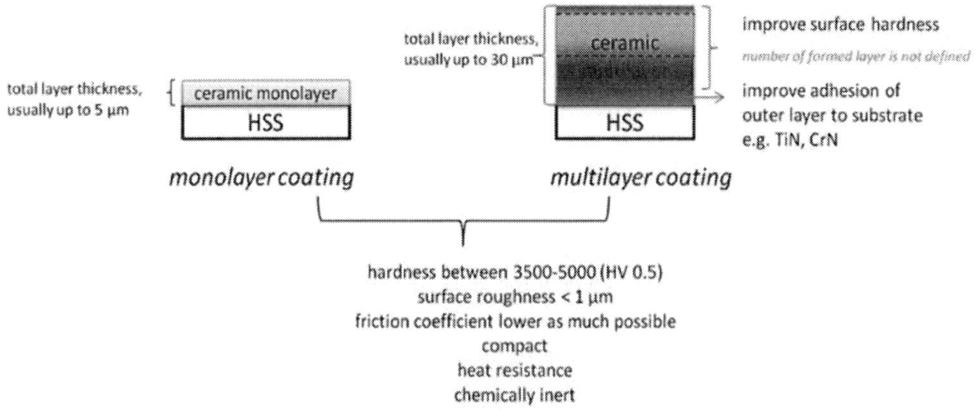

Fig. 1. Single layer and multilayer coatings.

The chemical composition of the single and multilayer coatings might be the same or very similar. However, the structure and phase composition of the layers can be strongly different. The final results of the metal surface modification depend on the number and type of applied techniques. Due to the condition during surface treatment e.g. temperature, the same chemical compounds may exist in various crystal structures due to phase transitions. In some cases the amorphous form of one phase might form and change the physical properties of the coatings. Due to the goal of the surface treatment - strong adhesion of the layers to the substrate and high hardness value - the number of the applied techniques for multilayer coatings formation might be limited. It is important to apply the appropriate instrumental techniques to control the properties of the multilayer coatings as they might be expensive and time consuming. Thus it's recommended that the formation of multilayer coatings should be correlated with the basic investigations as well as with the applied investigations which are desirable to be conducted in the industrial environment. Many techniques are appropriate for the formation of functional hard coatings. The parameters applied during surface treatment are important, and the quality of the used substance determines the final effects of the multilayer coating formation. As pointed out by [6] through the 4 phases identified within the dynamic analysis of the cutting process, the cutting conditions are very much different in continuous cutting than in interrupted cutting. Thus the coatings which are applied on the cutting tool must be adapted to the requirements of the cutting process besides to those of the substrate material. When it comes to turning processes, even though amongst the most important properties of the coatings to be focused on should be hardness when it comes to abrasion wear and surface roughness for adhesion wear, it's not enough to properly assess the quality and behavior of that particular coating during cutting. Each multilayer coating should be designed based on some general aspects of the cutting process and coating properties but the custom structure and properties of the coating should be devised after some real cutting tests.

The comparison of the coatings is complicated due to the various parameters applied during mechanical measurements. Thus, it is not possible to find the optimal method, parameters and the best phase composition for hard coating formation on high-speed stainless steel.

References

[1] K. Bobzin, CIRP J. Manufact. Sci. Technol. (to be published)
[2] V.P.Astakov, *Tribology of Metal Cutting*, (Elsevier, 2006)
[3] W. Grzesik, Inserts. Wear, 240, **9** (2000)
[4] J. Rech, J.L Battaglia, A. Moisan, J. Mater. Process. Technol. **159**,119 (2005)
[5] T.-H. Fanga, S.-R. Jianb, D.-S. Chuu, Appl. Surf. Sci. **228**, 365 (2004).
[6] J.-E. Stahl, *Metal Cutting Theories and Models* (Lund University Sweden 2012)
[7] D. P. Saini, M. Ali, ICME 9th International Conference on Manufacturing Excellence, 1 (2003)
[8] T. Aiso, U. Wiklund, M. Kubota , S. Jacobson, Wear **368-369**, 379 (2016)
[9] F. Saba, E. Kabiri, J.V. Khaki, M. H. Sabzevar, Powder Technol. **288**, 76 (2016)
[10] J. Zhang, Q. Xue, S. Li, Appl. Surf. Sci. **280**, 626 (2013)
[11] M.A. AL-Bukhaiti, K.A. Al-hatab, W. Tillmannb, F. Hoffmannb, T. Sprute, Appl. Surf. Sci. **318**, 180 (2014)
[12] T. Weirather, K. Chladil, B. Sartory, D. Caliskanoglu, R. Cremer, W. Kölker, C. Mitterer, Surf. Coat. Technol. **257,** 48 (2014)
[13] S.B. Abusuilik, Surf. Coat. Technol. **284,** 384 (2015)
[14] M. Yasuoka, P. Wang, R.-i. Murakami, Surf. Coat. Technol. **206,** 2168 (2012)
[15] K. Kawata, Thin Solid, Films **386,** (2001) 271-275

[16] A. F. Rousseau, J. G. Partridge, E. L. H. Mayes, J.T. Toton, M. Kracica, D. G. McCulloch, E. D. Doyle, Surf. Coat. Technol. **272,** 403 (2015)
[17] K. Valleti, C. Rejinb, Shrikant V. Joshi, Mater. Sci. Engin. A **545,** 155 (2012)
[18] C.-M. Shi, T.-G. Wang, Z.-L. Pei, J. Gong, C. Sun, J. Mater. Sci. Technol. **30**, 473 (2014)
[19] S. Liu, Y. Yang, R. Ji, X.T. Zeng, W.J. Clegg, Scripta Mater. **130,** 242 (2017)
[20] B. Warcholiński, A. Gilewicz, Z. Kukliński, P. Mysliński, Vacuum **83,** 715 (2008)
[21] A. Gilewicz, B. Warcholinski, Tribology Internat. **80**, 34 (2014)
[22] M. S. Kabir, P. Munroe, Z. Zhou, Z. Xie, Surf. Coat. Technol. **309,** 779 (2017)
[23] J.J. Roa, R. Rodríguez, V. Lamelas, R. Martínez, E. Jiménez-Piqué, L. Llanes, Surf. Coat. Technol. **308,** 414 (2016)
[24] W. Wu, W.Chen, S.Yang, Y. Lin, S. Zhang, T.-Y. Cho, G. H. Lee, S.-C. Kwon, Appl. Surf. Sci. **351,** 803 (2015) 803
[25] S. Yang, Kevin Cooke, Hailin Sun, Xiaoying Li, Kaijie Lin, Hanshan Dong, Surf. Coat. Technol. **236,** 2 (2013)
[26] S.Yang, E. Wiemann, D. G. Teer, Surf. Coat. Technol. **188–189,** 662 (2004)

Use of microscopy techniques in failure analysis of the plastic injection molded parts to prevent the risks of serial defects in the assembly processes

Ion Cristian Braga*, *Razvan* Udroiu, and *Anisor* Nedelcu

Transilvania University of Brasov, Manufacturing Engineering Department, Mihai Viteazu No.5, 500174, Brasov, Romania

Abstract. Nowadays, the plastic injected molded parts are components used in the automobile for both exterior and interior. A mechatronic device uses those molded parts for actuating in order to reach some effect; therefore they should be durable and aesthetically satisfying the customer requirements. The aim of this paper is to present the use of microscopy techniques in failure analysis of plastic molded parts as part of a mechatronic device in order to prevent the risks of cracking occurred in the serial assembly production. Due to the constraints, the thin walls are used in some small plastic parts and they can be broken in the next assembly steps, but the detection of the defects is not facile. At the validation of the batch, the sampling parts are tested to simulate the breakage and the tear zone is investigated using microscopy techniques. The information is used as a feedback to adjust the injection parameters, reference samples being created by experiments to support the team to reduce the time for adjusting the parameters.

1 Introduction

Year by year, the automotive industry is challenged to drastically reduce emissions. Speaking of Europe for instance, in regard to diesel engines, the emission control related cost increased more than three times for the EURO 6 limits compared with EURO 3 [1] as is presented in Figure 1. And the emissions are linked with the weight of the vehicle, as this is linked with the engine power. Therefore, the automakers are using more and more the alternative materials to metals in order to reduce the weight of the vehicles, so the plastic materials are one of the choices. In Europe, according to Plastics Europe, out of the demands of 47 million tons, 8.9% are used in the automotive industry [2].

* Corresponding author: jbc69mc@gmail.com

DOI: 10.1051/matecconf/201711204009

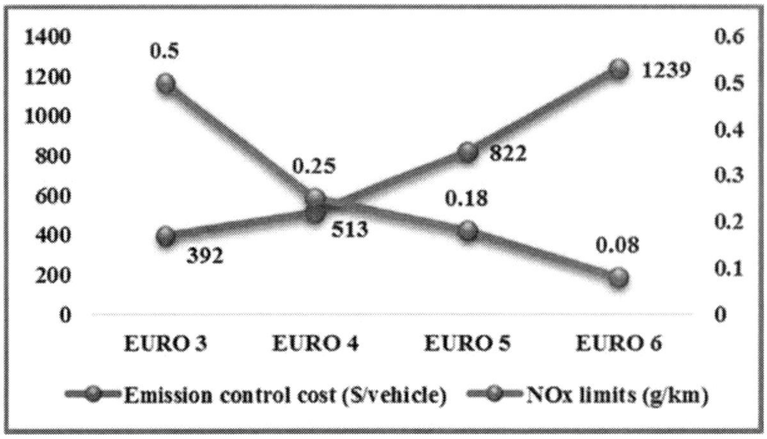

Fig. 1. NO_x limits of diesel cars related emission control cost.

The plastic parts are used in different areas of the vehicles, starting with the exterior parts as bumpers, headlight housings, side mirror housings, front grille, wheel covers, panoramic roof, etc. or interior parts as door panels, center console, and instrument panel.

In regard to the technologies used to manufacture the plastic materials as automotive components could be mentioned extrusion, blow molding, injection molding, compression molding, thermoforming, vacuum forming.

The purpose of this paper is to present the failure analysis of plastic part manufactured in the plastic injection molding, the part being a part of a mechatronic device. Nowadays, the mechatronic devices used in the vehicles become smaller and the haptic feedback requested is more sensitive, therefore the paper present the risk prevention of the cracks occurrence at assembly of the plastic molded parts by using the microscopy techniques at validation of the batches and use the feedback in the plastic injection molding department to adjust properly the injection molding parameters.

2 Plastic injection molding

Injection molding is a process consists of heating a plastic material until become soft enough and force it into a closed mold, then by cooling down the material will be solidified and will form the specific product (Figure 2).

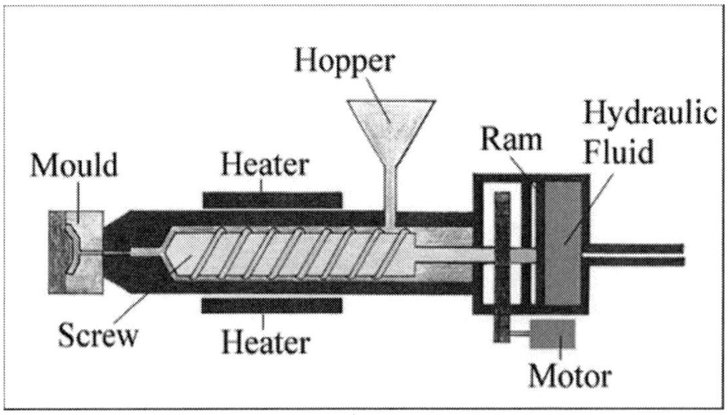

Fig. 2. Injection molding machine [3].

There are many parameters which must be controlled during the process in order to assure the quality level of the parts, but also the optimal cost, and these parameters could be split into four major categories: pressure, distance, time, and temperature. The best setting for these parameters depends on the material used in the molding injection and of the type of mold, as well as the status of the injection machines and, also the same importance, the environmental conditions [4].

3 Case Study

The study case is made on the cylindrical injection molded part which is assembled in a mechatronic device over the other cylindrical injection molded part. The assembly is made by pushing the studied part on the other part until the snap windows are secured in the three hooks positioned at 120 degrees. The thin wall is less than 1 MM, the load by snapping comes from "inside" (Figure 3).

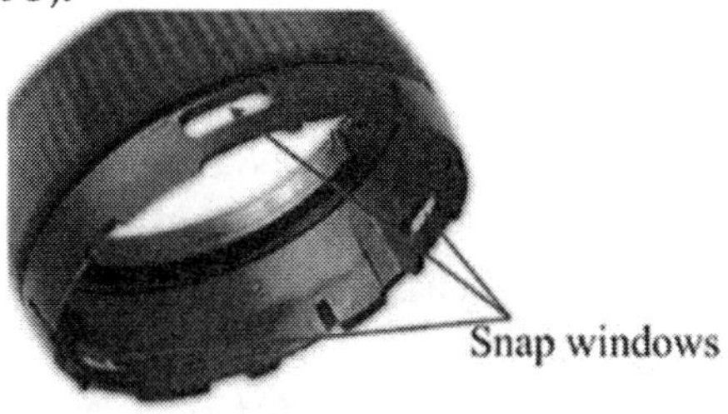

Fig. 3. Snap windows of the molded part.

The plastic material used is a composite material, an acrylonitrile-styrene-acrylate (ASA) and polyamide (PA) blend, which have a big resistance to scratches, impact resistant as well. As in the most of all plastic injection molding processes, the validation of the batches is made by measuring some dedicated dimensions established from the design phase, regular a diameter and a height, also the weight of the part. The parameters and the setting of these parameters are carefully established from the design phase and proofed by the capability of the processes and by the validation tests as well. As can be seen in figure 3, yellow marked is one of three injection points which are positioned at 120 degrees, so are equally arranged in relation to each snap window.

Currently, the start of production is made by measuring some specific dimensions plus the weight of 5 samples at the beginning of the batch and every 4 hours. The parts are taken by the quality shift responsible and measured in the laboratory; if the dimensions and the weight are according with the requirements then the process is released. But in the serial production, the variation of the parameters or the changes in the processes can lead to some risks, the small cracking of the material can occur on the thin wall of the areas near the snap windows. And this will occur at the time of the assembly when the detection of these cracks will be very low, therefore the risk to deliver these assemblies in the next operation or even to the vehicle assembly line will increase the cost of quality.

The root causes investigation is not easily to be made, so the microscopy technique was choosing to support in the analysis. In fact, the samples of the parts with small creaks of the thin wall were studied at microscope and base of the analysis of the images the brittleness of the material was detected (Figure 4). The measurements are done immediately after injection molding, the parts are kept some minutes in the laboratory conditions for the accuracy of the measurements, so there is not an influence of the moisture or temperature.

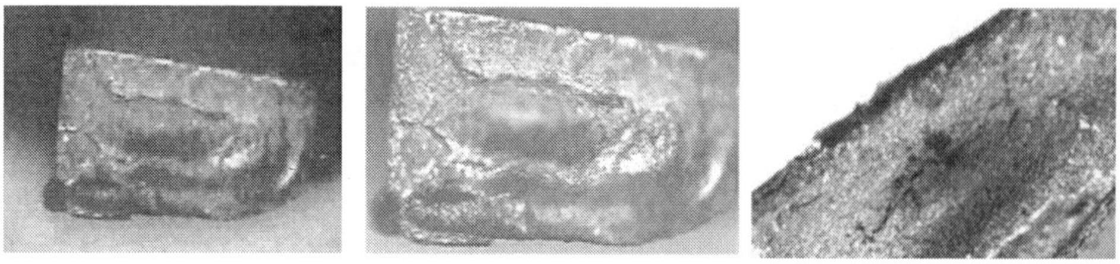

2x 5x 10x

Fig. 4. Microscopy images of the thin wall cracks with objective lens of 2x, 5x and 10x.

Based on the factors detected, as the material used was not changed, the dimensions measured and monitored with the statistical techniques shows not a significant variation, either on the studied part nor of the mating part, obviously, the combination of the parameters in injection molding are not the right one.

To implement this study in the batch validation, some experiments were developed to find the relation between the small crack occurrence and the value of the pull force in the middle of the thin wall area, as can be seen in Figure 5.

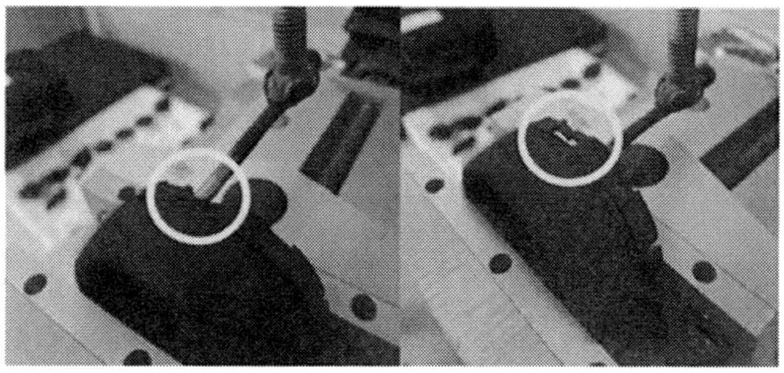

Fig. 5. Pull force measurement experiments.

The pull force measurements were conducted on each of three thin wall areas from each cavity out of the mold (four cavities for the studied part) until the thin wall was broken or until a certain maximum value of the force (for the cases when the part start to be deformed, but not broken at the maximum force). The number of the measurements was 5 shots from each cavity, and the average of the measurements can be seen in Table 1. The minimum value measured was 8.59 N and the maximum was the 21.16 N, but the mode was almost near the average of these 60 measurements. More, in the assembly processes were checked 50 pieces of sub-assemblies to detect the cracking of the thin wall due to snapping and the result was around 35% of broken parts (each sub-assembly have 3 injected parts used, so there were tested 150 pcs).

Table 1. Average of the pull force measurements of each cavity.

	Force (N)			
Thin wall area	Cav. 1	Cav. 2	Cav. 3	Cav. 4
Area 1	17.21	16.89	17.02	17.15
Area 2	16.34	16.32	16.96	17.54
Area 3	12.48	17.43	17.43	16.43

Using the knowledge from these experiments, next step developed was to find the optimal combination of the parameters in injection molding to prevent this brittleness of the

DOI: 10.1051/matecconf/201711204009

parts. At the beginning, were identified the factors with the most impact in the brittleness of the parts: the moisture of the resin was choosing as impact factor, because of the water droplets that will turn to steam when heated in the injection unit, nozzle temperature was another factor, as too hot temperature will overheat the material, then the injection pressure, as the low temperature will lead to weld areas and not good filling conditions, afterward the cycle time (short cycle will not allow the proper melting of the resin), back pressure, screw speed, and injection fill rate will lead to a thermal degradation of the plastic material through increased shear heat [5-6].

Thereby, a multi-approach team with the members from the injection molding setting, quality specialists, molding specialists, after deep analysis of the possible variations and the last changes in the processes, established six parameters and their levels to be part of the experiments. As the full factorial design of experiment was not an option, due to many factors, it means too many combinations and too much time consuming, the Taguchi method is preferred to be used for the optimization of the parameters. In the presented study, the higher value of the pull force is expected, so the signal-to-noise (S/N) ratio characteristic "the-higher-the-better" will be chosen and the orthogonal array L27 (Figure 6) [7].

Level	Factors						Level
min	290	270	8	2	35	20	1
median	300	280	12	5	55	30	2
max	310	290	16	8	75	40	3

	Hot runner	Melt temperature	Cooling time	Moisture	Holding presure	Injection speed	pull-out force average 3 snaps per cavity			
	Levels						Cav1	Cav2	Cav3	Cav4
Trial 1	1	1	1	1	1	1				
Trial 2	1	1	1	1	2	2				

Fig. 6. Parameters and their level for Taguchi method experiments.

After each experiment, the pull force was measured on one chosen area out of three thin walls for the cavities from 1 to 4 (based on the measurements and team analysis). The average of three measurements from each cavity was transposed in the experiments table, then using a software (in this study was used Minitab software) [7-8] the optimal combination was found [9]. After the production running, the sample parts were measured out of the optimal parameter setting, and the pull force obtained, as average, was increased to 35 N, in some cases the thin wall was deformed at 38N but did not present the cracks, finally no defects were found after changing of the parameters.

The cracking area was again studied under the microscope and the fractured areas were not observed anymore (Figure 7).

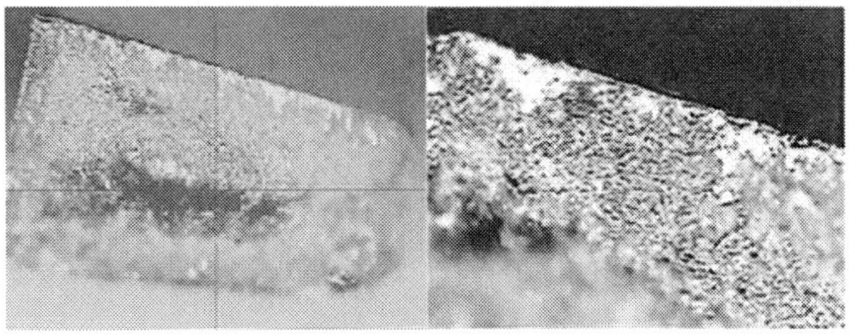

Fig. 7. Microscopy images of the thin wall cracks after optimization (with objective lens of 2x and 10x).

4 Conclusions

Using of the microscopy techniques in failure analysis of the plastic injection molded products as part of the validation control of the batches in the serial production will prevent the risks of occurrence of cracks or other similar defects in further assembly processes. Based on the type of the material and the desired shape and structure of the part, a failure catalog will support the inspector to detect if there is an unexpected effect and to react accordingly.

The paper presents the solution to reduce the poor-quality cost or even the possible complaints from the customer generated by the occurrence of small cracks in plastic molding injection parts only by implementing some supplementary tests made on the already established sample for validation of the processes in injection molding. The pull force measurements and microscopy techniques will lose way less time and cost as the waste generated by the non-conform product that can pass the assembly tests. Thus, the update of the control plan with the measurements to be made, also with the value of the characteristics and method, sampling, frequency, and reaction plan it is expected.

As the plastic materials are more often used in the automotive industry as an alternative to the heavy materials, new plastic material will bring also new challenges in serial production, therefore, the Taguchi method could be used to find the optimal combination of the parameters after the new variations or changes implemented in the injection molding processes.

Last but not least, the combination of the standard validation of the start of production batches by measuring of some dimensions and the weight of material with the pull tests and/or microscopy techniques will lead to lean manufacturing processes and to a high-quality level of the products.

References

1. F. Arena, D. Spera, W.-D. Hoppe, H. Suzuki, E. Kirstetter, *The future of diesel engines*, http://www.adlittle.com, (2016)
2. http://www.plasticseurope.org, (2016)
3. http://www.design-technology.org/.
4. D. M. Bryce, *Plastic Injection Molding...material selection and product design fundamentals*, IPLAS, (2016)
5. http://plastictroubleshooter.com, (2017) ;
6. D. V. Rosato, D. V. Rosato, M. G. Rosato, *Injection Molding Handbook,* **3rd ed.**, (Kluwer Academic Publishers, 2000)
7. https://www.minitab.com/en-us/products/minitab/, (2016)
8. G. Taguchi, *Introduction to quality engineering*, Mc Graw Hill, New York (1990)
9. I.C. Braga, A. Nedelcu, *Optimization of the Plastic Injection Molding Parameters by using the Taguchi Method to Prevent Jammed Components in the Assembly Process of Mechatronic Devices,* CONAT 2016, 575, (2016)

Analysis of some specular nano-oxide-layer spin valve used for digital applications

Daniela Ionescu*, and *Petre-Daniel* Matasaru

Gheorghe Asachi Technical University of Iasi, Department of Telecommunications, Carol I Blvd., No.11, 700506 Iasi, Romania

Abstract. Some peculiar spin valve structures with increased magnetic-resistance (MR), based on Co alloys, have been analyzed in this paper by simulation methods (with help of the HFSS 13.0 program) and their performances were determined for applications at digital sensors. Structures involve specular nano-oxide-layer spin valves (NOL-SV) presenting a NOL insertion inside the pinned layer. A controlled structure has been obtained, simulations indicating the parameters correlation in order to obtain the MR maxima. Values of around 5.8...16 % for the MR ratio have been obtained, respectively a variation of the resistance-area product of 0.7...9 mΩ•μm^2, for an individual layer thickness of around 2-3.2 nm, in an applied field of about 0.4...2.7 kOe. Our results illustrate better performances of the NOL spin valves in comparison with the results previously reported in literature.

1 Introduction

In a spin valve, two magnetic layers separated by a nano metric spacer present electrical resistance which can change between two values depending on the relative alignment of the magnetization in the layers. The resistance change is a result of the giant magnetoresistive (GMR) effect in the structure.

In the specular nano-oxide-layer spin valves (NOL-SV), nano-oxide layers are introduced inside the layer with pinned magnetization and above the layer with free orientable magnetization in order to obtain a specular scattering effect of the spin-polarized electrons, which has as a result the increasing of the magnetic resistance ratio (MR) of the structure, for thinner sense layers. Geometrical dimensions and physical nature of the layers in the spin valve influence significantly the magnetoresistive effect inside the material samples and their fine variation can determine resonant evolution of the MR in some specific domains [12], [14].

This type of spin valves are used in read head and related digital applications. Among these one can mention electric and electronic devices like: magnetic read heads, magnetic field sensors and GMR isolators, microelectromechanical systems (MEMS) and other devices [1], [3], [13], [15]. The magnetic field sensors are devices used to read data in hard disk drives, or like biosensors, etc.

* Corresponding author: danaity@yahoo.com

2 Spin valves characterization

2.1 Theoretical considerations

The configuration of the stack of layers in the NOL spin valve was represented in Figure 1. The main magnetic layers (with pinned, respectively free magnetization) are separated by a nanometric non-magnetic spacer. The theoretical support of the oscillatory antiferromagnetic coupling of the magnetic layers in the stack was considered: the RKKY (Ruderman – Kittel – Kasuya - Yosida) theory and the quantum well models. The intensity of the antiferromagnetic coupling depends strongly on the nonmagnetic spacer properties and in the case of the NOL-SV, on the NOL layers positions in the stack.

For estimating the MR of the considered spin valves using the simulation data, the following formula has been applied, using the sensing current which transforms the change in resistance in a readback voltage [1], [2], [12], given by simulation:

$$\frac{dR}{R} = \frac{R(H) - R(0)}{R(0)} \tag{1}$$

Where $R(H)$ is the resistance of the sample in a magnetic field H, and $R(0)$ is the resistance in null field.

The current-perpendicular-to-plane (CPP) geometry was considered, where the sensing current flows perpendicular to the ferro / nonferro-magnetic interfaces. The specific resistance, AR represents the product of the area A through which an uniform CPP current flows, and the sample resistance R. Variation of this parameter is:

$$\Delta AR = AR_{anti\parallel} - AR_{\parallel} \tag{2}$$

Which represents the difference between the specific resistances in the anti-parallel and parallel states (magnetizations of adjacent ferromagnetic layers in the stack) [4], [5], [10], [11]. On the context, the CPP magnetic resistance ratio was considered as [14], [15], [16]:

$$MR_{\mathrm{CPP}}\,[\%] = \frac{\Delta AR}{AR_{anti\parallel}} \tag{3}$$

2.2 NOL spin valves structure

Some peculiar spin valve structures with increased magnetic resistance (MR), based on Co alloys, have been analyzed in this paper by simulation methods (with help of the HFSS 13.0 program) and their performances were determined. Structures involve specular nano-oxide-layer spin valves (NOL-SV), presenting a NOL insertion inside the pinned layer and also above the free layer. Thickness of the oxidation layer is about 1 nm (0.5-2 nm for parametrical analysis).

The NOL-SV structure of layers (from bottom to above layers) is represented in Fig 1: Ta (3.7 nm) / NiFeCr (2.4 nm) / PtMn (5.5 nm) / $CoCr_{15}Pt_{12}$ (pinned, 1.8 nm) // oxidation (NOL1) // $CoCr_{15}Pt_{12}$ (pinned, 2.2 nm) / Ru (2 nm) (spacer) / $CoCr_{15}Pt_{12}$ (free, 2.8 nm) // oxidation (NOL2, Ta_2O_5) // Ta (3.5 nm) capping layer. Structure is placed on Si substrates.

Above a buffer layer (Ta or Ti) and an underlayer, the antiferromagnetic (AFM) PtMn layer is placed. This AFM layer represents a pinning layer, which fixes the magnetization of the bottom hard magnetic layer and raises its coercivity. The hard magnetic $CoCr_{15}Pt_{12}$ layer (in contact with the AFM PtMn layer) represents the pinned layer. It is divided in two parts, inside it a nano-oxide-layer (NOL1) is grown in an oxidation process. After a nanaometric non-magnetic spacer, a magnetic layer with free orientable magnetization is placed: the $CoCr_{15}Pt_{12}$ layer. The capping layer encloses the structure, below which an oxidation layer NOL2 (Ta_2O_5) is grown.

The hard magnetic material was considered: $CoCr_{15}Pt_{12}$ - $CoCr_{22}Pt_{14}$, respectively CoCrTa. For the oxide of the pinned layer we have considered the possibilities reported in literature [2], [6], [8]: in NOL1: CoO, CrO, $CoCr_2O_4$, Cr_2O_3, Cr_3O_4 , respectively Co_3O_4.

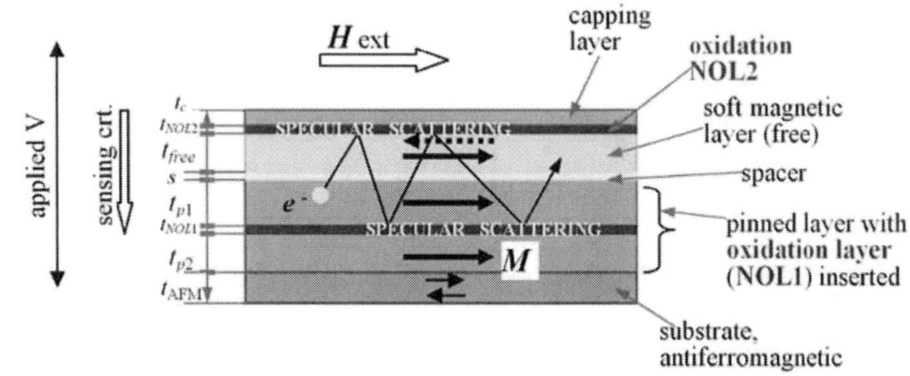

Fig. 1. Layer configuration of the considered specular nano-oxide-layer spin valve. The current-perpendicular-to-plane (CPP) geometry was considered, where the sensing current flows perpendicular to the ferro / nonferro-magnetic interfaces.

In the specular nano-oxide-layer spin valve, the sensing current flows perpendicular to the ferro / nonferro-magnetic interfaces, in the considered current-perpendicular-to-plane (CPP) geometry (Figure 1).

The role of the two nano-oxide-layers in SV is to enhance the specular reflection of spin-polarized electrons from the oxide interface, while conserving their spin direction [4], [5]. The above oxidation layer NOL1 allows the direct exchange coupling between the PtMn layer and the pinned CoCrPt layer. The mean free path of majority spin-polarized electrons is extended, due to specular reflection at metal and insulator interfaces. In the same time, the coercitive field H_c increases when an oxidation layer is inserted into the pinned layer [5], [8], [9].

The main effect is that the GMR ratio is enhanced by insertion of the oxidation layers inside the pinned layer and after the free magnetic layer [4], [8].

3 Results for structure magnetic resistance

We have used the structural simulation data to calculate the CPP magnetic resistance ratio for the considered spin valve structures, presenting oxidation layers. The individual layer thickness in the stack in Fig. 1 was around 1.8...5.5 nm, depending of the layer, in an applied field of about 0.4...2.7 kOe. A sense current, I, of 0...3 mA has been considered in the stack of layers. For parametrical representation, the NOL1 layer thickness t_{NOL1} was about 0.5...2 nm, while the free layer thickness t_{free} was of 1.8...3.5 nm.

For the 3D structural simulations, the mesh was set in the ANSYS HFSS program (High Frequency Electromagnetic Field Simulator) at 16.6 x 16.6 x 22.4 nm^3 per cell, chosen to overall the exchange interaction length for the magnetic layers in the stack, which is of nanometers order [13], [15], [19].

The resistance-area product and it's variation was also calculated for the NOL-SV with the two kind of hard magnetic material (CoCrPt, respectively CoCrTa).

Graphs for the MR ratio as a function of the free layer thickness t_{free} are given in Fig. 2, with the NOL1 layer thickness t_{NOL1} as parameter. The CPP magnetic resistance ratio presents an increasing with t_{free}, with a wide maximum which occurs around 2.9 nm for t_{free}, value imposed by the magnetic ion nature in the hard magnetic alloy.

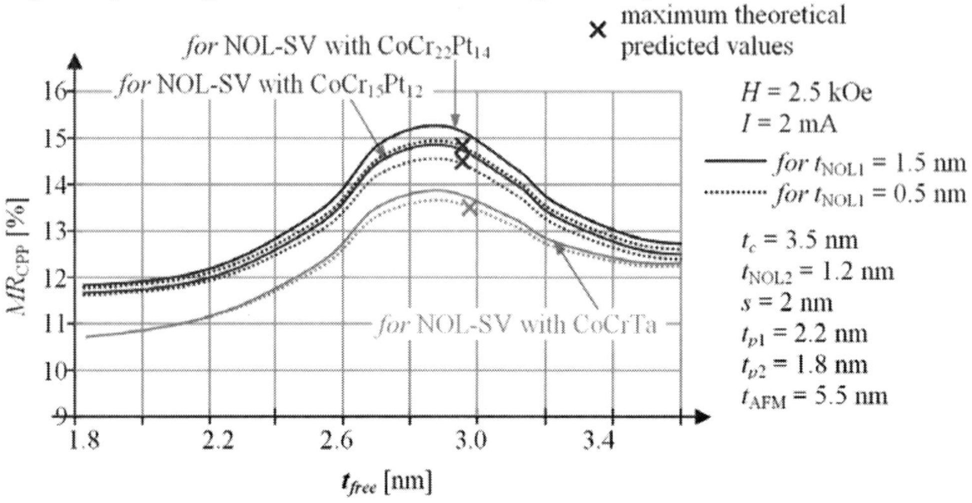

Fig. 2. The CPP magnetic resistance ratio of the considered specular nano-oxide-layer spin valves as a function of the free layer thickness t_{free}. Theoretical predicted maxima were represented on graphs.

One can observe that higher content of Cr in the hard magnetic alloy determines higher values for magnetic resistance. Concerning the materials conductivity, the Pt conductivity of $0.96 \cdot 10^7\ \Omega^{-1} \cdot m^{-1}$ being lower in comparison with Ta conductivity of $0.76 \cdot 10^7\ \Omega^{-1} \cdot m^{-1}$, generates a little bit fewer conduction electrons participating at scattering processes, which determines lower values of the magnetic resistance.

Theoretical predicted maximum on each graph occurs for thicker free magnetic layer (corresponding to a higher magnetization value in the layer) and displays a lower value. The oxidation layer thickness influences the curves, thicker oxidation layer being equivalent with a more accentuate specular scattering effect and avoiding the possible tunneling phenomena.

The distance effect is also reported in literature, the GMR effect increasing generally versus the thickness of individual magnetic layers, but the curves shape depends on the distance of the NOL to the magnetic layers [4], [7], [8].

Graphs for the variation of the resistance-area product as a function of spacer thickness s were given in Figure 3.

The thickness of above part of the hard magnetic layer t_{p1}, which represents also the distance in this layer to the below oxidation layer NOL1, was considered as parameter. The free layer thickness and thickness of the NOL1 layer were considered for obtaining the maximum GMR effect in the spin valve.

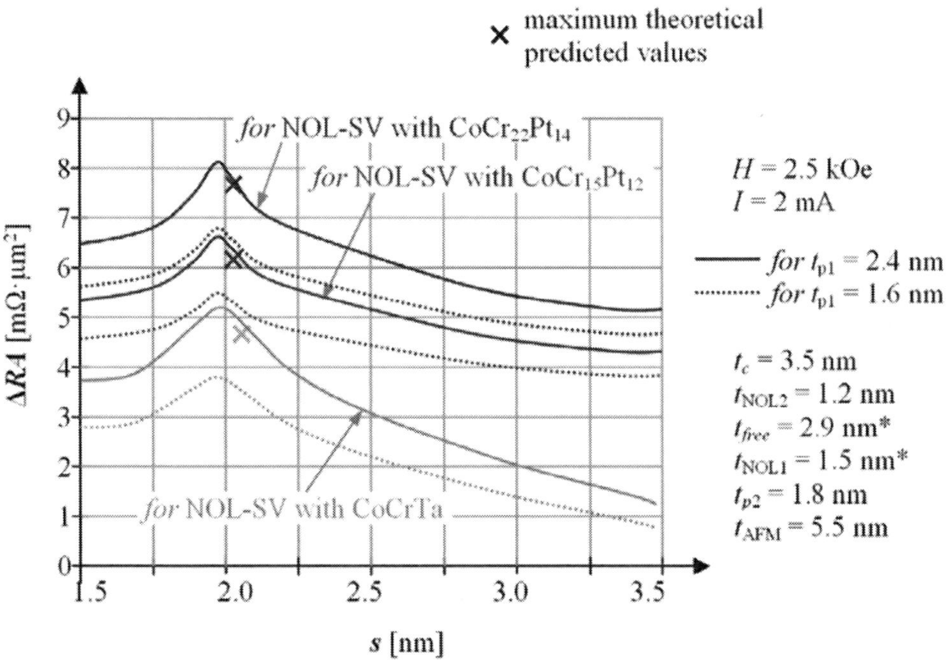

Fig. 3. Variation of the resistance-area product as a function of spacer thickness *s*, for the considered specular nano-oxide-layer spin valves. Theoretical predicted maxima were represented on graphs.

A variation of the resistance-area product of 0.7...9 mΩ•μm^2 was obtained, presenting a maximum around a value for spacer thickness of 2 nm. Even the *RA* values are decreasing when the nano-oxide-layers are inserted in the spin valve, the occurring maximum is of interest, implying higher values of the CPP magnetic resistance ratio (rel (1)), while the antiparallel specific resistance $AR_{anti||}$ also decreases.

4 Conclusions

Structures involve specular nano-oxide-layer spin valves (NOL-SV) presenting a NOL insertion inside the pinned layer have been analyzed in this paper. A controlled structure has been obtained, simulations indicating the parameters correlation in order to obtain the MR maxima.

Values of around 5.8...16 % for the MR ratio have been obtained, respectively a variation of the resistance-area product of 0.7...9 mΩ•μm2, for an individual layer thicknesses of around 1.8...5.5 nm, in an applied field of about 0.4...2.7 kOe.

Our results illustrate better performances of the NOL spin valves in comparison with the results previously reported in literature for the classic spin valves [13], [17], [18], [19].

The GMR ratio is enhanced by insertion of the nano-oxide-layers inside the pinned layer and after the free layer, thus creating a specular scattering space for the spin-polarized electrons.

Geometrical dimensions (especially active layers and spacer thicknesses) and physical nature of the layers in the spin valve influence significantly the magnetic resistance variation at device level and the obtained resonances of the magnetic resistance ratio in some specific domains are of great importance for particular applications.

Performances of the NOL-SV imply two great improvements: a higher MR ratio and thinner sense-layer inside the spin valve structure.

References

1. C. Reig ; D. Ramirez ; H.H. Li ; P.P. Freitas, IEE Proceedings - Circuits, Devices and Systems **152**, 4, 5 (2005)
2. L. Wang, J. Qiu, W.J. McMahon, Y.H. Wu, Physical Review B **69**, 21 (2004)
3. J.Dae Suh, C. A. Ross, Physica Status Solidi (B) **244**, 12 (2007)
4. F. Shen, Q. Y. Xu, G. H. Yu, W. Y. Lai, Z. Zhang, Z. Q. Lu, G. Pan, A. Al-Jibouri, Applied Physics Letters **80**, 23 (2002)
5. F. Shen, Q. Y. Xu, G. H. Yu, Z. Q. Lu, J.-W. Cai, Z. Zhang, Acta Physica Sinica **51**, 8 (2002)
6. S. Y. Yoon, D. H. Lee, J. Y. Lee, Y. S. Kim, D. H. Yoon, S. J. Suh, Journal of the Korean Physical Society **44**, 5 (2004)
7. C. Yongxiang, W. Haicheng, Y. U. Guanghua, Journal of University of Science and Technology Beijing **30**, 12 (2008)
8. C.L. Platt, K.W. Wierman, E.B. Svedberg, T.J. Klemmera, J.K. Howarda, David J. Smith, Journal of Magnetism and Magnetic Materials **247**, 2 (2002)
9. J. Ventura, J. Sousa, J. P. Araujo, P. P. Freitas, Journal of Applied Physics **101**, (2007)
10. Y. Xu, D. D. Awschalom, J. Nitta, Junsaku (Eds.), *Handbook of Spintronics* (Springer Netherlands, 2016)
11. T. Dietl, D. Awschalom, M. Kaminska, H. Ohno (Eds.), *Spintronics, Volume 82, 1st Edition*, (Academic Press, Elsevier, 2008)
12. D.D. Awschalom, D. Loss, N. Samarth (Eds.), *Semiconductor Spintronics and Quantum Computation* (Springer Science & Business Media, 2013)
13. B. Seongtae, H. Jack, *Spin Valves in Spintronics Applications: Materials, Device Applications, and Electrical* (VDM Verlag Dr. Müller, 2010)
14. K. Sato, E. Saitoh (Eds.), A. Willoughby, P. Capper, Safa Kasap (Series Eds.), *Spintronics for Next Generation Innovative Devices*, (Wiley, 2015)
15. G. C. Lombardi, G. E. Bianchi, *Spintronics: Materials, Applications and Devices* (Nova Science Publishers, 2009)
16. E. R. Hedin, Y. S. Joe (Eds.), *Spintronics in Nanoscale Devices* (CRC Press, 2016)
17. R. A. Lukaszew (Ed.), *Handbook of Nanomagnetism, Applications and Tools*, (Pan Stanford Publishing Pte. Ldt., Singapore, 2015)
18. H. Yu, L. Ni, Y. Wang, *Non-Volatile In-Memory Computing by Spintronics (Synthesis Lectures on Emerging Engineering Technologies)*, (Morgan & Claypool Publishers, 2016)
19. C. Felser, G. H. Fecher (Eds.), *Spintronics: From Materials to Devices*, (Springer Science & Business Media, 2013)

MATEC Web of Conferences **112**, 04011 (2017) DOI: 10.1051/matecconf/201711204011
IManE&E 2017

Study of the areal density in the read heads with spin valves with nano-oxide-layer insertion

Daniela Ionescu[1,*] and *Gabriela* Apreotesei[2]

[1]Gheorghe Asachi Technical University of Iasi, Department of Telecommunications, Carol I Blvd., No.11, 700506 Iasi, Romania,
[2] Gheorghe Asachi Technical University of Iasi, Department of Physics, Dimitrie Mangeron Blvd. 67, 700050, Iasi, Romania

Abstract. Read heads performances have been analyzed by simulation (HFSS) in the case of the devices designed with a category of spin valve structures with nano-oxide-layer insertion, based on Co alloys. Different structures of the stack of layers in the spin valve were considered in order to increase the magnetic resistance due to contribution of the insertion layer. A parametrical evolution of the resistance-area product variation ΔRA, and of the areal storage density have been obtained using the simulation and theoretical data. Results indicate areal densities up to 780 Gbit/in^2 for the system manufactured in practice, when the variation of the resistance-area product was greater than 7.4...9 mΩ•μm^2, strongly dependent on the nano-oxide insertion layer. These performances can be overlapped by parameters correlation, indicated by simulation. Consequently, our simulations offer solutions for improving the recording performances by controlling the spin valve structure.

1 Introduction

The read/write heads of the hard disk, using spin valves with nano-oxide-layer insertion (NOL-SV), have the role to transfer the information (the succession of bits) between the magnetic storage media and the electronic components which processes the digital signal. The read heads are converting the magnetic pulses (bits) into voltage levels and have to be able to work with a high speed and a great sensitivity. The case of the perpendicular recording was considered, where the bits point up or down perpendicular to the disk surface, in comparison with the parallel recording case, where the bits are moving in the disk plane. Higher fields are necessary at perpendicular recording to set the magnetization, but the advantage is that the magnetization is more stable.

A category of NOL-SVs has been studied in this paper, for the advantages presented by such a kind of structures: an increased GMR ratio by NOL insertion inside the pinned layer and after the free layer, while the sense-layer can be implemented thinner than in the case of classical SV. The NOL-SV are used in giant magneto resistive (GMR) read heads, for magnetic recording disk drives where requirements for higher density recording have to be fulfilled. High magnetic resistance and thinner layers are necessary for spin valves and the

* Corresponding author: danaity@yahoo.com

solution is represented by the specular reflection layers such as nano-oxide layers (NOLs) inserted in the spin valves.

2 Microstructures with discontinuous oxidation layer

2.1 Theoretical considerations

The NOL spin valves present a great magnetic resistance ratio (MR), which ensures a great sensitivity to the device used for applications [1], [2]. The theoretical model was based on the Valet-Fert theory [3], which represents the theory for giant magnetic resistance in the spin valves with current-perpendicular-to-plane (CPP) geometry. The theory applied in The NOL-SV case is known as the modified Valet–Fert model for the current-perpendicular-to-plane giant magnetic resistance spin-valve with a current-confined-path nano-oxide layer. The current which is passed perpendicular to the ferro / nonferro-magnetic interfaces it is assumed that presents a uniform current density uniform across the area. The electronic transport in magnetic multilayers depends on the electrons scattering phenomena due to the spin-orbit interactions.

The specific resistance *AR* has been computed in the case of the considered NOL-SV, like the product of the area *A* through which a CPP current flows transverse on interfaces and the sample resistance *R*. The difference between the anti-parallel and parallel states (when the magnetizations of adjacent ferromagnetic layers in the stack are anti-parallel, respectively parallel) leads us to the specific resistance variation:

$$\Delta AR = AR_{anti||} - AR_{||} \tag{1}$$

Which represents the difference between the specific resistances in the anti-parallel and parallel states [2], [4].The CPP magnetic resistance ratio was computed as [5], [6]:

$$MR_{\mathrm{CPP}}\,[\%] = \frac{\Delta AR}{AR_{anti||}} \tag{2}$$

The areal density at data storage was considered as the amount of data that can be stored in a given amount of hard disk platter and represents a measure of the number of bits that can be stored on the unit of area. Physically, the areal density can be calculated as the product of the track density and the linear or recording density. The track density indicates how many tracks can be placed down on an inch of radius on the platters. The linear or recording density indicates how tightly the bits are packed within a length of track (bits per inch per track, BPI). Every track on the surface of a platter has a different length, because they represent concentric circles, and different tracks are written with different densities. The maximum linear density used on each drive is indicated on datasheets.

2.2 Characteristics of the structure

In the case of the devices designed with a category of spin valve structures with nano-oxide-layer insertion, hard magnetic materials based on Co alloys have been considered ($CoCr_{15}Pt_{12}$ - $CoCr_{22}Pt_{14}$, respectively CoCrTa). Different structures of the stack of layers in the spin valve were taken into account, in order to increase the magnetic resistance due to contribution of the insertion layers.

DOI: 10.1051/matecconf/201711204011

The considered layers configuration in the spin valve read head is indicated in Figure 1. The sensing current flows perpendicular to the ferro / nonferro-magnetic interfaces, in the considered current-perpendicular-to-plane (CPP) geometry.

The magnetic resistance enhancement is due to the specular reflection effect inside the NOL-SV structure. The spin-polarized electrons scatter at the NOL surfaces, placed above the free layer and inside the pinned magnetic layer. The mean free path of majority spin-polarized electrons is consistently extended due to specular reflection at metal and insulator interfaces, on NOLs [7], [8], and [9]. In the same time, an improvement of the resistivity of the current-confined-path occurs [9], [10].

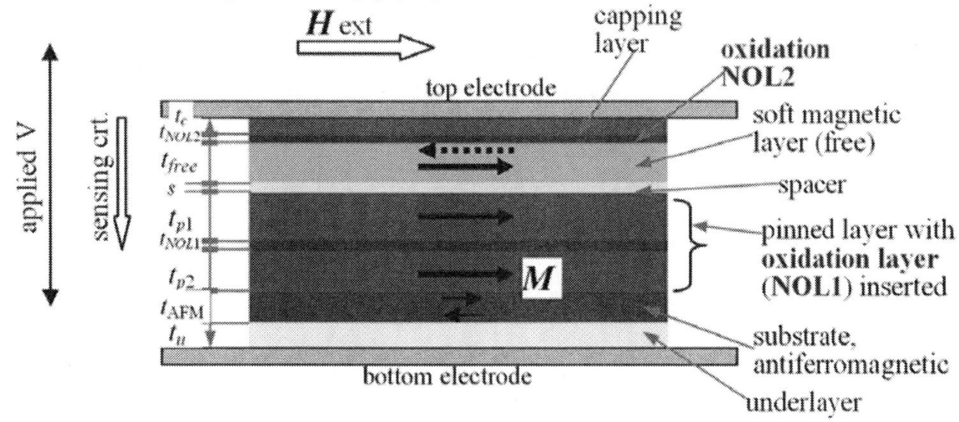

a)

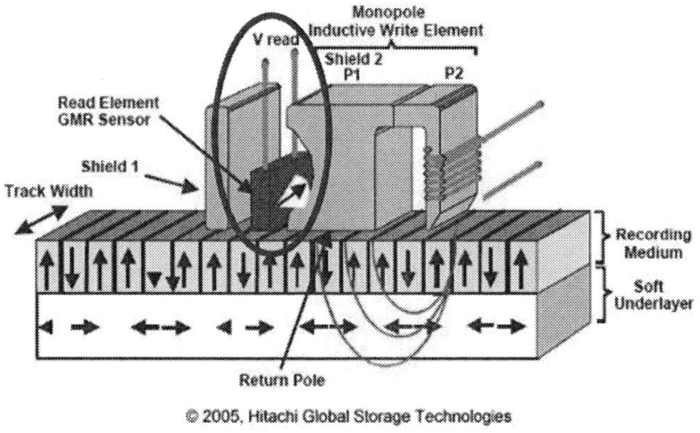

b)

Fig. 1. *a)* Layer configuration of the considered specular nano-oxide-layer spin valve. The current-perpendicular-to-plane (CPP) geometry was considered, where the sensing current flows perpendicular to the ferro / nonferro-magnetic interfaces. *b)* Read element with GMR sensor (inside the red oval), placed on a racetrack memory element, at perpendicular recording (*after* Hitachi Global Storage Technologies, 2005).

The NOL-SV structure of layers (from bottom to above layers) can be described: Ta (buffer, 3.7 nm) / NiFeCr (2.4 nm) / PtMn (antiferromagnetic, 5.5 nm) / CoCrPt (pinned, 1.8 nm) // oxidation (NOL1) // CoCrPt (pinned, 2.2 nm) / Ru (spacer, 2 nm) / $CoCr_{15}Pt_{12}$ (free, 2.8 nm) // oxidation (NOL2, Ta_2O_5) // Ta (capping layer, 3.5 nm). Structure is placed on Si substrates.

The oxidation layers present a discontinuous structure, oxidation occurring preferentially along the grain boundaries on columnar directions as it was reported in literature [7], [11],

 DOI: 10.1051/matecconf/201711204011

and [12]. The discontinuous texture of the oxidation layers enhances the specular scattering and, in the same time, maintains the strong direct exchange coupling between the PtMn pinning layer and the pinned hard magnetic material, increasing the GMR effect.

3 Results for the areal density

The performances of the read heads with NOL-SV have been analyzed by simulation, using the HFSS program 13.0. A parametrical evolution of the resistance-area product variation ΔRA, and of the areal storage density have been obtained, using the simulation and theoretical data.

In Figure 2 was represented the areal density of hard disk drives in function of the CPP magnetic resistance ratio of the considered spin valves. The external magnetic field and the sensing current for the spin valve have been indicated. The thicknesses of the first oxidation layer NOL1 place inside the pinned layer, t_{NOL1}, respectively the thickness of the pinned layer region situated above NOL1, t_{p1}, were considered as parameters and the influence of their modification on the results was illustrated on graphs.

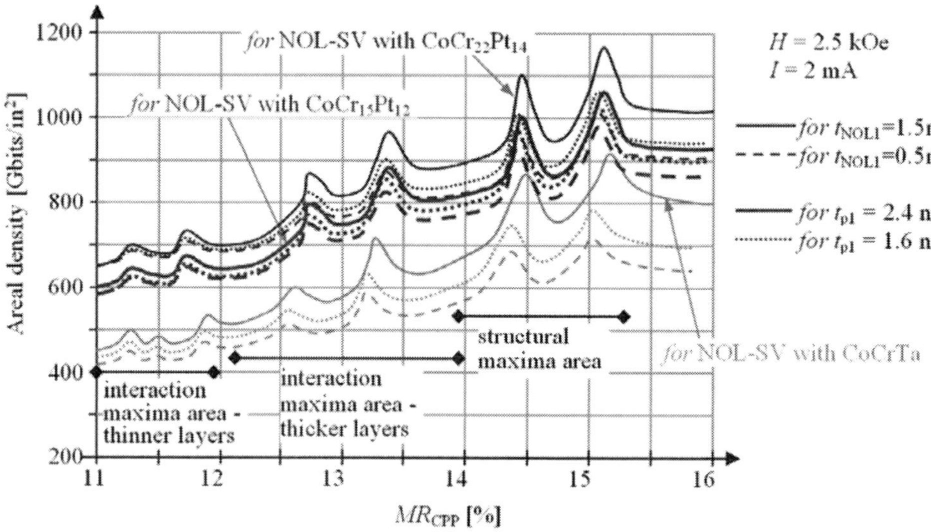

Fig. 2. Areal density of hard disk drives in function of the CPP magnetic resistance ratio of the considered spin valves. Structural, respectively interaction maxima were illustrated on graph. Internal parameters are correlated in order to obtain optimal results (correlated values are marked with $).

The presence of groups of single and multiple maxima can be noticed on graphs.

The main structural maxima of the areal density are imposed by the magnetic ion in the hard magnetic alloy and their characteristics depend on the internal interactions in the exchange coupled system, hard - soft magnetic. A few interaction maxima, generally less intense, are controlled by other factors, like the resonant coupling phenomena between structure, external field and sensing current.

Two zones can be remarked in the interaction maxima area, namely an area with maxima obtained for thicker magnetic layers (3 - 4 nm), respectively for thinner magnetic layers (2 - 3 nm), for which the magnetization is lower.

When different parameters of the structure have been modified, maxima with magnitude depending on the spin current were identified. Their originate in the magnetization dynamics, which is determined by the torques moving the spins, torques generated by the exchange interaction between conduction electrons and DW magnetizations, under the influence of external magnetic field.

The areal storage density modifications in function of the specific resistance variations were calculated and represented on graphs in Figure 3. Graphs are useful for the practical read heads design and setting the material manufacturing criteria.

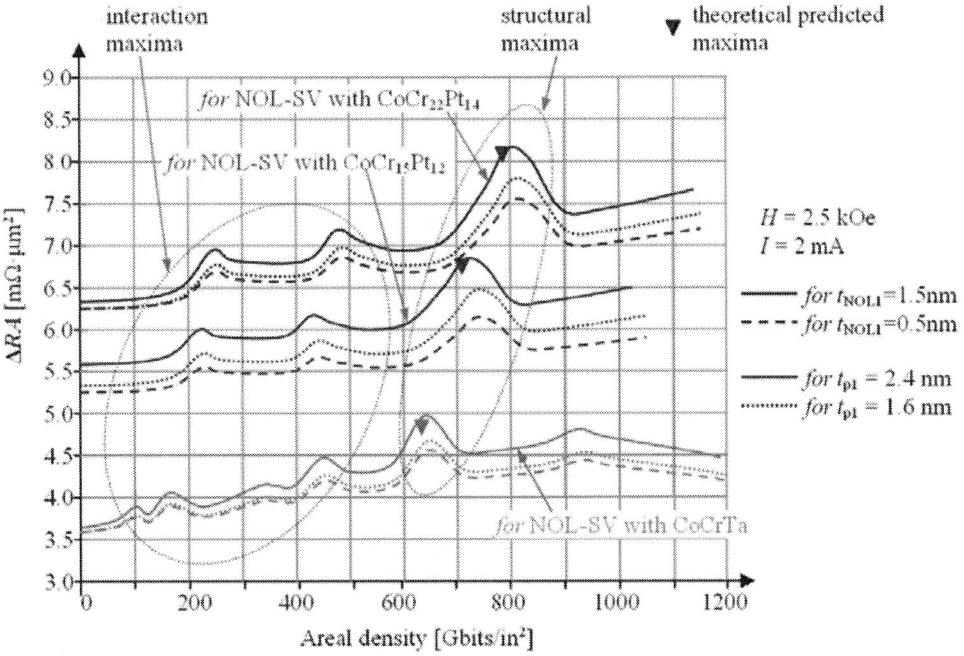

Fig. 3. Evolution of the areal density for a continuous modification of the specific resistance variation, for the considered spin valves. Structural, respectively interaction maxima of the ΔRA, obtained for different subdomains of the areal density values, are indicated on graphs. Internal parameters are not entirely correlated in order to obtain optimal results.

One can remark maxima on the areal density graphs, for different subdomains of the areal density values. These maxima can be characterized like structural and interaction maxima, in function of physical mechanism which generates them.

The main structural maxima of the specific resistance variation, which corresponds to the higher values area for areal density, are imposed again by the magnetic ion nature in the hard magnetic alloy. The maximum magnitude depends on the exchange interactions strength. If the alloy structure changes, the maximum shifts consistently and also its magnitude modifies.

For each type of magnetic material, interaction maxima occurs, less intense than the structural maximum. These maxima are present for materials with a common magnetic ion (Co in our case), and are controlled by the resonant phenomena occurring at structure level when structure interacts with external excitation. The magnitude of these maxima depends mostly on the internal structure of the alloy.

Internal parameters are not entirely correlated in order to obtain optimal results.

Superior performances of the spin valve read heads can be achieved by correlation of the magnetic material nature and geometrical parameters of the magnetic layers, spacer and oxidation layers in the spin valve, as simulations indicate.

4 Conclusions

Results indicate maximal areal densities up to 780 Gbit/in^2, when the variation of the resistance-area product was greater than 7.4...9 mΩ•μm^2, strongly dependent on the nano-oxide insertion layer. These represent the usually reached values for the SV systems manufactured in practice [1], [9], [13]. These values are lower that the maximal values indicated by the areal density - MR curves obtained by simulation with parameters correlation. On the latest mentioned curves, superior values for areal density (over 800 - 1000 Gbit/in^2) can be reached only in systems where the magnetic resistance ratio overlaps a threshold of about 14 %. An extremely careful correlation of all physical and geometrical parameters of the SV has to be practiced in order to obtain these results. Simulations can indicate the values of the necessary parameters which can ensure such a level of performance, in a non-invasive way.

Consequently, the main advantages of the analyzed NOL-SV structures can be pointed out as: higher MR ratio at not very high values of *RA*, corresponding to areal density maxima in the read heads.

Our simulations offer solutions for improving the recording performances by controlling the NOL spin valve structure.

References

1. C. Nam, Y. Jang, K. S. Lee, J. Shim, B. K. Cho, J. Nanosci Nanotechnol. **6,** 11 (2006)
2. J. O. Ventura, J. B. Sousa, P. P. Freitas, A. Veloso, Journal of Magnetism and Magnetic Materials 272–276 (2004)
3. T. Valet, A. Fert, Phys. Rev. B **48**, 7099 (1993)
4. L. Chen, Microsc Microanal **11**(Suppl 2) (2005)
5. A. Gupta, S. Mohanan, M. Kinyanjui, A. Chuvilin, U. Kaiser, U. Herr, Journal of Applied Physics **107**, 093910 (2010)
6. Teruya Shinjo (Ed.), *Nanomagnetism and Spintronics*, Second Edition (Elsevier, London UK, 2013)
7. D. H. Lee, S.Y. Yoon, J. H. Kim, S. J. Suh, Asian-European international conference on plasma surface engineering; Jeju City, Republic of Korea, 28 Sep. - 3 Oct. 2003, S0040-6090(04)00956-3 (2003)
8. H. D. Quang, N. T. Hien, S. K. Oh, N. H. Sinh, S. C. Yu, Journal of Physics D: Applied Physics **37**, 23 (2004)
9. Hideaki Fukuzawa, Hiromi Yuasa and Hitoshi Iwasaki, Journal of Physics D: Applied Physics **40**, 5 (2007)
10. Katsuaki Sato, Eiji Saitoh (Eds.), *Spintronics for Next Generation Innovative Devices* (John Wiley & Sons, UK, 2015)
11. F. Shen, Q. Y. Xu, G. H. Yu, W. Y. Lai, Z. Zhang, Z. Q. Lu, G. Pan, A. Al-Jibouri, Applied Physics Letters **80**, 23 (2002)
12. Y. Xu, D. D. Awschalom, J. Nitta, Junsaku (Eds.), *Handbook of Spintronics* (Springer Netherlands, 2016)
13. D. Weller, L. Folks, W. Lee, J.-U. Thiele, IEEE Transactions on Magnetics **36**, 1 (2000)

MATEC Web of Conferences **112**, 04012 (2017) DOI: 10.1051/matecconf/201711204012
IManE&E 2017

Assessment of metallic materials required for shock stress protection

Sorin V. Cristea, *Marius* Bibu, *Cristian* Deac*, *Alina* Gligor, and *Marinela* Inta

Lucian Blaga University of Sibiu, Romania, Department of Industrial Engineering and Management, Emil Cioran Str., 1, Sibiu, Romania

Abstract. Establishing which metallic materials are suitable to be used for the protection of operators and machinery, when subjected to stress impact in the drilling industry, for surface exploitation, mines and quarries, is a complex process. Due to the harsh operating conditions, a careful evaluation of the various materials used is of utmost importance. This paper addresses the issues relating to the methodology of determining the appropriate homogenous materials to be used, based on a correspondence between their characteristics and applications. Starting with a list of specific operating conditions, one has established a set of requirements and has assessed the market; the recommended materials for the job are alloy steels and aluminum alloys.

1 Introduction

The process of refining the requirements provided by a recipient of a technical system represents a first step, followed by a good management of the project to implement the system can guarantee its success. When the project concerns the technical endowment or, more precisely, when purchasing a system is expensive, choosing the optimum technical solution in terms of the beneficiary ensures an optimum ratio between costs and performances, for a punctual, specific situation.

The paper is divided into three chapters. The introduction presents the current state of theoretical developments of the concepts that led to the establishment of a rapid and simple methodology for linking certain requirements to the materials' features.

In the second chapter the applied methodology is presented. Therefore, it is proposed to quantify certain requirements of current and future beneficiaries of equipment for surface quarrying mining. The requirements were limited to a subsystem of this type of equipment, namely the protection plates at shock and a correlation with market demand was established.

The third chapter presents the results obtained, in the end being presented the conclusions in the approach proposed.

Human activities are often conducted in areas at risk. Risks exist for both the technology as well as for its operator. For quarrying machinery for resource exploitation, a separate category of risks is represented by those generated by a possible environmental physical aggression. In other words, the machines are in a position to be impacted by elements of the field. This is due not so much to the movement of the machinery, performed most often at speeds below 20 km/h, but especially of rocks and earth movements under the influence of environmental factors often unpredictable.

*Corresponding author: cristian.deac@ulbsibiu.ro

Stresses on impact are usually shock stresses. The choice of subsystems' materials for protection to these stresses, protection defined herein limiting, just as a set of technical solutions aimed at avoiding/limiting damage to technology and altering the health of the operator, requires a detailed risk analysis in relation to technical solutions available. The first step in developing means of protection and limiting the effects of shock stresses were made in the military. It has started with the leather shield, then wood, wood and metal, then homogeneous metal plates and non-homogeneous and layered structures followed.

In the field of quarrying machines the use of metal, homogeneous protective plates remains the basic constructive solution, primarily due to the low cost price. Increasing the performance of these protective plates involves the use of materials that can provide an optimal ratio between characteristics and costs.

In addressing the issue we began with the concepts of systems engineering and operational research. Systems engineering, widely represented in specialty papers, for example [1-7] is the set of integrated activities that controls a project as a whole, implements and integrates a comprehensive set of interacting components and systems in order to ensure satisfaction of all users and other stakeholders, within the restrictions due to dynamic operational environments in which the system works. From the operational research area [8-9] analysis and multi-criteria optimization elements were used.

2 Methodology

We have aimed to establish a practical and rapid methodology of choosing the necessary materials for a protection solution against shock stresses from the recipient's requirements and the destination of the machinery.

We started with a general description of the operational capabilities of the machinery in quarrying. The specific requirements for active subsystems were not evaluated: mills, hoes, blades, dippers, etc. We reviewed the threat to be countered and the threatening environment in which the machine must operate was evaluated. Two examples of some extreme situations are shown in Figure 1.

a) rock slides on the machinery. b) landslides covering the machinery.

Fig. 1. Extreme environmental situations of aggression on quarrying machinery.

Existing solutions to provide protection to identify disadvantages were examined. It has been found that the basic solution is the protection with homogeneous metallic plates. It was not taken into account to ensure total protection against any aggression because such solution leads to a total mass of equipment that would significantly affect mobility and progression in the field. Only the aggression of rocks with variable hardness and geometry was considered, often with sharp edges and impact energies of up to 10 kJ (the equivalent of a mass of 200 kg falling from a 5 m height).

Operational performance parameters of protective plates were identified.

For converting certain qualitative formulation into numbers, an index of the beneficiary's need i_{nb} was established, by comparing the number of references of a characteristic to the total number of characteristics to be assessed. The weighted level of confidence of i_{pi} requirement was calculated, depending on the degree of restriction (must be / is desirable) with the relationship:

$$i_{pi} = 1 / [(1 - i_{nb})(N-1)] \quad (1)$$

Where N is the number of opinions expressed.

The calculation is an adaptation after [10-12] and the results are interpreted as an affinity diagram.

The characteristics of some groups of materials were reviewed. The needs of some beneficiaries were identified. The market demand for homogeneous metal plates was evaluated. The suggested materials for making protective plates with good resistance to shock stresses were settled. Their characteristics were quantified and the material proper to the defined punctual situation was determined.

3 Results obtained

There were identified eleven characteristics of protection plates to shock stresses. Ten beneficiaries or potential beneficiaries of these types of equipment were identified, who were asked to specify the features they consider most important. The index of the beneficiary's need for this and the requirement's weighted level of confidence are shown in Table 1.

Table 1. The beneficiary's need index and the weighted level of confidence of the characteristic.

Crt. no.	Characteristic	i_{nb}	i_{pi}	Crt. no.	Characteristic	i_{nb}	i_{pi}
1	Density	0.36	0.16	7	Abrasion resistance	0.73	0.37
2	Hardness	0.91	1.00	8	Corrosion resistance	0.55	0.22
3	Thermal expansion	0.27	0.14	9	Machinability	0.09	0.11
4	Mechanical durability	0.45	0.18	10	Easy mounting/dismounting	0.27	0.14
5	Impact bending strength	0.64	0.28	11	Hardness	0.18	0.12
6	Chemical resistance	0.27	0.14				

It can be noticed that machinability issues are not particularly significant to the beneficiary. Moreover, the problem of hardness is significantly less important than the ease of mounting-dismounting operations. Last but not least, it can be noticed that hardness and abrasion resistance are absolutely needed characteristics, hardness being mentioned by all respondents.

Evaluation of results, given the large size of the protective plates in general and cost prices the field of materials related to ferrous and non-ferrous alloys was limited.

The market demand was analyzed. Medium hardness materials were considered, because they enable a better hardness/impact resistance ratio than materials with high or low hardness. Weldability and good machinability on high temperatures considerations have limited the scope to weak or alloyed temperable steels or aluminum alloys.

As the number of producers of homogeneous metal plates is large, the portfolio of customers was considered as an adequate selection criterion.

For the materials for which data was found, their composition was also considered, necessary to further, detailed studies.

In terms of Romanian steel, OB steels were identified, alloy steels with special destination. Their characteristics are shown in Table 2 and the compositions in Table 3.

From manufacturers of homogeneous metal plates for impact protection there were chosen: Swedish Oxelosund Steel AB - Sweden for laminate plates of alloy steel – Hardox and Kaiser Aluminium Corporation for aluminum alloy plates. There were also considered Aluminium Company of America - ALCOA, USA, ALCAN - England and Pekingese - France.

Table 2. Characteristics of alloy OB 2 – OB 6 ([13]).

Alloy	$R_{p\,0,2}$	R_m	A	Z	KCU	HB	HRC
	[daN/mm²]		[%]		[daJ/cm²]		
OB 2	100	110	15	55	6-12	285-341	28.3-35.0
OB 3	70-110	105-120	10-15	30-50	9-11	285-341	28.3-35.0
OB 4	90-110	100-120	12-15	45-55	7-11	285-341	28.3-35.0
OB 5	90-110	105-115	12-15	24-25	10-14	285-341	28.3-35.0
OB 6	74-78	91-94	17.5	34-39	5-6	262-311	24.8-31.5

Table 3. Characteristics of alloy OB 2 – OB 6 ([13]).

Alloy	Chemical composition [%]							
	C	Si	Mn	Cr	Mo	Ni	P	S
OB 2	0.25-0.31	0.18-0.35	0.30-0.55	1.80-2.3	0.25-0.35	Max.0.5	0.030	0.030
OB 3	0.28-0.34	0.18-0.35	0.30-0.55	1.40-1.9	0.25-0.35	1.00-1.5	0.030	0.030
OB 4	0.28-0.34	0.18-0.35	0.30-0.50	1.80-2.3	0.28-0.38	1.50-1.9	0.030	0.030
OB 5	0.27-0.33	0.18-0.35	0.30-0.50	1.80-2.3	0.28-0.38	1.80-2.3	0.030	0.030
OB 6	0.24-0.31	0.20-0.40	0.30-0.60	1.20-1.75	0.25-0.35	3.00-3.5	0.025	0.025

Hardness and tensile strength for several Hardox steels are shown in Table 4, and their composition in Table 5.

Table 4. Hardness and tensile strength of Hardox alloys ([14]).

Alloy	Hardness HBW	σ_r [MPa]
400	370 … 430	900 … 1100
450	390 … 475	1000 … 1300
450CR*	425 … 475	1400 … 1600
500	450 … 540	1250 … 1400
550	525 … 575	-
HITUF	310 … 370	850
Extreme	57 … 63 HRC	-

* With good resistance to abrasion.

Of the Kaiser Aluminium Corporation 7000 series aluminum alloys, alloys of Al-Zn-Mg type were identified. The chemical composition of two alloys is shown in Table 6.

Reviewing the properties of various aluminum alloys it has been observed that they have hardness between 40 HB and 70 HB. For a proper shock stress protection, we believe that the required hardness should be of at least 70HB.

Table 5. Chemical composition of certain Hardox alloys ([14]).

Alloy	Chemical composition [%]								
	C	Si	Mn	Cr	Ni	Mo	B	P	S
400	0.32	0.7	1.6	1.4	1.5	0.6	0.004	0.025	0.01
450	0.26	0.7	1.6	1.4	1.5	0.6	0.005	0.025	0.01
450CR*	0.18	0.25	1.3	0.1	0.1	0.04	0.003	0.015	0.004
500	0.3	0.7	1.6	1.5	1.5	0.6	0.005	0.02	0.01
550	0.37	0.5	1.3	1.4	1.4	0.6	0.004	0.020	0.01
HITUF	0.2	0.6	1.6	0.7	2.0	0.7	0.005	0.05	0.02
Extremes	0.47	0.5	1.4	1.2	2.5	0.8	0.005	0.015	0.01

Table 6. Chemical composition of some Al-Zn-Mg alloys ([15]).

Alloy	Chemical composition [%]							
	Al	Si	Fe	Zn	Mn	Mg	Bi	Cr
Cu Ni7439, 7440	Min 80	0.1-10	0.1-1.0	0.1	5.0	0.1-5.0	0.1-3.0	0.1-1.0
Cu Cr7439,7440	Min 80	0.1-15	0.1-5.0	0.1-10	0.1-5.0	0.1--5.0	0.1-3.0	0.1-5.0

The material's properties were evaluated in the order of the characteristic's weighted value of confidence i_{pi} (Table 1). Scores between 0.0 and 1.0 were given for the assessed characteristic's values, with 0.0 for the minimum value to 1.0 for the maximum amount for those materials. The first five features were analyzed, in decreasing order of the beneficiary's need i_{nb}, where data was willing. There were established the recommended materials for making protection plates with shock applications.

4 Conclusions

The beneficiary's need index is easier to interpret in terms of quality than the need's weighted level of confidence.

To achieve impact protection plates weldable steels are recommended, usuallylow alloyed and, in some cases, medium alloyed. The chemical composition of these steels is generally as follows: 0.23 ... 0.34% C; 0.18 ... 1.60% Si; 0.30 ... 1.60% Mn; 0.30 ... 2.30% Cr; 0.15 ... 0.38% Mo; 0.1 ... 3.50% Ni. Most shock-resistant weldable steels that are commonly employed are processed as profiles, usually plates with thicknesses from 8-160 mm. By applying this methodology OB2 steel was determined of the domestic production and Hardox 450CR steel from Swedish Oxelosund Steel AB.

If after the evaluation of the general organizing solution of the machine, a significant reduction of its total mass is chosen, even with additional costs, aluminum alloys from the 7000 series are recommended. By applying this methodology, the 7710 alloy was selected, having a hardness of 120 HB.

Selecting a material to meet a well-defined need is effectively achieved in design, every day by engineers, metallurgists, materials science and engineering specialists or designers. Choosing a material at the stage of establishing a generic system solution suitable to a beneficiary, according to a set of characteristics, other than those analyzed in the design stage, has the advantage of better cost control.

In the speciality literature [16-18] most frequent approaches refer to specific behaviors occurring in the material or class of material for a set of given stresses, the paper proposes a rapid and easy methodology for the appropriate selection both of real stresses and the needs of a beneficiary. Of course, further analysis of its behavior in operation cannot be excluded.

References

1. ***, *Systems Engineering Handbook*, NASA Headquarters, ISBN 978-0-16-079747-7, 2007
2. ***, *Systems Engineering Fundamentals*, D.A.U./D.S.M.C Press, Virginia, USA, 22060-5565 (2001)
3. ***, *Guide to the Systems Engineering Body of Knowledge (SEBoK)*, (Stevens Institute of Technology, 2014)
4. A. Kossiakoff A., W.N. Sweet, S.J. Seymour, S.M. Biemer, *Systems Engineering Principles and Practice*, 2nd ed. (Wiley, 2011)
5. Y. Bar-Yam, *When Systems Engineering Fails --- Toward Complex Systems Engineering* (IEEE International Conference on Systems, Man and Cybernetics, Washington DC, 2003)
6. S.N. Syed Salim, M. Zainon, *Control Systems Engineering* (Durian Tunggal Melaka, 2010)

7. T.I. Ören, L. Yilmaz, *Synergy of Systems Engineering and Modeling and Simulation (Proceedings of the SCS International Conference on Modeling and Simulation – Methodology, Tools, Software Applications, 2006)*
8. H.A. Taha, *Operations Research* (Pearson/Prentice Hall, 2007)
9. J.B. Goldberg, *Operations Research, Applications and Algorithms*, 4th ed. (Thomson, 2004)
10. *** *Materials Selection and Design*, ASM Handbook, **20** (ASM International, 1997)
11. J.M. Regan, F.M. Downey, *The TASCFORM Metodology: A tehnique for assessing comparative force modernization*, 5th ed. (TASC, 1993)
12. *** *Properties and Selection: Irons, Steels, and High-Performance Alloys*, ASM Handbook, **1**, 10th ed. (ASM International, 1993)
13. S. Cristea, *Blindaje omogene* (Editura Academiei Fortelor Terestre, Sibiu, 2010)
14. *** *High-strength steel – sheet, plate, coil, tube, profile* (Swedish Steel Oxelosund AB, 2016)
15. *** Kaiser Aluminum (Kaiser Aluminum, 2016)
16. L.C. Jones, R.J. Llewellyn, *Sliding abrasion resistance assessment of metallic materials for elevated temperature mineral processing conditions*, Wear, **267, 11** (2009)
17. W. Hoffelner, *Damage assessment in structural metallic materials for advanced nuclear plants*, Journal of Materials Science, **45** (2010)
18. H. Zhu, T. Wei, D. Carr, R. Harrison, L. Edwards, W. Hoffelner, D. Seo, K. Maruyama, *Assessment of Titanium Aluminide Alloys for High-Temperature Nuclear Structural Applications*, JOM, **64** (2012)

Experimental and numerical analysis of aramid fiber laminates with DCPD resin matrix subjected to impact tests

Miroslaw Bocian[1,*], *Joanna* Pach[2], *Krzysztof* Jamroziak[1], *Mariusz* Kosobudzki[3], *Slawomir* Polak[4], *Dariusz* Pyka[4], *Adam* Kurzawa[2], and *Janusz* Kurowski[5]

[1] Wroclaw University of Science and Technology, Department of Mechanics, Materials Science and Engineering, Smoluchowskiego 25, Wroclaw 50-370, Poland
[2] Wroclaw University of Science and Technology, Chair of Foundry, Polymers and Automation, Laboratory of Plastics, Lukasiewicza 5, Wroclaw 50-371, Poland
[3] General Tadeusz Kosciuszko Military Academy of Land Forces, Czajkowskiego 109, Wroclaw 51-150, Poland
[4] Wroclaw University of Science and Technology, Department of Metal Forming and Metrology, Lukasiewicza 5, Wroclaw 50-371, Poland
[5] The company IM/kompozyty, Topolowa 5, Stanowice 55-200, Poland

Abstract. Research for innovative composite material solutions for impact absorbers is focused on the analysis of their ability to absorb impact energy. For this purpose, aramid fiber laminate with DCPD resin matrix was prepared. Laminate samples were tested on the drop test and were subjected to ballistic loads. The laminate structure was modeled with ABAQUS software with the objective to optimize and estimate the ability to absorb impact energy in the range of high speeds (350m/s).

1 Introduction

Composite parameters depend on the properties of the constituent phases, their number, and volume of the composite, the orientation of the reinforcement in the matrix and the geometrical characteristics of these phases. For this reason, it is important to pay attention to the strength characteristics of the dispersed phase and the continuous phase (matrix). This refers to identifying various properties, such as mechanical, technological, physical and chemical [1-3].

Optimization in the laminate production, particularly ballistic laminates, is focused on the use of thermoplastic polymers [4, 5] and elastomers [6] for matrixes. The common use of these type of matrixes applies to reinforcements based primarily on aramid and polyethylene fibers, where the process of absorbing the energy of the impact loads (ballistic) is oriented mainly on the delamination of the matrix until it fractures. According to some studies [7, 8], the matrix is capable of dispersing up to 35% of the impact energy as a consequence of breaking. The mechanism of ballistic impact energy dissipation in case of a single layer will exceed the strength of the matrix in case of shearing and bending, and at

* Corresponding author: miroslaw.bocian@pwr.edu.pl

the layers edge of the laminate delamination is initiated, this process can occur with a specific progress of fractures [9]. The amount of energy causing delamination of the matrix is much lower than the critical amount of energy causing the destruction of the fiber [10]. This process is divided into two stages [11]. In the first stage, a high energy and velocity projectile on impact causes fiber shearing and matrix exfoliation in the composite. In the second stage, as the projectile decelerates, the remaining impact energy is involved in the process of delamination of the matrix. Composite fibers located underneath the concentration of the impact energy in the deeper layers are subject to stretching.

Authors have observed that, by replacing recently used and described in several papers hardened matrixes with a thermoplastic or an elastomer matrix, it is possible to perform optimization of the ballistic laminate in terms of both the volume efficiency, by the reduction of the reinforcement layers, and a local dissipation of impact energy. As an example, referred work [12] is focused on the analysis of using a styrene-butadiene-styrene (SBS) copolymer based matrix.

In this paper, the main focus is on the tests of dicyclopentadiene (DPCD) type of laminate matrix. Reinforcement of the laminate samples were prepared from aramid fibers. The laminates were subjected to tests for the impact energy dissipation in both ballistic tests and numerical analyzes.

2 Experiments

2.1 Materials

To prepare a DCPD matrix laminate sample, Twaron® aramid fabric of 280g/m^2 basic weight and plain weave structure was used as reinforcement. The fabric thickness of 0.4 mm has a fiber density of approx. 1.45g/cm^3, and a fiber diameter of 12 μm. For laminate production 12 layers of fabric were used, where the sample had a thickness of 4.4 mm and a weight of 224 g. Modified dicyclopentadiene (DCPD) was used as the matrix resin, which is a by-product secreted from processing crude oil. The polymerization process was carried out in the presence of a second generation Grubbs catalyst [13].

Applied in this paper matrix is characterized by improved parameters of static tensile and bending strength, high relative elongation at rupture and high compressive strength in comparison to other types of resins. Material after super saturation with the DCPD was subjected to pressure at room temperature (21°C) at a pressure of approx. 1.5 MPa and was left for 12 hours. Resultant laminate was cut into 100x100 mm pieces (used size was dependent on the coupling at the test stand) in order to perform energy-intensive testing on a drop tower and on a ballistic track.

2.2 Impact test

Perforation of used samples in the resilience test was performed using a Dynatup Instron 9250 drop tower equipped with a HV Instron Dynatup 1-800-I-U-SERVE impactor. The impact test stand is shown in Figure 1.

The geometry of the impactor, which allows executing concentrated loads, is similar to the geometry of the Parabellum 9×19 mm pistol bullet. The accessories used at the test stand allow registration of parameters such as the impact force, maximum deflection and impact velocity. The value of the energy absorbed by the laminate is determined, based on the force curve as a function of time. The value of kinetic energy of the impactor was determined experimentally and adjusted to 150 J. The impact tests were performed at 21°C for 6 samples.

Fig. 1. The impact testing stand: 1 – the impact machine, 2 – the striking head, 3 – the impactor, 4 – catch for mounting samples.

2.3 Ballistic test

Ballistic tests consisted of shootings Parabellum 9x19 mm ammunition into laminate samples using a Polish production pm-84P Glauberyt type machine gun. Tests were carried out according to the specifications of the European norm CEN EN 1522. Six shooting tests were conducted, for which the average muzzle velocity of the projectile was $v_0 = 349.6$ m/s. The speed of the projectiles was measured with a Doppler Weibel SL-525E radar and was mirrored by a Chronograph CED Millennium measurement gate.

Samples were mounted in a special holder, supported on the base with a ballistic plasticine. The plasticity of the base was verified by dropping three times a steel weight (cylinder with a diameter of ø44 mm with a spherical end) of 1 kg from a height of 2 m, where a cavity of 25 ± 3 mm should be the result.

2.4 Numerical analysis

Ballistic simulations performed in ABAQUS program were narrowed in order to simplify the numerical model of the projectile, namely, including in the analysis only the core of the projectile and the translation speed, that is, omission of the projectile rotation [14]. Kinetic energy of the projectile was realized by considering the initial speed $v_0 = 349.6$ m/s and the mass of the projectile as 8 g, thus the initial kinetic energy is as 488.9 J.

Homogenization of modeled object was made in order to determine the mechanical properties of the laminate. Material data used for the simulation are summarized in Table 1.

Table 1. Characteristics of the projectile material properties, aramid material and DCPD.

Specification	E	v	ρ	A	B	n	$SIGY$	FS
	[GPa]	[-]	[kg/m^3]	[MPa]	[MPa]	[-]	[N/m^2]	[-]
Core lead (alloy of Pb1 and antimony)	16	0.42	11300	24	300	1	-	-
Jacket (brass M90)	115	0.31	8940.9	206	505	0.42	-	-
Twaron	115	0.3	1450	-	-	-	3.6E+9	0.04
DCPD	3.1	-	980	-	-	-	-	-

Where: E - Young's modulus, v - Poisson's ratio, A - yield at zero plastic strain, B - hardening constant, n - hardening exponent, $SIGY$ - Yield strength, FS - Effective plastic strain, failure criterion.

DOI: 10.1051/matecconf/201711204013

3 Results and discussion

3.1 Responses under impact loading and ballistic

Based on carried out piercing drop tests of the laminate samples and firing on the shooting range (Figure 2) samples were subjected to a thorough analysis in terms of absorbed impact energy.

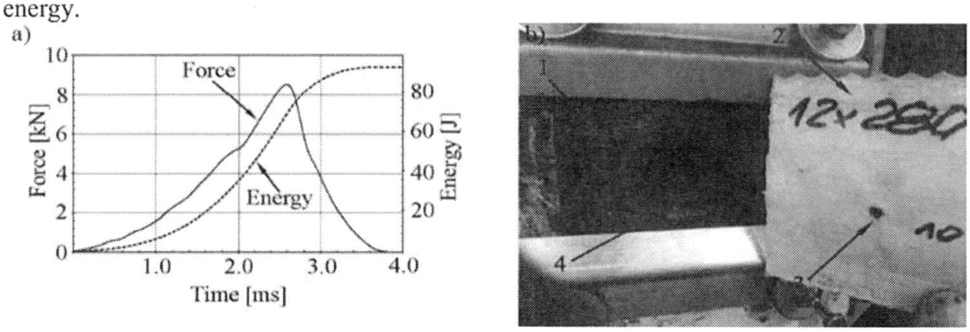

Fig. 2. Selected recorded results: a) after piercing on the drop test, b) after shooting into the laminate: 1 – clamping and the support base for the sample, 2 – shot sample, 3 – penetration point of the projectile in the laminate, 4 – bending region in the plasticine base.

In the drop test, the average energy absorbed by the laminate settles at 92.9 J with an average impact speed of the impactor at 8.8 m/s, the force which acted on a sample was 8.5 kN. Registered laminate deformation is 21.95 mm. During the ballistic test, samples absorbed the total kinetic energy (488.9 J), which was distributed on the deformation of the projectile and the deformation of the laminate. Dynamic deflection of 28 mm of the laminate and the local influence area of approx. 36 cm^2 were estimated based on the stamped in the plasticine cavity.

3.2 The SEM analyses of the composites

By using the Scanning Electron Microscope (SEM), an in-depth analysis of the matrix and the strengthening of the material as a result of the impact were carried out. Based on the SEM images it is shown that the laminate is properly saturated with the resin, as the resin is present between individual strands of fiber. As show in Figure 3, in the contact section between the impactor and the laminate, the reinforcing fibers in the composite are sheared and the damage of the matrix is related with compression and shearing of the fibers, which is observed as delamination.

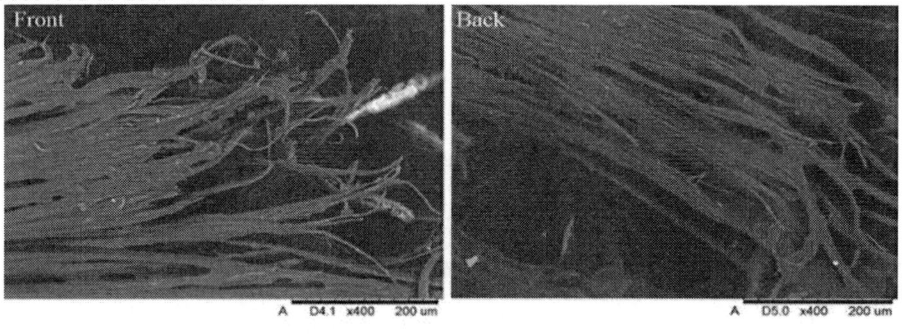

Fig. 3. SEM image of the damage from the drop test.

At the rear part of the laminate fibers are stretched by the impactor and their shearing occurs to a lesser degree. Visible damage to the matrix is a result of fiber delamination.

Damage to the reinforcement layer and the matrix from firing a 9 mm Parabellum bullet is show in Figure 4.

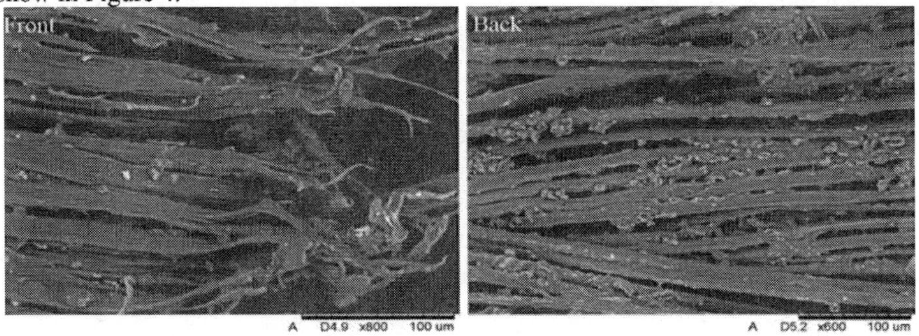

Fig. 4. SEM image of the damage in a sample after the ballistic test.

At the start, the difference in the damage of the reinforcement is visible. This difference is due to the fact that, for the ballistic testing, there is a deformation of the projectile, where the frontal surface is increased, and thus to a large extent the matrix takes an important function. It can be concluded that the matrix significantly affects the dissipation of the energy absorbed by the shot sample.

3.3 Energy absorption in simulation by FEM

Conducting an analysis while using the Finite Element Method (FEM), a model which enhances the dissipation of the impact energy was developed. As shown in Figure 5a, the dissipation of kinetic energy and the extent of deformation (Figure 5b) were correctly interpreted.

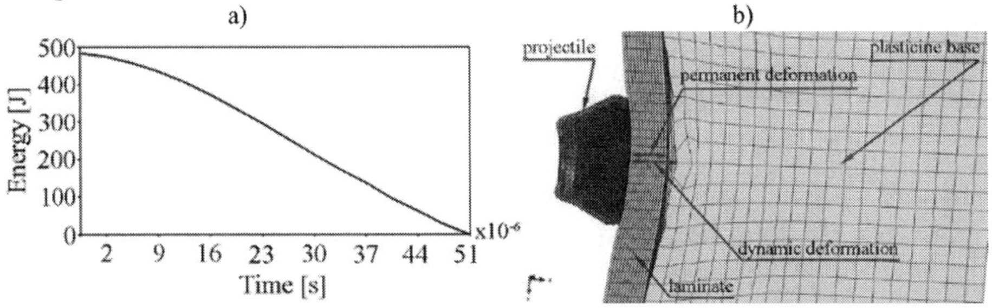

Fig. 5. Results of the numerical analysis: a) graph of the impact kinetic energy, b) deformation of the laminate.

4 Conclusions

This work is considered as an initial research of testing laminates composed of aramid fabric with DCPD resin matrix. Samples were subjected to tests, performed on a droptower and ballistic tests. Energy dissipation was compared at low (drop test) and high (ballistic test) load speeds. It was observed that there was a higher efficiency of the matrix at high loads. As a result of microscopic examination, correct saturation of the reinforcement in the matrix was concluded. Used DCPD resin satisfies the requirements in terms of energy absorption.

As observed in the SEM images, the resin saturates the reinforcement to such a degree that it does not cause the embrittlement of the fiber, which can be observed in the case of epoxy resins and glass fiber. In the following research, work of the authors will focus on various climatic conditions, taking into consideration the aging process, mainly the aging of the matrix.

Calculations have been carried out in Wroclaw Centre for Networking and Supercomputing (http://www.wcss.pl), grant No. 74

References

1. A. Abe, H-H, Hans-Henning Kausch, M. Möller, H. Pasch, *Polymer Composites – Polyolefin Fractionation – Polymeric Peptidomimetics – Collagens* (Springer-Verlag, Berlin Heidelberg, 2013)
2. R. F. Gibson, *Principles of composite material mechanics* (Third edition, Boca Raton, CRC Press: Taylor & Francis Group, 2012)
3. W. Barnat, P. Dziewulski, T. Niezgoda, R. Panowicz, Application of composites to impact energy absorption, Computational Materials Science **50**, 4, 1233-1237 (2011)
4. A. K. Bandaru Chavan, S. Ahmad, R. Alagirusamy, N. Bhatnagar, Ballistic impact response of Kevlar® reinforced thermoplastic composite armors, Int. J. Impact Eng. **89**, 1-13 (2016)
5. J. G. Carrillo, R. A. Gamboa, E. A. Flores-Johnson, P. I. Gonzalez-Chi, Ballistic performance of thermoplastic composite laminates made from aramid woven fabric and polypropylene matrix, Polymer Testing **31**, 512-519 (2012)
6. M. Bocian, K. Jamroziak, M. Kosobudzki, The analysis of energy consumption of a ballistic shields in simulation of mobile cellular automata, Advanced Materials Research **1036**, 680-685 (2014)
7. L. Iannucci, R. Dechaene, M. Willows, J. Degrieck, A failure model for the analysis of thin woven glass composite structures under impact loadings, Computers and Structures **79**, 8, 785-799 (2001)
8. R. H. Zee, C. Y. Hsieh, Energy loss partitioning during ballistic impact of polymer composites, Polymer Composites **14**, 3, 265-271(1993)
9. S. A. Salpekar, Analysis of delamination in cross–ply laminates initiating from impact induced matrix cracking, Journal of Composites Technology and Research **15**, 2, 88-94 (1993)
10. T. J. Kang, C. Kim, Impact energy absorption mechanism of largely deformable composites with different reinforcing structures, Fibers and Polymers **1**, 1, 45-54 (2000)
11. P. J. Hogg, Composites for ballistic applications, *Journal of Composites Processing* (CPA, Bromsgrove U K, March, 2003)
12. J. Pach, D. Pyka, K. Jamroziak, P. Mayer, The experimental and numerical analysis of the ballistic resistance of polymer composites, Composites Part B: Engineering **113**, 24-30 (2017)
13. A. Fürstner, L. Ackermann, B. Gabor, R. Goddard, C. W. Lehmann, R. Mynott, F. Stelzer, O. R. Thiel, Comparative Investigation of Ruthenium-Based Metathesis Catalysts Bearing N-Heterocyclic Carbene (NHC) Ligands, Chem. Eur. J. **7**, 15, 3236-3253 (2001)
14. M. Bocian, K. Jamroziak, M. Kosobudzki, Analysis of material punching including a rotational speed of the projectile, Solid State Phenomena **220-221**, 571-576 (2015)

MATEC Web of Conferences **112**, 04014 (2017) DOI: 10.1051/matecconf/201711204014
IManE&E 2017

Tribological properties evaluation of Ni-P coating manufactured with electroless plating on aluminum alloy substrate

Piotr Jabłoński*, *Kazimierz* Czapczyk, *Piotr* Siwak, *Łukasz* Furmański, *Paweł* Grobelny, *Piotr* Czajka, and *Adam* Patalas

Poznan Univeristy of Technology, Department of Mechanical Engineering and Management, Maria Curie Sklodowska pl. 5, 60-965 Poznan, Poland

Abstract. Results of tribological tests of Ni-P coating were presented in following paper. Tests were conducted with tribotester device using linear reciprocating wear test. Tribological wear test conditions, especially speed, load and sample geometry were described. Characteristics of coefficient of friction variation during wear testes were recorded. Conclusions included results interpretation especially focusing on influence of coating thickness in its properties.

1 Introduction

Aluminum alloys are nowadays widely used among all branches of industry, including automotive (engines, performance cars), marine (hulls), railroad vehicles (carriage structure), and aviation (frames, equipment). This increasing demand is connected with properties of these materials- especially low weight and corrosion resistance, while compared with steel. Among all aluminum alloys, EN AW 7075 (AlZn5.5MgCu) presents very satisfying mechanical properties. Still, lot of research is conducted on field of its weldability and other joining methods [1]. However, steel can be recognized as more common material, mainly thanks to its strength, lower material cost and resistance to abrasive wear. What is more, mechanical properties and wear resistance of both carbon and alloy steels can be easily increased by heat treatment or application of anti-wear coatings. Still, properties of several aluminum alloys can be modified by both heat treatment and coating application, in limited range. [2]. Unfortunately, it's impossible to adopt same coatings and methods that are commonly used for steel, to aluminum alloys. This is caused by significant difference in material properties, especially hardness and surface adhesion. Also, application of same technologies that are applicable for steel, is not valid for aluminum alloys because of lower processing temperatures.

However, among known engineering solutions, several coatings can be successfully used for steel and aluminum alloy substrates. Ni-P (nickel-phosphorus) coating manufactured with electroless, chemical reduction method can be applied to aluminum alloy substrate [3]. In contrast to electric manufactured coatings, electroless Ni-P coating can be applied not only

* Corresponding author: piotr.le.jablonski@doctorate.put.poznan.pl

for decorative reasons, because of better adhesion to substrate and higher resistance to abrasive wear [4].

Moreover, mechanical properties of these coatings can be improved by composition modification, e.g. with ceramic additives, what makes them composite coatings. Nano dispersive additives of PCD (poly-crystalline diamond), corundum, silicon carbide, silicon nitride, PTFE (Polytetrafluoroethylene) can be used, in order to prepare Ni-P coating for actual application. Both hardness increase and wear resistance can be achieved by proper composition modification with additives, what can be very promising, e.g. for applications were dry friction can occur [4-5].

Ni-P coating described in following paper should be considered as basic coating without any additives or modifications. Main aim of research presented above is to demine tribological properties of Ni-P coating manufactured on AW-0705 aluminum alloy.

Conducted tribological test were also aimed to determine and evaluate possibilities of Ni-P electroless coating application on aluminum alloy substrate in specific technical applications. What is more, very interesting results were achieved by combining such coatings with application of innovative lubricants with additives. Numerous records among state of art proved usefulness of Ni-P coating on steel substrates, as a successful method for increasing wear resistance [5-6]. This allows to make an assumption, that further research on field of application of electroless Ni-P coating on different substrates should be considered.

2 Overview of electroless Ni-P coating properties

Nickel- phosphorus coatings are amorphous, homogenous, hard, fragile, self-lubricating, vulnerable to soldering and corrosion resistant. Because of that, they can be applied on wide variety of fields. Ni-P coatings manufactured with electroless methods because of high hardness show major wear resistant directly after plating and also after heat treatment. Also because of its amorphous nature, Ni-P coating can be considered as good heat barrier [7].

While compared with nickel based coatings manufactured with galvanic techniques, electroless plated Ni-P coating presents following advantages [8-9]:

- Nickel coating tend to create coating of homogenous thickness, on numerous surfaces exposed to coating solution, regardless to shape and position of coated piece,
- Coating structure is homogenous and independent from substrate surface structure,
- Higher hardness that can be increased by further heat treatment,
- Higher corrosion resistance,
- Can be manufactured directly on aluminum substrate,
- Can be easily soldered.

As electroless plating widens spectrum of possible application of nickel based coatings, also presents some disadvantages, mainly connected with economic aspect [8.9]:

- Chemical compound used during electroless plating are far more expensive that these used in galvanic processes,
- Regeneration and preparation of coating solution is complicated and expensive process.

Basing on coating properties, presented advantages and disadvantages, and previous research, it can be stated that described Ni-P coating can be very promising solution for increasing wear resistance of parts made of AW-7075 aluminum, and should be subject of tribological tests.

3 Testing methodology

Samples consisting of AW- 7075 aluminum substrate with Ni-P coating manufactured with electroless chemical method was tested. Chemical composition of substrate alloy is presented in Table 1.

Table 1. AW-7075 chemical composition. Elements and percentage participation.

Al	Zn	Mg	Cu	Fe	Si	Mn	Cr	Zr	Ti	Other
Rest	5.1- 6.1	2.1- 2.9	1.2- 2.0	Max 0.5	Max 0.4	Max 0.3	0.18- 0.28	Max 0.25	Max 0.2	Max 0.05

20 mm x 20 mm x7 mm samples were cut of a 120 mm AW-7075 aluminum disc. Samples were machined to achieve perpendicular faces and desired roughness. Ni-P coating of various thickness was plated on samples. Good adhesion between coating and substrate was achieved thanks to surface etching, which helped in oxides removal. After surface preparation, Ni-P coating were produced. During described test, coating of 20 μm and 10 μm in thickness were applied.

Tribological test were conducted with UMT-2 Brucker device. Tool with bearing ball was used for testing. Ball was made of chromium stainless steel of 62 HRC hardness, according to PN-EN ISO 4527:2005. Diameter of balls used in testing was 6.32 mm (1/4 inch). Bearing ball was placed in holder, and fixed to avoid rotation. Ball was pushed towards tested sample. Reciprocating movement of tested sample was introduced. Sample was experiencing linear reciprocating movement of specific speed and distance. Thanks to this configuration of tested sample and tool (bearing ball), contact zone was circle (according to Hertzian theorem). Because of applied kinematics, wear trace was expected to be a single line. Whole test could be described as common pin-on-flat tribological test.

Sample with coating prepared earlier were loaded normally to its surface via bearing ball with force of 5 N. Simultaneously, reciprocating linear movement was introduced. Because of lubricant absence, test conditions are resized as dry friction. Single testing cycle consisted of two strokes on distance of 10 mm. Testing cycle started and stopped in same point. Speed of sample during test was set to 10 mm/s. Duration and number of cycles was different for each sample. Test was terminated after recognition of coating breakage.

Thanks to device design, applied force of 5 N stayed constant during test. Device was equipped with force sensor and regulator. If force measurement and regulation would be omitted, load force would decrease during test because of material wear. In described tests, force was regulated with precision of +/– 0.5 N. Such low value of force was assumed, as substrate was made of soft material. During test performed with hard substrates, greater forces can be applied. Device used in test is capable of applying axial forces of 200 N.

Before each test, tested samples were cleaned. Data acquainted during test covered: time [s], frictional force F_T [N], acoustic emission level AE [V] and also coefficient of friction- μ. Device was equipped with two dimensional force sensor, which allowed to measure force in direction of load and in direction of sample movement. Basing in this values, coefficient of friction could be calculated.

For each test, frictional force variation was recorded. Because of the reciprocating movement, frictional force was alternating. To allow further processing of results and values, signal for frictional force was periodically filtered in order to achieve positive values. As well, peaks occurring during movement direction change were filtered. In total, two samples were tested. Thickness of coating for sample I was 0.02 mm and for sample II- 0.01 mm.

4 Results presentation and discussion

Samples were evaluated regarding to recorded coefficient of friction and variations of acoustic emission during test. Data acquisition and processing were performed with UMT Brucker laboratory software. Frequency of data acquisition was 100 Hz. Precision of force measurement was 0N. Precision of acoustic emission measurement was 0.05 V. Figure 1 and 3 present variations of coefficient of friction during tribological tests. AE variation are shown in Figure 2 and 4, respectively. Interpretation of these results was described in following chapter.

DOI: 10.1051/matecconf/201711204014

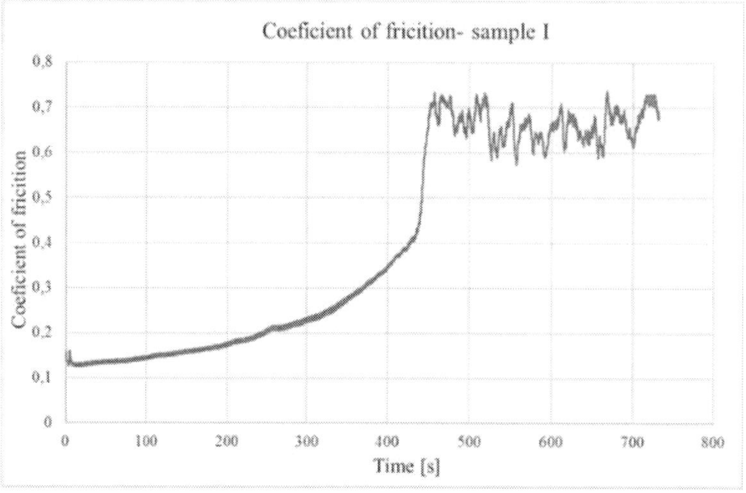

Fig. 1. Coefficient of friction for sample with 20 μm Ni-P coating.

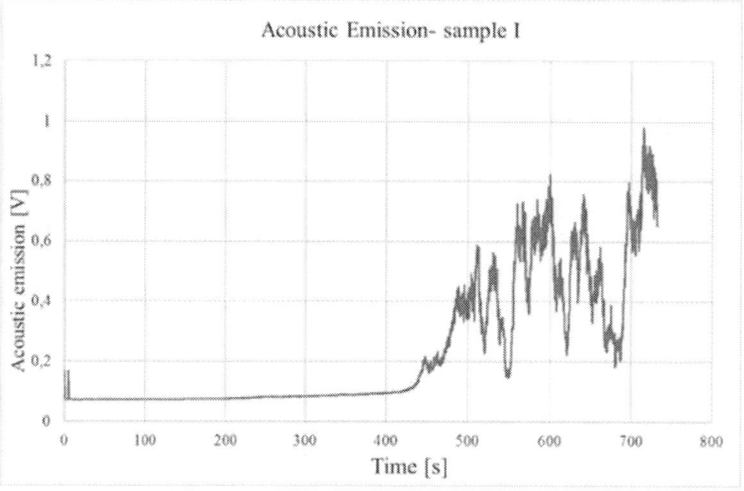

Fig. 2. Acoustic Emission for sample with 20 μm Ni-P coating.

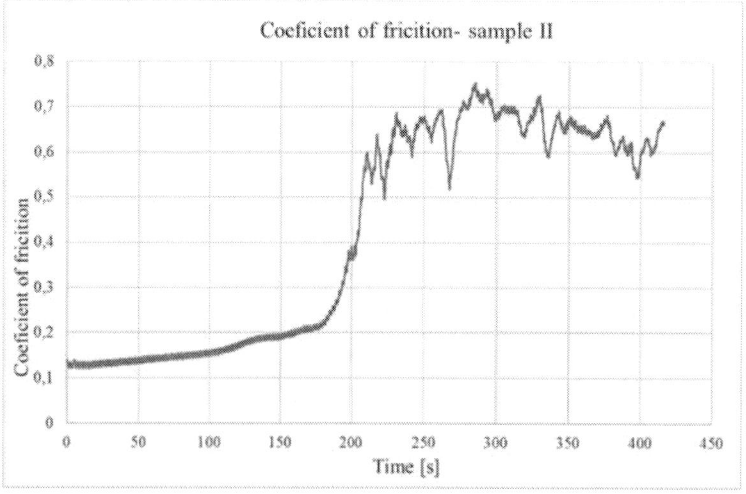

Fig. 3. Coefficient of friction for sample with 10 μm Ni-P coating.

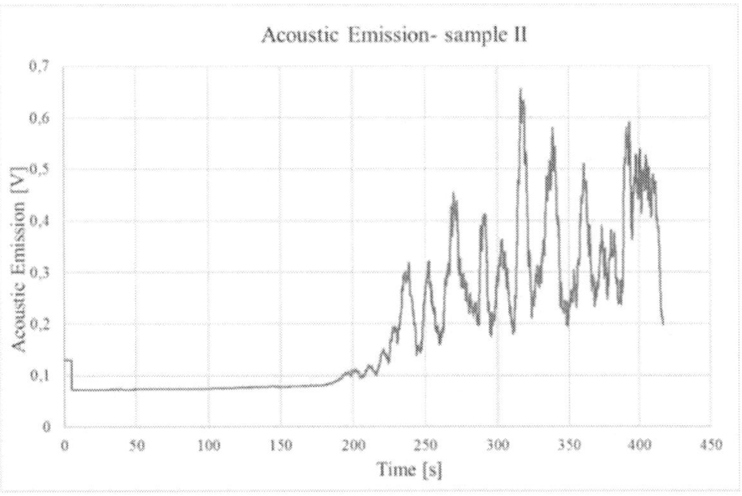

Fig. 4. Acoustic Emission for sample with 10 µm Ni-P coating.

5 Conclusions

Obtained results allowed to evaluate resistance to abrasive wear of Ni-P coated samples. Two samples were tested, with various coating thickness. All results presented by graphs in Figure 1-4 share similar characteristics. In case of coefficient of friction, monotonous rise of recorded magnitude is recorded in first phase of experiment. After certain number of cycles, rapid increase in COF magnitude occurred. After this peak, magnitude of COF starts to fluctuate around specific value. In case of coating of lower thickness, fluctuations present greater amplitude, when compared with results for thicker coating. Minor peaks in the beginning of test can be described as noise caused by initiative contact between testing tool and sample.

AE variations can be also considered as consisting of two phases. In first phase acoustic emission is constant and low. Minor monotonic increase can be observed shortly before end of first phase. Second phase, analogically as in case of coefficient of friction, begins with rapid increase in AE magnitude. After this peak, acoustic emission fluctuates at high levels. As well, minor peak of AE level is observed in beginning of test. Analogically to COF measurement, this peak is considered as noise caused by initial contact.

As point of rapid rise is coherent for both acoustic emission and coefficient of friction and in case of both samples, it can be stated that in that point breakage of coating occurs. Low values of AE during first phase of experiment may be connected with abrasive wear of coating. Rapid increase of both AE and COF is caused by breakage of coating and beginning of friction between substrate and testing tool. Great fluctuations and significant value of AE is caused by relatively high roughness of substrate surface, because of itching applied before coating plating.

Conducted tests allowed to evaluate influence of coating thickness on durability of Ni-P coating in pin-on-flat tribological test. Obtained results show that coating of thickness equal to 20 µm was able to carry load and friction for about 400 seconds before observable coating breakage. This value is two times higher, when compared with results obtained for 10 µm coating, which broke after 200 seconds from test beginning.

Value of kinetic coefficient of friction for first phase of experiment was varying from 0.1 to 0.3, and after coating breakage reached 0.6 which proves that applied coating helps in decreasing value of coefficient of friction. Stable and low AE signal in first phase of coating wear proves that wear processes that occurs during coating wear are stabilized and no intense destructive phenomena are present, like e.g. adhesive wear. This means that wear of Ni-P coating in described test is a predictable process, and final, destructive phase of wear can be anticipated. In case of machine maintenance this is considered as significant advantage, as lesser probability of unexpected and unpredictable failures means easier machine maintenance.

References

1. C.G. Rhodes, M.W. Mahoney, W.H. Bingel, R.A. Spurling, C.C. Bampton, Scr. Mater., **36**, 69-75, (1997)
2. J.N. Balaraju, K.S. Rajam, , Surf. and Coat. Tech., **195**, 154-161, (2005)
3. R. Elansezhian, B. Ramamoorthy, P. Kesavan Nair, Surf. and Coat. Techn., **203**, 709-712, (2008)
4. Y.Tsai, F. Wu, Y. Chen, P. Peng, J. Duh, S. Tsai, , Surf. and Coat. Techn, **146–147**, 502-507, (2001)
5. M.H. Staia, E.J. Castillo, E.S. Puchi, B. Lewis, H.E. Hintermann, Surf. and Coat. Techn, **86**, Pages 598-602, (1996)
6. P. Sahoo, S. Kalyan Das, Mat. & Design, **32**, 1760-1775, (2011)
7. Q. Zhaoa,Y. Liua, H. Müller-Steinhagenb, G. Liuc, Surf. and Coat. Techn, **155**, 279–284, (2002)
8. S. Alirezaei, S.M. Monirvaghefi, M. Salehi, A. Saatchi, Wear, **262**, , 978–985, (2007)
9. K. Hari Krishnan, S. John, K. N. Srinivasan, J. Praveen, M. Ganesan, P. M. Kavimani, Metal. and Mat. Trans., **37**, 1917–1926, (2006)

An investigation of the influence of a fiber arrangement of a laminate on the values of stresses in the composite panel of a modified freight wagon using the FEM method

Andrzej Buchacz, *Andrzej* Baier, *Krzysztof* Herbuś*, and *Piotr* Ociepka

Institute of Engineering Processes Automation and Integrated Manufacturing Systems, Faculty of Mechanical Engineering, Silesian University of Technology, Konarskiego 18a, 44-100 Gliwice, Poland

Abstract. In the work is presented the investigation of influence of fiber arrangement of a laminate on the values of stresses in the composite panel of a modified freight wagon using the FEM method. The conducted investigations concerned the reduced part of a side panel of a freight wagon. On this panel were mounted the tested composite panels. For this purpose, a special screw connection with a blind rivet nut was used. The first stage of works included manufacturing of a composite material consisting of four layers of a laminate. Each layer of the laminate is a composition of epoxy resin and carbon fibers with a plain weave in which fibers strands are woven at the angle of 90^0. In the next step it was changed the main angle of placement of fibers in all layers of the laminate. Wherein, all layers of the laminate were characterized with the same angle of fibers with respect to the global coordinate system used in the numerical test. As a result of the study it was observed the relationship between the main angle of fibers of the composite material and the values of stresses occurring in the composite plate.

1 Introduction

In recent years it is observed a dynamic growth in the use of composite materials in various technical fields. One of the major types of used composite materials is the composite material reinforced with fibers. In these materials the matrix acts as a binder for the fibers. This allows joining the fibers together within a composite layer as well as the individual layers within the multilayer composite material. In the fiber materials [1-2] as the reinforcing phase are used various kinds of fibers. The most commonly used fibers are: glass fibers, graphite ones and carbon ones. While, as the matrix various types of resins are used, among which it could be mentioned: epoxy resin, phenolic one, polyester one, nylon one and polycarbonate one.

The increased interest in composite materials implies the creation of tools that could be used, already in virtual testing phase, to verify the application of composite structures in the analyzed field of problems. One such tool is PLM Siemens NX software [3-11], which

* Corresponding author: krzysztof.herbus@polsl.pl

supports the process of designing and construction verification in terms of the motion dynamics and also stress analysis as well as production preparation.

Within the conducted investigations [12-15] realized by the research team and concerning the implementation of the project which objective is to extend the exploitation time of wagons between intervals associated with their regeneration it was decided to use, for this purpose, composite panels reinforced with fibers. The main task of the modernized freight wagons is to transport cargo in the form of loose or aggregate material (stone, hard coal, gravel). Both at the loading and unloading of transported cargo it comes to frequent damage of the inner shell of the wagon body. The resulting damage is of the mechanical nature. The main problem in this case is the phenomenon of decreasing in the thickness of the shell plate of the wagon body as a result of its abrasion. This attrition is of a stochastic nature. Another type of mechanical damage of the metal sheet is a result of the impact of an actuator of the loading (unloading) machine. In the body plating occur plastic deformations as the result of these actions. Another analyzed issue, considered with the plating damage, was the problem of an aggressive type of a cargo. The result of chemical interactions between transported materials and the plating of the wagon body are also local defects of the plate. The diagnosed dangers, in relation to freight wagons exploitation, were the basis for selection of composite material reinforced with fibers. As the matrix was used epoxy resin, and in the case of fibers there were used fiberglass and carbon fiber.

This paper presents studies on the application of the composite material which is a composition of epoxy resin and carbon fiber fabric. Such chosen composite material has a high resistance to abrasion, with high susceptibility and general chemical resistance. For the purpose of virtual testing the part of the modified wagon wall of the 418V type was extracted (Figure 1).

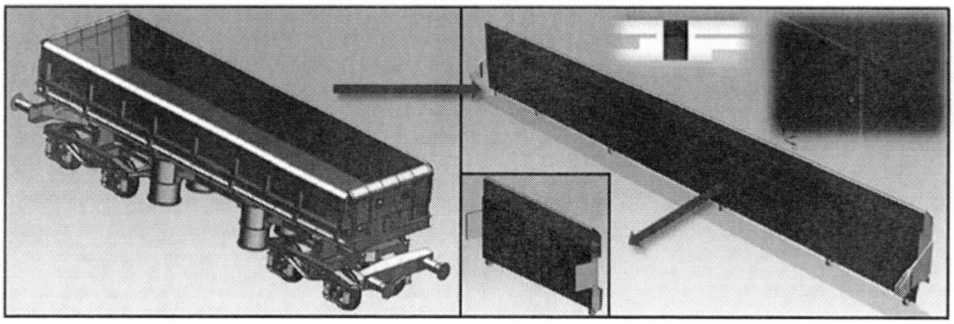

Fig. 1. Extracting of the part of the side wall of the freight wagon of the 418V type.

The method used for extraction was such determine to ensure the possibility of elaboration the physical test stand and to improve the process of performing the numerical stress analysis and of their validation [14]. The adjusting of the FEM numerical model of the side wall of the wagon consisted of such model modifications to obtain the matching of calculated results and with the results of experimental tests (strain gauge ones). The matched model of the side wall of the wagon was the base for numerical analysis, using the FEM method, of the modernized freight wagon. In this case, to the FEM model was added the plates of composite material and it was analyzed values of stresses and displacements in relation both to the entire side wall and the composite panels themselves.

2 Definition of the composite material

The used composite material was the composite one consisting of four layers. Each layer of the composite material was a composition of the epoxy resin and carbon fiber with a plain weave, in which the fibers are woven at the angle of 90^0. In Figure 2 is shown the structure of the used composite material.

The axes x, y, z represent the global coordinate system, in relation to which the analysis was performed using the finite element method. The axis 1 represents the orientation of fibers of the weft, and the axis 2 represents the orientation of fibers of the warp of the fabric woven of carbon fibers. Wherein the angle between the fibers of warp and of weft is always constant and is equal to 90^0.

The individual layers of the composite material are arranged along the y axis. However, as the main direction of the fibers was assumed the z axis. In the case, when the orientation of fibers of the weft of the fabric woven of carbon fiber (1) coincides with the main direction of fibers arrangement (z), the angle $\theta = 0^0$. In any other case the angle $\theta \neq 0^0$.

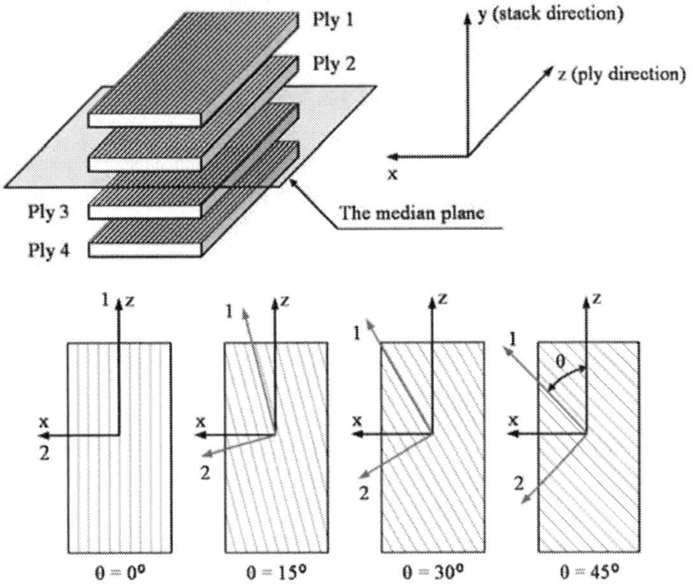

Fig. 2. The structure of the composite material.

It was assumed that all layers, within the composite material, are characterized by the same angle of fibers arrangement θ, in relation to the main direction of fibers arrangement (z). Therefore it was defined the four composite materials, which structure corresponds to the structure shown in Figure 1, with the angles θ respectively 0^0, 15^0, 30^0, 90^0 (all four layers, within a single composition of the composite material, have the same angle of fibers arrangement). To conduct the numerical tests, using the FEM method, it was implemented the presented structure of composite material in the PLM Siemens NX program.

The applied epoxy resin has the following basic (standard) properties [12]: mass density 1300 [kg/m^3], Young's modulus 3000 [MPa], Poisson's ratio 0.37. Whereas the used carbon fabric has the following basic features: mass density 1770 [kg/m^3], Young's modulus 240000 [MPa], Poisson's ratio 0.22.

One layer of the composite material has the following properties: matrix material – epoxy, matrix volume fraction – 0.66, warp fiber material – carbon fiber, weft fiber material – carbon fiber, fiber volume fraction – 0.34, balance coefficient – 0.5, weft fiber angle – 90 degrees, Young's modulus 1 – 44240 [MPa], Young's modulus 2 – 44240 [MPa], Young's modulus 3 – 3000 [MPa], Poisson's ratio 12 – 0.032, Poisson's ratio 13 – 0.344, Poisson's ratio 23 – 0.344, shear modulus 12 – 1649 [MPa], shear modulus 13 – 1047 [MPa], shear modulus 23 – 879 [MPa], mass density – 1460 [kg/m^3], thickness – 1 [mm].

3 FEM analysis of the modified side of the freight wagon

The conducted investigations concerned the reduced part of a side panel of a freight wagon. On this panel were mounted the tested composite panels. For this purpose, a special screw connection with a blind rivet nut was used.

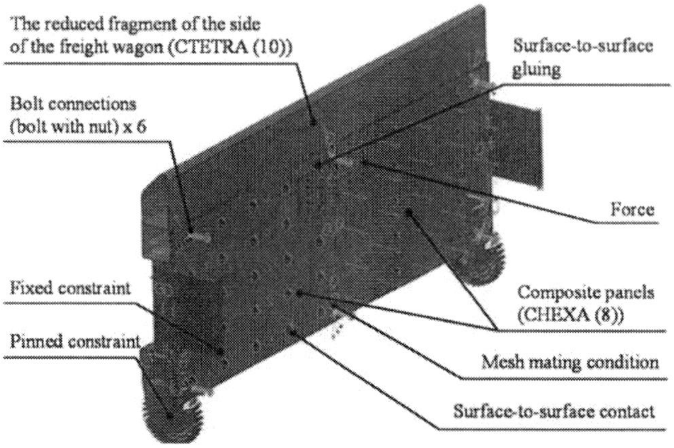

Figure 3. The model of the reduced fragment of the side of the freight wagon prepared to the FEM analysis.

The numerical investigations were performed in the PLM Siemens NX software. In the first step, it was created a mesh of finite elements of the steel elements of the side wall of the wagon. In this case was used finite elements of the CTETRA type (10 nodes tetragonal finite elements). Next was defined the mesh of finite elements with respect to the composite panels. In this case it was used finite elements of the CHEXA type (8 nodes hexagonal finite elements). To all elements of the plating of the freight wagon was assigned the same material steel.

While, with regard to the mounted composite panels, it was used, the previously defined and described, composite material. In the next step, the following boundary conditions were defined in relation to the analyzed issue (Figure 3): Fixed and pinned constraint – mapping the method of mounting of the of side wall of the wagon; Mesh mating condition – between elements of the wagon plating, which are permanently connected one to another; Surface-to-surface contact – between elements that are coming in contact under the load in the system (between the wagon plating and composite panels); Surface-to-surface gluing – between the composite panel and the beam, by which the load is transferred to the system and Bolt connections – mapping the method of joining the composite panels and the wagon plating.

On the basis of such prepared model it was conducted a series of strength tests using the FEM method. The particular, virtual experiment referred to a specific θ angle. However, in all tests it was assumed the constant value of the force acting on the analyzed system equal to 20 [kN].

In Figure 4 is shown the distribution of reduced stresses in the layer with the maximal effort of the composite material in relation to the θ angle of fibers arrangement and the distribution of displacements in relation to the entire system. On the Figure 4 are presented the maps of stresses and displacements in two extreme cases ($\theta = 0^0$, $\theta = 45^0$).

While, in the Table 1 are gathered the values of the maximal reduced stresses in the particular layers of the composite material. In the mentioned Table also are shown the maximal displacements in relation to the entire system.

According to the presented stress distribution it could be observed the phenomenon of stress concentration in the areas in which there is a screw connection of the composite panel with the wagon side wall plating (Figure 4).

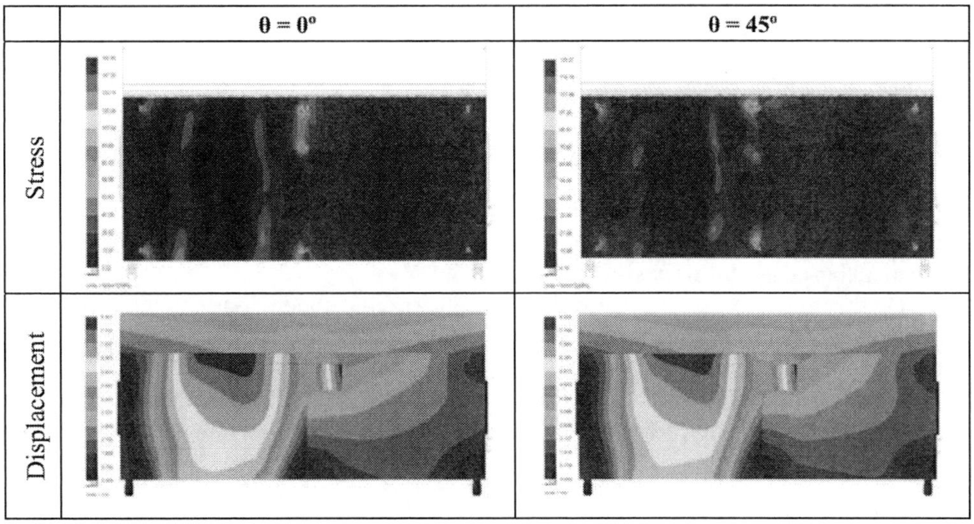

Fig. 4. Distribution of stresses in the layer, in which are the highest reduced stresses and distribution of displacements in relation to the entire system.

This phenomenon is very dangerous due to the possibility of mechanical damage of the outer layers of the composite panel under the influence of the axial force acting in the screw joint. This force is necessary to properly mount the composite panels to the side wall of the wagon.

Table 1. The maximal values of the reduced stresses (Von-Mises) in the ply of the composite panel and the value of the displacement.

Stress [MPa]	**θ = 0°**	**θ = 15°**	**θ = 30°**	**θ = 45°**
Ply1	160	163	145	130
Ply2	80	73	68	68
Ply3	97	95	97	96
Ply4	163	169	149	138
Displacement [mm]	8.4	8.45	8.5	8.5

On the basis of the obtained results it could be concluded that the highest values of reduced stresses are in the layer 1 and 4. While, much smaller values of reduced stresses are in the two middle layers (ply2 and ply3).

In accordance with the obtained values it could be said that there is a correlation between the values of the stresses occurring in the layers of the composite material and the angle of arrangement of the weft fibers of the fabric woven of carbon fiber. When the fibers of the weft in the fabric of carbon fiber are placed at the angle $\theta = 45^0$ in relation to the main direction of the fibers it was observed the decrease of reduced stresses in the extreme layers of the composite panel by 19%. However, in the case of ply2 the decrease of stress values was equal to 15%. In the case of ply3 there was no significant decrease in stress values.

The maximum absolute displacement measured on the composite plate is equal to 8.5 mm. The composite panel in the neutral position (without load) is spaced from the metal paneling of the wagon by the value of 5.2 mm. If the load is applied along the y axis, first the composite panel is pushed to the paneling of a wagon and then will move along together with the wagon wall (5.2 mm + 3.3 mm = 8.5 mm).

4 Conclusions

In the result of the conducted investigations it was observed the relationship between the θ angle of arrangement of the weft fibers of the fabric of carbon fibers and the values of stresses occurring in the particular layers of the composite panel. In the case of the composite material which is the composition of epoxy resin and carbon fibers with a plain weave, in which the fiber strands are woven at the angle of 90^0, the lowest values of stresses in the composite layers was observed when the angle $\theta = 45^0$. Wherein, all tests were carried out under identical boundary conditions.

On this basis it was determined the most preferred method of assembly (because of the angle of arrangement of the weft fibers of the fabric) the composite panels to the side panel of the wagon to minimize stresses occurring in the composite plate. Furthermore, the proposed arrangement of the fibers reduces the unfavorable influence of the mechanical reaction of the joining elements on the values of stresses in the outer layers of the composite material.

References

1. R. Martinsa, L. Reisa, R. Marat-Mendesa, Procedia Structural Integrity, **1**, 066–073 (2016)
2. J. Bienias, H. Dębski, B. Surowska, T. Sadowski, Computational Materials Science, **64**, 168–172 (2012)
3. W. Banaś, K. Herbuś, G. Kost, A. Nierychlok, P. Ociepka, D. Reclik, Advanced Materials Research, **837**, 537-542 (2014)
4. K. Herbuś, P. Ociepka, IOP Conf. Series: Materials Science and Engineering, **95**, 012096 (2015)
5. A. Dymarek, T. Dzitkowski, K. Herbuś, G. Kost, P. Ociepka, Advanced Materials Research, **837**, 351-356 (2014)
6. P. Ociepka, K. Herbuś, IOP Conf. Series: Materials Science and Engineering, **95**, 012101 (2015)
7. K. Herbuś, G. Kost, D. Reclik, J. Świder, Advanced Materials Research, **837**, 582-587 (2014)
8. K. Herbuś, P. Ociepka, IOP Conf. Series: Materials Science and Engineering, **95**, 012084 (2015)
9. P. Ociepka, K. Herbuś, IOP Conf. Series: Materials Science and Engineering, **95**, 012100 (2015)
10. K. Herbuś, P. Ociepka, IOP Conf. Series: Materials Science and Engineering, **145**, 042018 (2016)
11. K. Herbuś, P. Ociepka, IOP Conf. Series: Materials Science and Engineering, **145**, 052010 (2016)
12. A. Buchacz, A. Baier, K. Herbuś, M. Majzner, P. Ociepka, Applied Mechanics & Materials, **809/810**, 944-949 (2015)
13. M. Płaczek, A. Wróbel, A. Baier, IOP Conf. Series: Materials Science and Engineering, **95**, 012042 (2015)
14. A. Buchacz, A. Baier, J. Świder, M. Płaczek, A. Wróbel, K. Herbuś, P. Ociepka, W. Banaś, M. Sobek, Ł. Grabowski, M. Majzner, *Analytical and experimental tests and determination of characteristics of components working as assemblies of innovative structures of repaired freight cars* (Wydawnictwo Politechniki Śląskiej, Gliwice 2016)
15. A. Buchacz, A. Baier, K. Herbuś, M. Majzner, P. Ociepka, *Springer Proceedings in Mathematics & Statistics, Dynamical Systems: Modelling*, **181**, 97-105, (2016)

Specific heat of nano-ferrites modified composites

Cristian Muntenita[1], *Claudia* Ungureanu[1], *Vasile* Bria[1], and *Iulia* Graur[1,2,*]

[1]"Dunarea de Jos" University of Galati, Research and Development Centre for Thermoset Matrix Composites, Galati, Romania
[2]Diagnose and Measurement Group, Galati, Romania

Abstract. The specific heat of nano-ferrites modified composites was studied using differential scanning calorimeter (DSC) method in the temperature range of 30 to 150°C. Initially, nano-ferrites were introduced in epoxy systems in order to improve the electromagnetic properties of formed materials. Together with the changes in electromagnetic properties some modifications occur regarding thermal and mechanical properties. The materials were formed by placing 5g or 10g of ferrite into 250g polymer matrix leading to a very low weight ratio of modifying agent. At so low ratios the effect of ferrite presence should be insignificant according to mixing rule. Anyway there is possible to appear some chelation reaction with effects on thermal properties of materials. Three types of epoxy resins had been used as matrix and barium ferrite and strontium ferrite as modifying agents. The thermal analysis was developed on two heating-cooling cycles and the specific heat was evaluated for each segment of the cycle analysis.

1 Introduction

Composite materials are the first materials with internal structural layout conceived by man, both through of their molecular structure and through preferential directions, thus giving them increased resistance, superior to those of their constituents. The use of polymeric materials represent a number of advantages compared to traditional materials [1-2]. The properties of these materials are: low energy consumption during production, minimize the impact on the environment, corrosion resistance, tensile strength, wear resistance, resistance to high temperatures, the hardness of the surface, dimensional stability, resistance to the environment, the thermal conductivity or thermal insulation, electrical conductivity, and large usage spectrum: medicine, by making dental implants and bone, aerospace industry by developing various components of the vehicle, automotive industry by developing various structural components, the sports industry through the development of various vehicles: bicycles, boats and sport equipment, energy industry through by applications in the field of wind energy (the wind turbines whose blades are made of composite) [3-5].

The developing of the composite materials was based both on economic and technical considerations. Among them we can remember: the need to make materials with special

* Corresponding author: iulia.graur@ugal.ro

properties, impossible to attain by traditional materials, the need to increase the safety and reliability in operation, the need to reduce the consumption of materials deficient, expensive or precious, with the possibility to reduce the time for the technological manufacturing and man-hours and last but not least the extend of the lifetime of the materials [6-7]. To get superior properties of polymers one usual way is to modify it by placing another phase, for instance a nano-sized one, into the polymer volume [8]. Ferrites remain the best magnetic material and cannot be replaced by any other magnetic elements because they are inexpensive, more stable and have a wide range of technological applications in transformer cores, high quality filters, radio wave circuits and operating devices [9-10].

According to the literature, the formation of polymeric composites, does not allow discussing interest issues from a single point of view, more than that, polymers properties can be controlled by synthesis and adapted to applications. Starting from the idea that the composite materials have good mechanical properties and affordable forming techniques, these are increasingly used in elements repairing or even replace them. Here we can mention the current trend of replacing traditional structures with composite materials, which have a higher life expectancy, high corrosion resistance, good elastic properties than traditional materials (pipelines for transportation - gas, water) [11-13].

Specific heat is the quantity of heat required to raise the temperature of a unit mass of a substance one degree under specified conditions. This property is measured in a calorimeter. In practice, the specific heat at constant pressure or enthalpy, Cp, is the measured quantity, with values reported in J/(kgK) in SI units. The standard test method for measuring the specific heat of polymer matrix composites is ASTM E1269-95, and is based on Differential Scanning Calorimetry (DSC). The temperature range can be extended depending upon the instrumentation and specimen holders used. A quantitative measurement of energy imparted to a test specimen as a function of temperature must be obtained to determine specific heat. Thus, the instrument used for these measurements must be calibrated in both the heat flow and temperature modes. Application of the DSC test method to polymer matrix composites may also encounter difficulties due to specimen mass loss from either moisture evolution or material decomposition, but this problem can be overcome by taking proper precautions [14].

This work goal is to analysis the composite materials modified with nano-ferrites (barium and strontium ferrites) with matrix made from three different epoxy resins, in particular specific heat.

2 Experimental procedure

Three epoxy systems were chosen namely: Epiphen RE4020-DE 4020 (Bostik), (C) Epoxy Resin C (R&G Gmbh Waldenbuch), and (HT) Epoxy Resin HT-2 (R&G Gmbh Waldenbuch). Ferrites used in this study are commercially available ferrite: Strontium ferrite ($SrFe_{12}O_{19}$) and barium ferrite ($BaFe_{12}O_{19}$). All these systems were modified by mixing the main component of the respective epoxy system (the resin) with 5g of ferrite and 10g of both ferrites [15].

The codifications for the materials used in this study are: C1, E1, H1 - epoxy resin with 5g barium ferrite; C2, E2, H2 - epoxy resin with 5g strontium ferrite; C3, E3, H3 - epoxy resin with 5g strontium and 5g barium ferrite; C, E and H - unchanged resins.

In order to determine the specific heat a differential scanning calorimeter DSC Mettler Toledo was used. This device is equipped with a sensor that is designed to control the atmospheric conditions inside the cell and ensure proper symmetry between the final sample and the reference sample. The DSC operation and data acquiring are possible with the two elements of DSC Mettler Toledo: measuring chamber (cell) and STARe application for data acquisition and evaluation. The specific heat was determined from the heat absorption curve and / or the released heat for each type of material. Test method is defined by two cycles as follows: 30 minutes 30ºC isotherm heating from 30°C to 150°C with a ratio of 10K-min, 150°C isotherm and cooling from 150°C to 30°C with a ratio of 10K/min and the cycle is repeated.

Both heating and cooling DSC curves can provide information on the nature of tested material. When talking about polymeric materials, the curves provide information on the glass transition temperature and melting temperature.

3 Results and discussion

Regarding specific heat values of studied materials, it may be noted that measured results shows several values lower than expected. This is due to chemical interactions which lead to formation of a new polymer. Thermal analysis were carried out in different temperature ranges from 30°C to 150°C. Interest temperature interval in this study is 70-120°C. For these values comparative charts were made (Figure 1-3). In the case of C modified epoxy resin materials, specific heat values shows a significant increase compared with unchanged epoxy matrix materials. The highest specific heat values were obtained for C modified epoxy resin materials. This is caused by the type of modifying agent (baric ferrite) used alone or in combination with strontium ferrite for C1 and C3 materials. It can be seen that the modifying agent has a beneficial effect on the materials thermal properties (Figure 1).

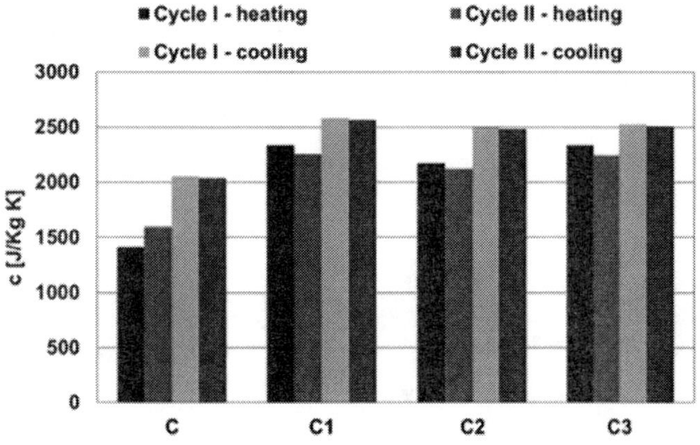

Fig.1. Specific heat values of C modified epoxy resin materials (70-120°C temperature interval).

In the case E type materials the same working protocol was kept, in order to determine the heat absorption. This test, like the others, has aimed to highlighting the thermal properties (specific heat) for the modified materials in comparison with unchanged epoxy resin. In the case of E modified epoxy resin materials it is noted that, when heated, the amount of modifying agent and also the combination of modifiers and matrix is a good one. Thus, can be noticed increased specific heat values for E3 material (Figure 2). For cooling there are not significant variations in specific heat values. Under the tested conditions it can be seen that baric ferrite improves the thermal properties of E type materials and C type materials.

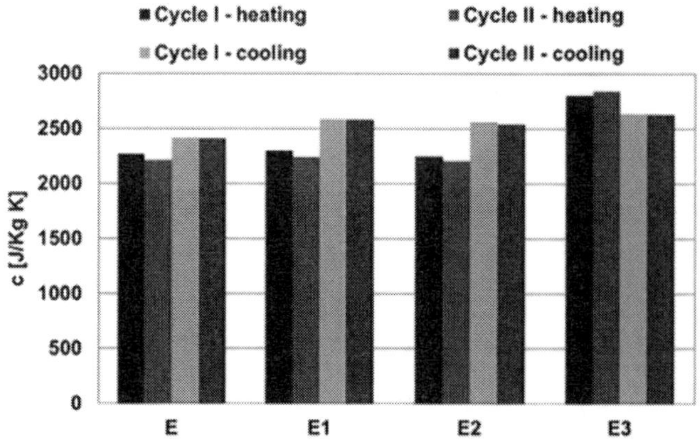

Fig.2. Specific heat values of E modified epoxy resin materials (70-120°C temperature interval).

HT type materials have shown variations in the specific heat values depending on the used modifying agent. Figure 3 shows, in graphical form, the specific heat values following the two cycles of heating and cooling. We can observe the behavior of the modified materials compared with epoxy resin and also the modifying agent behavior. The highest values of specific heat were recorded for HT modified epoxy resin materials (with baric ferrite) (Figure 3). For HT3 material specific heat values average in the studied temperature interval can be influenced by the amount of modifying agent in the polymer matrix. For cooling there are not significant variations in specific heat values for HT modified epoxy resin materials.

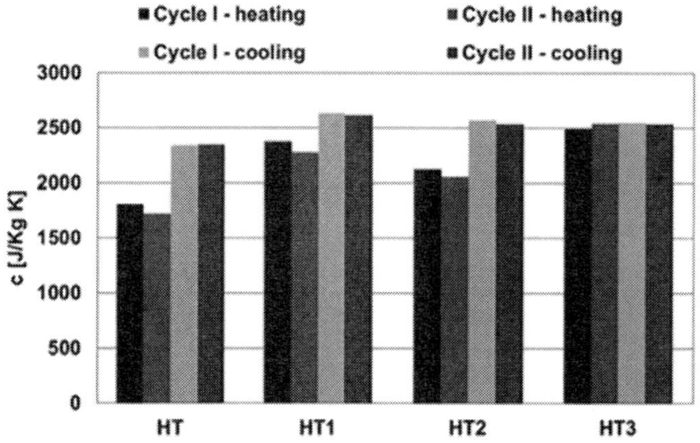

Fig.3. Specific heat values of HT modified epoxy resin materials (70-120°C temperature interval).

Before and after specific heat values tests, the samples were weighed to determine the mass loss (Figure 4). This loss is influenced by the presence of residual water and other volatile compound from the polymer. Generally the largest mass loss values were recorded for C modified epoxy resin materials while the lowest were recorded for E modified epoxy resin materials.

DOI: 10.1051/matecconf/201711204016

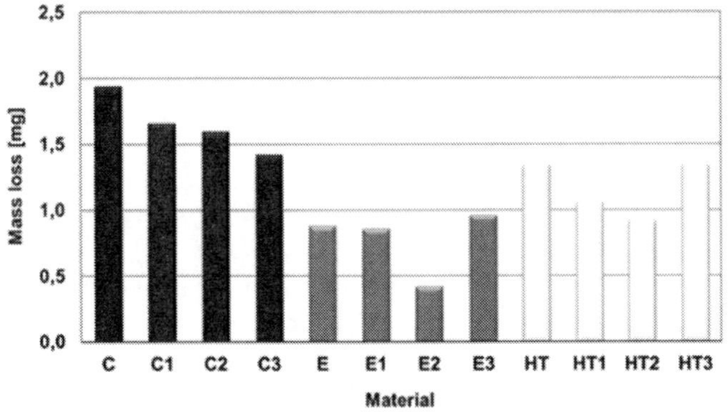

Fig.4. Mass loss of epoxy matrix materials.

Strontium ferrite produced significant variations in the mass loss depending on the type of used epoxy system. Thus the highest value is recorded for materials for E type materials and the lower for HT type materials. The combination of strontium ferrite and baric ferrite led to lower mass loss in the case of C type materials and higher in the case of HT type materials compared to homogeneous systems.

4 Conclusions

In the comparative analysis on heating cycles, different behavior and higher values of specific heat is due to chemical interaction between epoxy matrix and modifying agent. Also, the interaction between the two agents resulted in an improvement of the thermal properties of modified materials. Modified materials specific heat values are generally higher compared to unmodified epoxy systems. Better thermal properties were observed for those materials where both modifying agents were used. The mass loss is visible for all materials, regardless of matrix type or amount of modifying agent. Mechanisms that could explain the behavior of nano-ferrite modified materials can be explained by the nature and by the used amount of modifying agent.

The present study is part of research activity under the Project POSCCE 12P01.024/CD111. The research activity was developed at Research and Development Centre for Thermoset Matrix Composites, "Dunarea de Jos" University of Galati.

References

1. D. Craver Clara, C. E. Jr. Carraher, J. Appl. Polym. Sci., in *Introduction to polymer science and technology*, (2000)
2. A. Curt, *NM Epoxy Handbook,* Nils Malmgren AB, (2004)
3. J. P. Pascault, R. J. J. Williams, Epoxy Polym. , *New Materials and Inovations*, (2010)
4. L.V. Mc Adams, J.A. Gannon, *Encyclopedia of polymer science and engineering*, **6**, (1986)
5. P.D. Wallace, *Thermosetting resins*, Borden Chemical, (2011)
6. C. Muntenita, C. Ungureanu, V. Bria, I. Graur, *Bending Behaviour of Nano-ferrites Modified Epoxy Resins*, ICMSAV2016& COMAT2016, **6**, (2016)
7. C. Muntenita, V. Bria, C. Eni, A. Cîrciumaru, I. Graur, *Physical characterization of nano-ferrites modified epoxy resins*, Mat. Plastice, **53**, 3, (2016)

8. J.C.J. Bart, *Additives in Polymers. Industrial Analysis and Applications*, **1**, (2005)
9. M. Delgado, A. Lazaro, J. Mazo, B. Zalba, *Review on phase change material emulsion and microencapsulated phase change materiel slurries: materials, heat transfer studies and applications*, Renew.Sustain.EnergyRev., **16,** (2012)
10. B. A. Strong, *Fundamentals of Composites Manufacturing: Materials, Methods, and Applications*, **2**, SME, (2008)
11. A. Fallahi, L. Rajabi, F.A. Taromi, *DSC Analysis of Thermosetting Polyimides Based on Three Bismaleimide Resin Eutectic Mixtures*, Iranian Polymer Journal, **20**, (2011)
12. J. Le Blanc, *Filled Polymers. Science and Industrial Applications*, (2010)
13. C. Barreneche, A. Solé, L. Miró, I. Martorell, A. InèsFernández, L.F. Cabeza, *Study on differential scanning calorimetry analysis with two operation modes and organic and inorganic phase change material (PCM),* Thermochim.Acta., **553,** (2013)
14. Composite Materials Handbook, *Polymer matrix composites guidelines for characterization of structural materials*, **1**, (2002)
15. C. Muntenita, I. Graur, A. Boboc, C. Ungureanu, V. Bria, Compressive Behavior of Nanoferite Modified Epoxy Resins, The Romanian Review Precision Mechanics Optics & Mechatronics, **50**, (2016)

Tensile tests of specimens made of selected group of the filament materials manufactured with FDM method

Cezary Grabowik[1,*], *Krzysztof* Kalinowski[1], *Grzegorz* Ćwikła[1], *Iwona* Paprocka[1], and *Paweł* Kogut[2]

[1]Silesian University of Technology, Faculty of Mechanical Engineering, Institute of Engineering Processes Automation and Integrated Manufacturing Systems, Konarskiego 18A, Gliwice, Poland
[2]3DGence, Konarskiego 18C, Gliwice, Poland

Abstract. This paper presents results of tensile tests carried out for specimens made of the selected group of the filament materials. As a manufacturing technology FDM 3D printing method was chosen. The selected group of the filament materials involved the group of wood, PLA, ABS, PET, PMMA and ASA. Herein, it should be noticed, that technical data sheets that are delivered by filament materials producers include data that are valid for only one specific printing direction. This printing direction is deliberately selected, in such way that ensures the best material characteristics. Therefore, received during the research results allow to make comparison between a catalogue data and data obtained in the printing process. It aids both mechanical designer and product manufacturer at the stage of the printed product mechanical properties shaping.

1 The selection of the materials subjected to the tensile test

Currently, a vast variety of materials is available on the market. This group of materials is called filament. Filament is a thermoplastic material. It is wound on the spool in the form of a thin wire. Typically, the spool contains 1 kg of a material. In our case, 1.75 mm diameter filament was used. Manufacturers of the filament are constantly trying to surprise clients with new materials. A few years ago, only ABS or PLA were in the offer. Nowadays, these materials are the basis of FDM technology and they are the most popular. Newer and newer materials are commercially available. For example, materials with admixtures of wood or chalk, which were created for design applications. On the other hand, the impact of the industry has demonstrated a need for materials for engineering applications, such as ABS, ASA, Nylon, PET, PMMA or rubber. Above the basic properties of the selected materials are given.

3DGO Wood is a product name of the material made by 3DFilamenty Company. The original name of this filament is LayWoo-D3 or LayWood and it was created by Kai Parthy from CC-Products Company [1]. This filament consists of 60% of PLA which is the base of this filament and 40% of wooden powder which is an admixture. The material has a wood

* Corresponding author: cezary.grabowik@polsl.pl

color and it smells like wood during printing, that is why, it is ideal for printing elements imitating wood [2]. Printouts can be easily machined, sanded and painted. The manufacturer guarantees that the material is biodegradable, does not deform or warp during printing and is easy to print with low temperature of the hot end. The recommended temperature of the hot end is between 205 and 230°C and temperature change results in different shade of a printout. The admixture of wooden powder causes the decrease of the melting point (compared to the standard PLA). The tolerance of the filament diameter is equal to ±Ø0.05 mm and the roundness is 95%. Print settings are similar as in case of PLA, but the temperature of the hot end can be lower. In this case, 95% of material flow was used and temperature of print head was equal to 212°C on the first layer and 210°C on others. Initially, the cooling fans were turned off. From the third layer it was set at 50% and from fifth layer to the end of printout it was set at 100% of cooling. This material can be printed a little bit slower than 35 mm/s to obtain a much better quality of the printouts.

PLA – polylactide is a biodegradable thermoplastic material. PLA is one of the most popular materials in 3D printing (FDM). Its popularity is gained due to the low price, satisfactory strength and ease of printing as a result of material's low shrinkage. PLA made by INK3.de was used to prepare the samples. The density of PLA is equal to 1.24 g/cm3. The recommended temperature of the hot end is between 180 and 210°C, but the melting temperature is 210°C ±10°C. The tolerance of the filament diameter is equal to ±Ø0.05 mm and the roundness is 95%. PLA can be printed without using the heated bed but it is suggested to use it because this increases the adhesion of the print to the table. The typical value of MFI (Melt Flow Index) is equal to 6.0 g/10 min [3]. PLA has a low shrinkage factor so it does not deform or warp after strong cooling. That is why, the first three layers of the samples were made without cooling, whilst next two with 50% of cooling and others with 100% of cooling fans. The samples were printed with the print head temperature equaled to 215°C and temperature of the heat bed of 70°C. The 95% of material flow was used in the process.

ABS – acrylonitrile butadiene styrene is a popular thermoplastic polymer. In contrast to PLA, ABS has a much higher shrinkage factor. As a result, ABS requires much higher temperature on the heat bed. In the printing process ABS by Wolfix company was used. The density of ABS is approximately 1.08 g/cm^3. The tolerance of the filament diameter is the same as for PLA and equals ±Ø0.05 mm but the roundness is ±0.02 mm. The recommended temperature of the hot end is in the range 230÷270°C. ABS also requires using the heated bed and its temperature should be around 100°C [4]. ABS has a higher shrinkage factor than PLA so it requires low level of cooling. During the preparation of the samples, the percentage of cooling fans was set to 15%. The hot end temperature was 250°C for all layers and the temperature of the heat bed was 100°C for first layer and 95°C for other layers. The material flow was 99%. Due to the high shrinkage of the material, the heat chamber is found very useful to prevent deformation or warping.

PET – poly (ethylene terephthalate) is a common thermoplastic polymer. It provides outstanding resistance to weather conditions. When it is exposed to UV light it does not deform or lose its color. PET has unique properties of the adhesion. It adheres well to the heating table and has a low thermal shrinkage factor. PET is odorless during printing. In the printing process PET by EFilament3D Company was used. The density of PET is approximately 1.38 g/cm^3. The tolerance of the filament diameter and the roundness are not specified by the manufacturer. The only data given by the manufacturer are recommended temperatures. So, the temperature of the print head should be approximately 230°C and the temperature of the heat bed more than 60°C [4]. PET has an average shrinkage factor, that is why 50% of cooling was used starting from the fourth printed layer. The temperature of the print head was equal to 232°C on the first layer and 230°C for the others. The temperature of the heat bed was 75°C on the first layer and 73°C on others. 97% of material flow was used.

DOI: 10.1051/matecconf/201711204017

PMMA – poly (methyl methacrylate) is a transparent thermoplastic material which is also known as acrylic glass or Plexiglas. The shrinkage factor for PMMA is similar to that of ABS that is why PMMA requires the heated bed. In order to prepare the samples, PMMA made by Spectrum Filaments Company was used. The density of PMMA is approximately 1.18 g/cm^3. PMMA has scratch resistance, high temperature resistance, high weather resistance, the sustainability of printing process and high quality of the printout. It transmits UV light, and some sources report that PMMA can be a replacement of glass. The tolerance of the filament diameter is equal to ±Ø0.05 mm but the roundness is not specified by the manufacturer. The recommended temperature of the hot end is between 240 and 255°C. The temperature of the heat bed should be between 80 and 90°C [4]. PMMA has as high shrinkage factor as ABS so the percentage of cooling was set to 15%. The temperature of the print head was equal to 250°C and the temperature of heat bed was 95°C on the first layer and 90°C on other layers. 98% of material flow was used.

ASA – acrylonitrile styrene acrylate (ASA) is also a thermoplastic material. It is an alternative to ABS because ASA has better weather resistance than ABS. Because of UV light resistance, ASA is a very good option for printouts for outdoor use. It also has a low level of yellowing. This feature is very helpful when the printout needs to be used for a long time. In order to prepare the samples, ASA made by Fillamentum Company was used. The density of ASA is approximately 1.07 g/cm^3. The tolerance of the filament diameter is equal to ±Ø0.05 mm but the roundness is not specified by the manufacturer. The melt volume index is 6 cm^2/10 min. The melting temperature is between 220 and 250°C, but the recommended temperature of the hot end is between 250 and 255°C. The temperature of heat bed should be between 90 and 100°C. The shrinkage factor is between 0.3 and 0.6% [4]. ASA has similar printing settings as ABS. The temperature of the print head was 250°C and the temperature of heat bed was 100°C for first layer and 95°C for other layers. The percentage of cooling was equal to 15% and the material flow was at 97%. During the tests 3DGence ONE 3D printer was used. The open filament policy is the most important feature in 3D printers. This allows to use different materials which are manufactured an especially for FDM printers. 3DGence ONE is used in rapid prototyping, manufacturing small and batch production.

2 The tensile test results

In each case the five samples for each filament material type were subjected to the tensile tests. The tensile test results achieved for 3DGO Wood filament material both in the Figure 1 and Table 1 are presented. Two marginal samples were rejected. Other correct samples have the following numbers: 3, 4 and 5. The value of the peak stress is between 16.6 and 16.9 MPa. The value of the strain at peak stress is between 3.5 and 3.8%. The value of the Young's modulus is between 794 and 797 MPa.

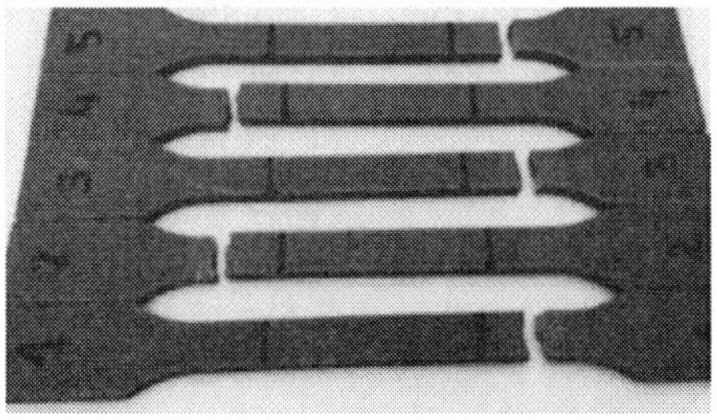

Fig. 1. 3DGO Wood – results of tensile tests.

Table 1. Tensile test data for 3DGO Wood.

No.	Peak load [N]	Peak stress [MPa]	Strain at peak stress [%]	Modulus [MPa]
1	646.487	16.2	3.2	793.015
2	685.611	17.1	5.3	812.715
3	664.899	16.6	3.5	794.841
4	674.762	16.9	3.8	797.109
5	671.824	16.8	3.5	794.333

For PLA all samples were printed with high quality. The tested samples are presented in Figure 2. The same as for 3DGO Wood, all the samples were broken out of the working section. None discoloration can be seen near the breaking points. All collected parameters for PLA are presented in Table 2. The two extrema results were rejected. Other correct samples have the following numbers: 3, 4 and 5. The value of the peak stress is between 38 and 38.4 MPa. The value of the strain at peak stress is between 5 and 5.3%. The value of the Young's modulus is between 1074 and 1100 MPa.

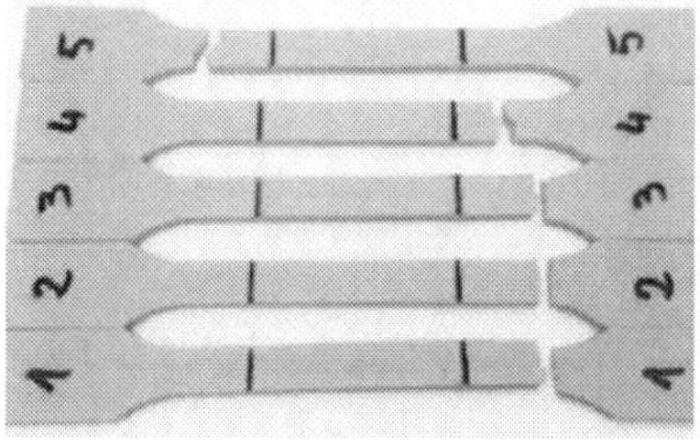

Fig. 2. PLA – results of tensile tests.

Table 2. Tensile test data for PLA.

No.	Peak load [N]	Peak stress [MPa]	Strain at peak stress [%]	Modulus [MPa]
1	1461.407	36.5	4.9	1081.730
2	1724.771	43.1	5.0	1228.285
3	1537.018	38.4	5.3	1086.438
4	1518.917	38.0	5.0	1100.184
5	1530,403	38,3	5,3	1074,699

ABS is the difficult material to print. All samples were printed with high quality. Tested samples are presented in Figure 3. Unfortunately, all the samples were broke out of the working section. Distinct discoloration can be seen near the breaking points. All collected parameters for ABS are presented in table 3. Two marginal samples were rejected. Other correct samples have the following numbers: 1, 2 and 5. The value of the peak stress is between 34.6 and 35 MPa. The value of the strain at peak stress is between 5.5 and 5.6%. The value of the Young's modulus is between 856 and 864 MPa.

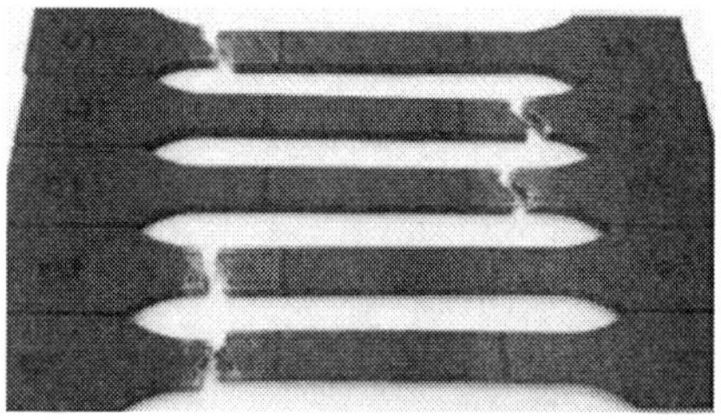

Fig. 3. ABS – results of tensile tests.

Table 3. Tensile test data for ABS.

No.	Peak load [N]	Peak stress [MPa]	Strain at peak stress [%]	Modulus [MPa]
1	1461.407	36.5	4.9	1081.730
2	1724.771	43.1	5.0	1228.285
3	1537.018	38.4	5.3	1086.438
4	1518.917	38.0	5.0	1100.184
5	1530,403	38,3	5,3	1074,699

PET is a little bit harder to print than PLA. All samples were printed with high quality. Tested samples are presented in Figure 4. Surprisingly, three samples broke in the working section. The cross-sections of the tested samples are very homogeneous. All collected parameters for PET are presented in Table 4. Two marginal samples were rejected. Other correct samples have the following numbers: 1, 4 and 5. The value of the peak stress is between 49 and 51.1 MPa. The value of the strain at peak stress is between 7.8 and 8.1%. The value of the Young's modulus is between 909 and 933 MPa.

DOI: 10.1051/matecconf/201711204017

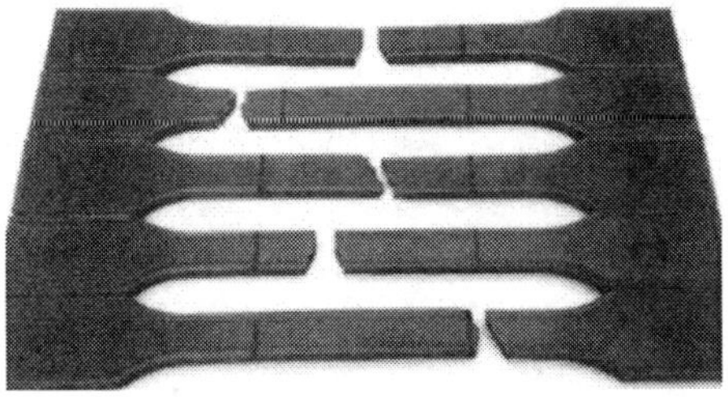

Fig. 4. PET – results of tensile tests.

Table 4. Tensile test data for PET.

No.	Peak load [N]	Peak stress [MPa]	Strain at peak stress [%]	Modulus [MPa]
1	1960.23	49.0	7.8	909.99
2	1697.86	42.4	6.9	846.17
3	2052.73	51.3	8.1	927.84
4	2033.92	50.8	8.0	928.32
5	2044.39	51.1	8.1	933.51

PMMA is one of the hardest materials to print. All samples were printed with good quality. Tested samples are presented in Figure 5. Surprisingly, sample no. 5 has three breaks and sample no. 2 has two breaks. Three samples broke in the working section. The cross-sections of the tested samples are not so homogeneous. All collected parameters for PMMA are presented in Table 5. Two marginal samples were rejected. Other correct samples have the following numbers: 1, 3 and 4. The value of the peak stress is between 54.3 and 56 MPa. The value of the strain at peak stress is between 5.3 and 5.5%. The value of the Young's modulus is between 1355 and 1376 MPa.

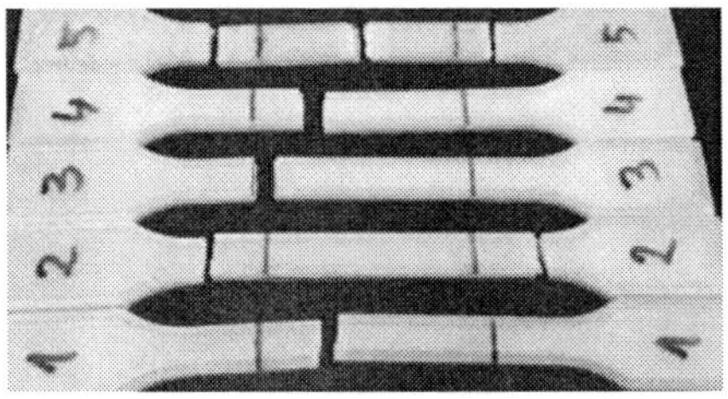

Fig. 5. PMMA – results of tensile tests.

Table 5. Tensile test data for PMMA.

No.	Peak load [N]	Peak stress [MPa]	Strain at peak stress [%]	Modulus [MPa]
1	2240.664	56.0	5.3	1376.651
2	2329.248	58.2	5.9	1366.475
3	2183.541	54.6	5.4	1360.051
4	2171.132	54.3	5.5	1355.035
5	2156.152	53.9	5.2	1349.742

ASA is as hard to print as ABS. All samples were printed with high quality. Unfortunately, all the samples broke out of the working section. Slight discoloration can be seen near the breaking points. All collected parameters for ASA are presented in Table 6. Two marginal samples were rejected. Other correct samples have the following numbers: 1, 3 and 5. The value of the peak stress is between 31.8 and 32.1 MPa. The value of the strain at peak stress is between 7.1 and 7.9%. The value of the Young's modulus is between 663 and 684 MPa.

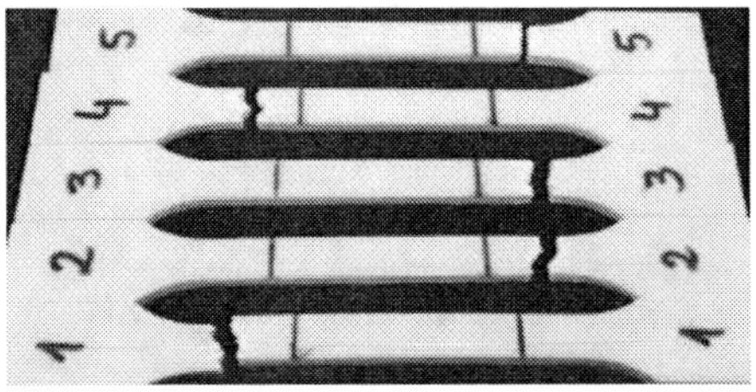

Fig. 6. ASA – results of tensile tests.

Table 6. Tensile test data for ASA.

No.	Peak load [N]	Peak stress [MPa]	Strain at peak stress [%]	Modulus [MPa]
1	1285.697	32.1	7.9	684.363
2	1299.721	32.5	8.3	678.889
3	1270.319	31.8	7.2	663.789
4	1260.193	31.5	7.2	667.619
5	1276.971	31.9	7.1	679.332

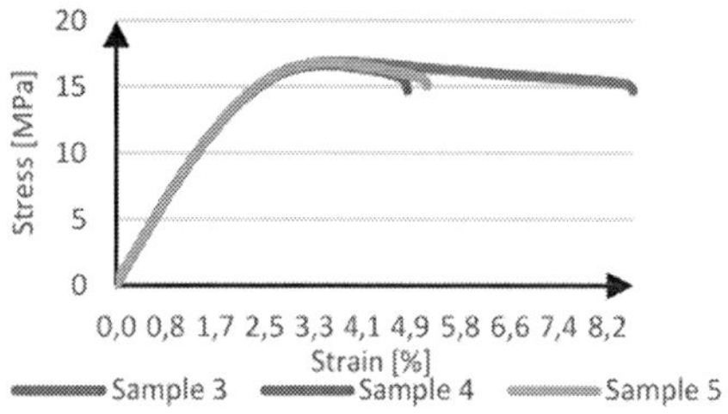

Fig. 7. Stress-strain diagram - Wood.

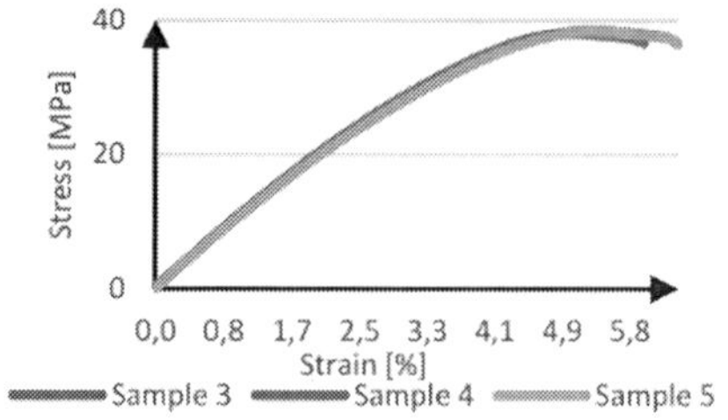

Fig. 8. Stress-strain diagram - PLA.

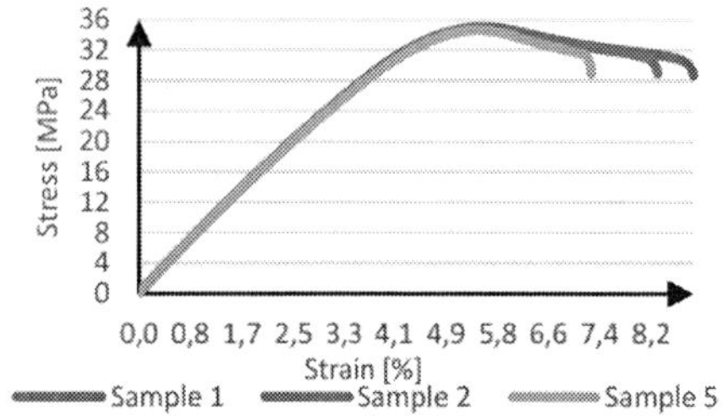

Fig. 9. Stress-strain diagram - ABS.

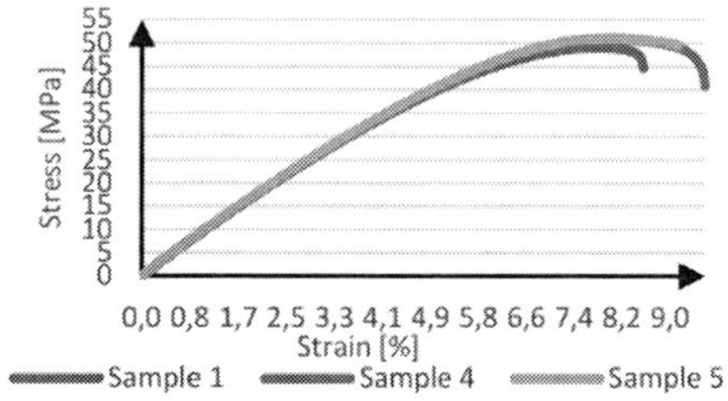

Fig. 10. Stress-strain diagram - PET.

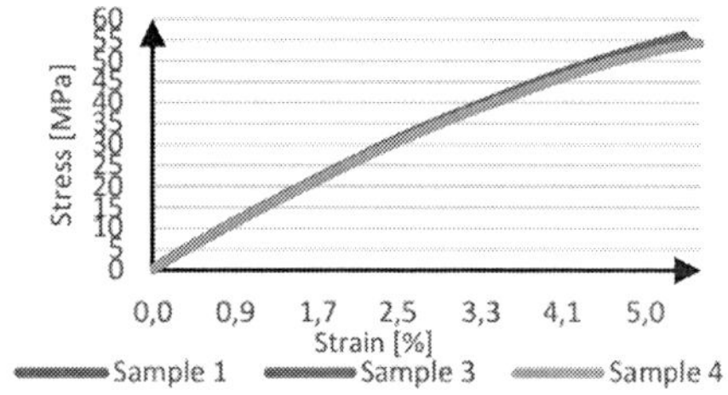

Fig. 11. Stress-strain diagram - PMMA.

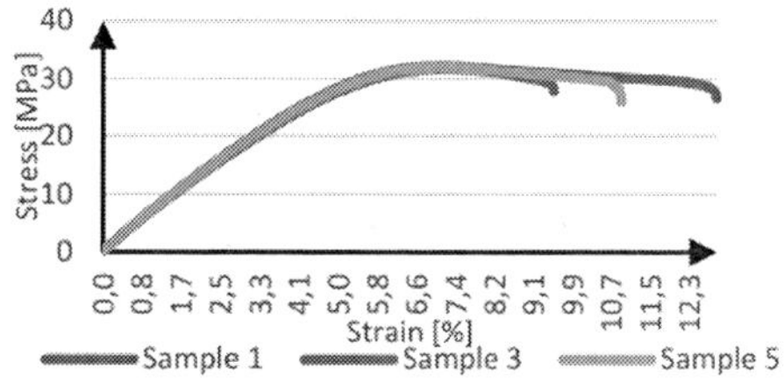

Fig. 12. Stress-strain diagram - ASA.

3 Conclusion

According to the research the weakest material is 3DGO Wood which has almost twice lower mechanical properties compared to ASA. Comparison of the most popular materials such as PLA and ABS is very interesting. It turns out that PLA has bigger average peak stress than ABS. On the other hand, ABS has a lower value of average Young's modulus which means that ABS is softer than PLA. ASA can be compared to a softer version of ABS, but the difference is small. The strongest material, presented in the ranking, is PMMA. Although PET is not as fragile as PMMA, it has a very good impact resistance. Also, due to the ease of printing, PET is the best choice for tough applications.

References

1. https://3dprintingindustry.com/news/journey-laywoo-d3-interview-wood-filament-inventor-kai-parthy-23357/
2. http://3dgo.pl/pl/p/3DGO-WOOD-500g-1%2C75mm/79
3. https://spectrumstore.pl/pl/p/Filament-PMMA-1.75mm/106
4. http://3dfilamenty.com/pl/c/Filamenty/19

A tribological studies of graphite solid lubrication effect on UHMW polyethylene in silicone oil lubricated conditions

Piotr Jabłoński*, *Adam* Patalas, *Piotr* Czajka

Poznan University of Technology, Institute of Mechanical Technology, Piotrowo 3, 60-965 Poznan, Poland

Abstract. Recent advances in surface engineering and solid lubrication technologies offer unique possibilities for better controlling friction and wear under boundary or marginally lubricated rolling, sliding or rotating contact conditions. The aim of the study was to evaluate influence of graphite solid lubrication addition on silicone oil tribological performance. Studies of graphite solid lubrication effect was carried out with using tribological testing station UMT-2 Bruker. Tests were performed against UHMW polyethylene samples and under various conditions, differing in relative speed and the load of interacting frictional pair.

1 Introduction

Knowledge about the material characteristics of moving parts is especially important for mechanical engineering, therefore conducting tribological studies is substantial in determination of the usefulness of frictional pairs in different applications. During this studies a variety of engineering materials properties like coefficient of friction, wear resistance and their behavior under lubrication condition might be verified.

It has been documented, that popularity of solid lubrication use in a field of wear resistance is increasing this day. Using of solid lubricant materials is an effective way to simplify design and prolong service life of the operating systems under harsh conditions [1-3]. The fields in which application of solid lubrication is more preferable than oil or grease lubrication are especially aerospace or food industry and also conditions in which mechanical parts works under high temperatures and pressure [4].

Example of solid lubrication material is ultra-high weight polyethylene (UHMWPE), which mechanical properties such as low coefficient of friction, high wear and chemical resistance in corrosive environment, high impact strength combined with low temperature of embrittlement ensure its stability and the potential for broad applications in various operating conditions at fields of engineering [5]. This material might be used in slide bearings, belt pulleys, tension belt rollers and protective coatings of mining silos interior surfaces and hoppers [6-7].

* Corresponding author: piotr.le.jablonski@doctorate.put.poznan.pl

It has been observed that combining multiple solid lubricants is a successful way in reduction of friction between moving parts [1, 9], therefore the aim of conducted research was to evaluate possible use of graphite as an addition to silicon oil in lubrication of friction pairs made of UHMWPE and steel.

2 Materials and Methods

The research presented in following paper were conducted with use of tribological testing station Tribotester Brucker UMT-2 in the laboratory at the Faculty of Mechanical Engineering and Management at the Institute of Mechanical Technology of Poznan University of Technology. This station might be used for tribological testing of ferrous, non-ferrous metals, plastics, ceramics, composites and various types of coatings, under "dry" and fluid or solid lubrication conditions. The tribological testing station can be programmed for motion in several axes. The ability to control of the trajectory of the sample and the counter-sample allow to use different methods of evaluation of wear, which are used for evaluation of engineering materials characteristics.

The research on tribological properties of UHMWPE in different lubricating conditions was performed with counter-sample made of martensitic stainless steel X105CrMo17. The trade name of used UHMWPE is TIVAR® manufactured by Quadrant EPP. According to Quadrant EPP catalog, used UHMWPE molecular weight is 5 * 10^6 g/mol. The samples of UHMWPE were prepared in form of plate 25 mm in length and 5 mm in thickness. Diamater of the counter-sample was 6.35 mm.

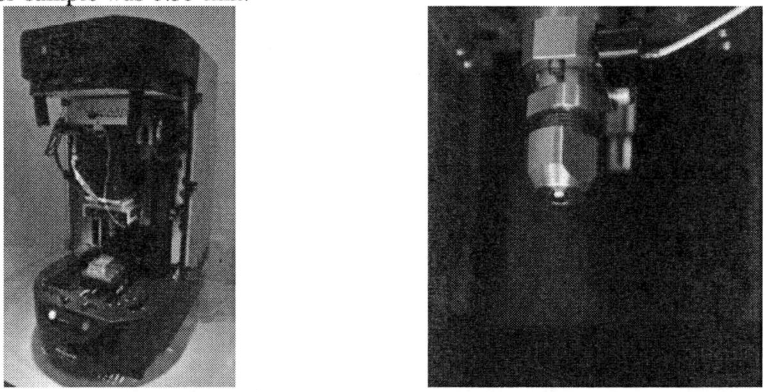

Fig. 1. a) Tribotester Brucker UMT-2 and b) counter-sample of martensitic stainless steel X105CrMo17.

The tribological studies were performed with use of a pin-on-flat method. The kinematic diagram of this method and parameters of the test are show in Figure 2.

The counter-sample performed a reciprocating motion on the fixed sample with an amplitude L = 20 mm and a linear velocity v = 10 mm/s. While the tests were performed, the counter-sample was set to apply the constant axial force F = 20 N.

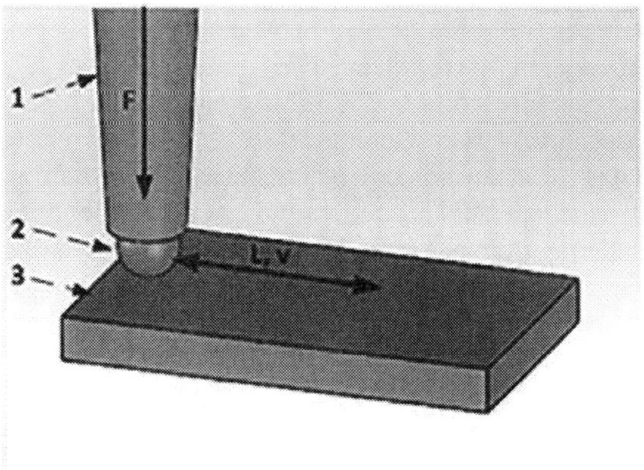

Fig. 2. Kinematic diagram of pin-on-flat method. 1 - griper of counter-sample, 2 – counter-sample, 3 – the sample of UHMWPE.

Using the equations (1) and (2) of the classic Hertzian contact stress theory value of contact pressure between the ball and polymer plate was calculated. In this equations (*a*) stands as radios of the contact area, (*υ*) is Poisson's ratio, (*E*) is Young modulus and (R_1) is ball radius (in this case flat radios (R_2) is understood as a sphere with an infinitely large radius) [8-9].

$$a = \sqrt[3]{\frac{3 \cdot F \cdot \left(\frac{1-\upsilon_1^2}{E_1} + \frac{1-\upsilon_2^2}{E_2}\right)}{4 \cdot \left(\frac{1}{R_1} + \frac{1}{R_2}\right)}} \, [mm] \quad (1)$$

$$p_{max} = \frac{3 \cdot F}{2 \cdot \pi \cdot a^2} \, [MPa] \quad (2)$$

Parameters which were used to calculations together with obtained results are presented in Table. 1.

Table 1. Parameters and obtained results of Hertzian contact pressure calculations.

	Steel X105CrMo17 ball-indenter	**UHMWPE plate**
F [N]	20	
R [mm]	6,35	-
E [MPa]	$2 \cdot 10^5$	763
υ	0,283	0,317
	Results	
a [mm]	0,48	
p_{max} [MPa]	40,92	

The maximum allowable contact stresses of UHMWPE is estimated to be equal 15 - 20 MPa and according to DIN 53 456 hardness of UHMWPE is equal to 35 MPa, therefore plastic deformation in the contact area between ball indenter and plate was expected. However, it must be remembered that due to high nonlinearity of UHMWPE calculated value of Hertzian contact pressure stress might be different in real [10-11].

The studies consisted of four tribological tests with different conditions of lubrication: dry, silicone oil lubrication and silicone oils suspension with 0.5 % or 1 % graphite addition. Silicone oils have excellent fluid properties and are commonly used as lubricants in systems designed to run under hydrodynamic and elastohydrodynamic conditions where bearings are made with non-ferrous materials. These oils have probably the best viscosity–temperature characteristics of any synthetic oils, with very low volatility and excellent oxidative and thermal stability. Their main disadvantage is the inability to miscible with petroleum oils [8, 12].

Graphite next to molybdenum disulfide is the predominant type of material which can be used as solid lubrication agents or dry lubricants. Graphite is known to be composed of a series of stacked parallel layers. Bonding forces within these layers between carbon atoms are significantly stronger than interatomic bonding forces between the layers. Thus, graphite's physical properties are characterized with layer like anisotropy. This kind of structure is called lamellar structure [4]. This structure is very favorable in sliding motion because the lamellae orient parallel to the surface in the direction of motion and slide easily over each other, resulting in low friction and preventing contact between sliding components even under high loads [12-14].

This study used synthetic graphite from Sigma-Aldrich and universal silicone oil, which was laboratory tested and calibrated to ensure the highest possible consistency and viscosity to $1000 MPa \cdot s$.

3 Results

During tribological test axial force, frictional force and coefficient of friction was recorded in both the time and cycles function. In Figure 3 are shown the curves of coefficient of frictions change in time for various lubrication conditions.

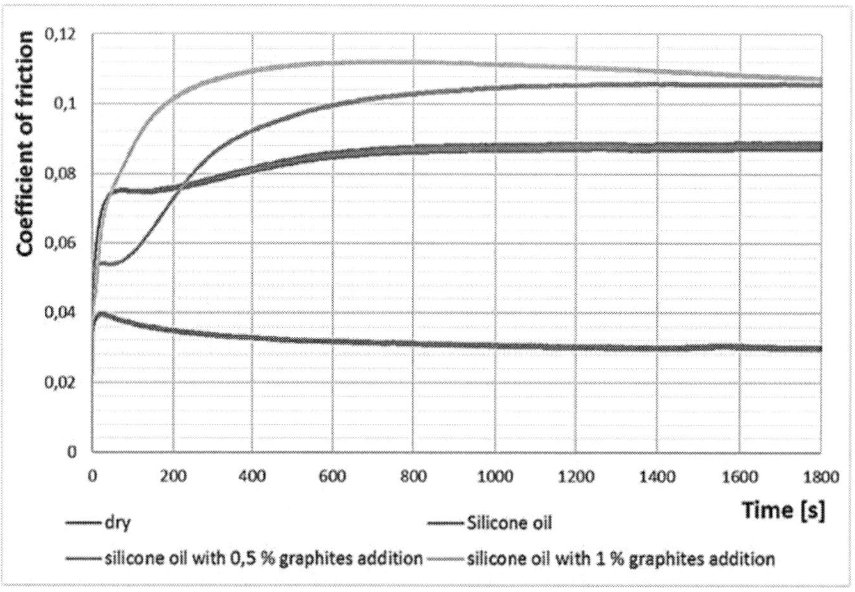

Fig. 3. Changes in the coefficient of kinetic friction for different lubricating agents.

In Table 2 the obtained results of tribological test for different lubricants are presented.

Table 2. The results of tribological tests for various lubricants.

	Dry	Silicone oil	Silicone oil with 0,5% graphite addition	Silicone oil with 1% graphite addition
The mean coefficient of friction after stabilization time	0,087	0,030	0,106	0,107
The maximal coefficient of friction	0,089	0,040	0,106	0,112
The time of stabilization	800 s	900 s	1100 s	-

The kinetic coefficient of friction measurements for dry samples and lubricated by silicone oil with/without 0.5 % and 1 % graphite addition, has been shown to increase the coefficient of kinetic friction in early time of test. Only in case of lubrication under pure silicone oil conditions are not shown pronounced changes.

During tribological test stabilization of measured value of coefficient of friction was observed. Time of stabilization is evident on curves in Figure 3 and are appointed. A stabilization of measured value of coefficient of friction indicates that used parameters – number of cycles, counter-sample, velocity, and its amplitude – were chose correctly.

In Table 2 the mean values of coefficient of friction that achieved after stabilization time are compared. There is noticeable, that the terminal coefficient of friction for silicone oil condition are nearly three times lower in comparison with dry samples. It is surprising that in tribological tests with silicone oil with graphite addition the highest value of the kinetic friction coefficient are reached.

4 Conclusions and Planned Further Research

The aim of this study was to perform tribological research for frictional pair made of UHMWPE and stainless steel X105CrMo17 under various conditions of lubrication. Conducted research allowed to obtain data about values of the coefficient of friction under: dry, silicon oil, and silicon oil with addition of graphite lubrication. Base on the research results (see table 1 and chart presented in Figure 3 the following conclusions were set out:

- Use of the silicon oil as a lubricant for the tested friction pair resulted in significant decrease of coefficient of friction value from 0,087 to 0,03 - approximately 3 times drop in value. This was caused by formation of oil film that separated surfaces of sample UHWMPE and steel counter-sample.
- Addition of graphite to silicon oil had negative synergy influence on lubrication of tested friction pair. As it is presented in Figure 1 addition of 0.5% and 1% graphite significantly affected measured value of the coefficient of friction which in both cases was higher than the coefficient of friction on dry surfaces. The probable cause was a graphite indentation into UHMW Polyethylene surface that resulted in in change of friction pair from UHMWPE-steel into graphite-steel.
- Another possible reason of coefficient of friction increase, was breaking of oil film caused by change in lubrication condition from fluid into boundary lubrication, as the result of graphite particles adherence to surface of the steel counter-sample and graphite indentation into UHMWPE. In this case friction pair was changed from UHMWPE-steel into graphite-graphite.
- Obtained data indicates that in this case addition of graphite to silicon oil is not recommended.
- In future research possible use of lubrication by silicon oil with addition of such solid lubricant as: expanded graphite with a highly developed specific surface area,

hBN or MoS_2 will be tested in application for friction pairs made of different engineering materials i.e. steel-steel, steel-ceramics.

References

1. J. Zhena, J. Chenga, S. Zhua, J. Haoa, Z. Qiaoa, J. Yanga, Trib. Inter., **110**, p. 52–56, (2017)
2. R.W. Maruda, E.E. Feldshtein, S. Legutko, et al., J. of Fri. and Wear, **36**, p. 548–553 (2015)
3. K. Adamczuk, S. Legutko, G. M. Krolczyk, R. W. Maruda, App. Mech. and Mat., **809-810**, p. 1055-1060, (2015)
4. K. Miyoshi, *Solid Lubrication Fundamentals and Applications,* CRC Press, (2001)
5. S. V. Panin, L. A. Kornienko, V. O. Alexenko, L. R. Ivanova, Key Eng. Mat., **712**, p. 155-160, (2016)
6. P. Siwak, A. Patalas, P. Jabłoński, K. Peta, Trib., p. 139 – 148, (2016)
7. Braskem – Brochure of produced Ultra High Molecular Weight Polyethylene https://www.braskem.com.br/Portal/Principal/arquivos/HRP_15-016_v02_UTEC_Brochure_iPad.pdfHigh-temperature tribological behavior of a nickel alloy matrix solid lubricating composite under vacuum, (access 20.02.2017)
8. G. W. Stachowiak, A. W. Batchelor, *Engineering Tribology,* Butterworth-Heinemann, (2005)
9. S. D. Mesarovic, N. A. Fleck, The Royal Society, **455**, pp. 2107-2128 (1999)
10. Z.M. Jin, D. Dowson, J. Fisher, Med. Eng. & Phys., **16**, p. 398-405, (1994)
11. GG Tech - http://www.ggtech.com.pl/12-polietylen-pe, (access 27.03.2017)
12. T. Hisakado, T. Tsukizoe, H. Yoshikawa, J. of Lub. Tech., **105**, 2, (1983)
13. B. Chen, Q. Bi, Y. Xia, J. Hao, Trib. Inter., **41**, 12, p. 1145-1152 (2008)
14. J. C. Spear, B. W. Ewers, J. D. Batteas, Nano Today, **10**, 3, p. 40-43 (2015)

Application possibilities of fuzzy networks concept in metallurgical industry

Adrian Ioana*, *Augustin* Semenescu, *Mihnea* Costoiu, and *Dragos* Marcu

University Politehnica of Bucharest, Spl. Independentei 313, Bucharest, Romania

Abstract. Our article presents possibilities of applying the concept Fuzzy Networks for an efficient metallurgical industry in Romania. We also present and analyze Fuzzy Networks complementary concepts, such as Expert Systems (ES), Enterprise Resource Planning (ERP), Analytics and Intelligent Strategies (SAI). The main results of our article are based on a case study of the possibilities of applying these concepts in metallurgy through Fuzzy Networks. There are presented the domains afferent to KIBS are defined complying with the standardized classification of industrial sectors according to European Monitoring Centre on Change (EMCC). It is also analyze important approaches in the specific scientific literature, the research methodology a study case. KIBS concept implementation to the Romanian metallurgical industry aims also to include performance of activities concerning the formulation and development of new specific concepts.

1 Introduction

The concept of Knowledge Intensive Business Services (KIBS) is basically defined as providing efficient services performed by specialized companies to other businesses interested in intensively using advanced knowledge [3-5].

KIBS has diverse areas of concern as specific activities. These include:

- Conception;
- Designing;
- Research-development;
- Training (learning – training);
- Implementing new technologies;
- Consulting in: management, marketing, advertising;
- Quality Assurance and Certification.

Figure 1 presents the algorithmic scheme of KIBS's design and specific correlations.

The beneficial aspects of KIBS concept application like: quality increase, economical efficient making, sales optimization are pointed out in many works of scientific literature.

The Romanian metallurgical industry suffered unfortunately an important and severe regression during last period. So, many production works of this industry branch have been

* Corresponding author: adyioana@gmail.com

closed or severely diminished production. Most relevant examples in this regard are the Steel works of Călan, Oţelul Roşu, Reşita, Târgovişte [2, 6].

Further on, are presented several possibilities to apply the KIBS concept to the Romanian metallurgical industry. It is also analyze important approaches in the specific scientific literature, the research methodology a study case.

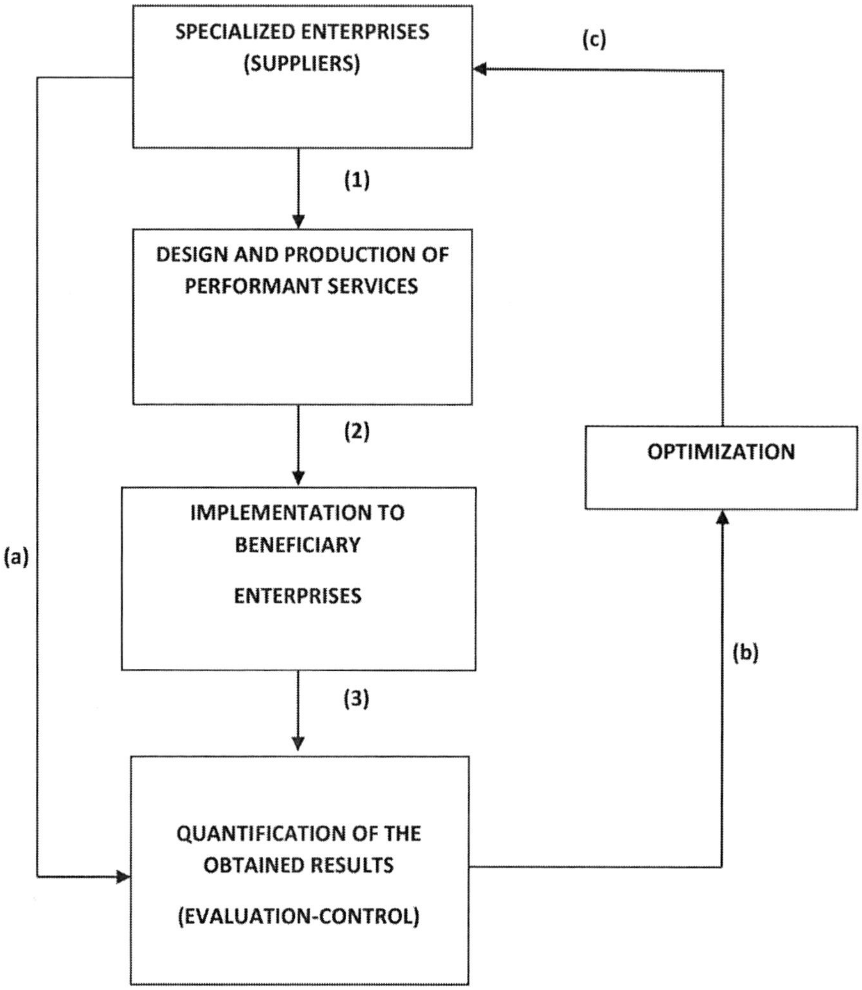

Fig. 1. Algorithmic scheme of KIBS's design and specific correlations. a – Suppliers evaluate the KIBS concept implementation on beneficiary's side. (b) + (c) – Supplier optimizes KIBS concept based on evaluation outputs. (1) - (3) – Time phasing of KIBS concept realization.

2 Possibilities and prerequisites to apply the KIBS concept to the Romanian metallurgical industry

According to European Monitoring Centre on Change (EMCC), the domains afferent to KIBS are defined complying with the standardized classification of industrial sectors [1, 7]. NACE Division 72: Computer activities and related:

- 72.1: Hardware consultancy
- 72.2: Software consultancy and supply

- 72.3: Data processing
- 72.4: Database activities
- 72.5: Maintenance and repair of office, accounting and computing machinery
- 72.6: Other computer related activities

- NACE Division 73: Research and experimental development:
    - 73.1: Research and experimental development on natural sciences and engineering
    - 73.2: Research and experimental development on social sciences and humanities
- NACE Division 74: Other business activities:
    - 74.3: Technical testing and analysis
    - 74.4: Advertising
    - 74.5: Labor recruitment and provision of personnel
    - 74.8: Miscellaneous business activities n.e.c.
    - 74.11: Legal activities
    - 74.12: Accounting, book-keeping and auditing activities; tax consultancy
    - 74.13: Market research and public opinion polling
    - 74.14: Business and management consultancy activities
    - 74.15: Management activities of holding companies
    - 74.20: Architectural and engineering activities and related technical consultancy
    - 74.81: Photographic activities
    - 74.84: Other business activities n.e.c.

KIBS concept implementation to the Romanian metallurgical industry aims also to include performance of activities concerning the formulation and development of new specific concepts as is (Dumitrana et. al, 2015, Ioana and Marinica, 2014, Ioana et. al. 2013, Ioana, Mirea and Bălescu, 2010): Expert system (ES), Enterprise Resource Planning (ERP), Fuzzy Network, Analytics Strategy and Business Intelligence (ASBI).

Unlike the majority computing programs, that require complete pieces of information to issue a decision, expert systems are designed to find the optimal solution based on available data the same way a human expert would act (Kallweit, Spreer and Toporowski, 2014, Semenescu, Ioana and Marcu, 2014).

An expert system (ES) is a computer program made of two components:

- A database (as big as possible) and
- A set of rules (program, algorithm) to search the given database aiming to find the optimal solution of the given problem

The database (DB) and the set of rules (SR) are set up by questioning the domain experts (Abrudan, Plăiaş and Dabija, 2015, Ioana and Semenescu, 2013, Ioana, Semenescu and Preda, 2012, Ioana, Mirea and Bălescu, 2009, 2010).

Figure 2 presents an expert system (ES).

The cost of producing similar products vary between different countries because certain resources, such a labor, raw materials and energy will be cheaper in some countries than others.

Multinational enterprise are able to take advantage of these cost differences by buying components or products from countries that have these advantage. For example, a motor-vehicle manufacturer in an economy with high labor costs may purchase certain components from another country with significantly lower costs.

An expert system is made of five components, rendered in Figure 3.

Knowledge Base (KB) serves to stock all pieces of knowledge (facts, rules, solving methods, heuristics etc.) specific to the domain of application retrieved from human experts or other sources.

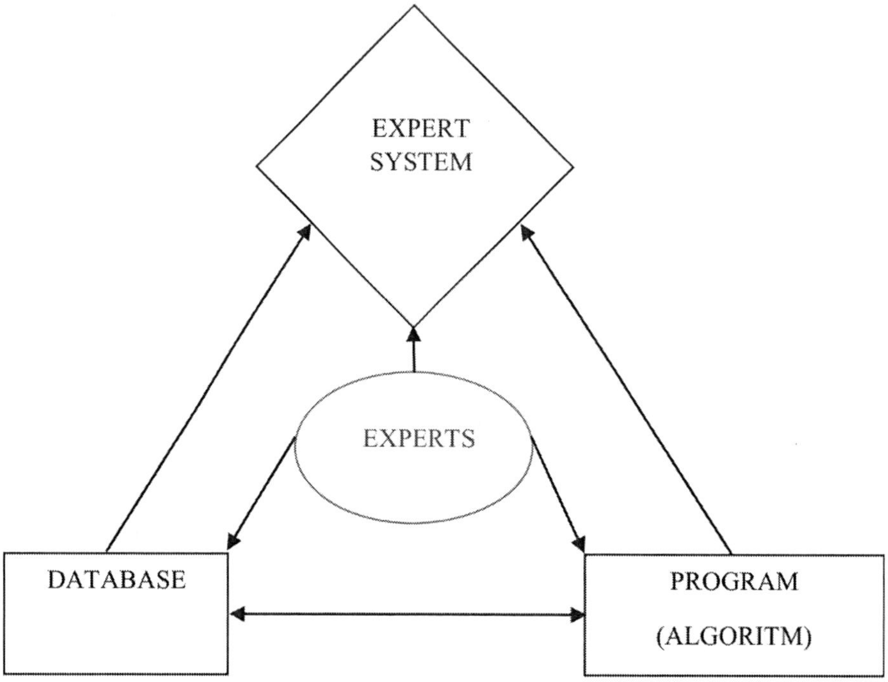

Fig. 2. Expert System Scheme.

Inference Engine (IE) is a program that contains checking, procedural and operational reasoning by means of which the database is explored applying reasoning in order to assert solutions, recommendations and conclusions.
Dialogue Interface (DI) permits communication with user during consulting sessions as well as access for users to facts and knowledge in the database for adding or updating thereof Oh, Teo and Sambamurthy, 2012).
Knowledge Acquiring Module (KAM) assists the expert user to introduce knowledge in a form recognized by the system and to update the knowledge base.
Explicative module (EM) is aimed to explain to users the knowledge level available in the system as well as the reasoning process that is performed or the solutions asserted during consulting sessions (Renko and Druzijanic, 2014, Sîrbu, Săseanu and Ghiță, 2015).
The explanations in such a system, if designed accordingly, improve the way the users perceive and accept the system.
[1)] Inference: a logical operation to pass from one statement to another while the following statement is derived from the previous.
An ES is based on two distinct and complementary components:
a. **Programming technologies** which allow a large volume of knowledge as well as the way to interfere therewith.
b. **Developed constructions** and methodologies that enable effective use of these technologies.

Reasoning and knowledge should not be treated separately in an ES since such a system requires harmonizing thereof.

The expert who understands the need for an expert system in its field can obtain help from the knowledge engineer (computer specialist), so following discussions with this one, certain knowledge can be reformulated to enable the use there of in a computer.

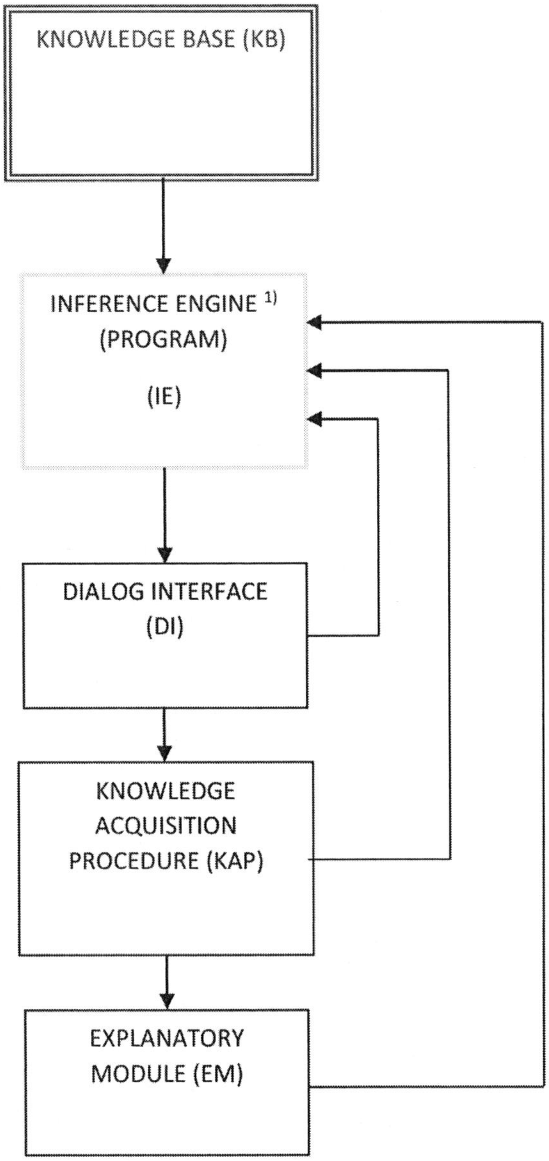

Fig. 3. Architecture of an ES.

An **Enterprise Resource Planning** (ERP) is a combination of business management practices with information technology, whereby company's business processes are integrated into the information system with the goal to achieve business specific objectives.

Equity finance represent finance raised from shareholders in the form of ordinary shares and reserves, as opposed to non-equity shares and to borrowing.

3 Conclusions

The Romanian metallurgical industry has unfortunately suffered major setbacks in recent years. Privatization strategies related to metallurgical production units were deficient, euphemistically said. In this negative context, the application of the KIBS concept to the metallurgical industry still left in Romania can represent a revival.

The following concepts complementary with KIBS concept are of particular importance: Expert System (ES) Enterprise Resource Planning (ERP), Fuzzy Networks, Strategy Analytics and Business Intelligence (SAI).

Absolute performance standard is a standard set at the theoretical limit of performance. For example, in manufacturing, the theoretical quality standard of "zero defects", which it is impossible to improve on, might be set as the absolute performance standard. Such standards may be achievable in practice or may form an ideal against which an organization may judge its progress.

An accounting process used in absorption costing in which the overhead of an organization is borne by the production of that organization by the use of absorption rates.

The cost accounting system in which the overhead of an organization are charged to the production by means of the process of absorption.

Cost are first apportioned to cost centers, where they are absorbed using absorption rates.

Equity capital is the part of the share capital of a company owned by ordinary shareholders, although for certain purposes, such as pre-emption rights, others classes of shareholders may be demand to share in the equity capital and therefore be entitled to share in the profits of the company or any surplus assets on winding up.

Equity dividend cover is a ratio that show many times the dividend to ordinary shareholders can be paid out, of the profits of a company available for distribution.

The higher the cover, the greater the certainty that dividends will be paid in the future.

References

1. N. Bharadwaj, R. Walker Naylor and F. Hofstede, IJRM, **26(3)**, 216-227, (2009).
2. M. Costoiu, A. Ioana, A. Semenescu, N. Constantin, B. Florea, V. Rucai, C. Dobrescu, M. Polifroni, L. Păunescu, EEMJ, **15, No. 10**, 2279-2284, (2016).
3. M. Costoiu, A. Ioana, A. Semenescu, D. Marcu, *IOP Conference Series-Materials Science and Engineering*, **161**, 012104, (2016).
4. M. Costoiu, A. Ioana, A. Semenescu, D. Marcu, *IOP Conference Series-Materials Science and Engineering*, **161**, 012105, (2016).
5. A. Ioana, RRSI, **43**, 169-179, (2013).
6. A. Ioana, A. Semenescu, JOM, **65, Issue 8**, 951-957, (2013).
7. A. Ioana, V. Mirea, C. Bălescu, JECECSR, **2**, 129-137, (2010).

Retail optimization in Romanian metallurgical industry by applying of fuzzy networks concept

Adrian Ioana*, *Augustin* Semenescu, *Mihnea* Costoiu, and *Dragos* Marcu

University Politehnica of Bucharest, Spl. Independentei 313, Bucharest, Romania

Abstract. Our article presents possibilities of applying the concept Fuzzy Networks for an efficient metallurgical industry in Romania. We also present and analyze Fuzzy Networks complementary concepts, such as Expert Systems (ES), Enterprise Resource Planning (ERP), Analytics and Intelligent Strategies (SAI). The main results of our article are based on a case study of the possibilities of applying these concepts in metallurgy through Fuzzy Networks. Also, it is presented a case study on the application of the FUZZY concept on the Romanian metallurgical industry.

1 Introduction

The Romanian metallurgical industry suffered unfortunately an important and severe regression during last period. So, many production works of this industry branch have been closed or severely diminished production. Most relevant examples in this regard are the Steel works of Călan, Oţelul Roşu, Reşita, Târgovişte [1-2].

An expert system is made of five components: Knowledge Base (KB), Inference Engine (IE), Dialogue Interface (DI), Knowledge Acquiring Module (KAM), Explanatory Module (EM) [3-7].

The explanations in such a system, if designed accordingly, improve the way the users perceive and accept the system.

Figure 1 shows the most important modules that make up a rule-based expert system. The user interacts with the system through a user interface which may use menus, natural language or any other style of interaction).

Then an inference engine is used to reason with both the expert knowledge (extracted from our friendly expert) and data specific to the particular problem being solved. The expert knowledge will typically be in the form of a set of IF-THEN rules.

The case specific data includes both data provided by the user and partial conclusions (along with certainty measures) based on this data. In a simple forward chaining rule-based system the case specific data will be the elements in working memory.

Almost all expert systems also have an explanation subsystem, which allows the program to explain its reasoning to the user.

Some systems also have a knowledge base editor which help the expert or knowledge engineer to easily update and check the knowledge base.

* Corresponding author: adyioana@gmail.com

One important feature of expert systems is the way they (usually) separate domain specific knowledge from more general purpose reasoning and representation techniques. The general purpose bit (in the dotted box in the figure) is referred to as an expert system shell.

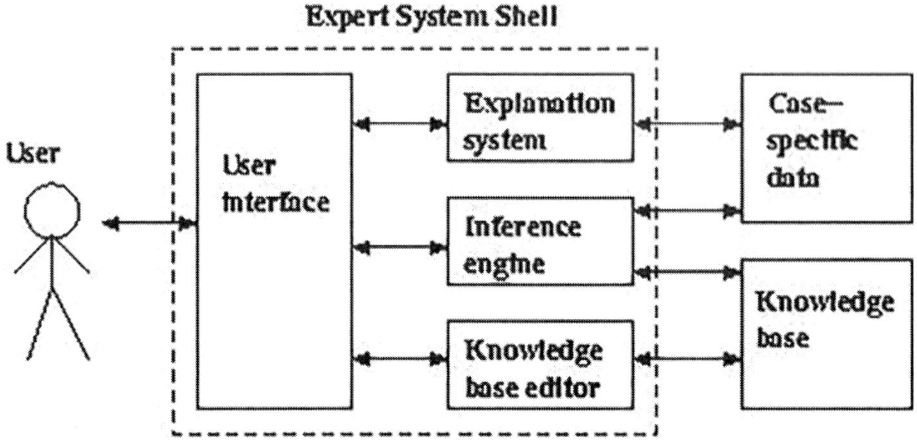

Fig. 1. Modules that make up a rule-based expert system.

As we see in the figure, the shell will provide the inference engine (and knowledge representation scheme), a user interface, an explanation system and sometimes a knowledge base editor. Given a new kind of problem to solve (say, car design), we can usually find a shell that provides the right sort of support for that problem, so all we need to do is provide the expert knowledge.

There are numerous commercial expert system shells, each one appropriate for a slightly different range of problems.

Expert systems work in industry includes both writing expert system shells and writing expert systems using shells. Using shells to write expert systems generally greatly reduces the cost and time of development (compared with writing the expert system from scratch).

Writing an expert system generally involves a great deal of time and money. To avoid costly and embarrassing failures, people have developed a set of guidelines to determine whether a problem is suitable for an expert system solution:

The need for a solution must justify the costs involved in development. There must be a realistic assessment of the costs and benefits involved.

Human expertise is not available in all situations where it is needed. If the ``expert'' knowledge is widely available it is unlikely that it will be worth developing an expert system. However, in areas like oil exploration and medicine there may be rare specialized knowledge which could be cheaply provided by an expert system, as and when required, without having to fly in your friendly (but very highly paid) expert.

The problem may be solved using symbolic reasoning techniques. It shouldn't require manual dexterity or physical skill.

The problem is well structured and does not require (much) common sense knowledge. Common sense knowledge is notoriously hard to capture and represent. It turns out that highly technical fields are easier to deal with, and tend to involve relatively small amounts of well formalized knowledge.

The problem cannot be easily solved using more traditional computing methods. If there's a good algorithmic solution to a problem, you don't want to use an expert system.

Cooperative and articulate experts exist. For an expert system project to be successful it is essential that the experts are willing to help, and don't feel that their job is threatened.

Also need any management and potential users to be involved and have positive attitudes to the whole thing.

The problem is of proper size and scope. Typically you need problems that require highly specialized expertise, but would only take a human expert a short time to solve (say an hour, max).

It should be clear that only a small range of problems are appropriate for expert system technology. However, given a suitable problem, expert systems can bring enormous benefits. Systems have been developed, for example, to help analyze samples collected in oil exploration, and to help configure computer systems. Both these systems are (or were) in active use, saving large amounts of money.

2 Case study on the application of the FUZZY concept on the Romanian metallurgical industry

Applying the FUZZY concept in metallurgy implicitly assumes the development and implementation of ERP systems as well. A positive and relevant example of effective implementation and development of ERP systems in the Romanian metallurgy is the SC Tenaris S.A. Calarasi. The relevance of this example resides in the leading position of this company in the Romanian metallurgical industry. So, the modern organization and planning system of this production unit ensures a high level of technological and economic efficiency. In this work, we will also present a case study applying Fuzzy networks on data characteristic for SC Tenaris S.A. Calarasi.

There are two types of motivations that determine to implement ERP, which are also characteristic for SC Tenaris S.A. Călărași [4-6]:

A. **Technological motivations:**

- Replacing a non-integrated information system (multiple applications, obsolete or not, which operate independently);
- Replacing one or more obsolete systems, integrated or not;
- Improving the quality and accessibility of information;
- Integrating business processes and supporting systems;
- Acquisition of a system able to support projected business increase;
- Simplification of the integration of new businesses (acquisition of other companies) into the current technological infrastructure.

B. Operational motivations

- Optimization of business processes;
- Reduction of structural costs;
- Improving response times to customer requests;
- Simplifying complex, but ineffective business processes;
- Implementation of new business strategies;
- Massive business expansion;
- Standardization of business processes at company level.

Ergonomics represent the study of the physical and psychological aspects of work. It is primarily concerned with designing jobs and work environments that are safe and efficient. For example, such human characteristics as size, reach and flexibility must match the machines and tools that people have to use.

Escalation clause is a clause in a contract authorizing the contractor to increase the price in specified conditions of all or part or goods he has contracted to supply. Escalation clauses are common in contracts involving work over a long period in time of high inflation.

In Figure 2 we set forth an algorithmic diagram of an Analytic and Intelligent Strategy (AIS).

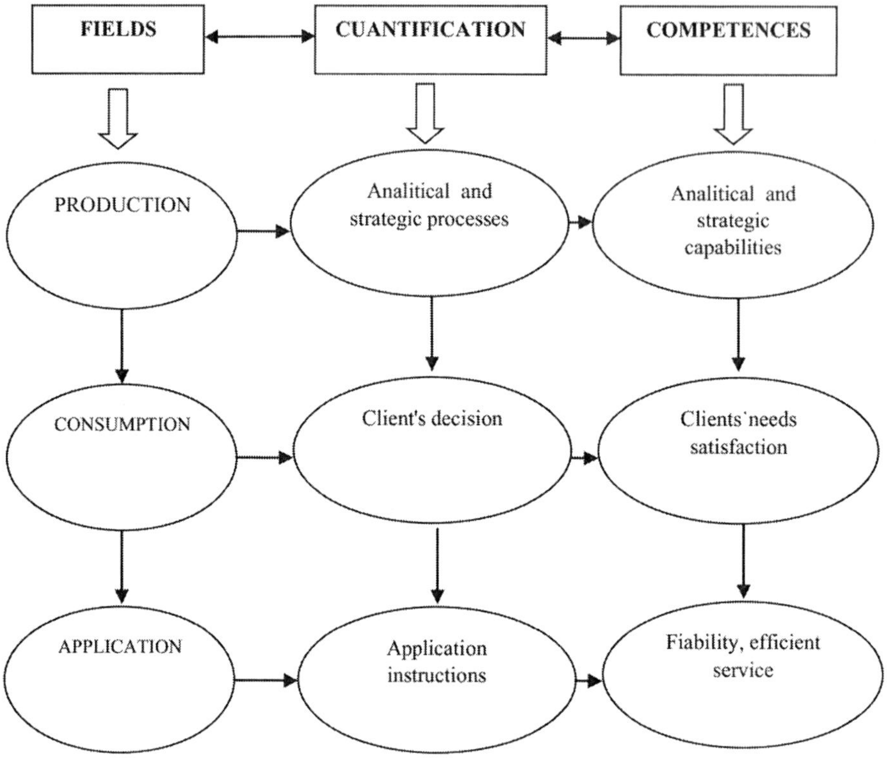

Fig. 2. Algorithmic diagram of AIS.

We have set up a case study applying Fuzzy networks based on AIS and on SAI and on ERP and ES concepts previously analyzed. Fuzzy logic is a powerful methodology for solving specific issues (including case studies), offering a multitude of applications in controlling and information processing. This powerful methodology gives us an effective way to draw precise conclusions stemming vague, ambiguous or imprecise pieces of information.

Fuzzy logic based applications can lead to any of its potential benefits, such as: performance, productivity, simplicity and lower costs (Figure 3).

The Fuzzy networks principle is to use neural networks and fuzzy set theory for modelling practical systems. The algorithm related to a Fuzzy network is implemented as an inference engine that can automatically deduce precise conclusions from certain facts (data).

Equity-linked policy is an insurance or assurance policy in which a proportion of the premiums paid are invested in equities.

The surrender value of the policy is therefore the selling price of the equities purchased. As more premiums are paid the portfolio gets larger.

Although investment returns may be considerably better on this type of policy than on a traditional endowment policy, the risk is greater, as the price equities can fall dramatically reducing the value of the policy.

With unit-linked policies, a much wider range of investments can be achieved and the risk is correspondingly reduced.

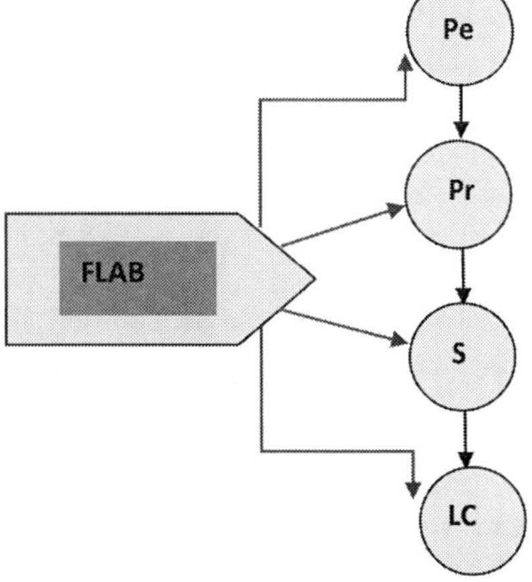

Fig. 3. Fuzzy Logic Application Benefits (FLAB).

Pe - Performance; Pr – Productivity; S – Simplicity; LC – Low Costs

Case study data rendering consumption (sales) analyze of an assortment of metallic products (high alloy, stainless and refractory steels) are set forth in Figure 4 (the variation of numerical results pertaining to Fuzzy networks specific for consumption (sales) analyze of metallic products).

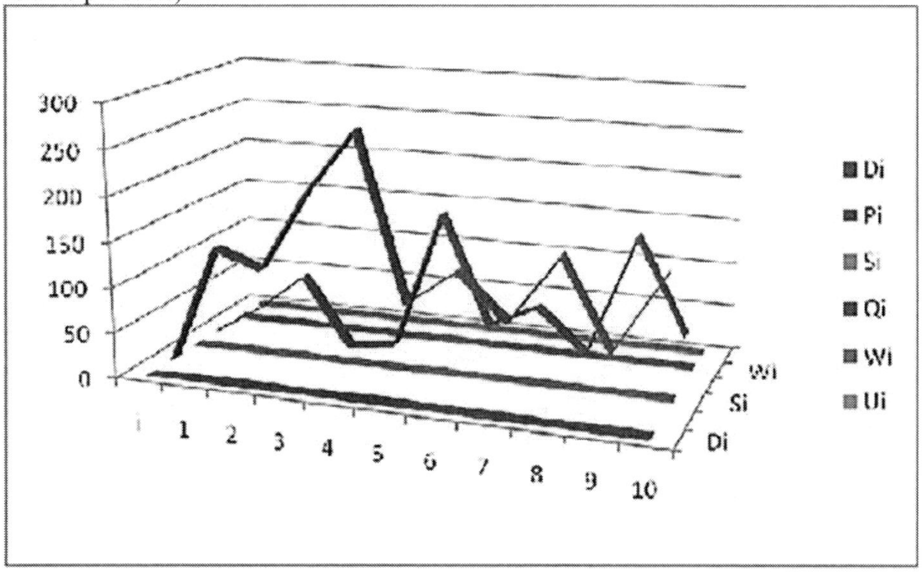

Fig. 4. Variation of numerical results pertaining to Fuzzy networks specific for consumption (sales) analyze of metallic products.

i – Analyzed type of metallic product
Di – Parameter for clients' subjective evaluation with respect to consumption (depreciation)
Pi – Price of product type
Si – Costs related to "i" type product (percentage)

Qi – Factor of subjective satisfaction quantification by consumers related to consumption value
Wi – Consumer's preference share
Ui – Utility function for each "i" product type.

3 Conclusions

Applying the fuzzy concept in the metallic materials industry has as main purpose and results to increase efficiency in producing and selling these materials. An important role in achieving this purpose is also played by complementary factors such as: Expert Systems (ES), Enterprise Resource Planning (ERP), Analytics and Intelligent Strategies (SAI).

Architecture of Expert Systems has the following elements: Inference Engine (IE), Dialogue Interface (DI), Knowledge Acquiring Module (KAM), and Explanatory Module (EM).

The case study of applying the fuzzy concept in the metallic materials industry was based on two sets of preliminary conditions motivations on one hand a set of technological (motivations) and on the other hand a set of operational motivations (dealing with the execution).

Algorithmic diagram of an Analytic and Intelligent Strategy (AIS) contains the fields: production, consumption, application.

Equity accounting is the practice of showing in a company-s accounts a share of the undistributed profits of another company in which it holds a share of between 20% and 50%.

The share of profit shown by the equity-holding company is usually equal to its share of the equity in the other company. Although none of the profit may actually be paid over, the company has right to this share of the undistributed profit.

Ethical dilemmas is the moral quandaries that can occur in running a business. While not confronting the law, most of these dilemmas arise as a result of conflict between what the businessperson sees as a necessary in the interests of the business and his or her personal ethical values.

References

1. N. Bharadwaj, R. Walker Naylor and F. Hofstede, IJRM, **26(3)**, 216-227, (2009).
2. M. Costoiu,A. Ioana, A. Semenescu, N. Constantin, B. Florea, V. Rucai, C. Dobrescu, M. Polifroni, L. Păunescu, EEMJ, **15, No. 10**, 2279-2284, (2016).
3. M. Costoiu, A. Ioana, A. Semenescu, D. Marcu, *IOP Conference Series-Materials Science and Engineering*, **161**, 012104, (2016)
4. M. Costoiu, A. Ioana, A. Semenescu, D. Marcu, *IOP Conference Series-Materials Science and Engineering*, **161**, 012105, (2016).
5. A. Ioana, RRSI, **43**, 169-179, (2013).
6. A. Ioana, A. Semenescu, Technological, JOM, **65, Issue 8**, 951-957, (2013).
7. A. Ioana, V. Mirea, C. Bălescu, JECECSR, **2**, 129-137, (2010).

Influence of electrical parameters on morphology of nanostructured TiO_2 layers developed by electrochemical anodization

Gabriela Strnad [1,*], *Zoltan* German-Sallo [2], *Laszlo* Jakab-Farkas [3], *Cecilia* Petrovan [4], and *Diana* Portan [5]

[1] "Petru Maior" University of Tirgu Mures, Faculty of Engineering. Department of Industrial Engineering and Management, Nicolae Iorga 1, Tirgu Mures, Romania
[2] "Petru Maior" University of Tirgu Mures, Faculty of Engineering, Department of Electrical Engineering and Computers, Nicolae Iorga 1, Tirgu Mures, Romania
[3] Sapientia University of Cluj Napoca, Faculty of Technical and Human Sciences, Tirgu Mures, Romania
[4] University of Medicine and Pharmacy of Tîrgu Mureş, Faculty of Dentistry, Oral and Maxillo-Facial Department, Gh. Marinescu 38, Tirgu Mures, Romania
[5] University of Patras, Department of Mechanical and Aeronautical Engineering, Patras, Greece

Abstract. Ti6Al4V alloy micro rough surfaces with TiO_2 self-organized nanostructured layers were synthesized using electrochemical anodization in phosphate/fluoride electrolyte, at different end potentials (5V, 10V, 15V, and 20 V). The current – time characteristics were recorded, and the link between current evolution and the morphology of developing oxide layers was investigated. On flat surfaces of Ti6Al4V alloy we developed TiO_2 layers with different morphologies (random pores, nanopores of 25...50 nm, and highly organized nanotubes of 50...100 nm in diameter) depending on electrical parameters of anodization process. In our anodization cell, in optimized conditions, we are able to superimpose nanostructured oxide layers (nanotubular or nanoporous) over micro structured surfaces of titanium based materials used for biomedical implants.

1 Introduction

Titanium and its alloys are nowadays widely used for manufacturing of dental and orthopedic implants manufacturing, due to their excellent biocompatibility, good mechanical strength, and high corrosion resistance. Ti6Al4V alloy is of major importance in hard tissue (teeth, bones) replacement and surgery, as it presents increased mechanical properties [1, 2].

The unique biocompatibility of titanium based materials is a result of the formation of a native TiO_2 compact layer, with a thickness of 2...10 nm, on their surface. The process of implants osseointegration of implants can be enhanced by modifying their surface at the micro level, in order to obtain a micro rough surface, with a large bone-to-implant contact

* Corresponding author: gabriela.strnad@ing.upm.ro

area. Recent results demonstrate that surface modification at nano scale level, by producing an ordered nanostructured surface, promotes enhanced bone apposition [3-12].

Electrochemical anodization method is used for surface modification of titanium and titanium based alloys, as it can control the morphology, structure and chemical composition of TiO_2 layers, leading to enhancement of the surface properties [13, 14]. By optimizing the process parameters of anodization, highly order and self-arranged nanostructured (nanoporous or nanotubular) surfaces can be developed [15-19]. These nanostructures combine very well with osseous tissue and can be the perfect basis for osteoblasts integration during the process of bone regeneration. Recent literature arguments the favorable osseointegration around TiO_2 nanotubes of 15...100 nm diameters modified surfaces by over-expression of genes related to high activity of osteoblasts and mesenchymal stem cells, and by fast kinetics of hydroxyapatite (basic bone component) formation, surrounding titanium implant.

Nanostructures formation is governed by a very complex mechanism that occurs in interdependent conditions. Phosphate-fluoride solutions used as electrolytes proved to be very effective on TiO_2 nanoporous/nanotubular self-organized structures development. Inside the anodization cell the formation of nanostructures is determined by the competition between anodic oxide formation and chemical dissolution of the oxide, and the presence of hydrofluoric acid is the key factor which causes nanoporous or nanotubular oxide layer development. The nanotubular TiO_2 layer develops in three stages: first is compact oxide formation, second is the initial porous structure formation, and third is self-organized nanotubes growth. A proper combination of electrolyte composition, anodization potential, anodization time, and a perfect balance between potentiodynamic stage and potentiostatic stage are required in order to develop uniform and highly-organized nanotubular oxide structures on titanium [13, 19, 20]. There are many results on development of titanium nanotubes on pure titanium (CpTi) surface. But the formation of self-ordered nanostructures on two phase ($\alpha+\beta$) titanium alloys, such Ti6Al4V is, is a more demanding process, due to the selective dissolution of the less stable phase and/or different reactions rate of the different phases of the alloy [13]. Moreover, the vast majority of results are obtained on extra polished surfaces.

In this context our major research direction is enhancement of surfaces of biomedical titanium based implants, at nano scale level, by developing TiO_2 nanostructures (nanotubes/nanopores) with 15...100 nm diameter, in order to improve implants osseointegration. Our recently reported research led to successful development, using electrochemical anodization, of self-ordered nanostructured oxide layer on the surface of two phase Ti6Al4V alloy by using electrochemical anodization in H_3PO_4/HF electrolytes [21]. Our results have a high degree of novelty, proving nanostructured TiO_2 layers deposition on surfaces with less demanding preparation prior to anodization, such are turning and milling, surfaces that presents a roughness R_a in 1...1,5 μm range. Also we developed nanostructured layers on sand blasted and acid etched surfaces (SLA) [22].

Present paper aims to investigate the influence of electrical parameters on the morphology of developing oxide layers on flat micro rough surfaces, of Ti6Al4V alloy, in order to optimize the conditions of electrochemical anodization.

2 Methodology

In present set of experiments our samples of Ti6Al4V titanium alloy were cut from a round bar, using a CNC turning machine. They have disc shape of 16 mm in diameter and 3 mm in height. Surtronic 25 (Taylor Hobson) roughness tester was used for roughness measurements.

Before the anodization all samples were subjected to a cleaning process in distilled water followed by ethanol, and after then were dried.

DOI: 10.1051/matecconf/201711204021

The electrochemical anodization process was performed in a custom built electrochemical cell having two electrode configuration. The samples were connected to the anode of the cell, while the cathode was in form of ∅16 x 3 mm disc, made of high purity copper (99.99%). For all experiments the distance anode-cathode was 15 mm. 1M H_3PO_4 + 0.5 wt% HF was used as electrolyte in anodization experiments. Electrolyte was prepared from reagent grade chemicals (Chemical Company) and deionized water. Experiments were carried out at room temperature.

For electrical parameters control and registration a direct current stabilized power supply, and voltmeter and ammeter were used. Several experiments were conducted with different anodization potentials: 5 V, 10 V, 15 V, and 20 V, applied with a sweep rate of 0.1 V/s. After the end potential was reached, we applied a potentiostatic stage by holding constant the end potential for an anodization time of 2 hours. After anodization, the samples were rinsed with deionized water, cleaned in ethanol and dried.

The surface morphology was evaluated using scanning electron microscopy (SEM), performed in a JSM 5200 (JEOL) scanning electron microscope, operated at 25 kV. We collected images at magnifications of 100X, 1500X, 10000X, 20000X, and 35000X, using secondary electrons and backscattered electrons. For the visualization of layers thickness we used a tilt angle of 45°. The dimensions of nanostructures developed on the modified surfaces were measured on SEM micrographs using an open source graphical image processing software (Gimp).

3 Results and discussion

Titanium alloy Ti6Al4V is a two phase (α+β) alloy, with 6 wt% aluminum stabilizing the α phase (hcp), and 4 wt% vanadium stabilizing the β phase (bcc).

Table 1 summarizes the electrical process parameters used in our experiments and results on morphology and diameters of self-organized TiO_2 nanotubes/nanopores that were synthesized by anodization.

Table 1. Electrical process parameters and anodization results.

Electrical parameters			Results	
Sample	Potential [V]	Current [mA] in potentiostatic stage	Morphology of TiO_2 layer	Nanostructures diameter D_i [nm]
A	5	2.2...4.4	random pores	< 10 nm
B	10	2.2...3.5	ordered nanopores	25...45
C	15	3.1...6.5	ordered nanopores	30...50
D	20	2.8...9.6	self-ordered nanotubes	60...100

After electrochemical anodization the surface of the samples is covered by TiO_2 oxide layer. The oxide layer grows over the entire surface, as the Figure 1.a, collected at low magnification, shows. At higher magnification the morphology of oxide layers can be examined in details. Figure 1.b presents the SEM micrograph of highly organized, self-arranged TiO_2 layer developed on sample D, using an anodization potential of 20 V. The nanotubes are fully developed, their walls being clearly split one of the other. The internal diameter of the nanotubes is in 60...100 nm range, the external diameter is 100...160 nm, and the wall thickness is 20...30 nm. The thickness of the nanostructured oxide layer, measured on SEM micrographs not shown here, is 200...350 nm.

 DOI: 10.1051/matecconf/201711204021

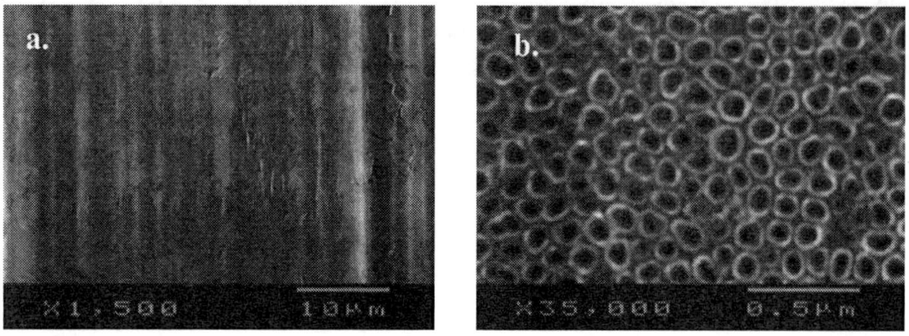

Fig. 1. SEM micrographs of TiO_2 nanotubular layer developed on micro rough surface of Ti6Al4V alloy by electrochemical anodization in 1M H_3PO_4 + 0.5 wt% HF, anodization potential 20 V, anodization time 2 hours (a – 1500X magnification; b – 35000X magnification)

Our results obtained by monitoring the current in the process of titanium layer development using an anodization potential of 5 V show that after the reaching of the end potential (50 seconds from the start of the process), period in which the current has an oscillated evolution between 0...8 mA (not shown in Figure 2), the current is continuously decreasing from 4.4 mA (at 5 min after the beginning of the process) to 2.2 mA (at the end of anodization). The SEM micrographs of the TiO_2 layer show the development of a compact oxide layer with superficial surface pores with openings less than 10 nm (Figure 3.a).

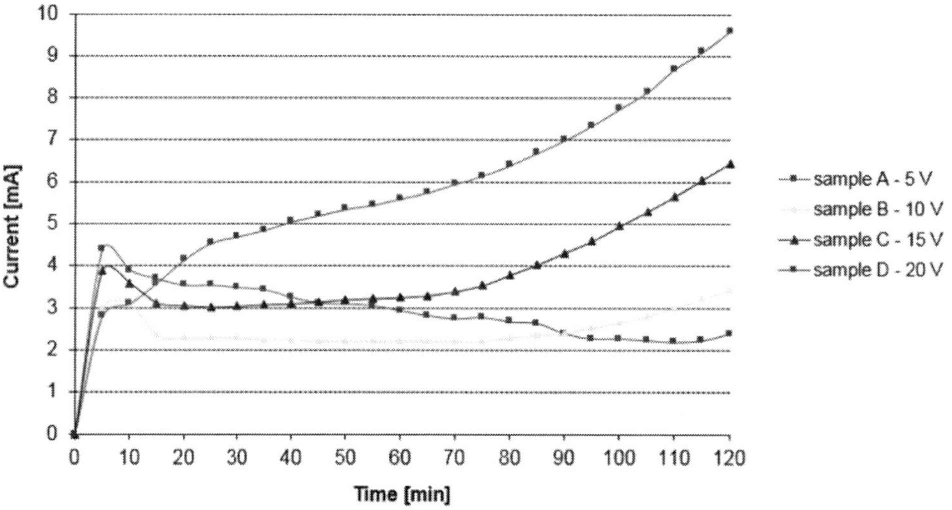

Fig. 2. Current-time characteristic for electrochemical anodization of Ti6Al4V samples in 1M H_3PO_4 + 0.5 wt% HF at different anodization potential, anodization time 2 hours

By using increased anodization potentials, 10 V and 15 V, in the potentiostatic stage, after an initial increase in current, a drop occurs. After that a stabilization of current appears (around 2.2 mA for U = 10 V, and around 3 mA for U = 15 V), followed by an increase in current, with 0.03 mA/min when anodization potential is 10 V, and with 0.06 mA/min when anodization potential is 15 V. SEM micrographs of developed layers show ordered nanoporous morphology with nanopores diameters in 25...50 nm range (Figure 3.b and 3.c).

By using an anodization potential of 20 V, the current evolution, after the first 200 s needed for reaching the end potential, shows a short stabilization phase (5 minutes) followed by a continuous increase in current with approx. 0.06 mA/min. The SEM micrographs taken from the TiO_2 layer reveal a well-developed nanotubular topography, with highly organized nanotubes of 60...100 nm in diameter. The tubes wall thickness is of 20...30 nm (Figure 3.d).

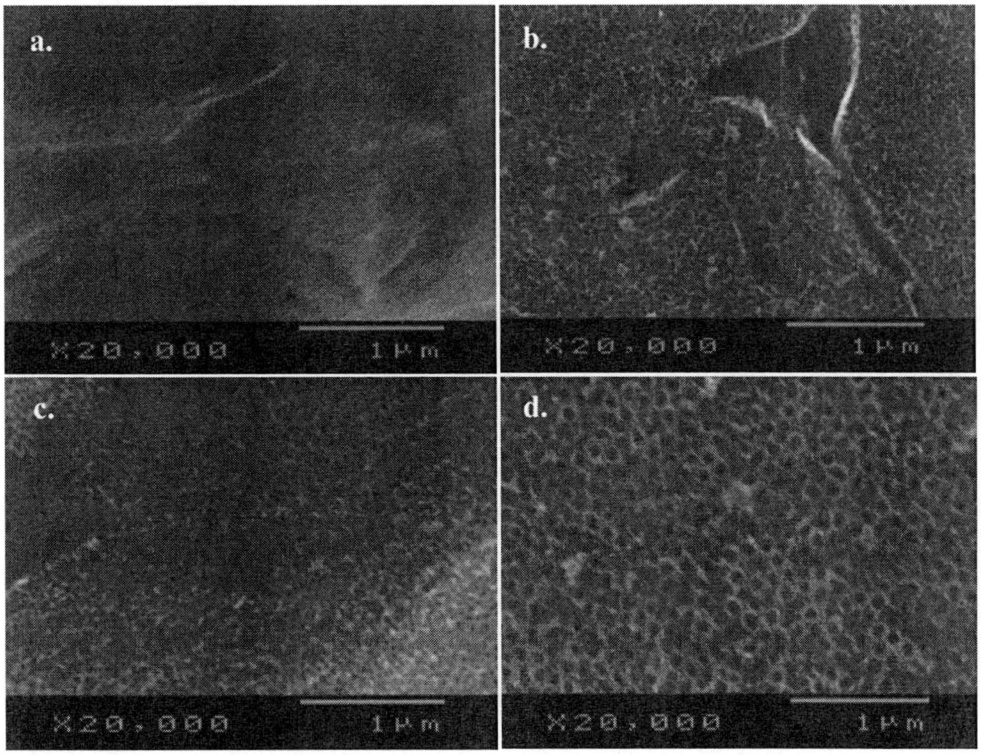

Fig. 3. SEM micrographs of TiO_2 layers developed on micro rough surface of Ti6Al4V alloy by electrochemical anodization in 1M H_3PO_4 + 0.5 wt% HF, anodization time 2 hours, 20000X magnification (a: anodization potential U = 5 V; b: U = 10 V; c: U = 15 V, d: U = 20V)

Analyzing the developed topographies in relation with the current evolution during the process, our results show that, for Ti6Al4V alloy anodized in phosphate/fluoride aqueous solution, when nanotubes are growing the magnitude of the current is higher than when a compact oxide or a nanoporous one is formed.

Our results show that, in optimized conditions, we are able to superimpose nanostructured oxide layers (nanotubular or nanoporous) over micro structured surfaces of titanium based materials for biomedical implants by using electrochemical anodization.

4 Conclusions

We monitored the growth of TiO_2 oxide layer on two phase ($\alpha+\beta$) Ti6Al4V alloy, with micro rough surface, using electrochemical anodization in 1M H_3PO_4 + 0.5 wt% HF solution, and recorded current-time characteristics.

For synthesis of the oxide layers we work by maintaining a constant potential (5, 10, 15, and 20 V) for duration of 2 hours. Our results show that when the current is low (<4 mA) and decrease continuously, the TiO_2 layer is a compact one with superficial nanopores with diameter less than 10 nm. A different evolution of current, an initial decrease, a steady state, and a final step when the current is increasing, leads to the formation of nanoporous structures, with nanopores of 25...50 nm. This ordered nanoporous oxide structures are developing at low currents (2.2 ...3.5 mA, and 3.1...6.5 mA). Higher currents in the anodization cell (2.8 ...9.6 mA) were recorded by us when highly organized, self-arranged nanotubular oxide layer is developing.

These results allow us to optimize and control the electrochemical anodization process in order to develop nanotubular TiO_2 layer, by means of current control.

This work was supported by a grant of the Romanian National Authority for Scientific Research and Innovation, CNCS/CCCDI-UEFISCDI, project number PN-III-P2-2.1-PED-2016-0142, within PNCDI III.

References

1. K. Gulati, S. Ivanovski, Expert Opinion on Drug Delivery, doi: 10.1080/17425247. 2017.1266332 (2016)
2. W. Wang, C. K. Poh, C. K. *Titanium alloys in orthopaedics* (*Titanium alloys – advances in properties control*) chapter 1, 1-20 (2013)
3. K. Gulati, S. Maher, D. M. Findlay, D. Losic, Nanomedicine, **11(14)** 1847-1864 (2016)
4. D. Karazisis, A. M. Ballo, S. Petronis, H. Agheli, L. Emanuelsson, P. Thomsen, O. Omar, Int J of Nanomed, **11** 1367-1381 (2016)
5. A. Mazare, Corrosion Science, **103** 215–222 (2016)
6. M. Kulkarni, A. Mazare, E. Gongadze, S. Perutkova, V. Kralj-Iglič, I. Milosev, P. Schmuki, A. Iglic, M. Mozetic, Nanotechnology, **26** 062002 (2015)
7. K. Narendrakumar, M. Kulkarni, O. Addison, A. Mazare, I. Junkar, P. Schmuki, R. Sammons, A. Iglic, Dental Materials, **31** 1460-1468 (2015)
8. E. Beltrán-Partida, A. Moreno-Ulloa, B. Valdez-Salas, C. Velasquillo., M. Carrillo, A. Escamilla, E. Valdez, F. Villarreal, Materials, **8** 867-88 (2015)
9. M. Kulkarni, A. Mazare, P. Schmuki, A. Iglic, *Biomaterial surface modification of titanium and titanium alloys for medical applications*, in *Nanomedicine*, chapter 5, pp 111-135 (2014)
10. S. Minagar, J. Wang, C.C. Berndt, E. P. Ivanova, C. Wen, Journal of Biomedical Materials Research A, **9999A** 1-14 (2013)
11. A. Ross, T. Webster, International Journal of Nanomedicine, **8** 109-117 (2013)
12. C. Frandsen, K. Brammer, S Jin, International Journal of Biomaterials, doi : 10.1155 /2013/ 513680 (2013)
13. 13. Krasicka-Cydzik, *Anodic layer formation on titanium and its alloys for biomedical applications (Titanium alloys – towards achieving enhanced properties for diversified application)*, chapter 8, pp 175-200 (2012)
14. K. Brammer, et al, *Biomaterials and biotechnology schemes utilizing TiO_2 nanotube arrays (Biomaterials science and engineering)*, chapter 9, pp 194-210 (2011)
15. M. Kulkarni, et al., Advanced Material Letters, **7(1)** :23-28 (2016)
16. K. Gulati, A. Santos, D. Findlay, D. Losic, Journal of Physical Chemistry C, **119** 16033-16045 (2015)
17. A. Mazare, et al, Surf. Interface Anal., **46** 186-192 (2014)
18. K. Lee, et al., Chemical Reviews, **114** 9385–9454 (2014)
19. N. A. Al-Mobarak, A. A. Al-Swayih, Int. J. Electrochem. Sci., **9** 32–45 (2014)
20. P. Roy, S. Berger, P. Schmuki, Angew. Chem. Int. Ed., **50** 2904 – 2939 (2011)
21. G. Strnad, C. Petrovan, O. Russu, L. Jakab-Farkas, IoP Conf. Series: Materials Science and Engineering, **161** doi: 10.1088/1757-899X/161/1/012051 (2016)
22. G. Strnad, L. Jakab-Farkas, C. Petrovan, O. Russu, O., Procedia Engineering, accepted, in process (2017).

A study of an influence of a fiber arrangement of a laminate ply on the distribution and values of stresses in the multi-layered composite material

Krzysztof Herbuś* and *Piotr* Ociepka

Institute of Engineering Processes Automation and Integrated Manufacturing Systems, Faculty of Mechanical Engineering, Silesian University of Technology, Konarskiego 18a, 44-100 Gliwice, Poland

Abstract. In the work are presented studies related with the influence of a fiber arrangement of a laminate ply on the distribution and values of stresses in the multi-layered composite material. For this purpose, the characteristics of the three-point bending test, according to the standard PN-EN ISO 7438, of specimens made from the composite material, where a single ply is a composition of epoxy resin and glass fibres, was mapped. The modelling process of the multi-layered composite material and its strength verification was performed in the PLM Siemens NX system. Based on the results of performed numerical studies, the relation between the value of the main angle of an arrangement of fibers in each plies of the laminate, and the distribution and values of stresses, occurring in the examined specimens has been determined.

1 Introduction

In recent years, a significant increase in the use of composite materials in engineering applications has been observed. One of the main groups of composite materials is multilayer composite materials. These materials are characterized by a distinctive structure within which the matrix and reinforcement can be distinguished. As the matrix material could be used a variety of resins that are intended to assemble the reinforcing layers. Hence one of the basic forms of matrix reinforcement is the use of different types of fibres. The reinforcement in the form of a fibre can take the form of a rowing where all the reinforcement fibres are laid in the same direction or a fabric in case the reinforcement fibres are woven in accordance with the determined weave. Reinforcement can also be discontinuous when it is in the form of staple fibres. Due to the continuous expansion of the areas of application of multilayer composite materials, research is conducted to determine their correct modelling [1-3] and verification through a virtual experiment using the computer-aided approach.

The modern systems of the CAD/CAE class enable conducting researches concerning the kinematic, dynamic, and MES analysis [4-17]. The paper presents computer aided researches on the influence of fibres orientation of a laminate layer on the distribution and values of stresses in a multi-layered composite material. Virtual tests were conducted in accordance

* Corresponding author: krzysztof.herbus@polsl.pl

with the recommendations of PN-EN ISO 7438. This standard refers to a three-point test of sample bending. In Figure 1 is shown the accepted scheme of the three-point test of composite samples bending reinforcement with glass fibres. According to the standard, the ratio of support spacing to the height of the tested sample should be equal to 16.

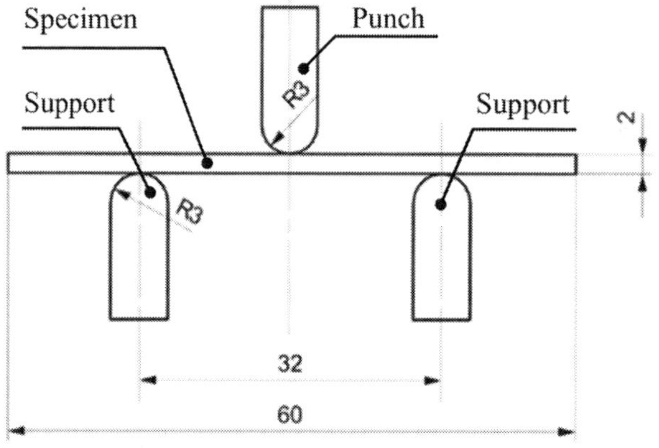

Fig. 1. The scheme of the three-point bending test.

In the accepted scheme of supporting and loading of the tested sample there are two fixed supports with a spacing of 32 mm and a punch that can move only in the vertical axis. The punch acts on the composite sample in the middle of its length. The tested sample has the following dimensions: length 60 mm, height 2 mm, width 10 mm.

2 A structure of the used composite material

The structure of the sample subjected to the cycle of virtual experiments consists of four layers. Each layer is a composition of epoxy resin and fiberglass. The used fiberglass reinforcement is in the form of a fabric. Within the fabric, the fibres of weft and warp are weaved, relative to each other, at the angle of 90° (Figure 2). In the created composite material the main direction of fibres orientation is determined according to the Y axis of the absolute coordinate system. The method of laying the following composite layers was defined, as being consistent with the direction of the Z axis. It was assumed that the layer 1 is the layer that directly contacts with the punch and the layer 4 is the layer that directly comes into contact with the supports. The other two layers are laid between them. Because of the nature of the carried out analyses, an important parameter is the principal angle of alignment of the fibres (θ) within the composition of a composite material. This angle is defined as the angle between the Y axis of the absolute coordinate system and the direction of the fabric weft.

To conduct the numerical tests, using the FEM method, it was implemented the presented structure of composite material in the PLM Siemens NX 10 program (Figure 2). The applied epoxy resin has the following basic properties: mass density 1300 [kg/m^3], Young's modulus 3000 [MPa], Poisson's ratio 0.37. Whereas the used glass fibres have the following basic features: mass density 2540 [kg/m^3], Young's modulus 70000 [MPa], Poisson's ratio 0.2.

 DOI: 10.1051/matecconf/201711204022

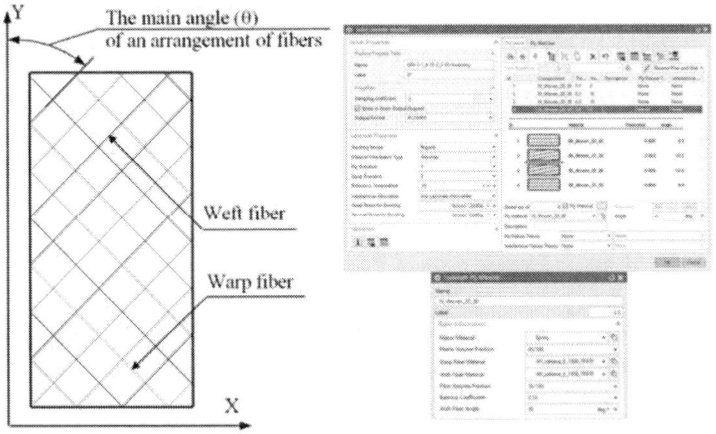

Fig. 2. Definition of the structure of the composite material.

One layer of the composite material has the following properties: matrix material – epoxy, matrix volume fraction – 0.45, warp fiber material – glass fiber, weft fiber material – glass fiber, fiber volume fraction – 0.55, balance coefficient – 0.53, weft fiber angle – 90 degrees, Young's modulus 1 – 24250 [MPa], Young's modulus 2 – 22230 [MPa], Young's modulus 3 – 3000 [MPa], Poisson's ratio 12 – 0.079, Poisson's ratio 13 – 0.32, Poisson's ratio 23 – 0.326, shear modulus 12 – 2329 [MPa], shear modulus 13 – 1118 [MPa], shear modulus 23 – 1443 [MPa], mass density – 1982 [kg/m^3], thickness – 0.5 [mm].

3 FEM analysis of the three-point bending of specimens

To model the form of displacement of the analysed composite sample during the three-point bending test, a model for FEM was prepared. This model includes the following types of boundary conditions: fixed constraint, surface-to-surface contact, enforced displacement constraint.

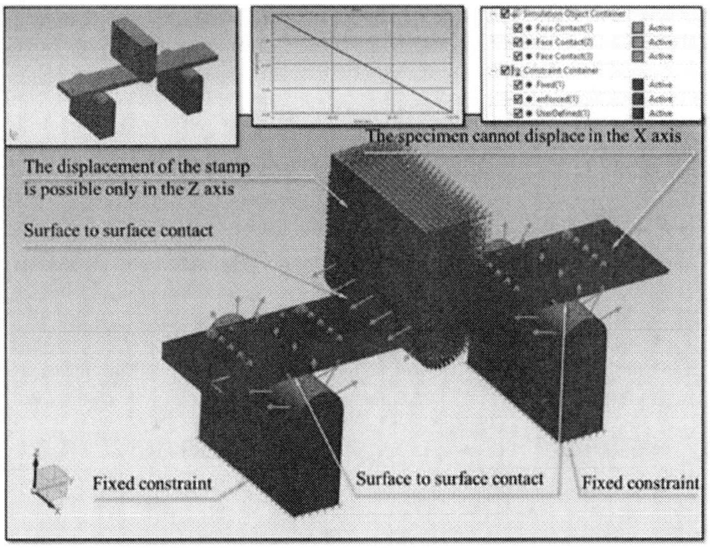

Fig. 3. The model prepared to the FEM analysis.

DOI: 10.1051/matecconf/201711204022

The model also includes an input force in the form of the punch displacement at the speed of 2 mm/min. This displacement was only possible along the Z axis of an absolute coordinate system.

The first series of numerical tests was carried out assuming that all the layers of the composite material composition had the same principal angular position of the fibres relative to the absolute coordinate system. The stress distribution, in this case, changes constantly up to the angle of $\theta = 45°$. When this value is exceeded, the obtained stress distribution is a mirror image with respect to the previous (Figure 4). This is because the impact factor equal to 0.5 is considered.

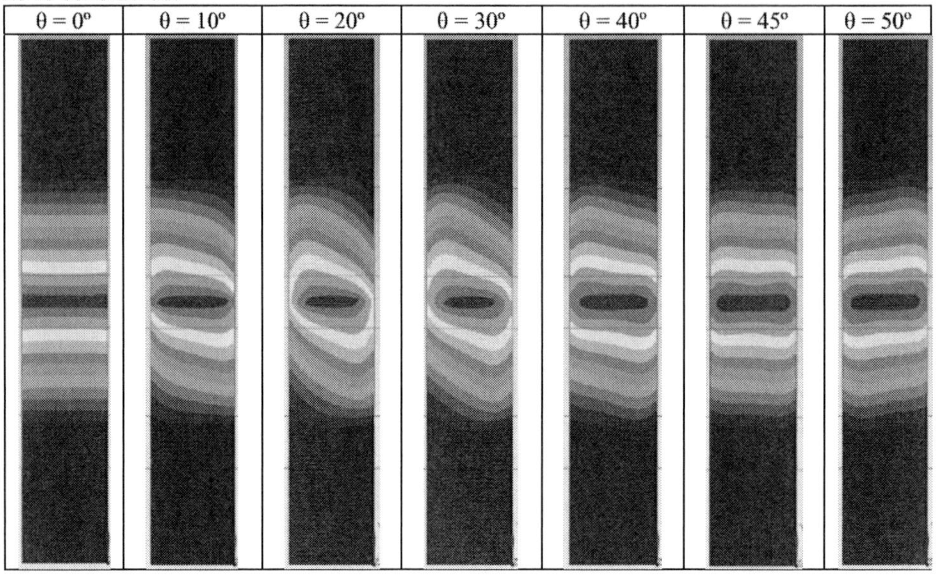

Fig. 4. Distribution of stresses depending on the angle of an arrangement of the ply (ply1)

In the Figure 5 is shown a graph presenting the relationship between the angle θ and the maximum values of reduced stresses in the particular layers of the composite material. It could be seen that as the angle θ increases up to 45°, the maximal values of reduced stresses in all layers are decreasing.

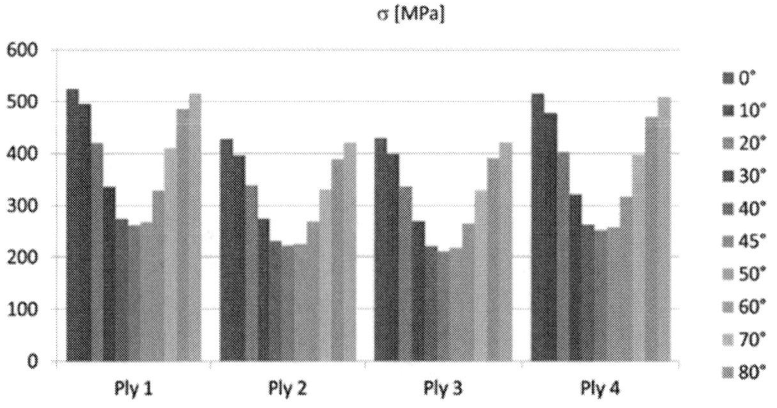

Fig. 5. The values of reduced stresses depending on the angle of an arrangement of the ply (all the plies arranged in the same way).

The next cycle of the numerical tests was conducted with the assumption that the angle θ with respect to the layers 1 and 4 is unchanged, while the angle θ with respect to the layers 2 and 3 varies at the same time between 0 and 90° (Figure 6a).

In this case are observed: a significant decrease of stresses in layers 2 and 3 and an increase in stress values in layers 1 and 4.

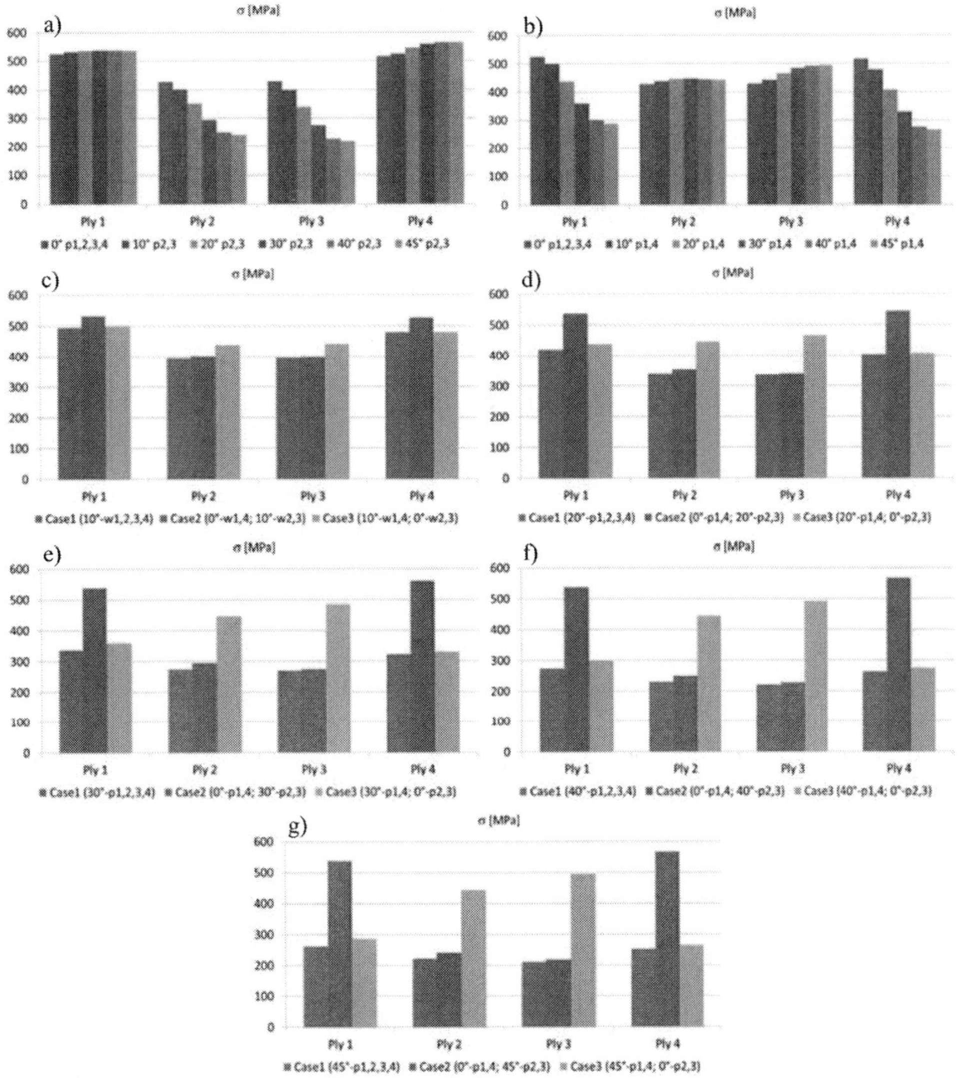

Fig. 6. The values of reduced stresses depending on the angle of an arrangement of the ply (the plies arranged in the differently way).

The last cycle of numerical tests was carried out basing on the assumption that the angle θ with respect to the layers 2 and 3 is unchanged while the angle θ with the respect to the layers 1 and 4 changes simultaneously between 0 and 90° (Figure 6b).

In this case are observed: a large decrease in stress values in layers 1 and 4 and an increase in stress values in layers 2 and 3. In Figures from 6c to 6g are shown summaries of the results obtained in these three cycles of tests to illustrate the effect of the change of the main angle of fibres orientation θ on stress values occurring in particular layers of the composite material.

4 Conclusions

The change of the angle of the fibres orientation causes the change of the stress distribution in particular layers of the composite material. If all layers of a composite material are uniformly oriented (have the same angle), the most uniform stress distribution, in the individual layers of the composite, is obtained. With the increasing difference of the angle of layers orientation in the composite material, could be observed greater differences between the stress values in the individual layers of a composite material.

The created FEM model allows virtual conducting the three-point bending of samples with respect to any fibre-reinforced composite material. It should be stressed, however, that the created model allows determining the relationship between stress values in individual layers of multilayer composite materials and the angle θ, only in terms of its nature or tendency. To determine the exact relationship between the stress values and the angle θ, the FEM model should be fitted to the real object.

References

1. R. Martinsa, L. Reisa, R. Marat-Mendesa, Procedia Structural Integrity, **1**, 066–073 (2016)
2. P. Folega, Journal Of Vibroengineering, **17**, 155-163 (2015)
3. W. Cecot, M. Oleksy, Computers and Mathematics with Applications, **70**, 1391–1400 (2015)
4. W. Banaś, K. Herbuś, G. Kost, A. Nierychlok, P. Ociepka, D. Reclik, Advanced Materials Research, **837**, 537-542 (2014)
5. K. Foit, G. Ćwikła, MATEC Web of Conferences **94**, UNSP 05002 (2017)
6. A. Gwiazda, Advanced Materials Research, **837**, 393-398 (2014)
7. A. Gwiazda, Applied Mechanics and Materials, **474**, 417-422 (2014)
8. K. Herbuś, P. Ociepka, IOP Conf. Series: Materials Science and Engineering, **95**, 012096 (2015)
9. A. Dymarek, T. Dzitkowski, K. Herbuś, G. Kost, P. Ociepka, Advanced Materials Research, **837**, 351-356 (2014)
10. P. Ociepka, K. Herbuś, IOP Conf. Series: Materials Science and Engineering, **95**, 012101 (2015)
11. K. Herbuś, P. Ociepka, IOP Conf. Series: Materials Science and Engineering, **95**, 012084 (2015)
12. P. Ociepka, K. Herbuś, IOP Conf. Series: Materials Science and Engineering, **95**, 012100 (2015)
13. K. Herbuś, P. Ociepka, IOP Conf. Series: Materials Science and Engineering, **145**, 042018 (2016)
14. K. Herbuś, P. Ociepka, IOP Conf. Series: Materials Science and Engineering, **145**, 052010 (2016)
15. A. Buchacz, A. Baier, K. Herbuś, M. Majzner, P. Ociepka, Applied Mechanics & Materials, **809/810**, 944-949 (2015)
16. A. Buchacz, A. Baier, J. Świder, M. Płaczek, A. Wróbel, K. Herbuś, P. Ociepka, W. Banaś, M. Sobek, Ł. Grabowski, M. Majzner, *Analytical and experimental tests and determination of characteristics of components working as assemblies of innovative structures of repaired freight cars* (Wydawnictwo Politechniki Śląskiej, Gliwice 2016)
17. A. Buchacz, A. Baier, K. Herbuś, M. Majzner, P. Ociepka, *Springer Proceedings in Mathematics & Statistics, Dynamical Systems: Modelling*, **181**, 97-105, (2016)

Nano graphite as additive in soybean oil

George Catalin Cristea, *Cosmin* Dima, *Dumitru* Dima, *Constantin* Georgescu, and *Lorena* Deleanu*

Dunarea de Jos University, Galati, Romania

Abstract. Vegetable oil-based lubricants have several disadvantages as compared to mineral and synthetic ones, including low viscosity that not encourage the generation of a continuous film when the tribosystem runs, consequently, implying a mixt or boundary. This is why the additivation of such oils is of great interests for researchers, producers and users. This paper presents results of testing the soybean oil additivated with nano graphite (0.25%wt, 0.50%wt and 1%wt) on a four ball machine. The friction coefficient is slightly increased for the additivated lubricants, but no evident dependency on concentration and test conditions was noticed. Supplementary tests will give the opportunity of a statistical approach of this parameter. Any concentration of nano graphite makes the wear parameter (wear scar diameter) to slightly increase as compared to the values obtained for the non-additivated soybean oil. The increase of the nano graphite concentration in this vegetable oil does not influence significantly the wear parameter, meaning that, at least for the tested regime, this is not an efficient anti-wear additive.

1 Introduction

A SWOT analysis of lubricants based on vegetable oils points out the non-polluting character and the renewable resources for these materials [1, 2] But vegetable oil-based lubricants have several disadvantages as compared to mineral and synthetic ones, including low viscosity that not encourage the generation of a continuous film when the tribosystem runs, consequently, implying a mixt or boundary lubrication. This is why the additivation of such vegetable oils is of great interests for researchers, producers and users.

Tang [3] presented a review of the mechanisms of friction reduction and anti-wear of nanoparticles in lubricants: rolling effect [4], protective film, mending effect and polishing effect.

Shahnazar et al. [5], based on Rudnick works [6, 7], presented a well-documented classification of nano additives in lubricants, those based on carbon being classified in four main allotropic classes: zero-dimensional (fullerene), one-dimensional (nanotubes, nanowires, nanorodes), two-dimensional (graphene), three-dimensional (graphite, nano-sized diamonds). Graphite has been used as micro particles in lubricants but adding it at nano level could enhance the friction and wear performance of the tribosystem.

* Corresponding author: lorena.deleanu@ugal.ro

Hwang [8] concluded that lubricants with nano additives improve the tribological behavior as compared to microaddition of the same base oils. The presence of nanoparticles between the frictional surfaces significantly reduced contact between the frictional plates by playing the effective role of ball bearings. Analyzing the influence of additive shape, the fiber-like nanoparticles agglomerated easier than spherical nanoparticles and thus, the dimensions of agglomerated fiber-like nanoparticles became larger than that of the lubricant film, resulting in an increase in surface roughness and inhibition of the inflow of lubricating oil at the inlet of the frictional surfaces.

Martorana et al. [9] reported that ethanol containing graphite for gear pump proved to be significantly resilience without undergoing scission and degradation during running, but without significantly increasing the viscosity of the lubricant.

Lee [10] studied the graphite as nano additive in a mineral transmission oil and concluded that graphite nanoparticles decreases the metal contact between sliding surfaces by acting as ball bearing spacers.

The four ball test allows for easy comparing the tribological characteristics of a lubricant [11], results given by Solea [12] for several vegetable oil (olive, rapeseed, corn and soybean] allowing for comparing their performance to the mineral and synthetic ones.

This paper aims to report the influence of nano graphite as additive in soybean oil on the tribological behavior of the formulated lubricants.

2 Testing methodology and lubricants

The lubricants formulated with soybean oil additivated and nano graphite in different concentrations (0.25%wt, 0.50 %wt and 1% wt) were tested on a four ball machine in order to evaluate the improvement that this recipe provides for lubricating characterization with the help of parameters as friction coefficient, wear scar diameter. Table 1 presents the typical composition of this oil [12].

The nano graphite was supplied by PlasmaChem and has the following characteristic [13] and was produced by chemical disintegration of graphite. The average particle radius is 400-450 nm.

Table 1. Typical Fat acid composition for the tested soybean oil

Fat acid	C14:0	C16:0	C16:1	C17:0	C18:0	C18:1	C18:2	C18:3	C20:0	C20:1	C22:0
	0.11	12.7	0.13	0.05	5.40	21.60	52.40	5.70	0.25	0.20	0.50

Table 2. Chemical composition of the steel the balls are made of (wt%)

Element	C	Cr	Mn	Si	S	P
Steel grade EN31	1.0	1.3	0.5	0.35	0.05	0.05

The degumming and refining process of soybean oil is done at Prutul SA Galati and prevents oil to form gum deposit and to ferment [14].

The four-ball machine (in the "LubriTest" Laboratory of "Dunărea de Jos" University of Galati, was designed by Schaeffer for a maximum load of 6000 N, in agreement with the new standard SR EN ISO 20623:2004. Petroleum and related products. Determination of the extreme-pressure and anti-wear properties of fluids. Four ball method (European conditions).

The test balls are lime polished, made of chrome alloyed steel balls (Table 2), having 12.7±0.0005 mm in diameter, with 64-66 HRC hardness. The sample oil volume required for each test was 8 ml ±1 ml. The test method for investigating the lubricating capacity was EN ISO 20623:2003 Petroleum and related products - Determination of the extreme-pressure and anti-wear properties of fluids - four ball method.

The test parameters were: three values for speed (1000 rpm, 1400 rpm and 1800 rpm corresponding to the following sliding speeds 0.383 m/s, 0.537 m/s and 0.691 m/s, respectively), three values for loading force (100 N, 200 N and 300 N), testing time 1 hour.

The issue that has to be solved with such anti-wear additive is its dispersion in oil. Thus, the authors, taking into account that the tested base oil is a mix of triglycerides of fat acids (Table 1) propose a method for obtaining a good dispersion. The formulated lubricants were obtained in a small quantity of 200 ml each. The steps followed in this laboratory technology were:

- weighting the additive and the dispersing agent with an accuracy of 0.1 mg,
- mechanical mixing of additive and equal mass of guaiacol (supplied by Fluka Chemica), chemical formula being $C_6H_4(OH)OCH)_3$ (2-methoxyphenol), for 20 minutes; this dispersing agent is compatible with both additive and vegetable oils,
- adding gradually the soybean oil, measured for getting 200 g of lubricant with the desired concentrations of additive,
- stirring with a magnetic homogenizing device for 1 hour,
- sonication + cooling of 200 g lubricant for 5 minutes with the help of sonicator Bandelin HD 3200 (Electronic GmbH & KG Berlin); the lubricants is heating at approximately 70 °C; the cooling time was 1 hour; this step sonication + cooling is repeated 5 times for obtaining a total sonication time of 60 minutes. The parameters of sonicating regime are: power 100 W, frequency 20 kHz ±500 Hz, continuous regime.

3 Results

3.1 Friction coefficient

Being used as additive in oil, solid lubricants, including carbon-based materials as graphite, black carbon, even nano carbon tubes and onion-like graphite [15], considerably reduce friction between surfaces and wear where liquid lubricants do not perform desirably [16, 17]. For the rolling effect, no chemical or mechanical reaction occurs, and the spherical nanoparticles are likely to roll between the friction surfaces and change the pure sliding friction to the mixed sliding–rolling friction.

The analysis of the friction coefficient (COF) takes into account the comment of professor Czichos [18] that distinguished, in general, four steps in the evolution in time of this tribological parameter. Graphics in Figures 2 to 4 are obtained by moving average of 200 successive values. The rate sampling for COF was 2 per second.

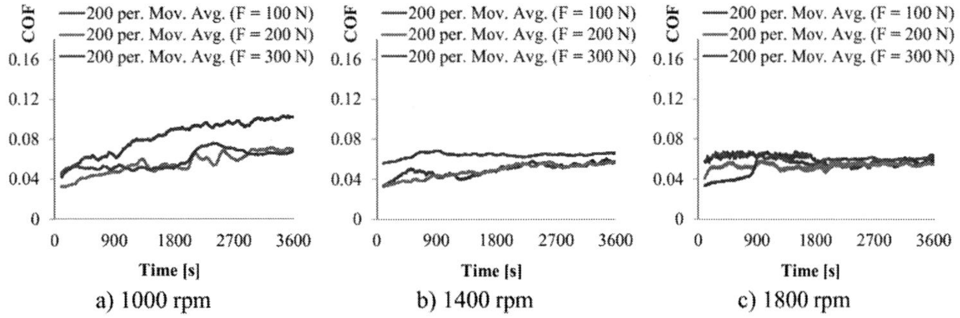

Fig. 1. Friction coefficient versus time for non-additivated soybean oil.

For the non-additivated soybean oil, the friction coefficient is less influenced by loading force at 1800 rpm and an increasing evolution was obtained for the lowest values of speed

DOI: 10.1051/matecconf/201711204023

and force (100 N). Stable period is obvious after several minutes but oscillations are more visible for low force and speed.

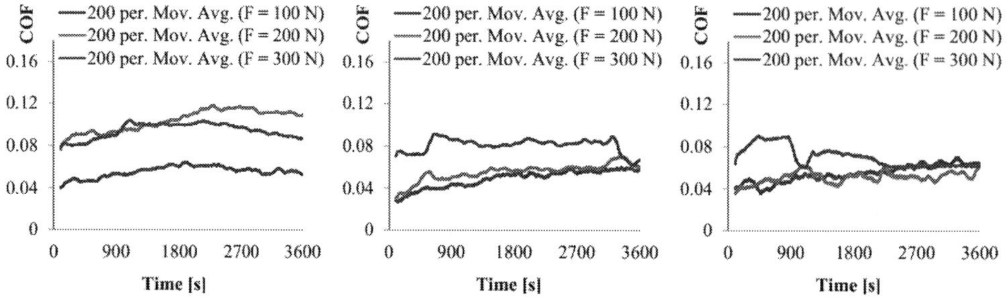

a) 1000 rpm b) 1400 rpm c) 1800 rpm

Fig. 2. Friction coefficient versus time for soybean oil+ 0.25% nano graphite.

Adding 0.25% nano graphite in soyben oil makes the friction coefficient to slightly increase, especially for low speed (Figure 2). For a concentration of 0.5% additive, there were obtained higher values for COF, especially for F=300 N, suggesting a mixt regime and dragging the nanoparticles along the contact (Figure 3).

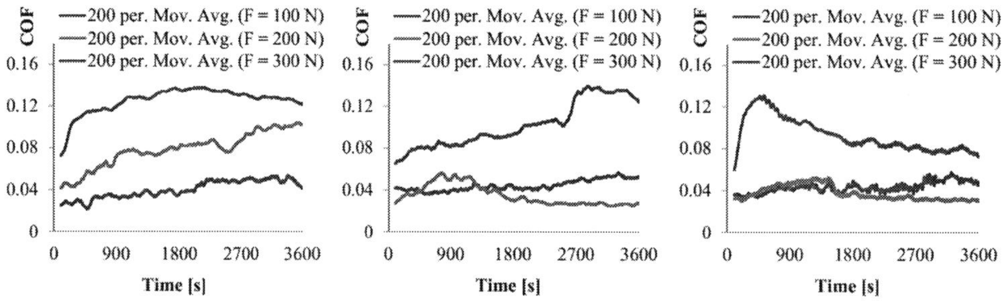

a) 1000 rpm b) 1400 rpm c) 1800 rpm

Fig. 3. Friction coefficient versus time for soybean oil+ 0.5% nano graphite.

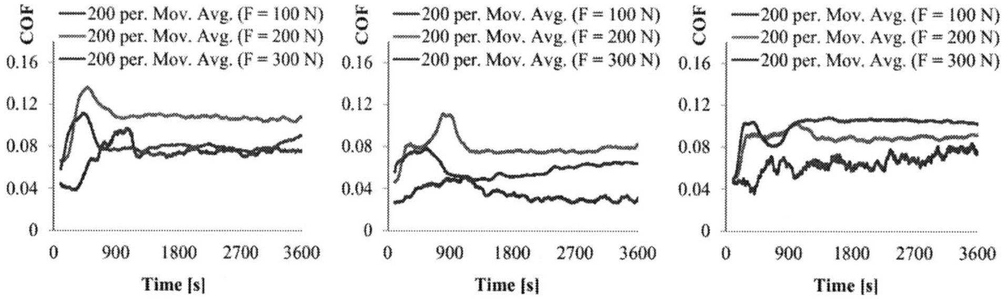

a) 1000 rpm b) 1400 rpm c) 1800 rpm

Fig. 4. Friction coefficient versus time for soybean oil + 1.0% nano graphite.

Except for several values of COF, the average of this parameter has a similar evolution with the additive concentration for both 1 hour and the last 10 minutes of running, meaning a stable evolution of this parameter (Figure 5). Values are spread on larger interval for the concentration of 0.5%wt graphite and are closer for the neat oil and the lubricant with 1%wt.

DOI: 10.1051/matecconf/201711204023

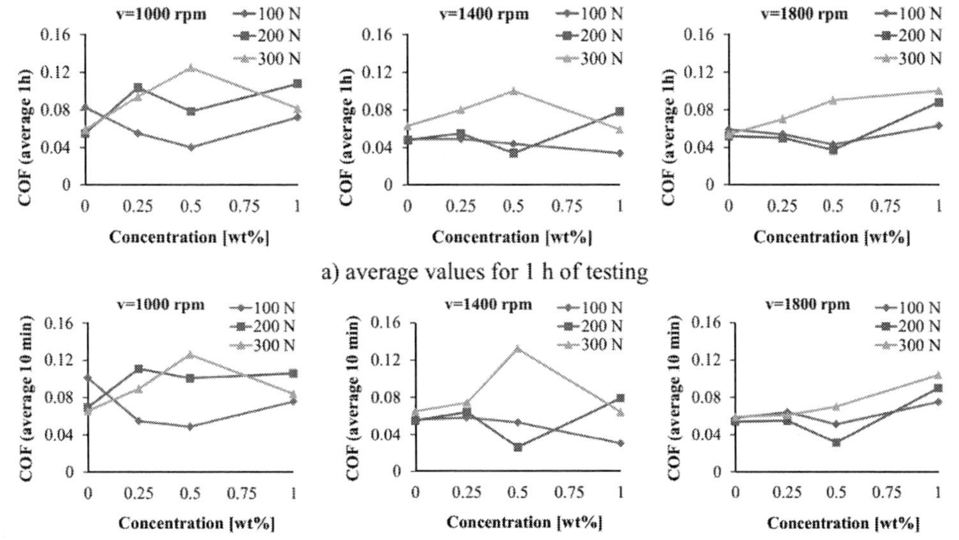

a) average values for 1 h of testing

b) average values for the last 10 minutes of testing

Fig. 5. Influence of nano graphite concentration on the friction coefficient (COF).

3.2 Wear

An optical microscope was used to measure the wear scar diameters (WSD) (an example being given in Figure 6). Each value in Figure 7 is the average of six measurements. Each ball diameter was measured in the sliding direction and perpendicular to it.

Fig. 6. WSD for soybean oil additivated with 1.0%wt nano graphite; test conditions: 100 N, 1000 rpm, 1 hour.

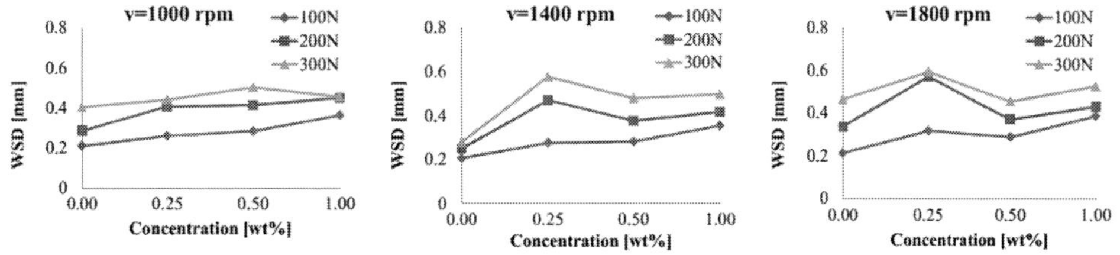

Fig. 7. Influence of additive concentration on the average value of WSD.

Adding nano graphite as additive in soybean oil, it is obvious that the wear is not reduced, but WSD seems to be less influenced by the testing regime. Further increase in additive concentration maintain the values of WSD in a narrow range (Figure 7), but for the investigated range of testing parameters (speed and normal force), there is not clear the tendency.

4 Conclusions

After testing the soybean oil additivated with nano graphite in different concentrations, the authors formulated the following conclusions.

The friction coefficient is slightly increased for the additivated lubricants, but no evident dependency on concentration and test conditions was noticed. Supplementary tests will give the opportunity of a statistical approach of this parameter.

Any concentration of nano graphite makes the wear parameter (wear scar diameter) to slightly increase as compared to the values obtained for the non-additivated soybean oil. The increase of the nano graphite concentration in this vegetable oil does not influence significantly the wear parameter, meaning that, at least for the tested regime, this is not an efficient anti-wear additive.

References

1. L. A. T. Honary, E. Richter, *Biobased lubricants and greases technology and products* (John Wiley & Sons, 2011)
2. J. C. J. Bart, E. Gucciardi, S. Cavallaro, *Biolubricants*, (Woodhead Publishing Ltd, 2013)
3. Z. Tang, S. Li, Curr Opin Solid State Mater Sci, **18**(3), 20 (2014)
4. Y. Y. Wu, W. C. Tsui, T. C. Liu, Wear, **262** (2007)
5. S. Shahnazar, S. Bagheri, S. B. A. Hamid, Intern. J. of Hydrogen Energy, **41**, 17 (2016)
6. L. R. Rudnick, *Natural oils as lubricants* (CRC/Taylor & Francis Group, 2006)
7. L.R. Rudnik, *Lubricant Additives. Chemistry and Applications* (Taylor&Francis, 2009)
8. Y. Hwang, C. Lee, Y. Choi, S. Cheong, D. Kim, K. Lee et al., J Mech Sci Technol, **25**(11), 5 (2011)
9. P. Martorana, I.S. Bayer, A. Steele, Industrial Eng Chem Res **49**(22), 6 (2010)
10. J. Lee, S. Cho, Y. Hwang, C. Lee, S. H. Kim, Tribology Letters, **28**(2) 6 (2007)
11. C. Spanu, M. Ripa, I. Stefanescu, L. Deleanu, The Annals of "Dunarea de Jos" University of Galati, Fascicle VIII, Tribology, **5** (2007)
12. L. C. Solea, PhD thesis, "Dunarea de Jos" University, Galati, (2013)
13. *** Nano materilas and related product, 1st ed. Catalogue PlasmaChem (2016)
14. E. M. J. Deffense, *From organic chemistry to fat and oil chemistry* (Oléagineux Corps Gras Lipides/OCL, 2009)
15. L. Joly-Pottuz, B. Vacher, N. Ohmae, J. M. Martin, T. Epicier, Tribol Lett, **30(1)** 12 (2008)
16. P. Nagendramma, S. Kaul, Renewable and Sustainable Energy Reviews, **16**, 11 (2012).
17. M. Gulzar, H. H. Masjuki, M. A. Kalam, M. Varman, N. W. M. Zulkifli, R. A. Mufti, R. Zahid, J Nanopart Res **18** (2016)
18. H. Czikos, *Tribology – A System Approach to the Science and Technology of Friction, Lubrication and Wear* (Elsevier, 1978)

Production of recyclates – compared with virgin Plastics – a LCA Study

Birgit Kjærside Storm*

Aalborg University Esbjerg, Niels Bohrs Vej 8, 6700 Esbjerg, Denmark
Plastix A/S, Gammel Landevej 3, 7620 Lemvig, Denmark

Abstract. Plastix A/S is a Danish cleantech company transforming discarded fishing trawls and nets into valuable green raw materials. Plastix' technology and processes solve a maritime waste problem and contribute to a more circular green economy and reduce landfilling, marine pollution, CO_2 emissions and especially loss of valuable resources. Plastix' recycling technology enables recovery of discarded fishing trawls and nets via mechanical and thermal processes transforming the waste into valuable recycles which can be converted into plastic products replacing virgin raw materials. The performance has been proved through a Life Cycle Assessment (LCA) study. The results from the LCA study are compared with the production of virgin materials. The results of the LCA show that especially the carbon footprint is remarkable better for Oceanix than for virgin plastics. Oceanix HDPE is 5 times better than virgin HDPE, when talking about the carbon foot print, and the results for Oceanix PP and Oceanix PA6 are 5 times and 20 times better compared with virgin PP and PA6. Also other environmental indicators are better for Oceanix compared with virgin plastics.

1 Introduction

Plastix A/S is a Danish cleantech company transforming discarded fishing trawls and nets into valuable green raw materials. Plastix' technology and processes solve a maritime waste problem and contribute to a more circular green economy and reduce landfilling, marine pollution, CO_2 emissions and especially loss of valuable resources. Plastix' recycling technology enables recovery of discarded fishing trawls and nets via mechanical and thermal processes transforming the waste into valuable recycles which can be converted into plastic products replacing virgin raw materials.
The materials collected and recycled in Plastix A/S are mainly high density polyethylene, HDPE, polypropylene, PP, and polyamide 6, PA6. They are the main part of the materials used for fishing trawls and nets used in the fishing industry. Plastix' cleantech recycling processes are as follows showing the four steps in the plant: 1. Gate control and storage; 2. Rough cleaning and cutting; 3. Cleaning and final separation; 4. Extrusion

After the extrusion all batched are controlled in the laboratory for measuring the main properties of the produced granulates.

* Corresponding author: bkstorm@bbsyd.dk

The input materials – the fishing nets and trawls – are collected from different ports in Europe. Plastix A/S wants an input with as low part of other types of waste. For obtaining a good quality of fishing gears a point system for ports and other suppliers has been developed. They can get a certificate for the quality. The higher quality the better certificate the supplier will receive. In Figure 1 an example of the certificates is shown. The certificates can be that one shown in Figure 1, or it can be a bronze, silver or a golden certificate.

In the gate control the received materials are weighed and visually inspected for the quality. If there are too much other things in the container than fishing gears, the container can be sent back to the supplier or the prize for the materials can be reduced to that Plastix A/S shall have a prize for taking the material instead of paying something for the "raw" material. The container receive a number, so it after it possible during the production to find back to input material.

Fig. 1. An example of a certificate for a supplier to Plastix A/S.

The raw materials are stored in the grey zone. The grey zone is the place, where the input materials are received and roughly sorted up. The grey zone is placed outdoors. In the grey zone the net and trawls are sorted in different materials. Other parts - as for example ropes and floaters - are cut off and stored for being either sold or also recycled. Totally Plastix A/S has a very small amount of waste, because all types of materials are regarded as a resource.

From the grey zone the materials are sent to the green zone, where it will be cut into pieces so most of the knots are cut and the materials are in pieces, so a washing of it can be made. Quite a big part of the sand in the fishing gears are removed in the grey zone as well. In the green zone it is a problem for the cutting equipment that there is sand left in the fishing gear, because it causes wear on the knives. The polyamide is a quite tough material, and it itself is a challenge to cut. That combined with the sand gives wear on the knives. Plastix A/S has its own mechanical workshop to develop and to maintain the machinery.

From the grey zone the materials are transferred to the blue zone, where it is washed and sorted out in different fractions, so only a minor part of polyamide will go into the polyethylene fraction, and a minor part of polyethylene will go into the polyamide fraction. A minor part of PP in HDPE and HDPE in PP can be accepted, because those materials are compatible if only a small amount of contamination part is present. In the blue zone the washed materials are dried, too, so dry materials are sent to the white zone, which is the extrusion zone. Here the cut, washed and dried materials will be made to granulate or pellets

Fig. 2. Raw materials of Polyethylene and granulates of polyethylene. (Photo made at Plastix A/S)

and stored in big bags before being delivered to the customer. In Figure 2 granulates can be seen together with a piece of the raw materials. The granules and the raw materials shown in the figure are polyethylene.

A laboratory control of all inputs and outputs is made. The input is control for the type of polymer. During the production the content of sand and water in the washed and dried material are measured before sending it to the extruder.

The output is controlled for strength, flexibility, density and melt flow index. All those results are used for the technical datasheets for the recycles.

Plastix A/S has decided to call the recyclates made from fishing gears for Oceanix, which then become Oceanix PA, Oceanix HDPE and Oceanix PP.

For improving that it is a clean tech product Plastix A/S makes a life cycle assessment has been made for the products.

2 Life Cycle Assessment

Life Cycle Assessment, LCA in the following, is a tool for systematic evaluation of the environmental aspects of a product or service system through all stages of its life cycle. LCA provides an adequate instrument for environmental decision support. Reliable LCA performance is crucial to achieve a life cycle economy. The International Organization for Standardization (ISO) has standardized this framework within the series ISO 14040 on LCA.

For making a Life Cycle Assessment ISO 14040 and ISO 14044 is used. A LCA is often called a "cradle-to-grave" or a "cradle-to-cradle" analysis depending on, how the analysis is made. A "cradle-to-gate" analysis can be made as well. In this case the analysis is called a "grave-to-cradle" analysis, because Plastix A/S use discarded materials as raw materials and convert it to a new raw material.

For evaluation of the LCA a functional unit needs to be decided for comparison of the results as well as system boundaries need to be defined. The stages in LCA are:

1. Raw material stage
2. Manufacturing stage
3. Use stage
4. Waste stages – or end of use stage
5. Transport stage.

For all stages the same functional unit needs to be regarded. The system boundaries are set up, so they will give the true picture – or a picture as true as possible – of the product regarded [1, 2].

3 Life Cycle Assessment for Oceanix recyclates

All the calculations in the LCA for Oceanix have been performed by Provice by Tomas Sander Poulsen [3]. The LCA has been performed as a part of a project co-funded by Eco-innovation Initiative of the European Union, Retrawl.

According to the standards mentioned above the functional unit in the study is defined as 1 ton of produced plastic granulates output of the three materials PA6, HDPE and PP.

The LCA calculations in the report are conducted by Simapro calculation tool and the Ecoinvent database [4, 5].

By defining the unit as stated above, a closed loop assessment of discarded trawl and net produced from recycled or virgin plastic can also be conducted.

In this case a comparison study of discarded materials and virgin materials are compared. For making that possible the used model is a "cradle-to-gate" analysis is performed for both type of materials, even that the right analysis for the recycles should be a "grave-to-cradle" analysis, but that would give a wrong picture for comparison with virgin materials.

The trawl and net can be delivered from different harbors situated 20 km away or 1500 km away from Plastix A/S manufacturing plant. In the report a distance for the transport and the transport system has been 1250 km, where 1000 km are covered with ship and 250 km with truck. In Figure 3 the system boundaries for the data collection are shown.

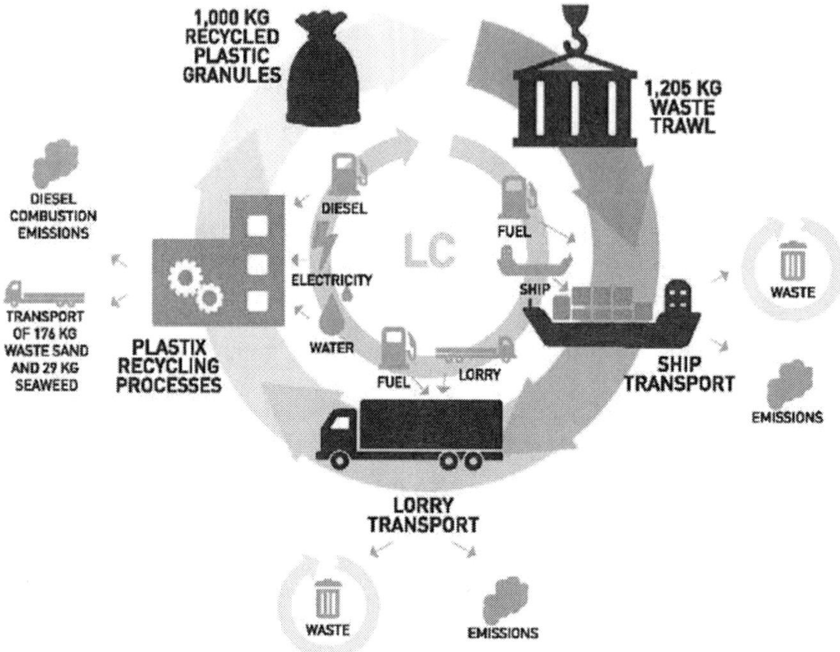

Fig. 3. System boundary of data collection for Plastix recycling scenario including the mass balance [3].

As illustrated in the above shown Figure 3, the process flow of the recycling scenario of PA6, PE and PP has a life cycle consisting of the following steps:
1. Transport of discarded waste to Plastix A/S manufacturing plant
2. The recycling processes at Plastix A/S to final recycles.

Plastix' cleantech recycling processes are as follows showing the four steps in the plant:
1. Gate control and storage
2. Rough cleaning and cutting
3. Cleaning and final separation
4. Extrusion

In the LCA it has been used that 1 ton of recycled plastic granulates will need 1,075 kg discarded net and trawl, because it is assumed that there will be sand and organic matter corresponding to approximately 75 kg per 1000 plastic. This number is set as an average part, because as mentioned above there are different amounts of sand and other materials in the received raw material.

The processes for cleaning, cutting, washing, separation, drying and extrusion use energy. The amount of energy used per ton plastic depends on the sort of plastic. Internal transport is also an energy consuming process and is calculated as well. In Figure 3 all

different contribution to energy consumption are listed together with the emissions to the environment and waste from the production.
In the production washing water is cleaned and recirculated. Those processes are calculated in the LCA.

All data related to the Life Cycle Inventory of the virgin plastic production are collected from the Ecoinvent database [5], and no site specific data have been added. Figure 4 shows the system boundaries of data collection and process flow for the virgin product scenario.

The LCA for the virgin products use non-renewable resources as naphtha and benzene for PE and PP and for PA6, caprolactame made from crude oil is used as raw material. The product system for making virgin plastic consists of the sub-processes:

1. Mining/production of raw materials
2. Refining of raw materials (producing monomers and other raw materials)
3. Additional processing
4. Polymerization
5. Transport

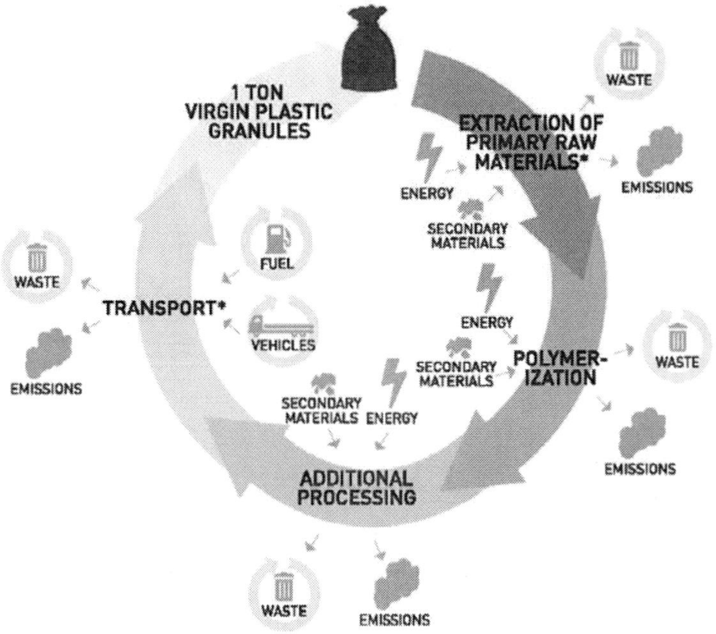

Fig. 4. System boundaries of the data collection and process flow for the virgin plastic production scenario [3].

The LCA calculations in the report are conducted by Simapro [4] calculation tool and the Ecoinvent database [5].

The impact categories included in the LCA are a shortlist of environmental impacts calculated in SimaPro. The impact categories are: Climate Change, Terrestrial acidification, Marine and fresh water eutrophication, photochemical oxidant formation, Particulate matter formation, Marine eco toxicity, Fossil depletion.

Climate change is mainly calculated as the contribution of CO_2 from the processes and is illustrated by the carbon foot prints in the processes. The CO_2 is calculated because it has a global warming potential, GWP. For CO_2 the potential is set to 1. Other matters have a higher GWP, but none of them are significant here.

Each single process emits different substances and they may or may not contribute to a category.

The environmental impacts of the processes at Plastix A/S are mainly related to electricity consumption, water consumption, landfill avoidance and transport activities. The results show that the recycling processes at Plastix A/S outperforms the virgin plastic scenario. The main part of the contributions is from the electricity consumption at Plastix A/S, more specifically the part of non-renewable coal energy sources (23%) that supply the electricity production. The electricity consumption of the recycling process is small compared to the energy consumption from virgin plastic processes. Even with these impacts from electricity use, the Plastix A/S process outperforms the virgin plastic scenario. If the electricity used by Plastix A/S came from renewable energy sources, the impacts from the recycling process would be close to none, but until now that's not the case.

Energy use is the largest input in the Plastix A/S recycling process. A comparison only on energy can be made, since PlasticsEurope [6-8] reports inform of the energy required to make the different virgin plastics.

Table 1 the different energy consumption levels are illustrated.

Table 1. Energy consumption per ton produced virgin plastic and Plastix plastic (Oceanix). Plastix total energy consumption is only 5 – 11% of virgin plastic energy consumption [3].

	Virgin PA	**Plastix PA**	**Virgin PE**	**Plastix PE**	**Virgin PP**	**Plastix PP**
Process energy Consumption per ton (kWh)	25,167	1,257	9,444	922	8,750	976
Plastix energy consumption in relation to virgin energy consumption %		5,0%		9,8%		11,2%

All the results of the calculations of the environmental impacts for as well Plastix A/S materials (Oceanix) as for the virgin plastics are collected in Figure 5. All results are normalized to a European citizen's person equivalent. Figure 5 determines that the Plastix A/S impacts are significantly lower than all the virgin plastic scenario in all the environmental categories except fresh water eutrophication and marine eco toxicity. Both impacts are related to indirect emissions from consumption of non-renewable coal-based energy.

The results of the LCA for Plastix A/S recycled plastics (Oceanix) and the comparison between the virgin plastics are shown in Figure 5.

Even when the results have been calculated for a worst case scenario for Plastix' products, the LCA shows better results than for virgin materials. The worst case scenario has been calculated for

- amount of input sand transported with waste trawl is 15% instead of 5%, handled in process and disposed toll landfill;
- transport scenario is 650 km by truck and 600 km by ship instead of 250 km by truck and 1000 km by ship.

The worst case scenario shows still reduction in the carbon footprint for being 17 times better for the recycled PA6 than for virgin PA6, and 4 times better for recycled PE and PP compared with virgin PE and PP.

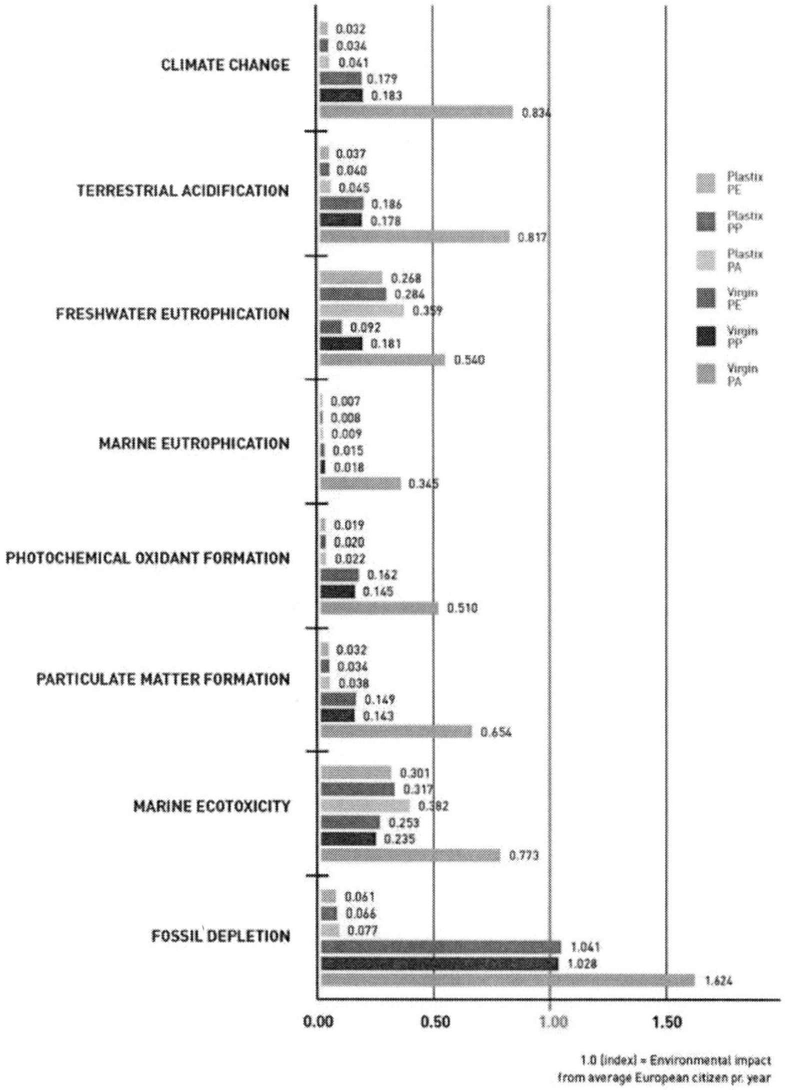

Fig. 5. LCA results of environmental impacts from one tones produced plastics with Plastix recycling technology and virgin plastic production (normalized to a European citizens person equivalent) [3].

Fig. 6. Carbon footprint for Plastix A/S recycled plastic, Oceanix, compared with virgin plastics [4].

The reduction in the carbon footprint, which has been regarded as being important for climate change considerations are as shown in Figure 6. The reduction in carbon footprint for PA6 is 20 times compared with virgin PA6, and 5 times for PE and PP compared with virgin PE and PP.

4 Conclusion

Plastix A/S use discarded fishing nets and trawls transforming tem to new raw materials for use in the plastic industry. The discarded nets and trawls are treated through a separation process, a cutting process, a washing process and an extrusion process to granulate called Oceanix. The polymers in the nets and trawls are HDPE and PA6 and in ropes PP.

The LCA results illustrates that the environmental impacts and here especially the carbon footprint also regarded as climate change of Plastix' cleantech technology and processes for recycling of discarded trawls and nets are substantially better than the production of virgin plastic of the same types.

With regard to fossil depletion the Plastix A/S recycling scenario performs by a factor 17 better than virgin PE, a factor 16 better than virgin PP and a factor 21 better than virgin PA.

With regard to the carbon footprint the recycling scenario performs by a factor 5.7 for PE, 5.5 for PP and 20.5 for PA compared with virgin PE, PP and PA.

References

1. ISO 14040, LCA principles and framework (2016)
2. ISO 14044, LCA requirements and guidelines (2016)
3. Plastix and Provice; Life Cycle Assessment of The Retrawl Project (2016)
4. Simapro LCA calculation tool version 8.2
5. Ecoinvent LCA database version 3.2
6. Plastic Europe: Ecoprofiles and Environmental Product Declarations Report, on PA6 (2014)
7. Plastic Europe: Ecoprofiles and Environmental Product Declarations Report, on PE (2014)
8. Plastic Europe: Ecoprofiles and Environmental Product Declarations Report, on PP (2014)

Possibilities and applications of smart textiles

Elena Rotari [1,*], and *Corina* Negara[2]

[1]Alecu Russo Balti State University, Faculty of Exact, Economic and Environmental Sciences, Department of Physical and Engineering Sciences;
[2]Alecu Russo Balti State University, Faculty of Exact, Economic and Environmental Sciences, Department of Mathematics and Informatics

Abstract. In the process of creating of new clothing products, the designers choose textiles which have: appearance, comfort, durability, shape retention, protection from bad weather, etc. These aspects not fully satisfy the new factors, such automatic regulation of body temperature; signs of heart attack; fever and others. Smart textiles possess such advanced properties. The functionality of smart textiles consists in informing, protecting and relaxing the wearer. In this research, we approached and revealed the application of e-textile materials and their importance in clothing. The research methodology consists in the efficacy of applying Cu wires to the fabric and the result obtained. The results obtained are positive and they are revealed in the research.

1 Introduction

With the development of personal computers (PCs), in the mid-70's together with the advent of the Internet (in 1989), an unprecedented technological explosion was recorded, with major implications in all areas of human activity. Everyday life, current requirements call for the use of phones, PDAs, technical means of which people cannot be deprived. An applicability of informational sources constitutes also the intelligent textiles: dresses, shirts, jackets and girdles, all being obtained by weaving high conductivity yarns together with cotton, nylon or polyester yarns [2]. The metal wires connect the sensors to the microcontrollers inserted in the garments and the special programs control the communications in the semi-fabric network.

To begin with, it is necessary and reasonable to give an explanation of the term „e-textile", which is, in fact, a combination of high tech apparel and fashionable apparel.

The notion of „e-textile" [1] was introduced in the late 90's in order to disseminate the materials that change their properties under the action of certain stimuli, where processes take place by introducing substances with specific properties (detection, processing, updating, etc.) in materials of different textile-types, plastic, metal or metal alloys, building constituents, etc.

The three processes (detection-processing-updating) are embedded separately in intelligent materials and must be joined by means of an automatic control system.

* Corresponding author: rotarielena81@yahoo.com

The concept of intelligent material derives from the intelligent forms of natural systems (fabrics/materials), that is, living organisms. As a result, Intelligent Materials are designed as a material that performs natural functions of sensing, actuation, control and intelligence. Intelligent materials ("smart / intelligent materials") have the ability to adapt themselves to external stimuli.

The use of the smart textile were a preoccupation for the researchers and some of the research results were published over the years.

Chan Vili studied the use of shape memory materials in developing high performance smart textiles, taking into consideration ways for enhancing aesthetics of woven interior textiles [2]. Such materials are capable of monitoring the temperature change and can react by the textiles shape.

Mondal showed that the use of phase change materials could be considered as advantageous in using the storing or releasing the latent heat, when such materials are incorporated in thermos-regulated smart textiles [3] by coating or encapsulation. He analyzed the use of some recent concepts and applications in the field of smart textiles and their extension in the market.

Diaconescu et al. described an experimental study aiming to measure and display the temperature by means of a commercial sensor [4]. An electrical scheme able to incorporate the sensor was designed and applied. The experimental study proved the possibilities of using such a solution to highlight the temperature inside and outside.

Massaroni et al. took into the use of fiber optic sensors incorporated in smart textiles to continuously monitor some physiological parameters [5]. They found that there are fields of application where the so called standard sensors cannot be employed. The smart textiles could be applied in respiratory and heart beat monitoring, offering good metrological performances.

The research presented in this paper aimed to present some results of using certain sensors in generating and analyzing the behavior of smart textiles.

2 Methodology of experimental investigation

Smart textile materials are made of intelligent fibers and yarns, using nanotechnologies, sensors, processors, devices for various usages: military equipment, medical articles (for determining the health condition, prophylaxis and treatment) protection equipment (for firefighters, workers operating in particular conditions, etc.), sports equipment (mountaineering, swimming, athletics, etc.), entertainment (apparel products with built-in music devices), household and industrial appliances, etc. [6].

Depending on the type of reaction to stimuli, smart textile materials [7] can be: passive (only sensing the external stimuli), active (sensing and responding to stimuli) and very smart (sensing, reacting and adapting to stimuli).

The concept of "smart materials" is based on outstanding properties in terms of durability, comfort adjustment, thermal protection, shape maintenance, scratch restoration/mending, reduced weight, low production costs, etc.

Other types of „e-textile" materials [4] are also electronic textiles, they combine the finishing processes of air and vapor permeable fabrics with the insertion of metallic materials.

The adding of nickel, copper or silver as a superficial layer having different thicknesses in the textile material gives it special physical and electrical properties, which can have many applications.

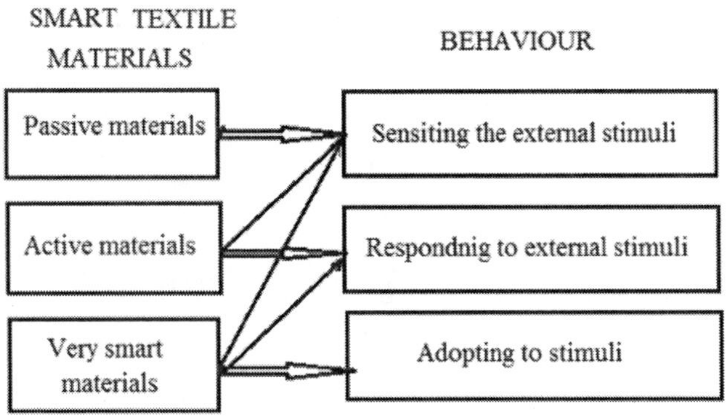

Fig.1. Classification of smart materials considering their behavior.

For example, the thermal conductivity of fabrics increases thousands of times in consort with the metal fiber, while ordinary fabrics offer a minimal isotherm.

Another category of conductive materials is represented by electronic textiles that allow the passage of electricity through them or incorporated portable devices, microphones, sensors, counters or integrated circuits.

Electronic apparel products [1] have a new and revolutionary property, namely sharing of information.

Clothes are now capable of recording, analyzing, storing, sending and displaying information, having many applications: in medicine, for example, smart bandages that monitor the health parameters of injuries, security-detection of danger and help requesting, sport-monitoring of parameters.

Smart textiles will only be successful in the long run if the sensors and their associated components are fully integrated into the product.

This represents a great challenge because, besides the technical aspects, concepts, materials, structures and treatments, we must primarily take into consideration the suitability of the product to the field of use in or as a textile material [8].

The technique of weaving and knitting of copper yarns inside the thread makes their incorporation into the suit lead to a skin particular contact so that the signals can be directly taken over, without the need for the use of the electrode.

Similarly, the girdle is considered comfortable, inoffensive, safe, wearable and having maintenance (washing) possibilities. Figure 2 shows a portion of the girdle, facilitating the visibility of the position of copper wires, the interconnections and the signal transmitting antenna.

Further down we present the proposed scheme for the experiment (Figure 2).

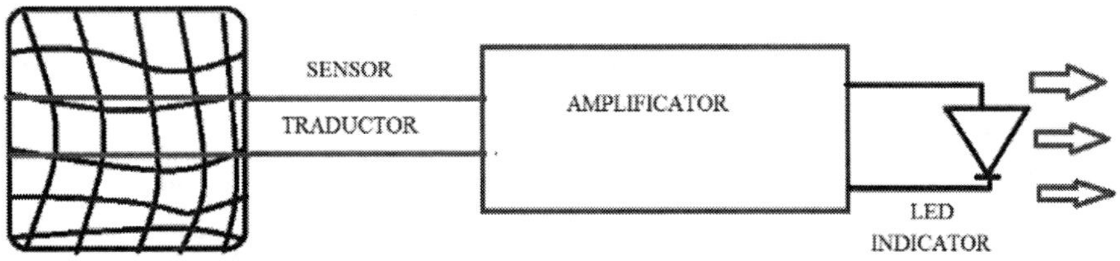

Fig. 2. Electrical operating diagram.

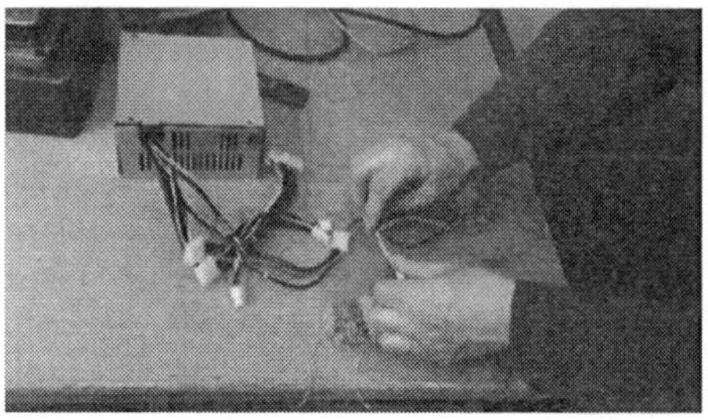

Fig. 3. Experimental method of sensor verification.

The girdle was made of natural fibers, having Lycra elastomeric yarns in their composition, copper wires with a diameter of 0.3...0.5 mm which are placed 2 cm one from another, because located at smaller distances, they are sensible and react even at the touch of the hand.

3 Results of experimental investigations

The permeability to vapors. In order to assess the vapor transmission through diffusion, direct permeability indicators are used Pv [g] and vapor permeability coefficient μ [g/m²h]. As an indirect indicator, the resistance to vapor transmission Rv [mm.m2.h/g] is used for calculations. The determinations are carried out according to STAS 9005-79.

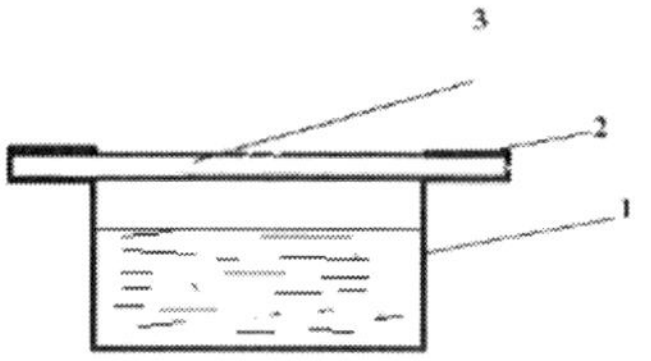

Fig.4. Herfeld glass with textile.

The method used to obtain the data needed to determine these indicators is the Herfeld glass method (Figure 4). According to this method, in a Herfeld glass (1), 50 ml of distilled water are placed and the top of the glass is covered with the test material (3) and it is fixed with a ring 2. This assembly formed of the glass is weight, water and textile material (M_i Mass), and then this assembly is introduced into an environment with relative humidity $\varphi = 0\%$ (desiccator), maintaining a certain time interval (t), after which it is rewound (M_f mass). The passage of water vapor is made, therefore, from the environment with high relative humidity to the environment with relatively low humidity (from the Herfeld glass to the environment).

We made a piece of girdle using the weaving method and the other one using the knitting method. The copper threads with a diameter of 0.3 ... 0.5 mm were placed together with the warp yarn having the same diameter at a distance of 20 mm between the rows. The free ends have been connected to the 4.5 V power supply. The section of the girdle works and is considered e-material, because it reacts to humidity and the result is indicated by a LED sensor. The sensor response time starts with 80%...100% with the increase of humidity above 80 %.

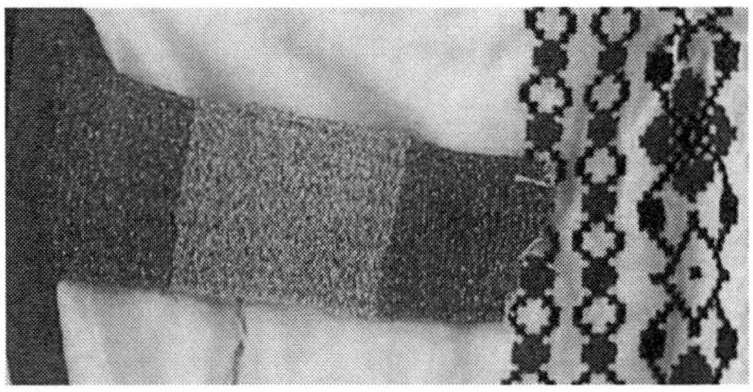

Fig. 3. Woven girdle pattern.

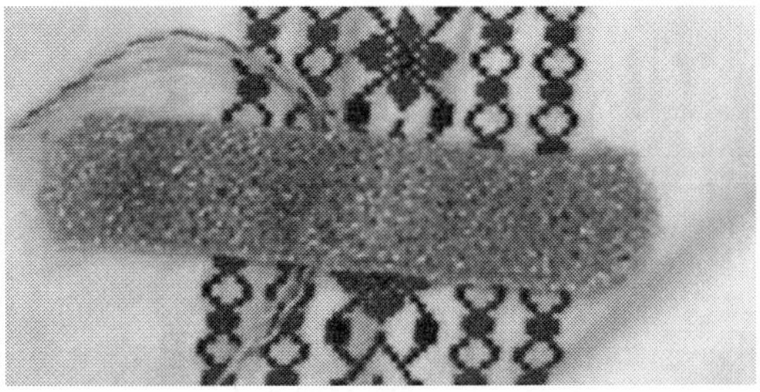

Fig. 4. The girdle pattern made using the knitting method.

a) b)

Fig. 5. Presentation of e-textiles on the blouse surface: a – obtained by the knitting method; b – obtained by the weaving method.

The conductive materials combine the finishing processes of the air and vapor permeable fabrics with the insertion of metallic materials.

The addition of nickel, copper, silver as superficial layers having thicknesses in the textile material, gives it special physical and electrical properties, which can have many applications.

Permeability Pv [g/ m²h mbar] to water vapors indicates the amount of water in grams, in the form of vapor, which passes within one hour through the surface of 1 m² of the analyzed textile material, at a partial pressure gradient of 1 mbar, oriented in the normal direction of the surface of the textile sample.

By means of the Herfeld glass method, the mass difference (M_i - M_f) the permeability to vapors is represented Pv [g]. If this mass difference is related to the initial mass M_i vapor permeability is obtained expressed as a percentage [%].

The relation (1) is applied for the calculation of vapor permeability at the passage of vapors P_v and the relations (2) and (3) are used to determine the vaporization coefficient μ that the resistance to the passage of vapors R_v.

$$Pv = M_i - M_f \; [g] \tag{1}$$

$$\mu = Pv/S.t \ [g/m^2h] \tag{2}$$

$$R_v = d/\mu \ [mm.\ h.\ m^2/g] \tag{3}$$

Where:
S $[m^2]$ - the vaporization surface;
t[h] - the residence time of the Herfeld glass, with water and fabric in the desiccator.
d [mm] - the thickness of the textile material.
μ $[g/m^2h]$ - the vaporization coefficient;
R_v $[mm.\ h.\ m^2/g]$- resistance to vapor transmission.

4 Conclusions

For the textile industry, the potential offered by combining the textile technology domain with the information technology one is immense, offering new valences to apparel products, both to normal and professional ones.

Smart clothes will certainly become the rule and not the exception in everyday life. The problem lies in the correct choice of the product market.

For the textile industry, the potential offered by the combination of these two areas of maximum and strict necessity is immense. Starting with medical products, going through the range of protective clothing and ending with interior design objects, we can say that the future is open to smart textile products, whatever they may be.

Over the years, the smart textiles materials were a subject of preoccupation for the researchers from different countries.

Taking into consideration the information found in the specialty literature, some proposals of using e-textiles obtained by the knitting method and by weaving method, respectively, were formulated.

In the future, there is the intention to investigate the behavior of this e-textiles in various experimental conditions.

References

1. European's technological platform for future industry of textiles and garments. A prospection for 2020. www.research.ro/img/.../1156747771textile%20 viziune.doc/
2. Y. Y. F. Chan Vili, Textile Research Journal, **77** (5) (2007)
3. S. Mondal, Applied Thermal Engineering **28** (2008)
4. V. D. Diaconescu, L. Vornicu-Albu, R. Diaconescu, I. Racea, A. Grigoriu, Buletinul AGIR **3** (2015)
5. C. Massaroni, P. Saccomandi and E. Schena, J. Funct. Biomater 6 (2015)
6. C.A. Roberts, Intelligent material systems – The dawn of a new materials age, Journal of Intelligent Material Systems and Structures, 4 (1993)
7. M.F. Ashby, Materials Selection in Mechanical Design (Butterworth-Heinemann, Oxford, 1995)
8. A. Yoshihito, Information processing using intelligent materials –Information-processing architectures for materials processors, J. Intell. Mater. Syst. and Struct., 5-May, 418-423 (1994)

MATEC Web of Conferences **112**, 05001 (2017) DOI: 10.1051/matecconf/201711205001
IManE&E 2017

The estimation with Artificial Neural Networks of some quality parameters for the surfaces processed by superfinishing

Daniel-Constantin Anghel[1,*], and *Alexandru* Ene[2]

[1]University of Pitesti, Department of Manufacturing and Industrial Management, Tg. din Vale Str., no.1, Pitesti, Romania Romania
[2]University of Pitesti, Department of Electronics, Computers, Communications and Electrical Engineering, Tg. din Vale Str., no.1, Pitesti, Romania

Abstract. This paper presents a study on the quality parameters obtained by superfinishing. The quality is characterized by the roughness. They are dependent on the following process parameters: circular feed, the contact pressure between the piece and the tool, frequency of oscillation of the tool, the coverage degree between the tool and the piece and the basic time. Because the dependence between inputs and outputs is a nonlinear one, in this paper we used an artificial feed forward neural network (ANN). The ANN is trained with the backpropagation algorithm, using as training patterns data measured from the mechanical process. The ANN is used to estimate some parameters from future experiments of the mechanical process.

1 Introduction

One of the basic demands for the new products launched on the market is represented by the quality level. Because the clients are more and more exigent with the products, the manufacturers must give a distinct attention to quality. A part of general quality is represented by the quality of the processing surfaces.

As presented [1, 2], the analyse of the manufacturing parameters effect over the quality indexes shows that only a part of them has a significant effect to be considered and studied.

Processing by superfinishing is part of the final processing of surfaces with severe quality requirements. The authors conducted a research on the quality parameters for the surfaces processed by superfinishing.

The quality is characterized by the roughness (Ra and Rz) and they are dependent on the following process parameters: Vs, p, f, Ga and tb:

- V_s is the rotation speed of the piece (circular feed);
- p – the contact pressure between the piece and the tool;
- f – the frequency of oscillation of the tool;

* Corresponding author: daniel.anghel@upit.ro

- G_a – the coverage degree between the tool and the piece defined as the ratio between the length of the contact arc and the length of the circle in the cross section of the piece [1];
- t_{be} – the basic time actually scheduled.

The material of the part is an Fgn 700 – 2 iron (EN-GJS-700-2 cast iron, according to SR EN 1563:1999 standard).

For the study presented in this paper, a number of 32 experiments was chosen. These experiments are showed in the Table 1.

Table 1. The experiments of superfinishing.

Independent variables					Dependent variables	
					R_{aISO}	R_{zISO}
v_s [m/min]	p [bar]	f [cd/min]	G_a	t_{be} [s]	[μm]	[μm]
4.1	1.5	1250	0.2	118	0.1	0.52
12.86	1.5	1250	0.2	15	0.36	1.66
4.1	2.16	1250	0.2	15	0.22	1.31
12.86	2.16	1250	0.2	118	0.17	0.89
4.1	1.5	1800	0.2	15	0.29	1.47
12.86	1.5	1800	0.2	118	0.16	0.93
4.1	2.16	1800	0.2	118	0.19	0.71
12.86	2.16	1800	0.2	15	0.35	1.6
4.1	1.5	1250	0.4	15	0.23	0.95
12.86	1.5	1250	0.4	118	0.15	0.75
4.1	2.16	1250	0.4	118	0.08	0.53
12.86	2.16	1250	0.4	15	0.23	1.04
4.1	1.5	1800	0.4	118	0.17	0.99
12.86	1.5	1800	0.4	15	0.19	0.96
4.1	2.16	1800	0.4	15	0.20	1.00
12.86	2.16	1800	0.4	118	0.15	0.76
2.31	1.8	1500	0.28	42	0.17	0.87
22.76	1.8	1500	0.28	42	0.19	1.00
7.261	1.25	1500	0.28	42	0.12	0.60
7.261	2.59	1500	0.28	42	0.16	0.77
7.261	1.8	1041.6	0.28	42	0.15	0.78
7.261	1.8	2160	0.28	42	0.17	0.80
7.261	1.8	1500	0.14	42	0.25	1.16
7.261	1.8	1500	0.56	42	0.11	0.65
7.261	1.8	1500	0.28	5.36	0.27	1.42
7.261	1.8	1500	0.28	330	0.10	0.65
7.261	1.8	1500	0.28	42	0.15	0.75
7.261	1.8	1500	0.28	42	0.14	0.77
7.261	1.8	1500	0.28	42	0.15	0.79
7.261	1.8	1500	0.28	42	0.14	0.77
7.261	1.8	1500	0.28	42	0.14	0.71
7.261	1.8	1500	0.28	42	0.14	0.77

Because the dependence between inputs and outputs is a nonlinear one, in this paper we used an artificial feed forward neural network (ANN).

2 Feed Forward Neural Networks

Feed forward neural networks are intelligent devices that are able to recognize images, to approximate non-linear functions or to estimate different parameters. For instance, they are widely used to recognize hand written letters or digits. A classical successful application of neural networks, used in some post offices is to classify the received letters using the postal code number, and this postal code is hand written on the envelope [3]. They recognize images in the same manner as we, human beings, recognize them. We do not recognize a hand-written digit using a mathematical model of that digit. We recognize it, because, in the past, we have seen many examples of that digit. We learn through examples. The feed forward neural network learns to recognize or to classify an image, through examples, too. These examples are called training patterns.

A feed forward neural network consists of layers of interconnected artificial neurons. There is an input layer, one or more hidden layers and an output layer. The neurons in the same layer are not interconnected, but each neuron from a layer is totally connected with all the neurons from the next layer. This organization of artificial neurons in layers is biologically inspired (we have a sensorial layer of neurons, layers of processing neurons in cortex and a layer of motor neurons). The strength of a connection between two artificial neurons, is simulated associating to that connection, a real value, positive or negative, called weight (in the biological model, we have excitatory connections between neurons, that correspond to the positive weights, and we have inhibitory connections that correspond to negative weights).

An input layer could have binary inputs (as is typically in pattern recognition) and/or analogue inputs.

In Figure 1 is presented a 3-2-2 feed forward neural network (3 neurons in the input layer, 2 neurons in the middle layer i.e. in the single hidden layer, and 2 neurons in the output layer).

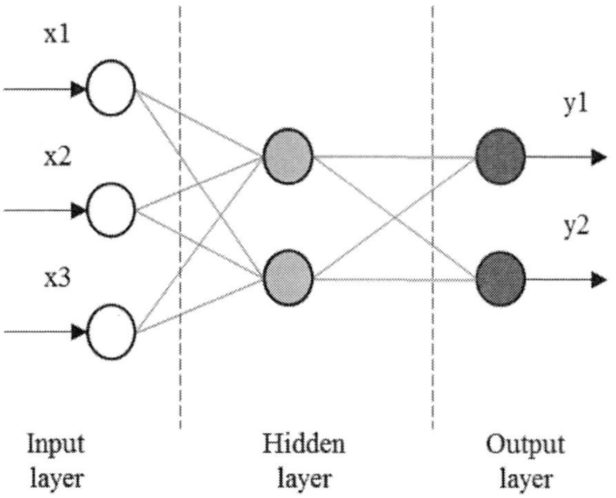

Fig. 1. A 3-2-2 feed forward neural network.

The neurons from the input layer have an identity transfer function (output=input), and they have the role to distribute the current input to all the neurons from the next layer.

The other neurons from the other layers have a sigmoidal transfer function, in which: output=1/(1+exp(-activation)).

The activation for an artificial neuron is defined as the weighted sum of all its inputs. This is a nonlinear transfer function and this nonlinearity makes the ANN capable to approximate nonlinear functions. And this is the second major application of feed forward neural networks. And for this reason, we used in this paper a feed forward neural network to model the nonlinear relationship between some mechanical parameters and the two quality parameters.

In [4] is presented a feed forward neural network that is able to learn a nonlinear process described by a single input and a single output. In this paper, we developed a neural network that is able to learn a nonlinear process, but it has more inputs.

A feed forward neural network can be implemented in many different ways:

-A hardware implementation, using electronic VLSI neural chips. This offers the greatest speed for processing, but is the least flexible.

- by software simulation, using parallel computers

-by software simulation using common programming languages, on ordinary computers This is the most flexible approach, but is not fast.

-software simulation using special neural software (for instance Matlab with its Neural ToolBox)

In order to use a neural network for a specific task (either pattern recognition or nonlinear function approximation), firstly, the architecture of the network has to be determined: i.e. the number of inputs, the number of hidden layers, the number of neurons from each hidden layer, and the number of neurons from the output layer.

The number of neurons from the input layer is set by the application. In our application we have five mechanical parameters, so we will use five neurons in the first layer.

The number of neurons in the output layer is also set by the problem to be solved.

The number of hidden layers and the number of neurons in each hidden layer is experimentally chosen. Typically, an ANN that is used for pattern recognition has one hidden layer, but an ANN that is used for nonlinear function approximation has two hidden layers. So, in our application we used two hidden layers, each having five artificial neurons.

After this stage (determining the architecture) we have to train the network. Through the training, all the weights from the network are computed. We can say that the "intelligence" of the network stays in its weights.

When the network is software simulated, the values of the weights are saved in a computer file.

After the network is trained, it can be used to solve the real problem for which it was trained. In the case of a pattern recognition problem, the image to be recognized is placed on networks' inputs (generally, each pixel from the image is connected to an input neuron) , and the network computes the outputs of all neurons from the hidden layer and after this, it calculates the outputs of all neurons from the output layer. The output neuron which has the biggest output corresponds to the class from which the input image belongs. In the case of our approximation problem, after the input data are placed on the inputs of the network, the network calculates the outputs of all neurons from the two hidden layers, and then the outputs of neurons from the output layer.

3 The software application

Firstly, we used for our application a 5 – 5 – 5 -2 feed forward neural network. But, through simulations, we decided that is better to use a 5 – 5 – 5 – 1 neural network, that is firstly trained to estimate the Ra parameter (the set of weights for Ra is calculated), and then, the same architecture 5 – 5 – 5 – 1 is trained to estimate the Rz parameter (a new set of weights is calculated). The network that we used, learns the training patterns faster than the network 5-5-5-2. The network is simulated with Java programming language.

An example of a training pattern for Ra:

Inputs: 4.1, 1.5, 1250, 0.2, 118
Output: 0.1

An example of a training pattern for Rz:

Inputs: 4.1, 1.5, 1250, 0.2, 118

Output: 0.52

These data, in order to be successfully processed by the neural network, had to be scaled, because all sigmoidal neurons of the network have an output in the interval (0, 1). So, for each from the five input parameters and also for the two output parameters, we calculate the maximum (max) and the minimum (min) from all the experimental training patterns. In order to scale a value x, from [min, max] interval to a value y in an [A, B] interval, we used a linear scaling, given by the formula:

$$y=(x*(B-A)+A*max-B*min)/(max-min) \tag{1}$$

In this application, good results had been obtained with A=0.1 and B=0.9.

The feed forward neural network is trained using the back-propagation algorithm. This algorithm iteratively calculates all the weights from the network, based on available training patterns.

We say that the network converges when it learns with a specified error, all the training patterns. In this application the learning error was 1%. So, in our case, the network converges when total learning error <0.01

The back-propagation algorithm starts initializing all the weights with small random values, real numbers positive and negative.

The back-propagation algorithm continues in the following manner:

WHILE the network did not converge REPEAT
FOR all training patterns REPEAT
begin
calculate the outputs of all neurons (using the current pattern)
compute the error of learning current pattern
modify all the weights using this error (the delta rule of the backpropagation algorithm)
end
calculate the total error of learning all the patterns (as the sum of individual learning error for each pattern)

A training epoch is done when all the training patterns had been presented on the inputs of the network, its weights were modified according to the delta rule and the total error had been computed.

After the training, the network is tested with data that it did not learn (data that are different from the training patterns), in order to see the ability of network to generalise. For the test, we take a set of values for the 5 input parameters: Vs, p, f, Ga and tb, scale them, and place them on the 5 inputs of the network, and the program computes the output of the network (using the weights calculated by the back-propagation algorithm). In order to compare the value obtained at the output of the network (y) with the real value, the output of the network must be re-scaled, using the formula:

$$yr=(y*(max-min)-A*max+B*min)/(B-A) \tag{2}$$

4 The experimental results

Our 5 – 5 – 5 – 1 feed forward neural network converges in tens of thousands of epochs, for a 1% learning error. (of course this number is variable, depending on the random values that had been used for the initialization of the weights). The speed of convergence for the network depends also of the scaling interval [A, B].

We tested the network with values that it did not learned. For example, for the following set of input values: 12.86, 1.5, 1250, 0.2, 15, the value measured for Rz was: 1.66. The value computed by the network was 1.71. The network converged in about 60,000 epochs, as illustrated by the following execution of the program:

THE TRAINING PHASE:
Epoch 1: total learning error=3.398574329350182
Epoch 2: total learning error=3.11323717907116
…………………
Epoch 59901: total learning error =0.010000067948527896
Epoch 59902: total learning error=0.009999888421416941
THE NETWORK COVERGED !
THE TESTING PHASE:
Input values: 4.1, 1.5, 1250, 0.2, 118
Correct value: 0.52, neural network computed value 0.5202859669811853

Input values: 12.86, 1.5, 1250, 0.2, 15
Correct value: 1.66, neural network computed value: 1.7115214289464546

The first set of input values, used in the testing phase was a pattern used for the training of the network. That is why the correct value and the computed value were quite the same. But the second set of input values were not used in the training, and still, the error is acceptable.

5 Conclusions

In this paper, we used a feed forward neural network, to approximate the parameter of quality of the surface, Ra and Rz. The networks have only analogue inputs.

The necessary data for training and for testing the neural network is half of necessary data than using the method of process functions (classical method), which is a considerable advantage in financial terms, in terms of availability of machine tools etc.

The approximation error could be improved if we had more available experimental data, in order to use much more training patterns.

References

1. Il. Banu, *PHD Thesis*, ("Transilvania" Univ. of Braşov, 1999)
2. Il. Banu, V. Nicolae, Int. Conf. TMCR 2006, Iasi, May, 25th – 27th (2006)
3. A. Ene, C.Stirbu, *Neural networks* (in Romanian), (Ed. Univ. Pitesti, 2008).
4. A. Ene, C.Stirbu, ECAI 2016, Ploiesti, DOI: 10.1109/ECAI.2016.7861197 (2016).

Automation of the evaluation process of technical condition and repair planning of freight wagons using computer tools

Marek Płaczek*

Silesian University of Technology, Faculty of Mechanical Engineering, Institute of Engineering Processes Automation and Integrated Manufacturing Systems, Konarskiego 18A, 44-100 Gliwice, Poland

Abstract. Freight wagons are verified about their technical condition at a certain time interval and during this qualification process a standard paper qualifying protocol is used. It is necessary to carry out a more accurate operational documentation in order to obtain more precise information about the process of damage of individual elements. This documentation should be created starting from the production of wagons. The automation of this process using computer tools is presented. A ready to use solution is proposed.

1 Introduction and aim of the research works

Development of all technical devices and technology using computer aided method is the main idea of research works conducted by scientist all over the world [1-5].

This paper contains a report on the part of works conducted in the research and development project entitled "Analytical and experimental studies and determination of the structural features of components and assemblies in innovative structure of repaired wagons". This project was realized within the Program of Applied Research by Institute of Engineering Processes Automation and Integrated Manufacturing Systems of Silesian University of Technology together with consortium partners: company DB Schenker Rail Poland SA and Germaz. The main objective of the project was to develop a technology of modernization of freight wagons for the transport of coal and aggregates, through the use of innovative materials and technologies to repair this type of wagons during periodic repairs. The main aim of the project was to improve the operating conditions of considered types of freight wagons. To increase their resistance to corrosion and freezes transported cargo to the shell of the body in the winter conditions, and thus an easier unloading as well as verification of strength of modernized carriages and an estimation of the possibility of reducing their weight, while maintaining or increasing the permissible load [6-8].

One of elements of the project was also to develop a system for diagnosing the technical condition of the modernized shell of wagon body during operation. For this purpose the use of non-destructive testing methods of technical state of constructions was used, including

* Corresponding author: marek.placzek@polsl.pl

DOI: 10.1051/matecconf/201711205002

methods that use the analysis of dynamic response of the object. Therefore research which examines the possibility of use of the foils as sensors in the system of vibration measurement of tested items were conducted [9]. Application of the composite panels to the freight wagon's body shell was proposed as the solution that can solve mentioned problems during exploitation of freight wagons [6-10]. The composite panels composed of fiberglass and epoxy resin were proposed. They were mounted on the body shell using rivet nuts. What is more the body shell of the modernized freight wagon was painted using an anti-corrosion agent, while the standard freight wagon body is not painted inside at all.

In the research project a statistical analysis of data from renovations of freight wagons was done. At this moment the qualifying protocol that is used in the DB Schenker Rail Poland SA Company in order to decide which elements of the freight wagon should be repaired or replaced by the new ones includes such data as:

- the type of the analysed freight wagon;
- drawings of the analysed type of wagons including views of all walls and the floor;
- the identification number of the wagon;
- the date and type of the previous repair and the date and type of the actual renovation;
- the table with listed elements of the wagon with the possibility to mark whether the item is suitable for repair, replacement, or whether it is in good technical condition;
- signatures of responsible persons and the place for other notes.

A statistical analysis of data from qualifying protocols were analysed in order to verify if some wagon's elements are the most exposed to damage during exploitation of freight wagons [10]. Qualifying protocols from years 2012 up to 2014 were analysed. The percentage of repairs of individual wagons elements was specified on the basis of protocols analysis.

The most important, taking into account the aim of the research work – the modernization of the freight wagons using composite panels mounted to their body shell, was the information about the damage of its elements but results of qualifying protocols analysis showed that there is no any part of the freight wagons body shell that may be indicated as the most exposed to damage during standard exploitation of the freight wagons.

It should be mentioned that in order to obtain more precise information about the process of damage of individual elements of the freight wagons body shell during their exploitation it is necessary to carry out a more accurate operational documentation starting from the production of wagons. This documentation should be created starting from the production of wagons and it should take into account operating conditions of carriage, such as the type of transported cargo, as well as all carried out repairs.

2 Automation of the evaluation process of freight wagons

The possibility of development of the process of qualifying freight wagons to repairs as well as the whole process of their repairing using computer aided systems was also an aim of this research project. The aim of the study was the use of selected virtual environments to support the process of repair of freight wagons and analysis of project-based approach to this issue, based on qualifying protocols shared by DB Schenker Rail Poland SA.

Assessment of the technical condition of a freight wagon is complex and time-consuming. It consists of several stages. The qualifying protocol must be filled in with information about the date and place of qualifying process, the type and number of the freight wagon. Then, data about freight wagon's elements are entered - data about wheelset, trolley, sheathing of the wagon's body and others. It is important that measured values of thickness of all elements must be written down. The number of the sheathing elements is about twenty-five depending on the type of the freight wagon that is being serviced. The thickness of each element is measured in two places - on the top and bottom of the plate of the sheathing. Each element should be qualified if it is in good condition or should be repaired or replaced by the new one. This is why the qualifying process can be considered as uncomfortable manual paper work and very often the documentation was filled carelessly. If someone wanted to introduce changes he had to write down the changes on paper, which unfortunately is often associated with the fact that the document was unreadable.

In order to improve the work of technicians during the qualifying process of freight wagons, to be able to quickly and easily enter data and to archive the data in one file (making possible a detailed analysis of repairs) two applications were created.

The mobile application for Android system in order to enter and send data is the first. The second one - a web application used for receiving data from the tablet. By using it, data can be presented in the form of a spreadsheet with the option to download in CSV format. Created protocol can be opened in programs such as Microsoft Excel, Open Office Calc or text editor. The aim of the study was to write applications for the Android device in conjunction with the possibility to write data to a spreadsheet. The application was written to create a separate file, accessible to all users, which will be the result of recording the data entered into the application. In order to use the application and record the results the device need to be connected to the Internet.

Due to the rapid development of new technologies and mobile devices such as tablets and smartphones, the application was written for the tablet with the most popular mobile operating system that is Android. The program works well with the version of the API level 15 - a version number 4.0.3. Below this version there is no guarantee that the application will run correctly. To write a mobile application the Java language was used with the tool developed to build such projects called Gradle. Therefore, the device do not need to store any data (collected only during fulfilling the sheets and forwarded) is not used there natively embedded in the Android SQLite data base [11-13].

When creating the application to store data between screens a mechanism called Shared Preferences was used. This mechanism allows keeping a small amount of simple data types such as integer or string. In the case of the collection of data fields a class object StringBuilder was used. For providing Internet connection and send data using POST method the class HttpURLConnection was used.

To create a web application the PHP programming language, the MySQL database, the library jQuery and the CSS Twitter bootstrap framework were used. The PHP programming language is one of the most popular languages for building web pages and frequently to cooperate with it the MySQL database is used [11-13].

The first step after starting the application is login of the user. To add a new user one must enter it to the list in the application code. This prevents the application from creating additional accounts by unauthorized persons. Additionally, on each screen and each card of the application in places where one can enter the text a standard keyboard "qwerty" is available. After a successful login to your account application goes to the main menu, which appears on the left side of the display. At this point the user of the application has a choice of sections such as data, wheelset, trolley, sheathing of the wagon body and others. Each section runs as a "checklist", so there is no possibility that while switching windows any data will be lost.

After selecting the appropriate section the user is able to supplement the data according to the physical condition of a classified freight wagon. It is necessary to introduce the type of wagon, wagon number, type/series, the last type of repair etc. also the user has the opportunity of note comments.

After completing all sections and addition of comments the user clicks on the icon in the lower right corner and the application automatically sends and saves generated qualifying protocol.

3 The project approach to the freight wagons renovation process

Technological processes and management in large enterprises are based on the project approach to problems. To achieve a higher degree of management process tools supported on computer systems are most commonly used [14-16].

Market demand for such services has resulted in a large amount of software in a broader or less degree supporting the processes of production management, web services, repair maintenance services and as a result, maintenance of external resources, internal resources and control of process lines. Depending on the size of the enterprise and approach to the management more or less developed and powerful tool can be applied. Its goal is to ensure increased productivity, control and reduce operating costs.

The most common systems supporting production processes and technology are systems CMMS (Computerized Maintenance Management Systems). These tools work with other systems via bus integration tools such as ERP (SAP) and surveillance systems of technological processes (SCADA).

In the research project the use of IBM Maximo Asset Management to manage the qualification process and repair of freight wagons was proposed [15,16]. It was assumed design approach to managing the process of repairing a freight wagons. The result of the performed simulation was the rise of a project in which all elements relating to the process flow of information related to the classification of repairing a freight wagon were introduced. In this process of the project approach to the issue of supervision a collective summary of information on the implementation of the objective pursued and the costs incurred at him were received.

From the project it is possible to perform multiple operational reports on project tasks, work orders, distributed cost, ordered materials and services, and the status of their advancement.

In Figure 1 assumption of the information flow of work orders in proposed project is presented. A scheme of the process of repair or maintenance of a freight wagon is presented in Figure 2.

In addition, the possibility of developing the system for classification of freight wagons to repair in a system SIMPLE.ERP was analysed. Key benefits that entail the use of the proposed classification system are:

- The ability to use the system SIMPLE.ERP on a mobile device such as tablet. The user can enter data on a regular basis during the inspection of the wagon. The entered information, at the same time can be seen in other places in the enterprise, for example in the planning department or in the warehouse. This gives the ability to quickly respond to the demand.
- Classification system created in SIMPLE.ERP deprives employees need to use multiple sheets of repairs card. It causes savings and no need to archive multiple documents. It also facilitates access to the history of performed operations that are archived in the form of data on the server.
- It gives the ability to quickly analyse data by exporting data to a Microsoft Excel file, what in the case of traditional repairs cards was impossible.
- With a single program, the operator is able to classify the data, at the same time, check availability of the part in the module "magazine" and in the case of their absence, order them using another module. All these activities, the operator can perform using a single application.

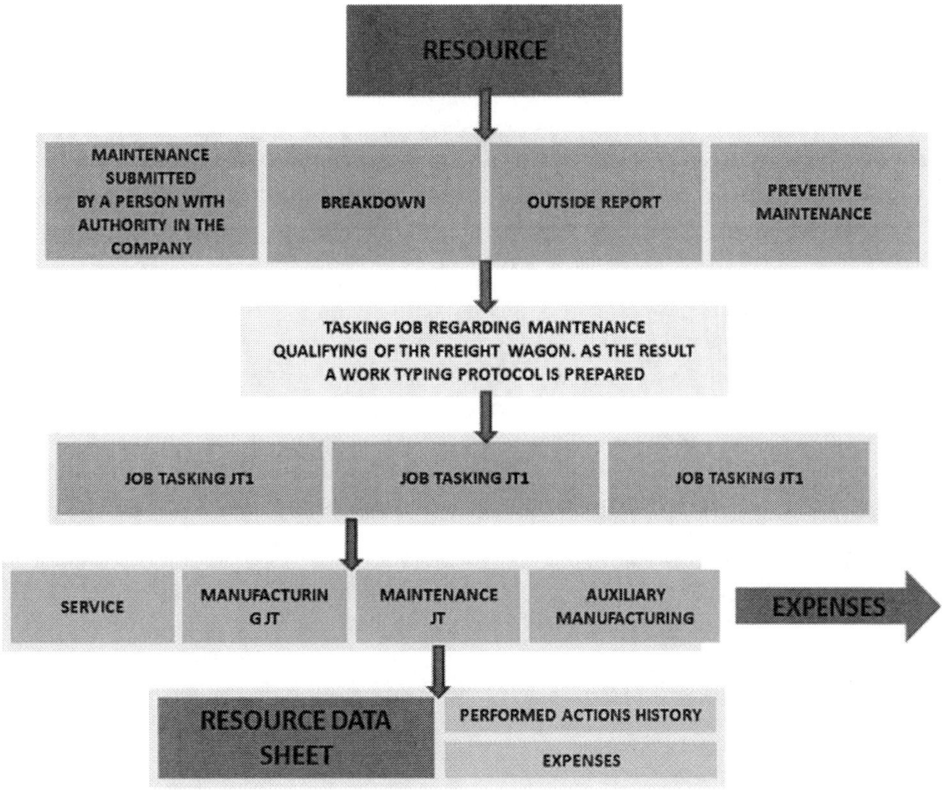

Fig. 1. The information flow of work orders.

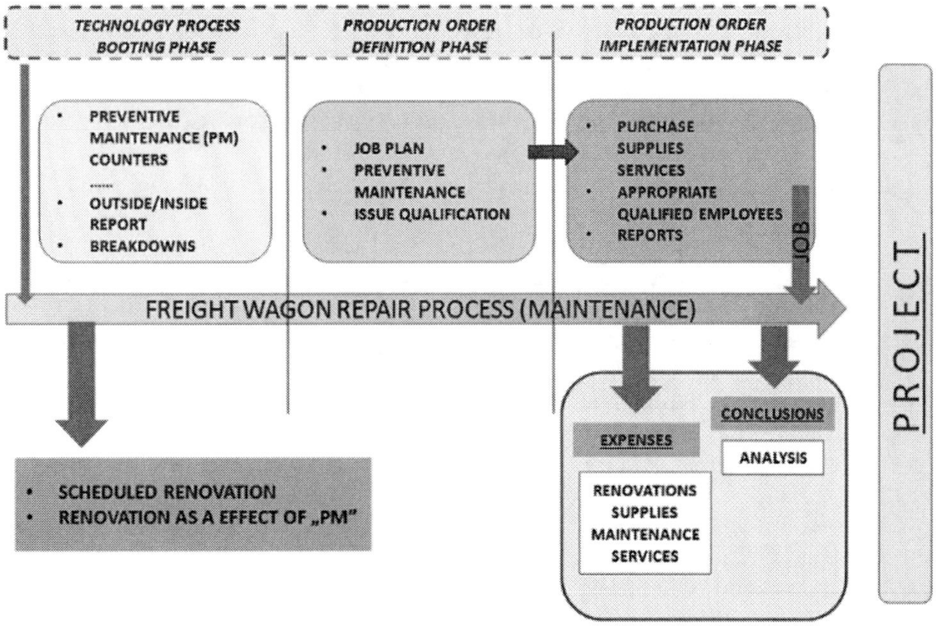

Fig. 2. A scheme of the process of repair or maintenance of a freight wagon.

4 Conclusions

As a response to the needs of the industry company the application to support the process of collecting and analysing data needed for planning of the freight wagons renovation process was created. With this application the process, which until now was entirely physically performed by employees can be automated. Entering data is divided into several stages, between which you can move flexibly and freely edit them - the blurring of the old data with new ones was removed - everything is clear. After all the fields were filled, just press a button, and all the data will be sent to the server.

The second application is installed on the server, which receives the data and presents it in the similar form as before - on the standard qualifying protocols, making it visually does not differ from what is known to employees. At the same time the computer data storage system was introduced - so far there was the need to store a lot of paper documents. Another advantage is no need to involve an additional employee to the time-consuming operations of rewriting data sheets (often illegible). With the introduction of this automated process analysis and qualifying of the wagons there is also a possibility of easily examination and comparing of the incoming data. It will allow for an even better future work and planning of repairs and the purchase of parts.

The project approach to the freight wagons renovation process was also proposed. An application of computer aided systems for supporting the processes of production management, web services, repair maintenance services and, as a result, maintenance of external resources, internal resources and control of process lines were considered. The proposed solutions can bring a lot of benefits to the process of freight wagons qualifying and their renovation process.

References

1. K. Jamroziak, M. Bocian, M. Kulisiewicz, Mechanics, **19(5)**, 549-553 (2013)
2. P. Ociepka, K. Herbuś, A. Gwiazda, Adv. Materials Research **1036**,1011-1016 (2014)
3. E. Rusiński, S. Dragan, P. Moczko., Pietrusiak D., Arch. of Civil and Mechanical Eng., **12**, 471-476 (2012)
4. A. Buchacz, A. Wróbel, Solid State Phenomena, **164**, 239-242 (2010)
5. A. Wróbel, J. of Vibroengineering, **14 (2)**, 534-537 (2012)
6. A. Wróbel, M. Płaczek, A. Buchacz, M. Majzner, Int. J. Materials and Product Technology, **50 (3/4)**, 259-275 (2015)
7. M. Płaczek, A. Wróbel, A. Baier, IOP Conf. Ser.: Mater. Sci. Eng. **145,** 042026 (2016)
8. M. Płaczek, A. Wróbel, A. Buchacz, IOP Conf. Ser.: Mater. Sci. Eng. **145,** 072012 (2016)
9. M. Płaczek, A. Buchacz, A. Wróbel, Eksploatacja i Niezawodnosc – Maintenance and Reliability **17**, 443–449 (2015)
10. M. Płaczek, A. Wróbel, A. Buchacz, IOP Conf. Series: Materials Science and Engineering **161,** 012107 (2016)
11. E. Hellman, *Android platform. New challenges* (Helion, Warsaw, 2014, In Polish)
12. J. Annuzzi jr., L. Darcey, S. Conder, *Introduction to Application Programming* (Helion, Warsaw, 2016), In Polish
13. R. Nixon, *PHP, MySQL and JavaScript. Introduction* (Helion, Warsaw, 2015)
14. K. Kalinowski, C. Grabowik, W. Kempa, I. Paprocka, Advanced Materials Research, **837,** 422-427 (2014)
15. S. Krajewski, Z. Banaszak, Wiadomosci Gornicze, **61 (2)**, 107-113 (2010), In Polish
16. W. Puchała, W. Biały, G. Bobkowski, Scientific Papers of Maritime Academy in Szczecin, 5(77), 403-412 (2015), In Polish

MATEC Web of Conferences **112**, 05003 (2017) DOI: 10.1051/matecconf/201711205003
IManE&E 2017

Multi-assortment production flow synchronization. Multiscale modelling approach

Bozena Skolud[1,*], *Damian* Krenczyk[1], and *Reggie* Davidrajuh[2]

[1]Silesian University of Technology, Gliwice, Poland
[2]University of Stavanger, Stavanger, Norway

Abstract. Increasing competition strongly influence a company that is looking for the efficient method of production control. Companies expand their production flexibility and control method in the whole value chain. In this paper, authors propose a new description of the production inspired by the multiscale modelling. In particular, the idea based on the modelling of concurrent cyclic processes and the ability to use this approach in complex production systems is presented. This paper also discusses modelling and simulation of multiscale production flows using the Activity-Oriented Petri Nets.

1 Introduction

Success of every firm depends on effective utilisation of its resources (machines, finances, employee skills, etc.) Recently the competitive advantages, however, are not in the monthly, quarterly and annual aggregated level of production flexibility but on a daily and weekly basis on the market–driven flexibility [1]. This approach is observed in the automotive industry, which adapts flexible manufacturing systems. Firms continue to expand their flexibility in the whole value chain.

Enterprises producing a variety of products tend to stabilise production on the one hand, but on the other hand, must satisfy the expectation of demanding customers, which is characterized by high dynamics in today's market. In many cases, the production recognised by the customer as diversified and multi-variant, from the manufacturer point of view is seen as stable, because manufacturing of subsequent diverse products requires neither presenting nor other interventions. With the standardisation of certain parts (construction) transport, manufacturing, manipulation, the assembly can be carried out fluently. In many cases, final products are manufactured by methods characteristic for mass production. Suppliers of sub-elements and raw-materials for production have to adjust their production to recipient needs in term of assortment diversity, quantity, timeliness and delivery data (just in time or even just in sequence).

The objective of this paper is the presentation of approach, which could be considered as multiscale modelling of a production system. It is combination of multiscale idea known from material science and multi-assortment rhythmic production elaborated by authors of this paper in recent years.

* Corresponding author: bozena.skolud@polsl.pl

DOI: 10.1051/matecconf/201711205003

2 Multiscale in production

There are phenomena in nature, which are observed in many scales. Such examples are atmospheric phenomena observed in the range of meters and on thousands of kilometres, and the chemical reactions that can last for seconds and hours. In chemistry, material science and mechanics a multiscale approach related to client expectation for material with specific properties is applied [2, 3, 4].

Multiscale modelling involves the determination of material characteristics or behaviour of the system at one level by the use of information or a model from another level.

In relation to the production systems, the multiscale issue was presented in [5]. This approach is seen as linking production with sales. It shows the use of methods based on knowledge for effective modelling and simulation of assembly systems, so as to present the strong links between their infrastructure and products that are to be made in it. Therefore, it was created a library of resources based on knowledge, which is used for modelling at the line (cell) or at the level of the plant. The authors propose the use of the idea of multiscale modelling to control the flow of concurrent cyclic processes implemented at various levels of production (Fig. 1).

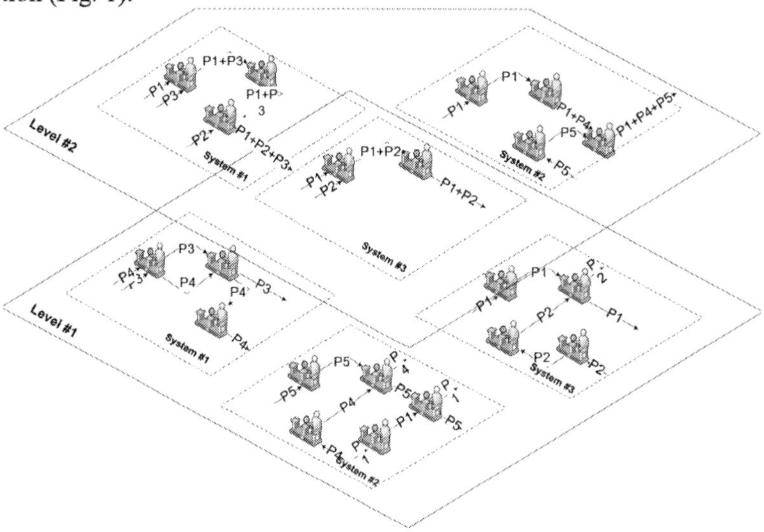

Fig. 1. Multiscale production system consist of multiple levels, where each level consists of many concurrent cyclic processes.

The proposed approach is associated with the control of production flow through the construction of a system of concurrent cyclic processes. This type of flow was chosen because for it at the planning stage an accurate evaluation of the functioning of the system is possible already.

It is assumed that sufficient condition guaranteeing deadlock-free and starvation-free process realisation in the system are satisfied. These conditions define a subset of the permitted dispatching rules controlling process execution.

This means, among other, that the sufficient conditions guaranteeing a cyclical process determined, provide the basis for an analytical determination of production flow parameters, such as:

- cycle (repetitiveness period) of the system,
- number of critical resources (resources fully occupied)
- storage capacity,
- timely realisation of production orders.

3 Rhythmic production modelling

The problem of planning the production flow in production systems belonging to the class of Cyclic Concurrent Processes Systems (CCPS) is considered (Fig. 2). In the CCPS production systems, cyclic discrete processes P_j are concurrently executed under mutual exclusions on the shared production resources M_i. The work cycle of the system is established by technological operations and the processes are sequentially carried out on the shared resource, in the order that is regulated with a local dispatching rules allocated on the resources R_i.

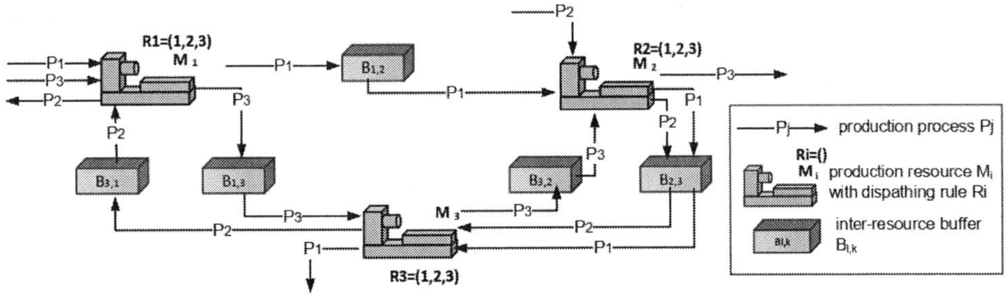

Fig. 2. System of concurrent processes.

The cyclic concurrent processes system with dispatching rules is defined as:

$$SC_{ls}=(M_{ls},PP,B,Cs,R), \quad (1)$$

where:

l – level of production realisation,

s – production system ID,

$M_{ls}=\{M_i,\ i = 1,2,...,m_{ls}\}$ – the set of resources of s-th producer on l-th prod. level,

$PP_{ls}=(P_{ls},MP_{ls})$ – the structure of the production processes of s-th producer on l-th production level, where:

$P_{ls}=\{P_j,\ j = 1,2,...n_{ls}\}$ – the set of processes of s-th producer on l-th level,

$MP_{ls}=\{MP_j,\ j = 1,2,...n_{ls}\}$ – the set of production processes matrix of s-th producer on l-th production level:

$$MP_j = \begin{bmatrix} mp_{1,1} & mp_{1,2} & \cdots & mp_{1,h} & \cdots & mp_{1,H_j} \\ mp_{2,1} & mp_{2,2} & \cdots & mp_{2,h} & \cdots & mp_{2,H_j} \\ mp_{3,1} & mp_{3,2} & \cdots & mp_{3,h} & \cdots & mp_{3,H_j} \end{bmatrix},$$ *j-th* production process matrix,

where:

h – operation sequence number (according to the order of route operations),

H_j - the number of operations in the *j-th* process route,

$mp_{1,h}$ – the resource number, on which the *h-th* operation is performed,

$mp_{2,h}$ – the cycle time of *h-th* operation,

$mp_{3,h}$ – the setup time of *h-th* operation,

N, – the batch size,

T– the period of batch entering the system,

$B_{ls}=\{B_{u,k},\ u = 1,2,...,m_{ls};\ k= 1,2,...,m_{ls};\ u\neq k\}$ - the set of inter-resource buffers allocated to the neighbouring resources (M_u, M_k),

$R_{ls}=\{Ri,\ i = 1,\ 2,...,m_{ls}\}$ – the set of dispatching rules allocated on the resources.

In [6,7] it has been proven that in the system when processes routes do not create closed loops a deadlock problem does not appear.

3.1 Balance condition

A system of repetitive production is characterised by a simultaneous realisation of processes; this approach is not limited to production line. However, one should remember that in the case when production routes create a closed loop, the deadlock can appear. To avoid this problem sufficient condition concerning a flow balance and buffers capacity should be satisfied for deadlock free and starvation free processes realisation. The balance condition in the system is satisfied when number of elements entering the system in one repetitive period (common multiple of system cycle *T*) should be equal to the number of elements leaving the system in the same period. The system balance is accomplished in the case when the following equations are satisfied for the shared machine [7, 8, 9]:

$$\chi_1 n_{1,1} = \chi_2 n_{2,1} = \ldots = \chi_i n_{i,1} = \ldots = \chi_m n_{m,1},$$
$$\ldots$$
$$\chi_1 n_{1,j} = \chi_2 n_{2,j} = \ldots = \chi_i n_{i,j} = \ldots = \chi_m n_{m,j}, \quad (2)$$
$$\ldots$$
$$\chi_1 n_{1,n} = \chi_2 n_{2,n} = \ldots = \chi_i n_{i,n} = \ldots = \chi_m n_{m,n},$$

where:
χ_i – the repetitiveness of the dispatching rule allocated on the *i-th* machine,
$n_{i,j}$ – the repetitiveness of the *j-th* process in the dispatching rule allocated on the *i-th* machine.

Additionally the balance must be satisfied between all the levels of production. The cycle time for these levels can differ (on a common multiple). The balance is satisfied when (3) is satisfied:

$$\sum_{s=1}^{S_1} \frac{T}{T_{1,s}} \chi_i^{1,s} n_{i,j}^{1,s} = \sum_{s=1}^{S_2} \frac{T}{T_{2,s}} \chi_i^{2,s} n_{i,j}^{2,s} = \cdots = \sum_{s=1}^{S_l} \frac{T}{T_{l,s}} \chi_i^{l,s} n_{i,j}^{l,s} = \cdots \sum_{s=1}^{S_L} \frac{T}{T_{L,s}} \chi_i^{L,s} n_{i,j}^{L,s} \quad (3)$$

where:
S_l - No. of producer (production systems) on *l*-th level,
T - common multiplicity of cycles,
$T_{l,s}$ - cycle time of *s-th* system on *l-th* level,
$\chi_i^{l,s}$ – the repetitiveness of the dispatching rule allocated on the *i-th* machine of *s-th* system on *l-th* level,
$n_{i,j}^{l,s}$ - the repetitiveness of the *j-th* process in the dispatching rule allocated on the *i-th* machine of *s-th* system on *l-th* level,
L - No. of levels.

3.2 Interoperation buffers allocation

Buffers allocation and their capacities are one from list of constraints which should be satisfied in order to ensure deadlock free and starvation free functioning of the system. This ensures a bottleneck resources utilisation according to theory of constraints (TOC) investigated and described in [10]. Constraints in the production system are divided into two groups, i.e. the producers constraints and the customers constraints.

Simultaneously with the balance condition, the condition of the buffers space allocation in the system should be satisfied. It means that the capacity of buffer $B_{l,k}$ allocated between machines M_l, and M_k for the *j-th* process should satisfy the following condition (4):

$$B^j_{l,k} = n_{l,j} \cdot \chi_l, \quad (4)$$

 DOI: 10.1051/matecconf/201711205003

Interoperation buffers capacity condition assures that capacity of buffer allocated between resources M_l, and M_k should not be less than the number of elements executed according given process during one cycle (see equation (4)).

Interoperation buffers capacity condition must be satisfied also between levels (5)

$$B^{j}_{1,k} \geq \frac{T}{T_{1,s}} \chi_i^{1,s} n_{i,j}^{1,s} \tag{5}$$

This condition provides a storage capacity that guarantees preservation of all the elements made in the common multiple of the work period (T) of the system.

4 Modelling with GPenSIM

Petri net model for a multiscale production flow can be obtained in three steps. Step-1: A system: a system within a level is modelled as a submodule (see fig.3a). Step-2: A level: the submodules that are representing the systems within a level are connected to form a module that represents a level of production flow, as shown in the fig. 3b. Step-3: Multi-level (multi-scale).

a)

b)

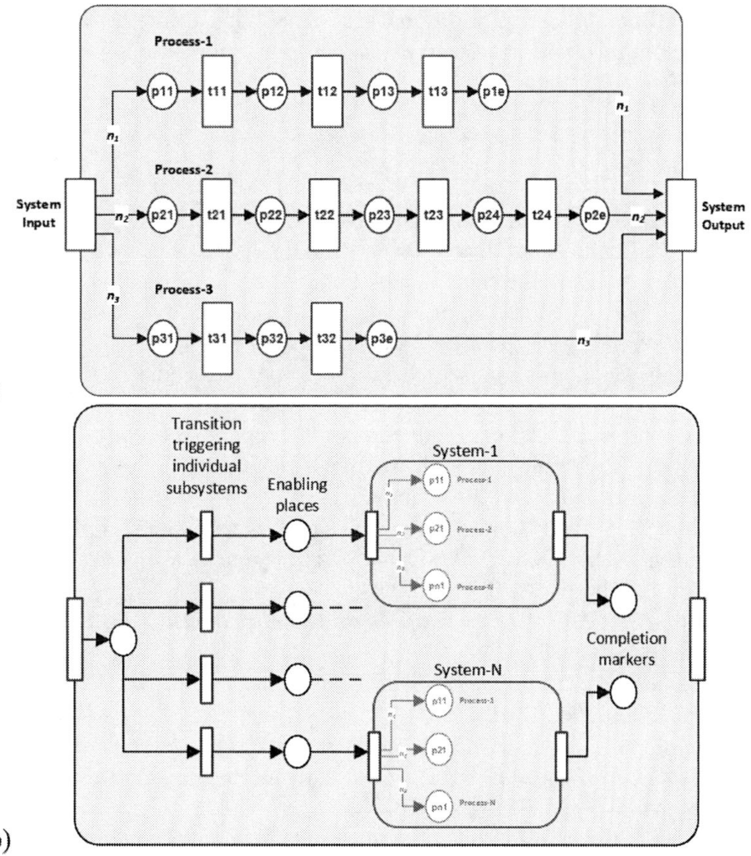

Fig. 3. a) The level module consisting of many system submodules; individual subsystems are enabled by the controlled firing of the triggering transitions; b) The system submodule, with three internal processes, each with 3, 4, and 2, operations, respectively.

Finally, the modules that represent the levels are connected to form the complete model for the multiscale production flow.

It may seem that the resulting model will be huge. However, by compact modules obtained by the AOPN approach [11, 12] and with the support for modular Petri net model building provided by the GPenSIM software [13], very useful Petri net model can be obtained for multiscale production flow

5 Conclusions

In the paper, the approach to the rhythmic production planning inspired by multiscale modelling is presented. The production flow is controlled by dispatching rules, which are repetitive and are allocated to the resources. The approach proposed in the article is based on the previous work of the authors, which is used for the propagation of constraints. The proposed solution containing the multiscale model of the production system with the formulated sufficient conditions of processes flow will allow to obtain a solution in the form of dispatching rules for different levels, without the need for use of exhaustive search methods. They allow also immediate determination of selected qualitative and quantitative indicators values of the system.

The continuation of this approach is to look at the production through the context of multiscale modelling and create additional rules involving different levels of production. The further work in this field will be an extension of the methodology to another parameter and more complicated systems.

References

1. T. Tomino, I. J. Production Economics **118**, pp. 375–386, (2009)
2. E. Weinan, B. Engquist, Z. Huang, Phys. Rev. **B 67**, 092101 (2003)
3. S.A. Hosseini, N. Shah, Interface Focus, **1(2)**, pp. 255–262. (2011)
4. I.S. Cole, C. Chu, M. Breedon, D.Winkler, 20th *International Congress on Modelling and Simulation (Modsim2013),* pp. 726-732, (2013)
5. M. Neumann, C. Constantinescu, E. Westkamper, Procedia CIRP, **3**, pp. 406–411, (2012)
6. B. Skołud, Journal of Machine Engineering. **8(2)**, pp. 5–13 (2008)
7. D. Krenczyk, B. Skołud, Hybrid Artificial Intelligent Systems, Lecture Notes in Computer Science, **6679**, pp. 397–404, (2011)
8. D. Krenczyk, B. Skołud, Applied Mechanics and Materials, **657**, pp. 961-965, (2014)
9. D. Krenczyk, A. Dobrzanska-Danikiewicz, Journal of Materials Processing Technology. **164** pp. 1388–1394, (2005)
10. E. Goldratt, *The Goal. A process of Ongoing Improvement*, North River Press, Great Barrington, Mass, (1992)
11. R. Davidrajuh, *2012 IEEE International Conference on Systems, Man, and Cybernetics (SMC)*, pp. 1201-1206. (2012)
12. B. Skolud, D. Krenczyk, R. Davidrajuh, *International Joint Conference SOCO'16-CISIS'16-ICEUTE'16. ICEUTE 2016.* Advances in Intelligent Systems and Computing, vol **527**, pp. 397-407, (2017)
13. R. Davidrajuh, *GPenSIM. A general purpose Petri net simulator for mathematical modeling and simulation of discrete-event systems in MATLAB*, url: http://www. davidrajuh. net/gpensim.

The network control system of high-bay warehouse

Julian Malaka, *Andrzej* Wróbel*, and *Marek* Płaczek

Silesian University of Technology, Faculty of Mechanical Engineering, Institute of Engineering Processes Automation and Integrated Manufacturing Systems, Konarskiego 18A, 44-100 Gliwice, Poland

Abstract. Presentation of developing a method of automation of the storage process using electric drives with frequency converters, logic control and communication in industrial networks was the main purpose of this article. A connection structure was proposed to exchange information between devices that are part of a high-storage warehouse. It was assumed that modern communication protocols are used to synchronize the drives and to create a central control and information center in the PLC. The results of theoretical considerations were applied in practice by performing a laboratory model of a high storage warehouse with a developed automatic control system. Benefits of the proposed solutions was shown in the conclusions.

1 Introduction

The work was focused on the aspects of control and automation of traffic in a high-bay warehouse. Before the research, the character of the devices was known. It was necessary to consider the automation of the process in question with the use of hardware solutions, which was the subject of research. It was expected that a high-storage warehouse model would be created, with the implemented control system developed during the study.

2 Choice of hardware and communication solutions

Taking into account the speed of processing and information exchange, functionality as well as device compatibility, hardware solutions and the structure of the control system were proposed to allow for the achievement of the intended objectives.

2.1 Drives

Linear motion in two perpendicular axes in the plane was the main assumption of motion. Horizontal displacement was achieved by a mechanism consisting of a wheelchair moving along a track coupled to a belt gear belt. The gear wheel is driven by a gearmotor with an electric three-phase asynchronous motor, equipped with an incremental encoder. The

* Corresponding author: andrzej.wrobel@polsl.pl

MOVIDRIVE® MDX61B frequency converter from SEW EURDRIVE was selected for motor control. On the carriage is mounted to the gear system comprising a screw-rolling, and an electrical actuator with single-phase motor. It corresponds to the vertical displacement of the gripper for the storage of items. The servo drive is controlled by B & R's ACOPOS 8V1016.50 2 frequency inverter. Frequency inverters have been selected for control of both drives to operate over a wide range of rotational speeds and with high positioning accuracy. In order to perform the intended tasks, both frequency inverters must be equipped with additional modules - encoder cards and communication cards (for MOVIDRIVE® MDX61B - for Modbus / TCP protocol, for ACOPOS 8V1016.50-2 - for POWERLINK protocol).

2.2 Master Control

It was assumed that the master control unit in the proposed system is to be a programmable logic controller. In addition to supporting various communication protocols, and input and output signals PLC allows to establish a connection with process visualization on the screen synoptic device HMI or SCADA system. An important aspect is collecting data in the control unit on the state of the drives that based on them to send signals to the actuators, triggering the corresponding movement sequences. The parameter transmitted by the network, which determines the possibility of positioning and synchronizing the motors to the current position, determined using the encoder.

2.3 Main idea of network connection

In Figure 1 diagram of the network connection, on which are marked the path and direction of flow of information is shown. Due to the configuration of the hardware to communicate with their inverters from various manufacturers, all with different network cards, it was selected two protocols: Modbus and POWERLINK. In both cases, the connection is realized via Ethernet. You can also use it to communicate with the HMI device driver, and SCADA system with a computer, thanks to the OPC Unified Architecture. Connection of industrial equipment via Ethernet becoming general practice because of switches, hubs or routers can be easily added to the network further objects participating in the exchange of data.

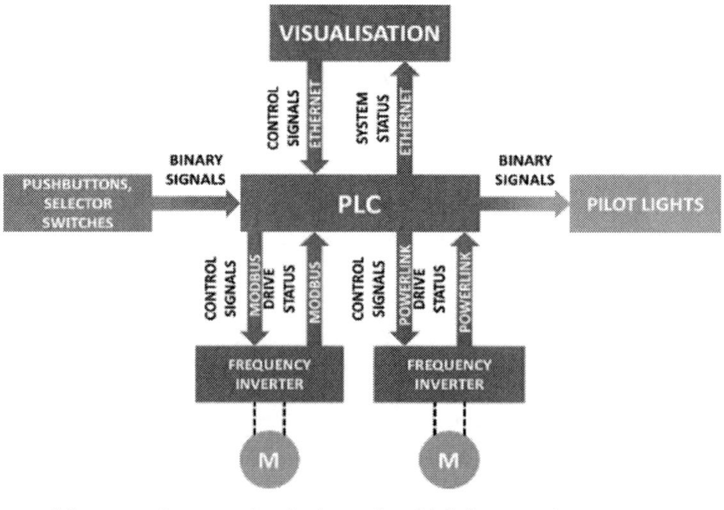

Fig. 1. Diagram of the network connection in the project high-bay warehouse.

3 Application of developed solutions in practice

In order to accomplish the intended tasks, the frequency inverters had to be equipped with additional modules - encoder cards and communication cards (MOVIDRIVE® MDX61B for Modbus / TCP protocol, ACOPOS 8V1016.50-2 for POWERLINK protocol).

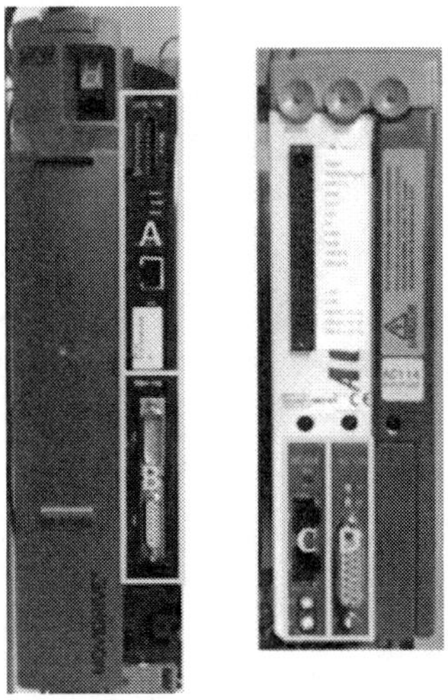

Fig. 2. View of the front panels used frequency converters with attached modules, where: A - communication card DFE11B; B - DEH11B Encoder card; C - communication card AC114; D - Card Encoder AC120.

The system uses an X20CP1584 B & R controller with two Ethernet connectors - one for POWERLINK communication and the other for configuring the device to exchange information using the Modbus protocol. A network switch was used to connect the PLC, MOVIDRIVE® MDX61B frequency inverter and computer [1, 2-4].

3.1 Hardware Configuration

Programming of X20CP1584 was carried out in a computer program Automation Studio. It also allows you to configure all the devices B&R connected to the controller. In this way, introduced among other things, information about the type of motor connected to the inverter or the address of the device having to exchange data with the controller via Modbus protocol. The frequency converter SEW EURODRIVE was configured using a computer program MOVITOOLS® MotionStudio. Description parameter aimed at properly communicated frequency converters and PLC are given in sections 3.3.

3.2 Configuring the connection of the controller with the drive ACOPOS 8V1016.50 2

Communication between the controller and devices connected to it shall be established by placing, connecting and editing components on the graphical diagram hardware that shown in Figure 3.

When entering drive to the hardware scheme in the Automation Studio must indicate in which additional modules are equipped. The presence of a network adapter allows you to create the schematic connection with the controller, which requires no additional configuration.

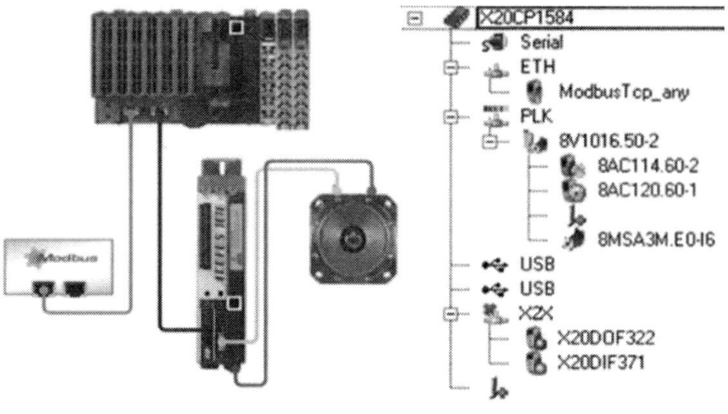

Fig. 3. Graphical representation of the hardware connections and the tree components created during the configuration of the PLC in the Automation Studio.

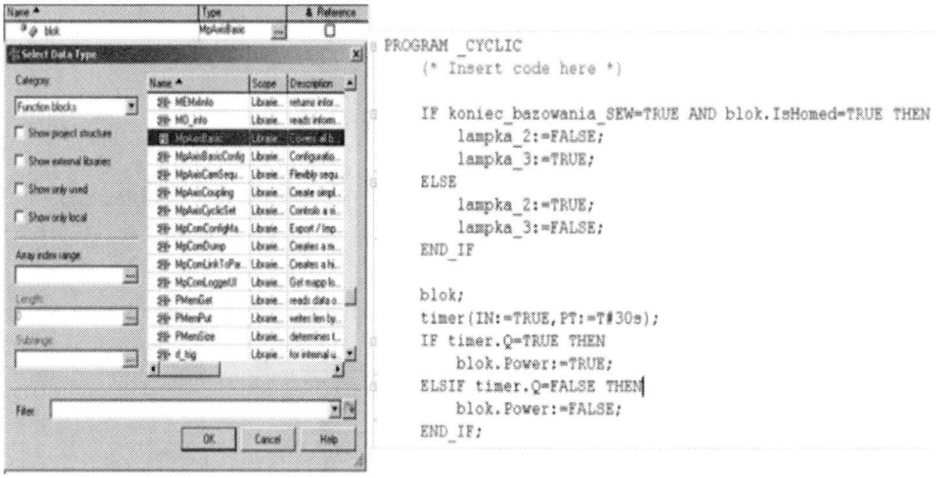

Fig. 4. The declaration of the function block MpAxisBasic and a part of the control program, which uses the unit for controlling the drive from B & R.

Send control information and parameter values to the drive system is done directly from the driver program. This is done by special function blocks to control the drive axle that is declared in the variables.

3.3 MOVIDRIVE® MDX61B Frequency Drive Driver Configuration

Controlling the frequency inverter from SEW-EURODRIVE can be done using the signals coming to it from the network [5-7, 8].

For this purpose it is necessary to set the parameters "Setpoint source" and "Control signal source" to "Fieldbus". The device will detect what network and what protocol is designed installed communication card. You must enter the data addressing device on the Ethernet network. MOVIDRIVE® MDX61B can receive from the controller three words process (marked in the parameter list as PO) and three process data words to the controller may send (marked in the parameter list as PI).

The data exchanged may include a selection from the list of commands, such as setpoint speed and control word to control the movement of the engine, and status data, eg. Current position read from the encoder (Figure 5, Figure 6).

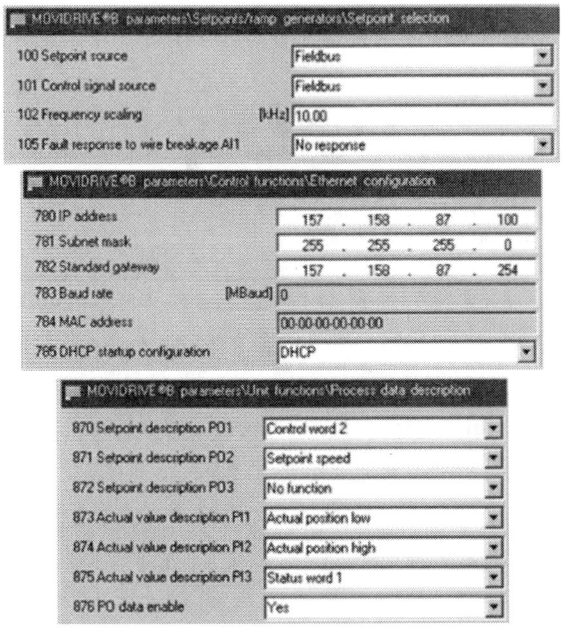

Fig. 5. Group of the frequency inverter in MOVITOOLS® MotionStudio.

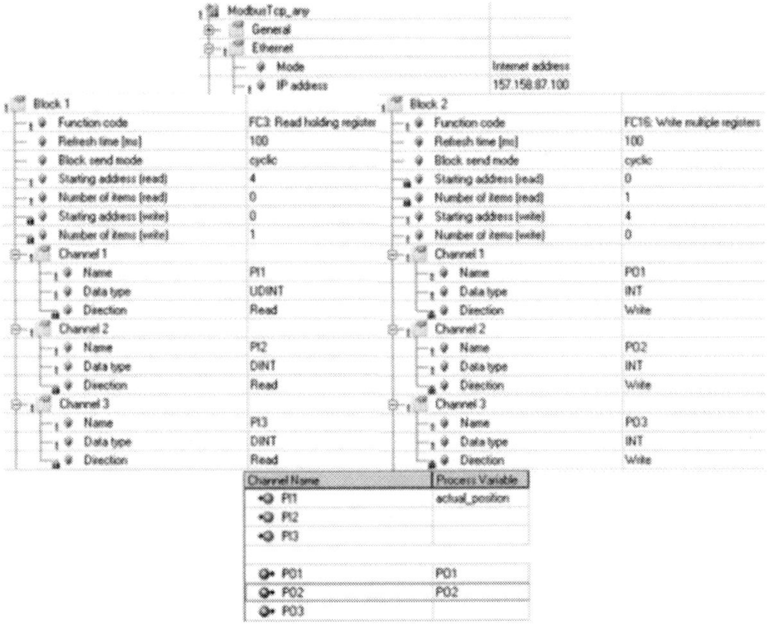

Fig. 6. Example of the window parameters of Modbus device in the Automation Studio.

The diagram hardware in the Automation Studio is a Modbus device, which in this case is a frequency converter SEW EURODRIVE.

In the configuration, enter the IP address, create blocks with the respective functions of Modbus (to read and write registers), indicate the initial address registers words process (in MOVIDRIVE® MDX61B they have value 4) and define the three channels of the appropriate data type in both blocks. These channels are then assigns a variable in the controller, which is used when programming the sequence of the drive.

4 Summary

Automation of a high-bay warehouse with electric drives with frequency inverters and industrial communication networks was developed and implemented.

The laboratory stand is working in a prescriptive manner, so it can be stated that the developed control system is working correctly. The use of the POWERLINK and Modbus networks has provided a fast, stable and easy to establish process information exchange within the automation system in question.

The study also shows the advantages of electric drives with frequency inverters. In addition to high accuracy and wide range of speed control, it is possible to exchange control and status data with other devices.

In addition, compared to pneumatic or hydraulic applications, the electric motor together with the whole transmission ratio is quite simple to operate, taking into account, for example, the supply of the supply medium. The frequency inverter can be equipped with a number of expansion modules, which allows you to connect the devices to the industrial network via the appropriate communication protocol.

References

1. A. Wróbel, W. Surma, *IOP Conf. Series: Mat. Sci. & Eng.* **145**, 052004 (2016)
2. B&R, *ACOPOS User's Manual* (2014)
3. B&R, *X20 system User's Manual* (2016)
4. B&R, *Automation Studio - Files of documentation*
5. D. Reclik, G. Kost, J. Świder, SPG - Fast Tracked Vehicles (in Polish) **2**(22), (2007)
6. Operating Instructions – MOVIDRIVE® MDX60B/61B, SEW-EURODRIVE (2010)
7. MOVIDRIVE® MDX61B Fieldbus Interface DFE11B Ethernet, SEW-EURODRIVE (2004)
8. A. Wróbel, *IOP Conf. Series: Mat. Sci. & Eng.* **95**, 1, 012090 (2015)

Distribution of time to buffer overflow in a finite-buffer manufacturing model with unreliable machine

Wojciech M. Kempa[1,*], *Iwona* Paprocka[2], *Cezary* Grabowik[2], *Krzysztof* Kalinowski[2], and *Damian* Krenczyk[2]

[1] Silesian University of Technology, Faculty of Applied Mathematics, Institute of Mathematics, 23 Kaszubska Str., 44-100 Gliwice, Poland
[2] Silesian University of Technology, Faculty of Mechanical Engineering, Institute of Engineering Processes Automation and Integrated Manufacturing Systems, 18A Konarskiego Str., 44-100 Gliwice, Poland

Abstract. One of the most important characteristic of each queueing model is time to buffer overflow. In the paper the distribution of that time in a single-channel manufacturing system, modelled by a Markovian finite-buffer queue with machine breakdowns, is studied. By using the analytical approach based on the idea of embedded Markov chain, total probability law and linear algebra, the formula for Laplace transform of the time to buffer overflow conditional distribution is found. Numerical illustration is presented.

1 Preliminaries

The problem of finite buffer (magazine) capacity is a typical one in manufacturing engineering. Jobs arriving to the system being saturated are being lost or must be redirected to another destination. The in-depth investigation of the loss process cannot be based only on the well-known coefficient as the loss ratio, defined as the part of the total number of jobs being directed to the system (manufacturing line), which are lost due to the buffer overflow. To analyze the phenomenon more precisely from the probabilistic point of view, it is necessary to know, e.g. the probability distributions of buffer overflow durations and times of reaching them. The problem becomes more essential in the case of unreliable machine, where successive failure-free times are followed by repair periods during which the processing of jobs is suspended.

In [1] and [2] the studies on the application of different-type queueing models in manufacturing engineering can be found. As one may notice, the analytical results for distributions of the time to the buffer overflow in one-channel queues modelling single-machine manufacturing lines, in fact, are restricted mainly to systems without machine breakdowns. In [3-4] the formulae for the time-to- buffer overflow period were obtained for

* Corresponding author: wojciech.kempa@polsl.pl

models with BMAP (Batch Markovian Arrival Process) and MMPP (Markov-Modulated Poisson Process) input flows, respectively. One can find another results related to buffer overflow period and the loss process, e.g. in [5] and [6]. Compact-form representation for the LT (Laplace transform) of time-to-buffer overflow distribution for the model with Poisson arrival stream and setup/closedown times was derived in [7]. In [8] (see also [9]) a similar formula was obtained for the GI/M/1/N-type queueing system (using the well-known Kendall's notation of queueing systems).

In the paper the conditional CDF (cumulative distribution function) tail of the time β to the buffer overflow, conditioned by the number $X(0)$ of jobs present in the system at the opening, is studied, i.e. the stochastic characteristic

$$B_n(t) \stackrel{\text{def}}{=} \boldsymbol{P}\{\beta > t | X(0) = n\}, \quad t > 0, \qquad 0 \le n \le N-1, \qquad (1)$$

where N stands for the system capacity. By applying the approach based on the memoryless property of exponential distribution and the formula of total probability, a system of integral equations for the conditional time-to-buffer overflow distribution is derived. The solution of the corresponding system written for LTs is obtained in a compact form using the linear-algebraic approach.

2 Model description

In the paper a model of a single-machine manufacturing line described by a finite-buffer one-channel M/M/1/N-type queueing system is investigated, in which the input flow of jobs is described by a Poisson process with rate λ, and the processing of a single job is provided in an exponential service time with mean μ^{-1}, according to the FIFO (First In First Out) service discipline. In the case if an arriving job finds the machine busy with processing, it joins the finite-capacity buffer (magazine) queue and waits for service. The maximum number of jobs present simultaneously in the system is bounded by N, i.e. we have $N-1$ places in the buffer and one place in the processing machine. If the arriving job finds the system being saturated, i.e. with N jobs present, then this job is lost.

The system starts the evolution with a machine being functional at time $t = 0$. The working (failure-free) period is assumed to be exponentially distributed with mean γ^{-1}. At the completion moment of the failure-free period a machine breakdown occurs and a generally distributed repairing time with CDF $G(\cdot)$ begins. After finishing the repair time, a new failure-free period begins, and so on. The sequences of random variables representing successive inter-arrival, processing, failure-free and repairing times are supposed to be totally independent. Additionally, it is assumed that the magazine (buffer) can contain some jobs before the start of the processing (before time $t = 0$).

3 Integral equations for time-to-buffer overflow distribution

Let us start with considering the case when the magazine is empty before the starting moment of the system.

Considering the situations in which the first job arrives before and after time t, we obtain:

$$B_0(t) = \lambda \int_0^t e^{-\lambda x} B_1(t-x)dx + e^{-\lambda t}. \qquad (2)$$

In the case if at least one job is accumulated in the buffer queue before the opening of the system, the probability distribution of the time to buffer overflow depends on times of occurrences of the first job arrival, first job departure and first machine breakdown.

Indeed, let us define the following random events:

- D_1- before time epoch t the first job arrives and precedes the first departure epoch and the first machine breakdown;
- D_2 - before time t the first job departure occurs, and precedes the first arrival epoch and the first machine breakdown;
- D_3 - before time t the first machine breakdown occurs, and precedes the first arrival and departure epochs;
- D_4 - before time t neither the first job arrival, nor job departure, nor the first machine breakdown occurs.

Due to the fact that for a fixed time t random events D_i, for $i = 1, 2, 3, 4$, are separable in pairs and $\boldsymbol{P}\{D_1\} + \boldsymbol{P}\{D_2\} + \boldsymbol{P}\{D_3\} + \boldsymbol{P}\{D_3\} = 1$, we can then write:

$$B_n(t) = \sum_{i=1}^{4} B_n^{(i)}(t) \stackrel{\text{def}}{=} \sum_{i=1}^{4} \boldsymbol{P}\{(\beta > t) \cap D_i | X(0) = n\}, \tag{3}$$

where $t > 0$ and $1 \leq n \leq N - 1$.

Suppose now that the buffer (magazine) is not empty at the starting time, respectively i.e. $1 \leq n \leq N - 1$. By applying the continuous version of the total probability law and by using the memoryless property of the exponential distribution, we get the following representations:

$$B_n^{(1)}(t) = (1 - \delta_{n,N-1})\lambda \int_0^t e^{-(\lambda+\mu+\gamma)x} B_{n+1}(t - x)dx, \tag{4}$$

$$B_n^{(2)}(t) = \mu \int_0^t e^{-(\lambda+\mu+\gamma)x} B_{n-1}(t - x)dx, \tag{5}$$

$$B_n^{(3)}(t) = \gamma \int_{x=0}^{t} e^{-(\lambda+\mu+\gamma)x} dx \left[\int_{y=0}^{t-x} \sum_{k=0}^{N-n-1} \frac{(\lambda y)^k}{k!} e^{-\lambda y} B_{n+k}(t - x - y) dG(y) \right.$$
$$\left. + \bar{G}(t - x) \sum_{k=0}^{N-n-1} \frac{[\lambda(t - x)]^k}{k!} e^{-\lambda(t-x)} \right], \tag{6}$$

$$B_n^{(4)}(t) = e^{-(\lambda+\mu+\gamma)t}, \tag{7}$$

where the notation $\delta_{i,j}$ is the Kronecker delta function and $\bar{G}(t) \stackrel{\text{def}}{=} 1 - G(t)$. Let us substitute now into (2) and (4)-(7) the LT on the argument. Define

$$b_n(s) \stackrel{\text{def}}{=} \int_0^\infty e^{-st} B_n(t)dt, \qquad \Re(s) > 0, \ 0 \leq n \leq N - 1. \tag{8}$$

Since $b_n(s) = \sum_{i=1}^{4} \int_0^\infty e^{-st} B_n^{(i)}(t)dt$ for $1 \leq n \leq N - 1$, then we obtain now from (2) and (4)-(7) the following corresponding system of equations written for LTs:

$$b_0(s) = \frac{1}{\lambda + s}(\lambda b_1(s) + 1), \tag{9}$$

DOI: 10.1051/matecconf/201711205005

$$b_n(s) = \frac{1}{\lambda+\mu+\gamma+s}\Bigg[\big(1-\delta_{n,N-1}\big)\lambda b_{n+1}(s) + \mu b_{n-1}(s) + \gamma \sum_{k=0}^{N-n-1} \big(g_k(s) b_{n+k}(s) + \bar{g}_k(s)\big) + 1\Bigg], \quad 1 \le n \le N-1, \tag{10}$$

where we denote:

$$g_k(s) \stackrel{\text{def}}{=} \frac{\lambda^k}{k!} \int_0^\infty e^{-(\lambda+s)t} t^k dG(t), \tag{11}$$

$$\bar{g}_k(s) \stackrel{\text{def}}{=} \frac{\lambda^k}{k!} \int_0^\infty e^{-(\lambda+s)t} t^k \bar{G}(t) dt. \tag{12}$$

4 Compact-form solution for LT

As one may notice, if we introduce the following sequences:

$$\alpha_0(s) \stackrel{\text{def}}{=} \frac{\mu}{\lambda+\mu+\gamma+s},$$
$$\alpha_{n+1}(s) \stackrel{\text{def}}{=} \frac{1}{\lambda+\mu+\gamma+s}\big(\gamma g_n(s) + \delta_{n,1}\big(1-\delta_{n,N-1}\big)\lambda\big), \quad n \ge 0, \tag{13}$$

$$\phi_n(s) \stackrel{\text{def}}{=} -\frac{1}{\lambda+\mu+\gamma+s}\left(\gamma \sum_{k=0}^{n-1} \bar{g}_k(s) + 1\right), \qquad 1 \le n \le N-1, \tag{14}$$

the equations (10) can be transformed to the form:

$$\sum_{k=-1}^{N-n-1} \alpha_{k+1}(s) b_{n+k}(s) - b_n(s) = \phi_{N-n}(s), \qquad 1 \le n \le N-1. \tag{15}$$

The next step consists in using the substitution $\beta_n(s) \stackrel{\text{def}}{=} b_{N-n}(s)$. In consequence, we get from (9) and (15) the following system of equations:

$$\sum_{k=-1}^{n-1} \alpha_{k+1}(s) \beta_{n-k}(s) - \beta_n(s) = \phi_n(s), \qquad 1 \le n \le N-1, \tag{16}$$

$$\beta_N(s) = \frac{1}{\lambda+s}\big(\lambda \beta_{N-1}(s) + 1\big). \tag{17}$$

The infinite-sized system of equations (for $n \ge 1$) of type (16) was considered originally in [10]. It was proved there that its general solution can be represented in form:

$$\beta_n(s) = M(s) R_n(s) + \sum_{k=1}^{n} R_{n-k}(s) \phi_k(s), \qquad n \ge 1, \tag{18}$$

where successive terms of the sequence $R_k(s)$ can be computed via recursive formulae, namely

$$R_0(s) \stackrel{\text{def}}{=} 0, R_1(s) \stackrel{\text{def}}{=} \frac{1}{\alpha_0(s)}, R_{k+1}(s) = R_1(s)\left(R_k(s) - \sum_{i=0}^{k} \alpha_{i+1}(s)R_{k-i}(s)\right), k \geq 1. \quad (19)$$

Due to the fact that the number of equations in (16) is finite, then the correspondence (17) can be used as a specific-type boundary condition required for finding the formula for $M(s)$. Indeed, by taking $n = N$ in (18), we obtain

$$\beta_N(s) = M(s)R_N(s) + \sum_{k=1}^{N} R_{N-k}(s)\phi_k(s). \quad (20)$$

Similarly, introducing (18) into (17), we get

$$\beta_N(s) = \frac{1}{\lambda + s}\left[\lambda\left(M(s)R_{N-1}(s) + \sum_{k=1}^{N-1} R_{N-1-k}(s)\phi_k(s)\right) + 1\right]. \quad (21)$$

By comparing now the right sides of (20) and (21), we eliminate $M(s)$ as follows:

$$M(s) = \frac{\sum_{k=1}^{N-1}\left(\frac{\lambda}{\lambda+s}R_{N-1-k}(s) - R_{N-k}(s)\right)\phi_k(s) + \frac{1}{\lambda+s}}{R_N(s) - \frac{\lambda}{\lambda+s}R_{N-1}(s)}. \quad (22)$$

Now, by having in mind the relationship $\beta_n(s) = b_{N-n}(s)$ and by referring to (18) and (21), we can formulate the following main result:

Theorem 1: *The LT $b_n(s)$ of the time-to-buffer overflow CDF tail, conditioned by the initial buffer state n, in the considered model of a single-machine manufacturing system with machine failures, can be written in the following form:*

$$b_n(s) = \frac{R_{N-n}(s)\left[\sum_{k=1}^{N-1}\left(\frac{\lambda}{\lambda+s}R_{N-1-k}(s) - R_{N-k}(s)\right)\phi_k(s) + \frac{1}{\lambda+s}\right]}{R_N(s) - \frac{\lambda}{\lambda+s}R_{N-1}(s)} + \sum_{k=1}^{N-n} R_{N-n-k}(s)\phi_k(s), \qquad 0 \leq n \leq N-1, \quad (23)$$

where the representations for $\phi_k(s)$ and $R_k(s)$ are given in (14) and (19), respectively.

Since for a non-negative random variable X we have $\boldsymbol{E}X = \int_0^\infty \boldsymbol{P}\{X > x\}dx$, then from Theorem 1 the following corollary follows:

Corollary 1. *The mean time to the buffer overflow $\boldsymbol{E}T_n$ in the considered manufacturing model, conditioned by the number $0 \leq n \leq N-1$ of jobs accumulated in the magazine before the starting epoch, is the following:*

$$\boldsymbol{E}T_n = b_n(0) = \frac{R_{N-n}(0)\left[\sum_{k=1}^{N-1}\left(R_{N-1-k}(0) - R_{N-k}(0)\right)\phi_k(0) + \frac{1}{\lambda}\right]}{R_N(0) - R_{N-1}(0)} + \sum_{k=1}^{N-n} R_{N-n-k}(0)\phi_k(0). \quad (24)$$

5 Numerical example

As an illustrative example, let us consider the system described by the following parameters: $\lambda = \mu = 4$ jobs/minute (equal arrival and departure rates), $\gamma = 1$ (mean failure-free time equals 1 minute), exponentially distributed repairing times with mean 0.2 minute and the system size $N = 6$. Three different initial levels of buffer saturation: empty buffer ($n = 0$), half-filled buffer ($n = 3$) and almost saturated buffer ($n = 5$) are considered in Figure 1, where transient evolution of conditional time-to-buffer overflow CDF tail is visualised. Moreover, average durations of time to buffer overflow equals, respectively: 3.985; 2.690 and 1.050.

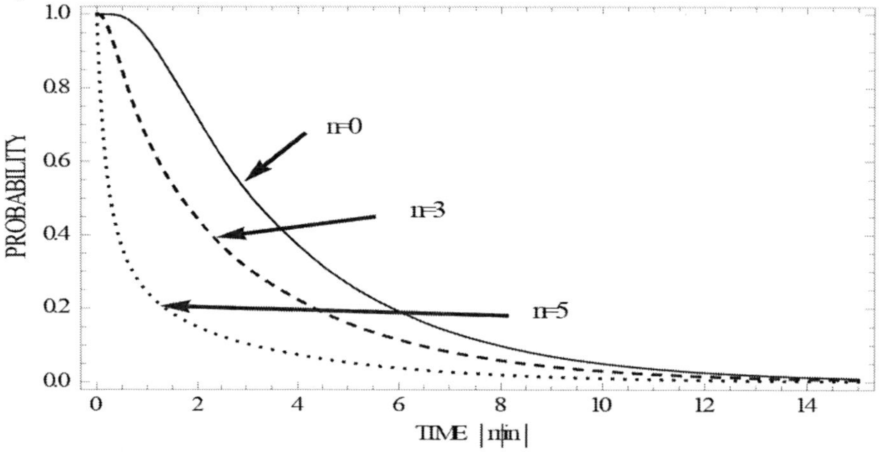

Fig. 1. Transient evolutions of time-to-buffer overflow conditional CDF tails.

6 Conclusions

A Markovian finite-buffer queue with machine breakdowns is investigated as a model of a single-channel manufacturing system. The compact-form formula for the Laplace transform of the time to the first buffer overflow conditional distribution is derived, by using the approach based on the paradigm of embedded Markov chain, total probability law and linear algebra. A numerical illustrative example is presented. The extension of the analysis is planned in future for the case of batch arrivals of jobs.

References

1. M.K. Govil, M.C. Fu, J. Manuf. Syst. **18 (3)** (1999).
2. H.T. Papadopoulos, C. Heavey, Eur. J. Oper. Res. **92 (1)** (1996).
3. A. Chydziński, Stoch. Models **23** (2007).
4. A. Chydziński, Lect. Notes Comput. Sci. **4479** (2007).
5. A. Al Hanbali, Queueing Syst. **67 (3)** (2011).
6. E.Y. Lee, K.K.J. Kinateder, Stoch. Proc. Appl. **90 (1)** (2000).
7. W.M. Kempa, I. Paprocka, Adv. Intel. Sys. **523** (2017).
8. W.M. Kempa, *Proc. of 35th the International Conference Telecommunications and Signal Processing* (TSP 2012), IEEE (2012).
9. W.M. Kempa, J. Adv. Telecommun. Electrotech. Signals Syst. **1 (2-3)** (2012).
10. V.S. Korolyuk, *Boundary-value problems for compound Poisson processes* (Naukova Dumka, Kiev, 1975).

MATEC Web of Conferences **112**, 05006 (2017) DOI: 10.1051/matecconf/201711205006
IManE&E 2017

Pneumatic muscle actuated parallel asymmetrical gripper system with one mobile jaw

Tudor Deaconescu*, and *Andrea* Deaconescu

Transilvania University of Braşov, Department of Industrial Engineering and Management, 500036, 29 Bd. Eroilor, Braşov, Romania

Abstract. The paper pertains to the field of the current global endeavours in industrial robot construction, the presented research being oriented towards identifying innovative constructive solutions for gripper systems. The utilisation of the linear pneumatic muscle as actuator of the gripper system ensures a construction that is light, highly compliant, and that meets the safety requirements related to interaction with humans. The paper further presents and discusses such a system of asymmetrical construction with a single mobile jaw.

1 Introduction

Besides the more or less evolved mechanisms mimicking human motions developed by the end of the 19th century, it was the 20th century industrialisation of global economy that brought about the invention of the modern means of computing and their deployment in the control of working equipment, eventually leading to the invention of robots.

According to the International Federation of Robotics (World Robotics 2016), worldwide robot sales peaked in 2015 with a total of 253,748 units, recording a 15% increase over the previous year. Figure 1 illustrates the 12-year dynamics of worldwide robot sales [1].

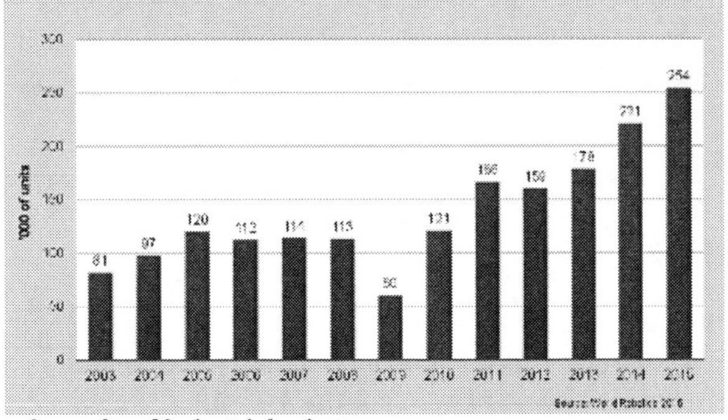

Fig. 1. Worldwide annual supply of industrial robots.

* Corresponding author: tdeacon@unitbv.ro

DOI: 10.1051/matecconf/201711205006

The main driver of the growth in 2015 was the general industry with an increase of 33% compared to 2014, in particular the electronics industry (+41%), metal industry (+39%), the chemical, plastics and rubber industry (+16%) [1]. A similar dynamic is forecasted for the next years, as shown by the prognosis in Figure 2 [1]:

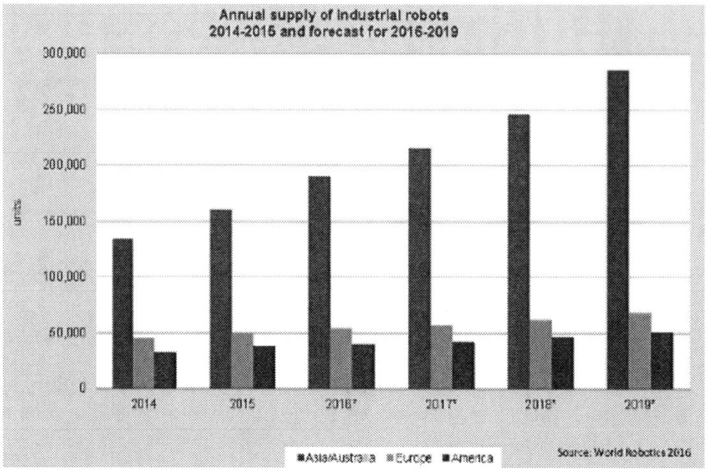

Fig. 2. Forecast of worldwide industrial robot supply.

The future increase in industrial robot demand is determined, *i.a.* by the implementation of the Industry 4.0 concept, linking the real-life factory with virtual reality, which is bound to play a key role in global manufacturing. From 2017 to 2019, robot installations are estimated to increase by at least 13% on average per year: 8% in the Americas and in Europe and 15% in Asia/Australia. Total global sales will reach about 413,000 units in 2019 [1].

Within this context, the gripper system, an essential component of any robot, needs to adapt to the new market requirements, such as to improve its energy efficiency, to be compact and easy-to-use – feature ensured by novel and light materials used for its manufacturing, and to allow optimum human-robot collaboration.

The construction of such gripper systems, satisfying the above requirements, is obtained by analogy-based design, a powerful tool meant to widen the horizon of inspiration in engineering. Nature is the obvious and generous source of inspiration for developing new products. Numerous scientific papers discuss the transfer of knowledge from natural sciences, and in particular from biology to engineering and coin the term of bionic design (bio-mimetics) emphasizing its enormous potential for developing new products and technologies [2].

This paper aims at presenting and discussing an innovative solution for such a bio-inspired, self-adaptive, light and environment friendly gripper system, endowed with a pneumatic muscle as the motion generator.

2 Construction of the gripper system

Grippers are those components of robot systems that facilitate the temporary contact with the manipulated object, ensuring its position and orientation during transport and assembly. The most important characteristics of an industrial gripper are the developed force and its structural rigidity – with a significant impact on the positioning accuracy. A small number of degrees of freedom allow a good reliability of such a gripper, and also contribute to its lower price [3].

Gripping is achieved by mechanical contact forces. While in natural systems such forces are developed by muscles, in artificial grippers it is motors that generate the forces. The motor proposed for the new gripper system discussed in this paper is the pneumatic muscle. The pneumatic muscle is a contracting membrane-based system that under the action of compressed air increases its diameter and shortens its length proportionally with the magnitude of the fed pressure. Due to the cumulated benefits of pneumatic muscle's special technical characteristics (shock absorbing capacity, light weight, small overall dimensions, reduced mass by power unit and compliance), they can be deployed successfully in the construction of gripper systems [4, 5].

With a pneumatic muscle as motor element and a power transmission mechanism based on gears, a parallel asymmetrical gripper system with one mobile and one fixed jaw was conceived, with the structural and block diagrams shown in Figure 3.

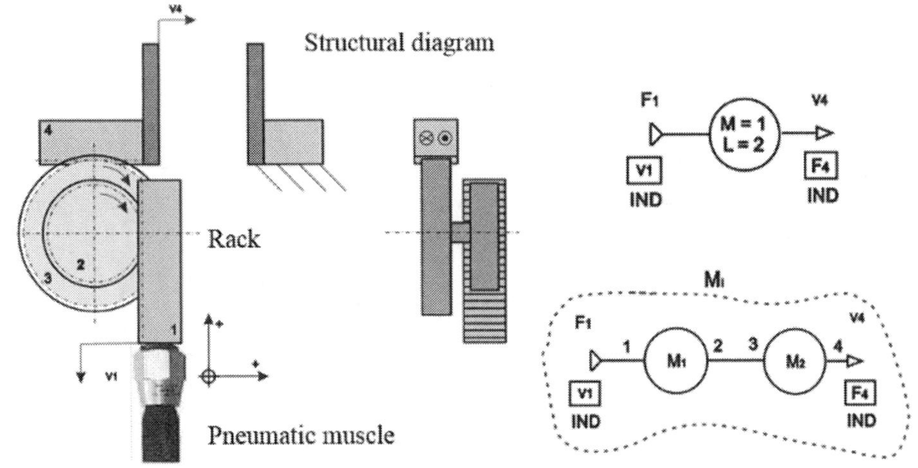

Fig. 3. Parallel asymmetrical gripper system with one mobile and one fixed jaw.

The analysed mechanism consists of a single linkage with series connected components, the number of external links of the system being L = 2. The system input is its connection to the motor (pneumatic muscle), and its output is the connection to the jaw support.

The system consisting of the two gear mechanisms that are not interconnected has the degree of mobility determined by equation (1):

$$\sum M_j = M_1 + M_2 = 1 + 1 = 2 \quad (1)$$

where j denotes the number of component gears.

Upon coupling the gears, the system's degree of mobility results as:

$$M = \sum M_j - L_c = 2 - 1 = 1 \quad (2)$$

The degree of mobility indicates that the studied fixed axes mechanism is characterised by an independent exterior motion (velocity v_1) and a dependent input force (F_1). Consequently the remaining exterior motion (v_4) is dependent, and the remaining exterior force is independent (F_4).

From the qualitative viewpoint a motion transmission function $v_4 = f(v_1)$ and a transmission function of the forces $F_1 = f(F_4)$ will be determined for the analysed gripper.

Determining the transmission forces for the velocities entails establishing the transmission ratio i_{14}:

DOI: 10.1051/matecconf/201711205006

$$i_{14} = i_{12} \cdot i_{34} = \frac{v_1}{\omega_2} \cdot \frac{\omega_3}{v_4} = \frac{\omega_2 \cdot R_2}{\omega_2} \cdot \frac{\omega_3}{\omega_3 \cdot R_3} = \frac{R_2}{R_3} = -\frac{z_2}{z_3} = -\frac{2}{3} \quad (3)$$

where z_2 and cu z_3 are the number of teeth of gears 2 and 3 ($z_2 = 20$; $z_3 = 30$).

The transmission function of the velocity will be:

$$v_4 = \frac{v_1}{i_{14}} = -\frac{3}{2} \cdot v_1 \quad (4)$$

If the friction forces in the system are not neglected and the inertial effects of the masses in motion are not considered, the transmission function of the forces is determined starting from equation (5):

$$F_1 \cdot v_1 \cdot \eta + F_4 \cdot v_4 = 0 \quad (5)$$

where η denotes the efficiency of the system and is calculated with equation (6):

$$\eta = \frac{1}{(\beta \cdot \eta_{14})^{-1}} \quad (6)$$

where η_{14} is the global efficiency, computed as: $\eta_{14} = \eta_{12} \cdot \eta_{34} = 0.97^2 = 0.941$. As this is a series aggregate with a single power branch, the output distribution coefficient is $\beta = 1$.

With these notations the transmission function of the forces will be:

$$F_1 = \frac{1}{\eta} \cdot \left(-F_4 \cdot \frac{v_4}{v_1}\right) = \frac{1}{\eta} \cdot \left(-F_4 \cdot \frac{1}{i_{14}}\right) = \frac{1}{\eta} \cdot \left(\frac{3}{2} \cdot F_4\right) \quad (7)$$

or

$$F_1 = 1{,}4115 \cdot F_4$$

Figure 4 shows the geometrical model of the gripper system. In the case of this constructive solution gripping is achieved by form and friction.

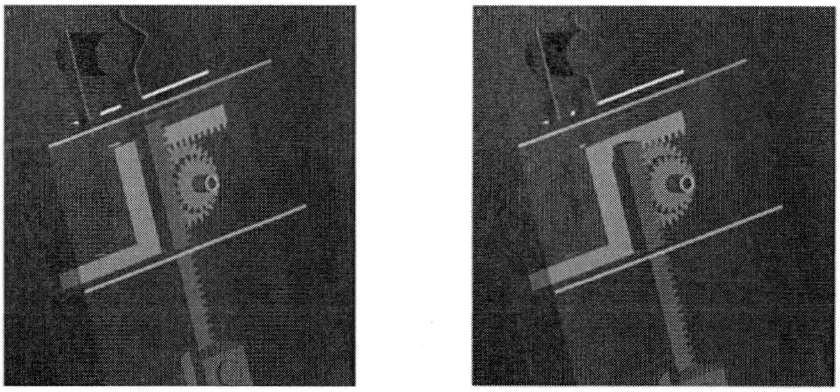

Fig. 4. The geometrical model of the gripper system.

The utilised pneumatic muscle is the smallest of the current product range of Festo AG & Co (MAS-10-45N-AA-MC-O-ER-EG). Its diameter is of 10 mm, and the length of the active part is of 45 mm. Upon being fed air at a pressure of 6 bar, the muscle shortens by 9 mm, representing 20% of its initial length.

The characteristic diagrams of the pneumatic muscles, as well as the MuscleSim application developed by Festo indicate that for a 4 mm axial contraction the muscle develops a force of 207 N [6, 7]. Under these circumstances the maximum mass of the object that can be seized by this gripper system is determined by equation (8):

DOI: 10.1051/matecconf/201711205006

$$m = \frac{\mu \cdot F_4}{(g+a_S)\cdot S} = \frac{\mu \cdot \frac{F_1}{1.4115}}{(g+a_S)\cdot S} \frac{0,2\cdot\frac{207}{1.4115}}{(9,81+10)\cdot 2,5} = 0,592\ kg \tag{8}$$

where: g = 9.81 m/s^2 is the gravity acceleration; a_S = 10 m/s^2 is the emergency stop deceleration; μ = 0.2 is a friction coefficient and S = 2.5 is a safety coefficient.

Figure 5 presents the functional model of the gripper system obtained by means of the MatLab SimMechanics module.

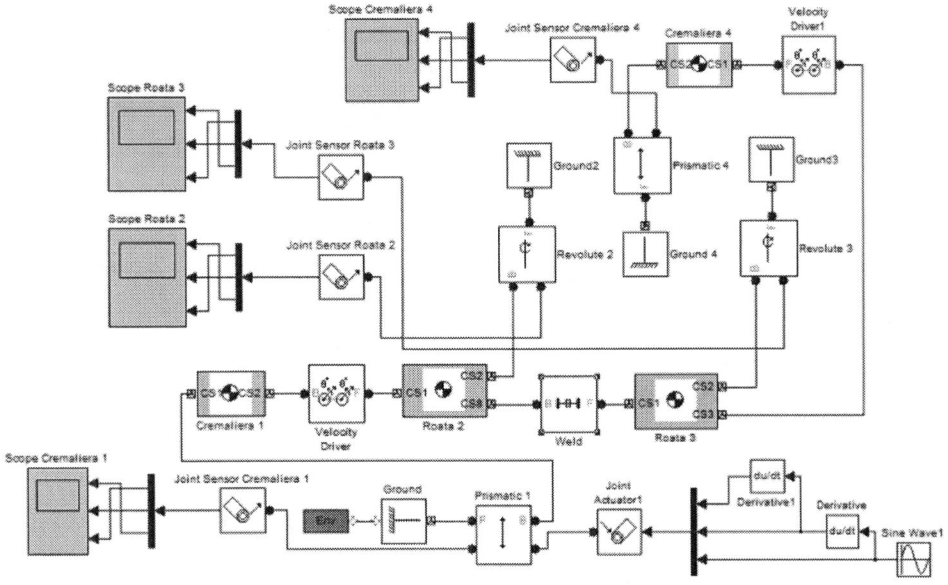

Fig. 5. The functional model of the gripper system.

The motion of the gripper system is generated by a Sine Wave block, which was selected for the discussed system because of the similarity of the pneumatic muscle's motion behaviour to a sine curve. The output quantity of this block is a linear displacement applied to rack 1 and is given by equation (9):

$$s(t) = A \cdot \sin(\omega t + \varphi) = 9 \cdot \sin\frac{\pi}{2}t \quad [mm] \tag{9}$$

where A is the amplitude of the motion, ω – its pulse, and φ is the initial phase.

Figure 6 shows the motion parameters of gears 2 and 3 upon applying the above described sine signal for 2 seconds.

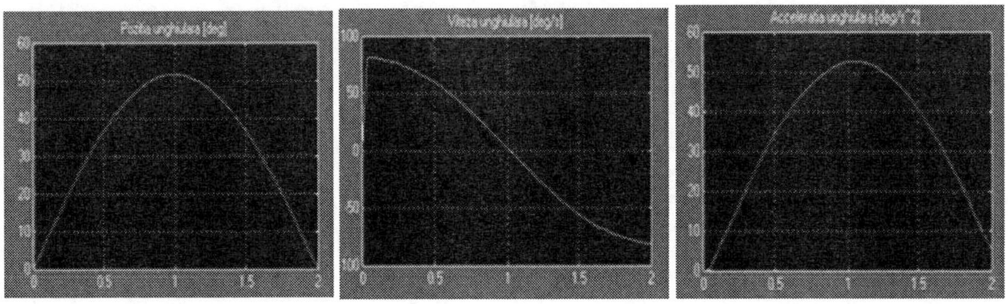

Fig. 6. Variation of the rotation angle, angular velocities and angular accelerations of gears 2 and 3 versus time.

It can be noted that for a 9 mm displacement of Rack 1, gears 2 and 3 turn synchronously by 51.56°.

Figure 7 shows the evolution of displacements [mm], velocities [mm/s] and acceleration [mm/s^2] of Rack 4 versus time.

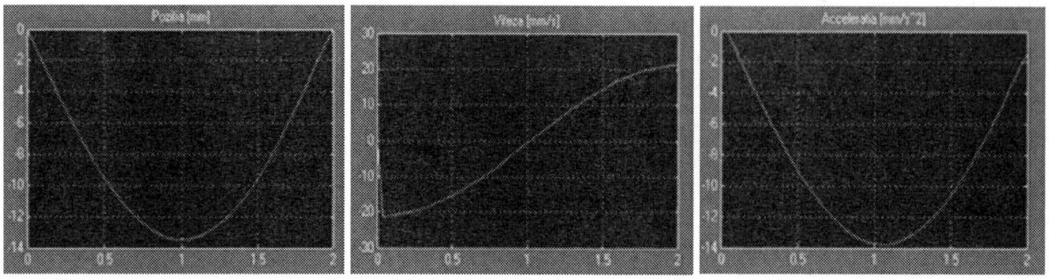

Fig. 7. Evolution of the position, velocity and acceleration of Rack 4 versus time.

The first graph reveals that a 9 mm displacement of the pneumatic muscle causes Rack 4 to move by 13.5 mm. Further it can be noticed that the beginning of the motion features a sudden leap of the velocity up to 21.2 m/s, after which the motion continues in a smooth manner, following the sine motion imposed by the pneumatic muscle. This sudden leap occurs at the beginning of the motion, before contact with the object is achieved, and does not influence the gripping process.

3 Conclusion

Within the context of a global increase in demand for industrial robots, their gripper systems need to undergo significant development such as to meet the requirements of the Industry 4.0 concept. In this respect the paper presents and discusses an innovative variant of a pneumatic muscle actuated gripper system. Characteristics of this gripper are constructive simplicity, low weight, and more importantly, the compliant behaviour specific to pneumatic drives. The latter characteristic of pneumatic muscles led to the concept of deploying this gripper system in applications requiring adaptability to concrete circumstances of object manipulation.

References

1. International Federation of Robotics. *Executive Summary World Robotics 2016 Industrial Robots*. http://www.ifr.org/fileadmin/user_upload/downloads/World_Robotics/2016/Executive_Summary_WR_Industrial_Robots_2016.pdf. Accessed: 12.12.2016
2. G. Fantoni, D. Gabelloni, J. Tilli, *How to design new grippers by analogy*, Dep. of Mech., Nuclear and Prod. Eng., University of Pisa, Report (2012).
3. D. Negrea, T. Deaconescu, A. Deaconescu, Applied Mechanics and Materials, **541-542**, 852-856 (2014).
4. A. Deaconescu, T. Deaconescu, *International MultiConference of Engineers and Computer Scientists Hong Kong*, **II,** 1872-1875 (2009).
5. A. Deaconescu, T. Deaconescu, *Annals of DAAAM, 19th International DAAAM Symposium Trnava, Slovakia*, 0365-0366, (2008).
6. Festo, Fluidic Muscle DMSP/MAS, https://www.festo.com/rep/en_corp/assets/pdf/info_501_en.pdf Accessed: 12.12.2016.
7. Festo MuscleSim v. 2.0.1.5., Festo MAS Fluidic Muscle Simulation Software (2005).

Linear pneumatic motors – a comparative study

Tudor Deaconescu*, and *Andrea* Deaconescu

Transilvania University of Braşov, Department of Industrial Engineering and Management, 500036, 29 Bd. Eroilor, Braşov, Romania

Abstract. The paper presents a comparative study of the performance of single-acting cylinders, diaphragm cylinders and pneumatic muscles, and offers users information that allows the selection of an optimum technical solution. Such a study was necessary, in view of the numerous papers on pneumatic muscle applications found in literature, that assert the superiority of pneumatic muscles over other pneumatic linear motors in relation to quantities like dimensions, mass, developed force or energy-to-mass ratios, however without offering concrete data.

1 Introduction

The increasingly higher requirements related to the performance parameters of manufacturing systems fuel the competition between the various actuation modalities: mechanical, electrical, or fluid. This generates the continued improvement of each type of actuation and their convergence towards performance maximisation.

Compressed air is one of the most efficient means of actuation and automation of manufacturing systems. Over time pneumatic actuations have undergone continuous development, with a gradual passing form the utilisation of individual components to the deployment of complex automation systems. New systems were conceived, adding mechanical, electronic and sensor elements to the classical pneumatic structures.

The increasingly larger scale of compressed air utilisation in industrial applications is due to its benefits, like easy generation and storage, non-flammability, minimum explosion risk, the minimal maintenance effort required by pneumatic systems, etc.

The linear pneumatic motors used in industry are rod or rodless cylinders, diaphragm cylinders, bellows cylinders, as well as pneumatic muscles.

Further on the paper presents a comparative study of the performance of rod cylinders, diaphragm cylinders and pneumatic muscles, offering users the necessary information for selecting the optimum solution for each application. For the same size of the three types of pneumatic motors (diameter, stroke) the comparison refers to dimensions, masses, developed forces and energy-to-mass ratios.

Such a study was called for, in view of the numerous papers on pneumatic muscle applications found in literature, that assert the superiority of these parameters in pneumatic muscles over other pneumatic linear motors, however without offering concrete data [1-3].

* Corresponding author: tdeacon@unitbv.ro

2 The analysed pneumatic motors

The aim of the study discussed in this paper is to highlight the performance of pneumatic muscles, as compared to linear motors whose developed forces are not constant over the entire stroke length. These are single-acting cylinders and diaphragm cylinders.

Single-acting cylinders are used when force needs to be applied in a single direction, typically for feed motions. The return of the piston is automatic, by means of a built-in spring. The force developed by such a cylinder is [4]:

$$F = p \cdot \frac{\pi \cdot D^2}{4} - F_{FR} - F_{SPR} \quad (1)$$

where p is the pressure of the compressed air, D – the piston diameter, F_{FR} – the friction forces, F_{SPR} – the force of the spring.

Diaphragm cylinders are used in gripping devices and transfer arms of industrial robots, clamping devices of machine tools, stopper devices, etc. [4]. By eliminating the conventional sliding piston seal, diaphragm cylinders produce nearly frictionless motion, except for the miniscule amount of friction at the rod seal [5]. The force developed by these cylinders is determined by equation (2) [4]:

$$F = S_e \cdot p - F_{SPR} \quad (2)$$

where S_e denotes the effective diaphragm area.

The pneumatic muscle is a motor that causes displacement by modifying its geometrical form under pressure. As compressed air is fed into the muscle a traction force is generated along its longitudinal axis, and the muscle shortens in proportion with the increase in internal pressure. The pneumatic muscle reaches the maximum developed force as soon as contraction starts, which force decreases to zero by the end of the stroke. The force developed by a pneumatic muscle is given by equation (3) [6]:

$$F = p \cdot \frac{\pi}{4} \cdot d^2 \cdot \left[\frac{3 \cdot \cos^2 \alpha - 1}{1 - \cos^2 \alpha}\right] \quad (3)$$

where p is the working pressure, d the interior diameter of the pneumatic muscle, and α is the wrapping angle of the synthetic fibre mesh that envelopes the interior tube.

From the three categories of actuators described above, the following models were selected for the comparative study of linear pneumatic motor performance:

- a single-acting cylinder model ESNU ISO 6432, manufactured by FESTO, Germany;
- a diaphragm cylinder model MM-1 (349-180-007) manufactured by TECO Pneumatic, USA;
- a pneumatic muscle, model MAS-20-N, manufactured by FESTO Germany [7].

In the single-acting cylinder and the pneumatic muscle the piston and tube diameter, respectively, is of 20 mm, while the diaphragm cylinder diameter is of 17.7 mm (0.7 in). The strokes of the former two types of motors are of 25 mm, and the stroke of the diaphragm cylinder is of 17.7 mm (0.7 in). Figure 1 shows the three discussed linear motors.

Single-acting cylinder   Diaphragm cylinder   Pneumatic muscle

Fig. 1. Linear pneumatic motors considered for comparison.

3 Comparative studies

The comparative analysis of the performance of linear pneumatic motors is based on a set of assessment criteria that take into consideration motor dimensions and mass, the developed forces and the energy-to-mass ratio.

From the viewpoint of *overall size* of the three types of motors, Table 1 centralises the main values, as indicated in the technical documentation of these devices.

Table 1. Dimensions of the analysed pneumatic motors.

Motor type	Total length, including assembly elements [mm]	Maximum exterior diameter [mm]
Single-acting cylinder	132	36
Diaphragm cylinder	71.374	25.4
Pneumatic muscle	333	30

It can be noticed that as to the axial dimensions the pneumatic muscle is the longest, and the diaphragm cylinder the shortest motor. As to the radial size, the three motors have comparable dimensions, with a slight advantage of the diaphragm cylinder.

The *masses* of the pneumatic motors depend on the lengths of their strokes. In single-acting cylinders the masses corresponding to a 0 mm stroke is of 0.167 kg, and increases by 0.0072 kg for each 10 mm of stroke. Equation (4) is the linear function that describes the dependency of the pneumatic cylinder mass on the stroke (s) [8]:

$$m_{ESNU}(s) = 0.0072 \cdot s + 0.167 \quad (4)$$

In the case of the analysed cylinder whose stroke is of 25 mm, the corresponding mass is of 0.185 kg.

In the case of the diaphragm cylinder, its mass as indicated by the manufacturer is of 0.162 kg.

While the mass of artificial muscles is evidently directly dependent on their length, it is smaller than that of a pneumatic cylinder of the same dimensions due to the deployed materials. Figure 2 presents the dependency of the analysed pneumatic muscle mass on its length. The graph was plotted based on the information provided by the MuscleSIM 2.0.1.5 application provided by the manufacturer.

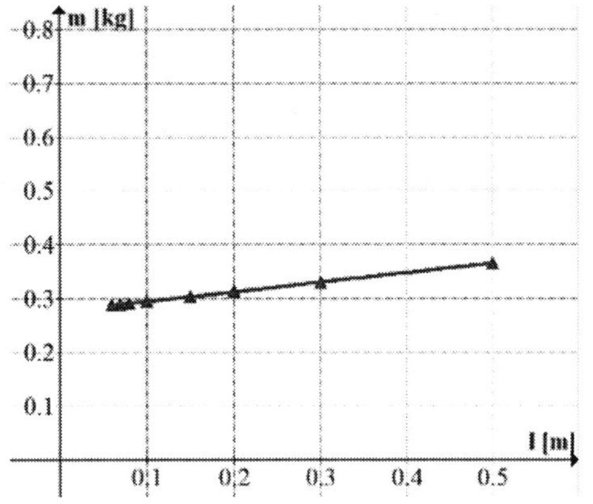

Fig. 2. Dependency of the pneumatic muscle mass on its length.

Equation (5) represents the linear function that describes best the dependency of the pneumatic muscle mass on its length [8]:

$$m_{20}(l) = 0.178 \cdot l + 0.277 \quad (5)$$

where l denotes the length of the active part of the pneumatic muscle (length of the elastic tube). For the analysed muscle whose active length is of 162.9 mm, the corresponding mass is of 0.306 kg.

The determined values of the masses of the three motors reveal the pneumatic muscle as the heaviest and the diaphragm cylinder as the lightest actuator.

Table 2 presents the *forces* developed by the three pneumatic motors. The maximum working pressure taken into consideration was of 6 bar.

Table 2. Forces developed by the analysed pneumatic motors.

Motor type	Maximum force F_{max} [N]	Minimum force F_{min} [N]
Single-acting cylinder	169	16.5
Diaphragm cylinder	139.74	31.15
Pneumatic muscle	1200	330 (at $\varepsilon = 0.2$)*

*ε = the specific axial contraction

Based on the data in Table 2, Figure 3 shows the plotted lines of the force variation versus stroke in the three analysed pneumatic motors.

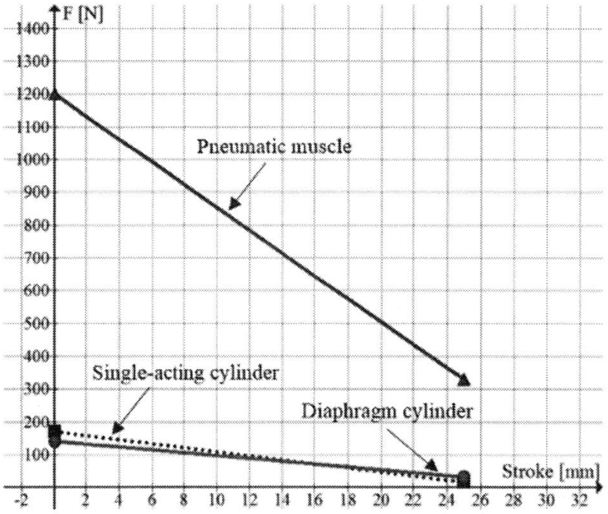

Fig. 3. Variation of the developed force versus stoke in the three linear motors.

The graph reveals clearly that the forces developed by the pneumatic muscles are 8 to 10 times greater than those generated by single-acting cylinders and diaphragm cylinders, respectively.

In the case of the pneumatic muscle however, the dependency of the force on the stroke s is not linear. Equation (6) [8] represents the computational relationship of the force developed by the pneumatic muscle in dependence on the specific axial contraction $\varepsilon = s/l$.

$$F_{20}(\varepsilon) = 1.678 \cdot 10^5 \cdot \varepsilon^4 - 1.087 \cdot 10^5 \cdot \varepsilon^3 + 2.924 \cdot 10^4 \cdot \varepsilon^2 - 8640.755 \cdot \varepsilon + 1490.71 \quad (6)$$

Another criterion of comparison of linear pneumatic motors concerning their energy efficiency is the *energy-to-mass ratio*. Various expressions of this indicator for pneumatic muscles can be found in [8] and [9]. Thus paper [8] puts forward a computational relationship of the energy-to-mass ratio that takes into consideration the specific axial contraction ε of the pneumatic muscle, its length l and mass m.

$$R_{\frac{E}{m}} = \frac{l \cdot \int F(\varepsilon) \cdot d\varepsilon}{m} \tag{7}$$

Starting from equation (6) and based on the graph of Figure 4, equation (8) computes the energy developed by the analysed pneumatic muscle:

$$E_{20}(l) = l \cdot \int_{0.038}^{0.25} F_{20}(\varepsilon) \cdot d\varepsilon = 130.6123 \cdot l \tag{8}$$

The lower limit of the integral above (0.038) represents the minimum specific axial contraction starting at which the muscle generates the force; the upper limit of the integral above (0.25) is the maximum specific axial contraction at that the muscle still generates a minimal force.

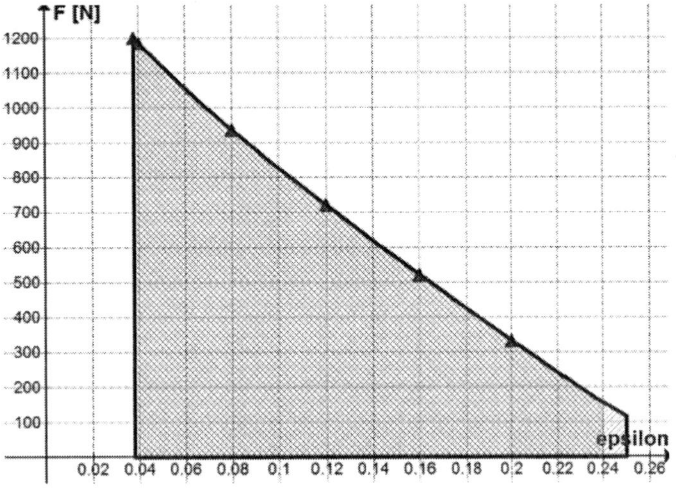

Fig. 4. Working range of the Festo MAS-20-N Fluid Muscle.

When the specific axial contraction is 0.25 ($\varepsilon = 0.25$), the force developed by the muscle is of about 120 N, smaller than indicated in Table 2 (F = 330 N for $\varepsilon = 0.20$).

Equation (9) obtained by inserting equation (8) into equation (7) calculates the energy-to-mass ratio of the studied pneumatic muscle:

$$R_{\frac{E}{m},20} = \frac{E_{20}(l)}{m_{20}(l)} = \frac{130.6123 \cdot l}{0.306} \tag{9}$$

For the two pneumatic cylinders the relationships for the energy-to-mass ratios are:

$$R_{\frac{E}{m},ESNU} = \frac{F \cdot s}{m} = \frac{(-6.1 \cdot s + 169) \cdot s}{0.185} \tag{10}$$

$$R_{\frac{E}{m},MM-1} = \frac{F \cdot s}{m} = \frac{(-4.3436 \cdot s + 139.74) \cdot s}{0.162} \tag{11}$$

Based on the above equations, Table 3 features the comparison of the calculated values of the energy-to-mass ratio for the three types of analysed linear pneumatic motors.

Table 3. Comparison of the energy-to-mass ratios.

Motor type	Energy-to-mass ratio
Single-acting cylinder	22.82
Diaphragm cylinder	15.26
Pneumatic muscle	69.46

The analysis of these data shows that the pneumatic muscle ensures an energy-to-mass ratio superior to other types of pneumatic motors of similar dimensions.

4 Conclusion

Table 4 presents a synopsis of the results yielded by the comparison of the three types of linear pneumatic motors.

Table 4. Synopsis of the results.

Motor type	Dimensions	Masses	Forces	Energy-to-mass ratio
Single-acting cylinder	😐	😐	😐	😐
Diaphragm cylinder	☺	☺	😐	😐
Pneumatic muscle	☹	☹	☺	☺

An overall assessment of the results reveals that pneumatic muscles, although currently insufficiently known and applied, offer numerous benefits. In future, based on detailed knowledge of their advantages, pneumatic muscles are bound to replace single-acting cylinders or diaphragm cylinders in an increasing number of applications.

References

1. N. Tsagarakis, D.G. Caldwell, *Proc. of the IEEE International Conference on Robotics & Automation*, San Francisco, USA, pp. 3641-3646 (2000).
2. N. Nakamura, M. Sekiguchi, K. Kawashima, T. Fujita, T. Kagawa, *Bath Workshop on Power Transmission & Motion Control*, (2002).
3. G.A. Medrano-Cerda, C.J. Bowler, D.G. Caldwell, *Proc. of the IEEE/RSJ International Conference on Intelligent Robots and Systems*, Pittsburgh, USA, 378-383, (1995).
4. E. Pashkov, Y. Osinskiy, A. Chetviorkin, *Electropneumatics in Manufacturing Processes* (Isdatelstovo SevNTU, Sevastopol, 2004)
5. Hydraulics & Pneumatics. Engineering Essentials: Cylinders, http://m.hydraulicspneumatics.com/200/TechZone/Cylinders/Article/False/6423/TechZone-Cylinders?page=2 Accessed: 26.12.2016
6. S. Hesse, *The Fluidic Muscle in Application* (Blue Digest on Automation, Esslingen, 2003).
7. Festo Fluidic Muscle DMSP/MAS, https://www.festo.com/rep/en_corp/assets/pdf/info_501_en.pdf Accessed: 26.12.2016
8. T. Deaconescu, A. Deaconescu, Int. J. of Mech., Aerospace, Ind., Mechatronic and Manuf. Eng., **10(10)**, 1598 – 1604, (2016).
9. D.H. Plettenburg, *Proc. of the 2005 IEEE 9th International Conference on Rehabilitation Robotics*, Chicago, IL, USA, 545-549 (2005).

Operational parameters estimation for a flexible manufacturing system. A case study

Adriana Florescu*, *Sorin* Barabaş, and *Flavius* Sârbu

Transilvania University of Braşov, Department of Engineering and Industrial Management, Mihai Viteazul 5, Braşov, Romania

Abstract. The complex structure of flexible manufacturing systems and rapid changes currently taking place with the evolution of the market and increase product diversity involves a detailed analysis of manufacturing processes. Manufacturing systems analysis can be done using methods based on analytical models and simulation. By applying these techniques one can view, analyze and forecast the performance of a dynamic system or of a new implemented system. Performance is one of the main factors affecting the design, development and configuration of any flexible manufacturing system. In this context, the present research will be directed to flexible manufacturing system structure analysis, to assess its performance, as well as simulating and validating these systems. It will generate and analyze key performance parameters, for the designed system by the use of specific software. Thus, it will study: the influence of the initial conditions or parameters on the behaviour of the flexible manufacturing system, the influence of the developing commands and system states on output or the influence of changes in the system structure on its behaviour.

1 Introduction

Flexible manufacturing systems (FMSs) are those manufacturing systems intended for manufacturing tasks of large typological diversity, high complexity, ensuring time delivery and minimal manufacture costs, while manufacturing is unpredictable , being held in small batches with frequent change. From the point of view of the global characteristics, flexible manufacturing systems are superior to manufacturing systems which they replace.

In their evolution [1] there can be noticed efficient flexible systems implementing for processing/assembly, but also many failures, determined most of the times by an over evaluation of technical solutions, by the impossibility of managing efficiently these systems which are extremely complex, by huge investment they need. It is very difficult a correct substantiation of the investment costs, which most of the times exceed the expectations. Current, flexible manufacturing systems are being increasingly implemented [2, 3]. Between them and standard machine tools or transfer lines it will be a coexistence with a tendency to increase the use of flexible manufacturing systems. Based on the many achievements of the FMS it is now possible to have fundamental data on the behaviour in service, the requirements for use and areas of utility, particularly with regard to the economic aspect.

* Corresponding author: fota.a@unitbv.ro

With regard to flexible manufacturing systems modeling operation, there are still several attempts, some recent, [4, 5] to define and analyse the performance indicators of the system. Achieving and introduction into service of a flexible manufacturing system involves a lot of expenses and the expected economic effect is hard to say. For this reason, both during the design and the operation of the FMS, modeling and simulation technique has been used [6, 7, 8].

In this context, the authors propose their analysis of the structure flexible manufacturing systems in order to assess their operational parameters and the knowledge of system behaviour through simulation before implementation. Using simulation we intend to clarify some issues regarding the description of the system operation, development of theories and hypotheses based on previous functioning, using the model to predict future functioning, namely the establishment of the effects of changes in the system.This paper is aimed at developing and applying optimization of flexible production systems in order to increase technical and economic performance of these systems, based on previous research conducted by the authors [9, 10].

2 Structure of the flexible manufacturing systems. Case study

The evolution in time of the condition of flexible systems for processing cylindrical parts, for dynamic systems could be described by algorithms or complex procedures [9], which in turn better describe the actual operation.

Figure 1 presents the structural scheme of FMS which will be used in the estimation of the operational parameters for cylindrical parts in real processes. Its validation was done by computer simulation for actual physical applications [10]. This system must implement a flow of operations as follows: cylindrical semi-automatic supply from a warehouse; the transport of the semi-finished on a conveyor to the drilling machine processing station; piece drilling operation, with control of the computer drilling depth and speed of advance; transport of the drilled piece towards the processing by turning simulation station; simulation of the turning process: clamping the piece between two tanks and turning it; checking the exterior odds of the piece via a measuring laser sensor (online control); displaying and recording of the measured data on the computer; transport the parts on the gear cutting station and grinding station and at final transport to the warehouse of finished goods; storing of the parts in the warehouse of finished products or ejection of rejects, which are not within the tolerances, settable via software SCADA.

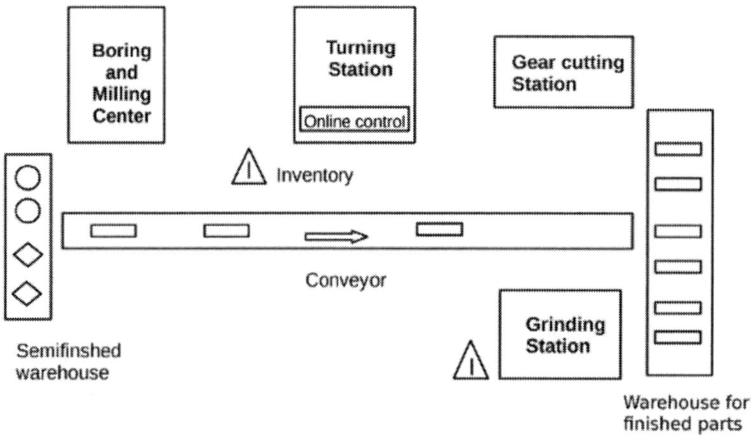

Fig. 1. Structural scheme of FMS.

3 Case study regarding estimation of operational parameters of FMS through simulation

The performance is one of the main factors affecting the design, development and configuration of FMS. Measurements, analytical models and simulation models are alternative performance analysis techniques. The models based on the simulation are preferable because they allow to obtain accurate results for complex and larger calculations, with fewer assumptions.

The methodology applied in the study of the processing of cylindrical parts in a FMS was by decomposing the system, detailing the internal structure of it, so the running processes of all its subsystems components could be algorithm-based and introduced step by step in the computer, observing accuracy and decision into account the interaction with the external environment, [9]. The evolution in time of the system state was described by algorithms or complex procedures that better express its actual functioning. Within the analysis through modeling, it will take into account the following categories of time: directly productive activities require working times (t_w); indirectly productive activities uses auxiliary times (t_a), which may be auxiliary times during the work cycle (t_{aw}) that are also called idle times and are composed of times for: supply / evacuation, positioning, waiting, change tools, grip / detaching in work device; auxiliary times off work cycle (t_{ao}): transport / transfer times, handling times, adapting times for manufacturing batch, waiting times. Major influences on system performance have waiting times which are generally unpredictable times, challenged by asincronysmus of the system. To improve performance of the system, during the design / manufacturing, activities that consume large auxiliary time are emphasized.

3.1 Input data

To study and analyse the performance of a cylindrical parts processing FMS, [10] will consider the following initial data of entry into model: current part in manufacturing (P_k, $1 \leq k \leq p$), p - number of types parts from typological diversified manufacturing task; the processing of parts P_r (k, $k = ct$). It considers serial production, manufacture executing on the size batches of the size of $n_b[pcs/batch]$, for $\forall P_k \in FP_k, k \in \aleph, k = \overline{[1,r]}$ with the frequency of occurrence of the lot $v_{bk}[no.batch/\tau]$. Any model is developed on a time interval τ (week, month, and year).Performance analysis simulation system will be made in relation to a number of operational parameters: the duration of manufacturing (T_m), the average manufacturing duration (T_{mk}) of a part on the whole system of machine, the production rate (R_p) on the part, on the machine (m_i), production capacity, average unit through production time - $\overline{T}_{mu}[h/pcs]$, the period of the machine cycle - $T_c[min/pcs]$, the degrees of utilization in time for workstation (η_{tu})and in days (η_{du}), production in processing (P_p). The developed model aims to know the system behaviour as well as it performance, and if possible, even before their physical realization.

3.2 Simulation of FMS

The simulation can intervene when designing the FMS to determine the static charge or for more elaborate validation solutions that rely on dynamic models. These techniques are used to substantiate decisions on the optimal exploitation of FMS. Selecting the programming language in which the simulation model will be coded has significant effects on the quality of results analysis through simulation of the systems.

This case study will use ARENA Simulation Software version 15.00.00001 (Arena Trial version), [11]. ARENA is built on the SIMAN simulation language. Alongside the encoded transcription of the simulation model, the simulation language performs a graphical animation, which allows to view on the display the behaviour of the simulation. ARENA is considered powerful software specialized in the analysis and evaluation of complex manufacturing systems [12, 13].

ARENA is software for dynamic simulation of discrete technical systems that includes FMS and uses terms like: *entity* is part of the system which constitutes a subsystem; *attributes* are characteristics of the entities; *activities* are dynamic processes that lead to modification of attribute values; *event* produces changing the value of an attribute. ARENA requires knowledge of random distributions, probability, and statistics.

3.2.1 Simulation modelling of proposed FMS. Results

For the realization of the imitative model of the FMS and its computer simulation were considered these dates: the number of the machines-tools from the system, the number and the type of vehicles and the transport speed, the number of deposits and their capacity, the type parts, the technological route, the cycle time of processing of each type of piece and each module of the flexible system, the matrix distances between machines; transfer time between transport means and machine-tools, the average adjusting time of each machine-tools. For analysis were used analytical and configuration models, previously designed FMS [9, 10].

Figure 2 presents the flexible system designed model and analyzed in the previous paragraph, model built using ARENA simulation software, which describes the logic of model. Arena is an easy-to-use, powerful modeling and simulation software tool that allows the user to construct a simulation model and run experiments on the model. Arena provides a simple method to setup the simulation parameters and the model input parameters [11].

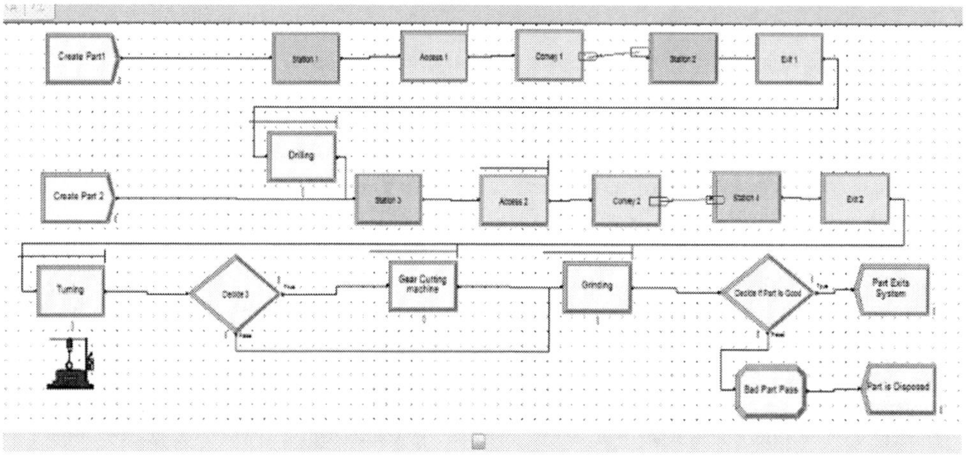

Fig. 2. Flowchart of the FMS model.

The software generates several reports as a result of a simulation run.

As a first result obtained by evaluating the flow of parts on machine and processing stations and examining conflicts regarding the application of limited resources, it could be evaluated and optimized the system layout, the selection of equipment and industrial logistic subsystem, as well as the operating procedures.

The analysis was performed for the following parameters required in the design and optimization of the production system: real manufacturing times for each task of production, production rate, laden workstations, analysis of bottlenecks, allocation of tasks on workstations, scheme of composing system.

After running of the simulation model show the result of throughput analysis such as average, minimum and maximum times for production.

Table 1 shows part of the Category by Replications report. For example, from this report the average VA time is 2.9776 minutes for Part 1 and 3.3240 for Part 2. The increase of the system performance was achieved by reducing the handling times of pieces, waiting time and duration of manufacturing.

Table 1. Category overview time in FMS.

SFF

Replications: 1 Time Units: Minutes

Entity

Time

VA Time	Average	Half Width	Minimum Value	Maximum Value
Part1	2.9776	(Insufficient)	2.6148	3.3871
Part2	3.3240	(Insufficient)	2.8805	3.8507

NVA Time	Average	Half Width	Minimum Value	Maximum Value
Part1	0.00	(Insufficient)	0.00	0.00
Part2	0.00	(Insufficient)	0.00	0.00

Wait Time	Average	Half Width	Minimum Value	Maximum Value
Part1	12.8531	(Insufficient)	0.8971	22.7689
Part2	12.2111	(Insufficient)	0.00	22.3008

Another important performance measure is also utilization of the workstations from the system. Table 2 shows part of the Resources report produced after the simulation run. The resource (Machine 2 - Turning and Machine 4 - Grinding, for example) utilization was 99.06% and 98.62% (average) and 100% (maximum). Analyzing the operational parameters of the system it aims to balance by raising the load Machines 3 (Gear Cutting) and Machine1 (Drilling). This will reduce the losses on the production line.

Table 2. The Resources report of the FMS model.

Resource

Usage

Instantaneous Utilization	Average	Half Width	Minimum Value	Maximum Value
Machine 1	0.7188	(Insufficient)	0.00	1.0000
Machine 2	0.9906	(Insufficient)	0.00	1.0000
Machine 3	0.6365	(Insufficient)	0.00	1.0000
machine 4	0.9862	(Insufficient)	0.00	1.0000

Figure 3 shows part of the Queues report. Arena provides a facility to generate or draw dynamic plots of a simulation run (Figure 3).

SFF

Replications: 1 Time Units: Minutes

Queue

Time

Waiting Time	Average	Half Width	Minimum Value	Maximum Value
Access 1.Queue	0.5340	(Insufficient)	0.00	4.7441
Access 2.Queue	9.9123	(Correlated)	0.00	17.7556
Drilling.Queue	0.02810185	(Insufficient)	0.00	0.4567
Gear Cutting machine.Queue	0.4779	(Insufficient)	0.00	4.3909
Grinding.Queue	1.9665	(Correlated)	0.00	4.9092
Turning.Queue	0.2418	(Correlated)	0.00	0.8231

Other

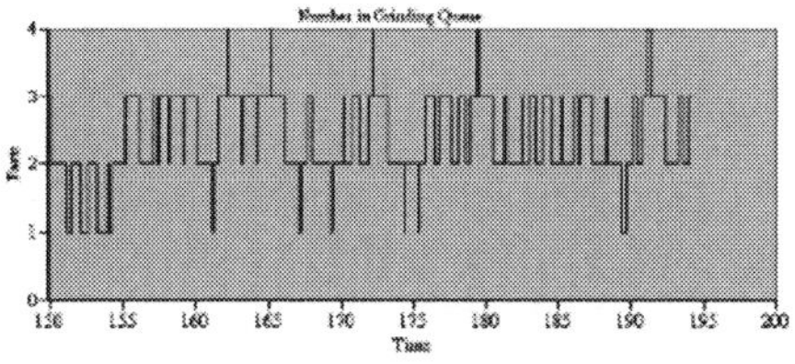

Fig. 3. The Queues reports of the FMS model and dynamic simulation run.

By knowing and evaluating these parameters it can be performed some optimizations on the process and the system by: determining the optimal type and number of workstations, choosing logistics system and developing the layout of the system. The parameters and performance indicators will be set in order to optimize the synthesis of the system in real-time environment. One obtained increasing production capacity and the degree of utilization. The reduction of production processing and the duration of manufacture have increased the production rate and thus to reduced production costs.

4 Conclusions

Modeling and simulation of a flexible manufacturing system involves creating a model of the system and its implementation in a computer utility program aimed verify system operation and its performance evaluation.

To achieve the research it was carried out verification and validation of the designed models, the stage by which their validity was confirmed by using existing software and modern equipment in the modelling and simulation laboratory of the production systems and the research centre. This work identified opportunities to shape the research and the application in industry. During the process of simulation cylindrical parts processing one were able to identify and eliminate errors, adverse situations and events.

In the simulation were used values of the transitions related to conditions from the real process so that the simulation better reflected the behaviour of the flexible line. Thus, they were able to study the complex interactions within the analysed flexible production system, which helped to increase the performances of the system. An important role in the FMS is knowing the number of processing subsystems, their layout and diversity. Layout-structure configuration of processing systems and logistics subsystem has a direct influence on the production capacity and adaptability through reconfiguration. In the simulation phase, assessing performance criteria allows the introduction of strategies for planning the system resources.

References

1. M.P. Groover, *Fundamentals of modern manufacturing, materials, processes and systems*, Edition: 4, pp. 935 (John Wiley & Sons, 2010)
2. B.S.P. Reddy, C.S.P. Rao, Int. J. Sim. Mod., **10** (2011)
3. Y. Koren, M. Shpitalni, J. of Manuf. Syst., **29 (4)**, 130-141,(2011)
4. G. Seebacher, W., Herwig, Int. J. Prod. Ec. **153** 340–351 (2014)
5. A. I. Khaleel, *Int. Conf. Ind. Eng. Op. Manag.*, Detroit, Michigan, USA, September 23-25 (2016)
6. R. Pisuchpen, E. J. of Assam. Autom., **32(1)**, 51-61 (2010)
7. E. Abdulziz, A. Mustufa, M. Hammad, A. Javed, J. Eng. Sci., **4**, 115-129 (2011)
8. F.T.S., Chan, I. J. Manuf. Tech., **21**, 328-338 (2009)
9. A. Fota, *Designing of the machines systems. Modeling and simulation* (Transilvania University of Brasov Publishing House, 2004)
10. A. Florescu, S. Barabaş, Int. J. Sci. Eng. Investig., **5 (59)**, (2016)
11. Arena User's Guide, Rockwell Automation Inc., USA (2014)
12. W.D. Kelton, R.P. Sadowski, N.B. Swets, *Simulation with ARENA. 5th ed.* (McGrawHill, New York, 2010)
13. J.M. Garridi, *Object oriented simulation. A modeling and programming perspective*, (Springer, 2009)

Laser scanning and CAD conversion accuracy correction of a highly curved engineering component using a precision tactile measuring system

Athanasios Giannelis[1], *Ioanna* Symeonidou[2], *Dimitrios* Tzetzis[2,*], *George* Kaisarlis[3], and *Panagiotis* Kyratsis[4]

[1]Qcontrol, GR 17675, Athens, Greece
[2]International Hellenic University, 14km Thessaloniki - N. Moudania, Thermi GR57001, Greece
[3]National Technical University of Athens, School of Mechanical Engineering, Athens, GR15780, Greece
[4]Western Macedonia University of Applied Sciences, Department of Mechanical Engineering and Industrial Design, Kila Kozani GR50100, Greece

Abstract. Impellers are used in various mechanical applications and they usually operate under severe conditions. Very often a replacement of an impeller is required after certain operation hours due to material failure. Since it is a critical and expensive spare part, accurate manufacturing, following the design requirements of the part, is necessary. In order to avoid any dimensional manufacturing defects, quality control procedures are used based on reverse engineering. The aim of the current paper was to evaluate the overall accuracy of the reverse engineering procedure, when a laser scanner is used for digital data capturing. The specified accuracy of the laser scanner was of ±0.084mm and the scanned data were converted to a Computer Aided Design (CAD) file using commercial design software. In order to assess the accuracy of the laser scanning and the CAD conversion procedure, a highly accurate coordinate measuring machine was used with a touch trigger probe and specified accuracy of ±0.006 mm. The results have shown that the deviation produced by the laser scanning and the associated reverse engineering methodology was up to 1mm, mostly observed at the edges of the freeform surfaces. It was concluded that such large deviations are caused from the inaccuracy of the laser scanner, possible errors created by the mesh – polygonal model creation, but most importantly from the errors when creating the B-splines.

1 Introduction

Reverse Engineering (RE) is the process of discovering the technological principles of an object or component through analysis of its structure and function [1]. It is very often used in mechanical parts for a variety of reasons, which may relate to the availability of a spare

* Corresponding author: d.tzetzis@ihu.edu.gr

part or in order to redesign and improve the function of the spare part [2]. There is a great range of techniques, which are all accompanied by their advantages and limitations. Most of the RE techniques include three main phases, namely a) data capturing, b) data processing (point cloud processing, triangulation, segmentation, alignment, meshing) and c) downstream applications i.e. CAD model, prototyping [3, 4].

The present research aims to analyse the advantages and the disadvantages of the procedures applied when creating a 3D model, measure and evaluate the errors formed during different stages of the process, study the accuracy of the reverse engineering procedures and detect any weaknesses in order to propose improvements of the process. An impeller is chosen for the evaluation since often it requires replacement after a certain number of operation hours or after material failure.

For the RE of the impeller a laser scanner was used and evaluated with respect to its accuracy and ease of use. For the evaluation, a coordinate measuring machine (CMM) with touch trigger probe, which is basically a tactile measurement method with a high accuracy, was used as well. With the CMM the part was re-measured and the measured data were compared to the created 3D CAD model with the use of specialized measuring software.

2 Laser scanning procedure

The laser scanner used was the ROMER Absolute Arm 7-Axis from the "SI" series, a configuration that requires no external controller box and has good characteristics, ergonomics and ease of use. Its automatic laser power control allows scanning of different surface types [5].

The operation of the scanner is manual and therefore it has to be manually driven over the surfaces which are to be scanned. The mounting of the part must safeguard that it shall not be moved during the scanning process. Through a set of absolute encoders the system recognizes the position of the scanner and the angle to the part. For the data collection and the data acquisition InnovMetric PolyWorks™–Standard was used, as it offers a range of benefits related to mesh generation, sketching and file format compatibility. The process was divided in certain steps using a variety of tools: a) IMAlign™ – collecting and editing the point cloud, b) IMMerge™ – creating the polygon file, c) IMEdit™ – edit the polygons and creating the curves & NURBS, d) IMInspect™ – measuring software.

3 Coordinate measuring machine

For the evaluation of the performed methodology, a system with a higher accuracy is required. With this system the deviations are measured between the created 3D model and the real object. For selecting an appropriate measuring system the commonly-used "Rule of Ten" was followed, i.e. the measuring instrument should be 10 times more accurate than the characteristic to be measured. The whole idea was to choose an instrument that was capable of detecting the amount of variation present in a given characteristic. Since the certified scanning system accuracy of the laser scanner was ±0.084mm, the measuring system used for the evaluation should offer a system accuracy of less than ±0.009mm. For measuring the object in order to perform a CAD to part comparison, the CMM DEA/Brown & Sharpe/Mistral with PH10M/TP200 Renishaw head was used (measuring range: 700mmx700mmx500mm, specified measuring accuracy according to ISO 10360–2: U(μm) =2.5+L(mm)/250, where L is the measured length). For the evaluation the measuring software PC-DMIS CAD++ was used, since it was state-of-the-art metrology software that effectively uses CAD models for the task of inspecting and evaluating parts.

4 Reverse engineering workflow

4.1 Scanning and data collection

When setting up the scanning workstation, the area of the part used for mounting was kept to a minimum. For the scanning procedure, the impeller was placed on the side of a granite table on which the scanner was also mounted and stabilized. Polyworks™ IMAlign™ was used for digitizing the part and for collecting the point cloud data. A plug-in tool enabled the direct interface with the laser scanner and the real time visualization of the digitalization process.

The laser density is automatically adjustable according to the part's surface colour and gloss. The point cloud is organized in a grid whose spacing was set by the parameters max edge length and interpolation step. The interpolation step used for this particular set-up was 0.3mm and the max edge length was set at the 10x the interpolation step. The maximum angle parameter was adjusted to 70deg. for providing sufficient points to reconstruct the surface. The outcome was several clouds of points (highlighted in different colours – see Figure 1). Since only the upper part of the impeller was totally accessible for the scanner laser stripe, the scanned data included several gaps that appear as holes. In Figure 1, the non-closed scans appear blue, as these are the scanned surfaces from the inside.

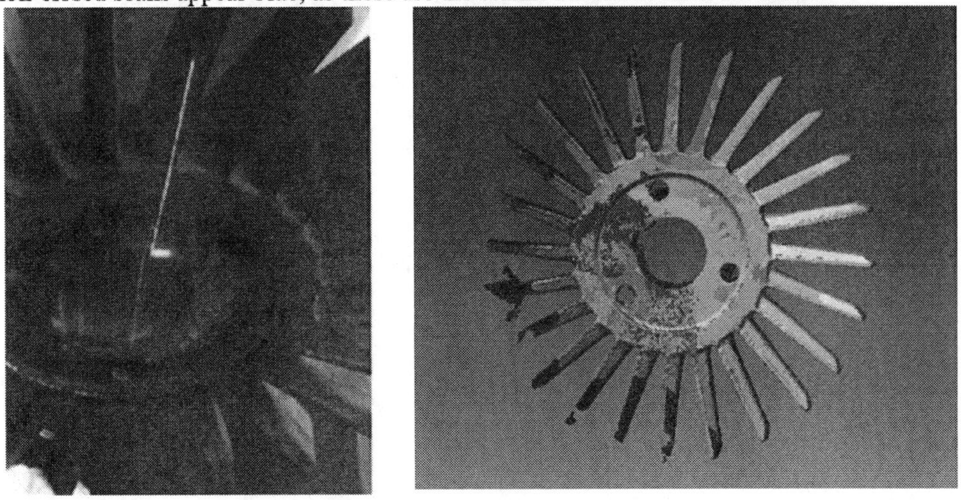

Fig. 1. The scanned object and the corresponding point clouds.

4.2 Processing of the point clouds

The created cloud of points is a cluster of different sub-clouds. Every time a new start point was used a new grid of points was created. This was used in order to log the different scan directions, which is helpful for further point processing. These several scans were overlapped in several areas (Figure 2). For the impeller used, 84 scans were performed.

The next action was to remove captured noise from reflection and fixtures, as well as points, which were not required for further processing. For creating the polygon- mesh and proceed with the CAD creation, it was recommended to reduce the size of the file. The IMAlign™ (Figure 3) reduced the cloud of points by reducing the overlapping points and by keeping only two overlapping clouds of points in areas where the scans were coincident.

DOI: 10.1051/matecconf/201711205009

4.3 Creating the polygons (stl) and the curves (B-splines)

The polygon mesh creation procedure was performed automatically and the parameters were predefined by the grid settings. For processing the polygon mesh, the IMEdit™ tool was employed. It was possible to optimize, smooth and reconstruct the polygon mesh and also to reconstruct surfaces in order to close the holes. In the process of transforming the point clouds into polygonal surfaces, holes appeared in the areas, where no points were captured. For optimizing the mesh, the polygons were reconstructed so that the angles, sizes and vectors were as uniform as possible. Smooth and reconstruct were not considered at this stage, as they would change the original surface data. Automatic hole filling was used for small holes, based on curvature calculation, while bigger holes were evaluated by the user via an interactive procedure. By completing the aforementioned steps, a watertight mesh was created, which was suitable for further digital fabrication.

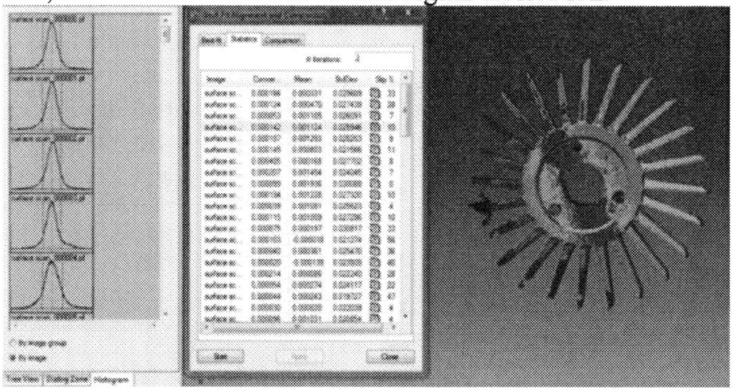

Fig. 2. A best fit alignment aligns the scans on the overlapping areas.

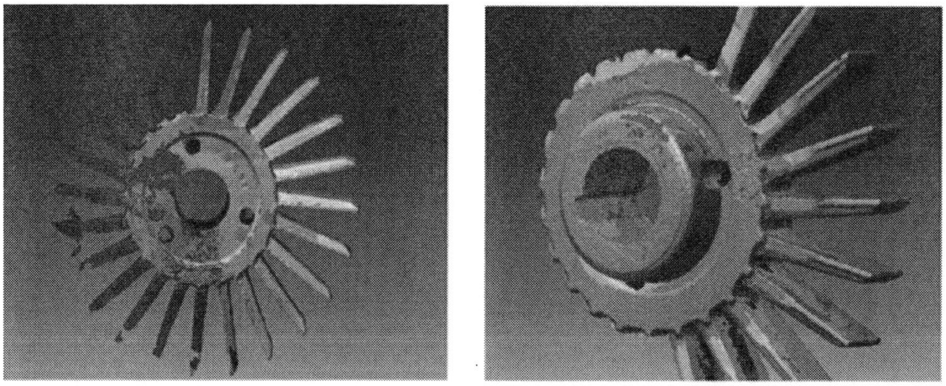

Fig. 3. Cleaning and reducing the cloud data.

Among the possible options for reconstructing a model, it was decided to create the curves and export them to a CAD system. The data could be imported in the CAD system and used directly to extract the features of the model. For the rapid reconstruction of the model one single blade was considered enough, as this could be circularly patterned. All the created sketches were exported as an .iges file, which is a universal file format and was imported in the CAD system (Figure 4). Planes were created based on sketches. Then these 2D features were converted into 3D features by using several of the available CAD tools. Geometric and freeform features were used to create the core base and the blades respectively.

DOI: 10.1051/matecconf/201711205009

4.4 Measurements with a tactile measuring system

Matching the CAD model coordinate system with the part coordinate system is a primary, yet rather delicate CMM task. For this work, a typical feature-based *(plane/circle/circle)* coordinate system was chosen and the potential errors were considered a priori. PC-DMIS software aligned CAD and part coordinate systems. The CMM was then set in DCC mode to automatically take over scanning procedure by: (i) generating the theoretical CMM paths for sets of points on repetitive scanning patterns, through the use of *"Linear Open Scan"* tool, (ii) capturing representative single points by the *"Auto Vector Point"* option. The CMM probe head was driven by the X, Y, Z coordinates of the selected CAD features and random points, which should be the same with the coordinates of the actual part. The real coordinates were recorded and the deviation was calculated (range: max=0.94mm, min=0.031mm). This range was confirmed after several experimental iterations, thus one cylindrical feature and 87 measuring points were considered representative.

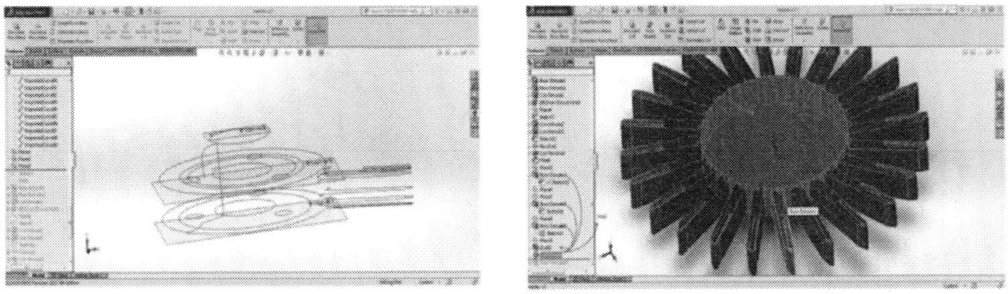

Fig. 4. Importing the data in SOLIDWORKS™ and creating the blades through circular pattern.

4.5 Measurements with a laser scanner

The second part of the evaluation of the accuracy of the reverse engineering procedure was to compare the convergence with the laser scanned data to the created CAD file. For this, the Polyworks™ software offers a measurement plugin named IMInspect™. The CAD file and the cloud of points were imported and aligned to each other. Then the deviation of the scanned points to the surfaces was measured. The software calculated automatically the deviation of all the measuring points to the nominal data; this was visualized on the CAD screen.

5 Data analysis

The aim of this research was to examine the accuracy of the reverse engineering procedure of creating 3D models from scanned data for mechanical parts with freeform surfaces. For this reason the measuring software's Polyworks™ and PC-DMIS were used. In both cases, the measured points with the tactile system and with the laser scanner were compared to the nominal values of the CAD data. In the examined case, the following measurements were applied: 1. Comparison of the created 3D Model within the used CAD system with the points measured with the laser scanner within the Polyworks IM™ inspect software tool and 2. Comparison of the created 3D Model in the CAD software with the points measured with the tactile system of the CMM in the PC-DMIS software. With both measurement procedures it was observed that the max deviations were up to 1mm. These deviations were mostly monitored near the edges of the freeform surfaces (Figure 5). Even though the RE technology of hardware and software is continuously improving, the design process steps are still error-prone. The entire procedure is not yet fully automated and therefore the experience of the operators and designers is of crucial importance. During the whole procedure and by evaluating the data with several systems it was observed that the errors have been determined by the following causes: 1. Laser scanning accuracy of ±0.084μm; 2. Unevenness of the impeller surfaces caused by the operation of the component; 3. Errors created by the mesh – polygonal model creation; 4. Errors created while generating the B-Splines; 5. Selected curves are not representative enough for the part; 6. Taking into account one single blade for the pattern and other errors made by the designer in the design software; 7. Alignment Errors of the CAD data with the measured points of the CMM.

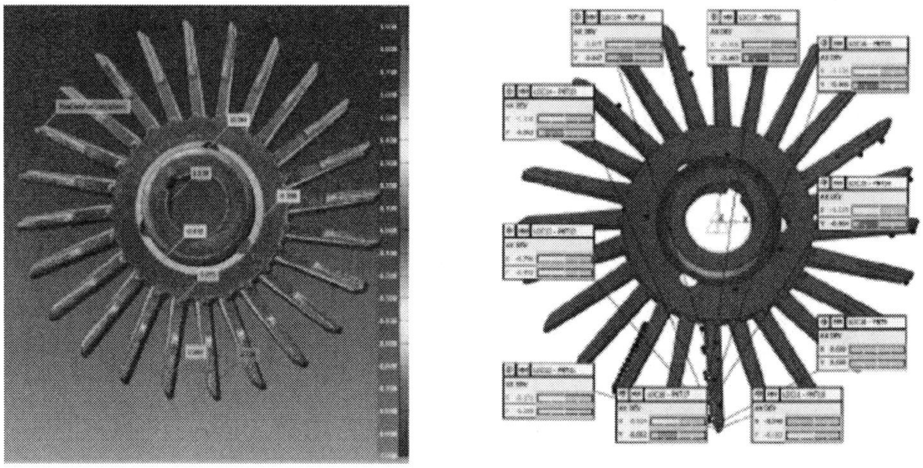

Fig. 5. The colour map scale showing the max and min values of the measured deviation.

6 Conclusions

The measurement results provide sound evidence that the technology is not ready yet to offer a fully-automated solution that safeguards the accuracy during reverse engineering of parts. Lack of user-friendly tools for the analytical, GUM-based expanded uncertainty estimation of the whole RE procedure is certainly a drawback. It could be possible to redesign the CAD model according to the measured deviations but the outcome is questionable and should be further examined. In any case, the technology provides certified accuracy of the measurement instrumentation which can, to a certain extent, ensure the designer about the quality of the RE procedure.

References

1. V. Raja, K. Fernandes, Eds., *Reverse Engineering - An Industrial Perspective* (Springer, 2010)
2. T. Varady, R.R. Martin, J. Cox, , Comput.-Aided Des., **29(4)**, 255–268 (1997)
3. K. Papadopoulos, D. Tzetzis, G. Oancea, P. Kyratsis, Academic Journal of Manufacturing Engineering, **14(1)**, 28-35 (2016)
4. G.R. Buican, G. Oancea, A. Manolescu, Applied Mechanics and Materials, **693**, 285-290 (2014)
5. "Romer Absolute Arm | SI." [Online]. Available: http://www.exactmetrology.com/metrology-equipment/romer-scanner/romer-absolute-arm [Accessed: 11-Feb-2017]

Determining workspace parameters for a new type of 6<u>R</u>SS parallel manipulator based on structural and geometric models

Lucian Milica, *Alexandru* Năstase and *Gabriel* Andrei*

Dunărea de Jos University of Galaţi, Department of Mechanical Engineering, Domnească 111800201, Galaţi, Romania

Abstract. Workspace geometric modelling of a new type of 6<u>R</u>SS parallel manipulator is described below. In the beginning, the researches undertaken in this area by other authors are highlighted and then a definition of this type of mechanisms is provided. The structural model of the 6<u>R</u>SS manipulator is briefly described. Inverse geometric model and translation subspace methods are used in order to determine the dimensions that define the workspace volume of the parallel manipulator. The reachable workspace is defined as a subset of the whole workspace in relation with the positions achieved by the characteristic point.

1 Introduction

In the last thirty years several studies on the applicability of parallel manipulators have been developed. Various applications of precision engineering such as lithography, microscopic assemblies, medical devices etc., require high-precision positioning in order to manipulate an object in different environments.

Regardless of the number of degrees of freedom, fewer [1-8] or more [9-12], the parallel mechanisms have proven their utility in various domains such as flight simulators, telescope position control, radar antennas, pointing devices [13].Parallel mechanisms have a robust construction and can move bodies of considerable mass and size at high speed, establishing a translational motion or rotation of the mobile platform. Compared to serial mechanisms, parallel manipulators have some special characteristics: high rigidity, high dynamic loading capacity, chain closed structure, high positional accuracy. Determining the workspace of a parallel manipulator is obviously an important issue in the design stage.

2 Structural model of the mechanism 6<u>R</u>SS

A mobile structure of a parallel mechanism, as illustrated in Figure 1, consists of n fixed curves (c_1) ... (c_n), n spatial actuators, b_1 ... b_n, linked to a mobile platform by n spherical joints, B_1 ... B_n. Links between n spatial actuators b_1 ... b_n and curves (c_1) ... (c_n) are achieved through n "point-curve"-type joints denoted by A_1 ... A_n.

*Corresponding author: gabriel.andrei@ugal.ro

DOI: 10.1051/matecconf/201711205010

According to the quantitative analysis of the structure mobility, the following parameters are defined:

- $m = n + 1$ (m is the number of mobile kinematic elements);
- $k_2 = n$ (number of A_j kinematic joints of second class);
- $k_3 = n$ (number of the B_j kinematic joints of third class).

The degree of mobility of the mechanism is given by Grübler-Kutzbach formula:

$$M = 6 \cdot m - \sum_{n=1}^{5} c \cdot k_n \tag{1}$$

where:

- c is the kinematic class of joints;
- k_n is the number of joints of nth class.

In the mechanical structure shown in Figure 1, the degree of mobility is:

$$M = 6 \cdot m - \sum_{n=1}^{5} c \cdot k_n = 6(n+1) - 2n - 3n = n + 6 \tag{2}$$

Present approach deals with the workspace of the mobile platform of a RSS manipulator with six degrees of freedom in different situations, when the constraints imposed on the platform are fully determined, limited or non-existent.

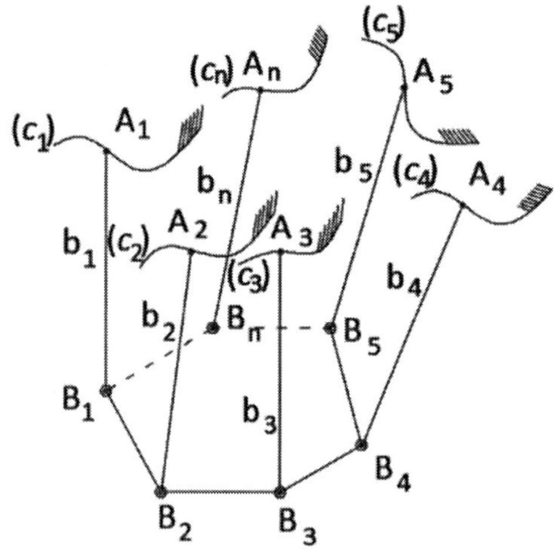

Fig. 1. Generic structural scheme of the parallel mechanism.

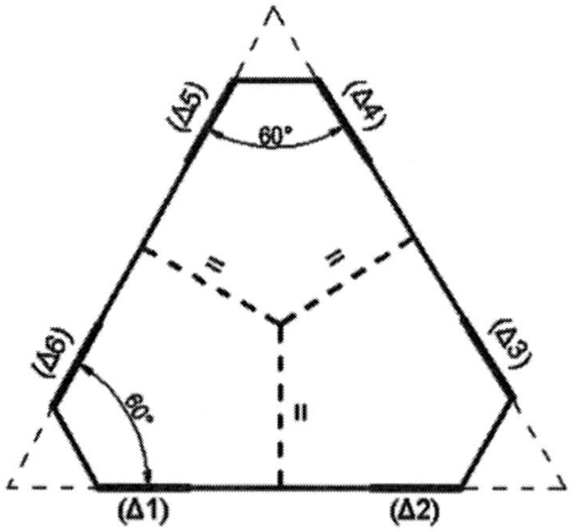

Fig. 2. Positioning of the axes of the rotational joints.

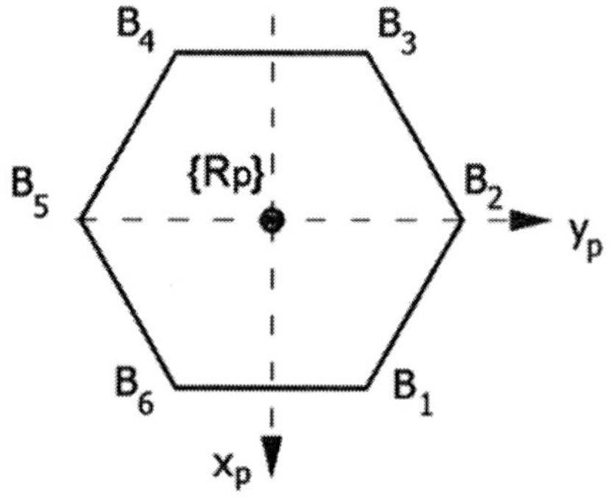

Fig. 3. Positioning of the spherical joints of the mobile platform.

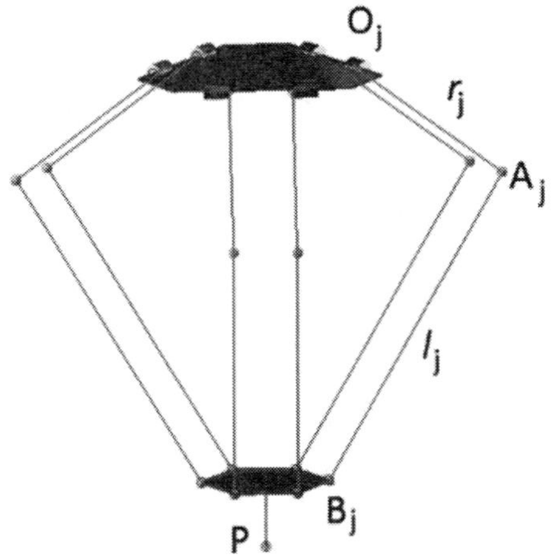

Fig. 4. Mechanism kinematic scheme.

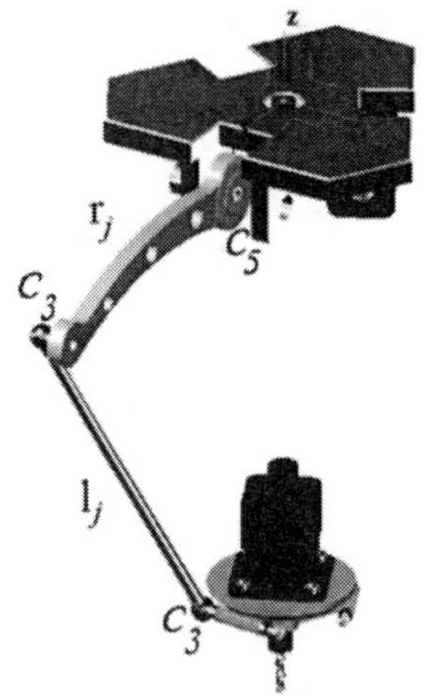

Fig. 5. New design version of 6RSS parallel manipulator (only a single actuating branch viewed).

The mechanism is characterized by the following features:
- the axis of rotational active joints Δ_j that generates circular trajectories (c_j) are coplanar, twos coincident and disposed central-symmetrically (Fig. 2);
- the spherical joints of the mobile platform are coplanar and their centres are located on the corners of a regular hexagon (Fig. 3);
- the actuating arms r_j are of equal length (Fig. 4);
- the rods l_j are of equal length (Fig. 4);
- the distance between the spherical joints A_j and A_{j+1} is equal to the distance between B_j and B_{j+1}, $j = 1, 3, 5$ (Fig. 4).

A new design version of 6RSS parallel manipulator consists of a fixed plate with six servomotors, and a mobile platform on which a seventh servomotor is fitted to drive a milling tool (Fig.5).

3 Inverse geometric model

In the case of inverse geometric model, parameters x, y, z, α, β, γ of $\mathbf{T}_{P,B}$ matrix - giving the mobile platform position - are known while parameters $\theta_1^k...\theta_6^k$ of each kinematic chain are to be determined.

In the following section, an algorithm for determining the angular parameter δ for one actuator will be described. A Cartesian coordinate system of versors (**v, u, k**) is attached to cylindrical joint C, where **u** is the unit vector of the joint axis, **k** is parallel to z_0 fixed axis of the reference system (x_0,y_0,z_0), and $\mathbf{v} = \mathbf{u} \times \mathbf{k}$.

Furthermore, from Figure 6 it can be noticed the projection B' of point B on the plane determined by the unit vectors **v** and **k** that contains point A.

The algorithm of inverse geometrical model includes the following steps (Fig. 7):

1. Calculating the coordinates of point B using the relation:

$$\mathbf{r}_B = \mathbf{T}_P^0 \cdot \mathbf{b} \tag{3}$$

where **b** is the vector of local coordinates of the point B in the platform reference frame.

2. Determining the distance d between point B and plan [**k, v**]:

$$d = (\mathbf{r}_B - \mathbf{r}_C) \cdot \mathbf{u} \tag{4}$$

 DOI: 10.1051/matecconf/201711205010

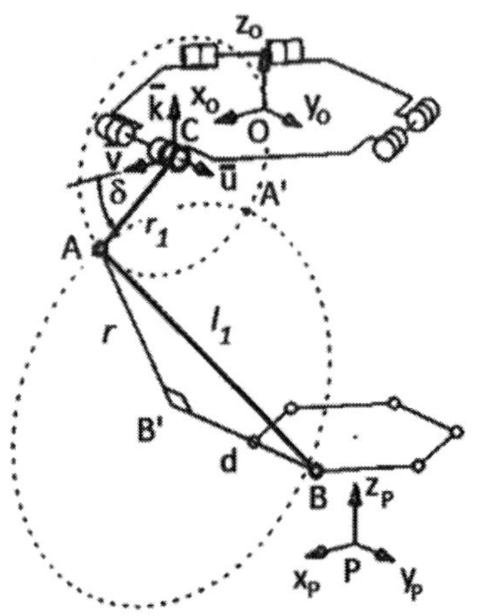

Fig. 6. Sketch of a single kinematic chain of the parallel manipulator 6RSS.

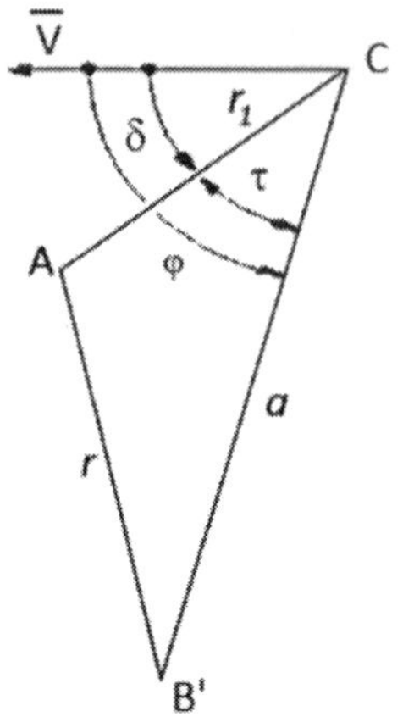

Fig. 7. Highlighting the angles φ, δ and τ.

If $|d| \geq l_1$, jump out to point 6.

3. Calculating the coordinates of point B':

$$\mathbf{r}_{B'} = \mathbf{r}_B - d \cdot \mathbf{u} \quad (5)$$

4. Calculating the distances a = B'C and r = B'A:

$$a = |\mathbf{r}_C - \mathbf{r}_{B'}| \text{ and } r = |\mathbf{r}_A - \mathbf{r}_{B'}| \quad (6)$$

If a $\geq r + r_1$ or a $\leq r - r_1$, jump out to point 6.

5. Calculating the value of angle φ.

Resolving *AB'C* triangle, and determining angle τ. Obtaining the solution $\delta = \varphi - \tau$.
End of algorithm.

6. There is no solution. End of algorithm.

4 Translation subspace

In a particular case when the mobile platform is parallel to the fixed platform and the axes of their associated coordinate systems are parallel to each other, then adjacent actuating arms are parallel ($\theta_1=\theta_2$, $\theta_3=\theta_4$ and $\theta_5=\theta_6$), at any time of movement. Based on the above observation, a method for determining the working space through a geometric demonstration about the locus of B_k points has been developed (Fig. 8). The method presented below is available only under specified conditions, applied just for a single linkage chain (e.g. $O_kA_kB_k$). The restriction imposed by the stiffness of the actuating arm A_kB_k, i.e. constant distance between points A_k and B_k, can be interpreted as a locus of A_k positions in relation to B_k, which is a spherical surface (Fig. 8a).

Then the locus of point B_k will be the volume generated by the sphere defined above, when it sweeps the arc of radius $r_k = O_kA_k$ and angle θ_h (Fig. 8b). Figure 9 shows a section through a plane normal to the axis of rotation of the arm. A_0 and A_1 are the centre of the sphere at initial and end position, respectively. Spherical surface was divided into two semi-spheres, S_1 and S_2. Figure 9a presents the section of the volume resulted after the displacement of the spherical surface. It is noticed that in this figure there is an empty space inside. The volumes generated by S_1 and S_2, whose intersection is the volume investigated, can be seen in Figure 9b. The reachable workspace is defined by the totality of space positions, orientations and locations that can be achieved by the characteristic point in a certain configuration of the actuating arms. This space is a subset of the whole workspace.

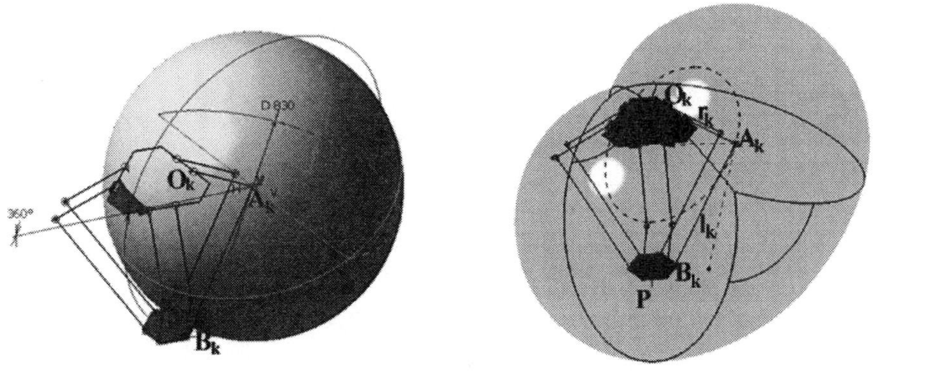

a **b**

Fig. 8. Spherical surface generated by locus of point B_k (a); the volume swept when the sphere is moving on the arc with radius r_k (b).

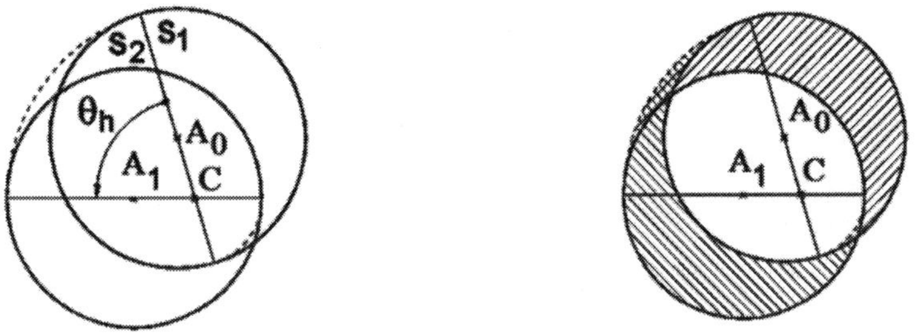

a **b**

Fig. 9. Sections through the locus of point B_k: (a) the section of the volume resulted; (b) projection of the intersection of volumes.

Actual determination of reachable workspace was done with a software application written in AutoLISP. Using the same AutoLISP code, it is possible to determine the reachable workspace of the parallel mechanism 6<u>R</u>SS, for different combinations of parameters a_1, b_1, a_2, b_2, r_1, l_1, h, θ_{oc}. Furthermore, a simulating approach is possible when certain positions must be reached during a particular manufacturing operation.

5 The volume of the translational workspace

In order to maximize the workspace volume, an algorithm for determining the boundaries of the operating space has been developed. The proposed algorithm consists of the following steps:

- cutting the reachable workspace with an horizontal plane of an arbitrary elevation, "z";
- in the section obtained, determining the vertices of an inscribed rectangle of maximum area, $x_{opt} \times y_{opt}$; the product between elevation "z" and rectangle area is the volume of a parallelepiped V_{opt}, which is maximum, related to the section plane (Fig.10);
- repeating previous steps for a number of "z" values and determining a volume chart.

The maximum-maximorum value of the volume chart will be the searched volume with entailed boundaries.

Further, an AutoLISP program to perform the above algorithm for a large number of equidistant plans was written (Fig.11). Using the program, an analysis for 20 elevation plans was carried out. Then, the chart of the maximum inscribed volume, depending on the elevation "z" of the current plan, was accomplished. The graph in Figure 12 shows the value of maximum-maximorum volume.

 DOI: 10.1051/matecconf/201711205010

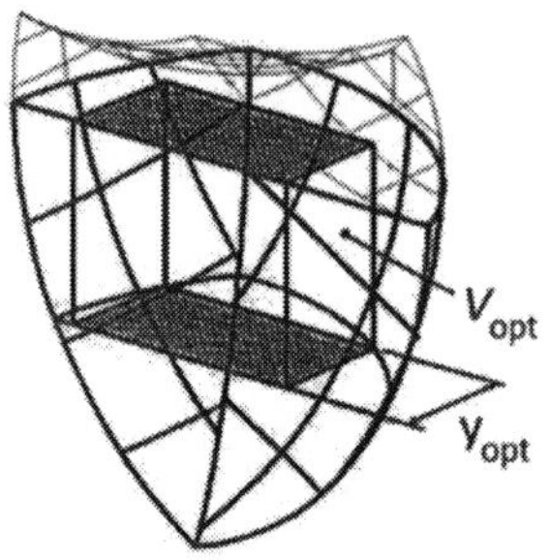

Fig. 10. Determining the rectangle of maximum area.

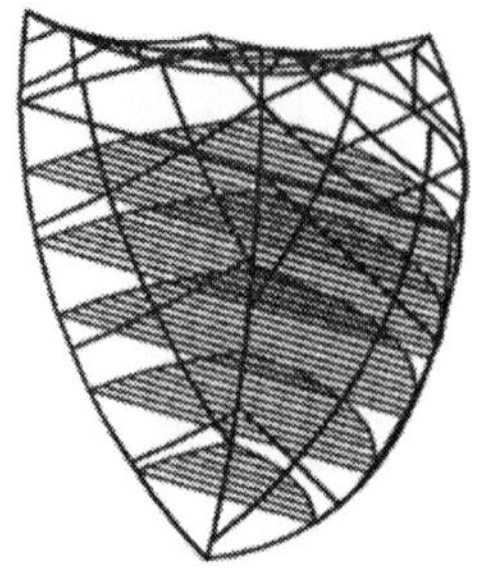

Fig.11. Equidistant sectioning of the reachable workspace.

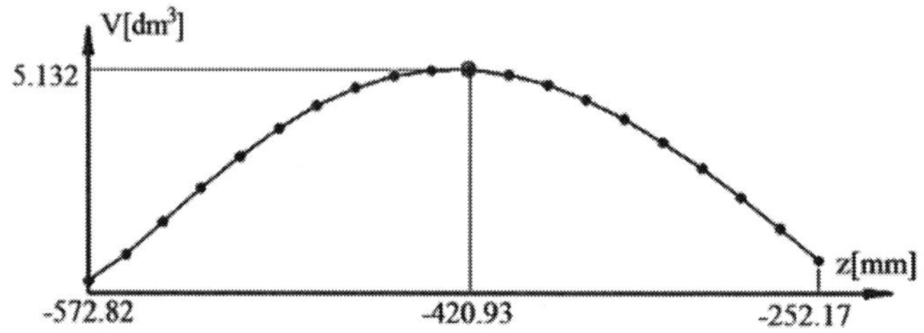

Fig. 12. Volume chart and the maximum-maximorum value.

6 Conclusions

The inverse geometric model and the subspace of translations were used in order to develop algorithms and software applications for determining optimal workspace of 6RSS parallel manipulator. The method described above is useful to assess the performance of parallel manipulators designed according to a certain structure and it can be extrapolated for other structural schemes. Using such algorithms, extreme values of optimal translational workspace may be inferred. Therefore a better understanding of the operation capability of parallel mechanisms could be achieved. Accurately defined dimensions for the translational workspace entails the range of kinematics parameters.

References

1. B. Roger, N. Scott, Mech. Mach. Theory, **56**, 138-155. (2012)
2. Y. Zhang, Q.Liao, H. Su,S.Wei, Mech. Mach. Theory, **52**, 47-58 (2012)
3. C. Lee, M. Jacques, Mech. Mach. Theory,**51**, 185-194 (2012)
4. R. Amir, A. Alireza, R. Mohammad, Mech. Mach. Theory, **51**, 195-216(2012)
5. S. Liu, Y. Yu, Z. Zhu, L. Su, Q. Liu, J. Cent. South. Univ. T., **17**, 323-331 (2010)
6. Q. Li., Q. Chen,C. Wu,Z. Huang, Mech. Mach. Theory, **65**, 46-57 (2013)
7. R. Srivatsan, B. Sandipan, Mech. Mach. Theory,**62**, 150-165 (2013)
8. Z. Du, R. Shi, W. Doung, J. Cent. South. Univ. T, **19**, 2482-2487 (2012)
9. H.S. Kim, L.W. Tsai, ASME J. Mech.Des. **125**, 92–97 (2003)
10. C.H. Liu, S. Cheng, ASME J. Mech. Des. **126**,1006–1016 (2004)
11. I.A. Bonev, J. Ryu,Mech. Mach. Theory, **36**, 1-13 (2001)
12. L. Milica, G. Andrei, MATEC web of conf.,**94**(2017)
13. D. Zhang, *Parallel robotic machine tools*, Springer(2010)

Algorithms of control parameters selection for automation of FDM 3D printing process

Paweł Kogut, *Krzysztof* Kalinowski*, *Cezary* Grabowik, *Grzegorz* Ćwikła, and *Iwona* Paprocka

Silesian University of Technology, Faculty of Mechanical Engineering, Institute of Engineering Processes Automation and Integrated Manufacturing Systems, Konarskiego 18A, 44-100 Gliwice, Poland

Abstract. The paper presents algorithms of control parameters selection of the Fused Deposition Modelling (FDM) technology in case of an open printing solutions environment and 3DGence ONE printer. The following parameters were distinguished: model mesh density, material flow speed, cooling performance, retraction and printing speeds. These parameters are independent in principle printing system, but in fact to a certain degree that results from the selected printing equipment features. This is the first step for automation of the 3D printing process in FDM technology.

1 Introduction

3D printing (also known as a spatial printing) belongs to additive manufacturing group of techniques, and opposite to the machining processes - three-dimensional physical objects, based on a computer model, are "formed out of nothing" [1-3]. In additive process object is created step by step by successive addition of small pieces of material, thin layers [4,5].

Originally, 3D printing was one of rapid prototyping techniques used to build moulds and prototypes themselves. However, the development of this technology that affects the achievement of better accuracy of printouts caused that 3D printing has become a method of receiving ready-made objects which are characterized by very good properties [6]. The possibilities of using spatial printers mainly depend on production methods [7], materials, which are available on the market and the volume of the printer. In practice, all 3D printing methods are similar because they form the object layer by layer in order to create complex shapes. The difference, however, is in the way of creation those layers and the material which is used in the process.

The popularity of FDM technology derives from its cheapness and wide availability. It is also the most popular due to the number of printers which are available on the market [8]. This method is based on building the model layer by layer. Each layer consists of thin lines. The lines are applied according to the generated G-Code commands. After applying all lines on the given layer, the head of the printer (hotend) raises the value of the next layer thickness or the heatbed moves down, depending on the type of the printer.

* Corresponding author: krzysztof.kalinowski@polsl.pl

2 Control parameters

The process of creating 3D objects is under computer control. The proper selection of parameters allows controlling the 3D printing process and guarantee obtaining good quality models [9-11]. Currently, 3D printing is not fully automated but it can be changed in a few years. The key to success, an algorithm for selecting parameters that most influence the process of 3D printing should be made, because such parameters control this process. The creation of algorithms improves further work on the automation of the 3D printing process and can also be helpful for the today's operators of 3D printers.

The main aim of presented research is to find the most responsible parameters affecting the 3D printing process in FDM technology. These parameters are examined in order to present how they affect the final printout. Subsequently, the method of choosing optimal values for these parameters should be defined after making test prints. During the tests 3DGence ONE 3D printer was used.

In this study the parameters that significantly influence the final form of the printout were presented in Figure 1.

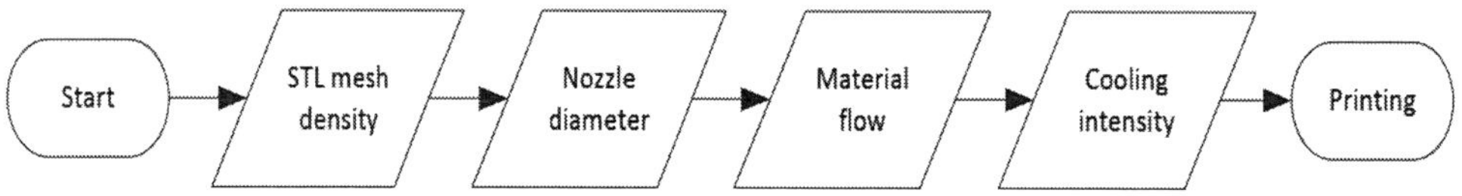

Fig. 1. The parameters' selection algorithm.

The parameters as material flow, percentage of cooling, retraction or printing speed can be customized in a slicer program designed to prepare the G-Code. The most important parameters are discussed below.

2.1 Model mesh density

The first step to print the object is its modelling in CAD software. For printing process, the model is translated to a grid triangle according to the STL (STereoLithography) standard. The main problem is choosing the proper mesh density for a model. It should be selected according to expected quality of final printout, with assumed level of details mapping [6,9].

Table 1 shows the three cases of mesh density: optimal, coarse and dense grid. The exampled test object consists of a cube, a pyramid, a cylinder with a hole and a ball and has 85 x 25 x 18 mm. Case 1 presents correctly generated mesh. As it is seen, the grid mesh density occurs on curved surfaces such as the cylinder with a hole, the ball and rounded edges of the stand. The case 2 was generated with insufficient accuracy. The small mesh density caused the round elements such as a ball or a cylinder more "angular". The last case (3) was generated with the greatest available accuracy. The density of the grid was so large that the edges of the elements, which were close to each other, made the colour of the model locally black.

Table 2 summarizes the main differences between these models.

The difference between coarse and optimal grid is noticeable. The number of elements in optimal grid is more than twice bigger than in coarse grid. Also, the file size is more than twice greater but it is still acceptable. The volume is also increased. The smallest element size and the angle tolerance available in the program were used in dense grid. The number of elements in dense grid was more than 200x larger than in optimal. Also, the increase of the file size was huge. The value of the volume did not change significantly in contrast to the size of the file, which was unacceptable in this case.

Table 1. Comparison of the mesh types.

Case	Grid	Model after meshing
1	Optimal	
2	Coarse	
3	Dense	

Table 2. Comparison of the mesh types.

Number	Mesh type	Mesh element	Angle tolerance	Number of elements	File size	Model volume
Case 1	Optimal grid	0,0415 mm	10°	2806	137 KB	10,37 cm^3
Case 2	Coarse grid	0,1076 mm	30°	1164	58 KB	10,33 cm^3
Case 3	Dense grid	0,0045 mm	0,5°	611350	21,9 MB	10,38 cm^3

2.2 Diameter of the nozzle hole

In most printer systems there's a possibility of use nozzles with different hole sizes. It is done manually by the operator but in the future can be automated. The appropriate diameter of the nozzle hole should be chosen for the specific project. The most popular hole sizes are between 0.1-0.5 mm. Smaller sizes allows achieving higher resolution and generally better surface quality of printed models. This is at the expense of much longer printing time due to the larger number of layers.

2.3 Material flow

The parameter of material flow enables correcting value of a fed filament during printing process. By increasing this flow the amount of extruded filament increases. The filament which is pushed by the extruder swells after leaving the nozzle. Each material behaves differently during printing. Therefore, proper value of the material flow must be chosen to get an accurate printout for each material [12-15].

An algorithm of a proper value selection of the material flow is shown in Figure 2. The first step is the measurement of the wall thickness in the calibration print. The value of the wall thickness should be equal to the diameter of the nozzle hole. The material flow can be also controlled during printing process for obtaining thickness of the wall smaller or bigger than the diameter of the nozzle hole.

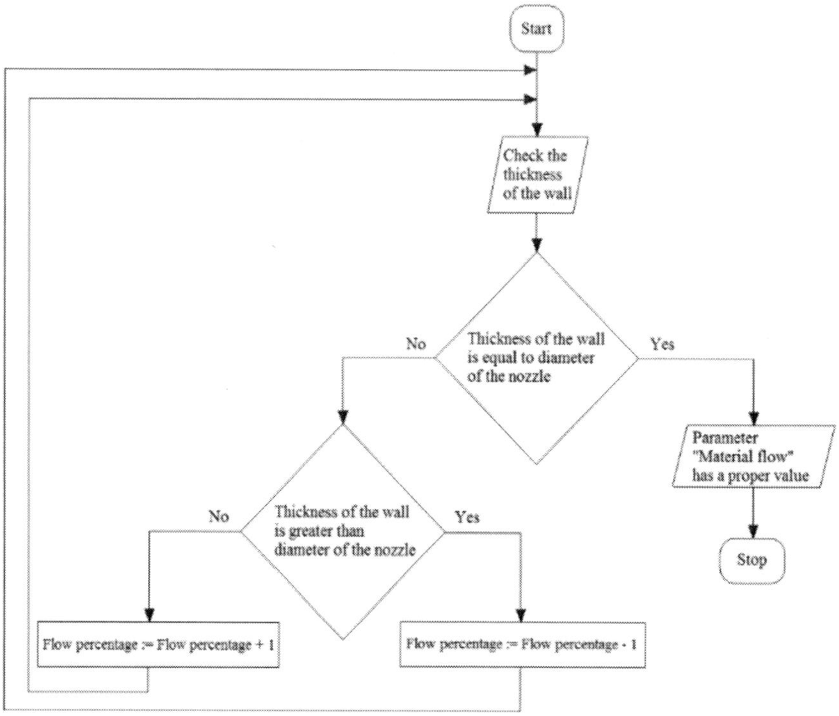

Fig. 2. The material flow algorithm.

2.4 Cooling

Cooling is one of the most important factors in FDM technology. Filament, heated in the printhead and extruded in the form of a plasticized line, requires cooling after laying. It would be seem that if the plastic path is not cooled properly it falls by gravity but insufficiently, cooled line tucks up (Figure 3). This is undesirable for two reasons: rolled up layers reduce the quality of the printout and the head can hit the rolled up layer during the application of the following layer.

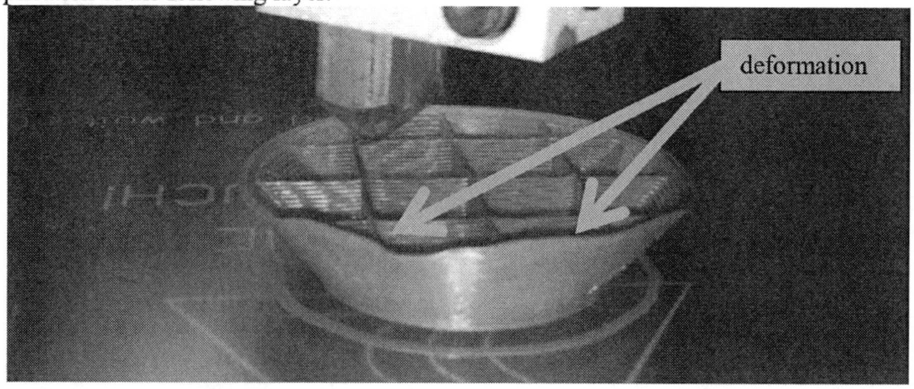

Fig. 3. The rolled up printout.

In extreme cases, it can cause the detachment of the printout from the heated bed. Sided cooling can cause one side of the model print correctly but the other side is deformed (Figure 4). The poor quality of the printout is usually caused by rolling up and badly arranged layers. The Monkeyfab Prime (a) and 3DGence ONE (b) printers were used for this two test printouts. The application of double fans on both sides of the model is very helpful to achieve good quality of the printout - their placement prevents deformation of the layers on each side of the printout.

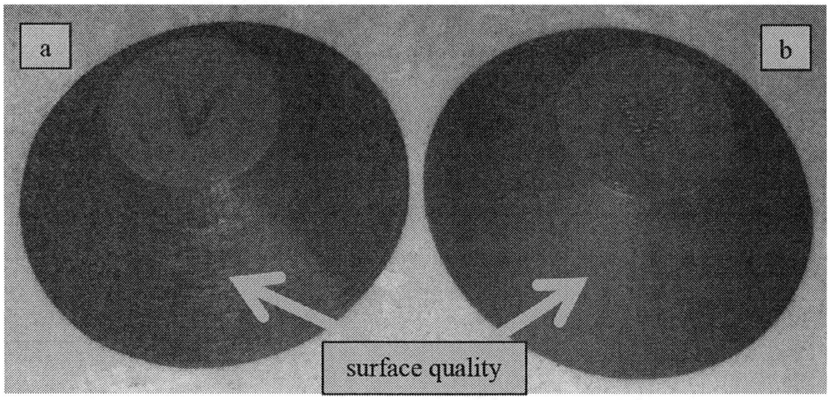

Fig. 4. The quality of surfaces using one side cooling and both sides cooling.

The proposed algorithm for selecting the cooling rate is presented in Figure 5.

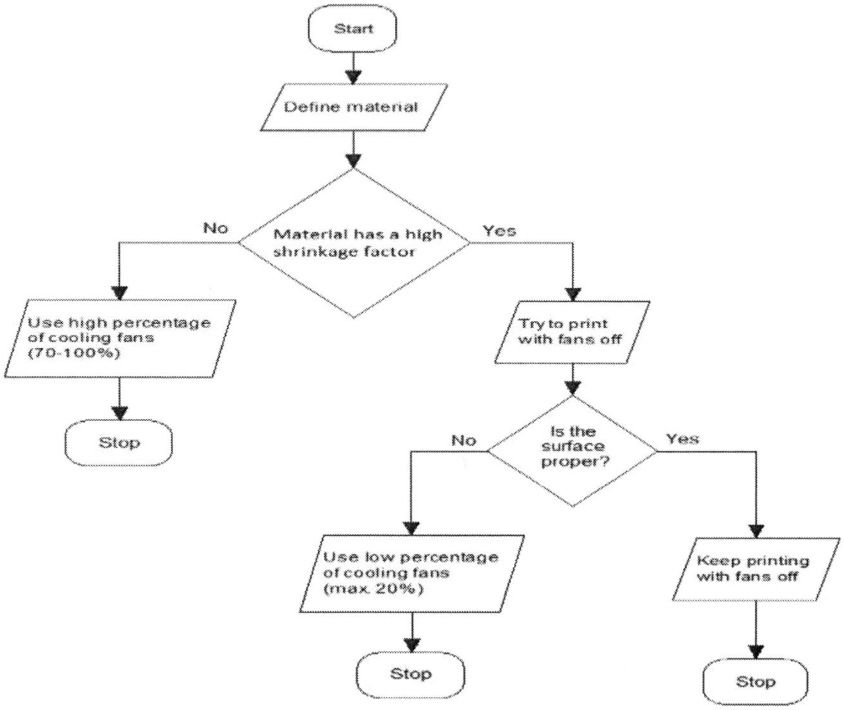

Fig. 5. The cooling rate selection algorithm.

Material characteristics strongly influence the need for the printout cooling. The first step is analysis of the filament material shrinkage factor [15]. In the case of a low value, the cooling should be more intensive. Depends on the cooling efficiency its value should be between 70 and 100% of nominal speed. However, if the material has a high shrinkage factor, the printing process should be executed without object cooling. Subsequently, the surface of the printout should be evaluated. Cooling should be turned off if the evaluation of the printout quality is satisfactory and acceptable. In the case of using ABS, printing without cooling provides a better adhesion between layers of the printout by which an increased strength can be achieved. On the other hand, a low percentage of cooling should be used if the quality of the printout is unacceptable. A dozen or so percent of cooling should be used in this case. 20% of cooling is the value of the upper limit. Above this value, the printout can be deformed during printing.

3 Summary

The proper selection of presented, most important parameters of the 3D printing process in FDM technology enables obtaining printouts with exact dimensions and excellent surface quality. Nowadays, FDM technology is not fully automated, but this may change in a few years. The creation of selection parameters algorithm is the first step to achieve this. The proposed algorithm consists of a selection of the optimal mesh density, appropriate diameter of the nozzle hole and the proper value of material flow and percentage of cooling. Because of the fact that the material type has a great influence on certain parameters (e.g. ABS in the case of percentage of cooling) all new materials should be examined separately.

References

1. 3D printing history, http://3dphoenix.pl/historia.html access: 02.2016.
2. A brief history of 3D printing, http://individual.troweprice.com/staticFiles/Retail/Shared/PDFs/3D_Printing_Infographic_FINAL.pdf, access: 02.2016.
3. P. Siemiński, G. Budzik, *Additive Techniques, 3D printing.* (in Polish), (Publishing House of Warsaw Polytechnic, Warszawa, 2015)
4. P. Brzeskot, E. Wąsek, Ł. Zatorski, *3d printing - innovations you can rely on* (in Polish). Mechanik, 12CD2, **88** 197-199 (2015)
5. H. Lipson, M. Kurman, *Fabricated - The new world of 3D printing*, (John Wiley & Sons, Inc., Indiana, 2013)
6. A. Kaziunas France, *Make: 3D Printing*, (Maker Media Inc., 2014)
7. A. Gwiazda, R. Knosala, J. of Mat. Proc. Technol. **64**/1-3, 181-188 (1997)
8. B. Evans, *Practical 3D Printers - The Science and Art. of 3D Printing*, Technology in action (2012)
9. J. Horvath, *Mastering 3D Printing*, Technology in action, (2014).
10. Description of 3D printing technology, http://www.allaboutlean.com/3d-printing/3d-printing-overview/ access: 08.2016
11. T. Dzitkowski, A. Dymarek, Journal of Vibroengineering **17**, 4, 1578-1592 (2015)
12. Opportunities in 3D printing, http://www.marsdd.com/wp-content/uploads/2014/04/MAR-CLT6965_3D-Printing_White_paper.pdf access: 02.2016.
13. Micromart and SB Ltd, *The Ultimate Guide to 3D Printing*, (2014)
14. S. Topolska, J. Łabanowski, Mater in Technol. **49**/4, 481-486 (2015)
15. Materialise's press materials, http://www.materialise.pl/przeglad-technologii-i-materialow, access: 02.2016.

The reliability problems of reading RFID tags in logistics warehouse systems

Pawel Zajac*, and *Stanislaw* Kwasniowski

Wroclaw University of Science and Technology, 27 Wybrzeze Wyspianskiego st. 50-370 Wroclaw, Poland

Abstract. The use of RFID (radio frequency identification) tags in logistic warehousing systems has been analysed. The research was carried out at the RFID measuring station at Wroclaw University of Technology. The test results for 125 kHz tags in dynamic write and read tags in one or several variants are presented. The applications presented in the article are universal and apply in widely understood logistic systems. In particular, they define the area of their use in logistic storage systems, in particular the identification of tagged and conveyed craft units, forklift trucks / stackers, or manually. Test results may improve the reliability of the automatic identification subsystem in logistic storage systems.

1 Introduction

Tags are used more and more frequently in logistics warehouse systems, automatic identification systems and other systems as well. The authors developed a warehouse rack overload sensor as a safety element of warehouse racking systems whereby an active tag was used (patent pending). Cooperating with logistics centers, the authors have noticed the need to increase efficiency of logistics warehouse systems. The paper [1-4] discussed those issues in relation to bar codes. Luggage RFID at airports has been known for a long time, albeit conditions of work of a passive tag write/read system are different than those in which the airport systems work. Unfortunately, increasing the efficiency of internal transport does not work in a way that would speed up a conveyor or a forklift truck, whereas an automatic identification system is capable of self-adaptation. Antennas, which are part of the identification system Figure 1 and are installed in the so-called gates, have to read reliably and as fast as possible all moved and stored shipping units along with piece items placed on top of them.

This problem is also encountered in the process of collection of piece items with the use of a mobile tag scanner in the case of a long and unsuccessful search of shipments in a warehouse (collection time gets longer which is a waste of time in accordance with the lean management rules) [5,6]. That is why fast and reliable identification of shipping units in loading subprocesses is very important. The research on the reliability of reading and writing tags, whose results were presented in this paper, was conducted at a laboratory station in the Laboratory of Automatic Identification in Logistics Systems at the Wroclaw

* Corresponding author: pawel.zajac@pwr.edu.pl

University of Science and Technology. The said results were of a universal nature and delineated the area of application of tags. Moreover, they were also checked during implementation work.

2 Logistics warehouse system vs. RFID

Warehouse operation can be divided into the following processes: receipt, storage, collection, release. According to the requirements of the processes, an automatic identification system is selected so as to be reliable [7]. In this case, this applies to the selection of the RFID tags (this issue was discussed in detail in [8, 9]), i.e. tag type/frequency, coded information vector and information read/write conditions in line with the WMS (Warehouse Management System) standards. Once such tags have been placed in a closed warehouse system (Figure 1), they have to work reliably.

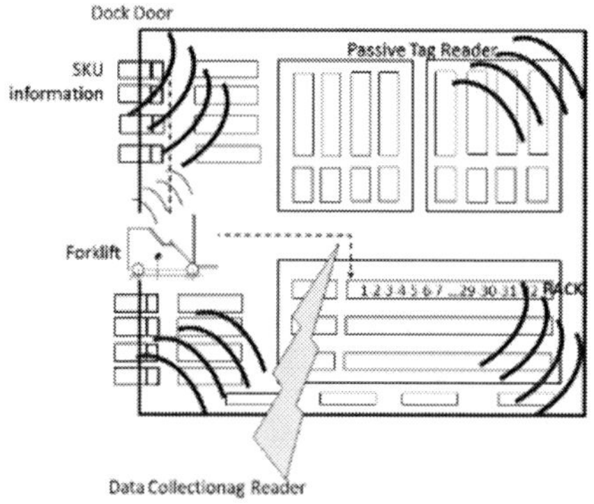

Fig. 1. RFID operation algorithm in the LSM (logistics warehouse system).

Activating a stationary antenna of an RFID reader or a mobile one in front of a full rack of found RFID pallets does not provide the answer to the question where the pallet sought for is – it is necessary to use functions and information collected in the WMS. However, it is not always possible, and it is certainly impossible in the case of a system failure. That is why, as part of safety improvements, non-collision read protocols and e-signatures are being introduced into the LSM. Issuing a read command in a place with a larger number of tags will cause many devices to send back a response to the base station at the same time. The individual pieces of information will overlap and the signal will eventually be unreadable. In order to avoid such situations, anti-collision algorithms are employed which are based on the mechanism of randomness. Waiting time of a base station is divided into so-called time slots. Transponders sending responses randomly choose a slot to transmit it in. This reduces the number of collisions that can occur, albeit it does not solve the problem completely. It may so happen that two transponders will draw the same slot. Philips has solved this problem using an 8-bit CRC (Cyclic Redundancy Check) algorithm. They are used in the WMS for navigating forklift trucks, e.g. determination of their whereabouts and operation status [10].

3 Factors affecting reliability

Experiments conducted under warehouse conditions allow spotting a problem with reliability of reading piece items. In the case of a cargo pallet with approx. 60–80 marked items, this reliability (or rather lack thereof) is not critical yet. However, when attempting to use RFID with a pallet with over a thousand items (e.g. books or pieces of clothing), the said reliability becomes critical. Currently existing tags and RFID readers are capable of reading approx. 300 tags per second. It would seem that it is enough then to stop under a gate with an antenna (a gate usually has several antennas: 2–5 pcs., this results from the radio wave coverage of the area of identification) for about 5 second in order for all tags to be read. Unfortunately, passive tags are powered by means of an query field of the antenna, there being no two-way communication between the two devices. Furthermore, a tag does not receive any information from the antenna that it can switch itself off already. In view of the above, tags closest to antennas will manage to response at least a dozen or so times, whereas tags further away sometimes do not response at all. The issue of determining electromagnetic characteristics of the packaging material becomes a major one as does the impact of placement and collection of items on pallets on its course. Actions aimed at increasing reliability can be expressed under three items: 1. Analysing the reliability of RFID systems, one comes across such terms as **application Id** (Id: identification). These are the terms which characterise the same mechanism consisting in communicating with a specific tag group only. A device activation command, sent from a base station, may contain a key which initialises communication only with those tags which have the same key saved in their memory. If a key received from a base station is different than that saved in the transponder, the device will remain silent. This type of tag identification is used in situations when items of various types are identified, i.e. items found in the same place (e.g. a warehouse) which are marked with identifiers containing different keys. Instead of reading all tags within the base station's coverage, only information from the tags that are of interest to us is received, i.e. from tags marked with appropriate keys; 2. Using authenticity checking mechanisms. Thanks to this, the data contained in a transponder are encrypted and can only be accessed by entering a correct password. Such mechanisms were employed, inter alia, in the TRIAS (triple radio automatic identification system) system. The principle of operation of this mechanism is based on a random number generator and an encryption key. A base station sends the number received from the random number generator to a transponder. Using the received number and its own identifier as the basis, the transponder generates a response. Subsequently, the transponder sends its own identifier and the generated response to the base station. On the basis of the previously generated number and the received identification number, the base station generates a series which should correspond to the response from the transponder. This completes the verification of authenticity of a tag; 3. Controlled access to information using tags with encryption of information. Two types of such devices are distinguished: tags with information protected during transmission and transponders with saved encrypted information. The principle of operation of those first ones consists in encrypting data before sending them from a transmitter and decrypting them once they have been received by the receiver. This technique is rarely used, since tags have to be equipped with encryption processors which raises the cost of devices and decreases the speed of their operation. The second group of transponders includes transponders with saved encrypted information. Once a base station has received encrypted information from a tag, it has to decrypt it using an appropriate key. This works similarly the other way round, i.e. in order for the base station to save information to the transponder, it first encrypts it with a key and the sends the data which are saved in the transponder's memory in encrypted form. Due to the necessity to perform encryption and decryption of data, the transmission gets slowed down.

4 Dynamic tests

The construction of the test station was shown in Figure 2a. A tag reader was connected with a computer. Tags were fixed to a rotating disc whose revolutions were controlled by means of an inverter. The inverter (T1) was connected with the rotating disc (T2) which transponders (T3) were mounted to. A straight edge was placed in close proximity to the rotating disc, in the tag's horizontal plane. The reader along with an antenna (reader) were placed over the rotating disc, but in such a way that only one tag was within the antenna's coverage. The placement of the reader relative to the tag was perpendicular. The position of the reader in relation to the rotating disc was being changed during measurements.

The set of alphanumerical characters used in writing tags corresponds to the set of MSI characters which was developed at MSI Data Corp. on the basis of the Plessey code. The MSI bar code, also called the modified Plessey code, is used mainly to mark racks with assortment. This code enables writing any number of digits in the 0–F hexadecimal or the 0–9 decimal system, albeit the version for writing in the decimal system is used more frequently. The data vector has the following structure: start sign, any number of digits, checksum, stop sign. MSI uses either one or two digits to write the checksum. It can be calculated in several different ways and it depends on the application which way will be selected. For the purposes of this article, "modulo 10" has been chosen.

UniqueMaker v1.2 is a programme used to communicate with an antenna for reading/writing passive tags working at the frequency of 125 kHz, through the RS 232 interface. Next, the antenna gets connected to a computer in the programme which will automatically establish a connection with the antenna. A tag is then placed within the coverage of the reader's antenna. The programme displays the "read code" which is written in the tag. Finally, we enter a code to be written in the "New code" field and then click on "Programme", thus programming the tag. Experimental research plan: reading distance - this test has been designed to check the maximum distance at which the transponder can be read by the reader; Mutual positioning (perpendicular or parallel) - this test has been designed to check the best positioning of the transponder relative to the reader's antenna ; Tag movement speed relative to the reader - this test has been designed to check the maximum linear speed at which the tag can be read.

4.1 Tag reading

The test was to verify to what extend readings of several transponders collided. Constant angular speeds were determined when conducting the tests. The possessed devices were being checked several times for approx. 30 s after setpoints had been changed in order to stabilize the measurement conditions. In order to avoid measurement interference, the system was calibrated. The first tag was placed on the rotating disc with a diameter of 150 mm. Linear speed was approx. 1 m/s. Reading distance was 40 mm. Subsequently, the second tag and the third tag were added. At the outset of the measurements, test revolutions were performed in order to determine the range of revolutions. One tag was fixed to the disc which was then set in motion. After several attempts, the maximum number of revolutions was successfully established for tags operating at the frequency of 125 kHz, this value having been 140 rpm. Next, three tags at a distance of 180 mm from one another were mounted on the disc.

The disc with tags was set in motion and then it was established that the rotational speed was equal to 140 rpm. The reader was put closer to the rotating disc so that the center of the reader's antenna and the tag's center were aligned. Reading angle of 90° was maintained. Each time no result was obtained, rotational speed was reduced. When a reading signal was obtained, the disc was stopped and the reading distance measured.

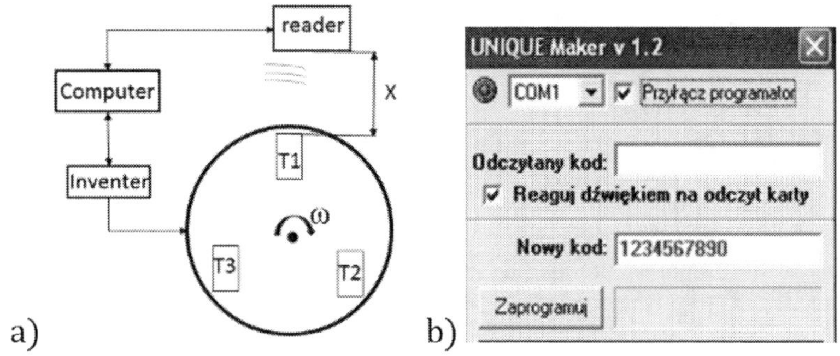

Fig. 2. Diagram showing the measurement station of the UniqueMaker v1.2 programme.

Tests were also conducted for one and two tags, but mounted on a smaller arm. In order to calculate the linear speed, the following formula $V = \frac{2\pi rn}{60}$ was used, where V – linear speed [m/s], r – disc radius [m], n rotational speed [rpm]. The obtained results were shown in Figure 3.

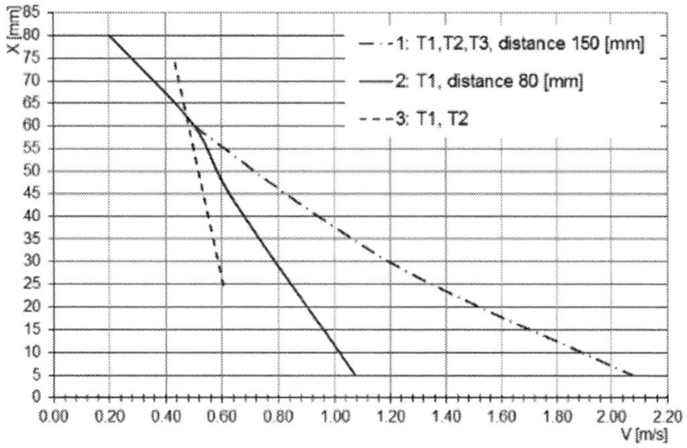

Fig. 3. Diagram for dynamic tests.

4.2 Tag writing

Stage of writing information in a tag, 1÷3 pcs., simultaneously, during movement. In this case, 10 alphanumeric characters can be written (in line with the definition of the information vector above). Tag writing time is approx. 1.5÷2.0 s. It is constant and does not change in the case of a change in distance of the tag from the programmer or the number of tags remaining within the antenna's coverage. **Measurement 1.** During programming, the transponder has to stay within the antenna's coverage. In order to ensure fully correct writing, the tag has to remain motionless. In the event of movement of the tag, writing may be incorrect. The programming time is approx. 1.5÷2.0 s. In the case of failure to perform a correct write operation, the transponder will not be "visible" for the reader at all. **Measurement 2.** If there are two tags within the coverage of the programmer's antenna, one tag will be read (the one with better coverage). During writing, the same new code is written in both tags. Reading or writing will not possible if two tags are placed in their axis and are stuck together. In the case of two tags in the axis, the distance between them has to be at least 10 mm. If the tags are stuck together, they have to be spaced in order to eliminate alignment. **Measurement 3.** Simultaneous writing of three tags within the antenna's coverage will take place if: three tags are placed axially, but perpendicularly, to the antenna and the distance between them is at least 15 mm.

5 Conclusions

The optimum linear speed for the tags (0.5 m/s) can be read from the graph in Figure 3. In this case, the number of tags or the mutual distance between them does not play any role. Furthermore, the smaller the speed, the higher the reading distance; and the higher the linear speed, the closer the reader has to be. This allows recommending these tags for use in logistics warehouse systems where it overlaps with the standard speed of roller conveyors. This is an important research proposal, universal in application.

The maximum reading distance is connected with the number of tags in their own vicinity. In the case of one tag, the reading distance X is 80 mm at a speed of 0.2 m/s, whereas with two tags, the maximum reading is 74 mm at a speed of 0.4 m/s. With three tags, the reading distance decreases to 60 mm at a speed of 0.5 m/s. The maximum speed at which the tags are read is 2.1 m/s. However, in order to make use of such a high speed, the distance between the reader and the moving tag has to be very short. Series 2 includes one tag on an 80 mm arm, whereas series 3 includes two tags on the same arm; tags are 170 mm away from each other. It can be observed that the reading speed of two tags at the same speed decreases. This phenomenon occurs only over the speed of 0.63 m/s. It is the other way round at lower speeds, i.e. two tags are read faster, despite that intuition could suggest that it would be one tag that is read faster.

References

1. H. Bhatt, B. Glover, *RFID Essentials* (O'Reilly, 2006)
2. D. M. Dobkin, *The RF in RFID. Passive UHF RFID in Practice* (Elsevier, 2008)
3. K. Finkenzeller, *RFID Handbook: Fundamentals and Applications in Contactless Smart Cards and Identification* (John Wiley & Sons, 2003)
4. A. Rizzi, R. Montanari, M. Bertolini, *Logistica e tecnologia RFID* (Springer, 2011)
5. A. Bujak, P. Zajac, *Communication in Computer and Information Science, Telematics in the Transport Environment **329**,* 199-210 (Springer, 2012)
6. A. Bujak, P. Zajac, Communications in Computer and Information Science, Activities of Transport Telematics **395**, 361-369 (Springer 2013)
7. A. Kierzkowski, T. Kisiel, J. Air Transp. Manag. (to be published)
8. S. Kwasniowski, T. Nowakowski, M. Zajac, *Intermodal transport in logistics networks*, (Publishing house of Wroclaw University of Technology, Wroclaw, 2008)
9. S. Kwasniowski, M. Zajac, P. Zajac, *Communication in Computer and Information Science, Transport System Telematics **104***, 391-399 (Springer, 2010)
10. P. Zając, *Progress in Automation, Robotics and Measuring Techniques*, 345-355 (Springer, 2015)

MATEC Web of Conferences 112, 05013 (2017) DOI: 10.1051/matecconf/201711205013
IManE&E 2017

Problems of modelling Proportional–Integral–Derivative controller in automated control systems

Anna Doroshenko*

Moscow State University of Civil Engineering, Yaroslavskoye shosse, 26, Moscow, Russia, 129337

Abstract. The actual task of modelling automatic control systems is to use control algorithms that include some function that limits the output signal. Proportional–integral–derivative (PID) controllers are used to control dynamic processes with variable at wide limits parameters and uncontrolled violations in automated process control systems. Despite the wide distribution, this type of controllers is not easy to configure. In article the problems of modelling proportional–integral–derivative controller are discussed. The methods of combating the effect of integral saturation are studied. The results of using algorithms of combating the effect of integral saturation are represented.

1 Introduction

The actual task of modelling automatic control systems is to use control algorithms that include some function that limits the output signal.

Proportional–integral–derivative (PID) controllers are used to control dynamic processes with variable at wide limits parameters and uncontrolled violations in automated process control systems.

The PID controller was invented in 1910 [1]. In 1942, Ziegler and Nichols developed a method for its tuning, and after the advent of microprocessors in the 1980s the development of PID regulators was taking place at an increasing rate [2].

The PID controller is one of the most common type of controllers. About 90 - 95% of the controllers currently in operation use the PID algorithm [3]. The reason for such high popularity is the simplicity of construction and industrial use, the clarity of functioning, the suitability for solving most practical problems and low cost.

The PID controller has received wide application in various in purpose and design systems of automatic control of the working parameters of equipment, which in turn makes it easy and easy to provide signal generation [4]. The ultimate goal is to obtain high data accuracy, as well as the quality level of the transient process itself [5-8].

The PID controller uses the proportional-integral-differential regulation law. The PID controller usually has additional service properties of automatic tuning, signalling, self-

* Corresponding author: pochta.avd@gmail.com

diagnostics, programming, unstressed switching modes, remote control, the ability to work in an industrial network, etc [9].

Some of the problems arise because of the complexity of operation. In many PID controllers, the differential component is disabled only because it is difficult to properly configure. Users neglect the calibration procedure, insufficiently thorough knowledge of the dynamics of the controlled process does not allow choosing the correct parameters of the regulator. As a result, 30% of the regulators used in the industry are set up incorrectly. Therefore, the main efforts of researchers are currently focused on the search for reliable methods for automatic adjustment of regulators, both integrated in the PID controller, and functioning on a separate computer.

After the emergence of cheap microprocessors and analogue-to-digital converters PID controllers use automatic parameter adjustment, adaptive algorithms, fuzzy logic methods, genetic algorithms are used. The structure of the controllers became more and more complicated [1-39].

2 PID controller parameters

Despite the wide distribution, this type of controllers is not easy to configure, since it is necessary to search by three parameters: proportional (kp), integral (ki) and differential (kd) components [10-12]. The transfer function of these regulators is usually represented as follows:

$$W(s) = k_p\left(1 + \frac{1}{T_i s} + T_d s\right) = k_p + k_i\frac{1}{s} + k_d s \tag{1}$$

At present, there is still no single method for calculating the parameters of the controller for the control loop. The most famous methods are the empirical Ziegler-Nichols method (the method of undamped oscillations), the damping oscillation method, the Kuhn method, the Schädel method, etc. Each of the methods has advantages and disadvantages and is applied under certain requirements to the regulatory system. The choice of method directly depends on the control object.

The actual task for modeling automatic control systems is to use control algorithms that include some function that limits the output signal.

The equation of a nonlinear element of saturation type is represented by the formula:

$$\bar{u} = \begin{cases} u_{min}, u < u_{min} \\ u, u_{min} \le u \le u_{max} \\ u_{max}, u > u_{max} \end{cases} \tag{2}$$

Integral saturation is the effect that is observed when the PI or PID controller has to compensate for a long time an error outside the range of the controlled variable, since the controller output is limited, it is difficult to reduce the error to zero.

If the control error persists for a long time, the value of the integral component of the controller becomes very large. This, in particular, occurs if the control signal is limited enough that the calculated output of the controller is different from the actual output of the actuator. Since the integral part becomes zero only a short time after the error value changed sign. Integral saturation can lead to a large overshoot.

In order to avoid this effect in the modeling of PID regulators in automated control systems in software complexes, algorithms for protection against integral saturation are laid down.

3 PID controllers modelling

In order to avoid this effect in the modeling of PID regulators in automated control systems in software complexes, algorithms for protection against integral saturation are laid down.

For example there are two built-in protection algorithms in Matlab Simulink: Back-Calculation (Figure 1) and Clamping (Figure 2).

Simulink is an interactive tool for modeling, simulating and analyzing dynamic systems. It provides the ability to build graphical block diagrams, simulate dynamic systems, investigate the operability of systems and improve designs. Simulink is fully integrated with MATLAB, providing immediate access to a wide range of analysis and design tools. These advantages make Simulink one of the most popular tool for designing control systems.

In the additional settings of the controller, you can set limits on the input signal. And when using this setting, you can configure protection against the integral saturation effect. This effect occurs when the actuator has a saturation limitation on the input. Because of this limitation, it is not possible to reach the set point, which results in a non-zero error value at the regulator input. During operation, the integrator continues to accumulate this error, the signal at its output increases, but does not participate in the regulation process (does not affect the object). As a result, there is a delay in the transient process. In order to avoid this effect, you must set the limit "Limit output" and choose one of the two methods of protection "Anti-windup method".

Algorithm "back-calculation". In this algorithm, the integral saturation effect occurs with the help of additional feedback, for the transmission of signals, the difference between the received control signal and taking into account the saturation at the integrator input is taken into account. Figure 1 shows the scheme of the "back-calculation" algorithm.

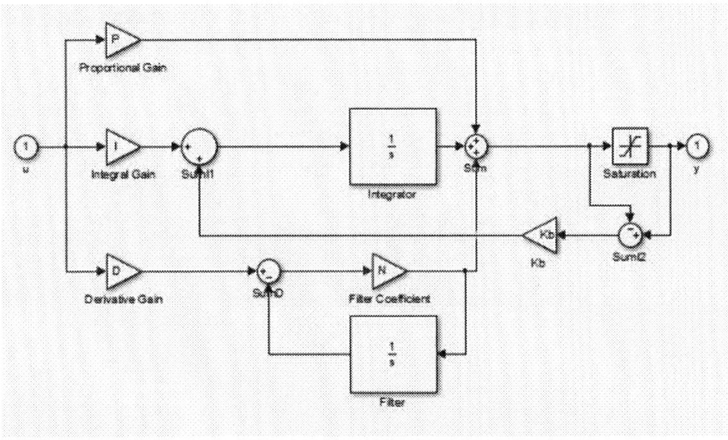

Fig. 1. Back-Calculation algorithm.

The back-calculation anti-windup method uses a feedback loop to discharge the PID Controller's internal integrator when the controller hits specified saturation limits and enters nonlinear operation.

Another commonly used anti-windup strategy is based on conditional integration.

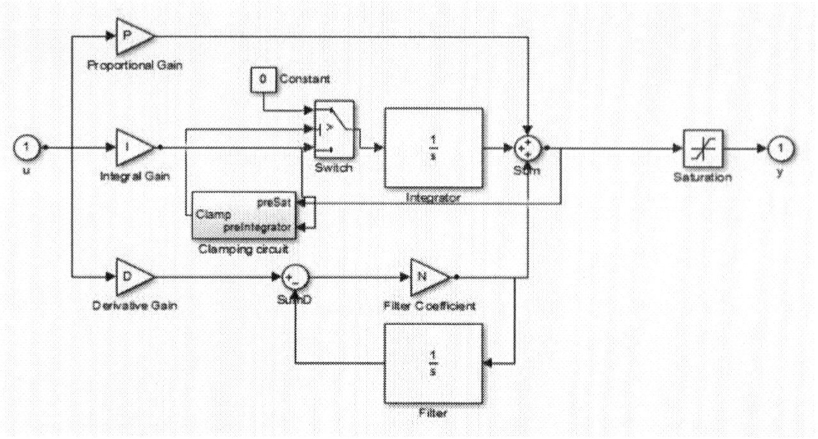

Fig. 2. Clamping algorithm.

When the algorithm changes, the structure of the PID controller changes. When the control action on the object reaches saturation, we programmatically disconnect the integrator from the control. The advantage of this method is that the nature of the transient process is much better, that is, it terminates faster and the regulator output does not exceed saturation.

Clamping will always work. It detects when there is an integrator overflow and sets the integral path of the PID controller to zero to avoid. Clamping is a commonly used anti-windup method, especially in the case of digital control systems. In serious applications, however, there is also forward clamping involved. This mechanism must be implemented manually.

Back Calculation highly depends on the back calculation coefficient Kb. If the user doesn't know how to actually calculate the parameter Kb don't use back-calculation. This method calculates the difference between the actual controller output and the saturated output and subtracts it from the I-Gain path, amplified by Kb. In most of cases the default value Kb=1 will lead to worse results than clamping, it is even possible that it has no effect at all. Kb should be calculated based on the sampling time or in case a D-Gain is involved, based on D- and I-Gain. Appropriate literature should be consulted to calculate the coefficient. Back calculation with a properly set coefficient enables better dynamics than clamping

Consider the results of using the back-calculation algorithm in the example with PID controller and without it (Figure 3). The results obtained demonstrate high efficiency of back-calculation algorithm use.

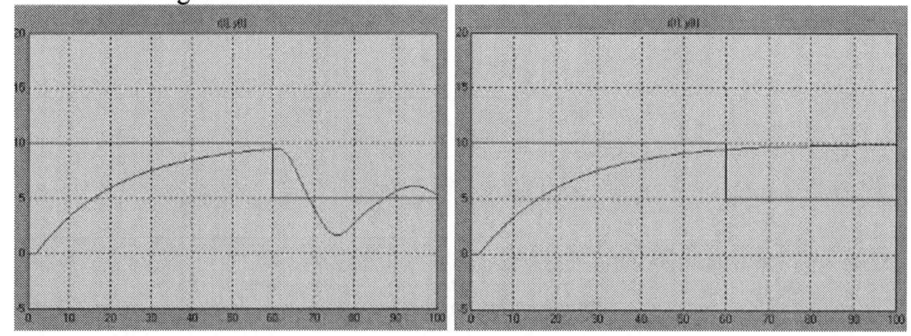

Fig. 3. The results of using the back-calculation algorithm – left figure, without using the back-calculation algorithm – right picture.

3 Discussion

With a properly tuned PID controller, in most cases, it is possible to ensure that all system requirements are met. Because of their simplicity, they have received the widest distribution. According to statistics, more than 90% of all industrial regulators are PID-controllers.

Constantly growing market requirements to reduce the time of regulation, to the quality of the transient process, to the degree of weakening of the influence of external disturbances and noise, simplification of the tuning procedure and the need to control objects with a large transport delay initiated the appearance of many modifications of PID regulators.

Integral saturation is the result of nonlinearities in the system associated with limiting the output control. The negative effect of integral saturation causes a deterioration in the quality of regulation. Increase in regulation time and large overshoot. Under certain conditions, integral saturation can lead to self-oscillations, even if in the linear zone the control system ensures a high quality of regulation. Therefore, various ways have been developed to eliminate the effect of integral saturation, which do not allow the control signal to reach the boundary values and, nevertheless, ensure a good quality of transient processes. These methods are included in various simulation complexes of automatic control systems.

4 Conclusions

The use of integral saturation compensation algorithms has the following advantages: implementation of the method does not require powerful computing facilities and the quality of regulation is significantly improved.

When developing automatic control systems it is necessary to take into account the presence of some non-linearities and use algorithms for taking into account the limitations, depending on the available automation equipment, the complexity of the process and the technological requirements for it.

The use of integral saturation compensation methods in self-tuning systems of automatic control using adaptation and identification algorithms is promising.

References

1. M. Bertocco, S. Cappellazzo, A. Flammini, M. Parvis, IMTC/2002, **2**, 1261 – 1264 (2002)
2. J. Ziegler, N. Nichols, Trans. ASME, **64**, 759–768 (1942)
3. K. Astrom, T. Hagglund, ISA, , 460p (2006)
4. S.A. Ajwad, , J. Iqbal, M.I.Ullah, A. Mehmood, Frontiers of Mechanical Engineering, **10 (2)**, 198-210 (2015)
5. N. Makisha, Pr. Eng., **165**, 1092-1097 (2016)
6. A. Volkov, E. Batov, Pr. Eng., **111**, 849-852 (2015)
7. A. Volkov, Advanced Materials Research, **838-841**, 2969-2972 (2014)
8. L. Shirokov, P. Chelyshkov, E. Romanenko, MATEC Web of Conferences, **86**, 04062 (2016)
9. A. Volkov, L. Sukneva, Applied Mechanics and Materials, **587-589**, 338-341 (2014)
10. M. Miranda, K. Vamvoudakis, *Proceedings of the American Control Conference*, 5443-5448 (2016)

11. Z. Yang, Y. Gao, D. Zhang, T. Huang, Key Engineering Materials, **238-239**, 375-380 (2003)
12. A.J. Winsby, S.E. Burge, M.B. Widden, International Journal of Electrical Engineering Education, **41** (4), 292-306 (2004)
13. S. Skogestad, Journal of Process Control, **13** (4), 291-309 (2003)
14. Z.L. Gaing, IEEE Transactions on Energy Conversion, **19** (2), 384-391 (2004)
15. K.H. Ang, G. Chong, Y. Li, IEEE Transactions on Control Systems Technology, **13** (4), 559-576 (2005)
16. M.V. Kothare, P.J. Campo, M. Morari, C.N. Nett, Automatica, **30** (12), 1869-1883 (1994)
17. P. Ge, M. Jouaneh, IEEE Transactions on Control Systems Technology, **4** (3), 209-216 (1996)
18. Z. Gao, *Proceedings of the American Control Conference*, **6**, 4989-4996 (2003)
19. R.W. Longman, International Journal of Control, **73** (10), 930-954 (2000)
20. C.C. Hang, K.J. Astrom, W.K. Ho, *IEE Proceedings D: Control Theory and Applications*, **138** (2), 111-118 (1991)
21. K.J. Åström, T. Hägglund, Journal of Process Control, **14** (6), 635-650 (2004)
22. A.I. Dounis, C. Caraiscos, Renewable and Sustainable Energy Reviews, **13** (6-7), 1246-1261 (2009).
23. Y.Q.Chen, I. Petráš, D. Xue, *Proceedings of the American Control Conference*, art. no. 5160719, 1397-1411 (2009)
24. H. Zhang, Y. Shi, A. Saadat Mehr, IEEE Transactions on Industrial Electronics, **58** (12), art. no. 5699920, 5396-5405 (2011)
25. M. Zamani, , M. Karimi-Ghartemani, N. Sadati, M. Parniani, Control Engineering Practice, **17** (12), 1380-1387 (2009)
26. F. Padula, A. Visioli, Journal of Process Control, **21** (1), 69-81 (2011)
27. Y. Li, Q. Xu, IEEE Transactions on Control Systems Technology, **18** (4), art. no. 5282525, 798-810 (2010)
28. W. Tan, IEEE Transactions on Power Systems, **25** (1), art. no. 5361327, 341-350 (2010)
29. Y. Luo, Y. Chen, Automatica, **45** (10), 2446-2450 (2009)
30. N. Makisha, E3sConf, **6**, 01002 (2016)
31. A. Biswas, S. Das, A. Abraham, S. Dasgupta, Engineering Applications of Artificial Intelligence, **22** (2), 343-350 (2009)
32. L. Guo, J.Y. Hung, R.M. Nelms, IEEE Transactions on Industrial Electronics, **56** (6), 2237-2248 (2009)
33. P.E.I. Pounds, D.R. Bersak, A.M. Dollar, *Proceedings - IEEE International Conference on Robotics and Automation*, art. no. 5980314, 2491-2498 (2011)
34. H. Zhang, Y. Shi, A.S. Mehr, International Journal of Robust and Nonlinear Control, **22** (2), 183-204 (2012)
35. D.Q. Truong, K.K. Ahn, Mechatronics, **19** (2), 233-246 (2009)
36. I. Pan, S. Das, A. Gupta, ISA Transactions, **50** (1), 28-36 (2011)
37. Y. Luo, Y.Q. Chen, C.Y. Wang, Y.G. Pi, Journal of Process Control, **20** (7), 823-831 (2010)
38. R.J. Wai, J.D. Lee, K.L. Chuang, IEEE Transactions on Industrial Electronics, **58** (2), art. no. 5437232, 629-646 (2011)
39. M. Kano, M. Ogawa, Journal of Process Control, **20** (9), 969-982 (2010)

Grouping of parts for co-processing on flexible rotor lines

Yevhen Bulenkov *, and *Alexander* Mikhaylov

Donetsk National Technical University, Department of Manufacturing Engineering, 58, Artyom Street, Donetsk, Ukraine

Abstract. The problem of parts combining for group processing on flexible rotary lines is considered in this article. The advantages of group processing in a single tool unit in terms of performance and metal consumption of lines are presented in this article. The technique of parts combining for group processing is proposed in this paper.

1. Introduction

Fasteners are one of the most widespread products of mechanical engineering. Increase in efficiency of their production is an important task, thus works on improvement of technology of their production are conducted. Cold forming of bolts, made of micro-alloyed steel, is investigated [1] to enable elimination of thermal treatment. In the study reported in [2], two forging approaches, traditional multistep forging and injection forging, were analyzed by using a finite element method in order to improve production efficiency in automobile fasteners production. The technological modernity of a machine used during the production process of screws is evaluated too [3]. These works allow developing manufacturing techniques of screws, but this technology doesn't consider feature of manufacturing machines and lines. Production of screws on flexible rotor lines is the most effective way of their processing [4-6]. The known design methods for these lines [4-5] allow creating multinomenclature rotor lines on which several different products are made. At the same time, the known rotor lines do not allow to make large number of different parts at the same time. They possess insufficient flexibility. Works in the field of grouping of products for their joint processing are conducted. Authors of paper [7] propose to apply a modified discrete firefly algorithm to the machine loading problem of the flexible manufacturing systems. Grouping parts into families which can be produced by a cluster of machine cells is the cornerstone of cellular manufacturing is proposed in [8]. Optimization of the integrated part type selection problem and machine loading problem is described in [9]. These works allow us to propose new type of rotor lines. The flexible rotor lines offered by us [6] mean processing of several screws in the flexible tool node. Grouping of parts for co-processing is one of the main problems for these lines. Thus, finding a technique for grouping of parts is an actual task for production of screws on flexible rotor lines.

* Corresponding author: bulenkov@ukr.net

The purpose of this article is finding a technique for grouping of parts for their co-processing.

To achieve this goal it is necessary to examine features of new flexible rotor lines, connection between structure of line and a structure of products and features of grouping of parts for co-processing.

2. Grouping of parts for their co-processing

For the solution of objectives the new design of the flexible rotor machine is offered, where in each flexible working position several different products are processed. It should be noted that the different structures of a flow of products in the flexible rotor line means transfer of products with different forms and sizes by the same position.

Figure 1 shows a sample of a flexible rotor line, which provides processing of several sizes of goods in each working position. In this case, the figure shows two flexible rotor machines (TrM1 and TrM2) for goods transportation and one machine for goods processing (TM). This sample of flexible rotor line is a part of line for producing 20 different types of screws. In processing, goods come from transport rotor TrM1 and processed in group tool module (FIN) by one of the tools. After completion of the technological cycle of processing goods are transferred to the transport rotor TrM2. During idling in group tool module tool is replaced for processing other nomenclature by turning tool drum.

The parts made on the flexible automatic rotor line differ in a head form, diameter and length of a rod, a step of a thread and a form of a groove for the screw-driver.

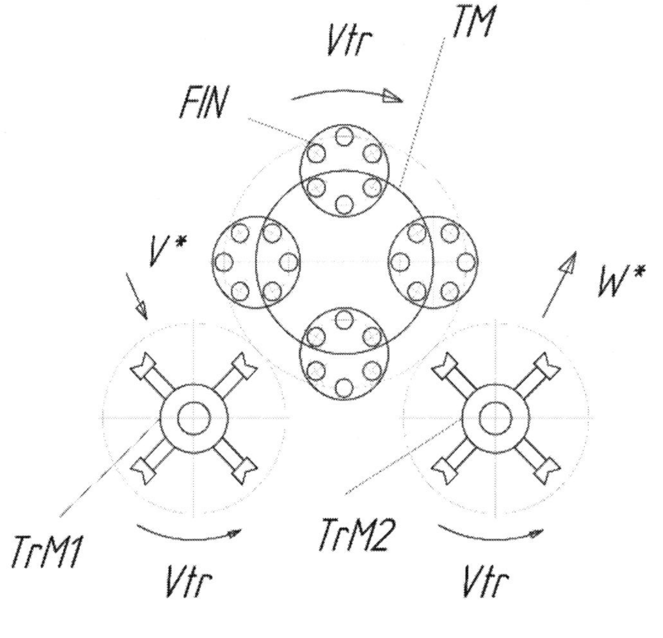

Fig. 1. A sample of a flexible rotor line.

During transportation products differ not only by the sizes, but also by a form of surfaces by which a transport rotor is holding them. At production of rod fasteners at different production phases their capture will be carried out as on cylindrical surfaces at stages of a shaping of a head, as on a threaded surface after knurl of a thread.

At the same time, cylindrical surfaces even for one screw will differ in the sizes: before thread rolling reduction of a rod is made. Creation of one gripping body for each screw will lead to unfairly big expenses and complication of transport devices design. In these conditions transfer of products with use of physical properties of material is proposed as more rational. These properties for all screws will be identical that will allow to unify designs of gripping bodies and to simplify designs of transport rotors. In this regard, we offered to use permanent magnets in a design of gripping bodies of transport rotors that will allow one gripping body of a transport rotor to transfer products of different standard sizes.

It should be noted that the problem of insufficient loading of flexible rotor line is closely connected with quantity of standard sizes of the products processed by this machine. If processing of a small amount of different products is supposed, then their annual programs have to be big (Figure 2a). Processing of a large number of different products on the contrary assumes their small programs of release (Figure 2b).

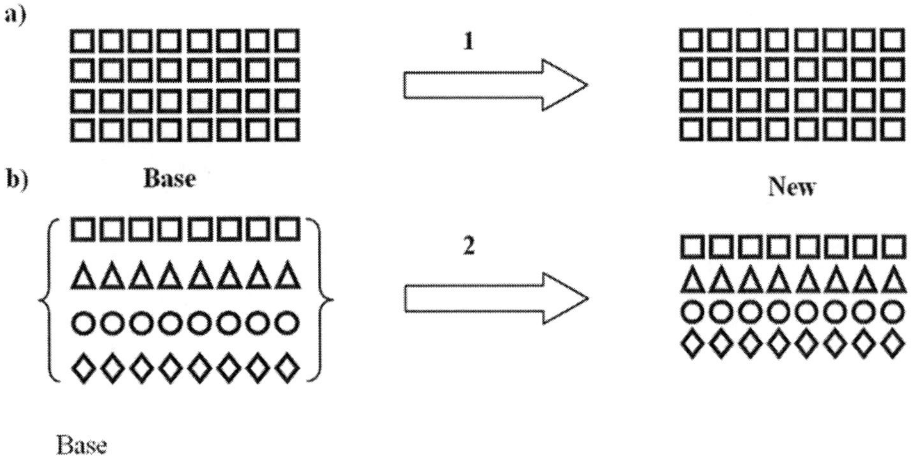

Fig. 2. The summary annual program of release of flexible rotor line.

In the conditions of the market relations the second way is represented as more attractive. It is necessary to choose such structures of a flow of products which will allow integrating the bigger quantity of different standard sizes of products with smaller annual programs of release. Thus, required loading of flexible rotor line will be provided due to merging of group of butch productions in mass.

The summary annual program of release of flexible rotor line is defined as the sum of annual programs of release of products with separate standard sizes:

$$Q = Q_A + Q_B + ... + Q_S, \tag{1}$$

where Q - the summary annual program of release of flexible rotor line;

Q_A, Q_B, ... Q_S - annual programs of release of products with separate standard sizes.

In flexible rotor lines the quantity of the products processed in each position will be defined by structure of a flow of products in the line. For simplification of rational flow structure search we will accept as an assumption that required annual programs of release of products are equal among themselves ($Q_A = Q_B = ... = Q_S$). Besides, time of a kinematic cycle Tk is a constant ($Tk = const$). The annual program of release of flexible rotor line can be determined by a formula:

DOI: 10.1051/matecconf/201711205014

$$Q = P \cdot a = \frac{Up \cdot a}{Tk}, \quad (2)$$

where P - cyclic productivity of flexible rotor machine, number of pieces/min.;
a - annual fund of operating time of the machine, min;
Up - quantity of flexible tool nodes, piece;
Tk - time of a kinematic cycle, min.

Let's consider influence of different parts flow characteristics on the size of annual program of release of product with separate standard size Q_A.

For the flows differing in a ratio of quantity of standard sizes of products and quantities of flexible tool nodes, the annual program of release of products of one standard size is determined by dependence

$$Q = Q_A + Q_B + \cdots Q_S = Z_O \cdot Q_A = \frac{U_p \cdot a}{T_k}, \quad Q_A = \frac{U_p \cdot a}{Z_O \cdot T_k}, \quad (3)$$

where Zo - quantity of instruments in flexible tool nodes, piece

At the same time, if $Zo < Up$, then the required annual program of release of products of one standard size is more than at equality of these values (Figure 3). When $Zo > Up$, the required annual program of one standard size products release is less than annual program of release of flexible tool node. Thus, it is necessary for ensuring loading of the flexible rotor line that the quantity of standard sizes of products was more than the number of flexible tool nodes in flexible rotor machine.

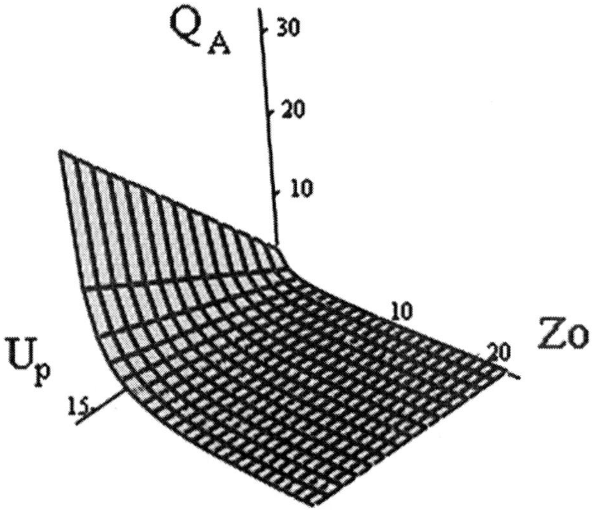

Fig. 3. Annual program of release of products with separate standard size.

The required annual program of release of products of one standard size in a case when flows differ in quantity of screws, processed at the same time, is found on similar expression:

$$Q = Q_A + Q_B + \cdots Q_S = Z_O \cdot Q_A = \frac{t \cdot U_p \cdot a}{T_k}, \quad Q_A = \frac{t \cdot U_p \cdot a}{Z_O \cdot T_k}, \quad (4)$$

where t - quantity of workpieces, processed at the same time.

Thus, with increase in quantity of standard sizes of products the required annual program of release of parts of one standard size remains without change. It is connected with the fact that in flexible instrumental node products of different standard sizes are processed. In case of single processing ($t = 1$) dependence will be similar represented in Figure 3. At the mixed processing with increase in quantity of screws, processed at the same time, the required annual program of release of parts of one standard size increases because in each flexible instrumental node identical products are processed.

Required annual programs of release of products of one standard size in the readjusted flexible rotor machines will be more, than in machines with simultaneous processing of products. It is connected with the fact that annual programs of production of products are invariable, and all pieces have to be processed in shorter time accordingly to readjustment. Thus, in this case the offered flexible rotor machines will be the most preferable as they provide a possibility of processing of the greatest number of standard sizes of products.

The structure of technology processes of products production is represented as set of elements - the operations of technology process (which are carried out in separate flexible rotor machine) and the relations between them (route compliance between parts and working positions). At the same time route compliance exerts very great influence on structure of flexible rotor machine (Figure 4), - at identical quantity of products different quantities of flexible tool nodes and intervals of shaping forces are possible.

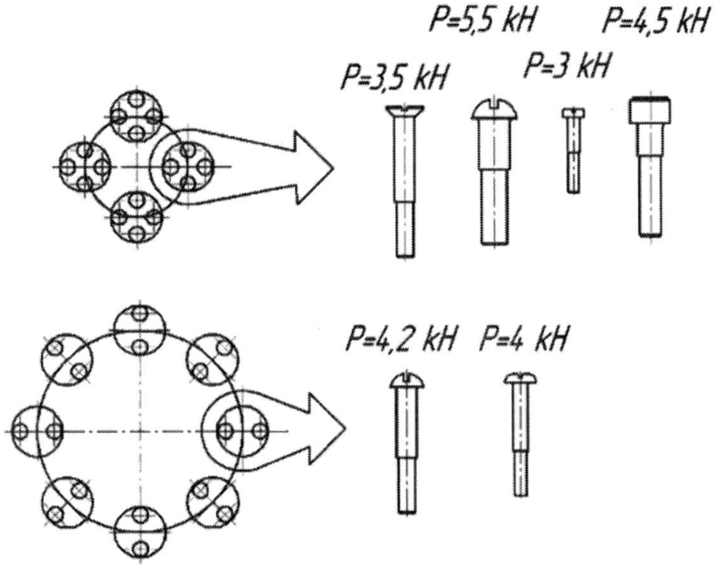

Fig. 4. Connection between structure of line and a structure of products.

Thus, the reasonable decision on grouping of parts for co-processing on flexible rotor lines can be made only after consideration of products routing in flexible rotor line and studying of products transportation.

Use of the theory of products routing as link between products manufacturing methods and structure of the line on which they will be made allows integrating development of technology processes of products production with structural synthesis of flexible rotor line.

The general sequence of development of technology processes of products production includes development of route technology process with the subsequent development of operations structure (Figure 5). It should be noted that process design of products production is carried out at the same time for all group of products.

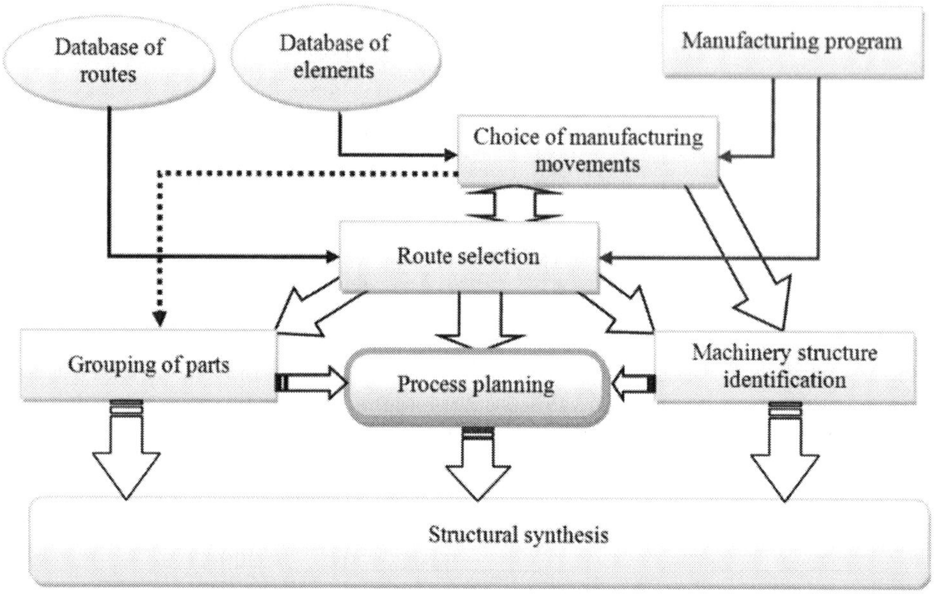

Fig. 5. The general sequence of development of technology processes of products production.

Design of technology process for products production is based on the choice of manufacturing movements and route selection. It is carried out parallel to machinery structure identification and grouping of parts.

3. Results

This work is dedicated to the development of design methods for flexible rotor lines that will combine several large-scale productions of products of different sizes in the mass diversified flow of goods by processing several sizes of goods in each group tool module. The developed method allows to group parts for co-processing on flexible rotor lines. The advantages of group processing in a single tool unit in terms of performance and metal consumption of lines are presented here.

References

1. L. Janíček, J. Petruška, B. Maroš, S. Rusz, J. of Mater. Proc. Tech., **125–126**, 341-346 (2002).
2. C. Senyong, Q. Yi, MATEC Web of Conf., **21,** 02007 (2015).
3. M. Ingaldi, K. Lestyánszka Škůrková, Tech. agricult., **17(4)**, 101–104 (2014)
4. L. Koshkin, *Rotary and rotary conveyor lines.* (Moscow, Machine building, 1986).
5. A. Mikhaylov, *Basics of synthesis for flow-spatial technology systems of continuous action.* (Donetsk, DNTU, 2002).
6. Y. Bulenkov, Progr. Tech. and Syst. for Mach.-Build. **46** 30-37 (2013).
7. E. Bottani, P. Centobelli, R. Cerchione, L. Del Gaudio, T. Murino, Int. J. of Ind. Eng. Comp., **8,** 363–372 (2017)
8. G. C. Onwubolu, The South Afr. J. of Ind. Eng., **9(1),** (2011).
9. W. F. Mahmudy, AENG Int. J. of Comp. Sci., **42(3),** 254-264 (2015).

Numerical simulation with the finite element using the contact between the soil and the actively working body

Dan Chiorescu[1,*], *Esmeralda* Chiorescu[1], and *Oana* Dodun[2]

[1]University of Agronomical Sciences and Veterinary Medicine, 3, Mihail Sadoveanu Alley, 700490, Iaşi, Romania
[2]Gheorghe Asachi Tehnical University, 59 A, Dimitrie Mangeron Blvd, 700050, Iaşi, Romania

Abstract. In this present paper, because of the complexity of the system soil – agricultural machine, we will use an analytical model which respects the geometry of the active element, realising a prediction of the forces which result at the dislocation of the soil. This study analyses the behavior of the working tool, part of the soil processing machine, using the Finite Element Method (FEM) in three different stages. In the pre-processing stage, the objective was to design a three dimensional model in CATIA V5, in keeping with the geometry of the active element, represented by the Cartesian coordinates, together with a portion of the soil rendered as a parallelepiped shape. The second stage followed the introduction of conditions both for the working part, through the fastening of the plowshare frame, the moving direction and velocity, and for the soil, through the action of the cohesion and internal friction forces. In the third stage, called the processing stage, there is the simulation of the process of soil displacement done in real conditions, for various degrees of refinement of the discretization network in finite elements.

1 Introduction

Because agriculture brings a major contribution to the sustainable development of the economy, there exists a major demand for innovative technologies, for durable systems of the agricultural plot, which prevents or minimizes the degradation of the soil. In general, the reason for the studies done by researchers is done to evolve agricultural machines which function autonomously for the benefit of the farmers [1].

The soil is a complex environment that can be considered, for simplification purposes, as a combination of solid particles, organic matter and porous spaces that allow air breakthrough and water accumulation and movement. The share of soil particles in its volume, for an ideal productive soil, should be: 45% mineral particles, 5% organic matter, 25% porous spaces filled with water and 25% porous spaces filled with air [2].

Research has shown that about half of the energy is used for cultivating the soil. This high energy consumption is caused by the large amount of soil to be engaged during

* Corresponding author: dan_chiorescu@yahoo.com

cultivation because of the inefficient method of transferring energy from the working part to the soil [3, 4].

The main research studies have focused on the interaction between the soil and the type of work, different soils (physical and mechanical traits of the soil), the working part (shape, angle) and working conditions (working depth, cutting width, velocity).

Because of the complexity of the soil-machine system, analytical models are used, limiting the study to the geometry of the active plowshare, making a prediction of the forces resulting from engaging the soil.

The continuous development of computers combined with the improvement of the mathematical techniques and methods have lead to the occurrence of new generations of very efficient software which have succeeded in simulating various agricultural operations. The determination model of the soil-tool system to be submitted to the finite element analysis consists in passing from a continuous structure having an infinite number of points to a discrete model with finite number of points called knots, covering the model in a discretization network [5, 6].

The finite element method (FEM) used in the current paper analyses the interaction between the soil and the working part by studying the stress – deformation function in relation to the real working conditions. By using the finite element method, the main changes in the material structure are highlighted (the material thinning rate, the stress field, the direction of the deformations) together with the soil deformation [7].

2 Material and method

The programming software Ansys was used for the simulation with the finite element method, allowing the study of the behaviour of the active working part, the normal stress being analysed in real conditions, at various depths and velocities for a soil with a clay-sandy texture [8, 9].

To analyse the behaviour of the working part within the soil-tool system, using the finite element method, a tridimensional model was constructed replicating exactly the geometry of the active element while the soil was rendered through a parallelepiped. In this paper the active working part under study is a plowshare made in 3d using the designing software CATIA, the file being saved with the .igs extension in order to be imported. The import of the basic design of the plowshare is made in the Explicit Dynamics module.

The experiments were made in the Western Plain of Romania on a soil of the non-carbonated chernozem (black earth) type with mainly clay-sandy texture, using the PP-4-30 plow having the following technical characteristics:

- Working width on the plough-body, b_{Tr}=30 cm; Number of plough-bodies n_{Tr}=2;
- Plow Working width $B= b_{Tr}\cdot n_{Tr}$ =30·2=60 cm=0.60 m; Total mass m=565 kg;
- Plow weight $G_{pl}=g\cdot m$=9.81·565=5543 N.

The physical – mechanical properties of the material of the plowshare OLC 45 under the Romanian standard STAS 880-80 and the European standard SR EN 10083-2:1995 symbolized as 1C45 in Table 1, are introduced in the pre-processing stage.

Table 1. Mechanical and physical characteristics of the 1C45 steel.

Steel brand	Tensile yield strength $Rp_{0.2}$ [MPa]	Ultimate tensile strength R_m [MPa]	Poisson Coefficient n_{med}	A [%]
1C45	410	800	0.3	14

Considering the nonlinear plastic behaviour of the Drucker Prager type of the clay-sandy soil, its characteristics are shown in Table 2.

Table 2. Soil properties.

Soil Properties	Corresponding Values
Soil Type	Sandy clay
Young's modulus of elasticity (Es)	19000 kPa
Posson's ratio (η)	0.3
Cohesion (C)	23kPa
Internal friction angle (Ø)	32

The simulation of the plowing process at depths between 15 and 40 cm is made tridimensional, taking into account only the active working part that is the plowshare. A system of coordinates is assigned to the soil, made of 3 axes: Ox, Oy and Oz, in which the plowshare movement direction is positive along the Ox axis (Figure 1).

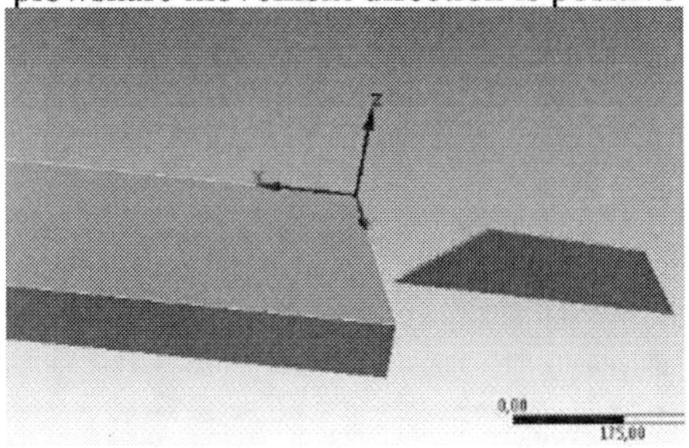

Fig. 1. The system of axes and the plowshare movement direction.

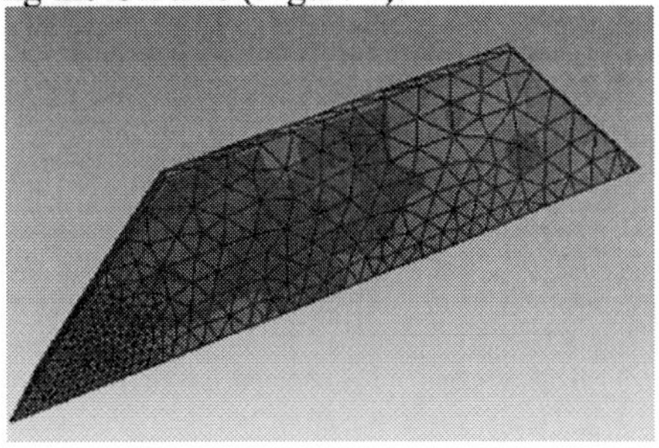

Fig. 2. The ddiscretization of the plowshare.

We are doing a tridimensional study on the unitary stresses in relation to the plowshare movement direction in the plowing process along the Ox direction for 3 velocity values (0.50; 1.00; 1.50 m/s) at different depths (0.18; 0.25; 0.36; 0.38 m) and the 12 values of the traction resistance. The stress resistance force was established experimentally using pin transducers. It was used a VISHAY - HS 100 MG7128 type force cell of 100 kN, invertor number FO21633 – USA, calibrated on 20.09.2015. For the measurements it was used a data system consisting of a laptop and the master unit – Traveller 1, model MUT – 1, 1016-S type with 8 SG-2 type tensometric amplifying channels with a transmission band of 1kHz. The results were obtained on a sandy clay soil at different depths in Table 3.

Table 3. Maximum and minimum normal stress at different velocities and depths.

No.	Velocity v [m/s]	Depth a [m]	Traction resistance R_{tr} [N]	Maximum normal stress [MPa]	Minimum normal stress [MPa]
1	0.50	0.18	4806	52.534	-98.763
2		0.25	6780	44.438	-93.213
3		0.36	8620	71.544	-94.100
4		0.38	9040	44.751	-106.560
5	1.00	0.18	6507	42.186	-92.926
6		0.25	8850	43.461	-95.585
7		0.36	9805	73.271	-146.03
8		0.38	10127	46.154	-93.029
9	1.50	0.18	6842	48.752	-92.345
10		0.25	7920	39.268	-93.258
11		0.36	11860	245.32	-270.71
12		0.38	12604	136.52	-362.46

 DOI: 10.1051/matecconf/201711206001

After introducing the mechanical characteristics of the plowshare material as well as the soil properties, we need, for the simulations set, to choose the optimum discretization for the working part (plowshare) as well as the constraints within the system, according to the importance during the working process. Thus, for the whole plowshare we choose a discretization of 1mm, and for the tip 0.5 mm (Figure 2). For the soil the discretization value is 3 mm, to shorten the time needed for cycles rolling.

One can notice following simulation that for a clay-sandy soil at a velocity v= 0.5 m/s, the maximum normal stress on OX direction is 52.534 MPa (Figure 3), for a depth of a=0.18 m, and 44.751 MPa for a=0.38 m (Figure 4).

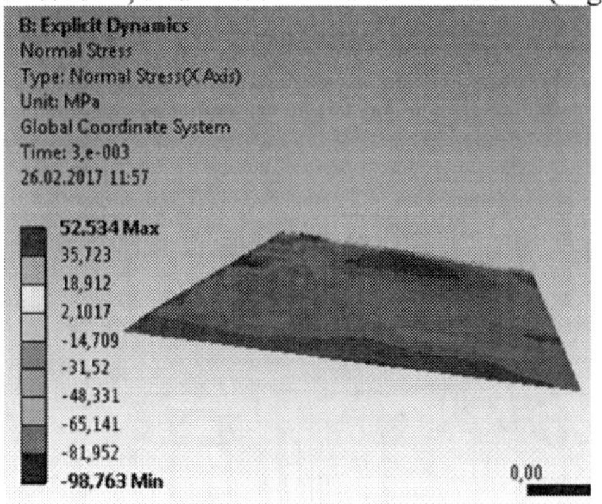

Fig. 3. Normal stress on movement direction x, at v=0.5m/s, a=0.18m, for clay – sandy soil.

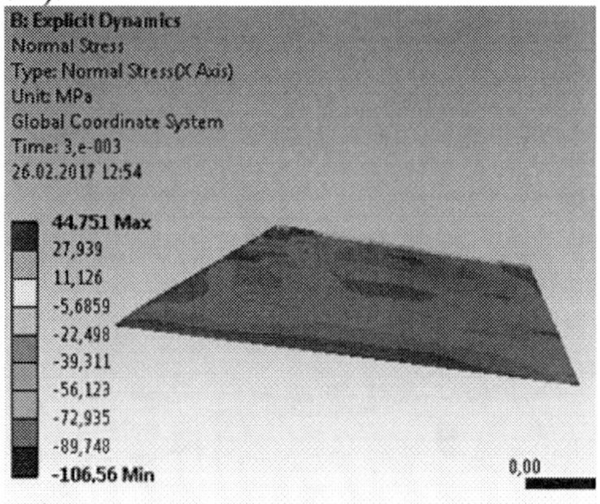

Fig. 4. Normal stress on movement direction x, at v=0.5m/s, a=0.38m, for clay – sandy soil.

To move the plowshare with the velocity of 1.0 m/s, the maximum normal stress is 42.186 MPa (Figure 5) at 0.18 m depth, while for 0.38 m depth, the maximum normal stress is 46.154 MPa (Figure 6).

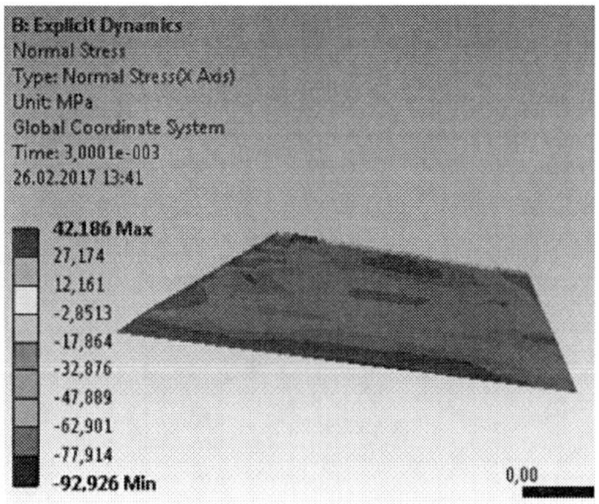

Fig. 5. Normal stress on movement direction x, at v=1.0 m/s, a=0.18m, for clay – sandy soil.

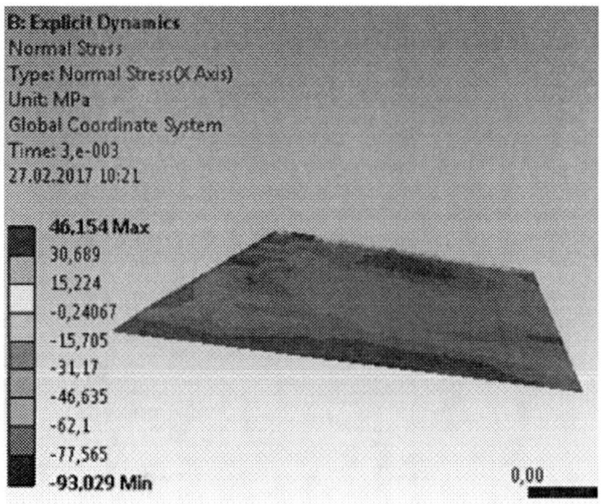

Fig. 6. Normal stress on movement direction x, at v=1.0 m/s, a=0.38m, for clay – sandy soil.

Following the simulation of the basic process (plowing) of the clay-sandy soil for a plowshare movement velocity of 1.50 m/s, the maximum normal stress is 48.752 MPa (Figure 7) at a depth of a=0.18 m and it has a value of 136.52 MPa for a depth of 0.38 m (Figure 8).

The study was based on the analysis of the maximum normal stress at a depth variation of a=0.18; 0.25; 0.36; 0.38, for three values of the plowshare mouvement velocity v=0.5; 1.0; 1.5 m/s (Figures 9, 10, 11).

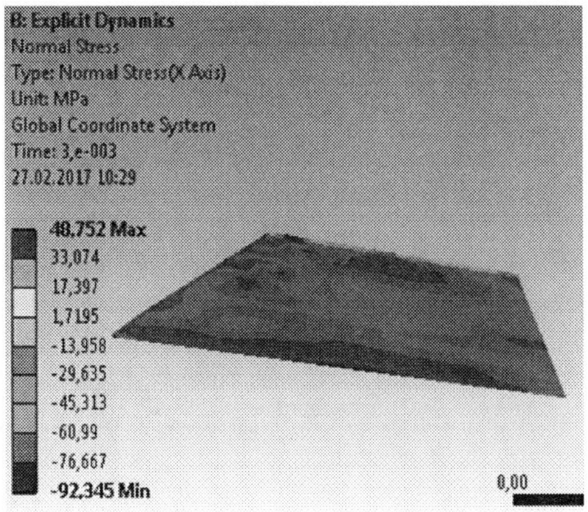

Fig. 7. Normal stress on movement direction x, at v=1.5 m/s, a=0.18m, for clay – sandy soil.

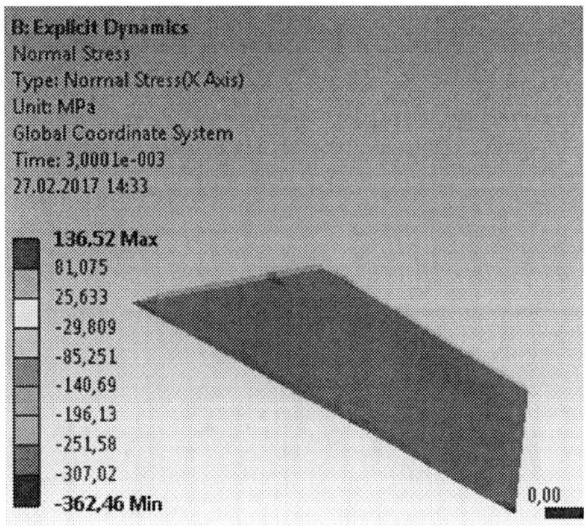

Fig. 8. Normal stress on movement direction x, at v=1.5 m/s, a=0.38m, for clay – sandy soil.

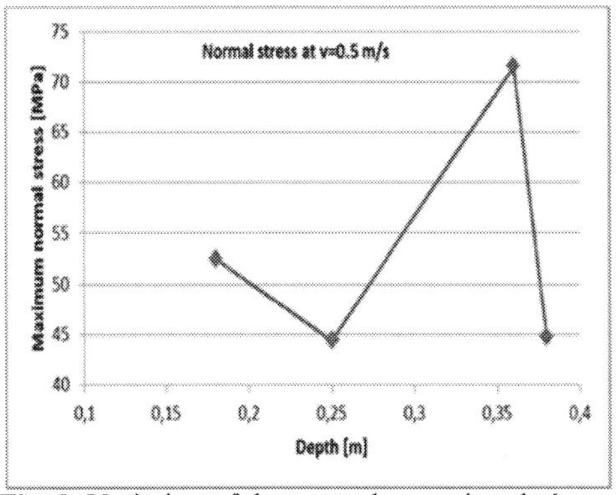

Fig. 9. Variation of the normal stress in relation to the working depth at velocity v=0.5m/s for clay-sandy soil.

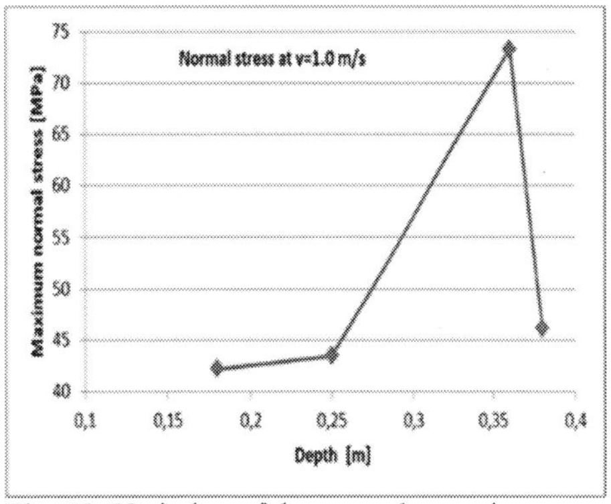

Fig. 10. Variation of the normal stress in relation to the working depth at velocity v=1.0 m/s for clay-sandy soil.

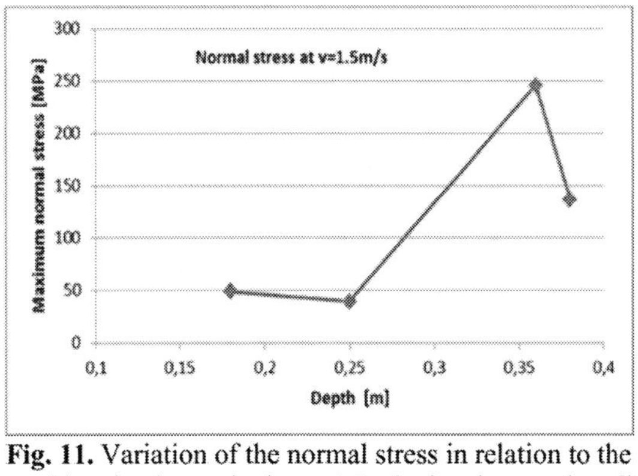

Fig. 11. Variation of the normal stress in relation to the working depth at velocity v=1.5m/s for clay-sandy soil.

3 Conclusions

A representative set of twelve problems was designed regarding the plowing process using the plowshare, differentiated through three values of movement velocity and four different depth values.

DOI: 10.1051/matecconf/201711206001

All twelve problems concerning the soil plowing were solved though numerical simulations using the Ansys software package, employing the same refinement degree of the discretization network with finite elements in Cartesian coordinates.

Caused by the large weight of plowing machines, the soil compaction, highlighted as well by the thinning of the superficial layer, has a significant influence, especially down to 0.36 m depth.

Hence, an increase in the traction resistance is noticeable translated into increased stress in the working part. Between 0.36 m and 0.38 m depth, there is a decrease of the normal stress due to better aeration and a more drilling circulation of the free water in the soil.

Following simulation, we can conclude that the clay-sandy soil can receive high tensile or shear stresses due to the linking forces between the particles or between the aggregates and the particles, achieving low deformations for this type of soil.

For the analysed type of soil, the wear of the working part is evident especially at the tip and the base of the plowshare blade, the main cause being the friction with the soil.

For this reason, special measures are required to increase the durability of the active parts through various methods like coating the surface with ceramic materials based on titanium compounds.

References

1. G. Jităreanu, I. Tenu, et.all., *Technologies and Machines for Mechanizing Soil Tillage in Order to Employ the Sustainable Agriculture Concept*, Edp. "Ion Ionescu de la Brad" (2007)
2. A. Canarache, *The physics of agricultural soil*, Edp. Ceres, Bucureşti, România (1990)
3. E. Chiorescu, D. Prepelita, St. Popescu, Envrironmental Engeneering and Management Journal, Vol.**8**, nr.3, 447-452, ISSN 1582-9596- Iaşi (2009)
4. E. Chiorescu, Şt. Popescu, D. Toma, Envrironmental Engeneering and Management Journal, Vol.7, nr. 6, ISSN 1582-9596, 809-813, Iasi (2008)
5. H. Bernacki, J. Haman, CZ. Kanafojski, **I**, *Naţional Technical Information Service*, Springfield, Ilinois, USA (1972)
6. H. Bentaher, A. Ibrahmi, E. Hamza, M. Hbaieb, G. Kantchev, A. Maalej, et al, *Soil Till*, Res. **134**, 11–16 (2013)
7. R. Jafari, Proceeding F-and-B Proceedings of the 1st WSEAS International Conference on Finite Differences-Finite Elements-Finite Volumes-Boundary Elements, 122–127 (2008)
8. C. Kosmas, S. Gerontidis, M. Marathianou, B. Detsis, T. Zafiriou, W. Nan Muysen, G. Govers, T. Quin., K. Vanoost, *Soil Till*, Res. **58**, 31–44 (2001)
9. R. Derpsch, M.A. Florentín, K. Moriya. Proyecto Conservación de Suelos MAG – GTZ, San Lorenzo, 40 (2000)

DOI: 10.1051/matecconf/201711206002

Products eco-sustainability analysis using CAD SolidWorks software

Luminița I. Popa[1,*], and *Vasile* N. Popa[2]

[1] Transilvania University of Brașov,Department of Automation and Information Technologies, Mihai Viteazul 5, Brașov, Romania
[2] Regional Department for Defense Resources Management Studie s, Mihai Viteazul 130, Brașov, Romania

Abstract. This article is focused on the analysis of environmental impact and Eco-sustainability of models designed using CAD SolidWorks software. We have evaluated the material it was made the whole ansamble, in terms of strength, durability and environmental pollution considering the carbon footprint, energy consumption, air acidification and eutrophication. We considered the whole product life-cycle management, from raw material extraction, processing it, piece production, assembly it, and use it until the end of his life, considering the mode of transport and the distance between these stages. The case study presents the virtual model of the product and Sustainability Report.

1 The concept of sustainability

The sustainability definition is the one proposed in 1987 by the Commission for Environment and Development (known as the Brundtland Report), according to: "the sustainable development is the type of economic development that meets the needs of present generations without compromising the ability of future generations to meet their own needs" .

One of the major challenges of sustainable development is to find ways to encourage environmentally friendly economic activities and discourage those causing damage to the environment (air, water and soil or subsoil).

2 Sustainable design

Behind the sustainable design concept there is a whole philosophy of economic, social and environmental responsibility that should follow each corporation, company, organization, agency or state institution.

Sustainable design must consider the environment and natural resources, which are accessible to the community, protecting the environment from the negative effects of production activities and using resources with responsibly.

* Corresponding author: mluminita2001@yahoo.com

Sustainable design looks at how your product's development, from cradle to grave, will affect four crucial *environmental factors*: *air acidification*; *carbon footprint*; *total energy consumed*; and *water eutrophication*. Measuring these impacts will help you better design for the environment.

2.1 Lifecycle analysis

Definition of lifecycle analysis is emerging from the content of ISO 14040 and aims to achieve the following goals: composition of a record with relevant inputs and outputs of the system; evaluating the potential environmental impacts associated with these inputs and outputs; interpretation of record results and impact phases in relation to the study objectives. There are three major components of the Lifecycle Analysis: analyzing evidence, assessing the impact and assess possible improvements.

LCA addresses the overall impact on the environment associated with the production, use and disposal of materials and identifies the key indicators of environmental pollution sources.

2.1.1 Environmental impact

Environmental impact includes expedition and use of limited natural resources, atmosphere, land and water pollution, environmental quality, ecosystem health and human health, as well as evaluating safety problems at every stage of the production cycle. An example of this is the Eco Index industry model, created by the collaborative efforts of over 100 producers and retailers and coordinated by the Outdoor Industry Association. The result of this model is a software application that guides its users through a series of questions for each of the six life cycle stages with seven key areas of impact. (Table 1)

Table 1. Eco Index industry model (made by the authors).

STAGES/ IMPACTS		**Before customer**					**Customer**	
				Manufacturig Operations				
		Material Aquisition	**R&D Opera-tions**	**Process**	**Pack-aging**	**Warehouse, Distribution Transport**	**Product, Use, Reuse, Mainte-nance**	**Recycle, Final Disposal**
		A	B	C	D	E	F	H
ENVIRONMENT								
Air	1							
Water	2							
Solid Waste	3							
ENERGY	4							
RESOURCE USE	5							
HEALTH								
Chemical	6							
Physical	7							
Biological	8							
SAFETY								
Chemical	9							
Electrical	10							
Mechanical	11							

2.1.2 The 5-Pillar Method

The 5-pillar methodology involves the evaluation of a novel chemical process and comparison to another comparable one (same functionally for final product) based on a set of key parameters [1- 4].These five parameters represent the technical performance of a scaled-up process, economic feasibility, environmental impact, human health, and systemic risks. For this method, the parameters that contribute to the final score , are shown in Table 2:

Table 2. Overview of 5-pillar assessment method.

Innovative new process		Conventional process			Total score
Economic constraint (EC) (0.3)	Environmental impact of raw materials (EI) (0.2)	Process costs and environmental impacts (PCEI)	EHS index (EHSI),(0.2) Environ. Health& Safety	Risk aspects (RA) (0.1)	1,00
Price ratio	Cumulative energy demand	Energy loss Index	**Environment**	Feedstock supply risk	
Practical yields	GHG (Green-house Gas) emissions	Product concentration	Persistency	Market risk	
Allocated raw material costs		Water content	Air hazard	Infrastructura (availability) risk	
Market price		Boiling point difference	Water hazard	Regional feedstock availability	
		Mass loss index	Solid waste	Application-technical aspects	
		Reaction energy	**Health**		
		No. of coproducts	Irritation		
		Pretreatment	Chronic toxicity		
			Safety		
			Mobility		
			Fire/explosion		
			Reactivity		
			Acute toxicity		

3 Products eco-sustainability analysis using CAD SolidWorks software

Fig. 1. The virtual model of plastic storage box.

Once completed the plastic box modeling, we will start the product sustainability study using the "Sustainability" menu (Figure 1). By accessing the "Set Material" tab, was launched the plastics materials database available. (Figure 2). Figure 2 is a database that contains class plastics, "Material Class". In the fields from the table head are shown the characteristics of each material (Specific Heat, Mass density, etc.).

Fig. 1. The product, "Sustainability" menu.

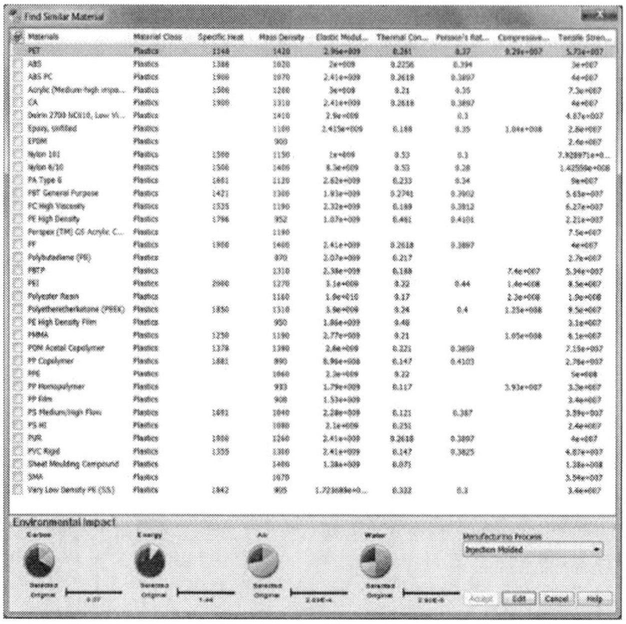

Fig. 2. The plastics materials database.

From the plastics database is chosen the material with "High Density PE", as indicated in Figure 1.

We use the world map is in this menu in order to choose where the product will be transported after production.The greater is the distance between the manufacturer and the user, the less ecological is the product.In this case, the final use of the product will be in North America (Figure 3) .

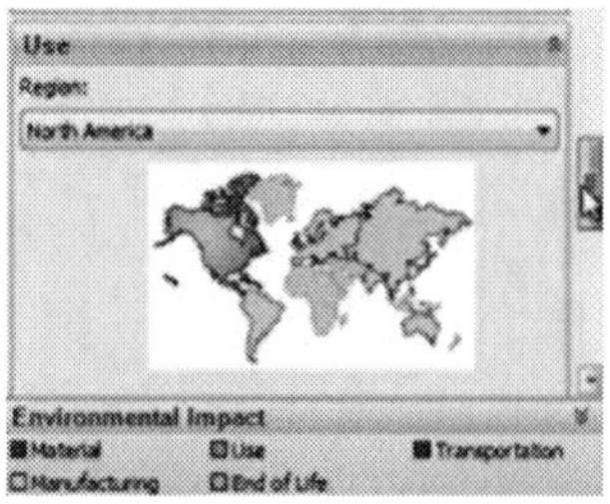

Fig. 3. User Menu.

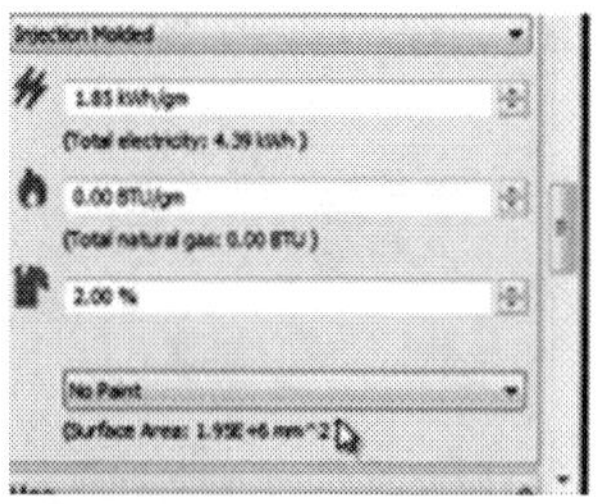

Fig. 4. Injection Molded Tab.

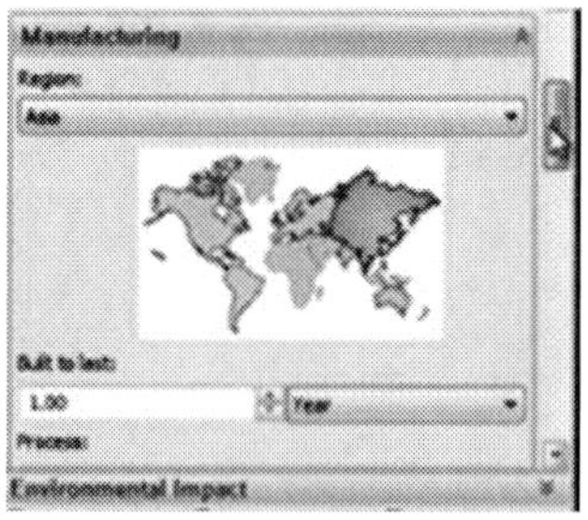

Fig. 5. Manufacturing Tab.

In the "Manufacturing" process, it was chosen, as the manufacturing technology, the injection molded process "Injection Molded" (Figure 4). In Figure5 is represented the manufacturing location –Asia. .Also, it was initially opted for option "No paint". In these conditions , environmental impact factors , are represented in Figure 6.

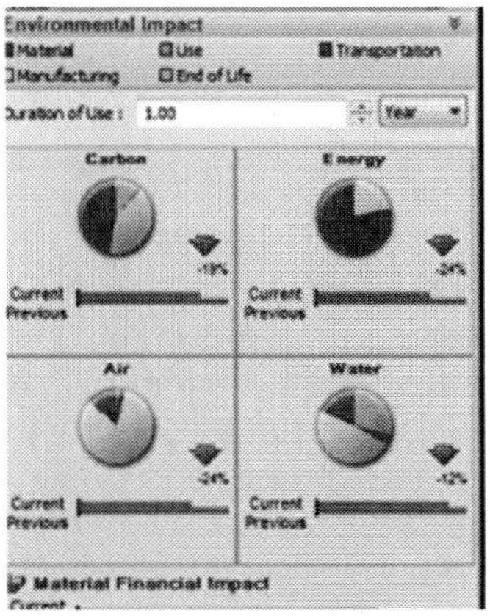

Fig. 6. Menu Environmantal Impact.

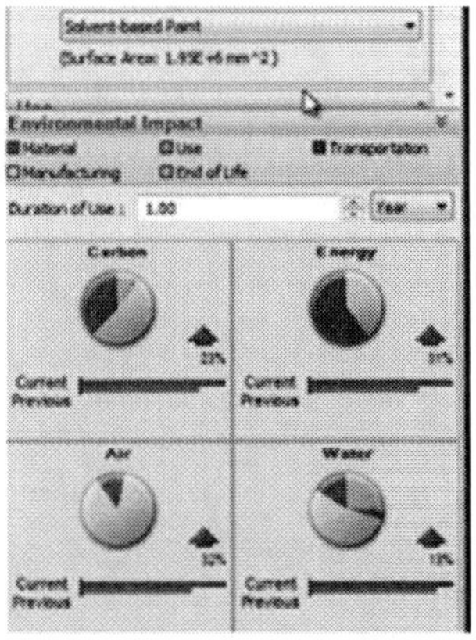

Fig. 7. Injection Molded Paint.

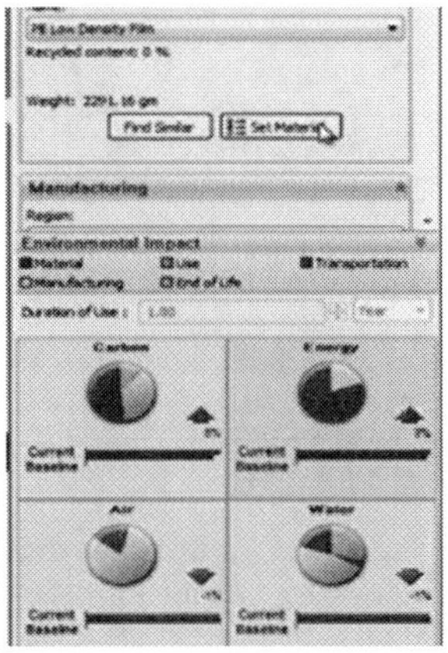

Fig. 8. Set material PE Low.

If was chosen the mold painting mode, in Figure 7 a can be seen the environmental indicators modification.If was chosen material change ("PE Low density Film"), this choice can be seen as another environmental indicators modification (Figure 8). Similar material PE Low could be seen in Figure 9.

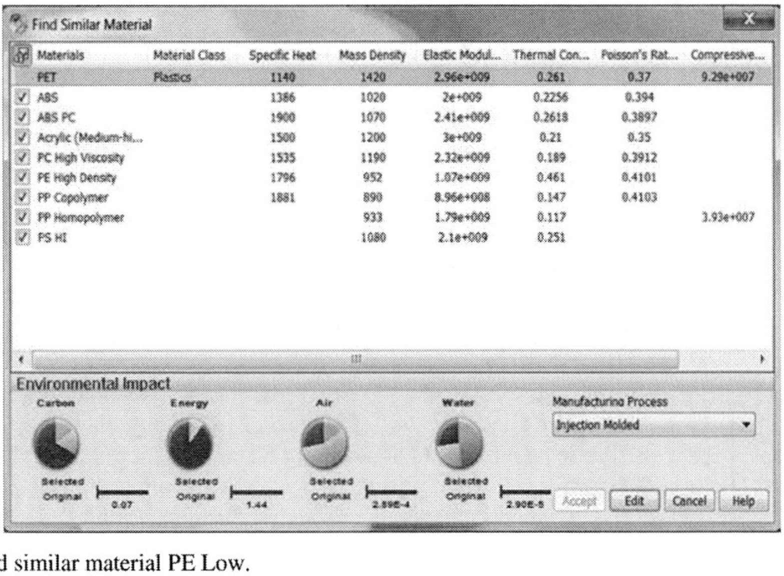

Fig. 9. Find similar material PE Low.

It was generated the Sustainability Report. The outcome is presented in Figure 10.

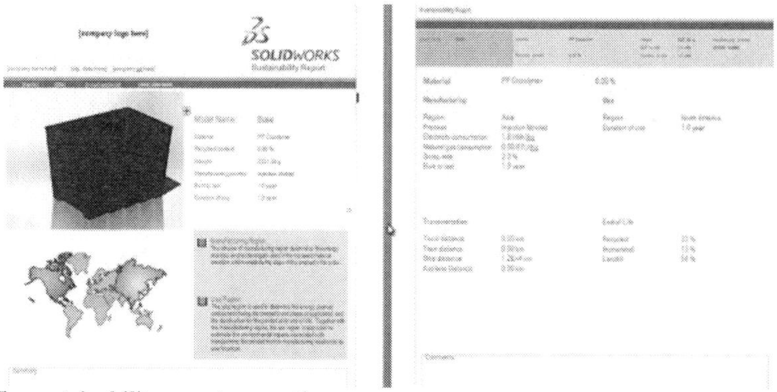

Fig.10. The sustainability report generation.

4 Conclusions

EcoDesign guidelines are a set of best design practices and represent a database of ecological questions and relative answers using a set of well-known standard design parameters [2]. The raw material and environmental impacts emissions (in the case of GHG, Table 2), are associated with bioethanol as a range of different values that can be found in literature [3]. Environmental indicators have become objective functions or constraints in the LCA models. LCA methodologies are becoming quite common because they allow both retrospective and prospective assessment of the environmental impact of a certain product or process (taking into account its entire life cycle) [4]. Solidworks Sustainability provides real-time feedback on environmental impact factors.

The results appear in the Environmental Impact Menu and could be updated dynamically with any changes. Solidworks Sustainability is useful to compare with experimental results or for better sustainability of designed products. Reducing the used material is made by optimizing the design full power. The less material is used for a product, the less resources are used in the production process.

References

1. A. D. Patel, J. A.Posada, L. Shen and M. K. Patel, *Early R&D Stage Sustainability Assessment:The 5 –Pillar Method* ,Wiley Series in Renewable resources, John Wiley&Sons, Ltd., (2016)
2. F. Cappelli, M. Delogu, M. Pierini, Proceedings of LCE (2006) http://citeseerx.ist.psu.edu/viewdoc/download?doi=10.1.1.128.9866&rep=rep1&type=pdf, accessed 13.01.2017
3. A. D. Patel, et al, ESI or Energy & Environmental Science,(2012) http://www.rsc.org/suppdata/ee/c2/c2ee21581k/c2ee21581k.pdf, , accessed 12.07.2017
4. Sustainability of Products, Processes and Supply Chains Theory and Applications, **36**, 1st Edition, Elsevier (2015) https://www.elsevier.com/books/sustainability-of-products-processes-and-supply-chains-theory-and-applications/you/978-0-444-63472-6, accessed 11.01.2017

Method for optimization of the orientation and fixing system of workpiece for the construction of control devices

Daniela-Monica Iordache*, *Alin-Daniel* Rizea, *Aurel* Costea, *Eduard-Laurențiu* Nițu, and *Alexandru* Babă

University of Pitești, Manufacturing and Industrial Management Department, Târgu din Vale Street No.1, Pitești, Romania

Abstract. The development and evolution of technological equipment for machining, assembly and control ensure the modernization of manufacturing processes. Devices as subsystems of technological system in the general context of the development and diversification of machinery, tools, workpiece and drives are made in a variety of sizes and constructive variants that create difficulties in their structure and improvement. Part of the research in recent years presented in this paper have as major objectives the increase of accuracy, productivity and flexibility of orientation and fixing devices for control operations. To this end there have been developed a mathematical model, a new method of working and an algorithm for optimizing the construction of the orientation and fixing system of a new type of control device.

1 Introduction

Technology, originally defined as the science of methods and means of work and control or as a set of processes used to achieve a product, has evolved in the last 25 years by attempts to scientifically systematize knowledge and to develop correct methods of construction for modern technological systems.

Complex optimization of technological processes presented in specialized works [1-4] requires well-reasoned decisions in the field of technological equipment, implicitly in the field of device construction which justifies a systemic approach to manufacturing problems for highlighting the multitude of factors in their interdependencies.

The need to optimize the activities of the stages of design, production and exploitation [1-5] requires an objective analysis of the technological processes in which the orientation and fixing device of the workpiece is a component of the system that also includes technical, technological components and human factor.

The research and achievements in recent years have had as major objectives to improve performance and flexibility of processing, assembling or control equipment, in general, and of device, in particular.

*Corresponding author: monica.iordache@upit.ro

The aim of this paper is to present original contributions of the authors on developing a mathematical model and a method of optimizing the orientation and fixing scheme of workpiece for the construction of work and control devices.

2 The device in the operational technological system

The device is considered in literature [1, 2, 6, 7] a subsystem of a technological system, which constitutes a unit in terms of technology, construction and functionality, which establishes and maintains the orientation of workspiece or tools, with the possibility to take over functions of machine-tool or operator.

Although there are a wide variety of constructive devices, determined by the multiple functions they perform, regardless of their type, having several common elements, the role of orientation being the most important.

In the design phase, the two categories of problems requiring a solution are:
- the theoretical study to establish the optimal orientation and fixing scheme of the semi-product;
- the organological study to establish the optimal variant of construction for the orientation and fixing device of the workpiece.

3 Methodology to optimize the orientation and fixing system of the workpiece

In the field of machine construction it is frequently noted that the interest to apply in industry optimal solutions did not develop enough because it appeals exclusively to the experience of designers and the problem of optimizing complex processes and technological equipment is not a simple one [1-4, 6, 8]. Devices in general, but especially those for orientation and fixing workpiece, have recorded the lowest progress as compared to other subsystems, due to their constructive diversity [1-3, 6].

3.1 Method to optimize the orientation and fixing system of workpiece to build devices

According to the general theory of systems, the technologist is an analyst of the technological system, well-informed on the components of this system, who sets the solution to orient and fix the workpiece.

Figure 1 presents a new method to optimize the orientation and fixing system of workpiece, by which is obtained the optimal orientation and fixing scheme (SOF-O), expressed through informational symbolization.

The general algorithm to design - optimize SOF-O involves the following steps:

E_{01} - Chart of the operation with simplified drawing, data and information from PO and detail or assembly drawing;

E_{02} - Identifying explicit and implicit quotas and conditions required in the operation analyzed;

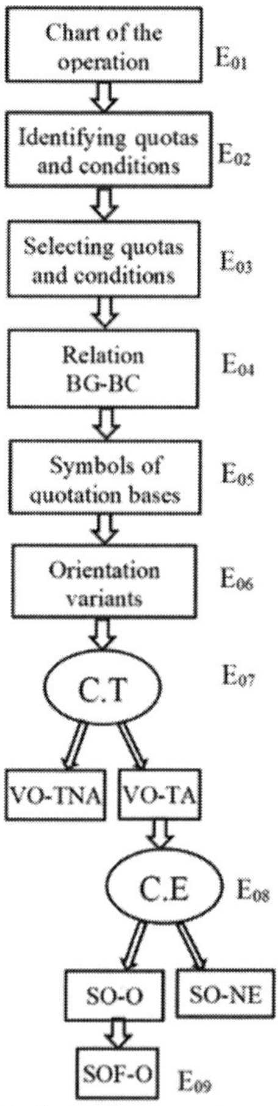

Fig. 1. Optimization scheme.

E_{03} - Selecting quotas and conditions according to specified criteria;
E_{04} - Establishing the relation of dependency between the bases of generation (measurement) and those of quotation;
E_{05} - Symbols of quotation bases;
E_{06} - Establishing technically possible orientation variants (VO-TP) ;
E_{07} - Application of the technical selection criterion (error calculation) abiding by the condition $\varepsilon or < \varepsilon oa$;
E_{08} - Application of the economic selection criterion to obtain optimal orientation scheme (SO-O) ;
E_{09} - Choosing the fixing system to establish the optimal orientation and fixing scheme (SOF-O).

3.2 Establishing technically possible orientation variants (VO-TP)

E_{01} Chart of the operation
On the simplified drawing of the part to control (or process) a system of axes XYZ is associated, the quotas and conditions to be controlled (or carried out) in the analyzed operation are given, the surfaces of the part (processed or unprocessed) are numbered, previously processed surfaces are marked with a continuous thick line and will be controlled or processed with an interrupted thick line; the other surfaces are drawn with a thin line.

Figure 2 presents the chart of the control operation for the assembly water pump, as a complex case study to establish the optimal orientation and fixing scheme (SOF-O). Checking the quotas and conditions in this case is difficult because some processed surfaces are assembled with other parts (shaft and turbine) and the other ones will be partially covered by elements of the orientation system.

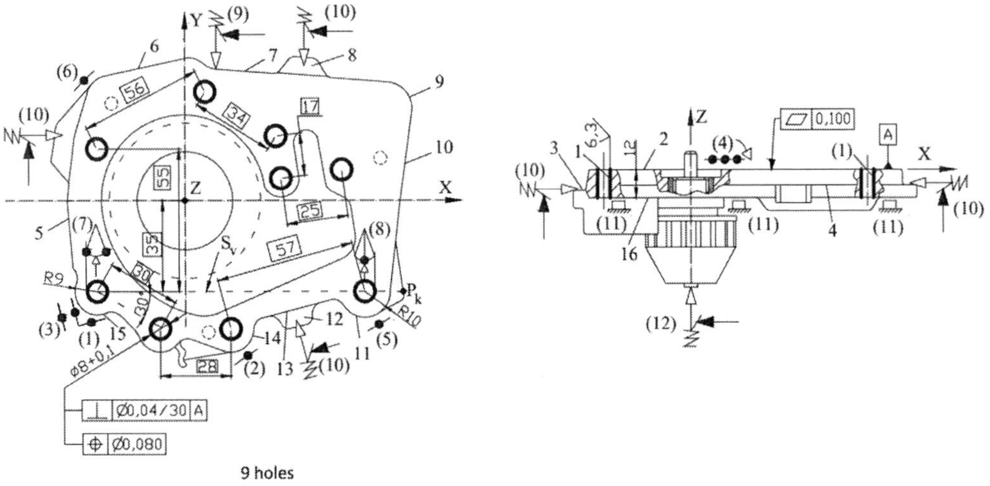

Fig. 2. Chart of the operation.

E02 Identifying the quotas and conditions imposed by the analysed operation
In Table 1 are specified the explicit or implicit quotas and conditions for the analyzed operation, the measuring or generating bases and quotation bases. Notations used refer to geometric elements of flat surface (S), right (D) and point (P).
E03 Selecting important and necessary quotas and conditions related to the orientation of the workpiece

The quotas and conditions are analyzed and value 0 is assigned to conditions that are not related to the orientation of the semi-product and value 1 to those conditions that are linked to its orientation.

When there are two conditions that express the same thing, the condition which includes the other one is maintained.

Table 1. Quotas and conditions.

Quotas and conditions to carry	Measuring or generating bases	Quotation bases	Observations
$C_1 - Ø8^{+0.1}$	Dz_1-z_1	-	
$C_2 - 30$	Dz_1-z_1	Dz_1-z_1	Notation Dz_1-z_1, is read:
$C_3 - R9$	Dz_1-z_1	Dz_{15}-z_{15}	D - right ;
$C_4 - 28$	Dz_1-z_1	Dz_1-z_1	z-z, is the direction of axis z ;
$C_5 - 57$	Dz_1-z_1	Dz_1-z_1	1- refers to surface 1 (noted on
$C_6 - 55$	Dz_1-z_1	Dz_1-z_1	the operation chart).
$C_7 - 56$	Dz_1-z_1	Dz_1-z_1	Notation S_2 is read:
$C_8 - 34$	Dz_1-z_1	Dz_1-z_1	S- is a flat surface;
$C_9 - 17$	Dz_1-z_1	Dz_1-z_1	2- refers to surface 2.
$C_{10} - 25$	Dz_1-z_1	Dz_1-z_1	For S_V, it is understood that it is
$C_{11} - \perp$	Dz_1-z_1	S_2	the imaginary (implicit) plane
$C_{12} - R10$	Dz_1-z_1	S_V	passing through axes $D_{Z1\text{-}Z1}$ of
$C_{13} - //$	Dz_1-z_1	Dz_1-z_1	the two holes in bosses R9 and
$C_{14} - ⏥$	S_2	S_2	R10.
$C_{15} - 12$	S_2	S_2-S_{16}	

If there are two contradictory conditions for a generating basis, the condition involving greater accuracy is maintained. According to these criteria the values are: C1=0; C2=0; C3=1; C4=0; C5=0; C6=0; C7=0; C8=0; C9=0; C10=0; C11=1; C12=1; C13=0 (is included in C11); C14=1; C15=0.

If the multitude of conditions imposed to the operation analysed is expressed as a product of a line matrix with the assigned values and a column matrix with the notations of conditions, a line matrix with the important and necessary conditions linked to the orientation of the workpiece is obtained.

$$(0\,0\,1\,0\,0\,0\,0\,0\,0\,0\,1\,1\,0\,1\,0)x(C_1C_2C_3\dots C_{15})^T = (C_3 + C_{11} + C_{12} + C_{14}) \qquad (1)$$

E04 Establishing the relation of dependency between measuring (generating) bases and those of quotation (obtaining quotation bases)

Conditions C_3, C_{11}, C_{12}, C_{14} refer to the dependency of the generating bases (BG) and quotation bases (BC), which can be expressed by relation: R (BG, BC), formulated as a function of the form:

$$Y = f(X)\cdot W, \qquad (2)$$

where Y-generating basis; X-conditioning elements; W- quotation basis.

The relation adapted for this relation is:

$$Ebg = f\,(cc)\cdot Ebc \qquad (3)$$

where: Ebg - elements of generation; f(cc) – function of quotas and conditions; Ebc – elements of quotation bases.

By customizing of the relations (2) and (3) for the analyzed case, there are obtain:

C_3, Dz_1-z_1= f(R9)·Dz_{15}-z_{15}

C_{11}, Dz_1- $z_1 = f(\perp)\cdot S_2$
C_{12}, Dz_1- $z_1 = f(R10)\cdot S_V$
$C_{14}, S_2 = f($▱$)\cdot S_2$
Dz_1-z_1, is the axis of the hole (holes) $Ø8^{+0.1}$;
S_2 is the stretched flat surface 2;
Dz_{15}-$z_{15}$is the axis of the surface S9;
Sv is the imaginary plane (implicit) passing through axis;
Dz_1- z_1 and axis Dz_1- z_1 of surface R10.

All these elements must be explicit, real and palpable.

Plane Sv, imaginary, will be made explicit by axis Dz_{15}-z_{15} and a fixed point P_k which have a real correspondent on the workpiece.

Therefore it can be written:

$$Dz_1\text{-}z_1 = f(R9)\cdot Dz_{15}\text{-}z_{15} + f(\perp)\cdot S_2 + f(R10)\cdot Dz_{15}\text{-}z_{15} / P_k + f(▱)\cdot S_2 \quad (4)$$

In its general form the relation becomes:

$$Dz_1\text{-}z_1 = f(R9, \perp, R10, ▱)\cdot(Dz_{15}\text{-}z_{15}, S_2, P_k) \quad (5)$$

The elements of the quotation bases are:

$$Ebc\{Dz_{15}\text{-}z_{15}, S_2, P_k\} \quad (6)$$

where: BC_I - (Dz_{15}-z_{15}); BC_{II} – (S_2); BC_{III} – (P_k)

E05 Symbolization of quotation bases

Using informational symbolization, on the operation chart will be attached the icons that materialize the quotation bases:

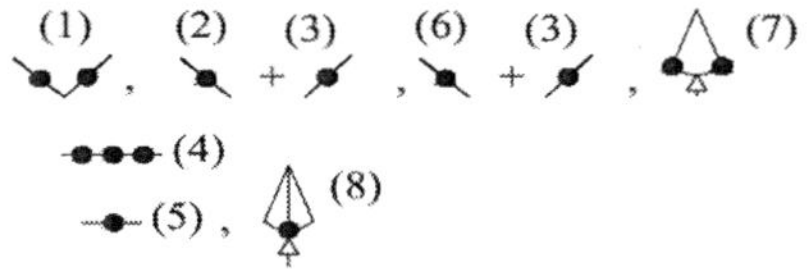

BC_I
BC_{II} (7)
BC_{III}

E06 Establishing technically possible orientation variants (VO-TP)

To establish technically possible orientation variants (VO-TP), condition of taking up to six degrees of freedom is:

$$\begin{array}{l} A_1(a_1, a_2, \ldots a_{m1}) \\ A_2(b_1, b_2, \ldots b_{m2}) \\ \ldots \\ A_n(l_1, l_2, \ldots l_{mn}), \end{array} \quad (8)$$

where:
A_1 - the set of symbols for the first quotation basis;
A_2 - the set of symbols for the second quotation basis;
...
A_n - the set of symbols for the last quotation basis.

To generate all orientation variants it will be used the cartesian multiplication of sets A_1, A_2, … A_n:

$$A_1 \text{ x } A_2 \text{ x} \ldots \text{x } A_n = \{(a_1, b_1 \ldots l_1); (a_1, b_1 \ldots l_2); \ldots (a_{m1}, b_{m2} \ldots l_{mn})\} \quad (9)$$

For the analyzed set, the combination of symbols of orientation variants is:

V_I – (1)+(4)+(5) ; V_{II} – (1)+(4)+(8) ;
V_{III} – (2+3)+(4)+(5) ; V_{IV} – (2+3)+(4)+(8)

V_V – (6+3)+(4)+(5) ; V_{VI} – (6+3)+(4)+(8) ;
V_{VII} – (7)+(4)+(5) ; V_{VIII} – (7)+(4)+(8)

In order to check the maximum number of degrees of freedom cancelled, a matrix will be conceived based on calculation, by entering the number of degrees of freedom cancelled for the symbols belonging to the variants:

$$\begin{matrix} V_I \\ V_{II} \\ V_{III} \\ V_{IV} \\ V_V \\ V_{VI} \\ V_{VII} \\ V_{VIII} \end{matrix} \quad \begin{pmatrix} 2\,3\,1 \\ 2\,3\,1 \\ 2\,3\,1 \\ 2\,3\,1 \\ 2\,3\,1 \\ 2\,3\,1 \\ 2\,3\,1 \\ 2\,3\,1 \end{pmatrix} \begin{pmatrix} 1 \\ 1 \\ 1 \end{pmatrix} = \begin{pmatrix} 2+3+1 \\ 2+3+1 \\ 2+3+1 \\ 2+3+1 \\ 2+3+1 \\ 2+3+1 \\ 2+3+1 \\ 2+3+1 \end{pmatrix} = \begin{pmatrix} 6 \\ 6 \\ 6 \\ 6 \\ 6 \\ 6 \\ 6 \\ 6 \end{pmatrix} \tag{10}$$

All variants meet the condition imposed and become technically possible orientation variants (VO-TP).

E07 Application of the technical selection criterion (CT)

For the analysis of the orientation accuracy of technically possible variants (VO-TP): it has to be calculated the allowable orientation error (ε_{oa}); the real orientation error (ε_{or}); checking at the end the accuracy of variants by condition $\varepsilon_{or}<\varepsilon_{oa}$.

Allowable orientation errors (ε_{oa}) are determined by relation [1, 3, 5, 8, 9]:

$$\varepsilon_{oa} = \left(\frac{1}{5} \cdots \frac{1}{10}\right) T_P \tag{11}$$

where T_P is tolerance imposed on the drawing. For the quotes or conditions not specified in the plan of execution, the values of standards or norms must be adapted.

By imposing $\varepsilon_{oa} = \frac{1}{8} T_P$ are obtained the following allowable orientation errors:

C_3 -$\varepsilon_{oa} = \frac{1}{8}\, 0{,}4 = 0{,}05mm$; C_{11} - $\varepsilon_{oa} = \frac{1}{8}\, 0{,}04 = 0{,}005\ mm$;

C_{12} - $\varepsilon_{oa} = \frac{1}{8}\, 0{,}6 = 0{,}075mm$;C_{14} - $\varepsilon_{oa} = \frac{1}{8} 0{,}1 = 0{,}012\ mm$

Real orientation errors (ε_{or}) are calculated as follows.

V_I – (1)+(4)+(5)

C_3 -$\varepsilon_{or} = \frac{T_d}{2sin\alpha/2} = \frac{0{,}4}{2sin45^\circ}$ =0,285 mm. This value cannot be considered since during the control operation in the orientation variant V_I, the surfaces of the part (S_{15}, S_2, S_{11}) come into contact with support elements of the control device which are identical to those on the drilling device and have an error of execution and installation of maximum 0.010 [mm].

In this case, the errors are:

V_I – (1)+(4)+(5); C_3-$\varepsilon_{or} = 0{,}010\ mm$; C_{11}-$\varepsilon_{or} = 0$; C_{12}-$\varepsilon_{or} = 0{,}010\ mm$; C_{14}-$\varepsilon_{or} = 0$

V_{II} – (1)+(4)+(8); C_3-$\varepsilon_{or} = 0{,}010\ mm$; C_{11}- $\varepsilon_{or} = 0$; C_{12}-$\varepsilon_{or} = 0$; C_{14} -$\varepsilon_{or} = 0$

V_{III} – (2+3)+(4)+(5); C_3-$\varepsilon_{or} = 0{,}148\ mm$; C_{11}-$\varepsilon_{or} = 0$; C_{12}-$\varepsilon_{or} = 0{,}010\ mm$;$C_{14}$- $\varepsilon_{or} = 0$

For C_3 it is considered that element (3) is the same semi-prism of the element (1) and semi-prism (2) influences with half of the tolerance field of the surface:

C_3 -$\varepsilon_{or} = \frac{0{,}01+0{,}2}{1{,}414} = 0{,}148\ mm$

V_{IV} – (2+3)+(4)+(8); C_3-$\varepsilon_{or} = 0{,}148\ mm$; C_{11}-$\varepsilon_{or} = 0$; C_{12}-$\varepsilon_{or} = 0$; C_{14}-$\varepsilon_{or} = 0$

V_V – (6+3)+(4)+(5); C_3-$\varepsilon_{or} = 0{,}219\ mm$; C_{11}-$\varepsilon_{or} = 0$; C_{12}-$\varepsilon_{or} = 0{,}010\ mm$; C_{14}-$\varepsilon_{or} = 0$

For C_3 it is considered that element (3) is the same semi-prism of element (1) and semi-prism (6) influences with half of the tolerance field of the surface:

$C_3 - \varepsilon_{or} = \frac{0,01+0,3}{1,414} = 0,219\ mm$

V_{VI} – (6+3)+(4)+(8); C_3-$\varepsilon_{or} = 0,219\ mm$; C_{11}-$\varepsilon_{or} = 0$; C_{12}-$\varepsilon_{or} = 0$; C_{14}- $\varepsilon_{or} = 0$

V_{VII} – (7)+(4)+(5); C_3-$\varepsilon_{or} = 0$; C_{11}-$\varepsilon_{or} = 0$; C_{12}-$\varepsilon_{or} = 0,010\ mm$; C_{14}-$\varepsilon_{or} = 0$

V_{VIII} – (7)+(4)+(8); C_3-$\varepsilon_{or} = 0$; C_{11}-$\varepsilon_{or} = 0$; C_{12}-$\varepsilon_{or} = 0$; C_{14}-$\varepsilon_{or} = 0$

The calculated values are written in Table 2 where the technically possible variants (VO-TP) matching relation $\varepsilon_{or} < \varepsilon_{oa}$ for all conditions C_3; C_{11}; C_{12}; C_{14} become technically acceptable orientation variants (VO-TA) and the others are removed as technically unacceptable orientation variants (VO-TNA).

Table 2. Synthesis of allowable and real orientation errors.

Crt. nb.	VO-TP	C_3 $\varepsilon_{oa} = 0,05$ mm	C_{11} $\varepsilon_{oa} = 0,005$mm	C_{12} $\varepsilon_{oa} = 0,075$mm	C_{14} $\varepsilon_{oa} = 0,012$ mm	VO-TA
		ε_{or} [mm]	ε_{or}	ε_{or} [mm]	ε_{or}	$\varepsilon_{or} < \varepsilon_{oa}$
1	V_I	0.010	0	0.010	0	X(Yes)
2	V_{II}	0.010	0	0	0	X(Yes)
3	V_{III}	0.148	0	0.010	0	-
4	V_{IV}	0.148	0	0	0	-
5	V_V	0.219	0	0.010	0	-
6	V_{VI}	0.219	0	0	0	-
7	V_{VII}	0	0	0.010	0	X(Yes)
8	V_{VIII}	0	0	0	0	X(Yes)

E08 The economic analysis VO-TA

In Table 3 the economic optimization obtained by the method of utilities proposed in paper [1-3, 7] allows establishing optimal orientation elements under the form of optimal orientation scheme (SO-O) based on coefficients of importance assigned to the orientation elements for the six economic criteria analyzed ($CE_1 \ldots CE_6$).

CE_1- number of elements composing the support symbolized;

CE_2 - complexity of elements;

CE_3 - cost of constructive elements associated to the informational symbolization;

CE_4 - productivity concerning the convenient use of the device;

Table 3. Economic optimization by the method of utilities.

No.	No VO-TA	Symbolization associated BC	Criteria and utilities						$\sum C_i$	Decision
			CE_1	CE_2	CE_3	CE_4	CE_5	CE_6	Max	Optimal locally
	$Nr._{VO-TA}$ = 2x1x2= 4	BC_I	Importance coefficients (Ci)							
1		 (1)	1	0.9	0.9	1	0.9	0.8	5.5	Optimal BC_I
2		 (7)	0.6	0.6	0.5	0.6	0.5	0.7	3.5	
		BC_{II}								
3		 (4)	1	0.9	0.9	1	0.9	0.9	5.6	Optimal BC_{II}
		BC_{III}								
4		 (5)	1	1	1	1	0.9	0.9	5.8	Optimal BC_{III}
5		 (8)	0.6	0.6	0.5	0.6	0.5	0.7	3.5	

CE_5 - wear and cost of maintenance (exploitation);

CE_6 - reuse of support elements for the construction of other elements.

The optimal orientation scheme (SO-O) is:

SO-O=  (12)

E09 Choice of the fixing system

A secondary adjustment force of the support with self-place and lock (9) brings the part that sits on the flat base of pre-orientation (11) into contact with the first short line (1) and support plate 5 (see Fig.2). The main flat base (4) flips in working position to bring the part under the action of the support (12) with self-place and lock, which originally has the function of a main adjustment force, afterwards becoming a fixing force.

For the measurements, the three special supports (10), with auto-position and blocking further come into contact with the part and define a flat substitute base, which preserves the original orientation of the part and accuracy of the operation, and then the flat base (4) flips.

The optimal orientation and fixing scheme is:

SOF-O : 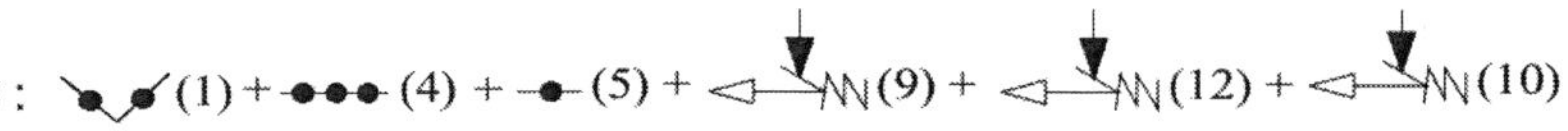 (13)

4 Conclusions

In the activity of research-design and realization of industrial applications it is necessary to have a well-reasoned multi-criteria approach of the problems analysed, leading to objective decisions and rational solutions to build devices and technological equipment. The systemic approach on issues to optimize processes and technological equipment is determined by the need to highlight the multitude of factors influencing it and their interdependency in order to establish rational solutions in the field of industrial engineering.

In this paper the original contribution of the authors refers to a mathematical model and a new method to optimize the orientation and fixing scheme of workpieces to build work and control devices. Based on this methodology, there was obtained a solution of a technological device as industrial application, with clear new elements from a kinematic, technological, structural and functional point of view, which is the subject of a patent.

References

1. A. Brăgaru, Proiectarea dispozitivelor. *Teoria şi practica proiectării schemelor de orientare şi fixare*, **1**, (Ed. Tehn. Bucureşti, 1998).
2. D. M. Iordache, A. Costea, Al. Babă, *Metode de calcul şi modele matematice pentru optimizarea proiectării dispozitivelor* (Ed. Univ. din Piteşti, 2016).
3. N.V. Ivan, T. Păunescu, R. Udroiu, M.C. Ivan, C. Găvruş, R. Pescaru, *Tehnologia Construcţiilor de Maşini - Teorie şi abordări inovative*, **I**, (Ed. Univ. Transilvania din Braşov, 2010).
4. G. H. Qin, W.H. Zhang, Int. J. Adv. Manuf. Tech., **29**,(2006).
5. A. Sturzu, *Bazele proiectării dispozitivelor de control al formei şi poziţiei relative a suprafeţelor în construcţia de maşini*. (Ed. Tehn. Bucureşti,1977).
6. J.J.X. Liu, D.R. StrongInt. J. Adv. Manuf. Tech., **21**, (2003).
7. T. Păunescu, H. Bulea, R. Păunescu, *Dispozitive modulare*, **II**, (Ed. Univ. Transilvania din Braşov, 2008).
8. B. Li. Int. J. Adv. Manuf. Tech.,**17**, (2001)
9. A. Raghu, S.N. Melkote, Procedings of IMECE'03 2003 ASME International Mechanical Engineering Congress & Exposition Wasington, D.C., (2003).

5-axes modular CNC machining center

Radu-Eugen Breaz*, *Sever-Gabriel* Racz, and *Octavian-Constantin* Bologa

Lucian Blaga, University of Sibiu, Department of Industrial Machines and Equipment, Emil Cioran 4, Sibiu, Romania

Abstract. The paper presents the development of a 5-axes CNC machining center. The main goal of the machine was to provide the students a practical layout for training in advanced CAM techniques. The mechanical structure of the machine was built in a modular way by a specialized company, which also implemented the CNC controller. The authors of this paper developed the geometric and kinematic model of the CNC machining center and the post-processor, in order to use the machine in a CAM environment.

1 Introduction

CNC machine-tools represent the backbone of nowadays machine-building industry. Consequently, using and programming CNC machine-tools are important subjects within the curriculum in technical universities. While 3-axes CNC machine-tools are widely used in industry, currently we are witnessing the introduction of 5-axes CNC machine-tools in more and more industry sectors.

In order to offer the students state of the art training with regards of CNC machine-tools, universities have to equip their laboratories with both 3-axes and 5-axes machine tools. While industrial 5-axes CNC machine-tools may be a rational choice where complex and highly accurate parts are required, for teaching purposes, this king of equipment may be too expensive. So, the problem tackled by this research work was to develop a 5-axes machining center, with similar technological capabilities as industrial state of the art solutions, at an affordable level of development costs and suitable to be used for teaching the future engineers how to use and program 5-axes CNC machine-tools.

1.1 Open architecture CNC machine-tools

Industrial CNC machine-tools are supplied as closed architecture systems. Their structure, kinematic and CNC controllers are designed to be used solely for their basic purpose, without any possibility to change it. Practically these systems have no re-configurability and cannot be adapted in order to perform other machining processes other the one they were designed for. The re-configurability is seen here as the ability to add or remove modules from the machine, to equip it with supplementary degrees of freedom (axes). Their CNC controllers have also closed architecture which cannot be modified.

* Corresponding author: radu.breaz@ulbsibiu.ro

As opposite to the closed architecture, researches were reported in the literature of open systems CNC architecture [1-2]. One of the most known approaches of developing such systems are OSACA (Europe), OMAC (USA) and OMEC (Japan) [3]. These systems are based on building a CNC system form modular components and standardized interfaces. Other approaches of developing open architecture CNC systems are relying on the STEP NC concept, which tries to develop a machine independent programming language [4].

2 The structure of the 5-axes CNC machining center

The workspace of CNC machine-tools is defined as a Cartesian one, with three translational axes (X, Y, and Z) and three rotational axes (A, B, C), as presented in Figure 1.

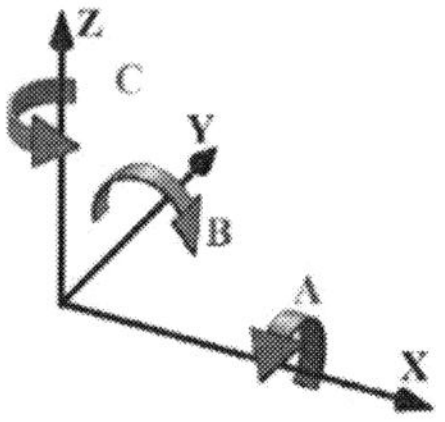

Fig. 1. The Cartesian workspace used for CNC machine-tools.

A large amount of the CNC machine-tools on the market can only perform translational movements on X, Y and Z axes and are called 3-axes CNC machine-tools. In order to machine complex parts sometime one or even two supplementary rotational movements in combination with translation along X, Y and Z axes are required. In order to perform such motions 5-axes CNC machine-tools are used.

A simplified diagram of a 5-axes CNC machine-tool, with X, Y, Z translational axes and A, C rotational axes is presented in Figure 2.

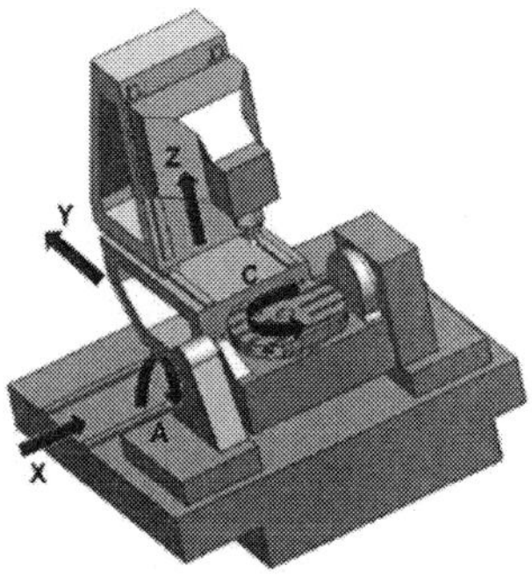

Fig. 2. Simplified diagram of a 5-axes CNC machine-tool with X, Y, Z, A and C kinematic.

In order to develop the 5-axes modular machining center, a requirements list regarding the machine's characteristics was build. The list is presented below:

- the kinematic if the machine has to allow simultaneous movements on 5-axes;
- the machine has to be built in a modular way, in order to allow its use both as 3-axes and 5-axes machine-tool;
- the machine has to be programmed using G-code (formalized as ISO 6983);
- the control loops on each axis has to be closed loop ones (using servomotors for actuation and encoders for feedback – no stepping motors were allowed);
- the overall developing costs of the system must not exceed 25.000 Euro.

2.1 Mechanical structure

The machine structure was designed modularly, by combining the following modules:
A. Custom built modules:
- mechanical frame; fixed table; movable portal unit for X-axis; movable slide for Y-axis; movable slide with Z-axis.

B. Special units from ISEL company:
- rotary tilting unit DSH-S combined with rotary unit RDH-S;
- spindle motor unit iSA 1500 (with manual tool exchange).

The overall view of the 5-axes CNC machining center is presented in Figure 3.

Fig. 3. Overall view of the 5-axes machining center.

The custom-built modules also include in their structure specific ISEL components such as linear guideways and linear slides. The general assembly of the machining center was made by company General Numeric, from Brașov, Romania, a company specialized in manufacturing CNC profiling machines (oxy-gas and/or plasma cutting) and CNC routers.

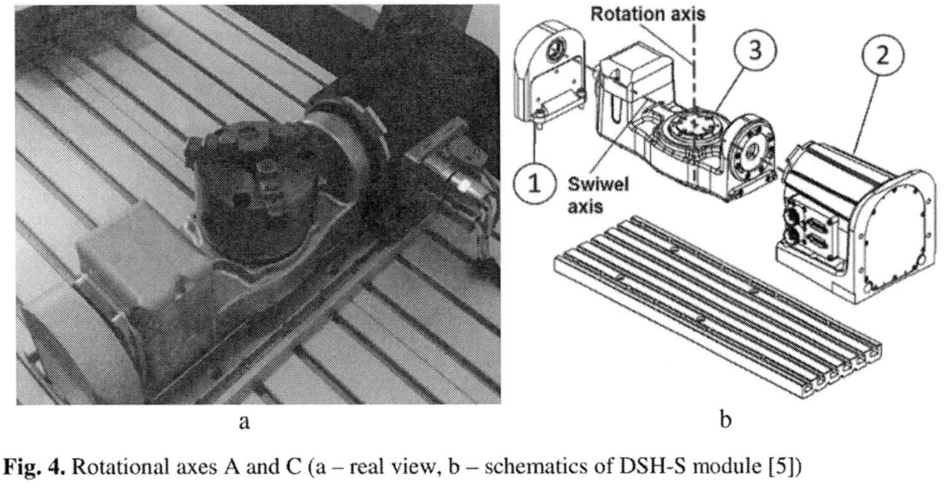

a b

Fig. 4. Rotational axes A and C (a – real view, b – schematics of DSH-S module [5])

Figure 4 a, b shows the combined modules DSH-S (tilting) and RDH-S (rotary) which embodies the rotational axes A and C respectively. The combined modules are based upon the ISEL harmonic drive and allow the rotational movements around the swivel axis (A) and the rotary axis (C) presented in Figure 4 b [5].

DOI: 10.1051/matecconf/201711206004

In Figure 4 b, 1 represents the counter bearing, 2 the swivel unit and 3 the rotary indexing table RDH-S.

2.2 The CNC controller

In order to provide open-architecture capabilities to the machining center the LinuxCNC software system was used as CNC controller. It can drive milling machines, lathes, 3D printers, laser cutters, plasma cutters, robot arms, hexapods, and more. LinuxCNC is also compatible with many popular machine control hardware interfaces and provides to the user several graphical user interfaces [6]. The integration of LinuxCNC as controller of the 5-axes machining center was also made by General Numeric company. The GUI used for this machine-tool is presented in Figure 5.

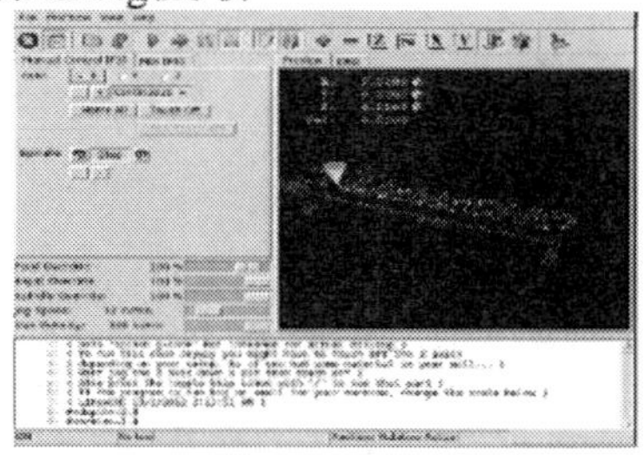

Fig. 5. Graphical user interface (GUI) of LinuxCNC controller.

3 The geometric and kinematic model

5-axes machining operations rises a series of problems which do not appear during normal 3-axes machining. While during 3-axes machining only reciprocal movements between tool and workpiece occur and have to be taken into consideration for collisions avoidance, things are very different for 5-axes machining. The number of movements are higher, because of the supplementary rotational movements, and collisions may occur not only between tool and workpiece but also between tool, workpiece and machine slides and platters and between machine slides and platters themselves.

The simulation interface from Linux CNC is only able the represent the toolpaths, without taking into consideration the workpiece and machine geometry and the potential collisions. Moreover, most of CAM software packages do not take into consideration the geometry and kinematics of the machine when simulating the machining process.

To address this problem, the authors of this work have developed a geometric and kinematic model of the 5-axes machining center, by performing the following steps [7]:

- building a 3D model of the machine-tool;

This step was performed by a joint modelling and import process. While the custom build modules of the machine were modeled, the 3D models of specialized ISEL modules were imported form the files provided by the manufacturer. Figure 6 shows the 3D model of the machining center, developed in CATIA.

- separating each kinematic axis of the machine form the 3D model and saving it as an individual module;

During these two first steps, some alterations of the geometry of the components are allowed in order to reduce the size of the files. The user has to keep in mind that only the components which may lead to collision are important, while inside components, or components which cannot interfere with others components may be simplified or even removed. For example, from Figure 6 it can be noticed that the height of the machine has been reduced, because the machine supports cannot collide with any moving parts.

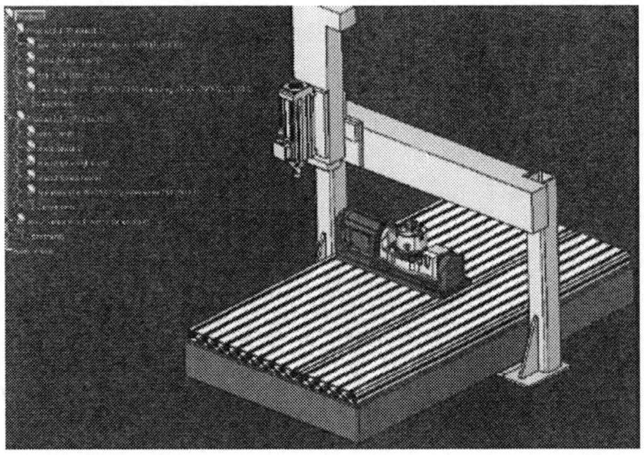

Fig. 6. The 3D model of the machine in CATIA.

- exporting the axis modules as igs file and re-importing them in a CAM software package;
- using a specialized module of the CAM software to build the kinematic model of the machine;

During this final step, the user has to define the hierarchies between each axis (for example, axis X supports both axes Y and Z, axis Y supports axis Z) and the kinematic dependencies between the axes (for example, axis X cannot be moved without moving also axes Y and Z). Also, the stroke length (and angular limits) for each axis have to be defined.

After building the geometric and kinematic model, the CAM software was able to realistically simulate the machining processes unfolded on the machine.

Figure 7 presents a screenshot form a simulation of a 3+2 axes indexed milling process (the rotational movements involved in the process are indexing movements, performed outside the milling process, used only for positioning the part).

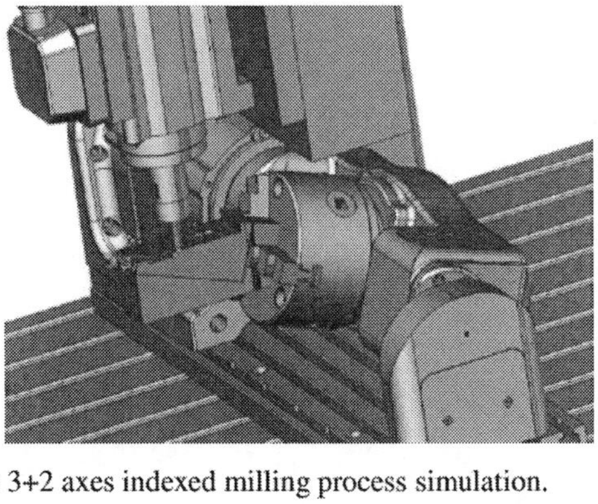

Fig. 7. Screenshot form a 3+2 axes indexed milling process simulation.

Figure 8 a, b presents screenshots form a simulation of a 5 axes continuous milling process. During this machining process simultaneous movements occur on all five axes of the machine.

4 Conclusion

During this research program, a modular 5-axes machining center was developed and

implemented. The machine was built by a specialized industrial company according to a requirements list made by the authors of this work. In order to be able to use the machine

DOI: 10.1051/matecconf/201711206004

in a CAM environment, a geometric and kinematic model of the machine was built, which allow the user to realistically simulate the machining process.

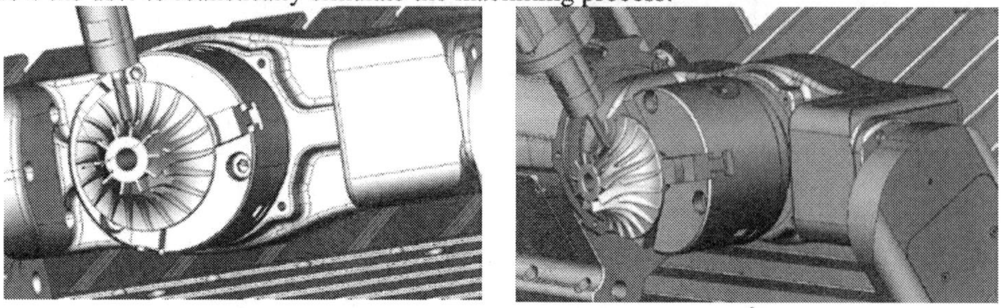

a b

Fig. 8. Screenshots form a 5 axes continuous milling process simulation.

By using the geometric and kinematic model, the CAM program is able to generate the complex toolpaths required for 5-axes machining operations, by taking into consideration the real kinematic capabilities of the machine (hierarchies and dependencies between machine axis). Also, the simulation process is able to detect not only tool-workpiece collisions, but also the collisions between the mobile elements of the machine.

The authors want to thank the students from the second study year (now graduates of 2016) of "CAD/CAE/CAM Techniques in Plastic Deformation" Master study programme from "Lucian Blaga" University of Sibiu, which built the 3D assembly model of the 5-axes machining center. Special thanks are also owed to Eng. Meszaros Mihai Dan, graduate of the above mentioned Master programme, which assisted the authors in the process of developing the kinematic and geometric model of the machining center.

References

1. H. Tianliang, Z. Chengrui, L. Riliang, L. Peng, Int. J. Adv. Manuf. Technol., **40**(5), 541–552, (2009)
2. S. Park, S.-H. Kim, H. Cho, Int. J. Adv. Manuf. Technol., **27**, 788-796, (2006)
3. W. Sperling, P. Lutz, Robot Manuf., **6**, 613–620, (1996)
4. C. Zhang, R. Liu, T. Hu, Int. J. Comput. Integr. Manuf., **19**(6), 508–515, (2006)
5. *** https://www.isel.com/en/downloads/dl/file/id/4268/assembly_instruction_rdh_dsh.pdf
6. *** http://www.linuxcnc.org/
7. A. L. Chicea, R. E. Breaz, O. Bologa, Applied Mechanics and Materials, **809-810**, 1004-1009, (2015)

MATEC Web of Conferences **112**, 06005 (2017) DOI: 10.1051/matecconf/201711206005
IManE&E 2017

Applicability of ANN models and Taguchi method for the determination of tool life in turning

Nikolaos E. Karkalos, and *Angelos* P. Markopoulos*

[1]National Technical University of Athens, School of Mechanical Engineering, Section of Manufacturing Technology, Heroon Polytechniou 9, 15780, Athens, Greece

Abstract. Tool life is an important parameter in machining processes, affecting directly the quality of machined components and the process cost. It is already shown that various parameters can affect tool life such as process parameters, i.e. depth of cut, cutting speed and feed, or material properties of cutting tool and workpiece. The determination of the effect of each parameter on tool life is of crucial importance when designing the manufacturing process of a product in order to select suitable process parameter values and tool types. Several empirical formulas for the determination of tool life exist in the relevant literature; especially in the case of CBN cutting tools for turning, a cubic polynomial formula was proposed to model the relationship between tool life and cutting speed. The determination of the polynomial parameters was performed by conducting cutting experiments for several cutting speeds, without the aid of a design of experiments (DoE) method in order to model properly this non-linear relationship. In this paper, the feasibility of determining this non-linear relationship by conducting experiments designed by Taguchi method and using artificial neural networks (ANN) is investigated for several cases and conclusions on the applicability of this approach are presented.

1 Introduction

The determination of the correlation between tool life (and tool wear) and various process parameters is essential for efficient manufacturing. Tool wear refers to the gradual deterioration of cutting tools' performance due to various factors, e.g. thermo-mechanical or chemical ones. Some of the types of tool wear include flank wear, crater wear and notch wear among others; their effects on the process, apart from increased cutting forces and temperature are the decreased dimensional accuracy and quality of produced parts and tool breakage, which eventually result in increased cost.

As far as tool wear models are concerned, apart from the empirical or semi-empirical approaches [1-2], firstly established from the early 1900s, there also exist some numerical

* Corresponding author: amark@mail.ntua.gr

and several soft computing models, capable of predicting cutting tool wear in various cases. Leone et al. [3] employed an ANN model for the prediction of tool wear progression during machining of a Nickel-based alloy. They found that it was superior from a regression model, as it was able to predict maximum flank wear (VB_{max}) for cutting speeds more than 50 m/min, which was the safe machining limit for the examined cases, with significant accuracy. Rao et al. [4] used an ANN model for the prediction of tool wear, surface roughness and vibration of workpiece during boring. They conducted experiments designed by Taguchi method with two levels of cutting speed, nose radius and feed values. The proposed model consisted of four layers and exhibited mean percentage error of 4.52%, 4.26% and 2.92% for surface roughness, amplitude of vibration and tool wear, respectively.

D'Addona et al. [5] developed an ANN model of tool wear during machining of nickel superalloys. For the model, data from experiments at five different cutting speeds were employed and the ANN contained three layers; hidden layer neurons number was determined by a "cascade learning" procedure in which new nodes are added to the model until no further reduction of error is observed. Root mean square error (RMSE) of the results up to the largest cutting speed value were acceptable, whereas an underestimation of the wear land size was observed for the cutting speed of 55 m/min when the wear value was over the limit value. D'Addona and Teti [6] created a predictive tool for crater wear using image data of the cutting tool and an ANN model. They conducted tests for different values of cutting speed, feed rate and depth of cut. After images corresponding to each studied cases were properly processed, a three-layer ANN was used to predict tool wear and comments were made on the correct strategies for tool wear recognition.

However, ANN models are often employed for on-line monitoring of machining processes as well. Srikant et al. [7] used a modified back propagation ANN for online tool wear prediction in cases of wet machining. They conducted various experiments with dry and wet conditions and various cutting fluids and obtained results concerning cutting forces, temperature, flank wear and surface roughness. The inputs for the ANN model were forces, temperature, surface roughness, machining time, viscosity and thermal conductivity of cutting fluid and workpiece hardness and the structure of neural network was optimized using simulated annealing method. Absolute error of prediction was lower than 5%, thus the results were considered satisfying. Corne et al. [8] created a real time tool wear and breakage prediction system using ANN. They conducted experiments of hole drilling and observed tool wear during the process. For the development of the ANN model, they used data from five different experiments with spindle speed between 400 and 600 rpm and feed of 0.05 and 0.1 mm/rev, for a total of 36 input/output data samples. Levenberg-Marquardt training algorithm and a single hidden layer of five neurons were chosen. They compared the performance of models with direct power and force measurement and observed that models with force input data were more accurate, but given power and force are correlated, measurement of power load during the process can lead to reliable prediction of tool wear.

Liu and Altintas [9] employed an ANN for the online monitoring of flank wear. The input variables of the model were cutting speed, feed rate, and cutting force ratio as well as past values of measured force ratio and estimated flank wear and the data covered ten different cutting conditions. The real time monitoring system was implemented and the performance of tool wear prediction model was found to be in significant agreement with test results on new turning experiments. Krishnakumar et al. [10] developed a tool wear predictive tool based on vibration signals monitoring for high speed machining of Ti alloy. They conducted six machining experiments with various cutting speed and spindle speed values for a total of 87 data samples and created an ANN model with standard error, kurtosis and median or vibration data as inputs. It was found that the ANN model was superior from a decision tree model for the classification of cutting tool condition. Other notable studies with neural network models include studies with neuro-fuzzy approaches [11, 12].

In the current work, the applicability of standard DoE methods and ANN models for the approximation of tool life-cutting speed relationship is investigated. The tool life curves derived from the ANN models are directly compared to the experimental ones in order to determine up to which degree tool life and tool life curve characteristics, such as minima and maxima, can be approximated.

2 Tool life models

Tool life models are often developed in respect to process parameters. Some of the most notable tool life equations include Taylor, Safonov, Wu and Granovski among others [2]. The simplest among them is the well-known Taylor equation for tool life which correlates cutting speed (v_c) with tool life (T) and tool-workpiece characteristics and can provide a preliminary approximation for tool life:

$$v_c T^n = c \tag{1}$$

where n and c are constants depending on cutting tool and workpiece. This equation can be generalized to include depth of cut (a_p) and feed rate (f) as well:

$$v_c T^n a_p^x f^y = c \tag{2}$$

where n, x, y, c are constants, which are experimentally derived. Moreover, some approaches consider other process-related parameters and even the value of VB_{max}. However, for the case of machining with CBN tools, Kundrák [13] created a tool life formula based on a rational function with a cubic polynomial as denominator, intending to describe tool life-cutting speed relationship for a wide range of cutting speed values:

$$T = \frac{C_{T_1}}{v_c^3 + C_{T_2} v_c^2 + C_{T_3} v_C} \tag{3}$$

where T represents tool life, v_c represents the cutting speed and C_{T1}, C_{T2}, C_{T3} are constants depending on process conditions. This formula will be employed in the current work for the comparison of predicted and experimental results.

3 Methodology

In the present work, an investigation of the applicability of DoE and ANN for tool life prediction of CBN tools for a wide range of cutting speeds is presented. For all cases, data from [2] are employed to develop ANN models for the prediction of tool life. After some trial runs were performed, the main investigation is conducted; two experimental designs with three factors at four levels, namely Taguchi L16 orthogonal array and 4^3 full factorial design are employed to provide the tool life determination experimental runs, for which values from the actual experiments are used, either directly or indirectly through equation (3). The number of levels was selected to be four, as it can provide the minimum sufficient number of points for the approximation of equation (3), which contains a cubic polynomial. The results from the simulated experiments conducted from these designs are then used to create a predictive model using ANN models. Finally, results from ANN models are compared to the results presented in [2] in order to determine at which degree experiments with standard experimental designs can "capture" the tool life curve behaviour for the whole range of cutting speeds.

In this work, the comparison of experimental with ANN predicted curves will be conducted in terms of correlation coefficient R, RMSE, mean absolute error (MAE) and mean percentage error (MPE) in order to assess the level of accuracy of the two approaches. All ANN models employ radial basis functions and are implemented in MATLAB environment with cutting speed v_c (m/min), feed f (mm/rev), depth of cut a_p (mm) as inputs and tool life T (min) as the only output. Spread parameter was determined after a trial-and-error procedure and was kept constant with value $\sigma = 0.5$ in every case. The value range for the input variables is 11-120 mm/min for the cutting speed, 0.025-0.125 mm/rev for feed and 0.05-0.25 mm for the depth of cut.

4 Results and discussion

The evaluation of results is conducted by comparison of actual values from the experiments, namely 6 cases, presented in [2], which are also displayed in Table 1. In Table 2, the performance of each of the two approaches is presented.

Table 1. Characteristics of the cases considered in this study.

No	f (mm/rev)	a_p (mm)	No	f (mm/rev)	a_p (mm)
1	0.025	0.1	4	0.050	0.05
2	0.075	0.1	5	0.050	0.15
3	0.125	0.1	6	0.050	0.25

Table 2. Mean values and range of values for R, RMSE, MAE and MPE for the two approaches.

Mean values				
Approach	R	RMSE	MAE (%)	MPE (%)
L16+ANN	0.9714	33.3731	26.8680	-21.4911
4^3 + ANN	0.9634	29.0291	23.7730	32.2893
Range of values				
Approach	R	RMSE	MAE (%)	MPE (%)
L16+ANN	0.9387-0.9827	16.749-53.996	18.909-48.552	-41.513-35.646
4^3 + ANN	0.9187-0.9870	15.919-39.638	19.897-33.230	20.936-58.110

For the cases where the L16 and ANN approach was used, it is observed that the results are not very satisfactory as error is around 25%; although it is exceeding the widely accepted value of 10% for prediction errors, the fact that only 16 input/output pairs were used renders this result not totally inacceptable and it will be quite useful for the purposes of the current study. Correlation coefficient value is quite high, indicating that there is strong correlation between predicted and experimentally derived data and RMSE value is not extremely high.

For the cases where the full factorial and ANN approach was used, similar results are generally observed. MAE is below 25% in all but one case whereas MPE values are quite large in all cases (close to 50%). Correlation coefficient values are above 0.9 in every case and RMSE mean value of 29.0291 indicates a moderately high degree of error between predicted and experimental curves.

Comparing the results from the two different experimental designs, it can be observed that there are several similarities but also some differences. At first, as can be seen from Table 2, the values of correlation coefficient and errors in each case are close enough, with the second approach exhibiting better mean values for RMSE and MAE, as well as narrower value range for these two quantities. These values are indicative of the results portrayed in Figure 1 and Figure 2, as it is evident that none of the two approaches can fully approximate the experimental curve.

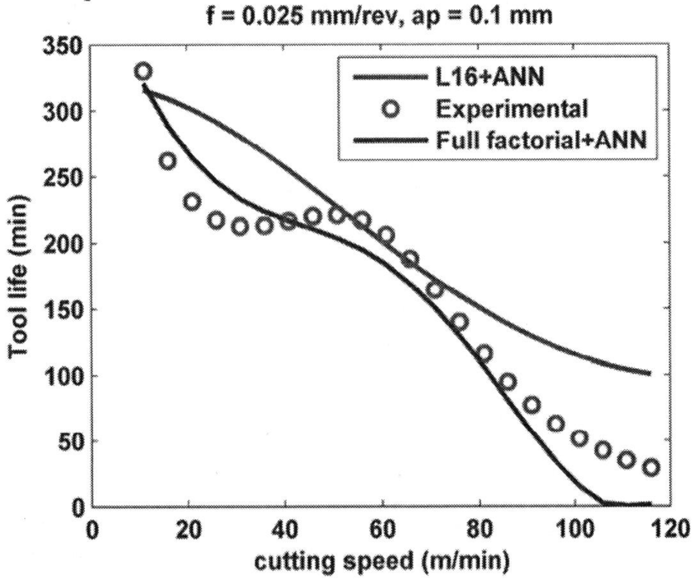

Fig.1. Experimentally and ANN derived curves for $f = 0.025$ mm/rev, $a_p = 0.1$ mm.

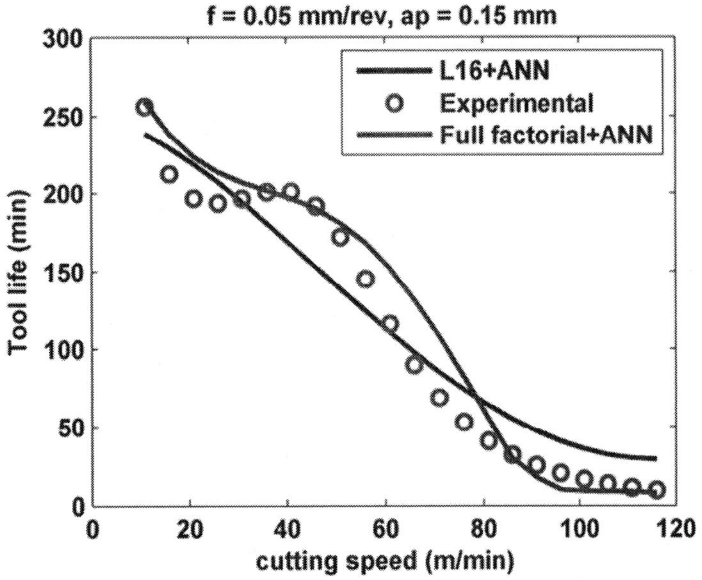

Fig.2. Experimentally and ANN derived curves for $f = 0.05$ mm/rev, $a_p = 0.15$ mm.

It seems that the full factorial approach, which used a lot more data than the L16 approach, i.e. 64 vs. 16, but the same numbers of levels, can provide more accurate estimations of the shape of the experimental curve. Furthermore, the curve derived by the first approach tends to overestimate tool life at low and high speeds, whereas approximates the experimental curve better at moderate cutting speeds. As for the curve derived using the second approach, it can approximate the experimental one adequately at some regions but still the minima and maxima of the experimental curve are not quite accurately predicted.

Considering that the experimental curve was derived by using 9 levels for the cutting speed instead of 4, the results are satisfactory for the prediction of tool life but not satisfactory for the prediction of the distinct regions of the tool life-cutting speed curve.

5 Conclusions

In this paper, an investigation for the applicability of DoE methods and ANN for the prediction of tool life of CBN cutting tools was conducted. The experiments for the derivation of tool life-cutting speed relationship were conducted using standard experimental designs such as Taguchi method and full factorial method, both at four levels for each factor. Several useful conclusions were drawn:

• Both approaches provided moderate results concerning RMSE, MAE, and MPE, whereas they exhibited high correlation coefficient values. The mean values for MAE and MPE are about 23-32%, which is above the acceptable limit of 10% but results are not very far from the experimental ones, as indicated by RMSE.

• The ANN trained with results from the full factorial design approach is approximating better the shape of the experimental curve but cannot predict the minima and maxima adequately. It can approximate the experimental curve better at moderate cutting speeds but generally it overestimates the tool life value at the other regions.

• In general, it can be concluded that the non-linear relationship of tool life and cutting speed for a wide range of cutting speeds is difficult to be derived using standard design procedures as it requires data for many cutting speed values to adequately approximate this relationship, something that is not usually the case with DoE methods. However, although the experimental curve was not sufficiently predicted, the deviations of the actual tool life values were not very high. The latter can be useful for preliminary examinations of tool wear and tool life of CBN tools.

In order to develop more accurate models for the approximation of tool life-cutting speed curve, several attempts can be conducted, such as to employ various experimental designs and performance functions for training of the ANN model or a different type of ANN model.

References

1. A.G. Mamalis, J. Kundrák, M. Horváth, Int J Adv Manuf Tech **20**, 475-479 (2002)
2. A.G. Mamalis, J. Kundrák, M. Horváth, J Manuf Sci E-T ASME **127**, 328-332 (2005)
3. C. Leone, D. D' Addona, R. Teti, CIRP J Manuf Sci Tech **4**, 327-331 (2011)
4. K. Venkata Rao, B.S.N. Murthy, N. Mohan Rao, Measurement **51**, 63-70 (2014)
5. D. D'Addona, T. Segreto, A. Simeone, R. Teti, CIRP J Manuf Sci Tech **4**, 33-37 (2011)
6. D.M. D' Addona, R. Teti, Procedia CIRP **12**, 252-257 (2013)
7. R.R. Srikant, P.V. Krishna, N.D. Rao, P I Mech Eng B-J Eng **225**, 1009-1018 (2011)
8. R. Corne, C. Nath, M. El Mansori, T. Kurfess, Procedia Manuf **5**, 1-14 (2016)
9. Q. Liu, Y. Altintas, Int J Mach Tool Manu, **39**, 1945-1959 (1999)
10. P. Krishnakumar, K. Rameshkumar, K.I. Ramachandran, Procedia Comp Sci **50**, 270-275 (2015)
11. S.S. Gill, R. Singh, J. Singh, H. Singh, Expert Syst Appl **39**, 4171-4180 (2012)
12. A.I. Azmi, Adv Eng Softw **82**, 53-64 (2015)
13. J. Kundrák, *The scientific principles of increasing the effectiveness of inner surfaces cutting with CBN tools*, (Kharkov, 1996)

MATEC Web of Conferences **112**, 06006 (2017) DOI: 10.1051/matecconf/201711206006
IManE&E 2017

Optimization of the mechanism from equipment for overcoming obstacles with design of experiments

Vlad Gheorghiță*, and *Cătălin* Gheorghiță

Politehnica University of Bucharest, Department of Manufacturing Technology, Splaiul Independeței, No. 313, 060042, Bucharest, Romania

Abstract. A data collection considers how the experimental factors, controlled and uncontrolled, fit together into a model that will meet the specific objectives of an experiment and satisfy the practical constraints of resource and time. Customizing any product to improve performance requires a reliable method to measure performance results, in order to accurately assess the effects that design changes have on them. This paper has the purpose to analyse the equipment for overcoming obstacles. A comprehensive approach to model equipment requires performance data related to the design parameters. Slipping, slanting, becoming unstable, sliding, overturning and blocking while climbing steps and stairs are common problems and may cause instability for the user of the equipment. The Taguchi Method is used to optimize several parameters of the proposed mechanism and to ensure the stability through horizontality of the equipment. This method provides an opportunity to reduce the number of required sample size and is implemented by an orthogonal experimental array which enables the effect of multiple parameters to be simultaneously assessed.

1 Literature review

The development of devices that can improve people's quality of life can be a task for engineering researchers. Over the past several years, research of motorized equipment was performed by several scientists who have analysed the types of mechanisms to access obstacles. There are many studies conducted in areas related to climbing equipment. These include crawler type, leg type, hybrid type and wheeled type. Ghani and others [1] investigated an equipment used indoor, evaluating different mechanisms for accessing stairs: track, legs and wheeled combined system, analysing various forces and couples acting. Quaglia [2] analysed aspects of developing equipment for climbing stairs, which can move in structured and unstructured environments that can go over obstacles and can go down or up the stairs. The device consists of a frame, seat and a linkage mechanism. The framework consists of a chassis built with two motorized units, support for two electric motors, two wheels and a battery. The seat has a tubular structure consisting of a frame and

*Corresponding author: vlad.gheorghita@upb.ro

a pivoting wheel. Linkage mechanism is responsible for the relative movement between the frame and the seat during operation. To climb the stairs, it is necessary to move the seat back, then refocus and finally, to raise swivel wheel. The mechanism is driven by a motor connected to a lead screw mechanism. When the seat reaches the desired position, the engine is off and needs no additional energy to maintain position.

Lawn [3] identified the developments of the mechanisms of mobility support, while studying the importance of the wheels. A proposed mechanism is based on the use of four wheels, the rear wheels are driven independently and the front wheels are free wheels. The influence of external factors such as cost, weight, aesthetics, range of operation, safety, operational efficiency or comfort were evaluated by overview of recent advances in mobility assistive devices available for curbs or stairs.

Yoneda [4] proposed a deformable track considering the most fundamental track based problem that of the high pressure exerted on the stair edges. The tracked vehicle can easily climb stairs by using two active arms in front. One arm is basically used to put in contact the crawlers with the first step and to assist the vehicle motion, while the other arm adjusts the payload position, changing the posture of the carrier in order to maintain a constant posture.

Yun et al. [5] have developed transmuting track equipment which has a dynamic track tension characteristic. They have used fuzzy logic control as an optimal estimation algorithm to estimate the proportion of various factors that affect the track tension. Frame of the equipment consists of fixed shelf, seat and track transmutation mechanism in which the main factors affecting tension are determined by fuzzy decision [5].

A hybrid solution combines the high effectiveness of wheel locomotion on flat ground with the performance of a track on stairs. The advantages of tracks are the stability and regularity of the trajectory during stair-climbing. The disadvantage is represented by the dimensions and weight involved in the use of a motorized track. This solution is one of the most effective, and it is used in most commercial stair-climbing devices. Equipment with pure leg locomotion has a high climbing capability, but it requires a complex structure for both the actuation and the control systems. The stability could be a critical aspect, and the seat oscillates due to leg movement during both stair-climbing and flat ground motion, thus causing an uncomfortable sensation for the user.

The stair climbing IBOT include 4 active wheels and 2 front small casters. Only the two rear wheels make contact with the ground using the front casters to provide free wheeled steering. Dual wheel cluster type stair climbing equipment uses two pairs of wheel clusters, the rear of which for drive and the front as freewheeling casters. The stair climbing motion is forward down the stairs and back up the stairs. The front cluster rotates passively during stair-negotiation. Scalevo device is a hybrid between a wheelchair and a Segway device. It is equipped with a track platform that rises and helps in moving across the stairs, managing to climb or to descend without the intervention of the user or others. The rotary technology is used in Scalevo design, thus the device can rotate on the spot and can change direction easily.

Morales et al. [6] designed a hybrid two decoupled equipment mechanisms in each axle, one to negotiate steps and the other to position the axle with regard to the device in order to accommodate the overall slope. Kinematic model was utilized to improve the trajectory planning.

2 Modelling of the mechanism

One of the most important features of the design of equipment for overcoming obstacles is its ability to surmount physical obstructions that may occur (e.g. threshold, stairs). The basic requirements for a product of this type are: to enable to move along various kinds of

DOI: 10.1051/matecconf/201711206006

surfaces (e.g. uneven surface, sand, snow); to have a small size; to enable the movement up and down the stairs.

A shape of the proposed mechanism is presented in Figure 1. The mechanism consists of: a drive wheel R_1, two power driven wheels R_2 and R_3 articulated through two arms. Parallelogram mechanism is placed in the centre and is operated by Link2 through a lever. System stability is achieved by wheel R_3 and Link1.

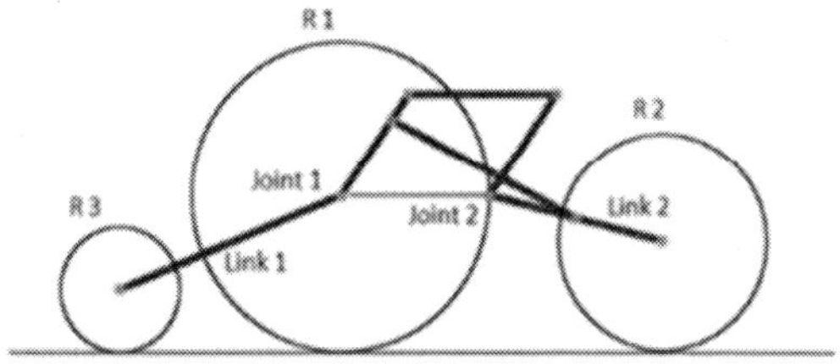

Fig. 1. Diagram of the mechanism.

The mechanism of the equipment presented in this paper should be able to overcome obstacles in a comfortable and safe way, and it must be stable during climbing operation. The way to overcome the obstacles for the type of scale chosen, in three steps is shown in Figure 2.

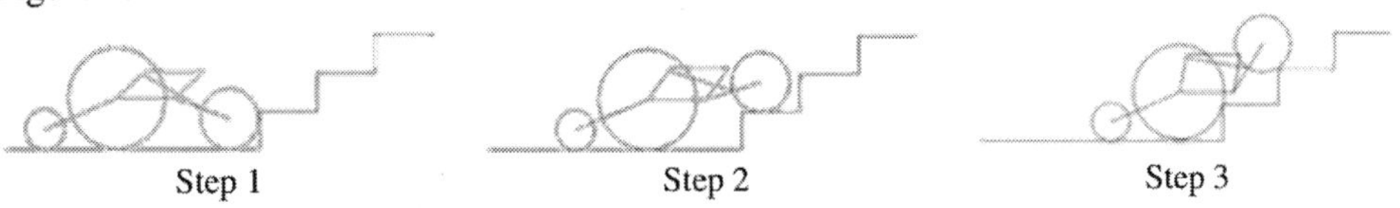

Fig.2. Steps for overcoming obstacles.

One type of stair (300mm X 190mm) is selected to determine a robust optimal solution. The result will show the variation of the horizontal position of the platform support. The kinematic model of the mechanism shown in Figure 3, is built and it is simulated in Matlab.

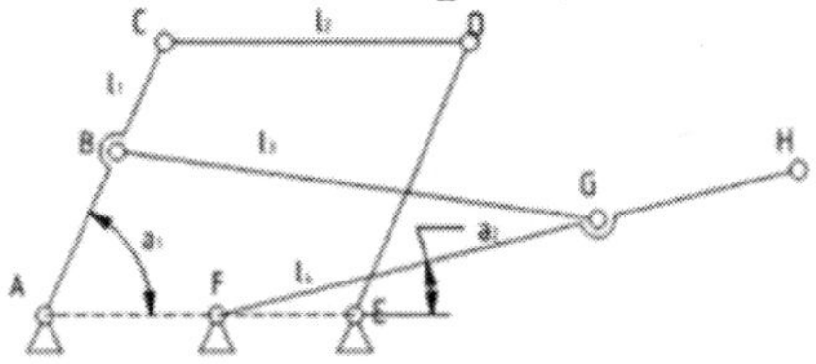

Fig.3. Analysis of the mechanism

A kinematic analysis of a connection is used to assess its performance keeping the design goal and to ensure that the mechanism has no flaws. A complicated structure will determine unnecessary energy losses [7]. Models and references to prior experimental work have identified important design parameters of similar devices, for ensuring the horizontal and vertical seat position. The seat connects with the mechanism in points C and D, and the wheel driven R_2 in H point. Changing the position of lever FG is transmitted to the mechanism ACDE through the arm BG. The angle α_2 is the angle made by FH link with the horizontal. The range of the angle α_2 was ± 30°, this variation being determined by the height of the staircase taken into consideration.

3 Optimization procedure

Experimental design can be used to reduce design costs by speeding up the design process, reducing late engineering design changes, and reducing product material and labour complexity. Designed of Experiments are a powerful tool to achieve manufacturing cost savings by minimizing process variation and reducing rework and the need for inspection [8].

Taguchi method shows the importance of studying the response variation using the signal to noise (SN) ratio, resulting in minimization of quality characteristic variation due to uncontrollable parameter. The experimental design proposed by Taguchi involves using orthogonal arrays to organize the parameters affecting the process and the levels at which they should be varied [8]. Instead of having to test all possible combinations like the factorial design, the Taguchi method tests pairs of combinations. In this study, four dimensions of the link (l_1, l_2, l_3, l_4) have been considered with their range as design parameters shown in Table 1 and were improved using the optimization procedure.

Table 1. Input parameters and their levels.

Levels	l_1 [mm]	l_2 [mm]	l_3 [mm]	l_4 [mm]
1	100	200	250	150
2	125	250	300	200
3	150	300	350	250

The parameters for design of experiment, based on Taguchi's L9 Orthogonal Array design have been selected as shown in Table 2.

Table 2. Design of Experiment Using L9 Orthogonal Array.

Exp. No.	l_1 [mm]	l_2 [mm]	l_3 [mm]	l_4 [mm]
1.	1	1	1	1
2.	1	2	2	2
3.	1	3	3	3
4.	2	1	2	3
5.	2	2	3	1
6.	2	3	1	2
7.	3	1	3	2
8.	3	2	1	3
9.	3	3	2	1

4 Results and discussions

The objective function was the minimizing of the inclination angle. To determine the effect each variable has on the output, the SN ratio needs to be calculated for each experiment conducted. The parameter $\bar{y}_i$ is the mean value of the performance characteristic for a given experimentand the parameter s_i is the variance and is given as:

$$\bar{y}_i = \frac{1}{N_i}\sum_{u=1}^{N_i} y_{i,u} \tag{1}$$

$$s_i^2 = \frac{1}{N_i - 1}\sum_{u=1}^{N_i}(y_{i,u} - \bar{y}_i)^2 \tag{2}$$

where: i = experiment number; u = trial number ;N_i = number of trials.

The observations for the inclination angle of the platform support are shown in Table 3. Three trials were performed for each experiment, and the informations below were collected.

The SN ratio measures how the response varies relative to the nominal or target value under different noise conditions. The Nominal is Best SN ratio is useful for analysing or

DOI: 10.1051/matecconf/201711206006

Table 3. Experimental data.

Exp. No.	l_1 [mm]	l_2 [mm]	l_3 [mm]	l_4 [mm]	Trial 1	Trial 2	Trial 3	Mean	Standard deviation
1.	100	200	250	150	5.75	5.66	5.70	5.70	0.045
2.	100	250	300	200	6.49	6.43	6.46	6.46	0.031
3.	100	300	350	250	8.28	8.19	8.18	8.21	0.055
4.	125	200	300	250	6.18	6.15	6.09	6.14	0.045
5.	125	250	350	150	7.65	7.57	7.61	7.61	0.041
6.	125	300	250	200	8.30	8.29	8.25	8.28	0.026
7.	150	200	350	200	6.26	6.24	6.19	6.23	0.036
8.	150	250	250	250	6.61	6.56	6.55	6.57	0.032
9.	150	300	300	150	7.40	7.34	7.35	7.36	0.032

identifying scaling factors, which are factors in which the mean and standard deviation vary proportionally. For Taguchi dynamic designs, it is used one SN ratio which is closely related to the Nominal is Best SN ratio for static designs [8].

$$SN_i = 10 * log\frac{\overline{y_i}^2}{s_i^2} \tag{3}$$

Below is shown the response table, where is calculated an average SN value for each factor.

Table 4. Experimental Observations and S/N Ratio for the inclination angle.

Exp. No.	l_1 [mm]	l_2[mm]	l_3[mm]	l_4 [mm]	S/Ni
1.	100	200	250	150	42.04
2.	100	250	300	200	46.66
3.	100	300	350	250	43.47
4.	125	200	300	250	42.54
5.	125	250	350	150	45.58
6.	125	300	250	200	49.90
7.	150	200	350	200	44.75
8.	150	250	250	250	46.21
9.	150	300	300	150	47.19

Compute the SN ratio for each experiment for the target value case, was created a response table and a response chart, and were determine with Matlab Software the parameters that have the highest and lowest effect.

Table 5. Ranking of parameters Response Table for Signal to Noise Ratios.

Level	l_1 [A]	l_2 [B]	l_3 [C]	l_4 [D]
1	44.05	43.11	46.05	44.94
2	46.01	46.15	45.47	47.11
3	46.05	46.86	44.60	44.08
Delta	2.00	3.75	1.45	3.03
Rank	3	1	4	2

The effect of factors is then calculated by determining the range: Δ_2 = Max – Min = 46.86 - 43.11 = 3.75 and Δ_3 = Max – Min = 46.05 - 44.60 = 1.45. Therefore, link with length l_2 has the largest effect on the mechanism stability and link with length l_4 has the smallest effect. From the experimental data we got better values for angle inclination at minimum dimensions of the links. Optimum parameters resulted from Taguchi analysis is: l_1 = 100 mm, l_2 = 200 mm, l_3 = 350 mm, l_4 = 250 mm.

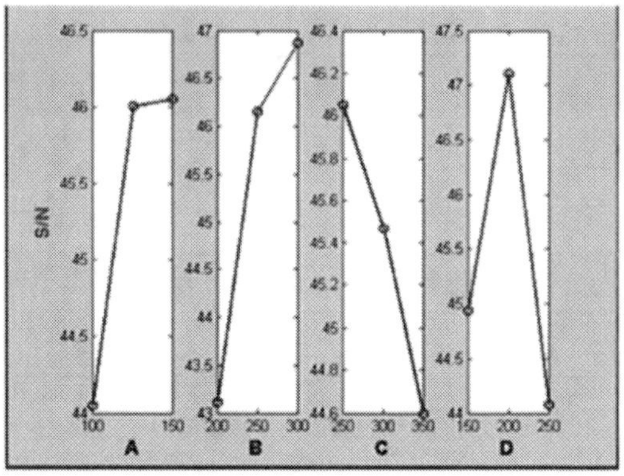

Fig.4. Response chart for Signal to Noise Ratios.

Conclusions

This paper has presented application of Taguchi method to determine the optimal dimensions for the components of a mechanism. The concept of SN ratio is used to determine the effect and influence of the dimensions links on output measure represented by the angle of the mobile platform. Matlab software is used for analysis the response graphs of average values and SN ratios. From the analysis results that l_2 and l_4 played a more important role on output measure than other dimensions. The model is useful for selecting the optimal values for the lengths of the elements considered in the study. On understanding the effect of each parameter, decisions can be made for getting a reliable and compact product, with reduced weight, small dimensions and low cost. Experiments at the optimum set of dimensions for inclination angle will be carried out in further research. For this purpose, the mechanism will be practically made and will be tested under field conditions to verify the results obtained by simulation. Future work can be done by including more parameters as the angles that the mechanism elements must be able to negotiate if other types of obstacles occurred in terms of ensuring the stability.

References

1. Ghani, *Control of a Multi-Function Wheelchair for Disabled/Elderly Mobility*, University of Sheffield (2003)
2. G. Quaglia, W. Franco, R. Oderio, Proceeding of IEEE International Conference on Robotics and Biomimetics, Guilin, China (2009)
3. J.M. Lawn, T. Sakai, M. Kuroiwa and T. Ishimatsu, Int. J. of HWRS-ERC (2001)
4. K. Yoneda, Y.Ota, Proceeding of 4th International Conference on Climbing and Walking Robots CLAWAR 2001, 569-576(2001)
5. T. Yun, W. Ting, Y. Chen, L. Xiaofan, Proceeding of 29th Chinese Control Conference, Beijing, China (2010)
6. R. Morales, V. Feliu, A. Gonzalez, Robotics and Autonomous Systems, **58**, 97-114 (2010)
7. A. Mallik, A. K.,Ghosh, *Kinematic analysis and synthesis of mechanisms*, CRC press, Inc.,Boca Raton, FL, USA, (1994)
8. Y.Wu, A.Wu, *Taguchi methods for robust design*, ASME Press, New York, (2000)

MATEC Web of Conferences **112**, 06007 (2017) DOI: 10.1051/matecconf/201711206007
IManE&E 2017

Aspects of derivative causality in bond-graph models

Radu Ibănescu[1,*], and *Mihaela* Ibănescu[2]

[1]Department of Theoretical Mechanics, "Gheorghe Asachi" Technical University of Iaşi, Bd. D. Mangeron, nr. 59 A, 700050, Iaşi, Romania
[2]Department of Structural Mechanics, "Gheorghe Asachi" Technical University of Iaşi, Bd. D. Mangeron, nr. 1, 700050, Iaşi, Romania

Abstract. The bond-graph method used in the analysis of system dynamics problems leads to a system containing a number of differential equations equal to the number of energy storing elements in integral causality and a number of algebraic equations equal to the number of energy storing elements in derivative causality. When the system has many energy storing elements in derivative causality, the system of equations becomes a very difficult differential-algebraic one. The paper proposes the simplest procedure for finding the appropriate form for numerical simulation of the equations system. The efficiency of the proposed solution is exemplified by using the bond-graph model of a mechanical system which contains nine energy storing elements in derivative causality.

1 Introduction

The modelling of system dynamics by using the bond-graph method presumes a profound analysis of the manner of transmitting and transforming the power from the driving element to the working element [1-3]. It represents a valuable alternative to other methods for solving dynamics problems [4,5]. More difficult problems in obtaining the motion equations and in performing the numerical simulation occur when the bond-graph model contains energy storing elements in derivative causality [6,7]. The system of equations which results in such circumstances is a differential-algebraic one, difficult to be resolved. In case of a numerical simulation with block diagrams, derivative blocks occur, which, because of the anticipative character of the derivative, can lead to computing errors and simulation stopping [8]. In order to render evident these difficulties and to propose and demonstrate the simplest solution, the paper presents the bond-graph model of a mechanism which contains two elements in integral causality and nine elements in derivative causality.

2 System description

The mechanical system, whose dynamics will be further modelled, is presented in Fiure.1. It consists of o bar of mass M_1 and length $2l$, whose ends can move away along two fixed

* Corresponding author: ribanesc@yahoo.com

mutual perpendicular directions. At the end of the bar which moves along the vertical direction, a slide of mass m is pinned, which is attached to a spring of stiffness k. When the bar is in the vertical position, the spring is undeflected. At the end of the bar which moves along the horizontal direction, a disk of mass M_2 and radius R is pinned. The disk rolls without slip along the horizontal direction. The two weights of mass M_4 are supported by ideal wires, which are attached at the centre of the disk of mass M_2. Each wire passes over a pulley of mass M_3 and radius R. The friction is ignored. The system oscillates if, for example, the bar is inclined with angle φ_0 with respect to the vertical direction at the initial moment and then it is let free.

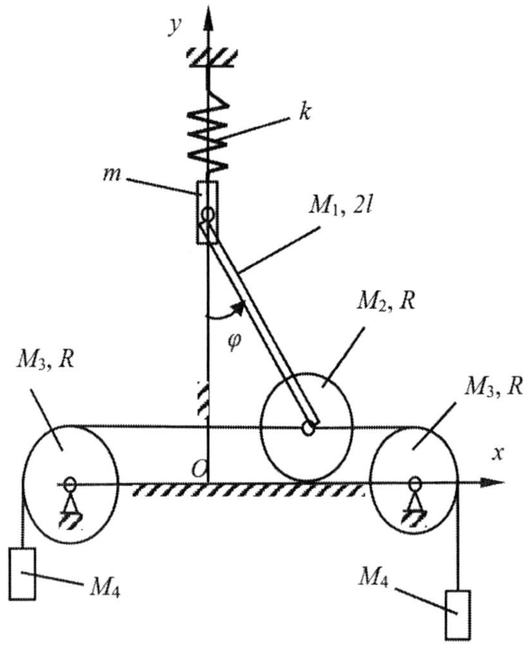

Fig. 1. The mechanical system.

3 The bond-graph model

The bond-graph model is presented in Figure 2. The element I that is assigned with integral causality, corresponds to the bar rotation (bond 10), after the causal assignment of sources. This fact leads to the causal assignment of all bonds. The element C, corresponding to the spring, results in integral causality and all the other nine elements I, corresponding to element translation and rotation inertia (bonds 9, 1, 13, 33, 16, 19, 22, 26 and 29) result in derivative causality. This means that the system of equations which describes the mechanical system dynamics will contain two differential equations and nine algebraic equations. These equations are:

$$\begin{aligned}\dot{p}_{10} = &-l\cos\varphi\dot{p}_{13} - 2l\cos\varphi\dot{p}_{16} - \frac{2l}{R}\cos\varphi\,\dot{p}_{19} - 2l\cos\varphi\dot{p}_{22} - 2l\cos\varphi M_4 g + \\ &+\frac{2l}{R}\cos\varphi\dot{p}_{26} - 2l\cos\varphi\dot{p}_{29} + 2l\cos\varphi M_4 g + l\sin\varphi\dot{p}_1 - l\sin\varphi M_1 g + \\ &+2l\sin\varphi\,\dot{p}_9 - 2l\sin\varphi mg + 2lk\sin\varphi\,q_7 + \frac{2l}{R}\cos\varphi\,\dot{p}_{33}\end{aligned} \quad (1)$$

$$\dot{q}_7 = -\frac{6\sin\varphi}{M_1 l}p_{10} \quad (2)$$

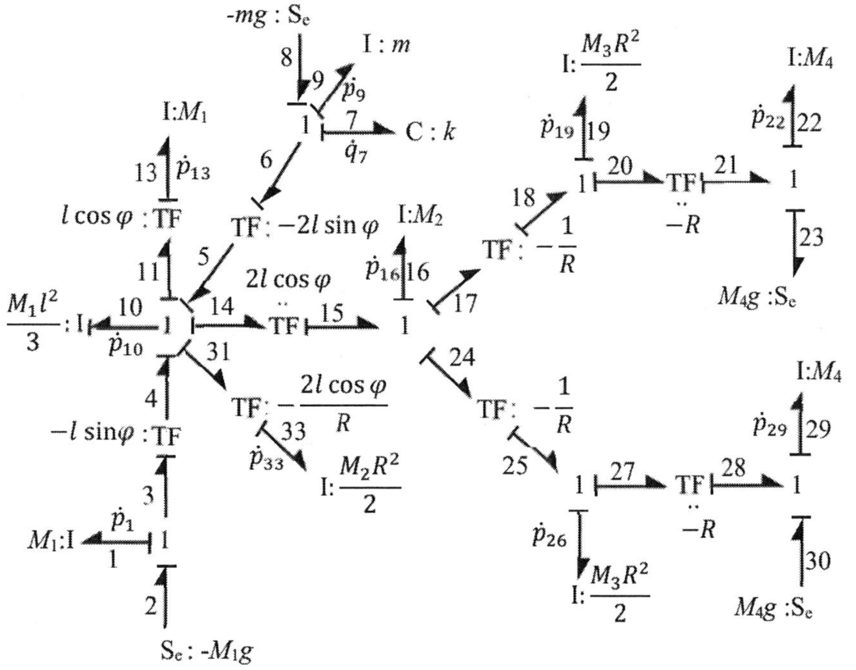

Fig. 2. Bond graph model

$$p_{13} = \frac{3\cos\varphi}{l} p_{10} \tag{3}$$

$$p_{16} = 6\frac{M_2}{M_1}\frac{\cos\varphi}{l} p_{10} \tag{4}$$

$$p_{19} = -3\frac{M_3}{M_1}\frac{R\cos\varphi}{l} p_{10} \tag{5}$$

$$p_{22} = 6\frac{M_4}{M_1}\frac{\cos\varphi}{l} p_{10} \tag{6}$$

$$p_{26} = -3\frac{M_3}{M_1}\frac{R\cos\varphi}{l} p_{10} \tag{7}$$

$$p_{29} = 6\frac{M_4}{M_1}\cos\varphi\, p_{10} \tag{8}$$

$$p_1 = \frac{3\sin\varphi}{l} p_{10} \tag{9}$$

$$p_9 = -6\frac{m}{M_1}\frac{\sin\varphi}{l} p_{10} \tag{10}$$

$$p_{33} = -3\frac{M_2}{M_1}\frac{R\cos\varphi}{l} p_{10} \tag{11}$$

where q_7 is the spring elongation and the other variables are linear momenta ($p_9, p_1, p_{13}, p_{16}, p_{22}$ and p_{29}) and angular momenta (p_{10}, p_{19}, p_{26} and p_{33}).

4 The use of equations system resulted from the bond-graph model

In order to perform a numerical simulation, it is necessary to obtain a system of differential equations in explicit form. They are found by differentiating the previous nine algebraic equations (the equations (3) to (11)) with respect to time. The derivatives have the form:

$$\dot{p}_{13} = -9\frac{\sin\varphi}{M_1 l^3}p_{10}^2 + 3\frac{\cos\varphi}{l}\dot{p}_{10} \tag{12}$$

$$\dot{p}_{16} = -18\frac{M_2}{M_1^2}\frac{\sin\varphi}{l^3}p_{10}^2 + 6\frac{M_2}{M_1}\frac{\cos\varphi}{l}\dot{p}_{10} \tag{13}$$

$$\dot{p}_{19} = 9\frac{M_3}{M_1^2}\frac{R\sin\varphi}{l^3}p_{10}^2 - 3\frac{M_3}{M_1}\frac{R\cos\varphi}{l}\dot{p}_{10} \tag{14}$$

$$\dot{p}_{22} = -18\frac{M_4}{M_1^2}\frac{\sin\varphi}{l^3}p_{10}^2 + 6\frac{M_4}{M_1}\frac{\cos\varphi}{l}\dot{p}_{10} \tag{15}$$

$$\dot{p}_{26} = 9\frac{M_3}{M_1^2}\frac{R\sin\varphi}{l^3}p_{10}^2 - 3\frac{M_3}{M_1}\frac{R\cos\varphi}{l}\dot{p}_{10} \tag{16}$$

$$\dot{p}_{29} = -18\frac{M_4}{M_1^2}\frac{\sin\varphi}{l^3}p_{10}^2 + 6\frac{M_4}{M_1}\frac{\cos\varphi}{l}\dot{p}_{10} \tag{17}$$

$$\dot{p}_{1} = -9\frac{1}{M_1}\frac{\cos\varphi}{l^3}p_{10}^2 - 3\frac{\sin\varphi}{l}\dot{p}_{10} \tag{18}$$

$$\dot{p}_{9} = -18\frac{m}{M_1^2}\frac{\cos\varphi}{l^3}p_{10}^2 - 6\frac{m}{M_1}\frac{\sin\varphi}{l}\dot{p}_{10} \tag{19}$$

$$\dot{p}_{33} = 9\frac{M_2}{M_1^2}\frac{R\sin\varphi}{l^3}p_{10}^2 - 3\frac{M_2}{M_1}\frac{R\cos\varphi}{l}\dot{p}_{10} \tag{20}$$

Parameter φ is determined from the constitutive equation of element *I*, which is connected by the bond 10 and is in integral causality:

$$\dot{\varphi} = \frac{p_{10}}{\frac{M_1 l^2}{3}} \tag{21}$$

The system of 12 differential equations containing the equations (1), (2) (12), (13), (14), (15), (16), (17), (18), (19), (20) and (21), is in implicit form and has the unknowns p_{10}, p_{13}, p_{16}, p_{19}, p_{22}, p_{26}, p_{29}, p_1, p_9, p_{33}, q_7 band φ. The system of equations should be resolved like an algebraic system with respect to derivatives of the unknowns, in order to obtain a system of differential equations in explicit form, that can be numerically integrated. This operation involves a difficult calculus.

A simpler solution is to substitute the derivatives given by equations (12), (13), (14), (15), (16), (17), (18), (19), (20) in equation (1). In the final shape of the resulted equation, the variable p_{10} and its derivative are replaced by:

$$p_{10} = \frac{M_1 l^2}{3}\dot{\varphi} \tag{22}$$

$$\dot{p}_{10} = \frac{M_1 l^2}{3}\dot{\omega} \tag{23}$$

where $\omega = \dot{\varphi}$ denotes the angular velocity of the bar. In this manner, it is much easier to obtain a system of three differential equations in explicit form:

$$\dot{\omega} = \frac{(-4m+(6M_2+4M_3+8M_4))\sin\varphi\cos\varphi\omega^2+2lk\sin\varphi q_7+\frac{2mg+M_1g}{l}\sin\varphi}{4m(\sin\varphi)^2+\frac{4}{3}M_1+(6M_2+4M_3+8M_4)(\cos\varphi)^2} \quad (24)$$

$$\dot{\varphi} = \omega \quad (25)$$

$$\dot{q}_7 = -2l\,\omega\sin\varphi \quad (26)$$

having as unknowns ω, φ and q_7.

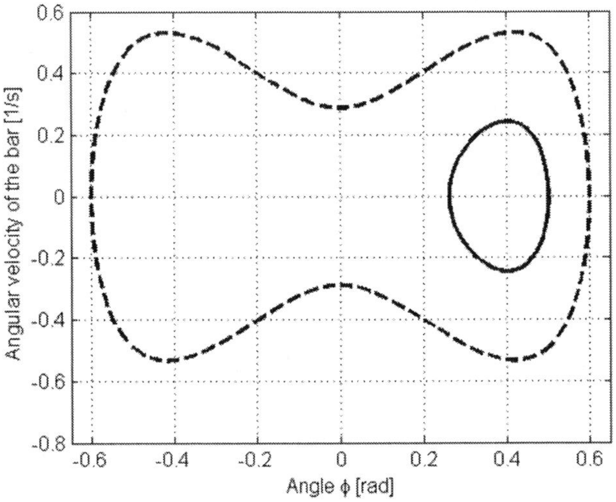

Fig. 3. The angular velocity of the bar with respect to the angle φ.

The following numerical values are considered: m=4kg, M_1=8kg, g=9.81m/s, k=1000N/m, L=0.5m, M_2=6kg, M_3=6kg and M_4=10kg.

Equations (24), (25) and (26) are numerically integrated in MATLAB, by using different initial conditions.

The variation of the angular velocity in terms of vertical bar position, given by angle φ, is shown in Figure 3.

The curve represented by the continuous line is obtained when the initial conditions are $\varphi(0) = 0.5$ rad and $\omega(0) = 0\ s^{-1}$.

The initial potential energy is not high enough to produce the passing of the bar to the left side of the vertical position, that is to have negative values of the angle φ.

The curve represented by the dashed line is obtained when the initial conditions are $\varphi(0) = 0.6$ rad and $\omega(0) = 0\ s^{-1}$.

The initial potential energy is now enough big to produce the passing of the bar to the left side of the vertical position, that is to have negative values of the angle φ. The angular velocity varies in a very interesting manner. It increases at the beginning and then decreases, when the bar approaches the vertical position. The evolution is then symmetric with respect to axis Oy.

Here there are some comments referring to the two curves represented in Figure 3. The system does not move if the initial conditions are zero. When the initial conditions are slightly different from zero, for example they are $\varphi(0) = 0.01$ rad and $\omega(0) = 0\ s^{-1}$, the maximum value for angle φ is 0.56779 rad. If the initial value of the angle φ is 0.56779 rad, then the initial potential energy will determine the bar to come very close to the vertical position, but it will not pass to the left side of the vertical position. The initial conditions for the angle φ chosen for the two diagrams are less and higher than the value 0.56779 rad. The bar remains at the right side of the vertical position in the first case because the initial potential energy is not high enough, so that the bar to pass to the left side of the vertical position. The bar passes to the left side of the vertical position in the second case, because the initial potential energy is high enough to make the bar to pass at the left side of the vertical position.

5 Conclusions

The bond-graph method used in the analysis of system dynamics problems leads to a system of equations, containing a number of differential equations equal to the number of energy storing elements in integral causality and a number of algebraic equations equal to the number of energy storing elements in derivative causality. After differentiating the algebraic equations, there are two possibilities for finding a system of differential equations in explicit form, suitable for numerical simulation. To get this explicit form represents a very tedious task in case of systems which contain a great number of elements in derivative causality. The easier way, revealed by the present paper, is to substitute the derivatives of the algebraic equations in the differential equations corresponding to the energy storing elements in integral causality. In this manner, the final system has a low number of differential equations, equal to the number of energy storing elements in integral causality.

This procedure is exemplified by using the bond-graph model of a mechanical system which contains two energy storing elements in integral causality and nine energy storing elements in derivative causality. There are also presented the results of the numerical simulation and some comments about the motion of the mechanism in different initial conditions, in order to demonstrate the efficiency of the proposed procedure.

References

1. W. Borutzky, *Bond Graph Methodology. Development and Analysis of Multidisciplinary Dynamic System Models* (Springer, 2010)
2. D. C. Karnopp, D. L. Margolis, R. C. Rosenberg, *System Dynamics. Modeling and Simulation of Mechatronic Systems*, (John Willey & Sons, 2006)
3. R. Ibănescu, *Graph Based Modelling in Engineering*, 3, (Springer 2017)
4. R. Ibănescu, M. Ibănescu, Materials Science and Engineering **161** (2016)
5. R. Ibănescu, C. Ungureanu, Applied Mechanics and Materials, **809-810**, 914 (2015)
6. L. Xiaotian, W. Anlin, Mech. Mach. Theory, **80**, 112 (2014)
7. V. Ragusila, M.-R. Emami, Simul. Model. Pract. Theory **40**, 132 (2014)
8. R. Ibănescu, International Conference ModTech Proceedings, **I**, 493 (2011)

MATEC Web of Conferences 112, 06008 (2017) DOI: 10.1051/matecconf/201711206008
IManE&E 2017

Hob mill for spline shaft, profiled by generating trajectories method. Graphical application in CATIA

Florin Susac, *Virgil Gabriel* Teodor*, *Nicuşor* Baroiu, and *Nicolae* Oancea

"Dunărea de Jos" University of Galaţi, Department of Manufacturing Engineering, Domnească 111, Galaţi, Romania

Abstract. The hob mill is a cutting tool with a wide application for generation in machining of splined shafts. A method for profiling the hob mill for generation of square splines is presented in this paper. The analytical development was done based on a complementary theorem of surface enwrapping — the method of generating trajectories. A numerical application for a standardized spline shaft is presented. Based on the same theorem is presented a graphical application, developed in the CATIA design environment, for profiling the hob mill which generated standardised square splines.

1 Introduction

The squared splines are ordered curls of surfaces with parallels flanks and rectilinear generation. The generation of these complexes surfaces is made by milling with hob tools. This imposes the profiling of the primary worm of this tool's type.

They were created graphical methods [1, 2], which allow approximating the profile of hob mill for generating such composed surfaces.

The cylindrical surface of the splines and the helical surface of the worm, approached as reciprocally enwrapping surfaces, represent issues of surfaces enveloping of second degree, according to the Olivier theorems [3, 4].

In this way, it is possible to analytically determine a rigorous form of the enwrapping worm of the splined shaft, as base for the hob mill. The analytical method, based on the second theorem of Olivier, assume handling of complicated equations, with obvious disadvantages.

There were developed complementary theorems for the surfaces enveloping, which, for real generating cases, has sentences which simplify the analytical effort for surface determining. Such complementary theorems are: the "minimum distance" method; the method of "substitutive circles family", the method of "generating trajectories" [5].

Also, there were presented graphical methods, developed in graphical design environments as AutoCAD, [6], which allows, by 3D modelling the generating of the surfaces in order to determine the solid model of the hob mill.

*Corresponding author: virgil.teodor@ugal.ro

2 Hob mill for splined shaft

They are defined the reference systems and shaft's crossing section presented in Figure 1.

The reference systems are: *xyz* is the global reference system, with *z* axis as the main axis of the shaft; *XYZ* — mobile reference system, joined with the C_1 centrode of the shaft. Initially, this system is overlapped to the global system; $\xi\eta\zeta$ — mobile reference system, joined with the C_2 centrode of the rack-gear tool.

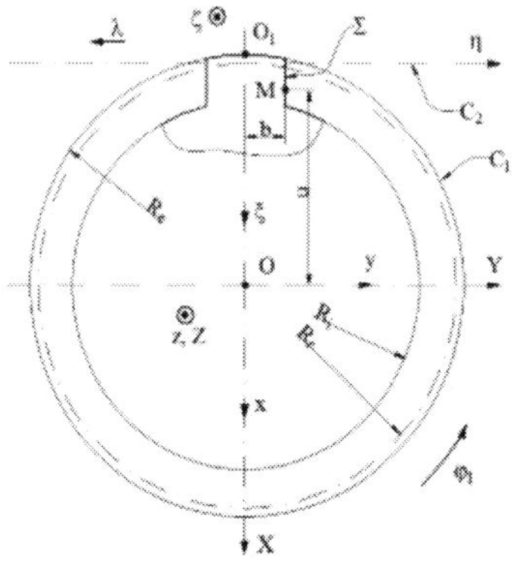

Fig. 1. Splined shaft. Reference system.

2.1 Generating rack-gear

In Figure 1, the crossing profile of the splined shaft and the reference systems are presented. The process kinematics assumes the movements: rotation of the C_1 centrode with φ_1 angular parameter and translation of the generating tool, with parameter λ. The link:

$$\lambda = R_r \cdot \varphi_1, \tag{1}$$

is defined as rolling condition. R_r represents the rolling radius of the splined shaft.

The analytical equations of the splined shaft are:

$$\Sigma \begin{vmatrix} X = -u; \\ Y - b; \\ Z = t, \end{vmatrix} \tag{2}$$

with u and t independent variables. Usually, the rolling radius is $R_r = \sqrt{R_e^2 - b^2}$. The generating kinematics assumes the motions:

$$\mathbf{x} = \boldsymbol{\omega}_3^T(\varphi_1)\mathbf{X};\ \boldsymbol{\xi} = \mathbf{x} - \mathbf{a};\ \mathbf{a} = \begin{pmatrix} -R_r \\ -R_r \cdot \varphi_1 \\ 0 \end{pmatrix}. \tag{3}$$

The relative motion is:

$$\boldsymbol{\xi} = \boldsymbol{\omega}_3^T(\varphi_1) \cdot \mathbf{X} - \mathbf{a}. \tag{4}$$

The Σ flanks family, in the rack-gear's reference system is:

$$(\Sigma)_{\varphi_1} \begin{vmatrix} \xi = -u\cdot\cos\varphi_1 - b\cdot\sin\varphi_1 + R_r; \\ \eta = -u\cdot\sin\varphi_1 + b\cdot\cos\varphi_1 + R_r\cdot\varphi_1; \\ \zeta = t. \end{vmatrix} \quad (5)$$

The enwrapping of this surfaces' family is the generating rack-gear flank. The enwrapping condition is determined based on the "generating trajectories" theorem. The direct parameters of the normal to the spline's flank are calculated. By partial derivative, from (2), results:

$$\vec{n} = \begin{vmatrix} \vec{i} & \vec{j} & \vec{k} \\ -1 & 0 & 0 \\ 0 & 0 & 1 \end{vmatrix} = \vec{j}, \quad (6)$$

so, the direction of the $\vec{N}_\Sigma$ normal on to the current point of the flank (2) has equation:

$$\vec{N}_\Sigma = -u\cdot\vec{i} + (b+q)\cdot\vec{j} + t\cdot\vec{k}, \quad (7)$$

where, q is a variable scalar value. The generating trajectory of the normal (7) is:

$$(\vec{N}_\Sigma)_{\varphi_1} \left| \xi = \omega_3^T(\mathbf{j}) \times \mathbf{N}_\Sigma + \mathbf{a}. \right. \quad (8)$$

It is imposed the condition that the $(\vec{N}_\Sigma)_{\varphi_1}$ trajectories pass through the gearing pole:

$$P \begin{vmatrix} \xi = 0; \\ \eta = R_r\cdot\varphi_1; \\ \eta = t. \end{vmatrix} \quad (9)$$

From (8) and (9) result the conditions:

$$-u\cdot\cos\varphi_1 - (b+q)\cdot\sin\varphi_1 + R_r = 0; \; -u\cdot\sin\varphi_1 + (b+q)\cdot\cos\varphi_1 + R_r\cdot\varphi_1 = R_r\cdot\varphi_1, \quad (10)$$

from which, eliminating the q parameter, the specific enwrapping condition results:

$$u = R_r\cdot\cos\varphi_1. \quad (11)$$

The assembly of equations (5) and (11) is the enwrapping of the rack-gear's family flanks (12). The rack-gear is a cylindrical surface with generation parallel with ζ axis.

$$C_{\varphi,t} \begin{vmatrix} \xi = -R_r\cdot\cos^2\varphi_1 - b\cdot\sin\varphi_1 + R_r; \\ \eta = -R_r\cdot\sin\varphi_1\cdot\cos\varphi_1 + b\cdot\cos\varphi_1 + R_r\cdot\varphi_1; \\ \zeta = t. \end{vmatrix} \quad (13)$$

2.2 Helical surface conjugated with generating rack-gear

In Figure 2, it is presented the axis position of the future hob mill, the axis $\vec{V}$, inclined regarding the front plane of the spline with angle ω. Let assume that the helical motion with

$\vec{V}$ axis and p helical parameter is decomposed in two motions: one translation along the rack-gear's cylindrical surface's generation, denoted with $T(\vec{t},v)$ and one rotation around an axis parallel with $\vec{V}$ and at distance a_0 from this, $(\vec{A},\varphi)$.

In this way, the decomposition of the helical movements becomes:

$$(\vec{V},p)\Box\, T(\vec{t},v)+(\vec{A},\varphi). \tag{14}$$

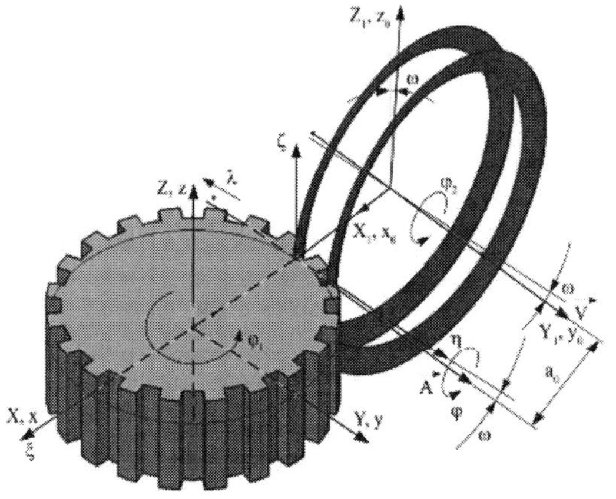

Fig. 2. Hob mill's axis and reference systems.

We have to notice that the $C_{\varphi,t}$ surface, in the motion $T(\vec{t},v)$ it's moving in the direction of its own generation. So, the characteristic curve of the $C_{\varphi,t}$ surface does not depend by this motion. The characteristic curve will depend only by the rotation around the $\vec{A}$ axis. The a_0 distance is defined as $a_0 = p\cdot\tan\theta$, where θ is the angle between the $\vec{V}$ axis and $\vec{t}$ generation.

2.3 Worm generation kinematics

There are defined the reference systems: $x_0y_0z_0$ is global reference system, with y_0 axis overlapped to the worm's axis; $X_1Y_1Z_1$ — mobile reference system, with Y_1 axis overlapped to the $\vec{V}$ axis and $\xi_1\eta_1\zeta_1$ — mobile reference system, with η_1 axis overlapped to the $\vec{A}$ axis.

The generating process kinematics assume: the motion I is the rotation of the shaft around its own axis; II — translation of rack gear along the η axis and III rotation of the hob mill. The relative position of fixed reference systems is given by:

$$\mathbf{x_0}=\boldsymbol{\beta}\cdot(\mathbf{x}-\mathbf{A});\ \mathbf{A}=\begin{pmatrix}-R_r-a_0\\0\\0\end{pmatrix};\ \boldsymbol{\beta}=\begin{pmatrix}1&0&0\\0&\cos\omega&-\sin\omega\\0&\sin\omega&\cos\omega\end{pmatrix}. \tag{15}$$

The rotation and translation of the rack-gear around and along the $\vec{V}$ axis is given by:

$$\mathbf{X_1}=\boldsymbol{\omega}_2(\varphi_1)\cdot\boldsymbol{\beta}\cdot(\mathbf{x}-\mathbf{A}). \tag{16}$$

The rotation and translation of the rack-gear around and along the $\vec{V}$ axis is given by:

$$\mathbf{X}_1 = \boldsymbol{\omega}_2(\varphi_1)\cdot\boldsymbol{\beta}\cdot(\mathbf{x}-\mathbf{A}). \tag{17}$$

By developing equation (18), see also (19), the following is obtained:

$$\left[C(\varphi,t)\right]_{\varphi_2} \begin{vmatrix} X_1 = (\xi + a_0)\cos\varphi_2 + (\eta - p\varphi_2\cos\omega)\sin\varphi_2 - \zeta\sin\varphi_2\cos\omega; \\ Y_1 = (\eta - p\varphi_2\cos\omega)\cos\omega - \zeta\sin\omega; \\ Z_1 = (\xi + a_0)\sin\varphi_2 + (\eta - p\varphi_2\cos\omega)\cos\varphi_2\sin\omega + \zeta\cos\varphi_2\sin\omega. \end{vmatrix} \tag{20}$$

The enwrapping of the (20) surfaces' family is obtained using the specific theorem of the generating trajectories. The direct parameters of the normal to the helical surface are given by(21). In the generating movement(4), the normal (22) describe a family. Considering that the normal family intersect the $\vec{A}$ axis, results the enwrapping condition.

$$\vec{n}_c = \dot{\eta}_\varphi\cdot\vec{i} - \dot{\xi}_\varphi\cdot\vec{j}. \tag{22}$$

3 Numerical application

A numerical application is presented for a hob mill to generate a splined shaft with: $b = 9$ mm; $R_e = 62.5$ mm; $R_i = 56$ mm; $R_r = 62.338$ mm. The graphical method presumes determination of the profile for the rack-gear cutting tool, using the method of "generating trajectories".

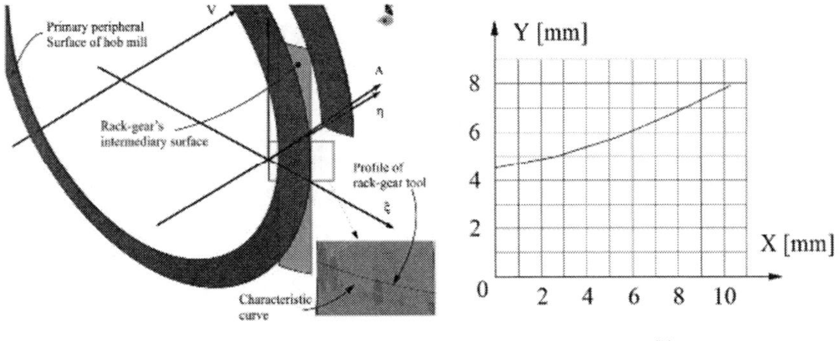

a). b).

Fig. 3. a). Characteristic curve; **b).** Axial section.

Consequently, the following algorithm is applied. The flank surface of the rack-gear is created extruding the profile of the rack-gear tool along the direction of ζ axis. This is the *I* surface. The point which represents the origin of the $X_I Y_I Z_I$ reference system is created.

A line which passes through the origin of the $\xi\eta\zeta$ reference system and is inclined with angle ω regarding the η axis is draw. This line represents the $\vec{A}$ axis. The support plane of this axis is the $\eta\zeta$ plane. The new $X_I Y_I Z_I$ reference system is created. The axis of this new system are $X_1 \equiv X$ and $Y_1 = \vec{A}$. Parallel with axis $\vec{A}$ is created the axis of future hob mill, the $\vec{V}$ axis. This will pass through the origin of the $X_I Y_I Z_I$ reference system. Onto the axis $\vec{A}$, a current point is generated. From this point, a line normal to the surface *I* is drawn.

According to the "generating trajectories" theorem, the intersection point between the normal and the surface *I* belongs to the characteristic curve too. This point is kept using the

ISOLATE command and the last two steps are resumed in order to find others points onto the characteristic curve.

In Figure 3.a is presented the form of the characteristic curve and in Figure 3.b is presented the form of the axial section for the hob mill. In Table 1, there are presented the coordinates of points form for the characteristic curve and axial section of hob mill.

Table 1. Coordinates of points form characteristic curve and axial section.

Crt.	Characteristic curve			Axial section	
no.	X [mm]	Y [mm]	Z [mm]	X [mm]	Y [mm]
1	0.000	4.500	-0.234	0.000	4.500
2	1.149	4.666	-0.587	1.192	4.674
3	2.325	4.905	-0.774	2.373	4.918
4	3.493	5.210	-0.905	3.538	5.225
5	4.649	5.568	-1.021	4.688	5.584
6	5.791	5.966	-1.133	5.825	5.983
7	6.922	6.399	-1.237	6.951	6.415
8	8.041	6.862	-1.329	8.063	6.877
9	9.148	7.357	-1.409	9.163	7.370
10	10.242	7.882	-1.481	10.249	7.894

4 Conclusions

The CATIA design environment allows, by software capabilities, to obtain a faster solution, very rigorous and, at the same time, very intuitive, for the helical surface which represents the primary worm of the hob mill.

An application of the "generating trajectories" method, developed for enwrapping surfaces with single point contact is proposed in this paper. An algorithm, developed in analytical form, is presented for a numerical application of the same square splined shaft.

The paper details a numerical example for which the results are presented in Table 1 and Figure 3. There is presented the form of characteristic curve at the contact with rack gear tool and, at the same time, the axial section of the peripheral surface of the worm.

The method presented in this paper can be also applied to other ordered curves of profiles (toothed parts), both for noninvoluted profiles and involute tooth. This method can be developed and integrated in a software product.

This work was supported by a grant of the Romanian National Authority for Scientific Research and Innovation, CNCS – UEFISCDI, project number PN-II-RU-TE-2014-4-0031.

References

1. S. Berbinschi, V. Teodor, N. Oancea, Int. J. of Ad. Man. Tec. **64** 291-304 (2013).
2. N. Baroiu, V. Teodor, N. Oancea, B. Pol. I. Jassy, ISBN 1011-2855, 27-36 (2015).
3. F. L. Litvin, **1212**, NASA, Sc. and Tech. Inf. Div., Washington D.C. (1984).
4. S. Radzevich, *Kinematic Geometry of Surface Machining*, CRC Press, ISBN 978-1-4200-6340-0 (2008).
5. N. Oancea, *Surfaces generation by winding*, Galati Uni. Press, ISBN 973-627-170-4, ISBN 973-627-107-2, ISBN 973-627-176-6, ISBN 973-627-239-7, **I-III** (2004-2008).
6. N. Oancea, Dunărea de Jos J., Galaţi, V, ISSN 1221-4566 48-53 (2002).

AUTHOR INDEX

AUTHOR INDEX

AUTHOR INDEX

AUTHOR INDEX

21st Innovative Manufacturing Engineering & Energy International Conference (IManE&E 2017)

MATEC Web of Conferences Volume 112 (2017)

Iasi, Romania
24 - 27 May 2017

Part 2 of 2

Editors:

L. Slatineanu
G. Nagit
O. Dodun
V. Merticaru
M. Coteata
M.I. Ripanu
A.M. Mihalache
M. Boca
R. Ibanescu
C.E. Panait
G. Oancea
P. Kyratsis

ISBN: 978-1-5108-4430-8

Printed from e-media with permission by:

Curran Associates, Inc.
57 Morehouse Lane
Red Hook, NY 12571

Some format issues inherent in the e-media version may also appear in this print version.

Printed by Curran Associates, Inc. (2017)

For additional information, please contact EDP Sciences – Web of Conferences at the address below.

EDP Sciences – Web of Conferences
17, Avenue du Hoggar
Parc d'Activité de Courtabœuf
BP 112
F-91944 Les Ulis Cedex A
France

Phone: +33 (0) 1 69 18 75 75
Fax: +33 (0) 1 69 28 84 91

contact-edps@webofconferences.org

Additional copies of this publication are available from:

Curran Associates, Inc.
57 Morehouse Lane
Red Hook, NY 12571 USA
Phone: 845-758-0400
Fax: 845-758-2633
Email: curran@proceedings.com
Web: www.proceedings.com

TABLE OF CONTENTS

PART 1

PART 2

MATEC Web of Conferences 112, 06009 (2017) DOI: 10.1051/matecconf/201711206009
IManE&E 2017

FEA analysis and design optimization for a multifunctional piece of furniture

Georgeta Haraga[1,*], and *Adrian Mihai* Goanţă[2]

[1] "Politehnica" University of Bucharest, Department of Engineering Graphics and Industrial Design, Spl. Independenţei No.313 Street, District 6, 060042, Bucharest, Romania
[2] "Dunărea de Jos" University of Galati, Engineering and Agronomy Faculty of Braila, Research Centre for Mechanics of the Machines and Technological Equipments, Calea Calăraşilor Street, 810017, Brăila, Romania

Abstract. The paper presents an original approach to the integration of computer aided design and finite element analysis for a multifunctional piece of furniture using CATIA Generative Structural Analysis workbench. Finite Element Analysis (FEA) application is an important engineering technique in the furniture industry. FEA is an accurate method for numerical solution of field problems. A major problem of mesh generation today is access to CAD geometry in a efficient and precise manner. Starting with a solid modelling for the creation of the desired piece of multifunctional furniture is used in this case, the automatic generation of the finite element meshes. Getting more accurate results for the mesh refinement process can be made by changing parameters. The article outlines the developed model for analysing the supporting structure of the armchair. This model provides simulating of the different loads and checking the raised stresses and deformation in the structures analysed. The main objective of this research is to evaluate strength of wooden cedar armchair for a person weighing 140 kg and for this the finite element analysis (FEA) has been used. Multifunctional furniture is the answer to the challenges contemporary lifestyle, creativity in design activities that require optimal space utilization and reducing costs.

1 Introduction

Furniture can be not only a functional product but also an artistic product and which can be associated with a form of decorative art. Furniture can contain different types of furniture pieces such as chairs, armchairs, tables, beds, cupboards and shelves. The pieces of furniture can be made using many materials like wood, metal and plastic. Thus, as an alternative to massive furniture, heavy and take up much space, the designers have brought into focus the multifunctional furniture, affordable, easy to install and operate.

Multifunctional furniture is considered a modern concept that can perform several functions simultaneously and which is based on recent knowledge of design, technology

[*] Corresponding author: g_haraga@yahoo.com

and ergonomics. A set of furniture contains several kinds of individual pieces of furniture with multiple functionalities.
Many multifunctional furniture pieces are created from the need to satisfy the several requirements of the most demanding users. For these reasons, the conceptual model proposed for this study is a multifunctional armchair, which was created by necessity to have a working personal space. The armchair was designed in environmentally conscious manner, being made out of wood-based material. This study has been structured into two main parts: in the first part - Modelling a piece of furniture and in the second part - Analysis of final model with FEA. The multifunctional piece of furniture proposed for the modelling and analysing is made in CATIA software, offering a realistic design that allows visualizing this model throughout its development. Produced of the company Dassault Systemes and distributed by IBM, CATIA offers many types modules with many applications in a variety of fields such as construction industry of machines, shipbuilding industry, automobile industry, aerospace, electrical engineering etc. [1]

2 Modelling process

Multifunctional furniture is known for extremely flexible design and that allows multiple uses. Furniture can be customized for the individual that needs to optimize space utilization.

The planning operation of space allocation is a specific optimization process, that involves allocating limited. The entire design process starts with the marketing research focused on several existing products and on potential user for the armchair [2]. Main criteria in furniture design are aesthetic, ergonomic and safety [3].

A significant number of benefits can be realized on during the development phase. The first part of this study involves designing of a multifunctional armchair with space optimization solutions for small room. Using CATIA Part module, the basic concept consists of two custom armrests of the chair with pockets of rectangular profile. The armchair is considered very durable being made from cedar wood, having all dimensions displayed in mm (W x H x D): 960 x 875 x 860. A better image in terms of colour to this model can be obtained by applying the Properties command. Custom tables can be inserted into each pocket to provide space optimization solutions. The armrests of this piece of

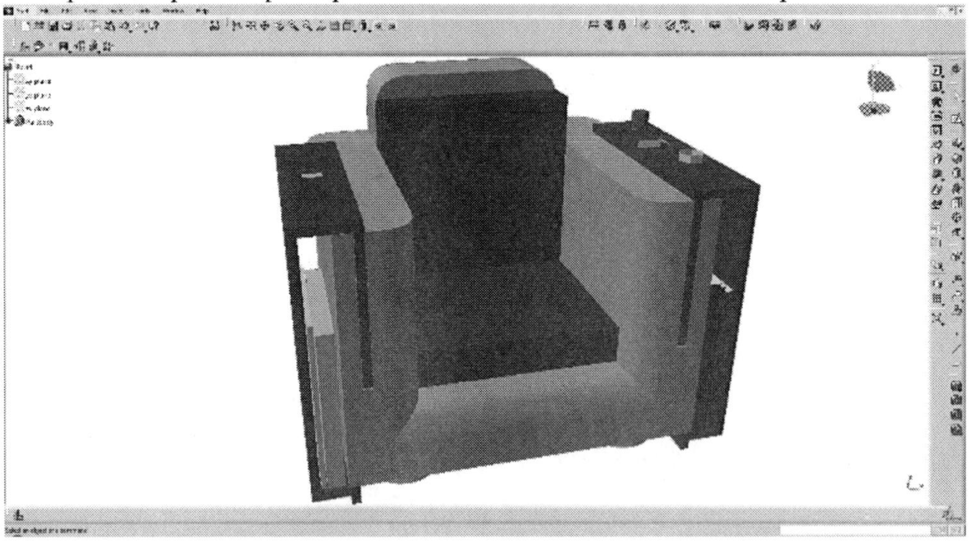

Fig. 1. Modeling of a multifunctional armchair.

furniture can be equipped with wooden tables on each side as can be seen in Figure 1. Wooden tables can be used for rest periods, but it also, can be an excellent support for multiple everyday objects such as digital tablet or laptop, newspapers or magazines, books, coffee cups etc.

3 Finite element analysis process

Finite element analysis (FEA) is a significant computerized numerical method that provides approximate solutions for structural and thermal analysis in different applications areas.

FEA for a multifunctional piece of furniture (with/without custom wooden tables) can be done by using Generative Structural Analysis workbench of CATIA software.
A case of static analysis involves analysing the static boundary conditions of a CATIA Analysis document and consists of three main stages: Pre-processing, Processing and Post-processing.

The Pre-processing step includes preparing to a specific part for Finite Element Analysis on the model and involves many stages such as: definition of type of analysis, definition of the material properties, generating mesh, applying restraints and loads.

The choice of material plays a crucial role in the design process and in this case cedar wood has been chosen. Cedar wood is often used for many applications in the furniture industry because it has a significant number of advantages, such as: lightweight, dimensionally stable, softwood and highly durable. Material properties are very important for any type of analysis and the mechanical properties of cedar wood such as: Young's modulus is 1.5e+010N_m^2 and Density is 750kg_m^3 can be seen in Figure 2.

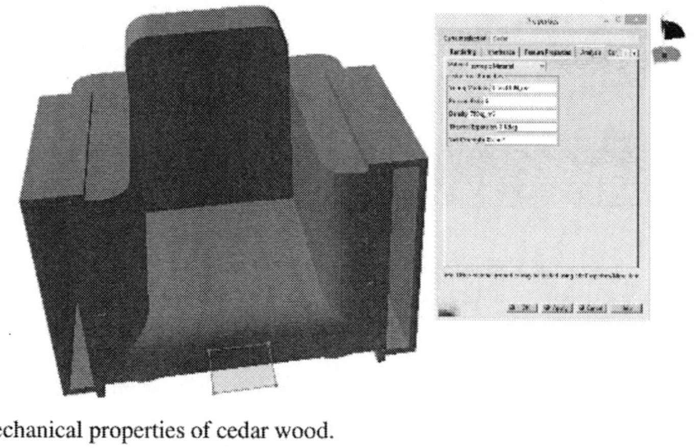

Fig. 2. The mechanical properties of cedar wood.

Generating Mesh represents meshing the part to create Finite Element model using elements and nodes. In both cases studied have been used “size” and “sag” and in the Octree Tetrahedral Mesh box have been entered the following values: Size -10 mm and Sag - 2 mm. “Size” refers to the elemental size of the mesh seen on the specific part and “sag” refers to the internal distance between the mesh size and the actual geometry of the part [4].

In the next step, there are established the restraints and loads on the model analysed. Any static analysis requires a linear static computation by referencing restraints and loads. The restraints can be achieved by the Restraints bar. According to the Clamp condition, the displacements are always zero in all three directions as shown in Figure 3.

The Processing is a second stage of the FEA process and includes the analysis for obtaining the correct solution (displacements, stresses etc.). This stage is an available

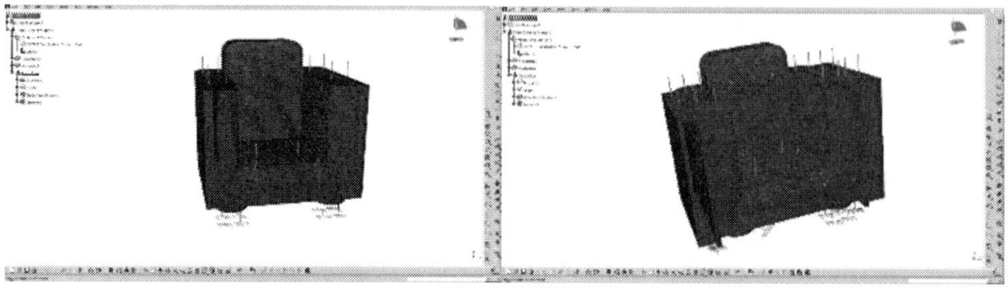

Fig. 3. Applying of the restraints and loads.

solution of the finite element problem provided by Pre-processing stage to find out the displacement values and to run the analysis and all these can be carried out by Compute command.

The Post-processing is the last stage of the FEA process and it refers graphical display of displacements and stresses for quick and easy interpretation, sorting and printing of results. The main toolbar in the Post-processing stage is called Image (Deformation, Von Mises Stress, Displacement). The Deformation option is used to view the deformed shapes according to the restraints and loads that were applied as can be seen in Figure 4.

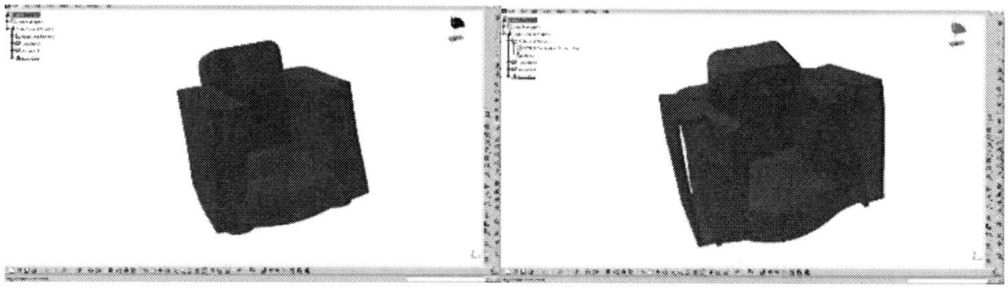

Fig. 4. Application of the Deformation command.

Stress analysis can be applied to entire structure or to a specific part [5, 6]. The state of stress is described by the six Cauchy stresses (the normal stresses - σx, σy, σz and the tangential stresses - τx, τy, τz) which can vary from one point to another [7, 8]. Usually, the Cauchy stress tensor is used if the stress analysis of the model proposed is experiencing with small deformations when the differences in the distribution of stresses can be neglected.
The Von Mises stress is a combination of these and calculates values using the formula [9]:

$$\sigma_{VM} = \sqrt{\frac{1}{2}\left[\left(\sigma_x - \sigma_y\right)^2 + (\sigma_x - \sigma_z)^2 + \left(\sigma_y - \sigma_z\right)^2 + 6\left(\tau_{xy} + \tau_{xz} + \tau_{yz}\right)^2\right]} \quad (1)$$

As a general rule, the Von Mises Stress is only a positive number and in terms of principal stresses, σ_{VM} can also be written as [10]:

$$\sigma_{VM} = \sqrt{\frac{1}{2}[(\sigma_1 - \sigma_2)^2 + (\sigma_2 - \sigma_3)^2 + (\sigma_3 - \sigma_1)^2]} \quad (2)$$

where σ_1, σ_2, σ_3 are namely principal stresses.

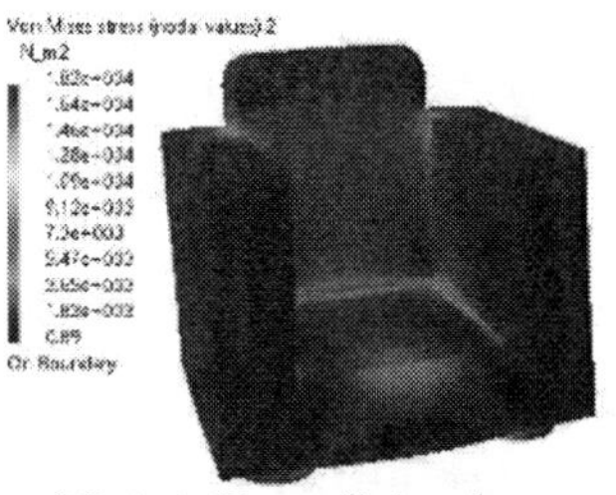

a. Mesh 1 (linear finite elements – Concept1)

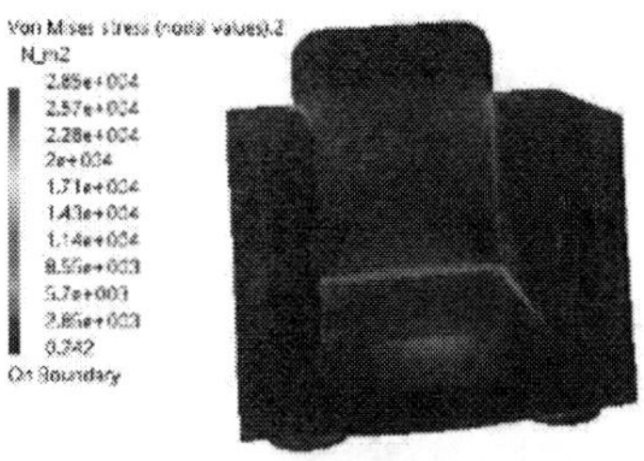

b. Mesh 2 (parabolic finite elements – Concept1)

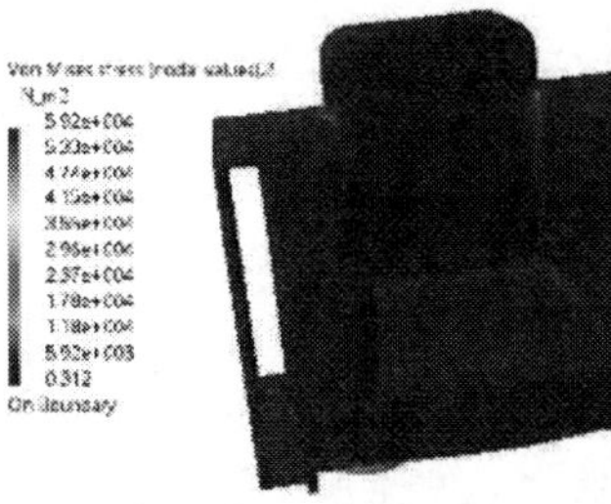

c. Mesh 3 (linear finite elements – Concept2)

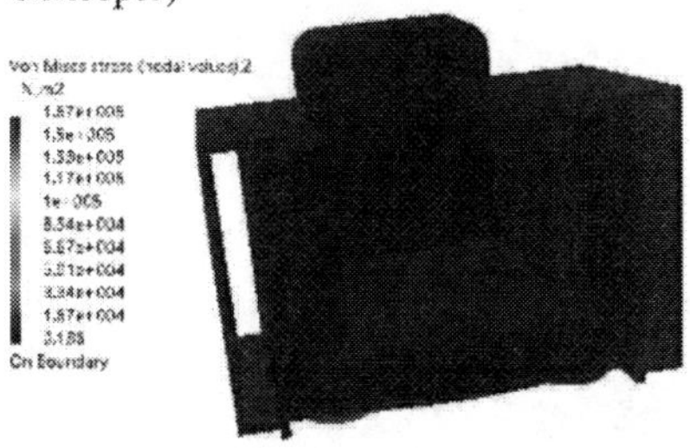

d. Mesh 4 (parabolic finite elements – Concept2)

Fig.5. The von Mises stress (Static Case Solution 1 - Mesh 1, 2, 3, 4 – a, b, c, d).

The Von Mises stress is the following icon from the Image toolbar. It is very important to understand that Von Mises stress is used by designers to check whether their design will withstand a given load condition [11].

The mesh refinement process can be made by changing global parameters can be seen in the Figure 5. In the first two cases analysed for Concept 1, Mesh 1 has the global mesh parameters (size =10, sag=2) with linear finite elements (145348 nodes, 651125 elements) and the global error rate of the analysis is $\varepsilon_{p\ initial}$=20.72%, respectively, Mesh 2 has the global mesh parameters (size =10, sag=2) with parabolic finite elements (1004303 nodes, 651125 elements) and the global error rate of the analysis is $\varepsilon_{p\ final}$=4.17%. In the second two cases analysed for Concept 2, Mesh 3 has the global mesh parameters (size =10, sag=2) with linear finite elements (227259 nodes, 984167 elements) and the global error rate of the analysis is $\varepsilon_{p\ initial}$=27.25%, respectively, Mesh 4 has the global mesh parameters (size =10, sag=2) with parabolic finite elements (1541050 nodes, 984167 elements) and the global error rate of the analysis is $\varepsilon_{p\ final}$= 12.02%.

The Displacement option allows viewing translational displacement of the model based on the loads that were applied upon it [12]. Displacements, Precision, Principal Stress distributions options are used along with a rich Colour Palette as shown in Figure 6.

4 Conclusion

The aim of the proposed in this paper is to design an armchair which has a study space optimization solution and make finite elements analysis for the supporting structure of an armchair with multifunctional facilities, taking into account the characteristics of its components. This paper is focused on the finite element analysis (FEA) of a multifunctional armchair having the following main objectives: choosing an the eco-friendly material that is quite frequently used in the furniture industry; adaptive mesh refinement; visualizing structural analysis results; reducing the global error rate, providing reasonable estimates.

DOI: 10.1051/matecconf/201711206009

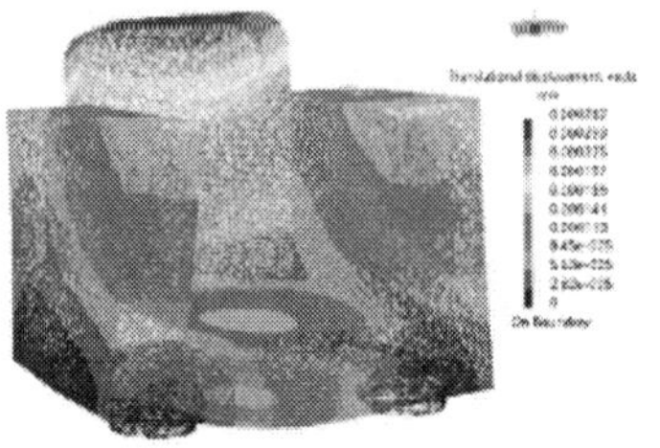

a. The Displacement option – Concept 1

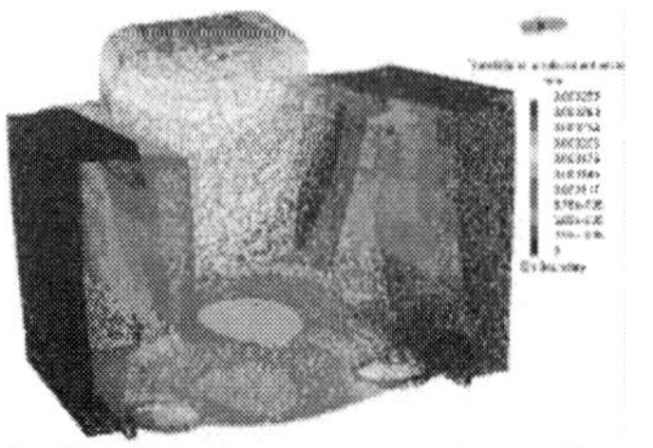

b. The Displacement option – Concept 2

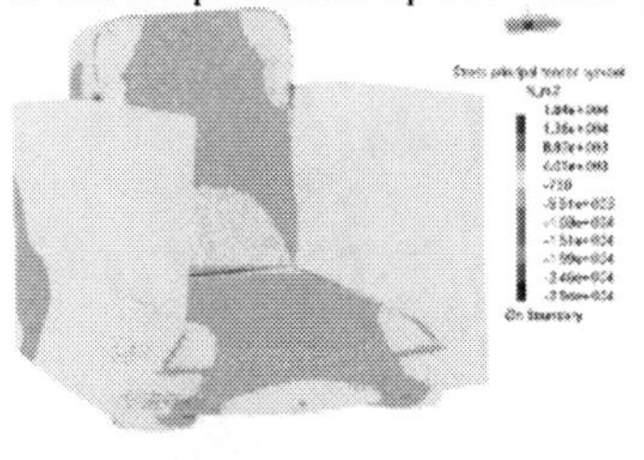

c. The Principal stress option – Concept 1

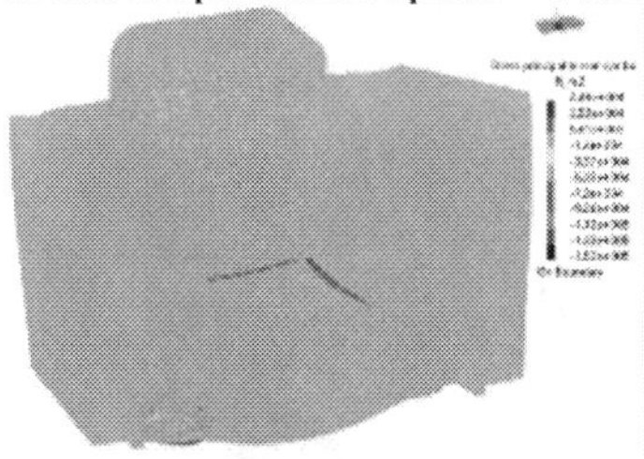

d. The Principal stress option – Concept 2

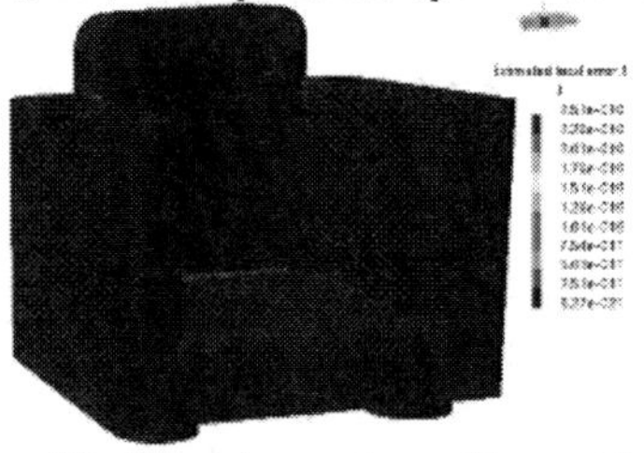

e. The Precision option – Concept 1

f. The Precision option – Concept 2

Fig.6. The Displacement, the Principal stress and the Precision options.

Making high-quality computations in the most automatic way possible was one of the main goals of the present paper. The final results of stress distribution and deformation are submitted for the analysis for this piece of multifunctional furniture. In FEA, there is an interpret the results from the color-coding. The red color shows the maximum deformation areas and the blue color shows the minimum deformation areas.

The mesh control is an important step in the Post-processing stage. The mesh parameters (geometric parameters - size and absolute sag and type of the finite element – linear or parabolic) have a significant influence on the accuracy of numerical analysis.

The initial results for the global error rate of the analysis obtained in these case studies, are: Concept 1: $\varepsilon_{p1initial}$=20.72% (Mesh 1); Concept 2: $\varepsilon_{p2initial}$=27.25% (Mesh 3). For 3D models discretized with linear finite elements, the global error rate are in the range 27% - 20%, percent error that cannot be accepted. Under these circumstances, our goal is to reduce the percent (%) the global error rate of the analysis. Reducing of the global rate error can be done by changing the global parameteres and type of element. For Concept 1, the error has been reduced to $\varepsilon_{p1final}$=4.17% (Mesh 2) that is substantially less than the 20.72% and for Concept 2, the error has been reduced to $\varepsilon_{p2final}$=12.02% (Mesh 4) that is a significant less than the 27.25%.

Increasing the number of nodes can improve the accuracy of the results, but at the same time, it increases the solution time and cost. In recent years, have made a considerable progress in this direction, but there are still many problems remaining.

References

1. A.M., Annals of Dunărea de Jos, Univ. Galaţi, XIV Mech. Eng., **1** 9-12 (2013)
2. A. Fantu, L. Bârsan, Bulletin of Transilvania Univ. Braşov, România, **4 (53)** 2 I 39-44 (2011)
3. I. Ntintakis, V. Iakovakis, G. Ntalos, J. Kechagias, Conference on Current Issues in Global Furniture 14-22 (2013)
4. G. Haraga, D. Ghelase, L. Daschievici, Computer Eng. and Appl. (CEA '16), ISI Proceding, Barcelona, Spain, 217-220 (2016)
5. CATIA V5 Fundamentals, EDU-CAT-E-V5R15 (2005)
6. D., Zahariea, Internat. J. of Modern Manufact. Techn. **3** No.2 117-122 (2011)
7. http://www.cadcamlab.org/courseware/PDFs/Stress%20Analysis.pdf
8. https://www.capinc.com/2014/02/12/frequently-asked-questions-on-von-mises-stress-explained
9. http://www.sdcpublications.com/pdfsample/1-58503-227-1-2.pdf
10. http://www.learnengineering.org/2012/12/what-is-von-mises-stress.html
11. https://en.wikipedia.org/wiki/Cauchy_stress_tensor
12. http://www.cadcamlab.org/course/images/Example_Manuals_2006/Stress%20Analysis%20R14.pdf

MATEC Web of Conferences **112**, 06010 (2017) DOI: 10.1051/matecconf/201711206010
IManE&E 2017

Manufacturing ontology through templates

Vlad Diciuc[1,*], *Mircea* Lobonţiu , and *Adrian* Petrovan

Technical University of Cluj Napoca, North University Center of Baia Mare, Romania

Abstract. The manufacturing industry contains a high volume of know-how and of high value, much of it being held by key persons in the company. The passing of this know-how is the basis of manufacturing ontology. Among other methods like advanced filtering and algorithm based decision making, one way of handling the manufacturing ontology is via templates. The current paper tackles this approach and highlights the advantages concluding with some recommendations.

1 Introduction

The lifecyle of a certain product, in its completeness from idea and design, project, technology, manufacturing planning, manufacturing and exploitation to removing the product from its use implies activities that rely on a series of principles, specific logic, structures and technological regimes, manufacturing structures, commercial activities, etc. which are basically based on knowledge. The cost of developing a product is highly influenced by the quantity and quality of the knowledge used. The cost, quality, training and service conditions within warranty and post warranty period are fundamental elements which sustain market competition. The manufacturing cost, if correctly estimated in the design and product development stages leads to a high rate of success and to a controlled competition. Despite the fact that 60% to 80% of the final product cost – raw materials, manufacturing costs, capital investments, auxiliary costs – is accessed starting the technological and functional design, a lot of managers consider the design as an unproductive and costly activity, even if a lot of companies have design costs not more over 5% of the project's budget [1].

When launching a product, two aspects are to be considered starting the design stage: the estimation of the manufacturing costs, especially on grounds of technological performance and the duration of the product's manufacturing activities. The duration of the activities that lead to the development of the product is estimated to be 80%-90% of the time to market, being influenced by the complexity of the product, the manufacturing technology and by the duration of the technological preparation of manufacturing.

Regarding the issues with product quality, Saaksvuori and Immonen [2] stated that 40% of these issues are caused by a deficient design activity. A design activity considered to be deficient may have multiple causes: the lack of knowledge regarding the quality of the furnishing, component incompatibilities, the dispersion of knowledge and so on.

[*] Corresponding author: vlad.diciuc@cunbm.utcluj.ro

2 On the concept. Manufacturing ontology

The concept of "ontology" has been used in the last years in the field of knowledge management and computer assisted collaborative engineering.

The term "ontology" is borrowed from philosophy and it has been introduced in 1613 by the German philosopher Goclenius and popularized especially by Chr. Wolf. The idea of such a philosophical discipline has been defined in Aristotle times who called it "the first philosophy" and later, for this the term "metaphysic" has been used. In the pre-marxist and contemporary idealist philosophy ontology represented that speculative theory about existence as a state of fact, meaning common features and principles to any existence. Merriam-Webster dictionary defines ontology as "a branch of metaphysics which deals with the nature and the relationships of the existence".

The term ontology has been adopted by researchers in the field of Artificial Intelligence (AI) which have recognized the applicability of mathematic logic [3] arguing that they can develop ontologies as computational models which allow for certain types of automatized reasoning. In the 80's the AI community has proposed the use of the term "ontology" to address a different theory of the modelled world as well as a component of knowledge systems. Some researchers inspired by the philosophical ontologies have seen in the applicable computational ontologies aspects of the "applied" philosophy.

At the beginning of the 90's in an effort to create interoperability standards, an entire set of technologies which recognize the value of ontologies as a standard in the knowledge based systems has been created [4]. Following this stage the ontology became a technical term widely used especially in the field of computer science.

Ontologies are now used to capture knowledge of a certain domain of interest. Ontology describes the most important concepts and relationships of a certain area of interest. A widely accepted definition for ontology is given by Gruber, i.e. "an explicit specification of a conceptualization"[5]. According to Gruber [5], the conceptualization is an abstract notion, a simplified view of the world that we want to represent to serve a certain purpose. Regardless of the community type which will change and share knowledge through ontology, be it human or non-human agents (within AI), ontologies establish a common terminology and for this reason the conceptualization asks for a proper formalization.

As an application in the technical domain, Lobonţiu and Petrovan [6] have proposed a definition of ontology in product development: "The ontology represents that set of concepts, hierarchically structured which describes a certain knowledge domain which can be used to create a knowledge base. The ontology of a technical domain contains concepts, a structure, a subordinated hierarchy, arbitrary relationships between concepts, axioms, functions and other constraints".

Lemagnan [7] presents a simple example for the perception of the manufacturing ontology concept: "let's say my ontology defines the concept of aluminium as a sub-concept of raw material. The concept of drill as a sub-concept of tool and drilling as a kind of operation are also defined. Properties drill diameter and drill speed are added to these concepts to refine them. The ontology finally binds these concepts through the relationships canBeMachinedBy (between a raw material and an operation) and uses (between an operation and a tool). Once these concepts and relationships are established, rules may be written (like "for aluminium and drilling diameter smaller than 5 mm, the drill rotation should be 3000 rpm").

In the manufacturing field, ontology is still in its research stage [8], lacking a clear definition of the working area. This working area can include small areas belonging to the product's manufacturing cycle like design or machining or it can include the whole manufacturing process.

Studies regarding the manufacturing ontology in relation to the PLM can be found in papers like [8-12], each aiming for a certain stage of the PLM like Design/Development, machine selection, assembly, etc.

Neither of the mentioned studies tackles the ontology of machining. These are all treated under the umbrella of CAM software and manually written NC code but the actual ontology of the machining process is not treated. Although it's a branch of PLM, the number of variables and the nature of this branch have a significant influence over the final product and as well over the costs because of the selected machine, tools, cutting strategies, etc.

In the following chapters the development of ontology aspects that emerge in CAM controlled machining processes and the implementation of these aspects through templates is discussed. These aspects will form the so-called manufacturing ontologies of parts, divided and organized on part families, hierarchically structured within the manufacturing process with their own axioms, functions and constraints thus generating the manufacturing ontology of the whole product.

3 Methodology. Interdisciplinary

As well mentioned in the paper [13] manufacturing ontology involves interdisciplinary. Here we include the machining specialist who synthesizes the input variables which influence the cutting, the project manager which forces economic efficiency constraints and the IT specialist which, depending on the situation, generates the interface based on the data received or it conceives a way to input and adapt the received/collected data into existing systems/interfaces according to the variable relationship and their weight over the final result (Figure 1).

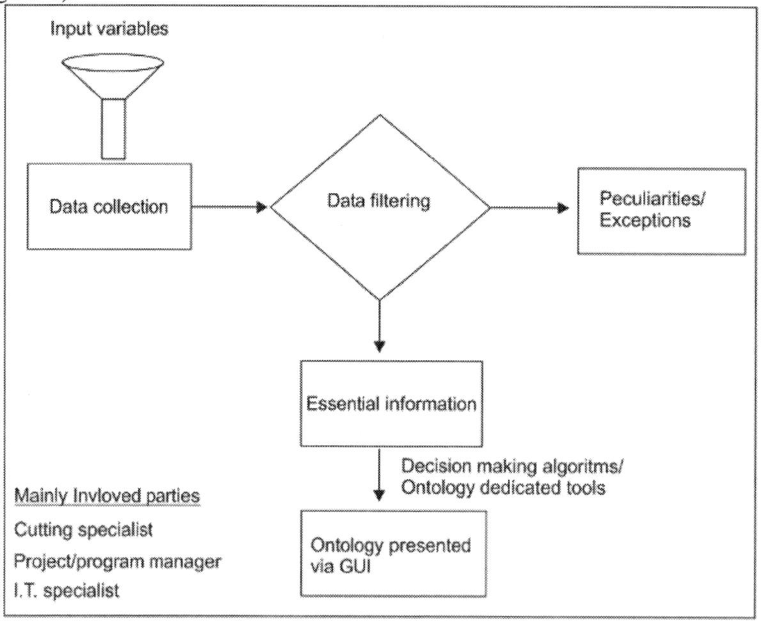

Fig. 1. Manufacturing ontology methodology.

The stages of creating the ontology presented in Figure 1 will not be discussed during this paper as they were very well described in the papers referenced here. It will however be underlined the implications each stage has over the machining domain.

Data collection initially implies the identification of all the input variables which influence the machining process. Within these variables the following are the most important:

- the cutting regime (a_p – axial depth of cut [mm], a_e – radial depth of cut [mm], V_c – cutting velocity [m/min], V_f – feed velocity [mm/min]);
- the cutting pattern: raster, zig-zag, offset, profile, etc.;
- the type of cutting process: turning, milling, boring, etc.;
- the machining method: roughing, semi-finishing, finishing;
- the order of the operations: according to the access to each surface, to the conditions within the cutting process, to the used tools, etc.;
- the order of the cutting within the same operation: level first, depth first, custom;
- the definition of the working and clearance planes;
- the workpiece material, the tool and insert material, the tool geometry;
- the machine type used (power, number of axes, maximum travel on each axis);
- the cutting environment (dry, wet, mist, etc.).

Data filtering is dependent on the CAM software used and which determines if all or just part of the above mentioned variables can be included in the piece machining technology. These represent the core of that manufacturing ontology. If some of the data mentioned above cannot be included in the software, updates or plug-ins will be devised to improve the application as much as possible. The rest will pass into the Peculiarities/Exceptions area. In this area one may also find the information that cannot be digitized and it depends on the machining specialist to implement them on the workshop floor as good as possible.

Essential information includes: cutting regime, cutting pattern, tool geometry (as dimensions) and cutting operation succession. Based on this information the machining ontology can be devised for a certain surface or for a certain part.

Peculiarities/exceptions – as it was explained previously, this category includes either variables that cannot be defined within the CAM software either uncommon aspects that are very particular to the context.

Manufacturing ontology presented via GUI – it represents the crystallization of the manufacturing ontology via a GUI (Graphical User Interface) which is driven by decision making algorithms and ontology dedicated tools. This GUI is either custom created for a much more intuitive and user-friendly experience or as a template made by using the software's available resources in this case the appearance being the one with which the user is accustomed from using the standard software.

In the following chapter an example is being presented of a manufacturing ontology implemented in NX CAM from Siemens via templates.

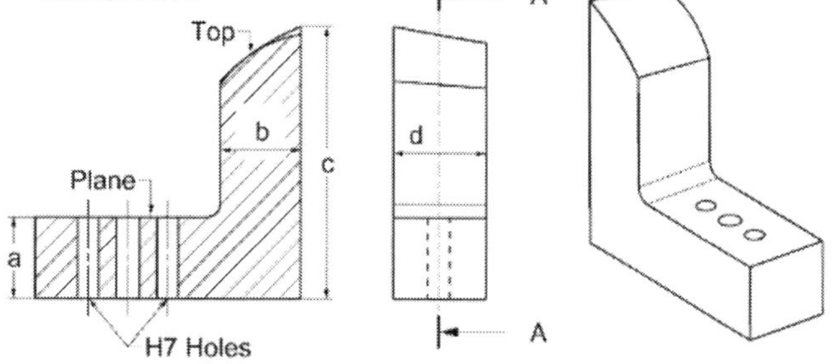

Fig. 2. Piece defined through dimensional parameters and surfaces.

4 Example of manufacturing ontology implemented through templates in NX

The piece presented in Figure 2 is often met in the automotive industry as part of an assembly which holds different car body parts to be welded.

It is defined through dimensions a, b, c, d and most of the times by 2 H7 precision holes for the dowel pins and one hole for the screw. The part must sustain the body part of the car this being laid on the Top surface. Considering the shape and the type of the surfaces, all these parts are identical which makes them ideal for a template because they can be manufactured using the same technology. In NX CAM, when building a template, the main philosophy is: "start with a process that works and reuse it for similar geometry". For this we need to have the manufacturing technology done in NX CAM for one such a piece and using it as a starting point to configure the template within NX CAM.

The operation sequence, the used tools and the parameters of the cutting for each case are presented in the figure below (Figure 3).

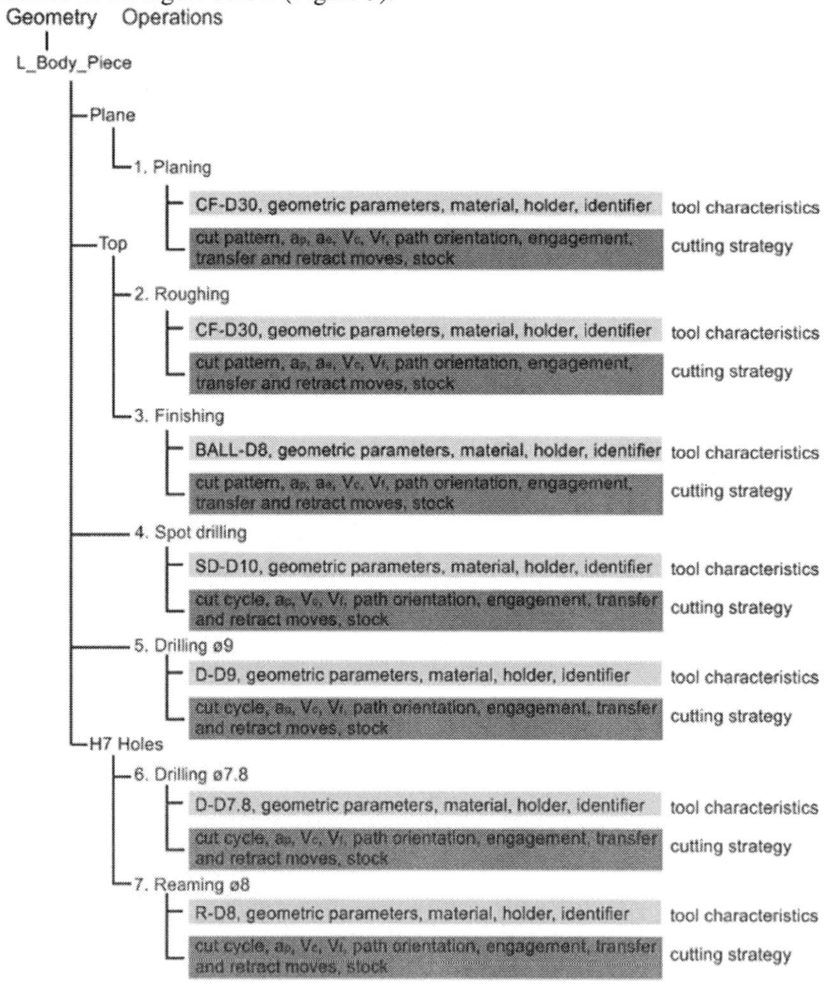

Fig. 3. Levels of subordination within the template.

The template will be created based on the parent geometry, i.e. in this case we named it L_BODY_PIECE. When opening the new part from the same part family to which we want to apply the known technology, the user must insert by using the NX CAM menus the L_BODY_PIECE geometry. Once this geometry inserted, everything regarding the manufacturing technology will also be inserted (tools, cutting operations, etc.). The user will then only have to click and define the surfaces referenced (Top, Plane, H7 Holes and screw hole). Everything else is being done by the template (creation of tools with all the defined parameters, creation of the cutting operations following the known order of execution and the given cutting parameters). Additionally to limit the access of the user to certain sensitive information of the manufacturing technology areas of the menu can be hidden thus also limiting the possibility of error/mistake to a minimum. Both interface customization and knowhow information input by using NX CAM is extremely easy and it does not require coding. Everything is done via mouse clicks.

It can be seen that in such template using cases, the end user (not the user who configures the template) must meet the minimum requirements of using the software tool regarding the CAM module. Basically this is a manufacturing ontology template for a certain type of pieces.

5 Conclusions

In the present paper the authors have tried to bring up front the concept of ontology in the technical domain by defining its basic structures. Through a working template we have created a manufacturing ontology of a typical part for the automotive industry.

Through the manufacturing ontology a specific database system is being created with usage procedures similar to those included in product development ontologies. The data will be referencing the whole process (technology, cutting data, time estimation, tooling, fixtures, probing tools, etc.). In this way the following advantages will be created for the industry: the time to manufacturing is being shortened, the certification of an in-use technology, the knowledge used, tooling and fixture solutions, known providers. On the other hand, if a fluctuation of personnel emerges, the replacement employees will use ontology models which will by then be certified.

References

1. W.G. Sullivan, J.A. Bontadelli, E.M. Wicks, *Engineering Economy*, Eleventh Edition, Prentice Hall, New Jersey (2000)
2. A. Saaksvuori, A. Immonen, *Product Lifecycle Management*, Second Edition, Springer (2005)
3. J. McCarthy, "Circumscription - A Form of Non-Monotonic Reasoning", *Artificial Intelligence* 5(13) 27-39 (1980)
4. R. Neches, R.E. Fikes, T. Finin, T.R. Gruber, R. Patil, T. Senator, W. R. Swartout, , "Enabling technology for knowledge sharing", *AI Magazine* 12(3) 16-36 (1991)
5. T.R. Gruber, *Knowledge Acquisition*, 5(2) 199-220 (1993)
6. M. Lobonțiu, A. Petrovan, , Management and Electrical Eng. J., ISSN(print)1583-624X, ISSN (online) 2360-2155 43-56 (2012)
7. S. Lemaignan, S. Siadat, J-Y. Dantan, A. Semenenko, Proceedings of the IEEE Workshop on Distributed Intelligent Systems: Collective Intelligence and Its Applications (DIS'06) (2006)
8. D. Penciuc, A. Durupt, F. Belkadi, B. Eynard, H. Rowson, Towards a PLM interop for a collaborative design support system, Proc. CIRP 25 369-376 (2014)
9. G. Bruno, D. Antonelli, A. Villa, A reference ontology to support product lifecycle management, Procedia CIRP 33 41-46 (2015)

10. B.A. Kendrick, S.T. Newman, V. Dhokia, A novel product representation to highlight cross-assembly dependencies and product robustness, Procedia CIRP 25 46-52 (2014)
11. A. Felic, B. Koenig-Ries, M. Klein, Process-oriented semantic knowledge management in product lifecycle management, Proc CIRP 25 361-368 (2014)
12. G. Rehange, J. Gausemeier, Ontology-based determination of alternative CNC machines for a flexible resource allocation, Proc CIRP 31 47-52 (2015)
13. M. Landherr, C. Constantinescu, Intelligent management of manufacturing knowledge: foundations, motivation scenario and roadmap, Proc CIRP 3 269-274 (2012)

MATEC Web of Conferences **112**, 06011 (2017) DOI: 10.1051/matecconf/201711206011
IManE&E 2017

Finite element simulation of the C70 orthogonal cutting process

Vlad-Bogdan Tomoiagă[1,*], *Marcel Sabin* Popa[1], *Stefan* Sattel[2], *Glad* Conţiu[1,*], and *Marius* Bozga[2]

[1] Technical University of Cluj-Napoca, Faculty of Machine Building, Department of Manufacturing Engineering, Bulevardul Muncii 103-105, Cluj-Napoca, Romania
[2] Gühring KG, Herderstraße 50-54, Albstadt, Germany

Abstract. Finite element simulations are intended to predict the effects of the technological parameters on the cutting process. The process simulation may reveal many aspects of the cutting process before testing it in laboratory: thermal transfer, elastic and plastic strains, hardening levels and force levels. Simulation is also used for predicting some process issues that are hard, and even impossibly noticeable. This paper analyzes the influence of the K factor on the cutting of C70 material, from the point of view of chip formation and thermal transfer - elements considered important in the evolution of tool wear on the rake and flank surfaces. The chip formation is also studied and considered as a deciding factor in choosing the right geometry of the tool. The obtained results allow to make observation under the optimal geometry that was tested.

1 Introduction

The simulation goal is to predict what might happen in a real cutting process. FEM (finite element method) is used for crash tests, fluid flows, heat transfer, and plastic or elastic strain predictions. Cutting simulation allows seeing how chip is formed or the heat is transferred between the chip and workpiece. Due to the numerous advantages as well as the large availability of commercial FE programs, these phenomena were studied for cutting edges with different K factors. From this point of view, Denkena [1] performed a relevant study with variable K factor (Figure 1).

In Denkena's study [1], the K factor is defined as follows in equation (1):

$$K = S_\gamma / S_\alpha \quad (1)$$

The study conducted an orthogonal cutting process, where S_γ and S_α were variable. The study focused on the influence of the K factor on the separation point, where it was observed that in the case of a symmetric cutting edge the size of the stagnation point increases with increasing of S_γ and S_α and for asymmetrical cutting edge only S_α has a significant influence on the stagnation point.

* Corresponding author: vlad_tomoiaga@yahoo.com

DOI: 10.1051/matecconf/201711206011

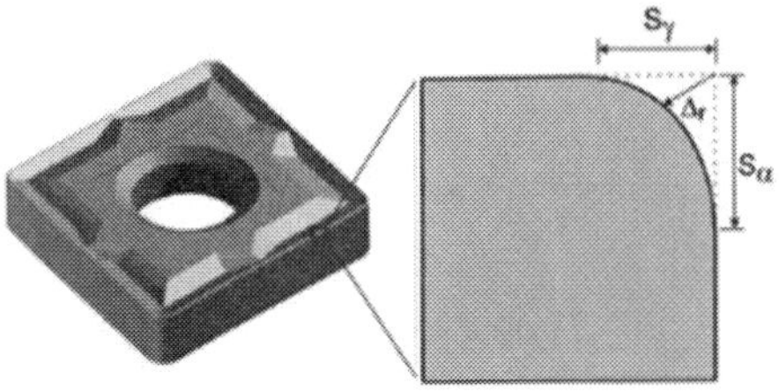

Fig. 1. K factor [1].

S_α - cutting edge segment on the flank face
S_γ - cutting edge segment on the rake face
Δr - profile flattening

In terms of the cutting forces, it was noticed that they increase with the increasing value of S_γ and S_α. This cannot be confirmed in the case of thrust forces F_t, where it appears that these forces grow only when S_α is variable. When S_α is constant and S_γ is variable, the force F_t increases to a specific point, and after that it stabilizes. The temperature during the cutting process has the same value even if S_γ and S_α change. The effect of S_γ and S_α variation on the rake and flank face temperatures was studied by the authors. The main conclusion is that the temperature increases proportionally with S_γ and S_α when K = 1. Segebade [2] has studied the influence of factors K: 0.2; 0.3; 2 and 5, using different cutting parameters during a broaching process. The author studied the influence of K factors on the quantitative grain size and the accumulated plastic strain. Denkena [3] studied the stagnation zone in front of the cutting edge, having asymmetric micro geometries (K~1.7 and K~0.6), observing how the stagnation zone increases when K~1.7, is higher than the stagnation zone for factor K~0.6. This can also be found in paper [4], where the K factors used in the simulation were 0.5 and 2.

This paper deals with the simulation of the C70 orthogonal cutting with the same radius of the cutting edge but different values of the K factor. The orthogonal cutting simulation was chosen due the fact, that in this case it easier to see the influence of the K factor. This material has been chosen, because experiments will be made in the future using this material to compare the simulation with the experimental part. All simulations should lead to an optimal K factor that has to be used in the case of C70 material. This paper aims to emphasize the need of controlling the microgeometry in the machining process.

2 Chip formation and heat transfer during cutting

Three zones of chip deformation can be distinguished during a cutting process (Figure 2): primary deformation zone (1), secondary deformation zone (2), and third deformation zone (3). As the name says, the primary deformation zone is the main deformation zone of the cut material, which is characterized by angle Ø [5]. The secondary deformation zone is the area where the chip moves along the rake face of the cutting tool, in this area being recorded the highest temperature during a cutting process [5]. This is the most sensible problem that has to be studied. As it can be seen in the Figure 2, from the microgeometry of the tool, the cutting edge has no influence on the first deformation zone. The secondary zone is mainly influenced by the macro-geometry (rake angle). The secondary zone is also influenced by the shape of the rake face (concave, convex or plane), the friction coefficient and the usage of cooling fluids. This secondary zone mainly determines the wear of the rake face. Depending on the material and on the work tool, different chip configurations are

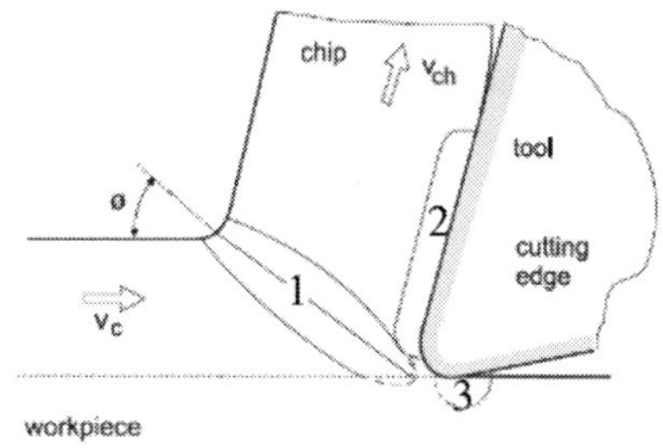

Fig. 2. Cutting mechanism [5].

obtained. The third deformation zone results in material deformation under the flank face of the cutting tool.

From the point of view of the chip, four types of chips can be distinguished: continuous, lamellar, segmented and discontinuous [5] (Figure 3).

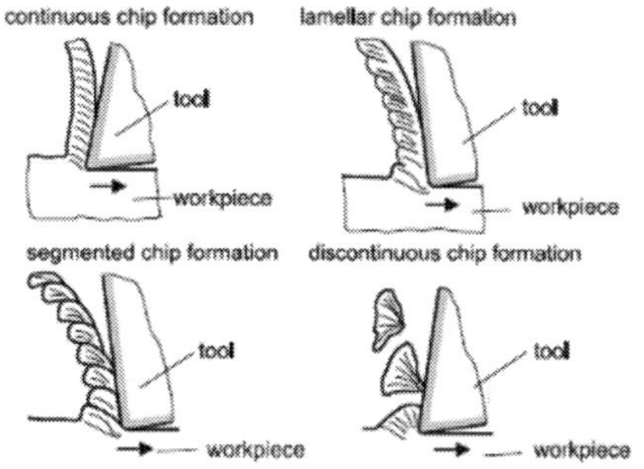

Fig. 3. Types of chips [5].

Heat is released during cutting in different environments such as: chip, tool and workpiece. In order to simplify the FE model, heat transferred to the environment is neglected.

Figure 4 presents the thermal transfer during an orthogonal cutting process. In the primary deformation zone ($Q_{primary\ deformation\ zone}$) (eq. (2)), heat is transferred to the chip ($Q_{deformation\ chip}$), and material ($Q_{deformation\ workpiece}$) that is going to be deformed.

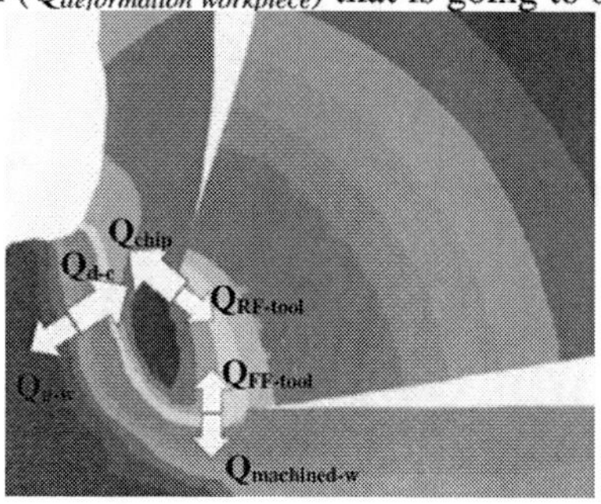

Fig. 4. Heat transfer during cutting.

In the case of the secondary deformation zone ($Q_{secondary\ deformation\ zone}$) (eq. (3)), heat is generated due to the friction between the chip (Q_{chip}) and the rake face of the cutting tool ($Q_{RF\text{-}tool}$). The last zone, where a significant amount of heat is relieved, is the third deformation zone ($Q_{third\ deformation\ zone}$) (eq. (4)), where the heat is emitted by the flank face tool ($Q_{FF\text{-}tool}$) respectively in the machined workpiece ($Q_{machined\text{-}workpiece}$).

$$Q_{primary\ deformation\ zone} = Q_{deformation\ workpiece} + Q_{deformation\ chip} \tag{2}$$

$$Q_{secondary\ deformation\ zone} = Q_{chip} + Q_{RF\text{-}tool} \tag{3}$$

$$Q_{third\ deformation\ zone} = Q_{machined\text{-}workpiece} + Q_{FF\text{-}tool} \tag{4}$$

3 Simulation and discussion

The simulations were performed using the Advant Edge software. The same macro geometry of the cutting tool was used in the orthogonal cutting, as well as the same radius, while the microgeometry K factors were set to the following values: 0.5; 1 and 1.4. All components of the simulation model were meshed with triangular elements of various sizes, with the mention that in the contact zone the elements were smaller with the tendency of

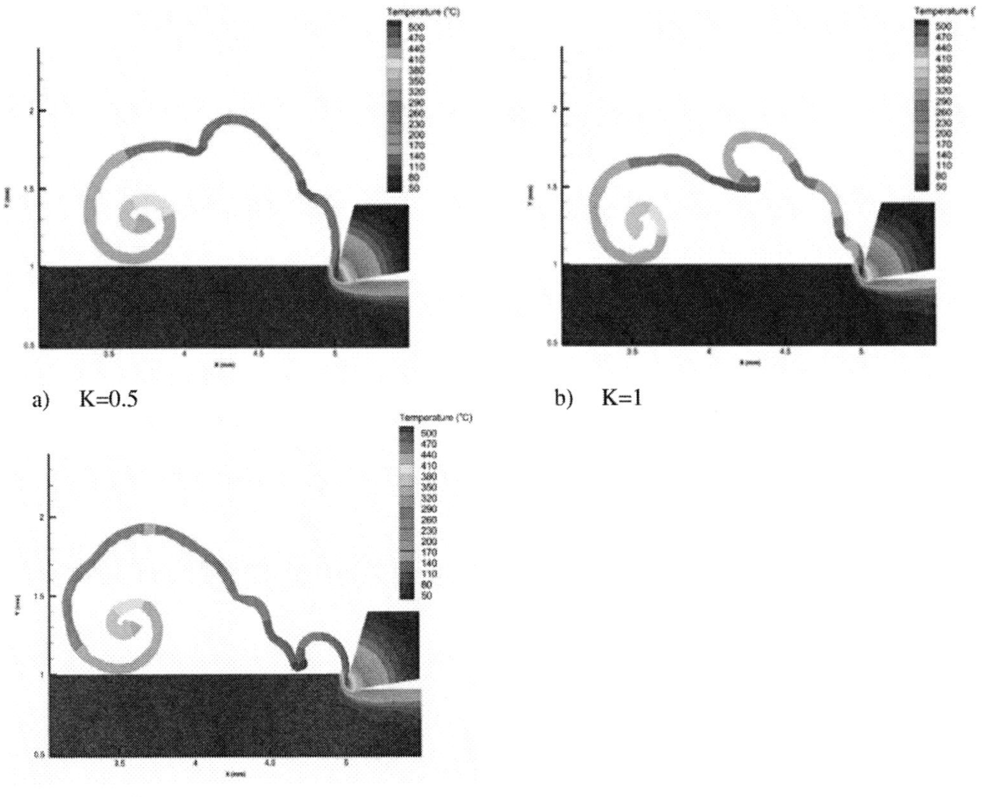

a) K=0.5

b) K=1

c) K=1.4

Fig. 5. Chip formation and temperature in orthogonal cutting.

increasing after losing contact with the tool. In case of the cutting tool made from carbide, with a minimum size of 0.03 mm and a maximum of 0.3 mm of the meshed element. C70 was used as machined material, with the size of the elements in the range of 0.02 mm and 0.1 mm. The cutting parameters of the simulation are as follows: v_c=80 m/min, t=0.1 mm, and an initial temperature of 20°C. The friction coefficient between the cutting tool and material is μ=0.5.

The chip formation was studied in the first stage. As one may notice in Figure 5a), in the case K=0.5, the chip is more compact than in the other two cases (K=1 and K=1.4).

In this case, after the chip makes the first loop, it begins to increase in size, not being restrained like in the beginning and at one point reaching the greatest height (continued chip) of all simulated chip formation. In some cases, this aspect can lead to difficulties in the evacuation of the chip (when machining the interior of cylindrical surfaces).

In the case of orthogonal cutting using K=1 (Figure 5b)), one may notice how the chip is fragmented. The chip presents the most fragmentations, facilitating its evacuation and a lack in the tendency of getting congested. The fragmentation in the case of symmetrical microgeometry is more pronounced than in the case of K=0.5, but similar to K=1.4. The last simulation, where K=1.4 (Figure 5c)) was used, shows how the chip increases a lot more than in the other two cases (K=0.5 and K=1). Unlike the other two simulations, one may notice that the fragmentation of the chip appears a lot later, but after that fragments are shorter.

Another important aspect of these simulations is heat dissipation in the chip, machined material and tool. A large amount of dissipated heat during the orthogonal cutting is

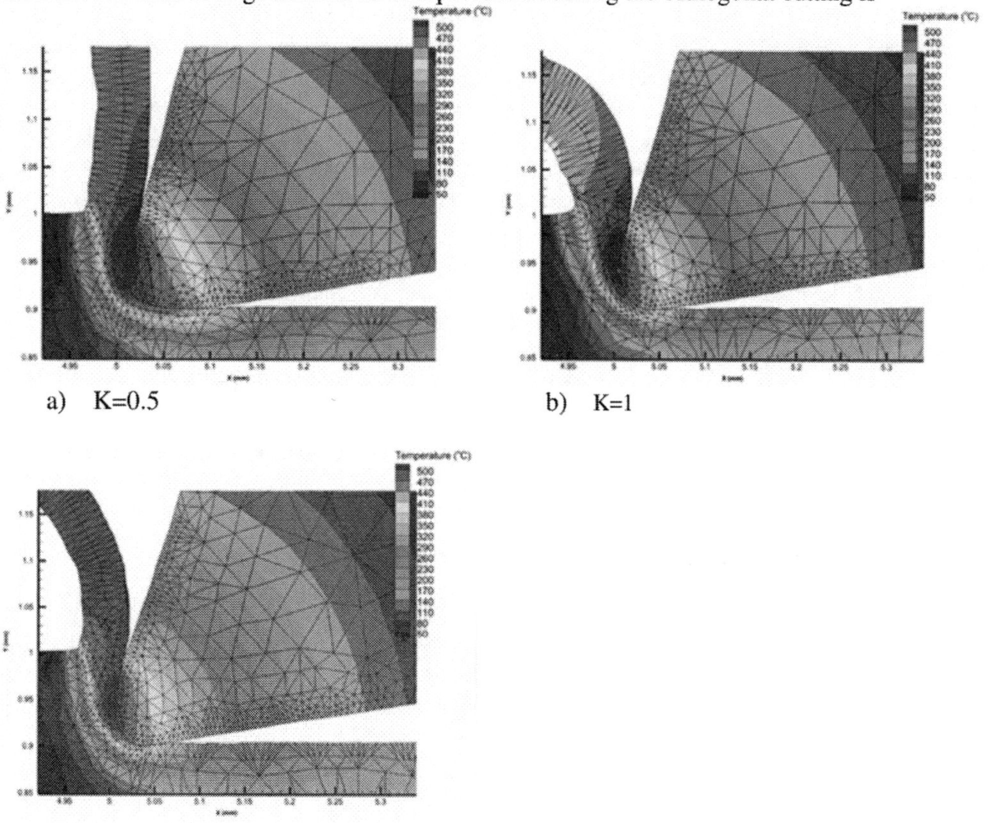

a) K=0.5 b) K=1

c) K=1.4

Fig. 6. Temperature distribution in orthogonal cutting.

transferred to the chip. The chip temperature in case of the simulation with a micro-geometry with K=0.5 is higher than in the simulations with K=1, reaching temperatures over 490°C in some areas. By comparison, one may notice in Figure 6 that in the case of cutting with K factor 0.5, the primary and secondary zone generates a higher heat quantity than in the case of utilizing a K factor 1.0. The cutting edge also heats up to temperatures of

500°C, but again one may notice that the heat released in the tool is higher in case of K factors of 1.0 and 1.4. The highest temperature is found on the rake face, confirming the hypothesis issued by Denkena in [5]. This situation can be explained by the fact that the friction of the chip on the rake face is severe due to the contact surface.

According to Figure 6, the deformation zones of the machined material can be differentiated. Case K=0.5 (Figure 6a)) has a smaller primary zone than K=1.0 (Figure 6b)), but the second and the third zone are clearly more extended - fact that leads to a greater amount of heat release. In the case of K=1.4 (Figure 6c)), the third deformation zone is smaller than in the other two cases, resulting in less heat being transferred to the part (chip and tool), despite the fact that the primary deformation zone generates more heat than in the case of K=1.0.

An advantage that can be observed in Figure 5 and Figure 6 is the reduced amount of heat transferred to the part. For the flank face, K factors of 1.0 and 1.4 represent an advantage; despite the fact these two factor levels cannot be really differentiated. Case K=0.5 is inferior from the point of view of thermal stresses induced in the cutting edge. The interrupted chips from case K=1.0 are due to the plastic deformations at lower temperatures than in the other two cases. Due to the high temperatures in the chip (case 0.5 and 1.4) the material becomes ductile and consequently the chip will be of lamellar type or even continued. The lowest temperature at the edge can be found at the cutting tool with K=1.4. During the machining process, the least amount of heat is released in the machined material. The highest temperature following the cutting process can be found in the simulation with K=0.5. In this case, the temperature is higher on the length of the machined workpiece but also in its depth. Here, a high quantity of heat is transmitted in the workpiece and the highest temperature is achieved. For the K=1.4, the temperature of the machined material is higher than in the case K=0.5, but lower than for K=1 (Figure 5).

In terms of the finite element mesh used in the simulations, one may notice how the meshing of the cutting tool does not modify. The elements of the cutting surface following the cutting process have the same shape. The only meshed elements which modify during the cutting process are the ones of the chip and the ones of the material that will be cut in the area of the edge. In all three simulations it can be observed how, in the area of the edge and in the secondary deformation zone, the meshed elements have the same dimensions, having the same variation of the size. In the primary deformation zone, the dimensions of the meshed elements do not vary much in the secondary deformation zone.

4 Conclusions

The research aimed at establishing an optimal microgeometry for machining the C70 material. One may conclude that the K factor is a very important element in the cutting process (but not only in this case). A K factor of 0.5 will lead to generating a large amount of heat and so, to a pronounced wear of the cutting tool both on the flank face and on the rake face. It is not recommended to the drilling processes because that does not determine the adequate deformation of the chip for obtaining a lamellar or segmented chip. The microgeometry with a K factor=1.4 determines the formation of chips with lamellar shape and tending towards a segmented shape, but having an advantage over the 0.5 micro-geometry, of generating a large amount of heat transmitted towards the chip and not towards the preform.

The K factor 1.0 is the most advantageous from the thermal point of view. The amount of heat is the lowest in this case. The chip takes the majority of the heat, but the types of the chips that are obtained are lamellar which can be segmented and so are easy to evacuate. The tendency of heat generation in the third area is small, so the flank face is protected against thermal stresses. The secondary area of deformation is the most reduced.

The future research will be focused on obtaining different types of microgeometries on different types of cutting tools in order to validate the theory that has been deduced form FE simulation and implementing it in the industry.

References

1. B. Denkena, J. Kohler J, M. Mesfin, *Prod. Eng.* **6** 329–338 (2012)
2. E. Segebade, F. Zanger, V. Schulze, Influence of different asymmetrical cutting edge microgeometries on surface integrity, *Procedia CIRP* **45** 11-14 (2016)
3. B. Denkena, D. Biermann, Cutting edge geometries, *CIRP Annals – Manuf. Tech.* **63** 631-653, (2014)
4. E. Bassett, J. Kohler, B. Denkena, On the honed cutting edge and its side effects during orthogonal turning operations of AISI1045 with coated WC-Co inserts. *CIRP J. of Manuf. Sc. and Tech.* **5** 108–126 (2012)
5. H.K. Toenshoff, B. Denkena, Basics of cutting and abrasive processes, *Springer-Verlag Berlin Heidelberg* 21-36 (2010)

DOI: 10.1051/matecconf/201711206012

Balancing the manufacturing lines through modelling and simulation using Tecnomatix Plant Simulation

Florin Blaga[1,*], *Iulian* Stanăşel[1], *Voichiţa* Hule[2], and *Alin* Pop[1]

[1]University of Oradea, Industrial Engineering Department, Universităţii 1, Oradea, Romania
[2]University of Oradea, Mechanical Engineering and Automotive, Universităţii 1, Oradea, Romania

Abstract. A frequently problem found in case of serial production is the balancing of flow lines. This paper shows how using modelling and simulation procedures can put in evidence the bottleneck in the manufacturing flow. These situations occur to the case where the corresponding times of the technological operations differ very much. Using the program TECNOMATIX PLANT SIMULATION can be highlighted those dysfunctions that may appear during of the manufacturing system operation. It is also possible to identify the solutions to allow equilibration of the lines of the manufacturing process. The program TECNOMATIX PLANT SIMULATION offers a variety of information useful for/in the decision makers in the management of the manufacturing process.

1 Introduction

An objective of management activities characteristic manufacturing system organized as manufacturing flow is their balance, and the removal of the bottleneck [1, 2]. Also the *TECNOMATIX PLANT SIMULATION* software allows the modeling and simulation of manufacturing systems. The program is oriented object, allowing easily the change and the maintenance of complex models. The model is built from a collection of objects (block) which interact with each other [3, 4].

To each object it can be assigned initial information (default settings). During the simulation these information is processed and, ultimately, can be used as data in decision-making processes [3].

The efficiency of *TECNOMATIX PLANT SIMULATION* software was put in evidence in the flow lines applications [5].

The paper presents the use of *TECNOMATIX PLANT SIMULATION* software for functional optimization of the manufacturing line of the sliding gear part type. The arrivals in the system and processing times are considered random variables.

The paper highlights the way in which can be identified the bottleneck places using modelling and simulation techniques, and how they can be removed.

Also in the paper are presented solutions for increasing productivity.

[*]Corresponding author: 1. florin_blaga2000@yahoo.com

2 Overview of the product

The part for which will be designed the manufacturing line is a sliding gear whose drawing is presented in Figure 1.

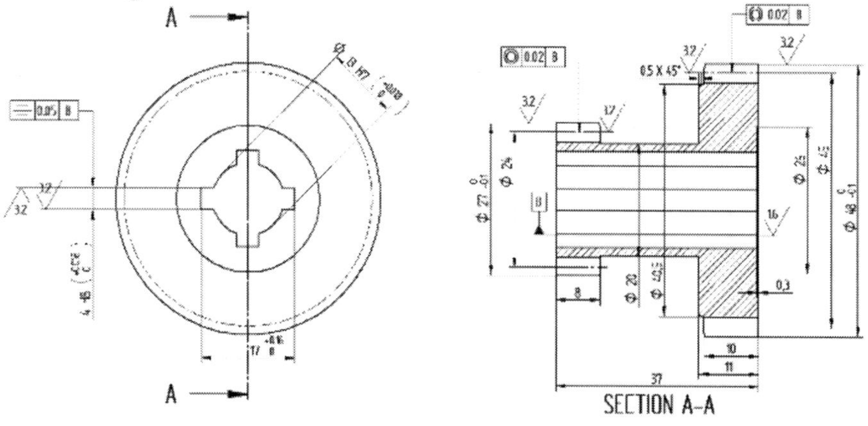

Fig. 1. The drawing of sliding gear part.

In Table 1 are presented the sequencing of operations for workpiece manufacturing

Table 1. The necessary operations of sliding gear manufacturing process.

Operation	Time (min)	Surface Speed (m/min)	Feed (mm/rot)	Workstation version 1	Workstation version 2
Face1 rough	21	40	0.25	F1_R	F1_R_v2
Face 1 finish	7	60	0.16	F1_F	F1_F
Center drill	1	15	0.12	CD	CD
Drill D11	7	15	012	D_11	D_11_v2
Finish ID 13	2	50	0.16	F_ID13	F_ID13
Rough OD_28	17	40	0.25	R_OD28	R_OD28_v2
Finish OD_27	6	60	0.16	F_OD27	F_OD27
Groove OD_20	14	30	0.1	GR_OD20	GR_OD20_v2
Cut Off	13	30	0.1	CUT	CUT_v2
Face2 rough	14	40	0.25	F2_R	F2_R_v2
Face 2 finish	4	60	0.16	F2_F	F2_F
Broaching straight sided spline	8	2	0.05 mm/tooth	BSSS	BSSS
Gear hobbing z30	40	40	3	GH_z30	GH_z30_v2
Gear shaping z16	28	4	0.07	GS_z16	GS_z16_v2
Gear teeth rounding z30	35	12	0.3	GTR_z30	GTR_z30_v2
Gear teeth rounding z30	30	12	0.3	GTR_z30p	GTR_z30p_v2
Gear shaving z30	32	35	0.2	GSv_z30	GSv_z30_v2
Gear shaving z16	26	35	0.2	GSv_z16	GSv_z16_v2
Gear running z30	25	-	-	GR_z30	GR_z30_v2
Gear running z16	25	-	-	GR_z16	GR_z16_v2
Gear control	26	-	-	GC	GC_v2
Grinding ID_13	20	-	-	G_ID13	G_ID13_v2
Final control	34	-	-	FC	FC_v2

process. To each operation it is associated the following technological parameters: name, time, surface speed, feed, workstation in the first model version, workstation in the second model version.

3 Modeling and simulation of manufacturing system activity

The model of manufacturing line of sliding gear is made by using the *TECNOMATIX PLANT SIMULATION* software. The first version of the model is characterized by the fact that to each operation of manufacturing process one block of *SingleProc* type. This is presented in Figure 2.

The arrival modeling of the blank in the system is realized through the type *Source* block (Figure 2). This block enables parameter setting the interval between two successive inputs (*Interval*). Considering this parameter a random variable, it will be associated a function of Poisson distribution [6]:

$$f(x)=\frac{\lambda^x}{x!}e^{-\lambda}\ ;\quad x=0,1,2,... \tag{1}$$

In our case x is a random variable (the time between two inputs) and λ is the medium value of the random variable. In the analyzed case Stream has the value 1 and λ=300 seconds (Figure 2.a).

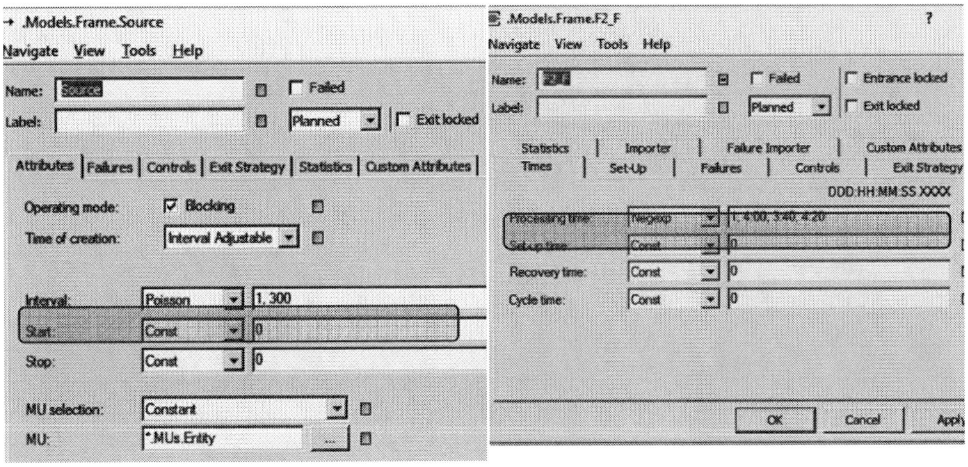

(a) (b)

Fig. 2. Version 1 of manufacturing line model for the sliding gear production.

Regarding the workstation, the most important issue is the processing time change. When the processing time is a random variable, a density of probability will be associated to this one (probability function). In many cases the density of probability of processing time is negative exponential function [6] like this:

$$f(t)=\beta e^{-\beta t} \tag{2}$$

where, t is the random variable (processing time) and β is the average value of processing time. In the case of *Face 2 finish* (*F2_F*) operation, *Stream* has the value of 4 minutes, *Lower Boud* has value of 3 minutes and 40 seconds, *Upper Bound* has the value of 4 minutes and 20 seconds (Figure 2.b).

In the same way the processing times are defined in the other operations. Considering this manufacturing time like random variables is an approach much more closely to what is happening in the real manufacturing system.

After the model is made, for passing to the next step that is simulation, it has to be defining the simulation process parameters. This can be done thought the *EventController* block (Figure 3). So, the time interval in witch will be simulate the manufacturing system activity is 8 hours (Figure 3).

After the simulation is made, using the *BottleneckAnalyzer* option, it can be showed the status that every workstation and their weight related to the reference time of 8 hours. It can be seen the work period (green colour), the waiting period (gray colour). During the 8 hour interval, the workstations can be found in pause (blue colour), can be broken (red colour), or they are not provided in the working program (blue colour). Bottlenecks that may occur in the operation of workstations are represented with yellow colour.

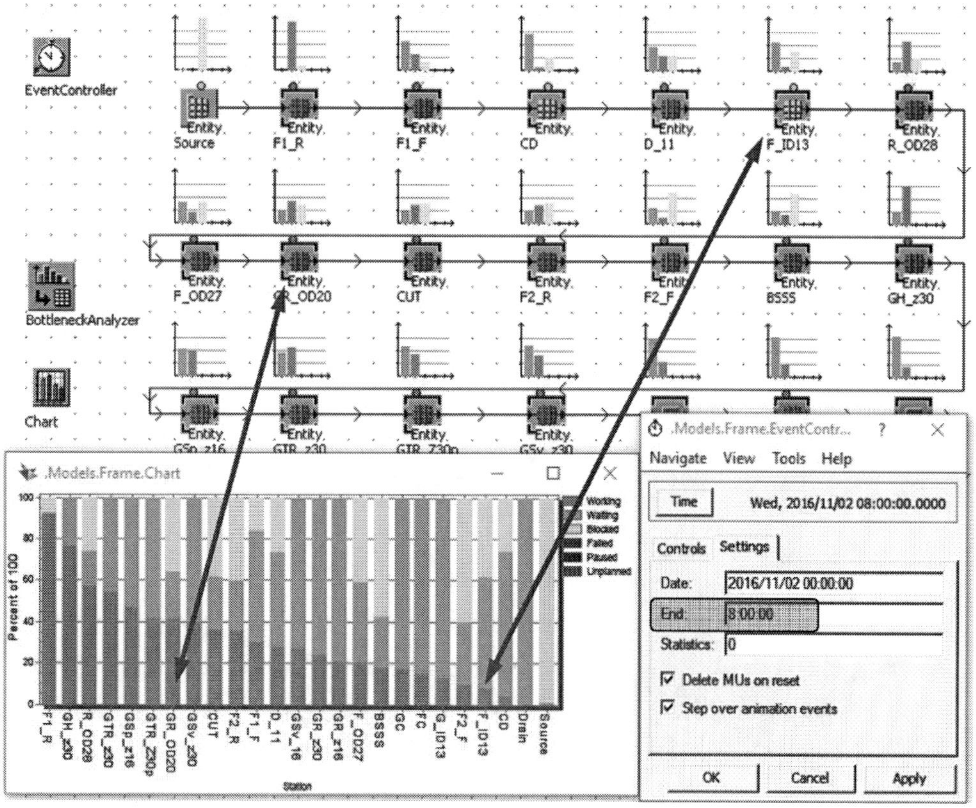

Fig. 3. The simulation system activity result. Version 1.

Through the *Chart* options it's generate a diagram (Figure 3) that shows the system components status during the simulation process (8 hours). In this case it can be seen the bottleneck frequency on the manufacturing flow. The reason of these bottlenecks is the difference between the processing times of different operations.

After the simulation, the performance indicators of every workstation can be analysed. In the case of *Frame block* associated to *Finish ID 13* (*F_ID13*) operation, by selecting the

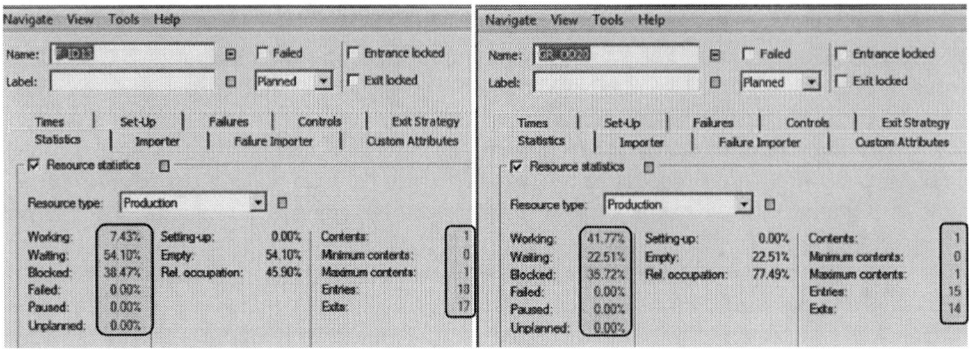

(a) (b)

Fig. 4. The system simulation results. Version 1: (a)- *F_ID13*; (b)- *GR_OD20*.

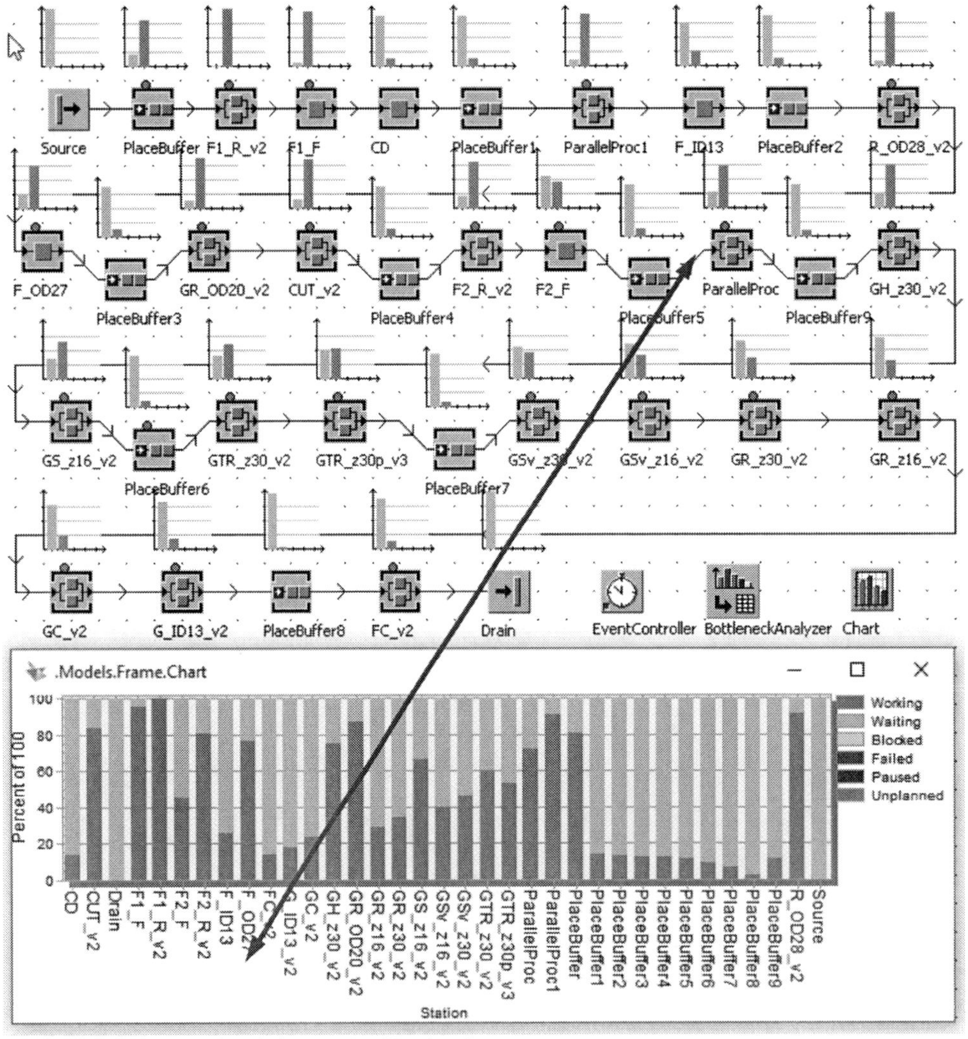

Fig. 5. The system simulation results. Version 2.

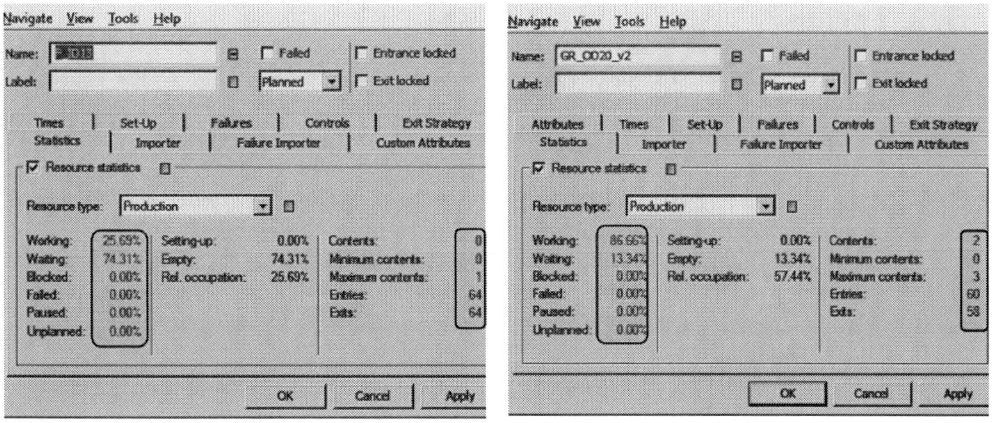

(a) (b)

Fig. 6. Simulation results. Version 3: (a)- *F_ID13*; (b)- *GR_OD20_v3.*

Statistics button, it can be seen that during the 8 hours, the workstation is working 7.43% of total time, 54.10%it's waiting, 38.47% it's blocked (Figure 4.a).

In the case of *Groove OD_20* (*GR_OD20*) operation it can be seen that the workstation is working 41.77% from total time of 8 hours, 22.51% was waiting, 35.72 % is blocked (Figure 4.b).

To ensure a continuous manufacturing flow and increase in productivity it will adopt two solutions: involves buffers installation between the workstation with smaller processing time and those with processing time bigger and allocated more workstation for operations with processing time bigger than the operation time that precede them.

In the Table 1 are found the workstations name for the version 2 of model. In the version 2 of model it was used the *ParallelProc* type frame (Figure 5).

The second version of model highlights the significant increase in productivity. In this way on the *F_ID13* operation is made 64 parts (Figure 6.a), and the *GR_OD20_v2* operation are made 58 parts (Figure 6.b).

4 Conclusions

In the conditions of serial production can appear imbalances during manufacturing system operations, organize as flow lines. These imbalances are caused by differences in the duration of the various operations. In these conditions can occur bottlenecks and increase waiting times. Such disturbances can be highlighted through modeling and simulation.

Modelling and simulation of manufacturing systems using *TECNOMATIX PLANT SIMULATION*, offers the possibility to evaluate their performance for various production tasks (orders).

The different model versions that may be associated with a system that can provide useful information decision makers, responsible for management activities.

The case of study presented in the paper showed a gradual improvement of production system performances. Version 2 of model allowed bottlenecks removing and highlight a significant increase in productivity.

DOI: 10.1051/matecconf/201711206012

References

1. I. Abrudan, D. Cândea, *The engineering and management of manufacturing systems* (in romanian), (Ed. Dacia, Cluj Napoca, 2002)
2. E. E. Adam, R. Ebert, *The management of production and operations*(in romanian),(Ed. Teora, Bucureşti, 2001)
3. M. Iosip, E. Oprea, D. Boricean, *The achievement of digital product manufacturing using virtual prototype* (in romanian), (Ed. Qual Media, Cluj-Napoca 2010)
4. S. Bangsow, *Manufacturing Simulation with Plant Simulation and SimTalk- Usage and Programming with Examples and Solutions*, Springer-Verlag, Berlin Heidelberg, (2010)
5. F. Blaga, C. Bungău, V. Hule, *R. Man. Ec. Eng.*, **12**, 25 2013)
6. I. Abrudan, *Flexible manufacturing systems. Concepts of design and management* (in romanian), Ed. Dacia, Cluj- Napoca, (1996)

MATEC Web of Conferences **112**, 06013 (2017) DOI: 10.1051/matecconf/201711206013
IManE&E 2017

Using noncircular gears for the unloading door kinematics modification

Mircea Niculescu*, and *Laurenţia* Andrei
"Dunărea de Jos" University of Galaţi, Faculty of Engineering, Department of Mechanical Engineering, str. Domnească nr. 111, 800201, Romania

Abstract. The paper introduces a new kinematics of the billet reheating furnace unloading door, by using a noncircular gear train in the drive mechanism. The main objective of the kinematics modification is to reduce the heat loss by increasing the unloading door speed, during the stroke, while kept it at the conventional value at the beginning and end of stroke. In order to modify the door kinematics, the drive mechanism output motion is controlled by noncircular gears whose transmission ratio is defined as a hybrid multiparameters function. A study is developed to analyse the influence of the defining parameters on the variation of the gears transmission ratio and on the discharging door speed, respectively. Based on the analytical study, a certain law of variation was chosen to describe the mechanism kinematics; in accordance to the kinematics variation, the mating noncircular centrodes/pitch curves were modelled and the gears teeth are generated, considering constant circular pitch and constant pressure angle, respectively.

1 Introduction

Due to nowadays ease of design and manufacture, compactness and low costs, noncircular gears have become an efficient solution for the driving mechanisms of closing/regulation industrial applications, where velocities are subjected to variations or motions of driven elements should obey imposed trajectories [1]. For example, Y. Takahiro proposes a vehicle window regulator that varies the window speed, increasing it as the window gets near to its fully closed position [2]. The regulator is based on the meshing of two noncircular gears of circular eccentric, elliptical and logarithmically helical shapes. J. P Nolta proposes a noncircular gears pair defined by a transmission ratio that varies from a linear function to a non-linear one, proportional to an exponential function, in order to influence the closing element of the motor vehicle climate control system [3]. G. Vonhausen proposes noncircular gears as driving mechanism for an actuation valve, in order to adjust the liquid flow [4], while H.C Buchanan uses noncircular gears to adjust the windshield wiper speed [5].

The authors of this paper are taking into account the usage of noncircular gears for the

*Corresponding author: mircea.niculescu@yahoo.com

furnace discharging door speed with the aim of reducing the heat loss during the specific stroke.

2 Modified kinematics of the unloading door drive mechanism

In hot-rolling process, the flow first operation is the billets heating, at 1250°C temperature, into specialised furnace. The furnace is successively fed with billets, one by one, by means of Push Machine. The billets are discharged through the front side of the furnace, at a rate of one piece at every 2-3 minutes; every time, the discharge door is opened, in order to enable the fall of billets on the roller table, and then is closed. Taking into account that the billets discharging period is shorter (1-2 seconds) than the time required for the door opening and closing process (12 seconds), a large amount of heat is lost through the open door space. In order to reduce this energy loss, the door kinematics could be varied by introducing a noncircular gears train within the door driving mechanism; the authors intend to keep at constant designed value the door velocity while opening and closing the door and to increase the door velocity during the stroke.

2.1 The analysis of the actual door mechanism kinematics

The discharge door of the billets heating furnace is driven by an asynchronous motor (1) and a mechanism consisting of a worm gear (2) and a driving chain (4), as shown in Figure 1 [6]. The door (8) is lifted by means of chains (7) that are wound on the wheels (6).The total transmission ratio is 140 assures a door lifting velocity of 0.08 m/s. This velocity is required when the furnace door is opened and closed, in order to avoid the mechanical impacts on the furnace frontal structure. To open the door, the motor shaft is clockwise rotated; to close the door, the motor shaft rotational motion is reversed. The door stroke is electrically limited by a controller.

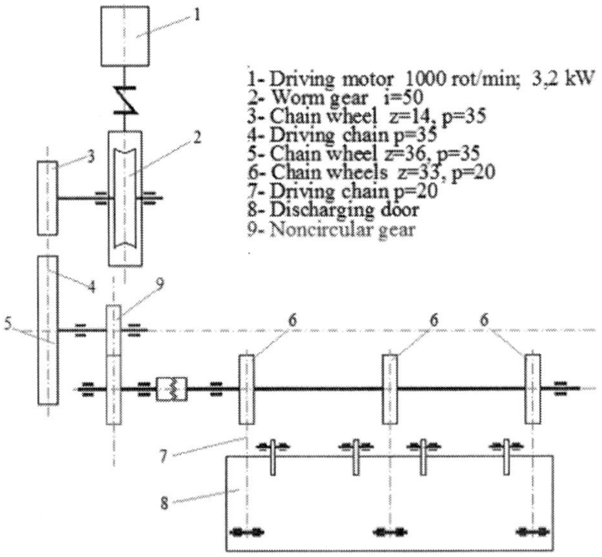

Fig. 1. The modified driving mechanisms of the discharging door, by inserting noncircular gears.

2.2 The modified kinematics of the door driving mechanism

In order to increase the door angular velocity, a noncircular gears train is placed within the door driving mechanisms, between the driving chain (4) and the driving shaft of the wheels (6), as shown in Figure 1. The noncircular gears (9) perform less than one complete rotation as long as the total rotational angle of the driven gear should be $\theta_{2t} = 308.3°$ (corresponding to the opening door angle of 45°). The gears geometry consists of a combination of circular and noncircular areas, in correspondence to the desirable transmission ratio variation.

In order to achieve smooth opening and closing motion for the discharging door, the transmission ratio variation is divided into three phases: constant conventional value of 1:1 at the beginning of stroke, nonlinear increase to a maximum value during the opening process and constant value again, at the high rate of 2:1, till the discharging door is completely opened. Reversing de motor shaft rotational motion, the door is closed by a high constant speed for a while, and then the speed is gradually reduced to the conventional value and kept at this value while the door is clos

2.3 The influence of noncircular gears defining parameters, on the door mechanism kinematics

In order to modify the kinematics of the door driving mechanism, the noncircular gear transmission ratio is defined as a multiparameters hybrid function:

$$i_{21}(\varphi_1) = \begin{cases} 1, & \varphi_1 \in [0, \varphi_u] \\ \dfrac{i_{max}+1}{2} + \dfrac{1-i_{max}}{2} \cdot \sin\left(\dfrac{\pi}{\varphi_v - \varphi_u} \cdot \varphi_1 + \dfrac{\pi(\varphi_v - \varphi_u)}{2(\varphi_v - \varphi_u)}\right), & \varphi_1 \in [\varphi_u, \varphi_v] \\ i_{max}, & \varphi_1 \in [\varphi_v, \varphi_{1t}] \end{cases} \quad (1)$$

where i_{max} is the maximum transmission ratio achieved during the door stroke; φ_u – the driving gear rotational angle while the motion is transmitted by the constant conventional ratio (1:1); φ_v – the driving gear rotational angle where the variable motion turns into an uniform one; φ_{1t} – the driving gear rotational angle when the door is closed, corresponding to the defined angle of the driven gear $\varphi_{2t} = 308.3°$. The variation of the transmission ratio is chosen so that a continuous, positive and derivative function is defined.

The rotation angle of the driven gear is expressed by [1]:

$$\varphi_2(\varphi_1) = \int_0^{\varphi_1} i_{21}(\varphi_1) d\varphi \quad (2)$$

Imposing that $\varphi_2 > 0$, $\varphi_{2t} = 308.3°$ and considering the variation of the gear transmission ratio from equation (1), the total rotation angle of the driving gear could be calculated by equation (3)

$$\varphi_{1t} = \frac{\varphi_{2t}}{i_{max}} + \frac{i_{max} - 1}{2 i_{max}} \cdot (\varphi_u + \varphi_v) \quad (3)$$

Figure 2 illustrates the results of an initial study that is developed to analyse the influence of the i_{max}, φ_u and φ_v defining parameters on the transmission ratio variation. As noticed in Figure 2, the transmission ratio variation, from $i_{min} = 1$ to its maximum value, depends on the angular difference φ_v - φ_u; increasing the angular difference, as shown in Figures 2a, b, the transmission ratio variation gets smoother, the slope and the door acceleration being reduced.

Also, the increase of the maximum transmission ratio, as seen in Figure 2c, increases the driving gear total rotation angle; this increase is not an issue for the mechanism

DOI: 10.1051/matecconf/201711206013

kinematics, taking into account that the driving gear doesn't perform a full rotation; Choosing a higher ratio, i_{max}, the technological conditions are improved as the time when the door is open and the heat is lost, respectively, are reduced.

Figure 3 illustrates the influence of the transmission ratio defining parameters on the discharging door speed. The parameters φ_v and φ_u, especially their difference φ_v - φ_u, have a similar influence on the door velocity, v_u, as they have on the transmission ratio. The door velocity was calculated based on the angular velocity of the driving motor and considering the total value of the transmission ratio from the driving mechanism shown in Figure 1.

Considering the above analysis, the following initial parameters were chosen: $\varphi_u = 10^\circ$, $\varphi_v \in [60^\circ, 90^\circ]$ and $i_{max} = 2$.

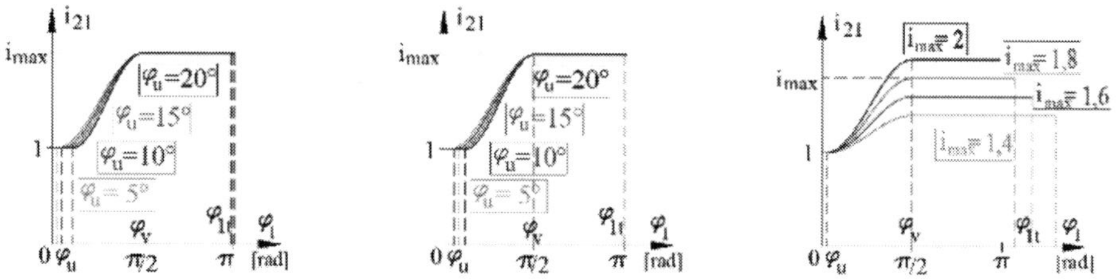

a) i_{max}=2; φ_u=5° b) i_{max} = 3; φ_v = 90° c) φ_u=5° ; φ_v=90°

Fig. 2. The influence of the transmission ratio defining parameters φ_u (a), φ_v (b) and i_{max} (c) on the transmission ratio variation.

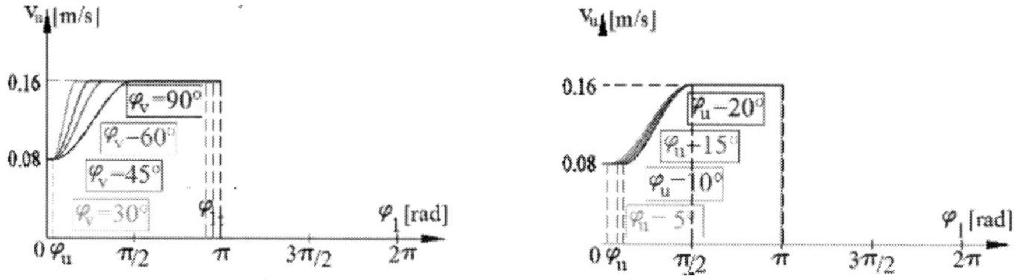

a) i_{max}= 2 ; φ_u= 5° b) i_{max}= 2 ; φ_v= 90°

Fig. 3. The influence of the transmission ratio defining parameters φ_u (a), φ_v (b) and i_{max} (c) on the door lifting velocity.

3 Non-circular gear generation

The main steps required within the noncircular gears generation process are (i) the modelling of the mating noncircular centrodes/pitch curves and (ii) the teeth generation.

3.1 Noncircular pitch curves modelling

The noncircular gears pitch curves that assure the transmission ratio defined by the equation (1), have the following polar equations [1], [7]:

$$r_1(\varphi_1) = \frac{D}{1 + i_{21}(\varphi_1)} \tag{4}$$

$$r_2(\varphi_2(\varphi_1)) = D\frac{i_{21}(\varphi_1)}{1 + i_{21}(\varphi_1)} \tag{5}$$

where φ_1 is the driving centrode polar angle, uniformly increased; φ_2 – the driven centrode polar angle, varying in accordance with equation (2); *D*- the gears center distance.

Based on the equations (4) and (5), two pairs of noncircular pitch curves are modelled for the initial data: the functioning cycle dividing angles φ_u = 10°, φ_v = 60° and 90° respectively, the transmission ratio is varied within the limits i_{min}= 1 and i_{max}= 2, and the gear center distance is D = 250 mm. The pitch curves illustrated in Figure 4 are modelled in AutoCAD environment, using an AutoLISP code.

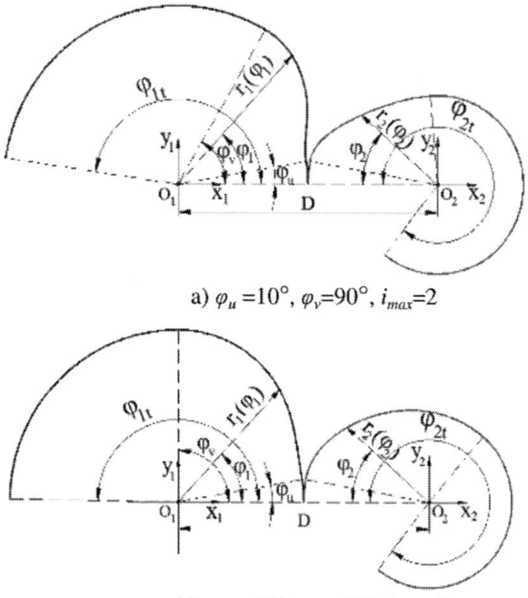

a) φ_u =10°, φ_v=90°, i_{max}=2

b) φ_u =10°, φ_v=60°, i_{max}=2

Fig. 4. Pitch curves of noncircular gears.

As illustrated, the chosen transmission ratio parameters leads to open noncircular centrodes, with favourable geometry, consisting of circular arcs and convex noncircular arcs, with a reduced area. The driving centrode shape exhibits a concave zone whose curvature has shown to be permissive for the further tooth flanks generation.

3.2 Gear teeth generation

For the gear teeth flanks generation, an analytical procedure is applied considering the local geometry of the noncircular pitch curve and the profile of a "single tooth rack cutter" that rolls over the pitch curve. Following the theory presented in [7], the coordinates of the tooth flank points are expressed by equation (6)

$$\begin{aligned} x_{1ij} &= r_1(\varphi_{1ij}) \cdot \cos\varphi_{1ij} \pm s_{ij} \cdot \cos\alpha \cdot \cos(\mu_{ij} + \alpha + \varphi_{1ij}) \\ y_{1ij} &= r_1(\varphi_{1ij}) \cdot \sin\varphi_{1ij} \pm s_{ij} \cdot \cos\alpha \cdot \sin(\mu_{ij} + \alpha + \varphi_{1ij}) \end{aligned} \tag{6}$$

where r_{1ij}, φ_{1ij} are the polar coordinates of the instantaneous center of rotation; s_{ij} – the rolling distance along the current tangent line; μ_{ij} – the angle of the current tangent relative to the positioning vector; α - the standard pressure angle (20°); (-) sign is for the gear tooth addendum and (+) sign is for the deddendum segment, when the rolling is counter clockwise performed [7]. Similar equations are written for the driving gear tooth opposite flank points.

 DOI: 10.1051/matecconf/201711206013

Based on the driving gear tooth flanks profiles, the driven gear tooth conjugate profiles are analytically expressed considering the gears meshing [7]. Using the coordinate transformation, the driven gear conjugate tooth flanks are expressed, in its fixed coordinate system, by equation (7), where φ_{2ij} is the driven gear rotational angle, defined by equation (2).

$$\begin{bmatrix} x_{2ij} \\ y_{2ij} \end{bmatrix} = \begin{bmatrix} \cos\varphi_{2ij} & \sin\varphi_{2ij} \\ -\sin\varphi_{2ij} & \cos\varphi_{2ij} \end{bmatrix} \cdot \left(\begin{bmatrix} -D \\ 0 \end{bmatrix} + \begin{bmatrix} \cos\varphi_{1ij} & \sin\varphi_{1ij} \\ -\sin\varphi_{1ij} & \cos\varphi_{1ij} \end{bmatrix} \cdot \begin{bmatrix} x_{1ij} \\ y_{1ij} \end{bmatrix} \right) \tag{7}$$

AutoLISP codes automatically generated the noncircular gears teeth flanks, as illustrated in Figure 5. Constant circular pitch and pressure angle have been considered during the teeth generation. In Figure 6 the noncircular gears sections are completed by additional editing operations, i.e. the teeth addendum and dedendum were considered at *m* and 1,25*m*, respectively (*m* – the gear modulus), and the tooth root is filleted by a 0,25*m* radius.

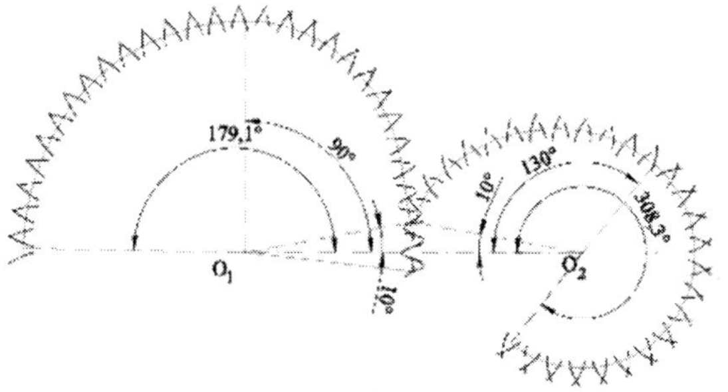

a) φ_u =10°, φ_v =90°, i_{max} =2

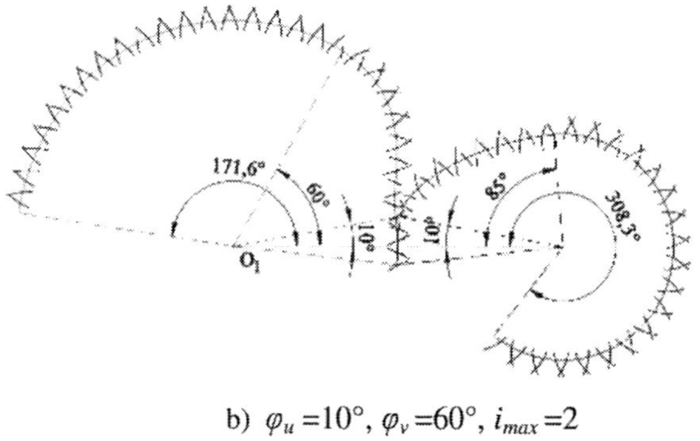

b) φ_u =10°, φ_v =60°, i_{max} =2

Fig. 5. The teeth flanks generation of the noncircular gear.

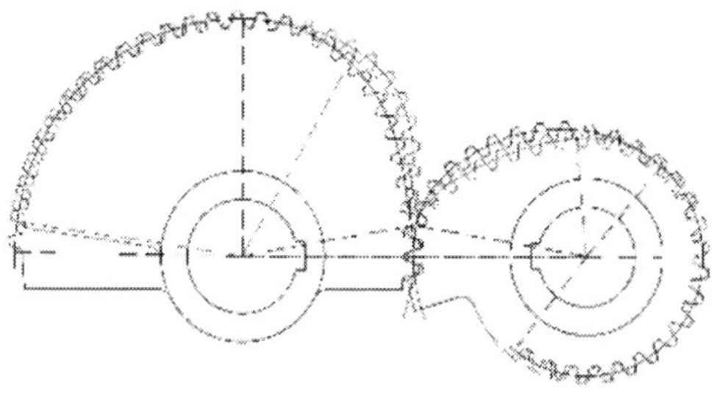

Fig. 6. Comparison of the noncircular gears, in both designed cases.

4 Conclusions

The paper presents a modified kinematics for the unloading door drive mechanism of the billets reheating furnace, in order to reduce the heat loss; the modified kinematics is achieved from a noncircular gear train whose transmission ratio is defined as a hybrid function with multiple parameters; the flexibility of the gears design is therefore increased and enabled to consider the technological requirements. Analysing the influence of the transmission ratio parameters on the door relative velocity, proper values were chosen and introduced into the further noncircular gear design. Original AutoLISP codes were used for the analytical studies and geometric modelling of the noncircular gears.

References

1. F.L. Litvin, *Noncircular Gears Design and Generation*, Cambridge University Press (2009)
2. Y. Takahiro, et al., United States Patent 4998379 Mar 12 (1991)
3. J. P. Nolta, United States Patent US 2014/0000397 A1 (2014)
4. D.Vonhausen, United States Patent 4638977 (1987)
5. H.C. Buchanan, et al., United States Patent 4543839 Oct 1 (1985)
6. IPROLAM project. no 505.30-012 (1977)
7. M. Niculescu, L. Andrei, 7th International Conference on Advanced Concepts in Mechanical Engineering, Iasi (2016)

On the vertical interaction between a three-axle bogie and track

Traian Mazilu[1,*], and *Marius* Gheţi[2]
[1]University Politehnica of Bucharest, Department of Railway Vehicles, 313 Splaiul Independenţei, Bucharest, Romania
[2]Alstom Transport, 32 Răzoare, Bucureşti, Romania

Abstract. In this paper, the 'moving wheel' model is used aiming to analysis new aspects of the interaction between three-axle bogie and track due to the harmonic roughness of the rail. The three-axle bogie of a locomotive is considered and its model is reduced to three loaded wheels running along a rail supported on a continuous elastic foundation. The track model consists of a Euler-Bernoulli beam resting on elastic foundation including two elastic layers (Winkler model) representing the rail pad and ballast and one inertial layer for sleepers. The rail deflection due to the moving dead loads of the wheels is calculated and the frequency response functions of the wheels, rail in the contact points and contact forces are analysed, including the three resonance frequencies.

1 Introduction

The small-scale undulations (roughness) of both rolling surfaces - the wheel and the rail - are among the common causes of the wheel/rail vertical vibration. The study of the vibration generated by a wheel while rolling on a rail in the presence of the small-scale undulations is critical in predicting the short-pitch rail corrugation [1] and the rolling noise [2].

Usually, the wheel-rail vibration behaviour is studied using a model in which only a single wheel is present [3]. This is not the case in practice, where multiple wheels roll on the rail [4, 5].

The issue of the wheel/rail interaction may be solved in two different ways - using either the model of a 'moving irregularity' between a stationary wheel and rail, or the model of a 'moving wheel' along the rail. The model of the moving irregularity is much easier to use and it is currently utilized for the frequency-domain analysis. However, the 'moving wheel' model is more realistic but requires much more effort to apply [6, 7].

This paper deals with the forced vertical vibrations of a three-axle bogie running along a track with small-scale undulations aiming to point out some of main features of the bogie/track interaction. To this end, the 'moving wheel' approach is applied for a simplified bogie/track model due to the symmetry. Equations of motion are solved using the Green's function method and the rail deflection under moving dead loads and frequency response

*Corresponding author: trmazilu@yahoo.com

functions are calculated and analysed. It is shown that the three-axle bogie/track system exhibits three resonance frequencies, but these emerge in one due to the damping.

2 Equations of motion

The mechanical model applied in this section is presented in Figure 1, where it can be seen three loaded wheels of mass M uniform moving with the velocity V along a continuous elastically supported rail. The three loaded wheels represent the model of a three-axle bogie, while the continuous elastically supported rail is idealised representation of a track, both considered as symmetric structure. Each wheel is considered rigid body and is loaded with the force Q_0. The distance between the front wheel and the second wheel is a_{12} and the distance between the second wheel and the rear one is a_{23}, respectively. Wheels are only coupled through the rail and the coupling via suspension can be neglected when the frequency is higher than the natural frequency of the bogie.

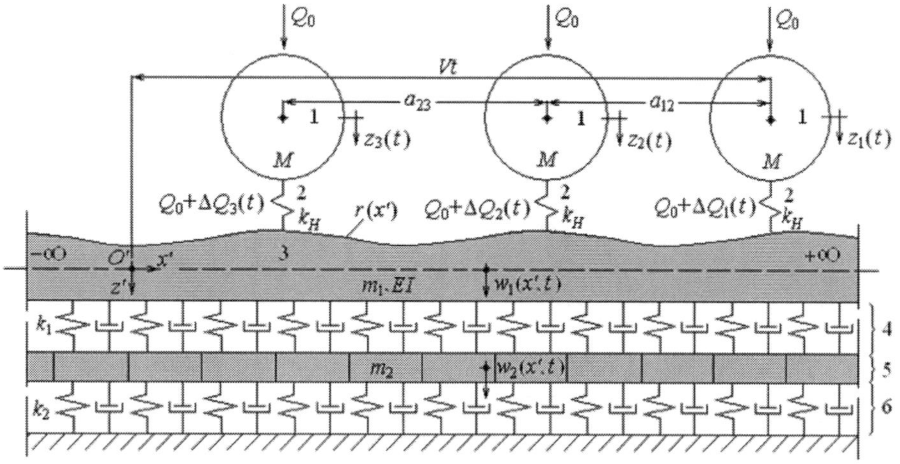

Fig. 1. Mechanical model: 1. wheel; 2. wheel-rail contact; 3. rail; 4. railpad; 5. sleeper; 6. ballast.

The rail is modelled as an infinite uniform Euler-Bernoulli which is continuous supported on two elastic layers, representing the viscoelastic characteristics of both railpad and ballast. The sleeper's inertia is modelled using an intermediate inertial layer between the railpad layer and the ballast layer. Practically, following this manner, the influence of the equidistant support given by sleepers is neglected; this approach can be applied up to 6-700 Hz [3].

The parameters for the track model are: EI – the bending stiffness of the rail, m_1 – the rail mass per unit length, k_1 – the railpad stiffness per unit length, c_1 – the damping constant per unit length of the railpad, m_2 – the sleeper mass per unit length, k_2 – the ballast stiffness per unit length and c_2 – the damping constant per unit length of the ballast.

Vertical displacement of the i wheel is $z_i(t)$ with i = 1 to 3, the rail deflection is $w_1(x',t)$ and the displacement of the sleeper layer is $w_2(x',t)$. The interaction between the wheels and the rail is the result of the presence of the roughness $r(x)$ on the rail surface. Considering that the dynamic component of the contact force ΔQ_i is much smaller than the static load Q_0, the dynamic of the wheels/rail contact is in the limits of the linear model.

Applying the mechanical laws, we can write the equations of motion:

- for wheels

$$M\ddot{z}_i = -\Delta Q_i \tag{1}$$

- for the track

$$EI\frac{\partial^4 w_1}{\partial x'^4}+m_1\frac{\partial^2 w_1}{\partial t^2}+c_1\left(\frac{\partial w_1}{\partial t}-\frac{\partial w_2}{\partial t}\right)+k_1(w_1-w_2)=\sum_{i=1}^{3}(Q_0+\Delta Q_i)\delta(x'-Vt-a_i)$$
$$m_2\frac{\partial^2 w_2}{\partial t^2}+c_1\left(\frac{\partial w_2}{\partial t}-\frac{\partial w_1}{\partial t}\right)+c_2\frac{\partial w_2}{\partial t}+k_1(w_1-w_2)+k_2 w_2=0 \quad (2)$$

- for the wheels/rail contact

$$z_i-w_1(Vt+a_i)-r(Vt+a_i)=\frac{\Delta Q_i}{k_H},\ \text{i=1 to 3}, \quad (3)$$

where $a_1 = 0$, $a_2 = a_{12}$, $a_3 = a_{12} + a_{23}$ and k_H – the stiffness of the wheel/rail contact.
New coordinate is introduced

$$x'=Vt+x, \quad (4)$$

to solve the equations of motion. Subsequently, the equations of motion of for the track are transformed by applying the following transformations

$$\frac{\partial^n}{\partial x'^n}\to\frac{\partial^n}{\partial x^n},\ \frac{\partial^n}{\partial t^n}\to\left(\frac{\partial}{\partial t}-V\frac{\partial}{\partial x}\right)^n. \quad (5)$$

It obtains

$$EI\frac{\partial^4 w_1}{\partial x^4}+m_1\left(\frac{\partial^2 w_1}{\partial t^2}-2V\frac{\partial^2 w_1}{\partial t\partial x}+V^2\frac{\partial^2 w_1}{\partial x^2}\right)+c_1\left(\frac{\partial w_1}{\partial t}-V\frac{\partial w_1}{\partial x}-\frac{\partial w_2}{\partial t}+V\frac{\partial w_2}{\partial x}\right)+$$
$$+k_1(w_1-w_2)=\sum_{i=1}^{3}(Q_0+\Delta Q_i)\delta(x-a_i)$$
$$m_2\left(\frac{\partial^2 w_2}{\partial t^2}-2V\frac{\partial^2 w_2}{\partial t\partial x}+V^2\frac{\partial^2 w_2}{\partial x^2}\right)+c_1\left(\frac{\partial w_2}{\partial t}-V\frac{\partial w_2}{\partial x}-\frac{\partial w_1}{\partial t}+V\frac{\partial w_1}{\partial x}\right)+ \quad (6)$$
$$+c_2\left(\frac{\partial w_2}{\partial t}-V\frac{\partial w_2}{\partial t}\right)+k_1(w_1-w_2)+k_2 w_2=0.$$

Also, the contact equation becomes

$$z_i-w(a_i)-r(a_i)=\frac{\Delta Q_i}{k_H}, \quad (7)$$

Considering the linearity of the mechanical model, the issue of the wheels/rail interaction can be divided in two: the stationary response and steady-state harmonic behaviour. To solve these issues the Green's function method is applied [6].

First, we occupy by the steady-state harmonic behaviour if all variables are harmonic

$$\overline{w}_{1,2}(x,t)=\overline{W}_{1,2}(x)e^{i\omega t},\ \overline{z}_i(t)=\overline{Z}_i e^{i\omega t},\ \Delta\overline{Q}_i(t)=\Delta\overline{Q}_i e^{i\omega t}\ \overline{r}_i(t)=r_o e^{i\omega(t-a_i/V)}. \quad (8)$$

Inserting these variables in equations of motion for the track (6), it reads

$$\mathbf{L}_{x,\omega}\mathbf{p} = \mathbf{q}\,, \tag{9}$$

where

$$\mathbf{L}_{x,\omega} = \begin{bmatrix} L_{11} & L_{12} \\ L_{21} & L_{22} \end{bmatrix}, \mathbf{p} = \begin{bmatrix} \overline{W}_1 & \overline{W}_2 \end{bmatrix}^{\mathrm{T}}, \ \mathbf{q} = \begin{bmatrix} \sum_{i=1}^{3} \Delta Q_i \delta(x - a_i) & 0 \end{bmatrix}^{\mathrm{T}}$$

with

$$L_{11} = EI\frac{\mathrm{d}^4}{\mathrm{d}x^4} + m_1 V^2 \frac{\mathrm{d}^2}{\mathrm{d}x^2} - V(c_1 + 2\omega m_1 \mathrm{i})\frac{\mathrm{d}}{\mathrm{d}x} + k_1 - \omega^2 m_1 + c\omega \mathrm{i}$$

$$L_{12} = L_{21} = c_1 V\frac{\mathrm{d}}{\mathrm{d}x} - k_1 - c\omega\mathrm{i}$$

$$L_{22} = m_2 V^2 \frac{\mathrm{d}^2}{\mathrm{d}x^2} - V(c_1 + c_2 + 2\omega m_2 \mathrm{i})\frac{\mathrm{d}}{\mathrm{d}x} + k_1 + k_2 - \omega^2 m_2 + \omega(c_1 + c_2)\mathrm{i}.$$

The complex amplitude of the rail can be written as

$$\overline{W}_1(x) = \int_{-\infty}^{\infty} G(x,\xi)\sum_{i=1}^{3} \Delta\overline{Q}_i \delta(\xi - a_i) = \sum_{i=1}^{3} G(x, a_i)\Delta\overline{Q}_i\,, \tag{10}$$

where

$$G(x,\xi) = m_2 V^2 G^o_{\xi\xi} - V(c_1 + c_2 + 2\omega m_2 \mathrm{i})G^o_{\xi} + \left(k_1 + k_2 - \omega^2 m_2 + \omega(c_1 + c_2)\mathrm{i}\right)G^o\,,$$

with $G^o = G^o(x,\xi)$ is the Green's function of the differential operator

$$L_{x,\omega} = L_{11}L_{22}\,.$$

The contact forces are obtained from the following set of equations

$$\begin{aligned}
(G_w + G_{11} + G_H)\Delta\overline{Q}_1 + G_{12}\Delta\overline{Q}_2 + G_{13}\Delta\overline{Q}_3 &= -r_0 e^{-\mathrm{i}\omega a_1 / V} \\
G_{21}\Delta\overline{Q}_1 + (G_w + G_{22} + G_H)\Delta\overline{Q}_2 + G_{23}\Delta\overline{Q}_3 &= -r_0 e^{-\mathrm{i}\omega a_2 / V} \\
G_{31}\Delta\overline{Q}_1 + G_{32}\Delta\overline{Q}_2 + (G_w + G_{33} + G_H)\Delta\overline{Q}_3 &= -r_0 e^{-\mathrm{i}\omega a_3 / V}
\end{aligned} \tag{11}$$

where $Gij = G(a_i, a_j)$ and

$$G_w = -\frac{1}{\omega^2 M}\,,\ G_H = \frac{1}{k_H}$$

are the wheel receptance and the contact elasticity, respectively.

The complex amplitude of the wheels and the rail in the contact points are

$$\overline{w}_i = \sum_{j=1}^{3} G_{ij}\Delta\overline{Q}_j\,,\ \overline{z}_i = G_w \Delta\overline{Q}_i\,. \tag{12}$$

Finally, the response function for contact forces, wheel and rail displacements in the contact points can be derived

$$\overline{H}_i^{\Delta Q} = \frac{\Delta \overline{Q}_i}{r_0}, \ \overline{H}_i^{w} = \frac{\overline{z}_i}{r_0}, \overline{H}_i^{r} = \frac{\overline{w}_i}{r_0}. \tag{13}$$

The stationary behaviour can be solved following the same way with $\omega = 0$ and Q_0 instead of ΔQ_i.

3 Numerical application

This section shows some numerical results derived from the above model, considering the following parameters: $M = 1000$ kg, $a_{12} = 2.10$ m, $a_{23} = 2.25$ m, $EI = 6.42$ MNm^2, $m_1 = 60$ kg, $c_1 = 92.95$ kNs/m^2 (damping ratio = 0.3), $k_1 = 400$ MN/m^2, $m_2 = 129$ kg, $c_2 = 92.95$ kNs/m^2 (damping ratio = 0.3), $k_2 = 400$ MN/m^2, $Q_0 = 100$ kN and $k_H = 1.5$ GN/m.

Figure 2 shows the rail deflection under dead load of the wheels for $V = 60$ m/s. At that velocity, the rail deflection is like the case of the stationary loaded wheels. However, the deflection is smaller than the one of a single loaded wheel due to relative long distances between wheels.

The frequency response function of the wheel's exhibit three peaks corresponding to the resonance frequencies at 43.1, 44.2 and 44.9 Hz as can be seen in Fig. 3, where the result from the undamped case is shown.

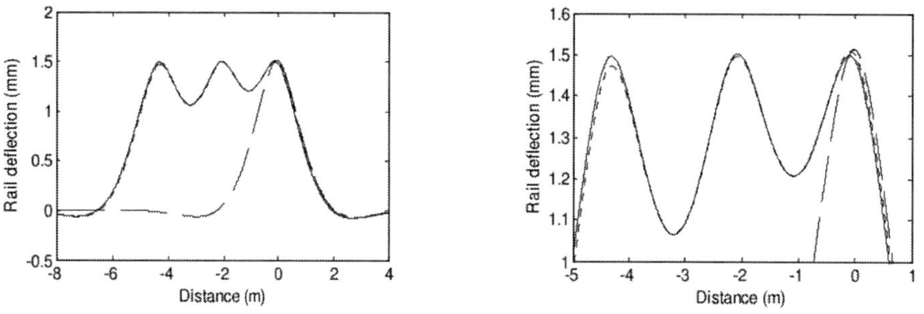

Fig. 2. Rail deflection: —, $V = 60$ m/s; - - -, $V = 0$; – – –, wheel alone at $V = 60$ m/s.

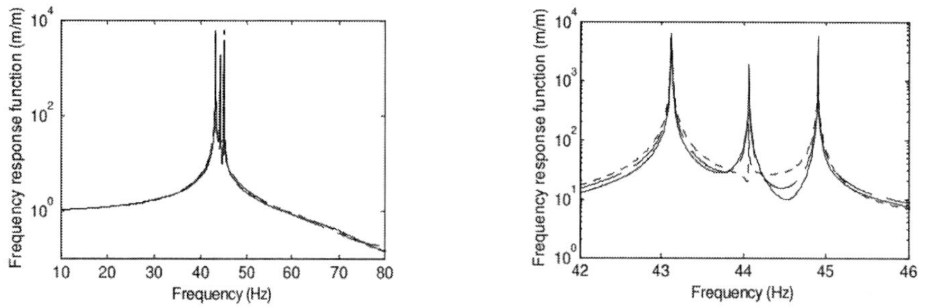

Fig. 3. Frequency response function of the wheels: —, front wheel; - -, second wheel; – –, rear wheel.

Figure 4 displays the frequency response functions of the wheels, rail in the contact point and contact forces. The three resonance frequencies are included into a single peak due to the damping of the track. At high frequencies, the response of the three-axle bogie/track system exhibits alternant peaks and deeps due to the rail bending waves reflected between wheels.

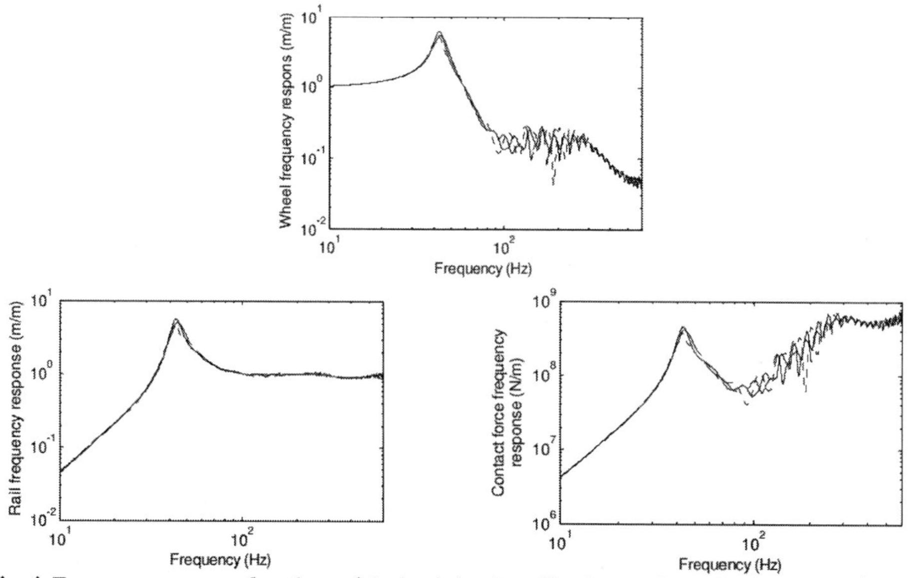

Fig. 4. Frequency response functions of the bogie/track at 60 m/s: —, front wheel; - -, second wheel; – –, rear wheel.

4 Conclusions

Wheel/rail vibration generated by the roughness of the rolling surfaces represents an interesting topic from many reasons, particularly the rolling noise and wear of the rolling surfaces. This paper is focused on the interaction between a moving three-axle bogie and track due to harmonic roughness. Mix 'moving wheel' model and the Green's function method is applied to calculate the rail deflection under dead loads of the wheels and the frequency response functions of the wheels, rail in the contact points and contact forces.

Three-axle bogie/track exhibits three resonance peaks (undamped track) which shift in one due the track damping. As the knowledge authors, this aspect never has been revelled in the past. Contact forces have alternate peaks/dips which can contribute or not to the growing of the undulations on the rolling surfaces. However, at the damped resonant frequency, all three wheels have the similar behaviour, meaning the all three exhibit the same dynamic stress.

This work has been funded by UEFISCDI, project number 76PED/2017, code PN-III-P2-2.1-PED-2016-0748.

References

1. Y. Soeta, R. Shimokura, Applied Acoustics **74**, 1160-1166 (2013)
2. N. Correa, E. G. Vadillo, J. Santamaria, J. Herreros, Wear **352-353**, 196-208 (2016)
3. S. L. Grassie, R. W. Gregory, D. Harrison, J. Mech. Eng. Sci. **24**, 77-90 (1982)
4. T. X. Wu, D. J. Thompson, J. Sound Vib. **239**, 69-97 (2001)
5. T. X. Wu, D. J. Thompson, J. Acoust. Soc. Am. **108**, 1046-1053 (2000)
6. T. Mazilu, J. Sound Vib. **239**, 4597-4619 (2013)
7. X. Sheng, T. Zhong, Y. Li, J. Sound Vib. **395**, 160-186 (2017)

DOI: 10.1051/matecconf/201711206015

CADOProSys Software. Process selection

Catrina Chivu*, *Cătălin-Iulian* Chivu, and *Cristina* Găvruş

Transilvania University of Braşov, Department of Engineering and Industrial Management, Mihai Viteazul 5, Braşov, Romania

Abstract. CADOProSys was developed by a group of teachers from Engineering and Industrial Management Department of Transilvania University of Brasov-Romania. Present paper is focused on module design for process selection that allows undergraduates to choose the appropriate technology based on product's specification, to obtain route sheets. Based on these route sheets the users will be able to determine the equipment requirements, materials requirements (type and quantity), standard times (computed based on theory or estimated based on firm's standards), and layout requirements. The Process selection module is interconnected with logistic module (sub-modules oriented on suppliers selection and facility layout design).

1 Introduction

In a business plan, designing and planning the new production system is one of the most important parts of the start-up.

Today, on the market, there is a lot of ERP (Enterprise Resource Planning) software, focused on company's management. These packages offer management for both discrete and repetitive manufactures in the automotive. There are managed accounting, business analytics human resources, inventory control, invoicing, quality control & inspection, supply chain, repair & maintenance, etc. Each of this software may be used for complete enterprise resource planning and manufacturing execution functionality including generating shipping labels and packing slips and then tracks shipments online to multiple shipping addresses [1]. Thus, there is software for assembly process simulation (e.g. DELMIA) [2]. One of the most powerful and complex instrument is offered by Siemens, through Tecnomatix software [3]. This software allows designing innovative systems based on product engineering, manufacturing engineering and production. It is a very complex instrument that may be used to increase productivity through planning and to manage manufacturing by optimising subsystems. The most disadvantageous aspect of this product is the complexity, which make it usable only by specialists, not beginners in the field of manufacturing system design.

Analysing the market's offer it can be underlined that a few of such software are oriented on designing a manufacturing system, and less than that for teaching future engineers what are the rules that they should follow when they need to design new system or modernize an existing one (Figure 1).

* Corresponding author: catrina.c@unitbv.ro

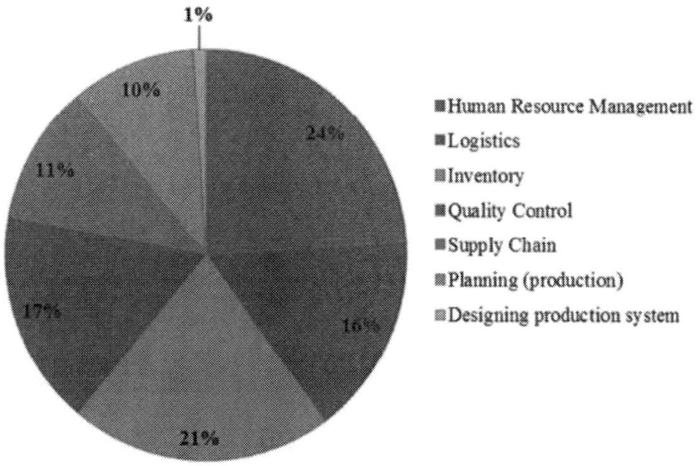

Fig. 1. Manufacturing software, based on field of interest (processed data from [1]).

2 CADOProSys. Manufacturing design module

2.1 General concept

Product and service choices, capacity planning, process selection, and layout of facilities are among the most basic decisions because they have long-term consequences for business organizations.

The impact of process selection goes beyond operations management, affecting entire organization and its ability to achieve its mission. So process selection choices very often have strategic significance. Different process types have different capacity ranges, and once a process type is functioning, changing it can be difficult, time consuming, and costly.

Table 1. Process types [4].

	Job Shop	**Batch**	**Repetitive/ Assembly**	**Continuous**	**Project**
Description	Customized products	Semi-standardized products	Standardized Products	Highly standardized products	Nonrepetitive set of activities
Advantages	wide variety of work	flexibility; easy to add or change products	Low unit cost, high volume, efficient	Very efficient, very high volume	Competiti-veness on market, variety of work
Disadvantages	Slow, high cost per unit, complex planning and scheduling	Moderate cost per unit, moderate scheduling complexity	Low flexibility, high cost of downtime	Very rigid, lack of variety, costly to change, very high cost of downtime	High cost per unit, complex planning, high variety of equipment and skills for employees

Obviously, long-term forecasts as well as an organization's mission and goals are important in developing a process strategy [4]. Process selection defines the strategy of the firm, having impact on capital intensity and process flexibility, an appropriate selection affecting directly facility and equipment choices, layout, and work design.

Before starting process selection it should be considering desired: variety in products, flexibility of equipment and volume output.

There are five basic process types: job shop, batch, repetitive, continuous, and project (Table 1).

To operate effectively, a manufacturing firm must have systems that allow it to efficiently accomplish its type of production. Production systems can be divided into two categories: (1) production facilities and (2) manufacturing support systems. Production facilities refer to the physical equipment and the arrangement of equipment in the factory. Manufacturing support systems are the procedures used by the company to manage production and solve the technical and logistics problems encountered in ordering materials, moving work through the factory, and ensuring that products meet quality standards [5].

To design educational software focused on production system, it should be considered both types of manufacturing operations: processing operations and assembly operations. It also should be included both stages of system: new production system or an existing one.

CADOProSys Software was developed based on the production systems design principles and correlated with QFD (Quality Function Development) chart. Software allows the students to both design new production systems and develop an existing system through implementing new product, or increasing the production [6, 8-12].

Each module is a combination of requirements and results that interconnect to other module or direct with the conclusion page. Because CADOProSys Software is an educational one, after each selection done by students are synthetized the generated restrictions.

2.2 Module design

CADOProsSys module oriented on manufacturing the product has, as main goal, to obtain the route sheets, one sheet per part. The outputs of the manufacturing module will be used in the logistics module: to select de suppliers, to design the facility, to select employees.

CADOProSys is educational software, thus, almost all selections done in each step is followed by some theoretical details regarding the selection, with rules, data required to resolve de problem.

A typical route sheet contains the following information [7]:

1. **Heading of Route Sheet:**
   - component name and part identification;
   - drawing reference number;
   - lot size
2. **Body of Route Sheet:**
   - operational sequences;
   - equipment requirements;
   - standard times;
   - machine capacity/ hour;
   - material requirements (type and quantity)

The general logic diagram of the module (Figure 2) is, in fact, the diagram of the steps mandatory in selecting of manufacturing processes.

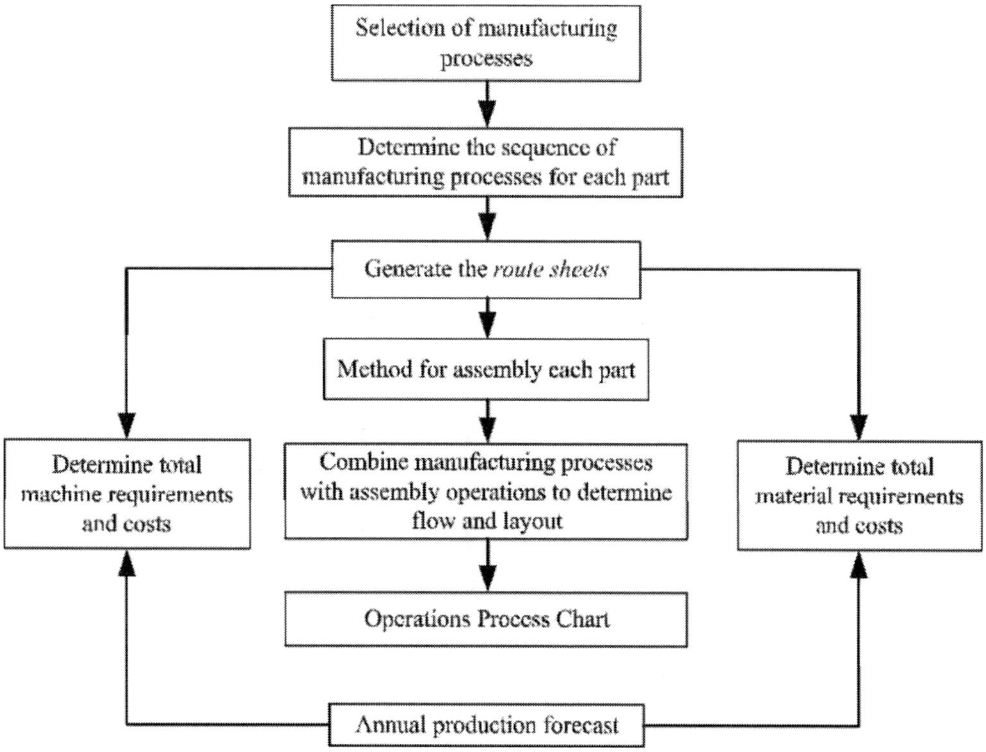

Fig. 2. Logic diagram of the module.

Thus the input data, grouped on stages of manufacturing are presented in Table 2.

Table 2. Input data.

Steps	**Input data**
Product analysis	parts list; assembly drawings/ part drawings; assembly chart (some of the activities are only if manufacturing process is an assembly one).
Manufacturing process	- manufacturing the type: changing shape/ machining part to fixed dimension/ obtaining a surface finish/ joining parts or materials/ changing physical properties; - machine cost database; - material costs database
Route sheet	part name and code; drawing reference; lot size; operation code; operation description; jigs, fixtures, tools; machine type.

In the present paper there are shown the steps and software facilities in the case of an existing manufacturing system (Figure 3), with continuous process (Figure 4).

According to the manufacturing process's type, the interface will ask for input data and explain requirements (according to Table 2). Machine and material costs database are automatically generated at the end of the suppliers' module of CADOProSys software. The end of that module gives also the space required for each device, equipment of machine-tools, according to the technical data sheets, information used in layout module of the software.

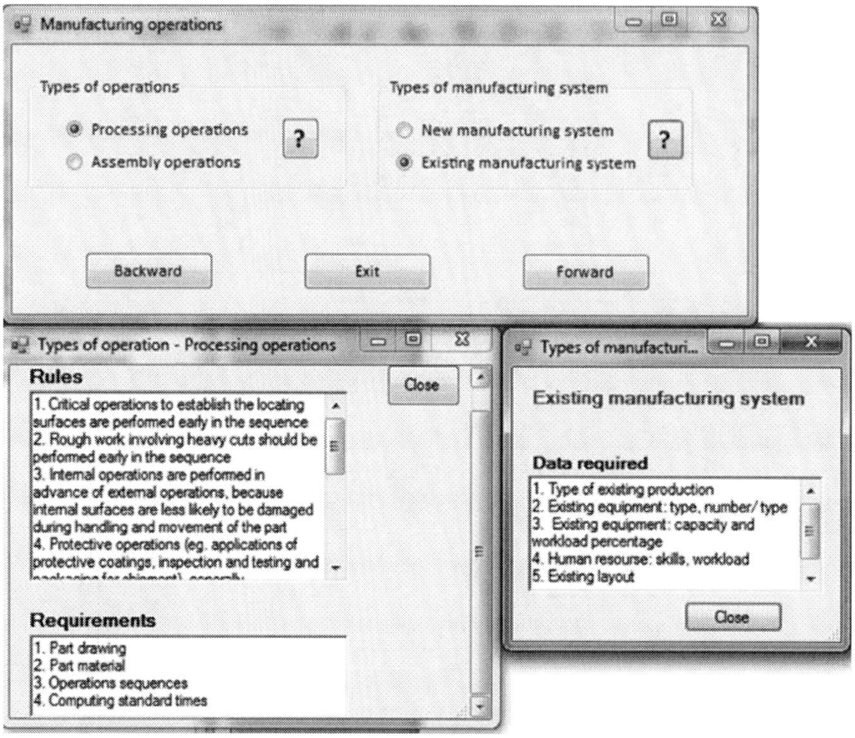

Fig. 3. Defining processing system.

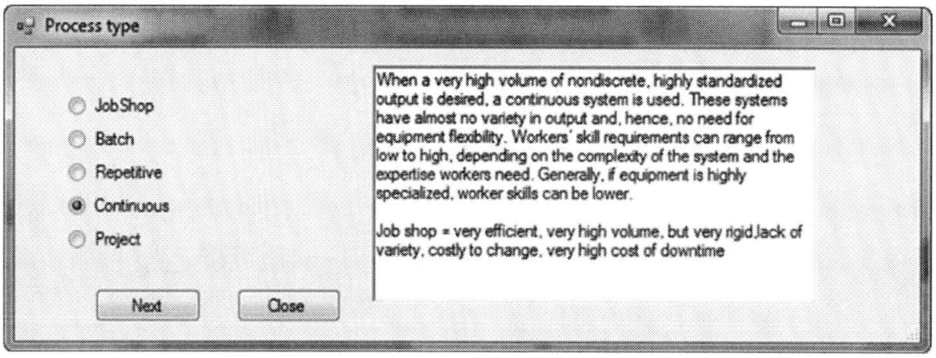

Fig. 4. Process type selection.

3 Conclusions and results

CADOProSys – process selection is able to generate de route sheet, to compute materials and equipment required quantity and, for existing system to determine if the desired lot size is able to be manufactured considering the data from the system. To show the results, in the following is considering the following situation: to implement, in an existing manufacturing system, the flow of linear guides. The product is manufacturing together with other 18 products (lot size is 760 pieces/ month). Manufacturing flow contains: complex cutting, treating, washing, control and assembly. For the complex cutting firm uses two CNC machines. Data are introduced according to interface from Figure 5.

DOI: 10.1051/matecconf/201711206015

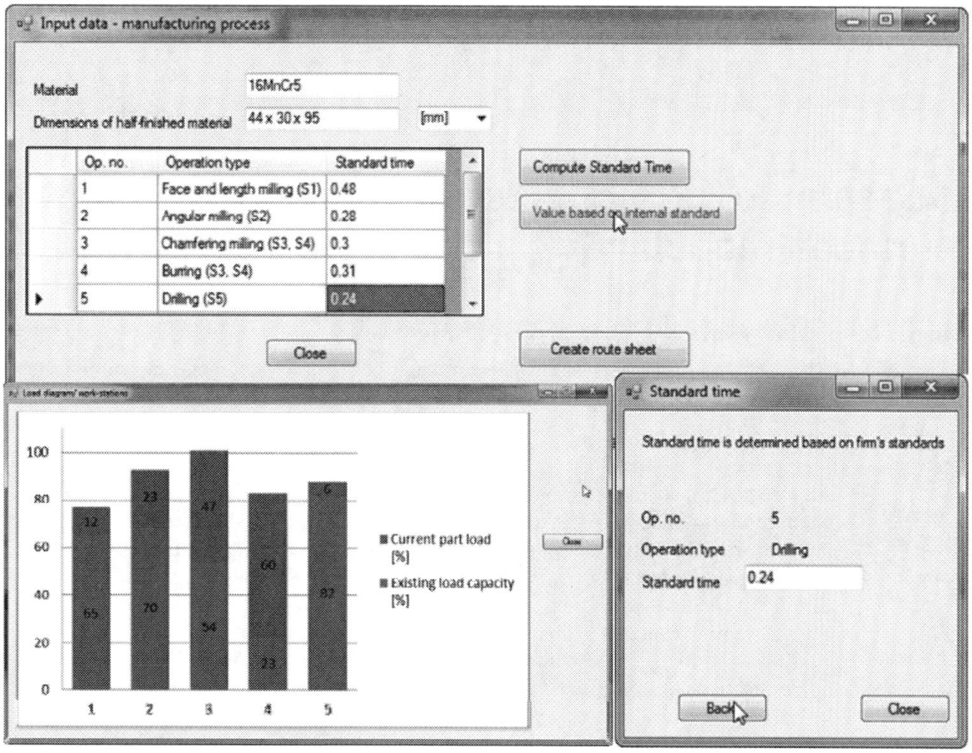

Fig. 5. Input data.

Based on these data, it may be generated route sheet for each part of the final product, which is an Excel file, particularized for the firm and product. Using the data it can be computed load for each work-station (values and graphical representation – Figure 5).

CADOProSys is educational software, designed, as a main purpose, to change the students' focus: from calculus (that is automatically done by the software) to understand the essence of the manufacturing system design. Using CADOProSys students will be able to feel better the problems of manufacturing systems, to follow the design steps, to understand interconnections between requirements of such design, essentially to answer to "why?", "when?", "what?" and "what for?" questions regarding manufacturing systems design.

References

1. Software Advice. http://www.softwareadvice.com/manufacturing
2. Digital Manufacturing Solutions. DELMIA. http://4dmsco.com/
3. Siemens Tecnomatix www.plm.automation.siemens.com/en_us/products/tecnomatix/
4. W. Stevenson, *Operations Management* (McGraw-Hill Irwin, 2012)
5. M. Groover, *Fundamentals of Modern Manufacturing* (John Wiley & Sons, Inc., 2010)
6. N. Slack, A. Brandon-Jones, R. Johnston, *Operations Management* (Pearson, 2013)
7. A. Garcia-Diaz, J. McGregor Smith, *Facilities Planning and Design* (Pearson, 2008)
8. E. Klimczak. *Design for Software: A Playbook for Developers.* (Wiley P.H., 2013)
9. P. Tchounikine. *Computer science and educational software design* (Springer, 2011)
10. D. Unger, S. Eppinger. CJEN, **22**, 689 (2011)
11. D. Cochran, A. Chu, M.U. Jafri, Z. Bi. JIII, **2**, 65 (2016)
12. A. Thecle, Y. Benama, N. Perry. Adv. PMS, **440**, 1 (2014)

Aspects to be considered in case of variable surfaces modelling

Andrei Mihalache[*], *Gheorghe* Nagit, and *Vasile* Merticaru

Gheorghe Asachi Technical University of Iasi, Department of Machine Manufacturing and Industrial Management, Blvd. Dimitrie Mangeron no. 59A, Iasi, Romania

Abstract. Proper triangulation will bring benefits regarding the numerical conditioning of nodes within the point cloud but will also allow certain nodes to be able to move around those that are neighboring them. In this way, surface features as chained regions or curves will have a certain degree of freedom which will allow them to slide one towards another. At a more careful evaluation there is a major drawback in terms of lack of possibility to take into consideration the operations which depend upon approximate shapes inside the ideal shape evaluation process. In order to be able to control the accuracy of the process there is an impetuous need for all surfaces to be described functionally. Any triangulation procedure has to be able to allow control over the analyzed surface topology and any other after modifications: dividing the surface along an incorporated curve in order to define a new edge or sticking two surfaces together in order to form a mutual boundary edge.

1 Introduction

This paper's goal is to present an identification procedure of points within a point cloud that represent the boundaries and in such way to be able to define the entire surface based on those points. This procedure would move the nodes inside the point cloud around the neighboring already considered points. This technique is somehow similar to general triangulation procedures that are presented in the scientific literature. The proposed procedure will not only lead to a noise reduction inside the point cloud and point randomly distributed to be ordered but will also maintain a higher quality of the triangulation process of nodes from the point cloud as they are identified and thus considered. A proper triangulation will bring benefits regarding the numerical conditioning of nodes within the point cloud but will also allow certain nodes to be able to move around those that are neighboring them. In this way, surface features as chained regions or curves will have a certain degree of freedom which will allow them to slide one towards another. As other authors state it has been noticed that by combining a Laplacian procedure in finishing of neighboring nodes with a Delaunay triangulation we may achieve satisfying results [1]. In order for the proposed procedure to work we will assume that instead of considering a surface being described by a single empty circumference circle the authors will consider it

[*] Corresponding author and co-author(s): andrei.mihalache@tuiasi.ro

as described by a sum of empty circumference spheres. In such way we will maximize the minimal value of the considered angle inside triangulation process (measured in 3D space). The procedure proposes to triangulate in an optimum way any type of surface through a series of edge flip operations. Those operations should keep and maintain the initial topology of any surface in such a way that will always be a valid mesh which would allow re-triangulations of the same surface as the process advances [2].

2 The triangulation procedure

In order to be able to approximate a free surface, as others have mentioned, we will need a very precise description of the surface's 3D limits within the point cloud. Also a very significant reason to succeed is to have a rigorous view upon the surface boundaries no matter the chosen approach: either a point interpolation which will take into account the curve radiuses of the considered boundaries or an approach based on the consideration that limits are used for cutting through the overlapped and contour fitted surface [3]. The method proposes a procedure for point detection that is representing the boundaries. It aims to fit a cylindrical section along all boundaries of the surface that has been defined by the point cloud in such a way that the cylindrical surface would touch at any time two points within the point cloud between two rotations and the direction of the cylinder's axis to be parallel at the normal of the point cloud (Figure 1).

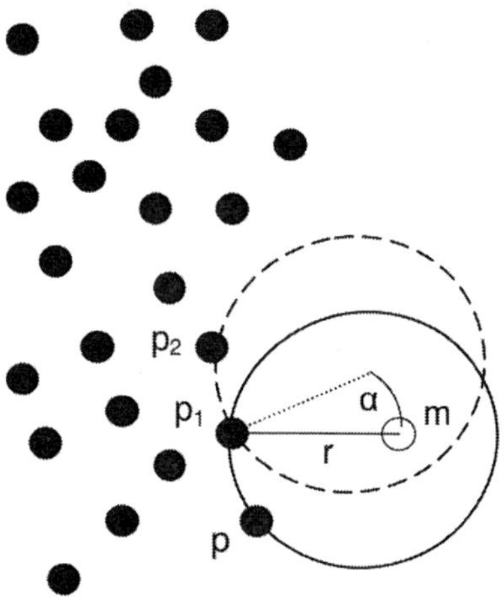

Fig. 1. The rolling process of a cylindrical surface in order to detect the boundaries of a surface described by a point cloud.

An express requirement of the proposed method is that the interior of the cylindrical surface to be free at any time. That implies that it should not contain any point from the point cloud which will lead eventually to the description of a polyline especially drafted for the boundaries of the analyzed surface if the order of connecting the boundary points is respected. The procedure will identify a new boundary point when two points already are in touch with the cylindrical surface as depicted in Figure 2.

The point that we are looking for, touching the cylindrical surface must be in an area that would equal the $2r$ radius of a sphere that would in take the last point that has been identified in such a way [4]. To ease the computation process the authors have appealed to a geometric description that allows a quicker identification of points that are in the neighbor by using a 3D grid. The grid will have a search radius of $2r$. As illustrated in Figure 2 we will rotate the cylinder around the last identified boundary point which has the direction of the cylinder axis defined as rotational.

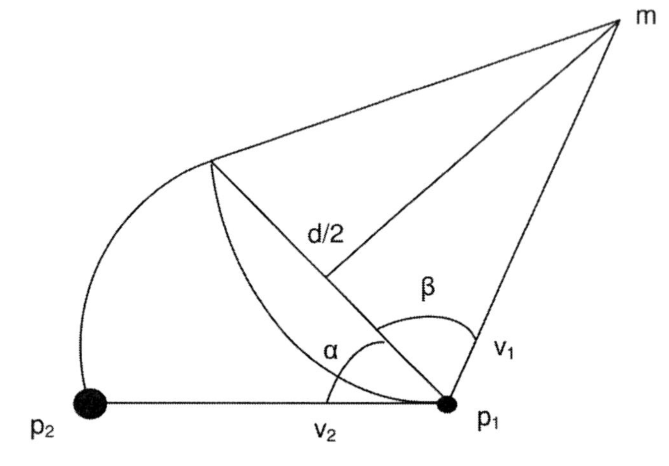

Fig. 2. The computing procedure of α in order to connect p_2 with the cylindrical surface.

Thus, we will consider p_1 as the last identified boundary point and n_1 as the estimated normal to the surface in p_1. The m point is arbitrary chosen along the cylinder axis. In order to achieve the connection of a p_2 point with the surface of the cylinder the method supposes that the cylinder is virtually rotated around p_1 having as rotational axis the n_1 [4].

The angle is determined as following: on the plane that contains the n_1 normal we project the line $m - p_1$ that will receive a v_1 badge and for the $p_2 - p_1$ from the plane that contains the normal we will have v_2 [5]. The β angle may be computed as follows:

$$\beta = \arccos(\tfrac{\|v_2\|}{2r}), \text{ where } \|v_2\| = \sqrt{v_2 \times v_2} \quad (1)$$

In this case the returning angle α may be computed as follows:

$$\alpha = \begin{cases} \sphericalangle(v_1, v_2) - \beta \text{ for } det(n, v_1, v_2) > 0 \\ \sphericalangle(v_1, -v_2) - \beta + \pi \text{ for } det(n, v_1, v_2) < 0 \end{cases} \quad (2)$$

In this manner the procedure will ensure that for all the points inside a sphere of a $2r$ radius around the p_1 previously identified boundary point will result in α turning point. The point which will result with the smaller positive value for α will be considered as the next point along the boundary line p_2 [5].

In this way we are able to approximate the surface normal for the newly identified point. In consequence the cylinder axis is being adjusted till it gets identical with the new surface normal from the p_2 point in such way that the cylinder surface will continue to remain defined in p_1 and p_2 points [6].

The whole procedure repeats itself until a loop is identified or up till it reaches the situation where there is a need for a α having a bigger value than $\frac{4}{3}\pi$ in order to identify a new boundary point. In order for the proposed procedure to work the user must specify the r radius, the first point and eventually the initial direction in the case that edge detection is the main concern.

The dimension of the geometrical description that has been used to identify all points that are contained inside a r radius sphere around the last identified edge point is given by the known r radius of the cylinder [6].

3 Operations on surface

The authors will present some of the base operations performed in surface modelling. The goal is that of illustrating some significant aspects that may appear in case of variable surfaces.

3.1 Control curve incorporation

A control curve of a surface may be described if we choose to consider a large number of points which may be or not neighbors on the surface which would allow the control curve to pass through. The facet dividing process is used to allow such points to be placed arbitrary on the surface. The operation adds a new node in the middle of a triangular facet and then connects it with every tip of the triangle. As we identify and then insert in the mesh these anchor points the process will result with a sequence of nodes and edges at our disposal that may be attached to the already designed ones.

As the literature states one way to describe a curve defined by two points is to identify a sequence of neighboring facets that are interconnected by these nodes [7]. By combining them we shall get a polygon which contains the curve that connects the two points. As soon as the curve has been identified another similar one is being created inside the 3D space which has its points constrained in the original positions anchored in such a way that it would allow the previously identified curve to approximate it.

3.2 Topology change

Any triangulation procedure has to be able to allow control over the analyzed surface topology and any other after modifications: dividing the surface along an incorporated curve in order to define a new edge or sticking two surfaces together in order to form a mutual boundary edge [8].

The authors take into account the fact that inside 3D space there is the possibility to coexist large discontinuities in the distribution of the points from the point cloud that are produced by occlusions. Those areas won't be described explicitly and thus small creases will appear throughout the surface. That is why the proposed algorithm has to take into account all the areas that are contained within the analyzed perimeter when combining or dividing points that are describing edges. The imposed condition in this case is that those described curves are forced to coincide.

3.2.1. Curve correspondence

In order to merge two surfaces along a pair of edges the nodes that belong to the two curves must be reordered in such a way that they will correspond. A simpler procedure may be that which takes into account the possibility to refine incrementally the curve considering lesser edges. This would be possible by dividing the longest edge up till both curves will reach approximately the same number of nodes [6]. The proposed procedure would be very useful in case the user neighbors the curves in order to merge them.

3.2.2. Surface division and merging

The premise that stands at the basis of our proposed procedure states that a surface is considered as a collection of nodes and that each one of those is randomly neighboring a series of other neighboring nodes. The circle's radius that encapsulates a certain number of neighboring points may be bigger or smaller as the described area requires it but also depending on the topological and geometrical complexity of the analyzed surface. In order to merge two curves that are situated at the surface's boundary after they have suffered operations that made them correspondent, the procedure will move the inner edges of a vertex towards another for each pair of nodes in the order they have been identified. This will transform the last nodes from edge points into mapped points and thus the previous edge curve will remain totally disconnected in respect with the resulted merged surface. This will determine a wipe out of its nodes and edges.

Dividing a surface along of a closed curve which has already been mapped represents the opposite process: edges that belong to a node are being classified as part of the interior, exterior or some region's boundary line. We then copy the boundary nodes and edges and then delete the inner edges connections from a copy of the surface's limits and also the outer edges connections from the other copy [7].

3.3 Surface intersection

A solution may be the one considering the intersection curves as self-considered edges of the resulted composed surface. In order to intersect two varying surfaces we firstly consider a curve that approximates the shape of the intersection curve of those two considered surfaces. Then those surfaces are redefined in such a manner that they would retain or interpolate this independent curve as a new edge. This manner of combining surfaces will lead to simplification in the implementation processes of common operations that are based on intersections.

3.4 Control tools for outer shapes

In order to be able to control the accuracy of the process there is an impetuous need for all surfaces to be described functionally. But this tedious work will only harden even more the process by accumulating an even larger number of operations in the case of free form surfaces [8].

That is why the authors consider that by controlling outer shapes that were previously represented we may also be able to control surface's merged regions shape. The nodes that are along edges inside an externally controlled region and those from an already mapped region will represent a blend of both neighboring and free nodes constrained topologically speaking.

This would allow that the shape of the surface from the controlled area to influence the shape of the mapped area in such a manner that the information regarding tangents or its curvature's propagation along edges.

3.5 Structured models construction

The process of combining some parameterized shapes depending on the interlaced and merged regions leads to the construction and definition of models that contain free form surfaces. The control tools types of these shapes may be organized into libraries which would allow a hierarchical distribution of shapes depending also on their ability to deform [6].

This study has led to the conclusion that control tools resulted collection serves at the construction of a similar mesh that clothes the skeleton. As the model gets refined its edges are recomposed in order to accommodate the imposed changes with respect to keeping the same general shape.

4 Conclusions

The procedure proposes an algorithm for triangulation. Applying it takes us to some considerations about the required operations that may be needed in surface modeling. Considering each surface as a collection of modes that are neighboring randomly other nodes allows us to proper close the gaps between surfaces. Also by considering previous mapped forms or primitives we are able to analyze surfaces in such a way that many may be appropriate with the ones already considered thus making it easier to process large volumes of information.

Our approach considers the surface as a sum of hollow circumference spheres. From the triangulation's point of view it's easier to map the surface by means of known algorithms as it is more accurate. It also allows more nodes to migrate towards each other thus making it possible for them to unite inside the procedure. As we rotate the theoretical cylinder around considered points from the surface we are able to approximate the surface normal making it possible to move the cylinder until its revolution axis becomes identical with the normal.

An immediate application of the proposed procedure may be the one that identifies points that are randomly scattered inside a point cloud by using a cylindrical surface which may be rolled out in order to describe in space a curve. In such a case instead of considering the normal by means of the proposed procedure we may estimate tangents by aligning the cylinder axis with a perpendicular arbitrary direction to the considered tangent one.

At a more careful evaluation upon the proposed method we may observe that a major drawback would be the lack of possibility to take into consideration the operations which depend upon approximate shapes inside the ideal shape evaluation process. For example we cannot base ourselves on the points that are at the intersection of two approximated surfaces in order to be able to formulate a conclusion regarding their ability to vary despite the fact that they are intersected.

Taking into account also the introduced error by the meshing process we cannot say for certain that the way two approximated surfaces do intersect or not and also the manner in which they do it gives us any insights about the true intersection area topologically speaking. From this point of view it appears that the intersecting surfaces are being cut along edges when they are merged along their intersection curves.

References

1. M.Y. Li, W.B. Ding, Procedia CIRP, The 9th International Conference on Digital Enterprise Technology – Intelligent Manufacturing in the Knowledge Economy Era, **56**, 178-181 (2016)
2. T. Sovany, Z. Tisler, K. Kristo, A. Kelemen, G. Regdon Jr. , European Journal of Pharmaceutics and Biopharmaceutics, 7th International Granulation Workshop: Granulation across the Length Scales, **106**, 79-87 (2016)
3. G. Trummer, C. Marte, P. Dietmaier, C. Sommitsch, K. Six, Wear, **352-353**, 136-145 (2016)
4. A. Needleman, Procedia IUTAM, Mechanics for the World: Proceedings of the 23rd International Congress of Theoretical and Applied Mechanics, ICTAM2012, **10**, 221 – 246 (2014)
5. G. Xu, Q. Zhang, Computer Aided Geometric Design, **25**, Iss. 3, 181–202 (2008)
6. P. Zhang, Z. Liu, Measurement, **94**, 37–45 (2016)
7. A. Chida, H. Masudan, Journal of Computational Design and Engineering, **3**, Iss. 4, 322-329 (2016)
8. W. Peng, C. Xu, Z. Feng, Neurocomputing, **179**, 228–237 (2016)

MATEC Web of Conferences **112**, 06017 (2017) DOI: 10.1051/matecconf/201711206017
IManE&E 2017

Analysis of cylindrical roller bearings design in order to optimize the classical process using FEM

Alin Marian Puşcaşu*, *Octavian* Lupescu, and *Ana* Bădănac

Technical University "Gheorghe Asachi" of Iasi, Department of Machine Manufacturing and Industrial Management, B-dul Mangeron, 59A, Iaşi, 700050, România

Abstract. This paper presents the structural and functional improvement of cylindrical roller bearings using finite element method. This method has a widely application and enjoys by an extended use in structural areas thermal analysis, mechanical and vibration. The advantages of the finite element method (FEM-Finite Element Method) are numerous and important. In case of conception a design it can be studied and modelled the structure behaviour in various media load in real time, therefore, based on the obtained results, the model can be modify before creating the final drawings for execution. Once is developed the CAD model, using FEM method it can be analyse the design structure in detail. Therefore using FEM can save time and money by reducing the required number of prototypes.

1 Introduction

As it is known, the bearings are essentials part of all rotating machines whose function is to support the machines. More than 90% of use rolling element bearings and their failure may lead to a catastrophic failure of the machine. Radial bearings are considered as the most critical components in industrial applications, [1]. Given the fact that the bearings are designed to withstand high loads and are mounted on complex and expensive cars and machinery, nothing can be left to chance in the bearing selection process suitable for the application in question [2]. Strictly related to design technologies, developed concern for research and development relating to the materials used, the heat treatment of surfaces, machining, surveying, contacts, profiles, analysis and simulation software's, leading to high levels of knowledge crucial to continuously improve technological conditions incorporated in the final product. The bearing types currently used in practice are: cylindrical roller bearings, thrust bearings, tapered roller bearings and spherical bearings [3]. In the case of cylindrical roller bearings they are manufactured in a wide variety of construction types and sizes, especially on single row rollers, and in some situations two or more rows of rollers, with a roll cages or side by side rollers. In the cylindrical roller bearings, the rollers are guided by the side of one of the rings collars.

* Corresponding author: puscasu_alin@yahoo.com

In the case of bearing with cage, ring collar and cage with retained rollers can be extracted from the other ring; this means that the bearings are removable. With this, mounting and dismounting of bearings becomes much easier, especially in places where due to loading conditions, tight fits are required for both rings.

2 Method

To optimize the design cylindrical roller bearings was done first resistance analysis benchmarks. Analysis was prepared using the finite element method. This method consists of a continuous body mesh and finished in several finite element (FEM - Finite element method). The mesh structure is meant in subdivision mathematical model with simple geometric form, which does not overlap, called finite elements. If the answer to each finite element simulations is expressed on a finite number of degrees of freedom that represent the values of unknown function (movement function) in a number of crucial points. Thus the answer to the mathematical model will result as an approximation of the response obtained by assembling discrete model answers all model elements. This method has wide application and enjoys widespread use in areas of structural, thermal analysis, mechanical and vibration. If we can conceive a design study model and structure behaviour in various media load in real time, based on the results, we may change the model before creating the final drawings. Once the CAD model is developed, using FEM method we can analyse in detail a structure. FEM programs, in our case ANSYS software found at S.C. Rulmenti S.A. Barlad has a library of tens of types of finite elements, from which the user must choose the most effective type for bearing structure. According to [4] ANSYS simulation software allows the designer to predict with confidence how the product will work under real conditions. Simulation is a powerful modular system. ANSYS Mechanical is a complete analysis FEM for structural type analysis, including linear and nonlinear dynamic studies.

The product offers a complete set of elements, material models and equation solving a wide range of mechanical design problems. FEM contains no "indications " or restrictions in this regard. However to ensure efficient model requires that the types of finite elements are chosen depending on numerous factors and conditions, the first of which is functionality, distribution of loads and tensions, to assure the DOF and limit certain others. By selecting ANSYS Mechanical simulation as a tool, S.C. Rulmenti S.A. Barlad has a large gain in terms of high computing power and response pattern of its products. It covers a wide range of properties, such as tension, strain, vibration characteristics, thermal characteristics, and durability.

3 Analyse and Results

For the achievement of experimental research was used a cylindrical roller bearing type NJ with brass cage. Components for this type of bearing is used in CHM speed reducer shaft, was first designed in Creo Parametric software, (Figure 1). Then the geometry for finite element analysis was exported Ansys 17.0, (Figure 2). After importing the geometry, we have conducted a series of optimizations on the components of the geometry in order to simplify and to reduce the possibility of calculation errors.

As a first result of importation, the bearing surfaces are cut symmetrically, so the first stage is to use MERGE command and merge the surfaces (Figure 3). In the second phase were carried out a number of sections for each element of the ensemble, using commands such as; slice, delete, flat, extruded, the result being represented by a set cut in as simple bodies, in order to simplify our analysis (Figure 4). In the third step was placed in a shaft assembly located on the inner ring (Figure 5).

DOI: 10.1051/matecconf/201711206017

Fig.1. Geometry in CREO 3.0.

Fig.2. Geometry in ANSYS 17.0.

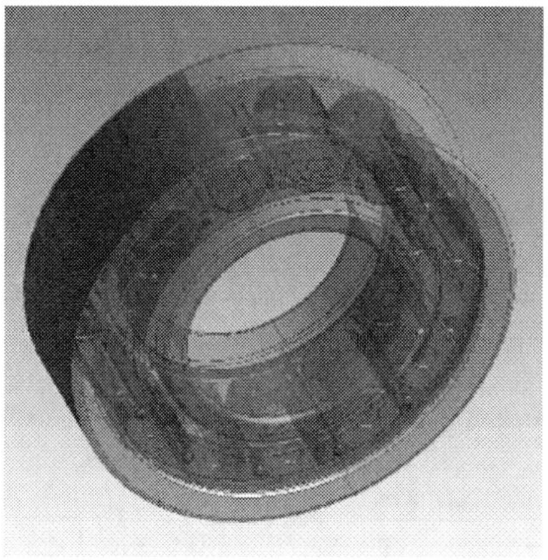

Fig.3. Merging separated faces.

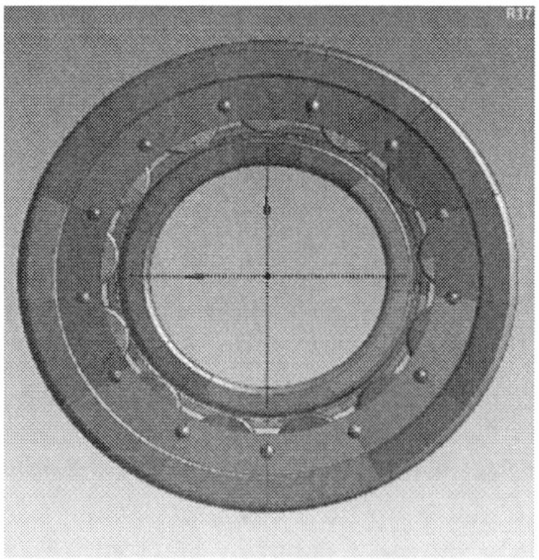

Fig.4. Dividing bodies.

Fig.5. Shaft on the inner ring.

After completing the operations to simplify geometry, we passed to the next stage, namely the introduction of mechanical model, at this stage were established the contacts between parts, couplings, meshing (Figure 6).

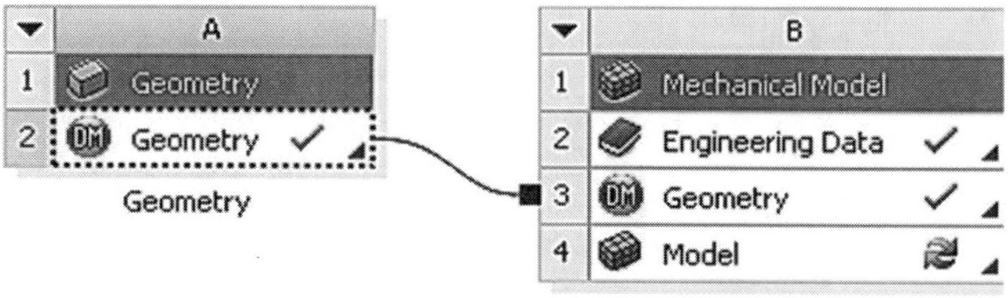

Fig.6. Mechanical Model.

It has achieved a uniform bearing mesh, with a number of nodes and elements as small as possible, to ease the process of calculating and also to shorten the settlement period. Optimization of the mesh was carried out in order to obtain a continuous mesh and a correlation between the number of finite elements and the number of associated nodes, Figure 7. I chose a simple type generally fixed mesh, a mesh type Hex roller type and a mesh cage and tetrahedrons for rivets. It chose a maximum size of items of 2 mm and a minimum allowable size of 0.5 mm with a growth rate of 1.2 mm. As shown in Figure 8, the resulting mesh is top quality with a relatively small number of nodes respectively elements in view of the structure and complexity. Also for rings was established a custom mesh.

For this present study it was chosen a MultiZone mesh with features recommended sizing, Hexa with each element of 2 mm mesh, free method of uniform surface type and concentration of each element type nodes Kept. For the rollers I selected a method hex dominant node with a concentration on each element type dropped. For cage and cage cover was admitted a mesh patch type and method conforming tetrahedrons.

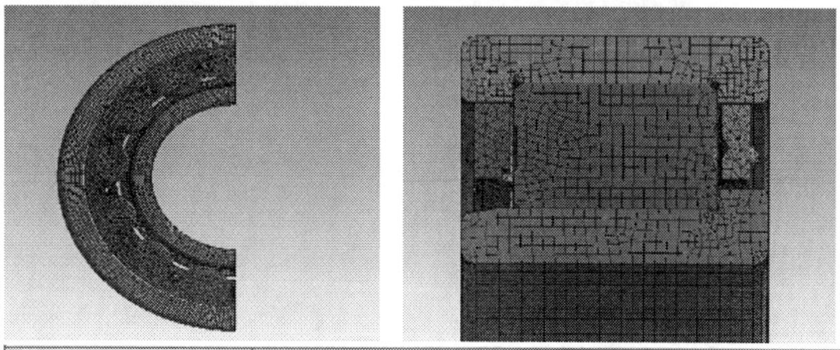

Nodes	1249193
Elements	720805
Mesh Metric	None

Fig.7. Mesh 1.

If was chosen the rivet, as a method hex dominant surface mesh with a quad / tri. This is given the low number of nodes and elements and how uniform meshing for analysis was chosen the first method of meshing. At the end of the round mesh was introduced a condition that would be provided contacting as many knots possible between rollers and rings, this command is going node.

DOI: 10.1051/matecconf/201711206017

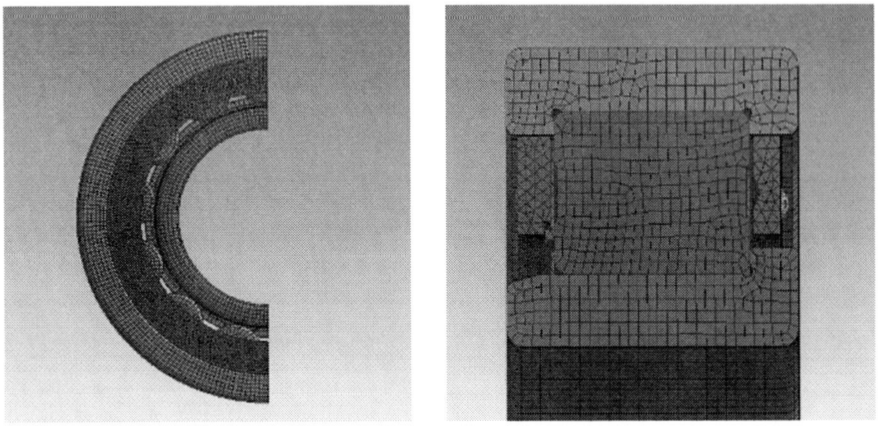

Statistics

Nodes	865133
Elements	447160
Mesh Metric	None

Fig.8. Mesh 2.

At this stage, were established limits to which the bearing is subjected, the moving part, the fixed part and axial displacements. The fixed part of the bearing is outer ring, was imposed displacements constrains and rotations in all 3 axes (fig.9).

A load was applied on the inner ring of the bearing, radial load limit request the catalogue is 376 Kn, but for the analysis was elected a calculated value of 80 Kn. The charge is applied on the inner ring through the shaft, (fig.10).

In the Figure 11 are presented the results of static analysis for contact pressure and respective in Figure 12 for bearing life.

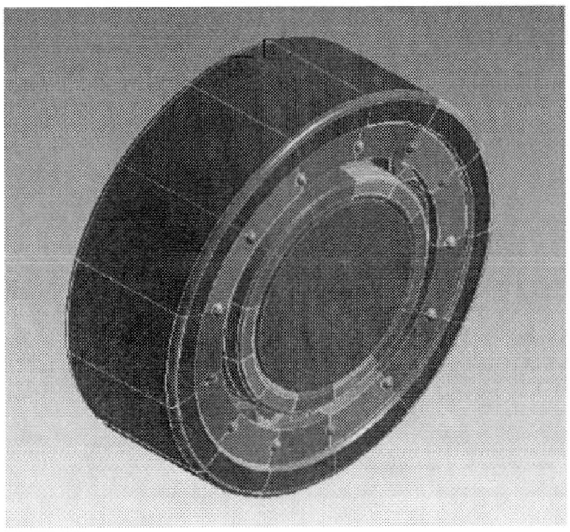

Fig.9. Outer ring fixed.

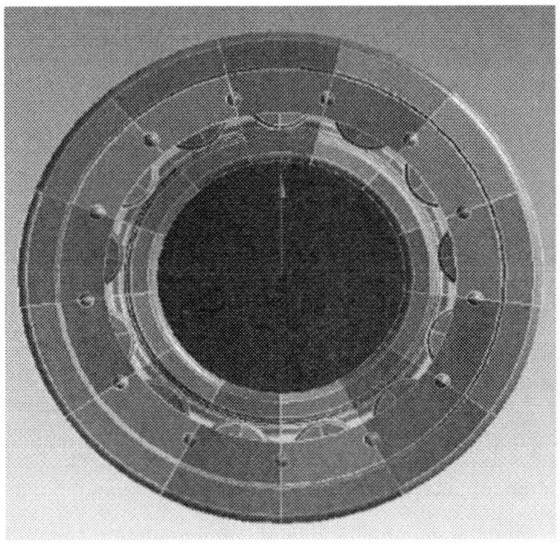

Fig.10. Shaft loaded.

DOI: 10.1051/matecconf/201711206017

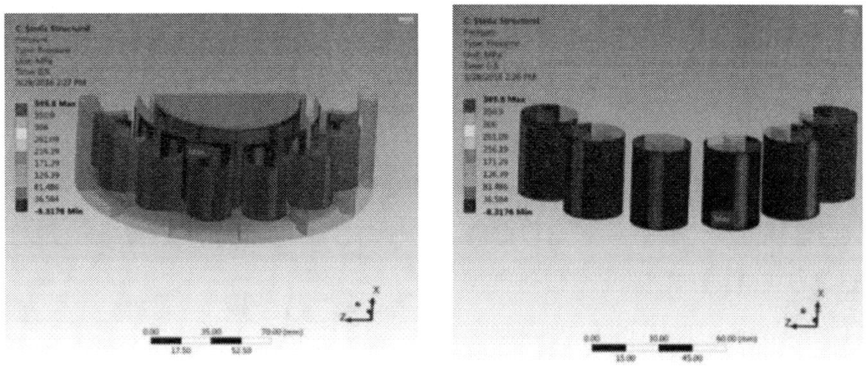

a) Element contact b) Contact surface of the rollers

Fig.11. Contact pressure.

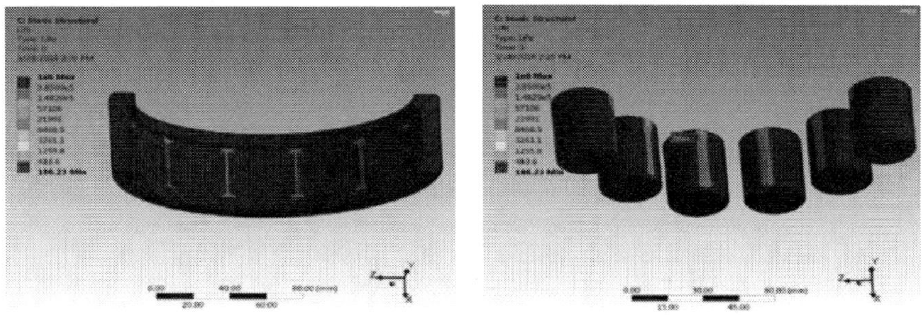

a) The life time of the outer ring b) The life time of the rollers

Fig.12. Bearing life.

4 Conclusion

Analysis of the results led to the following conclusion: finite element analysis software Ansys allowed through static structural analysis the accumulation of information's about stress and deformation of the structure rollers and rings, the life of the bearing was improved and load distribution on the rollers changed disappeared from the discharges end of the rollers. They reveal spikes on the edges of the rollers. The value of the bearing life is similar to the amount recorded in the S.C. Rulmenti S.A. Barlad catalogue.

References

1. D. R. Pradhan, Mohan Handbook of Research on Manufacturing Process Modelling and Optimization Strategies, (2014)
2. Y. Zhang, T. Qingchang, Z. Kuo1, L. Jiangang, Analysis of Stress and Strain of the Rolling Bearing by FEM method, international conference on applied physics and industrial engineering. Physics procedia 24, pp. 19-24, (2012).
3. P. Chunjun, Thesis of Static Analysis of Rolling Bearings Using Finite Element Method, (2009).
4. E. Suli, Lecture Notes on Finite Element Methods for Partial Differential Equations, Mathematical Institute University of Oxford, (2012).

Ant colony optimisation for scheduling of flexible job shop with multi-resources requirements

Krzysztof Kalinowski[1,*], *Damian* Krenczyk[1], *Iwona* Paprocka[1], *Wojciech M.* Kempa[2], *and Cezary* Grabowik[1]

[1] Silesian University of Technology, Faculty of Mechanical Engineering, Institute of Engineering Processes Automation and Integrated Manufacturing Systems, Konarskiego 18A, 44-100 Gliwice, Poland
[2] Silesian University of Technology, Faculty of Applied Mathematics, Institute of Mathematics, Kaszubska 23, 44-100 Gliwice, Poland

Abstract. The paper presents application of ant colony optimisation algorithm for scheduling multi-resources operations in flexible job shop type of production systems. Operations that require the participation of two or more resources are common in industrial practice, when planning are subject not only machines, but also other additional resources (personnel, tools, etc.). Resource requirements of operation are indicated indirectly by resource groups. The most important parameters of the resource model and resource groups are also described. A basic assumptions for ant colony algorithm used for scheduling in the considered model with multi-resources requirements of operations is discussed. The main result of the research is the schema of metaheuristic that enables searching best-score solutions in manufacturing systems satisfying presented constraints.

1 Introduction

Efficient organization of production flow, setting achievable plans and efficient use of available resources is one of the basic assumptions of a successful enterprise. Detailed scheduling of production processes improves coordination of activities in this area. In particular, it allows assessing the capacity of production resources for given constraints, precisely define the amount and dates of realization of the orders received, etc. However, the growing level of complexity of the issues in the production planning and control, and the dynamics of changes resulting from the interaction with the environment significantly hinder making the right decisions and motivate the use of efficient methods and software tools.

There are many methods for supporting the process of scheduling; among them we can distinguish methods which guarantee finding the optimal solution and approximate methods, heuristics, which do not give such a guarantee. Optimal methods are dedicated for

* Corresponding author: krzysztof.kalinowski@polsl.pl

precisely defined tasks structure and system configuration, with specific constraints and criteria. Hence, only a small percentage of problems can apply them in practice.

More attractive in practical applications are approximate algorithms and methods using heuristics, metaheuristics [1] and fuzzy logic [2,3] and simulation tools [4,5]. Abandoning the determination of optimal solutions for sufficient solutions provides an acceptable time for determination of a satisfactory solution and the ability to analyze more complex models [6,7]. Among metaheuristics, promising results are obtained by using swarm intelligence methods. Swarm intelligence (SI) deals with the implementation of rules of collective behaviour of small beings from biological systems. In this paper we focus on the ant colony and their behaviour when searching for a food – characterized by decision-making decentralization and self-organization.

The implementation of ant colony algorithm in scheduling domain is the subject of many works [8, 9, 10, 11]. In many cases authors proposes some extensions dedicated for a special kind of scheduling task's model: flow configuration, constraints etc. In [12] a simple resource levelling approach used for scheduling projects with multi-mode execution activities is proposed. Modified ACO approach for a multi-constraint (precedence and resources constraints) project scheduling problem is described in [13]. In [14] they implement Elitist Ant System algorithm with Job Shop Scheduling Problem. Improving the basic ant system is also proposed in [15], where a pheromone alteration strategy that utilizing the behaviour of artificial ants to perform local search is introduced. ACO algorithms can also effectively solve the scheduling problems with employee allocation [16] and multiple objectives [17,18]. An ant colony optimization approach (ACO) is also widely used for the shop scheduling [19-23] resource-constrained scheduling (RCS) and resource-constrained project scheduling problems (RCPSP) [24]. Novel operation scheduling algorithms using the ant colony optimization approach for both timing-constrained scheduling (TCS) and RCS is proposed in [25]. The combinations of two pheromone evaluation methods are discussed in [26].

2 Issue description

The considered scheduling problem can generally be formulated as follows: given is the flexible job shop type manufacturing system. The set of available resources production R is given by: $R = \{R_i, i=1,\dots, m\}$, where: R_i - i-th resource m - the number of all available resources. Resources may have a different level of specialization (machines and equipment), skills and abilities (human resources) as presented in [27]. The tasks for the system are represented by a set of orders. A production order is described by a set of operations, with given precedence relationships. An operation requirements regarding to resources are formulated indirectly, through resource groups:

$$GRz = \{GRz_s, s = 1,\dots,mg\}, \tag{1}$$

where: GRz – set of resource groups, mg – size of this set, GRz_s – s-th resource group.

The resource group is understood as a set of unrelated, parallel resources, enabling the execution of a certain number of treatments of a particular operation. The need for the required number of competences groups and the required number of resources from these groups are defined separately for each operation. A single production resource Ri (renewable, discrete) is described by:

$$R_i = (c_i, kj_i, LwGR_i, Ka_b), \tag{2}$$

where: c_i - capacity, kj_i – unit cost, $LwGR_i$ – list of resources groups, Ka_b – b-th working timetable.

The capacity of the resource indicates the maximum number of concurrently executed tasks. A resource with a capacity greater than 1 is a resource that can process more tasks simultaneously. Unit cost of resource is a parameter used in determining the cost of the in the given route. List of groups determines the ability to perform treatments/operations or groups of operations. The level of resource competence in a given group is determined by priority factor. Working timetables describe availability of the resources to perform operations. Resources can operate according to the same or different working timetables. In presented model, the problem of finding a feasible schedule that meets all hard constraints and best meets time and cost based criteria (as soft constrains) is considered.

3 Ant colony optimisation algorithm

Proposed method of determining solution is based on ant colony optimisation metaheuristic. Proposed algorithm attempts to imitate natural, collective behaviour of ants in the anthill environment when they search and collect food [10,22,26]. In particular, the following characteristics of this kind of swarms are taken into account: pheromone trail and the strength of its impact for ants move, evaporation of the pheromone over time and trend for finding the easiest path to food.

A general scheme of ACO algorithm is shown in Figure 1. The base of searching is a directed and/or type graph of operations, describing precedence relationships between all planned operations [7,27]. This graph is created from given production orders and enables defining complex processes, containing machining and assembling/disassembling operations and their multi-routes structures. At the beginning, initial values of parameters affecting the way of searching for solutions are set: the pheromone trail τ, pheromone evaporation factor ρ, and α, β – parameters representing importance of pheromone (global heuristic) and importance of calculated distance between given node and allowable successors (local heuristic), respectively. The number of iterations and/or stop conditions should be also defined. Population of ants may be fixed or variable in subsequent iterations. Each ant generates one solution by passing through the graph of operations. Beginning from the first node, the set of eligible successor nodes (*E*) is updated. Given operation is assigned to this set only if all its predecessors are visited. The ant chooses the next operation probabilistically. The value of the probability of the v node selection starting from the node u is calculated as follow [8,10]:

$$p_k(u,v) = \frac{[\tau(u,v)]^\alpha [\eta(u,v)]^\beta}{\sum_{w \in S_k} [\tau(u,w)]^\alpha [\eta(u,w)]^\beta} \quad (3)$$

where: S_k – the set of eligible nodes, w – the id of node, $\tau(u,w)$ – the current amount of pheromone of transition $u \rightarrow w$, $\eta(u,w)$ – the distance (local attractiveness) of transition $u \rightarrow w$.

The distance is calculated separately in each decision situation, depending on selected parameters connected with an operation:

$$\eta(u,w) \leftarrow f(e_t d_t + e_{Tb} d_{Tb} + e_{Te} d_{Te} + e_c d_c) \quad (4)$$

where: d_x – distance based on processing time (d_t), beginning time (d_{Tb}), end time (d_{Te}), cost (d_c), e_x – weights, respectively.

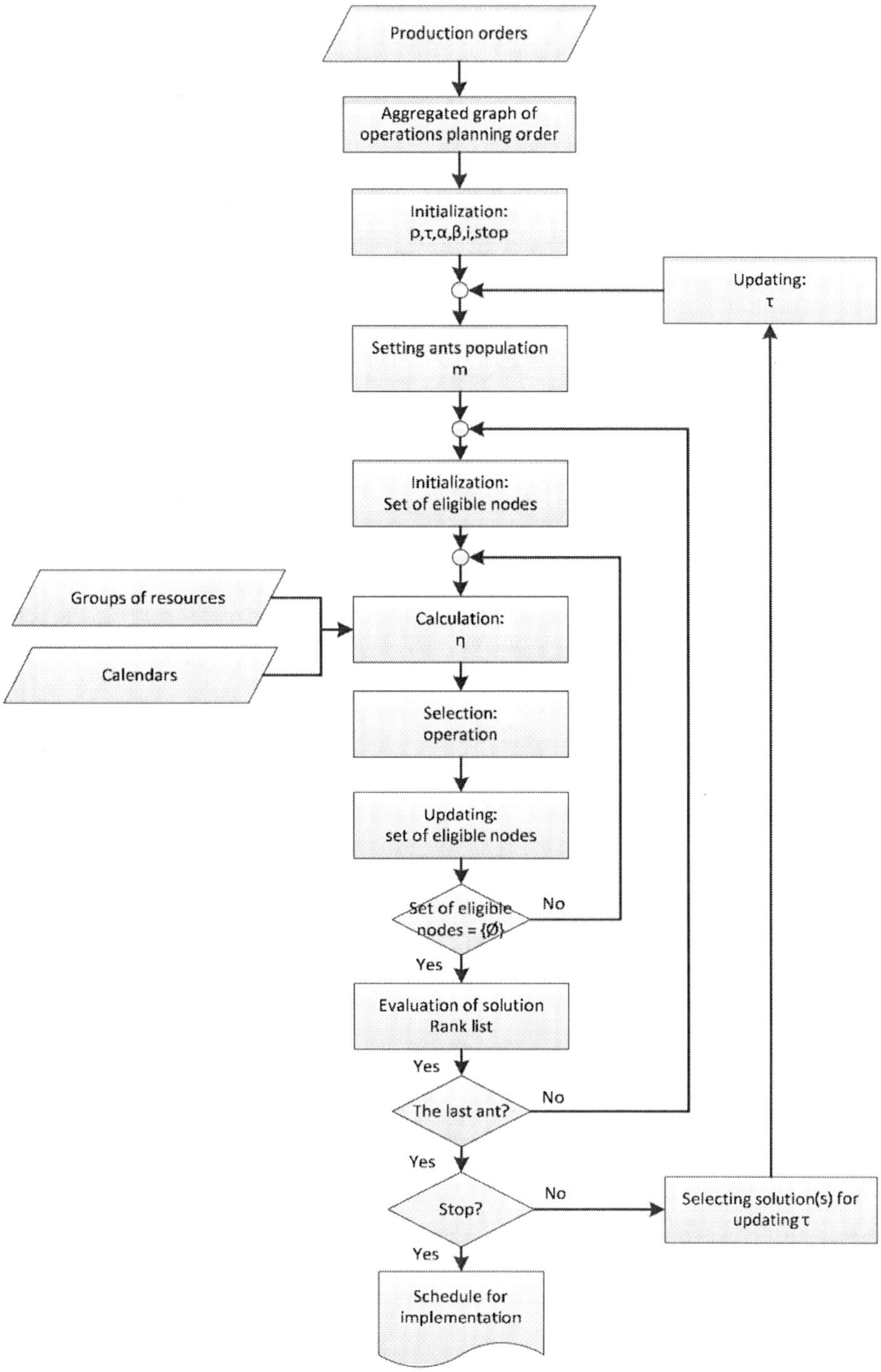

Fig. 1. The scheme of ACO algorithm.

These parameters are used for searching resources in required operations groups using local heuristic. An example of a single operation scheduled with requirements of resources groups is presented in Figure 2.

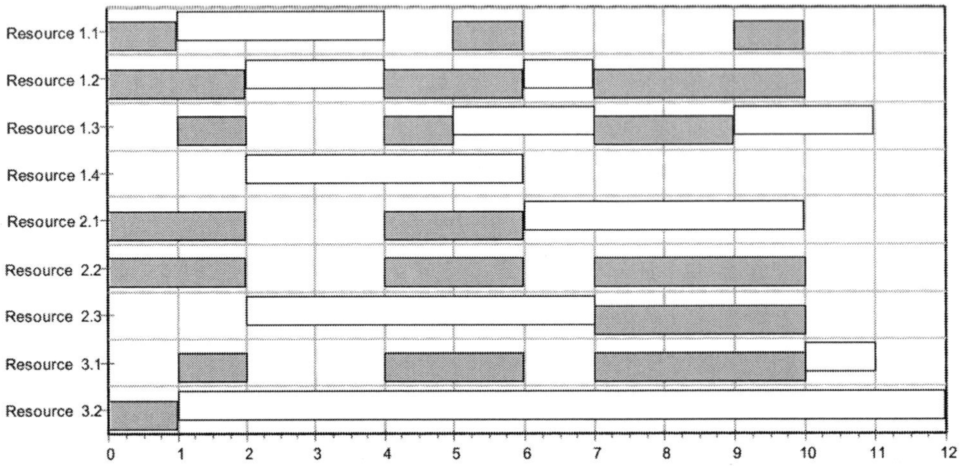

Fig. 2. The distribution of the single operation between resources groups.

The operation has following characteristics: processing time t=7, divisibility = true, resource requirements (resources group, required amount of resources): (1,2), (2,2), (3,1). Resources have limited availability, white bars in Gantt chart means that a resource is busy. Because of lack of required number of two available resources from the group 1 in 2-4 time interval and from the group 2 in 6-7 operation should be temporarily suspended.

After selecting an operation the set E is updated (selected operation is removed and new eligible are added). Obtained solution is positioned in the ranking list. When all ants from a given iteration have built complete schedules the *Stop* conditions are checked. Both quality and time based criteria can be used there. If the *Stop* conditions are not satisfied the subset of schedules are selected for updating pheromone. The rule of updating pheromone is as follow [10,12,17]:

$$\tau(u,w)_{i+1} \leftarrow (1-\rho)\tau(u,w)_i + \Delta\tau_b(u,w) \tag{5}$$

where: $\Delta\tau_b(u,w)$ - indicates the amount of pheromone adding to the edge *(u,v)* calculated proportionally to the value of solution in the set of best solutions.

Summary

The article presents the ant colony optimization algorithm dedicated for extended scheduling model related to distribution of operations on greater number of resources. Although ACO do not guarantee finding optimal solutions, this kind of metaheuristics, classified to widely understood swarm intelligence, is an effective tool for searching feasible solutions in computationally hard tasks, such as NP-complete problems.

The presented method can be used for scheduling both machining and assembly operations of production processes, in which the limited resources belong to different groups of resources (machines, human resources with different competences, tools etc.) and require simultaneous assignment to the operation. It is also be useful when resources have to be exchanged during long-time operations, e.g. in shift work or in a hazardous environment. The area of potential applications of the proposed approach is also in operations where the number of resources from particular resources groups does not have to be constant and The duration of the operation may be vary depending on the number of working resources.

The results obtained are a promising alternative for exploring the space of possible solutions in scheduling of manufacturing systems. Software implementation, as an extension of

KbRS scheduling system [28] functionality, and performance testing of the presented algorithm is provided for further research.

References

1. R.-H. Huang, Ch.-L. Yang, Int. J. Production Economics **115**, 163–170 (2008)
2. M. Diering, K. Dyczkowski, A. Hamrol, Advances in Intelligent Systems and Computing, **368**, 415–425 (2015).
3. M. Diering, K. Dyczkowski, IEEE International Conference on Industrial Engineering and Engineering Management, **2016**, 228–232 (2016)
4. D. Krenczyk, M. Jagodzinski, Advances in Intelligent Systems and Computing, **368**, 451-461 (2015)
5. A. Kampa, G. Gołda, I. Paprocka, Computers 2017, **6**, 10, 1–12 (2017)
6. D. Lei, Int J Adv Manuf Technol **43**, 926–938 (2009)
7. K. Kalinowski, M. Zemczak, Advances in Intelligent Systems and Computing, **368**, 475-483 (2015)
8. B. Chandra Mohan, R. Baskaran, Expert Systems with App **39,** 4618–4627 (2012)
9. X. Li, X. Shao, L. Gao, W. Qian, Int. J. Production Economics **126,** 289–298(2010)
10. M. Dorigo, V. Maniezzo, A. Colorni, Proceedings of ECAL91 – European Conference on Artificial Life, Elsevier, 134-142, (1991)
11. C.W. Leung, T.N. Wong, K.L. Mak, R.Y.K. Fung, Computers & Industrial Engineering **59** 166–180 (2010)
12. M. Garmsiri, M. R. Abassi, J. of Industr. Eng. Intern., **8**:7, 1-9 (2012)
13. R.-M. Chen, S.-T. Lo, IJCSNS Int. J. of Com. Sci. and Net. Sec., **6**,11, 75-84 (2006)
14. E. Flórez, W. Gómez, L. Bautista, International Journal of Artificial Intelligence & Applications (IJAIA) **4**, 4, (2013)
15. M. Ventresca, B. M. Ombuki, Tech. Rep. N° CS-04-04, Department of Computer Science, Brock University (2004)
16. W.-N. Chen, J. Zhang, Trans. on Soft. Eng. **39**, 1, 1-17 (2013)
17. B. Yagmahan, M. M. Yenisey, Expert Systems with App **37,** 1361–1368 (2010)
18. R.H. Huang, Ch.L. Yang, Int. J. Production Economics **115**, 163– 170 (2008)
19. S.G. Ponnambalam, N. Jawahar and B.S. Girish, New Advanced Technologies, InTech (2010)
20. G. Ćwikła, Manufacturing 2014: Contemporary Problems of Manufacturing And Production Management, Poznan Univ Technol, Springer, 44-56 (2014)
21. A. Rossi, G. Dini, Robotics and Comp.-Integrated Manufacturing **23** 503-516 (2007)
22. Ch. Blum, M. Sampels, J of Mathematical Modelling and Alg. **3**: 285–308 (2004).
23. L.N. Xing, Y.W. Chen, P. Wang, Q.S. Zhao, J. Xiong, Applied Soft Computing **10** 888–896 (2010)
24. Ch.-W. Chiang, Y.-Q. Huang, Third International Conference on Innovations in Bio-Inspired Computing and Applications (2012)
25. G. Wang, W. Gong, B. DeRenzi, R. Kastner, IEEE Trans. on CAD of Integr. Circ. and Syst., **26**, 6, 1010-1029 (2007)
26. D. Merkle, M. Middendorf, and H. Schmeck, Journal IEEE Transactions on Evolutionary Computation **6**, 4, 333-346 (2002)
27. K. Kalinowski, C. Grabowik, I. Paprocka, W. Kempa, IOP Conference Series; Materials Science and Engineering **95** 1757-8981 (2015)
28. http://imms.home.pl/kbrs - KbRS scheduling system [2017.03.30]

An attempt of CNC machining cycle's application as a tool of the design feature library elaboration

Cezary Grabowik[1], *Krzysztof* Kalinowski[1], *Damian* Krenczyk[1], *Iwona* Paprocka[1], and *Wojciech* Kempa[2]

[1]Silesian University of Technology, Faculty of Mechanical Engineering, Institute of Engineering Processes Automation and Integrated Manufacturing Systems, Konarskiego 18A, Gliwice, Poland
[2]Silesian University of Technology, The Faculty of Applied Mathematics, Institute of Mathematics, Konarskiego 23A, Gliwice, Poland

Abstract. This paper presents a novel approach to a problem of the design feature library elaboration. As a tool of the design feature library development CNC machining cycles were proposed. Because of the great number of commercially available CNC machine controllers, with different CNC machining cycles definitions, it was necessary to make a decision about a research methodological framework, it is the selected CNC machine controller. Taking into account the criterion of popularity as the research framework the selected group of Sinumerik CNC machine controllers was chosen. Presented in the paper idea of the feature library development is based on an assumption saying that it is possible to find a relationship between a particular CNC machining cycle and the simple design feature or even compound design features. Identified, thanks to this assumption, set of the design features could be the base for elaboration of the design feature library. This solution, it is the feature library next gave opportunity for elaboration of the feature based design modelling module (FBDMM) working in the SIEMENS NX system environment. Hence, the FBDMM module can support both a designer and CNC machine programmer which is possible due to received in the module modelling paradigm. In FBDMM module the removal feature based modelling technique is received.

1 Methods of the design modelling with design features

Whilst considering design modelling methods with application of the design feature library, which is implemented in any CAD/CAM/CAE system, it is possible to distinguish the following three modelling methods:

- socalled removal method (technological) which mimics removal manufacturing processes like turning, milling, drilling, reaming and so on. In this method a design is modelled by performing subsequent Boolean's operations, so in each single operation a particular design feature is subtracted from the blank in order to get a finished part model. The schema of modelling process with technological method in the Figure 1a is shown. One of the most important drawbacks of this method is lack of possibilities of the design feature library preparing only by means of the design features with a simple design shape structure. In most cases it is needed to work out at least a part of the library with compound design shape features. The biggest advantage of the technological method is its similarity to real manufacturing processes, and it gives possibility of working out a CAPP system with really simple inference mechanism. Well-designed structure of the original modelling module implemented in a CAD/CAM/CAE system environment with the properly worked out a set of the design rules allows to a certain degree to control a design modelling process. Consequently, a designer during the modelling process is supervised in order to follow given modelling standards. In the considered modelling standard the particular design process stages should respond to the subsequent manufacturing process stages. It means that consecutively performed Boolean's operations would respond to certain manufacturing process operations or cuts. As a result, at the first stage of modelling process the "rough" shape of the model, which corresponds to the product stage after roughing, is achieved. At the second stage of modelling process the shaped features of the model are made – the stage after profiling, but taking into consideration that in CAD systems product model is always modelled taking nominal dimensions into account, in principle, after the second modelling stage the final shape of the product is achieved.

- so called additive method (constructional) in opposite to removal modelling method a model design shape come into being as a result of performing of the subsequent unite operations in the sense of the Boolean's operations. A product model is built by adding of the design features together. From manufacturing technology point of view it corresponds to additive manufacturing, in which the product is manufactured with one of the 3D printing – rapid prototyping methods or classic manufacturing technologies like bonding, welding etc. The schema of constructional modelling process in the Figure 1b is shown. Like in case of the technological method the biggest drawback of constructional method is necessity of creation of the design features with a compound design shape. This problem can be illustrated by the example of the shaft pin with the splineway manufactured on its surface. In case of the constructional method it would be necessary, at the design feature geometrical shape identification stage, to create a compound shape. As a result, in the considered case it would be the difference between the pin geometrical shape and splineway shape in the sense of the Boolean's operation. As a consequence, it is easy to notice that it is almost impossible to work out the design feature library including all possible combinations of the compound design feature structures.
- so called hybrid method (design-technological), the product model is created by uniting or subtracting particular objects from each other. The schema of modelling process with hybrid method in the Figure 1a is shown. Looking for similarities in the manufacturing realm it corresponds to manufacturing using both removal and bonding manufacturing techniques.

Taking into consideration drawbacks and merits of the particular methods the biggest usefulness in the process of the CAPP system creation shows the first above mentioned method, it is technological modelling method. It comes from fact that this method is in its sense similar to manufacturing processes being realized in real industrial conditions. In addition, as a consequence of deployment of the technological method it would be possible to work out such structure of the CAPP system user interface that the manufacturing process would be designed in the background somewhat.It means that a designer making a design and its design structure would make a process plan, simultaneously. In such case the order of the design features in the design structure would correspond to the order of the manufacturing operations or cuts in the process plan structure. This assumption let us to make the CAPP knowledge based system with simple inference mechanism. Consequently, the inference mechanism is built of a set of inference rules that interlink a certain design feature with strictly ascribed manufacturing feature. Despite unquestionable advantages of such solution its effectiveness is in principle limited to a group of products such as shafts, sleeves and disks.

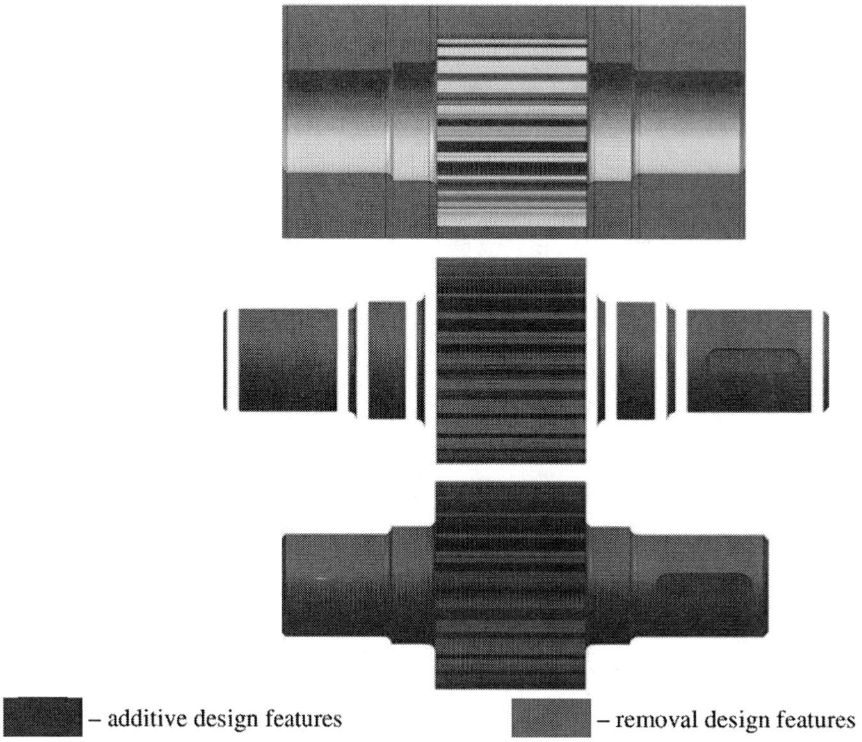

Fig. 1. Design modelling methods with design features, a) removal, b) additive, c) hybrid.

2 Machining cycles as base for the design feature library elaboration

Machining cycles are commonly considered as the important element of CNC control programmes. They allow to automise the CNC programming process, and they also limit the programmes size. Thanks to them it is possible to change manufacturing parameters in an easy and quick way. The concept of machining cycle is understood as a constant parameterized subprogram stored in the control system memory used for programming typical manufacturing operations such as turning, drilling, milling, threading etc.

Machining cycles programming process is very often supported by the dialog programming module like in case of Sinumerik control system. In such case a CNC programmer is visually assisted in order to make them familiar with the particular machining cycle parameters and their permissible values [5-6]. Taking into account they characteristic the machining cycles can be roughly grouped in to the three groups it is: (i) drilling cycles, (ii) milling cycles and (iii) turning cycles. As it was mentioned in the abstract paragraph because of the great number of commercially available CNC machine controllers, with different CNC machining cycles definitions, it was necessary to make a decision about a research methodological framework, it is the selected CNC machine controllers group. Taking into account the criterion of popularity,in the considered case in Polish industry conditions, as the research framework the selected group of Sinumerik CNC machine controllers was chosen. Although, presented in the paper result can be generalized for other CNC machine controllers like these offered by Fanuc, Heidenhain or Okuma producers. In our work we were focused on the group of Sinumerik 810D, 840D and 840Di CNC controllers. The idea of the feature library development is based on a general assumption saying that it is possible to find a relationship between a particular CNC machining cycle and the simple design feature or a compound design features.

2.1 An overview of the machining cycles for Sinumerik 810D/840D/840Di

From definition machining cycles are motion sequences defined for drilling, boring/reaming, tapping etc. according to standard DIN 66025. These machining cycles are called in the form of subroutine with the unique cycle name and appropriate set of the parameters. In Sinumerik 810D/840D/840Di control systems there are five drilling and five boring/reaming machining cycles available. The short description of these cycles in the Table 1 is given [1,2].

Table 1. Drilling and boring machining cycles in Sinumerik 810D/840D/840Di [1].

Machining cycle type	Cycle name	Cycle function and description
Drilling, centering	CYCLE81	The tool drills at the programmed spindle speed and federate to the programmed final drilling depth.
Drilling, counterboring	CYCLE82	The tool drills at the programmed spindle speed and federate to the programmed final drilling depth. A dwell time can be allowed to elapse when the final drilling depth has been reached.
Deep-hole drilling	CYCLE83	The tool drills at the programmed spindle speed and federate to the programmed final drilling depth. Operation is performed with a depth infeed of a maximum definable depth executed several times. The tool can be retracted in order to swarf remove or for chip breaking.
Rigid tapping	CYCLE84	Rigid tapping with programmed spindle speed and federate to the programmed thread depth.
Tapping with compensation chuck	CYCLE840	Taping with compensation chuck operation with programmed spindle speed and federate to the programmed thread depth.
Boring 1	CYCLE85	Different feedrates for boring and retraction.
Boring 2	CYCLE86	Oriented spindle stop, definition of retraction path, retraction in rapid traverse, definition spindle direction of rotation.
Boring 3	CYCLE87	Spindle stops M5 and program stop M10 at drilling depth, continued machining after NC Start in rapid reverse, definition of spindle direction of rotation.
Boring 4	CYCLE88	As for CYCLE87 plus dwell time at drilling depth
Boring 5	CYCLE89	Boring and retraction at the same speed.
Row holes	HOLES1	Drilling of a row holes i.e. a number of holes that lie along a straight line or a grid of holes.
Hole circle	HOLES2	Drilling a circle of holes. The type of holes must be determined by the drilling cycle.

Turning cycles in most CNC machine controllers are used for programming of the following manufacturing cuts: turning, boring, grooving, threading or manufacturing undercuts. In addition, there are some hole manufacturing cycles in Sinumerik

810D/840D/840Di control systems. They are derived from above mentioned boring/reaming cycles excluding specific milling manufacturing cycles for example machining of the holes patterns. There are seven turning machining cycles available in Sinumerik 810D/840D/840Di control systems. The short description of these cycles in the Table 2 is given.

Table 2. Turning machining cycles in Sinumerik 810D/840D/840Di [1].

Machining cycle type	Cycle name	Cycle function and description
Grooving	CYCLE93	Turning of symmetrical and asymmetrical groves for longitudinal and travers machining on straight contour elements.
Undercut	CYCLE94	Machining of undercuts of E and F forms in accordance with DIN509 with the usual load on a finished part.
Stock removal	CYCLE95	With this cycle, contours can be machined in the longitudinal and facing directions, inside and outside. This cycle is freely selectable for roughing, finishing, complete machining).
Thread undercut	CYCLE96	Cycle used for thread undercuts machining of A, B, C, and D forms in accordance with DIN 76 on parts with a metric ISO thread.
Thread cutting	CYCLE97	Turning of threads.
Thread chaining	CYCLE98	It allows to produce several concatenated cylindrical or tapered threads with constant lead in longitudinal or face machining, all of which can have different thread loads.
Extended stock removal	CYCLE950	The same as for CYCLE95, but the finished part contour profile cannot have any relief cuts must be continuous. For roughing the programmed infeed is set precisely, the last two roughing steps are divided equally. Roughing it performed to the programmed to final machining allowance. Finishing is performed in the same direction as roughing.

Because, at the current stage of our research we are focued only on rotational symmetric parts such as shafts, sleeves and shilds, the machining cycles analisis was limited to gruops of boring/reaming and turning cycles. The analisys of the usebillity of milling cycles for design features identification was omitted. Taking into account the set of available machining cycles, in the consiedered case these of the Sinumerik 810D/840D/840Di machine contollers it can be stated that is possible to work out a set of the desing features that can be next treated as the base for feature based design modelling module elaboration.

In the Figure 2 aproposed structure of design feature classes worked out on the basis of the usuablity of the machine cycles is shown. It could be easily noticed thatthis structure is strictly oriented on machining cycles. Moreover, the characteristic feature of it is its simplicity compere to other structures for example presented in [3]. This simplicity results from fact that the internal structure of the particulars classes is more complex than in case of the complex class structures but with less complex structures of the classes. The base class in the proposed solution is *Design* class that represents the product itself. The rest of classes are derived from this class. Design features such as pins – cylindrical, conical and profile external surfaces are represented by object of the *Profile* class. Holes both circumferential and travers are represented by instances of the *Drilling* class and its derivative classes it is *Centering*, *HoleDrilling*, *DeepHoleDrilling*, *Counterboring*, and *Boring* classes respectively. Design features of threads by means of the *Tapping*, *Thread* class, and their derivative classes are represented in the model, whilst undercuts by the *Undercut* class and its derivative classes such as *UndercutFormE*, and *UndercutFormF*

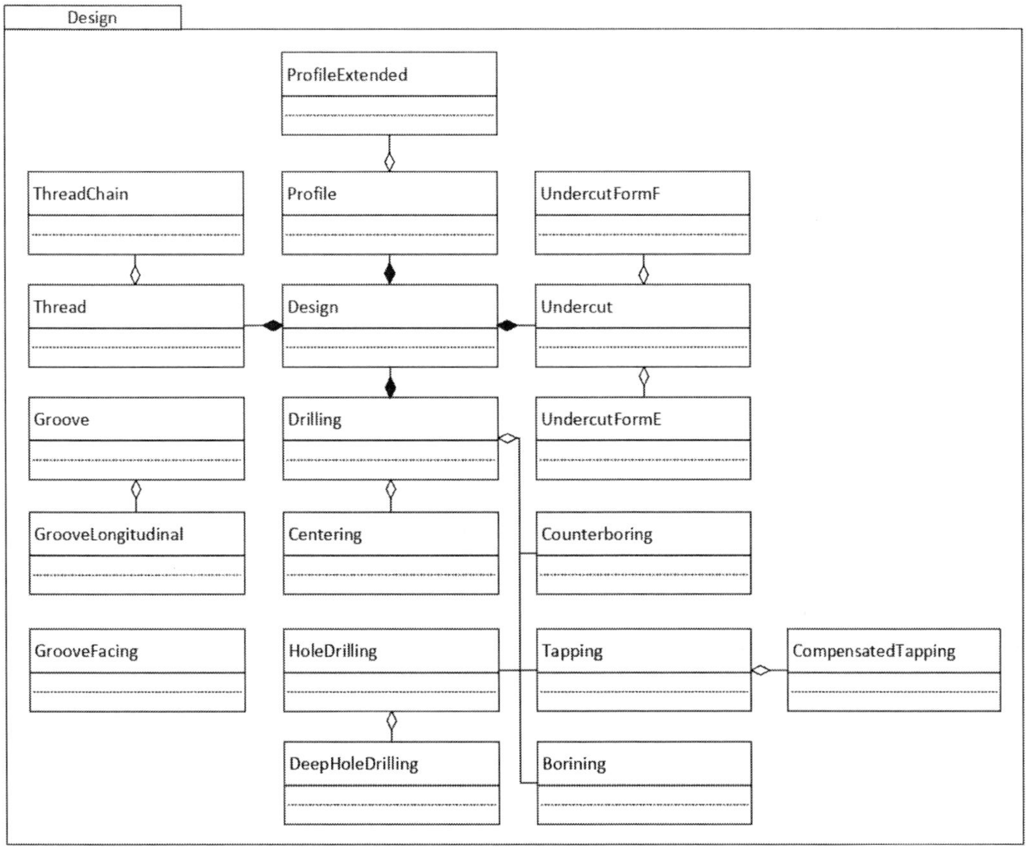

Fig. 2. The proposed structure of the design features.

classes. Grooves in the design features class structure are represented by the *Groove* class and its derivative classes *GrooveLongitudinal*, and *GrooveFacing*.

Conclusions

In the paper an attempt of application of the machining cycles in order to build the design feature library is presented. At the current state of the research the proposed structure of the design features is strictly oriented on manufacturing operations that can be performed with a lathe and milling machines equipped with Sinumerik 810D/840D/840Di CNC controller, but the achieved result are promising enough to say that they can be applied for any other CNC controller. This result will be used for creating of the prototype of the CAPP system.

References

1. Siemens, *Sinumerik 840D/840Di/810D/FM-MC programming guide* (Siemens, 2000)
2. C. Grabowik, K. Kalinowski, D. Krenczyk, Lect. note. in arti. inte. 7209 (2012)
3. A. Kampa, G. Gołda, I. Paprocka, Comp. 6, 10 (2017), doi:10.3390/computers6010010
4. A. Gwiazda, R. Knosala, J. of Mat. Proc. Technol. **64** 1-3 (1997) doi: doi:10.1016/S0924-0136(96)02566-6
5. A. Gwiazda, W. Banaś, A. Sękala, K. Foit, P. Hryniewicz, G. Kost, IOP Conf. Ser. Mater. Sci. Eng. 95 (2015) doi:10.1088/1757-899X/95/1/012104
6. K. Foit, A. Gwiazda, W. Banas, A. Sekala, P. Hryniewicz, IOP Conf. Ser. Mater. Sci. Eng. 95 (2015) doi: 10.1088/1757-899X/95/1/01210

MATEC Web of Conferences **112**, 06020 (2017) DOI: 10.1051/matecconf/201711206020
IManE&E 2017

From 3D layout to dynamic simulation model

Alin Gavriluță[1,*], *Ana* Gavriluță[1], *Eduard-Laurențiu* Nițu[1], and *Jérémie* Schutz[2]
[1]University of Pitești, Department of Manufacturing and Industrial Management, Pitești, Romania
[2]Université de Lorraine, LGIPM, EA 3096, Metz, F-57045, France

Abstract. The design of a supply system for an assembly line possess many challenges. A tool that helps the decision making and the adjustment of the supply system parameters is the dynamic simulation. This article has the aim of showing on a given case, the road from workstation layout, to its simulation model and an example of parameter that can vary and its effect of workstation performance. Starting from the 3D model of the workstation, varying the number of diversities present in the workstation, by using the simulation we can see the effect of this variation on the workstation cycle time.

1 Introduction

The globalization and strong competition in currentmarketplaces have forced companies to change theirways of doing business [1]. Presently, high productivity and a quick response to customers are essential for most producers [2]. Reducing the cycle time and inventories are objectives in managing production systems.

To enhance the system performance when designing layouts for assembly line workstation, the positions of the manufacturing system components must be appropriately determined by considering the sequence of tasks the operator conduct during the assembly process.

The purpose of this research is to present a methodology of crossing from layout to dynamic simulation for reducing the cycle time. To describe the methodology we used IDEF 0. Functional analysis methodology used in the study – IDEF0 [3] is based on abstraction and consists of the decomposition of the line activities in as fine details, until reaching the level of detail wanted.

2 Methodology of crossing from layout to dynamic simulation

To describe the working methodology of crossing from layout design to dynamic simulation and its functional links, it was used the IDEF0, functional modelling method.

Specific of this modelling method is the levelling of description from a higher point of view (A0) to as much details as needed. We consider the layout design and the simulation study as part of the overall design of the workstation and will be presented as so.

The workstation design has the following elements: input – process and environmental data; resources – software, tools of design and analysis, specialist, "know how"; controls –

*Corresponding author: gavrilutaalin@yahoo.com

rules, norms and standards and as output – workstation design and operating mode, Figure 1.

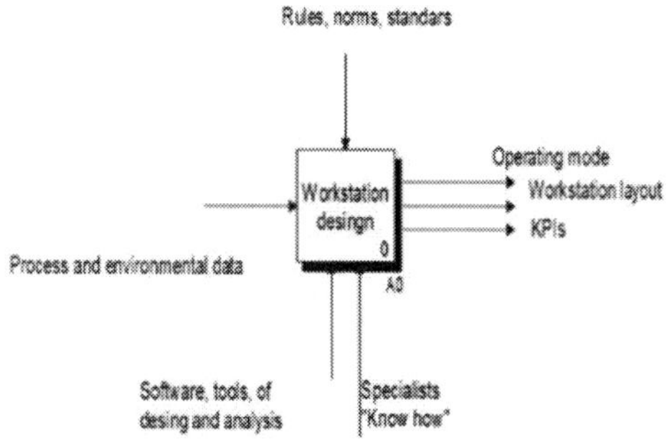

Fig.1. Workstation design model.

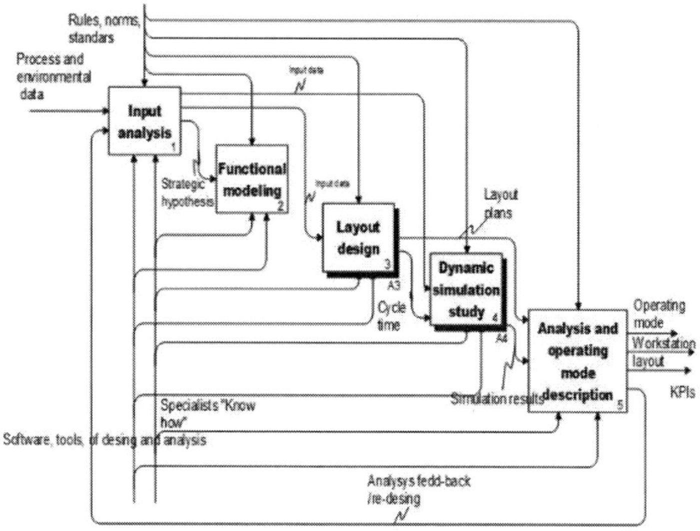

Fig.2. The steps of workstation design.

The first level of decomposition, Figure 2, has as elements:
-A1 – Input analysis: the entry data are processed and structured as to comply with the needs of design. Results are: strategic hypothesis that will guide the design steps and input data (cycle time, variables etc.) that will be the base of layout design and simulation;
- A2 – Functional modelling: as in this example, it helps in choosing elements as supply method, operation mode guidelines and explains the links between the system elements. The results are needed for layout design and dynamic simulation model construction;
- A3 – Layout design: uses the entry data and functional modelling guidelines to construct a layout and to determine the cycle time characteristics of it. Its results are used in dynamic simulation study and in the analysis step;
- A4 – Dynamic simulation study: starting from the functional model, this interactions are transformed in a dynamic model that will mimic the behaviour of a real life system, estimating the impact of strategic hypothesis, layout and variables from the design phase.
- A5 – Analysis and operating mode description: compiling the results of previous steps, the designers can compare performance KPIs concluding in the need of redesign or

validation of results. As end products the designer delivers: workstations layouts, operating modes, KPIs (key performance indicator) estimations etc.

Giving a closer look at layout design, Figure 3, has as elements:

- A31 – Design of dependent elements: this elements are dependent of the supply method (conveyer, dynamic racks, trolleys etc.), of the method of workstation loading/unloading (manual, automatic, semiautomatic), of transportation of parts between workstations (on trolleys, conveyers, hangers etc.), or other dependency factors. The result of this step are elements as: support for components, dynamic racks, cranes, transport platforms;
- A32 – design of independent elements: this elements are given by process characteristics and are the same no matter what layout characteristic or supply method are used. The results are: tool stands, HMI (human machine interface) design, machining or assembly automatic workstations etc.
- A33 – workstation layout construction: taking all the designed elements, the layout designer builds the workstation variants and comparing the specific KPIs, choses the more performant ones.
- A34 – layout plan construction: this step consists in the construction of the system graphic design. Taking all workstations design, the designer places them in interdependency and ensures the correct flows. The results are layout plans and distance measurements between the elements of the layout. The distances and the positioning of elements in the 3D model is the base of time measurement and ergonomic conclusions to evaluate the layouts.
- A35 – distances and cycle time impact estimation: using the layout plan results the designer can make time measurements, ergonomic studies and variable estimations. This will be used in the dynamic simulation study.

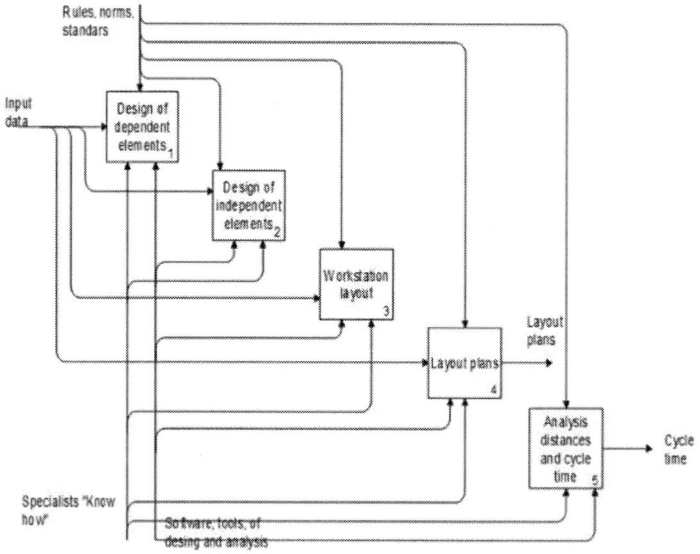

Fig.3. The steps of layout design.

This functional modelling can be used as base for further system design description upward and inward, depending on the needs.

3Application of methodology on a given case

3.1 Workstation and operating mode description

The presented example is an automotive assembly line workstation [4], Figure4. The work process is mainly manual and it contains dependent and independent elements. The assembly is made on a transport pallet that travels between workstations on a roller conveyer.

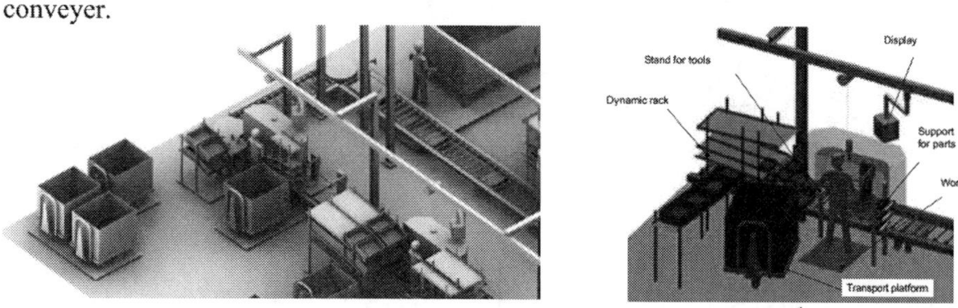

a b

Fig.4. Workstation 3D model.

The core of the workstation is the strike point area. This is an area where the operator has the minimum effort and the maximum speed in movement. The aim of the designer is to keep the operator at maximum in this area in order to reduce the Non added Value due to operator movements.

The workstation assembles 4 types of products. At each cycle there are 9 parts (small, medium and large) assembled in a specific sequence. Of each part there have to be 2 hour of autonomy, this giving the capacity of storage on place (racks for small and medium parts, transport platform for large parts). The sequence of operation is concentrated in strike zone, exception being the movement to get the parts.

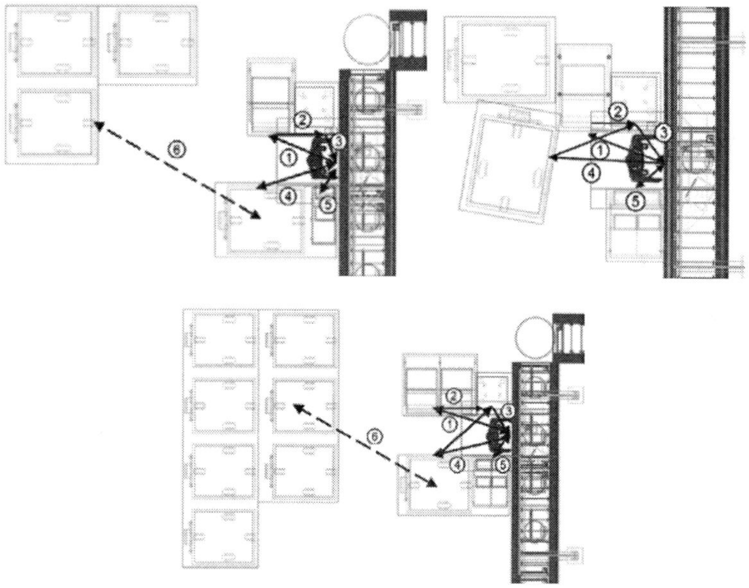

Fig.5. Sequence of operations impact in movement (1, 2 and 3 diversities of parts).

At each setup (batch change), the operator has to change the platform for large parts with the other type that is in the vicinity.

3.2 Cycle time impact

The cycle time needed to the assembly process is 1.2 min. This we consider constant and independent of the layout design.

As example, the details of the time impact of the layout for 2 diversities of parts are presented in Table 1 and Table 2. The operations that appear with a specific frequency are very important, them giving the perturbation that influence the workstations efficiency and have an impact in the rest of the line (upstream and downstream).

Table 1. Sequence of operations impact in operator movements (2 diversities of parts).

Sequence no.	Distance	No. of tours	No. steps	Time (min.)
1	1.5	1	3	0.048
2	1.4	1	2	0.032
3	0.6	1	1	0.016
4	1.2	2	2	0.064
5	0.5	4	1	0.064
6 (setup)	*3.7*	*2*	*6*	*0.192*
Total	5.2	-	15	**0.224**

Table 2. Packaging specificity and the impact of it on cycle time.

Parts	Coef.	No. parts/ pack.	No. of packs	Freq. of change (hours)	Time (min.)
R1	2	2900	1	29	0.053
R2	1	30	4	2.4	0.053
R3	1	100	1	2	0.053
R4	1	80	2	3.2	0.053
R5	6	1500	1	5	0.053
R6	1	1020	1	20.4	0.053
R7	1	250	1	5	0.192
R8	1	1500	1	30	0.053
R9	6	1500	1	5	0.053

It can be observed that we have movements that are made at each cycle time and ones that are done depending on the supply method (autonomy, no. of parts in the packaging, setup at batch change), Figure5.

If it is to transpose this situation in a dynamic simulation model, it will be as following:

- process cycle time – invariable or has small variations that can be with no impact. This is given by tools, machines, control equipment, etc.
- movement cycle time – this is specific and dependent of the layout characteristics. It can be input in simulation as a variable cause of the high variations due to operator influence. This will also vary from one diversity to another due to different distance implied;
- change of packaging cycle time – this is specific and dependent of the supply method and the diversity number, that gives the distances in the workplace. It can be input in simulation with a cyclic currency (ex. at each 250 work cycles, the capacity of packaging for the component R7, at 2 diversities variant, the operator has a 0,192 min. packaging change time);
- batch change cycle time due to process – the time needed for tools change, machines setup, etc. This cycle is independent of the layout;

- batch change cycle time due to layout: considering the method of supply and the position of the alternative parts that have to be assembled, this cycle time can be with great influence on the line flexibility. For example: on an assembly line were the process setup is "0", the supply setup is the only responsible of line flexibility at batch change.

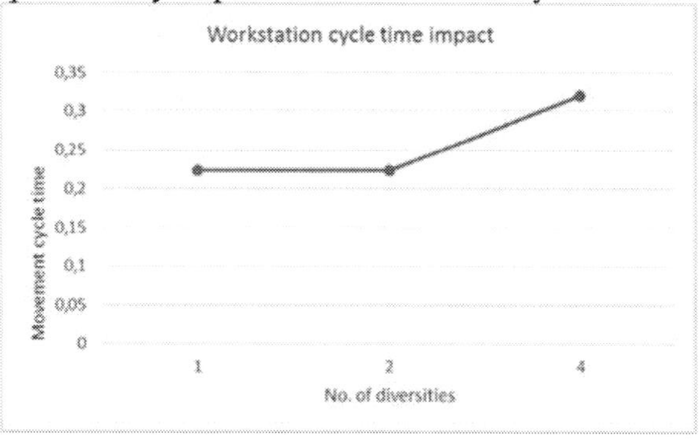

Fig.6. Results of analysis of movement on the workstation cycle time.

3 Conclusions

The layout design has a strong impact on the dynamic simulation analysis. The tendency can be to take into account the direct process time and setup. As started to show in this article, the layout has an import influence, if not more of an importance than the breakdown.

In our example the cycle time direct linked with the process is 1.2 min. Making the sum of all movement done per cycle the cycle time ends to be from 1.42 to 1.54 min, concluding that the impact of the layout is between 15 to 22% from the workstation cycle time (fig 6).

Also it can be viewed that in a line where the setup at batch change can be 0, which theoretically can be completely flexible, due to logistic reconfiguration of workstation, leads to a decrease of 13% the line productivity.

The production line system, and more specific the assembly line system is more and more a concept of batch of 1, reduction of WIP (work in process) (small if no buffers between workstations), inline conveyer transport. These concepts imply that any cycle time variation has a strong influence in the overall system. To detect this influence, it is discovered the need of dynamic simulation of flow, and in it the analysis of the impact of workstation layout.

The next step of the study is to extend the analysis from one workstation to a full assembly line.

References

1. R.S. Chena, K.Y.Lua, S.C.Yua, H.W.Tzenga, C.C. Chang, Information and Management **41** 25-37 (2003)
2. J.D. Wells, W.L. Fuerst, J. Choobineh, Information and Management **35**53-62(1999)
3. A. Rotaru, A. Gavriluţă, E. Niţu, Applied Mechanics and Materials **371** 116-120(2013)
4. A. Gavrilută, E. Niţu, A. Rotaru, Proceedings of the 3th Review of Management and Economic Engineering – International Management Conference "A new dilemma: between East and West" 263-274 (2012)

DOI: 10.1051/matecconf/201711206021

Analysis on the influence of supply method on a workstation with the help of dynamic simulation

Alin Gavriluță[1,*], *Eduard-Laurențiu* Nițu[1], *Ana* Gavriluță[1], and *Jérémie* Schutz[2]
[1]University of Pitești, Department of Manufacturing and Industrial Management, Pitești, Romania
[2]Université de Lorraine, LGIPM, EA 3096, Metz, F-57045, France

Abstract. Considering the need of flexibility in any manufacturing process, the choice of the supply method of an assembly workstation can be a decision with instead influence on its performances. Using dynamic simulation, this article wants to compare the effect on a workstation cycle time of three different supply methods: supply on stock, supply in "Strike Zone" and synchronous supply. This study is part of an extended work that has the aim of compering by 3D layout design and dynamic simulation, different supply methods on an assembly line performances.

1 Introduction

By Blackstone and Cox [1], an assembly line is an assembly process were the equipment and the workstations are so arranged that the sequence of assembly is followed.

A workstation is the place where the processes of assembly have place. It is defined by a precise area and contains known tasks. The parts that have to be assembled need a entire sequence of operation in the workstation [2]. In an assembly line, the method of supply with parts is of great importance, it represents one of the basics of the performance of the system and also makes a transition between two different processes: logistic and assembly operations.

The supply of an assembly line represents the transport of components needed for each workstation where the products are assembled. The supply must consider two main actions: to ensure the needed quantity and to ensure all the types of components (diversity) for all types of products.There are more types of supply methods: supply based on stock, supply in Strike Zone and synchronous supply.

The supply on stock is the most often met. Wänström and Medbo [3] say that in this type of supply the parts are placed in storage inside the assembly line. The parts stocks are refilled periodically after they are depleted. Caputo and Pelagagge [4] considering that comparing to other methods of supply of type JIT, this type is one discontinuous. It is considered that this method increases the effort of the operator due to the parts handling and makes it harder for him to search for the needed components.

The supply in Strike Zone means a reorganization of the workstation and its supply, allowing the operator to concentrate only on the tasks of assembly process, reducing the

*Corresponding author: gavrilutaalin@yahoo.com

wasted time with the movement to choose parts. The operators' activities can be more focused on ensuring the quality of its work. To do that, the operator can receive the needed parts in a rack or on conveyers that are placed on the side or in front of the operator, in an optimum ergonomic area, known as Strike Zone. This is easier to use with an integrated management system of the assembly line which makes the link between the production program and the flow of supply.

The synchronous supply can be used only if exists an integrated management system of the assembly line. The information taken from this system are used to synchronize the flows of supply with the flow of production, the parts arrive in the workstation exactly in the moment and order they are needed.

The simulation is the process of creating a model after a real life system (existing or in design phase) and making a series of experiments on it, with the purpose of understanding its behavior and to evaluate new optimization strategies. The simulation means the representation of a system or a phenomenon, basing on a model that allows the analysis of its characteristics and its distinct behavior in different hypothetical situations [5].

With the help of simulation the researcher can generate scenarios that in real life can't be observed and more importantly is the possibility to test conditions and scenarios before implementing them, reducing considerably the costs and the degree of risk taken. Also another strong advantage of the simulation is that it gives a better understanding of the complex interaction of certain variation factors and the interdependency between them. The results are more as tendencies and help to formulate or test analytical expressions.

The dynamic simulation, next to the layout design, is the main tool used in this study and analysis of the methods of supply of the production system.

2 Research methodology

At the base of the dynamic simulation study is the functional model. This modelling method was used to explain the methodology of simulation study, Figure 1:

- A41 – Dynamic simulation model construction: using the functional model and the results of layout design, the model that mimics the real life system is built. The model must copy the real life operating mode and has to include the variations that will give the "randomness" specific to real life situations. This model is further transferred to validation

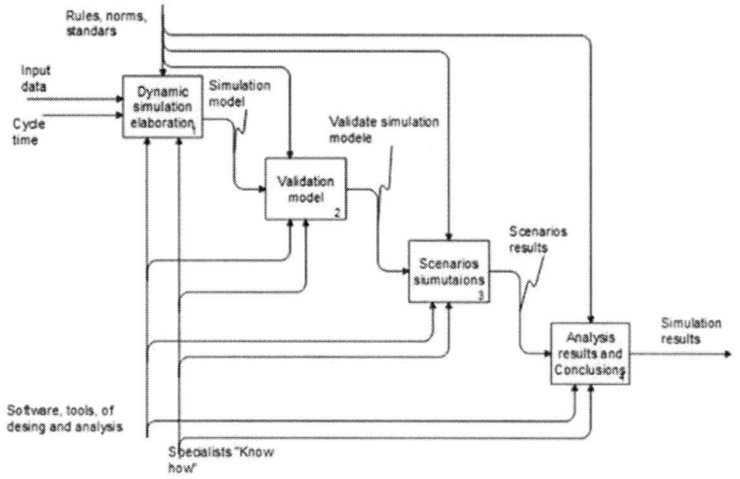

Fig.1. Dynamic simulation study.

step;
- A42 – Model validation: has as aim to ensure the correctitude of the simulation model from operation point of view and variable interdependence. The best way to validate is to compare with a real life reference scenario. The result is a valid model that will be used in scenarios simulation.
- A43 – Scenarios simulation: starting from the validated model, the designer has to implement all the foreseen scenarios exiting performance results as system capacity, efficiency, OEE etc.
- A44 – Analysis of results and conclusions: considering the art of dynamic simulation, this step is the most challenging. Starting from scenarios results, the designer has to make conclusions and to detect bottlenecks or tendencies.

As entry data, the model has the input that is common with layout design, and data that come from the analysis of the layout.

By analysing this input, the designer has to separate events that are common for each cycle time and events that appear with a frequency. This frequency can be variable, as a workstation breakdown or exterior perturbation, or invariable, as packaging change when it gets empty or programed tools changes or preventive maintenance.

Another factor that generates events is the setup time at batch change. This can be of two types: setup time due to process (tools change, program change etc.), or setup time due to workstation supply (change of packaging, change of parts location etc.).

In Figure 2, are represented as events the setup time at batch change and the change of packaging when empty.

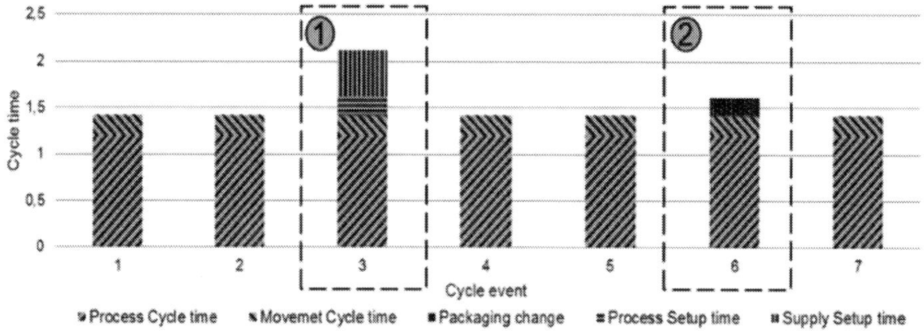

Fig.2. Example of events that appear in the workstations (cycle time in minutes).

The events have constant frequency and variable frequency. The dispersion of the variables must be determined or chosen from the research literature, according to its specificity.

An example can be the triangularvariable distribution. This distribution expression is used frequently to represent the variation of manual activities. Considering a mean manual cycle time of 0.224min., a dispersion of the variable can be ± 10 %. This dispersion copies the accepted variation of an operator manual cycle time that has a normal degree of dexterity.

Other variables must be determined by analysing the historic of events and extracting a low that represents them. As example, the logistic system doesn't always give the right quantity of parts at the needed moment, so the workstation can have stoppagesdelays due to lack of parts.

3 Application of methodology on a given case

3.1 Workstation supply and cycle time impact

Starting from the influence of the supply method on the workstation, Figure 3, each configuration has to be converted in time influence.

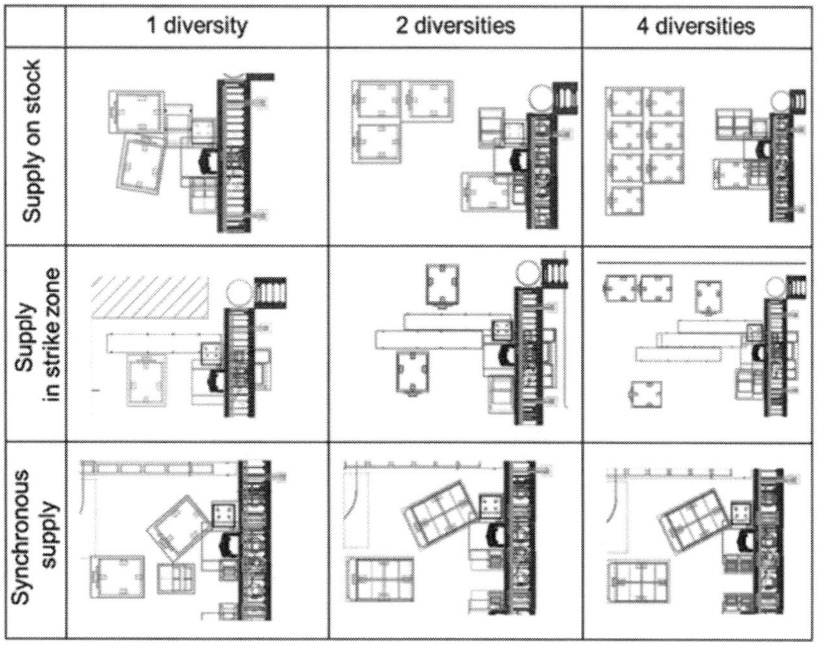

Fig.3. Different workstation configuration, depending on the supply method.

Considering the process cycle time a variable not influenced by the method of supply, this will be the same in all the simulation scenarios.

The operator movement differs from one configuration to another. In the case of supply on stock the operator receives the parts directly from logistic storage area. There are no other handling operations on the flow. In the case of strike zone supply, the parts are loaded on the supply conveyors and racks that transport the parts more concentrated in the working area. For the synchronous supply, the parts are sequenced in a preparation area (picking, kitting), this operation adding one more handling on the flow.

The impact of the layout on the cycle time is presented in the table 1. We can observe there that for the supply on stock, beside the cycle time, the workstation has a setup time due to workstation rearrangement.

In Strike Zone, beside the workstation cycle time, appears the operation of preparation with its on time added. For synchronous supply, the sequencing also means a handling that also comes with a time impact. All this time added to the end product will be included in the dynamic model and will be transposed in influence on production and costs.

Beside this durations, the model also includes the random generated events like: breakdown or external influences.

A factor that is not taken into account is the link of the workstation with the other workstations, cause of the limit of the analysed perimeter.

Table 1. Sequence of operations impact in operator movements.

	Workstation (cycle time impact - min.)			
Supply method	Product diversity	1	2	4
On stock	Workstation	1,424	1,424	1,540
	Preparation	-	-	-
	Setup	-	0,192	0,224
In Strike Zone	Workstation	1,232	1,25	1,27
	Preparation	0,064	0,064	0,096
	Setup	-	-	-
Synchronous	Workstation	1,344	1,36	1,36
	Preparation	-	-	-
	Setup	-	-	-
	Kitting	0,098	0,197	0,296
	Picking		0,138	0,194

3.2 Dynamic simulation model

Starting from the functional modeling of the industrial system, it is built a dynamic model. To simplify the work and scenarios implementation, the model is as simple as possible, and complex enough to take into account all the considered influences on the final parameters, Figure 4.

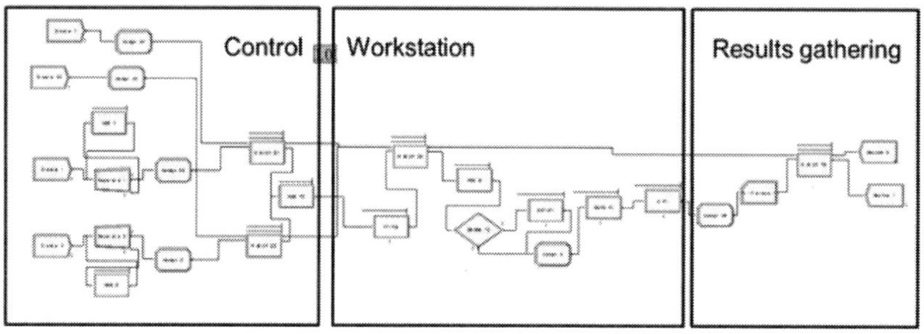

Fig.4. The dynamic simulation model

As for this model, there are 3 distinct sub-models: control, workstation and results gathering.

The control module is responsible of: order analysis, production grouping and attributes allocation. This module also manages the system conduction a pull or push type system, a certain flow or another depending of the type of supply. The workstation module is the one that mimics the operating mode of the workstation and its characteristics as: setup at product change, workstation supply, and the logistic preparation stage. The result gathering module defines and gathers the results as needed for analysis.

To compare the 3 types of supply methods it were considered a set of scenarios varying the number of products manufactured on the line. To compare the 3 hypotheses of supply and the impact of diversity, it was used the production per hour and the cost per part (due to the operations of process and handling). The results are shown in Figure 5.

If we consider the production per hour we can observe that the higher production is at the workstation supplied in strike zone. Also this type of supply is the most stable at variation of number of products. The synchronous supply is the second most productive supply method and on this area of variation of number of products has also a stable reactivity. The supply on stock has the lowest productivity and it is seen a strong decrease at the increase of number of products (more than 7% decrease from 1 diversity to 4 diversities).

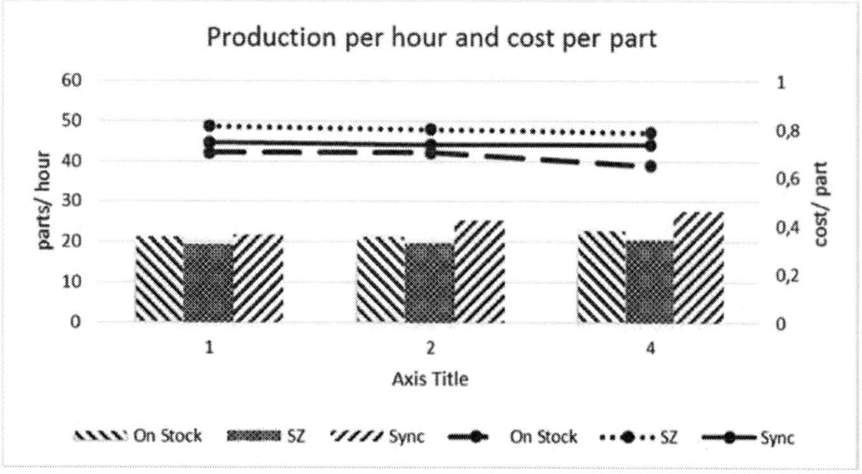

Fig.5. Results of simulation in: production per hour and cost per part.

Considering the cost per product it can be observed that the lowest cost also for the supply on strike zone. Also the increase of cost on this type of supply is of only 5%.The second lest costly type of supply is the supply on stock with a cost increase at number of products variation of 8%. The most costly supply is the sequenced supply. Also the increase of number of products leads to an increase of 28% of cost.

4 Conclusions

The conclusions of this layout and dynamic simulation model are very important, mostly because the results are contra intuitive. Considering more than one factor of influence and including a larger view of the subject can only increase the correctness of the conclusions of the studies.

As example, considering the number of handlings on the flow, it could be concluded that the less costly is the supply on stock, but due to the high impact of the layout (operator movement), the production decreases, and the cost per unit increases. Also it can be seen that two methods of supply by a logistic preparation area (strike zone and synchronous) have different impact on production and cost.

The next step to make, in order to have a better result is to build all the industrial system (workstations, logistic preparation areas and also to include de movement of the logistic operators). Gathering all this data and transposing them in the dynamic model we will have an overall impact of the supply method.

Another conclusion is that we can observe that it doesn't exist a solution valid in all cases. What the designer can do is to construct a flexible system, which can be easily adapted to variations and can be easily changed in case of need.

References

1. J.H. Jr. Blackstone, J.F. Cox, *APICS Dictionary* (Twelfth ed.) APICS 2008
2. Y.A. Bozer, L.F. McGinnis, Int. J. of Prod. Ec. **28** 1 (1992)
3. C. Wänström, L. Medbo. J. of Manuf. Tech. Manag. **20 - 1** 30 (2009)
4. A.C. Caputo,P.M. Pelagagge, Ind. Manag.& Data Syst. **108 - 6** 726 (2010)
5. A.E. Blanes, *Simulasion y Reingineria. Gestion de Sistemas avanzados de fabrication*, Polytechnic University of Valencia (2010)

MATEC Web of Conferences **112**, 06022 (2017) DOI: 10.1051/matecconf/201711206022
IManE&E 2017

Modelling and arrangement of composite panels in modernized freight cars

Marek Płaczek[1,*], *Andrzej* Wróbel[1], and *Maciej* Olesiejuk[2]

[1]Silesian University of Technology, Faculty of Mechanical Engineering, Institute of Engineering Processes Automation and Integrated Manufacturing Systems, Konarskiego 18A, 44-100 Gliwice, Poland
[2]IBS Poland Sp. z. o. o, PLM Support Department, Jagiellońska 4, 44-100 Gliwice, Poland

Abstract. A process of modelling in a CAD system and designing of arrangement of composite panels used for freight cars' body shell protection against corrosion and for easier unloading of transported cargo in winter conditions is presented in this work. Arrangement of used composite panels was designing in order to fulfil assumed criteria and thus to improve the process of freight cars modernization during periodic repairs.

1 Aim of the research work and state of the art

The main problem during exploitation of freight cars designed for coal transport is corrosion of the body shell. The goal of the work was to elongate the period between periodic repairs (by better corrosion protection) and improve conditions of exploitation of modernized freight wagons (for example easier unloading during winter conditions – no freezes of the charge to the freight wagon body shell). The proposed solution is application of composite panels to the freight car's body shell. The main causes of degradation of freight wagon body are acid rains, sulphur precipitated from coal and mechanical damage occurring during loading and unloading Figure 1. The repair of those wagon must be conducted every few years and the process itself is time consuming and expensive.

New materials and technology are solutions that can develop technical devices [1-5]. Also in case of railway transport there are a lot of research works carried out in order to develop this mean of transport. Some solutions proposed as a development of freight cars are applications of aluminium alloy or stainless steel [6,7]. Application of aluminium because of the high specific strength to weight ratio of aluminium enables increased payload and fuel savings as well as corrosion protection [6]._Applications of special types of stainless steel can also extend financial planning horizons when they consider maintenance and replacement as proposed authors in paper [7]. As they explain the true cost of ownership comparisons between several candidate materials - mild steel, conventional stainless and specialty stainless steel have repeatedly shown that replacement periods can be extended by as much as ten years with specialty stainless steels such as Cromgard. It can result in a very significant reduction in overall lifecycle cost. They

[*]Corresponding author: marek.placzek@polsl.pl

claimed that during inspection of coal wagons built in specialty stainless steel it was proved that it can provide a projected ultra-long service life and require minimal maintenance. They inform that during inspection after years of exploitation of a typical vehicle from the fleet, which was completely built in specialty stainless steel, the hopper doors were found to be in a bright, smooth condition and no discernible metal loss was detected. The wear rate was consistent with the 40-year life expectancy required. The proposed stainless steel provides a very good resistance to corrosion and abrasion. It also allowing the construction of robust structures at relatively low weight what is very important in case of freight cars [7].Developments in composite materials for the rail industry is presented also in [8] where authors present current status and options for further cost effective weight savings in rail vehicles obtained by application of composite materials. Less weight and as a result less energy consumption is presented as one of the benefits brings by the composite materials application. This advantage can also be used in freight cars.

In the presented research work an application of the composite panels to the freight wagon's body shell was proposed as the solution that can solve mentioned problems during exploitation [9-14]. The overall thickness of the composite was 5 mm and the maximum width could not exceed 1200 mm due to manufacturing restriction. Composite panels were mounted on the body shell using rivet nuts. Each panel was fastened with twelve ϕ8 rivet nuts with the assumption that there were five rivets on each of the longer side. Due to corrosion of the wagon's shell, rivets were to be mounted on reinforced parts of wagon (elements of the frame - for example to posts). What is more the body shell of the modernized freight wagon was painted using an anti-corrosion agent, while the standard freight wagon body is not painted inside at all.

Fig. 1. EAOS 1415-A3 freight wagon (*a*) and 418V Dumpcar freight wagon (*b*).

2 Modelling of composite panels

Composites modules were designed for two widely used in Poland freight wagons EAOS 1415-A3 and Dumpcar 418V. For the needs of this work, a 3D documentation of those wagons was prepared by Silesian University of Technology. The documentation was originally made in NX Siemens 8.5 software and then via STEP 214 ISO standard exported into CATIA V5-6R2014 software, which was the default work environment. The design process began with determining the actual interior dimensions of both freight wagons.

Additionally initial number of used composite modules taking into account criteria of their width and thickness was estimated. During this estimation manufacturing processes were also considered to provide most productive fabrication and limiting diversity of composite plates' sizes. Finite number, distribution and dimensions of plates were calculated with analytical method. Prior to the design process a material library was defined containing information such as Young's modulus, Poisson's ratio, density, cured thickness and width about used materials such as fiberglass and epoxy resin [Table 1]. Library was created in CATIA Material Library application. The design itself was conducted in dedicated application - Composite Design.

Table 1. Materials specification.

	Fiberglass	Epoxy resin
Young's Modulus [GPa]	8	3,5
Poisson's ratio	0,31725	0,4
Density [kg/m^3]	2600	1170
Cured thickness [mm]	0,46	0,088
Material width [mm]	1200	1220

At next stage sketches were made in Generative Shape Design application, which were then used to create planar surfaces. Those planar surfaces along their contour were to be used as composite basis. In Composite Design application defined a Laminate value which determined composite structure. In Composite Parameters dialog box materials to be used were defined as well as their order and orientation.

Following composite layers were defined:

I. Epoxy resin – thickness 0,088 mm; orientation 0°
II. Fiberglass – thickness 0,46 mm, orientation -45°
III. Fiberglass – thickness 0,46 mm, orientation 45°
IV. Epoxy resin – thickness 0,088 mm; orientation 0°
V. Fiberglass – thickness 0,46 mm, orientation -45°
VI. Fiberglass – thickness 0,46 mm, orientation 45°
VII. Epoxy resin – thickness 0,088 mm; orientation 0°
VIII. Fiberglass – thickness 0,46 mm, orientation -45°
IX. Fiberglass – thickness 0,46 mm, orientation 45°
X. Epoxy resin – thickness 0,088 mm; orientation 0°
XI. Fiberglass – thickness 0,46 mm, orientation -45°
XII. Fiberglass – thickness 0,46 mm, orientation 45°
XIII. Epoxy resin – thickness 0,088 mm; orientation 0°
XIV. Fiberglass – thickness 0,46 mm, orientation -45°
XV. Fiberglass – thickness 0,46 mm, orientation 45°
XVI. Epoxy resin – thickness 0,088 mm; orientation 0°

As result total composite thickness added up to 5,128 mm. Laminate value created in previous steps was then defined on a base surface. Result of defining the area where laminate will be applied was obtaining a surface-based composite plate design. Next stage assumed creating an associative solid from surface-based design. This solution provides an ability to modify the base surface, composite structure as well as its thickness, which would change the solid after an update. Although the resulting solid is associative with composite parameters it is created as object without history of its design and information about materials from which it was made of. The purpose of creating solid-based composite modules was to be able to create technical documentation of those object as well as ability to position them in the context of the wagon.

Provided freight wagon models were opened in Assembly Design application and mounting holes in wagon floor and plating were made. Holes were created using "Hole" function and their diameter was set to 11mm as stated in their specification. In next step mounting holes were duplicated using "Rectangular Patter" to fulfil the requirement of twelve instances per composite plate. More holes were placed in a manner to provide 3 mm clearance between each plate.

Similarly, holes in the composite modules were created, which were to be used as engineering connection references in later stages of the design. Then "Multi Instantiation" function was used to duplicate and position identical composite plates in the context of the wagon. During described process of multi instantiation 3 mm clearance between models was also taken into consideration. Comparable actions were made in terms of symmetrical objects in the context of wagon. Additional symmetry plane were made and "Assembly Symmetry Wizard" was used to instantiate objects in the wagon. At the next stage engineering connections of coincidence and contact were defined between the composite plates and the body of the wagon to provide stability to the assembly.

At the final stage of the design process rivet nut model was created in Part Design application which was instantiated in wagon context. Rivet nuts were positioned using coincidence and contact engineering connections. Composite plates in EAOS 1415-A3 wagon were mounted using 462 rivet nuts which position were defined using 924 engineering connections. Similarly, 168 rivet nuts were used in context of 418V Dumpcar assembly positioned using 336 engineering connections. After design process was completed technical documentation was made in Drafting application. Each composite plate was dimensioned according to ISO 1101 standard.

3 Number and arrangement of composite panels

Technological processes which describe the manufacturing of composite plates define design requirements which should be taken in consideration. One of the most critical requirements was maximum width of the plate. Although it is possible to obtain composites of greater width using the same materials the process itself require using transition zones which makes it more expensive and time consuming.

Another restriction connected with possible number of composite panels was corrosion of wagon plating. What is more creating mounting holes in already severely damaged wagon plating could cause significant reduction of durability that is why size of composite modules had to be similar to distance between reinforcing posts. Described solution provided ability to create mounting holes in the least fragments of wagon body.

Last but not the least requirement was defined by repair processes which are usually done in groups of two people. That is why dimensions of composite plates should allow workers to handle and mount them in the interior of the freight wagon.

Taking into consideration all project requirements concerning composites plates' size and needed clearance between them following configurations were obtained:

- 418V Dumcar – to cover interior of the wagon body eighteen composite plates were used in four different sizes:

1. Four plates were used to cover middle of side walls, two plates per side (longitudinal cross-section of the 418V Dumpcar wagon – Figure 2*a*);
2. Four plates were used to cover the ends of side walls, two plates per side (Figure 2*a*);
3. Two plates were used to cover the front and back walls, one plate per side (top view of 418V Dumpcar wagon – Figure 2*b*);
4. Eight plates were used to cover floor of the wagon (Figure 2*b*).

- EAOS 1415-A3 – to cover interior of the wagon body forty three composite plates were used in eight different sizes:

1. Eleven plates to cover the floor of the wagon (top view – Figure 3*a*);
2. Four plates to cover front and back walls, two plates per side (transverse cross-section of EAOS 1415-A3 wagon – Figure 3*b*);
3. Eight plates to cover the ends of side walls, four plates per side (longitudinal cross-section of EAOS 1415-A3 wagon – Figure 3*c*);
4. Eight plates to cover the middle of side walls, four plates per side (Figure 3*c*);
5. Four plates to cover the middle of side wall, two plates per side (Figure 3*c*);
6. Four plates to cover plating above doors, two plates per side (Figure 3*c*);
7. Four plates to cover left door wing, two plates per side (Figure 3*c*);
8. Four plates to cover right door wing, two plates per side (Figure 3*c*).

DOI: 10.1051/matecconf/201711206022

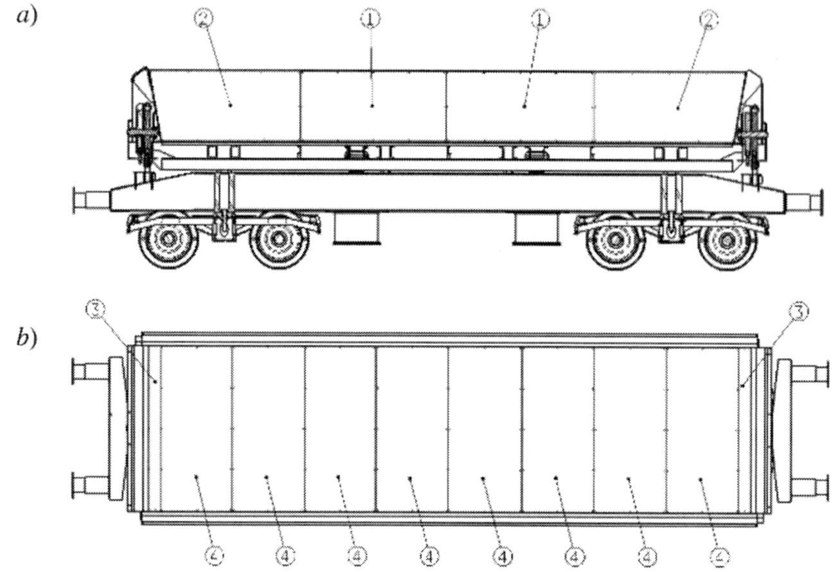

Fig. 2. The arrangement of composite plates - 418V Dumpcar wagon: *a*) longitudinal cross-section, *b*) top view.

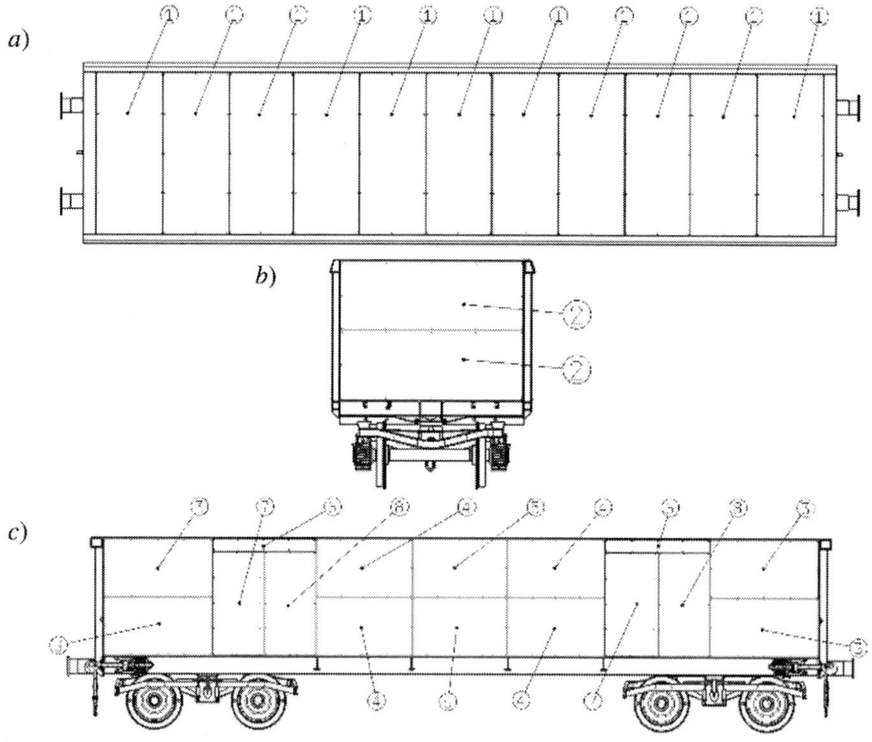

Fig. 3. The arrangement of composite plates - EAOS 1415-A3: *a*) top view, *b*) transverse cross-section, *c*) longitudinal cross-section.

4 Conclusions and discussion

As a response to the needs of the industry company concept of freight wagon modernization described in this paper could contribute to major savings. Developed solution could influence on service time as well as limiting costs of their repairs. CAD software used in this project along others available on the market can providesimplifying design and modification processes due to historical-based design. Complex software such as CATIA V5-6provide tools for effortless modification owning to design parametrization giving the ability to re-design components and assemblies by changing a small number of values. In this paper only a short report of research works concerned with composite modelling and solid model creation using dedicated application, assembly creation with the use of engineering connections, technical documentation creation according to ISO 1101 standard was mentioned.

The proposed solution was verified by a number of laboratory tests as well as test on the real objects during their standard exploitation and results of this works are presented also in other authors' publications [9-11]. Obtained results proved that it can extend the period between repairs of freight cars and improve their maintenance. On the other hand the proposed solution with application of composite panels to the body of the standard freight car during its periodic repair, without necessary to replace the wagon plating, is cost effective juxtaposing with other proposals with application of aluminium alloy or stainless steel [6-8]. The proposed technology can be applied not only in new freight wagons but also in renovated ones in order to prolong the period of exploitation.

References

1. A. Sękala, G. Kost, A. Dobrzańska-Danikiewicz et al.,IOP Conf. Ser.: Mater.Sci. Eng. **95** 012134 (2015)
2. K. Jamroziak, M. Bocian, M. Kulisiewicz, J. of Theoret. and Appl. Mechanics, **51(4)** 827-835 (2013)
3. P. Ociepka, K. Herbuś, A. Gwiazda, Adv.Materials Research **1036**1011-1016 (2014)
4. J. Tuma, J. Simek, J. Skuta et al., Springer Proceedings in Physics **139** 619-624 (2011)
5. K. Klarecki, D. Rabsztyn, M. Hetmańczyk, Ekspl. i Niezaw. – Maintenance and Reliability, 17(3), 338–344 (2015)
6. Skillingberg M, Green J. (2007) Aluminium Applications in the Rail Industry, Light Metal Age [http://www.aluminum.org/resources/electrical-faqs-and-handbooks/marine, access 31.03.2017]
7. AZO Materials, Applications of Cromgard Stainless Steel in the Railway Industry (2005) [http://www.azom.com/article.aspx?ArticleID=2929, access 31.03.2017]
8. AZO Materials: Developments in Composite Materials for the Rail Industry from TRB Lightweight Structures (2014) [http://www.azom.com/article.aspx?ArticleID=11061, access 31.03.2017]
9. A. Wróbel, M. Płaczek, A. Buchacz, M. Majzner, Int. J. Materials and Product Technology **50 (3/4)** 259-275 (2015)
10. M. Płaczek, A. Wróbel, A. Baier, IOP Conf. Ser.: Mater. Sci. Eng. **145** 042026 (2016)
11. M. Płaczek, A. Wróbel, A. Buchacz, IOP Conf. Ser.: Mater. Sci. Eng. **145** 072012 (2016)
12. M. Płaczek, A. Buchacz, A. Wróbel, Eksploatacja i Niezawodnosc – Maintenance and Reliability **17** 443–449 (2015)
13. M. Płaczek, A. Wróbel, A. Buchacz, IOP Conf. Series: Materials Science and Engineering **161** 012107(2016)
14. A. Wróbel, M. Płaczek, A. Buchacz, Acta Mech. et Automatica, **10** (1) 17-21 (2016)

Simplified simulation of impact bullet - stratified pack for restraining ballistic tests

Catalin Pirvu[1,*], *Traian Florian* Ionescu[2], *Lorena* Deleanu[2], and *Simona* Badea[3]

[1]National Institute for Aerospace Research "Elie Carafoli" INCAS, Bucharest, Romania
[2]"Dunarea de Jos" University, Galati, Romania
[3]Scientific Research Center for CBRN Defense and Ecology, Bucharest, Romania

Abstract. This paper presents a macro FEM isothermal model for simulating the impact of a 9 mm FMJ bullet on a stratified packs in order to restrain the preliminary tests for actual stratified packs made of layers of Twaron LFT SB1plus (Teijin), taking into account the friction between layers and between bullet and layers, in order to obtain the number of layers that arrest the bullet. The layers were considered as an orthotropic material, with maximum equivalent plastic strain of 0.06. The materials are modeled as bilinear isotropic with hardening. The analyzed parameter to validate the model was the number of broken layers. The average value of back face signature for a 12 layer pack was of 19.44 mm recommending it for protection level of II and IIA (NIJ Standard-0101.06-2008).

1 Simulation results

1.1 The model

Today, woven or multi-directional fabrics, made of fibers with high impact resistance [1, 2, 3], are used for individual protection armors [4, 5, 6]. Reports on ballistic impact simulation are published, from micro-scale [7, 8] to macro one [9-14], including simplifying hypotheses as considering a rigid projectile [15].

The aim of this paper is to run a simplified simulation of the impact bullet - stratified pack for restraining one or more parameters involved in the pack testing.

The isothermal model has 14 solid bodies: 12 identical layers considered a group of bodies (with the option *multiple materials)*, overlapped and rigidly fixed on their contour, and 2 bodies for the bullet (also *multiple materials)* with bonded connections (Table 1, Figure 1). The analysis is of structural type and solved with Lagrangean method, in Explicit Dynamics, using AutoDyn [16, 17, 18]. One layer dimensions are 6.10^{-2} m x 6.10^{-2} m x $0.6 \cdot 10^{-4}$ m. As the impact direction is the same to the symmetry axes of the model, the simulation is run for a quarter. The contact among bodies take into account the friction [19]: the friction coefficient between layers is 0.4, characteristic for sliding polyethylene against itself and the friction coefficient between a layer and the bullet jacket is 0.3.

* Corresponding author: pirvu.catalin@incas.ro

DOI: 10.1051/matecconf/201711206023

The meshing (Figure 1) is presented for the pack with 12 layers. The maximum side of an element is $5x10^{-4}$ m. Initial condition refers to the bullet velocity just before impact (here 400 m/s). Many recent models accepted the rigid clamp on the contour [15], even if for body armor, the pack is simply supported on visco-elastic material. The maximum number of cycles is 10^7 and the maximum error for energy was 0.9. Materials are modeled as hardening bilinear isotropic. This simplifying hypothesis could be justified by the arrangement of long aramid fibers, long at four different angles (0, 90, 45, -45) [20]. After consulting the literature [21, 22], the failure criterion was the ultimate plastic strain, set for 0.05.

Table 1. Characteristics of the model.

Body	**Nodes**	**Elements**
Layer	1250	576
Pack with 12 layers	18816	8748
Bullet (jacket + core)	1432	6289
Bullet jacket	739	2211
Bullet core	996	4078

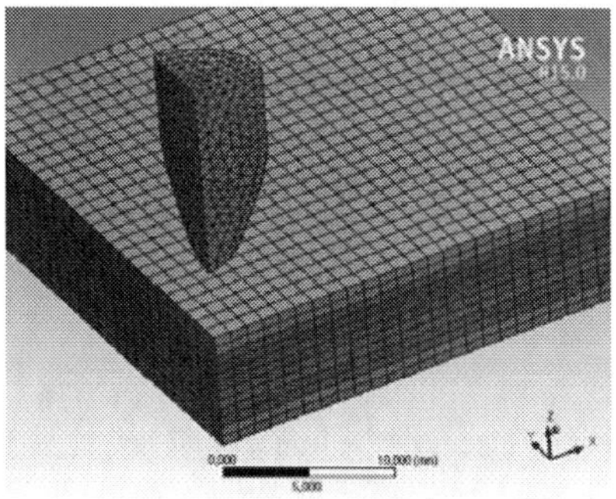

Fig. 1. Model meshing, 12 layers.

Table 2. Material characteristics used for modeling the impact.

Material	Young, Pa	Poisson ratio	Bulk modulus, Pa	Shear modulus, Pa	Yield limit, Pa	Modulul tangent, Pa
Layer	$7x10^{10}$	0.35	$7.77x10^{10}$	$2.59x10^{10}$	$6.3x10^8$	$1.9x10^{10}$
Copper alloy	$1.1x10^{11}$	0.34	$1.14x10^{11}$	$4.10x10^{10}$	$2.8x10^8$	$1.15x10^9$
Plumb	$1.6x10^{10}$	0.3	$4.44x10^{10}$	$5.55x10^9$	$3x10^7$	$1.1x10^8$

2.2 Results

Figure 2 presents time steps of the impact for the pack with 4 layers. Simulation was run for $2x10^{-4}$ s. After $1x10^{-4}$ s, the first layer is broken and the stress concentration is not exactly under the bullet tip, but in circular sector of the layer edge, situated at approximately 1/4 of the projectile radius.

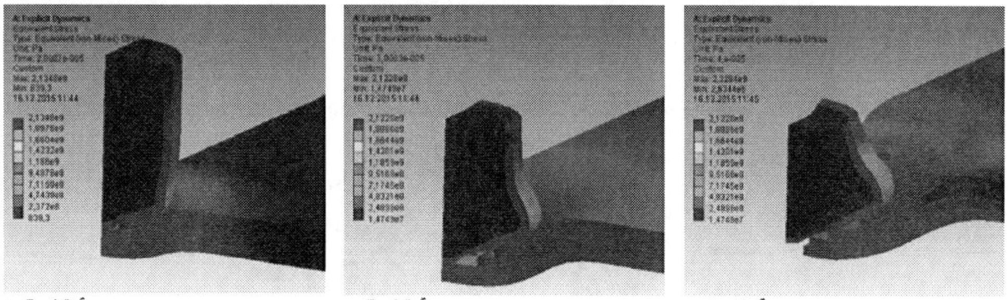

t=$2x10^{-5}$ s   t=$3x10^{-5}$ s   t=$4x10^{-5}$ s

Fig. 2. Imagine of the impact, at different time steps, for a pack made of 4 layers.

At t=2·10^{-5} s, the maximum stress (2058 MPa) occurs on the bullet jacket and within the first layer, on the impact axes, but also in the volume of the first layer, near the broken ridge that continues to be compressed by the bullet. The first layer has been already broken.

At t=3·10^{-5} s, the second layer is already broken and the first layer is laterally pushed, towards the bullet peripheral zone, producing the jacket thinning (actually, tests revealed the petal broke of the jacket) [23]. The bullet has already had the mushroom shape.

At t=4x10^{-5} s, 4 layers have been already failed. Stress concentrators are noticed on the bullet and near the edges of the broken layers that are still under compression. Layer 3 has high stress value just under the bullet tip (on the model axes). Till t=19.10^{-4} s, no other layer is broken.

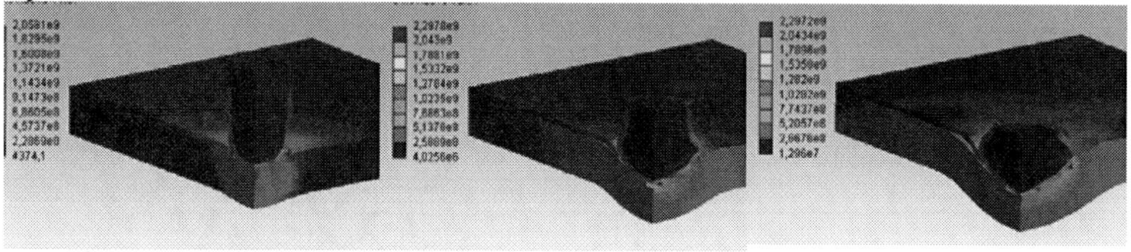

t=2·10^{-5} s t=3·10^{-5} s t=4·10^{-5} s

Fig. 3. Images of the impact of 4 layers pack.

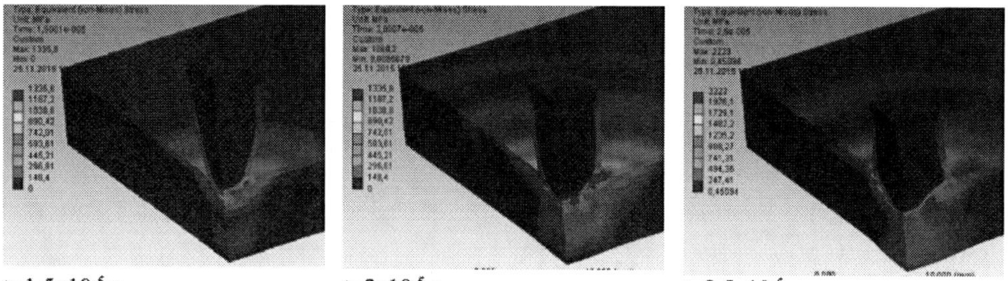

t=1.5x10^{-5} s t=2x10^{-5} s t=2.5x10^{-5} s

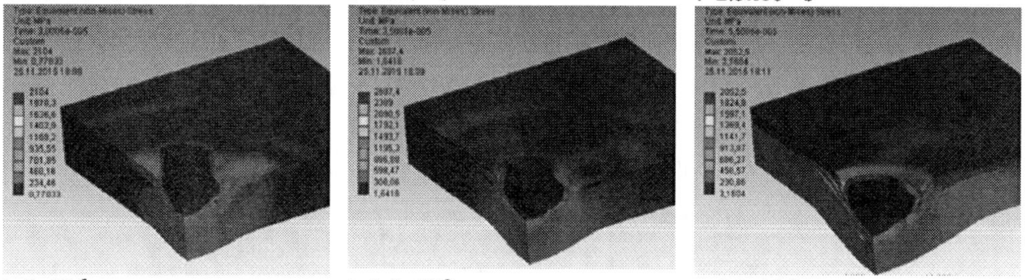

t=3x10^{-5} s t=3.5x10^{-5} s t=5.5x10^{-5} s

Fig. 4. Distribution of von Mises stress for different moments during the impact (12 layers).

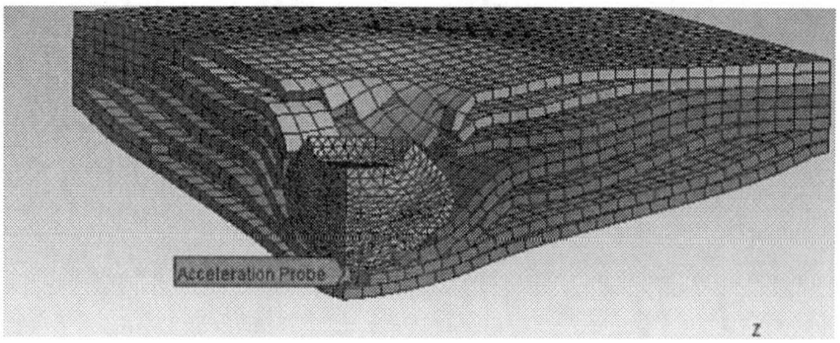

Fig. 5. 8 layers pack with 7 broken layers (t=10^{-4} s).

The simulated deformation of the central point of the pack is 17 mm, close to the measured BFS, for 12 layers packs (see Table 5), even if the actual packs have 500 x 500 (mm).

After shutting, the packs made of 12 layers LFT SB1plus had 4...5 broken layers (Figure 6), similar to the model (Figure 4). For the pack of 8 layers (Figure 5), only the last layers resists. Thus, from running the model, it results a conclusion that tests should begin with packs with 8 or more layers in order not to have total penetration.

3 Experimental work

These tests were done taking into account the American standard NIJ 0101.04/2004 [24, 25]: the projectile was a bullet of 9 mm FMJ, initial velocity being measured in the range of 400-420 m/s, and the target was fixed at 3 m. Fires were done in the laboratory of CCSACBRNE, by specialized personnel. The impact velocity was measured with the help of a chronograph Oehler model 43. Other devices included: rigid support for ballistic barrel; a ballistic barrel for bullets of 9 mm FMJ; box for the ballistic clay 610 x 610 x 140 ± 2 (mm); firing table with compensated kick [26]. Test conditions were: 19...23 ± 5°C; relative humidity: 50 - 70%; atmospheric pressure: 760...764 ± 1 mmHg. The clay grade was Roma Plastilina no. 1. The backface signature (BFS) was measured with a depth calipers, having the accuracy of ±0.1 mm. After each measurement, the calipers were cleaned for avoiding eventually adhered clay on the active measuring element. Actual layers were fixed with the help of a sewing line (200 mm on two opposite sides, at 40 mm from the edge).

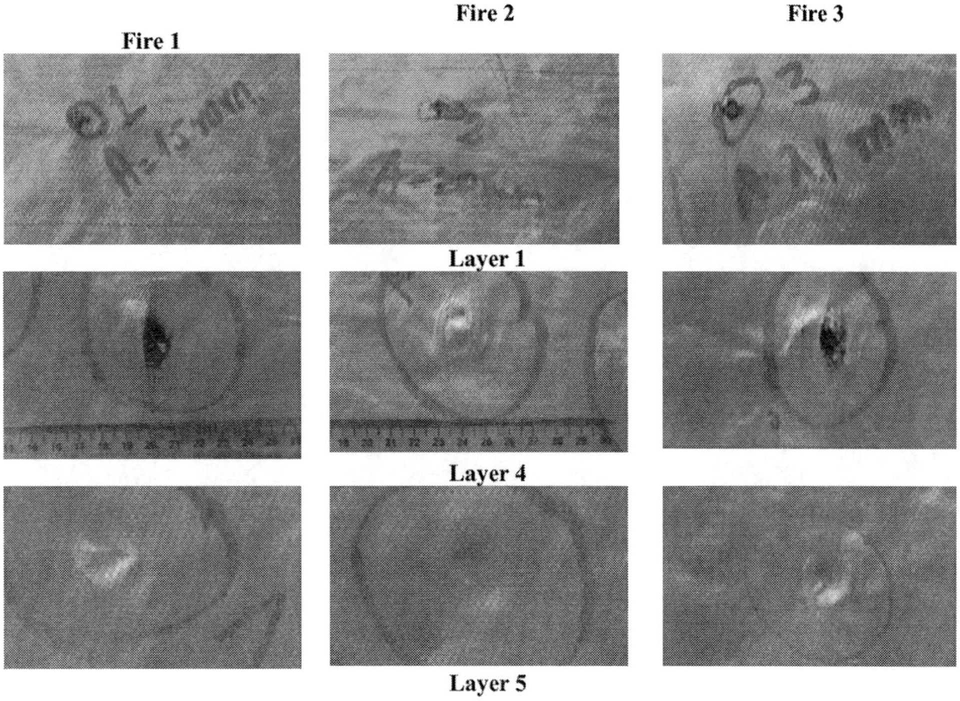

Fig. 6. Frontal view of a sample made of LFT SB1plus [23].

Figure 7 presents the values of backface signature (BFS) for the tested packs. The bold line for 44 mm is the accepted limit according to [24]. The values suggest a further decrease of the layer number without overpassing this limit, but tests are necessary to confirm this it. The average value for BFS, $Y_{average}$, for a number of fire N=18, is

$$Y_{average}=(1/N)\cdot SUM(Y_i)=19.444 \text{ mm} \quad (1)$$

Y_i being the value of each BSF, i=1...N [23]. The results recommend the packs made of 12 layers of LFT SB1plus for a protection level II and IIA. The standard deviation is s=4.6554.

The upper limit of tolerance, Y_u [24], for packs made of 12 layers of LFT SB1plus is:

$$Y_u = Y_{average}+K_1\cdot s=25.9354 \text{ mm} \quad (2)$$

and this value is much less than the permissible one of 44 mm.

Table 5. BFS for the packs made of LFT SB1plus layers.

Number of layers/ test number	Fire 1	Fire 2	Fire 3
	BSF [mm]		
12/1	15	20	21
12/18	29	24	21
12/12	23	25	25
12/19	18	12	21
12/13	14	15	17
12/9	16	14	20
8/8	26	TP	TP
8/9	21	TP	TP
8/6	30	33	31

TP - total penetration

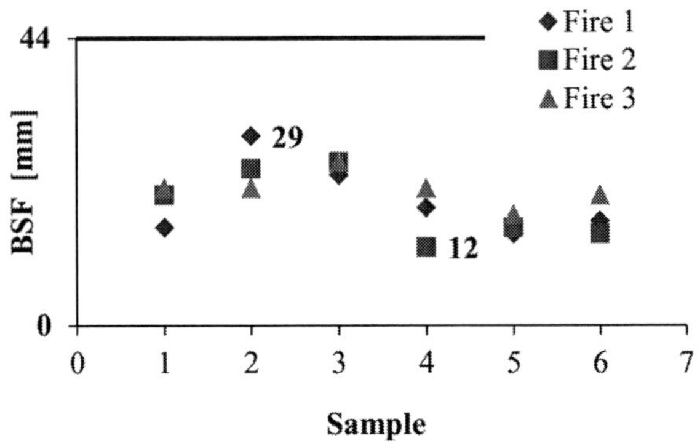

Fig. 7. Measured BFS for the packs made of LFT SB1plus layers.

4 Conclusions

The simulation pointed out the total penetration for the pack with 4 layers and 7 broken layers in the pack with 8 layers; the pack with 12 layers has only 4 broken ones (Figure 8). Analyzing the evolution of the velocity of the bullet head (Figure 8) and the kinematic energy of the bullet (Figure 9), one may notice the difference between total and partial penetration.

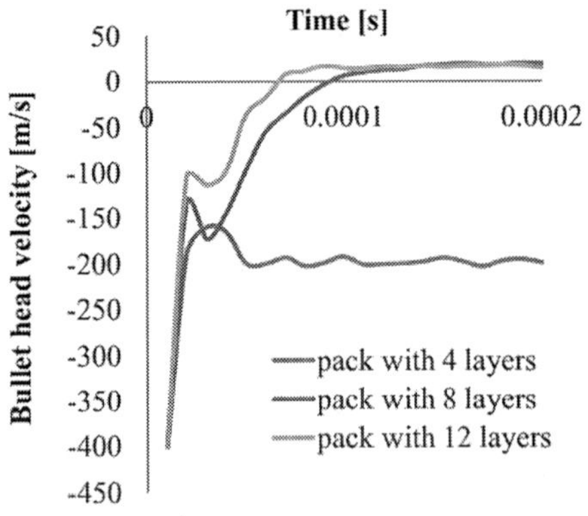

Fig. 8. Velocity of the bullet top.

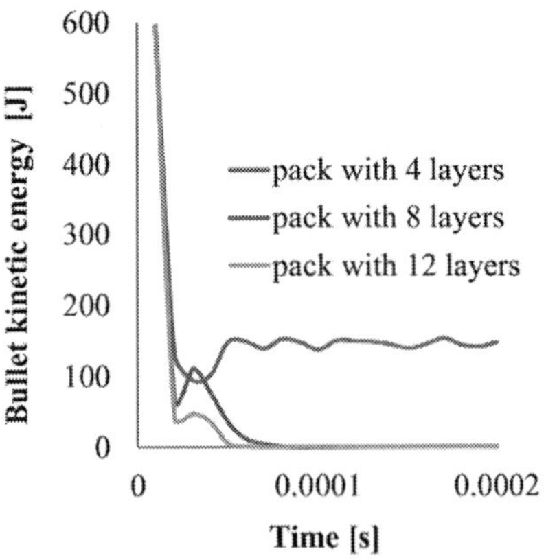

Fig. 9. Bullet kinetic energy during simulation.

Thus, the model was validated as for the pack made of 8 layers, there were obtained total penetration for more than half of the fires, but there were bullets arrested between the

8th and 7th layers. But such a risk is not acceptable and tests were done for 12 layers of LFT SB1plus.

The presented simulation on ballistic packs for individual armor with 12 layers, even if it was simplified, estimated the failure of 5 layers, and experiments validate this. Thus, a simulation at macro level may be useful in a rough estimation of the thickness (or number of layers) of the pack. Obviously, the simplifying hypotheses, the material properties and the conditions during simulation (isothermal and with friction) have to have realistic values and the results may shorten the range of some parameters, here the number of layers supposed to resist to a certain threat.

References

1. P. M. Cunniff, M. A. Auerbach, 23th Army Science Conference, Orlando, SUA, http://web.mit.edu/course/3/3.91/OldFiles/www/slides/cunniff.pdf (2002)
2. R. R. Kumar, K. S. Raghuram, Review of Research, **3** (2014)
3. S. Mohan, S. Velu, Intern. J. of Impact Engineering, **63** (2014)
4. A. Bhatnagar, *Lightweight ballistic composites* (CRC Press, New York, 2006)
5. A. Purushothaman, G. Coimbatore, S. S. Ramkumar, J. of Engineered Fibers and Fabrics, **8** (2013)
6. C. W. Ong, C. W. Boey, S. R. Hixson, J. O. Sinibaldi, Intern. J. of Impact Engineering, **38** (2011)
7. M. Grujicic, J. Snipes, S. Ramaswami, V. Avuthu, C.-F. Yen, B. Cheeseman, Intern. J. of Structural Integrity, **7/4** (2016)
8. R. Barauskas, A. Abraitiene, Composite Structures, **104** (2013)
9. S. Chocron, I. Benloulo, J. Rodrigez, V. Sanchez-Galvez, J. de Physique, **IV**, 07 (C3), C3-821-826 (1997)
10. M. El-Habti, PhD thesis, Queen Mary University of London (2010)
11. B. A. Gama, Jr. J. W. Gillespie, Intern. J. of Impact Engineering, 38 (2011)
12. P. Soltani, M. Keikhosravy, R. H. Oskouei, C. Soutis, Applied Composite Materials, **18** (2011)
13. G. Nilakantan, E. D. Wetzel, T. A. Bogetti, Jr. J. W. Gillespie, Composite Structures, **95** (2013)
14. D. Zhu, A. Vaidya, B. Mobasher, S. D. Rajan, Composites, Part B, **56** (2014)
15. G. M. Zhang, R. C. Batra, J. Zheng, Composites, Part B, **39** (2008)
16. *** AutoDyne composite modeling (ANSYS Inc., 2012)
17. T. Delpero, G. Lepoittevin, A. Sanchez, *Finite Element Modeling with ANSYS*, (Strukturlabor, 2010)
18. V. Năstăsescu, A. Ştefan, C. Lupoiu, *Analiza neliniară prin metoda elementelor finite. Fundamente teoretice şi aplicaţii* (Military Technical Academy, Bucharest, 2001)
19. Y. Duan, M. Keefe, T. A. Bogetti, B. A. Cheeseman, B. Powers, Intern. J. of Impact Engineering, **32** (2006)
20. *** *Ballistics Teijin Aramid*, Ballistics material handbook 38-1405/2012 (2012)
21. M. J. Hinton, A. S. Kaddour, P. D. Soden, *Failure Criteria in Fibre Reinforced Polymer Composites: The World-Wide Failure Exercise* (Elsevier Ltd., 2004)
22. D. Kreculj, B. Rašuo, Tehnicki vjesnik, **20**, 3 (2013)
23. C. Pirvu, PhD thesis, "Dunarea de Jos" University (Galati University Press, 2015)
24. *** Ballistic Resistance of Body Armor, NIJ Standard-0101.06, U.S. Department of Justice Office of Justice Programs National Institute of Justice (2008)
25. *** *Penetration Testing. Procedures and Methodology* (EC Council Press, 2011)
26. I. Safta, PhD thesis, Military Technical Academy, Bucharest (2011)

Aspects of modelling classical or synchronous modelling with Solid Edge ST 9

Adrian Mihai Goanta[1,*], and *Georgeta* Haraga[2]

[1]"Dunarea de Jos" University of Galati, Engineering and Agronomy Faculty of Braila, Research Centre for Mechanics of the Machines and Technological Equipments, Calea Calarasilor Street, 810017, Braila, Romania
[2]"Politehnica" University of Bucharest, Department of Engineering Graphics and Industrial Design, Splaiul Independentei No.313, Street, District 6, Postal Code 060042, Bucharest, Romania

Abstract. The current situation of the design activity is dependent on both the level of training of the human resources and the financial resources of companies required purchasing the design software packages and complex calculation equipment. Consequently, the situation is very diverse in the sense that there are design cases using only drawing software but also classical 3D or synchronous modelling situations, simple or integrated into software packages that meet the Product Lifecycle Management (PLM) principles. The natural tendency in modelling and design is primarily to the high computing power integrated software or somewhat simplified versions that, however, allow at least FEA modelling, simulation and the related 2D documentation. The paper presents some aspects of modernity in synchronous modelling as compared to the classic one, made with 2016 version of Solid Edge software from SIEMENS. Basically there were studied and analysed aspects of modelling ease, speed of changes and also optimization of commands in the modelling process of the same piece in the two versions mentioned: classic and synchronous. It is also presented the alternative path from one method to another within the same process of piece modelling, depending on the advantages provided by each method. In other words, the work is based on a case study of modelling a piece under the two modelling versions of which some aspects were highlighted and conclusions were drawn.

1 Evolution of Traditional Design Techniques and State of the Art

In small and medium business, traditional design before 1980 was done on board and paper followed by translation of both assembly and execution drawings on vellum transparent vellum/tracing paper which could be further multiplied in many copies and distributed in workshops for subsequent making and processing of the product. Original drawings on transparent tracing paper were archived in special libraries and archiving was based on the name and number assigned to each project. Any changes applied to older projects were

* Corresponding author: goanta_a_m@yahoo.com

made by adding new drawing boards, either as assembly drawing or execution drawing and, when too many changes were necessary, the entire documentation of the initial project was made again and eventually another identification number was assigned. With the development and expansion of computer technology, meaning the emergence of PC computer systems and their more powerful version, called graphics workstations, work could move on gradually from the design to the drawing board to computer aided design in the sense of computer – based drawing. In high technology sectors with adequate funding, graphic stations began to be used for three-dimensional modelling to achieve a photorealistic visualization close to the real designed products. Graphics station hardware resources of that time were well above personal computers, which used to determine very high costs of acquisition for most design companies. One should not forget that the acquisition costs of computers were added to the costs of commercial licenses of the software. If we compare the computing systems of today with performance of the graphics workstations of those times, it is hard to believe how was possible to work 3D at that historical times at the beginning of three-dimensional designing, considering the fact that the processors had maximum 250 MHz frequency and graphics cards, particularly important for 3D visualizations, did not exceed 4-8 MB AGP memory. Compared with those graphics workstations, personal computers of the time managed with great difficulty to generate a Wireframe of 3D solid and when they did it, they did it gradually giving the impression of a lengthy animation. At this point, it must be said that the situation is extremely varied in terms of the design method applied. This is because companies have extremely varied sizes and financial capacities. Thus at the current moment, small firms still use the method of computer –aided design while large business with extended activity and equally significant financial resources have implemented 3D design which implicitly allows getting automatic orthogonal or axonometric projections, the designer having just to add the necessary sizes for manufacturing, shape and position deviations, technical notes and all the other documents required by the rules of the technical drawing. It should be noted that at this moment there are companies that have managed to overcome this level of design complexity, meaning that they use integrated PLM software's (Product Lifecycle Management) which in a single file introduce information about both the 3D model, the execution drawing of the model, and information about the physical properties of materials, finite element analysis results, information on manufacturing simulation, etc.

2 The classic method with tree-history (ordered)

The tree - history method is also called parametric modelling method, where the 3D model is based on two-dimensional parameters and two-dimensional constraints that help the part keep its form according to the related parts family. At this point it is also called the classical method [1] because it tends to be generalized for all design firms. The moment for launching this method was 1989 when the company T-FLEX and a later the company ProEngineer introduced, as a working method in the field of CAD (Computers Aided Design), those 3D parameterized models , which implicitly kept the history of all operations that have led to obtaining three-dimensional model. Synthetically, the parametric modelling assumes, in addition to 3D modelling, the existence of defined parameters within a certain timeline. These parameters can be changed both during the design / modelling and at the end, but cause chain propagation of all the other parameters that are in a determination relationship with the changed one. Stages to be completed when using this method are: defining the sketch plan, sketch achievement in the drawing plan chosen, making parameterized listing, imposing 2D restrictions, defining 3D characteristics and combining them. Regardless of the method chosen (classical or synchronous) the design continues with orthogonal and axonometric projections, projections sizing, adding shape and position deviations, and adding technical notes. For Solid Edge this tree- history method is called ordered. Basically, in the area of the modelling screen, superimposed on it there is an area where a number of information are written about the history of all operations performed, which is called *PathFinder*. For example Figure 1 illustrates the 3D models related to two pieces with different tree histories, but both contain at *PathFinder* level the same categories of information: file name, references of the base plans, the label "Ordered" and then a set of sketches followed by designation of a 3D feature. Mention must be made that in the files chosen it was considered that sketches should not be visible not to make difficult the understanding of the final geometry.

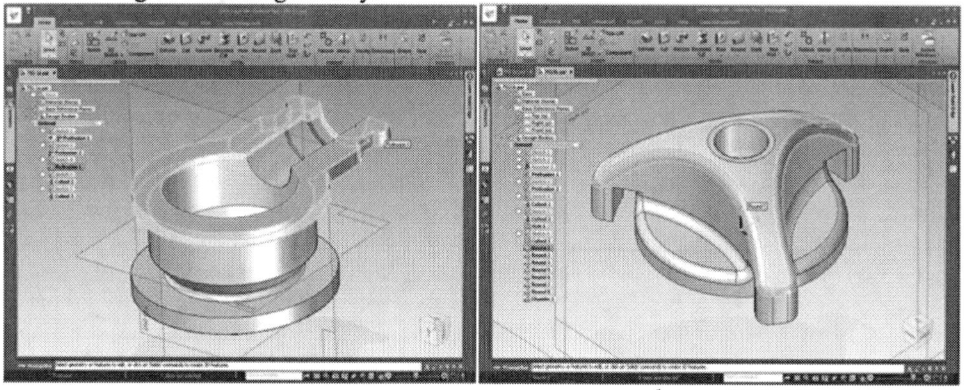

a b

Fig. 1. Parts made with (ordered) tree history.

Also it can be seen in each of the examples in Figure 1 that selecting a 3D feature from the tree *PathFinder* involves the allocation of a certain colour (shade of gray) to those portions of the three-dimensional model which relates to the feature selected. The tree structure of the PathFinder also observes the chronology of generating sketches or local 3D feature. Changing any element of the *PathFinder* structure means to keep in the modelling space those items inside the selected item for editing and not showing temporarily the subsequent ones [2].

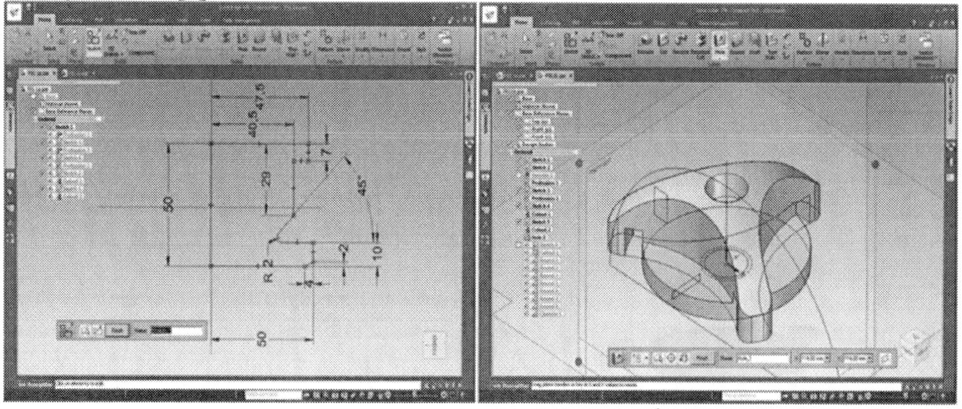

a b

Fig. 2. Editing components from the tree structure.

As seen in Figure 2, the sketch editing involves its viewing in the modelling space and temporary not showing all the elements of the tree structure achieved subsequently. Upon completion of editing by clicking "Finish", the sketch will close and will involve changes related to the characteristic "Protrusion1" while updating all the characteristics not shown during editing. Figure 2b shows that editing the feature „Hole1" involves viewing all previous features and temporary concealment of those subsequent to the edited one. In other words in the chronology of the completed operations there are elements of "parent" type and elements of "child" type. In some cases, an item "parent" may have subordinated several elements of "child" type. During editing a "parent" item, all the elements of "child" type will be temporarily hidden. For decades modelling based on history ruled the CAD for various reasons among which: the undeniable superiority compared to the previous stage of drawing assisted by computer, making quick changes by editing parameters, applying the concept "Rolling back", the special utility of the "parent-child" relationship, etc.

3 The Synchronous method

The emergence of the Synchronous technology is in year 2008, when Siemens launches the Solid Edge version called ST1 (Synchronous technology) designed to achieve a true revolution in the design field as, along with the method of history modelling , introduced a new method and modality of editing solids. This new method has proved from the very beginning extremely useful especially in case of importing [3] in Solid Edge native files of other competing software. Until now continuing the modelling on such files imported from other software was difficult and time consuming. Synchronous technology is the combination of the best features based on CAD history and best functions of Direct Modelling, along with some special features, with the ultimate goal to speed up the completion of the model and making changes. This idea is suggested in Figure 3.

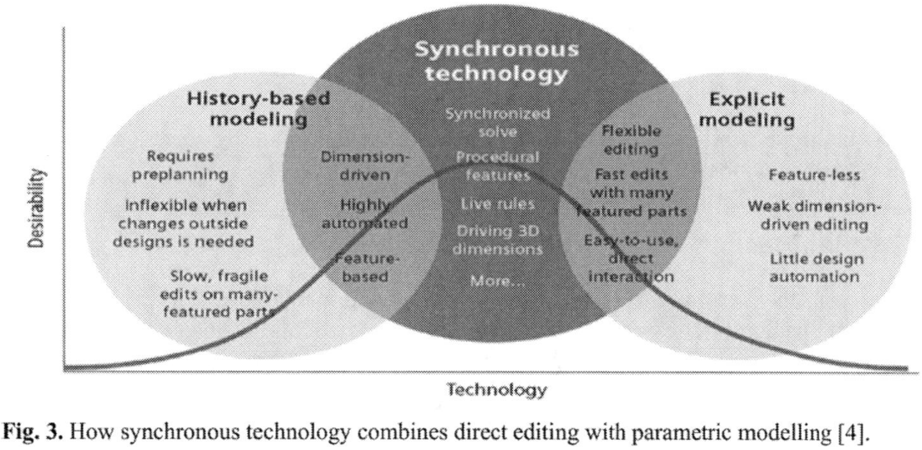

Fig. 3. How synchronous technology combines direct editing with parametric modelling [4].

Unlike the technology based on history, where things happen in turn, the synchronous version brings a total novelty according to which several processes occur simultaneously. The instruments that initialize these simultaneous processes are: PMI 3D Dimensions, Steering Wheel editing interface, Face Relations and Design Intent.

3.1 PMI (Product Manufacturing Information's) 3D Dimensions

In the classical variant with tree-history, changing the 3D feature is based on the link between "parent" and "child", meaning it acts on the parent who is actually a sketch and when the sketch is completed, the change is made on the "child" of the 3D feature type. This process is a sequential reconstruction. Unlike this case, the synchronous method offers the designer a new tool called PMI (Product Manufacturing Information) particularly strong and fast because it causes the change through direct action on the dimensions displayed on the screen, due to the fact that they no longer are to be found at the sketch, they are actually moved in the three-dimensional solid. Thus, the synchronous method causes the parent-child link to break but also add this new changing instrument called PMI that, when selecting a size of the solid, enables the designer to intervene directly on the solid by requiring application of the modification to the left, centre or right side, as shown in Figure 4.

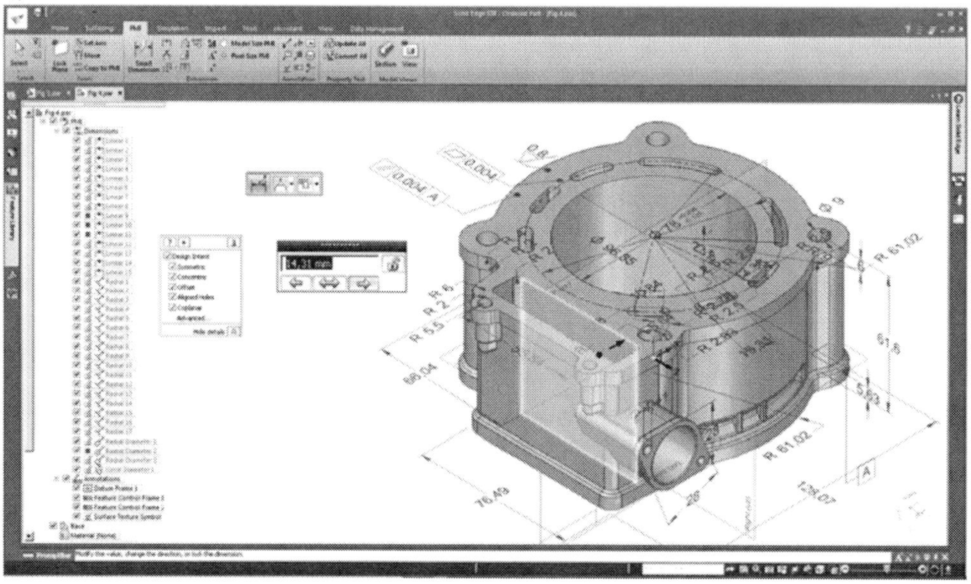

Fig. 4. Solid Edge PMI editor dialog.

At the same time it enables blocking the dimensional value by selecting the lock in the new working window.

3.2 Steering Weel

The steering wheel is an extremely powerful tool made available to the designer by the synchronous technology, by which one may select a particular surface, followed by its orientation in space and the choice of how to implement the change, meaning selecting a travel axis or the plan containing the tor axis of rotation. This tool determines the movement or rotation of the following elements: Reference planes (except the base reference planes) Coordinate systems (except the base coordinate system), Sketches, Sketch elements, Curves, Faces, Features and Design Bodies. The appearance of the steering wheel on the screen in the full version is a progressive one, meaning that from the first click on any of the model face it will be selected the surface and in that place only the origin and normal axis will appear, while the second click performed on origin will cause full appearance which will allow commencement of the orientation process in 3D space.

Progressive display of the wheel depends on the type of surface selected and the displayed elements within it are sufficient to achieve all the modifications possible for that surface.

3.3 Face Relations and Design Intent

 DOI: 10.1051/matecconf/201711206024

In synchronous technology relationship between adjacent faces is no longer related to the parent-child relationship. This allows the system to simultaneously solve the model and eliminates the need for a sequential history reconstruction. If, for instance, a line in space is tangent to an arc, and the two graphic elements are coplanar, then the tangent relationship is inflicted on the two sides as well. In case of importing geometry from other 3D design software, the synchronous technology allows the user to add additional information to imported geometry. Design Intent and Advanced Design Intent panel are two tools that together reveal the true power of the synchronous technology. Design Intent dialog box displays the current rules which were associated with a selected geometry. The second level of help is Advanced Design panel. Here, users have full access to see what rules the system has applied for the selected geometry and all the rules on which can intervene to select them.

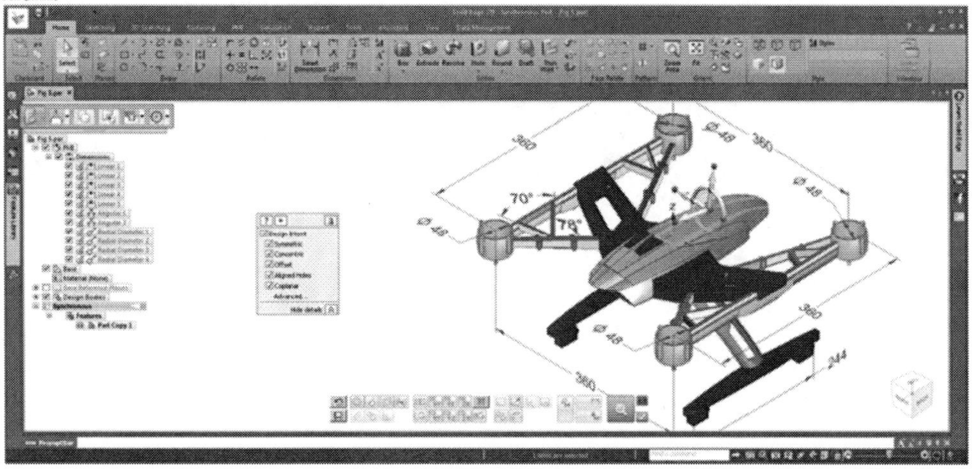

Fig. 5. Design Intent.

Rules in the lower region of Figure 5 may be enabled or disabled. There are several different options available to users, including control deactivation or stopping all controls.

4 Conclusions

Analysing the references mentioned in the end and those presented in this paper we can draw the following conclusions about the two methods studied, namely the ordered (with history) type and the synchronous type:

- The classic method (with history) has been applied for more than 25 years in the field of aided design and therefore has created stronger dependent relations among its users, who because of habit, prefer to apply a very well-known method rather than to test and then apply a superior method as the synchronous one.
- The classical method characterized by parent-child relationships that determine the sequential reconstruction of the model with every change of the parent is very much like the instruction lines chronologically ordered within any program.
- Both methods have strengths but also weaknesses, which is why it is recommended to be used with the aim of creating as robust models as possible. ST9 Solid Edge allows switching from one working mode to another whenever the user wants to.

- The synchronous method has several extremely powerful tools such as: PMI 3D Dimensions, Steering Wheel editing interface, Face Relations and Design Intent, which have no correspondence with the history based method.
- The synchronous methods involve working with regions and require greater computing power from the hardware.

References

1. P. Dumitrache, JIDEG, **10/1,** 5 (2015)
2. G. Musca, *Solid Edge the complete solution for the projection mechanic*, (Edp. PIM 2008)
3. A.M.Goanta, IOP Conf. Series: Mat. Science and Eng., *Modern Technologies in Industrial Engineering*, **145**, 1 (2016)
4. https://www.plm.automation.siemens.com/en/products/solid-edge/st9/ , accessed by 10.01.2017

Modelling of just-in-sequence supply of manufacturing processes

Tamás Bányai*, and *Ágota* Bányai

University of Miskolc, Institute of Logistics, 3515 Miskolc-Egyetemváros, Hungary

Abstract. The customer oriented production led to the growth of complexity of manufacturing and connected logistics processes. In many production companies one of the largest asset on balance sheet is inventory. To avoid inventory problems and to be the winners of today's market situation manufacturing companies try to decrease heavy inventory levels through just-in-time based supply strategies. The aim of this research work is to analyse these supply strategies. The first part of the paper describes the just-in-time based supply and summarises the most important characteristics of them. The second part focuses on the modelling of just-in-sequence based in-plant supply. The models makes it possible to determine different in-plant supply strategies.

1 Introduction

Logistic processes integrate the supply chain of production companies from the purchasing of raw materials required for the final products through in-plant materials handling to the recycling/reuse of used products covering the four functional areas of logistics: purchasing, manufacturing/production, distribution, and recycling. Today the just-in-time (JIT) based material supply strategies of manufacturing processes are gaining more and more importance as, being flexible and reliable, they significantly increase the cost efficiency.

Supply strategies are part of logistic strategies. The most important tasks of logistic strategies are the followings: increase of capacity utilisation of manufacturing and logistic resources; reduction of throughput time; reduction of in-process inventory without increasing supply risk; reduction of operation costs of manufacturing and logistic processes; increase of flexibility to answer customer's demands; enhanced transparency of processes to support the efficiency of lean tools; integration of manufacturing processes into the whole enterprise process through ERP. The operation management of in-plant supply should seek to improve the cost structure of manufacturing related logistic processes, reduce transportation, warehousing and material handling costs. Probably one the most popular tool of lean and operation management is the JIT supply strategy and the newer, so-called just-in-sequence (JIS) supply strategy. According to the literature, the just-in-sequence supply strategy is based on JIT philosophy [1], except that the goal is not only to fulfil the 7R rule, but also to ensure the requested sequence of products.

* Corresponding author: alttamas@uni-miskolc.hu

2 Literature review

Just-in-time became a very popular operational strategy because of its success, especially in the field of automotive industry and mechatronic assembly plants. Today just-in-time philosophy is used both in production and service industry [2, 3].

The development of business environment for just-in-time based supply is one of the core problems, because the roles of suppliers, customers, 3PL partners must be defined to ensure improved efficiency of related processes. The business environment can be defined as a full business cycle, where call-offs contain all information to feed production [4].

Supply models of manufacturing are mostly described as manufacturer-vendor models. One of the first works investigated the manufacturer-vendor model [5]. In a later study the importance of time was underlined and generalise the time varying demand rate [6, 7]. Another topic that has received significant attention in the literature is the multi-level supply of manufacturing processes, where demand and manufacturing rate, delivery and material handling time are both as deterministic and stochastic parameter are taken into consideration [8].

There are three main streams of just-in-sequence literature, which are focusing on purchasing and supply chain, just-in-time production and supply strategies and optimisation of just-in-sequence supply. One of the first milestones of supply chain research was the appearance of cost-benefit models [9] and the researches in the field of selection and evaluation of suppliers [10-12]. The supply chain complexity has increased so much that their planning cannot be carried out only on the base of analytical tools of operations research, but in many cases heuristic methods and simulation techniques are required. The laws of nature may be as relevant sources of inspiration for heuristics as living bodies or human-depending phenomenon [13]. Living bodies inspired heuristics are for example ant colony optimisation [14], firefly optimisation, cuckoo search [15] or bat algorithm [16]. simulated annealing [17], gravitation search [18], intelligent water drops [19] are inspired by physical laws, while harmony search [20] is based on a strongly human-depending attitude.

Several models, can be found in the literature, defining conditions to support the decision: when is it required to change the JIT supply to JIS [21]. These researches focus on the whole supply chain and highlight the macrologistic aspects of JIS supply [22].

Production companies need to manage their supply chain efficiently, reduce cost and improve the accuracy and efficiency of delivery and in-plant supply through production strategies like JIT and JIS [21]. There is a growing interest in the field of automotive industry for JIS supply. Researches showed the following advantages of JIS supply: increased effectiveness of in-plant transportation in automotive industry [22], up-to-date automatic identification tools, like RFID wristband supports JIS processes [23], supermarket based assembly line supply can be based on JIS, especially in the case of line-integrated supermarkets [24], JIS supply can be supported by up-to-date IT and telecommunication solutions [25].

The aim of this paper is to investigate the most important JIS supply strategies. The contribution of this paper is the description of models of JIS based manufacturing supply systems, like ship-to-sequence (StS), build-to-sequence (BtS) and pick-to-sequence (PtS) supply.

This paper is organized as follows. Section 2 presents a literature review, which systematically summarizes the research background of supply chain, JIT and JIS supply. Section 3 describes the model framework of JIS philosophy based in-plant and external supply. Section 4 presents the three major models of JIS supply, by the aid of which it is possible to determine the most important in-plant supply solution methods. Conclusions and future research directions are discussed in Section 5.

3 Model framework of just-in-sequence supply

The advantages of JIS supply can be summarised as follows: easier application of lean tools and solutions through enhanced transparency; increased operational efficiency through high inventory turnover rate; increased level of control over manufacturing, assembly and logistics; reliable respond on customer's demands; better utilisation of resources including human resources through better supply chain management; lower costs through avoiding excess inventory (high rate of inventory turnover, decreased inventory holding costs, lower inventory damage costs). The different solutions of JIS supply can be used in different cases. Within the frame of this section authors are showing a general model framework, by the aid of which it is possible to define a wide range of JIS based supply strategies.

Figure 1 demonstrates the model framework of JIS supply. This general model makes it possible to define the different models of JIS based supply strategies. The model framework includes the following participants of the supply chain: JIS supplier, intermediate storage (traditional warehouse, consignment store or cross docking facility) manufacturing plant with production and assembly processes.

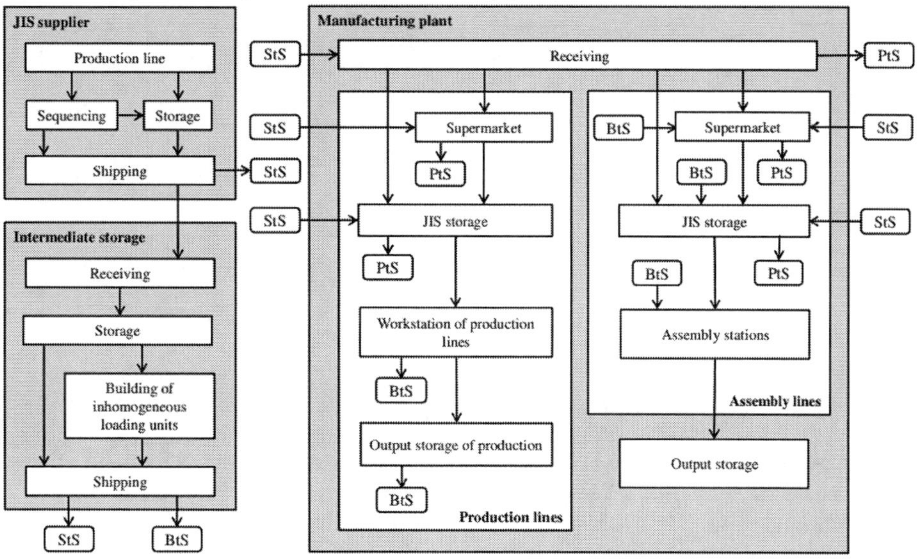

Fig. 1. Model framework of just-in-sequence supply including possible relations of pick-to-sequence, ship-to-sequence and build-to-sequence supply.

In the case of build-to-sequence, the required parts are produced according to the required sequence of the customer's assembly plant. In the case of pick-to sequence supply the required parts are sequenced and picked from a pre-production storage. It means, these solutions of supply can be defined as in-plant supply, where no external suppliers are involved. In the case of ship-to-sequence supply the supply chain represents a vertical cooperation among networking partners, like suppliers, intermediate storages and production plants as customers.

Suppliers can be involved into JIS based supply, if they have integrated IT solutions, segmented production and their purchasing is synchronised with the production processes. Intermediate storages required up-to-date identification solutions and an integrated warehouse management system to avoid failures through warehousing and related materials handling, like packaging, sequencing, building of homogeneous or inhomogeneous loading units, loading and unloading.

4 Modelling of just-in-sequence supply

Within the frame of this section the major directions of just-in-sequence based supply strategies will be defined. The strategies are described with the sequence of material handling and manufacturing processes, required to supply products from JIS suppliers to customers.

4.1 Ship-to-sequence supply

In the case of ship-to-sequence strategy, the required products are sequenced outside of the manufacturing plant. The required products are sequenced either in the plant of JIS supplier or in the intermediate storage or cross docking facility. In this case, the inventories of the customers are decreased and shifted back to the intermediate storage of JIS supplier.

Table 1. Ship-to-sequence strategies to supply assembly plants.

JIS supplier	1	2	3	4	5	6	7	8	9
Outsourced production	♦	♦	♦	♦	♦	♦	♦	♦	♦
Storage	♦	♦	♦	♦	♦	♦	♦	♦	♦
Sequencing	♦	♦	♦	♦	♦	♦			
Shipping to intermediate storage				♦	♦	♦			
Shipping to customer	♦	♦	♦				♦	♦	♦
Intermediate storage									
Receiving products from suppliers				♦	♦	♦	♦	♦	♦
Storage				♦	♦	♦	♦	♦	♦
Sequencing							♦	♦	♦
Shipping				♦	♦	♦	♦	♦	♦
Manufacturing plant									
Receiving parts in the input storage			♦			♦			♦
Supermarket for assembly		♦	♦		♦	♦		♦	♦
JIS storage for assembly	♦	♦	♦	♦	♦	♦	♦	♦	♦
Assembly	♦	♦	♦	♦	♦	♦	♦	♦	♦
Output storage	♦	♦	♦	♦	♦	♦	♦	♦	♦

4.2 Pick-to-sequence supply

The pick-to-sequence supply is not in any case a just-in-sequence supply for the whole supply chain because the loading unit is sequenced for the customer's demands in the manufacturing plant. The shipping from JIS supplier through intermediate storage to the manufacturing plant can be either just-in-time or just-in-case.

Table 2. Pick-to-sequence strategies to supply manufacturing plants.

JIS supplier	1	2	3	4	5	6	7	8	9	10	11	12
Outsourced production	♦	♦	♦	♦	♦	♦	♦	♦	♦	♦	♦	♦
Storage	♦	♦	♦	♦	♦	♦	♦	♦	♦	♦	♦	♦
Building loading units	♦	♦	♦	♦	♦	♦	♦	♦	♦	♦	♦	♦
Shipping to intermediate storage	♦	♦	♦	♦	♦	♦	♦	♦	♦	♦	♦	♦
Shipping to customer	♦	♦	♦	♦	♦	♦						
Intermediate storage												
Receiving products from suppliers							♦	♦	♦	♦	♦	♦
Storage							♦	♦	♦	♦	♦	♦
Building loading units												
Shipping							♦	♦	♦	♦	♦	♦
Manufacturing plant												
Receiving parts in the input storage	♦		♦	♦	♦	♦	♦		♦	♦	♦	♦
Supermarket for production			♦		♦				♦		♦	
JIS storage for production			♦	♦	♦	♦			♦	♦	♦	♦
Production for assembly line			♦	♦	♦	♦			♦	♦	♦	♦
Output storage for production			♦	♦	♦	♦			♦	♦	♦	♦
Supermarket for assembly	♦		♦	♦			♦		♦	♦		
JIS storage for assembly	♦	♦	♦	♦	♦	♦	♦	♦	♦	♦	♦	♦
Assembly	♦	♦	♦	♦	♦	♦	♦	♦	♦	♦	♦	♦
Output storage	♦	♦	♦	♦	♦	♦	♦	♦	♦	♦	♦	♦

4.3 Build-to-sequence supply

In the case of build-to-sequence supply of the manufacturing plant, products are produced according to the required sequence of the customer's assembly plant. The just-in-sequence production of required parts can be insourced or outsourced and should be realized based on just-in-time philosophy.

Table 3. Build-to-sequence strategies to supply manufacturing plants.

JIS supplier	1	2	3	4	5	6	7	8
Outsourced production			♦	♦	♦	♦	♦	♦
Sequencing to customer's needs			♦	♦	♦			
Storage						♦	♦	♦
Building loading units						♦	♦	♦
Shipping			♦	♦	♦	♦	♦	♦
Intermediate storage								
Receiving products from suppliers						♦	♦	♦
Storage						♦	♦	♦
Sequencing for customer's demands						♦	♦	♦
Shipping						♦	♦	♦
Manufacturing plant								
Receiving parts in the input storage			♦	♦	♦	♦	♦	♦
Supermarket for production								
JIS storage for production								
Production for assembly line	♦	♦						
Output storage for production		♦						
Supermarket for assembly		♦	♦	♦		♦	♦	
JIS storage for assembly	♦	♦	♦	♦	♦	♦	♦	♦
Assembly	♦	♦	♦	♦	♦	♦	♦	♦
Output storage	♦	♦	♦	♦	♦	♦	♦	♦

The cost function of the analysis of just-in-sequence strategies is based on the costs of production, warehousing, sequencing and transportation.

$$C=min(C_{P,out}, C_{P,in})+ min(C_{W,Su} +C_{W,Ma}, C_{W,Su}+C_{W,St}+C_{W,Ma})+$$

$$+min(C_{Se,Su}, C_{Se,St}, C_{Se,Ma})+ min(C_{T,Su} +C_{T,Ma}, C_{T,Su}+C_{T,St}+C_{T,Ma}) \tag{1}$$

The optimization of just-in-sequence strategies can be based on the above mentioned objective function. For production capacities the following constraints have to be taken into consideration (2):

- the outsourced production is less or equal to the production capacity of JIS supplier;
- the internal production is less or equal to the production capacity of manufacturer;
- the sum of outsourced and internal production is equal to the product demand of assembly plant.

$$q_{P,out} \leq q_{P,Su} \text{ and } q_{P,in} \leq q_{P,Ma} \text{ and } q_{P,out} + q_{P,in} = q_P \tag{2}$$

For logistic capacities the following constraints have to be taken into consideration (3):

- the required transportation capacity is less or equal to the available transportation capacities among JIS suppliers, intermediate storages and manufacturers;
- the required warehousing capacities are less or equal to the available capacity in JIS suppliers, intermediate storages and manufacturers.

$$q_{T,Req} \leq q_{T,Av} \text{ and } q_{W,Req} \leq q_{W,Av} \tag{3}$$

The just-in-sequence problems can be solved depending on the complexity of the problems either with LP solution methods or in the case of NP-hard problems with heuristics.

5 Summary

Sourcing strategies are very important for a successful supply chain. It is important to understand how just-in-time based sourcing of manufacturing companies works to choose the best solution. Just-in-sequence solutions support the decrease of inventory levels, but the design and operation of these supply processes can be expensive. Within the frame of this paper authors described a model framework of just-in-sequence supply and defined the most important JIS based supply strategies: ship-to-sequence, pick-to-sequence and build to sequence.

This project has received funding from the European Union's Horizon 2020 research and innovation programme under grant agreement No 691942. This research was partially carried out in the framework of the Center of Excellence of Mechatronics and Logistics at the University of Miskolc.

References

1. S. Werner, M. Kellner, E. Schenk, G. Weigert, Robot. Comp. Int. Manuf. **19**, 107 (2003)
2. A. S. Aradhye, S.P. Kallurkar, Procedia Eng. **97**, 2232 (2014)
3. H. Vass, Zs. K. Szabo, Procedia Econ. Fin. **32**, 479 (2015)
4. C. O. Brien, M. Head, Int. J. Prod. Econ. **42**, 41 (1995)
5. S. K. Goyal, Oper. Res. Quart. **28,** 865 (1977)
6. M. Omar, D. K. Smith, Comput. Ind. Eng. **42**, 35 (2002)
7. M. Omar, Int. J. Prod. Econ. **118**, 185 (2009)
8. A. Banerjee, S. L. Kim, Int. J. Oper. Prod. Manag. **15**, 237 (1995)
9. J. L. Cavinato, J. Bus. Log. **13**, 285 (1992)
10. T. Y. Choi, J. L. Hartley, J. Oper. Manag. **14**, 333 (1996)
11. M. Segura, C. Maroto, Expert. Syst. Appl. **69**, 87 (2017)
12. M. Tavana, A. Fallahpour, D. D. Caprio, F. J. Santos-Arteaga, Expert. Syst. Appl. **61**, 129 (2016)
13. A. P. Piotrowski, J. J. Napiorkowski, P. M. Rowinski, Inf. Sci. **246**, 191 (2014)
14. M. Dorigo, L. M. Gambardella, Biosystems **43**, 73 (1997)
15. V. Bhargava, S. E. K. Fateen, A. Bonilla-Petriciolet, Fluid Phase Equilibr. **337**, 191 (2013)
16. T. Niknam, S. Sharifinia, R. Azizipanah-Abarghooee, Energ. Convers. and Manage. 76. pp. 1015-1028 (2013)
17. A. R. McKendall, J. Shang, S. Kuppusamy, Comput. Oper. Res. **33**, 2431 (2006)
18. S. K. Saha, R. Kar, D. Mandal, S. P. Ghoshal, J. King Saud. Univ. Eng. Sci. **26**, 69 (2014)
19. P. R. Srivastava, Comput. Electric. Eng. **43**, 270 (2015)
20. P. Veres, T. Bányai, B. Illés, Procedia Eng. **100**, 46 (2015)
21. S. M. Wagner, V. Silveira-Camargos, Int. J. Prod. Res. **49**, 5713 (2011)
22. J. Kaneko, W. Nojiri, J Transp Geography, **16**, 155 (2008)
23. A. N. Dwivedi, *Innovative Solutions for Implementing Global Supply Chains in Emerging Markets* (Elsevier B.V. 2016)
24. A. Lorenc, M.Szkoda, ICIL 168 (2016)
25. M. Kirch, O. Poenicke, Smart Systech Article number 7156013 (2015)
26. N. Boysen, S. Emde, Eur. J. Oper. Res. **239,** 820 (2014)
27. G. Heinecke, S. Lamparter, R. Lepratti, A. Kunz, MIM 1902 (2013)

Optimisation problems of networking manufacturing processes

Péter Veres*, *Tamás* Bányai, and *Béla* Illés

University of Miskolc, Institute of Logistics, 3515 Miskolc-Egyetemváros, Hungary

Abstract. The globalization of economy and market led to increased networking in the field of manufacturing. The processes of these manufacturing including logistics became more and more complex. This paper aims to report design aspects of networking manufacturing focusing on NP-hard optimisation problems. The authors describe general models of logistic processes of networking manufacturing systems and present heuristic algorithms to solve the optimisation problems focusing on the solution of knapsack and traveling salesman problems with harmony search and firefly algorithms.

1 Introduction

The increased diversity of customer's needs has led to the increase of the complexity of manufacturing processes. Complex, networking manufacturing processes required complex logistics and supply chain solutions both in-plant and outside of the manufacturing plant.

Heuristic algorithms offer suitable tools to support the design and control of these networking systems. A wide range of networking problems can be defined as knapsack and traveling salesman problems. Some special and generalized algorithms were created and published in different research works: the first exact solution methods focused on either bounded or unbounded problems [1, 2]; the new solution methods are based on heuristic: assessing solution quality of biobjective 0-1 knapsack problem using evolutionary and heuristic algorithms [3], bounded set-up knapsack problem [4], image knapsack problems [5]. Much literature has been published in the field of optimization of large scale networks and complex engineering systems by the aid of Harmony Search Algorithm (HAS). There are several applications for harmony search algorithms from the sizing optimization of truss structures [6] to the transport energy modeling [7] or nuclear energy production [8].

The harmony search algorithm can be used to create hybrid algorithms to develop more effective solution methods for large systems [9-11]. Harmony search algorithms make it possible to realize global harmony search [12, 13] and local-best harmony search [14].

This paper is organized as follows. Section 1 presents a short literature review. Section 2 describes two manufacturing related NP-hard optimization problem with heuristic solutions. Conclusions and future research directions are discussed in Section 3.

* Corresponding author: altveres@uni-miskolc.hu

 DOI: 10.1051/matecconf/201711206026

2 Manufacturing optimization in the case of networking companies

There are many problems in the field of logistics. Most of these problems are so complicated that a simple model can't specify them. When this occurs several model and methods needed to get solutions, and we have to decide it which we will choose. The appropriate model and approach, can significantly reduce the computational time, human resources and save money.

2.1 Knapsack problems in manufacturing

The optimization of logistic systems and processes is the key factor of the economical operation of a company. Capacity problems occur every day in every field of manufacturing and logistics. It has the biggest effect on purchasing, distribution and production, but the problem can appear in services and recycling too. Just to mention some activities the problem affects scheduling of transportation; transportation; loading and unloading of products; storage in the warehouses or internal storages (buffering); building of homogeneous and inhomogeneous loading units; handling of production resources.

There are lots of ways to improve the capacity of a company. The easiest way is to buy more space, building, machine or transport vehicle. But the companies don't want to spend millions of dollars on huge investments, which they don't necessary need, so the thing to do is to overview the capacity management. Most of the time capacity problem can be represented as a knapsack problem (Fig.1.). The one-dimensional knapsack problem focuses on one question: which products should be chosen to maximize the profit within a limited space (or another parameter)?

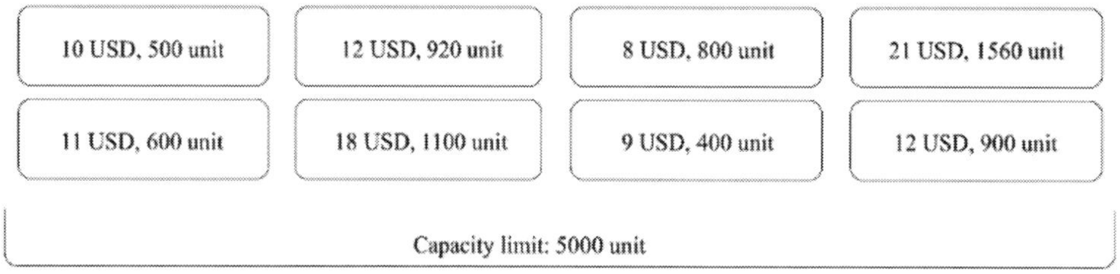

Fig. 1. Knapsack problem.

The one-dimensional knapsack problem can be solved by basic combinatory or linear programming tools, like brute-force search, but unfortunately most of the cases one parameter is not enough to describe a problem in a company, and certainly won't do in a supply chain. Lots of factors, limits, parameter connections and constraints affect the solution and this gives us a NP-hard problem. From this point of view, linear programming can't be used effectively, but heuristic algorithms can give us answers very quickly if we use the good ones with proper settings.

There are several predefined knapsack problems, and all of them have some kinds of solution in the literature: unbounded knapsack problems; bounded set-up knapsack problems; image knapsack problems. We only create a little application to solve the basic knapsack problem, but with some work, it can be used for the most difficult ones too.

One of our previous works was a modification of a meta-heuristic algorithm, that can easily solve the knapsack problem [15]. The base of the method called Harmony Search Algorithm, and it's a music-based swarm type algorithm. It has 3 types of variable modification; copying, slightly changes or total randomization; which chosen randomly affected by weights. These 3 modes with the proper settings can be so robust that it can get out of a local optimal point, but it also can refine the solution greatly if it has time.

$$f(x) = \sum_{j=1}^{n} p_j \cdot x_j \rightarrow max \quad (1)$$

$$\sum_{j=1}^{n} w_j \cdot x_j \leq W \quad (2)$$

$$0 \leq x_j \leq r_j \quad and \quad x_j \in Z \quad and \quad j = 1 \ldots n \quad (3)$$

where p_j is the cost, w_j is the weight, x_j is decision variable, W is the weight or capacity limit. The basic knapsack problem doesn't need a complicated model as we can see above. It only contains an objective function, that maximizes the profit and a limit of the space we have, which cannot be exceed. We can make it more complicated, if we add another limit, like weight; or we need to give a formation or sequence for the things inside.

HSA is a hybrid algorithm which means it contains several separated parts from other algorithms, so we made some modifications too. First a memory function was added, because when the search space is too flat or chaotic (noisy) it's easily lose the optimal point. Then a solution refining mechanism took place inside the method, that gives us more punctual solution if the running has more time.

2.2 Traveling salesman problem

This problem always occurs when somebody or something must reach at least two places and we must decide in which order. Every company that does some sort of transporting meets this problem on daily basis. There are lots of solutions in the literature and on the market too, they called route planning applications. We are working on a basic one too, but we focused on some extra functions, that can save money.

The basic TSP is based on distances between the reachable points. We try to find the shortest summarized distance as we reach all the points we require. The points and distances are fixed, so the only thing that can be modified is the sequence. In this case the basic TSP is just a permutational problem, and it's not hard to handle. But most of the time minimize the traveled distant is just not enough. There are some parameters, and requests that must be fulfilled: energy consumption, fastest route, emission, deadlines, driving and resting hours, occupancy, empty space, utilization, and we are only speaking about solo transportation. If we place parallel transporting and multiple warehousing in the equation, they just make the computing even harder. With these parameters, the problem gets too hard to be solved by standard linear programming tools, we must use heuristics again.

We create a basic route planning application with HSA, and it worked perfectly, but we only counted with the distances. For the next step, we leave the Matlab language and create a basic route planning application with C# for better visualizing the results. After the language, the base of the application is also changed. A new algorithm has been introduced and we want to try it. It's called Firefly Algorithm (FA). It's also a nature based method, but it simulates the movement of a group of fireflies.

The next update for the application was a specific function, where we not only measure the distance, but to examine the stock levels, and create multiple routes for solutions. A new visual and mathematical model was created. For these new parameters had to be added: costs, efficiency, volumes, weights, etc.
Summarized route length:

$$L = \sum_{t=1}^{\tau} \sum_{\alpha=1}^{\delta(t)} \left(l_{f(t,\alpha),p_{t,\alpha(t),1}}^{KR} + \sum_{s=1}^{m(t,\alpha)-1} l_{p_{t,\alpha(t),s},p_{t,\alpha(t),s+1}} + l_{p_{t,\alpha(t),m,f(t,\alpha)}}^{KR} \right) \rightarrow min. \quad (4)$$

 DOI: 10.1051/matecconf/201711206026

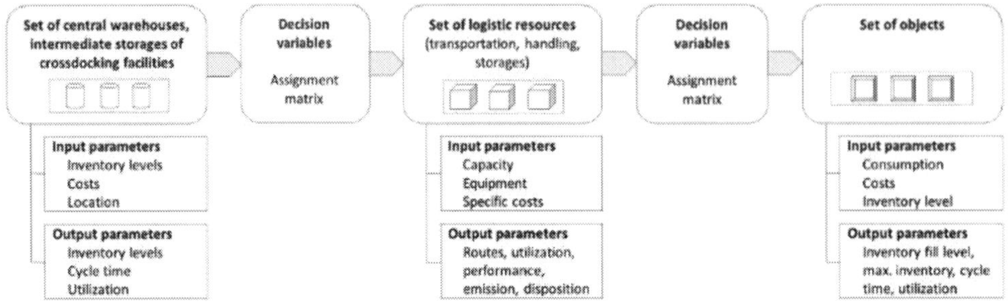

Fig. 2. TSP flowchart.

where: τ is the number of cycles of the investigation period, α is the route ID, $\delta(t)$ is the number of routes in the tth investigation period, $f(t,\alpha)$ is the central storage ID in the tth investigation period visited by the αth route, is the objects number in a given route, $p_{t,\alpha,\beta}$ is the objects ID number in the tth investigation period visited by the αth route in the βth place, $m(t,\alpha)$ is the number of objects in the tth investigation period visited by the αth route.

The following objective function is determining the routes occupancy:

$$\eta^J = \sum_{t=1}^{\tau}\sum_{\alpha=1}^{\delta(t)} \frac{\sum_{\beta=1}^{m(t,\alpha)} K^{max}_{p_{t,\alpha,\beta}} - K^{akt}_{p_{t,\alpha,\beta}}}{K^J_\alpha} \rightarrow max. \tag{5}$$

where $K^{max}_{p_{t,\alpha,\beta}}$ is the maximum storage capacity of the βth object, $K^{akt}_{p_{t,\alpha,\beta}}$ actual filled capacity of the βth object, K^J_α is the αth route assigned vehicle capacity. To minimize the investments costs it's important to optimize the routes number. The central storage plays an important role in the system, therefore during the optimization of the supply chain transport and storage costs should be minimized.

$${}^{KR}K^{MF} = \frac{\tau}{T^C} \cdot c^{SZ} \tag{6}$$

where T^C is the central storage replenishment cycle time, c^{SZ} is the order-processing cost related to the central storages refill process. The storage costs in the full time horizon can also be calculated.

$${}^{KR}K^{T} = \sum_{t=1}^{T^C}\sum_{\alpha=1}^{\delta(t)}\sum_{\beta=1}^{m(t,\alpha)} K^{max}_{p_{t,\alpha,\beta}} - K^{akt}_{p_{t,\alpha,\beta}} \tag{7}$$

Based on these two costs, the objective function for the central storage can be written in the following form.

$${}^{KR}K = {}^{KR}K^{MF} + {}^{KR}K^{T} \rightarrow \min. \tag{8}$$

For the routes we can also create an important limitation, which expresses the fact that the volume served by the route can't be more than the vehicles capacity.

$$\sum_{\beta=1}^{m(t,\alpha)} K^{max}_{p_{t,\alpha,\beta}} - K^{akt}_{p_{t,\alpha,\beta}} \leq K^J_\alpha \tag{9}$$

The application was created for ATM refilling, but it can be easily modified for other route planning purposes with inventory management: milkrun, vending machine management, smart recycling bins, smart container refilling or emptying, etc. The biggest problem was that the algorithm and the model serve different purpose. The algorithm works with floating points, and tries to find the global minimum by changing the distance between points, but the model needs a simple integer and permutational solution. We have hard time to discretize the algorithm, but in the end it worked greatly.

For our last two applications we change the algorithm again, because the Firefly algorithm was too specific for these tasks, too many modifications were required to work properly. We find a novelty algorithm, that called Black hole, and it works brilliantly. The method it's use is quite simple, and easy to discretize.

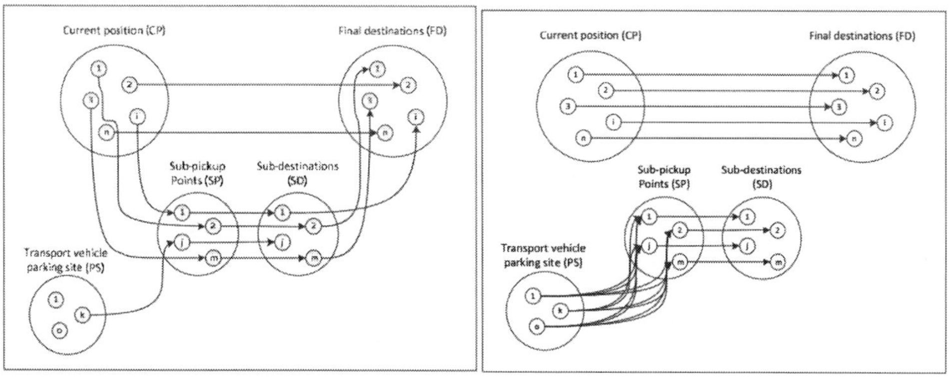

Fig. 2. TSP flowchart.

The first application with the new algorithm analyses the running transport vehicles if they can fulfill another request, within their original goal. The objective in this case is the money that can be saved by running extra errands.

$$CS = \sum_{l=1}^{p} \left[CT_1 * L_{FDi}^{CPi} + CT_2 * \left(L_{SPj}^{PSk} + L_{SDj}^{SPj} \right) \right] -$$

$$-CT_1 * \left(L_{SPj}^{CPi} + L_{SDj}^{SPj} + L_{FDi}^{SDj} \right) + \sum_{s=1}^{q} CO_s + \sum_{t=1}^{m-q} CP_t \Rightarrow max. \quad (10)$$

$$i = 1 \dots n\,; j = 1 \dots m\,; k = 1 \dots o \quad (11)$$

$$0 \leq q \leq m\,;\ n \leq p \leq n + m \quad (12)$$

$$CAP_{Ti}^{V} - V_{TGi} \geq \sum_{r \in \theta_i} V_{TDrj} \quad (13)$$

$$CAP_{Ti}^{W} - W_{TGi} \geq \sum_{r \in \theta_i} W_{TDrj} \quad (14)$$

$$TC_i + T_{SPj}^{CPi} + T_{SDj}^{SPj} + T_{FDi}^{SDj} \leq TR_i \quad (15)$$

where CT_1 is the original transportation cost, which is proportional with the distance, CT_2 is the transportation cost of a new transport vehicle, CO_s is the cost of the sth order processing and assigning a new vehicle, CP_t profit from fulfilling the tth demand, L_{FDi}^{CPi} is the distance between the current position of the ith vehicle and the ith final destination, L_{SPj}^{PSk} is the distance between k^{th} vehicle parking site and the jth sub-pickup point, L_{SDj}^{SPj} is the distance between the jth sub-pickup point and j^{th} sub-destination, L_{SPj}^{CPi} isthe distance between the current position of the ith vehicle and the jth sub-pickup point, L_{FDi}^{SDj} is the distance between the jth sub-destination and the ith final destination, n is the number of original transports and vehicles that are on road, m is the number of incoming transport demands, o is the number of transport vehicles parking sites, q is the number of new vehicles needed to fulfil the transport demands, p is the number of vehicles will be transporting including the new demands, θ_i is the set of elements of j, that the ith vehicle can transport due to capacity restriction, $CAP_{Ti}^{V}/CAP_{Ti}^{W}$ is the volume/weight capacity of the ith vehicle, V_{TGi}/W_{TGi} is the volume/weight of the transported goods in the ith vehicle, V_{TDj}/W_{TDj} is the volume/weight of the jth transport demand, T_B^A is the time needed to go from A to B, TC_i is the current time of the ith vehicle at the CP point, TR_i is the requested arrival time of the ith vehicle to its original destination.

The above mentioned models make it possible to support the design and operation processes of manufacturing related supply chain problems, while a huge number of in-plant supply chain problems can be defined as a special type of knapsack or TSP problems.

3 Conclusions and future research directions

Manufacturing systems are more and more complex. Heuristic optimization offers powerful tools and methods to design networking manufacturing systems. Within the frame of this paper authors described two possible optimization methods of knapsack and traveling salesman problem with harmony search and firefly algorithm to support the optimization of various manufacturing related logistic problems. As a result of the demonstrated research it can be expected that in the future years more and more optimization application will be developed in the field of supply chain. The results of this study can be used to improve the whole logistics process of companies from the purchasing through production logistics to distribution.

This research work has received funding from the European Union's Horizon 2020 research and innovation programme under grant agreement No 691942. This research was partially carried out in the framework of the Center of Excellence of Mechatronics and Logistics at the University of Miskolc.

References

1. S. E. Elmaghraby, Eur. J. Oper. Res. **38**(2), 42 (1989)
2. S. Martello, P. Toth, Oper. Res. Lett. **9**(1), 15 (1990)
3. R. Kumar, P. K. Singh, Appl. Soft. Comput. **10**(3), 711 (2010)
4. L. A. McLay, S. H. Jacobson, Discrete Optim. **4**(2), 206 (2007)
5. B. Zeng, J. P. Richard, Discrete Optim. **8**(2), 277 (2011)
6. S.O. Degertekin, Comput. Struct. **92–93,** 229 (2012)
7. H. Ceylan, H. Ceylan, S. Haldenbilen, O. Baskan, Energy Pol. **36**(7), 2527 (2008)
8. M. Aghaie, T. Nazari, A. Zolfaghari, A. Minuchehr, A. Shirani, Ann. Nucl. Energy **57,** 1 (2013)
9. B. Wu, C. Qian, W. Ni, S. Fan, Comput. Math. App. **64**(8), 2621 (2012)
10. M. Fesanghary, M. Mahdavi, M. Minary-Jolandan, Y. Alizadeh, Comput. Method. Appl. M. **197**(33-40), 3080 (2008)
11. F. Amini, P. Ghaderi, Appl. Soft. Comput. **13**(5), 2272 (2013)
12. D. Zou, L. Gao, S. Li, J. Wu, X. Wang, J. Syst. Software **83**(10), 1678 (2010)
13. C. Cobos, D. Estupiñán, J. Pérez, Appl. Math. Comput. **218**(6), 2558 (2011)
14. Q.-K. Pan, P. N. Suganthan, J. J. Liang, M. F. Tasgetiren, Expert Syst. Appl. **38**(4), 3252 (2011)
15. T. Bányai, P. Veres, Adv. Log. Sys. **7**(1), 13 (2013)

Experimental investigation of noise characteristics for HVAC silencers

Carmen Bujoreanu*, and *Marcelin* Benchea

Technical University "Gheorghe Asachi" Iaşi, Mechanical Engineering Faculty, Department of Mechanical Engineering, Mechatronic&Robotics, 700050 Iaşi, Romania

Abstract. Heating, ventilating and air-conditioning (HVAC) systems represent a major source of inside buildings noise with a negative effect on the acoustical environment. Therefore, it is necessary to add attenuation to the system. This is provided by sound attenuators or silencers manufactured in different geometrical configurations and with various absorptive lining materials. The paper deals with the insertion loss, as a performance criterion for in-duct noise, of some dissipative silencers types. The measured data are compared in order to establish the efficiency of every tested silencer type. Third-octave band analysis of random noise of an air handling unit (AHU) from a HVAC system is realized without and with silencers (ducts fitted). Insertion loss measurements are made in an anechoic room using a test facility that agrees the requirements of ISO 7235:2009. The results highlight the effects of the silencers geometry, pressure-air flow and loading relations upon the sound attenuation level. The silencers selection could be a difficult task for the designers which have to consider all these parameters.

1 Introduction

Silencers, also called sound attenuators, absorbers or mufflers, are designed to reduce the noise transmitted from a source to a receiver. In the HVAC field, the noise attenuation in ventilation ducts becomes of major interest as the acoustic comfort requirements have become more stringent. The European Standards EN ISO 7235:2009 (EN ISO 2009a) and EN ISO 11691:2009 (EN ISO 2009b) and the American Standard ASTM E477-06a (ASTM 2006) set the experimental measurement procedures in order to determine the insertion loss (attenuation), in frequency bands, of ducted silencers and air-terminal units with and without airflow.

For HVAC applications, the most common silencers are duct silencers, installed on the intake and/or discharge side of a fan or air handler. Also, they may be used on the receiver side of other noise generators such as terminal boxes, valves, dampers, etc. There are several types of silencers used in the sound attenuation: reactive, active and dissipative [1].

For the wide frequency bands, the dissipative silencers are recommended. They attenuate the sound especially through the presence of the sound-absorbing materials with a certain

* Corresponding author: carmen.bujoreanu@gmaill.com

flow resistivity. These silencers are the most widely used for noise control where fan noise, flow induced noise and engine noise must be attenuated [2].

The duct silencers are available in various shapes, including rectangular, round, elbow ones, and different sizes in accord to the duct design. Usually, a commercial duct silencer's outer appearance is similar to a piece of duct. It consists of a sheet metal casing with length commonly ranging from (1-3 m) [2]. Sound-absorbing material lining one or more of the silencer sides, in a usual configuration. Other silencer's configurations include a centrally located bar or several splitters mounted horizontally or vertically inside it [2].

Some comparisons with experimental data revealed noticeable difference between practice and theory and emphasize the limitations of theoretical approaches for the silencer design [3, 4, 5].

The paper is an experimental work and it deals with the insertion loss, as a performance criterion for in-duct noise, of some dissipative silencers types, intended to be commercialized, from the same producer. The measured data are compared in order to establish the efficiency of every tested silencer type. Third-octave band analysis of random noise of an air handling unit (AHU) from a HVAC system is realized without and with silencers (ducts fitted). Insertion loss measurements are made using the test facility that agrees the requirements of ISO 7235:2009 [6]. The results highlight the effects of the silencers geometry, pressure-air flow relation and loading upon the sound attenuation level and lead to a better silencer design.

2 Experimental methodology

The lined silencers used in specific HVAC systems need to consider some parameters as: the air-flow direction and velocity, the pressure drop after the silencer insertion and the most important, the insertion loss in decibels (dB) in each octave band. The use of silencer naturally produces the fan loading increasing and both the acoustic and air flow performances must be evaluated when they are designed [7, 8].

ISO 7235: 2009 specifies the methods for determining the insertion loss, in frequency bands, of ducted silencers with and without airflow, the sound power level of the flow noise generated by ducted silencers, the total pressure loss of silencers with airflow, and the transmission loss of air-terminal units.

The test facility for measurements of insertion loss in dynamic conditions, with airflow, is presented below.

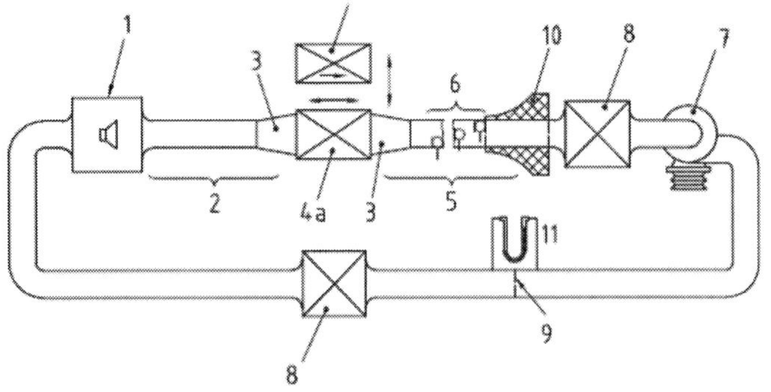

Fig. 1. Test facility for measurements of insertion loss with airflow and/or flow noise Arrangement for in-duct measurements (ISO 7235: 2009).

The elements of the test facility presented in Fig.1 are: 1 - loudspeaker unit in the source chamber ; 2 - test duct in front of the test object ; 3 – transitions ; 4a - test object ; 4b -

substitution duct ; 5 - test duct behind the test object ; 6 - wind-shielded microphones ; 7 - fan (installed to generate either forward or reverse flow relative to the direction of sound propagation ; 8 - fan silencers ; 9 - nozzle for measurement of flow rate (alternatives: orifice or Venturi nozzle) ; 10 - anechoic termination (we have used an anechoic room) ; 11- static pressure measurement

The experimental setup for our measurements consists of: centrifugal fan providing a variable airflow (from 1V to 10V with a ratio of 1V), device for measuring the flow rate (anemometer), device for measuring the pressure drop (pressure differential meter), special sound-source equipment (consists of a white noise generator B&K, an amplifier and a broad band loudspeaker), the test object (the dissipative silencers), the transition elements on either side of the test object, special receiving-sound equipment (soundmeter B&K connected with the microphone and NIDAQ board, laptop with LabVIEW soft compatible with National Instruments DAQPad for data processing).

The insertion loss D_I of the test object represents the reduction in the sound power level in the duct behind the test object, due to the insertion of the test object into the duct in place of a substitution duct. It is given by the equation:

$$D_I = SWL_{II} - SWL_I \text{ [dB]} \quad (1)$$

where: SWL_I - the level of the sound power in the frequency band considered, propagating along the test duct when the test object is installed; SWL_{II} - the level of the sound power in the frequency band considered, propagating along the test duct when the substitution duct replaces the test object.

The insertion loss, D_I, must be measured from spatially energy-averaged sound pressure levels *SPL* at identical points or paths. In one test series, SPL_I is determined with the test object installed and in a further series, SPL_{II} is determined with the test object replaced by the substitution duct [6].

The insertion loss measurements were realized according to the procedure for different airflows variations of the centrifugal fan (from 1V to 10V with a ratio of 1V) and the white noise generated from sound-source equipment.

The experimental setup and the tested silencers are presented in Fig.2.

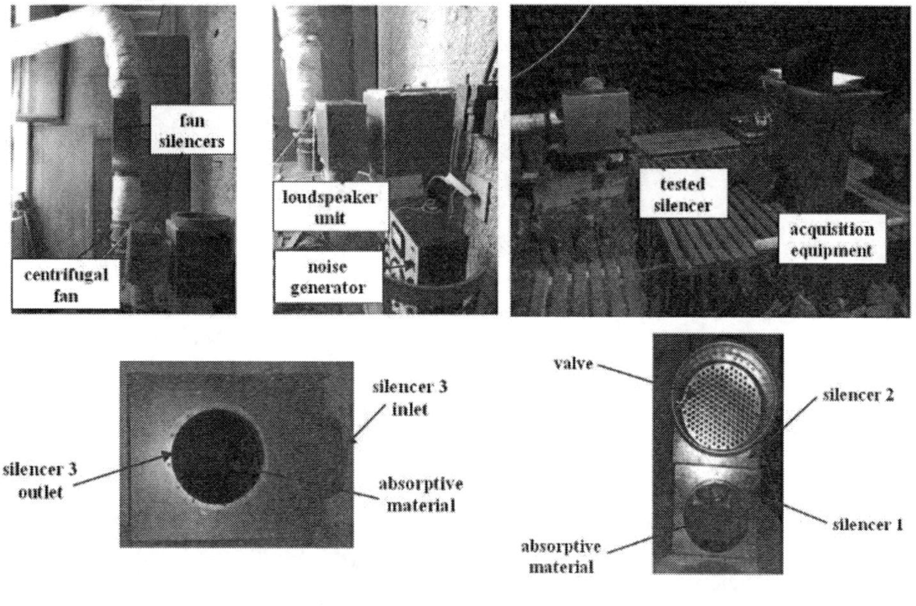

Fig. 2. Experimental setup outside and inside the anechoic room.

The tested silencers, with different configurations, are noted with:

- *"silencer 1"*- the straight silencer with valve (Φ125/Φ125 mm and inlet-outlet air flow direction is 180^0); the valve can be open or closed.
- *"silencer 2"* – the elbow silencer with valve (Φ125/Φ125 mm and inlet-outlet air flow direction is 90^0 rotated); the valve can be open or closed.
- *"silencer 3"* - the straight silencer without valve (Φ125/Φ125 mm and inlet-outlet air flow direction is 90^0 rotated).

The absorptive material lining the silencers 1 and 2 is different from that lining the silencer 3, so their acoustical properties, in terms of sound absorption, are not the same.

First of all, we have measured the noise level dB(A), A weighted, by means the LabVIEW soft, using the third octave analysis of the sound pressure levels, dB(A), for nominal frequencies of 63, 125, 250, 500, 1000, 2000, 4000, 8000 Hz and total band power (within any specified frequency band). Then, we have calculated, according to relation (1), the insertion loss for every silencer tested.

3 Results and discussions

We have selected some experimental results to be presented, comparatively, in the following figures. Figure 3 reports the pressure drop versus the air flow and Figs. 4 and 5 express the sound attenuation (in terms of sound pressure) in the specified silencers. It should be noted that, according to the manufacturer, the most used regime in HVAC exploitation corresponds to 4-6 V loading. The sound attenuation for this loading is of interest. However, we also present some results acquired in the extreme regime of 10 V.

Figure 3 compares the pressure-air flow evolution for the specified silencers of the same inlet-outlet dimensions, Φ125/Φ125, but different air flow direction, during the most unfavorable loading range (10 V).

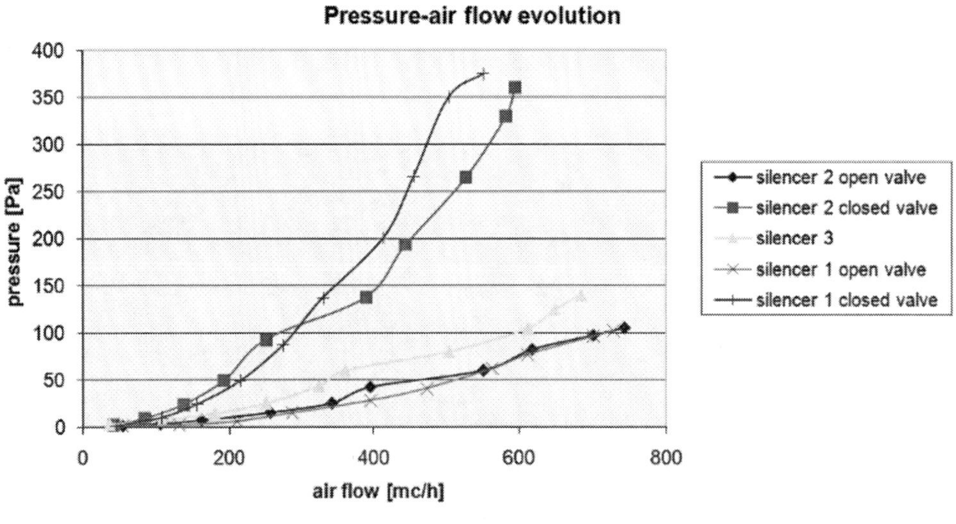

Fig. 3. Pressure versus air flow.

Note that we record the same evolution using *silencer 1 open valve* as for *silencer 2 open valve*, even the air flow direction is 90^0 rotated. Instead, for approximately the same pressures, the air flow values are greater for *silencer 2 closed valve* than those for *silencer 1 closed valve*, therefore the rotated direction of air flow is important in the closed valve configuration. Obviously, there are noticeable differences between open and closed variants, in terms of pressure. Also, the loading increasing leads to the increase of the pressure-air flow curve slope. There are little differences between the *silencer 3* (another absorptive material used inside) and *silencer 1 open valve* values on the pressure-air flow curve, so the lining has no influence on these parameters evolution, as it was expected.

Figures 4 and 5 report the sound attenuation versus the octave band frequencies, for the loading of interest in HVAC exploitation. We compared the sound attenuation for the three silencers types with the same inlet-outlet, but different flow direction and lining. As a general observation, we emphasize the attenuation decreasing with the loading increasing for all silencers analyzed.

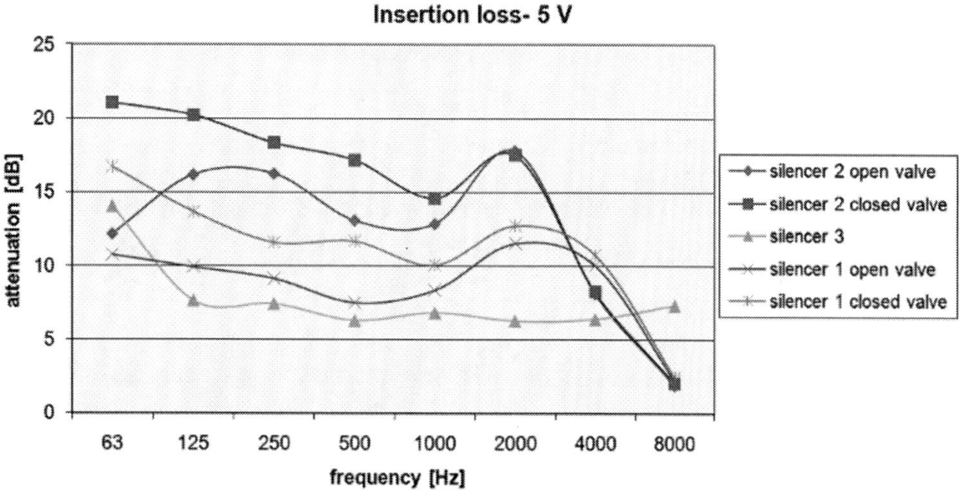

Fig. 4. Insertion loss for 5 V loading.

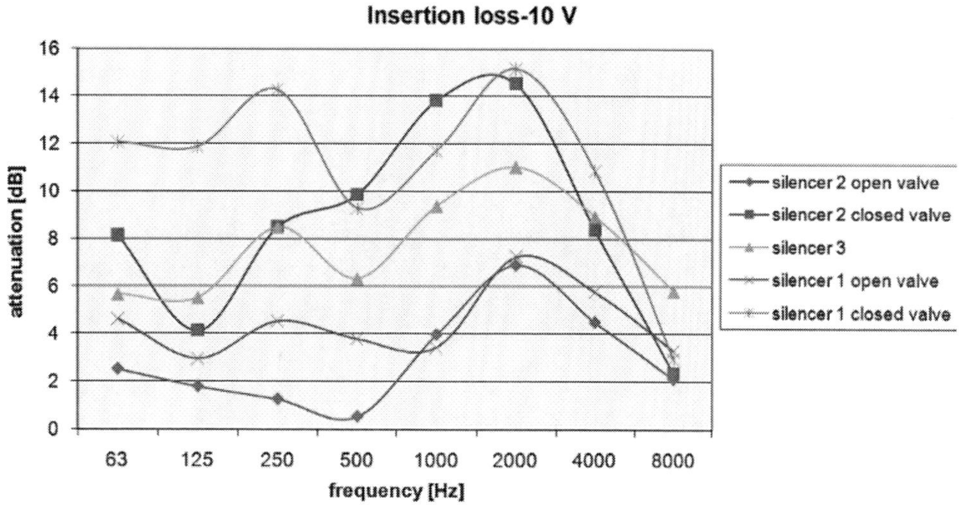

Fig. 5. Insertion loss for 10 V loading.

With the loading increasing, at 5 V (Fig. 4) - considered the usual exploitation of HVAC - we remark the good attenuation at lower frequencies, under 125 Hz, and a poor one above 4000 Hz. *Silencer 2* ensure the best sound attenuation from the analyzed types and we already observed the *closed valve* advantage in this case. The attenuation values decrease by 5-6 dB in 250-500 Hz frequencies area, compared with those from 1 V loading, but they are growing with 5-6 dB for lower frequencies under 250 Hz. *Silencer 3* (different absorptive lining) offers a weaker attenuation in the middle frequencies, compared to the other discussed silencers. However, we observe that at low (under 250 Hz) and high frequencies (over 4000Hz), *silencer 3* has a good sound attenuation behaviour.

Figure 5 presents the extreme situation, less used by potentially beneficiaries, when the loading reaches its maximum (10 V). The best sound attenuation behaviour (14-15 dB) belongs to the *silencer 1 closed valve*, and the second place is occupied by the *silencer 2 closed valve*. We remark the *silencer 3* for its satisfactory sound attenuation at 8000 Hz and certain constancy in attenuation behaviour along all the loading stages.

We can summarize all the results above presented for the tested silencers:
- silencer 2 (both open and closed valve configurations) provides a good attenuation in the middle frequencies range at every loading stage.
- silencer 3 offers a better attenuation in low and high frequencies and a moderate one in the middle frequencies range at every loading stage.
- silencer 1 closed valve is effective in sound attenuation at the maximum loading.

4 Conclusions

The paper is an experimental investigation on sound attenuation performances of some silencers types used for HVAC systems. The measurements were realized according to the standard and a test facility was implemented in an anechoic room. The results have emphasized the attenuation decreasing trend with the loading increasing of the tested silencers, as well as the important influence in attenuation of the absorptive material lining. The silencer lining influences the sound attenuation more than the geometry in certain situations. The loading increasing leads to the increase of the pressure-air flow curve slope. Also, the flow direction influences the acoustic level and improves the attenuation (90^0 rotation of flow direction, as in silencer 2). The inlet-outlet size of the silencer, the length, and the acoustical properties of the lining must be well correlated, in accord to the purpose and needs to be covered through its use in a specific application. It's a difficult task for the silencer designer. Future work may concern a theoretical investigation on sound attenuation in a silencer, with precautions and knowledge gained through this experimental investigation.

References

1. ASHRAE, *Handbook-Sound and Vibration Control* **48**, 279 (2011)
2. D. Borelli, C. Schenone, I. Pittaluga, *Proc. of Meetings on Acoustics* **14**, 040004 (2012)
3. K.S. Peat, *J. Sound Vibration* **146,** 353-360 (1991)
4. M. L. Munjal, B. K. Behera, P. T Thawani, *Appl. Acoust.* **54**, 229-238 (1998)
5. T. W. Wu, C. Y. R Cheng, P. Zhang, *J. Acoust. Soc. Am.* **111**, 2566-2572 (2002)
6. EN ISO 7235 : 2009, *Acoustics- Measurement procedures for ducted silencers- Insertion loss, flow noise and total pressure* (2009)
7. A. Bhatia, *PDHOnline* 1-72 (2012)
8. W.T.W Cory, *Fans & Ventilation: A Practical Guide* (Elsevier, 424, 2005)

Testing the assumption of linear dependence between the rolling friction torque and normal force

Stelian Alaci[1,2,*], *Luminita* Irimescu[1,2], *Marian-Adrian* Popescu[1], *Daniel* Fodorcan[3], *Dumitru* Amarandei[1,2] and *Florina-Carmen* Ciornei[1,2]

[1]Stefan cel Mare University, Department of Mechanics and Technologies, 13 University Street, Suceava, Romania
[2]Integrated Center for Research, Development and Innovation in Advanced Materials, Nanotechnologies, and Distributed Systems for Fabrication and Control (MANSiD), 13 University Street, Suceava, Romania
[3]Aalborg University, Design of Mechanical System, Fredrik Bajers Vej 5, 9220 Aalborg, Denmark

Abstract. Rolling friction is present in all nonconforming bodies in contact. A permanent topic is the characterization of the moment of rolling friction. A number of authors accept the hypothesis of linear dependency between the rolling torque and the normal force while other researchers disagree with this assumption. The present paper proposes a method for testing the hypothesis of linear relationship between rolling moment and normal pressing force. A doubly supported cycloidal pendulum is used in two situations: symmetrically and asymmetrically supported, respectively. Under the hypothesis of a linear relationship, the motions of the pendulum should be identical.

1 Introduction

The interaction between two solid bodies can be accomplished in two ways: by direct contact or via a field. In mechanical engineering applications, the first manner of interface, by direct contact, is met in the majority of cases. Most recently, according to the requirements imposed by diminution of energy utilization and reduction of wear, modern solutions were adopted, specifically the direct contact between the elements of a mechanical system was eliminated. Thus there can be mentioned the magnetic bearings, pneumatic actuators and so on. Even with all these advancements, a mechanical system within which there is no direct interaction (mechanical contact) between its elements is implausible. The scheme of a contact between two solid bodies is presented in Figure 1. It is assumed that the boundary surfaces are coming into contact in point P, where at least one of the two surfaces admits a unique normal $\boldsymbol{n}$. One of the bodies is considered as reference body, therefore its action upon the other body is materialized by the reaction force $\boldsymbol{R}$ and reaction moment $\boldsymbol{M}$. The relative motion between the two bodies is characterized

* Corresponding author: alaci@fim.usv.ro

by the sliding velocity $\boldsymbol{v}_{al}$ situated in the common tangent plane and by the angular velocity ω, which in its turn decomposes into the spin angular velocity ω_s, parallel to the normal $\boldsymbol{n}$ and the rolling angular velocity ω_r, placed in the tangent plane. Sequentially, the components of the reaction torsor are projected as follows: the reaction $\boldsymbol{R}$ into normal reaction N and tangential (friction) reaction $\boldsymbol{T}$ which is parallel but in opposite sense to the sliding velocity; the reaction moment having as components the spin torque $\boldsymbol{M}_s$, parallel and opposite to the spin velocity, and the rolling torque $\boldsymbol{M}_r$ parallel and opposite to rolling angular velocity ω_r. The characterization of the components specific to friction, $\boldsymbol{T}$, $\boldsymbol{M}_r$ and $\boldsymbol{M}_s$ is currently an open subject, still.

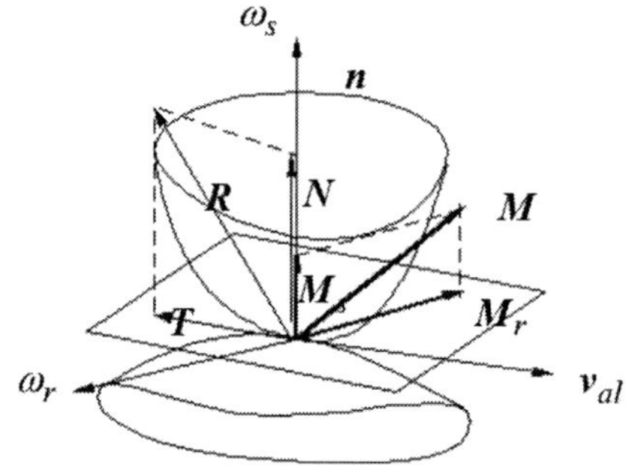

Fig. 1. Reaction forces and reaction torques from a mechanical contact.

Another remark to be considered concerns the dimensions of the contact region. Two situations are observed: first, when the contact zone has dimensions comparable to those of the bodies and therefore the contact point cannot be established - the case is stated as *conform contact*; second, when the contact between two nonconforming bodies is theoretically made in a point - the situation of *nonconform contact* or *Hertzian contact*. It is to be noted that while the friction force $\boldsymbol{T}$ and spin torque $\boldsymbol{M}_s$ can be met in both types of contact, the rolling torque $\boldsymbol{M}_r$ is met only in the contacts of nonconforming bodies. For an actual contact situation, two or even all three components can co-exist. If the aim is to obtain pure rolling motion between two bodies, this assumes that sliding velocity and spin angular velocity are simultaneously vanished ($v_{al} = 0$; $\omega_{sp} = 0$).

In the technical literature there are presented expressions of the three components of friction torsor depending on the normal pressing force N and on the characteristics of the relative motion between the two bodies. It is unanimously accepted that in dry friction conditions the friction force is proportional to the normal pressing force. The spin torque depends on normal pressing force at a power law, the exponent depending on the pressure distribution on contact area. Regarding the rolling friction torque, there are works that accept the hypothesis of proportionality to normal pressing force [1-3] while other papers disagree with this assumption, and in most of them there are proposed dependencies of power law type between rolling friction moment and normal force [4]. Cherepanov [5] reaches the conclusion that for the contact between a ball and an elastic half-space the exponent is $4/3$. In the present paper an answer is sought concerning the linear dependency between the rolling friction torque and normal force. An affirmative answer to this question would substantially simplify the model of dynamical behaviour of a mechanical system where rolling friction is encountered since the linear dependency should conduct to linear differential equations.

2 Principle of method

The employment of a mobile body that makes two identical contacts with an immobile body is proposed in order to verify the principle of linearity between rolling friction torque and normal force. All the components of reaction forces torsor $N_{1,2}$, $T_{1,2}$ $M_{r_{1,2}}$ occur in the contact points. Assuming the pure rolling motion hypothesis, the velocities of the contact points from the mobile body are zero (or equal, in this particular case) and therefore the angular spin velocity ω_{sp} is identical to zero. Under these conditions, the motion of the body is a parallel motion. A mobile system in translational motion is considered, as presented in Figure 2.

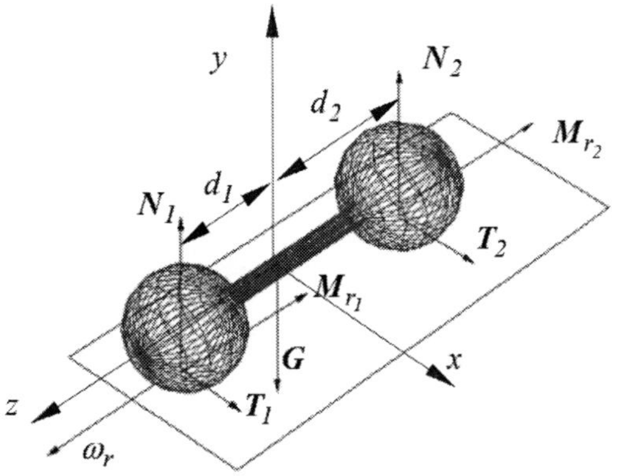

Fig. 2. Scheme of a body with two contact points in pure rolling motion.

The motion of the centre of mass projected on the vertical axis has the equation:

$$N_1 + N_2 - G = 0 \tag{1}$$

The body does not have rotation motion with respect to x axis, so it results:

$$N_1 d_1 = N_2 d_2 \tag{2}$$

where $d_{1,2}$ is the distance from the contact point to the weight's support. With the notation:

$$d = d_1 + d_2 ; \tag{3}$$

it results:

$$N_1 = (1-\xi)G,\ \ N_2 = \xi G \text{ where } \xi = d_1 / d \tag{4}$$

If the rolling friction torque depends on the normal pressing force and obeys the law:

$$M_r = s_{r\beta} N^{\beta} \tag{5}$$

where $s_{r\beta}$ is a proportionality factor, the total rolling friction torque is:

$$M_r = M_{r1} + M_{r2} = s_{r\beta}[(1-\xi)^{\beta} + \xi^{\beta}] = s_{r\beta} f(\xi,\beta) \tag{6}$$

where $s_{r\beta}$ is a constant and $f(\xi,\beta)$ is:

$$f(\xi,\beta)=(1-\xi)^{\beta}+\xi^{\beta} \quad (7)$$

In Figure 3 the variation of the function (7) that appears in the structure of the rolling friction torque is represented for four values of parameter β . For pure rolling hypothesis, assuming that an external torque M_a , parallel to the support axis, acts upon the body, the moment of momentum theorem written with respect to the instantaneous rotation axis is:

$$J_G + mr^2 = M_a - M_r = M_a - s_{r\beta}[(1-\xi)^{\beta} + \xi^{\beta}] \quad (8)$$

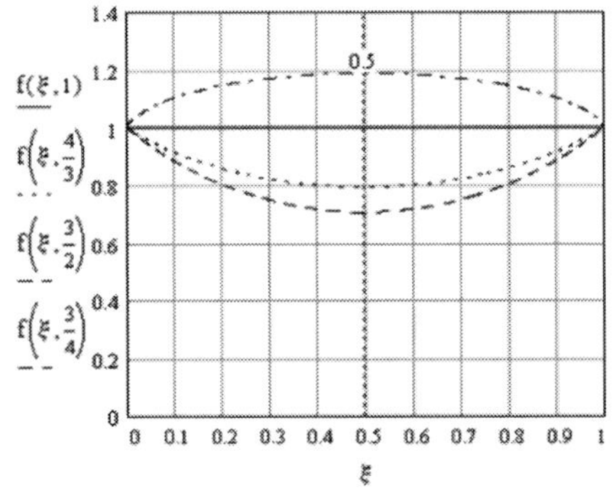

Fig. 3. Friction torque variation with respect to the position of the centre of mass.

Figure 3 and relation (8) show that the motion of the body is independent on the position of the centre of mass of the body only for $\beta = 1$ when linear dependency between the rolling torque and normal force exists.

3 Experimental set-up

The experimental device presented in Figure 4 was used for experimental validation. On a cylindrical rod 1 a vertical rod 2 is fixed.

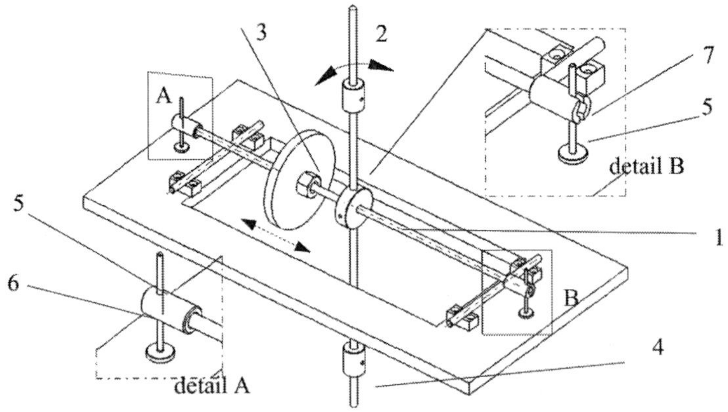

Fig. 4. Principle of experimental device.

A disc 3 can slide along the horizontal rod. The rod 1 has two points of contact with two identical parallel rods, placed in a horizontal plane. The system configuration materializes a cycloidal pendulum. When the pendulum is brought out of the equilibrium position and then set free, it will oscillate, and the oscillation motion is damped due to rolling friction torque. The oscillation period of the pendulum can be varied using the mass 4. A laser stick device is mounted at the end of the rod and thus the pendulum motion can be registered via a non-contact technique. To ensure the same launching position of the pendulum, when the position of the disc is changed, the horizontal shaft is blocked using the stud 5 and bushings 6 and 7, fixed on the shaft. The experimental test consists in the study of pendulum motion for different positions of the centre of mass (accomplished by different positions of the disc 3). In the case when the torque depends linearly on normal pressing force, according to Figure 3, the rolling friction torque is the same, irrespective of the position of the disc 3, and therefore the motion of the pendulum should be the same, regardless of the position of the mobile disc. In an opposite situation, the rolling friction torque depends on the position of the centre of mass, attesting nonlinear dependence between the rolling friction torque and the normal force. Based on the design presented in Figure 4, the experimental test-rig was built and it is presented in Figure 5.

Fig. 5. Actual experimental device.

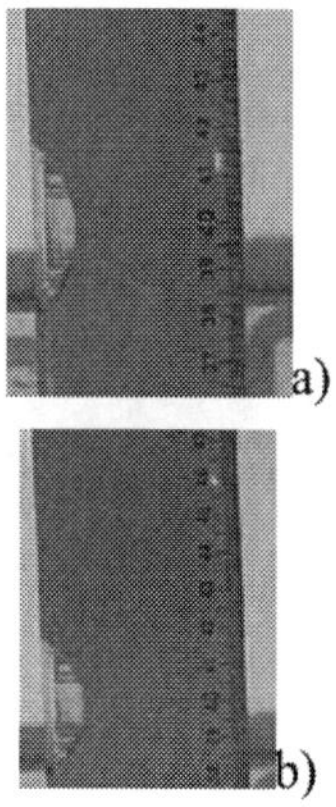

Fig. 6. Spot image.

A series of tests was carried out on the experimental device for certain pairs of materials, namely steel-aluminium, steel-rubber and steel-steel. An experimental analysis consists of establishing the pair of tested materials, positioning the disc in the lateral or central position on the rod, launching the oscillatory motion, and recording the motion of the spot along the scale with a digital camera.

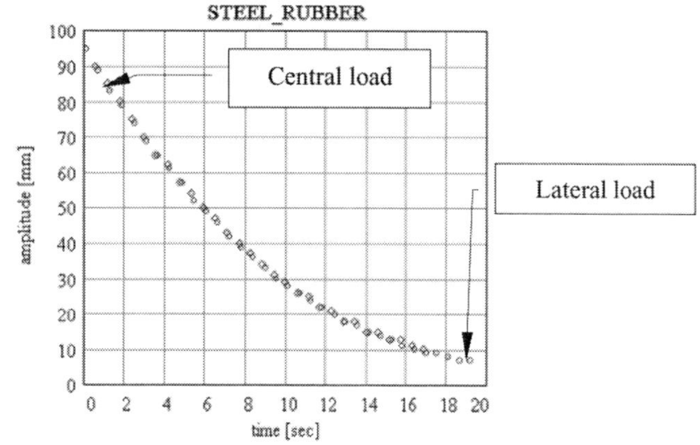

Fig. 7. Steel-rubber amplitude decrease.

The video is then split into frames and the amplitude of the motion and corresponding instants are found. The amplitude can be determined accurately, as shown in Figure 6, since at maximum elongation the spot is immobile (Figure 6b). The experimental data are plotted as follows: on the same graph, the amplitude decrease is compared for the two positions of the disc on the rod, for the same pair of materials. In Figure 7 the behaviour of damping from steel-rubber contacts is presented, and it can be noticed there is a coincidence between the points from the two plots. In Figure 8 there are plotted the data obtained for the pair steel- aluminium, and the decrease in amplitude differs - in the case of the central disc there is a stronger damping of amplitude.

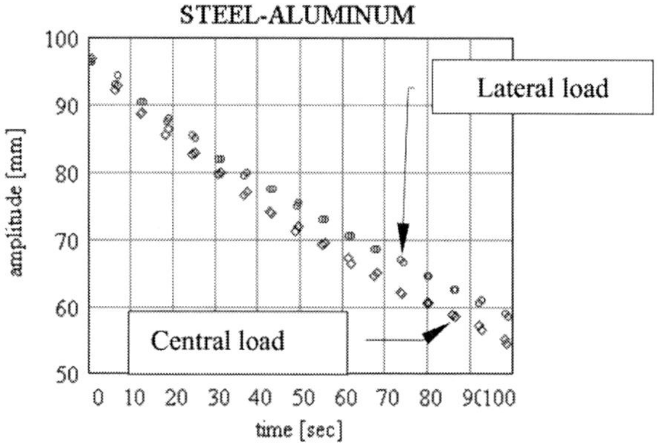

Fig. 8. Amplitude decrease for steel-aluminium pair.

4 Conclusions

The paper presents a proposed method and device for the study of linear dependence between the rolling friction torque and normal load of a rolling contact. The principle of operation of the test rig consists in damped motion of a cycloidal pendulum with double contact. Testing dissimilar materials like rubber-steel and aluminium-steel, one can state that there is not a general tendency for the pairs of materials that is a linear dependency between rolling friction torque and normal load. As a general conclusion, for any pair of materials it is necessary to find the particular law expressing the dependence between rolling torque and normal load. The paper regards only incipient research in applying the proposed method.

The infrastructure used for this work was partially supported by the project "Integrated Center for research, development and innovation in Advanced Materials, Nanotechnologies, and Distributed Systems for fabrication and control", Contract No. 671/09.04.2015, Sectoral Operational Program for Increase of the Economic Competitiveness co-funded from the European Regional Development Fund.

References

1. R. Voinea, V. Ceaușu, D. Voiculescu, *Mechanics* (in Romanian), (EDP, 1983)
2. P. P. Teodorescu, *Mechanical systems, Classical Models*, *Vol. 1*, (Springer, 2007)
3. D. Mangeron, N. Irimiciuc, *Rigid body mechanics, Vol. 1.* (in Romanian), (Ed. Tehnică Bucuresti 1978)
4. A. V. Korolev, A. A. Korolev, Journal of Friction and Wear, **37**, 2, 119 (2016)
5. G. P. Cherepanov, J. Appl. Mech. Techn. Physics, **55**, 1, 182 (2014)

Damped periodic motions used in the study of the drag coefficient

Stelian Alaci[1,3,*], *Constantin* Filote[2,3], *Daniel* Fodorcan[4], *Erwin-Vasile* Alexandru[1], *Dumitru* Amarandei[1,3] and *Florina-Carmen* Ciornei[1,3]

[1]Stefan cel Mare University, Department of Mechanics and Technologies, 13 University Street, Suceava, Romania
[2]Stefan cel Mare University, Computers, Electronics and Automation Department, 13 University Street, Suceava, Romania
[3]Integrated Center for Research, Development and Innovation in Advanced Materials, Nanotechnologies, and Distributed Systems for Fabrication and Control (MANSiD), 13 University Street, Suceava, Romania
[4]Aalborg University, Design of Mechanical System, Fredrik Bajers Vej 5, 9220 Aalborg, Denmark

Abstract. The paper proposes a method and the related experimental setup for finding the drag coefficient for a body. The device consists in a pendulum materialized by a hinged rod initially in horizontal position. A spherical body immersed in a liquid is attached to the free end of the rod. The restoring force is created by a helical spring. The inertial characteristics of the pendulum can be changed using an additional mass. The differential equation characteristic to the motion of the pendulum is deduced and subsequently integrated numerically. The coincidence between experimental and theoretical data allow for viscosity evaluation.

1 Introduction

Real fluids can transmit shear stresses between two surfaces between which relative motion exists, due to the property of the fluid - *viscosity*. The experimental finding of viscosity is still an actual problem. There are a large number of works dedicated to the subject, starting with classical monographs of Stokes [1-2] and ending with the most recent papers and thesis. The subject was approached both from a fundamental sciences point of view [3] and by researchers from applied sciences [4]. A device used in viscosity measurements must satisfy the following condition: to permit the finding of the magnitude of interaction between two rigid surfaces between which there is a liquid when there is imposed motion between the surfaces or, to allow relative motion between the two surfaces when the interaction between them is imposed. The relative motions imposed between the surfaces can be periodical or non-periodical. The shape of the surfaces and the relative motion are the main criteria for viscometer classification [5]. Landau and Lifshitz, [3] present analytical expressions of the calculus of hydrodynamic friction for different combinations of surfaces and relative motions.

* Corresponding author: alaci@fim.usv.ro

2 Principle of proposed device

The idea of the proposed test rig started from the Nagarajan paper [6]. He uses a viscometer presented in principle in Figure 1, where one of the surfaces is the boundary surface of a gold sphere 1 immersed in an unbounded fluid. The ball is connected to a helical spring 3 and the dynamic system ball-spring is stimulated into oscillation on vertical direction. An electronic transducer 2 registers the motion of the ball and the damping phenomenon is revealed. The causes of damping the author identifies are the drag force and the correction term of inertial force due to the dislocated liquid by the sphere. From the calculus of logarithmic decrement of damping he determines the viscosity of the liquid. Besides the forces met above, other forces are responsible of motion damping [7] like the force describing the history of the motion of the fluid or the correction term for the drag force due to the effect of finite boundaries of the liquid.

The schematic of the new proposed viscometer is presented in Figure 2. It consists of an aluminium horizontal pipe 1 of square cross-section, which can rotate about the joint O. A vertical shaft 2 is firmly attached to the rod, having a sphere 3 fastened at the inferior end that can be immersed into the liquid to be studied, 4. At the other end of the vertical shaft, a body 5 can be fixed eccentrically with respect to the horizontal rod. A supplementary body can be fixed on the rod at a specific distance, on the opposite side of the joint O with respect to the vertical shaft. The function of this body is to change the moment of inertia JOz of the pendulum with respect to the axis of oscillation and the position of the centre of mass. The gliding body 5 from the vertical shaft has the task to get the centre of mass of the pendulum on the axis of horizontal rod. Additionally, a series of holes with vertical axis are made in the rod and thus the position of the point where the spring 7 is tied up can be modified. The upper end of the spring is fixed in a bushing that can translate vertically. This ensures the possibility that after the spherical body is immersed into the liquid the pendulum will be brought into horizontal position by modifying the spring pre-stressing force. Therefore the moments of weight, pre-stressing and buoyancy forces are balanced. The motion of the pendulum is investigated via a non-contact method. To this purpose, at the end of the rod a small laser stick 8 is fixed and the spot can be watched on a vertical screen E having a scale S. After the pendulum is set into oscillation, the motion of the spot on the scale from the screen is video-captured and after that the recording is split into frames for identification of extreme values of spot displacement and the index of the frame.

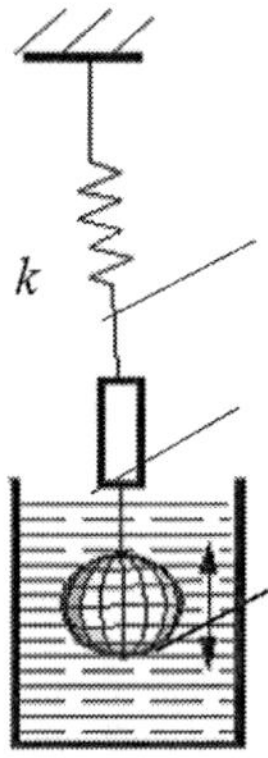

Fig. 1. Viscometer with sphere [6].

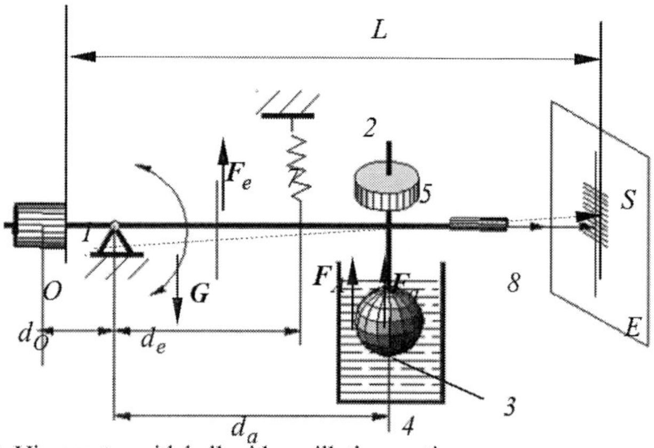

Fig. 2. Viscometer with ball with oscillating motion.

Compared to the viscometer presented in Figure 1, the proposed device has the advantage that the effects of forces acting upon the pendulum can be controlled, diminished or amplified, by varying the distances of the points of application of forces acting upon the pendulum. The principle of viscosity evaluation is similar to the one used by Nagarajan [6] and refers to the comparison between the damping of theoretical model and the experimental damping. The inertial forces acting upon the ball are [6]:

$$M = m_b + m_c + m_s / 3 \quad (1)$$

Where m_b is the mass of the ball, m_c is the mass of the transducer core and m_s is the mass of the spring. In our case, $m_c = 0$ and $m_s \cong 0$ since the mass of the spring is insignificant compared to the mass of the ball. The mass of the displaced liquid by the oscillating ball must be considered; Stokes [1] introduced the correction of inertia force:

$$F_b = \left(\frac{1}{2} + \frac{9}{4} \mu \right) m' \frac{d^2 x}{dt^2} \quad (2)$$

where m' is the mass of the liquid displaced and μ is a dimensionless parameter having the expression:

$$\mu = [\eta T / \pi \rho a]^{1/2} \quad (3)$$

where η is the dynamic viscosity, ρ is the fluid density, T is the period of oscillation and a is the radius of the sphere. The total inertial force is given by:

$$F_i = -\left[m_b - m' \left(\frac{1}{2} - \frac{9}{4} \mu \right) \right] \frac{dx^2}{dt^2} \quad (4)$$

Next it is assumed that the period of oscillation of the ball is not affected by the viscous damping. After the ball is completely immersed into liquid, the device is balanced and the pendulum is set to horizontal position using the additional mass 6. The elastic force acting on the ball is proportional to the spring deformation x and elastic constant of the spring k_e:

$$F_e = -k_e x \quad (5)$$

The damping force on the sphere that oscillates vertically in a liquid is given by:

$$F_d = -\frac{9}{2} \frac{\pi \mu M'}{T} (1 + \mu) \frac{dx}{dt} \quad (6)$$

and being valid in the domain of velocities that situate the Reynolds number in the range $Re = [40 \div 200]$. The equation of motion of the pendulum is obtained by applying the moment of momentum theorem about the joint centre:

$$J_O d^2 \theta / dt^2 = -F_i d_a - F_d d_a - F_e d_e = 0 \quad (7)$$

where J_O is the moment of inertia of the pendulum with respect to the joint O.
Since the inertial force F_i depends on the acceleration, this is taken into account by replacing J_O with a new moment of inertia J'_O that considers the effects of the liquid

displaced and set into motion by the sphere, as well. With the assumption of small oscillations, $\theta \leq 5^{\circ}$, after some manipulation the equation 7 becomes:

$$J'_O \frac{d\theta^2}{dt^2} + \frac{9}{2}\frac{\pi\mu m'}{T} d_a^2 (1+\mu) \frac{d\theta}{dt} + k_e d_e^2 \theta = 0 \quad (8)$$

The equation 8 is a linear homogenous differential equation with constant coefficients, written under the standard form:

$$A\ddot{\theta} + 2B\dot{\theta} + C\theta = 0 \quad (9)$$

where:

$$A = J'_O \cdot d_a^2, B = \frac{9}{4}\frac{\pi\mu m'}{T} d_a^2 (1+\mu), C = k_e d_e^2 \quad (10)$$

With the notations:

$$\alpha = B/A, \quad \beta = \sqrt{C/A - (B/A)^2} \cong \sqrt{C/A} = T \quad (11)$$

one can attest that between the logarithmic decrement of the model and the experimental damping exponent α the following relation subsists:

$$ln(\theta_0 / \theta_{0n}) = \alpha n T \quad (12)$$

where T is the period and θ_0, θ_n are two angular amplitudes corresponding to the instants 0 and nT, respectively. Using relation 11 in equation 12, the final relation on which the evaluation of viscosity is based is obtained:

$$ln\frac{\theta_0}{\theta_{0n}} = \frac{9}{4T}\pi\left[\frac{\eta T}{\pi\rho a}\right]^{1/2} m' d_a^2 \left(1 + \left[\frac{\eta T}{\pi\rho a}\right]^{1/2}\right)\frac{nT}{J'_0}. \quad (13)$$

3 Experimental device. Results and discussions

The experimental set-up employed in laboratory tests is presented in Figure 3 where the liquid used is glycerine. The oscillations of the pendulum are 5^0 visualised on the screen via the laser spot and the geometrical restriction $\theta \leq 5^{\circ}$ can be controlled. The angular oscillations are mapped as linear displacements of the spot on the scale and recorded with a camera in real time. The decrease of the angular amplitude is established on computer, by analyzing the frames of the movie. To eliminate the errors due to the internal friction of the pendulum, the test was also performed in air. The decrease of the amplitude of the pendulum for the two experimental cases was fitted by exponential function of the form $\theta_0 \, exp(-\alpha t)$. The experimental data and the interpolation curves obtained for air and glycerine are given in Figure 4. To eliminate the effect of internal friction, assuming that the values of the damping exponents are small, the following approximation is used:

$$exp(\alpha x) \cong 1 + \alpha x \tag{14}$$

and it results:

$$\alpha_g = \alpha_{ga} - \alpha_a \tag{15}$$

where α_{ga} is the exponent for global experiment, α_a is for pendulum oscillating in air and α_g is the exponent deduced for the damping produced only by glycerine.

a)

b)

Fig. 3. Experimental set-up: assembly (a) and detail (b).

The theoretical curve due to damping produced only by glycerine is also plotted in Figure 4. For the oscillations in air and liquid, the periods are practically the same $T_{exp} = 1.167\,sec$, as resulting from Figure 5, and the hypothesis that the period is insensitive to the effect of viscosity is confirmed.

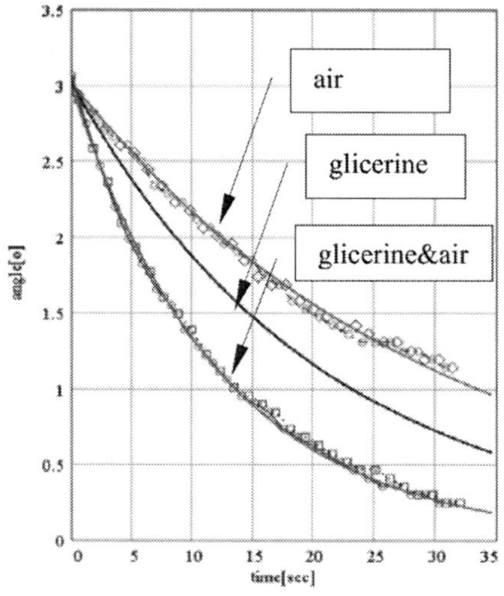

Fig. 4. Amplitude variation-experimental and interpolation curves.

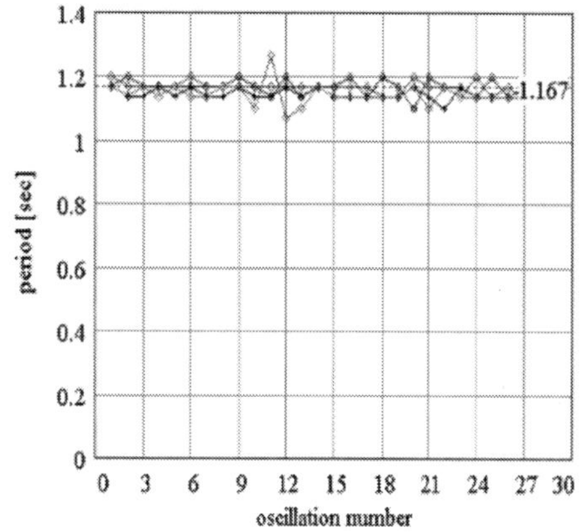

Fig. 5. The period of oscillation is not affected by viscosity.

From the second equation 11 it results that the moment of inertia can be found J'_O imposing the condition that the period of the model equals the experimental period T_{exp}.

$$J'_O = k_e d_e^2 T_{exp} / (4\pi^2) \tag{16}$$

Knowing J'_O and using the first equation 11, with the condition that the logarithmic decrement of the model is the same as the experimental damping exponent α and taking into account the equation 3, a transcendental equation that allows finding the viscosity is obtained:

$$\alpha_g = \frac{\frac{9}{4}\pi\left(\eta\frac{T_{exp}}{\pi\rho' a}\right)^{1/2}\frac{m'}{T_{exp}}d_a^2\left[1+\left(\eta\frac{T_{exp}}{\pi\rho' a}\right)^{1/2}\right]}{k_e d_e^2 T_{exp}/(4\pi^2)}. \tag{17}$$

The experimental parameters were: $a = 0.02\,m$; $\rho' = 930\,kg/m^3$; $J'_0 = 0.128\,kg \cdot m^2$; $k_e = 70\,N/m$; $d_e = 0.25\,m$; $d_a = 0.45\,m$; $T_{exp} = 1.167\,sec$;. The damping interpolation exponent $\alpha_g = 0.048\,sec^{-1}$ results from $\alpha_{ga} = 0.081\,sec^{-1}$ and $\alpha_a = 0.033\,sec^{-1}$. With these values, the value obtained for the experimental dynamic viscosity is $\eta = 0.977\,Pa \cdot s$, in very good agreement with the values from literature.

4 Conclusions

The present paper proposes a method and device for finding the viscosity of a fluid. The method is based on the viscometer with vertical oscillations of a ball and spring system. The effect of damping forces of the fluid upon the oscillations of the ball is characterized by logarithmic decrement of oscillations, accepting thus an exponential function for the decrease of oscillation amplitude. The proposed viscometer has oscillations about a fixed point. Based on the perfect similitude between translational motion and rotation motion around a fixed axis, the method of finding the drag force is applicable to the new viscometer. The progress brought by the new proposed method is that the equation of the model is described by a series of parameters that allow the amplification/ attenuation of the effect of specific forces occurring in the dynamic system, and thus the model is more flexible.

The infrastructure used for this work was partially supported by the project "Integrated Center for research, development and innovation in Advanced Materials, Nanotechnologies, and Distributed Systems for fabrication and control", Contract No. 671/09.04.2015, Sectoral Operational Program for Increase of the Economic Competitiveness co-funded from the European Regional Development Fund.

References

1. G. G. Stokes, Trans. Cambridge Phil. Soc., IX, 8 (1850)
2. H. Lamb, *Hydrodynamics* 6th ed. (Cambridge University Press1994)
3. L. D. Landau, E. M. Lifshitz, *Fluid mechanics*, 2-nd Ed. (Pergamon Press 1987)
4. A. Cameron, *The principles of lubrication*, (John Wiley & Sons Inc 3rd edition 1981)
5. S. V. Gupta, *Viscometry for Liquids* (Springer Int. Publ. Switzerland 2014)
6. M. K. Nagarajan, Canadian J of Chemistry, **46**, 1969 (1968)
7. R. A. Nelson, M. G. Olsson, Am J Phys **54** (2), 112 (1986)
8. F. C. Ciornei, S. Alaci, Fasc. Manag. Tech. Eng. **XIII-1**, 67 (2014)

Investigation of geometric parameters' influences on the measurement uncertainty by various measuring sensors

Rauf Oezden* and *Marcel Sabin* Popa

Technical University of Cluj-Napoca, Department for Manufacturing Technology, Romania

Abstract. The following paper was written in cooperation with the TU of Cluj-Napoca and provides an overview on the possibilities of calculating measurement uncertainty in the industrial environment of geometrical measurement. Scheduling effective influences reveals measurement uncertainties. Furthermore, experimental approaches, which prove that measurements with different sensors arrive at difficult to reproduce data, are presented. Experimental methods, which reveal inevitable errors in measurement because they violate the laws of mathematics, are presented as well.

1 Introduction

Geometrical measurement is an integral part of the industry. Component evaluation and process control require great reliability on the part of the measurement results.

Often there is no clear statement on a measurement's accuracy and the uncertainties it is fraught with. Reason for this is commonly poor knowledge of the origin of measurement uncertainty of the influences that affect the results and their scope. Furthermore, there is a lack of experience on how to deal with individual influence factors.

Today, a measurement result, which was determined in a single measurement, is often accepted as the true value of a test object, but the measurement result is always an estimator, which is synthesized from several measurements. The true value of an item is imperfect and not accurately determinable in its nature. Each measurement is subject to influences, either by the environment, in which it is conducted, by the gauge, by the operator, by the measurement strategy or by the type of measurement subject.

Measurement uncertainty is considered to be a quantitative dimension for the quality of the respective measurement result. It is associated with the measurement result and characterizes the area of values, which can be ascribed to the measurement parameter through the conducted measurement. It answers the question of how well the determined result mirrors the measurement parameter's true value. Furthermore, it facilitates the assessment of measurement results' reliability, enabling comparisons of other measurements with the same measurement parameter or comparisons with reference values. Trust in the comparability of measurement results is very important in international trade. It

* Corresponding author: rauf.oezden@outlook.com

helps to remove trade barriers. Since production engineering constantly evolves, driven by international competition, tolerances decliner more and more. Components become increasingly smaller and delicate. Production measurement technology settles immediately in or close to production. Correspondingly, the requirements for gauges in terms of speed, accuracy, reliability and sensitivity permanently increase. Thus, the effort to determine reliable measurement results increases. In order to understand and adapt properties such as influences, which affect measurement results, the calculation of measurement uncertainties through modeling or the assessment of causes, it is crucial consider the standards of product specification. Standards guarantee for a consistent approach in reducing inaccurate assessments and in creating global comparability of results. The particular advantage is the international confirmation of standards.

2 Measurement errors and measurement uncertainty

Measurement results can only comparable when they have been obtained under reproducible conditions. Measurement results shall become comprehensible through the designation of reference conditions [1]. Among the reference conditions is the reference temperature, which is set to 20°C in accordance with the standard [2].

Furthermore, the complete measurement sequence (measurement under repeatability conditions or under comparability conditions) and the influential variables, which themselves are not part of the measured variables, but influence the measured value, have to be constituted. In the measurement of geometrical quantities, the temperature of the measured object and its environment are important influential variables, which can entail measurement deviations.

The measuring sensors' user strives to keep measurement deviations to a minimum with the help of suitable measurement strategies, favorable measurement sequences and shielding from disturbances.

Figure 1 illustrates the path via the measurement deviations towards the measurement results and the occurring measurement uncertainties.

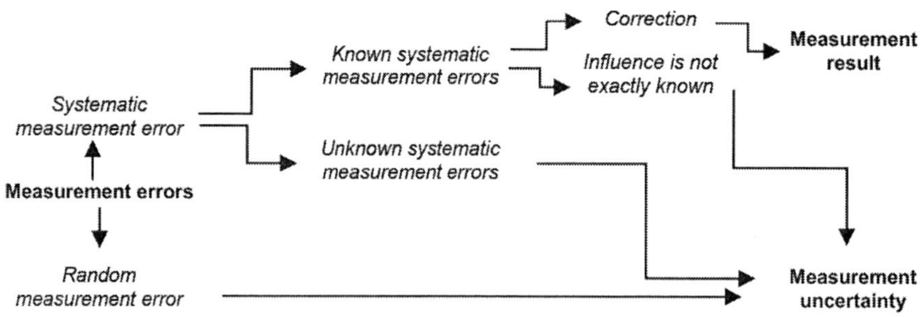

Fig. 1. Influences that describe a complete measurement result.

As far as the systematic measurement deviations are determined under comparability conditions, they are considered to be known and can influence the measurement result als corrections. The absolute value of the correction complies with the systematic measurement deviation, although with opposite portents.

Random measurement deviations can only partially be estimated via measurement sequences under repeatability conditions, because they still contain unknown fractions of the systematic measurement deviations. It is not known here, with which portent the measurement uncertainty is afflicted [1]. Therefore, it is included in the measurement result as portent undetermined value (±).

2.1 Determination of measurement uncertainty

2.1.1 Fundamentals

The definition of measurement uncertainty renders it a quantitative measure of the respective measurement result's quality. It answers the question of how well the obtained result mirrors the measurement`s value. It enables the user to evaluate the reliability of the measurement result, e.g. to compare the results of different measurements with the same measurement values to each other or to reference values.

Often, a measurement value is to be compared to threshold values, which are defined in a specification or normative provision. In this case, the measurement uncertainty can yield whether the measurement result is clearly within the predefined limits or whether it hardly meets the demands. When the measured value is close to a limit value, there is a high risk for the measurement value not to meet the required demands.

2.1.2 Important policies

An overview of relevant standards and policies for the determination of measurement uncertainty in geometrical measurement technology is listed in the following:
- DIN EN ISO 14253 Inspection by measurement of workpieces and measuring equipment
- DIN EN ISO/IEC 17025 Requirements for the competence of testing and calibration laboratories
- DIN V ENV 13005 Guide to the Expression of Uncertainty in Measurement (GUM)
- VDA5/ISO22514-7 Test suitability of measurement/Capability of measurement processes
- VDI/VDE 2617 Accuracy of coordinate measuring machines

Today's high-tech economy with worldwide quality standards demands safe and particularly precise measurements. Only thus, extensive division of labor, precision fit, long life, product and consumer protection are possible. The basis for this are international standards and the tracing of any measurement values with known measurement uncertainty. An internationally accepted approach to measurement uncertainty determination is provided in the GUM (Guide to the Expression of Uncertainty in Measurement). In order to apply this comprehensive guide in accordance with the standard, a simple procedure for the determination of measurement uncertainty on the basis of GUM is developed in the international standardization committees [3]. In the process, a cost-optimized solution is developed for a certain measurement problem via an iterative method in consideration of measurement uncertainty. This approach is called the PUMA method (Procedure for Uncertainty Management). In accordance with the standards, methods for measurement evaluation, which are based on Gum, are compiled within this standard. These also constitute the basis for the subsequent mathematical representation [4, 5].

2.1.3 Mathematical process

According to the afore mentioned standards, the evaluation of a measurement including the assessment of measurement uncertainty can be conducted in difference steps: (The approach compliant with GUM is subdivided into eight steps)
- Establishment of a model, which describes the measurement values' connection to all other involved values mathematically.
- Calculation of the measurement result and the measurement uncertainty for the measurement variable from the prepared data.
- Specification of the complete measurement result and determination of further uncertainty.

Basis for the assessment of measurement uncertainty is a model, which describes the connection between measurement value y as result variable and the input variables $x_1, x_2, \ldots, x_n$ in the form of a mathematical function:

$$y = f(x_1, x_2, \ldots, x_n) \quad (1)$$

Provided that the measured variable has been measured multiple times, the arithmetic average has to be calculated from the individual measurement results:
$\overline{x}$ from characteristic values:

$$\bar{x} = \frac{1}{n}\sum_{i=1}^{n} x_i \quad (2)$$

It denotes the unwarranted measurement result, which is directly available in a measurement. If several measurements are conducted, the standard deviation from the average u has to be determined. It is calculated from the standard deviation of the attention value s, whereas n denotes the quantity of attention values:
s from characteristic values:

$$s = \sqrt{\frac{1}{n-1}\sum_{i=1}^{n}(x_i - \bar{x})^2} \quad (3)$$

The influences that take effect on a measurement evoke systematic and random measurement deviations. The known systematic deviation is used to correct the measurement result.

The random and unknown systematic deviations account for the measurement uncertainty. Their influence has to be estimated, unless it has been considered by the repeated measurement [6]. Thereto, data from earlier measurements, manufacturer's data or knowledge on the behavior and the characteristics of measurement objects and measurement sensors can be used.

According to the standard the assessment of uncertainty percentage via repeated measurement is designated method A and the assessment via alternative means is designated method B [7]. It is of crucial importance for the subsequent roundup of all measurement uncertainty components that the individual components' employed scatter parameters are equivalent and therefore transmittable.

This is achieved through interpreting the measurement uncertainty as standard deviation. In order to obtain a complete measurement result, the correction has to be added. The result provides the best estimate for the measurement parameter, which is determined via this measurement. For the determination of measurement uncertainty via method B, the standard deviation as well as the standard uncertainty have to be calculated from the available information. Following the determination of standard uncertainty u:

Method type A (Normal distribution): Statistical analysis (series of measurements)

$$u = s \quad (4)$$

Method type B (Rectangular distribution):
- Other information (Value is between a lower limit -a and an upper limit +a)
- All values in between are equally likely (no further information)
- Specification in a calibration certificate [8]

$$\bar{x} = \frac{(a_+ + a_-)}{2} \quad (5)$$

$$u = \frac{\Delta a}{\sqrt{3}} \quad (6)$$

In accordance with [1], determination B (method B) is based upon earlier measurements, experience, specifications in calibration certificates or in manufacturer catalogues.

Catalogue specifications are then interpreted as extended measurement uncertainty U. Typically, catalogue specifications do not list information on probability distribution and level of reliability.

In these cases, normal distribution and a reliability level of 95.45% are assumed and an amplification factor of k=2 is set. Hereby, a measurement uncertainty specification from a catalogue can be ascribed to the standard uncertainty u:

$$u = \frac{U}{k} \tag{7}$$

This becomes necessary in cases, in which several measurement uncertainty components $u_1,\ u_2,\ u_3, \dots\ u_n$ are to be combined to standard uncertainty u_c :

$$u_c = \sqrt{u_1^2 + u_2^2 + u_3^2 + \cdots + u_n^2} \tag{8}$$

The extended measurement uncertainty U is calculated from the combined standard uncertainty with the amplification factor k, whereby an amplification factor of k=2 is usually assumed for a reliability level of 95.45%.

$$U = k \cdot u_c \tag{9}$$

3 Experimental determination of avoidable influences from practice

Avoidable influences must be prevented in a geometrical measurement's early phase and should not be considered in the calculation of a measurement uncertainty. The following study demonstrates a classic case within the industries, which is often predetermined as an inspection criterion in technical requirements and results in the misjudgement of a measurement uncertainty.

Circle measurements are among the most common measurement assignments in product development. In accordance with [9], operators often ask themselves, with how many measuring points a workpieces' circles or cylinders are to be measured. The difficulty in measuring the radii and the associated circle segment is, among others, the identification of the discharge radius, which is hard to identify on a tangible workpiece.

In accordance with [10] the radius of a circle can be prefixed in order to eradicate this problem.

When, for example, the circle is calculated from measurement points, the circle center and the circle diameter normally have to be calculated. As an alternative side condition, the diameter can be prefixed and the center has to be calculated.

In accordance with the standard, at least 90° of a circle segment have to be existent as measureable contour to specify radii [11]. However, this is seldom arranged in the technical drawings. Although this provision (at least 90°of a circle segment existent as measureable contour) applies to synthetic material, different circle segments are measured with visual- and tactile sensors in the following experiment, which features the example of a calibrated testing pin (maximum permissible deviation of ± 2 µm), revealing fluctuation in metallic materials as well.

Table 1 and Figure 2 provide an overview of deviations using the example of a calibrated testing pin under the provision of different circle segments.

DOI: 10.1051/matecconf/201711207004

Table 1. Evaluation on a calibrated Pin Gauge.

Tested on a calibrated Pin Gauge *Reference Diameter = 10,000 mm*			
Circular Segment [°]	Scanpoint O/T	Optical Radius [mm]	Tactile Radius [mm]
15°	8/12	4,847	4,846
30°	17/15	4,890	4,892
45°	36/31	4,953	4,956
60°	47/43	4,965	4,963
75°	64/61	4,972	4,971
90°	109/107	4,995	4,996
120°	132/124	5,000	5,001
150°	181/169	5,001	5,001
180°	234/219	5,001	5,000
270°	327/322	5,000	5,001
360°	378/367	5,001	5,001

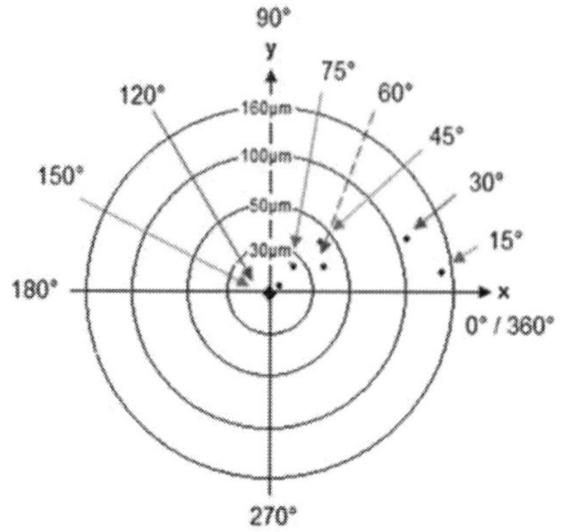

Fig. 2. Uncertainty to the reference middle point.

For stability reasons, measurements were repeated fivefold per circle segment following the Gauss-strategy. The result for the radius thus complies with the average of five measurements. In a first step, 15° of the ideal contour were taken in and measured with the calibrated testing pin. Figure 2 reveals that the circle center shows a measurement deviation from the reference focus larger than 15 µm. It becomes clear that measurements under an available circle segment of 90° reveal different or rather fluctuating deviations. Measurements over 90° show comparable measurement values, merely an inevitable random measurement deviation comes into effect here. It is also important to note, that the measurement results become more stable with an increased number of recorded measuring points.

4 Conclusions

This article describes measurement deviations and measurement uncertainty with the help of a clear concept in detail. Influences which take an effect on measurement results and hence lead to measurement uncertainties were highlighted experimentally. The idea was to show that not all influences, which are known prior to an experiment, actually have a substantial effect on the measurement result. Furthermore, non-avoidable influences were described, which always occur under certain conditions and steadily afflict the measurement result with a measurement uncertainty. Through the establishment of a measurement uncertainty record, small influences can be analyzed and further avoided. Future research should focus on the observation of measurement uncertainties in workpieces with alternating properties, such as color, surface etc.

References

1. W. Dutschke, *Manufacturing Metrology*, (Teubner Verlag, 2002)
2. ISO 1 (Standard), *Reference temperature for geometric product specification* (2016)
3. ISO 14253-2 (Standard), *Guidance for the estimation of uncertainty* (2011)
4. DIN 1319-3 (Standard), *Evaluation of measurements of a single measurand* (1996)
5. DIN 1319-4 (Standard), *Evaluation of measurements* (in German) (1999)
6. ISO 22514-7 (Standard), *Capability of measurement processes* (2012)
7. ISO/IEC 17025 (Standard), *General requirements for calibration laboratories* (2005)
8. T. Hausotte, *Manufacturing Metrology* (in German) (Erlangen University, 2015)
9. F. Cebulla, G. Smith, S. Midha, *NMS* (in German), QZ Hanser Verlag, **56** (2004)
10. A. Weckenmann, *Coordinate measuring technique* (in German) (Hanser Verlag, 2012)
11. DIN 16742 (Standard), *Tolerances and acceptance conditions* (2013)

Rigidity versus deformability hypothesis in impact dynamics

Florina-Carmen Ciornei[1,3,*], *Stelian* Alaci[1,3], *Dumitru* Amarandei[1,3], *Constantin* Filote[2,3], and *Ioan* Tămăşag[1,3]

[1]Ştefan cel Mare University, Department of Mechanics and Technologies, 13 University Street, Suceava, Romania
[2]Ştefan cel Mare University, Computers, Electronics and Automation Department, 13 University Street, Suceava, Romania
[3]Integrated Center for Research, Development and Innovation in Advanced Materials, Nanotechnologies and Distributed Systems for Fabrication and Control (MANSiD), 13 University Street, Suceava, Romania

Abstract. The complexity of the study concerning an impact phenomenon depends strongly on the accepted simplifying hypothesis for the model. One of the most important assumptions in the study of mechanical impact refers to the absolutely rigid or compliant character of colliding bodies. It is anticipated that the employment of hypothesis of deformable body should lead to a better modelling of the process. The present paper presents a qualitative comparison between the results obtained via the two methods, considering one of the simplest impact models, namely the collision with dry friction between a metallic ball and an immobile flat body.

1 Study methods of impact phenomenon

Two main directions in approaching the impact phenomena have been outlined in technical literature. The first method considers the impact between any two bodies as an instantaneous event. The second approach regards bodies as deformable ones and the impact happens during a finite time. So, every impact phenomenon is characterized by the existence of two distinctive phases: compression and restitution. The compression phase lasts the moment the first points of the two bodies are contacting until the instance the normal approach between the two bodies reaches the maximum. The restitution phase begins from the moment corresponding to the maximum approach and lasts until the last points of the two bodies separate from contact. No matter the considered hypothesis, the coefficient of restitution (COR) is a parameter characteristic to both methods. Newton [1] describes cinematically the coefficient of restitution e as the ratio with changed sign between the normal components of relative velocities of initial contact points:

$$e_{Newton} = -(\boldsymbol{v}''_1 - \boldsymbol{v}''_2)\cdot\boldsymbol{n} / (\boldsymbol{v}'_1 - \boldsymbol{v}'_2)\cdot\boldsymbol{n} \tag{1}$$

* Corresponding author: florina@fim.usv.ro

In (1), $\boldsymbol{n}$ represents the unit vector of the normal to the bodies' surfaces. For the present work, the symbol $(')$ is used for the parameters characteristic to initial collision time and the symbol $('')$ is used for the parameters corresponding to the final moment. The definition (1) for the COR was useful until Kane [2], analyzing the plane impact with friction for a double pendulum reaches the conclusion that accepting the definition (1) for the COR leads, when the geometry and initial kinematical state are conveniently chosen, to disobedience of the law of conservation of energy. To surpass this intricacy another definition for the COR is required. To this end, the hypothesis of finite continuous variation of impact force must be accepted. The new definition of the COR is due to Poisson [6]. Thus:

$$e_{Poisson} = \int_{t_c}^{t_f} \boldsymbol{F} \cdot \boldsymbol{n}\, dt \Big/ \int_{t_i}^{t_c} \boldsymbol{F} \cdot \boldsymbol{n}\, dt = P_r / P_c \tag{2}$$

Considering that:

$$\int_0^t \boldsymbol{F} dt = \boldsymbol{P} \tag{3}$$

represents the percussion, according to Hibbler [3] the definition states that the COR is the ratio between normal percussions corresponding to the restitution (r index) and compression phases (c index), respectively. There were denoted t_i, t_c t_f the moments corresponding to contact initiation, maximum approach and impact ending, respectively.

2 The plane of percussions method (Routh)

The models mentioned above are applicable where the friction force presents a continuous variation with velocity. In the case of dry Coulomb friction, the condition is not obeyed due to the fact that the friction force is characterized via inequalities, the dry friction forces being unilateral constraints [4]. A reference work in the impact with friction domain is owing to Wang and Mason [1]. For the study of two-dimensional impact with dry friction, they apply the plane of percussions method, an extremely intuitive method, proposed by Routh [5]. As a principle, as shown in Figure 1, for the two bodies denoted 1 and 2, contacting in point O, an axis system is defined, with the axes directed along the normal and tangent to the surfaces of the bodies, and the centers of mass of the two bodies, C_1 and C_2 are established according to it. For both bodies, the Newton-Euler dynamic equations [3] are written and the final kinematical parameters are found, as function of tangential P_t and normal P_n components of interaction percussion, Figure 2. Finding the relative velocity of impact points, in the plane of percussions [5], the geometrical locus of the points where the normal component of the velocity $\boldsymbol{v}_{21}$ (relative velocity of the two contacting points [1]) is zero is represented by the line of maximum compression (C), Figure 3, and the geometrical locus of the points from percussions plane where the tangential component of the velocity $\boldsymbol{v}_{21}$ is zero is represented by the stiction line (S); in addition to these two straight lines, the straight line of limit friction is traced, defined as:

$$P_t = \mu P_n \tag{4}$$

and the ending straight line (T), parallel to (C), on which the impact finishes according to Newton's hypothesis.

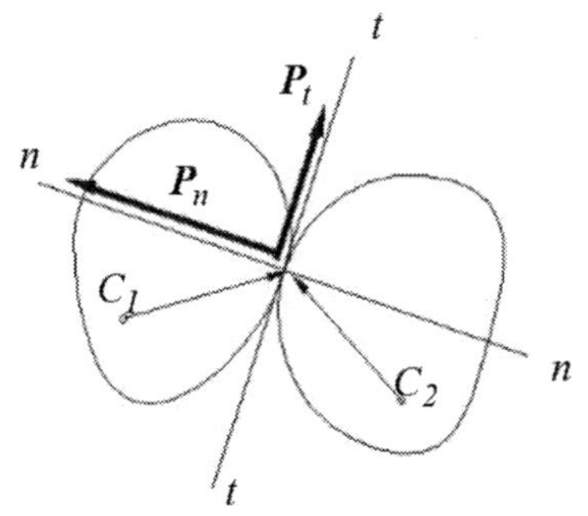

Fig. 1. Plane impact with dry friction.

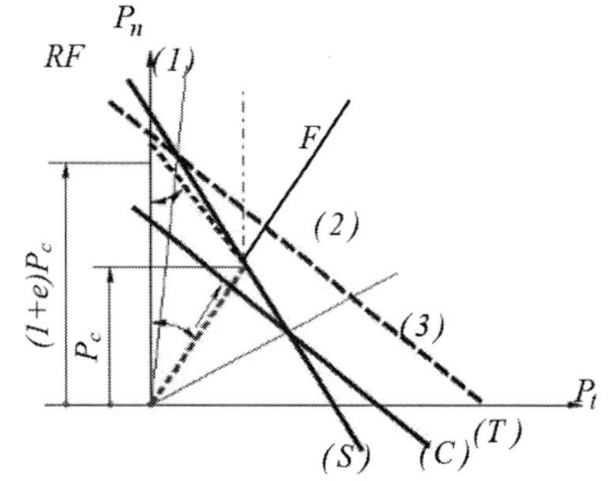

Fig. 2. Plane impact with friction for two bodies.

For the study of a plane impact with friction, the hypothesis is made that during the entire impact process the normal percussion increases monotonically. Thus, at the beginning of impact, the characteristic point starts from origin and moves along the limit friction line LF. When the compression line is reached, the approaching phase ends. The impact ends when (knowing that $P_f = P_c + P_r$) according to equation (2), the next relation is fulfilled:

$$P_f = (1+e)P_c \tag{5}$$

The motion state of the system modifies only once, when the characteristic point reaches the stiction line (S). The characteristic point will move along the steepest of the stiction lines (rolling relative motion) or along the reverse friction line (RF) when the sliding reverses its sign. At the intersections between the stiction line and compression line, and stiction line and terminal line, two points are defined. These points, together with the origin, describe two half-lines crossing the plane in three regions, 1, 2 and 3, Figure 3. Only in the domain 1 the same result is obtained with both definitions of COR.

3 Percussions plane for the impact between a dropping ball and a tilted immobile plane – rigid bodies hypothesis

For the impact between a ball in free fall and an inclined immobile plane tilted at ϕ angle with respect to the horizontal, the percussions plane is presented in Figure 3. Applying the methodology described by Goldsmith [6] concerning the manner proposed by Routh for the impact between a mobile sphere and a fixed plane, the following conclusions become known: the stiction line has the equation of a vertical line, while the maximum compression line, as well the termination line, become two horizontal lines. From Figure 4 one can notice two domains, depending on the relation between the friction angle $\beta = atan\,\mu$, and the angle δ, namely smaller or greater than it. The angle δ is defined [6] by:

$$\delta = atan[2\tan\phi / 7(1+e)] \tag{6}$$

If $\mu \leq tan\,\delta$ the characteristic point moves only along the limit friction line $LF1$ (sliding impact) and the impact ends in the E_1. If $\mu > tan\,\delta$ the characteristic point moves along the limit friction line LF_2 (sliding) until it reaches the stiction line in A, after that it moves along the stiction line until the end of impact, in E_2 (rolling motion, reverse sliding being impossible for this case) where the percussions have the values:

$$P_{t_2} = (2/7)mv'\sin\phi;\quad P_{n_2} = (1+e)mv'\cos\phi \tag{7}$$

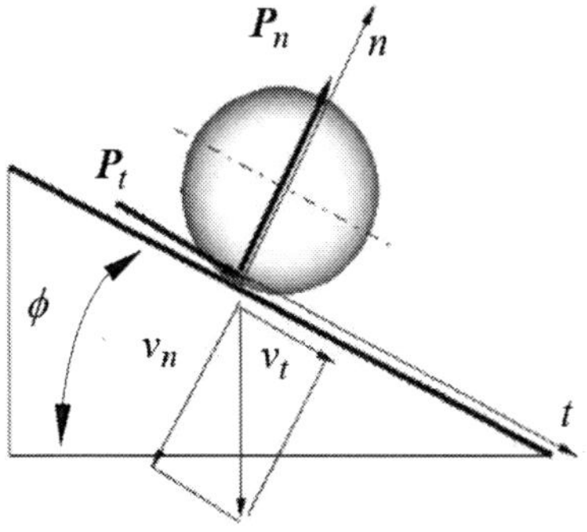

Fig. 3. Plane impact between a dropping bal and a tilted plane.

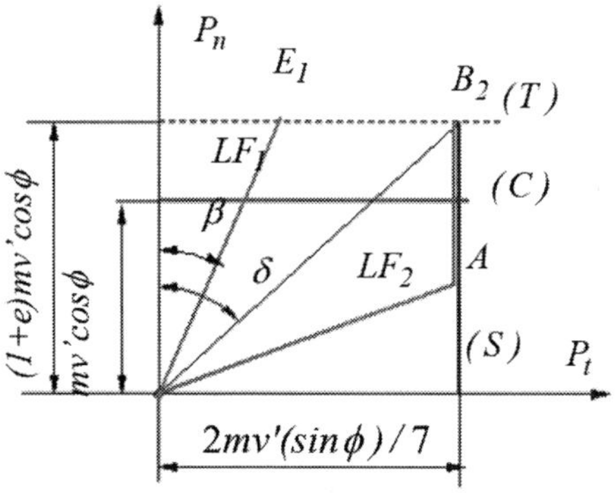

Fig. 4. Percussion plne for impact between a ball and a tilted plane.

The certainty of two different behaviours of the ball stated by the tilting angle of the plane was proved in a recent work [7] where, considering the impact between a ball and a metallic disc that rotates about a vertical axis, Figure 5, it is shown that the mentioned impact case can be assimilated as the impact between a ball and an inclined plane; the space run by the ball after impact is found and there is an obvious rotation velocity value delimiting the linear increase of the space from the space decrease, Figure 6.

Fig. 5. Test-rig for the study of ball-disc impact.

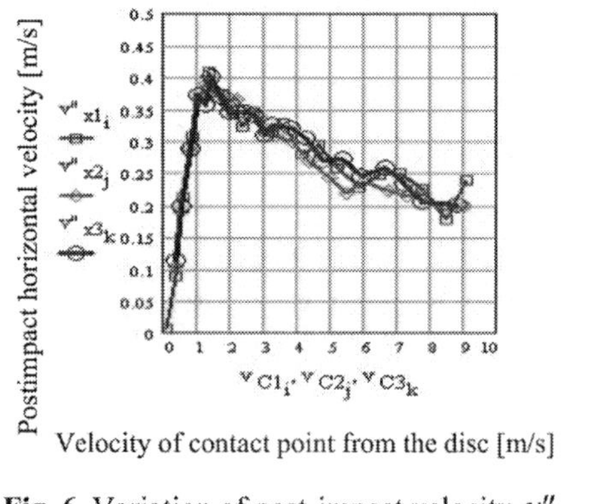

Fig. 6. Variation of post-impact velocity v''_x.

4 Percussions plane for the impact between a dropping ball and a tilted immobile plane – deformable bodies hypothesis

The problem of experimental study impact assumes finding both the variation of normal approach and impact force with time. For the case of elastic centric impact, Timoshenko [8] determines analytical expressions for maximum impact force, maximum normal approach and impact time, based on the assumption that during the whole impact period, the impact force has the same relation as in Hertzian case. The equations show that the period of a collision is quite short, of the order of milliseconds, while the accelerations are of the order $(1000-10.000)g$. The significant values of impact force and reduced time of collision are the main impediments in the study of collisions. In a recent paper, Garland and Rogers [9] present the experimental study of the plane impact with friction between a ball and an inclined plane. The schematic of the principle [9] is shown in Figure 7, using a mathematic pendulum consisting in a wire and steel ball. A body 2 having great mass is placed eccentrically with respect to the plane of oscillation of the pendulum. On the surface of the body a tri-axial shock sensor 3 is mounted, the axes of the sensor being oriented on the following directions: Oz normal to the plane surface, Oy vertical in the impact point and Ox tangent in the impact point. The ball is launched from B' position, strikes the sensor 3 and reaches the post-impact position B''. The launching amplitude is small enough as to consider the trajectory of the centre of the ball as a plane one. This approximation is confirmed by the experimental values of vertical force Fy which are negligible compared to normal force Fz. In [9] there are presented graphs of the variations of normal and tangential forces, for different incident angles ϕ of the ball. For a tilt of $\phi = 15°$ the data obtained by Garland and Rogers [9] were used in the present work as a set of experimental points, interpolated with fourth degree polynomials and then in the variations of normal and tangential impact forces were obtained as presented in Figure 8.

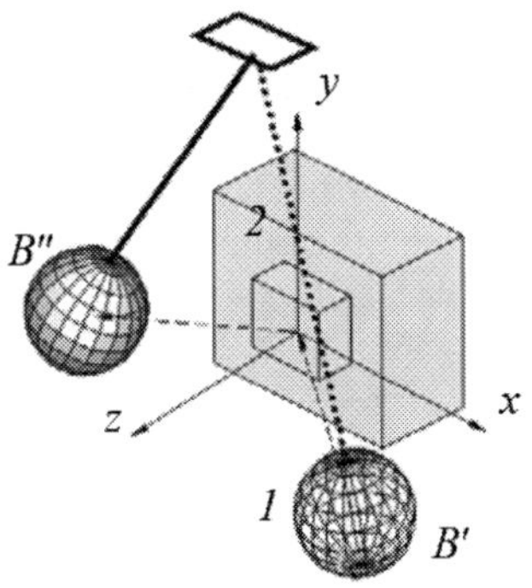

Fig. 7. The Garland experiment principle.

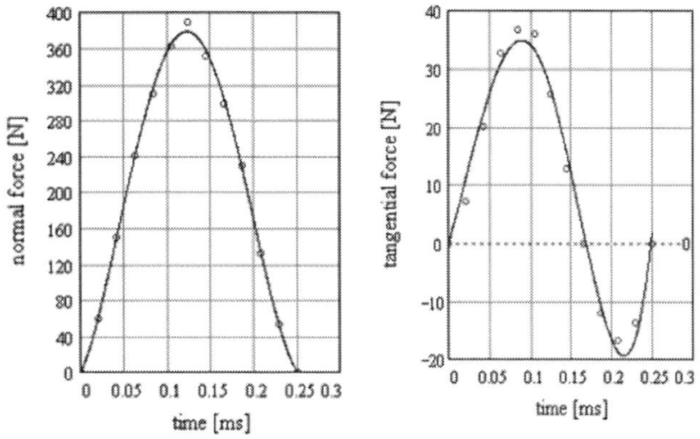

Fig. 8. Normal and tangential impact forces (interpolations of experimental point data from [9]).

With known dependencies for $Fn,t = Fn,t(t)$, using the relation 3, the variations in time of tangential and normal percussions were established, $Pn(t), Pt(t)$, and the plane of percussions was traced using these percussions, as presented in Figure 9; also represented is a broken line consisting in an inclined straight line and a vertical line, according to Figure 4. To accept the hypothesis of rigid bodies and to compare the results with the ones under the hypothesis of deformable bodies is equivalent to accepting that the broken line form Figure 9 (obtained as in explained in Figure 4) is an approximation of the continuous curve from the same figure. But, as one can remark, there is a significant difference between the two graphs. It can be considered that, for the first stage of impact, the broken tilted line is a good approximation of the continuous curve, but for the restitution phase this equivalence is hardly plausible. One of the most significant differences between the two methods consists in the fact that while the rigid body model estimates that tangential percussion increases or at least is constant during the last phase of collision, the deformable body model assumed the occurrence of a maximum for tangential percussion during collision, and afterwards a decrease of it. Given the variety of applications in the domain of multibody dynamics, [10-13], the results could be a base for future research.

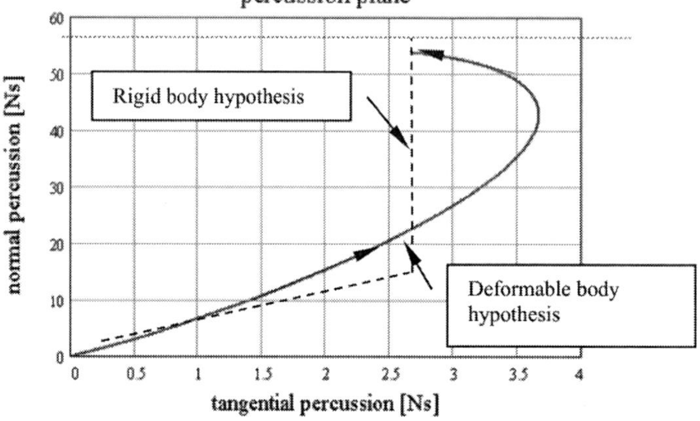

Fig. 9. Percussions plane for deformable bodies compared to rigid body.

5 Conclusions

The paper analyses one of the simplest collision cases, namely the dry friction impact of a ball with a tilted plane. There are two main methods of approaching the problem: the rigid bodies assumption and the compliant bodies method. For the ball-inclined plane impact, the authors used both data from research literature and experimental results from own research work, and a qualitative comparison between the two methods is presented. Differences between the two methods are noticeable even from qualitative analysis. The results of the paper are a starting phase for future work aiming to correlate the results of the two methods, considering that the broad fields of applications, from granular and powder materials phenomena, grasping in robotics, landing trains in aircraft to percussions in sports. The infrastructure used for this work was partially supported by the project "Integrated Center for research, development and innovation in Advanced Materials, Nanotechnologies, and Distributed Systems for fabrication and control", Contract No. 671/09.04.2015, Sectoral Program for Increase of the Economic Competitiveness co-funded from the European Regional Development Fund.

References

1. Y. Wang Y, M. T. Mason, J. Appl. Mechanics **59**, 635 (1992)
2. T. R. Kane, Stanford Mech. Alumni Club Newsl. 6-10 (1984)
3. R. C. Hibbler, *Engineering Mechanics: Dynamics* (Prentice Hall, 2012)
4. F. C. Pfeiffer, C. Glocker, *Multibody Dynamics with Unilateral Contacts* (Wiley 1996)
5. E. T. Routh, *Dynamics of a system of rigid bodies* (Macmillan, London, 1905)
6. W. Goldsmith, *Impact, The Theory and Phys. Behav. Collid. Solids (*Dover Pub, 2001)
7. S. Alaci, F. C. Ciornei, C. Filote, J. Balkan Tribol. Assoc. **22** (2), 1552 (2016)
8. S. P. Timoshenko, J. N. Goodier, *Theory of elasticity* (McGraw-Hill, New York, 1970)
9. P. P. Garland, R. J. Rogers, J. Appl. Mechanics, ASME **76,** 031015 (2009)
10. O. A. Bauchau, A. Laulusa, ASME J. Comput. Nonlin. Dyn. **3** (1), 011, 005 (2008)
11. M. Baumann, R. I. Leine, Int. J. Robust Nonlin. Control **26**(12), 2542 (2016)
12. S. Antoniuk et.al., Granular Matter **12,** 15 (2010)
13. C. E. Anderson, Holmquist T.J., Int. J. Impact eng. **56**, 2 (2013).

Approaches for reducing structural vibration of the carbody railway vehicles

Mădălina Dumitriu* and *Cătălin* Cruceanu

[1]University Politehnica of Bucharest, Department of Railway Vehicles, 313 Splaiul Independenţei, 060042, Bucharest, Romania

Abstract. Reducing the weight of the railway vehicles stands as a decisive rule in their design, entailed by higher velocities, the need to consume less energy and lower the manufacturing costs, along with the maximization of the use of loads on the axle. Once complied with this rule, the vehicle flexibility increases and leads to an easy excitation of the structural vibrations in the carbody, with an impact upon the ride comfort in the railway vehicle. For a better ride comfort in lightweight railway vehicles, both vibration isolation approaches and structural damping approaches have been introduced. The paper herein submits a brief review of the main structural damping approaches aiming to reduce the amplitude in the carbody structural vibrations, based on the use of the piezoelectric elements in passive control schemes. The paper outcomes show the potential of the presented methods concerning the reduction of the flexible vibrations in the carbody and the ride comfort improvement.

1 Introduction

A lower weight is a basic criterion in designing the railway vehicles with respect to reaching higher velocities with the lowest energy consumption possible. The weight decrease of the railway vehicles also contributes to the reduction of ground vibration and construction-cost saving [1]. Similarly, the demand from the transportation service providers to maximize usage of the limited axle loads compel the railway industry to adopt the lightweight carbody designs [2, 3]. While the requirements of strength and crashworthiness are complied with, the reduction in the carbody weight is usually done by a significant decrease of the structural stiffness [2]. The lighter the vehicle carbody, the higher its flexibility, thus allowing an easier excitation of the carbody structural vibrations. The structural vibration leads to carbody fatigue which affects the dynamic performances and service life of the vehicle [4]. There are likewise numerous theoretical and experimental studies to validate that the level of carbody vibrations can be strongly influenced by the flexible modes of the carbody vibrations, mainly for the high speed vehicles [2, 4 - 6]. In some cases, this carbody structural flexibility accounts for half of the perceived vibrations, while the rest is due to rigid modes [7].

The value of the frequencies for the first structural modes of the carbody vibration often fall within the frequency range of high human sensitivity to vibration. As a result, structural

*Corresponding author: madalinadumitriu@yahoo.com

vibrations of the vehicle carbody have a great impact on the ride comfort, both for high speed trains [2, 8] and for the commuter or metro trains [1, 3, 9]. In spite of the fact that the carbody structural vibrations are rather complex [7], the greatest influence upon the ride comfort comes from the first carbody natural bending mode [6, 10] whose frequency usually ranges from 6 to 12 Hz, an interval where the human body shows a higher sensitivity to vertical vibrations.

The conventional solution to reduce the structural vibrations of the vehicle carbody involves an increase in the carbody stiffness. Different solutions can be adopted to come to this situation [11]. One is to have a carbody as short as possible but this solution contradicts the desire to maximise the number of passengers in relation to the number of bogies, as an example. Another one aims the carbody cross-section, which should be the largest possible. However, the cross-section must not interfere with the prescribed carbody gauge profile. Furthermore, crosswind stability problems may arise. Stiffer yet heavier carbodies can be designed but this is nonetheless in opposition to the lightweight design targets. Similarly, the necessary reinforcements for the carbody stiffening often lead to a more complicated and therefore more expensive manufacturing of the carbody [3].

More concepts on the reduction of the comfort-relevant vibrations of the carbody vibrations have already been introduced. They can be classified into two complementary approaches, namely the vibration isolation approaches to include passive, semi-active and active concepts with the role of stopping the excitation propagation through the suspension and structural damping approaches - passive or active - with the aim at reducing the carbody structure vibration amplitudes [9].

This paper will provide a brief review of the main carbody structural damping approaches, based on the use of the piezoelectric elements in passive control schemes. The piezoelectric elements are used in passive control schemes to suppress the flexible vibration modes of the carbody and, hence, to improve the ride comfort by benefiting from the ability of the piezoelectric elements to convert the mechanical energy into electrical energy.

2 Passive damping structural approaches

The concepts regarding the passive structure damping of the railway vehicle carbody rely on the ability of the piezoelectric elements to convert the mechanical energy, i.e. the vibration energy in the structure, into electrical energy.

Takigami [12] applies this concept in order to develop a new technique to suppress the bending vibration of railway vehicle carbodies that use piezoelectric elements which are electrically shunted by an external circuit (shunt circuit). When the vibration occurs as shown in Fig. 1, a strain is produced in the piezoelectric elements attached to the carbody which is deformed. As a result, a voltage is generated. When a shunt circuit including a resistor is connected to the piezoelectric elements, the voltage generated by the elements causes a current to flow to the circuit. As the current flows through the resistor in the shunt circuit, the electric energy is dissipated in the form of Joule heat, triggering a loss of energy. The energy loss has a damping effect on the carbody, thereby suppressing the bending vibration. The suggested theory is tested on a 1:5 scale model of Shinkansen carbody model, with eight piezoelectric elements - four on each of the lateral side beams - and to a stationary vibration test. The results of the vibrations test in Fig. 2 show that the acceleration amplitude at the carbody centre has lowered to less than half of its value.

The same method for improving the ride comfort through reducing the flexible vibrations in the railway vehicles carbody by means of the piezoelectric elements attached on the carbody is also submitted by Hansson [13].

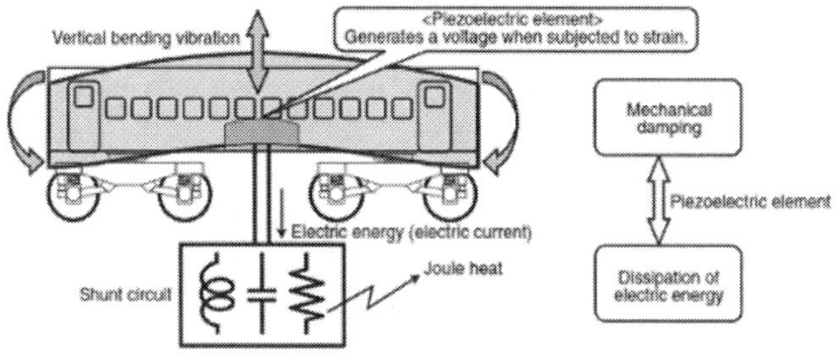

Fig. 1. Scheme of bending vibration suppression technique [12].

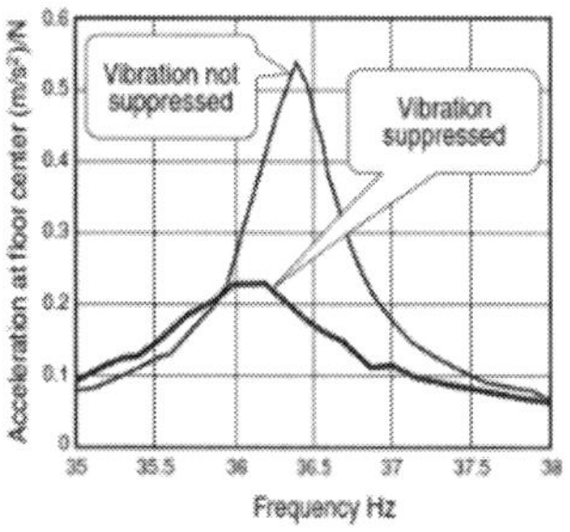

Fig. 2. Vibration test results [12].

The efficiency of the method is both theoretically and experimentally researched. Two types of electric shunt circuits are studied; a single-mode shunt circuit which is tuned for a single mode damper, and a multi-mode shunt circuit for two-mode damping. The piezoelectric material (lead zirconate titanate) (LZT) is bonded to a host structure and the electrodes of the piezoelectric element are shunted with some electrical impedance.

Should the shunt consist of an inductor and resistor in series, the shunt combined with the capacitance of the piezoelectric element creates a damped electrical resonance. The resonance can be tuned so that the piezoelectric device acts as a damped vibration absorber. A damped vibration absorber replaces a single structural mode with two highly damped modes. (Fig. 3, a). To simultaneously suppress multiple modes, one multi-resonant shunt circuit is used (Fig. 3, b).

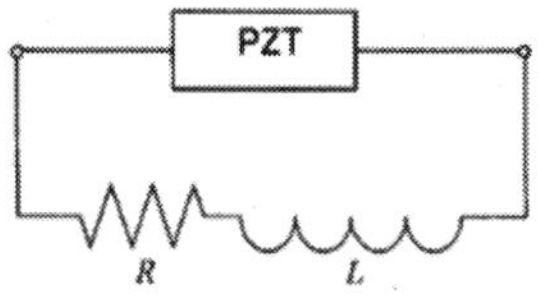

a. Single mode damping.

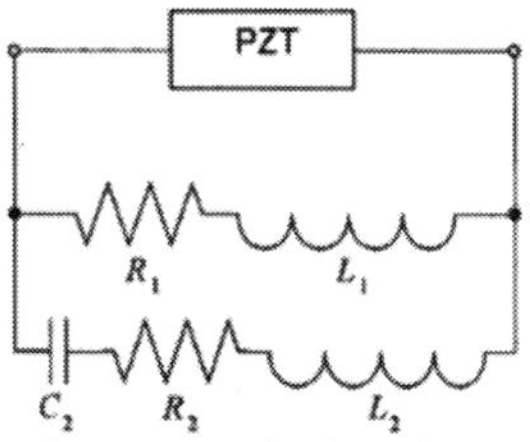

b. Two-mode damping.

Fig. 3. Shunt circuits [13].

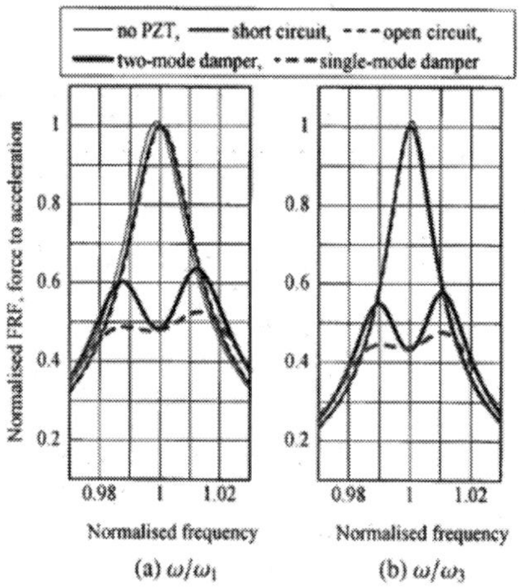

Fig. 4. Simulated FRF peak reduction at the first (a) and third (b) bending modes, near the centre of the carbody [13].

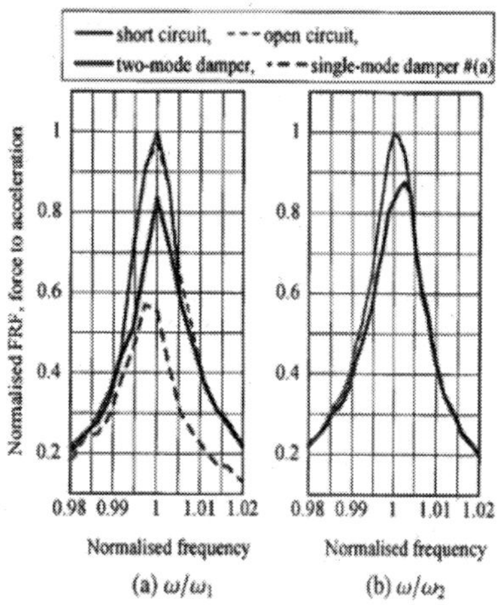

Fig. 5. Measured FRF peak reduction at the first bending mode (a) and at a local mode (b) at 76.2 Hz near the centre of the carbody [13].

To estimate the effectiveness of the piezoelectric damping, a 1:5 scale model of a suspended Shinkansen vehicle is investigated experimentally and numerically, both with and without piezoelectric elements. The results in Fig. 4 correspond to the numerical simulations. The frequency response function (FRF) value is shown to be reduced at the first bending mode peak by 48% for the single-mode damper and by 37% for the two-mode damper. The two-mode damper simultaneously reduces the frequency response value at the third bending mode peak by 43%. In case the single-mode damper is targeted at the third bending mode instead of the first, the peak value can be reduced by 53%. Figure 5 shows the measurement test results. When applying the single-mode damper, the FRF peak value at the first bending mode is reduced by 43%. The two-mode damper reduces the FRF peak value at the first bending mode by 17% and the peak value at the local mode at 76.2 Hz by 12%, at the same time.

The study shows that the effectiveness of the vibration suppression depends on the placement, size and amount of piezoelectric elements, where the placement should be as far away from the neutral axis of the car body as possible. The multi-mode damper is useful when the suppression of more vibration modes within the comfort-sensitive frequency interval is desired. Alternatively, the multi-mode damper cannot exceed the performance of the single-mode damper to suppress a single vibration mode.

Kamada et al. [14] proposed new control methods for shunt damping using stack type piezoelectric transducers (PZTs) and the effectiveness and the robustness of the designed circuits were confirmed through 1:6 scale model experiments that have almost the same eigenfrequencies of the first five modes as the Shinkarisen light carbody.

The railway vehicle is modelled as a simple beam supported on two points by a spring in parallel with a damper (see. Fig. 6). The stack type PZTs are attached at the centre of the beam. For the study, the first three bending modes in the total system were taken into account.

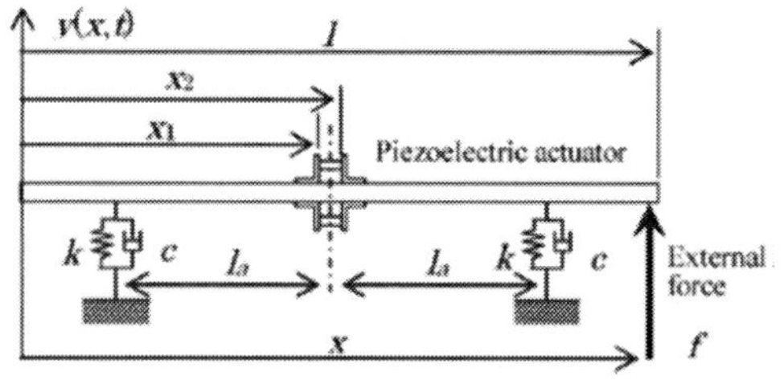

Fig. 6. Analytical model [14].

Two shunt circuits have been investigated, where one is the LR–LRC shunt and the other is the negative capacitance shunt. A single vibration mode is reduced to a simple LR circuit. The multiple damping is done by parallel connection of a RLC circuit to the original LR circuit. The LR–LRC circuit was designed to suppress the first and third bending modes of the beam.

The tests results are incorporated in Fig. 7 - 9. Fig. 7 shows a frequency response of LR shunt. On the left, a third vibration mode is efficiently suppressed since the circuit has been thus designed, which is not valid for the fifth vibration mode. Similarly, the LR shunt used on the right is only efficient for the fifth vibration mode. Fig. 8 includes the experimental results for the LR –LRC shunt. Both first and third bending modes are noted to have been effectively suppressed. The reduction rates are 29% for the third mode and 69% for the fifth one, which were a bit worse than the individual shunt damping by a single LR circuit. The results for the negative CR shunt are shown in Fig. 9, confirming that they are efficient for the suppression of both vibration modes and that the performance is better than for other circuits.

DOI: 10.1051/matecconf/201711207006

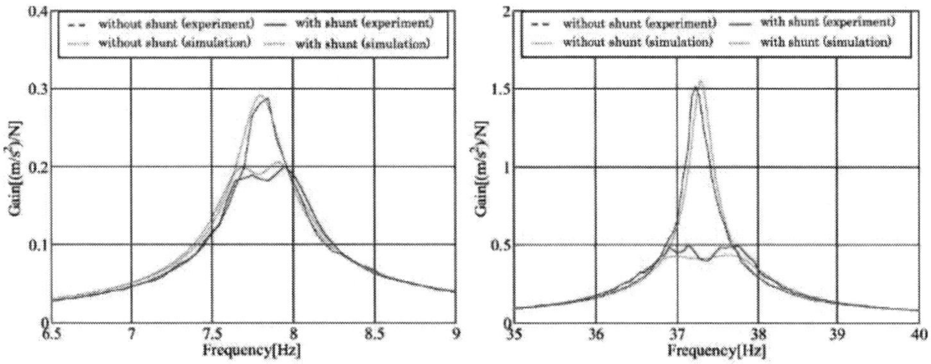

(a) Tuned for the 3rd mode (b) Tuned for the 5th mode

Fig. 7. Frequency responses of the beam acceleration by single LR shunt [14].

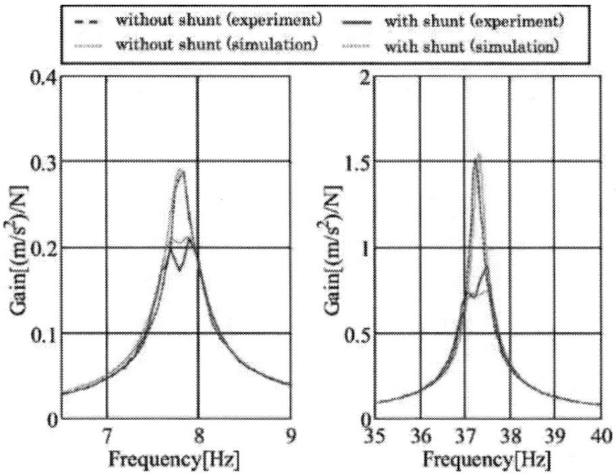

Fig. 8. Frequency responses by LR–LRC shunt [14].

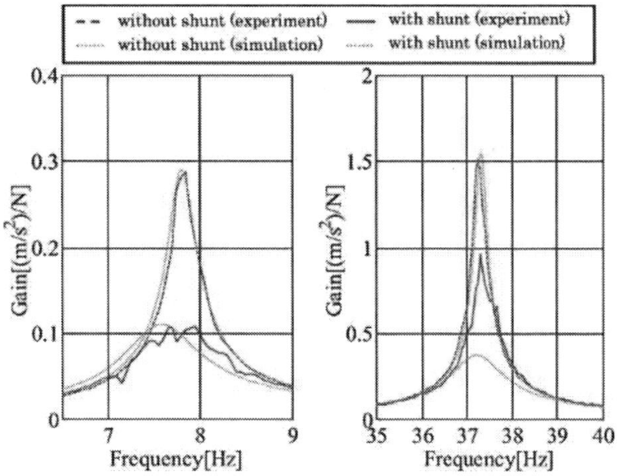

Fig. 9. Frequency responses by negative CR shunt [14].

Based on the results above, the conclusion is that both LR shunt and the negative capacitance shunt are efficient for the suppression of a single vibration mode of the carbody in the railway vehicles. The multiple vibration modes can be suppressed by parallel connection of a LR with a LRC circuit.

3 Conclusions

The concern for lighter railway vehicles so as they are able to travel at higher speeds with less energy consumption and reduced manufacturing costs while having a maximum use of the loads on the axle has taken shape in a less heavy vehicle carbody. But the weight decrease is usually done by a significant decline in the structural stiffness that fosters the easy excitation of the flexible vibrations in the carbody. At this rate, the level in the carbody vibrations goes up and the ride comfort will be significantly affected if the increase occurs in the frequency range where the human body has a higher sensitivity to vibrations.

For the ride comfort, the literature of review introduces more concepts regarding the reduction in the carbody vibrations, which can be classified into vibration isolation approaches and structural damping approaches.

The paper herein submits the main structural damping approaches based on the use of the piezoelectric elements in the passive systems to control the carbody flexible vibrations. As a rule, the passive systems of controlling the carbody flexible vibrations rely on the ability of the piezoelectric elements attached to the carbody to convert the mechanical vibration energy into electrical energy that is later dissipated via a shunt circuit.

The studies described in this paper are a good evidence for the potential of the suggested methods concerning the reduction of the flexible vibrations in the carbody and the ride comfort improvement. The passive systems of the structural damping in the carbody herein featured can turn into an alternative to the active suspension that, in most cases, cannot be applied to improve the ride comfort due to the high costs of implementation and maintenance.

This work has been funded by University Politehnica of Bucharest, through the "Excellence Research Grants" Program, UPB – GEX - 2016. Research project title: *Research on developing mechanical and numerical models for the virtual evaluation of the dynamic performances in the railway vehicles* (in Romanian). Contract number: 48/26.09.2016.

References

1. T. Tomioka, T. Takigami, Y. Suzuki, Veh. Syst. Dyn **44**, Suppl., 272-285 (2006)
2. J. Zhou, R. Goodall, L. Ren, H. Zhang, Proc. IMechE, Part F: J. of Rail and Rapid Transit **223**, 461-471 (2009)
3. G. Schandl, P. Lugner, C. Benatzky et al. Veh. Syst. Dyn. **45**, 9, 835-847 (2007)
4. G. Yang, C. Wang, F. Xiang, S. Xiao, Chinese J. Mech. Eng. **29**, 6, 1120-1126 (2016)
5. M. Dumitriu, Archive of Mechanical Engineering **LXI**, 4, 609-625 (2014)
6. G. Diana, F. Cheli, A. Collina et al., Veh. Syst. Dyn **38**, 3, 165-183 (2002)
7. P. Carlbom, *Carbody and Passengers in Rail Vehicle Dynamics* (Doctoral Thesis, Stockholm, 2000)
8. H. Shi, P. Wu, J. Mech. Sci. Technol. **30**, 1, 55-66 (2016)
9. A. Schirrer, M. Kozek, *Proceeding of the 6th Mediterranean Conference on Control and Automation Congress Centre*, 481-486 (2008)
10. M. Dumitriu, J. Engineering Science and Technology Review **8**, 4, 32-39 (2015)
11. A. Orvanäs, *Methods for reducing vertical carbody vibrations of a rail vehicle* (Report in Railway Technology Stockholm, 2010)
12. T. Takigami, Railway Technology Avalanche **9** (2005)
13. J. Hansson, M. Takano, T. Takigami, T. Tomioka, Y. Suzuki, JSME Int. J., Ser. C **47**, 2, 451–456 (2004).
14. T. Kamada, R. Kiuchi, M. Nagai, Veh. Syst. Dyn **46**, Suppl., 561–570 (2008)

Modelling of structural flexibility of the railway vehicles carbody

Mădălina Dumitriu* and *Camil* Crăciun

University Politehnica of Bucharest, Department of Railway Vehicles, 313 Splaiul Independenţei, 060042, Bucharest, Romania

Abstract. The accuracy in the models of railway vehicles in the ride comfort evaluation mainly derives from the modelling of the carbody structural flexibility. The former part of the paper herein reviews the 'flexible carbody' type methods used to represent the carbody structural flexibility in the approaches targeting the identification of certain methods to reduce the flexible vibrations in the carbody so as to improve the ride comfort of the railway vehicles. What seems to apply are the simple beam models such as Euler - Bernoulli or Timoshenko, structure type models and detailed multi-body, finite-element models or combinations of multi-body and finite-elements models. The latter part will feature a contrastive analysis regarding the comfort index coming from the numerical simulations based on 'rigid carbody' and 'flexible carbody' original models. It is highlighted that the comfort performance of a railway vehicle evaluated on the basis of a 'rigid carbody' mode is overrated.

1 Introduction

While running, the railway vehicle is subjected to steady-state vibration, mainly from the track irregularities, which they are transmitted from via the bogies and the carbody to the passengers. The carbody vibration exhibits simple vibration modes - rigid modes, and complex vibration modes - vibration due to the carbody structural flexibility. In some cases, this carbody structural flexibility accounts for half of the perceived vibrations, the rest being due to rigid modes [1].

Given the fact that the reduction in the weight of the railway vehicles stands as one of the most influential designing criteria, called for by the increase in velocities, need for lower energy consumption, manufacturing costs and maximization of the use of loads on the axle, the issue of the carbody structural flexibility is extremely relevant. A lower structural stiffness usually triggers a smaller weight in the carbody. The lighter the carbody, the higher its flexibility, which leads to an easier excitation of the carbody structural vibrations. The structural vibration causes carbody fatigue which affects the dynamic performance and service life of the vehicle [2, 3]. But the most important effect of the carbody structural vibrations is visible in the ride comfort of the railway vehicles [4 - 6]. This action occurs due to the first flexible vibration modes of the carbody whose

*Corresponding author: madalinadumitriu@yahoo.com

eigenfrequencies are found within the frequency interval where the human body shows a higher sensitivity to vibrations. More exactly, it is about the first mode of carbody vertical bending with the frequency usually between 6 and 12 Hz that has the highest influence on the ride comfort.

In the last years, numerous theoretical studies have been conducted in order to identify certain methods of reducing the ride comfort-relevant flexible vibrations of the carbody. The analytical evaluation of the comfort performance in the railway vehicles involves the use of the numerical simulation applications as an investigation instrument that are developed on the basis of certain theoretical models of the vehicle [7]. The precision of the models largely depends on the carbody modelling. The 'rigid carbody' simple model can be used to explain some basic phenomena of the vibration behaviour in the railway vehicle. This model, however, becomes unsuitable when the ride comfort is being evaluated, as will seen in this paper. The results derived from such models prove that the critical positions in terms of ride comfort are found at the ends of the carbody. This fact does not always reflect in reality, since there are situations in which the level of vibrations at the carbody centre is comparable with the level recorded at its ends [4, 8].

The former part of the paper herein reviews the models used to represent the carbody structural flexibility in the approaches targeting the identification of certain methods to reduce the flexible vibrations in the carbody so as to improve the ride comfort of the railway vehicles. The latter part will include an evaluation of the ride comfort in high speed trains, by using the results from numerical simulations based on both 'rigid carbody' and 'flexible carbody' original models. These have been developed by the authors starting from the basic models by improving the secondary suspension model which includes pitch and longitudinal elements. It is therefore highlighted that the comfort performance of a railway vehicle evaluated on the basis of a 'rigid carbody' mode is overrated.

2 Models of 'flexible carbody' type

The models of 'flexible carbody' type vary from relatively simple Euler-Bernoulli or Timoshenko beams, to a structure type models consisting of plates and beams or multi-body, finite-element detailed models or combinations between multi-body and finite-element models.

A model widely utilized to analyze the flexible vibrations in the carbody bending or their influence upon the ride comfort consists in a free-free equivalent beam, of Euler-Bernoulli type. Such a model was used by Zhou et al. [5] to investigate the influences of the carbody vertical flexibility on the ride quality in the passenger railway vehicles. Dumitriu [8] studied the position of the carbody critical point in terms of the ride comfort. Tomioka and Takigami [9] employed the model of Euler-Bernoulli elastic beam for presenting a theory based on the use of the longitudinal vibration in bogies as a dynamic vibration absorber (DVA) to reduce the vertical bending vibration of railway vehicle carbodies. In a study on the dynamic analysis of the influence of carbody flexibility upon the dynamic response of the train-bridge system, Kunpeng et al. [10] established a three-dimensional (3D) model of the vehicle based on flexible multi-body dynamics, where the carbody is simulated as an Euler-Bernoulli beam.

The same beam model is adopted to represent the carbody in more papers, in order to examine the influence of the equipment suspended from the carbody chassis upon the vibration of the flexible carbody for high-speed electric multiple unit (EMU). For instance, Dumitriu [11] evaluated the influence of the suspended equipment and of its fastening manner (rigid or elastic), in correlation with the speed behaviour upon the regime of vertical vibrations in the carbody at the high speed trains. Shi et al. [12] applied the theory of dynamic vibration absorber (DVA) to restrain the vibration of carbody for high-speed electric multiple unit. For this purpose, the carbody was modelled as an Euler-Bernoulli beam with the equipment mounted on the chassis regarded as a DVA. Relying on this model, the influence of the suspension parameters of equipment on the flexible vibration reduction of carbody is studied, while including the suspension frequency, damping ratio, mounting position, and mass of equipment.

The use of the Euler-Bernoulli beam is also suitable for the elaboration of certain methods to reduce the flexible vibrations in the carbody of the railway vehicles, based on the piezoelectric elements in passive control and active control systems. Hansson et al. [13], for example, modelled the vehicle carbody as an elastically supported Bernoulli-Euler beam in their study for a method to improve the riding comfort by reducing vertical flexural vibrations in railway carbodies via piezoelectric elements. These elements are attached on the carbody to convert the vibration energy into electrical energy, which can be dissipated in a shunt circuit. Kamada et al. [14] submitted a method for the active control of the vibrations in the carbody of the railway vehicles based on the combined use of the piezoelectric actuators and linear actuators. Thus, elastic vibrations are suppressed by piezoelectric actuators and rigid-body vibrations are reduced by linear actuators.

The theory of Timoshenko beam is frequently applied for the track modelling [15] and more rarely for the same action on the vehicle carbody. The carbody is modelled via a Timoshenko beam when taking into account both bending deformation and also shear deformation. Yang et al. [2] applied a method based on Timoshenko beam theory to infer an analytical formula for the calculation of the first frequency of the carbody vertical bending. To study the vertical vibrations of a vehicle running on a track with irregularities, Young and Li [16] modelled the carbody as a Timoshenko beam with limited length. With the help of a numerical procedure to simulate the vibrations in the vehicle, the effect of the carbody flexibility upon the vertical vibrations in the vehicle/track system is investigated.

The analysis of the 3D flexible carbody vibrations of the railway vehicles uses structure type models, multibody models (MBMs), finite element models (FEMs) or MBM with the FEM combinations. Tomioka et al. [17] proposed an analytical method to investigate the 3D flexural vibration of railway vehicle carbodies in which the carbody is modelled as a box-type structure consisting of plates and beams. To investigate the improvement in the ride comfort of lightweight railway vehicles by means of an active vibration reduction system using piezo-stack actuators, Schandl et al. [6] used a MBM of the vehicle. Diana et al. [4] developed a complex FEM of the vehicle, suitable for reproducing its dynamic behaviour in the 0 ... 50 Hz frequency range, especially with respect to the ride comfort problem. Carlbom [1] combined the multibody system theory with the FE method so as to examine the dynamic response of the railway vehicle.

3 Evaluation of the ride comfort

This section features the evaluation of the ride comfort at the vertical vibrations for a railway vehicle, based on the partial comfort index (N_{MV}) obtained from numerical simulations. To this end, the vehicle is represented by both a 'rigid carbody' and also a 'flexible carbody' type models (see Fig. 1). The vehicle is running on a perfectly rigid track with vertical irregularities that are described against each axle by the functions $\eta_{1...4}$. The track vertical irregularities are considered to represent a stationary stochastic process that can be described by means of the power spectral density [18].

In the 'rigid carbody' model, the carbody is described by a rigid body with two vibration modes, i.e. bounce z_c and pitch θ_c. For the 'flexible carbody' type, the carbody is represented by a free-free equivalent beam, a constant section and uniformly distributed mass, of Euler-Bernoulli type.

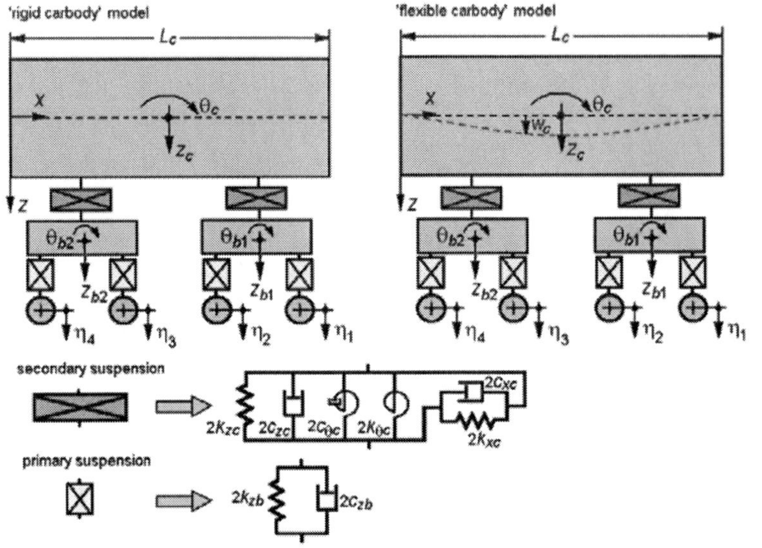

Fig. 1. The vehicle model: 'rigid carbody' model; ,flexible carbody' model.

The carbody displacement $w_c(x, t)$ is the result of overlapping between the rigid modes of vibration - bounce z_c and pitch θ_c and the first mode of carbody structural vibration - vertical bending,

$$w_c(x,t) = z_c(t) + (x - L_c / 2)\theta_c(t) + X_c(x)T_c(t)\,, \tag{1}$$

where L_c is the carbody length, $T_c(t)$ is the coordinate of the carbody vertical bending and $X_c(x)$ represents the natural function of this vibration mode, described in the equation

$$X_c(x) = \sin\beta_c x + \sinh\beta_c x - \frac{\sin\beta_c L_c - \sinh\beta_c L_c}{\cos\beta_c L_c - \cosh\beta_c L_c}(\cos\beta_c x + \cosh\beta_c x) \tag{2}$$

with
$$\beta_c = \sqrt[4]{\omega_c^2 \rho_c /(EI)}\,,\ \cos\beta_c L_c \cosh\beta_c L_c - 1 = 0\,, \tag{3}$$

where ω_c is the natural pulsation of the carbody bending.

For both models, each bogie is represented by a rigid body which is able to perform two motions in a vertical plan, namely bounce z_{bi} and pitch θ_{bi}, with $i = 1, 2$.

The model of the secondary suspension to a bogie includes two Kelvin-Voigt systems; one of them is for vertical translation, with stiffness $2k_{zc}$ and damping constant $2c_{zc}$, and another one for rotation, with the pitch angular stiffness $2k_{\theta c}$ and damping constant $2c_{\theta c}$. The model of the secondary suspension also includes a Kelvin-Voigt system which models the transmission system of the longitudinal forces between the carbody and bogie. This model has an elastic constant $2k_{xc}$ and a damping constant $2c_{xc}$. The primary suspension corresponding to an axle is modelled via a Kelvin-Voigt system, with the elastic constant $2k_{zb}$ and the damping constant $2c_{zb}$.

The motion relations of the carbody and bogies come from the implementation of the laws of mechanics. While the 'rigid carbody' model has six motion equations with ordinary derivatives, the 'flexible carbody' acquires four such equations - the motion equations of the bogies and the carbody, written as below

$$EI\frac{\partial^4 w_c(x,t)}{\partial x^4} + \mu I\frac{\partial^5 w_c(x,t)}{\partial x^4 \partial t} + \rho_c\frac{\partial^2 w_c(x,t)}{\partial t^2} = \sum_{i=1}^{2} F_{zci}\delta(x - l_i) + \sum_{i=1}^{2}(M_{ci} - h_c F_{xci})\frac{\mathrm{d}\delta(x - l_i)}{\mathrm{d}x}, \tag{4}$$

where δ(.) is Dirac's delta function and F_{zci}, F_{xci} and M_i stand for the forces and moments due to the secondary suspension of bogie i. To solve this equation, the modal analysis method is applied, well known in vibration theory [19]. The answer will be three motion equations, corresponding to the vibration modes of the carbody.

Table 1. The parameters of the vehicle numerical model.

$m_c = 34.0 \cdot 10^3$ kg	$EI = 3.16 \cdot 10^9$Nm^2	$2k_{zc} = 1.20$ MN/m	$2c_{xc} = 50.0$ kNs/m
$m_b = 3.20 \cdot 10^3$ kg	$L = 26.4$ m	$2k_{xc} = 4.00$MN/m	$2c_{\theta c} = 2.00$ kNm
$m_{mc} = 35.2 \cdot 10^3$ kg	$2a_c = 19.0$ m; $2a_b = 2.56$ m	$2k_{\theta c} = 256$ kN/m	$2c_{zc} = 34.3$ kNs/m
$J_c = 1.96 \cdot 10^6$ kg·m^2	$h_c = 1.30$ m	$4k_{zb} = 4.40$ MN/m	$4c_{zb} = 52.2$ kNs/m
$J_b = 2.05 \cdot 10^3$ kg·m^2	$h_b = 0.20$ m	$k_{mc} = 89.0$ MN/m	$c_{mc} = 53.1$ kNm/s

Table 3. The natural frequencies of the vehicle vibration modes.

Carbody vibration mode	Frequency	Bogie vibration mode	Frequency
Carbody bounce	1.17 Hz	Bogie bounce	6.61 Hz
Carbody pitch	1.46 Hz	Bogie pitch	9.63 Hz
Carbody bending	8.00 Hz		

The results of the numerical simulations derived from the models above are featured in Fig. 2 and 3. The same parameters of numerical simulation have been considered, where their selection is congruent with a high speed train (Table 1). The natural frequencies of the vibration modes in the carbody and bogie are featured in Table 2.

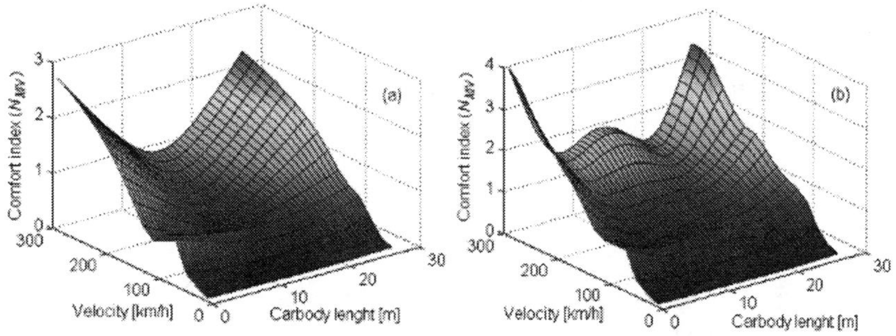

Fig. 2. Comfort index: (a) 'rigid carbody' model; (b) 'flexible carbody' model.

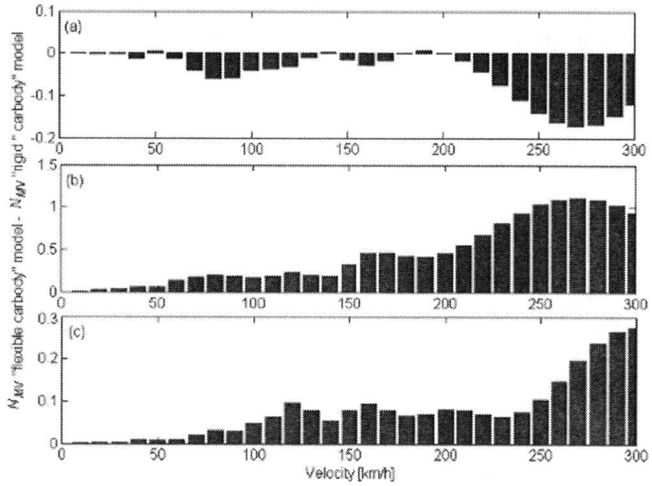

Fig. 3. N_{MV}'flexible carbody' model - N_{MV}'rigid carbody' model: (a) above the front bogie; (b) at the carbody centre; (c) above the rear bogie.

Fig. 2 shows the N_{MV} calculated along the carbody, in dependence on the velocity, according the standard for evaluation of ride comfort for rail passengers [20]. Differences between the results coming from the two models are noticed and they become more obvious along with the speed increase. As a rule, it is about an overestimation of the ride comfort evaluated by means of the ‚rigid carbody' model. The same thing is pointed out at in the diagrams in Fig. 3, displaying the difference between N_{MV} calculated on the basis of 'flexible carbody' model and 'rigid carbody' model at the carbody centre and against the bogies. The highest difference between these results occur at the carbody centre, where the level of vibrations increases with the velocity, due to the vertical bending of the carbody. These results are consistent with the ones derived from other research [4].

4 Conclusions

The representation of the carbody flexibility is still an essential element for the accuracy of the railway vehicle model which is used in the studies targeting the identification of certain methods to reduce the carbody vibrations and improve the ride comfort. The part one in this paper reviewed the 'flexible carbody' models used in such studies and the part two evaluated the ride comfort for a high speed train, based on the comfort index from numerical simulations. The results from 'flexible carbody' and 'rigid carbody' models are compared and an overestimation in the comfort performance is demonstrated with the latter model. The differences are quite visible mainly at the carbody centre, where the level of vibrations increases with the velocity, due to the vertical carbody bending.

This work has been funded by University Politehnica of Bucharest, through the "Excellence Research Grants" Program, UPB – GEX - 2016. Research project title: *Research on developing mechanical and numerical models for the virtual evaluation of the dynamic performances in the railway vehicles* (in Romanian). Contract number: 48/26.09.2016.

References

1. P. Carlbom, *Carbody in rail vehicle dynamics* (Ph.D Thesis, Stockholm, 2000)
2. G. Yang, C. Wang, F. Xiang, S. Xiao, Chinese J. Mech.l Eng. **29**, 6, 1120-1126 (2016)
3. C. Hui, Z. Weihua, M. Bingrong, Int. J. Veh. Struct. Syst. **7**, 2, 55-60 (2015)
4. G. Diana, F. Cheli, A. Collina, R. et al., Veh. Syst. Dyn **38**, 3, 165-185 (2002)
5. J. Zhou, R. Goodall, L. Ren, H. Zhang, Proc. IMechE, Part F **223**, 461-471 (2009)
6. G. Schandl, P. Lugner, C. Benatzky, et al., Veh. Syst. Dyn. **45**, 9, 835-847 (2007)
7. G. Schupp, Mech. Based Des. Struct. Mach. **31**, 3, 297–314 (2003)
8. M. Dumitriu, Archive of Mechanical Engineering **61**, 4, 609-625 (2014)
9. T. Tomioka, T. Takigami, Veh. Syst. Dyn. **48**, Suppl., 467–486 (2010)
10. W. Kunpeng, X. He, X. Man, J. Mech. Sci. Tech. **29**, 9, 3571-3580 (2015)
11. M. Dumitriu, Archive of Mechanical Engineering **63**, 1, 25-44 (2016)
12. H. Shi, R. Luo, P. Wu, J. Zeng, J. Guo, Sci. China Tech. Sci. **57**, 7, 1425–1438 (2014)
13. J. Hansson, M. Takano, T. Takigami, et al., JSME Int. J., Ser. C **47**, 2, 451–456 (2004)
14. T. Kamada, T. Hiraizumi, M. Nagai, Veh. Syst. Dyn. **48**, 73–87 (2010)
15. T. Mazilu, Veh. Syst. Dyn. **48**, 9, 1021-1042 (2010)
16. T.H. Young, C.Y. Li, Veh.Syst.Dyn. **40**, 5, 329–349 (2003)
17. T. Tomioka, T. Takigami, Y. Suzuki, Veh. Syst. Dyn. **44**, Suppl., 272–285 (2006)
18. ORE C 116/RP1, *Interaction between vehicles and track* (Utrecht, 1971)
19. A.A. Shabana, *Theory of vibration* (Springer-Verlag, 1991)
20. ENV 12299, *Railway applications ride comfort for passengers measurement and evaluation* (1997)

 DOI: 10.1051/matecconf/201711207008

Experimental modal analysis of an electric locomotive body

Ion Manea[1,*], *Ioan* Sebesan[2], *Gheorghe* Ghita[3], *Mihai* Matache[4], *Sorin* Arsene[2], and *Gabriel* Prenta [1]

[1]SC Softronic SRL, R&D Department, Craiova, No.40, Dolj, Romania
[2]Polytechnic University of Bucharest, Sector 6, Splaiul Independentei, No. 313, Romania
[3]Institute of Solid Mechanics of Romanian Academy, Bucharest, Constantin Mille, No.15, Romania
[4]INMA Bucharest, Sector 1, Blvd. Ion Ionescu de la Brad, No. 6, Romania

Abstract. The study of the structural dynamics is essential for understanding and assessing of any engineering product performance, knowledge of dynamic structural response being fundamental to ensuring a sustainable and safe product functioning. From all techniques for structural response assessment, the experimental modal analysis (EMA) provides one of the most comprehensive tools for collecting the experimental data, relevant for structure investigation and assessing of the structural response under normal or imposed operating conditions. Once identified the structure in terms of modal model, it can begin to its optimization to face new challenges or functioning tasks. The article presents an experimental modal analysis application, performed on a LEMA 6000kW electric locomotive body on the production cycle of SC Softronic Craiova. The application was made to validate the dynamic finite elements model realised in order to certify the crashworthiness locomotive design and to evaluate the body interaction with bogie and track to avoid the locomotive structural resonance phenomena.

1 Background of experimental modal analysis

Experimental modal analysis (EMA) is the process of building the mathematical model of a system based on experimental determined modal parameters in view to investigate the structural response to the given load conditions, to evaluate the effect of controlled structural changes, or to evaluate the structure response to the required conditions in new applications.

Usually, by using the numerical analysis techniques it is done the analysis of structure vibration state in terms of frequency response function (FRF). A set of FRF measurements carried out in well-defined conditions and points on the structure, can provide a comprehensive response of the structure to specified conditions of excitation.

Through experimental modal analysis are determined the modal parameters of the system in terms of: modal frequencies, modal damping and modal shapes, which together provide a

* Corresponding author: ion.manea@softronic.ro

complete description of the intrinsic dynamic characteristics of the structure, independent of the structural excitation.

Modal analysis can be achieved through both analytical techniques and experimental techniques, the two types of analysis leading practically to the same modal parameters. A complete modal analysis assumes utilization of the combined techniques, analytical and experimental, in this case the experimental model serves as a validation tool of analytical model. Modal models are not the models of the system itself, but rather are models of dynamic behavioural of the system constrained by a set of hypotheses and boundary conditions.

The analytical model is based on equations associated with the system mass and stiffness distribution, calculated by imposing a given set of boundary conditions. The analytical model, usually, is prepared using the finite elements method (FEM) and associated mathematical model involves a set of complicated coupled differential equations that can be solved only by using advanced computing techniques.

The experimental model is based on vibrational response data measured on the system in well-defined conditions, which is a mirror of the system under measurement. Associated mathematical model consists of a limited set of differential equations, one for each degree of freedom. This model is often referred as "experimental modal model".

Experimental modal analysis techniques are used in a wide range of applications, including: structural optimization; prediction of dynamic behaviour; assessment of the excitation forces from only vibration response; prediction of structural response under the effect of complex excitations; validation of analytical finite elements model by experimental data; detection, location and evaluation of structural defects.

2 Mathematical models used in experimental modal analysis

The most general mathematical model is one in which the matrices of mass, stiffness and damping forces are estimated based on measurements of vibrational excitation and response.

The mathematical model is based on a system of differential equations adapted to the calculation method, which can be in time domain or frequency domain. Equation (1) shows the general form of the moving equation for the time domain, and equation (2) shows the general form of the moving equation for the frequency domain.

$$[M]\{\ddot{x}(t)\}+[C]\cdot\{\dot{x}(t)\}+[K]\{x(t)\}=\{f(t)\} \qquad (1)$$

$$-\omega^2[M]\{X(\omega)\}+i\omega[C]\cdot\{X(\omega)\}+[K]\{X(\omega)\}=\{Q(\omega)\} \qquad (2)$$

If the equations (1) and (2) are used as a model to estimate the modal parameters, the unknown matrices elements must first be estimated from experimental data, through measurements carried out in time or frequency domains. Once the matrices elements were estimated, the modal parameters are found as solution of a eigenvalues and eigenvectors problem.

Because of the truncated data in terms of content, frequency range, number of degrees of freedom, or due to errors in measurements, the matrices in the equations (1) and (2) are not generally directly comparable with the matrices determined by an finite element approach. Instead, the matrices which are estimated by such procedures provide valid relationships between input-output quantities and correct modal parameters. This is because there are an infinite number of sets of matrices of mass, damping and stiffness which leads to the same modal parameters in the same frequency range and in the same measurement range. The most mathematical models describing the dynamics of an elastic system with N degrees of freedom, can be arranged in a suitable form, in such a way as to be described by the following relationships defined in the time domain (relation 3) or frequency domain (relation 4) :

$$h_{pq}(t) = \sum_{r=1}^{N} A_{pqr} e^{\lambda_r t} + A_{pqr}^{*} e^{\lambda_r^{*} t} \quad (3)$$

$$H_{pq}(\omega) = \sum_{r=1}^{N} \frac{A_{pqr}}{i\omega - \lambda_r} + \frac{A_{pqr}^{*}}{i\omega - \lambda_r^{*}} \quad (4)$$

$$A_{pqr} = Q_r \Psi_{pr} \Psi_{qr} \quad (5)$$

$$A_{pqr} = \Psi_{pr} \mathrm{L}_{qr} \quad (6)$$

In relations (1) ... (6) have been used the following notations [1]:
ω – angular frequency variable (rad/sec); p – measured degree-of-freedom (response); q – measured degree-of-freedom (input); r – modal vector number; N – number of modal frequencies; A_{pqr} – residue for mode r; Q_r – complex modal scaling coefficient for mode r; Ψ_{pr}– modal coefficient for measured degree-of-freedom p and mode r; L_{qr}– modal participation factor for reference degree-of-freedom q and mode r; λ_r – system pole.

The great majority of modal parameter estimation techniques assume that the system is linear and time invariable. In fact, depending on the analysed system and on conditions under which excitation was carried out, these assumptions are valid in a proportion greater than or less than, but almost never true at all. The analytical model represented by equations (4) is a nonlinear model in terms of modal parameters, λ_r. The difference between the various modal parameter estimation algorithms consists in the modality of solving the nonlinearities of mathematical models. There are a wide variety of algorithms for determining the modal parameters, but the most used techniques are: Rational Fraction Polynomial-Z (RFP-Z) and Polyreference Time (PDT). Both techniques works with complex modes and have the great advantage that can process data input from multi-reference imputes. In this way, any situations of real modes or unireference excitation can be treated as a particular case.

3 Experimental modal analysis applications to railway vehicles

3.1 Excitation and measuring equipment and software

-Piezoelectric accelerometers type 355B03, 15 pieces, sensitivity: 10.19 mV/(m/s^2); measuring range: ±490 m/s^2; frequency range: 2-5000 Hz;

-Force transducer type 208C03, sensitivity: 2248 mV/kN; measuring range: ±2.224 kN; frequency range: 36 kHz;

-Electrodynamic shaker type 11077, force range: ±100N, frequency range: 3Hz ÷ 20kHz;

-Power amplifier type LV103, power: 100VA, frequency range: 3Hz ÷ 20kHz;

-Data acquisition system type SoftronicDataAcquisition, carrying case including 1xLAN-XI type 3160A042 and 1xLAN-XI type 3053B120, with the following main characteristics: inputs: 16(voltage and CCLD/ICP); outputs: 2 (voltage outputs from generator); energy independence with all acquisition systems in operation: 5 hours;

-PULSE LabShop, software module for data acquisition control and real-time multi-channel processing;

-Structural Dynamic Test Consultants, software module working under PULSE LabShop, providing a dedicated environment for modal analysis tests driving using the classical excitation technics of impact hammer or shaker. Drives the measurements through simple steps, providing an intuitive graphical interface which connects the test configuration, geometry of test structure and information concerning the degrees of freedom of excitation and response. At the final stage, the measured FRF together with the structure geometry are transmitted to PULSE Reflex Modal Analysis, which is automatically launched.

-PULSE Reflex Modal Analysis, post-processing mode for classical modal analysis. Uses a wide variety of modal identification modules, including PTD and RFP techniques.

-PULSE Reflex Correlation Analysis, correlation analysis module that provides the verification and validation of analytical models obtained by finite element methods through the experimental modal analysis tests.

3.2 Experimental modal analysis test on an electric locomotive body

During the tests the locomotive body was only equipped with the two obstacle deflectors, without caps doors and other equipment. To ensure a correct application of EMA techniques, it was ensuring the conditions of pseudo-freedom vibrations of the body by suspending it on four winches, which, by their lifting axle and sole support, provides enough elasticity to consider the locomotive body as a free body, for the low level vibration displacements, so not restricting the degrees of freedom of the body [2].

It has used the method of excitation in one point by sinusoidal excitation sweep in the frequency range of 2-200 Hz with electrodynamic shaker attached to the strength structure of sideframe, under the driver's cab, near the mounting point of obstacle deflector. The excitation force transmission was carried out by means of a spring steel rod with a diameter of 5 mm, to avoid the effects of parasitic coupling between the shaker and body. The mounting schemes for EMA tests is presented in Fig. 1.

Fig. 1. Experimental modal analysis test on a locomotive body using the shaker excitation method.

Data acquisition was carried out under SDTC in which it was achieved the body geometry and it has been developed the test program, being planned eight sequences for measuring of dynamic response. For the 8 measuring sequences, the structure excitation was maintained at said point, point 67 as shown in Figure 2. At one measuring sequence, the dynamic response was determined in 5 measuring points, for each point being determined the vibratory response on the vertical, longitudinal (along the body) and transverse (across the body) directions. So for each sequences were measured 15 degree of freedom of response and 1 degree of freedom of reference (excitation). The global EMA test involved a total of 120 measuring response for one reference point 67, resulting in a total of 120 frequency response functions.

Figure 2 presents the panel driving for one measuring sequences under SDTC. On the geometric model it is positioned the shaker excitation in 67 point and frequency response functions of points 36, 49, 75, 46 and 53 in the 7th measuring sequence. The distribution of the measurement points was carried out in such a way as to be representative for dynamic response of locomotive body and to be enough distanced in order to achieve a correct body animation in the natural vibration modes. To achieve a high accuracy of FRF, has selected a frequency resolution of 62.5 mHz and a linear mediation over 237 FRF sequences.

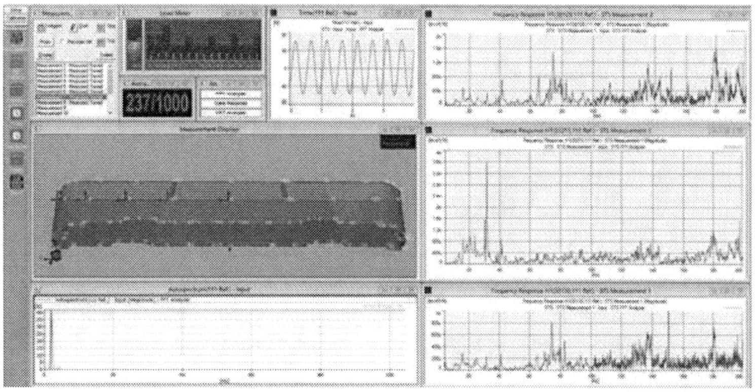

Fig. 2. The driving panel of EMA tests under Structural Dynamic Test Consultants.

3.2 Modal analysis of electric locomotive body

Modal analysis of recorded data was performed with PULSE Reflex Modal Analysis. In the first stage it was done the validation of acquired data by analysing the noise in excitation and response, as well as by structure animation at different frequencies in the test domain, as in Figure 3. In this step can be eliminated the aberrant data.

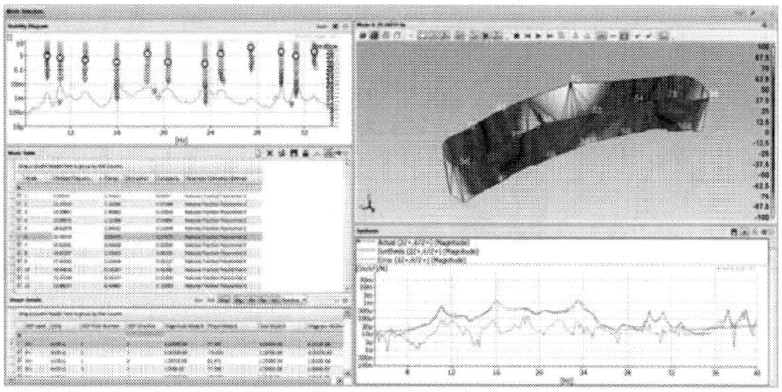

Fig. 3. The validation panel of PULSE Reflex Modal Analysis.

In this paper have been used two methods to estimate the modal parameters, Rational Fraction Polynomial-Z (RFP-Z) and Polyreference Time (PDT), both the higher order methods, recommended for working in wide band frequency, being able to working with large number of modes. Given the purpose application, validating the analytical model with finite elements, the frequency domain was limited to 5 ... 35Hz.

Figures 4 ... 11 presents the modal shapes of locomotive body in their first 8 vibration eigenmodes in frequency domain of 5 ... 35Hz.

Modal shapes can be exported in a tabular form providing: damped frequency (Hz), damping (%), and mode complexity. For each of modes are transmitted separately for each measured degree of freedom (DOF) the following parameters needed for the modes rebuilding: DOF direction, mode magnitude, mode phase. These parameters are transmitted separately for: residue, unity vector length and unity modal mass.

Mode shapes determined by EMA serve as a validation tool for analytical model realised by finite elements analysis. Validated FEA model was used for evaluating the collisions resistance of LEMA locomotive and to estimate the body interaction with the bogie and track to avoid the structural resonance phenomena. This analysis are not the subject of this article.

DOI: 10.1051/matecconf/201711207008

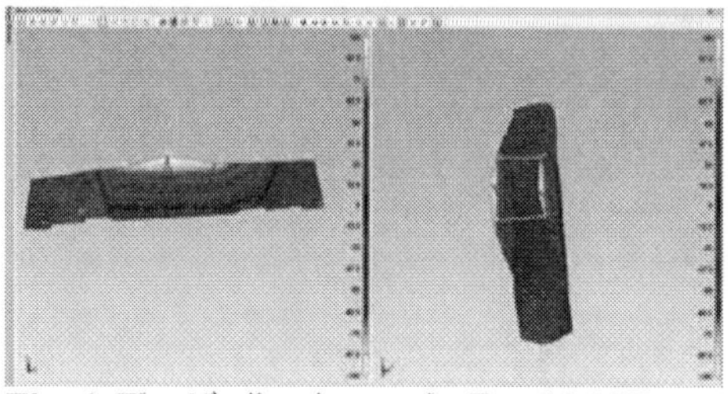

Fig. 4. The 1st vibration mode, Fq= 11.10Hz

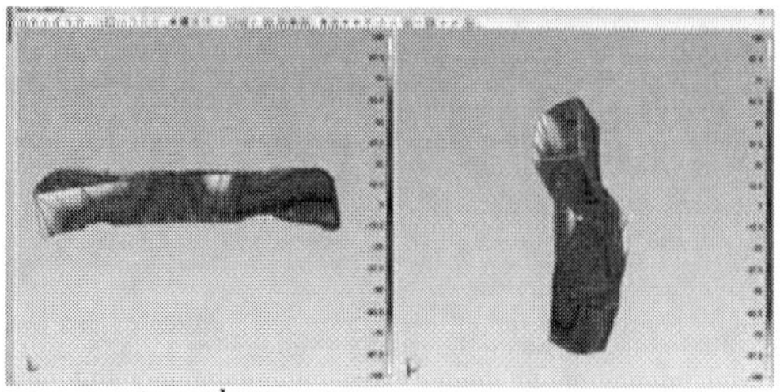

Fig. 5. The 2nd vibration mode, Fq= 15.98Hz

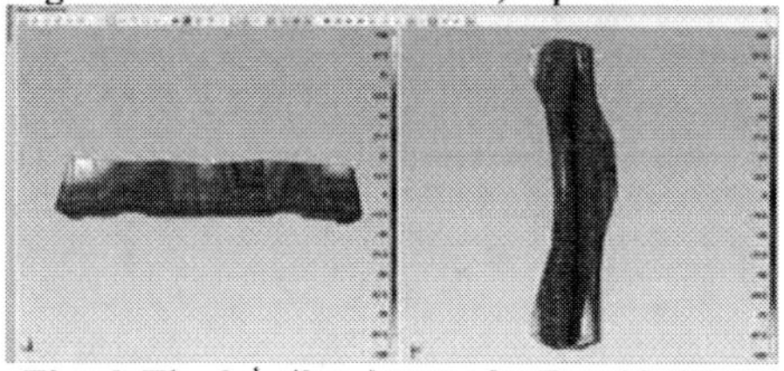

Fig. 6. The 3rd vibration mode, Fq= 18.62Hz

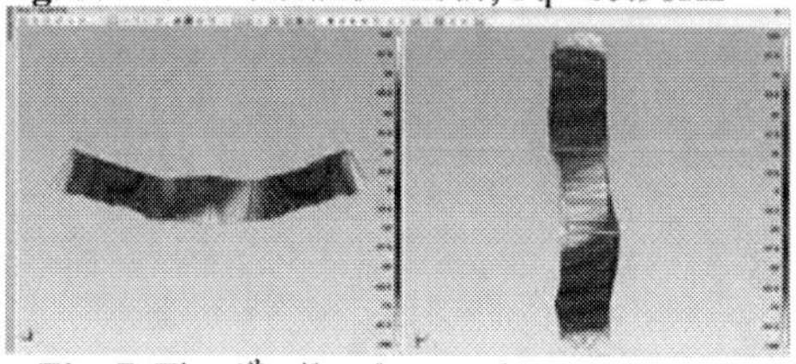

Fig. 7. The 4th vibration mode, Fq= 20.38Hz

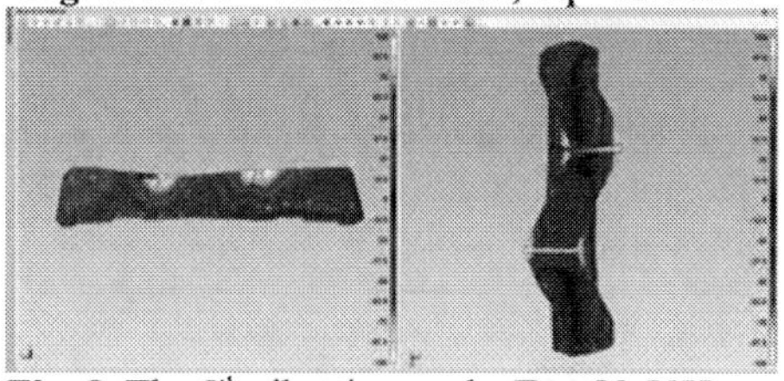

Fig. 8. The 5th vibration mode, Fq= 23.52Hz

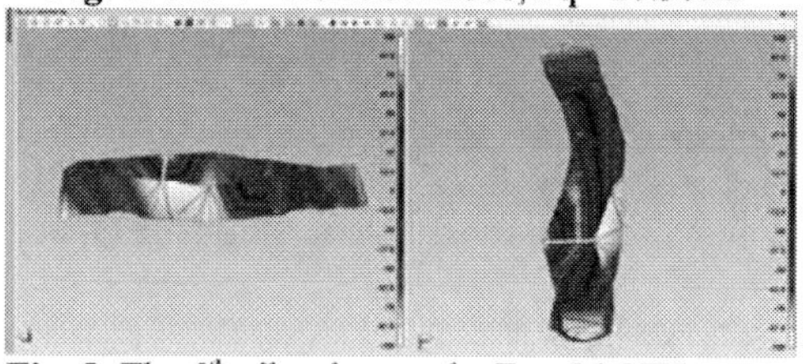

Fig. 9. The 6th vibration mode, Fq= 24.87Hz

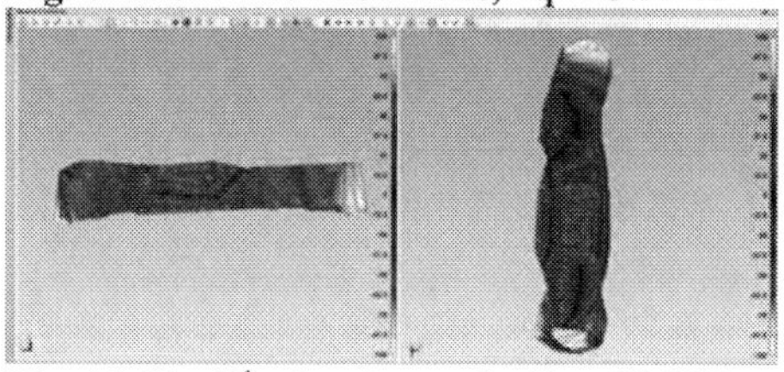

Fig. 10. The 7th vibration mode, Fq= 27.43Hz

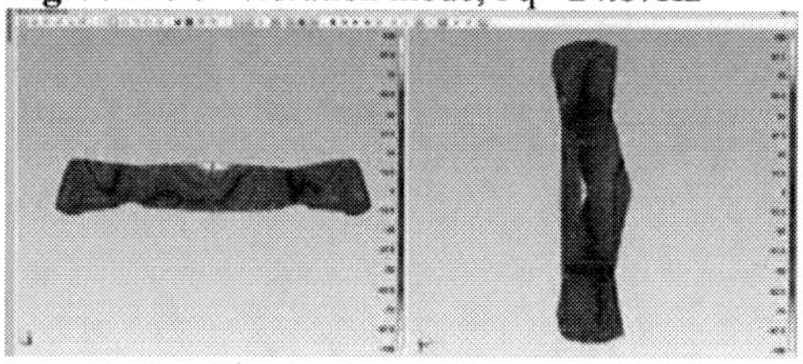

Fig. 11. The 8th vibration mode, Fq= 30.04Hz

4 Conclusions

The article presents an EMA application performed on an empty locomotive body in order to achieve the modal model which shall be later used as a validation tool for analytical finite element model. The modal model is described through: damped frequency (Hz), damping (%), and mode complexity. It can be successfully used for prediction of the body structural response under the effect of complex excitations like as that from bogie and track.

This work was supported by a grant of the Romanian National Authority for Scientific Research and Innovation, CNCS–UEFISCDI, project number PN-III-P2-2.1-PTE-2016-0008, within PNCDI III

References

1. R. J. Allemang, D. L. Brown, *Experimental Modal Analysis and Dynamic Component Synthesis - Modal Parameter Estimation*, vol3 (1987)
2. I Manea, I Sebesan, Gh. Ghită, *Modal analysis of theoretical and experimental. Application to railway vehicles*, Bucharest, Matrix (2016)

Integration of complementary methods for monitoring stress/strain of wind turbine blades structures

Adriana Savin[1,*], *Rozina* Steigmann[1], *Nicoleta* Iftimie[1], *Mariana Domnica* Stanciu[2], *Narcis Andrei* Danila[1,3], and *Paul-Doru* Barsanescu[3]

[1] National Institute of R&D for Technical Physics, NDT Department, Iasi, Romania
[2] Transilvania University of Braşov, Faculty of Mechanical Engineering, Brasov, Romania
[3] Technical University "Gheorghe Asachi" of Iasi, Faculty of Mechanical Engineering, Iasi, Romania

Abstract. The most important part of wind turbine is the blade that must be tested during the fabrication and during the functioning when can be damaged by moisture absorption, fatigue, wind gusts or lightning strikes. The novelty of the paper is represented by the employing of wireless sensors embedded in a scalable wind turbine blade made from Glass Fiber Reinforced Plastics together with Fiber Bragg Gratings in order to effectuate structural health monitoring in static conditions. The sensors are placed on critical location on blade determined by FEM simulation and a comparison between the complementary methods is done.

1 Introduction

The EU is currently studying the possibility of updating the air pollution standards for power plants based on coal, because they are the largest source of emissions of sulfur and mercury dioxide, nitrogen oxides, arsenic, lead and cadmium. Thus, electrical power market obtained by harvesting wind power is expanding. In these conditions, to increase the power of turbines, wind turbines (WT) manufacturers use glass-fibers reinforced plastic (GFRP) for WT blades (WTB) allowing construction, due to low weight, of increased size blades. The monitoring of WTB made from GFRP have been reviewed in [1], proposing either statistical pattern recognition using simulation or experimental data, or in [2] considering mechanical property testing and full scale testing as well as nondestructive testing methods. Condition monitoring can be carried up using thermography [3] and fatigue crack detection performance can be done [4]. Implementing a process of fault detection and characterization of structures is refers as Structural Health Monitoring (SHM) [5]. The SHM process involves the observation of a system over time using periodically sampled dynamic response measurements from an arrangement of sensors, the extraction of damage-sensitive features from these features to determine the current state of system health [6]. In SHM domain, 7 axioms are accepted [7]. In our case, the most important are:
Axiom I: All materials have inherent flaws or defects.

* Corresponding author: asavin@phys-iasi.ro

Axiom II: The assessment of damage requires a comparison between two system states.
Axiom IVa: Sensors cannot measure damage. Feature extraction through signal processing and statistical classification is necessary to convert sensor data into damage information.

The nondestructive testing (NDT) techniques for in-service inspection and determination of regions with high degradation risks are developed, (function on the type and the size of the WT) [8]. In order to decrease the fabrication and maintenance costs, as well as for avoiding unproductive time, NDT is required both during fabrication and in-service.

In Romania, over 3% of annual electrical power is produced by WT [9], the majority being GE1.5sle or similarly [10]. The most frequently flaws in turbine blades are microscopic and mesoscopic flaws, appeared in matrix, undetected by classical NDT procedures (as phased array ultrasound), broken fibers can appear and develop under moderate loading, cracks and delamination due to low energies impacts, etc. Starting from the scheme presented in Figure 1, flaws mentioned above must be detected and evaluated by NDT methods with good probability of detection (POD) under high reliability coefficient.

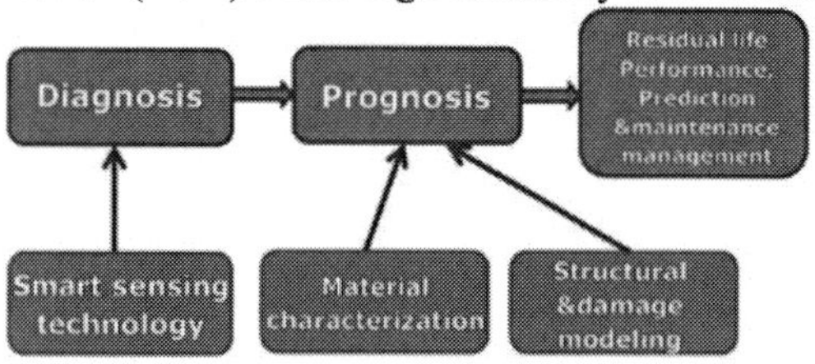

Fig. 1. SHM principle.

The paper presents the results of testing of a WTB, in static conditions, using wireless sensor (WRS) and Fiber Bragg Gratings (FBG) embedded within in order to monitor possible damages that further can be transformed into flaws. To determine the righteous of solutions, reliability of correct diagnosis probability, prognosis, and evaluation of residual lifetime and maintenance management, a scalable WTB has been constructed, that had the scaled dynamics of a full scale blade as well as enough sensors for measurements as well as monitoring. The maximum stress zones [11] and damage evolution and remnant stress [12] have been determined using Finite Element Modelling (FEM). The results of complex mechanical tests, effected on scalable WTB models (in our case a blade of 1750mm length) are used to give efficiency to monitoring strategy.

2 SHM based on transducers

The SHM of WTB is characterized by sensors network around the blade to detect and locate any occurring damage. Taking into account the efficacy of the method and *a priori* knowledge about WTB, the sensors are distributed according to the most expected damage area, aiming the minimizing of the sensors number (Figure 2).

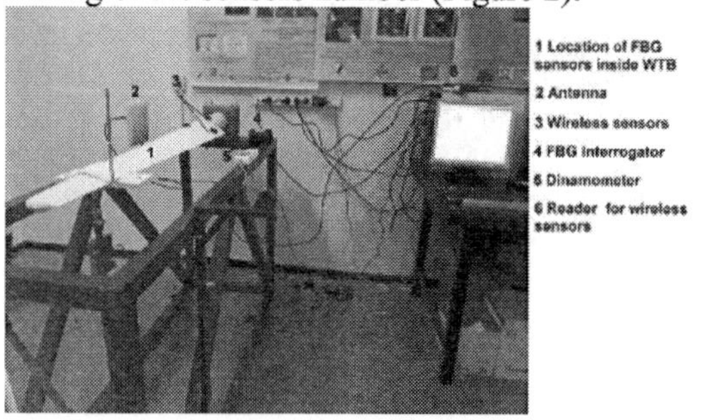

Fig. 2. Testing stand.

The blade has been realized from E-glass/epoxy EPIKOTE Resin MGS LR 385 composite. The leading edge is straight and the trailing edge conical for an easy construction. The profile follows NACA airfoil. For increased structural strength and stiffness at 0.286R (R- the total length of distance between the center of rotor and the tip of blade), the same NACA has been applied both for the upper surface and lower surface, keeping the aerodynamically performances of the blade's tip. The profiles between 0.268R and tip were linear interpolated. A compromise has to be found between high-resolution and long propagation distance. The results are obtained at preliminarily testing of a scalable WTB, in laboratory, using optical fiber (OF) sensor technology and passive radiofrequency identification (RFID) WRS as uni-dimensional displacements sensor.

2.1 Sensors with Fiber Bragg Gratings (FBG)

SHM includes the use of FBG for detecting delamination in composite laminates [13] and monitor impact event occurrence [14]. [15]. The embedded sensor can monitories the structure in critical regions. The central wavelength of this signal, called Bragg wavelength λ_B is related to the physical parameters of the grating according to $\lambda_B = 2n\Lambda$ where n is the effective refractive index of the fundamental mode propagating inside the fiber, Λ is the pitch of FBG. When the fiber containing FBG is submitted to strain, the central wavelength is displaced to higher or smaller values [16]. The direction and the magnitude of displacement is proportional with the modification of strain or temperature. The strain axial sensitivity is

$$\frac{1}{\lambda_B}\frac{d\lambda_B}{d\varepsilon} = \left(\frac{1}{n_{eff}}\right)\frac{\partial n_{eff}}{\partial \varepsilon} + \left(\frac{1}{\Lambda}\right)\frac{\partial n_{eff}}{\partial \Lambda} \approx (1 - p_e) \tag{1}$$

where $\partial\lambda_B$ is the FBG wavelength shift; ε<<1, p_e is photo-elastic coefficient of the fiber $p_e \approx 0.22$, FBG sensors whose gauge lengths are about 10mm were used for SHM.

$$\frac{\partial\lambda_B}{\lambda_B} \approx (1 - p_e)\varepsilon; \quad \frac{\partial\lambda_B}{\lambda_B} \approx 0.78\varepsilon \tag{2}$$

The FBG strain sensing can be expressed as $\varepsilon = \frac{1}{0.78\times10^{-6}}\frac{\Delta\lambda_B}{\lambda_B}$.

FBG can assure and improve the functioning of WTB due to the detection properties, small dimensions and reduced weight, allowing the embedding them into the base material of the blade (being embedded at intersection of longeron with NACA profile, inside the WTB). In this structure, 3 FBG were embedded along the central longitude of the blade, placed into critical points determined by FEM of blade under vibration. Figure 3a presents the division of the blade in critical regions, in Figure 3b is presented the result of simulation under a force F_y=200N, producing the displacement of the tip with 1.59 mm. The FEM model shown below takes into account the presence of the reinforcement structure, who's mass cannot be neglected, especially if one considers that the distance from the axis of rotation increase the inertia and can reduce the frequency associated with the first mode of vibration of WTB. Fortunately, the glass fibers employed makes the structure very rigid.

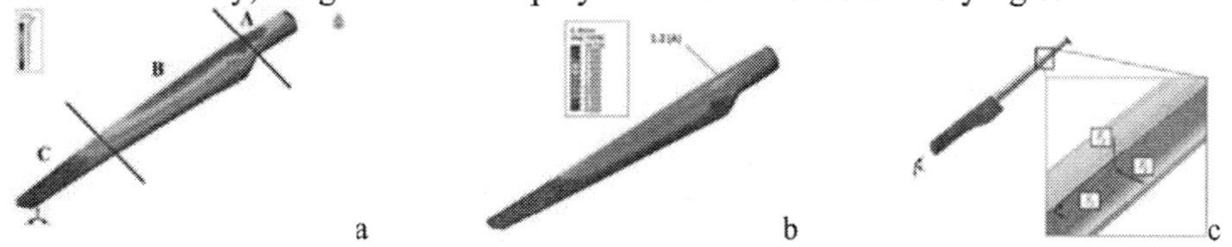

Fig. 3. FEM simulation of WTB: a) division of blade in critical regions; b) Von Mises stress; c) loading forces disposition.

FBGs sensors were placed in region B at 307 mm, 362 mm and 406 mm from the hub fixing in stand. All the loadings are static. The WRS were placed in region with maximum critical points established by FEM.

2.2 Wireless sensors (WRS)

The passive WRS designed to monitory stress/strain status have as sensitive element a special type of metamaterial (MM), that follows the relative displacement (compression or tension) of its components due to an impact and can develop slows or either fast. The measurement of strain provides critical information on the force presented on material [17]. In specific construction as 2D geometry on flexible support, copper split ring resonators can be used as stress/strain sensors [18], [19]. In function of type and geometrical dimensions, the sensor has the resonance frequency in the range of RF and microwaves. The quality factor of the equivalent circuit has value in the range of tens or hundreds. MM [20] presents resonant properties [21] and around resonance frequency displays unusual properties. The inductance and the capacitance are given by [22]

$$L = 0.2\mu_0\left(-\frac{w}{2}\sinh^{-1} l + \frac{w}{2}\sqrt{2} + \left(l-\frac{w}{2}\right)\sinh^{-1}\left(\frac{l-\frac{w}{2}}{\frac{w}{2}}\right) - \sqrt{\left(l-\frac{w}{2}\right)^2 + \left(\frac{w}{2}\right)^2}\right); C = \varepsilon_0\varepsilon_r\frac{w_1 t}{g} \quad (3)$$

where w_1 is the distance between strips, w is the width of strip, l is the length of strip, g is the gap and t is the thickness of copper layer strip. The two circuit elements who constitute the sensor are considered as serial connected. The series resistances modifies the value of quality factor of the circuit element and, implicitly, the frequency band. The resonance frequency is

$$f_r = \frac{c}{4(L+\Delta L)\sqrt{\varepsilon_r}} \quad (4)$$

where f_r is the estimated resonance frequency, c is the speed of light in vacuum, L is the length of the copper layer, ε_r is the dielectric permittivity of the substrate, ΔL is the additional length who compensate the effect due to thickness, width and dielectric constant of the substrate. For the developed sensor, the resonance frequency is f_0=834 MHz. Parallel with capacitance C, a capacitive element sensitive to stress/strain is connected, with the capacitance varying linear with the strain as $C_{sensitive} = \frac{\varepsilon_r\varepsilon_0 w_1 g}{u}(1+\varepsilon_s)$ where u is the thickness of dielectric layer and ε_s is strain in μm/m. For the capacitive element, sensitive to stress/strain, to function upon a law closer to this, it is imposed that the Poisson ratio of dielectric support shall be as high as possible. For polyimide, the layer supporting copper strips, the Poisson ratio is 0.4. If ε_s=10 μm/m, the connection between the two capacitance make C increases with 10%, inductance L remaining unchanged and the resonance frequency become 780MHz. In order to obtain the best sensor with MM for monitoring the glass fiber composite structure, a sensor which allow the determination of resonance frequency displacement at the changing of dielectric properties from the sample has been developed. The system consist in a RFID reader and a RFID tag [11], the tag including the sensor and the integrate circuit. When the sensor detect a modification of the strain ε, the frequency is

$$f_r' = \frac{c}{4(1+\varepsilon)(L+\Delta L)\sqrt{\varepsilon_r}} = \frac{f_r}{1+\varepsilon} \approx f_r(1-\varepsilon) \quad (5)$$

Thus, when the load is small, the resonance frequency modifies almost linear with load, the load can be determined if the resonance frequency is measured. The RFID tag antenna assures that the interrogation frequency *f* shall be equal with the one of RFID tag to obtain perfect matching of impedance between antenna tag and IC chip. The smallest amount of energy must be transmitted toward reader to activate RFID tag, the transmitted power threshold (measured through the reader) reach minimum value at resonance frequency.

3 Experimental results

GFRP currently used in construction of WTB are usually long fiber-enforced composite materials. The fatigue test was performed with 50 kN steps of tensile loading. The FBG are connected at optical system Spectral Eye 600 [23] coupled with PC via RS 232-USB2 connector. The data processing has been made in Matlab 2014b, the temperature correction being made, the equipment indicating the temperature of fiber, too. The central wavelength is 1527 nm and strain sensitivity 7.8x10^{-7}με. Figure 4a presents the raw data recorded by the device for different loadings. The relative variation of Bragg wavelength was determined in function of loading is presented in Figure 4b showing that the relative variation of Bragg wavelength is linear. The dependency strain-load for WTB in a loading-unloading cycle for three sets of experimental measurement is presented in Figure 5. The same linear dependency strain-load can be shown, even the existence of a remnant stress at force removal, indicating an accumulation of energy in WTB composite structure, preponderant in the resin.

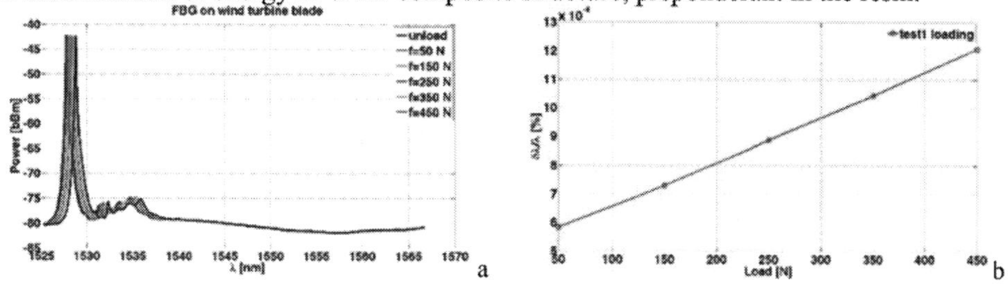

Fig. 4. FBG measurements: a) response to successive loading; b) relative variation of Bragg wavelength.

These stresses are relieving in relatively short time once with the material relaxation, so that the new tests are not influenced. The measurements using WRS were carried on the same time with those using OF sensors. The reader antenna (Figure 2) is fixed on the upper part of the stand, at 30cm from the middle area of WRS placement locus. The antenna is connected to a logger reader coupled to PC by USB. The loading force is configured to have an increment of 243με at each step. The mean threshold of interrogation power, in dBm, is read by logger reader at each loading step. Figure 6 plots the experimentally dependency strain-load for WTB into a cycle loading for the three WRS placed in different critical regions.

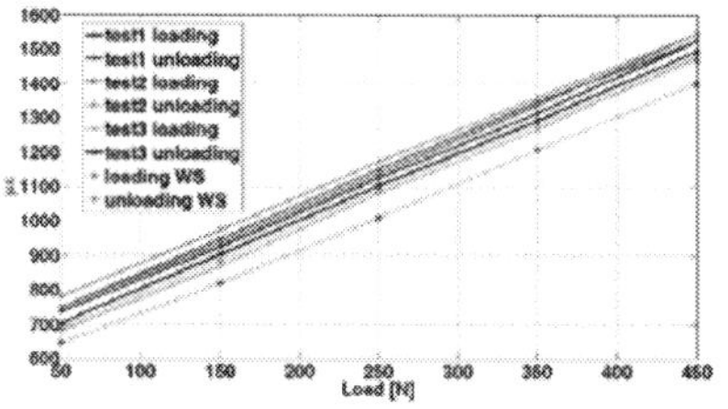

Fig. 5. The dependence of strain vs. loadings.

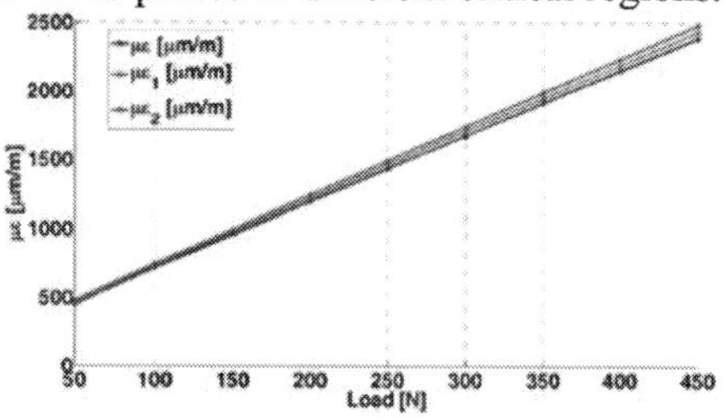

Fig. 6. Signal delivered by the 3 wireless sensors for loading of WTB.

It can be observed that the strain has a practically linear dependency on to the applied load on all WRS with small difference on slopes. Although the sensors are not located at exactly the same location, a great resemblance has been found. The measurement results of the FBG interrogation system was in excellent agreement with those obtained by wireless sensors measurements. The FBG sensor system and the wireless sensors offer many advantages over traditional resistive strain gauges. These include remote sensing, easy installation, non-corrosive and lower maintenance cost. This shows that complementary technologies are a good alternative for structural strain monitoring.

4 Conclusions

Monitoring complex structures as WTB has become a requirement. Practical applications demonstrates that, in order to avoid environmental disasters, it is necessary to establish diagnosis and prognosis methods, based on using information obtained from sensors constructed on known physical principles. The monitoring is close related with nondestructive evaluation and the trend is to obtain real time information. Scalable WTB have been constructed and tested to loadings using FBG and WRS embedded in the maximum concentration stress zones. The sensors responses obtained in static conditions were analyzed following a linear tendency, even they were not placed in the same regions. Both types of measurements shows advantages as remote sensing, simple embedding, and easy interpretation. The tests were carried on scalable models because in the frame of the project that sustains the paper, the blades will be employed into a demonstrator to show the righteous of solutions, reliability of correct diagnosis probability, prognosis, and evaluation of residual lifetime and maintenance management.

This paper is partially supported by Romanian Ministry of Research and Innovation under project PN-II-PCCA-2013-4-0656 and Nucleus Program PN 16-37 01 01.

References

1. D. Adams, J. White, M. Ramsey, C.Farror, Wind Energy **14**, 4 (2011)
2. B. Yang, D. Sun, Renew. Sust. Energ. Rev. **22** (2013)
3. B. Gao, W.L. Woo, Y. He, G.Y. Tian, IEEE T Ind. Inform. **12**, 1 (2016)
4. S.G. Taylor, K. Farinholt, M. Choi, H. Jeong, J. Jang, G. Park, J.R. Lee, M.D. Todd, J. Intel. Mat. Syst. Str. **25**, 5 (2014)
5. D. Balageas, C.P. Fritzen, A. Güemes, *Structural health monitoring,* **90**, (John Wiley & Sons, 2010)
6. C.R. Farrar, K. Worden, Philos. T. Roy. Soc. A. **365**, 1851, (2007)
7. K. Worden, C.R. Farrar, G. Manson, G. Park, P. R. Soc. A, **463**, 2082 (2007)
8. Wind turbine accident compilation, http://www.caithnesswindfarms.co.uk/fullaccidents.pdf
9. Wind Energy Targets for 2020 – 2030, EWEA (2011) http://www.ewea.org/fileadmin/files/library/publications/reports/Pure_Power_III.pdf
10. www.gerenewableenergy.com/wind-energy/turbines.html
11. R. Grimberg, G.Y. Tian, A. Savin, R. Steigmann, G.S. Dobrescu, Stud. Appl. Electromag. **39** (2014)
12. S. Antaya, Z. Ahmed, Eng. Fail. Anal. **35** (2013)
13. S. Takeda, Y. Aoki, Y. Nagao, Compos. Struct. **94** (2012)
14. K. S. C. Kuang, R. Kenny, M. P. Whelan, J. W. Cantwell, P. R. Chalker, Smart Mater. Struct. **10**, 338 (2001)
15. J.M. López-Higuera, *Handbook of Optical Fibre Sensing Technology*, (Wiley and Sons: Chichester, West Sussex, UK, 2002).

16. C.A.J. Gouveia, J.M. Baptista, P.A.S. Jorge, *Current Developments in Optical Fiber Technology*, Ch.13 (Dr. Sulaiman Wadi Harun Ed. 2013).
17. A. Deivasigamani, A. Daliri, Mod. Appl. Sci. **2** (2013)
18. T. Chen, S. Li, H. Sun, Sensors **12** (2012)
19. J.B. Pendry, A.J. Holden, D.J. Robbins, IEEE T. Microw. Theory Tech. **47**, 11(1999)
20. D. Sievenpiper ; Lijun Zhang ; R.F.J. Broas ; N.G. Alexopolous ; E. Yablonovitch IEEE T. Microw Theory Tech., 47, 11, (1999)
21. J.B. Pendry, A.J. Holden, W.J. Stewart, I. Youngs, Phys. Rev. Lett., **76**, (1996)
22. A. Savin, R. Steigmann, G.S. Dobrescu, 12th Biennial Conference on Engineering Systems Design and Analysis (ESDA2014) V002T07A027
23. FOS SpectralEye 600 Instruction Mannual, v.2.02, FOS&S, Belgium

Comparative study on the lateral run-out of friction surfaces measurement of brake discs using a brake roller tester and a dial gauge

Marius Toma, *Cristian* Andreescu, and *Dan* Micu*

University Politehnica of Bucharest, Faculty of Transports, Automotive Engineering Department, Splaiul Independentei 313, Romania

Abstract. Brake system diagnosis is one of the most common and necessary technical operations applied to the car, regardless of its type and operating phases. Measuring the diagnostic parameters on a roller brake tester is a fast operation with no disassembly necessary. Measuring the run-out of friction surfaces of brake discs with a dial gauge is an action that requires more extensive preparatory operations but it offers a high accuracy of the results. The paper aims to analyze the correlation between the dial gauge measured values and the diagnostic obtained using the brake roller tester.

1 Introduction

The continuous improvement of vehicles performances and the growth of traffic intensity make the vehicle braking system one of the most important safety elements in the road traffic. The braking system must fulfil its specific functions throughout the entire vehicle life. Considerable effort is spent in the design and testing of disk brakes of modern passenger cars [1]. Diagnosing technical condition of this system is one of the most common and necessary technical operations applied to the car, regardless of its type and stage of use and therefore special attention is given to it [2].

National and international regulations on braking performances has been improved in time, changing in correlation with the evolution in the conception of new cars. Regulations on testing and diagnostic parameters of the braking system are set out in Regulations 13 and 78 UNECE and European Community Council Directive no. 71/320 / EEC. In Romania, the standard STAS 11960-84 was conceived according to UNECE Regulation 13. Also technical regulations on road safety are approved by RNTR [3].

The most used method to check the brake system is using a roller brake tester. Usually, roller testers measure the braking force at each wheel of the car, the force applied on the brake pedal and the weight of the car. Based on these measurements, the following parameters are calculated:

* Corresponding author: dan.micu@upb.ro

Ovality – relative difference between minimum and maximum brake force, measured on the same wheel:

$$O = \frac{Max(F_{br}) - Min(F_{br})}{Max(F_{br})} \cdot 100\ [\%]. \quad (1)$$

Relative imbalance between the braking forces on the same axle:

$$D = \frac{|F_{br.l.} - F_{br.r.}|}{Max(F_{br.l.}, F_{br.r.})} \cdot 100\ [\%]. \quad (2)$$

The effectiveness of the braking system:

$$E = \frac{\sum_{i=1}^{n}(F_{br})_{max.i}}{G} \cdot 100\ [\%], \quad (3)$$

where $\sum_{i=1}^{n}(F_{br})_{max.i}$ is the sum of maximum braking forces from all the wheels, and G is the vehicle weight during the test; $Max(F_{br})$ – maximum brake force; $Min(F_{br})$ – minimum brake force; $F_{br.l}$ – braking force at the left wheel; $F_{br.r}$ – braking force at the right wheel; $Max(F_{br.l}, F_{br.r})$ – the maximum value between $F_{br.l}$ and $F_{br.r}$, G – the car weight.

The caliper housing may not fully retract after the brake is released. The remnant contact between the pad and the disc affects the fuel efficiency and the life of the pad friction lining and increases the temperature and wear of disc and pads. This off-brake calliper drag is greatly affected by the lateral run-out, which is defined as the side-to-side motion of the disc as it turns on the hub [4].

„Regulation No 90 of the Economic Commission for Europe of the United Nations (UN/ECE) — Uniform provisions concerning the approval of replacement brake lining assemblies, drum brake linings and discs and drums for power-driven vehicles and their trailers" limits the values for lateral run-out friction surface of the disc brake to 0.05 mm [5]. Limit values set by the manufacturer for this parameter depend on constructive peculiarities of the system that contains the brake discs and on the user's exigency. Run-out limit values imposed by some manufacturers are shown in Table 1 [6, 12]:

Table 1. Lateral run-out limit values [6, 12].

Vehicle	Fiat Tipo 1.4 L/16V	Chrysler 300-3.0L V6 24V V.V.T.	Lancia Grand Voyager 2.8L TD	Jeep Cherokee 1.4 16V	Opel Astra H
Value [mm]	0.05	0.035	0.05	0.04	0.05
Vehicle	Jeep Grand Cherokee 3.0 V6 TD	Jeep Wrangler 2.8L 4cyl	Jeep Renegade 1.4 L	VW cars	Škoda Octavia
Value [mm]	0.05	0.025	0.02	0.05	0.05

Table 1 shows that the specified limits are lower or equal to the limits in accordance with European legislation [5].

Determination of lateral run-out is done using a dial gauge and following the procedure indicated in the maintenance and repair manuals.

When a roller brake tester is used, the lateral run-out is estimated considering that the variation of the braking force at one wheel (the so-called „Ovality") is proportional to the run-out of the brake disc of the respective wheel. Thus, a value expressed as a percentage, is computed according to the relation (1).

The ovality limit measured on the roller brake tester is not regulated by legislation, but it may be specified by the manufacturer of the tester; e.g. BOSCH indicates a maximum acceptable value of 25% for its BSA 43XX tester [7].

A braking system which has good effectiveness and good relative imbalance is considered in good condition according RNTR1, no matter the value of ovality.

2 Preparing and conducting experimental research

The study presents the results of the lateral run-out of friction surfaces measurement of brake discs using a brake roller tester and a dial gauge. The use of the brake roller tester is a fast operation, part of the braking system diagnostic program. This measurement represents an indirect estimation of the run-out. The use of a dial gauge is an action that requires more extensive preparatory operations but it provides accurate results.

Both methods have been used in order to show if it is possible to establish a correlation between the values measured by these two methods.

Experimental research has been conducted using three cars: Opel Astra, VW Passat, Škoda Octavia. Some of their technical data are listed in Table 2.

Table 2. Specifications of tested vehicles.

No.	Vehicle	Curb weight [kg]	Kilometres	Year of manufacture
1	Opel Astra	1395	171158	2007
2	VW Passat	1579	96850	2007
3	Škoda Octavia	1385	174637	2008

2.1 Lateral run-out measurement of brake discs using a brake roller tester

In the first stage, tests were conducted on Bosch BSA 43XX roller tester located in University Politehnica of Bucharest, Faculty of Transports, Automotive Engineering Department.

Before starting, the vehicle has to be driven 5 km on an urban route, followed by intense heating of the brakes on the roller tester. Tire pressure was adjusted at nominal value. Each car has been tested three times successively.

During the test, the pressure on the brake pedal rises gradually. When the braking force at the wheel reaches a reference value memorized by the computer of the roller tester, a message appears on the tester's display which indicates that the operator must keep constant the pressure on the brake pedal for a few seconds, until the message disappears. The roller tester computes the ovality during this period and displays it in percent.

2.2 Lateral run-out measurement of brake discs using a dial gauge

In the second phase of the experimental research, the test consisted in measuring the brake discs lateral run-out using a dial gauge. The procedure indicated in the maintenance and repair manuals for ovality measuring was followed. It was similar for all the three tested cars.

After removing the wheels from the vehicle, the wheel brake assembly components were blown with compressed air to remove dust particles. Each disc brake was tight with its studs using the helpful parts (struts) as suggested in Figure 1.

The dial gauge was fitted with a magnetic base on the shock absorber (Figure 1). The spindle of the dial gauge was positioned 10 mm towards the interior from the edge of the disc, as recommended in [12].

For the wheels of the front axle, measurements were made on both surfaces of the brake disk. No correspondence has been established between the measured values on the two surfaces but only the maximum measured values were retained. For the wheels of the rear axle, the lateral run-out was only measured on the outer surface of the disc.

Fig. 1. Mounting the dial gauge on the front brake assembly.

3 Experimental results

In the case of the lateral run-out measurement using a brake roller tester, the braking forces developed at each wheel have been measured and used to calculate the relative imbalance, the effectiveness and the ovality. The rolling resistance when the brakes were not applied was also measured.

In the case of the periodically testing of the car's technical condition, the regulations only specify acceptable limits for effectiveness, relative imbalance and rolling resistance when brakes are not applied. Taking into consideration the effectiveness and the relative imbalance, for which all values did not exceed the limits, it results that the three tested cars were in good operation condition. In the case of the rolling resistance, only few the tests revealed higher values than normal for two of the tested cars, the arithmetic means of the three values being under limits. Therefore, the cars operation condition may also be considered acceptable from this criterion point of view.

If all three cars are considered in good or acceptable condition for effectiveness, relative imbalance and rolling resistance criteria, the ovality shows that one vehicle is in good condition, one is at the limit and the other is defective.

The operator did not notice improper behaviour of the braking system during preparatory operations (heating brakes) nor during the tests themselves for VW and Opel cars. But while driving Škoda there were excessive vibrations and pulses in the brake pedal during braking. This behaviour is confirmed also by the roller tester measured values.

For Škoda and Opel it has to be specified that during the last year they have run on short distances followed by long periods of immobilization which can cause deposits on the brake discs (areas heavily oxidized, uneven brake pistons sliding) that affect the reproducibility of results (due to rapid and permanent evolution of the surfaces quality during measurements).

The lateral run-out values obtained using the dial gauge are listed in Table 3.

Table 3. The lateral run-out [mm] values obtained using the dial gauge.

Vehicle	Front Right Wheel			Front Left Wheel			Rear Right Wheel			Rear Left Wheel		
	Int.	Ext.	D.T.	Int.	Ext.	D.T.	Int.	Ext.	D.T.	Int.	Ext.	D.T.
Opel Astra	0.03	0.028	25	0.04	0.05	24.7	*	0.03	9.6	*	0.07	9.6
VW Passat	0.015	0.03	24.2	0.01	0.028	24.8	*	0.04	12.4	*	0.05	12.6
Škoda Octavia	0.11	0.09	19.4	0.67	0.19	19.9	*	0.025	9.1	*	0.04	9.4

where: Int. = interior surface of the brake disc; Ext. = exterior surface of the brake disc; D.T. = brake disc thickness.

Comparative results concerning the ovality measurement on the roller tester (rectangles) and lateral run-out measurement using dial gauge (circles) are shown in Figure 2.

Analyzing the results, it is observed that it cannot be establish a correlation between the ovality value obtained by using the roller tester and the lateral run-out measured using a dial gauge. This is because of differences between the influences acting in these two methods. Measuring the ovality using roller tester is an indirect method based on the braking forces measurement, while measuring the lateral run-out using the dial gauge is a direct, high precision action.

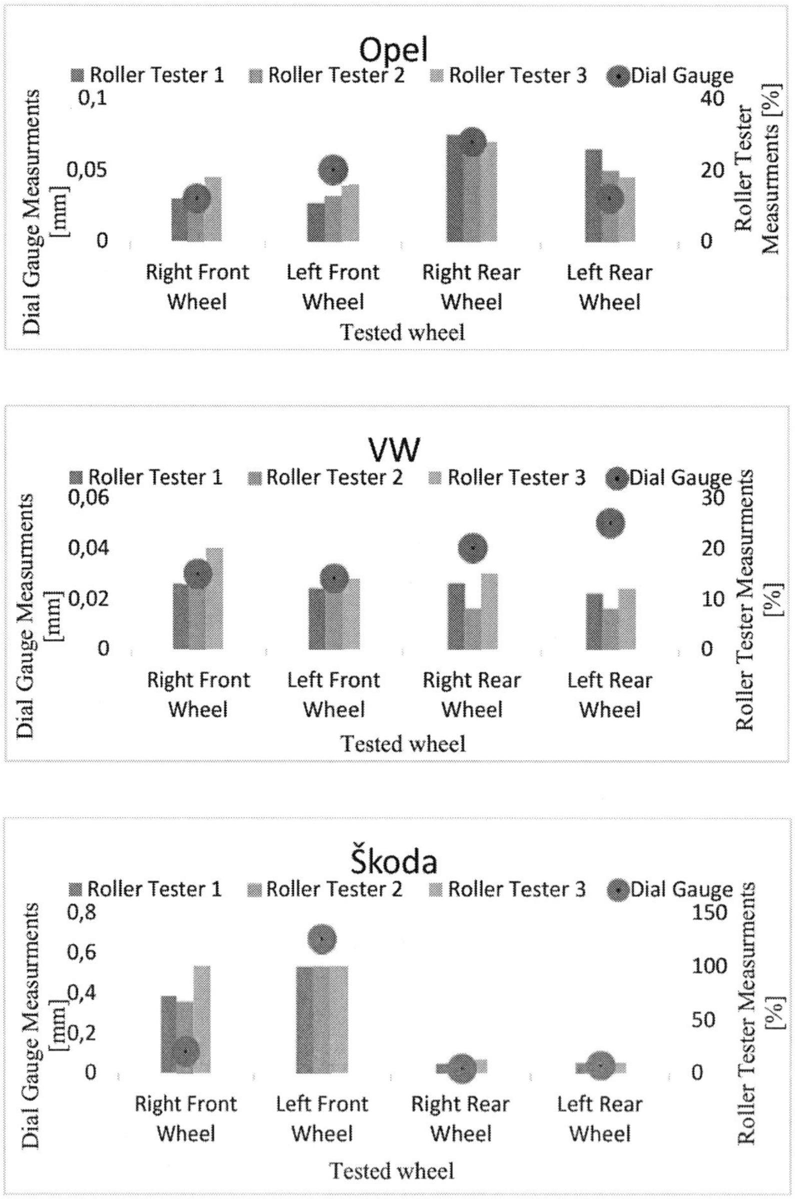

Fig. 2. Comparative results of lateral run-out using roller tester and dial gauge.

Disc replacement imposed by an inadequate value of the lateral run-out is decided by using a dial gauge. Most of the times, the roller tester measurement grabs attention and suggests the direct measurement using dial gauge. Identification by the driver of an inappropriate behaviour of the car is a subjective action and depends on the readiness and understanding of phenomena that occur in the braking process of the wheel.

4 Conclusions

Analyzing the results, it is observed that it cannot be establish a correlation between the ovality value obtained by using the roller tester and the lateral run-out measured using a dial gauge. The variations of the braking force noticed during the test on a roller stand (which defines the value of ovality) may be produced not only by the run-out of the brake disc, but also by other causes that can modify the friction coefficient, such as damaged friction surfaces, hesitation in caliper movement, seizing of piston in the wheel braking-cylinder, etc.

These influences are more important when the car is operated with large periods of immobilization which can cause deposits on the brake discs (areas heavily oxidized, uneven sliding pistons brake).

However, if the acceptable limit of ovality is overpassed, a direct check of the brake disc run-out must be complete using the dial gauge and the final decision of replacing the brake disc may be only taken on the basis of this test.

References

1. T. Jearsiripongkul, P. Hagedorn, KKU Engineering Journal, **31**, 299 (2004)
2. C. Andreescu, M. Oprean, Gr. Danciu, G. Anghelache, *Automobile Diagnosis - Practical Activities*, (PRINTECH Publishing House, 2002)
3. *** Order of the Romanian Transport Ministry No. 2133/2005– RNTR 1
4. S.H. Masood, L. Singh, *5th Australasian Congress on Applied Mechanics, (*2007)
5. *** Regulation No. 90 of the Economic Commission for Europe of the United Nations (UN/ECE)
6. ***Technical Specifications: Fiat Tipo, Jeep Cherokee, Grand Cherokee, Wrangler, Renegate, Lancia Grand Voyager, VW Passat
7. *** Technical Specifications of BOSCH Brake Stand, Model BSA/SDL43XX
8. M. Stratulat, C. Andreescu, *Automobile Diagnosis*, (Societatea Ştiinţă şi Tehnică S.A., 1998)
9. *** Vehicle registration certificate of VW Passat, 2007
10. *** Vehicle registration certificate of Škoda Octavia 2008
11. *** Vehicle registration certificate of Opel Astra 2007
12. *** Maintenance and repair manuals of Opel Astra, Škoda Octavia, VW Passat 2007

Parametric study of the distribution of longitudinal dynamic forces developed in the train body using hysteretic characteristics of Ringfeder buffers

Camil Crăciun*, *Mădălina* Dumitriu, and *Cătălin* Cruceanu

University Politehnica of Bucharest, Department of Railway Vehicles, 313 Splaiul Independentei, Romania

Abstract. During braking, due to the interaction between the vehicles, longitudinal dynamic forces, characterized by different values and an irregular distribution along the train, develop in the train body, affecting the passengers` comfort or even the traffic safety. The paper is a parametric study having as objective to determine the influence of the number of coaches, vehicle weight and delay time of the development of braking forces on the size and distribution of the longitudinal dynamic forces developed in passenger trains during braking. The first analysis is based on the change in the number of vehicles which are part of the train. The second case analyzed refers to the total mass of the vehicles, which increases by 50% compared to standard train mass. In the third situation, there are taken into account variations of ± 20% of the length of the train overall pipe that are able to generate delays related to the occurrence and development of the braking forces at each vehicle of the train body.

1 Introduction

In exploitation, the appearance and development of longitudinal dynamic forces are due primarily to the principle of operation of the basic brake, exploitation characteristics of the vehicles forming the train, its length, but especially to the characteristics of buffer and draw-gear devices that equip these vehicles.

Taking into consideration the way the basic brake works (indirect compressed air brake) [1-4], the appearance of longitudinal dynamic forces is influenced by the time when the distributors that equip each rail road vehicle come into action, being commanded by pressure drop along the general pipeline and the propagation speed of the brake wave. Distributors are devices that control the pressure in the brake cylinders according to the lower of the pressure in the general pipe of the train, so, the size of the brake force for each train vehicle. Thus, in the train body, at a certain time, vehicles located in the first half begin to brake, while those from the back of the train do not have the brake force or they have just begun to develop it. This causes collisions between vehicles in the train (longitudinal dynamic forces). Another

* Corresponding author: craciun_camil@yahoo.com

aspect that cannot be neglected is the length of the train that influences, on the one hand, the general pipe length with implications on the delay time in the development of brake force and, on the other hand, it contributes to increase the train weight with influences on the maximum values of the longitudinal forces. The effect of the delayed development of brake forces on the longitudinal dynamic behaviour of the freight trains is approached by Nasr and Mohamadi in the work [5].

The collision devices used in rail vehicles are designed to protect the structure of the vehicles, to diminish the forces between them during braking and to dissipate a part of the energy during collisions [6]. Characteristics and the constructive type influence the development and/or reduction of the longitudinal dynamic reactions with profound implications on the running stability of the vehicles [7]. Also, to study the longitudinal dynamic of the trains in brake, equipped with manual hook-screw-type coupling, the elastic characteristics of these devices must be known.

In this paper, the study of the distribution of longitudinal dynamic forces (forces of compression and tensile) in the body of passenger trains found in brake is done by changing three parameters: the length of the train – by increasing or decreasing the number of wagons (*Nw*), the train weight (*Wt*) –increasing it by 50% and the change of the braking delay time (*Td*) by 20%, with the main purpose to highlight the influences of longitudinal forces.

2 Dynamic System

2.1 Vehicle model

Knowing that the train consists of a locomotive and a finite number of coaches, the model used in the software calculation consists of *n* rigid bodies representing the vehicles entering in the train composition, connected by elastic and damping elements representing buffer and draw-gear devices.

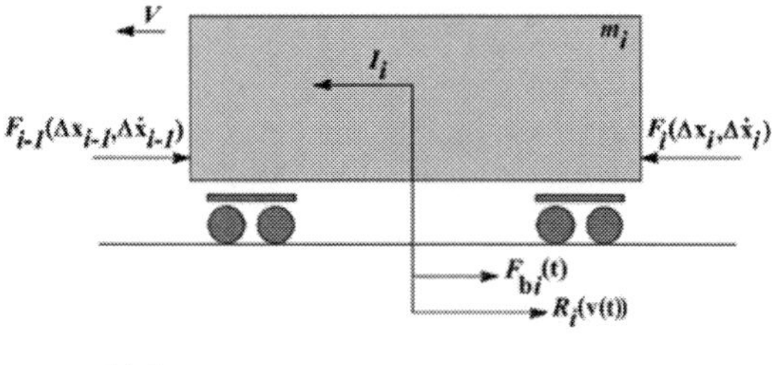

Fig. 1. Forces acting on a vehicle.

At the level of a vehicle in the train body (excepting the first and the last vehicle) the forces that act are: forces of inertia Ii, braking forces $F_{bi}(t)$, resistances to march $R_i(v(t))$ and the forces from buffer and draw-gear devices (from the front vehicle $F_{i-1}(\Delta x_{i-1},\Delta\dot{x}_{i-1})$ and from the rear vehicle $F_i(\Delta x_i,\Delta\dot{x}_i)$), for i = 1÷ n, where $\Delta x_i = x_i - x_{i+1}$ represents the relative movement from the buffer and draw-gear devices (see Fig. 1).

For the first vehicle there are not forces on buffer and draw-gear devices found in the front of the train, but only on those who come into contact with the next vehicle in the train and at the last vehicle, there are not forces on buffer and draw-gear devices from the back of the train.

Thus, applying the laws of the mechanics, we obtain *n – 1* nonlinear equations, each describing the movement between two consecutive vehicles.

 DOI: 10.1051/matecconf/201711207011

For a group i of buffer and draw-gear devices (coaches i and $i+1$) the relation that characterize the movement is [8]:

$$\ddot{y}_i = \frac{F_i(y_i,\dot{y}_i) - F_{i+1}(y_{i+1},\dot{y}_{i+1}) + F_{f,i+1}(t) + R_{i+1}(v(t))}{m_{i+1}} + \\ + \frac{F_i(y_i,\dot{y}_i) - F_{i-1}(y_{i-1},\dot{y}_{i-1}) - F_{f,i}(t) - R_i(v(t))}{m_i} \tag{1}$$

It is specified that, in order to simplify the above equations, the notation $y_i = \Delta x_i$ was introduced.

2.2 Model of buffer and draw-gear devices

Mathematical modeling of buffer and draw-gear devices has as main aim the achievement of a functional mathematical model with the help of which to be able to properly evaluate the progress and the size of longitudinal dynamic reactions in the train body during braking.

Consequently, the calculation of the forces for the traction, collision and coupling devices follows the below equation [9, 10]:

$$F(x,\dot{x}) = \begin{cases} k_e \cdot x + k_f \cdot |x| \cdot \tanh(u \cdot \dot{x}) & for \quad x < 0, \\ 0 & for \quad x = 0, \\ k_{ec} \cdot x + k_{fc} \cdot |x| \cdot \tanh(u \cdot \dot{x}) & for \quad x > 0 \end{cases} \tag{2}$$

where x is the stroke and $\dot{x}$ the speed of such devices, k_e and k_f - specific constants for elastic and friction forces that develop in the collision device, k_{ec} and k_{fc} represent specific constants for elastic and friction forces that develop in the traction device, and u is a scaling factor.

2.3 Calculation model of the braking force

To calculate the maximum braking force developed by each vehicle, putting the condition to avoid, under normal circumstances, the lock of the axles during braking actions [7 - 11] it results that the adhesion force between the wheel-rail should not be exceeded, thus:

$$F_{f,i} \le F_a = \mu_{a,i} \cdot m_i \cdot g \tag{3}$$

where μ_a represents the wheel-rail adhesion coefficient, and m_i represents the weight of each vehicle in the train body.

Taking into account the evolution of the pressure from the brake cylinder in addition to the maximum pressure stabilized in it, we can determine the brake force thus [8, 11]:

$$F_{f,i}(t) = \mu_a \cdot m_i \cdot g \cdot \frac{p_{cf,i}(t)}{p_{cf\max}} \tag{4}$$

In the previous relation, $p_{cfi}(t)$ represents the evolution in time of the pressure in the brake cylinder, and p_{cfmax} is the maximum stabilized pressure.

2.4 Application

In order to point out the influence of parameters *Nw*, *Wt* and *Td* regarding the evolution of longitudinal dynamic forces, as well as the distribution of maximum compression and stretching forces that appear in the buffer and draw-gear devices, the programme created in Matlab was run [8 - 10] where the data from Table 1 were used. It should be specified that the train undergoes emergency braking from top speed of 180 km/h on a railway in alignment and landing so that the only influences that lead to differentiation of results should be introduced only by required parameters.

In the study it is considered that the evolution of pressure in the brake cylinders of all vehicles entering the train composition is made by the same feature and filling time.

Table 1. Numerical model parameters.

The main parameters	Values
The locomotive weight	$m_l = 120$ t
A wagon weight	$m_v = 40 \dots 60$ t
Constant specific to collision devices	$k_e = 2.8 \cdot 10^6$ N/m
	$k_f = 1.4 \cdot 10^6$ N/m
Constants specific to traction and binding devices	$k_{ec} = 5.46 \cdot 10^6$ N/m
	$k_{fc} = 2.43 \cdot 10^6$ N/m
Maximum pressure in the brake cylinder	p_{cfmax}=3.8 bar
Wheel-rail adhesion coefficient	μ_a=0.1

The value of the pressure developed in the brake cylinder necessary to calculate the brake forces was experimentally determined in laboratory specialized stand, using distributors from passenger vehicles used in mining.

3 Results and discussions

In order to start, it is analyzed the influence of *Nw* parameter on the distribution of longitudinal forces of compression and tensile in the train body. The results are shown in Figures 2 and 3.

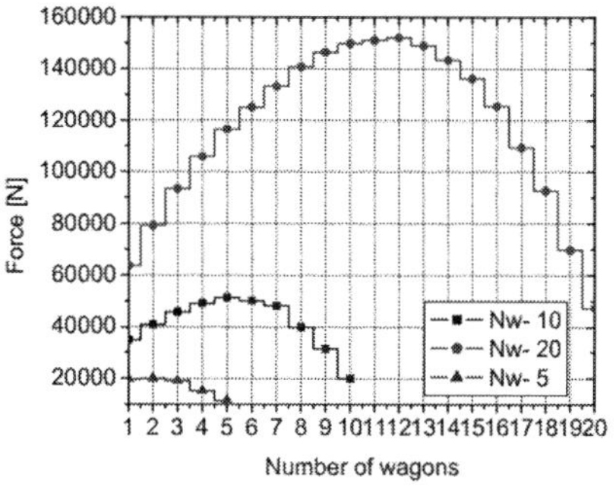

Fig. 2. Distribution of compression forces according to *Nw* variation.

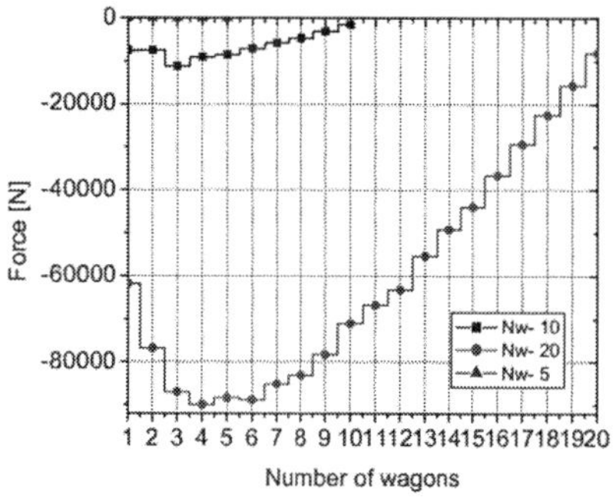

Fig. 3. Distribution of tensile forces according to *Nw* variation.

Doubling the number of coaches, starting from the model train composed often wagons and an engine [8], leads to triple compression forces (see Fig 2). The maximum force has a displacement towards the second half of the train (coupling nr.12) in relation to the standard train, where the maximum force is on coupling no. 5 at the half of the train. The tensile forces grow very much highlighting the phenomenon of recoil after equalizing the brake forces on the vehicles (see. Fig. 3).

Reducing *Nw* to half shows a significant decrease of compression forces by about 60%, totally modifying the allure of distribution curve. The maximum compression in this case is registered on the first two collision devices. Tensile forces, in this case, are practical void the train shows no rebound (Fig.3).

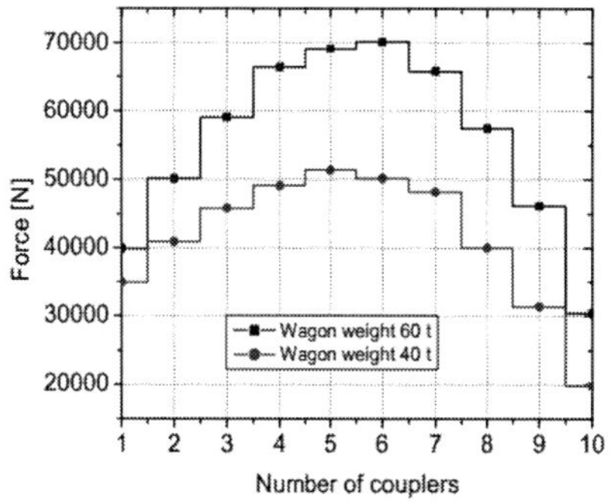

Fig. 4. Distribution of compression forces to increase the weight of a wagon by 50%.

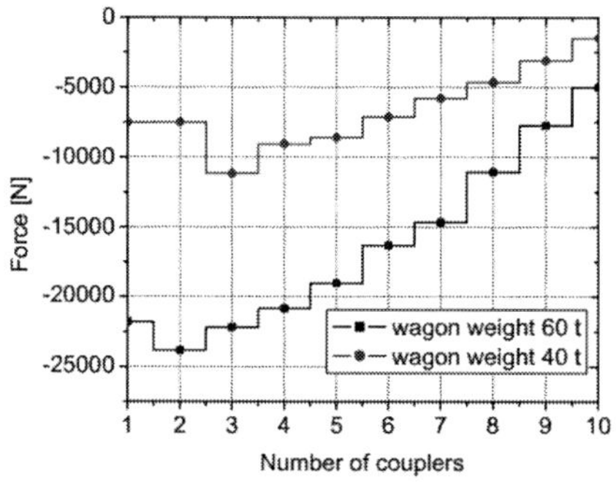

Fig. 5. Distribution of tensile forces to increase the weight of a wagon by 50%.

To observe the influence of the weight on the size and distribution of longitudinal dynamic forces, the study is done on the standard train, changing the weights of all contained wagons. So, a weight increase by 50% leads to marked increase of the compression forces (Fig. 4) approximately by 40% and of tensile forces for about 2.2 times (Fig. 5).

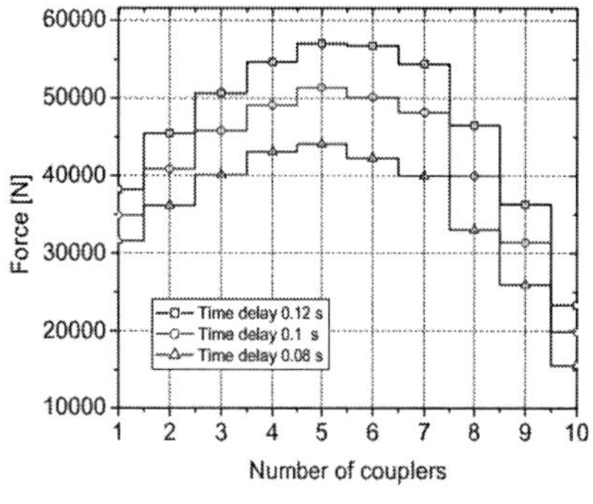

Fig. 6. Distribution of compression forces according to the delay time.

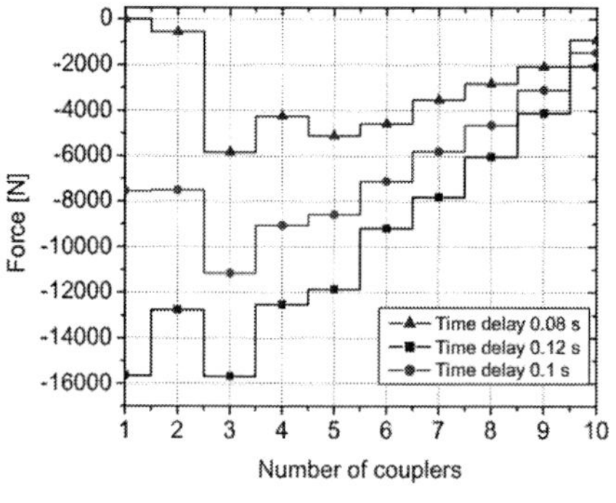

Fig. 7. Distribution of tensile forces according to the delay time.

Changing the delay time regarding the evolution of brake forces directly influences the size of the longitudinal dynamic forces (Fig. 6 and 7). Interestingly, though the values of tensile or compression forces grow depending on the value of *Td*, the distribution of these forces remains practically unchanged. If *Td* reduces, the brake forces between the vehicles of the train equalize faster due to the filling time of the brake cylinder and the forces between the vehicles are small. Increasing the delay time, the time increases until all forces between the vehicles equalize, so that, at a certain time, because of the differences of high brake forces, longitudinal dynamic forces (of compression or tensile) increase.

Conclusions

Longitudinal dynamic forces, that develop in the draw-gear devices when the train brakes, represent the consequence of the successive entry into action of the brakes of each vehicle of the train.

As outlined in the paper, the size of these forces and their distribution are influenced by many parameters, the current study being focused on three of them, namely: the train length, towed weight and the delay time in braking.

Increase or decrease of number of wagons (*Nw*) leads to increases or reductions in the values of the longitudinal forces, but essential changes, as expected, appear to the distribution of these forces referred to the length of the train. Train weight (*Wt*) contributes as long as the vehicles are identical and equal loads only on the values of longitudinal forces.

Variations of the braking delay time (*Td)* parameter change the distribution of tensile forces during stretching, the compression forces having the same distribution, recording only increases or decreases depending on how *Td* changed. Therefore, if a train is made up of vehicles whose lengths vary by 20%, then the forces developed between the vehicles of the train would behave totally differently compared to the train made up of wagons with identical lengths.

This work has been funded by University Politehnica of Bucharest, through the "Excellence Research Grants" Program, UPB – GEX - 2016. Research project title: *Research on developing mechanical and numerical models for the virtual evaluation of the dynamic performances in the railway vehicles* (in Romanian). Contract number: 48/26.09.2016.

References

1. L. Pugi, D. Fioravanti & L. Rindi, *12th IFToMM World Congress*, (2007)
2. P. Belforte, F. Cheli, G. Diana, S. Melzi, Veh. Syst. Dyn., 19 (2008)
3. R. C. Sharma, M. Dhingra, R. K. Pathak, Int. J. Eng. Res & Tech (IJERT), **4**, 6 (2015)
4. C. Cole *Longitudinal train dynamics, chapter 9 in Handbook of Railway Vehicle Dynamics* (Ed. Taylor & Francis Grup, 2006)
5. A. Nasr, S. Mohammadi, *Proc. ImechE,* J. Rail and Rapid Transit, **224,** 11 (2010)
6. W. Gamon, The Archives of transport, **33**, 8 (2015)
7. C. Cruceanu, *Brakes for rail vehicles* (in Romanian) (Ed. MatrixRom, Bucharest, 2009)
8. C. Crăciun, M. Dumitriu, *Appl. Mech. Mater.*, **809-810**, 6 (2015)
9. C. Crăciun, T. Mazilu, Ann. Fac. Eng. Hunedoara – Int. J. Eng., **12**, 8 (2014)
10. T. Mazilu, C. Crăciun, Ann. Fac. Eng. Hunedoara – Int. J. Eng., **13**, 6 (2015)
11. C. Cruceanu, *Train Braking, chapter in Reliability and Safety in Railway* (Ed. Xavier Perpinya, InTech, 2012)

About the influence of wheel slide protection devices action on longitudinal dynamic of trains

Cătălin Cruceanu*, *Mădălina* Dumitriu, and *Camil* Crăciun

University Politehnica of Bucharest, Department of Railway Vehicles, 313 Splaiul Independenţei, 060042, Bucharest, Romania

Abstract. Wheel slide protection devices (WSPD) are destined to avoid important sliding of wheels during braking actions in case of temporarily impaired wheel-rail adhesion by correspondent reductions of air pressure in brake cylinders. Accordingly, longer braking distances and higher longitudinal in-train forces occur, potentially affecting the traffic safety and the passengers comfort. A general analyse and evaluation of these effects are the main targets of the present study. The theoretical considerations are sustained by simulations of braking process, considering situations of normal, respectively diminished wheel-rail adhesion determining random actuations of WSPD. The air pressure evolution in brake cylinders of was experimentally determined on a computerized brake system test stand and adequately used as input in simulations. The results indicate an increase of in-train forces and a more complex evolution of the longitudinal dynamic actions between the vehicles in the case of degraded wheel-rail adhesion, when the random actuation of WSPD have major influence in the whole braking process.

1 Introduction

Train braking is a complex process, influenced by numerous operational, technical and constructive factors. Specific aspects contribute to generate in-train forces that, in certain situations, may affect the train integrity and safety. Hence, it is to notice a growing interest on longitudinal dynamics of trains with the aim to define appropriate recommendations regarding both operating and design of railway vehicles, in order to recommend improvements of safety and energy efficiency of this transport mode, e.g. [1-5]. Most of the studies are focused on long heavy freight trains featured with brake blocks and UIC brake system, operating in slow-action mode. In comparison, passenger trains are shorter, lighter and more uniform in consist. Destined to run with higher speeds, the braking system require more sophisticated equipment, determining particular braking characteristics that consequently have major influences on the longitudinal dynamics of trains.

For railway transport, the deceleration ability is limited by the adhesion between wheel and rail. As an open system, the wheel–rail contact is exposed to environmental factors and contaminants that randomly may result in lowering in a very large extent the adhesion on the railhead. Braking actions on slippery track increase the braking distance and the blockage or

* Corresponding author: c_cruceanu@yahoo.com

partial blockage of wheelset conducts to wheel flats, affecting the safety of the traffic, the vehicle, the track and the passengers comfort [6-7].

In order to provide high braking power to satisfy the safety and operational requests, for passenger vehicles, systems are usually designed at the limits of wheel-rail adhesion. This results in a pronounced sensitivity of the braking system to degraded adhesion [8]. Consequently, all passenger vehicles are featured with WSPD that, in case of important sliding, rapidly diminish the air pressure in brake cylinders until the affected wheelset returns to normal rotational speed according to the translational speed of the vehicle.

In such situations, important random variations of the vehicles' braking forces affect the whole longitudinal dynamics of the train. Supplementary longitudinal oscillations are generated in the train body, while the magnitude, time evolution and disposition of compressive and tensile forces between adjoined vehicles are completely modified.

The main aim of the study is to evaluate in certain particular situations the effects of WPSD during braking in low adhesion conditions on the in-train forces. The approach towards the subject is by simulations, using an original program in which measured air pressure variations in brake cylinder while actuating a mechanical WPSD are used as input.

While, to the authors' knowledge, no such studies have been published before, the investigations are focused on a very simple composition of the train and the simulation program involves certain assumptions to avoid, as much as possible, the interference of other parameters in the results obtained.

2 Theoretical aspects

Mechanical and pneumatic modelling of trains has been discussed in an extensive of publications, see for instance [1-5, 8-12].

In the present work, the train is classically considered an elastic-damped lumped system of n individual rigid masses m_i representing each vehicle, connected by couplings denoting well defined elastic and damping equivalent characteristics (Fig. 1).

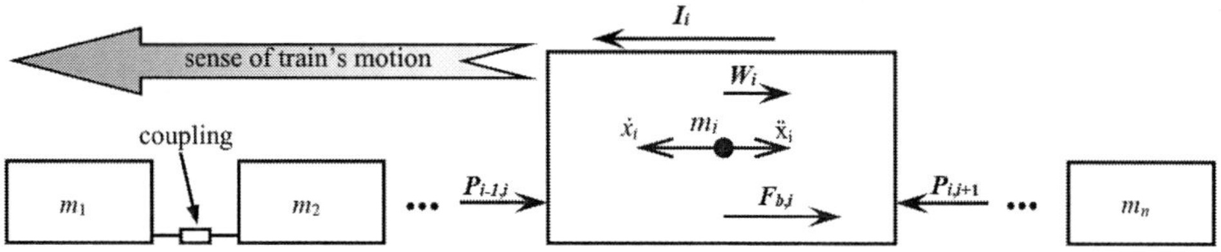

Fig. 1. Mechanical model of the train.

Assuming a track with no curves or slopes and considering the multibody formulation, the displacement x_i of the generalized "i"-th vehicle of the train is governed by the following forces: the braking forces $F_{b,i}$ the resistances W_i, the in-train forces $P_{i-1,i}$ and $P_{i,i+1}$ acting in the connections with neighboured vehicles and the inertial forces I_i (see Fig. 1). The equation of motion is:

$$m_i \cdot \ddot{x}_i + P_{i,i+1} - F_{b,i} - W_i - P_{i-1,i} = 0 \tag{1}$$

Applied to the vehicles of the train, considering $i \in [1, n]$ and attaching the boundary conditions $P_{0,1} = P_{n,n+1} = 0$, eq. (1) defines the differential equation system of second order describing the movements of the system presented in Fig. 1.

In the present study the forces in couplers were evaluated assuming an algebraic sum of elastic and friction forces, considering RINGFEDER-type buffers built in Romania by ICPVA-SA featuring trailed passenger vehicles. There were determined [11]:

$$P_{i,j}(\Delta x_{i,j},\Delta\dot{x}_{i,j}) = \begin{cases} k_b \cdot \Delta x_{i,j} + c_b \cdot |\Delta x_{i,j}| \cdot \tanh(u \cdot \Delta\dot{x}_{i,j}), \text{ for } \Delta x_{i,j} < 0; \\ 0, \text{ if } \Delta x_{i,j} < 0; \\ k_t \cdot \Delta x_{i,j} + c_t \cdot |\Delta x_{i,j}| \cdot \tanh(u \cdot \Delta\dot{x}_{i,j}), \text{ for } \Delta x_{i,j} > 0 \end{cases} \quad (2)$$

In eq. (2) k and c are equivalent constants depending on the elastic, respectively friction characteristics between the metallic rings inside, indexes "b" and "t" are referring to buffer and traction devices, $\Delta x_{i,j}$ is the stroke, $\Delta\dot{x}_{i,j}$ the relative speed for the considered element of the couplers between i and j adjacent vehicles of the train, $u >> 1$ is a scaling factor.

Considering the vehicles designed for a specific running speed $V_{max} \geq 160$ km/h equipped with disc brakes, the brake force is described accounting the dependency on instantaneous air pressure in brake cylinders p_{BC} and the wheel-rail adhesion coefficient in normal conditions [8, 13]:

$$F_{b,i}(t) = \begin{cases} 0, \text{ if } p_{BC,i}(t) < 0.4 \text{ bar, else} \\ \dfrac{p_{BC,i}(t)}{p_{BC,max}} \cdot m_i \cdot g \cdot \dfrac{0.33}{1+0.011 \cdot V_{max}} \end{cases} \quad (3)$$

In eq. (3), $p_{BC,max}$ is the maximum pressure in brake cylinders for emergency braking and $g = 9.81$ m/s^2 is the gravitational acceleration.

Regarding the pneumatic processes, the main aspects involved were approached as follows: an average transmission speed for brake actuation of 250 m/s in long of the train and experimentally determined data, adequately implemented into the simulation program, for the air pressure evolution in the brake cylinders.

The operating principle of WSPD is illustrated in Fig. 2. Once the braking force become greater than the wheel-rail adhesion force (*1*), important sliding occurs. The system experience differences between the translational speed of vehicle and the peripheral speed of wheels. When a certain threshold value is exceeded (*1'*), the air from the brake cylinders is rapidly evacuated (*1''*), resulting a decrease in braking force. When becomes lower than the actual exploitable adhesion (*2*), the wheelset accelerates to return to normal rotational speed (*2'*). This process engenders for the air pressure in brake cylinders (*2''*) a prompt return to the initially commanded level, unless no other overbraking tendency.

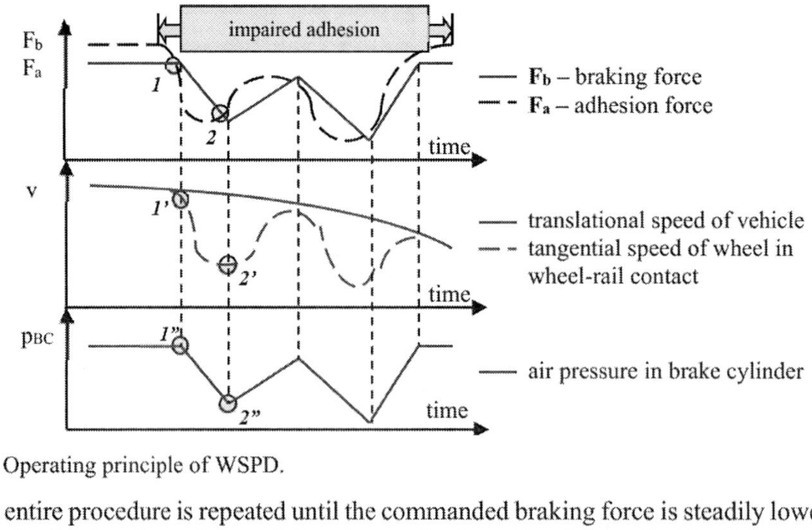

Fig. 2. Operating principle of WSPD.

The entire procedure is repeated until the commanded braking force is steadily lower than wheel-rail adhesion force, i.e. until the slippery track section is travelled by the vehicle. As parts of a train, the vehicles reach successively such areas and the random specific operation of each WSPD conduct to important instantaneous differences in braking forces. These definitely influence the longitudinal dynamics of the whole train.

3 Input data and assumptions

In order to get an image on the effects of WSPD actions on in-train forces in case of low wheel-rail adhesion, several simulations were performed. As a primary stage in our studies, certain constraints and simplifying hypotheses were assumed. The most relevant are outlined as follows: short passenger train of four identical vehicles equipped with disc brakes, KE 1 air distributor and mechanical WSPD type M2; vehicles have classical couplers consisting of RINGFEDER type buffers and traction devices; screw couplings tightened up, with no clearance between the buffer discs; the train is submitted to an emergency braking performed at 160 km/h on a straight track with no slopes; the vehicles reach successively the low adhesion area, the moment of WSPD actuation depending on the running speed and position of each vehicle in the train composition; identical filling characteristics of brake cylinders; identical response of vehicles' WSPD after actuation.

The main input data are presented in table 1.

Table 1. Main input data.

Vehicle weight [t]	50	
Vehicle length [m]	25	
Specific constants for couplings in eq. (2) [N/m]	$k_b = 2.8 \cdot 10^6$	$c_b = 1.4 \cdot 10^6$
	$k_t = 5.46 \cdot 10^6$	$c_t = 2.43 \cdot 10^6$
Maximum brake force/vehicle [kN]	58.65	
Maximum air pressure in brake cylinders [bar]	3.837	
Filling time of brake cylinders [s]	3.4	

The air pressure evolutions in brake cylinders for emergency braking, with and without WSPD actuation, were experimentally determined on the computerised system in our Brakes Laboratory. Detailed information about it and equipment can be found in [14]. The simulation program [10-11] was improved by adequately implementation of the acquisitioned pressure data. It was considered a length of about 205 m for the impaired adhesion section, after reaching the maximum pressure in brake cylinders (Fig. 3).

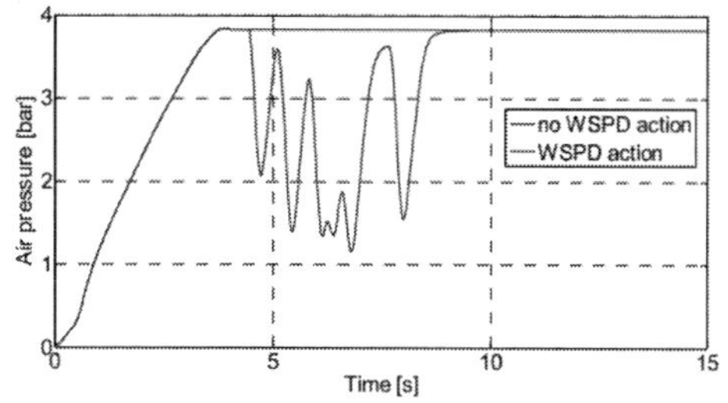

Fig. 3. Experimentally determined air pressure evolutions in brake cylinders, emergency braking.

For the sake of illustration, it is to notice that the stopping distances resulted (following [13]) by simulations for individual vehicle are 915 m, respectively 988 m.

4 Results, comments and conclusions

The main outputs achieved by simulations are the maximum values of buff and draft forces in couplers (see table 2) and the diagrams presenting time evolution of in-train forces. In Fig.

4 are presented some of the most representative: case 0000 (no WSPD action) – considered as baseline, cases 1101 and 0111 (maximum tensile, respectively compression forces between adjacent vehicles) and case 1111 (all vehicles affected by low adhesion).

Table 2. Main input data.

	Maximum buff forces [kN]			Maximum draft forces [kN]		
case*/coupler	1-2	2-3	3-4	1-1	2-3	3-4
0000	*8.2*	*13.0*	*15.3*	*-2.0*	*-3.7*	*-4.1*
1000	14.1	13.0	15.5	-39.1	-38.9	-35.2
0100	19.1	13.0	15.3	-7.4	-26.5	-23.5
0010	18.0	23.2	15.3	-10.4	-12.1	-20.1
0001	29.0	36.0	32.1	-14.3	-26.8	-24.8
1100	23.6	13.0	15.3	-50.0	-45.1	-23.9
1010	25.4	35.2	15.3	-46.4	-40.5	-37.2
1001	32.7	44.0	44.4	-44.6	-36.8	-29.3
0110	35.2	34.7	15.3	-16.7	-36.0	-33.2
0101	37.5	36.5	38.8	-20.8	-29.4	-30.6
0011	24.2	49.1	38.8	-17.5	-17.9	-39.1
1110	27.6	19.8	15.3	-57.9	-54.9	-27.3
1101	33.4	37.9	38.3	**-59.7**	-48.8	-38.6
1011	42.6	51.9	31.9	-45.1	-40.5	-33.6
0111	52.1	**52.6**	39.7	-22.3	-32.6	-22.5
1111	48.7	47.4	37.1	-57.9	-53.7	-29.6
* 0 – vehicle with no WSPD action during braking; 1 – vehicle with WSPD actuation						

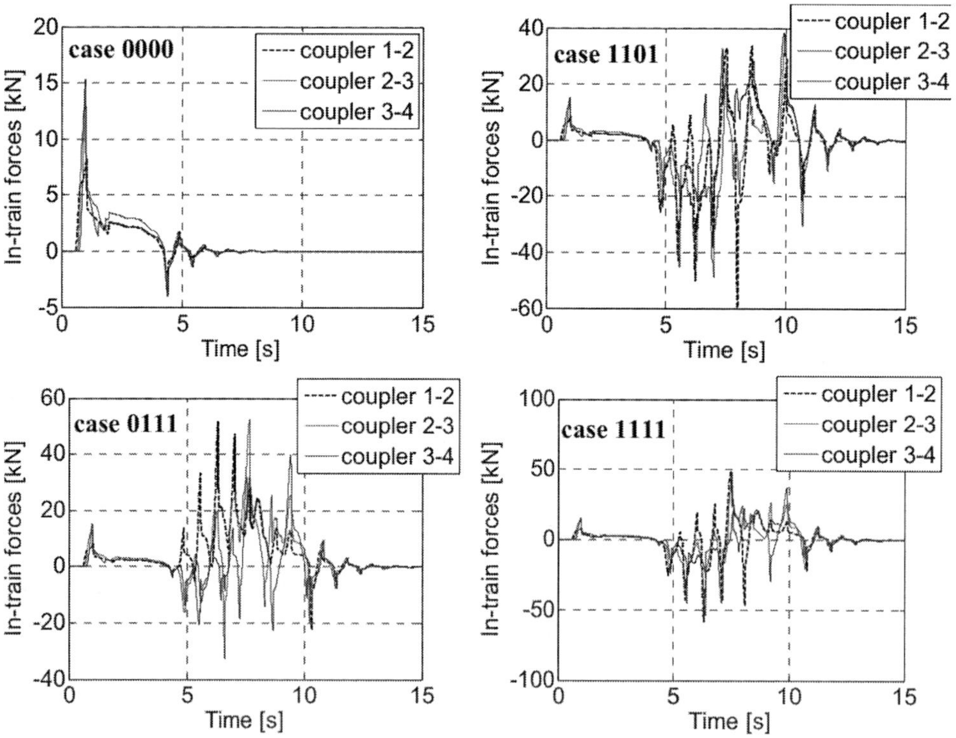

Fig. 4. Time history of in-train forces evolution (buff forces positives, draft forces negatives).

It can be seen that during the first 4.44 s (until the train reaches the low adhesion sector) the braking process unfolds identically in all cases. Beyond this point, the diagrams analyse highlight both substantial modification of the whole longitudinal dynamics of the train and important increase of in-train forces. In the presented case study, considering the particular situations and assumptions presented in section 3, whenever WSPD actions are involved, the magnitude of buff and draft forces increase. These growths can exceed in certain cases more than 3, respectively 14 times the dynamic longitudinal forces between adjoined vehicles in train resulted in the baseline case (good wheel-rail adhesion, no WSPD actuations). Even so, these forces are low enough as not to endanger the traffic safety.

It is also interesting to notice that the highest levels of in-train forces do not correspond to the case of actuation for all WSPD, but on only three vehicles of the train.

One still may note that for simulations was selected a mechanical type of WSPD, known to operate simultaneous on all wheelsets of the vehicle and to engender high variations of pressure in brake cylinders. In case of featuring the vehicles with electronic WSPD, lower effects are expected on in-train forces in the same adhesion conditions.

In conclusion, WSPD are necessary for the traffic safety and mandatory for passenger and traction rail vehicles, requirement imposed by international regulations. Our study highlights certain complementary aspects on the complexity of the train braking process. The actuations of WSPD during braking actions performed in low adhesion conditions engender longer stopping distances, considerable modifications in longitudinal dynamics of the train and significant increase of in-train forces. Obviously, for a more complete picture of effects, more work is required but, at the moment, our approach represents at least a starting point and a source of interesting aspects in passenger trains braking studies.

This work has been funded by University Politehnica of Bucharest, through the "Excellence Research Grants" Program, UPB – GEX - 2016. Research project title: *Research on developing mechanical and numerical models for the virtual evaluation of the dynamic performances in the railway vehicles* (in Romanian). Contract number: 48/26.09.2016.

References

1. C. Cole, Longitudinal train dynamics, in *Handbook of railway vehicle dynamics*, S. Iwnicki (Ed), Taylor & Francis, 239–278 (2006)
2. Q. Wu, S. Luo, C. Cole, J. Mod. Transport. **22**, 3, 127–136 (2014)
3. L. Pugi, D. Fioravanti, A. Rindi, 12th IFToMM World Congress, (2007)
4. A. Nasr, S. Mohammaddi, Proc. IMechE, **224**, Part F: JRRT, 523-534 (2010)
5. M.J. Arsahd, Attic-ur-Rehman, A.I. Durrani, H. Ahamad, JFET, **20**, 1, 38-53 (2013)
6. O. Arias-Cuevas, *Low adhesion in the wheel-rail contact*, (Doctoral Thesis, Delft, 2010)
7. Y. Zhu, *Adhesion in the wheel-rail contact*, (Doctoral Thesis, Stockholm, 2013)
8. Q. Wu, C. Cole, S. Luo, M. Spiryagina, Veh Syst Dyn **52**, 6, 733–758 (2014)
9. C. Cruceanu, Train braking, in *Reliability and Safety in Railway*, X. Perpinya (Ed), InTech, 29-74 (2012)
10. C. Cruceanu, R. Oprea et al., *Proc. of Int. Conf. on Computer Science WORLDCOMP'09*, 58–64 (2009)
11. C. Crăciun, T. Mazilu, Ann. Fac. Eng. Hunedoara – Int. J. Eng., **12**, 3, 19-26, (2014)
12. C. Cruceanu, C. Crăciun, AMM, **378**, Trans Tech Publications, 74-81 (2013)
13. C. Cruceanu, C. I. Crăciun, I. C. Cruceanu, AMM, **809-810**, 1085-1090 (2015)
14. C. Crăciun, C. Cruceanu, *Brakes for railway vehicles – Applications* (in Romanian), Matrixrom (2011)

On the stick-slip vibration in the suspension of a freight wagon

Traian Mazilu[1,*], *Vlăduţ Marian* Dinu[2]

[1]University Politehnica of Bucharest, Department of Railway Vehicles, Splaiul Independenţei 313, Bucharest, Romania
[2]S.N.T.F.C. 'CFR CĂLĂTORI' S.A. Blvd. Dinicu Golescu 38, Bucharest, Romania

Abstract. Damping based on the dry friction is frequently in suspensions of the freight wagons. The paper presents some aspects of vertical vibration of freight wagons. It uses a simple model considering that the vehicle was reduced to a mass suspended on a wheel, moving at constant speed on a rigid track with harmonic irregularity. Stick-slip vibration can occur due to the friction and it is characterized by sudden changes in the wheel acceleration affecting the ride quality. The paper shows the influence of stick-slip vibration on the wheel-rail dynamic force.

1 Introduction

Railway vehicle suspension is equipped in many cases with damping elements based on the dry friction [1]. We refer mainly to the suspension of freight wagons which almost all are fitted with such damping elements: leaf springs (two-axle wagons or the bogies HOURS or H), Lenoir devices (Y 25 bogie wagons) and damping with friction wedges (Diamond bogie wagons).

The presence in the suspension systems of the damping with dry friction is likely to cause vibration stick-slip which could affect the ride quality and induce excessive wheel-rail contact forces which result in reducing the fatigue strength of both running gear of the vehicle and track superstructure, as well as increased wear of the wheel/rail rolling surfaces.

Modelling dry friction damping elements is mainly based on the model of Coulomb friction [2]. Friction does not depend on sliding speed, but has its opposite. Coulomb's model can be improved to a certain extent admitting as in reality that static friction is greater than sliding friction.

Mathematical model of the Coulomb friction uses the signum function. Although it is very simple, this model raises serious numerical integration problems when used for modelling the suspension systems of the vehicles. It's mainly the so-called numerical instability whose effect consists of high-frequency vibration that do not has physical correspondent. From the mathematical point of view, this numerical instability is given by nature signum function which is discontinuous at zero crossing. To overcome this, the method is applicable regularization or harmonic balance method. If the first approach, signum function is replaced by another which has a similar shape, but is still in the home. In the case

*Corresponding author: trmazilu@yahoo.com

of the second approach, it is considered an equivalent to a harmonic series that is modulated both in phase and in amplitude [3]. Another method is the so-called switching method that considers the integration of equations of motion distinct phases with changing the number of degrees of freedom depending on the outcome of checks on the conditions of slipping and adhesion [4].

This paper deals with the forced vibrations of a system with dry friction elastic consists of two rigid bodies arranged vertically connected by two elements, one linear elastic characteristic and the other is a dry friction damping. The system is in permanent contact with a rigid base undulating through the lower body. It is considered that the system moves along the rigid base, and its form is sinusoidal type. Interesting aspects of the stick-slip vibrations and their influence on dynamic contact force between lower body and rigid base. The described model is a simplified representation of a freight wagon suspension friction as dry as, for example, wagon equipped with bogies Y 25.

2 Equations of motion

The mechanical model is presented in Figure 1, comprising two rigid bodies connected by a spring element that works together with a dry friction element. This system can simply model 1/8 freight wagon moving at a constant speed *V* on a track with vertical irregularity η. Indeed, the mass rigid body *M* is the suspended mass of the vehicle corresponding to a wheel and the rigid body of mass *m* is the wheel. Suspension stiffness is *k* and friction force developed in the damping element is F_f. Displacement of the body *M* is *z*, and η is the displacement of the wheel.

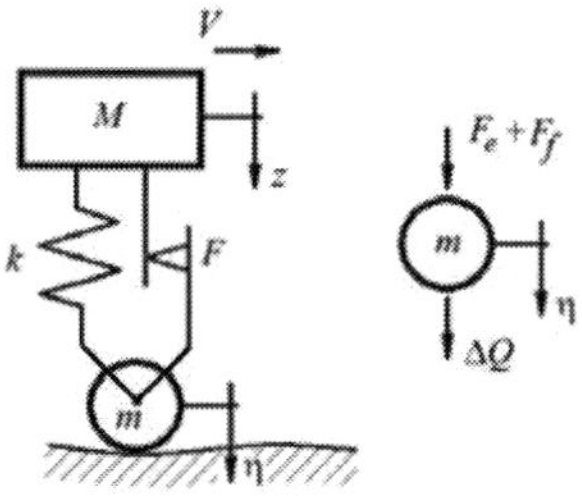

Fig. 1. Mechanical model.

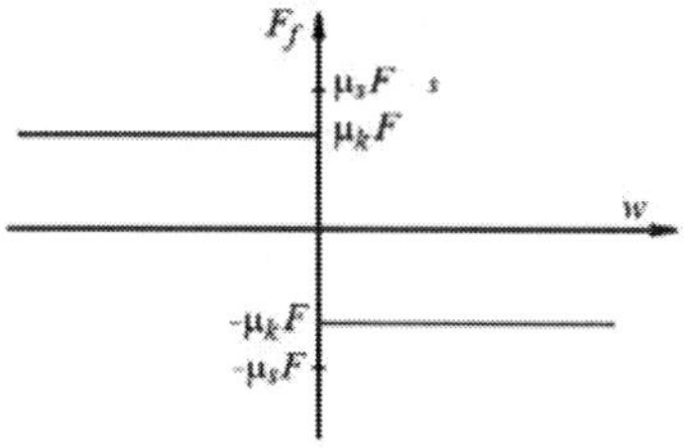

Fig. 2. Friction force versus slip velocity.

Friction force has the following form, depending on the slip velocity (see Fig. 2)

$$F_f = -\mu_k N \mathrm{sign}(w) \text{ for } w \neq 0 \tag{1}$$

$$|F_f| \leq \mu_s N \text{ for } w = 0, \tag{2}$$

where μ_k is the kinetic friction coefficient, μ_s - static friction coefficient, *N* – normal force and *w* is the slip velocity.

Equations of motions are

$$\begin{aligned} M\ddot{z} + k(z-\eta) + F_f &= 0 \\ m\ddot{\eta} + k(\eta - z) - F_f &= \Delta Q, \end{aligned} \tag{3}$$

where ΔQ is the dynamic force between wheel and track.

New variable is introduced in order to solve the equations of motion

$$u = z - \eta . \tag{4}$$

Equations of motion become

$$\begin{aligned} M\ddot{u} + ku + F_f &= -M\ddot{\eta} \\ m\ddot{\eta} - ku - F_f &= \Delta Q. \end{aligned} \tag{5}$$

The relative motion between the two bodies exhibits two distinctive alternating sequences: the stick and the slip. During the slip sequence, the friction force is given by the equation (1), where the slip velocity results from Eq. (4).

The following conditions are accomplished during the stick sequence

$$u = \text{constant}, \; \dot{u} = 0, \; \ddot{u} = 0, \tag{6}$$

and the friction force is obtained from the first equation (5)

$$F_f = -ku - M\ddot{\eta} \tag{7}$$

as long as the Eq. (2) is valid.

There is a symmetric lock zone due to the friction, where the motion stops when the stick conditions are accomplished. The lock zone length, z_l, can be calculated with

$$z_l = \mu_s N / k . \tag{8}$$

Next, considering the track irregularity of harmonic form

$$\eta = \eta_o \sin \omega t , \tag{9}$$

where η_o is the amplitude, ω is the angular frequency and t stands for time, we are looking for the solution to the equations of motion. To this end, the analytical solution for each stick/slip sequence of the motion is established.

For the slip sequence, the following equation has to be solved

$$M\ddot{u}(t^*) + ku(t^*) + \mu_k N \text{sign}[\dot{u}(t^*)] = \omega^2 M\eta_o \sin[\omega(t_k + t^*)] \tag{10}$$

where t_k is the start time moment of the k slip sequence and t^* stands for time during the k slip sequence, and $0 \leq t^* \leq t^*_k$ with t^*_k the time period of the k slip sequence.

Considering the initial conditions $u(t^* = 0) = u_{ak}, \dot{u}(t^* = 0) = 0$, the solution to the equation (10) is

$$\begin{aligned} u(t^*) = &\left\{ u_{ak} + \Delta u_k \text{sign}[\dot{u}(t^*)] - \frac{\Omega^2}{1-\Omega^2}\eta_o \sin(\omega t_k) \right\} \cos(\omega_o t^*) - \Delta u_k \text{sign}[\dot{u}(t^*)] - \\ &- \frac{\Omega^3}{1-\Omega^2}\eta_o \cos(\omega t_k) \sin(\omega_o t^*) + \frac{\Omega^2}{1-\Omega^2}\eta_o \sin[\omega(t_k + t^*)], \end{aligned} \tag{10}$$

where $\Delta u_k = \mu_k N/k$ and $\Omega = \omega/\omega_o$.

The time period of the k slip sequence is obtained from the zero velocity condition

$$\omega_o\left\{u_{ak}+\Delta u_k \text{sign}\left[\dot{u}(t_k^*)\right]-\frac{\Omega^2}{1-\Omega^2}\eta_o\sin(\omega t_k)\right\}\sin(\omega_o t_k^*)+$$
$$+\omega_o\frac{\Omega^3}{1-\Omega^2}\eta_o\cos(\omega t_k)\cos(\omega_o t_k^*)-\omega\frac{\Omega^2}{1-\Omega^2}\eta_o\cos\left[\omega(t_k+t_k^*)\right]=0. \quad (13)$$

which can be solved by iteration method.

3 Numerical application

This section shows some numerical results derived from the above model, considering the following parameters corresponding to 1/8 empty freight wagon fitted with Y 25 bogie: M = 2325 kg, m= 650 kg, k = 830 kN/m, μ_k = 0,17, μ_s = 0,204, $N = Mg\text{tg}(\alpha)$ = 9126 N, where g is the gravitational acceleration and α is the Lenoir device angle in respect to vertical axis (α = 21°43′) [1]. It obtains the lock zone length z_l = 2.243 mm and the natural frequency of the vehicle of 3.0 Hz.

Figure 3 shows the free vibration when the initial displacement is 14 z_b and zero velocity condition. It observes 3 periods in the first second of the motion according to the natural frequency of 3 Hz. The wave attenuation is linear and the motion stops after 8 semi-periods.

Figure 4 displays the free vibration in the presence of a high frequency vibration (dither [5]) of 90 Hz and the amplitude of 10 μm. This dither can be originated by the wheel/rail interaction due to the roughness. When the free motion starts, its attenuation is linear, but then it becomes exponential like linear damping case. The motion never stops within the lock zone which has no effect.

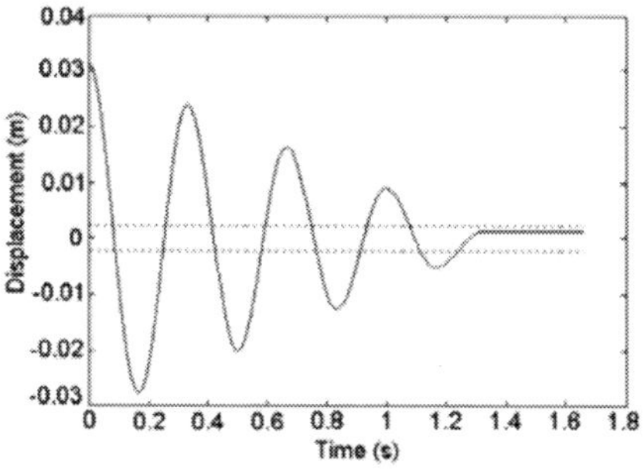

Fig. 3. Free vibration:
—, vibration, ---, lock zone.

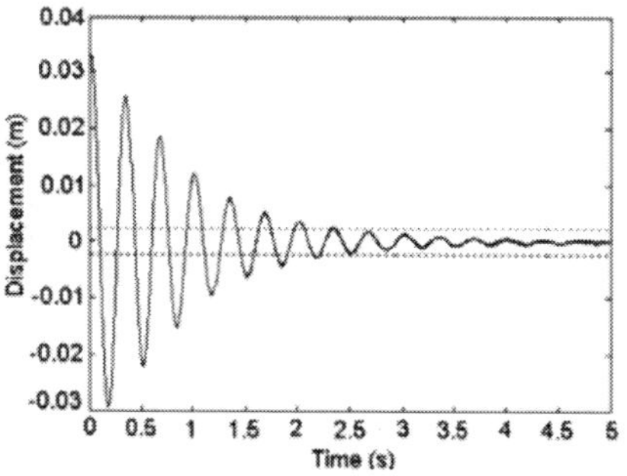

Fig. 4. Free vibration in the presence of dither of 90 Hz and 10 μm amplitude:
—, vibration, ---, lock zone.

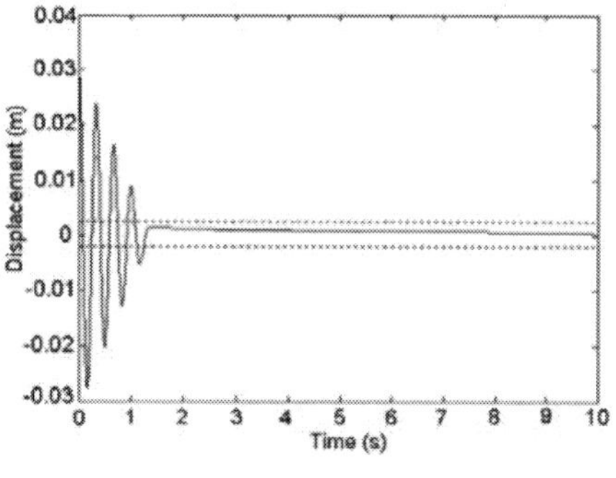

Fig. 5. Free vibration in the presence of dither of 90 Hz and 1.75 μm amplitude:
—, vibration, - - -, lock zone.

It must be underlined that the dither amplitude should be sufficiently high to maintain the motion, alternatively the lock zone stops the motion. For instance, Fig. 5 presents such case, where the amplitude of the dither is only 1.75 μm. Figure 6 shows the forced vibration due to a low-frequency harmonic excitation (6 Hz) with amplitude of 566 μm. Vibration exhibits alternately stick-slip sequences. Two stick/slip sequences occur during a time. Acceleration of the suspended mass and the contact force have sudden variations and high peaks which could affect the ride quality and the wear of the wheel/rail rolling surfaces.

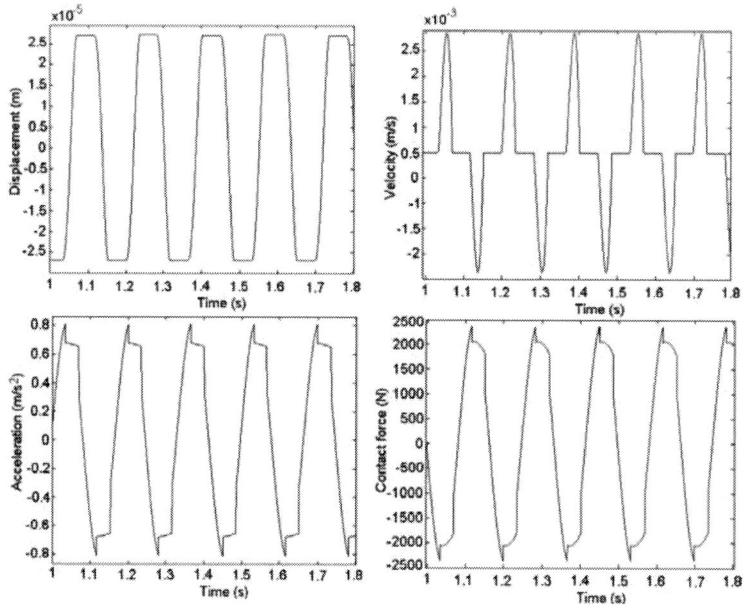

Fig. 6. Forced vibration due to a harmonic irregularity (η_o = 560 μm, frequency 6Hz).

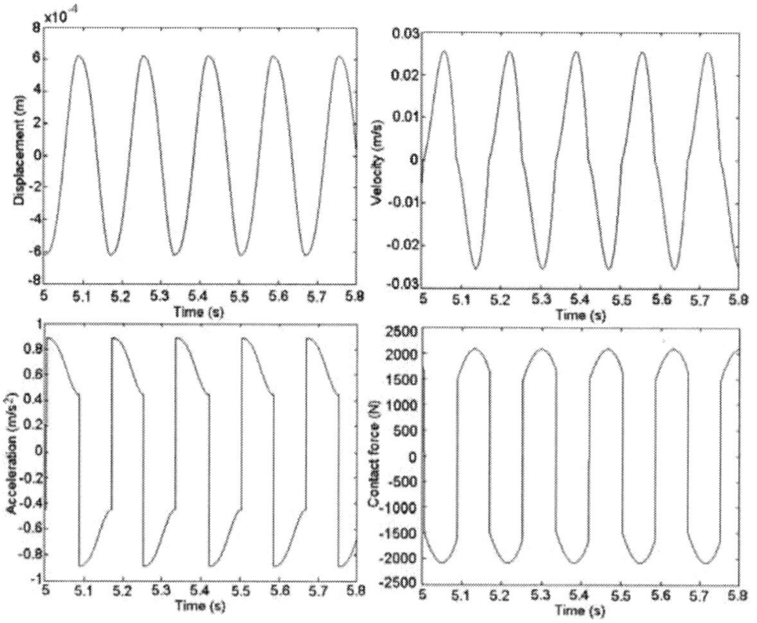

Fig. 7. Forced vibration due to a harmonic irregularity (η_o = 840 μm, frequency 6Hz).

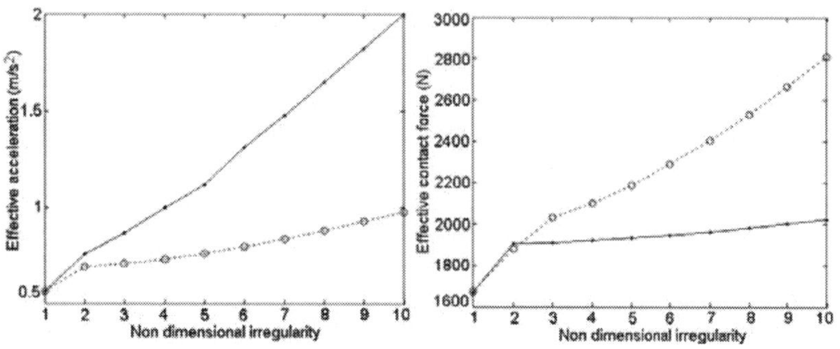

Fig. 8. Acceleration of the suspended mass and the contact force:
—•—, excitation 6 Hz; - - -○- - -, excitation frequency 9 Hz.

When the amplitude of the excitation is higher, the vibration is different. For instance, the alternately stick/slip sequences disappear when the excitation amplitude is 841 μm (Fig. 7). Figure 8 shows the acceleration of the suspended mass and the contact force versus the excitation amplitude considering two frequencies, 6 Hz and 9 Hz, respectively. The non-dimensional amplitude equals the ratio between the amplitude and $z_b(\omega_o/\omega)^2$. The suspended mass acceleration is higher at 6 Hz, while the contact force, at 9 Hz.

4 Conclusions

Damping based on the principle of dry friction is a technical solution applied to almost all types of freight wagons. To improve the dynamic performance of the freight wagons it is essential to understand the dynamic behaviour in the presence of the dry friction based on the outcome from the theoretical models.

The paper presents a simple model for a freight wagon equipped with bogies Y25 to highlight several important aspects of the free and forced vibration, due to excitation from a track with harmonic irregularity.

Free vibrations are influenced by the existence of high-frequency vibrations such as those induced by wheel-rail interaction in the presence of the rolling surfaces roughness.

If low-frequency excitation amplitude is relatively small, the forced vibration is a jerking motion. At higher amplitudes of track harmonic irregularity, jerky character could disappear.

Vehicle behaviour has nonlinear characteristic which is reflected in the relation between the suspended mass acceleration and contact force, on the one hand, and the excitation amplitude on the other hand.

Outcomes of this paper can be taken as reference to validate more sophisticated models.

References

1. C. Burada, M. Buga, A. Crăsneanu, *Elements and Structures of Railway Vehicles* (In Romanian), (Ed. Tehnică, Bucureşti, 1980)
2. B. Brogliato, *Nonsmooth Mechanics Models, Dynamics and Control* (Springer International Publishing, 2016)
3. G. J.Stein, R. Zahoranský, P. Múčka, J. Sound Vib. **311**, 74-96 (2008)
4. D. Karnopp,J. Dyn. Syst. Meas. Contr. **107**, 100-103 (1985)
5. J. Piotrowski, Veh. Syst. Dyn. **48**, 675–703 (2010)

Theoretical aspects concerning working fluids in hydraulic systems

Irina Ţiţa[1,*], *Irina* Mardare[1] and *Dorin Emil* Husaru[2]

[1]Technical University "Gheorghe Asachi" Iaşi, Department of Fluid Mechanics, Fluid Machinery and Fluid Power Systems, Blvd. D. Mangeron, 59A, 700050, Iasi, România
[2]Technical University "Gheorghe Asachi" Iaşi, Department of Mechanical Engineering, Mechatronics and Robotics, Blvd. D. Mangeron, 43, 700050, Iasi, România

Abstract. Among the properties of working fluid, viscosity is the most important as it regards especially to pumps. In order to study the behavior of hydrostatic transmission it is important to create a reliable research instrument for dynamic simulation. Our research expertise being in SimHydraulics consequently this instrument is the suitable block diagram. The purpose of this paper is to present the possible ways to customize the properties of the working fluid in the block diagram.

1 Introduction

In fluid power systems the working fluid has to transmit energy. It must transmit the prime movers energy to the actuator so it can perform work. The fluid must flow easily to reduce power losses and make the circuit respond quickly [1, 2].

Among the properties of working fluid, viscosity is the most important as it regards especially to pumps. One can use either dynamic viscosity η [N·s/m^2, Pa·s in SI or Poise in CGS] or kinematic viscosity υ [m^2/s in SI or St in CGS] knowing that dynamic viscosity is the quantitative expression of fluid's resistance to flow, while kinematic viscosity is the ratio of the fluid's viscous force to the inertial force and kinematic viscosity is the ratio of dynamic viscosity to density. In this paper viscosity will be used for kinematic viscosity.

$$\upsilon = \eta / \rho \tag{1}$$

Most manufacturers specify viscosity limits for their pumps and it is recommended to stay within the suggested limits. Pressure drop in the pump suction line generally is low and if the oil is too thick the pump will be damaged due to cavitation. On the other end, if fluids are too thin, pump bypass wastes energy and generates extra heat.

All other components in the circuit could operate on any viscosity fluid because they only use what is fed to them. However, thicker fluids waste energy because they are hard to move [3, 4].

There are two critical aspects regarding working fluid: the resilience of the hydraulic fluid's viscosity across a wide range of operating temperatures and the shear stability of the hydraulic fluid – sufficient to withstand the stress of demanding operating conditions.

* Corresponding author: iddtita@yahoo.com

Considering that the main interest in our research laboratory is on hydrostatic transmissions used in wind turbines and the fact that the average range of operating temperature over a year is of 60°C, viscosity of the working fluid is very important.

In order to study the behavior of hydrostatic transmission it is important to create a reliable research instrument for dynamic simulation. Our research expertise being in SimHydraulics consequently this instrument is the suitable block diagram. The purpose of this paper is to present the possible ways to customize the properties of the working fluid in the block diagram.

2 Working fluid properties

Viscosity is the most important property of the working fluid in hydrostatic transmissions. Viscosity of every hydraulic fluid depends on its operating temperature Fig. 1 [5].

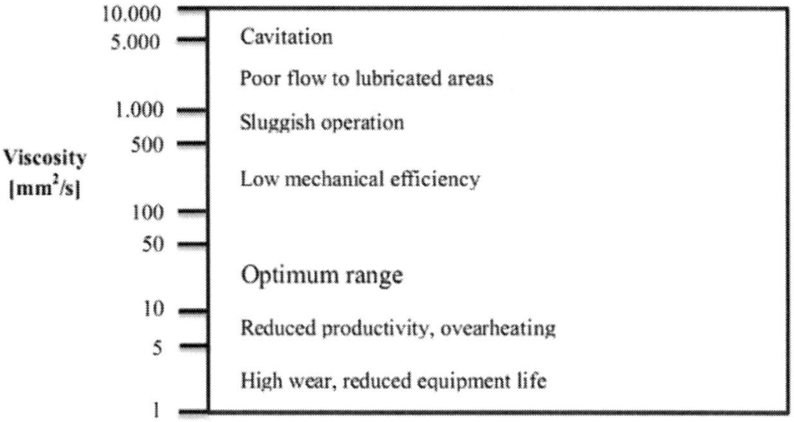

Fig. 1. Kinematic viscosity domains.

The ideal viscosity range for a fluid – and thus its highest efficiency – is usually between 10 and 100 mm²/s, depending on the application. If a hydraulic fluid is too thin this leads to increased wear, lower power transmission, lower volumetric efficiency and overheating. A fluid of too high a viscosity leads to a loss of power, decreased mechanical efficiency, and in the worst cases, cavitation and mechanical failure.

The operable range of temperatures in which a fluid is effective is called its "temperature operating window - TOW" [4] (Fig. 2).

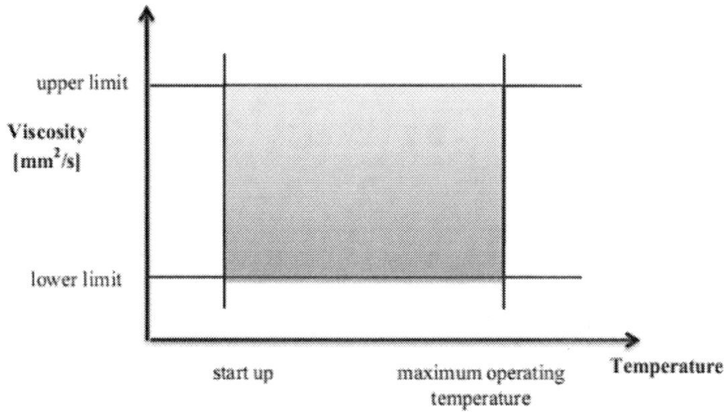

Fig. 2. Temperature operating window – TOW.

The position and the width of the window depend largely on the characteristics of the pump and the hydraulic fluid itself. The operating temperature requirements are defined by the application. Within these limits, from start up to emergency situations, the fluid has to protect the hydraulic equipment. A measure of viscosity change from one temperature to another is viscosity index - VI. Most industrial hydraulic oils run in the 90 to 105 -VI range and are satisfactory for most applications. Viscosity modifiers enhance the oils ability to remain at a workable viscosity through a broad temperature range [1, 4].

Viscosity has to be high enough to protect the equipment and to guarantee pump efficiency. On its way from the tank to the pump, the fluid has to pass several hydraulic resistances. Friction within the fluid itself and friction on the walls and compression in the pump creates heat, heat that reduces the viscosity.

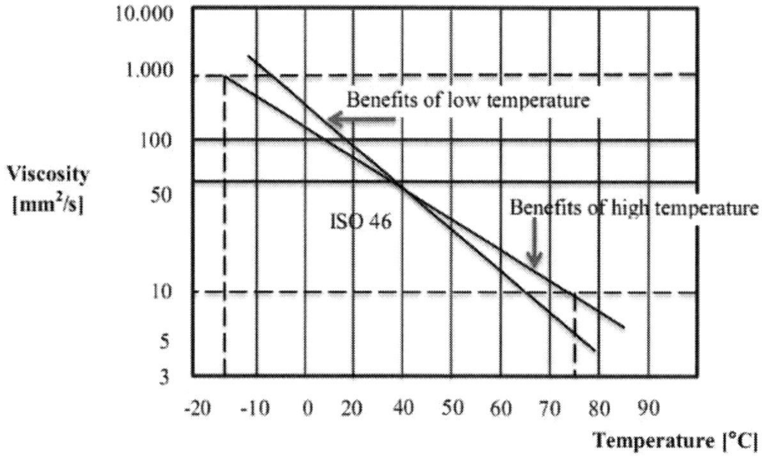

Fig. 3. Case study for two oils meeting ISO 46 Viscosity [5].

The diagram in Fig. 3 shows the viscosity of two fluids across a range of operating temperatures. While both fluids meet pump manufacturer requirements, the fluid with the higher viscosity index (VI) permits a broader Temperature Operating Window (TOW). One may see the comparatively high viscosity of the High VI fluid when reaching high temperature (higher volumetric efficiency) as well as its comparatively lower viscosity at low temperatures (higher mechanical efficiency) [6].

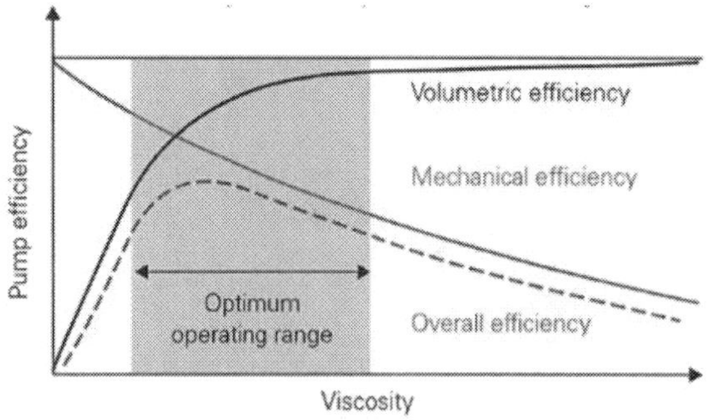

Fig. 4. Pump efficiency according to kinematic viscosity [7].

While the amount of conveyed oil grows with increased viscosity, the mechanical efficiency of the pump is decreasing Fig. 4. The overall efficiency of a pump is the product of both curves. Only with a suitable viscosity for its hydraulic fluid will a pump work efficiently. If a hydraulic fluid can be maintained within an efficient viscosity range throughout its entire operating temperature spectrum, the pump will operate more efficiently.

Given that petroleum oil can pollute drinking water, with the agricultural and mining industries being major consumers of hydraulic fluid, bio based hydraulic fluids are used in environmentally sensitive applications when there is the risk of an oil spill. Vegetable oils from Canola used as base stocks for fluids where biodegradability is considered important. [8]. Some of the significant advantages of using biodegradable hydraulic fluids are: adequate viscosity and viscosity index, proper viscosity to minimize internal leakage low temperature fluidity, cleanliness and filterability, corrosion control, shear stability, low volatility etc.

3 Hydraulic fluid in SimHydraulics

In order to consider properties of working fluid Matlab / SimHydraulics has two blocs: Custom Hydraulic Fluid –Simscape/Foundation Library/Hydraulic Utilities and Hydraulic Fluid Simscape/SimHydraulics/Utilities [9] (Fig. 5 and Fig.6).

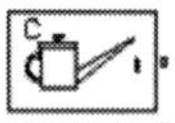

Fig. 5. Custom Hydraulic Fluid Block.

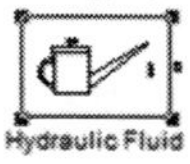

Fig. 6. Hydraulic Fluid Block.

In Custom Hydraulic Fluid block one can specify, according to producer's specifications: fluid density, kinematic viscosity, bulk modulus and relative amount of air trapped, as one can see in Fig. 7.

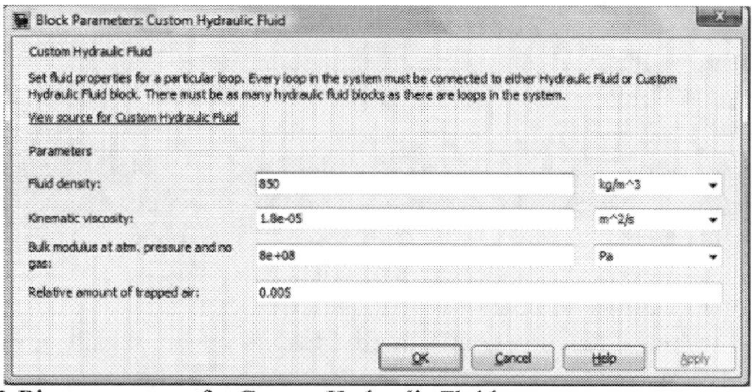

Fig. 7. Bloc parameters for Custom Hydraulic Fluid.

Hydraulic Fluid block offers a drop-down list of some fluids (Fig. 8).

Fluids in the list can be grouped in the following categories: a) Skydrol eco-friendly, fire resistant intended for lubrication components, aircraft use [10]; b) HyJet synthetic oils, fire resistant, stable at temperature variation, intended for aircraft use [11]; c) Fluid MIL oils are military specifications, synthetic hydrocarbon base, fire resistant for aircraft use [12]; d) Oil 10W is synthetic engine oil, high VI, for heavy vehicles use; e) Oil SAE is a synthetic motor oil for heavy vehicle use also [11, 13]; f) Transmission fluid ATF (Dexron III) is synthetic, high VI, for cars automatic transmission or any other hydraulic transmission [11, 13]; g) ISO VG oils are industrial mineral oils with good anti-foaming, anti-oxidant and anti-rust characteristics, ISO VG 22 generally used in airlines for air tools, ISO VG 32 ideal for use in high-powered machine tools, ISO VG 46 is normally required for industrial plant working under high-pressure etc. [14]; h) Brake fluids are synthetic oils with additives to improve performances in extreme functioning conditions, for vehicles braking systems [11]; i) Gasoline is a product fuel gas resulting from the liquefaction; j) Diesel fuel is a hydrocarbon mixture useful for Diesel engines; k) Jet Fuel is a hydrocarbon mixture useful for aircraft engines [11]; l) Water-Glycol 60/40 is mix with frost protection up to a temperature of 51^{0}C, nontoxic fluid and consequently used for hydrostatic transmission or cooling systems in pharmaceutical or food industry

Selecting working fluid parameters: type, relative amount of trapped air and temperature, corresponding fluid properties: density, viscosity and bulk modulus will be automatically selected from the characteristics.

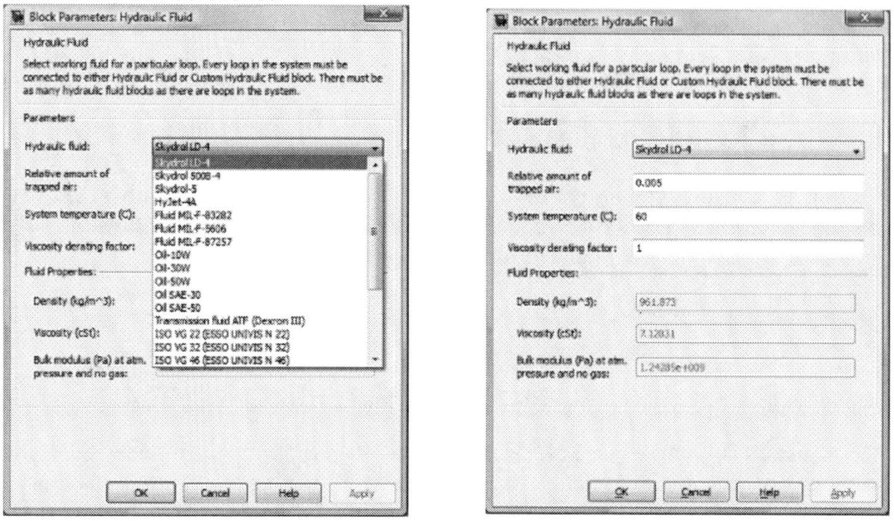

Fig. 8. Bloc Parameters for Hydraulic Fluid.

One can get information on the fluids and their properties considered in the literature used in MATLAB® list of fluids. Information is given as plots for viscosity, density and bulk modulus with temperature.

For researches in our laboratory it has been identified some problems regarding a simulation diagram: how could one consider the influence of the temperature variation; how to consider a variation with temperature for a property (e.g. viscosity) founded to be different from the one considered in MATLAB including vegetable fluids and how to consider the possible change of a property after a number of working hours.

Our laboratory offers technical resources and expertise to determine viscosity curve for a fluid. For example for SAE 10W-60 dynamic viscosity η variation obtained is presented in Fig.9 [15].

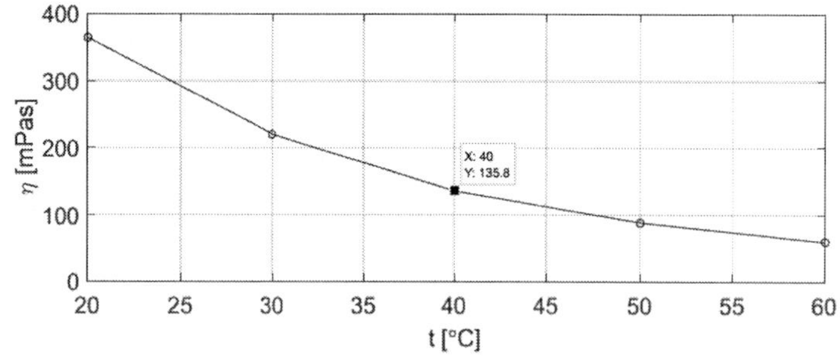

Fig. 9. Approximated dynamic viscosity.

In order to solve the identified problems we intend to evaluate the fluid properties at the beginning and after a measured number of working hours. For the simulation diagram we intend to create a new component which will consider the properties as a Lookup Table depending on temperature variation. In the table will be considered the properties of the fluid obtained with RheolabQC.

4 Conclusions

If a hydraulic fluid is too thin this leads to increased wear, lower power transmission, lower volumetric efficiency and overheating. A fluid of too high a viscosity leads to a loss of power, decreased mechanical efficiency, and in the worst cases, cavitation and mechanical failure. It is of high importance to choose the working fluid.

In order to study the behavior of hydrostatic transmission it is important to create a reliable SimHydraulics block diagram as a research instrument for dynamic simulation.

In this paper was revealed the actual SimHydraulics possibilities for using a working fluid.

Considering the final purpose of the research which consists in hydrostatic transmissions for wind turbines, the main specific problems regarding the working fluid was identified and also the future researches.

The authors would like to acknowledge the technical resources offered by the Laboratory of Fluid Properties Measurement, from the Department of Fluid Mechanics, Fluid Machinery and Fluid Power Systems, "Gheorghe Asachi" Technical University of Iasi, Romania. The Laboratory of Fluid Properties Measurement has been equipped with technical resources with the financial support of the grant ENERED, POSCCE-A2-O2.2.1-2009-4, ID 911.

References

1. J. Watton, Springer Verlag, London, (2007).
2. J. Watton, Cambridge University Press, London, (2009).
3. S. Lunt, T. Meindorf, 9 IFK, **3**, 11, (2014)
4. Hydraulics & Pneumatics http://hydraulicspneumatics.com/fluid-power-basics/hydraulics (accessed 18.02.2017)
5. DynaVis http://formulators.dynavis.com/cms/formulators/en/hydraulic-fluid-performance/viscosity/pages/default.aspx, (accessed 25.02.2017)
6. H. Murenhoff, Proceeding of the 12th International Conference on Tools: ICT-2007, Miskolc, (2007).
7. Machinery Lubrication http://machinerylubrication.com/Read/776/maximum-efficiency-hydraulic-fluids (accessed 17.02.2017)
8. D. Placek, (edited by L. Rudnick ed., CRC Press, 2006).
9. MATLAB/Simscape https://uk.mathworks.com/products/simscape.html (accessed 10.02.2017)
10. Skydrol http://www.eastman.com/Brands/EAS/Skydrol/Pages/Home.aspx accessed 22.02.2017
11. Viscopedia www.viscopedia.com (accessed 20.02.2017)
12. QC Lubricants http://qclubricants.com/hydraulic_fluid_mil.htm
13. Castrol http://www.castrol.com/en_us/united-states/other-fluids/automatic-transmission-fluid-atf.html (accessed 25.02.2017)
14. Crownoil https://www.crownoil.co.uk/lubricants/all-you-need-to-know-about-hydraulic-oil/ (accessed 22.02.2017)
15. D. Zahariea, D. E. Husaru, *IManEE,* 6 (2017).

Tooth contact analysis of spur gears. Part 1-SAM analysis of standard gears

Spiridon Crețu*, *Nicolae* Pop, and *Stelian* Cazan
Technical University of Iași, Mechanical Engineering Department, 63 D. Mangeron St., Iași, Romania

Abstract. The involute gears are sensitive to the misalignment of their axes which determines transmission errors and perturbations of pressures distributions along the tooth flank. The concentrated contacts in gears are no longer as Hertz type. A semi-analytical method was developed to find the contact area, pressures distribution and depth stresses state. The matrix of initial separations is found analytically for standard and non-standard spur gears. The presence of misalignment as well as the flank crowning and flank end relief are included in the numerical analysis process.

1 Introduction

Under real running condition the meshing is altered by various deviations of the machine parts involved [1], but also by elastic deformations [2], which determine transmission errors and perturbations of pressures distributions along the flank, especially at its end sides, Figure1a. The pressures concentrations accelerate the fatigue and wear phenomena resulting a much shorter life for the machine parts subjected to rolling contact [3]. The concentrated contacts in gears meshing are no longer of Hertz type [4-6], and no straight analytical solutions is available. Semi-analytical methods have been developed to find the contact area, pressures distribution and depth stresses state [7-9]. For spur and helical gears some techniques as flank crowning or flank relieving, Figure 1b, are commonly used to modify the initial line contact to a localized bearing contact [5, 6]. The normal load acts along the contact line and a common contact area develops. The shape and size of the common contact area, as well as the pressures distribution on this area, are unknown.

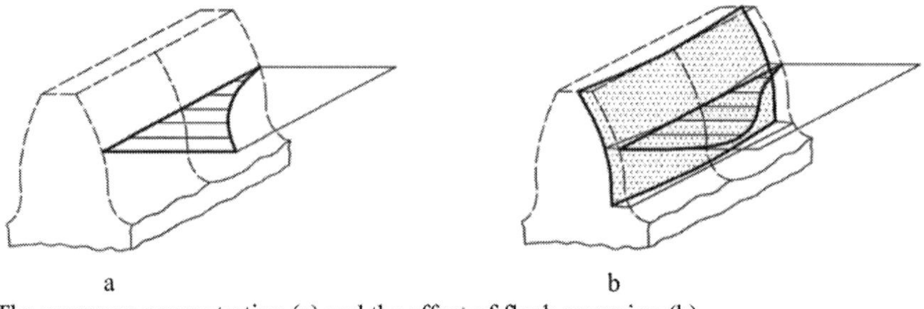

Fig.1. The pressures concentration (a) and the effect of flank crowning (b).

* Corresponding author: spcretu@tuiasi.ro

2 SAM modelling of non-Hertz concentrated contacts

The semi-analytical method uses the initial (no load) separations between surfaces of the conjugate teeth that sustain the meshing process. A virtual rectangular contact area, denoted A_v, is built on the common tangent plane, around the initial contact point, and a Cartesian system *(x, y, z)* is introduced, the *x-O-y* plane being the common tangent plane, Figure 2.

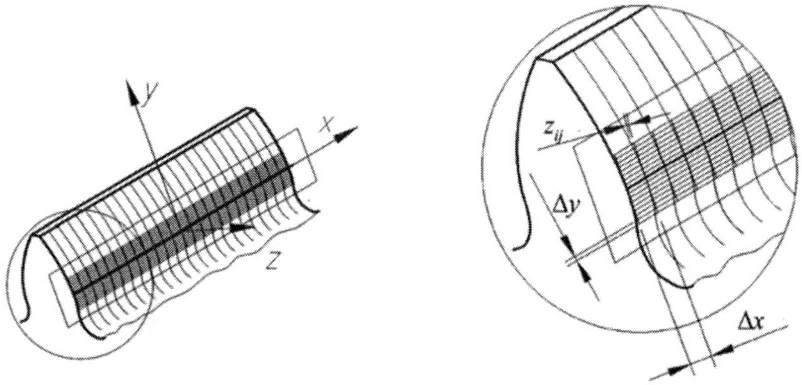

Fig. 2. The virtual rectangular contact area.

A uniformly spaced rectangular array is built on the virtual rectangular contact area with the grid sides parallel to the *x* and *y*-axes. The nodes of the grid are denoted by *(i, j)*, where indices *i* and *j* refer to the N_x grid columns and N_y grid rows, respectively. The real pressure distribution is approximated by a virtual pressure distribution, characterized by unknown constant values p_{ij} inside of each (i, j) patch [8, 9]. The surface deformation is modeled by the following six linear algebraic equations:

a) geometric equation of the elastic contact:

$$g_{ij} = h_{ij} + w_{ij} - \delta_0 \tag{1}$$

where: g_{ij} is the gap between the normal loaded surfaces, h_{ij} is the separation between unloaded surfaces, w_{ij} is the elastic deformation of the two surfaces, measured along the normal load, and δ_0 is the rigid displacement of the contacting bodies;

b) equation of the normal surface displacement:

$$w_{ij} = \sum_{k=0}^{Nx-1} \sum_{j=0}^{Ny-1} (K_{i-k,j-l} * p_{kl}) \tag{2}$$

c) load balance equation

$$\Delta x \Delta y \sum_{i=0}^{Nx-1} \sum_{j=0}^{Ny-1} p_{ij} = F \tag{3}$$

d) constraint equation of non-penetration:

$$g_{ij} = 0, \xrightarrow{yields} p_{ij} > 0, \ (i, j) \exists A_r \tag{4}$$

e) constraint equation of non-adhesion:

$$g_{ij} > 0, \xrightarrow{yields} p_{ij} = 0, \ (i, j) \nexists A_r \tag{5}$$

f) *elastic-perfect plastic* behavior of the material:

$$p_{ij} \geq p_Y \Rightarrow p_{ij} = p_Y \tag{6}$$

where p_Y is the value of the pressure able to initiate the plastic yield.

The influence function K_{ij} represents the value of the surface deformation created in the point *(i, j)* by the unit pressure acting in the elementary rectangle *(k, l)* [9].

The components of the stress tensor are obtained by superposition·

$$\sigma_{ij}(x,z,y,) = \sum_{k=0}^{Nx-1}\sum_{l=0}^{Ny-1}\left(C_{ijkl} * p_{kl}\right) \quad (7)$$

where the influence function $C_{ijkl}(x,y,z)$ describes the stress component $\sigma_{ij}(x,y,z)$ due to a unit pressure acting in the patch *(k, l)*, [10]. A single-loop conjugate gradient method was used to solve the mentioned algebraic system of equations. To increase the efficiency of the numerical algorithm, a dedicated discrete fast Fourier transform routine for 3D contact problems was used to solve the convolution products. The entire algorithm has been applied in a C++ computer code, named Non-Hertz [9, 10].

3 Matrix of separations

Spur involute gears are in line contact at every instant and consequently the virtual rectangular contact area is built having the contact line as median line. The half-width of the virtual rectangular area has a value b_v related to the value b_H obtained considering the Hertz hypothesis. The target is to determine the value h_{ij} of the initial separation in each mesh point *(i, j)* of the virtual rectangle. Two fixed coordinate systems, $O_1v_1w_1$ and $O_2v_2w_2$, are attached to pinion and gear Figure 3a, and two coordinates transformations are operated.

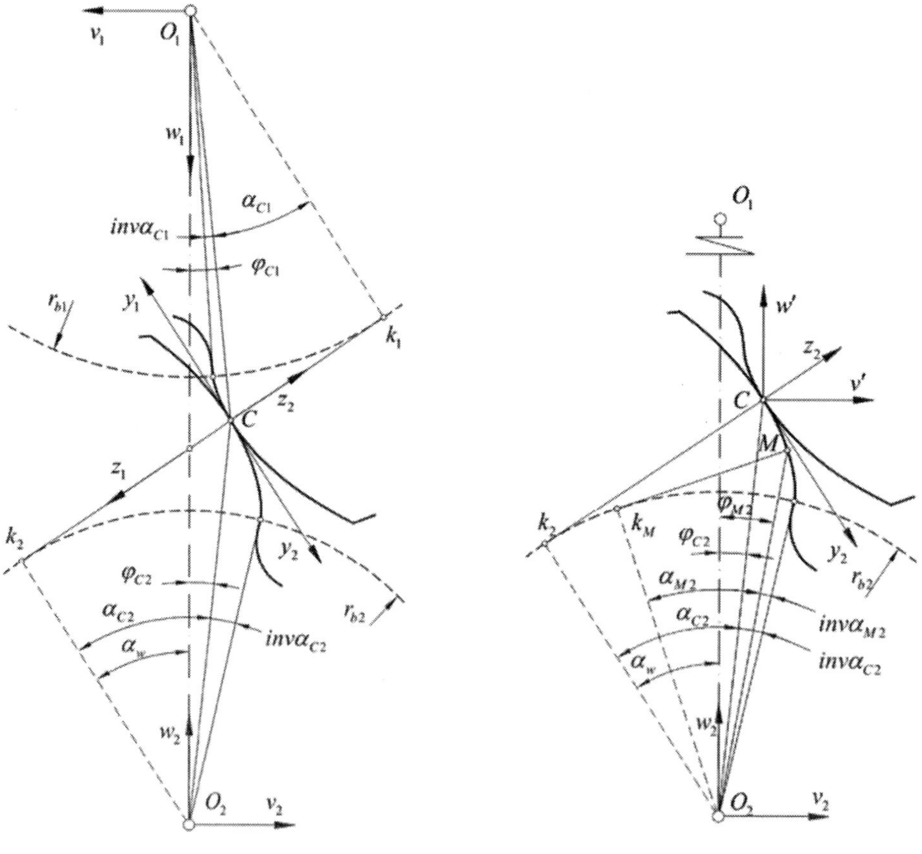

a b

Fig. 3. The main geometric elements and coordinate systems.

The first one is a translation from the fixed system $O_2v_2w_2$ to the mobile coordinate system $v'Cw'$, Figure 3b. The second coordinate transformation is a rotation with the angle θ, needed to obtain the Cz axis along the common normal of the teeth surfaces and Cy axis in the tangent plane of the contact point, Figure 3b. In the mesh grid of virtual rectangular area the point *(i,j)* is located at the distance y_j from the contact line. A mobile point *M,* with the position angle φ_{M2}, Figure 3b is considered on the flank of gear 2:

$$\varphi_{M2} = \varphi_{C2} - (\mathrm{inv}\alpha_{M2} - \mathrm{inv}\alpha_{C2}) \tag{8}$$

For gear 2 and point $M_2(x_i, y_j)$ the pressure angle α_{M2} is found as solution of the equation:

$$f(\alpha_{M2}) = y_{M2} - y_j = \frac{r_{b2}}{\cos\alpha_{M2}}\sin(\theta + \varphi_{M2}) - \frac{r_{b2}}{\cos\alpha_{C2}}\sin(\theta + \varphi_{C2}) - y_j = 0 \tag{9}$$

The Newton-Raphson iterative method provides the solution α_{M2}.

For point *(i,j)*, situated on virtual contact area, the equations (8, 9) give the value of initial separation $h2_{ij}$ between the tangent plane and flank surface of gear 2:

$$h2_{ij} = z_{M2[i,j]} \tag{10}$$

The same technique is used to obtain the initial separation $h1_{ij}$ [11]:

$$h1_{ij} = z_{M1[i,j]} \tag{11}$$

The normal separation between the surfaces of the meshing teeth is:

$$h_{ij} = h1_{ij} + h2_{ij} \tag{12}$$

The misalignment as well as the flank crowning or end flank relief are further included in the computing process for the matrix of initial separations data. The Non-Hertz software uses the separations matrix as initial data in an iterative process to obtain the pressures distribution, contact area and stresses states in the shallow layers of loaded teeth [9, 10].

4 TCA analysis of standard spur gears

The numerical analysis has been performed using the data for standard spur gears: $\alpha_0 = 20^0,\ h^*_{ao} = 1,\ c^*_0 = 0.25,\ \rho^*_0 = 0.38$ [11].
$z_1 = 25,\ z_2 = 51,\ m_n = 4\ \mathrm{mm},\ x_1 = 0,\ x_2 = 0,\ B = 50\ \mathrm{mm}.$
The nominal power parameters were: $T_2 = 30$ kW, $\varpi_2 = 52.35$ rad/s.

4.1 Standard spur gears - the edge effect

The phenomenon of pressures concentration at the end sides of the line contact with finite length is called *edge effect* [5, 13, 14]. To be in agreement with experimental evidence of the edge effect Guilbault [13] introduces a correction factor ψ into the 3D computational algorithm. In case of a flank with sharp edges the high values of pressures developed in the end areas of the flank might induce plastic deformations able to change locally the contact geometry and to attenuate the edge effect [14].To avoid the cumbersome calculations needed to evaluate the edge effect in standard spur gears, the present algorithm considers, for both sides of the flank, a chamfer with the radius $R_{ch}=0.2$ mm.

4.2 Numerical analysis

Standard spur gears, having a straight line lead (SL-lead), obtained with good manufacturing practice and operating in ideal conditions, were first considered.

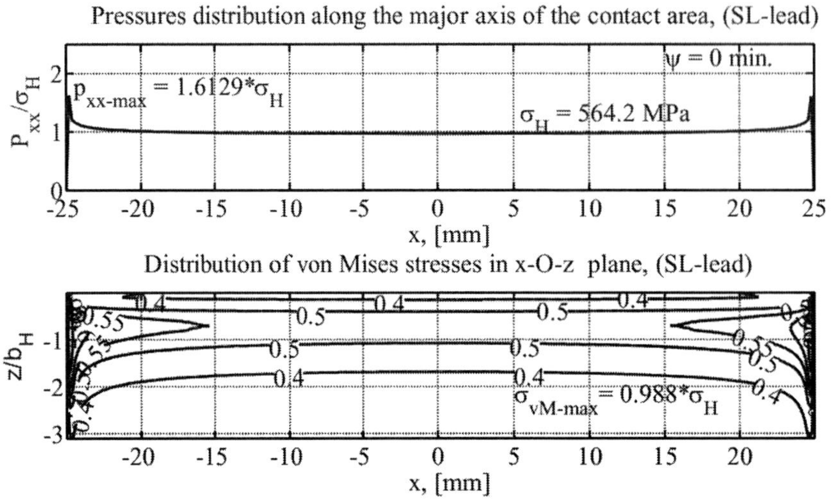

Fig. 4. Distributions of pressures and von Mises stresses for operating without misalignment.

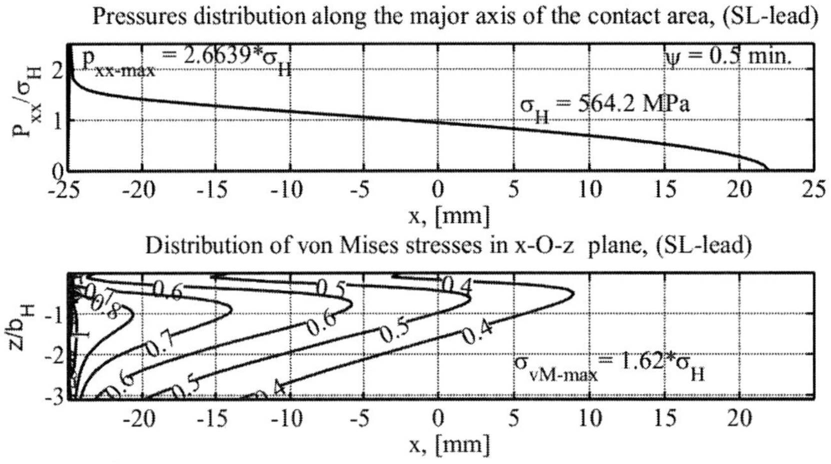

Fig. 5. Distributions of pressures and von Mises stresses for operating with misalignment, $\psi =$ 0.5 min.

The pressures distribution is presented in Figure 4 for the case when the contact point C is in the meshing pole, $\alpha_{C1} = \alpha_{C2} = \alpha_w = 20^o$. Along the tooth flank the maximum pressure is very close to the Hertz pressure except end sides areas where the edge effect manifests, Figure 4.

The edge effect becomes much damaging when even a small misalignment exists. Figure 5 depicts the pressures distribution obtained using the same operating data added with a misalignment of $\psi = 0.5$ min. As effect of this small misalignment, the p_{max}/σ_H ratio between the maximum value of pressure and Hertzian stress, increased from 1.61 to 2.66. Some of plasticity criteria consider the value of von Mises stress as responsible for starting the material yielding inside the stressed volume [9], and for the initiation of the rolling contact fatigue [3, 15]. In this respect the σ_{vM-max}/σ_H ratio between the maximum value of von Mises stress and Hertz stress, increased from 0.99 to 1.62.

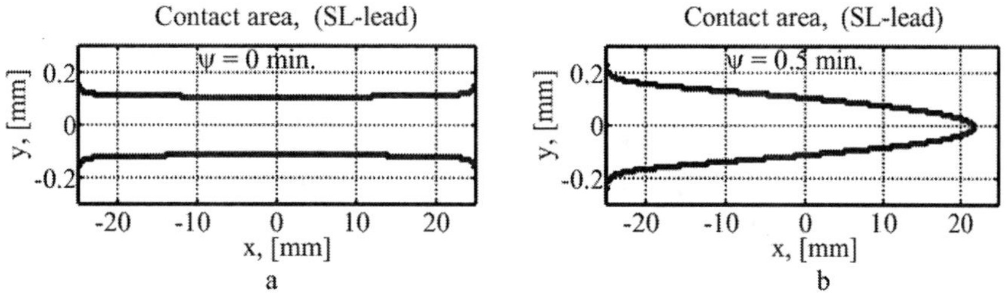

Fig. 6. Contact areas: a- operating in ideal conditions; b- operating with misalignment.

The shape and size of the contact area suffer major modifications when there is misalignment between axes, Figure 6. Modified spur gears are commonly used to avoid the harmful edge effect.

The main direction of the modified spur gears is to replace the line contact by a point contact. This means to localize their bearing contact by a proper modification of the surface of one of the mating gears, the pinion being the preferred one.

5 Conclusions

A robust and very fast semi-analytical method, Non-Hertz, is presented to solve the concentrated contacts achieved between conjugate teeth during meshing.
An analytical method was developed to find the matrix of separations between the surfaces of the conjugate teeth.
The harmful edge effect has been pointed out and the need for flank modification and its analysis was revealed.

References

1. S. Li, Mechanism and Machine Theory, **42**, 6 (2007)
2. Y. Siang-Yu, T. Shyi-Jeng, Mechanism and Machine Theory, **97**, 190-214 (2016)
3. F. Sadeghi, B. Jalalahmadi, T. Slack, N. Raje, N. Arakere, ASME J. Tribol., **131**, 2 (2009)
4. K. Johnson, *Contact mechanics*, Cambridge University Press (1985)
5. F. Litvin, A. Fuentes, *Gear geometry and applied theory*, Cambridge University Press (2004)
6. ISO 6336-1, *Calculation of load capacity of spur and helical gears*, Switzerland (2006)
7. I. Polonsky, L. Keer, ASME J. Tribol. **122**, 1 (2000)
8. S. Creţu, E. Antalucă, O. Creţu, Annals Univ. Galaţi Romania **24**, 26-35 (2003)
9. S. Creţu, Bull. Inst.Polit. Iaşi, **LI (LV)** 1-2, 1-31 (2005)
10. S. Creţu, Elastic plastic concentrated contact (in Romanian), Ed. Politehnium, Iaşi, Romania (2009)
11. N. Pop, S. Creţu, A. Tufescu, App. Mech. Mat. **658**, 531-356 (2014)
12. STAS 821, Romania (1982)
13. R. Guibault, ASME J. Tribol. **133**, 2 (2011)
14. S. Creţu, M. Benchea, A. Iovan-Dragomir, J. Balkan Tribol. Assoc. **22**, 1 (2016)
15. N. Popinceanu, E. Diaconescu, S. Creţu, Wear, **71**, 265-282 (1981)

Tooth contact analysis of spur gears. Part 2- Analysis of modified gears

Nicolae Pop, *Spiridon* Crețu*
Technical University of Iaşi, Mechanical Engineering Department, 63 D. Mangeron St., Iaşi, Romania

Abstract. The misalignment between gears axes or direction deviations of teeth leads accentuate the edge effect, Modified spur gears which localize the initial bearing contact by proper modifications of the lead profile are used to avoid the development of the edge effect. A semi-analytical method was involved to find the contact area, pressures distribution and depth stresses states for the standard gears and gears with modified lead profile. The lead modifications concerned the crowning of the flank surface and the end relieving of the tooth flanks, both being applied to the pinion teeth only.

1 Introduction

The spur and helical involute gears are running in line contact at every instant and therefore they are very sensitive to the misalignment of their axes, elastic deformations or various deviations of parts involved in the power transmission process, which determines transmission errors and pressures concentration at the ends of the contacting lines [1, 2].

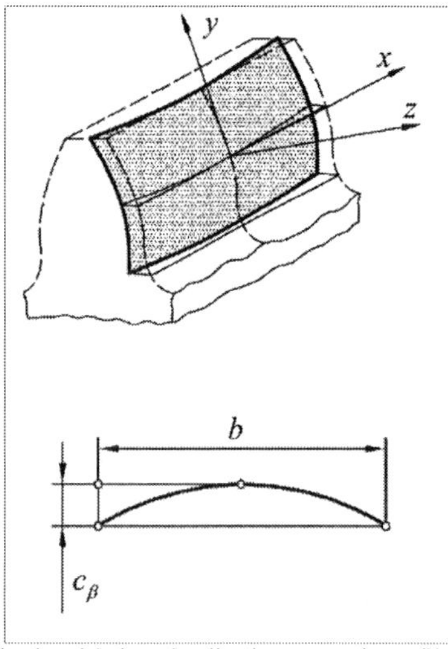

Fig. 1. Flank with longitudinal crowned profile.

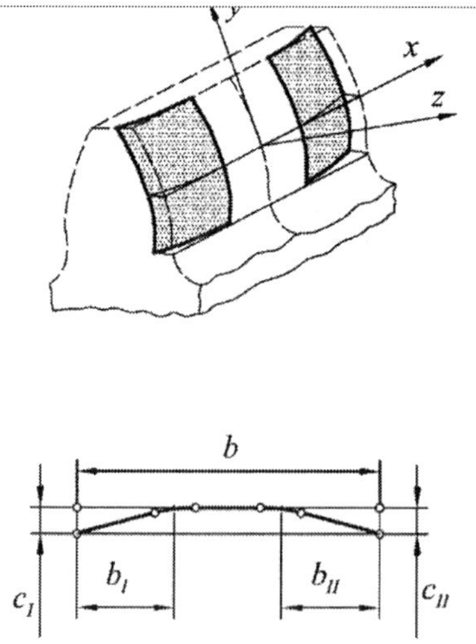

Fig. 2. Flank with end relief profile.

* corresponding author : spcretu@tuiasi.ro

The concentrated contacts in gears meshing are no longer of Hertz type [3], and semi-analytical methods (SAMs) have been developed to find the contact area, pressures distribution and depth stresses state [4-7]. For spur and helical gears some techniques as flank crowning Figure 1, or flank end relief Figure2, are frequently used to modify the initial line contact to a localized bearing contact [1, 2, 8-10].

2 Pinion modification by longitudinal crowning of flank profile

All simulations have considered the same nominal loading and the contact point C in the meshing pole, $\alpha_{C1} = \alpha_{C2} = \alpha_w = 20^o$ resulting the Hertzian stress $\sigma_H = 564$ MPa, and half-width of Hertzian contact area $b_H = 0.113$ mm. Different values for both, the misalignment angle ψ between gears axes and the crowned parameter C_β, have been taken into account.

2.1 Modified spur gears with longitudinal crowned flank profile, case $\psi = 0$

For $C_\beta > 0$ and $\psi = 0$ the contact between the mating teeth is transformed into a point contact type, located evenly along the flank. The contact area is an elongated ellipse with the greater half-axis along the flank. The pressure distribution is a quasi-ellipsoid and no edge effect develops if the value C_β was correlated with the normal load, Figure 3. A l higher value for C_β reduces the contact area and considerably increases the maximum pressure.

2.2 Modified spur gears with longitudinal crowned flank profile, case $\psi > 0$

2.2.1 Pressures distribution

When the operating conditions included a misalignment $\psi = 0.5$ min, the usage of a rather small value for the lead parameter $C_\beta = 5$ µm, led to a skewed pressures distribution with an emerging edge effect and a smaller contact area, Figure 4. The detrimental effect of misalignment is revealed by stress analysis.

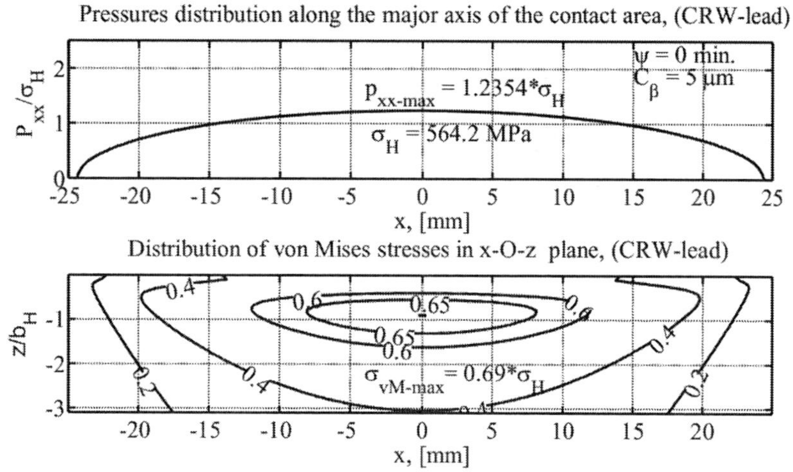

Fig. 3. Distributions of pressures and von Mises depth stresses, ($\psi = 0$, and $C_\beta = 5$ µm).

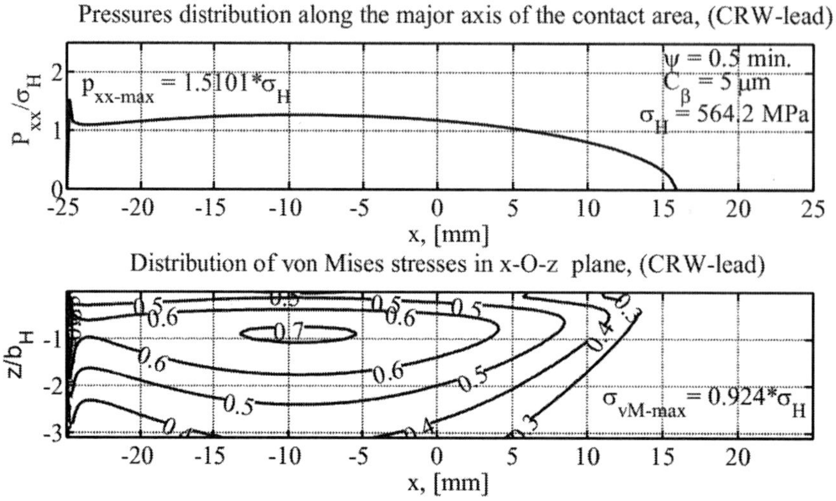

Fig. 4. Distributions of pressures and von Mises depth stresses, (ψ = 0.5 min., and C_β = 5 µm).

The maximum value of pressure increased, from p_{max} = 697 MPa in middle of the quasi-ellipsoid of pressures, Figure 3, to p_{max} = 853 MPa at the edge, Figure 4.

The maximum value of the von Mises stress increased from 389 MPa, Figure 3, to 521 MPa, Figure 4. However, a good correlation of the value for crowning parameter C_β with operating values of misalignment and radial load, assures more favorable distributions for both the contact pressures and von Mises depth stresses.

For a large misalignment, as ψ = 4 min., a crowned flank with the parameter C_β = 30 µm changed the location for pressures and von Mises stresses distributions Figure 6, but preserved the shape and values provided by the case C_β = 30 µm and ψ = 0 min.

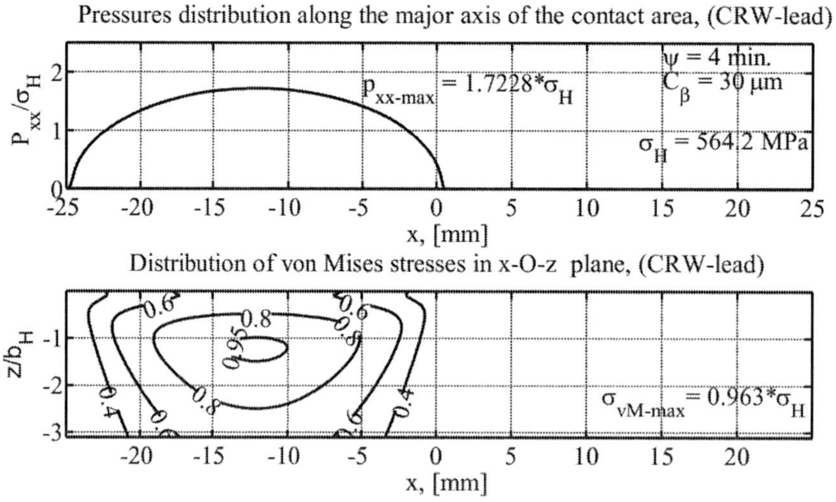

Fig. 5. Distributions of pressures and von Mises depth stresses, (ψ = 4.0 min., and C_β = 30 µm).

 DOI: 10.1051/matecconf/201711207016

3 Flank modification by end relief

The end relief is defined by parameters $b_{I(II)}$ and c, Figure 2, representing the length of the modified flank and end relief of profile, respectively [2]. In this numerical study, the parameter $b_{I(II)}$ was maintained constant, $b_{I(II)} = 0.1 * B$ and the parameter c was variable.

3.1 Modified spur gears by end relief of flank profile, case $\psi = 0$

The contact area is a rectangle with little wider ends, (*dog bone* shape). The pressures distribution is close to an elliptic cylinder distribution and no edge effect develops when the value of parameter c was accurate correlated with the normal load, Figure 6.

Pressures distribution along the major axis of the contact area, (ER-lead)

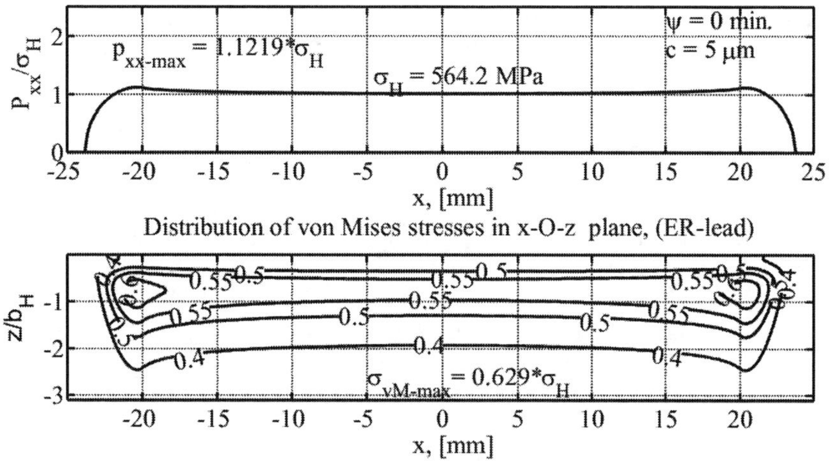

Fig. 6. Distributions of pressures and von Mises depth stresses, ($\psi = 0$ min., and $c = 5$ μm).

For operating condition without misalignment between gears axes the longitudinal modification of tooth flank profile by end relief provides smaller values for the maximum pressure in comparison with the values provided by the longitudinal crowning of the tooth flank profile.

3.2 Modified spur gears with end relief of flank profile, case $\psi > 0$

3.2.1 Distribution of pressures

When the operating conditions included a misalignment $\psi = 0.5$ min., the pressures shown a skewed distribution with a slight edge effect, Figure 7.

The harmful effect of misalignment is revealed by stress analysis. When for parameter c the value $c = 5$ μm was used, the maximum value of pressures increased, from $p_{max} =$ 633 MPa, Figure 6, to $p_{max} = 854$ MPa, Figure 7.

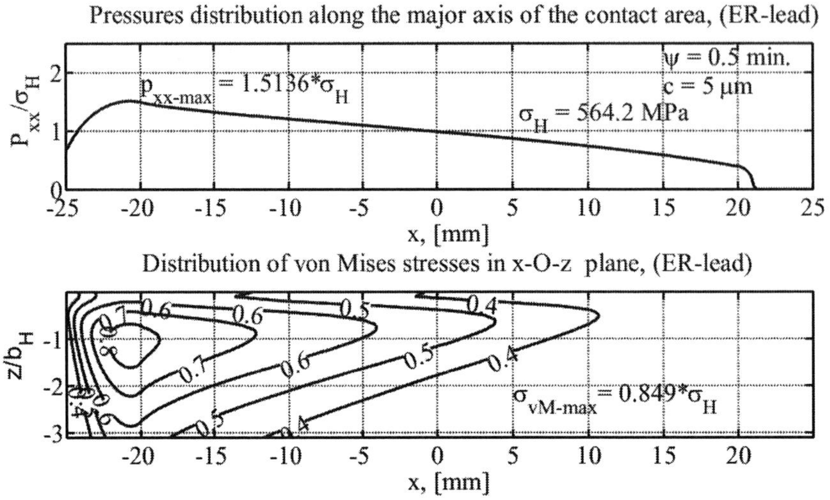

Fig. 7. Distributions of pressures and von Mises depth stresses, (ψ = 0.5 min., and c = 5 μm).

The skewed distribution of pressures generates a lop-sided depth distribution of von Mises stresses, with the maximum value in the vicinity of maximum pressure, below the contact surface, Figure 7 and Figure 8.

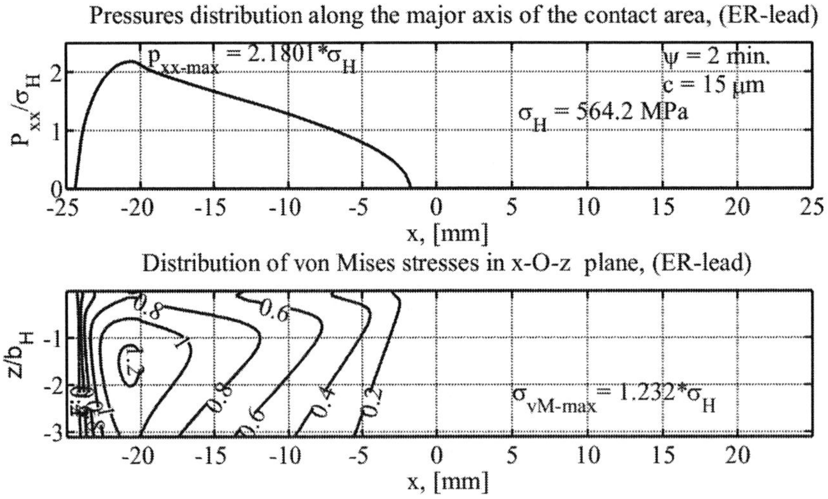

Fig. 8. Distributions of pressures and von Mises depth stresses, (ψ = 2 min., and c = 15 μm).

3.2.2 Contact areas

When gears operate with a small misalignment, $\psi = 0.5\ min.$, the contact area is changed from quasi-rectangular shape to a trapezoidal shape, Figure 9a. For greater values of misalignment, the contact area has a triangle shape, the contact length representing only a portion of the flank width, even less than half-width, Figure 9b.

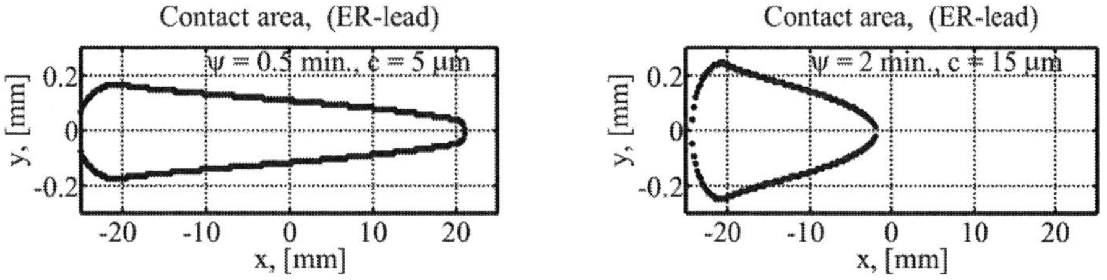

a   b

Fig. 9. Contact areas: a - small misalignment, $\psi = 0.5$ min. and small end relief $c = 5$ μm ; b - large misalignment, $\psi = 2$ min. and average value for the end relief $c = 15$ μm.

4 Conclusions

Modified spur gears able to localize the initial bearing contact have been analyzed using the semi-analytical method Non-Hertz presented in part 1of the paper.

Spur gears with pinion teeth having a longitudinal crowned flank profile make an ellipsoidal distribution of pressures on an elliptical contact area. These gears have indicated a good capacity to avoid the end edge effect when the operating conditions included the misalignment between gears axes.

For operating conditions without misalignment, the spur gears modified by end relief of flank longitudinal profile generate a pressures distribution close to an elliptic-cylinder and a quasi-rectangular contact area. The presence of misalignment creates a skewed pressures distribution with severe pressures concentrations at one end of the tooth flank.

For given operating conditions, a tooth contact analysis (TCA) is needed to find the optimum values for parameters that define the modified spur gears.

References

1. F. Litvin, A. Fuentes, *Gear geometry and applied theory*, Cambridge University Press (2004)
2. ISO 6336-1, *Calculation of load capacity of spur and helical gears*, Switzerland (2006)
3. S. Creţu, E. Antalucă, O. Creţu, Annals Univ. of Galaţi, Romania **24** 26-35 (2003)
4. S. Creţu, Bull. I.P. Iaşi **LI(LV)** 1-31 (2005)
5. K. Johnson, *Contact mechanics*, Cambridge University Press (1985)
6. D. Nelias, V. Boucly, E. Antălucă, S. Creţu, ASME J. Tribol. **129** 791-800 (2007)
7. S. Creţu, Elastic plastic concentrated contact (in Romanian) (Ed. Politehnium, Iaşi, Romania, 2009)
8. N. Pop, S. Creţu, A. Tufescu, App. Mech. and Mat. **658**,531-356 (2014)
9. Y. Siang-Yu, T. Shyi-Jeng, Mec. Mach. Th.,**97**, 190-214 (2016)
10. S. Creţu, N. Pop, S. Cazan, IManee17, Iaşi, Romania (2017)

Design algorithm for generatrix profile of cylindrical crowned rollers

Spiridon Crețu*

Technical University" Gheorghe Asachi" of Iaşi, Department of Mechanical Engineering, 63, D. Mangeron St., Iaşi, Romania

Abstract. The cross-section of roller profile controls the pressure distribution in the contact area and radically affects the roller bearings basic dynamic load rating and rating lives. Today the most used roller profiles are the logarithmic profile and cylindrical-crowned (ZB) profile. The logarithmic profile has a continuous evolution with no discontinuities till the intersection with the end fillet while ZB profile has two more discontinuities at the intersections points between the crowning circle and straight line generatrix. Using a semi-analytical method, a numerical study has been carried out to find the optimum ZB profile for rollers incorporated in cylindrical rollers bearings. The basic reference rating life (L_{10_r}) has been used as optimization criterion.

1 Introduction

Design changes which reduce material stresses appeared to be the main possibilities to improve roller bearing lives within the same envelope dimensions. The cross -section of roller profile controls the pressures distribution in the contact area and radically affects the bearing basic dynamic load rating and rating lives, [1-3]. Consequently, the primary target was to reduce stresses in the roller raceway contact by modifying roller profile.

The diversity of crowning profiles includes: the single straight line with end fillets, single arc or a combination of multiple arcs, cylindrical-crowned (ZB), Figure 1, and logarithmic. Reviewing papers [4-7] have pointed out that the contact pressures tends to be significantly higher around the profile's discontinuities than the pressures attained in the middle of the contact area. The edge effect is considerably amplified when the bearing operates with a misalignment angle ψ between axes of the outer and inner rings [2-4].

2 End relief set up

The size of cylindrical roller is described by diameter D_w and length L_w. The cylindrical-crowned profile is defined by two parameters: the value ZK of the end relief, and value b representing the length of circular profile, Figure 1. For a particular size (D_w*L_w) of cylindrical roller, the end relief Z_K of ZB profile is established considering the values provided by the chosen logarithmic profile.

* Corresponding author: spcretu@tuiasi.ro

DOI: 10.1051/matecconf/201711207017

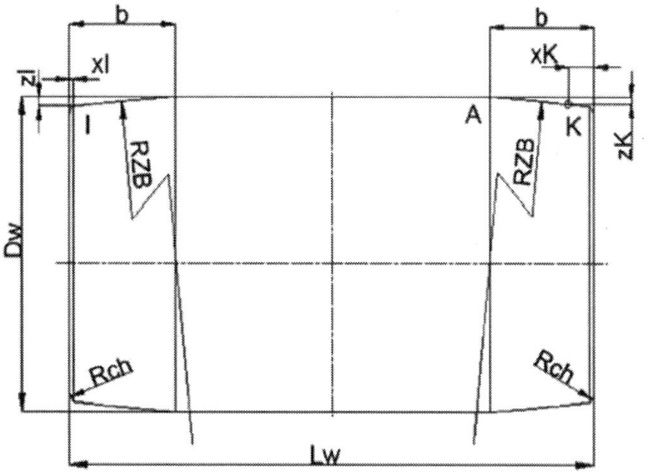

Fig. 1 Cylindrical crowned roller.

For usual applications, and rollers made of steel the profile functions $P(x)$ are presented in international standards [3] and, with some modifications, in technical documentations of bearings manufacturers.

For rollers having a length, $L_{we} \leq 2.5 * D_{we}$, the international standard ISO 16281 [3] gives:

$$P(x) = 0.00035 D_{we} \ln\left[\frac{1}{1-(2x/L_{we})^2}\right] \quad (1)$$

Other commendation aims the minimum value Z_{T_min} for the end relief in a given point T of the minimum function $P_{min}(x)$:

$$P_{min}(x_T) = Z_{T-min} \quad (2)$$

Using function $P_{min}(x)$ the profile functions $P_{nom}(x)$ and $P_{max}(x)$ are established for nominal and maximum profiles, repectively. In this paper was considered: $P_{nom}(x) = 1.5 * P_{min}(x)$ and $P_{max}(x) = 2.0 * P_{min}(x)$, Figure 2.

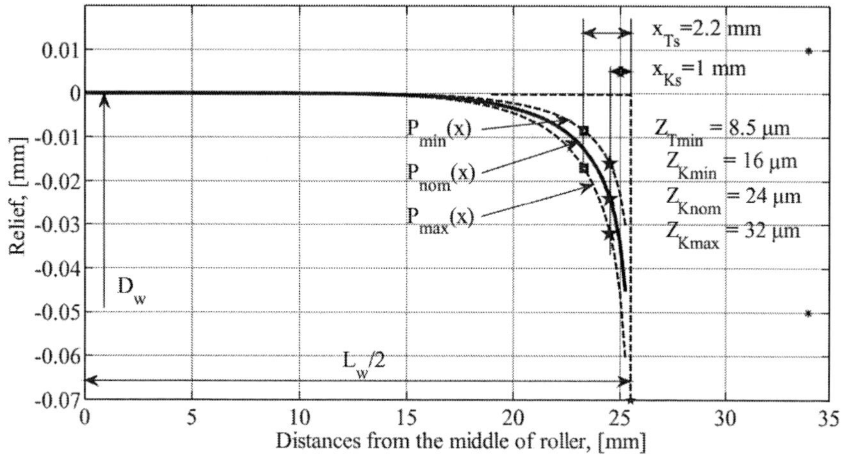

Fig. 2. The nominal profile shape $P_{nom}(x)$, function curves $P_{min}(x)$, $P_{max}(x)$ and end relievs for logarithmic profiled cylindrical roller components of NJ2320E bearing.

3 Length of circular profile

The value of b is established to maintain high values for the basic reference rating lives when the operating parameters might significantly change: from medium loads and reduced misalignments ($F_r = 0.1*C_{r}$, $\Psi = 1$ min.) till very high loads and severe misalignments ($F_r = 0.3*C_r$, $\Psi = 3.0$ min). A numerical study has been accomplished to point out the dependence of the basic reference rating lives L_{10_r} on length b of roller generatrix with circular profile.

3.1 Rating Lives

3.1.1. Modified rating life

To evaluate the modified rating life of bearings ISO 281 [8] uses the equation:

$$L_{10\,\mathrm{m}} = a_1 a_{\mathrm{ISO}} \cdot L_{10} = a_1 \cdot a_{\mathrm{ISO}} \cdot \left(\frac{C_r}{F_r}\right)^p \tag{3}$$

where P is the dynamic equivalent radial load, p=*10/3* and $a_1 = 1$ when a 90% reliability is admitted. The life modification factor a_{ISO} is:

$$a_{\mathrm{ISO}} = f\left(\frac{e_C \cdot C_u}{P}, \kappa\right) \tag{4}$$

where C_u is the fatigue load limit, whereas the parameters e_C and κ take into account the contamination and lubrication conditions. For mineral oil lubrication and bearing raceway surfaces machined with *good manufacturing quality*, the condition of lubricant separation is described by the complex parameter κ , defined as the ratio of the actual kinematic viscosity ν to the reference kinematic viscosity ν_1 .

3.1.2 Basic reference rating life. The lamina technique

Additional to the influencing parameters described by ISO 281 [8], the standard ISO 16281 [3] takes into account three more operating parameters: the misalignment, operating internal clearance and internal load distribution on rolling elements. The main steps to obtain the basic reference rating life L_{10_r} [3] are mentioned in the following:

Pressures distributions and distribution of contact stresses. The roller-raceway contacts of length L_{we} are divided into a number of n_s laminae each of width w, and $L_{we} = n_s \cdot w$. A number of laminae $n_s = 256$ was used in calculations. The distribution of contact stresses $q_{j,k}$ over the length of the j-roller is established using the pressure distribution provided by Non-Hertz semi-analytical method described in [9, 10]. The virtual rectangular contact area was discretized with 256*64 patches, each of them supporting a different unknown pressure. More details are in [11, 12]. The influence of misalignment on pressures distributions achieved in the contact between the most loaded roller and inner ring of NJ2320 bearing is comparatively exemplified for operating conditions without misalignment $\psi = 0$ Figure 3, and with a misalignment $\psi = 1 minute$, Figure 4.

Functions for stress concentrations stresses. Approximated functions of the stress concentration stresses over the roller length are further developed for the inner ring raceway $f_i\,[j,k]$, and outer ring raceway $f_e[j,k]$ [3].

Dynamic equivalent load on a lamina. The dynamic equivalent load on a lamina k of a rotating inner ring is:

$$q_{kei} = \left[\frac{1}{Z_w} \sum_1^{Z_w} \left(f_i[j,k] \cdot q_{j,k}\right)^4\right]^{1/4} \tag{5}$$

DOI: 10.1051/matecconf/201711207017

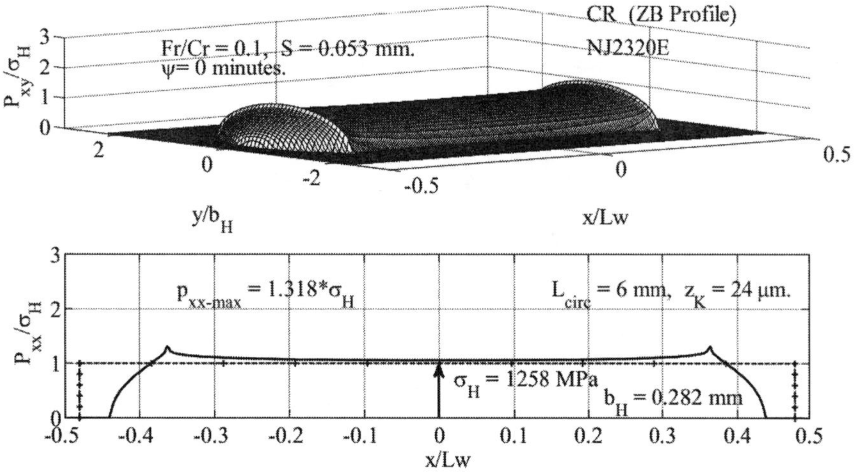

Fig. 3. Pressures distribution for operating conditions without misalignment.

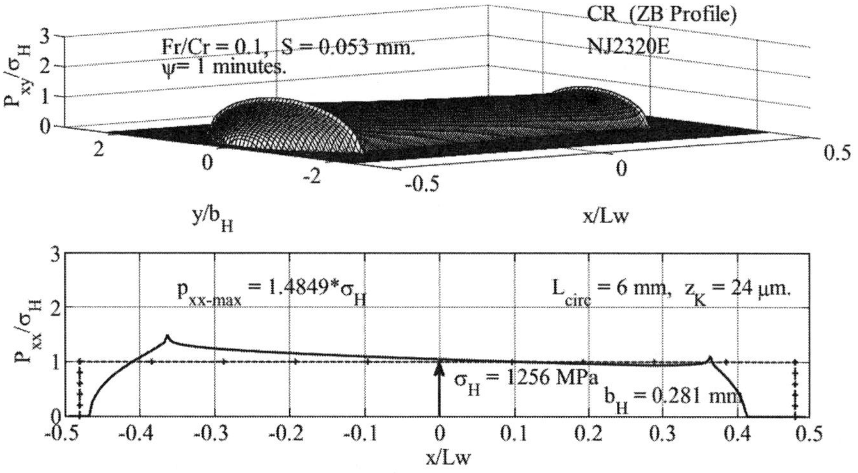

Fig. 4. Pressures distribution for operating conditions with $\psi = 1minute$ misalignment.

The dynamic equivalent load on a lamina k of a stationary outer ring is:

$$q_{kee} = \left[\frac{1}{Z_w} \sum_1^{Z_w} \left(f_e[j,k] \cdot q_{j,k}\right)^{4.5}\right]^{1/4.5} \tag{6}$$

Fatigue lives of roller-raceway contact laminae. Considering a fourth power load-life dependence for the line contact, the fatigue lives of roller-raceway contact laminae is:

$$L_{ijk} = \left(\frac{q_{cij}}{q_{ijk}}\right)^4, \; L_{ejk} = \left(\frac{q_{cej}}{q_{ejk}}\right)^{9/2} \tag{7}$$

The basic reference rating life L_{10_r} is calculated using the primary life formula:

$$\boldsymbol{L_{10\_r}} = \left\{\sum_{k=1}^{m} \left[\left(\frac{q_{kci}}{q_{kei}}\right)^{-4.5} + \left(\frac{q_{kce}}{q_{kee}}\right)^{-4.5}\right]\right\}^{-8/9} \tag{8}$$

3.1.3 Modified reference rating life L_{nmr}

The modification factor a_{ISO}, calculated using equations (34) to (36) from [8], is determined, for each lamina and considered in equation of modified reference rating life:

$$L_{nmr} = a_1 \left(\sum_{k=1}^{n_s} \left\{ \left[a_{ISO} \left(\frac{e_C \cdot C_{ur}}{P_{ks}}, \kappa \right) \right]^{-9/8} \cdot \left[\left(\frac{q_{kci}}{q_{kei}} \right)^{-4.5} + \left(\frac{q_{kce}}{q_{kee}} \right)^{-4.5} \right] \right\} \right)^{-8/9} \quad (9)$$

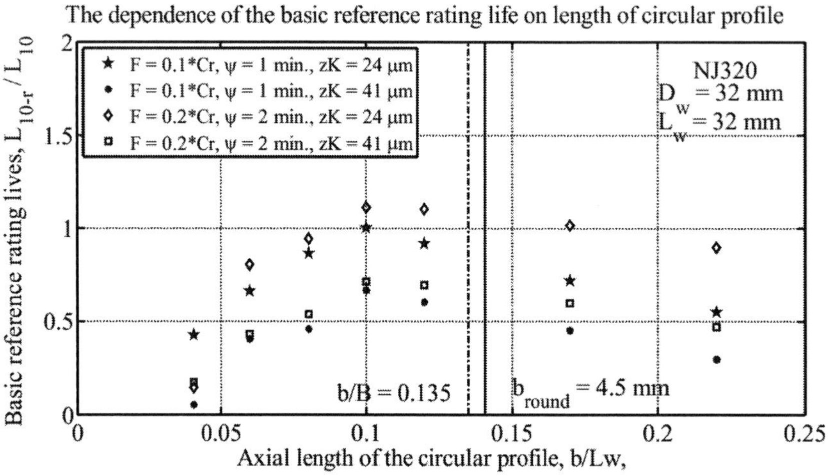

Fig. 5. The basic reference rating lives as function of length of circular profile, NJ320 bearing.

The following formulae has been derived for the minimum length of t circular profiles:

$$b_{adim} = 0.125 + 0.01 * \frac{L_w}{D_w} \quad (10)$$

$$b_{min} = \text{round}(b_{adim} * L_w,\ 0.5) \quad (11)$$

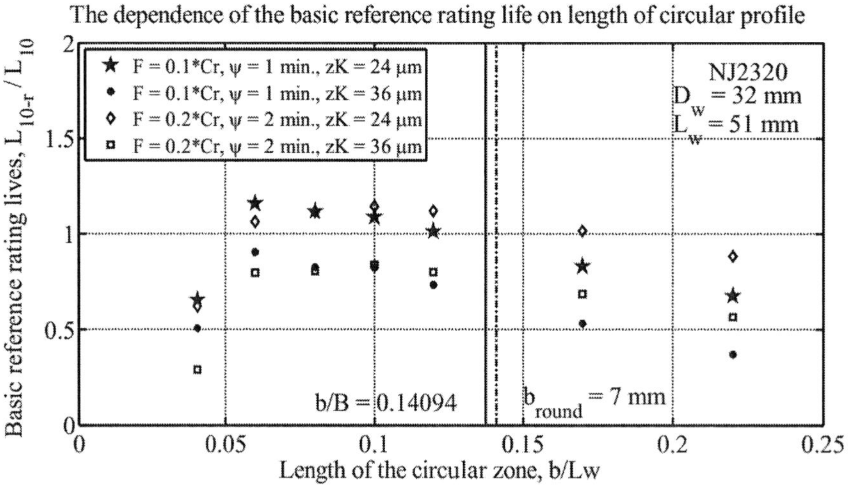

Fig. 6. The basic reference rating lives as function of length of circular profile, NJ2320 bearing.

3.2 Basic reference rating lives and the length of circular profile

The various operating conditions were considered as the input data in the computations of basic reference rating life L_{10_r}.

The values for length b have been selected in order to maintain high values of basic reference rating lives for a large field of operating parameters.
Some results of computations are exemplified in Figure 5 for NJ2320 bearing and in Figure 6 for NJ320 bearing. The positions of the minimum length b_{min} inside the lives dependencies plots are also presented in Figure 5 and Figure 6.

4 Conclusions

The size of cylindrical roller is determined by diameter D_w and length L_w. The cylindrical crowned profile is defined by two parameters: the value ZK of the end relief, and the length b of circular profile.

For any particular roller size (D_w*L_w), the end relief Z_K of cylindrical crowned profile is established considering the values provided by preferred logarithmic profile.
A numerical study has been accomplished to reveal the dependence of the basic reference rating life, L_{10_r} on the length b of roller generatrix with circular profile.

The value of b was established to maintain the high values for the basic reference rating lives for a large field of operating conditions: from usual loads and misalignment ($F_r = 0.1*C_r$, $\Psi = 1$ min.) till very high loads and severe misalignment ($F_r = 0.3*C_r$, $\Psi = 3.0$ min).

References

1. E. Ioannides, G. Bergling, A. Gabelli, *Acta Scandinavica, Me* **137** 1-80 (1999)
2. T. Harris and M. Kotzalas, *Rolling bearing analysis-Advanced concepts of bearing technology* (Boca Raton: CRC Press, Taylor & Francis Group, 2007)
3. ISO 16281, *Rolling bearings-methods for calculating the modified reference rating life for universally loaded bearings,* Switzerland (2008)
4. S. Crețu, M. Benchea, A.Iovan-Dragomir, J. Balkan. Tribol. Assoc., **22** 1 (2016)
5. H. Reusner, Ball Bearing Journal. **230** 2-10 (1987)
6. J. de Mul, J. Kalker, B. Fredriksson, ASME J. Tribol., **108** 140-148 (1986)
7. H. Fujiwara, T. Kobayashi, T. Kawase, K. Yamauchi, *Tribol. Trans.* **53**, 910-916 (2010)
8. ISO 281, *Rolling bearings—Dynamic load ratings and rating life*, , Switzerland, (2007)
9. I. Polonsky, and L. Keer, ASME J. Tribol. **122,** 1 (2000)
10. S. Crețu, E. Antalucă, O. Crețu, Annals Univ. Galați Romania **24**, 26-35 (2003)
11. S. Crețu, Bull. Inst. Polit. Iași, **LI (LV)** 1-2, 1-31 (2005)
12. S. Crețu, *Elastic plastic concentrated contact* (in Romanian) (Ed. Politehnium, Iași, Romania, 2009)

Theoretical research on the issue of heat stress that appears in mechanism of automotive braking system and their influence

Tudor Ulian *, *Vlad* Vornicu, *Sorinel* Talif, and *Edward* Rakosi

"Gheorghe Asachi" Technical University of Iasi, Faculty of Mechanical Engineering, Department of Mechanical Engineering and Automotive Engineering, Romania

Abstract. The braking system is one of the most important car systems with application in traffic safety. The thermal stress in this system, especially when the brake is intensively used, is affecting the efficiency of the system, its response and therefore the road safety.

1 Introduction

Efficiency of the braking system and good functionality, in correlation with his main roles, improves the average travel speeds of car, but also the passenger safety. The studies show that much of the vehicle's kinetic energy is converted into heat, which leads to poor functionality of the entire braking system.

As negative effects of brake system warming can be mentioned: rapid wear of the brake pads or brake disc, cracks or deformation of the brake disc, occurrence of the "fading" phenomenon.

In this context, one of the important challenges of the automotive industry is to improve the braking performances. To obtain a reliable constructive solution have been done a series of theoretical and experimental research based on numerical modeling, finite element analysis and laboratory studies.

2 Modeling of brake disc - pads assembly

In this paper, the numerical modeling of disc brake – pads assembly was performed, customized to the braking system of the car. The model has considered the constructive solution, based on an unventilated disc. To perform the simulation, firstly the components of the braking system were modeled, in order to reproduce as faithfully as possible the geometry and their position during operation. This modeling was performed using CATIA V5 R19 design environment, widely used in the automotive design.

In Figure 1-5 are shown the 3D models of the unventilated brake disc, caliper, the piston inside the caliper, brake pads and whole disc pads.

* Corresponding author: uliantudor@yahoo.fr

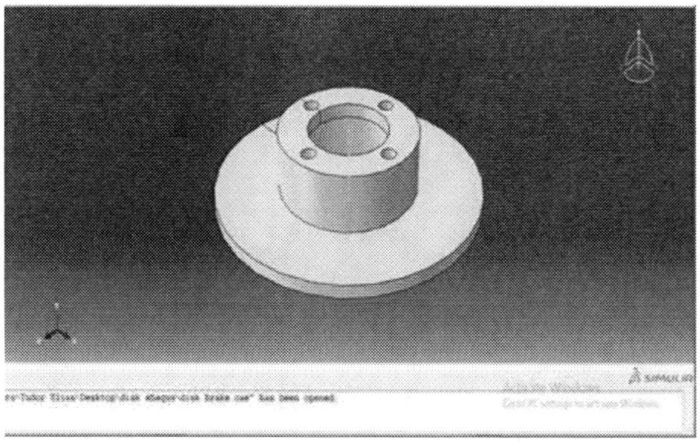

Fig. 1. Brake disc.

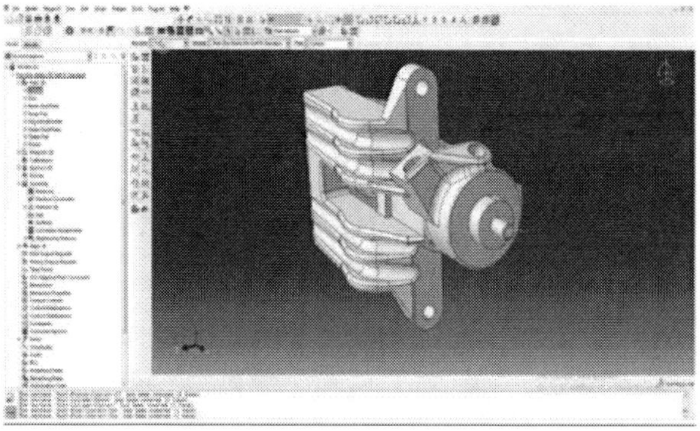

Fig. 2. Caliper.

Fig. 3. Piston.

Fig. 4. Brake pads.

DOI: 10.1051/matecconf/201711207018

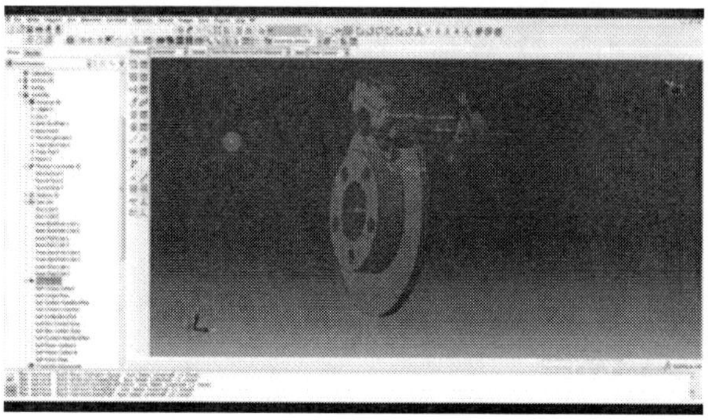

Fig. 5. Brake disc – brake pads assembly.

Modeling of the disc and brake pads, in discretized form is shown in Figure 6 and Figure 7.

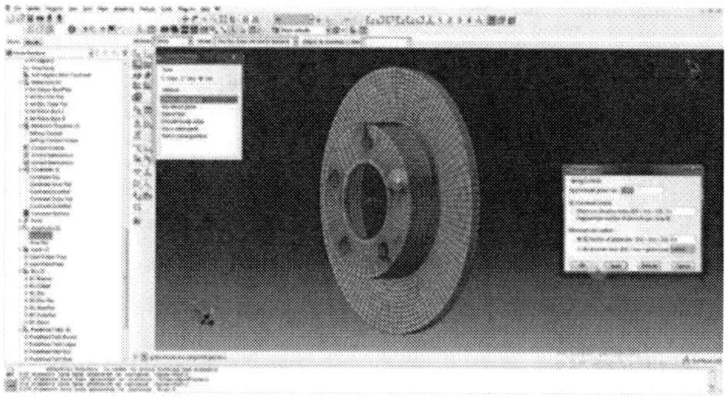

Fig. 6. Meshed brake disc.

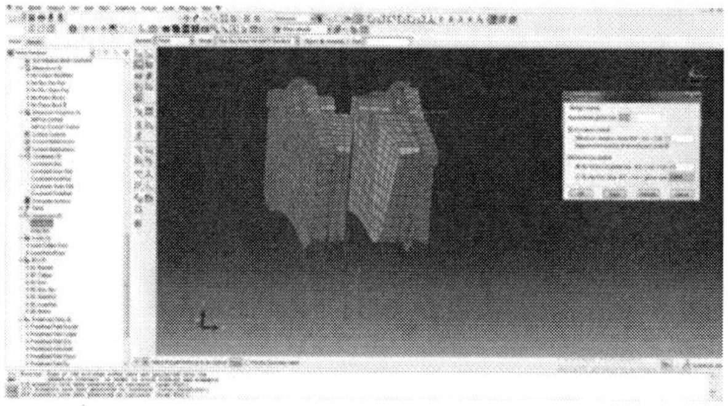

Fig. 7. Meshed brake pads.

3 Simulation and study of the temperature distribution

The analysis of temperature distribution on the friction surface between the brake disc and brake pads was done by using finite element analysis (FEA), based on ABAQUS 6.14 software, produced by Dassault Systèmes Company. This program provides a very accurate analysis, being able to identify areas with critical mechanical and heat stress, in order to offer viable technical solutions.

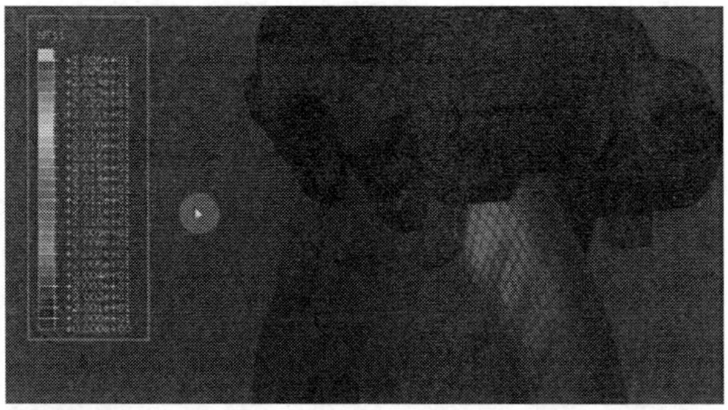

Fig. 8. Distribution of temperature of the brake disc at intense braking (50 km/h) - achieved at the start of braking.

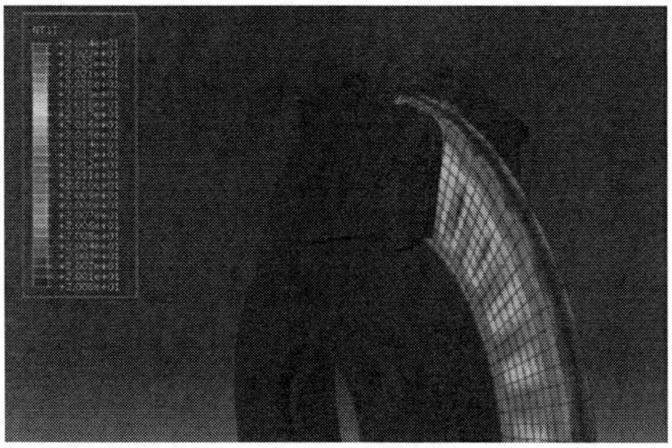

Fig. 9. Distribution of temperature of the brake disc at intense braking (50 km/h) - achieved at the end of braking.

For this analysis, it was considered the following situation: an intense braking from two representative values of speed, 50 km/h and 100 km/h to 0 km/h. This analysis is based on the following assumptions:
- caliper, support of caliper and the piston are considered rigid bodies, which will not conduct heat, generated during braking;
- disc initial temperature is 20 °C;

In order to realize a complex and relevant investigation it was considered useful an analyze on achieving the maximum temperature of the disc starting from intermediate moments, characterized by lower temperatures. This approach is aimed to obtain an evolution in the temperature distribution on the brake disc.

Fig. 10. Distribution of temperature of the brake disc at intense braking (100 km/h) - achieved at the start of braking.

DOI: 10.1051/matecconf/201711207018

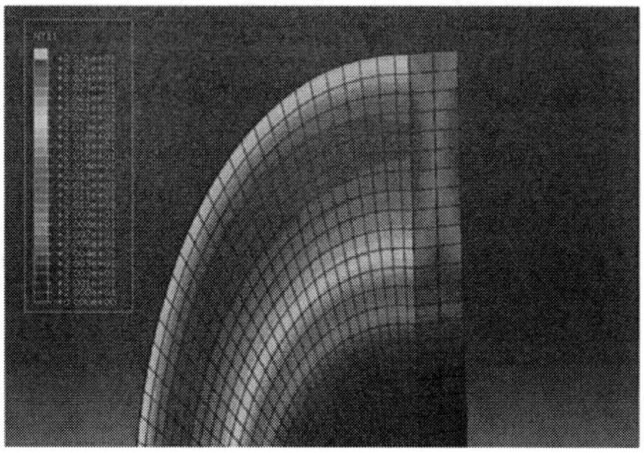

Fig. 11. Distribution of temperature of the brake disc at intense braking (100 km/h) - achieved at the end of braking.

Distributions of temperature obtained in this way are shown in Figure 8, Figure 9, Figure 10 and Figure 11. The analysis highlights the thermal storage areas exposed to extreme heat stresses.

4 Results

The analyses can take a few theoretical results, summarized below. Thus, during the braking from the speed of 50 km/h until 0 km/h, the disc can reach temperatures up to 150 °C. Also, from the speed of 100 km/h until 0 km/h, the disc may reach temperatures up to 360 °C, which represents a high risk and may cause cracks and premature wear of the braking system. In Figure 12 and Figure 13 are presented two diagrams which show the evolution of the temperature of the disc brake in time, during braking and cooling after stopping the vehicle. The temperatures were considered on a network node placed in the maximum temperature.

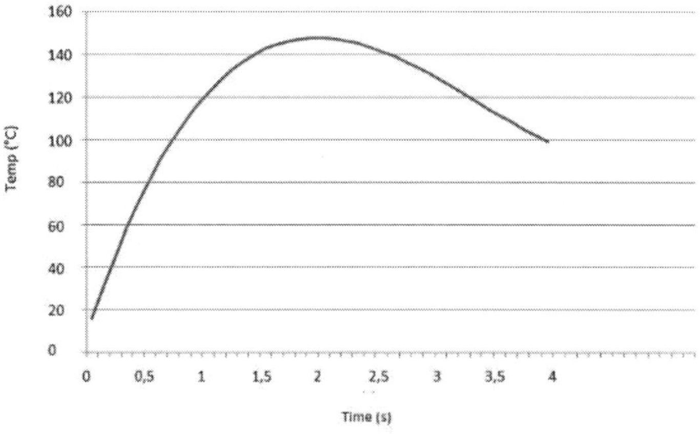

Fig. 12. Time-Temperature diagram for a particular node (50km/h).

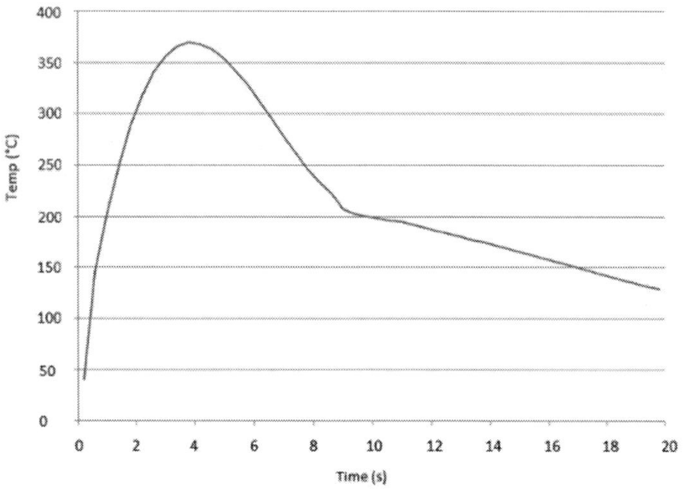

Fig. 13. Time-Temperature diagram for a particular node (100km/h).

5 Conclusions

The performed simulation leads to the possibility of viewing the phenomenon of braking from thermal point of view. The evolution of the temperature distribution in the specific braking cases is also highlighted.

These researches provide the temperature distribution on surface of the disc, thus identifying the areas with high heat stress and high mechanical stress. Based on these, we can determine the areas considered with high risk.

Is envisaged, in order to develop the study, an extension of simulation by considering a higher number of brake modes, with the expansion of the input data (materials, forces etc.), so that the results will cover a larger area of issues.

References

1. G. Frăţilă, *Automotive knowledge, maintenance and repair* (EDP, Bucuresti, 2005)
2. C. Spulber, Ş. Voloacă, *Aspects of simulation of thermal stress in disc brake using Infrared Thermography*, *OPTIROB* (Sinaia, Romania, 2011)
3. M. Untaru, G. Frăţilă, G. Poţincu, N. Seitz, I. Tabacu, G. Pereş, T. Macarie, *Automotive Design* (EDP, Bucuresti, 1982)
4. Ş. Voloacă, G. Frăţilă, *Uncertainty of the Disc Brake Thermal Stress Measured by Thermography*, Annals of the Academy of Romanian Scientists Series on Science and Technology of Information, **4** (2011)

Analysis of reaction forces in human ankle joint during gait

Daniel Ganea, *Elena* Mereuta, *Silvia* Veresiu, *Madalina* Rus, and *Valentin* Amortila*
"Dunarea de Jos" University of Galati, Department of Mechanical Engineering, 111 Domnească Street, Romania

Abstract. The paper aims to analyze the human ankle joint reaction forces with a multibody system approach, using MSC Adams software and its features. The paper presents the results provided by the simulation of a spatial mechanical structure, which replicates the human's locomotion system in terms of bone structure. The mechanical structure used for determining the human ankle joint reaction forces consists of 7 kinematic elements linked by 6 spherical joints constrained to 1 DOF each.

1 Introduction

The study of joints forces in human body joints is extremely important for physiotherapists, coaches and athletes. It is often difficult to determine these internal forces using noninvasive methods.

However, the inverse dynamics analysis can provide with good approximation the forces in human joints.

While direct dynamics is used for determining how human body actuators are used, the inverse dynamic analysis is widely used in estimating joints moments and reaction forces [1, 2].

Using the inverse dynamics approach in human body motion analysis the computational effort is considerable reduced by avoiding the numerical integration of state equations. Unfortunately, the approach is poorly explored in biomechanics [2, 3].

The same method was use by some authors to estimate the force developed by human body actuators [4]. They have used a multibody model with 5 degrees of freedom.

Other authors [5] have analyzed the reaction force developed in the knee joint and also the force developed by human body actuators, using the inverse dynamic method combined with an MBS model.

The ankle is the anatomical structure of the human leg that connects two main kinematic elements that composes the human locomotor system. The first kinematic element is the shank formed by the tibia and the fibula. The second kinematic element is the foot, which is more complex, having 26 bones and 33 joints (Figure 1, 2).

This kinematic chain is configured to ensure the stability and mobility of the human body on various surfaces during running, walking and even during orthostatic position [6].

* Corresponding author: valentin.amortila@ugal.ro

Human gait is a semi-automatic action. This highly complex activity is unpredictable and automated. The aim of this article is to determine the reaction forces in the ankle joint, during a complete gait cycle, under three different loading cases. It is important to determine these forces not only for prosthesis design, but also for physiotherapists.

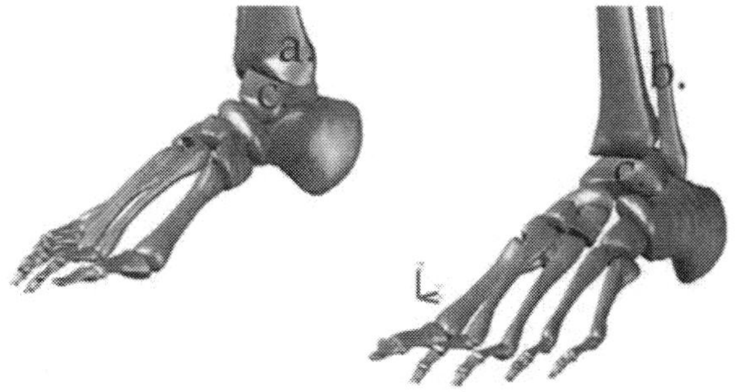

Fig. 1. Ankle joint: a. tibia bone; b. fibula bone; c. astragal bone.

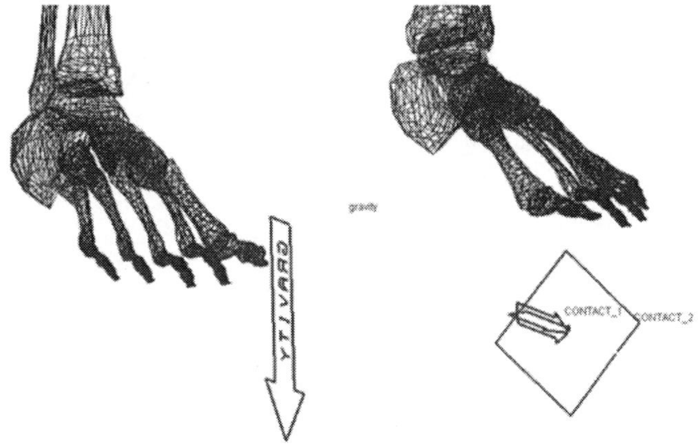

Fig. 2. Human ankle joint multibody system.

2 Method

The methods used in this study are mainly related to modelling and computer simulation. Due to the complexity of lower limb, some simplifying assumptions are required in order to facilitate the MBS modelling process. However, the simplifying assumptions have influence on the accuracy of the output data.

In order to get a more accurate model of the lower limb, we have imported a 3D geometric pattern available at 3Dlancer virtual library (http://3dlancer.net/en/3dmodel-human-skeleton-4792.html).

The imported 3D mesh model (Figure 3) was transformed into 3D solid using the Mesh to Solid feature of AutoCAD.

The solid thus created was imported in MSC Adams and the mechanical properties were associated to each kinematic element.

The kinematic structure is analyzed considering the following forces: the weight, the contact forces, the friction forces and applied forces that represent the supplementary load (Figure 4).

As input data, we have considered also the time variation of six angular parameters, previously determined using a depth camera (Kinect).

A specially created computer application analyzes the images taken by the Kinect sensor and determines the coordinates of the main joints. It is also important to perform a kinematics analysis, which provides angular velocities and accelerations.

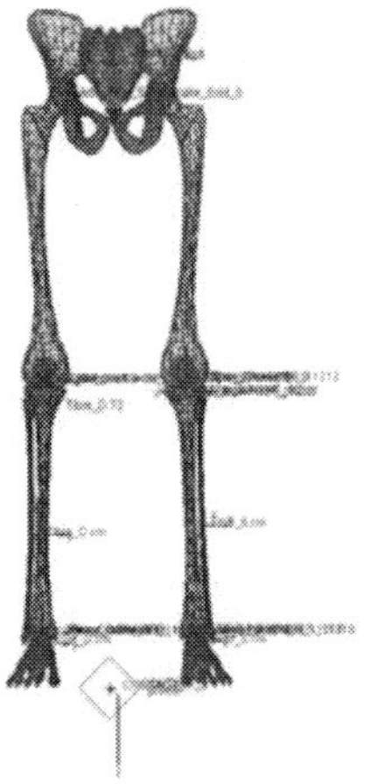

Fig. 3. 3D Mesh lower limb model.

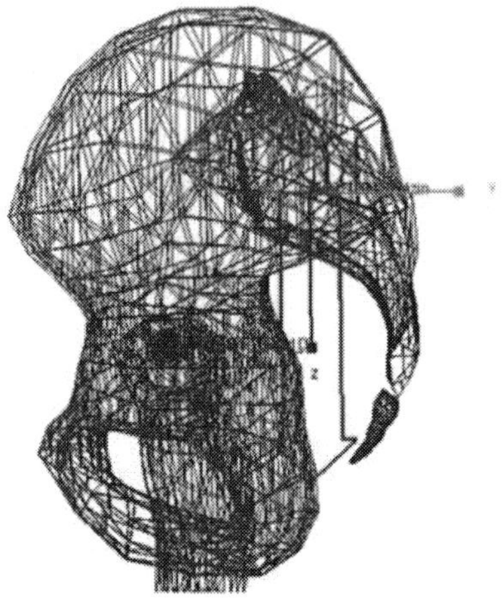

Fig. 4. Graphical representation of added weight.

The analysis was performed for a complete gait cycle, consisting of two steps, in 3 loading cases:

- First case - a normal gait cycle;
- Second case - a gait cycle with 34.46% of his body mass overload;
- Third case - a gait cycle with 64.93% of his body mass overload.

The gait cycle was divided into four stages (Figure 5). The first stage named start simulation includes two phases. The second stage is the support stage and includes 4 phases. The last two phases are the swing phase and stop simulation phase [6].

The dynamic analysis provides, besides the reaction forces the angular velocities and the angular accelerations for right ankle joint (RAJ) and left ankle joint (LAJ).

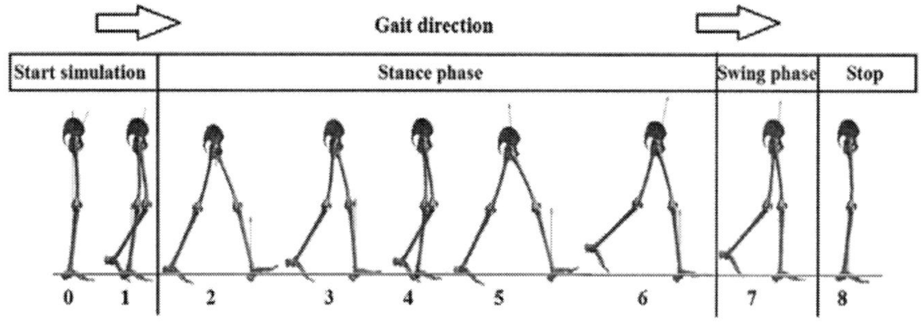

Fig. 5. Gait phases.

3 Results

The right ankle joint (RAJ) angular velocity has amplitude near 5 deg/s. The right lower limb kinematic chain reaches this value during the sub – step of initial contact. For the second

 DOI: 10.1051/matecconf/201711207019

loading case, the right ankle joint (RAJ) angular velocity shows an increasing of amplitude, from 0.32 deg/s to 0.34 deg/s representing 6.25%. For the third loading case the amplitude decreased by 20.5%.

Relatively to right ankle joint (RAJ), the left ankle joint (LAJ) time variation of angular velocity reaches a smaller amplitude (of 1.87 deg./s), towards the end of sub – phase called support completion.

The left ankle joint (LAJ) average angular velocity during the first and second loading cases is 0.32 deg/s. During the third loading case, the data reveals a decreasing of 15.62%.

The right ankle joint (RAJ) time variation of angular acceleration amplitude is reached when the multibody passes from phase 1 to phase 2 (Figures 6, 7, 8).

For the first loading case, the angular acceleration amplitude is 6.46 deg/s^2, for the second loading case the amplitude is 10.68 deg/s^2, which increases with 95% during the third case (20.83 deg/s^2). The average angular acceleration data shows an equal value in all three loading cases (0.33 degrees/s^2).

The LAJ angular acceleration variation with respect to time reaches its amplitude during the sub – phase called support completion (Figures 6, 7, 8).

During the first loading case the angular acceleration time variation curve with respect to time reaches a maximum value of 2.91 deg/s^2, during the second case a rather small increasing to 3,01 deg/s^2 (3.4%) and a significant increasing by 27.5% (3,84 degrees/s^2). The average angular acceleration shows a slight increasing from 0.12 deg/s^2 to 0,138 deg/s^2, during the second and third loading cases.

The simulation also provides information on the reaction forces, revealing that the ankle joints are stressed.

During the gait cycle, the reaction force time variation reaches its amplitude at the end of phase 2 of the first studied case (Figures 9, 10, 11).

During the second loading case, the reaction force reaches an amplitude increased by 90% (1397.27 N), while in the third loading case it increased by 12.26% (1568.6 N).

The average reaction force recorded in the first loading case was 107.04 N, 180.03 N during the second one and 233.92 N during the third loading case.

The data reveals the fact that the most stressed joint is the LAJ. The amplitudes are the biggest during the beginning of the swing phase.

The LAJ reaction force is 852,154 N in the first loading case and the average is 204.05 N. During the second loading case the reaction force has a maximum value of 1114 N and an average value of 315.09 N.

The data resulted from the last loading case reveals that in this case the left ankle joint takes the biggest mechanical stress (1631.6 N with an average of 410.39 N).

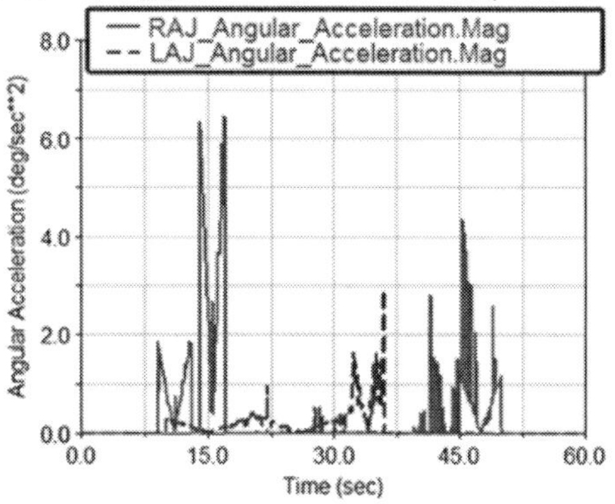

Fig. 6. RAJ and LAJ angular acceleration with respect to time (first loading case).

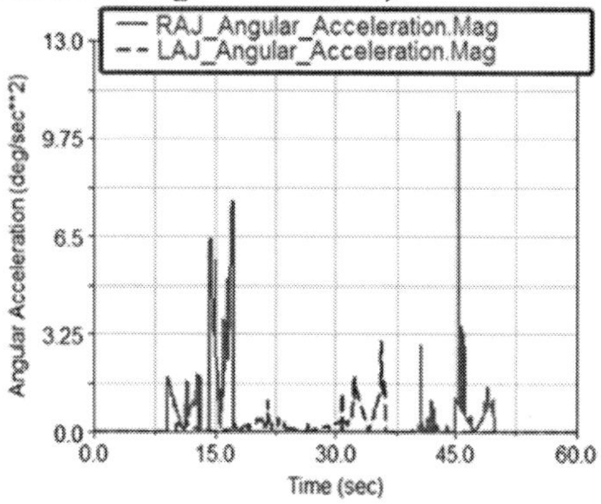

Fig. 7. RAJ and LAJ angular acceleration with respect to time (second loading case).

 DOI: 10.1051/matecconf/201711207019

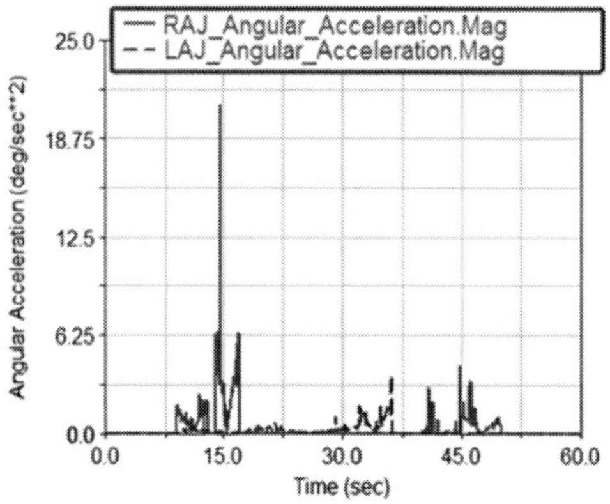

Fig. 8. RAJ and LAJ angular acceleration with respect to time (third loading case).

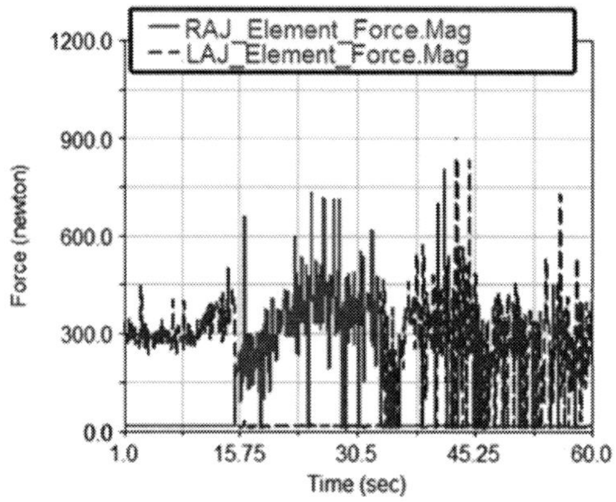

Fig. 9. Reaction force with respect to time recorded in RAJ and LAJ (first loading case).

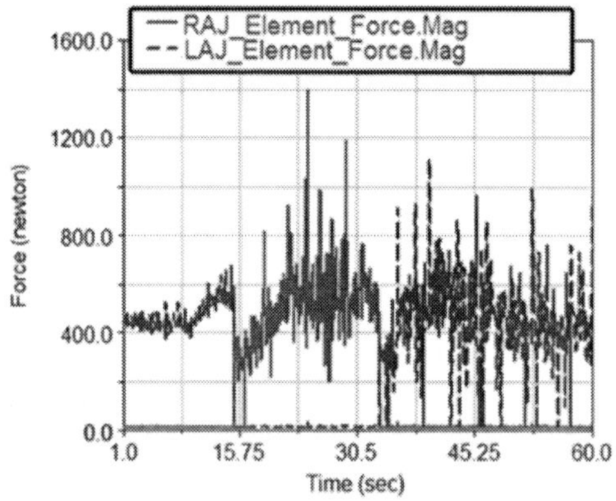

Fig. 10. Reaction force with respect to time recorded in RAJ and LAJ (second loading case).

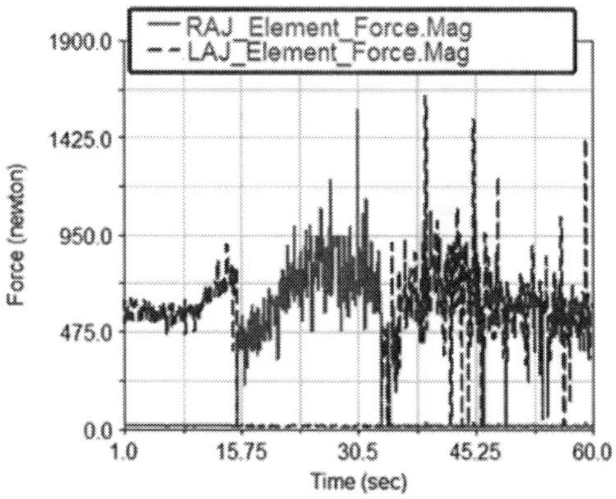

Fig. 11. Reaction force with respect to time recorded in RAJ and LAJ (third loading case).

3 Conclusions

The study of joints forces in human body joints is extremely important for physiotherapists, coaches and athletes. It is often difficult to determine these internal forces using noninvasive methods. However, the inverse dynamics analysis can provide with good approximation the forces in human joints.

Using the inverse dynamics approach in human body motion analysis the computational effort is considerable reduced by avoiding the numerical integration of state equations.

Simulating the model using laws of motion for the variation of kinematic parameters, we have been able to determine the reaction forces variation with respect to time for left and right ankle joints (LAJ, RAJ).

The simulated data revealed that the amplitude of the reaction force with respect to time for the right and left joint is asymmetrical.

For the first loading case, it can be observed that the reaction force has different amplitudes. Due to the applied forces, during the second loading case it is noticed that the mechanical stress rise with 90.04 % for the RAJ and with 30.72% for the LAJ.

For the third loading case the data reveals that the reaction force increased significantly, to 1568.60 N (133.34%) for the RAJ and to 1631.60 N (91.46%) for LAJ.

To conclude, we were able to set a model which can be used in the study of forces that are developed during gait. This model can provide information on the internal forces which are important in prosthesis design. Also, the amplitudes of these forces are significant for optimal selection of more efficient endurance exercises for a better recovery of injured lower limbs.

References

1. S. Chowdhury, N. Kumar, Journal of Rehabilitation Robotics, **1**, 2, (2013)
2. M. Ackermann, *Dynamics and energetics of walking with prostheses*, (2007)
3. D. Ganea, E. Mereuta, C. Mereuta, J AMM, **555** (2014)
4. F. Moissenet, L. Cheze, R. Dumas, J Multibody Syst Dyn, **28** (2012)
5. C.Y. Lin, P.J. Walter, S.A. Banks, G.M. Pandy, B.J. Fregly, J Biomech, **43** (2010)
6. C. L. Vaughan, B. L. Davis, J. C. O'Connor, *Dynamics of Human Gait*, (2nd Edition, Kiboho Publishers Cape Town, South Africa, 1999)

Atmospheric air density analysis with Meteo-40S wind monitoring system

Dănuț Zahariea[1,*], and *Dorin Emil* Husaru [2]

[1]"Gheorghe Asachi" Technical University of Iași, Department of Fluid Mechanics, Fluid Machinery and Fluid Power Systems, D. Mangeron 59A, Iași, România
[2]"Gheorghe Asachi" Technical University of Iași, Department of Mechanical Engineering, Mecatronics and Robotics, D. Mangeron 43, Iași, România

Abstract. In order to estimate the wind potential of wind turbine sites, the wind resource maps can be used for mean annual wind speed, wind speed frequency distribution and mean annual wind power density determination. The general evaluation of the wind resource and the wind turbine ratings are based on the standard air density measured at sea level and at 15°C, ρ_S=1.225 kg/m^3. Based on the experimental data obtained for a continental climate specific location, this study will present the relative error between the standard air density and the density of the dry and the moist air. Considering a cold day, for example on Friday 10th February 2017, on 1-second measurement rate and 10-minute measuring interval starting at 16:20, the mean relative errors obtained are 10.4145% for dry air, and 10.3634% for moist air. Based on these results, a correction for temperature, atmospheric air pressure and relative humidity should be always considered for wind resource assessment, as well as for the predicting the wind turbines performance.

1 Introduction

Wind turbine energy generation involve complex aspects such as: wind energy conversion [1], planning and building permits (regional framework, building permit, assessment studies of environmental impact, connecting turbines to the electrical grid), transportation problems, erection on the site, grid connection, operating and monitoring, safety aspects, maintenance and repair [2, 3], as well as economics of wind energy, [4]. All these aspects are covered by technical, environmental and economic studies, standards, regulations, researches all over the world, and represent after all, work done by human resources.

However, the most important aspect, which represent the starting point of the wind turbine energy generation process is the wind resource assessment, [5], which allows to obtain important data about the wind potential, like mean annual wind speed, wind speed frequency distribution and mean annual wind power density. The wind power density, P [W/m^2], which represent the power in the upstream wind, is defined by the relationship, [6]:

$$P = 1/2 \cdot \rho V^3 \quad (1)$$

where ρ [kg/m^3] is the air density, and V [m/s] is the mean annual wind speed.

* Corresponding author and co-author(s): dzahariea@yahoo.com

DOI: 10.1051/matecconf/201711207020

Analyzing the equation (1), it can be observed that the wind speed has an exponent of 3, so even small change in wind speed leads to major changes in wind power density. It is noted also, the presence of the air density, and even if its exponent is 1, it plays an important role in the evaluation of the wind power density, which will be further discussed in this paper.

The general evaluation of the wind resource on the large scale should be performed using the wind resource data stored in the so-called wind atlases. Based on the meteorological data gathered mainly from all national weather stations, and using numerical wind models, national wind resource estimates have been obtained in different forms: tables, charts and maps. Collecting wind data from many countries allows obtaining the regional wind resource data, such as the European Wind Atlas [7], or the Global Wind Atlas, [8]. All these online wind resource data are available, generally, at three heights 50, 100 and 200 m. Further, wind speed can be extrapolated to the wind turbine rotor height, using different models of the wind speed distribution with altitude in the atmospheric boundary layer, like the logarithmic and power laws, [2]. After the general evaluation of the wind resource has been successfully completed and a potentially suitable wind turbine site has been selected, a more deeply analysis should be performed, this time on a small scale in order to obtain more precise information on the site-specific wind resource. The first approach is the analysis of the site topography looking after microscale effects such as ridges oriented perpendicular to the prevailing wind direction, highest elevations, and areas where local winds can funnel, [5]. However, the most important analysis is the on-site wind monitoring for at least one year, using specialized wind measurement systems.

It is very important to observe that the general evaluation of the wind resource is based on the standard air density measured at sea level and at 15°C, ρ_S=1.225 kg/m^3, [6]. However, there are regions around the world where the air density has very different values from the standard air density, due to high fluctuations of temperatures, humidity and atmospheric air pressure. For example, in [9] has been found that the monthly air density is at most 6% less than the standard air density for the tropical climates of Trinidad and Tobago. Measuring the temperature, atmospheric air pressure and relative humidity in Gyor, Hungary from 2004 to 2006, it has been found that, even if the average air density (1.229 kg/m^3) is rather close to the standard air density, the estimation of the wind power production was 16% more accurate, [10]. Another parameter that modifies the air density is the site elevation; the impact of the air density variation with altitude (up to 3000 m) on functional performances of a three-bladed small wind turbine has been investigated in [11], and it has been established that the performance of a wind turbine designed to work at sea level decrease with increasing altitude.

In this study, the influence of the measured values of temperature, atmospheric air pressure and relative humidity on the air density will be comparatively analyzed, with respect to the standard air density by computing the relative errors between the standard air density and the density of the dry, respectively the moist air, for a continental climate specific location with latitude 47,155876 and longitude 27,60095. The experimental data will be obtained with the Meteo-40S wind monitoring system which is composed by Ammonit Meteo-40S data logger [12], Vaisala barometric pressure sensor [13], Galltec hygro/thermo sensor [14], and Thies ultrasonic 2D anemometers, [15].

2 Wind monitoring system description

The wind monitoring system about we will discuss in this paper, is located in the Wind Turbine Field Test Laboratory from the Department of Fluid Mechanics, Fluid Machinery and Fluid Power Systems, Faculty of Machine Manufacturing and Industrial Management at "Gheorghe Asachi" Technical University of Iaşi, Romania. The coordinates of the wind monitoring system are: latitude 47,155876 and longitude 27,60095. Using the Google Earth application, the wind monitoring system location can be observed in Fig. 1.

The wind monitoring system is identified with the name MFMAHP Data Logger selected by the user and the data logger code D153031 provided by the producer.

Fig. 1. Location of the wind monitoring system.

There are 4 sensors connected with the Meteo-40S data logger: 1 barometric, 1 hygro/thermo sensor, and 2 anemometers:

- The atmospheric air pressure is measured with the PTB110.2 barometric sensor produced by Vaisala, [13]. The principal technical characteristics of the barometric sensor are: pressure range 600÷1100 hPa, total accuracy ±0.3 hPa for +15÷+25°C, and ±1.5 hPa for-40÷+60°C.
- The air temperature and relative humidity are measured with a Galltec hygro/thermo sensor, [14]. The hygro/thermo sensor is placed inside a weather and radiation shield. The principal technical characteristics of the hygro/thermo sensor are: temperature range -30÷+70 °C, temperature accuracy ±0.2 K, relative humidity range 0÷100 %, relative humidity range ±2%.
- The 2D ultrasonic anemometers are produced by Thies, [15], and can measure two velocity components (u and v), as well as the wind direction. The principal technical characteristics of ultrasonic anemometers are: wind speed range 0÷75 m/s, accuracy ±0.1 m/s ≤5 m/s, ±2% of meas. value > 5 m/s, resolution 0.1 m/s; wind direction range 0÷360 °, accuracy ±1°, resolution 1°.

The data logger, the barometric sensor, the power supply and the electrical connectors are placed inside a steel cabinet. The sensors configuration is performed using the data logger, for example in remote control system, through a Web browser application. After the physical connection, the data logger is able to detect the type of sensors. There are two important parameters that user should carefully select: the measurement rate (selected 1-second) and the file statistics interval (selected 10-minutes).

3 Experiment

The experimental data are stored in the data logger internal memory at the sample rate of 1 second for all sensors, averaged on a 10-minute basis and saved in a data file on a daily basis. The daily data file, which is saved in a CSV file format, can be downloaded locally using a USB connection, or remotely using an Ethernet connection and the same Web browser application. Moreover, the data source, stored in the data-logger internal memory at the measurement rate, can be analyzed by downloading on different intervals: 10-minute, 1-hour, 1-day, and 1-week. The Meteo-40S wind measurement system uses digital signature and encryption methods in order to protect the experimental data integrity and authenticity.

For this paper, the experimental data measured on Friday, 10th February 2017, on a 10-minutes interval starting at 16:20 will be analyzed. The experimental data are presented in Fig. 2, a) (for the temperature, t [°C]), Fig. 2, b) (for the atmospheric air pressure, p_a[hPa]), and Fig. 2, c) (for the relative humidity, φ [%]).

DOI: 10.1051/matecconf/201711207020

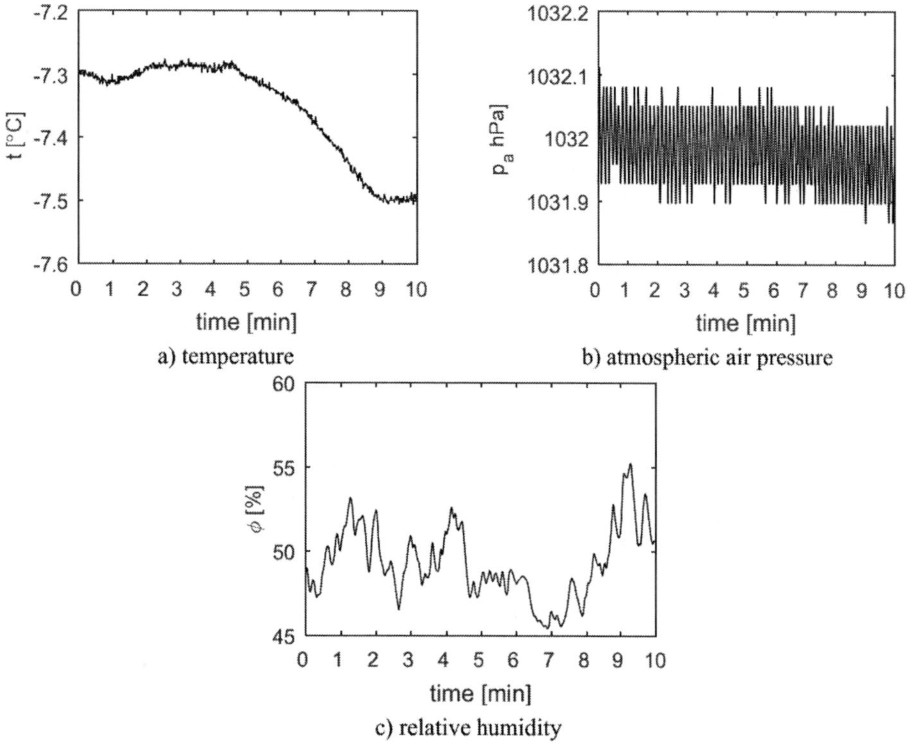

Fig. 2. Experimental data.

4 Discussion

Based on the experimental data presented in Fig. 2, the air density has been computed and presented comparatively for dry air (ρ_d), and moist air (ρ_m), in Fig. 3.

For the dry air, the ideal gas law has been used:

$$\rho_d = \frac{p_a}{RT} \tag{2}$$

where R [J/kg·K] is the specific gas constant for dry air, and T [K] is the absolute temperature.

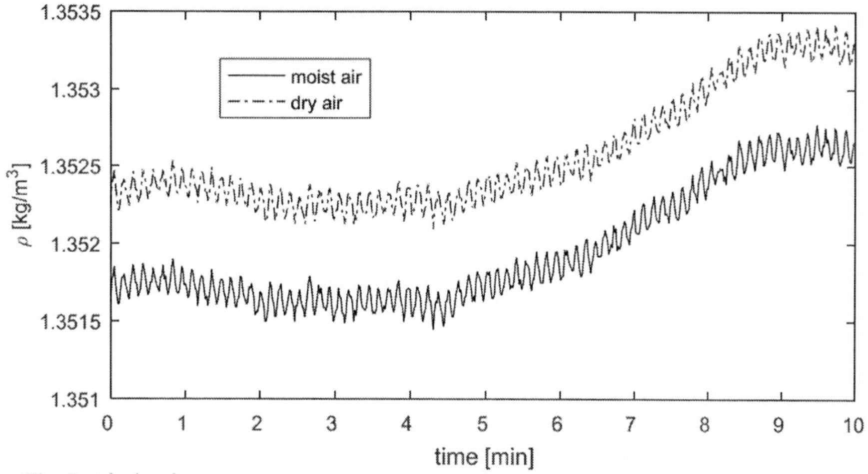

Fig. 3. Air density.

For the moist air the following relationship has been used, [12]:

$$\rho_m = \frac{1}{T}\left[\frac{p_a}{R} - \varphi \cdot p_v\left(\frac{1}{R} - \frac{1}{R_v}\right)\right] \tag{3}$$

where p_v and R_v are the water vapor pressure and the gas constant of water vapor.

Considering the standard air density value ρ_s=1.225 kg/m^3 as the reference value, the relative errors for dry air density ε_{sd} [%], and for moist air density ε_{sm} [%] will be computed with the equations (4) and (5). Similarly, one can calculate the relative error between the dry air density and the moist air density, ε_{md} (6).

$$\varepsilon_{sd} = \frac{|\rho_s - \rho_d|}{\rho_s} 100\ [\%] \tag{4}$$

$$\varepsilon_{sm} = \frac{|\rho_s - \rho_m|}{\rho_s} 100\ [\%] \tag{5}$$

$$\varepsilon_{md} = \frac{|\rho_m - \rho_d|}{\rho_m} 100\ [\%] \tag{6}$$

The relative errors are presented in Fig. 4, a) for the dry air density, in Fig. 4. b) for the moist air density, and in Fig. 4. c) for the relative error between the dry air density and the moist air density. The mean values of the relative errors are: 10.4145% for dry air, 10.3634% for moist air, and 0.0462% for the relative error between the dry and the moist air density.

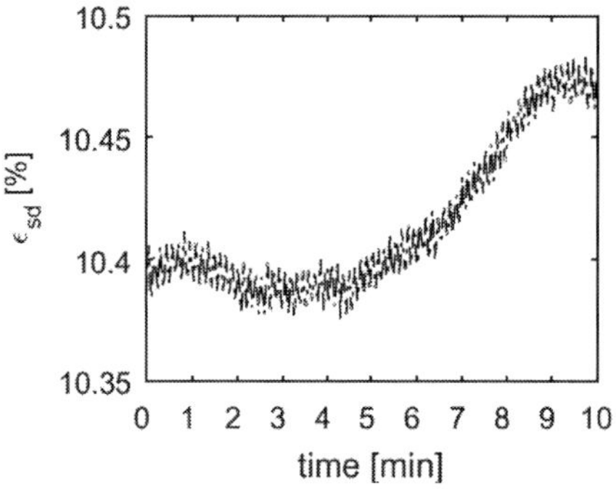

a) dry air, mean value 10.4145%

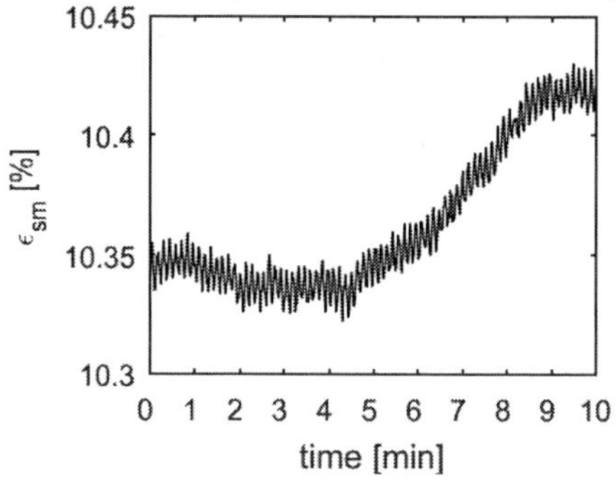

b) moist air, mean value 10.3634%

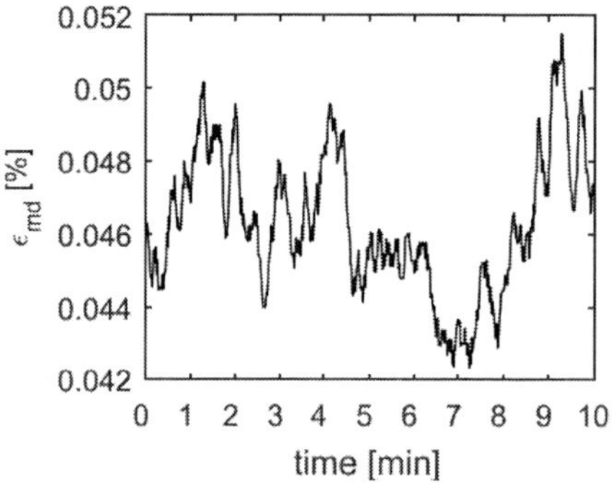

c) dry air with respect with moist air, mean value 0.0462%

Fig. 4. Relative error.

5 Conclusions

Measuring the temperature, the atmospheric air pressure and the relative humidity on 1-second measurement rate and 10-minute measuring interval, starting at 16:20 on Friday 10th February 2017, for the continental climate location defined by latitude 47,155876 and longitude 27,60095 (in Iaşi, Romania), computing the density for the dry and the moist air, and the corresponding relative errors with respect to the standard air density, it has been found that these errors are 10.4145% for dry air, and 10.3634% for moist air. In addition, the relative error between the dry and the moist air density is 0.0462%. Based on these results, a correction, at least for temperature and atmospheric air pressure, should be always considered for wind resource assessment, as well as for predicting the wind turbines performance. For better accuracy, the relative humidity should also be considered, thus, the moist air density should be used, instead of standard air density. The high measurement quality of sensors combined with a high-performance data logger, recommended the Meteo-40S wind measurement system as one of the most advanced product for wind resource assessment and wind farm monitoring, as well as for basic meteorological data recording and processing.

The authors would like to acknowledge the technical resources offered by the Laboratory of Wind Energy 1 – Wind Turbine Field Test, from the Department of Fluid Mechanics, Fluid Machinery and Fluid Power Systems, "Gheorghe Asachi" Technical University of Iaşi, Romania. The Laboratory of Wind Energy 1 – Wind Turbine Field Test has been equipped with technical resources with the financial support of the grant ENERED, POSCCE-A2-O2.2.1-2009-4, ID 911.

References

1. M. Cheng, Y. Zhu, Energ. Convers. Manage., **88**, (2014)
2. E. Hau, *Wind Turbines. Fundamentals, Technologies, Application, Economics* (Springer, Berlin, 2006)
3. K. Dai, A. Bergot, C. Liang, W.-N. Xiang, Z. Huang, Renew. Energ. **75** (2015)
4. S. Krohn, P.-E. Morthorst, S. Awerbuch, *The Economics of Wind Energy. A report by the European Wind Energy Association*, http://www.ewea.org/fileadmin/ewea_documents/documents/publications/reports/Economics_of_Wind_Main_Report_FINAL-lr.pdf, accessed 9.02.2017
5. B.H. Bailey, S.L. McDonald, D.W. Bernadett, M.J. Markus, *Wind Resource Assessment Handbook*, http://www.nrel.gov/docs/legosti/fy97/22223.pdf, accessed 9.02.2017
6. J.P. Hennessey, J. Appl. Meteorol. **16** 2 (1977)
7. I. Troen, L.E. Petersen, *European Wind Atlas*, (Riso National Laboratory, Roskilde, Denmark, 1989)
8. Technical University of Denmark, Department of Wind Energy, *Global Wind Atlas*, http://globalwindatlas.com/map.html, accessed 9.02.2017
9. X.T. Chadee, R.M. Clarke, ISRN Renew. Energ., ID 378027 (2013).
10. Z. Farkas, *Considering Air Density in Wind Power Production*, Cornell University Library, arXiv:1103.2198, https://arxiv.org/pdf/1103.2198.pdf, accessed 31.03.2017
11. A. Pourrajabian, M. Mirzaei, R. Ebrahimi, D. Wood, J. Wind Eng. Ind. Aerod. **126** (2014)
12. Ammonit Measurement GmbH, *Wind Measurement for Site Assessment*, http://www.ammonit.com/en/wind-solar-messsysteme/windmessung, accessed 01.02.2017.
13. Vaisala GmbH, PTB110 Barometer for Industrial Use, http://www.vaisala.com/Vaisala%20Documents/Brochures%20and%20Datasheets/PTB110-Datasheet-B210681EN-C-LoRes.pdf , accessed 9.02.2017
14. Galltec Mess- und Regeltechnik GmbH, *Humidity and Temperature Sensors*, https://www.galltec-mela.de/product/humidity-and-temperature-sensors-for-hvac-l-series/114/EN, accessed 01.02.2017
15. Thies Clima GmbH, *Wind Ultrasonic Anemometer*, https://www.thiesclima.com/en/Products/ Wind-Ultrasonic-Anemometer/, accessed 01.02.2017

MATEC Web of Conferences **112**, 07021 (2017) DOI: 10.1051/matecconf/201711207021
IManE&E 2017

Design and manufacturing front hood for electric vehicle by carbon fibber

Paul Bere[1], *Calin* Neamtu [2,*], and *Cristian* Dudescu[3]

[1] Technical University of Cluj-Napoca, Manufacturing Engineering, Blvd. Muncii 103-105, Romania
[2] Technical University of Cluj-Napoca, Design Engineering and Robotics, Blvd. Muncii 103-105, Romania
[3] Technical University of Cluj-Napoca, Mechanical Engineering department, Blvd. Muncii 103-105, Romania

Abstract. The polymeric reinforced composite materials (FRP) are a new materials group which have a lot of research reported in the last years. Our paper present the results obtained in case of use FRP for manufacturing an electric vehicle front hood in order for lightweight chassis and increase the autonomy. The design stage and 3D modelling of the mould, material design and manufacturing front hood from Carbon Fibber/Epoxy are presented.

1 Introduction

In the last time, the new materials as composite materials are thoroughly researched for a lot of researchers' team. They must find solutions to scientific challenges, and practical applications ever greater. These materials are used in high performance began, in areas such as aerospace, military or aviation. Step by step these advanced materials are used in our day by day life in different domain. Such materials are also fibre-reinforced polymer which has an accelerated development. FRP is composite materials made by plastic (polymer) matrix and reinforced fibres. These two components put together are very impressive mechanical properties in condition of low density.

Today FRP are in a continues interest and development and are intensive applied in automotive manufacturing industry, medicine, aircraft, motorsport, manufacturing machines industry or building construction. The McLaren Company used these materials in 1981 and built the first Formula One prototype car. Revolutionary car considered for that time, it brought a significant change in the car performance. In 2002, the first car made by carbon fibre polymer (CFRP) was produced.

The increasing demand on reduced emissions, lower fuel consumption and higher safety in the automotive industry requires not only a development of alternatively powered vehicles, but also consistent lightweight design. The use of lightweight materials such as glass or carbon FRP is a possible approach to achieve these goals. In contrast to metal materials FRP have very high values for specific stiffness and strength [1]. The usage of FRP structures within new vehicle concepts beneficially leads to a further enhancement of structural safety while lowering the vehicle's mass. The use of FRP for the Body-in-White (BIW) shows a

* Corresponding author: calin.neamtu@muri.utcluj.ro

weight saving potential of up to 60% to 70% compared to metal materials [2-3]. This potential has been exploited for many years in motorsport and comes more and more into focus for conventional cars. In particular, for future alternatively powered vehicles, weight saving is directly linked to increased range which brings FRP-materials into such cars in greater extent (e.g. BMW i3 [4]).

In addition to the weight saving potential FRP also show a very high ratio of energy absorption per weight, compared to metal structures (CFK~ 100kJ/kg vs. Al ~50kJ/kg [5]). This makes it a very interesting material for the crash-structures in a car. Actual applications can be found predominantly in sports cars but as soon as some shortcomings of those materials (cost, recycling, etc.) are solved, it is very likely that they are applied also in more vehicle types in the future. In [6-7] the authors presented an important application of FRP in a car door vehicle. Adaptive crash structures show potential for further increase of safety and lightweight performance. Different approaches with varying design goals and realizations have been discussed in recent publications. The main goals for the use of adaptive structures can be summarized as described in [8].

The numerical simulation of the FRP vehicles structures are a very important step in the design of components. [9-11]. In [12] the authors presented a methodology to design and manufacturing of the component from Formula One car. This can be applied in all the development CFRP car components.

In this paper, the authors presented the design and the manufacturing process for CFRP front hood in the electric car prototype case. The methodology for design of the component and the composite material layers' distribution were indicated. The vacuum assisted resin transfer mould (VRTM) to obtain the CFRP front hood are applied. The analysis of the typical load cases and computation of lateral, transversal and torsional stiffness of the front hood were done using finite element (FEA) commercial software.

2 Design and manufacturing method

2.1 Design

The front hood has been cut from general assembly shapes of an electric prototype vehicle. The design of this it was done using CATIA V5 software program using surface design and composite design modules. It was performed behind it based on the shape of the external surface of the model (Fig. 1. a, b). The hood are some fixing points and reinforcement zone for its hardening, this element was modelled in according with similar element from actual cars.

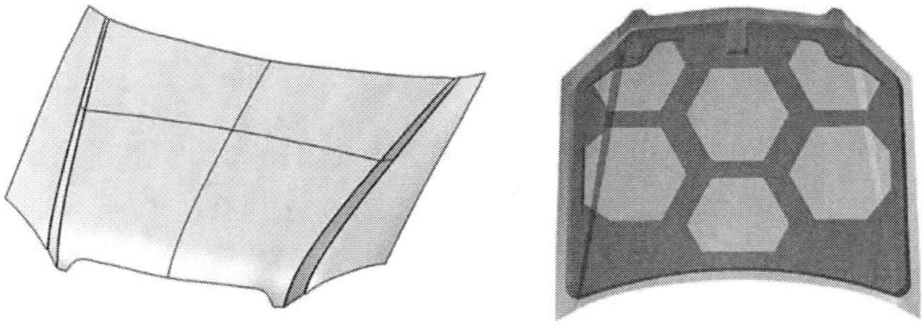

a. b.

Fig. 1 Front hood CAD prototype.

2.2 Numerical simulation

The finite element model contains all geometrical components of the above described hood, the simulation was realized using ANSYS Workbench software [13] and the ACP tool for layup modelling of the composite materials. The composite material used in the simulation was defined as in the manufacturing process, the orthotropic material proprieties of the fabric epoxy carbon woven being those provided by the software material database (Epoxy Carbon Woven (230 GPa) Wet). The shell elements are used where the number of the elements was 21600. Connection between the inner and outer part is made by mesh connection in the contact areas. The bases for the analysis are the established static load cases [2-7] to determine the longitudinal, transversal and torsional stiffness. The hood is bended in cases of the lateral and transversal stiffness tests and distorted in the torsional stiffness test.

In all three tests, the hood is mounted in design position and constrained in the supporting points, the hinges in the rear hood area and the buffer points in the front area, respectively. The loads are applied in vertical direction and have for each load case the intensities shown in Figs. 2-4.

The assessment criterion is the total deflection of the load application point on the outer panel. Figs. 2-4 presents the total displacement distribution and its maximum value for investigated load cases. The obtained values give an indication about the hood stiffness in comparison with a similar one made from steel and ensure in this stage the proper composite and structural design. Comparison between experimental and numerical stiffness of the structure is necessary to be performed in the next step and consists future work.

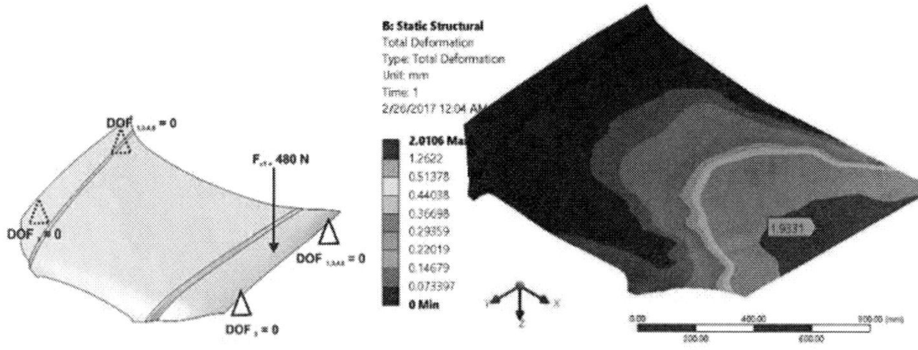

Fig. 2. Lateral stiffness of the composite hood: a) load case and b) total displacement.

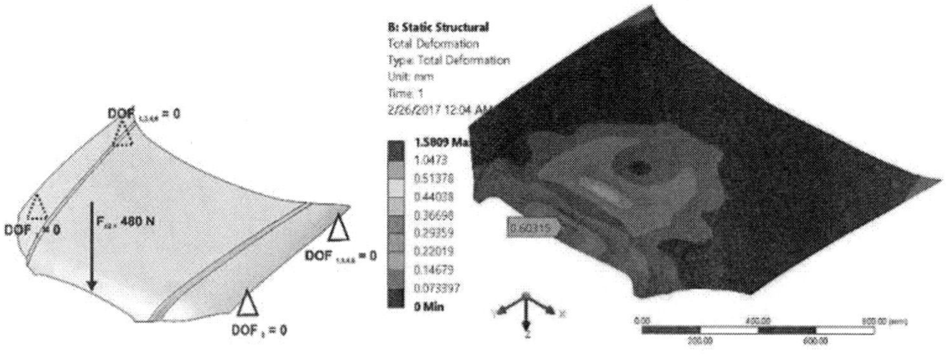

Fig. 3. Transversal stiffness of the composite hood: a) load case and b) total displacement.

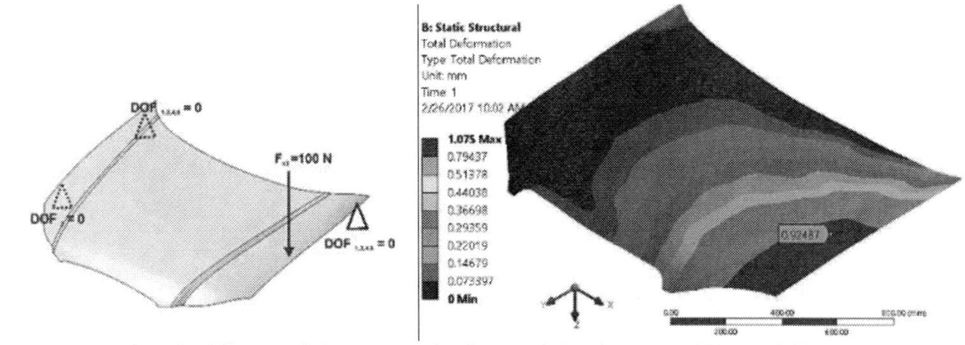

Fig. 4. Torsional stiffness of the composite hood: a) load case and b) total displacement.

2.3 Materials and manufacturing method

The materials used for manufacturing of front hood it was carbon fibre/epoxy. The balanced woven fabrics Twill 2X2 by 200g/m^2 were used. The distribution layers were [90/0 $\pm 45_2$/90/0].

For the matrix L285 epoxy resin and hardener 286 from Lange-Ritter (Germany) was used. The mixing ratio according to the company prescription was 100:40 parts by weight.

The epoxy blocks were used to obtain the front hood mould. There was milling by CNC machines. To have a sufficient sealing surface of the vacuum bag the borders of the mould were extended by 100 mm. Two moulds were manufacturing, one for exterior of the hood and one for back part (Fig. 5). The surface of the moulds was mechanical processing by glass paper sanding by 1000 grit.

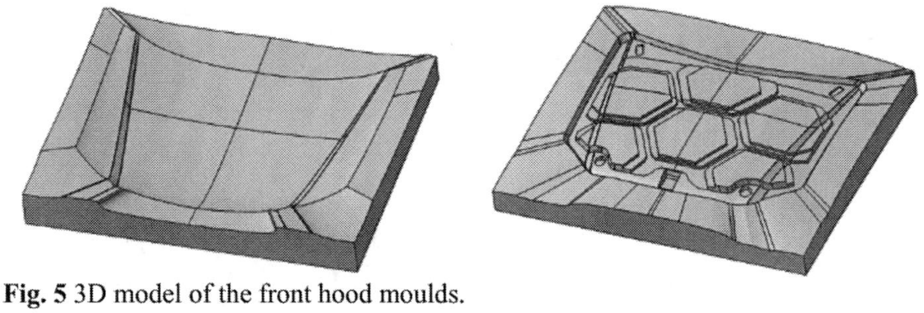

Fig. 5 3D model of the front hood moulds.

The surface was traded with several layers of Mould Sealer S31 agent supplied by the Jost Chemicals Company. This procedure closes the porosity of the mould's material. To realize a good thermal stability of the moulds a thermal treatment in an oven at 80° C for 8 hours' time were applied. To prevent the stick the CFRP laminating by the mould surface, several times of mould sealer tip Frekote B15 Aero by Loctite were applied.

The VRTM technology was applied to obtain the front hood by CFRP. The distribution of the carbon fibre layers was done according to the previous presented. The reinforced material was applied in the mould cavity. The CFRP reinforced material was covered by peel ply layer and breather fabric. On the border of the reinforced material the spiral tubing (wrapped in peel ply) were applied. All the system was covered by vacuum bag foil (Fig. 6). On the border of the mould the vacuum bag foil were seals by vacuum sealing type. The resin transfer was made on centrum of the mould and the vacuum valve was applied on the border.

On the bag has been applied a vacuum pressure of -0.9 Bar 30 minutes tines. In this timer, the epoxy resin was mixing with hardener. To eliminate the air bubbles from mixing process the resin was vacuumed. After 30 min., the epoxy resin was transferred in the carbon layers by five vacuum valves. The resin transfer was made progressive from a valve to another. The process was stopped when the resin covered all the composite materials and arrives on the border of the mould. The vacuum pressure eliminates all the resin excess in a trap filter.

Composite materials polymerization was done at 20° C for 12 hours. Supplementary thermal treatments of the composite were applied in the oven at 80° C for 8 hour. After complete polymerization, the composite pieces were released and the excess border materials were cut (Fig. 7). It was manufacturing two CFRP separate pieces. One for the top of the hood and one for back (reinforced structure). This two parts were bonded by a structural adhesive type Scotch Weld, EC-9323 B/A.

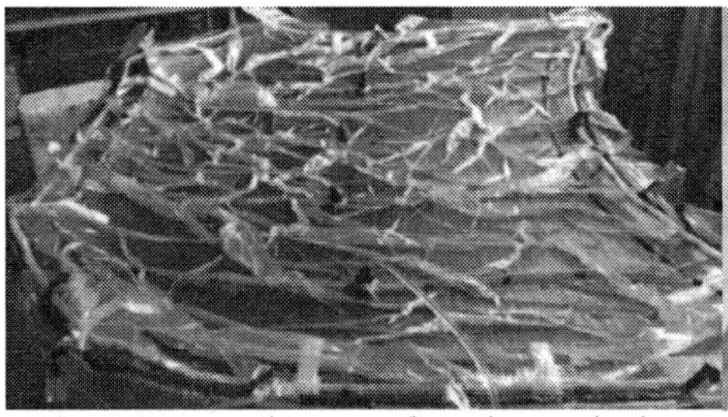

Fig. 6. Vacuum bag manufacturing method.

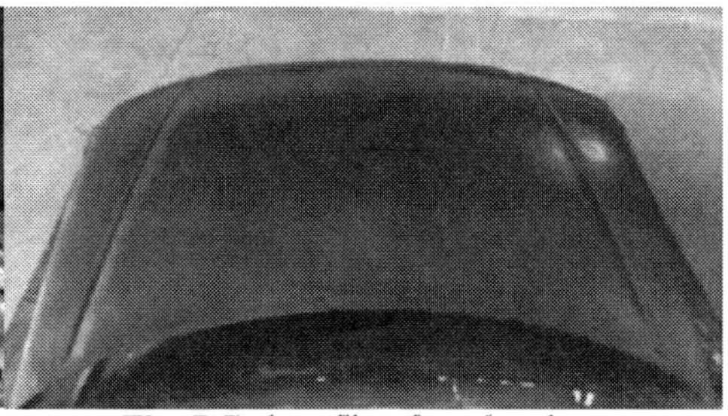

Fig. 7 Carbon fibre front hood.

3 Results and discussions.

A very good surface of the CFRP pieces on the mould part was obtained. This copies the mould surface and has a very good accuracy. On the back part of the carbon piece the surface is dimply typically for VARTM technology. The obtained degree of reinforcement was 64% in the weight fraction.

The design of the front hood was well done applied for CFRP components. The shape of the hood respected typical design for composites. The edges and ridges are adapted and the surfaces are connected by radius. No air bubbles or damage of reinforced materials are observed in a critical edge. The carbon fabric respected the initial disposal on the mould.

Comparatively with a similar hood designed of steel with metal sheets thickness of 0.5 mm the composite material hood has an increase in lateral stiffness of 52%, in transversal stiffness of 8% and in torsional stiffness of 30% (simulation results). The density of CFRP material from front hood obtained is 1450kg/m^3. The CFRP front hood presented in Fig. 4 has 3880g weight. Comparative front steel hoods by 0.5 mm have 9400g and from aluminium alloy by 0,8 mm have 8120g weight. If we compare the weight of the hood realize form this metal pieces we obtained a weight decrease to 69% in a steel case and 53% in aluminium alloy case.

4 Conclusions.

The composite materials used to manufacture of the electric vehicle front hood have a significant weight reduction. This has a substantial contribution of the mass properties reduction of the electric vehicle and increases the autonomy of the car. The manufacturing technology to obtain the front hood is commonly used in high performance CFRP domain.

The composite structure obtained is homogeneous well pressed, and the degree of reinforcement was 64%. The exterior of the carbon/epoxy piece looks good without porosity. The weight reduction between steel front hood and presented one is 2,42 times and from aluminium alloy is 2,09 times lighter.

The FE analyses of the mechanical structure of the vehicle front hood indicates that the design of the composite material fulfil the requirements of the standard static tests.

This work was supported by a grant of the Romanian National Authority for Scientific Research and Innovation, CNCS/CCCDI–UEFISCDI, project number PN-III-P2-2.1-BG-2016-0210, within PNCDI III.

References

1. A. Ahmed, L. Wei, Thin-Walled Structures, **99** 97-108 (2016).
2. M. Hamacher, R. Wohlecker, L. Ickert, Abaqus Users' Conference, May 19-22, Newport, Rhode Island, USA, 1-16 (2008).
3. C.E. Torricelli, L. Splendi and M. Pettazzoni, in Proceedings of the World Congress on Engineering WCE 2011, July 6 - 8, London, U.K (2011).
4. N. Bhaskar, P. Rayudu, International Journal of Current Engineering and Technology, **5** (2015).
5. S. M. Darwish, S. M. Elseufy, A. Ahmad, Journal of Management & Engineering Integration, **7** 62 (2014).
6. S. Darwish, H. M. A. Hussein, A. Gemeal, International Journal of Engineering & Technology IJET-IJENS **12** (2012).
7. W. Tan, Y. Wang, L. Li, Y. Zhang, in Proceedings of the 2012 International Conference on Computer Application and System Modeling. Atlantis Press, (2012).
8. M. Grujicic, G. Arakere, V. Sellappan, J. C. Ziegert, F. Y. Koçer, D. Schmueser, Multidiscipline Modeling in Materials and Structures, **5** 1-28 (2009).
9. Q. Liu, Y. Lin, Z. Zong, G. Sun, Q. Li, Composite Structures **97** 231-238 (2013).
10. A. Zinno, E. Fusco, A. Prota, G. Manfredi, Composite Structures **92** 2208–2219 (2010).
11. G. Gstrein, C. Kurzböck, J-M. Opelka, J. Krollmann, in Proceedings of 24th International Technical Conference on the Enhanced Safety of Vehicles, Göteborg, Sweden, 8-11 June, 1-12 (2015).
12. P. Bere, C. Neamtu, in Proceedings of 3rd International Conference on Quality and Innovation in Engineering and Management, Cluj-Napoca, Romania, 21-26 (2014).
13. ANSYS Workbench, Release 16.0, ANSYS, Inc., Canonsburg, USA (2015)

DOI: 10.1051/matecconf/201711207022

Influence of hauling on railway car vertical load

Marius-Adrian Spiroiu*

University POLITEHNICA of Bucharest, Department of Rolling Stock, Splaiul Independenței 313, Bucharest, Romania

Abstract. The aim of this paper is to investigate the effects of pitch phenomenon induced by traction process on railway cars. There are identified and analysed the vertical load variations occurring both on bogies and on axles as a result of pitch motion. It is investigated the influence on vertical supplementary loads of various parameters such as vehicle mass, speed, acceleration, vehicle dimensional data.

1 Introduction

The present paper investigates the effects of pitch phenomenon induced by traction process on hauled railway cars. The pitch motion - rotation about axis y - of the car body of railway vehicles is caused by the couple (torque) produced as a result of the height difference between the point of application of the traction force of the locomotive and the bogie bolster. In a similar way occurs bogie pitch - because of the height difference between the bogie bolster and the axle journal.

Pitch motion induces load variations both on vehicle's bogies and axles. These variations (loading and unloading) are undesirable on railway vehicles from the point of view of ride quality and traffic safety and have negative impact on adhesion forces, thus affecting traction (for motor vehicles) and braking processes.

Interest in studying this phenomenon is quite old - see [1] - and it still is a subject of research nowadays – see, e.g., [2], but studies have focused only on motor rail vehicles, for which the magnitude and the impact of pitch is more important – affecting mainly the traction force. Regarding the pitch effect on railway cars, it may be mentioned only the paper [3], where is evaluated the influence of braking process on vertical load variation.

In these circumstances, the present paper aim is to analyse the vertical load variations on bogies and axles caused by hauling in the case of a 4-axle passenger car with two suspension levels. The analysis takes into account the vehicle traction force, drag (resistance force) and acceleration.

2 Vertical load variations induced by hauling process

In Figure 1 are shown the forces acting on the railway car as a result of hauling process; there are also indicated the main geometrical parameters of interest.

* Corresponding author: marius_spiroiu@yahoo.com

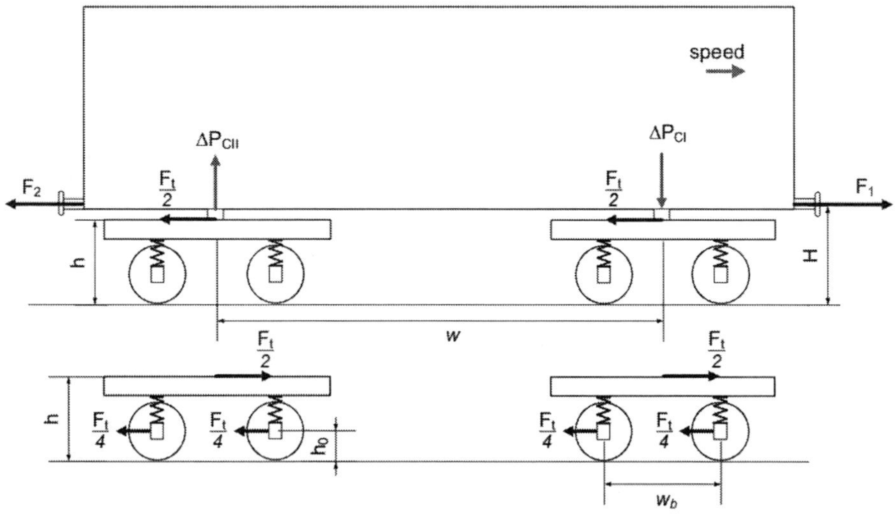

Fig. 1. Forces acting on vehicle carbody and bogies during the hauling process.

In the case of a hauled car, traction forces are provided by motor vehicle(s). The car "receives" a force F_1 from the vehicle in front and transmits a force F_2 to the vehicle behind. The difference between these two forces is the fraction of the train total traction force that is used for hauling the car:

$$F_t = F_1 - F_2 . \tag{1}$$

The equation of motion for the wagon can be written as:

$$F_1 - F_2 - R_v = m(1+\gamma)a , \tag{2}$$

where m, a and R_v are the vehicle mass, acceleration and resistance force, respectively and γ is a mass factor modelling the effect of accelerating rotational components of the vehicle. Therefore, the car hauling force is

$$F_t = F_1 - F_2 = R_v + m(1+\gamma)a . \tag{3}$$

Due to the difference of elevation (H versus h) between the point of application of the hauling force and the connection point between carbody and bogie, respectively, occurs a force couple (torque) which rotates the vehicle carbody about y axis, given by:

$$C_{yc} = F_t(H-h) \tag{4}$$

or, using equations (1) and (3),

$$C_{yc} = (F_1 - F_2)(H-h) = [R_v + m(1+\gamma)a](H-h) \ . \tag{5}$$

The pitch angle of the carbody caused by the action of the torque in (5) is [4]:

$$\varphi_{yc} = -\frac{C_{yc}}{K_{xc}} , \tag{6}$$

K_{xc} being the carbody suspension longitudinal angular stiffness, given by [4]:

$$K_{xc} = \sum_{i=1}^{2}\sum_{j=1}^{2} k_{cij} \cdot (x_{cij} - x_{c0})^2 \,. \tag{7}$$

where x_{c0} is the x-coordinate of the center of motion of the vehicle carbody; x_{cij} and k_{cij} are, the x-coordinates and the stiffness of the bogie-carbody suspension points, respectively. Considering $k_c=k_{cij}$, $i=1,2; j=1,2$, and $x_{c11}=x_{c12}=x_{c1}, x_{c21}=x_{c22}=x_{c2}$, the variation of vertical load on the bogies given by the rotation of vehicle carbody are:

$$\Delta P_{CI} = 2k_c(x_{CI} - x_{c0})\varphi_{yc}; \quad \Delta P_{CII} = 2k_c(x_{CII} - x_{c0})\varphi_{yc} \,. \tag{8}$$

For $x_{c0} = {}^{w}\!/_{2}$, $x_{c11} = x_{c12} = 0$, $x_{c21} = x_{c22} = w$, where w is the vehicle wheelbase, the load variations (8) can be written:

$$\Delta P_{CI} = 2k_c\left(0 - \frac{w}{2}\right)\left(-\frac{F_t(H-h)}{w^2 k_c}\right) = F_t\frac{H-h}{w} = [R_v + m(1+\gamma)a]\frac{H-h}{w} \tag{9}$$

$$\Delta P_{CII} = 2k_c\left(w - \frac{w}{2}\right)\left(-\frac{F_t(H-h)}{w^2 k_c}\right) = -F_t\frac{H-h}{w} = -[R_v + m(1+\gamma)a]\frac{H-h}{w} \,. \tag{10}$$

Notice that the first bogie in the travelling direction is loaded, while the second one is unloaded.

Due to the difference in elevation (h versus h_0) between the carbody-bogie connection point and the axle guiding system, respectively (see Figure 1), occurs a force couple (torque) which rotates the bogie about y axis, given by:

$$C_{yb} = \frac{F_t}{2} \cdot (h - h_0) \tag{11}$$

under the effect of which the bogie frame is rotated by the angle

$$\varphi_{yb} = -\frac{C_{yb}}{K_{xb}} \tag{12}$$

where K_{xb} is the bogie suspension longitudinal angular stiffness, given by

$$K_{xb} = \sum_{i=1}^{2}\sum_{j=1}^{2} k_{bij} \cdot (x_{bij} - x_{b0})^2 \tag{13}$$

where x_{b0} is the x-coordinate of the center of motion of the bogie; x_{bij} and k_{bij} are, respectively, the x-coordinates and the suspension stiffness corresponding to each axle journal.

Because of the combined effect - on each bogie - of torque C_{yb} and of load variations ΔP_{CI} and ΔP_{CII}, the vertical load variation for the j journal of axle i can be written as:

$$\Delta P_{ij} = k_{bij}\left[\pm f_z + (x_{bij} - x_{b0}) \cdot \varphi_{yb}\right], \tag{14}$$

where f_z is the vertical deflection of axle suspension caused by load variations ΔP_{CI} and ΔP_{CII} which is of opposite sign for the two bogies – see equations (9) and (10):

$$f_z = \frac{\Delta P_{CI}}{4k_b} = -\frac{\Delta P_{CII}}{4k_b} = \frac{F_t}{4k}\frac{H-h}{w}. \tag{15}$$

Assuming that bogies are elastically and geometrically symmetric, i.e. $k_{bij}= k_b$, i=1, 2, 3, 4; j=1, 2; $x_{b11}= x_{b12}=x_{b1}=0$, $x_{b21}= x_{b22}=x_{b2}=w_b$, $x_{b31}= x_{b32}=x_{b3}=0$, $x_{b41}= x_{b42}=x_{b4}=w_b$ (w_b - bogie wheelbase) and $x_{b0}=w_b/2$, the axle journals supplementary vertical loads given by equation (14) can be written as:

$$\Delta P_{11} = \Delta P_{12} = k_b\left[\frac{F_t}{4k_b}\frac{H-h}{w} - \left(0-\frac{w_b}{2}\right)\frac{F_t(h-h_0)}{2w_b^2 k_b}\right] = \frac{F_t}{4}\left(\frac{H-h}{w}+\frac{h-h_0}{w_b}\right) \tag{16}$$

$$\Delta P_{21} = \Delta P_{22} = k_b\left[\frac{F_t}{4k_b}\frac{H-h}{w} - \left(w_b-\frac{w_b}{2}\right)\frac{F_t(h-h_0)}{2w_b^2 k_b}\right] = \frac{F_t}{4}\left(\frac{H-h}{w}-\frac{h-h_0}{w_b}\right) \tag{17}$$

$$\Delta P_{31} = \Delta P_{32} = k_b\left[-\frac{F_t}{4k_b}\frac{H-h}{w} - \left(0-\frac{w_b}{2}\right)\frac{F_t(h-h_0)}{2w_b^2 k_b}\right] = -\frac{F_t}{4}\left(\frac{H-h}{w}-\frac{h-h_0}{w_b}\right) \tag{18}$$

$$\Delta P_{41} = \Delta P_{42} = k_b\left[-\frac{F_t}{4k_b}\frac{H-h}{w} - \left(w_b-\frac{w_b}{2}\right)\frac{F_t(h-h_0)}{2w_b^2 k_b}\right] = -\frac{F_t}{4}\left(\frac{H-h}{w}+\frac{h-h_0}{w_b}\right) \tag{19}$$

Denoting $\Delta P_{ij}=\Delta P_i$, i=1...4; j=1,2 and taking into account equation (3):

$$\Delta P_1 = \frac{R_v + m(1+\gamma)a}{4}\left(\frac{H-h}{w}+\frac{h-h_0}{w_b}\right) \tag{20}$$

$$\Delta P_2 = \frac{R_v + m(1+\gamma)a}{4}\left(\frac{H-h}{w}-\frac{h-h_0}{w_b}\right) \tag{21}$$

$$\Delta P_3 = \frac{R_v + m(1+\gamma)a}{4}\left(-\frac{H-h}{w}+\frac{h-h_0}{w_b}\right) \tag{22}$$

$$\Delta P_4 = \frac{R_v + m(1+\gamma)a}{4}\left(-\frac{H-h}{w}-\frac{h-h_0}{w_b}\right) \tag{23}$$

It can be seen from equations (9), (10) and (20) … (23) that supplementary loads on bogies and axles journals induced by hauling process are directly proportional to vehicle mass, acceleration and resistance. On the other hand, it is noticeable that suspension stiffness no longer appears in the expressions of vertical forces variation. The load variation on the bogies are equal in magnitude and of opposite sense. The same situation (equal and opposite forces) occurs in the case of vertical loads on axles 1 and 4, respectively 2 and 3. The supplementary forces also depend on the geometrical characteristics of the vehicle. In the case of the bogies, loadings are amplified by the height difference H-h and are decreasing when the vehicle wheelbase w increases. For the axle journals loads, there are more influences, involving bogie and vehicle wheelbases and differences in elevation H-h and h-h_0. The magnitude and sense of the supplementary forces are given by the sign of the two terms in parentheses – see equations (20) … (23). It can be seen that the first axle has the higher loading, while the fourth has the higher unloading. Taking into account the specific dimensions of railway vehicles, it follows that $(H-h) < (h-h_0)$ and $w > w_b$, so the second term in parentheses is more important as magnitude.

3 Numerical application

In this section is considered the case of a railway passenger car with the following characteristics: m=52000 kg, w=18 m, w_b=2.5 m, H=1.065 m, h=0.96 m, h_0=0.46 m, k_c=560 N/mm, k_b=680 N/mm. The car resistance is given by

$$R_v = mr_v = m\left(1.65 + \frac{V^2}{4000}\right), \tag{24}$$

where V is the vehicle speed in km/h and the mass factor is considered to be $\gamma = 0.03$.

It is assumed that the vehicle accelerates from 0 to 100 km/h with a constant acceleration of 0.3 m/s^2.

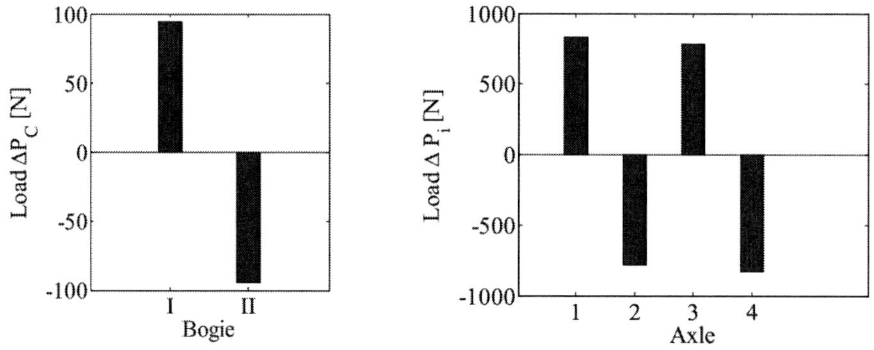

Fig. 2. Load variations on bogies (left) and on axle journals (right) for v=20 km/h, a = 0.3 m/s^2.

In Figure 2 are presented the load variations on bogies and axles for a speed of v=20 km/h. It can be observed that graphical representation confirms the considerations made in the previous section. Load variations are not very important as magnitude; also, it is to notice that loads on axle journals are significantly higher than the bogie ones.

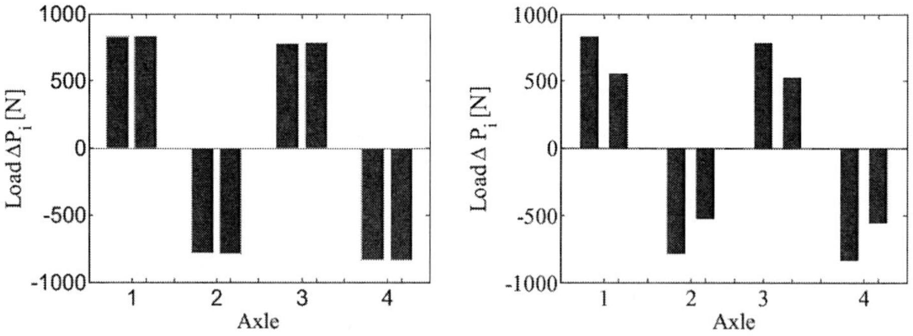

Fig. 3. Influence of vehicle speed (left) and acceleration (right) on axle journal load variations.

In Figure 3 can be observed that the influence of vehicle speed on axle journal vertical loads is negligible: increasing speed from 20 to 100 km/h leads to an increase (in absolute value) of the supplementary forces by approximately 1%. On the other hand, it can be seen that reducing the vehicle acceleration from 0.3 to 0.2 m/s^2 leads to a proportional reduction of all journal vertical loads. These influences are found equally in the case of bogies load variations.

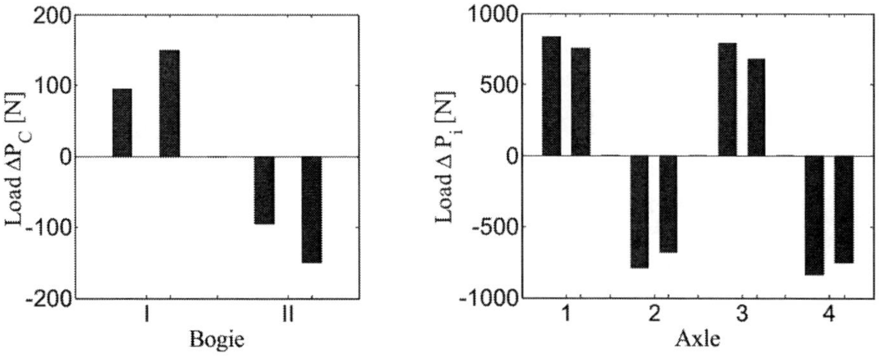

Fig. 4. Influence of carbody-bogie connection point height on bogie (left) and axle journal (right) load variations.

Geometric parameters of the vehicle have an important influence on bogie and axles load variations. A modification of the height of carbody-bogie connection point from 0.96 to 0.9 m has an important effect on al load variations– see Figure 4. The supplementary loads on the bogies are increasing - in absolute value, while those on the axles are reduced. This reduction is not uniform, being more accentuated in the case of axles 2 and 3.

4 Conclusion

The aim of this paper was to investigate the effects of pitch phenomenon induced by hauling process on railway cars. The analysis carried out revealed that pitch motion leads to the occurrence of load variation both on bogies and on axles. The axle supplementary loads are higher than the bogies ones. The load variation on the bogies are equal in magnitude and of opposite sense, a similar situation occurs in the case of vertical loads on axles 1 and 4, respectively 2 and 3. The first axle has the higher loading, while the fourth (last one) has the higher unloading.

The supplementary loads magnitude is not very important. For the considered usual vehicle parameters and common hauling regime, the magnitudes of load variation are below 1% of the static axle load. The supplementary loads are strongly influenced by vehicle mass and acceleration, increasing with it. On the other hand, the vehicle speed has very little impact. Of great influence are also the geometric parameters of the vehicle – wheelbases and levels of points of transmission of the traction force from the chassis to the axles. The supplementary loads on the bogies and on axle journals are increasing - in absolute value - with the increase of the corresponding height differences.

References

1. G. Borgeaud, Bulletin de L`A.I. du Congrès des Chemins de Fer, 691 (1967)
2. I. Sebeşan, M. M. Călin, *Civil Engineering and Urban Planning III*, 493 (CRC Press 2014)
3. C. Crăciun, C. Cruceanu, M. Spiroiu, Acta Tehnica Corviniensis – Bull. of Eng., **8(2)**, 17 (2015)
4. G. Popa, B. Ţăruş, *Supporting structures for rail vehicles* (in Romanian), (Matrixrom, Bucharest, 2005)

MATEC Web of Conferences **112**, 07023 (2017) DOI: 10.1051/matecconf/201711207023
IManE&E 2017

An analysis of the train–turnout interaction

Mihaela Cristina Tudorache*, *Răzvan Andrei* Oprea

Railway Vehicles Engineering Department, University Politehnica of Bucharest, Romania

Abstract. A distinctive feature of the application proposed in this work is the occurrence of simultaneous contact patches on a single wheel, i.e., the multi-point contact case. A difficulty in the study of this phenomenon is that the contact areas depend on the deformation of the bodies, whereas these deformations are functions of the contact loading and the contact area. This is contrary to what is usually the case in finite element analysis, where the load and displacement are prescribed at different (fixed) parts of the boundary.

1 Introduction

Crossings points and switches are designed to offer a continuous support for the wheels throughout the crossing, giving a smooth transition. However, high impact loads may arise [1,2], leading to damage, fracture and wear of both the crossing and the wheels and to high maintenance costs. Therefore, investigations of dynamics and damage in a crossing are of great interest. The basis for the prediction of component life is an adequate determination of stresses and strains.

The UIC (International Union of Railways) and railway researchers have been accomplished several computer applications for the study of train–turnout interaction [3]. The common feature is the employment of commercial multi-body or finite element (FE) software.

Multi-body simulation packages like SIMPACK or GENSYS have been used to simulate the vehicle track interaction due to passing over a turnout [4,5,6,7] and in some cases the results were validated through full-scale field tests [8].

The strip method has been used to determine the wheel–rail contact forces in a turnout with the aid of the Hertzian method to model the wheel–rail contact [9]. The rails have also been modeled in the turn-out zone using several Euler-Bernoulli beams [10].

Flexible track model and real-time wheel-rail penetration contact model in NUCARS was employed for the crossing design [11].

Elastic and elastic–plastic finite element models of the contact are investigated with FE software like ABAQUS. The finite element method is not limited by the half-space assumption or applicable to a linear-elastic material model only [8,12,13]. The subject regarding the phenomena resulted from the contact between the road wheel and the railway

* Corresponding author: cristi_tudorache2003@yahoo.com

represents the base for assuring the safety on guidance traveling and the main focus in the development of railway vehicles and it is approached from different directions in the research field.

It is important to be mentioned that up until recently the problem regarding the influence of these contact phenomena resulted from the contact between the wheel and the railway over the safety guidance has been done using simplifying hypothesis, approximation methods which lead to illustrative methods. Thus, in the present paper it will be presented a three-dimensional analysis at the level of the point of contact between the wheel and the turn-out switch from the point of view of the tip of the switch, using the Kalker's Contact algorithm. This is an iterative algorithm which offers an output with an extremely good accuracy in the field of railway transportation [14].

By analyzing the literature, it can easily be remarked the necessity for thorough documentation regarding the study of the influence that the wheel-railway contact phenomena has upon the guidance safety in the sense of reducing the wear of the running surface and optimizing the parameters of the wheel-railway contact surfaces.

2 2. The basic principles of Kalker theory

The theory of the linear contact developed by Kalker starts from the De Pater [17] idea, based on the fact that for small values of the pseudo-slip effect the slip area is very small, which considers that the entire contact surface is assimilated with an adhesion area. According to this hypothesis, the condition at the extremity is given by the following equations:

$$0 = w_{x,y} = V \cdot (v_x - \varphi \cdot y, v_y + \varphi \cdot x) - V \frac{\partial u(x,y)}{\partial x} \text{ - within the contact ellipse.} \quad (1)$$

$$0 = p(x,y) \text{ - outside the contact ellipse} \quad (2)$$

By integrating the relation (1) results:

$$g(y) = -V \cdot u(x,y) + V\left(v_x x - \varphi \cdot x \cdot y, v_y x - \frac{1}{2}\varphi \cdot x^2 \right) \quad (3)$$

where *g(y)* represents an arbitrary function.

The arbitrary function *g(y)* is determined from the condition that the traction force must be constant along the border, where the railway and the wheel come into contact. The final result does not respect though the limit value of the friction force at the end limit of the contact surface [15]. This thing is justified by the fact that in the exterior vicinity of the contact limit the bodies materials are free of internal stress.

The traction increases up to the rear edge of the adhesion limit then the normal load is removed and the traction suddenly drops to 0. Thus, on the edge, the friction index shows a discontinuous, infinite fluctuation. Calculating the *g(y)* function will lead to a linear expression displaying the interdependence between forces and pseudo-slip:

$$\begin{aligned} &F_x = -a \cdot b \cdot C_{11} \cdot v_x,\ F_y = -a \cdot b \cdot G\left(C_{22} \cdot v_y + \sqrt{a \cdot b} \cdot C_{23} \cdot \varphi\right) \\ &M_\varphi = -\left(\sqrt[3]{a \cdot b} \cdot G \cdot C_{32} + (a \cdot b)^2 \cdot G \cdot C_{33} \cdot \varphi\right) \end{aligned} \quad (4)$$

where: *a, b* – semi-axes of the contact ellipse, v_x, v_y, φ - longitudinal, transversal and spinning pseudo-slip movements in the contact points, C_{11}, C_{22}, $C_{23,32}$, C_{33} – Kalker parameters which depend on the value of Poisson index and being determined based on the following equations [15,16]:

$$C_{11} = -\frac{\partial F_x}{\partial v_x} / (a \cdot b \cdot G),\ C_{22} = -\frac{\partial F_y}{\partial v_y} / (a \cdot b \cdot G)$$
$$C_{23} = -\frac{\partial F_y}{\partial \varphi} / \left(\sqrt[3]{a \cdot b} \cdot G\right)_{v_x = v_y = \varphi = 0} \tag{5}$$

Kalker defines the pseudo-slip parameters as following:

$$k_x = \frac{a \cdot b \cdot G \cdot C_{11} \cdot v_x}{\mu \cdot N},\ k_y = \frac{a \cdot b \cdot G \cdot C_{22} \cdot v_y}{\mu \cdot N},\ k_\varphi = \frac{\sqrt[3]{a \cdot b} \cdot G \cdot C_{23} \cdot \varphi}{\mu \cdot N} \tag{6}$$

The longitudinal, transversal and spinning pseudo-slip movements are established based on the following equations:

$$v_x = \frac{w_x}{V},\ v_y = \frac{w_y}{V},\ \varphi = \frac{r \cdot \omega_s}{V} \tag{7}$$

where: ω_s is the spinning angle speed. The pseudo-slip speeds $w_{x,y}$ are established based on the (1) relation.

3 Numerical application

Within this section, we will be analyzing the behavior of an axle equipped with wheels having the S1002 profile (the nominal rolling radius is of 460 mm) over a simple railroad siding made of a rail track profile type UIC 60, 1:20 and 1:40. The speed with which the axle will pass over the railroad siding is of 35 m/s. The contact point is analyzed in a three-dimensional reference frame. The normal load applied on the wheel is considered to be of 10 kN. For evaluating the behavior between wheel and the rail tongue an iterative program has been developed within the Kalker CONTACT frame, applied for the dual-contact analysis. As a result, the calculus stages which mention the input data for the CONTACT program are the following, [17, 18]:

1. Determining the coordinates of the first point of contact through the method of the minimum height difference.
2. The calculus of the penetration within the hypothesis of a single contact point.
3. Verifying the possibility of the development of a new contact area. If this is possible the previous two stages are re-iterated.
4. Determining the penetrations within the contact areas.
5. Re-compute the values for the vertical components (through the equations shown below). If the values obtained here show differences compared to the values obtained at step 4, differences which are not within the accepted margins, then the calculus at step 4 is re-performed again with vertical components determined at step 5 up until the results obtain are within the imposed limits.

By analyzing the influence that an inclination of the tip of the turn-out switch has over the phenomena from within the contact are we can observe that the grip area is considerably enlarged in the case of an inclination of 1:20 (see Fig. 1a), when comparing that with an inclination of 1:40 (see Fig. 1b). This effect appears only in the contact point from the edge of the wheel, where the guidance is achieved. At the point from the rolling surface or the leaning point the effect is insignificant; here only the sliding surface counts.

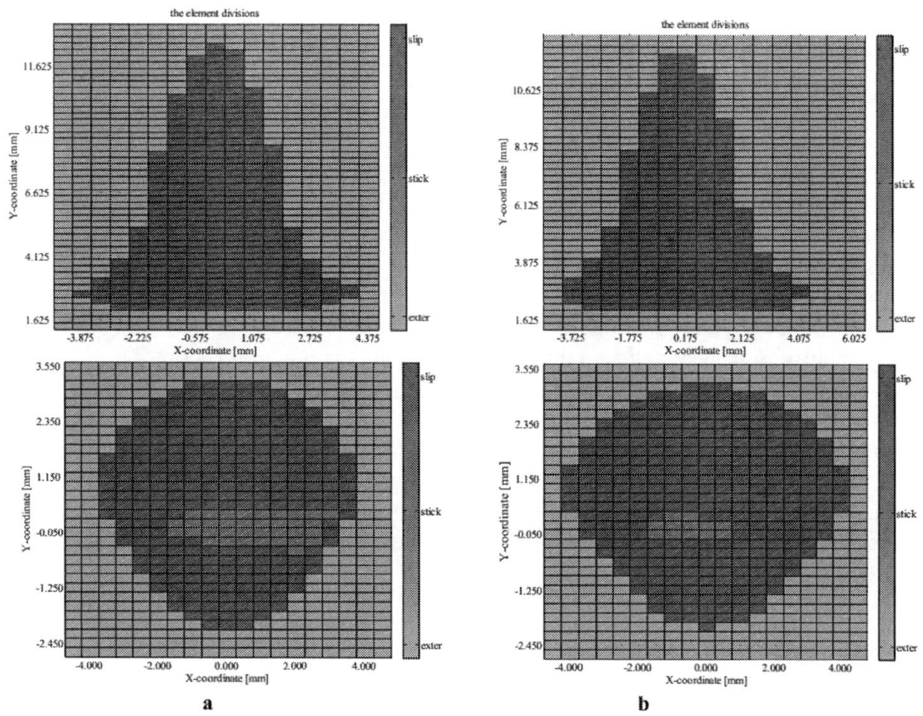

Fig. 1. The distribution of the adherence and sliding zones in the two contact points, for different inclinations of the turn-out switch tip: 1 - 1:20, b - 1:40.

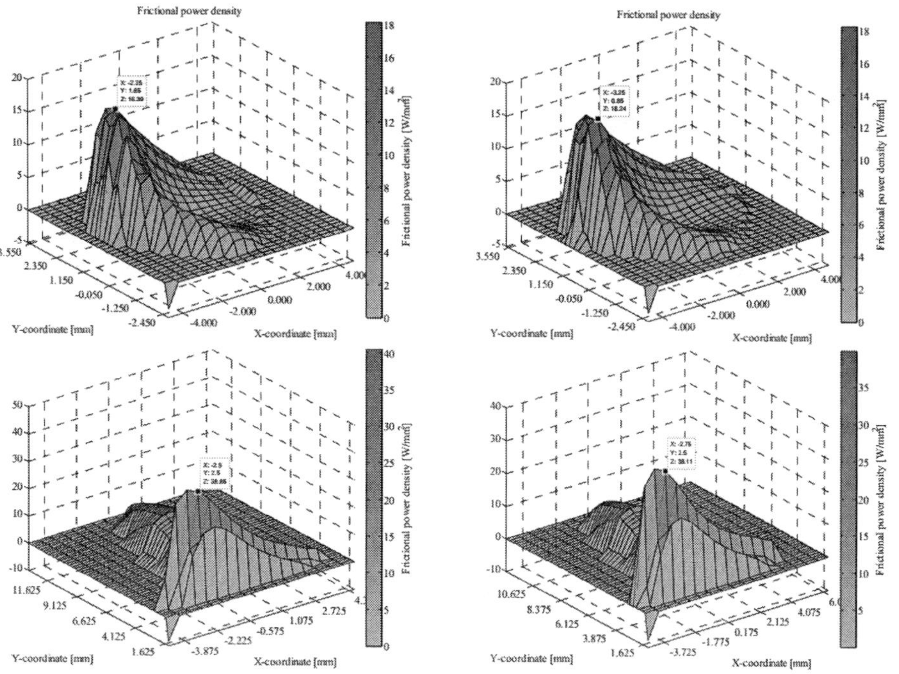

Fig. 2. The density of the power of friction, for different inclinations of the turn-out switch tip: left - 1: 20, right - 1:40.

We can assert that there is an elongation of the shape of the contact surface for the 1:20 inclination, a fact clearly visible in both points and justified by a better guidance. This assertion is reasoned by the fact that for the wear profiles the larger half-axis of the contact surface is orientated perpendicularly on the railway.

According to Fig. 2 the wear in the contact points differs from one inclination to the other. The density of the friction power is increasing in the fulcrum (rolling surface) for the 1:20 inclination and it is decreasing on the wheel's edge, thus assuring a good reliability of the wheel and of the turn-out switch tip.

In the end the effect resulting from the inclination of the turn-out switch tip is reflecting over the guiding transverse forces; for the 1:20 inclination a smaller value is obtained in comparison with a 1:40 inclination 1:40, (458N / 498N).

The application is designed for frictional concentrated contact problems. The solution is numerically efficient and it has been previously validated for the general case of the wheel-rail contact.

Multi-point contact is a common situation for the studied system. In the approach proposed, the bodies deformations and the contact areas can be determined from the contact loading. Further, the number and position of the contact points can be calculated either on-line or off-line. The scheme proposed also allows the computation of the loading and stress state starting from the local deformations.

References

1. E. Kassa, Simulation of Dynamic Interaction between Train and Turnout, Lic. Thesis, Chalmers University of Technology (Department of Applied Mechanics, Gothenburg, Sweden, 2004)
2. M. Wiest, W. Daves, F.D. Fischer, H. Ossberger, Deformation and damage of a crossing nose due to wheel passages, Wear **265**, 9–10 (2008)
3. Y.Q. Sun, C. Cole, M. McClanachan, The Calculation of Wheel Impact Force Due to the Interaction between Vehicle and a Turnout, Proceedings of the Institution of Mechanical Engineers, Part F: Journal of Rail and Rapid Transit **1** (2010)
4. G. Schupp, C. Weidemann, and L. Mauer, Modelling the contact between wheel and rail within multibody system simulation, Veh. Syst. Dyn., **41** 5 (2004)
5. G. Schupp, Simulation of railway vehicles: necessities and applications, Mech. Based Des. Struct. Mach. **31** 3 (2003).
6. E. Kassa, C., Andersson, and J.C.O. Nielsen, Simulation of dynamic interaction between train and railway turnout, Veh. Syst. Dyn. **44** 3 (2006)
7. E. Kassa, J.C.O., Nielsen, Dynamic interaction between train and railway turnout: full-scale field test and validation of simulation models, Veh. Syst. Dyn. **46** 5 (2008)
8. E. Kassa, J.C.O. Nielsen, Stochastic analysis of dynamic interaction between train and railway turnout, Veh. Syst. Dyn. **46** 5 (2008)
9. E. Kassa, J.C.O. Nielsen, Dynamic train–turnout interaction in an extended frequency range using a detailed model of track dynamics, J. Sound Vibr. (2009)
10. Z. Ren, S. Sun, W. Zhai, Study on lateral dynamic characteristics of vehicle/turnout system, Veh. Syst. Dyn., **43** 4 (2005)
11. X. Shu, N. Wilson, C. Sasaoka, J. Elkins, Development of a real-time wheel/rail contact model in NUCARS and application to diamond crossing and turnout design simulations, Veh. Syst. Dyn., **44** (Suppl) (2006).
12. C. Andersson, T. Dahlberg, Wheel/rail impacts at a railway turnout crossing. Proc. IMechE, Part F: J. Rail and Rapid Transit, **212**(F2) (1998)
13. M. Wiest, W. Daves, J.C.O. Nielsen, H. Ossberger: Assessment of methods for calculating contact pressure in wheel-rail/switch contact, Wear **265** 9–10 (2008)

14. E.A.H. Vollebregt, User guide for CONTACT, Vollebregt & Kalker's rolling and sliding contact model, VORtech Computing the Scientific Software Engineers (2012)
15. J.J. Kalker, Wheel-rail rolling contact theory, Wear **144** (1991)
16. I. Sebeşan, Railway Vehicles Dynamics (MatrixRom, 2011)
17. C. Tudorache, R. Oprea, A method for studying the interaction of wheel-turnout multi-point contact, Scientific Bulletin - University Politehnica of Bucharest, Series D: Mechanical Engineering **77** 1 (2015)
18. M. C. Tudorache, The influence of wheel-rail phenomena upon guiding safety for railway vehicle, PhD Thesis (Bucharest 2013)

 DOI: 10.1051/matecconf/201711207024

Experimental research regarding static rigidity in transversal direction of "normal" module

Dragoş-Florin Chitariu*

Technical University "Gheorghe Asachi" of Iasi, Prof. Dr. Doc. D. Mangeron Str., 67, Iaşi, Romania

Abstract. The paper presents some experimental results regarding the total deformation of "normal" modules found in the structure of modular sets. The study presents the main aspect regarding the experimental methodology used for the determination of static rigidity in the transversal direction, the measuring scheme and test stand used. Modules deformations were analysed, under the action of transversal static force, in the presence of axial pretension force, using advanced equipment. The research conducted showed that modules with symmetric structure present increased stiffness compared to the asymmetric modules and T cross channels.

1 Introduction

In a technological system fixtures are used for orientation-positioning-clamping/tightening of workpieces during machining, measuring, assembly. Modular fixtures are characterized by the use of components that are normalized, standardized. The notion of modularity is related to determining a dimensional correlation based on module and by using multipliers various dimensions can be set. By non-permanent assembly, various fixtures can be built for clamping various workpieces, in various machining, measuring, assembly conditions [1-2]. Modular fixtures have the advantage of high flexibility, reduced assembly times. The main disadvantages: lower technical performance than special permanent fixtures and high cost caused by high geometrical accuracy imposed to modules that are comprise in a modular fixture and the assembled fixture rigidity is lower as a result of a high number of joints [3-6].

The notion of system rigidity (stiffness), in general, therefore of technological systems also, represents the system capacity to oppose deformation under the action of exterior forces [5], [7].

The rigidity (stiffness) of a regular or a modular fixture can be defined as the ratio of the force (or torque) acting on the fixture, in a particular direction, and the deformations (linear or angular) of the structure, in the same direction or not [5].

The rigidity is an important fixture requirement; in order assure the dimensional/ geometrical precisions, during machining, inspection, assembly, especially for difficult machining and severe geometric conditions [1, 2, 3, 5]

Static stiffness offers information for evaluating the accuracy of machining process and for evaluation of "resistance" to vibration. Both static and dynamic rigidity/ stiffness, of a

*Corresponding author: chitariudragos@gmail.com

modular fixture, is determined by the number, arrangement and static stiffness of the modules within the fixture structure [4, 8].

Based on the forces acting on the modular fixture-workpiece system during clamping and machining and decomposition of the structure the following joints may be considered:

- joints axially tensioned (on normal direction);
- joints transversal tensioned, with axial pretension (on normal direction);
- joints torque tensioned, with axial pretension (on normal direction).

Static stiffness analytic determination is difficult due to multiple joints present in the structure and limited information regarding the static behaviour of these joints and, in addition, the surfaces of the modules that compose the modular structure are accompanied by a series of geometrical deviations (dimensional, shape, relative position, waviness, roughness) which can lead to a series of deviations of orientation-positioning and clamping deviations during clamping and machining [1, 3, 5, 8, 9].

In the case of joints transversal tensioned, with axial pretension (on normal direction) the information in the Machine-tools and Fixtures fields are limited, compared to the information regarding joints axially tensioned (on normal direction) in the form of values for contact rigidity between the workpiece and supports/ clamping elements without values for the total (overall) rigidity (consisting of contact rigidity and intrinsic rigidity) of fixtures components which compose the structure [1, 2, 8, 9, 10, 11].

In the paper a work methodology is proposed in order to experimentally determine the values for the total deformation of modules structures, found in modular fixture, which can lead to geometrical deviations caused by static transversal forces in the presence of pretension static forces on axial (vertical) direction of modules and modular structures - assembly forces. Also in this paper, the proposed modules (Figure 3), will be associated with a base plate (Figure 2).

In the case of modular fixtures, frequently, modules that form the fixture body are used for orienting and positioning of workpieces, thus justifying the study of this kind of modules.

2 Method

The developed research methodology is based on the knowledge from the Machinery construction field [7, 11], Machine tools field [9], and fixture field [1, 2, 3, 4, 5, 10].

The research methodology contains the following stages:

1. Identification of the influence factors on modular fixture deformation on transversal direction in the presence of assembly forces (on normal direction). Among the factors that may influence the transversal deformations of modular devices we considered: 1-the acting forces (assembly forces, clamping forces, machining forces); 2-module material; 3-clamping fittings; 4-surface roughness of the joint contact surfaces, 5-number of joints found in the structure; 6-orientation and positioning of the module in the modular structure of the fixture related to the base plate; 7-number of clamping points of the base plate on the machine tool table; 8-module type and size; 9-clamping force [2, 3, 4, 5].

2. Selection of constant and variable factors. The constant factors, which can influence the transversal deformations of modular devices in the presence of axial assembly forces, considered are: axial assembly forces, module type; orientation and positioning of the module in the modular structure of the fixture. The variable factors considered are: transversal force: type of modules, surface on which the transversal force acts.

3. Development of force and deformations measuring scheme and the selection of the force and displacement measuring method. In order to determine the rigidity on transversal direction, in the presence of axial assembly forces, the methodology requires the measurement of the acting force and the fixture displacements.

 DOI: 10.1051/matecconf/201711207024

Figure 1 presents an illustration of the experimental stand designed developed and used to determine the stiffness of modular fixtures [4].

Fig. 1. Test stand designed and developed to determine the stiffness of modular fixtures [4].

The acting force is measured with the force transducer and simulates the effect of the clamping and machining forces. The force transducer is positioned horizontally at 16 mm to the top surface of the module. The force is applied horizontally by a hydraulic cylinder; deformations were measured at 16 mm to the top surface and at 16 mm near the base plate of the modular structure as presented in Figure 1.

The measuring equipment consists of: force transducer CTL model 1000 produced by Laumas Elettronica srl; inductive displacement transducer type Wl/10mm produced by Hottinger Baldwin Messtechnik; Spider 8 multi-channel electronic PC measurement unit produced by Hottinger Baldwin Messtechnik; computer [3, 4, 5].

Data processing was performed by filtration using the "moving average" filter. The static force considered was composed from: assembly forces, tightening forces and the static component of the machining forces.

The tested modular structures consisted of modules (elements) from the modular kit SEM – 64 DISROM, made in Romania,. General characteristics of the SEM – 64 DISROM modular kit are: basic dimensions (distance between channels) – 64 mm, channel width – 14 mm diameter of threaded holes – M12 and M6, diameter of screws and bolts – M12 and M6, holes – Ø 13 mm [3, 4, 5]. Histogram 6 presents dimensions and shape of tested modules.

The "normal" module means that the modules section dimensions are 64 × 76 mm. The "narrow" module means that the modules section dimensions are 64 × 38 mm, Fig. 2. These modules generally fulfill (most common) the functions of body components and orientation-positioning of workpieces.

The material for the modules is 38MoCrAl09, which after heat treatment and nitrating treatment, acquires a surface hardness of 880-1100 HV. Active surface roughness *Ra* is between 0.4 and 0.8 μm [3, 4, 5].

Measurements of the displacements and forces were made on the transversal direction at a force of 550 daN (5.5 kN). Data processing and plotting of diagrams was conducted in NI Diadem software. The axial force used for assembly force had the value 36.140 N (3.641 daN).

DOI: 10.1051/matecconf/201711207024

3 Results

Diagrams regarding the evolution of deformations (stiffness) are presented in Fig. 2, the results refer to modules associated with the base plate.

Considering the asymmetric construction of modules the acting forces were applied on the 64 mm and 76 mm side, resulting different sides that oppose the transversal forces.

In order to correctly orientate and position the modules on to the baseplate guide keys were used during testing.

Figure 2 presents the measured maximum deformation for support "normal" module CSN — 003 module (filtered experimental result), at maximum height and minimum height for the second set of determination.

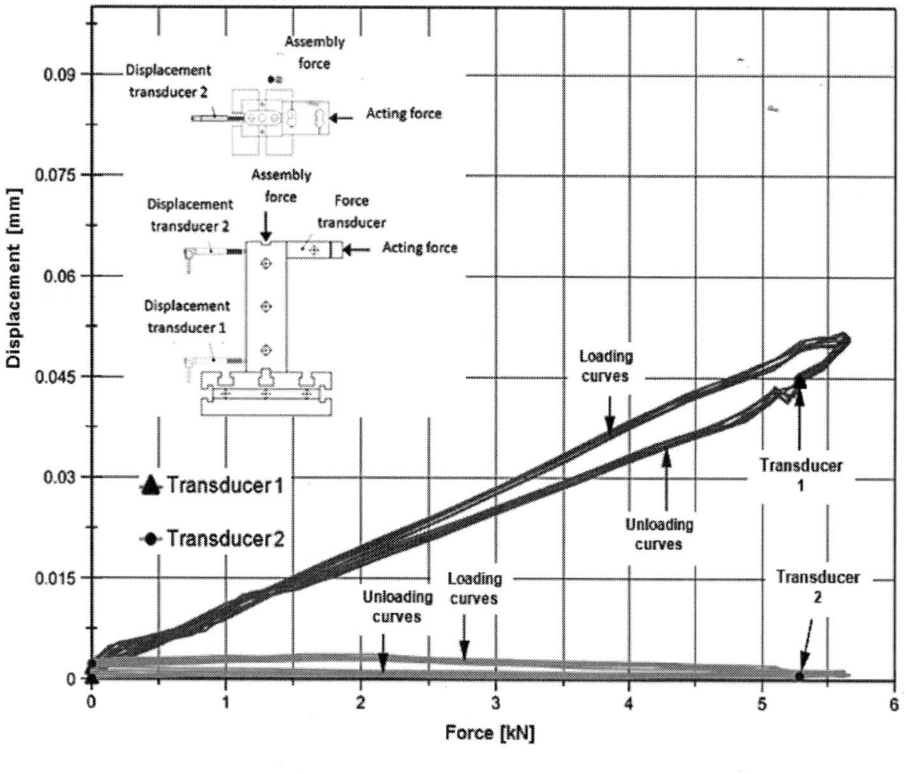

Fig. 2. The deformation (rigidity) curves for support "normal" module CSN — 003 module (filtered experimental result).

The maximum measured displacement is 0.0519 mm for an acting force of 5.56 kN.

The maximum measured displacement at the base of the modular structure is 0.0031 mm for the same force.

The curves shapes are non-linear, aspect specific to joint structures, as presented in literature [9, 10, 11]. The allure of the deformation curves show inflection points, specific to the application of hydraulic actuation force, respectively, the "hydraulic shock" similar to the behaviour presented in [3, 4, 6,].

Figure 3 presents, synthetically, as a histogram, the maximum deformation of modular structures considering acting force of 5.5 kN, on transversal direction and a 36140 N assembly force on axial direction. The acting force was on the 64 mm and the 76 mm side.

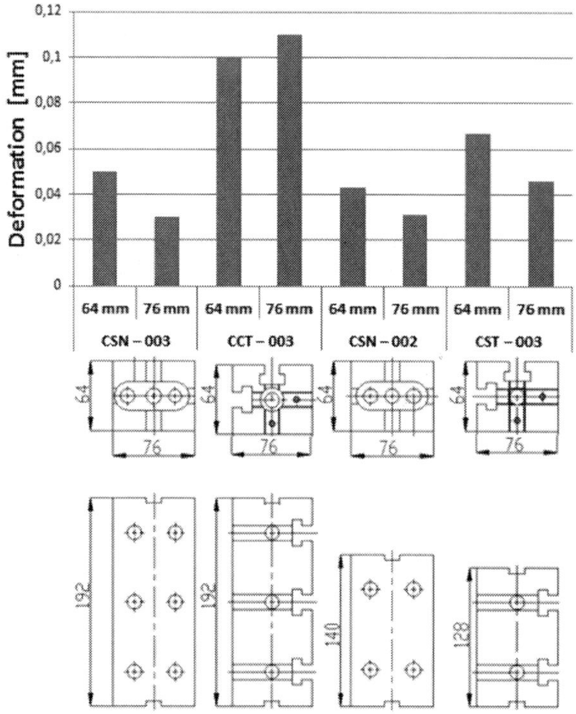

Fig. 3. Maximum deformation considering a 5.5 kN transversal force for the considered structures.

The presented results are after the reduction of the gap caused by the clearance in the wedge assembly. In order to reduce the gap first loading-unloading cycle is performed.

After the first loading cycle and the reduction of gap/ clearance, in order to correctly evaluate the elastic deformations, the measuring equipment was reset to zero a second cycle was performed, without interventions on the system, resulting the graph from Figure 2.

Table 1. Rigidity values calculated for the top of the structure (measured by transducer 2).

Module type	Rigidity value at the top of the module [daN/mm]	
	76 mm side	64 mm side
CCT – 003	4.867	5.500
CCT – 002	12.613	7.746
CSN – 003	11.120	18.333
CSN – 002	17.741	12.883

From Figure 3 analysis it is observed that the measured deformations on 64 mm side (the 76 mm side is opposing deformation) are reduced compared to deformations when the force acts on 76 mm side (the 64 mm side is opposing deformation). Measured deformations are lower by 10 -40 % when the acting force is applied on 76 mm side.

The rigidity values calculated based on Figure 2 values indicate high rigidity value at the base of the module around 110000 [daN/mm], for all modules.

During testing guide keys were used to orientate and position the modules on the base plate. The absence of guide keys can lead to rotations of the modules under transversal acting force.

4 Conclusions

The research conducted has focused on determining the transversal stiffness of modular structure modules constructed from "normal" module from SEM - 64 DISROM modular set. A series of deformation (stiffness) curves were drawn.

Higher stiffness values have been measured for modular structure with modules which do not have cross – channels and which have a symmetric structure. The experimental results indicates a tilting/ leaning of the "normal" module, after the displacements of the entire modular structure alongside the base plate, corresponding to the play between the wedge and the T slot, similar to the results presented for "narrow" [4].

The highlighted displacements and deformations of the modular structure can lead during clamping and machining to "dimensional"/ position deviations from the intended position of the workpiece affecting the machining accuracy.

The research will be extended in the future for more complex structure, typical structures found in this field, in order to asses modular fixture rigidity in common working conditions.

References

1. N. Gherghel, *Fixture Design and Exploitation* (in Romanian), **1** (I.P. Iaşi, 1981).
2. N. Gherghel, N. Seghedin, *Conception and Design of Supports of Technological Fixtures* (in Romanian), (Tehnopress, Iaşi, 2006).
3. D.F. Chitariu, App. Mec. and Mat. **657**, 514-518 (2014)
4. D.F. Chitariu, App. Mec. and Mat. **809-810**, 754-759 (2015).
5. D. Chitariu, N. Gherghel, Bul. Inst. Polit. Iasi **LIX (LXIII)**, Fasc. 1, 101 – 110 (2013).
6. D. Chitariu, N. Gherghel, Bul. Inst. Polit. Iasi **LIX (LXIII)**, Fasc. 1, 111 – 118 (2013).
7. V.S. Korsakov, *Machining Accuracy* (in Romanian, translation from Russian) (I.D.T., Bucharest, Romania 1963).
8. Y. Rong, Y. Zhu, *Computer Aided Fixture Design*, (Marcel Dekker Inc., New York, 1999).
9. Y. Ito, *Modular design for machine tools* (The McGraw-Hill Companies, New York, 2008).
10. V. Tache, A. Brăgaru, *Fixture for machine-tools, Design of Orientation and Positioning Schemes of Workpieces,* (Ed. Tehnica, Bucharest, Romania, 1976).
11. C. Picoş et. al., *Design of Machining Technologies, Design Manual*, **1** (in Romanian), (Ed. Universitas, Chisinau, Moldova, 1992).

Conceptual design of jewellery: a space-based aesthetics approach

Vaia Tzintzi[1], *Athanasios* Manavis[1], *Nikolaos* Efkolidis[1], *Christos* Dimopoulos[2], *Konstantinos* Kakoulis[1] and *Panagiotis* Kyratsis[1,*]

[1]Western Macedonia University of Applied Science, Department of Mechanical Engineering and Industrial Design, Kila Kozani GR50100 Greece
[2] European University Cyprus, School of Sciences, Nicosia, Cyprus

Abstract. Conceptual design is a field that offers various aesthetic approaches to generation of nature-based product design concepts. Essentially, Conceptual Product Design (CPD) uses similarities based on the geometrical forms and functionalities. Furthermore, the CAD-based freehand sketch is a primary conceptual tool in the early stages of the design process. The proposed Conceptual Product Design concept is dealing with jewelleries that are inspired from space. Specifically, a number of galaxy features, such as galaxy shapes, wormholes and graphical representation of planet magnetic field are used as inspirations. Those space-based design ideas at a conceptual level can lead to further opportunities for research and economic success of the jewellery industry. A number of illustrative case studies are presented and new opportunities can be derived for economic success.

1 Introduction

The conceptual design inspired from different science fields can become a straightforward activity for the industrial designers. Specifically, several designers from different backgrounds (i.e., furniture design, fashion, transportation and architectural design) are using nature and its evolution as an inspiration for their product designs [1]. The main reason of that are the opportunities created for innovation and creativity (Figure 1). Essentially, conceptual design is a field that offers various aesthetic approaches to generation of nature-based product design concepts. The term nature can be defined as both the earthly nature and at the same time the outer space nature.

Jewellery design has been around for thousands of years. It has changed and further developed over the years. Valuable metals and precious stones are designed, processed together for jewellery creations. Jewels are based on aesthetics but can incorporate different concepts related to religion. Jewellery is the peak where beauty, even passion, is combined with immortality [2]. Contemporary jewellery is an extremely exciting field which explores

* Corresponding author: pkyratsis@teiwm.gr

many new and innovative materials and techniques (i.e., CAD systems, rapid prototyping and 3D scanning methods, laser cutting and engraving systems etc.).

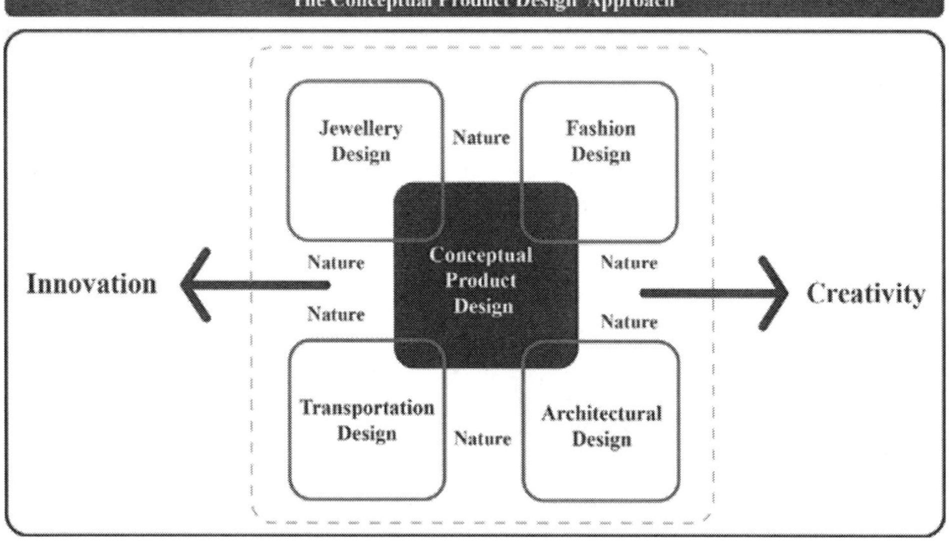

Fig. 1. The Conceptual Product Design approach.

2 Nature-based design approach

Humans have always been inspired by nature capabilities in problems solving and innovation [3]. There is a great number of nature-inspired design approaches that were developed in the recent years, i.e. structural design and morphology, material and texturing. The main advantage of bio-inspired design methods consists in offering designers an organized process for attaining a model that may be applied in design [4]. Final creations could be based on the relations between function, ergonomic features and aesthetics of nature.

Structural design is one of the most important components of any entity in the universe, which has a physical dimension. The term structural design means: the art and science of creating a structural system in nature or in the artificial world [5]. The structural systems in nature have two directions: a) structural systems in living systems and b) structural systems in inanimate entities.

Material selection for a clear-cut nature-based application is a thorough, lengthy and expensive process. There are many constraints to be considered in selecting materials. There are some situations that certain criteria for a material are defined at the beginning of the design project. Designer selects a material and he/she must consider fulfilling the three basic requirements: a) service requirements, b) fabrication requirements and c) economic requirements. On the other hand, the designers might be very attentive about the aesthetics criteria of material, such as colour and texture [6].

2.1 Conceptual Product Design Framework

Conceptual Product Design (CPD) is more of an art than an actual science. The framework followed from industrial designers incorporates several sessions of creative processes (i.e. brainstorming, mind-map and mood boards) with an aim to find alternative ideas for each product. Each idea inspired from nature was associated with the final products in order to inspire further the process of ideas generation.

CAD-based sketches were used so as to outline each idea at a conceptual design level and all of them were presented to an audience (designers, artists, engineers and marketers). The team of all these specialists was able to make comments about the ideas by completing a short questionnaire. An additional outcome was that these possible future customers would be willing to pay more for purchasing a product with a specific concept inspired from nature (Figure 2). The audience consisted of 18 students of the Master's program in Product Design and Systems.

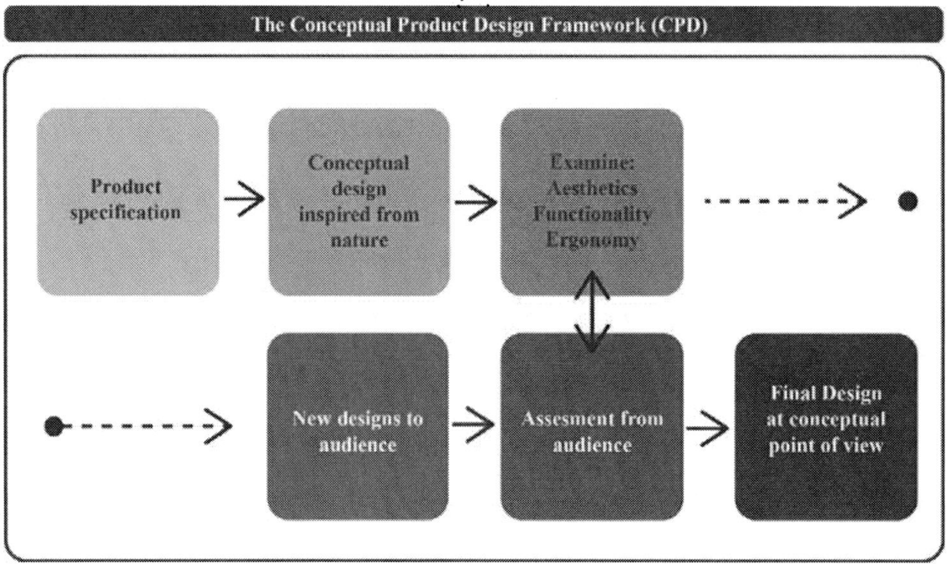

Fig. 2. The conceptual product design framework.

2.2 Jewellery industry

Jewellery is a universal form of adornment. They are made of shells and stone from prehistoric times [7]. The main themes of the first jewelleries were the flowers and the floral motifs. It is clear that the nature was the first inspiration for the jewellery industry.

Today, the boundaries of jewellery have been continually redefined. Conventions have been challenged by successive generations of industrial designers, often educated at art and design schools and immersed in radical ideas. New technologies, innovative design concepts and modern materials have overturned the notions of status traditionally implicit in jewellery.

2.3 Inspirations from space

The cosmic theme always awakens insatiable curiosity. This specific nature (planets, asteroids, stars, galaxies, black holes, and satellites) is not often used as a design concept for product designers. Moreover, the term inspired from nature design is considered more related to earth based nature. For this reason, the space-based aesthetics design approach offers a great advantage for the industrial designers. The specific inspiration, at a conceptual level, can lead to further opportunities for research and economic success. Innovative and novel ideas can be created and offer a great deal of satisfaction to the appropriate customer groups.

3 Conceptual case studies for the jewellery industry

The current paper presents a series of products, at conceptual level, inspired from nature. The proposed CPD concept is about jewelleries that are inspired from space. A number of galaxy features, such as galaxy shapes, wormholes and graphical representation of planet magnetic fields are used as an inspiration.

• Ring design based on a wormhole visualisation. A ring is a round band, usually of high quality and made of precious metals, worn as an ornamental piece around the finger. There are different types of rings, depending on their shape, material, use or meaning. Some of these rings have a symbolic function such as the wedding band. In this case, a ring proposal is presented based on the form of a wormhole.

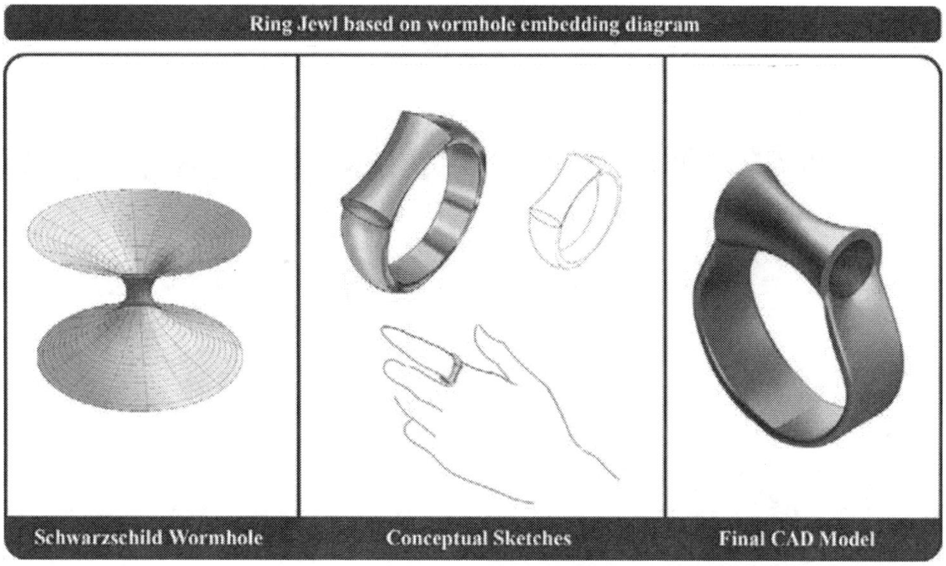

Fig. 3. A ring design based on the wormhole embedding diagram.

A wormhole is a hypothetical shortcut between two distant regions of a space-time. A three dimensional wormhole is impossible to be fully visualized. However, there is a human-friendly approach of the wormhole visualization. It is known as Schwarzchild embedding diagram, which is the wormhole analogue for a static, non-rotating, Schwarzchild black hole. The proposed product of the ring incorporates all these special geometries together with the aesthetics elements that excuse the wormhole concept (Figure 3).

• Earring design based on the Earth's magnetic field. An earring is a piece of jewelry attached to the ear via a piercing in the earlobe or another external part of the ear. They are the most essential jewelry item for every woman, they sparkle and shine and add much glamor to their appearance. There are different types of earrings. The categorization is based on different types of closures: posts, fish hook, lever backed, kidney and hoop. In this case, the presented earrings are inspired from the geometrical characteristics of the earth's magnetic field (Figure 4). The earth is considered a large size magnet, generating a great deal of magnetic field around itself. The view of the magnetic field used, at a conceptual level, is based on real science observations that have been made since the beginning of the space age. The magnetosphere bulges on one side and tapers off on the other, because it is pushed on the side facing the sun by the solar wind and stretched out in the earth's shadow. These specific lines were used as the main inspiration for the earring design.

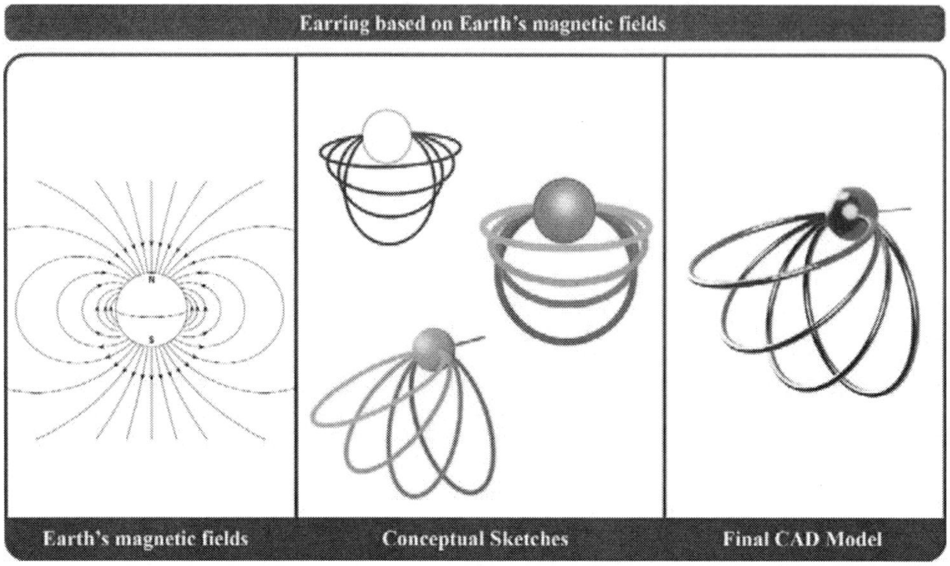

Fig. 4. Earrings design based on the geometrical characteristics of earth's magnetic field.

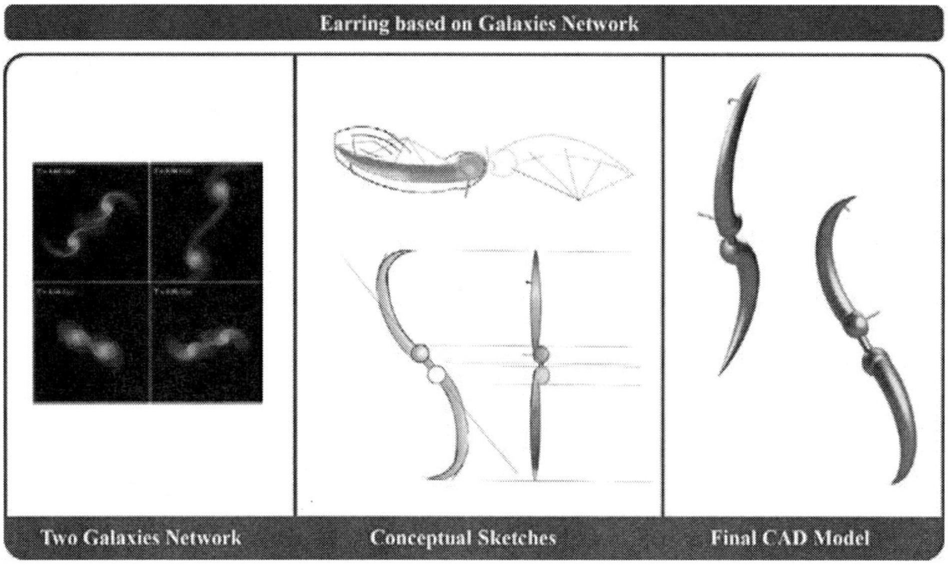

Fig. 5. Earring design based on galaxies network.

• Earring design based on galaxies network. Galaxies interact with each other and are considered connected with other galaxies by gravity. Galaxy groups can be small (two galaxies orbiting each other), or large (cluster of thousands of galaxies extending for more than ten million light years). Clusters are made up of two basic types of matter: luminous matter (i.e. stars, hot gas) and dark matter. Dark matter does not shine on its own, and the only way we know it exists is because of its gravitational effect on luminous matter [8]. The final result of earring has all these specific aesthetics elements that are described for the small group of two galaxies (Figure 5). The key features of the proposed design are based on the orbital geometry between two galaxies.

4 Conclusions

A series of conceptual product designs were presented based on the design inspired from space.

The CPD framework provides a stable process for assessing the proposed products. Following the Conceptual Product Design framework industrial designers are able to follow a structured way for presenting novel ideas for products that are able to offer commercial success and communicate their messages to the future customers.

Those space-based design ideas at a conceptual level can lead to further opportunities for research and economic success for jewellery industry.

The methodology is considered to be an extension of the design inspired from nature.

References

1. P. Kyratsis, E. Dimou, A. Manavis, N. Bilalis, *AMM*, **760,** 33 (2015)
2. L. Ulusman, Ç. Bayburtlu, Procedia Soc. Behav. Sci., **46,** 5928 (2012)
3. Y. Bar-Cohen, Int. J. Aeronaut. Space, **13(1)**, 1 (2012)
4. C.A.M. Versos, D.A. Coehlo, *Biologically inspired design: Methods and validation*, (Industrial Design-New Frontiers, 2011)
5. A. Mosseri, WIT Trans. Ecol. Envir., **73**, 589 (2011)
6. E. Karana, P.Hekkert, P. Kandachar, *Smart Design: 1st International Conference Proceedings,* **29**, 1081 (2008)
7. L.C. Molinari, M.C. Megazzini, E. Bemporad, Gold Technol., **23**, 3 (2008)
8. https://imagine.gsfc.nasa.gov/science/objects/clusters.html Accessed 22/02/2017

Experimental investigation on wear resistance of a new cutting fluid using a four ball tribometer – optimization of additive percent

Viorel Paleu*, *Viorel* Goanţă, and *Bogdan* Istrate

"Gheorghe Asachi" Technical University of Iaşi, Department of Mechanical Engineering, Mechatronics and Robotics, 61-63 D. Mangeron Blvd., Iaşi, Romania

Abstract. The cutting fluid manufacturers aim to obtain the best products at the lowest possible price. To achieve such a compromise experimental work is needed, trying to get the lowest allowed content of additive in the base oil. The paper presents the results of standard tests carried out on four ball machine. This tribometer is used to analyze the load-carrying properties of cutting fluids. The results are compared with the standard requirements, related to the measured wear scar on the three lower balls of the four ball device. An optimum of sulphur additive content into the base oil was found around 10%. According to the actual standards and following the results of carried-out tests, for this specific case the lowest acceptable content of additive is around 4%.

1 Introduction

A major problem in the metal cutting process is the condition of the cutting tools, influencing the quality of the machined surfaces and the overall machining costs. The dry cutting can generate heat and high temperatures, intensifying the tool wear, increasing the cutting forces and diminishing the surface accuracy of the workpiece.

Three main categories of the cutting fluids are used today: cutting oils, water soluble fluids, and gases. A cutting fluid reduces the friction at tool-chip and tool-workpiece interfaces enhancing the tool life, cools the tool and the chip and facilitates the removal of the chip [1]. To minimize the wear of the tool the cutting fluid have to be a good lubricant, to provide a separating film in the cutting zone.

A good cutting fluid must be able to rapidly eliminate the heat from the tool - workpiece – chip system, especially by convection. For a good flow the cutting fluid is requested to be less viscous, but for high wear resistance it requires to be additived with extreme pressure (EP) additives.

The production cost is an important aspect that interests the cutting fluid manufacturers. It can be minimized by choosing cheap low viscosity base oils and optimizing the quantity of good but expensive EP additives. To achieve such a compromise experimental work is needed, trying to get the lowest allowed content of additive into the base oil.

* Corresponding author: vpaleu@yahoo.com

There are different methods to experimentally asses the performances of the cutting fluids. According to ASTM D2783-3, various lubricants performances can be determined on a four ball machine: load - wear index, the weld point, last non-seizure load etc.

A review of the tribological properties of cutting fluids is presented on large in [2]. Some results on four ball machine testing of oil based additived cutting fluids and cutting nanofluids are reported in literature [3-5].

This paper presents the results of standard tests carried out on a four ball machine to analyze the load-carrying properties of cutting fluids. The results are compared with the standard requirements, related to the measured wear scar on the three lower balls of the four ball device.

2 Materials, testing machine and test conditions

2.1 Materials

The main properties for the base oil, high-sulphur content polysulfide EP additive, and tested balls are indicated in Table 1.

Table 1. Properties of the base oil, EP additive, and bearing steel balls.

Material	Property	Value	Units
Base oil	Kinematic viscosity at 40^0C	13.9	mm^2/s,
	Density at 20^{0C}	853	Kg/m^3
EP additive	Kinematic viscosity at 25^0C	87	mm^2/s
	Kinematic viscosity at 40^0C	39	
	Kinematic viscosity at 100^0C	6	
	Pour Point	-48	^0C
	Sulfur	36.9	% Weight
Steel balls	Material	AISI 52100 bearing steel	
	Diameter	1/2	inch
	Density	7800	$Kg/\ m^3$
	Elasticity modulus	208	GPa
	Hardness, HV10	700	$Kg/\ mm^2$
	Roughness, R_a	0.018... 0.035	µm

2.2 Four ball machine

A four ball machine is a tribometer designed to test lubricants behaviour in a tribosystem composed by one rotating ball pressed against 3 stationary balls firmly held together and immersed in lubricant (see Figure 1). The four balls touching each other form a pyramid, between the upper and the three lower balls obtaining point contacts.

The four ball machine is composed by an electrical motor and a belt transmission (1), a spindle (2), the four balls testing device (3), the load lamellar spring (4), and a frequency static converter (6).

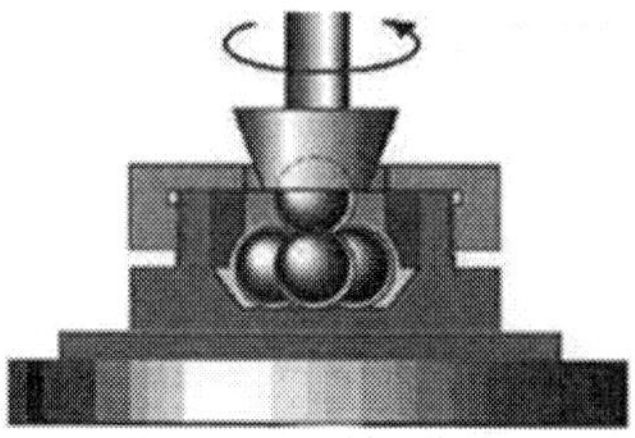

Fig. 1. The testing device of the four ball machine.

Figure 2 presents a general view of the testing machine used in this paper.

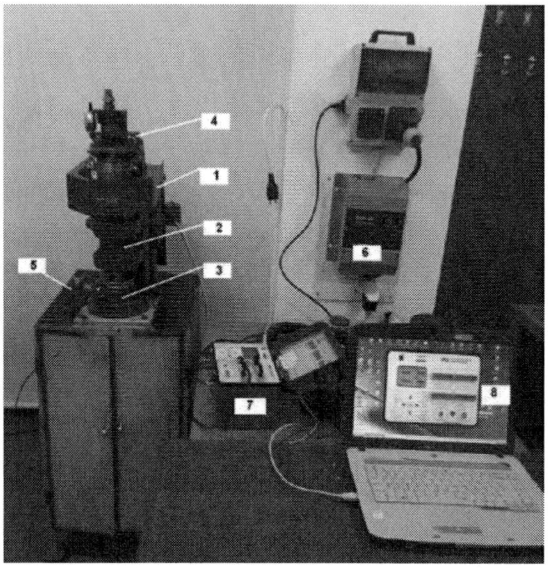

Fig. 2. General view of the four ball machine.

2.3 Data acquisition system

The monitored parameter is the friction torque between the upper rotating ball and the three static lower balls. The friction torque created by the friction forces acting on the point contacts between the rotating upper ball and the three stationary balls tries to rotate the testing device. The utilized sensor is resistive, being a metallic elastic leaf (5), with two resistive strain gages brazed on it, attached to the testing device (3). A pin from the frame of the machine blocks the elastic leaf, producing an elastic deformation proportional with the friction torque within the testing device.

The data acquisition system is composed by a P3 – Vishay resistive bridge (7) with four channels, and its software application. The data acquisition system is calibrated by dead weights method. The friction torque values during tests are saved on the internal hard disk of a notebook (8).

The load is maintained constant during tests, by the lamellar spring (4), which was previously calibrated on a compression standard machine.

2.4 Testing conditions

The properties of the tested materials are indicated in Table 1. The cutting fluid is obtained by mixing 2 up to 20 percent of EP additive into the base oil, samples of 1 liter being prepared.

DOI: 10.1051/matecconf/201711207026

For each test new four unworn steel balls are used, and after their mounting in the testing device about 10 ml of additived oil is added, till the top ball is completely covered. The top ball is clamped to the end of the rotating spindle.

The tests are carried-out according to a specific standard for cutting fluid testing on four ball machines, equivalent to ISO 6743/7 L-MHD and ISO 6743/7 L-MHF, which imposes a mean wear spot measured on the three lower balls of maximum 2 millimeters. The test conditions are: speed of the shaft (rotating upper ball):1500 rpm; constant axial load during tests: 1500 N; time of each test: 1 minute; concentration of the EP additive in the base oil: 2% … 20 %. In fact, this is a specific test from the general ASTM D2783-3 standard, called compensation scar diameter.

After each test the wear scar diameters on the lower balls were measured and inspected on a microscope at a high magnitude scale.

3 Results and discussions

3.1 Compensation scar diameter

The compensation scar diameter is defined by ASTM D2783-3 standard as the average diameter, in millimeters, of the wear scar measured after each test on the stationary lower balls. This wear scar is created by the rotating upper ball, pressed against the stationary balls by an axial force, in the presence of a lubricant, without causing scuffing or welding of the balls. Figure 3 presents the compensation scar diameter evolution versus EP additive percent.

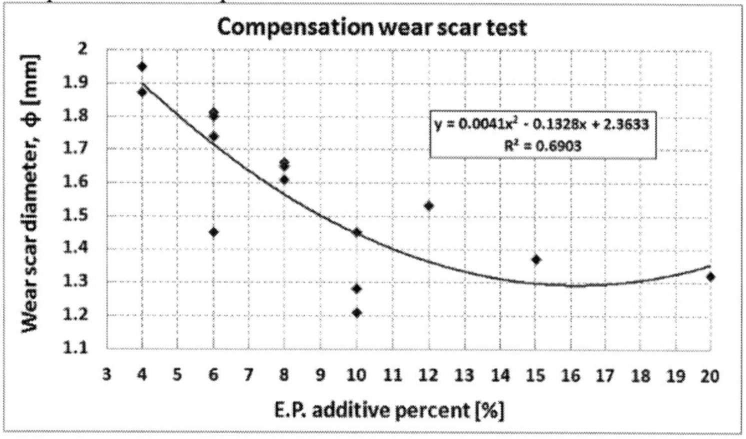

Fig. 3. Mean wear scar diameter versus E.P. additive percent into the base oil.

The wear scar diameter becomes almost constant if the additive percent is increased above 10%. According to ISO 6743/7 L-MHD and ISO 6743/7 L-MHF standards, the minimum accepted additive percent into the base oil is 4%, the mean wear scar on the three lower balls reaching the limit accepted value of 2 millimeters.

When the obtained results were spread, we repeated up to four times the tests. The variation of the wear scar diameter, Φ, versus EP additive percent, C, can be interpolated with a polynomial function of the second order, see equation (1). The R^2 parameter is 0.69, better than for logarithmic (R^2=0.65), power law (R^2=0.62), linear (R^2=0.53), or exponential (R^2=0.52).

$$\Phi = 0.0041 \cdot C^2 - 0.1328 \cdot C + 2.3633 \tag{1}$$

DOI: 10.1051/matecconf/201711207026

3.2 Microscopy analysis of the wear scars

Analyzing the surface of the wear scar by microscopy, it can be observed that no catastrophic wear occurred for the additived samples of oil during the imposed test condition (duration 1 minute, speed 1500 rpm, axial load 1500 N), for EP additive percent between 4 and 20%. Figure 4 shows some images for different percent of additive into the base oil.

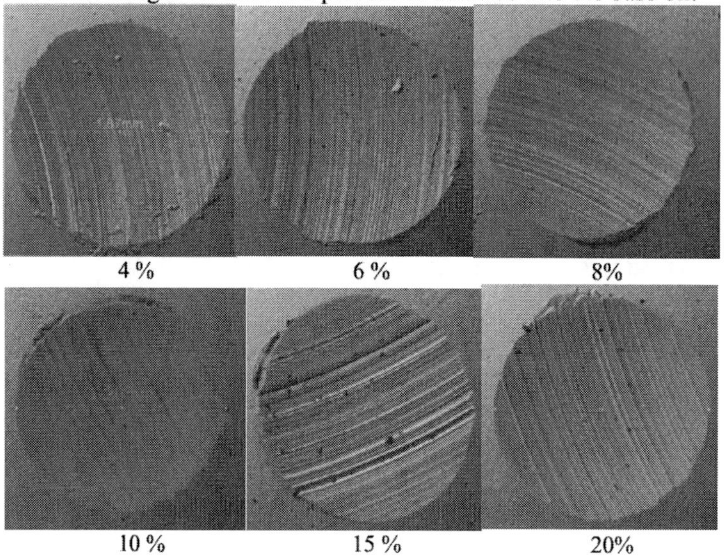

Fig. 4. Images of the wear scars for different percent of additive.

The visual inspection of wear scars shows some wear particles at the border of the scars, and also some lubricant and EP additive scraps.

3.3 Friction torque evolution

Friction torque evolution during tests versus EP additive content in the base oil is represented in Figure 5. The friction torque values are distinct at the end of the tests, when the balls lubricated by a sample of oil with 2% EP additive failed by scuffing. The balls welded together when scuffing appeared (for 2% of additive), and the friction torque almost doubled its value. The moment when scuffing produced is indicated on Figure 5 by a strong increase of the friction torque. For the rest of the samples the values are very near, being between 2 and 3 N·m. The monitoring of the four ball machine tests by friction torque evolution is a confident method, allowing the prediction of the occurrence of scuffing in just few seconds.

Analyzing the results, one can be observed that the lowest friction torque and the best surface quality was obtained for 10% of EP sulphur additive into the base oil. Increasing the content of additive from 10% to 20% or decreasing it to 4%, the friction torque augments. But from economical viewpoint the most advantageous case accomplishing the existing standards for cutting fluids wear resistance the optimum is obtained for 4% of additive. For this percent of additive into the base oil the requirements of the standards are still accomplished, that is the mean wear scar on the lower balls is less than 2 millimeters.

Important savings can be obtained by cutting oil manufacturers by adding just the minimum necessary of additive.

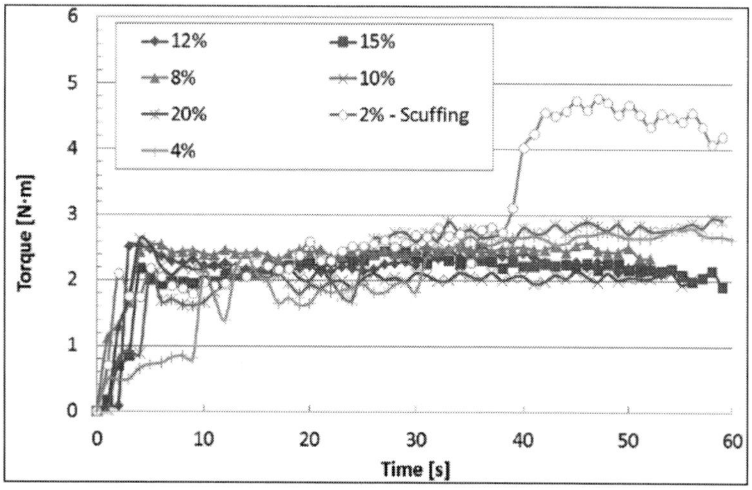

Fig. 5. Friction torque evolution versus EP additive content.

4 Conclusions

The tribological properties are very important when developing a new cutting fluid. A good carrying-load of the cutting fluid assures a reduced wear of the tool. The cutting fluid manufacturers must minimize the costs, while accomplishing the requirements of the actual standards. To establish the minimum allowed EP additive concentration into the base oil for new oil based cutting fluid, the compensation wear scar test is developed on four ball tribometer. Analyzing the results, one can be observed that the lowest friction torque and the best surface quality was obtained for 10% of EP sulphur additive into the base oil, but from economical viewpoint the most advantageous case accomplishing the existing standards for cutting fluids wear resistance the optimum is obtained for 4% of additive. The monitoring of the four ball machine tests by friction torque evolution is a confident method, allowing the prediction of scuffing occurrence in just few seconds.

Interpolating the obtained experimental results, an empirical formula was found for the dependence between the mean wear scar diameter, measured on the three lower balls, and the EP additive concentration into the base oil. For samples of oil with different percent of additive, the research is intended to be continued with standard incipient seizure load tests.

This work was supported by VIPOIL Company from Iaşi, Romania. The authors also acknowledge Mr. Igor Plugaru, manager of VIPOIL Company, for supplying the technical data on base oil and additives.

References

1. Y. Isik, Strojniški vestnik, J. Mech. Eng. **56** 3 (2010)
2. A. Anand, K. Vohra, M.I. Ul Haq, A. Raina, M.F. Wani, Tribol. Ind. **38** 4 (2016)
3. A. D. Dongare, A. J. Gite, IOSR J. Eng. **04** 8 (2014)
4. A. D. Dongare, G. J. Vikhe Patil, Int. J. Eng. Resear. Dev. (IJRED) **4** 6 (2012)
5. M. Mosleh, M. Ghaderi, K.A. Shirvani, Jh. Belk, D.J. Grzina, J. Manuf. Proc. **25** (2017)

DOI: 10.1051/matecconf/201711208001

Physiotherapy devices able to generate ethical dilemmas

Nadinne Roman*

Transilvania University of Braşov, Department of Basic, Prophylactic and Clinical Disciplines, Nicolae Balcescu 56, Braşov, Romania

Abstract: Physical therapy is a medical specialty where the professionals help restore movement and function when someone is affected by injury, illness or disability. This paper wishes to establish the connection between ethics, physiotherapy and bioengineering. The research method was achieved using academic database searches based on specific keywords. A SWOT analysis of the physiotherapy devices utilization and design was made, for extracting ethical considerations. The main results suggest that physiotherapy devices are able to generate ethical dilemmas, classified in 4 main items: (1) Bioengineering in physical therapy, ethical and clinical standards for manufacturers; (2) Social impact of physical therapy devices and ethical issues; (3) Inter-professional lack of communication and ethical concerns; (4) Bioengineering ethical research and education. As conclusions, for the physical therapy or electrotherapy research equipment development, a multidisciplinary team is needed. The equipment used in rehabilitation must fulfil specific technical and scientific requirements drafted by the professionals.

1 Introduction

Physical therapy and psychiatry are medical specialties where the professionals helps restore movement and function when someone is affected by injury, illness or disability. A physiotherapist specialist provides medical health care services in a prophylactically manner for patients likely to develop musculoskeletal disorders. The physiatrists and physical therapists' activity are centered on patient`s treatment personalization, seeking to improve their patient's quality of life. The rehabilitation area treats a broad scope of cases, from the brain and spinal cord injuries, to post traumatic sequelae, congenital locomotors disorders, and stroke and burn victims. Usually, these specialists coordinate a multidisciplinary team of professionals, like physical therapists, occupational therapists and collaborate with the specialized clinicians like neurologist, podiatrist, pediatrician, and other specialized doctors [1]

Physical therapy is a medical field where devices and forms of therapy used varies depending on several factors such as the type of disease, the affected tissue, body`s area or segment affected, the patient's age or other associated pathologies like respiratory diseases, cardiovascular diseases, blood coagulation, diabetes, rheumatoid or psychiatric disorders.

* Corresponding author: nadinneroman@gmail.com

This paper wishes to establish the connection between ethics, physiotherapy, and bioengineering, considering the multifactorial elements involved in the field of physiotherapy.

2 Method

Research method was achieved using academic database searches and academic engines based on specific keywords like: ‘physiotherapy”, “medical devices”, “bioengineering”, “ethics”, “rehabilitation”, “psychiatry”, “moral”, “electrotherapy”, “bioethics”. We developed search criteria for each database: Isi Web of Science, Scopus, Elsevier and Google Scholar, but the main papers resulted from one particular search engine-Google Scholar. The literature in this field is poor developed so the papers analyzed have heterogeneous structures. The main physiotherapy devices suppliers’ user manuals were searched too.

We have analyzed the data from the papers and user’s manuals selected performing a SWOT analysis for physiotherapy devices, highlighting their strengths, weaknesses, opportunities, and threats. After analyzing the weak and strong elements concerning physiotherapy and electrotherapy devices, we have extracted four major ethical considerations concerning physiotherapy devices.

3 Results

The main results suggest that physiotherapy devices are able to generate ethical dilemmas in design and use. The ethical issues were distributed in four items.

3.1 Bioengineering in physical therapy, ethical and clinical standards for manufacturers

Bioengineering is a multidisciplinary framework, with multiple application domains, starting with skin grafting, tissue replacement, bone reconstruction and further continuing with implantable devices, biomedical micro electro-mechanical systems, prosthetics or limb replacement, but also concerning neuro robotics or physical therapy and electrotherapy devices. With the expansion of the bioengineering field, it also occurred the approaches concerning ethical issues in this branch, technology that interact with human life made in order to solve problems in medicine and biology.

With the expansion of the bioengineering field also growth the concerning of ethical issues regarding technologies that interact with human life [2]. The robotics and bioengineering in physical therapy area are less developed comparing with industrial robotics. The robots’ specification within these fields is different because the user interface is the human being compared to industrial robots which usually do not permit human interaction. Secondary, in rehabilitation medicine robots or devices, are often items used by patients and not necessarily by medical professionals [3]. The need for specific rules or legislation caused many institutions to create frameworks and guidelines for regulating the ethical standards concerning bioengineering development and human beings’ safety. In this case, ethics is not only a philosophical branch, but it is practical and serves to human race beneficence [2].

For bioengineers is important to have ethical knowledge concerning device or materials manufacturing and to be able to make an ethical decision. An ethical dilemma arises when two major ethics principles are conflicting with each other. The practical aspects of ethics it is disputed in an engineering student’s research. The research revealed that using case study interactive education, and implementing before the basic ethical information "plays a significant role in learning moral reasoning skills" [4]. There is a poor boundary line concerning the interface of physiotherapy medical devices towards the therapeutic or prophylaxis utilization. The manufacturers and the medical devices suppliers are developing new technologies and some of them can be used both on ill or healthy subjects. There are many forms of electrotherapy or physical therapy devices in which the end user is the simple patient or client. In this case, the user doesn't have the required knowledge to be able to assess if the devices use might harm. There are physiotherapy medical devices that are not recognized by insurance companies due to lack of sufficient clinical randomized trials background. In this scenario, it cannot be proved the manufacturer's device effectiveness or lack of contraindications. This is a major ethical concerning among the physiotherapy devices users, to apply a therapy which proves decades later to have harmful effects unknown at the right time [5].

A recent physical therapy device proves that usually is insufficient medical scientific study background in this field. A well-suited example is extracorporeal shock wave therapy (ESWT) which was developed almost 17 years ago, whose manufacturer user's guide does not specify the cavitation process through effects. A 2016 study proves that ESWT generates cavitation and produce side effects including blood vessel rupture, not mentioned in the device's patent or other medical suppliers' manufacturers [6-7].

Organizational and regulatory institutions like World Health Organization (WHO) and The European Economic Community have some regulations and guidelines concerning medical device approval.

One of WHO statements refer to medical devices safety and it argues for "shared responsibility among the stakeholders". There are no special conditions regarding physiotherapy devices or concerning the ethical aspects of bioengineering or medical devices manufacturing [8-9]. Ethical responsibilities should be included in patents and manufacturers activity, including those with a prophylactically purpose.

3.2 Social impact of physical therapy devices and ethical issues

Physical therapy devices and bioengineering products used in rehabilitation represent an interface between the physical therapists, doctors, engineers and patients [2]. The physical therapy equipment is produced and utilized with the scope to improve patient's quality of life and regain movement trough activities of daily living. Lately, the physical therapy industry and neuro robotics developed. The utilization of these devices is highly expensive and involves a lot of ethical challenges.

The educational level and performance of the physical therapists, the patient's education and rehabilitation emotional distress period are key factors in ethical issues emerged. Some of the physical therapists even asking themselves if their profession is in danger to disappear, according to the manufacturer and patents development. But shortly studying the robots' usefulness and interface, they proved to be only another instrument in the hands of physical therapists, having more the role to fulfill the neuromuscular disorders are more prevalent due to precarious lifestyles. And here comes a new ethical issue about how much such equipment (neuro robots or exoskeletons) justifies the costs of manufacturing and how much it actually reflects in the cost of equipment the idea of novelty. This kind of ethical problem balances the good of a small group of people with high financial standards, with frustration and poor health of poor people. The electrotherapy devices used in physical therapy are different types of electromagnetic waves, ultrasound, lasers, continuous current, high voltage currents, and low frequency or high-frequency currents. These types of electrical therapies are noninvasive treatment methods which may interfere each odder and cause patient's adverse effects. The ethical implications of manufacturing and using these devices is a gap in this field. The professional community should reconsider the ethics approach of these devices use and manufacture. The reconsiderations shall be applied during the prototyping and patenting phase and also after placing on the market.

Fig. 1. SWOT analysis for physiotherapy devices.

motor rehabilitation, not the functional recovery of the patients, with the advocacy for the inability of robots to understand the patient's needs and the proprioceptive feedback considered superior to visual feed-back [10].

And in terms of costs, physical therapy equipment reported at the level of neuro robots and exoskeletons is extremely expensive. Especially for developing countries where

Another important ethical concern is raised by highly expensive physical therapy devices. The new equipment developed for gait rehabilitation are a well-suited example. This equipment is unaffordable for public health institutions and affordable only in private medical practice. Therefore, may put in ethical dilemmas the patients and also the physiotherapists.

The patients mostly striving beyond their ability to access those services and the physiotherapists to have their professional work hampered, for not being able to provide full medical services. Some of these ethical challenges and dilemmas may be avoided by rules and cooperation of manufacturers, sellers, and national government.

3.3 Inter-professional lack of communication and ethical concerns

The analyze of the manufacturer's user manuals and physiotherapists professional associations guidelines for physiotherapy equipment use revealed that in some cases there are major discrepancies regarding the therapies contraindications.

Some of the associations and manufacturers permit using different types of electric agents like ultrasound on composite or metal implants, other don't.

It is quite a major mismatch in terms of contraindications regarding electrical implant carriers. Also from this perspective, certain associations and manufacturers allow the use of electrical agents to a patient with a pacemaker, but not in the chest area.

Most of the therapies are contraindicated to pregnant women, but not because of the physical agents' effects, but due to lack of studies in the area. Which is confirmed by some manufacturers in the user manuals [11-14]. Ethical reasoning does not mean a person should be confided into its personal boundaries in terms of thinking and application of codes of conduct or rules of moral behavior.

The ethical problem in this case appears when a patient might be harmed by a physiotherapist or a doctor, after the use of one these therapies, although the protocols of use were strictly followed. Whose responsibility is then? Is the medical professional to be blamed, the patentees, the bioengineers or the manufacturers?

And they often create ethical disputes and problems in some medical centers where physicians or physiotherapists disputed new knowledge about contraindications. There are cases where some professionals take into account the new permissiveness of technology, and others refuse to put them into practice for fear of injuring patients. Thus, the evolution and treatment of a patient may suffer or be gained by the absence or presence of a physical procedure that can cause it to recover faster or better.

Worldwide, in most countries, physiotherapists are independent medical professionals and they apply these medical procedures without being supervised by other medical specialists. In Romania and in several other countries there is an exception, due to physical medicine and rehabilitation doctors.

They make the medical prescription, they set the treatment parameters, the length and they decide what is indicated and what is contraindicated. This way they are also sharing the responsibilities with physical therapists, but shouldn't be shared also with medical engineers? The lack of communication between professionals like bioengineers, doctors, physiatrists and physical therapists can give rise to situations that endangered patient health.

Another problem that may arise from the use of physiotherapy equipment is related to the poor training of physiotherapists in using new equipment. In most cases, training is implemented by a sales representative of the company providing the equipment. And most of the time, these medical representatives do not have enough basic knowledge to get an appropriate training. Very often, not knowing answers to questions asked by the people they need to train.

3.4 Bioengineering ethical research and education

The misconduct arose in research lately played an important role in ethical research frameworks. Federal or governmental agencies, academic institutions, professional association or professional communities have struggled to create specific rules for biomedical scientific research. The majority of research misconduct have emerged from the biomedical world and researches that prove that many medical pharmaceutical agencies that fund medical studies tend to overestimate the beneficial effect of their products [15].

Concerning physical therapy research, is very difficult to obtain accurate results because of two main factors: 1) the patient illness- which in most cases doesn't exist similarities that can provide scientific research results, and 2) the physical therapy manufacturers use different types of software design, with different electro physical parameters that sometimes are not disclosed in technical books [13-14].

The development of engineering process faces ethical issues related to principles like Justice, Beneficence, and Respect for Autonomy. There are many cases when bioengineers are somehow forced to accept to make compromise settings being vulnerable in front of the organizational management. The reflexive principium approaching for bioengineers is a useful tool for helping both specialists and students to face their ethical obligations, in the international context and cross-cultural bioengineering development. The conflict of interests that might arise in the engineering field is an ethical problem that needs to be discussed and solved because it affects also the engineers whose informed consent is not taken adequately and violating their autonomy.

The ethical education must be provided in a manner to determine the ethical reasoning skill. It is not enough just to offer information concerning what is ethic to do and what is not or knowledge about normative, descriptive and meta ethics. The process by which arises ethical reasoning skill needs more then information, needs applied ethics, problem case solving issues and the development of a specific ethical personal characteristic for the professionals in every field, but especially in the sciences. Ethical reasoning doesn't mean limiting personal thinking and application of codes of conduct or rules of moral behavior. Ethical reasoning is an It is a main component of cognitive and moral development process that must keep pace with the evolution of technology to prevent disasters affecting human life and to protect life, human dignity and human rights.

4 Conclusions

Physical therapy professionals, bioengineers, and doctors should communicate and create a team for offering quality medical services and respecting human rights and dignity.

A multidisciplinary team is needed for the physical therapy or electrotherapy research equipment development. There is a need also for a good communication channel between the physical therapist professionals and medical engineers for preventing the ethical dilemmas arising from the physical therapy equipment use.

There is a major need for ethical practice of problem based cases for biomedical engineers which perform their activities in the biomedical industry field. Codes of conduct and guidelines are very important instrument for ethical challenges raised in biomedical field, but they don't always approach each problem directly to equipment authors or engineers, one of these issues rarely delineated it concerns the authors' responsibility for controlling the translation process through the moment when the humans benefit from their creation. The ethical education must be provided in a manner to determine the ethical reasoning skill. This means the need for information, applied ethics, problem case solving issues and development of a specific ethical personal characteristic. Ethical reasoning doesn't mean limiting personal thinking and application of codes of conduct or rules of moral behavior. Ethical reasoning is the main component of cognitive and moral development process. It must keep pace with the evolution of technology, to prevent disasters affecting human life and to protect life, human dignity, and human rights.

As it regards the national government regulations it requires a multidisciplinary team, to realize a normative framework for the safe utilization of these devices. The framework must supply minimal technical and scientific devices requirements, include guidelines regarding the ethical implications and shared responsibility [16].

References

[1] J. Weiss, T. Pobre, L. Weiss, *American Handbook of Physical Medicine & Rehabilitation*, Ed. New York: Oxford University Press, (2010).
[2] L., Rogozea, G.Sechel, A. Fleancu , *Ethical aspects in bioengineering research.,* WSEAS International Conference. Proceedings. Mathematics and Computers in Science and Engineering, 139-143, (2009)
[3] W., Vela, L., Blanco, A., Gonzalez, J., Alococer, Major Trends In The Development Of Ankle Rehabilitation Devices, Dyna Rev.Fac.Nac.Minas, vol. **79**, Dec(2012).
[4] L., Gloeckner, C., Fortunato, A., Pinkus, The Role of Professional Knowledge in Case-Based, Sci Eng. Ethics ,. 767-787,(2015)
[5] Aetna., Aetna Insurance Company. [Online], http://www.aetna.com/cpb/medical/data/300_399/0325.html, (2017)
[6] BTL, Shock Wave Users' Manual, (2008)
[7] N.B.M., Császár, N.B., Angstman, S., Milz, C.M., Sprecher, P., Kobel, M., Farhat, J.P., Furia, C., Schmitz, Radial Shock Wave Devices Generate Cavitation, Plos One, (2015.)
[8] World Health Organization, Medical Device Regulations-Global overview and guiding principles, Geneva, (2003)
[9] European Council, Council Directive 93/42/EEC of 14 June 1993 concerning medical devices, http://eur-lex.europa.eu/legal-content/EN/TXT/?uri=CELEX%3A01993L0042-20071011
[10] B.A., Hagiu, The Physiotherapist Will Be Replaced By Robot? Sp Soc Int J Ph, vol. **16**, 53-57, 2016
[11] Physiotherapy Canada. ELECTROPHYSICAL AGENTS-Contraindications and Precautions: An Evidence-Based Approach to Clinical Decision Making in Physical Therapy, Physiother Can, (2010)
[12] T., Watson. Electrotherapy, Contraindications, http://www.electrotherapy.org, (2007)
[13] Chatanooga, Djoglobal., User Manual, (2008)
[14] BTL-BTL 4000 Users' Manual, (2007)
[15] P., Herxheimer, A., Lexchin, J., Mansfield, P., Norris., Drug promotion what we know, what we have yet to learn-Reviews of materials in the WHO/HAI database on drug promotion WHO. WHO/EDM/PAR/, (2004)
[16] L., Rogozea, F., Leasu, A., Repanovici, M., Baritz, *Ethics, robotics and medicine development,* Proceedings of the 9th WSEAS international conference on Signal processing, robotics and automation,264-268, (2010)

Application of some ideas from the axiomatic design principles for construction of a learning management system in Romanian higher engineering education

Dragoş Iliescu[1,*], and *Felicia Veronica* Banciu[2]

[1] University Politehnica of Bucharest, Faculty of Engineering and Management of Technological Systems, Doctoral School, 313 Splaiul Independentei, 060042, Bucharest, Romania
[2] Politehnica University of Timişoara, Mechanical Faculty, IMF Department, 300222 Mihai Viteazu Bd., Timişoara, Romania

Abstract. In education, the communication processes are critical. The result of education process depends on a significant manner by the quality of the incurred communication. To enhance learning in higher engineering education, an application of axiomatic design for the construction of a Learning Management System is proposed. The clients of such a system are identified, and their expectations were gathered as well. Functional requirements and design parameters to be designed are compiled regarding the two principles of axiomatic design. Finally, we investigate four design options to select the optimal design solution.

1 Introduction

Learning is a complex cognitive process. It can be under study in different extends: individual, group or organizational learners. Learning and teaching are often presented with a certain amount of indistinguishability. Nevertheless, the two are different by their meaning, and a particular relation occurs in between [1-5]. Learning can be under study from a theoretical or an experiential perspective, but the main characteristic - learning is a process - claims the process of learning to be under control. There is a particular granularity involved, aligned with the interested parties classifying the objectives of a higher education organization, and the learning management system concept is introduced as a method to control the related process. Work focus on identifying the needed information, constraints, modules and interfaces to develop an effective Learning Management System (LMS).

The social dimension of learning is well known and accepted [6-7]. The interested parties are the representatives of universities, committees and education beneficiaries [8]. Based on social network theory, the concept of bridge tie creates a beneficial impact [6, 9] by clear and structured information. All interested parties are equally represented.

* Corresponding author: dragos.iliescu@ieee.org

2 Interested parties

There are multiple interested parties in connection with the LMS benefits. Two major actors are implied: *Society* and *University*, both as *Agentive Social Object* class [10].

The *Society* represents a generalization of the *Student*, *Professor* and *Other* organizations. The *University* got the function of *education*, where *Professor* and *Student* act the specified roles of *Teacher, Assessor* and *Learner*. The major actors, *Society* and *University*, both generalize *Professor*, *Student* and *Other* organizations actors. The generalization goes into details with granularity if taking account of part-whole relations. In the case of the *University*, all entities involved are agentive, and the relation is classified by the functions of *education*, *teaching*, *learning* and *assessment*.

For the case of *Society*, the part-whole relation can be agentive to the same extent as for *University*. However, in the present work, it is under study only the metonymy. A member relation type [7] relates the *Professor*, *Student* and *Other* organizations to the *Society*. In the same manner, the *University*, as a metonymic part of *Society*, got the function of *education*, and communication would act between *University* and *Society*, and also, between *Professor*, *Student* and *Other* [8, 11].

2.1 Voice of customer

The interested parties play specific roles. The actors have specific requirements they have to fulfil. The *University* having the function of *education* will call the *Professor*, with the function of *teaching* (*Tch*), as the source of knowledge (term knowledge apply here for the tuple of {knowledge, behavior, attitude} [4]) to be transferred to the *Student*. In addition, the *Professor* got the function of *assessment* (*Asm*).

The *Student* deploys the function of *learning* (*Lrn*). The function of *education* is deployed as an aggregation (*Agg*) of *teaching*, *learning* and *assessment*. We have by consequence a knowledge transfer, from *teacher* to *student*, using certain defined channels. The *assessment*, by another hand, is a metrological part (proper part) of *education* (*Edu*) and represents the evaluation of transferred knowledge effectiveness. The *University* assign function of *assessment* to the *Professor* [8].

2.2 Interaction specificity in teaching-learning environment

Education claims for a communication process across involved actors. Each connection branch has own features supplying a particular identity over specified interface. The communication between teacher and students, composing the *knowledge transfer* function, and between student and teacher forming the *feedback* [12], is here for exemplification.

The communication channels are bidirectional, and there is a two-way relationship between teaching and learning [1]. The process is controlled by the LMS, with the objective of consistency preservation in teaching-learning [1, 11, 13, 14], and to enhance the education environment. The criteria used in this construction are those defined by Entwistle [1], Kolb and Kolb [13], Doignon & Falmagne [14] and Iosif et al. [2].

3 Learning process modelling

Learning process model considered for this study, illustrated in figure 1 is based on the Experiential Learning Theory [5], embedding Entwistle [1] and Iosif et al. [2] results as well. Key features to retain here are (i) learning is a holistic process and (ii) relearning is an integral part of learning.

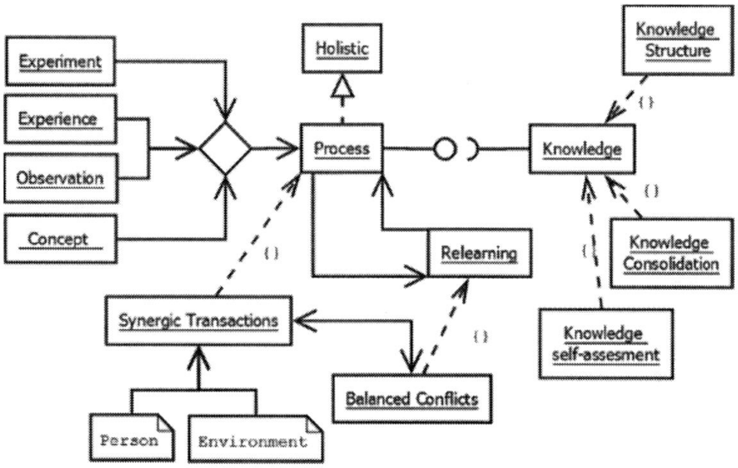

Fig. 1. Learning process model.

The learning model reveals the expectations expressed under the form of management of interactions for (i) teacher-to-teacher, (ii) student-to-student and (iii) teacher-to-student [15-17]. Finally, the *University* expectations come regarding the management of dedicated facilities as requested by quality assurance for education [8]. The four sources the information flows from (see figure 1) emphasize the formation of a new knowledge depend on more than one information source, but, on another hand, there is a specific relation across these sources. It is an equilibrium described by ELT [5, 13], under the control of *Professor*, and perceived by *Student*. Depending upon the equilibrium proposed by *Professor* and perceived by *Student*, the later will finally adopt a more conceptual development path, or he/she will become a practitioner focusing on action side. It is obvious that we refer here to knowledge in the tuple of {knowledge, behavior, attitude} (§2.1), but we observe too the equilibrium depicted in figure 1 acts as a regulator for behavior and attitude in the same tuple, as perceived by *Student*.

3.1 Interested parties

The axiomatic principles [18] guide the process of LMS design, and the defined goal is to propose an informational structure able to facilitate the fulfilment of interested parties expectations presented in Table 1.

Table 1. Professors, students and university expectations.

Interested party	Identification	Expectation
Professor	CR1	Management of interactions (teacher-to-teacher, student-to-student and teacher-to-student)
Student	CR2	Knowledge structure and content
	CR3	Knowledge consolidation
	CR4	Knowledge self-assessment
University	CR5	Management of dedicated facility

The definition of *education*, with the meaning: x is the knowledge for transfer, by teacher T to student A. Transferred knowledge effectiveness would be under asses by university Y as a quantifiable measure of supplied education, which is formalized herein by equation (1).

$$Edu(x, \Upsilon) = Agg[Tch(x, T), Lrn(x, \Lambda)] \wedge Asm(x, \Upsilon) \quad (1)$$

There are two possible perspectives we note in (1): (i) transfer efficiency and (ii) transferred knowledge effectiveness. The implementation of LMS pose both aspects [3, 19].

3.2 Customer and Functional Requirements mapped to Design Parameters

Four options were analyzed against the fulfilment rate of functional requirements (FR) by the Design Parameters (DP). These options describe an LMS with different degree of integration and interactivity in relation with named FR.

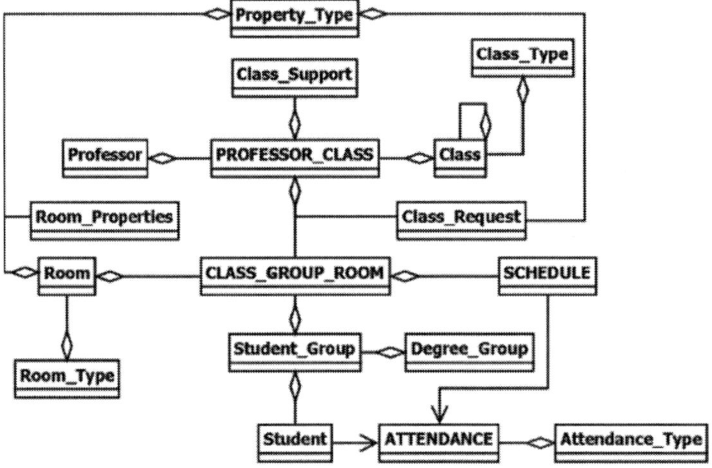

Fig. 2. Learning management system model for option one (baseline).

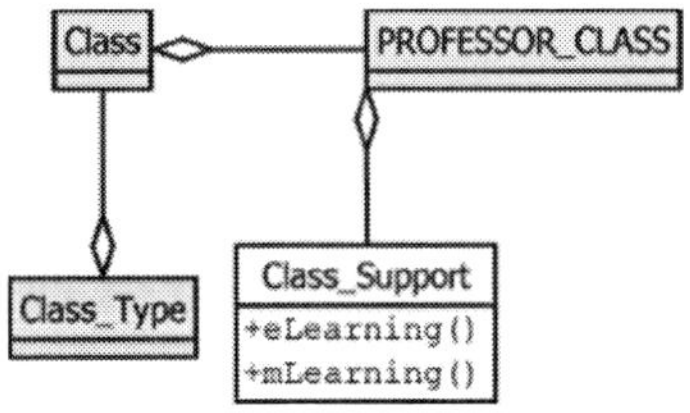

Fig. 3. Enhancing method option two - eLearning

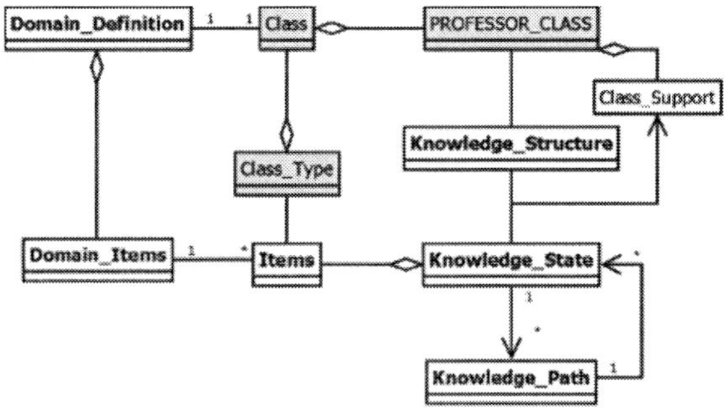

Fig. 4. Enhancing method option three – Knowledge Space

Design option one (see figure 2) focus on the definition of *Professor* and *Student*. Class, classroom, related facilities and their properties are defined in the informational view. By defining a joining entity of class-professor-group, the *University* should be able to manage in a satisfactory manner the available resources and to monitor the *Student* attendance (the efficiency and effectiveness of it) to specified class activities.

Option one supplies the necessary management level and is currently applied in higher education. The knowledge transfer from *Professor* to the *Student* is based on the *Professor* charge. Supplying of electronic data (class support information) to students is present, but is constraint by the in charge professor depending upon this information is made available.

Design option two (see figure 3) enlarge the option one above and introduces the technology of eLearning and mLearning. Knowledge transfer to student preserves the features shown by option one, but enter the mobile availability of information transfer by student request, improving the relearning as an integral part of the process.

Design option three (see figure 4) implements the concept of knowledge space (KS). The knowledge structure becomes part of the curricula. By applying the concept, the knowledge scope and utility apply also, helping the students to get a better understanding over the specified curricula. The knowledge path gets an extensional definition, adding a proper contribution to the degree the student master the transferred knowledge.

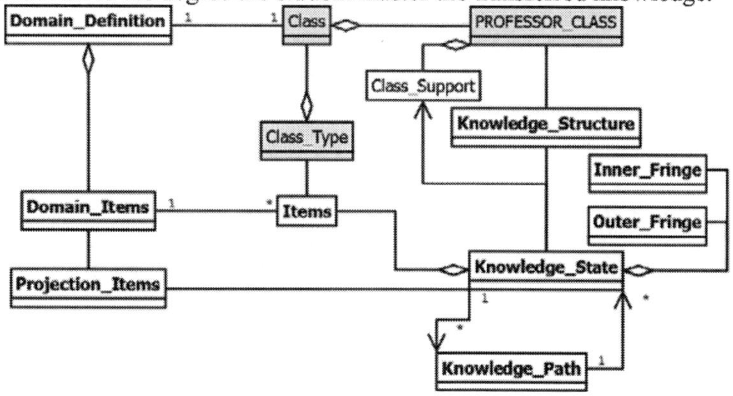

Fig. 5. Enhancing method option four – Learning Space

Design option four (see figure 5) implements the concept of learning space. It offers benefits of accuracy in the definition of how the knowledge is related one to another within the same knowledge domain. The design option four has the most interactive features among all considered designs, supplying not only the knowledge transfer but curricula understanding as well. It brings the benefits of knowledge space concept by an improved knowledge structure.

One particular challenge occurs in the implementation of option four. There should be a mechanism in place to capture the student's feedback, to evaluate it, and later, *Professor* to apply these observations into practice, thus creating a new and improved learning experience. Inner and outer fringe can develop such a mechanism but it is outside of present scope (see figure 5).

The calculus summary and formulae for information content are presented in detail on [20].

4 Conclusion

One particular aspect we should highlight against the optimal design solution. The particularity of the options proposed above is that each of them is based on the previous model. By considering each option being a knowledge state, overall, the four options come in respect with the KS.

By the particularity noted above, the university can decide progressively on the design option to implement. This strategy will allow the *University*, *Professor*, *Student* and *Other* entities to adapt themselves to the described communication platform requirements. As the integration and interactivity are intensively present, the *University* can progressively create the needed steps to gain proper skills for his members.

The design of an LMS has studied in this paper. The applied methodology for the stated design is complying with the axiomatic principles with the goal of bridge tie implementation. The customers of the designed system were identified, along with their specific functions and roles, to establish the expectations that each involved actor would pose against the developed LMS.

The identification of customer expectations, related functional requirements and design parameters, conducted to a design proposal with the identified configuration option four as the optimal solution. The optimal solution, as the informational content principle point out based on the expected degree of fulfilment, is proposed as an evolving configuration, by a modular information platform.

The work presented in this paper was partly performed by the corresponding author during the PhD program with the support of University Politehnica of Bucharest, Faculty of Engineering and Management of Technological Systems.

References

1. N. Entwistle, *Researching Higher Education: International Perspectives on Theory, Policy and Practice* (Routledge, Abingdon, 2016)
2. G. Iosif, A. Marhan, I. Juvina, S. Trausan–Matu, Rev. Inf. Ec. **3** 135 (2015)
3. B.S. Dias, A.J. Diniz, Educ. Technol. Soc. **17** 307 (2014)
4. P. Ernest, J. Educ. Teaching **15** 13 (1989)
5. A.Y. Kolb, D.A. Kolb, Acad. Manag. Learn. Edu. **4** 193 (2005)
6. S.P. Borgatti, D.S. Halgin, Organ. Sci. 2011; **22** 1168 (2011)
7. J. Krogstie, G. Sindre, H. Jorgensen, Eur. J. Inform. Syst. 2006; **15** 91 (2006)
8. ***, *METHODOLOGY of External Evaluation, Standards, Reference Standards and List of Performance Indicators of the Romanian Agency for Quality Assurance in Higher Education* (Romanian Agency for Quality Assurance in Higher Education, Bucharest, 2006)
9. D. Iliescu, I. Ciocan, M. Dijmărescu, M. Gheorghe, *Proceedings of 11th International Scientific eLearning and Software for Education Conference* (Carol I National Defence University Publishing House, București, 2015)
10. C. Masolo et al., *WonderWeb: Ontology Infrastructure for the Semantic Web* (Laboratory for Applied Ontology- ISTC-CNR,Trento, 2003)
11. D.L. Eisler, J. Comp. High. Educ. **13** 3 (2001)
12. S. Robson, D. Leat, K. Wall, R. Lofthouse, *Cross Cultural Teaching and Learning for Home and International Students: Internationalisation of Pedagogy and Curriculum in Higher Education* (Routledge, London, 2013)
13. A.Y. Kolb, D.A., *The SAGE Handbook of Management Learning, Education and Development* (SAGE, London, 2005)
14. J. Doignon, J. Falmagne, *New Handbook of Mathematical Psychology* (Cambridge Handbooks in Psychology, 2016)
15. C. Kyriacou, R. Kunc, Teach. Teach. Educ. **23** 1246 (2007)
16. J.R. de Jong et al. Teach. Teach. Educ. 2012; **28** 948 (2012)
17. A.I. Friedman, Soc. Psychol. Educ. **7** 435 (2004)
18. N.P. Suh *Research in Engineering Design* (Springer-Verlag, London, 1998)
19. V. Hubert, K. Rob, *Proceedings of 6th International Conference on Computer Supported Education CSEDU-2014* (SCITEPRESS, Barcelona, 2014)
20. D. Iliescu, F.V. Banciu, http://dx.doi.org/10.13140/RG.2.2.10526.13125

Complex approach in telecommunication engineering education: develop engineering skills by a team project

Luminița Scripcariu*, and *Petre-Daniel* Mătăsaru

Technical University "Gheorghe Asachi" of Iasi, Faculty of Electronics, Telecommunications and Information Technology, 11 Carol I Blvd., Iasi, Romania

Abstract. This paper provides an overview of the educational process of telecommunication engineering students by presenting the preparation of a team project focused on information security. Our educational approach combines basic knowledge such as mathematics with specialized engineering notions and various skills. The project theme is to design, implement and test an encryption algorithm. Students are provided with online courses, specific software programs and Internet access. They have to choose an encryption algorithm, to study its details and to write the script of the encryption algorithm in MATLAB program. The algorithm is implemented in C/C++ programming language and tested. Finally, a concurrent team tries to break the algorithm by finding the decryption key. It is an interactive approach which combines various education methods including gaming concepts. The covered topics provide students professional outcomes such as knowledge and use of specific mathematical tools and software environments (C/C ++ programming languages, MATLAB), abilities to design, develop, implement and test software algorithms. The project also provides transversal outcomes such as ability to team work, skills of computer use and information technology and capability to take responsibilities. Creativity is also encouraged by extending the algorithm to other encryption key lengths than the usual ones.

1 Introduction

The project is proposed to students, from "Telecommunication Systems and Technologies" bachelor program, during the "Introduction to Communication" class, in order to apply the theoretical notions presented in the course unit, focused on information security and encryption algorithms.

The objective of this activity is to offer thorough knowledge of encryption techniques to the students and to develop their programming skills.

Students can choose between different encryption algorithms, both with public or private encryption key.

* Corresponding author: luminita.scripcariu@gmail.com

Usually the team is composed by three students and one of them is the team leader which coordinates all the activities of the project.

The team securely accesses bibliographical references through Moodle application, provided by the faculty servers (http://edu.etti.tuiasi.ro).

The basic mathematical notions about Galois Fields must be known. Students have programming skills from previous courses. Matlab operating skills are also required. The students are solving the project tasks during six hours. Finally, a student workshop is organized and each team presents the project to the others. Competitiveness is stimulated. Like in a gaming environment, a challenge is launched to the other teams, to find the secret decryption key.

In this way, the educational process has the main goal to develop different cross-disciplinary professional skills such as knowledge and use of specific mathematical tools and software environments (C/C ++ programming languages, MATLAB), abilities to design, develop, implement and test software algorithms and to develop student's transversal skills like team working, creativity and sense of responsibility. Educational methods and techniques are adapted to students, with different levels of motivation and attitudes about learning and working [1].

2 Cross-disciplinary requirements

Encryption algorithms are intended to ensure information security [2]. Telecommunication engineers must have knowledge of **security techniques** used by communication systems and networks.

Different encryption algorithms can be used:

a. private-key algorithms: DES (Data Encryption Standard), TDES (Triple DES), AES (Advanced Encryption Standard) etc.

b. public-key algorithms: RSA (Rivest-Shamir-Adleman), El-Gamal etc.

The difficulty of implementing these algorithms consists in working with very large numbers and also in working with different data structures.

Understanding these algorithms is an essential requirement for the students as future specialists on telecommunication systems.

The **mathematical aspects** of the algorithm must be well known by students.

DES and TDES are binary block algorithms. Students must know Boolean algebra notions. The encryption key is a 64-bit sequence for DES and 128-bit sequence for TDES [3].

AES works on the Galois Field GF (256) with 8-bit input values, with 128, 192 or 256-bit keys [4]. In this case, the operations are made algebraically based on the Galois Field rules.

First the students read the standard and analyze the encryption algorithm, by studying all the steps.

It is important that the actual cryptography standards are freely available on the Internet [5].

Numerical data is used to exemplify the algorithm. Different input sequences are chosen and encrypted by hand in order to finally test the implemented algorithm. Numerical examples are essential for understanding how the algorithm works. In fact, it is confirmed the fact that some details are lost by reading only the theoretical description of a standard. For example, students do not always pay attention to the data type until a numerical value is required as an example.

After well-understanding the algorithm, students write the Matlab scripts of the encryption algorithm and the decryption one.

This activity requires **programming** skills and **knowledge of Matlab commands** [6]. This is a good opportunity for students to learn and use the Matlab tools which are required by various employers.

After writing the Matlab scripts, students test them with different input data.

This is a laborious process because different errors occur and students must solve all the problems before using the script. They must have patience and work systematically. In this time, the student working style is shaped.

The finality of the project consists in a software program, written using C/C++ language, which encrypts and decrypts a file. It is a difficult operation but working as a team is crucial and efficient. Some students could not complete this task in time without the help of the other colleagues. It is a good opportunity for each student to learn from the experience of the others.

Presenting the project is made by all the members of the team. The presentation itself is a challenge for students. Using a specific presentation program implies different competences on software operation and highlights their experience in promoting a project in front of a large audience. Communication skills are also developed by this activity.

The project requires many cross-disciplinary capabilities from mathematical, programming and technical areas, and educational methods are adapted to the students' needs [7].

3 Project Description

3.1 Project theme

The theme of the project is "Design, implement and test an encryption algorithm as Matlab scripts and as a C/C++ program". The project of a team working on AES algorithm is described as a relevant example in the following paragraphs.

3.2 Project tasks

The project tasks consist in:

- Reviewing the Galois Fields' properties.
- Studying AES standard and making a synthesis of it.
- Applying AES on numerical data as an exercise.
- Writing the Matlab scripts of AES encryption and decryption algorithms.
- Testing the Matlab scripts with numerical data.
- Implementing AES algorithms as a C/C++ programs with a friendly graphical user interface (GUI).

Each task requires special abilities and students are trained to complete these tasks as a team. Teacher is supervising the team and provides technical information at the request of students.

3.2.1 Reviewing Galois Fields properties

Galois Fields (GFs) are algebraic finite fields with internal operations of addition and multiplication [8].

Special rules must be followed in order to operate in these fields.

Students learn about Galois Fields as an introductory lecture and then they use it to implement an encryption algorithm.

Let us denote by m the number of bits used to express a symbol from GF and by $p(x)$ the m-degree primitive polynomial used to multiply numbers in GF (2^m). For example, AES algorithm uses GF (256) = {0, 1, 2 … 255} and 8-bit values.

The elements of a GF can be expressed as decimal values, as binary sequences and as polynomials:

$$a = [a_0 \; a_1 \; a_2 \; \ldots . a_{m-1}] \tag{1}$$

$$a(x) = \sum_{i=0}^{m-1} a_i x^i \tag{2}$$

Conversion of decimal values into binary form and knowledge of Boolean algebra are prerequisites of this course unit and of this project.

The addition of two numbers from GF is made modulo-2 bit-by-bit:

$$s = a + b \Leftrightarrow s_i = a_i \oplus b_i, \;\; for\, i = 0 \;\; to \;\; m-1 \tag{3}$$

The multiplication of two elements is made using the polynomials associated to each number and the chosen primitive polynomial $p(x)$:

$$c(x) = \{a(x) \cdot b(x)\} \bmod p(x) \tag{4}$$

Numerical exercises are done by the team to become familiar to GF's algebra. Students have to pay attention to operate according to the rules specified by the GF and avoid of adding or multiplying numbers as usual decimal values. This is a good exercise of attention and concentration. It develops abstract understanding and exercise solving skills.

3.2.2 AES Synthesis

FIPS PUB 197 standard is studied by the team and the steps of the algorithm are described synthetically:

- Each data block is written as a sequence of 16, 24, 32 or 64 bytes that become the elements of the data-matrix having four rows. For example, the data matrix used by AES-128 is presented below:

$$S = \begin{bmatrix} s_{00} & s_{01} & s_{02} & s_{03} \\ s_{10} & s_{11} & s_{12} & s_{13} \\ s_{20} & s_{21} & s_{22} & s_{23} \\ s_{30} & s_{31} & s_{32} & s_{33} \end{bmatrix} \tag{5}$$

- Each byte is substituted according to the following affine transformation containing the bits from the binary form of the byte [$b_i \; b_{i+1} \; b_{i+2} \; b_{i+3} \; b_{i+4} \; b_{i+5} \; b_{i+6} \; b_{i+7}$] and a constant byte ***c*** = 0110.0011 = 0x.63. This step is called *SubByte* and it is described by the following relation:

$$b'_i = b_i \oplus b_{(i+4)\bmod 8} \oplus b_{(i+5)\bmod 8} \oplus b_{(i+6)\bmod 8} \oplus b_{(i+7)\bmod 8} \oplus c_i \tag{6}$$

- Next step, called *ShiftRows*, consists of a permutation of the elements. Each row of the state-matrix is shifted with a different shift value: 0 for the first row, 1 for the second one, 2 for the third and 3 for the last:

$$S' = \begin{bmatrix} s_{00} & s_{01} & s_{02} & s_{03} \\ s_{13} & s_{10} & s_{11} & s_{12} \\ s_{22} & s_{23} & s_{20} & s_{21} \\ s_{31} & s_{32} & s_{33} & s_{30} \end{bmatrix} \quad (7)$$

- The so-called step of the algorithm *MixColumns* is a simple multiplication of each column of the data matrix with a given matrix A, according to the algebraic rules of GF.
- The final step, *AddRoundKey*, consists in adding the key matrix to the data matrix.

The algorithm has a different number of rounds (10, 12 or 14), depending on the length of the key. A key schedule algorithm is applied to generate the sub-keys for these rounds.

Numerical examples of AES algorithm are calculated by students in order to fully understand the details of the algorithm which requires in-depth mathematical knowledge.

Making calculation with numerical data is an essential way for understand the algorithm, to be aware of the type of data used for each step of the algorithm (binary, decimal, octal etc.), to be aware of the size of the data block processed by the algorithm (8-bit word, 4-by-4 matrix, 4-by-1 column vector) or of how the algorithm operates (based on bitwise logical AND, exclusive-OR, multiplication of two arrays using the rules of addition and multiplication of the Galois Field with 8-bit elements etc.).

3.2.3 Implementing AES as Matlab scripts and C/C++ programs

To accomplish the task of writing the encryption and decryption scripts in Matlab program, students must have previous knowledge of this software program widely used in engineering. Matlab scripts are written for the encryption and the decryption algorithms.

Students must use their knowledge about the syntax used to write Matlab scripts or to access the Matlab manual [6]. Graphical User Interfaces (GUIs) are designed by the students based on their knowledge of how to make a GUI in Matlab program or accessing the documentation.

Then both algorithms are implemented using the C/C++ language as two independent applications. This is a difficult task which requires advanced programming abilities, a lot of time, a systematic thinking and an efficient human resource management. Strong leadership abilities are also required for at least one member of each team.

Tests are performed by students to validate their applications.

This task develops the programming skills of students and enforces the spirit of teamwork.

3.3 Project results

The selected team implements AES in Matlab and as C/C++ programs, with some easy-to-use GUIs.

The team writes the Matlab scripts for 128, 192, 256 and 512-bit key sequences. The last specified key length is suggested by the teacher and it is presented as an extension of the classical AES algorithm, sustaining the creativity and adaptability of students to execute special tasks.

The algorithms can be tested by the team itself and other colleagues, too, in order to identify possible weaknesses or non-functional aspects of the applications.

Testing the algorithms by others is an interesting and useful exercise for all students. All of them get used with the algorithms, applying it on a particular data set. They become aware of the details of the application and therefore, some changes are made after testing it, in order to optimize different aspects. Testing becomes an intermediary step of the project itself.

The difficulty of the project has been overcome by the team spirit and by the perseverance of the students. By doing this project, students get used to tackle complex tasks and show that they are capable of making a complex project as a team.

3.4 Project Presentation

The presentation of projects is done in a large room, in front of the audience consisting in about 60 students and two teachers. A video-projector and a computer are available to expose the prepared presentation.

The team has 15 minutes to present the project and five minutes to answer the questions from the audience. Students must have a suitable outfit and control their emotions. They must speak clearly and fluently, without reading from different other helpful materials.

The audience can address questions to the team and the answers are evaluated among other aspects of the project.

The implementation of the algorithm is the contribution of the team and it was well appreciated by the audience during presentation.

The presentation of the project is a valuable experience for the students, useful for future job interviews in which they will participate.

4 Conclusions

This paper illustrates the efficient concept of engineering education by developing cross-disciplinary team projects. We illustrate the way of making a project on an information security theme, by a team of students enrolled in the third study year of the bachelor study program "Telecommunications Systems and Technologies". Students improved their professional and transversal skills working on a complex project and the results were above expectations. Educational methods are adapted to the students with different attitudes and learning needs. We consider that the educational objectives are accomplished and all the intended skills of the students are formed by doing a complex specialized project.

References

1. R.M. Felder, R. Brent, *Understanding Students Differences,* J. of Engineering Education (JEE) **94,** 1 (2005)
2. B. Schneier, *Applied cryptography, second edition* (NY: John Wiley & Sons, Inc., 1996)
3. *** FIPS PUB 46-3 (Federal Information Processing Standards,1999)
4. *** FIPS PUB 197 (Federal Information Processing Standards, 2001)
5. E. Barker, *NIST Special Publication 800-175B, Guideline for Using Cryptographic Standards in the Federal Government: Cryptographic Mechanism* (NIST, 2016)
6. B.D. Hahn, D.T. Valentine, *Essential MATLAB for Engineers and Scientists, 4e,* (Academic Press, 2010)
7. W.Z. Khan, S. Al Zubaidy, *Engineering design approach in Marine Engineering: A bridge between Training Need Analysis (TNA) and Engineering Education*, Intern. J. Of Engineering And Science (IJES) **5**, 3 (2016)
8. L. Scripcariu, P.D. Mătăsaru, *On the Substitution Method of the AES Algorithm, Proceed. of the Intern. Symposium on Signals, Circuits and Systems, ISSCS 2013*, Iaşi, Romania, **1** (2013).

MATEC Web of Conferences 112, 08004 (2017) DOI: 10.1051/matecconf/201711208004
IManE&E 2017

The interaction between personal and situational factors in developing future engineers' creativity

Aurel Mihail Țitu[1,*], *Claudiu* Pirnau[2], and *Mironela* Pirnau[3]

[1]Lucian Blaga University of Sibiu, 10 Victoriei Blvd, Sibiu, Romania
[2]Politehnica University of Bucharest, 313 Splaiul Independentei, sector 6, Bucharest, Romania
[3]Titu Maiorescu University, 187 Vacaresti Ave, sector 4, Bucharest, Romania

Abstract. The purpose of this paper is to analyze the interaction between the technological and the social field. Growth and development require a continuous source of ideas. Their manufacture is efficient only by stimulating the creativity development in the knowledge economy. This essential skill will lead to making smart products and technologies improvements. This will also be the result of the interactions between the new IT-cloud technologies and other various factors, such as: knowledge, skills, approaches, attention, motivation and intellectual property.

1 Introduction

In the context of higher education, the benefits of learning and personal development arise from defining the objectives of each student. The correlation of these objectives with the inner values and personal priorities leads to the creative and technical formation of future specialists. The target-setting trends arise from the interaction of the factors which contribute to the creativity and personality development. The factors influencing the future engineers' development fall into two categories: personal and situational. The main personal factors are represented by knowledge, skills and attitudes. Among the situational factors we mention attention, state of mind, motivation and reward [1].

A coherent and creative educational process, associated with innovative practical experience, results in maximizing the effect of the interactions between the two factors. Processing and implementing the knowledge gained by the engineering students are based on the interaction between personal and situational factors. According to the specialized literature [2], personal creativity, denoted by the letter "c" is specific to the individual, involving ideas or products new to a particular person. Creativity 'C' implies the existence of new elements that receive social validation, being sufficiently valuable to be added to the culture. The interaction between the above mentioned factors is analyzed from two perspectives: technical higher education organizations and social media. In the four chapters hereof, this paper addresses interactions specific to the future engineers' professional development.

* Corresponding author: mihail.titu@ulbsibiu.ro

2 The interaction between cognitive science and engineering

Cognitive sciences represent an interdisciplinary research field that integrates disciplines such as psychology, artificial intelligence, linguistics, philosophy, education, anthropology and neuroscience. The understanding level and the employment of the psychological knowledge specific for the people executing technical activities are very high. This fact is firstly generated by the volume and the complexity of the specific interactions, according to Table 1.

Table 1. Specific interactions related to the engineer profession.

No.	Type of interaction
1.	Engineer – Engineer
2.	Engineer – Managers
3.	Engineer – Specialists in Economics and Finance
4.	Engineer – Subordinate staff
5.	Engineer – Systems (educational systems, cognitive systems etc.)
6.	Engineer – Collaborators
7.	Engineer – IoT (The Internet of things, also entitled The Industrial Internet)

The number of interactions is proportional to the size of the organization. The volume and the quality of the technical and psychological knowledge of a certain organization increase along with the number of employees and the used systems. Some of the latter are: calculation systems, operational systems, communication systems, data base systems, transport systems, production systems, cooling systems, construction site organization systems, ventilation systems etc. Regarding collaborators, they may be of a variety of professions such as doctors, when the engineers aim the medical institutions as beneficiaries. Within a type of economy based on knowledge, the knowledge engineer uses the knowledge data for two purposes: for the development of the artificial intelligence and for the regional intelligent development.

The new regional clusters contain within their hub four types of managers: the knowledge manager, the transverse manager, the facility manager and the senior sustainable development manager. The organization performance that learns is proportional to the performance of the intellectual capital. The engineers would be the first beneficiaries of the cognitive systems: the interactions between the artificial intelligence and Big Data (also known as Smart Inputs/Entries) would generate processes (called Smart Outputs) some of the self-learning (Machine Learning), all within a new science entitled Data Science. (For example: various universities offer Data Science master degree programs). The main domains that can be found at the intersection between the artificial intelligence and cognitive sciences are: psychology, linguistics and philosophy (Fig. 1). The interactions of many human experts and knowledge engineers lead to the effectiveness of the process of knowledge transfer and to the increase of the creativity level. Due to the fact that production systems are coordinated by specialists in technology, they will be analyzed in terms of features specific to engineers [3-4]. ***Example***: A computational model for metaphors understanding is initially based on the representation of the meaning of the word **"engineer"**.

The representation of the word can be **localized** (see Figure 2), corresponding to a single structure (concentrated in a single memory location) or **distributed** (see Figure 3), corresponding to a tree whose root is the meaning of the word and whose leaves are its individual features.

For these representations, the following main features of an engineer are used: analytical (centered on native curiosity), meticulous, communicative, creative (search for new, innovative ideas) logical, perfectionist, good mathematician, effective in finding crisis solutions, good team worker (integrates and works well in a team) and possessor of technical knowledge [5-6].

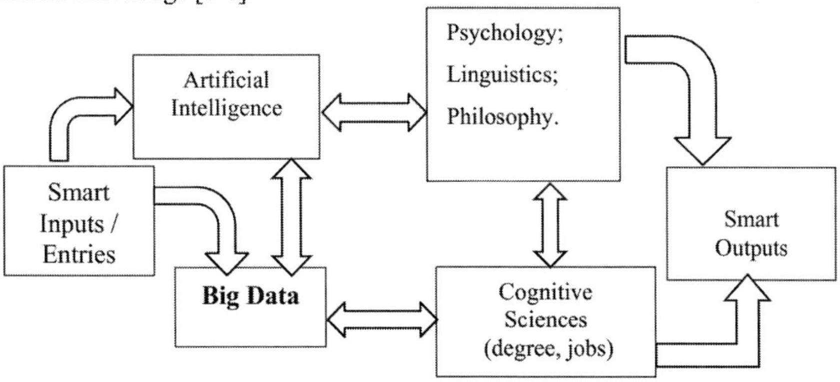

Fig. 1. Specific interactions of the cognitive systems.

The representation of a sentence such as "The engineer sketches products in the enterprise" can be achieved based on an edge-labeled tree as shown in Figure 4: the sentence is the root of the tree, the leaves are the words and the edges indicate the thematic role of each word. Each leaf represents a meaning of a word which may be represented in any of the forms shown in Figures 2 and 3. Thus, by applying cognitive sciences, models are performed and developed in order to represent the general knowledge of the subjects involved in the engineering field.

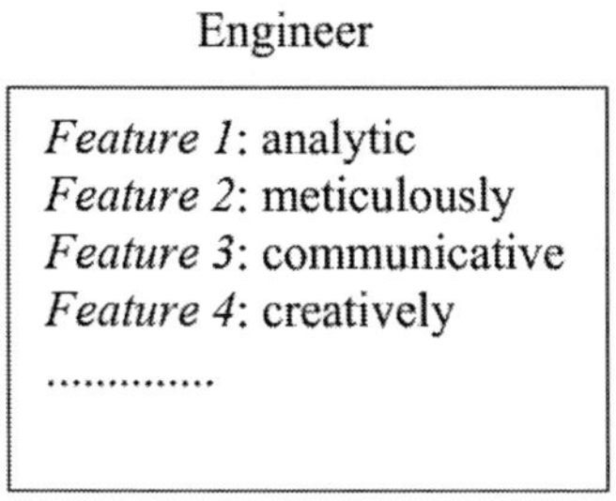

Fig. 2. Localized representation.

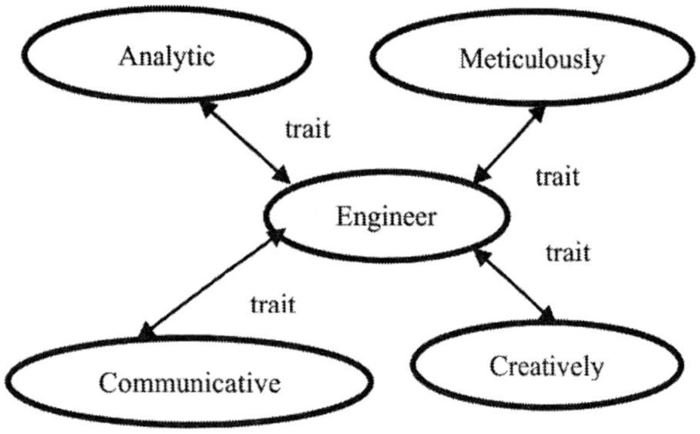

Fig. 3. Distributed representation.

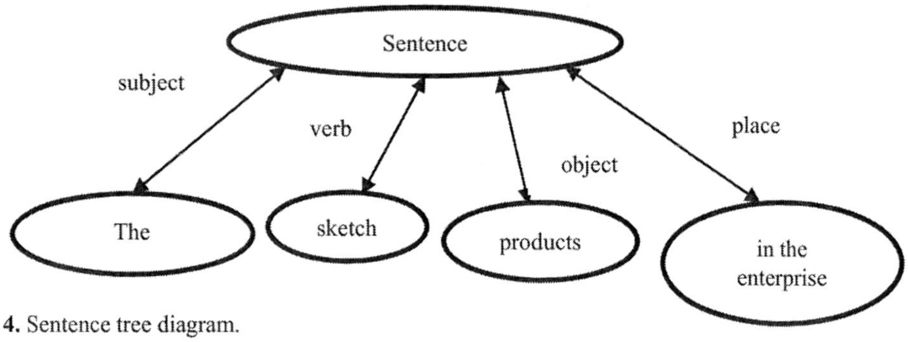

Fig. 4. Sentence tree diagram.

Since the majority of the models use interactions through the Internet services, nowadays it is even more important to emphasize the intellectual property role regarding the protection of the acquisition process and the knowledge transfer.

3 The role of intellectual property protection in engineering and related fields

Ensuring that the policy on copyright, related rights and the protection of industrial property provides strong incentives for creativity, while promoting innovation in the knowledge-based economy is an important task of the actors involved in the technical higher educational process [7].

The copyright legislation is constantly adapting to technological changes (digital technology and networks), trying to find a balance between protecting the creative works and maintaining the advantages of the free flow of information.

The interaction between copyright and target audience (knowledge which contributes to the increase of the intellectual capital creativity) occurs through the instruments existing in the digital environment. The effectiveness and legality of these instruments may lead to the development of new business models, implicitly creating new jobs, so necessary in the context of forecasting the increase of the number of unemployed people (at least in Europe) due to the evolution of ITC [8].

Most interactions between intellectual property legislation and other factors that contribute to the sustainable development of the knowledge-based economy, take place on the Internet (cloud, social commerce, blogs etc.).

The objectives of the copyright policies at national level and policies on the use of Internet services worldwide can and should work in tandem. The sustainable Internet governance must rely on transparency, cooperation and strengthening the multilateral governance, which allows the development of creativity and innovation while simultaneously building trust and protecting certain important rights and interests [9].

Intellectual property laws (everything related to industrial property, copyright and related rights) interact mainly with: competition laws; norms regulating the consumption of medicines and drugs (implicitly their purchase by means of Internet services); legislation on foreign direct investment; standardization procedures (highly important in industrial and education activities in the engineering field); monopoly legislation (the interaction between intellectual property and competition); technology transfer; commercial law; consumer protection; human rights; innovation and diffusion of green / clean patents and technologies (design rights); human capital; biodiversity and knowledge (technical, traditional, economic etc.); Smart Sustainable Development (learning companies performance); international conventions in force etc. The status of engineering in the context of Intellectual Property Law is shown in Figure 5.

The intellectual property management is an important task for the teams of engineers. The usage of the social networks within various tasks represents an efficient method generated at the intersection between the digital economy and the one based on knowledge. [10-11].

4 The interaction between social media and engineering

The mass-media social engineering has different aims, depending on the enterprise and its domain, on the social network and its characteristics and on the interactions between the two entities, according to Table 2 [12].

Promoting the engineering field through social media must be included in the general marketing plan of the enterprise. In order to increase the impact two main elements are needed: creativity and authenticity. Whenever some beneficiaries are addressed, it is mandatory for the used social media channels to be analyzed.

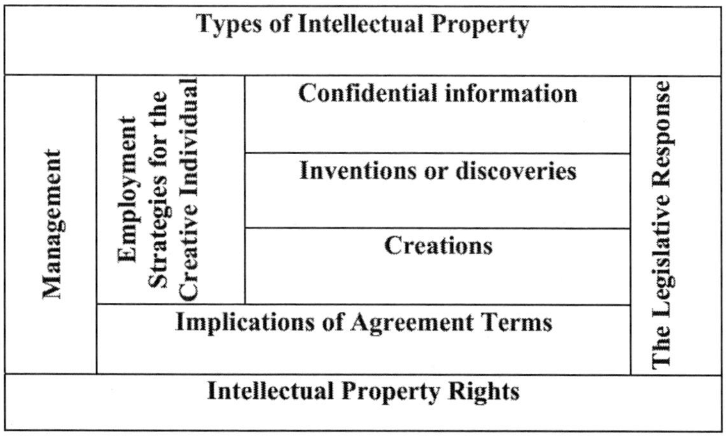

Fig. 5. Engineering in the context of Intellectual Property Law.

Table 2. Social Media Engineering Strategies.

No.	The name of the social network	Strategies / Type of article
1.	LinkedIn	How to create a profile that changes "heads"; The perfect resume; The presentation of specific tendencies of certain industrial (the division of a relevant content).
2.	Instagram	Promoting the gender equality in specific engineering domains by presenting female employees and their roles.
3.	Facebook	Presenting the gender differences depending on the engineering domain; The importance of the courses regarding the ongoing professional training.
4.	Twitter	The promoting campaigns of messages including logos and photos (their main consequence is to increase the employees' loyalty); Promoting certain events through strategies such as: "seize the day". Example: photos from the hackathon, paintball or softball championship; Promoting certain specific holidays. Example: International Labor Day.

Thus, an accurate choice of the social media channel can be insured. The content of the messages must be addressed to the people themselves, and not to enterprises. Those who write and read messages are people who must be persuaded to repeat their involvement. In order to build a brand time is needed. This implies consistency at all levels: production, quality, marketing, management, social media etc. According to some studies performed by Cadenas Part Solutions LLC, 60% of the engineers use social media to find new products, and 53% of them use social media for product recommendations. A number of 8.522.177 of engineers are present on LinkedIn, and from those, a number of 3.054.315 are from SUA. The companies that accomplish design activities for CAD-CAM prefer social media such as Facebook, Twitter, LinkedIn, Google+ and Pinterest [13]. The instantaneous character of social media involves the fact that a potential client may contact the enterprise by asking questions regarding various posts, a procedure that cannot be performed in case of an e-mail address or a phone number. Example: the Twitter lists allow the separation and the supervision of messages of distinct categories (clients, perspectives, competitors and professionals in engineering), thus these to be used whenever necessary.

The evaluation of the impact of social media engineering can be accomplished through certain coefficients and indicators. Their complex characteristic may be analyzed in a study of "Knowledge, creativity, innovation and promoting - KCIP" type. The KCIP acronym will become a symbol of specific interactions of the knowledge based economy: Social Media Engineering, IoT, Data Science, Intellectual Property, Artificial Intelligence etc.

5 Conclusions

The feedbacks generated from responses to specific questions will be constituted in new sets of knowledge and ideas. The interactions between cognitive science and engineering, as in the example shown, allow the creation and development of computational models. These models are based on artificial neural networks (self-learning intelligent systems) or production systems (designed and developed by means of the engineering specialists' creativity).

According to studies carried out herein, the intellectual property laws can be defined as a global interaction between organizational capital and the three dimensions of the intellectual capital: human, structural and relational. This interaction should be coordinated and driven in two directions: online and offline. The interactions map prioritizes online interactions based on the quantity and quality of existing and shared knowledge. The interactions in social media have highlighted the interdependence between knowledge discovery, knowledge engineering and knowledge management. The main reasons for the engineering companies to have a mass media activity are as follows: the need of marketing capacity growth, the diversification of the methods of communication to customers, the development of methods regarding targeting new clients and the possibility of creating new online professional community networks (Example: LinkedIn Mechanical Engineering Forum – 178.461 members).

Through new restructured concepts and knowledge, social media contributes to the development of loyalty of all types of employees, thus to the engineers. The use and growth of social media based on the interactions among knowledge, creativity, innovation and promotion also implies the promotion of creativity by validating creative thinking of students from the engineering domain.

References

1. R.M. Felder, North Carolina State University, CEE, **22(3),** 120–125 (1988)
2. *** Encyclopedia of creativity, **1. A-H,** AP, U.S.A. (1999)
3. F. Bouz, H. Dave, IBM Redbooks, NY, (2016)
4. L. Candy, Z. Bilda, ACM C&C UC Berkeley CA, (2009)
5. T. Barrett, R. Donnelly, University College Dublin, DIT, In B. Higgs & M. McCarthy (Eds.), **EI II,** (2008)
6. F.J. King, L. Goodson, F. Rohani, CALA, Florida State University, (1998)
7. M. Ţîţu, C. Oprean, LAP Lambert Publishing House, ISBN-13 978-3-659-79332-5, ISBN-10 3659793329, Germany, (2015)
8. L. Bently, B. Sherman, 4th Edition, Oxford University Press, (2014)
9. ***The Department of Commerce. IPTF. USA, (2013)
10. K. Adams, National Center on Education and the Economy for the New Commission on the Skills of the American Workforce, (2006)
11. R. Miller, P. Heuzenroeder, S. Tso, A. Ruhrmann, Engineers Australia, Spruson & Ferguson, (2013)
12. M.A. Storey, C. Treude, A. van Deursen, L.T. Cheng, Delft University of Technology, SERG, TRS, The Netherlands, (2010)
13. R. Zesch, Digital Parts Catalogs 101, **SICM**, 1-13, Cincinnati, USA, (2014)

The use of creative techniques in the engineering human resources development

Claudiu Pirnau*, *Daniel* Ghiculescu, and *Ion Niculae* Marinescu

Politehnica University of Bucharest, 313 Splaiul Independentei, sector 6, Bucharest, Romania

Abstract. Starting from the role of new technologies in the development of engineering innovation, this paperwork has two main research directions with respect to creativity and innovation in engineering education. These are based on a combination of creative techniques with traditional engineering education methods. The first component of our research presented herein aims the use of creative methods such as Creative Training and Image Streaming, and the second component of the research statistically analyses the Social Media posts regarding the use of Cloud Computing technology as a support for innovation. The aim of combining such lines of research will lead to the emergence of specific knowledge needed to identify and implement new creative techniques in engineering human resources. Applying such creativity techniques within groups of students can result in innovative ideas usable in ongoing research projects, in certain deadlock situations on finding critical solutions to overcome blockages.

1 Introduction

The future engineers' education is based on two fundamental factors: creativity and innovation [1]. The parameters needed in developing the creativity and the innovation can be analyzed from two perspectives: the individual who learns (intellectual capital) and the creative community (Creative & Connected Community, Creative Bio - Community etc.) [2-3]. The introduction of ECED diagrams (Engineering Creativity Enterprise Diagram) reflects the degree of industrialization and the technological level of any community, correlated with the one of creativity and innovation [4]. This approach can result in courses and specific activities, which would increase the creative and innovative skills and capacities of the future engineers.

This paperwork arose from the need to improve the students' ability with respect to the long term creative thinking, thus contributing to the improvement of the future industrial products design. The creative thinking skills enable students to generate hypotheses, to generate and extend ideas, to exercise their imagination and to seek innovative alternatives for solving problems [5]. Combining ideas can lead to "another way" of doing things, to the continuation of "the same way" of doing things or to ideas about "how not to continue" working that way [6-7].

* Corresponding author: claude.pyr@gmail.com

2 Encouraging the engineering students' creativity

Case study - a group of 20 engineering students (group 1) using computer-aided design by means of Inventor program, attended six lectures on creativity, according to the following schedule: three lectures throughout 15 days, 7 days of pause, followed by the last three lectures during another 15 days. At the end of each lecture, the subjects solved a creativity test. At the end of each of the two periods, competitions were organized for the design of a piece having a high degree of complexity. The contestants were divided into two groups: one group consisting of 20 students randomly enrolled (group 2) and the group trained in creativity (group 1). The projects carried out by the creative students were much more elegant and more correctly performed, so the creative training can be associated not only to some behavioral changes but also to practical activities (see Figure 1).

The same method was applied by an automotive company to the students of an engineering faculty. In this case, students were required to come up with some variants of Electrical Discharge Machining (EDM) of some stamping molds needed for the performance of some bodywork elements. From the technical, economic and creative perspective, the results were significantly superior for the group that was trained in creativity (in this case the Group 3). The randomly created group was named Group 4. The comparative results are shown in Table 1.

Table 1. Comparative results

Criterion / Group	Execution Time (T)	Design (D)	Correctness (C)	Materials Economy (E)	Novelty Degree (N)
Group 1	4	5	5	4	5
Group 2	2	3	3	4	4
Group 3	5	4	5	5	4
Group 4	3	3	3	4	4

For each criterion taken into account, the grades from 1 to 5 were given as follows: 1 (poor), 2 (satisfactory), 3 (good), 4 (very good), 5 (excellent).

A code according to its name was assigned to each criterion: time-T, design-D, correctness-C, economy-E and novelty-N. The evaluation of the creativity was performed by calculating the determinants of square matrices corresponding to the groups 1 and 2, 3 and 4 or 1 and 3, leading to the diagram from Figure 2.

The evaluation of the creativity can be integrated into the chaos theory. Small fluctuations of the creativity will lead to different results in the engineering complex system. Their analysis will result in various conclusions on the use of creative education in higher technical education. The chaos equation would be simple if the students' creativity increased indefinitely, but the effect generated by individual resources, the quantity and quality of the accumulated education, make this equation incorrect [8].

The simplest equation, which takes into account the above factors, might be as follows:

$$ncg_1 = r \cdot ncg_0 \tag{1}$$

Where: ncg_1 is the number of creative graduates in the next year; ncg_0 - the number of creative graduates in the current year; r - the growth rate, considered between 1 and 5.

DOI: 10.1051/matecconf/201711208005

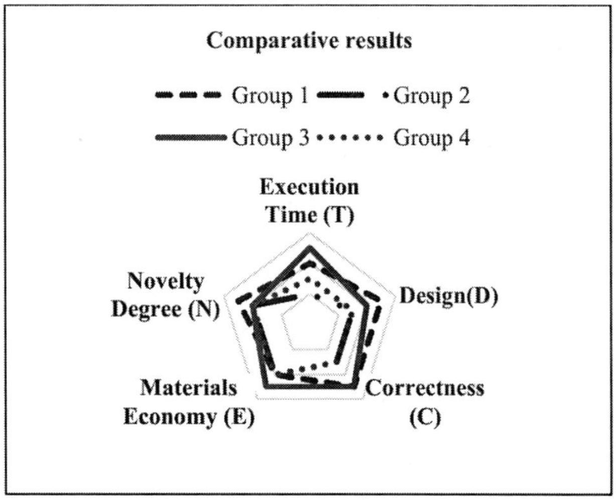

Fig. 1. Practical results of engineering creative education.

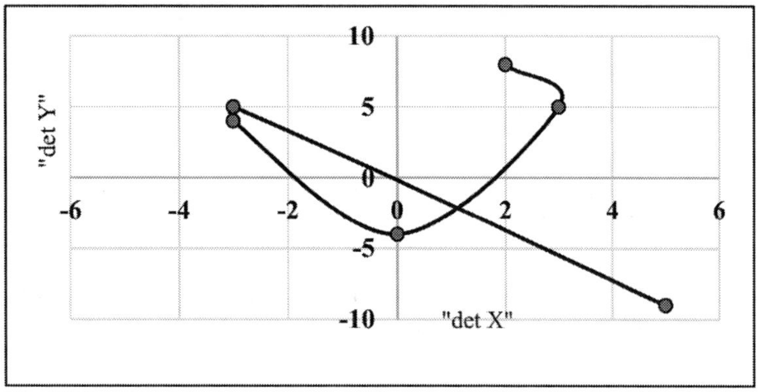

Fig. 2. Creativity Evaluation Diagram.

3 Boosting the engineering students' intelligence and creativity with the aid of Image Streaming technique

The use of Image Streaming technique based on creative thinking can have positive results, especially in the internships that students do in companies and organizations that perform engineering activities [9-10]. The stages required in this technique are as follows:

Stage 1 – expecting the completion of an internship of at least 7 days in order to increase the impact of the new activities and technologies on students. In this stage the accumulated personal factors such as knowledge, skills and attitudes lead to the development of the creativity.

Stage 2 - in a quiet office for 30 minutes maximum (this meeting can last between 10 and 30 minutes), each student describes aloud the reminded images (concentrating on the inner screen of the mind) related to a certain technological process (realizing that different frequencies of the thoughts received through the stream of consciousness is an "input" for the brain, as consciousness cannot be perceived as an object). The descriptions are recorded using any type of device that can record human voice (useful in stage 3). Subsequently, the thoughts are converted into electrical energy, which is amplified and sent to the central nervous system, so as the understanding phenomenon to take place. During the concentration process, along with images, other sensorial items may be involved such as: smell (examples: the smell of oil, coolant, dielectric fluid used in EDM etc.) or the noise specific to a particular processing activity. Focusing takes place in a complex of situational factors (attention, mood, motivation, reward etc.), with a significant impact on the creativity development process [11]. Thus, the evolution of an image description in conjunction with various other senses acquires a fluid nature (the description flows logically and coherently).

Stage 3 - during the seminar courses, each group of students listens to audio recordings and has discussions on specific topics of interest, developing methods of image decoding. The method is considered as a kind of "brainstorm" based on descriptions, not on ideas and answers, as in usual brainstorming, but with rules similar to this.

Stage 4 - the most important outcome of the discussions is the feedback, an element that stimulates intelligence and creativity. This technique is used both as a method to increase the creativity, as well as in order to acquire certain accelerated learning techniques. Since cloud technology replaces IT local systems, the data obtained are stored in the cloud, to be used in satellite or terrestrial analogue networks, particularly in best-effort IP digital networks (for example: Internet).

4 Social Media, the innovation support of creative engineering education

As result of our concern to analyze the way in which new IT technologies are perceived in the online environment, we collected tweets by streaming methods from July 10 to July 22, 2016 [12]. The used words flow was: "cloud computing, virtualization." The total number of collected tweets was 184,000, of which only 79,752 have unique content, the difference being retweets. The collected data were grouped according to the criteria: General information on cloud computing (Table 2); Cloud Computing Providers (Table 3). For these collected data we determined statistical indicators on the calculation of means, synthetic position indicators and scattering indicators.

Table 2. General Cloud.

Words group searched in tweets	Number of tweets
Cloud Engineer	223
Cloud Engineer jobs	129
Cloud risk	656
Cloud encrypted	36
Cloud Developing	21
Cloud- Machine Learning	453
Cloud- virtualization	430
Mean indicators	
Arithmetic mean	278.26
Harmonic mean	74.71
Geometric mean	160.79
Position indicators	
Median	223
Scattering indicators	
Index of dispersion	635
Standard deviation	222.47

Table 3. Cloud Providers.

Words group searched in tweets	Number of tweets
Cloud IBM	5368
Cloud Google	6215
Cloud Amazon	2373
Cloud Azure	3679
Cloud Oracle	3385
Cloud Microsoft	9666
Google Microsoft Amazon	630
Mean indicators	
Arithmetic mean	4473.71
Harmonic mean	2312.83
Geometric mean	3465.57
Position indicators	
Median	3679
Scattering indicators	
Index of dispersion	9036
Standard deviation	2720.18

Unless the set of numbers provided contains duplicates, the returned modal value is # **N/A.** Cloud and virtualization are technologies that have been developed in order to reduce costs and maximize the use of IT resources.

Starting from the actual situation where virtualization can be achieved without cloud - whereas cloud without virtualization is hardly doable - and analyzing the data in Table 2, we note that the "Cloud-virtualization" appears twice as much as the "Cloud Engineer" relationship.

Furthermore, the "Cloud Engineer" is twice more frequent than the "Cloud Engineer jobs" group. However, according to Figure 2 and the data in Table 2, the "Cloud risk" pair has priority in the analyzed tweets. The main cloud suppliers, according to analyzed data, are shown in Figure 3.

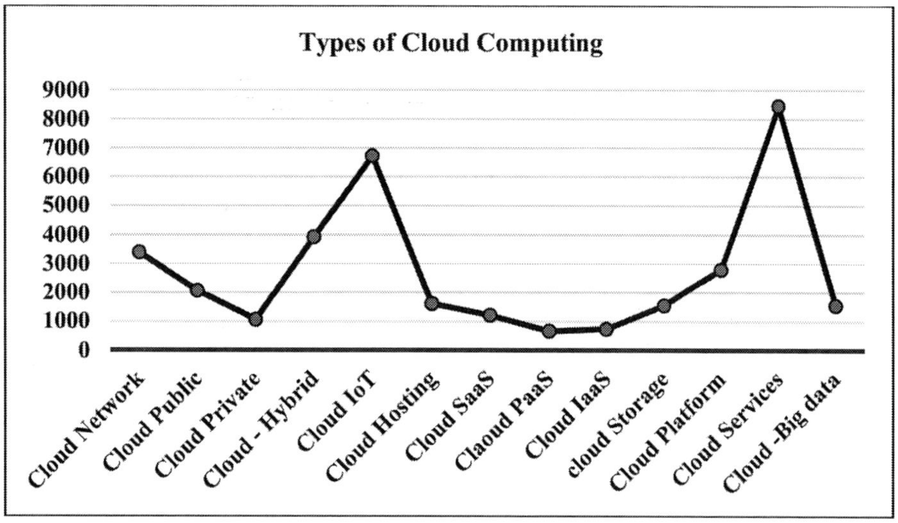

Fig. 3. Frequency of tweets pairs containing cloud.

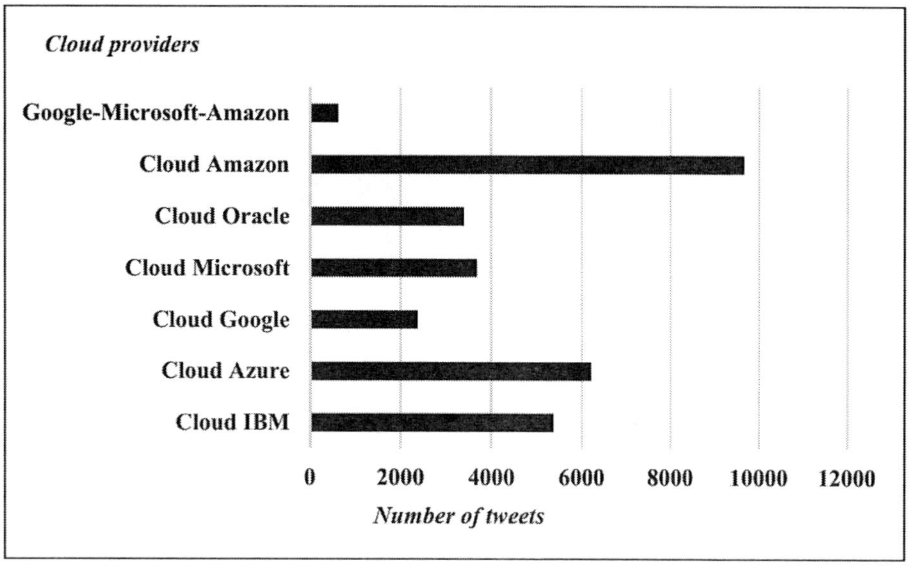

Fig. 4. Cloud provider's distribution.

Researches financed through the projects: Joint Applied Research Project supported by MENUEFISCDI, project no. PN-II-PT-PCCA-2013-4-0236, Contract no. 222/2014.

5 Conclusions

Creativity and innovation are the most important drivers of the economic growth sustainability and the competitive achievements in knowledge-based economy. The engineering departments of the technical universities should develop new disciplines and programs to develop creativity, based on an original and valuable implementation of creativity increase techniques, providing practical solutions in order to include creativity in the engineering programs.

The activity of a cloud engineer through a creative and innovative action must be carried out in conjunction with the security elements and taking into account the risks the data in the cloud may be exposed to. At present there are technologies that revolve around cutting edge security platforms that protect the resources in the cloud.

The social media data can also be analyzed in terms of unconventional processing by means of ultrasounds in order to identify new specific knowledge. It is known that the involvement of those people, who are not part of the project team as they work in another field, tends to be more creative than that of those who worked for a long time in the same field.

Based on subsequent research, the results obtained by teams of students will be analyzed, following the above mentioned stages and implementing the possible solutions, relative to construction of the end part of ultrasonic horn, where the tool for ultrasonically aided electrical discharge machining is clamped, in the frame of the PN-II-PT-PCCA-2013-4-0236 project, “High productivity and precision technology and equipment for micro-electrical discharge machining aided by ultrasonic”.

References

1. C.V. Doicin, S. Chircu, B.G. Coteț, M.E. Ulmeanu, M.S. Sălvan, AMM, **834**, 5, (2016)
2. J.M. Eger, San Diego State University, (2012)
3. W. Sarkissian, D. Hurford, C. Wenman, Earthscan publishes in association with the International Institute for Environment and Development, London, UK, (2010)
4. I. Badran, EJEE, **32, Issue 5**, 13, (2007)
5. B. Muirhead, IJITDL, **1, No. 12**, (2004)
6. R.F. Verderber, D.D. Sellnow, K.S. Verderber, *Student Edition. Cengage Learning*, USA, **2**, (2013)
7. N.P. Kirillov, E.G. Leontyeva, Y.A. Moisenko, International Conference of Research Paradigms Transformation in Social Sciences, Elsevier Ltd. (2015)
8. H.N. Teodorescu, IJGS, **41, No. 8**, 20, (2012)
9. W. Wenger, *Project Renaissance*, Singapore, **2**, (1992)
10. D. McKeag, School of Electrical and Mechanical Engineering, University of Ulster at Jordanstown, UK, (2008)
11. L.D. Ghiculescu, N.I. Marinescu, O. Alupei, N. Căruțașu, Key Engineering Materials, **651-653**, 6, (2015)
12. C. Pirnau, M. Titu, L. Rosca, M. Pirnau, *Proceeding The 2nd International Scientific Conference SAMRO 2016, "News, challenges and trends in management of knowledge-based organizations"*, Paltinis, Sibiu County, Romania, 8, (2016)

Innovative mechatronic system for testing laterality

Claudiu Mereuta*, *Daniel* Ganea, and *Elena* Mereuta

Dunarea de Jos University of Galati, Romania

Abstract. The paper presents the construction and operation of a mechatronic system for determining and improving the reaction rate and the dominance of cerebral hemispheres, in order to reduce imbalances between left and right laterality and increase the selective reaction rate to visual stimuli. The mechatronic system consists of an experimental platform and tailor made software that runs on a personal computer (PC) or notebook. The mechatronic system is an innovative one, as there aren't any devices to determine laterality and improve the reaction rate on the left / right side.

1 Introduction

Testing laterality is a challenge for researchers, because there aren't devices or equipment for it. Most tests are based on observation, on the execution of current activities or on questioning subjects. It seems that being left-handed or right-handed has biological and physical causes. The human brain consists of two hemispheres, the left and right. The right hemisphere is the place where emotions, intuitions are emerging. It controls the motor functions of the left side of the body. The left hemisphere is the space of language and logic but manages the right side of the human body. The latter is prevalent in everyday life, in relation to the right hemisphere functions.

Left- handed can be educated to use the right hand, but the question is if the benefits of such action outweigh the disadvantages that may occur. All these manifestations are usually called the opposite of laterality [1].

The laterality opposite phenomena is relatively common in sports since most branches of sport necessarily require the development of sensory-motor skills for both dominant half and for the contralateral side. The laterality issue has been studied by many authors, but there wasn't a way to estimate, to explain or test it to be universally accepted. There are tests based on observation, tests based on questionnaires, exercises for laterality education like Ozeretski - Guillmain spatial orientation test - Piaget Head, Harris laterality test, laterality battery of Galifret - Granjon [2-7].

The mathematical models designed to explain the laterality prove to be narrowed, because a lot of empirical factors and coefficients were considered [8, 9].Predilections for left hand and its consequences have been the subject of research [10]. Some authors have attempted to discover the gene responsible for the dominance of cerebral hemispheres [11,

*Corresponding author: cmereuta@ugal.ro

12], others have tried to reveal if there is a connection between laterality and cognitive development [13], while others were concerned about the advantages and of asymmetry [14, 15].

The frequency of polymorphisms in relation to the use of the right hand or left hand was another topic of scientists [15]. Some authors have studied by different means the evolution of brain lateralization [16], or have tried to estimate the life expectancy of left-handed and right-handed [17], together with the differences in performance of some usual movements [18].

2 Description of the mechatronic system

The idea of a mechatronic system for assessing and improving the speed rate and the dominance of the cerebral hemispheres, emerged from an existing system called Batak used to improve the reaction and hand eye co-ordination [19].

A presentation of the mechatronic system was made by the first author of the paper, in his habilitation thesis [20]. So, as it is presented in the habilitation thesis, the use of this device will render the possibility to reduce imbalances between left and right laterality and to increase the selective reaction rate to visual stimuli, thus becoming an important tool for testing laterality and for the rehabilitation of disabled persons.

The mechatronic system consists of experimental platform and specially designed software that runs on a personal computer or notebook. A USB cable connects the platform and the computer.

The mechatronic system consists of the following elements:

a) A set of 13 smart buttons. Each button contains a comeback contact (operated by touch), a light source with selectable intensity and color (red, green, blue or combinations thereof) and an electronic circuit that automatically turns off the light source when the button is pressed;
b) A mechanical structure for positioning and support frame for the smart buttons. The support structure is in the shape of a regular octagon with 0.6m apothem. The smart buttons are placed according to a suitable schema;
c) An electronic circuit with microprocessor that connects the smart button to the computer. The microprocessor contains a tailor-made program for this application which performs the following functions: allows multiple pre-programmed sets of exercises, some containing a fixed sequence (repeated) of buttons and colors, some with completely random order; receives the code of the test to be run from PC; allows offline exercises for demonstration purposes, or as individual training or as warming game; activates a light source for each smart button; measures the reaction time since light activation to pressing the button; disables the button and moves to the next combination of buttons and lights, if a button has not been pressed for 2.5seconds;transmits the following information to the application: code button, the hue of light used and the reaction time measured in milliseconds.

3 The operating principles of the mechatronic system

The operating principles are described by the microcontroller and computer software flow-charts. The flowchart for the microcontroller application that will run on the experimental platform is shown in figure 1.

The software is a Java application that provides a user friendly graphical interface which will help to assess the reaction time to visual stimuli of different frequencies and wavelengths.

The PC application flow-chart is shown in figure 2, while sequences marked with*A - *B are presented in detail, see figure 3. The computer software application is based on three major processes: preprocessing, serial communication and processing (Fig. 4).

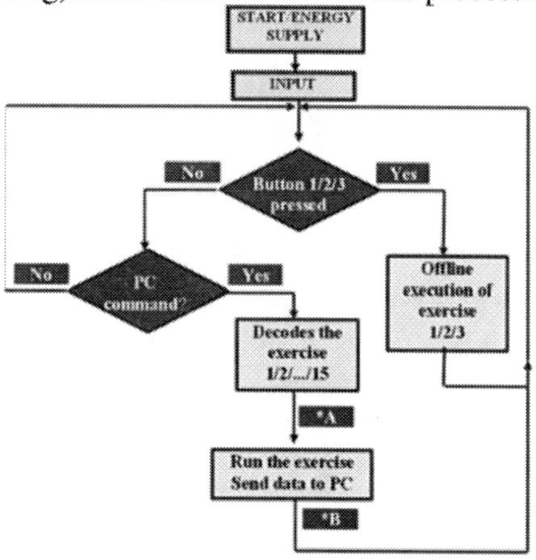

Fig.1. Microcontroller flow-chart.

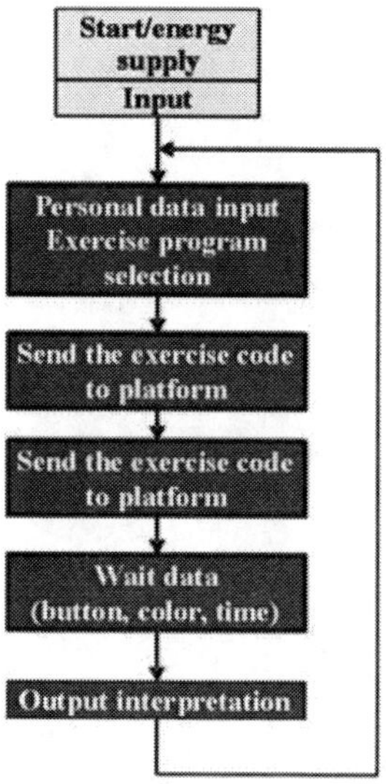

Fig.2. PC application flow-chart.

4 The implementation of the mechatronic system

The purpose of testing using the mechatronic system is to reduce imbalances between the cerebral hemispheres, to increase the reaction speed to visual stimuli, and to increase the selective response speed to stimuli of different colors.

Several categories of athletes, children and people with neuromotor deficiencies are targeted. As general population is right-handed, imbalances between the two hemispheres of the brain are widening and some activities controlled by nerve centers of a hemisphere can be performed more slowly or cannot be performed simultaneously with other activities [21]. The complexity of everyday life requires the softening of these imbalances and therefore such a test is appropriate only if followed by an individualized training program.

For the athletes category different meanings on laterality are developed, but all of them take into account all the sports [22].

It is possible to design programs to improve performance, especially for athletes, people with slight disabilities caused by psychomotor disorders and children. Thus, it appears that some left-handed athletes are more advantaged during direct competition (i.e. boxing, sword, floret, tennis and table tennis etc.).

There are also sports at which being left-handed is an advantage (i.e. darts, snooker, golf) [22]. A clear division of sports in which being left-handed is an advantage cannot be made. What is important is that during competitions the athlete reacts appropriately, anticipates the actions of opponents and takes decisions quickly. Therefore, it is important for athletes to develop their skills on both arms, thus being able to handle many tactical positions on the competition field, in order to adapt to specific situations during competition.

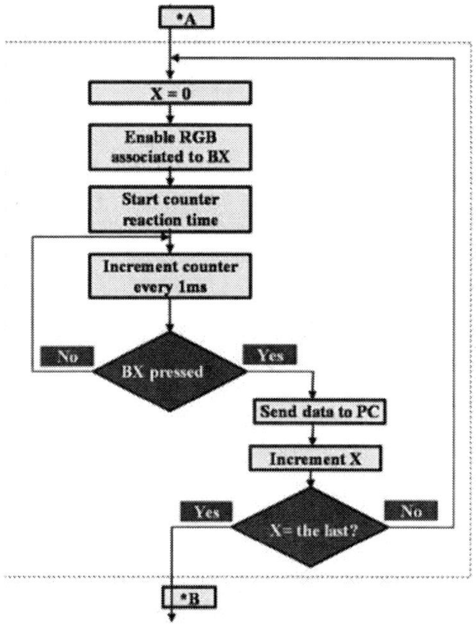

Fig.3. A*-B* sequences flow-chart.

Regarding the children, it is possible to assess laterality only after 5-6 years old, and skills can be educated and directed using games towards the efficient use of both arms and thus to the development of both cerebral hemispheres and the stimulation of all areas of the cerebral cortex. People with slight disabilities can attend training programs to increase the reaction rate.

A friendly interface was designed. See figure 5 which presents the participant's details and the testing program that is applied. See figure 6 which presents the results of the testing procedure.

It is possible to design programs to improve performance, especially for athletes, people with slight disabilities caused by psychomotor disorders and children. Thus, it appears that some left-handed athletes are more advantaged during direct competition (i.e. boxing, sword, floret, tennis and table tennis etc.).

There are also sports at which being left-handed is an advantage (i.e. darts, snooker, golf) [22]. A clear division of sports in which being left-handed is an advantage cannot be made. What is important is that during competitions the athlete reacts appropriately, anticipates the actions of opponents and takes decisions quickly. Therefore, it is important

 DOI: 10.1051/matecconf/201711208006

for athletes to develop their skills on both arms, thus being able to handle many tactical positions on the competition field, in order to adapt to specific situations during competition.

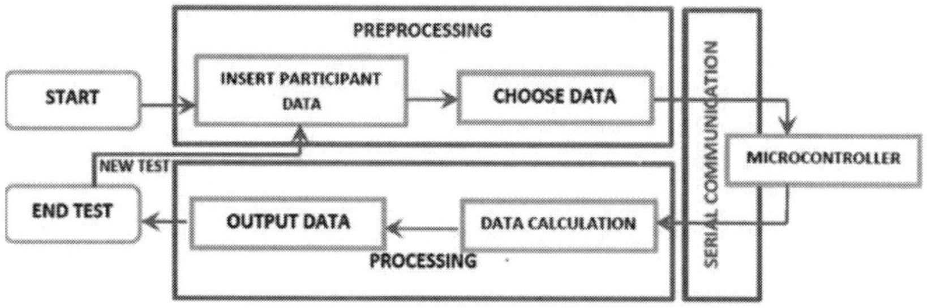

Fig.4. Major processes of PC application.

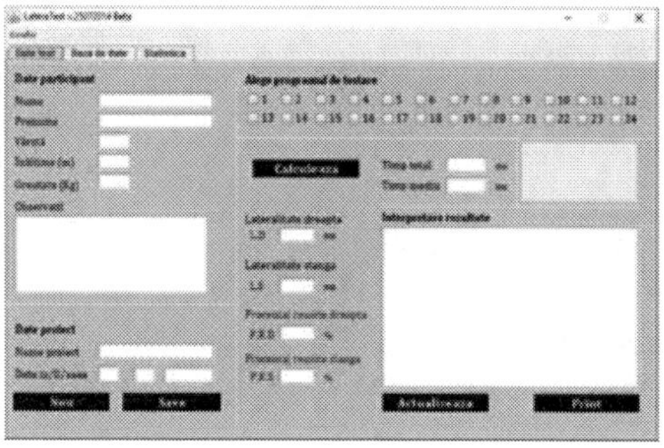

Fig.5. Interface for participant's details.

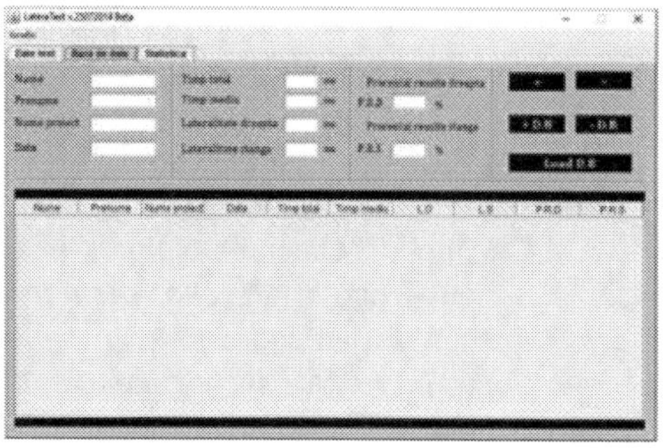

Fig.6. Interface with the testing results.

Regarding the children, it is possible to assess laterality only after 5-6 years old, and skills can be educated and directed using games towards the efficient use of both arms and thus to the development of both cerebral hemispheres and the stimulation of all areas of the cerebral cortex. People with slight disabilities can attend training programs to increase the reaction rate. A friendly interface was designed. Fig. 5 presents the participant's details and the testing program that is applied. See figure 6 which presents the results of the testing procedure. A friendly interface was designed. Fig. 5 presents the participant's details and the testing program that is applied. See figure 6 which presents the results of the testing procedure.

5 Conclusions

The experimental mechatronic system is an innovative one, as there are not such devices to determine laterality and improve the reaction rate on the left/right side. The system can be successfully used by athletes, by physiotherapists for recovering individuals which have temporarily lost the upper limb mobility, and for stimulation the children's brain hemispheres, from an early age.

The system implementation will open new research directions in the evaluation of laterality and study of the neuronal processes that determine the response to stimuli. It is possible to combine the mechatronic system with other devices, in order to determine brain nerve stimuli and muscle reactions controlled by brain nerve centers.

Regarding the experimental research, this will assume a series of tests that will be performed on different groups of subjects: athletes, children and individuals with slight neuromotor disabilities. For each of them, random testing programs will be generated and thus the left / right speed rate to white light stimuli will be assessed.

The software allows the randomly generation of programs with different colored lights (red, blue, white, green). Subjects will be asked to press the buttons with a certain colored light (e.g. only red buttons, or just green and blue buttons, etc.).

The software will allow the generation of reports for these cases, evaluating the left/right speed rate, analyzing in this way also the responses to colored visual stimuli. The data will be statistically analyzed for each testing protocol and initial status reports will be automatically generated. The next step will consist in the development of different training programs using the same mechatronic system.

The training will be customized and adapted to each participant in the experiment. The next step will be the final testing, aiming to track the progress of each participant in the experiment. The participants will be tested in temperature, brightness and noise level limits according to the lab conditions. Subjects will be aware of the experiment protocols and they will be asked to sign an agreement for their participation in the experiment tests. The experiment is non-invasive and does not cause trauma to participants.

References

1. M.Ciucurel, C. Ciucurel, Citius, Altius, Fortius, 2:15-22 (2002)
2. S. Lungu – Nicolae, *Psychological tests*, Bucureşti, (1991)
3. I. Muşu, A.Taflan(coord.) *Integrated educational therapy*, Ed. Pro Humanitate, (1999)
4. E. Verza, F. Verza, *Psychogenetic and psychodynamic landmarks in the knowledge and evaluation of the child,* Bucureşti, (2000)
5. E. Vlad, *Evaluation in the educational-therapeutic act*, Ed. Pro HumanitateBucureşti, (1999)
6. E. Vrăşmaş, *Pre-school child education,* Ed. Pro Humanitate, Bucureşti, (1999)
7. V. Oprea, E. L. Niţu, D. Chiriacescu, P. Lungu, *Set of tools and tests for the educational assessment of children with disabilities,*Bucureşti, EdituraMarlink, , ISBN 973-8411-06-8 (2003)
8. D.M. Abrams, M.J. Panaggio, , J. R. Soc. Interface 9, 2718–2722, (2012)
9. S. Ghirlanda, E. Frasnelli, G. Vallortigara, Phil. Trans. R. Soc. B 364, 861 –866, (2009)
10. V. Laurens, M. Raymond, C. Faurie, Phil.Trans. R. Soc. B 364, 881 –894 (2009)
11. C. Francks, et al. Mol. Psychiatry 12, 1129–1139 (2007)
12. I.C. McManus, et al. Editorial commentary: is LRRTM1 the gene for handedness? Laterality 14, 1–2. (2009)
13. M. Magat, C. Brown, Proc. R. Soc. B 276, 4155–4162. (2009)
14. L.J. Rogers, P. Zucca, G. Vallortigara, G. Proc. R. Soc. Lond. B 271, S420–S422 (2004)
15. G. Vallortigara, L.J. Rogers, Behav. Brain Sci. 28, 575 –589 (2005)
16. S. Billiard, C. Faurie, M. Raymond, J. Theor. Biol. 235, 85–93 (2005)
17. S. Ghirlanda, G. Vallortigara, Proc. R. Soc. Lond. B 271, 853 – 857 (2004)
18. D.F. Halpern, S. Coren, Nature 333, 213 (1988)
19. http://www.batak.com accessed on 10/19/2013
20. C. Mereuta – Habilitation thesis available at: http://www.unitbv.ro/Portals/31/ Abilitare/Teze/Teza/05-MEREUTA-Teza%20abilitare%20ENG.pdf
21. M.S. Hardie, L. Wright, Front. Psychol., (2014)
22. F. Loffing, F. Soelter, N. Hagemann, PLOS ONE Volume:9 Issue:8 Article Number:e105800, (2014)

Innovative method for training students to develop enterprising, decision-making and cooperation skills through complex computer-networks team-design

Petre-Daniel Mătăsaru*, and *Luminița* Scripcariu

Technical University "Gheorghe Asachi" of Iasi, Faculty of Electronics, Telecommunications and Information Technology, 11 Carol I Blvd., Iasi, Romania

Abstract. Our paper presents a short overview of the educational methods used in the learning process in telecommunication domain, targeting skills and competences students acquire and develop during the semester and focusing on the ones that are most valuable and appreciated on the labor market after graduation. Our research of the market shows that besides basic engineering knowledge, skills like initiative, team-playing, business analysis, decision-making, marketing and creative project presentation are considered very valuable assets by HR recruiters. We implement and test an improved method that combines classic techniques with the use of modern digital tools, emphasizing on specific tasks that coach the student how to deal with real markets, extract valuable data through analysis, design up-to-date computer-networks, make correlated decisions based also on economic arguments, team working and assume responsibilities. This is achieved through a real-case study and project themes for computer-networks that involve real situations with technical and budgetary challenges, market analysis and research through online facilities and dealing with specialized software for network design and simulation with a creative presentation. The presented approach is intended to prepare faculty staff to implement innovative and self-improving teaching methods in engineering educational process and thus make an educational reform happen [1].

1 Introduction

Most instructional methods used in the learning process in the electronics, telecommunication and information technology domain are consisting of a combination of core elements of active, collaborative, cooperative and problem-based learning [2]. In engineering literature, collaborative learning may be referring to any type of educational method that makes the students work together in small teams in order to reach a common goal [3], cooperative learning may be a particular form of collaborative learning where students also work in teams to achieve common goal, but they are assessed individually [4-

* Corresponding author: dmatasaru@etti.tuiasi.ro

5] and problem-based learning may represent an educational method where specific problems are used to provide context and motivation for the learning that follows, involving significant amounts of self-oriented learning effort on the student side [2].

While the above mentioned three techniques are universal and represent a common ground for all university staff with a technical background, active learning makes a difference, engaging students in the learning process by requiring them to do meaningful activities, get quantitative and qualitative results and acknowledge their contribution. The presented training process aims improving skills of both students, and teaching staff.

2 Quick overview on the graduate employ-ability skills

Many studies on Higher Education in Europe reveal the fact that both employers and future employees graduate identify several key soft skills and competences integral to the employment process.

According to Andrews and Higson [6], among these is the ability to plan and think strategically, the capability to communicate and interact with others, either in teams or through networking, good written and verbal communication skills, Information and Communication Technology skills, creativity, good self-management and time-management skills. Furthermore, these key features may form the basis and main values of current employment interviews.

Others [7] emphasize on the need of engineering schools to strengthen their coverage of fundamentals, teach a more realistic and up to date engineering design and operations, including quality management, provide student training in critical and creative thinking skills, with good engineering ethics within the development of technology.

Without regard to the identified types of skills, three important themes stand up, each of them focusing on various components of graduate employability: Business Specific Issues referring to Hard business-related knowledge and skills, Interpersonal Competencies referring to Soft business-related skills and Work Experience and Work-Based Learning. [6]

From the employer point of view, the added-value of employing an engineering graduate with enterprising skills may be that a superior level business approach represented more than just simple evidence of getting and applying knowledge, it also reflected the individual's ability to think in a critical and applied manner to the economic environment. A plus may be given also by the fact that engineering graduates with good decision making skills require less supervision and mentoring, are more mature and fit sooner into their working environment.

From the graduate point of view, a mixed approach of studying both in terms of engineering and economics may be considered positive in achieving more business-focused skills for their portfolio that would provide better employment chances. However, studies point out for the graduates' perspective that even though many of them feel confident in respect of knowledge focused abilities, they perceive themselves as lacking the necessary level of presentation skills [6].

3 Improved training method: designing, testing and presenting a real-case computer network as a student-team enterprise

The instructional method that we apply comes as a response to the labor market needs according to various studies and our own research made through meetings, discussions and collaborations with representatives of national and international companies in the electronic, telecommunications and Information Technology domain. It is, in fact, a solution that emphasizes on the interaction of students and the business environment in groups with focus on the practical issues instead of the classical individual theoretical learning. Thus, besides the basic engineering knowledge that most of the students/graduates get through classical educational process, skills like initiative, team-playing, business analysis and decision-making, marketing and creative results presentation are also targeted.

Our method combines classic instructional techniques with the use of modern digital and online tools, focusing on specific tasks that coach the student on how to deal with the real market, extract valuable data through analysis and data-mining, design up-to-date computer-networks, make correlated decisions based not only on technical issues but also on economic arguments, work in a team and assume responsibilities and finally, present in an attractive manner the results of the teamwork.

This is achieved through a real-case study and project themes for computer-networks that involve real situations with technical and budgetary challenges, market analysis and research through online facilities and dealing with specialized software for network design, network simulation and creative presentation.

Project evaluation is made in an innovative way. Students must evaluate the projects of their colleagues from the entrepreneurial point-of-view as possible beneficiaries. They have to select the best project as a compromise between technical and economic aspects. This contest motivates students to improve their entrepreneurial and argument-based decision-making skills.

3.1 Creating work groups and choosing the theme of the project

Cooperative learning models involve the creation of working groups by the coach and our method does not cross the line in this matter. Then group members choose a team-leader by consensus. What we are doing different is to choose the theme of the project based on the common background of the students that are part of the working group. Thus, students are more responsible with their project since they are involved in the decision process and are better motivated to study and work harder as a team because they feel like belonging to the project and the results of their work will be closer to the real technical and economic environment on a real-case study.

For example, student work groups' choices are computer network design, testing and implementation for their home high-school, public libraries in their home cities, companies where they had internships, various types of private businesses like newspapers, internet-cafes, call-centers and restaurants where they had part-time jobs as students.

3.2 Changing perspective from a school project to a student enterprise

After setting the working groups and the theme of the project, the next step is to approach the project from the business point of view, students becoming "virtual entrepreneurs" and forced this way to get acquainted to concepts like market research, data mining, human resources cost, material costs, project budget, communication and PR, besides the technical aspects of a computer-network design.

Within the project students are required to create a "costumer chart" with general information about the type of activity, company estimated number of employees, offices and organizational chart, current and future development plans. Based on this "costumer chart", the design of the project begins.

The next step is to effectively design the computer network, establishing the location of the computers in the building, the connections diagram, the physical and logical topology, the necessary sub-nets number depending on the departments of the beneficiary company (see figure 1), the hosts' IP addresses per each sub-net (see table 1), and the required security levels.

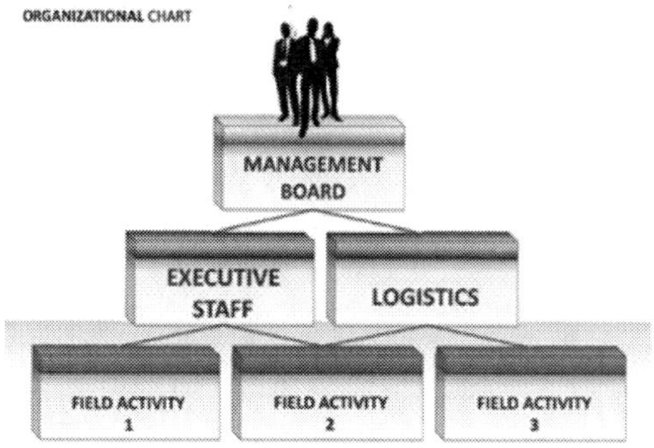

Fig. 1. Organizational chart of a "project beneficiary".

Decisions have to be made in order to meet the optimal solutions for all problems, so the students in the working group not only have to collaborate, but also to face their colleagues' suggestions, analyze them and, if necessary, sustain with arguments the better outcome.

Table 1. Example of IP allocation table

Sub-net	Sub-net Mask	IP		Default Gateway
S#1	255.255.255.224	H#1	192.168.1.34	192.168.1.33
		H#20	192.168.1.53	
		H#30	192.168.1.62	
S#2	255.255.255.224	H#1	192.168.1.66	192.168.1.65
		H#20	192.168.1.85	
		H#30	192.168.1.94	
S#3	255.255.255.224	H#1	192.168.1.98	192.168.1.97
		H#20	192.168.1.117	
		H#30	192.168.1.126	
S#4	255.255.255.224	H#1	192.168.1.130	192.168.1.129
		H#20	192.168.1.149	
		H#30	192.168.1.158	
S#5	255.255.255.224	H#1	192.168.1.162	192.168.1.161
		H#20	192.168.1.181	
		H#30	192.168.1.190	

Once every technical aspect is determined, the following step is to implement, configure and test the computer network using specialized engineering software.

In the end, after the completion and validation of all required tests concerning technical issues, an estimation of costs per chapters and overall budget has to be made.

For this, students are asked to search individually the web for network components and other material prices and calculate, from an economic point of view, a reasonable close to reality cost of the entire project. Other decisions, this time, economical, have to be made, after a comparison between the prices at working team level.

An example of overall budget calculus is given below:

UTP Cable *CAT5e: 1000m * 0.99 RON/m = 990 RON;*
Switch *(Switch Allied Telesis AT-GS950/24):4 * 1,630 RON = 6,520 RON;*
Router *(Router Linksys Cisco RV082-EU): 2 * 2,024 RON = 4,048 RON;*
RJ-45 Connectors*: 200 * 0.2 RON = 40 RON.*
Cable channels *(80 x 60, 200 m): 1,500 RON.*
Server *(Server HP Tower ML110 G7): 2,259 RON.*

Students also calculate the labor costs based on an estimation of working hours and workers' qualification. For example, for this project, the **labor costs** are evaluated to 10,000 RON.

Thus the **TOTAL AMMOUNT** results as 26,257 RON.

3.3 Improving presentation skills

A final objective of our instruction technique is to improve the student's presentation skills by requesting that the entire project to be presented in front of all the other teams in a modern manner and a time limit.

The presentation should be made using an online tool like Prezi [8] and every member has to present a part of the group project. Thus, cooperation and presentation skills are improved; focus is put also on other relevant features for the employer like creativity, responsibility, good time-management and the opening to new technologies. Project presentation includes the network simulation (see figure 2) and testing (see figure 3).

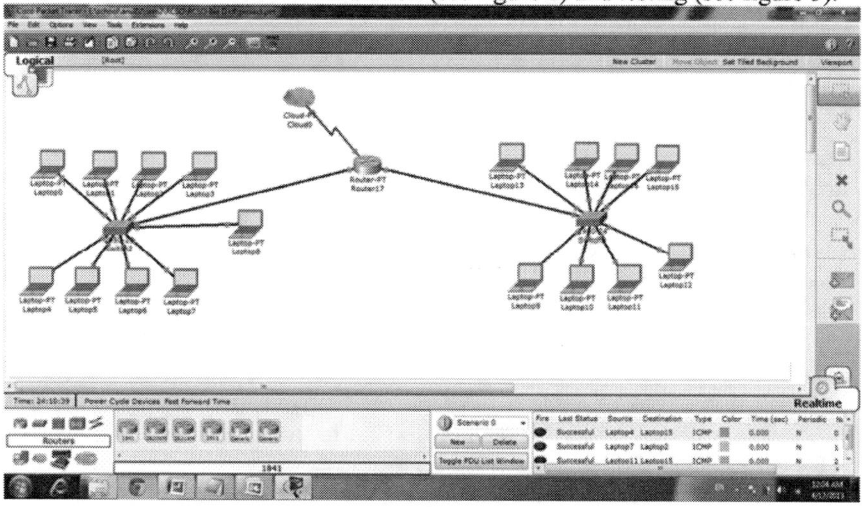

Fig. 2. Implementation, configuration and testing of the computer network design.

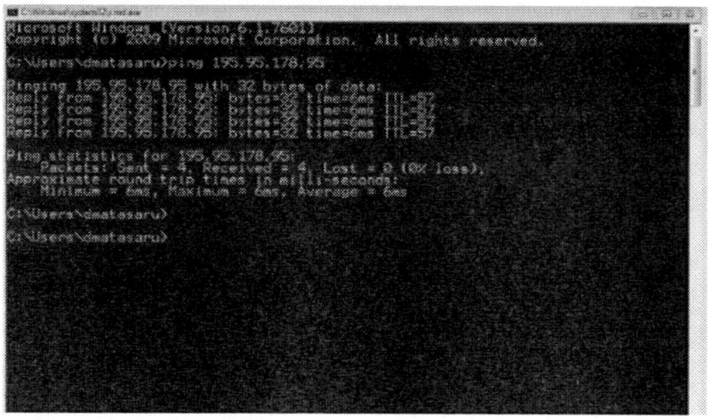

Fig. 3. Connectivity Testing in Command Prompt.

3.4 Developing entrepreneurial skills

After all projects are presented, each student evaluates them, from the investor's point-of-view, and chooses three projects that he appreciates to be very well done and economically justified. Comparing their project with the others, students become aware of the strong points and the weak points of their own project.

3.5 Project evaluation

A polling session is organized and after the count of votes, one project is declared as eligible to be financed and communicated to the auditorium.

This final activity imposes another kind of responsibility to students. They are not implicated as technical staff but they act as investors so their entrepreneurial skills are developed.

The teacher acts as a neutral moderator during the presentation of the projects, leaving the decision on the best project to be made by the students.

In the same time, the projects are technically evaluated by the teacher himself without knowing the student polling result.

Finally the teacher evaluation results and the polling result are confronted. This is the moment when the righteousness of students' decision is analyzed.

4 Conclusions

Our model incorporates both classic instruction techniques and business-oriented methods with the use of digital tools and could be categorized as active learning. The final objective is to enable, in an innovative way, the graduated student to be properly prepared for the labor market and the employer to be satisfied with the quality of the future employees simultaneously with training skills improvement of the training staff.

Both technical and economic knowledge, as well as initiative, strong decision-making abilities and the capacity to properly present the results of the work represent an important asset for any engineering company.

Last, but not least, the professional experience that students get while working on the computer-network design represents an added value for anyone's portfolio when applying for a job and sustaining a job interview.

The overall experience is valuable for both students and teacher.

References

1. R.M. Felder, J.E. Stice, A. Rugarcia, *The future of engineering education, VI. Teaching methods that work*, Chem. Engr. Education, **34**, 3, 208–215 (2000)
2. M. Prince, *Does Active Learning Work? A Review of the Research*, Journal of Engineering Education, **93**, 3, 23–231 (2004)
3. *Online Collaborative Learning in Higher Education*, http://clp.cqu.edu.au/glossary.html (accessed 12/3/2003)
4. B. Millis, P. Cottell, Jr., American Council on Education, *Cooperative Learning for Higher Education Faculty* (ORYX Press, 1998)
5. P. Feden, R. Vogel, *Methods of Teaching: Applying Cognitive Science to Promote Student Learning* (McGraw Hill Higher Education, 2003)
6. J. Andrews, H. Higson, Higher Education in Europe, *Graduate Employability, 'Soft Skills' Vs 'Hard' Business Knowledge: A European Study*, **33**, 4, 411-422 (2008)
7. R. M. Felder, D. R. Woods, J. E. Stice, A. Rugarcia, Chem. Engr. Education, *The future of engineering education. II Teaching methods that work*, **34**, 1, 26–39 (2000)
8. http://www.prezi.com

Forecast of a quality indicator in academic activity using statistical methods

Silvia Veresiu*, *Elena* Mereuta, *Madalina* Rus, *Daniel* Ganea, and *Valentin* Amortila

"Dunarea de Jos" University of Galati, Faculty of Engineering, Romania

Abstract. The paper presents the time evolution of an indicator for quality, which is related to self-evaluation of academic staff in technical faculties. The study is focused on two major issues that compose the academic activities: teaching and research activity on one side, and managerial activities on the other side. The aim of the study is to reveal the forecast of self-evaluation using statistical methods and to assess if possible, the minimum accepted value of the quality indicator.

1 Introduction

In the recent years, quality has become a higher education priority in universities all over the world.

The changes on the labor market and the rise of the institutions that provide higher education (private universities, increasing number of public universities, new study programs and specializations) determine the employers to ask for official validated information about the quality of education in the targeted institutions.

The official recognition is reflected by the accreditation obtained by the university itself. The national agency licensed to grant temporary authorization or accreditation performs the external evaluation of to the quality of the educational process by complying with the existent criteria.

The criteria of accreditation encourage the institutions of higher education to try innovative programs and procedures and to adopt them when they prove to be successful.

The higher education all over the world suffered major changes due to the permanent character of the changes occurred in the social, political and economic environment.

In order to fulfil the valid standards, the institutions of higher education must develop policies and procedures of quality assurance of the educational process. These must commit themselves explicitly to develop an education that recognized the importance of quality in their activity.

According to "The Methodology of External Evaluation, Standards and the List of Performance Indicators of ARACIS" [1], during the process of education's quality assurance, three issues are under focus: the institutional capacity, the educational efficiency and the management of quality.

The institutional capacity refers to: the university's organization on the basis of an integrated and well-structured management and to: the organization of the teaching -

*Corresponding author: silvia.veresiu@ugal.ro

learning processes focused on the students, methods and techniques that allow the assimilation of knowledge measured by the outcomes learning and research; evaluation criteria of the students, the existence of a qualified teaching staff, the existence of available resources and learning facilities, of the services offered to the students.

The quality's management is focused on activities, strategies, techniques, structures, policies by means of which the university proves the permanent administration system; the existence of an infrastructure, of proper equipment, and well prepared and capable human resources able to achieve the aims.

The educational efficiency refers evaluation of the performances obtained in ensuring and improving the quality of the education they offer.

The evaluation of the three issues is performed using criteria and standards expressed by performance indicators. The criteria represent the basis of organizing and functioning of the educational institution. The performance indicators are tools for measuring the achievement of the work carried out by the university in question, in relation to the standards. These indicators can vary from the minimum level to the maximum one.

2 Teaching staff assessment

The quality of teaching staff is regularly assessed through the expertise, capacity for transmitting knowledge to students and professional ethics.

The evaluation criteria mentioned above, are summarized in an evaluation sheet that includes information from four topics: self-evaluation, peer review, assessment of students and evaluation of head department.

This paper refers only to the first issue, which is conducted annually, based on self-assessment forms.

The self-assessment sheet is approved by the Senate, is drawn up annually and represents a multi-criteria assessment of all teaching and research activities covered on the previous calendar year.

The self-assessment sheet includes two parts:

A. Criteria for assessing the teaching and research activities and promotion criteria,

B. Other criteria (professional, administrative, cultural).

The A set of criteria represents the elementary performance criteria according to ARACIS standards, the performance indicators to measure the level of scientific research, according to National Council of Scientific Research, and criteria for annually financing the higher education institutions. The A set of criteria form the basis of professional promotion [2]. The sets of criteria A and B are the basis for granting the performance rewards.

Completion and submission of the self-evaluation sheet in compliance with deadlines shows that the entire teaching staff has acquired and developed a culture of quality according to the requirements of a total quality management.

The paper aims to study the evolution of the scores recorded for self-assessment criterion A in the schedule for all teaching positions in two technical faculties. Also, considering the results statistical methods will be used for converting a quantitative indicator into a qualitative one.

3 Research methods

Two statistical instruments were used as research methods: the moving average [3-4] and the stochastic oscillator [5-6].

The first method was used to analyze the collected data between 2008 and 2016. The annual scores were systematized and thus time series were generated.

The moving average method consists of calculating the seasonal component of time series by dividing the trend to total successive values of the series. The method is used when a time series shows fluctuations, in order to smooth the evolution. The moving average method aims at highlighting the central tendency of evolution of a phenomenon. The method is characterized by elasticity, the ability to easily adjust to new conditions. Using regression the basic trend of evolution of the studied phenomenon is established, by cancelling large fluctuations, which may distort real evolution. [7]

The averages are called mobile because constantly the first term of the previous average is deleted, while the next term is added.

Thus, based on average annual recorded scores from self-assessment files, criterion A, seasonal component, irregular component and the trend component were calculated and the combination of them determined the forecast for the next three years. [8]

Using stochastic oscillator is imported from financial analysis. It can provide information on the stability of the market, determining the actions of those operating in the market. In this paper, the stochastic oscillator is used as an indicator of the equilibrium state, providing stability range, with which one can determine the maximum and minimum values of the self-assessment scores.

The stochastic oscillator is represented by two lines: %K and %D, where %K is the main line, and %D is a moving average of %K.

The %K line is computed using the formula (1):

$$\%K = \frac{C - L_n}{H_n - L_n} 100 \tag{1}$$

Where: C is the annual score;
L_n is the minimum score from the last three years;
H_n is the maximum score of the last three years.

The %D line is computed using the formula (2):

$$\%D = \frac{H_3}{L_3} 100 \tag{2}$$

Where: H_3 is the sum of the value C-L_n for the last three years;
L_3 is the sum of the H_n-L_n for the last three years.

4 Results and discussions

For this study the average scores of each teaching staff category from two technical faculties (criterion A) were considered (table 1). The scores were normalized so that the number of teachers (time-varying) should not distort the study.

Table 1. Average scores of self-evaluation records.

Year	Technical faculty 1				Technical faculty 2			
	Prof.	**As.prof.**	**Lecturer**	**Assist.**	**Prof.**	**As.prof.**	**Lecturer**	**Assist.**
2008	993.62	840.65	316.49	151.22	546.16	252.59	110.00	120.16
2009	1304.87	825.00	543.39	289.98	1068.76	337.38	182.17	173.73
2010	2021.07	822.75	610.28	457.68	1902.62	925.18	334.43	466.60
2011	1849.88	923.13	732.14	456.47	1962.90	1116.30	324.80	145.50
2012	1613.29	1000.40	708.81	712.82	2137.18	976.29	548.15	670.64

2013	1139.90	462.64	474.38	327.47	1844.47	770.24	687.98	297.19
2014	1228.53	762.35	492.82	381.20	2193.01	606.93	782.28	455.18
2015	1074.19	791.55	485.31	437.24	2307.80	968.69	806.82	274.09
2016	1184.28	784.38	484.82	515.00	2307.99	1230.81	768.41	350.53

The data were computed using the moving average method. The figures 1-4 present the historical evolution of average scores, the center moving average (CMA) (the baseline) and the forecast for the following three years, for all categories of teaching staff.

The intermediate results showed that it is possible to emphasize the seasonality irregularity components, to reveal the seasonality components and to predict the future trends for the average score, getting rid of irregularities and seasonality. For example, for the average score of professors from the first technical faculty, the seasonality irregularity was in 2009 24% below the baseline (CMA), while in 2012 was 6% above the baseline (fig.1 a). The seasonality components are below the smooth outline (CMA) (fig.1a). For the second technical faculty, the results revealed that the seasonality irregularities components were 14% above the baseline (in 2016), and 18% below the baseline, in 2009 (fig. 2b). Regarding the seasonality components, they were 8% above the baseline in 2010 and 2% below the smooth outline in 2011 (fig.1b). The same analysis can be performed for all categories of teaching staff: associate professors, lecturers, assistants (fig. 2-4).

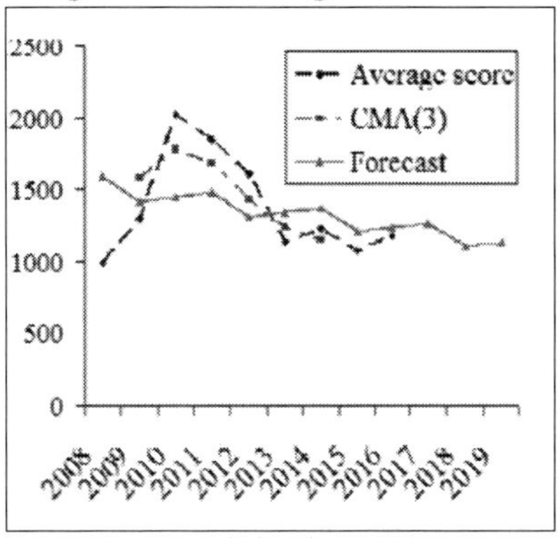

a) faculty 1

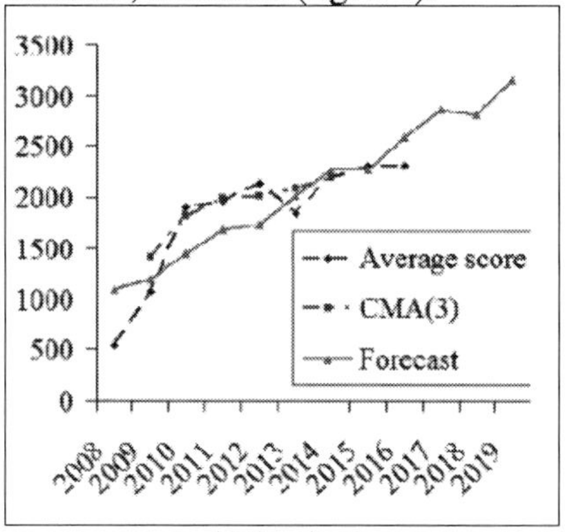

b) faculty 2

Fig. 1. Forecast of professors 'average scores.

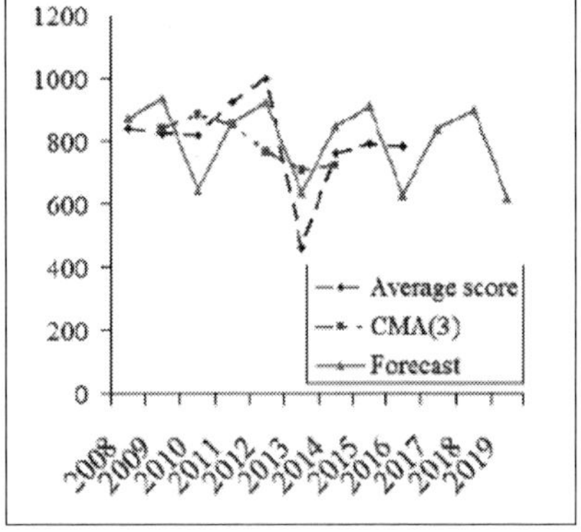

a) faculty 1

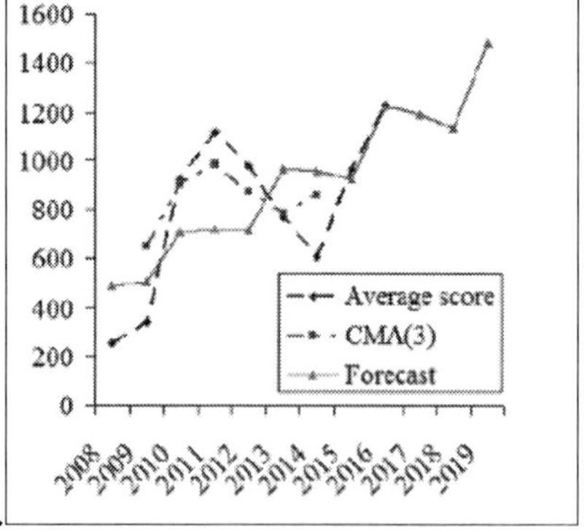

b) faculty 2

Fig. 2. Forecast of associate professors 'average scores.

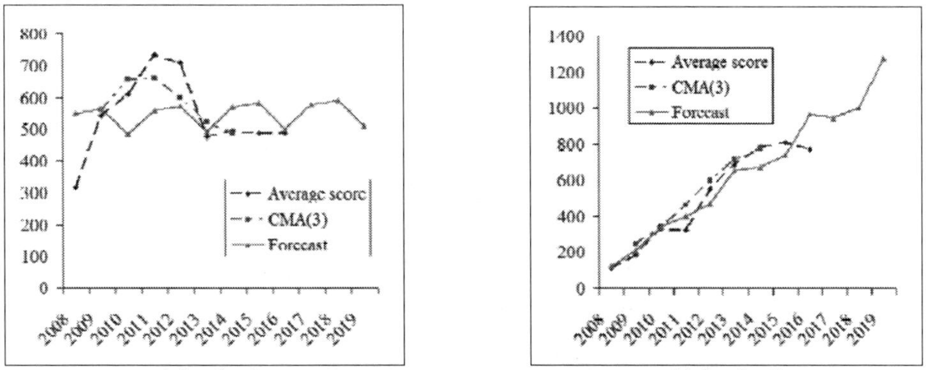

a) faculty 1 b) faculty 2

Fig. 3. Forecast of lecturer 'average scores.

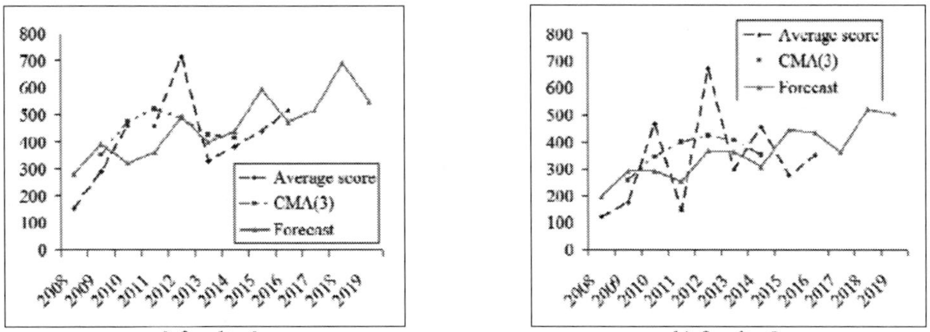

a) faculty 1 b) faculty 2

Fig. 4. Forecast of assistants 'average scores.

After determining the forecast of average scores, the next step consists of determining the stability interval for the average score, for the next three years. In order to achieve this goal, the stochastic oscillator method is used (fig. 5). This method provides the maximum and minimum values of the average score, for the next cycle of three years. These values are the extreme values from the stability range [20%, 80%]. Thus, the %K line is the taken into account and all the values outside this range are considered extreme and neglected. The analysis revealed the fact that for the professors' in faculty 1, the range for average scores is: [331, 674], while for the professors in faculty 2, the range is [915, 2162] (fig.6).

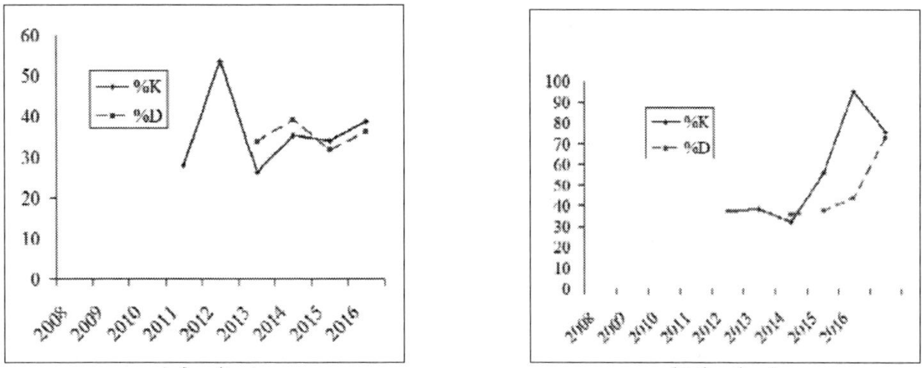

a) faculty 1 b) faculty 2

Fig. 5. Stochastic oscillator for the professor's average score.

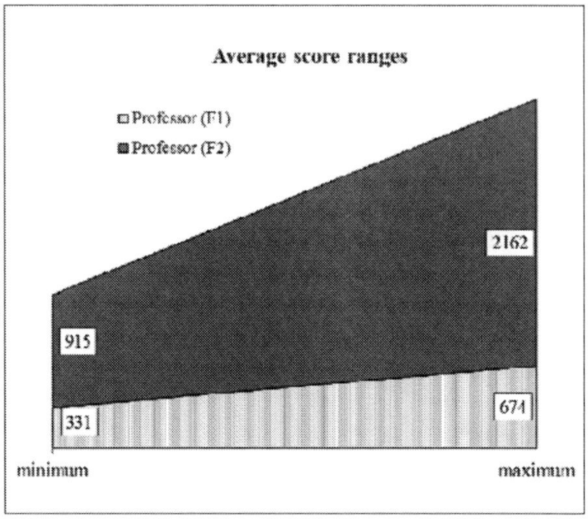

Fig.6. Average score ranges for professor.

The same analysis can be performed for all categories of teaching staff. The teaching and research activities can be assessed now using ratings. If the score is below the minimum value, the rating is insufficient. For a score between the minimum and maximum value, the rating is good and for the score above the maximum value, the rating is very good.

5 Conclusions

Based on the forecasts provided by the moving average method and the stochastic oscillator, it is possible to set limits for rating each category of teaching staff.

This approach led to a qualitative assessment, starting from a quantitative expression of the annual performances.

It is important to perform the same analysis for each faculty or department, even if they are technical ones.

The study can be extended to all faculties, humanities, arts, economic, medical etc.

References

1. *** *External evaluation methodology, standards, reference standards and list of performance indicators,* (www.aracis.ro/proceduri/)
2. *** *Annual evaluation procedure for teachers* (http://www.qa.ugal.ro/index.php?page=evaluare)
3. C. Anghelache, *Theoretical and economical statistics guide*, Ed. Economica, Buc., (2008)
4. E. Biji, et al., *Statistics for economists*, Ed. Economică, Bucuresti, (2012)
5. G. C. Lane, *Getting Started With Stochastics*,3 (1998)
6. D. Zapodeanu, D. Popa, Annales Universitas Apulensis, Univ. "1 Dec. 1918" Alba Iulia, **issue. 8**, vol. 2, 316-321 (2006)
7. C. Anghelache, G. C. Anghelache, F. P. C. Lilea, A. G. Baltac, Romanian Journal of Statistics **issue. 8**, 4-15 (2012)
8. https://www.oanda.com/forex-trading/learn/technical-analysis-for-traders/stochastic/lines

Experimental tests for foot pressure analysis during orthostatic position and gait

Daniel Ganea, *Elena* Mereuta, *Silvia* Veresiu, *Madalina* Rus, and *Valentin* Amortila*
"Dunarea de Jos" University of Galati, Department of Mechanical Engineering, 111 Domneasca Street, Romania

Abstract. Human body postural deviation during normal activities such as gait or orthostatic position can cause injuries. The paper presents an experimental study regarding the distribution of contact pressures during gait and orthostatic position, aiming to evaluate the load distribution between forefoot (FF) and rear-foot (RF), the projection of center of gravity and the support area variation while conducting stability metering and feet load distribution tests.

1 Introduction

The musculoskeletal system is a complex one and the asymmetry of motion stresses the human foot, which provides support and flexibility during weight transfer while standing or walking.

Posture ability is a neurological control activity dependence by nervous system that provides the optimal spatial framework for body adjustment. The input data for posture adjustment are received by the brain through various peripheral channels, processed and send to different systems for deployment. High plantar pressure generates posture abnormalities. Thus, an inadequate distribution on the plantar region occurs due to various factors such as eyesight deficiency, hearing deficiency, musculoskeletal system pathologies and mental states [1, 2].

Human body postural deviation during normal activities such as gait or orthostatic position can cause malfunctions or injuries not only to the foot but also in different areas of the body. Some studies highlight the fact that posture and equilibrium problems can be associated with rheumatology and orthopedic diseases such as back pain, neck tension and ankle instability [3]. Postural deviation is a state of imbalance that leads to functional overload, degeneration and dysfunction [1] and can be studied using photography, magnetic resonance imaging and x-ray [4-7].

The anatomical structure which is basic for support and propulsion for any bipedal activity is the foot [8].

Due to its complex structure, the foot is responsible for maintain body balance in statics and dynamics, by symmetrical distribution of plantar pressure. Structurally, the human foot is divided in forefoot (FF), mid-foot, and the rear-foot (RF). Pressure distribution analysis for this anatomical structure implies sophisticate techniques such feet pressure distribution

* Corresponding co-author: valentin.amortila@ugal.ro

and stability metering tests. Using these analyzing techniques, we can assess the malfunctions of the postural system by mapping the plantar pressure distribution.

The aim of this paper is to present an experimental study regarding the distribution of contact pressures during gait and orthostatic position, aiming to evaluate the load distribution between FF and RF, the projection of center of gravity and the support area variation while performing feet load distribution and stability metering tests.

2 Method

In this study, we conducted an analysis of foot pressure variation during standing and orthostatic position while performing feet load distribution and stability metering tests. All tests were performed in laboratory conditions without the influence of any external factors.

The feet pressure distribution test implies a static and a dynamic analysis. During the static test, the participant is asked to stand straight on the static area of the platform touching the specific alignment tool with the heel. The participant must be in a natural comfortable position. The software needs 5 seconds to determine the mean values of small oscillation. Using this method, the body weight distribution on the plantar surfaces and the geometry of both legs while standing straight are determined.

Dynamic analysis was used to determine the ground reaction evolution while walking. The participant performed a cycle of four completed steps equivalent with 12 footprints.

In order to measure the stability of a participant that stands a stability metering test was conducted. The test lasted 5 seconds while the participant stood with eyes closed.

For this experimental tests, we have used Diasu technologies. More exactly, the tests were performed by using a Podoscanalyzer and an UltraSensor LAC platform both combined with Milletrix Software capabilities.

3 Results

3.1 Feet pressure distribution analysis

3.1.1 Static analysis

The static test results indicates that the contact plantar surface is 182 cm^2 (Fig. 1) with an average pressure of 450.5 g/cm^2. It can be noticed that the contact pressure is uneven distributed on both legs. Static analysis test revealed that the body center of pressure (CoP), corresponding with the body's gravity center (letter C in Fig. 2), is projected in the left of the support polygon. According to normal values, this deviation can indicate a moderate mono lateral deviation, with a hyper load. Thus, the left foot contact surface is 99.5 cm^2 carrying 44.2 kg load, about 54% of body mass. The difference of 37.7 kg is distributed on the right foot on a smaller surface of 82.5 cm^2.

In addition, it can be noticed that the projection of CoP is located towards back, indicating a more RF support, 64.20% for the left RF and 82.70% for the right RF. This was caused by the fact that the axis of the legs are not placed at the same angle. The right leg axis forms with the sagittal plane an angle of 26.2^0, while the left leg an angle of 13.9^0 (Fig. 3). Due to difference in positioning of both axis it can be seen that the center of gravity of the left foot (the letter L) is placed toward the front of the foot to the right (point R). In addition, this variation results to the fact that center of each foot (CoF, letters L = left and R=right) reveals a misalignment with respect to CoP.

The maximum points of pressure (red colored) were detected on both rear feet, more on the left. These aspect have pointed out a mono lateral load. The maximum pressure of 1141 g/cm^2 (letter M) was detected in the right RF. The load distribution on each leg highlights an adequate bodyweight distribution between left and right legs, according to normal values (Fig. 4, 5). The load distribution between FF and RF for the left and right legs was comprised within the physiological values. The plantar surface was uneven distributed. Speaking of FF of both legs were highlighted an excessive surface's difference, more on the left. Between the RF a moderate difference of surface, more on the left was highlighted.

During static analysis, the foot geometry was divided in 6 areas: the FF (medial and lateral, including toes, if detected), the RF (medial and lateral) and the mid-foot (external/internal). The pressure distribution in this analysis, during orthostatic position was also uneven. It can be noticed that the RF of the human legs is more stressed as compared to FF. The mid-foot areas of both legs have almost no contact with the ground especially in the medial area.

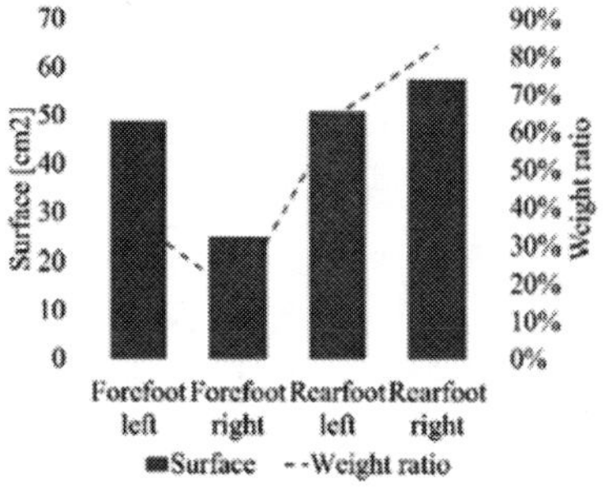

Fig. 1. Surface - weight ratio comparison for left foot and right foot division areas.

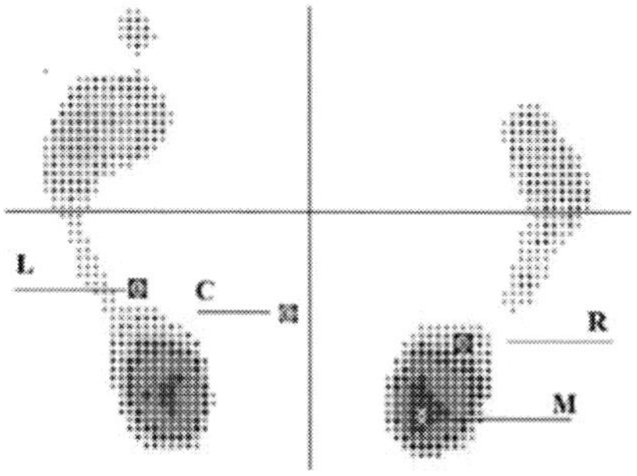

Fig. 2. Static analysis comparison between FF and RF divisions for left and right foot.

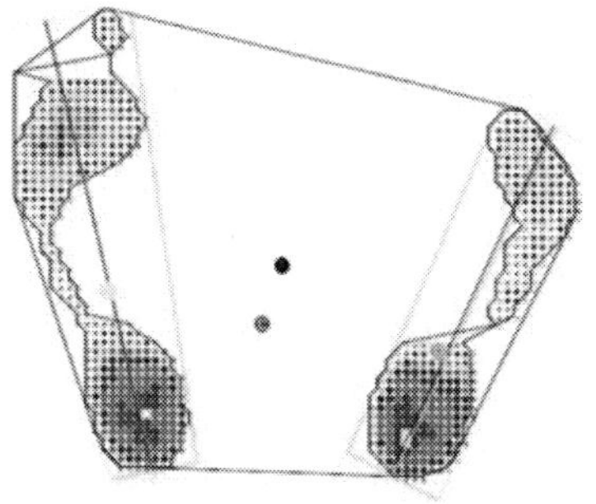

Fig. 3. The support polygon.

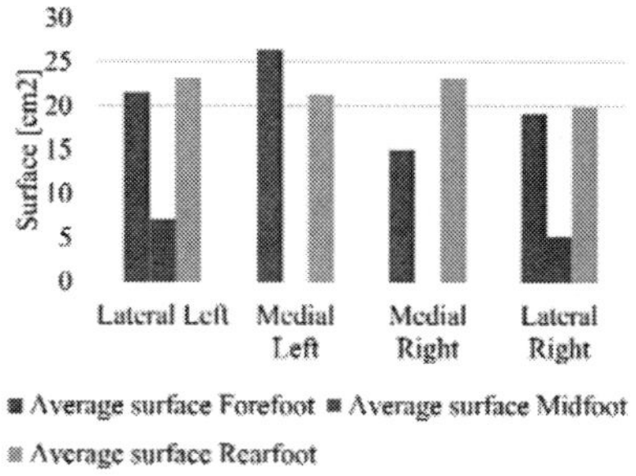

Fig. 4. Plantar average surfaces for medial and lateral areas.

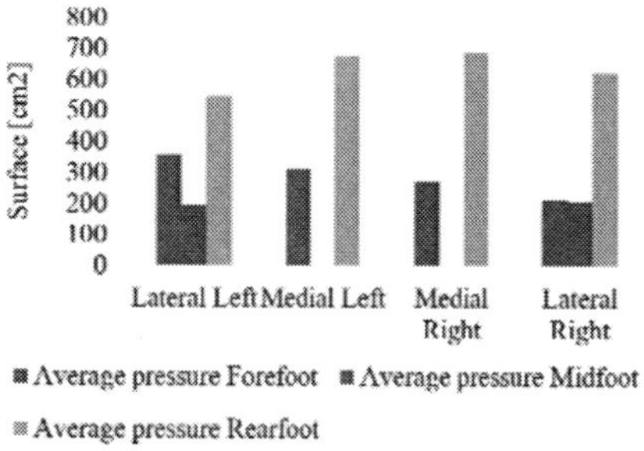

Fig. 5. Analysis of average static pressure.

3.1.2 Dynamic analysis

During this test, the participant performed four completed steps equivalent with 12 footprints (Fig. 6).

The aim of this test was to analyze the ground reaction evolution within plantar region, during bipedal gait. The test was performed in laboratory conditions without the influence of any external factors, such as obstacles. Even so, the participant expressed a rather unstable gait generating an unstable posture.

By analyzing each image and dividing the footprint into FF, mid-foot and RF it can be noticed that the maximum point of pressure and the ground reaction varies from footprint to footprint. This is due to the fact that the geometry of each footprint varies.

During gait or standing, the plantar pressure should be distributed rather even and the maximum point of pressure should be applied within the RF surface.

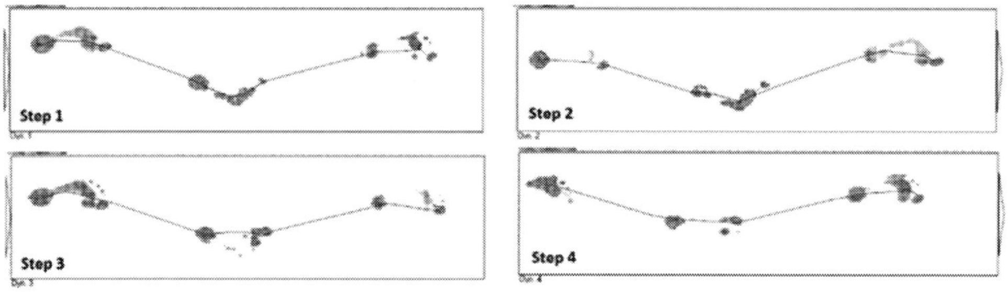

Fig. 6. Ground reaction evolution within plantar region, during bipedal gait.

By analyzing the data from the feet pressure distribution analysis, the data reveals an imbalance in the prevalence of weight load.

It can be noticed that during this test, even if the participant executed a controlled gait, the contact pressures and surfaces for both legs were variable (Fig. 7). By analyzing the gait cycle, the data reveals that both legs are stressed uneven.

By analyzing figure 7 we have noticed that during the dynamic test the maximum point of pressure for three of the footprints (1, 2, and 9) was smaller due to a more adequate plantar surface. Thus, a more efficient body mass distribution is emphasized. In this case, the plantar surface is ranging from 71.25 cm^2 to 113.25 cm^2. The body weight distribution generated a maximum point of pressure that varies from 1335.7 g/cm^2 to 1472 g/cm^2.

The average surface difference between FF and RF was 32% in the case of 18% loading.

In contrast are footprints number 4 and 12. The load is similar or greater that in the previous case and due to a very small plantar surface, 18 cm^2, the maximum point of pressure exceeds 5000 g/cm^2.

By dividing the footprint in two areas, the RF and the FF, and by analyzing the mass distribution we can observe that not all surfaces are stressed in the same manner (Fig. 8). Normally, the body weight must be more concentrated in the heel area, known as the RF. This means that the heel joint is more in extension phase rather than flexion when approaching to ground. This pattern is notable for footprints number 2, 4, 5, 10 and 11. A particular case is footprint 6. During this step an increased imbalance is noticeable. The maximum point of pressure is located within phalanges regions. More precisely, the foot approached the ground with a high heel angle of flexion rather than extension angle.

By analyzing figure 9, we have noticed that the left foot mean plantar surface, on average was twice as high as the right foot. For the left foot, it can be noticed that the full contact surface is completed in the middle of the step while for the right foot towards the end.

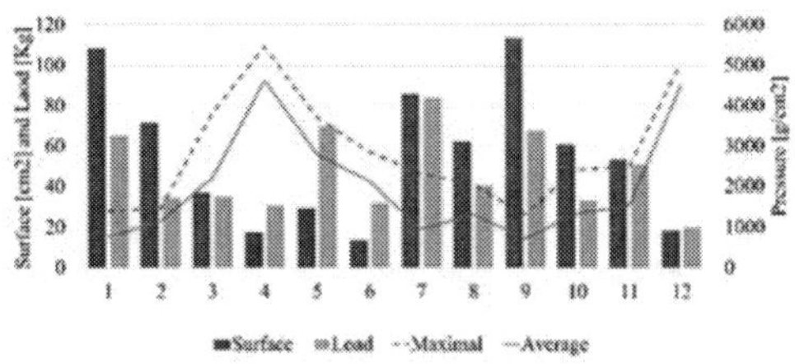

Fig. 7. Comparison between some dynamic analysis parameters for each of the 12 footprints.

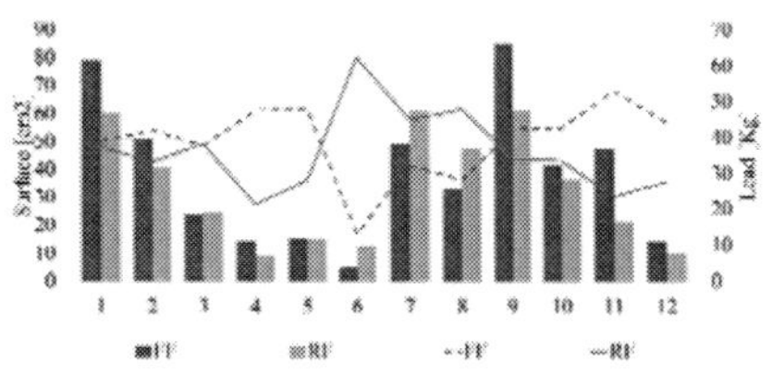

Fig. 8. A parallel between surface and load evolution while dividing each 12 footprints into FF and RF.

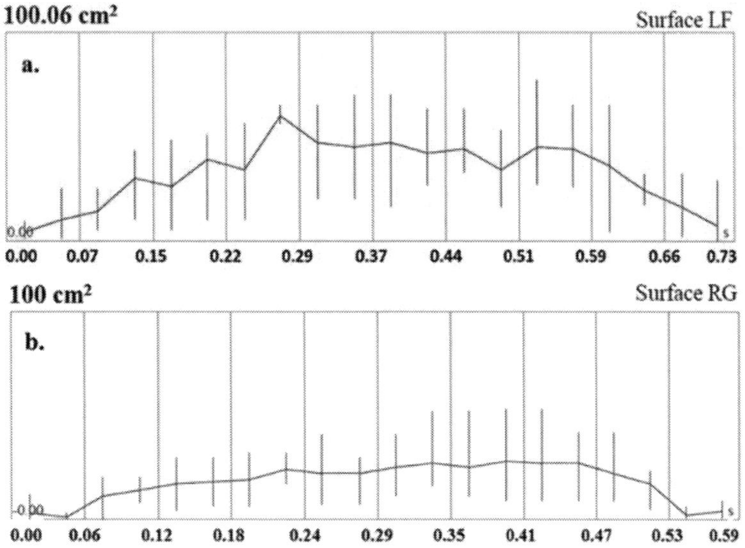

Fig. 9. Mean plantar surface; a. left foot; b. right foot.

3.2 Stability metering

The stability metering test objective was to study the body involuntary swing during standing. The test was performed in the absence of any external perturbations and any voluntary movements. Thus, the postural control system function such as steady – state can be analyzed by collecting information about the mean position of center of pressure (CoP). Normally CoP displacement is within 1 cm^2. The standing still position is inherently unstable due to small perturbation and fluctuations. By analyzing the data (Fig. 10), we have noticed that the participant posture has a small deviation towards left. This is noticeable because the support area is not equal divided on both legs. Thus the maximum point of pressure (red area) is located on the left RF. Due to a deviated towards left we noticed that the projection of CoP is shifted to left. The confidence ellipse (Fig. 11) with an area of 0.78 cm^2 (major half axis of 0.85 cm and minor half axis of 0.29 cm) and an eccentricity of 0.93 contains the projection of CoP during sway, described as the area that covers the center of a pattern with a given probability of 90%.The ellipse major axis indicates the main direction of sway, -31.1^0.

The peaks of sway are represented in figure 12. By analyzing the radar balance graph we can easily evaluate the sway most important direction. The graph shows that body's involuntary swing direction while standing is oriented mainly towards right side of the body.

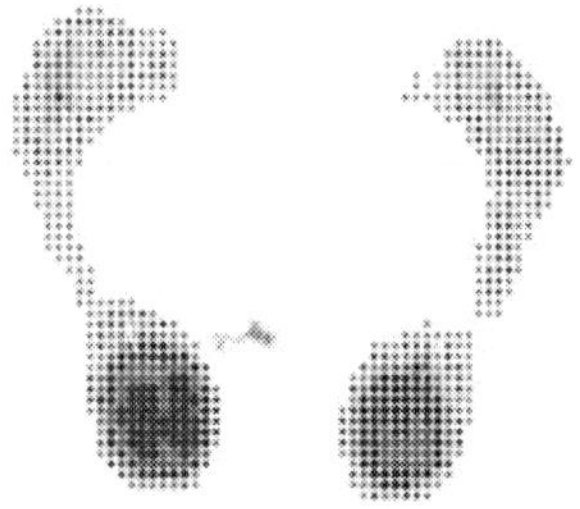

Fig. 10. The evolution of plantar stress while stability metering.

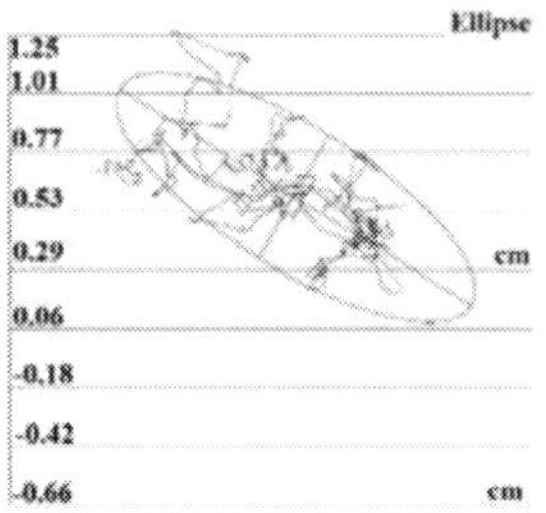

Fig. 11. The confidence ellipse.

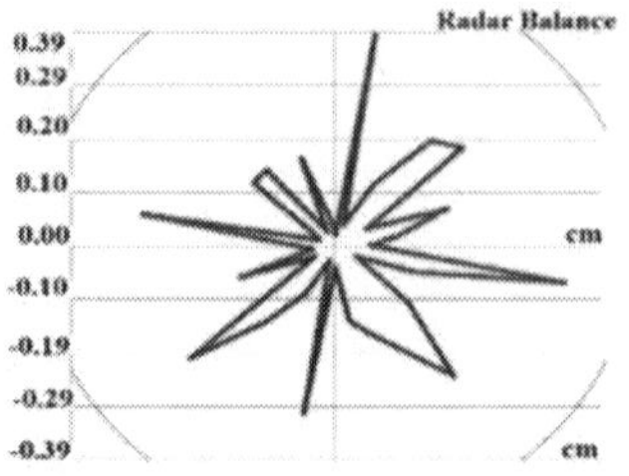

Fig. 12. Radar balance.

4 Conclusions

The foot is one of the most complex structure of the human body. By combining 26 bones connected with 33 joints and more than 100 muscle and tendons the human foot is a complex mechanical structure characterized by structural strength. Due to its complex geometry the human foot role is to absorb socks, to support the body weight offering balance and mobility.

Body posture is a neurological dependent ability. Also for an adequate posture, the human body needs a free of abnormalities mechanical structure. Thus, abnormalities that occurs in other parts of the body can influence the foot dynamics.

High plantar pressure may occur do to posture abnormalities. By using sophisticate technologies, we can evaluate the geometry and the dynamic of the plantar surfaces. By conducting stability metering and feet pressure distribution analysis, we can evaluate the area that usually receive the maximum pressure. Anatomical speaking, these areas are the heel and metatarsal head.

By analyzing pressure and plantar surfaces, we can conclude that the participant has no abnormalities like flat foot or high arch and hyperextension of the toes. The high arch and hyperextension of the toes can be indicated by high pressure associated with a decreased support within the midfoot area. On the other hand, flat foot represents low stress with an increased plantar surface. In our study, the data indicates that the participant does not present any of these abnormalities. However, tests like feet pressure distribution and stability metering have an excellent potential to provide top research in the field of human body posture.

References

1. M. Bellizzi, R. Rizzo, G. Bellizzi, M. Ranieri, M. Fanelli, G. Megna, U. Procoli, Strabismus.**19**, 1,(2011)
2. B. Bricot, *Total Postural Reprogramming* (Dux Lucis Books, Santa Monica, 2008)
3. B. Missaoui, P. Portero, S. Bendaya, O. Hanktie, P. Thoumie, Clin. Neurophysiol.**38**, 6 (2008)
4. J.L.P. Rosário, I.Y.Nakashima, K. Rizopoulos, D. Kostopoulos, A.P. Marques, Clinical Chiropractic.**15**, 3-4 (2012)
5. J.S. Steffen, I. Obeid, N. Aurouer, O. Hauger, J.M. Vital, J. Dubousset, W. Skalli, European Spine Journal.**19**, 5 (2010)
6. E. Berthonnaud, J. Dimnet, R. Hilmi, Comput. Med. Imaging Graphics.**33**, 8 (2009)
7. H. Suzuki, K. Endo, J. Mizuochi, H. Kobayashi, H. Tanaka, K. Yamamoto, Eur. Spine J. **19** (2010)
8. J.L. Rosário, J Bodyw, Mov Ther.**18**, 2 (2014)

MATEC Web of Conferences **112**, 08010 (2017) DOI: 10.1051/matecconf/201711208010
IManE&E 2017

Mechanical design of an electronic control unit using axiomatic principles

Vlad Cazacu*, *Cristina* Savu, *Claudiu* Gurita, and *Gheorghe* Nagit

"Gheorghe Asachi" Technical University of Iaşi, Romania

Abstract. If the engine of the car can be considered as the heart, then the E.C.U's represents the brain of the car. Electronic control units (E.C.U's) are electronic devices which control the way different components of a car (engine, windows, airbags, etc.) react in some situations (overheating, button pressed by a passenger, crash, etc.). Axiomatic design is a set of principles that theorizes the act of conceiving a new project. Based on two axiom this method comes into designers help, giving them the option to reach in a short period of time a fully functional and compliant product without supporting the design of the product on chance, past experiences or "try and fail" principle.

1 Introduction

In the past years, with the overwhelming products that came to market a lot of enterprises put more and more money in research and development departments so the new products that will came to live will be different and attractive for the clients.

The main problems when designing a new product are the cost and the time needed for developing. Both these problems can be solved by using axiomatic design which creates a complete product lifecycle and simplifies the stages through a product has to pass from conception until recycle.

Considering that in automotive industry electronic component became used more and more, the necessity to apply a successfully design method is obvious.

In this paper we will try to apply the axiomatic design for a mechanical design of an ECU which will control the sunroof of a car and the windows, so basically the main input that this ECU will have is provided by the passengers.

2 Axiomatic design applied to mechanical design of an ECU

Axiomatic design is considered a pretty new concept, even if there have been several years since professor Nam Pyo Suh presented it. As its founder states, axiomatic design helps in taking "correct decisions", "shorten lead time " ; "improves the quality of products " , "deal with complex systems", "simplify service and maintenance", "exchanges creativity". The ease of use is given by the fact that axiomatic design is based on matrix relationships, so every information given or assumed is connected in an easy to access and review way. Even

* Corresponding author: v.cazacuconstantin@yahoo.com

more, for simplifying the means of conceiving, axiomatic design theorizes the act of conceiving by two axioms: The independence axiom and the information axiom. [1-3]
First axiom imposes the maintenance of the functional requirements independent.
Considering the functional requirements (FR) of a product, design parameters (DP) have to be defined so the product gets from the conception state to physical domain. Every functional requirement detected or imposed can be connected with one or more design parameters. Depending the way functional parameters are connected to design parameters, we can create connection matrixes (between FR and DP) which can be diagonal, triangular (superior or inferior) (see figure 1) or any other kind of matrix.

$$[A] = \begin{matrix} A1 & 0 & 0 \\ 0 & A2 & 0 \\ 0 & 0 & A3 \end{matrix}$$

Diagonal matrix – uncoupled design

$$[A] = \begin{matrix} A11 & 0 & 0 \\ A21 & A22 & 0 \\ A31 & A32 & A33 \end{matrix}$$

Triangular matrix – decoupled design

Fig. 1. Axiomatic design resulting matrix types- A1, A2, A3, A11...A33 elements of the resulting matrix.

Each matrix type defines a type of design as it follows: the diagonal matrix defines an ideal design named uncoupled design, triangular matrixes define a decoupled design and for any other types of matrix the resulting design is considered to be a coupled design [4].
In order to create a proper design, it is needed that the resulting design to correspond to first axiom and also it is needed that the matrix A to be at least a triangular matrix if not a diagonal matrix.
For creating a robust design, it is needed that each functional requirement starting with the intention of design which is considered the zero-hierarchy level to be decomposed into lower and lower functional requirements by respecting the hierarchy and also a zig-zag principle presented in the figure below. [5- 7]

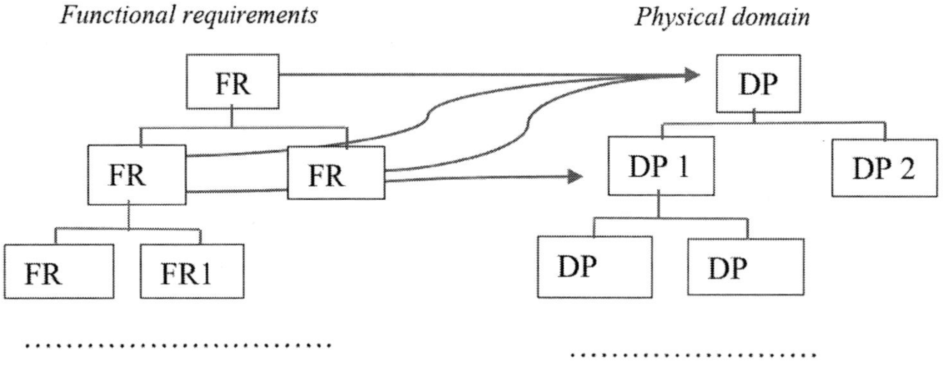

Fig. 2. Functional requirements and conception parameters dependence decomposition.

After the decomposition took place and several designs of the product have been established applying the second axiom will lead to the final design that will be chosen.
Second axiom is used for determining the optimal solution from the multiple solution that have been determined after applying the first axiom.
It defines the quantity of the information (I) of a functional requirement as the probability (P) of the functional requirement to be satisfied [6]

 DOI: 10.1051/matecconf/201711208010

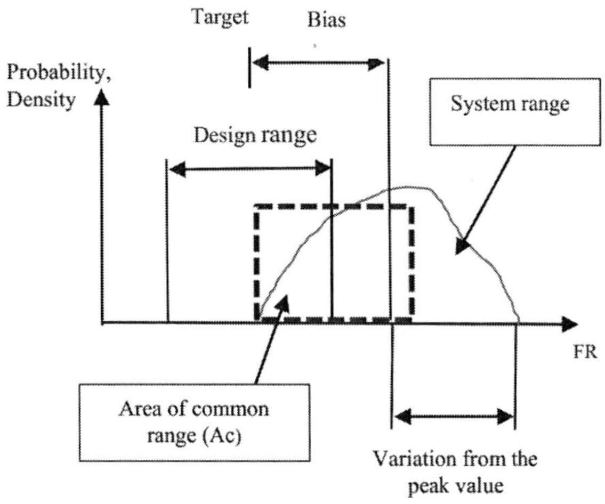

Fig. 3. System range and common range representation of functional requirements in relation with process probability.

$$I = log_2 \frac{1}{P} = -log_2 P \qquad (1) [4]$$

For *k* functional requirements for an uncoupled design *I* can be expressed as:

$$I = \sum_{k=1}^{i} log_2 \frac{1}{P_k} = - \sum_{k=1}^{i} log_2 P_k \qquad (2) [4]$$

For determining the probability of success it is necessary to determine the System range and the Common range.

System range can be determined as the area under the system range (A_{sr}) graph and the common range is represented by the area under the common range area (A_{cr}) (Fig. 3).
The probability is equal to:

$$I = log_2 \frac{A_{sr}}{A_{cr}} \qquad 4]$$

3 Practical use

Practical use will be explained using a mechanical ECU design.
Following the theory given above firstly we have to think the functionality of the product and find the functional requirements. An important contribution in establishing the functional requirements come from the client which states some of the needs that the product will have to accomplish during its production, mounting and functioning.
Client specification:

- Assembly of the parts has to be done with a minimum number of tools [8];
- Maximum dimensions are : L=140 mm ; l=100 mm ; h =45 mm;
- Allow the connection with the main computer by two connectors;
- Number of components involved in the assembly have to be less than 10 (the printed circuit board (PCB and its components are considered as 1 item);
- Total weight of the assembly has to be less than 200 g;
- Functionality between -40 °C and 110°C;

- Mechanical protection IP5KX/IPX1 [9].

Analyzing the client specifications, besides the requirements specified by the client we can establish several more functional requirements which we will centralized in the table below:

- Considering the low weight that the assembly need to have, we will use as material for the housing and cover a plastic material;
- Designing a demountable assembly;
- Material used : plastic, aluminum;
- Assembling features: screws, self-fastening plastic features (snap fit);
- Sealing for mechanical protection: realized by the shape of the assembled parts, introducing supplementary sealing feature (gasket);
- Organizing the PCB inside of the assembly: one or two PCB's;
- Designing a two-piece assembly (cover and housing).

Table 1. Establishing the hierarchy of the functional requirements.

ECU									
Maximum dimensions				Number of components		Working environment			
Length	Width	Height	Weight	Connectors	Assembled	Mechanical protection		Temperature range [°C]	
140 mm	100 mm	40 mm	200 g	2	As few as possible	IP5KX	IPX1	-40	- 100

Housing	Cover	PCB	Assembling features	Dust protection	Vertical water drips

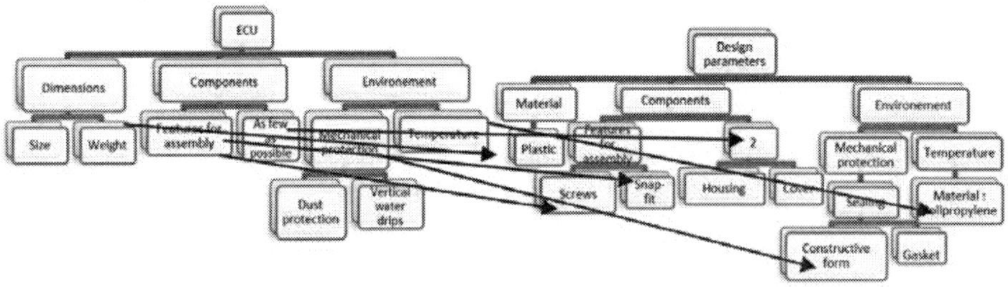

Fig. 4. Functional requirements and design parameters dependences.

After establishing the dependencies between functional requirements and design parameters in the diagram above we created the matrix A (presented in Table 2) bellow which presents the ideal, uncoupled design of the assembly.

As we can see there are many dependencies between single elements related to the ECU and Design Parameters as well as between elements of both. The aim is to obtain a functional constructive form taking into account as many as possible elements and variables which may lead to the desired result.

Table 2. The ideal, uncoupled design of the assembly.

	Use of plastic	Snap-fit assembly	Design with 2 components	Sealing: by constructive form	Use of polypropylene
Weight	X	O	O	O	O
Features for assembly	O	X	O	O	O
Use as few possible features	O	O	X	O	O
Mechanical protection	O	O	O	X	O
Temperature range	O	O	O	O	X

From the diagram above (see figure 4) we can observe it results a coupled design because weight and temperature have the same design parameter and also, number of features of assembly is satisfied by two parameter designs: snap-fit assembly and designing only with two components: housing and cover.
From all this we can observe now that our design at this state doesn't respect the first axiom. By reviewing and decomposing the functional requirements we can split the number of components that the client asked for our design into using as few features as possible (thus we chose to have basically two components created by us: housing and cover) and using the snap-fit design for realizing the assembly (this way we remove the need for screws, which will make our overall cost higher).

Also, clearly specifying the material as Polypropylene (PP) we have decomposed the material section and now the resulting matrix drives us to a decoupled design. As we can see from the illustration presented in figure 5 just a few elements from the original setup were taken into account, such as: the snap fit design, the cover, the PCB, the sealing specific to the constructive form, the connectors as well as the housing of the entire assembly itself.

This certifies that this design suits most of the client's requirements.

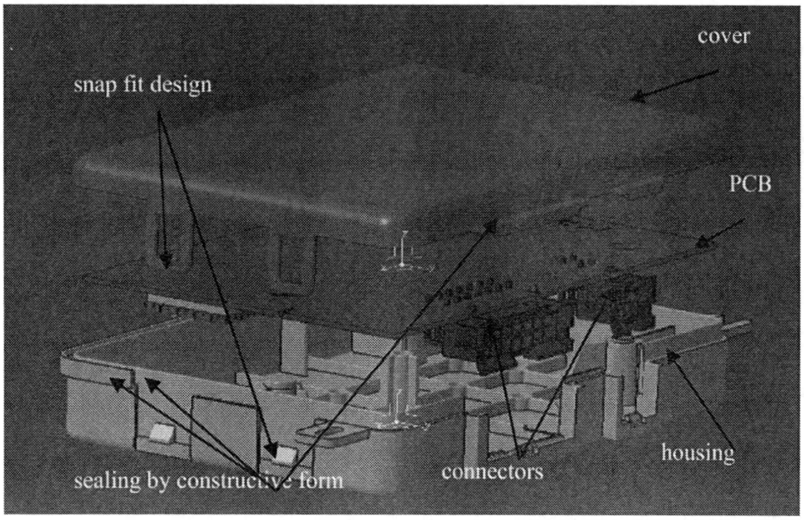

Fig. 5. Final design of the mechanical features of an ECU.

4 Conclusion

Even if we didn't manage to reach the second axiom, we realized a fully functional design (see figure 5) by respecting the first axiom, so our goal of applying the axiomatic design for the mechanical features of an ECU was met. As we hoped we manage to create a compliant design of the ECU which respect all the client requirements and also the requirements imposed by the actual chosen design that derived from the main demands such as the use of plastic materials that involved using the molding injection technology and thus creating draft angles for all the surfaces of the assembly.

Even if we had initially a coupled design, applying decomposition method we were able to reduce the dependences between the functional requirements and the design parameters and we got to the final design which fully respect the first axiom.Described as a facile way of getting to the final goal design, axiomatic design can create difficulties if the designer does not understand the fully functional needs of the product or if he does not initially establish all the possible solutions for the design problem this way, not having enough design parameters to create the design matrix.

In this paper, we demonstrated that axiomatic design is a method of conceiving a product which can drive to a good result by respecting the customer needs, defining the solutions for the requirements, create a solution, analyze it and compare it with the client requirements. Many times the client based specifications are more or less suitable for a proper design in terms of its mechanical properties, cost-performance ratio or even its suitability in the desired market. Nevertheless as it is the case with many other great designs no element or variable should be left aside because it could mean the difference between a great product or just another fail design.

The authors feel that the proposed designs meets all functional parameters as it is already in the production cycles.

References

1. C. A. Brown, 1st CIRP Conference on Surface Integrity (CSI), Procedia Engineering 19 53 – 59 (2011);
2. Gollapudi Arun Manohar, International Journal of Research in Engineering and Technology, 430-436 (Apr-2015);
3. Max Fuchs, Michael Eckrich, Olaf Müller, Jan Philipps, Peter Scholz, Advanced designing and validation techniques for electronic control units, 1998 ;
4. Ion Grozav, Quality improvement using axiomatic design, Buletinul AGIR nr. 1-2, (2008);
5. Dym, C. L., Agogino, A. M., Eris, O., Frey, D. D. and Leifer, L. J., Teaching and Learning. Journal of Engineering Education, 103–120 (2005);
6. Q. Cheng, C. Xiao, L. Cai, G. Zhang and P. Gu, Proceedings of the 2012 IEEE 16th International Conference on Computer Supported Cooperative Work in Design (CSCWD), Wuhan, pp. 282-288 (2012);
7. M. Dastagiri and M. M. Prasad, Frontiers in Automobile and Mechanical Engineering, Chennai, pp. 222-228 (2010);
8. System Design and Analysis based on AD and Complexity Theories, https://ocw.mit.edu/courses/mechanical-engineering/2-882-system-design-and-analysis-based-on-ad-and-complexity-theories-spring-2005/lecture-notes/lec202.pdf, retrieved 7.02.2017 ;
9. ISO 20653 :2013, Road vehicles - Degrees of protection (IP code) - Protection of electrical equipment against foreign objects, water and access

 DOI: 10.1051/matecconf/201711208011

Effect of the pitch angle and of the number of blades on depression created under impellers

Corina Băbuţanu[1], and *Lucian* Mândrea[2,*]

[1]National Research Institute for Electrical Engineering ICPE-CA, Efficiency in Energy Conversion and Consumption Department, 313 Splaiul Unirii, Bucharest, Romania
[2]University Politehnica of Bucharest, Department of Hydraulics, Hydraulic Machinery and Environmental Engineering, 313 Splaiul Independentei, Bucharest, Romania

Abstract. This work presents a set of experiments concerning the behavior of several types of impellers regarding the depression formatted and operating in a specific range of rotations. Various mixer designs including a four pitched blades impeller, a six flat blades impeller and four two flat blades impeller were compared. Experimental data confirmed the analytical results previously obtained. The results showed that there are similarities regarding the effects generated by four pitched blades impeller and one of the two flat blades impeller. Also, it is found that two short-blade impeller produces the lowest depression of all tested impeller. The maximum depression produced by all type of impellers do not exceeds 120 mmH_2O at 400 rpm, which is a small value to realize the vaporization of the fluid and to reach the cavitation. The equipment is not in danger to be destroyed. The numerical results obtained by experiments can be used in various practical applications.

1 Introduction

From extractive and food industry and up to newest industries, like aerospace industry or pharmaceutical, impellers, stirrers or mixers are widely used for homogenization multiphase mixtures, acceleration of chemical reactions or movement of substances through devices and equipment. Beside issues related to energy and/or environmental impellers use raises also issued related to depression appeared behind the moving blades. Depending on the physical and chemical characteristics of the substances involved in industrial processes, depression which occurs at impellers operation and its dimension are factors with negative potential effects, like slowing of movement, vortices that tend to lower the degree of homogeneity, etc. Given the practical importance of the issue, there are many studies which approach both theoretical and experimental study of the phenomenon, in order to remove possible undesirable effects [1-4]. Analytical approaches are quite rapidly but not very accurately, while numerical solutions can be more precise but are more complicated and time consuming.

The authors of this work previously obtained an original analytical solution used to calculate the depression which appears during operation of the impellers. It was obtained

*Corresponding author: mandrea_lucian@hotmail.com

the radius variation in the depression area, the pressure inside the depression zone and the depression height [5]. The problem was also studied in [6], the author obtaining a formula that calculates the height of the depression area.

The maximum radius of the depression zone at the contact level between the impeller and the water is [5, 7, 8]:

$$R = \sqrt{\frac{2}{\rho\varepsilon^2 t^2}(p_e - p_v)} \tag{1}$$

In the space under the impeller in rotation, the pressure inside the depression zone becomes smaller than the exterior pressure (the light blue area). The exterior pressure of the water is the p_e in the blue area. ε is the tangential acceleration and ρ the water density. t is the time variable.

Using variable radius r, one can calculate the pressure inside the depression zone:

$$p = p_e - \frac{\rho\varepsilon^2 t^2}{2} r^2 \tag{2}$$

The height of the rotation paraboloid from figure 1 is:

$$H = \frac{\varepsilon^2 t^2}{2g} R^2 \tag{3}$$

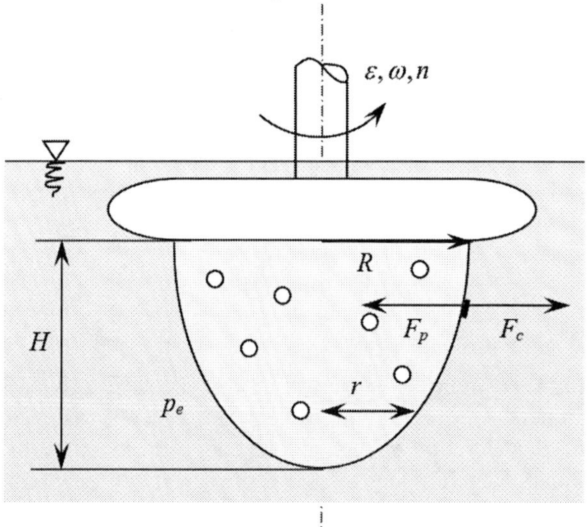

Fig. 1. Characteristics of the depression zone created under impeller.

2 Materials and methods

In order to obtain supplementary data concerning depression caused by impeller operation, a new set of experimental measurement was made. The aim of these measurements was to determine the influence of the type of impeller on the depression size. Experiments were conducted in an experimental setup that is shown in Figure 2 a. It consists of a transparent cylindrical vessel with the ratio D/H of about 1.5 and a capacity of 0.126 m^3. Equipment is endowed with a variable speed motor on whose shaft can be fitted impellers of different types and sizes. Also, the setup is equipped with electronic speed control, and a panel which displays rotational speed and momentum produced by the studied impellers. To measure the depression caused by the impellers, a pressure plug is mounted at a distance of 120 mm from the vessel bottom.

The seven types of impellers are shown in Figure 2 b.

The experiments were made utilizing tap water under normal pressure and temperature.

a.

b.

Fig. 2. The first experimental setup (a) and the tested impellers (b).

Seven types of impellers (four impellers with two flat blades, an impeller with four flat blades, an impeller with four pitched blades, and an impeller with six flat blades) were tested in a speed range between 100 and 500 rpm. All the impellers had a hub diameter of 20 mm and blades size as follows:

- two thin flat blades impeller - 100 mm length and 10 mm width;
- two flat blades impeller - 100 mm length and 20 mm width;
- two wide flat blades impeller - 100 mm length and 40 mm width;
- two short flat blades impeller - 60 mm length and 20 mm width;
- four flat blades impeller - 100 mm length and 20 mm width;
- four pitched blades (45° pitch angle) impeller -100 mm length and 20 mm width;
- six flat blades impeller - 100 mm length and 20 mm width.

The depression ΔH (expressed in mmH_2O) was measured in five different positions of impellers who were located vertically at 10, 20, 30, 40 respectively 50 mm from the pressure plug. Its evolution for each type of impeller is shown in figures 3-7.

3 Results and discussion

Analyzing the diagrams obtained from testing the seven propellers some observations regarding their behavior can be made. Thus, it appears that for the impellers with blades having the same dimensions, as the number of blades increases, the greater is the depression (two flat blades impeller, four flat blades impeller and two flat blades impeller). It also notes that, for the impeller with the same diameter the depression increases with an increased width of impeller (two thin flat blades impeller, two flat blades impeller and two wide flat blades impeller). Similarly, for the impeller of the same width, diameter increase lead to increased depression (two flat blades impeller and two flat short blades impeller).

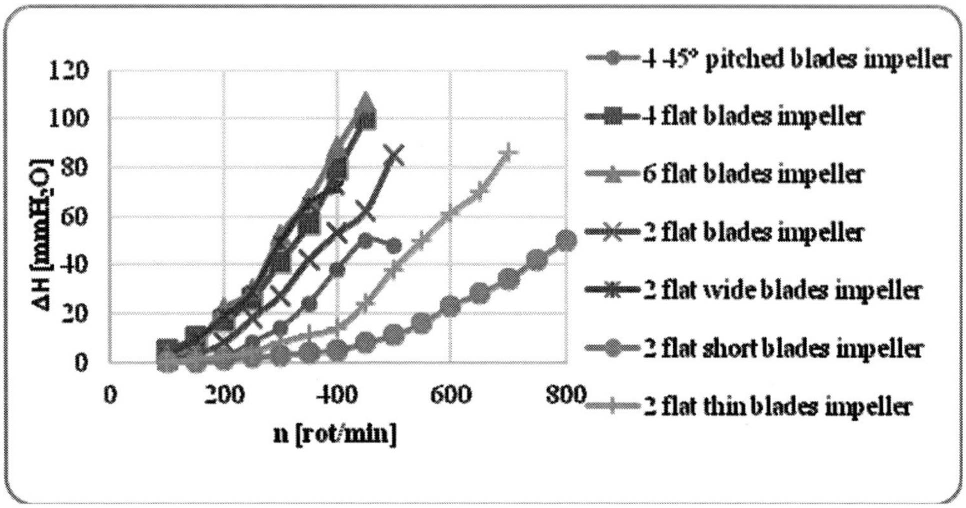

Fig. 3. Variation of depression produced at 10 mm from bottom impeller.

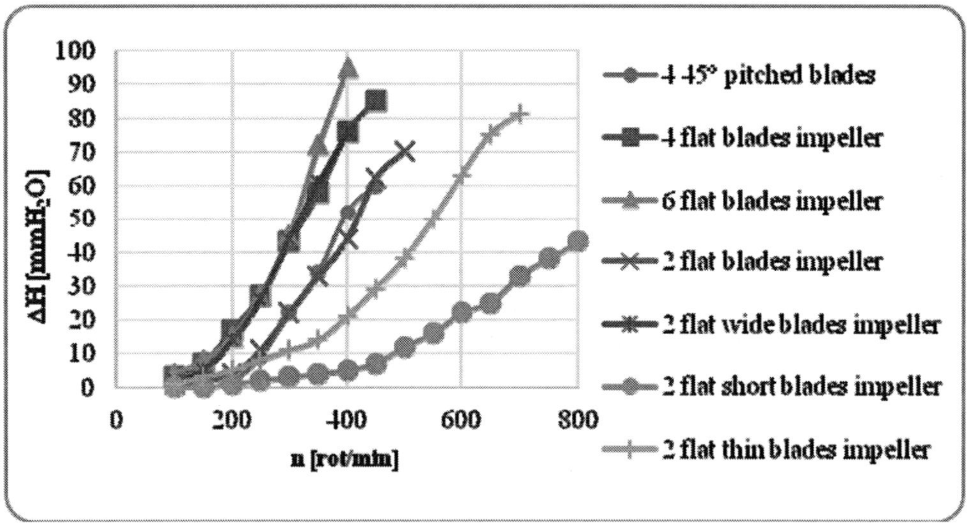

Fig. 4. Variation of depression produced at 20 mm from bottom impeller.

DOI: 10.1051/matecconf/201711208011

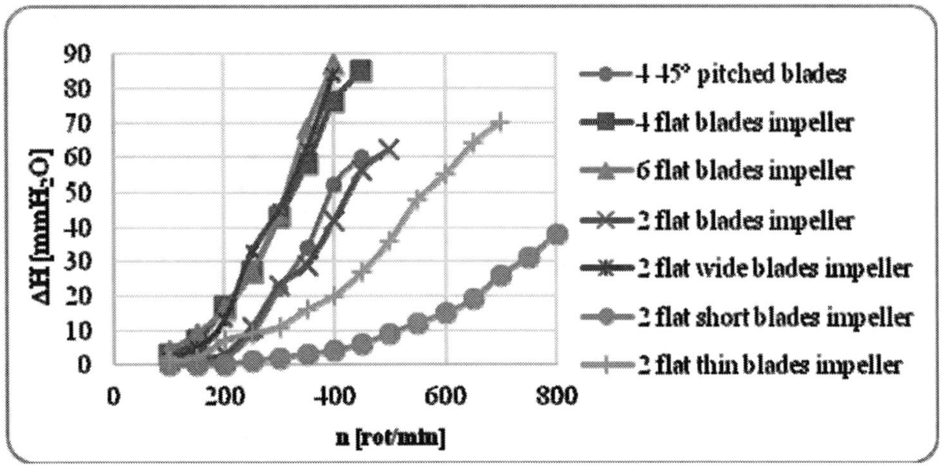

Fig. 5. Variation of depression produced at 30 mm from bottom impeller.

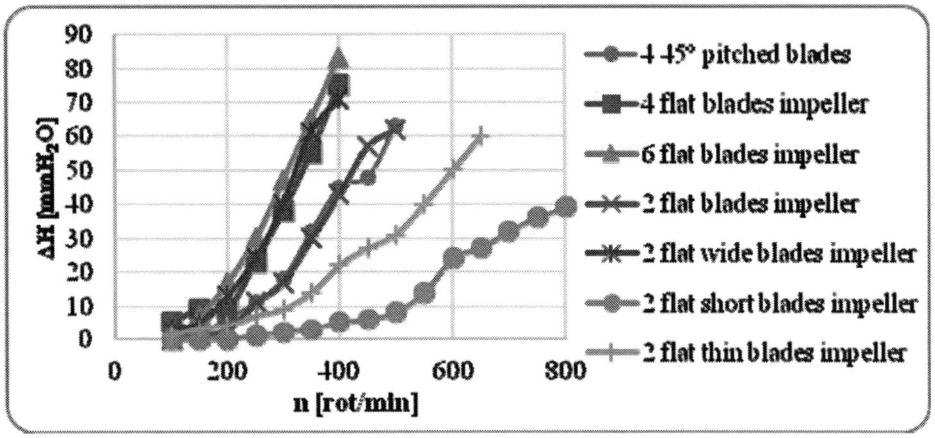

Fig. 6. Variation of depression produced at 40 mm from bottom impeller.

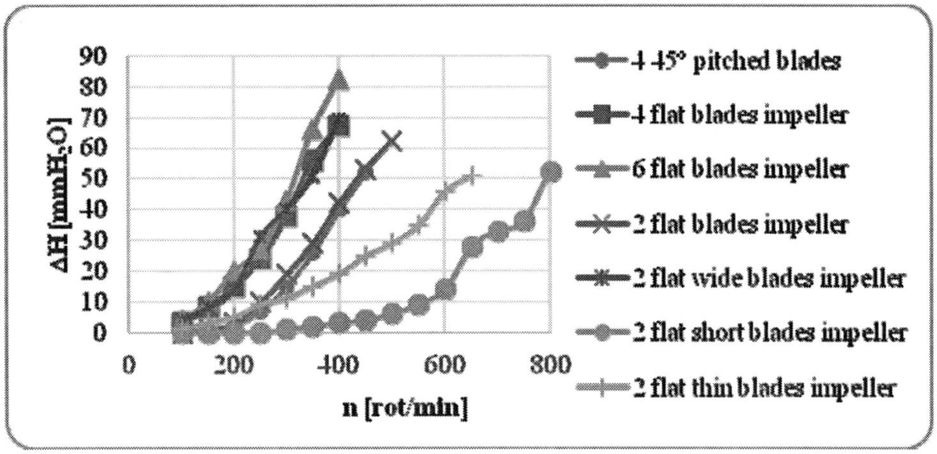

Fig. 7. Variation of depression produced at 50 mm from bottom impeller.

In all of the five measuring points it was found that the depression caused by the two flat blades impeller is quasi identical with the one caused by the four 45° pitch blades impeller.

Also in the range between 100 and 400 rpm is observed that the depression created by the six flat blades impeller, four flat blades impeller and four 45° pitch blades impeller is almost similar. We also note that short-bladed impeller produces a depression much lower than all the other types.

Regarding the distance between the impeller and the measuring point, there is a decrease of depression with increasing distance, which confirms theoretical approach (see Figure 1).

4 Conclusions

The present work shows that the impeller type plays an important role on the depression size it can create. Although many of the conclusions were expected results, these experiments have quantified the depression created by the operation of these propellers. Depending on the reaction mode on different operating conditions (pressure, viscosity, temperature, etc.) tested equipment can be used in various practical applications.

From the figure 3 to figure 7 it is very clear that the depression expressed by ΔH increases as the rotating speed increases, thing in correspondence with the theoretical part (formula number 2).

The maximum depression produced by all type of impellers do not exceeds 120 mmH_2O at 400 rpm in our case, which is in fact a small value to realize the vaporization of the fluid and to reach the cavitation. The equipment is not in danger to be destroyed.

The research was conducted within NUCLEUS Program, contract no. 14N/2016, *Innovative solutions regarding the design and implementation of hydrokinetic turbines* financed by the Romanian Executive Agency for Higher Education, Research, Development and Innovation Funding.

References

1. K. Souidi, A. Mardaru, M. Roudet, A. Marcat, D. Della Valed, G. Djelveh, Che. Eng., Sci., **74**, 9, (2012)
2. C. Hsiao, A. Jain, G.L. Chahine, *Proc. of 26th Symposium on Naval Hydrodynamics Rome*, **26**, 12 (2006)
3. Z. Trad, C. Vial, J.-P. Fontaine, C. Larroche, Che. Eng. Sci. (to be published)
4. J. Bin, L. Xian-wu, W. Yu-lin, L. Shu-hong, X. Hong-yuan, O. Akira, *J. of Hydrodynamics* **22**(5), 6 (2010)
5. L. Mândrea, G. Oprina, C. Băbuţanu, D. Marin, G. Negreanu, Applied Mechanics and Materials **841**, 6 (2016)
6. P.R.N. Childs, *Rotating Flow*, (Butterworth Heinemann Elsevier, 2010)
7. L. Mândrea, C. Cipu, S. Cănănău, World Renewable Energy Congress Bucharest Romania (2015)
8. L. Mândrea, C. Băbuţanu, G. Oprina, G. Militaru, *Proc. of 20th Innovative Manufacturing Engineering and Energy Conference IManEE 2016*, IOP Conference Series Materials and Engineering, **161**, 6 (2016)

MATEC Web of Conferences 112, 08012 (2017) DOI: 10.1051/matecconf/201711208012
IManE&E 2017

Improving the education in the field of patient autonomy in rehabilitation doctors working with engineers

Roxana Elena Bliuc [1,*], *Vasile* Astarastoae [2], and *Sorin Ioan* Stratulat [3]

[1] Grigore T Popa University of Medicine & Pharmacy, Iasi, Romania
[2] Grigore T Popa University of Medicine & Pharmacy, Iasi Inst Legal Med, Iasi, Romania
[3] Grigore T Popa University of Medicine & Pharmacy, Romanian Railrd Clin Hosp, Rehabil Phys Med & Balneoclimatol Clin, Iasi, Romania

Abstract. The present study analyzes the use of bioethical expertise of Romanian rehabilitation doctors working in a hospital for engineering professionals, the Romanian Railways Clinical Hospital Iasi. The knowledge of the specific legislation by the medical personnel, proper communication, shared decision making and the use of informed consent are essential for effective healthcare provided to engineers, a group of professionals with a great contribution to the development of rehabilitation robotics and medical technology.

1 Introduction

Rehabilitation is a medical field in which engineering plays an important role. It is used for devices, wheelchairs and prostheses that replace functional abilities. Robots have begun to be used in physical therapy for repetitive and straining tasks that are usually performed by the therapist [1]. Conversely, in engineering, a profession in which physical requirements are sometimes minimal, including mostly planning, designing, modeling, analyzing, rehabilitation was successful in disabled patients with stroke and spinal cord injury who return to work [2].

However, to reduce the high prevalence of musculoskeletal diseases in the sedentary work of engineers, these professionals should be instructed by the rehabilitation doctors on how to activate their lumbar muscles through breaks for walking and alternating office work with activities in standing position, since ergonomic chairs are not effective as long as the sitting position itself causes the atrophy of lumbar muscles [3].

In this way, engineers could assume a more active role in their healthcare. Therefore, the aim of our study was to investigate to what extent Romanian doctors are ready to establish a relationship that respects the bioethical principle of patient's autonomy, i.e. the patient right to confidentiality, privacy, truth-telling, access to essential medical assistance, but also the right to make autonomous decisions about healthcare, including the right to refuse treatment and to have access to a correct informed consent [4-5].

* Corresponding author: roxana.bliuc@gmail.com

In order to assess this bioethical principle in a population of engineers, we have conducted our study mainly at the Romanian Railways Clinical Hospital Iasi. The knowledge of the specific legislation by the medical personnel, proper communication, shared decision making and the use of informed consent are essential for effective healthcare provided to engineers, a group of professionals whose scientific innovations in medicine were remarkable: the development of X-rays, cardiac catheterization, angiography, prosthetic devices, artificial organs and nanomaterials used in tissue engineering, nuclear medicine, diagnostic ultrasound, implantable drug delivery systems, computers for monitoring patients in intensive care units, computed tomography, magnetic resonance imaging, positron emission tomography, implantable defibrillators, stents, neuronal pacing systems, the study of the genome, engineering of stem cells [6].

Respecting patient's autonomy may be a little more challenging in medical rehabilitation, considering some specific aspects that may impact the patient's decisions regarding their health care and overall can impair their degree of autonomy. In this matter, in rehabilitation clinics, patients relate to a multidisciplinary team, not to a single doctor, frequently suffer from physical limitation, chronic pain and can have a chronic course of the illness. Also, patients in rehabilitation services are vulnerable to develop depression and anxiety. A study done in this matter found that among patients that came for chronic treatment in a rheumatology clinic, almost half of them were also suffering from depression or anxiety [4].

The considerable duration of the recovery process encourages the development of a close relationship between the doctor and the patient, the therapeutic relation being the foundation of the medical rehabilitation [5]. The present study intends to analysis the view of Romanian doctors in regard to the doctor-patient relationship with focus on autonomy of the patients in medical rehabilitation units. The main aspects followed in our study are referring to the knowledge on the specific legislation by the medical personnel, correct communication with the patient, involvement of the patient in the medical decision and also aspects concerning informed consent.

2 Material and methods

2.1 Sample group

The study group included 50 doctors selected from Iasi (located in North-East Romania) and was conducted during October 2014 – December 2015. The group surveyed comprised 20 doctors from the Rehabilitation Clinic of the Railway University Hospital Iasi, 10 doctors from the Rehabilitation Hospital Iasi and 20 doctors that work in private clinics. 48 doctors are specialized in medical rehabilitation, and 2 doctors are surgeons. All personnel that was implicated had to sign copyrights forms as well as other agreements in order for the authors to be able to publish the results.

Table 1. Characteristics of the sample group.

Characteristics	Sex (F)	Age group (30-39 years)	Marital status (M)	Medical experience (>20 years)	Individual medical practices	Field of rehabilitation medicine
N	44	20	28	7	27	48
Percent	88	40	56	14	54	96

DOI: 10.1051/matecconf/201711208012

2.2 Survey

All the subjects responded to a survey composed of questions that concern basic aspects related to doctor-patient relation such as communication and giving information to the patient, being informed of medical legislative framework or patient's involvement in making decisions regarding medical acts, but also aspects regarding informed consent.

2.3 Statistical analysis

The data were centralized in SPSS 18.0 data bases and statistically analyzed by using the T Student test; $p < 0.05$ was considered as statistically significant.

3 Results

3.1 Socio-demographic data

Analyzing the *gender* in the study group revealed a predominance of the female gender (88%), with a sex ratio of 7.3/1 female/male. Regarding *age groups*, a frequency peak was recorded at the 30-39 age group (40%).The distribution of doctors according to their *marital status* shows a slightly higher percentage of married individuals (56%).

3.2 Medical practice data

Type of medical practice was evaluated and revealed that 54% of the doctors surveyed conduct their activity in individual medical practices, 10% of the subjects being involved in both systems, state and private. Regarding *medical experience,* 50% of the surveyed doctors have less than 5 years of working experience in the medical system and only 14% have over 20 years of experience. In conjunction with the socio-demographic elements, the duration of the medical working experience longer than 20 years was observed more frequently in female doctors (p=0.06), with ages over 50 (p=0.001) and married (p=0.131).

3.3 Legislative framework

On the group surveyed it can be noticed that, in general, there is a partial knowledge (52%) or lack of knowledge (8%) of the legislative framework that regulates medical activity, with only 40% of the doctors being informed (16%) or fully informed (24%) regarding this aspect. The legislative framework is significantly frequently known by doctors younger than 50 (p=0.032).

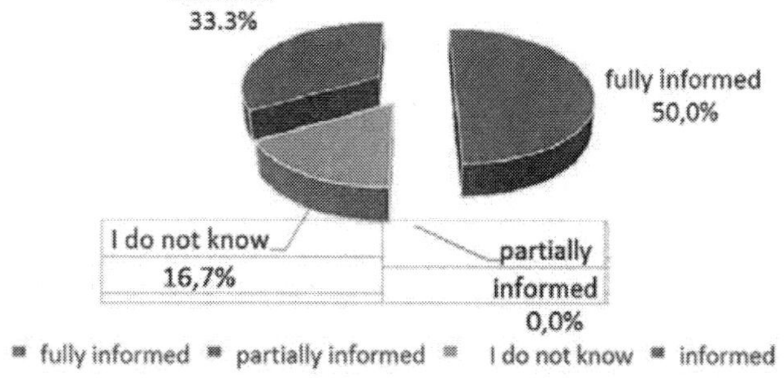

Fig 1. Structure of study group according to knowledge of legislative framework.

3.4 Self-assessment of communication with patients

All the doctors surveyed evaluate their communication with patients as good (40%) and very good (60%). A very good self-assessment of communication with patients was recorded more frequently among female doctors (87.5%; p=0.725), with ages under 50 (72.7%; p=0.086), married (66.7%; p=0.123). While only 88% of the doctors surveyed evaluated the quality of their medical activity as high and very high, but all of them claim they have a good and very good communication with patients.

3.5 Patient involvement in making decisions regarding medical investigations and treatment

The doctors surveyed reported patient involvement in making decisions regarding investigations and treatment as follows: 20% in all situations and 56% in majority of situations. Patient involvement does not present significantly different percentages in relation to the socio-demographic characteristics of the doctors (p>0.05).

Fig 2. Structure of sample group in relation to patient's involvement in making decisions regarding medical acts.

3.6 Information regarded as confidential

Among the types of information obtained by doctors the most frequently regarded as confidential are personal data (80%), patient diagnostic and result of investigations (68%) as well as prognosis for the condition (64%). Approximately half of the doctors surveyed (52%) consider the state of the patient confidential. The types of information regarded as confidential were significantly more frequent in the answers of doctors with ages under 50 (p<0.05).

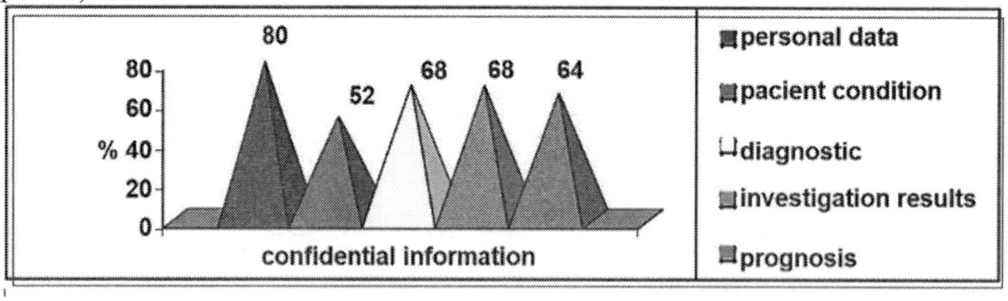

Fig 3. Structure of sample group in relation to the information regarded as confidential.

 DOI: 10.1051/matecconf/201711208012

3.7 Transmitting information about patients

Doctors reported that they transmit information about the patient to law enforcement (84%) or in situations in which the patient consents to this (88%). The information considered confidential, under the conditions of patient consent, is transmitted by 99.5% of the female doctors (p=0.001), with ages under 50 (84.1%; p=0.042) and married (50%; p=0.028). When family caregivers require, information about patients is offered significantly more frequently by young doctors (62.5%; p=0.009) and married doctors (62.5%; p=0.024). Transmitting information about patients to law enforcement is more frequently performed by doctors with ages under 50 (81%; p=0.034).

3.8 Components of informed consent used in medical practice

Most commonly, patients are informed in connection to the risks of treatment and existent alternatives, description of treatment and its effects. The components of informed consent most commonly used in medical practice by the subjects surveyed were informing the patient about the purpose of medical investigations (96%) and informing them about the cost of treatment and the necessity for connected care (88%). In relation to the socio-demographic characteristics of the doctors surveyed, with only one exception, no significantly different percentages were noticed, the components of the informed consent are used more frequently by females, with ages under 50, married (p>0.05).

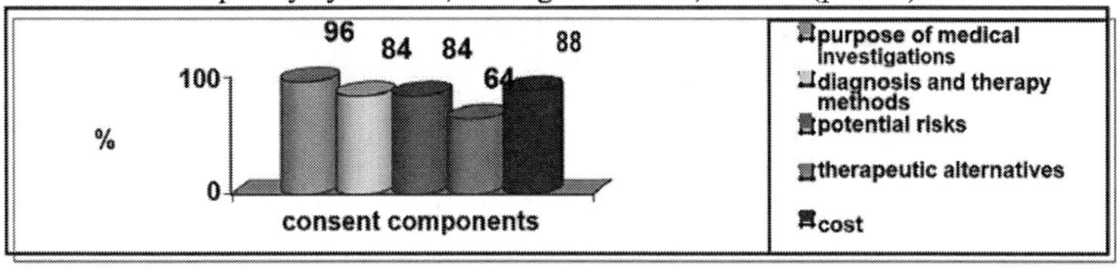

Fig 4. Structure of sample group in relation to the components of informed consent used in medical practice.

4 Discussion

The principle of autonomy requires encouraging the patients to participate in taking medical decisions regarding their own healthcare. It is known that empowering the patient to assume a more active role in taking medical decisions may improve the doctor-patient relation and ultimate may bring an improvement in the therapeutic act. The principle of autonomy has to be sustained by the medical personnel through a deep respect of the ethical values and a good understanding of bioethical principles and legislation that protects the patient [7].

In our study we found that generally there are gaps in information regarding the knowledge of the legislative framework under which medical care is practiced. The majority of the individuals surveyed responded that are partially or uninformed in this matter. Also, the lack of knowledge regarding medical legislation may have an overall negative impact on the quality of the medical act. These findings encourage improving awareness and education in medical legislation in the Romanian healthcare system.

Efficient communication with patients is the foundation of a good therapeutic relation and a key element in empowering the patient to be actively involved in medical decision [8]. It is known that failing to communicate with the patient may decrease the patient compliance and have a negative outcome in the medical response [9].

Most commonly, patients in our study were found to be informed about the purpose of medical investigations and also about the cost of treatment and the necessity for additional care. The rehabilitation medicine faces some specific difficulties comparing to other fields of medicine including the unpredictable character of the disease evolution, treatment efficiency, but also the disability risk, reason for which the discussions with the patient should be periodically resumed.

Another important aspect that strengthens the doctor-patient relation is respecting the confidentiality of the information about the patient. Violating confidential information may have both ethics but also legal consequences [10]. In rehabilitation, the medical act implies obtaining a great amount of private or intimate information, by means of physically examining the patient, through anamnesis and medical analyses. Sharing information with other people should happen only with patient consent or in situations in which breaking of law is detected.

5 Conclusions

Improving performance of the medical rehabilitation team can be also achieved by increasing the medical personnel's awareness on the bioethical aspects of medical practice, which concern patient autonomy, ensuring wellbeing, doing-no-harm and equity.

The doctors from Iasi who participated in this study generally know the legislative framework, inform patients about the evolution of disease, consult them regarding the application and cost of treatment and respect the confidentiality of patient information. On the other hand, our study highlights a general deficit of information regarding medical legislation and also a decreased involvement of the patients in their healthcare decisions.

We consider that it is necessary to clearly establish the role of the medical rehabilitation team, so that the involvement in the rehabilitation medical act is conducted for the patient's benefit and in compliance to the ethical and legal framework. In conclusion, our study draw attention on the necessity to improve knowledge concerning legislative framework and bioethical norms especially in rehabilitation services and patients to be fully informed and encouraged to take decisions in order to improve the quality of the medical care in Romania.

References

1. S. Xie, *Advanced Robotics for Medical Rehabilitation. Current State of the Art and Recent Advances* (Springer, Cham, 2016)
2. L. Smith, Workplace Health Saf. **64**,84-88 (2016)
3. F. Mörl, I. Bradl, J Electromyogr. Kinesiol. **23**,362-368 (2013)
4. R. Chadwick, *Encyclopaedia of Applied Ethics Vol. 3 (M-R)*. (Elsevier, Amsterdam, 2012)
5. V.A. Entwistle, S.M. Carter, A. Cribb, K. McCaffery, J.Gen.Intern.Med. **25,** 741-74 (2010)
6. J.D. Bronzino, D.R. Peterson, *Medical Devices and Human Engineering* (CRC Press, Boca Raton, 2015)
7. Law no 46 of 21 January 2003 regarding patient rights. Official Monitor no 51 of 29January2003. URL: http://legislatie.just.ro/Public/DetaliiDocument/41483
8. J. Fong Ha, N. Longnecker, Ochsner J. **10**, 38-43 (2010)
9. L. Kourkouta, I.V. Papathanasiou, Mater Sociomed. **26**, 65-67 (2014)
10. K. Blightman, S.E. Griffiths, C. Danbury, Contin Educ Anaesth Crit Care Pain. **14**, 52-56 (2014)

Development of a web based instrument on higher education structures of industrial engineering

Ioan - Cristian Tarba*

Manufacturing Engineering Department, Politehnica University of Bucharest, Splaiul Independentei 313, Romania

Abstract. The research and development of assisted operational instruments on higher education structures of industrial engineering represent a continuous and complex process. The present paper contributes to the building up of support elements and an assisted operational instrument on higher education structures of industrial engineering, with focus on the specific curricula. The use of tested and validated constructive solutions from other projects, as base for the new design, reduces the design time.

1 Introduction

The quality and the reliability of the computing systems are essentially determined by the software applications. The complexity of the software was and still is a challenge to those who design, implement, maintain, run and develop the systems, even the simple ones. Thus, as a new instrument, the methodology for software development consists in rigorous approach, as a plan, on the work style, technique of code writing, within the teams, between which the coordination and communication become essential [1].

Within methodologies for software development, there are traditional (formal) and agile approaches [1-3]. During the present research, the following methodologies have been applied: cascade methodology, evolutionary methodology, spiral methodology and the agile methodologies - SCRUM [4] and Extreme Programming [5].

The technologies used for the development of software applications are selected based on a various number of criteria, because of a large variety of programming languages [6, 7], database systems and solutions for the user interface [8].

The modular web applications are more and more developed [9, 10].

The algorithms' building is an important part of web applications development [11].

2 Objective and method of research

The main objective of the present work is to configure a modular web application as assisted design instrument to support the operational actions on curricula, especially in the framework of higher education industrial engineering structures, with the main aim to offer a tool for the

* Corresponding author: cristian.tarba@imst.pub.ro

managers of the departments to elaborate the curricula needed for education process and to reduce the time for getting the curricula done. During the research there were discussions about the input data and expected results of the developed application. The actually tools used for elaborating the curricula were spreadsheets and the time for completing them were critically affected by other factors. Thus, the time for elaborating the curricula using spreadsheets method cannot be a reference and it is obvious that the developed tool will reduce this time considerably.

The research activity has been structured with regard to the following reference elements: selection of the programming languages; building an algorithm, the structure and content of the application; conducting a case study in a real higher education institution.

3 Design of the objective modular web application

The objective modular web application is named Integrated Operational Instrument – Web Application, in short IOI.AW.01.

The development of the IOI.AW.01 application is based on an open source platform provided by a Linux distro, CentOS 7, and the selected solution for the persistence of data which is MongoDB database system. As development technologies, the followings have been applied: Python and JavaScript, as programming languages; Flask framework [12] with its 2 modules: Werkzeug, which is a development web server and powerful debugger, and Jinja2, which is a language for creating templates; Bootstrap framework used for the user interface design.

The IOI.AW.01 application provides with the possibility for the user to access the process of adding new curriculum, modifying the existing ones, adding/ modifying/ deleting the disciplines which belong to curriculum. Because the decisions regarding the content of a curriculum are to be taken by the persons in charge within the higher education institution, it was developed a routine for managing the users and the roles assigned to them. Thus, when the user login in he/she will have access only to those collections in the database which they have permission assigned to.

The IOI.AW.01 application is developed structured on 4 sections: View Curriculum, Modify Curriculum, Add New Curriculum and Copy New Curriculum, presented in Fig. 1.

Each of the defined section within the application can be accessed in its homepage, as presented in figure 1.

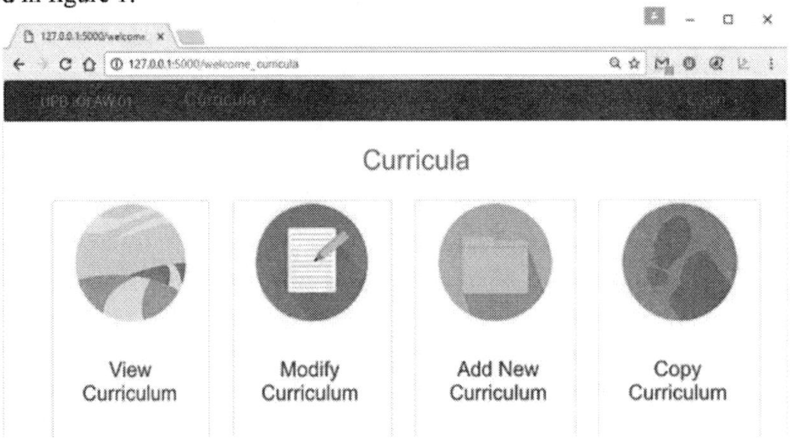

Fig. 1. Homepage of the application.

The scheme block of the software system as well as the work flow of the data through it are presented in figure 2.

 DOI: 10.1051/matecconf/201711208013

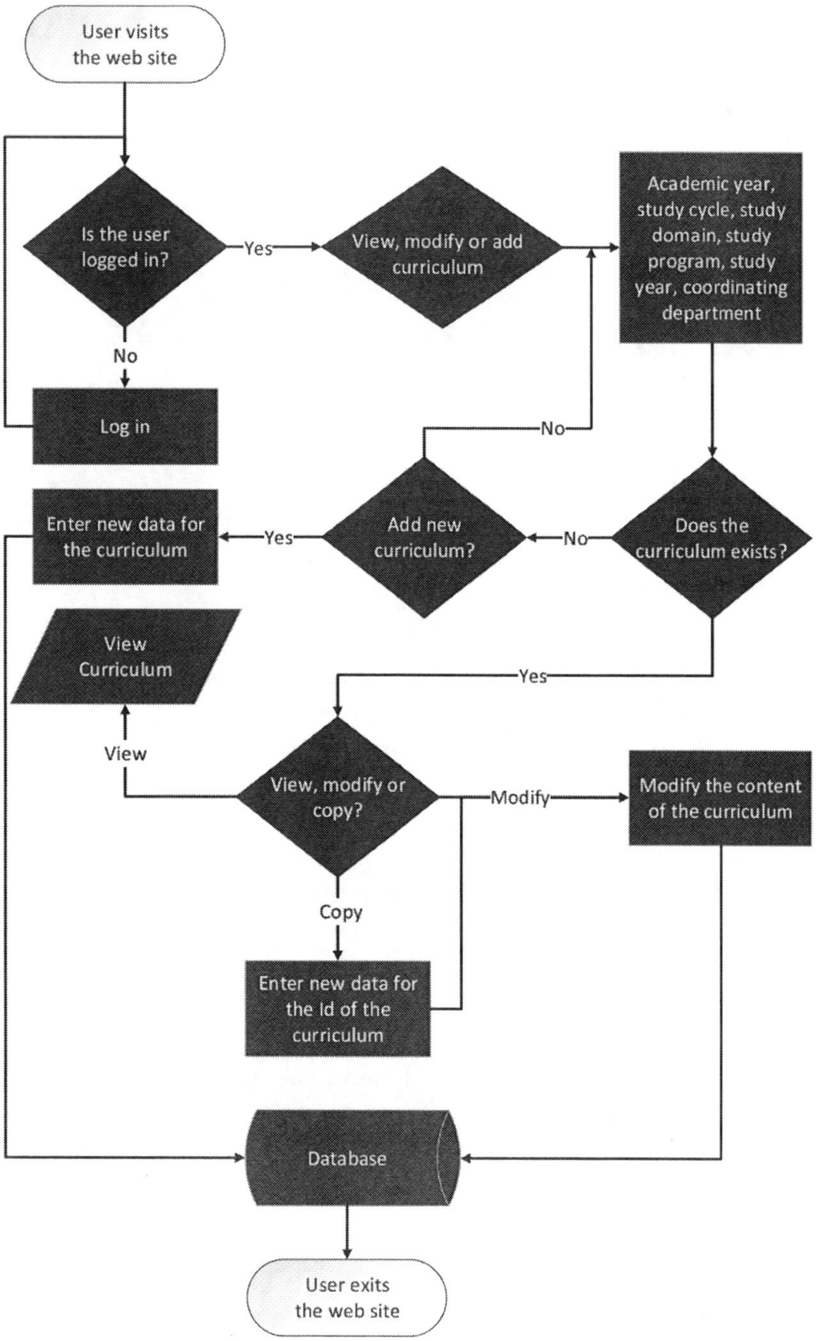

Fig. 2. Scheme block of the software system.

3.1 View Curriculum section

The purpose of the View Curriculum section and of the algorithm developed was to return to the user the view of a curriculum of which the identification data was entered in a web form. There were defined the route, /search_curriculum, the view function, search_curriculum()

and the html template, filter_curricula.html, which is used to generate a form in order to get the data from the user, as presented in figure 3, and the route /view_curriculum/<curriculum_id>, the appropriate view function, view_ curriculum(curriculum_id), and template curriculum.html, which leads to displaying the curriculum, as presented in figure. 4.

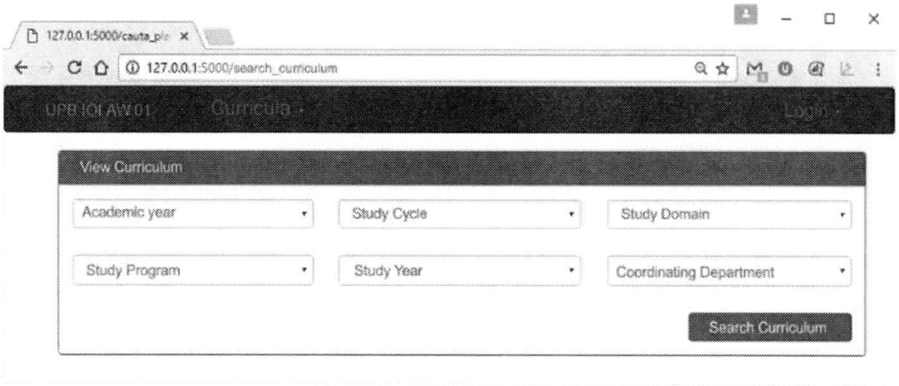

Fig. 3. Curriculum search form.

The steps the algorithm performs are the following:

1. The route /search_curriculum is accessed in the web browser;
2. It is called the view function search_curriculum():
   a. A connection to the database is created and retrieved the appropriate collections: curricula, study_cycles, study_domains, study_programs, departments, faculties;
   b. There are created the lists for populating the form elements in the html template, such as: yearslist, cycles_list, domains_list, programs_list, study_years_list, departments_list. Any combination of the 6 elements in these lists could be valid because they are already in the database collections. They are also called the general identification data for a curriculum;
3. The selected data in the web forms will form the id of the curriculum and used to search the database collections for an existing curriculum identified by this calculated id;
4. The database will be searched for the existence of the calculated id:
   a. If the id was not found in the database, the system will raise an error and will redirect the user to the search form page to modify the general identification data for a curriculum in order to calculate another id;
   b. If the id was found in the database, the user is redirected to the route /view_curriculum/<curriculum_id> and the view function view_ curriculum(curriculum_id) is called:
      i. There are retrieved from the database the needed collections: curricula, disciplines, departments;
      ii. The documents of the collections curricula and disciplines are filtered used curriculum id as filter.
      iii. There are created lists (study_programs, program_name, study_domains, domain_name, coordinating_dep, dep_name, academic_year, study_year, disciplines, semester, opt_disciplines) that are passed to the html template, named vieplan.html to appropriate display the data, as presented in figure 4.

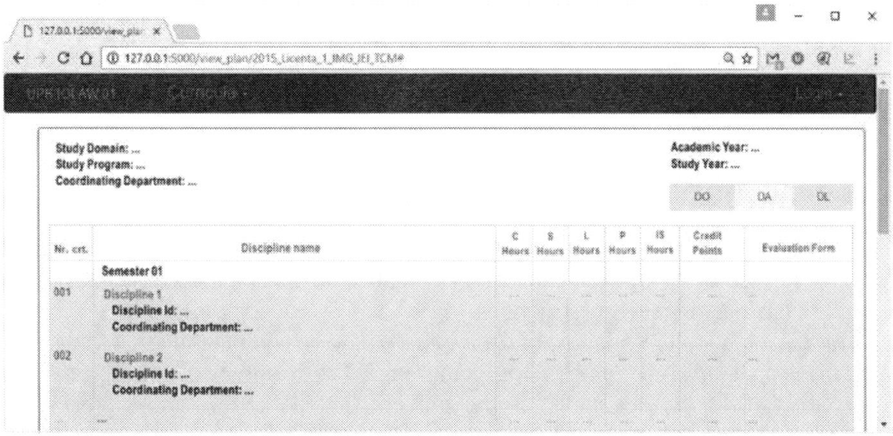

Fig. 4. View curriculum.

3.2 Modify Curriculum section

The section *Modify Curriculum* refers to the process of altering the content of a curriculum, but not the general identification data of a curriculum. In fact, modifying the curriculum consists in adding, editing, deleting the disciplines and their characteristics. The application administrator will have the task of deleting the general identification data of a curriculum.

There were defined the route, /search_edit_curriculum, the view function, search_edit_curriculum() and the html template, filter_edit_curricula.html, which is used to generate a form in order to get the data from the user, and the route /edit_curriculum/<curriculum_id>, the appropriate view function, edit_ curriculum(curriculum_id), and template edit_curriculum.html, which leads to displaying the curriculum, as presented in figure. 5.

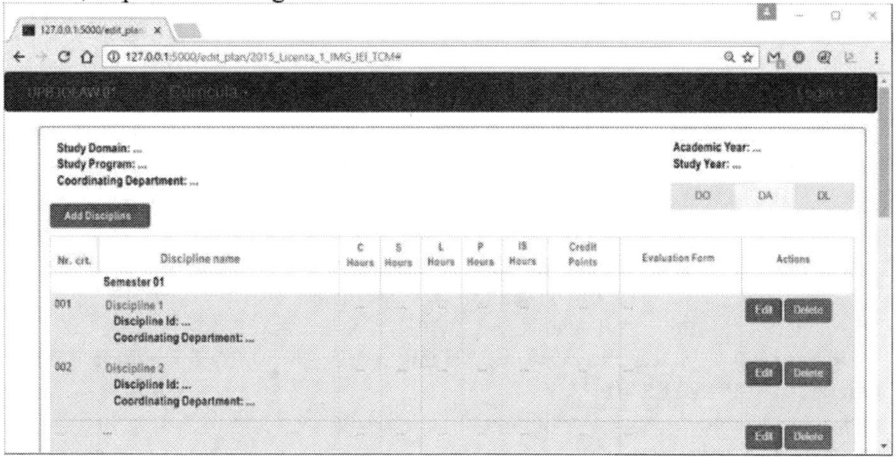

Fig. 5. Modify a curriculum.

The algorithm for searching and displaying a curriculum for editing is similar to the algorithm presented for View Curriculum section. In addition there were defined the route /add_discipline/<curriculum_id>, the view function add_discipline(curriculum_id) and the html template add_discipline.html which will offer the user possibility to add new disciplines to an existing curriculum identified by curriculum_id. Also, for each discipline there were added actions to edit or delete it. Thus, for the edit discipline action were defined: the route

/edit_discipline/<discipline_id>, the view function edit_discipline(discipline_id) and the html template edit_discipline.html. For the delete discipline action were defined: the route /del_discipline/< discipline_id >, the view function del_ discipline(discipline_id) and the html template del_discipline.html.

4 Conclusions

The IOI.AW.01 application has been developed and implemented as assisted design instrument to support the operational activity on curricula development associated to higher education industrial engineering structures.

As for the methodology adopted in the research, it was a combination of a traditional one (waterfall) and an agile one (extreme programming). The requirements of the final product were established at the beginning of the process, but later, the whole process of the development of the software system was according with the four basic activities described by extreme programming methodology: coding, testing, listening and designing.

The algorithm was developed with possibilities of future improvements such as validating the curricula against the nationwide regulations, building statistics regarding different elements of a curriculum, implementing the application in more fields than industrial engineering field.

The chosen solution for the data persistence is a NoSql database, which uses rules defined by the developer to deal large amounts of unstructured data

The low working time and high performing quality represent relevant performance indicators of IOI.AW.01 application.

This work is supported by the Sectorial Operational Program for Human Resource Development, financed from the European Social Fund and the Romanian Government under the contract number POSDRU/159/1.5/ S/137070.

References

1. D. E. Avison, G. Fitzgerald, Information Systems Development: Methodologies, Techniques and Tools (1995)
2. J. Highsmith, Agile Software Development Ecosystems (2002)
3. J. Highsmith, Agile Project Management: Creating Innovative Products (2010)
4. K. Schwaber, M. Beedle, Agile Software Development with Scrum (2002)
5. C. Novac, Extreme Programming - a Challenge for Software Developers, Revista Informatica Economică, **31**, no. 3: pp. 80 – 83 (2004)
6. V. Loganathan, PySide GUI Application Development (2013)
7. D. Ascher, A. Martelli, A. Ravenscroft, Python Cookbook (2005)
8. J. Anderson, J. McRee, R. Wilson, Effective UI: The Art of Building Great User Experience in Software (2010)
9. D. Esposito, Modern Web Development: Understanding domains, technologies, and user experience (2016)
10. I. Maia, Building Web Applications with Flask (2014)
11. D. Baldwin, Algorithms and Data Structures: The Science of Computing (2004)
12. M. Grinberg, Flask Web Development: Developing Web Applications with (2014)

Transformed objects: the influence of unruly product design

Athanasios Manavis[1], *Georgios* Pliatsios[1], *Eva* Dimou[1], *Apostolos* Korlos[2], and *Panagiotis* Kyratsis[1,*]

[1] Western Macedonia University of Applied Science, Department of Mechanical Engineering and Industrial Design, Kila Kozani GR50100 Greece
[2] Alexander Technological Educational Institute of Thessaloniki, Department of Automotive Engineering, P.O. Box 141, Thessaloniki, GR57400, Greece

Abstract. Nowadays, the image of a product becomes more important than the product itself. Unruly design is defined as the methodology of all objects that are designed with the intention to undermine the existing design-paradigm of the functionalists. At the same time, the term surreal object is based on the magical, transformative concept of the juxtaposition of diverse elements usually found in everyday objects. The present paper outlines the general principles of the unruly design and surrealism art movement, together with their influence on styles of the conceptual product design. A number of illustrative case studies are depicted, through an ergonomic approach, in order to ensure an appropriate interface relationship between art and product design. The products' transformations have been selected in order to stress the idea that using unruly design offers additional advantages in creating concepts from a different point of view. Innovation and unusual product design can create the basis for a complete product design transformation culture.

1 Introduction

Product design is about making things that can lead to change. Designers, through specific acts, bring into life not only solutions that help people to overcome problems, but also artefacts that offer opportunities for innovation in the field of aesthetics. More specifically, product aesthetics defines the object that gives pleasure and delight, through the incorporation of aesthetic features [1-3].

Nowadays, the image of the product becomes more important than the product itself. Furthermore, the relationship between beauty, usability and functionality shows the importance of aesthetics in product design (Fig. 1) [4]. A modern theory that describes the relationship between the image of the product and the product itself is the unruly design. The core of unruly design theory is based on both the postmodern and surrealism art movements (Fig. 2).

* Corresponding author: pkyratsis@teiwm.gr

 DOI: 10.1051/matecconf/201711208014

2 The Three Factors

2.1 Unruly Design

Unruly design is defined as the methodology of all objects that are designed with the intention to undermine the existing design-paradigm of the functionalists [5]. In this case, the definition of the product functionality is extended with emotional and cultural aspects, at the cost of usability, functionality and beauty. An unruly designed product follows five basic principles, the: a) combination of different interest domains, b) inspiration from popular culture, c) incorporation of a form complexity, d) use of ready-mades and e) use of uncommon materials [6].

The main influence of unruly design in product development is that, it can support designers to understand the implementation of meaning into demand driven design practice, and therefore extends the possibilities for making meaningful objects.

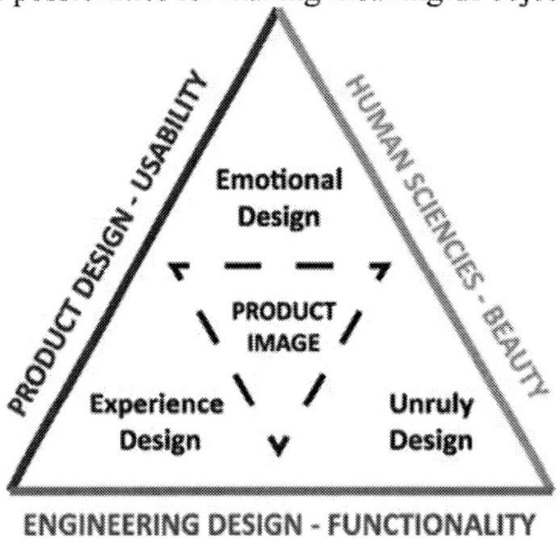

Fig. 1. The product image triangle.

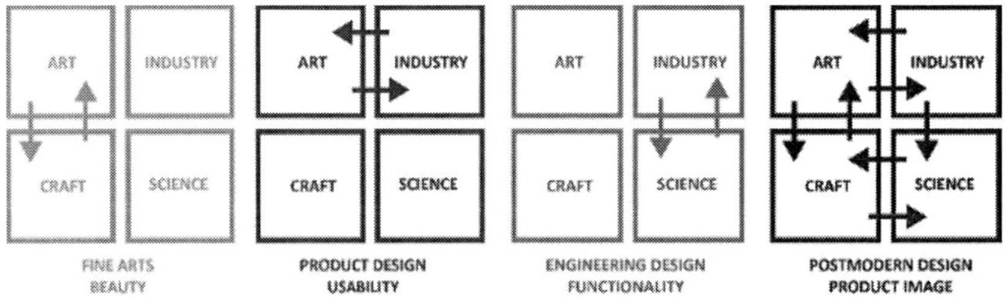

Fig. 2. Design theories and relationships between art, industry, craft and science fields [2].

2.2 Surreal Object

Surrealism occupies a unique place within the history of the 20th century avant-garde because of its international influence and longevity. The Manifesto of Surrealism explains this art movement as a primarily tributary creation of an interior model, where dreams, the subconscious and automatism in creation, inspired a poetry designed to deny reality and throw it into turmoil [7-9]. The most well-known surrealist artist Salvador Dalí gave an initial definition to objects with a symbolic function. These objects led themselves to a minimum of mechanical functioning and they are based on the fantasies and representations that can arise from the performance of subconscious acts.

2.3 Ergonomics

Ergonomics scopes to build a bridge between design parameters and human compatibility factors. It makes design more successful for user's acceptance. It is more than just product & user relationship tool. Design ergonomics is also an advancement in its journey, crossing the concept of the user, the product and function, the look at the user, while offers a product relationship in totality. A human user of a product is not a physical and cognitive processor. His needs are to be addressed in tune with his emotions, values, hopes, fears, and anxiety over new adaptations in life [10-12].

3 Transformed Objects

Transformed objects are a combination of all three factors unruly design, surreal object and ergonomics (Fig. 3). In this paper, several illustrative case studies are presented, through an ergonomic approach, in order to ensure an appropriate interface relationship between both the art and product design.

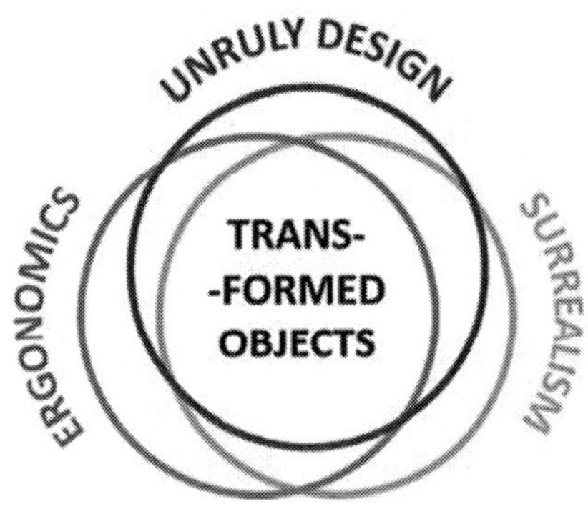

Fig. 3. The three factors of transformed objects.

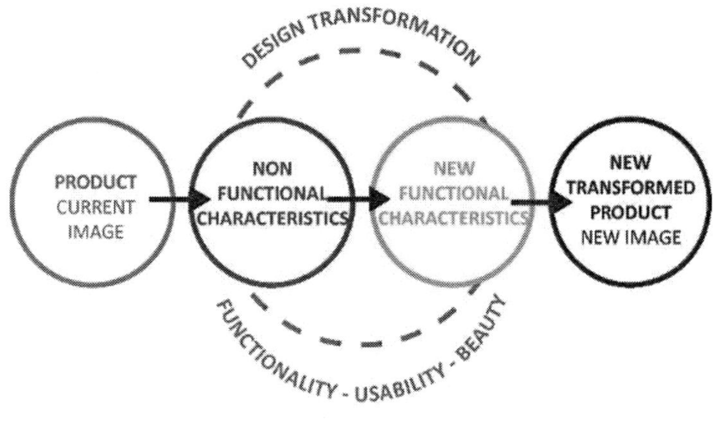

Fig. 4. The design process of transforming objects.

Figure 4 depicts the whole process used when the case studies were prepared. Initially, the designer chooses a full functional, useful and "beautiful" product. The next step is him/her to decide what product characteristics might be cancelled or manipulated from the function and the ergonomic point of view. The third stage relates with a conceptual product design, using design sketches for the development of the final idea of the novel transformed object. All these new criteria (functional, ergonomic and aesthetics) discovered at previous stages, are added at the new object. Finally, a transformed object is defined when following this creative design process. The aim is to build a new product image in addition to the current one. The whole framework of transformed objects offers opportunities to designers for developing creativity and experimentation opportunities. This provides an additional path for innovation and unusual product design that can offer a solid basis for a complete product design transformation culture.

3.1 Wearable media player design

The watch as a product is something personal. Everyone has its own taste and preferences in design and style. Watches are small timepieces intended to be carried or worn by a person. They are designed to keep working, despite the motions caused by personal activities. The five features that describe the ideal watch are: a) the material of the product (a titanium or ceramic watch is considered of higher value), b) the kind of watch glass used (a quality watch always has sapphire glass in addition to crystal glass), c) the design of the watch mechanism (a good, affordable watch is based on the quartz principle or uses a certified mechanical mechanism), d) its water resistance and e) the quality of its strap. All these features are selected with an aim to manipulate the users' perception from the conceptual design point of view (Fig. 5).

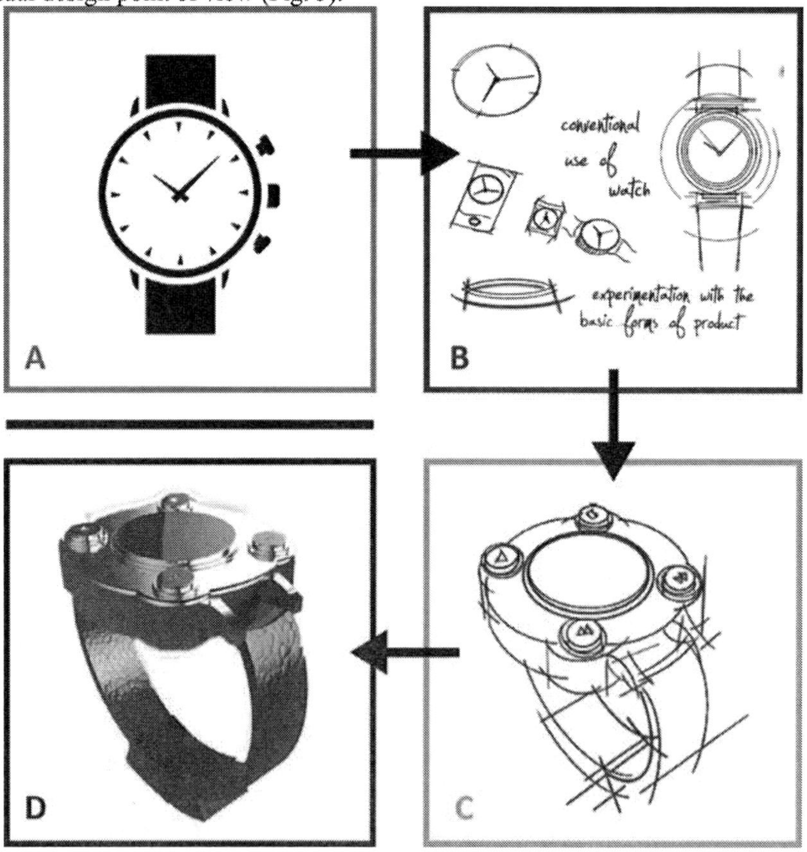

Fig. 5. A watch is transformed to a wearable media player: a) current image of the watch; b) non-functional characteristics; c) new functional characteristics; d) new transformed product.

 DOI: 10.1051/matecconf/201711208014

This procedure showed that a watch without ergonomics could be used i.e. with a cut strap, a blind screen and nonfunctional buttons. This specific design transformation guides to fundamental questions for the new product image, such as "what is the new role of this object?", "what features are to be replaced for the novel image of this product?"
At the third stage of the design process, the novel identity of a wearable media player with updated features is presented. The cut strap is replaced by a magnetic clip, the blind screen is substituted by a high-resolution monitor (like smartphones) and the nonfunctional buttons are redesigned to be fully ergonomic and functional keys. Finally, the main purpose of the initial watch is replaced from the final concept that promotes music. In addition, it underlines the difference between an anxiolytic watch and an entertainment media player.

3.2 Data Storage Device design

A key is a device that is used primary to operate a lock. A typical key is a small piece of metal consisting of four parts: a) tip, b) cuts, c) shoulder and d) bow. In this case, a typical key is transformed into a USB data storage device by the transformed object design process proposed (Figure 6). The experimentation with the basic forms of the product, guides the rotation of the door key upper part. Then an extrusion of the surface leads to a different geometry. During this step, the addition of the USB geometry should be highlighted.

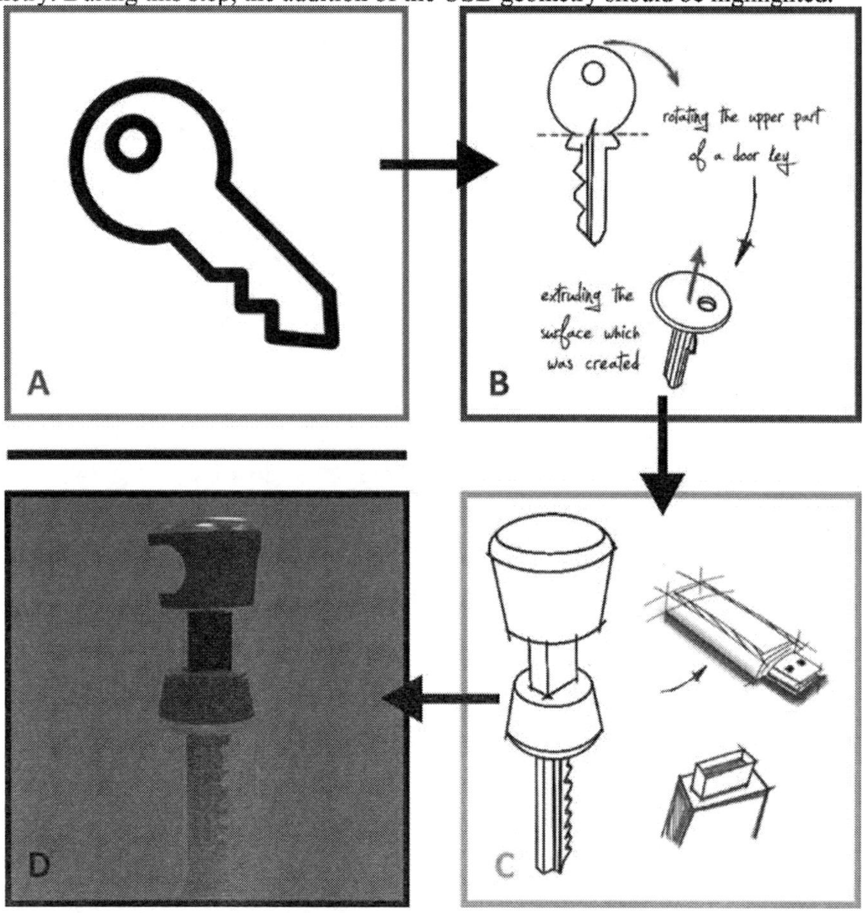

Fig. 6. A key is transformed to a USB stick.

New functional characteristics are explored and the role of the cap leads to the data storage device geometry. Finally, a new image of this transformed product appears as a novel data storage device that is included in a fully functional, ergonomic and "beautiful" key. Two functions were incorporated in one ergonomic object. A unique approach of using all the three factors: unruly design, surrealism and ergonomics.

4 Conclusions

A new methodology was presented, based on the creation of transformed objects. The proposed framework could offer a series of great design and marketing advantages to enterprises that would be willing to follow this path. Transformed objects are a combination of three factors: unruly design, surrealism object and ergonomics. This specific approach makes use of shapes and functions that customers associate with a number of different original products and produces unusual items.

The relationship between beauty, usability and functionality shows the importance of aesthetics in product design. The proposed conceptual designs incorporate not only innovative creative design, but the creation of novel "product images". A new chain of products based on this unusual approach could trigger further new product designs and design methodologies.

References

1. P. Moutaftsi, P. Kyratsis J APPL PACKAGING RES, **8(3)**, 1 (2016).
2. P. Kyratsis, N. Efkolidis, Ath. Manavis, St. Peristeri, ACAD J MAN ENG, **11(4)**, 56, (2013)
3. P. Hekkert, H. Leder, Product Aesthetics: Product Experience (Elsevier Science, 2008)
4. M. Hassenzahl, Product Experience: Aesthetics in interactive products: correlates and consequences of beauty (Elsevier Science, 2008)
5. W. Eggink, A. Reinders, 5th International Congress of International Association of Societies of Design Research, Tokyo (2013)
6. W. Eggink, 4th World Conference on Design Research. The International Association of Societies of Design Research (IASDR) (2011)
7. A. Breton, *Manifestoes of Surrealism* (University of Michigan Press, 1969)
8. R. Spiteri, *A Companion to Dada and Surrealism*, 110 (2016)
9. E.H. King, *A Companion to Dada and Surrealism*, 416 (2016)
10. L. Rungtai, J.G. Kreifeldtb, INT J IND ERGONOM, **27**, 259 (1999)
11. L. Rungtai, Proceedings of the Human Factors Society (1992)
12. L. Rungtai, J.G. Kreifeldt, Proceedings of the Human Factors Society (1992)

The necessity of a dual education system for Romanian automotive industry. Case study: The German dual vocational training

Ramona Pîrlog[1,*], *Ioan-Vlăduț* Nuțu[2], *Anca Alexandra* Purcărea[1], and *Aurelian* Vlase[2]

[1]Politehnica University of Bucharest, Faculty of Entrepreneurship, Business Engineering and Management, Bucharest, Romania
[2]Politehnica University of Bucharest, Faculty of Engineering and Management of Technological Systems - I.M.S.T., Bucharest, Romania

Abstract. According to the official data, Romania has more than 600 companies involved in the automotive industry, with more than 203,600 employees. However the automotive industry is facing a real problem when it comes to finding skilled work force. In the west part of the country, the unemployment rate is negative and companies are forced to find solutions for their lack of personnel. This study presents the solutions that were taken by Germany in order to solve the matter. A dual education system was implemented which combines trainings in a company with vocational education. This kind of system is practiced in countries such as: Germany, Austria, Switzerland and South Korea. As of 2005 young German people can learn one of 356 apprenticeship occupations through this program. In France, dual education has gained a lot of popularity since the 1990s. In order to gain more experience and have multilateral development, in Germany the students are also send to workshops run by guilds or chamber of commerce. The advantages of the dual training system are numerous and this program can help the industry by providing skilled workers with flexible qualifications, solution which may as well be suitable for the Romanian Automotive Industry.

1 Introduction

In recent years, the private sector showed a growing interest for a qualified and well trained work force with good technical skills. In the automotive sector, the counties in the West of the country are experiencing negative unemployment and the private companies are forced to find solutions for their lack of personnel.

Because of this shortage of human resources, companies are losing business and turnover. Representatives of car manufacturers in Romania consider that the scarcity of trained personnel is due to the migration of the well trained work force to more developed markets in Western Europe or the Middle East. The significant increase in investments in

*Corresponding author: ramona.pirlog@gmail.com

new facilities and distribution service is not directly proportional to the increase number of specialists, which leads to a lack of personnel. [1]

A relatively new trend for Romania is recruiting students in the final years or recent graduates and offer them numerous specialized trainings through their own education centers. However, this approach is suitable for the large companies, the small and medium size enterprises have difficulties in finding the skilled workers.

2 Consideration regarding the Dual Education System

The main objective of vocational training: to produce skilled workers with flexible qualifications who are mobile and capable of working in their chosen fields.

According to the International Centre for Technical and Vocational Education and Training (UNESCO-UNEVO) the Dual Education System is called "dual" because it combines apprenticeships in a company and vocational education at a vocational school in one course. In the company, the apprentice receives practical training which is supplemented by theoretical instruction in the vocational school.

The European Centre for the Development of Vocational Training CEDEFOP adds that the dual education system combines periods in an educational institution or training center and in the workplace. The scheme's alternation can take place on a weekly, monthly or yearly basis. Depending on the country and applicable status, participants may be contractually linked to the employer and/or receive a remuneration. [2]

This system is practiced in several countries, notably Germany, Austria, Hungary, Bosnia and Herzegovina, Croatia, Serbia, Slovenia, Macedonia, Montenegro and Switzerland, but also Portugal, Denmark, the Netherlands, France, Australia and Egypt, and for some years now in China and other countries in Asia.

The German Dual Education System is one of the cornerstones of the successful brand '**Made in Germany**' due to which a small nation of only 80 million people is able to be the world champion in exports for several years. The system is widely practiced in Germany for over 350 officially-recognized training occupations. [3]

In 2011, about 68% of all apprentices received an employment contract after completing their training. Almost 1.4 million young people are currently on a vocational training course in Germany accordingly to the economic development agency of the Federal Republic of Germany, The Germany Trade & Invest (GTAI). [4]

Germany has Europe's lowest youth unemployment rate, has thereby increased the competitiveness of Germany's economy and enjoys an excellent reputation as a result of the dual system. [5]

As stated by the Indo-German Training Center, the main features of a dual education system are:

- **Joint educational responsibility** meaning that the training institution and the private companies work together in order to create a unique synergy between learning and working. It also leads to a constant modernizing of the curricula and guarantees an up-to-date learning that follows the developments of the economy.
- **Practice integration**. Theory imparted in the classroom is complimented with hands-on experience at the workplace, so that real-life situations immediately test the effectiveness of classroom theory and vice versa. The system seeks to impart structured knowledge and active competence, in their proper context.
- **Best future perspectives**. In choosing the dual system, the student chooses a safe future. 100% of the students receive a job placement immediately after completion of the program.

In 2012 was launched the dual education programmer at Kecskemet College in Hungary for 25 vehicle engineer students in partnership with Mercedes-Benz Manufacturing Hungary and Knorr-Bremse Fékrendszerek Kft and the dean of the college stated that "these students are the drivers of the labor market as their achievements are significantly better than those of regular students". [6]

3 Costs and Benefits of the Dual Educational System

„The costs of the dual educational system are shared by regional governments, private companies, and the apprentices themselves. The government pays for the costs of the public education side of training, which amounts to about one percent of the country's gross domestic product" [4], while the companies pay for all of the costs associated with the on-the-job training.

As a result from a survey did by the Federal Statistical Office in 2013, the annual net costs for companies per apprentice (2007) in Germany was 3.596 euros. However the companies will receive engineers with working experience and commitment towards the company by the end of the training. [7]

As shown in figure 1 the apprentice receives practical training in the company for three or four days which is supplemented by theoretical instruction in the vocational school for another day or two.

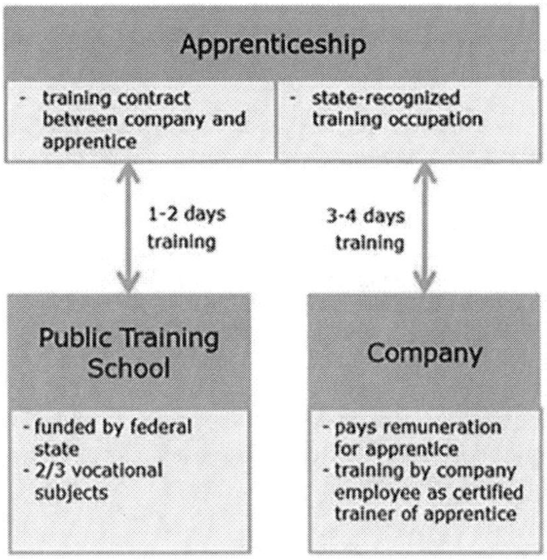

Fig. 1. Dual System of Apprenticeship. [7]

Prof. Dr. Dieter Euler states the following benefits of the dual education system:

- Training is designed to meet the practical needs of the labor market
- Occupational skills enable individuals to work in their chosen fields
- Skills are applicable to a wide range of settings within the field, which allows individuals to be flexibly employed in a variety of businesses
- Skill profile makes trained workers more mobile
- In some cases skills are developed in a fixed process, without separate steps or interruptions
- Skill development may in some cases be combined with aspects of career and occupational development

- A responsible agency (usually the chamber of commerce or crafts) administers a centralized examination to assess trainees' skills, adhering to the principle that the teacher and the examiner should not be one and the same
- Benefits: skills enhance workers' flexibility and mobility, reduce the danger of social marginalization and raise educational levels in a non-academic context

In figure 2 it is shown the supply and demand in the dual training employment in Germany between 1995 and 2011.

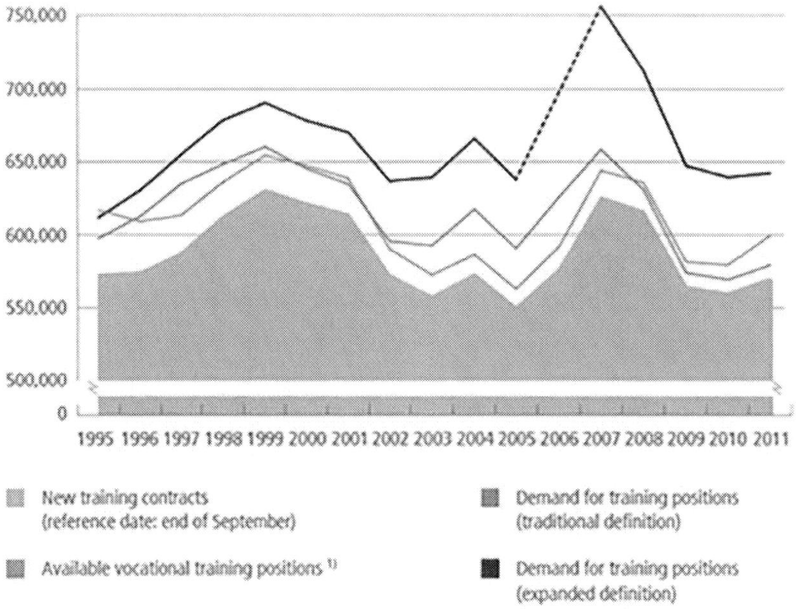

Fig. 2. Supply and demand in training in Germany 1995-2011. [8]

The Federal Institute for Vocational Education and Training (BIBB) conducted a survey in 2004 in order to establish the reasons that drove the companies to take part in dual education and 94% responded that „trainees fit company's needs", 93% said that „acquisition of specialists was not available on labor market" and 72% consider this kind of program as a „possibility to choose best candidate when considering employment after end of apprenticeship".

4 Automotive Industry in Romania

According to the official data, Romania has more than 600 companies involved in the automotive industry, with 203,600 employees. Approx. 20% of all Romanian exports are from the automotive industry.

The total turnover of the automotive industry in 2014 was estimated to 18 billion Euro, a 7% increase from 2013. [6] As regards the turnover of the automotive parts sector, it increased by 8.7% in 2014 as compared to 2013. The entire automotive industry is estimated to have contributed around 12% to the Romanian GDP in 2014. The main drivers of this sector remain the Renault Group via the international "entry" program and the export of components, directly or through integrators.

In 2015, the Romanian automotive industry showed a solid, steady growth. According to the estimates of the Association of Automotive Manufacturers of Romania (ACAROM) in the first 5 months of 2015, automotive production increased by 9%, the export being 1.2% lower than in the same period of last year. Internal automotive sales increased by 35%, while the ratio of second hand cars vs. new cars produced domestically and sold in Romania is 7.3/1 (the average age of second hand cars is 10.6 years). The Association of Automotive Manufactures of Romania - ACAROM is a professional body of companies involved in the automotive industry, associated in order to represent the interests of the automotive sector. ACAROM, a non-profit organization, has 144 members and was set up in 1996. [6]

In figure 3 it is presented the evolution in turnover for automotive companies, both for automobiles and spare parts. In 2014, the turnover was 13% more than in 2013.

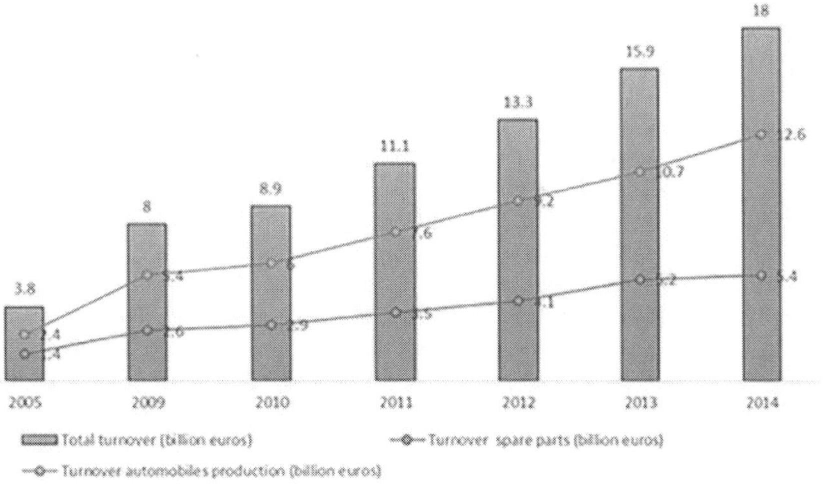

Fig. 3. Turnover for automotive industry in Romania 2005-2014. [6]

The Association of Automotive Manufactures of Romania ACAROM represents the interests of more than 150 companies from the automotive industry. In the last years they tried to be the connection between the private sector and government. Among their requests there is one regarding human resources. The private companies are asking for the following : a) developing partnerships between public educational institutions and private employers to better prepare students, b)equipping school laboratories with modern and up to date equipment, c) initiate training courses for the automotive industry, d) the need to review the study programs taking into account the requirements of the industrial environment and advances in the field of high technologies, e) he need for review of educational material in order to fold the requirements of employers, f) Improvement in terms of quality, quantity and skill set developed by the educational system in Romania in certain professional fields.

5 Conclusions

At the meetings of the automotive industry associations, the representatives of the companies involved in car manufacturing or auto parts are asking the Ministry of Education to implement a dual education system. The involved parties see this system as a solution to the lack of well trained personnel. The charts presented above show that the automotive industry is in a constantly development in our country thus requiring work force. The main advantage of the program is that the student starts as an entry-level employee who can take tasks according to his growing abilities and at the end of the program, the company will have an employee who already knows the workflow and who, during a considerable amount of time, has benefited from the experience of more experienced co-workers. The student also has his share of benefits by earning money from the beginning and not after exams.

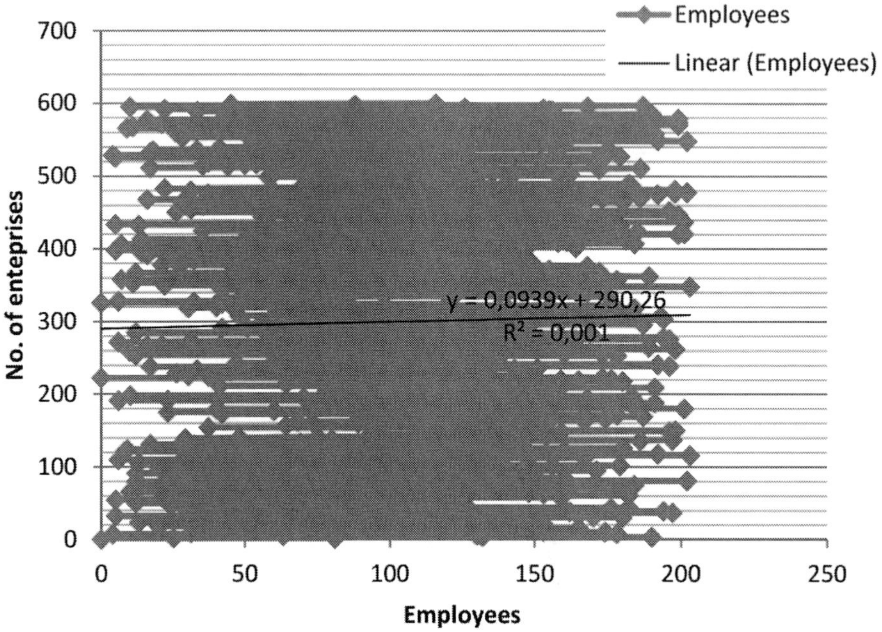

Fig. 4. Romanian enterprises versus number of employees.

As we can see although there is a lack of highly skilled personnel in the Romanian enterprises (over 600) the trend is ascending as more and more students are applying to these companies as they are more involved and invest constantly in skills hunting strategies. Thus we see that educational dual systems strategies pay off in long term projects.

The success of the German dual educational system has understandably attracted much attention from many other industrial nations. However the acceptance is affected by economic, social and pedagogical aspects, as well as by the cultural setting, which we will turn to now.

In general, it can be stated that the acceptance of vocational training in countries with a high level of dual vocational training is higher than in those countries in which dual vocational training does not exist at all or only exists to a very limited degree.

References

1. C. Wieland, E. Lezcano, (2016)
2. D. Euler, Bertelsmann Stiftung (2013)
3. Germany Trade & Invest (2014)
4. Clark, Charles S. "Youth Apprenticeships." The CQ Researcher (1992)
5. J. Petrosky (1996)
6. ACAROM research and studies (2013-2016)
7. Germany Trade & Invest (GTAI) - http://www.gtai.de
8. AutorengruppeBildungsberichterstattung (2012)

Lean engineering in the design process: An industrial application

Toufik Boudouh*, and *Samuel* Gomes

ICB UMR 6303, CNRS, Univ. Bourgogne Franche-Comté, UTBM, Belfort, France

Abstract. The objective of this paper is to present a methodology used to streamline the product development process by introducing Lean Engineering concepts. We are interested specifically in routine design which represents in some companies more than 80% of the design activity. A project management model was developed and implemented in a PLM tool. This allows faster initiation of projects in which project planning is automatically performed. Hence, non-added value tasks in project planning were identified and reduced.

1 Introduction

To increase their competitiveness, companies are required to increase the turnover of their products. They should continuously develop attractive new products efficiently in order to increase sales volume and ensure sustainability in globalized markets. Streamline of routine engineering processes becomes necessary to improve quality and productivity in design and will allow more time to develop new products.

We must reduce the length of the non-added value tasks to free up time for innovative projects. Engineering design is a source of competitive advantage for manufacturing companies. Reducing the development time is a key factor in the successful completion of the product development process in a time based competition context. During the last decades, the design process is subject to new methodologies and tools to make it more and more efficient. As what was done in manufacturing, lean engineering techniques were introduced into the design process. The objective is to reduce the non-added value tasks of the design process.

Design is an information intensive activity. Because of the complex information dependencies that exist between design tasks, we cannot perform the design process as a once-through procedure. Therefore, iterations are necessary to resolve design problems. These iterations are often seen as non-added value tasks. Understanding and controlling iterations can improve the design process, and reducing them would have positive effects for the product development cycle time [1]. The purpose of lean engineering is to introduce new methodologies helping the designer approach the ideal design solution as rapidly as possible.

In this paper, we describe a project management methodology, implemented using a PLM system. The objective of this methodology is to streamline the product development process by introducing Lean Engineering techniques in project management. We are interested

* Corresponding author: toufik.boudouh@utbm.fr

specifically in routine design which represents in some companies more than 80% of the design activity [2-3]. Thus, reducing these tasks will allow more time for innovative design. The methodology was applied in a SME in which two types of projects where identified: long projects and short projects. For each type, a project management model was developed and implemented in the PLM tool. This allows faster initiation of projects in which project planning is automatically performed. Hence, non-added value tasks in project planning were identified and reduced. The use of the PLM tool allows also taking into account the manufacturing issues since the design phase.

2 Methodology

The use of a PLM (Product Lifecycle Management) tool for coupled with a streamlined design methodology, simplifies the achievement of the project, document archiving, and improves communication between actors involved in the project and product lifecycle. ACSP is an organizational tool for collaborative project management providing these features. It was developed at the University of Technology of Belfort-Montbéliard, France. This tool provides a software environment for distributed collaborative design [3], centralizing different data related to the design project of the product and its manufacturing process. Positioned in the context of concurrent engineering [4], the goal of such a tool is to support the design process in terms of cooperation between the different actors of the design: engineers, technicians, decision makers, users, manufacturing operators, etc.... It also allows a better definition of the project (planning, project team definition, workload rates, etc…), and a better visualization of tasks to be performed by each project member. In addition to the basic function of document management and communication specific to PLM systems, the tool must allow to manage the product development process through considering interrelated tasks. Standardization of some project templates, grouping already predefined tasks (according to various project models in the company), the standardization of the duration of tasks, the standardization of documents and intelligent inter-connections of tasks, saves considerable time in developing the project plan. Visualization of the workload profile of each project member also allows to create realistic schedule estimates, taking into account the implication of these members in other projects. Therefore, from standard projects, the creation of a new project including project scheduling is facilitated by a semiautomatic formatting of the project.

3 Industrial case study and application

Our methodology was tested and implemented in a SME company. Despite the presence of a design office and a process planning office, we noticed in the beginning of the project the absence of a standard design methodology. There is no formal method to capture and reuse the knowledge gained at the completion of each project. We have also identified several barriers to the integration of a PLM tool in such a company. These barriers range from the employees' resistance to change, to the number of ongoing projects as well as the amount of data to be added to the PLM by each project member. As actors in the implementation of the PLM tool, we felt it necessary to provide a clear and easily usable tool. It was also necessary to go through several steps before a full deployment of the tool.

The first step in PLM implementation was to map the current way of working by listing all the tasks and deliverables of any project. Indeed, the company has no standard design procedure for its projects and project activities can be performed in any order. In this case, information may be lost and the communication is poor among the different project members. Once the tasks identified, the next step was to define standard deliverables, using a model that should be shared by all the actors involved in the different product development projects. A standard schedule of the project tasks is necessary to get homogenous and streamlined project information. A Gantt chart was then constructed to organize these tasks. Each task was assigned a standard average duration. Using the information gathered, it was possible to construct and setup project tasks in the PLM tool (ACSP).

The main characteristics of the studied company are the number, diversity, and complexity of projects:

- 150 projects are undertaken simultaneously.
- Project durations are very different from one project to another.
- Project complexity is variable from one project to another (from simple to complex).

For these considerations, during the creation of the WBS (Work Breakdown Structure) for the identification of standard activities, two project templates have been created: short project model and long project model.

3.1 Long project template

The long project template is a project model including all phases necessary for the proper achievement of a project, from pre-study to concept validation (Figure 1). In this model, market and technical environments are unknown. So, this will require further research activities, such as functional analysis or marketing analysis. Long projects involve all the departments of the company from the sales department to the production shop floor. Generally, long projects are mainly initiated by the innovation department, or after a comparative study, or for alignment to standards.

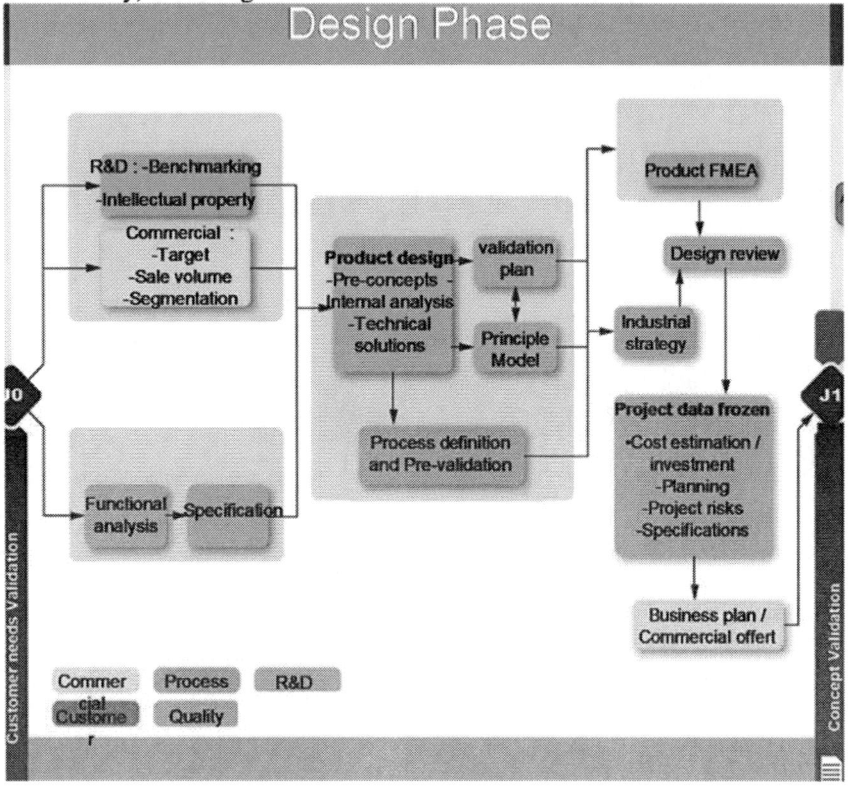

Fig. 1. Long project template: design phase.

3.2 Short project template

The short project template includes various types of short projects. These projects could be classified into two categories. Projects with lower risks in the design phase where the risks in terms of reliability and consistency are controlled. The technical solution is generally obvious; the validation of this solution is performed by evaluation of its cost, time and performance using analogy or the experience of project members. Projects with strong risks in the design phase, where design risks are vague and require further investigations using product and process FMEA. The difference with a long project is characterized by a better mastered environment because it's much more known, so a phase of information research much more short and simple (Figure 2). Reflection on the choice of a path to high or low risk is determined in advance by experts and managers. In case of doubt, the project risk is considered as high and a feasibility study is performed including a product and process FMEA. In the case of short projects with low risk, a solution is evaluated by various experts involved in the product development and the management relying on the company's knowledge and experience. Capitalization of this knowledge is then necessary to facilitate and accelerate the evaluation of future projects. Generally, short projects consist of component modifications, or creating a product variant on the basis of a known model in order to respond to a customer demand. These tasks, which are routine task, mainly occupy 60% of the company's studied projects. Thus, in a short project, the customer quotation is delivered much more quickly than in a long project. Reactivity is the most important factor for the customer.

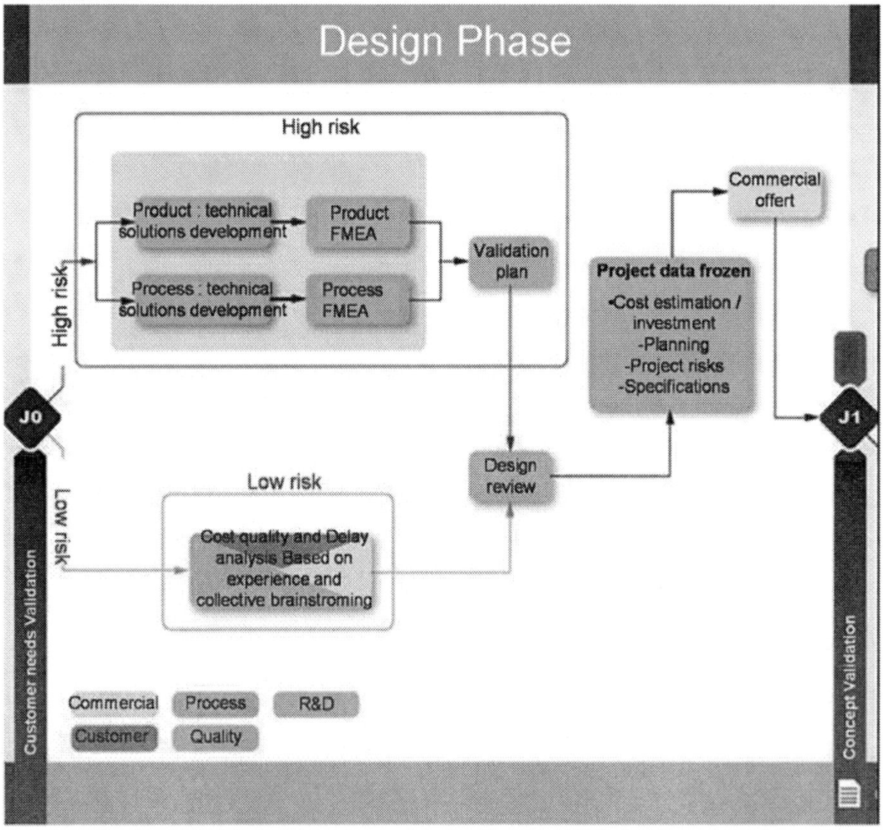

Fig. 2. Short project template: design phase.

3.3 Automating the implementation of projects

A module for creating projects allows the quick and easy initiation of projects. Based on performance indicators which allow knowledge capitalization of the company, the project templates (long and short) are continuously optimized. This adjusts the standard duration of tasks, the actual workload, and the management of standard documents.

This automated implementation allows having a look ahead of the whole project, thereby improving the timeliness, the organizational quality of the project, and also being more reactive to customer needs in terms of quality, cost, and time.

Moreover, by using a module for project duplication, projects of different lengths can be created from standard project through automated coefficients, thus setting the length of the tasks automatically for new projects.

Thus, the user avoids performing an update of the duration of each task of a new project and can extract quickly a schedule of the project (Figure 3).

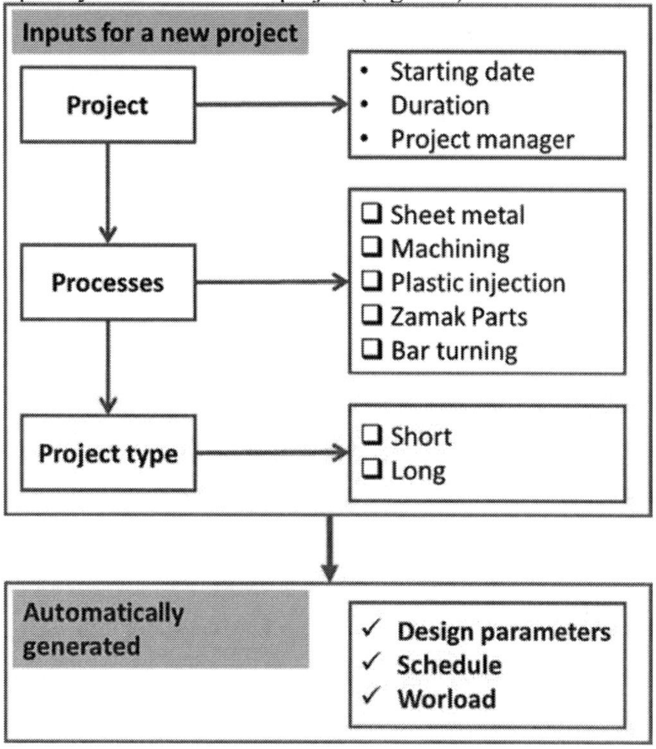

Fig. 3. Initiation of a new project.

4 Conclusion

Lean engineering deals with the reduction of waste in the product development process. Waste is considered as any task without a significant added value. Reducing these tasks will free up valuable time that could be used for innovation.

Lean engineering concepts should be first applied for routine design which account for nearly 80% of the time in a design office.

We proposed in this paper to add semi-automated project management functionalities in a PLM to generate significant time savings in the implementation of new projects.

Resource management and automatic verification of workloads allow for project initiation with realistic durations and avoid resource overload.

Once the PLM tool is set up in the company and used by all team project members, the prospects for a highly productive engineering design can be deployed in the product development process.

Indeed, after company knowledge capitalization by coupling the ACSP tool with KBE (Knowledge Based Engineering) techniques [5], it becomes possible to automatically generate engineering applications. The goal being always the reduction of the time dedicated to routine tasks.

References

1. T.R. Browning, IEEE Trans. Eng. Manag., **48** (2001)
2. T. Boudouh, C. Vernier, N. Lebaal, S. Gomes, *14th International DSM Conference*, Kyoto, Japan, (2012)
3. S. Gomes, D. Monticolo, V. Hilaire, M. Mahdjoub, Int. J. Prod. Lifecy. Mgt, **2** (2007)
4. G. Sohlenius, Ann. CIRP, **41** (1992)
5. S. Gomes, P. Serrafero, D. Monticolo, B. Eynard, *International Conference in Product Lifecycle Management*, Lyon, (2005)

Product development process improvement: A review of Design Structure Matrix methods

Toufik Boudouh[1,*], and *Larbi* Bendada[2]

[1]ICB UMR 6303, CNRS, Univ. Bourgogne Franche-Comté, UTBM, Belfort, France
[2]Université Larbi Ben M'hidi, Oum El Boughi, Algeria

Abstract. Iteration is an important characteristic of the design process. The Design Structure Matrix (DSM) is a powerful method allowing modelling, analysis, and optimisation of iterations in the design process. Boosted by the concepts of concurrent engineering, it has been widely developed during the decades by researchers from both academia and industry. The objective of this paper is to provide a review of different DSM based techniques used to improve the performance of the product development process.

1 Introduction

Engineering design is crucial for the competitiveness of manufacturing companies. An optimal timing and scheduling of the design process is a critical factor in the successful time-to-market completion of the product development process. Design is a multidisciplinary process and then requires a large variety of approaches to be understood. Therefore, the existence of a great number of design process models should not be surprising, and we have to consider them as complementary rather than competing. Design is also a dynamic process because of to the continuous changes in the flow of information among its tasks.

In this work we will focus on the iterative aspect of the design process. Iteration is defined as the process by which a solution is approximated step by step [1]. It is also defined to be the repetition of design tasks due to the arrival or discovery of new information [2]. Iteration has been used as an important element in several research works for developing models of the design process. These models are used to make important managerial decisions in planning design processes.

The design structure matrix (DSM) is one of the methods used to model and analyse iterations in the design process. DSM is a network modelling tool used to model the constituting elements (tasks, components, people...) of a system (design project, product, organization…) and their interactions. In the last decades, an important research work has been undertaken to develop and apply DSM to a wide range of engineering management problems, including product design, systems engineering, project planning, and organization design. This led to the emergence of different types of DSMs. Browning [3] proposes a classification of DSMs into four categories: time-based DSMs with (1) activity-based DSMs and (2) parameter-based DSMs; and static DSMs with (3) component-based DSMs and (4) people-based DSMs. We propose to provide a review of the main DSM models found in the

* Corresponding author: toufik.boudouh@utbm.fr

literature. We present also some techniques used to improve the design process by manipulating the DSM.

The remainder of the paper is organized as follows. We provide in the next section a brief description of iterations in the design process and a review of models accounting for iterations. We present then the basic DSM model and its extensions. Some methods used to optimize the DSM are then presented.

2 Iterations in the design process

Design is an information processing activity. Because of the complex information dependencies that exist between design tasks, we cannot achieve the design process using a straightforward procedure. Therefore, iteration is fundamental to resolve design problems. We can observe two types of iterations in the design process: planned iterations, caused by coupled tasks or uncertainty; and unplanned iterations, caused by design errors or the introduction of new information. Understanding and controlling iterations can improve the design process, and reducing them would have positive effects for the product development cycle time. In many papers it was reported that iterations represent up to 70% of total development effort in a design process [4]. It is then important to consider the iteration aspect of design tasks when developing models of the design process. The introduction of concurrent engineering concepts, in order to reduce the product development time, has drawn attention to consider thoroughly the iterative aspect of the design process.

Several models of the design process that account for iterations are derived from the DSM method. One of these models is the sequential iteration model developed by Smith and Eppinger [2], in which they use linear systems theory to identify controlling features of iteration in a coupled development process. The authors developed also the work transformation matrix model [5], which has a fully parallel structure. In this model, the tasks are re-sequenced and an expected duration for each sequence is determined. While both of these models are useful in characterizing the two extreme cases of product development (parallel and sequential iteration) for any number of tasks, they do not model intermediate scenarios where overlapping might be more appropriate.

Carrascosa [6] developed a mathematical model based on characterizing the information exchanged between tasks in terms of probability of change and impact. DSM has been used in this model as an interface to graphically represent the information flow. The model provides a basis for determining the appropriate task sequence and degree of concurrency to minimize development time and cost.

Belhe and Kusiak [7] proposed a different matrix based methodology to define and sequence tasks by mapping their technical relationships. The model developed includes probabilistic OR and XOR (exclusive OR) relationships between tasks. In this model, the probabilities of executing one or more of the OR/XOR paths are dependent on the iteration number, and these probabilities are fixed in advance. In this model, the length of each task is fixed and deterministic and does not change as the iteration process progresses.

In addition to these models, other models of the iteration process in design do not use the DSM approach. The model developed by Boudouh et al. [8] presents the effects of parallel scheduling. Doing tasks in parallel allows the completion of the design process more quickly. On the other hand, doing tasks in series leads to less tasks requiring repetition, and therefore the development cost is lessened. The objective of the model is to determine how much parallelism is desirable, and whether minimizing development time justifies an increase in development cost.

Krishnan et al. [9] developed a model for overlapping nominally sequential tasks in order to reduce development lead time. In their model, the downstream activity begins with preliminary upstream information and incorporates subsequent upstream design changes in future iterations. They present a framework to determine how to disaggregate design information and overlap consecutive stages based on the evolution and sensitivity properties of the information exchanged.

3 DSM models

3.1 Basic model

The original DSM was first developed in the 1970's by Steward [10] to analyse information flow. The objective was to apply the square matrix to represent a network of design variable interactions in order to solve large systems of equations. The order in which the variables are solved is determinant to minimize the number of iterations needed to solve the problem. DSM was then picked up in the early 1990's at MIT by Eppinger [11] and applied to a number of industry projects in the manufacturing industry.

The DSM is a matrix representation of the elements of a system and their interactions. If the considered system is the design process, the corresponding DSM will be a square (N * N) matrix, mapping the interactions between the set of N design tasks. In the DSM method, we assume that each design task can be modelled as an information processing task, using information supplied by other tasks, and creating information for other ones. The original DSM matrix is square with one row and one column per task. The tasks are listed in a roughly chronological sequence of execution. In a simple DSM, diagonal elements of the matrix are not used. Off-diagonal elements indicate information flows between design tasks. DSM was used to model different types of systems and not only the tasks of a design process. Hence, DSM can be used to model the product's architecture. In this case, the elements of the matrix would be the components of the product and the interactions would be the interfaces between the components.

Using DSM, we can represent four types of interactions between design tasks:

- Serial (or independent): Output of the upstream tasks is necessary to execute the downstream tasks, so they are executed sequentially.

- Parallel (or independent): Tasks can be executed simultaneously. Notice that in case where the tasks use the same resource, even if they are independent, they should interact to share the same resource.

- Coupled (or interdependent): Each task needs the information supplied by the other task. In this case, the tasks must be iterated until they converge to an acceptable solution.

- Conditional (or contingent): The execution of downstream tasks depends on the result of upstream tasks. The DSM representation of conditional processes needs the adoption of specific notations.

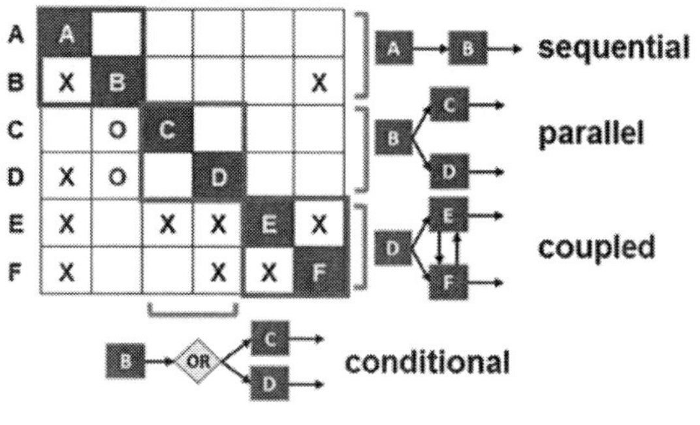

Fig. 1. Relationships types in a DSM.

Figure 1 shows the four relationship types and their DSM representations. Reading across a row indicates all of the tasks whose output is required to process the task corresponding to the row, whilst reading down a column indicates all of the tasks which receive output from the task corresponding to the column. Upper diagonal elements of a DSM indicate the existence of cyclic information flows.

3.2 DSM models

The original DSM method proposed by Steward [10] does not contain quantitative information about the strength of interaction between the matrix elements. Several extensions have been therefore developed to allow quantitative analysis. Over recent years, DSM was used in different domains related to the design process, including product design, systems engineering, project planning, and organization design. This led to the emergence of different types of DSMs. A classification of the different DSM models was proposed by Browning [3].

3.2.1 Activity-based DSM

In this type, DSM is used to model the activities (or tasks) of a process and their interactions. Each task transforms one or more inputs into one or more outputs. This type is well suited to model the activities of the design process. The main objective of the model is to analyse the process architecture and determine the best schedule of the design tasks that will minimise iterations. The DSM elements are sequenced by using manipulation techniques called partitioning.

3.2.2 Parameter-based DSM

This type of DSM is used to model low level design tasks. The elements of the DSM are the design parameters that should be determined in a design task. For complex products, the determination of the value of one parameter depends on several other ones. So, the modification of one parameter will influence many others. The use of DSM will help in determining the sequence of parameter computations that allow minimum need for iterations. As in activity-based DSM, we use partitioning techniques to sequence the elements of the DSM.

3.2.3 Component-based DSM

This type of DSM is used to model the interactions between the different components of a product. Many interactions can be depicted through interfaces between the components. The analysis of component-based DSM results in defining clusters which are sets of components grouped on the basis of certain relationships. The manipulation techniques of the DSM used in this case are called clustering. The results of the analysis could be for example the determination of modular designs.

3.2.4 Team-based DSM

This type of DSM is used to model the interactions among units in an organisation. Organisations represent a network of people sharing a common objective such as the development of a new product. Analysing a team-based DSM highlights interteam relationships which leads to a better understanding of interfaces between teams and improve coordination and communication. Clustering is the manipulation technique used to analyse team-based DSMs.

4 Reducing iterations in the design process

In this section, we present an example of the DSM analysis. We focus on time-based DSMs and especially task-based DSMs. After modelling the design process using the DSM technique, we obtain the as-is model in which the iterations are highlighted and visible in the matrix. It is then possible to reduce them by applying a technique called partitioning or sequencing. This is an operation that consists in reordering the DSM rows and columns such that the feedback marks are eliminated or minimized in the new DSM. Design activities or tasks are then rescheduled accordingly. Figure 2 represents an initial DSM of a design process with 11 activities.

	A	B	C	D	E	F	G	H	I	J	K
A				X							X
B				X							
C				X			X		X		
D											
E							X				
F								X		X	
G			X		X						
H											
I			X	X			X	X			
J											
K				X		X					

Fig. 2. Example of an initial DSM.

Upper diagonal elements of the matrix represent the feedbacks or the iterations. In this example, we have 9 elements in the upper diagonal part of the matrix. They should be eliminated or minimized in order to improve the performance of the design process. By applying a partitioning algorithm, we obtain a new DSM (Figure 3 left) in which the number of upper diagonal elements is reduced to only 3.

There are several heuristics used to partition a DSM. They generally differ just in the way they identify the loops in the DSM. The heuristics attempt then to find the sequence of the design tasks so that each task do not begin until it receives all the input information needed to its execution from its predecessors. We can find in the following references the details of different techniques used to partition a DSM [12] [13].

	D	H	J	B	C	E	G	I	F	K	A
D											
H											
J											
B	X										
C	X						X	X			
E							X				
G					X	X					
I	X	X			X		X				
F		X	X								
K	X								X		
A	X									X	

	H	J	F	D	E	G	C	I	B	K	A
H											
J											
F	X	X									
D											
E						X					
G					X		X				
C				X		X		X			
I	X			X		X	X				
B				X							
K			X	X							
A				X						X	

Fig. 3. Partitioned DSM.

The use of partitioning heuristics does not always yield an optimal partitioned DSM. For our example of Figure 2, we can find another solution (Figure 3 right) that seems better than the one presented in Figure 3 left. When there is no quantification of the strengths of the relationships between tasks in a DSM, we should then try to have the remaining upper diagonal elements of the DSM closer to the diagonal. In this case the iteration process will be reduced. We have developed in [14] a new approach to partition a DSM. It is formulated as an integer problem and an optimal solution is possible to obtain.

5 Conclusion

We presented in this paper a quick review of DSM models. DSM is a powerful method and tool to capture the main interactions among the elements of a system. In the field of product development, DSM has been identified to be useful because it can generate and represent design process architecture, and, by manipulating operations (partitioning and clustering) it can improve the process, reduce cost, duration, and risk.

References

1. G. Pahl, W. Beitz, *Engineering Design: A systematic approach,* (Springer-Verlag, 1996)
2. R.P. Smith, S.D. Eppinger, Manage. Sci., **43** (1997)
3. T.R. Browning, IEEE Trans. Eng. Manag., **48** (2001)
4. T. Boudouh, C. Vernier, N. Lebaal, S. Gomes, *14th International DSM Conference*, Kyoto, Japan, (2012)
5. R.P. Smith, S.D. Eppinger, CERA, **6** (1998)
6. M. Carrascosa, S.D. Eppinger, D.E. Whitney, *ASME Design Automation Conference*, Atlanta, (1998)
7. U. Belhe, A. Kusiak., J. Mech. Des., **118** (1996)
8. T. Boudouh, D. Noyes, Int. J. Mech. Prod. Syst. Eng., **6** (2002)
9. V. Krishnan, S.D. Eppinger, D.E. Whitney, Manage. Sci., **43** (1997)
10. D. Steward, IEEE Trans. Eng. Manage., **28** (1981)
11. S.D. Eppinger, T.R. Browning, *Design Structure Matrix Methods and Applications*, (MIT Press, Cambridge, 2012)
12. S.H. Cho, S.D. Eppinger, *Proceedings of DETC2001/DTM-21691 conferences, ASME*, (2001)
13. S.H. Cho. *An integrated method for managing complex engineering projects using the design structure matrix and advanced simulation*, MS thesis, (MIT, 2001)
14. T. Boudouh, O. Grunder, App. Mech. Mater., **760** (2015)

The effect of cushioning system on impact attenuation of athletic footwear

Stergios Maropoulos[1,*], *Giannis* Korakidis[1], *Dimitris* Fasnakis[1], *Sophia* Papanikolaou[1], *Maria* Papagiannaki[2], and *Fotini* Arabazti[2]

[1]Department of Mechanical Engineering and Industrial Design, Technological Education Institute of Western Macedonia, Koila, 50 100 Kozani, Greece.
[2]School of Physical Education & Sport Science, Aristotle University of Thessaloniki, Ippokratous 22 Ag. Ioannis, 62 122 Serres, Greece.

Abstract. Several studies sought to determine impact attenuation of shoes through the evaluation of force platform or accelerometer measurements. Recent literature however points out methodological flaws of "subject inclusive" experimentation, as this approach is unable to evaluate neither the foot/ground interface nor the absorbed energy allocation. This study provides a holistic evaluation of the biomechanical response of athletic footwear to individual runner characteristics through advanced mechanical testing.

1 INTRODUCTION

As a periodic motion, running exposes the human musculoskeletal system to intense impacts amounting up to 2.5 times the runner's weight. The repeated nature of these impacts over long distances has been documented to induce joint overloading and a variety of other running related injuries. Cushioning systems, found in technical footwear, provide protection to recreational and professional athletes in mitigating these transient forces. Lieberman et al. [1] showed that the specific properties of these midsole systems o can change individual gait patterns. These changes are provoked by sensory mechanisms at the foot-ground interface (during foot-strike), which the body perceives as a source of potential injury. Creagh et al. [2] also suggest that the terrain plays a crucial part in the running style and observed that stride length and stance phase increases both, at higher speeds and on soft terrain.

A holistic understanding of kinematics is required to evaluate human locomotion and performance during running. When a moving object (e.g. the foot) collides with a non-moving one (e.g. the ground), there is an exchange of energy and momentum between the 2 objects. This periodic phenomenon re-occurs during each foot-strike and results in a transient force that is transmitted, throughout our musculoskeletal system. Whittle [3] showed that the magnitude of the transient forces and the subsequent transient shock waves are highly dependent on the material between the food-ground interface, suggesting the use of viscous-elastic material for the construction of midsole systems of athletic shoes. With a

* Corresponding author, email: maropou@teiwm.gr

probability of 1 injury for every 1000 hours of training [4], the shock mitigation is arguably the most important aspect of technical footwear.

The ground reaction forces recorded during running have a distinctive pattern, which recent studies have associated to both, body mass and the running speed [5]. Although in theory the more flexible the cushioning system, the less percussive the forces, impact mitigation is a non-linear phenomenon, complicating the prediction of a shoe's absorbing capacity. The ground reaction, during foot-strike, is distributed by the midsole in multiple dimensions (space, time and a subject specific frequency spectrum). As a result, the effect of the cushioning material might be concealed (to some extent) in low frequency components of the active swing phase, inducing forces to other foot parts. Furthermore, most studies [6] assume that the ground reaction force is exercised in only one direction, accelerating the center of mass, whereas gait patterns are more accurately described by multi-body systems [7].

The human body compensates for the transient forces created during the foot-ground strike through both, suitable joint alignment and the presence of viscous-elastic tissue in the heel bed and the joints. Athletic footwear provides additional protection, as its main purpose is to alter the perception of the different environmental conditions in favor of the runner [8], by dissipating the transient shock waves created during the foot-ground interface. However, individuals identify environmental stimuli and comfort in a different way, thus rendering the optimal footwear choice subjective. The choice of the type of footwear cushioning can be affected by the morphological characteristics of an individual foot [9] and other criteria that are not always bound to the shoe quality and performance.

Over the past decade several studies sought to determine impact attenuation of shoes through the evaluation of force platform or accelerometer measurements ([3], [10], [11]). Recent literature however [5] points out methodological flaws of "subject inclusive" experimentation, as the former approach is unable to evaluate neither the foot/ground interface nor the absorbed energy allocation.

Despite the abundance of different running shoes, most of the available cushioning systems can be assigned to one of three materials: foam, gel and air based. Ideally a shock-absorbing midsole, regardless of material type, should dissipate the kinetic energy of the impact, while transmitting a constant plateau force to the runner [12]. It is therefore essential to determine which cushioning system accommodates individual athletes better, in order to render running-related injuries less prevalent.

The primary hypothesis of this investigation is that the shock attenuating capacity of different midsole systems, is strongly related to impact energy and thus, each shoe type should be optimized for a specific body mass. This study provides an in-depth evaluation of the biomechanical response of athletic footwear to individual runner characteristics (weight and impact speed) through advanced mechanical testing and compares the shock absorption capacity of the three types of cushioning systems (foam, gel and air).

2 EXPERIMENTAL METHODS

ASTM standards suggest the evaluation of running shoes through guided impacts ranging from 5 to 7 joules (ASTM F1614–99, [13]; ASTM F1976–06, [14]). Even though it is adequate to evaluate the shock absorption capacity of athletic footwear destined for a 50-percentile male (175cm height and 78 kg body mass) running at a high pace, it is not sufficient to cover a wide spectrum of sub-elite runners, which are expected to significantly vary in body mass and thus impact energy. According to literature, a 50-percentile male should have a shoe size of 42 [15], but anthropometric data suggest that this shoe size would also cover individuals varying by more than ±30 kg in body mass. This indicates that the impact energy would exceed or fall short of the test limits by more than 40%, an error that might increase even further if gender-specific criteria are considered.

As the aim of this investigation was to determine the dependence of impact attenuation properties of athletic footwear by the runner's body mass, only the conditions occurring at the end of the swing phase (foot strike) were simulated. A polymer (ABS) heel was prototyped on an open source 3D printer (MendelMax 2) to match both, the impact pattern of a classic heel-strike runner and the midsole geometry of the tested shoes. This was achieved by texturing the impact module's lower surface [5]. The shoe-heel was mounted on a guidance unit, providing rigid support to the shoe while ensuring the application of the impact load in the desired direction. The assembly consisted of two linear rails, guiding the axial ball bearings on which the heel was mounted [5]. Shock attenuation was measured using an INSTRON CEAST 9350 drop-weight testing machine.

The experimental assembly along with one shoe specimen at a time, was inserted into the device. The load, corresponding to a runner body mass range of 45–70 kg, was propagated onto the midsole through the artificial heel. 5kg impact load variations towards a 10kg under- and 20kg over-weight subject were also considered. Two different strike velocities were tested, 0.4 and 0.5 m/s. Additionally, the mass of the heel-guidance assembly, placed between the force transducer and the shoe, was considered during the drop test calibration to avoid interference with the force and acceleration data.

In order to achieve the body mass range of 45-70 kg some adjustments were made taking into consideration the different strike velocities. For 0.4 m/s the 45-70 kg range translated to 3.6 to 5.8 J of strike energy and equivalent mass of 7.014 to 11.014 kg, while for 0.5 m/s the strike energy was in the range of 5.6 to 8.75 J and the equivalent mass was in the range of 11.014 to 17.514 kg. Three type of cushioning systems (foam, gel and air based) were evaluated for loads corresponding to a female runner of average body mass. For each type of shoe the test was conducted 3 times, to ensure the good repeatability of the results. All specimens were new, as prior impact conditioning would significantly alter the shock attenuation capacity of footwear [5]. All the tests were conducted at 25 °C and normal humidity. Several parameters were registered during each test. The on shoe (cushioned) impact force was registered along with the maximum vertical displacement and the % force attenuation.

3 RESULTS

The test results for both speeds, where mild running corresponds to 0.4 m/s and fast running to 0.5 m/s, and for three types of midsole systems are shown in Figure 1. For the gel based cushioning almost no difference was observed among the two test speeds. Foam based cushioning lead to slightly varying impact attenuation, when considering the 0.4 and 0.5 m/s impact velocity, whereas air based cushioning exhibited rather significant differences. Specifically, for the air technology fast running produces 8% less impact force compared with mild running across the whole range of body masses examined. As for the foam cushioning fast running develops 2-3% more force, revealing a less loading-rate dependent behavior when compared to air.

Further examination of the results reveal a strong relation of the impact force, generated during foot-strike, to the runner's body mass. This was notable for both, air and gel cushioning, while the foam technology was not greatly affected by the weight, at least at the range tested in this study. For the gel cushioning the difference in the force dissipation performance between the highest and the lowest point is 250N and for the air cushioning is 140N. At this point it must be considered that these are impact forces that transmitted to the musculoskeletal system at every foot-strike, thus indicating the significance of footwear choice in terms of fatigue phenomena over long distance running.

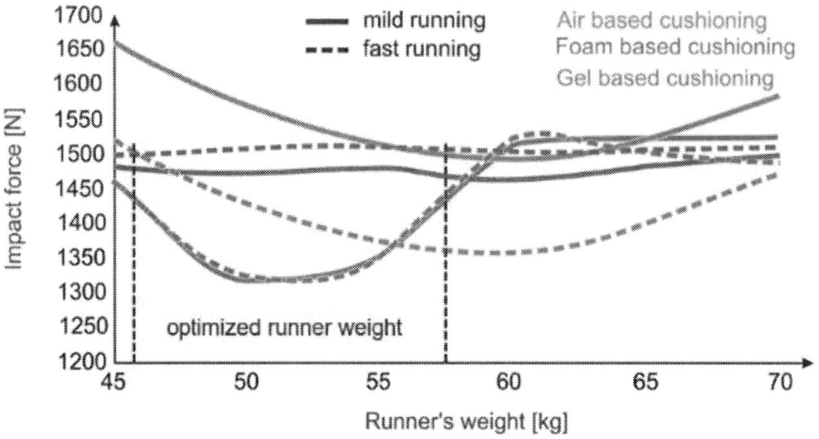

Fig. 1: Measured impact force for all types of cushioning.

Gel based cushioning exhibited a superior performance when compared to the other midsole systems, at least at a body mass range of 45-56 kg, as is shown in Figure 1. For heavier athletes, any cushioning might be suitable depending the body mass and the type of training. Specifically, for fast running the air cushioning systems accommodates athletes in the range 56 to 70 kg better, as the impact force transferred to the foot is significantly lower throughout this span. However, the performance of air cushioning for mild running is the worst of all three in the whole range tested in this study, excluding the 60-65 kg span. For mild training the foam based technology is superior to the other two for runners with body mass from 58 to 70, making it the best choice in this body mass range.

4 DISCUSSION

The investigation provided refined insight on running induced impact transmission to the human body. The results showed that the impact attenuation properties of the shoe are weight depended for the two out of three cushioning types tested in this study.

The pressure relief (responsible for the comfort) in the area of the foot is dependent of the elastic properties of the material used for the footwear midsole, while the dissipation of the transient shock wave (injury prevention) during foot strike depends on properties such as viscosity [11]. If the midsole is too elastic energy will be rebound to the foot, which will result in a higher impact force. Viscous midsoles, on the other hand, will result in lower forces, but the reduction in elasticity may slow down the shape recovery of the material [3]. This is the reason why technical footwear usually employs viscous-elastic materials as cushioning media. Viscous-elastic materials are however strain-rate dependent, a response that was affirmed by the catalogued results. The loading rate, which is strongly dependent on activity type (e.g. running will result in higher impact velocities than walking or even jogging), is considered an important factor for midsole materials. At higher strain-rates, materials are expected to display increased stress values [16], which may influence the activation of secondary absorption mechanisms, often found in cushioning structures.

The results of this investigation are in line with this. Gel based cushioning is generally the result of a composite sole material, comprising of a gel surrounded by a foam. As a result, these cushioning types absorb energy by dissipating the load though the gel phase, while evoking a secondary response once the gel is deformed beyond a threshold that forces it against the foam. This induces a bi-directional strain that is only activated at higher strain-rates, thus providing superior support during running. This contributed to the superior performance of the gel-based midsole, when compared to the other two cushioning systems. It should however be mentioned that this was observed for a specific loading range and rate, while other limitations include the use of just one shoe size and gender consideration.

Nonetheless, optimal footwear choice remains an important issue for millions of recreational and professional runners. According to the results of this investigation, the impact dissipation capacity of footwear systems depends on both, body mass and activity type (loading range). As a result, the choice of the athletic footwear must dependent on the anthropometric characteristics of the athlete, as a less optimal shoe will lead to increased impact loads that might prove strenuous if exposed over long distances.

5 CONCLUSION

The principals of momentum and energy preservation would suggest the impact and reaction force, are a product of body mass and acceleration/deceleration. In these terms, athletes with increased body mass should be exposed to higher impact forces when compared to lighter subjects.

The results of this investigation clearly state that this is not the case for the midsole cushioning systems tested. Runners with lower body mass may subject to significantly increased impact forces, comparing to heavier ones, as a midsole may be underperforming if out of its optimal range.

Although modern footwear is capable of properly attenuating the shock occurring during foot strike, athletic shoes should be carefully chosen in order to avoid muscle and joint related injuries.

REFERENCES

1. D. E. Lieberman, M. Venkadesan, W. A. Werbel, A. I. Daoud, S. D'Andrea, I. S. Davis, R. O. Manq'eni, Pitsiladis, Nature, **463**, 531–535, (2010)
2. U. Creagh, T.Reilly, A. Lees, Ergonomics, **41**, 7 ,1029-1033, (1998)
3. M. W. Whittle, Gait and Posture, 264-275, (1999)
4. W. van Mechelen, **14**, 5, 320–335, (1992)
5. A. Tsouknidas, M. Pantazopoulos, D. Sagris, D. Fasnakis, S. Maropoulos, F. Arabatzi, N. Michailidis, article ID 7136238, (2017)
6. M. R. Shorten, International conference on the Engineering of Sport, (2002)
7. N. Pandey, S. Roshan, R. Kharate, M. Sonawane, V. Bhivate, N. S. Ujwal, **3**, 5, 66–70, (2014)
8. J. J. Gibson, Houghton Mifflin, (1979)
9. M. Kouchi, M. Mochimaru, H. Nogawa, and S. Ujihashi, Paper presented at the 7th Footwear Biomechanics Symposium, (2005)
10. M. F. Bobbert, H. C. Schamhardt, B. M. Nigg, J. Biomechanics, **24,** 1095-1105, (1992)
11. H. A. Gruber, K. A. Boyer, T. R. Derrick, J. Hamill, Journal of Sport and Health Science, 113-121, (2014)
12. R. A. McNeil, American Scientist, **72**, (1984)
13. ASTM F1614–99, 8 Applied Bionics and Biomechanics ASTM International, (2006), https://www.astm.org
14. ASTM F1976–06, ASTM International, (2006), https://www.astm.org
15. R. H. Miller, B. R. Umberger, J. Hamill, G. E. Caldwell, Proceedings of the Royal Society, 1498-1505, (2012)
16. K.D. Bouzakis, G. Maliaris, A. Tsouknidas, Computational Materials Science, **59**, 133-139 (2012).

Decision support in maintenance activity by using modern techniques for equipment condition monitoring

Ionel Paunescu, *Paul Liviu* Paunescu, and *Ştefan* Velicu*

University Politehnica of Bucharest, Romania

ABSTRACT. As the complexity of the equipment and embedded technologies grow, the decision to take action for keeping the normal operation condition is increasingly challenging. The development of methods and instruments meant to improve availability, reliability and maintainability of equipment involves research on the of the related subassemblies and components in various conditions of operation. To maximize technical system performance, to ensure the operational safety and to decrease functioning costs, the authors of this paper propose the implementation of some monitoring techniques, using technical indicators specific to the system (markers for maintenance) and an innovative process based on the electricity footprint method for monitoring the operational behavior of the critical items in a kinematic chain. The analysis of the monitoring techniques presented in this paper leads to an effective act for keeping the normal operation condition of a technical system.

1. Introduction

Starting with the 9th decade of the 20th century, many industrialized countries used the specialized software products for technical systems maintenance. In our country, it took a long time to show that the maintenance of technical systems is included in the category of computer-based activities.

The paper [1] specifies that the management of maintenance works takes into account several approaches, namely:

- proactive scheduling and planning;
- team work;
- contractors-oriented productivity;
- continuous improvement of maintenance process;
- continuous improvement of materials and services purchase process.

* Corresponding author: velstefan@hotmail.com

To maintain the operational status of a technical system it is necessary to consider the existing types of maintenance, namely: scheduled maintenance and unscheduled maintenance.

These two large types of maintenance are classified into groups and subgroups as follows:

A) Group 1- unscheduled: - reactive.
B) Group 2- scheduled: - preventive,
- proactive
- by monitoring;
- scheduled.

The first group (A) has a short path to follow, but also a major drawback difficult to control during the production process, namely the action taken when a component of the technical system fails. Although the costs of maintenance operations in this variant are reduced to labor costs and to the equivalent value of the defective component or group of components, the overall costs for the re-commissioning of the technical system are increased with the costs of machine down-time and also the costs of the damages caused by stopping the manufacturing line (when there is no backup equipment) [1] [4].

The second group (B) has a more complex path to follow, the costs for maintenance operations are higher because of monitoring equipment and staff specialization, but the moment of intervention in the equipment can be controlled [1].

Regarding these techniques that currently use modern but expensive instrumentation (Figure 1), this research paper puts particular emphasis on the implementation of sensors on certain critical components of the technical system, the monitoring of preset parameters by means of cheap and user-friendly instrumentation that reduces the operational risks [2].

Fig. 1. Monitoring instrumentation (Fluke 430 Series, power quality analyzer).

For maintaining the equipment operating costs at the value of proper operation period it is effective to use the method of monitoring some technical indicators specific to the system (maintenance markers) which lead us to balance the result obtained with the sustainable costs (Figure 2).

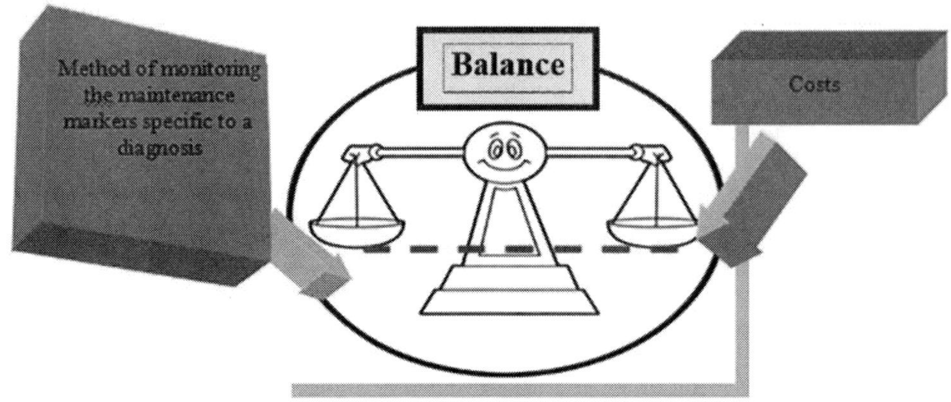

Fig. 2. Balance of maintenance.

2. Monitoring system of kinematic chain faults based on electricity footprint method as maintenance marker

To simulate the operation of a kinematic chain it has been developed the dynamic model of this one (Figure 3). This model basically describes the relationship between the instantaneous angular position of each shaft and the torque of the electric motor reflected in the motor shaft and electric parameters (voltage, current). Any disturbance of the rotational movement of the kinematic chain is transmitted to the motor shaft. By monitoring the current absorbed by the motor and the rotational speed of a shaft it is possible to determine accurately which is the mechanism that generates disturbances (gearing, bearings, misalignments etc.) and the evolution of degradation over time.

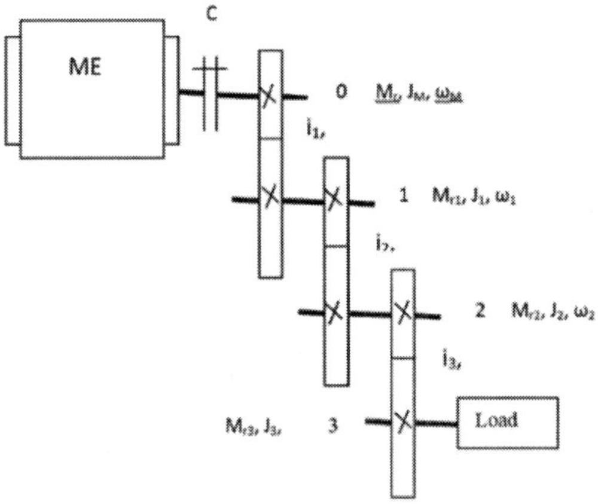

Fig. 3. Dynamic model of the kinematic chain.

Let us assume that a certain constraint to a rotation of the main shaft, starting by the angle θ_3 , results in the variation of the speed, namely of the angular velocity ω over a time Δt.

This variation of the angular velocity determines an angular acceleration ε_3, where:

$$\varepsilon_3 = \frac{d\omega_3}{dt} \quad (1)$$

which shall be propagated in the motor shaft:

$$\varepsilon_M = \frac{d\omega_3}{dt} \cdot i_1 \cdot i_2 \cdot i_3 \tag{2}$$

In this case, the dynamic torque in motor shaft will be:

$$M_{din} = J \cdot \varepsilon_M \tag{3}$$

The angular velocity can be expressed depending on speed:

$$\omega = \frac{2 \cdot \pi \cdot n}{60} \tag{4}$$

In which case the dynamic torque will be:

$$M_{din} = J \cdot \frac{dn_M}{30} \tag{5}$$

This dynamic torque will entail a variation of the motor shaft power:

$$\Delta P_{din} = M_{din} \cdot \Delta\omega_M \tag{6}$$

Given the constant voltage (Uconst), the power variation will be seized by the current variation ΔI. The change of current intensity size measured at the main electric motor is a consequence of changes in the operational parameters of some elements that compose the technical system. In this case, the aforementioned equations are changed accordingly. Going on reverse logic we can reveal the components that disturb the rotational movement (bearings, gearing, wear, eccentric masses etc.) and thus a maintenance decision can be taken in order to keep the normal operating parameters of the equipment [5].

3. Experimental researches

Using these data, the authors propose a method to measure the parameters of the electric motor for kinematic chain drive (current, rotational speed and temperature) and to determine a relationship between the variation of these parameters and the proper operation of movement transmission mechanisms. The hypothesis was verified on an experimental stand so designed to be able to simulate different conditions of failure of kinematic chain components: uniform and nonuniform wear of a bearing, coaxiality error of motor coupling, an eccentric mass on a shaft, gearing error or combinations of these ones. The system includes a module for data acquisition, represented by an interface NI-DAQ 6251 (b), connected to a computer(a) via USB port, an extension module(c) which contains sensors for measuring the current, voltage and temperature of the motor, and also a LabVIEW software application for processing and representing the information recorded by the acquisition module [3].

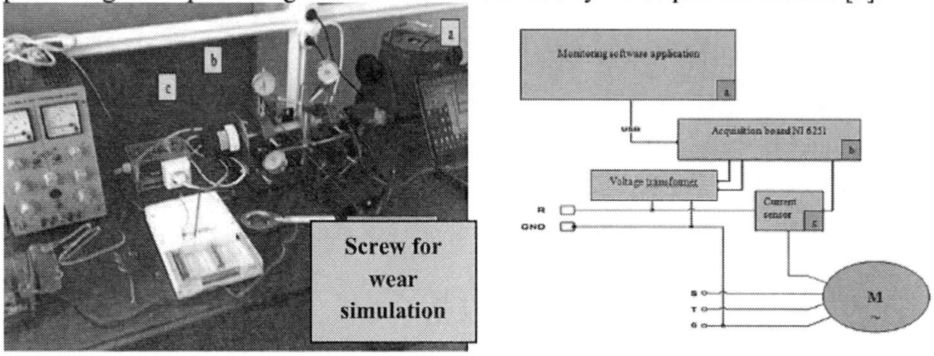

a) b)

Fig. 4. Monitoring system of electricity footprint (a – research stand, b - block diagram).

The research paper focuses on studying the thermal and mechanical behavior of the electric motor, transmitted by the elements of a kinematic chain in normal status and in wear conditions too. The electric signals obtained by using LabVIEW software, adapted to this research paper, were analyzed and compared with the standard signals obtained for "NORMAL" status - MT (machine tool) operation within the initial parameters, according to the Instructions Book [3].

The shape of the signals obtained is structured as follows:

a) constant amplitude of the signal displayed (with determined values, up to the value of 800 mA), corresponding to the stage of kinematic chain without fault;
b) increase of signal amplitude over the value of 800 mA, accompanied by an increase of motor housing temperature over 60°C, corresponding to the stage of worn out element in the kinematic chain(damaged bearing).

The parameters of electric motors (rated current and power consumption) and the increased temperature of motor housing can be indicators that signal the wear of an element of the kinematic chain during motor operation. In the results of our research, the increase of the value of current is associated with the beginning of an abnormal operation caused by the wear of some components that can be of critical nature.

This size can be controlled; therefore when a maximum value is reached, the decision-maker (specialist of maintenance) can stop the system for adjustment/change of components (Figure 5).

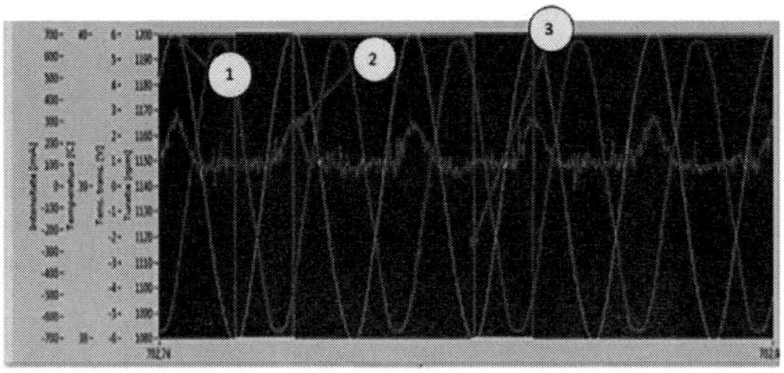

Fig.5. Virtual instrument of monitoring (current intensity- 1, temperature- 2, motor speed- 3), LabVIEW application.

As the operating hours number increases, the shape of current changes its amplitude. This size can be controlled and if there is a maximum value, the system can be stopped by the decision-maker (specialist of maintenance) for adjusting/replacing a component.

Table 1 shows the values of current, temperature and rotational speed. The increased values of current and temperature are obvious, also the decrease of the rotational speed entailed by the equipment running duration. We notice that during a normal operation the values are constant and the equipment operation keeps within the set scope - of normal operation.

By monitoring the system, we observed and recorded increased values of markers associated with the increase of the number of operation hours.

The research paper highlights the advantage of monitoring the parameters by means of markers set for the vital components of the equipment and helps the specialist to decide when to take action.

Table 1 Measured values (normal , fault).

No	Time (h)	Amperage (A)	Speed (RPM)	No	Time (h)	Amperage (A)	Speed (RPM)	Temp C°
1	1	0.125	1440	1000	10	0.7	1200	41
2	1 h 30′	0.125	1440	1001	10h 30′	0.7	1380	41
3	2	0.125	1440	1002	11	0.71	1380	43
4	2 h 30′	0.125	1440	1003	11h 30′	0.72	1380	43
5	3	0.125	1440	1004	12	0.72	1380	43
6	2 h 30′	0.125	1440	1006	12h 30′	0.72	1380	43
7	4	0.125	1440	1007	14	0.73	1400	44
..	....	....	....	n	n+1	0.85	1400	60

Figure 6 shows the analysis of monitored markers values. This analysis helps the maintenance department to:

- make maintenance decisions at the most appropriate time;
- achieve a strategy of maintenance and to develop the responsibility of the involved staff;
- preserve the technical system reliability according to the recommendations of its manufacturer.

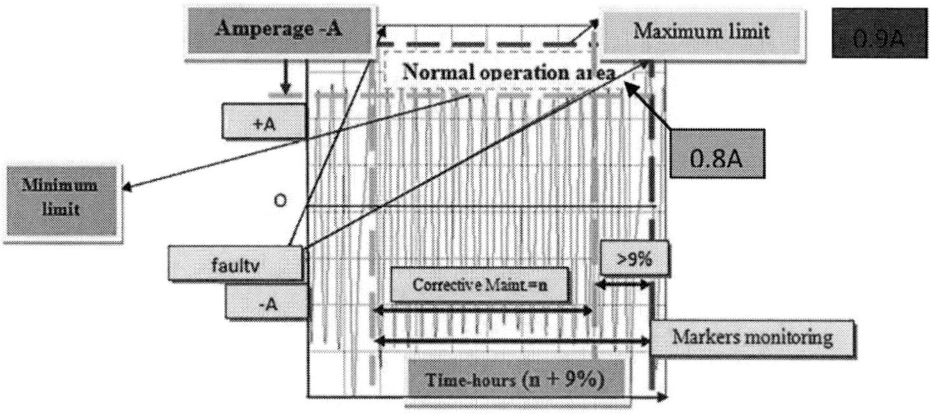

Fig.6. Monitoring of the operation area by electricity footprint.

4. Conclusions

The proposed research method allows obtaining valuable information that lead to the diagnosis of equipment/kinematic chains by using the monitoring and modern analysis methods. The thorough knowledge of equipment components enables to find the mechanisms that explain the response of the material to the action, generally destructive, of some exterior factors such as the mechanical, chemical or thermal ones. Also, the establishment and analysis of some suitable markers of maintenance can technically justify certain decisions. The selection of the most appropriate parameters to be monitored and the improvement of the analysis algorithms were the targets of the research whose results were presented in this paper. Regarding the scientific perspectives that this paper presents: a dynamic mathematical model can be determined in the future that can simulate the functioning of the equipment till the maintenance intervention.

References

[1] C. Bîrleanu, Student Scientific Session (May, 2015)
[2] http://en-us.fluke.com/site/fluke-sitemap/
[3] I. Păunescu, P. Păunescu, St. Velicu, Fourth International Conference MIDIS 2015-Sibiu (2016)
[4] St. Velicu, I. Păunescu, D. M. Paraschiv, ICMAS **9,** 1 (2014).
[5] Al. Velicu. *Research on integrated maintenance of technological systems* (UPB, 2010)

Some aspects regarding the relationship between consumer and Industrial Design

Andrei Dumitrescu*

Politehnica University of Bucharest, Department of Manufacturing Engineering, 313 Splaiul Independentei, Bucharest 060042, Romania

Abstract. The product manager, as the leader of product-related activities, has the task of identification the product's main features that will distinguish it on the market. Some of these will be related to the industrial design. This paper presents the results of an experiment designed to explore the relationship between consumer and industrial design. The experiment addressed issues like the consumer's awareness of industrial design, its influence on product assessment, etc.

1 Introduction

The globalization has made a world in which the competition is tougher than ever. When approaching this subject, academic authors and speakers use to quote brand guru David A. Aaker [1]. He discussed with thousands of CEOs and almost all of them complained about the high competition in their field. Only one was the exception: the director of Panama Canal. But it seems that even that one will have competition: Nicaragua's National Assembly has approved plans to make a connecting channel through their country [2].

Because of large funds invested in technological research, the technology is progressing at a high rate and new improved products appear on the market at shorter intervals of time. In order to cope with quality issues, the vast majority of companies implemented quality management systems (ISO 9000, etc.). So, the cases of poor quality products manufactured by regular companies are very rare.

The result is that the market is flooded with very similar products, possessing almost identical functions with parameter values in a very close interval and at the same (more or less) price. More, the quality is guaranteed for years. The consequence is the impasse of the consumer. How a consumer visiting a virtual shop on internet can easily and reliably choose among hundreds of very similar products?

There are two main strategies that today consumers are using: they rely on well-known and trusted brands or they let themselves "seduced" by the product's industrial design, assuming that a company that invested in product aesthetics also invested in practical functions, usability and quality.

The experts studied the consumer responses to products and discovered that cognition and affect influences the buying behaviour [3, 4]. Behavioural responses are: approach and avoid. The approach response triggers further investigation of the product, leading to a

* Corresponding author: andrei.dumitrescu@upb.ro

possible purchase, while the avoid response directs to the ignoring of the product. More in-depth, it was shown that consumers observing a product attain pleasure from reflecting about its quintessence and not just from pure sensorial inputs [5].

However, the consumer literature generally presents the consumer as a simple observer of the products, who ignores the fact that these products were conceived accordingly to a design specification; and especially that the products were designed to be regarded by the consumer in a particular way. Just few works highlight this particular aspect [6, 7, 8, 9].

Nathan Crilly [9] considers that in some situations consumers can be unaware of the fact that products have been designed, but in other situations they are aware of the design aspect and this influences their reaction. Consumers can inference from the product features and the designer's intentions. These intensions might be: eliciting attention; emphasising the manufacturer's brand; creating attraction towards the product; attribution of meanings; etc.

Crilly [9] chose the mobile phones as the subject of his investigations and the interview as method. By asking questions like: "What do you like or dislike about your mobile phone?", "Why do you think your phone is the way it is?" and "When you are looking at, or using, products such as mobile phones, how aware are you that those products have been designed?", he found evidence for the inference process related to the designer's intentions.

The topic of knowledgeably and not-knowledgeably persons was approached by Odette da Silva, Nathan Crilly and Paul Hekkert [10]. They used students from an industrial design programme and product images & designer intention statements from student projects. Among other conclusions, they discovered that knowing designer intentions enhanced the positive assessment of products. Going further with the reasoning, the researchers assumed that consumers can learn about the designers' intentions from statements made by designers, press releases, marketing campaigns, critical reviews, etc. So, the designers, manufacturers and publicity agencies should directly advertise their intentions in order to get a favourable consumer assessment.

2 Design of experiment

Analysing carefully the results obtained by other researchers, some significant aspects has emerged. In one investigation, mobile phones were used, but it should be mentioned that mobile phones are very personal and involving products, so consumers are more sensitive to all information related to them, including the fact that they have a "designed" status. So, an experiment based on more mundane products should lead to more relevant results.

Another important aspect is related to the use of products (and design statements) from student projects in an experiment with design students as participants. Inevitably, a student project is a by-the-book endeavour, meaning that all design phases are fully completed (including a design statement). In the practical world, some design phases are skipped or partially completed. So, a design statement could be missing for an average product. Also, the design students are definitely more sensitive to design statements than the average consumer. Finally, as indicated above, the consumer is usually overwhelmed by the number of the products available on the market and an associated design statement for each of them (more likely very similar) would make no difference. The design statement would be really relevant for sought-after products (very personal or very expensive etc.).

Without disregarding the results obtained from the researches mentioned above, the author of the present paper considers that the results of a new experiment aimed at average consumers and using a very mundane type of product will complete the knowledge in the field of the complex relationship between consumer and industrial design.

The selected type of product was the set of salt and pepper shakers. In experiment, there were used images of four sets of salt and pepper shakers, considering the third set (the most simple and mundane) as reference. These sets are displayed in Figures 1 – 4.

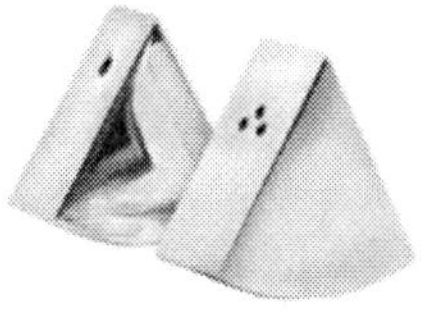

Fig. 1. Set 1.

Fig. 2. Set 2.

Fig. 3. Set 3.

Fig. 4. Set 4.

It was decided that the participants will be students from a technical university. This narrowed the addressed population, but increased the relevance of experiment for the younger population. Also, this eased conducting of experiment.

After reviewing the similar experiments, it was decided to use the following questions:

1. Indicate the aesthetic value of the product. [1 ... 7]
2. What do you like at the product? [open question]
3. What do you dislike at the product? [open question]
4. Indicate your impression regarding the product using only one adjective. [open question]
5. Why do you think that the product had been designed with the features indicated by you? [open question]
6. Indicate again the aesthetic value of the product. (You can indicate the same value as before or a different one.) [1 ... 7]

The reason for repeating the first question similar at the end was to determine if the participant is changing her/his assessment after thinking about the product features and the "designed" character of the product.

3 Experimental results

The experiment was performed with 161 participants (81 female and 80 male participants). All participants were students enrolled at a large technical university in Romania. The participants filled the same questionnaire with 6 questions per product.

The open responses were analysed individually and the essence of the response was recorded. The essence was a feature or a quality. Some features and qualities were general (shape, colour, etc.), others were product specific (antithesis, etc.).

Table 1. Products' features and qualities liked by participants.

Like [%]	Product 1	Product 2	Product 3	Product 4
Shape	51.06	27.84	16.99	32.45
Colour(s)	15.43	42.61	NR	22.34
Texture	5.85	5.68	NR	NR
Material	7.98	2.84	NR	NR
Design	13.30	7.95	7.84	13.83
Utility	NR	NR	38.57	NR
Significance	NR	NR	NR	19.15
Simplicity	NR	NR	22.22	NR
Antithesis	NR	NR	NR	5.32
Other	6.38	13.07	14.37	6.91
Total	100	100	100	100

Note: NR – not relevant values

The products' features and qualities liked by the participants are displayed in Table 1. (Note that a set of shakers was considered as a single product.)

Analysing the results from Table 1, it can be observed that shape and colour were the most used features in the assessment of liked features. Texture and material were used only when they possess on outstanding character (as for products 1 and 2). Design (as an integrative feature for product aesthetics) was entailed seldom. Only when the industrial design was poor, participants were considering qualities like utility and simplicity (product 3). When the product's message was eloquent, people were more inclined to reflect upon the significance circulated by the design object and considered even possible "figures of speech", like antithesis. Concluding, participants were assessing the industrial design in the first instance, taking utility as granted and, only when design was poor, they took into account the utility.

When analysing the disliked features and qualities, it was observed that people were very precise and analytic about what they do not like, more than in the case of liked characteristics. It was like they were afraid to choose the wrong product. The products' features and qualities disliked by the participants to experiment are displayed in Table 2.

Table 2. Products' features and qualities disliked by participants.

Dislike [%]	**Product 1**	**Product 2**	**Product 3**	**Product 4**
Shape	18.87	27.85	22.54	23.59
Colour(s)	23.90	27.85	7.04	15.73
Texture	NR	5.06	NR	NR
Material	11.32	6.33	7.75	6.74
Design	NR	4.43	4.93	2.24
Size	3.77	6.33	NR	7.86
Holes	8.81	8.86	4.93	13.48
Utility	10.07	NR	NR	NR
Triviality	NR	NR	29.58	NR
Other	23.27	13.29	23.23	30.36
Total	100	100	100	100

Note: NR – not relevant values

Again, the shape and colour were used as assessment criteria, but in a lesser degree comparing to the liked features. Texture was an insignificant feature, but material had a greater importance than previously. Design was less considered than before, but now even less. Size was an assessment criterion, but considered only by few participants. Utility was ignored with the notable exception of product 1, which raised the fear of instability. The triviality of "poor" design (product 3) was the most annoying characteristic. If in the previous assessment the holes were seldom mentioned, the holes (especially position and size) were indicated by several participants. A special mentioned should be made about product 4 which was totally liked by 54% participants (no disliked features).

Even if the question regarding the adjectives was aimed primarily to make participants to think more about products' design, also this question lead to significant results. The adjective used by participants are displayed in Table 3.

Because it was an open question, the participants used a large range of adjectives and the individual weight of almost all adjectives was relatively low (lower than 6.8%). The notable exceptions were "simple" (15.53%) for product 3 and "funny" for product 4. Actually, these two adjectives were obvious when looking at the two products. Reasonable choices for adjectives were "beautiful", "nice", "simple", "funny" and "playful". When some participants were in doubt what adjective to use, they chose "interesting".

Table 3. Adjectives used by participants.

Product 1 [%]		Product 2 [%]		Product 3 [%]		Product 4 [%]	
Modern	9.32	Interesting	5.59	Simple	15.53	Funny	14.29
Interesting	5.59	Beautiful	4.97	Trivial	9.94	Beautiful	6.21
Simple	5.59	Nice	4.97	Classic	9.32	Interesting	6.21
Beautiful	4.97	Weird	3.73	Mundane	6.83	Playful	5.59
Elegant	3.73	Modern	3.73	Useful	5.59	Modern	5.59
Nice	3.73	Cute	3.11	Nice	4.35		
Cold	3.73			Beautiful	3.73		
Other	63.36	Other	73.91	Other	44.72	Other	62.11
Total	100	Total	100	Total	100	Total	100

The responses to the open question "Why do you think that the product had been designed with the features indicated by you?" are synthesised in Table 4.

Table 4. Intentions attributed to designers.

Product 1 [%]		Product 2 [%]		Product 3 [%]		Product 4 [%]	
To be sold well	16.15	To grab attention	18.13	For utility	42.24	To grab attention	22.36
To grab attention	14.29	To be distinguished	11.25	For simplicity	7.45	To be sold well	16.77
For utility	13.66	To be sold well	10.63	To be sold well	7.45	For amusement	12.42
To look modern	6.83	For utility	10.00	For lowering costs	6.21	To be distinguished	6.21
To be distinguished	5.59	To look nice	7.50	To grab attention	3.11	For utility	6.21
To match the market needs	3.73	To match the market needs	5.00			To look nice	4.97
Other	39.76	Other	37.50	Other	33.54	Other	31.06
Total	100	Total	100	Total	100	Total	100

Among the relevant intentions attributed to designers, there should be made a distinction between objective intentions and subjective intentions. The former category includes "for utility", "to match the market needs", "for simplicity", "to look nice", "for amusement" and "to look modern" and the latter – "to grab attention", "to be sold well" and "for lowering costs". Unfortunately, the participants felt that the subjective intentions are prevalent when the design is more elaborate.

Odette da Silva, Nathan Crilly and Paul Hekkert [10] discovered, in an experiment briefly described in introduction, that people who are aware of designer's intentions are more inclined to enhance their positive assessment of products. Because of the reasons indicated in "Design of experiment" section, the designer's intentions were not used, but participants were persuaded to think about the "designed" character of products in the Crilly's manner [9]. They were asked to assess the aesthetic value of the products at the beginning and at the end of the questionnaire. The mark averages are displayed in Table 5.

Table 5. Average aesthetic values – before and after [max. 7].

Product 1 [%]		**Product 2 [%]**		**Product 3 [%]**		**Product 4 [%]**	
Before	After	Before	After	Before	After	Before	After
5.09	5.11	4.46	4.44	4.01	4.02	5.78	5.88

It can be easily observed that only in the case of elaborate design with a clear message (product 4), the awareness of the "designed" character determined participants to increase their assessment marks. In the case of mundane design or even neat design, thoughts about the designed character did not improve the assessment of industrial design.

So, reasoning about the product's aesthetics, its "designed" character, and probable designer's intentions does not improve the results of aesthetic assessment of usual products, with the exception of products with an elaborate design. This exception partially confirms the results obtained by researchers that used products designed by-the-book and students enrolled at a design programme as participants to research.

4 Conclusions

An experiment was conducted using usual products (four sets of salt and pepper shakers). The participants to the experiment were university students. Because university students can easily be consumers of salt and pepper shakers, the experiment conclusions can be drawn for consumers. The conclusions are:

Shape and colour are the main features assessed by consumers when asked about what they like or dislike about a certain product. Texture and material are considered only when these features possess an outstanding character. The overall design is considered as an assessment criterion by few consumers. Utility is assessed only when the industrial design is disappointing or when there are doubts about the product's proper functioning. Significance is considered only if the message was really eloquent. Triviality of poor-conceived design can be the most annoying characteristic.

The choices for adjectives to describe product's design can be from a large palette, if the design is not striking in a certain way. "Interesting" is used by consumers as an escape option when judging design.

Consumers consider that an elaborate design indicates rather subjective design intentions like grabbing attention and increasing the sales. Consumers associate a professional designer contribution only with products with an elaborate and neat design.

Consumers enhance their aesthetic assessment of products as a result of the awareness of the "designed" character only in the case of elaborate design.

References

1. D.A. Aaker, *Building Strong Brands* (Simon & Schuster, 2010)
2. https://en.wikipedia.org/wiki/Nicaragua_Canal
3. P.H. Bloch, J. Marketing, **59**(3), 16 (1995)
4. J. O'Shaughnessy, *Explaining buyer behavior* (Oxford University Press, 1992)
5. P. Bloom, *How pleasure works: Why we like what we like* (Vintage, 2011)
6. N. Crilly, J. Moultrie, P. J. Clarkson, Design Studies, **25**(6), 547 (2004)
7. M.E.H. Creusen, J. P. L. Schoormans, J. Prod. Innov. Manag. **22**(1), 63 (2005)
8. P.M.A. Desmet, P. Hekkert, I. J. Design **1**(1), 57 (2007)
9. N. Crilly, I. J. Design, **5**(3), 1 (2011)
10. O. Da Silva, N. Crilly, P. Hekkert, I. J. Design, **9**(2), 21 (2015)

Method for energy-efficient planning of the industrial processes

Gabriel Frumuşanu [1,*], *Nicolae* Badea[2], *Cezarina* Afteni[1], and *Alexandru* Epureanu[1]

[1] Dunărea de Jos University of Galaţi, Department of Manufacturing Engineering, Domnească Street 111, 800201 - Galaţi, Romania

[2] Dunărea de Jos University of Galaţi, Department of Automatic Control & Electrical Engineering, Ştiinţei Street 2, 800210 - Galaţi, Romania

Abstract. In industry, process planning has a major impact on manufacturing performance. This paper proposes a method aiming to increase the efficiency in process planning, more specific in selecting the set of plans with high potential of performance. The method works on the base of processes classification after their energy efficiency. It is applied for processes grouped in the same utility area, and involves the definition of energy classes. A case study, having as purpose to prove the feasibility and effectiveness of the proposed method is also presented. Energy labelling may further become an efficient tool enabling the manufacturing process optimization from energetic point of view.

1 Introduction

The numerous dramatic events from recent years (floods, droughts, devastating storms), due to environment changes issuing from the negative impact of human activities, have brought in front of human kind the new challenge synthesised in the concept of sustainable development, one of its key-features being energy efficiency. The industrial activity, deployed through industrial processes, involves by its nature the use of a consistent share from the energy consumed at global level. Hereby, the energy efficiency of the industrial process is a top priority subject.

The energy needed to perform an industrial process has two main components: the energy required for underlying the process and the energy consumed inside the process, for producing its utile result (e.g. machining a surface, welding two parts etc.). Although the machine and the industrial process have no reason of existing one without another, the existence of two energy components leads to the possibility of speaking about energy efficiency at machine level and process level, respectively. Furthermore, separately rating the energy efficiency of machine and industrial process makes sense, because the same process may be performed on different machines, as well as the same product may be obtained using different processes. In the first case, one might need to choose a machine for running a process (having the energy efficiency as criterion), while in the second, the process for obtaining a product is the choice to be made.

* Corresponding author: gabriel.frumusanu@ugal.ro

The researches presented in literature dedicated to energy efficiency in manufacturing approach the problem at two distinct levels. First one is the level of the natural process, which may be included in different industrial processes. For example, [1] starts from the energetic components of material removal in the cutting process and consider the consumed energy as a function of five parameters (cutting speed, feed rate, tool nose radius, cutting edge radius and rake angle). A similar approach in energetic modelling the cutting process is presented in [2, 3]. The second level is the one of the machine – process couple. In this case, the energy components consumed by both machine and process are put together and the efficiency is globally assessed when milling [4], turning & grinding [5], grinding [6]. More comprehensive approaches are detailed in [7, 8].

In this paper, the energy efficiency problem is approached at an intermediary level, addressing the ensemble formed by the natural process and consumables needed for running the process. By consumables we mean, for example, the cutting tool and the cooling liquid, in the case of a cutting process, the forming die and the lubricant (forming process), the electrode and the inert gas for shield (arc welding), etc. Paper target is to develop a method for energy-efficient planning of industrial processes. Without generality restraining, for sake of discourse clarity we will further address to manufacturing processes (one of the most relevant category of industrial processes). Because any product can be manufactured by following diverse process plans, in practice one have to find the optimal solution, by choosing between the plans with highest performance, which should apply processes with optimality potential. As energy efficiency is addressed here for performance assessment, the tool we propose in order to find the appropriate processes consist in grouping them in energy efficiency classes, defined for procedures utility areas. Next section presents the method proposed for energy-efficient planning, while the third presents a case-study for energy labelling the industrial process. Last section is for conclusion.

2 Proposed method for processes planning

2.1 Planning stages

In manufacturing companies, the first action to be followed after receiving an order for a certain product is processes planning, which has a major impact on manufacturing performance. Processes planning is supposed to cover three successive stages, namely selection, programming, and analysis.

Selection starts by decomposing the order in jobs (at first level) and, subsequently, each job in tasks (at second level), see Figure 1. Job means manufacturing a product component, while task means an elementary manufacturing process (e.g. manufacturing a hole, a key slot or a thread). After order decomposition, procedures available in the considered workshop are assigned to the tasks needing to be achieved. Selection purpose is to generate set of sub-optimal process plans. Here, by *industrial process* we mean the actual deployment of a natural processes set, artificially provoked and taking part correlated, conducted towards an expected result (generically meaning *task accomplishment*). At process input there are four items: materials, energy, time and information. Any process deploys on a logistic bed, here designated as *workstation*. The methodological algorithm according to which the process takes part is called *procedure*.

Programming consists in establishing the actual conditions for running the selected plans, namely values for all parameters of the involved procedures, as well as the setup of workstations assigned in this purpose.

Analysis aims to choose the optimal plan, after the considered performance criterion. This can be done once and for all, before manufacturing beginning, or along the order accomplishment, depending on actual circumstances. Second option takes into account possible modifications of manufacturing conditions, which may lead to the situation when another sub-optimal plan becomes optimal, instead of the previously chosen one.

The proposed method distinguishes by the particular manner for performing the first planning stage, namely selection.

2.2 Selection

As jobs decomposition in tasks is not unique and procedures assignment to tasks either, it follows that more variants of processes planning can be generated. A process plan variant result after selecting from a task pool the tasks required for jobs achievement, the assigned manufacturing procedures from a procedures pool and the support workstations from a workstations pool, and after combining tasks and procedures in processes (Figure 1). As it can be noticed, a procedure may be used in more processes and a workstation may support different procedures.

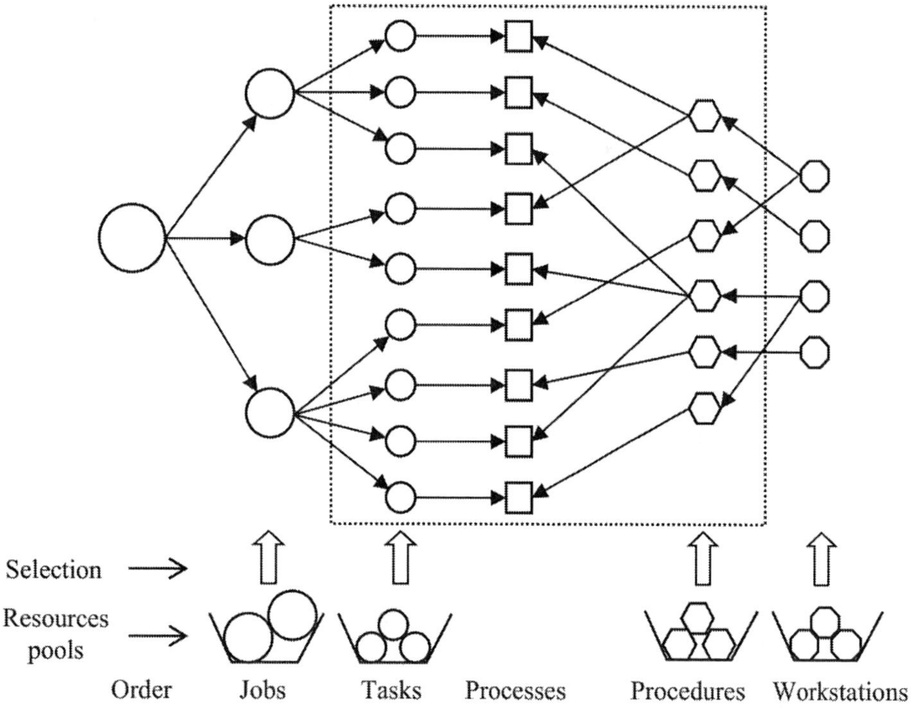

Fig. 1. Graf illustrating a generic variant of process planning.

The final action from selection stage is the assessment of all available plan variants, followed by choosing the set of sub-optimal plans, having high potential of performance. For finding the sub-optimal plans, an appropriate assessment tool is necessary. In our vision, such tool is a system of performance classes. In this paper we will refer, in particular, to energy efficiency classes.

The assessment is proposed to be realized based on energy labelling, which is a *classification* (which highlights the similitude between processes) and not by their *measurement* (which highlights the difference between them), as currently practiced. In our vision, classification is a more appropriated tool than measurement, because it works faster and is more adequate when addressing to process plans potential.

2.3 Processes classification

The proposed classification supposes efficiency assessment for the processes from the same utility area, based on processes similitude and reflecting the potential performance they can more likely reach. Processes classification consists in identifying the energy class to which a given process belongs.

2.3.1. Utility area definition

In order to build an energy classes system, a precursory action needs to be accomplished. It consists in defining utility area, materializing the connection between procedure and task the last one being addressed in general sense. For example, *task* might mean manufacturing a certain type of surface (plane exterior, cylindrical interior, thread etc.), manufacturing with a given precision, or manufacturing surfaces with specified roughness.

One should notice that the same task can be achieved by different procedures (e.g. a cylindrical interior surface can be machined by drilling, boring, turning, grinding), while the same procedure may cover different tasks (milling can be used for generating plane surfaces, profiled surfaces or threads). Hereby, the *utility area* may be formed by *i)* the tasks which can be achieved by applying a given procedure (e.g. the tasks t_1, t_2 and t_4 for the procedure P_1, Figure 2), or *ii)* the procedures covering a certain task (e.g. the procedures P_2, P_3 and P_4 for the task t_3, Figure 2). First of the two utility area definitions has application in product design, when adjusting the tasks to procedure potential. The second is necessary in product manufacturing, if choosing among the procedures covering a task; as this paper is about process planning, from now-on this will be the considered definition of utility area.

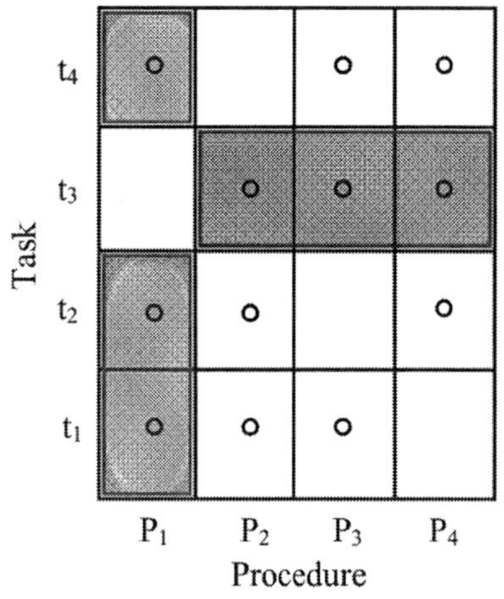

Fig. 2. Utility areas definition.

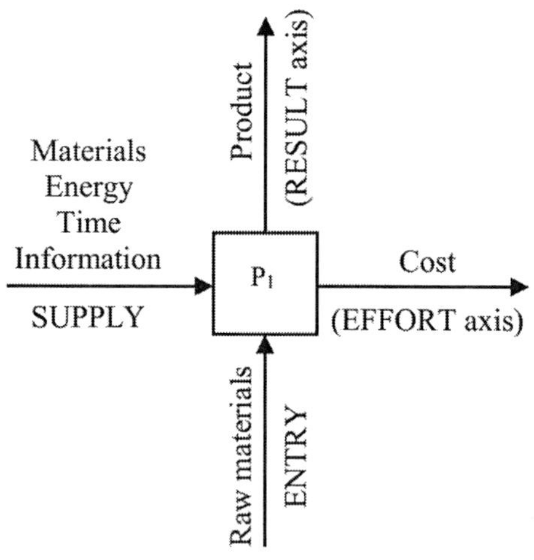

Fig. 3. Procedure flowcharts.

2.3.2. Energy classes definition

If looking now to a procedure (for sample P_1), then we can assign it two axis, namely *Effort* axis and *Result* axis (see Figure 3). Along the first one, as input there is the supply (consisting in materials, energy, time and information) and as output – the corresponding cost. As second axis entry there are raw materials (the workpiece but also materials with technologic role), which are necessary to obtain the product by applying the processing algorithm.

Manufacturing processes classification after their energy efficiency is approached based on the general efficiency definition:

$$Efficiency = \frac{Result}{Effort} \tag{1}$$

The manner of defining and representing the energy classes for the procedures covering a task (for sample t_3) is presented in Figure 4.

In the depicted diagram, each generic process P is represented as a point. The effort involved to run the process and the result issued from this effort, both measured through appropriate indicators, are giving P coordinates. Because procedure and task parameters may vary, P position may also change. The locus of points corresponding to all possible values of parameters for a given procedure defines a close domain assigned to it in the graphic representation. By introducing a third coordinates axis, circular, the energy efficiency of a process is reflected through the angle made by OP segment with *Effort axis*, as suggested on the picture. The energy efficiency classes can be defined now by dividing in several sectors the angular domain comprised between the lines corresponding to extreme efficiencies (inferior/superior), as they can be revealed at classification moment.

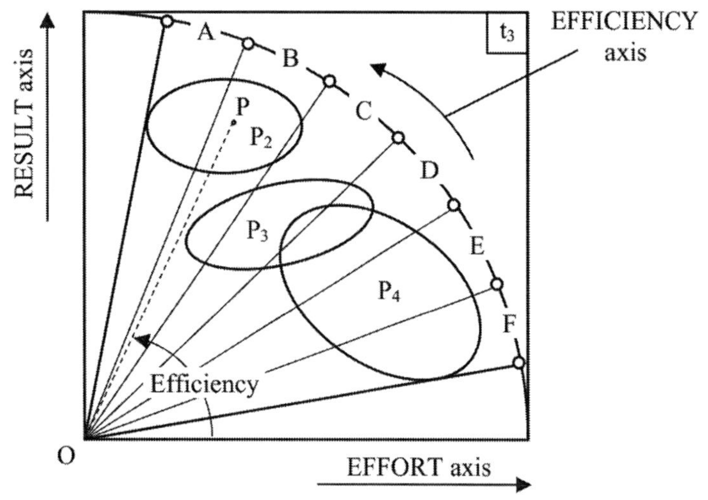

Fig. 4. Effort – Result – Efficiency diagram for t_3 task.

By similitude with products energy-labelling, according to current EU directive [9], the efficiency classes for manufacturing processes are labelled by letters from *A* (the most efficient) to *F* (the least efficient, if considering six classes). The label of a certain procedure is given by classes which its domain covers. Some remarks have to be made:

- The assigned efficiency class is not necessarily unique; actually it may vary depending on procedure parameters values. For example, the efficiency class of P_3 process (Figure 4) is $B - D$.
- The same process might belong to several utility areas, having different efficiency labels in each case. Moreover, in the same efficiency class one may find more procedures, from different tasks.
- Despite the efficiency labelling is a local classification, it has the same structure, no matter of the considered task or procedure.

- The possible enhancement of procedures efficiency, subsequent to the moment when the efficiency classes were defined, may lead to occurrence of new classes, above *A*, which will be denoted by *A*+, *A*++ etc., until the entire classification will be updated.

3 Case study

As case study, we present the application of the algorithm for developing the energy classes system for the utility area resulted by associating the task "Cut-off from bar" with five procedures supposed to be available for achieving it, as specified in Figure 5.

Here by cut-off we mean to detach a material piece from remaining bar by removing the layer of material previously existing between two plane parallel transversal sections of the bar. For simplicity, we accepted to characterize the task and each of the considered procedures by two features, as mentioned in Figure 5. For sample, in task case, chosen features are material and dimensions (more specific, hardness and transversal dimension, respectively).

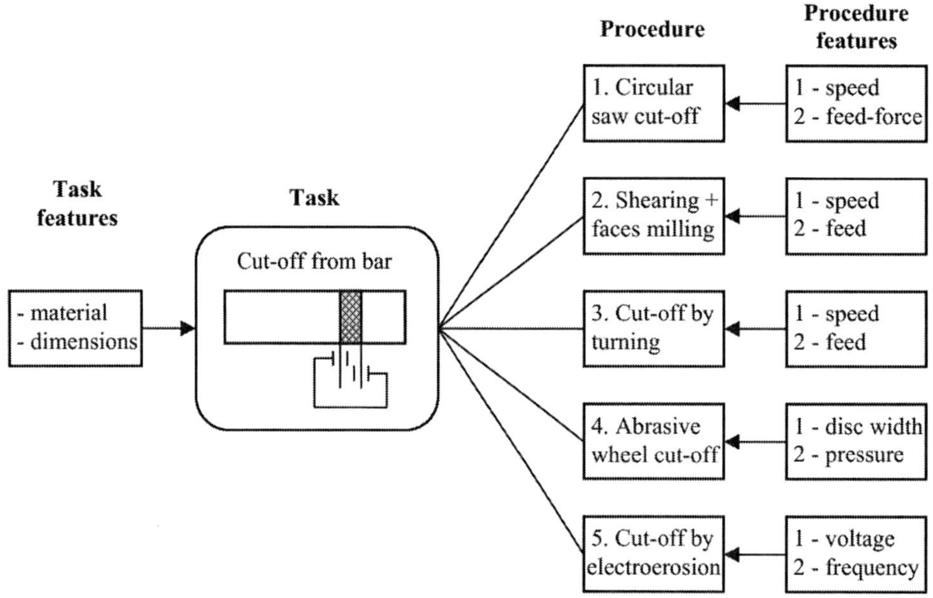

Fig. 5. Utility area for iron bar cut-off.

If speaking in general, after selecting the available procedures for achieving the considered task, the table of the *characteristic processes* need to be built. Each characteristic process results by combining the task (with actual features) with one of the procedures (also with actual features). The number of representative values for task and procedures features depends on the extension of their variation domains. The table's number of rows is equal to characteristic processes' number.

The process energy efficiency *EE* is defined as:

$$EE = \frac{SA}{EC}, \tag{2}$$

where *SA* means the area of the frontal surface delimiting the detached piece of material, expressed in cm^2, while *EC* is the energy consumed in cut-off process, in Wh, resulting in general, as sum of four components:

$$EC = EC_p + EC_{pm} + EC_{am} + EC_t \,. \tag{3}$$

In relation (3) EC_p means the energy directly consumed in the process while EC_{pm}, EC_{am} and EC_t are the amounts of energy embedded in processed material, auxiliary materials and tool share consumed in the process, respectively.

Excerpts from the representative processes table, generated as artificial dataset for the addressed case study, are presented in Table 1.

Table 1. Energy efficiency of representative processes.

Crt. no.	Task features: material / HB [MPa]	Task features: dimension [mm]	Procedure number	Procedure feature value: Feature 1	Procedure feature value: Feature 2	Sectioned area [cm^2]	Energy consumed [Wh]
1	steel / 1700	40	1	1400 m/min	1000 N	12.56	2.56
2	brass / 650	60	2	150 m/min	1.2 mm/rot	28.27	1.4
3	steel / 2000	50	3	70 m/min	0.2 mm/rot	19.63	8.17
4	steel / 1700	55	4	1.6 mm	115 MPa	23.76	3.2
5	steel / 2300	25	5	40 V	4600 Hz	4.91	4.1
6	bronze / 900	50	1	1400 m/min	800 N	19.63	4.36
7	steel / 2000	60	2	145 m/min	1 mm/rot	28.27	8.83
8	brass / 650	70	3	120 m/min	0.25 mm/rot	38.48	9.24
9	steel / 2300	50	4	2.5 mm	130 MPa	19.63	6.54
10	steel / 2500	35	5	40 V	4600 Hz	9.62	6.87

By graphical representing the characteristic processes in (*EC*, *SA*) coordinate we obtain the general diagram from Figure 4, particularised in the case of cut-off utility area (Figure 6). If splitting in six the angular domain comprised between the points giving the most efficient process and the less efficient one, then the efficiency domain for classes A ... F results (see Table 2). After inscribing the points corresponding to the same procedure in a closed curve we can further assess the energy class(es) for each procedure (see Table 3).

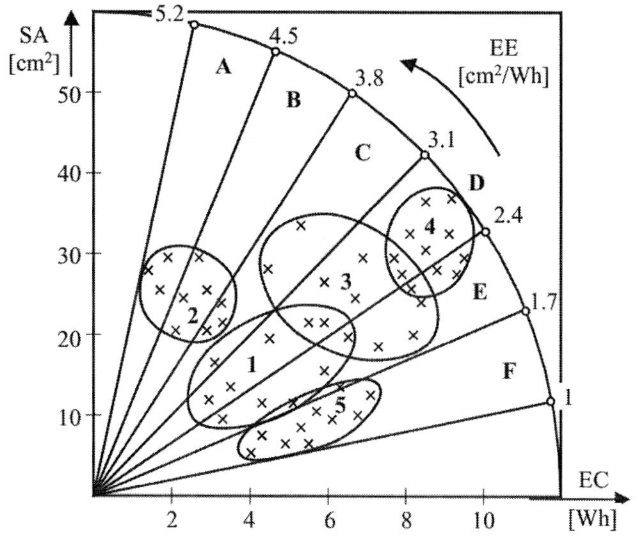

Fig. 6. Energy classes for cut-off process.

It should be noticed that, due to harmonisation between general interest (sustainability) and particular interest (economic efficiency) through legal regulations (e.g. EU directives), the use of manufacturing procedures corresponding to lower energy efficiency classes might become prohibitive.

Table 2. Energy classes domains.

Energy class	*EE* domain [cm^2/Wh]
A	4.5 ... 5.2
B	3.8 ... 4.5
C	3.1 ... 3.8
D	2.4 ... 3.1
E	1.7 ... 2.4
F	1.0 ... 1.7

Table 3. Procedures energy classes.

Procedure	Energy class
1. Circular saw cut-off	C - E
2. Shearing + faces milling	A - C
3. Cut-off by turning	C - E
4. Abrasive wheel cut-off	D - E
5. Cut-off by electroerosion	E - F

4 Conclusion

This paper proposes a method aiming to increase the efficiency in process planning. The method works on the base of processes classification after their energy efficiency. It is applied for processes grouped in the same utility area, and involves the definition of energy classes with the help of a newly imagined graphical representation: the Effort – Result – Efficiency diagram.

The performed case study shows that:

- The method has a wide potential of applicability.
- Both method implementation and building of energy classes system afferent to a utility domain are feasible.
- The method is effective, the differences between procedures being significant from energy efficiency point of view.
- Method application may lead to an enhancement of processes performance by facilitating process planning optimization.

The authors acknowledge the use of the research facilities developed by RO-054 Project, financed through EEA Grant 2009 program, in performing the research presented in this paper.

References

1. H. S. Park, T. T. Nguyen, J. C. Kim, Transactions of FAMENA, **XL-1**, 1 (2016).
2. E. Abele, S. Braun, P. Schraml, Procedia CIRP **29**, 251 (2015).
3. J. Ma, X. Ge, S.I. Chang, S. Lei, Int. J. Adv. Manuf. Technol., **74**, 1701 (2014).
4. H. S. Yoon, J. Y. Lee, M. S. Kim, S. H. Ahn, Journal of Cleaner Production, **78**, 54 (2014).
5. Y. Guo, J. R. Duflou, B. Lauwers, Int. J. Adv. Manuf. Technol., **75**, 503 (2014).
6. P. C. Priarone, Int J Adv Manuf Technol, **86**, 2107 (2016).
7. K. Salonitis, P. Ball, Procedia CIRP – Manuf. Systems, **7**, 634 (2013).
8. L. Zhou, J. Li, F. Li, Q. Meng, J. Li, X. Xu, Journal of Cleaner Production, **112**, 3721 (2016).
9. Directive 2010/30/EU on the indication by labelling and standard product information of the consumption of energy and other resources by energy-related products.

MATEC Web of Conferences **112**, 09004 (2017) DOI: 10.1051/matecconf/201711209004
IManE&E 2017

A peeling DEA-game cross efficiency procedure to classify suppliers

Corrado lo Storto*

[1]University of Naples Federico II, Department of Industrial Engineering, Piazzale V. Tecchio 80, 80125 Naples, Italy

Abstract. This paper proposes a non-parametric procedure to classify suppliers. The procedure adopts a peeling approach and Data Envelopment Analysis-game cross efficiency (DEA-GXE) method to group a set of suppliers into classes. In DEA-GXE each supplier is seen as a player competing in an uncooperative environment, seeking to maximize efficiency under the condition that the other suppliers'cross-efficiency does not deteriorate. The procedure is useful when there is a long list of suppliers that should be assessed by the customer, providing an agile decision-making tool.

1 Introduction

Supplier assessment and classification is an important step of the purchasing decision-making process in a manufacturing company. Selecting high quality suppliers provides competitive advantage with the management of the supply chain. Choosing good suppliers as partners along the supply chain contributes to reducing operational costs, improving process and product quality. A number of scholars focused on the supplier selection problem, i.e. the analysis and measurement of the performance of a set of suppliers in order to generate a relative rank and select one or more of them. Literature includes a large number of approaches and models that are based on multi-objective optimization (MOP) [1,2], analytic hierarchical process (AHP) [3,4], simple multi-attribute rating techniques (SMART) [5], and data envelopment analysis (DEA) [6,7]. However, even though they provide systematic frameworks that purchasing managers can utilize, most of them are not easily implementable and imply a large amount of subjectivity to weight individual criteria. Vice versa, frameworks based on DEA are more effective to reduce the evaluators' subjectivity as weights are endogenously determined from data and no assumptions relative to relationships among variables are necessary.

This paper proposes a non-parametric DEA based procedure to classify suppliers. The procedure adopts DEA-game cross efficiency (DEA-GXE) method to rank a dataset of suppliers and a peeling procedure to group them into classes. In DEA-GXE every supplier to be ranked is seen as a player competing in an uncooperative environment, seeking to maximize efficiency under the condition that the cross-efficiency scores of the other suppliers do not deteriorate. The procedure is particularly useful when there is a large number of

* Corresponding author: corrado.lostorto@unina.it

suppliers that should be preliminarily assessed by the customer and classified, providing an agile decision-making tool. The paper is organized as follows. Section 2 describes the DEA-GXE method and the peeling procedure. Section 3 reports the result of the implementation of the method. Finally, section 4 presents some concluding remarks.

2 Method

2.1 Supplier classification procedure

The procedure to classify suppliers implements DEA game-cross efficiency (DEA-GXE) to generate efficiency measurements relative to suppliers and the DEA peeling approach to classify them into different classes.

2.2 DEA game-cross efficiency

DEA is a non-parametric linear programming based technique that calculates the relative efficiencies of a number of decision making units denominated DMUs. The efficient units are identified from this set and combined together to construct an efficient frontier used as a benchmark to measure the efficiency of inefficient units [8].

In the proposed approach, each DMU is an individual supplier S_i in the dataset that can be associated to a specific production function producing $\boldsymbol{Y}=(x_{ij})\in\Re^{s\times n}$ outputs by combining $\boldsymbol{X}=(x_{ij})\in\Re^{m\times n}$ inputs. The DMU ability to produce a given amount of outputs with the minimum consumption of inputs or the maximum amount of outputs by consuming the same amount of input independently of any input and output price set is referred to as technical efficiency (TE).

The original model proposed by [9] classifies units into two groups, e.g. the efficient (scoring TE=100%) and the inefficient units (scoring TE<100%) and provides a relative rank for these latter only, having a weak discriminating power. Scholars have proposed different approaches and methods to improve the discriminating capability of DEA and generate full rankings among alternatives [10,11]. The approach based on the calculation of the cross-efficiency scores of DMUs is effective to improve the discriminating capability of DEA and does not need introducing any subjectivity in the ranking procedure as a consequence of weighting schemes provided by expert judgment [12, 13].

Assume that there are n DMUs and every DMU j (j=1,...,n) produces s outputs y_{rj} (r=1,...,s) consuming m inputs x_{ij} (i=1,...,m). The cross-efficiency of each DMU j, θ_{kj}, can be easily calculated using the optimal values of weights computed for DMU k , ω_{rk} and μ_{ik}, as follows

$$\theta_{kj}=\frac{\sum_{r=1}^{s}\omega_{rk}^{*}y_{rj}}{\sum_{i=1}^{m}\mu_{ik}^{*}x_{ij}} \qquad j\neq k \text{ and } j=1,...,n \tag{1}$$

The DMUs cross-efficiencies are calculated in two steps. First, by solving model (2) n times basic DEA is performed to calculate the optimal weights assigned to DMU inputs and outputs as suggested by [9]. Here an input orientation and constant returns to scale are assumed.

$$\begin{aligned}
&\text{Max } \sum_{r=1}^{s} \omega_{rk} y_{rk} \\
&\text{s.t. } \sum_{i=1}^{m} \mu_{ik} x_{ik} = 1 \\
&\sum_{r=1}^{s} \omega_{rk} y_{rj} - \sum_{i=1}^{m} \mu_{ik} x_{ij} \leq 0 \quad j = 1,...,n \\
&\omega_{rk} \geq 0, \mu_{ik} \geq 0, \ r = 1,...,s \text{ and } i = 1,...,m
\end{aligned} \tag{2}$$

Second, the $n \times n$ cross-efficiencies matrix of DMUs is generated. In the matrix, the off-diagonal entries show the cross-efficiencies, while the diagonal entries are the basic DEA efficiency scores. Finally, for each DMU k, cross-efficiencies are aggregated to obtain a mean cross-efficiency measurement.

Liang et al. [14] have proposed a DEA game cross-efficiency measurement. In DEA game, the DMU cross-efficiency is measured as the pay-off of a non-cooperative game that the DMU aims at maximizing. Scholars also suggested a robust iterative algorithm to generate the best average game cross-efficiency measurements, and proved that the suggested algorithm originates a set of unique Nash equilibrium DEA weights. In order to calculate the cross-efficiency of DMU k related to DMU j, a set of weights is found to maximize the efficiency of DMU k with the additional restriction that the efficiency of DMU j does not diminish. The efficiencies of the other DMUs should be known to calculate the efficiency of each DMU, and vice versa.

The cross-efficiency of DMU k related to DMU j is denoted as

$$\alpha_{kj} = \frac{\sum_{r=1}^{s} \omega_{rj}^{k} y_{rj}}{\sum_{i=1}^{m} \mu_{ij}^{k} x_{ij}} \qquad k = 1,...,n \tag{3}$$

where ω_{rj}^{k} and μ_{ij}^{k} are optimal weights. The following linear programming model (4) is utilized to implement the proposed procedure [14, 15]

$$\begin{aligned}
&\text{Max } \sum_{r=1}^{s} \omega_{rj}^{k} y_{rj} \\
&\text{s.t. } \sum_{i=1}^{m} \mu_{ij}^{k} x_{il} - \sum_{r=1}^{s} \omega_{rj}^{k} y_{rl} \geq 0 \qquad l = 1,...,n \\
&\alpha_k \times \sum_{i=1}^{m} \mu_{ij}^{k} x_{ik} - \sum_{r=1}^{s} \omega_{rj}^{k} y_{rk} \leq 0 \qquad j = 1,...,n \\
&\sum_{i=1}^{m} \mu_{ij}^{k} x_{ij} = 1 \qquad j = 1,...,n \\
&\omega_{rj}^{k} \geq 0, \mu_{ij}^{k} \geq 0, \ r = 1,...,s \text{ and } \ i = 1,...,m \\
&\alpha_k \leq 1
\end{aligned} \tag{4}$$

2.3 The peeling approach

The idea of "peeling the DEA onion" was firstly suggested by [16]. In the peeling approach, DEA is applied to variable datasets that are extracted from the main dataset at different steps [17]. In the first step, DEA is applied to the whole dataset to find efficient DMUs. At this step, efficient DMUs compose the first layer and accordingly are classified in the first class. Efficient DMUs are removed from the whole dataset and a smaller dataset is extracted from the first one, including only previously inefficient DMUs.

In the second step, DEA is applied to the reduced dataset and DMUs resulting efficient compose the second layer, assigned to the second class, and finally removed from the dataset to move to the next step. This process continues until all the DMUs in the dataset have been assigned to a layer.

3 Example

The dataset for this example is taken from [18]. It contains data relative to 23 suppliers which are assessed according to 11 attributes, respectively clustered into 6 inputs and 5 outputs (see Table 1).

Table 1. Original inputs and outputs.

inputs		ouputs	
variable	**attribute**	**variable**	**attribute**
QMP(X1)	quality management practices and systems	QUALITY(Y1)	quality
SA(X2)	documentation and self audit	PRICE(Y2)	price
PMC(X3)	process and manufacturing capability	DELIVERY(Y3)	delivery
MF(X4)	management of the firm	CRP(Y4)	cost reduction performance
DD(X5)	design and development capabilities	OTHER(Y5)	other
CR(X6)	cost reduction capabilities		

To meet the strictly positivity requirement of DEA, zero values were preliminarily eliminated by adding a constant scalar equal to 0.05 to all measurements. Additionally, because of the large number of input and output variables, in order to increase DEA discriminating capability, Principal Component Analysis (PCA) was used to reduce the dataset dimensionality. After running PCA, variables were reduced to 2 inputs and 2 outputs whose measurements were generated as weighted averages of original measurements using PCA factor loadings as weights. The new set of inputs and outputs considered for the subsequent analysis are showed in Table 2.

Table 2. Inputs and outputs used in the analysis.

new inputs and outputs	original inputs and outputs
I_1	QMP(X1), PMC(X3), MF(X4), DD(X5), CR(X6)
I_2	SA(X2)
O_1	QUALITY(Y1), DELIVERY(Y3), OTHER(Y5)
O_2	PRICE(Y2), CRP(Y4)

Table 3 reports the output of the implementation of the classification procedure. Without loss of generalizability, the procedure was stopped after the generation of the first 4 layers (layer 0,…,layer 3) and aggregation of suppliers in 4 classes (1,...,4). Particularly, table provides information relative to the game cross-efficiency score, rank for every layers, and suppliers excluded from the dataset at each step. In the first step, only one supplier was evaluated as 100% efficient. In the second step, the DEA game cross-efficiency algorithm did not identify any 100% supplier. Differently from conventional DEA, the best cross-efficiency score may be less than 1 because it is finally calculated as the mean value of different efficiencies. Thus, as a decision-making rule adopted in this example, suppliers scoring more than 0.95 were excluded from the dataset to move to the next step. In the same way, layers 2 and 3 were generated by excluding the 100% efficient suppliers.

Table 4 displays how suppliers have been classified. Class 1 contains one supplier only (S22), class 2 contains five suppliers (S4, S6, S7, S15, and S23), class 3 includes one supplier (S20), and finally class 4 includes one supplier (S12).

Table 3. Results (first 4 layers).

Supplier	Layer 0		Layer 1		Layer 2		Layer 3	
	DEA-GXE	rank	DEA-GXE	rank	DEA-GXE	rank	DEA-GXE	rank
S1	0.463	19	0.464	18	0.660	11	0.661	11
S2	0.378	21	0.370	20	0.527	14	0.534	14
S3	0.862	8	0.857	7	0.962	4	0.982	3
S4	0.981	3	0.998	1	-	-	-	-
S5	0.676	11	0.707	10	0.912	5	0.922	4
S6	0.945	6	0.971	5	-	-	-	-
S7	0.958	4	0.986	3	-	-	-	-
S8	0.564	15	0.581	15	0.779	8	0.786	7
S9	0.333	22	0.345	21	0.435	16	0.445	15
S10	0.886	7	0.864	6	0.971	3	0.989	2
S11	0.576	14	0.589	13	0.727	10	0.743	9
S12	0.736	10	0.761	9	0.998	2	1.000	1
S13	0.606	12	0.633	11	0.817	6	0.826	5
S14	0.562	16	0.589	14	0.758	9	0.768	8
S15	0.985	2	0.988	2	-	-	-	-
S16	0.411	20	0.419	19	0.525	15	0.540	13
S17	0.483	18	0.495	17	0.607	13	0.620	12
S18	0.316	23	0.320	22	0.373	17	0.380	16
S19	0.589	13	0.612	12	0.797	7	0.809	6
S20	0.775	9	0.796	8	1.000	1	-	-
S21	0.530	17	0.541	16	0.656	12	0.669	10
S22	1.000	1	-	-	-	-	-	-
S23	0.947	5	0.976	4	-	-	-	-

Table 4. Classified suppliers (first 4 classes).

Class	Classified Suppliers				
class 1	S22 (1.000)				
class 2	S4 (0.998)	S6 (0.971)	S7 (0.986)	S15 (0.988)	S23 (0.976)
class 3	S20 (1.000)				
class 4	S12 (1.000)				
...	...	...	...	...	...
class j	...	...	...	...	...

4 Conclusion

This paper has presented a computational procedure to classify suppliers. The procedure is based on the implementation of DEA game cross-efficiency (DEA-GXE) and a peeling approach to associate suppliers to classes. While DEA-GXE is considered an effective procedure to generate rankings, a difference in the ranking position of suppliers may be due to small differences of their efficiency scores (i.e., the third or even the fourth digit). Thus, the resulting ranking can be rather aleatory because of measurement uncertainty, implemented algorithm, and numerical approximation. Combining both DEA-GXE and the peeling procedure allows having a more robust and refined classification than using only the classical peeling approach as suggested by [16]. The innovative procedure has been applied to classify a dataset of 23 suppliers. Results indicate that the proposed procedure can be easily implemented and is effective to generate classes of suppliers. Additionally, given its flexibility, it can be adapted to context by modifying exclusion and selection rules.

References

1. S. Deng, R. Aydin, C.K. Kwong, Yun Huang, Comput. Ind Eng., **70**, 150 (2014)
2. R. Narasimhan, S. Talluri, S.K. Mahapatra, Decision Sci., **37**, 577 (2006)
3. F.T.S. Chan, Int. J. Prod. Res. **41**, 3549 (2003)
4. V. Yadav, M.K. Sharma, Journal of Modelling in Management, **11**, 326 (2016)
5. S.H. Huang, H. Keska, Int. J. Prod. Econ., **105**, 510 (2007)
6. R.C. Baker, S. Talluri, Comput. Ind. Eng. **32**, 101 (1997)
7. J. Seydel, Ind. Manage. Data. Syst., **106**, 81 (2006)
8. W.W. Cooper, L.M. Seiford, K. Tone, *Data Envelopment Analysis: A Comprehensive Text with Models, Applications, References and DEA-Solver Software*, (2nd Ed. Springer-Verlag, Berlin, Germany, 2006)
9. A. Charnes, W.W. Cooper, E. Rhodes, Eur. J. Oper. Res., **2**, 429 (1978)
10. N. Adler, L. Friedman, Z. Sinuany-Stern, Eur. J. Oper. Res., **140**, 249 (2002)
11. L.A. Meza, M.P.E. Lins, Ann. Oper. Res., **116**, 225 (2002)
12. J.R. Doyle, R.H. Green, J. Oper. Res. Soc., **45**, 567 (1994)
13. C. lo Storto, Sustainability, **8**, 124 (2016)
14. L. Liang, J. Wu, W.D. Cook, J. Zhu, Oper. Res., **56**, 1278 (2008)
15. J. Wu, L. Liang, Y. Chen, Omega, **37**, 909 (2009)
16. R.S. Barr, M.L. Durchholz, L. Seiford, Southern Methodist University Technical Report, **5**, 1 (1994)
17. M. Bougnol, J. Dul'a, Ann. Oper. Res., **145**, 339 (2006)
18. R. Narasimhan, S. Talluri, D. Mendez, The Journal of Supply Chain Management: A Global Review of Purchasing and Supply, **Summer**, 28 (2001)

Organisational strategy in industry

Cristina Fenişer[1,*], *Arik* Sadeh[2], and *Florin* Lungu[3]

[1]Technical University of Cluj Napoca, Department of Management and Economic Engineering, Cluj Napoca, Romania
[2]Holon Institute of Technology, Faculty of Technology Management, Holon, Israel
[3]Technical University of Cluj Napoca, Department of Management and Economic Engineering, Cluj Napoca, Romania

Abstract. Organizational strategy is a comprehensive topic debated in the literature and is widely recognized the critical role that the survival of their enterprises regardless of the type. Information on the strategic behaviour of enterprises in industry is limited.
A study was conducted in order to explore the types of strategy adopted by managers of small enterprises in the industry. The analysis is based on the typology of strategies undertaken by Miles and Snow. Three dimensions are covered: product, customers and competition. The obtained data was analysed using quantitative techniques to find out whether the independent variables chosen are predictors of how managers fulfil the role of strategist.

1 Introduction

The strategy is one that describes desirable image of the Organization in the future and is designed to target towards what aims to make the organization. Strategic thinking is the key to achieving success in a competitive environment which is constantly changing.

The authors describe the different typology for the strategic behaviour of the company. It ensures the care is good in respect of the external website.

In their work about Organization Strategy, Structure, and Process, Raymond E. Miles and Charles Snow [1] mentioned that companies develop strategies depending on the way in which companies decide to address three fundamental issues: entrepreneurial, technical and administrative. Entrepreneurial problem relates to how they should be managed. A technical problem is involved with how the company implements enterprise and solutions to the problem. The third issue is how to take into account the administrative structure of the company for the implementation of solutions to the first two problems. Although the companies are choosing different solutions to these problems, Miles and Snow suggests that more companies analogous solutions. So, they identify four types of organizational strategies: prospective, defensive, analyser and reactive. [2]

* Corresponding author: cristina.feniser@mis.utcluj.ro

Organizations that adopt prospective strategy address the problem of identification and exploitation of entrepreneurial new products, as well as the opportunities available in the market [3]. These organizations are successful in business environments and have an element of peril. They succeed by examining the market constantly in search of new opportunities. These companies have large production lines and often promote creativity more than efficiency. It prioritizes the development of new products to meet new needs and demands of consumers and changeable. The problem of these companies is to coordinate various business activities and promote innovation. Prospective companies solve this problem through decentralization, by encouraging cooperation between the different departments and units, they hire generalists (specialists).

Defensive strategies organizations are faced with the problem of maintaining entrepreneurial stable positions on the market and carry out their mission in the best business environments. These organizations achieve success by specializing in certain areas and through the use of standardized technical processes that maintain low costs. Organizations have aimed mainly at defensive efficiency, and the problem focuses on ensuring administrative efficiency through centralization and formal procedures. Because their environment is changing slowly, they can rely on long-term planning [4].

Organizations have common display characteristic analysers with the prospective and defensive, so their entrepreneurial problem is maintaining position in existing markets and identify new markets and products. Their problem is to maintain the operational effectiveness of established products but also to be flexible at the same time with new business opportunities. Accordingly, it seeks to maintain low costs, as well as identifying and developing new products to be competitive when the market changes. Administrative problem is to manage both aspects. Just like the prospective, the analyser organizations promote collaboration between departments.

As the name implies, reactive organizations, it does not have a strategy or a systematic structure [5]. They are not prepared to deal with changes in the external business environment. When you have a strategy and a well-defined structure, they no longer correspond to the external environment. The development of new products will fluctuate depending on how it perceives the Environment Manager [6]. These organizations do not have long-term plans, because for them the environment is changing too quickly, making them useless.

Various researchers have tried to validate the Miles and Snow typology [7]. So, Shortel and Zajac [8] found that firms are more likely to prospective who adopt new products; the analysers are firms that adopt new management procedures and the defensive are the first to adopt new production technologies [9].

The authors describe the behaviour patterns of stable typology for strategic business behaviour which gives it a good adaptation to the external environment.

2 The study

Based on the characteristics of each type of strategy we formulated twelve affirmations concerning the Organization's behaviour product, customer and competition. These statements represent items in a questionnaire asking respondents to indicate on a Likert scale is likely to adopt certain behaviour. They were included in the research of industrial firms in the County of Alba. Sampling method was simple random, each company was assigned a number and the selection was done by the method "used to make."

Each of the twelve items of the questionnaire corresponds to one of four types of strategies as indicated by Table 1 below.

DOI: 10.1051/matecconf/201711209005

Table 1. The twelve items in the study and the corresponding strategy.

ITEM	Strategy
1. Trying to be the first to provide products to your customers	Prospective
2.Focus on a limited line of products	Defensive
3. Carefully develop new products	Analyzer
4. Develop new products in response to what is currently available	Reactive
5. Strive to be the market leader without disregarding competition	Prospective
6. Emphasize quality and better prices for customers	Defensive
7. Develop marketing strategies in line with other companies	Reactive
8. Rarely are the market leader but carefully analyze market trends	Analyzer
9. Your company operates in a market type "niche" to protect themselves from competition	Defensive
10. Analyze competition actions, learn from their mistakes	Analyzer
11. Strive to stay ahead of the competitions	Prospective
12. Try to keep up with competitors	Reactive

From the analysis of the responses of managers of firms in Alba County to the questions in the questionnaire were obtained the following results:

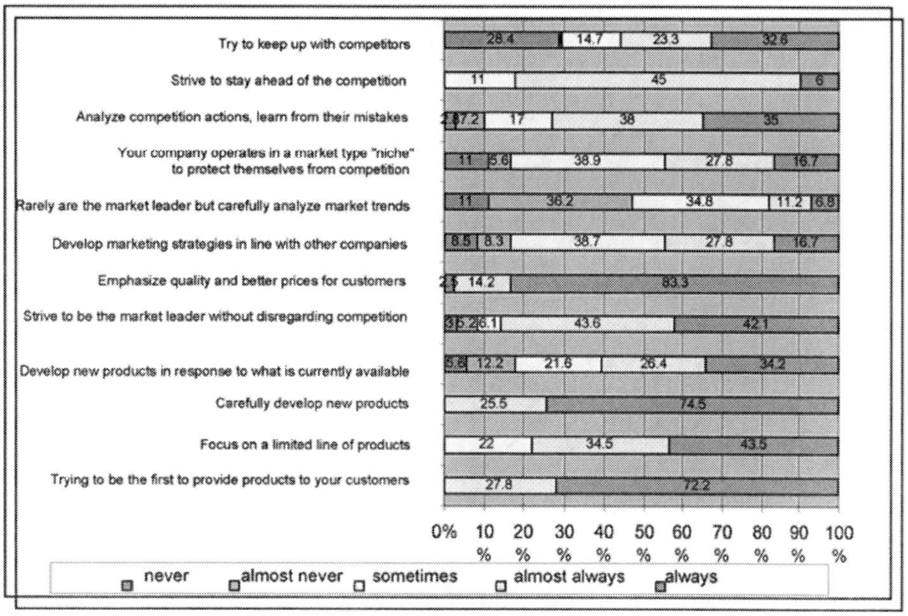

Fig. 1. The frequency of responses to the allegations regarding the conduct of the company.

Prospective strategy is adopted by the majority of the managers and holding them with analysers shall be regarded as leaders in the market that analyses market trends (Table 2). Prospective strategy has the highest frequency of replies "always" or "almost always."

Table 2. The strategies adopted by the managers of companies in Romania, Alba.

	Never	Almost never	Sometimes	Almost always	Always
prospective			11%	45%	6%
prospective				27.8%	72.2%
prospective	3%	5.2%	6.1%	43.6%	42.1%
analyzer				25.5%	74.5%
analyzer	2.8%	7.2%	17%	38%	35%
analyzer	11%	36.2%	34.8%	11.2%	6.8%
defensive	2.5%			14.2%	83.3%
defensive	11%	5.6%	38.9%	27.8%	16.7%
defensive			22%	34.5%	43.5%
reactive	5.6%	12.2%	21.6%	26.4%	34.2%
reactive	28.4%	1%	14.7%	23.3%	32.6%
reactive	8.5%	8.3%	38.7%	27.8	16.7%

The answers "always" and "almost always" are considered as the ones that indicate the presence of the strategy referred to in the statement. For each type of strategy were formulated three allegations, resulting in 12 questions.

From the analysis of the responses on the strategy depending on the number of employees, by size of firm, noted the trend medium and large enterprises to adopt strategies and defensive analyzers. The explanation for this result is that in companies with a large number of employees change occurs in a long time, it is necessary for maintaining stability and centralization is an important cost control while small firms are more flexible and at the same time more sensitive to changes in the external environment. In the case of micro enterprises, the percentages do not indicate a type explicitly, the difference between the answers were not significant.

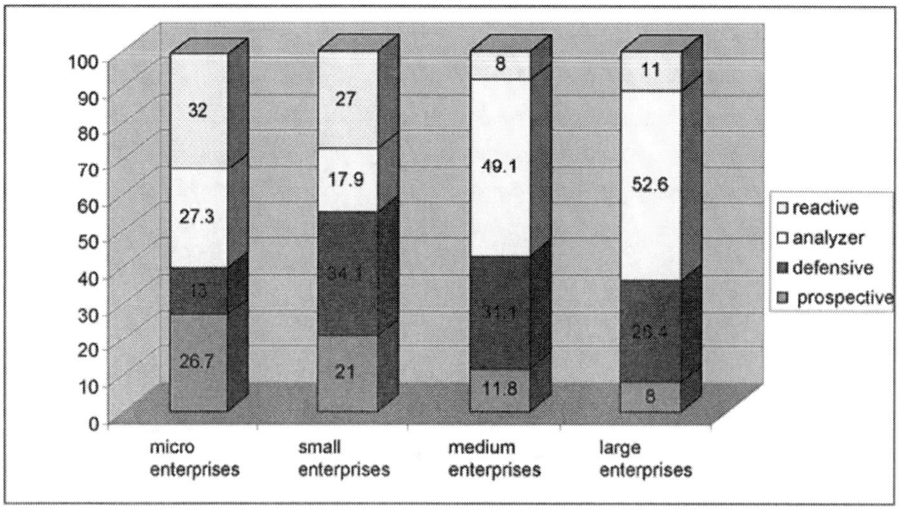

Fig. 2. The type and size of company strategy.

The variable sales results show us that the companies with the lowest turnover recorded last year are predominantly those who adopt the strategy analyzers, a sign that the company is oriented toward lowering costs and at the same time are turning to new products and is attentive to the behaviour of the competition. For firms with turnovers, and for those with larger turnovers have not identified a predilection for specific strategies.

The experience of the manager is a variable that influences organizational behaviour as well as the strategic default. Firms are run by managers with long experience tend to choose prospective and analyser strategies. We observe that firms with less experience, up to 5 years shall adopt analyzers and defensive strategy, trend that remains for the next category, that of firms between 5 and 10 years. It is worth mentioning is that the strategy is present analyzers with a considerable percentage, between 29.9% and 44.4% regardless of the years of experience of the company.

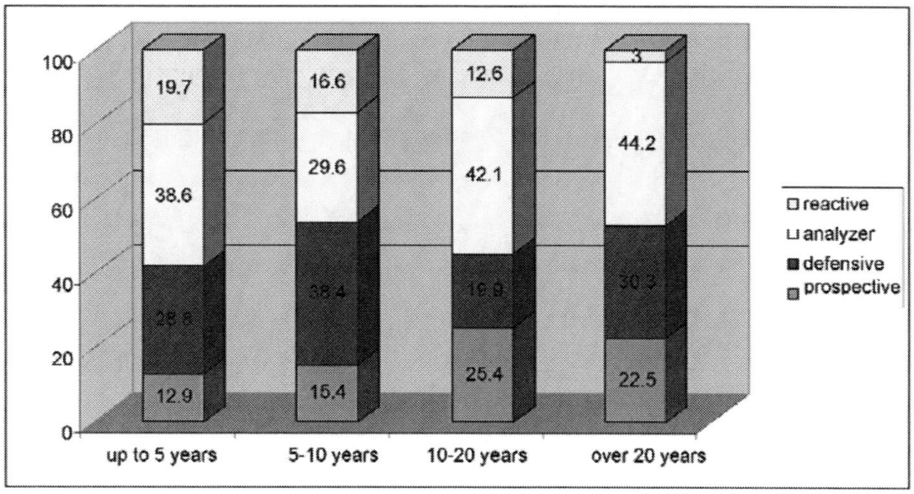

Fig. 3. The type of exercise indeed and firm age .

3 Conclusions:

The objective of this study was to test the validity of the typology of organizational strategy in firms. The study was conducted in firms in the County of Alba Romania. The model shows that firms have looked a lot like the etymologized strategies by the two authors, regardless of size or managerial experience. So, we conclude that the four strategies are not exclusive but are stages or phases in the development of the company strategy, because as Miles and Snow, mentions the adoption of a strategy is manager's option in function of how it perceives and market competition.

The attractiveness of this kind is given by the fact that specifies the relationship between strategy and structure of the company and treats the company as a whole in interaction with the environment. It is also to note that there has been major criticism levelled over time.

References

1. R. E. Miles, C. C. Snow - *Organizational Strategy, Structure, and Process* (McGraw-Hill, 1978)
2. M. Abd M. Dincer M. Yildirim E. Dil, Procedia Social and Behavioral Sciences **24**, 601 (2011)
3. K. Kylaheiko A. Jantunen K. Puumalainen, P. Luukka, Int. J. Production Economics, **131**, 273 (2011)
4. M. A. Koseoglu, C. Topalogluc, J. A. Parnelld, D. L. Lestere, International Journal of Hospitality Management, **34**, 81 (2013)
5. A. Thomas, K. Ramaswamy - British Journal of Management, **7** (1996)
6. G. Johnson, K. Scholes, R. Whittington, *Exploring Corporate Strategy: Text & Cases* (Prentice Hall, 2009)
7. V. Isoherranen, P. Kess, Modern Economy, **2**, 575 (2011)
8. S. M. Shortell, E. J. Zajac, Academy of Management Journal, **33**(4), 817 (1990)
9. D. N. Angwin, M. Meadows, Long Range Planning, **48**, 235 (2015)

Current state in the research on electronic monitoring systems for the security and flow of objects and individuals

Marcela Man Dietrich[1,*], and *Marcel Sabin* Popa[2]

[1]Helix Software and Engineering GmbH, Automation Engineer, 60326 Frankfurt am Main, Germany
[2]The Technical University of Cluj-Napoca, Machine Building, 400020 Cluj-Napoca, Romania

Abstract. This paper sets forth the current state of security systems in prisons from Romania and around the world, particularly aiming electronic systems of monitoring the flow of people, materials control and perimeter security, focusing on the research results concluded by motion detection tests and devices. The currently used systems were observed in order to put an analysis of the methodology together and implement and perfect these systems in protected areas. The protection of citizens must be performed to the extent that is allowed by the legislation.

1 Introduction

As a result of a petition that I addressed to the Romanian National Administration of Penitentiaries, it was learnt that nowadays, the majority of the prison units do not have electronic systems of monitoring the flow of people. Prisons that have integrated electronic access control devices have in endowment the following: video monitoring and dispatching systems; systems with electronic access control that are based on turnstiles and personalized magnetic cards. For the control of materials placed inside the prison units, the following electronic systems are used: port metal detectors ports and handheld metal detectors; baggage scan x-ray devices.

2 Methods of ensuring the security perimeter of premises

To ensure the perimeter security of prisons (The figure 1 illustrates the inside area of a prison), indoor and outdoor surveillance is realized through collaboration of the following physical facilities and electronic devices [1]:

2.1 Video surveillance subsystems

These are made with closed circuit, and are composed of cameras installed on segments covering all due perimeter, as well as interior and exterior spaces, connected to the central command system and control type dispatch.

* Corresponding author: marcela_man@hotmail.com

Fig. 1. Barbed wire used around the area allowed to prisoners.

On the 20th of January 2009, Romania had legalized wiretaps, recordings of SMS's and e-mails. The law was passed by the Romanian Parliament. Article 28 of the Constitution provides: "Secrecy of the letters, telegrams and other postal communications, of telephone conversations and other legal means of communication is inviolable."

2.2 Installation for the detection of intrusion in perimeters, made of metal wires

This installation of detection of the penetration in the perimeter is mainly against escalation or destruction of the interior fence (Fig. 1), consisting of detectors vibration and shock in yarn (sensitive burglar detectors climbing ring, cutting and deterioration) installed on the fence of metallic wires mounted on supportive poles on the inner fence that separates the forbidden area of the ownership sector.

2.3. Protection systems with microwave beams

These are some trespassing detection installations in exclusion zones (interior forbidden zone) by means of a beam of microwave protection. Microwaves are radio waves; whose wavelength is between 1 mm (300 GHz) and 1 m (0.3 GHz). [2] Microwave applications are of interest in relation to their propagation through the transmission lines and waveguides as well as electromagnetic resonators, instead of the conventional resonant circuit.

3 Safety in penitentiaries

The electronic central units, as in the central electronic unit of detected system evaluation and their transmission to the control computer and the control system computer play a big role in the security in prisons and are a main component of the running systems.

Fig. 2. The secured cells of a penitentiary.

3.1 Security in penitentiaries

In this chapter, an issue of interest to this article will be addressed, the current situation of the security in prisons in Romania and abroad [3]. The relevance of psychological behavior of detainees is very high, in order to implement a security system properly and analyzing existing ones.

Investigations made into the correctional environment:

• Environment in prisons:

It fluctuates, it is complex and difficult to manage within the contravention population; increasing operational requirements due to illicit drug trafficking, and the activities of band members; old infrastructure and security equipment, as per Fig. 2.

• Strategic objectives: [4]

Improving safety and security, infrastructure, capacity and operations by: a more rigorous control of entry into institutions perimeter inspection, searches and visits of offender will be under attention; detector-dog expansion teams; more thorough information security; better training of correctional officer; better equipment, better performing electronic products to support safety and security initiatives.

3.2 Safety in prisons in case of fire

• Monitoring the performance :

The person responsible must have procedures for employees monitoring the performance of safety systems in case of fire in their areas of control [4] [Fig.3]. European safety, health and counseling committee in case of fire through the founding the committee of staff in health and safety is responsible for the quality assurance inspections of fire safety and to identify any remedial action that may be necessary to make after an inspection.

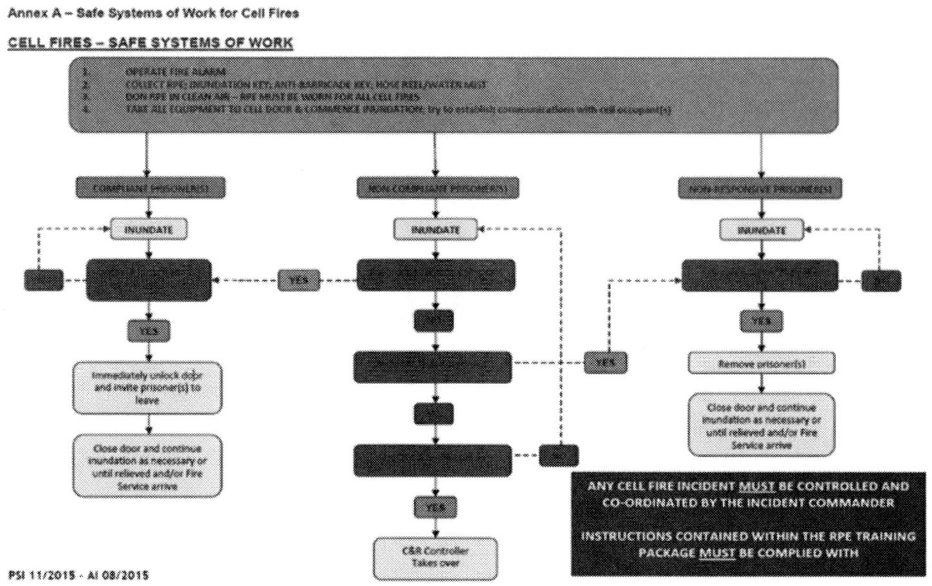

Fig. 3. The schematic cells used in the technique of reaction control in case of a fire.

4 Motion Detection

The research and the study of this paper is also focused on movement detection, which in the video surveillance systems is an important tool for the security of all. It warns when someone is on their own property, without authorization. The understanding of the way this technology works, helps with the configuration of the best regions for motion detection and alerts, but also officials the understanding of detection and movement.

To understand how motion detection in video surveillance systems works, we must first understand how a surveillance camera works. The current study indicated that inside the camera there is an image sensor and camera lenses directs light - when light hits the image sensor, each pixel is there for getting as much light possible. The fact that the pattern of bright and dark areas on the pixel becomes full video image is closely related to the "pictures and videos".

When configuring motion detection, select a region or area to monitor a door for example. The way it works is by comparing sequential images frame by frame of video images; if enough pixels have changed between these frames, the camera software determines something changed and will send an alert. The degree of change is determined by the level of sensitivity and level settings affect the percentage of motion detection. [5]

It depends on the level of sensitivity to changing light / dark levels of pixels. Higher sensitivity will rise to more changes while lower sensitivity will require a high level of change, such as lights occurring in a dark room. The percentage of change, on the other hand, is the percentage of pixels that have recorded a change. If the door is monitored, establishing the proportion of 50% will warn when something big enters the room, such as a person, but it ignores something smaller.

As a basic concept, the best balance in motion detection settings to prevent too many false alerts, it may take some time. During the camera configuration, the detection of someone´s motion in the concerned region can be tested; the settings being adjusted as needed. It may take some time, but the effort is well worth it when notification occurs a burglary at a store or vandalizing a parking zone. Fig. 4 plots this process.

Fig. 4. Edge detection in motion [9].

4 AVIAN – Heart Beat Detector

A device that was deeply studied and used for this research is AVIAN [6]. Advanced interrogation and notification system for vehicles detects the presence of people hidden in them. Using data from special sensors found AVIAN shock wave generated by the beating heart (fig. 5), that plugs into any surface or object with which the body is in contact. AVIAN collects and analyses data using signal processing algorithms to detect a hidden person in less than 1 minute.

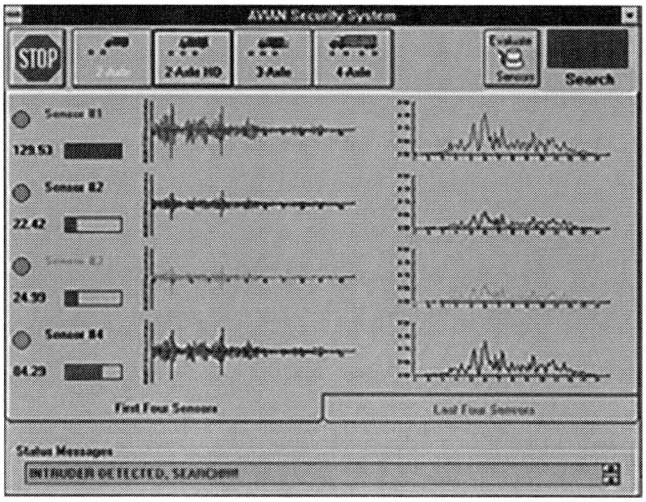

Fig. 5. The display of the Avian Heart beat detector [9].

AVIAN system consists of an industrial personal computer loaded with specially developed software, a touch-screen monitor and special sensors. The operator uses the touch screen to select the type of vehicle, the vehicle before it starts scanning. The processed data provides the operator with concrete information or evidence search.

The results of the study show that a complete search lasts less than 14 seconds from the placement of sensors on the vehicle. The system was found to be 100% effective and reliable in the detection of human presence in ideal conditions (strong to moderate wind protection is necessary to eliminate false positive indications). [8]

AVIAN is a cost-effective method to accurately and quickly search large or small vehicles for hidden people, regardless of content. Operation with a screen with icons makes users′ training a simple matter that can be accomplished in less than an hour. The computer system doesn′t need maintenance.

The cost of an Avian is less than the cost of one correction officer or servant of mobile security. It is accurate and reliable and provides the ability to quickly search the officer's fully loaded vehicle without moving or disturbing the contents of the vehicle.

5 Conclusions

This paper has presented the start of a research in security prison institutions, with some basic concepts, and the purpose is to implement this knowledge in the area of security systems of protected buildings. It has brought an improved classification of the monitoring systems, as well as a highlighting of the main aspects specific to the monitoring systems used in prisons. In this sense, it is to achieve a viable and applicable method to ensure the protection of persons and materials security (e.g. in case of fire, to prevent escape, or to detect weapons and illegal substances) as well as detailed analysis and deepening the current situation of the psychological part involved, as well as to compare the methods used in our country to the outside.

The possible directions of the future authors` research could be the analysis of the methodology of the placing of surveillance systems in a correctional institution, as well as a complete classification of the identified monitoring systems relevant groups and highlighting the main characteristics of each group.

This research was carried out with the support of the Technical University of Cluj-Napoca.

References

1. The Ministry of Justice, the Department of Homeland Security and the Penitentiary Regime, Answer ANP – following the correspondence with Marcela Man (in Romanian).
2. Associated Press (2001), Prison escape probe to focus on lax security. Retrieved January 28, 2003, http://www.clickonsa.com/ant/news/stories/news-20010108-085202.html
3. American Jail Association (1993). American Jail Association Code of Ethics, http://www.corrections.com/aja/resolutions/index.html
4. AI 08/2015, PSI 11/2015, 2015, „Fire Safety in Prison Establishments", NOMS Agency Board
5. http://ittrends.ro/
6. Federal Bureau of Prisons (1999). Introducing AVIAN. Retrieved February 5, 2003, from http://www.heartbeat-detector-avian.com
7. www.mesa.org
8. http://www.geovox.com/
9. http://netsecurity.about.com/

The method of using current regulations and standards in designing management and technologies of construction

Sergey Sinenko*

Moscow State University of Civil Engineering, Department of Construction Technology and Organization, 129337 Moscow, Russian Federation

Abstract. Economic efficiency of using funds in construction of buildings and structures begins from an effective, proper design, which is based on modern, cutting-edge, advanced equipment, streamlined organization and process solutions of construction operations. In the light of this it is considered application of "Self-Organization and Technology of Building" multifunctional software package capable of solving various engineering tasks on designing management of construction operations in accordance with the applicable rules and regulations in view of using software for automatic generation of workflow. Implementing this software in the construction management processes may help to solve management tasks at the construction site.

1 Normative study of designing management of construction operations

Designing management of construction operations involves choosing the most effective management and process scheme of construction of buildings and structures, as well as application of contemporary process solutions for construction and installation which on the one hand should reduce estimated cost of construction and installation, as well as construction period of a facility, and on the other hand should provide high-quality construction in compliance with safety requirements, occupational safety, fire and environmental safety, in accordance with applicable rules, regulations and standards [1].

In the current context of present-day competition improvement the principal participants of investment and construction activities shall effectively arrange construction operations to receive maximum profit at minimal cost of time and resources without any failures as to construction deadlines causing expenses increasing in the course of projects implementation and loss of profit for all participants of construction and investment activities.

At present the market offers a majority of universal software packages that we are used to subdivide into professional and "desktop" (amateur). They mainly differ in the labour intensity when applying since professional software is quite time-consuming however it

* Corresponding author: 6622040@mail.ru

DOI: 10.1051/matecconf/201711209007

may provide more flexible outcome, and amateur packages may be used just once, and spending little time and few efforts you shall receive quite an informative result.

Further, designing process of construction operations involves choosing the most efficient process solutions and the use of up-to-date devices that improve quality of construction products while reducing the time and cost of construction. Software market offers various products that can solve such tasks, but often they are not comprehensive enough [2, 3].

2 The method of designing management

2.1 Justification of choosing the software

It's offered a contemporary way to use current standards and specifications to meet the challenges of construction management and technologies: using the "Self-Organization and Technology of Building" ("SOTB") software. This software allows independent solving tasks in construction process management with the use of up-to-date technologies.

Self-Organization and Technology of Building is a multifunctional software package for solving tasks of construction operations management with the use of updated regulations. The software is also intended for the design of construction technologies: "SOTB" allows monitoring the progress of construction in real time.

The software is multifunctional and involves the following: financial planning; automatic generation of calendar and resource plans; maximizing the level of construction management; possibility to avoid unplanned costs, thereby increasing the profitability of construction companies [4].

2.2 Components of the software package

"SOTB" software consists of several components that interact with each other. Currently, there are 4 elements represented at the Figure 1, each of which has its own characteristics and meets certain requirements, which generally allow the software working smoothly and effectively.

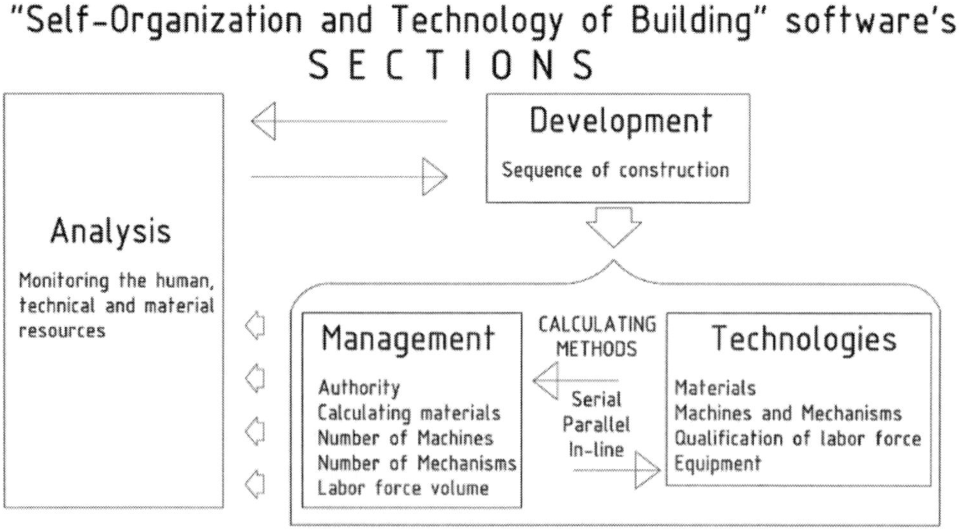

Fig. 1. "Self-Organization and Technology of Building" software's sections.

Self-Organization and Technology of Building is, first of all, proper management and technology of construction operations. The management section of the software is a table into which we enter a construction site data.

"SOTB" displays the necessary building codes and state standards, while, in addition to the regulatory framework, the software automatically opens a new tab with authorities where a company should register documents. Also, "Management" section has an additional function for calculating the necessary materials, the number of machines and mechanisms, and finding the labour force volume needed.

"SOTB" software package offers construction methods being the most appropriate for a given amount of work. "SOTB" independently calculates three methods of construction and installation works (serial, parallel and in-line), thus allowing the customer choosing the most suitable and profitable one.

In Development section "SOTB" offers approval of construction operations sequence. Since the sequence can be attributed both to management, and to technologies, the software developers found it appropriate to make "Development" section and to make a sequence of construction and installation a separate item.

"SOTB" users can make any changes in this section, and their proposals will be automatically updated in "Technologies" and "Management" sections. Changing a sequence may for example changing the materials to be more exact, timing of their delivery to a construction site or workers' involvement may partially change or the use of machines and equipment.

The main and more important aspect is the ability of "SOTB" to identify equipment, construction machinery downtime, or non-rational employment of workers. For this purpose, the software contains the fourth element the "Analysis".

Analysis section is on a paying basis, but, despite this, the majority of users buy this optional item to have the whole software. "Analysis" allows monitoring the use of human, technical and material resources. Also, this section includes a part of financial policy, thus clearly shows the costs and revenue of a company from using each of the methods of construction and installation works [5, 6].

The software is multifunctional and involves the following: financial planning; automatic generation of calendar and resource plans; maximizing the level of construction management; possibility to avoid unplanned costs, thereby increasing the profitability of construction companies [4].

2.3 Focus areas of the software package

"SOTB" allows preparing construction operations in two areas: preparation for a facility construction and preparation of construction and installation management.

2.3.1 Management preparation

The Management Preparation includes justification, approval, obtaining a building permit which is regulated for example in Russia by "The order of development, coordination, approval and composition of studies of investments in construction of enterprises, buildings and structures..." [7].

The Management Preparation includes engineering research, development of design estimates, ordering equipment. "SOTB" will allow developing design & estimate documentation, choosing the most efficient option of using the equipment, and also functions of the software include a list of authorities that will provide the necessary material and technical resources within the given timeframe.

Technical preparation is divided into:

- off-site preparation;
- on-site preparation;
- facility preparation.

Off-site preparation includes providing access roads, supply lines and networks, in some cases, this includes creating a base. On-site preparation includes creating a geodetic control network, arrangement of roads, construction site fencing, and providing communication. Facility preparation is a breakdown of axes of buildings, fixing, delivery of materials, cranes installation.

"SOTB" software is directly responsible for technical preparation. After entering certain data (overall size of the building, of the construction site and location of the building relative to the construction site), it designs temporary supply lines to the building, parameters of which have already been entered into the software.

Also, Self-Organization and Technology of Building offers recommended location of temporary roads, places for storage of materials, equipment installation location (the software designs all these based on the documents responsible for compliance with the rules at the facility and at the construction site generally). The software also has a "Preferences" feature, where you can manually change one or more parameters; in this case, "SOTB" will offer a different design of the facility, taking into account the customer's preferences [8].

2.3.2 Process preparation and realization

Process preparation is to implement the solutions contained in the Work performance projects (in the Process charts).

Self-Organization and Technology of Building software develops operations and financial plan for a company in its key activities and based on the following:

- Plan of the construction facility;
- Plan of the construction operations;
- Technical development plan;
- Mechanization plan;
- Labour plan;
- Procurement plan;
- Profit and cost plan;
- Income and expenses plan.

All of these plans are based on the main principles of term scheduling and supervising [9].

2.3.3 Term scheduling and supervising

Timeline analysis procedures provided by "SOTB" enable to estimate the project's total duration, any possible earlier date of completing thereof, dates of the project's goals start and finish, time reserves and to identify critical tasks.

To perform the timeline analysis, it's required the estimations of the project goals duration. Such parameter may be directly set for work or may be calculated. In certain spheres, for instance, in construction the project shall be initially planned not subject to the works duration but subject to the scope thereof.

In "SOTB" software such principle shall be realized by estimating the duration subject to assigned resources or an algorithm set by a user. It may be used probability estimates received as the result of the risks analysis. Reflecting the project's critical path may be made possible by several means including colouring on the screen or receiving reports providing the information only regarding the critical path goals.

2.3.4 Resources planning

One of the key differences of any system for scheduling is the type of resources to be supported and the way the resources management process is implemented in a multi-project. All the majority of resources types in the context of pattern description of scheduling goals reasonably are subdivided into two key types (renewable – reproduced, non-stock or being stored – resources distinguished by an ability to be compensated in case of any idle time; nonrecoverable – those being wasted completely changing their actual form and excluding repeated use).

"SOTB" is capable of considering such resources types and may take into account the kind of "perishable resources" that we may be storing for a fixed period, and upon expiration thereof unused odds shall not be transferred to the following intervals and are simply lost.

"SOTB is also capable of considering the resources with various qualifications list including selecting the least loaded resource in the course of the work fulfilment [10].

2.3.5 Analysing the resources provided for the project's works

"SOTB" offers powerful features to analyse the extent of the project's coverage by resources by means of comparing their demand's profiles and availability thereof in the course of the project implementation. Such analysis may be made both for a separate resource and for any group thereof.

"SOTB" As opposed to many other scheduling packages program enables to perform analysis and rescheduling not only for renewable resources but also for physical ones providing for this purpose two basic methods for dates estimation in the course of resource planning: in the context of limited time and limited resources [11].

2.3.6 Costs planning and supervising

Supervising any monetary assets expended in the course of the project implementation is one of the key goals in the project being no less important than resources management. Though the functions of costs supervising are provided by the majority of packages for projects management using thereof quite seldom gives satisfactory results: mainly this is due to the fact that the offered cost functions disagree with a wide range of methods applied in a company for costs planning, tracking and statements making.

"SOTB" operations functions regarding the costs have been developed to secure basic procedures of cost supervision and appear to be a flexible mechanism that managers and schedulers shall as close as possible adapt for the environment they face [12].

2.3.7 Analysing the works cost based on the actual output

It is deemed not practical to track only the difference between the amounts planned for the works fulfilment and the actual amounts since the simple comparison of the two figures upon the works completion shall be of no help for management and supervising. The very actual and planned costs do not provide a manager enough information if the non-fulfilled work has exceeded the budget.

Processing the costs data together with the actual achievements in "SOTB" is entitled "earned value cost control system" being analysing the works cost on the basis of actual output. As a rule, the aforementioned three parameters are depicted as aggregate costs curves in graphic form. The three parameters taken together, being the Planned cost of Scheduled works, the Planned cost of the Fulfilled works, the Actual cost of the Fulfilled works, shall form the core of the costs control system based on actual output [13, 14].

3 Software package application

Self-Organization and Technology of Building software greatly facilitates solving management tasks at the construction site. Implementing this software in the construction management processes may help to avoid many complexities and details, among the house erecting could be planned using the newest materials and technologies.

The great advantage of the software is the regulatory framework, being the basis of "SOTB" including the basic regulations for management and technologies of construction operations. It makes it possible to implement the most rational construction process.

After the beta-release the "SOTB" software will be also applied on the professional practice level to the educational process and scientific research activity in Moscow State University of Civil Engineering at the Department of Construction Technology and Organization.

4 Conclusion

From this it could be concluded that every type of work has its own management and process aspects and therefore using Self-Organization and Technology of Building software with functional ability of managing each of construction process automates the search for the construction regulatory framework.

The "SOTB" software was designed and developed by Sergey Sinenko with the aim of making possible to solve various engineering tasks on designing management of construction operations by dint of creating an independent algorithm of works by entering input data and parameters of the facility, to obtain the documents mentioned in "Management" section, and follow automatic creation of workflow.

References

1. S.O. Novikov, A.A. Sapukhin, Proc. of MSUCE **7**, 96 (2013)
2. A.A. Volkov, V.O. Chulkov, R. Kazaryan, M.A. Fachratov, O.N. Kyzina, R.K. Gazaryan, AMM **580**, 2281 (2014)
3. V.S. Semenov, E.V. Tkach, S.A. Tkach, T.A. Rozovskaya, Proc. Eng. **111**, 763 (2015)
4. B.V. Zhadanovsky, S.A Sinenko, Science Review **9-2**, 435 (2014)
5. S.O. Novikov, B.V. Zhadanovsky, Science Review **21**, 341 (2015)
6. B.V. Zhadanovsky, S.A. Sinenko, AMR **838-841**, 280 (2013)
7. The Russian Government, *Set of Rules 11-101-95* (Russian newspaper, 2011)
8. P.P. Oleinik, Proc. Eng. **153**, 208 (2016)
9. V.O. Chulkov, R.K. Gazaryan, V. M. Fakhratov, A.V. Belyaev, Proc. of VGASU **23**, 91 (2012)
10. E.K. Zavadskas, T. Vilutienė, Z. Turskis, J. Šaparauskas, ACME **14**, 114 (2014)
11. D. Jato-Espino, E. Castillo-Lopez, J. Rodriguez-Hernandez, J.C. Canteras-Jordana, AutoCon **45**, 151 (2014)
12. S.O. Novikov, Matec Web of Conf. **86**, 04032 (2016)
13. A.A. Lapidus, Proc. of MSUCE **7**, 96 (2013)
14. P. De Wilde, D. Coley, Build. and Env. **55**, 1 (2012)

Estimation of total time on test for large samples

Marius-Adrian Spiroiu*

University POLITEHNICA of Bucharest, Department of Rolling Stock, Splaiul Independenţei 313, Bucharest, Romania

Abstract. The aim of the present paper is to identify total time on test (TTT) estimators which require only partial information on sampling units. This allows an important simplification of the testing or observation procedure. The approach is based on the division of the testing/observation period into intervals so that TTT estimators to be determined using only classes numerical attributes. Division is made based either on failure rank or on time. A case study based on real field empirical data is carried out in order to assess the effectiveness of TTT estimators.

1 Introduction

The total time on test (TTT) is a concept introduced in the second half of the last century, developed and popularized in the seventies especially by Barlow and Campo [1]. Since then, it has been a subject of interest to many researchers and different new applications of the concept have been defined and studied (see, e.g., [2-5]). Many of these developments have proven to be very useful in the area of reliability analysis, particularly those concerning the identification of reliability mathematical model and the characterisation of failure rate.

Although the main use of the TTT concept is the field of reliability analysis, recently it has found its utility also in other areas such as economics, risk assessment, maintenance scheduling, etc. [3,5].

When several units are simultaneously put under test (or observed in real operation), the units failure will occur at different times. Therefore, at each moment of time during the test it will be a number of units still in operation and a number of units that have failed. The TTT statistic is given by the sum of all units life lengths, either completed (the case of units that have failed) or incomplete (units without failure).

The most common use of TTT is to obtain information about the failure rate, based on the shape of the so-called TTT transform plot. An important property of this plot is that the slope of the graphical representation at any point (i.e. any moment of time) is the inverse of the failure rate at that point. This type of analysis can be performed for both complete and incomplete data and it typically requires evaluating the TTT for each time-to-failure, i.e. for every tested/observed unit failure time. Taking into account that it is increasingly difficult to obtain complete information about the lifetime of products, the present paper investigates

* Corresponding author and co-author(s): marius_spiroiu@yahoo.com

the estimation of TTT in the case of large samples, assuming that not all the units time-to-failure are known.

2 The total time on test concept

Given a non-negative random variable X and $F(x)$ its distribution function, the total time on test (TTT) transform of X is defined as

$$T(u)=\int_{0}^{F^{-1}(u)}[1-F(x)]dx,\quad u\in[0,1] \tag{1}$$

where

$$F^{-1}(u)=\inf\{x/F(x)\geq u\} \tag{2}$$

is the quantile function of X.

The scaled TTT transform is given by the ratio

$$T_s(u)=\frac{T(u)}{T(1)}=\frac{\int_{0}^{F^{-1}(u)}[1-F(x)]dx}{\int_{0}^{F^{-1}(1)}[1-F(x)]dx} \tag{3}$$

and the plot of $T_s(u)$ versus u is the scaled TTT transform curve on the basis of which the characterisation of failure rate can be done. The shape of the curve allows to decide whether the failure rate is constant, increasing or decreasing with time.

In most of the cases, the random variable X consists of the lifetime (time of operation without failure) of units whose reliability is analyzed. In this case, the TTT transform in equation (3) can be expressed as [6]:

$$T_s(u)=\frac{\int_{0}^{t_u}R(t)dt}{\int_{0}^{\infty}R(t)dt} \tag{4}$$

where $R(t)$ denotes the reliability function and $t_u=F^{-1}(u)$.

If $t_1\leq t_2\leq\ldots\leq t_n$ are the successive failure times of a sample of n units put to test at time 0, the empirical TTT up to the ith order statistic is given by [6]

$$TTT(n,i)=\sum_{j=1}^{i}t_j+(n-r)t_i \tag{5}$$

which represents the total time on test of the n simultaneously tested units accumulated up to the time of the ith failure.

The TTT transform graph is obtained by plotting the points (i/n, $TTT(n,i)/TTT(n,n)$) - in the case of complete data - or (i/k, $TTT(n,i)/TTT(n,k)$) in the case of incomplete data, when the test ends after the kth failure (with $i=1\ldots k$).

3 Estimation of TTT for large samples

As it could be seen above, plotting the TTT transform curve requires the calculation of TTT value at each failure time, so it has to be known the moment of failure for every tested or observed unit. This is not very convenient in the case of units observed in real operation, especially if it comes to large samples. For this reason, the aim of the present paper is to investigate algorithms for estimating TTT in the case of large data samples.

The idea is to get an estimation of TTT without being necessary to know the failure-times of all units observed or tested. This would both decrease the amount of calculations required and, most of all, would simplify the procedure of obtaining data during the entire testing or observation period.

The natural solution to simplify the procedure is to divide the testing period into intervals and to evaluate TTT only for the limits of these intervals instead of calculating it for each failure time of tested units. Under these circumstances, the questions that must be answered are:

- On what basis should be the observation period divided into classes ?
- How (i.e. using what formula) should be estimated the TTT for each class ?

Regarding the first question, two solutions may be considered for distributing the data into classes:

- the quantiles method: in this case each class contains an equal number of failed units;
- the equal intervals method: in this case each class has the same duration.

In other words, the first method is based on failure rank while the second is based on time. It is to be noticed that if the first method is used, the TTT transform plot points will have equidistant abscissae.

Regarding the second question, two solutions may be considered for estimating TTT for each interval:

- to evaluate TTT using the upper limit of the class interval;
- to evaluate TTT using the midpoint of the class interval.

In either case, equation (5) can be used to estimate TTT, with the observation that this time t_j are no longer failure times of individual units but class intervals parameters (upper limit or midpoint, as the case may).

If the intervals upper limit is used, TTT estimation is given by

$$\hat{TTT}(n,i) = \sum_{j=1}^{i} r_j t_{sj} + (n - \sum_{j=1}^{i} r_j) t_{si} \, , \tag{6}$$

where r_j is the number of failures in class j and t_{sj} is the superior limit of class interval j.

If the intervals midpoint is used, TTT can be estimated by a similar equation, except that it uses in the first term the class intervals midpoint (t_{mj}) instead of its upper limit t_{sj}:

$$\hat{TTT}(n,i) = \sum_{j=1}^{i} r_j t_{mj} + (n - \sum_{j=1}^{i} r_j) t_{si} \, . \tag{7}$$

Equations (6) and (7) can be further particularized according to each considered case: if the quantiles method is used, equations (6) and (7) become, respectively:

$$\hat{TTT}(n,i) = r_c \sum_{j=1}^{i} t_{sj} + (n - i r_c) t_{si} \tag{8}$$

and

$$\hat{TTT}(n,i) = r_c \sum_{j=1}^{i} t_{mj} + (n - ir_c)t_{si}, \tag{9}$$

where r_c denotes the number of failures in each class (the same for all classes). Furthermore, for a complete test, when $n=pr_c$, where p is the number of classes, equations (8) and (9) can be written in the form

$$\hat{TTT}(n,i) = r_c \left[\sum_{j=1}^{i} t_{sj} + (p-i)t_{si} \right] \tag{10}$$

and

$$\hat{TTT}(n,i) = r_c \left[\sum_{j=1}^{i} t_{sj} + \left(p - i - \frac{1}{2} \right) t_{si} \right]. \tag{11}$$

It can be seen from equations (10) and (11) that, whether the class intervals upper limits or midpoints are used, the only data necessary to estimate TTT (taking into account that the number of failed units per class is previously set and assuming that the total number of classes is known) are the time intervals limits.

If the "equal intervals" method is used, equations (6) and (7) become, respectively

$$\hat{TTT}(n,i) = t_c \left[\sum_{j=1}^{i} jr_j + \left(n - \sum_{j=1}^{i} r_j \right) i \right] \tag{12}$$

and

$$\hat{TTT}(n,i) = t_c \left[\sum_{j=1}^{i} \left(j - \frac{1}{2} \right) r_j + \left(n - \sum_{j=1}^{i} r_j \right) i \right], \tag{13}$$

where t_c denotes the length of each class interval (same duration for all intervals), hence $t_{sj}=jt_c$, for $j=1 \dots i$.

It can be seen from equations (12) and (13) that also in the case of "equal intervals" method, whether the class intervals upper limits or midpoints are used, the only data necessary to estimate TTT (taking into account that the size of the class interval t_c is previously set) are the number of failed units in each class.

Estimators given by equations (10) … (13) are covering the four possible combinations of the considered cases - concerning division into classes and data used to evaluate TTT. Based on any of which can be set up a plan for testing or observing in real operation a set of sampling units. Choosing a suitable plan is depending on the specific conditions related to statistical units features, possible testing/observing procedures, available resources, etc.

4 Case study

A case study is carried out in this section, in order to assess the effectiveness of the previously presented estimators. The study is based on real field empirical data, observed and collected in actual operations of freight wagons [7]. The sample size is of 1802 units, of which in the case of n =190 have been registered failures over the observation period of 350 days.

As the 190 failure times are precisely known, the TTT value can be accurately calculated at any time. This (real) value can be used as a basis of comparison for the estimated values of TTT, obtained using the expressions derived in the previous section.

In tables 1 and 2 are presented the accurate and the estimated values of TTT in the two hypotheses concerning the distribution of data into classes: equal time and equal number of failed units, respectively. In each scenario, calculation was made using both the upper limit and the midpoint of the class interval. TTT was calculated as the total operation time accumulated by the 190 units failed during the observation period, which was divided in 10 intervals (35 days or 19 failed units each).

Table 1. TTT estimation – equal time classes.

i	t_i [days]	r_i	$TTI(n,i)$ [days]	$\hat{TTT}(n,i)$ [days]	
				upper limit	*midpoint*
1	35	21	6294	6650	6282.5
2	70	23	11823	12565	11795
3	105	27	16419	17675	16432.5
4	140	14	20282	21840	20352.5
5	175	22	23574	25515	23642.5
6	210	20	26079	28420	26197.5
7	245	14	28047	30625	28157.5
8	280	24	29322	32340	29452.5
9	315	8	30020	33215	30187.5
10	350	17	30347	33810	30485
			RMSE:	2277.82	100.57
			CV [%]:	10.25	0.452

Table 2. TTT estimation – equal failure number classes.

i	t_i [days]	r_i	$TTI(n,i)$ [days]	$\hat{TTT}(n,i)$ [days]	
				upper limit	*midpoint*
1	33	19	5953	6270	5956.5
2	61	19	10484	11058	10478.5
3	84	19	13794	14554	13756
4	115	19	17591	18677	17584.5
5	157	19	21977	23465	21973.5
6	182	19	24138	25840	24111
7	225	19	26997	29108	26970.5
8	257	19	28568	30932	28490.5
9	298	19	29718	32490	29659
10	350	19	30347	33478	30153
			RMSE:	1866.27	70.77
			CV [%]:	8.90	0.337

To overall assess the deviation of the proposed TTT estimators in relation to actual values, the root mean square error (RMSE) and the coefficient of variation (CV) are computed using the relationships:

$$RMSE = \sqrt{\frac{\sum_{i=1}^{p}\left[TTT(n,i) - \hat{TTT}(n,i)\right]^2}{p}} \tag{14}$$

$$CV = \frac{RMSE}{m(TTT)} \tag{15}$$

where p is the number of classes and $m(TTT)$ is the mean of actual TTT values. Notice that CV is expressed as a percentage in tables 1 and 2.

It is obvious that using the midpoint of the class interval is clearly preferable to the alternative of using the upper limit of the class interval (see the values of CV and RMSE in tables 1 and 2). Also, division into classes according to quantiles method – equal number of failures per class – seems to be preferable.

An additional analysis (table not included) made in the case of 19 classes of 10 failed units each resulted in the following values of CV: 4.153 % (using the upper limit) and 0.684 % (using the midpoint). By making a comparison with previous results (see table 2), 8.90 % and 0.337 %, respectively, it does not necessarily result that an increase of the number of intervals (a more detailed analysis) has a positive impact in any situation.

5 Conclusion

Given the difficulty to get complete information about the lifetime of products, the aim of the present paper was to identify TTT estimators which require only partial information. This is convenient especially for large samples, implying a major simplification of the procedure of obtaining data during the entire testing or observation period.

The approach was based on the division of the testing period into intervals so that TTT estimators to be determined using only classes numerical attributes. Division was made based either on failure rank or on time and TTT was estimated using either the class upper limit or class midpoint. The TTT estimators were obtained for each considered case, the only necessary data being the classes number of failed units or time intervals limits (according to the case). On the basis of this data required can be established plans for observing in real operation or testing a set of sampling units.

A case study based on real field empirical data was carried out in order to assess the effectiveness of the TTT estimators. The root mean square error (RMSE) and the coefficient of variation (CV) were used for an overall evaluation. The numerical results showed that using the midpoint of the class interval is more suitable and that division into classes according to quantiles method seems to be advantageous.

However, choosing a suitable testing plan is depending on the specific conditions related to statistical units features, possible testing/observing procedures, available resources, etc

References

1. R. E. Barlow, R. A. Campo, *Reliability and Fault Tree Analysis*, 451 (SIAM, Philadelphia, 1975)
2. M. Csörgö, H. Yul, Stochastic Processes and their Applications **68**, 229 (1997)
3. N. Unnikrishnan Nair, P. G. Sankaran, Statistical Methodology **10**, 93 (2013)
4. F. Belzunce, X. Li, J. F. Pinar, J. M. Ruiz, J of Stat. Plan. and Inf. **133**, 111 (2005)
5. A. M. Franco-Pereira, M. Shaked, Statistical Methodology **18**, 32 (2014)
6. M. Spiroiu, *Reliability and Maintenance of Railway Vehicles* (in Romanian), (Matrixrom, Bucharest, 2006)
7. M. A. Spiroiu, Applied Mechanics and Materials **809-810**, 1097 (2015)

Analyzing the functional capabilities of modern systems, methods and means for scheduling

Sergey Sinenko*

Moscow State University of Civil Engineering, Department of Construction Technology and Organization, 129337 Moscow, Russian Federation

Abstract. This work deals with the approaches for scheduling the process of territory development with building complexes of any objects being erected or reconstructed in sequence based upon up-to-date software products, provides the analysis of the functional capabilities of systems, methods and means for scheduling and gives an overview of modern software for scheduling.

1 Significance of long-term scheduling in construction

At present major megapolises worldwide are actively applying the method of territory development with building complexes including the city centres development, often using reconstruction and restoration resulting in adapting for residence or commercialization any premises in the existent erections [1].

In the current context of present-day competition improvement the principal participants of investment and construction activities shall effectively arrange construction operations to receive maximum profit at minimal cost of time and resources without any failures as to construction deadlines causing expenses increasing in the course of projects implementation and loss of profit for all participants of construction and investment activities.

Building a complex of objects shall require long-term scheduling and consequently may be faced by the goal of generating long-term investment construction spreads including the matter of optimizing the sequence of building complex objects development. Hence, it requires the tools enabling to select the construction appropriate sequence being at the same time resistant to perturbing factors. At present this matter may only be settled at the level of methodological foundation of rather general nature indicating that requires to address the goals focused on modern tolls basis for effective scheduling [2].

2 Effective scheduling

2.1 Goals for scheduling, basic definitions

Scheduling refers to the fundamentals of management and is applied in various sectors: power generation sector, oil and gas sector, metallurgical industry, construction, etc. The

* Corresponding author: 6622040@mail.ru

reason for scheduling development is the projects complexity increasing and necessity of raising the efficiency of implementing thereof. Construction companies face the goal of optimal distribution of limited resources in time. Investment and construction project is a complex system shifting from one stage to another one in the course of its whole life cycle. Such stages are referred to as the project phases, and each of them has its specific features and risks, so projecting methods for scheduling at various stages shall be different [3, 4].

2.1.1 Stages of an object building process

At the first (pre-investment) stage of an object building since the deficiency of information about the project the scheduling shall be projected taking into consideration any available consolidated indices addressing the following goals: estimating the project's efficiency and feasibility, fixing the project's term and cost thereof, developing the process of investing into the project.

An investment stage is the main project's stage including negotiations holding and contracts making, projecting, construction, marketing. In the course of bidding scheduling enables to receive the most preferable tender bid addressing the goals as follows: scheduling the milestones for the works to be performed, building a financing schedule, executing statements for personnel and physical resources required.

For the purposes of scheduling it may be divided into 2 steps: the process of the project development and its implementation supervising. The first step includes the goals as follows: defining the scope of work, executing schedules for the works and costs, timelines development, resources allocation, estimating and optimizing the milestones, constructing the charts for needs in personnel resources and machines, arranging the schedule of materials procurement and supply, estimating the costs at various stages and risks evaluation [5].

2.1.2 Modern systems for scheduling

In accordance with the goals faced by scheduling at present there is a number of modern systems that shall have capabilities as follows:

- organizational and technological diagram creation including describing the global parameters of the project scheduling, describing the structure of the scope of works, supporting the project calendar;
- settlement means subject to the critical path method (being the derivative of organizational and technological diagram);
- resource planning means (supporting the resources calendars, resources destination, calendar scheduling with limited resources, making the list of resources and costs items);
- supervising means for the project implementation process (recording targeted parameters of the project's timelines, entering actual parameters of the set goals, comparing scheduled and actual parameters, forecasting the outstanding works fulfilment);
- means for preparing reports and graphical presentation of the project's structure (by means of Gannt Chart, network diagram, PERT diagram);
- means for costs analysis [6].

2.2 Scheduling with the help of software packages

At present the market offers a majority of universal software packages that we are used to subdivide into professional and "desktop" (amateur). They mainly differ in the labor intensity when applying since professional software is quite time-consuming however it may provide more flexible outcome, and amateur packages may be used just once, and spending little time and few efforts you shall receive quite an informative result.

Except for universal programs, in more recent times extra software packages are widely distributed enabling to treat each project much more individually. They address the goals as follows:

- risks analyzing, planning the best strategy for risks management (@RISK by Palisade Corp.);
- timeline scheduling subject to the goals, salary accounting (HMS Software, Time Control);
- timeline scheduling in the course of limited resources (Best Schedule for Project by Parsifal System Inc.);
- integrating the management system into corporate systems (Project Gateway by Marin Research Inc., Project Management Integrator by Time Line Solutions Corp., etc.) [7].

The most applicable universal professional software packages with the cost of more than 1000$ are Artemis Project View, Open Plan Professional, Primavera Project Planner.

Among the amateur universal software packages the most frequently used are OpenPlanDesktop, Project 98, Project Scheduler, Super Project, Time Line.

3 Applying Open Plan Professional when projecting scheduling

In respect that the sectioning the software depending on the outcome that is required let consider the key features of Open Plan Professional Software Complex as a system being among the leaders in the sphere of professional systems for projects management (Figure 1) [8].

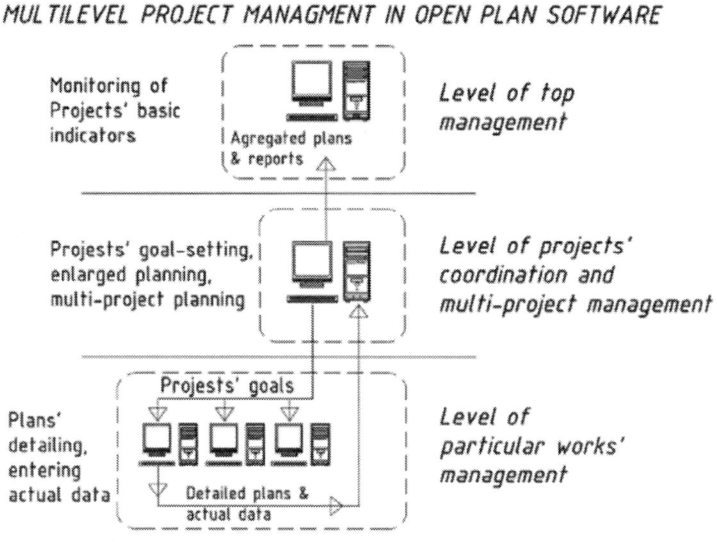

Fig. 1. Projects' management system in Open Plan Software.

3.3.1 The software package primary function and its capabilities

Open Plan Professional refers the category of professional software packages and correspondingly possesses powerful features of resource and cost planning. In this context it's recommended to apply it for major projects requiring detailed scheduling and for integration with other organization automated systems.

3.3.2 Software functional capabilities

The software enables to create a resource breakdown structure and enables to automatically allocate resources among the goals subject to qualification thereof. It also enables to analyze risks, calculate the probability thereof and any consequences subject to any possible delays in the materials delivery, budget deficit.

3.3.3 Further capabilities for resources management

Open Plan's powerful features for resources planning enable to properly manage any types of resources: human, equipment, materials, finance (Figure 2) [9].

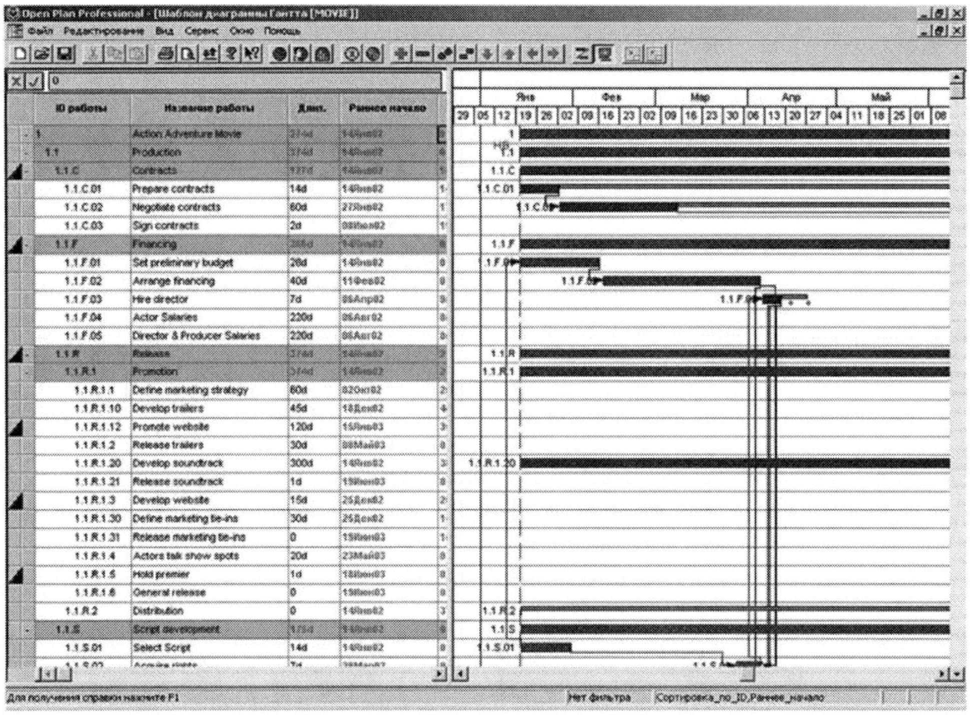

Fig. 2. Resources management in open plan professional.

3.3.3.1 Costs analyzing

The software enables to create a resource breakdown structure and enables to Open Plan provides for special procedures for costs scheduling and supervising particularly noting the means of analyzing and building reports for actual output. The facilities for "memorizing" several options of any project implementation and inputting the actual data regarding costs for work and resources workout enable to analyze the project costs both forecast and actual [10].

3.3.3.2 Risks Analysis

Analytical tools built in Open Plan enable to evaluate any probable risks in estimating the term of separate works, whole stages and the whole project completion, using the facilities as follows:
- procedures of inputting probability distribution for any selected or all project's works;

- analyzing the risks subject to Monte Carlo Methods to estimate the probabilities input into the project's dates;
- providing the reports being used to analyze the uncertainties influence on the project's implementation [11].

3.3.3.3 Multi-project analysis

Open Plan's standard facilities enable to integrate independent projects enabling to manage resources and finance taking into consideration the projects' priority being implemented by any company and receiving the consolidated statements (Figure 1).

3.3.3.4 Means for creating the project's model

The project's structuring is one of the key concepts of the management concepts due to which the project's hierarchic subdivision into components is provided that are much easier to schedule and supervise. Comparing to other professional systems of the projects scheduling and management, Open Plan provides the most powerful present-day project's models based on the following basic concepts:
- goals hierarchic structure (WBS -Work Breakdown Structure);
- network model (PERT Diagram represented at Figure 3);
- resources hierarchic structure;
- hierarchic system of activities coding [12].

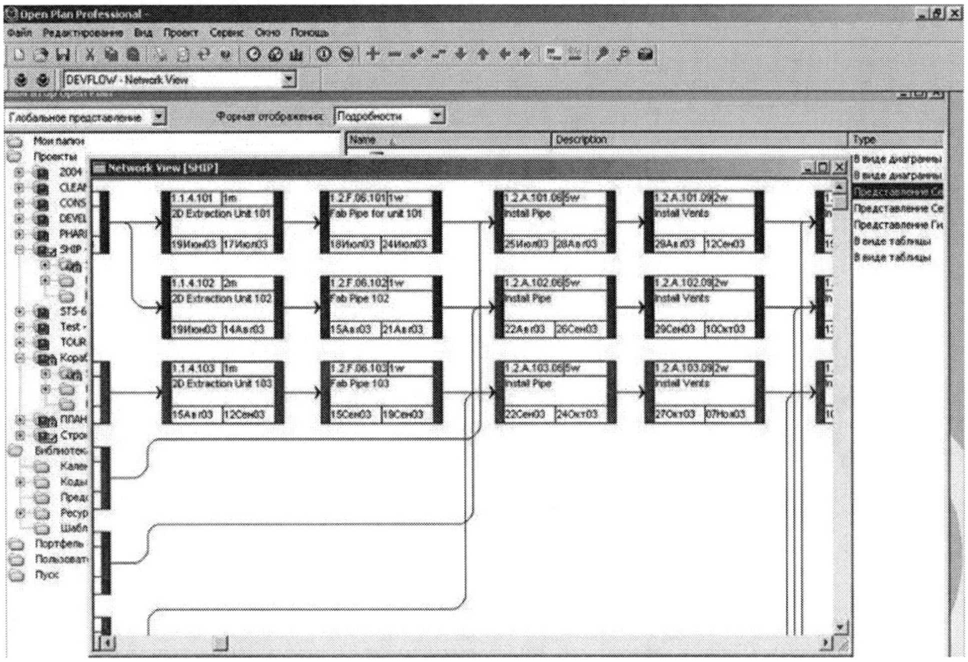

Fig. 3. Network model view in Open Plan Software.

3.3.3.5 Multi-project planning

Operations opportunities in Open Plan in multi-project mode enable users to consider one major project as consisting of much smaller sub-projects providing thereby flexible management at various levels.

Working in multi-project mode provides facilities to supervise and distribute unified company's data among all projects being implemented by it. Uniting projects in such a way serves two purposes at once: firstly, analyzing the resources loading on a scale of the whole company, secondly, providing the environment for integrated software-based control of large complex projects subdivided into sub-projects [13, 14].

4 Conclusion

Following the results of the study it's reasonable to make the following conclusions:
1. Building a complex of objects shall require long-term scheduling and consequently there arise the goals of forming long-term construction spreads including the matter of optimizing the priority of developing the objects being the part of a building complex.
2. There is a number of modern ways, methods and means for scheduling enabling to settle various goals and aiding the development of scheduling methods; at the same time there are universal software packages, separate programs and applications enabling to settle general and more specific goals of scheduling. In this angle Open Plan is deemed one of the leaders in the sphere of professional systems of projects management.
3. Open Plan Software greatly facilitates solving management tasks at the construction site. Implementing this software in the construction management processes may help to avoid many complexities and details, among the house erecting could be planned using the newest materials and technologies. The great advantage of the software is the regulatory framework based of including the basic regulations for management and technologies of construction operations. It makes it possible to implement the most rational construction process.

References

1. B.V. Zhadanovsky, S.A Sinenko, Science Review **9-2**, 435 (2014)
2. B.V. Zhadanovsky, D.G. Dragan, S.A Sinenko, Tech. and org. of build. prod. **2**, 38 (2014)
3. S.O. Novikov, B.V. Zhadanovsky, Science Review **21**, 341 (2015)
4. B.V. Zhadanovsky, S.A. Sinenko, AMR **838-841**, 280 (2013)
5. V.S. Semenov, E.V. Tkach, S.A. Tkach, T.A. Rozovskaya, Proc. Eng. **111**, 763 (2015)
6. A.A. Volkov, V.O. Chulkov, R. Kazaryan, M.A. Fachratov, O.N. Kyzina, R.K. Gazaryan, AMM **580**, 2281 (2014)
7. S.O. Novikov, A.A. Sapukhin, Proc. of MSUCE **7**, 96 (2013)
8. V.O. Chulkov, R.K. Gazaryan, V. M. Fakhratov, A.V. Belyaev, Proc. of VGASU **23**, 91 (2012)
9. S.O. Novikov, Matec Web of Conf. **86**, 04032 (2016)
10. P.P. Oleinik, Proc. Eng. **153**, 208 (2016)
11. E.K. Zavadskas, T. Vilutienė, Z. Turskis, J. Šaparauskas, ACME **14**, 114 (2014)
12. D. Jato-Espino, E. Castillo-Lopez, J. Rodriguez-Hernandez, J.C. Canteras-Jordana, AutoCon **45**, 151 (2014)
13. A.A. Lapidus, Proc. of MSUCE **7**, 96 (2013)
14. P. De Wilde, D. Coley, Build. and Env. **55**, 1 (2012)

How Axiomatic Design can promote creativity in the design of new products

António Gabriel-Santos[1,*], *Alberto* Martinho[1], *João* Fradinho[1], *Miguel* Cavique[2], *Pedro* Sousa-Santos[3], *António* Gonçalves-Coelho[1], and *António* Mourão[1]

[1] UNIDEMI & Faculty of Science and Technology, NOVA University of Lisbon, Campus of Caparica, 2829-516 Caparica, Portugal
[2] UNIDEMI, CINAV & Naval School, Naval Base of Lisbon – Alfeite 2810-001 Almada
[3] DEMI, Faculty of Science and Technology, NOVA University of Lisbon, Campus of Caparica, 2829-516 Caparica, Portugal

Abstract. In product development, creativity is the driving force for doing something that leads to innovation. A typical ideation background has three subsystems: inspiration, dematerialization and recombination. The most basic concepts of Axiomatic Design, *i.e.* domains, hierarchies and zigzagging, as well as the two design axioms, provide a powerful framework to implement ideation, as to make creativity easier when developing a new product. The result of dematerialization is in the functional domain, which is the place where the customer needs are presented by functional requirements and purged of any kind of physical bias. The recombination creates the design parameters, at the physical domain, which are the set of elements of the design object that have been chosen to satisfy the functional requirements, into a new materialization profile of a new product. In this paper, a concrete case illustrating the above-mentioned concepts is presented.

1 Introduction

The success of an innovative new product highly depends on creativity during the conceptual stage of the design process.

On the one hand, creativity can be defined as the ability to produce work that is both novel and appropriate. The economic viability is not an important aspect when looking for new ideas, new concepts and/or even new approaches to deal with existing or emerging problems. On the other hand, innovation refers to the entire process by which an organization generates new ideas and converts them into novel, useful and viable commercial products, services, and business practices.

A typical ideation background is composed by three subsystems: inspiration, dematerialization and recombination [1]. The relations between dematerialization, recombination and the basic concepts of Axiomatic Design will be analysed in this paper.

Section 2 of this paper introduces the concept of ideation framework. Section 3 explains the principles of Axiomatic Design. Section 4 succinctly presents the deployment of a machine aimed to tilt the revolution axis of steel reels from the vertical to a horizontal

* Corresponding author: agms@fct.unl.pt

 DOI: 10.1051/matecconf/201711209010

alignment, which is presented as an example. At last, Section 5 contains the conclusions of the present work.

2 The Lean Ideation Framework

The "Lean Ideation Framework" proposed by Marques *et al.* and depicted in Figure 1 represents the aforesaid dematerialization and recombination subsystems that physically behave as being just one [2].

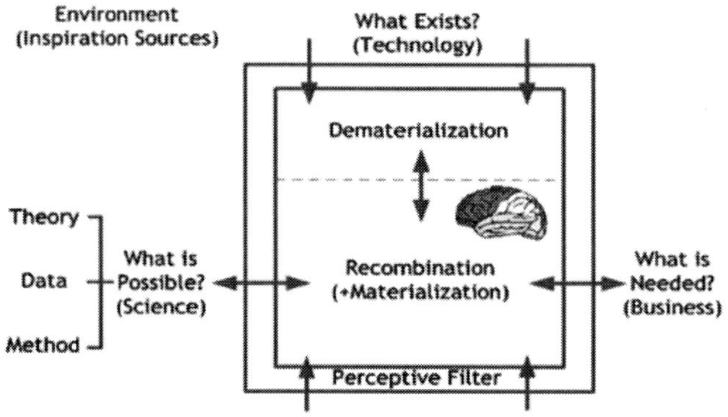

Fig. 1. The Lean Ideation Framework (adapted from [1]).

The inspiration subsystem consists on the available knowledge that exists in the designing environment, which is composed by scientific findings and technological achievements, as well as specific opportunities that depend on the business and the market surroundings. The dematerialization subsystem dissociates by abstraction the physical aspects or the functions embodied in an existent product into elementary functions to be performed by the new product for fulfilling the customer needs. The subsystem recombination is the complement of dematerialization, and is performed when the designer generates, applies and changes the content of the domains of knowledge for a new materialization profile in a new product.

Essentially the designers use the dematerialization phase of the model to release, by abstraction, from the existing concepts. The inspiration phase is driven by necessity, knowledge and technology. In the recombination step the association of new concepts drives the final solution.

3 Axiomatic Design

Axiomatic design (AD) is a theory created by N.P. Suh that mathematically describes the whole design process, according to the mode the human mind operates in the development of the designing activities.

Therefore, AD provides a systematic approach to design that is based on scientific principles, by introducing axioms and theorems, as well as the concepts of domains, zigzag decomposition and design equations, which can be used at the development of all the levels of the design process [3].

According to Suh, "Design, as the epitome of the goal of engineering, facilitates the creation of new products, processes, software, systems, and organizations through which engineering contributes to society by satisfying its needs and aspirations" [3].

3.1 The Design Environment

In the AD terminology, the depiction of the world of any design object is made in any one of four design domains: the customer, the functional, the physical, and the process domains [3, 4]. These domains are shown in Figure 2, and their contents can be described as follows:

"Customer Domain": contains the Customer Needs (CNs), i.e. the attributes that the customer seeks in the product or in the system that must be designed;

"Functional Domain": contains the Functional Requirements (FRs) of the design object. In a good design, they are the minimum set of independent requirements that completely describe the functional needs of the design solution, and which must be defined in a solution-neutral manner;

"Physical Domain": contains the Design Parameters (DPs) of the design solution. The DPs are the elements of the design solution that are chosen to satisfy the specified FRs;

"Process Domain": contains the Process Variables (PVs) that characterise the production process of the design solution, i.e., the variables that allow attaining the specified DPs.

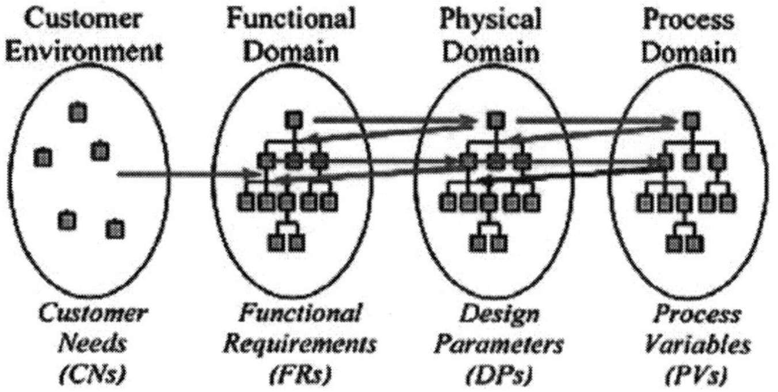

Fig. 2. Axiomatic design domains, their contents and relationships (adapted from [5]).

As shown in Figure 2, for each pair of adjacent domains, the left domain represents 'What is required to achieve, i.e. the goals', while the right domain represents 'How the goals are achieved, i.e. the proposed design solution' [3, 4].

Design is attained by interactions between the goals and the way to achieve them. The goals of the design are specified in the functional domain (where only immaterial items exist), and the way of achieving them is proposed in the physical domain (where the possible real solutions able of performing the specified functionalities are represented).

Constraints are the bounds for the acceptable solutions, and are classified as "input" and "system" constraints [3, 4]. The input constraints are part of the customer needs, while the system constraints are being discovered along the design process.

The mapping between the FRs and DPs can be summarized by equation (1), the "design equation", where {FR} is the "FR vector", {DP} is the "DP vector", and [A] is the "design matrix"

$$\{FR\} = [A]\{DP\} \tag{1}$$

Where $A_{ij} = \frac{\partial FR_i}{\partial DP_j}$. If DP_j disturbs FR_i, then the corresponding design matrix element A_{ij} is non-zero. Otherwise A_{ij} is zero.

3.2 Hierarchies and Zigzagging

Another important concept in AD is the hierarchical decomposition through zigzagging between contiguous domains, as for example zigzagging between the functional domain and the physical domain, in a top-bottom way, beginning at the system level and continuing through levels of more detail, as it is also shown in Figure 2.

After finding the top-level FRs and DPs that appropriately satisfy the general goals of the object under design, one should decompose the design object until a sufficient level of detail is attained at the so-called leaf-level, which means that DPs should not need either redesigning or further decomposition, so that the design could be implemented. The hierarchy established between FRs and DPs represents the design structure, which is also known as the system architecture.

An example of a possible zigzagging path between the functional and the physical domains of the design of an engine is shown in Figure 3.

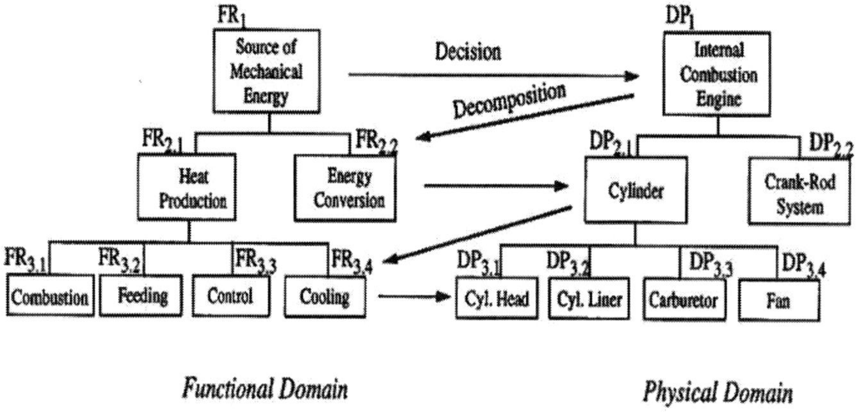

Fig. 3. The zigzag path of the hierarchical decomposition of an internal combustion engine [6].

3.3 The Principles of Design

The underlying hypothesis of AD is that there are two fundamental principles that govern good design practice [3, 4]:

The Independence Axiom (the first axiom):

Maintain the independence of functional requirements.

This means that each FR should be such that adjusting each DP just disturbs the accomplishment of one of the FRs.

The Minimum Information Axiom (the second axiom):

Minimize the information content of the design.

The purpose of this axiom is to help in finding out the alternative design solution that has minimum information, which is the one that has highest probability of achieving the FRs.

The independence axiom and the specific constraints should be applied to the design equation during the decomposition, as to ensure that an "uncoupled" or a "decoupled" design solution is obtained at each level of the design process. Uncoupled and decoupled solutions are characterized by diagonal and triangular design matrices, respectively [3, 4]. Any other shape of squared design matrix corresponds to a "coupled" design solution that should be avoided. Since the design process does not lead to a unique solution, the information axiom should be used to compare the alternative solutions that were previously found and select the alternative design solution with the highest probability of achieving the FRs. [7].

 DOI: 10.1051/matecconf/201711209010

4 An Application Example

An example of a real case where AD was used to the development of a machine design and check the soundness of the selected design solution is as follows.

According to AD, the design main goal must be carefully defined at the onset of the design process. Only after they are clearly stated, one can proceed detailing the appropriate design solutions [8].

In the designing process of a plant to make steel products it was required to rotate the symmetry axle of steel sheet reels from the vertical to a horizontal alignment, as to allow using an existing crane to move away the reels (see Figure 4).

Fig. 4. The design task as defined in the Customer Environment.

The design process starts with the definition of a set of FRs such that the CNs are satisfied. In this case, we have found the following 2nd-level FRs, which are depicted in Figure 5.

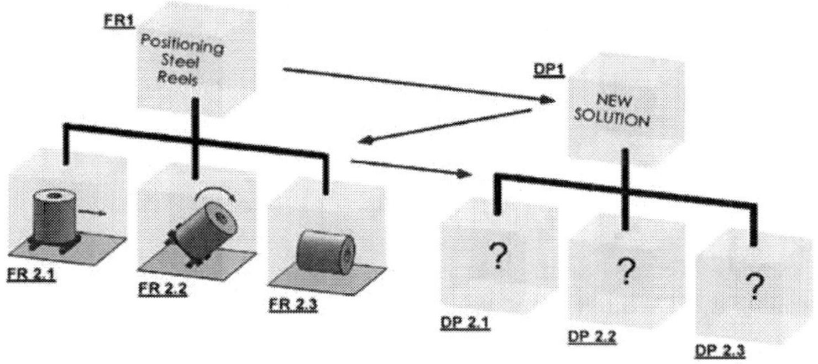

Fig. 5. Functional Requirements that satisfy the Customer Needs.

$FR_{2.1}$ – Receive the steel coils that come on motorized roller conveyors.
$FR_{2.2}$ – Rotate the axle of the steel reels from the vertical to an horizontal alignment.
$FR_{2.3}$ – Keep the axle of the steel in a horizontal alignment.

The design process continues creating solution concepts, at the physical domain, which are capable to perform the required functions.

For each one of the FRs there are several possible solutions that can be defined through different DPs. Figure 6 depicts some possible alternative solutions.

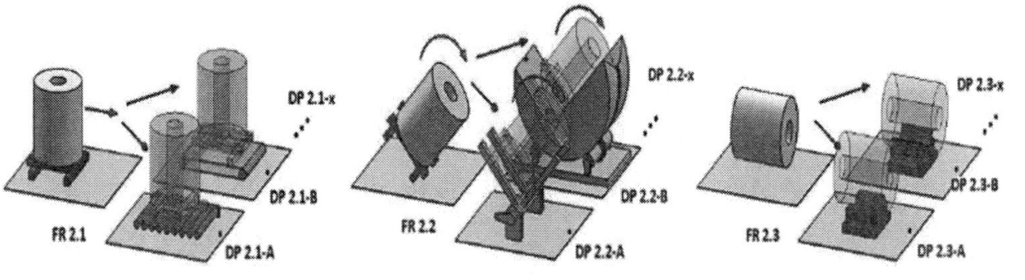

Fig. 6. Alternative solutions that satisfy $FR_{2.1}$, $FR_{2.2}$ and $FR_{2.3}$.

In some specific design solutions, it can happen that the number of DPs is larger than the number of FRs, which makes them "redundant" designs. In such cases, one can use the method followed by Gabriel-Santos *et al.* [9], in which the utilization of system constraints consisting in relationships between some of the DPs, allows clustering the DPs, as to match the number of DPs to the number of FRs.

From the several partial alternative solutions that were generated to satisfy the FRs, one selects the ones that compose the global solution with the higher probability of success. Figure 7 depicts the first higher-level solutions that result from the zigzag decomposition of the new product to positioning the steel sheet reels.

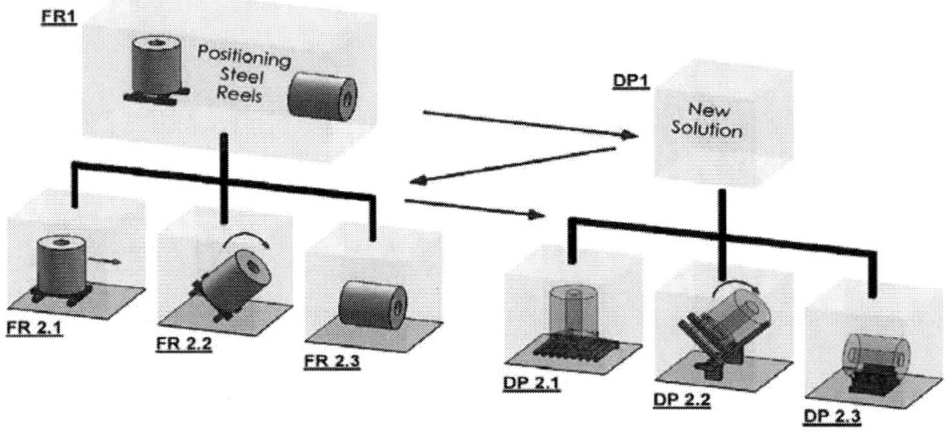

Fig. 7. The first two higher levels of the Functional Domain and Physical Domain of the new product under design.

The design process continues by repeating the procedure described above at the subsequent hierarchical levels, in a top-down sequence, until a level that contains sufficient detail to clearly describe the design object.

After reaching this level, the composition of the selected sub-solutions is performed, in a bottom-up sequence, until reaching the overall solution. In the Simplified Ideation Framework, this procedure is called Recombination, which combines the selected design parameters at the physical domain, thus attaining the materialization profile of the new product.

In this case, we are considering only the first two decomposition levels, so that the global solution is obtained by the composition of the selected sub-solutions, as shown in Figure 8.

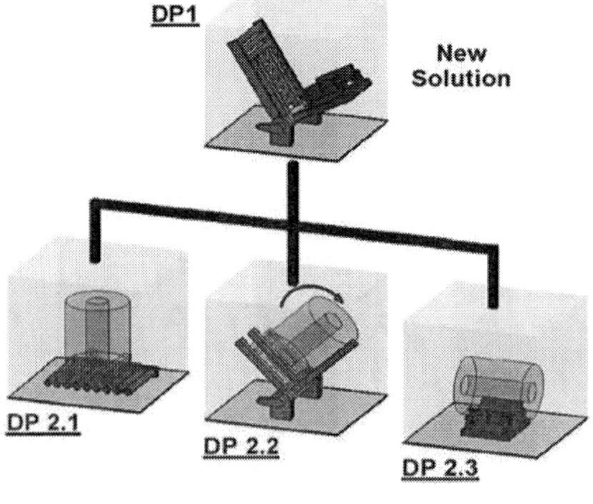

Fig. 8. The global solution resulting from the composition of the selected sub solutions.

DOI: 10.1051/matecconf/201711209010

Figure 9 shows the new product after being manufactured.

Fig. 9. Three different views of the designed new product rotating steel sheet reels.

As one could see, AD is a powerful tool for evaluating the generated design solutions, through the independence axiom, as well as for selecting the best one among alternative solutions through the information axiom. Additionally, the AD's unique zigzag decomposition technique allows efficiently deploying compound design solutions. However, Axiomatic Design does not offer any specific tool for synthesizing new design solutions.

The synthesis of new design solutions requires creativity, which is related to inspiration, both topics belonging to the scope of the cognitive sciences, as explained, for example, by de Bono [10] and much more extensively by Belous [11, 12].

Therefore the creativity techniques should be used to quickly generate several new design solutions, while AD should be used to quickly and objectively evaluate their efficiency.

5 Conclusions

The main benefits of using AD for promoting creativity in the design of new products are the following:

- It allows to organize and to prioritize the CNs, thus facilitating the minimization of the number of the independent FRs that characterizes the object under design.
- It allows defining the FRs in a solution-neutral environment, i.e., in a way that is not biased by predetermined ideas or existing solution.
- It provides a method to help focussing on the creative aspects of design, while also ensuring functionality, performance, testability, and feasibility of the systems.
- It emphasises an often-weak area of the design process — the complete understanding of the FRs and of their inter-relationships at all the design levels.

Axiomatic Design helps promoting creativity:

- In the problem definition, which is done in terms of solution-neutral FRs, thus creating conditions that facilitate the deployment of the dematerialization subsystem.
- In the separation of the elements of the design process by grouping the design object elements in different design domains according to their nature.
- In defining of the design structure, through the zigzag decomposition that hierarchizes de definition of the design details and facilitates ensuring independence, as prescribed by the First Axiom.
- In systematically dividing compound design problems into several sub-problems, thus facilitating the search for alternative solutions for each of the sub-problems at the different hierarchical levels.
- In selecting the alternative sub-solutions with the higher probability of success, as prescribed by the Second Axiom.
- In quickly evaluating the output of the creativity techniques, which do not provide the means for objective evaluation.

Acknowledgments

The authors gratefully thank the sponsorship of Fundação para a Ciência e Tecnologia through the Strategic Project UID/EMS/00667/2013 – UNIDEMI.

References

1. P. Marques, E. Henriques, A. Gabriel, International Journal of Academic Research, **2**, No 1,160, (2010)
2. P. Marques, A. Silva, E. Henriques, *RPD 2008*, Oliveira de Azeméis, Portugal (2008)
3. N. P. Suh, *The principles of design* (New York: Oxford University Press,1990)
4. N. P. Suh, *Axiomatic Design: Advances and Applications* (New York: Oxford University Press, 2001)
5. B. Gumus, A. Ertas, D. Tate, I. Cicek, J. Eng. Design, **19**, No. 3, 185 (2008)
6. A. M. Gonçalves-Coelho, J. F. Mourão António, J. J. Pamies-Teixeira, *The Vision for the Future Generation in Research and Applications*, J. Cha *et al.* (eds), Swets & Zeitilinger, Lisse, 419, (2003)
7. T. Derrick, *International Mechanical Engineering Congress and Exposition IMECE2015*, Houston, Texas. IMECE2015-51282 (2015)
8. M. Nordlund, T. Lee, S.-G. Kim, *International Mechanical Engineering Congress and Exposition IMECE2015*, Houston, Texas. IMECE2015-52893 (2015)
9. A. Gabriel-Santos, J. Rolla, A. Martinho, J. Fradinho, A. Gonçalves-Coelho, A. Mourão, *Procedia CIRP* **53** 193-197 (2016)
10. E. de Bono, *Lateral Thinking: A Textbook of Creativity* (London: Penguin Books 1977)
11. V. Belous, *La Performantique — L'ingénierie de la performance humaine* **I** (in French), (Iaşi: Editura "Gh. Asachi", 1995)
12. V. Belous, *La Performantique — L'ingénierie de la performance humaine* **II** (in French), (Iaşi: Editura "Gh. Asachi", 1995)

Study of iterations in the design process of a product for automotive industry

Daniel-Constantin Anghel[1,*], *Nadia* Belu[1], *Laurentiu-Mihai* Ionescu[2], and *Nicoleta* Rachieru[1]

[1] University of Pitesti, Department of Manufacturing and Industrial Management, Str. Tg. Din Vale, nr. 1, Pitesti, Romania Romania
[2] University of Pitesti, Department of Electronics, Computers, Communications and Electrical Engineering, Str. Tg. Din Vale, nr. 1, Pitesti, Romania

Abstract. This paper present an experiment realized in order to observe the iterations in a design process. The main objective of this work is to study the iterations during the design process using a laboratory experiment, in order to understand how and why iterations occur. The different forms of iterations as they occur in practice are identified. This study will help us in the classification of iterations in order to distinguish useful iterations from undesirable ones. The results of the study might be used to improve the manner of working in the field of engineering design.

1 Introduction

In this paper, a laboratory experimental was used to make observations about the iterations in a design process. Iteration is a fact of life in any project. The larger, more novel and more interconnected a project, the more of an issue it can be [1, 2]. Iteration can be defined to be the repetition of design tasks due to the arrival or discovery of new information [3].

For the industries, engineering design is a source of competitive advantage. One of the key factor for the success for a corporation is the reducing the development time of the product development process.

For product developers, it is very important to understand iterations, so you can manage them through the design process.

But, the design process is a complex and dynamic one. In order to understand the design process, the experts have used a big number of methods and models. Most models are complementary each to other, the approaches used vary depending on the context, on the vision of experts and on the scope. Unfortunately, many of the models are developed based on intuition and practice.

The experiment presented in paper aims at observing and understanding iterations production, the effects of these iterations, how they can be anticipated etc.

This paper is organised as follows: a literature review of design process models integrating the iteration aspect of engineering design is presented. After this literature review, the research method used in this study and the experiment are presented. The next section

* Corresponding author: daniel.anghel@upit.ro

presents the experimental data and observations, and their analysis. The conclusion of the paper present a summary of our work and a presentation of future developments.

2 Literature review

A way of understanding and acting on the design process is through its modelling. The work of developing a model provides a better understanding of the functional and behavioural characteristics of the design process. The use of the model makes it possible to define, test and improve strategies for the acquisition and engagement of design activities. The literature offers us a wide range of models. These models can be classified into two main families: prescriptive models and descriptive models [4]. Most of the models encountered and which can be used to study iterations are prescriptive models.

Smith and Eppinger [3] propose a model for the description of sequential processes with the presence of iterations. The model is used to find an initial order of activities leading to a minimal duration of the design process. Due to the assumptions adopted concerning the duration of the activities and the probabilities of questioning which are considered as constant and previously known.

The model developed by Krishnan [5] is used to manage the paralleling of two design activities. Downstream activity starts with partial information from upstream uncompleted activity. This can lead to poor premature design choices leading to loss of quality or causing long and costly iterations. The authors identified two characteristics of design process activities, called sensitivity and evolution, that allow us to choose the optimum degree of overlap to be adopted by determining information that can be frozen early in the design process and which information is to be used in a design process. Preliminary form. The limitations of this model are that it considers only the problem of overlapping only two activities. Moreover, it does not take into account the feedback from the downstream activity to the upstream activity and only downstream activity is reiterated in the process.

Yassine [6] also developed a mathematical model of two design activities that can be sequential, parallel, or partially overlapping. The duration of the iterations and their number are represented by random variables with a known distribution. The model calculates the total duration of the process.

Numerous research papers have approached the problem of iterations in the design process from close and distant. Indeed, these contribute greatly to the extension of development times. Osborne [7] reports that iterations account for between 13% and 70% of the total development time for Intel's semiconductor design activities. It also reports that the variation in development time is mainly due to iterations.

In their work, Pahl and Beitz describe the design process as a succession of phases [8]. They define iterations as the process by which a solution is approached step by step. They take place between the different phases and often within each phase. In this case, the iterations make it possible to refine a design solution. This procedure can be assimilated to that used in mathematics for solving an equation or a system of equations starting from an initial solution used to calculate another more precise solution which will in turn be used until convergence Towards an acceptable solution. This type of iteration is often encountered in interdependent design activities. For example, in order to carry out the activity C, it is necessary to know the value of the parameter y provided by the activity B. To calculate this parameter, the activity B in turn needs the value of the parameter x supplied by the activity A, but this needs the result of the activity C represented by the parameter z. In this case, an iterative process is required to compute all the parameters.

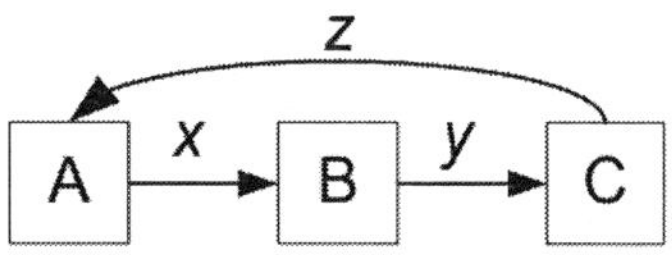

Fig. 1. A design iteration.

As a brief conclusion, in the literature, the iteration:

- is needed to solve complex problems [9];
- may help to deal with a changing context [9];
- may be reduced by a focus on central information-consuming/generating nodes [2];
- is strongly influenced by small changes in task time if close to capacity [10].

3 Design Experiment

The activity of design is a social one. In order to understand all the aspects of the design process, it is useful to use experimentations in order to observe how the designers interact, and how to process progress. Design could be accomplished in different situations: by an individual designer, or by a design team or several teams. The members of a design teams could work synchronously or asynchronously; they also could be geographically distributed [11].

The main objective of observations is to understanding cognition, creativity and innovation in the design process.

The role of our study is to observe iterations in a design activity performed by a multidisciplinary team. The observational techniques were used to record the design process. The records were used to make different study of the activity of participants.

The experiment described in this paper was developed at the University of Pitesti. The theme of the experiment was to design a roof lighting system for an SUV for intervention in situations of disasters or for the frontier service. The system must be able to adapt to all the atmospheric condition: clear sky, snowing, raining, fog, smog, pollution.

The experimental subjects were 4 PhD, three from the Department of Manufacturing and Industrial Management and one from Department of Electronics, Computers, Communications and Electrical Engineering. Each participant was given a distinct role. There were roles for a mechanical designer (Catia V5), an electrical and programmer engineer (Arduino), a risk analyst, and a mechanical specialist. The design team was led by a leader.

The experiment lasted for two weeks and there were 4 design sessions. All the sessions were recorded in order to fully capture the activity of design. In addition to these sessions, the designers performed their own tasks individually and asynchronously, but they could change information with others.

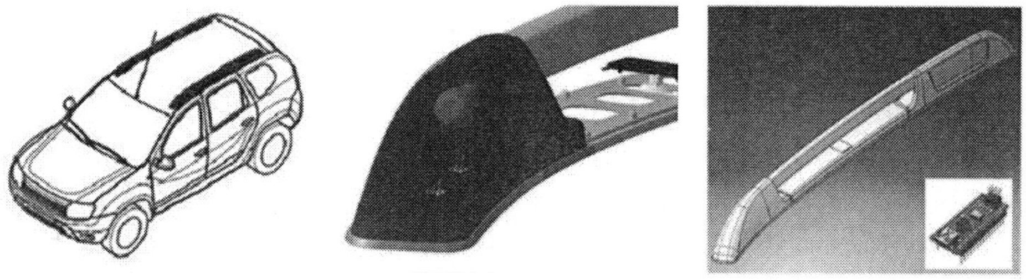

Fig. 2. The product - a roof lighting system.

4 Observations and results

In this paper, it is presented the third working session, which corresponds to the maturity of the product "design embodiment." In this session, they were identified a great number of iterations.

The design tasks performed by the designers were established and then were identified all the iterations based on the links between the tasks, by use of DSM method, figure 3.

Task Name	1	2	3	4	5	6	7	8	9	10	11	12	13	14	15	16
1																
2	1		1											1		
3	1					1										
4		1	1													
5			1	1												
6				1						1						
7		1		1										1		
8				1					1							
9			1	1				1		1						1
10									1							1
11		1		1		1			1							
12											1					
13				1		1			1							
14		1										1				
15		1										1		1		
16															1	

Legend: *Presentation of the problem; Analysis of solution for the roof bar; Analysis the location for the projectors; Choice of type of the projectors; Study of mechanical link projectors – bar; Electrical connection projectors - the command module; Detailing solutions for the fixture elements; Analysis of the command for the projectors; Solution development of the platform for analysis of atmospheric conditions; Developing the software for the platform; The risk analysis; The choice of materials; Evaluating the product in terms of ambient environmental conditions; Detailing solution for the roof bar; Rapid prototyping of the roof bar; Testing the solution for the system.*

Fig. 3. The DSM representation of the third working session of design.

Design Structure Matrix (DSM) based models [3,6] have been used to represent the iterative structure of engineering design. DSM uses a matrix representation of the design process. The DSM matrix is square with one task by column and by row. Information flows between tasks are indicated in the off-diagonal elements of the matrix. Two types of information flows are distinguished: feed forward (lower diagonal elements) and iterations (upper diagonal elements). With this representation, cyclic information flows are easily captured and the need for iterations is identified.

4.1 The sources of the iterations

After the analysis, we identified three main sources of iterations:

The change of objectives. During the design process, some initial data or solutions already proposed can be changed at any time for different reasons, leading to a repetition of a number of design tasks.

Interdependence between the tasks. These are mutually dependent tasks, in this case several iterations are necessary to arrive at an acceptable solution. In this case, the total delay of the design process often depends on the initial scheduling of the interrelated design tasks.

Design errors. These errors are all the more important as the design process is more and more complex and involves more and more people, especially in the context of simultaneous engineering, where several different trades are taken together.

4.2 The typology of iterations

The iterations can be classified according to a multi-dimensional typology. We propose a typology with four classification criteria.

Voluntary / involuntary iterations. Voluntary iterations are due to interdependent activities and sometimes changes in objectives. Involuntary iterations are mainly due to design errors.

This classification makes it possible to better predict the actions to be taken to reduce the development time of the product. The reduction of iterative processes can be achieved by better structuring the activities of the design process: decoupling interdependent activities, better scheduling of design activities, reduction of coordination time between design teams, and so on. However, the reduction of iterative processes can increase the risk of failure of the solutions developed and must therefore be considered with caution.

Short/long iterations. An iteration can be considered short or long. This classification makes it possible to evaluate the impact of iterations on the time and/or the overall cost of the activities involved. This evaluation requires a form of quantification linked to the attributes of the activities (duration, cost) or the number of iterated activities. In this case, we must study the compromise between several short iterations and a limited number of long iterations.

Positive/negative iterations. Some iterations produce value, but not all of them. The iteration that can be eliminated without loss of value is considered a negative iteration. This classification makes it possible to specify how to handle the iterations (elimination, better control, reduction, etc.).

Fast/slow iterations. In this case, it is the speed at which a design task or part of a task is repeated to be in conformity with others or to correct errors.

In figure 4 the iterations are classified according to their sources.

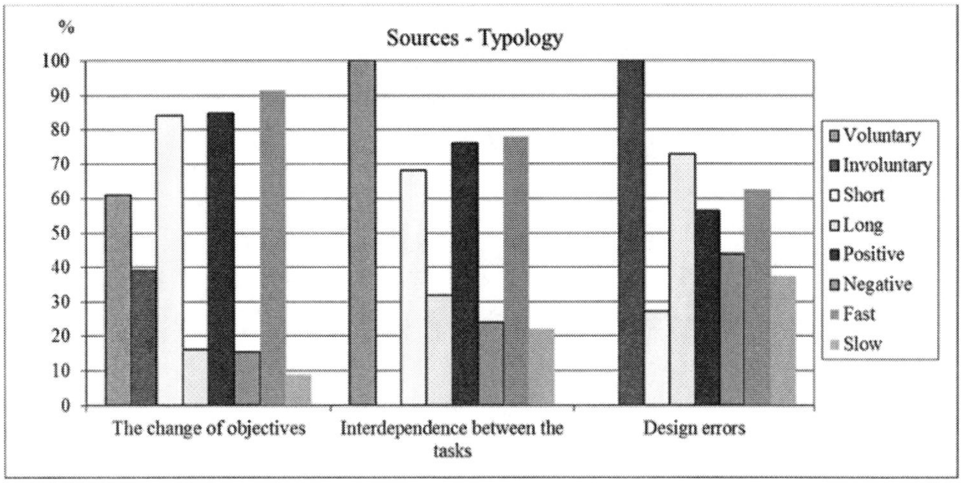

Fig. 4. Classification of iterations according to sources.

Table 1. Sources/typology of iteration.

Sources/Typology	The change of objectives	Interdependence between the tasks	Design errors
Voluntary	61	100	0
Involuntary	39	0	100
Short	84	68,2	27,1
Long	16	31,8	72,9
Positive	84,7	76,2	56,4
Negative	15,3	23,8	43,6
Fast	91,3	77,9	62,6
Slow	8,7	22,1	37,4

5 Conclusions

Experiments provide data needed to understand the design process. In this study, we presented an iterations analysis on a design experiment.

Regarding changes of objectives, the iterations voluntary is more numerous because it is needed to explore the space of solutions in order to meet the new requirements of the design. Thus, a large number of short iterations, positive, is engaged to improve the chosen solution. A short iterations character makes iterative process to be done quickly.

It is very interesting that in the case of interdependence between the tasks, the involuntary iterations do not occur, which seems natural in theoretically. All these iterations are entered voluntarily, aiming to us "closer" to the final stage of the product.

In terms of design errors, it generates long iterations, affecting the duration of the design process. Unfortunately, these sources of iterations cannot be predicted, their appearance being made randomly.

This type of analysis of a design experience aids the project manager to build and distribute design teams, to optimize the design of a product in the new context of globalisation.

References

1. J. Mihm, C. Loch, A. Huchzermeier, Manage. Sci. **46**(6), 733 (2003)
2. D. Braha, Y. Bar-Yam, Manage. Sci. **53**(7), 1127 (2007)
3. R. P. Smith, S. D. Eppinger, Manage. Sci. **43**, 1104 (1997)
4. S. Finger, J.R. Dixon, Research in Eng. Design. **1**, 57 (1989)
5. V. Krishnan, S.D. Eppinger, Journal of Mech. Design **117**, 491 (1995)
6. A. Yassine, K. Chelst, , D. Falkenburg, IEEE Trans. on Eng. Manage. **46**, 2, 144 (1999)
7. S.M. Osborne, (M.I.T. 1992)
8. G. Pahl, W. Beitz (Springer-Verlag, 1997)
9. R.D. Moen, C.L. Norman, Qual. Prog. **43**(11), 22 (2010)
10. B.A. Huberman, D.W. Wilkinson, Comput. Math. Organ. Theory. **11**, 307 (2007)
11. T. Boudouh, D-C. Anghel, O. Garro, *Advances in Design*, 377 (Springer London, 2006)

Methodology for risk analysis of projects

Maria Magdalena Roşu*, *Rodica* Rohan, and *Elena - Cristina* Juganaru

[1]Politehnica University of Bucharest, Department of Manufacturing Engineering, Splaiul Independenţei, 313, Sector 6, Bucharest, Romania

Abstract. The risk in the organization activity as an economic and social system open, adaptive, with varying degrees of permeability to the influences from the current business environment, which is increasingly unpredictable and in which the only constant is the change, refers to the probability of not complying with the objectives set in terms of performance, execution and cost. The insufficient application of the recognized project management methodologies can be one of the main causes of projects failures in the organization with major influences on the activity efficiency and the performance recorded. Therefore, the methodology proposed in the paper, wants to be an effective tool, a formalized risk management tool, considered as a cyclical process, with several distinct phases, indispensable to the current organizational practice which should contribute to optimizing the project performance and its successful completion.

1 Introduction

In the specialty literature the concept of project is presented in a variety of ways, but a jointly accepted definition is that *a project is an ensemble of actions or works designed to achieve a single and measurable objective* [1].

Thus, the main features of a project are the following: time limits (well set start and end date) and well-defined location in space; it has well-defined objectives and a unique purpose that must be fulfilled in response to an identified problem; it has an approved budget within which must fall; it goes through several phases (design, planning, execution, assessment, completion); it consumes resources (human, material, time, money, which are known from the beginning of the activities); it is an ensemble of activities that involve specific risks; it has a responsible – the leader or project manager, and responsibilities of each team member participating in achieving the project aim are well defined; it has a certain degree of risk that must be well managed for the project to be realized in the timeframe set; it creates a new value - product / service / idea / structure - unique and with a high degree of novelty; it requires interdisciplinary collaboration within a particular organizational structure.

The goal of project management is to prevent or predict as many of the problems that arise and to plan, organize and control the activities so that the project can be completed as effectively as possible and at the time set. This process starts before moving to the allocation of resources and should continue until the whole activity ends.

* Corresponding author: magdalenarosu@yahoo.com

According to ISO Guide 73:2009, Risk management — Vocabulary, risk is defined as "*effect of uncertainty on objectives*" where "*an effect is a deviation from the expected — positive and/or negative*". Usually, risk magnitude is expressed as the combination of likelihood and consequences of an event [2].

Risk management is a "*method of managing that concentrates on identifying and controlling the areas or events that have a potential of causing unwanted change*" (Caver 1985), cited by [3]. But, in the same time, risk management is an integral part of project management and should be considered a component of any project management methodology [3].

2 Process of risk management

During the project, the risk management will be treated as an iterative process based on the continuous improvement principle being permanent monitored.

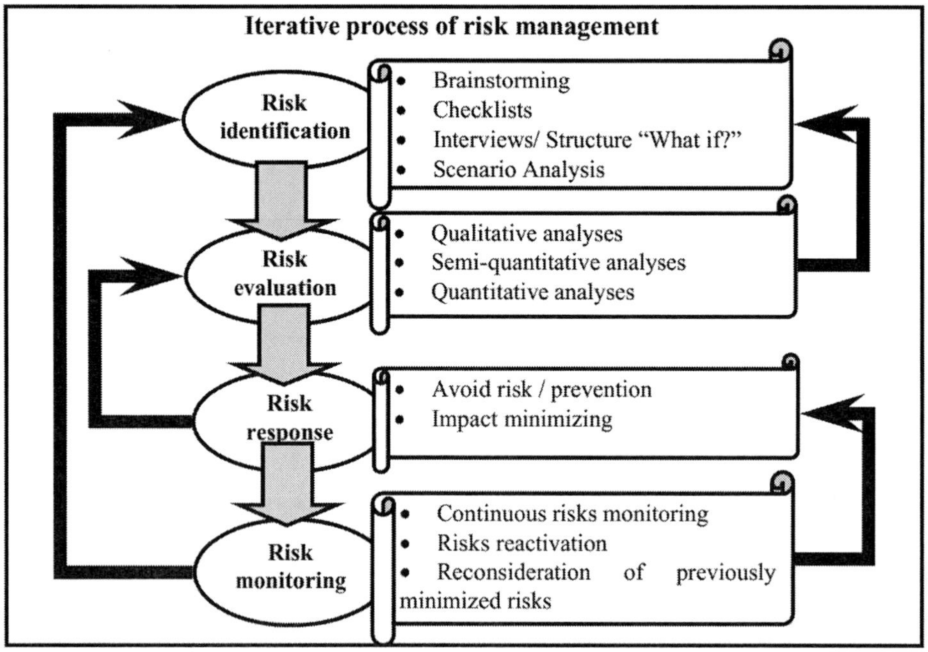

Fig. 1. Process of risk management - adaptation of [6].

All risks which can arise during the project implementation will be taken into account, including those encountered in the preparatory phase of the project proposal. Because some risks are unknown in the proposal stage or emerging, this process will be iterative, see Figure 1.

2.1 Risk identification

The first step in this iterative process is to identify as many risks as possible that can be recognized and associated with the project objectives. In order to manage and complete the project successfully risk identification is a critical step and the entire project team should be involved from the beginning of the project.

For risk identification the authors recommend techniques and methods such as: Brainstorming, Checklists, Structured or Semi-Structured Interviews and Scenario analysis.

Table 1 presents a number of risks that may emerge during a project and we refer mainly to research projects thanks to the authors experience in this field.

Table 1. Identified risks.

Risk number	Risk description
1.	Lack of experienced personnel
2.	Project team modification
3.	Poor experience of the project manager in the relation with the team members and with the persons responsible from the target company
4.	Risks related to the methodology and the development technologies
5.	Lack of interest from potential users for project outcome
6.	Bankruptcy risk of target company

2.2 Risk evaluation

The risks identified in the previous step should be assessed in a process of analysis based on the previous experience of the project manager and implementation team. The results of this process of assessment must be considered during the whole project implementation.

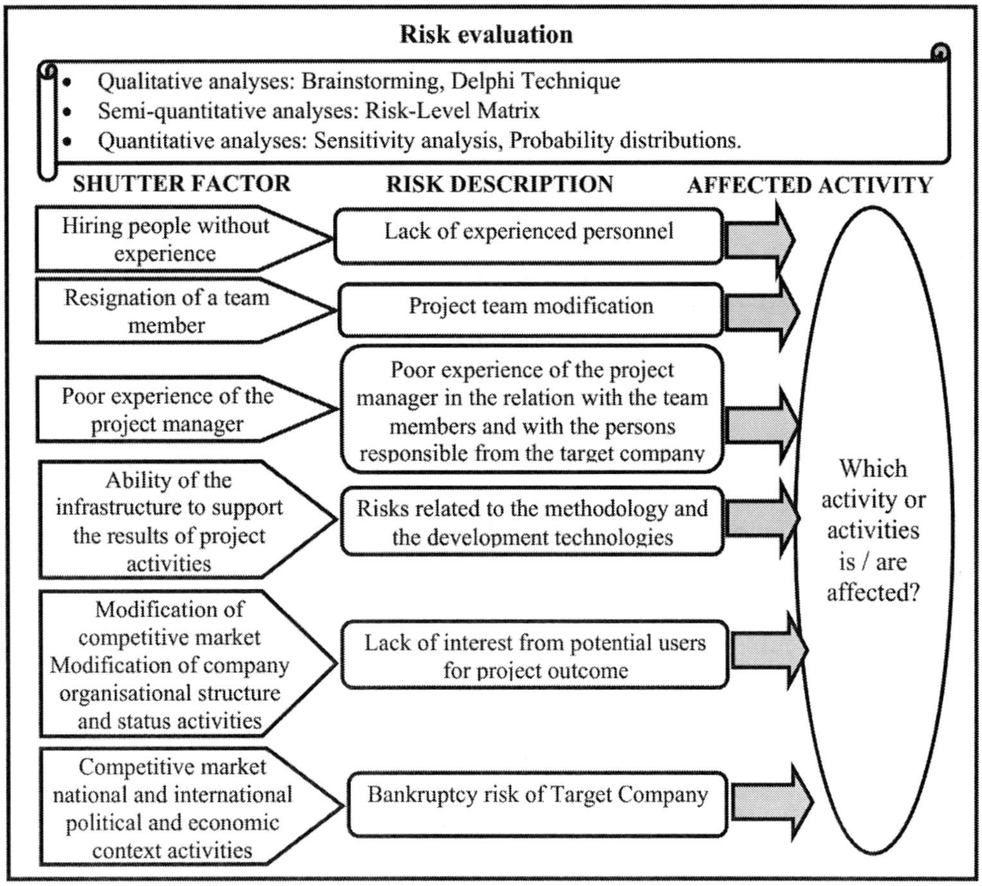

Fig. 2. Process of risk evaluation.

Risk evaluation can be done by qualitative, semi-quantitative and quantitative methods [5]. Thus, it is estimated the probability of the risk emergence, the impact it can have negative or positive on the project's activities and trigger factor of the risk emergence. Risk evaluation should consider the affected activity, the control you have over the risk, potential losses and any benefits or opportunities presented by the risk. In prioritization process the project manager should be the ultimate decision maker by establishing which risks are the most worrisome [3].

For the risks identified in the previous stage and presented in Table 1, the trigger factors and the affected activities will be identified. This step is summarized in Figure 2.

2.3 Risk response

Risk response development is a critical element in the risk management process and requires finding the best solutions to prevent or response to the identified threats. In this stage are established if any and what action will be taken to address evaluated risks.

Since some risk events have potentially positive outcomes, it's recommended to consider responses for both threat and opportunity [2].

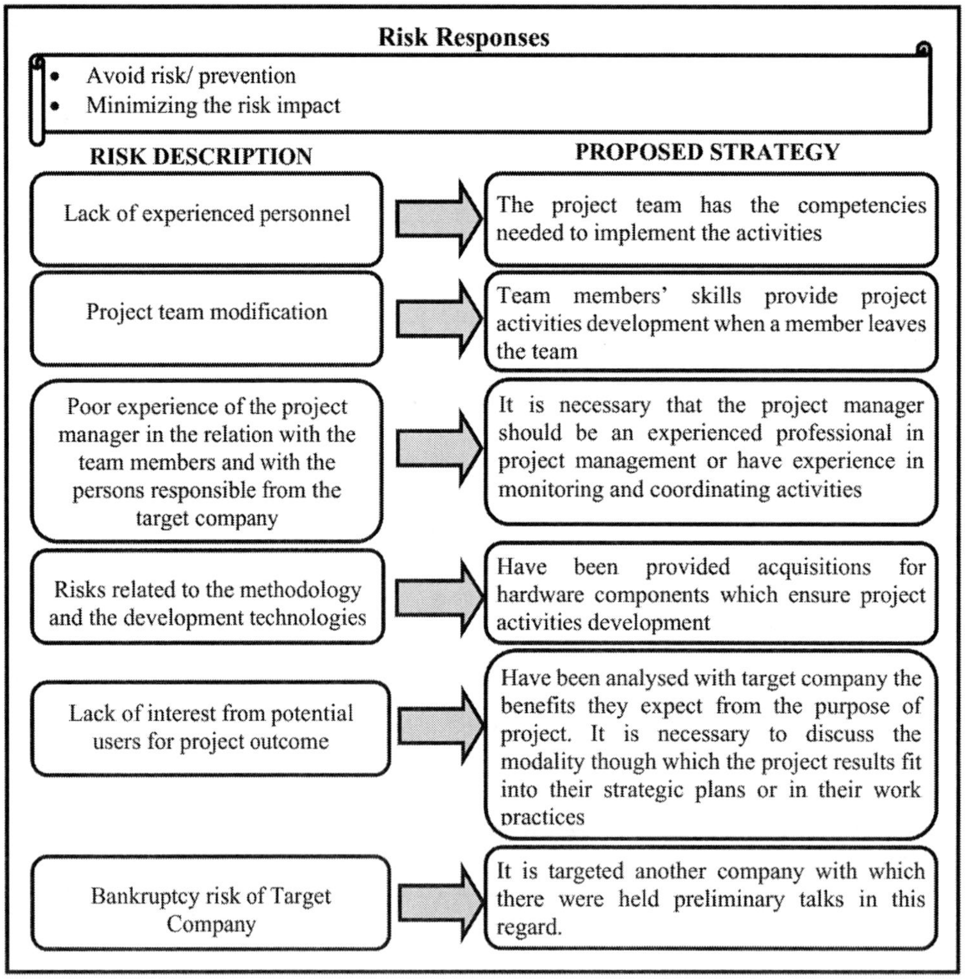

Fig. 3. Process of risk response.

The literature [3], [4], [10] classifies response strategies for threats as follows: *avoidance* – the threats are eliminated by removing the cause or chancing the approach; *transference* - the impact and management of the risk are transferred to a third party (insurers, subcontractors, and partners); *mitigation* – it is the process of taking specific actions to reduce the probability and/or reduce the impact of a risk so that if it does occur, the consequences are smaller and easier to fix; *acceptance* - is the decision to deal with the consequences if a risk event occurs.

The implementation of measures aimed at attenuating the risk consists in applying strategies to eliminate the trigger factors that can generate a risk. Figure 3 shows the proposed strategies for eliminating, attenuating or accepting risks for the case analyzed.

2.4 Risk monitoring

For an adequate monitoring of the risks that may emerge during the development and implementation of a project it is necessary to develop a risk management plan as presented in Table 2. It is a synthetic picture of *most* risks that must be taken into account during the development / implementation of a project.

The *Risk Management Plan* is a response to the possible events arising from the risk and to the way in which action should be taken to eliminate the potential risks that may emerge during the projects implementation.

Table 2. Risk management plan.

Risk description	Shutter factor	Affected activity	Responsibility	Proposed strategy
Lack of experienced personnel	Hiring people without experience	Will be indicated project activities	Who is responsible for that?	The project team has the competencies needed to implement the activities
Project team modification	Resignation of a team member	Will be indicated project activities	In general project manager is responsible for that	Team members skills provide project activities development when a member leaves the team
Poor experience of the project manager in the relation with the team members and with the persons responsible from the target company	Poor experience of the project manager	Will be indicated project activities (coordinated of all project activities)	Project Manager	It is necessary that the project manager should be an experienced professional in project management or should have experience in monitoring and coordinating activities
Risks related to the methodology and the development technologies	Ability of the infrastructure to support the results of project activitie	Will be indicated project activities	Who is responsible for that?	Have been provided acquisitions for hardware components which ensure project activities development
Lack of interest from potential users for project outcome	Modification of competitive market Modification of company organizational structure and status	Will be indicated project activities	Who is responsible for that?	Have been analyzed with target company the benefits they expect from the purpose of project. It is necessary to discuss the modality though which the project results fit into their strategic plans or in their work practices
Bankruptcy risk of target company	Competitive market national and international political and economic context	Will be indicated project activities	Who is responsible for that?	It is targeted another company with which there were held preliminary talks in this regard.

For a more accurate analysis of the effects generated by the risks we should consider possible scenarios for the project evolution depending on the emergence of the identified risks, on the reactivation of the risks that have been attenuated and even on reconsidering the risks which have not been given particular importance during the evaluation.

Therefore, the process of identifying and monitoring risks in the development and implementation of projects is an iterative process that may suffer major changes from one stage to another.

3 Conclusions

In this paper we have proposed our own methodology for identifying and attenuating risks that may emerge during the implementation of projects. This methodology was based on the authors experience on coordinating projects and, in particular, research projects and can be apply for any type of projects in the same mode.

So, the risk management of projects is essential to all organizations who implement all types of projects. Many projects fail because in the initiation phase there are not identified all the risks that can emerge during the development and implementation of the project activities. Some risks may have a negative effect on the project evolution and some of them can have positive effects.

Thus it is very important: to integrate the risk management in all the projects; to develop a methodology that identifies *"almost all"* the risks that may emerge during the project implementation, the risk elements identified must be quantified in order to assess and measure the risk and finally appropriate measures will be provided to attenuate or eliminate the risk.

This methodology can be adapted to any kind of project by identifying all the risk elements that can influence negatively the projects implementation.

References

1. M. M. Rosu, C. Neagu, *Project Management* (in Romanian) (Ed. Printeh, Bucharest, 2014)
2. *** ISO Guide 73:2009, Risk management — Vocabulary risk
3. C. L. Pritchard, *Risk management – Concepts and Guidance* (CRC Press, 2014)
4. P. S. Royer, *Project Risk Management: A Proactive Approach* (Management Concepts, Viena, 2000)
5. *** ISO/IEC 31010:2009, Risk management — Risk assessment techniques
6. D. Benţa, *Software development and risk management in complex projects* (in Romanian), (Babeş-Bolyai University, Cluj-Napoca, PhD Thesis, 2011)
7. ***ISO 9001, *Quality management systems – Requirements* (2015)
8. S. VR. Vilarinho, R*isk Management Model in ITIL*, (https://fenix.tecnico.ulisboa.pt/downloadFile/395144242579/Risk%20management%20on%20ITIL%20(Article).pdf)
9. N. D. Ignat, *Dependability analysis in risk management and critical infrastructure vulnerability* (POLITEHNICA University of Bucharest, Bucharest, PhD Thesis, 2014)
10. I.D. Filipoiu, C. Rânea, *Project Risk Management,* (in Romanian), http://www.omtr.pub.ro/didactic/indrumare/mdpp/mpdp_ii_7+8+9.pdf

Reducing the noise pollution level along main roads in Sibiu, Romania by optimizing the traffic flow

Cristian Deac*, *Valentin* Petrescu, and *Alina* Gligor

[1] Lucian Blaga University of Sibiu, Romania, Department of Industrial Engineering and Management, Emil Cioran str., 4, Sibiu, Romania

Abstract. Noise pollution is one of the most important negative effects of road traffic, especially in cities. The current paper presents an attempt to reduce the noise level created by the road traffic along some main roads in the city of Sibiu, Romania by suggesting ways to optimize the traffic flow. The number of cars and the resulting noise levels were recorded in a key area of Sibiu. The noise levels were compared to standards and to measurements taken in areas with less traffic. Then, they were fed, together with road geometry data, into a specialized simulation software that allowed the computer-aided reproduction of the traffic flow, but also the modifying of conditions such as the road geometry to achieve, coupled with a multicriterial analysis, an optimization of the pedestrian and road traffic flow in view of a reduced noise pollution.

1 Introduction

The rapid increase of the world population in the last few decades has led to an increased crowdedness especially in the larger cities, which generally translates into larger delays for the travel to, within or from those cities, but also into more pollution, both in terms of noxious gaseous emissions and in terms of noise [1].

The city of Sibiu is located in central Romania and has a stable population of around 150000 inhabitants. It must be added however, that the city is an important industrial and commercial centre and also has a rich culture and history, so there is also a significant temporary population – tourists, businessmen, workers etc. All this contributes to a significant pressure on the road structure of the city, which, despite sustained efforts over the years, continues to lead to difficulties in driving through the city, but also leads to an increased pollution.

In the current paper, the authors analyzed the traffic conditions in a key area from the central part of Sibiu, Romania and, with the help of specialized software instruments, offer some solutions for the decrease of the road traffic levels and also, at the same time, for the decrease of the noise pollution level.

* Corresponding author: cristian.deac@ulbsibiu.ro

DOI: 10.1051/matecconf/201711209013

2 The analyzed area, means and method

For the current paper, the authors have focused on the area around the Unirii square in the centre of Sibiu, which represents an important crossroads area, where the north-south road traffic intersects with the east-west road traffic through the city (Figure. 1).

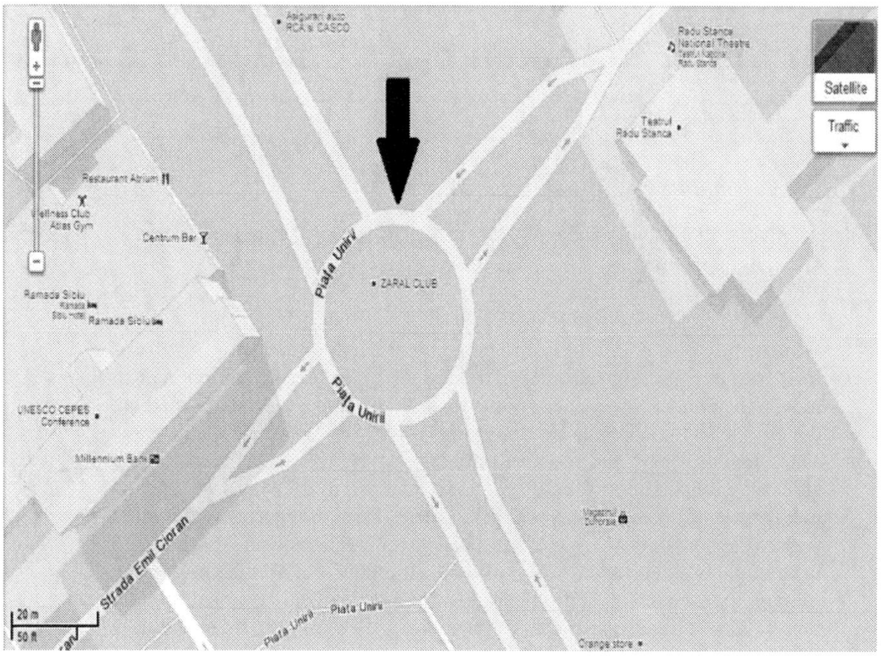

Fig. 1. The Unirii square and crossroads in central Sibiu, targeted by the present study.

In this area, both the road traffic and the pedestrian traffic are very busy, for following reasons:

- It is located close to the city's historical centre;
- It is an area of economical and administrative interest (in or near to it there are residential areas, but also important institutions such as banks, hotels, shopping centres, the local Police headquarters, a hospital etc.)
- The area is crossed or bordered by a series of main roads that connect the various neighborhoods in the city but also funnel the traffic towards the exits from the city and towards the surrounding cities;

Even if the geometry of the crossroads seems to be well organized, during peak hours their traffic capacity is visibly surpassed and there is an excessive sound level pressure. Especially in the morning there result long waiting queues on some directions.

The intended analysis seeks to determine the precise traffic and noise pollution conditions and to find the optimal solutions for reducing or possibly eliminating the problems.

The main activities that were carried out for this study included:

- analysis of the studied area
- collection of data regarding the traffic conditions
- simultaneously, collection of data regarding the noise pollution conditions
- comparison of the noise pollution data with the relevant national standards
- modelling of the crossroads area in a specialized software and simulation of the traffic conditions
- determining optimization solutions for the traffic and for the pollution.

The data regarding traffic conditions can be collected using one of the following methods [2]:

- manual collection using a team of human observers
- analysis of video recordings
- collection using specialized equipment (sensors, GPS etc.)

When assessing the collected data, there need to be taken into account also factors such as the meteorological conditions, the traffic intensity and the traffic's complexity. Also, it is important to select appropriate time periods for the counting.[2, 3]

For the current study, the data on the number of vehicles passing through the crossroads from various directions and heading to various directions was collected by means of video recordings, on several days, using time periods of 15 minutes each. After that, the recordings were reviewed and the vehicles were counted manually.

Beneath the actual car counting on each individual lane, several other elements were recorded, in order to facilitate the subsequent computer-aided modelling of the traffic in the crossroads area:

- the crossroads' geometry and equipment
- the traffic light cycles and the corresponding durations of each colour
- the pedestrian traffic
- the length of waiting queues
- frequency of the public transportation means etc.

Given that the traffic flow through the crossroads contained different types of vehicles (small cars, motorcycles, trucks, buses etc.), the authors have applied the coefficients indicated in the standard SR 7348-2001 [4] for obtaining numbers of so-called equivalent vehicles using the formula:

$$N_{equiv} = N_1c_1 + N_2c_2 + \ldots + N_ic_i \tag{1}$$

where:

N_{equiv} – number of equivalent vehicles for the considered traffic flow and time period;

$N_1, N_2, \ldots, N_i$ – number of vehicles of a certain type in the flow;

$c_1, c_2, \ldots, c_i$ – correction coefficients corresponding to each type of vehicle. For example, for cars, $c = 1$, for motorcycles $c = 0.5$ and for buses $c = 3.5$.

Since there have been counted only very few motorcycles and also very few trucks (most of the heavy traffic uses the ring road around Sibiu without entering it), these vehicle categories have been neglected in the actual vehicle count.

Also, in parallel there has been determined the noise level in the crossroads for the same time ranges, thus allowing a correlation of the results.

The noise level can be determined for individual vehicles (useful for assessing the noise produced by each category of vehicles) or for the entire road flow.

The main noise sources in the road traffic are [5, 6]:

- noises emitted by the car engines;
- noises emitted by the contact between wheels and roads;
- noises made by the air currents when interacting with the vehicles;
- other sources (brakes, door slamming, loading/unloading etc.);

The noise levels were measured using a sound level meter model 2100 made by Quest Technologies, from the endowment of the Faculty of Engineering of the "Lucian Blaga" University of Sibiu, Romania. In order to obtain a clear signal and to eliminate disturbance factors, the sound level meter was equipped with a band-pass filter.

The traffic data were then input into the specialized software package Synchro Studio 7. For this, the authors first modelled the geometry of the crossroads and then the traffic-related data were added. The main element determined using this traffic simulation was the so-called crossroads usage capacity. This was then used as base for a comparison with a proposed optimization solution.

The optimization of the traffic in the analyzed crossroads was based on modifications to the crossroads' geometry, even if the existing geometry does not allow to carry out major changes.

The adopted optimization solution consisted in the adding of a supplemental lane on one direction (Unirii square, descending) where only right turns are allowed and also one supplemental lane each on two other roads (Unirii square, ascending and Corneliu Coposu Blvd.) where only left turns were allowed.

In the end, considerations were made regarding the noise level in the optimized solution.

3. Results and discussion

The data obtained after counting the vehicles passing through the crossroads coming from the Unirii square (descending branch), according to their type and direction in which they were heading, is presented in Table 1.

Table 1. Number of vehicles counted on the Unirii square (descending branch) according to their type and the direction in which they were heading.

Type of vehicle	Cars			Bus0es			Trucks			Vans			Motor-cycles		
Movement direction / Time range	←	↑	→	←	↑	→	←	↑	→	←	↑	→	←	↑	→
7:30 - 7:45	80	231	35	4	3	-	-	3	-	6	28	-	2	1	-
7:45 - 8:00	119	447	44	3	3	-	-	4	-	4	21	1	2	-	-
8:00 - 8:15	48	219	28	3	3	-	1	3	-	3	16	-	-	-	-
8:15 - 8:30	34	156	21	3	3	-	2	1	-	-	18	2	-	1	-

Since the data corresponding to all 4 branches of the crossroads would lead to a very large table, in Table 2 these data have been presented after applying the equivalation formula 1 for all vehicles except buses, which are indicated in a special row.

Table 2. Number of vehicles determined for all branches of the studied crossroads.

Road / Time range	1 Emil Cioran str.			2 Unirii square (descending)			3 Corneliu Coposu Blvd.			4 Piața Unirii (ascending)		
Direction of movement	←	↑	→	←	↑	→	←	↑	→	←	↑	→
7:30 - 7:45	64	43	71	88	278	35	174	90	135	37	283	155
7:45 - 8:00	78	65	95	125	486	44	263	124	183	75	386	242
8:00 - 8:15	41	38	65	55	249	28	139	81	74	58	229	129
8:15 - 8:30	21	28	27	41	182	21	113	63	85	31	163	86
Number of buses (#/h)	-	-	-	13	12	-	12	-	13	-	12	12
Overall number of equivalent cars (7:30-8:30)	204	174	258	355	1237	128	731	358	523	201	1103	654

The number of equivalent vehicles indicated in the last row of Table 1 was determined using equation 1 for each particular flow on a street and in a certain direction. For example, for the traffic coming from Unirii square (descending) and heading straight ahead, the number is:

$N_{2.2} = (278+486+249+182)*1 + (12*3.5) = 1237$

The noise levels recorded in the crossroads in parallel with the video recording of the traffic are indicated in Table 3.

Table 3. Noise level determined in the crossroads for various time periods.

Time period	Determined noise level (dB)
7.30-7.45	79.56
7.46-8.00	93.40
8.01-08.15	86.32
8.16-8.30	83.93
8.31-8.45	82.50
12.00-12.15	76.80
12.16-12.30	77.30
12.31-12.45	78.01
17.00-17.15	85.80
17.16-17.30	91.25
17.31-17.45	92.71
17.46-18.00	86.20

As can be seen, there is a correspondence between the time of the maximal number of cars passing through the crossroads and the maximal noise level recording, which indicates that the solution for reducing the noise level in this crossroads resides with optimizing the traffic in the area as a whole.

The simulation of the traffic in the crossroads in the initial state, carried out using the software Synchro Studio 7 lead to the calculation of the crossroads usage capacity as visualized in Figure 2.

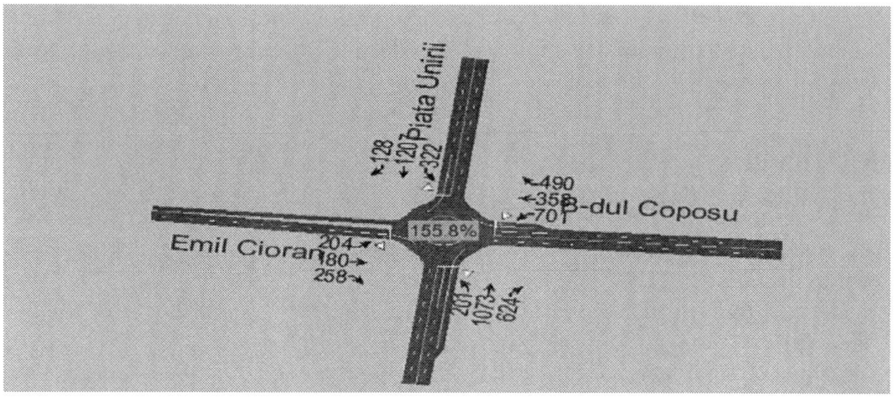

Fig. 2. Calculation of the crossroads usage capacity for the initial state of the Unirii crossroads using Synchro Studio 7.

As can be seen in Figure 2, the crossroads usage capacity exceeds 100% by far, indicating that traffic is not unfolding under normal conditions.

After applying the optimization proposal detailed in the previous paragraph, the crossroads usage capacity has decreased somewhat, as indicated in Figure 3.

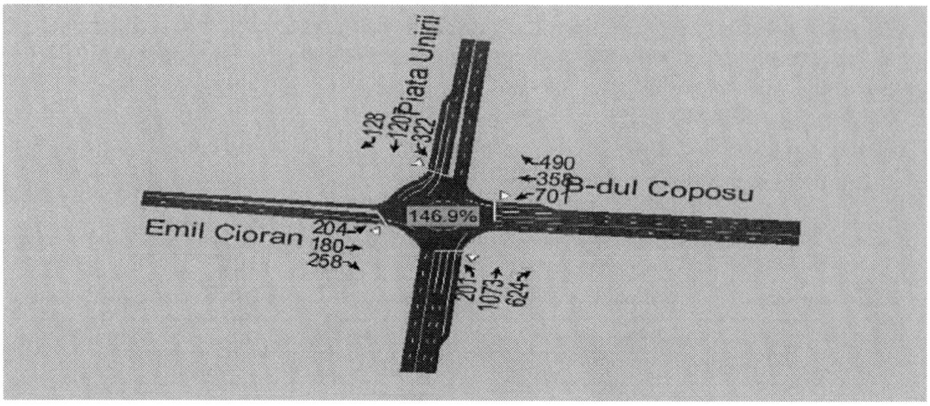

Fig. 3. Calculation of the crossroads usage capacity for the optimized Unirii crossroads, using Synchro Studio 7.

It can be thus concluded that an optimization solution of the crossroads, even if limited by the existing infrastructure, can still lead to a decrease of the noise level by about 9%, i.e. by some 8-10 dB.

4. Conclusions

The current paper has presented a study on the traffic conditions in a busy crossroads area in the centre of Sibiu, Romania, on the corresponding noise level and also suggested a possibility for optimizing the traffic and reducing the noise level.

The proposed optimization solution succeeded in reducing the crossroads usage capacity and the associated noise level by some 9%. However, it is obvious that the resulting values are still significantly higher than those allowed by the current standards. Unfortunately, the current infrastructure conditions in the area do not allow bolder changes and thus better traffic and noise conditions.

In future, the authors intend to work together with the local authorities and with other specialists in order to find solutions for a larger reduction of the crossroads usage capacity in the area.

References

1. M. Wiering, J. van Veenen, J. Vreeken, A. Koopman, *Intelligent traffic light control* (Technical Report uu-cs-2004-029, Institute of Information and Computing Sciences, Utrecht University, July 2004)
2. J. Barceló, M. Delgado, G. Funes, D. García, J. Perarnau, A. Torday, SIAM News, 40, **9** (2007)
3. D. Popovici, V. Boboc, I. Găluşcă, Transport Systems and Urban Traffic (in Romanian), (Editura Societăţii Academice „Teiu Botez", Iaşi, 2004)
4. *** SR 7348-2001, Road works. Equivalation of vehicles for determining the traffic capacity. ASRO, Bucharest, 2001
5. W. Burghout, Technical report, PREDIKT Project, (2005)
6. A. Sachelarie, E. Golgoţiu, Traffic and Road Safety (in Romanian) (Casa de Editură Venus, Iaşi, 2002).

Study on the design of industrial control valves with manual drive

Andrei Pascariu*

"Gheorghe Asachi" Technical University, Iaşi, România

Abstract. The paper presents the main components and the operating mode of the manually operated control valves used in natural gas transport and distribution installations. Main parameters are specified: inlet-outlet pressure (P), temperature(T), flow rate (Q), which are taken into considerations from the design phase and beyond. Noise is another parameter, in close liaison with the pressure and flow. As their value increases with both the inlet noise levels downstream of the valve is greater. By optimizing internal components, especially cages, noise can be limited below the maximum allowed by regulations. Also, there are presented control valve constructive types so that the optimal variant can be chose.

1 Introduction

Control valves are equipments which are used in natural gas transport and distribution installation for flow rate / outlet pressure control. The movement of piston modifies the flow area through the valve and the control of flow rate/ outlet pressure is made.

Are valves with valve plate, flanges and straight body. These are built on a cast carbon steel body; The size of the control valves can be from DN 25 up to DN 400 and nominal pressure from PN 16 up to PN 100.

Actions shall be taken so the pressure won't rise above the maximum allowable pressure (PS). For short periods it is allowed a pressure equal to 1,1 x maximum allowable pressure (PS).

2 Systemic analysis of manually operated control valves constructions and functioning

The main components of a control valve are presented in Figure1.

The body 1 is the basic component parts to be assembled which constitute the control valve. Inside the gas flow direction change and gas expansion take place. At the bottom of ***the inspection cover 2*** is mounted, which allows access to the ***seat 3***. This is a designated location by passing gas and to realize that closing and regulation of gas. The inside diameter is equal to the nominal diameter of the control valve (ex. $Ø_{scaun}50$ DN = 50 mm). Effective adjustment is performed by the ***shutter member or piston 4***, which performs a linear movement equal to 1/4*DN. The ***silencer 5*** is designed to reduce the noise emitted by the

* Corresponding author: andrey_octavian@yahoo.com

DOI: 10.1051/matecconf/201711209014

fluid flow through the valve and to strengthen the assembly formed with ***piston guide 6.*** This serves to guide and center ***piston 4***, the interior is fitted with sealing, through the sealing between the piston guide and piston the force generated by the pressure on the piston is cancelled. Linear movement of the piston 4 is caused by the rotation of the ***stem 7. Connection part 8*** is support element for ***stem guide 9*** on the valve body 1. ***Handwheel 10*** is the actuating system of the valve. Through the seal between the stem 7 and the guide stem 9 does not occur gas emissions in the atmosphere. This sealing can be done in several versions.

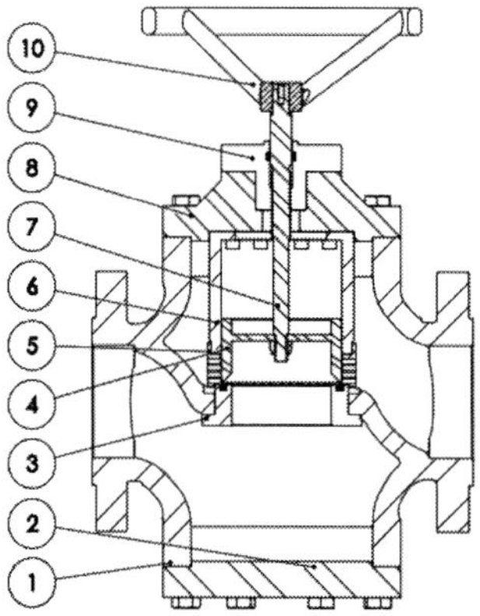

Fig .1. Manually operated control valve.

In Figure 2 are shown, the existing solutions of components that compose a control valve. Among these is proposed a new solution *silencer slotted in one cage.* Going through the scheme, may result in a control valve body with 2 port, with the captive O-ring seat, high pressure piston, slotted silencer and sealing the stem with two rings. For each component can develop a new improved version.

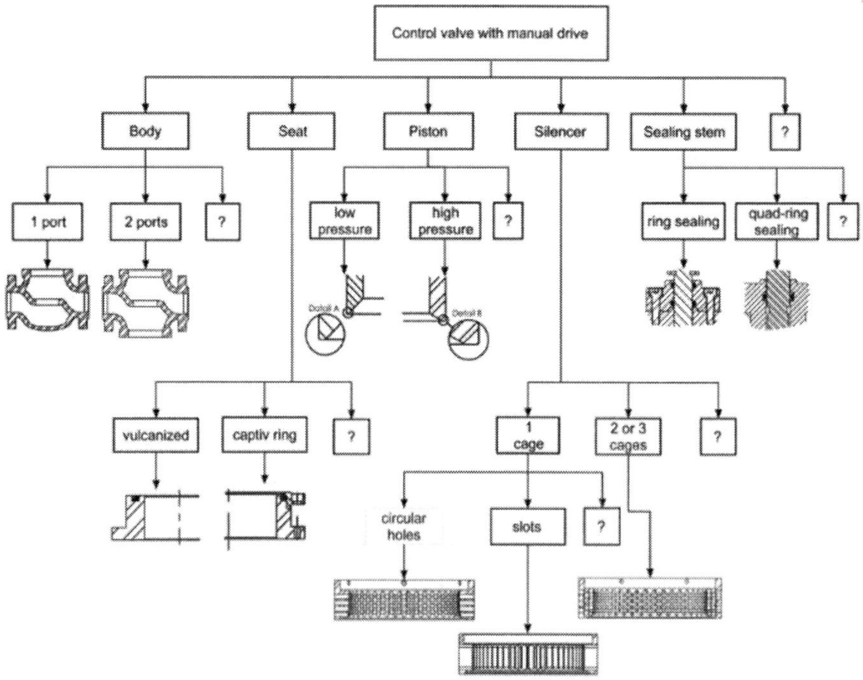

Fig. 2 Chart ideas with main components of control valves with manual drive.

DOI: 10.1051/matecconf/201711209014

The functioning of the valve: the gas enters the valve through the inlet connection, passes through the seat and into the piston, (flow to open-FTO). The gas amount which will pass to the outlet connection varies depending on the distance of the piston from the seat (fig.3). Expansion is made when the gas passes through the space between the piston and seat. If the control valve is equipped with silencer expansion is made when the gas passes through the holes of the silencer.[1]

The piston movement compared to the seat is realized by the manual actuator (with screw). The piston stroke, from 0% to 100% is determined by multiplying the piston diameter with 0,25 (0,25 x Ød piston). In this way, the passing section area through seat is equal to the "cylinder" area generated by the piston movement from 0% to the 100% opening.

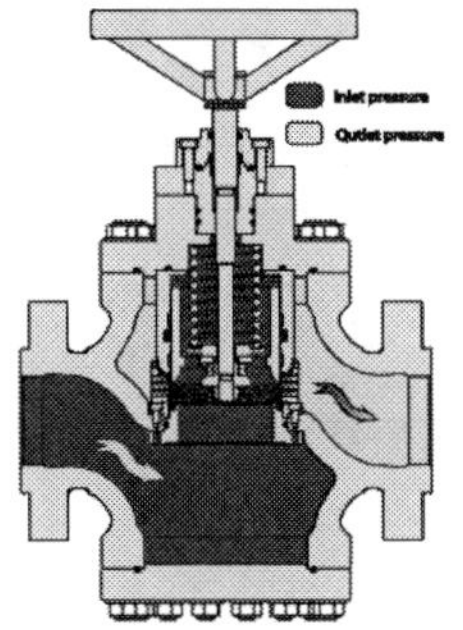

Fig. 3. Manually operated control valve. [1]

The main parameters of the gas at the valve inlet are: flow rate pressure and temperature

The control valve operating mode is determined by the three main parameters: input and output gas pressure(P), mass flow rate (Q) and temperature(T)[2]. The flow rate is the physical parameter which is measured in m^3, which can be normal (N) or standard (S), per hour (h) or per day (d). The flow speed must not exceed, at outlet flange, 20 m/s.

The flow rate control is made also by pressure variation, for this reason we cannot have a low flow at high pressure (at control valve outlet).

3 Analysis of the components

Control valve bodies are manufactured by casting, from normative accepted materials (WCB vs LCC) [3], depending on minimum functioning temperature. Body can be made in two constructive types: with 1 and 2 ports.

The body with two ports (fig. 4.a) has the advantage that allows access to the underside of the seat, easy maintenance, it can be equipped with safety device, the actuator mechanism can be mounted on both ports. The disadvantage is that it requires inspection cover involving semiproduct processing and additional fasteners.

The advantage of body with 1 port (fig. 4.b) is that it has fewer parts, but has the disadvantage that cannot be equipped with safety device.

Fig. 4. Control valve body: a) with 2 ports, b) with 1 port.

 DOI: 10.1051/matecconf/201711209014

Coupling connections are body parts and serve as connection to the sections and equipments from upstream and downstream. They can be manufactured in 2 constructive types:

- Standardized flanges PN or Class (RF, RTJ) (fig. 5.a);
- Weld ends (fig. 5.b);

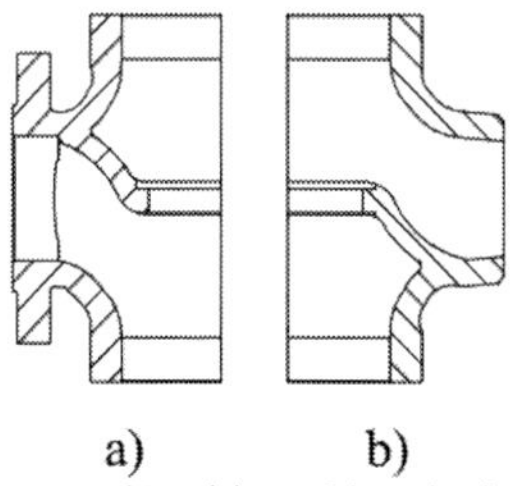

Fig. 5. Control valve body: a) with flanges, b) with weld ends. [4]

The seat, at this valve type can be realized in 2 types, vulcanized (fig. 6.a) or captive ring (fig. 6.b). NBR or VITON ring are used with a 70 ShA or 90 ShA hardness. The rings with the hardness of 90 ShA are used for valves working at pressure above 60 bar.

Fig. 6. Seat variants: a) vulcanized, b) captive ring.

The silencer is an accessory which can be mounted on the valve if the noise level is high and above the limitation imposed by the normative. It can be manufactured from a single cage with circular holes (fig.7.a) or slots (fig.6.c), or by 2 or 3 cages assembled into each other with a deviation between the holes made in it (fig.6.b); in this way, the noise reducing capacity is increased.

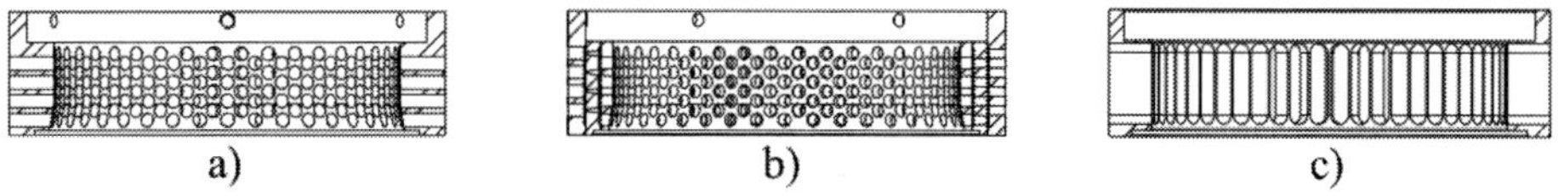

Fig. 7. Versions silencer: a) with 1 port, circular holes; b) with 3 cages, circular holes; c) with 1 cage with slots.

The design of this style flow cage gives the added benefits of flexibility of size, pressure class, materials of construction, rangeability and attenuation [5].

The piston, by its movement, modifies the fluid flow area. Depending of closing pressure value, the form of the piston sealing part is sharp (fig.8.a) or flat (fig.8.b).

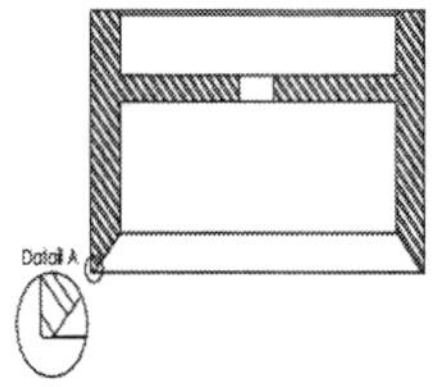

Fig. 8. a. Low pressure piston.

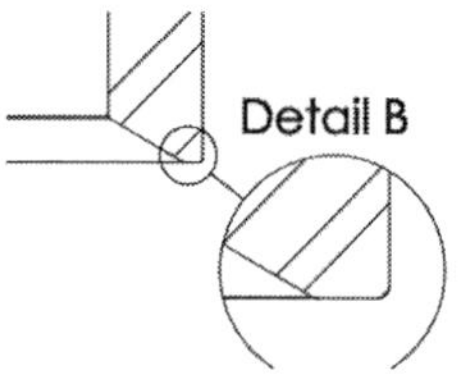

Fig. 8. b. High pressure piston.

The stem guide is placed at the upper side of the valve and helps the stem to perform the linear movement and with it, the piston. This component also seals the stem. The stem sealing can be made by several means available on market. One way is sealing with two rings, the first one realize the sealing and the second one serves as a reserve if the first one brakes (fig.8.a).

Other mode of sealing is with quad-ring (fig.8.b). The manufacturers recommend using quad-rings for dynamic sealing. The stem performs two movements, rotation and translation, and the sealing is considered dynamic. For the surfaces in contact with sealing elements are prescribed special machining conditions. In normative SR EN 15848 *Measurement, test and qualification procedures for fugitive emissions* the stem emissions, methods and means of determination for these oversights are regulated [6].

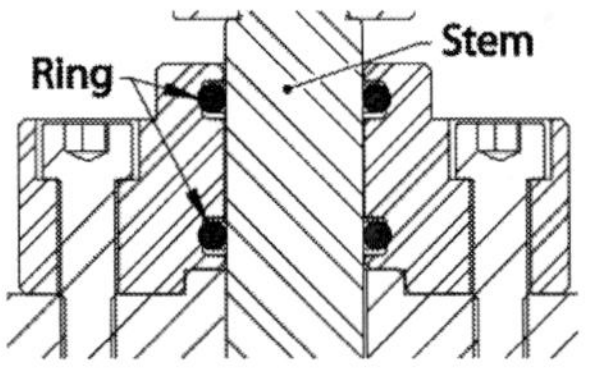

Fig. 8. a. Ring sealing. [1]

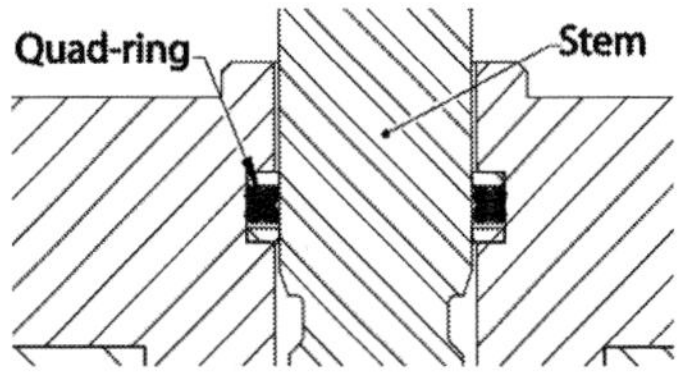

Fig. 8. b. Quad-ring sealing.

Schematically in the Figure 9 it is presented the detector probe (sniffing) method, allows the measurement of the local emission of the stem sealing system . The measured concentration is expressed in parts per million volume (1 ppmv = 1 mil m3 = 1 cm3 1m3).

This procedure is intended to locate and classify leaks only, and is not used as a direct measure of mass emission rates from individual sources [6].

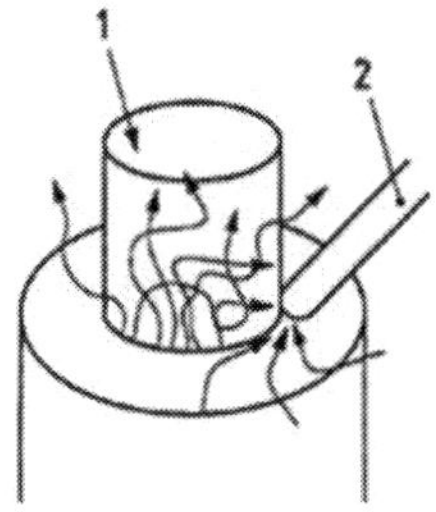

Fig. 9 Local measurement sniffing. [6]

In the next figure (Figure 10) constructive variants of the control valve components are presented. From this image, depending on the parameters, you can define the constructive solution of such equipment.

4 Conclusions

This paper shows the main components of the control valve manually operated. Each component can be designed in many variants. To achieve a control valve assembly pairs are chosen variations or part that would best fit in terms of functionality and cost, application for which is being executed.

The new elements presented in this paper are stem sealing with quad-ring and slotted silencer.

For optimal solution of low pressure control valve is composed of body with two ports, with vulcanized seat mounted, sharp active piston edge, slotted silencer and the stem sealing quad-ring.

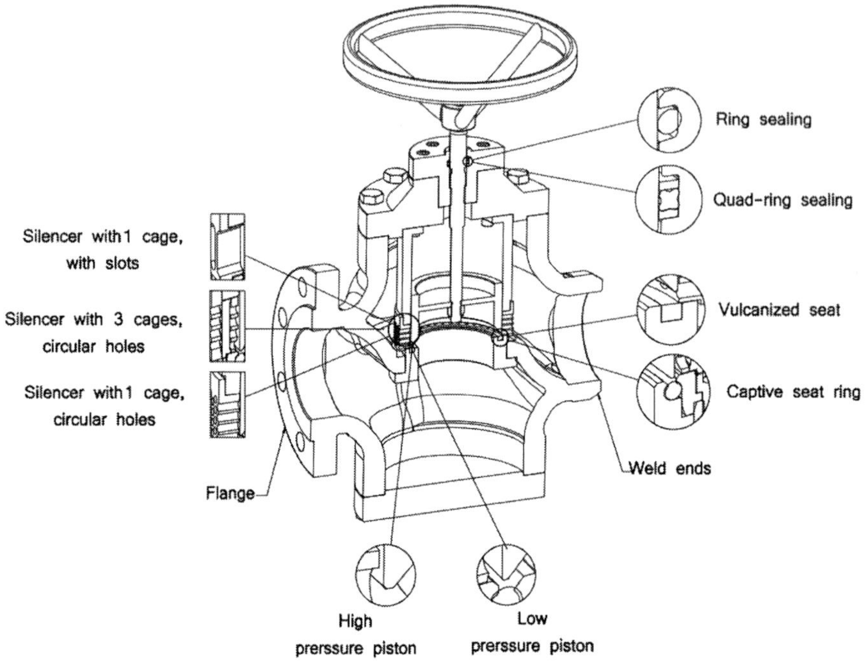

Fig. 10. Constructive solutions for the components of manually operated control valve.

With control valve manually operated with working a high pressure, on the body 2 port is recommended to mount the seat ring captive, flat active edge piston (not to damage the O-ring due to force high press), silencer with 3 cages and circular holes (for low noise) and the stem sealing quad-ring.

References

1. TOTALGAZ INDUSTRIE, Techical manual (in Romanian), MT_RR 931_M, nr. 1181
2. M.V. Balyaba, M. A. Ermilov, A. N. Kryuchkov, Procedia Engineering, **176**, 577 (2017)
3. SR EN 334+A1:2009, Gas pressure regulators for inlet pressures up to 100 bar
4. SR EN 60534, Industrial process control valves
5. http://www.documentation.emersonprocess.com/groups/public/documents/articles_articlesreprints/ce_control_vlvs_evolution_desi.pdf
6. SR EN 15848, Measurement, test and qualification procedures for fugitive emissions

Production planning and scheduling with material handling using modelling and simulation

Damian Krenczyk[1,*], *Wojciech M.* Kempa[2], *Krzysztof* Kalinowski[1], *Cezary* Grabowik[1] and *Iwona* Paprocka[1]

[1] Silesian University of Technology, Faculty of Mechanical Engineering, Konarskiego 18A, 44-100 Gliwice, Poland
[2] Silesian University of Technology, Faculty of Applied Mathematics, Institute of Mathematics, Kaszubska 23, 44-100 Gliwice, Poland

Abstract. Increase flexibility of manufacturing requires implementation of effective planning IT tools. Planning and scheduling problems traditionally consider the production resources as the only constraints. Since the raw material and intermediate transferring times are relevant in relation to the production times, it is required to model the material handling equipment and its activities. The problem of scheduling subject to manufacturing and logistical constraints based on data-driven models generation method is presented. A KbRS and FlexSim systems are used as a scheduling tool.

1 Introduction

Increase automation and flexibility of manufacturing systems which allow to take measures to adapt production systems to produce many variants of products in variable production batches, requires implementation of innovative and intelligent technologies and effective planning IT tools and techniques. The production planning and scheduling problems traditionally consider the production resources as the only constraints in the system [1]. Since the raw materials and intermediate products transferring times are relevant in relation to the production times, it is required to model the material handling equipment (especially in the automated material handling systems) and its activities in a scheduling process in the real industrial applications. The handling vehicles operations should be planned and synchronized with production resources operations. Thus, the production flow plans, configuration of manufacturing systems, scheduling and control, particularly at the operational level, should be developed in a way that ensures the optimal implementation of production orders in the context of constraints associated with both the manufacturing system as well as material handling equipment [1, 2, 3, 4]. Their mutual planning allows to determine the type, quantity, place and time of raw material and semi-finished products use, thus determining at the same time the transport, handling and storage operations, together

* Corresponding author: damian.krenczyk@polsl.pl

with the accompanying information flow. Thus, the required changes in the area of organizing production preparation make it necessary to improve the flow of information and integration of decision support systems.

In the problem addressed in this paper, a scheduling and simulation systems integration based on data-driven semi-automatic simulation models generation method is presented. The proposed approach allow to eliminate the disadvantages associated with the traditional - expensive and time-consuming process of creating the simulation models. Methods of data mapping, data transformation and data exchange, between heterogeneous IT systems in the process of the simulation models generation have been used.

The presented method can become an efficient and effective tool to support the planning and verification of the production schedule subject to manufacturing and material handling constraints in the production system. Developed integration software module RapidSim allows for fast simulation model generation and conduct simulation experiments (with the visualization of the production processes flow) for rapid analysis of prepared schedule with regard to available quantity and the capacity of the material handling vehicles, the timetables, and the location and capacity of the material handling systems as well as.

A commercial software KbRS and FlexSim were used as a scheduling and simulation tool for testing the integration module. Developed RapidSim software has been applied to carry out the verification simulation experiments. Due to the use of the intermediate neutral data model, the integration module can be commercialized with any scheduling and simulation system which provide interfaces for data exchange (open system) and internal programming languages. Data and simulation models used in the verification experiments are provided as well.

The paper is organized as follows. Section 2 is devoted to presenting methods for semi-automatic generation of simulation models. Section 3 is devoted to the presentation of the practical implementation of the proposed solution. The production scheduling system KbRS used to support the process of generating production schedules and a brief description of the FlexSim simulation system with a practical example of its application are also presented. Finally, Section 4 presents our conclusions.

2 The integration of scheduling and simulation systems

The integration of scheduling and simulation systems is performed using the method of automating the process of creating simulation models based on the concept of Data-driven simulation model generation, which are now the largest group among the developed concepts [5, 6, 7, 8, 9] The simulation model is created automatically simultaneously with its parameterization based on specific data instances obtained automatically from the planning and scheduling system. Generating (creating) the models could therefore be performed by users without knowledge of how to build the simulation model. In this approach to build a simulation model does not require the use of standard computer simulation tools (usually specific modules of simulation systems), which are used in conventional "manual" approach, because the model is built automatically using data analysis methods and algorithms processing [5, 6, 10, 11].

The semi-automatic simulation model generation method based on parametric hybrid approach is proposed in this paper. It combines data mapping and data transformation methods with the concept of a neutral intermediate data model. This approach allows to generate a model for any method of production flow in the manufacturing system, based only on the data taken from scheduling system, for which between the manufacturing resources (deriving from the technological routing process flow in system) standard objects related to material handling equipment are created automatically.

Due to the fact that in the scheduling system (which is the source of the data used in the present method), no data regarding the specific parameters of internal transport are available (type, speed, acceleration, capacity, location, etc.), the objects imitating the behaviour of the material handling system must be generated with default parameters. A characteristic feature of the proposed approach to generate models in a semiautomatic way, is that the missing data and parameters for created material handling system objects can be introduced by the operator/planner (already within the target simulation computer software) in the last step before performing verification simulation experiments (Figure 1).

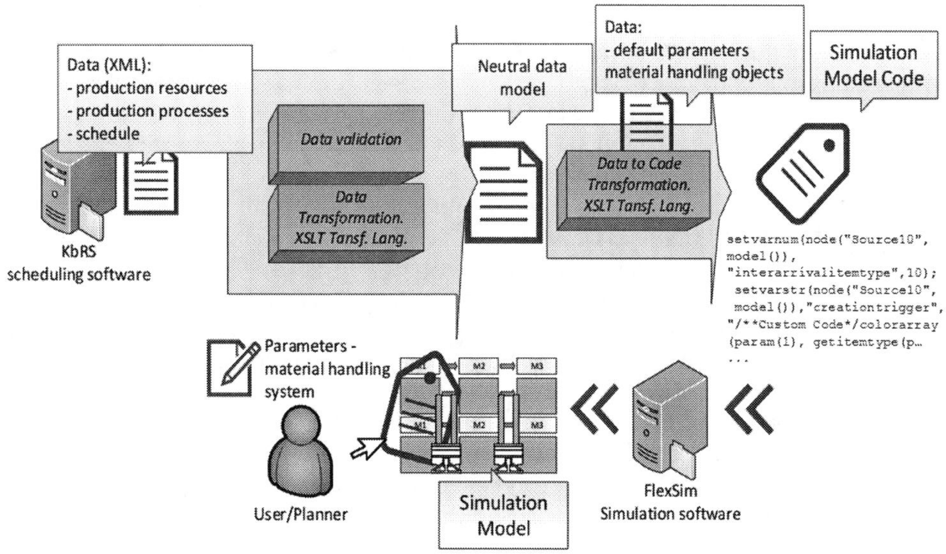

Fig. 1. Semi-automatic simulation model generation method.

The XML schema developed for the method of automating the process of creating of simulation models contains definitions of the XML document structure. The data structure defines the manufacturing system resources, i.e.: machines, interop, input and output warehouses, products, details of the production processes, i.e. technological routes, setup times and cycle times, as well as execution of production operations sequence data (production plans and schedules obtained from scheduling software). These data are obtained directly from scheduling system. The process of data transformation into input file for simulation systems is performed using XSLT processors (Extensible Stylesheet Language Transformations).

At the stage of transformation the data from the intermediate model are converted into script code using the data transformation method in the internal programming language of simulation systems (e.g. FlexScript for FlexSim system). The code contains script functions for creating a component objects of the simulation model, along with connections between them - and thus a complete simulation model. The instructions contained in the code also create information objects, i.e. tables contain data on the processing time of operations executed on resources, tables with resources work schedule, data on processes routes as a function of performing the connection between the respective atoms in the model, together with the corresponding objects of transport resources or other material handling system components. It also contains the parameters for the scenes and scripts enable the implementation of simulation and visualization processes, according to data contained in the information resources.

3 Practical implementation

Practical implementation of the presented method was carried out using a commercial scheduling and re-scheduling KbRS system and FlexSim simulation software. The process of data transformation (system integration) was carried out using RapidSim software, which is computerized implementation of the previously described methods.

3.1 KbRS scheduling system

The Knowledge based Rescheduling System (KbRS) is a production scheduling system dedicated for multiassortment production flow together with the best possible utilization of production resources and taking into account accepted evaluation criteria. The basis for determining production schedules are the production system data set, technological and organizational data (data describing the production processes) along with the data of production orders – routes, setup time, processing time, due dates, priorities, and the volume of production [1,11]. These data could come from many (often heterogeneous) sources of MRPII/ERP/CAPP systems. Due to the frequent gaps in information in existing databases, a part (or all) of the data can be entered into the system manually through a graphical user interface. After entering all the required data and defining input parameters the searching of available schedules is executed by two basic scheduling strategies: forward and backward. Schedules are generated by processes scheduling algorithms according to established common priority rules i.e. EDD (Earliest Due Date first), LDD (Latest Due Date first, LPT (Longest Processing Time first), SPT (Shortest Processing Time first) or Random, and set of the best solutions are saved. Production schedules are presented in form of the Gantt chart, an overview of evaluation indicators and analysis of individual schedules by other reports can be provided. The final decision on the selection and application of a schedule is taken by the planner.

3.2 FlexSim simulation software

FlexSim provides extensive support for building, analysis and verification simulation models developed by the FlexSim Software Products, Inc. FlexSim is fully customizable, graphically-oriented simulation software that integrates modelling, simulation and 3D visualization and animation in 3D object oriented drag-drop-connect environment. In FlexSim, discrete or fluid pre-defined modelling object are the most basic building element of a simulation and allows users to change the behaviour of objects, which can be parameterized at the stage of creating the model. The discrete objects (objects that create, send, store items awaiting processing, move items through the model, group and perform operations on products, etc.) are used to develop discrete-event simulation models where the model behaviour results from events that occur at discrete points in time [12]. Objects are available in the libraries of objects the extent corresponding to the possible uses of the various versions. Further provides mechanisms that import and export data and facilitate the analysis of data.

3.3 Illustrative example

The process of data transformation representing the computer implementation of the previously described integration methods was realized in the RapidSim software [8, 11]. The data transformation documents (XSLT) can be developed independently, so the software can be used for any scheduling and simulation systems. Input files to the RapidSim are XSLT documents constitute the definition of transformation functions for data, also positioning RapidSim work parameters for data retrieved from computer systems which are subject to integration - in this case KbRS and FlexSim (Figure 2).

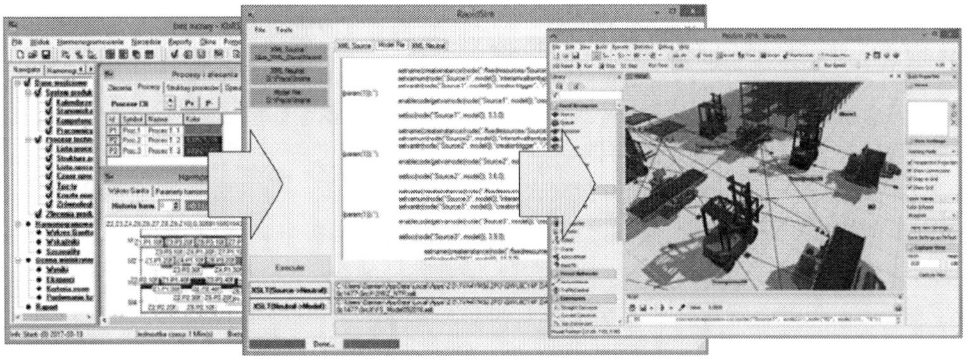

Fig. 2. RapidSim integration software.

Implementation of the process of simulation model creation consists of the following steps:

- loading the an XML document containing input data (KbRS - the results of scheduling),
- validation of the loaded XML document based on the definition of data models in the form of XML schemas,
- transformation of the XML document using a XSLT processor,
- generating a document (.mod) containing code in FlexScript (simulation model for FlexSim software),
- validation of the generated document.

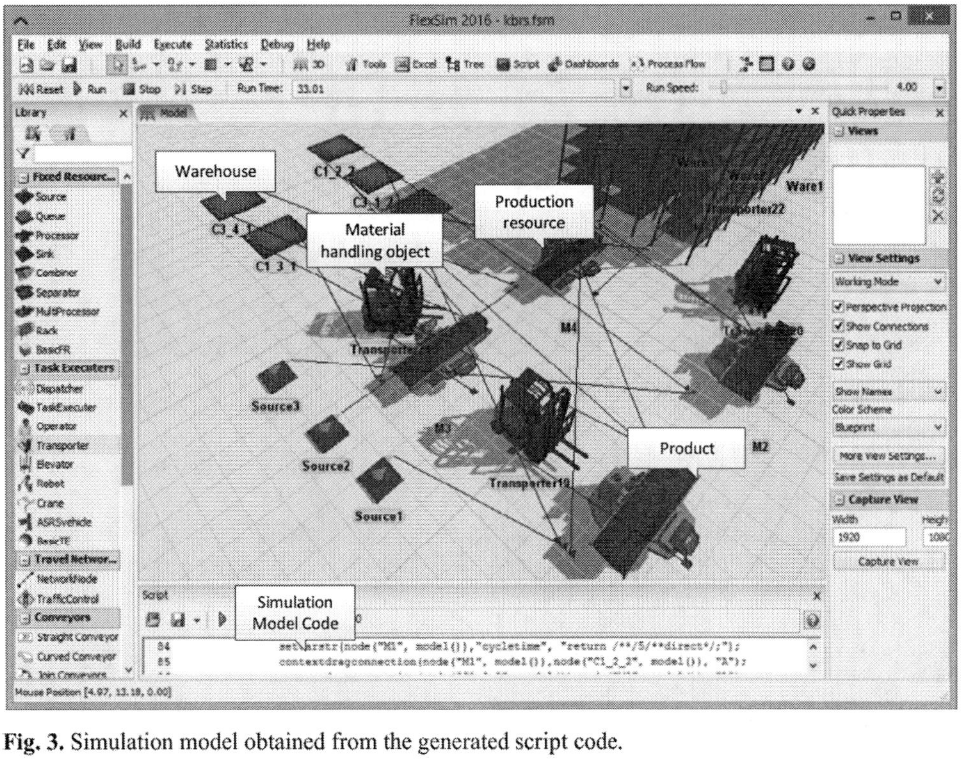

Fig. 3. Simulation model obtained from the generated script code.

Loading the resulting output document (MOD) in FlexSim software allows for automatic creation of a complete simulation model (Figure 2 and Figure 3), which is ready to carry out verification experiments, including the constraints related to material handling objects - analysis of available quantity and capacity of means of transport, timetables and potential hazards collisions.

Generated model, after setting the specific parameters for material handling objects, is ready to carry out simulation experiments. Performed simulations have confirmed the correctness and completeness of the connection in the generated automatically model.

4 Conclusions

The method of integration of scheduling and simulation systems proposed in the paper constitutes also data driven method of automatic generation of models for simulation systems, by eliminating the disadvantages associated with cost- and time-consuming process of building simulation models, can become an efficient and effective tool to support the planning and verification of production schedules subject to the material handling limitations.

The proposed RapidSim software solution enables, among others, to perform rapid analysis with regard to the available quantity and capacity of means of transport, timetable, and the location and storages. Thanks to its openness it may be commercialized with practically any scheduling system or simulation software. The subject of further work in this area will involve the expansion of the intermediate model data possible to obtain information concerning the other subsystems and manufacturing systems constraints.

References

1. K. Kalinowski, M. Zemczak, Advances in Intelligent Systems and Computing, **368**, 475 (2015)
2. A. Burduk, K. Musiał, Lecture Notes in Computer Science, vol. **9842**, 197 (2016)
3. M. Diering, K. Dyczkowski, *2016 IEEE International Conference on Industrial Engineering and Engineering Management*, 228 (2016)
4. G. Ćwikła, D. Krenczyk, A. Kampa, G Gołda, IOP Conf. Ser.: Mater. Sci. Eng. **95**, 012153, (2015)
5. J. Wang, Q. Chang, G. Xiao, N. Wang and S. Li, Computers in Industry, **62(7)** 765, (2011)
6. Y. J. Son and R.A. Wysk, Computers in Industry, **45(3)**, 291 (2001)
7. J. Wy, S. Jeong, B.I. Kim, J. Park, J. Shin, H. Yoon, S. Lee, Computers and Industrial Engineering, **60(1)**, 138 (2011)
8. D. Krenczyk, M. Jagodzinski, Advances in Intelligent Systems and Computing, **368**, 451 (2015)
9. U. Ghania, R. Monfared, R. Harrison, International Journal of Computer Integrated Manufacturing, Vol. **28**, No. 8, 844 (2015)
10. D. Krenczyk, B. Skołud, Applied Mechanics and Materials, **657**, 961 (2014)
11. D Krenczyk, I Paprocka, W M Kempa, C Grabowik, K Kalinowski, IOP Conference Series: Mater. Sci. Eng., **145**, 022018 (2016)
12. M. Beaverstock, A.G. Greenwood, E. Lavery, B. Nordgren, *Applied Simulation: Modeling and Analysis Using FlexSim* (FlexSim Software Products, Inc. 2012)

Hydraulic calculation and design of no-dig recovered pipelines

Roman Hurgin*

Moscow State University of Civil Engineering, Yaroslavskoye shosse, 26, Moscow, 129337, Russia

Abstract. The research of hydraulic properties of pipes to be relayed in repair bays or to be no-dig recovered is a constituent part of renovation approach. The designers do not have any information on hydraulic properties of new protective materials and have to use the hydraulic calculations procedures, strength and hydraulic data submitted by pipe manufacturers that may not be regarded as objective arguments towards selection of pipeline material. Role of experimental hydraulic studies is increasing therewith. One must have information on hydraulic properties of both pipes and repair materials (e.g. polymer covering, new types of polymer pipes, etc.) to analyze the situation. Two polymer pipes that can be used for trenchless renovation are considered and analyzed: corrugated polypropylene pipe DN 98 and bare pipe DN 105 made from unplasticized polyvinyl chloride. The experimental results to fix relative roughness factor are presented and analyzed in this paper.

1 Introduction

Current trends in wide application of new pipes made from different materials (including the synthetic ones) during repair and modification of old networks being destroyed result in using different pipes of materials for drainage system from year to year thereby making it difficult to evaluate hydraulic data and operation thereof, because each pipeline section is to be appropriately served (e.g. pipe cleaning, etc.) [1-5].

Nowadays there are no strict hydraulic dependencies of Chezy coefficient C and pipe friction number λ for pipelines made from new materials. Furthermore, actually each pipe manufacturer gives its own nonobjective evaluation criteria for different pipe compatibility. The problem is more aggravating with increase of new materials usage, with each of them being applied in repairing non-pressure network, thus resulting in some sort of a network with "patches". It may also result in hydraulic unbalance i.e. possible negative tendencies due to flooding in pipe junctions or in certain distance from junction points [6-10].

Many scientists (Prandtl L., Chezy A., Bazen A., Manning R., Pavlovsky N.N., Altschul A.D., Yakovlev S.V., Dobromyslov A.Ya., Kalitsun V.I., Kurganov A.M., Lukinych N.A., Fedorov N.F. etc.) were engaged in problems of hydraulic calculations and designing non-pressure networks in the last years. They put forward a number of formulas to be applied in hydraulic calculations of pipelines made from appropriate materials. However, nowadays

* Corresponding author: hurgin@mgsu.ru

both SP 32.13330.2012 [11] and other technical documents contain no strict unified procedures and recommendation for hydraulic calculation of gravity pipelines to be made from different materials. Application of different materials makes certain disorder both in theory and practice in calculating and designing non-pressure networks during repair thereof especially in the course of their no-dig repair, and thereby it requires the improved approach in unification and optimization of pipe calculation made from alternative materials [12-18].

The objective of the present work was to analyze the main approaches to the hydraulic calculation of straight sections of non-pressure pipelines and obtain the experimental hydraulic characteristics of two polymer pipelines used for trenchless renovation of gravity sewerage networks.

2. Theory of hydraulic calculations of gravity pipelines

The basis for hydraulic calculations is a condition for observance of uniform sewage water flow over the pipes under two basic formulae: flow continuity (1) and Chezy (2):

$$q = \omega \cdot V \tag{1}$$

$$V = C \cdot \sqrt{R \cdot i} \tag{2}$$

where q – fluid flow, m^3/s; ω – discharge area, m^2; V – mean velocity, m/s; R – hydraulic radius, m; i – hydraulic slope (to be equal to pitch of pipe under constant uniform flow); C – Chezy coefficient depending on hydraulic radius and roughness factor of pipeline wetted area, $m^{0,5}/s$.

Main problem in making hydraulic calculations is in finding Chezy coefficient C. A number of scientists proposed their own universal formulae (empirical and semi-empirical dependences) more or less describing dependence of Chezy coefficient C on hydraulic radius, roughness value of pipeline walls and other factors:

a) N.N. Pavlovsky formula [19]:

$$C = \frac{1}{n} \cdot R^y \tag{3}$$

where n – relative roughness of the pipe wall.

To determine y, formula is used:

$$y = 2,5\sqrt{n} - 0,13 - 0,75\sqrt{R}(\sqrt{n} - 0,1) \tag{4}$$

b) A. Manning formula [20]:

$$C = \frac{1}{n} \cdot R^{1/6} \tag{5}$$

c) To determine y, A.D. Altschtul and V.A. Ludov formula is used [19]:

$$y = 0,57 - 0,22\lg C \tag{6}$$

d) A.A. Karpinsky formula [19]:

$$y = 0,29 - 0,0021C \tag{7}$$

Hydraulic calculation tables and nomograms enabling the mechanical engineers to make hydraulic calculations for non-pressure networks and conduits made from some materials are charted under the indicated and other similar dependences.

Non-pressure pipelines are recommended to calculate by [11] using Darcy-Weisbach formula [20]:

$$i = \frac{\lambda}{4R} \cdot \frac{v^2}{2g}, \tag{8}$$

where λ – pipe friction number; g – gravitational acceleration, m/s^2. Chezy coefficient C may be determined by formula (9) under combining formulae (2) and (8):

$$C = \sqrt{\frac{8g}{\lambda}} \tag{9}$$

A number of scientists note that in practice the calculations under the formulae indicated above and other formulae describing dependence of Chezy coefficient C on different values give 2 – 9% discrepancies in results [21] and even up to 20% [19] thereby indicating possibility in applying thereof in engineering calculations but with some errors. Such discrepancies are explained by the fact that each calculation formula to determine Chezy coefficient C has its own roughness factor, in particular formulae (3) and (4) – non-dimensional coefficient n.

Among above indicated formulae the formulae proposed by N.N. Pavlovsky (3) and Manning (5) are mostly approved and better agreed with the experimental results. These formulae are approved and tested in engineering practice and may be easily applied in hydraulic calculations of non-pressure network made of ceramics, concrete and bricks, i.e. for materials where roughness factor n is approximately 0.013-0.014. Recognizing formulae (3) and (5) one must admit that they do not consider fluid viscosity and therefore are active in quadratic area of hydraulic resistance only [19]. When applied in the transition area of hydraulic resistance where most of drain pipes are used, formulae (3) and (5) are to be adjusted due to lower roughness factors n that is typical for new generation polymeric pipes.

Obtaining of universal dependences for hydraulic calculation of pipes made from different polymer materials resulted in adjustment of roughness factors n, with Manning formula (5) being applied as the basic hydraulic dependence to determine Chezy coefficient to simplify the calculations.

3. Materials and methods

Two kinds of Wavin pipes were tested in the hydraulic test bench (Figure 1): corrugated polypropylene pipe DN 98 and bare pipe DN 105 made from unplasticized polyvinyl chloride.

The following procedure was used for making the experiment:

– reading of pressure values from piezometers and Pitot tubes in two points of fluid located at 10 m in wide range of pipeline pitch (0.001-0.03);

– metering of flow running by ultrasonic flow meter with further calculation of average values of hydraulic radius for different operations of pipeline with appropriate diameters and determining values of factors C and λ by plotting empirical dependences $C = f(R)$;

– application of obtained empirical dependences $C = f(R)$ and formula (5) to get relative roughness factor values n for pipeline appropriate material.

Fig. 1. Hydraulic test bench with non-pressure pipes to be tested (the first and second from the right).

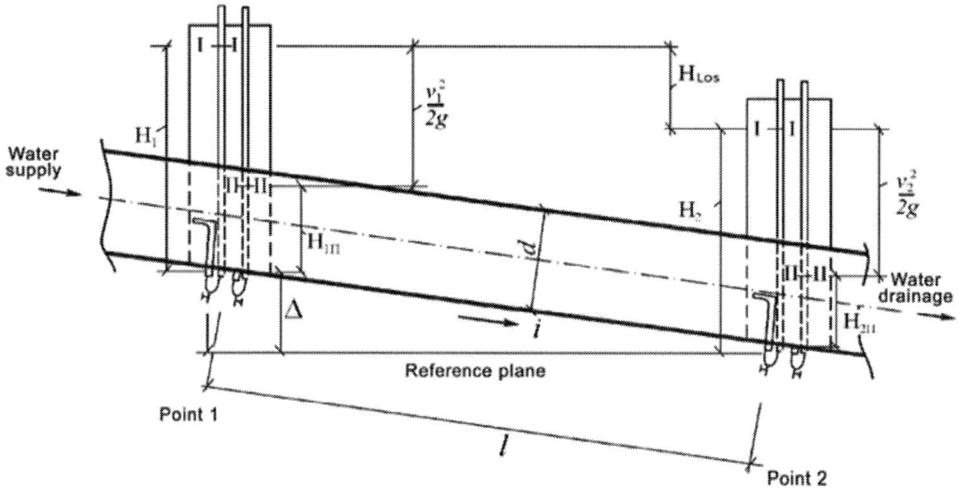

Fig. 2. Outline of hydraulic test bench.

4. Results and discussion

Full-scale experiment results to determine empirical dependences $C = f(R)$ are presented in Figure 3 and Figure 4.

Comparing the obtained dependences of $C = f(R)$ with formula (5) we calculated values of relative roughness n for three kinds of pipes under the calculated filling 0.6: $n_{1\ pipe}$ = 0.00998; $n_{2\ pipe}$ = 0.00939; if we compare the obtained values n with ceramic pipe relative roughness values $n_{cer.}$ = 0.0134, we may see the roughness of the pipes being tested to become significantly lower.

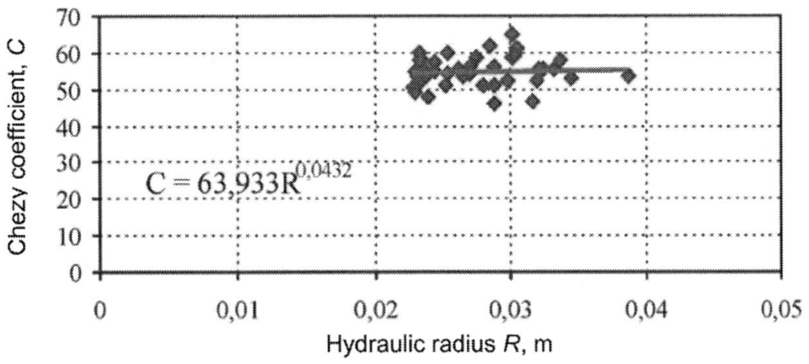

Fig. 3. Dependence curve of Chezy coefficient C on hydraulic radius R for the first pipe.

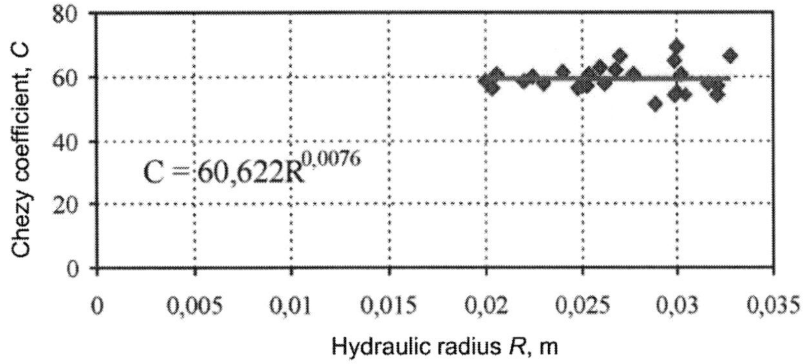

Fig. 4. Dependence curve of Chezy coefficient C on hydraulic radius R for the second pipe.

Therefore, experimentally found universal formulae (10 and 11) may be used by the design engineers to make hydraulic calculations to determine Chezy coefficient C under the calculated fillings. These formulae are adjusted to classical formulae of N.N. Pavlovsky (A. Manning) as:

$$C1\ pipe = 1/\ 0{,}00998R1/6 = 100{,}2R1/6 \tag{10}$$

$$C2\ pipe = 1/\ 0{,}00939R1/6 = 106{,}5R1/6 \tag{11}$$

5. Conclusions

1. The right choice of the repairing material enabling hydraulic compatibility between old and new pipeline sections made from different materials and of different diameters is the most important thing under no-dig restoration of worn pipelines.

2. Dependence of Chezy coefficient C on hydraulic radius, and calculated values of relative roughness factors for three kinds of polymer pipes that may be applied as the repairing material to be installed in available pipelines are obtained under hydraulic experiments in straight sections of pipelines.

3. High precision of calculating relative roughness factors n obtained for three functional dependences: linear, power and logarithmic is registered for all experiments in the appropriate flow range; these dependences may be applied as the calculated ones within the range of the tested filling h/d (similar to the calculated) and as hydraulic radii R.

4. Absolute values n calculated by Pavlovsky and Manning formula differ slightly (from 4 to 14% at the average). Such discrepancies may be considered as accepted for the engineering calculations because A. Manning and N.N. Pavlovsky formulae may be actually interchanged in further hydraulic calculations of non-pressure pipelines.

5. Negative aspects due to flooding of some sections of the network and surging of pressure head (water level) are experimentally found on curved pipelines in wide slope range (0.01-0.03). Under the results of analyses made it may be basically concluded that experimental developments confirm the intertubular space is to be filled to exclude any pipeline irregularities due to temperature differences and stabilization of hydraulic values.

6. The results of the research will be useful for choosing an optimal new pipe materials when considering the problems of ensuring the hydraulic compatibility of old and new sections of the drainage network that are renovated by polymer pipes.

References

1. S.V. Khramenkov, O.G. Primin, V.A. Orlov, *Reconstruction of pipeline systems* (DIA, 2008)
2. V. Chukhin, A. Andrianov, V. Orlov, *Proc. of International No-Dig 2014. 32nd International Conference and Exhibition,* **1B-1**, 1 (2014)
3. A. Andrianov, V. Chukhin, V. Orlov, *Proc. of the 2nd International conference of advanced materials, mechanical and structural engineering*, 81 (CRC Press, 2016)
4. R. Ishmuratov, V. Orlov, A. Andrianov, *31st International No-Dig Conference and Exhibition Sydney, Australia,* **2.16**, 1-7 (2013).
5. N. Maleeva, I. Averkeev, V. Orlov, *30th International No-Dig Conference and Exhibition, Sao Paulo* (2012)
6. V.A. Kharkin, *Systematization and optimization in operation of non-pressure drainage networks (Author's abstract)* (MSUCE, Moscow, 2003)
7. V. Orlov, A. Andrianov, Appl. Mech. and Mater. **580-583**, 2398 (2014)
8. V. Orlov, I. Averkeev, Appl. Mech. and Mater. **580-583**, 2384 (2014)
9. E. Risch, O. Gutierrez, P. Roux, C. Boutin, L. Corominas, Water Res. **77**, 35 (2015)
10. M. Le Quiniou, P. Mandel, L. Monier, Procedia Engineering **89**, 710 (2014)
11. SP 32.13330.2012, *Sewage. Public utilities (Actualized edition of SNiP 2.04.03-85)* (Moscow, 2012)
12. N. Makisha, V. Scherbakov, A. Smirnov, E. Scherbina, Int. J. of Appl. Eng. Res. **10**, 24, 44347 (2015)
13. V. Scherbakov, E. Gogina, T. Schukina, N. Kuznetsova, N. Makisha, E. Poupyrev, Int. J. of Appl. Eng. Res. **10**, 24, 44353 (2015)
14. N. Makisha, Procedia Eng. **165**, 1087 (2016)
15. N. Makisha, Procedia Eng. **165**, 1092 (2016)
16. N. Zaletova, Yu. Voronov, N. Makisha, Int. J. of Appl. Eng. Res. **10**, 21, 42454 (2015)
17. A. Santiago, M. Durango, *30th International No-Dig Conference and Exhibition 2012, Sao Paulo* (2012)
18. K. Laffrechine, D. Breysse, Y. Le Gat, P. Bourgogne, Tech. Sci. Meth. **6**, 61 (1999)
19. A.Ya. Dobromyslov, Pipelines and Ecology **2**, 21 (2000)
20. K. Subramanya, *Flow in open channels. Second edition* (McGraw-Hill Publishing Company Limited, New Delhi, 1997)
21. V.I. Kalitsun, *Hydraulic calculation of drainage networks* (Stroyizdat, Moscow, 1988)

Research on importance of cleanliness in manufacturing reliable products for automotive

Alexandra Dzetzit*, and *Gheorghe* Nagit
"Gheorghe Asachi" Technical University of Iasi, Department of Machine Manufacturing Technology, Romania

Abstract. Quality is ultimately defined by the client and is near the result of requirements and customer expectations. Currently the molten aluminum has known a huge increase in demands. Once with the enlarged volumes is also expected a high purity of the raw material which starts with the initial state of aluminum alloy. In the presented paper one of the methods that can assure the necessary quality is presented, with focus on particles and how this final result can be improved. What is even more important, in addition for the detection method is the effective and appropriate method of reducing them. The final results, the beneficial effects of filtering die casting metal with the bonded particle filter media, are presented in different aspects in order to achieve reduction of machining defects that in the ends achieves financial and casting marketability realizations.

1 Introduction

The automotive industries is growing fast and during the last year the technical development had inquired special classes for technical cleanliness. Nowadays, the cleanliness had become vital to the automotive "freedom of a control-or function area of critical impurities coming from the production process and environment". The problems with the product's cleanliness had always been there but now, once with the customer's growing demand for safety and comfort in driving, cleanliness it is necessaries to ensure quality and reliability in respect of the technical cleanliness standard [1]. Due to that, the tolerances of assemblies build into the vehicles are lower and lower and that for the parts are compiled to support higher and higher stress and therefore them are more sensitive to particle contamination. The standards are compiling to smaller classes and without regard to, imagine that a chip just a few microns in length can compromise an entire complex hydraulic system or even the completed fail of the system.

The particles can create the breakdown of diesel injection system than can reach the pressure of 2500 bars, a PCB (printed circuit board) that can cause an overstressed or short circuit, nozzles become blocked, valves congest, the bearing are damaged, this are only some of the reasons that can happen. In order to prevent the contamination, it is important to standardize the environmental production, for example the clean room's classes, particle monitoring, clothes, air conditioning. Development department is responsible for creating the

* Corresponding author : dzetzit.alexandra@yahoo.com

product specification, since technical cleanliness is a design feature. In general, cleanliness specification for a part, component or product are given to secure the function and/or to fulfill a cosmetic specification. State of the art is to distinguish between 3 classes by a light microscope with polarization filter (one of the measuring methods): metallic shiny, non-metallic shiny and fibers. Therefore, will be presented some methods for monitoring the inclusions inside the molten aluminum in the following paper, but in different aspects than concluding with the benefits that can be taken out of this method of measure. The system is capable of monitoring in raw material, the concentration and size of particles in the range of 20-300 μm in the absence of micro bubbles in the melt of aluminum [3].

2 Cleanliness determining tolerances and producing parts

2.1 Main ideas

A cleanliness specification should be taken in account where there is a significant risk because the system's function or functionality of a component could be serious damaged by particles contamination. But, if this risk it is not present, a cleanliness specification should not be made. Cleanliness is a characteristic that should be taken an account from the first phase of the product- design. Before establishing cleanliness limiting values, it is necessary to check the automotive standard [1] and other standards, in order to increasing the compactability of the system. A good example could be the installation of filters, magnets, air filtration or a clean room and if it is established that the component has a safety reliability characteristic. Other methods for realizing this proactive, feasible and money saving is also the most convenient and safely, reducing the metallic chips and non-metallic into the manufacturing process [2].

2.2 The importance of cleanliness classes in automotive

Particles inside complex systems are generally mobile, this "weakest link in the chain" usually determines the degree of cleanliness necessary for the whole assembly. Examples of typical fluid systems found in the automotive industry include: the cooling system, the air-conditioning and the exhaust system. A particle determined to be critical will only cause a malfunction if is located in the sensitive area of the system at the right moment in time and possibly if is orientated in a specific direction. For that the aim should be to determinate accurate the cleanliness standard in order to avoid:

–Damaging mechanisms/damaging effect (e.g. blocking, jamming, electrical shorting, etc.)

–Probability of a defect occurring, e.g. during a changeable state such as a valve closing;

–Large particle, high current lines – rapid corrosion

In Figure 1 are presented some possible failures from Jomesa showroom, failures produced by inclusions. This type of failures can cause leakages, blocker filters, blocking injectors, failures on pressure contacts or high contact resistance on electronic devices, bearing damages by hard particles, etc.

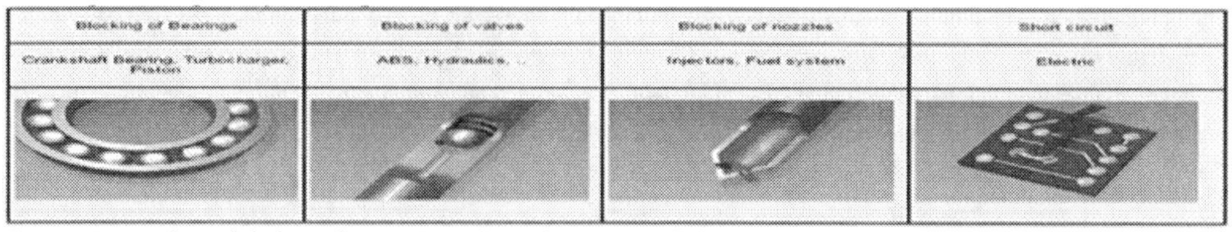

Fig. 1. Examples of defects from JOMESA microscope. [5]

3 Measuring cleanliness process

Due to quick development and specially requirements from the customers, it is necessary to evolve and this has been driven by finding lightweight materials with strong mechanical qualities. For the moment, the most conceivable, but least productive method is to expect for the customer to accuse the part for an incident and in this matter to diagnose an issue. For sure, the best manner to handle the problem is not the one in question.

It is possible that the defect is not caught on time and creates an incident that could cause loss of human life. Maybe this is an exaggerate effect but the reality is the things can occur. In case of aluminum it could be said that one of the most important parameters is metal cleanliness defined like size and per solid inclusions per unit volume. That is why during the paper it will be discussed a method, maybe not new on the market, but continuously developed, using the history behind.

By this way, was created Limca III, for detecting inclusions in molten aluminum during processing. It is known that these particles can degrade the mechanical proprieties and also can damage the finished surfaces. Defects like small holes, tears, whiskers, visual defects on polished surfaces can occur. Casting defects that are related to molten metal cleanliness result from three main categories:

1.General inclusion level and oxide films present in the bulk liquid;
2.Local clusters of inclusions and oxide films caused by poor metal treatment;
3.Formation of oxide films during mold filling itself [3];

It is much known in the literature [3] that the first two problems caused many incidents among casting's suppliers. Due to the critical high quality demands, products like thin gauge sheet for body stock, foil or plates and designed parts with complicates bodies, make the presence of non-metallically and metallically particles to be detected and eliminated. There were developed several techniques based on pre-concentration / filtration step, followed by metallographic analysis such as the PoDFA from Alcan and the LAIS from Union Carbide, like K-Mold or Metal Vision with the aim of monitoring the concentration of particles in the melt and in the product [3]. Due to mentioned fact those methods take several days for getting any results and this means the involvement of resources that were required to study other solutions.

Besides, for this type of process procedure are mandatory the human resources with strong background on metallurgical process in order to complete all the analyses. The final results can have an extreme impact on the output, but cumulating everything the cost is not worthy enough, especially if is done on a regular frequency. Investigation on the market for building up filters during the pressure filtration test itself had been running. But the most effective and quite new on the market is the one who is cleaning the molten alloys by removing from the beginning the solid inclusions, an instrument named Limca III. Over the years, from the '80 until now, several Limca instruments were developed. In the raw material, in molten aluminum, the solid inclusions are mainly non-metallic compounds, such as oxides, borides, carbides, nitrides, chlorides, fluxing salt [3].

Limca instruments is based on the fact that this type of particles are not such eagerly conductive that the molten aluminum, mainly using resistive pulse / electric anticipating area technique (ESZ) [8]. A non-conducting glass pipeline overflowed with a 300μm hole and couple of tungsten electrodes, one inside the tube and one outside, are inundate the molten aluminum during measurement, like the figure below (Figure2).

The particles non-metallically have an electrical dynamism different from the fluid that they are included. In this way, they are compress into an electrically bumper capsule. Achieving to pass through the gap, the volume of the influencing fluid is dismissed causing the temporally shifting in the orifice of the electrical resistance.

The actual pick is observed as a voltage pulse. In this way, the pulses are intensified, counted and classified by their magnitude. By this route, it is provided an exact number (it is important to mention that is expressed in thousands of inclusions per kilogram of metal (k/kg)) and also the size distribution of the particles. The inclusion recognition is here disturbed in the presence of micro bubbles due to their non-conducting performance. Effort was spent to separate non- metallic inclusions and gas bubbles, but no final solution is available to measure proximate degasser units [4].

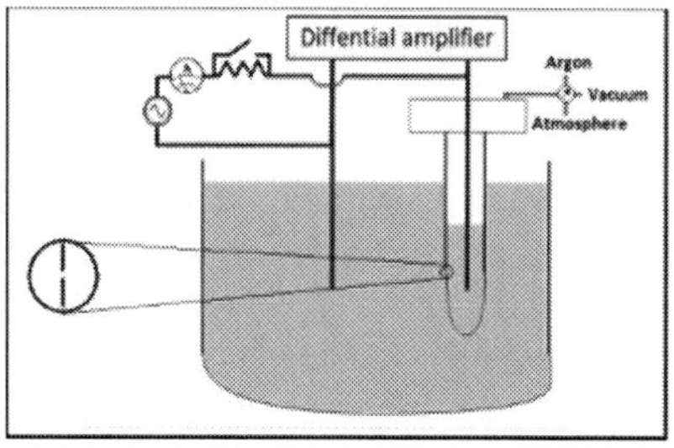

Fig. 2. Limca III functioning system.

In the latest research, it has been discovered that it was feasible ranging 20 microvolts signals, comparable to 20 microns inclusions in the molten aluminum during the process at 700°C in the power-driven rowdy climate of aluminum processing focal plant. The non-metallic particles are detected in the molten aluminum due to limited-conductivity. This is realized by Limca using a vacuum to drag the molten aluminum into an aluminosilicate glass probe through 300 micron's loop. How also was described above, the electrical current is the same to all the opening [6].

Main difference between other methods and Limca III can be observed in the decisional tree (Figure 3.) and providing major improvements on casting process knowledge, with main influence on environment, productivity, recycling, flexibility, quality measurement.

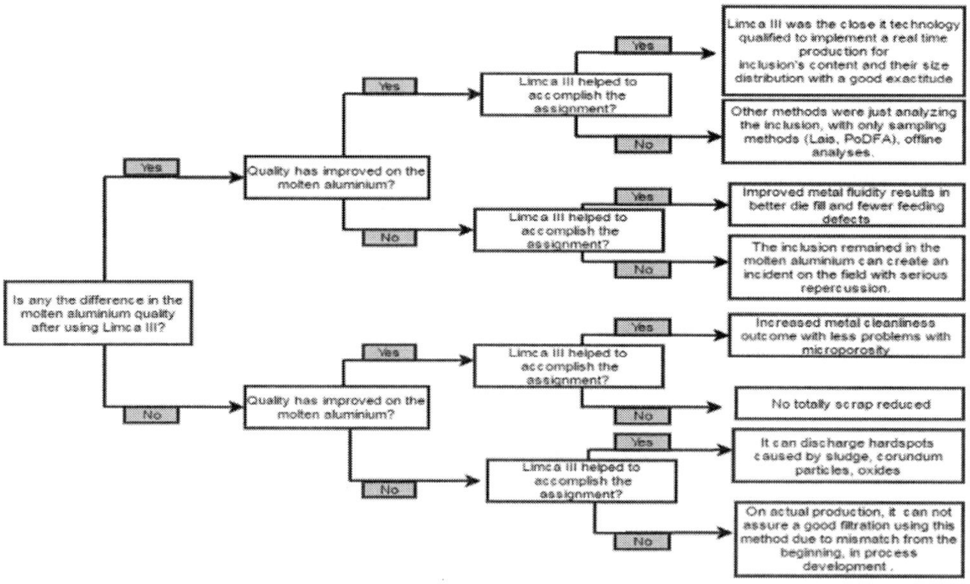

Fig. 3. Decisional tree in the benefits of using Limca.

DOI: 10.1051/matecconf/201711209017

4 Case study on Limca III's results

Using all the terms involved in Figure 4, Isikawa diagram, the results can show the necessary step for introducing filtering particles inside molten aluminum (Figure 4).

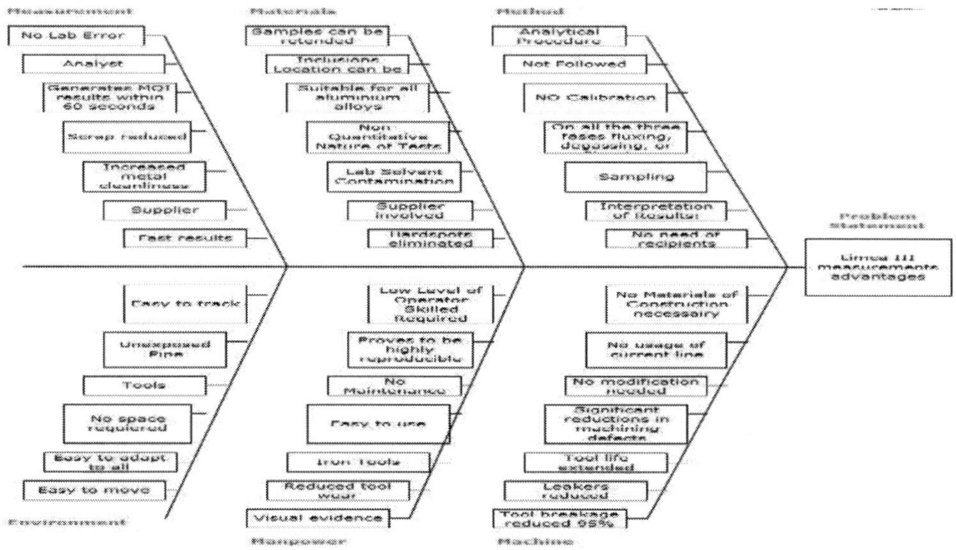

Fig. 4. Ishikawa- using Limca in the process line.

The results taken out from this measurement contain data that shown the actual cleanliness of the material used. Results are replicated but long-term metallurgical elements will depend on the standard procedures and starting stocks employed in the manufacturing process. In agreement with the length to which the material assessed in this study is exemplified, the inclusion content has a certain repeatability. In the following pictures, we can detect how distinct it is the process from the quantities detected. It can be seen in the Figure 5 that the values are from different stages of the casting: exit furnace, after degasser and CFF (ceramic foam filter). Background that had side effects on the measuring we can add: all tests were carried out at 685 °C and a pressure of 12.1 psi. The tests were carried out between 680 and 690 °C and a pressure of 12.1 psi. The material used was A380 and the refined melt weight samples were 1000 g weight.

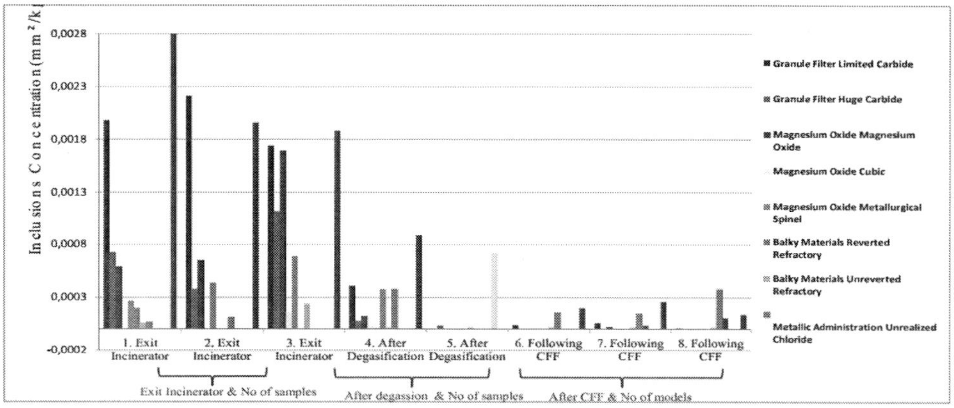

Fig. 5. Metallographic Analysis Results from a Limca measurement.

Once the measurement is done, it is very visible in the graph below, (Figure 6), a significant improvement over unfiltered material and by this manner, the fluidity is improved with normal industry data.

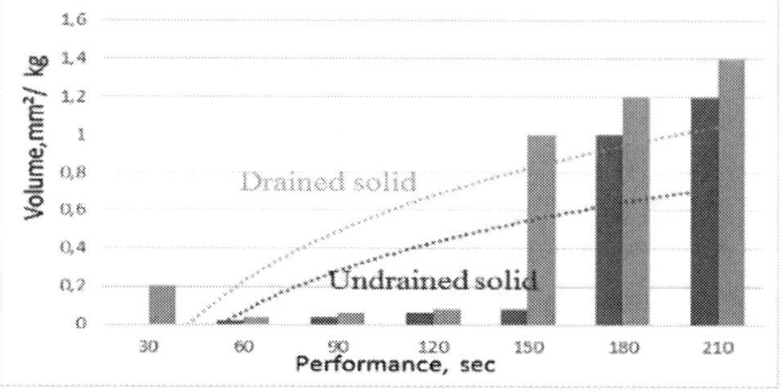

Fig. 6. Benefits of filtration of molten aluminum.

From the pictures, it can be seen that the metallurgical spinels (MgAl2O4) can cause big aversion in the aluminum process because of their large size and cohesion. The rate of structures increases as deliquesce is heating up. A hard glassy minerals consisting of an oxide of magnesium and aluminum formed in a cube-shape are generally formed from the reaction between MgO chunks and aluminum oxides [7]. Because of their large dimensions, both types of this formations can easily be dragging out of the metal. Most chloride inclusions derive from fluidity's equipment using chlorine gas as an inclusion hollow agent and in the end, is acting as a metal cleanliness process. Another process that is enlarging in time is that after downgrading in contact with magnesium in which alloys are included, the obstinate particles become similar in composition and color as a metallurgical formation, but due to their more increased size, they are easily detectable. Usually, refractory inclusions are large solid particles which may be exceptionally destructive to the practice.

5 Conclusions

By the end of this paper it can be seen that there is a clear difference between the quality of the metal in the melting and casting furnaces, as well as after the in-line treatment processes. Aluminum alloys and treatment applied on it, in order to meet the specification, especially in castings processes are significant in the cleaning process. Results are reproducible but long-term metallurgical quality will depend on the standard procedures and stocks built in the manufacturing processes. That is why utilizing Limca III has developed a new generation of treatment equipment in order to reduce and control the particles. Due to that, also the progress of consistent turnout specialty materials wouldn't be now in production without this continuous check on inclusion's dimension.

References

1. VDA 19, Inspection of Technical Cleanliness, **2**nd (Revised Edition) (2014)
2. M. Rochowicz, M. Bargende, Intern. Stuttgarter Symposium, 641 (2015)
3. Alain A. Simard, *The Minerals, Metals & Material Society*, 1090, (2000)
4. M. Badowski, S. Instone, *The Minerals, Metals & Materials Society,* 1071, (2012)
5. http://www.jomesa.com,
6. T. L. Buijs, D. Gagnon, C. Dupuis, *The Minerals, Metals & Materials Society*, 973, (2015)
7. J. Smith, *Metallographic Analyses*, (Int. Al. Casting, Quebec, Canada, 2006)
8. R.I.L. Guthrie, *Metallurgical and Material Transactions B*, **31B**, 767 (2000)

Analysis of the performance management within industrial systems: a case study

Flavia Fechete*, *and Anişor* Nedelcu

[1] Transilvania University of Braşov, Department of Manufacturing Engineering, Mihai Viteazu No.5, Braşov, Romania.

Abstract. Improving performance has become one of the critical issues for gaining competitive advantages for companies. Monitoring and improvement of the industrial systems' performance has become an increasingly complex task. In order to improve performance, companies have to implement a performance management system. This paper aims to analyze the performance management system within a multinational company. There will be analyzed the performance indicators monitored by the company and the models used to evaluate the performance. The aim of the study is to identify potential causes that led to getting poor performance; that will allow highlighting areas where goals were not achieved in order to apply the proper measures for improvement.

1 Introduction

Industrial system's performance is the relative size by which it can be characterized the activity of the production systems, using specific indicators [1].

Improving the strategic performance, the operational, team or the individual performance is a major objective of any company.

Performance management is the mode in which the entity ensures a sustainable success in their activity by improving the performance of the organization, of the team or the individual performance [2, 3].

Performance is measured and evaluated using criteria or performance indicators. In order to obtain a true and fair view of the performance, entities must implement a system of indicators. Indicators are used for proper evaluate, report and improve the organizational performance [4, 5].

This paper aims to examine the manner that performance management is achieved in an industrial system. Thus, there will be analyzed the key performance indicators monitored by the company in order to propose improving measures for the identified risk areas. Performance improvement is proposed to be achieved by identifying and analyzing root causes using 5 Whys technique. Root cause analysis is a method used by the performance management within developed companies. The method is considered a primary-method to be used in the early stages of specific analysis of the process management [6].

* Corresponding author: fechete_flavia@yahoo.com.

2 Industrial system's performance management. A case study

The experimental research was conducted in the Automotive division of a multinational company and it was concentrated on the product "outer ring" for roller cam followers.

Processing technology for pieces like "outer ring" involves achieving the following main steps: turning, surface grinding, centerless grinding, outer diameter grinding by vibrating, interior grinding and super finishing.

Performance management in the segment Z of the multinational company which manufactures parts like "outer ring" is achieved by daily monitoring of a set of nine key performance indicators and by taking immediate measures to improve where it is necessary (where the goal was not reached or which exceeded the goal set in a negative manner).

The key performance indicators that are monitored and evaluated in the multinational company are shown in Table 1.

Table 1. Key performance indicators tracked by the multinational company.

Indicator	Definition	Measurement units	Frequency of calculation	Response
Non quality costs	Expenses incurred if the product does not meet quality requirements.	Percentage	Daily	Taking measures in case of exceeding the goal
Complaints	Non-compliance of the product delivered to the client.	Pieces	Daily	Taking measures in case of complaints
Delivery	Cumulation of the previous day deliveries.	Percentage	Daily	Escalation
Backlogs	Cumulation of the previous day backlogs.	Percentage	Daily	Escalation
Stocks	Accumulation of the quantity of goods available as backup.	DOH	Daily	Long-term plan to reduce inventory
Productivity	The ratio between the manufactured products and workload.	EUR/h	Daily	Taking special measures and immediate action
Work accidents	Number of workplace accidents.	Number	Daily	

Table 1 presents the main key performance indicators assessed daily by the multinational company: non quality costs, customer complaints, performed deliveries, backlogs, stocks, work productivity and work accidents; each indicator has a proper definition, a unit of measurement, a calculation formula and a method of action in case of not achieving the established goal.

As the quality of industrial production is a prerequisite for achieving performance [3], special attention will be given to non quality costs indicator.

Non quality costs relate to expenses incurred if the product does not meet quality requirements. It can be seen a downward trend in this indicator towards the end of the review period (Fig.1).

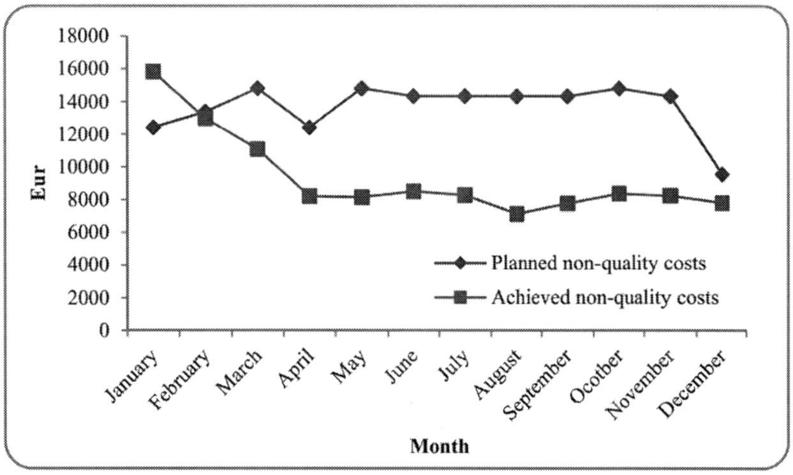

Fig.1. Evolution of non-quality costs in the period January to December 2015 for the multinational company.

In order to identify the distribution of non-quality costs, a Pareto chart was conducted (shown in Figure 3), which notes that over 50% of the non-quality costs in 2015 are represented by scrap parts.

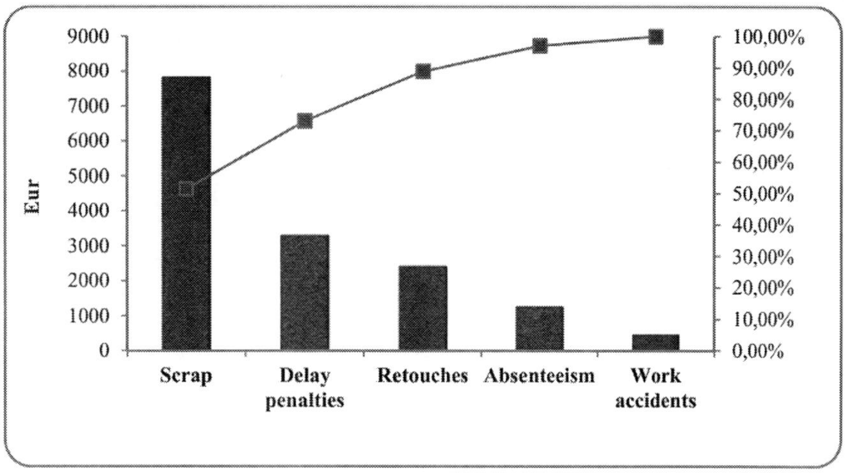

Fig.2. Pareto chart of the distribution of the non-quality costs in the multinational company.

In order to identify the poor performance on the part of scrap in the process of bore grinding it would be used the 5 Whys technique (Table 2). Asking "why" five consecutive times is an approach that many Six Sigma practitioners use to help identify the "root cause" of a problem or a defect [7]. Using a series of questions "why?", this technique is meant to explore the cause and effect relationships or work process, highlighting a particular problem. The goal is to find the essence of the problem in order to correct it [8, 9].

Table 2. 5 Whys analysis applied for the bore grinding process.

<table>
<tr><th>Category</th><th>1st Why</th><th>2nd Why</th><th>3rd Why</th><th>4th Why</th><th>5th Why</th></tr>
<tr><td rowspan="6">Scrap parts after set-up of the machine</td><td rowspan="3">The machine get's cold (the broach, guides, grinding stone)</td><td rowspan="3">Long down times of the machine

Lack of raw material</td><td>Machine break down that is not repaired for long time</td><td>Because there is only one setting person on the shift</td><td rowspan="3"></td></tr>
<tr><td>The planning was inadequate and it wasn't planned enough raw material</td><td rowspan="2">The production is not monitored</td></tr>
<tr><td>The planning of the production is not respected</td></tr>
<tr><td rowspan="3">The tools needs to be changed often</td><td>Re-adjustment of the machine</td><td></td><td></td><td></td></tr>
<tr><td rowspan="2">Machine break down</td><td>Because of ends of raw material got in the bore grinding</td><td>Ends of raw material not detected</td><td></td></tr>
<tr><td>Mixed up parts from the turning, heat treatment, calibration</td><td>After changing the part-time at turning/calibration remain parts from the previous operation</td><td>The instruction for changing and cleaning the machine is not respected</td></tr>
<tr><td>Tools are not changed on time</td><td>There is not enough personnel to change the tools</td><td></td><td></td><td></td><td></td></tr>
</table>

	on time				
Grinding stone broken	Parts with degree	Lack of chamfer at inner diameter	The mill is broken	Because of an incorrect adjustment	There is not a standard way to adjust the machine
	Parts with variation at the outer diameter	Because of the tolerance at exterior grinding			
	The machine did not removed chips at grinding	Lack of material on inner diameter	Because the drill could be bigger at turning operation	They mixed the tools when they took them from the tool preparation	
			Because of the mill that cuts inappropriate	Because of the incorrect adjustment	There is not a standard way to adjust the machine
Parts with black spots or roundness after bore grinding	Because the parts from the lathe have run out	Because at the turning they don't work with covered tools			
		The drill from the turning operations it is not well sharpened	Sharpened internally with different method from the supplier		
Too much material to process at bore grinding	Change of the calibration process with a smaller ball				

5 Whys analysis allowed the identification of detailed root causes that led getting a large percentage of scrap in the process of bore grinding. Further analysis using statistical methods such as regression analysis will allow identifying the exact causes and the immediate intervention action to reduce the percentage of the scrap obtained.

Table 3 shows the six main causes that led to an increasing number of scrap, according to the high coefficient of confidence given by the team of specialists.

Table 3. Identified main root causes for scrap parts.

	Identified root cause	Conclusion
1	Not enough personnel at bore grinding (for corrections)	In the days with only 3 persons on the line the scrap rate was higher than the days with 4 persons on the line.
2	The spread of exterior diameter at bore grinding	The operator does not make corrections on time
3	Setting parts with different value from the one on the label	46% of the setting parts were not ok
4	Too much added material at bore grinding	When changing the calibration ball with one bigger, the scrap rate was significant smaller than usual
5	Black spots at bore grinding	There are many parts that have runout at turning and when reaching bore grinding black spots appear on the surface
6	Scrap parts after dressing	After each dressing of the machine approx. 5 parts are scrap.

3 Conclusions

This paper aims to analyze the manner that performance management is achieved in an industrial system. The research revealed that the industrial system uses a total of nine key performance indicators that are monitored daily, also being taken necessary measures to improve the obtained results.

As quality is the prerequisite to achieve performance, special attention was paid to the indicator of non-quality costs. Pareto analysis showed that over 50% of non-quality costs are the scrap parts.

This paper aims to identify potential causes that led to getting the high percentage of scrap, which will allow applying the appropriate improvement measures in order to increase the performance of the analyzed industrial system. The application of the 5-whys analysis provides a fact based and structured approach to problem identification and correction that focuses on not only reducing defects but also in eliminating them.

References

1. H. Aguinis, J. Harry, B.H. **55,** 385 (2012).
2. M. Armstrong, *Management of human resources* (Codex Published House, Bucharest, 2001).
3. A. Draghici, A. Popescu, L. Gogan, P.S.B.S **124**, 544 (2014).
4. F. Osmani, G. Maliqi, P.S.B.S. **41**, 434 (2012).
5. J. Cai, X. Liu, D.S.S **46**, 512 (2009).
6. U. Dombrowskia, T. Mielkea, P.C.I.R.P. **7**, 569 (2013).
7. A. Ayad, J.M.D **29**, 556 (2010).
8. E. D. Arnheiter, J. Maleyeff, T.Q.M.M **17**, 5 (2005).
9. U. Murugaiah, S. J. Benjamin, M. S. Marathamuthu, S. Muthaiyah, I.J.Q.R.M **27**, 527 (2010).

MATEC Web of Conferences **112**, 10001 (2017) DOI: 10.1051/matecconf/201711210001
IManE&E 2017

Energy evaluation at a winery: a case study at a Portuguese producer

João Correia[1], *António* Mourão[2], and *Miguel* Cavique[3,*]

[1] Faculdade de Ciências e Tecnologia, Universidade NOVA de Lisboa, Campus de Caparica, 2829-516 Caparica, Portugal
[2] UNIDEMI & Faculdade de Ciências e Tecnologia, Universidade NOVA de Lisboa, Campus de Caparica, 2829-516 Caparica, Portugal
[3] UNIDEMI & CINAV & Escola Naval, Base Naval de Lisboa – Alfeite 2810-001 Almada, Portugal

Abstract. The introduction of cooling systems in the wine industry to control the fermentation has allowed the oenologist to produce more and more excellent wines. In this regard, the alcoholic fermentation is a target for various studies that aims at explaining the chemical reactions involved in the release of energy. The aim of this paper is to evaluate the energy consumption of a winery and to discuss and understand the main parameters involved in the process of fermentation. The weather profile during fermentation and the schedule of charging the tanks with freshly affect strongly the needs of cooling power, and the energy use. The study conducted at the Adega da Ervideira in the South of Portugal allowed to define a model for the computation of the cooling power and the electricity consumption. The heat gains from outdoor in convection mode and the heat released during maturation and fermentation phases are the main contributors for the cooling requirements at a winery. As a result of the real fact study, it will allow an oenologist to estimate the cooling power and energy for a winery as well as to produce other types of wines.

1 Introduction

In the last thirty years the wine production changed from an empirical activity to a scientific based approach, what made possible to increase the quality of the wine produced. Many factors interfere in the production of the wine, which need to be taken into account in order to get a wine with a certain quality. The quality of a wine starts on the vineyards, depending on the type of soil, on the solar exposition, on the watering and on the care activities especially in what concerns to the grapes.

After the growing process, the oenologist needs to define the date to harvest the grapes checking the maturity of the grapes. The process of harvesting may be mechanical or human based, depending on availability of resources or on the quality of grape required for special wines. Special attention needs to be taken into account regarding the transportation, in order to avoid a preliminary fermentation. Many producers tend to transport the grapes in refrigerated vans and harvest the grapes during the night to maintain the bunches at

* Corresponding author: cavique.santos@marinha.pt

temperatures lower than 20 °C.

Thus, inside the winery the collected grapes undergo a process of maceration that allows creating the must. The must may be filtered, and sometimes undergoes centrifugation or flotation processes to get rid of solid parts, activities usual in the production of white wines. In this phase, the oenologist may ask to correct the acidity and the sugar content of the must. Next, the must is transferred to large wine tanks, where the fermentation takes place at precise temperatures according to the type of wine. During this phase most of the ethanol is produced.

The maturation process allows warranty the stability of the wine colour and the reduction of the astringency of the wine. This phase develops in oak barrels or stainless steel vats during a period of time that varies from six months to two years. The maturation continues in bottles, a process that should occur at a stabilized temperature. All of these activities interfere in the organoleptic characteristics of a wine, allowing to define the flavour, the astringency, the colour, the acidity and the aromatics of a wine.

The aim of this paper is to establish the energy balance of a winery, allowing to define the energy needed to run it. Most of the energy needed comes from the cooling process, which is responsible for about 54% of the energy use of the winery. Therefore, changing the fermentation temperature to alter the organoleptic characteristics impacts the energy use of a winery.

The energy model developed in this study takes into account the energy released during the fermentation process and the transfer of energy between the tanks and the atmosphere. The fermentation process starts by a natural decomposition of the glucose, ending on the formation of ethanol [1]. During the process, energy of circa 550 kJ/kg of glucose is released [2-5] in the fermentation process, which has to be removed from the tanks in order to avoid increasing the temperature of the must. The energy released transfers by natural convection between the must and the atmosphere, making the calculation of energy needed to depend on the local weather condition.

Thus is possible to calculate the cooling power needed in a vineyard, as well as to estimate the energy used in cooling and to evaluate the energy use for different operation conditions. The application to a winery of Ervideira, Alentejo, Portugal shows the value of the model briefly described above.

2 Energy in the fermentation process

The fermentation process was first exposed by Guy-Lussac in the equation of decomposition of glucose $C_6H_{12}O_6 \rightarrow 2CH_3CH_2OH + 2CO_2$, showing that the reaction produces ethanol and carbon dioxide while releasing energy. Nowadays, the wineries control the starting of the fermentation by introducing yeast in the must after filling the tanks with fresh squeezed must. During the fermentation, the must is usually in contact with solid parts of the grapes, in order to get special organoleptic characteristics developed in the maceration process [6]. Moreover, moving the must avoid the development of bacteria on the surface of the must in contact with the ambient air. The old maned chewing process was able to allow the aforementioned contact, but now a pump moves the must from downwards to upwards of the tank. A similar technology rolls the must in a rotary vat with horizontal axis, allowing the carbon dioxide released to stay in contact with the must, periodically discharged into the ambient [7].

Increasing the temperature of fermentation to the range of 25 °C to 30 °C, being 30 °C the better for the development of the yeast, gives rise to more coloured wine attained by a faster formation of ethanol. The wineries tend to ferment red wines at about 25 °C achieving an almost complete fermentation. For white and rose wines the fermentation is controlled at lower temperatures to get fruity wines by avoiding the evaporation of aromatics [8].

The production of the glucose in the grapes starts in the photosynthesis process, an oxi-reduction process that allows the energy from sunlight to supply the ADP (adenosine diphosphate) and other phosphate molecules. These molecules thus form ATP (adenosine triphosphate) and NADPH (Adenine Nicotinamide Dinucleotide), used to produce glucose, a second product of photosynthesis [1].

During the fermentation process, the yeasts allow producing alcohol and carbon dioxide from the glucose in an exothermic process. Glycolysis is the set of reactions that degrade glucose in pyruvic acid, or pyruvate, latter reduced in a set of reactions leading to produce alcohol and carbon dioxide.

The reduction of pyruvate or pyruvic acid ($C_3H_4O_3$) is anaerobic, done by the action of NADH formed during glycolysis. The pyruvate give origin to Acetaldehyde (C_2OH_4) and later reacting with the molecules of NADH generates ethanol (C_2OH_6) [1,2].

S. Yonsel et al. range the enthalpy released in the decomposition of glucose from -450 to -550 kJ/kg, a value similar to the one attained by Dubrunfaut in 1856 who stated a value of -527 kJ/kg [4]. In 1979, Boulton defined it to be -651.28 kJ/kg for broad sugars [5, 9]. Recently in 2005, Von Stockar et al. defined the enthalpy of decomposition of glucose to be -100 MJ/kmol or -555.1 kJ/kg of glucose [10]. The book "2006 ASHRAE-Refrigeration Manual (SI)" presents in the "Beverages" section similar values for the enthalpy of decomposition of glucose, varying between -83.7 to -100.5 MJ/kmol most used to be 99.6 MJ/kmol [11]. Hence, a sugar content of 220 kg/m3 of must decomposed to 98.5% releases an energy per cubic meter of must of 119.2 MJ.

Therefore, estimating the energy of fermentation from a certain quantity of must is a question of knowing the moles of CO_2 released in a steady state process. The set of fermentation reactions changes the glucose to ethanol, a product with about half the density of glucose. El Haloui et al [12] correlated the energy release to CO_2, the CO_2 released to the change of glucose, and the sugar content of the must to the density of the must as well as to the energy released. Based on the aforementioned work, A. López and P. Secanell [13] estimate the rate of heat released during fermentation, from the variation of the density of the must.

Being Z the sugar concentration of the must in g/L, Z_0 an initial sugar concentration, and d the density of the must, A. López and P. Secanell found:

$$Z = 0.99109 \cdot Z_0 - 2096.3 \cdot d + 2078 \qquad (1)$$

The rate of energy released, Q [kJ], follow the rate of sugar consumed Z as:

$$\frac{dQ}{dt} = 0.55 \frac{dZ}{dt} \qquad (2)$$

And the total energy released from a state Z0 to Z:

$$Q_e = 0.5451 \cdot Z_0 - 1152.97 \cdot d + 1142.9 \qquad (3)$$

Equation 3 allows to compute the total energy released until time t, knowing the initial sugar content of the must and the density at time t.

3 Heat Transfer

The Ervideira winery is open to the exterior on the North and East bounds and covered by a non-insulated white metal sheet. The tanks are therefore at atmospheric conditions, subjected to radiation from the cover and to direct radiation from the East area.

Hence, it is necessary to calculate the convection heat-transfer coefficients, inside and outside, of the tanks in order to determine the heat flow at any hour. Inside the tanks the convection is natural convection as well as in the outside of the tanks, due to low wind velocities found in the winery.

In natural convection the convection heat-transfer coefficients is governed by the Grashof number, Gr, that in turn varies with the difference of temperature between the wall temperature and the fluid temperature. The Nusselt number, Nu, allows determine the heat convection coefficient, h, for a characteristic length, x, and fluid thermal conductivity, k. Nu varies with the geometry and with the Rayleigh number, Ra, according to:

$$Nu = \frac{h \cdot x}{k} = C \cdot Ra^{m} \tag{4}$$

In Equation 4, C and m vary with geometry and with the Rayleigh number, this being $Ra = Gr \cdot Pr$, where Pr is the Prandtl number of the fluid slightly dependent on temperature. Therefore, Nusselt number varies mostly with the temperature of the surface.

For each type of surface of tanks and vats, determining the surface temperature involves an iterative method, as the h value varies with temperature. The method starts by determining the convection heat-transfer coefficients based on an estimated surface temperature, thus calculating the heat flux, and then the new surface temperatures. Achieving a new surface temperature a new iteration is performed, until reaching convergence to the value for h. The heat balance needs to determine h for each surface, at each hour of the energy calculation, as the outside temperature varies during the period of wine production.

Figure 1 depicts a vertical and a horizontal tank, showing on the horizontal vat picture the possibility of rolling the vat allowing to transfer the must from the downside to the upside.

Fig. 1. Vertical (left panel) and horizontal (right panel) vats

Using Equation 4 with the appropriate coefficients it is possible to calculate the heat-transfer coefficients from the inside and outside convection on each vat. Figure 2 depicts the type of heat-transfer coefficients computed for each vat.

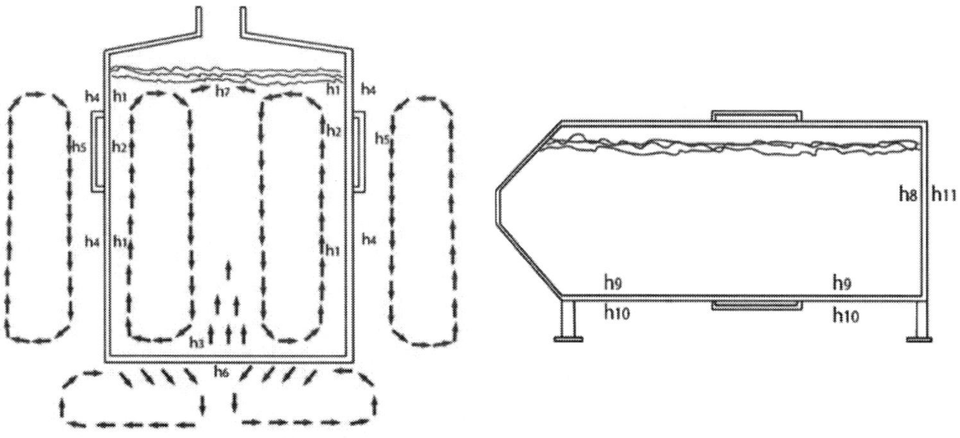

Fig. 2. Heat-transfer coefficients for vertical vats (h1 to h7) and for a horizontal vat (h8 to h11)

The heat-transfer coefficients by convection vary depending on the direction of the heat flux, making it necessary to compute the heat-transfer coefficients when the room temperature is lower and when it is higher than the must temperature. Table 1 shows an exert of the computed heat-transfer coefficients for two vertical (25 CV, 5 CV) and one horizontal (10 V) vat in the situation when room temperature is higher than the must temperature.

Table 1. Convection heat-transfer coefficient [W/m^2·°C] for some VATs for room temperature higher than must temperature

Winery reference of the vat	h1	h2	h3	h4	h5	h6	h7	h8	h9	h10	h11
25 CV	109	118	177	8	19	3					
5 CV	134	139	218	12	12	4	1				
10 V								111	111	4	2

The tanks exchange heat by radiation with the metal sheet that covers the winery and to the East facade of the winery directly opened to outdoors. The direct and diffuse radiation at each hour of the day entering the east façade, Ee, allows determining the solar gains of the tanks nearby. The heat flux gained per square meter is $q = \alpha \cdot E_e$, where α is the absorptivity of the material taken as 0.5.

The metal sheet introduces a radiation shield to the tanks whose temperature is determined based on the heat balance between the solar radiation received by the roof and the convection heat transfer on upper and lower sides of the cover. The temperature of the roof, Tr, allows determining the radiation heat transfer per unit area to each surface of the tank at temperature T_{ti}, after calculating the view factors, $F_{i\text{-}r}$, of the surfaces to the roof, by:

$$q = \alpha \cdot F_{i-r} \cdot \sigma \cdot (T_r^4 - T_{t_i}^4) \quad (5)$$

4 Energy Model for Ervideira

The Ervideira winery, located nearby Évora, Portugal, has 160 ha of vineyard and produces about 800.000 bottles of different wines each year. The following study reports to the period of August 11 to October 12, 2015, when data for tank scheduling and energy use was collected. The winery has 30 tanks for wine fermentation with a total volume of 385 m3 and 16 more for wine maturation. The temperature during the maturation phase might not overcome 20 °C, and the fermentation might occur at 25 °C for red wines and at 15 °C for white wines. A refrigeration water network with a central chiller allows removing the heat gained by the tanks due to the chemical process and to the heat transfer. The chiller has 149 kW of refrigerating power, fluid R410a, overall seasonable efficiency of 2.58, with auxiliary pumping system of 2 kW.

The INETI weather data file for Évora, TRY format, have the data for temperature, solar radiation and wind that allows defining the outdoor conditions during the period of wine production. Évora is a hot place during the summer, with design temperature of 35.2 °C at 97.5% of probability of occurrence. Anyway, the average outdoor temperature during the period of fermentation is about 24 °C, having a low impact on the energy introduced in the vats allocated for the fermentation of red wines.

Regarding the data of the must, the winery has daily data for each tank concerning temperature and density of the must. Therefore, Equation 3 allows determining the daily energy removed from each tank, as depicted on Figure 3 (right panel) for a white wine tank. A real tank receives many different charges of maturated must from different types of grapes, making the density to increase anytime a new must is added. Figure 3 depicts a real evolution of sugars against density, for a white grape variety. On the right panel it shows the energy needed to be removed from a white wine tank, calculated by the variation of density, averaged by the red line tendency.

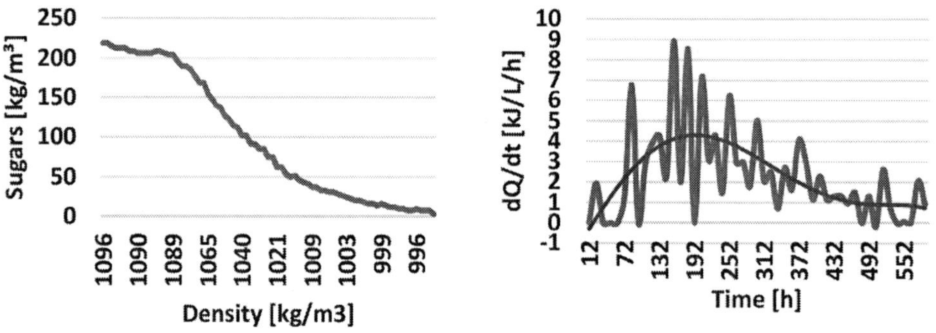

Fig. 3. (left panel) Sugar content and density (data collected daily) for a white grape variety; (right panel) real energy released in a tank due to fermentation and charging

The total heat removed by the chiller from the fermentation tanks is the sum of heat released during fermentation processes and energy transferred by radiation and convection from the environment.

The hourly weather data for Évora and Equation 4 allow determining the hourly convection coefficients for each surface of each tank. Equation 5 allows computing the energy transferred by radiation.

Therefore, knowing the production stage at each vat during the year of 2015 it is possible to calculate the energy transfer by convection and radiation to each vat, and accounting the energy needed to remove in the fermentation process.

Figure 4 depicts the total thermal power transfer into the fermentation tanks during the season of August to October by convections and radiation, not including the fermentation energy. It does not account for the thermal energy of the fermentation process. Anyway, it accounts for the different temperatures of white and red fermentation vats.

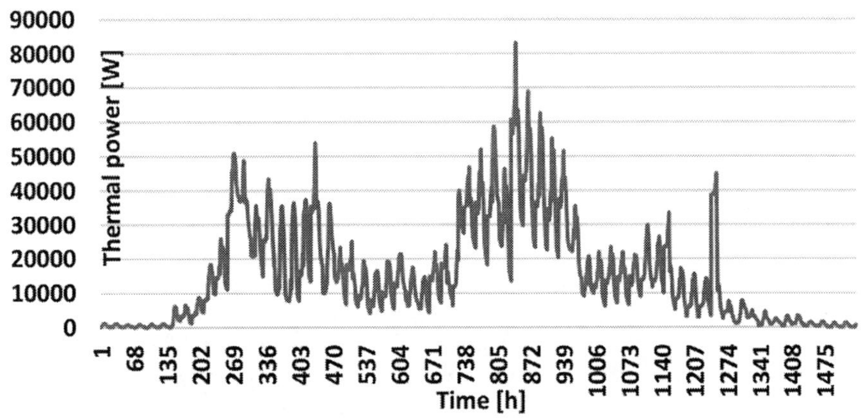

Fig. 4. Total convection and radiation thermal power for all fermentation tanks during the season

In a similar way, Figure 5 shows the thermal energy needed to be removed from all maturation tanks between May and October.

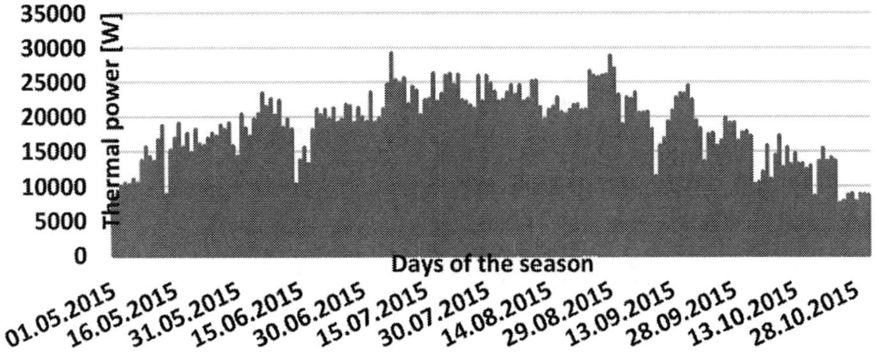

Fig. 5 Total convection and radiation thermal power for all maturation tanks during the season

Accounting the complete season of maturation and fermentation, the energy for maturation is about 77% of the all energy. Concerning the period of about two months when maturation and fermentation happens the total energy for maturation is similar to the one for fermentation tanks. The fermentation of red wines is achieved at a constant temperature of about 25 °C, making the contributions of the heat exchange by radiation and convection to be below 15 %. On the other hand, white wines has lower periods of fermentation at lower temperatures than the red wines, making convection and radiation to attain about 65% of the needs of energy. The effect of radiation accounts for less than 5%.

Figure 6 depicts the complete amount of the thermal energy use during the wine production from May to October 2015. The picture shows in the bottom a white area from 11 August to 12 October regarding the energy for fermentation, and in the top area the thermal energy regarding the convection and radiation. The figure displays a maximum of 136 kW of cooling power, for a maximum production of the chiller installed of 149 kW, being a reference of the quality of the model. During the wine production period the cooling needs are 96.2 MWh, corresponding to about 37.3 MWh of electricity use in the chiller. Considering the pumps, the electricity use is 43.8 MWh, about 54% of the total electricity consumption of the winery, which is within the range of energy consumption for this type of applications [14]. The electricity invoices shows an increase for the same period of 37 MWh, making the model to be in an acceptable margin of error of 15%, common on energy evaluations.

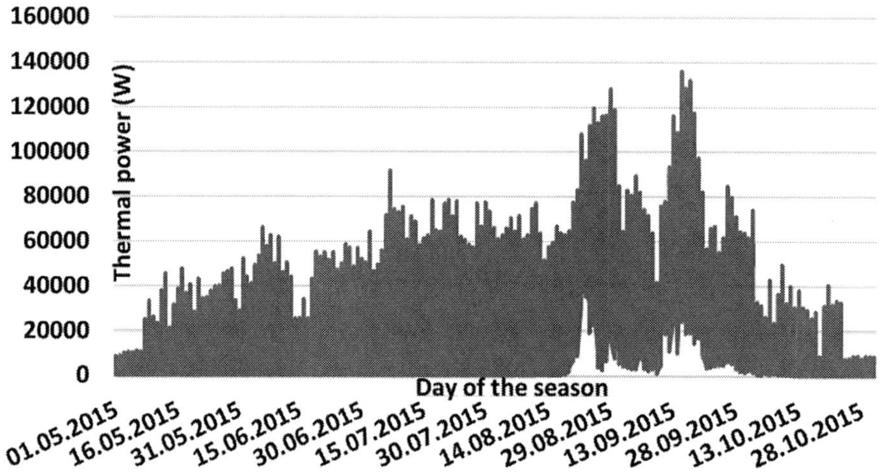

Fig. 6. Total thermal power for all fermentation and maturation tanks including the fermentation power in white.

The model allows testing various improvements in the winery, the best being the insulation of the tanks and the insulation of the metal sheet of the roof. The former allows a reduction of electrical energy use of 18.3 MWh and the roof insulation only 1.1 MWh, both improvements influencing the stability of the temperature in the tanks. Due to the stability of temperature in the tanks, the insulation improvement is more than an energy improvement: it is a quality improvement of the production of wine.

5 Conclusion

The Adega da Ervideira is a winery located near Évora, South of Portugal, producing 800.000 bottles of wine a year. The winery has 46 tanks, 30 of it dedicated to the fermentation process with a total volume of 385 m^3. Évora is a location with hot temperatures during the summer, what makes important to reduce the heat transfer to the tanks.

The electric energy use during the wine production represents about 54% of the annual electricity consumption of the winery. The electricity for wine production totals 43.8 MWh/a, therefore about 5.5 MWh/a per 100.000 bottles.

Regarding the thermal cooling power, the winery needs a maximum cooling power of 136 kW, accomplished by a chiller of 149 kW, or less than 20 kW to produce 100.000 bottles.

In what concerns to improvements in energy, the insulation of the tanks is the most important improvement due to the hot climate of the location. This improvement would reduce the electricity use by about 40% during the wine producing period, or roughly 20% of the electricity paid in a year.

The authors gratefully thank the sponsorship of Fundação para a Ciência e Tecnologia through the Strategic Project UID/EMS/00667/2013 – UNIDEMI.

References

1. N. G. Hopkins, William; P. A. Huner, *Introduction to Plant Physiology* (NJ John Wiley and Sons, 2008)
2. O. Matias and P. Martins, *Biology (in Portuguese)*, (Areal Editores, 2009)
3. P. L. O. Volpe, *New Chemistry (in Portuguese)* **20** (5), 528-534 (1997)
4. S. Yonsel, G. Ç. Bülbül-Caliskan, M. Köni and L. Dagasan, Chem Biochem Eng Quart **21** (4), 395–408 (2007)
5. M. I. Zenteno, J. R. Pérez-Correa, C. A. Gelmi, and E. Agosin, J Food Eng **99** (1), 40–48 (2010)
6. J. M. Sablayrolles, Food Res. Int. **42** (4), 418–424 (2009)
7. A. D. Cardoso, *Wine, from grape to the bottle (in Portuguese)*, (Âncora Edi., 2007)
8. http://extension.psu.edu/food/enology/wine-production/an-introduction-on-low-temperature-fermentation-in-wine-production
9. R. Boulton, Am J of Enol Vitic **30** (2), 152-156 (1979)
10. U. Von Stockar, T. Maskow, J. Liu, I. W. Marison, R. Pati, and C. U. M, , *J of Biot* **121**, 517–533 (2006)
11. American Society of Heating Refrigerating and Air-Conditioning Engineers, "Beverages," in *ASHRAE Handbook—Refrigeration (SI)*, ASHRAE (2006)
12. N. El-Haloui, D. Picque and G. Corrieu, J Food Eng **8**, 17–30 (1988)
13. A. López and P. Secanell, Inter J of Refrig **15** (5), 276–280 (1992)
14. S. Nordestgaard, *Improving Winery Refrigeration Efficiency*, Edition of the Australian Wine Research Institute (AWRI) (2011)

Turning process monitoring of internal combustion engine piston's cylindrical surface

Pawel Twardowski[1], *Maciej* Tabaszewski[2], and *Szymon* Wojciechowski[1,*]

[1] Poznan University of Technology, Faculty of Mechanical Engineering and Management, Institute of Mechanical Technology, Poznan, Poland
[2] Poznan University of Technology, Faculty of Mechanical Engineering and Management, Institute of Applied Mechanics, Poznan, Poland

Abstract. In this paper the monitoring method of turning process of combustion engine piston's cylindrical surface is proposed. During this process, the PCD diamond tool cuts the piston's cylindrical surface with the 3 grooves. The first groove is made in the cast iron insert. In case, when the machining allowances are selected inappropriately, the tool cuts the cast iron insert and consequently generates the reject. The proposed monitoring system based on vibration signals analysis enables the detection of these critical situations and react, in order to maintain the production quality.

1 Introduction

The chip material's removal during machining processes is accompanied by plastic-elastic deformations of work material induced by the ploughing mechanism [1, 2], as well as the thermal phenomena in the cutting zone [3-5]. The appearance of these phenomena is strictly correlated with the generation of cutting forces and vibrations, which in turn can significantly affect the deterioration of the machined surface quality [6, 7]. On the other hand, the force and vibration signals can be applied for the machining process monitoring.

The literature survey shows the examples of vibration signals' application to the monitoring of cutting tools' condition [8]. However, Scheffer et al. [9] proposed the machining process monitoring on the basis of the acoustic emission and cutting force signals. The application of this kind of systems is important, especially during the automated manufacturing, where the human's interference is limited. In above specified conditions, the problem related to the cutting tool's control becomes essential [10].

2 Experimental details

In this paper the monitoring method of turning process of combustion engine piston's cylindrical surface with the application of mechanical vibrations' measurement is proposed.

Figure 1 shows an image of the IC engine's piston with the visible grooves intended for the piston rings.

* Corresponding author: sjwojciechowski@o2.pl

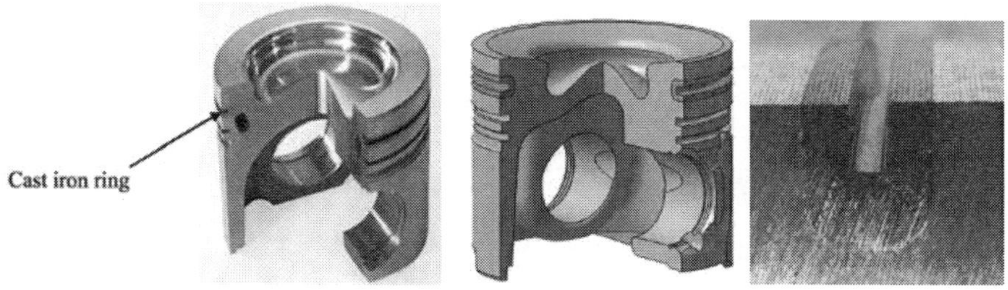

Fig. 1. View of piston internal combustion (IC) engine.

The first groove contains the cast iron insert, in order to strengthen its structure. This groove is fitted to the compression ring, which is responsible for the maintaining the appropriate pressure in the combustion chamber. The subsequent two rings are: scraper ring (the middle one) and oil control ring.

The investigated problem is connected with the positioning accuracy of cutting tool's working part in the special toolholder. In this instance, the cutting edge's positioning tolerances are contained in the narrow range equaled to ± 0.01 mm. Exceeding these values during the machining process can lead to the collision between the PCD tool and cast iron insert. Furthermore, the additional factor which can contribute to the collision is the relatively low accuracy of the cast iron insert submerged inside the aluminum body (dimensional deviation higher than 0.01 mm).

The first groove (together with the cast iron insert), intended for the compression ring is being machined before carrying out the preliminary turning of the piston's cylindrical surface. The longitudinal turning of the cylindrical surface is carried out with the application of PCD (polycrystalline diamond) tool. As it is well known, the PCD tool cannot be applied to the cutting of the ferritic materials, because of carbon diffusion into the iron alloys. Therefore, the machining allowances has been selected in order to avoid the tool's contact with the cast iron ring (which has been previously machined in the finishing conditions). Nevertheless, the instant contact of PCD tool with cast iron ring during machining is possible when the cutting depth is selected inappropriately, or the ring has been manufactured incorrectly. The piston manufacturing process is completely automated, thus the PCD tool's condition is not controlled. In case when the PCD tool will start to cut the cast iron ring, then the rapid growth of tool wear will occur, which can lead to the rupture of the intermetallic bonding between the aluminum alloy and cast iron ring. Consequently, the production shortcoming is being generated.

Therefore, the primary task of the proposed monitoring system is the detection of situation in which the PCD tool cuts the iron ring, and subsequently the immediate stoppage of the process, together with sending the appropriate message to the machine's control system. The primary subsystem of the developed monitoring system is the module intended to measurement and analysis of mechanical vibration signals (Figure 2).

The turning of the piston's cylindrical surface is carried out in the following steps:

- PCD tool is located outside the work piece,
- activation of rotational speed and feed motion,
- initiation of aluminum turning with the selected parameters (AB section–Figure 3),
- approach of tool to the cast iron ring (point B),
- if the machining allowance (depth of cut a_p) is appropriately selected, then tool is not contacting with the section BE (Figure 4),
- if the machining allowance is inappropriately selected, the tool cuts in the areas: BC and DE (tool cuts the cast iron – Figure 5),
- further machining of aluminum (section EF) until the tool's output (point F).

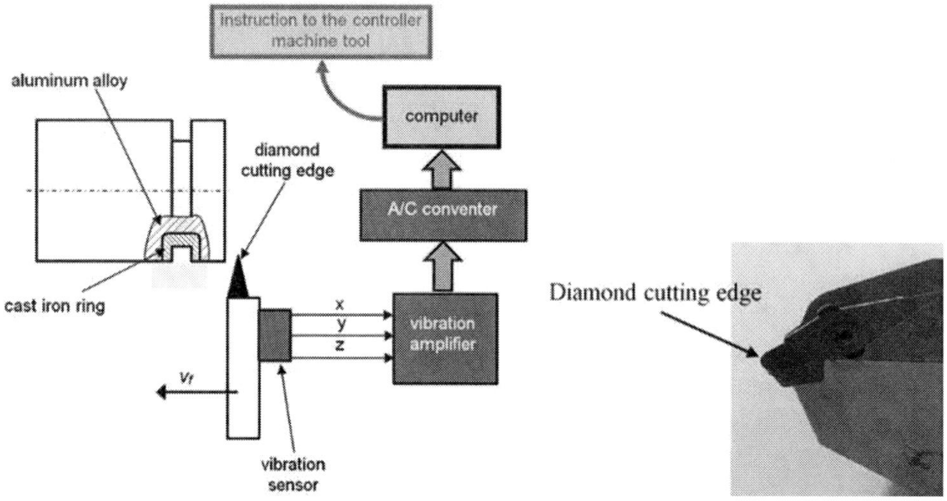

Fig. 2. The scheme of the proposed turning process monitoring system.

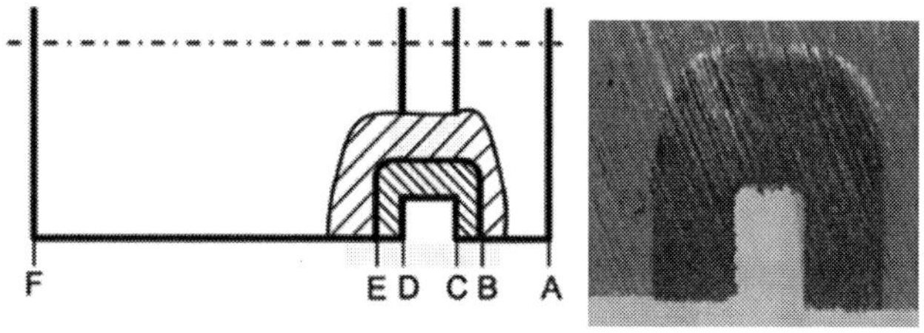

Fig. 3. The phases of turning of the piston's cylindrical surface.

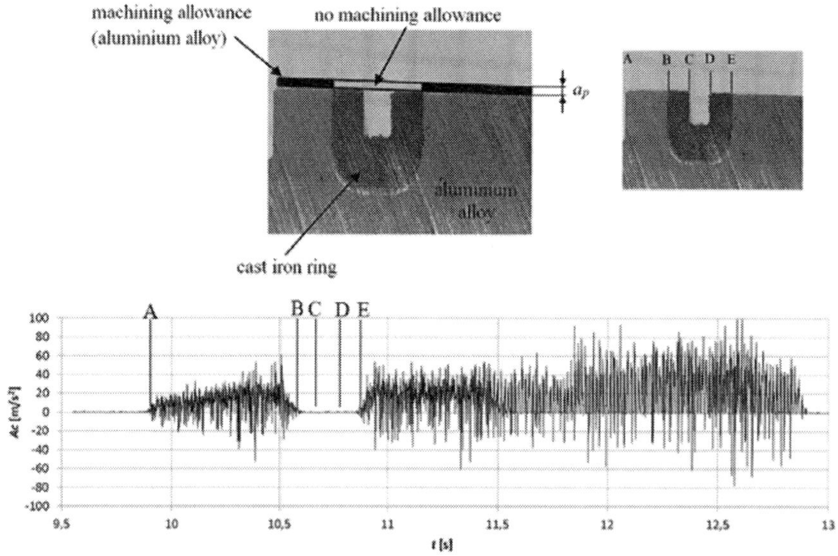

Fig. 4. The collocation of machining allowance and typical vibrations' time course
(n = 2000 rev/min, f = 0.6 mm/rev, a_p = 0.5mm).

 DOI: 10.1051/matecconf/201711210002

During the machining of cast iron (BC and DE sections) with the PCD tool, the rapid growth of vibration amplitude is observed, which is further applied to the evaluation of the process condition. The differences in the vibrations level are large enough to carry out the real-time process monitoring.

Figure 6 depicts the control algorithm of the turning correctness, on the basis of the measurement of vibration signals generated during machining. The control process is carried out in the real-time and cyclically for the each manufactured piston.

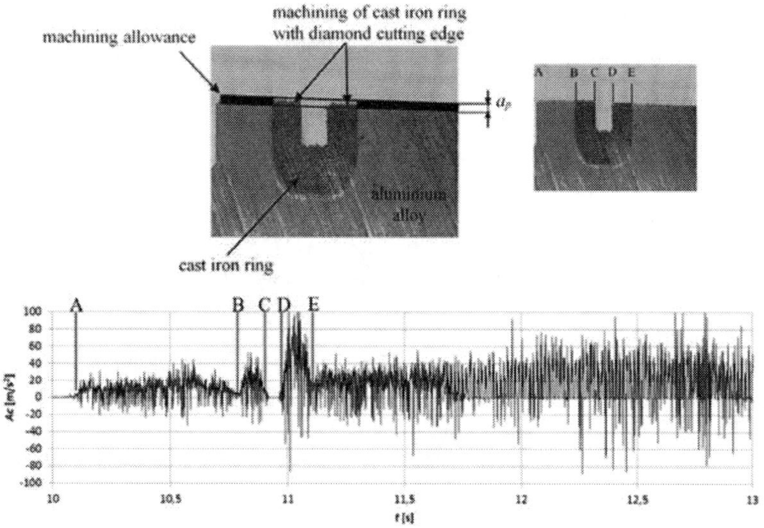

Fig. 5. The incorrect collocation of machining allowance and typical vibrations' time course (n = 2000 rev/min, f = 0.6 mm/rev, a_p = 0.5mm).

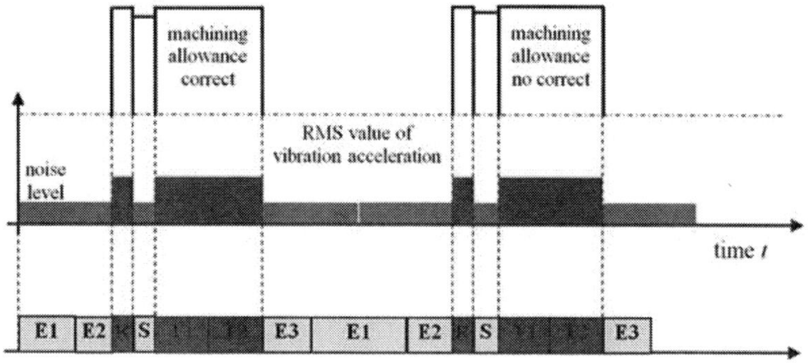

Fig. 6. The control algorithm of the piston's cylindrical surface turning correctness.

The control algorithm is conducted in the following stages:

E1 – The phase of background level detection – the average value of background level is set in the narrow configurable window, moving along the samples, in the time range including the time of the specific phase. The absolute value is determined from the digital signal available in the window. On the basis of this information, the threshold value is being calculated. This value has to be exceeded by the signal's average value in order to recognize this signal as one accompanying the machining process.

E2 – The phase of awaiting the trigger. The signal contained in the specified time window is being preliminarily processed: the removal of constant component and calculation of

signal's absolute value. The average value of signal is set in the narrow configurable window, moving along the samples, in the time range including the time of the specific phase.
R – During this stage, the program passes into the awaiting phase of tool's approaching the area of cast iron ring.
S – Analogically as in E2 phase, the continuous measurement of average vibration values in the set window is conducted. During this stage, the correctness of the piston's production process is evaluated. In case of vibrations with the values averaged in the specified time interval, which significantly exceed the previously set background level, the piston is graded as inappropriately manufactured, otherwise (lower vibration values) the piston is categorized as correctly manufactured.
T1 – The program passes into the awaiting phase during the time needed to the tool's output.
T2 – The signal contained in the specified time window is being preliminarily processed: the removal of constant component and calculation of signal's absolute value. The average value of signal is set in the narrow configurable window, moving along the samples, in the time range including the time of the specific phase.
E3 – During this stage the program is paused for the short time, before the initiation of the next cycle. This pause is needed to the movement of the tool to the starting position, and collecting the next element.
After the execution of the full cycle, the subsequent initiation of the E1 cycle is conducted.

3 The verification of the proposed system and conclusion

The proposed monitoring system has been tested for the various conditions: longitudinal turning with fluids, rotational speed n = 1600 – 2523 rev/min, feed f = 0.6 – 0.8 mm, depth of cut a_p = 0.5 – 1 mm. Figures 7 and 8 depict two characteristic situations, i.e. when the tool is not cutting in the BC, DE areas, and when the cutting occurs in these areas.

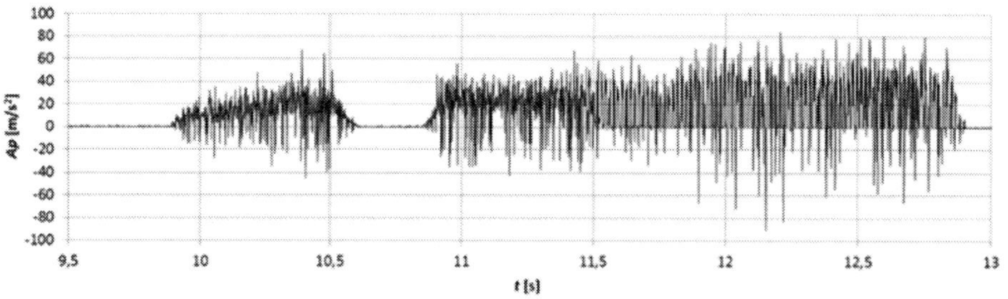

Fig. 7. The exemplary vibration time course, when tool is not cutting in the BC and DE areas (n = 2000 rev/min, f = 0.6 mm/rev, a_p = 0.5mm).

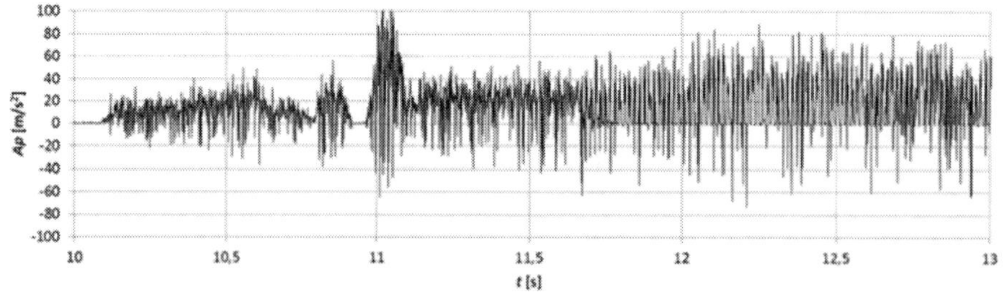

Fig. 8. The exemplary vibration time course, when tool is cutting in the BC and DE areas (n = 2000 rev/min, f = 0.6 mm/rev, a_p = 0.5mm).

The presented courses are significantly different. The observation of vibration signals enables the evaluation of machining allowance selection correctness. Turning with the PCD tool in the BC and DE areas – besides the negative effect on tool life – very often results in the rupture of the intermetallic bonding between the aluminum alloy and cast iron. In this case the piston is graded as a reject and is consequently intended to the re-casting.

In the each of the analyzed cases, the system correctly identified the condition in which the PCD tool contacted with the cast iron ring. Thus, the expensive rejects have been avoided. Therefore, this system can be successfully applied in the automated production of the IC engine pistons.

References

1. S. Wojciechowski, R.W. Maruda, P. Nieslony, G.M. Krolczyk, Int J Mech Sci **119**, 360–369 (2016)
2. S. Wojciechowski, P. Twardowski, Procedia CIRP **1**, 289 – 294 (2012)
3. D. Przestacki, P. Szymanski, S. Wojciechowski, Composites Part A **91** (1), 370–379 (2016)
4. D. Przestacki, M. Jankowiak, Journal of Physics: Conference Series **483** (2014)
5. D. Przestacki, Procedia CIRP **14,** 229–233 (2014)
6. S. Wojciechowski, P. Twardowski, M. Pelic, R.W. Maruda, S. Barrans, G. Krolczyk, Precision Eng **46,** 158-165 (2016)
7. S. Wojciechowski, P. Twardowski, M. Pelic, Procedia CIRP **14,** 113 – 118 (2014)
8. P. Gierlak, A Burghardt, D. Szybicki, M. Szuster, M. Muszyńska, Mechanical Systems and Signal Processing, (On-line)
9. C. Scheffer, H. Kratz, P.S. Heyns, F. Klocke, Int J Mach Tools Manuf **43,** 973–985 (2003)
10. C. Scheffer, P.S. Heyns, Mechanical Systems and Signal Processing **18,** 1219–1242 (2004)

MATEC Web of Conferences 112, 10003 (2017) DOI: 10.1051/matecconf/201711210003
IManE&E 2017

Experimental investigations of LPG use at the automotive diesel engine

Cristian Nutu[1,*], *Constantin* Pana[2], *Niculae* Negurescu[2], *Alexandru* Cernat[2] and *Ionel* Mirica[2]

[1]University Politehnica of Bucharest, Department of Transportation, Splaiul Independentei 313, București, Romania
[2]University Politehnica of Bucharest, Department of Thermotechnics, Engines, Thermal Equipments and Refrigeration Instalations, Splaiul Independentei 313, București, Romania

Abstract. The liquefied petroleum gas has a great potential to improve energetically and pollution performance of compression ignition engines due to its good combustion properties. This paper presents results of the researches carried on a car compression ignition engine with a 1.5 dm^3 displacement, fuelled with diesel fuel and liquefied petroleum gas by diesel-gas method at the operating regimens of 70% and 55% engine load, engine speed of 2000 rpm and for substitute ratios between (6-19)%. A specific objective of this paper is to establish a correlation between the optimum adjustments and the substitute ratio of the diesel fuel with liquefied petroleum gas for the investigated regimens to limit the maximum pressure and smoke level, knock and rough engine functioning and having regard to decrease the fuel consumption and the level of the pollutant emissions.

1 Introduction

In the last years, the internal combustion engines and especially compression ignition engines increased the energetic performances either because of the researcher's continuous work or because of the need of fossil fuels saving, whose resources are exhaustible. The smoke and the smell of exhaust gases imposed reticence of many users to use vehicles equipped with compression-ignition engines. This behavior of the compression-ignition engine has been greatly improved by solutions adopted to limit the environmental pollution and by the adoption of legislative rules. Tougher legislation on emission level imposed the adoption of solutions to reduce it to the legislated levels. One of the recommended solutions to reduce emissions from compression ignition engine is the use of alternative fuels. Of these, liquefied petroleum gas (LPG) is a clean and efficient alternative fuel, its use being possible without major structural changes of the engine.

Liquefied petroleum gas is a fuel which generally consist of a mixture of 2 hydrocarbons, propane and butane, in different ratios, depending on season. Because of its good burning properties and because of the price liquefied petroleum gas is a very good

*Corresponding author: cristi_cmt@yahoo.com

alternative fuel for the compression ignition engine. The LPG properties, comparative with the diesel fuel properties are presented in the Table 1.

Table 1. LPG properties, compared to diesel fuel properties [1].

Properties		Diesel fuel	LPG	
			Propane	Butane
Density	[kg/m³]	800-840	503	-
Vaporization heat	[kJ/kg]	465	420	-
Self ignition temperature	[°C]	355	481	544
Inflammability limits	[%]	0.6-5.5	2.1-9.5	-
A/F ratio	[kg/kg]	15	15.71	15.49
Flame temperature	[°C]	2054	1990	-
Lower heating value	[MJ/m³]	36000	23000	-
Lower heating value	[MJ/kg]	42.5	46.34	45.55
Cetane number CC		40-55	-2	-

Liquid LPG density is lower than diesel fuel density, 503 kg/m³ for LPG and 800-840 kg/m³ for diesel fuel [1], thus the mass of the same volume of fuel is lower, leading to a lower autonomy for the vehicle fuelled with LPG.

The vaporization heat of LPG is lower than diesel fuel vaporization heat (420 kJ/kg to 465 kJ/kg for diesel fuel [1]), allowing to vaporize faster and to consume less local heat in the case of direct injection in the combustion chamber.

The LPG self ignition temperature is higher than the diesel fuel self ignition temperature, 481 °C – propane, 544 °C –butane, 355°C diesel fuel [1], which underlines the worsening of self ignition properties. Therefore fuelling a diesel engine with LPG requires the use of specific methods.

The flame temperature of LPG lower than the diesel fuel flame temperature leads to an important reduction in nitrogen oxides emissions..

The LPG lower heating value higher than diesel fuel lower heating value provides an increase in the amount of heat released during the combustion of fuel for the same fuel quantity.

The extremely low cetane number of LPG underlines its very low self ignition properties. Therefore to fuel a diesel engine with LPG involves specific methods. In this paper the authors chose to fuel the engine by Diesel-Gas method.

1.1 The Diesel-Gas method

The Diesel-Gas method consists of gaseous LPG injection in the intake manifold of the engine, using a conventional LPG fuelling system. Thus in the combustion chamber burns LPG, prior injected in the intake manifold on the intake stroke and ignited by the flames occurred in the pilot diesel fuel sprays. The diesel engine is equipped with another fuelling system for LPG injection. The LPG quantity injected in the intake manifold is tuned with the load and limited by the occurrence of knock (because of the higher compression ratio), and by the smoke emission, which at higher substitute ratios is in high concentrations in exhaust gases. Also the durability of the engine can be affected at higher substitute ratios.

1.2 The state of the research art

Results of a compression ignition engine fuelled with LPG are presented by Qi et. al. in the work [2]. The authors experimented direct injection of a LPG-diesel fuel mixture (with the help of a nitrogen tank) with different proportions: 0, 10, 20, 30, 40 %, leading to a

DOI: 10.1051/matecconf/201711210003

decrease in the pollutant emissions of the engine. In the work [3] the authors decreased the level of the nitrogen oxides emission fuelling the diesel engine with LPG. Although the level of nitrogen oxides emission decreased, the level of unburned hydrocarbons increased. To reduce this emission the authors used a glow plug [3]. The same solution was found by [4], but in this case the engine was fuelled with methane. An increase of the level of unburned hydrocarbon emission was obtained also in [5, 6]. To reduce the level of the unburned hydrocarbons emission the authors used exhaust gas recirculation. In the work [7] by fuelling a four cylinders diesel engine with LPG an increase in the engine efficiency with 4% was obtained, when the engine functioned at full load. At partial loads the efficiency of the engine increased with increasing substitute ratio of the diesel fuel with LPG [7]. Also the smoke emission level was with 40-60% lower than in the case of the standard engine, fuelled with diesel fuel [7]. The reduction of the smoke emission level by fuelling a compression ignition engine with LPG is presented also in the paper [8].

2 Experimental investigations

Experimental investigations were carried out on a K9K 792 1.5 dCi diesel engine, equipped with a LPG fuelling system, corresponding to Diesel-Gas method, at 70% and 55% engine load and 2000 rpm. Laboratory equipments (Figure 1) are: 1 – 1.5 dci diesel engine; 2 – engine cooling system; 3 – engine water cooler; 4 – intercooler fan; 5 – engine angular encoder; 6 –AVL piezoelectric pressure transducer; 7 – diesel fuel injector; 8 – LPG injector; 9 – Turbocharger; 10 – intake air drum; 11 – intake air flow meter; 12 – exhaust gas recirculation; 13 –Schenck E90 dyno; 14 – dyno-engine coupling; 15 –Schenck E 90 dyno cooling water pump; 16 –dyno cooling system; 17 –AVL Dicom 4000 gas analyzer; 18 –AVL Dicom 4000 Opacimeter; 19 –AVL charge amplifier; 20 – PC + AVL data acquisition system; 21 –Schenck E 90 dyno controller; 22 – temperatures displays: a) – exhaust gas; b) – intake air; c) – engine oil; d) – engine cooling liquid; e) – engine oil pressure; 23 – diesel fuel and LPG injection control Laptop; 24 – diesel fuel tank; 25 – diesel fuel mass flow meter; 26 – fuel filters; 27 – high pressure pump for common Rail; 28 –Common Rail; 29 – LPG tank; 30 – LPG mass flow meter; 31 – LPG vaporizer; 32 –LPG ECU; 33 –diesel engine ECU; 34 –intercooler; 35- supercharge pressure measuring system; 36- throttle actuator.

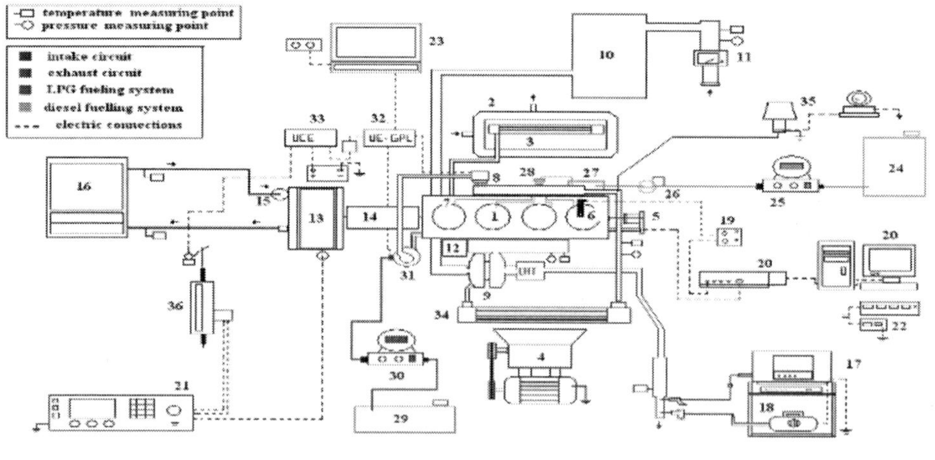

Fig. 1. Test bed diagram.

The engine parameters are presented in the Table 2.

Table 2. Engine parameters

Bore	76 mm
Stroke	80.5 mm
Connecting rod length	134 mm
Maximum power/speed	52 kW @ 3900 rpm
Maximum torque/speed	156 Nm @ 2000 rpm

2.1 The working procedure

First was determined the reference, fuelling the engine with diesel fuel, than the diesel fuel cyclic dose was decreased, and the LPG cyclic dose was increased. The engine power was maintained the same like in the case of standard fuelling with diesel fuel. The energetic substitute ratio has the mathematical relation presented in equation 1.

$$x_c = \frac{m_{LPG}H_{i_{LPG}}}{m_{LPG}H_{i_{LPG}}+m_{dieselfuel}H_{i_{dieselfuel}}} \cdot 100 \ , \qquad (1)$$

where: m_{LPG} [kg]- the LPG dose; $m_{dieselfuel}$ [kg]-the diesel fuel dose; H_i [kJ/kg]- the caloric heating value.

The investigated energetic substitute ratios of the diesel fuel with LPG was between [0-11.85] % for 70% engine load and [0-19]% for 55% engine load. The substitute ratio of diesel fuel with LPG was limited in both cases (70% engine load and 55% engine load) to limit the maximum pressure and smoke level, knock and rough engine functioning. Also to limit the nitrogen oxides emission level and to proper position the combustion near top dead center the injection timing advance was reduced with [1.1-6.7] °CRA, depending on the regimen.

3 Results

The experimental and theoretical investigations led to the following results.

3.1 In cylinder pressure

The pressure inside the cylinder decreased for all the investigated substitution ratios of diesel fuel with LPG. This can be explained by the reduction of the diesel fuel injection advance, to limit the nitrogen oxides emissions level. The figure 2 shows the measured in cylinder pressure for the investigated cases.

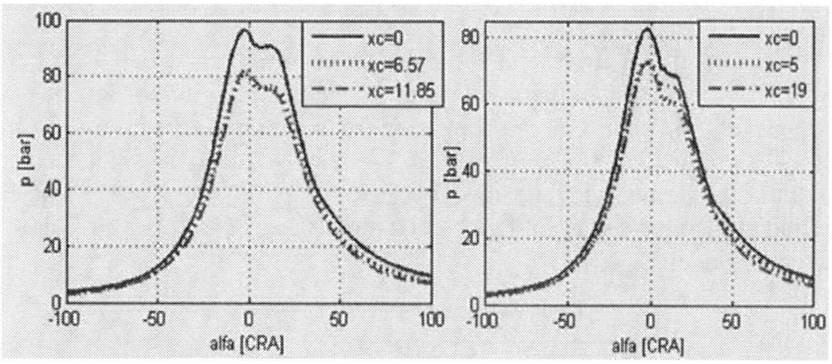

(a) (b)

Fig. 2. The in cylinder pressure for 70% engine load, (a) – 70% engine load, (b) – 55% engine load.

3.2 The maximum rate of pressure rise

The combustion noise, evaluate by the maximum rate of pressure rise decreased for all the investigated substitute ratios of diesel fuel with LPG either for 70% engine load or 55% engine load because the diesel fuel injection timing advance was reduced in order to limit the nitrogen oxides emissions level. The maximum rate of pressure rise is presented in the figure 4.

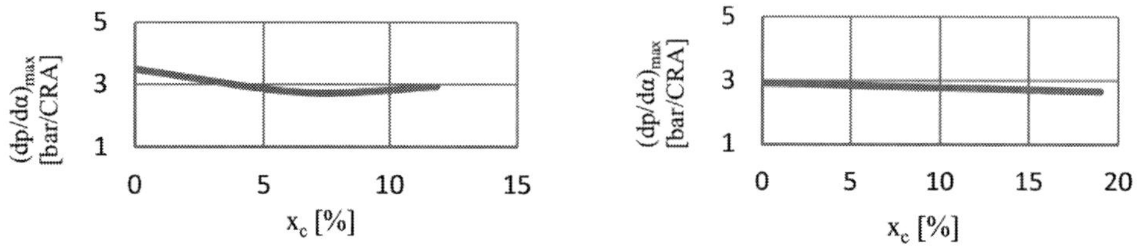

(a) (b)

Fig. 4. The maximum rate of pressure rise, (a) – 70% engine load, (b) – 55% engine load.

3.3 The nitrogen oxides emission level

The nitrogen oxides emission level was lower than in the case of fuelling only with diesel fuel for all the investigated substitute ratios x_c. This was possible because in the case of engine fuelling with LPG by Diesel-Gas method, the cylinder global temperature decreases due to good LPG burning properties and due the reduction of injection timing advance, which has a major role in temperature decreasing. Also the reduction of injection timing advance leads in reducing the time required for forming reaction of nitrogen oxides. The nitrogen oxides emission level is presented in the Figure 5, for all the investigated cases.

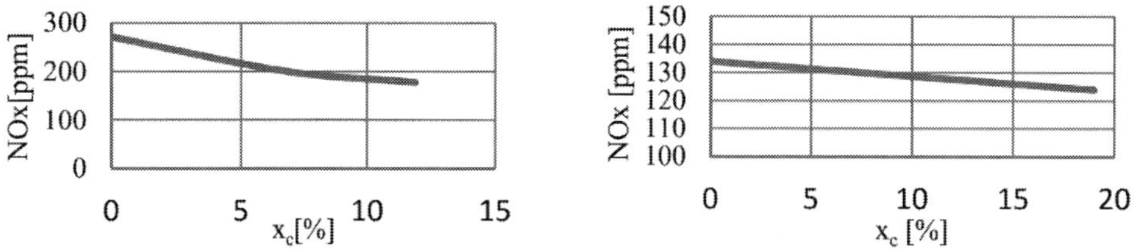

(a) (b)

Fig. 5. The nitrogen oxides emission level, (a) – 70% engine load and (b) – 55% engine load.

3.4 The smoke emission level

The smoke emission, measured with the opacimeter and evaluated by the coefficient of absorption k, had a lower level than in the case of standard engine for low substitute ratios of diesel fuel with LPG and a higher level for the maximum substitute ratio investigated (11.85 for 70% engine load and 19 for 55% engine load). The smoke emission level increases with increasing substitute ratio because of the intake process worsening, LPG being injected in the intake manifold and replacing a part of the intake air. Table 3 presents the smoke emission level.

Table 3. The smoke emission level.

70% engine load		55% engine load	
xc[%]	k[1/m]	xc[%]	k[1/m]
0	0.38	0	0.58
6.57	0.36	5	0.42
11.85	0.91	19	1.13

3.5 The fuel consumption

The energetic specific fuel consumption (Table 4) decreased for all the investigated substitute ratios of diesel fuel with LPG.

Table 4. The energetic specific fuel consumption.

70% engine load		55% engine load	
xc[%]	ce[kJ/kWh]	xc[%]	ce[kJ/kWh]
0	11163	0	13486
6.57	8980	5	13300
11.85	9083	19	12813

4 Conclusions

The experimental investigations led to the following conclusions:

1. The level of the nitrogen oxides emission decreased with ~34% for 11.85% substitute ratio of the diesel fuel with LPG for 70% engine load and with ~7% for 19% substitute ratio and 55% engine load.
2. The level of the smoke emission decreased with ~5% for 70% engine load and with ~27% for the 55% engine load;
3. The level of the maximum pressure decreased when the diesel fuel was substituted with LPG;
4. The maximum rate of pressure rise decreased for all the investigated cases, the maximum value of 2.65 bar/°CRA being recorded for the maximum substitute ratio. When the engine was fuelled only with diesel fuel the maximum rate of pressure rise was ~3 bar/°CRA.
5. The brake specific energetic consumption decreased with ~19% when the diesel fuel was 11.85% substituted with LPG for 70% engine load and with ~ 5% when the diesel fuel was 19% substituted with LPG for 55% engine load.

The authors would like to thank to AVL List GmbH Graz, Austria, for providing the possibility to use the research equipments.

References

1. M. G. Popa, N. Negurescu, C. Pană, *Diesel Engines* **I, II** (in Romanian), (Matrix Rom, Bucureşti, 2003)
2. D. H. Qi, Y. Z. Bian, Z.Y. Ma, C. H. Zhang, S. Q. Liu, J. of Energy Conv. Man. **48** (2), 500-509 (2007)
3. P. Vijayabalan, G. Nagarajan, Jordan J.l of Mech. and Ind. Eng. **3** (2), 105-110 (2009)
4. V. Aesoy, H. Valland, SAE Transactions, 960767 (1996)
5. S. Salman, C. Çinar, C. Hasimoglu, T. Topgul, M. Civiz, TEKNOLOJİ **7**, 3, 455-460 (2004)
6. J. Cao, Y. Bian, D. Qi, Q. Cheng, T. Wu, Proc. of the IMECHE Part D J. of Auto. Eng. **218** (5), 557-565 (2004)
7. C.V. Sudhir, H. Vijay, S. Desai, Y. Kumar, P. Mohanan, SAE Transactions, 2003-01-3087 (2003)
8. C. Pana, N. Negurescu, M. G. Popa, Al. Cernat, U.P.B. Sci. Bull., Series D, Vol. **72** (1), 25-34 (2010)

Investigation of pressure retarded osmosis power production

Nikolaos Taousanidis[1,*], and *Konstantinos* Gavros[2]

[1]Mechanical Engineering & Industrial Design Dpt, Western Macedonia University of Applied Sciences, Koila Kozani, Kozani, 50100, Greece
[2]Electrical Engineering Dpt, Western Macedonia University of Applied Sciences, Koila Kozani, Kozani, 50100, Greece

Abstract. A major source of energy exists where there is mixing between aqueous solutions of different salinities. This energy source is particularly concentrated where fresh water rivers flow on to the ocean. The power, represented by the osmotic pressure difference between fresh water and salt water, may be called salinity gradient power. In this study the pressure retarded osmosis method for the extraction of salinity gradients' energy is investigated, main problems and difficulties are pointed out and finally the whole subject is justified with experimental results.

1 Introduction

After decades of anticipation we may now be close to the ability to extract substantial amounts of energy from the sea. Serious efforts are underway in several countries to use the "free" energy of tidal oscillations, surface waves and mainly that resulting from thermal gradients. Even though a small portion of the solar radiative flux is directly deposited in the sea, the total energy budget of the ocean is large compared to man's needs. Some of the incident solar energy establishes and maintains thermal gradients. Water is evaporated, condensed and precipitated nonuniformly and establishes salinity differences.

This energy source exists at the interface between waters of differing salinities and is particularly concentrated where fresh water rivers flow on to the ocean. The power, represented by the osmotic pressure difference between fresh water and salt water, may be called salinity gradient power. Salinity power has certain features that make it attractive as a source of energy. It is large and untapped. Its employment could have little environmental impact. Compared with other sources of ocean energy, its density is high. It is renewable due to water evaporation by the sun and subsequent precipitation.

Experiments indicate that the equivalent pressure head between 0.5 molar seawater and fresh water is about 24atm. This pressure is equivalent to a 240m water head. One may conceptualize it is a 240m waterfall at the mouth of every river. Various techniques for the extraction of this form of energy has been proposed or developed. In this study we deal with the pressure retarded osmosis (PRO) technique [1].

*Corresponding author: taousan@teiwm.gr

2 Principle of pressure retarded osmosis

Osmosis is spontaneous transport of solvent molecules from a dilute solution to a more concentrated solution across a semipermeable membrane. In 1976, Sidney Loeb proposed pressure retarded osmosis (PRO) as a novel application for osmotic membranes [1]. The oil prices at that time were low and therefore the interest in further development of this technology was low. However, in 1997 Thorsen and Holt started a project together with the Norwegian company Statkraft [2, 3] and the world's first PRO pilot plant started in 2009.

PRO utilizes the osmotic pressure difference between two sources of different salinity to perform work and thus produce energy. The relationship between Reverse Osmosis (RO), Pressure Retarded Osmosis (PRO) and Direct Osmosis (DO) is illustrated in Figure 3.1 for an ideal semipermeable membrane. In DO, the membrane separates two solutions of unequal concentration, typically a salt solution from pure water. A flow of water from the water side into the salt solution takes place due to the difference in osmotic pressures between the two solutions. If an increasing pressure is applied to the salt solution, the magnitude of the water flow will decrease until no flow occurs when the applied pressure equals the osmotic pressure difference across the membrane. The regime in which there is a flow of water into a pressurized salt solution is known as pressure retarded osmosis. If the water volumetric permeation rate through the membrane, $\Delta\dot{V}$, is divided by membrane area, we have the water permeation flux, J_w. The relation between water permeation flux and applied forces, in the ideal case, is given by [8]:

$$J_W = A\ (\Delta\pi - \Delta P)\ , \tag{1}$$

where A is the membrane water permeation coefficient , $\Delta\pi$ the osmotic pressure difference across the membrane and ΔP is the hydrostatic pressure difference across the membrane.

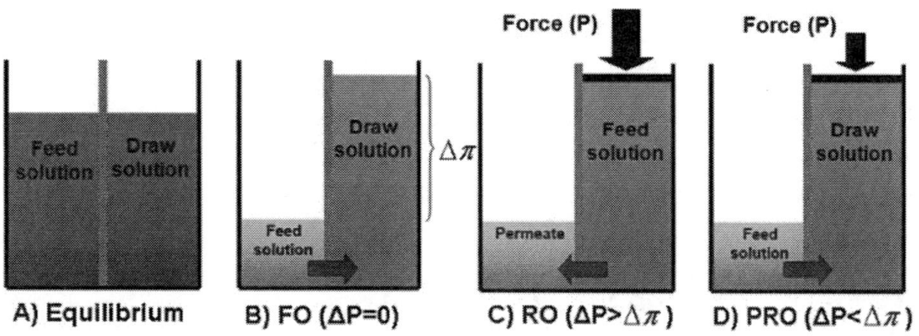

Fig. 1. Classifications of osmotic pressures: a. osmotic equilibrium, b. DO ($\Delta P = 0$), c. RO ($\Delta P > \Delta\pi$) and d. PRO ($\Delta P < \Delta\pi$) [5]

It is clear from equation (1) that if ΔP=0 we have direct osmosis (DO), for which $J_w = A\ \Delta\pi$. However if $0<\Delta P<\Delta\pi$ we have Pressure Retarded Osmosis (PRO), so defined because the direction of permeation flux is still the same as in direct osmosis but the flux is decreased as the hydraulic pressure increases (in contrast to reverse osmosis (RO), a well developed technology, for which $\Delta P>\Delta\pi$ and in which an increase in ΔP increases the flux). PRO water permeates against the hydraulic pressure gradient, i.e., the flux is "uphill". The subsequent depressurization of the permeate through a hydroturbine-generator set would produce power by what may be described as an "osmotic waterfall".

The maximum power per unit flux is obtained at the maximum hydrostatic pressure under which PRO takes place, i.e., the osmotic pressure difference $\Delta\pi$. However, when ΔP is close to $\Delta\pi$ the transmembrane water flux is very small, and a very large membrane area would be required, resulting in a high capital cost, being preferable, to operate PRO systems under maximum power per unit membrane area conditions.

The power per unit membrane area, W that can be generated in PRO is the product of the water flux across the membrane and the hydrostatic pressure of the salt solution [8]:

$$W = J_w \, \Delta P = A\,(\Delta\pi - \Delta P)\,\Delta P \tag{2}$$

By differentiating equation (2) with respect to ΔP, it can be shown that W reaches a maximum when $\Delta P = \Delta\pi/2$. Substituting this value for ΔP in equation (2) yields

$$W_{max} = A\,\Delta\pi^2/4 \tag{3}$$

According equation (3) the maximum power in a PRO system is directly proportional to water permeability coefficient A, and thus high flux membranes are preferred. It is also proportional to the square of the osmotic pressure difference. This arises because increasing the osmotic pressure of the salt solution increases both the optimum pressure at which the system operates (i.e. $\Delta\pi/2$) and the water flux through the membrane at that pressure.

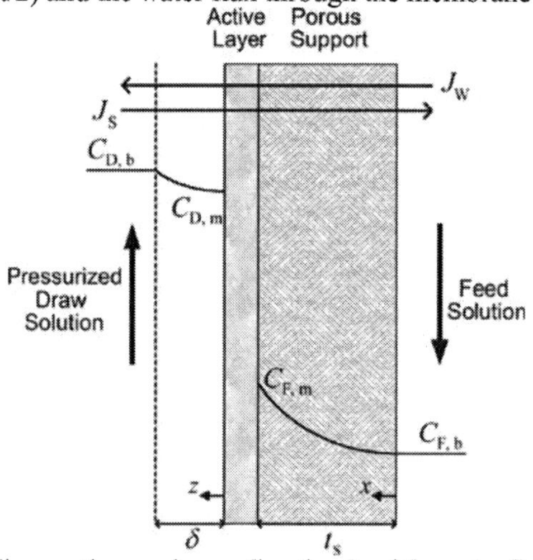

Fig. 2 Concentration profile over the membrane, direction J_w of the water flux and the salt flux Js are also shown [7]

The case is particularly different in real membranes as a series of reducing effects exist. They include salt leakage due to salt permeation in the 'opposite' direction and concentration polarization. While the first one is not considered a major problem, although reduces the effective osmotic pressure difference, the latter one is more severe. Concentration polarization occurs externally on the two sides of the membrane due to accumulation or depletion of solutes (external polarization) and internally in the necessary bulky porous substrate (internal polarization). Disturbing the boundaries on the two sides of the membrane we can minimizes external polarization. However this does not affect internal concentration polarization. Figure 2 schematically represents the results of these reducing effects upon concentrations and consequently upon osmotic pressures.

Equation (1) can then be modified to

$$J_w = A\,(\Delta\pi_{eff} - \Delta P) = A\,(\pi_{D,m} - \pi_{F,m} - \Delta P) \quad . \tag{4}$$

Equation (4) is not a useful expression for the water flux across a PRO membrane because the osmotic pressure is not known. As Lee et al. [8], outline the water flow can be accessed by the equation.

$$J_W \approx A\left(\frac{\pi_2}{1+BK} - \Delta P\right) \tag{5}$$

where B is the salt permeation coefficient and K is a measure of the resistance to salt transport in the porous substrate. In this model, used as a basis also by other researchers [6], [7], membranes parameters A and B could be determined from RO while K can be obtained from DO measurements where $\Delta P = 0$ and pure water is used on one side of the membrane, i.e. $C_{F,m} = 0$. In this study the above model will be used and proved experimentally for novel design membranes, in the direction to overcome the commercial viable limit of $5 W/m^2$ [4]. To date, no commercial membrane has demonstrated a power density larger than $3.5 W/m^2$.

3 Experiment

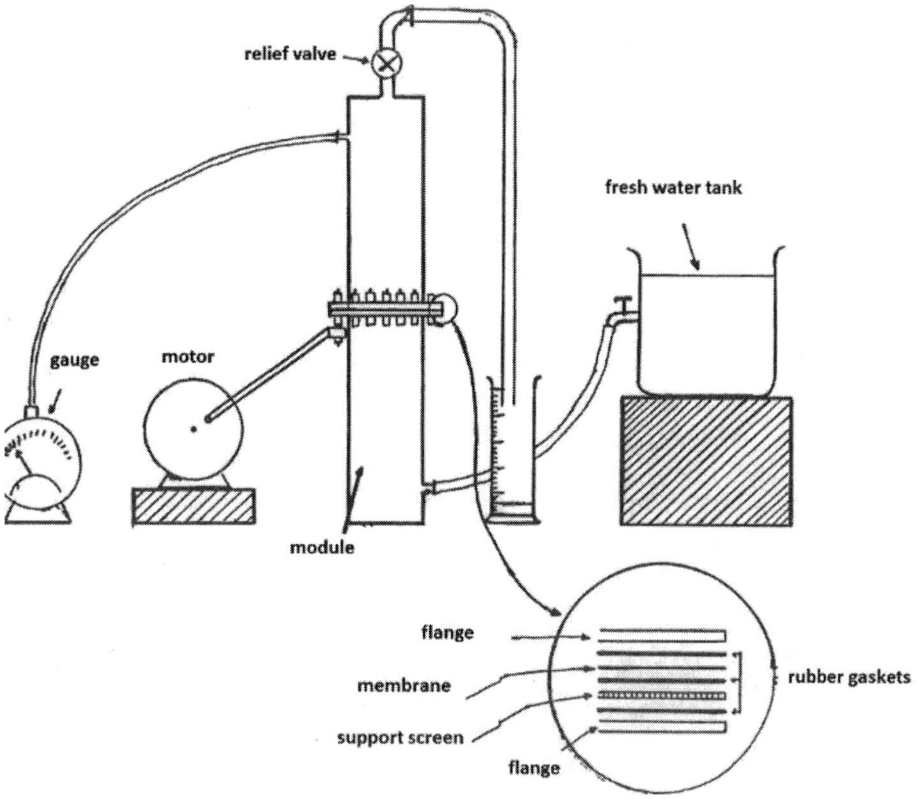

Fig. 3. Schematic drawing of the apparatus and circuit used for the experiment.

The basic apparatus used for the experiment was a cylindrical stainless steel pressure vessel. It contains the membrane, a support stainless steel screen for the membrane and three rubber gaskets, as shown in Figure 3, to make the whole assembly leak - proof. All of them (membranes, screen and gaskets) are between the two flanges, each one of each chamber, the whole module consisting of the upper and lower chamber. The upper chamber is the salt water chamber and the lower chamber is the fresh water chamber. The salt water chamber is provided with a pressure gauge. Also the outlet of the diluted solution from this chamber is provided with a relief needle valve. The outlet we have just named, lead the diluted solution to a graduated glass tube. Thus, we could measure the permeate volume and with the aid of a simple chronometer to calculate the permeate rate $\Delta\dot{V}$ (ml/min) of fresh water through the membrane into the salt water chamber. The fresh water chamber is connected via a plastic tube to a fresh water tank, which continuously supplies water to the module to achieve a continuous operation.

Considering the external polarization, we put glass marbles in the upper chamber and plastic balls in the lower. The former because of their weight and the latter because of the water buoyancy come both very close to the membrane. We connected the apparatus via an eccentric aim to a low speed electric motor. When the power is on, the vessel is shaken vigorously by the arm resulting to elimination of the boundary layers which assist to the concentration polarization, moving both the glass marbles and the plastic balls.

Two HTI-sheet Cellulose Triacetate membranes of a very thin asymmetric cellulose acetate salt rejecting skin and an open - structured polysulfone porous substrate, were investigated. Both have salt rejection index of 95,6% for a 0,5% NaCl aqueous solution and operating conditions: pH range 1-11, temperature range 0°C-85°C, operating pressure 0-80 bar for the HTI-CTA-011 and 0-50 for HTI-CTA-013 and water capacity is 45 l/m^{2}h for the first and 70 l/m^{2}h for the second. As the latter proved much better we present the results just for the HTI-CTA–013.

The system was fresh water – 3.5 % wt. Aqueous NaCl solution. The experiment was undertaken, after the membrane was freed from the conservative medium (glycerine) by emerging it in the water for about 60 minutes and the original backing was carefully removed. That, improves the short - term efficiency of the membrane (time ranges of minutes) but causes a quick deterioration of the rest of the membrane.

The membrane has been used for a period of not more than two hours every day for about one week. With these time constraints we tried to minimize salt leakage and concentration polarization. The disturbance caused by the stirring effect of the combination of the low speed motor and the plastic and glass balls, interferes the growth of any boundary layer of different concentration than that in the chamber. The membrane requires about 50 minutes to reach the steady state flux, because there is not salt leakage when we start the experiment and the salt - water solution has not yet been diluted.

The effects of concentration polarization on water flux are illustrated in Table 1.

Table 1. Fluxes for both cases of continuous and interrupted agitation and power outputs

	INTERRUPTED AGITATION			CONTINUOUS AGITATION			
Time (min)	Readings (flow in ml)	flow rate (ml/min)	flux of water (ml/m^2 min)	Readings (flow in ml)	flow rate (ml/min)	flux of water (ml/m^2 min)	Work watts x 10^{-6}
0	-			-			
10	10	1.00	0.02325	10	1.00	0.02325	2.49567
20	30	2.00	0.0465	30	2.00	0.0465	9.98268
20	58	2.80	0.0651	58	2.80	0.0651	19.56606
40	92	3.40	0.07905	92	3.40	0.07905	28.84996
50	130	3.80	0.08835	130	3.80	0.08835	36.0375
60	170	4.00	0.093	170	4.00	0.093	39.99307
70	202	3.20	0.0744	208	3.80	0.08835	36.0375
80	227	2.50	0.0581	248	4.00	0.093	39.99307
90	249	2.20	0.05115	288	4.00	0.093	39.99307
100	272	2.30	0.05347	329	4.10	0.09533	41.95664
110	300	2.80	0.0651	368	3.90	0.0906	57.8964

When the stirring - shaking was stopped, the water flux decreased markedly. This is the result of a higher salt concentration at the interface between the salt solution and the membrane than in the bulk solution and the development of a diluted salt solution layer near the membrane at the fresh water side. Resumption of stirring led to a sharp rise of water flux, surpassing the steady state value. This surge was real and reflected the condition in which virtually the entire osmotic pressure gradient was instantaneously imposed across the salt rejecting layer before the salt concentration profile could again develop within the membrane. However, the stirring cannot prevent the decline of the flux after a certain time that is the concentration polarization cannot be avoided ultimately.

Results show that the maximum flux is 0.09533 ml/cm²min. For NaCl concentration of about 3.5%, molality is around 0.6 and for 15°C, the theoretical osmotic pressure is about 26.4 bar. Thus, the water permeation coefficient can be calculated as follows:

$$A_{013} = 0{,}09533/26{,}4 = 0{,}00361 \text{ ml/cm}^2 \text{ min bar}. \tag{6}$$

Knowing the A values and the fluxes from Table 1 we could form last column with the estimated values of power output from the following equation:

$$W = J_w \Delta P = J_w J_w/A = J_w^2/A. \tag{7}$$

Thus the maximum power output is 41,95664 pWatts. These values are not the representative of the expected output from a salinity gradient source but illustrate the great significance and potential of these renewable energy forms.

4 Conclusions

1. Membrane properties under PRO conditions can be calculated using a model taking into account concentration polarization. Required parameters - water permeation coefficient A, salt permeation coefficient B, and resistance to salt transport in the membrane substrate K - can be obtained from RO and DO experiments.
2. Concentration polarization is a major problem in PRO. External concentration polarization can be minimized by stirring to reduce the thickness of boundary layers.
3. Internal concentration polarization can only be reduced to an acceptable level by using membranes with an open substrate. Without due regard for internal concentration polarization, it is unsafe to project PRO performance from RO performance.
4. Because PRO operating pressures with seawater/fresh water salinity gradient can be lower than those typically used in RO, compaction of the membranes due to the hydrostatic pressure gradient can be smaller in PRO, and PRO fluxes in the absence, of concentration polarization can be higher than high pressure RO data would suggest.
5. There is a serious prospect in energy extraction from salinity gradients if membrane problems could be resolved, attaining a production of minimum 5W/m².

References

1. S. Loeb, , J Membrane Sci **1**, 49–63 (1976)
2. K. Gerstandt, K.-V. Peinemann, S.E. Skilhagen, T. Thorsen, T. Holt, Desalination **224**, 64–70 (2008)
3. T. Thorsen, T. Holt, J Membrane Sci **335**, pp 103–110 (2009)
4. S.E.Skilhagen, Desalin Water Treat **15**, 271-278 (2010)
5. I. L. Alsvik and M. Hägg, Polymers **5**, pp 303-327 (2013)
6. A.Achilli, T.Y.Cath, A.E.Childress, Journal of Membrane Science **343**, 42–52 (2009)
7. N.Yip, A.Tiraferri, W.Phillip, J. Schiffman, L.Hoover, Y.Kim, and M.Elimelech, Environ. Sci. Technol. **45**, 4360–4369 (2011)
8. K.L. Lee, R.W. Baker, H.K. Lonsdale, J Membrane Sci **8**, 141–171 (1981)

Energy efficiency regulation for industrial products and manufacturing

George-Vlad Badea[1], *Gabriel* Frumuşanu[2,*], *Nicolae* Badea[3], and *Alexandru* Epureanu[2]

[1]BGV Energy Consultancy, Pascal Aristide Street 49-51 Bucharest, Romania
[2]Dunărea de Jos University of Galaţi, Department of Manufacturing Engineering, Domnească Street 111, 800201, Galaţi, Romania
[3]Dunărea de Jos University of Galaţi, Department of Automatic Control & Electrical Engineering, Ştiinţei Street 2, 800210, Galaţi, Romania

Abstract. The paper deals with the energy efficiency of industrial products or manufacturing as compared to the framework legislative measures implemented by EU through the Eco-design and Energy Labeling Directives. The Eco-design implementing measures such as taking into account all phases of the life cycle (manufacturing, transport, use, disposal), as well as the essential environmental aspects (consumption, materials, emissions, waste, etc.) for each phase, are considered. The implementing measures should have no significant negative impact on the functionality, health and safety, affordability and industry's competitiveness, as well as they should not impose proprietary technology on manufacturers and not be an excessive administrative burden for them. In this paper a method for implementing Legislative measures concerning the Eco-design and Energy labeling of industrial product is proposed. It grounds on the analysis of particular interest versus general interest relation, for each product case. Method application consists in products classifying relative to the two types of interest, followed by a voluntary agreement between manufacturers operating on market and EU. Finally, the paper presents the limits and possibilities for Eco-design of industrial products and manufacturing industry.

1 Introduction

At European level, the right policies are in place, developing on three main pillars: security of supply (energy security), environment (sustainable development), and internal market (competition). In developing these competitive, sustainable, and secure energy systems throughout Europe [1], EU has adopted and is implementing a short-term Europe 2020 policy. In order to achieve the short-term goals covered by Europe 2020 (the 20/20/20 climate and energy targets – 20% reduction of GHG emissions, 20% share of RES and 20% reduction of primary energy use by 20% improvement in energy efficiency), the energy legislative framework has been updated and designed for effective implementation. Currently in force, the general framework for promoting energy efficiency is set by the

* Corresponding author: gabriel.frumusanu@ugal.ro

Eco-design and Energy Labeling for consumer equipment and industrial products, the EPBD Directive [2], as well as the latest regulatory development promoting energy efficiency - the EED Directive [3] from 2012.

The Directive on Eco-design [4, 5] has the purpose of reducing the environmental impact of products on their entire life cycle. It establishes a framework for the European Commission to elaborate compulsory standards regarding the environmental impact, on energy efficiency, for different products groups whose use has impact on energy consumption. The products covered by the directive are labeled in the following way:

- *Energy-using products (EuPs)*: products that use energy to fulfill their function as well as industrial products that transfer, use and generate energy
- *Other energy related products (ErPs)*: products that do not use energy to fulfil their function but have a significant impact on the energy consumption and can therefore contribute to saving energy (such as windows, insulation materials, etc.).

The directive on Eco-labeling provides two types of compulsory products requirements:

- *Specific requirements*: setting the limit values such as maximum energy consumption or minimum quantities of recycled material
- *General requirements*: which do not set specific consumption limits but may contain *information requirements*, that the products are "*energy efficient*" or "*recyclable*".

The Eco-design Directive is meant to be used together with the Energy Labeling Directive complementing the requirements for product energy labeling and implementing energy labeling measures. The energy labeling process runs in parallel to the establishment of mandatory requirements.

2 Legislative measures concerning the Eco-design and Energy labeling of industrial product

If with regard to the manufacturers of consumer goods the eco-design regulations require them to decrease the energy consumption of their products by setting minimum energy efficiency standard, in the *industrial sector* the European Commission promotes mostly *voluntary agreements* with manufacturers in order to reduce the energy consumption of their products. These are self-regulatory agreements signed by industry in order to meet specific criteria of the Directive on Eco-design and are further assessed and monitored by Commission.

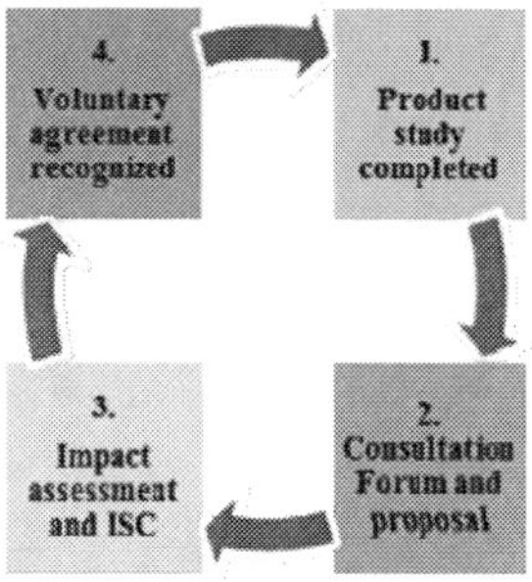

Fig. 1. The voluntary agreement process

The process of implementing voluntary agreements (Figure 1) implies as first phase that every group of products undertake a market study, the state of technology and other relevant aspects. Then the voluntary agreement or the self-regulatory initiative is subject to consultations on the implementation measures. After the consultation with the members of the Consultation forum, the Commission evaluates the impact and if the group of products is included in the Eco-design Working Plan (thus fulfilling the above-mentioned requirements), then the proposal is submitted to the other internal groups of the Commission for consultation (Inter-Service Consultation). Depending on the feedback, the proposal is accepted or reviewed. The last step is to recognize the voluntary agreement if the requirements of the Eco-design Directive are met.

3 Proposal for energy efficiency regulation technique

Although energy is one of the most important cost factors in industry, *in products manufacturing there is no legislative measure for energy labeling so far.* Reducing energy demand in product manufacturing is a challenge because the requirements for electricity in the production and processing industry depend on the energy need of the machines that are run and energy required for the processing of materials [6].

The objective of energy efficiency concept is to reduce the energy consumption of machines and processes continuously and extensively, for all production process – from product design to production planning and engineering, all the ways to production itself and the related services. Maximum energy efficiency in production is possible only by a good choice of machines [7] and when all components interact perfectly one with another [8, 9].

Energy labeling of machines used in production processes and process optimization for working machines can constitute an important step in promoting sustainability in manufacturing, in order to reduce electricity demand.

The regulation technique for energy efficiency that we propose starts from the above-mentioned prerequisites (energy is an important cost factor in industry and there is no legislative measure in this area yet). The regulation technique is designed to be used starting from the first step of the voluntary agreement process, namely the *product study completed* stage. It also has the purpose of creating a database necessary to be able to go easily through the other three stages of the process (consultation forum and proposal, impact assessment and ISC and voluntary agreement and recognition).

The regulation technique of energy efficiency is referring to the following aspects:

1. Establishing the limits that separate the acceptable products from the unacceptable products on the market from an energy point of view
2. Implementing the regulation

These aspects are addressed in detail below.

3.1 Delimitation of the acceptable products from the unacceptable products

Setting the limit that separates the products acceptable on the market from the products unacceptable on the market starts from the following observations:

- Energy efficiency is not the single criteria for assessment of a product. Indeed the product can be appreciated by multiple criteria, such as costs (acquisition, operation, and maintenance), productivity, quality, waste recycling levels, etc. Each criterion presents a certain interest.
- Certain criteria represent the interests of market actors, generically called *particular interest.* Even the actors not acting in the market (inhabitants of a city, population of a country) may have interests in terms of product. This is called *general interest.*

The line that separates acceptable from unacceptable products on the market is determined by the extent to which these interests can be harmonized. The boundary can be established only at the level of a certain utility area, as only at this level the products are comparable. For instance, let us consider that energy efficiency represents the general interest and economic efficiency the particular interest. In the diagram from Figure 2, each dot represents a certain product, and the coordinates of the dot are the average values of the two criteria corresponding to that product. The shape of the dotted line that separates the cloud of points (PQRS line, Figure 2) highlights the level of coherence of the two interests and shows if there is a need for regulation or not. In each given case, the regulatory line splitting the cloud of points is MN. This line splits the cloud of dots in two domains, acceptable and unacceptable (i.e. PMNQ is the acceptable and MNRS the unacceptable one). The shape and position for the regulation line is the objective for going through the last three stages of the voluntary agreement process: Consultation forum and proposal, Impact assessment and ISC, and Voluntary agreement and recognition.

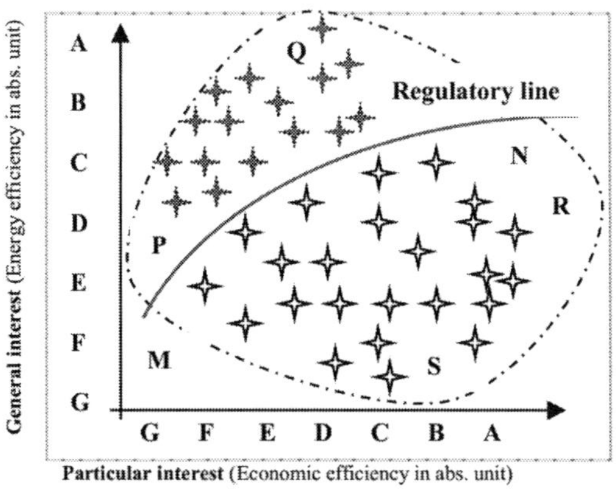

Fig. 2. General interest (energy efficiency) vs. Particular interest (economic efficiency) and regulatory line: Acceptable domain – MNQP; Unacceptable domain – NRSM

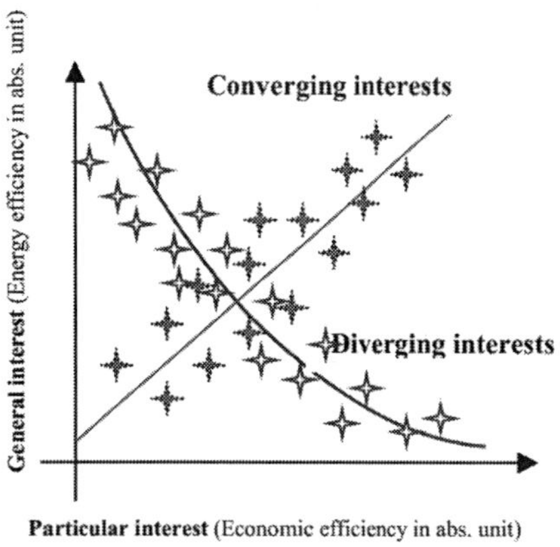

Fig. 3. Converging and diverging interests

Figure 3 presents the two possible scenarios: the case in which the interests are converging (depicted in green), thus, a regulation is not necessary, and the case in which the interests are diverging and a regulation is required (depicted in red).

3.2 Regulation implementing

Firstly, it should be noted that for the practical implementation of the regulation, it is more convenient to know the class to which the product belongs, after each considered criterion. Thus, both enouncement of the regulation and verification of its application become easier.

The classification algorithm from below is a particularization of the classification algorithm proposed in [10]. According to this, it covers two steps, namely the *identification of the utility area* and the *delimitation of the classes*.

In order to build a classes system, a precursory action needs to be accomplished. It consists in defining utility area, materializing the connection between *product* and *task* the last one being addressed in general sense. For example, *task* might mean the excavation of a quantity of earth, obtaining a certain microclimate in a residential building or manufacturing a certain type of surface (exterior plane, cylindrical interior, thread etc) with a given precision and a specified roughness.

If looking now to a product, then we can assign it two axis, namely *Effort* axis and *Result* axis (Figure 4). Along the first one, as input, there is the supply (consisting in materials, energy, time and information) and as output – the corresponding cost, for example. As second axis entry there are task ordered, while the result is the task accomplishment (by product operation).

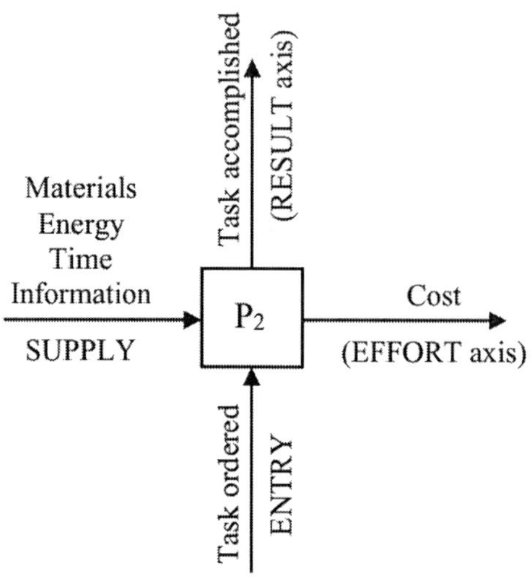

Fig. 4. Product flowcharts

Products classification after their energy efficiency is also approached based on the general efficiency definition, as ratio between *Result* and *Effort*.

The manner of defining and representing the classes for the products covering a given task is presented in Figure 5. In the depicted diagram, each case in which this task was accomplished is represented as a point *M*. *The effort* involved during this task accomplishment and *the result* issued from this effort, both measured through appropriate indicators, are giving *M* coordinates. The locus of points corresponding to all possible values of parameters for a given product defines a close domain assigned to it in the graphic representation (for example, see the domains of products *P2*, *P3*, and *P4*, in Figure 5). By introducing a third, circular coordinate axis, the efficiency is reflected through the angle made by *OM* segment with *Effort axis*, as suggested on the picture.

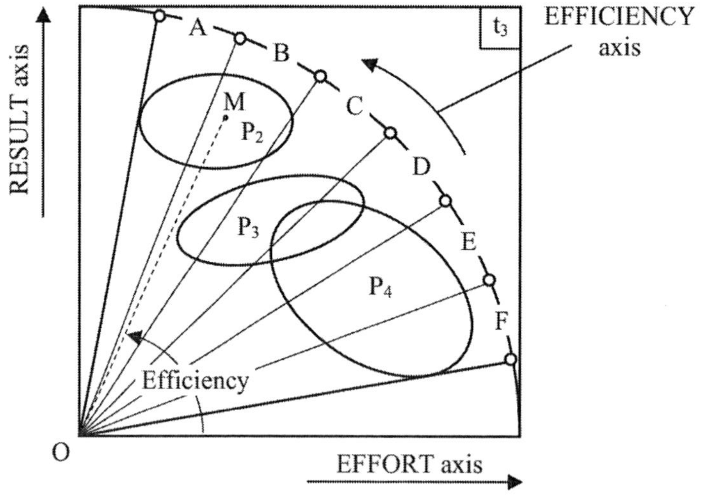

Fig. 5. Effort – Result – Efficiency diagram for a given task

The efficiency classes can be defined now by dividing in several sectors the angular domain comprised between the lines corresponding to extreme efficiencies (inferior/superior), as they can be revealed at moment when the product study is completed.

In order to generate the diagram in Figure 5, the numerical values of the two criteria for each product (from the products that belong to the same utility area) should be known.

For regulation, implementation is necessary to extend the classification to both criteria. For instance, regarding the products represented in Figure 5, the classification has to be extended to economic efficiency as well as to energy efficiency. As result, the cloud of dots could be framed in a rectangle with dashed lines, where each side is given by the extreme values of the criterion. Then, each side is divided into classes, denoted A, B, etc. This way, the products could have a double tag and the regulation statement & verification of its application could be exclusively made based on labeling.

5 Conclusions

This paper proposes a technique for development and implementation of regulation on eco-design for industrial products, applied on the basis of products classification by two criteria: particular interest of market actors and general interest of non-market actors.

The proposed technique could bring the following benefits:

- It can be applied to any industrial product and/or process in which the respective product participates, including product manufacturing processes.
- The development and implementation of regulation is eased because it is based on double labeling of products (energy and economic labeling). This way precludes the situations where one of the interests with regards to the product is disregarded.
- The technique is efficient because the differences between the products are revealed with sufficient accuracy if the number of classes is large enough. The increase in the number of classes can highlight finer performance differences between the products.

The proposed technique application may lead to improvement of products performance, avoiding market saturation with underperforming products.

The authors acknowledge the use of the research facilities developed by RO-054 Project, financed through EEA Grant 2009 program, in performing the research presented in this paper.

References

1. European Commission, An energy policy for Europe, COM 1 final, Brussels, 2007
2. EPBD Directive - Directive 2010/31/EU of the European Parliament and of the Council of 19 May 2010 on the energy performance of buildings, OJ L 153, 13-35
3. EED Directive - Directive 2012/27/EU of the European Parliament and of the Council of 25 October 2012 on energy efficiency, OJ L315, 1-56
4. http://ec.europa.eu/enterprise/policies/sustainable-business/documents/eco-design/legislation/framework-directive/index_en.htm#h2-1
5. http://www.eceee.org/ecodesign/
6. Recommendation 2016/2125 EU, on guidelines set by industry self-regulation
7. K. Tanaka, Energy Policy **36**, 2887– 2902 (2008)
8. K. Tanaka, Energy Policy **39**, 6532–6550 (2011)
9. A.P. Kuznetsov, P. Blau, H.-J. Koriath, M. Richter, Procedia CIRP **46**, 340 – 343 (2016)

Study of the cycle variability at an automotive diesel engine fuelled with LPG

Liviu Nemoianu[1], *Constantin* Pana[1], *Niculae* Negurescu[1], *Alexandru* Cernat[1,*], *Dinu* Fuiorescu[1], and *Cristian* Nutu[2]

[1]University Politehnica of Bucharest, Department of Thermotechnics, Engines, Thermical and Frigorific Equipment, Splaiul Independentei 313, Bucharest, Romania
[2] University Politehnica of Bucharest, Department of Automotive Engineering, Splaiul Independentei 313, Bucharest, Romania

Abstract. Liquid Petroleum Gas is a viable alternative fuel for diesel engines due to its ability of emissions and fuel consumption reduction. Combustion variability at LPG diesel engine is analysed for maximum pressure, maximum pressure angle and indicated mean effective pressure. Combustion variability is influenced by the increase of LPG cycle dose which lead to the increase of the cycle variability coefficients values, but without exceeding the admitted values that provide diesel engine reliability. Analysis of COV values establishes the maximum admitted values of LPG cycle.

1 Introduction

Nowadays, in the global content regarding the impact on the environmental of the gas emissions resulted from the fossil fuels combustion, including the impact of diesel engine on environment, special measures are taken in order to reduce as much as possible the pollution especially in the urban areas. This issue was discussed at the 21st Session of the Conference of the Parties (COP21) from the 2015 Paris Climate Conference. Moreover, since the C40 Events-C40 Mayors Summit 2016 (in December 2016) pollution issue becomes a priority for some capital cities (like Paris, Madrid, Athens and Mexico City) which want to be free from automotive diesel engines by 2020 and for the whole France by 2025. The progressive diminution of the oil reserves all over the world highlights the necessity of using alternative energy derived from durable and renewable resources, [1]. The use of the alternative fuels, even in partial substitution, may represent a viable solution to reduce the pollutant emissions, but it's important to maintain the normal engine manoeuvrability for the new fuelling solution, [1]. The Liquid Petroleum Gas (LPG), an alternative fuel used on large scale at spark ignition engines, offers the advantage of an already existing distribution infrastructure in many European countries, [1]. The LPG can be use as alternative fuel at diesel engines with remarkable reductions of soot and NO_X emissions, [1]. Ying Wang [2] studied the combustion in a diesel dual-fuel premixed charge compression ignition combustion engine with LPG as ignition inhibitor showing the

[*] Corresponding author: cernatt@gmail.com, cernatalex@yahoo.com

influence on the appearing positions of the maximum cylinder pressure and on pressure peak values when LPG mass fraction in DME (dimethyl ether)/LPG mixture increases. At the same time the NO_X emission decrease slightly, [2]. Surawski [3] shows that the use of LPG leads to the pre-mixed combustion of LPG-air mixture followed by the in-cylinder maximum pressure and pressure rise rate increase. Polk [4] limits to maximum the propane substitutions so the NO_X and smoke emissions decrease. Boretti [5] converts a heavy duty truck diesel engine to operate with diesel fuel-LPG, showing the reduction of particulate and CO_2 emissions. Tira [6] uses the LPG (100% propane) for diesel engine fuelling and calculate the coefficient of variability (COV) of Indicated Mean Effective Pressure (IMEP) to evaluate the combustion stability from a minimum of 200 cycles. Tira observes an accelerated combustion of the mixture and an increased in-cylinder pressure [6]. At low load the LPG addition causes a reduced combustion duration, being established the maximum fraction of LPG addition with acceptable cycle combustion variability, COV_{IMEP} < 5%, [6]. Elnajjar [7] uses a single cylinder, naturally aspirated, four strokes with indirect injection, water cooled Ricardo E6 diesel engine. Different composition of LPG has lower impact on the engine efficiency and has a significant influence on the combustion noise level [7]. Tira [8] analyses the running of a LPG-diesel dual fuelled engine operated with rapeseed methyl ester and gas-to-liquid diesel fuels, proofing that the combustion cycle variability remains acceptable in running conditions at which 60% of liquid fuel was substituted by LPG. Selim [9] presents the performance of dual fuel engine running on natural gas/LPG by using pilot fuel derived from jojoba seeds. Selim uses 100 cycles for cyclic variability evaluation and determinates the combustion noise defined by the in-cylinder maximum pressure and maximum pressure rise rate, [9]. The use of jojoba fuel and LPG leads to diesel engine performance increases, the combustion noise and the cyclic variability of the combustion being reduced, [9]. Ashok [10] shows the performance, combustion and emission characteristics of the LPG diesel dual fuel engine by using intake gaseous fuel. The use of LPG fuel at a high compression ratio engine leads to the increase of combustion noise and of the cyclic variation of the combustion, [10]. Renald [11] shows that at the fuelling with 60% of diesel and 40% of LPG the thermal efficiency of the diesel engine increases with about 5% when and the mechanical efficiency was increased about 5%. Poonia [12] use a single cylinder diesel engine fuelled with LPG at the load regime of 3.7 kW at speed of 1500 min^{-1}, recording 100 in-cylinder pressure diagrams for cyclic variation evaluation. At diesel fuelling the COV of IMEP is below 6% and the COV of maximum pressure is under 2% comparative to dual fuelling. Once with the rise of engine load, up till 80%, the COV values rise too for diesel-LPG fuelling [12]. For all engine loads, the COV of peak pressure is higher than diesel mode and is influenced by the irregular combustion process of air-LPG mixture, [12]. For engine loads bigger than 60% and exhaust gas recirculation (EGR) use raised values of COV are registered because the combustion velocity is reduced by the EGR [12].

In this paper the combustion cycle variability of a K9K diesel engine (from Dacia Logan MCV van) fuelled with LPG by diesel gas method is analyzed. At LPG fuelling important reductions of soot (with 60%), NO_X (with 40%) and HC (with 80%) emissions were obtained, [1]. The LPG injectors are connected to the inlet manifold and electronic controlled by a special ECU connected back to back with the main engine ECU (Engine Control Unit).

2 Methodology of the experimental research

The LPG cycle quantity is evaluated by an energetic substitute ratio x_c taking into consideration the lower heating values of LPG and diesel fuel. The value of x_c=0 [%] is set up as reference regime (engine fuelled with only diesel fuel). The coefficients of cycle variability (COV) used to evaluate the intensity of cycle-to-cycle variation are calculated with the relation (1):

$$(COV)a_i = \frac{\sqrt{\dfrac{\sum_{i=1}^{n}\left(a_i - \dfrac{\sum_{i=1}^{n} a_i}{n}\right)^2}{n-1}}}{\dfrac{\sum_{i=1}^{n} a_i}{n}} \cdot 100\% \qquad (1)$$

where n is the number of cycles, a is the parameter of which variability is study and is defined by indicated mean effective pressure IMEP, maximum pressure p_{max} and the angle where maximum pressure occurs, α_{pmax} in the cycle number "i", [1, 13]. The COV of maximum pressure is suitable at cycle-by-cycle variability evaluation for regimes with injection timing closer to the value of timing for MBT (Maximum Brake Torque). The COV of maximum pressure angle, α_{pmax}, reflects the combustion variability during the first phase of combustion. The $(COV)_{IMEP}$ shows how the engine responds to the combustion variability [1, 13]. The accepted limit value of $(COV)_{IMEP}$ defines the limit of in-cylinder mixture leaning and reflects the variability of flame development during the first phase of combustion. The COV values for maximum pressure and IMEP are taking into consideration when the limit value of x_c is established. If $(COV)_{pmax}$ and $(COV)_{IMEP}$ don't exceeded 10% the normal automotive manoeuvrability is assured [1, 13]. Thus, the cycle variability study has a significant relevance when alternative fuels, like LPG, are used to diesel engine in order to evaluate the response of the engine to the variability of the LPG combustion. To establish if the LPG is a suitable alternative fuel for diesel engines and can assure the normal mechanical operation of the engine and good drive ability, not only the pollutant emissions decreasing, such a study is mandatory.

3 Results

The regime of 2000 min^{-1} and 85% load, a regime of full load and maximum torque brake, offend used in the operation of the commercial vehicles, was investigated at dual fuelling. For a correct cycle variability study, a number of 150 consecutive pressure diagrams have being measured and registered by AVL Indimodul 621 data acquisition system and analyzed in terms of maximum pressure, maximum pressure angle, IMEP and angles of 5% mass fraction burned (MFB) for different substitute ratios, x_c. The cycle dispersion registered between values of maximum pressure, in consecutive cycles, is shown in the Figure 1. Next to the values find it in the average cycles, the minimum and maximum values registered in the individual cycles are shown. The COV values for maximum pressure are: 0.56 at x_c=0; 0.57 at x_c=2.5; 0.65 at x_c=6.7 and 1.44 for x_c=28. The coefficient $(COV)_{pmax}$ don't exceed values of 1.44%, but the increase tendency of $(COV)_{pmax}$ with the rise of x_c values can be take into consideration, especially for x_c=28%. The value registered for x_c=28% is 2.5 times bigger comparative to the value of standard engine. The increase of LPG cycle dose leads to the increase of maximum pressure, maximum pressure cycle dispersion and of the dispersion of successive running cycles. The angle of maximum pressure, α_{pmax}, is registered closer to the TDC, with is related to a rapid, brutal combustion, Figure 2.

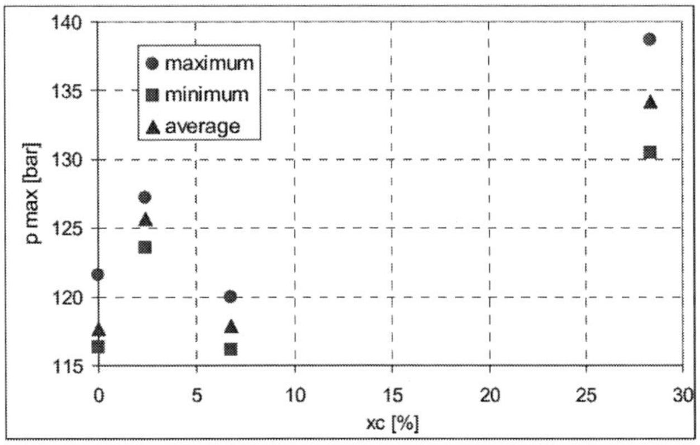

Fig. 1. Cyclic dispersion of maximum pressure values versus x_c at 2000 min^{-1} speed and 85% load.

The COV values for angle of maximum pressure are: 12.5 at x_c=0; 9.7 at x_c=2.5; 12.4 at x_c=6.7 and 12.7 for x_c=28.

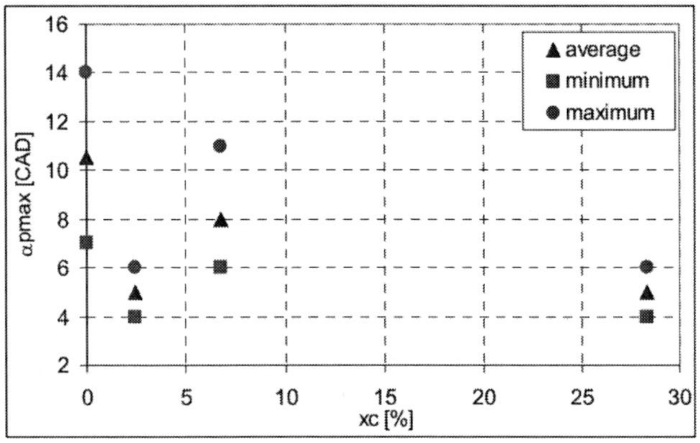

Fig. 2. Cyclic dispersion of angles of maximum pressure versus x_c at 2000 min^{-1} speed and 85% load.

The variability of maximum pressure angle slightly decreases comparative to reference regime for the maximum x_c, but for averaged x_c values the dispersion is similar to diesel fuelling. For x_c=2.5% the $(COV)\alpha_{pmax}$ is improved comparative to diesel fuelling. The increasing tendency of $(COV)\alpha_{pmax}$ for some x_c can be related with the increased cycle variability during the initial phase of combustion, COV of α5%, Figure 4. The COV values for IMEP are: 1.1 at x_c=0; 1.57 at x_c=2.5; 0.7 at x_c=6.7 and 1.3 for x_c=28. The limitation of $(COV)_{IMEP}$ leads to the reduction of the variability of flame development during the premixed stage of combustion. Variability of the flame development during combustion premixed phase may be affected by the increase of x_c. For x_c=28% $(COV)_{IMEP}$ is 1,4 times bigger comparative to reference value, the limitation of x_c to 28% can maintain the normal engine running at this regime. If the admitted limit is $[(COV)_{IMEP}]_{max}$=10%, then appears no limitation issue at x_c =28% for this operating regime. If the substitute ratio is limited to x_c=28% the maximum pressure and maximum pressure rise rate are limited at p_{max}=138 bar and $(dp/d\alpha)_{max}$=6.68 bar/CAD, respectively, values measured in individual cycles. This phenomenon is related with a combustion process closer to TDC (Top Dead Centre) and leads to the increasing of the maximum pressure rise rate during the combustion process.

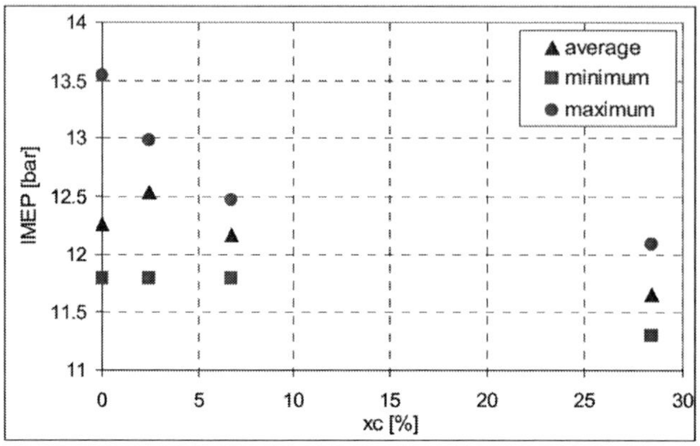

Fig. 3. Cyclic dispersion of IMEP values versus x_c at 2000 min^{-1} speed and 85% load.

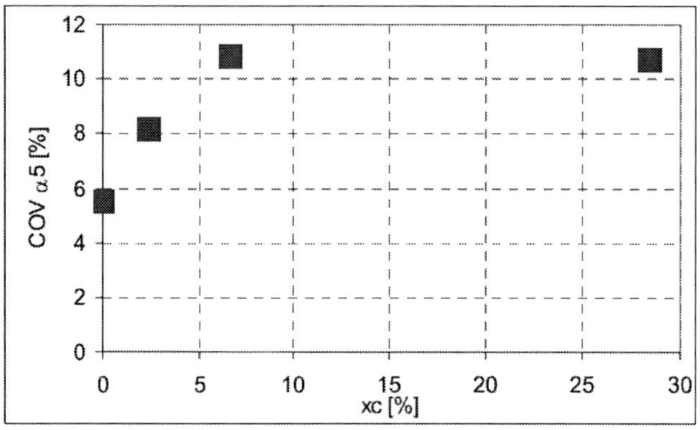

Fig. 4. Coefficient $(COV)_{\alpha 5}$ of MFB versus x_c at 2000 min^{-1} speed and 85% load.

The initial very lean mixtures of air-LPG achieved in-cylinder before combustion starts will influence the engine mechanical running and the cycle variability at the beginning of the combustion process, evaluated by variability of the angle of 5% of mass fraction burned. The $COV\alpha_{5\%}$ increases from 5.6, for diesel fuelling to 10.7 at maximum LPG dose, Figure 4. This increase is in correlation with the raised cycle variability of maximum pressure and of IMEP and in order to assure the normal manoeuvrability of the engine at LPG fuelling the x_c value must be limited at 28% for the investigated regime.

4 Conclusions

The experimental study of dual fuelled diesel engine leads to the following conclusions:
The diesel and LPG engine fuelling leads to the increase of in-cylinder maximum pressure and of its cycle variability coefficient $(COV)_{pmax}$. The increased values of $(COV)_{pmax}$ at x_c=6.7…28% are related with the increased dispersion between the angles of maximum pressure and upper values of $(COV)\alpha_{pmax}$. This aspect may reflect the increase of the combustion cycle variability during the premixed combustion phase at the increase of x_c over 28%. The LPG fuelling leads to the increase of maximum pressure rise rate $(dp/d\alpha)_{max}$ and of the cycle variability, evaluated by $(COV)_{IMEP}$. In order to assure a good drive ability of the dual fuelled diesel engine the LPG dose, x_c, is limited by the maximum values or by the maximum variation interval of same running characteristic parameters of the engine, as

is follows: if the maximum admissible level of the cycle variability is $(COV)_{IMEP}$=1.3%, then the substitute ratio will be limited to x_c=28%; if the maximum pressure rise rate is limited to the value $(dp/d\alpha)_{max}$=6.68 bar/CAD, then the LPG substitute ratio will be limited at x_c=28%, value that will leads to a maximum acceptable in-cylinder pressure level of 138 bar that assure the normal engine operation. LPG fuelling leads to the increase of the cycle variability of the beginning of combustion process, with the rise of $(COV)\alpha_{5\%}$, fact that influence the cycle variability of IMEP, $(COV)_{IMEP}$. The cycle variability study shows that the influence of LPG fuelling on combustion process and on engine mechanical operation are minimal if the limited x_c values are respected . Thus, the study relives that the LPG is a viable alternative fuel for diesel engines and the benefits of pollutant emissions decreasing can be also assured for a normal engine running. The values of the calculated COV parameters show a good manoeuvrability of the LPG fuelled diesel engine and define the LPG as a viable alternative fuel for diesel engines.

Partial, the work has been funded by the Sectoral Operational Programme Human Resources Development 2007-2013 of the Ministry of European Funds through the Financial Agreement POSDRU/159/1.5/S/134398. This work has been founded by University Politehnica of Bucharest through the "Excellence Research Grants" Program, UPB-GEX. Identifier: UPB-EXCELENTA-2016 Experimental Researches of hydrogen use at diesel engine. Contract number 38/26.09.2016. The authors address special thanks to AVL GmbH Graz Austria for providing the necessary equipments.

References

1. C. Pana, N. Negurescu, M. G. Popa, A. Cernat, UPB Sci. Bull., Series D, **72** (1) (2010)
2. Y. Wang, H. Liu, Z. Huang, et al., Energy **96**, 278–285 (2016)
3. N.C. Surawski, B. Miljevic, et al., Fuel **133**, 17–25 (2014)
4. A. C. Polk, C. D. Carpenter, K. K. Srinivasan, et al., Fuel **132**, 135–148 (2014)
5. A. Boretti, F.P.T. **113**, 97–108 (2013)
6. H. S. Tira, J. M. Herreros, et al., Fuel **118**, 73-82 (2014)
7. E. Elnajjar, M. O. Hamdan, et al., Renew. Energy **56**, 110–116 (2013)
8. H.S. Tira, J.M. Herreros, et. al., Energy **47** (1), 620-629 (2011)
9. M. Selim, et. al., Renew. Energy **33**, 1173-1185 (2008)
10. B. Ashok, S. Denis, C. Ramesh, C., Kumar, Alexandria Engineering Journal **54**, 105–126 (2015)
11. R. Thomas, P. Somasundaram, Energy Procedia **14**, 1427 – 1438 (2012)
12. M.P. Poonia, Y. B. Mathur, J.E.R.D. **7** (2), 746-755 (2012)
13. J. B. Heywood, *Internal Combustion Engines Fundamentals* (McGraw-Hill Book Company, 1988)

Theoretical calculation model of the propulsion engine admission parameters for different functional conditions

Vlad Vornicu*, *Tudor* Ulian, *Edward* Rakosi and *Sorinel* Talif

"Gheorghe Asachi" Technical University of Iasi, Faculty of Mechanical Engineering, Department of Mechanical Engineering and Automotive Engineering, Romania

Abstract. Inlet pressure as an engine parameter has a decisive role in the effective power and torque. Spark ignition engines achieve the load adjustment by changing the shutter position, which represents a quantitative adjustment. Obtaining a theoretical model to demonstrate the dependence between the engine speed and load will make possible an optimization of the combustion process.

1 Introduction

The pressure at the end of the intake stroke, currently denoted p_a, represents the pressure of the mixture between the fresh load and residual gasses from the cylinder at the end of the intake stroke when the cylinder gasses volume is maximum. This pressure value affects other parameters of the characteristic operating cycle for internal combustion engines, with an impact on the overall performance, by influencing the effective power, effective torque and the specific fuel consumption. This article presents a calculation model for p_a pressure, developed by the authors, which takes into account, different from other known models, the load and engine speed modification, during a variation of the powered vehicle operating mode. In this way, the vehicle speed, fuel consumption or engine economic regime can be more precisely estimated [1].

2 Proposed Theoretical Model

In the context of this model, it is considered that the efficiency of cylinder filling, commonly referred by many authors as filling ratio, defined according to the literature as the charge amount of fresh mixture ratio effectively retained into the engine cylinder and also by considering the gasodynamic and thermal losses, it's not affected by the engine speed variation, but it is affected by the engine load, main consideration for spark ignition engine.

This consideration can be revealed by the following expression:

$$\eta_v \neq f(n) . \tag{1}$$

* Corresponding author: vladd.alex@yahoo.ro

By taking into consideration the recommended values for η_v on full load [2], and also experimental results [3], it was subsequently adopted a filling coefficient $\eta_v = 0.87$.

Further, the situation of p_a intake pressure variation was developed, at the same time with the engine load modification. In order to do this, one of the p_a intake pressure expressions, recommended by some authors, was adapted, by introducing the amendment of filling coefficient modification, dependent on the engine load, denoted with x.

In this way:

$$p_a(\chi) = \frac{p_0}{T_0} \cdot T_a \cdot (1 - \varepsilon^{-1}) \cdot (1 - \varphi_{af} + \gamma_r) \cdot \eta_v \quad . \tag{2}$$

By introducing further $\Phi_0(p_0, T_0) = \frac{p_0}{T_0}$, as a parameter which defines the environmental conditions, regarding pressure and temperature, the equation (2) becomes:

$$p_a(\chi) = \Phi_0 \cdot T_a \cdot (1 - \varepsilon^{-1}) \cdot (1 - \varphi_{af} + \gamma_r) \cdot \eta_v \ . \tag{3}$$

In this expression, φ_{af} quantifies the influence of cylinder additional filling, having the well-known designation of additional filling degree, and γ_r represents the residual gasses coefficient, being a direct criterion of cylinder residual gasses degree of cleaning evaluation.

In order to simplify the development of the proposed model, it is considered $\varphi_{af_{med}} \cong 0.16$ as a positive average and constant value of the additional filling degree, not affected by the engine operating mode changing.

Regarding the dependence between the inlet pressure and engine load calculation, a linear function is introduced in the model:

$$\eta_v(\chi) = \alpha_n \cdot \chi + \beta_n \quad , \tag{4}$$

which shows, for spark-ignition engines, the filling coefficient variation in case of engine load modification, for different engine speeds.

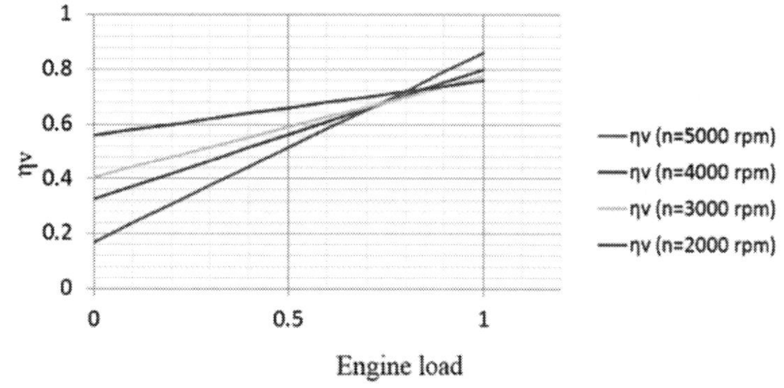

Fig. 1. Dependence between filling coefficient and engine load

Establishment of α_n and β_n coefficients from expression (4) was made by analyzing a data series from the literature for η_v, both in terms of, as described above, the maximum corresponding value of the total engine load and the minimum engine load.

The representation of this variation, for an engine speed interval which ensures a stable operationality is shown in Figure 1, as a family of lines with different tendencies.

In this way, for the proposed model it is obtained the mathematically expression which highlights the intake pressure variation together with the engine load modification, for different values of the engine speed:

$$p_a(\chi) = \Phi_0 \cdot T_a \cdot (1 - \varepsilon^{-1}) \cdot (1 - \varphi_{af} + \gamma_r) \cdot (\alpha_n \cdot \chi + \beta_n) \,. \tag{5}$$

With α_n, β_n values determined this way, the intake pressure expression $p_a(\chi)$ becomes:

$$p_a(\chi) = \Phi_0 \cdot (1 - \varepsilon^{-1}) \cdot (0{,}84 + \gamma_r) \cdot T_a(\chi) \cdot \begin{cases} 0.69 \cdot \chi + 0.17, & n = 5 \cdot 10^3 \\ 0.475 \cdot \chi + 0.325, & n = 4 \cdot 10^3 \\ 0.368 \cdot \chi + 0.406, & n = 3 \cdot 10^3 \\ 0.2 \;\; \cdot \chi + 0.56, & n = 2 \cdot 10^3 \end{cases} [rpm] \,. \tag{6}$$

Graphical analysis of Figure 1 highlights that for low engine loads, η_v has high values in low engine speeds domain, the lower rate of airflow contributing on gasodynamic loss lowering. By increasing η_v, the filling duration is contributing in case of small engine speeds, contribution which is higher for these low engine speeds. In case of high engine loads, the gasodynamic resistance is decreasing, irrespective of engine speed value, which leads to η_v increasing for high engine speeds domain, when the inertia of intake air flow is higher, encouraging in this way the filling. It can be noted a situation of quasi-equilibrium for an engine load of 80%, when η_v is kept almost constant for all the considered engine speeds.

In expression (6), T_a represents the temperature of fresh mixture at the end of intake stroke.

To ensure a better approximation of the theoretical model to the reality, this article proposes an analytical modeling of intake temperature T_a, depending on the engine load, for different engine speed values. In order to do this, it is considered that T_g, the temperature of exhaust gasses highlights very accurate the engine operating regimes and therefore the engine thermal regime.

Because of this reason, it is considered that the level and variation of this temperature will influence the heat transfer of the fresh mixture, taking into consideration the engine components surfaces, as intake manifold, cylinder head, pistons, cylinder walls, and also the residual gasses from the combustion chamber, determining considerable the level and evolution of T_a temperature of the fresh mixture at the end of the intake process.

On this line, the start consists in adopting a variation of the exhaust gasses temperature, T_g, dependent on the engine load, by using a quadratic function, whose coefficients are calculated by using collected and processed data experimentally. The results of this processing are summarized in Figure 2.

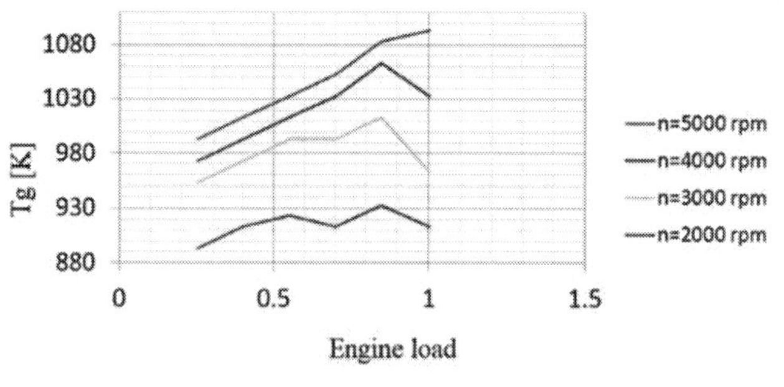

Fig. 2. Exhaust gasses temperature variation dependent of engine load, experimentally determined

The coefficients of this parabolic function of the exhaust gasses temperature, T_g, were introduced according to the minimum and maximum values of T_g temperature and also according to the quadratic function mathematical properties.

By customizing the quadratic function in the present situation, we can define the general function which shows the exhaust gasses temperature variation, $T_g(\chi)$, as following:

DOI: 10.1051/matecconf/201711210007

$$T_g(\chi) = a_{nT_g} \cdot \chi^2 + b_{nT_g} \cdot \chi + c_{nT_g}\,, \qquad (7)$$

where a_{nT_g}, b_{nT_g} şi c_{nT_g} represents the coefficients of $T_g(\chi)$ function.

As already shown in Figure 2 is highlighted the variation of exhaust gasses temperature, T_g, experimentally obtained, depending on the engine load, for different engine speed values. T_g temperature variations for four different engine speed values were mathematic assimilated with four parabolic functions, having the coefficients determined by the method described above.

In these conditions, $T_g(\chi)$ function is expressed as following:

$$T_g(\chi) = \begin{cases} -278 \cdot \chi^2 + 473 \cdot \chi + 893, & n = 5 \cdot 10^3 \\ -167 \cdot \chi^2 + 284 \cdot \chi + 913, & n = 4 \cdot 10^3 \\ -28 \cdot \chi^2 + 48 \cdot \chi + 943, & n = 3 \cdot 10^3 \\ -56 \cdot \chi^2 + 95 \cdot \chi + 873, & n = 2 \cdot 10^3 \end{cases} \quad [rpm]. \qquad (8)$$

The graphical representation of the exhaust gasses temperature, by considering the parabolic function $T_g(\chi)$ analytically determined is highlighted in Figure 3.

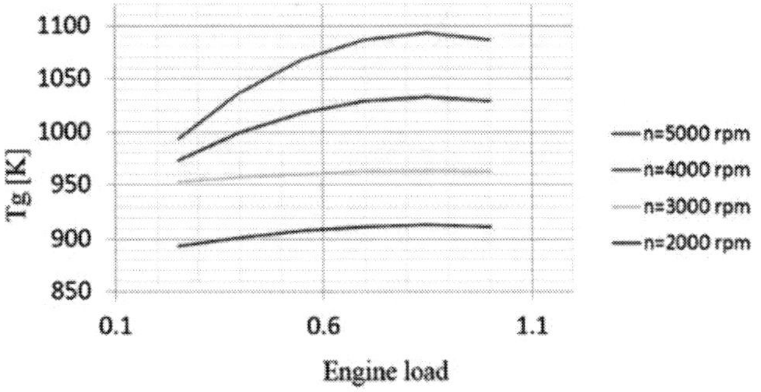

Fig. 3. Parabolic variation of exhaust gasses temperature T_g depending on the engine load

Based on the reasoning exposed above, sill considering for T_a temperature the same type of variation, we can write the analytical expression used to determine the variation of temperature depending on engine load, for different engine speed values $T_a(\chi)$:

$$T_a(\chi) = a_{nT_{ga}} \cdot \chi^2 + b_{nT_a} \cdot \chi + c_{nT_a} \qquad . \qquad (9)$$

Considering the evolutions of temperatures T_a and T_g with the same difference $\Delta T = T_{max} - T_{min}$ and adopting values for $T_{a_{min}}$ temperature, corresponding to the minimum engine load, on every engine speed value, the coefficients of the function were determined, according to expression (10). Using this coefficients, function $T_a(\chi)$ becomes:

$$T_a(\chi) = \begin{cases} -278 \cdot \chi^2 + 473 \cdot \chi + 225, & n = 5 \cdot 10^3 \\ -167 \cdot \chi^2 + 284 \cdot \chi + 263, & n = 4 \cdot 10^3 \\ -28 \cdot \chi^2 + 48 \cdot \chi + 312, & n = 3 \cdot 10^3 \\ -56 \cdot \chi^2 + 95 \cdot \chi + 299, & n = 2 \cdot 10^3 \end{cases} \quad [rpm]. \qquad (10)$$

In this conditions, the expression of $p_a(\chi)$ becomes:

$$p_a(\chi) = \Phi_0 \cdot (1 - \varepsilon^{-1}) \cdot (0.84 + \gamma_r) \cdot \begin{cases} -278 \cdot \chi^2 + 473 \cdot \chi + 225 \\ -167 \cdot \chi^2 + 284 \cdot \chi + 263 \\ -28 \cdot \chi^2 + 48 \cdot \chi + 312 \\ -56 \cdot \chi^2 + 95 \cdot \chi + 299 \end{cases} \cdot$$

$$\begin{cases} 0.69 \cdot \chi + 0.17, & n = 5 \cdot 10^3 \\ 0.475 \cdot \chi + 0.325, & n = 4 \cdot 10^3 \\ 0.368 \cdot \chi + 0.406, & n = 3 \cdot 10^3 \\ 0.2 \cdot \chi + 0.56, & n = 2 \cdot 10^3 \end{cases} [rpm] \tag{11}$$

Coefficient γ_r, direct and particular criterion to estimate the quality of purification of the engine cylinder by the residual gasses is defined by the following expression:

$$\gamma_r = \frac{M_r}{M_a} \quad , \tag{12}$$

where: M_r – quantity of residual gasses [kmol], M_a – quantity of fresh mixture [kmol].

The influence of the functional factors on the residual gasses coefficient is complex, the proposed model being focused on this aspect. In this way, the modification of this coefficient depending on the engine speed value has a minimum contribution [2, 4].

A second influence is given by the engine load. Thus it was demonstrated experimentally that increasing the engine load implicitly leads to an increase of the residual gasses quantity, which remains inside the engine cylinder after the evacuation process. On the other hand, as highlighted by some authors, it may be noted a sharp increase of the residual gasses coefficient γ_r while the spark-ignition engine's load is reduced. In the same time, the residual gasses acts in a chemical way, by introducing a kinetic consequence, because being chemically inert substances, leads to a reduced fuel burn rate. This inhibitor action interferes with a large share of the spark-ignition engines, in partially load regimes.

Based on this brief related analysis, in this article developed by the authors a variation of residual gasses coefficient, depending only on the engine load is adopted, considering that the variation depending on the engine speed is insignificant. Considering this analysis, and also data from the literature, a variation depending on the engine load according to the following expression is established:

$$\gamma_r(\chi) = a_{\gamma_r} \cdot \chi + b_{\gamma_r} \tag{13}$$

in which the coefficients of this function are: $a_{\gamma_r} = -0.3$ şi $\mathrm{b}_{\gamma_r} = 0.35$. In this situation, function $\gamma_r(\chi)$ becomes:

$$\gamma_r(\chi) = -0.3 \cdot \chi + 0.35 \tag{14}$$

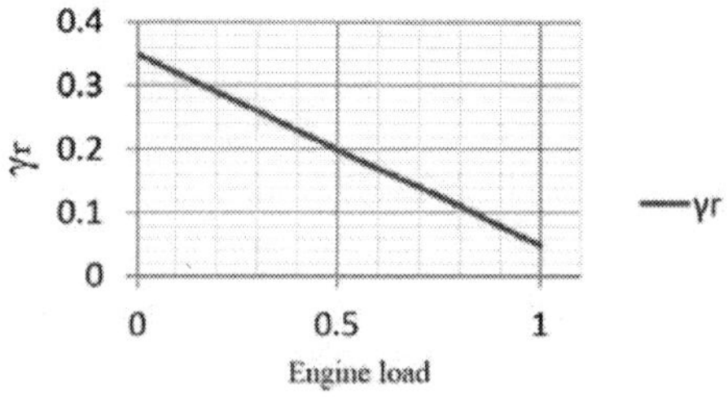

Fig. 4. Variation of residual gasses coefficient depending on the engine load

Finally, by using these functions, we can define the $\prod$ parameters at the end of the intake process, obtained through this developed model:

$$\Pi:\begin{cases} T_a(\chi)=\begin{cases}-278\cdot\chi^2+473\cdot\chi+225, & n=5\cdot10^3\\ -167\cdot\chi^2+284\cdot\chi+263, & n=4\cdot10^3\\ -28\cdot\chi^2+48\cdot\chi+312, & n=3\cdot10^3\\ -56\cdot\chi^2+95\cdot\chi+299, & n=2\cdot10^3\end{cases}\ [rpm]\\ p_a(\chi)=\Phi_0\cdot(1-\varepsilon^{-1})\cdot[0{,}84+(-0{,}3\cdot\chi+0{,}35)]\cdot\begin{cases}-278\cdot\chi^2+473\cdot\chi+225\\ -167\cdot\chi^2+284\cdot\chi+263\\ -28\cdot\chi^2+48\cdot\chi+312\\ -56\cdot\chi^2+95\cdot\chi+299\end{cases}\cdot\\ \cdot\begin{cases}0.69\cdot\chi+0.17, & n=5\cdot10^3\\ 0.475\cdot\chi+0.325, & n=4\cdot10^3\\ 0.328\cdot\chi+0.406, & n=3\cdot10^3\\ 0.2\cdot\chi+0.56, & n=2\cdot10^3\end{cases}\ [rpm]\end{cases} \quad (15)$$

3 Theoretical results. Simulations

Based on these simulations, the variation of temperature T_a and pressure p_a were highlighted, depending on the engine load, for different engine speed values, as shown in Figure 5 and Figure 6. In this way, the parameters which establish the engine performance can be easily obtained, especially the fuel consumptions of the engine.

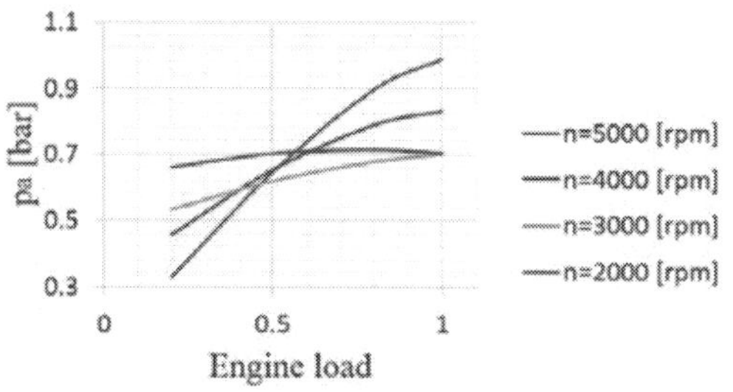

Fig. 5. Intake pressure p_a variation dependent of engine load

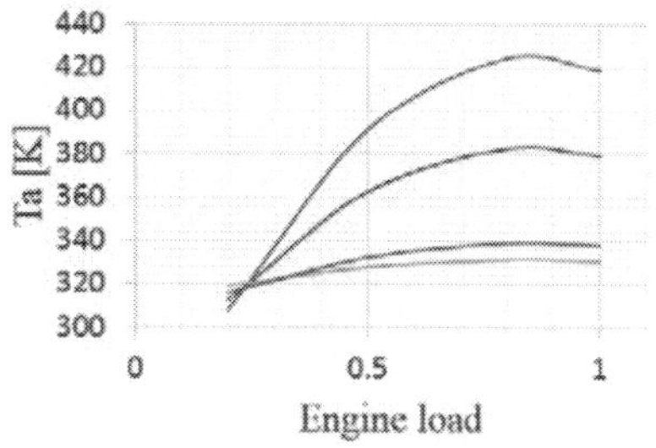

Fig. 6. Intake temperature T_a variation dependent of engine load

4 Conclusions

It is estimated that the developed model allows a performance evaluation and a behavior study of the spark-ignition engines, used in automotive propulsion. In this way, a contribution in the cooperation between the engine and the vehicle optimization can be offered, still in the design phase, shortening the introduction of the manufacturing cycle for new models. The result of this study will lead in the end to cost reductions.

References

1. V. Raimbault et al., *16. Internationales Stuttgarter Symposium*, 721-735 (Springer, 2016)
2. B. Grünwald, *Theory, calculation and construction of motor vehicles engines* (in Romanian), (EDP, București, 1980)
3. S. Buhl et al., Flow Turbul Combust J. **98** (2), 579-600 (2017)
4. D. M. Dogariu at al., *CONAT 2016 International Congress of Automotive and Transport Engineering*, 187-194 (Springer, 2016)

Aspects of experimental investigations of laser plug ignition use at spark ignition engine

Bogdan Done*

University POLITEHNICA of Bucharest, Faculty of Mechanical Engineering and Mechatronics, Department of Thermotechnics, Engines, Thermical and Frigorific Equipment, Splaiul Independentei no.313, Bucharest, Romania

Abstract. Nowadays, the necessity of pollutant emissions reduction brings to the fore new technologies developed for a better control of the combustion process. Ignition is the process of starting radical reactions until a self-sustaining flame has developed and strongly affects the formation of pollutants. Among the new technologies used for ignition and combustion control, the Laser Plug Ignition system is defined as an innovative technology which overcomes several limitations of conventional spark ignition. Laser ignition technology presents many advantages for engine operating control, engine performance improvement and pollutant emissions reduction. The objective of the paper is the experimental research of the laser ignition used in the spark ignition engine. The laser plug ignition system was mounted on an experimental spark ignition engine and tested at the regime of 90% load and 2800 rpm and dosage, $\lambda=1$. The experimental results present the influence of laser ignition on different engine parameters compared to electric spark ignition. The influence on in-cylinder pressure, heat release rate, engine efficiency and pollutant emissions level is analyzed. Compared to a conventional spark plug, a laser ignition system assures efficient engine operation at lean dosages.

1 Introduction

The negative effects of environmental pollution with gas emissions produced by the combustion of fossil fuels, an interesting aspect discussed also at the 21st Session of the Conference of the Parties from the 2015 Paris Climate Conference in November 2015, brings to the fore more than ever before the necessity of developing new technologies for the control of the combustion process in thermal machines. New technologies can be applied to automotive internal combustion engines in order to improve their performance, efficiency and pollutant emissions by applying a direct control on combustion process, [1-2]. Among the newly developed technologies for combustion management, laser technology is one of the newest and it can be successfully applied in classic spark ignition engines. Laser technology can be used to develop a new Laser Plug Ignition system (LPI) with the ability to control the ignition and combustion processes that will lead to the improvement of engine performance in terms of pollutant emissions and fuel efficiency, [3]. Phases of spark

*Corresponding author: bogdan.done@yahoo.fr

formation in correlation with breakdown intensity and the subsequent ignition, which depends on minimum ignition energy (MIE) must be established for LPI use, [1]. From this point of view the issue may be represented by the fact that the delivered energy is sufficient for ignition but its intensity is very low, so the spark cannot be formed; also, the spark formation may occur but the energy for combustion may be insufficient, [1, 3]. The mechanisms through which laser radiation can ignite a mixture of air-fuel are: thermal initiation (TI), non-resonant breakdown (NRB), resonant breakdown (RB) and photo-chemical ignition (PCI) [3]. During the thermal initiation (TI) phase the laser energy is taken by the in-cylinder mixture, the mixture being heated over the limit ignition temperature. Also, by heating a specific target surface inside the combustion chamber the TI phase can be established [3, 4]. Analogous to the electric spark ignition, during the non-resonant breakdown (NRB) phase the laser beam creates an electric field of high intensity which leads to the dielectric state, [1]. At one or more specific wavelengths, the resonant absorption at atomic level is specific to the resonant breakdown phase (RB). Comparative to NRB, during the RB the free electrons required for the breakdown process are created. There are two preceeding stages to create free electrons: the stage of non-resonant (NRPD) photo-dissociation of a molecules and the resonant photo-ionization of the atom created during the NRPD stage. The essential stage for processes of non-resonant breakdown (NRB) and resonant breakdown (RB) is the non-resonant photo-dissociation of a molecules (NRPD) by absorption and ionization. The non-resonant breakdown is a laser ignition mechanism that produces and emits light, heat and a shockwave, [5, 6]. The photo-chemical ignition (PCI) is a process that requires the absorption and dissociation of a single photon in ultraviolet radiation [5, 6]. Two-protons or non-resonant photo-dissociation of a molecule absorption in matter can lead to energy release in the form of a single high-energy photon, when the intensity is high. The advantage of this aspect is that the short wavelengths can be used for resonant absorption based on the action of longer wavelength laser light, [5, 6]. Regarding the mechanism of non-resonant breakdown Dearden et. al. used laser-induced sparks that offer advantage of small size design, but shorter in duration and which operates at higher temperatures [5, 7].

Comparative results regarding the operation of an automobile engine that was ignited with classical spark plugs as well as with laser spark plugs were presented recently by research groups from National Institute for Laser, Plasma and Radiation Physics (INFLPR) and Renault Technologie Roumanie (RTR), Bucharest, Romania [8]. After measurements made for the K7M 812k engine, lower CO and HC emissions level were identified, using the Laser Plug Ignition. At the speed regime of 1500 min^{-1}, due to improvement of the combustion process using laser ignition, the CO emissions decreased by 18...25% and the HC emission decreased by 14...17%. For higher engine speeds, over 2000 min^{-1}, the decreases are limited to ~3%. For laser use the NO_X emission increases by 8% for a speed of 1500 min^{-1} and by 2% for a speed of 2000 min^{-1}, the researchers from INFLPR and RTR explained that the higher flame temperature reached in the first part of combustion leads to the production of significant quantities of NO_X, [8]. As a solution for NO_X emission reduction the INFLPR and RTR propose the increase of EGR rate (Exhaust Gas Recirculation), [8]. The researchers of INFLPR explain the CO_2 emission increases on laser ignition use by stating that the carbon content entering into combustion and resulting from the combustion process is constant and is a compromise between HC and NO_X when the internal combustion engine calibration is necessary, [8]. Measurement results show that the power of the engine ignited by laser system increases by ~3% in comparison with classical ignition system. For INFLPR the most critical problem of laser ignition (LI) use during the experimental investigation was the possibility of destroying the optical element coatings. The optical element is designed to construct the focusing line and during engine operation damage of the lenses from the pump line can sometimes appear, [5, 7, 8]. Usually, the laser lenses are purchased from the standard market, and are designed with no special coatings. This technical issue is eliminated by the special coating technology applied to the lenses using high-damage threshold layers, [15]. Also, to eliminate this disadvantage the uncoated lenses at critical points in the laser beam (for high intensity laser beam) can be used. The second solution, proposed by many researchers, has already been used for different engines but the deposit of the combustion products on the sapphire window was also present. H. Ranner et. al. [7] investigated and proposed for this problem the solution of self-cleaning, a method that cleans the window with the laser beam itself, and can be done in three steps. As for the first method Ranner [7] explains that the initial part from the higher laser pulse energy, Ep= 4 [mJ], was able to clean the laser window, although just partially. The second method is the possibility of using two laser pulses due to the increased duration of the pump pulse. To clean the lenses the first laser pulse was used, this cleaning method being the most efficient. A third cleaning method was applied by Ranner [7] who used a four stroke engine which admits double triggering per cycle. Thus, in order to clean the window of cylinder no. 4 before a new ignition started, the laser pulsses were applied to cylinder no. 4 in the exhaust stroke while the ignition was being initiated in cylinder no.1, [7]. The window of laser plugs for cylinder no. 2 and 3 were cleaned in the same way. Due to the fact that Ranner used a water compact cooling system for each laser plug the engine was able to operate for a few hours without coating issues for windows or for optical elements [9, 10]. Mullett and Dearden [5, 11] studied the LI system performance on a Ford Zetec engine at a speed regime of 1500 min^{-1} and 36° before Top Dead Centre (TDC) and calculated several laser energy values between 4…92 [mJ], [5].

Regarding the emissions control, the reduction of pollutant emission, especially the NO_X emission, can be successfully assured by the use of exhaust gases to dilute the inlet fresh charge. The use of high ratios of exhaust gas recirculation (EGR), especially in partial load operation regimes, can significantly improve the engine efficiency, knowing that in urban traffic, automotives consume 30% of their fuel in idle conditions [12, 13]. R. D. Fruechte et al. reach the conclusion that a reduction in idle speed from 800 min^{-1} to 650 min^{-1} could assure fuel savings up to 24% at idle regime [12, 14]. With no electrodes to quench the flame kernel, like in classical spark plugs, the LPI could offer improved cold start performance. Laser energies could be readily increased during cold start to improve the ignitability of the fuel. The experiments made at the Technical University of India, in Lonere-Raigad-Maharashtra, show that LPI has advantages compared to the conventional spark plug ignition (SPI), [15]. The laser ignition system used leads to the attenuation of fuel consumption and to a pollutant emissions decrease by almost 20% compared to the classic spark plug ignition system, [15]. Swapnil Harel et.al. [15] affirm that it is important for these advantages of LI use to be achieved for the same engine noise level and same smoothness of running level. Harel [15] use a frequency-doubled Nd:YAG laser to study the influences of the wavelength on laser ignition process [15], finally finding out that there are no influences. Harel shows that the fuel consumption and the emissions from the exhaust gases can be significantly improved by the use of laser ignition within the fuel spray, [15]. Technology of classic spark plug ignition within the fuel spray cannot be used because the conventional spark plug will not operate normally, the laser plugs not being affected by this issue, [15]. Harel also studied the reliability of the laser ignition system trying to find out if the deposits resulting from the combustion process can affect the laser beam window and the normal running of the laser ignition system, [15]. Using the classic spark plugs ignition system and maintaining the installed the laser beam window in the combustion chamber, the engine ran around 20 hours, at different loads, with simulated regimes of cold start and missfiring. After 100 running cycles the engine was stopped, the laser window was dismounted, inspected and Harel noticed that the deposit from the combustion had been eliminated by the laser beam, [15]. Harel shows that when using a certain threshold intensity at the laser beam entrance window, the engine can operate normally, without misfiring, [15]. Harel used the high pulse energy of the first few laser pulses only to assure normal engine running at cold start, a fact that was also beneficial for the cleaning process of the beam entrance window [15].

The present paper shows some aspects of the experimental investigation of laser ignition used on an experimental engine equipped with laser plug ignition system. The effects of LI use on in-cylinder pressure, heat release rates and heate release laws versus classic spark plug ignition are analyzed.

2 Experimental investigation

The experimental research was developed on an experimental single cylinder SI engine, equipped with Laser Plug Ignition. The engine operating regime was 2800 rev/min, 90 % load. The experimental single cylinder engine was mounted on the test bed adequate instrumented, its schema being presented in Figure 1.

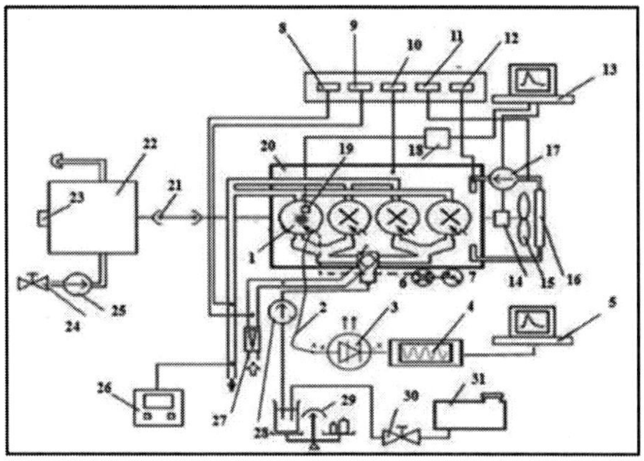

Fig. 1. Experimental test bed schema.

1-laser plug ignition, 2-optical fibre, 3-laser diode, 4-laser power supply, 5-PC with soft laser, 6,7-the ensemble breaker distributor (cam with one corner), 8-inlet air temperature measurement indicator, 9-exhaust gas temperature measurement indicator, 10-engine oil temperature measurement indicator, 11-engine oil pressure measurement indicator, 12-cooling liquid temperature measurement indicator, 13-PC equipped with AVL acquisition board, 14-crank angle encoder, 15-cooling fan, 16-cooler,17-engine water pump, 18- Kistler charge amplifier , 19-piezoelectric Kistler pressure transducer, 20-spark plug ignition, 21- coupling, 22-Schönebeck B4 hydraulic dynamometer, 23-mechanical snuff speed, 24-air flow meter, 25-hydraulic dynamometer water pump, 26-AVL DiCom Analyzer 4000, 27-air flow meter, 28- gasoline fuel pump, 29-gravimetric fuel flow meter, 30-gasoline consumption tap, 31-tank.

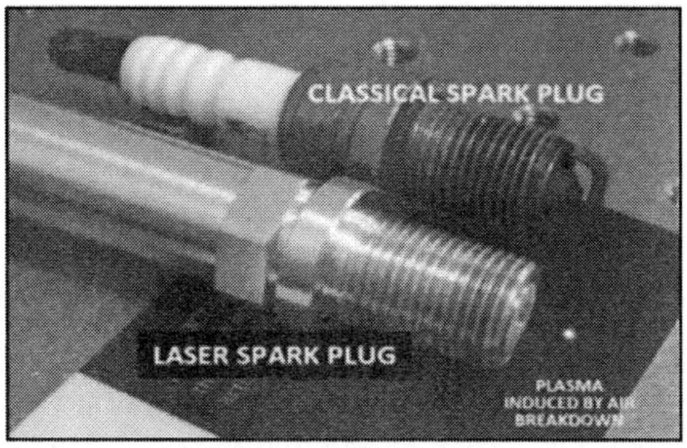

Fig. 2. A photo of a laser spark plug is shown in comparison with a classical spark plug. The plasma induced in air by optical breakdown is visible

The laser spark plug used in the experiments was provided by INFLPR, Laboratory of Solid-State Quantum Electronics, Magurele, Romania. A photo of the laser spark is shown in Figure 2.

The laser medium was a Nd:YAG/Cr4+:YAG ceramic structure (Baikowski Co., Japan) that consisted of a 8.0-mm long, 1.0-at.% Nd:YAG ceramic, optically-bonded to a Cr4+:YAG ceramic with saturable absorption (SA) [8, 16]. The initial transmission of Cr4+:YAG SA was around 40%. Monolithic configuration of the resonator was obtained by coating the high reflectivity mirror at lasing wavelength, λem= 1.06 μm on the Nd:YAG free side and the outcoupling mirror with reflectivity R= 50% at λem on the Cr4+:YAG free surface. The Nd:YAG side was coated for high transmission (T> 0.98) at the pump wavelength, λp= 807 nm. The optical pump was performed with a fiber-coupled diode laser (JOLD-120-QPXF-2P, Jenoptik, Germany) that was operated in quasi continuous-wave mode; the pump pulse duration was 250 μs and repetition rates up to 100 Hz were used. Typically, the laser yielded pulses with energy of 3.8 mJ at 1.06 μm for the pump with pulses of ~35 mJ at 807 nm; the laser pulse duration was around 1 ns.

3 Results

The experimental research was carried on the SI engine initally equipped with Spark Plug Ignition system (SPI), defined as reference and secondly on the engine equipped with Laser Plug Ignition system (LPI) . The operating regime was 2800 rev/min, 90 % load and air-fuel ratio λ=1. For the investigated regime, it was measured with the resolution of 1 CAD and registered with the AVL data acquisition system.

Figure 3 presents the averaged in-cylinder pressure diagrams measured for Spark Ignition System and for Laser Ignition System. The peak pressure rose from 43 bar up to 46 bar. The increase of maximum pressure, by almost 7%, is in correlation with the increase of the maximum pressure rise rate registered for laser ignition versus classic ignition, from 1.7 bar/CAD to 1.9 bar/CAD. Also, the maximum pressure is achieved sooner per cycle for laser ignition compared to spark ignition, the angle of maximum pressure being reached with 2 % closer to TDC for Laser Ignition, Figure 3. The initial phase ends ~ 4 degrees sooner for laser ignition compared to the spark ignition system.

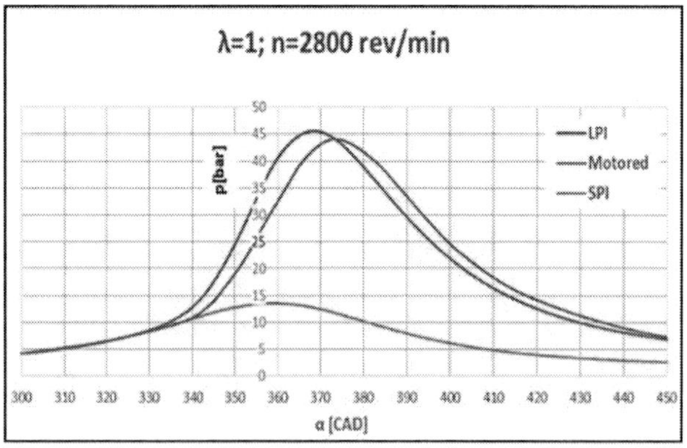

Fig. 3. Pressure diagrams registered for SPI and LPI

Heat release rate is presented in Figure 4. The peak values registered per cycle heat release rate are comparable for laser and classic ignition system, but the maximum is attained 5 CAD sooner per cycle for laser ignition. Also, the heat release rate for laser ignition starts sooner,

4 CAD quicker versus spark ignition and decreases faster, in a much more significant way compared to the tendency registered for classic ignition.
These aspects are in corellation with the heat release allure, Figure 5. For laser ignition system the heat release starts 5 CAD earlier compared to the classic ignition system.

The initial phase of the combustion is reached 6 CAD sooner per cycle and has a shorter duration compared to the classic ignition system, almost 22% shorter, the value decreasing from 9 CAD down to 7 CAD. The reduction of the duration of the initial phase of combustion is due to a much higher energy developed by laser ignition system versus spark ignition system. Also this affects the main phase of the combustion which is registered 6…7 CAD sooner per cycle compared to classic ignition, Figure 4.

The heat release law rises faster for LPI compared to SPI and the 10% of conventional mass fraction burned is reached almost 5 CAD sooner per cycle for LPI system compared to classic ignition system. The 50% mass fraction burned per cycle is achieved before TDC at 355 CAD for laser ignition and at 360 CAD for spark ignition system. The conventional end of the combustion process, reflected by 90% of cycle heat release, appears 8 CAD sooner per cycle for laser ignition system, Figure 5. The main influence of laser ignition is reflected on the combustion process which starts earlier and ends earlier compared to the combustion phenomenology registered for the classic ignition system.

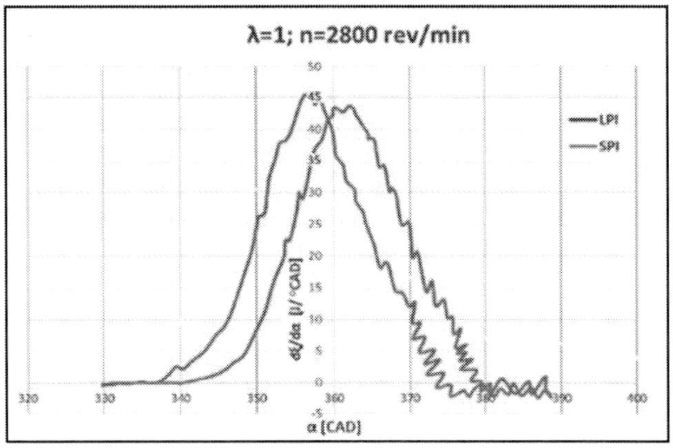

Fig. 4. Heat release rate diagrams evaluated for SPI and LPI

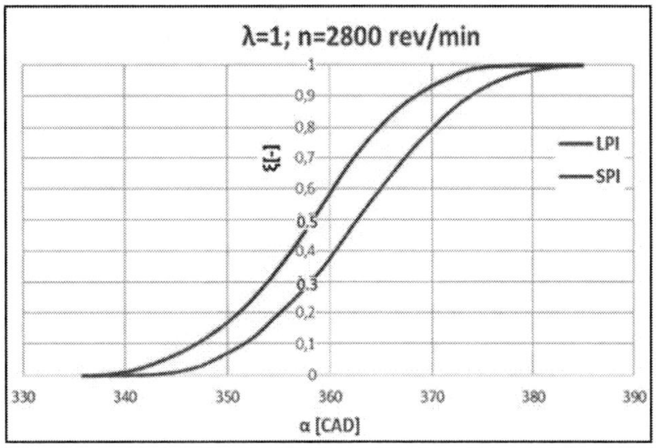

Fig. 5. Heat release law diagrams evaluated for SPI and LPI

4 Conclusions

The experimental results obtained from the laser ignition used in the spark ignition engine compared to the classic ignition system have led to several important conclusions presented as follows:

The in-cylinder maximum pressure increases with almost 3 bar on laser use. Also the maximum pressure rise rate registered for classic ignition increases from 1.7 bar/CAD up to 1.9 bar/CAD for laser ignition. The maximum pressure appears earlier per cycle on laser use, the angle of maximum pressure being reached 2 % closer to TDC.

When the laser ignition system is used, the initial phase ends around 4 CAD earlier per cycle comparative to spark ignition system.

The maximum values of heat release rate and heat release law for SPI are comparable to the values registered for LPI, but the heat release rate and the heat release law are reached almost 5 CAD earlier per cycle for laser ignition.

The initial phase of the combustion decreases by 22% at laser ignition use versus the classic ignition system. The stage of 10% heat release per cycle is reached 6 CAD sooner per cycle in the case of the laser ignition system. The reduction of the duration of the initial phase of combustion is due to a much higher energy developed by laser ignition system versus the spark ignition system. 50% of mass fraction burned per cycle is achieved before TDC, 5 CAD earlier for laser ignition.

The experiments were performed with a laser spark device developed at National Institute for Laser, Plasma and Radiation Physics, Laboratory of Solid-State Quantum Electronics, Magurele, Ilfov, 077125, Romania. The authors would like to thank to Mr. Pavel Nicolaie, Mr. Dinca Mihai and Mrs. Croitoru Gabriela for their help and assistance during the experiments. Also, the authors address special thanks to the AVL GmbH for providing the necessary eqiupments.

References

1. P. D. Ronney, Opt. Eng. **33**, 510–522 (1994)
2. N. Negurescu, C. Pana, M.G. Popa, *Internal Combustion Engines. Processes* (Matrixrom Bucharest, 2009)
3. C. Morgan, Rep. Prog. Phys. **38**, 621–665 (1975)
4. S. S. Vorontsov, V. N. Zudov, P. K. Tretyakov, and A. V. Tupikin, Thermophys. Aeromech. **13**, 615–621 (2006)
5. G. Dearden and T. Shenton, Opt. Express **21** (S6 Suppl 6), A1113–A1125 (2013)
6. Y. L. Chen, J. W. L. Lewis, and C. Parigger, J. Quantitative Spectrosc. Radiative Transf. **67**, 91–103 (2000)
7. H. Ranner, P. K. Tewari, H. Koefler, M. Lackner, E. Wintner, A. K. Agarwal, and F. Wintner, Opt. Eng. **46**, 104301 (2007).
8. N. Pavel, T. Dascalu, G. Salamu, M. Dinca, N. Boicea, and A. Birtas, Opt. Express **23**, 33028-33037 (2015)
9. S. Lorenz, M. Bärwinkel, R. Stäglich, W. Mühlbauer, and D. Brüggemann, Int. J. Engine Res. **17**, 139–150 (2016)
10. H. Chen, V. Page, Z. Kuang, E. Lyon, G. Dearden, and T. Shenton, 3rd Laser Ignition Conference (LIC'15), Argonne National Laboratory, USA, W2A.2 (2015)
11. J. D. Mullett, *Laser-Induced Systemsfor Gasoline Automotivr Engines,* PhD Thesis, University of Liverpool (2009)
12. P. B. Dickinson, A. T. Shenton, J. D. Mullett, G. Dearden, and A. Scarisbrick, Proc. 10th Int. Symp. on Advanced Vehicle Control (AVEC10), 22–26 (2010)
13. R. Jurgen, *Automotive Electronics Handbook* (McGraw-Hill, 1995)

14. R. D. Fruechte, F. E. Coats, and C. H. Folkerts, IEEE Proc. 17th ISECE Conference, 467–472 (1983)
15. S. Swapnil. Harel, Mohnish Khairnar, Vipul Sonawane, IJSR **3** (7), 1551-1560 (2014)
16. T. Dascalu, G. Salamu, O. Sandu, M. Dinca, and N. Pavel, Opt. & LASER Techn. **67**, 164-168 (2015)

Assessment of small hydropower potential by software. Case study

Alexandru Moldoveanu [1], and *Daniela* Popescu [2,*]

[1]National Institute of Hydrology and Water Management, Water Management Department, 97 Bucuresti - Ploiesti, Bucharest, 013686, Romania
[2]Technical University Gheorghe Asachi of Iaşi, Department of Fluid Mechanics, Fluid Machinery and Fluid Power Systems, B-dul Dimitrie Mangeron nr.59A, Iasi, 700050, Romania

Abstract. The rivers hydropower potential is considered one of the oldest renewable energy source used in the electricity production process. A method to investigate the possibility to construct a micro hydropower system on a small river is presented. The analysis and the hydropower assessment were done by using Vapidro-Aste software. Results point out that micro-hydro units can be implemented in remote locations or hybrid renewable energy systems, while the environmental flow is guaranteed.

1 Introduction

The European Union vision regarding the policy in the field of energy is defined by the sustainable development concept. This concept represents a major challenge for the European Union and especially for ensuring energy security with clean energy at competitive prices for the population and also taking into consideration the increasing global energy demand. By the ratification of the Kyoto Protocol, the European Member States committed to achieve the objectives on reducing the greenhouse gas emissions. One of the first measures performed at the European level was the development of a unique energy trade market and the promotion of electricity produced from renewable energy sources.

In this frame, the European Commission issued the Directive 2009/28/CE on the promotion of the use of energy from renewable sources and its main objective is that 20% from the total internal gross consumption of energy for each Member State must be produced from renewable energy sources until 2020 [1]. In the Directive 2009/28/CE acceptance, only the small hydropower potential developed in hydropower plants with an installed capacity smaller than 10 MW can be considered a renewable source of energy for generating electricity.

New energy policies boosted the study of micro-power (up to 100kW) and pico scale hydropower (up to 5 kW), as a research direction worthy of investigation. Micro or pico hydro turbines can be used as standalone installations or as part of small hybrid renewable energy systems (HRES), that combine wind turbines, photovoltaic modules, micro-hydro and biomass units [2], [3]. They can be installed with minimum construction costs on rivers

*Corresponding author: daniela_popescu@tuiasi.ro

that have low heads and fluid flow rates. In this situation, only several types of turbines are recommended, the cross flow turbine, the Archimedean screw and the Kaplan micro-turbine.

In this paper a method to investigate the possibility to construct a micro hydropower system on a small river is presented. A case study, the Topolog river, is in focus. The method is based on the Vapidro-Aste software to identify the best way to use the hydraulic potential, characterized by low head and low fluid flow rates. The method takes into consideration an important restriction; part of the fluid flow rate cannot be used to produce electricity, because it must be discharged to protect the river ecosystem. The environmental flow takes into account the natural variability of the flow regime and describes the quantity and quality of water flows required to protect aquatic ecosystems of the rivers. In this study, it was taken into consideration according to the Water Framework Directive 60/2000/EC provisions for the environmental objectives achievement regarding environmental flow downstream the hydropower plant [4].

The results of the complex analysis indicate that a cross-flow turbine is recommendable in the studied case.

2 Vapidro-Aste software presentation

Vapidro - Aste is a software for the assessment of small hydropower potential. The Vapidro-Aste was developed at the same time as the GIS (Geographic Information System) spreading utilization all over the world, preserving the software developers trend to use GIS as an interface. It was developed by the Italian Lead Partner RSE S.p.A., within the SEE HYDROPOWER project [5-6], financed in the frame of the South East Europe Programme.

The main steps for applying Vapidro – Aste are: loading a DTM (Digital Terrain Model), introducing the hydrological and technical data, introducing the financial and economic data and performing the analysis of the river selected for the assessment of the small hydropower potential [7].

The main hydrological and technical input data are the following:

- the mean flow of the analyzed river sector (in one to four cross sections);
- the environmental flow;
- the withdrawals and restitutions along the analyzed river sector;
- hydropower efficiency coefficients.

After introducing the DTM (Digital Terrain Model) the Vapidro – Aste is able to generate the river network on the selected area. The following steps consist of introducing the required data mentioned above and performing the technical analysis.

The main economic and financial data are the following:

- the updating interest rate;
- the plant lifetime;
- the price of the energy produced;
- the price of the green certificates needed.

After introducing the economic and financial data, the Vapidro – Aste is able to perform the economic analysis.

The main outcomes of the Vapidro – Aste are the following:

- the automatic river network generation and sub-basin computation;
- the gross and net heads distribution along the analyzed river sector;
- the hydropower potential distribution (the maximum installed capacity) along the analyzed river sector;
- the distribution of the mean annually energy production corresponding to the mean annually operating hours of the small hydropower plant;

- the optimal number of small hydropower plants for the analyzed river sector;
- the most suitable locations for the water abstractions and the small hydropower facilities placed on GIS maps for the analyzed river sector development scheme.

3. Case study. Topolog River

The Topolog River is a tributary of the Olt River and has a length of 111 km with a river basin area of 543 km^2 [8]. The Topolog River springs are located under the Negoiu Peak (2535 m) and the river presents the typical characteristics of the Southern Carpathians Rivers. The river forms after the confluence of two streams springing from the glacial cirques, the Negoiu and Scara streams, the first being considered its ancestor (1880 m). In the upper part of the river or upstream of the Capul Bulzului, the river has slopes of 45 m/km and the valley was being shaped in the crystalline schists with frequent injections of gneiss, which leads to the development of the selective erosion [9]. In figure 1, the cartographic representation of the Topolog and Olt River Basins is presented.

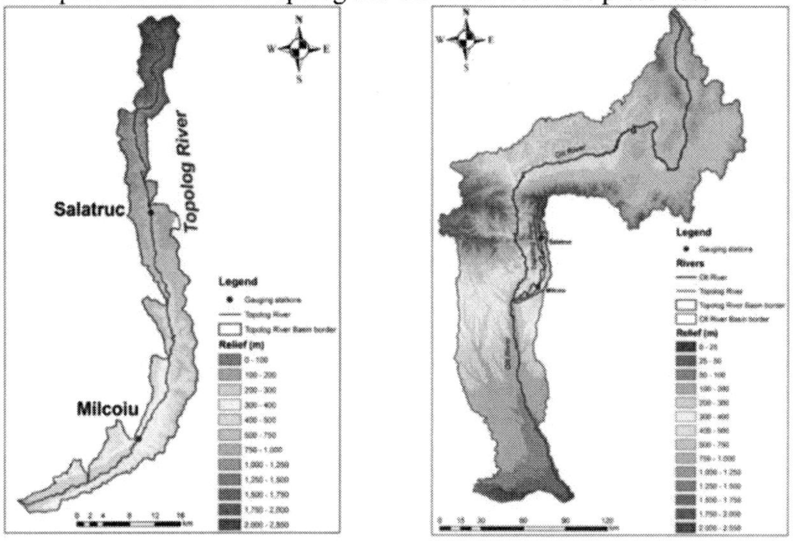

Fig. 1. The cartographic representation of the Topolog and Olt River Basins.

3.1 The selected methodology and the input data used

Along the Topolog River, a sector with the length of 34 km was selected by Vapidro - Aste and represents the river sector from its springs to the Salatruc gauging station.

The selected methodology, a single small hydropower plant with diversion, was considered for a single environmental flow downstream of the water works corresponding to a single hydrological mean flow and without abstractions and restitutions along the selected river sector.

In the next step the following input data were used:

- the mean flow (Q_{mean}= 1.27 m^3/s) at the Salatruc gauging station;
- the environmental flow (Q_{env}= 0.2 m^3/s);
- the hydropower efficiency coefficients required by the Vapidro – Aste main menu.

In this hypothesis, the economic and financial data was not taken into consideration and the economic analysis was not performed, due to the fact that the future micro hydropower unit is planned to be used only for private purposes and will not be connected to the national energy grid.

3.2 The results of the Vapidro – Aste application on the selected river sector

After performing calculations, the software associates a series of river sub-basins to the river sector analyzed that represent the river catchment. Then, the Vapidro – Aste correlates these results with the data extracted from the digital terrain model (DTM) resulting head (the height difference between two sections of calculus) which is presented in Figure 2 as a map and as a graph.

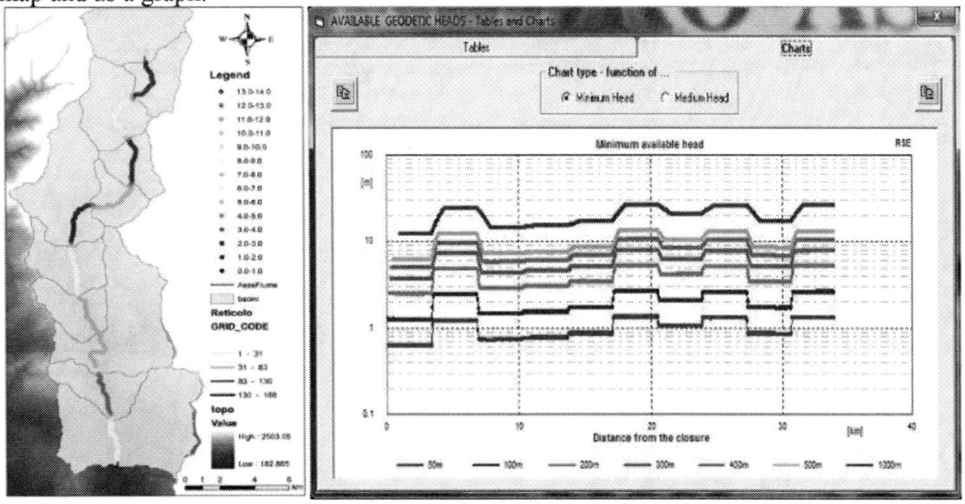

Fig. 2. The representation of the heads distribution along the analyzed river sector

The hydropower potential assessment of the analyzed river sector was performed based on the heads and the optimum installed capacity which was calculated based on the input data mentioned above. The distribution of the maximum installed capacity is presented in Figure 3 as a map and as a graph.

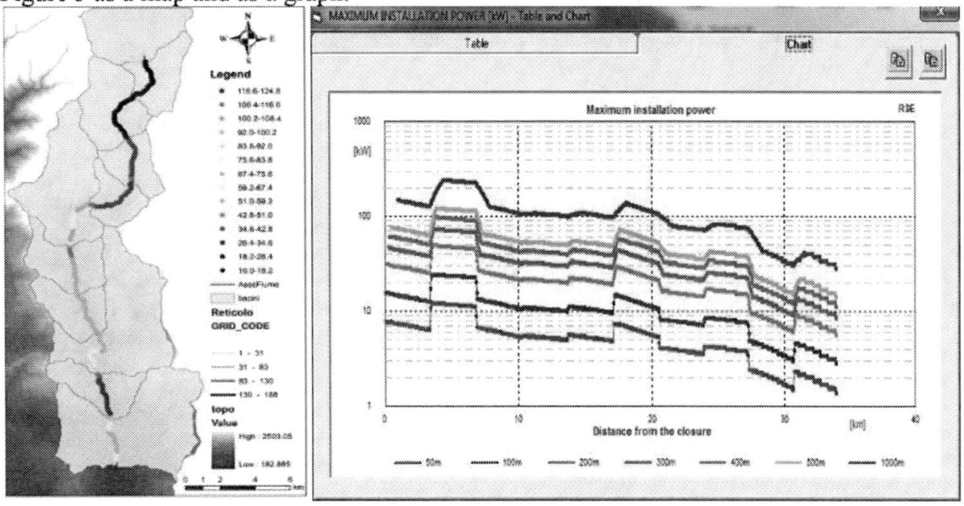

Fig. 3. The representation of the installed capacity distribution along the analyzed river sector

In the presented case study, the optimum installed capacity on the analyzed river sector resulted from calculus is 22 kW and the mean electricity production computed for an operation time of the hydropower unit of 4632 hours/year is about 102 MWh/year. The distribution of the mean electricity production corresponding to the operation hours of the hydropower system within a year is presented in Figure 4 as a map and as a graph.

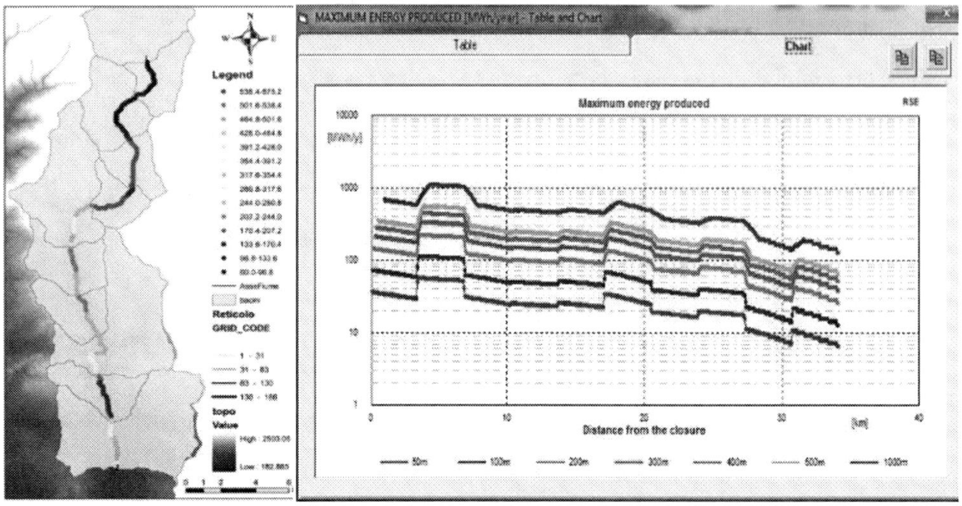

Fig. 4. The mean electricity production distribution representation along the analyzed river sector

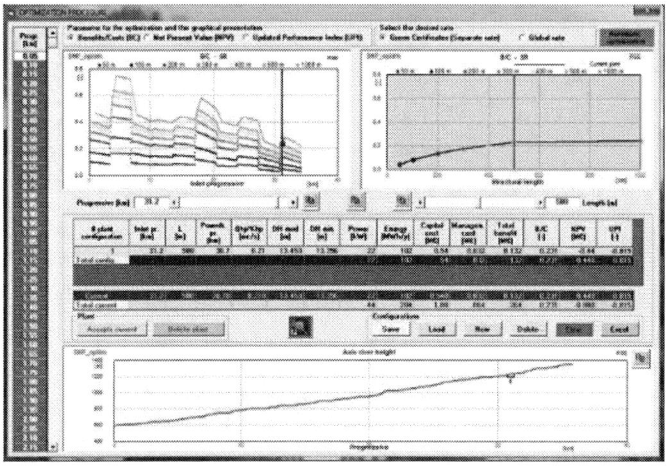

Fig. 5. The result of the automatic/manual function of the Vapidro – Aste

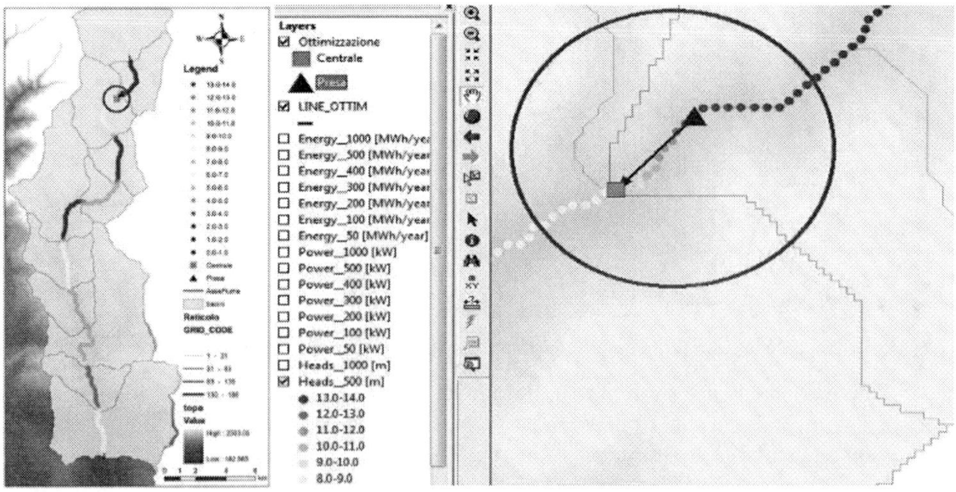

Fig. 6. The cartographic representation of the optimized location of the micro - hydropower unit.

The optimal distance between the water intake and the micro-hydro facility location is the result of the automatic/manual procedure of the Vapidro – Aste software. For this study, the resulted length of the manual procedure is about 500 meters and the optimal locations for the water intake and the micro-hydro power unit, from the technically point of view are presented above in Figure 5 and in Figure 6.

4 Conclusions

The optimal distance between the water intake and power plant location and their optimal positions along the river sector, represent relevant information for potential investors in the field of small hydropower plants construction.

Every potential site must be analyzed case by case and the Vapidro – Aste application represents only a preliminary stage for the suitable sites identification in order to build a new micro hydropower system.

In the present case study, the results revealed that the optimum installed power on the analyzed river sector is 22 kW, which corresponds to a distance of 500 meters from the water intake to the micro turbine and for a mean head of 13-14 meters. In this case, a cross flow turbine is recommended. The mean electricity production is about 102 MWh/year and is computed for an operation time of the hydropower unit of 4632 hours/year.

The results took into consideration the environmental flow to the hydropower assessment of the analyzed river and were presented both as GIS maps, as well as graphics.

Concluding, micro-hydro units can be implemented in remote locations or hybrid renewable energy systems, without a significant risk of impact on the river ecosystem.

The paper contributes to achieving the objectives and highlights the results of the research project "Cross Flow Hydro Turbines for Pico Renewable Energy Systems", contract 45/1.07.2014, financed by the Romanian Executive Agency for Higher Education, Research, Development and Innovation Funding (UEFISCDI) in the frame of the research program Partnerships in Priority Domains – PN II, project code PN-II-PT-PCCA-2013-4-1901.

The VAPIDRO-ASTE software has been used. It has been produced by RSE S.p.A in the frame of the SEE HYDROPOWER project financed by the South East Europe Program and the Research Fund for the Italian Electrical System [5-7].

References

1. European Council, *Directive 2009/28/EC on the promotion of the use of energy from renewable sources, amended by Directive 2013/18/EU* (2009)
2. S. Meshram, G. Agnihotri, S. Gupta, JRSE **6**, 013140 (2014)
3. A. Afshin, M. Angelo, A. Pedrasa, *TENCON 2012- IEEE Region 10 Conference*, 10.1109/TENCON.2012.6412230 (2012)
4. European Council, *Water Framework Directive 2000/60/EC* with further amendments (2000)
5. M. Peviani, J. Alterach, A. Danelli, *SEE HYDROPOWER Project, targeted to improve water resource management for a growing renewable energy production, HYDRO 2011*, Praha October 17-19 (2011)
6. http://www.seehydropower.eu/
7. J. Alterach, A. Elli, M. Vergata, SEE Hydropower -*VAPIDRO ASTE 4.0 User Guide. WP 5 – Common strategies to improve SHP Implementation* (2012)
8. Ministry of the Environment, *Cadastral Water Atlas of Romania* (in Romanian) (1992)
9. I. Ujvary, *Geography of Romania's waters* (in Romanian) (Editura Ştiinţifică, Bucuresti,1972)

Renewable energy technologies: enlargement of biofuels list and co-products from microalgae

Nadezhda I. Chernova[1], *Sofya V.* Kiseleva[1,2], *Mikhail S.* Vlaskin[2], and *Yulia Yu.* Rafikova[1,*]

[1]Lomonosov Moscow State University, Moscow, Russia
[2]Joint Institute for High Temperatures of Russian Academy of Science, Moscow, Russia

Abstract. Microalgae is a perspective feedstock for producing a wide variety of biofuels and co-products with high added value. An alternative to the traditional technology of biodiesel from algae by the transesterification is the technology of hydrothermal liquefaction (HTL). The article presents the results of promising strains screening and directed cultivation of microalgae for the processing by means of variety of technologies and production of valuable co-products. An algorithm for selecting suitable areas for industrial plantations of algae is presented.

1 Introduction

Among renewable energy sources biomass is one of the most popular: there is a significant increase in the scope and types of biomass use, recycling technology development and its application in the energy sector. Therefore, an urgent task is to find and study the use of new types of biomass, as well as the integrated technologies of its processing. In recent years the microalgae biotechnology which gives the opportunity to receive biofuels from microalgae are considered as a promising trend of bioenergy. Microalgae are very efficient solar energy converter that organizes the transformation of carbon dioxide into the complex energy consuming biomolecules comprising carbohydrates, proteins, lipids, which can be subjected to further transformation.

At the heart of research and engineering developments aimed on microalgal biofuel production lies the microalgae culture, which should grow rapidly, synthesize and accumulate a significant amount of lipids in the form of triacylglycerols and liquid carbohydrates [1, 2]. Currently the whole number of microorganisms collections, particularly of microscopic algae, were created; in Russia collections of microalgae for energy purposes still are not created. The search of the perspective microalgae strains for energy purposes and creation of collections and efficient cultivation systems are certainly actual.

Development and testing of microalgae large-scale cultivation methods (MACs), as well as the methods of biomass processing into the suitable products, including energy products, was the topic of the authors study over the last years. The objects of the study were representatives of green and blue-green MACs: *Chlorella, Dunaliella, Spirulina, Haematococcus*, etc. [3, 4]. Due to uniqueness of the *Spirulina/Arthrospira (A. platensis*

[*] Corresponding author: ju.rafikova@gmail.com

(Nordst.) Geitl.) biochemical composition, it was also studied as a biologically active food and feed additives. The tests of spirulina as a feed additive in poultry farming, cattle breeding and fur farming were carried out, guidelines for its application were developed and the patents of the Russian Federation were obtained.

Obviously, the most important factor in determining the profitability of the fuel production from biomass in general, and from microalgae - in particular, is the situation at the world oil market. A possible way to reduce the cost of biofuel from algae and to make it competitive on the market consists of using them to simultaneously obtain biofuels and high-value products, like nutraceuticals (BAA), phycocyanin, phycoerythrin and allophycocyanin, astaxanthin, chlorophyll, vitamin B12, polyunsaturated fatty acids omega-3 and omega-6, and others. In case of using algae for sewage cleaning it is advisable to convert the biomass of MACs to methane or biofuel.

2 Methodology and data analysis

The introduction of advanced lipid strains into the biofuels production is complicated by following well-known fact: the simultaneous achievement of high productivity in biomass and lipids is impossible under the microalgae cultivation. Induction of lipid biosynthesis can be achieved by creating a physiological stress: limitation or starvation for nitrogen and/or phosphorus; maximum or minimum value of the light intensity, pH, temperature; ultraviolet irradiation; effects of heavy metals and other chemicals, osmotic stress, and others. One of the possible way of maximizing the microalgae biomass accumulation with optimal lipid-content is the conducting cultivation in two stages. The first stage includes a maximum accumulation of biomass under standard growth conditions. At the second stage the production of biomass in stressful conditions to induce lipid synthesis is carried out.

As one of the most famous and expensive outdoor farmed species of microalgae/cyanobacteria remains *A. platensis*; it was used in the experiments under the two-step-cultivation and production of high lipid content in the algae. The object of the study were two clonal cultures of *A. platensis: strain rsemsu 1/02-P* (direct) and *strain rsemsu 1/02-T* (relaxed) as well as the culture of *H. pluvialis rsemsu Hp-1/11* resulting from the natural variability of morphological features. Testing of screening methods and selection of lipid-stress-conditions for lipid induction in microalgae have been conducted on *spirulina* [5]. As a result, samples of biomass with a high content of lipids were obtained (Table 1).

Traditionally, processing of algae biomass into biodiesel comprises drying, extraction of lipids, transesterification to give methyl esters of fatty acids constituting the biodiesel. At the same time the energy costs for drying and extraction is high. An alternative to this method is the HTL technology [6]. The main advantages of this method include the absence of pre-drying the feedstock and a relatively high yield (the ratio of product mass to the mass of raw material) due to the fact that all components of microalgae (lipids, proteins and carbohydrates) are converted into biofuel. In this case the main aim is to produce high biomass productivity of strains, rather than high lipid-content. However, the bio-oil produced from algae by HTL, is characterised by large amounts of nitrogen. This is due to high content of proteins in the original biomass. Therefore, the use of highly productive MAC strains with low-protein and, accordingly, high-lipid and/or carbohydrate still is important.

Table 1. The biochemical composition of investigated strains microalgae at different growing conditions and the possible appropriate processing technology, %

Genus/species/ strain	Cultivation conditions	Proteins	Neutral lipids	Total carbohy drates	The optimal processing technology	The products
A. platensis rsemsu 1/02-P	standard conditions	60.7	12.1	7.1	1) extraction co-products	phycocyanin, β-carotene, chlorophyll, B12, nutraceuticals (BAA)
					2) HTL	bio-oil
	stress conditions 1[1]	39.2	26.1	34.7	transesterification / acetone-butanol-ethanol fermentation/ HTL	biodiesel / biohydrogen, biobutanol, ethanol/bio-oil
	stress conditions 2[2]	24.3	30.9	24.3		
	stress conditions 3[3]	38.7	32.0	29.3		
A. platensis rsemsu 1/02-T	standard conditions	65.6	13.5	5.3	1) extraction co-products	phycocyanin, β-carotene, chlorophyll, B12, nutraceuticals (BAA)
					2) HTL	bio-oil
	stress conditions 4[4]	30.5	36.1	33.4	transesterification / acetone-butanol-ethanol fermentation / HTL	biodiesel/ biohydrogen, biobutanol, ethanol/bio-oil
	stress conditions 5[5]	32.6	47.1	20.3		
Haematococcus pluvialis rsemsu Hp-1/11	standard conditions	27.2	4.0	60.8	HTL/ acetone-butanol-ethanol fermentation	bio-oil/ biodiesel/ biohydrogen, biobutanol, ethanol
	stress conditions 6[6]	13.3	16.8	60.9	1) extraction co-products	astaxanthin
					2) transesterification/acetone-butanol-ethanol fermentation/ HTL	biodiesel/ biohydrogen, biobutanol, ethanol/bio-oil

[1] stress conditions 1- Zarrouk's medium [7] that does not contain nitrogen and phosphorus; cultivation in open ponds of the volume V=20.0 l including the air barbotage (containing 2% CO_2); steady-lighting conditions I=(160±5), μE/(m^2s); T=25±1°C; growing and photosynthesis (duration) – 15 days.

[2] stress conditions 2 – Zarrouk's medium that does not contain nitrogen and phosphorus; cultivation in open ponds of the volume V=20.0 l including the air barbotage (containing 2% CO_2); steady-lighting conditions I=(250±5), μE/(m²s); T=25±1°C; growing and photosynthesis (duration) – 14 days.
[3] stress conditions 3 – Zarrouk's medium that does not contain nitrogen and phosphorus; cultivation in open ponds of the volume V=20.0 l, including the air barbotage (containing 2% CO_2); steady-lighting conditions I=(215±5), μE/(m²s); T=25± 1°C; growing and photosynthesis (duration) – 15 days.
[4] stress conditions 4 – ½ Zarrouk's medium; cultivation in closed bioreactors of the volume V=1.0 l including the air barbotage (containing 2% CO_2) tank in the reactor; lighting I=(55±5), μE/(m²s); lighting requirements: light cycles (night/day)=16:8; T=21±1°C; growing and photosynthesis (duration) – 10 days.
[5] stress conditions 5 - Zarrouk's medium that does not contain nitrogen and phosphorus; cultivation in open ponds of the volume V=1.0 l; steady-lighting conditions I=(400±5), μE/(m²s); T=21±1°C, growing and photosynthesis (duration) – 2 days.
[6] stress conditions 6 - Optimal Haematococcus Medium (OHM) [8] that does not contain nitrogen and phosphorus; cultivation in open ponds of the volume V=1.0 l; steady-lighting conditions I=(450±50), μE/(m²s); T=25±1°C; growing and photosynthesis (duration) – 1 day.

3 Discussion

As indicated above, the profitability of fuel production from biomass is determined by competitive prices for hydrocarbon resources. Therefore, at low prices for oil, coal, natural gas, it is economically justified the use of microalgae to produce an expensive by-products (Table 1), and converting of waste to biofuels (e.g., HTL method).

It is shown that as a result of the selection of methods for the microalgae cultivation the biomass with the desired properties can be produced. Thus, the biochemical composition of *A. platensis* and other MACs showed in the experiments a high lability in relation to environmental conditions. If biomass *A. platensis st. rsemsu 1/02-P* and *st. rsemsu 1/T-02* under standard conditions contained a large amount of protein (60.7±2.5% and 65.6±2.7%), and lipid and carbohydrate contents are low (12-13% and 7-5%, respectively), in the stress conditions the biochemical composition of both strains becomes more balanced (content of proteins is reduced to 39-24%; content of neutral lipids is increased to 36-47%, and the carbohydrate content - up to 20-34%). *H. pluvialis* in response to stress (within 1 day of exposure) reached a significant increase in lipid content (4.2 times), which testifies to its considerable potential. Biosynthesis of the expensive antioxidant - astaxanthin - in the biomass of algae significantly increases the profitability of biofuel production [9].

Thus, the potential biomass of the MAC after the stress may be universal, suitable for producing biodiesel by transesterification of lipids and biodiesel by HTL technology, and to produce biohydrogen, bioethanol and biobutanol. However, first of all, to improve the economic attractiveness of using microalgae it is important to extract from the biomass the most valuable components - phycocyanin, β-carotene, chlorophyll, B12 (spirulina), astaxanthin, β-carotene (haematococcus). In case of spirulina rich in protein biomass should be used as a nutraceuticals (BAA).

As noted above, the bio-oil derived from algae has high nitrogen content. Increasing lipid content in cells by means stress and preliminary extraction of proteinaceous compounds may provide substantial reduction of nitrogen content in bio-oil.

For Russia, the production of biofuels from algae is important in connection with the need to achieve a European standard fuel, i.e., for production of gasoline additives. Our experience in large-scale microalgae cultivation provides the basis of organizing industrial production in the Russian Federation [3, 10]. In connection with the above conditions of microalgae cultivation the task of evaluating territories in terms of their suitability for large-scale microalgae production can be formulated. The list of the most important climatic factors includes the following:

- Illumination of 4.0 kWh/m^2/day considered sufficient for the microalgae production;
- Preferred ratio of light and dark phases during growth shall be not less than 6:18, as days with less than 6 hours of sunshine are considered to be insufficient for algae growth;
- Cultivation temperature 12°C. As shown by numerous experiments, spirulina/arthrospira stops growth at temperatures below 12°C. Various types of microalgae grow over a wide temperature range, but usually, they are sensitive for freezing and therefore cannot survive lowering of the temperature below 0°C.

For mapping of these factors we used regular spatial grid data from the database NASA Surface meteorology and Solar Energy. Comprehensive analysis of the spatial distribution of the above climatic factors allowed us to identify territories at the South of Russia for more detailed study, which include the Southern, North Caucasian and Crimean Federal Districts, where the climatic conditions are optimal for the cultivation of MACs (Figure 1). The following infrastructural factors were identified for mapping on regional and local scale: 1) availability of water, 2) CO_2 sources, 3) low-temperature waters, 4) the available lands for commercial microalgae cultivation (Figure 2).

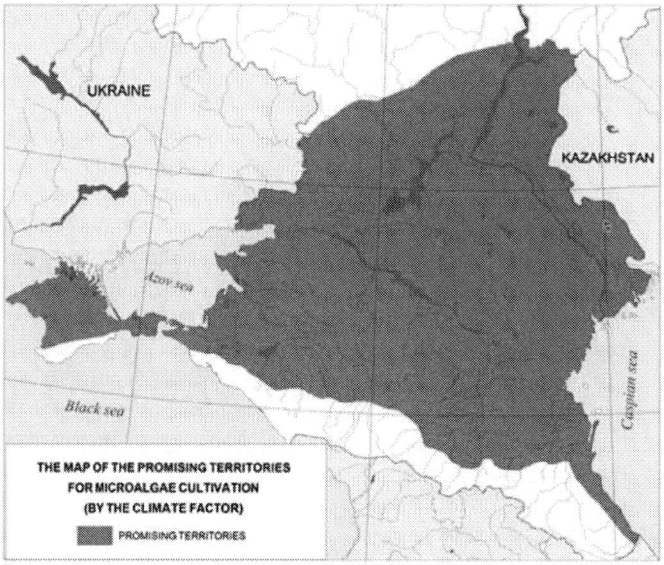

Fig. 1. The map of the promising territories.

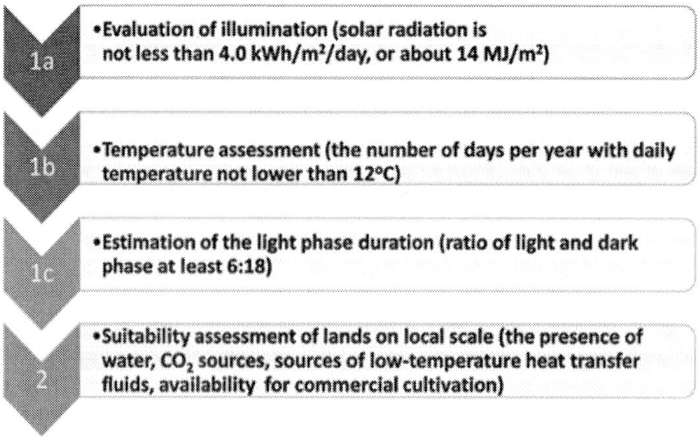

Fig. 2. The sequence of the territory assessment for the large-scale microalgae production by the climatic (1a, 1b, 1c) and infrastructural factors (2).

4 Conclusions

The experimental results of the stressful conditions impact on microalgae strains prove the ability of direct target products obtaining. This allows us to expand the range of biomass processing technologies with appropriate ones for MACs use (transesterification-acetone-butanol-ethanol fermentation-HTL) and a list of relevant energy and by-products (biodiesel, biohydrogen, bioethanol, biobutanol, bio-oil, phycocyanin, beta-carotene, astaxanthin, chlorophyll, B12, feed, food additive). Thus, the sensibility of the biomass to the effects of environmental factors enhances the opportunity for cost-effective microalgae use.

On the basis of own long-term experimental data and literature we stated the list and ranges of climatic and infrastructural factors for selecting suitable territories for large scale microalgae production. Analysis of the spatial distribution for the climatic factors allowed to identify territories of Russia, where the cultivation of MACs could be effective.

This research has been supported by Russian Science Foundation (project № 17-19-01617). The analysis of optimal processing technology applied to investigated strains microalgae was carried out by Vlaskin M.S. and Kiseleva S.V. in Joint Institute for High Temperatures of the Russian Academy of Sciences.

References

1. L. Rodolfi, G. Chini Zittelli, N. Bassi, G. Padovani, N. Biondi, G. Bonini, M. R. Tredici. Biotechnol. bioeng. **102**, 100-112 (2009)
2. L. Raslavičius, V. Semenov, N. Chernova, A. Kersys, A. Kopeyka. Renew. Sust. Energ. Rev. **40**, pp. 133-142 (2014)
3. N. Chernova, T. Korobkova, S. Kiseleva. Therm. Eng. **57**, 937–945 (2010)
4. N. Chernova, T. Korobkova, S. Kiseleva, S. Zaytsev, N. Radomskii. *Sust. Man. Selig. G.* (Springer-Verlag, Berlin, 2012)
5. N. Chernova, S. Kiseleva, S. Zajcev. Altern. Energ. Ecol. **12**, 70-83 (2014)
6. D. Barreiro, W. Prins, F. Ronsse, W. Brilman. Biom. and Bioen. **53**, 113-127 (2013)
7. *Katalog cultur microvodoroslei v collektsiyakh SSSR* (in Russian) (IPPRAS Publ., Moscow, 1991)
8. J. Fábregas, D. Domínguez, M. Regueiro. Appl. Microbiol. Biotechnol. **53**, 530-535 (2000)
9. M. Huntley, D. Redalje. Mitig. Adapt. Strateg. Glob. Chang. **12**, 573–608 (2007)
10. N. Chernova, S. Kiseleva., O. Popel'. Therm. Eng. **61**, 6 (2014)

Autonomous power supply using solar energy in Russian Far East regions

Nina A. Komarova[1], *Yulia Yu.* Rafikova[1,2,*], *Alexey B.* Tarasenko[2], and *Sofya V.* Kiseleva[1,2]

[1]Faculty of Geography, Lomonosov Moscow State University, Moscow, Russia
[2]Joint Institute for High Temperatures of Russian Academy of Science, Moscow, Russia

Abstract. The most useful application of PV solar power in Russia are autonomous power systems in regions with high costs of organic fuels (due to transportation problems). Effective application of PV power needs comprehensive data analysis for solar energy resources, electric and heat load graphs and fuel costs. This paper is devoted to climate and power load data analysis for Russian Far East regions (in connection with PV application) and PV-based power plant mathematical modeling for several locations.

1 Introduction

Due to the large area, complex and heterogeneous climatic and geographical conditions of Russia, as well as the extremely non-uniform distribution of fuel and energy resources (FER), a significant part of the country is characterized by high production cost for electricity and heat. Therefore, one of the most important socio-economic problem of the country is to ensure reliable electricity supply in the territory of the Russian Federation.

Autonomous power sources are also needed, especially for consumers at the dead-end power lines, in some regions of the country covered by the centralized power grid, due to restriction of joining the grid and runout of generating and transmitting equipment. In areas with a developed centralized power grid, there is also a number of local consumers with a temporary need for power supply from independent sources. Electrical equipments, temporary buildings and structures used during construction of roads and other infrastructure could become such a consumers.

In decentralized area main energy consumers are local settlements. The island settlements automatically fall into this category. Isolated consumers are receiving electricity from the local diesel gensets, or in some rare cases, using local energy resources such as wood, coal, peat, etc. Usually transport expenses share in cost of energy received from gensets is quite high

For decentralized zones the combined or hybrid power plants, which include diesel generators and systems based on renewable energy sources (solar and wind power plants) could be effective. The most important direction in the development of decentralized energy supply is photovoltaic application in these areas. Despite relatively high costs, photovoltaic is less sensible to low local renewable potential, than wind power. Also non-

*Corresponding author: ju.rafikova@gmail.com

uniformity of solar energy potential throughout the region is much less, than for wind. However, in this case, due to sharp seasonal fluctuations in solar radiation, there is a problem of generated electricity storage, including the complex systems of energy storage.

It is common, that main contribution at the cost of energy generated by diesel power plants make fuel costs. Therefore, the ratio of investment in new capacity and cost of fuel displaced are the main factors, which determine the choice of an autonomous source of power generation. Selection of the optimal composition of autonomous power plants with different types of power generators and batteries due to specific geographical conditions is a difficult task and is carried out on the basis of reliable data on the load of consumers, the potentials of renewable energy resources, climatic, geological data, the cost of fuel oil, the degree of assurance of power and a variety of other factors.

2 Analysis of the typical autonomous energy consumers in the decentralized zone of Russian energy supplies

The main parameters that determine the character of electrical load of large autonomous consumers, are the type of load (in general case, municipal, social or industrial) and the number of consumers [1]. Determination of typical electrical load for decentralized consumer is a complex task due to climatic, geographic and technical characteristics of each particular region and locality. In addition, the obtaining and analysis of real load curves are complicated by the absence of detailed monitoring of energy consumption data (summer and winter days regime). Data of the magnitude of the peak power consumption of electrical energy in the region are more available.

The data, that characterize the potential consumers of photovoltaic power plants in different regions of Russia, was collected and processed, allowing to analyze the typical energy consumer in decentralized areas. Two types of these areas were identified based on their consumers specific features: a) the decentralized energy supply areas (Far East Federal District, the Arctic zone of the Russian Federation (AZRF), island territories); b) the areas of centralized power supply with the limitations of electric power supply and the presence of niche consumer (Moscow and Krasnodar region).

Data systematization and analysis allowed to identify the following groups of electricity consumers in the Russian Arctic and Far Eastern Federal District as a whole:
0.1-1 kW - autonomous lighting for road, sea and river transport, telecom repeaters, weather stations, signal units (buoys, beacons, lights airfield), including objects of ground navigation and communication of Northern Sea Route (NSR);
1-10 kW - individual settlements, isolated farms, border posts, telecommunications systems, large navigation and communication systems airfields and of NSR for year-round use;
10-100 kW – island villages, mining and gold mines, tourist camps;
100-500 kW - coastal and island sites of service of the Northern Fleet and NSR, small production and processing enterprises, settlements with a population of 100-500 people;
500-1000 kW - reloading base, fishing farms, large deer farms, shift villages and towns (up to 1000 people);
1000 kW - large settlements (more than 1000 people), port facilities; military facilities, villages, enterprises of forestry, mining and oil and gas industry.
The main types of consumers are: 1) navigation and connections facilities; 2) mobile consumers; 3) port complexes; 4) local settlements.

The data array, characterizing the indicated types of consumers, has been created on the basis of statistical data on energy consumption for isolated local settlements in Far Eastern Federal District and the Russian Arctic (Kamchatka Territory, Altai Territory, Republic of Buryatia, Irkutsk Region, Republic of Sakha (Yakutia)) and are different by the municipal districts. Currently, power supply is based on diesel generators for the most part of these consumers.

As an example, Figure 1 shows the load curves (summer and winter regime day) for a group of villages in Kamchatka. The power consumption is normalized to the maximum power value among all settlements for each hour (for June and January).

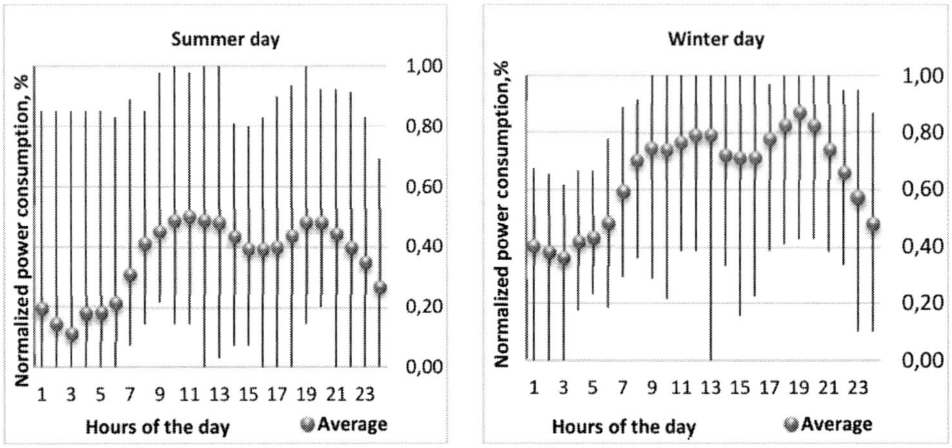

Fig. 1. Normalized load curves (average values and dispersion) for isolated local settlements, Kamchatka region.

Because of the correlation between population and consumption of electric power the maximum power frequency consumption diagrams were built for all isolated local settlements groups (Figure 2). It can be seen from the figure, that the logical shift of the maximum power consumption varies from 0-100 kW (population up to 500 people, Republic of Sakha (Yakutia) and Irkutsk region) to 400-2000 kW (population up to 1000-5000 people, Kamchatka).

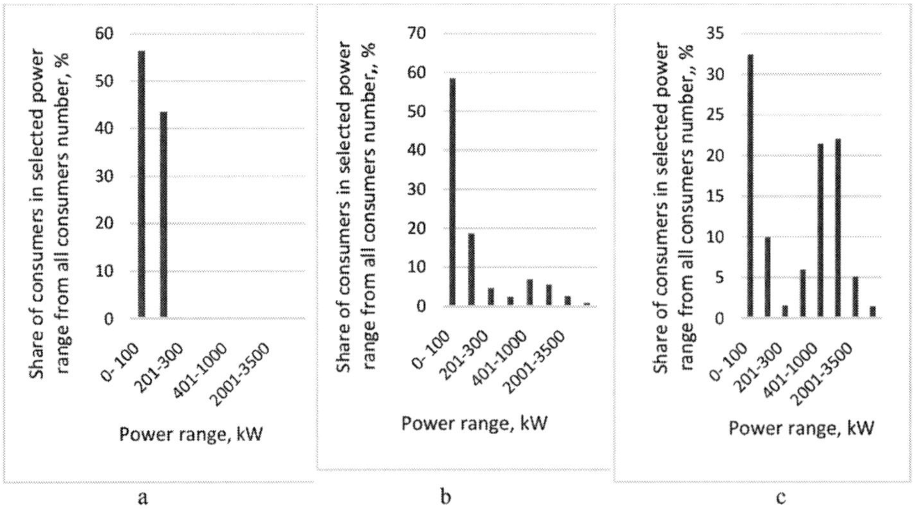

a b c

Fig. 2. Maximum power share distributions for Irkutsk region (a), Kamchatka (b) and the Republic of Sakha (Yakutia) (c).

All of the identified samples, as well as standard load curves allow to simulate the concrete load curves (summer and winter days regime) even in the absence of detailed monitoring from human settlements.

In the presence of typical load schedules for optimization of power supply in remote decentralized settlements of the Russian Federation Far Eastern region, including using renewable energy sources, it is necessary to carry out modeling, solar-diesel or wind-diesel electrical installations with accumulation. However, due to the complexity of both the modeling itself and the conduct of feasibility studies at the first stage of research, we carried out optimization of energy supply for the first group of consumers discussed above – for navigation facilities.

3 Optimization calculation for autonomous consumer powered from photovoltaic and batteries using load graph and climate data

Optimization calculations were performed for autonomous lighting-signal unit powered from PV array (Figure 3) using mathematical model based on hour, month and year energy balance calculation.

$$W_{PV} = W_{acc} + W_{con} \tag{1}$$

W_{pv} – energy, produced by PV array, W_{acc} – stored energy, W_{con} – energy, consumed by lighting-signal unit.

Minimal cost at 100% covering of given load graph was taken as optimization criteria. Leading lighting-signal marks on Tarantseva island (42.05°N 131.08°E) to south of Vladivostok and Rechanya Matuga (61.68°N 160.23°E) in Kamchatka were selected for investigation. First location is operated during the whole year, but the second – only in summer navigation (from April to October, [2]), so load graphs and climate conditions are quite different. Such autonomous consumers can be characterized by low power (10-100 W) which doesn't allow efficiently use diesel genset, stable energy consumption during the year, absence of operative staff in the location. Last circumstance demands very high reliability.

Multi-crystalline PV module RZMP 155 (R'azan factory for metal-ceramic devices JSC) parameters with peak power of 155 W were used in both calculations. Tilt angle of latitude+15° was selected to increase winter energy output and prevent PV array snow covering.

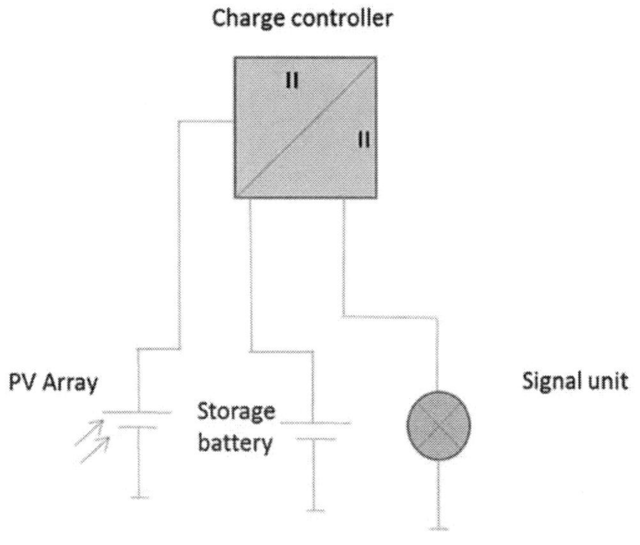

Fig. 3. One-line electric scheme for autonomous lighting-signal unit powered from PV

Mathematical model includes calculation of hourly rows for temperature and solar radiation data based on approach described in [5] and using NASA Power data [6], calculation (for every hour) of PV array energy production, depending on solar radiation, temperature and charge controller efficiency (2)-(4):

$$T_{pv} = T + (\frac{G}{800} \times (NOCT \quad - 20)) \tag{2}$$

$$P_{max} = P_{max}^{STC} \times \frac{G}{1000} \times (1 + \gamma \times (T_{pv} - 25)) \tag{3}$$

$$P_{op} = P \times \eta_c \tag{4}$$

Here T is environment temperature (average between maximal and minimal value from NASA Power), T_{pv} – operational temperature of PV array, G – solar radiation, obtained from NASA Power and transformed into hour data for tilted surface using approach from [5], NOCT – PV cell operation temperature at 800 W/m^2, 20°C, P_{max} – maximal available power output of PV module at T and G conditions, $P_{max}{}^{STC}$ – at standard test conditions (1000 W/m^2, 25°C), γ - PV module power temperature coefficient, P – minimal power value between power consumption for battery charging and signal unit feed and P_{max}, P_{op} – operational hour power value of PV module, including charge controller efficiency dependence on input power. For most types of available charge controllers this dependence is characterized by large efficiency drop for very low input power. Approximation for this dependence was based on controller passport data and experiments and is close to

$$\eta_c \sim \frac{1}{(P/_{Pmax})^2} \tag{5}$$

Storage battery based on lithium-ion ($LiC_6||LiFePO_4$) accumulators was integrated into model using its efficiency (84-86%) measured experimentally for LFP200 AHA (Winston Battery Co) on lower charge-discharge currents, close to those in solar-powered power unit. Energy, indicated as W_{acc} in (2), was multiplied by 0,85 according to losses on storage battery efficiency before further usage for covering night or low-insolation days load graph. Calculation results are shown in Table 1.

Table 1. Optimal combination of PV and batteries for PV powered autonomous leading marks.

Location	Geography	PV peak power, W	Battery capacity, kWh	Capital costs, USD
Rechanya Matuga	61.68°N 160.23°E	465	10,2	9900
Tarantseva island	42.05°N 131.08°E	465	10,8	9000

It's worth mentioning, that in two different regions the power units for leading marks have similar structure. The problem is not only in climate conditions (though heavy rains season in summer is characteristic for Tarantseva island, increasing need in batteries to run through it and PV modules to charge the batteries), but also in load graph – absence of consumption in winter allows to store energy in batteries to use it mostly in April for Rechnaya Matuga. Assuming the whole year operation of leading mark in Rechnaya Matuga one will need 930 W of peak PV power with 15,4 kWh battery and will have capital costs of 15380 USD. So load graph has the same importance as climate conditions.

4 Conclusions

The most promising consumers of energy from renewable sources in the Russian Federation are located at the Far Eastern region. On the basis of statistical information processing, the following types of consumers were identified: 1) navigation and connections facilities; 2) mobile consumers; 3) port complexes; 4) the inhabited localities. For settlements in the three regions of the Russian Far East (Kamchatka, Irkutsk region, Yakutia) we constructed typical load curves for the summer and winter regime days (average values and dispersion at each hour of the day) normalized for the maximum power consumption among the villages of each region. By the presence of maximum power consumption for each group of localities during the day and at least one data on the power consumption during the day it is possible to restore a typical load schedule for a particular village. This makes it possible for further modelling and optimization of the energy supply for these settlements, including use of renewable energy sources

The paper presents the results of optimization modeling for an autonomous lighting-signal unit, where power supply can be carried out using a solar system with accumulation. The modeling has been carried out for two points in the Primorsky Territory (the Tarantsev Island) and the Magadan Region (the River Matuga), which are characterized with very complicated climatic conditions, PV peak power - 465 W, battery capacity - 10 kWh and capital costs - near 9000$.

This research has been supported by Russian Science Foundation (project № 16-19-10659). The analysis of the typical autonomous energy consumers and optimization calculation for autonomous consumer were carried out by Tarasenko A.B., Kiseleva S.V. and Rafikova Yu.Yu. in Joint Institute for High Temperatures of the Russian Academy of Sciences.

References

1. E.N. Sosnina, A.U. Kechkin, D.A. Filatov. Proceed. of the NNTU **107**, 100-105 (2014)
2. O.S. Popel, S.V. Kiseleva, M.O. Morgunova, T.S. Gabderahmanova, A.B. Tarasenko. The Arctic: ecol. and econ. **1** (17), 64-69 (2015)
3. N.N. Melnikov, V.P. Konuhin, V.A. Naumov, S.A. Gander. Vestnik MSTU **18**, 198-208 (2015)
4. www.gisre.ru
5. J. A. Duffie, W. A. Beckman. *Solar Engineering of Thermal Processes* (John Wiley and Sons, Inc, Hoboken, New Jersey, 2006)
6. https://power.larc.nasa.gov

The scientific studies on smart grid in selected European countries

Serhat Orkun Tan[1,*], *İlker* Türker[2], and *Türker* Toku[3]

[1]Karabuk University, Department of Electrical and Electronics Engineering, Karabuk, Turkey
[2] Karabuk University, Department of Computer Engineering, Karabuk, Turkey
[3] Karabuk University, Department of Information Processing Management, Karabuk, Turkey

Abstract. Smart grid is a power system consisting of many transmission and distribution systems subjected to an automation which are efficient, reliable and coordinated with each other. As a nature friendly technology, Smart grid come into prominence due to the increasing energy consumption and limited renewable energy sources around the world. In the near future, the use of renewable energy sources is not expected to grow rapidly; but the transmission and distribution systems will be enhanced by Smart grid technologies. Considering these significant benefits, the studies have been increased on Smart grid technologies to meet the energy requirement in each country. Herewith, the aim of this study is to analyse the scientific studies in developed European countries such as Italy, Germany, United Kingdom, France and Spain to find out the increment rate of the importance devoted to the Smart grid technologies in academicals manner. The scientific researches on Smart grid are achieved from the Web of Science database and the statistical analysis have been made by utilizing proper SQL queries in combination with Excel Power Pivot for these countries. The correlation between the scientific studies on smart grid and the virtual smart grid applications are also outlined for each selected country.

1 Introduction

The predictions for future energy consumption and the calculations for the amounts of natural reserves remaining in the world and their consumption times are made in seminars and symposiums which are organized all over the world due to increasing energy demand. According to the scenario prepared by the International Energy Agency (WEO2016), the world's energy needs continue to grow and a 30% rise in global energy demand to 2040 means an increase in consumption for all modern fuels. But global aggregates mask various trends and significant changes between fuels. Additionally in 2040 hundreds of millions of people will be left without basic energy services [1]. From this point of view, this future concerns require the application of methods such as smart grid to achieve safer and sustainable energy [2-3].

Smart grid (SG) networks include smart meters and monitoring systems to monitor and update the electricity networks and provide user safety while ensuring consistently reliable

* Corresponding author: serhatorkuntan@karabuk.edu.tr

and high quality distribution [4-6]. Adaptation to existing technologies, commercial appealingness, continuous developments and investments are significant reasons that deployed SG all over the world at different rates. Implementing SG across countries that faced huge demand and aging electrical infrastructure is crucial to make the energy grids more reliable, efficient and sustainable in these countries. In addition, SG is a significant factor in promoting the utilization of technologies including electric vehicles and renewable energy sources such as geothermal energy, wind energy and solar heat and power [3, 7-10]. The importance given to the applications on SG is necessary for countries when taking into account economic and environmental interests for a near future [11-14]. The modernization of the electricity network with new installations is also a reason for the rapid development of SG especially in various regions of the world [15, 17]. For the most efficient use of energy, the need for SG all around the world is increasing day by day in terms of energy management and integration with renewable energies [18].

Considering these important contributions to the energy sector, many researchers around the world make scientific studies that shed light on SG applications. A study on SG technologies in developing countries obtained the data from ISI Web of Knowledge and presented the published items on SG for 20 years up to 2012 [3]. As can be seen from Figure 1, there were fewer than 50 academic studies before 2008. After this year smart grid technology has become a hot issue and a very rapid increase in publications has been observed in 2011. As a result, SG technology is a widespread topic, and studies, applications and investments are expected to increase rapidly in the future.

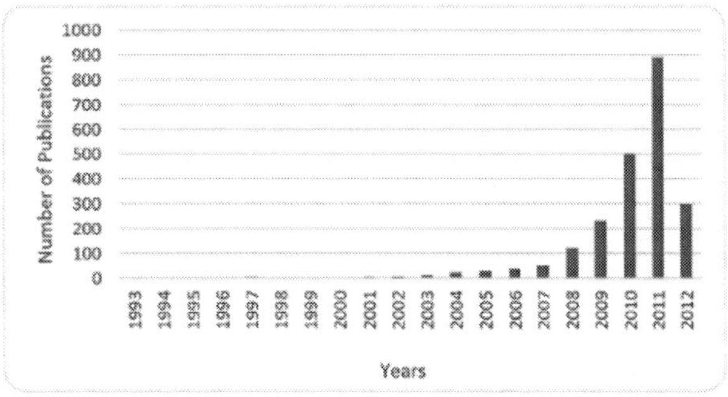

Fig. 1. Published items on smart grid for 20 years up to 2012.

2 Methodology

Depending on the data at Web of Science database, automatic retrieval of information was made by a data retrieval method which used a data retriever program to read the clipboard periodically if any new data has been placed therein. Through the program, each new data search is checked whether the pre-determined criteria can be met. In addition, if the program meets the criteria, it initiates data retrieval by performing a web based search based on the data. If the criteria are not met, the program will not perform data retrieval. [19]. Data extraction from the Web of Science database and statistical analysis with the SQL server management studio is done through the operation logic of the Automatic Data Collection Method. As keywords to focus our study, we entered the words "smart grid" and also as a compound word of "smart grid" to the Web of Science database for not to skip any publications in this area. The statistical distributions of the researches made on the smart grid and the analysis values of the data were transferred to Excel Power Pivot and the analysis results were visualized.

DOI: 10.1051/matecconf/201711210012

Fig. 2. The data retrieval schematic diagram of selected countries from Web of Science

3 Results and discussion

Initiatives on SG have been attracted considerable attention in many countries around the world. The production and distribution of energy varies due to the increase in energy demand, reliability, and environmental concerns. Consequently, organizations around the world are searching for more effectual and reliable methods to energy production and distribution. The "smart grid" has revealed as an operative way to manage the peak load of energy, power supply integration from renewable energy sources and increase consumer control over domestic usage of energy [12, 20-22]. UK, Germany, Italy, Spain and many other developing countries are adopting SG technology to prevent energy pilferage.

In recent years, many countries have invested significant amounts in SG technology and the global market is expected to grow continuously for ten years. The cumulative data for the budget of SG projects in Germany, Italy, France, Spain and UK is presented in Figure 3. The data in the figure is assumed as all the budget of related countries is allocated in the starting year. The total budget is calculated by the sum of research and development (R&D) and demonstration and deployment (D&D) budgets of smart grid projects for the countries.

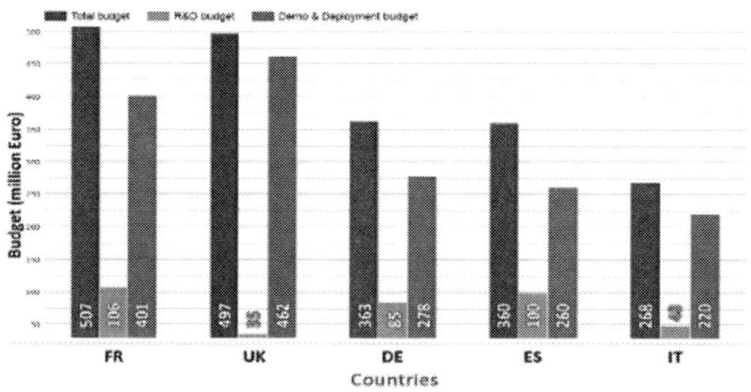

Fig. 3. The cumulative data for the budget of Smart grid projects of the selected countries [23].

As seen from Figure 3, France and the UK are the countries with the highest investment in SG projects. The highest budget was allocated by France and Spain to the R&D and especially Spain comes first in ratio by allocating nearly 30% of its total budget to these projects. On the other hand, UK is leading by allocating more than 90% of its budget to the D&D but it takes also the last place by 35 million euro budget to the R&D. Generally, Figure 3 can be evaluated as, all countries give priority to allocate most of their budgets to the D&D where France, Spain and almost Germany give more importance to the R&D for Smart grid projects. Considering the budgets allocated to SG projects, Figure 4 shows the linkages of cooperation in European multinational projects. Approximately 170 projects carried out by multinational consortia and collaboration of European countries on these projects is presented in Figure 4 [23].

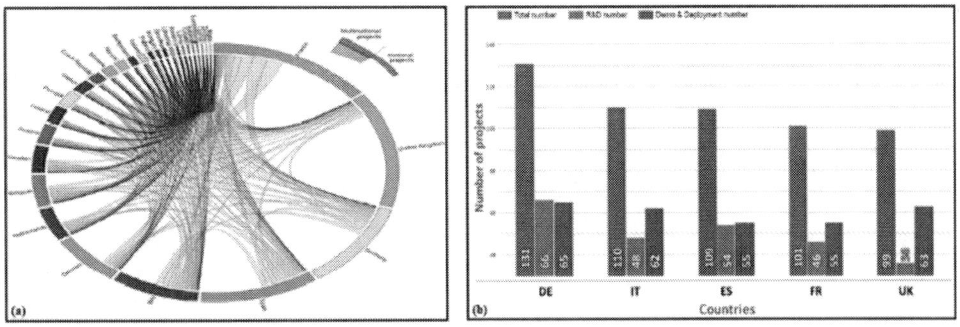

Fig. 4. a) The linkages of cooperation in European multinational projects, **b)** The total smart grid project numbers of selected countries [23]

The most striking point in Figure 4.a is the multinational projects region that, French, Germany, Spain and Italy allocated nearly 50% of their budget to the multinational projects where UK allocated a shoestring budget to these projects. As mentioned in Figure 3, UK allocated 90% of its budget to the D&D and this can also be explained in Fig.4.a as UK allocated many of its budget to the national SG D&D projects. This situation is also verified in Fig.4.b which indicates the total SG project numbers of selected countries. As seen Figure 4.b, apart from UK the other selected countries have almost the same number of R&D and D&D projects. The budget of multinational projects on SG for these countries are also proportional to the project numbers on SG projects. Considering R&D projects on SG, UK has least project numbers and this can be explained by its budget on R&D projects and also its budget on multinational projects. All in all, the budget allocated for D&D projects is more than the budget for R&D projects when considering the total budget and number of projects together. It is important to reveal the relation between the academic and scientific studies on the SG technology and the budgets and project numbers that countries allocate to this technology in order to see how the applications interact with the scientific studies in this area. From this point of view, Figure 5 indicates the number of publications that each selected countries have been made for 8 years. As seen from Figure 5, the average number of publications increase by years and this is also in direct proportion to the developments, budgets and projects on the SG technology. Considerable number of publications have been made during 2009-2016 period as SG technology has become a hot issue around the world.

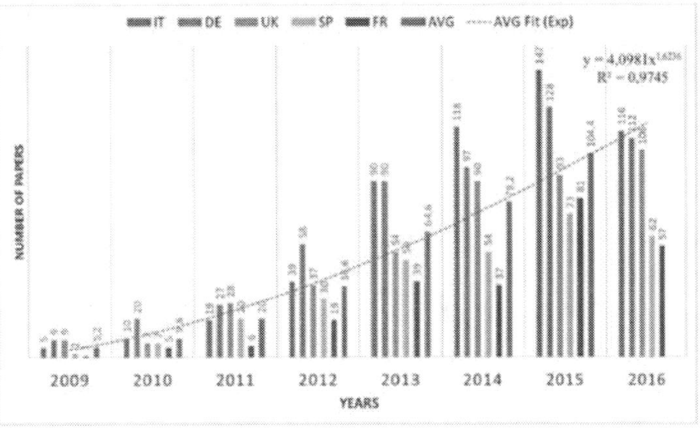

Fig. 5. The number of publications that each selected countries by years.

It is clearly seen in Figure 6.a that the same rate of increment is observed in this time period in the worldwide publications. 2015 was a very efficient year for both Europe and

 DOI: 10.1051/matecconf/201711210012

world when taking into account the number of publications on SG. It was also observed that there were very few publications on SG before 2008, when the publications made around the world were evaluated since 1975. Figure 6.b presents the number of publications of the most publishing countries in the world from 1975 to this time. USA and China are far ahead of other countries in terms of the number of publications as shown in Figure 6.b.

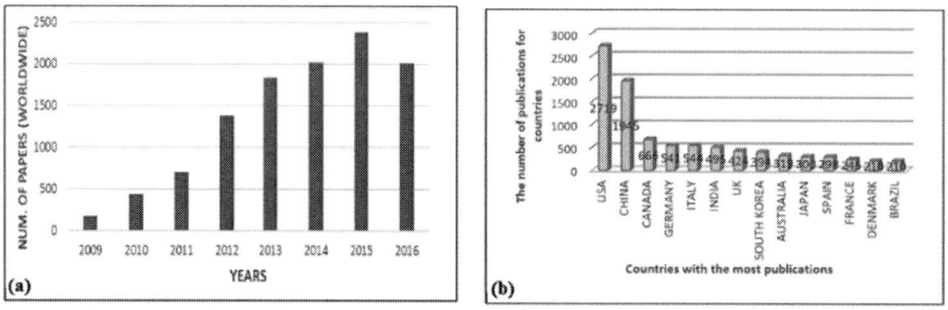

Fig. 6. a) The total number of publications all over the world by years, **b)** The distribution of the top publishing countries worldwide

In general, it can be assumed that the numbers of the publications of these five prominent European countries are similar to the project numbers and budgets allocated to this area. Especially when the numbers of publications and project numbers examined, it can be seen as, the publication numbers of the UK are slightly ahead of their project numbers and this also changes the ranking in this sense. However, the fact that the numbers are close to each other makes it possible to assume that the scientific studies or publications increase with increasing applications or projects in countries basis. A quantitative way to abstract the relation between data series is extracting correlations between them. Considering the total numbers of budgets (B), projects (PR) and publications (PU) for the selected countries, we calculate the correlations as: B-PR: -0.537, B-PU: -0.617 and PR-PU: +0.603. Here, the only positively correlated pair is projects and publication counts. But elaborating the components of the first two parameters as R&D and D&D subsections, the correlations are extracted as following: BRD-PRRD: +0.536, BRD-PU: -0.621, PRRD-PU: +0.305. As seen, for R&D data, budget and project pair becomes positively correlated, while the correlation for projects-publications decrease and budget-publication remains same. Lastly, evaluating the data for only D&D sections, we find correlations as: BDD-PRDD: -0.046, BDD-PU: -0.42, PRDD-PU: +0.914. These results show that project and publication counts are very strongly correlated for the D&D data. Such a strong correlation close to 1 is an indicator of a possible relation between publications funding between the D&D budget.

4 Conclusions

The governments of many developed countries have significant critical objectives as economic growth, environmental sustainability and energy security. The SG is one of the critical way to achieve these objectives. Thus, European developed countries such as Germany, UK, France, Spain and Italy have regional and multiregional projects and allocate budgets to SG. In addition to this SG technology in these countries is also supported by scientific or academic studies. In this study, the correlation between the scientific studies and the projects and allocated budgets have been investigated by comparing the number of studies, the number of projects and total allocated budgets of the related European countries. It has been observed that the allocated budgets, projects and scientific studies increase years by years and vary in direct proportion to each other. This increment is also similar in countries all over the world. Consequently, it is clear that the importance given to SG technology have increased each passing year by the projects and the allocated budgets and this increase is supported by the increment of scientific or academic studies made in the selected leading European countries.

We thank to Scientific Research Projects (BAP) Coordinatorship of Karabuk University for funding this study with project code of KBÜBAP-17-YD-230.

References

1. www.iea.org, "*World Energy Outlook 2016*", International Energy Agency (IEA)
2. C. Cecati, C. Citro, P. Siano, IEEE Trans. Sustain. Energy **2** (4), 468-476 (2011)
3. M. Fadaeenejad, A. M. Saberian, M. Fadaee, M. A. M. Radzi, H. Hizam, M. Z. A. AbKadir, Renew Sust Energ Rev **29**, 828–834 (2014)
4. S. Kaufmann, K. Künzel, M. Loock, Energy Policy **53**, 229–239 (2013)
5. H. Gharavi, R. Ghafurian, Proceedings of the IEEE **99**, 6, 917–921 (2011)
6. G. Rietveld, J. P. Braun, R. Martin, P. Wright, W. Heins, N. Ell, P. Clarkson, N. Zisky, IEEE T Instrum Meas **64** (6), 1355-1363 (2015)
7. N.C. Batista, R. Melício, J.C.O Matias, J.P.S. Catalão, Energy **49** (1), 306–315 (2013)
8. D.S. Markovic, D. Zivkovic, I. Branovic, R. Popovic, D. Cvetkovic, Renew Sust Energ Rev **24**, 566–577 (2013)
9. A.S. Malik, M. Bouzguenda, Energy **54** (1), 365–371 (2013)
10. M. Alonso, H. Amaris, C. Alvarez-Ortega, Expert Systems with Applications **39** (55), 13–22 (2012)
11. J. Gao, Y. Xiao, J. Liu, W. Liang, C. L. P. Chen, Future Generation Computer Systems **28**, 391–404 (2012)
12. https://www.smartgrid.gov/files/Global_Smart_Grid_Federation_Report.pdf
13. M. M. Abdullah, B. Dwolatzky, Proceedings of IEEE AFRICON (2009)
14. https://www.eia.gov/analysis/studies/electricity/pdf/intl_sg.pdf
15. M. G. Simões, IEEE Transactions on Industry Applications **48** (2), 1154-1162 (2012)
16. D. Mah, J. Vleuten, P. Hills, J. Tao, Energy Policy **49**, 204-216 (2012)
17. J. Yuan, J. Shen, L. Pan, C. Zhao, J. Kang, Renew Sust Energ Rev **37**, 896–906 (2014)
18. N.S. Nafi, K. Ahmed, Mark A. Gregory, M. Datta, J Netw Comput Appl **76**, 23–36 (2016)
19. S. E. Waggoner, *US8762401 B2* (2010)
20. N. IqtiyaniIlham, M. Hasanuzzaman, M. Hosenuzzaman, Renew Sust Energ Rev **67**, 776–790 (2017)
21. http://ses.jrc.ec.europa.eu/sites/ses/files/documents/smart_grid_projects_in_europe_lessons_learned_and_current_developments.pdf
22. H. Farhangi, IEEE Power Energy Mag **8** (1), 18-28 (2010)
23. http://ses.jrc.ec.europa.eu

Elemental and phase characterization of quartz samples from Meliti (Florina, Greece)

Georgios Charalampides*, *Anastasia* Liana, *Konstantinos* Vatalis, and *Nikolas-Ploutarch* Benetis

Department of Environmental Engineering, Technological Educational Institution of Western Macedonia, Koila Kozani, 50131, Greece

Abstract. In this work, a geological, phase and elemental characterization of quartz samples, deriving from Meliti area, is presented. The bedrock of the investigated area belongs to Pelagonian Nappe (Middle Triassic – Lower Jurassic) and consists predominantly of metamorphic basement including mainly mica schists where biotite dominates in relation to muscovite. Specifically, the quartz samples investigated in this study were collected at Moutsouna (Roudina) site about 3,5 km NNW of Meliti village. SEM-EDS, ICP-MS and XRD results indicate that quartz mineral originating from Meliti area is free from other microcrystalline phases. Hence, the quartz mineral found is highly pure and can be used for production of ultra-high purity quartz for special applications, as long as the deposits are found to be exploitable.

1 Introduction

Quartz is the second most abundant mineral in the Earth's continental crust, after feldspar. It is found in nature in varying purities and forms. Commercially, quartz is traded in various qualities for different industrial applications. It can be used either for high added value applications such as silicon-metal wafers, optical glass or PV panels or for ordinary applications such as foundry sand for metal castings or as a filler for adhesives and grouts [1-3]. Quartz deposits have been found in many regions in Greece. Most important-occurrences of quartz in Greece can be found in Lavrio and Penteli (Attiki), Soufli (Thrace), Petsofas (Lesvos), Avysallos (Serifos), Ios (Cyclades) and Meliti (Florina, Western Macedonia). The latter was the region that was investigated in the present study. The occurrence of quartz in each of the aforementioned places may have different forms and colors [4-6].

2 Geological setting

In general, the bedrock of the investigated area (Roudina near Meliti village, close to city of Florina, Western Macedonia Region, Greece) belongs to Pelagonian Nappe (Middle Triassic – Lower Jurassic) and consists predominantly of metamorphic basement including the following: (a) Schists of various mineralogical composition. They are mainly mica schists

* Corresponding author: chara@teiwm.gr

where biotite dominates in relation to muscovite (garnet schists, chlorite schists, epidote schists). Muscovite schists are also present in a smaller extent (chlorite – chloritoid schists, chlorite – garnet schists). (b) Basic metamorphic igneous rocks within the schists of the metamorphic system (region of the village Skopos) and in less extent, these igneous rocks are intercalated (amphibolites gneisses, metagabros, greenstones). (c) Orthogneisses consisting mainly of mica orthogneisses, augene gneisses and gneisses of schistone structure.

2.1. Sampling

Specifically, the quartz samples investigated in this study were collected at Roudina site about 3,5 km NNW of Meliti village (Figure 1) from Pleistocene aged conglomerate series consisting predominantly of various calcareous pebbles, with mainly calcitic matrix. Heavily bedded with a thickness up to some tenths of meters. The alternate with sandstones and grit and they develop in the areas of the villages Kella, Klidi and Vevi. Also the conglomerate series include sandstones, sand and red clays of river – torrential origin; the thinner elements gradually dominate towards the center of the basin. Even more, lenses of Neocene limestones and marbles are surrounding the small quartz occurrences of the Roudina site. These lenses consist of white-yellow to white-gray, compact, fossiliferous mainly limestones, marls alternating with gray-greenish, clayey marls and clays (Geological Map, Vevi sheet, IGME).

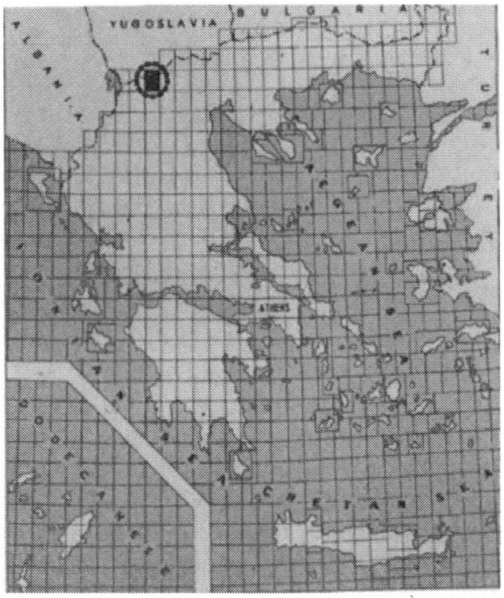

Fig. 1. The geological setting 40° 50' 0" North, 21° 35' 0" East of Meliti (Florina, Greece) in the map of Greece

The present geological survey is focused on the quartz mineral deposits observation in Roudina area. Samples from three different positions were collected. Area photographs of the sampling areas are shown in Figure 2. From these three positions, five samples were examined. Due to the non-uniformity of the terrain, more samples (three) were examined from Position 1 in order to obtain better sampling conditions of the area in this position. From the other two positions one sample per position was examined (Table 1). The quartz occurrences mainly consist of milky quartz. Quartz fractions of spheroidal shape (with a diameter up to 40 cm) were found scattered over long distances covering the entire survey area. It seems that these quartz occurrences in the study area constitute mechanical secondary concentrations of weathering and transportation of a pre-existing deposit. They might have been created mainly by the transposition, sorting and deposition process of clastic weathering products. Generally, these alluvial (fluvial) quartz concentrations could be an important source of secondary mechanical deposits.

Fig. 2. The three sampling positions in the setting of Roudina (Meliti, Florina).

3 Characterization Techniques

X-Ray Diffraction (XRD) was carried out in order to identify the mineralogical phases which are present in the quartz samples. Prior to the XRD measurements, the quartz samples were mechanically crushed and sieved through a 45μm stainless steel sieve in order to obtain fine powder. XRD analysis was performed, using a 3003 TT Rich. Seifert Diffractometer (Ni filtered, with Cu Ka irradiation, λ=0.154 nm). The 2θ scan range varied from 5° to 80°. Phase identification and quantification was performed by matching the acquired XRD spectra with the COD-AMCSD database. The phase identification and the XRD measurement procedure includes an instrumental error of approximately 2% wt, regarding the phase composition.

Morphological examination was conducted by means of Scanning Electron Microscopy (SEM), using a JEOL 6610LV. The elemental analysis, by means of Energy Dispersive Spectroscopy (EDS), was carried out using an X-Max 80 (Oxford Instruments) detector attached to the SEM. Images, elements maps and spectra were acquired and analysed using AZtech Nanoanalysis software. Prior to SEM observation the samples were coated with carbon in order to eliminate charging under the electron beam.

Trace-element analysis was carried out by inductively coupled plasma mass spectrometry (ICP-MS). A Perkin Elmer ELAN 6100 ICP-MS device was used for the trace-element detection in the quartz samples. The standard deviation of the SEM-EDS and ICP-MS measurements was calculated via the manufacturer's software.

Table 1. Sampling positions and sample names.

Position	Sample Name
1	Q1
	Q2
	Q3
2	Q4
3	Q5

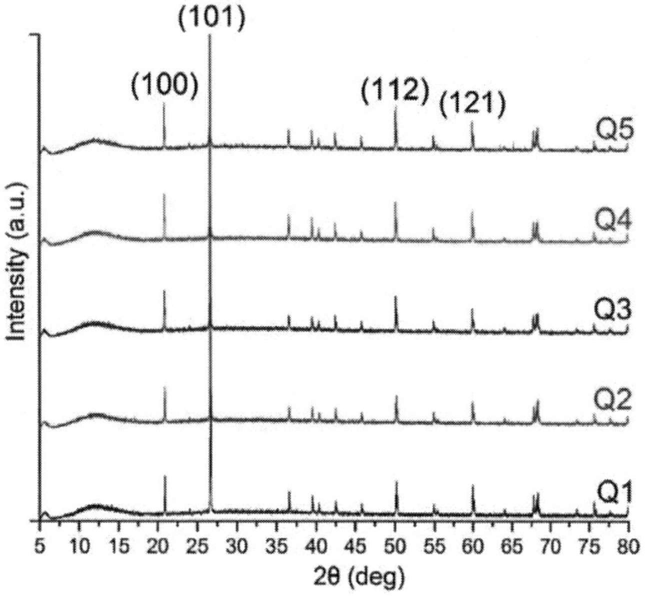

Fig. 3. XRD patterns of the five samples examined in this study. It is obvious from these patterns that, except pure quartz, no other accessory microcrystalline phases were detected in the samples.

4 Results and Discussion

The XRD results are given in Figure 3. The XRD patterns of all samples exhibited only the distinct peaks that correspond solely to quartz. Specifically, the peaks at 20.9°, 26.6°, 50.1° and 59.95° 2θ correspond to the crystallographic planes (100), (101), (112) and (121) of the mineralogical phase of quartz. Moreover, SEM-EDS analysis was performed in order to gain insights into the main elements of the investigated mineral samples. In Figure 4, indicative SEM-EDS results of the Q1 sample are shown. As it is shown form Figure 4, the EDS spectrum denotes the presence of only Si (52.2 % wt) and O (47.8 wt %). A Si:O ratio 1:2 is calculated by converting the %wt compositions of the elements into atomic composition (at %) of the oxide (SiO_2). Furthermore, on the right side of Figure 4 an elemental map of Si and O is shown. Si and O appear to be evenly distributed on the surface of the sample, which indicates the absence of other major elements or impurities. Rather similar spectra and elemental maps were obtained from the rest of the samples examined.

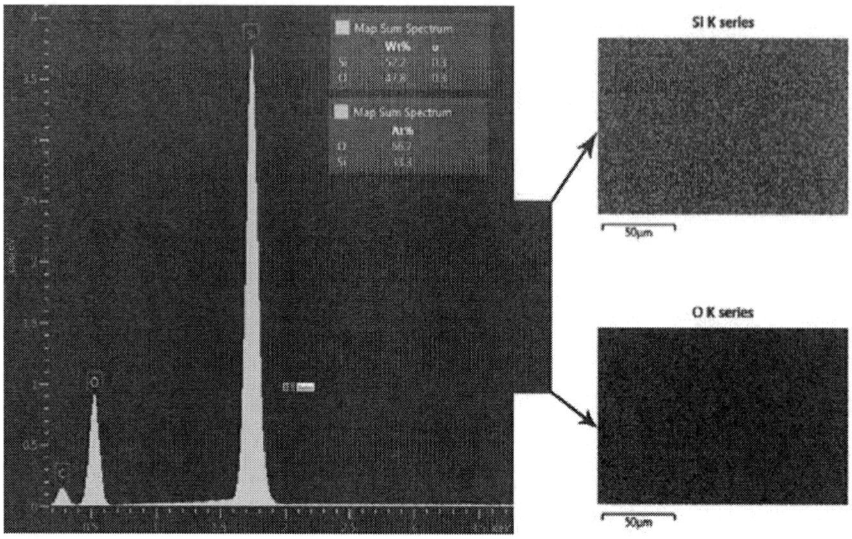

Fig. 4. SEM-EDS analysis and elemental mapping of the main chemical elements Si and O of the quartz sample (Q1) taken from position 1.

Nevertheless, the ICP-MS trace element analysis (Table 3) revealed the presence of five trace-elements (Li, Na, Al, K, Fe) in the quartz samples examined. The Q5 sample, deriving from position 3, exhibited the higher trace-element concentrations. Loss of Ignition (%LOI) results for all samples was quite low (lower than 0.2%) as expected for quartz mineral. Our analyses results are in good agreement with the literature findings [7, 8].

Table 2. EDS results for all samples investigated. The standard deviation is indicated with ±.

Sample Name	% wt elemental composition		% wt elemental composition	
	Si	O	Si	O
Q1	52.2±0.3	47.8±0.3	33.3±0.3	66.7±0.3
Q2	51.7±0.3	48.3±0.3	33.3±0.3	66.7±0.3
Q3	51.7±0.5	48.3±0.5	33.3±0.5	66.7±0.5
Q4	52.1±0.4	47.9±0.4	33.3±0.4	66.7±0.4
Q5	50.3±0.4	49.7±0.4	33.3±0.4	66.7±0.4

Table 3. ICP-MS trace-element analysis and Loss on Ignition (%LOI) results.

Sample Name	mg/kg (ppm)					%LOI
	Li	Na	Al	K	Fe	
Q1	<0.50±0.01	44±0.88	245±4.9	18±0.40	430±8.6	0.10
Q2	<0.50±0.01	47±0.94	135±2.7	<10±0.20	165±3.3	0.15
Q3	1.80±0.04	25±0.50	135±2.7	<10±0.20	70±1.4	0.13
Q4	0.50±0.01	<20±0.40	135±2.7	23±0.46	215±4.3	0.07
Q5	1.6±0.03	<20±0.40	7700±154	550±11	530±10.6	0.04

5 Conclusions

In this work quartz samples originating from Meliti area (Florina, Greece) were analyzed. According to the XRD analysis the quartz mineral originating is free from other major or even accessory microcrystalline mineral phases. According to semi-quantitative chemical analysis (SEM-EDS) the major elements detected in the samples were only Si and O in the ratio 1:2 (SiO_2). On the other hand, trace-element analysis, carried out by means of ICP-MS indicated the presence of several trace elements in the quartz interstitial crystal sites. Consequently, the quartz mineral found in the investigated area is undoubtedly very pure and therefore this could be used for the production of ultra-high purity quartz, as long as the deposits are found to be exploitable.

The authors would like to thank the Specific-Account Research Committee (ELKE) of TEI of Western Macedonia for financing the research work presented here under the research grant 80161 Moreover, the contribution of Apostolos Baklavaridis (Materials Science Engineer, PhD) to the SEM-EDS and XRD analyses is gratefully acknowledged.

References

1. S. Platias, K.I. Vatalis, and G. Charalampides, Procedia Econ. Financ. **14**, 491 (2014)
2. K.I. Vatalis, G. Charalambides, and N.P. Benetis, Procedia Econ. Financ. **24**, 734 (2015)
3. K.I. Vatalis, G. Charalampides, S. Platias, N.P. Benetis, Procedia Econ. Financ. **14**, 624 (2014)
4. T. A., *The mineral wealth of Greece (In Greek)* (Giachoudis Publications, Thessaloniki, 2005)
5. N. Arvanitidis, S. Kilias, *Vein Quartz Deposits of Northern Greece – Exploitable Resources for Industrial Uses, Mineral Deposits* (Balkema, Rotterdam, 1997)
6. N. Arvanitidis, J. Geochem. Explor. **62**, 217 (1998)
7. A.H. Damman, S.M. Kars, J.L.R. Touret, E.C. Rieffe, J. Kramer, R.D. Vis, I. Pintea, Eur. J. Mineral. **8**, 1081 (1996)
8. H. Naka and H. Kurayasu, ISIJ Int. **33**, 1252 (1993)

Temperature-dependent viscosity analysis of SAE 10W-60 engine oil with RheolabQC rotational rheometer

Dănuț Zahariea[1,*], and *Dorin Emil* Husaru[2]

[1]"Gheorghe Asachi" Technical University of Iași, Department of Fluid Mechanics, Fluid Machinery and Fluid Power Systems, D. Mangeron 59A, Iași, România
[2]"Gheorghe Asachi" Technical University of Iași, Department of Mechanical Engineering, Mecatronics and Robotics, D. Mangeron 43, Iași, România

Abstract. The purpose of this work was to determine a viscosity-temperature relationship for SAE 10W-60 engine oil. The rheological properties of this engine oil, for a temperature range of 20÷60 °C, were obtained with RheolabQC rotational rheometer. For the first reference temperature of 40 °C, the experimental result was obtained with a relative error of 1.29%. The temperature-dependent viscosity was modelled, comparatively, with the Arrhenius and the 3rd degree polynomial models. Comparing the graphs of the fits with prediction bounds for 95% confidence level, as well as the goodness-of-fit statistics, the preliminary conclusion was that the 3rd degree polynomial could be the best fit model. However, the fit model should be used also for extrapolation, for the second reference temperature of 100 °C. This new approach changes the fit models order, the Arrhenius equation becoming the best fit model, because of the completely failed to predict the extrapolated value with the polynomial model.

1 Introduction

The liquids viscosity is very sensitive to the temperature, especially for different kind of engine oils. Generally, for any fluid viscosity analysis, the corresponding temperature should be mentioned. There are many temperature-dependent viscosity models, many of these for a specific class of liquids; one of the most well-known being the Arrhenius equation. For a certain liquid, it is recommended to obtain by experiment the viscosity at different temperatures and then, to find the values of the parameters from the Arrhenius equation which fits the experimental data. This methodology was used in many researches, for analysis of different liquids: biodiesel fuels [1], vegetable oils [2], vegetable oil-diesel fuels blends [3], and pollock oil [4]. Different other temperature-dependent viscosity models were also used, for example 3rd degree polynomial models for blends of soybean oil, soy biodiesel and petroleum diesel oil [5], Vogel equation, 6th degree polynomial and Gaussian equation for 10W-40 engine oil [6], 3rd and 4th degree polynomial for 0W-40, 5W-40, 10W-40, 15W-40 and 20W-40 engine oils [7].

* Corresponding author: dzahariea@yahoo.com

One of the most important measurement method of rheological properties of fluids is the rotational method which consists, in principal, in two parts: one volume calibrated cup which will be filled with a precise volume of sample fluid and a measuring bob having a certain geometry. The Searle rotational method assume stationary calibrated cup and rotational bob. On the contrary, the Couette rotational method assume rotational cup and stationary measuring bob, [8]. For the Searle rotational method, the measuring bob will be driven with a controlled rotational speed n [rpm], as consequence the shear rate $\dot{\gamma}$ [1/s] will be known. The measured parameter is the shaft torque M [Nm], as consequence the shear stress τ [Pa] will be known, and finally the dynamic viscosity η [Pas] will be compute with the formula $\eta = \tau/\dot{\gamma}$. This is the so called controlled shear test, CSR. There is also another type of Searle rotational method, known as the controlled shear stress test, CSS, which consists in setting the torque/shear stress, measuring the rotational speed/shear rate, and finally obtaining the dynamic viscosity, [9]. Depending on the geometry, the rotational rheometers may be: concentric cylinder measuring system (CC MS); double-gap measuring system (DG MS); combined cylinder and cone-and-plate measuring system (Mooney/Ewart measuring system, ME MS); cone-and-plate measuring system (CP MS) and parallel-plate measuring system (PP MS), [9].

In this study, the RheolabQC rotational rheometer with concentric cylinder (CC MS) [10] will be used for performing a CSR test in order to evaluate the rheological properties of the Castrol EDGE SAE 10W-60 engine oil for a temperature range of 20÷60°C. The experimental data will be validated considering the viscosity presented in the Castrol EDGE SAE 10W-60 data sheet for the first reference temperature of 40°C, [11]. Two temperature-dependent viscosity models will be used: the Arrhenius model, and the 3rd degree polynomial. In order to obtain the best fit model a comparatively analysis will be performed from the point of view of the graphs of the fits with prediction bounds for 95% confidence level, as well as of the goodness-of-fit statistics. However, the final decision will depend also on the viscosity obtained by extrapolation for the second reference temperature of 100°C.

2 Rheometer description

RheolabQC is a rotational rheometer composed by the main components: 1-basic unit, 2-coupling for connecting measuring systems, 3-temperature control unit, 4-measuring systems, and 5-computer with Rheoplus software, Figure 1.

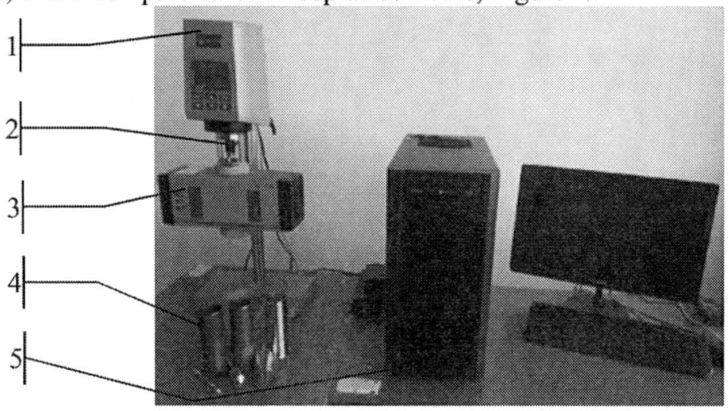

Fig. 1. RheolabQC rotational rheometer-general view.

The basic unit contain the rheometer principal components: mechanics, electronics and control systems and can be directly operated using a keypad and a LCD display from the front view of the basic unit, Figure 2, a). On the rear view of the basic unit there are different connectors: power supply, RS232 for PC or printer, Ethernet, PS2 for keyboard, Pt100 temperature sensor, Figure 2, b). The coupling for connecting the measuring systems is presented in Fig. 2, c). The coupling includes also an automatic component recognition and configuration system, named Toolmaster, for taking into considerations and transferring to the Rheoplus software all the relevant information about the connected measuring system. The relevant data is stored in a transponder chip in each measuring system. The air counter-cooling Peltier temperature control unit is presented in Figure 2, d). The temperature can be adjusted in the range of 0÷180°C. Three measuring systems can be used, Figure 2, e): two concentric cylinder measuring systems CC10, Figure 2, f) and CC27, Figure 2, g); and one double-gap measuring system DG42, Figure 2, h).

The RheolabQC principal technical specifications are: viscosity measuring range 1÷10^9 mPas; rotational speed 0.01÷1200 rpm; torque 0.25÷75 mNm; shear stress 0.6÷17800 Pa; shear rate 0.02÷4000 1/s.

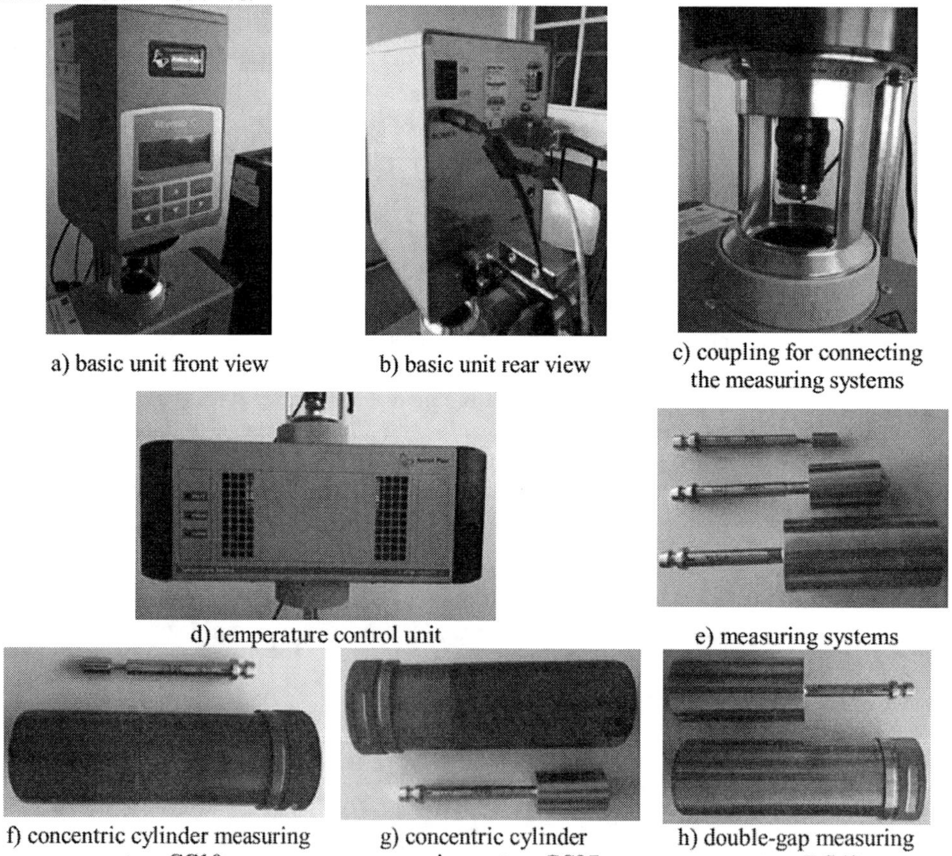

a) basic unit front view b) basic unit rear view c) coupling for connecting the measuring systems

d) temperature control unit e) measuring systems

f) concentric cylinder measuring system CC10 g) concentric cylinder measuring system CC27 h) double-gap measuring system DG42

Fig. 2. RheolabQC rotational rheometer-main components.

3 Setting up the experiment

For this experiment, the CC27 concentric cylinder measuring system has been selected. The sample volume for this measuring system is 19 ml, thus the measuring system has been filled with this volume of SAE 10W-60 oil. Then, the sample container was push up through the Peltier temperature control unit and locked up with the locking screw, meanwhile the measuring system was connected with the RheolabQC coupling system. The experiment was controlled using the Rheoplus software with a CSR flow curve default workbook. The measuring system (CC27-SN38994) has been detected using the Toolmaster automatic component recognition and configuration system. The principal characteristics of CC27-SN38994 measuring system are: maximum shear rate 1935.96 s^{-1} and maximum shear stress 952.311 Pa. For the first experiment, the setup temperature is 20°C.

The measuring profile is defined by the linear ramp shear rate range $\dot{\gamma}$=1÷100 s^{-1}, with 50 measuring points and a linear variable measuring point duration starting from 5 seconds and ending to 10 seconds, thus the total experiment interval being 375 seconds. Before starting the first test, a temperature control process was defined with the setup temperature of 20°C and the temperature tolerance of ±0.1°C, with 2-minute checking time and 5-minute timeout.

4 Experimental results

The first experiment has been carried out at the temperature of 20°C. The experimental data is presented in Figure 3, a) for the flow curve and Figure 3, b) for the viscosity curve. The flow curve suggests the newtonian character of the sample. In order to aproximate the experimental viscosity curve, the Newton model function (linear regression method) has been used, which represent the recommended method for idealviscous flow behavior. The experimental data and the Newton regression line are presented in Figure 3, c); the approximated value of dynamic viscosity being 364.59 mPas. Several new experiments have been performed using the same methodology and the same parameters, except the temperature for which the following values have been considered: 30°C, 40°C, 50°C and 60°C. The approximated values of dynamic viscosity obtained with the same Newton reggresion method are presented in Figure 4.

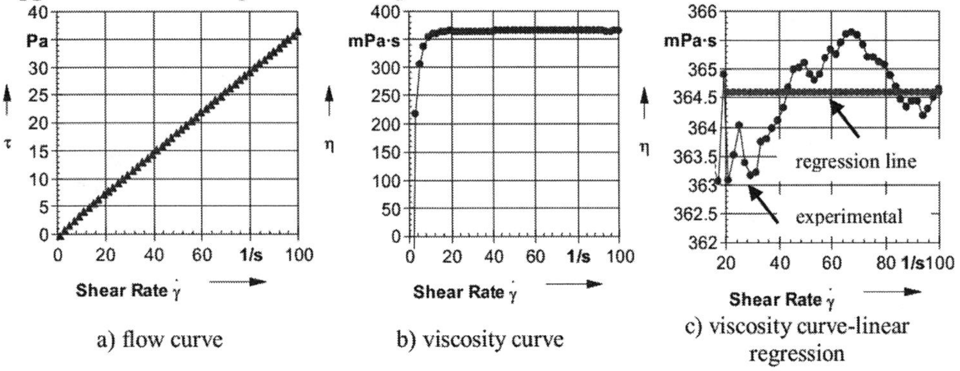

Fig. 3. Experimental data and Newton regression approximation for t=20°C.

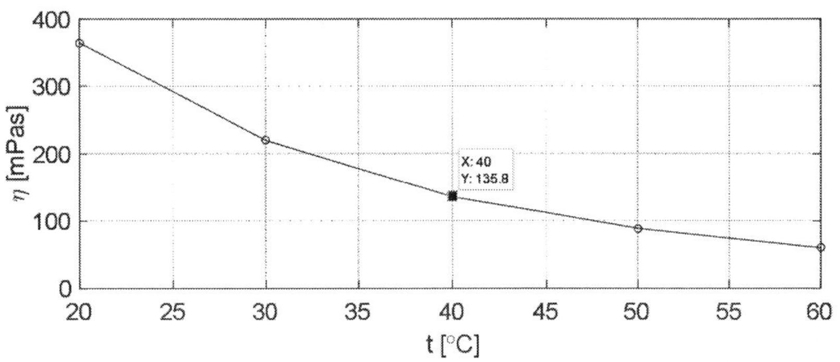

Fig. 4. Approximated dynamic viscosity.

DOI: 10.1051/matecconf/201711210014

In order to validate the experimental data obtained with RheolabQC rheometer and presented Figure 4, the information provided in the SAE 10W 60 product data sheet has been considered, [11]. According to this data sheet, at 40°C the kinematic viscosity is 160 mm^2/s. At the same temperature, considering the density 0.838 g/cm^3 from [8] and the dynamic viscosity 135.81 mPas from Figure 4, the kinematic viscosity will be 162.06 mm^2/s, for which the relative error respect to the product data sheet value is only 1.29 %.

5 Discussion

In order to obtain a temperature-dependent viscosity relationship, the Arrhenius model has been used comparatively with the 3rd degree polynomial model. The equations, the coefficients and the goodness-of-fit statistics (the sum of squares due to error-SSE, R-square, Adjusted R-square, the root mean squared error-RMSE) are presented in Table 1. The graphs of the fits with prediction bounds for 95% confidence level are presented in Figure 5 (Arrhenius model), and Figure 6 (polynomial model). From Table 1, it is obvious that the polynomial model leads to better goodness-of-fit statistics for all parameters; thus it would seem that the polynomial model would be the answer. Moreover, for the experimental temperature range 20÷60°C, there are not significant discrepancies related to the shape of the fit curves, including the prediction bounds. Outside of the experimental temperature range, however, is something else; the polynomial curve change dramatically his shape, together with the prediction bounds which are strongly diverging. The estimated viscosity for the second reference temperature is -302.2 mPas; the polynomial model completely failed to predict this value (which is 18.17 mPas, [11]).

Table 1. Temperature-dependent viscosity model's characteristics.

No.	Model	Coefficients	Goodness-of-fit statistics				Viscosity at 100°C
			SSE	R^2	Adj. R^2	RMSE	
1.	Arrhenius $\eta = Ae^{B/(t+273.15)}$	A=8.674x10^{-5} B=4471	5.869	0.9999	0.9999	1.3987	13.85
2.	3rd degree polynomial $\eta = p_1 t^3 + p_2 t^2 + p_3 t + p_4$	p1=-0.003501 p2=0.6122 p3=-38.38 p4=915.4	0.336	1	1	0.579	-302.2

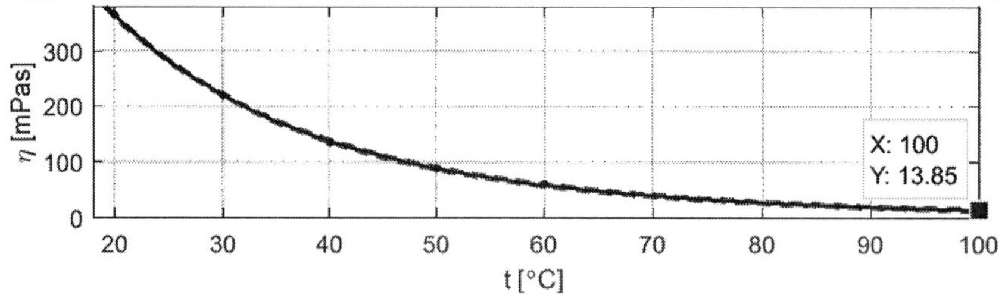

Fig. 5. Fit curve with prediction bounds, confidence level 95%-Arrhenius model.

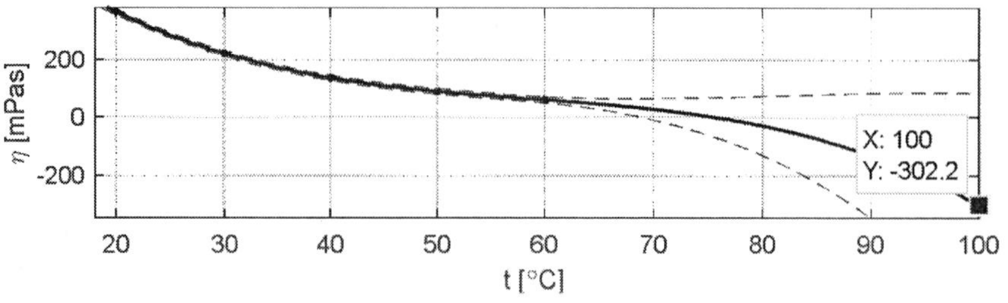

Fig. 6. Fit curve with prediction bounds, confidence level 95%-3rd degree polynomial model.

On the other side, for the Arhenius model, a very good behaviour can be observed outside of the temperature range, the prediction bounds are quite uniform, and the estimated viscosity for 100°C is 13.85±4.55 mPas, with confidence level of 95%, which is a very good result, considering the large temperature range 60÷100°C without experimental data.

6 Conclusions

RheolabQC rotational rheometer can be successfully used for temperature-dependent viscosity experimental analysis. The fluid viscosity is extremely sensitive at the temperature; thus, a high-performance temperature control unit (like the Peltier unit C-LTD180/QC) is mandatory for rheological experiments. The rheological test performed on the Castrol EDGE SAE 10W-60 engine oil shows a good agreement with the product data, with a relative error of 1.29% at the first reference temperature of 40°C.

Comparing multiple fit models should be done not only by comparing the goodness-of-fit statistics; in fact, for this analysis all these coefficients indicate that the polynomial model should be preferred. However, considering the ability of the model to estimate outside of the experimental temperature range, it has been proven that this first conclusion was wrong; the Arrhenius model being capable to estimate the product data sheet viscosity value of 18.17 mPas with the interval 13.85±4.55 mPas, with confidence level of 95%, while the polynomial model with the estimated value of -302.2 mPas represents a total failure. Improving the accuracy of the fit model can be done by increasing the number of experimental data points, inside the initial temperature range 20÷60°, but also, if it is possible, for higher temperatures.

The authors would like to acknowledge the technical resources offered by the Laboratory of Fluid Properties Measurement, from the Department of Fluid Mechanics, Fluid Machinery and Fluid Power Systems, "Gheorghe Asachi" Technical University of Iasi, Romania. The Laboratory of Fluid Properties Measurement has been equipped with technical resources with the financial support of the grant ENERED, POSCCE-A2-O2.2.1-2009-4, ID 911.

References

1. S. Kerschbaum, G. Rinke, Fuel **83** (3) 287-291 (2004)
2. S.G.E. Giap, J. Phys. Sci. **21** (1), 29-39 (2010)
3. Z. Franco, Q.D. Nguyen, Fuel **90** (2), 838-843 (2011)
4. S. Sathivel, J. Huang, W. Prinyawiwatkul, J. Food Eng. **84** (2), 187-193 (2008)
5. O.A. Aworanti, S.E. Agarry, A.O. Ajani, Adv. Chem. Eng. Sci. **2**, 444-452 (2012)
6. L. Severa, M. Havlicek, V. Kumbar, Acta Univ. Agric. Silvic. Mendel. Brun. **4**, 95-102 (2009)
7. V. Kumbar, A. Sabaliauskas, Acta Univ. Agric. Silvic. Mendel. Brun., **6**, 95-102, 1763-1767 (2013)
8. http://www.viscopedia.com
9. T.G. Mezger, *The Rheology Handbook. For users of rotational and oscillatory rheometers* (Vincentz Network, Hannover, 2006)
10. http://www.anton-paar.com/us-en/products/details/quality-control-rheometer-rheolabqc/
11. http://msdspds.castrol.com/bpglis/FusionPDS.nsf/Files/33478B845E8AACDD80257FDB004B749E/$File/BPXE-ABN7DR.pdf

Researches regarding primary control in hydraulic systems

Irina Tița*, and *Irina* Mardare

Technical University "Gheorghe Asachi" of Iași, Department of Fluid Mechanics, Fluid Machinery and Fluid Power Systems, Bvd. D. Mangeron, 59A, 700050, Iasi, Romania

Abstract. The technology in wind turbines has developed very rapidly but there are still a lot that can be improved also regarding new technologies. One example is wind turbine with hydraulic transmission. At the beginning low power wind turbines are in view. First of all the wind energy is meant to be used by isolated users for household and garden equipment or pumping water. Later, if results will be as expected, and wind potential satisfactory, such systems could be connected to electric grid. In our research laboratory we must build an experimental setup. The simulation for wind turbine and fixed displacement pump coupled to it will be realized using a variable displacement piston pump. As the variable wind speed has as a result variations of the pump flow, the variable displacement pump from the test rig may reproduce a similar variation law. In this paper some aspects regarding the variable displacement pump are detailed. This study is necessary for the future development of the research.

1 Introduction

Although the technology in wind turbines has developed rapidly there are still a lot that can be improved and new technologies emerge frequently. One example is wind turbine with hydraulic transmission. The long term goal is to see if a mid-size wind turbine with a hydraulic transmission is suitable for the commercial market.

In recent years mathematical models of fluid power transmission systems in wind turbines were developed and validated at the TU Delft. Laboratory experiments were successfully carried out at scales of 600W (validation of control method with water hydraulics & rotor in wind tunnel) and 10kW (coupled oil and water circuits). Another set of experiments was carried out using the 1MW test rig (oil) at the Institute for Fluid Power Drives and Controls (IFAS) at RWTH Aachen University [1, 2].

Considering the case of low power wind turbines, the structure of such a system is presented in Figure 1. As one can see there is a positive displacement pump coupled directly to the turbine. For the protection there is a brake. The pump is coupled hydraulic at hydraulic motor through the high pressure line and the low pressure line. At the hydraulic motor axle may be coupled in order that isolated users to extract wind energy and use it for household and garden equipment or pumping water.

* Corresponding author: iddtita@yahoo.com

It is important that if possible to configure the fluid power transmission system in such a way that no form of active control is required for the hydraulic transmission. In these conditions passive control is an inherent characteristic of the system.

Considering the future development of wind turbines with hydraulic transmissions one must try to answer some justified questions:

- Will a hydrostatic transmission be able to control the speed of the generator, so that the generated electricity satisfies the electric grid standards during normal operational conditions, without the use of a power converter?
- Can the drivetrain be realized by using already existing standard components?
- Can the system have a good overall efficiency?
- Is it possible to place some of the powertrain components in the nacelle and some of the components close to the ground to reduce weight in the nacelle and enable easier maintenance of some components? [3].

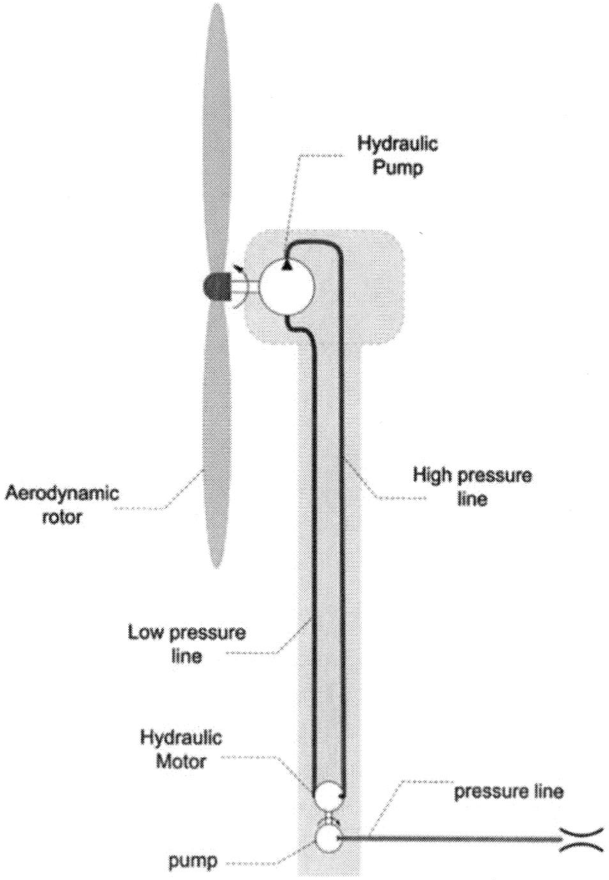

Fig. 1. Structure of wind turbine with hydraulic system [1].

In order to study such a system it must be configured an experimental rig. The simulation for wind turbine and fixed displacement pump coupled to it will be realized using a variable displacement piston pump. As the variable wind speed has as a result variations of the pump flow, the variable displacement pump from the test rig may reproduce a similar variation law [4].

 DOI: 10.1051/matecconf/201711210015

2 Primary control in hydraulic systems

The hydraulic system consists of a hydrostatic transmission built of a variable-displacement pump, a fixed - displacement motor, and fixed displacement pump as a booster pump, pressure relief valves and check valves [5, 6]. The schematic of closed circuit hydrostatic transmission system is presented in Figure 2.

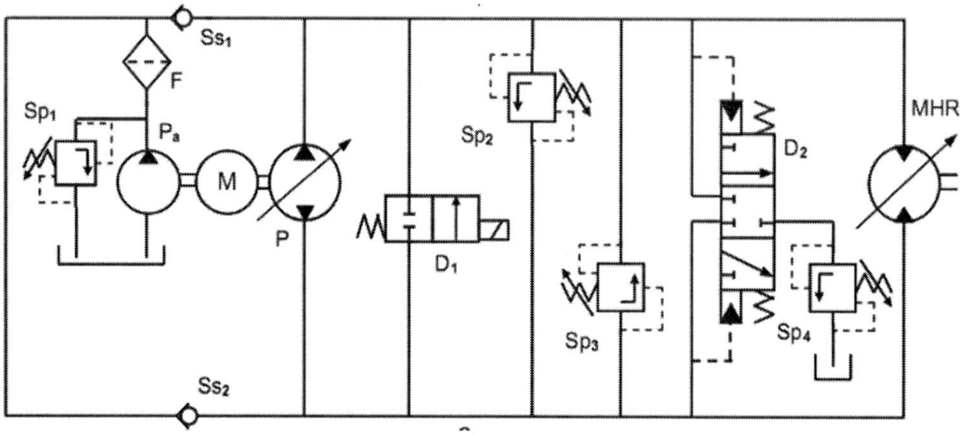

Fig. 2. Hydraulic system with variable displacement pump.

Equations for the system are as follows [7 - 9].

The pump outlet flow:

$$Q_{1p} = V_p \cdot \omega_p - \frac{p_1}{R_p} + \frac{p_2}{R_i} \tag{1}$$

where: V_p is displacement of hydraulic pump [$m^3 \cdot rot^{-1}$], ω_p is rotational speed at pump axle [$rad \cdot s^{-1}$], p_1 is pressure at pump outlet [$N \cdot m^{-2}$], p_2 is pressure at pump inlet [$N \cdot m^{-2}$], R_p is output resistance for the pump [$N \cdot m^{-2}/m^3 \cdot s^{-1}$], R_i is internal resistance [$N \cdot m^{-2}/m^3 \cdot s^{-1}$].

The motor inlet flow:

$$Q_{1m} = V_m \cdot \omega_m + \frac{p_1}{R_m} - \frac{p_2}{R_i} \tag{2}$$

where: V_m is displacement of hydraulic motor [$m^3 \cdot rot^{-1}$], ω_m is rotational speed at motor axle [$rad \cdot s^{-1}$], p_1 is pressure at motor outlet [$N \cdot m^{-2}$], p_2 is pressure at motor inlet [$N \cdot m^{-2}$], R_m is output resistance for the motor [$N \cdot m^{-2}/m^3 \cdot s^{-1}$], R_i is internal resistance [$N \cdot m^{-2}/m^3 \cdot s^{-1}$].

The motor outlet flow:

$$Q_{2m} = V_m \cdot \omega_m + \frac{p_1}{R_i} - \frac{p_2}{R_m} \tag{3}$$

The pump inlet flow:

$$Q_{2p} = V_p \cdot \omega_p - \frac{p_1}{R_i} + \frac{p_2}{R_p} \tag{4}$$

Considering:

$$\frac{1}{R_p} = \left(\frac{1}{R_e} + \frac{1}{R_i}\right)_p \text{ and } \frac{1}{R_m} = \left(\frac{1}{R_e} + \frac{1}{R_i}\right)_m \tag{5}$$

$$\frac{1}{R_{ep}} = \left(\frac{V_p \cdot \omega_p}{E} + \frac{1}{R_{ext}} + \frac{1}{R_{tim}}\right)_p \text{ and } \frac{1}{R_{em}} = \left(\frac{V_m \cdot \omega_m}{E} + \frac{1}{R_{ext}} + \frac{1}{R_{tim}}\right)_m \tag{6}$$

where: R_p is output resistance for the pump [$N \cdot m^{-2}/m^3 \cdot s^{-1}$], R_m is output resistance for the motor [$N \cdot m^{-2}/m^3 \cdot s^{-1}$], R_i is internal resistance [$N \cdot m^{-2}/m^3 \cdot s^{-1}$], R_e is external resistance [$N \cdot m^{-2}/m^3 \cdot s^{-1}$], R_{ep} is external resistance for the pump [$N \cdot m^{-2}/m^3 \cdot s^{-1}$], R_{em} is external resistance for the motor [$N \cdot m^{-2}/m^3 \cdot s^{-1}$], R_{ext} is external leakage resistance [$N \cdot m^{-2}/m^3 \cdot s^{-1}$], R_{tim} is timing groove resistance [$N \cdot m^{-2}/m^3 \cdot s^{-1}$], E is bulk modulus of hydraulic oil [$N \cdot m^{-2}$].

From the equations of the pump outlet flow rate and the motor inlet flow rate:

$$V_p \cdot \omega_p - \frac{p_1}{R_p} + \frac{p_2}{R_i} = V_m \cdot \omega_m + \frac{p_1}{R_m} - \frac{p_2}{R_i} \tag{7}$$

$$V_m \cdot \omega_m = V_p \cdot \omega_p - p_1 \cdot \left(\frac{1}{R_p} + \frac{1}{R_m}\right) + \frac{2p_2}{R_i} \tag{8}$$

The torque equations for the pump and the motor:

$$M_p = V_p \cdot (p_1 - p_2) + c_V \cdot \omega_p + M_{scp} \tag{9}$$

$$M_m = V_m \cdot (p_1 - p_2) - c_V \cdot \omega_m - M_{scm} \tag{10}$$

where: M_p is torque at pump axle [N·m], M_m is torque at motor axle [N·m], c_V is viscous friction coefficient [$N \cdot m/rad \cdot s^{-1}$], M_{scp} is torque embracing striction and coulomb friction at pump axle [N·m], M_{scm} is torque embracing striction and coulomb friction at pump axle [N·m].

The transmission may be evaluated with:

$$\frac{1}{R_1} = \left(\frac{1}{R_p} + \frac{1}{R_m} - \frac{2}{R_i}\right) \text{ and } \frac{1}{R_2} = \left(\frac{1}{R_p} + \frac{1}{R_m}\right) \tag{11}$$

$$\frac{\omega_p}{\omega_m} = \frac{\frac{V_p}{V_m} - \frac{M_m}{\omega_p \cdot R_2 \cdot V_m^2} - \left[\frac{p_2}{\omega_p \cdot R_1 \cdot V_m} + \frac{M_{scm}}{\omega_p \cdot R_2 \cdot V_m^2}\right]}{\left[1 + \frac{c_V}{V_m^2 \cdot R_2}\right]} \tag{12}$$

$$p_1 = p_2 + \frac{M_m}{V_m} + \frac{c_V \cdot \omega_p}{V_m} \cdot \frac{\omega_m}{\omega_p} + \frac{M_{scm}}{V_m} \tag{13}$$

$$\frac{M_m}{M_p}=\frac{\left(p_1-p_2\right)-\frac{c_v\cdot\omega_p}{V_m}\cdot\left(\frac{\omega_m}{\omega_p}\right)-\frac{M_{scm}}{V_m}}{\left(p_1-p_2\right)\cdot\left(\frac{V_p}{V_m}\right)+\frac{c_v\cdot\omega_p}{V_m}+\frac{M_{scp}}{V_m}} \tag{14}$$

The transmission efficiency is:

$$\eta_{tran}=\frac{M_m\cdot\omega_m}{M_p\cdot\omega_p} \tag{15}$$

In the laboratory of Wind Turbine Control Systems, from the Department of Fluid Mechanics, Fluid Machinery and Fluid Power Systems it is used a variable displacement axial piston pump PV023L1K1T1NUPK with maximum displacement V_p = 23cm^3/rot = 3,66·10^{-6}m^3/rad, nominal pressure of 350·10^{5}N/m^2, maximum pressure of 420·10^{5}N/m^2, speed between 400 and 3000 rot/min, moment of inertia 0,0017 kg·m^2. Positive displacement motor has V_m = 4·10^{-6}m^3/rad, the rotational speed of the pump is ω_p=1440rpm, external and internal resistances are R_e = 0,5·10^{12} N·m^{-2}/m^3 s^{-1} and R_i = 2·10^{12} N·m^{-2}/m^3·s^{-1}. Viscous friction coefficient is c_υ = 0,02 N·m/rad·s^{-1}.

Using same kind of hydraulic machine torque embracing striction and coulomb friction at pump axle is considered to be equal with torque embracing striction and coulomb friction at motor axle M_{scp} = M_{scm} = 6N·m. The pressure from the auxiliary pump set at pressure valve Sp_l is 6·10^{5}N/m^2.

On the basis of equation (5) R_p = R_m = 0,48·10^{12} N·m^{-2}/m^3 s^{-1} and equation (11), there result R_1 = 0,35·10^{12} N·m^{-2}/m^3 s^{-1} and R_2 = 0,24·10^{12} N·m^{-2}/m^3 s^{-1}.

On the above considerations equation (12) is

$$\frac{\omega_p}{\omega_m}=\frac{\frac{V_p}{V_m}-\frac{M_m}{578{,}8}-0{,}0132}{1{,}0052} \tag{16}$$

Equation (13) gives pressure p_l [N/m^2]:

$$p_1=6\cdot10^5+2{,}56\cdot10^5\cdot M_m+7{,}54\cdot10^5\cdot\frac{\omega_m}{\omega_p}+15\cdot10^5 \tag{17}$$

Equation (14)

$$\frac{M_m}{M_p}=\frac{\left(p_1-6\right)\cdot10^5-7{,}54\cdot10^5\cdot\left(\frac{\omega_m}{\omega_p}\right)-15\cdot10^5}{\left(p_1-6\right)\cdot10^5\cdot\left(\frac{V_p}{V_m}\right)+7{,}54\cdot10^5+15\cdot10^5} \tag{18}$$

From (16), for V_p/V_m=0,915 we obtained graphic variation in Figure 3 for speed ratio, and from (17) considering $p_{l\,max}$=300·10^{5}N/m^2 graphic variation for $p_l/p_{l\,max}$.

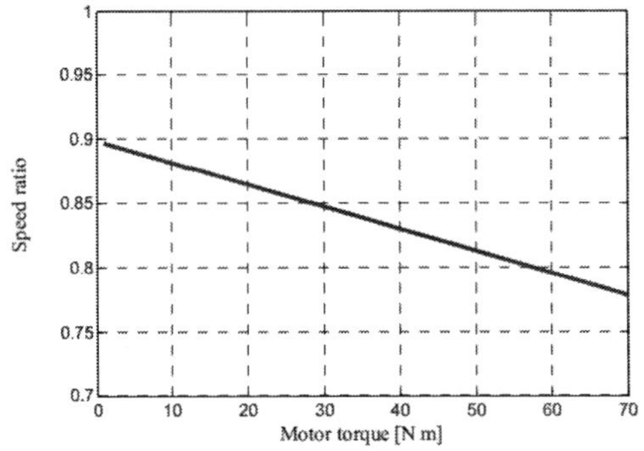

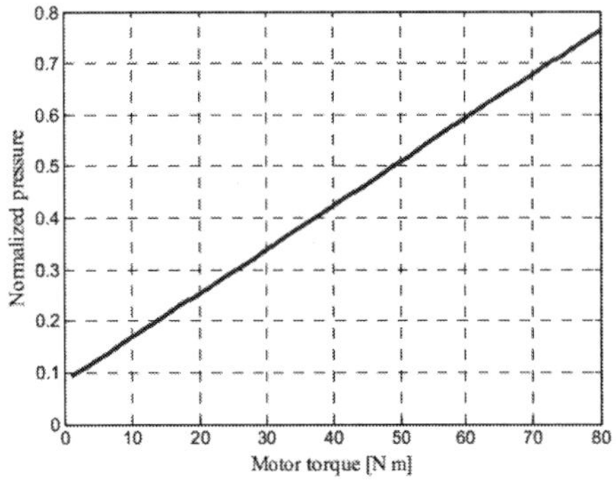

Fig. 3. Normalized variation speed.

3 Conclusions

Although the technology in wind turbines has developed rapidly there are still a lot that can be improved and new technologies emerge frequently. One example is wind turbine with hydraulic transmission.

In order to study the behavior of hydrostatic transmission it is important to design a test rig including the hydrostatic transmission and the components used to simulate wind turbine and fixed displacement pump coupled to it which will be realized using a variable displacement piston pump .

Considering the final purpose of the research which consists in hydrostatic transmissions for wind turbines, the main specific problems the way to configure the wind turbine and the connected pump was identified.

The authors would like to acknowledge the technical resources offered by the Laboratory of Wind Turbine Control Systems, from the Department of Fluid Mechanics, Fluid Machinery and Fluid Power Systems, "Gheorghe Asachi" Technical University of Iasi, Romania. The Laboratory of Wind Turbine Control Systems has been equipped with technical resources with the financial support of the grant ENERED, POSCCE-A2-O2.2.1-2009-4, ID 911.

References

1. N. Diepeveen, A.J. Laguna, 9IFK, **3**, 12, (2014)
2. J. Schmitz, G. Matthiesen, H. Murrenhoff, , 9IFK, **3**, 9, (2014)
3. Z. Jiang, L. Yang, Z. Gao, T. Moan, Energy Procedia, **53**, 12 (2014).
4. A. Nikranjbar, A.N. Sharbabaki, Majlesi Journal of Energy Management, **2**, 6, (2013).
5. K.E. Rydberg, *The 11-th Scandinavian International Conference on Fluid Power*, SICFP'09, June 2-4, Linköping, Sweden, (2009).
6. H. Murenhoff, *Proceeding of the 12th International Conference on Tools: ICT-2007*: September 6-8, Miskolc, (2007).
7. J. Watton, *Fundamentals of Fluid Power Control* (Cambridge University Press, London, 2009).
8. J. Watton, *Modelling, Monitoring and Diagnostic Techniques for Fluid Power Systems*, (Springer Verlag, London, 2007).
9. D. Zahariea, Applied Mechanics and Materials, **656**, 5, (2014)

Congestion effect on renewable energy production efficiency in Europe

Corrado lo Storto*

University of Naples Federico II, Department of Industrial Engineering, Piazzale V. Tecchio 80, 80125 Naples, Italy

Abstract. In the last decade the renewable energy facilities have become very important for generating electricity in Europe. Estimating the production efficiency of renewable energy industry is thus a major concern of governments. This paper implements Data Envelopment Analysis (DEA) to measure efficiency in the generation of electricity from renewable sources during 2002-2011 in 31 European countries. As input congestion may introduce bias in the efficiency analysis, an improved DEA model is used to correct and avoid efficiency underestimation.

1 Introduction

In Europe, the industry of the electricity generation has experienced profound changes since the beginning of the 2000s. The slowdown of energy demand, the need for climate change mitigation, the search for greater efficiency, security and affordability of energy provision, and the decrease of costs thanks to technological innovation have put pressure on the industry to change the mix of energy sources. In this new landscape, renewable energy sources have largely contributed to this change. Indeed, renewable energy sources have a higher mitigation potential of greenhouse gas emissions and lower health and environment impact than fossil fuel sources [1]. Furthermore, as they allow reducing fossil fuel imports and stimulating the creation of jobs, skills and technological innovation, renewable energy sources contribute to economic and social growth, too. In Europe, contribution to fossil fuel import savings was about €16 billion in 2015, while forecasts predict that savings will be €58 billion in 2030 [2]. According to statistics released by EUROSTAT, between 2004 and 2014 the relative weight of renewable energy sources in the generation of electricity in the EU-28 countries grew from 13.5% to 24.9% [3]. In 2014, the EU-28 countries had the largest solar-PV and wind power capacity in the world.

Contrarily to investment to increase electricity generation through conventional sources, investment leveraging on the utilization of renewable energy sources requires considerable amount of money per energy generated to build plants, but low operating expenditures. Thus, investment decisions to add new electric power capacity from renewable sources should take into account a number of factors, such as investment risk, cost of capital, future demand, technological innovation trends. National governments have an important role to promote investment by designing policy and industry regulation mechanisms. In particular,

* Corresponding author: corrado.lostorto@unina.it

DOI: 10.1051/matecconf/201711210016

policymakers need to address the choice of the renewable sources to invest on in order to meet longer terms objectives and achieve the higher efficiency in the generation of electricity. Thus, an important issue in the energy policy is to understand what renewable sources contribute more efficiently to the aggregate production of electricity in a country.

This paper adopts Data Envelopment Analysis (DEA) to measure efficiency in the generation of electricity from major renewable sources from 2002 to 2011 in 31 European countries. As input congestion may introduce bias in the efficiency analysis, an improved DEA model is used to measure it to correct and avoid efficiency underestimation.

The rest of the paper unfolds as follows. Section 2 illustrates the method implemented to calculate efficiency and evaluate the influence of congestion. Section 3 provides information relative to dataset and variables used. Section 4 and section 5 respectively present major results and conclusion.

2 Method

DEA is a well known non-parametric technique utilized to conduct benchmarking analyses and investigate how to improve performance, specifically efficiency. DEA has been applied in a wide range of industries and contexts, i.e. drinking water management, banking, transportation, healthcare, manufacturing, automotive, energy, education, high-performance computing urban management, etc. [4-14]. In the DEA approach a benchmarking frontier is generated as a set of linear combinations of inputs and outputs of a number of units to be evaluated denominated DMUs. The relative efficiency of DMUs is calculated by running the same number of linear programming models, one for each DMU [15]. A DMU k lying on the envelopment frontier is 100% efficient, and it is unable to increase one of the outputs without increasing at least one input or decreasing at least one of the other outputs. Vice versa, if DMU k is placed under the frontier it is considered inefficient.

In this study, each country C_k is a DMU k in the dataset that is associated to a production function which converts $\boldsymbol{X}=(x_{ij})\in\Re^{m\times n}$ inputs (i.e., maximum power generation capability of power plants using a certain renewable source in the country) into $\boldsymbol{Y}=(x_{ij})\in\Re^{s\times n}$ outputs (i.e., electricity generated by all power plants in the country using the same renewable source). Technical efficiency (TE) of DMU k denotes its ability to produce the maximum amount of outputs by consuming the same amount of inputs or to produce a given amount of outputs with the minimum consumption of inputs, independently of any input and output price set.

Assuming costant returns to scale (CRS), the linear programing model to compute technical efficiency in an input-orientation is as follows

$$
\begin{aligned}
& TE^{CRS} = DF^{i}\left(x_k, y_k\right) = \min_{\theta,z} \\
\text{s.t.}\quad & Y'z \geq y_k \\
& X'z \leq \theta x_k \\
& \theta \geq 0, z \geq 0
\end{aligned}
\tag{1}
$$

where $\theta=DF_i(x_k,y_k)$ is the radial input efficiency measure of DMU k, and z is an intensity variable.

The CRS assumption can be easily changed to variable returns to scale (VRS) by adding the constraint $\boldsymbol{I}z=1$, where $\boldsymbol{I}$ is a $n\times1$ unity vector.

Usually increases in inputs cause increases in outputs. Conventional DEA provides a measure of efficiency that is based on the assumption of strong (or freely) disposability (SD) of inputs. Inputs are strongly disposable if an input increase does not decrease outputs. Under this assumption, all energy power capacity inputs are radially decreased when an inefficient country is projected onto the efficient envelopment frontier.

However, there are situations where increasing one or more inputs may reduce one or more outputs. In such situations there is congestion in inputs of production process and increasing one or some inputs to produce more outputs is undesirable [16]. Weak disposability (WD) is the term used to refer to such situation.

In order to investigate if there is congestion in the production process, Färe and Grosskopf [17] suggest to assume both strong and weak input disposability at the same time. Technical efficiency TE can be decomposed into the product of three components as follows

$$TE^{CRS}=TE^{VRS}\times SE\times CN \quad (2)$$

where

$TE^{CRS}=DF_i(x_k,y_k|\text{CRS,SD})$ measures the total efficiency of DMU k under the assumption of strong input disposability

$TE^{VRS}=DF_i(x_k,y_k|\text{VRS,WD})$ measures pure technical efficiency of DMU k under the assumptions of VRS and weak input disposability

$SE=DF_i(x_k,y_k|\text{CRS,SD})/DF_i(x_k,y_k|\text{VRS,SD})$ measures scale efficiency of DMU k that is related to the size. If SE<1 there are scale inefficiencies.

$CN=DF_i(x_k,y_k|\text{VRS,SD})/DF_i(x_k,y_k|\text{VRS,WD})$ measures congestion of DMU k as a consequence of lack of strong disposability of its inputs. If CN<1 congestion occurs.

3 Data, variables and model specification

The dataset includes 31 European countries. Data relative to both the installed power capacity and electricity generation from renewable sources between 2002 and 2011 were retrieved from the European Commission statistical database. The production function model specification includes 4 inputs and 1 output. Particularly, the following inputs were used: the total installed capacity of the country relative to energy generation from hydro, wind, solar-PV and waste sources. Other forms of renewable sources have been excluded because of their scarce diffusion across sample countries. The total yearly electricity generated from the renewable sources considered in the study was used as the unique output. Measurement units are in Mw.

All inputs and output have been divided by their respective means to reduce computational scaling problems.

4 Results

Table 1 presents individual measurements relative to technical efficiency (TE) from 2002 to 2011. Scores were calculated under the assumptions of constant returns to scale and input strong disposability. Figures show an extremely varied picture. Countries largely differ with respect to their achieved efficiency. For instance, technical efficiency of total installed capacity in Italy is between 0.300 and 0.469, while in Ireland it is between 0.173 and 1.000. The behavior of technical efficiency over years was generally very unsteady. However, for some countries more evident patterns can be identified. Efficiency scores remained rather unchanged for Italy, except in 2009 and 2010. Vice versa, technical efficiency increased in Belgium, Spain, Lithuania, UK. Romania is the only country in which technical efficiency achieved the maximum value and remained stable over the observation period.

Table 1. Technical efficiencies between 2002 and 2011.

Country	TE (CRS, SD)									
	2002	**2003**	**2004**	**2005**	**2006**	**2007**	**2008**	**2009**	**2010**	**2011**
Belgium	0.223	0.204	0.235	0.261	0.310	0.351	0.385	0.394	0.418	0.540
Bulgaria	0.322	0.378	0.329	0.417	0.421	0.456	0.465	0.418	0.334	0.374
Czech Republic	0.341	0.190	0.266	0.241	0.257	0.212	0.214	0.259	0.322	0.507
Denmark	1.000	1.000	1.000	1.000	1.000	1.000	1.000	1.000	1.000	0.466
Germany	0.702	0.688	0.810	0.824	0.790	0.858	0.830	0.757	0.841	0.429
Estonia	0.490	0.967	0.597	1.000	1.000	1.000	1.000	1.000	1.000	0.360
Ireland	0.514	0.432	0.501	0.655	0.819	0.908	1.000	1.000	1.000	0.173
Greece	0.221	0.674	0.333	0.572	0.708	0.267	0.321	0.801	0.547	1.000
Spain	0.328	0.502	0.447	0.400	0.453	0.493	0.496	0.557	0.773	1.000
France	0.440	0.424	0.426	0.378	0.402	0.425	0.456	0.427	0.468	0.461
Croatia	0.632	0.602	0.800	0.691	0.665	0.617	0.734	0.781	1.000	0.358
Italy	0.393	0.364	0.404	0.355	0.353	0.323	0.387	0.440	0.469	0.300
Latvia	0.320	0.291	0.376	0.419	0.324	0.345	0.392	0.431	0.750	0.515
Lithuania	0.150	0.185	0.177	0.155	0.147	0.194	0.202	0.231	0.398	0.511
Luxembourg	0.150	0.137	0.132	0.140	0.142	0.145	0.149	0.131	0.216	0.491
Hungary	0.672	0.832	1.000	1.000	1.000	1.000	1.000	1.000	1.000	0.288
The Netherlands	1.000	1.000	1.000	1.000	1.000	0.927	1.000	1.000	1.000	0.708
Austria	0.609	0.504	0.542	0.538	0.523	0.511	0.494	0.522	0.506	0.680
Poland	0.448	0.347	0.350	0.382	0.300	0.348	0.366	0.408	0.535	0.608
Portugal	0.318	0.588	0.360	0.221	0.426	0.417	0.369	0.472	0.722	1.000
Romania	1.000	1.000	1.000	1.000	1.000	1.000	1.000	1.000	1.000	1.000
Slovenia	0.578	0.495	0.685	0.586	0.568	0.519	0.633	0.705	0.599	0.622
Slovakia	0.584	0.352	0.359	0.441	0.409	0.475	0.426	0.512	0.359	0.400
Finland	0.642	0.600	0.716	0.597	0.604	0.713	0.886	0.651	0.695	0.412
Sweden	0.678	0.544	0.575	0.669	0.554	0.588	0.624	0.568	0.651	0.759
United Kingdom	0.388	0.350	0.423	0.498	0.543	0.549	0.588	0.635	0.616	1.000
Iceland	1.000	1.000	1.000	1.000	1.000	1.000	1.000	1.000	1.000	0.761
Norway	0.770	0.619	0.637	0.793	0.668	0.755	0.762	0.684	0.794	1.000
Switzerland	0.461	0.453	0.432	0.407	0.392	0.441	0.450	0.443	0.440	1.000
F Y R of Macedonia	0.283	0.504	0.444	0.454	0.483	0.299	0.242	0.367	1.000	0.512
Turkey	1.000	1.000	1.000	1.000	1.000	1.000	1.000	1.000	1.000	0.580

Figure 1 displays the plot of TE(CRS,SD), TE(VRS, WD), SE and CN means from 2002 to 2011. The graph clearly shows that on average technical efficiency calculated assuming strong input disposability continuously increased till 2009, but after 2010 it drastically diminished, probably as a consequence of the downward trend of the European economy. However, the trend of the SE curve suggests that the improvement of the scale size operations had a not negligible weight in the increasing of efficiency scores. Indeed, the steady increase of the SE measurements indicates that values of both TE(CRS,SD) and TE(VRS,SD), on average, get closer and closer together over the years (as SE is closer to 1). Finally, the curve trend related to CN emphasizes that the effect of input congestion compensate for the efficiency improvement determined by scale economies, particularly from 2006 to 2009.

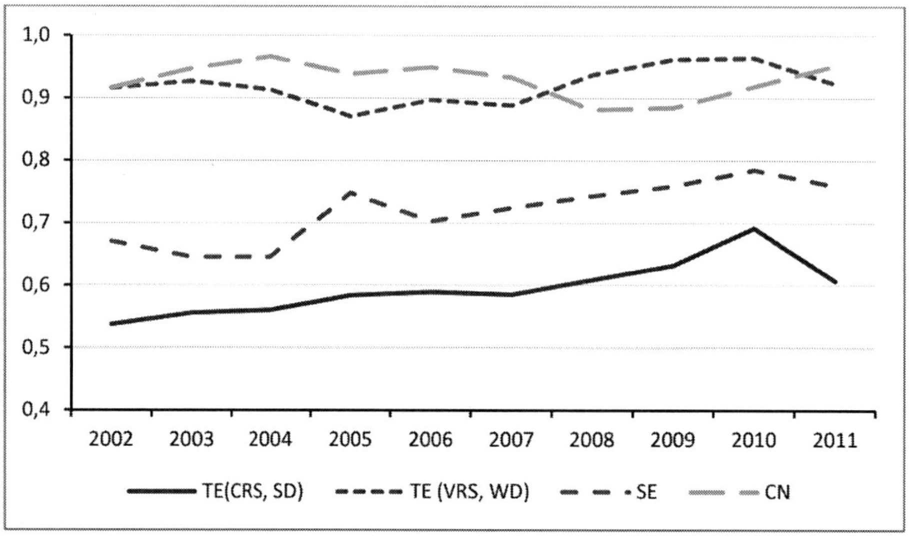

Fig. 1. Means plot of TE(CRS,SD), TE(VRS), SE and CN from 2002 to 2011.

5 Conclusion

This paper has illustrated the results of benchmarking study which has measured the production efficiency relative to the generation of electricity from renewable sources in 31 European countries between 2002 and 2011. The study has implemented Data Envelopment Analysis to compute efficiency. Additionally, the effect on efficiency of both input congestion and scale economies have been evaluated.

Results indicate that countries largely differ with respect to their achieved efficiency. The behavior of technical efficiency measurements over years was generally very unsteady. Finally, to a large extent efficiency of installed renewable power capacity was influenced by both scale economies and input congestion.

These findings suggest that in designing market regulation and incentives mechanisms, policy-makers should take into account not only efficiency, but also congestion and scale economies issues in order to identify the optimal mix of renewable sources.

This study considered only the country energy generation capacities from renewable sources as inputs and the amount of electricity generated from these latter as output of the electricity production model. Consequently, it has not provided a comprehensive picture of the generation of electricity in the countries of the sample. Future research might include further variables in the efficiency model, such as the electricity generation capacity from fossil fuels, the total generation of electricity in the country, the amount of CO_2 emissions. Additionally, a two step DEA analysis might be purposefully performed to correct efficiency and congestion measurements by taking into account the influence of external variables, i.e. national GDP, country size and economic structure, etc. [18].

References

1. http://fs-unep-centre.org/sites/default/files/publications/globaltrendsinrenewableenergyinvestment2016lowres_0.pdf
2. EEA, *Renewable energy in Europe 2016 – Recent growth and knock-on effect*, (European Environment Agency Report no. 4/2016, 2016, Copenhagen)
3. http://epp.eurostat.ec.europa.eu/statisticsexplained
4. T. Agasisti, Education Economics **19**, 199-224 (2011)
5. J.R. Barth, C. Lin, Y. Ma, J. Seade, F.M. Song, J Bank Financ **37**, 2879-2892 (2013)
6. A.G. Goncharuk, C. lo Storto, Energ Policy **101**, 456-466 (2017)
7. B. Hollingsworth, Health Econ **17**, 1107-1128 (2008)
8. C. lo Storto, B. Capano, Ener Educ Sci Tech – A **32**, 3061-3072 (2014)
9. C. lo Storto, B. Capano, ARPN J Eng Appl Sci **11**, 6234-6239 (2016)
10. C. lo Storto, Water **5**, 2058-2079 (2013)
11. C. lo Storto, Appl Mech Mater **309**, 335-341(2013)
12. C. lo Storto, Inz Ekon **25**, 320-332 (2014)
13. C. lo Storto, Cities **51**, 52-63 (2016)
14. R.C. Marques, C.P. Barros, Appl Econ Lett **18**, 29-37 (2011)
15. A. Charnes, W.W. Cooper, E. Rhodes, Eur J Oper Res **2**, 429-444 (1978)
16. R. Fare, L. Svensson, Econometrica **48**, 1743- 1753 (1980)
17. R. Fare, S. Grosskopf, Zeitschhrift für Nationalökonomie **43**, 257-271 (1983)
18. L. Simar, P.W. Wilson, J. Econ **136**, 31-64 (2007)

Influence of the material used to build the blades of a wind turbine on their starting conditions

Eugen-Vlad Năstase*

Technical University Gheorghe Asachi of Iaşi, Department of Fluid Mechanics, Fluid Machinery and Fluid Power Systems, Iaşi, Romania

Abstract. Wind energy has been shown to be one of the most viable sources of renewable energy. Hydraulic machines that convert the energy of a fluid into mechanical energy are called turbines. A wind turbine is a device which extracts kinetic energy from the wind. With increasing energy demands is necessary to increase the size of wind turbines. Under these conditions the turbine will start only at high wind speeds. On the other hand, the control of high speed is more difficult and the reduction of the inertial forces becomes mandatory. This study presents an analysis of the material influence on the wind turbine starting conditions.

1 Introduction

Global energy consumption in the last decades has increased at a very rapid rate. Present trends in global population growth, rapid industrialization, and urbanization in major population centers of the world suggest that the world energy demand will continue to increase in the next years. A solution to this problem could be producing energy from renewable sources [1-3]. The main types of renewable energy is wind power, solar power, geothermal energy, hydropower and various forms of biomass. These sources have been coined renewable due to their continuous replenishment and availability for use over and over again. Compared to many other conventional energy sources, wind power's environmental effects are fairly minor [4]. Wind power does not release any toxic gas pollution into the atmosphere and does not consume any amount of fuel. The only energy that is used due to wind power is the energy used in actual construction of the wind power plant. The increase in power capability is produced by the increase in the rotor diameter [5], and hence the increase in the turbine inertia which in turn increases its response time to changes in wind speed [6]. This study is structured on 5 sections and presents an analysis of the material influence on the wind turbine starting conditions. Section 2 will present the diagram of energy transformation in a turbine and the main components for a wind turbine.

In Section 3, the design methodology of blade will be shown. In section 4 will be analyzed the drive train for a wind turbine and will be reported a modal analysis for three materials used in blade construction. Section 5 will present the conclusions of this study.

*Corresponding author: nastase_eugenvlad@yahoo.com

2 General consideration

A turbine is a hydraulic machine that transform hydraulic energy of a fluid into mechanical energy. The block diagram which shows the transformation of hydraulic energy (load H and flow Q) into mechanical energy (torque M and angular velocity ω) is shown in Figure1.

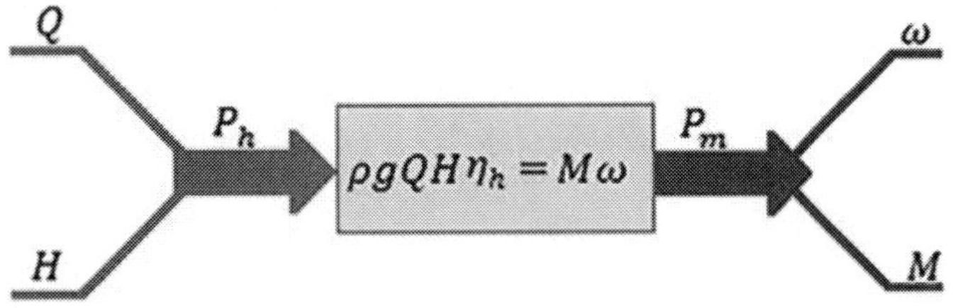

Fig. 1. Block diagram of the energy transformation in turbine.

Most modern wind turbines are built with a horizontal-axis similar to the one seen in the Figure 2. The hub and the blades of the wind turbine are constructed with an aerodynamic design and when interacting with the wind the rotor begin to spin. The turbine shaft spins with the blades and is the mechanism that transfers the rotational/mechanical energy of the blades towards the electrical generator. Generally for increase the rotational speed of the turbine shaft is used a gearbox.

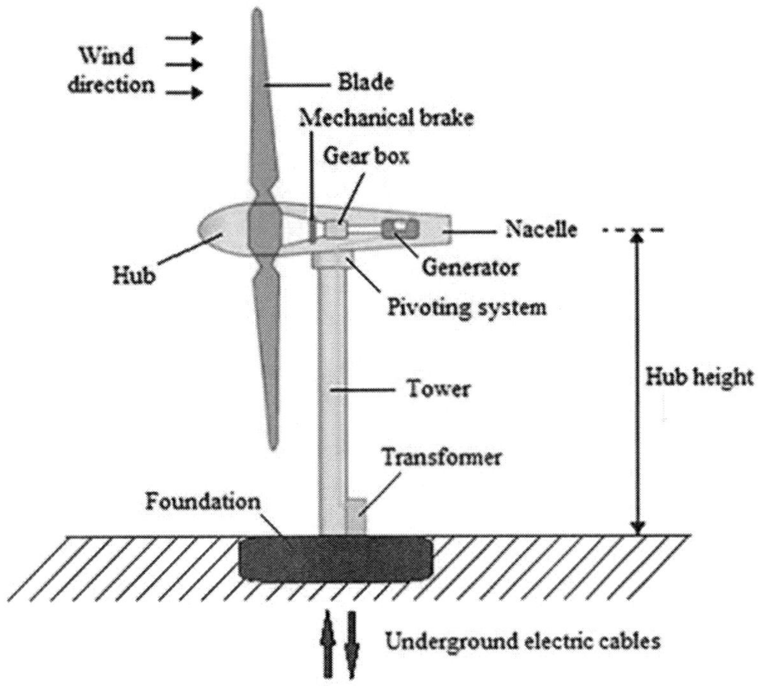

Fig. 2. Overview of main components for a wind turbine.

The generator converts the mechanical energy from the wind into electrical energy. The turbine tower contains wiring so the generator can send electricity into a transformer or a battery which will eventually distribute usable electric power. A brake is installed to prevent mechanical failure from high wind [7].

DOI: 10.1051/matecconf/201711210017

3 Design methodology of blade

Aerodynamic performance is fundamental for efficient rotor design. Aerodynamic lift is the force responsible for the power yield generated by the turbine and it is therefore essential to maximise this force using appropriate design. A resistant drag force which opposes the motion of the blade is also generated by friction which must be minimised.

The methodology used for aerodynamic calculations is the Blade Element Momentum Method. The fundamental components of the rotor are the blades [1, 8, 9]. They are the devices that convert the force of the wind into the torque needed to generate useful power.

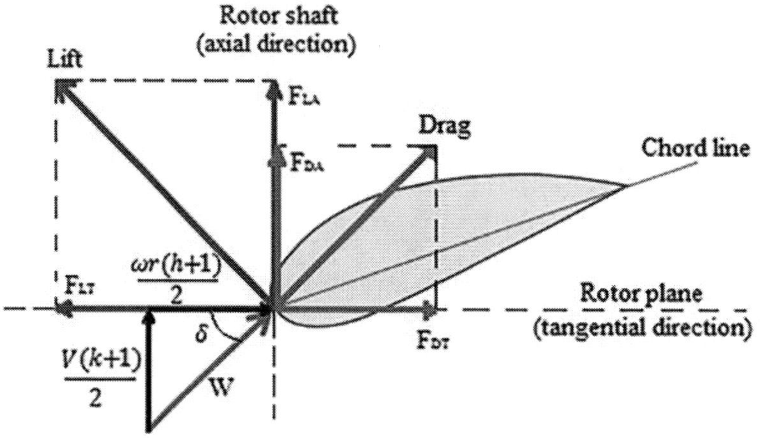

Fig. 3. Components of force and velocity in the blade cross-section.

The essence of this method is to divide the turbine blade into a finite number of cross-sections. The finite blade length between the two nearest cross-sections is called the blade element. The relative angles of attack and consequently the induction factors and incidences are found for each blade cross-section and averaged for the elements. The found measures allow the calculation of the aerodynamic efficiency, lift and drag forces and coefficients [10]. By designing the lift force and drag force on axial and tangential direction are obtained elementary forces (Figure 3):

$$F_A = F_{LA} + F_{DA} \quad (1)$$

$$F_T = F_{LT} - F_{DT} \quad (2)$$

Where F_A, is the elementary force on the axial direction and F_T is the elementary force on the tangential direction. Tangential component is that which gives elementary torque. In Figure 3 also can be seen velocity triangle for a wind turbine. This triangle is formed by the decomposition of W velocity in the horizontal direction and in the tangential direction. The axial velocity and the tangential velocity are affected by h and k coefficients. These coefficients take into account the induced rotation of the airflow which goes through the rotor provoked by the sheet of vortices originated at the tip and the hub of the finite blade.

For a given situation, specifically for an average wind speed, V = 7m/s and a rotor diameter D = 5,8 m, was made aerodynamic blade computing. In these conditions it was calculated the blade in ten sections, as can be seen in Figure 4.a. 3D model of blade, presented in Figure 4.b, was obtained uniting the ten sections using Catia software. Other method to change dynamic and mechanical properties of wind turbine is modifying the material, which the blade is made of [9].

DOI: 10.1051/matecconf/201711210017

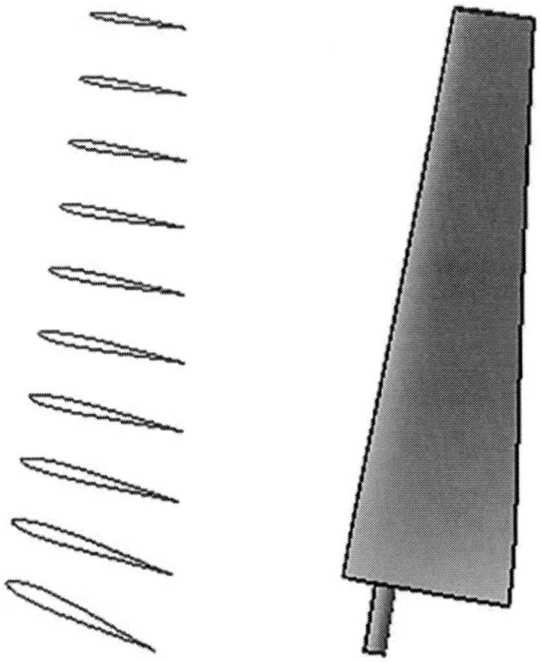

Fig. 4. The blade of wind turbine (a -profile layout in sections; b – 3D model).

4 Kinematic chain for wind turbines

The energy conversion chain is organised into four subsystems (as can see in Figure 5): aerodynamic subsystem, consisting mainly of the turbine rotor, which is composed of blades, and turbine hub, which is the support for blades; drive train, generally composed of: low-speed shaft – coupled with the turbine hub, speed multiplier and high-speed shaft – driving the electrical generator .

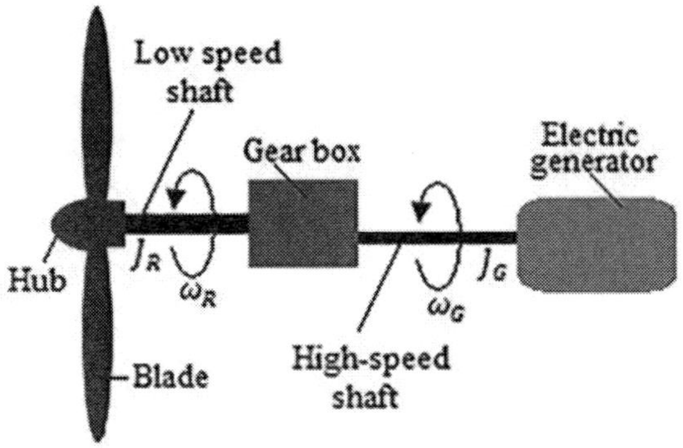

Fig. 5. The diagram of the drive train.

After the analysis of the diagram presented in Figure 5, we can write the equations of equilibrium:

$$M_R - M_{GBi} = J_R \frac{d\omega_R}{dt} + B_R \cdot \omega_R \tag{3}$$

$$M_{GBe} - M_G = J_G \frac{d\omega_R}{dt} + B_G \cdot \omega_G \qquad (4)$$

Where:

- B_R and B_G are the constant friction in the rotor and generator system;
- M_R is the rotor torque;
- M_{GBi} is the gear box torque to the start;
- M_{GBe} is the gear box torque to the end;
- M_G is the generator torque to the end;
- J_R is the moment of inertia of the rotor;
- J_G is the moment of inertia of the generator;
- ω_R is the rotational speed of the main shaft;
- ω_G is the rotational speed to the start of generator.

From (3) result:

$$\frac{d\omega_R}{dt} = \frac{M_R - M_{GBi} - B_R \cdot \omega_R}{J_R} \qquad (5)$$

The moment of inertia of the rotor for a wind turbine with *n* blades [11] is given by:

$$J_R = n \cdot m \cdot r_i^2 \qquad (6)$$

Where, *n* is the number of blades, *m* is the mass of one blade, and *r* is the radius of weight centre of blades. After analysis of relations (5) and (6) we can draw the following conclusion : importance of decreasing moment of inertia of rotor hence its inertial mass, will lead directly to increase rotational speed of main shaft so will enable faster start at low speed wind of power plant.

One of possible way to reduce the moment of inertia is given by reducing weight of blades. Wind turbinie blades have been made from a variety of materials that range from wood to metals to composites. The key structural design requirements for the turbine's blades are: must be strong to resist the extreme loads, must be stiff to prevent collision with the tower, must be as light as possible to minimize the cost and the starting, and should be stiff and light to avoid resonance.

These requirements require the use of materials with high stiffness properties to withstand a long time of service and extreme loads and respectively a low density is needed to reduce centrifugal and gravitational forces. For the modal analysis, we considered three materials, more exactly blade made of steel, aluminum and epoxy-carbon. After designing the blade in the Catia software, this is imported into Ansys and based on the geometry and properties of the material, the mass of the blade is computed and values for the three materials considered is given in Table 1.

Table 1. Mass of the blade for the three materials.

Material	Steel	Aluminum	Epoxy carbon
Mass of blade (Kg)	685.37	241.85	139.69

The modal analysis for the blade is executed by ANSYS Workbench [12] and the results is presented in Figure 6.

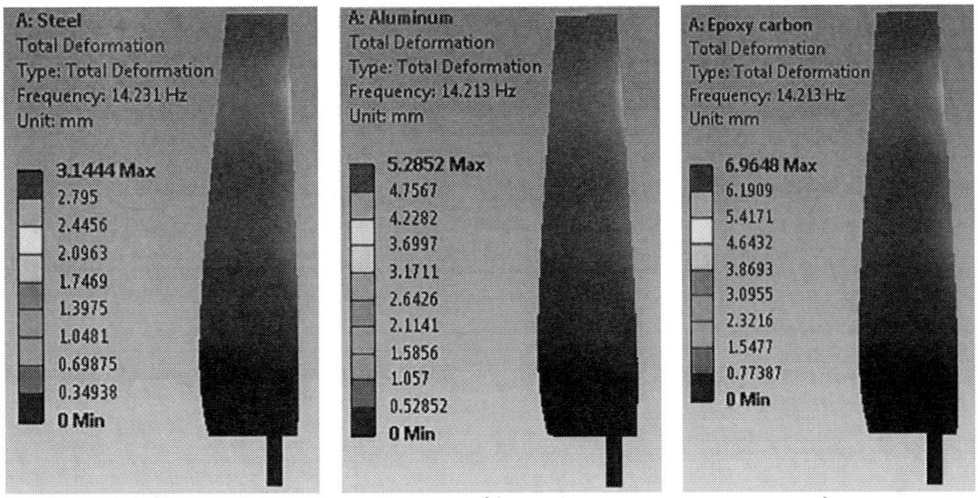

a) b) c)

Fig. 6. The diagrams of the modal analysis.

5 Conclusions

Wind power is an important source of renewable energy and plays a significant role in the world energy policy.

With increasing demand for renewable energy it is necessary to increase the dimensions of the rotors, which leads to the increase of the inertial masses.

Importance of decreasing moment of inertia of rotor hence its inertial mass, will lead directly to increase rotational speed of main shaft so will enable faster start at low speed wind of power plant.

After building the blades using the three materials and calculating the masses, one can conclude that the optimal solution is for the Epoxy-carbon blade.

Taking into account the inertial mass and after modal analysis performed on the blade in the three situations it can be concluded that the situation most favorable is for Epoxy-carbon.

Composites materials are ideal for producing wind turbine blades because of their low weight and the mechanical properties needed for any blade design.

References

1. T.M. Asif, P. K.Yadav, IJESRT **4** (8), 326-332 (2015)
2. M. Chandrala, A. Choubey, B. Gupta, IJERA **2**, 1244-1248 (2012)
3. W.T. Chong, K.C. Pan, S.C. Poh, A. Fazlizan, C.S. Oon, A. Badarudin, N. Nik-Ghazali, Renew. Energy **51**, 388-397 (2013)
4. V. Anbarasan, S. Eswaran, IJMERR **4** (1), 286-291 (2015)
5. A.S. Rathore, S. Ahmed, IJEST **3**, 7975-7980 (2011)
6. U. Jassmann, M. Reiter, D. Abel, IFAC **6**, 10107-10112 (2014)
7. V. M. Kumar, B. N. Rao, Sk. Farooq, IJAER **11** (6), 4491-4499 (2016)
8. T. Chuamvarasart, C. Chantharasenawong, S. Wongkittirat, ACEM14, 10 (2014)
9. T. Nitin, N.D. Mittal, A. Siraj, IJAER **1**, 500-507 (2010)
10. I. Curtu, I.Teşulă, M.D. Stanciu, P. Piscoi, A. Savin, Pro Ligno **11**, 157-164 (2015)
11. S. Barabas, F. Sarbu, B. Barabas, A. Fota, WASET **9**, 878-881 (2015)
12. P. Lengvarský, J. Bocko, M. Hagara, AJME **1**, 271-275 (2013)

Peculiarities of clarifiers' reconstruction at waste water treatment plants

Nikolay Makisha[1,*], and *Artem* Kulakov[2]

[1]Moscow State University of Civil Engineering (MSUCE), Yaroslavskoye Shosse, 26, Moscow, 12933, Russia
[2]Vologda State University (VSU), Lenina Street., 15, Vologda, 160000, Russia

Abstract. The article reveals the ways in which the reconstruction of clarifiers for wastewater treatment plant can be performed. One of the most common method is bioflocculation that means use of corresponding properties of excess activated sludge or biofilm, which both have in its composition so called extracellular biopolymers that determine the spatial structuring and bioflocculation of cell formations. Multiple researches showed that use of surplus sludge as bioflocculant provides increased efficiency of sedimentation in any waste water up to 85-90 % and reduce biochemical oxygen demand (BOD) in the clarified water by 40-50 %. The other method to be implemented for reconstruction of clarifying facilities is thin-layer modules to raise the efficiency of settling. Thin-layer modules can be used in a wide range of capacity of sewage works. The possibility of application depends on the structural size and condition of building structures, the amount of suspended solids and other parameters.

1 Introduction

Sedimentation is the easiest, cheapest and least energy intensive method to remove out of wastewater suspended impurities with a density different from the density of water. Under the action of gravity the particles of dirt settle on the bottom of the tank or float to the surface.

The relative ease of settling facilities determines their wide application in various stages of wastewater treatment and further handling of the produced sludge. Depending on its purpose and location in technological schemes of wastewater treatment clarifiers may be classified into the following structures: settling tank - primary, secondary and tertiary (contact tank); sludge thickeners.

2 Types of clarifiers

Primary settling tanks are located in the technological scheme of wastewater directly after a sand traps and used for separation of suspended solids from waste water. Achieved effect 40-60% leads to a 20-40% decrease of BOD in the clarified wastewater to the original value

* Corresponding author: nmakisha@gmail.com

[1,2].

In order to avoid the increased growth of activated sludge in the aeration or biofilm in the biofilters residual concentration of suspended solids in the clarified effluent after primary clarifiers should not exceed 100-150 mg/l. Influent concentrations of suspended solids in the waste water that are normally of 200-500 mg/l, that determines the choice of the most reasonable technology for primary clarification and the desired duration of settling.

Secondary clarifiers are the second part of the biological treatment facilities that is located in the treatment sequence after the phase of biological oxidation and serves to separate the activated sludge from the biologically treated water discharged from the aeration tanks, or to hold the biofilm supplied with water from a biofilter. The efficiency of the secondary clarifiers determines the final effect of water purification from suspended solids [1-3].

In technological schemes of biological treatment in the aeration tanks secondary clarifiers may also determine the volume of the aeration facilities, that depends on the concentration of the return sludge and the degree of recycling, the ability of the clarifiers to effectively separate the high-concentrated sludge mixture.

Sludge thickeners. The sequence of slaudge treatmetn determines how the thickening may be applied: for residues from primary clarifiers; for excess activated sludge; for mixture of residue and sludge; for flotation sludge; for sludge after stabilization.

The growth of activated sludge depends on the content of dissolved and suspended (mainly organic) media in the purified water and the efficiency of the primary clarifiers. The better primary sedimentation tanks work, the smaller is the amount of excessive active sludge [4,5].

3 Methods of intensification

Wide scope of studies conducted through years in MSUCE and VSU helped to develop and test different methods of intensification of processes of sedimentation of sewage and the thickening of the produced sludge. The most popular known method of intensification of primary settling for the treatment of municipal wastewater is associated with the use of bioflocculation properties of excess activated sludge and biofilm, which has in its composition extracellular biopolymers that determine the spatial structuring and bioflocculation of cell formations.

Excess activated sludge and biofilm are the natural bioflocculation agents to be appeared during biological wastewater treatment. Use of their bioflocculation properties is advisable as one of the most efficient methods that has certain physical and chemical effects on the formation of agglomerations of fine suspension in the process of its sedimentation [1, 2, 5-7].

Research in the field of sedimentation and hydrodynamic conditions of its implementation helped to develop and optimize technology of primary clarification of wastewater using surplus sludge as bioflocculant, which provides increased efficiency of sedimentation in any waste water up to 85-90 % and reduction of BOD in the clarified water by 40-50 %. Possible scheme of implementation of this technology in radial primary sedimentation tank is shown in Figure 1.

Area of bioflocculation located in the central part of the radial clarifier to ensure waste water effective contact between the particles and fine suspension of active sludge during 20-minutes stay.

Air input flow supplemented by the aerator device in the form of perforated pipes, which together provides the area of bioflocculation with required gradient of mixing speed of 50-60 sec (Figure 2, curve 1).

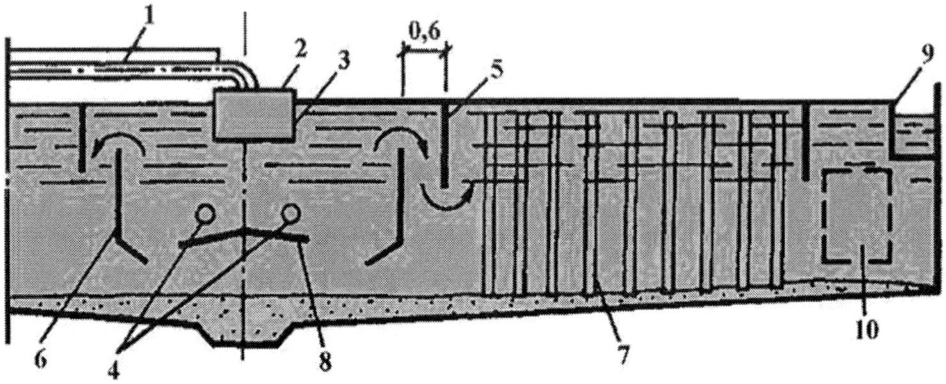

Fig. 1. Scheme of primary clarifier with bioflocculation chamber: 1 — inlet of waste water and sludge; 2 — distribution chamber; 3 — bioflocculation zone; 4 — perforated aerators; 5 — semi-submerged partition wall; 6 — submerged walls; 7 — low-rate mixer; 8 — protection shield; 9 — collection weir; 10 — thin-layer blocks

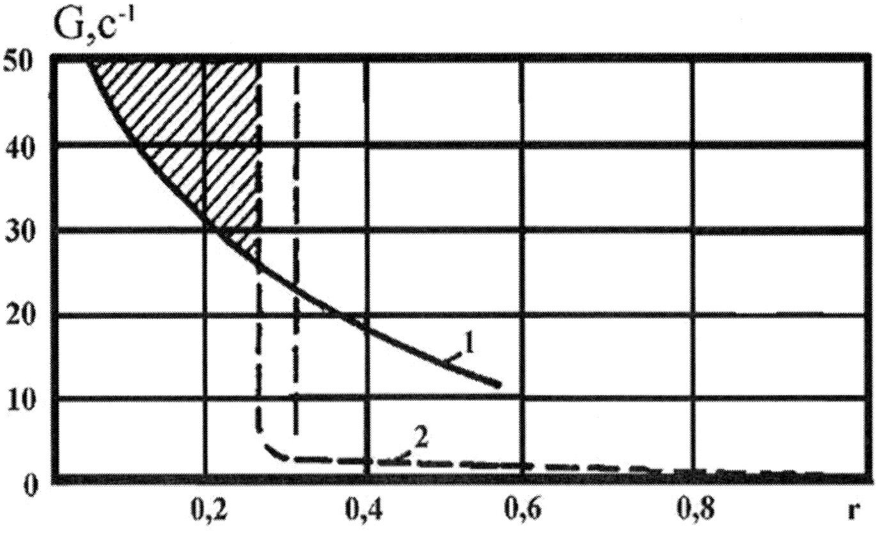

Fig. 2. Distribution of rate gradients before (curve 1) and after the reconstruction (curve 2)

After the bioflocculation waste water passes under the baffle zone of air separation, where the adhering air bubbles separated to prevent possible degradation of the sedimentation conditions [8].

In the settling zone of the clarifier sedimentation is stimulated by discordantly mixing, which for G=1-2 sec^{-1} provides optimal conditions for deposition of suspended particles and compaction of sediment formed. Cross-scheme thin-layer blocks located on the periphery of the sedimentation tank clarify water in the final stage before it reaches the collection tray. Acquired experience of MSUCE and VSU researchers in operation of the primary clarifiers, modified under this scheme, showed its high effectiveness to settle suspended solids with 60 to 80% and reduce BOD in the clarified water by 40-70%. The disadvantage of this method of intensification is the material demand for thin-layer blocks.

Under optimal supply of activated sludge 160-200 mg/l, the efficiency of the clarification of suspended solids and BOD were respectively 75-80 % and 50-70 %, while the moisture content of the mixture of sludge and excess sludge discharged from the clarifier was 96.0-96.5 per cent [9, 10].

Intensification of the primary clarifiers is also possible through the use of continuous sediment pumping with its subsequent thickening. The advantages of this technology is keeping of almost zero (not more than the height of the scrapers) layer of sediment on the bottom of the settler, thereby enhancing the effect of water clarification. Quick removal of drop-down sediment, by means of bottom scrapers, allows avoiding deposits of sludge with subsequent anaerobic decay and input of hardly settled compounds into the treated water.

It is necessary to improve the construction of devices that removes floating materials from the water surface of the radial settling tanks (the most typical solution for wastewater treatment plants). Swinging entrance bunkers, when truss scraper passes over them, collect a floating substance along with a significant amount of water to provide humidity of the mixture at about 97 % [11-13].

4 Sedimentation with thin-layer modules

The essence of the method, advantages and disadvantages. A thin layer (tubular or plate) clarifiers used to increase the efficiency of settling. At shallow depth sedimentation occurs rapidly, which helps to reduce the size of the tanks. The efficiency of tubular and plate sedimentation tanks is nearly similar. The height of the thin-layer space is recommended to be of 1-2 m, the distance between the plates — 25-200 mm, length — 0.6-1 m Thin-layer blocks can be installed in case of conventional clarifiers (Figure 3) [5, 14, 15].

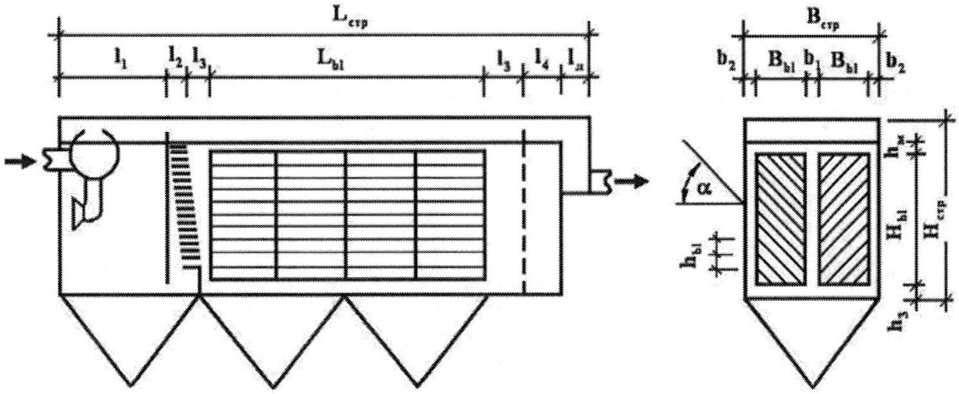

Fig. 3. Clarifier with thin-layer module

Application of thin-layer elements can significantly reduce the duration of settling and consequently the volume of the tanks (the duration of the treatment is 4-10 min). Thin-layer sediment tanks allow intensifying of suspended solids sedimentation process, together with 60 % reduction of necessary volume and 25-30 % increase of the purification effect if compared to the commonly used clarifiers. As with conventional clarifiers, they do have zones of water distribution, sedimentation and collection and zone of accumulation of sediment.

The settling area of shelf sections or tubular elements is divided into a number of shallow layers (to 15 cm). Shelf sections are assembled from flat or wavy plates, easy to operate. The tubular sections have greater rigidity, which provides stability of dimensions

DOI: 10.1051/matecconf/201711210018

throughout its length and higher flow speed than the shelf section. On the other side tubular modules could be faster bulked with sediments (and hardly cleaned) and require greater material consumption [4, 6, 16].

The hydraulic mode of operation of the clarifiers has a significant impact on the effect of their work. The better the design of the clarifier, the higher the efficiency of the detention of suspended solids that is connected to the way how water enters the sedimentation tank i.e. the speed of water inlet and depth of partition wall in a horizontal settler. Hydraulic mode of operation is estimated by the coefficients of volumetric use and useful actions of the settlers.

The other advantage of thin-layer sediment tanks is that parallel plates in cross-section of the clarifier provides more regular distribution the incoming water flow and maintains this distribution throughout its length. Therefore, volumetric use is much higher in multilevel clarifiers [17, 18].

Thin-layer sedimentation tanks are classified according to the following criteria: design of blocks – tubular and shelving; mode of operation – periodic (cyclic) and continuous; mutual movement of clarified water and with displaced sediment — direct-flow, counter-flow or combined motion.

The cross section of the tubular sections may be rectangular, square, hexagonal and round. Shelf sections assembled from flat or corrugated sheets have a rectangular cross-section. They are made of steel, aluminum and polymers (polypropylene, polyethylene, fiberglass etc.). Normal flow rate for shelf elements is 5-10 mm/s and up to 20 mm/sec for tubular. Tube can be installed with low (to 5°) and large (up to 45-60°) slope. Tubular elements with a slight tilt work intermittently. The sedimentation is conducted first following by residue removal.

Thin-layer modules (Figure 4) can be used in a wide range of capacity of sewage plants. The possibility of application depends on the structural size and condition of constructions, the amount of suspended solids and other parameters [19, 20].

Fig. 4. Thin-layer module

The best results can be reached if the method of thin-layer settling applied for the secondary clarifiers that helps not only to improve significantly suspended solids removal efficiency but also to make a slight reduction contamination of BOD.

When using thin-layer blocks they should be regularly regenerated by means of mechanical, hydraulic, pneumatic or combined methods, as they can accumulate particles of suspended solids on the modules' surface when operating. The frequency of regeneration of thin-film modules is determined in the process of operation, but not less than 1 time in 10 days.

5 Conclusions

1. Within the process of clarifiers reconstruction the most important properties of these facilities, which largely depends on the hydraulic factors and their peculiarities should be considered.
2. Violation of waste water flow speed, flow distribution, settling velocity and other hydraulic parameters of clarifiers may eventually lead to malfunction of the biological treatment facilities and sludge treatment.
3. Methods of reconstruction such as bioflocculation and thin-layer modules application help to achieve a significant reduction of suspended solids and BOD concentration in waste water

References

1. A.G. Pervov, A.P. Andrianov, T.P. Gorbunova, A.S. Bagdasaryan, Petr. Chem. **55** (10), 879-886 (2015)
2. I. Gulshin, A. Kuzina, IJAER **10** (21), 42618–42623 (2015)
3. E.S. Gogina, O.V. Yantsen, O.A. Ruzhitskaya, AMM **580-583**, 2354-2357 (2014)
4. A. Volkov, V. Chulkov, R. Kazaryan, M. Fachratov, O. Kyzina, R. Gazaryan, AMM **580-583,** 2281-2284 (2014)
5. A.G. Pervov, A.P. Andrianov, E.B. Yurchevskiy, Petr. Chem. **55** (10), 871-878 (2015)
6. E. Gogina, A. Pelipenko, MATECCONF **73,** 03007 (2016)
7. V.N. Varapaev, S.A. Doroshenko, A.Y. Trotsko, A.V. Doroshenko, IJAER **10** (21), 42588-42592 (2015)
8. O. Kuzina, E. Pankratov, V. Tkachev, MATECCONF **86,** 05023 (2016)
9. V. Orlov, A. Andrianov, AMM **580–583**, 2398-2402 (2014)
10. V.N. Varapaev, A.V. Doroshenko, I.Y. Lantsova, Pr. Eng. **153**, 816-823 (2016)
11. E.S. Gogina, O.A. Ruzhitskaya, O.V. Yantsen. AMR **919–921**, 2145–2148 (2014)
12. N. Makisha, E3sconf **6**, 01002 (2016)
13. A.G. Pervov, A.P. Andrianov, DWT **35** (1-3), 2-9 (2011)
14. E. Gogina, I. Gulshin, AMM **580-583**, 2367-2369 (2014)
15. A. Volkov, O. Kuzina, Pr. Eng. **153**, 838-843 (2016)
16. N. Makisha, Pr. Eng. **165**, 1087-1091 (2016)
17. A.A. Volkov, A.V. Sedov, P.D. Chelyshkov, D.A. Lysenko, A.V. Doroshenko, IJAER **10** (22), 43269-43272 (2015)
18. N. Makisha, Pr. Eng. **165**, 1092-1097 (2016)
19. E. Gogina, I. Gulshin, Pr. Eng. **153**, 189-194 (2016)
20. A.G. Pervov, A.P. Andrianov, V.A. Chukhin, R.V. Efremov, IJAER **10** (22), 43517-43525 (2015)

Removal of nitrogen and phosphorous from domestic wastewater

Artem Kulakov[1,*], and *Nikolay* Makisha[2,*]

[1]Vologda State University, Lenina Street, 15, Vologda, 160000, Russia
[2]Moscow State University of Civil Engineering, Yaroslavskoye Shosse, 26, Moscow, 129337, Russia

Abstract. This work is aimed at providing a deep cleaning from nitrogen and phosphorus at the requirements level for a small capacity waste water treatment plants with active sludge. The wastewater treatment technology includes: mechanical screen, anaerobic reactor, anoxic reactor, two aerobic reactors with floating media, aerobic reactor, clarifier, bioreactor, purification, filter, filter of tertiary treatment with the sorbent, ultraviolet disinfection facility. The sludge treatment includes aerobic stabilizer and facility of mechanical dewatering. The developed technology ensures deep purification of household waste waters from nitrogen and phosphorus. The degree of purification in the ammonium-ion reaches of 99.5 %. Efficiency of total nitrogen removal (the sum of ammonium, nitrite and nitrate nitrogen) is 85 %. The efficiency of biological purification of phosphates is 97.8 percent that is achieved without the use of coagulant. Requirements are met in most cases.

1 Introduction

Implementation of advanced wastewater treatment at small treatment plants is a difficult task due to dimensions of the facilities, the irregularity of the composition and volume of sewage [1, 2]. There are different effective technologies of removal of nitrogen and phosphorus to major stations [3-7]. However, their adaptation to small structures may be sometimes difficult [1, 8].

At the stations of biological purification with active sludge the main task is to ensure effective mixing, aeration and/or mechanical devices. The creation of anaerobic conditions for the biological dephosphotation and anoxic conditions for denitrification processes is the basis how to achieve environmental standards.

Performance improvement of small objects (Figure 1) is required in the formation of biomass in condition of regular load on the facilities by organic and biogenic substances as well as by volume. This work is aimed at providing a deep cleaning from nitrogen and phosphorus at requirements level at a small capacity waste water treatment plants with active sludge.

* Corresponding authors: artem.a.kulakov@gmail.com, nmakisha@gmail.com

Fig. 1. Waste water treatment plants of small capacity

2 Methods

The analyses of wastewater were performed according to standard procedures on certified equipment in the laboratory sewage treatment plant.

The main indicators characterizing the efficiency of domestic wastewater treatment, is as follows:

- Suspended solids (SS).
- Chemical oxygen demand (COD).
- Biochemical oxygen demand (BOD_5).
- Ammonium-ion (NH_4).
- Nitrite-ion (NO_2).
- Nitrate-ion (NO_3).
- Phosphate-ion (PO_4).

A total the three nitrogen forms was calculated to assess changes in the content of nitrogen:

$$\Sigma N = 0,78*NH_4+0,3*NO_2+0,23*NO_3,\ \mathrm{mg/l} \quad (1)$$

Technology correction was based on the methods of process simulation to achieve maximum use facilities' capacity [9-12].

3 Technology

Facilities designed for purification of household sewage water of industrial site and residential complex. The volume of treated wastewater varies from 10 to 24 m^3/day., an average flow is 15 m^3/day.

Wastewater is pumped to the treatment facilities by means of sewage pumping stations (Figure 2). Wastewater goes through mechanical screens to remove suspended particles of size greater than the size of the grid opening from wastewater, as well as long fibers, sand and grease. Partially cleaned from mechanical impurities wastewater received in the intermediate tank, from which it is evenly pumped to treatment in the biological treatment unit [13-18].

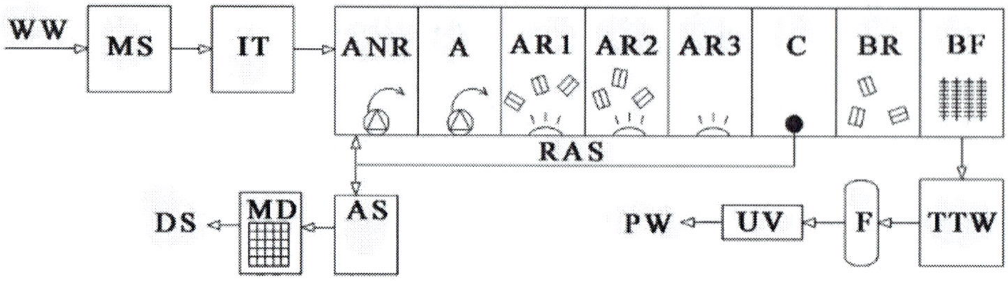

Fig. 2. Technological scheme: MS – mechanical screen, IT – intermediate tank, ANR – anaerobic reactor, A – anoxic reactor, AR1, AR2 – aerobic reactor with floating feed, AR3 –aerobic reactor, C – clarifier, BR – bioreactor, BF – biofilter, TTW – tank for biologically treated water, F – filter, UV – UV-disinfection , AS – aerobic stabilizer, MD – mechanical dewatering, WW – waste water for treatment, PW – purified waste water, RAS – return active sludge

The first stage of biological treatment is anaerobic reactor (ANR), where it is mixed with circulating activated sludge (CES). Then the sludge mixture enters anoxic tank (A). The liquid mixing in the anaerobic and anoxic tanks is performed by submersible pumps. After anoxic tank the sludge mixture sequentially passes through the three aerobic reactor (AR), equipped with a system of fine-bubble aeration. The first and second AR is equipped with a floating feed for immobilization of biofilms (Figure 3), thus treatment goes by a combination of biomass: free-floating and attached [8, 9, 19-24].

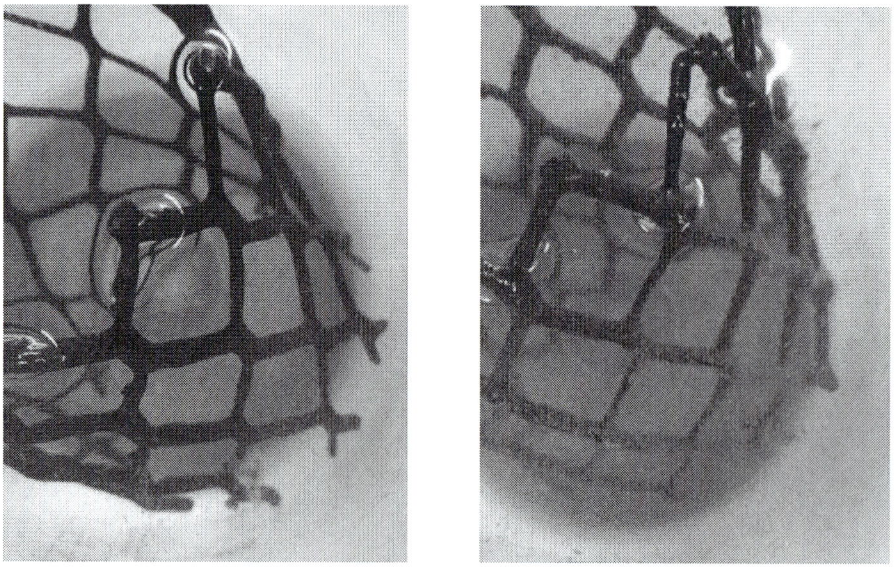

Fig. 3. New bioreactor media (left), bioreactor media with biomass (right)

After aerobic reactors the sludge mixture enters the clarifier where the water is separated from sludge, which is pumped by airlift pumps back to the anaerobic reactor. The clarified water passes through the bioreactor of tertiary treatment with floating load (the same load in aerobic reactors) and biofilter with brush feed to remove fine particles.

The biologically purified waste water flows into the tank and it is pumped to the filter with sorbent and ultraviolet disinfection. Removal of phosphates can be made by means of coagulant (if necessary). Excess activated sludge is treated in aerobic stabilizer and facility of mechanical dewatering.

3 Results

Long observation of the facility work confirmed the stable treatment of wastewater with high efficiency for removal of nitrogen and phosphorus. The concentration of dissolved oxygen determined at different points to evaluate the oxygen regime structures (Table 1).

The conditions with low oxygen is maintained in the anaerobic and anoxic reactors to provide the processes of denitrification and biological phosphorus removal. The concentration of dissolved oxygen in the aeration tanks that is above 2 mg/l has positive effect on the processes of nitrification [25-28].

The analysis of the content of pollutants in the wastewater showed a high degree of purification on nitrogen and phosphorus and compliance with requirements on them in most cases (Table 2) [29-33].

Table 1. Concentration of dissolved oxygen

Tank	Value [mg/l]
Anaerobic reactor	0,2
Anoxic reactor	0,05
Aerobic reactor 1	2,2
Aerobic reactor 2	4,9
Aerobic reactor 3	7,2
Clarifier	2,5
Bioreactor	4,5
Biofilter	6
Tank for treated water	7,5

Table 2. Concentration of pollutants in waste water

Indicator	Concentration [mg/l]		Efficiency [%]	Limits
	In	Out		
Suspended solids	135-154	2-4	98	3
COD	455-515	40-55	90	-
BOD_5	92-115	1,5-3,4	97	2
Phosphate-ion	6,9-10	0,11-0,3	97,8	0,6
Ammonium-ion	82-90	0,21-0,46	99,5	0,5
Nitrite-ion	0,03-0,08	0,12-0,39	-	0,08
Nitrate-ion	0,53-0,71	37-46	-	40
ΣN	64,1-70,4	8,7-11	85	-

The analysis of activated sludge sedimentation and structure of activated sludge was carried out (fig.4). The activated sludge has a well-defined structure and a high rate of sedimentation.

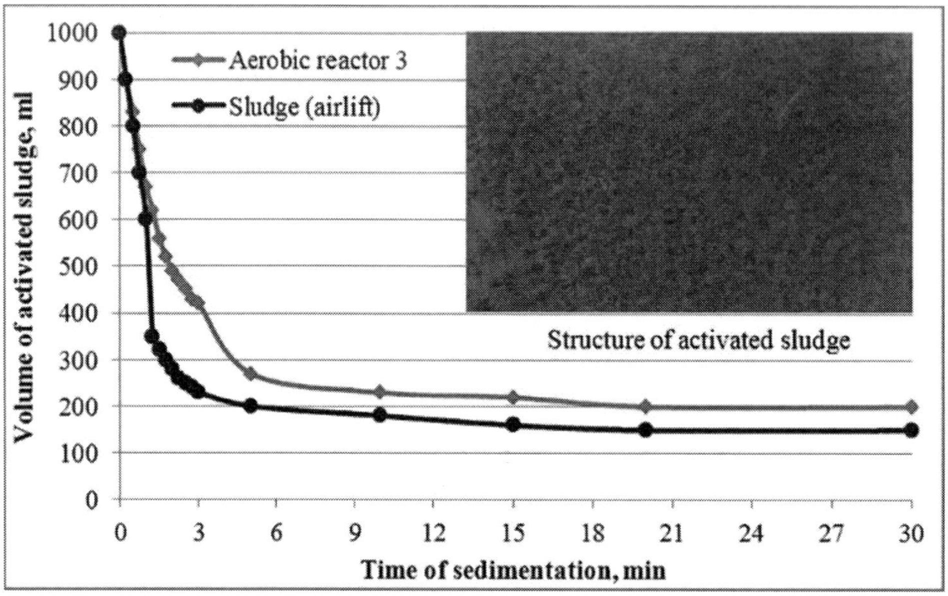

Fig. 4. Sedimentation of activated sludge

Cleaning efficiency in the total of the three nitrogen forms (ammonium, nitrite and nitrate) is 85 %, which is also fairly high, that can be obtained by means of strict following of operation procedure (Table 3).

Table 3. Operation procedure

Action	Regularity
Visual analysis of immobilized biomass	Daily
Visual estimate of mixing efficiency in ANR, A and ARs	Daily
Visual control of sludge level in clarifier	Daily
Volumetric sludge dose estimate in ANR, A and ARs	Daily
Dry compound sludge dose estimate in ANR, A and ARs	Weekly
Sludge index calculation	Weekly
Determination of dissolved oxygen concentration	Daily
Cleaning of tray and walls	Weekly
Aeration of biofilter (5-10 min)	Weekly
Pumping of excessive sludge	Weekly

4 Conclusions

1. The developed technology allows providing a deep purification of household waste waters from nitrogen and phosphorus. The degree of purification in the ammonium-ion reaches of 99.5 %. In the process of nitrification there is a slight increase in nitrites and nitrates that characterizes the occurrence of denitrification. The effectiveness in the total of the three nitrogen forms (ammonium, nitrite and nitrate) is 85 %. The efficiency of biological purification of phosphate make up 97.8 percent, the result is achieved without the use of coagulant. Requirements are met in most cases.

2. Identified areas for further improvement structures. For increase of efficiency of preliminary mechanical cleaning of large debris and improve the reliability of the pumps required extra grate (sieve) with opening size of 4 mm.
3. Maintaining a high cleaning efficiency is only possible while ensuring proper operation of facilities.

References

1. A.A. Kulakov, ViST **5**, 25-30 (2013)
2. A.A. Kulakov, A.F. Shafigullina, Vodoochistka **8**, 28-36 (in Russian) (2016)
3. N. Makisha, V. Scherbakov, A. Smirnov, E. Scherbina, IJAER **10** (24), 44347–44349 (2015)
4. V. Scherbakov, E. Gogina, T. Schukina, N. Kuznetsova, N. Makisha, E. Poupyrev, IJAER **10** (21), 44353–44356 (2015)
5. E. Gogina, I. Gulshin, Pr. Eng. **117**, 107–113 (2015)
6. A.G. Pervov, A.P. Andrianov, T.P. Gorbunova, A.S. Bagdasaryan, Petr. Chem. **55** (10), 879-886 (2015)
7. I. Gulshin, A. Kuzina, IJAER **10** (21), 42618–42623 (2015)
8. E.S. Gogina, O.V. Yantsen, O.A. Ruzhitskaya, AMM **580–583**, 2354–2357 (2014)
9. O.A. Ruzhitskaya, E.S. Gogina, AMR **919–921**, 2141–2144 (2014)
10. A. Volkov, V. Chulkov, R. Kazaryan, M. Fachratov, O. Kyzina, R. Gazaryan, AMM **580-583,** 2281-2284, (2014)
11. A.A. Kulakov, Wat.Ecol **1** (61), 26-40 (2015)
12. A.G. Pervov, A.P. Andrianov, E.B. Yurchevskiy, Petr. Chem. **55** (10), 871-878 (2015)
13. O.A. Ruzhitskaya, E.S. Gogina, AMR **919–921**, 2153–2156 (2014)
14. Kulakov A.A., Lebedeva E.A., Umarov M.F., EiPR **11**, 33-36 (2011)
15. E. Gogina, A. Pelipenko, MATECCONF **73,** 03007 (2016)
16. Kulakov A.A., Lebedeva E.A., Vodoochistka **12**, 10-19 (2011)
17. V.N. Varapaev, S.A. Doroshenko, A.Y. Trotsko, A.V. Doroshenko, IJAER **10** (21), 42588-42592 (2015)
18. N. Makisha, O. Yantsen, AMM **587–589**, 640–643 (2014)
19. O. Kuzina, E. Pankratov, V. Tkachev, MATECCONF **86,** 05023 (2016)
20. V. Orlov, A. Andrianov, AMM **580–583**, 2398-2402 (2014)
21. N. Makisha, Y. Voronov, E. Poupyrev, V. Volshanik, IJAER **10** (21), 41919–41922 (2015)
22. V.N. Varapaev, A.V. Doroshenko, I.Y. Lantsova, Pr. Eng. **153**, 816-823 (2016)
23. N. Zaletova, Y. Voronov, N. Makisha, IJAER **10** (21), 42454–42455 (2015)
24. E.S. Gogina, O.A. Ruzhitskaya, O.V. Yantsen. AMR **919–921**, 2145–2148 (2014)
25. N. Makisha, E3sconf **6**, 01002 (2016)
26. A.G. Pervov, A.P. Andrianov, DWT **35** (1-3), 2-9 (2011)
27. E. Gogina, I. Gulshin, AMM **580-583**, 2367-2369 (2014)
28. A. Volkov, O. Kuzina, Pr. Eng. **153**, 838-843 (2016)
29. N. Makisha, Pr. Eng. **165**, 1087-1091 (2016)
30. A.A. Volkov, A.V. Sedov, P.D. Chelyshkov, D.A. Lysenko, A.V. Doroshenko, IJAER **10** (22), 43269-43272 (2015)
31. N. Makisha, Pr. Eng. **165**, 1092-1097 (2016)
32. E. Gogina, I. Gulshin, Pr. Eng. **153**, 189-194 (2016)
33. A.G. Pervov, A.P. Andrianov, V.A. Chukhin, R.V. Efremov, IJAER **10** (22), 43517-43525 (2015)

AUTHOR INDEX

AUTHOR INDEX

AUTHOR INDEX

AUTHOR INDEX